Sixteenth European Photovoltaic Solar Energy Conference

Proceedings of the International Conference held in Glasgow, United Kingdom

1 – 5 May 2000

Edited by:

H. SCHEER
Eurosolar
Bonn, Germany

B. McNELIS
IT Power
Eversley, United Kingdom

W. PALZ
European Commission

H.A. OSSENBRINK
European Commission
DG JRC,
Ispra

P. HELM
WIP-Renewable Energies
Munich, Germany

Volume III

Conference organised by:
WIP-Renewable Energies
Sylvensteinstr. 2
D-81369 München
Germany
Tel: +49 89 720 1235
Fax: +49 89 720 1291
Email: wip@wip-munich.de
Web: www.wip-munich.de

Proceedings published by:
James & James (Science Publishers) Ltd
35-37 William Road
London NW1 3ER
UK

A catalogue record for this book is available from the
British Library.

ISBN 1 902916 18 2

Printed in the UK by Alden Press

Cover photos: (left to right) Detail of Photowatt's
automated production line in Bourgoin (Lyon), France;
photo: Photowatt International. Grid connected PV facade
on low-rental apartments in Surieux, France; photo: Total-
Energie. Installation of roof mounted PV system; Siemens
press photo.

Contents of Volume I and Volume II

For the reader's convenience, the contents of Volume I and Volume II are repeated below; the contents of Volume III start on page v.

Volume I

Volume II

Contents of Volume III

PV Modules and Components of PV Systems

VISUAL PRESENTATION VD2 PV Modules and Components of PV Systems

Implementation, Strategies, National Programs and Financing Schemes

PLENARY SESSION PB1 Implementation, Strategies, National Programs and Financing Schemes
Chairpersons: R. Vigotti, ENEL, Rome, Italy
P. Helm, WIP-Renewable Energies, Munich, Germany

Market Deployment in Developing Countries

ORAL PRESENTATION OB7 Market Deployment in Developing Countries
Chairperson: W. Palz, European Commission, Brussels

ORAL PRESENTATION OB8 Market Deployment in Developing Countries
Chairperson: A. Wilshaw, IT Power, Eversley, United Kingdom

ORAL PRESENTATION OBX Market Deployment in Developing Countries
Chairperson: G. Cadonau, Arbeitsgemeinschaft Solar '91, Zurich, Switzerland

Author Index (at the end of each volume)

Subject Keyword Index (at the end of each volume)

PV Modules and Components of PV Systems

Solar resource and photovoltaic generators for airships

Werner Knaupp

Zentrum für Sonnenenergie- und Wasserstoff-Forschung (ZSW)
Hessbrühlstrasse 21c, D- 70565 Stuttgart, Germany
e-mail: werner_knaupp@compuserve.com

ABSTRACT: Irradiance measurements on the envelope during airship flights as well as validated calculations of irradiance distribution with consideration of anisotropic diffuse irradiance on the basis of flight attitude angles and ground data for determination of solar resource at airship envelope at low altitude were carried out. With consideration of PV module specific properties electrical energy output from airship envelope segments covered with solar cells can be determined with good quality. Requirements regarding integrated photovoltaic modules for airships are identified. The feasibility of lightweight flexible photovoltaic modules and achievable power values was studied on the basis of possible substrate materials and deposition techniques for flexible CIGS thin film photovoltaic modules. Several metallic and non-metallic substrates are suitable for flexible CIGS thin film PV modules with a potential of up to 700 W/kg and 400 g/m².

KEYWORDS: Airships - 1: PV Module - 2: Solar Resource - 3

1   INTRODUCTION AND OBJECTIVES

Airships meet again with great interest worldwide due to advantageous new technologies and materials as well as new application areas. Airships have a great potential for cost efficient freight transports of very heavy or oversized goods (Figure 1) over large distances and in regions with low air traffic infrastructure.

Figure 1:   Transport of heavy or oversized goods could be realized by airships (Photo: Baumann, Bonn).

In high altitude airships could act as a stratospheric platform for telecommunication services [1] (Figure 2) and scientific research purposes like atmospheric monitoring tasks.

Figure 2:   High altitude airship serving as communications relay platform (Source: Airship Technologies Group, Bedford, UK).

For all applications the energy supply system is one of the key issues. Especially long enduring flight tasks need an inexhaustible, reliable and lightweight energy source. In this context photovoltaic generators play an important role. Design and layout of PV based energy systems require knowledge of power and energy characteristics of solar resource.

Main objectives presented here deal with the theoretical and experimental determination of solar resource at airship envelope and examinations regarding the requirements and the feasibility of lightweight flexible PV generators for airships.

2   DETERMINATION OF SOLAR RESOURCE

Important characteristic parameters for photovoltaic energy converters are the irradiance in the plane of array and the resulting operation temperature. For determination of solar resource at airship envelope irradiance measurements at different points on the envelope during airship flights at low altitude as well as theoretical calculations of irradiance distribution with consideration of diffuse irradiance on the basis of flight attitude angles and ground data were carried out.

2.1   Measurements

Measurement of irradiance at the airship envelope was realized on the basis of small monocrystalline silicon solar cells. For temperature correction of solar cell short circuit current the cell operation temperature was additionally measured with PT100 foil resistors on the back side of the cell. All sensors with their pre-amplifiers were calibrated against a thermopile pyranometer under real outdoor conditions. With that scaling functions for conversion of digital flight data into irradiance values could be derived. Global horizontal (GHI) and diffuse horizontal (DHI) irradiance data at ground were recorded in parallel to the measurement flights. Figure 3 shows the schematic placement of eight sensors (S1...S8) at the envelope for the measurement flights.

The measurement flights were carried out with the airship *LOTTE* (Figure 4) [2] by the Institut für Statik und Dynamik der Luft- und Raumfahrtkonstruktionen (ISD) of University of Stuttgart.

2.2   Simulation

The goal of the simulation is the determination of irradiance and electrical energy values from envelope segments on the basis of irradiance ground reference measurements and flight attitude angles. Irradiance in the plane of array can be calculated with the effective angle of incidence θ between normal of area segment and sun and with additional consideration of diffuse and ground reflec-

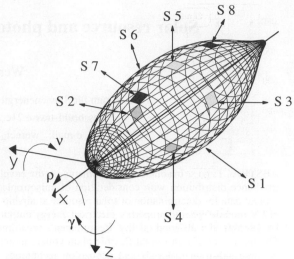

Figure 3:   Placement of irradiance sensors during measurement flights.

ted irradiance. The calculation requires assumptions regarding diffuse irradiance distribution over the sky dome and their parametric description. In this context different descriptions [3]-[6] were examined regarding their suitability for given geometry, accuracy and mathematical efforts. A simplified description with differentiation between circumsolar and isotropic portion [4] was found as a good compromise.

During the measurement flights operation data of the different sensors were recorded every two seconds. Figure 5 shows the time series of measured and calculated irradiance values and Figure 6 shows the scatter plot of calculated vs. measured irradiance for sensor 6. Calculated and measured values coincide basically. Inertia of flight attitude sensors and uncertainty in registration of applicable global and spectral albedo values during the flight limit the achievable accuracy. Figure 4 shows also that the dynamics of the irradiance for a certain segment area can be very high. This should be considered during layout process.

Figure 4:   Airship LOTTE during measurement flight   (Stuttgart, 20.06.1998).

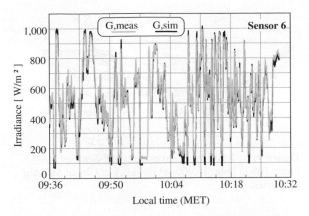

Figure 5: Measured and calculated irradiance profiles during measurement flight (Stuttgart, 20.06.1998).

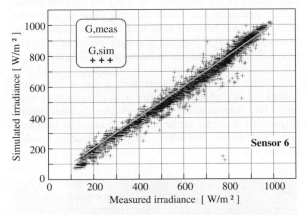

Figure 6: Calculated vs. measured irradiance during measurement flight (Stuttgart, 20.06.1998).

3   ELECTRICAL POWER AND ENERGY OUTPUT

On the basis of solar resource and with consideration of photovoltaic module characteristics electrical power and energy values can be determined. As an example for maximum achievable values calculations were carried out for a fully solar cell covered airship in LOTTE size at operation during daytime (Stuttgart, June 20th) at different stationery yawing, pitch, and roll angles.

Figure 7 shows the power profiles for four different positions I-IV with a solar cell efficiency η of 15 % at STC (≡ Standard Test Conditions) and disregarding electrical mismatch losses. The overall average efficiency $\overline{\eta}$ is lower mainly due to reflection losses at the three dimensional curvature.

As a second example for achievable power and energy values calculations were performed here for two airhip sizes and different solar cell coverage ratios (full coverage of the envelope and coverage of the upper half of the envelope).

Case I :   Full daytime operation of an airship in LOTTE size (16 m long, 4 m diameter) at June 20th, stationary position southwards, $\eta_{solar\ cell}$ = 15 % (STC).

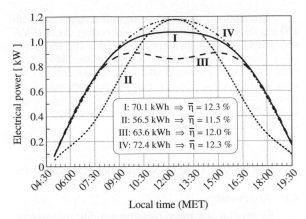

Figure 7: Calculated power output from photovoltaic generator for LOTTE size airship at different stationery positions {γ= yawing, ν=pitch, ρ= roll angle ; see Fig. 2} (operation during daytime, Stuttgart, June 20th, low altitude, fully covered with solar cells, $\eta_{solar\ cell}$ = 15 % STC, no electrical mismatch losses):

I   γ / ν / ρ = 180° / 0° / 0°
II   γ / ν / ρ = 90° / 0° / 0°
III   γ / ν / ρ = 180° / 20° / 0°
IV   γ / ν / ρ = 180° /-20° / 0°

Case II :   Full daytime operation of an airship in CL size (260 m long, 65 m diameter) at June 20th, stationary position southwards, $\eta_{solar\ cell}$ = 15 % (STC).

Table 1 lists the calculated power and energy values. In real configurations there will be some mismatch losses within the photovoltaic generator due to different solar cell orientations on top of the curved surface. There is an optimum between mismatch losses, voltage level and safety considerations. With detailed system engineering mismatch losses can be minimized well below 5 %.

Table 1: Solar resource and electrical output of two different airship sizes and two degrees of solar cell coverage   (operation during daytime, Stuttgart, June 20th, low altitude, stationery position southwards, $\eta_{solar\ cell}$ = 15 % at STC)

	100 % *of envelope*	*50 %* *of envelope*
LOTTE size **(16 m long)**		
Area	133.4 m²	66.7 m²
Radiant Energy	569.5 kWh	414.9 kWh
Electrical Energy	70.1 kWh	56.2 kWh
Peak Power	6.4 kW	5.7 kW
CL size **(260 m long)**		
Area	39,600 m²	19,800 m²
Radiant Energy	170.0 MWh	123.9 MWh
Electrical Energy	20.9 MWh	16.8 MWh
Peak Power	1.9 MW	1.7 MW

Conventional photovoltaic module technologies with 350 μm thick c-Si cells and thick embedding sheets (EVA and Tedlar) have a high a specific weight. The realization of a photovoltaic generator according to Table 1 in standard technology would take a significant share of the total airship mass. This is one of the motivations for investigations in the realization of flexible and lightweight photovoltaic generators on the basis of crystalline silicon cells or advanced thin film solar cells.

4   PHOTOVOLTAIC GENERATORS FOR AIRSHIPS

4.1   Requirements

In general the main requirements for photovoltaic generators for airships are:

- High power values per weight

- Easy current-/ voltage level adaptation

- Reliability

- Environmental resistance

    Especially for applications in high altitude the environmental conditions for the airship as well as the photovoltaic generator are very severe. Ambient temperature is below -55°C. There are also higher concentrations of UV radiation and ozone with implications on the choice of materials.

- Shape adaptation and mechanical load capacity

    Besides the demands from the three dimensional curvature there could occur other mechanical loads of the photovoltaic generator depending on the operation condition and construction type of the airship. Figure 8 shows as an example wrinkling of the envelope of a semi-rigid keel airship body resulting from an internal pressure drop [7]. In this case the PV module should tolerate a small bending radius.

Figure 8:   Wrinkling of envelope of a semi- rigid keel airship body after an internal pressure drop from 160 Pa to 85 Pa (Source: [7]) as an example for possible mechanical loads of photovoltaic modules at the surface.

For reduction of airship overall weight it would be desirable to build the photovoltaic generator as integral part of the envelope. With that there are additional requirements regarding tensile strength and gas impermeability of the sheets. In general there is no single material even for the envelope alone fulfilling all requirements.

Therefore airship envelopes are realized with several layers of strong but lightweight fibres (see example in Figure 9). In a similar way production of integral PV generators will be realized in a multi layer configuration.

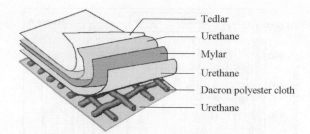

Figure 9:   Multi layer envelope of the *Sentinel 1000* blimp {68 m long, 10,000 m³ Helium} of Global Skyship Industries.

4.2   Crystalline silicon solar cells

The first approach during the construction of the airship LOTTE for realization of a slightly flexible photovoltaic generator was based on crystalline silicon solar cells embedded between very thin layers of plastic foils. With such a straight forward technique one could achieve a remarkable specific weight in the range of 1.7 kg/m².

Static and dynamic tests with a linear load were carried out at embedded samples with 350 μm thick solar cells for determination of mechanical characteristics (Figure 10). Analysis of the break diagrams after load tests give indications regarding minimal bending radius (Figure 11).

Figure 10:   Submodule sample with two mono c-Si cells during mechanical load tests (Photo: ISD, Stuttgart).

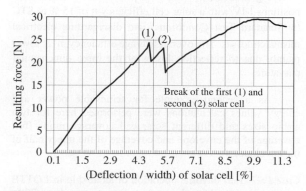

Figure 11:   Break diagram of a submodule sample with two embedded mono c-Si solar cells (5cm x 10cm each, 350 μm thickness). Minimal bending radius here is in the range of 20 cm [8]

The unintentional proof of the robustness of these "semi-flexible" photovoltaic modules occured with a crash of the airship LOTTE (see Figure 12). Only two module strips out of 110 pieces had a total failure. Several pieces were broken with a slight decrease in their efficiency (Figure 13). The airship was rebuilt again with the not broken modules.

Figure 12: Parts of solar airship LOTTE after crash of the temporary hangar. The black strips are the individual foil modules on top of the airship (Photo: H.Neupert).

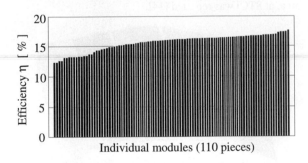

Figure 13: Measurement of all foil modules after the crash of solar airship LOTTE. Initial efficiency values were in the range of 16.5 %.

In the last years new results regarding ultralight photovoltaic modules for unmanned aerial vehicles have been reported. The use of thin 125 µm monocrystalline silicon cells (16 % STC efficiency) together with 51 µm EVA embedding and 23 µm Tefzel® cover sheets lead to an excellent value of only 0.481 kg/m² [9]. Better values have been reported only from aerospace applications with GaInP$_2$/GaAs concentrator cells.

Besides excellent specific power values of crystalline silicon solar cells unfortunately they don't offer the potential of "maximal" flexibility and total integration in the envelope.

4.3 Thin film solar cells

Driven by aerospace applications like power supply for low earth orbit satellites lightweight thin film photovoltaic generators are under development since many years. Thin film solar cell configurations based on amorphous silicon alloys or chalkopyrite films of Cu(Ga,In)Se$_2$. are currently most advanced for these applications.

Amorphous silicon cells have the big advantage, that there are very low temperature deposition processes available. This fact increases the number of possible substrate materials and with that the potential for direct integration in the airship envelope. From laboratory experiments with small scale multi-bandgap, triple junction a-Si alloy solar cells deposited on Kapton or stainless steel (schematic drawing in Figure 14) values were reported in the range of 0.170 kg/m² [10]. Currently available are heavier stainless steel configurations (Figure 15) for leisure applications which show already a quite good flexibility. Unfortunately the area related power value of large area modules (> 0.5 m²) is still very low in the range of 40 to 60 W/m².

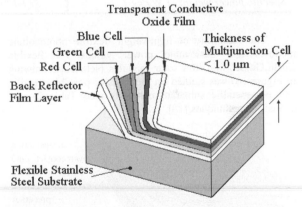

Figure 14: Schematic profile of flexible thin film solar cells based on amorphous silicon in a triple junction configuration on a stainless steel substrate [11].

Figure 15: Flexible amorphous silicon PV modules on a stainless steel substrate (Photo: W. Knaupp).

Table 2:   Material properties of different conductive and non-conductive substrates. Specific power weight is calculated with a substrate thickness of 100 μm and a max. power density of 100 W/m² at STC [12].
*) Polybenzoxazole (not commercially available);   **)   Organic fillers vaporize at 250...280°C

a) *Conductive Substrates* ⇒	Alu-minium	Stainless Steel	Molyb-denum	Nickel Alloys	Titanium	Graphite
Density [g/cm³]	2.7	8	10.2	8.9	4.5	0.7-1.2
Young's modulus [GPa]	70	193-200	320	214	108	?
Tensile strength: Rp0.2 [MPa] Rm [MPa]	35-115 65-135	230 540-750	620 780	150-600 500-1200	270 390-540	- >4
Thermal expansion coefficient [ppm/°C]	23-24.2	15.6-17.3	5.8	1.5-17	8.6	1
Thermal stability: Melting point [°C] Rp0.2 [MPa] after 400°C / 1h	660 10	1,400 220	2,610 600	1,450 100-500	1,668 250	3,000 ?
Specific power weight [W / kg]	*370*	*125*	*98*	*112*	*222*	*830-1,400*

b) *Non-conductive Substrates* ⇒	Glass	Polyimide	PBO *)	Ceramics
Density [g/cm³]	2.5-2.7	1.42	1.5	3.5-7
Young's modulus [GPa]	66-73	2.5	2.7	?
Tensile strength: Rp3 [MPa] Rm [MPa]		69 230	? 100-300	? ?
Thermal expansion coefficient [ppm/°C]	4.5-7.2	20-40	50-60	?
Thermal stability up to [°C]	550...650	≈ 400	≤ 500	**)
Specific power weight [W / kg]	*370-400*	*700*	*670*	*140-285*

The feasibility of lightweight flexible photovoltaic modules and achievable power values for flexible Cu(In,Ga)Se$_2$ thin film solar modules (schematic drawing in Figure 16) was studied on the basis of possible metallic and non-metallic substrate materials (see Table 2) and deposition techniques [12].

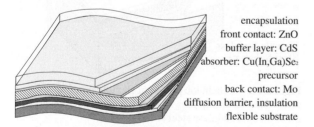

encapsulation
front contact: ZnO
buffer layer: CdS
absorber: Cu(In,Ga)Se₂
precursor
back contact: Mo
diffusion barrier, insulation
flexible substrate

Figure 16:   Schematic profile of flexible thin film solar cells based on Cu(In,Ga)Se$_2$ (Source: ZSW Stuttgart).

In general metallic substrates are on the one hand very interesting due to their thermal stability within the deposition process (see also Table 2) and their low cost potential, on the other hand all conductive subtrates need for realization of modules with interconnected cells an additional insulating layer. Realization of layers with good dielectric strength and adhesion behaviour on partly very rough metal surfaces especially after cell structuring is still a challenge. This applies even more to those conductive substrates like aluminium with a quite different thermal expansion coefficient comparing to Cu(In,Ga)Se$_2$ films (thermal expansion coefficient of CIS/ CIGS ≈ 8 ppm/°C). Recent results are given in [13].

Non conductive substrates offer the advantage of an insulating layer, however the thermal stability in usual high temperature deposition processes is a critical issue. Currently only polyimide substrates offer sufficient thermal stability during the high temperature production process. For small CIGS cells grown on a polyimide sheet (Figure 17) a very promising efficiency of 12.8 % (total area, at STC) was reported [14].

Figure 17:   High efficient single CIGS solar cells grown on a polyimide sheet (Source: Institute of Quantum Electronics, ETH Zürich).

The examinations of possible substrates can be summarized as follows. Under the criterion of "high power per mass" the use of aluminium as metallic substrate would lead to a specific power weight potential of flexible Cu(In,Ga)Se$_2$ photovoltaic modules in the range of 350... 400 W/kg. The use of polyimide in the group of non-metallic substrates increases the potential up to 700 W/kg.

With main emphasis on the criterion of "easy handling during production process" (similar thermal expansion coefficients of absorber and substrate, sufficient material strength at high temperature, ...) other materials like titanium, nickel alloys and thin glass are better suited. Use of titanium offers the potential of 200... 250 W/kg and thin glass of 350...400 W/kg specific power weight. With consideration of solar cell efficiencies an achievable mass per area of 400 g / m² seems to be possible in the future.

4.4 System Engineering

The basic questions of system engineering are at the first sight similar to those at conventional photovoltaic plants. However the fact that the generator is possibly changing its position relative to the sun in every moment according to wind situation makes the optimzation more complex.

Introduction and use of thin film solar cells will allow further optimizations in the field of system engineering due to increased possibilities for realization of complicated geometric module shapes, integrated complex interconnection schemes and due to the flexible choice of current and voltage levels even for small area subgenerators.

5 SUMMARY

Airships are at the entrance to a new era due to well adapted new application areas as well as advanced materials for all components. For most of the applications a reliable, inexhaustible and lightweight energy supply system is one of the key issues. In this respect photovoltaics and especially thin film technologies have a great potential.

The developed models for determination of solar resource and electrical output from photovoltaic modules on the airship envelope provide the basis for further detailed system optimizations.

The feasibility of lightweight flexible photovoltaic modules and achievable power values was studied on the basis of possible conductive and non-conductive substrate materials and deposition techniques for CIGS solar cells. Several metallic and non-metallic substrates are suitable for flexible thin film solar cells with a potential of up to 700 W/kg and 400 g/m².

For full development of solar energy potential as energy source for airships there are further efforts necessary for mass decrease, increase of robustness, flexibility and efficiency not only of the photovoltaic generator but of the whole energy system.

6 ACKNOWLEDGEMENTS

Parts of this work were conducted at ZSW within the Research Group "Airship Technologies" at the University of Stuttgart with support from the DFG, the national research council for academic research in Germany. The author also gratefully acknowledges the excellent cooperation with the institute *ISD* for carrying out the measurement flights.

7 REFERENCES

[1] http://www.skystation.com/

[2] Knaupp, W.; Schäfer, I.; *Solar powered airship -Challenge and chance-*, Conf. Records of the 23rd IEEE PV Specialists Conference, Louisville, 1993 (IEEE, New York, 1993), p. 1314-1319.

[3] Liu, B.Y.H.; Jordan, R.C.; *The interrelationship and characteristic distribution of direct, diffuse and total solar radiation*, Solar Energy Vol.4, 1960, p. 1-19.

[4] Hay, J.E.; Davies, J.A.; *Calculation of the solar radiation incident on an inclined surface*, Proceedings of the First Canadian Solar Radiation Data Workshop, Toronto, 1978, p. 59-72.

[5] Klucher, T.M.; *Evaluation of models to predict insolation on tilted surfaces*, Solar Energy Vol.23, 1979, p. 111-114.

[6] Perez, R.; Ineichen, P.; Seals, R.; Michalsky, J.; Stewart, R.; *Modeling daylight availability and irradiance components from direct and global irradiance*, Solar Energy Volume 44, Nr.5, 1990, p. 271-289.

[7] Wiedemann, B.; *Untersuchungen zum Tragverhalten halbstarrer Luftschiffe*, Proceedings of DGLR Workshop III "Flight Systems Lighter-Than-Air", Bremen, 28.05.1999.

[8] Rehmet, M.A.; Knaupp, W.; *Integration von Solarzellen in die Hülle von Luftschiffen*, ISD-report 94/1, Stuttgart.

[9] Nowlan, M.J.; Maglitta, J.C.; Darkazalli, G.; Lamp, T.; *Ultralight photovoltaic modules for unmanned aerial vehicles*, Conf. Records of the 26th IEEE PV Specialists Conference, Anaheim, 1997 (IEEE New York, 1997), p. 1149-1152.

[10] Guha, S.; Yang, J.; Banerjee, A.; Glatfelter, T.; Vendura Jr., G.J.; Garcia, A.; Kruer, M.; *Amorphous silicon alloy solar cells for space applications.* Proceedings of the 2nd World Conference on Photovoltaic Energy Conversion, Vienna, 1998 (EC JRC, Ispra, 1998), p. 3609-3613.

[11] http://www.ovonic.com/unisolar.html

[12] Herz, K.; *Flexible substrates for deposition of CIS thin film solar cells -Aeronautical applications*, Internal ZSW-Study, Stuttgart, 1997.

[13] Kessler, F.; Herz, K.; Powalla, M.; Baumgärtner, K.-M.; Schulz, A.; Herrero, J.; *CIS thin-film solar cells on metal foils*, 16th European Photovoltaic Solar Energy Conference, Glasgow, May 1-5, 2000, to be published.

[14] Tiwari, A.N.; Krejci, M.; Haug, F.-J.; Zogg, H.; *12.8% efficiency $Cu(In,Ga)Se_2$ solar cell on a flexible polymer sheet*, Progress in Photovoltaics: Research and Applications, Volume 7, Issue 5 (1999), p. 393-397.

THE 480 kW$_P$ EUCLIDES^{TM}-THERMIE POWER PLANT: INSTALLATION, SET-UP AND FIRST RESULTS

G.Sala, I.Antón, J.C. Arboiro
A.Luque
Instituto de Energía Solar
UPM
28040 Madrid – Spain
Ph/Fax:+3491 5441060 / 6341

E.Camblor, E.Mera,
M.Gasson
B.P.Solar Ltd POB 191
Sunbury –on Thames
TW167AX,UK
Phone/Fax 44 1932 76594

M.Cendagorta, P.Valera,
M.P.Friend, J.Monedero,
S.González , F.Dobón
ITER, Pol. Ind Granadilla
San Isidro –Tenerife - SP
Phone/Fax: 34922391000 /1

I. Luque
INSPIRA
C/ Chile 10
28230 Las Rozas
Madrid-Spain
Phone: +34 91 6304534

ABSTRACT: The D.C. section of the 480 kW$_p$ EUCLIDES-THERMIE demonstration plant, the world largest using concentration and the first subsidised by the European Commission, was completed on November 27th, 1998. One inverter, connected to a pair of arrays and injecting power into the grid, was already operative by that time. Currently a monitoring station and seven inverters, one per each pair of arrays are operative.
The EUCLIDES ^{TM} plant consists of 14 arrays, 84 meters long, each with 140 linear parabolic mirrors and 138 receiving modules series connected. The collectors are parabolic troughs with one axis tracking, oriented North/South and parallel to ground. The receiving modules were manufactured by BP Solar Ltd., and the rest of the sub-systems were developed by IES in a previous JOULE IV project. ITER was responsible of the mirror and structure set-up, overall installation, site preparation, inverter fabrication and grid connection.
The objective of this paper is to describe the installation of the whole plant, to present the initial performance and to asses the cost per W$_{peak}$ of this technology. The cost analysis promises that 3.84 ECU/W, all included for grid connection, is achievable at 10 MW/year production.
Keywords: Concentrator –1: Large Grid-connected PV systems –2: Parabolic Trough –3

1. INTRODUCTION AND OBJETIVES

This plant, based on the technology of the EUCLIDES™ concentrator, developed in the JOULE programme, was proposed to and subsidised by the THERMIE programme in 1996. The nominal power of the concentrator plant is 480 kW$_p$, using parabolic troughs and is shared in 14 arrays, each 84 m. long. (Fig. 1, 2 and 3.). The goal of this project was: (a) to demonstrate this technology in a real size plant; (b) to identify the most adequate sub-contractors for developing the tools and manufacturing the components of this technology; (c) to learn with accuracy the costs of the components in order to probe the cost reduction potential of the technology; (d) to asses the components and plant performance. The EUCLIDES™ components are: the concentration cells encapsulated into receiving modules, the passive heat sink, one axis tracking structure, transformer-less inverters and the mirrors.

The energy produced will be sold to the local utility at 0.216 EURO/kW.h, according to the current Spanish regulation.

At present, the project is in the phase of commissioning, and we hope to start the monitoring and continuous production in the coming months.

2. SHORT DESCRIPTION OF THE PLANT AND ITS SUB-SYSTEMS

The RTD JOULE project named EUCLIDES^{TM} led to the development of a full technology of PV concentration in 1994-95. A prototype 24 meters long was manufactured, and its performance was tested since September 1995 in Madrid. The efficiency obtained was 14.42 % (at 800 W/m^2 and 25ºC) and 10.95 % at yearly averaged operating conditions in Madrid [1-4].

Figure 1: Aerial view of the EUCLIDES Plant installed in the grounds of ITER in Tenerife (Canary Islands, Spain)

Figure 2: View of an array, with the tracking control box

Figure 3. A partial view of the PV field during initial operation set-up.

Figure 4: The seven TEIDE inverters for the fourteen arrays.

Based on this technology, the Instituto Tecnológico y de Energías Renovables (Tenerife, Spain), the Instituto de Energía Solar (Madrid, Spain) and BP Solar Ltd. (Sunbury on Thames, UK) have carried out the project of the world largest PV concentration grid-connected power plant: the EUCLIDES-THERMIE plant. It has been installed in the south of Tenerife (Canary Islands) in the grounds of ITER. This Institute co-ordinates the project and will own and monitor the plant. BP SOLAR acts in this project as supplier of the whole plant and is recipient of the EUCLIDES™ technology.

Figure 5: Close view of the wheel and the linear tracking mechanism.

The arrays, with 250 m² collector aperture, are North/South oriented and close to the ground. The geometric concentration ratio is X38.2, 1.2 times the one in the prototype (However only 9% of light collection increase can be expected by theoretical reasons). The mirror technology is based on metallic reflective sheets shaped with ribs to the optimum parabolic profile. Three different materials have been tested to be used as reflective material. The fully encapsulated receiving modules are made of 10 concentration LGBG BP Solar cells, also series connected. The modules are cooled with a passive heat sink.. Every two contiguous arrays are connected, in parallel, to one inverter sized 68 KVA. The output voltage at standard operating conditions is 750 Volts. The inverter, without intermediate transformer, was designed and manufactured by ITER [5].

3. OVERVIEW OF THE INSTALATION PROCESS AND LEASONS LEARN

3.1. The structure design was improved, simplified, with respect to that of the prototype and checked to withstands normal and extreme wind loads. [6]. It was built by JUPASA (Toledo, Spain) that arranged the elements to be shipped from the port of Valencia to Santa Cruz de Tenerife. Once arrived to the site all the beam elements were assembled, some steel reinforcements welded. Then sandblasted and painted. Then the beams were mounted on the previously installed central table were is located the driving mechanism.(Fig.5)

The operation took as much as 4 days to install the first but the 2 last arrays last were erected in 1 day each. All the work was carried out just with a truck provided with a 6 Tons crane. In spite of the weight and size of the steel array tracking structure (160Kg/m), the alignment of the axis, as measured at the central table, is better than 0.0003 radians, that is sufficient for correct operation and it was achieved using simple optical devices. The maximum bow was 12 cm. as predicted by the theory in value and location.

The effect of stationary wind (up to 30 Km/h) on the power output has not been observed, probably because other mismatch effects are currently larger than the wind one. The driving mechanism, a new version developed for this project, converts a powerful (200 kNewton) but smooth linear movement into a circular one by means of a pair of automatically equilibrated steel cables. Although the array structure of the prototype EUCLIDES (Madrid) must be deployed horizontally, the current design permits to tilt the axis up to 12 degrees. The tilt of the arrays in

Tenerife progress from 0.5 degrees at the most eastern to 5.5 degrees at the most western.

3.2 *The tracking control system*, fabricated by INSPIRA, adopting IES designed tracking strategies [7], was operative immediately after its connection, because it was previously checked at the Madrid prototype. The control unit has been operating in a continuous way in the last 8 months with a maximum pointing fixed to 0.2 degrees, but can be reduced to 0.05°, if required. The tracking control unit generates along all the system life a table of small angle values that corrects the theoretical (astronomical) position to a closer one providing more output power. From those small correcting angle values (shown in Fig 6) the errors in the calibration constants or a deviation of the array axis with respect to the N/S line can be known.

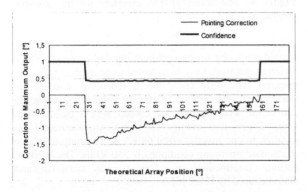

Fig 6 : Correction to calculated tracking angles to get maximum power output.

3.3. The Receiving Module

A second generation of receiving modules were developed for the EUCLIDES plant based on the experience of the Madrid Prototype modules that led 17 % efficiency at Standard Test Conditions. The new cells are larger, 11, 6 cm long and only ten were included in each module. The current capability of the by-pass diode was adapted to the larger cells and mirrors. Before the fabrication was initiated the cells and module interconnects were modified to decrease the series resistance from 13 to 10 mohm, and the sheet thermal resistance was reduced to 3-4 °C.cm^2/W between the cell and aluminium substrate.

There was also some time before the fabrication to check the quality of 10 modules for this new application, testing the module performance before and after several stressing processes. Humidity and temperature cycling, hi-pot at 2600V, normal operation and hot spot conditions under natural concentrated light during 800 hours. The fabrication was initiated recording the 1 sun I-V curve , the I$_{sc}$ at 27,5 suns with flash lamps, and predicting the performance at concentration with an I-V curve recorded at dark conditions [8]. The rejection limit was fixed at 215.64 W at STC. Fig 7 shows the distribution of the fabricated modules. The connectors of the module were greatly improved respect to the first design, but still more development is required on this subject, not only for reliability but for easy field receiving mounting & dismounting.

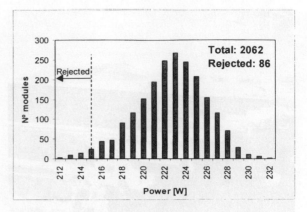

Figure 7: Power histogram of receiving modules at STC.

The modules were glued to heat sinks with a "thermal" acrylic adhesive that cures at room temperature. The module heat sink assembly were installed on the array and connected before the mirrors were. To avoid high voltage risks at the connection the modules were joined first in groups of 8 (about 40 volts)

3.4 The Mirrors

The shaping of the mirrors at the site longs several months because the cure time required for adhesive was 40 minutes at high temperature. Now, fast room temperature adhesives have been tested reducing the time by 5 times.The yield of this process was very high (99,5%), demonstrating that it can be done at the site, by not specially skilled personnel.

The reflective materials used in this plant are not optimised, because they were not previously tested for long outdoor exposure. The assumption of that risk was forced by the unavailability of the film 3M ECP305, the one tested during 2 years in Madrid and previously selected for the project. The mirrors cannot be installed as they were shaped, The storage caused some damage at the sealing film at the mirror edges. If the array was already tracking, the mirrors were aligned to focus during the mounting process. The trimming process was visual, locating the spot line on the receiver. Later, we verified that such visual method yield Isc values close to the maximum available current.

The installation of the mirrors on the arrays was a fast and simple task that consumed only 6 ManHours

The cleaning of the mirrors and receivers is a task that can be carried out in a simple way with workers located on the ground floor, thanks to the EUCLIDES structure concept.

3.5. The inverters

The solution adopted for the DC/AC conversion is not the most conventional for power plants, one big inverter, but for the sake of modularity, every EUCLIDES™ concentrator unit were delivered provided with its own inverter, Although such an approach guarantees the maximum modularity of the system and also gives very good reliability (the failure of an inverter affects only one unit), economical considerations and marketing advised that two arrays should share the same inverter. Modularity is then kept at the level of 68kWp. Therefore, this EUCLIDES™ plant has seven modular inverters 68 kWp, for every two arrays (Fig.4). The seven inverters are connected in parallel to the primary of the medium voltage transformer (380V/20KV). Avoiding the intermediate transformer saves around 4% of the overall energy of the plant. In the following Table I and Fig 8 the technical specifications of the inverters are shown.

Table I. Technical characteristics of the TEIDE inverter

Manufacturer	ITER
Rating	68.5 kWp x 7
Input (min)	600 V
Output	380 V
Eff. at rated power	97%
Eff. at 10% of rated power	95%
Total harmonic distortion	<1%

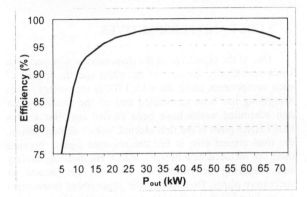

Figure 8: Efficiency curve of the inverters

Every inverter has the possibility to regulate its total, active and reactive power output as well as some other operating parameters, which can be either controlled externally, over the field bus, or by an internal programmable strategy based on MPP tracking algorithms.

3.6. Central Control Weather and Monitoring Station

It stands nearby to the plant and is used to monitor external magnitudes as outside temperature, wind speed and direction, direct solar radiation, global radiation, modules temperature, etc. The electrical parameters of the plant are also recorded at the inverters input and output, located in the control building with the 380V/20kV grid transformers (Fig. 9).

Fig. 9: Control building and visitors centre with the meteorological station at the top.

The magnitudes will be recorded and distributed following the recommendations of ISPRA (CODIGO DEL PAPEL), but some of them are specifically defined for this project because the lack of standards for concentration PV systems.

4. COMPONENTS AND PLANT PERFORMANCE

As the plant is still in commissioning,, there are not yet recorded continuous reliable operational data but several experiences carried out can inform about the initial performance and the trimming still required to optimise the output. Up to date, we can report that:

The structure withstood winds up to 90 Km/h twice last year without any sensible damage (just 20 mirrors were partially unglued from its ribs). The reliability of the overall tracking system has been demonstrated after 6 moths of continuous operation. The windy and salty conditions at the site reduce the optical efficiency 20 % per week, due to deposits on the mirrors and receivers, much if compared with the 2-3 % in Madrid.

The measured optical efficiencies are 85 % for one mirror, and 77 %, too low, for an array. The early array IV curves show a mismatch effect along the array that seems connected with the low optical efficiency. The problem is being investigated (Fig. 10).

Several arrays have been consecutively connected to the inverters allowing injection of power to the grid up to 17.2 kW at 850 W/cm2. This figure is justified by the low optical efficiency and a higher than expected operation cell temperature. After a few days under operation, several modules lost insulation, grounding the active cell line of its array. It prevented the power injection of the array to the grid and activated the safety alarm. This kind of failure has later occurred to 3 % of the modules (i.e. 58 receivers) but spread in all the arrays. This failure is quite surprising because the module insulation was tested, as said above, at 2600V. and the previous development laboratory tests reach more than 4000 V. Unhappily the identification and the substitution of the grounded modules is a difficult and very time consuming task.. BP Solarex will manufacture additional modules to substitute those presenting leakage in the plant once the failure reasons are identified. Then the the arrays will inject power again across the 7 inverters already installed.

During the off-grid periods we measured several figures of merit of the plant and the effect of defects and trimmings. The parameters investigated were: the electrical characteristics of arrays, the optical performance of the overall system, the thermal parameters in load, the optical aperture, the methods for mirror alignment, the mirror cleaning requirements, the possible components degradation, etc.

The effects of the array optics on the power output and Isc of the EUCLIDES system is presented in detail in a paper in this Conference to which you are addressed [9].

The whole IV curve of an array has been measured with a transient load developed for the EUCLIDES array voltage, close to 1000 V. The curves obtained in one array is the one shown in Fig 10 The curve shows a significant tilt (false shunt resistance) due mainly to optical mismatch along the array caused by misalignment of the mirrors, excessive gap between them, dispersion of mirror quality and of module performance. (However the modules were selected to be installed in series with very low electrical mismatch, within +/- 1%).

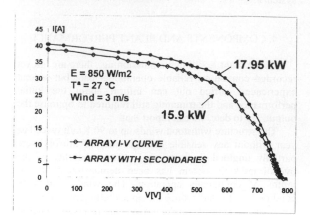

Figure 10: IV curve of an array with 138 modules in series.

To reduce the loses due to a the possible misalignment we installed in one array a secondary concentrator to increase the power output.(Fig. 11) The second IV curve in Fig. 10 demonstrated an increase in power output of 11 %, This amount was due to a major collection of light, valued as 3 % in tests of individual modules, and the rest, about 8 % must be attributed to mirror misalignment. The rest up to 21 % will be caused by mirror discontinuities or errors running parallel to the array axis, and not affected by the secondary. The current overall efficiency is 8.4%.

The loses can be understood with Fig 12. where the array and module currents versus the sun misalignment are plotted. The figure shows that the angle of aperture of the array regarding the short circuit current is much wider than the aperture regarding the maximum power output (point B) and also that the available Isc for the maximum power is only 83 % the maximum Isc of the array. This effect due to small misalignements of mirrors between themselves, requires a tracking accuracy of +/- 0.05, while the system is now operating at +/-0,2 degrees.

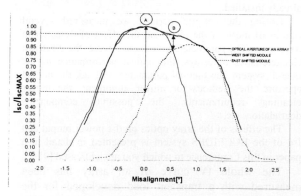

Figure 12: Angular aperture of an array in comparison module with the aperture of the most shifted modules.

This demonstration plant allows for the first time to study the real problems associated to pre-industrial concentrators, because all previous works were carried out at laboratory or prototype level.

Figure 11: Array equipped with V-secondary concentrators

5.COST AND PERSPECTIVES

One of the objectives of the demonstration project is to know realistically the cost of the plant and the impact of each components using the EUCLIDES technology, once everything has been terminated and all the necessary but non scheduled works have been carried out. But as this EUCLIDES plant is the first erected, one of demonstration, the total project cost is not the required figure, because there are included new designs, R&D, engineering, tooling and managing activities that will not be present in successive plants. Discounting the expenses of those items it was found that the Toledo demonstration plat cost was 7.66 EURO/W$_p$ (or 7.34 \$/W$_p$) and that this EUCLIDES-Tenerife plant has cost 5.14 EURO/Wp (or 4.93 \$/Wp), including in both cases the inverter and the grid connection. We guess that the cost reduction is significant, taking into account that in Toledo the know how of flat module installation was already acquired, but in EUCLIDES everything was new. In consequence the objective of 3,84 EURO/Wp (or 4.03 \$/Wp) for a production of 10 Mw/year seems very possible. In Table II we present the cost of every subsystem and other associated costs and forecast.

6. CONCLUSIONS

The EUCLIDES array concept, as one close to ground, is advantageous regarding the wind loads, the mounting requirements, the cleaning and the maintenance. Also it has been proved to be a singular solution because the BP Solar concentrator cells are the only competitive today. The industrialisation of the components has required very small investments in tooling (in the range of 800000 EURO), far from the usual investments on other promising cost reduction PV systems (like thin films). On site manufacturing of structure, mirror shaping and receiver assembly has been demonstrated.

Regarding the dissemination of this technology is advisable to fix standard rules for qualification, specification and rating of PV concentrators.

Regarding the initial problems found in some modules operation at reduced voltage will be checked in the next months. Investigation of the causes of optical mismatch must be issued. The use of secondaries as already tested in one array on alternative series/parallel connections inside the modules are immediate alternatives. The continuous

Table II: Cost of the EUCLIDES-THERMIE power plant (discounted tooling, engineering, and management cost)

Sub system	Cost (EURO)	Cost Wpeak (EURO)	Cost Wpeak ($)	Projection (EURO)
STRUCTURE & TRACKING	703845	1,466	1,404	1,10
MIRRORS	252425	0,526	0,504	0,45
HEAT SINKS	159395	0,332	0,318	0,33
RECEIVING MODULES	390658	0,814	0,779	0,81
Total Sub-Systems	1506323	3,138	3,005	2,69
MATERIALS & SUBCONTRACTORS	405310	0,844	0,809	0,80
TRANSPORTATION	93145	0,194	0,186	0.15
ON SITE MANUF & INSTALLATION	128708	0,268	0,257	0,20
TOTAL PLANT	2133486	4,445	4,256	3,84

monitoring along two years will provide the required information to reach maturity of this technology. According to the project experience and the manufacturing engineers, we expect that the most significant cost reductions will come from the structure-tracking and also from the operating efficiency. The cost reduction of the EUCLIDES technology has been demonstrated in comparison with previous flat panel grid connected plants. Analysis of each component cost shows that power plants in the range of 3.5 to 4.0 EURO/Wp at 10 MW/year production are achievable in short term with this technology.

REFERENCES

[1] G. Sala, J.C. Arboiro, A. Luque, J.C. Zamorano, J.C. Miñano, C. Dramsch, T. Bruton, D. Cunningham. "The EUCLIDES prototype: An efficient parabolic trough for PV concentration". Proc. 25th IEEE Photovoltaic Specialist Conference, Washington D.C (1996) 1207 – 1210.

[2] G. Sala, J.C. Arboiro, R. García, A. Luque, T. Bruton, D. Cunninghan. "Description and performance of the EUCLIDES concentrator prototype". Proc. 14th European Photovoltaic Solar Energy Conference, Barcelona (1997) 352-355.

[3] A. Luque, G. Sala, J.C. Arboiro. "Some results of the EUCLIDES photovoltaic concentrator prototype" Progress in Photovoltaic **5, 3** (1997) 121-195

[4] A. Luque, G. Sala, J.C. Arboiro. "Electric and thermal model for non-uniformly illuminated concentration cells", Solar Energy Materials and Solar Cells 51 (1998) 269-290

[5] G. Sala, J.C. Arboiro, A. Luque, I.Antón, M. Gasson, N. Mason, K. Heasman. 480 kWpeak Concentrator Power Plant Using the EUCLIDES[TM] Parabolic Trough Technology, Proceedings. 2nd WCPVSEC, Vienna, (1998).

[6] J.C. Arboiro, G. Sala, I.Molina, L. Hernando, E. Camblor, The EUCLIDES[TM] Concentrator: A Lightweight 84m Long Structure for Sub-Degree Tracking. 2nd WCPVSEC, Vienna (1998).

[7] J.C. Arboiro et al. "Self-Learning Tracking: a New Control Startegy for PV Concentrators" Progress in Photovoltaics **5** (1997) 213-226

[8] G. Sala, I. Antón, J.C. Arboiro, A. Flouquet, K.C. Heasman, N.B. Mason, M.P. Gasson, Acceptance Test For Concentrator Cells And Modules Based On Dark I-V Characteristics, 2nd WCPVSEC, Vienna (1998)

[9] I. Antón, G. Sala , J.C. Arboiro, J.Monedero and P.Valera . "Effect of the Optical Performance on the Output Power of the Euclides[tm] Array" 16th European PVSEC, Glasgow (2000)

The project has been subsided in the framework of the JOULE-THERMIE IV Programme.

PROGRESS AND PERFORMANCE OF NEW ENCAPSULATING MATERIALS

A. K. Plessing*[1], P. Pertl[1], S. Degiampietro[1], H.-C. Langowski[2], U. Moosheimer[2]

1)ISOVOLTA AG, Austria

Vianovastrasse 20, A-8402 Werndorf, Tel.: +43 3135 54314-0, Fax: +43 3135 54314-82

eMail: plessing.a@isovolta .com

*to whom correspondence should be addressed

2) Fraunhofer Institut Verfahrenstechnik und Verpackung, Germany

Giggenhauserstrasse 35, D-85354 Freising, Tel.: +49 8161 491 500

eMail: langowski@ilv.fhg.de

ABSTRACT: Today the encapsulation of solar cells is made by glass on the front side and different types of laminated film materials e.g. PVF-PET-PVF; PVF-Al-PVF; PVF-Al-PET-PVF for backside protection. In the centre of the PV-module EVA is used for shock absorption.

This paper is concentrating on new encapsulating materials and systems, which include new, high performing, dielectric barrier layers for front and backside application. Different substrate-barrier systems and their performance are described. Outstanding features of the new materials are very low moisture permeability, UV-protection and promotion of adhesion. The performance of those new materials meets long term stability of PV-modules and overcomes electrical insulating problems. Cost saving by cost effective components and process, weight saving, increased productivity and a variety of new designs and applications make the new materials and systems excellent candidates for new generation PV-modules.

Keywords: Encapsulation – 1: Moisture Barrier – 2: Reliability - 3

1. INTRODUCTION

The typical construction of today's PV-modules is a multilayer encapsulation system using polymer films and glass to protect the solar cells against environmental impacts.

In Figure 1 for backside protection, a composite material consisting of polyvinylfluoride (PVF) and polyethyleneterephtalate (PET) and aluminum (Al) and polyvinylfluoride (PVF) is used, for the frontside glass and for the embedding of the solar cells and as shock absorber ethylenevinylacetate (EVA) is the central encapsulant.

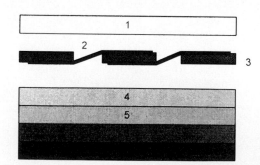

1: glass; 2: ethylenevinylacetate (EVA);
3: solar cells; 4: polivinyl fluoride (PVF);
5:polyethyleneterephtalate (PET);
6: aluminum (Al); 7: polyvinylfluoride (PVF)
Figure 1: Typical construction of a PV-module

In this example the solar cells are excellently protected against moisture by using the Al foil at the backside and the glass at the front side. The disadvantage of the Al foil is the electrical conductivity, which could create short circuits and a capacitor effect. The glass on the front side has excellent protective characteristics but it is heavy in weight [1].

This paper describes new encapsulating materials for front and backside, which include dielectric oxide layers with a high barrier against water vapour [2]. The composition of different materials is presented and their performance is demonstrated. To improve the UV protection the barrier layers for front side application can be adjusted. New applications by means of the new encapsulating materials are shown.

2. EXPERIMENTAL WORK

2.1 Barrier Layers

For barrier layers against water vapour stoichiometric aluminum oxide and non stoichiometric silicon oxide had been evaluated. The barrier layers had been produced by means of a vacuum web coater. The coating material had been evaporated by electron beam from a water cooled crucible and coated onto different polymer films.

The thickness of the barrier layers had been varied from 40nm to 400nm by changing the web speed of the roll to roll coater.

There are similar products published [3], [4], but the disadvantage of the detailed processes are on the one hand the cost of CVD-process with low productivity and on the other hand the direct application of the barrier coating onto the solar cells. This direct coating always results in some defects of the barrier layer, which reduces long term stability. Using polymer substrate films [5], but still in a CVD-process, improves the quality but not the efficiency. Therefore the described PVD-process in this paper for the coating of barrier layers on polymer substrate films results in an efficient, low cost and highly performing process for the production of encapsulating materials.

Substrate films had been selected according to mechanical stability, wheatherability, low cost and longterm stability.

The most interesting polymer substrate films are polyethyleneterephtalate (PET), polyethylenenaphtenate (PEN) and ethylenefluoroethylenecopolymer (ETFE). The significant features of the polymer films can be seen from Table 1.

Substrate Film	Characteristic
PET	cost effective, commonly used
PEN	improved barrier, costly
ETFE	highly transparent, costly

Table 1: Substrate films for barrier layers

The stability of the barrier layers on different substrate films had been investigated by aging in damp heat climate of 85°C and 85% r.h. The test specimen had been prepared by vacuum lamination of the substrate film with the barrier layer and EVA and glass or of the substrate film with the barrier layer and EVA and a composite of primer-PET-Al-PVF. The qualification criterion for the stability had been no delamination, no bubbles occured in the specimen and high adhesion strength between the layers.

2.2 Encapsulating Materials

For the investigation of encapsulating materials the substrate-barrier systems PET-SiOx, PEN-SiOx and ETFE-SiOx had been used.

For front side encapsulation the highly transparent ETFE-SiOx, using a 50μ thick ETFE-film, had been directly laminated to EVA.

For backside encapsulation laminated samples had been produced consisting of PVF-SiOx-PET-primer and PVF-PET-PEN-SiOx. From these composite materials test specimen had been prepared with EVA and glass or a composite of primer-PET-Al-PVF. These specimen had been aged in damp heat climate of 85°C and 85% r.h..

The classification of the test results had been made according to occuring delamination and bubbles.

2.3 Test Modules

The different encapsulating materials, including SiOx-layers, had been used for the production of test modules as can be seen from table 2.

Front side Encapsulant	Backside Encapsulant
ETFE-SiOx	glass
ETFE-SiOx	SiOx-PET-PVF
glass	Primer PET-SiOx-PVF
glass	SiOx-PET-PVF
glass	SiOx-PEN-PVF

Table 2: Encapsulants for test modules

The test modules had been aged in damp heat climate, had been tested in thermal cycles from 25°C to 80°C and in high voltage tests.

The criterion for the module performance had been no delamination, no bubbles, no discolouration and low leakage current.

3. RESULTS

3.1 Barrier Layers

Polymer films made of aromatic polyester or polymethylmethacrylate could not be used because of their amorphous structure. The coating of polyvinylflouride did not succeed in higher barrier values. Therefore the most interesting substrate films are PET, PEN and ETFE.

It was found that SiOx layers result in the higher moisture barrier compared to Al_2O_3. Therefore SiOx layers had been favoured for further work. Very low water vapour permeability of $0,1g/m^2d$ could be achieved by SiOx-layers coated on PET, PEN and ETFE. But the most important question was the long term stability in damp heat climate of 85°C and 85% r. h. The best laboratory results on moisture barrier and climatic stability can be seen from Table 3.

With the water vapour permeability of 0,6g/m²d for ETFE-SiOx, 0,5g/m²d for PET-SiOx and 0,1g/m²d for PEN-SiOx climatic stabilities of 2000 hours could be achieved at damp heat climate of 85°C/85% r.h.

Sample	Damp heat stability [hours] at 85°C/85% r.h.	Water vapour permeability [g/m²d] at 23°C/85% r. h.
ETFE-SiOx	2000 hr	0,6
PET-SiOx	2000 hr	0,5
PEN-SiOx	2000 hr	0,1

Table 3: Laboratory scaled performance of barrier layers

For industrial application laboratory samples do not represent the real potential of new materials. Therefore pilot scaled trials of SiOx-coating on different polymer substrates had been made. It was found, that the results are very close to the laboratory results, as can be seen from Table 4.

Sample	Damp heat stability [hours] at 85°C/85% r.h.	Water vapour permeability [g/m²d] at 23°C/85% r.h.
ETFE-SiOx	2000	0,6
PET-SiOx	2000	0,5
PEN-SiOx	1750	0,15

Table 4: Pilot scaled performance of barrier layers

For ETFE-SiOx the climatic stability of 2000 hours could be achieved at a water vapour permeability of 0,6g/m²d, for PET-SiOx the same stability of 2000 hours at 0,5g/m²d of water vapour permeability had been achieved and for PEN-SiOx a stability of 1750 hours in damp heat climate had been

achieved with a water vapour permeability of 0,15g/m²d.

The adhesion strength of the test laminates was only reduced by less than 20 % after 2000 hours of aging. The final adhesion values of 30 N/cm to 45 N/cm are still on a very high level. The results are summarised in Figure 2, which shows the adhesion strength after aging in damp heat climate.

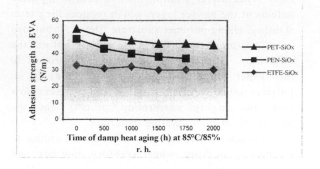

Figure 2: Adhesion strength after aging in damp heat climate

3.2 Encapsulating Materials

For the protection of solar cells it is not only necessary to achieve good protection against water vapour but also electrical insulation, mechanical protection and weatherability are essential requirements. For front side application the mechanically and electrically strong ETFE-film is considered to be the right encapsulant. For backside application composite materials had been produced, consisting of polyvinylfluoride film for weatherability, PET-film for electrical insulation and mechanical stability and barrier layers of SiOx on PET- and PEN-films. The performance of these materials had been tested in the water vapour permeability and in the stability in damp heat climate of 85°C/85% r.h. As can be seen from Table 5, the climatic stability achieved 1250 hours up to 2000 hours.

Today´s results demonstrate that the SiOx-barrier layer between PET and PVF film achieve significantly higher climatic stability compared to PEN-SiOx-layers, where the SiOx is directly laminated onto the EVA.

The light transmission of ETFE-SiOx is still on a very high level of more than 92% as can be seen from Figure 3.

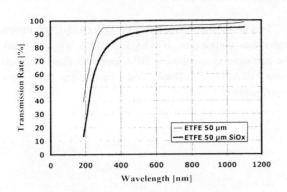

Figure 3: Light transmission of ETFE-SiOx

By reducing the oxygen content of SiOx the light transmission can be reduced and consequently the transmission of UV light can be reduced. This is a possible solution to improve UV-stability. The optimum ratio of silicon to oxygen for front side application still has to be worked out, which will result in an high light transmission for visible light and consequently in an high efficiency of the solar cells and in a good UV-protection of the inner encapsulation layers.

Encapsulating Material	Laminated to	Initial WVP [g/m²d] at 23°C/85% r.h.	Dampheat stab. [hrs] at 85°C /85% r.h.
ETFE-SiOx	EVA-glass	0.5	2000
ETFE-SiOx	EVA-primer-PET-PVF	0.5	2000
PVF-SiOx-PET-primer	EVA-glass	0.5	2000
PVF-SiOx-PET-primer	EVA-primer-PET-PVF	0.5	2000
PVF-PET-PEN-SiOx	EVA-glass	0,15	1250
PVF-PET-PEN-SiOx	EVA-primer-PET-PVF	0,15	1750

Table 5: Performance of encapsulating materials

With these very encouraging results the next step of producing test modules started.

3.3 Test Modules

With different new encapsulating materials, including barrier layers of SiOx, test modules had been produced and tested in damp heat climate, thermal cycles and high voltage test. The tested encapsulating systems can be seen from Table 2.

The central encapsulant had been standard curing EVA.

The results for the test modules consisting of ETFE-SiOx or glass on the frontside and SiOx-PET-PVF on the backside did not succeed in climatic stability. Further optimisation work will be done for the better understanding of the failing.

The best results could be obtained with ETFE-SiOx on the frontside and glass on the backside, with primer-PET-SiOx-PVF on the backside and glass on the frontside as can be seen from table 6.

Front side encapsulant	Back side encapsulant	Material shrinkage	Discolouration	Climatic stability	High voltage	Leakage
ETFE-SiOx	glass	-	no	+	+	+
glass	primer-PET-SiOx-PVF	+	slight	+	+	+
glass	SiOx-PEN-PET-PVF	+	yes	+	+	+

Table 6: Performance of test modules

It seems that for the frontside ETFE-SiOx and for the backside primer-PET-SiOx-PVF will be the first candidates for commercialisation.

The climatic tests included thermal cycles from −25°C to 80°C, 50 cycles and 200 hours in hot water of 90°C and simultanously 200 hours at 90°C/100%r.h.. The high voltage test had been made at 6000 V for 1 minute. Leakage current should be lower than 5µA. Material behaviour and discolouration had been evaluated by visual inspection.

Continued optimisation of the materials resulted in excellent climatic stability of ETFE-SiOx and of primer-PET-SiOx-PVF.

4. NEW APPLICATIONS

Using ETFE-SiOx for front side encapsulation as an alternative to glass, it is possible to produce very light weight PV-modules. The mechanical stability then has to be created by rigid backside material, which can be done by thick layers, e.g. 350 μm, of PET in the backside encapsulant.

For standard type PV-modules the improved protection against moisture means increased liability of PV-modules in damp heat climate, without using composites including aluminum, which creates electric insulating problems. These new types of encapsulants have the chance for successful encapsulation of sensible thin film solar cells.

5. CONCLUSION

New, dielectric moisture barrier layers, based on silicon oxide, can significantly improve the protection of solar cells. These barrier layers offer high barriers against water vapour, but without creating electrical insulating problems.

For front side application ETFE-SiOx with a light transmittance of 92% is of great interest, which offers the chance for very light weight PV-modules without glass.

Although the climatic stability of all composite materials had been on a very high level, the stability of the test modules did not always follow the same way. Further investigations have to find out the reasons of failing and the chances for improvement.

For backside application the composite material primer-PET-SiOx-PVF shows a very high level of performance, which should encourage commercialisation.
For future application the described new encapsulants have the possibility for the successful encapsulation of sensible thin film solar cells.

REFERENCES

[1] A.K. Plessing, S. Degiampietro, P. Pertl, "Laminated film material for solar cell encapsulation and their influence on PV-module production and development "Proc. 2nd World Conference on PV, Vienna Austria 1998, published by EC, JRC, Ispra Italy 1998, p 1915

[2] U. Moosheimer, H.-C. Langowski, C. Bichler, "New cost efficient components for the encapsualtion of photo-voltaic modules", Proc. 2nd World Conference on PV, Vienna Austria 1998, published by EC, JRC, Ispra Italy 1998, p 2272

[3] W. Stetter, V. Probst, H. Calwer, "Laminated structure which is stable with respect to climate and corrosion ", PCT Patent application WO 97/96334, 1997

[4] COCOSOL, JOR3-CT97-0140

[5] S. Okuno, "Transparent composite film", Japan, Patent application No. Hei-10(1998)-25357.

ACKNOWLEDGEMENT

This work was funded by the European Commission under the contract No. JOR3-CT97-0155. We also thank the participating companies BP Chemicals PlasTec, Shell Solar Energy, Fabrimex and Eurosolare for their contributions.

MONOLITHIC MODULES INCORPORATING STRING RIBBON SILICON SOLAR CELLS WITH WRAPAROUND CONTACTS

A. M. Gabor, D. L. Hutton, and J. I. Hanoka
Evergreen Solar Inc.
211 Second Ave, Waltham, MA 02451, USA
email: *gabor@evergreensolar.com*

ABSTRACT: Building on technology developments in the areas of continuous Si ribbon growth (String Ribbon) and polymeric materials for PV modules, a method is described for forming wraparound contacts on these wafers and then interconnecting the cells in a monolithic fashion. A significant reduction in processing steps and cost can then be realized, and thinner Si ribbon can be processed with higher yields. Modules with 84 cm^2 and 120 cm^2 String Ribbon wraparound cells and printed interconnect patterns were formed using this monolithic module procedure. Accelerated reliability testing of these modules is encouraging.

Keywords: Module Manufacturing: - 1: Ribbon Silicon - 2: Qualification and Testing - 3

1. CONVENTIONAL VS MONOLITHIC INTERCONNECT STRUCTURES

In an effort to lower costs in all steps of manufacturing photovoltaic modules, Evergreen Solar has developed a method for the continuous production of silicon ribbon directly from the melt [1] and novel polymeric materials to replace the conventional materials now used as an encapsulant [2] and as a backskin [3]. In this work, a method to form wraparound contacts on String Ribbon substrates has been combined with some of the unique properties of the backskin material to allow for a low cost module assembly process.

The conventional technique used by industry to interconnect single crystal or multicrystalline Si solar cells involves the soldering of flat wires from busbars on the front of the cells to contacts on the backs of the cells (see Fig. 1a).

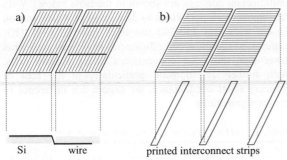

Figure 1: Cell interconnection techniques for a) conventional silicon wafer cells, b) Evergreen Solar's monolithic interconnects for silicon wafers.

An example of conventional module assembly production flow is shown in Figure 2.

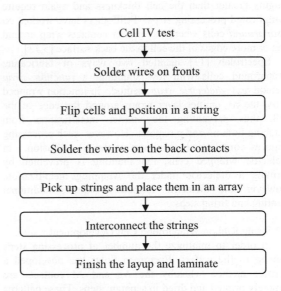

Figure 2: Module assembly production flow using the standard interconnect structure.

This technique has several disadvantages:
- The cells experience excessive handling and thermal shock which can lead to yield loss in terms of cell breakage.
- The handling of already soldered strings is difficult and can lead to broken solder bonds.
- The equipment used to automate these processes is expensive and difficult to maintain, or alternatively, the labor costs for unautomated processes are high.
- Front busbars block light and are considered aesthetically unattractive for some architectural applications.

Instead, we would prefer a simpler process with fewer cell handling steps. As an alternative, some PV technologies use processes that create so-called *monolithic* interconnects. This term implies that the connections between cells form as a consequence of processing steps applied to the whole module structure as opposed to the case here where cells are connected one by one and then are placed onto the module structure. In this paper we present a type of simple monolithic interconnect structure (see Fig. 1b) wherein all interconnections between cells occur on the back side of the cells.

2. WRAPAROUND CELLS

2.1 Background

Several groups have worked on back contact silicon cells. Many have used *wrapthrough* structures where an array of holes are laser cut, sawed, or etched through the wafer, and later the emitter and/or front contacts are wrapped through the holes to the back side. [4-10]. However, such processes do not match our low-cost goals since they involve a large number of processing steps, and since we wish to completely eliminate any silicon etching. Other groups have worked on back junction cells [11,12], but such cells require silicon with minority carrier diffusion lengths greater than the cell thickness and again require complicated processing steps. Still others have worked on *wraparound* cells where the emitter contacts wrap around one or more edges of the cell to the back surface [5,13].

Spectrolab [13] detailed two ways of fabricating wraparound cells with low shunting: a *junction wrap* method and *dielectric wrap* method. In junction wrapped cells, the n+ emitter layer wraps around the edge of the cell, thus preventing the emitter metallization from touching both n+ and p regions. However, such processing requires complicated patterning of the n+ diffusion. In dielectric wrapped cells, the shunting is prevented by forming a dielectric under the wrapping metallization. However, the dielectric formation requires additional printing and firing steps.

2.2 String Ribbon Wraparound Cell Development

In order to minimize the number of processing steps applied to the wafers, Evergreen Solar has developed a transfer method whereby the Ag and Al contacts are remotely printed and dried in separate steps. These patterns are then applied to the wafers in a gentle manner with low yield loss, even with thin wafers. In addition, any yield or quality problems with the printing steps are effectively decoupled from the cell yield, and large screens may be used for exceptional throughput. This indirect printing method is well suited to wraparound cell fabrication, since the dried Ag emitter metallization paste is somewhat flexible in its green state.

The shape of the edge was found to be important. In the processing of String Ribbon silicon, the ribbon is drawn from the melt in widths of either 5.6 or 8.0 cm. The ribbon is cut with a laser into wafers 15 cm in length. We found that the long string edges of the ribbon were naturally rounded, and that by adjusting the string material parameters and the Si growth parameters we could further control the edge shape to achieve good quality wrapped Ag fingers.

Finally, the Ag drying parameters and paste composition affect the flexibility of the fingers since the plasticity of the organic vehicle in the paste is controlled by how much solvent/plasticizer remains after the drying step.

By optimizing these parameters, cells were fabricated with low series conductivity loss due to the wrapped fingers.

To optimize the front grid design, we entered the commonly used iterative power loss equations [14] into a spreadsheet. For this exercise we used a grid design where the fingers wrapped around only one edge, as is discussed below in the module section. We considered both tapered fingers and fingers with a constant width of 150 μm. We empirically tested the theoretical optimization by incorporating several patterns with different numbers of fingers into a single screen. We then fabricated cells from the different transfer patterns. Patterns with untapered fingers gave the best results, and the wider cells required closer finger spacing. The chosen patterns for the 5.6 and 8.0 cm wide cells had front Ag coverages between 6 and 7%.

The front side of an 8.0-cm wide cell is shown in Figure 3. Most of the back side is covered by Al paste. These cell metallization patterns are currently transferred by hand to the wafers, but in future work we will automate the pattern transfer step.

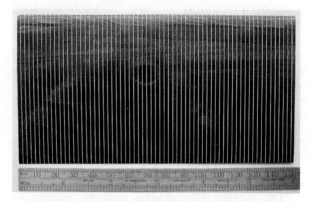

Figure 3: Front side of a 15 x 8 cm wraparound cell.

The performance of cells with TiO_2 and Si_xN antireflection coatings are shown in the table below. With the limited work performed to date, most of which was on 5.6-cm wide and 250-300 μm thick material with TiO_2 coatings, wraparound cell efficiencies have been somewhat lower than those for conventional String Ribbon cells. With further development, particularly in the area of reducing shunt conductance, we expect this difference to narrow significantly.

Table I. Performance of 15-cm long wraparound cells.

Width	AR	Voc	Jsc	FF	Eff.
5.6 cm	TiO_2	0.579	27.4	67.8	10.8
5.6	Si_xN	0.576	29.3	68.4	11.7
8.0 cm	TiO_2	0.546	28.4	66.4	10.3

3. MONOLITHIC MODULES

3.1 Production Flow

The proposed production flow for monolithic modules is shown in Figure 4. As compared to the production of standard modules using cells sawed from ingots, the number of processing steps and the cost are significantly reduced. In comparison to the conventional tabbing and stringing procedures shown in Figure 2, the module layout segment of the operation is greatly simplified.

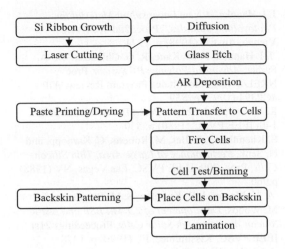

Figure 4: Future production flow for monolithic modules.

3.2 Interconnect Patterns

We considered two different interconnect patterns (see Figure 5), depending on whether the fingers wrap around one or two edges of the cell. The double wrap approach has the advantage in that fewer fingers are needed on the front side (less shadowing loss) since the I^2R power losses in the Ag are less, but the advantages and elegance of the single wrap approach are overwhelming. This pattern is similar to the one proposed by Gee for interconnecting *emitter wrap through* cells in his vision of *monolithic module assembly* [15]. In the single wrap structure, the current needs only to travel a few mm's in the interconnect metallization as opposed to several cm's in the double wrap structure. Thus, the conductivity requirements of the pattern are greatly reduced. In addition, a much smaller total area of interconnect metallization is needed, and wrapping the transfer pattern around one edge of the cell is easier than wrapping it around two edges.

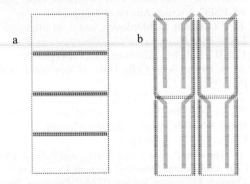

Figure 5: Interconnection pattern for cells with a) single and b) double wrap patterns.

The properties of Evergreen's novel backskin material allowed for several possibilities of accomplishing the interconnection process. Metal foil strips can be directly bonded to the backskin, or a conductive material can be printed directly onto the backskin. With a sufficiently large printer, all the patterns in a module may be printed in one step, thus ensuring perfect alignment between successive interconnect strips. Several different conductive materials were evaluated in terms of their conductivity, the contact resistance with fired Ag/Al or Ag paste on the cells, and in terms of the performance of modules or minimodules constructed with these materials under accelerated testing.

A module usually contains more than one column of cells. The connections between these columns is usually established with wide pieces of Sn or solder-coated Cu ribbon. In these interconnect pieces, the current travels along the length of the structure, not across the short width dimension, so the conductivity requirements are much higher than for the connections between cells. We used Cu ribbon strips for these interconnections in the prototype modules and minimodules, but printed conductive materials may work well for these cross column connections as well.

3.3 Module Layout

Following cell testing, a vision guided robot can merely pick up the cells and place them on the patterned backskin. The backskin itself is a modified thermoplastic which strongly bonds to the backs of the cells during the lamination cycle, thus eliminating the need for an encapsulant layer between the cells and backskin. Frameless modules have also been made using this material [16].

After the cells are placed on the backskin, the external contact leads are incorporated, and a sheet of Evergreen's novel encapsulant and the coverglass are placed on top. The entire structure is then placed in a laminator.

Photos of standard and monolithic frameless modules are shown in Figure 6.

Figure 6: Frameless modules with a) standard and b) monolithic interconnects.

3.4 Module Testing

We placed small modules in a humidity/freeze chamber and exposed them to accelerated testing where one 24 hour cycle comprised at least 30 minutes at $-40^{\circ}C$ and 20 hours at $85^{\circ}C$ with 85% relative humidity. As is shown in Figure 7a, the relative output power reduction of the module is less than 0.5% after 100 such cycles, indicating good module stability.

We also exposed larger modules to 100 purely thermal cycles. One 4 hour cycle goes from -40 to $90^{\circ}C$ and back down. Good stability was also seen for these modules as is shown in Figure 7b. We will expose the modules to larger numbers of cycles and other reliability tests in future work.

a)
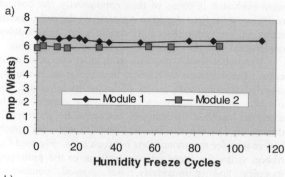

b)
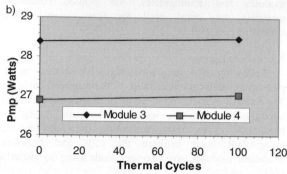

Figure 7:   Monolithic module performance under accelerated a) humidity/freeze and b) thermal cycling.

4. CONCLUSIONS

We have demonstrated a module manufacturing process with a greatly reduced number of processing steps as compared to conventional module processing. In particular, several steps that involve handling of the wafers have been eliminated. Evergreen's goal is to shift production toward 100-μm ribbon. The reduction in processing steps will facilitate the transition toward thinner wafers in the future where the sensitivity to yield loss from handling wafers is increased. Many of the problems in fabricating cells with wraparound contacts have been addressed, but more work is needed to bring cell efficiencies up to the level of standard cells, particularly in the area of reducing shunt conductance. These wraparound cells were successfully connected by a conductive interconnection pattern printed on the module backskin. Module reliability tests show promise for this monolithic interconnect structure.

5. ACKNOWLEDGEMENTS
We thank the Commonwealth of Massachusetts and the U.S. Department of Energy (DOE) for their support of this work through the Commercialization Ventures Program.

REFERENCES

[1] R. L. Wallace, R. E. Janoch, J. I. Hanoka, *String Ribbon – A New Silicon Sheet Growth Method*, Proc. 2nd WCPSEC, Vienna, Austria (1998) p. 1818.

[2]   J. I. Hanoka, *A New Encapsulant Material for Photovoltaic Modules*," Proc. 2nd WCPVEC, Vienna, Austria (1998) p. 1974.

[3]   J. I. Hanoka, P. M. Kane, R. G. Chleboski, M. A. Farber, *Advanced Polymer PV System*, Proc. NREL/SNL Photovoltaics Program Review, AIP #CP394 (1997) p. 859.

[4]   R. N. Hall, T. J. Soltys, *Polka Dot Solar Cell*, Proc. 14th IEEE PVSC (1980) p. 550.

[5]   S. Khemthong, P. Iles, M. Roberts, C. Karnopp, and N. Senk, *Performance of Large Area, Thin Silicon Cells*, Proc. 20th IEEE PVSC, Las Vegas, NV (1988) p.960.

[6]   A. V. Mason, J. R. Kukulka, S. M. Bunyan, and L. M. Woods, *Development of 2.7 Mil BSF and BSFR Silicon Wrapthrough Solar Cells*, Proceedings 21st IEEE PVSC, Kissimimee, FL (1990) p. 1378.

[7]   J. M. Gee, W. K. Schubert, and P. A. Basore, *Emitter Wrap-Through Solar Cell*, Proc. 23rd IEEE PVSC, Louisville, KY (1993) p. 265.

[8]   A. Schönecker, H. H. C. de Moor, A. R. Burgers, A. W. Weeber, J. Hoornstra, W. C. Sinke, P.-P. Michiels, and R. A. Steeman, *An Industrial Multi-Crystalline EWT Solar Cell with Screen Printed Metallisation*, Proc 14th European PVSEC, Barcelona (1997) p. 796.

[9]   E. Kerschaver, R. Einhaus, J. Szlufcik, N. Nijs, R. Mertens, *A Novel Silicon Solar Cell Structure with Both External Polarity Contacts on the Back Surface*, Proc. 2nd WCPSEC, Vienna, Austria (1998) p. 1479.

[10] A. Kress, P. Fath, G. Willeke, E. Bucher, *Low-Cost Back Contact Silicon Solar Cells Applying the Emitter-Wrap Through (EWT) Concept*, 2nd WCPSEC, Vienna, Austria (1998) p. 1547.

[11] R. A. Sinton et al., *Silicon Point Contact Concentrator Solar Cells*, Proc. 18th IEEE PVSC, Las Vegas, NV (1985) p. 61.

[12] D. L. Meier et. al, *Self-Doping Contacts and Associated Silicon Solar Cell Structures*, Proc. 2nd WCPSEC, Vienna, Austria (1998) p.1491.

[13] B. T. Cavicchi, N. Mardesich, and S. M. Bunyan, *Large Area Wraparound Cell Development*, Proc. 17th IEEE PVSC, Kissimee, FL (1984) p. 128.

[14] M. A. Green, Solar Cells, Univ. of New South Wales Ed (1992) p. 153.

[15] J. M. Gee and T. F. Ciszek, *The Crystalline-Silicon Photovoltaic R&D Project at NREL and SNL*, Proc. NREL/SNL PV Program Review Meeting, Lakewood, CO (1996) p. 1.

[16] J. I. Hanoka, P. E. Kane, J. Martz, and J. Fava, *Low Cost Module and Mounting Systems Developed Through Evergreen Solar's PVMaT Program*, Proc. 26th IEEE PVSC, Anaheim, CA (1997) p. 1081.

ENERGY RATING OF PHOTOVOLTAIC MODULES

D. Anderson, J. Bishop, E. Dunlop
European Commission Joint Research Centre
Renewable Energies - Environment Institute
European Solar Testing Installation (ESTI) TP450; Ispra (VA) 21020; Italy
Tel. +39-0332-789289; Fax. +39-0332-789268; Email; david.anderson@jrc.it

ABSTRACT: Electrical performance data measured under laboratory conditions (e.g. Standard Test Conditions) are felt to be of limited use in the design of a photovoltaic (PV) system. This work proposes a method for characterising the behaviour of PV modules using "performance surfaces" of electrical power output as a function of incident irradiance and ambient or module temperature. Performance surfaces were modelled from measured outdoor data and compared to surfaces generated from indoor laboratory testing. The suitability of this method for various PV technologies is investigated (crystalline silicon, amorphous silicon, Cd-Te etc.)
Keywords: Energy Options -1: Performance - 2: Modelling - 3

1. INTRODUCTION

The power output of a PV module depends on the intensity and spectral content of the incident radiation, the electrical operating point and the average cell temperature. Various methods of predicting the energy production of a PV module in the field have been proposed, such as the current working draft IEC standard 1853 [1], which includes factors such as incident angle and the definition of "standard days". This paper instead proposes a new approach to energy rating, based on the reduction of power output to performance surfaces of irradiance and ambient temperature. These surfaces can be combined with distribution surfaces of the conditions at the installation site to provide an estimate of the energy production. Performance surfaces are investigated using indoor and outdoor data, and compared to surfaces generated from existing standard laboratory data for Poly-Si, Cd-Te, tandem and triple junction technologies.

2. OUTDOOR PERFORMANCE DATA

An outdoor measurement system was built for the dual purpose of Nominal Operating Cell Temperature (NOCT) and Energy Rating (ENRA) studies (see Figure 1).

There are currently 6 KEPCO bipolar power supplies used to monitor the electrical performance of each PV module throughout the course of the measurement period. The power supplies are connected to the PV modules using a 4-wire connection, with a precision shunt resistor inserted on the negative line to measure the current. As the power supplies are driven by the controlling software, a Hewlett Packard datalogger is used to trace the current-voltage (I-V) characteristics of the modules. Each I-V curve is associated with the instantaneous irradiance, ambient and module temperature and saved with a time stamp. I-V curves are measured every 5 minutes throughout the day. All temperatures are measured with PT100 resistor temperature detectors (RTDs), and irradiance is recorded by both a pyranometer and a crystalline-silicon ESTI sensor. The modules featured in this paper have been monitored outdoors since mid-February 2000.

3. MODELLING OUTDOOR PERFORMANCE

Data collected over a period of weeks or months can be represented by a 3-dimensional plot of maximum power output against irradiation and ambient temperature (see Figure 2).

Figure 1: Outdoor measuring site at Ispra. The set-up can measure up to 12 modules continuously (in NOCT or Energy Rating mode) including thin-film and AC modules.

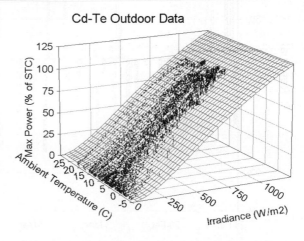

Cd-Te Outdoor Data

Figure 2: An example of a performance surface for a Cd-Te module, using data collected over one month. The surface can be modelled analytically to predict power output in various ambient conditions.

These surfaces were modelled to obtain a function for power output in terms of irradiance and ambient temperature. The measurement data was then fed back into this model to confirm the validity of this approach. The results are shown in Table 1 and Figure2.

Module	Test Period (days)	Total Energy			Mean Efficiency (%)	Error of Energy Estimate (%)
		In (kWh/m2)	Out (kWh/m2)	Predicted Out (kWh/m2)		
Poly-Si	31	133.83	14.19	14.31	10.6%	+0.8%
Cd-Te	31	134.23	7.08	7.08	5.3%	+0.1%
Tandem	38	172.06	8.40	8.45	4.9%	+0.6%
Triple	27	124.75	8.28	8.53	6.6%	+3.0%

Table 1: Summary of the outdoor measurements and prediction of energy using a model of power as a function of irradiance and ambient temperature. The model was determined from outdoor data.

Table 1 summarises the performance of the Poly-Si, Cd-Te, tandem and triple-junction modules over the initial test period. The performance surface models of irradiance and ambient temperature were used to confirm the validity of using such a model on outdoor data. Using the irradiance and ambient temperature data acquired during the course of each day, a prediction of the energy output was made and compared to the measured energy output. As can be seen, over the first month the difference between the predicted and measured energy output is +3% for the triple junction module and less than +1 % for the other three module types. However, these values do not tell us anything about the accuracy of the energy prediction on a day to day basis. Hence, Figure 3 shows the daily error (% residual) of the

energy prediction for each of the module types.

At this point, the validation of the irradiance and ambient temperature model was limited by the lack of measurement data. As the modules continue to be monitored, it will be possible to investigate how a model extracted from a month of data in (say) spring, can be used in the middle of summer. Similar analysis was done on the outdoor data using irradiance and module temperature as the dependant variables, but no significant increase in accuracy of the energy estimate was achieved.

4. ENERGY ESTIMATE FROM INDOOR PERFORMANCE MEASUREMENTS

Before the modules were mounted outside, they were tested indoors at the ESTI laboratory using a PASAN large area pulsed solar simulator (LAPSS) and a thermally controlled chamber. For each module, a series of measurements were made by connecting the module to a fixed resistor and recording current and voltage during the duration of the flash. A 26-position resistor box was used so that the value of the fixed resistor could be easily changed and the measurement repeated (see Figure 4). This entire process was also repeated at different temperatures, thus producing a data-set of I-V curves at continuous irradiance values over the range 25 to 60 °C.

Using the data from the indoor laboratory measurements, it was possible to create performance surfaces similar to the one shown in Figure 1, and compare these to the performance surfaces measured outdoors for the Poly-Si, Cd-Te, tandem and triple-junction modules. Figure 5 shows

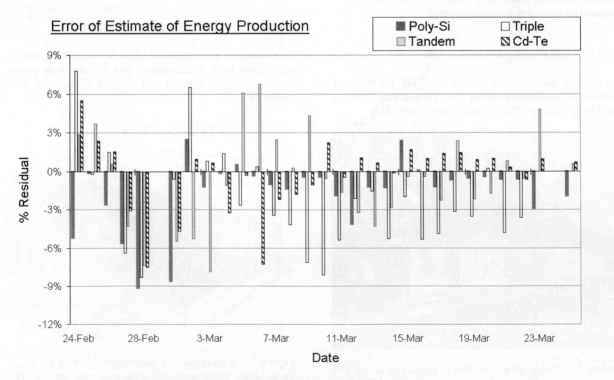

Figure 3: The daily error in the prediction of the outdoor energy production (Wh/m2/day) using a model of power as a function of irradiance and ambient temperature, determined from outdoor data. The error is defined as the difference between the measured and the predicted energy divided by the measured energy.

the indoor and outdoor surfaces on the same graph for the Poly-Si module. Figure 6 shows the difference between the two surfaces, and we can see that the two surfaces are generally coplanar, with the indoor surfaces lying about +2% above the outdoor measured surface.

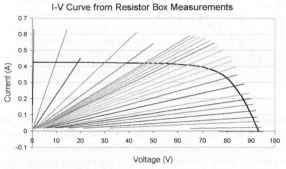

Figure 4: Resistor-box measurements for a module at constant temperature. Each of the straight lines (constant resistance) represents one of the 26 positions on the resistor-box. Along each of these lines lie points measured as the pulse from the solar simulator decayed. Since each point contains simultaneous values of irradiance, current and voltage, the corresponding I-V curve can be extracted by interpolation.

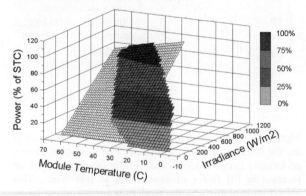

Figure 5: Indoor laboratory surface (plain) and outdoor measurement surface (shaded) for the Poly-Si module. The difference between the two surfaces is shown in Figure 6.

Using the indoor model of irradiance and module temperature, we can predict the outdoor power output of the module and verify the estimate. Figure 7 shows the results for the Poly-Si module. The results show that the outdoor energy production of the poly-Si module can be modelled successfully (to within 2%) by using the indoor performance surface model.

A similar analysis was made for the Cd-Te module, and the results are shown in Figure 8. In this case, the indoor model is significantly (8%) underestimating the power output of the module. Further analysis revealed that the increase in outdoor power was due to the fill factor deviating dramatically (+10%) from the indoor measurements. Although the short circuit current (Isc) and open-circuit voltage (Voc) can be predicted very accurately from the indoor data, the module seems to be performing quite

differently in terms of fill factor and power output in the field. Figures 9, 10 and 11 show the behaviour (indoor and outdoor) of the Cd-Te module in terms of current, voltage and fill factor respectively.

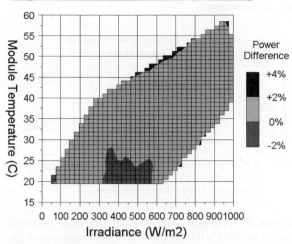

Figure 6: Plot of the outdoor performance surface subtracted from the indoor one. The majority of the indoor surface is 2% above the outdoor one, which can be attributed to the calibration limits of the irradiance measurements.

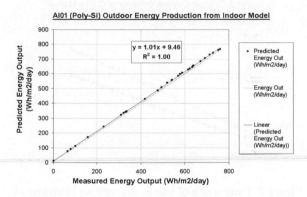

Figure 7: Predicted energy plotted against measured energy for the Poly-Si module. Each point on the chart corresponds to the total energy produced during one day. The energy prediction was made using indoor data only, and shows excellent agreement of less than 2%.

At this point of the analysis, the generic problem of characterising the response of a thin-film module in the laboratory became evident. For the Cd-Te module, the spectral mismatch factor successfully translated [2] the Isc and Voc points of the indoor measurements. However, the increase of fill factor outdoors results in a multiplication factor for our indoor power estimate of about 8%. Furthermore, spectral mismatch data for the tandem and triple junction modules was unavailable at the time of this publication, and therefore a meaningful comparison between indoor and outdoor power surfaces is not possible. The uncorrected indoor surface of the triple-junction module generally showed a ±3% variation from the observed outdoor surface. The tandem-junction module showed the worst deviation from the observed performance, with the uncorrected indoor power surface varying by +3 to -6% from the outdoor measurements, and falling off to nearer 10% at higher irradiances. Clearly, more analysis is needed on the performance of these special technologies.

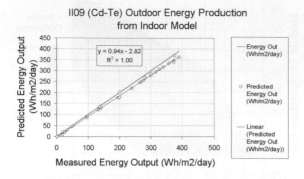

Figure 8: Predicted energy plotted against measured energy for the Cd-Te module. Each point represents a day of energy production. The energy prediction, made from the indoor surface model, underestimates the actual field performance by about 8%. This difference in power output is due to the increase in fill factor of the Cd-Te module when operating outdoors (see Figure 11).

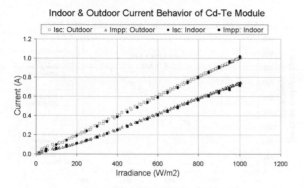

Figure 9: Comparison of indoor and outdoor behaviour of the current characteristics of the Cd-Te module. After the spectral mismatch correction, there is an excellent agreement between the indoor and outdoor values of short-circuit current (Isc) and current at maximum power (Impp).

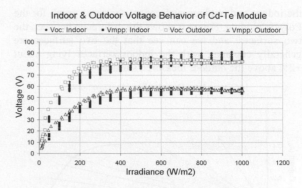

Figure 10: Comparison of indoor and outdoor behaviour of the voltage characteristics of the Cd-Te module. The indoor measurements appear at regular intervals of irradiance and are spread along the voltage axis as the module temperature was increased during the indoor measurements. The outdoor voltage values show good agreement with the indoor data. In the region 200-400 W/m2, the outdoor voltage values are lying above the indoor ones due to the low module temperatures of between 5 and 15 °C.

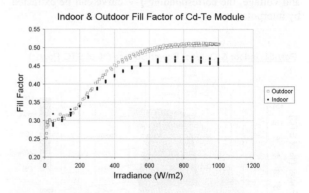

Figure 11: Comparison of indoor and outdoor behaviour of the fill factor of the Cd-Te module. The module is clearly performing better outdoors than in the laboratory, with an increase in fill factor of about 10% at 1000W/m^2. This behaviour explains why the model for energy output based on indoor data in Figure 8 was underestimating the actual field performance by about 8%.

5. PERFORMANCE FROM STC MEASUREMENTS

After modelling the performance of the modules from the indoor resistor-box measurements, it was desirable to recreate the indoor surface using the simplest of laboratory measurements. Hence, using the STC power as a starting point, a performance surface was generated using only temperature coefficients of current and voltage (α and β), the series resistance (Rs) and curve correction factor (K). The surface was expanded along irradiance and temperature axis using standard translation equations [3].

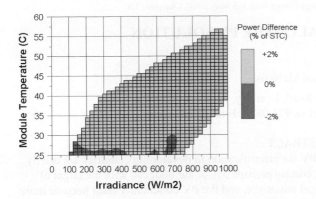

Figure 12: Poly-Si: outdoor performance surface subtracted from the indoor surface determined from the STC measurement and temperature coefficients, series resistance and curve correction. The surfaces are within 2% of each other, hence simple laboratory measurements can effectively predict outdoor behaviour for this module type.

When the surface was compared to the outdoor surface for the poly-Si module, the results were surprisingly good. The two surfaces were again generally within 2% of each other (see Figure 12). This result means that STC is a useful measurement and can be used, at least for poly-Si, to predict the outdoor performance with great accuracy.

6. USING PERFORMANCE SURFACES FOR ENERGY RATING

Once the performance surface for a PV module has been effectively determined, it is proposed that this surface is combined with a distribution surface of the environmental conditions to obtain an energy rating for that module. Figure 13 shows an example of a distribution surface of irradiance and ambient temperature, where the surface corresponds to the duration (as a percentage of the whole time period) that these conditions were observed.

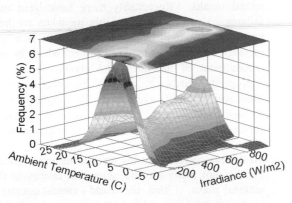

Figure 13: Distribution of environmental conditions as normalised frequency plot of irradiance and ambient temperature. An energy rating is obtained by multiplying the frequency of occurrence of a particular condition by the appropriate power output at that condition, and summing over the entire surface. If the above distribution represented the conditions during one month, then multiplying by the performance surface would result in an energy prediction for that month.

Multiplying the power surface by the conditions distribution and summing over all the conditions yields a value of energy which could be used as an absolute energy rating in units of energy per time (e.g. kWh/month). Alternatively, the total energy could be divided by the energy one would expect if the module operated continuously at STC power over the time period, to achieve an energy rating as a performance ratio, e.g. 0.63 of STC.

The advantage of this method is that the module electrical properties can be used as an independent unit from the local environmental conditions. This allows each user to modify the energy rating to suit the particular conditions of the installation site, thus achieving a more realistic estimate of the energy production. By reducing the model to the simple parameters of irradiance and ambient temperature, the rating is understandable and easy to implement for all PV users, since local environmental data is generally available, and in some cases documented for well over 50 years.

7. CONCLUSION

Outdoor performance of PV modules can be modelled successfully over time using performance surfaces of irradiance and ambient or module temperature. These surfaces can be recreated in the laboratory using simple standard measurements, although for some module technologies a correction procedure is necessary.

It is proposed that the use of these performance surfaces in conjunction with distribution surfaces of environmental conditions will result in an accurate energy rating of a particular module at a particular installation site.

8. REFERENCES

[1]   IEC 1853D, "Performance Testing and Energy Rating of Terrestrial Photovoltaic (PV) Modules

[2]   IEC 904-7, "Computation of Spectral Mismatch Error Introduced in the Testing of a Photovoltaic Device"

[3]   IEC 891, "Procedures for Temperature and Irradiance Corrections to Measured I-V Characteristics of Crystalline Silicon Photovoltaic Devices"

ELECTRICITY FOR ALL: THE PV SOLUTION

Bernard McNelis

IT Power, The Warren, Bramshill Road, Eversley, Hants. RG27 0PR. UK
Tel: +44 (0)118 973 0073, Fax:+44 (0)118 973 0820, E-mail: bmcn@itpower.co.uk

ABSTRACT

The Author's views on the success and failures of PV for electrification in developing countries are briefly summarised. The case for a global programme, costing perhaps 5% of aid budgets, with the target of *Electricity for All*, is presented. This requires political initiatives, and the PV community must become more aggressive in selling the PV solution for a better world.

INTRODUCTION

Electricity is the form of energy which can transform the way people live and work. In towns and cities, electricity provides light and power at home, in the school and in the office; it powers industry, communications and hospitals and allows entertainment from film, television and radio. Yet 40% of the world's population, more than two billion people (or 4 million villages) in the developing world do not have electricity[1].

In rural areas of developing countries, the population is dispersed and generally not connected to the electricity grid; living standards are poor in comparison to the main urban centres, with seemingly little prospect of improvement. The result is a general migration towards the cities and towns, resulting in overcrowding and poverty in urban areas, and a drain on human resources in rural areas.

The World Bank reports that over the past 25 years 1.3 billion people in developing countries have been connected to the electricity grid - an impressive achievement, but over the same period, the world's population has increased by 2 billion so there are 700 million more people without electricity[2]. The cost of installing new electricity generating capacity and extending the grid to bring quality energy services to many of the most needy rural locations is prohibitively high; the distances over which transmission cables have to travel are large and the electricity consumption levels once the grid reaches its rural destination are relatively low. This is not the most attractive business scenario for electric utilities, especially those being transferred to the private sector.

However, the use of PV systems to provide electricity is an increasingly important solution which is making a real and positive difference to the lives of tens of thousands of people, although the sustainability of many of these installations is in doubt. There is no need to build the expensive infrastructure that accompanies the generation of power from oil, gas or nuclear, as PV systems can be installed at the point of use. Using PV systems allows services and brings benefits which would otherwise be denied. The sectors affected include healthcare, agriculture and fisheries, education, micro-enterprise and community development. Increased effectiveness of social programmes, and improvement in productivity are a few of the important changes which PV can facilitate. For the 'regular' readers of EPVSEC Proceedings, an apology is due as this has all been said and written before! But, the purpose of this paper as will be seen, is to draw the attention of policy makers and politicians, not PV scientists!

TRADITIONAL PV CO-OPERATION

Various UN agencies and bilateral donors have over the past 20 years supported a huge range of small PV projects, throughout the developing world, and with mixed results. Regrettably there have been many failures. Even more regrettably mistakes are being repeated with more failures. But there is a clear message that a well-conceived project which includes commitment of the users and choice of the most appropriate hardware, delivers remarkable results. Moreover, on a life-cycle basis PV supplies services at lower cost then the conventional alternatives, eg. candles, kerosene lamps, small gasoline generators.

Most developing countries have some experience of PV, but not many include renewable energy in their national plans. China, the world's largest country, is starting to take renewable energy very seriously. The UNDP, together with GEF, Australian and Dutch funds, is helping develop the capacity for the rapid commercialisation of technologies including PV. The Chinese Renewable Energy Industries Association

(CREIA) has been set up as a trade association based on industrialised country models. CREIA is establishing an Investment Opportunities Facility (IOF) to bring together project developers, industry (both Chinese and foreign), government donors and banks to ensure that renewable energy makes a major contribution to China's expanding demand for energy[3]. But still the level of investment is very small in comparison with the scale of the market.

THE INDUSTRY

Today PV is a global business with an annual turnover of $1.5 billion. But this is minuscule compared with its potential. Manufacture of PV modules in 1999 of around 200 MWp is still less than a single "small" fossil fuel power station. Both leading multinational companies and innovative small enterprises are committed. The world's largest PV manufacturer is BP Solar. BP has set the ambitious target of a turnover of $10 billion by 2007 for BP Solar alone, and is making the investments necessary to achieve this. Other major PV producers are Kyocera, Siemens, Sharp, Photowatt, Astropower and Sanyo, and these "top 7" produce 75% of the world's PV modules. Shell is a small producer but has also set ambitious growth targets and is investing $500 million in its "5th core business", renewable energy. Globally, the industry is making investments, but profitability is poor, or in some cases, negative.

EU SUPPORTED PROJECTS

Applications of PV which can make the difference between life and death include water pumping and vaccine refrigeration. PV water pumps provide water on demand from boreholes - a huge improvement on open wells. In the countries of the Sahel, PV pumps have transformed the quality of life and improved health in villages. The European Union financed a $40 million regional solar programme with the Comité permanent Inter-Etats de Lutte contre la Sécheresse dans le Sahel (CILSS) which installed 1270 PV systems in the mid 90's in Burkina Faso, Chad, Gambia, Niger, Mali, Mauritania and Senegal. This was a large PV programme, but it only reached about 1% of the 67,000 villages in the CILSS countries[4]. The success of this initiative has now led to a second phase of the programme with around $80 million from Europe, which is currently in preparation. The European Union has recently provided $17 million for the use of PV to electrify 1,000 rural schools in South Africa[5]. Projects to introduce PV for household electrification in South Africa, are underway, including an ambitious one by Shell, but a serious barrier is the erroneous perception that solar electricity supply targeted at the rural (black) population is "inferior" to the grid-electricity used by the (white) city dwellers. The rural schools project should go some way to addressing this problem.

CAN GEF GRANTS MAKE A DIFFERENCE?

The availability of Global Environment Facility (GEF) funds to support serious scale investments has enabled the World Bank to develop projects which combine GEF grants with Bank loans to build the markets and the infrastructure needed to make markets sustainable. Renewable energy projects including PV in Argentina, Bangladesh, Benin, Cape Verde, China, Dominican Republic, India, Indonesia, Kenya, Lao, Morocco, Sri Lanka and Togo total almost $1.5 billion with $150 million of GEF grants. Projects are listed in Table 1. Most projects are at an early stage, so their degree of success cannot yet be assessed. The World Bank (more specifically, a small group of enlightened people within the Bank!) has undertaken a very large amount of work to overcome problems experienced with earlier projects. For example a guide to best practices for household PV electrification was published in 1996[6]. An extensive review of all the GEF PV projects is underway and results will be published later this year. This should provide very important insight to whether GEF funds can open sustainable markets[7].

Table 1.

Country	Funding ($million)				PV component
	GEF	World Bank	Other	Total	
Argentina	10	30	81	121	66,000 SHS
Bangladesh, Dominican Republic, Vietnam (IFC/SME Programme)	1.5				Investments in firms in the three countries
Benin	1	2	3	6	5,000 SHS
Cape Verde	5	18	25	48	4,000 SHS
China	35	100	309	444	10 MWp of SHS and hybrid systems
Global (Solar Development Group)	10	6	34	50	Finance for PV businesses
India (IREDA)	26	185	239	450	2.5 MWp: various applications
India, Kenya, Morocco (PV-MTI)	30		90–110	120–150	Investment in off-grid PV projects

Indonesia	24	20	74	118	200,000 SHS (disrupted by crises)
LAO PDR	1	1.5		2.5	20 battery charging stations (demonstration)
Sri Lanka	6	24		30	30,000 SHS
Togo	1	2	3	6	5,000 SHS

The IFC/GEF Photovoltaic Market Transformation Initiative is using $30 million of GEF funds to lever around $100 million additional finance for investments in India, Kenya and Morocco. Loans are provided to local entrepreneurial companies who know how to deal with real customers and have a vested interest in developing sustainable business. The first PVMTI solicitation, released in 1998, resulted in 45 proposals. The first investment of $2.2 million was made in late 1999 in Shri Shakti Alternative Energy, which is owned by India's largest privately-held distributor of liquefied petroleum gas. Shri Shakti is building a network of Energy Stores for the sale of consumer products based on photovoltaic and other alternative energy technologies, which will be sited in South India, Orissa and major towns in North India[8].

The level of funding from GEF is impressive, but it too early to determine the success of the projects, listed in Table 1. It is important to note that all of these projects are in reality "experiments" and many may fail.

THE IEA - LOOKING OUTWARD

The International Energy Agency (IEA), which is the energy forum for 24 industrialised countries, has been co-ordinating joint research and exchange of information between its member countries for the past ten years. The focus has been on improving the technology, (90% of world PV production is in IEA countries) as well as applications in members countries - ie grid feeding systems, both building-integrated and centralised PV power stations. Most recent growth in demand for PV has come from developed countries, in response to concerns about global climate change. PV does not yet compete economically with the existing electricity generation infrastructure. But it is widely accepted that increased demand will drive costs down sufficiently for this to be the case. Several governments therefore provide subsidies to encourage use of PV to build the

industry. Most notable is Germany where interest free loans are available to households and companies for PV installations, and in addition, from February 2000, the electric utility to which the building is connected must pay the "consumer" about $0.5 (DM0.99) per kWh generated. This policy is part of Germany's plan to meet its Kyoto target. The PV contribution to this will amount to 100,000 PV façades and rooftops totalling 350 MWp.

Recently, the IEA has launched a new project ("Task IX") to focus on PV market deployment in developing countries. This is using the expertise available from the main PV producers and users to help develop improved practices and build the infrastructure necessary for PV to become a significant renewable energy option in the developing world. PV is a technology where multi-national companies, governments, donors and NGOs can work together for their mutual benefit, as well as that of the 2 billion + unelectrified[9].

ELECTRICITY FOR ALL = POWER FOR THE WORLD

Wolfgang Palz presented his Global Photovoltaic Action Plan *Power for the World* at the 12th EPVSEC in Amsterdam in 1994[10]. So the ideas in *Electricity for All* are nothing new. Palz called for an ambitious programme to provide 10 Wp per capita for 20% of the world's population. This would total 10 GWp - a very attractive business for the industry, yet this would amount to only 0.5% of the electricity consumed by the other 80% (ie us). He pointed out that the cost of the programme, $3 billion per year over 20 years, would be less than 1% of global military expenditures.

It is the view of the author that another comparison is very interesting. Total world aid (Official Development Assistance, ODA) is more than $60 billion annually. So less than 5% of this could finance *Power for the World*. Moreover, the European Union and its member states provide more than 50% of total world aid. So a target of 10% of European aid could also achieve the goal. What is needed is a huge and concerted effort to convince politicians and decision makers of the global benefits of providing small amounts of electricity to "poor people", while at the same time proving that PV can be a solution in a sustainable way. The PV community should use up all their powers of persuasion to this end.

This conference is an important start. With a politician, Hermann Scheer, as Chairman, we must now convince others to share his vision. Including a World Bank/European Union Workshop in the Conference Programme is also a move in the right direction. Let us take this forward and deliver good results at the 17th EPVSEC in Munich in October 2001.

CONCLUSIONS

PV can make the world a better place. A global PV electrification programme will be a win-win! The unelectrified will receive light and other services, only available in towns, the PV industry will expand, employing people, not only in manufacture, but in installation, after sales service etc.

REFERENCES

(1) World Bank, Rural Energy and Development: Improving Energy Supplies for 2 billion people, Washington (July 1996)

(2) J. Bond, World Bank (1998), Opening statement, World Bank Energy Week: Extending the Frontiers of the World Bank's Energy Business, Washington D.C.

(3) W. Zhongying, Z. Junsheng, W. Wallace, L. Junfeng, F. Asseline, B. McNelis, J. Gregory, M. Mendis, Establishing a Renewable Energy Industries Association in China. 16th European Photovoltaic Solar Energy Conference, Glasgow, May 2000.

(4) CILSS/PRS, Regional Solar Programme, Lessons and Perspectives, European Commission.

(5) J.R Bates & J.P. Loineau, C, Purcell, W. Mandhlazi, H. Van Rensurg, Programme for the Electrification of 1,000 Schools in Eastern Province and Northern Cape in the Republic of South Africa. 16th EPVSEC, Glasgow, May 2000.

(6) A. Cabraal, M. Cosgrove-Davies, L. Schaeffer. Best Practices for Photovoltaic Household Electrification Programmes: Lessons from experiences in selected countries. World Bank Technical Paper No. 324, Washington D.C (1996)

(7) E. Martinot, Global Environment Facility, Washington DC. Monitoring and Evaluation of GEF PV projects. To be published 2000.

(8) Derrick, I. Simm, V. Widge, Progress with the IFC/GEF Photovoltaic Market Transformation Initiative (PVMTI), Proceedings of the 16th EPVSEC, Glasgow 2000.

(9) J. R. Bates, B. McNelis, A. Arter, W. Rijssenbeek, Deployment of Photovoltaic Technologies: Co-operation with Developing Countries: Task IX of the International Energy Agency's Photovoltaic Power Systems Programme. 16th EPVSEC, Glasgow, May 2000.

(10) W. Palz, Power for the World: A Global Photovoltaic Action Plan, Proceedings of the 12th European Photovoltaic Solar Energy Conference, Amsterdam (1994)

A NEW PV SYSTEM TECHNOLOGY – THE DEVELOPMENT OF A MAGNETIC POWER TRANSMISSION FROM THE PV MODULE TO THE POWER BUS

I. Weiss, A. Hänel, A. Sobirey, P. Helm
WIP, Sylvensteinstrasse 2, D-81369 Munich, Germany
Tel. +49-89-7201235, Fax +49-89-7201291, eMail: wip@wip-munich.de

H. Oppermann
Sun Power Solartechnik, Marktplatz 2-4, 61118 Bad Vilbel, Germany
Tel. +49-6101-58 45 50, Fax +49-6101-58 45 60, eMail: sales@sunpower.de

R. Tölle, T. M. Bruton, BP Solarex, 12 Brooklands Close, Sunbury-on-Thames, United Kingdom
Tel. +44-1932-765947, Fax: +44-1932-765293, eMail: ToelleR@bp.com

ABSTRACT: The New PV System Technology (NST) comprises the development of the several components such as the magnetic power transmitter and the magnetic power receiver, interface of PV module with fully integrated magnetic transmitter, the magnetic power bus and the central power processor. This innovative PV system technology concept removes the major weaknesses of state-of-the-art PV system technology such as varying lifetime expectancy of the PV system components, low degree of integration of electronic devices, sensitivity against shading, high cabling costs, significant mismatch losses and degradation of the PV generator. The magnetic coupling devices (transmitter and receiver) are very small devices which is possible due to the high transmission frequency of 40 kHz from the PV module to the power bus. This allows also an optimized integration of the high-frequency electronics and NST components and will lead to a significant increase life-time of the system because of the use magnetic material. The development of the prototypes will lead to a significant reduced PV system installation costs and reduced safety needs due to the use of a magnetically induced power bus with fully encapsulated magnetic power receivers. The economic benefits at the system level (reduced cabling and installation cost, reduced cost due to minimised safety needs which also accounts for the PV module allow for a total reduction of system cost of 15-20% (over the PV system lifetime) which will create an essential push for the take-up this technology of a large scale and open up new markets for PV. The project has been co-financed by the European Commission in the framework of the Non Nuclear Programme JOULEIII.

Keywords: PV Systems - 1: Magnetic Energy Coupling - 2: Magnetically Induced Power Bus - 3

1. OBJECTIVES AND INTRODUCTION

The objective of this project is to develop an New PV System Technology (NST) concept which aims at removing the major weaknesses of state-of-the-art PV system technology such as varying lifetime expectancy of the PV system components, low degree of integration of electronic devices, sensitivity against shading, high cabling costs and significant mismatch losses [1].

The development of the innovative PV system concept will support to overcome these problems. The project includes the development, manufacture and test of the NST component prototypes and the installation and monitoring of a NST pilot plant.

The following components are developments for the NST:

- Magnetically Induced Power Bus (MIPB) Technology with fully integrated Magnet Power Receiver (MPR) for energy transmission from the PV module to the Power Bus

- Innovative PV modules with fully integrated Magnetic Power Transmitter (MPT).

- Magnet Power Receiver (MPR) for contact free magnetically induced coupling of the PV modules to the electric circuit.

- Central Power Processor (CPP) which links the NST system to the public grid and which performs the overall system control.

- NST pilot plant realised at the facilities of one of the project partners

The prototypes were tested and operated in a test facility and results are presented in the following

2. OVERALL PROGRESS AND RESULTS

The following summarizes the overall progress of the relevant project period.

2.1 Design and Development of the Magnetic Power Transmitter (MPT) and the Magnetic Power Receiver (MPR)

The MPT and MPR were developed, designed and first prototypes were manufactured. The main results of the design of the MPT and MPR are as follows:

- To optimise the energy transfer at the air gap between transmitter and receiver (<1mm) different winding configurations were studied and assessed. An optimised winding configuration was shown which reduces the stray field and increases the coupling to up to a transmission power efficiency of 97%. Other configurations allowing for further improvements up to 98-99% will be investigated.

- For the selection of magnetic materials the electromagnetic compatibility aspect needs to be considered and that the transmission frequencies has been optimised. It is absolutely necessary to achieve a true sine-wave shape of the voltage output with very low harmonic distortion.

- The receiver and transmitter units will be mounted outdoors and integrated into the PV modules. The operating temperature of the transformer material is in the range of -25 to +85°C. There will be no influence on the quality of the transmission by the PV modules because they are not operating out of this range, even not in high temperature areas.

- The size of the prototypes of the transmitter and receiver are smaller than the junction box of the PV module.

2.2 Laboratory Testing of the Magnetic Power Transmitter (MPT) and the Magnetic Power Receiver (MPR) Prototypes

A test loop in the laboratory in order to verify system concepts, identify power transmission capabilities and improvement needs. A selection of the best technical solutions in terms of cost, reliability, power conversion and MPPT efficiency was made. The main results of the laboratory testing are the following:

- The laboratory tests which were performed on the MPT and MPR achieve an efficiency value of 97%. Other configurations allowing for further improvements up to 98% will be investigated.

- Re-design of the components on the basis of the results of the first test sequence for the two power switches of the MPT trench mosfets have been selected when re-designing the circuitry.

- Compared to the first approach the two electrolyte capacitors have been replaced by foil capacitors of little losses.

- The discretely designed control electronic of the first test circuits has been comprised in a highly integrated control circuit with level shifter ICs.

- The design of the air gap has been optimised from two coils mounted into each other to the more simple solution of two matched even surfaces. This solution has been possible by applying a very thin isolation foil inside the air gap between the two half coils.

- Re-test of optimised hardware prototypes:

  The tests of the prototype showed a power transfer capability via the air gap of 100W. The transmission efficiency achieved a value of 97 %. The operating frequency has been doubled, i.e. from 20 kHz to 40 kHz.

Figure 1 shows the developed Magnetic Power Transmitter (MPT) and the Magnetic Power Receiver (MPR) Prototypes.

Figure 1. MPT and MPR Prototypes for the PV System.

2.3 Development of the PV module/MPT interface and prototype manufacture

The electronics are integrated in an adequate casing, to adapt the electronics accordingly and to identify solutions to integrate the MPT into the PV module.

Several approaches have been made to find an appropriate case for MPT and MPR. Different kind of standard cases for DIN rail mounting have been checked for suitability. Two neighboured cases should contain MPT and MPR and the power be transferred via the touching sides of the two cases. The precise matching of the magnetic transfer areas of the MPT and MPR cases has to be considered .

The MPT has been integrated in the PV module connection box while the MPR shall be an integral part of the power bus. The PV module has 3 connections – minus, plus and centre tap for the bypass diodes. The PV modules are no longer connected in series and therefore the bypass diodes and corresponding terminals can be omitted. After removing these terminals and diodes the connection box of the solar module showed sufficient space for integrating the MPT. As the optimized MPT only requires one third of the connection box a new design of the cases is underway.

The electronic circuit of the MPT has then been built on a PCB fitting into the empty module connection box.

The magnetic transmitter has been integrated into the cover. For coupling the box of the MPR (Magnetic Power Receiver) two permanent magnets have been positioned at the right and left side of the connection box cover. While the MPT is built inside the standard connection box of the PV module, the MPR has a separate small case and is positioned and attached to the connection box via two permanent magnets.

The MPR has been integrated into the energy bus wire and is thus a fixed part of this wire. At the receiving stations no contacts are required. See Figure 2.

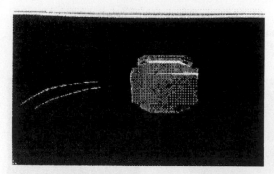

Figure 2. Coupled power transmitter and receiver

2.4 Development of the MI-Power Bus with integration Magnetic Power Receivers (MPRs)

The main results of the realization of the MI-Power Bus with integrated MPR are as follows:

The power bus consists of a two-pole wire with integrated MPRs. The power magnetically transferred via the air gap between MPT and MPR is converted into current and voltage inside the MPR. It is then fed into the power bus.

The mechanical adaptation between MPR and power bus has been solved by leading the two-pole wire through the MPR case. Inside the case the wires are interrupted and connected to terminals.

The final mechanical adaptation between MPR and power bus shall have in mind the facilitated PV system mounting and low cost production. The aim is therefore to automatically contact the MPRs on the bus wire in fixed distances. With this pre-mount "MPR-wires" a fast solar generator wiring can be achieved.

2.5 The Central Power Processor (CPP)

The requirements of a Central Power Processor controlling the power bus based on the state-of-the-art technology of the inverter technology additional design is required to activate the energy bus and to control the minimum and maximum voltage of the supplied power.

Further the output transformer of the Sun Profi shall be omitted since the MPT/MPR already provides isolation between solar generator and grid. The transformerless feeding into the grid shall increase the all-over efficiency and reduce the cost/W of installed power.

2.6 Design and Installation of a NST Pilot Plant

A valuable data source for a future development are the NST components and systems and introduce them into market ready products following the successful execution of this project. The Design of the NST system and building/system interfaces

Figure 3 shows the layout of the pilot plant:

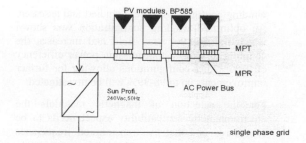

Figure 3. Pilot Plant Layout

Based on the prototype MPT/MPR the pilot plant units have been built. The size and shape had to follow the requirements to be placed inside the connection boxes of the solar modules. The MPT and MPR have been built on standard PCBs. A layout will follow upon results of the pilot plant operation and possibly further modifications.

The first test pilot plant has been installed with a set of 4 MPT/MPR at BP PV modules type BP585 each. The dc power bus consists of integrated MPRs. A standard Sun Profi Type SP1200 operates as CPP.

The PV modules are mounted at a protection railing at the end of the roof terrace of a building.

The MPTs are placed inside the PV module connection boxes as shown before. The MPRs are fixed on the cover of the boxes alike the prototype configuration described before. The Sun Profi SP1200 has been placed outside.

The final pilot plant will consist of a total of 12 PV modules that shall be completed with MPT/MPR upon further evaluations and modifications. The energy will be magnetically coupled to an ac power bus.

2.8 Performance Analyses

The complete NST pilot plant will be subject to a detailed monitoring phase and performance analysis. The objective of this task is to monitor and evaluate the PV system and further visualise the plant performance. The NST pilot plant is placed in such a way that the shading effect occurs and the advantages of the system can be demonstrated.

The monitoring phase will last one year to receive sufficient data and also compare them with the simulation of a conventional system.

2.7 Building Integration of PV system using NST Technology

This technology requires the identification of new requirements of PV building integration when applying NST. The objective is to reduce costs and increase acceptance of PV building integration. Various applications of the use of the NST are shown in Figure 4. The NST is applicable, especially, where shading effects occur and for PV modules with high voltages [2].

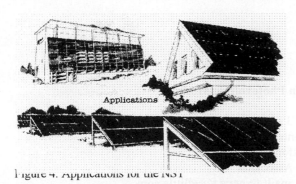

Figure 4. Applications for the NST

2.9 Installation of the NST

The magnetically induced power bus with the fully encapsulated magnetic power receiver is pre-assembled and therefore allows fast installation and less safety needs. PV modules can be installed or replaced without closing down the system and electricians are working with the lowest risk compared to the conventional PV systems. Figure 5 shows the illustration for the installation of the NST.

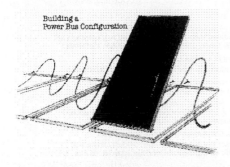

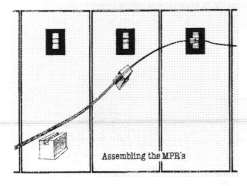

Figure 5. Installation of the NST.

3. BENEFITS

This new innovative technology will remove the weaknesses of the state-of-the-art technology of conventional PV system and consequently results in a significant cost reduction. More than 20% cost reduction of power conditioning devices (from 1.4 ECU/Wp to 1 ECU/Wp over 25 years lifetime) is the estimated cost-reduction potential of the PV system. This net reduction of cost will be achieved through:

- a significantly increased lifetime of components of more than 25 years and an MTBF of more than

100,000h which perfectly matches with the PV module lifetime. As a consequence, replacement of power conditioning equipment during plant lifetime will be obsolete with NST.

- The utilisation of high-frequency electronics will reduce the size of components and subsystems and allow for a full integration of individual components such as the MPT into the PV module. A high degree of integration of electronic components will be achieved with ideally leads to the "one-chip-inverter"

- The avoidance of direct electrical connections leading to increased contact resistance and corrosion but instead using magnetically induced coupling which is not susceptible to corrosion.

- Significant reduced system installation cost and reduced safety needs due to the use of a MI-Power Bus with fully encapsulated Magnetic Power Receivers. PV modules can be installed or replaced without closing down the system. Electricians working on the system have lowest risk compared to conventional PV systems.

Taking the economic benefits at system level into account (reduced cabling and installation cost, reduced cost due to minimised safety needs which also accounts for the PV module (protection class 2 will no longer be required)), they allow for an assumed total reduction of system cost of 15-20% (over the PV system lifetime) which will create an essential push to take up this technology at a large scale and open up new markets for PV.

4. CONCLUSIONS

The development of the NST will result in the following advantages:

- Lifetime of balance of system components are matching the PV module lifetime
- Increase of the Mean Time Between Failures (>100.000h)
- High degree of integration of electronic circuits
- Miniaturisation of electronic components
- Reduced losses
- Avoidance of isolation problems
- Cost Reduction

The innovative NST concept will open up new markets for PV building integration sector.

5. REFERENCES

[1] Report on the New PV System Technology performed in the framework of the Non Nuclear Energy Programme JOULE III co-financed by the European , Commission February 2000.

[2] A. Hänel, 'State of the Art in building integrated PV', European Directory of Sustainable and Energy Efficient Building 1999', 1999 James & James (Science Publisher) Ltd.

Opportunities for Cost Reductions in Photovoltaic Modules

Lisa Frantzis and Evan Jones, Arthur D. Little, Inc., 35 Acorn Park, Cambridge, MA USA
Charles Lee, Cambridge Consultants, Cambridge, UK
Madeline Wood, ETSU, Oxfordshire, UK
Paul Wormser, Solar Design Associates, US
Tel: 617-498-5688   Fax: 617-498-7007 email: frantzis.l@adlittle.com

The purpose of this paper is to examine the current and potential future module costs of five major photovoltaic (PV) technologies using a common modeling methodology. The five technologies examined were CZ silicon; cast silicon; amorphous silicon; cadmium telluride (CdTe); and copper indium diselenide (CIS). This work was sponsored in part by the UK Department of Trade and Industry.

Keywords: Module Manufacturing - 1: PV Module - 2: Cost Reduction - 3

Approach

A series of production cost models were used.

- Direct manufacturing costs:
  - all materials and manufacturing labor, including first-level supervision
  - depreciation for manufacturing capital equipment and facilities.
- Other typical business costs (management staff and SG&A) were not considered
  - they are a function of corporate strategy rather than technology
- Input data was gathered from PV manufacturers, manufacturing equipment providers, raw material suppliers and researchers active in PV technology development.

Approach

Processes were modeled step-by-step and linked to a common set of cost assumptions: labor, utilities, facilities and raw materials.

- The models consists of a series of linked spreadsheets
  - following the product down the manufacturing line step-by-step
- Modeled based on known existing processes as well as our knowledge of technology development programs across the world
- Inputs at each step include the number of units produced, throughput, operating uptime, labor loading, utilities consumption, floor space, direct materials added or consumed, and yield
- Yield losses include both broken and out-of-spec product
  - adjusted for the fact that some out-of-spec product can be sold either as a low efficiency product or for other applications at a discount.
- The manufacturing plants are assumed to operate 24 by 7 for 50 weeks a year.

Approach

A state-of-the-art 10 MW/yr production facility was modeled for each technology as the "current" base case.

- For crystalline technologies, there are multiple real-world examples at 10 MW/yr and above
- No thin-film facilities are currently producing 10 MW/yr creating some uncertainty in the model inputs
  - a-Si is modeled after a double junction device
    - has the most commercial experience of thin-film technologies
    - yield/efficiency are still lower than crystalline technologies. This may improve as production lines mature.
  - Although they have promise and are receiving significant investment, no large-scale CdTe or CIS facilities are producing commercial products today.
    - the base case represents a new facility in start-up mode
    - yields are low, leaving significant room for improvement
    - average aperture area stable efficiency is assumed to be 8% for CdTe and 9% for CIS

Module Direct Manufacturing Cost Summary of Existing Cost Structures

Module Direct Manufacturing Costs for 10MW Plant in 2000

	CZ Silicon	Cast Silicon	a-Si	CdTe	CIS
Average Efficiency*	15%	14%	6%	8%	9%
Yield*	85%	85%	75%	65%	65%

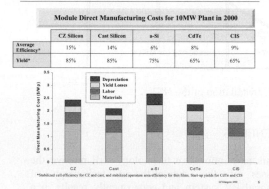

*Stabilized cell efficiency for CZ and cast, and stabilized aperature area efficiency for thin films. Start-up yields for CdTe and CIS

Module Direct Manufacturing Cost   Summary of Existing Cost Structures

Cast silicon has the lowest direct manufacturing cost structure in 2000.

- Nearly all new crystalline PV manufacturing facilities over the last five years have been cast
  - Cast is lower cost than CZ because it uses less silicon and can utilize a lower grade, lower-cost silicon
- CZ silicon has been the workhorse of the PV industry
  - will continue to be produced in fully amortized production facilities and higher efficiency is valued in some applications
- a-Si cost structure is high on a $/Wp basis even though it is the cheapest on a $/ft² basis.
  - low efficiency is one reason a-Si has been losing market share over the last decade

Module Direct Manufacturing Cost   CZ Single Crystal Silicon

The current direct manufacturing cost of CZ single crystal silicon modules is ~$2.45/Wp, dominated by material costs.

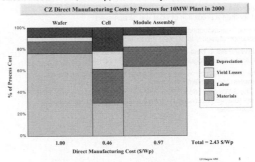

CZ Direct Manufacturing Costs by Process for 10MW Plant in 2000

Module Direct Manufacturing Cost   Cast Polycrystalline Silicon

The direct manufacturing cost of cast polycrystalline silicon modules is ~$2.10/Wp.

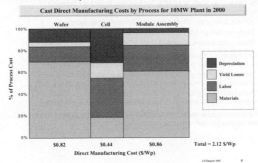

Cast Direct Manufacturing Costs by Process for 10MW Plant in 2000

Module Direct Manufacturing Cost   Amorphous Si

The direct manufacturing cost for a-Si modules is ~ $2.70/Wp.

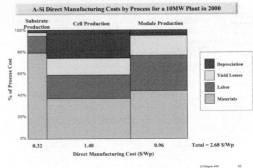

A-Si Direct Manufacturing Costs by Process for a 10MW Plant in 2000

Module Direct Manufacturing Cost   CdTe

The direct manufacturing cost for CdTe modules is ~$2.30/Wp.

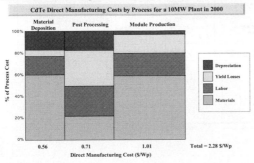

CdTe Direct Manufacturing Costs by Process for a 10MW Plant in 2000

Module Direct Manufacturing Cost   CIS

The direct manufacturing cost for CIS modules is ~$2.25/Wp.

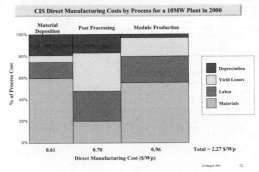

CIS Direct Manufacturing Costs by Process for a 10MW Plant in 2000

Module Cost Reduction Opportunities Overview

Four major cost reduction opportunities were examined for each technology: efficiency, yield, scale and material utilization.

- Cost reductions represent what is possible in the next 10 years, rather than a prediction of what will be achieved
- Multiple cost reduction pathways need to be pursued in each cost reduction category to achieve these aspirational goals
- If investment in both R&D and new manufacturing facilities is low for some technologies, these goals will not be met
- The cost reduction % listed represents the relative opportunity between the different categories
  - The $/Wp cost reductions displayed in the waterfall graph depict cumulative cost reductions; they should not be compared with each other

LFGlasgow 4/00   13

Module Cost Reduction Opportunities CZ Silicon

Direct manufacturing cost for CZ silicon could fall from $2.45 to $1.45/Wp over the next 10 years.

	Cell Efficiency	Yield	Scale (MW)	Material Utilization
Achievable Improvement	15% ➡ 18% stable	85% ➡ >95%	10 ➡ 100	20% ⬇ silicon usage
Improvement Pathway	• Process control • Front contact shadowing • Back surface field	• Eliminate gross failure of sawing • Prepare for thin wafers	• Vol. purchasing • Larger equip. • Higher throughput	• Improved slicing • Vapor etching • Improved packaging • Lower cost silicon
Cost Reduction	14%	5%	21%	9%

LFGlasgow 4/00   14

Module Cost Reduction Opportunities Cast Silicon

Direct manufacturing costs for cast modules could fall from ~$2.10 to $1.15/Wp in 10 years.

	Cell Efficiency	Yield	Scale (MW)	Material Utilization
Achievable Improvement	14% ➡ 17% stable	85% ➡ >95%	10 ➡ 100	20% ⬇ silicon usage
Improvement Pathway	• Process control • Hydrogen passivation • Back surface modification	• Eliminate gross failure of sawing • Prepare for thin wafers	• Vol. purchasing • Balanced line • Larger equip. • Higher throughput	• Improved slicing • Vapor etching • Improved packaging • Lower cost silicon
Cost Reduction	17%	4%	25%	8%

LFGlasgow 4/00   15

Module Cost Reduction Opportunities Thin Films

Thin-films have not yet reached their promise on a commercial scale, posing production cost modeling challenges.

- The base case plant is 10 MW (no thin film facility is operating at that capacity today)
  - 10 MW base case plant was used to maintain consistency
  - Scale up has proven to be a significant challenge
  - Base case assumes current state of technology
- The trade-offs between yield, throughput and efficiency have been and most likely will continue to be significant issues for thin films
  - Increasing efficiency through more complicated structures adds process steps, decreasing yield and throughput (especially a-Si).
  - Increased throughput can help to amortize the large capital investment, but tends to put pressure on yields and efficiency
- The following cost reduction analysis assumes that thin films meet their targets for both efficiency and yield in 10 years.

LFGlasgow 4/00   16

Module Cost Reduction Opportunities Amorphous Silicon

If a-Si overcomes significant challenges, direct manufacturing costs could drop from ~$2.70 to $1.40/Wp over the next 10 years.

	Aperture Area Efficiency	Yield	Scale (MW)	Material Utilization
Achievable Improvement	6% ➡ 8% stable	75% ➡ 90%	10 ➡ 100	Unquantified
Improvement Pathway	• Process control • Material quality	• Process definition and control	• Vol. purchasing • Balanced line • Larger equip. • Higher throughput	• Improved packaging • Improved germanium utilization
Cost Reduction	23%	7%	23%	5%

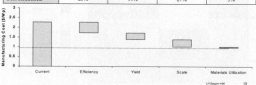

LFGlasgow 4/00   17

Module Cost Reduction Opportunities CdTe

Progress for CdTe modules could bring direct manufacturing costs from $2.30 to $0.95/Wp over the next 10 years.

	Aperture Area Efficiency	Yield	Scale	Material Utilization
Achievable Improvement	8% ➡ 12% stable	65% ➡ 90%	10 ➡ 100	Unquantified
Improvement Pathway	• Process control • Process definition • Contact enhancement	• Process definition and control	• Vol. purchasing • Balanced line • Larger equip. • Higher throughput	• Improved packaging • Improved tellurium utilization
Cost Reduction	25%	19%	27%	5%

LFGlasgow 4/00   18

Module Cost Reduction Opportunities CIS

CIS is similar to CdTe and could drop from $2.25 to $1.00/Wp over the next 10 years if significant progress is made.

	Aperture Area Efficiency	Yield	Scale (MW)	Material Utilization
Achievable Improvement	9% ➡ 13% stable	65% ➡ 90%	10 ➡ 100	Unquantified
Improvement Pathway	• Process control • Process definition • Contact enhancement	• Process definition and control	• Vol. purchasing • Balanced line • Larger equip. • Higher throughput	• Improved packaging (EVA/ZnO interaction) • Improved indium utilization
Cost Reduction	24%	19%	25%	5%

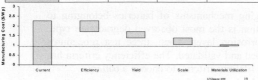

LFGlasgow 4/00   19

Module Cost Reduction Opportunities Range of Possible Costs Achievable by 2010

Thin film technologies could overtake cast silicon, but have greater risk.

- Cast silicon will likely remain the near-term dominant technology

- CdTe and CIS appear to have the largest upside potential

- CZ silicon can only achieve the cost reduction indicated if investment in new/larger production facilities occurs
  - CZ manufacturers are currently making incremental improvements to existing facilities

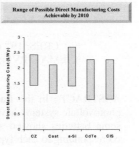

LFGlasgow 4/00   20

Module Cost Reduction Opportunities Range of Possible Costs Achievable by 2010

Emerging thin films, such as CdTe, will need to achieve an attractive combination of efficiency and yield to match cast Si.

- Due to the lack of commercial experience of CdTe and CIS, it is uncertain what efficiencies and yields are achievable.

- However, in order for CdTe to match current cast module direct manufacturing costs, CdTe will need to achieve total yields of 70% and aperture area efficiencies of 9%

- CIS will need to achieve similar types of targets

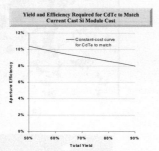

LFGlasgow 4/00   21

QUICKER ASSESSMENT OF LIFETIME OF PV BATTERIES (QUALIBAT)

Ph. Malbranche[1], S. Métais[1], F. Mattera[1], J. L. Martin[1], D. Desmettre[1],
C. Protogeropoulos[2], N. Vela[3], F. Fabero[3].

[1]GENEC, CEA Cadarache, 13108 Saint-Paul-Lez-Durance, FR. Email : philippe.malbranche@cea.fr.
phone : 33 4 42 25 66 02, fax : 33 4 42 25 73 65.
[2]CRES, 19th km Marathonos Av., 190 09 Pikermi, Athens, GR. Phone 301 603 99 00
[3]CIEMAT, PVLabDER, Avda. Complutense, 22. 28040 Madrid, ES. Phone : 34 91 346 6745

ABSTRACT : In the frame of the Qualibat project, damaging mechanisms of batteries belonging to photovoltaic systems have been studied. Irreversible sulphation is the most observed degradation type, some batteries showed shedding and the presence of a corrosion layer. Then, three test procedures corresponding to each type of degradation have been developed and validated. The efficiency criteria has been studied, it seems to be an interesting criteria, beyond the lifetime criteria, to choose a battery.
Keywords : Glasgow Conference - 1: Storage - 2 : Batteries - 3 : Test procedure

1. PURPOSE OF THE WORK

The Qualibat project concerns the development of test procedures of batteries for PV systems. This project involved three European laboratories (GENEC-France, CIEMAT-Spain, CRES-Greece) and two industrial partners (OLDHAM-France and CHLORIDE-England) during thirty months.

The main purpose of this project is to develop appropriate testing methods which reduce the number and the duration of the experimental tests to be performed and consequently to enable the wisest possible selection of the most appropriate batteries for PV applications.

1. APPROACH AND RESULTS

1.1. Photovoltaic batteries degradations

The first task was to study the damaging mechanisms of the batteries. Therefore, seven batteries were accurately characterised by electrical measurements, chemical titration in several location of the active mass, electrolyte analysis and finally S.E.M. observations of the grid and the active mass.

The main observed defect is the irreversible sulphation of active material with higher lead sulphate rates in the positive active material [1]. Moreover, stratification of the electrolyte is developed by the absence of significant gassing and this phenomenon is linked with the presence of irreversible sulphation (Figure 1a).

Loss of connection (shedding) between active material and the grid is observed, characterised by the presence of active material in the bottom of the container (Figure 1b).

The presence of an insulating barrier of lead oxide in the grid - active mass interface which could hinder the exchanges between the collector and the active mass can be observed (Figure 1c).

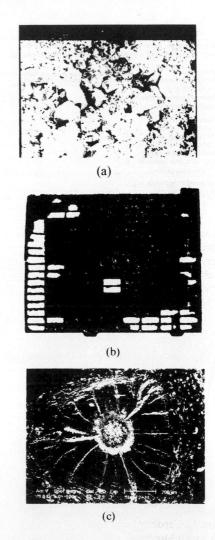

(a)

(b)

(c)

Figure 1 : Observation of positive plates.

(a) : S.E.M. photography of positive active material (Lead sulphate crystals).
(b) : Photography of a positive flat plate (Shedding).
(c) : S.E.M. photography of positive plate grid-active mass interface (Loss of connection of active mass).

1.2. Test procedures and results

The development and the validation of the test procedures which allows quicker degradations have been studied.

The work has been focused on four issues : shedding, irreversible sulphation and the recovery from deep discharges, corrosion and efficiency.

① The cycling or shedding test is based on high currents and quick cycles (three per day) (Figure 2).

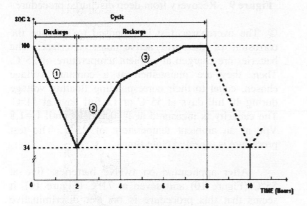

Figure 2 : Shedding test procedure.

For flooded batteries :
Discharge (1) : 2h at 0.33 C/10
Recharge (2) : 2h at 0.25 C/10
         (3) : 4h at 0.071 C/10
Overcharge ratio : 1.2 1.25

For VRLA batteries :
Discharge (1) : 3h at 0.22 C/10
Recharge (2) : 10h at 0.19 C/10 up to U=2.28 V
         (3) : 3h at 0.01 C/10 up to 2.28 V
Overcharge ratio : 1.05-1.25

A validation on four different batteries show that this procedure is around three times quicker than the usual ones, Figure 3 show the comparison of the results obtained with Qualibat test procedure with the ones obtained with a procedure developed in Genec, called "Moroccan procedure".

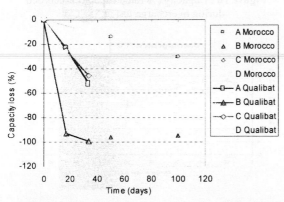

Figure 3 : Capacity loss changes of four identical batteries with Moroccan and Qualibat test procedure.

The use of this procedure on more than fifteen types of batteries results in an easy discrimination of the behaviour of the batteries in few months (Figure 4, Figure 5 and Figure 6).

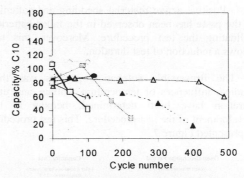

Figure 4 : Capacity value changes recorded during cycling tests performed on flooded flat plates batteries

The flooded flat plate designs show variable performance with most dropping below 70% capacity after 100 cycles (Figure 4).

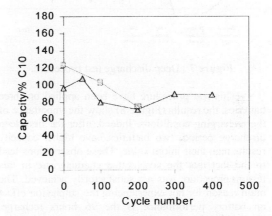

Figure 5 : Capacity value changes recorded during cycling tests performed on flooded tubular plates batteries

The flooded tubular designs are giving stable capacities at up to 400 cycles. One battery shows a gradual reduction in capacity over 200 cycles, this was identified as a short due to a paste lump in one cell.

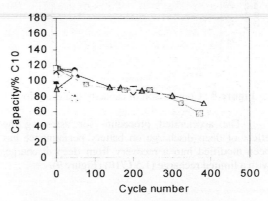

Figure 6 : Capacity changes during cycling tests performed on VRLA batteries.

Some of the valve regulated designs have reached 70% of rated capacity after 350 cycles.

Positive active material shedding and softening of the paste has been observed in the failed batteries, validating the test procedure. Moreover, this test allows a reduction of test duration.

② The irreversible sulphation process was then studied. Influences of temperature, voltage and time duration have been determined, helping in the development of the test procedure. This test procedure is presented Figure 7.

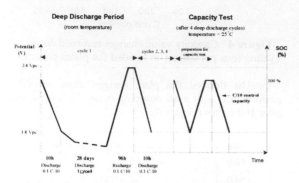

Figure 7 : Deep discharge test procedure.

This test procedure have been applied on three batteries, the results (Figure 8) show the importance of the recharging conditions. Indeed, after the first deep discharge period, two batteries have higher capacity results than their initial value. These observations lead to the fact that the accelerated characteristic in this testing procedure was not satisfactorily achieved. The main reason for not experiencing the sulphation effect on battery performance is the 96 hours recharge. During the charge, lead sulphate was transformed into lead dioxide, because the charged capacity was in some cases more than three times the nominal C/10 capacity value of the battery.

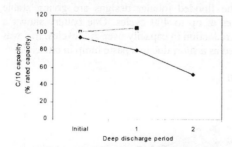

Figure 8 : Comparison of capacity test evolution

The accelerated procedure for assessing the effect of deep discharge on battery performance has been modified into a recovery from deep discharge, with a limited recharge (1.5 C/10) (Figure 9).

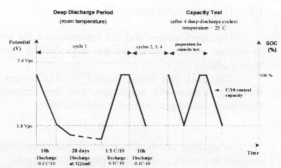

Figure 9 : Recovery from deep discharge procedure

③ The overcharge test is proposed to enhance the corrosion of the grid active mass interface. The batteries are charged at ambient temperature of 25°C. Then, they are maintained at a constant voltage chosen, equal to their corresponding floating voltage during 40 full days at 55°C or 12 full days at 71°C. The capacity is measured at I= 0.1 C/10 until U=1.8 Vpc at an ambient temperature of 25°C. The test procedure is repeated eight times.

After application on twelve batteries, five at 55°C (Figure 10) and seven at 71°C (Figure 11), it seems that this procedure is not yet discriminative enough and has to be strengthened.

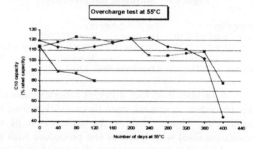

Figure 10 : Capacity (% rated capacity) recorded during overcharge test at 55 °C.

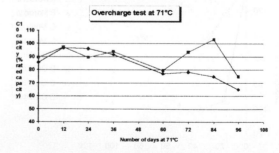

Figure 11 : Capacity (% rated capacity) recorded during overcharge test at 71 °C

④ The efficiency test (Figure 12) gives an additional information on the behaviour of the battery under specific PV conditions, namely cycling in low state of charge. Batteries are cycled 5 times between 0 % and 50 % of SOC. Two types of efficiency are then calculated, the faradic efficiency (Ah efficiency) and the energetic efficiency (Wh efficiency).

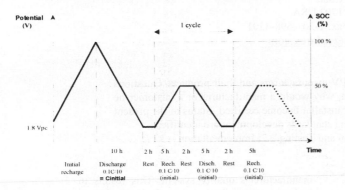

Figure 12 : Efficiency test procedure

Fiveteen batteries of different types were compared with this test procedure. An average value (between the 4th and the 5th cycle) has been calculated for each battery (Figure 13a and Figure 13b).

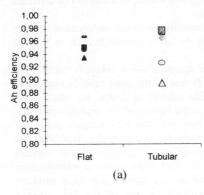

(a)

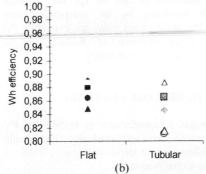

(b)

Figure 13 : Efficiency average values (between the 4th and 5th cycle) of flooded flat and tubular plates batteries
(a) : Faradic efficiency ; (b) : Energy efficiency.

Tubular plates batteries show higher Ah efficiency values dispersion (9 %) than flat plate batteries (5 %). The energy efficiency value dispersion is nearly equivalent around 8 % for both types of batteries.

Flat plates batteries and tubular plates batteries have similar faradic efficiency values. Energy efficiency values are higher for flat plates batteries (around 0.88) than for tubular plates batteries (around 0.85).

CONCLUSION

Results of this study will permit us to propose new test methods to choose the best suited battery for PV application with shorter time and less expensive test procedures than the existing ones.

The test procedure designed for quick cycling is above our expectations : three times quicker than the quickest one we have used so far. And the comparison made with another cycling procedure test validated the results obtained.

A result which was not expected is the one obtained with the deep discharge test procedure : six months at almost 0 V is not very damaging to some batteries, provided that the recharge is not limited, in order to recover from all the irreversible sulfation obtained. Therefore, this test is re-oriented from "resistance to deep discharge" to "recovery from deep discharge, with a limited recharge".

Regarding the further investigations carried out in order to obtain additional criteria to facilitate the battery selection process, the efficiency test seems to be interesting. A repeatable test procedure was developed, which makes interesting results achievable in four cycles (in less than two days) for most of the batteries. Discrepancies of more than 10 % are visible between different types of battery.

REFERENCE

[1] F. MATTERA, Ph. MALBRANCHE, D. DESMETTRE, J.L. MARTIN, "2nd World Conference and Exhibition on Photovoltaic Solar Energy Conversion", Wien (Austria) July 1998.

QUALIFICATION AND CERTIFICATION:
KEY ELEMENTS IN THE SUCCESSFUL IMPLEMENTATION OF PV

John H. Wohlgemuth
BP Solarex
630 Solarex Court
Frederick, MD 21703 USA
Phone: 1-301-698-4375   Fax: 1-301-698-4201
E-mail: jwohlgemuth@solarex.com

ABSTRACT: This paper discusses the need to certify PV manufacturers and their products. Questions answered include, what does certification actually mean, why would a manufacturer or a customer be interested in certification and how can a manufacturer or installer become certified. Status of the present certification program is reviewed and those areas requiring standards development are identified.
Keywords: Certifications and Standards – 1; Qualification and Testing – 2; Implementation – 3

1. INTRODUCTION

Every time a PV system fails to meet customer expectations, it is a set back for the PV industry. It is critical that PV systems perform as specified and as a minimum, meet the warranty lifetime. To achieve this all of the components in the system must be built with quality components to the appropriate design. This is accomplished by using a certified quality system and product qualification testing. The quality of the product is proven via qualification testing to an accepted standard and via control of the manufacturing process using a certified quality system.

The requirement to meet customer expectations and specifications is true for almost all products. Why then is it so important for PV? Because PV systems:

1. Have a very high up front capital cost. PV systems have to operate for many years in order to provide a return on the investment.
2. Are often deployed at remote sites where little or no maintenance is available.
3. Are often used to power critical loads that are worth much more than the PV itself.
4. Are often financed by governments or NGO grants and loans or with bank mortgages, which require a greater degree of assurance that the system will work as intended.

2. WHY SHOULD I WANT TO DO THIS?

Is such a system of certification and qualification going to be unnecessarily expensive? The answer to this is "No" it should not be. In reality it should save money for both the PV industry and its customers because:

1. Every PV system that does not meet specifications and the customer's expectations costs both the customer and the PV industry. These costs may include loss of future business for the seller and the entire PV industry.
2. Today some large customers require the PV manufacturer to implement such a system for their particular job. They require product qualification and do their own review of the manufacturer's quality system. This results in higher costs on both sides. The customer is paying all of the costs for qualification testing and a quality system review. The manufacturer has to go through qualification testing and the review for each such customer.
3. Today PV component manufacturers have to meet multiple standards and may have to pay for qualification testing at a number of different laboratories.

So an internationally accepted certification system should actually be an economic benefit to the PV industry.

The other question often asked is "aren't PV modules already one of the most reliable sources of electricity?" The answer of course is "yes". If we look carefully at the history of PV module reliability, we find that they got that way because many PV module manufacturers have really followed the formula for certification already. These module manufacturers are ISO 9001 certified, meaning they have a certified quality system, and their products have been qualified to one of the IEC module qualification standards by an independent test laboratory. The main issues today are extending the system to cover Balance of Systems (BOS) components and systems themselves.

3. HOW TO BECOME CERTIFIED

How would a manufacturer or installer of PV products become certified? Basically they would follow the following four steps.

1. Develop a quality system.
2. Have their quality system audited and certified, probably to ISO 9001 or 9002.
3. Have their products qualified to the appropriate international standard by an accredited testing laboratory.
4. Have the first three steps reviewed by a certifying agency.

This sounds very simple, but in reality there are a number of organizations and standards involved. The following subsections attempt to explain how they all interrelate before providing specifics on working through this process.

IEC – The International Electrotechnical Commission writes standards for electrical components including PV. TC-82 is the IEC technical committee that prepares standards for PV. TC-82 has prepared PV module qualification documents IEC 61251 for crystalline silicon modules and IEC 61646 for thin film modules.

IECQ – IEC Quality Assessment System for Electronic Components. IECQ has been retained by PVGaP to certify PV products. The IECQ system is implement by audit bodies called National Supervisory Inspectorates (NSI's). The NSI's assign the people who actually perform the final review and approve granting of a certification from IECQ and then from PVGaP.

PVGaP – Global Approval Program for PV. PVGaP has established a product certification system for PV components and systems. However, today the only PV component that can be certified is a PV module. The required international standards are not available for BOS or systems, so there is nothing to qualify/certify products to. PVGaP is preparing interim standards to get the process going, while IEC works on consensus international standards in these areas.

PowerMark – PowerMark is a non-profit Corporation that has established a PV product certification program in the United States. PowerMark runs its own certification program in the US. This program has led to the accreditation of one PV testing laboratory and the certification of two module manufacturers. PowerMark is also the US representative for PVGaP.

Accredited PV Testing Laboratories – These are laboratories that have been accredited specifically for PV testing by a third party laboratory accrediting agency. At least three laboratories (JRC, the PTL at Arizona State University and TUV) are certified to perform PV module tests to IEC standards. Tests performed at any one of these laboratories should be accepted anywhere in the world.

Blank Detailed Specification – This is a template for completing a detailed specification of a product. This is a critical part of the certification process, since the certification is actually done to the Detailed Specification prepared by the manufacturer of the product using the guidelines of the Blank Detailed Specification. The certification states that the products tested met the Detailed Specification and that manufacturer has an appropriate quality system to insure that future production continues to meet the Detailed Specification.

4. STATUS

There is a system in place today to certify PV components and systems. However, it is only complete for PV modules. If a PV module manufacturer has ISO 9001 or ISO 9002 for their factory and has qualified their modules to IEC 61215 or IEC 61646 at JRC, ASU or TUV, they are ready for certification. Depending on location, they should contact either PVGap or PowerMark to undergo a review of their ISO documentation and of the qualification testing and then pay a fee for certification.

So why are few PV module manufacturers certified today? There doesn't appear to be a compelling need because customers are not requiring or even requesting certification. Most PV modules are very reliable and come with long warranties. Many PV module manufacturers already have ISO certification and qualified modules. There is no competitive advantage for a module manufacturer to be certified under this process today.

Most customers today are worried more about the quality and reliability of the BOS components and of the system itself. Is there a system in place today to certify BOS components and systems? The system is there, but the lack of performance and qualification standards means that there is nothing to test to. This is rapidly changing. PVGaP has done an excellent job of fostering the development of BOS and systems standards. Soon interim PVGaP standards and IECQ Blank Detailed Specifications will be published. These will be submitted to IEC for development of international standards.

Performance and qualification standards for BOS and PV systems will be available shortly. There will soon be a need for accredited test laboratories to perform the BOS and systems performance and qualification tests. It appears that several laboratories including JRC and Florida Solar Energy Center (FSEC) will met this need.

Soon everything will be in place to begin certifying BOS component manufacturers and systems integrators. Such certification should provide a commercial advantage and is likely to be required by some large customers such as the World Bank. Since certified systems will require the use of certified components, it should then be a commercial advantage for PV module manufacturers to certify their products.

5. SUMMARY

A lack of standards has hurt the BOS and systems area. However, too many standards may be worse. A proliferation of national standards that are different than IEC standards will cause chaos and hurt the whole industry. Use of international, standards, preferably IEC if available or interim PVGaP, and full reciprocity of accredited laboratory testing are critical for this process to work and help PV become a successful worldwide business.

PV customers should begin asking their PV product vendors about component and system certifications now to show their interest. Both customers and PV manufacturers should provide their inputs to the standard writing activities underway in IEC and PVGaP. It is also critical that those in the PV business work with their trade associations and national committees, letting them know that you favor the use of international standards and testing reciprocity. This is the only way that PV will become a large, profitable business.

ENHANCEMENT OF PV MODULE EFFICIENCY USING REDUCED REFLECTION-LOSS SURFACE

Masami Tachikawa, Yousuke Nozaki, Takashi Nishioka, and Takeshi Yamada
NTT Telecommunications Energy Laboratories
3-1, Morinosato Wakamiya, Atsugi-shi, Kanagawa Pref., 243-0198 Japan

ABSTRACT: When a photovoltaic module is placed on a vertical surface the amount of power generated when the altitude of the sun is high is less than expected. We think that this problem arises from reflection loss at the glass surface. We used a structured glass surface to reduce this reflection loss and measured the power generated using a solar simulator. More power was generated by the module with the structured surface than by one with a conventional flat surface. We also estimated that the annual power output would be 10% greater than that from a module with a conventional surface. The theoretically calculated limits on the increased efficiency of modules with structured surfaces are 12.5%. Therefore, the module with structured surface is highly efficient. We also evaluated the field performance of a module with a structured surface on a sunny day and found that, for south-facing modules, 12.7% more power is obtained from one with a structured surface. When the modules were facing north, 7.13% more power is obtained from the one with the structured surface. Because the theoretically calculated limits on the increased efficiency of south-facing and north-facing modules with structured surfaces are respectively 15.1% and 10.1%, the module with the structured surface is effective not only with direct beam irradiance but also with diffuse or reflected irradiance.
Keywords: Facade - 1: PV Module - 2: High-Efficiency – 3.

1. INTRODUCTION

A PV (photovoltaic) module is sometimes put on a vertical surface in order to make the best use of limited space (Fig. 1), and the use of such modules has been increasing recently. The problem with them is that they generate less power than expected when the altitude of the sun is high [1]. We think this problem arises from the reflection loss at the glass surface of the module because at noon during the summer solstice in Tokyo, the incident angle of sunlight is nearly 80° and the reflection loss at a vertical air-glass surface is about 40%. The reflection loss at the surface of the module can be reduced by using a low-refractive index material and by using a structured surface. In the work reported here, we explored the use of a structured glass surface.

2. ESTIMATION OF REFLECTION LOSS

Figure 2 shows the cross-sectional structure of a typical PV module consisting of glass, encapsulant (EVA, ethylene vinyl alcohol), and a Si PV cell. Reflection occurs at the interface of two materials having different refractive indexes. The refractive index of EVA is almost equal to that of the glass ($n2=n3=1.5$), so there are only two interfaces where the refractive indexes are significantly different. One is the air/glass interface (refractive index $n1/n2 = 1.0/1.5$), and the other is the EVA/PV cell interface ($n3/n4 = 1.5/3.4$). There have been many attempts to reduce the reflection at the latter interface by using antireflection coating on the cell or making the cell surface a textured one. The glass surface, on the other hand, is generally flat. We therefore think the problem that arises when the incident angle of the sunlight becomes high is due to this surface.

The dependence of the reflectance of circularly polarized light on the angle of incidence at an air/glass interface ($n1/n2=1.0/1.5$) is shown in Fig. 3. Reflection increases as the incident angle increases and becomes about 10% at 60°, after which it increases rapidly. The arrows indicate the incident angle of sunlight under conditions in Tokyo, when the glass is assumed to be vertically oriented and facing south. The glass receives the most direct sunlight when the altitude of sun is low at the winter solstice. So, the reflectance is the lowest. As we move through the vernal equinox towards the summer solstice, the altitude of the sun increases, making the incident angle more oblique and the reflectance higher. Of course, the incident angle also depends on the time of day. It is lowest at midday, and highest at sunrise and sunset. For example, on an equinox, it is about 80 degrees before 7 a. m. and after 5 p. m., and the reflectance is over 40%. At the beginning of this work, it was pointed out that the problem

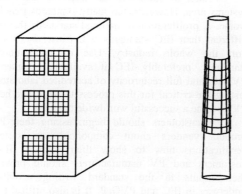

Figure 1: Examples of PV modules on vertical surfaces: facade modules on the wall of a building; cylindrical module wrapped around a pole.

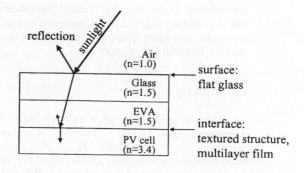

Figure 2: Conventional PV module structure.

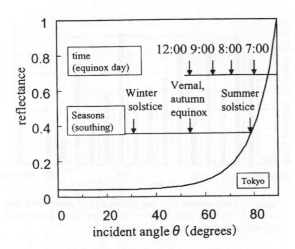

Figure 3: Relation between reflectance and incident angle at the air/glass interface. Arrows indicate the typical incident angles in Tokyo for a vertically mounted PV module.

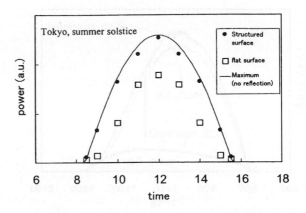

Figure 5: PV module power output as a function of the time of day.

occurs when the altitude of the sun is high. But it would be more correct that it occurs when the incident angle of the sunlight is high.

We used Fig. 3 to calculate the amount of power generated by a PV module over the course of a year. We found that, if there was no reflection loss, we could get 12.5% more power. This value is the upper limit on how much PV module efficiency can be increased by using a surface that reduces reflection loss. We also estimated the reflection loss for diffuse and reflected irradiance. We assumed the diffuse and reflected irradiance is the same for all directions and altitudes. We found that, if there was no reflection loss, we could theoretically get 10.1% more power for the diffuse or reflected sunlight.

3. EXPERIMENTS

We used a PV module with a structured glass surface, which is an array of vertical V-grooves (Fig. 4). The cross-sectional view of the horizontal surface is a triangular waveform; that is, the surface consists of two different azimuth planes. The incident angle and/or reflectance become low for one plane and high for the other. However, sun irradiance increases for the former plane and decreases for the latter. This reduces the apparent reflectance.

Furthermore, the light reflected at the latter is incident again to the glass at the former plane (multiple reflection and incidence). These effects reduce the reflection and increase the incidence of the light. Since dust tends to collect on surfaces with grooves, in this work, we oriented the V-grooves vertically to avoid this problem.

We used an AM1.5 solar simulator to measure how the power generated depends on angles of the two axes of a PV module. By adjusting the angles so that they correspond to a altitude and direction of the sun, we can simulate the amount of direct sunlight the module would receive at any time on any given day.

4. RESULTS

We evaluate the dependence of the PV module power output on the time of day on a given date. Figure 5 shows the results for the day of the summer solstice. Circles and squares show measured values, and the curve shows the power calculated assuming a conventional flat glass surface and that the reflection at the air/glass surface is zero. The power from the module with a structured surface was always greater than that from the module with a conventional flat surface and was very close to the theoretical limit. That means that our new design sufficiently reduces the reflection loss at the surface, at least on the summer solstice. By integrating the PV module power with the time of day, we can obtain the daily power output. We can estimate the annual power output by adding

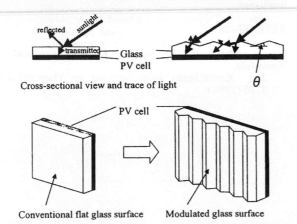

Figure 4: PV modules with conventional and structured surfaces. The incident angle becomes low for one surface. At another surface, the reflected light is incident again to the glass. These effects reduce the reflection of the light and increase PV module efficiency.

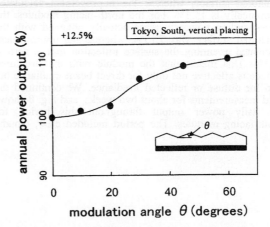

Figure 6: Relation between the annual power output and the modulation angle. Zero degrees corresponds to a conventional flat surface.

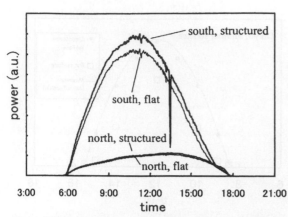

Figure 7: Relation between time of day and the output power of PV modules irradiated with real sunlight.

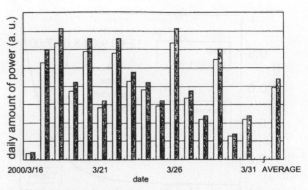

Figure 8: Daily amount of power obtained on 16 consecutive days using a PV module with a structured surface (gray bars) and a conventional flat glass surface (white bars).

up the daily power output for a year. We use these values as indexes of module efficiency.

Figure 6 shows how the annual power output depends on the grade of the surface modulation. As seen in the inset, we use the V-groove angle as the index of the grade of modulation. The power from the module increases as the modulation angle increases. At the practical grade of the angle, we achieved a 10% improvement in the annual power output. As mentioned before, the theoretical reflection loss at the air/glass surface is 12.5%. The module structure we used is therefore highly efficient.

5. FIELD PERFORMANCE

We measured the power output of PV modules with structured and conventional flat surfaces under the real sunlight conditions in Tokyo. The power output dependence on the time of day is shown in Fig. 7 for March 18th, 2000. It was a relatively clear day. We set the modules vertically and facing either south or north. When the modules faced north, the power output due to direct beam irradiance was almost zero and almost all of the power was due to diffused or reflected irradiance. When the modules faced south, the power was due both to direct beam irradiance and to diffused or reflected irradiance. For both south-facing and north-facing modules, more power was obtained from the modules with the structured surface. By integrating the output power over time, we calculated the amount obtained for the entire day. For south-facing modules, this power was 12.7% greater for the one with the structured surface, whereas the improvement calculated theoretically is 15.1%. For the north-facing modules, the corresponding power increase actually obtained with the structured surface was 7.13%, whereas the improvement calculated assuming the surface reflection loss is zero is 10.1%. This means that the module with the structured surface is effective not only for direct beam irradiance but also for diffuse or reflected irradiance. We continued the field measurements for about two weeks, and Fig. 8 shows the daily power output throughout this period for south-facing modules. The period included clear, cloudy,

and rainy days. The daily power output of the module with the structured surface was always greater than that of the module with the conventional flat surface. The advantage of the structured surface over the conventional flat one did not change systematically during the two-weeks period, so we can conclude that dust is not a problem in this time range.

6. SUMMARY

We have evaluated the use of a PV module with a structured glass surface for reducing reflection loss. The amount of power generated was measured using a solar simulator and found to depend on the grade of the surface modulation. At the practical grade of the angle, we achieved a 10% improvement in the annual power output. Since the theoretical reflection loss at the air/glass interface is 12.5%, the module with a structured surface is highly efficient. We also tested the field performance of a module with a structured surface on a sunny day and found that the amount of power obtained during the entire day when this module faced south was 12.7% greater than that obtained from a south-facing module with a conventional flat surface. When the modules faced north, 7.13% more power was obtained from the one with the structured surface. Because the theoretically calculated limits on the increased efficiency are respectively 15.1% and 10.1% for south-facing and north-facing modules, we conclude that the module with the structured surface is effective not only for direct beam irradiance but also for diffuse or reflected irradiance.

REFERENCE

[1] N. Matsuzaki, Y. Nozaki, K. Akiyama and T. Yamashita,
2nd Korea-Japan Joint Seminar on Photovoltaics, T-20 (1998)

LIGHT SOAKING OF AMORPHOUS SILICON, CdTe AND Cu(In,Ga)Se2 SAMPLES

Solveig Roschier, Ewan D. Dunlop
EC DG JRC Ispra
Via E. Fermi 1, I-21020 Ispra (VA), Italy
solveig.roschier@jrc.it, ewan.dunlop@jrc.it

ABSTRACT: In this study single junction amorphous silicon (a-Si), CdTe and Cu(In,Ga)Se2 (CIGS) photovoltaic (PV) material samples were exposed to controlled light soaking. Controlled light soaking is used as a preconditioning step in the type approval testing of thin film photovoltaic modules. The controlled light soaking procedure is based on test experience gained from amorphous silicon modules. In this study two series of a-Si, CdTe and CIGS samples were exposed to light. The samples were measured for their electrical performance frequently during the exposure. As a result the a-Si degraded in performance, as expected from the literature. The CdTe samples degraded in maximum power but all the samples did not behave consistently during the test. The CIGS samples mostly gained in maximum power although there were inconsistencies in the behaviour also within these samples. As a conclusion because the behaviour of CdTe and CIGS materials differs radically from that of a-Si under light exposure, the description of the controlled light soaking procedure in the qualification standard requires some revision in order to be applied to these different thin film materials.
Keywords: light soaking - 1: thin films - 2: qualification testing - 3

1. INTRODUCTION

In this study single junction amorphous silicon (a-Si), CdTe and Cu(In,Ga)Se2 (CIGS) photovoltaic (PV) material samples were exposed to controlled light soaking. Controlled light soaking is a preconditioning step in the type approval test for the qualification of a-Si modules to IEC standards [1]. The application of the controlled light soaking test to other thin film materials such as CdTe and CIGS has been assumed in the absence of other thin film qualification standards. However, a considerable difference in behaviour has been reported elsewhere for CdTe ([2], [3]) and CIGS ([4], [5], [6]) than that known for a-Si modules (e.g. [7], [8], [9], [10], [11], [12]). Thus the validity of light soaking in the qualification tests applied to thin film technologies other than a-Si has been investigated.

Two series of samples of a-Si, CdTe and CIGS modules and prototypes were tested in controlled light soaking. The electrical performance of the samples was measured frequently during the exposure. As a result the a-Si degraded in performance, as expected from the literature. The CdTe samples degraded in maximum power but all the samples did not behave consistently during the exposure time. The CIGS samples mostly gained in maximum power although there were inconsistencies in the behaviour also within these samples. As a conclusion because the behaviour of CdTe and CIGS materials differs radically from that of a-Si under light exposure, the qualification standard demanding controlled light soaking preconditioning step requires some revision to meet the specific needs of different thin film materials.

2. EXPERIMENTAL

2.1 Controlled light soaking equipment used

The controlled light soaking was performed in a chamber equipped with xenon mercury vapour lamps. The UV part of the spectrum in the light was filtered out with plain glass sheet which cuts off the wavelengths below 330 nm. The sample temperatures were regulated using two air fans. The samples were kept at open circuit and continuously monitored for their V_{oc} and the temperature. The irradiance during the light soaking was approximately 1000 W/m^2 and the irradiance was continuously monitored with a pyranometer.

The electrical performance of the first set of samples was measured with SpectroLab Large Area Pulsed Solar Simulator (LAPSS) before the test and then after every 24 hours of testing for the first 72 hours and then after every 48 test hours until 216 test hours. After 216 test hours the samples were stored in dark for a couple of weeks and then remeasured for electrical performance. The controlled light soaking was continued and the modules were then left under light exposure for another 187 hours before measurement, then for a further 48, 72 and 48 hours. The total testing time was 571 hours for the CdTe and the CIGS samples of this set. The a-Si samples finished the test after 523 hours of light soaking.

The second set of samples was also measured with SpectroLab LAPSS before the test and then infrequently first after 24 hours of test, then after next 115, 48, 72 and 48 test hours. The total testing time for these samples was 307 hours.

The samples tested are listed in Table 1 below. The CdTe samples in different series were by different manufacturers and they were prepared with different manufacturing technologies. The CIGS samples in both series came from the same source but the second series samples were of later, modified production.

Table 1. Sample description.

Quantity	Type	Size	Description
1st series			
2	a-Si	1x1ft^2	Framed module
2	CdTe	30x30cm^2	Framed module
2	CIGS	10x10cm^2	No frame, prototype
2nd series			
1	CdTe	5x5cm^2	No frame, prototype
2	CIGS	10x10cm^2	No frame, prototype

3. RESULTS

3.1 Amorphous silicon

The average degradation patterns for a-Si modules exposed to controlled light soaking are shown for short circuit current and open circuit voltage in Figure 1 below and for maximum power and fill factor in Figure 2 below. In average by the end of the test the I_{sc} had degraded by 17 %, the V_{oc} by 4 %, the P_{max} by 35 % and the fill factor by 18 %. Most of the degradation, 25 % in P_{max}, occurred during the first 48 hour of light soaking after which the modules degraded slowly 10 % more. Towards the end of the test the module degradation approached stability.

The interruption after 216 test hours and storing in dark had practically no effect on the a-Si module degradation. The difference after light soaking of 216 hours and after being thereafter in dark is only 0.5 % increase in maximum power. Both a-Si modules tested showed the same degradation pattern.

3.2 CdTe

The average behaviour of short circuit current and open circuit voltage during controlled light soaking of the first series CdTe modules is shown in Figure 1, and the behaviour of maximum power and fill factor is shown in Figure 2 below. The average changes in IV parameters for these modules at the end of test were 6 % degradation in I_{sc}, 3 % in V_{oc}, 10 % in P_{max} and 1 % in fill factor.

The two modules included in the controlled light soaking behaved very differently. The final decrease in P_{max} and in V_{oc} was about the same for both modules. However, in I_{sc} one module degraded 10 % and in fill factor no changes were observed. The other module degraded both in I_{sc} and fill factor 3 %. The oscillating behaviour of I_{sc} shown in Figure 1 and that of P_{max} in Figure 2 for these CdTe modules is in fact due only to one of the modules, illustrating the different behaviour of the samples.

The interruption in the light soaking after 216 hours affected only the module showing the oscillation pattern. This module was recovered during the time in dark almost to the original performance values. After continuing the light soaking, the module degraded quickly and stabilised there until the end of the test. The degradation pattern of damped oscillation was observed during this part of the test in I_{sc} and P_{max} for this module. The interruption did not seem to affect the other CdTe module exposed.

The degradation pattern of the CdTe sample in the second set of light soaking is shown for short circuit current and open circuit voltage in Figure 3 and for maximum power and fill factor in Figure 4 below. The sample degraded in I_{sc} 4 %, in V_{oc} 4 %, in P_{max} 11 % and in fill factor 5 %.

3.3 Cu(In,Ga)Se₂

The average behaviour for CIGS samples in the first set of testing in short circuit current and open circuit voltage are shown in Figure 1 and for maximum power and fill factor in Figure 2 below. On average the CIGS samples degraded in I_{sc} 1 % and increased in V_{oc} 3 %, in P_{max} 25 % and in fill factor 23 %.

The behaviour of the two modules exposed to controlled light soaking was the same. As a general pattern the maximum power and fill factor increased mainly during the first 24 hours of testing after which only slight changes

were observed. During the interruption in the testing after 216 hour of light exposure, the fill factor and P_{max} dropped sharply from an increase of 23 % and 25 % to 12 % when compared with the original values. No significant changes in I_{sc} or in V_{oc} where observed over the whole testing period.

The average degradation for the CIGS samples of the second set of light soaking is shown for short circuit current and open circuit voltage in Figure 3 and for maximum power and fill factor in Figure 4 below. On average the CIGS samples degraded in I_{sc} 3 %, in V_{oc} 1%, in P_{max} 1 % and increased in fill factor 3 %.

The two samples behaved very differently from one another during the second set of testing. The first sample had increased by 11 % in P_{max} and fill factor after 139 testing hours but then dropped sharply after 187 testing hours. The sample then increased again by 8% in P_{max} and 11 % in fill factor after 259 hours of testing but for the final measurement dropped to an overall increase of 4 % in P_{max} and 7 % in fill factor with respect to the initial values.

In the other sample P_{max} decreased after 187 testing hours 6 % having been stable around the initial value for the beginning of the test. In this sample the fill factor increased first by 2 % but dropped after 187 test hours to a degradation of 2 % and oscillated around that degradation value throughout the rest of the testing time. The I_{sc} and V_{oc} values of both CIGS samples had the same degradation patterns.

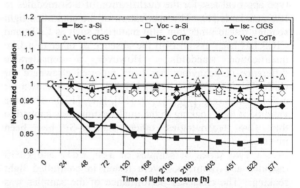

Figure 1. Relative or normalised average degradation of the first series of samples in light soaking. Degradation in short circuit current and open circuit voltage for amorphous silicon, CdTe and Cu(In,Ga)Se₂ samples are presented as a function of light soaking time.

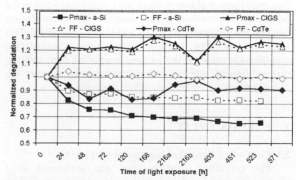

Figure 2. Relative or normalised average degradation of the first series of samples in light soaking. Degradation in maximum power and fill factor for amorphous silicon, CdTe and Cu(In,Ga)Se₂ samples are presented as a function of light soaking time.

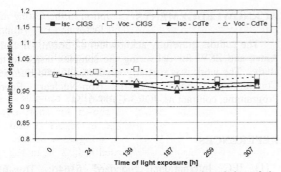

Figure 3. Relative or normalised average degradation of the second series of samples in light soaking. Degradation in short circuit current and open circuit voltage for CdTe and Cu(In,Ga)Se₂ samples are presented as a function of light soaking time.

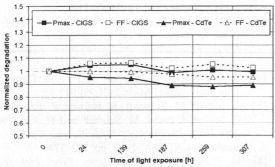

Figure 4. Relative or normalised average degradation of the second series of samples in light soaking. Degradation in maximum power and fill factor for CdTe and Cu(In,Ga)Se₂ samples are presented as a function of light soaking time.

4. DISCUSSION

4.1 Amorphous silicon

The behaviour of amorphous silicon thin film solar cell material under light is used as basis for the thin film qualification testing standard [1] which recommends light soaking to be used as preconditioning of the modules before the actual qualification testing. The amorphous silicon solar cell degradation behaviour under light soaking or the so called Staebler-Wronski effect was first described in 1977 [13]. There are several recent references describing this behaviour when both single junction and multi-junction amorphous silicon solar cells and modules have been tested in controlled indoor and uncontrolled outdoor light soaking conditions (e.g.[12], [11], [10], [9], [7], [8]). A possible cause for the degradation behaviour of a-Si under light was first suggested already in [13] and is discussed in more detail in [14], i.e. the compensation of dangling bonds by hydrogen atoms is playing an important role in the photoinduced metastability of a-Si material.

In our controlled light soaking study light induced degradation was observed mainly in short circuit current and maximum power and somewhat in fill factor. These results agree well with the experiences described in the various references listed above. Also as expected, the degradation stabilised towards the end of the test time. Thus the known light soaking behaviour for amorphous silicon photovoltaic material was confirmed.

4.2 CdTe

The CdTe samples did show some degradation especially in the maximum power when exposed to light soaking. However all the three samples tested in two series behaved very differently. One sample in the first series showed a strange oscillation pattern in maximum power and in short circuit current that however dampened towards the end of the test time. The interruption of the test and keeping the module in dark for some time affected this module such that the in light degraded parameters recovered nearly to their original values. The other CdTe sample in this series had a relatively stable behaviour of slight degradation throughout the entire testing period. The degraded parameter values recovered slightly during the interruption but in this module the changes were minor. The CdTe sample of the second series of testing showed in the end of the testing quite similar degradation values than the modules in the first series of testing. However, the degradation for this sample was smooth without any oscillation behaviour. The most common feature for all the CdTe samples was the about 10 % degradation in maximum power. However, it is difficult to give any general behaviour pattern for CdTe photovoltaic material under controlled light soaking based on the results of our study.

There are only few references previously describing the light soaking effects of CdTe photovoltaic modules [2], [3]. In [2] as a final result of 1500 hours of light soaking of CdTe modules no degradation was observed. However the figures in [2] show slight changes in IV parameters during the testing. In [3] it is argued that the stability behaviour of CdTe devices subjected to light soaking depends on the bias state. In their study the devices tested under resistive load were quite stable relative to devices tested at open circuit. They also suggest that increased temperature during light soaking does lead to increased rate of degradation. According to these findings the test method we used, as described in the qualification testing standard, that the samples are kept in open circuit under light soaking preconditioning step, maybe has enhanced the instability of the samples in our test. The temperature was around 50 °C all the time for all the samples so that should not have affected the degradation patterns. In future work further investigation of the effect of controlled light soaking with a resistive load would be useful.

4.3 Cu(In,Ga)Se₂

In the controlled light soaking study for the first series of CIGS samples there was no degradation observed but on the contrary large increases in maximum power and fill factor were noticed. The open circuit voltage and short circuit current were very stable throughout the entire testing period. The two CIGS samples in the first series behaved identically. For the second series of CIGS samples, which were delivered from the same source as the first series of samples but were of modified production, the changes in the light soaking were totally different compared with the first series. Also the two samples within the second series behaved differently from one another. In both of them the open circuit voltage and short circuit current were quite stable throughout the entire testing period. In one sample of this second series of testing the fill factor was also stable but the maximum power decreased slightly. For the other sample increases in maximum power and fill factor were observed but not to such an extent as in the first series. The decrease after 187 testing hours might be due to measurement done not straight after cooling of the samples but after a somewhat longer waiting period.

For CIGS material photovoltaic modules there are several references describing the material light induced behaviour and its possible causes, e.g. [6], [5], [4]. These references all observe increase in open circuit voltage and fill factor due to light exposure and the effect relaxes when kept in dark again. The same effect was observed when keeping the cells in dark with forward bias applied. In [4] it was claimed however that only the forward bias voltage is increasing the V_{oc}, illumination in short circuit did not have effect on the open circuit voltage. The opposite is claimed in [5] and [6].

The results of our light soaking study correlate well with the findings in literature when considering the increase in fill factor and relaxation of the effect when kept in dark again. The effect of increase in open circuit voltage (mentioned in the above references) was not however observed in this study. In [6] it is stated that the poorer quality devices show largest increases in V_{oc} and fill factor when illuminated. This is also consistent to our study since the second series samples, which showed fewer changes in fill factor and maximum power under illumination than the first series of samples, were of later and improved production.

5. CONCLUSION

For single junction amorphous silicon solar cell material the results of this controlled light soaking study confirmed what has previously been stated in the literature. That is, the a-Si photovoltaic modules degraded in light soaking mostly in short circuit current and in maximum power and somewhat in fill factor. Also as expected the degradation stabilised towards the end of the light exposure. In the CdTe a decrease in maximum power independent of the sample or manufacturer was observed. For other parameters no clear behaviour pattern could be observed during the light exposure. The dependence of bias state during the exposure to CdTe module behaviour should further be investigated. All the CIGS samples except one showed increase in maximum power and fill factor. However, the result differs between the first and second series samples tested, supporting the claim that the observed behaviour may be material quality dependent.

As a general conclusion, even though no clear general trend for CdTe and CIGS material behaviour under light exposure was observed, the behaviour differed significantly from that of a-Si. Since the qualification standard demanding controlled light soaking preconditioning step is based on the behaviour of a-Si material, revisions have to be considered for other thin film materials in order to meet their specific needs.

ACKNOWLEDGEMENTS

The authors gratefully acknowledge Fortum, BP Solar, Antec and ZSW for delivering the samples tested in this study. The work was partially supported by the European Commission, under DGXII Non-Nuclear Energy Programme, contract nr. JOR3-CT97-0154.

REFERENCES

[1] IEC International Standard 61646: Thin-film terrestrial photovoltaic (PV) modules - Design qualification and type approval, Geneva, Switzerland, 1996

[2] V. Ramanathan, L. A. Russell, C. H. Liu, and P. V. Meyers, *Solar Cells* **28** (1990) 129.

[3] R. C. Powell, R. A. Sasala, G. Rich, M. Steel, K. Bihn, N. Reiter, S. Cox, and G. Dorer, *Proc. 25th IEEE Photovoltaic Solar Energy Conference*, Washington D. C., USA, 1996, pp. 785-788.

[4] A. E. Delahoy, A. Ruppert, and M. A. Contreras, *Thin Solid Films* **361-362** (2000) 140.

[5] T. Meyer, M. Schmidt, R. Harney, F. Engelhardt, O. Seifert, J. Parisi, M. Schmitt, and U. Rau, *Proc. 26th IEEE Photovoltaic Solar Energy Conference*, Anaheim, CA, USA, 1998, pp. 371-374.

[6] D. Willett and S. Kuriyagawa, *Proc. 23th IEEE Photovoltaic Solar Energy Conference*, Louisville, KY, USA, 1993, pp. 495-500.

[7] L. Mrig, J. Burdick, W. Luft, and B. Kroposki, *Proc. 1st World Conference on Photovoltaic Energy Conversion*, Hawaii, USA, 1994, pp. 528-530.

[8] Y. Nakata, A. Yokota, H. Sannomiya, S. Moriuchi, Y. Inoue, K. Nomoto, M. Itoh, and T. Tsuji, *Japanese Journal of Applied Physics* **31** (1992) 168.

[9] A. Kholid, H. Okamoto, F. Yamamoto, and A. Kitamura, *Japanese Journal of Applied Physics* **36** (1997) 629.

[10] L.-F. Chen, F. Willing, L. Yang, Y.-M. Li, N. Maley, K. Rajan, M. Bennett, and R. R. Arya, *Proc. 25th IEEE Photovoltaic Solar Energy Conference*, Washington, D.C., USA, 1996, pp. 1137-1140.

[11] D. E. Carlson, R. R. Arya, M. Bennett, L.-F. Chen, K. Jansen, Y.-M. Li, J. Newton, K. Rajan, R. Romero, D. Talenti, E. Twesme, F. Willing, and L. Yang, *Proc. 25th IEEE Photovoltaic Solar Energy Conference*, Washington, D.C., USA, 1996, pp. 1023-1028.

[12] W. Luft, B. von Roedern, B. Stafford, and L. Mrig, *Proc. 23th IEEE Photovoltaic Solar Energy Conference*, Louisville, KY, USA, 1993, pp. 860-866.

[13] D. L. Staebler and C. R. Wronski, *Applied Physics Letters* **31** (1977) 292.

[14] D. Redfield, R. H. Bube, *Photoinduced Defects in Semiconductors*, Cambridge University Press, USA, 1996.

CONTACTLESS CARDS POWERED BY LIGHT

J.C. Zamorano and C. Algora
Instituto de Energía Solar. U.P.M.
Ciudad Universitaria s/n. 28040-Madrid. Spain
Phone: (34) 915441060; Fax: (34) 915446341; E-mail: zamora@ies-def.upm.es

ABSTRACT: The work described in this paper was developed a system, which allows the use of smart cards as a contactless cards, without physical contact with any terminal. This system is composed of two blocks: a card control terminal and a special wallet, where you may put in the smart card when contactless operation is necessary. This wallet is powered by a solar cell array, when illuminated from a light source into the card control terminal. Contactless operations are especially suitable for quickly data transactions, hands-free operations and applications in high security areas. The system works as a bi-directional infrared data-interface between the smart card and the card control terminal. The operational range is of up to one meter, with an angular mismatch within emitter and receiver of ± 30° in both horizontal and vertical planes. The data rate is 9.6 Kbauds, although the data interface is able to work up to 115 Kbauds. The actual operational range is voluntary limited. We are developing new strategies to improve this limitation.
Keywords: Power-by-light – 1. PV Array – 2. Power Conditioning – 3.

1. INTRODUCTION

Since the early fifties, magnetic credit cards have been used countless times. By using the magnetic properties of certain materials, data can be loaded permanently onto a magnetic strip.

Smart cards arose in the seventies and improve on the performance of magnetic cards by the inclusion of both a microprocessor and a memory. The transference of information is performed through contacts made on the surface of the card. Both these two kinds of cards need to make physical contact between themselves and a terminal reader.

There is another class of cards, which are able to work without this limitation. These cards are powered by radiowaves and data are sent also by this medium. They run without making physical contact with any terminal and therefore may be called contactless cards.

The simplest version is a read-only device, which is normally employed in access control. The most sophisticated of these cards include a memory to write and read with. It allows them to perform the same data transactions as both magnetic and smart cards. Nevertheless, in most cases the operational range of cards powered by radiowaves is no greater than a few centimetres.

We aim to demonstrate that this limitation could be overcome by using light to power the cards. This concept was first presented in a Patent [1] in which the authors claimed a codified card, powered by light coming from a special beacon.

A later Patent [2] claims to develop a system, powered by photovoltaic energy, by which any card can be used as a contactless card. As in the patent above mentioned, data transference was made by means of non-guided optical devices.

Based on this work, our objective was to develop a system where a smart card, accepted by a special apparatus, can transfer data to a control terminal and vice versa. The operating range can be up to one meter and data transference has been achieved using infrared wavelengths. Both the smart card as card writer and reader and data interface are powered by solar cells. This is a new low power PV application system that has been previously called as power-by-light.

By using this system, smart cards can be employed as contactless cards with an operating range greater than contactless cards powered only by radiowaves.

2. SETUP

The system, depicted in Figure 1, is composed of two blocks, called in the diagram "Card Control Terminal" (CCT) and "Photovoltaic Card-Wallet" (PCW). In essence, the photovoltaic card-wallet is a writer and reader card. The user must wear externally the PCW when contactless operations are necessary.

This block contains no batteries. A DC-DC converter adapts the electricity produced by the solar cells to the load.

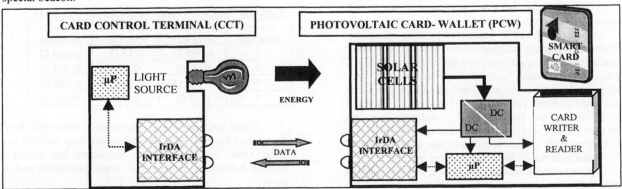

Figure 1: Blocks diagram of the system

This converter supplies IrDA interface, microprocessor, card writer & reader and also the smart card.

In the smart card acceptor a little microprocessor controls data to and from the card and sends it, normally by a serial interface, to a main processor where a card control program has been implemented. In this case, a microprocessor does the same function, but here we use a wireless interface based on an infrared device according to IrDA standards. A detailed description, excluding solar cells and power arrangement is shown in Figure 2

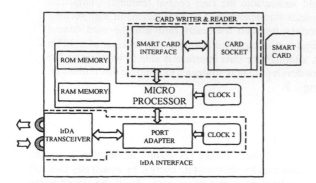

Figure 2: Functional diagram of the PCW

The card control terminal works like a smart card conventional acceptor, where data transmission to and from the card is made by an infrared link. Therefore, the CCT has another IrDA interface. A microprocessor runs the card-control application.

The light source from the CCT guaranties enough level of irradiance on the solar cell array housed within the PCW, regardless of ambient illumination.

A microprocessor could be included in the CCT in order to avoid data collision when several cards try to access the CCT simultaneously. However, it is possible to use a special program in the main processor, instead of the microprocessor, to control card access to the system.

3. DEVELOPMENT

The system can be considered as an inconstant load, which is powered with photovoltaic energy. This question will be the key part of our analysis.

3.1 Sizing of the photovoltaic card-wallet.

In this system the concept of Loss of Load Probability (LLP), very used in stand-alone PV systems [3], is meaningless because we must give enough electrical power to the load every time it is needed.

We consider the PCW composed by three main parts: electronics, including microprocessor and the card writer & reader; the smart card itself and the IrDA interface. All circuits are powered with 3.3 volts and we may distinguish four different situations: idle (I), sending data to the Card Control Terminal (SDTC), reading card and sending data to the CCT (RSDTC), and receiving data from the CCT and writing data onto the card (RDFCW).

During each situation, the average consumption of parts included in the PCW is summarised in Table I

Once we know consumption values, it is necessary to decide both the maximum current value allowed by the photovoltaic system and the size of the solar array.

Table I: Measured average consumption (at 3.3 V) of the PCW parts, for different operating phases.

PHASES \ DEVICES	I	SDTC	RSDTC	RDFCW
ELECTRONICS	6.2 mA	7.2mA	7.2mA	7.2mA
CARD	-	-	12mA	17mA
IrDA COMPONENTS	0.1mA	14mA	14mA	0.5mA
TOTAL	6.3mA	21.2mA	33.2mA	24.7mA

As a starting point, we can assume a maximum current value of 34 mA. Besides, we choose a PV array size of 54 cm^2 (9 cm x 6 cm) (no much bigger than the standard card size that is 8.5 cm. by 5.4 cm). This PV array would produce 540 mW when an irradiance of 100 mW/cm^2 falls on it (considering a pessimistic PV efficiency of about 10 %). This is approximately five times the maximum average power the system needs.

3.2 Configuration of solar cell array.

In this section we will study arrays with the same surface (54 cm^2), but which are composed of different numbers of solar cells. The active area is almost the same (48.4 cm^2) in all configurations, due to the special way for soldering cells without tabs, developed by the Instituto de Energía Solar in the frame of an European Project, called MENHIR [4] y [5].

We have made measurements in standard 1-AM1.5G. The Table II shows the results.

Table II: Electrical measurements carried out about arrays composed by different number of cells

Number of cells \ Parameters of the array	P_{MAX} (W)	η_{ce} (%)	FF	I_{SC} (A)	I_m (A)
2	0.713	14.7	0.73	0.845	0.783
3	0.73	15.1	0.74	0.568	0.512
4	0.7	14.5	0.76	0.412	0.372
7	0.681	14.1	0.77	0.224	0.208

For every array configuration, we have measured both individual cell and array series resistance by the method proposed in reference [6]. With these figures, we can calculate the cell- to-cell soldering resistance.

Table III: Series resistance measured for each array

No .of Array cells	Cell series resistance	Array series resistance	Soldering resistance
2	0.058 Ω	0.18 Ω	0.064 Ω
3	0.09 Ω	0.4 Ω	0.065 Ω
4	0.128 Ω	0.67 Ω	0.052 Ω
7	0.28 Ω	2.2 Ω	0.040 Ω

These soldering resistance values (Table III) are lower than the cell series resistance and they indicate a good result for the soldering procedure but cannot explain differences between the output powers of each array.

In Figure 3, we represent series resistance values and the square of the current in the maximum power point for each array

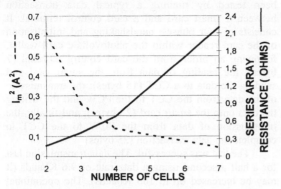

Figure 3: Square of the current at the point of maximum power and series array resistance in function of de number of cells, which compose the array

The product of both is represented in Figure 4, which shows, for each array, both maximum output power and power losses in the maximum power point

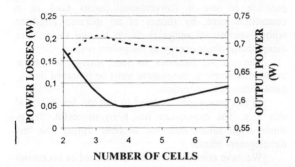

Figure 4: Power losses and output power vs. the number of cells

By looking at these results, there are different corners for both output power and power losses. It is possible to make a model to explain this behaviour, but this is out of the scope of this paper. Empirically, we can choose either three or four cell array to made the photovoltaic receiver.

3.3 Adaptation of electrical energy

The Photovoltaic Card-wallet operates at a nominal voltage of 3.3 V. Two, three and four cell array achieve an output voltage lower than 3.3 V. Then, a step-up converter is necessary to match the array output voltage to the nominal voltage. Seven-cell array achieves a voltage higher than the nominal and requires a step-down converter or a linear regulator.

Three types of DC-DC converters were studied: a regulated charge pump, a switched step-up regulator and a linear regulator. Several combinations of arrays and converters were made. Of course, the linear regulator was tested only with seven-cell array.

Results are shown in Figure 5, which depict the relation between conversion efficiency and the output current for the charge pump converter, MAX 1759, connected to arrays of three and four cells.

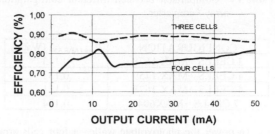

Figure 5: Efficiency vs. output current for MAX 1759

Figure 6 depicts the results of a switched step-up regulator, such as the IC MAX 1678, connected to two, three and four cell arrays.

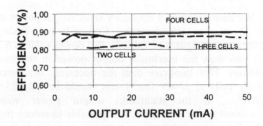

Figure 6: Efficiency vs. output current for MAX 1678

For the configuration composed of two cells and this converter, the maximum output current available was less than 30 mA.

A third type of converter, a linear regulator MAX 604, was also tested connected to a seven-cell array. The efficiency is shown in Figure 7.

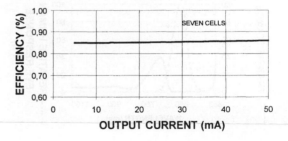

Figure 7: Efficiency vs. output current for MAX 604

3.4 Optimisation of the photovoltaic system

Once we have measured several combinations between arrays and converters, we need to decide which is the most adequate to power the photovoltaic wallet.

The two-cell array was rejected because it cannot provide enough current when connected to a step-up MAX 1678.

To compare arrays of three, four and seven cells, we fixed a realistic value of load current consumption, as is 30 mA. For all suitable combinations of the array-converter, we looked for the power transfer efficiency at this particular output current. Then, we calculated which array output power would be necessary to supply the load at this level of current. The results are summarised in Table IV.

Table IV: Comparison between different configurations array / converter

CONFIGURATION	$P_{out\ (ARRAY)}$ (Watts)
3 CELLS +MAX 1678	0,113
4 CELLS +MAX 1759	0,131
4 CELLS +MAX 1678	0,111
7 CELLS +MAX604	0,114

To power the photovoltaic wallet, a four cells array connected to a step-up MAX 1678 has been chosen, but every solution in Table IV is acceptable. We can only reject the second configuration, which needs 15 % more power in the same conditions.

4. THE LIGHT SOURCE ON THE CCT

We have obtained a necessary array output power of 0.111 W. to allow a load consumption of 30 mA. Taking into the account figures of Table II, this power value is 6.3 times less than the maximum output power in standard conditions. This implicate than the necessary irradiance onto the cells to power the system would be only 15.9 mW/cm^2 , if the maximum output power were proportional to the irradiance. It is possible to reduce this level of irradiance by using monochromatic radiation. This is due to the increase in efficiency the solar cells experience when illuminated by this class of radiation [7]. By combining a low-pass red filter (wavelength corner of 650 nm.), and a cut-off infrared filter (wavelength corner of 750 nm.) we have developed a set of filters for the ligth source of this system. The transmissivity of this filter is shown in Figure 8.togheter with the relative response of human eye, shown for comparison.

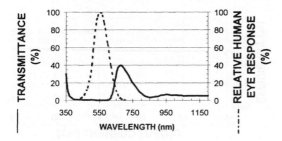

Figure 8: Transmittance of the set of filter used in the light source and relative human response vs. wavelength

We have made measures on the four-cell array illuminated by the above-mentioned light source. With a level of irradiance of 33 mW/cm^2 falling upon the array, the maximum power produced is 365 mW (for an V_m= 1.86 V and I_m= 0,196 A). Using data from Table II, this power efficiency is 56 % higher than in standard conditions.

Considering the maximum electrical power required for the system (0,111 W), the efficiency of the array (14,5 %), and the increment in efficiency due to quasi-monochromatic radiation and the active surface area of the solar array, the irradiance necessary to operate the system is 7 mW/cm^2. We have constructed a light source by using two 100 W halogen lamps and a set of filters described above. Using this together we have obtain a level of irradiance of 22,3 mW/cm^2 at a distance of 1,2 meters.

5. CONTACTLESS OPERATIONS

The assembly of all the parts of the system has been tested by running a typical data transaction between a smart card and a card control terminal. It consists of four phases: handshaking and initialisation of the card reader within the photovoltaic card wallet, following an order from the card control terminal (7 bytes); the reading of data from the smart card and sending the data to a CCT (15 bytes); the transmission of the data from the CCT to the PCW and the writing the data onto the smart card (25 bytes); and finally, the transmission of data from the PCW to the CCT to complete the data transaction (25 bytes).

The test was successful. The data transmission last for a half a second using a data rate of 9.6 Kbauds (it may be increased up to 105 Kbauds). The operational range is of up to 1 meter (with an angular mismatch of ± 20° in both horizontal and vertical planes). The potential infrared transceiver performance can reach a range up to three meters.

6. CONCLUSSIONS

We have developed a system by which it is possible to use a conventional smart card as a contactless card, by means of an infrared interface whose operational range is of up to one meter, and using photovoltaic energy as the power source.

The use of batteries to load energy has been avoided. Therefore, the system must be powered during each contactless operation.

In this prototype, we tried to test the feasibility of this idea. The experience has been succesful. Safety limits were taken into the account during the last development phase.

We have confidence in the smart card as becoming a common way for loading data either onto identification card, medical files, or bankcards. It should be possible to use only one card for all of these purposes. Our system could improve on the smart card by allowing it to be used as a contactless card itself.

7. ACKNOWLEDGMENTS

First, we give special thanks to ISOFOTON, Málaga, Spain, for the provision of solar cells. Our thanks go also to the G.U.T.I group from U.P.M., Spain for their contributions to smart cards and other devices. Also, the Photonic Technology group, from the same University, for their help in infrared communications.

8. REFERENCES

[1] Universidad Politecnica de Madrid- C. Algora, G. L. Araujo and J. de Dios, Patent 2080678 B1 (Spain).
[2] Universidad Politecnica de Madrid – C. Algora, F. Lopez and J.L Zoreda, Patent 2118048 B1 (Spain).
[3] E. Lorenzo, Solar Electricity. ISBN 84-86505-55-0
[4] MENHIR. Contract JOR3-CT95-0094. JOULE III
[5] Patent Pending P9802147. UPM-OTRI
[6] K. Krebs, Standard Procedures for terrestrial PV performance measurements. JSR Ispra.1981
[7] Spectral effects on PV-device ratings. Sol. Energy Materials.**27** (1992) 189

PV-THERMAL HYBRID LOW CONCENTRATING CPC MODULE

M. Brogren[1], M. Rönnelid[2], B. Karlsson[3]

[1]Dept. of Materials Science, Uppsala University, P.O. Box 534, S-751 21 Uppsala, Sweden
Phone: +46-(0)18-471 3134, Fax: +46-(0)18-500 131, E-mail: maria.brogren@angstrom.uu.se
[2]Solar Energy Research Centre, EKOS, Dalarna University College, S-781 88 Borlänge, Sweden
[3]Vattenfall Utveckling AB, S-814 26 Älvkarleby, Sweden

ABSTRACT: In this paper, the performance of a water-cooled PV-thermal (PV/T) hybrid system with low concentrating aluminium compound parabolic concentrators (CPCs) is discussed. The advantage of hybrid systems is their high total efficiency. By using concentrating hybrid systems, the cost per kWh is reduced due to simultaneous heat and electricity production and a reduced cell area. The thermal energy delivered from the module will fully compensate for the cost of the concentrating element and more electricity per cell area is produced. This technology is promising for future reduction of the cost of PV electricity.

During 1999, a concentrating hybrid system with a geometric concentration ratio of 4X and 0.5 kW$_p$ electric power was built at Vattenfall Utveckling, Älvkarleby, Sweden (60.5° N, 17.4° E). The yearly output is 250 kWh of electricity per square meter solar cell area, compared to 125 kWh for a conventional module, and 800 kWh of heat per square meter solar cell area. By using optimised reflectors and cells with lower series resistance, the yield could be further increased by 20%.

Keywords: Hybrid - 1: Concentrator - 2: Cost reduction - 3

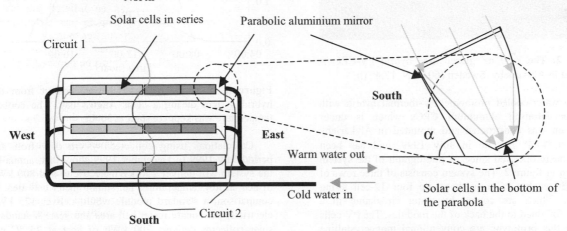

Figure 1. Schematic picture of the PV-thermal hybrid CPC module under evaluation. In each of the three CPC rows there are four 12-cell monocrystalline Si string modules. The system has two separate current circuits in parallel, each with 72 cells. The PV modules are cooled by water circulating in a cooling fin glued directly to the back of the modules.

1. INTRODUCTION

The current high cost of PV modules makes the use of reflectors or concentrators desirable. Advanced concentrator designs, such as compound parabolic concentrators (CPCs)[1], can significantly increase the electric power yield. Co-generation of heat and electricity increases the total efficiency of the system. The generation of heat thus contributes to the installation cost, giving better economy for the electrical fraction. In the future, solar thermal and solar electric systems may also compete about suitable areas for installation and then hybrids are an excellent solution.

In order to make the use of PV technology extensive the costs have to be radically reduced. For example, if PV electricity are going to play a role in the Swedish energy system in the future an integration of the modules in buildings and grid connection is necessary. However, this is not economical until the cost of PV technology has been reduced. Therefore, development of techniques for making PV electricity cheaper should be given high priority. The most important measure is to make the module itself cheaper. However, this is not enough, since the costs of installation and additional electrical equipment is a major part of the total system cost.

In this project, which is part of a larger building integrated PV programme, the main idea is to use cheap low concentrating elements in order to increase the output per cell area with a factor of 2 to 3. The objective is to halve the investment cost per yearly delivered kWh for the PV module with reflectors, compared to a conventional system.

Low concentration demands little sun tracking and makes facade integration possible. The irradiation distribution in Sweden is uniquely suitable for utilisation of east-west oriented concentrating systems with little sun tracking. This implies that systems with east-west oriented concentrating elements can be constructed with a concentrating factor of 2.5 without any need for sun tracking.[2] If the absorber is designed to be irradiated from both sides, i. e. substituted by a tandem cell, the effective concentration factor will be 5 times the solar irradiation.

2. THE ÄLVKARLEBY PV-CPC HYBRID SYSTEM

Figure 2. The 7.20 m² glazed area hybrid PV-CPC is installed in Älvkarleby, Sweden (60.5° N, 17.4° E).

The water-cooled prototype PV-thermal hybrid with low concentrating aluminium CPCs which is under evaluation was constructed and mounted in Älvkarleby (60.5° N, 17.4° E) late in July 1999, and has been continuously operated since. A photograph of the system is shown in figure 2. The system consists of three rows of CPCs. In each CPC row there are four 12-cell string modules which are cooled by water circulating in a cooling fin glued to the back of the modules. The PV cells used in this prototype are conventional monocrystalline 10×10 cm Si cells from Solartec, SC2166. The cells have been laminated onto aluminum profiles at Gällivare Photovoltaic and assembled to a hybrid system at Vattenfall Utveckling AB in Älvkarleby.

The geometric concentration factor of the symmetrical CPCs is 4X and the acceptance half angle, θ_a, is 12°. The CPCs are truncated to a height of 0.45 m and extended 0.2m on each side of the PV module, in order to avoid end effects.

The CPCs are covered with a newly developed anti-reflection glazing[3] for protection of the reflectors and the modules and for reduction of heat losses. The total glazed collector area is 7.2 m² and the total cell area is 1.5 m² The modules are connected in two separate series, as indicated in figure 1. Each circuit consists of six modules with altogether 72 cells. The nominal power for one circuit, given by the manufacturer, is 117.6 W. The CPC rows are east-west oriented and their inclination angle can be varied using a hydraulic system. The chosen acceptance angle requires that the tilt angle is changed four times per year. The hydraulic system was constructed only to facilitate measurements at different tilt angles and is, due to its high cost, not proposed to be a part of a future full-scale system.

3. RESULTS

Generated heat, water temperatures, voltages, currents, and peak power of the system are continuously monitored. Some results are shown in tables 1a and 1b. Figure 3 shows the total electric and thermal power during a clear August day. The maximum output is 330 W electric power and 2300 W heat at low temperatures per square meter cell area, corresponding to a 0.5 kW$_p$ electric output of the prototype system. A conventional module with the same cell area would have delivered 225 W electric power.

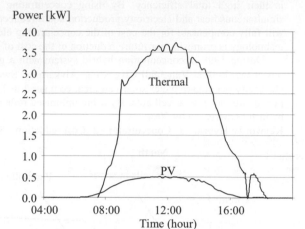

Figure 3. Total electric and thermal output from the hybrid system during a clear August day. The coolant temperature was kept constant at 30°C.

Calculations using collected system data from the period July 1999 to December 1999 show that, annually, the system will deliver 250 kWh electricity and 800 kWh of heat at low temperatures per square meter cell area. In comparison, a standard module would deliver 125 kWh electricity per square meter cell area and year. A standard solar collector delivers 700 kWh of heat at 25 °C per square meter glazed area and year. Examples of the electric and thermal efficiencies of the PV-CPC hybrid system are shown in figure 4.

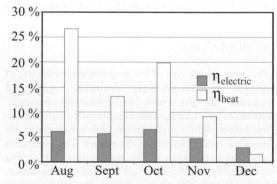

Figure 4. The electric and thermal efficiencies of the PV-CPC hybrid system during five months 1999.

3.1 Optical efficiency

There are several factors that limits the optical efficiency and thus the electrical efficiency of the hybrid system. The glass transmittance, weighted by the spectral distribution of the solar radiation and the cell sensitivity, has been determined by spectrophotometric measurements to 94%. The specular reflectance of the aluminium reflectors, determined the same way, is 77%. A reflector with higher reflectance, e. g. evaporated silver reflectors with a reflectance of 95%, will considerable increase the optical performance and hence the output of the system. Spectrophotometric measurements gives that the PV cell absorptance at near normal incidence is 93%. In total, this gives an optical efficiency of 67%. This means that the effective concentration ratio of the PV-CPC is reduced to 0.67×4=2.7.

In table 1b, the efficiency is given both per cell area and per glazed area. The short circuit current, I_{sc}, is proportional to the irradiation and the optical efficiency, η_{opt}, of the CPC can thus be calculated from the short circuit current with corrections for the geometrical concentration ratio, C, and the irradiated intensity, I.

$$\eta_{opt} = \frac{I_{sc, measured}}{I_{sc, s\tan dard\ mod ule}} \cdot \frac{I_{STC}}{C \cdot I} \quad (1)$$

Tables 1a and 1b show that fill factor and efficiency decrease with increasing temperature and irradiation. This is due to a combination of series resistance losses and a reduced open circuit voltage. These effects causes, peculiar, as it may seem, a higher electric efficiency at high angles of incidence where the optical losses are larger. The low efficiency at low intensity is explained by the fact that the sun was partly outside the acceptance area of the CPC. This is also the case in the I-V curve at low irradiation in figure 5.

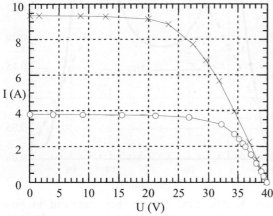

Figure 5. I-V curves at high irradiation (concentration and 950 W/m^2), x, and low (770 W/m^2, little concentration) irradiation, o.

Figure 5 shows that I_{sc}=9.35 A at I=950 W/m^2. Since I_{sc} is proportional to I, the expected I_{sc} at 1000W/m^2 irradiation is 9.84 A. The module manufacturer states that I_{sc}=3.72 A for a module at I_{STC}=1000 W/m^2 and a temperature of 25 °C. Since the concentration ratio is 4X, the expected I_{sc} at 100% optical efficiency would be 4×3.72 A=14.9 A. From this, the true optical efficiency for electricity production, η_{opt}, of the PV-CPC can be determined, using equation 1, to 9.84/14.9=0.66, which is in very good agreement with the result given by the spectrophotometric measurements.

3.2 Voltage drop

Voltage drop due to high cell temperatures and high series resistance limits the electrical performance of the PV-CPC combination. V_{oc} decreases from 41.5 V to 37.2 V and the electric efficiency decreases from 33% to 29.5% when the temperature is increased by 30 °C. The temperature dependence of V_{oc} is shown in figure 6. The temperature coefficient for the electric efficiency of the CPC system has been determined to -0.3% per °C. The fill factor decreases both with increased temperature and increased irradiation, as shown in table 1a. PV modules with lower series resistance are therefore desirable.

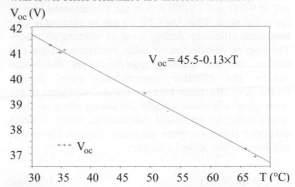

Figure 6. Voltage drop due to high cell temperatures limits the electrical performance.

Table 1a. Data from the manufacturer (at STC) and measured data for 6 modules with 72 cells in total.

Measure-ment	I_{sc}(A)	I_{mp}(A)	V_{oc}(V)	V_{mp}(V)	FF
Manu-facturer's	3.72	3.42	44.6	34.4	0.71
Low intensity	3.80	3.50	40.0	31.0	0.71
Moderate intensity	6.30	6.02	41.5	29.0	0.67
High intensity	9.35	7.72	39.8	27.4	0.57
High temp.	7.22	6.38	37.2	25.7	0.61

Table 1b. Data from the manufacturer and measured data for 6 modules with 72 cells in total.

Measure-ment	Cell temp. (°C)	I_{global} (W/m^2)	P(W)	η_{opt}	$\eta_{electric}$ per cell area/ glazed area
Manu-facturer's	25	1000	117.6	1	0.156/not given
Low intensity	42-46	770	108.5	0.32*	0.187/0.047
Moderate intensity	33-38	689	174.6	0.61	0.33/0.084
High intensity	43-45	950	211.5	0.66	0.297/0.074
High temp.	66	740	164.8	0.65	0.295/0.073

*Sun outside the acceptance area, only direct sun onto the cell.

3.3 Thermal performance

The U-value of the system is high and thermal losses limits thermal performance. Improved thermal insulation is required. These aspects are treated elsewhere[4].

3.4 CPC geometry

As shown in figures 7 and 8, the tilt angle has a significant impact on the shape of a curve of electric efficiency versus hour of the day. The CPC geometry is not perfect in the system under evaluation and there is a pronounced line of higher concentration that traverses the string module, in the north-south direction and back, during the day. When the line is in the middle of the cell, between the two fingers, there is a maxima in the efficiency. This movement of the line of highest concentration is the cause of the two maxima in figure 8. During the day shown, the line was in the middle of the cell at 9.00h and 15.00h and in the southernmost area of the cell at noon. Thus, the CPC tilt can be adjusted to give maximum efficiency during noon or morning/afternoon. The uneven distribution of solar irradiation on the cell surface, and the resulting uneven current distribution reduces the effective concentration ratio, though. Furthermore, the paste between the different parts of the cover glass will cause a slight shadowing effect on some of the cells. An optimised AR coating, which is being developed at the Ångström laboratory, Uppsala University, will increase the optical efficiency.

4. CONCLUSIONS AND ECONOMIC POTENTIAL

Voltage drop due to high cell temperatures and high series resistance limits the electrical performance and high thermal loss limits the thermal performance of the PV-CPC combination. PV modules with lower series resistance are required. With such cells and an optimised AR coating or silver reflectors, the yearly electric output can be increased by 20%.

Recently, an extremely long bifacial stationary CPC thermal collector has been developed in Sweden.[5] If the PV-CPC collector discussed here would be constructed with the same large-scale technology and with the standard thermal fin replaced by a special hybrid fin, with the PV cells glued onto a solar absorber, the system cost will significantly decrease compared to the prototype system. The thermal efficiency will increase as well. The expected investment cost for such a hybrid system would be limited to around €220 per m² glazed area or €1 per annually delivered kWh, electricity and heat included. This corresponds to €0.05 per kWh heat and €0.20 per kWh electricity, which is significantly below the cost of PV electricity today.

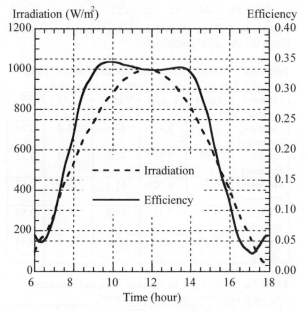

Figure 7. The electric efficiency of the PV-CPC. The weak minimum at noon is caused by voltage drop due to high series resistance.

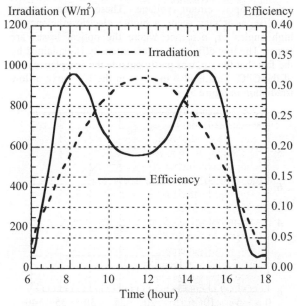

Figure 8. The CPC tilt has been adjusted to give maximum efficiency during morning and afternoon.

ACKNOWLEDGEMENTS

This work was supported by Elforsk AB, Sweden, and is carried out within the framework of the national research program Energisystem.

REFERENCES

[1]   W. T. Welford, R. Winston, High Collection Nonimaging Optics, Academic Press, San Diego (1989).

[2]   M. Rönnelid, Optical design of stationary solar concentrators for high latitudes, Thesis Acta Univ. Ups. 353, Uppsala, Sweden, 1998.

[3]   G. K. Chinyama, A. Roos, B. Karlsson, Solar Energy, V50 (2), p 105-111 (1993).

[4]   M. Brogren, B. Karlsson, Optical Efficiency of a PV-Thermal Hybrid CPC Module, Abstract submitted to the Eurosun 2000, Köpenhamn, 19- 22 June 2000.

[5]   B. Karlsson, G. Wilson, MaReCo – A Large Assymetric CPC for High Latitudes, Presented at ISES 1999 Solar World Congress, Jerusalem, 1999.

ANALYSIS OF LOW POWER STAND ALONE HYBRID PHOTOVOLTAIC-WIND SYSTEMS FOR TELECOMMUNICATIONS

M. Alonso-Abella, F. Chenlo, L. Arribas, L. Ramírez and J. Navarro

CIEMAT

Departamento de Energías Renovables

Avda. Complutense, 22. 28040 Madrid Spain

Tel.: 34-91-3466492 Fax.: 34-91-3466037 e-mail: abella@ciemat.es

ABSTRACT: This work shows the convenience or not of the inclusion of wind energy systems operating in junction with photovoltaic systems, oriented to power small stand alone telecommunication systems. A PV generator, a wind generator and a battery system compose the basic hybrid system analyzed. Simple models are used to predict the power generation for the wind and PV generators. An hourly measured year of global horizontal irradiance and wind speed (at 10m height) in 6 different locations in Spain representative of different irradiance and wind speed correlations were used. Simulations have been performed to obtain the size ratios: Battery - PV array generator - Wind turbine, that lead to a predefined loss of load hours (LOLH). For a defined LOLH there are an infinite number of potential ratios battery-PV-Wind. Only the economic criteria (including the battery lifetime dependence with the SOC evolution) will lead to obtain the optimum system size. For some locations, with a considerable wind resource (i.e. good solar-wind correlation), the use of a small wind generator can reduce not only the number of a PV system faults, but also the PV array peak power to be installed. Nevertheless economic aspects, as the ratio price/W for wind, photovoltaic or battery, can modify these results. System costs can be analyzed in function of the different parameters: location (irradiance and wind speed correlation), load profile, PV/wind/battery costs ratios. This work shows that the use of small wind turbines in photovoltaic stand-alone installations for telecommunications can reduce the final system cost in function, among others, of the correlation wind speed/irradiance for a given location and the different technology prices.

Keywords: Hybrid– 1: Sizing – 2: Economic Analysis– 3

1. INTRODUCTION

In some locations, with a defined wind speed-irradiance correlations, the use of a wind turbine in a PV system can reduce the final price of a *low power* stand alone installation. It is a very interesting application for professional telecommunication power supply installations where the allowed number of hours when the systems is not able to supply the load (system faults), loss of load hours (LOLH), should be even lower than 1 hour a year.

The main objective of this work is to supply a tool to give an answer to the question: What is the size ratio PV generator - Wind generator - Battery that minimizes the system cost for a given location and load profile?.

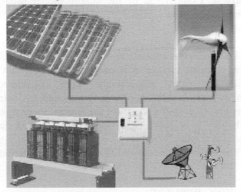

The basic hybrid PV-wind stand-alone system analyzed is composed by:

➢ A PV generator
➢ A wind turbine
➢ A battery
➢ A charge controller

Three different daily load profiles, with a daily energy consumption of 2.88 kWh/day (120 Ah/day x 24V), equal for all days during the whole year (constant, only in the night and only during the day load consumption) were used to perform hourly simulations:

(P1) Constant load profile
(P2) Load only in the night time
(P3) Load only during the daytime

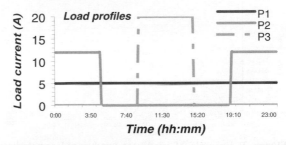

Figure 1: Load profiles: (P1) Constant-demand, (P2) Night Load and (P3) Day Load profiles.

2. IRRADIANCE AND WIND SPEED DATA

Hourly data for one year of global horizontal irradiance and wind speed (at 10m height) measured in six different locations in Spain were used for sizing procedure. These locations are representative of different irradiance and wind speed correlations (see Table I). These correlations will have a influence in the PV-Wind-battery sizes obtained in the sizing procedure. Not only the monthly and yearly values (Table II) are relevant, but also the low wind or/and low irradiance availability periods and its relation.

Irradiance on tilted surfaces was calculated using standard procedures.

Table I: Wind speed and irradiance correlations for the selected locations.

Location	Irradiance	Wind speed
Location 1	High	High
Location 2	Low	High
Location 3	Low	Low
Location 4	High	Low
Location 5	Medium	Medium
Location 6	High	Very Low

Table II: Wind speed(m/s) and irradiance (Wh/m²/day) for the selected locations.

	L1	38.3°	L2	37.6°	L3	37.7°	L4	37.7°	L5	38.5°	L6	40.4°
Jan	5.1	1815	6.3	2152	2.8	1952	3.5	1920	4.6	1853	1.0	1815
Feb	5.5	3810	6.8	3573	3.7	3151	3.0	3866	6.3	3917	1.1	3810
Mar	4.8	5267	6.2	4164	2.9	3942	3.0	5137	4.2	5358	1.0	5267
Apr	5.7	5784	5.3	4394	2.7	5224	3.5	5754	4.0	4658	1.1	5784
May	5.8	6916	6.0	5545	2.8	6237	3.7	6790	4.0	5878	1.2	6916
Jun	5.8	7789	6.0	6235	2.8	6655	3.0	7649	3.7	6610	1.2	7789
Jul	6.0	7327	6.2	6441	2.8	6488	3.1	7253	3.8	6396	1.2	7327
Aug	5.4	6664	5.2	6289	2.8	5714	3.0	6507	3.5	6979	1.1	6664
Sep	5.7	4702	4.8	4524	2.6	4082	3.6	4489	4.1	4805	1.1	4702
Oct	5.6	3845	4.1	3800	2.2	3416	3.2	4008	3.5	4085	1.1	3845
Nov	6.5	2504	5.9	2696	2.8	2426	4.1	2740	4.7	2461	1.3	2504
Dec	5.3	1701	6.0	1692	2.5	1668	3.6	1876	4.1	1664	1.1	1701
Mean	**5.6**	**4844**	**5.7**	**4292**	**2.8**	**4246**	**3.4**	**4832**	**4.2**	**4555**	**1.1**	**4844**

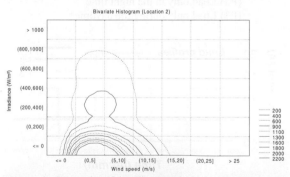

Figure 2: Example of irradiance-wind speed correlations (location 2).

3.- MODELS FOR COMPONENTS

Simple models have been used for the system components: The PV generator is assumed to produce a current directly proportional to the in plane irradiance. This is only valid when using a PV generator for charging a battery with good Fill Factor PV modules of more than 36 cells in series, that is the case in most of professional installations. The power delivered by the PV generator to the battery will be assumed as the PV array short circuit current multiplied by the battery voltage.

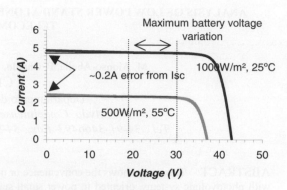

Figure 3: I-V curves for a PV array of 72 cells.

An equivalent wind turbine rotor area, A_e, has been defined in such a way that the power, P_w produced by the wind turbine is related with the wind speed, v, as

$$\begin{cases} P_w = \dfrac{1}{2}\rho A_e v^3 & v_o > v > v_m \\ P_w = 0 \, \forall v \le v_0 & P_w = P_m \, \forall v \ge v_m \end{cases}$$

ρ is the air mass density.

Comparing with wind power generator characteristics it was found that the equivalent rotor varies in the range of 0.2 to 0.5 the real rotor area. A starting charging wind speed threshold of v_0=3.5 m/s and a maximum allowable wind speed of v_m=15 m/s were considered for the calculations.

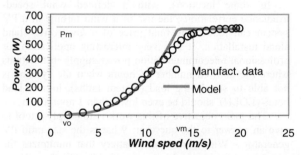

Figure 4: Comparison of the simple model used and a manufacturer power curve of a wind turbine of 400W nominal power.

The battery is characterized by the capacity (C20), a given constant charge efficiency and a maximum DOD (90% and 70% respectively were considered for calculations. A serial regulator is assumed to control the battery SOC: it will disconnect the generators charge when the battery is at 100% SOC and will disconnect the load when it is at Max DOD.

The system here considered will be of 24V nominal.

4. SIZING RESULTS

Hourly simulations have been performed for the different locations and load profiles in order to obtain the PV generator, wind turbine and battery sizes for a defined LOLH. Mainly, we will focus on 1 hour of system faults (LOLH=1) a year, that is the minimum acceptable for telecommunication systems.

Results can be presented as *iso-reliable* curves for each location, as i.e. the showed in Fig. 5, where the points in x-axis represent configurations with PV generator equal to zero (only wind).

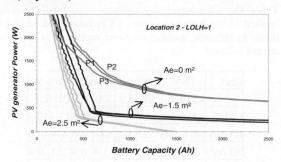

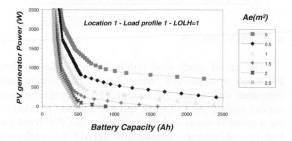

Figure 5 : Example of the obtained Iso-reliable curves for two locations.

When increasing the LOLH we can decrease the sizes of battery, PV generator and wind generator. Also when a small wind generator is introduced in the system (A_e=0.4 in Fig. 6) the PV and battery sizes ratio is also reduced, if there are enough wind resource in the given location.

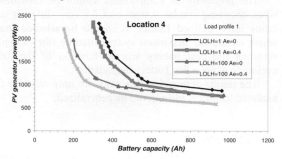

Figure 6: Example of system size reduction when introducing a wind turbine in a given location. Data for 1 and 100 LOLH.

The location (i.e. the wind-irradiance correlation) and load profile influence the ratios PV-Wind-Battery sizes (shape and slope) of the iso-reliable curves. Fig. 7 shows an example for locations 1 and 2.

5. ECONOMIC ANALYSIS

The total system cost, Ct, is calculated as:

$$C_t = C_{PV} PV + C_W A_e + C_B B + C_0$$

Where C_{PV} is the cost/Wp of PV, C_W is the cost/m² of the

equivalent turbine area, Ae, C_B is the cost/Ah (24V) of the battery capacity, B, and C_0 is a fixed cost. C_{PV}, C_W, C_B and C_0 include system installation, wind and PV support structures, charge regulator, maintenance and replacement over a 30 years period (battery replaced 3 times and 2 times the wind turbine). C_B also considers the cycling influence in the battery life time based on the generators/battery ratio sizes. The base case prices for the components were taken as: PV (3.75 €/Wp), battery(24V) (2.35 €/Ah), and wind turbine (1.25 €/W ≈ 937.5 €/m² Ae).

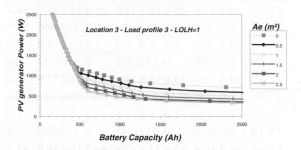

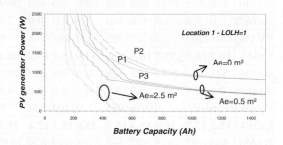

Figure 7: Iso-reliable curves (LOLH=1) for different load profiles and wind turbine equivalent areas.

There are many PV-Wind-Battery combinations to produce a desired LOLH. The optimum system size should be found considering economic criteria. The minimum system cost can be calculated as the minimum value of the system costs for different ratio sizes, as indicated in Fig. 8 for location 1, profile 1. The minimum system cost value defines the optimum PV-Wind-Battery size.

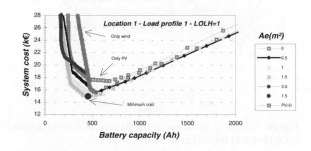

Figure 8: Determination of the optimum system size based on minimum system cost.

Table III presents the sizing results for all the locations and load profiles for a LOLH=1. Results for the hybrid, only PV and only wind configurations are presented, in order to compare. There are not locations, from the selected ones, whose optimum size is only wind, but we can note a large price decreasing (locations L1, L2 and L5) when a small wind turbine is introduced. Load profiles influence the component sizes, P3 (daily load) favour PV increasing in most cases, L2 is not affected because of its specially favourable irradiance-wind speed correlation .

Table III: Sizes and minimum system cost (LOLH=1) for the different locations and load profiles.

		P1			P2			P3		
		Hyb	PV	Wind	Hyb	PV	Wind	Hyb	PV	Wind
L1	Ct(k€)	**14.9**	**17.5**	**15.6**	**15.3**	**18.0**	**15.8**	**14.5**	**16.7**	**16.3**
	B (Ah)	457	681	538	525	703	557	430	566	605
	Ae (m²)	1.5	0.0	2.3	1.8	0.0	2.3	0.5	0.0	2.3
	PV (W)	384	1104	0	192	1152	0	816	1152	0
L2	Ct(k€)	13.3	18.7	23.6	13.8	19.0	23.9	13.2	17.9	24.0
	B (Ah)	590	703	1816	645	742	1847	586	625	1719
	Ae (m²)	0.4	0.0	1.2	0.4	0.0	1.2	0.4	0.0	1.6
	PV (W)	432	1248	0	432	1248	0	432	1248	0
L3	Ct(k€)	16.9	16.9	52.6	17.4	17.4	53.2	15.9	15.9	52.3
	B (Ah)	530	530	2935	605	605	3008	430	430	2891
	Ae (m²)	0.0	0.0	7.8	0.0	0.0	7.8	0.0	0.0	7.8
	PV (W)	1248	1248	0	1200	1200	0	1248	1248	0
L4	Ct(k€)	16.2	16.2	37.7	16.6	16.6	37.6	15.2	15.2	38.1
	B (Ah)	581	581	2354	625	625	2334	459	459	2402
	Ae (m²)	0.0	0.0	4.4	0.0	0.0	4.4	0.0	0.0	4.4
	PV (W)	1056	1056	0	1056	1056	0	1104	1104	0
L5	Ct(k€)	15.2	17.7	35.0	15.8	18.1	35.6	15.1	17.1	35.1
	B (Ah)	464	703	2915	527	747	2988	449	674	2930
	Ae (m²)	0.7	0.0	2.1	0.8	0.0	2.1	0.4	0.0	2.1
	PV (W)	768	1104	0	720	1104	0	912	1056	0
L6	Ct(k€)	17.5	17.5	368.6	18.0	18.0	368.6	16.7	16.7	368.5
	B (Ah)	681	681	56311	703	703	56311	566	566	56299
	Ae (m²)	0.0	0.0	0.0	0.0	0.0	0.0	0.0	0.0	0.0
	PV (W)	1104	1104	0	1152	1152	0	1152	1152	0

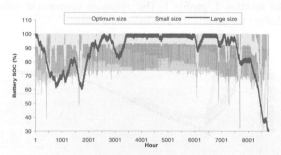

Figure 9: Hourly variation in the battery SOC over 1 year (LOLH=1;P1) for different battery sizes.

Finally, to decide the *optimum* system size from the economically minimum cost, while battery life time was already considered in the battery cost calculations, additional considerations are necessary. For example, if similar prices are obtained for very different battery sizes, it should be selected the one that is in accordance with the daily battery charge and discharge cycling in order to avoid a battery SOC for long periods of time or degradation and life time reduction for strong deep cycling.

Table IV indicates the total system cost reduction in different scenarios, where each technology will reduce 50% the base case cost.

Table IV: Cost sensibility analysis. Costs reduction are applied to the base case costs. Ct is calculated adding to these: installation, maintenance and replacement costs that are not reduced.

Cost	Base		50% PV		50% B		50% W	
	L1	L2	L1	L2	L1	L2	L1	L2
Ct(k€)	14.9	13.3	12.7	11.8	12.8	10.8	13.4	12.9
B (Ah)	457	590	527	584	488	590	474	590
Ae (m²)	1.5	0.4	0.1	0.3	1.4	0.4	2.1	0.4
PV (W)	384	432	1152	528	384	432	144	432

6. CONCLUSIONS

The use of small wind turbines in stand-alone hybrid PV-Wind systems can reduce the total system cost compared with a pure PV system. The cost reduction will depend mainly on the wind speed-irradiance correlations for a given location and the cost ratios for PV-Wind and Battery subsystems. Simulations should be performed (based on hourly data as maximum) to obtain the iso-reliability curves, and the economic criteria will define the minimum system cost for a given LOLH. Monthly or averaged wind speed values are not enough to define the system size: locations with the same amount of wind in average can lead to a very different PV-Wind sizes (i.e. locations 1 and 2), because of the size for a given LOLH depends on the low wind speed periods and low irradiance days.

The reliability of components should be considered when obtaining similar system cost for very different size ratios for a defined LOLH. It should be selected those which ensure the maximum system lifetime (for example, selecting the best battery size in function of the daily cycling to avoid a fast battery degradation).

The results obtained in this work are only valid for the analysed locations and can not be generalized.

7. REFERENCES

[1] F. Avia, I. Cruz. *Small Wind Turbine Technology Assessment,* Ciemat Technical Reports N° 875, Feb. 1999.

[2] G. Notton et al. *What hypothesis used for an economic study of electric generators for rural area? Literature survey and new suggestions.* Proc. 14th European Photovoltaic Solar Energy Conference, Barcelona (1997).

MODELLING AND OPTIMISATION OF THE CONTROL STRATEGIES FOR STAND–ALONE PV–DIESEL SYSTEMS

Christian Dumbs   Didier Mayer

Centre d'Energétique - Ecole des Mines de Paris

B.P. 207 - 06904 Sophia-Antipolis Cedex - France

Tel.: +33 4 93 95 75 75   Fax.: +33 4 93 95 75 35

ABSTRACT: Renewable energy sources are often the most economical option for electrification in rural areas where the grid connection is expensive or even impossible. The use of different energy sources such as photovoltaic panels, diesel generators, etc. allows to improve the system efficiency and the reliability of the energy supply compared to systems comprising only one single renewable energy source.

A point of general concern is the high cost of renewable energy-based stand-alone systems. A main objective consists therefore in finding a control scheme which minimises the waste of renewable energies while maximising the system efficiency and utilisation of the installed capacities.

Keywords: Hybrid – 1: Simulation – 2: Stand-alone PV Systems – 3

1. INTRODUCTION

Stand-alone photovoltaic hybrid systems are designed to be totally self-sufficient in generating, storing and supplying electricity to local electrical loads. The management of the energy flows within the system is performed through a regulation process. Energy flows are to a large extent defined by the meteorological conditions, i.e., the solar radiation, and the user load demand. Any control optimisation procedure is therefore limited to the following two control actions:

- the limitation of power supplied by the PV array
- the control of the auxiliary generator.

Optimising the system control means therefore to find an optimal and least-cost operating strategy, while evaluating its short-, medium- and long-term impact on operation, maintenance and component replacement costs.

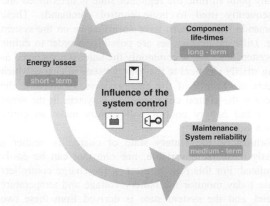

Figure 1: Influence of the system control on the overall system cost. Classification based on different time scales.

Figure 1 illustrates the influence of the control strategies on the overall system. 3 different time-scales are distinguished:

- Short-term effects of a given control strategy can be expressed in terms of energy balances and efficiencies characterising the system operation (unused or dumped PV energy, energy storage losses, energy conversion losses, necessary auxiliary energy) and associated costs can be evaluated.

- Medium-term effects are related to the need for maintenance and repair of the system components depending on the control strategy. Furthermore, loss-of-load related costs may be included.

- Long term effects of system control have to be evaluated with respect to the life-time of the system components. The time after which a battery replacement becomes necessary depends on the battery discharge/overcharge protection, the charge control during diesel generator runs, etc. Auxiliary generator wear depends on the start/stop frequency of the engine, maximum run time and the output power control.

2. APPROACH

In order to optimise the behaviour of a charge controller within a hybrid system, a simulation approach has been taken. We make use of our Hybrid Power System Simulation program (*HYPSIM*) which is based on detailed component modelling [1,2]. The program has been limited to the simulation of PV-diesel systems with electrochemical energy storage. A later extension to systems comprising wind turbines or any other additional energy source should be easily possible. Different connection schemes and control strategies allow for the study of various system configurations. The modular structure of the utilised software package (Matlab Simulink) allows for the analysis of various hybrid system configurations and the effects of different control strategies on their performance. Components are represented by separate simulation blocks which are connected to the controller via a defined set of control variables. The graphical interface makes the program user-friendly and different system configurations can be rapidly implemented.

In parallel to the simulation program development, a PV-diesel test bed was installed at the Centre d'Energétique in Sophia Antipolis. The test bed's design and component sizes (3 kWp PV, 3.2 kW diesel, 20 kWh battery, see figure 2) correspond to a typical mini-grid stand-alone system. The algorithms of various battery charge controllers can be easily integrated via the central PC of the installation and be used to define the control actions during the system operation. The system is fully operational since early 1999 and has been extensively used for the validation of the simulator.

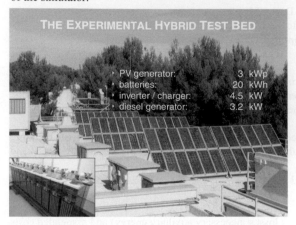

THE EXPERIMENTAL HYBRID TEST BED

- PV generator: 3 kWp
- batteries: 20 kWh
- inverter / charger: 4.5 kW
- diesel generator: 3.2 kW

Figure 2: View of the 3 kWp PV-Diesel hybrid test bed at the Centre d'Energétique, Sophia Antipolis.

Figure 3 illustrates the development of the two above-mentioned tools: the development of a computer code for hybrid system simulation (*HYPSIM*) as well as the installation of the test facility (experimental tool).

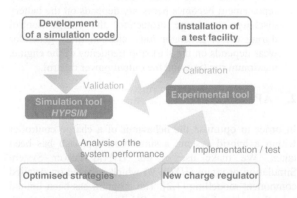

Figure 3: Development of two analysis tools for the optimisation of the control strategies of hybrid energy systems: the simulation tool HYPSIM and the experimental test facility at Sophia Antipolis

The work presented here is part of a research programme aiming at the optimisation of the hybrid system control and the development, implementation and test of a new charge regulator on the experimental facility.

3. MODELLING OF THE HYBRID SYSTEM CONTROL

The actions performed by a charge controller depend entirely on the input information received at a certain time. A first step in modelling the control problem consists therefore in choosing a well-adapted set of input variables for the charge controller. What we call charge control in this context is the compilation of this incoming information, the so-called events, to the output of the regulator, i.e., a set of binary variables. With respect to the

overall system, these variables determine the position of the connected relay contacts (open or closed), in other words, they control the connection, disconnection and operating mode of the system components. Any non-binary control schemes (such as PWM) can be easily reduced to a binary one (e.g. PWM enabled or disabled).

We can then represent the charge control problem by a finite state machine (FSM) where steps and transitions form the basic building blocks of the system.

We make use of a statechart diagram, which is the *Grafcet* [3], to represent the system behaviour. An example of a *Grafcet* representation is given in figure 4, showing a 3-stage ON/OFF regulator for the PV generator.

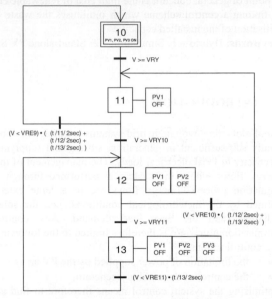

Figure 4: A Grafcet representation of a voltage-based 3-stage ON/OFF PV charge controller.

At any point in time the regulator state is determined and subsequently used to issue control commands. These commands will in return have an influence on the system state. Different approaches are possible in order to define the system state. Most commonly the latter is considered as being strictly identical to the battery's state of charge which reflects the history of the energy flows. Effectively, the battery is the central component of a system in the sense that a bad battery control will rapidly make an entire system fail.

Unfortunately, the battery state of charge is neither a directly measurable variable, nor one that can be easily calculated. For this reason, most of the charge controllers in use today monitor the battery voltage and temperature instead, and the system state is derived from these two variables.

4. OPTIMISATION OF THE HYBRID SYSTEM CONTROL

The potential for system control optimisation is inherently linked to the character and number of the different energy sources and sinks within a stand-alone hybrid system. PV-diesel systems with battery storage possess one degree of freedom with respect to the choice of the energy source to use, which leads to the necessary definition of a dispatch strategy and the corresponding possibility for optimisation.

Influence of the control strategy on energy losses

In the aim of outlining the potential for dispatch optimisation, those energy losses have to be identified which are a function of the adopted control strategy:

- *Unused or dumped solar energy*
  When the battery is fully charged, the system controller limits the PV generator's output power to avoid a battery overcharge. Consequently, a certain amount of PV energy remains uncaptured and has to be considered as an energy loss with respect to the potential maximum power output of the generator.
- *Battery storage losses*
  During the battery charge process, especially when approaching a high battery state of charge, part of the supplied energy is lost in parasitic reactions and cannot be recovered.
- *Energy conversion losses*
  There's no energy conversion without energy losses: typical DC/AC inverter efficiencies are as high as 95%, whereas AC/DC battery charger efficiencies are generally some percent lower.
- *Auxiliary energy generation losses*
  Due to the generally high no-load consumption of fossil fuel generators (about 30% of the maximum fuel consumption for diesel engines), the engine efficiency of the auxiliary generator is the highest at power levels close to the engine's rated output power. Running the engine at lower loads leads to non-negligible energy losses.

We took the standard voltage-based charge control as a reference case and evaluated the performance of other controllers with respect to the former. The typical features of a **voltage-based charge control** shall be shortly summarized hereunder:

- Over-charge protection for the battery
  If the battery reaches its maximum state of charge and there is excess power generation on the DC side, a voltage control is used to provide an over-charge protection for the battery. The controller monitors the actual battery voltage and limits the output of the power generators in order to keep the battery voltage below a pre-defined voltage threshold.
- Low battery protection
  When the actual battery voltage falls below the lowest admitted battery voltage for a pre-defined period of time, automatic diesel generator starting is initiated. In the same way, the auxiliary generator is shut down when the battery has been charged to a maximum voltage setting.

The voltage-based charge control is probably the most commonly used technique for charge regulation due to its simple use (directly measurable input variables: voltage + temperature) and robust character. However, since control actions are performed solely as a function of the actual voltage measurement, "undesired" operating modes as the one illustrated in figure 5 may occur.

In the example of figure 5, the normal operation of a PV-diesel hybrid system with voltage-regulation is shown for a

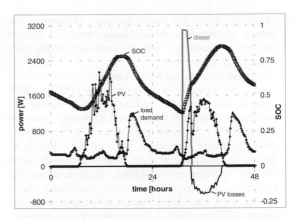

Figure 5: Power balance for the example of a standard voltage-based charge control.

duration of 2 days (48 hours). In the morning of the second day, the battery voltage reaches a critically low value, initiating automatic diesel starting and recharge of the batteries until a bulk charge voltage setting. The latter is reached at a time when the PV generator delivers a high net power (= generated PV power - load demand). Consequently, right after the diesel shutdown, the battery overcharge protection becomes active and disconnects part of the PV array (dumped energy, negative power values in figure 5)

In order to prevent operating modes as the one described above, various other dispatch strategies have been studied and the corresponding algorithms have been implemented in *HYPSIM*. An interesting discussion and classification of various hybrid system control strategies can be found in [4]. Depending on the type of the used control variables, we classified a certain number of strategies as shown in the table below.

Regulators using directly **measurable input variables**	Regulators using **calculated control variables**
- Battery voltage	- Battery state of charge
- Battery voltage + user load demand	- Energy balances: solar radiation and load forecast techniques ("perfectly known future", persistence, etc.)
- Battery voltage + time-based control	

This paper presents a critical analysis of a selection of 4 different control strategies which are based on:

1. the **standard voltage control**
   This case serves as a reference case, representing most of today's installed charge regulators.
2. the **voltage + time-based control**
   where the battery over-charge protection is entirely voltage-based (as above), whereas a diesel start-up is initiated at a predefined time of the day, most commonly at dusk when the generated PV power tends towards zero. Hence, the diesel generator is started systematically every day, unless the battery voltage was reaching a pre-defined upper threshold during the day, indicating that the battery is at a sufficiently high state

of charge. When the diesel generator is running, the battery undergoes a boost charge until the battery reaches a maximum voltage threshold which will initiate a generator stop.

3. the **persistence-based SOC control**

where the assumption is made that both the load demand and the solar radiation profile during the coming 24 hours will be identical to the one observed during the last 24 hours (persistence). In this strategy, diesel generator starting is initiated at a low SOC threshold, preferably at dusk, whenever this helps to prevent a diesel start during sunshine-hours the following day (persistence-based prediction). In a similar way, the diesel engine is stopped once the battery SOC has reached a level that allows a complete recharge of the battery using PV energy during the next 24 hours (persistence-based prediction).

4. the **SOC control based on a perfectly known future**

where future measurement data of the coming 24 hours of both the load demand and the solar radiation serve as an input to the simulator. The perfectly known future can be seen as a theoretical reference case for all forecast techniques. Obviously, such a strategy is of purely academic interest and cannot be implemented in a real charge controller. The charge control strategy is identical to the persistence-based SOC control where the persistence-based prediction has been replaced by the perfect knowledge of the future load and solar radiation data over the coming 24 hours.

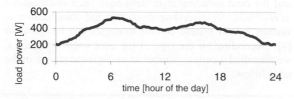

Figure 6: Daily load profile for the case study

In a case study using a hybrid system configuration according to the one of our test bed and a daily load profile as shown in figure 6, we have evaluated the performance of the 4 different control strategies described above. The simulation results which are shown in figure 7 are based on

input data of one month duration. The numbers on the x-axis of the bar-graph correspond to the respective control strategies, whereas the bars show two normalised values, the amount of lost PV energy due to an overcharge regulation (PV disconnection), and the quantity of diesel fuel that was consumed during one month. Moreover, the performance ratio for the overall system has been calculated for each of the four configurations.

When analysing the simulation results, we state that the standard voltage regulation (strategy 1) leads – as expected – to the highest PV energy losses and fuel consumption. Better results can be observed for strategies 2 and 3 whereas the strategy with the highest performance ratio, the "perfect prediction", yields fuel savings of up to 30% compared to the standard voltage regulation.

5. CONCLUSION

We present an optimisation approach based on both simulation and experimental tools, designed for the optimisation of the control strategies of PV-diesel hybrid systems. In this paper, we present an example of how the tools can be applied to the analysis of different control strategies. In the present case, optimisation criteria were defined in the aim of reducing energy losses during the system operation. The tools have shown to be extremely useful to evaluate energy balances for a given system under various dispatch strategies.

Ongoing works aim at including medium- and long-term optimisation criteria in the system simulation. Field experience has shown, that the minimisation of energy losses can only be one amongst several optimisation criteria. The impact of the control strategy on component lifetimes has furthermore to be considered in order to find truly optimised strategies taking into account all maintenance and replacement costs during the system operation.

Battery lifetime investigations were outside the scope of this paper. However, they are of uttermost importance when the overall goal is the minimisation of operation and life-cycle costs. Extensive battery testing is still required to identify appropriate battery charge cycles, regulation voltages, absorption charge times and time intervals between equalisation charges.

REFERENCES

[1] C. Dumbs, N. Maïzi, F.-P. Neirac, D. Mayer. Sizing and optimised control of PV-Diesel energy systems – Design and performance analysis. *In 2ⁿᵈ World PV Solar Energy Conference*, pages 245-248, Vienna, 1998

[2] C. Dumbs. Development of analysis tools for photovoltaic-diesel hybrid systems, PhD thesis, Ecole des Mines de Paris, 1999

[3] Sequential function chart, IEC 1131-3. International Electrotechnical Commission Publication, 1995

[4] C.D. Barley. Modeling and optimization of dispatch strategies for remote hybrid power systems. PhD thesis, Colorado State University, 1996

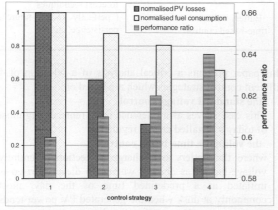

Figure 7: Comparison of different control strategies (numbers on the x-axis correspond to the different strategies as described above, PV losses and fuel consumption are normalised to 1)

GRADUAL PENETRATION OF PHOTOVOLTAICS INTO ISLAND GRIDS AND GRID MASTER CONTROL STRATEGIES

S. Tselepis, T. Romanos
Center for Renewable Energy
Sources,
19th km. Marathonos Ave.,
GR-190 09, Pikermi, Greece,
Tel. +301 6039 900,
Fax +301 6039 905,
E-mail : stselep@cres.gr

Barutti, W. Bohrer, L. Sardi, A. Sorokin,
A.N.I.T. Via N. Lorenzi 8, Palazzina 75 – 2o
Piano, 16152 Genova, Italy

P. Pinceti, M. Giannettoni, F. Poggi,
Genoa University - Electrical Engineering
Department, Via all'Opera Pia 11A,
16145 Genova, Italy,

F. Raptis, P. Strauss
ISET, Konigstor 59
D-34119, Kassel,
Germany

G. Olivier
Total Energie,
12-14 Allee du
Levant, 69890,
La Tour de
Salvagny, France

ABSTRACT: The aim of this work is to clearly indicate a way to transform island grids powered by diesel generators into hybrid ones with the main power contribution coming from distributed PV inverters. The daily production and consumption cycles will be balanced with battery storage allowing further the evolution to a 100% renewable energy system. The development of modular components (hardware and software) needed for this purpose is required as well.

Throughout the evolution process of a typical island system, the Grid Master Control System (GMC) goal is to achieve relevant improvements in terms of power availability, frequency and voltage regulation, fuel and maintenance savings, air emission and noise reduction.

The work was partially supported by the European Commission, under the Research Directorate General, Non-Nuclear Energy Programme, contract JOR3-CT97-0158. The paper focuses on the objectives, implementation and the Grid Master Control System
.

Keywords: Hybrid - 1: Grid-Connected - 2: Expert-System - 3

1. INTRODUCTION

The aim of the project is to clearly indicate a way to transform island grids powered by diesel generators into hybrid ones with the main power contribution coming from distributed PV inverters. The daily production and consumption cycles will be balanced with battery storage allowing further the evolution to a 100% renewable energy system. The development of modular components (hardware and software) needed for this purpose is required as well.

Throughout the evolution process of a typical island system, the Grid Master Control System (GMC) goal is to achieve relevant improvements in terms of power availability, frequency and voltage regulation, fuel and maintenance savings, air emission and noise reduction.

The GMC strategies have been defined in order to obtain the above mentioned improvements with the lowest impact over existing systems and to assure maximum easiness of the implementation. Distributed roof-top PV inverters will be dealt as autonomous devices, with no centralised control, equipped with local control logic that lead to a globally co-ordinated behaviour, maximising the solar energy production and the overall efficiency. Such a solution is preferred to standard centralised control to avoid the need of installing communication cables. Frequency and voltage regulation will be performed by the GMC through diesel power genset's voltage regulators and a Central Energy Buffer System's (CEBS, developed by A.N.I.T. for this project) with bi-directional power modulation capability. The power modulated PV inverters have been prepared by Total Energie. Different control strategies and operating modes have been defined in order to cope with system's evolution. GMC's tasks include configuration management, control strategies management and charging methods management.

Two solutions will be presented for the implementation to the GMC, one is based on the use of a fuzzy logic [1] and the second one on predictive control strategy [2]. The test runs will take place in a hybrid pilot plant at CRES. The supervision and automatic control of all the components installed will be performed by the central supervisory control unit (CSC), that sets the reference state of the power supply system and receives information on the actual state of each component. The exchange of information is performed by RS-485 cable, with a communication Fieldbus, serving as the medium to transfer information both ways. The information received in the central supervisory control station allows automatic operation.

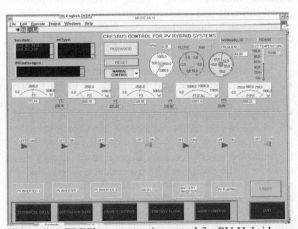

Figure 1 : CRESbus automatic control for PV Hybrid Systems

The visualisation and operational strategy of the PV hybrid system is achieved with the use of LABView, as presented in Figure 1.

The work advances a clean technology to be integrated in a wide variety of small utility grids, enhancing the quality and the reliability of the energy delivered to customers, contributing to energy savings, and furnishing employment opportunities both in industry and on islands where the systems will be implemented.

2. IMPLEMENTATION

2.1 Application area

The application area for the technology developed throughout the Project is in the multitude of small islands having the following characteristics:

- population up to a few thousand permanent residents; in these islands there is a significant increase in population during the tourist period. In some cases this is largely over 50% of the permanent residents;
- installed power capacity of diesel generators ranging from about 100 kW to up to about 10 MW. The yearly electric energy consumption may range from about 150 MWh to up to 40,000 GWh and its increase ranges from 5% to 8% per year;
- production cost of electricity ranging from about 12 cEuro/kW, for large size islands, to up to about 1 Euro/kW for the small ones.

Even though already the cost is competitive in some islands, photovoltaics are not introduced by utilities on a large scale due to the following main concerns:
- stability of the local utility grid with high PV penetration
- performances of PV systems with low grid power quality
- uncertainty regarding the real diesel fuel saving
- reliability of PV systems

2.2 Electricity costs and Power Quality in Islands

Data from Greek, Italian and French islands were collected and analysed for project purposes.

Even if some differences have been pointed out among the different utilities (energy costs, development policies, actual structure and renewable penetration), many common problems can be highlighted:
- electricity production cost is very high compared to interconnected electric systems, thus enabling PV systems to be competitive as fuel saving devices;
- large variations in grid frequency, up to 6% to 10%, are usual;
- large variations in grid voltage , up to 20%, are usual;
- frequent grid failures caused by grid weakness;

It should be noted that such a poor power quality is an obstacle to the diffusion of most commercial grid tied PV systems, since they usually require much more narrow frequency and voltage limits for continuous operation. This obstacle is overcome by the development of an appropriate grid tied tolerant inverter, suitable for the interface between PV systems and the island grid.

2.3 Problems related to high PV penetration

Major problems related to the high penetration of PV systems in island grids concerns grid stability and PV power excess.

As regards to grid stability, it has been noted that commercial inverters for grid tied PV systems require frequency and voltage limits that are very often exceeded in island grids. Such constraint violations are responsible for inverter disconnection from the grid, thus emphasising stability problems. In fact, if an overfrequency limit is exceeded due to a load reduction, as an example, this would cause the simultaneous disconnection of most of the PV systems. If PV power contribution is significant, such an event leads to sudden load excess that could evolve to a system blackout, or at least to greater frequency oscillations.

Another important issue concerns PV power excess. As previously mentioned, the population of the considered islands is variable during the year, and this implies a great variation of the load profile between winter and the tourist season. If the installed PV power is a considerable percentage of the peak load occurred throughout the year (in summer), let's say 30% to 40% or more, then it is expected that the PV power will exceed the load demand for several hours during the year. The conventional approach to solve this problem is to remotely control the PV inverters, with high costs and complexity increase. An alternative solution that does not require centralised control is presented in the paper.

2.4 PV inverter Maximum Power Point tracking

The algorithm of the Maximum Power Point tracking is as follow. The inverter supplies 100% power available from the solar array for the range of frequencies included between Fmin and Fmax. Should the grid frequency be higher than Fmax, because the load is low and we can not reduce anymore the power of the Diesel generator and the batteries are full, then the inverter will reduce gradually its output power to keep the grid frequency below Fmax. The inverter will supply again 100% of the energy available from the solar array once the frequency will drop below Fmax (this drop will result from a load increased in the grid). This algorithm has to give always the priority to the PV power production.

Table I: Transfer of the energy generated according to the grid frequency

GRID FREQUENCY	49 H$_z$	50 H$_z$	51 H$_z$
DIESEL GENSET POWER		100 %	20 %
PHOTOVOLTAIC INVERTER POWER		100 % available	0 %

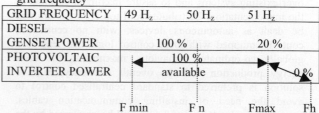

The different thresholds (Fmax, Fmin, Fh) can be configured with specific software.

2.5 Technical approach

The strategy for the gradual penetration of photovoltaics in island grids can be summarised into four phases (Figure 2):
Phase 1: the total PV contribution is significantly lower than the instantaneous load demand of supplied loads.

Tolerant PV systems are required to allow for economic operation of PV.

Phase 2: PV contribution on the island approaches the instantaneous load demand in some periods of the year. The technical limitations to additional introduction of PV are overcome by appropriate control strategies involving distributed PV power modulation.

Phase 3: Economic losses associated to growing PV power wastes due to additional introduction of PV are overcome by a centralised energy buffer.

Phase 4: Further increase in PV penetration and energy buffer capacity allows for temporary switch off of diesel generators.

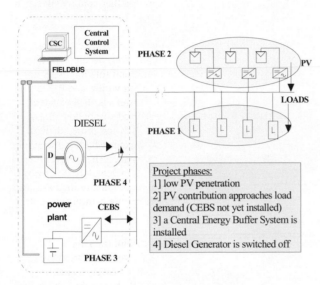

Figure 2: Block diagram of hybrid system.

A Grid Master Control unit (GMC) is required to manage diesel, energy buffers and PV generators. The Grid Master Control has to solve the following problems:

- frequency and voltage regulation;
- interaction between the centralised energy buffer, the diesel generators and distributed
- photovoltaics generating units;
- daily operating modes policy;
- alarms and emergency management;
- black-start procedure

The CEBS is able to perform voltage and frequency regulation functions on a passive grid, like any industrial UPS, but it also allows bi-directional power flow whenever the grid might become an active load. The CEBS should be considered as kind of unusual UPS with recovery capability. This particular feature will be crucial to the practical feasibility of phase IV of the project. In addiction, the CEBS will accept real and reactive power input signals from the GMC, while the Diesel Generator is running.

The paper is focused on the first two points in order to define grid voltage and frequency regulation policy and to define interfaces between the GMC and the various generating units.

3. GMC OBJECTIVES

The basic idea of the design is not to centralise the control of the distributed PV inverters in order to reduce costs and to increase system simplicity and reliability. Such a non-conventional solution requires some special control actions that are shortly described in the paper.

The Grid Master Control goal is to achieve the following improvements for the typical island systems:

3.1. Power Quality

- Power availability; the GMC is expected to bring considerable improvements in terms of stability of the local utility grid, reducing faults disturbances due to small isolated systems' intrinsic weakness
- Frequency regulation; to fully satisfy standards' requirements for frequency range in non-interconnected electrical systems, the grid frequency must be kept within rated frequency 2% over 95% of a week, and rated frequency 15% over 100% of a week. The target is to reach better results with a narrow frequency deviation.
- Voltage regulation; according to international standards, the system voltage, in normal operating conditions, must be kept within VR 10%, where VR is the system rated voltage.

3.2 Costs Savings

- Fuel savings; appropriated design of the GMC and its strategies should lead to diesel fuel savings
- Maintenance and machinery life cycles
- the GMC implementation will positively affect the rotating generating units' availability and will reduce the maintenance costs.

3.3 Environmental issues returns

- Air emission reduction;
- Noise reduction

A must of the project is to allow the transformation of existing conventional grids into hybrid grids to be gradual, in order to gain acceptance from local utility companies. Therefore, great importance is given to technical solutions, which guarantee easiness, adaptability and low impact over existing systems.

4. SYSTEM DESCRIPTION

When the large-scale penetration of PV systems is reached, in such a system the main actors are identified as follows:

- loads, to be considered for the absorbed power;
- tolerant grid tied PV inverters (PV), to be considered for their power injection;
- Diesel Generator(s) (DG), which are able to perform frequency and voltage regulation or power control,
- the Central Energy Buffer System (CEBS), which allows for a bi-directional power flow according to the control strategy implemented by the GMC

Some of these actors are present from the very first phase of the project, while some others will be added according to the system evolution described in the following paragraph. It has to be noted that while DG(s) and CEBS are located in the power plant, and therefore easily accessible from the control room, PV systems and loads are distributed along the island, and not subject to any remote control.

Two solutions will be implemented on the GMC through the Central Control System, one is based on the use of a fuzzy logic and the second on predictive control.

The specific skills of fuzzy controllers in dealing with uncertain information, of not clearly shaped constraints or limits and the ability to work on systems for whom mathematical models are not available or are difficult to be developed, handled or modified. A control approach based on adaptive scheme will be developed and tested at the pilot plant. The advantages of adaptive control in hybrid systems are the optimisation of fossil fuel consumption with respect to the load, maximizing the use of renewable energy sources.

The GMC, together with the various components developed for the project, have been installed in a hybrid pilot plant at CRES, in Greece. The tests will be performed in the following months.

5. CONCLUSION

The project indicates a way to transform island grids powered by diesel generators into hybrid ones with the main power contribution coming from distributed PV inverters and ultimately to establish 100% renewable energy electrified, isolated communities. The Grid Master Control System (GMC) goal is to achieve improvements in terms of power availability, frequency and voltage regulation, fuel and maintenance savings, air emission and noise reduction.

REFERENCES

[1]   M. Giannettoni, P. Pinceti, F. Poggi, Barutti, W. Bohrer, L. Sardi, A. Sorokin, Proceedings of 7th POWER-GEN Europe '99 Conference and Exhibition, June 1 to 3/1999, Frankfurt, Germany.

[2]   T. Romanos, S. Tselepis, G. Goulas, Poster VD2.2, Proceedings of 16th European Photovoltaic Energy Conference, 1-5 May 2000, Glasgow, UK.

SYSTEM OPTIMISATION AND POWER QUALITY ASSESSMENT OF GRID-CONNECTED PHOTOVOLTAICS – AN EXPERIMENTAL ON-SITE APPROACH

Achim Woyte [1], Ronnie Belmans [1] and Johan Nijs [2,1]

[1] Katholieke Universiteit Leuven, Kard. Mercierlaan 94, B-3001 Leuven
Phone: +32-16-32 10 30, Fax: +32-16-32 19 85, E-mail: woyte@esat.kuleuven.ac.be

[2] IMEC v.z.w., Kapeldreef 75, B-3001 Leuven
Phone: +32-16-28 12 84, Fax: +32-16-28 15 01, E-mail: nijs@imec.be and

ABSTRACT: A widely discussed question about grid-connected PV is the grade of modularity of a PV system. The 5 kWp installation described here includes a central inverter array, a string inverter, and an array of AC modules. In order to avoid additional electrical losses due to shading, the particular subsystems have been arranged corresponding to the differing sensitivities of the particular configurations to partial shading.
Utility companies often associate PV generation in the first place with questions of power quality. Results show that PV inverters generally spoken provide a much "cleaner" output than most electronic equipment.
Influences of system design and components on the installation expense are also addressed. A user friendly design of the module terminals for instance can have a direct impact on installation time and safety. Another question in this context is whether the still rather expensive AC modules can compensate for the expense that is necessary for a professional short circuit proof and ground circuit proof DC installation.
Keywords: Shading - 1: System - 2: Harmonics - 3

1. INTRODUCTION

In Belgium, up to the late nineties, there was little experience with grid-connected photovoltaics. For this reason, K.U.Leuven in close collaboration with the Interuniversity Microelectronics Center IMEC, has launched a long-term research project on the optimisation of grid-connected photovoltaics (PV) two years ago. A widely discussed question is the grade of modularity of PV system configurations. In order to assess the different system designs currently being discussed, K.U.Leuven has recently set up a 5 kWp grid-connected PV installation.

The points of interest are yield optimisation and cost reduction of grid-connected PV. Another important question for the grid operator is the impact of PV supply on quality and stability of the grid voltage. Finally, safety issues such as the islanding phenomenon are being investigated at K.U.Leuven.

2. YIELD OPTIMISATION

2.1 Electrical System Design

In order to study different design approaches, the implemented installation is split in three independent subsystems. A 480 Wp array applying various module inverters is integrated as well as a transformerless string inverter, operating at a relatively high DC voltage of 260 V and being supplied by a 1800 Wp array. The third subsystem with a power of 2880 Wp can be considered as the "classical" photovoltaic system for residential applications. Four module strings are connected in parallel to a 100 V DC bus feeding one central inverter. This system is switchable to a DC voltage of 150 V in order to enable performance tests on further inverters [1].

2.2 Partial Shading

The PV generator is situated on a flat roof, set up in three successive rows. It is shaded by a nearby obstacle as well as by the front rows themselves. The visible horizon is

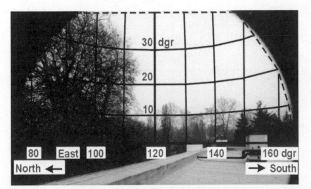

Figure 1: Fish-eye photograph of the site designated for the PV generator, taken horizontally from 90 cm above the roof plane, superposed by co-ordinates of the sky dome; taken into south-east direction

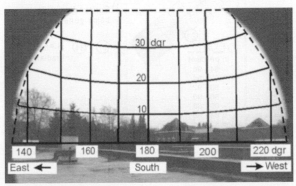

Figure 2: Fish-eye photograph as in figure 1, taken into south direction

reduced by vegetation and a neighbouring building. Looking at the designated site through a fish-eye lens gave an impression of the reduction of the visible horizon during the design phase (fig. 1 and 2). In the south-east there is an alone-standing tree that covers the sun for azimuth angles east of 110 °. Hence, this position will not receive any direct irradiance before 8 o'clock solar time, all over the year. With some more photographs an experienced designer can conclude that on this site roughly eight to fifteen percent average losses in annual irradiation will have to be expected.

More detailed calculations based on geometric figures of the shading obstacles have been carried out by means of the program STASOL [2]. Based on these simulations, the 43 PV modules have been arranged in order to minimise the losses due to shading by nearby obstacles. A spacing of 5.60 m between the module rows has been chosen. The modules are south oriented and 30 ° tilted. The annual irradiation loss by mutual shading of one row by another can then be estimated to be about 7 % [3]. Larger spacing between the rows would lead to intolerable shading of the middle array by the obstacle on the roof plane. For the final arrangement, the annual average irradiation on the total array surface has been calculated to 90 % of the unshadowed irradiation, distributed on the three arrays as shown in figure 3.

In order to avoid overproportional losses in electrical energy, only neighbouring modules, receiving approximately equal irradiation, should be connected in series. For the central inverter system, this issue has been realised as practically possible. The AC modules that are generally considered of being more shadow tolerant, are placed at the local minima of irradiation as indicated above. Long term observations will have to prove the suitability of the different system configurations for a partially shaded site like this. Some results can already be derived from the first months of operation.

2.3 Operational Results

The system has been connected to the public grid in December 1999. Since February 2000 it is monitored analytically following the guidelines of the European Joint Research Centre in Ispra [4]. Electrical performance data and module temperatures are logged for each inverter down to the individual string currents. As the generator site is partially shaded, in-plane irradiance is measured at four differently shaded positions of the generator site as indicated in figure 3 and at an unshaded position as well. As a reference, global and diffuse irradiance on the horizontal plane are recorded by pyranometers. Power analysing equipment with data logging facilities is also applied.

Some initial conclusions can be drawn comparing daily irradiance distributions measured at two winter days that were characterised by different weather conditions. February 11, 2000 was an almost cloudless day with low ambient temperatures and thus a very clear sky. February 29, 2000 was a cloudy day with almost no direct irradiance.

On the clear day, shading by obstacles becomes evident. In figure 4 we can observe about 5 % losses being constant during the day and thus originating from a reduced diffuse irradiance portion. Superposed there are further losses in the morning and evening hours that are much more significant. The nearer a position is situated to the south-east, the later it receives direct solar irradiance in the morning. The last one to receive direct irradiance is position 11 as had to be expected from the photograph in figure 1 and the simulation results as presented in figure 3. The evening losses are caused by the chimney of a neighbouring building that is situated west of the field of vision of figure 2.

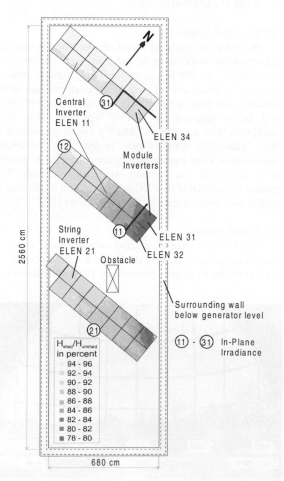

Figure 3: Generator site, array arrangement, and simulated annual in-plane irradiation normalised on unshadowed irradiation, height level of the lower side of the module frame: 45 cm above the roof plane

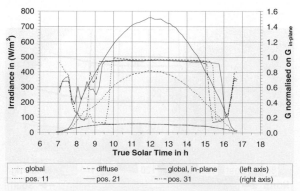

Figure 4: Left axis: global, diffuse and global in-plane irradiance (unshaded); right axis: in-plane irradiance normalised on unshaded value for different positions on the PV generator, clear day: February 11, 2000

The performance ratio (PR) has been calculated for both sample days and for all electrical subsystems (fig. 5). The module inverter at ELEN 34 had failed at that time and has now been replaced. The subsystem ELEN 32, in February 2000 also returned a very poor yield which was due to a loose terminal connection in one of the AC junction boxes.

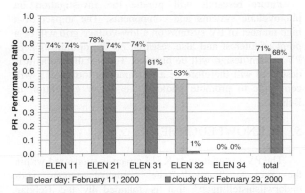

Figure 5: PR for two sample days, reference yield from unshadowed irradiation

In the above figure 5, PR is calculated by applying the reference yield as recorded at the unshaded reference location. The subsystems with central inverter (ELEN 11) and the one with string inverter (ELEN 21) perform approximately equally. However, the PR of the module inverter (ELEN 31) is much lower, especially on the cloudy day. With an overcast sky, with only diffuse irradiance, the irradiance on the partially shaded generator varies only very little in time. The reason for the low PR must thus be a poor performance of the inverter itself under such partial load conditions.

In order to split electrical losses from irradiance losses caused by shading, PR has also been calculated by applying for the reference yield the irradiation values that were recorded on the four partially shaded positions on the generator (fig. 6). The PR of central and string inverter are still comparable. No difference in performance can be found here, neither at clear sky conditions nor on a cloudy day.

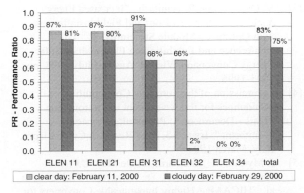

Figure 6: PR for two sample days, reference yield from irradiation values on the partially shaded generator

However, the outcomes for the module inverter at ELEN 31 are remarkable. The high PR of 91 % could be a consequence of short cabling connections that become more significant with high irradiance levels, and of the

absence of mismatch losses between several PV modules. In general, all the calculated values for PR are above 85 %. These values might be slightly, in the order of maximum 5 %, overestimated. Namely, three of the four positions on the generator whose irradiance values have been applied in order to calculate the particular reference yields, are situated on the lower edge of the particular module frames. Thus, they are shaded slightly harder than the corresponding modules themselves.

More recent results from spring 2000 indicate that shading losses on a year basis will become less significant than had to be expected from evaluating the site by means of the fish-eye photograph (fig. 7). Since shading in spring becoming less significant, the AC modules perform worse due to their lower partial load efficiencies. On the other hand, the string inverter now reaches an average PR of 78 % as a consequence of high voltage and transformerless design.

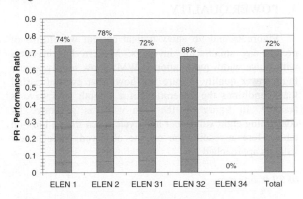

Figure 7: PR for the period from April 1 to April 21, 2000

In Belgium about 60 % of annual global irradiation are of diffuse nature, and not causing any cast shadows. Hence, with regard to their worse partial load performance, small module inverters are not necessarily the better choice from the energetic point of view. However, in order to work out more general conclusions, more data, at least describing several months of operation, will be necessary.

2.4 Practical Aspects

In practice, the decision for a particular inverter and system design will never be taken purely based on yield considerations. System configuration and components can have a direct impact on the installation costs. For the system installed at K.U.Leuven, namely the applied PV modules were shipped without bypass diodes. Installing the diodes manually took about 1.5 man hours per kWp and it contains an additional safety hazard if the modules' screw terminals are badly reconnected.

Figure 8 shows that the prices for small inverters are up to twice as high as for a central inverter. On the other hand, short circuit proof and ground circuit proof DC installation increases the global costs of a central system. With further development in AC module technology, module integrated inverters can still be expected to become cheaper [5]. Therefore, a trend towards modular systems may be expected even if long term reliability and the often stated positive impact on the energy yield with inhomogeneous irradiance distributions may still be doubted.

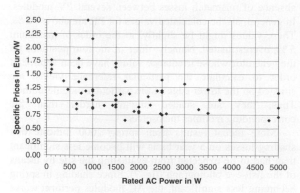

Figure 8: Specific costs of PV inverters as a function of rated power (Source: "Marktübersicht Wechselrichter", PHOTON 3/99, Aachen/Germany, and dealer information)

3. POWER QUALITY

Figure 9a and 9b show typical current waveforms of present-day small-scale inverters. Modern PV inverters with pulse width modulation (PWM) show mostly a very high power quality. Figure 9a shows that even at partial load conditions the generation of a smooth sine wave is feasible. In figure 9b the sine wave is generated by controlling eight values per half cycle taken from a look-up table. Here an oscillation can be observed while a new value is approached.

a) 90 W inverter

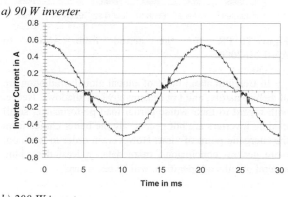

b) 200 W inverter

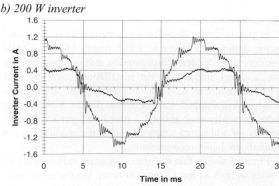

Figure 9: Current wave form of two module inverters at P_r and $0.3 \cdot P_r$

When compared to other electronic equipment, the above example shows that distributed PV generation units mean no serious hazard for the quality of the public grid voltage. However, in order to evaluate the impact with large grid penetrations, further research is still necessary. Future activities will pursue such issues. It might be feasible to implement measures for supporting or even improving the quality of the grid voltage into PV units.

4. CONCLUSIONS AND OUTLOOK

Future research will pursue the investigation on photovoltaic systems and components. The impact of a high number of PV systems on a low-voltage grid will be a point of interest and also research on the optimisation of PV system design will be pursued. The installation as it is described will therefore be a fundamental tool, and it will contribute to promote photovoltaic power generation in Belgium.

5. ACKNOWLEDGEMENTS

The PV installation at K.U.Leuven is sponsored within the project "Fotovoltaïsche zonnecelsystemen voor onderwijsinstellingen" that is founded by the Belgian utility companies Electrabel and SPE. The installed system is also supported by the Flemish regional government within the programme "50 % subsidie voor fotovoltaïsche zonnepanelen".

Acknowledgements also are paid to Soltech nv in Leuven/Heverlee for their competent advice in PV system design as to my colleagues from the Electrical Energy research group of K.U.Leuven for their physical assistance in setting up the installation as it is described above.

REFERENCES

[1] A. Woyte, R. Belmans, J. Nijs, et. al., "Practical Implementation of a 5 kWp Grid-Connected Photovoltaic System", Presentation C3-B.2, Power-Gen International 1998, Orlando, Fl, USA, December 9-11, 1998

[2] J. Grochowski, "STASOL - Ein Programm zur Simulation der Solarstrahlung", 12. Symposium Photovoltaische Solarenergie, Staffelstein, Februar 1997, in German

[3] V. Quaschning, R. Hanitsch., "Increased Energy Yield of 50 % at Flat Roof and Field Installations with Optimized Module Structures", 2nd World Conference and Exhibition on Photovoltaic Solar Energy Conversion, Vienna, Austria, July 1998, pp. 1993-1996

[4] Commission of the European Communities, "Guidelines for the Assessment of Photovoltaic Plants - Document A: Photovoltaic System Monitoring", Issue 4.3, Joint Research Centre, Ispra, Italy, March 1997

[5] J. Myrzik, M. Meinhardt, B. De Mey, C.F.A. Frumenau, H. Hofkens, Th. Krieger, P. Zacharias, et al., "HICAAP – Highly Integrateable Converters for Advanced AC-Photovoltaics, Study of Topologies, Principle Design", 2nd World Conference and Exhibition on Photovoltaic Solar Energy Conversion, Vienna, Austria, July 6-10, 1998, pp. 2146-2149

DEPLOYMENT OF A ROOFTOP, GRID-CONNECTED LINEAR FOCUS PV CONCENTRATOR SYSTEM

W. R. Bottenberg, P. Carrie, K. Chen, D. Gilbert, N. Kaminar, D. Lehmicke, C. Sherring
Photovoltaics International, LLC
171 Commercial Street
Sunnyvale, CA 94086, USA

ABSTRACT: PVI has installed a 30.5 kW AC concentrator array on the roof of a warehouse building at the Sacramento Municipal Utility District (SMUD). This system is the first installation of an advanced, new generation of line-focus, Fresnel lens concentrator modules. A system description and performance data is presented. The data show that the system performs as expected based on a new performance model.
Keywords: Concentrator – 1: Grid-Connected – 2: Rooftop – 3

1. INTRODUCTION

PVI has been developing a next generation, linear focus, one-axis tracking PV concentrator system during the last two years [1,2,3]. The modules employ an extruded plastic, Fresnel lens to focus the direct beam radiation on a string of 36 solar cells in series. The solar cells are mounted on an aluminum receiver that functions as a support and heat sink. The solar cells are encapsulated in ethylene vinyl acetate with a fluoropolymer cover. A bypass diode protects the solar cells in groups of nine. The module is assembled as a rigid framework for mounting the lens by attachment of two roll-form aluminum sides and two cast aluminum end caps.

The effective concentration ratio of the system is 8.5 at standard conditions. This allows the use of specially designed cells manufactured on a standard one-sun cell line. This has a strong advantage in lowering the cost of concentrator solar cells.

The modules are mounted on a frame support that provides the tracking function. A sun-following sensor actuates an electric motor/gear box that drives a linkage to rotate the twelve modules on the frame from east to west.

Figure 1: 30.5 kW AC PVI Array at SMUD

2. SYSTEM DESCRIPTION

2.1 General Description

The system consists of 240 SunFocus linear concentrator modules mounted on 20 tracker panels. The modules are wired into 10 source circuits of 24 modules each in series. The inverter is a Trace Technology PV40-GTI rated at 40 kW AC.

2.2 System Disposition

The SMUD 59th Street warehouse is oriented 15 degrees to the west because of its alignment to an adjacent set of railroad tracks. To optimize summer peaking performance the array structure is tilted at 29 degrees, which is about 9 degrees less than the latitude of 38.5 degrees.

3. INSTALLATION

3.1 Roof Attachment and Frame Installation

Each panel frame is attached to the roof through a mounting system that penetrates the roof and is securely attached to the building frame. The mounting system provides a 'hat' shield. This shield allows roofers to easily re-roof the building without having to remove the array. Each panel frame requires eight mounting points, four in the front and four in the back.

Frame feet are attached to the roof mounts. The frames are erected in place and module hardware is attached.

3.2 Module Assembly and Installation

Modules are shipped from the factory as 'tubs' stacked in groups of six. The tubs consist of the module receiver, sides and end cap. The lenses are included in a separately shipped stack. The modules are assembled on the roof by mounting the lens on the tub and applying clips that secure the assembly. A butyl sealant tape is applied to the mounting surfaces to provide dust protection. This assembly procedure dramatically reduces shipping costs.

The modules are mounted on the frame after pivot pins are attached to the end cap.

3.3 Tracker Assembly Installation

The motor/gear box tracker assembly mounts on the frame. A linkage bar connects the modules to the motor drive assembly. A short alignment procedure is used on the modules to maximize current output.

3.4 Electrical Installation

The modules are connected in series in the source circuit employing 'quick-connect' connectors. Electrical conduit brings the source circuit conductors to a system source circuit combiner box with a set of convenient switches and test points. The DC mains then are brought to the power-conditioning unit installed on the main floor. DC disconnect switches are provide on the roof and at the inverter.

3.5 Data Acquisition

The system installation included a data acquisition system (DAS), which measures DNI and POA irradiance, Vmp, Imp, Pac, Tambient, Tmodule, wind speed and wind direction every 10 minutes during the day.

4. SYSTEM PERFORMANCE

4.1 Expected Performance

Table 1: Expected array performance characteristics

System Parameter	Value Units
Rated Panel Output	1750 Watts DC
Panels	20
Available Power at Panel	35000 Watts DC
DC System	
DC Wiring Losses	2.0%
Panel Mismatch Losses	2.0%
Net DC Rating at Combiner Box	33614 Watts DC
AC System	
AC Wiring Losses	2.0%
Inverter Losses	5.0%
Net AC Rating	31295 Watts AC

4.2 Modeling

We modeled the performance of the array using a modification of the method developed at Sandia by King et al. [4]. The modified rating method used here is based on a more detailed methodology described in a separate paper in this conference [5]. Since Isc and Voc were unavailable, we used Imp as the indicator for system response to irradiance. This trial model is a first attempt to extend the PVUSA method that uses the global irradiance to calculate power. The PVUSA method does not include angle of incidence or spectral effects, which makes it inappropriate for tracking systems.

In this method, measured values of Imp are used to calculate the angle of incidence function and the response to a reduced air mass function. An estimate of the effective irradiance is calculated from the Imp values as shown in equations (1)-(6). The power response is modeled using the effective irradiance, ambient temperature and wind speed.

$$I_{mp}(Tc, E_{DNI}, E_{DIF}, amz, AOI) = \qquad (1)$$
$$[(E_{DN}I/850)*f_1(amz)*f_2(AOI) + C_6 E_{DIF}/150] *$$
$$\{[I_{mpo}/(1+C_6)] + \alpha_{imp}(T_{amb} - T_{ref})\}$$

$$E_e(E_{DNI}, E_{DIF}, amz, AOI) = \qquad (2)$$
$$I_{mp}(T_{ref} = 20°C, EDNI, EDIF, amz, AOI) / I_{mpo}$$

$$P_{mp}(Tc, E_{DNI}, E_{DIF}, amz, AOI) = \qquad (3)$$
$$a_0 + a_1*E_e + a_2*E_e^2 + b_1*WS + b_2*(T_{amb} - T_{ref})$$

$$f_1(amz) = 1 + c_{11}*amz + c_{12}*amz^2 \qquad (4)$$

$$amz = AMa(calculated) - 1.5 \qquad (5)$$

$$f_2(AOI) = 1 + c_{22}*(AOI/90)^2 + c_{24}*(AOI/90)^4 \qquad (6)$$

where:

I_{mp}	array current at maximum power, amps
I_{mpo}	standard max power current, amps
E_{DNI}, E_{DIF}	solar irradiance, direct and diffuse, w/m^2
E_e	effective irradiance
C_6	diffuse irradiance coefficient, amps
$f_1(amz)$	array air mass response function
$f_2(AOI)$	array angle of incidence response function
α_{imp}	temperature coefficient, max power, amps/°C
AMa	absolute air mass
amz	reduced air mass
AOI	angle of incidence
T_{amb}	ambient temperature, °C
T_{ref}	standard conditions reference temperature, °C
P_{mp}	array maximum power, watts
a_0	max power irradiance coefficient, watts
a_1	max power irradiance coefficient, watts
a_2	max power irradiance coefficient, watts
b_1	max power irradiance coefficient, watts/m/s
b_2	max power irradiance coefficient, watts/°C
WS	wind speed, m/s

T_{ref} is assumed to be 20°C following common standards. The absolute air mass, AMa, is calculated using time of day, seasonal time, and site location data.

4.3 Results

Figures 2 and 3 show the comparison of the measured response for Imp versus the calculated response for consecutive sunny and cloudy days in November.

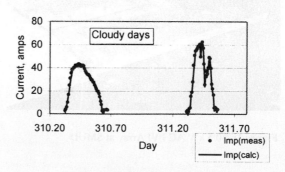

Figure 2: Array current at max power, cloudy days

The fall-off in the afternoon is due to the expected response to high angle of incidence in the afternoon winter hours. As mentioned previously, the building is tilted 15° to the west. During the winter the output is maximum in the morning. The panel will peak in the afternoon during the months from March through September.

The model adequately accounts for all of the features of performance: shadowing by clouds, varying irradiance, and angle of incidence. Table 2 shows the coefficients for the non-linear curve fit to the data for several days of data in November for the array at SMUD.

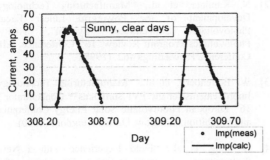

Figure 3: Array current at max power, sunny days

Table 2: Curve Fit Coefficients for Array Current

Parameter	Coefficient	Sigma
Imp_{oo}	88.3	1.0
C_6	0.0	
a_{22}	-7.23	0.08
a_{24}	0.00	
a_{12}	-0.057	0.016
a_{12}	0.0	
α_{isc}	0.0	

4.4 Power Modeling

The array power was modeled using equation (3) and the coefficients in Table 2 were obtained

Table2: Curve Fit Coefficients for Array Maximum Power

Parameter	Coefficient	Sigma	
a_0, irradiance	-1.0	0.8	kW AC
a_1, irradiance	40.7	3.2	kW AC
a_2, irradiance	-8.8	3.2	kW AC
b_1, wind speed	0.17	0.03	kW/°C
b_2, temperature	-0.04	0.02	kW/m/s

Figure 4 shows the good agreement between measured Pmax data and calculated values.

The standard conditions maximum power can be estimated by evaluating equation (3) with measured coefficients and by setting $E_e = 1$, WS = 1 m/s, and Tamb = 20°C. These conditions correspond to the common PVUSA Test Conditions (PTC) rating. For the array on the SMUD roof this value corresponds to 30.9 kW AC, which corresponds closely to the expected power shown in Table1. Figure 5 shows the measured and

calculated values for Pmax as a function of the effective irradiance, E_e.

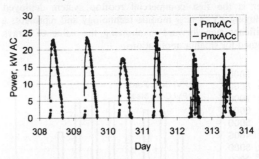

Figure 4: Calculated and measured array maximum power

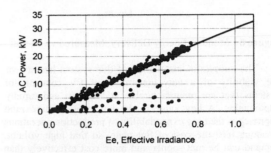

Figure 5: Estimation of System Power, Pmax as a function of effective irradiance, E_e.

5. SYSTEM ENERGY DELIVERY

We have developed a system prediction model to estimate PV linear concentrator performance using solar resource data, for example as found in the Solar Radiation Data Manual [6]. This model uses a more complete version of the model described in the previous section and is described in a separate paper at this conference [5].

Module response data was measured on a two-axis tracker for Isc, Imp, Voc and Vmp. Response functions for angle of incidence, effective air mass, diffuse irradiance and temperature were developed. Figure 4 shows the measured system AC energy delivered compared to the expected energy delivered. The expected monthly values are calculated using an array size of 30.5 kW and the monthly capacity factor evaluated using solar irradiance data for Sacramento [6]. The agreement is found to be very good, indicating overall consistency in the methodology.

6. CONCLUSIONS

A 30.5 kW AC array based on a next generation linear concentrator module has been deployed as a rooftop system on a SMUD warehouse in Sacramento, California. Initial measurements of system performance indicate that the system is performing to expectation and progress has been made on validating a useful performance model for linear focus, Fresnel lens concentrators.

Demonstration of the applicability of PV concentrator systems for rooftop, grid-connected systems is important for validating a wide range of applications such as grid support and peak-shaving. The system described in this paper is the first commercial rooftop system deployed employing PVI's new module technology and represents a significant step forward in showing the value of arrays designed for smaller applications.

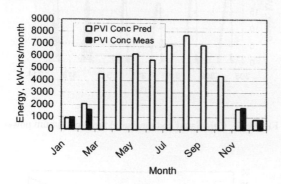

Figure 6: Monthly Energy Delivery, Measured vs. Model

The successful installation of this system has shown the viability of a new generation of PV concentrator systems that are manufacturable at low cost and rapidly deployable. A key value in the low concentration ratio approach is the easy expandability of production operations at about 1/10 the cost of flat plate, so that high volume demand can be met sooner and more cost effectively than other technologies.

REFERENCES

[1] The authors wish gratefully to acknowledge support from the Department of Energy through the PVMaT Program under subcontract ZAF-6-14271-11. This support enabled the development of manufacturing technology for significant reduction of manufacturing cost.

[2] N. Kaminar, et al., "Manufacturing Technology Development of the Powergrid Linear Focus Photovoltaic Concentrator System", NCPV Photovoltaics Program Review, 16th Conference, AIP Conference Proceedings 462, (1998).

[3] W. Bottenberg, et al., "Manufacturing Technology Improvements for the PVI SunFocus™ Concentrator", 16th European Photovoltaic Solar Energy Conference and Exhibition, Glasgow, United Kingdom (2000).

[4] D. King, et al., "Field Experience with a New Performance Characterization Procedure for Photovoltaic Arrays", 2nd World Conference and Exhibition on Photovoltaic Solar Energy Conversion Proceedings, Vienna, Austria (1998) 1947-1952.

[5] W. Bottenberg, "Measurement and Performance Prediction Method for PV Concentrating Systems", 16th European Photovoltaic Solar Energy Conference and Exhibition, Glasgow, United Kingdom (2000).

[6] "Solar Radiation Data Manual for Flat Plate and Concentrating Collectors", NREL/TP-463-5607 (1994).

DEVELOPMENT OF LOW COST FASTER-CURING ENCAPSULANTS FOR TERRESTRIAL PV MODULES

J.P. Galica, N. Sherman
STR (Specialized Technology Resources, Inc.)
10 Water Street, Enfield, Connecticut, 06082, USA
Ph. +01 860-749-8371, FAX +01 860-749-8234, E-mail: james.galica@strus.com

ABSTRACT: This research focuses on development of "faster-curing" EVA-based encapsulants. The implications of a successful development are significant, as reductions in process time increase a PV module manufacturing plant's rate of throughput and overall capacity. Allowing for lower-cost processing and gains realized by avoided cost of capital equipment, floor space, labor, etc., thereby reducing PV module manufacturing costs. Quantitative analysis of cure rate kinetics measured by Moving Die Rheometry determined that cure time reductions of ~70% were accomplished at 150°C, achieving full cure (crosslinking in excess of 80%), versus STR's commercial "fast-cure" 15295P EVA-based encapsulant. Prototype 60-watt polycrystalline silicon modules were manufactured using 5 and 6 minute lamination cycles in a Spire 240A laminator. Qualitative testing of EVA-based encapsulant samples processed during those trials suggests cure levels exceeding 80% crosslinking. The results to date, which are quite encouraging, are detailed including processing conditions, and current results from accelerated aging and qualification testing.
Keywords: Module Manufacturing – 1; Cost reduction –2; Encapsulation –3

1. INTRODUCTION

Under sponsorship of the Department of Energy in the late 1970's and early 1980's, efforts at Jet Propulsion Laboratories (JPL) and the former Springborn Laboratories, Inc., now Specialized Technology Resources, Inc. (STR), led to the development of a crosslinked ethylene vinyl acetate copolymer (EVA) formulation as encapsulant to protect delicate silicon solar cells in terrestrial PV modules.[1] In 1986, STR commercially introduced "fast-cure" EVA as an innovation that substantially increased laminator throughput efficiency by reducing the encapsulant's cure time, thereby reducing lamination process time by approximately 75% versus the prior art, "standard-cure" EVA encapsulants. Typical commercial "fast-cure" lamination cycles range between 12-18 minutes.

While several PV module manufacturers have since embraced the "fast-cure" technology, others continue to utilize the "standard-cure" EVA encapsulants. Most of these manufacturers have improved their own throughput rates through utilization of a two-stage lamination/cure process. This technique employs the laminator to evacuate air from within the PV module, melt the encapsulant, and press a defect-free uncured PV module. It is followed by oven post-curing to induce crosslinking (curing of the module) in an air-circulating oven.

On the basis of moving forward with the objectives detailed in the U.S. Photovoltaic Industry Roadmap, as conceived in 1999, PV module manufacturers concluded that they need to achieve a 5-fold reduction in module manufacturing costs by 2010 and a 10-fold reduction by 2020,[2] an ambitious objective requiring continual product and process improvement.

Focusing on those objectives, STR contends that increasing the curing-rate of a new generation encapsulant versus that of the present generation "fast-cure" EVA-based encapsulants will benefit the entire worldwide PV community by reducing PV module manufacturing costs. A significant reduction in the encapsulant's cure time, i.e., 50% or greater, will provide for lower-cost module processing and gains realized by avoided cost of capital equipment, floor space, labor, etc., thereby reducing overall PV module manufacturing costs. Those manufacturers employing post-curing ovens will be afforded the opportunity to discard that practice, recovering floor space and plant personnel. Module manufacturers processing "fast-cure" EVA by the single-stage lamination/cure process will be immediately afforded an estimated doubling of their lamination capacity, creating a tremendous opportunity for cost savings.

Throughout this development effort, STR has remained cognizant of the industries' requirement for a 30-year lifetime encapsulant. Capitalizing on STR's prior experience in developing UV stabilized, non-discoloring EVA-based encapsulants, every effort was made to select formulation constituents that were felt to have negligible affect on the encapsulant's long-term UV stability.[3]

Based on the work presented in this paper, STR concludes that significant progress has been demonstrated towards development of the next generation of "faster-curing" EVA-based encapsulants. Quantitative characterization of newly formulated experimental encapsulants formulated proved cure time reductions of 70 percent were obtainable, which led to the fabrication of prototype BP Solarex MSX-60 modules using conventional vacuum lamination equipment. The results of this work to date demonstrated that encapsulant cure levels of 80% were attained with total lamination/cure cycles as short as 5 minutes.

2. EXPERIMENTAL

2.1 Information Search / Materials Selection

Resulting from a literature review and dialog with industry experts, a number of approaches were conceived as possible means for acceleration of the encapsulant's cure rate: (1) Replacement of EVA with an alternative low-cost resin that demonstrates better thermal stability at higher temperatures; (2) Increasing the lamination processing temperature, thereby, reducing the curative's half-life and time required to cure the encapsulant (STR's 15295 "fast-cure" EVA curative is reported to have a half-life of 3.8 minutes at 150°C and less than 30 seconds at 171°C, an approximate 8-fold increase in rate); (3) The incorporation of acid scavengers and/or peroxide co-agents to accelerate the cure rate at processing conditions; (4) The incorporation of a more reactive (less stable) curative package, having a shorter half-life than the present curative system, to accelerate curing; (5) The incorporation of photo-initiators to make the encapsulant UV-responsive so that upon pre-treatment, the curative initiation temperature may be reduced.

Materials were secured and experiments were devised to pursue each of the described strategies over the course of this development.

2.2 Evaluation of Alternative Base Resins as Encapsulants

Resulting from the information search, twenty-five polymeric candidates were selected for consideration as candidate base resins for the "faster-curing" encapsulant system. For screening purposes, each was evaluated for optical clarity versus STR's 15295P "fast-cure" EVA-based encapsulant. Alternative candidates were comprised of other EVAs ranging from 10-33% vinyl acetate, ethylene butylacrylate copolymers ranging from 7-35% butyl acrylate; ethylene methyl acrylate copolymers ranging from 10-35% methyl acrylate, and ethylene octene copolymers ranging from 20-28% octene. Test specimens were prepared at a thickness of 0.46 mm via compression molding and extrusion processing, then evaluated for their average percent light transmission characteristics over the wavelengths of 360 – 900 nm using a Varian UV/Visible spectrophotometer.

2.3 Laboratory Compounding and Sample Preparation

A broad matrix of experimental formulations was designed to pursue reduction of the encapsulant's cure time. All experimental formulations were prepared using standard laboratory-scale plastics compounding equipment via batch processes. All formulation constituents were charged to the molten (fluxed) base resin using an oil heated/cooled 2-roll differential speed rubber mill. Once completely homogenized, the encapsulant was sheeted from the mill, then compression molded into sheet using a heated platen press. Processing conditions were monitored and maintained to prevent premature crosslinking of the encapsulants.

Subsequently, each experimental encapsulant was vacuum laminated using a laboratory vacuum laminator, having a 1200 cm^2 platen, into nominal 7 x 7 cm and 7 x 14 cm coupon-sized laminates. The construction consisted of the encapsulant sandwiched between two Teflon® impregnated release films, with glass positioned between the platen and sandwiched encapsulant. The construction was selected to closely simulate the heat transfer experienced by the encapsulant during commercial lamination of a module constructed with a flexible backsheet. A maximum lamination cycle of six minutes or less, including 105 seconds evacuation time, was employed.

2.4 Measurement of Cure Kinetics

Initial screening for viability was accomplished utilizing the coupon-sized vacuum-laminated samples to determine each formulation's and/or additive's impact on encapsulant cure time. Determinations were accomplished by measuring the degree of cure by standard gel content tests[5] employing hot toluene and/or hot chloroform. Due to the chemical bond formation of some of the final crosslinked compositions, it was necessary to substitute chloroform for toluene to achieve accurate gel content values.

Based on the results from gel testing, further reformulation was conducted to further improve upon the encapsulant's cure kinetics. Encapsulant formulations were compounded and compression molded into nominal 2.54 mm sheet, then characterized for their curing kinetics by Moving Die Rheometry versus a control.

The determination of cure rate was accomplished using a Monsanto MDR 2000 Moving Die Rheometer in accordance with ASTM D 6204-97, Standard Test Method for Rubber – Measurement of Unvulcanized Rheological Properties Using Rotorless Shear Rheometers. The rheometer measures the cure characteristics of encapsulants through use of a sealed rotorless moving die system. The encapsulant sample is sealed in a cavity formed by directly heated dies. The lower die oscillates at 1.66 Hz (100 cycles per minute); the reaction torque, measured at the upper die, correlates with the degree of crosslinking (change in viscosity) as a function of cure time.[4] Determinations were carried out at 150°, 155° and 160°C over 12-minute exposures. Torque versus time was recorded.

2.5 Prototype PV Module Lamination Trials

Multiple lamination trials led to the fabrication of full-scale prototype BP Solarex MSX-60 PV modules as qualification for each viable experimental encapsulant.

Initial trials involved fabrication of laminates using an SPI-LAMINATOR™ 350 at platen temperatures of 150°, 155° and 160°C. The laminates employed 510 mm x 1090 mm x 3.18 mm non-tempered float glass and BP Solarex-supplied 11.4 x 11.4 cm polycrystalline silicon wafers in place of actual solar cells. No solder, flux or ribbons were introduced during the initial trials. Craneglas 230 nonwoven glass scrim was incorporated between the substrate-side encapsulant and wafers, and a Tedlar™ backsheet was employed. The configuration is depicted in Figure 1. Sixteen wafers were used per mock-up, populating approximately 40% of the module with cells, and the remaining 60% was cell-free.

Figure 1: Lay-up of the Prototype Laminates

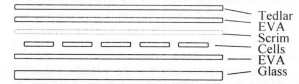

Subsequent lamination trials were conducted utilizing a Spire 240A laminator and actual BP Solarex module components to fabricate prototype MSX-60 PV modules. A broad range of lamination temperatures was studied based on a total laminator process time of 5 and 6 minutes. Cure levels were determined by Gel Content tests using chloroform.

3.   EXPERIMENTAL RESULTS

3.1 Average Light Transmission Values of Alternative Base Resins

Light transmission measurements revealed only five (out of twenty-five) resins having average light transmission values of 90% or greater, the minimum level felt to be acceptable. All five were ethylene-based copolymers. Twenty-nine to 33% EVAs had the highest light transmission values of up to 93%. Other ethylene copolymers containing 35% BA or 35% methyl acrylate, exhibited average light transmission values of 90%. All other resins evaluated ranged between 89-67% average light transmission. On an optical performance and cost-per-pound basis, 29-33% vinyl acetate containing EVAs were found to offer the most advantage over similarly priced resins. On that basis, EVA was selected as the primary choice resin for development of faster-curing encapsulants.

3.2 Results of the Cure Kinetics Measurements

Laboratory formulating and testing demonstrated that EVA-based encapsulants could be cured to gel content levels exceeding 70% in 4 minutes, and 80% based on a 5-minute lamination cycle processed at 140°C. The data is presented below in Table I.

Table I: Cure Level of Laboratory Laminate Specimens Prepared @ 140°C

Formulation	Cure Time	Gel Test Solvent	
		Toluene	Chloroform
X33561P	6 min.	44%	49%
X33562P	6 min.	63%	70%
X34646-4P	4 min.	70%	78%
X34646-4P	5 min.	80%	84%
X34646-4P	6 min.	85%	87%
X34646-10P	4.5 min.	72%	77%
X34646-10P	5 min.	74%	77%
X34646-10P	6 min.	83%	85%
X34646-14P	5 min.	75%	80%
X34646-14P	6 min.	80%	88%
X34646-16P	5 min.	72%	78%
X34646-16P	6 min.	78%	84%

Quantitative determinations of torque versus time at 150°, 155° and 160°C were accomplished by MDR for eleven EVA-based experimental formulations versus the 15295P control. The graphic representation of torque values measured versus time at 150°C is depicted in Figure 2. At 150°C, the control's maximum torque was 3.15 in-lb, requiring 12 minutes to achieve that level. Testing at 155°C and higher, resulted in a 2% increase in the control's maximum torque to 3.21 in-lb, but only requiring 8.3 minutes and 6.0 minutes to achieve that level at 155° and 160°C, respectively. This suggests that increasing the encapsulant's cure temperature by 10°C, from 150° to 160°C, results in a >50 percent decrease in cure time. Of course, this is only useful if the material can tolerate the increase in process temperature.

Figure 2: Moving Die Rheometry Results @ 150°C

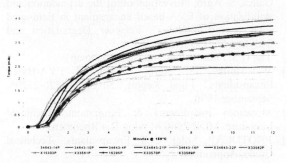

Table II compares the maximum torque values, as well as the rate and degree of cure, for 11 conceived formulations versus the control. Each demonstrated varying cure rate advantages, though X33569P and X33570P cured to lower torque values. This suggests that both cure time and cure efficiency are equally important criteria. X34643-10P and -14P resulted in maximum torque values approximately 45% higher than the control, suggesting a substantial increase in the number of crosslinkage sites, though not necessarily resulting in a higher measured gel content. The peak rate, a measure of the torque versus time relationship, indicated a 76% gain in cure rate based on X34643-14P at 150°C.

Table II: Moving Die Rheometry Results

@ 150°C, 12 Minutes				
	Torque (in-lb)			Peak Rate
Formulation	Min.	Max.	TC 90[1]	(Torque/Time)
15295P Control	0.03	3.15	7:40	1.10
X15303P	0.03	3.54	7:35	1.22
X33561P	0.03	3.19	7:42	1.04
X33562P	0.03	3.88	7:44	1.26
X33569P	0.03	2.27	5:11	1.38
X33570P	0.03	2.51	5:21	1.29
X34643-4P	0.03	4.22	7:37	1.38
X34643-10P	0.03	4.47	7:21	1.75
X34643-14P	0.03	4.54	7:05	1.94
X34643-16P	0.04	3.91	7:15	1.91
X34643-21P	0.03	3.97	7:35	1.56
X34643-22P	0.03	3.86	7:35	1.54
@ 155°C, 12 Minutes				
	Torque (in-lb)			Peak Rate
Formulation	Min.	Max.	TC 90[1]	(Torque/Time)
15295P Control	0.03	3.21	5:36	1.62
X15303P	0.03	3.65	5:37	1.84
X33561P	0.03	3.31	5:36	1.60
X33562P	0.03	4.13	5:40	1.95
X33569P	0.03	2.23	3:30	1.96
X33570P	0.03	2.49	3:36	1.77
X34643-4P	0.03	4.41	5:27	2.13
X34643-10P	0.03	4.74	5:13	2.72
X34643-14P	0.03	4.69	4:58	2.90
X34643-16P	0.04	4.01	5:04	2.78
X34643-21P	0.03	4.06	5:32	2.32
X34643-22P	0.03	4.04	5:24	2.27
@ 160°C, 12 Minutes				
	Torque (in-lb)			Peak Rate
Formulation	Min.	Max.	TC 90[1]	(Torque/Time)
15295P Control	0.03	3.21	3:50	2.36
X15303P	0.03	3.60	3:43	2.72
X33561P	0.03	3.20	3:44	2.32
X33562P	0.03	4.19	3:40	2.89
X33569P	0.03	2.10	2:20	2.53
X33570P	0.03	2.46	2:30	2.43
X34643-4P	0.03	4.44	3:34	3.17
X34643-10P	0.03	4.72	3:22	4.02
X34643-14P	0.03	4.70	3:11	4.32
X34643-16P	0.04	3.97	3:16	3.67
X34643-21P	0.04	4.04	3:36	3.27
X34643-22P	0.04	3.92	3:36	3.19
(1) TC 90 = Time to achieve 90% level of max. torque				

Table III indicates the cure time necessary for each candidate formulation to achieve a similar level of torque or cure versus the control at 150°, 155° and 160°C. The greatest advantage was noted with X34643-14P, accomplishing the control's 12-minute torque level in just 3.6 minutes at 150°C. Other formulations offered varying levels of advantages, as indicated. Further, the results indicate that processing temperature has a dramatic effect on the cure rate.

Table III: Time To Achieve Full Cure of Control @ Process Temperatures *(3.15 in-lb. torque)*

Formulation	150°C	155°C	160°C
15295P Control	12.0 min.	8.3 min.	6.0 min.
X15303P	7.3 min.	4.8 min.	3.3 min.
X33561P	10.8 min.	6.7 min.	5.7 min.
X33562P	5.6 min.	3.5 min.	2.3 min.
X33569P	NA	NA	NA
X33570P	NA	NA	NA
X34643-4P	4.8 min.	4.2 min.	2.0 min.
X34643-10P	4.0 min.	2.5 min.	1.5 min.
X34643-14P	3.6 min.	2.4 min.	1.5 min.
X34643-16P	5.0 min.	3.2 min.	2.2 min.
X34643-21P	5.3 min.	3.5 min.	2.4 min.
X34643-22P	5.6 min.	3.5 min.	2.5 min.

3.3 Lamination Trials Using Candidate Faster-Curing EVA-based Encapsulant Formulations

As indicated in Table IV, gel contents as high as 88-91% were measured on samples representative of actual PV modules processed in 6 minutes using PV module components and standard production laminators. Table V indicates that cure levels as high as 80% were achieved using 5-minute lamination cycles.

These results were accomplished by laminating BP Solarex MSX-60 PV modules using a Spire 240A vacuum laminator. Total evacuation time was held at 1 minute and 45 seconds (105 seconds) with the balance of the lamination cycle comprising press time. Gel specimens were prepared by laminating the encapsulant sandwiched between two Teflon® impregnated release films, with glass positioned between the platen and encapsulant. Those modules of good integrity and cure level are presently undergoing thermal cycling per IEC-1215 as an initial qualification of the encapsulant cure and durability.

Table IV: Lamination cycle = 6 minutes (105 sec pump, balance press)

EVA Type	Lamination Temp. (°C)[1]	Temp. Dot (°C)	Avg. Gel Content (%)	Comment
15295P Control	143	132	53	Good
X33561P	143	143	13	Good
X33562P	143	132	59	Good
X33569P	143	132	33	Few small bubbles
X33570P	143	132	71	Good
X33578P	143	132	45	Good
X34634-4P	143	132	91	Good
X34634-10P	143	132	71	Good
X34634-16P	143	132	63	Good
15295P Control	148	143	46	Few small bubbles
X33561P	148	143	43	Blister & bubbles
X33562P	148	143	64	Good
X33569P	148	143	66	Blisters & bubbles
X33570P	148	143	89	Blisters & bubbles
X33578P	148	148	80	Good
X34634-4P	148	143	84	Good
X34634-10P	148	143	88	Good
X34634-16P	148	143	88	Good

(1) Lamination temperature measured on top surface of release covering on the platen. It is, typically, significantly lower than the laminator set point due to the insulating quality of the release layer.

Table V: Lamination cycle = 5 minutes (105 sec pump, balance press)

EVA Type	Lamination Temp. (°C)[1]	Temp. Dot (°C)	Avg. Gel Content (%)	Quality
15295P Control	153	143	74	Bubbles
X33561P	153	143	71	Bubbles
X33562P	153	143	70	Good
X33569P	153	143	73	Bubbles & blisters
X33570P	153	143	71	Bubbles
X34634-10P	153	143	56	Good
X34634-16P	153	143	52	Good
X33578P	157		76	Gel Specimen Only
X34634-4P	157	143	47	Good
X34634-10P	157	143	75	(1) Blister
X34634-16P	157	143	80	Good
X34634-4P	167	149	NA	Blisters & bubbles

(1) Lamination temperature measured on top surface of release covering on the platen. It is, typically, significantly lower than the laminator set point due to the insulating quality of the release layer.

4 CONCLUSIONS

Based on the data generated to date, STR concludes that significant progress has been demonstrated towards development of the next generation of "faster-curing" EVA-based encapsulants that can offer tremendous cost savings advantage to PV module manufacturers. Quantitative laboratory characterization of the experimental encapsulants demonstrated that cure time reductions of 70% were achievable. Based on those results, limited quantities of BP Solarex MSX-60 modules were produced using conventional vacuum lamination production equipment. Based on preliminary gel test results and 2 weeks of thermal cycling in accordance with IEC-1215, the results to date are very encouraging. The data demonstrates that encapsulant cure levels of 80% were attained with total lamination/cure cycles as short as 5 minutes with good quality modules produced. All of the data generated regarding module fabrication is quite preliminary, as module production thus far has been minimal, and is presented without a statistical basis. Certainly, continued development and lamination trials are warranted to demonstrate the PV module's encapsulant's longevity as well as further optimization.

ACKNOWLEDGEMENTS

This work was supported by BP Solarex under DOE/NREL Subcontract ZAX-8-17647-05 entitled "PVMaT Improvements in the Solarex Photovoltaic Module Manufacturing Technology". Further STR would like to thank Spire Corp. and DuPont Packaging and Industrial Polymers for their in-kind laboratory support of this program.

REFERENCES

[1] P. Klemchuk, M. Ezrin, G. Lavine, W. Holley, J. Galica, S. Agro, "Investigation of the degradation and stabilization of EVA-based encapsulant in field-aged solar energy modules", Polymer Degradation and Stability, 55, 1997, 347-365

[2] The U.S. Photovoltaic Industry Roadmap

[3] W. Holley, S. Agro, "Advanced EVA-Based Encapsulants" Final Report, NREL/SR-520-25295, September 1998

[4] Monsanto Instruments & Equipment, Technical Literature, MDR 2000 Moving Die Rheometer

[5] PHOTOCAP Solar Cell Encapsulants, Technical Guide, Appendix

EVALUATIONS OF AN AUTOMATED PHOTOVOLTAIC MODULE TEST SYSTEM

M.J. Nowlan, J.L. Sutherland, E.R. Lewis, and S.J. Hogan
Spire Corporation
One Patriots Park
Bedford, MA 01730-2396

ABSTRACT: Performing electrical tests, including handling and probing, have been manual tasks in Photovoltaic (PV) module production lines up to this time. Evaluations were recently completed on a new automated system developed at Spire Corporation for performing tests on PV modules. The system performs tests for electrical isolation (hi-pot), ground continuity, and electrical performance (I-V measurement with a solar simulator) of PV modules. The equipment also includes automation for transporting, aligning, and probing modules. A number of tests were done to validate the performance of the integrated test system. System cycle times were measured to indicate the throughput of the automated system. As the PV industry grows and production levels increase, the ability to perform these tests with an automated system offers a means for reducing production cost and increasing throughput.

KEYWORDS: Photovoltaics -1: Measurements -2:Automation -3

1. INTRODUCTION

This paper reports on evaluations of a new automated system for comprehensive electrical testing of photovoltaic (PV) modules, designated the SPI-MODULE QA™ 350. The system, shown in Figure 1, integrates tests for high voltage isolation, ground continuity, and electrical performance of PV modules. The equipment also includes automation for transporting, aligning, and probing modules. The system was developed under funding by the U.S. Department of Energy PV Manufacturing Technology (PVMaT) program.

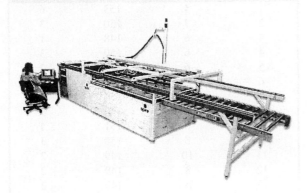

Figure 1 SPI-MODULE QA 350

2. INTEGRATED TEST SEQUENCE

The SPI-MODULE QA 350 automatically tests the ability of a photovoltaic module to maintain high voltage isolation (hi-pot) and acceptable ground continuity on exposed conductive surfaces. The unit also measures the I-V characteristics of a module with a multi-flash system which is part of Spire's proven series of Sun Simulators.

An input conveyor delivers and aligns a module, then a test carriage lifts the module above the conveyor. Test probes make contact to the + and - module output connections and the four frame sections to perform the hi-pot testing. The test probes are then configured to perform the ground continuity tests.

Upon completion of the safety testing, the carriage transports the module and places it on the SPI-SUN SIMULATOR™ 350i test plane. The carriage returns to its initial position and a second carriage moves over the module on the simulator. Four-point probes on the second carriage contact the module terminals and an I-V curve is measured. At this point, the first test carriage can pick up and simultaneously test another module (hi-pot and ground continuity tests) while the sun simulator measurement is being made on the first module. When the tests are done at both stations, the module on the sun simulator is transported to the output conveyor at the same time that the following module is transported to the simulator.

Pneumatic actuators, electronics, and software are provided for transporting, probing, testing, and switching the probe configuration for each test. A safety light curtain protects personnel from the high voltages applied during the hi-pot and ground continuity tests. While the SPI-MODULE QA 350 is limited to modules with a maximum dimension of 102 cm x 162 cm, systems with larger module size capacity can be produced.

3. TEST SYSTEM EVALUATIONS

3.1 Hi-Pot Test Measurements Evaluations

Testing was done on the SPI-MODULE QA 350 to determine the accuracy and reproducibility of the hi-pot test. Siemens Solar SP75 modules were used for these tests. The same formula is used by the IEC 1215,[1] ASTM E 1462,[2] and UL 1703[3] standards for calculating the hi-pot test voltage, V_t:

$$V_t = 2 \times V_s + 1000 \text{ V} \qquad 1$$

where V_s is the maximum system voltage rating for the module type. All three standards specify that V_t must be held for 60 s, during which time the leakage current must not exceed a specified value. Recognizing that the 60 s hold time limits the throughput of the hi-pot test, UL 1703 also provides a production line test method[4] that significantly increases the throughput of the test. In this test, V_t is increased by 20% over its usual value:

$$V_t = (2 \times V_s + 1000 \text{ V}) \times 120\% \qquad 2$$

The increased test voltage allows the hold time to be reduced from 60 s to 1 s. Thus the UL production line test method (Equation 2 with 1 s hold time) was selected for the SPI-MODULE QA 350 evaluations. The maximum system voltage rating, V_s, for SP75 modules is 600 V, so V_t was set to 2640 V on the hi-pot tester. The maximum leakage current allowed during the test is 50 μA.

V_t was applied between the frame (tester negative connection) and the shorted module output terminals in the junction box (tester positive connection). Initial tests showed very low leakage current levels of 0.0 to 0.1 μA on four different SP75 modules, which is at the resolution limit of the hi-pot tester. A shunt resistor was then wired between the module output terminals and the frame of an SP75 module to create a fixed (and artificially high) leakage current path that would allow the automated hi-pot test system to be evaluated. The hi-pot tests were repeated manually, with test leads terminated with alligator clips, and automatically, with the module being transported, aligned and probed by the QA 350.

The manual test results with the external fixed resistor were very repeatable, with leakage currents consistently measured at 284.3 mA. No variation was detected after 15 measurements.

Fifteen automated hi-pot measurements indicated an average leakage current of 284.4 mA, with a standard deviation of only 0.1 mA. This data shows good reproducibility and good agreement with the measurements obtained in the manual tests.

3.2 Ground Continuity Test Measurements Evaluations

Testing was done with the SPI-MODULE QA 350 to determine the accuracy and reproducibility of the ground continuity test. Ground path continuity tests were done manually and automatically using the same SP75 module used for the hi-pot tests. In the manual case, a clip lead (current return) was attached to the module frame next to a hole intended for attaching a ground lug. A second clip lead (current source) was attached to the other three frame members in turn, and ground continuity tests were done at each position. The anodized coating on the frame was scraped away under the clip leads to obtain a good electrical connection. A test current of 10 A was applied, the voltage was measured, and the resistance was calculated (R = V/I).

The results of the manual ground continuity tests are listed in Table I. The test was repeated 15 times on the same module, and the average ground path resistance was 10.0 mΩ. Some variation in the results was observed, probably due to differences in contact resistance between the clip leads and the frames. However, all of the results are well below the 100 mΩ maximum limit required to pass IEEE 1262[5] and UL 1703[6] standards. Note that the UL 1703 production line standard for ground continuity[7] only requires that electrical continuity be demonstrated.

The automatic ground continuity tests were done by running the SP75 module through the SPI-MODULE QA 350 fifteen times. After the module is aligned and lifted off of the roller conveyor, four sharp cone-shaped probes are pressed against the four frame members with small air cylinders. The probe that contacts the frame member with the hole for the ground stud is the current return probe. Three measurements are made in each test, using each of the three remaining probes in turn as the current source probe. The maximum resistance of the three measurements is listed in Table I for each test.

The average resistance of the 15 automatic measurements was 149.9 mΩ, significantly higher than the 10 mΩ measured in the manual case. However, the test leads are much longer than the leads used for the manual test, due to the length of cable needed to reach the test carriage. When the manual test leads were shorted out, the test lead resistance (R_{tl}) measured 0 mΩ, indicating that the tester was set up to zero out the resistance of these leads, which were supplied with the tester. However, when the automatic test leads were shorted out, an R_{tl} of 147 mΩ was measured. When this value is subtracted from the measured data, an average ground path resistance of only 2.9 mΩ is obtained. In production, the actual lead resistance can be zeroed out in the tester so that only the effective module ground path resistance is measured.

Table I Ground path resistance measurements done manually and automatically. R_{tl} = test lead resistance

Test Number	Ground Path Resistance (mΩ)		
	Manual	Automatic	
	R_{tl} = 0 mΩ	R_{tl}=147 mΩ	R_{tl} = 0 mΩ
1	4	153	6
2	7	150	3
3	6	148	1
4	6	150	3
5	6	149	2
6	13	152	5
7	16	150	3
8	8	150	3
9	6	151	4
10	8	150	3
11	10	149	2
12	8	149	2
13	19	149	2
14	23	149	2
15	10	149	2
Average	10.0	149.9	2.9
Std. Dev.	5.5	1.3	1.3

3.3 Sun Simulator Spectrum

Spire developed a new size inverted sun simulator, the SPI-SUN SIMULATOR™ 350i, that tests modules face down at conveyor height, eliminating the need to turn modules over in an automated production line. The simulator is designed to test modules up to 102 cm x 162 cm (40 inch x 64 inch). The SPI-SUN SIMULATOR 350i uses a pulsed xenon lamp and spectral filters to produce simulated sunlight for module testing.

The spectrum of the light produced by the 350i simulator was measured with a spectroradiometer from

400 nm to 1100 nm. The spectrum was compared to the standard spectral distribution of irradiance for the Air Mass 1.5 Global spectrum, defined in ASTM E 927,[8] as shown in Table II.

Table II Spectral distribution of irradiance for the SPI-SUN SIMULATOR 350i vs. AM1.5 Global standard

	Relative Light Intensity		
Wavelength	Spi-Sun Simulator 350i	AM.1.5 Global	Ratio
400-499	17.2%	18.5%	0.93
500-599	19.9%	20.1%	0.99
600-699	18.1%	18.3%	0.99
700-799	15.9%	14.8%	1.08
800-899	15.1%	12.2%	1.24
900-1100	13.5%	16.1%	0.84

Since the ratio of the simulator spectrum to the AM1.5 Global spectrum falls within the range of 0.75 to 1.25 of the reference intensity value for each wavelength interval, as shown in the last column in Table II, the 350i simulator has a Class A spectral match, according to ASTM E 927.

3.4 Sun Simulator Spatial Uniformity

Light spatial uniformity measurements were done to characterize the performance of the sun simulator. The entire test plane irradiance was measured at 40 locations (in a 5 x 8 matrix) with a 103 mm square single crystal silicon solar cell. The uniformity of total irradiance, in percent, was calculated according to the method defined in ASTM E 927:[9]

$$\text{Uniformity} = \pm 100 \times \left(\frac{\text{Irradiance}_{\text{Max}} - \text{Irradiance}_{\text{Min}}}{\text{Irradiance}_{\text{Max}} + \text{Irradiance}_{\text{Min}}} \right) \quad 4$$

Uniformity was measured under four conditions: 1) with normal ambient fluorescent room lighting, 2) with the room lights turned off, 3) with a large sheet of heavy black paper placed over the module and test plane, and 4) with the module transport carriage positioned over the test plane. The uniformity test results ranged from ± 2.11% to ± 2.34%, well within the simulator's design specification of ± 3% under all four test conditions.

3.5 Sun Simulator Irradiance Repeatability

Sun simulator irradiance repeatability data was obtained by placing a 103 mm square solar cell in a fixed location on the test plane and making 40 measurements. This test was done at five locations on the test plane, at the center and at locations near each corner. Repeatability, in percent, was calculated by the same formula as uniformity:

$$\text{Repeatability} = \pm 100 \times \left(\frac{\text{Irradiance}_{\text{Max}} - \text{Irradiance}_{\text{Min}}}{\text{Irradiance}_{\text{Max}} + \text{Irradiance}_{\text{Min}}} \right) \quad 5$$

The results of the repeatability tests are listed in Table III. The repeatability at all five test plane locations was ±0.11% or better, well within the ±1% design specification.

Table III Irradiance repeatability measurements, SPI-SUN SIMULATOR 350i

Test Plane Location	Irradiance Repeatability
Center	± 0.05%
Top left	± 0.09%
Bottom left	± 0.09%
Top right	± 0.11%
Bottom right	± 0.10%

3.6 Sun Simulator Calibration

The SPI-SUN SIMULATOR 350i uses a Spire electronic load and measurement circuit to obtain current and voltage data from the module under test. This circuit was electronically calibrated before installation in the sun simulator.

The short circuit current (I_{sc}) of a reference module (No. QC75) measured at NREL was used to calibrate the light intensity of the SPI-SUN SIMULATOR 350i. I-V curve data from this module was then measured on the 350i simulator. This data reported at 100 mW/cm^2, 25°C and AM1.5 Global spectrum is compared to data measured at NREL in Table IV. The module data measured at Spire agrees with the NREL data to within 1%.

Table IV Comparison of I-V curve parameters for module QC75 measured at NREL and at Spire.

Parameter	NREL	Spire	Δ(%)
V_{oc} (V)	21.78	21.62	-0.73
I_{sc} (A)	4.486	4.487	0.02
P_{max} (W)	73.5	73.6	0.14
V_{pm} (V)	17.47	17.41	-0.34
I_{pm} (A)	4.112	4.099	-0.32
FF (%)	71.83	71.35	-0.67

3.7 Sun Simulator Measurements Repeatability

Four Siemens Solar model SP75 modules were tested on the SPI-SUN SIMULATOR 350i under the following conditions:

- **Manual** - Four-wire (+V, -V, +I, -I) test leads with alligator clips were connected to the module junction box terminals. Ten complete I-V curves were measured for each module.
- **Automatic** - Each module was automatically transported and the junction box terminals were probed with spring-loaded four-point probes mounted on the simulator carriage. Each module was transported through the system ten times to obtain ten I-V curves.

The data obtained from each module under these two measurement conditions was averaged and repeatability was calculated as the standard deviation divided by the average. Calculations were done for open circuit voltage (V_{oc}), short circuit current (I_{sc}), maximum power (P_{max}), voltage and current at maximum power (V_{pm} and I_{pm}), and fill factor (FF). Data for the manual and the automatic test conditions of one representative module are listed in Table

V. Repeatability under both conditions was ±0.43% or better. Both conditions are within the ±1% repeatability specification for the sun simulator.

Table V Repeatablility of I-V data for manual and automatic conditions.

	Automatic Average	Automatic Repeatability (%)	Manual Average	Manual Repeatability (%)	Δ Average (%)
V_{oc} (V)	22.17	0.31	22.15	0.30	0.09
I_{sc} (A)	4.56	0.37	4.56	0.41	0.00
P_{max} (W)	73.77	0.43	73.49	0.35	0.38
V_{pm} (V)	17.93	0.43	17.90	0.33	0.17
I_{pm} (A)	4.11	0.26	4.11	0.22	0.00
FF (%)	73.06	0.40	72.80	0.43	0.36

The difference between the average values measured under the manual and the automatic test conditions is also listed in Table 3 (Δ Average column). The maximum difference is 0.38%, which confirms that data obtained by automated probing of the module terminals with the four-point probes is in good agreement with data obtained using clip leads in the manual case. These measurements are consistent with the other three modules tested.

3.8 System Throughput

The integrated test system was set up to run Siemens Solar SP75 modules. These modules have a power rating of 75 W and overall (frame) dimensions of 1200 mm x 527 mm x 34 mm. The simulator is a multi-flash system and the number of I-V points is programmable. For throughput analysis the I-V test was configured to measure 50 data points per curve. (Tests have shown that module measurements made with 50 versus 200 data points are equivalent). System throughput was measured as four SP75 modules were cycled continuously through the system. Modules were concurrently aligned at the input conveyor, tested at the hi-pot/ground continuity test station, and tested at the I-V test station as the cycle time was measured.

The cycle time, defined as the period from the time one module exits the system to the time the next module exits the system, was consistently measured at 52 seconds per module. This cycle time is equivalent to a throughput of 69.2 modules/hour. If a 10% downtime allowance is made for adjustments and maintenance, the throughput is 62.3 modules/hour, or 129,600 modules per year on an 8 hour/day, 5 day/week, 52 week/year basis. This throughput is equivalent 6.5 MW of 50 W modules or 9.7 MW of 75 W modules per year per operating shift. Annual throughput per shift is a linear function of module power rating.

4. CONCLUSION

Spire has successfully developed an automated test system based on proven simulator technology and incorporating methods to measure standard safety requirements of modules. Throughput of the measurement system was verified at 52 seconds per module, equivalent to almost 10 MW of 75 W modules per year on a single shift basis. This throughput includes complete testing for hi-pot, ground continuity, and module IV performance.

The measurement of module hi-pot withstand and ground continuity was compared with manual testing and found to agree within measurement error. Use of the proven SPI-SUN SIMULATOR technology for the IV characterization resulted in a Class A spectral match per international equipment standards, and measurement repeatability within 0.5%. The new SPI-MODULE QA equipment will allow the PV industry to easily measure both safety and performance of all modules produced. The demonstrated throughput and measurement capability will assist the industry in continued growth.

5. ACKNOWLEDGEMENTS

Spire gratefully acknowledges funding support from the United States Department of Energy (DOE) and National Renewable Energy Laboratory (NREL) through the PVMAT program for this work. We also thank Siemens Solar Industries for the supply of modules, and our PVMAT partners ASE Americas, AstroPower and Siemens for their evaluations and input.

REFERENCES

[1] IEC 1215 International Standard, "Crystalline silicon terrestrial photovoltaic (PV) modules - Design qualification and type approval," International Electrotechnical Commission. Section 10.3, Insulation Test (1993) 21.
[2] ASTM E 1462, "Standard Test Methods for Insulation Integrity and Ground Path Continuity of Photovoltaic Modules," Annual Book of ASTM Standards, Vol. 12.02, American Society for Testing and Materials (1997) 751.
[3] UL 1703 Standard for Safety, "Flat-Plate Photovoltaic Modules and Panels," Underwriters Laboratory, Inc., Section 26, Dielectric Voltage-Withstand Test (1995) 27.
[4] *Ibid*, Section 40, Factory Dielectric Voltage-Withstand Test, 52.
[5] IEEE Std. 1262, "Recommended Practice for Qualification of Photovoltaic (PV) Modules," Institute of Electrical and Electronics Engineers, Section 5.3, Ground-continuity test (1995) 10.
[6] UL 1703, *op cit.*, Section 25, Bonding Path Resistance Test, 27.
[7] *Ibid*, Section 42, Grounding Continuity Test, 53.
[8] *Ibid*, 502.
[9] ASTM E 927, "Standard Specification for Solar Simulation for Terrestrial Photovoltaic Testing," Annual Book of ASTM Standards, Vol. 12.02, American Society for Testing and Materials (1997) 501.

20 cm x 30 cm amorphous silicon solar modules on plastic film fabricated with the VHF-technology at high deposition rates for integration onto building panels

D. Fischer, D. Ciani, A. Closset, P. Torres, VHF-Technologies SA,

c/o EICN, Av. Hôtel de Ville 7, CH-2400 Le Locle,

Tel 41 32 930 39 99, Fax 41 32 930 30 50, E-mail: info@vhf-technologies.com

U.Kroll, M.Goetz, P. Pernet, X. Niquille, S. Golay, J. Meier, A.Shah, IMT, Breguet 2, CH-2000 Neuchâtel,

H. Keppner, University of Applied Science, Av. Hôtel de Ville 7, CH-2400 Le Locle;

A. Haller, H. Althaus, Ernst Schweizer AG, CH-8908 Hedingen

Abstract: In this paper, we report on the development of amorphous silicon solar cells on flexible plastic substrates, aiming at a future large scale production of low cost solar cells for building integration. The developments include in particular the application of VHF-plasma deposition to the fabrication of tandem and triple stacked cells, as well as the integration of such cells onto metal building materials.

As a result, at 80 MHz plasma excitation frequency, triple cell modules of 20cm x 30cm area and an active area efficiency of 4.1% have been successfully produced at i-layer deposition rates of 6.5 Å/s on polyimide film substrate, with a total silicon deposition time of only 23 min. In parallel, similar single junction solar modules have been successfully laminated onto aluminum building panels, confirming the fundamental feasibility of combining amorphous silicon foil modules with standard low-cost building materials.

Keywords: a-Si - 1 : Polymer Film -2 : Building Integration - 3

1. INTRODUCTION

It is the building integrated photovoltaïcs that is generally expected to bring about a substantial contribution to the electricity production in Europe in the course of the next century. For this application, amorphous silicon cells keep being of key interest, despite their comparatively modest efficiency figures. This is for the following reasons:

- amorphous silicon solar cells have the potential to become available as freely shapeable, low cost solar cell foils, which can be easily integrated in many ways onto standard building materials, in particularly onto metal sheets that are employed in large quantities in facades and roofs throughout Europe
- the corresponding solar cell building materials can be light-weight and unbreakable, and hence easily installed in a standard construction environment
- a low temperature coefficient of amorphous solar cells allows for high energy yields under unventilated integration conditions
- amorphous silicon solar technology is highly attractive due to its excellent energy payback and eco-compatibility figures

A pre-requirement for the desired massive deployment of amorphous silicon PV technology in Europe's building environment is however the availability of high quality amorphous silicon foil materials at moderate prices. Here, in the past, the industrial development of amorphous silicon PV production has been hampered by its comparatively high initial investment costs and risks (with respect to standard poly- and monocrystalline silicon solar cells). However, the VHF high rate silicon deposition process[1], introduced by the University of Neuchâtel, holds the promise of reducing this burden by enabling clearly higher throughputs and, thus, a substantial reduction of investment costs and risks. On the other hand, the continuous fabrication of amorphous silicon on flexible foil by a roll-to-roll process equally allows to reduce system complexity and cost.

Therefore, the combined application of roll-to-roll deposition on plastic foil and the VHF-high rate deposition is expected to be a wining strategy on the way to foster a widespread PV application for energy production. Figure 1 shows the outline of the envisaged fabrication of glass-less PV-building elements. By this strategy, it should become possible to realize in the future stabilized module efficiencies of about 7-8% at production costs of around 100 $/m². In the following paper, we describe the achievements so far made in moving towards this goal.

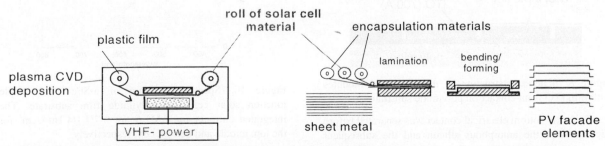

Figure 1: Projected fabrication of PV-building elements based on amorphous silicon solar cells on film substrate

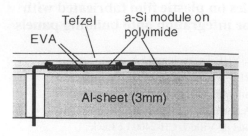

Figure 6: Schematic structure of test facade elements.

As a result, a facade element prototype including 4 series connected sub-modules of 46 cm² each was realized (Figure 7); it showed the following electrical initial state performance:

V$_{oc}$	12.9 V
I$_{sc}$	0.12 A
FF	56.6 %
Efficiency (aperture)	5.2 % (initial)
Illumination	91.2 mW/cm²
Module temperature	50°C approx.

Figure 7: Facade element prototype with four series connected sub-modules of 46 cm² each. The four sub-modules are directly laminated on an aluminum sheet, similar to those commonly used in metal facades.

5. LARGE AREA TRIPLE CELL MODULES ON POLYIMIDE FILM

In a final step, the VHF-large area deposition and the polyimide substrate module technology were merged. To reduce the module complexity and cost, an entirely different module structure was used (Figure 8):

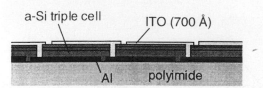

Figure 8: Schematic structure of the triple cell modules with thin ITO front contact on polyimide substrates.

First, the bottom electrical contact was separated <u>before</u> deposition of the amorphous silicon and the subsequent layers, thereby avoiding the application of an insulating paste (conventional approach). Secondly, to avoid the thick ZnO front contact and the Ag-finger-grid, those were replaced by a sole 700 Å thick sputtered ITO layer. For a <u>single junction</u> amorphous silicon cell, the use of such a thin TCO layer without finger-grid leads to unacceptably high series-resistance losses. A simple calculation [5] however shows that for amorphous tandem and especially for triple cells, the current density is decreased to such a degree, that the use of a thin, high impedance TCO becomes acceptable. Figure 9 shows the result of these calculations, demonstrating that the loss for a tandem and triple cell are reduced by a factor of 4 and 9, respectively, with respect to those of a single junction cell. With a 50 Ω/square sheet resistance, this enables a series-connection spacing of up to about 1cm width without a finger-grid based, while the losses remain lower than 10%.

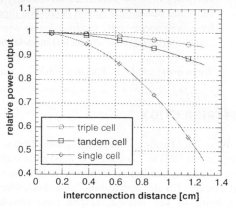

Figure 9: Calculated relative power output as a function of interconnection spacing of single, tandem, and triple cell amorphous silicon modules [5]. For the ITO layer, a 50 Ω/square sheet resistance is assumed. Further, all cells are assumed to have the same efficiency of 7.2%, but with a maximum power point voltage of 0.6V, 1.2V, and 1.8V, for single, tandem and triple cells, respectively.

As a consequence, triple stacked solar cells (a-Si/a-Si/a-Si) were developed on polyimide substrates. Figure 10 shows the spectral response of these triple cells. They yield a total cumulated AM1.5 current of 12 mA/cm².

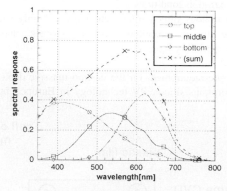

Figure 10: Spectral response of a-Si/a-Si/a-Si triple junction solar cells on polyimide film substrate. The integrated currents for AM1.5 are 3.7/4.1/4.1mA/cm² for the top, middle and bottom cell, respectively.

This relatively modest current is explained with the use of a specular back-contact and with still high absorption losses in the first tunnel junction. Back contact texturing, and an optimized tunnel junction should enable increase the current to about 14-15 mA/cm².

With the triple cell technology, 1cm wide test cells yielded a fill factor of 68% in a unilateral contacting scheme, demonstrating that a thin ITO top can be sufficient for current collection in an amorphous silicon series connected module. Based on this result, triple cell large area modules were fabricated in the reactor described in section 2. On a total area of 540cm² (28cm x 19cm) modules with 11 strings, each consisting of 19 series connected triple cells were realized (Figure 11). The series interconnection distance was 0.9 cm.

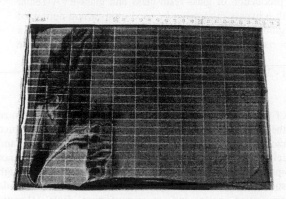

Figure 11: Triple junction amorphous silicon solar cell module (540 cm²) deposited in a large area 30 cm x 30 cm VHF-GD reactor at a plasma excitation frequency of 80 MHz. For this module, the i-layer deposition rate was of 6.5 Å/s, and the total triple cell deposition time was of 23 minutes. The top contact is a 700 Å thick ITO layer.

Figure 12 shows the resulting JV curve of one string out of such a module, showing a fill factor of 48%, and an active area initial efficiency of 4.1%. The evident remaining series resistance problem in the V_{oc} region we attribute to a still to be solved problem at the back-contact interface.

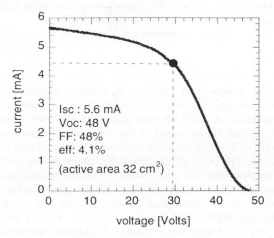

Isc : 5.6 mA
Voc: 48 V
FF: 48%
eff: 4.1%
(active area 32 cm²)

Figure 12: Initial state JV-curve of a string from a large area module as depicted in Figure 11. The active area efficiency is of 4.1%.

6. CONCLUSIONS AND OUTLOOK

In an approach to develop a PV technology option well suited for future energy production in Europe, amorphous silicon cells on flexible foil and their application on glass-less, metal based building materials are developed and tested. In this context, important cost reduction potentials are expected from:

(1) the roll-to-roll fabrication of high performance a-Si solar modules on plastic film substrate,

(2) the high deposition rate VHF-process for a-Si deposition, and

(3) from the application of plastic film substrate based a-Si PV-modules onto metal facade elements.

In this paper, it is shown that VHF-high rate deposition can yield homogeneous high quality a-Si layers and solar cells over substrate areas of 30 cm x 30 cm. Deposition rates for i-layer material of 6.5 Å/s are demonstrated at 80 MHz plasma excitation frequency. At this frequency, a-Si/a-Si/a-Si triple junction structures are deposited in a total deposition time of only 23 minutes. Based on this solar cell structure, series connected modules of 540 cm² total area, employing a low cost thin 700 Å ITO front contact without metal finger-grids, were fabricated. So far, with this technology, a maximal active area initial efficiency of 4.1% is obtained at an interconnection spacing of 9 mm.

Further, plastic film substrate based a-Si modules are successfully laminated on aluminum building-panels, demonstrating the fundamental feasibility of the envisaged facade integration of glass-less and series connected amorphous silicon modules.

Based on these results, a roll-to-roll pilot fabrication of amorphous silicon modules on plastic film is at present being implemented by VHF-Technologies SA together with the University of Applied Science of Le Locle, Switzerland.

Acknowledgements

This work was supported by the PSEL (Projekt und Studienfonds der Elektrizitätswirtschaft) under project No.88, by the CTI (Commission pour la technologies et l'innovation) under grant No. 3522.1, and by the Swiss Federal Department of Energy BEW/OFEN, under grant 19431.

References

[1] H. Curtins et al, Chem. Plasma Processing 7, p. 267 (1987)

[2] L. Sansonnens et al, Plasma Sources Science Technology 6, p. 170 (1997)

[3] U. Kroll et al, MRS Spring Meeting, San Francisco, 1999, to be published

[4] P. Pernet et al, 2nd WPVSEC, Vienna (1998) p.976

[5] according to A. Ricaud, "Photopiles solaires", 1997, Presses polytechniques et universitaires romandes, ISBN2-88074-326-5, p.219

DEVELOPMENT OF A HIGH CURRENT AND HIGH POWER TERMINAL SYSTEM WITH BYPASS DIODES FOR LARGE AREA MODULES INTEGRATED INTO BUILDINGS

K. Wambach
Flabeg Solar International GmbH
Haydnstr. 19 D-45884 Gelsenkirchen
Germany
Fax +49 (0)209 9134-120, Email wambach@pilksolar.de

ABSTRACT: Several terminal designs with bypass diodes were studied at high currents and poor ventilation of the junction boxes. Strategies to limit the thermal load of bypass diodes were examined. A new compact plug and socket circuit board was developed to allow an automated low cost production of glass-resin-glass and glass-EVA-Tedlar modules. The test results showed that small junction boxes with 6 bypass diodes can be used even in non-optimized installations with poor ventilation and partial shading. Design guidelines were revised for large modules proposing a revision of published solar cell specifications to reduce hot spot risks and thermal damage of the bypass diodes in the junction box.
Keywords: PV Module - 1: Building Integration - 2: Fabrication - 3

1. INTRODUCTION

The photovoltaic market is growing rapidly. The efficiencies and currents of modern solar cells are improved continously.

It is state of the art in PV system designs that bypass diodes have to be applied to prevent the modules from hot spot damage, but new technical solutions have to be developed to meet future system demands with high system voltages, high cell currents and high system relyability and fault tolerances. The risk of thermal module damage increases with ambient temperature and thermal load of the module from reversed biased cells and bypass diodes. High power general purpose modules with a large number of high current cells need tailored bypass diode solutions to protect the module from hot spot defects or overheating in the junction box in case of partial shadings.

The size of solar cells increased from 10 x 10 up to 15 x 15 cm. The currents of such modules vary from 2.7 to 9A, 5 A are very frequent today. At high currents special care has to be taken to cool the diodes, the terminals and the junction box during partial shading of the module. Results from module temperature measurements in the field, laboratory experiments and numerical calculations were reviewed and an improved automatically placable contact system was developed to reduce manufacturing costs and allow high performance at compact size.

2. FUNDAMENTAL CONSIDERATIONS

The hot spot test according to IEC1215 is a worst case test at normal operating temperature of the module at a well-defined irradiation and a function of cell and module type. At high environmental temperatures and module temperatures due to poor convection of the backside (warm facades, roof installations, wintergardens) much higher thermal loads well above a tolerable limit of 0.2 – 0.3W/cm² can be observed compared to the test conditions. The hot spot risks depend on the cell types and the distribution of the cell properties in production of the wafers and solar cells. Therefore the complete I-Vcurves in both quadrants and the distribution of breakdown voltages

of the reverse biased cells have to be known to minimize the number of bypass diodes necessary and thus reduce module production costs without sacrifying safety and relyability. Forward and reverse I-V curves are not specified in the data sheets of solar cells at present, though such information is of high importance to optimize module production costs and to manufacture fault tolerant modules. The use of bypass diodes requires additional terminals, more strings and interconnections as well as additional tests and maintenance in the junction box, all increasing the production costs significantly. Defect bypass diodes can short circuit several module parts or lead to permanent damage of the module if not conducting.

Recent measurements at TÜV and Ciemat [1-3] showed a broad distribution of reverse breakdown voltages of 7 – 18V for multicrystalline cells with dark currents of up to some amperes at –10V. Mono crystalline silicon cells in most cases have higher breakdown voltages of even more than 20V with currents below 1A at –10V. The slope at the breakdown voltage generally is higher compared to multicrystalline silicon cells. There was no indication that the cell sizes correlate with the breakdown voltage.

Using results from several hot spot tests of glass-resin-glass and glass-EVA-Tedlar laminates and numerical calculations with the programm PVTHERMAL [4-6] in a first estimation the hot spot risk increases at a given module design as a function of ambient temperature, irradiance and module temperature. Even at ambient temperatures of 20°C hot spot temperatures of more than 150 °C can be forecasted if the convection behind the module is low (e.g. in warm facades). As it was shown at JRC and ISE [4, 7, 8] the power dissipation can be even higher than 30W/dm² with irradiation without any damage detectable after the hot spot test. Double glass modules heat up slower than glass tedlar laminates and reach somewhat higher final temperatures. The module and junction box temperatures increase if backside ventilation is poor. The combination of high module temperatures, partial shadings and high temperatures of the operating bypass diodes can cause severe damage to the terminals, the junction box or the whole module.

4. EXPERIMENTAL SETUP

3.1 Heat Insulated Junction Boxes

Glass-resin-glass modules were assembled with two 4mm glass panes with a 2mm resin layer. Each module contained at least two solar cells. The sample sizes were 20 x 30 to 40 x 40 cm. A junction box (95 x 95 x 50 mm, PC, with up to 4 diodes or 95 x 95 x 30mm, PPO, with 3 diodes) was fixed on the back pane of each specimen. The diodes were installed in Wago spring clamps in case of the PC box and in MC terminal clips in the PPO box. The PPO box was used to test the new electronic circuit boards with 6 diodes during the development as well. Thermocouples were continously pressed on the package surface of a diode in the assembly. The modules were placed on top of a 10 cm PU insulation foam with a sufficient cutout for the junction boxes. Currents of 1 to 9 A were applied in forward direction through all diodes simultaneously via 2.5mm² cables to simulate the highest possible thermal load of the junction box. The voltage drop through the diodes and the current were measured to calculate the thermal power. The temperature changes were recorded until equilibration of the setup was observed.

3.2 Accelerated Aging Tests

The new contact system and the electronic circuit boards were certified according to IEC1215 and passed several internal tests. The contact system passed an internal UV irradiation test corresponding to 25 to 30 years in a middle European climate (5000h, 70°C) successfully. A humidity freeze test according to HF test of IEC1215 was carried out. During this test a continous current of 4.5 A passed through all contacts and a permanent 120VDC voltage was applied between all terminals on a second sample (50 cycles). A 500 cycles thermal shock test of the circuit board assembly showed no changes (–40°C/+85°C, equilibration time 30 min, temperature change within 30s). The circuit board with schottky diodes passed the impulse voltage test according to EN61000-4-5 and was grouped into class 2 of the norm.

4. EXPERIMENTAL RESULTS

4.1 Compact Junction Box With Terminal Clips

Currents from 1 to 5.5A were applied through 3 diodes (V_f=0,8V) installed with terminal clips in the PPO junction box. At a power of 5W (<2.5A) the temperature was above the upper limit of the diode specification (170°C). At 8W all polymer clips were melting. At currents above 3.5 A even through one diode only the specification limits were exceeded and changes of the contact clips could be detected. The surface temperatures of the diode package are shown in fig. 1.

4.2 Junction Box With Spring Clamp Terminals

In the PC junction box 4 diodes (V_f=0,8V) were installed in spring clamp terminals. A diode package surface temperature of 170°C was measured at 8W with changes of the materials. At 14 W the terminal polymers melted. The results were as expected from the first test. To avoid any damage of the junction box an air gap of about 2mm between the hot parts and the box should be ensured.

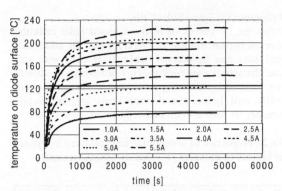

Figure1: Surface temperatures of three operating bypass diodes (terminal clips) in the PPO junction box placed in a heat insulation at different current levels.

4.3 Diode Selection

If the junction box is well ventilated the maximum thermal power should be < 10W (PC, PPO). If the ventilation is poor this has to be reduced < 5 W. This means that only one diode with a forward voltage drop of 0.8V at 4.5 A can be installed in the juncton box without exceeding the 5 W limit, with 3 diodes the power is > 10 W. The maximum junction temperature will be out of the specification anyway. As can be seen from table 1 it is better to reduce the currents through the diodes, .e.g. by parallel connection of selected diodes, or to use Schottky types with very low forward voltage drops though these are more expensive and have smaller reverse blocking voltages. The numbers printed in bold in table 1 indicate off specification operation of the diodes at different currents. At currents > 3A standard solder alloys on electronic boards melted if diodes with > 0.7V forward voltage drops were used. Recrystallisation and rapid aging of the solder joints was observed.

Table 1: Power dissipated in thermally insulated junction boxes (bold numbers: critical temperatures of diode junction)

Current [A]	No. of diodes	Thermal power at forward voltage 0.8V [W]	Thermal power at forward voltage 0.3V [W]
3	2	**4.8**	1.6
3	3	**7.2**	2.4
3	4	**8.6**	3.2
3	5	**9.6**	4.5
3	6	**14.4**	5.4
4.5	2	**7.2**	2.7
4.5	3	**10.8**	4.1
4.5	4	**14.4**	5.4
4.5	5	**18.0**	6.8
4.5	6	**21.6**	8.1
6.75	2	**10.8**	4.1
6.75	3	**16.2**	6.1
6.75	4	**21.6**	8.1
6.75	5	**27.0**	10.1
6.75	6	**32.4**	**12.2**

4.4 Properties Of Some Solar Cells

From the IV curves of reversed biased solar cells the number of bypass diodes in a module and the additional thermal load can be derived (table 2). If cells with soft reverse charactistics are used the current through the diodes can be reduced by allowing some reverse current through the cells well below the thermal limits of the module presumably all relevant cell data is provided.

Table 2: Properties of some solar cells relevant to bypass diode layouts of modules [1, 3 and own measurements].

Cell	Crit. breakdown voltage [V]	Current at -10V [A]	Power at -10V [W]
ASE/poly	-14	0.8 – 1.8	8 – 18
ASE/EFG	-10.4	1.3 – 2.0	13 – 20
ASE/ mono	-25	0.3 – 0.5	3 – 5
Solarex poly	-16	0.3 – 0.8	3 – 8

4.5 Terminal System Development

The new plug and socket terminal system developed was specified to meet the demands of several module types of both assemblies, glass-glass and glass-tedlar: The specification is shown in table 3. Some solutions how the specification could be met are listed in table 4. Most of them were used to reduce heat in the junction box.

Table 3: Terminal and board specification

Power	< 300Wp at 6 A or 120V
Temperatures in junction box	<120°C, < 170°C short time
Diode temperature	< 125°C, short time < 170°C
Power in junction box	< 10 W ventilated else <5W
Air gap between hot parts and junction box	> 2mm
Diodes	max 7 schottky diodes, max 6A/diode
Maximum operating current	10A at each of 8 terminals
Diameter of socket carrier	25 mm, thickness 3mm,
Glass thicknesses	3 – 9 mm, EVA-Tedlar possible
wire bridges for internal connection	8 positions, plug or solder terminal for cables of 4mm²
Junction box	MC JB/2 and bigger
automatic placement in module production possible	
ambient temperature	–40 to +100 °C
easy installation and service	
electrical safety class II	closed junction box

Table 4: Some methods to reduce heat in the junction box (methods printed in italics were not used in this case)

Applicable Method
Thick copper pads and double layer board
Optimize space between diodes
Solder melting point > 210°C
Allow heat transfer to module front
Allow heat transfer to thick wiring
Heat conductive resin on board
Use diodes of low forward voltage drop
Air gap to junction box 2mm
Allow some reverse current through cells to distribute currents through diodes and cells (soft reverse characteristics)
Sort out solar cells with low reverse blocking voltage breakdown
Improve convective and radiative cooling
Parallel connection of selected diodes
Maximize area for heat transfer
Apply active cooling

A micro plug and socket system was developed to fit in the biggest standard size of glass drill holes and to be on the same level as a 3mm glass pane. Up to 8 sockets can be used to contact up to 144 cells and 12 strings (fig. 2 and 3). A punched contact carrier welded to the micro-sockets is fixed with a double adhesive gasket at the drill hole of the back pane and insulated electrically from the solar cells in the module. The 8-pole plug is soldered to an electronic circuit board carrying the terminals and 6 bypass diodes. Plug and socket are fixed together with a 3 mm screw. The electronic circuit board is optimized to allow an excellent heat distribution. In addition a heat conducting resin is applied across the Schottky diodes of low forward voltage drop to reduce their surface temperature by another 20°C to about 125 °C at 7 A thus giving 70% higher performance compared to the results in fig. 1 at half of the size and lower costs (fig. 4 - 6).

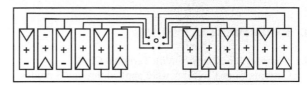

Figure 2: Interconnection scheme of two separate submodules in a laminate using the encapsulated sockets.

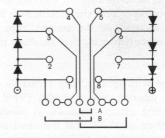

Figure 3: Schematic of electronic circuit, wire bridges for serial (A) or parallel connection (B) of two submodules can be applied

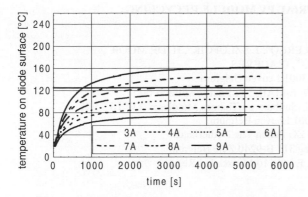

Figure. 4: Improvement with new circuit board. Notice the higher currents and twice as much diodes compared to fig. 1. PPO junction box in a heat insulation, 6 diodes on circuit board encapsulated in a heat conducting resin.

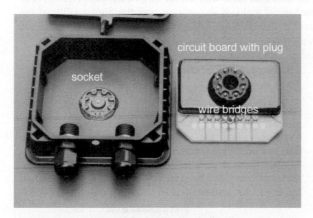

Figure 5: Socket encapsulated in glass-EVA-Tedlar module and circuit board

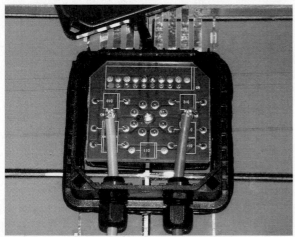

Figure. 6: Compact electronic circuit board connector with up to 7 diodes after installation in junction box

5. CONCLUSION

The forward voltage drops of conducting bypass diodes can lead to high thermal loads in a junction box. The thermal power has to be limited to about 10W in standard PV junction boxes or a maximum temperature of 125°C on all standard electronic parts. Higher temperatures cause accelerated aging of the materials and fatigue after several cycles. The use of long strings of high power cells reduces specific module costs, but requires a new design of terminals and bypass diode layouts. The individual properties of high current cells have to be considered during system design, the reverse voltage behaviour and reverse breakdown voltages should be published in the cell data sheets. Cells should be classified not only to maximum power or current I_{mpp} gradings but to their reverse currents I_{rev} at constant voltage, e.g. -12 V. Solar cell manufacturers should consider the integration of bypass diodes into the cells during new developments.

A new design of general purpose terminals and diode integration in junction boxes was presented with 70% higher performance at reduced size.

6. REFERENCES:

[1] W. Herrmann, M. Adrian, W. Wiesner, „Operational Behaviour of Commercial Solar Cells Under Reverse Biased Conditions", 2nd World Conf. On Photovoltaic Solar energy Conversion, Proc. Int. Conf. Vienna, Austria 6-10 July 1998 Vol II, 2357

[2] M.C. Alonso, F. Chenlo, „Experimental Study Of Reverse Biased Silicon Solar Cells", ibid,. 2376

[3] W. Herrmann, TÜV, Cologne, private communication

[4] J.J. Bloem, R. van Dijk, JRC, Ispra, experimental results to be published, see also M. van Schalkwijk, F. Leenders, J.J. Bloem, B. Cross, M. Sandberg, R.J.C. van Zolingen, „Prescript-Prestandardisation Activities For The Certification Of Roofs And Facades With Integrated Photovoltaic Modules", Final Report, JOR-CT97-0132, 1999

[5] Pilkington Solar International GmbH, „Development Of Manufacturing Techniques Of New PV Modules For Use In Facades And Sound Barriers", final report, BMB+F project 0329615, Germany 1998

[6] C. Bendel, T.M. Bruton, H. Müller, P. Niess, K. Wambach, J. Woodcock, „Color Variations Of Crystalline Silicon Solar Cells And Encapsulation Into PV Façade/Cladding Modules For Innovative Stylistic And Architectural Building Designs", Final Report, Jou2-CT94-0395, 1996

[7] H. Laukamp, K. Bücher, S. Gajewski, A. Kresse, A. Zastrow, „Grenzbelastbarkeit von PV-Modulen", 14. Symposium Photovoltaische Solarenergie, Staffelstein 1999

[8] H. Laukamp, A. Kresse, W. Leithold, „PV-Module ohne Bypass-Dioden?" 13. Symposium Photovoltaische Solarenergie, Staffelstein 1998

Acknowledgment

This work was funded by the government of North Rhine Westfalia, Germany

RECENT IMPROVEMENTS IN INDUSTRIAL PV MODULE RECYCLING

L. FRISSON, K. LIETEN, T. BRUTON*, K. DECLERCQ°, J. SZLUFCIK°, H. DE MOOR[+],
M. GORIS[+], A. BENALI[∇], O. ACEVES
SOLTECH Kapeldreef 60 B-3001 Leuven Belgium
TEL.: 32-16-298442 FAX : 32-16-298319 E-MAIL : Soltech@soltech.be
*BP Solarex 12 Brooklands Close Windmill Road UK -Sunbury on Thames Middlesex TW 16 7 DX
° IMEC Kapeldreef 75 B-3001 Leuven Belgium
[+] ECN Westerduinweg 3 NL - 1755 ZG Petten The Netherlands
[∇] Seghers Machinery Gentse Steenweg 311 B-9240 Zele Belgium
TFM Pol. Ind. Pla d'en Coll Gaià 5 E - 08110 Montcada i Reixac

ABSTRACT : The considerable growth of the PV market that started 30 years ago, will lead to a fast growing number end of life modules. If good solutions for recycling are developed, a huge accumulation of this end of life and rejected PV modules can be avoided. The aim of this work was to develop and to evaluate different recycling processes. Finally two methods have given acceptable results namely the pyrolysis in a conveyer belt furnace and the pyrolysis in a fluidised bed reactor. Especially for the fluidised bed reactor process, the development has reached an industrial level with the set-up of a big pilot reactor. The cost effectiveness of the process is demonstrated by the high mechanical yield of the process and the high quality of the reclaimed wafers, as proven by a high cell efficiency after reprocessing. The ecological impact of recycling is very high and the energy pay back time decreases drastically due to the avoided high energy consumption of the reclaimed silicon wafer.
Keywords : Recycling - 1: environmental effect - 2: cost reduction -3:

1. INTRODUCTION

Related to the PV module shipments 20 to 25 years ago, the accumulation of end of life PV modules will increase very fast. The producers of solar energy, a clean energy source, cannot afford a landfill destination for their end of life PV modules. Due to the relative high value and energy content of the silicon wafers, a dismantling operation for used and rejected PV module can be cost effective and is environmentally justified by a very positive life cycle analysis. Industrial recycling processes are optimised followed by a silicon wafer reclaiming and an optimised silicon solar cell and module process. The high quality of the reclaimed silicon wafers is proven by the solar cell efficiency results after reprocessing using an adapted solar cell process.

2. OPTIMISATION OF RECYCLING PROCESSES

Different potential recycling processes were proposed in the beginning of the project. Some of them seemed not to be promising and further research was stopped in an early stage.

Pyrolysis with microwave heating failed due to the non-uniform temperature distribution resulting in a considerable cell breakage.

Also dissolving the modules in a chemical reactor with tri-ethylene glycol at temperature between 220° C to 290° C resulted in negative results. The EVA swells up and does not release from the module. Tests with other solvents came to the same conclusion [1].

Immersion in hot nitric acid had already shown its potential [2]. New tests based on earlier experience showed that, though the process works, it is unlikely to become a viable industrial process due to huge amounts of nitric acid needed. Disposal of this chemical waste in a

responsible manner and the treatment of the toxic NO_2 gases would undoubtedly increase the cell recovery complexity and the energy involved as well as the financial cost significantly.

The thermal approach seems to be favourable to a chemical one [3]. Therefore, pyrolysis in a conveyer belt furnace looks promising as an industrial recycling process.

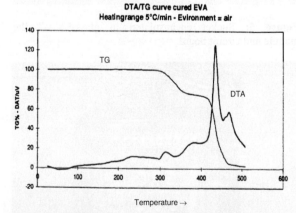

Figure 1a DTA/TG-curves from EVA in air

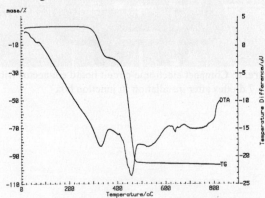

Figure 1b DTA/TG-curves from EVA in nitrogen

The EVA is burned away in the air atmosphere or decomposed under nitrogen at temperatures of 450°C resp. 480°C. The DTA/TG curves presented in fig. 1 shows this decomposition under ambient air or under nitrogen.

The tests with the thermal treatment under air resulted in a poor mechanical yield. This can be caused by the considerable temperature increase at the silicon surface due to the exothermic reaction leading to cracks.

Changing the furnace atmosphere from air to nitrogen avoids this exothermic reaction. With an optimised nitrogen flow and conveyor belt speed this reclaim process results in mechanical yields higher than 80%.

The most promising technique especially for industrial implementation is pyrolysis in a fluidised bed reactor [4].

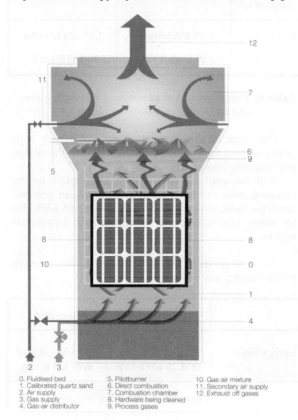

0. Fluidised bed	5. Pilotburner	10. Gas-air mixture
1. Calibrated quartz sand	6. Direct combustion	11. Secondary air supply
2. Air supply	7. Combustion chamber	12. Exhaust off gases
3. Gas supply	8. Hardware being cleaned	
4. Gas-air distributor	9. Process gases	

The fluidised bed reactor (fig.2) is filled with very fine sand that has a narrow particle size distribution. Due to an optimised air stream this sand is in a hot boiling fluidised state. In this fluidised state the sand takes the physical properties of a liquid. The modules are loaded in a basket and immersed in the fluidised bed.

Gasification of the EVA and back side sheet of the modules, sustained by the mechanical action of the silica, takes place. The off gases emerge from the reactor surface and pass immediately through the flame shield serving simultaneously as postcombustion and as a heat source for the reactor. The even temperature profile throughout the bed, the perfect mixing and the intense contact with the reactor recovers the silicon wafers and the glass in the optimal conditions.

Different parameters have been investigated. The process temperature has been optimised around 450°C. At this temperature a process time of 45 min. seems to be optimal.

A very important parameter to optimise is the fluidisation velocity of the sand particles. This velocity is directly related to the average diameter of the particles. For the recycling of crystalline silicon modules the velocity is kept extremely low at 1 cm/sec with a particle diameter of 100 micron.

A better collection yield of the reclaimed wafers was obtained by putting each module in a netting envelope. Also the positioning of the module in the loading basket seemed to be very important. An angle of 60 degrees gives the best results but leads to a worse occupation of the useful surface and a decreased capacity. New fixing structures with vertical positioning are under study now and have already given satisfactory results.

All the tests done up to now are with prepared samples with different module encapsulation technologies. The tests started with small 8-cell modules and are gradually going to the 36-cells modules. The project target of 80% mechanical yield of the wafers is already obtained, but with the new fixing structures now under study better results are very likely. A yield of almost 100% is achieved for the glass sheets.

3. SILICON WAFER RECLAIM AND CELL REPROCESS OPTIMISATION

The reclaimed solar cells have very low efficiencies as they were heated to 450°C for at least 20 minutes. Screen printed cells are shunted after such a heat treatment and cannot be re-used as such but have to be cleaned, etched and reprocessed into new cells.

Lifetime mapping with an MFCA scan on cleaned monocrystalline wafers indicate a preservation of the material quality after the recycling process. On the contrary, even a higher lifetime was measured on reclaimed wafers from solar cells that had an aluminium back side metallisation. The high material quality of the silicon wafer is confirmed by the very good cell efficiencies after reprocessing.

A general cleaning and etching sequence of the metallisation for most of the reclaimed wafers was a 15% HF treatment followed by a $H_2SO_4 : H_2O_2$ 4 to 1 solution at 80°C and finally followed by a 40% HNO_3 at 80°C. An emitter etch in 20% NaOH at 85°C is needed before starting the new cell process.

For reclaimed monocrystalline wafers the best process sequence is given in the following process flow chart.

Metal etching
I
Emitter and BSF etching
I
Surface texturisation
I
$POCl_3$ emitter diffusion
I
Parasitic junction removal
I
PE CVD SiN_X
I
Screen printing of metal contacts and cofiring

Despite a slightly worse texturisation almost the same results were obtained for reclaimed and for virgin wafers. Average results are presented in table 1.

	Jsc (mA/cm²)	Voc (mV)	FF (%)	Eff (%)
Reclaimed wafer	34.8	618	76.1	16.4
Virgin wafer	34.9	617	75.8	16.3

Table 1: Average illuminated IV parameters for reclaimed and virgin monocrystalline wafers.

The reprocessing of reclaimed multicrystalline wafers is somewhat more difficult. The emitter etch in 20% NaOH increases the step height at the grain boundaries, causing a lot of interruptions in the screen printed metallisation. The consequence is poor fill factors.

An optimised isotropic texturing process step overcame these problems. The PECVD SiNx process appeared again to be the best approach for reprocessing reclaimed wafers. The average cell results on reclaimed and virgin wafers are presented in table 2. Thanks to a better starting material the results for reclaim wafers were somewhat better.

	Jsc (mA/cm²)	Voc (mV)	FF (%)	Eff (%)
Reclaimed wafer	33.6	618	76.6	15.9
Virgin wafer	32.9	611	76.1	15.3

Table 2: Average illuminated IV parameters for reclaimed and virgin multicrystalline wafers.

Modules are made with these recycled wafers without any noticeable difference in mechanical yield. Accelerated ageing with the damp heat test does not show any additional degradation.

4. ECOLOGICAL IMPACT AND COST ANALYSIS

A life cycle analysis is made based on a module with 125 x 125 mm multicrystalline silicon cells. A standard module compared to a module using recycled wafers resulted in a 40% reduced energy consumption per generated kWh. The power generation is assumed in a sunny region for 20 years resulting in a total generation of 33 kWh/Wp or 71.9 kWh/waf. Table 3 shows the results.

Consumed energy of PV module		
	Standard	Recycled
Silicon production	7,55 kWh/wafer	-
Solar cell production	0,65 kWh/wafer	0,65 kWh/wafer
Module production	1,12 kWh/wafer	1,12 kWh/wafer
Recycling at end of life	-	0,4 kWh/wafer
Total	9,32 kWh/wafer 0,129 kWh/kWh$_{Gen}$	2,17 kWh/wafer 0,030 kWh/kWh$_{Gen}$

Table 3: Energy content of a module using virgin wafers and a module using recycled wafers

The reuse of the recycled silicon wafer with high energy content for a second life time improves the energy pay back time considerably.

With the small additional energy consumption for the recycling, solar cell and module process we can generate again the same amount of energy namely 1,65 kWh/Wp for sunny regions and 0,86 kWh/Wp for continental regions.

Table 4 presents the calculations of the strongly reduced energy pay back time.

	Standard	Recycled
Energy input	9,32 kWh/wafer or 4,26 kWh/Wp	2,17 kWh/wafer 0,99 kWh/Wp
Energy pay back time		
Sunny regions	2,58 years	0,6 years
Continental regions	4,92 years	1.14 years

Table 4: Energy pay back time for a module using virgin wafers and a module using recycled wafers

A second driving force behind the recycling process is the cost benefit. The high value of the silicon wafer is a significant cost benefit. The silicon wafer is recuperated in good condition and at relatively low cost. The cost calculations of the recycled silicon wafers are based on the presently available knowledge of the fluidised bed reactor process.

The following assumptions are made for this process:
- Mechanical yield of 80%
- Cycle time of 1 h gives capacity of 576 wafers/h
- Total investment cost of 575,000 Euro
  for a total line consisting of:
  fluidised bed reactor,
  a wet bench for etching and cleaning and
  a demi water installation.

The direct cost per recycled wafer as summarised in table 5, shows clearly the cost effectiveness of the process. The cost for the collection of the end of life modules is not included yet.

	Direct cost per recycled wafer
Investment	**0.047 Euro**
Energy	**0.011 Euro**
Etching, cleaning	**0.037 Euro**
Labour	**0.120 Euro**
Total	**0.215 Euro**

Table 5: Cost breakdown to recycle silicon wafers

5. CONCLUSIONS

Effective recycling processes were developed with very good results. The fluidised bed reactor process reached the industrial level by the construction of a pilot machine to demonstrate the exploitation capability.

The reclaimed silicon wafers preserved their initial high quality resulting in high efficiency recycled solar cells.

So, the PV industry has now a cost effective industrial recycling process available that can strengthen their positive image as clean energy producers.

6. ACKNOWLEDGEMENT

This research work is supported by the European Commission in the frame of a Brite Euram project under contract number BRPR - CT98- 0750

REFERENCES

[1] K. Sakuta et al : Attempt to recover silicon PV cells from modules for recycling 2nd world conference on Photovoltaïc Solar Energy Conversion July 1998 Vienna

[2] T. Bruton et al : Recycling of high value, high energy content components of silicon PV modules, 12 th European Photovoltaic Solar Energy Conference April 1994 Amsterdam

[3] K. Wambach : Recycling of PV modules 2nd World Conference on photovoltaïc Solar Energy conversion July 1998 Vienna

[4] L. Frisson et al : Cost effective recycling of PV modules and the impact on environmental, life cycle, energy payback time and cost 2nd World Conference on Photovoltaïc Solar Energy conversion July 1998 Vienna

LITHIUM-METAL BATTERIES FOR SMALL PV SYSTEMS

Andreas Jossen[1], Volker Späth[1], Harry Döring[1], Christos. Protogeropoulos[2], Nick Hatzidakis[2], S. Cofinas[2], Pnina Dan[3], Piero Calcagno[4], Gert Laube[4]

1: Center for Solar Energy and Hydrogen Research (ZSW), Helmholtzstrasse 8, 89081 Ulm, Germany, Tel.:+49-731-9530-0, e-mail: Ajossen@huba.zsw.uni-ulm.de

2: Center for Renewable Energy Sources (CRES), 19th km Marathonos Av. Gr-190 09 Pikerni Tel.: +30 1 6039900, Fax.: +30 1 6039905, e-mail: cprotog@cres.gr

3: TADIRAN Batteries Ltd., P.O. Box 1 Kiryyat Ekron, Israel 70500, Tel.: +972-8-9444366, Fax.: +972-8-9413062, e-mail: pninadan@netvision.net.il

4: WINSOL Energy Systems S.r.l., Via Enrico Fermi 23, 27100 Brindisi, Italy, Tel.: +39-831-546878, Fax.: +39-831-546879, e-mail: calcagno@iqsnet.it

Abstract: Small photovoltaic (pv) systems are normally autonomous systems and need batteries for energy storage. Up to now mainly lead-acid or nicd batteries are used. Within the described project a new type of battery, a lithium-metal battery was tested under solar operation conditions. For comparison other battery technologies (NiMH, lead-acid, RAM and LiIon) were tested too. Therefore special test regimes for parameter and cycling tests were developed.
The results show that lithium-metal batteries have high efficiencies, a very good low temperature performance and high shelf life. However an optimised management system is necessary to achieve a high lifetime.
The results of the battery tests were used to develop an optimized battery management system. Additionally the lithium-metal battery is optimized for solar operation. The optimized cells have a energy density of 158 Wh/kg. This parameter make them interesting for portable pv systems.
Keyword: Lithium metal battery – 1: Small pv systems – 2: Battery testing -3

This project is under financial contribution of the European Commission (JOR3 CT98-0305)

1 Introduction

Up to now in nearly most cases of stand alone pv systems lead-acid batteries are used. The reason therefore are mainly the low price, the low selfdischarge and the available system technology for lead-acid batteries. Only in a very few cases Nickel-Cadmium (nicd) batteries are used, but the high price, the high selfdischarge and the small energy efficiency are significant disadvantages of nicd systems. Nicd batteries are only used for very small applications and for applications at very low temperatures (arctic systems). However in a couple of applications lead-acid and NiCd Batteries do not fulfil the requirements. In this cases new battery technologies, i.e. Lithium-metal-Batteries, could be used.

However, this type of rechargeable battery is in an early state of development and there are some parameters that should be optimized for photovoltaic applications. Additionally the battery management system (BMS) of this battery type is different from BMS systems of lead acid or nicd batteries, normally used for these type of application.

The aims of this project are the determination of relevant battery parameters, the development of an optimised Lithium-Metal battery and the design of a battery-management-system (BMS).

The state of the art Lithium-metal (technical data see table I) cells were tested within the first project phase. Parameter tests and solar cycling tests were carried out with single cells. The results are used for the development of optimised Lithium-metal cells (AA size) for photovoltaic use. The specific energy, the cycle life, the self-discharge, and the shelf life are the key parameters to be improved.

Technical data of state of the art LiMe cells

Table I: Technical data of the investigated LiMe cells

Chemical System:	Li/Li_xMnO_2
Size:	AA, 8ml
Weight:	17gr
Nominal Voltage:	3.0V
Nominal Capacity:	800mAh
End of Discharge Voltage:	2.0V
Nominal Discharge Current @ RT:	250mA
Maximum Discharge Current (pulsed):	2A
Internal Resistance:	0.08Ω
Nominal Charge Rate:	80mA (to 3.40V)
Maximum Charge Rate:	250mA (to 3.45V)
Number of Cycles to 65% of Nominal Capacity @ RT	
@ nominal charge rate, DOD=100%:	300 to 350
@ nominal charge rate, DOD=50%:	600
Capacity retention after 1 year storage :	85% @ RT
Operating Temperature Range:	-30^0C to $+85^0C$
Storage Temperature:	0^0C to $+30^0C$

Additionally a low cost battery management system for this cells was developed. The development was attended by numerical simulation. Therefore a mathematical model of the lithium metal cell was developed and tested.

The improved Lithium-metal cell and the developed battery management system will be integrated at the backside of a photovoltaic module. Special consideration will be taken at the thermal behaviour of the battery. The developed system will be tested by field tests.

Taking the today high costs of Lithium-metal batteries into account, this battery type is not interesting for large pv systems, but it is an interesting candidate for small photovoltaic outdoor applications.

Within a market study a large number of applications were analysed. The most promising are shown in figure 1.

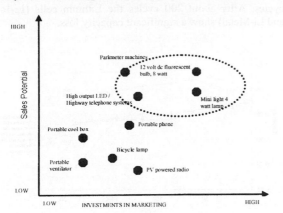

Figure 1: Small pv applications that are interesting for Lithium-metall storage systems

2 Battery testing

Within the project, laboratory tests were carried out. Additionally field tests will start soon.

For laboratory tests 800 mAh LiMe state of the art cells (Table I) were used. The parameter test and the simulated solar test take a wide temperature range into account.

In order to compare LiMe cells with none-advanced batteries, a small number of other battery types was tested under equal conditions (only cycling tests).

2.1 Laboratory test methods

Test procedures were defined to determine battery parameters as they are important for photovoltaic use. Unfortunately there exist only an IEC draft and a French Standard describing cycling regimes. Both standards describe cycling regimes for systems like solar home systems where deep cycling is required. The target applications are of a total other type. There the battery is mainly cycled shallow, with a daily DOD of approximately 10 to 20%. So it was necessary to define a adequate test procedure for cycling tests. For parameter tests the IEC draft 61960 part 1 and part 2 are used for orientation. Based on this standards a total parameter test procedure was developed.

2.1.1 Parameter tests

Within the parameter tests, well defined parameters were determined by constant current (potential) operation. The parameters that were measured are:

- relationship between temperature, discharge current and capacity
- relationship between temperature, charge current and capacity
- charge retention
- self discharge
- charge and energy efficiency
- deep discharge recovery

The test procedures are based on the draft IEC 61960. However the test procedures were adapted to LiMe batteries and some test procedures are oriented to solar operation. In example the charge and energy efficiency test procedures can be used to determine the efficiencies of the batteries at different state of charge (SOC) areas.

The detailed description of the test procedures are available at the author.

Parameter tests were carried out only with LiMe cells.

2.1.2 Cycling tests

For cycling two different regimes were developed, whereas the cycling depth is 10% or 20%.

The battery is daily cycled. The battery is charged during day and discharged during night. This cycling profile is generated by a superposition of a simulated insolation profile and a load profile. The load profile consists of a stand by load and the operation load. For example the stand by load is $0.02 I_{20}$ and the operation load is $2 I_{20}$.

The insolation profile was generated by measured insolation data.

Cycling tests were carried out at different constant temperatures (-20°C, RT, +40°C) but also with a day-periodic temperature profile. Figure 2 shows the state of charge during cycling.

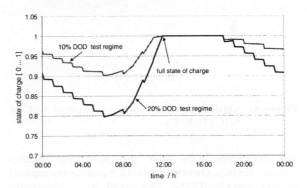

Figure 2: Calculated SOC for the two cycling regimes (10% DOD and 20% DOD regime)

2.2 Results

2.2.1 Parameter tests

The results show that the investigated cell shows only a small influence between charge- (discharge-) current and capacity. The influence of the temperature (0 - 60 °C) is small too. However at temperatures of -20°C and high discharge currents (> I_{10}=80mA) the capacity decreases below 50% of the nominal capacity (see figure 3). This should not be a problem in case of small stand alone photovoltaic systems where currents are normally below I_{10}. Maybe one exception could be applications with pulse-current discharge profiles (i.e. transmitters).

The charge and energy efficiencies are very high, especially at high state of charge. The Ah-efficiency is 99.5% and the energy efficiency is up to 97.6%. At lower state of charge the energy-efficiency is reduced from about 97.6% to 94%

The deep discharge recovery tests show that the LiMe cells are resist again deep discharge operation. The capacity loss is about 0.33% per week of deep discharge operation (see figure 4).

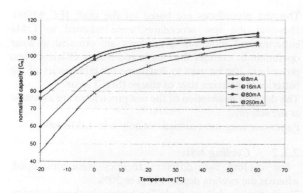

Figure 3: Battery capacity v.s. temperature for different discharge rates.

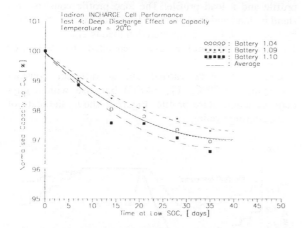

Figure 4: Capacity loss due to prolonged operation at low SOC (T=20°C)

The overcharge tests show that overcharge reduces capacity slightly. Overcharge for 12 h reduces the capacity of approximatly 0.25%. This could be a life-limiting criteria in case of small outdoor photovoltaic systems where the SOC is normally very high.

2.2.2 Cycling tests

Table II shows the efficiencies of the investigated batteries at the 20% DOD cycling test. The efficiencies are measured for a duration of two cycles. The LiMe cells show very high efficiencies, especially the Ah efficiency is approximately 100%. The energy efficiency is approximately 6% higher in comparison to lead-acid and 18 - 24% higher in comparison to RAM and NiMH batteries.

Cycling efficiencies

Table II: efficiencies during solar cycling with the 20% DOD test regime

	Lead-Acid	NiMH	RAM	LiMe
η-Ah	98.0%	77.7%	90.1%	99.9
η-Wh	91.5%	74.1%	80.0%	98.0%
η-U	93.4%	95.4%	87.9%	98.1%

The capacity development (normalised values) during cycling with 20% DOD is shown in figure 5. The capacity

of lead-acid batteries increase within the first 50 cycles. This is typical and is caused by formation processes. The Lithium cells do not show this behaviour. They show a slow continuos capacity decrease during the first 150 - 200 cycles. After about 200 cycles the Lithium cells (Li-Ion and Li-Metal) show a significant capacity loss.

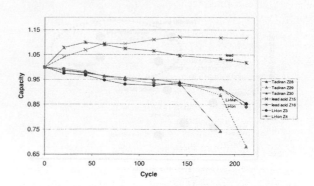

Figure 5: Capacity (normalised) v.s. cycle (20% DOD solar cycle) at 20 °C

Cycling tests at other temperatures than RT show that higher temperatures lead in a significant shorter lifetime of LiMe cells. For example at 40°C a cycle life of about 100 cycles was found. Another disadvantage is that the lifetime is influenced by the current rate during charge and discharge. Additionally it was found, that none constant current operation reduce battery lifetime. To overcome this disadvantage, a special battery management system is necessary.

3   System technology

Within the project an optimised battery management system (BMS) was designed and prototypes were developed. The BMS takes the characteristic of the Lithium-metal batteries into account.

To achieve the necessary flexibility and the battery capacity, the battery is organised in a serial-parallel connection of up to 5 strings with four cells. The basic structure is shown in the following figure.

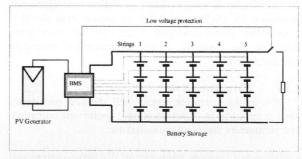

Figure 6: Schematic diagram of the BMS control in a small PV system

Up to now, it is not known if cell equalisation is necessary. The developed design of the BMS gives the possibility to do equalising or to disable the equalisation feature.

The costs of the total BMS are estimated to 15 - 30 Euro, depending on the equalisation feature. One target of the field tests is to determine the lifetime with and without charge equalisation.

The key features of the BMS are:

- Charge control
- Overcharge protection (single cell)
- Cell equalisation
- Single cell voltage monitoring
- Under-discharge protection
- Short-circuit protection
- Outdoor operation
- Incorporation of laboratory-derived characteristics
- Programmable characteristics using a notebook PC
- External visual indication of system status

The following figure shows the prototype of the BMS.

Figure 7: PCB of the developed management system

3.1 Numerical simulation

For system analysing and system design, a numerical battery model for the Li-Me cell was developed. Within the literature there was no model found fitted to the Li-Me battery cell. Therefore an existing model for lead-acid batteries [1] was modified for the Li-Me cell. The structure of the model is shown in Figure 8.

Parameters and equations for the components of the model were identified for different temperatures. The model has a average error of below 1%.

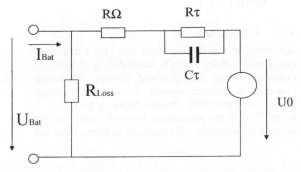

Figure 8: Schematic drawing of the battery model

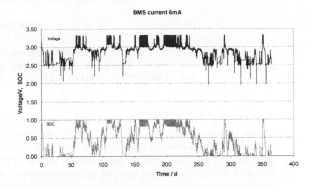

Figure 9: Result for a one year simulation of a system with Lithium-metal batteries

4 Optimization of Li-metal cells

One tasks within the project is to optimise the existing Lithium-metal cells, taking the PV applications into account. The key parameters are:

- Increasing of the energy density from to 140 to 175 Wh/kg (1Ah AA size cell)
- Increasing of lifetime to 2500 cycles at 10% DOD
- Increasing of shelf life
- Decreasing of self-discharge

Up to now an energy density of 158 Wh/kg is reached and shelf life of 3 years was established.

5 Conclusions

The results show that LiMe batteries are interesting for small solar applications. The high efficiencies, the good low temperature performance, and the high energy density are better in comparison to batteries used today, i.e. lead-acid and nicd. However the operation differs from the kind standard batteries are operated and a special battery management system is necessary to guarantee the lifetime. Such management systems were developed and field tests will be carried out soon.

The developed test procedures for parameter and cycling tests can also used for other battery technologies. The description of the test procedure are made available on request.

The optimisation of the LiMe cells will go on. The optimized cells will improve the overall performance of small pv systems with LiMe cells.

6 References

[1] Shepherd, C.M.: Design of Primary and Secondary Cells – II: An Equation Describing Battery Discharge, Journal of the Electrochemical Society, Vol. 112, No.7, 1965, p. 657.

BATTERY TESTING FOR SOLAR HOME SYSTEMS

K. Wolters, C. Helmke

European Commission, DG Joint Research Centre
European Solar Test Installation (ESTI), I-21020 Ispra, Italy
Tel.: +39 – 0332 – 789289, FAX: +39 – 0332 – 789268, E-mail: karsten.wolters@jrc.it

ABSTRACT: The European Solar Test Installation (ESTI) is involved in EU projects to increase the quality and reliability of Solar Home Systems (SHS's). Beside efforts to define standard procedures for performance tests of SHS's under indoor and outdoor conditions, a battery test facility for the conditioning and analysis of lead acid batteries is essential. As a service unit for SHS testing, it ensures homogeneous starting conditions for different batteries. Before and after the tests a detailed analysis of battery parameters, such as the dependence of capacity on discharge current and maximum capacity available as well as the internal resistance, allows one to observe subtle differences in battery performance. This report describes the design of the battery test facility and initial results.

Keywords: Batteries – 1: Capacity – 2: Internal Resistance – 3

1 INTRODUCTION

In many applications, such as for industrial uses or as starter batteries for vehicles, lead acid batteries have a satisfactory service life. In PV applications the requirements are different and field experience shows that the battery turns out to be the weakest component with the shortest service life.

In order to characterise the performance of SHS's and their components, an additional battery test facility for the conditioning and analysis of lead acid batteries is essential. As a service unit for SHS's tests, precise measurements of battery voltage, current and temperature under controlled charge and discharge conditions, ensures homogeneous starting conditions for different batteries. Before and after the tests a detailed analysis of battery parameters, such as the dependence of capacity on discharge current and maximum capacity available as well as the internal resistance, enables to detect subtle differences in battery performance. Moreover, the possibility for performing and developing test procedures to predict the long-term performance of batteries is a useful and important accessory for future research work.

Initial results have shown that some batteries that are in use in SHS's do not always meet the system requirements. Indicated capacities were not reached and were in fact sometimes below a justifiable tolerance of 80% of the rated capacity. Using existing standards, these batteries would have to be changed with such low capacities. Obviously the high battery requirements defined by international standards departs in some cases substantially from the reality.

2 SET-UP

A fully automated set-up has been designed and realised to charge and discharge batteries synchronously under controlled conditions like temperature and current, to calculate available capacities and to measure the internal resistance over a range of States Of Charge (SOC). The batteries are placed in a container filled with water in order to provide a uniform temperature. Using high precision bipolar power supplies constant currents from +20A to –20A are applied which covers adequately the range of currents found in Solar Home Systems. A data acquisition system measures current, voltage, battery temperature and ambient temperature at appropriate time-intervals.

The 400W bipolar power supplies can operate in voltage control or current control mode and can be connected in series or parallel as required. This enables to apply constant currents during discharge for precise capacity measurements and for preconditioning, to fully charge batteries either in constant voltage or constant current mode. Externally programmed current pulses in a time interval of few milliseconds are also possible for internal resistance measurements.

A Digital Multi Meter (DMM) with two plug-in 20-channel multiplexer cards is used to measure the voltage, the current (as a voltage over a calibrated 10-mOhm resistance), the temperature of each battery and the ambient temperature.

A program written in LabView, with a user-friendly and easy to read front panel, controls these instruments and saves the data in an easily accessible file system.

Beside capacity and resistance measurements, the high flexibility of the set-up, due to simple program modifications, also enables one to perform any test procedure.

3 DETERMINATION OF CAPACITY AND RESISTANCE

3.1 Capacity test

According to the IEC standard 896 [1], a battery is considered as fully charged, when during charging at constant voltage the observed current, or during charging at constant current the observed voltage shows no appreciable change during a period of 2 hours. This is the preparation for each battery to undergo a capacity test. The capacity is then calculated out of the discharge with a constant current until a final voltage of 1.80V/cell or 1.85V/cell and the discharge time.

When the initial average temperature T is different from the reference temperature (20°C), the uncorrected capacity C shall be corrected by means of the following equation to obtain the actual capacity C_A:

→ $\quad C_A = C / (1 + \lambda (T - 20))$

The coefficient λ shall be taken as 0.006 unless otherwise specified by the manufacturer.

3.2 Resistance test

The open circuit voltage V_{OC} divided by the short circuit current I_{SC} gives the internal resistance R_I of a battery. Since the short circuit current is hard to determine, a voltage drop ΔV, divided by a current drop ΔI, can also be used to calculate the internal resistance [2]. A current pulse of ca. 100ms is established (Figure 1). Recording voltage and current within an interval of $\Delta t = 6$ms, ΔV and ΔI can be determined. It follows:

→ $\quad R_I = \Delta V / \Delta I$

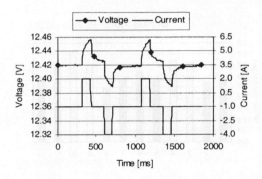

Figure 1: Sharp increase of voltage due to a current pulse. Polarisation effects cause continuous increase of voltage during the 100ms pulse.

Within the 100ms charge or discharge, polarisation effects are also measured and affect the result. A pulse of 1ms is recommended [3].

However, this influence is minimised by:

→ $\quad R_I^{corrected} = (V (t_1) - V (t_0)) / \Delta I$

t_0 = start of current pulse; $t_1 = t_0 + \Delta t$

To determine the resistance during charge or discharge periods over a range of State of Charge, the mean value of five measurements in positive and negative direction gives an adequate reliable result. The comparison with measurements using a 50Hz impedance meter validates the results that were obtained.

4 CURRENT DEFINITION

To compare currents at which batteries are discharged/charged the concept of the discharge or charge rate is normally used [4]. Here the indicated or rated capacity (C_R) refers to a discharge occurring at a constant current for a defined hour rate. For example, a 90Ah battery might be indicated as follows:

→ $\quad C_{100} = 90$Ah to 10.8V at 20°C

This would be the rated capacity C_R (= C_{100}). The index refers to 100h discharge at a constant current, in this case 0.9A, which is defined as the I_{100}. Since the capacity in Ampere-hours divided by a time in hours is a current in Ampere, the I_{100} is also written as $C_{100}/100$ or simply $C/100$.

→ $\quad I_{100} = C_{100}/100 = C/100 = 0.9$A

A current of 4.5A would be the C/20 and of 9A the C/10. However, if the same battery is discharged with a higher current, the capacity will decrease. Some manufactures indicate their batteries as well for the 10-hour and the 20-hour discharge rate. The same battery might have the additional indication:

→ $\quad C_{10} = 70$Ah to 10.8V at 20°C

The related current is:

→ $\quad I_{10} = C_{10}/10 = 7$A

To avoid any ambiguity as to whether a C/10 is 9A or 7A, the index specifics to which capacity the current refers. Within our measurements and results, the adjusted currents as discharge or charge rates always refer to the C_{100} (= C_R) capacity indicated by the manufacturer.

5 RESULTS

Capacity tests of 14 batteries, all intended to be used in PV applications, provided remarkable results.

- Most of the batteries did not achieve the rated capacity
- For some batteries significant variations in capacity with sequential cycles were monitored

All batteries had been prepared in accordance with the manufacture instructions and had been charged usually with $C_R/25$ either in constant voltage (CV) or constant current (CC) mode. After these preparations many discharges, mostly with $C_R/100$ and $C_R/10$, had been performed. Figure 2 shows two typical constant current discharge curves to 1.8V/cell with $C_R/100$ and $C_R/10$.

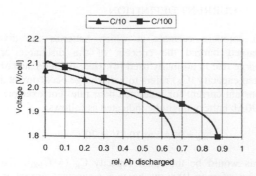

Figure 2: Capacity test. Constant current discharge curves at 20°C with C/10 and C/100. The relative Ah discharged refers to C_R

Eleven batteries could not be discharged to more than 90% of C_R. Two Batteries achieved the rated capacity and one battery was discharged to ca. 110% of C_R.

Typical charge curves are seen in Figure 3 and Figure 4. At around 2.3V/cell the gradient of the voltage curve increases and a gassing phase begins. Charging with constant current leads to a final Voltage of 2.7V/cell and more, depending on the charge current. A corrosion reaction at the positive plate might be accelerated by a dangerous high potential. However, short times at higher voltage will produce vigorous gassing that stirs up the acid and helps to remove stratification.

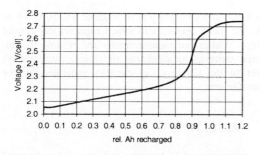

Figure 3: Constant current charge curve at 20°C with C/25. The relative Ah recharged refers to the preceded discharge

The relative Ah recharged in Figure 3 and Figure 4 refers to the Ah of the preceded discharge and not to the rated capacity. Therefore the Ah efficiency can be directly determined. This battery, charged with constant current, has a Ah efficiency of ca. $\eta_{Ah} = 83\%$ and when charged with constant voltage of ca. $\eta_{Ah} = 96\%$. Since some of the charge current goes into the gassing process, the lower efficiency with constant current charge is to be expected. However, after charging with constant current the available capacity was generally higher than after constant voltage charge. Therefore the determination of the Ah efficiency is incomplete. It is more appropriate to summarise the efficiency over a large number of cycles and to calculate the mean

value. A constant current or boost charge is necessary to keep stratification low.

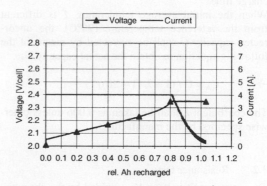

Figure 4: Constant voltage charge curve at 20°C. The relative Ah recharged refers to the preceded discharge

The cycling endurance of some batteries was insufficient. For the worst battery, a difference of ca. 15% of the achieved capacity with the same discharge current over sequential cycles was monitored (Figure 5). This battery was normally charged with constant voltage. A constant current charge between cycle 5/6 and 9/10 led to a remarkable increase of the capacity. This might indicate effects of stratification.

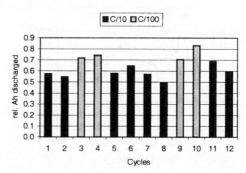

Figure 5: Significant variations in capacity due to a poor cycling endurance. Constant current charge between cycles 5/6 and 9/10. The relative Ah discharged refers to C_R

On the other hand, batteries, which had been charged with constant current for several consecutive cycles, sometimes intentionally for longer time periods at a high voltage, had losses in capacity of up to 10%. Significant losses in capacity were noticed after an excessive charge at around 2.8V/cell. Corrosion effects could have caused these degradations. A planned extension of the test facility is to open batteries for detailed examination of the plates and their degradations.

The results of the internal resistance measurements show that it can not be used for a sufficient SOC determination. The resistance remains almost stable over a wide range of SOC (Figure 6). The slight increase from 100% to 20% of SOC during continuous discharge is too small to be a reliable indicator. Although the resistance at the end of a discharge with C/100 is higher than after a discharge with C/10, no clear dependence on the delivered capacity could be recognised. However, a resistance measurement is used to detect serious defects of a battery, such as a short circuit between the plates. The continuous monitoring over many charge and discharge cycles could enable to identify aging effects, such as corrosion or sulphation, and might be therefore a criterion for the determination of the long time performance of a battery.

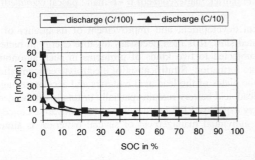

Figure 6: Internal resistance over a range of State of Charge (SOC) during continuous discharge with $C_R/100$ and $C_R/10$.

6 CONCLUSION

According to international standards, batteries are considered as "dead" when the actual capacity is less than 80% of the rated capacity. Our results showed that the capacity of several batteries was from the beginning below 90% and sometimes in fact below 80% of the rated capacity. However, this does not directly mean that the battery has also a bad performance, sometimes a good cycling endurance was monitored for such batteries. Therefore the initial measured capacity would have to be defined as the new rated capacity.

With the battery test facility, a strong initial charge or a defined state of charge can be applied to different batteries before conducting a SHS test, which ensures homogenous starting conditions. Also tests on available capacities, the range of the internal resistance over SOC and the charge efficiency, are performed. These are standard procedure for every battery before or after a SHS test. Additional, particular characteristics, such as the maximum capacity available depending on the maximum discharge current present in the intended SHS, as well as the determination of recommended thresholds for the charge regulator, can be defined. Finally, it is possible to perform any test procedure, which will help in future to obtain more information about the service time of a battery.

REFERENCES

[1]   International Standard IEC 896 – 1(2); Stationary lead-acid batteries – General requirements and methods of test; International Electrotechnical Commission (IEC); 1996

[2]   Linden, D.: Handbook of Batteries and Fuel Cells; McGraw-Hill Book Company; USA 1984

[3]   Zhang, J. / Canady, W.J.: Using High Speed Pulse as an Analysis Tool, In: The Twelfth Annual Battery Conference on Applications And Advances; California State University; California 1997

[4]   Spiers, D. / Royer, J.: Guidelines for the use of Batteries in Photovoltaic Systems; Canadian Energy Diversification Research Laboratory (CEDRL) of CANMET / Neste Advanced Power Systems (NAPS); Sipoo/Finland 1998

MULTI-BATTERY-MANAGEMENT,
AN INNOVATIVE CONCEPT OF PV STORAGE SYSTEM

Jean Alzieu & Guy Schweitz
Électricité de France, Division R&D, CIMA 8, Les Renardières,
77818 Moret-sur-Loing Cedex, France
Tel : 33-1-60-73-60-73 – Fax : 33-1-60-73-74-77
e-mail : jean.alzieu@edf.fr - e-mail : guy.schweitz@edf.fr

Pascal Izzo & Patrick Mattesco
Université de Montpellier II, LEM, Place Eugène Bataillon,
34095 Montpellier Cedex 5, France
Tel : 33-1-60-73-60-18 – Fax : 33-1-60-73-74-77
e-mail : patrick.mattesco@edf.fr - e-mail : pascal.izzo@edf.fr

ABSTRACT: Half of humanity is yet awaiting electricity, economical development and improvement of its quality of life. Renewable energy systems may be used for supplying electricity to isolated sites. It is then necessary to store energy in batteries. Vented lead acid batteries, which are generally used, are particularly subjected to two kinds of problems in such autonomous systems: stratification of electrolyte concentration and irreversible sulphate formation. Stratification of electrolyte concentration is provoked by incomplete charges; it leads to heterogeneous discharge of the plates of the lead acid batteries and accelerated ageing of the lower parts of these plates. Irreversible sulphate is produced when lead acid battery remains partially or totally discharged for a prolonged period. This process may lead to sensible reductions of battery capacity. EDF has developed an innovative concept of PV plant architecture and energy management. This concept, based on individual management of several batteries, is aimed at solving electrolyte stratification and irreversible sulphate formation.

Keywords: lead acid battery – 1: stratification – 2: battery management – 3

1. Introduction

Photovoltaic production profiles do not fit with consumption profiles. Hence, for stand-alone PV systems, excess of energy produced during periods of sunlight must be stored for when there is a deficit in production (at night for example). Furthermore, this storage enables to satisfy power requirements, which may be significantly higher than the instantaneous power supply of the photovoltaic panels.

Storage batteries most often used are vented lead acid batteries, which remain the best compromise for performances, cycle life and price. The management of these batteries is a sensitive issue. In stand-alone PV systems traditional management devices tend to protect the battery against damaging conditions: principally over-charges and over-discharges. But these are not the only damaging conditions. Incomplete charges lead to heterogeneous functioning of the plates and irreversible sulphate formation, which are the two main causes of premature failures of lead acid battery in such autonomous systems.

2. A few characteristics of vented lead acid batteries

The full charging of vented lead acid batteries consists of two steps: a main charging phase and a complementary phase known as gassing phase. These two steps are shown in Fig. 1.

2.1 Main charging phase
During this first phase, the electrolyte undergoes a heterogenisation of its concentration called «stratification» [1-3]. We know the electrolyte is involved, in the lead acid batteries, in the reactions of charging/discharging. Its concentration falls when discharging occurs and is re-established when charging occurs. The fall of concentration during discharging is relatively homogeneous throughout the electrolyte. Inversely, the charging reaction frees sulphate ions in the electrolyte and then increases its concentration. This takes place in the porous part of the active material of the electrodes i.e. a fairly limited volume, generally lower than 20% of the electrolyte volume. In this limited volume, the electrolyte attains a high concentration, corresponding to high densities. This leads to it oozing to the surface of the electrodes then, in the case of vented lead-acid batteries, to slide, under the effect of gravity, along their surfaces, towards the bottom of the container. This results in strong increases in the concentration of the electrolyte in the lower part of the battery elements with little change in the upper part (Fig. 1.b). This process called stratification increases throughout the charging. At the end of this phase, that is to say just before the gassing phase takes place, the accumulator is in a critical state: the electrolyte concentration in contact with the lower part of the electrodes can reach the double of that at the top of the electrodes. Maintaining this state to engage a discharge could be detrimental to the accumulator and could lead to a significant reduction in its lifespan.

2.2 Gassing phase
During this phase, the total current is made up of a charging current and an electrolysis current producing oxygen to the positive electrodes and hydrogen to the negative electrodes. The gassing phase has two functions: to approach in an asymptotic way a fully

charged state and, thanks to the mixing induced by the release of gases, to homogenize the concentration of the electrolyte.

This homogenisation of the electrolyte is the main function [4]. It is necessary for each charging. Without it in the discharging, which will follow, there is a heterogeneous functioning of the electrodes leading to significant reduction in their lifespan. Discharging under stratification leads to intense activity of the lower parts of the electrodes, which leads to their rapid destruction. This process is often the cause of premature failures in photovoltaic installation batteries.

The use of a high voltage threshold in traditional control systems leads to the homogenisation phase being reduced and even absent with as a direct consequence premature failures batteries.

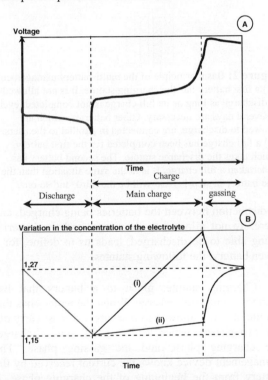

Figure 1: Voltage variation (Figure 1-A) and variation in the concentration of the electrolyte expressed by its density (Fig. 1-B) for a vented lead acid battery undergoing a cycle of discharging, then charging at constant current. The curves (i) and (ii) of Fig. 1-B correspond respectively to the measurements of density made at the bottom and at the top of a cell, i.e. under the stack of plates and at the surface of the electrolyte. Just one curve appears on discharge, it corresponds to the superposing of the high and low measurements, which are similar. The values indicated are given as an example.

2.3 Irreversible sulphate

The battery can remain in a weakly state-of-charge for extensive periods. But lead acid battery dislikes remaining long periods without going back to a fully charge state. On one hand due to the phenomena of stratification, described above, but equally due to the phenomena of recrystallisation. The lead sulphate, insoluble product of the discharge, is where recrystallisation phenomena occur: the smallest lead sulphate crystals tend to dissolve to the advantage of the largest, more thermodynamically stable crystals. In time this process leads to a reduced reactivity of the lead sulphate and to a reduction of its specific surface (large crystals). This leads to a reduced kinetic of dissolving during the following charging. Hence to widening distances of diffusion between the dissolving sites and the sites of reaction. Consequently, the capacity of the battery can be significantly reduced in time [5].

Increasing the size of the battery is sometimes used to reduce the adverse effects. The lifespan of the battery is then longer than that of a smaller battery, however, still reduced to what could be expected.
On large plants such as photovoltaic back-up power station of the network, mechanical mixing systems of the electrolyte can be used. These systems are relatively efficient to solve the stratification of the electrolyte, but they remain inefficient concerning the sulphate recrystallisation.

3. Traditional management of energy

The battery charges and discharges along with changes in sunlight and consumption. The charge and discharge are controlled in general by four threshold voltages [6] programmed in the regulator:

• A high voltage threshold is supposed to correspond to the full charge of the battery which, when reached, triggers the disconnection of the PV array.
• The battery voltage then decreases until it reaches a new threshold, which allows the reconnection of the PV array.
• A low voltage threshold associated to a depth of discharge not to be exceeded. When reached battery is then disconnected from the users.
• The arrival of a new charging current increases the voltage of the battery. When the battery is charged sufficiently, corresponding to a new voltage threshold, the battery is reconnected to the users.

Example of threshold values for a 2V lead cell:
- 2.4V: disconnection of panels
- 2.2V: reconnection of panels
- 1.9V: disconnection of the battery from the users
- 2.1V: reconnection of the battery to the users

Such a system of command has some drawbacks. Firstly, there does not exist in practice a strict correlation between the instantaneous voltage of the battery terminals and its state of charge. In particular, it is possible that the voltage is high during a charge although the state-of-charge of the battery is very low. For example when the battery has remained discharged for a long time.

Moreover, traditional managements of energy by voltage thresholds are not able to solve all the problems outlined above, inherent to the functioning of vented lead acid batteries. In particular:

- The homogenisation of the concentration of the electrolyte at the end of the charging is often badly achieved for two reasons. Firstly the sun can set before the charging is completed; the discharge, which will follow in the evening, will happen in a stratified electrolyte. The second reason is that even if the end of the charging is achieved during a sunny period, the disconnection of the panels on the threshold voltage leads generally to an insufficient gassing phase to ensure homogenisation, especially when there is no significant over sizing of the photovoltaic panels.
- A period of bad weather results often, for a traditional photovoltaic plant, in maintaining the battery incompletely charged. This results in the formation of irreversible sulphate and the corresponding loss of part of the capacity.

4.   Multi-battery management

EDF has developed a new principle of designing and managing battery storage of stand-alone PV systems. This principle is aimed at avoiding incomplete charges. With this intention we have imposed on ourselves a simple rule: every charging process, when started, must be completed. This means until a quantity of electricity judged sufficient has been supplied to the battery during the final gassing phase.

An immediate consequence of this rule is that it is not possible to function with just one battery. In fact, as a battery being charged is forbidden to discharge, if a user must be supplied, it is necessary to have another battery allowed to discharge to supply the current. Our basic rule obliges the use of at least two batteries.

Computer simulations corresponding to varied experimental conditions have shown that four batteries is a good number for this type of management. This we call multi-battery management [7]. Figure 2 depicts its basic principle.

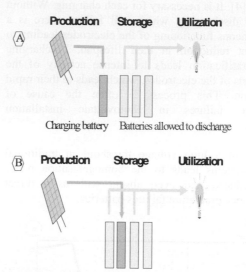

Figure 2: Basic principle of the multi-battery management: (A) a first battery has the charging statute. It is not allowed to discharge as long as its full charge is not completed, even if several days are necessary. Other batteries, which are allowed to discharge, are connected in parallel to the users. (B) a full charge has been completed for the first battery, which takes the discharge statute. The second battery has undertaken a full charge. It is in the same situation than the first battery in step 1. Next will be the third, and so on...

A distinction between the batteries being charged, and therefore not allowed to discharge, and the batteries being able to be discharged, leads us to define, for a given battery, the following statutes.

- Charging statute: given to a battery that has started a complete recharge. When the battery has this statute, it is not allowed to discharge. It must carry out the two phases, which lead to a full state-of-charge: the charging phase and the gassing phase. The management device knows the current received by the battery from the beginning of the charging phase. It detects, by analysing the current and the voltage of the battery, the beginning of the gassing. It calculates the amount of electricity, which must be received by the battery during the gassing phase in order that the battery reaches a full state-of-charge (Figure 3). Its value depends on the depth of discharge of the battery at the moment it starts it's charging. At the end of this step, the electrolyte is homogenous in the battery, which can then be discharged without danger of heterogeneous functioning.

- Discharging statute: a battery which does not have a charging statute. This statute allows discharging the battery. Batteries having this statute discharge in parallel. A diode type device does not allow the flow of current from one battery to another. A battery having the discharging statute is not normally allowed to be charged.

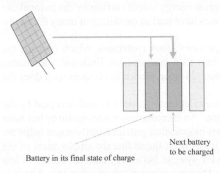

Battery in its final state of charge

Next battery to be charged

Figure 3: The gassing phase is necessary to ensure homogenisation of the electrolyte. When a battery is in the final stage of its charge, i.e. the gassing phase, it may be necessary to limit its charging current. Excess current is then directed to the next battery to be charged.

The multi-battery principle, based on the coexistence of these charge and discharge statutes, authorizes some associated functionalities. It allows periodic controls of the capacities of the batteries without exhausting stored energy. It is indeed possible to specifically discharge only one battery among four. This measurement is useful both for the state of charge evaluation and for state of health watching.

State of charge evaluation
In PV systems, when storage autonomy is at least 2 or 3 days, discharge rates are too low to have a very significant effect on the battery capacity. A simple amp-hour algorithm may be used to obtain state of charge evaluation then. Maximum state of charge is associated to full charges, which are systematic with multi-battery management. Minimum state of charge is more difficult to evaluate: its prediction is associated to the knowledge of the actual battery capacity that can vary sensibly during its life. Then, these periodic measurements of actual capacities provide accuracy to the state of charge evaluation.

State of health watching
Capacity loss is the principal parameter that is needed to define the state of health of a battery. Periodic actual capacity measurements are memorized. This allows evaluating capacity loss in term of percentage

of the initial capacity. Moreover, variations of actual capacity can also give additional information, for example on the amount of formed/recovered quasi-irreversible sulphate.

Other associated functionalities
To transform quasi-irreversible sulphate, it's necessary to charge the battery with a very reduced current for long periods. After sunset, a single battery may be put in charge of supplying this reduced current to the other batteries. Each battery, in turn, provides this service for the following nights.
In case of a cell failure, it is also possible to put out of service the corresponding battery, and continue to work with, for example 3 batteries in place of 4. In a general way this principle facilitates maintenance operations, watching of state of charge and of state of health.

5. Conclusions

Most stand-alone PV systems mainly use vented lead acid batteries to store the produced electricity. Management of these batteries is a sensitive issue. Multi-battery Management allows to overcome the two main causes of premature failures of photovoltaic plant vented lead acid batteries:

- Heterogeneous functioning of the electrodes, due to discharging in presence of a stratified electrolyte.
- Formation of quasi-irreversible sulphate due to prolonged periods without any full charge.

Moreover, some advantages are awaited:
- Increase of the lifespan of the batteries,
- Limitation of the over sizing of the installations,
- Maintenance operations made easier,
- Watching of state of charge,
- Watching of state of health.

REFERENCES

[1] L. Apateanu, A.F. Hollenkamp, M.J. Koop,
J. Power Sources, 46 (1993), pp. 239-250.
[2] D. Desmettre, F. Mattera, J. Alzieu,
2nd World Conference and Exhibition on Photovoltaic Solar Energy Conversion, 6-10 July 1998 Vienna Austria.
[3] D.J. Spiers, A.D. Rasinkoski,
J. Power Sources, 53 (1995), pp. 245-253.
[4] C. Armenta-Deu,
Renewable Energy, Vol. 13, No. 2 (1998), pp 215-255.
[5] R. Wagner, M. Schroeder, T. Stephanblome,
E. Handschin, *J. Power Sources*, 78 (1999) 156-163.
[6] T.D. Hund,
AIP Proceedings 394, Nov 1996, pp. 379-394.
[7] Patent No. 9808531 – 03/07/1998
EDF - Gestionnaire d'Énergie Renouvelable.

ENERGY LIMITATION FOR BETTER ENERGY SERVICE TO THE USER:
FIRST RESULTS OF APPLYING "ENERGY DISPENSERS" IN MULTI-USER PV STAND-ALONE SYSTEMS

Xavier Vallvé, Jens Merten
Trama TecnoAmbiental S.L.
Ripollés, 46 - 08026 Barcelona - Spain
Tel. 34 93 450 40 91 - Fax 34 93 456 69 48
tta@retemail.es

Klaus Preiser, Martin Schulz
Fraunhofer Institut für Solare Energiesysteme ISE
Oltmannsstr. 5 – 79100 Freiburg - Germany
Tel. 49 761 4588 216 – Fax 49 761 4588 217
preiser@ise.fhg.de

ABSTRACT: One of the difficulties of PV stand-alone village micro-grids is the adequate distribution of the limited energy among the users. The fact that several users are connected to a common energy source can hinder the rational use of energy with the adequate responsibility by the individual user. These issues have lead to the failure of many PV multi-user systems for rural village electrification in the past.

To overcome this problem, TTA developed an electronic *Energy Dispenser / Meter* (patented), which controls the energy consumption behaviour of each user and incentivizes appropriate user load management. Each user can choose from several tariffs. So an assigned amount of energy assures the equivalent 24 hour service to all users as it does the conventional electricity grid.

First results of social studies performed indicate that in general the *Energy Dispenser / Meter* is well accepted by the users and is thought to be a very useful device for improved energy service. An increased user responsibility has been detected, as they are motivated to care about their consumption. The system-information provided furthermore helps the users to manage their loads in a favourable way. On the technical side, it has been found that the employment of the Energy Dispenser / meter" improved the reliability of the operation of the PV system: No energy cut-off because of low batteries have been detected. The experience with the installed "Energy Dispensers/meter" shows that this device is indispensable for the flawless operation of PV stand-alone systems for local minigrids.

Keywords: Villages –1, Energy distribution – 2, Rural electrification – 3

1. INTRODUCTION

Lack of electrical services in rural regions inhibits any economic activities and leads to migration. The aim of rural electrification programs is to give an electrical energy service equivalent to the grid [1]. Experiences made during the design and execution of large-scale rural electrification programs revealed, that one key issue for the successful implementation is to guarantee each user a defined amount of energy *EDA* (*Energy Deliverability Assurance*). For a village, however, the energy service should also include the possibility to operate other appliances of common interest (such as machines for manufacture) as well as public lighting.

It is clear that in the case of wide spread of individual houses, a multi user system not viable because of high costs of the minigrid to be installed. In many cases, however, the houses of the village are so clustered that the installation of a centralized system becomes viable. Multi user systems with centralized PV-hybrid systems have some specific advantages which reduce the cost for each user and which are discussed in the following section.

However, such systems have a specific problem, which is the distribution of energy between the users. This problem has lead to the failure of village electrification projects in the past and is discussed in Section 3. In Section 4, we briefly overview the energy dispenser solving this problem, and we show up, how it is integrated in the layout of a multi-user system in Section 5. Finally, we describe first operation results made with the energy dispenser in rural villages in Spain.

2. Pros and Cons of multi user systems

In this section, we discuss the installation of stand alone PV systems for all users with respect to the installation of individual systems for each users.

2.1. Reduction of system costs

In rural regions, some users are often absent, or have periods with reduced energy consumption due to traditional activities such as shepherds, outdoor activities, etc. These periods may occur with weekly or seasonal frequency, and lead to increased capture losses of the individual system. This is shown in Figure 1, where there are several users not using all the available energy of their individual systems. In such situations, it would be desirable to allow a neighbour to use the energy.

In a multi-user system, these periods of reduced consumption are distributed between all N users, and the energy provision EDA_{tot} of the overall system may be smaller than the energy EDA_i guarantied for each user i:

$$EDA_{tot} = F_E \sum_N EDA_i,$$

where F_E is the energy utilisation factor which is less than one. This allows the system designer to reduce the PV power and the battery capacity of the multi-user system, which leads to reduced investment costs.

Other factors reducing system costs are [2]:

- The €/kWp-cost decreases with the size of the system.
- Centralised systems lead to reduced maintenance costs. For example, only one battery bank has to be checked
- Centralised systems are easier to be expanded, when additional energy requirements occur

2.2. Increase of performance

Another advantage of multi-user systems is the possibility to increase the service provided to the user. Only centralised systems permit the installation of common services in the village, such as public lighting, a public washing machine, or electrical tools for craft manufacture.

Another point concerns the availability of power for the users. Figure 2 shows the histogram of the power

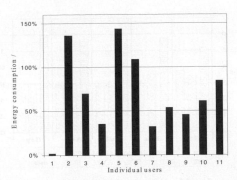

Figure 1: The ratio of energy consumed by the individual user and the energy deliverability assured by contract (*EDA*). Data shown are averaged during six months. There are many users which consume less than *EDA* or are absent, and there are others which consume slightly more.

consumption of typical individual PV systems in Spain (about 1kW$_p$ PV). As the typical operation period of user appliances is more than one hour, we can assume, that the long averaging period of the data (one hour) represent the continuous power demands on site. For a typical individual system, a continuous power of $P_{i,cont} = 400W$ would be sufficient. However, the high start up currents of refrigerators, pumps, etc. represent a high instantaneous power $P_{i,peak}$. This requires a relatively high inverter power, which are very rarely fully used by an individual user. Users of a smallest scale PV systems often do not have sufficient inverter power to start electrical machines like water pumps etc.

This situation is different in a multi user system. The probability, that all users require their guarantied maximum power $P_{i,cont}$ for continuous use at once, is low. This probability is even lower for instantaneous peak power $P_{i,peak}$ for start-up. Hence, the power provision Ptot of the overall system may be smaller than the power $P_{i,cont} + P_{i,peak}$ required for each individual user:

$$P_{tot} = F_{P,cont} \sum_N P_{i,cont} + F_{P,peak} \sum_N P_{i,peak},$$

where $F_{P,cont}$ and $F_{p,peak}$ are less than one. The inverter size of a multi user system is usually determined by the

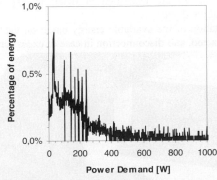

Figure 2: Power demand of single user systems. Data shown are from six different sites and hourly averaged. (40000 hours analysed, class-width 5W). As a typical load is always operative more than one hour, values shown are similar to the continuous power demand $P_{i,cont}$ of the users. There is a strong consumption below 10W (stand-by devices, single lights) and around 30W.

continuous power demand. Once covered this demand, there is still plenty of peak power available for each user. Therefore, in multi user system, a single user has more inverter peak power available than he would have in an individual installation.

2.3. Problems

Multi user systems have specific problems, which do not occur with single user systems. One of these is the enhanced effort necessary in the financial management for such installations, as all users to be connected have to agree at once to finance a multi user system. There are cases, where only a part of the village neighbours wishes to contribute and benefit to the multi user system. The solutions lays in systems with an layout which can be easily expanded.

Another problem is the energy distribution between the users. This aspect is addressed in the following sections.

3. THE ENERGY DISTRIBUTION PROBLEM OF MULTI USER SYSTEMS. EXAMPLE: A MINIGRID SYSTEM WITH TWO USERS

Besides these advantages of multi user systems, there are still specific challenges to be addressed. So the distribution lines of the minigrid increase costs and are not very beautiful.

The mayor challenge to be solved, however, for multi user systems with limited energy sources is the correct distribution of the energy between all parties connected. Sharing a common resource without limitations leads to the abuse by individual users. The fact that several users are connected to the same energy source hinders the rational use of energy with the adequate responsibility by the individual user. These issues have lead to the failure of many PV multi-user systems for rural village electrification in the past. The responsibility for such failures cannot be attributed to the user. Users are often not conscious about the energy consumption of the appliances they connect. A typical example is the washing machine with resistance heater. The lack of control animated the user to consume more energy than belongs to him.

This problem led to the development of a simple device that meters and limits energy consumption of each user. This device, the *TApS - energy dispenser / meter* enforces

Figure 3: Public lighting increases the living standard in rural villages and is often considered a priority by the users. (San Felices).

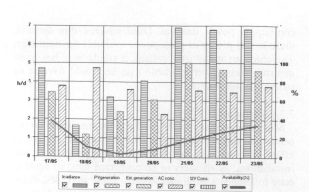

Figure 4: Daily monitoring data such as irradiance (left bars), generation (middle bars), and consumption (right bars). This graph is a standard output of the GIFA software developed at TTA for the management of stand alone PV systems. The data shown are normalised with the peak power of the PV modules installed (i.e. they are yields). They are expressed in units of equivalent sunshine hours (hours with $1000W/m^2$) per day. The absence of load management tools leads to consumptions exceeding energy generation. This leads to a poor battery availability index (line).

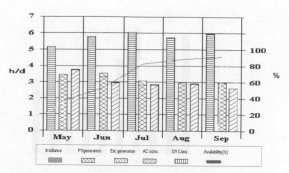

Figure 5: Monthly monitoring data. The high consumption of the two users resulted in a poor battery availability index in May 1999 (see Figure 1). The installation of the TApS-energy dispenser / meters on 6th of June resulted in a controlled energy consumption and normal battery availability index.

a balanced distribution of energy between all the households of the village. It has been found to be the only means to assure for each user the deliverability of a fixed amount of energy in multi-user systems.

As an example, how the energy dispenser may solve the energy distribution problem, we show here the simplest case of a multi user system with two parties only. They live in two houses of the same property, decided to implement a PV system with about 2kW, and to share this system. However, after a short time of operation, the users became cut off by low battery state of charge *SOC*. This is shown in Figure 4, where the battery availability index *AI* reaches zero. (This index is determined by the intelligent state of charge algorithm of the power conditioning unit *TApS - Centralita*: *AI*=0% ⇔ *SOC*=20%, *AI*=100% ⇔ *SOC*=100% [3]). After several discussions with the users it became clear, that this situation was caused by the abuse of one user only. With an individual system, this user would have been the only one affected by the lack of energy. In the multi user system, however, it is the <u>two</u> of them sitting in the dark.

The reason for this cut-off can be found in the fact that the energy consumption exceeds the energy generation on 17 - 19.5.1999. This situation occurred, because one user does not know, how much energy the other user is consuming. On the 19.5.1999, both users are finally disconnected by the system, as there is no more energy available in the batteries. In the following days, the users were more conscious about the limited energy

supply and moderated their consumption. However, after some days, similar situations occurred, as there were no means available, which could have allowed the users to coordinate their energy consumption. Frequent cut off due to low battery led to an unsatisfying situation for the users among themselves, and, as shown in Figure 5, to chronically low battery availability index in May 1999.

In order to resolve this unsatisfying situation, *TApS - energy dispensers* have been installed on 6th of June. This immediately lead to a control of the consumption of each user. As shown in Figure 5, the battery recovered in June and reached a satisfying battery availability index. This demonstrates, that the introduction of energy dispenser definitively solves the distribution problems of energy in multi user systems.

4.   THE ENERGY DISPENSER / METER

The *TApS - energy dispenser / meter* contains is a simple device which is installed in the house of each user (Figure 6). It contains an operation algorithm, which has been developed from TTA's experience in the execution of rural electrification programs. This algorithm and the operation principles of the *TApS - energy dispenser/meters* are described in detail in [4]. Main features of the device are:

- Limitation of the available energy based on the tariff contracted, and disconnection in case of excess energy

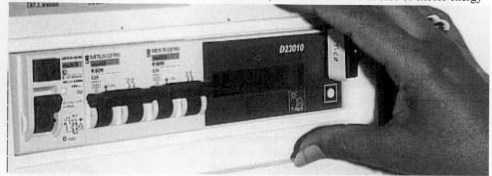

Figure 6: The *TApS energy dispenser / meter* mounted on a DIN-rail together with usual electrical circuit breakers. The hand just placed the individual user key *FEDI*, which contains the energy reserve of the user.

Tariff	Energy Deliverability Assured (EDA)		Power limit
	kWh/month	Wh/day	kW
TD-8/0,5	8,4	275	0,55
TD-17/0,5	16,7	550	0,55
TD-25/0,5	25,1	825	0,55
TD-33/1	33,4	1.100	1,10
TD-50/1	50,2	1.650	1,10
TD-67/2	66,9	2.200	2,20
TD-84/2	83,6	2.750	2,20
TD-100/2	100,3	3.300	2,20
TD-134/2	133,8	4.400	2,20

Table 1: Actual tariffs for the user to choose from. Example: with the tariff TD67/2, 2.2kWh may be used per day. These tariffs may be adapted to local requirements.

consumption (see Table 1 for possible tariffs).

- Measurement of the total energy consumed; memory with autonomy of long duration.
- Limitation of power, disconnection in case of excess power.
- Indicators and user advice.
- Adaption of operation algorithm to the state of charge of the battery: The bonus mode and restriction mode incentivizes adequate user load management. This optional feature is only available in the case of implementation with the power conditioning unit *TApS - centralita* C-8648, which transmits the system state via a domotic bus [3].

The dispenser and meter continuously measures the power and energy consumed. The programmed tariff (see Table 1) determines the maximum power and the amount of energy available for the user in the *regular energy reserve*. This available energy is increasing virtually according to the tariff contracted, when there is no

Figure 7: The 10kW$_p$ PV-plant in Escuaín. The structure of the PV system is also used as a housing of plant facilities (batteries, power conditioning, back-up generator).

consumption. The tariff is stored in an individual user key (called *FEDI, ficha electrónica de energía disponible individual*). This key can be taken off and used in another *TApS - energy dispenser / meter*. It also contains all the information about the energy available and the total energy consumed. By this way, it is possible to guarantee the user, that he may consume the contracted energy. This energy is called *EDA (Energy Deliverability Assurance)*.

5.   VILLAGE SYSTEMS WITH ENERGY DISPENSER / METER

Actually, nine multi user systems with *TApS - energy dispenser / meter* have been realised in Spain [5]. They are shown in Table 2. Figure 7 shows an example.

The architecture of the systems is based on a

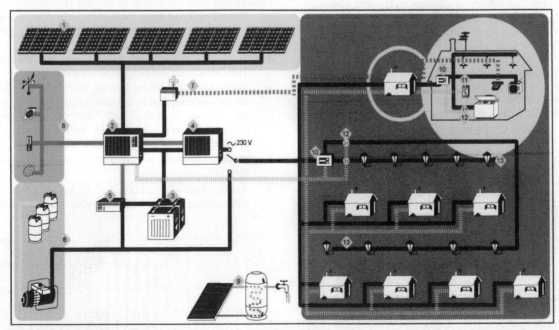

Figure 8: System architecture of village systems with *TApS* products. The energy dispenser / meter is located in each house (see upper right corner). The components in detail are: **1)** PV-array, **2)** *TApS - centralita* with MPP battery charger units, battery control and data acquisition, **3)** Batteries, **4)** *TApS - centralita* with Inverter unit, **5)** TApS - Battery charger, **6)** Emergency back-up generator, **7)** Cellular phone, **8)** Meteorological station (connected to **2**), **9)** Solar thermal collector for hot water supply (controlled by **2**) , **10)** *TApS - Energy dispenser / meter*, **11)** *TApS - Remote visualisation panel* for user information, **12)** *TApS - Remote controlled relay*, **13)** Public lighting with TApS ballast (electronic, two stages).

standardized layout, which is schematically shown in Figure 8. As outlined in Section 2, the use of a centralized system instead of individual systems reduces the need of PV power and battery capacity to be installed. Furthermore, the power conditioning (battery charger and inverter) can be concentrated in one single location on site. These facts lead to drastic improvements of system and maintenance costs.

The central power station contains a set of two *TApS-Centralitas*, which contain MPP-trackers, battery supervisor and monitoring system, and inverters.

Besides the AC grid for the energy supply, an information bus line is installed to each user. This *TApS-BUS* carries information about the system state, which is used by three different types of devices:

1. *Automatic load management relays* (N° 12 in Figure 8), which permit the operation of low priority loads as a function of the system state. This relay is typically used in each house for deep freezers, and is used for the public lighting. They are key tools for an automated load management
2. *Remote visualisation panel* (N° 11 in Figure 8), which give the complete system information to the system manger on site,
3. *Energy dispenser / meter* (N° 10 in Figure 8), which controls the energy consumption of each user.

6.   OPERATING EXPERIENCE

6.1.   Reliable System Operation

A typical daily load profile on a Saturday in *La Rambla del Agua* is shown in Figure 9. Most of the consumption occurs in the evening hours, for lighting and TV. The month of August is the month with most consumption on site. However, the battery availability index still shows satisfying values. This demonstrates that it is indeed possible to operate a stand alone PV system with 40 users with $10kW_p$ PV.

Figure 10 shows the histogram of hourly energy consumption data in *La Rambla del Agua* in August 1999. The very low power demands of single user systems disappeared. There is no continuous power demand beyond 5.5kW. This shows, that the inverter size of 7.5kW is sufficient for the users of *La Rambla del Agua*.

Village	Province	PV power	N° of users	Year
Escuaín	Huesca	10,2 kW	15	1997
La Rambla del Agua	Granada	10,2 kW	40	1997
Moriello de SamPietro	Huesca	2,25 kW	5	1997
San Felices	Huesca	10 kW	8	1998
Mipanas	Huesca	2,70 kW	4	1998
Aldea de Mirabal	Huesca	1.8kW	2	1999
Aldea de Estaronillo	Huesca	2.5kW	3	1999
Ascaso	Huesca	1.1kW	2	1999
Biscarra y Salamanca	Huesca	2.0kW	2	1999

Table 2:   Multi user systems realized with *TApS - energy dispenser / meters* in Spain.

6.2.   User acceptance

The idea of the multi user system has been developed together with the users of *La Rambla del Agua*, which wished a low cost solution of the electrification of their village [6]. Already in the design phase, it became clear, that the current and the energy of each user has to be limited. This lead to the development of the first generation of energy dispensers. After the successful operation, additional requirements occurred. These requirements have been then incorporated into the *TApS - energy dispenser / meter*:

• Use of surplus energy when the batteries are full. Solution: The benefaction mode for this case discounts the energy at half speed only.

• Use of the "neighbours" energy should be possible, when he is absent. Solution: The energy is stored in the user key FEDI, not the dispenser. So absent users may borrow FEDI to the present inhabitants.

• Use of community services from energy account of individual users. Solution: When a user want to use the communal television set, he takes the FEDI from his house and plugs it into the *TApS - energy dispenser / meter* of the community house. He also has the possibility to leave some energy "at home".

The *TApS - energy dispenser / meter* taking into account these remarks from the users has been installed in other villages.

6.3.   Different user behaviours

Figure 11 shows the readings of the *TApS - energy dispenser / meter* during August 1999. The consumptions are presented as a percentage of the guarantied energy *EDA* contracted by each user. Also shown are the levels of the regular reserve and the emergency reserve as percentages of their maximum possible values. One may observe four different classes of user behaviours:

• There is a broad majority of users consuming not all, but some of the energy contracted. They have good reserves in the regular and emergency reserve.

• There are many users without any consumption. They are absent and only sometimes in the village.

• There are other users, which have adapted their energy consumption to the state of the system (using the bonus mode of the *TApS - energy dispenser / meter*). So user 18 achieved a higher consumption than the contracted energy, and still has energy available in the reserve and emergency reserve.

• Another type of user is the one that consumes energy without taking care of the limitations of a stand-alone

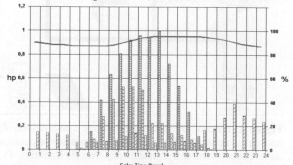

Figure 9: Hourly monitoring data of *La Rambla del Agua* on 14.8.1999. See Figure 4 for the legend. Note the high level of battery charge.

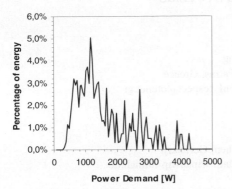

Figure 10: Power demand in the rural village *La Rambla del Agua* in August 1999. Data shown are hourly averaged.

system. They use all their energy contracted (User 2, 23) and therefore suffer low energy reserves.

6.4. User training seminars

In the beginning, the users had difficulties in understanding the concepts of the energy reserve and the emergency energy reserve. Also the different modes (bonus / normal / restriction) produced question marks in the faces of the users.

This shows that it is clearly necessary to explain the operation principle of the PV plant, but also the concepts of energy and power. A user who does not understand the reason of the energy and power limitation, will be always complaining, because asks why it is not possible to use a hairdryer. He has to know what is energy, why there is the need of energy efficient appliances, and the basic principals how a PV plant works. All this knowledge is essential for the understanding of the *TApS-energy dispenser / meter*, especially the differences of the bonus and restriction mode. We have to draw the same conclusion as for individual stand alone PV systems:

Only a trained user will be a satisfied user.

7. CONCLUSIONS

Multi user systems for the electrification of rural villages have specific advantages compared to individual systems. The centralisation of the system allows for reduced investment and maintenance costs, but also for increased performance and better service. Such systems favour the installation of common electrical appliances such as public lighting or special machines.

The main problem of multi users systems is the energy distribution between the users. This problem, leading to failures of village electrification projects with PV, is now addressed by a simple electronic device installed in each household. The first operating experiences with this *TApS-energy dispenser / meter* have shown, that the users accept the device. However, for a satisfying operation, it is necessary that the users do understand the operation of the PV system and the energy dispenser.

8. REFERENCES

[1] X Vallvé, J Serrasolses, *PV stand-alone competing successfully with grid extension in rural electrification: a success story in Southern Europe*. Proceedings of 14th E.

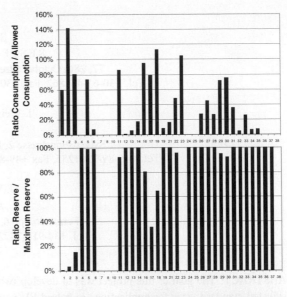

Figure 11: Analysing the dispenser / meter readings of *La Rambla del Aqua*. The top figure shows the total energy consumption of each user, as a percentage of the energy *EDA* guarantied by the tariff. Note the intelligence of user 18, who managed to consume 113% of his contracted energy. The middle figure shows the level of the energy reserve in %. The bottom figure shows the energy level of the emergency reserve in per cent.

P. S. E. C., Barcelona, june-july 1997, pgs. 23-26.

[2] Rehm, M. et al., *Influence of electricity demand profiles on the design of hybrid systems. A case study for South Africa*. Proc. 14th E.P.S.E.C., Barcelona, jul. 97, pp. 1110-1113.

[3] Vallvé, X. et al., *Stand-alone PV modular power conditioning station*, Proc. 12th E.P.S.E.C., Amsterdam, apr. 94 , pp. 1651-1654.

[4] X Vallvé, J Merten, E Figuerol, *Innovative load management for multi-user PV stand-alone systems*. Proceedings of 2nd World C.P.S.E.C. (15th E.P.S.E.C.), Vienna, july 1998, pgs.

[5] Boletín de SEBA nº 30, invierno 2000, Barcelona

[6] X Vallvé, JM Ruiz, J Serrasolses, *Organisational scheme for multi-user stand alone PV/diesel systems implemented in villages in Spain*. Proceedings of 2nd World C.P.S.E.C. (15th E.P.S.E.C.), Viena, july 1998, pgs. 3445-3448.

AN INNOVATIVE UNIVERSAL PHOTOVOLTAIC MANAGEMENT SYSTEM (UPMS) FOR STAND-ALONE HYBRID SYSTEMS

S. Pressas
R.E.S.P.E.C.T. Engineering
77, 28th October Street, GR-26223 Patras, Greece
Tel. +30-61-431862, Fax +30-751-6291, eMail: respect@otenet.gr

I. Weiss, A. Sobirey
WIP
Sylvensteinstrasse 2, D-81369 München, Germany
Tel. +49-89-7201235, Fax +49-89-7201291, eMail: wip@wip-minich.de

J. V. d. Bergh
ATERSA
Fernando Poo 6, E-28045 Madrid, Spain
Tel. +34-91-4747211, Fax +34-91-4747467

ABSTRACT: The goal of this project was to develop two *Universal Photovoltaic Management System (UPMS)* based on selected standard sizes for applications in hybrid PV power plants. R.E.S.P.E.C.T. Engineering and ATERSA developed and built the prototypes with a standard ac output rating of 3 kVA and 15 kVA in cooperation with WIP. The basic aims was to achieve substantial cost reduction, increased system efficiency, user friendliness and reliability improvements over the past conventional balance of system (BOS) configurations. This UPMS concept combines in one unit all power electronics, energy management unit, and their cabling necessary to control, regulate power, and monitor a complete hybrid PV plant. Cost reduction, increased system efficiency, user friendliness and reliability improvements are possible because all the critical BOS hardware and wiring are integrated in one box, resulting in use of common parts, cheaper transportation, installation, testing, and maintenance. Both UPMS prototypes have been installed and tested in two pilot plants and operate successfully. The inverter of the prototypes achieve a value of 92% and the cost reduction of the compact systems is 33% and 16 %. This project was co-financed by the European Commission in the framework of the Non Nuclear Energy Programme JOULE III.
Keywords: Photovoltaic Management System - 1: Stand-alone PV plants - 2: Cost Reduction - 3

1. INTRODUCTION

The principal goal of this project was to develop two prototypes of a *Universal Photovoltaic Management System* (UPMS) based on two selected standard sizes of 3-kVA and 15-kVA ac power output for stand-alone applications. The basic aim was to achieve significant cost reduction and reliability and efficiency improvements over previous balance of system (BOS) configurations.

The development of the universal PV management systems are considerably reducing costs in comparison with a conventional system configuration. The units are faster to install, and the simplicity of testing, initial check-out and maintenance reduces labour costs.

2. TECHNICAL DESCRIPTION

2.1 General Description

The two basic types of Universal PV Management System are briefly described below.
- Low-Power UPMS prototype: This modular standardized size will be used in low-power hybrid

PV power plant, installed near the user's or owner's residence, with a standard ac output rating of 3 kVA, PV array sized at 1.5 kWp, and battery capacity of 800 Ah [1]. All power electronics, power distribution switching and protective devices, monitoring capability, and energy management systems are completely packaged inside one container. The power sources, one solar irradiance sensor and the residential loads are the only connections to this UPMS. The prototype is installed and tested in the R.E.S.P.E.C.T. Engineering Testing and Demonstration Hybrid Power Plants located in Kourtaki, Argos, Greece.

- High-power UPMS prototype: This unit is also modularized and standardized but designed for the use in larger power applications such as centralized village power sources. Standard ratings selected are ac output rating of 15 kVA, PV array sized up to 10kW, and battery storage capacity of 1,200 Ah. The packaging of the UPMS is the same as for the low-power unit. The prototype unit is installed and tested in Gran Canaria, Spain.

2.2 Application Types

The application types which were identified for the 3-kVA UPMS are power supply systems for family houses, weekend/holiday buildings, water pumping systems, electrification of telecommunication systems for mobile telephones, light houses, fisheries, mountain refuges, monasteries, archaeological sites and tourist places in remote areas where the grid extension is too expensive and/or not possible in the near future. The low-power units can also be used as an uninterruptible power supply (UPS).

The 15-kVA UPMS will be used as central power supply for villages and high-power energy demand buildings. The unit will be able to supply electricity to facilities and buildings such as houses, schools, communication centres, health centres, police stations, street lights and water pumping systems.

2.3 Requirements for Both Prototypes

Both UPMS prototypes, 3kVA and 15kVA ac power output,. have to meet general requirements to achieve a high reliability, improved performance especially on the BOS components and to decrease costs due to the compact power management systems.

The power sources of both systems include a PV generator, wind generator and an ac generator (Diesel or gas). The selected storage systems are lead-acid battery cells with tubular positive plates for storage of the generated energy.

The typical loads to be powered by the PV system are ac loads which require a voltage of 230 Vac and a frequency of 50 Hz.

The energy demand for a typical family household was calculated and is illustrated in Figure 1.

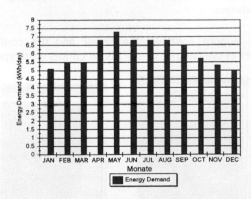

Figure 1. Energy Demand for a 3-kVA PV Application

The daily average energy demand for the high-power PV plant is 39 kWh/day. This includes houses, schools, communication centres, police station, hospital and water pumping systems.

The key system features of both UPMS prototypes are listed in Table 1.

Table 1. System Features for the UPMS Prototype Units

System Features	3-kVA	15-kVA
System voltage:	24 Vdc	120 Vdc
AC output voltage:	230 Vac	230 Vac
System output current:	125 A	125 A
Max. AC loads:	3 kW	15 kW

2.4 Design of the UPMS unit, 3kVA ac power output

The UPMS prototype is intended as a modular standardised size to be used in low-power hybrid PV power plant, installed near the users' or owners' residence, with a standard ac output power of 3 kVA, The system supposed to be powered with a PV array, wind generator and for emergency situations an auxiliary generator can be connected. The battery storage system are lead-acid cells. All power electronics, power distribution switching and protective devices, monitoring capability, and energy management systems are completely packaged inside one container. The power sources, one solar irradiance sensor and the residential loads are the only connections to this UPMS. Figure 2 shows the simplified block diagram.

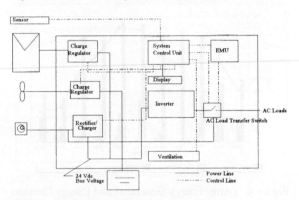

Figure 2. Block diagram of the low-power PV hybrid UPMS prototype with 3-kVA ac power output.

The UPMS prototype has been implemented and tested in a PV hybrid PV plant. The power sources and storage systems have been sized according the requirements [2].

The nominal power rating of the PV array is 1.5 kWp, the battery capacity rating is 800 Ah, the nominal power rating of the wind generator is also 3kW. For emergency purposes a petrol generator is connected to the unit and supplies power to charge the battery and directly to the ac loads.

In Figure 3 the designed 3-KVA UPMS prototype is shown. A minimum of display is integrated into the cover which allows immediate control of the UPMS unit.

Figure 3. 3kVA UPMS Prototype

2.5 Performance Evaluation Results of the 3kVA UPMS

The hybrid with the implemented UPMS has been continuously monitored since one year. Figure 4 shows the load demand and power supply of the UPMS system. The average energy load demand is about 5kWh per day as predicted and is very well matched by the power supplied by the PV generator. The auxiliary generators (wind and petrol) are required only for emergency situations.

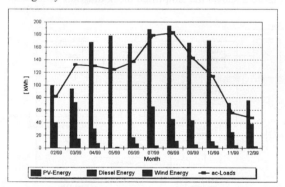

Figure 4. Yearly Energy Generated and Energy Demand

Figure 5 shows the inverter efficiency at partial ac loads. The inverter concept is able to achieve a value of 92% with 24V system voltage.

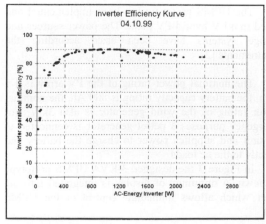

Figure 5. Inverter Efficiency Curve

2.6 Design for UPMS prototype, 15kVA ac power

This UPMS prototype unit is also modularised and standardised but designed for use in larger power applications such as centralised village power sources. The UPMS unit has been implemented and tested in PV hybrid PV plant in Gran Canaria, Spain. The sizes of the power sources and the storage system are as follows: PV array - 10.8 kWp, Wind generator - 3 kW, auxiliary generator - approx 12 kVA, and the battery system with lead-acid cells will have a capacity of 1,200 Ah/120Vdc. Figure 6 shows the simplified block diagram.

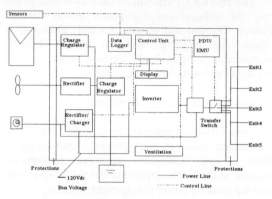

Figure 6. Block diagram of the high-power PV hybrid UPMS with 15 kVA ac power output.

The UPMS prototype has been connected to a lead-acid battery with tubular plates which was selected to use as storage system in the PV plant. To increase reliability and life-time of the battery operational techniques were defined such as 50% DOD.

The battery life-time is predicted to increase up to seven years instead of four years due to improvement of operational techniques of the battery. To assure full recharge of the battery systems a controlled gassing will be allowed.

Figure 7 shows the power and the control units of the cabinet of the 15kVA UPMS prototype.

Figure 7. Power and Control Unit of the 15kVA UPMS prototype

2.7 Performance Evaluation Results of the 15kVA UPMS Prototypes

The high-power prototype unit is installed and tested in Gran Canaria, Spain. In Figure 8 the inverter efficiency at partial load is shown which achieves a value of 92%.

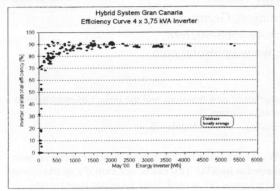

Figure 8. Inverter Efficiency at Partial Power Load

The charge regulation of the 15kVA has an intelligent charge and discharge regime implemented. At certain battery conditions it allows boost charging of the battery cell to assure full charge of the battery (Figure 9).

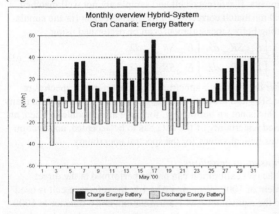

Figure 9. Charge and Discharge of the Battery System

3. COST REDUCTION

The cost reduction of the UPMS units achieve a value of 33% and 16% in comparison of previous conventional systems. This results from the reduction of devices and parts such as power transformers, input inductance capacitors, power semiconductor switches, fuses, control switches, cabling. The units will be faster to install, and the simplicity of testing, initial check-out and maintenance will reduce labour cost.

4. CONCLUSIONS

The development of the universal PV management systems considerably reduces costs in comparison with conventional configurations due to the compact design. The units will be faster to install, and the simplicity of testing, initial check-out and maintenance will reduce labour costs.

REFERENCES

[1] M.S.Imamura, I.Weiss, J. Garcia Martin, A.de Julian, 3- and 6-kW stand-alone PV systems for Utility Service, Final Report, JOULE II Programme, 1995.

[2] M.S.Imamura, P. Helm, I. Weiss, M. Grottke, Systems Technology Development for the Design of PV Plants, Final Report, JOULE II Programme, 1996.

PV MODULE AND CELL CALIBRATION PROCEDURES, ACCURACY AND PROBLEMS INDUCED BY RECENT TECHNOLOGICAL CELL AND MODULE IMPROVEMENTS

K. Bücher

Optosolar GmbH

Rittgasse 18, D 79291 Merdingen, Germany

Tel. +49 7668-902699, Fax. + 49 7668-902698

Email: klaus.buecher@gmx.de, www.optosolar.com

ABSTRACT

After a short discussion of the calibration of PV devices, specific properties of new materials and the impact of measurement routines, solar simulator performance and reference cells on calibration accuracy is given.

keywords: Calibration – 1: Qualification and Testing – 2: PV Module – 3.

1 INTRODUCTION

Recently, a variety of developments have made module calibrations more difficult:

- Due to building integration, module sizes have become very large (up to 1.5 x 3 m^2 and more).
- Large area cells lead to very high current production (up to 8 A)
- New crystalline solar cells have been developed with very high lifetimes and cell capacities.
- Crystalline silicon cells show an initial degradation of efficiency within the first 10 kWh of light absorption after their production
- Some thin film materials "remember" their pre-treatment with respect to open or short circuit condition, as well the light condition of storage
- For some newer materials, light soaking is necessary to start the photovoltaic conversion (i.e., response times to intensity variations are in the range of seconds).

In the following, calibration methods for PV cells and modules and the impact of the measurement equipment will be described.

2 CALIBRATION OF PV DEVICES

2.1 Calibration of standard cells and modules

Internationally, it has been agreed upon to report solar cell efficiencies or module power delivery at standard test conditions (STC, [1]), i.e. at an irradiance of 1000 W/m^2, a cell/module temperature of 25°C and the standardized spectrum AM 1.5. Actual measurement conditions, on the other hand, obviously deviate from these standardized conditions. Therefore, an IV curve translation and a spectral mismatch correction is necessary to transfer measured data to the standard test conditions.

For "most" cells and modules, the IV characteristic (IV) is only a function of cell/module temperature and the short circuit current of the device, independent of the light spectrum used to generate this current.So, if

$$IV – characteristic = f(I_{sc}, T_{mod}) \quad (1)$$

holds, the light source in use for the measurement does not need to conform exactly to the spectrum AM 1.5. In this case, any source (even a tungsten light source, which is cheap and can easily be stabilized) can be used.

Obviously, one has to find a simulator setting "equivalent" to 1000 W/m^2, AM 1.5. We use the term equivalent, because all PV devices have a spectrally varying sensitivity. Now, if the spectrum is not AM 1.5, we need a simulator setting that generates just as many electric carriers (or short circuit current) as the STC do. The physical light intensity (measured in W/m^2) need not be

1000 W/m^2 for this setting, but the same amount of current must be generated as by STC. This can be achieved if a spectrally well matched reference cell is used. If the simulator is set for this reference cell to deliver its STC calibration value , the subsequent calibration of a similar test object will give the correct IV at 1000 W/m^2, AM 1.5 (even if the simulator is physically set to some other W/m^2). To arrive at the STC performance, only a temperature transfer of the measured IV characteristic has to be done (usually with linear temperature coefficients for I, V resp.P).

This (implicit) approach to spectral mismatch correction using a well matched reference cell is easy to perform in practical experiments. If reference cell and sample are not well matched, a spectral mismatch correction of the measurement (or the simulator setting before measurement) can be performed using

$$MM = \frac{\int E_{sim} SR_{ref} d\lambda \int E_{ref} SR_{sample} d\lambda}{\int E_{ref} SR_{ref} d\lambda \int E_{sim} SR_{sample} d\lambda}, \quad (2)$$

after the simulator spectrum E_{sim}, the reference spectral response SR_{ref} and the sample SR_{sample} have been measured. In many cases, this is not an option (e.g. for large area modules). Then, an increased uncertainty of the data has to be accepted, and section 4.3 gives typical values for the spectral uncertainty.

2.2 Translation equations

If the cell calibration can not be performed at the direct equivalent of 1000 W/m^2 and AM 1.5, the reference cell is used to read the equivalent AM 1.5 irradiation, and a transfer to 1000 W/m^2 has to be performed. Many different procedures have been discussed in the literature, and an intercomparison of procedures for modules has been published in [2]. One possible way is the use of simulation with the 2 diode model. Other procedures correct each current point with the deviation of device photocurrent ΔI_{ph} between STC and measurement condition, and each voltage point by the additional voltage drop at the series resistance:

$$I_{STC} = I + \Delta I_{ph} \quad (3)$$
$$V_{STC} = V – (I_{STC} – I) R_s \quad (4)$$

Also, the logarithmic variation of open circuit voltage in the case of large irradiance differences between measurement and STC should be incorporated (ΔV_{oc} prop. ln (G_1/ G_2)). Again, other procedures correct the current directly by the ratio of the intensities.

All these translation equations may not be valid for thin films, and several respective studies are under way. Also, cell behavior may be dependent on the deposition technology of the thin films. For example, for some (but not all) CdTe deposition technologies, it has been shown that the open circuit voltage at low light levels reduces much slower than with ln G [3] (high V_{oc} is obtained even at low light levels).

2.3 Error sources in the calibration of new technologies

For newer technology cells, the assumptions as given above (e.g. eq. (1)) have to be checked. This must be done for each device (not only a material class), as often, deviations from solar cell theory may be true for one cell (type of material, type of impurity) but not for others.

Especially, the following effects must be considered:

- Amorphous silicon shows a spectral response that is a function of the cell voltage. The IEC standard 61646 [5] demands to use the SR at the MPP for the setting of the simulator.

- Thin film materials can show light soaking effects. IV characteristics can be different, whether the device has been stored in the dark or light before measurement, and can depend on the measurement direction (V_{oc} to I_{sc} or vice versa). Also, cells can be strongly nonlinear with respect to light intensity. This is of consideration for outdoor measurements where IVs may be translated from medium intensity to 1000 W/m^2. In some cases, e.g. some CdTe cells, the bulk resistivity might even be dependent on the light spectrum in the IR [4].

- Dye solar cells, but also modern high lifetime crystalline Si solar cells, can show long time constants. For certain dye cells, they are such that SR measurements with chopped light is not possible. For Si cells, high capacities may obstruct flashlight measurements of such cells.

- Deposition processes can influence cell calibration. It has been observed that some CIS cells/modules can easily be flashlight calibrated, whereas other deposition technologies do not (yet?) render to this analysis method. To a much smaller extend (a few % differences), such effects are observed for CdTe.

- Translation of IV has been mentioned in the previous section.

2.4 Specific Problems of module calibrations

Additional problems are caused by today's state of the art module constructions. Modules for building integration have sizes up to 2x3 m^2, and it is not possible to measure module SR (but at ESTI). For accurate calibrations, the production of small size samples in an equivalent optical encapsulation is necessary.

Also, the thermalisation, especially of glass-glass encapsulated modules, is difficult. Due to the large temperature coefficient of power, the temperature distribution must be uniform across the module. This may be difficult to achieve in outdoor measurements or DC simulators with high thermal load, but also in indoor flash measurements, if modules have been stored vertically. In many laboratories, there may be a layering of temperature between ground and ceiling leading to module temperature nonuniformities of up to 5K in winter.

3. SOLAR SIMULATORS FOR DEVICE CALIBRATION

Three different types of light sources are currently used to measure module power, in addition to natural sunlight in outdoor calibrations:

- steady state (DC) solar simulators with HMI or tungsten lamps (multilamp systems)
- long pulse flashlight simulators with a pulse plateau length of 2-3 ms and acquisition of the IV curve during one flash
- short pulse flashlight simulators using stroboscope type lamps (multiple flashes per second, 0.2 ms plateau), acquiring one IV datapoint per flash

Steady state solar simulators are rarely used (only in calibration laboratories), due to their extensive electrical power consumption (up to 80 kW for 2x2 m^2 module area) and their implicit heat radiation into the module. Also, it is difficult to achieve good spatial uniformity for large area modules. Calibration laboratories use this approach to excellent results (e.g. TÜV in Cologne with a sophisticated heat management and cooling systems for the modules). Nevertheless, the throughput can be small and inappropriate for use in module production facilities.

Up to now, short pulse multiflash simulators are most widely used for the production line calibration of PV modules. Nevertheless, this simulator type has shown fillfactor- and cell capacity related measurement problems in the past [6,7] due to its very short plateau time of the peak (0.2 ms out of a 1 ms pulse) and a dependence of the data on the sampling position during the pulse [7]. As a solution, a peak detector does not measure the module voltage simultaneously with the irradiation and module current, but measures the peak voltage reached over the entire period of the light pulse [7]. Of course, this approach gives certain demands on the amount of tolerable transient effects (charging capacities) or light soaking processes (thin films) within the cells which need not be obeyed by all modules.

To overcome these problems, long pulse flashlight simulators have been developed. Originally, this type of simulator has always been used in calibration laboratories (ESTI, ISE, JQA, NREL, ...) for its high precision that can be obtained. Even these simulators can not deal with all possible devices (e.g. light soaking in certain thin films), but the longer measurement times will avoid or reduce transient effects. Contrary to other systems, the measurement results can directly tell if the data is expected to be accurate due to the specifics of gathering the IV. This is because possible transient processes can be measured and observed during the pulse. Whereas these long pulse simulators originally needed long lamp/sample distances of 10 – 14 m and recharging times of minutes, these systems have recently been redeveloped to be usable for production line testing.

Fig. 1 shows a drawing of a respective instrument for module calibrations with an illumination area of 70 x 150 cm^2, as manufactured by our company, Optosolar. Here, the illumination is from below, well suited to modules leaving a thin film deposition line. Top illumination models are available with module sizes up to and above 2x2 m^2. Using plateau times of 2-3 ms and measurement times of up to 40 ms, these systems solve recent measurement problems with high capacity and transient solar cells.

4.0 SOLAR SIMULATOR REQUIREMENTS

Solar simulator requirements have been published in an international standard, IEC 60904-9 [8]. In detail, this standard classifies the spatial uniformity of the light, the temporal stability and the spectral distribution. For each property, three classes (A,B,C) are defined. In addition, the respective ASTM standard also specifies the light view angle.

Albeit desirable that solar simulators always fulfill the (highest) requirements of class A, for large area building integrated PV modules this is no longer possible. Here, one has to keep in mind that the standard was agreed upon at a time when module sizes were still smaller.

4.1 Light uniformity

The spatial uniformity of the illumination is of paramount importance for the accurate measurement of module power. Class A simulators may show nonuniformities up to +-2%, B up to +-5%, class C up to +-10%. As solar cells are series connected in a PV module, the corresponding current is dictated by the cell delivering the smallest current. If this is caused by (unwanted) variation of light intensity in the measurement area, errors will be unavoidable. The open circuit voltage will still be measured correct, but the module short circuit current and power can show errors of up to 10% in short circuit current, 6% in power at a +-5% light nonuniformity. Exact values depend on the uniformity distribution of the light, the averaged and total intensity in the module area (i.e. the number of cells receiving too high/too low irradiance) and the uniformity distribution of the cell characteristics.

Percentage of non uniform illuminated area	Class A (2% uniformity)	Class A	Class B (5% uniformity)	Class B
	I_{sc} deviation	P_{max} deviation	I_{sc} deviation	
10%	1.9%	0.5%	7%	1.7%
20%	2.6%	0.7%	8%	2.6%
30%	3.5%	1.5%	9.1%	4.3%
50%	3.8%	2%	9.5%	5.8%
Irradiation	960-1000W/m²		900 – 1000 W/m²	

Table 1 Deviation of I_{sc} and P_{max} from their true STC values, if different percentages of the test area are underilluminated..

Now, possible measurement errors not only depend on the highest/lowest irradiance value in the test plane, as classified by the standard IEC 60904-9, but also on the percentage of the test area with lower/higher irradiance. Table 1 gives possible measurement errors for a class A respective class B simulator. Here, in a computer simulation, a PV module has been suscepted to various intensity distributions. The illumination area with an irradiation of 4% below maximum (class A) or 10% below maximum (class B) has been varied between 10% and 50% of the total area. All these are possible representations of class A and B simulators. The errors increase directly with the nonuniformity and with the relative area of wrong light intensity. An algorithm has been developed during this investigation to correct these large uncertainties to the order of 2% even for large area modules [9].

4.2 Temporal stability and transient effects

The temporal stability of the illumination is classified identical to the spatial uniformity (A,B,C,+-2%, +-5%,+-10%). Nevertheless, it is no longer a major issue for most solar simulators, as light intensity can be measured parallel to the acquisition of each IV data point. Only fluctuations during this acquisition period influence the accuracy of the data. For outdoor measurements it is rather an issue that module temperature may not follow directly shortterm intensity fluctuations due to longer thermal time constants. Also, for measurements with a large intensity deviation from STC, care has to be taken with respect to translation equations to STC. For certain nonlinear thin film materials, such translations induce additional errors.

A much higher relevance must be given to the total duration of the measurement process. For steady state solar simulators or natural sunlight, heating of the module will cause measurement errors. Here, the measurement direction (open circuit to short circuit or vice versa) of the IV characteristic is of relevance, as the temperature coefficient is different for voltage, fill factor and current. Voc and MPP should be measured close in time to the temperature measurement. If the module was thermalised to the ambient by shading, the IV should be measured from V_{oc} to I_{sc}.

For some new materials, the photovoltaic process is dependent on the pre-treatment with light. Sometimes, thin film modules can not be measured using flashlight (e.g. certain, but not all CIS modules). For other thin film materials, it has been observed that the IV characteristic depends on the measurement time (e.g. max. power of a module 41W for a 5 ms scan, 45W for a 90 ms scan of the IV in steady light). Another example is given in table 3. In this experiment, the irradiance of steady light has been switched

off to analyze short time effects in thin film modules. It can be seen from table 3 that the short circuit current of the module does not follow the reduction of the light intensity as measured by the reference cell on time scales of 0.2 seconds, but rather reduces at a slower rate. This is of importance for IV measurements if the change of intensity just before the acquisition of a data point is in a similar time frame.

Similar effects (on a shorter time scale) can be found for high carrier lifetime, high capacity Si solar cells and modules. For solar cells, it has been possible to convert transient IV's to static IV's, using the capacity of the solar cells [10]. Similar attempts for full size modules have shown varying results in their translation accuracy.

Thus, in addition to the temporal stability demanded in the IEC standard, the illumination time in the case of flashlight simulators is of high importance to the accuracy of the data.

4.3 Spectral match of the simulator

Spectral errors of the light sources can be corrected by appropriate measurement procedures, as given in section 2.1. Anyway, the classification given by the standard is not very restrictive. Even outdoor sunlight fulfills class A nearly all day (but visibly changes its spectrum). Fig. 2 gives an example of two extreme outdoor spectra (global tilted (AM 1.5) and the direct beam component of AM 1.5). Although they distinctly differ in the blue and red component, both spectra represent class A. Thus, the choice of the right reference cells influences the measurement accuracy to a high extend. Using an unmatched Si cell as reference for other silicon cells will cause spectral uncertainties of typically 2% (up to 4%) for reasonable class A spectra. Using a pyranometer, for example, to calibrate a CdTe module in either spectrum of fig. 2 will lead to differences in the spectral match of about 5% due to the different spectral distribution.,.Up to 10% differences can occur for a very water rich spectrum (Mid Latitude Summer as described in [11]). Even a silicon cell will cause errors, as the water vapour absorptions will only be "seen" by one of the cells. As one does not want to use (possibly unstable) thin film reference cells, a better reference choice for the reference might be GaAs. For some CdTe technologies (but not all!), an even better match will be obtained with filtered GaAs cells, reducing the spectral uncertainty by about 1%. To summarise, even measurements in a class A simulator can impose high spectral measurement errors if inappropriate reference cells are used.

If module power at STC is to be specified, a pyranometer is especially unsuited, even for crystalline silicon cells. This is due to the wide infrared response of a pyranometer and its sensitivity to light energy in water absorption bands (for outdoor measurements) or infrared radiation (e.g. tungsten lamps) that are not seen by solar cells.

4.4 View angles

View angles can become a matter of concern if very large area PV modules are measured in setups using single or multilamp fields and a short distance between lamp and module. This is due to increased reflection off the glass surface at high incidence angles. The effect develops primarily near the edge of the module, where cells will see a lower light intensity. Also, effects of surface texturisation, light piping or multiple reflection in the module compound will be possible, leading to a redistribution of light intensity among the cells.

4.5 Choice of simulator

As shown in section 2, the main "ingredient" for accurate measurement is not only the solar simulator, but the equipment (solar simulator, measurements electronics and appropriate reference cells) combined with good measurement routines. Thus, special drawbacks of a certain simulator designs can be compensated by good practice.

To summarise, table 2 gives a survey of the various simulator types and their properties and their ratings, as personally seen by the author. Often, long and short pulse simulators will be favored due to negligible module heating.

5 CONCLUSIONS

The measurement accuracy for new technologies depends primarily on good practice. Especially, many crosschecks between steady state measurements, translation relations, linearity conditions have to be performed to arrive at accurate results. New technologies are still widely varying so that no general rules can be given to specific correction procedures.

6 REFERENCES

[1] IEC standard 61215
[2] S. Coors et al., Proc. 14th PVSEC 1997, p. 220
[3] Derk Baetzner, Diplomarbeit (physical diploma thesis), University Cologne, Germany
[4] D. Bonnett, priv. communication
[5] IEC standard 61646
[6] H.A.Ossenbrink et al., Proc. 23rd IEEE PVSC, 1993, p. 1194
[7]S.J. Hogan et al., Proc. 11th PVSEC, Montreux, 1992, p. 1374
[8] IEC standard 60904-9
[9] K. Bücher, to be published
[10] Frank. Lipps, Diplomarbeit (physical diploma thesis), K. Haas, Diplomarbeit (physical diploma thesis), both University Freiburg, Germany
[11] C. Gueymard, SMARTS 2, FSEC report FSEC-PF-270-95

7 ACKNOWLEDGEMENTS

Part of this work has been supported by the Deutsche Bundesstiftung Umwelt (DBU)

Table 2 In a DC light source, the light has been switched off, leading to a reduction of the light intensity within about 0.3 seconds (filament lamp). It can be seen from the table that the short circuit current of this module reduces at a slower rate as the light intensity. This is of importance for short time (e.g. flashlight) measurements of certain thin film technologies, but also to generator measurement systems using a capacitor as load. The last column gives the ratio of short circuit current and intensity. It should be constant for linear modules (this has been expected in this case from linearity measurements at low speeds).

Time after switching off the lamp	Relative intensity	Relative module short circuit current	Relative ratio of I_{sc} and intensity
0s	100%	100%	1
0.05 s	70%	85%	1.22
0.075 s	39%	45%	1.15
0.2 s	20%	20%	1

Fig. 1 Optosolar flashlight module tester Sol 7x15 for illumination from below, e.g. for the testing of thin film modules on a production line

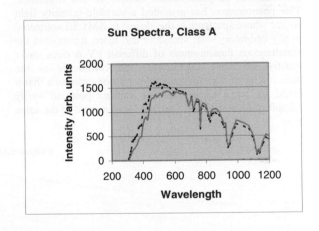

Fig. 2 The global tilted (AM 1.5) and direct beam spectra differ in the blue and infrared, both are class A.

	Sun	Multilamp	Short Pulse	Long Pulse
Spectrum	++	HMI + Tungsten --	+	+
Modul temperature control	-	o	++	++
Uniformity	++	o to +	+	+
Transient cell/module effects	++	++	o to +	(o to) +
Light fluctuations	o to ++	+	+	+
Power consumption	++	80kW	4kW	2kW
Measures everything	++	++	o to-+	(o to) +

Table 3 Comparison of various solar simulator types under the aspect of their use for large area encapsulated PV modules. ++: very good, ..., --: very bad

THE SOLAR ECLIPSE: A PERFECT VARIABLE IRRADIANCE SOURCE. AN INVESTIGATION OF THE LINEARITY PROPERTIES OF PHOTOVOLTAIC DEVICES

Authors: G. Agostinelli, J. Bishop, E.D. Dunlop, W. Zaaiman, D. Anderson, and B. Ebner

European Commission, Joint Research Centre, Renewable Energies Unit, TP 450, via E. Fermi 1, I-21020 Ispra, Italy

TEL.: +39–0332–789951, FAX: +39–0332–789268, E-MAIL: guido.agostinelli@jrc.it

ABSTRACT: A detailed study of photovoltaic devices and their response to natural sunlight is difficult since the experimental conditions, i.e. illuminated area, uniformity, spectral stability and distribution of the light source, can be very demanding to reproduce. Conventional laboratory approaches such as varying the intensity of the lamp or using neutral density filters do not provide satisfactory repeatability of the above conditions. Additionally, outdoor characterisation maintains ideal and consistent conditions only within a limited irradiance level. Total irradiance varies during the day between a large range but unfortunately so does the air mass. A rare event such as a solar eclipse combines two remarkable characteristics: the total irradiance varies while the spectral distribution remains virtually the same, and it exhibits an excellent spatial uniformity. These characteristics have been exploited to perform a study on photovoltaic devices and irradiance sensors of different technologies and their relative mismatch.

Keywords: Reference Cell - 1: Pyranometer - 2: Solar Eclipse - 3

1.INTRODUCTION

The solar eclipse 11th of August 1999 was, other than a fascinating event, an exceptional occasion to test outdoor performances of photovoltaic devices at low irradiance. This phenomenon has provided a variable intensity light source whose spectrum is the closest to AM1.5G compared to any laboratory solar simulator and more, it permitted the simultaneous measurement of different PV devices under exactly the same conditions. The confidence in the maintaining of the experimental conditions was such that it could be taken as a reference starting point to verify reliability of indoor and outdoor measurement of the same type.

2.EXPERIMENTAL SETUP

Different filtered and non-filtered thermal detectors, spectrum analysers, photovoltaic devices have been employed during the measurement, namely:
- 2 pyranometers for Global Total radiation monitoring
- 2 pyranometers for Diffuse Total radiation monitoring
- 1 pyranometer for UVA (315-400 nm) global radiation monitoring
- 1 pyranometer for UVB (290-315 nm) global radiation monitoring
- 1 Silicon PD optical spectrum analyser (OSA) for direct spectral (250-1100 nm, 5 nm resolution) irradiance monitoring

Eclipse Spectral Evolution

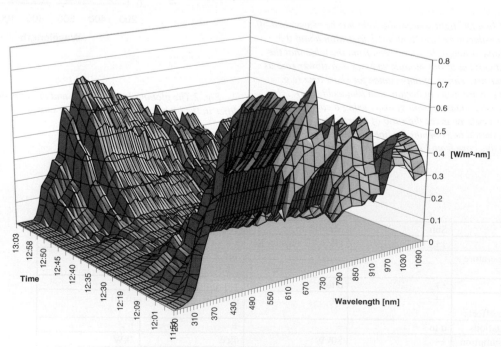

Figure 1: Solar spectrum evolution during the eclipse

- 1 optical spectrum analyser for global UVB spectral (290-325 nm, 0.5 nm resolution) irradiance monitoring
- 1 WPVS silicon reference cell
- 1 Calibrated silicon reference cell
- 1 Calibrated GaAs reference cell
- 1 ESTI sensor reference detector
- 1 CdTe Module
- 1 monocrystalline silicon module
- 1 polycrystalline silicon module
- 1 Tandem amorphous silicon module

All devices under test and the direct irradiance OSA were mounted on solar trackers so that direct sunlight would illuminate them at normal incidence. Ambient temperature, devices temperatures, and meteorological data have been monitored and recorded.

3.AIRMASS AND SUN ELEVATION

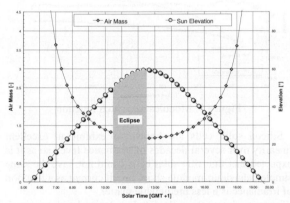

Figure 2: Airmass and sun elevation

Measurements were performed at ESTI outdoor test site in Ispra (Italy). Here the eclipse would reach at his maximum a covering factor of 90%, at AM1.16. The falling irradiance edge of the eclipse has been ruined by the presence of clouds, but clear sky during the rising part granted a window of 30 minutes of measurements in ideal conditions. During this period the total global irradiance varied from 50 to 400 W/m², ambient temperature varied from 12 to 24°C, module temperatures in average varied form 27 to 36°C, with a spread of a couple of degrees. Slight changes occurred in AirMass, which will be counted for in the analysis.

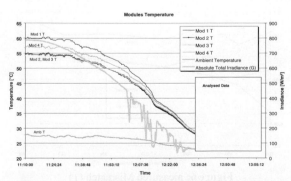

Figure 3: Module temperatures and Global irradiance

A detailed spectral evolution of direct radiation of the eclipse, in the range 250-1100 nm is illustrated in figure 1. The correlation index between each of the spectral lines and the total irradiance is practically unity on the whole Si-sensible range, fact that would lead to the expectation of minimal Mismatch correction deviation between different irradiance sensors. Direct irradiance monitoring is however not sufficient to establish the stability of the (normalised) spectrum, since about 10% of the total radiation is given by the diffuse component of sunlight. A prediction of its evolution during the eclipse was not a trivial issue, so as its monitoring at such low irradiances is particularly demanding with a standard OSA. The problem has been overcome by the analysis of other related irradiance parameters: data of diffuse versus global total radiation show a constant (within experimental errors) component of total diffuse radiation. As for the spectral content, the use of UV dedicated instrumentation confirmed a constant component of UVA, UVB and total UV radiation, and a detailed correlated evolution of the whole global UVB spectrum. As a conclusion, we can assume that the eclipse was a very good variable irradiance source.

4.MISMATCH ANALYSIS

The prime scope of this study was to examine linearity properties of PV modules and cells, and compare different technologies as best suitable absolute or relative (suitable for a particular class of devices) irradiance sensors. Under such a request second order effects given by minor changes in the spectrum could not be neglected and needed to be counted for. All devices are normally measured and calibrated at standard test conditions (AM15G spectrum, 1000 W/m², 25° C) while this paper aims to examine their behaviour over a range which reaches a lower decade of irradiance. Deviation from linear behaviour could not be evidenced by a bare linear correlation; an excellent tool to derive many properties of a device is given by the analysis of the evolution of its mismatch factor with respect to a 'reference' sensor.

We define the relative spectral mismatch factor of a device with reference to another, relative to a test and a reference spectrum [1] as:

$$M_{dev}^{ref} = \frac{\int E_{test}(\lambda)SR_{dev}(\lambda)\cdot\partial\lambda}{\int E_{test}(\lambda)SR_{ref}(\lambda)\cdot\partial\lambda} \cdot \frac{\int E_0(\lambda)SR_{ref}(\lambda)\cdot\partial\lambda}{\int E_0(\lambda)SR_{dev}(\lambda)\cdot\partial\lambda}$$

(eq. 1)

where E_{test} is the test spectrum, E_0 the reference spectrum, SR_{dev} the spectral response of the considered device, SR_{ref} the spectral response of the reference device; all parameters are dependent on wavelength λ.

M relates the measured short circuit current at a given test spectrum to the short circuit current at the reference conditions by means of the expression:

$$Isc_{dev}^{E_0} = Isc_{dev}^{Etest} \cdot \frac{1}{M} \cdot \frac{Isc_{ref}^{E_0}}{Isc_{ref}^{Etest}}$$

(eq. 2)

where Isc indicates the short circuit current and subscript and superscript define respectively the device and the spectrum. Assuming AM15G as reference spectrum, and knowing both the (external) absolute spectral responses of the devices and their calibration constants (Isc at STC), M can be calculated from the measured direct spectra with less than 0.5% of error by means of the approximations

$$Isc_{dev}^{Etest} \approx \int E_{measured}(\lambda) SR_{dev}(\lambda) \cdot \partial\lambda \quad \text{(eq. 3)}$$

$$Isc_{ref}^{Etest} \approx \int E_{measured}(\lambda) SR_{ref}(\lambda) \cdot \partial\lambda \quad \text{(eq. 4)}$$

$E_{measured}$ being the spectrum measured by means of the OSA; currents normalised to unit area. We define now the experimental 'Mismatch Function' M as the ratio of the measured signals normalised to their calibration values:

$$M_{dev}^{ref}(E_{test}) = \frac{Isc_{dev}^{Etest}(E_{test})}{Isc_{ref}^{Etest}(E_{test})} \cdot \frac{Isc_{ref}^{E_0}}{Isc_{dev}^{E_0}} \quad \text{(eq.5)}$$

Direct comparison between the calculated value for M and the one derived from the actual measurement of the short circuit current of the 'reference' and 'test' devices during the eclipse will immediately evidence deviation from predicted, irradiance proportional, current signal.

5.RESULTS

The first relevant result of the analysis is a sensible discrepancy of the irradiance readings between PV and thermal detectors at low irradiances. In fact that the minimum reading of the CM11 pyranometer is about 50 W/m² while all PV devices have a reading of about 60 W/m². Figure 5 shows the measured $M_{pyranometer}^{ESTIsensor}$ function of the pyranometer with reference to ESTI sensor reference detector; as it can be observed with decreasing irradiance the ratio between the two irradiance readings is changing and reaching lower values, which indicate a lower irradiance detected by the CM11. A virtually constant value of about 1.01 would instead be expected from analogous measurements at AM1.2 for crystalline Silicon Technologies [2]. Similar qualitative behaviour is observed versus all the other PV devices.

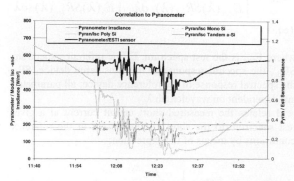

Figure 4: Mismatch (relative and absolute) to pyranometer

The CM11 is a thermal detector and three thermal mechanisms are active; some non-linearity can be expected.

-The CM11 has a 1% higher sensitivity at 1000 W/m² than at 100 W/m². This error increases rather linearly with irradiance.

-In clear sky conditions the CM11's zero point is somewhere below -5 W/m². This zero-offset is caused by IR radiation from the black sensor to the colder inner dome. (The tops of the domes are cooler due to IR-emission to the universe). This zero-offset can explain the discrepancy at 60 W/m² solar radiation. Apparently at 800 W/m² the zero-depression is compensated by the non-linearity error of about +8 W/m².

-A second type of zero-offset occurs at cooling down and heating up a pyranometer but the CM11 is very insensitive for these latest "thermal gradients".

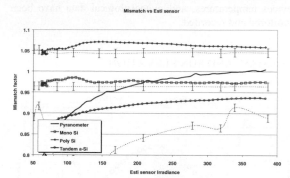

Figure 5: Modules mismatch function

Silicon solar cells are known to exhibit an excellent linearity over several decades of irradiance level, so as it is expected that thin film solar cells could not have the same simple behaviour; module short circuit current as a function of irradiance is a less predictable matter.

Figure 6 shows the calculated and simulated mismatch parameters relative to a Silicon sensor for three modules of different technology. Neither of the modules satisfies the theoretical M function. But while the Mono crystalline Silicon module exhibit a qualitative agreement, apart for a constant offset, Both the polycrystalline and particularly the Tandem a-Si module show a substantially different behavior with the one predicted. Mismatch relative to a pyranometer exhibits an even more accentuated deviation from the theoretical values. It has to be stressed that all the three devices gives very satisfactory results (<1% difference between predicted and actual mismatch) at irradiance levels above 800 W/m².

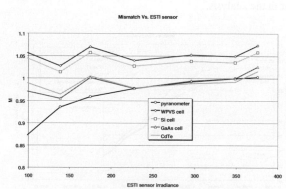

Figure 6: measured Mismatch (1)

Solar cells relative M function and comparison between measured and predicted values leads to some unexpected results: Si devices mismatch should be minimum because of the broadband sensitivity and spectral match. It is observed instead that their mismatch function is substantially the same when multiplied by a constant factor (not more of 1% greater than unity). Broader band on one hand and indoor calibration on the other seem to be the parameters that relate the various Si devices to minor outdoor response. This could be well explained by the combination of angular dependence of their quantum efficiency [3] and the subsequent difference between their STC calibration value (when measured indoor) and their outdoor performance.

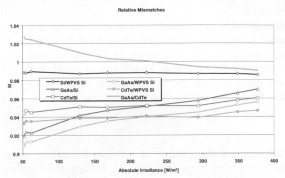

Figure 7: measured Mismatch (2)

Relative mismatch functions of PV devices of different classes (e.g. a GaAs vs. a Si device) show the same substantial behaviour and a good agreement with the predicted behaviour apart from the multiplication constants mentioned above. A good agreement is shown between the GaAs reference cell and the CdTe module theoretical and actual values; both the devices have been calibrated outdoor. Surprisingly, The CdTe module short circuit current exhibits an excellent linearity towards any of the other reference sensors. Another unexpected result is that the actual mismatch between CdTe and GaAs devices is not necessarily better than the one with Silicon devices.

6.CONCLUSION

The solar eclipse of august 11th, 1999, has been a fascinating laboratory for test and validation of some of the irradiance dependent parameter of photovoltaic devices. Modules, particularly of non-crystalline Si technologies, have some non-linear behaviour at irradiance levels below 400 W/m², not easy to model under such a circumstance. Thermal detectors might not necessarily be the best suitable irradiance sensors for PV application, as already mentioned as result of several application studies [4]. Furthermore, several uncertainty sources contribute to systematic measurement errors [5]. In our case combined non-linearity response and Zero-offset effects have been observed at irradiance level up to 800 W/m². Cell Mismatch analysis suggests that outdoor vs. indoor calibration can lead to different calibration constants, and confirm Si devices as the most reliable PV irradiance sensor. Other studies [6] have suggested the use if a GaAs reference device as irradiance detector for CdTe devices. This paper shows that this might be strongly dependent on the test spectrum.

GaAs and CdTe typical spectral response curves are strongly asymmetric with respect to the blue, which can contributes up to 10% of the total short circuit current of a GaAs devices. Outdoor characterisation might favour Si reference devices because of their broader response, which would lead to a systematically higher but more constant under changing methereological condition, mismatch correction factor. A suggested approach is that of using a GaAs device combined with a longpass filter in a similar but specular way to that of filtered Si [7] devices for a-Si, where light below 400 nm would be cut off.

ACKNOWLEDGEMENTS

The Authors wish to acknowledge cooperative discussion with Julian Gröbner (EI DG JRC) and Leo van Wely (Kipp & Zonen R&D)

REFERENCES

[1] ASTM E 793, "Determination of the Spectral Mismatch Parameter between a Photovoltaic device and a Photovoltaic reference cell"

[2] D.L. King; J.A. Kratchovil and W.E. Boyson, "Measuring Solar Spectral and angle of incidence effects on photovoltaic modules and solar irradiance sensors", 26th IEEE PV Specialist Conference (1997)

[3] G. Agostinelli and E. J. Haverkamp, "Angular dependancy of external quantum efficiency on high efficiency solar cells", 2nd World conference on PVSEC (1998)

[4] H. Haeberlin et al., "Comparison of pyranometer and Si Reference solar cell solar irradiation data in long term PV plant Monitoring", 13th EU PV conference (1995)

[5] T.L. Stoffel et al, "Current Issues in terrestrial Solar Radiation Instrumentation for Energy, Climate and Space application", New RAD '99 (1999)

[6] S. Roschier, W. Zaaiman and E.D. Dunlop, "Calibration and measurement techniques for thin film modules" 2nd World Conference on PVSEC (1998)

[7] E.D. Dunlop, W. Zaaiman, H.A. Ossenbrink and C. Helmke, "Towards a Stable Standard Reference Device for the Measurement and Monitoring of Amorphous Silicon Modules and Arrays"14th European Photovoltaic Solar Energy (1997)

HYBRID SYSTEMS – EASY IN CONFIGURATION AND APPLICATION

B. Burger, P. Zacharias
Institut für Solare Energieversorgungstechnik (ISET) e.V.
Königstor 59, 34119 Kassel, Germany, Tel. +49 561 7294-142, Fax: -100

G. Cramer
SMA Regelsysteme GmbH
Hannoversche Straße 1- 5, 34266 Niestetal, Germany, Tel. +49 561 9522-0, Fax: -100

W. Kleinkauf
Universität Gh Kassel, Institut für Elektrische Energietechnik (IEE)
Wilhelmshöher Allee 71-73, 34109 Kassel, Germany, Tel. +49 561 804-6344, Fax: -6512

ABSTRACT: The electricity supply in remote areas without public utility is very important worldwide, particularly in developing and threshold countries. This is an ideal application for isolated or "off grid" hybrid power supply systems. ISET developed a completely new bi-directional battery inverter with a rated power of 3.6 kW for such systems in cooperation with SMA Regelsysteme GmbH. Power ranges from 3.6 kW to 33 kW can be established with the parallel connection of inverters in single phase and three phase systems due to the modular design of the battery inverter. All power producers and consumers are coupled at the AC line in these modular systems.
Keywords: Hybrid - 1: Stand-alone PV Systems - 2: Inverter - 3

1. INTRODUCTION

The supply of small, peripheral consumers in the power range from 2 to 30 kW, which cannot be attached to a public grid, is worldwide, in particular in the developing and threshold countries, from large relevance. This is an almost ideal application for isolated or "off grid" photovoltaic power supply systems. The experiences with such systems have shown that these systems should be not only very reliably, economical and robust, but above all modularly structured and therefore easily subsequently expandable [1]. Also the connection of diesel generator sets and small wind energy systems should be possible in a simple manner. Only a simply structured and flexible system design for these photovoltaic power supply systems will enable a wide spread application.

On the basis of these requirements, the ISET in cooperation with SMA and with promotion of the German Federal Ministry BMWi developed the completely new battery inverter "Sunny Island" with a nominal output power of 3.6 kW. The usage of advanced microprocessor technology in combination with new power electronic circuit concepts provides a simply applicable and expandable system engineering for the power supply of remote areas.

2. THE BATTERY INVERTER

The central component of such a modular supply system is a battery inverter with the product name Sunny Island [2]. The AC output of the inverter must provide constant voltage and frequency for the consumers. A lead acid battery is used as energy buffer. Intelligent management and control algorithms integrated in the device enable it to supply not only different consumers but also to connect different generators, e.g. string inverters, small wind energy converters or diesel generator sets. The battery inverter must therefore be able to operate it in all four quadrants.

This requires the control of voltage, frequency, active power and reactive power of the AC voltage of the battery inverter. With appropriate coupling of three devices a three phase power supply is possible and by the direct parallel connection of several inverters at a phase an increased power will be achieved (in development). For simple configurations the battery inverter is able to take over the battery management and load management.

The DC voltage is controlled to provide a best possible battery handling, with respect to temperature dependent and current dependent voltage limits, the execution of regular total charge cycles and the adaptation of the charging algorithms to the battery type and the application conditions. Additionally the state of charge of the batteries is calculated and displayed.

The following requirements for battery inverters in modular structured island systems can be fulfilled:

- Operating modes: Voltage control - current control - parallel operation,
- Modular expandability,
- Extendibility for 1 and 3 phase island grids,
- Intelligent battery management for longest battery life: Charging and discharging control, regular full charging
- State of charge display
- Load management for simple basic configurations of small systems and
- High efficiency, also in the partial load range with low stand by losses.

To create a flexible applicable device, closed loop control and local management system (operational control) are taken over by its own processor. This allows the integration of a fast closed loop control and a complex management into the new battery inverter. The fast control allows the required operation modes and a parallel connection of inverters. The management system takes over the battery management, enables a limited load management and makes communication interfaces available for optional management devices. Figure 1 shows a block diagram of the battery inverter. The battery and the AC output are connected via circuit breakers. There are 8 relay contacts available for the following tasks:

- Starting a generator and connecting it to the island grid
- Switching of wind energy, consumers, utility and dump load
- Automatic control of the fan for the battery room and an optional electrolyte circulation pump.

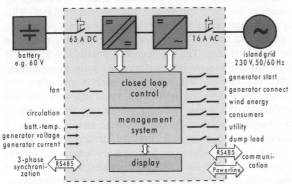

Figure 1: Block diagram of the battery inverter

2.1 Circuit of the static inverter

Figure 2 shows the power electronic circuit of the battery inverter. A bi-directional Cuk converter changes the battery voltage, which depending upon number of cells and charge can be between 40 V and 80 V, into a regulated DC-link voltage of 380 V. The HF transformer provides a electric separation between battery and grid, so that the battery is potential free. By the high frequency of 16.6 kHz the transformer is substantially lighter and smaller than a comparable transformer for 50 Hz. To the DC-link a single phase inverter with L-C-L filter is connected, which generates the sinusoidal voltage for the island grid. Since both, the Cuk converter and the inverter operate bi-directional, the static inverter can charge and discharge the batteries.

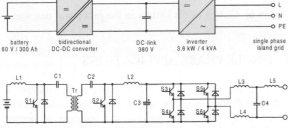

Figure 2: Circuit of the static inverter

2.2 Closed loop control

For the control of the Cuk converter a state control with overlaid closed loop PI control is used for DC-link voltage control. The digital closed loop control operates at the half clock frequency of the hardware, i.e. with 8.3 kHz.

The voltage control of the inverter is implemented as cascaded control. Figure 3 shows the structure of the closed loop control. The hardware components, which can be controlled, consist of the static inverter as dead time element with gain, the filter choke as PT1-element and the filter capacitor as I-element. With the measurement unit the capacitor current, the capacitor voltage and the offset of the capacitor voltage are measured. The capacitor current is controlled by the subsidiary control with proportional

reaction. It is relieved by a feed forward of the voltage desired value. The overlaid mains voltage closed loop control is a generalized PI controller [3], which controls both, the RMS and the phase position of the voltage exactly. The offset closed loop control is implemented as an integrator and controls the DC voltage offset of the output voltage to zero.

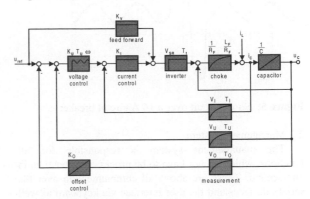

Figure 3: Simplified block diagram of the closed loop control of the inverter

Figure 4 shows the behavior of the closed loop control during a load branch with 2 kW. The voltage V_{grid} (channel 3) has only a short drop after switching and reaches already after approximately 1 ms again its desired value. A further improvement of the dynamics would be possible only with a higher DC-link voltage, since the inverter here already operates as control in the delimitation. This would increase however the losses of the power electronic components, so that the efficiency would be reduced.

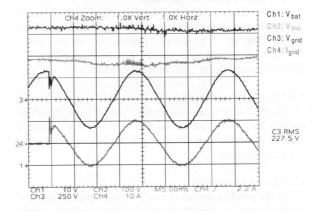

Figure 4: Load branch with 2 kW

If the output of the inverter is short circuit, an additional closed loop control for the choke current limits the output current of the inverter to the maximum current, which the power semiconductors can process. This current is higher than 60 A resulting in the fact that the inverter is able to trip normal circuit breakers of Class A with a rated current of 16 A. Figure 5 shows a measured short circuit. During the short circuit the voltage V_{grid} is nearly zero and the current I_{grid} rises to the current limit. After approximately 15 ms the circuit breaker trips and the voltage rises again on its desired sinusoidal value. Thus a selective protection is possible by circuit breakers in the island system just as in the public grid.

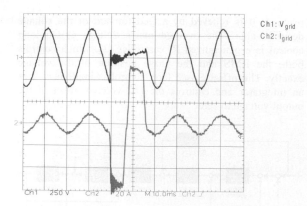

Figure 5: Short circuit over a 16 A circuit breaker

2.3 Management system

The management system is responsible for all functions, which do not have to be processed faster than in one second. These are above all communication over the serial interfaces and the user interface via keyboard as well as the graphic display. Additionally the management determines the operating mode. The following operating modes are implemented:

2.3.1 Voltage controlled operation

In the voltage controlled operation mode the output voltage of the inverter is regulated to its RMS value. The output frequency can be defined with a resolution of 10 mHz, the RMS value of the voltage with a resolution of 100 mV. A change from 50 Hz to 60 Hz can be made easily by software.

2.3.2 Current controlled operation

In the current controlled operation mode the inverter synchronizes to an external voltage supply. This can be a public grid or a generator. Depending on the given direction of current, the battery can be charged in this operating mode or the grid can be supported.

2.3.3 Three phase operation

In the three phase operation mode three inverters operate with 120° offset, so that a three phase current supply system is established. Synchronization is done with a digital interface.

2.3.4 Parallel operation

In the parallel operation mode several inverters operate synchronized on one phase, so that the available output power is increased to a multiple of 3.6 kW.

2.4 Battery management

The battery management is responsible for the charging and discharging control of the battery. It calculates the desired value for the charging voltage and is able to start an additional generator over relay outputs, in order to charge the batteries additionally to the PV power. When the battery has a low state of charge, a low priority load can be disconnected from the grid by another relay. In the case of a fully charged battery, a dump load will be connected or the PV power will be disconnected or short circuited. Additional relays are available for controlling a fan for the battery room and for the pump of an electrolyte circulation system.

For the calculation of the state of charge of the battery an algorithm is integrated, which uses a balance of ampere-hours combined with a calculation of losses and multilevel full charge recognition. Also an adapting current-voltage model of the battery is used to recalibrate the state of charge when the battery is not fully charged [4]. For the current-voltage model the linear correlation between the open-circuit voltage and the state of charge is used. The correlation is determined in phases after a full charge by means of the then well known ampere-hours balance. This is important, since the correlation between open-circuit voltage and state of charge for different battery types can be very different. Thus it becomes possible in most PV systems, the ampere-hour balance not only to recalibrate after full charges but practically each night, when the battery is discharged only with small currents. By the definition of a lower open-circuit voltage for the end of discharging additionally the ampere-hours capacity of a battery can be measured. For the determination of the battery capacity therefore no capacity test is necessary. Figure 6 shows the measured values of charge and discharge current, cell voltage and battery temperature as well as the calculated state of charge and desired value of the charging voltage over one week. The desired value of the charging voltage was not achieved here, since the battery was charged only with solar energy.

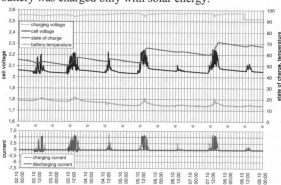

Figure 6: Charge of the battery in the process of one week

3. SINGLE PHASE APPLICATIONS

A simple single phase island grid can be established with one battery inverter and a lead acid battery. The closed loop control will enable the increase of output power by parallel connection of up to three battery inverters at one phase. In order to feed solar electricity into the island grid, conventional PV inverters e.g. string inverters from the Sunny Boy series from SMA can be used. Furthermore the integration of wind or hydroelectric power plants is possible with static inverters or single phase generators. For the increase of security of supply usually still a backup generator (e.g. Diesel generator) is used. Often these generators are already installed and can be integrated into the hybrid system. If a public grid is available from time to time as in many developing countries, then it can be attached also to the inverter. It operates then like an UPS and supplies the consumers in the case of power failures. Figure 7 shows the structure of a single phase island grid with the battery inverter Sunny Island and with different generators and consumers.

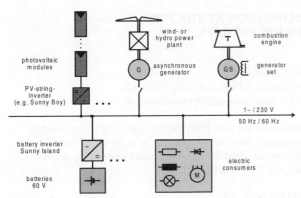

Figure 7: Example of the structure of a single phase modular island grid

4. THREE PHASE APPLICATIONS

The smallest three-phase system is a 11 kW power supply consisting of three Sunny Island each connected to a different phase. The three phases are synchronized via RS485 while the operating data is additionally sent over this link. Three phase systems make it much more easy to connect diesel or wind generators as these mostly are only available in three phase versions. Larger island systems consist of 6 or 9 inverters with two or three connected to each phase, all in all resulting in a total output power of 33 kW. The Sunny Islands can be freely connected to any battery set, i.e. several Sunny Island can use one or several sets of batteries. Although it is recommended to establish one single battery set for three phase systems. Figure 8 shows the structure of a three phase island system in principle and Figure 9 the prototype of a three phase hybrid system in the DeMoTec centre of ISET. First demonstration plants will be built on the Greek island Kythnos [5].

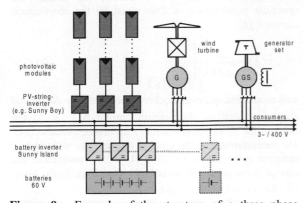

Figure 8: Example of the structure of a three phase modular island network

5. PERSPECTIVES

Due to the modular conception of the battery inverter, power ranges from 3.6 kW to 33 kW can be achieved by the parallel connection of inverters in single phase and three phase systems. Additionally it is possible to extend existing systems after the unit construction or to extend single phase systems to three phase systems. The application of this modular battery inverter will reduce planning and system costs for hybrid island grids.

Figure 9: Prototype of a three-phase hybrid system in the DeMoTec centre of ISET

The use of stand alone hybrid grid systems on the basis of the battery inverters "Sunny Island" will also enable a power supply in remote areas without mains connection for the first time and therefore will reduce the consumption of resources for the electric energy production. So even lower social classes have access to electricity for lighting, household and for small workshops.

The consistent modularity of this new system oriented concept allows a commercial use and self supporting retail structures (leasing of the systems or sale of the produced electricity), since the components (battery inverter, batteries, diesel sets...) do not have to be adapted for individual applications, but are universally applicable.

The ISET and SMA thank the German Federal Ministry BMWi for the promotion of the project "modular battery inverter: development of a battery for the modular system technology in PV systems" and the European Commission for the promotion of the projects "PV-MODE", "MORE" an "HYBRIX".

REFERENCES

[1] W. Kleinkauf, J.Sachau: Components for Modular Expandable and Adaptable PV Systems, 12th European PV Solar Energy Conference, Amsterdam, April 1994

[2] B. Burger, G. Cramer, A. Engler, B. Kansteiner, P. Zacharias: Battery Inverter for Modularly-Structured PV Power Supply Systems, 2nd World Conference and Exhibition on Photovoltaic Solar Energy Conversion, Hofburg Congress Center, Vienna, Austria, July 1998

[3] B. Burger: Transformatorloses Schaltungskonzept für ein dreiphasiges Inselnetz mit Photovoltaikgenerator und Batteriespeicher, Dissertation, Universität Karlsruhe, 1997

[4] M. Rothert, B. Willer: Möglichkeiten und Grenzen der Ladezustandsbestimmung von Bleibatterien in PV-Anlagen, 13. Symposium Photovoltaische Solarenergie, Kloster Banz/Staffelstein, 1998

[5] P. Strauss, D. Mayer, C. Trousseau, S. Tselepis, P. Romanos, F. Raptis, J. Reekers, M. Ibrahim, R.-P. Wurtz, F. Perez-Spiess, M. Bächler: Stand-Alone AC PV Systems and Micro Grids with New Standard Power Components, 16th European Photovoltaic Solar Energy Conference and Exhibition, Glasgow, United Kingdom, May 2000

PRIMARY REFERENCE CELL CALIBRATION AT THE PTB
BASED ON AN IMPROVED DSR FACILITY

S. Winter, T. Wittchen, J. Metzdorf
Phys.-Techn. Bundesanstalt, Bundesallee 100, D-38116 Braunschweig, Germany
Tel.: +49 531 5924100, Fax: +49 531 5924105, eMail: Juergen.Metzdorf@ptb.de

ABSTRACT: Based on several hardware and software improvements of the first DSR facility replacing also a number of components, both the typical operating times and the uncertainties have been reduced by a factor of two. The modifications of the DSR facility are described and the improvements are illustrated. An important aim was to improve the PTB's capability of carrying out the second WPVS recalibration in 2001 more efficiently.

Keywords: Calibration – 1: Spectral Response – 2 : WPVS World PV Scale – 3

1. INTRODUCTION

The PTB is one of the four qualified WPVS laboratories and the only European one that was included in establishing and presently maintaining the World Photovoltaic Scale (WPVS) used as a world-wide reference value for the calibration of primary reference solar cells and terrestrial PV performance measurements [1]. Such a global reference scale is the basic requirement for the mutual recognition of PV calibration certificates.

In the calibration of (terrestrial) solar cells and PV modules, the short-circuit currents I_{STC} and finally the conversion efficiencies both weighted according to the reference solar spectrum have to be determined under Standard Test Conditions (STC); i. e., $E_{STC} = 1000$ Wm^{-2} total irradiance, 25 °C cell temperature, AM1.5 reference solar spectral irradiance distribution $\partial E_{STC}(\lambda)/\partial \lambda = E_{STC,\lambda}$ of IEC 904-3, which is given in tabular form [2]. AM is the relative optical air mass. All the results in this paper are also valid if AM1.5 is replaced by, e. g., AM0.

The special difficulties of accurate solar cell calibrations are due to the high level of irradiance and large areas of the devices up to more than 100 cm² (cells) or 1 m² (modules) in connection with (i) the different spectral distributions of solar cell responsivities, (ii) the different and varying spectra of natural sun or solar simulator and the reference solar spectrum (double spectral mismatch) and (iii) the potential non-linearities of the short-circuit or photo-current at high irradiance levels [3].

Primary calibrations are carried out against radiometric standards (standard detectors and in addition standard lamps if necessary) traceable to SI units, while secondary calibrations against another reference cell are non-primary and non-independent and, therefore, correlated calibrations. The primary calibration methods applied by the four qualified WPVS laboratories are variously traceable to radiometric standards [1,3]. While the PTB uses a spectral method (see below), the other three laboratories (JQA/ETL in Japan, NREL in USA, TIPS in China) use integral methods with spectral mismatch correction based on the measurement of the relative spectral responsivity without determining the absolute spectral responsivity in Equation (1) directly.

I_{STC} is the required calibration value given by

$$I_{STC} = s(\lambda_0, E_{STC}) \int_{s(\lambda)\neq 0} s(\lambda, E_{STC})_{rel} E_{STC,\lambda}(\lambda) \, d\lambda \qquad (1)$$

Due to practical measuring conditions in radiometry (it is almost impossible to produce monochromatic radiation fields that are perfectly uniform at all wavelengths), relative spectral responsivity $s(\lambda)_{rel}$ covering the whole spectral range on the one hand and absolute spectral responsivity $s(\lambda_0)$ of a solar cell (or detector) at discrete wavelength(s) λ_0 on the other hand are measured separately. The variable E_{STC} indicates that the spectral responsivity of the (non-linear) cell has to be determined at the required high bias irradiance level $E_b = E_{STC}$ (i. e., not at low levels).

2. IMPROVED DSR FACILITY

The differential spectral responsivity (DSR) method is a spectral calibration method based on the measurement of the DSR as a function of wavelength λ in the presence of steady-state solar-like bias irradiance E_b setting the operating point which is determined by the short-circuit current $I_{sc}(E_b)$:

$$\tilde{s}(\lambda) = \frac{\Delta I_{sc}}{\Delta E(\lambda)} \bigg|_{I_{sc}(E_b)} \qquad (2)$$

with modulated quasi-monochromatic irradiance $\Delta E(\lambda)$ and photo-generated ac short-circuit current ΔI_{sc}. $\tilde{s}$ is the slope of the short-circuit current / irradiance characteristic. The measurement of the spectral responsivity without bias radiation according to Equation (1) is only acceptable and correct if the solar cell to be calibrated is perfectly linear ($\tilde{s}(\lambda, E_b) = s(\lambda)$ independent of $E_b \leq E_{STC}$).

In the PTB's DSR calibrations with varying bias radiation $E_b \leq E_{STC}$ including a linearity test, both the relative and the absolute spectral DSRs of a reference solar cell under test are determined. A dual-beam optical arrangement is used to measure relative but equally normalized DSR spectra of a reference cell at a series of discrete operating points that are set with a steady-state bias radiation at levels between $0.001 E_{STC}$ and $2E_{STC}$ (see Fig. 1). The chopped monochromatic radiation behind a double grating monochromator is measured with lock-in technique where the monitor photodiodes fixed within the spectroradiometer are calibrated against calibrated standard detectors (Si and InGaAs photodiodes) not exposed to the bias radiation (!) while substituting the test cell. To scale

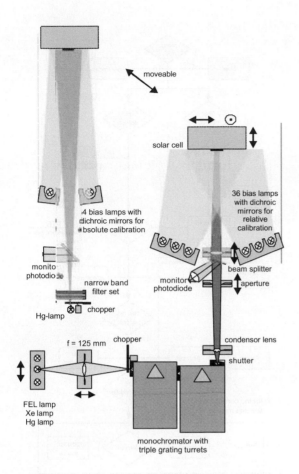

Figure 1. New version of the PTB's DSR Calibration Facility combining relative DSR spectroradiometer and absolute DSR filter monochromator. Monitor photodiodes are used to improve stability and reproducibility. The radiation shields are not included in this figure.

the relative DSR, the absolute DSR $\tilde{s}(\lambda_0, E_0)$ at one (or more) appropriate discrete wavelength(s) is measured at one low (but well-defined) bias level E_0 (about 20 Wm^{-2}) using a filtered medium-pressure Hg lamp. In contrast to the first DSR facility operated since 1987, the optical arrangement for the absolute DSR calibration is integrated into the new DSR calibration facility producing uniform spectral irradiance all over the cell area of up to 200 cm^2 without imaging optics. The modulation frequencies (about 135 Hz) for the relative and absolute DSR measurements are identical. More absolute calibrations at various wavelengths can be performed as a consistency check [4] (see also below). If a cell is linear and the (spectral) responsivity independent of (bias) irradiance, the calibrated responsivity is simply the mean of all the measured responsivities and Equation (1) is used to calculate I_{STC}. For the case of a non-linear cell, a simple iterative integration procedure is used to obtain I_{STC} [5]:

In the general case of non-linear cells, Equation (1) is not used but the AM1.5-weighted differential responsivity is calculated from the DSR data

$$\tilde{s}_{AM1.5}(E_b) \cdot E_{STC} =$$

$$\tag{3}$$

$$\tilde{s}(\lambda_0, E_0) \cdot \int\limits_{s(\lambda) \neq 0} \tilde{s}(\lambda, E_b)_{rel} \cdot E_{STC}(\lambda) d\lambda$$

The calibration value I_{STC} of non-linear cells is obtained with the aid of a simple integration over the short-circuit current $I_{sc}(E_b)$ by approximating the unknown upper integration limit I_{STC} within a few steps (see in Fig. 2 the non-reciprocal weighted differential responsivity versus short-circuit current)

$$E_{STC} = \int\limits_{0}^{I_{STC}} \tilde{s}_{AM1.5}^{-1}(I_{sc}) dI_{sc} \tag{4}$$

with $s_{STC} = s_{AM1.5}(E_{STC}) = I_{STC} \cdot E_{STC}^{-1}$.

Applying the efficient DSR method applicable to all kinds of reference solar cells independent of technology, the unit of spectral responsivity is transferred to high levels of irradiance including a linearity test covering the spectral range from 250 nm to 1600 nm without any gap. The standard detectors used are traceable to SI units via laser and broadband cryogenic radiometers. Fig. 3 shows the calibration chains in the PTB relating the calibration of solar cells and PV modules to national radiometric standards. Grey symbols indicate the direct calibration chain, while white symbols are used for the additional chain needed to determine spectral irradiance data for spectral mismatch corrections. The chains to the standard detectors and lamps are identical to the PTB's basic radiometric chain for the realization of the radiometric and photometric units. The main traceability chain from primary to secondary reference cells and to reference modules requires calibrations under natural or simulated solar radiation using steady-state or pulsed simulators. Spectral mismatch correction is in general required to reduce the relative calibration uncertainty to $< 5\%$. Spectroradiometers and filter sets [6] in combination with standard lamps are therefore used to correct the spectral mismatch between the PV devices on the one hand and the reference and solar/simulator spectra on the other hand.

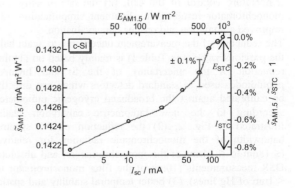

Figure 2. Weighted differential responsivity of a mono-crystalline Si solar cell as a function of short-circuit current.

3.   PERFORMANCE

Applying the DSR method for the calibration of primary reference cells at the PTB's first DSR facility since 1987 until quite recently, expanded uncertainties (k=2) of 1% (95% level of confidence) were obtained [3,4]. However, the DSR calibration is a rather time-consuming and thus costly business. Although 1% uncertainties of primary reference cells are reasonable, a markedly reduced uncertainty is desirable allowing a corresponding reduction of the uncertainty of secondary reference cells (see the calibration chain in Fig. 2). Therefore, the modifications of the improved DSR facility aimed at the reduction of both the operating time and the uncertainty focusing first on the improvement of the calibration of WPVS cells. The improvements are based on recent progress in radiometry and on the experience of operating the first DSR facility for more than 12 years.

The most important improvements of the new DSR facility are summarized as follows (see Fig. 1): extension of the spectral range (250 nm to 1600 nm) by using a double monochromator with triple grating turrets covering the whole wavelength range without any gap using programme-controlled selection of the gratings; integration of the setup for the absolute calibration at discrete wavelength(s) into the relative calibration; lens optics instead of mirror optics to avoid spherical abberation whereas the chromatic errors of the lenses are compensated by automatic moving the lenses as a function of wavelength; one double monochromator with triple grating turrets in subtractive dispersion instead of two monochromators with single grating turrets in order to (i) cover the whole spectral range and (ii) improve uniformity within the optical beam (up to 120 x 120 mm²); monochromatic light source (Hg lines produced by a medium-pressure Hg lamp with narrow-band filter set) for the absolute calibrations; mapping option for spatially resolved spectral response measurements (see below); improved automation and on-line data evaluation.

As a first result, both measurement time (operating time when the programme-controlled DSR facility is busy) and labour (actual input of staff and thus calibration fee charged) were reduced to about half the previous times. This improvement is mainly based on (1) the programme-controlled integration of the absolute DSR calibration procedure without extra alignment and without a second temperature control of the cell, (2) the use of only one monochromator resulting in alignment simplification, (3) improved automation and on-line data evaluation.

The reduction of the measurement uncertainty to again half the previous value (see Table I) is mainly based on (1) the reduction of the uncertainty of the Si and InGaAs photodiodes used as standard detectors which can directly be calibrated against the broadband cryogenic radiometer according to the new radiometric calibration chain illustrated in Fig. 2, (2) the reduction of the spectral bandwidth of the monochromatic radiation for relative (≤ 18nm to ≤ 11nm of the monochromators) and absolute DSR measurements (10nm of the filter monochromator to < 1nm of Hg lines), (3) better temporal stability and spatial uniformity ($\delta\lambda$ < 0.4 nm over cell area) of the well-defined centre wavelength of the monochromatic radiation, (4) doubled spectral irradiance and improved uniformity of the monochromatic radiation, especially for the calibration of WPVS cells, (5) improved image of the aperture on the

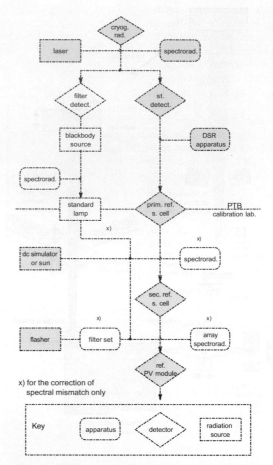

Figure 3. PTB's traceability chain for the calibration of PV devices.

solar cell behind the monochromator due to reduced image defects (spherical aberration), (6) improved electronics and signal-to-noise ratio, especially due to digital lock-in amplifiers, (7) improved temperature control.

Table I. PTB's DSR calibration: Uncertainty budget optimized for the calibration of WPVS cells.

Type A uncertainty: Uncertainty due to unstable cell temperature (± 0.5 K)	< 0.05 %
Type B uncertainties: Uncertainty of the standard detector(s)	< 0.1 %
Uncertainty due to nonlinear and/or narrow-band cells	< 0.05 %
Transfer uncertainties (repeatability) due to relative spectral responsivity	0.05%
absolute spectral responsivity at discrete wavelength(s)	0.05%
spectral mismatch between bias radiation and reference solar spectrum; non-uniformity of bias radiation; non-uniformity of monochromatic radiation; mismatch of cell area and irradiated area (image of the aperture); spectral bandwidth (≤ 11 nm) of the monochromatic radiation; nonlinearity of the amplifiers	0.2 %
Combined standard uncertainty	0.25 %

It was not necessary to improve the spectral match and the uniformity of the bias irradiance [7] and of the simulated solar radiation for the determination of fill factor and open-circuit voltage [4]. Moreover, it has been checked by varying the image A^* of the aperture on the cell (cell area A) within a wide range $0.25A \leq A^* \leq 0.98A$ that the uncertainty due to the mismatch $A^* \neq A$ is $< 0.05\%$ in practice if A^* is always less than A (see also [7]). Strictly speaking, the relative DSR measurement is a calibration with respect to spectral radiation power (using a optical beam smaller than the active area of the device), while the absolute DSR measurement is a calibration with respect to spectral irradiance (using a large uniform radiation field).

In addition, it has been proven that absolute DSR measurements can be carried out at every wavelength behind the monochromator (see Fig. 1) by using a mapping procedure: the solar cell is moved two-dimensionally resulting in a scan of the monochromatic optical beam of almost any diameter. Thus, flexibility of the DSR facility and the possibility of characterizing solar cells is increased.

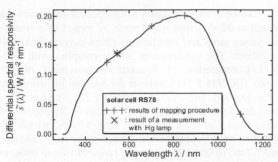

Figure 4. Scaling of the relative DSR by absolute DSR measurements (see text).

Fig. 4 shows the results of the absolute DSR calibration at the 546.1 nm Hg line compared with five mapping measurements that can be carried at any one wavelength to obtain also absolute DSR calibrations. The standard deviation of the five absolute DSR data (based on the mapping procedure) to the absolute DSR (relative DSR combined with absolute DSR at one Hg line) was found to be 0.15%. Moreover it has been shown, that the data of the mapping measurements are independent of the diameter of the scanning optical beam within wide limits.

4. CONCLUSION

Based on several hardware and software improvements of the first DSR facility and replacing a number of components, both the typical operating times and the uncertainties have been reduced to half their previous values. Expanded uncertainties of 0.5% (95% level of confidence) are obtained now. Thus, the new DSR facility is well-equipped for the second WPVS recalibration campaign to be carried out at the PTB in 2001 (see also [1,8]).

Acknowledgement. This work was supported in part by the BMBF.

REFERENCES

[1] C.R. Osterwald, S. Anevsky, A.K. Barua, K. Bücher, P. Chaudhuri, J. Dubard, K. Emery, B. Hansen, D. King, J. Metzdorf, F. Nagamine, R. Shimokawa, Y.X. Wang, T. Wittchen, W. Zaaiman, A. Zastrow, J. Zhang, Progr. Photovolt: Res. Appl. **7** (1999) 287, and The Results of the PEP'93 Intercomparison of Reference Cell Calibrations and Newer Technology Performance Measurements: Final Report, National Renewable Energy Laboratory Tech. Rep. NREL/TP-520-23477 (1998) 224 p. (available from the U.S. National Technical Information Service), and references therein.

[2] IEC 904-3, Measurement Principles for Terrestrial Photovoltaic Solar Devices with Reference Irradiation Data, Genf. (1989).

[3] J. Metzdorf, S. Winter, T. Wittchen, Metrologia. (2000) in press.

[4] J. Metzdorf, T. Wittchen, B. Nawo, W. Möller, Technical Digest of the Intern. PVSEC-5, Kyoto (1990) 705.

[5] J. Metzdorf, Appl. Optics **26** (1987) 1701.

[6] S. Winter, J. Metzdorf, Proc. 2nd World Conf. PVSEC, Wien (1998) 2312.

[7] J. Metzdorf, T. Wittchen, K. Möstl, Proc. 4^h Intern. PVSEC, Sydney (1989) 315.

[8] K. Emery, NREL / TP-520-27942 (2000) 16p.

AUTOMATED ABSOLUTE SPECTRAL RESPONSE CHARACTERISATION FOR CALIBRATION OF SECONDARY STANDARDS

Author: B. Ebner, G. Agostinelli, E. Dunlop

EC DG-JRC, Environment Institute / Renewable energies, TP 450, via E. Fermi 1, I-21020 Ispra, Italy

TEL.: +39–0332–789289, FAX: +39–0332–789268, E-MAIL: burkhard.ebner@jrc.it

ABSTRACT: This paper describes the features of the new automated Spectral Response Characterisation system built at European Solar Test Installation (ESTI) laboratories and illustrates the improvement obtained in resolution and precision of the measurements and its suitability to measure Silicon as well as non-Silicon devices. Furthermore the paper analyses typical problems faced in the execution of the measurement and the realisation of the set-up, supplying useful guidelines for the reader interested in carrying out SPR measurements.

KEYWORDS: Spectral Response – 1: Reference Cell – 2: Characterisation – 3

1.   INTRODUCTION

The spectral response of a photovoltaic cell is required to interpret laboratory measurements on devices and is useful for theoretical calculations. The procedure to measure the spectral response of a test cell is well known, and it is here briefly resumed for the use of the reader and clarity of exposition. A reference cell with known spectral response and a test cell are illuminated with perpendicular monochromatic light. The performance of the cell under test is then calculated relative to the reference cell by means of the relation

$$SR(\lambda)_{Test} = \frac{A_{\mathrm{Re}f}}{A_{Test}} \cdot \frac{I_{SC}(\lambda)_{Test}}{I_{SC}(\lambda)_{\mathrm{Re}f}} \cdot SR(\lambda)_{\mathrm{Re}f}$$

where A is the cell surface and $I_{SC}(\lambda)$ the short circuit current at a given wavelength, being Ref the indices for the reference cell and $Test$ for the test cell. Figure 1 shows schematically the set-up. A high-pressure xenon lamp produces the illumination on the measurement area. To insure that the device under test is operating in its linear response region (ignition conditions nearer to those of standard operating conditions of solar cells) a bias light system with variable intensity (0-100 mW/cm²) is used. The whole instrumentation is computer controlled. The control software is LabVIEW.

The new set-up calibrates the spectral response of a PV-cell at 32 different wavelength by means of a steady-state xenon lamp and interference filters ranging from 300 to 1200nm. The measurement set-up complies with IEC-904-7 Computation of Spectral Measurement of PV Device, IEC-904-8 [1] Guidance for Spectral Measurement of PV Devices and ASTM E1021-84 [2] Standard Test Methods for measuring the Spectral Response of Photovoltaic Cells. It uses two filter wheels, which are controlled by a stepper motor, each containing 16 narrow-band filters, to cover the response range of silicon cells, in wavelength steps not exceeding 50nm and a bandwidth of 10nm. Calibration is done against a WPVS (World PhotoVoltaic Scale) reference cell. The system is fully automated. The only required user intervention is to change the filter wheels and the devices under test. The rest is done automatically by the measurement software. The set-up is being expanded to 3 wheels (48 filters); 43 filters will be used to measure the spectral responsivity in wavelength steps not exceeding 20nm in the range between 300 and 1000nm, 5 filters are available between 1000 and 1200nm.

The measurement system improves the precision of the measurement in the typical critical ranges for spectral response measurements and overcomes most of the problems encountered in measuring SPR of non-crystalline Silicon devices. Variable intensity bias light (0-100 mW/cm²) produces the desired ignition condition. Monochromatic illumination can be roughly varied from a few mW up to 10W/m^2 (10nm windows). Isc measurement errors are reduced by means of a custom IV converter. Typical errors in the measurement of short circuit current in the range between 300 and 400nm and 1000 to 1200nm are now lowered to a max of ~1% (compared to former >6% of other set-ups) because of the increased signal to noise ratio achieved by means of the new design and the digital lock-in filtering capabilities. Though the mentioned results do not affect in a sensible way the overall calculation of Isc in AM1.5G, the precision achieved in single - EQE (External Quantum Efficiency) measurements is very important for the device characterisation.

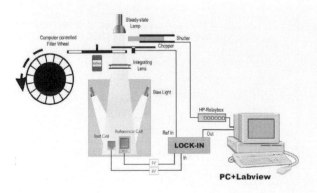

Figure 1 Set-up scheme

2. ERROR ANALYSIS

To analyse the performance of the spectral response measurement set-up many aspects have to be taken into consideration. For the error analysis of the calibration procedure of PV-cells, it is necessary to measure the accuracy of each instrument involved in the experimental set-up. We shall first consider intrinsic error sources, i.e. due to the instrumentation, and other error sources due to the settings and the way the system is operated.

2.1. Intrinsic error sources

The light source, its spectral distribution and its stability are of course crucial to the precision of the measurement. At regime, the intensity fluctuations and the temporal stability of the lamp are a considerable error source for the spectral response calculations. Contrary to what it might be expected, even when the total temporal stability of a lamp shows a stabilising trend, the detailed evolution of the spectra is likely to compromise the measurements. In the case of the considered set-up, only after 4 hours the average signal deviation of the intensity (over 30 min) is less than ±0.3% and only a few signals are exceeding ±0.5% (Figure 2).

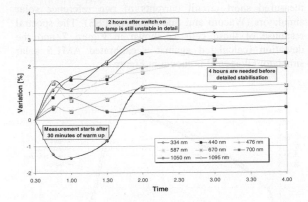

Figure 2 *Spectral evolution of an old Xe lamp irradiance*

The illumination uniformity on the test plane is another critical parameter when measuring cells of different sizes or non-homogenous characteristics. ASTM designation E1021-84 indicates ±2.5% as the required uniformity for spectral response measurements. There is no such explicit indication according to IEC standards. According to the IEC 904-2 [3] standard, a Class A simulator shall be used for the measurement and calibration of reference solar cells. The illustration shows that a Class A simulator (IEC 904-9 [4] illumination uniformity ±2%) could be achieved on a square of about 10x10cm² by ideal adjustment of the lens-system and the diaphragms. By positioning the reference cell in the region of the indicated square the spectral response of mini modules bigger than 20x20 cm² can be measured within the ASTM 1021-84 standard. (Figure 3). The bias light system consists of 24 adjustable 50W halogen lamps mounted in two circles above the test plane. According to ASTM E1021-84 the spatial uniformity of the bias light system should be at least ±10%. By adjusting all 24 lamps, according to the size and the position of the test device, the uniformity of bias light system matches this specification without problems.

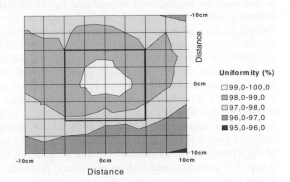

Figure 3 *Uniformity of the monochromatic light beam. Class A Simulator specifications (IEC 904-9) achieved at 10x10cm²*

2.2. Settings-depending error sources

The error of the lock-in amplifier is influenced by a few intrinsic noise signals produced by external resistors and the input noise of the lock-in amplifier. The measured noise of a resistor at the lock-in input, which is in this case the shunt of the I-V converter or the external shunt of the PV-device, can be expressed (in nano Volts) as

$$V_{noise}(rms) = 0.13\sqrt{R} * \sqrt{ENBW} .$$

The Electrical Noise Bandwidth (ENBW) is determined by the time constant (TC) and the filter slope of the lock-in. Hence the lock-in input noise is dependent on the parameter settings during the measurement and the selected shunt value of the I-V converter (or external shunt). Figure 4 gives an indication of the influence of the time constant and the shunt value.

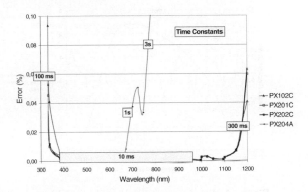

Figure 4 *Lock-In amplifier output errors [%] (calculated out of the measurement of four different solar reference cells measured with 24dB/oct (Slope) and 10-ohm shunt*

Figure 4 illustrates the percentage output noise of the lock-in amplifier acquired by measuring four different PV- reference cells (PX102C, PX201C PX202C, which are mono-crystalline silicon and PX204A a filtered silicon solar cell). The illustration shows that the output error of the lock-in amplifier in general is lower than 0.1%. Only in the wavelength ranges where the signals are very small and the response values of the PV-device is almost zero the error of the lock-in contributes to the error calculations.

In this case the biggest error is produced by the PX204A reference cell, which has no spectral response above 740nm. In the wavelength range where the response of the PV-device is not near to its limit the error of the lock-in amplifier can be neglected. Other errors could be induced by fluctuations of the signal and parasite noises. The program is set not to tolerate errors in the measured signal greater than 1% (standard deviation of a multiple series of measurements). This is the maximum assumed error of the lock-in data acquisition. The final error of the spectral response system is calculated as the square root of the sum of the squares of the single errors and is equal to ±1.12% (lamp, I-V converter and lock-in amplifier). It should be stressed that the margin of 1.0% error for the lock-in amplifier is an upper limit imposed a priori as a compromise between precision of the measurement and its speed.

3.   PERFORMANCE

3.1.   General Set-up

To check the performance of the system 3 procedures have been chosen: the repetition of the measurement, the cross comparison of Reference cells, and the comparison of the derived short circuit current. In the first case, one reference cell is measured against itself. In this way the percentage random errors of the measurement set-up can be evaluated. The reference cell should give a repeatable performance under identical test conditions. The average random deviation of the whole system is below 1%. The maximum errors are +1.3% and –0.9% and the standard error deviation is ±0.5% (see Figure 5). Even in the usually more critical parts of the spectrum the measurement set-up doesn't cause bigger errors.

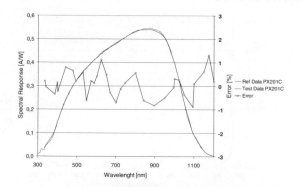

Figure 5*: Random system error derived by measuring a WPVS reference cell against itself*

In the second case, a reference cell is measured against a WPVS primary reference cell and the results are compared with the calibration value of the device under test. Figure 6 shows the spectral response measurements of 4 different ESTI solar reference cells, measured against the WPVS primary reference Cell PX201C. The short circuit currents of the reference cells were calculated out of the integral of the spectral response of each device and the AM 1.5 solar spectrum. The resulting short circuit currents were compared with the calibration value of the devices.
The deviation to the calibration value is indicated in the graph (Figure 6). Additional a CdTe thin film device, which is not exactly calibrated, is plotted. It is not possible to indicate a deviation to its calibration value.

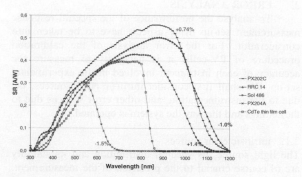

Figure 6*: Spectral Response of 5 Reference Cells (deviation to the calibration value indicated [%]) and 1 CdTe Thin Film Cell*

The third way is to calculate the short circuit current of the device under test and to compare it with the value of the I-V characteristic measurements. In this case the measured spectral response of two ASPIRE (Absolute Silicon Primary Irradiance Reference) cells [5] measured with and without quartz glass was compared with the measured short circuit currents of two reference solar simulators (Wacom and Spectrolab LAPSS). The spectral response of these cells is shown in Figure 7 and the deviation (calculated against a calibrated AM1.5 solar simulator (Wacom)) is shown in Figure 8.

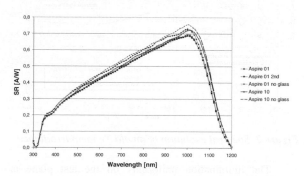

Figure 7*: Spectral Response of ASPIRE 01 and ASPIRE 10 measured with and without quartz glass*

The maximum deviation to the reference solar simulator is around 1.3% and four of the five measurements are below 1%.

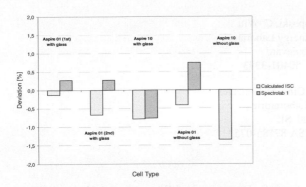

Figure 8: *Calculated Isc (out of the SPR Results of the ASPIRE 01 and 10 cells) and measured Isc (Spectrolab LAPSS) compared with the measured Isc value of the WACOM AM1.5 solar simulator*

3.2. Bias Light System

The bias light system performance was checked by measuring an ESTI 2 sensor with and without bias light. This sensor was selected because it was measured in the PEP (Photovoltaic Energy Project) 93 intercomparison by JQA (Japan Quality Assurance Organisation) [6] and a remarkable influence of the bias light was seen in this measurement. Therefore the same sensor was measured to control the performance of the bias light. The normalised plot of the spectral response with and without bias light is shown in Figure 9. The same influence of the differences between these two measurements was measured by JQA. A slightly deviation in the values of the differences might be due to differences in the bias light illumination intensity. The difference in the calculated was in this case about 3%.

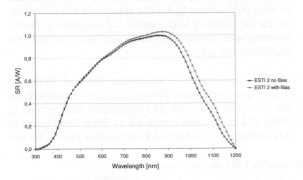

Figure 9: *Normalised spectral response of an ESTI 2 irradiance sensor measured with and without ~500W/m² bias light*

4. RESULTS:

The result of the error calculation and the performance evaluations proved the reliability of the measurement set-up. A system with a standard error deviation of about ±0.5% in the whole wavelength range and an illumination uniformity of a class A simulator on a square of 10x10cm² is achieved. The spatial simulator uniformity conforms to ASTM E1021-84 (±2.5%) and allows the measurement of the spectral response over an area greater than 20x20cm².

5. CONCLUSIONS:

The measurements performed up to now give very satisfactory results. The measurement of the Absolute Spectral Response of a WPVS Silicon Reference Cell leads to a calculated overall AM1.5G Isc current which is only 0,74% different from its calibration value. The new spectral response measurement facility is a good addition to the PV-device measurement facilities of the European Solar Test Installation. It will be used for reference cell calibrations and the characterisation of PV mini-modules and cells.

6. REFERENCES:

[1] IEC-904-8 Guidance for Spectral Measurements of a PV Device
[2] ASTM 1021-84 Standard Test Methods for Measuring the Spectral Response of Photovoltaic Cells
[3] IEC-904-2 Requirements for Reference Solar Cells
[4] IEC 904-9 Solarsimulator Performance Requirements
[5] Absolute Silicon Primary Irradiance Reference Claas Helmke JRC Ispra Italy 1996
[6] Report on PEP'93 Intercomparison of WPVS Series Samples F. Nagamine, R. Shimokava Japan Quality Assurance Organisation April 1994

TESTING TO SUPPORT IMPROVEMENTS TO PV COMPONENTS AND SYSTEMS

H. Thomas, B. Kroposki, C. Witt
National Renewable Energy Laboratory
1617 Cole Boulevard
Golden, CO USA 80401-3393

W. Bower, R. Bonn, J. Ginn, S. Gonzales
Sandia National Laboratories
1515 Eubank SE
Albuquerque, NM USA 87185-0753

ABSTRACT:
The National Photovoltaic (PV) Program is sponsored by the U.S. Department of Energy, and includes a PV Manufacturing Research and Development (R&D) project conducted with industry. This project includes advancements in PV components to improve reliability, reduce costs, and develop integrated PV systems. Participants submit prototypes, pre-production hardware products, and examples of the resulting final products for a range of tests conducted at several national laboratories, independent testing laboratories, and recognized listing agencies. The purpose of this testing is to use the results to assist industry in determining a product's performance and reliability, and to identify areas for potential improvement. This paper briefly describes the PV Manufacturing R&D project; participants in the area of PV systems, balance of systems, and components; and several examples of the different types of product and performance testing used to support and confirm product performance.
Keywords: Inverter- 1: Balance of Systems- 2: Components- 3

1. INTRODUCTION

The U.S. Department of Energy (DOE), in cooperation with the U.S. photovoltaic (PV) industry, has the objective of enhancing the U.S. PV industry leadership in the manufacture and commercial development of PV products. To further this objective, the Photovoltaic Manufacturing R&D project was initiated in 1990 between the DOE and the U.S. PV industry to help improve PV manufacturing processes and substantially reduce associated manufacturing costs. The project accomplishes these objectives by conducting competitive solicitations inviting proposals that address both technology-specific and generic problems in manufacturing R&D identified by industry. Work is implemented through cost-shared contractual agreements between the federal government, through the DOE and the National Renewable Energy Laboratory (NREL), and individual members of the U.S. PV industry. Technical teams from the National Center for Photovoltaics (NCPV,

which includes NREL and Sandia National Laboratories) manage the work.

The Manufacturing R&D efforts are divided into two parts, with the majority of the work in the area of PV Module Manufacturing. Results of this work are described in other publications [1,2]. The contracted efforts in PV System and Components are more fully described in the paper, "Progress in Manufacturing R&D in Photovoltaic Components and Systems," presented at this conference and in other publications [3,4]. The purpose of this paper is to review the variety of testing conducted in support of the PV systems and component R&D included in the PV Manufacturing project.

There have been two solicitations for work in the PV System and Component portion of the PV Manufacturing project. Work under the first solicitation was initiated in 1995, with most work completed by 1998. Participants in this solicitation and their work are listed in Table I. As part

Table I: PV System and Component Manufacturing subcontracts initiated in 1995 and now completed.

Company	Title
• Ascension Technology, Inc.	Manufacture of an AC Photovoltaic Module
• Advanced Energy, Inc.	Next-Generation Three-Phase Inverter
• Evergreen Solar, Inc.	Advanced Polymer PV System
• Omnion Power Engineering Corp.	Three-Phase Power Conversion System for Utility-Interconnected PV Applications
• Solar Design Associates, Inc.	The Development of Standardized, Low-Cost AC PV Systems
• Solar Electric Specialties	Design, Fabrication, and Certification of Advanced Modular PV Power Systems
• Trace Engineering	Modular Bi-directional DC-to-AC Power Inverter Module for PV Applications
• Utility Power Group, Inc.	Development of a Low-Cost Integrated 20-kW AC Solar-Tracking Sub-Array for Grid-Connected PV Power System Applications

Table II: System and Component Manufacturing R&D subcontracts initiated during 1998.

Company	Title
• Ascension Technology, Inc.	Cost Reduction and Manufacture of the SunSine™325 AC Module
• Omnion Power Engineering Corp.	Manufacturing and System Integration Improvements for One- and Two-kilowatt Residential PV Inverters
• PowerLight Corp.	PowerGuard® Advanced Manufacturing
• Utility Power Group, Inc.	Development of a Fully-Integrated PV System for Residential Applications

of a later PV Manufacturing solicitation, new participants were competitively selected and new contracts negotiated in 1998. This work is still in progress. These participants are listed in Table II. Participants in both solicitations included a variety of tests to support the planned product advancements, validate product performance to specifications, and adherence to safety standards as part of their subcontract agreements.

2.0 TESTING AND CHARACTERIZATION

System and component testing is conducted in collaboration with the NCPV, through independent testing laboratories, and at listing agencies. Testing at NCPV laboratories, including NREL and Sandia, is a collaborative effort, with specific tests defined by agreement between the labs and the manufacturer. Often, special tests are included to evaluate a particular aspect of interest to the manufacturer or to meet a special requirement from a potential customer. The NCPV laboratories test, but do not certify products.

Engineers at the NCPV laboratories conduct tests on production prototypes to identify needed improvements, test final products to compare them to the planned specifications, and assess long-term performance and reliability. Industry engineers assist in defining the types of tests and are often present during the testing. Typically, the manufacturer also applies to a qualified laboratory for a product listing for safety, or in the case of utility-interactive products, to determine if the product meets interconnection guidelines. Industry also sends products, prototypes, and samples to a variety of independent laboratories for failure-mode testing and specialized design evaluations. All these tests are in addition to the manufacturer's own specialized, in-house testing. The tests summarized in the following sections are only a few examples of the variety of tests conducted to establish a component's or system's performance and reliability.

2.1 Characterization and Evaluation at Sandia

Sandia evaluates prototypes and production components and systems in collaboration with the Manufacturing R&D project. Sandia evaluates the product relative to its performance specifications and assesses the design to meet certain safety and performance standards. The Sandia PV Photovoltaic Systems Evaluation Laboratory (PSEL) is located in Albuquerque, New Mexico. Products that have been tested include: a 60-kW hybrid inverter and a 30-kW grid-tied inverter from Advanced Energy, Inc.; a 15-kW-rated Integrated Power Processing Unit (IPPU) from Utility Power Group, Inc.; a 4-kW single-phase, packaged hybrid system from Solar Electric Specialties (now part of Applied Power Corp.); a 250-W SunSine module-scale inverter from

Ascension Technology, Inc. (now a division of Applied Power Corp. [APC]); and a 250-W module-scale inverter from Advanced Energy, Inc., under a Solar Design Associates subcontract. Several subcontracts are still in progress. Of these, Ascension will send Sandia their pre-production prototype and production versions of the new SunSine™ module-scale inverter, which incorporates their new soft-switching technology. Utility Power Group, Inc., will be sending their 12.9-kW grid-interactive inverter and energy storage unit for evaluation. The evaluation of the Omnion Model 3300 (100-kVA) grid-tied inverter is an example evaluating one production prototype from the Manufacturing R&D project.

As part of their contract, Omnion Power Engineering Corporation (now a division of S&C Electric Company) delivered the pre-production prototype Model 3300 power conversion system (PCS) inverter, developed as part of this agreement. Sandia operated the inverter with a PV array (up to 25 kW$_{dc}$) and a battery bank (over 100 kW) to conduct a series of tests. Test results confirmed that the unit met planned operating specifications, and several results are highlighted here. Omnion anticipated an inverter efficiency of 95%, without the transformer. Sandia measured 93% to 94%, with the transformer, at input power levels above 15 kW. Because a transformer requires some power, these results support the Omnion figure. Total harmonic distortion (THD) of the output current, expressed as a fraction of the inverter's rating, was below 4% for all power levels above 5 kW, supporting the company's specification of <5%. The voltage THD of the utility line at the test facility remained below 2% for all conditions tested. The maximum-power-tracker circuitry accurately extracted the maximum available from the test PV array. Acoustic noise was 63 dB, considered extremely low for a 100-kW inverter. The cooling fans effectively maintained the heat-sink temperature below 50°C, with a 25°C ambient operating temperature. As would be expected for prototype testing, Sandia also identified areas needing improvement. The conducted radio-frequency interference (RFI) exceeded the maximum allowed by the Federal Communications Commission (FCC), and additional filtering was recommended. However, the radiated RFI was negligible in the frequency spectrum relevant to FCC Part 15. Sandia noted the high-frequency trip point was set at 60.5 Hz, in accordance with an earlier draft version of Institute for Electrical and Electronic Engineers (IEEE) P929. Omnion will adjust the unit to 61.0 Hz, to comply with IEEE 929-2000. Sandia also noted that the low-AC-voltage trip point was set at 3% below nominal. The lab noted that this is much more conservative than the 10%-below-nominal specification. Investigators recommended an adjustment because the higher trip point could result in unnecessary interruption in inverter operation. Overall, Sandia reported that no operational difficulties were encountered. Omnion made the indicated modifications to the revised PCS Model 3300 and is working to obtain an Underwriters Laboratories (UL) listing for product safety [5]. An added benefit of testing this unit was that it provided the first opportunity to evaluate a large, grid-tied inverter for anti-islanding with a resonant resistive-inductive-capacitive (RLC) local load (i.e., part of the new IEEE 929-2000 interconnection standard). The inverter disconnected under a variety of loads within the time required by IEEE 929-2000. These test results are part of Omnion's Annual Technical Progress Report [6] and are also reported on the Sandia website [5]. An illustration of the Sandia testing setup is shown in Figure 1.

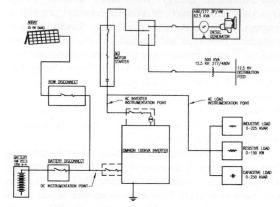

Figure 1. Diagram of test setup at Sandia's PSEL for the Omnion PCS Model 3300.

2.2 Accelerated Environmental Testing and Long-Term Performance Evaluation at NREL

NREL conducts accelerated environmental testing and long-term performance evaluations of PV modules and components at its Outdoor Test Facility (OTF), located in Golden, Colorado. NREL conducts performance testing and evaluation of products and systems in progress or resulting from the project. Products and components that NREL evaluated for the project include: a Modular Autonomous PV Power Supply from Solar Electric Specialties (now a part of Applied Power Corporation), the 250-Wac SunSine™300 AC PV Module from Ascension/APC; a 250-Wac module-scale inverter from Advanced Energy; and PowerCurb™ components from PowerLight Corp. Products still planned for testing include two Trace Engineering 2.5-kW Model PS inverters and the new 275-Wac SunSine™ AC PV Module from Ascension/APC. NREL also provided testing facilities for on-site valuations by UL. NREL's National Wind Center tested a prototype 60-kW hybrid inverter resulting from a subcontract with Advanced Energy. The unit was connected to a generator, a wind simulator, and a battery bank to evaluate its ability to manage hybrid operating conditions. The evaluation of Ascension/APC's SunSine™ 300 AC PV Module is an example of long-term performance testing at NREL's OTF. Figure 2 illustrates the test apparatus for the SunSine™300 AC PV Module and Advanced Energy's module-scale inverter.

As part of the 1995 subcontract, Ascension/APC delivered pre-production and production versions of their SunSine™ 300 AC PV Module for accelerated testing and long-term performance and reliability tests. This product consists of a 250-Wac module-scale inverter factory-

mounted to an ASE Americas large-area PV module. NREL conducted accelerated environmental testing by cycling the entire AC PV Module between -40° to +90 °C for 30 cycles of 24 hours per cycle. Measurements under Standard Test Conditions were made before and after testing. NREL reported the inverter operated at full power (250Wac) over the testing period, and no changes in performance were detected. Ascension/APC requested this test to establish a level of confidence that the SunSine™ 300's inverter would pass the same tests required for the PV module.

Figure 2: The SunSine™300 AC PV Module and Advanced Energy's module-scale inverter under test at NREL's Outdoor Test Facility.

Test results also provided an indication of the expected reliability of the inverter. Pre-production prototypes and production versions of the SunSine™300 AC PV Module were also tested for long-term performance and reliability. The products were mounted on outdoor racks at the OTF, connected to the grid, and instrumented to monitor the unit's performance under prevailing weather conditions in Colorado. Data collected include irradiance, wind speed, AC power output, PV module temperature, SunSine™300 inverter temperature, and capacitor temperature. Power output from the product and temperatures are collected every 5 seconds and averaged every 15 minutes. These SunSine™300 AC PV Modules are still under test, and based on the data, they have operated without anomaly since installation in early 1999. Initial results are more fully described in Ascension/APC's Final Report for this contract [7]. NREL is also conducting specific subcomponent tests for Ascension/APC's new product, a 275-Wac module-scale inverter now termed the SunSine™. Ascension/APC requested this test to provide data for estimating this new product's lifetime and to begin to evaluate its long-term performance.

2.3 Failure Modes and Reliability Testing by Independent Test Laboratories

Manufacturers have also used a variety of independent testing laboratories to evaluate their products. One of these laboratories, QualMark Corporation, has a testing approach that several manufacturers have used, or plan to use, to evaluate the failure modes of their products. QualMark's testing plans include two specialized tests: a HALT (Highly Accelerated Life Test) and a HASS (Highly Accelerated Stress Screening). The QualMark HALT is a test regime typically requested by the manufacturer for pre-production prototypes. The testing rationale is that when the product is stressed beyond its design limits and parts begin to fail, the result can be a list of design corrections to improve the product's reliability. The tests are not pass/fail, but are used to optimize the design. Two participants in the

Manufacturing R&D project, Advanced Energy and Ascension/APC, submitted their inverters to these tests, with Sandia's support [8]. Both companies discovered several areas for improvement. Ascension/APC notes, "The HALT supported by Sandia and NREL was very useful in weeding out potential failure modes in our SunSine™300 AC PV Module. We used HALT to achieve the highest possible reliability in the electronic design of our product. In the first 300 units built, only one unit had a failure in the electronics." Recently, Ascension/APC submitted the revised SunSine™ inverter for the HALT, and Utility Power Group will be submitting their 12.9-kW inverter later this year as part of their contractual agreements.

Other types of testing are also important. One manufacturer, PowerLight Corporation, included wind testing as part of their product validation. PowerLight's product, termed PowerGuard®, is a PV system that is mounted on a roof without roof penetrations. The unique product has two options for orienting the PV panels on the roof, either parallel to the roof or sloped, to better capture the available sunlight at higher latitudes. PowerLight had several versions of their angled-product design evaluated for wind resistance. They also evaluated scale models to test other methods of attaching PV laminates to their unique substrate. The product design was required to meet several constraints in addition to wind resistance, including low cost and an ability to collapse flat for a high packing-density for shipping. Results of these wind tests helped PowerLight engineers evaluate the lift and turbulence created by different design options, modify test designs and to then verify their final choice.

2.4 Listing and Certification at Recognized Laboratories

2.4.1. Underwriters Laboratories Inc.®
As part of the subcontract agreements, manufacturers have investigated obtaining UL, ETL (by Intertek Testing Services), or other recognized listing as an indication of product safety. Most companies in the project chose to apply for an UL-listing of their products, as the UL Mark is often the most readily recognized by building inspectors and other local authorities. A safety listing is obtained when samples of the product have been tested and evaluated by a certified listing agency, and they comply with appropriate standards. Products may then carry the listing mark. UL makes their listing determination based on product testing, and if the product has a novel design, listing may be based on a design review compared to known criteria. Several novel products have resulted from the PV Manufacturing R&D project and have required additional testing. These products ranged from the Ascension 250-W SunSine™ up to the Omnion 100-kW PCS Model 3300.

2.4.2 CE (Conformite Europeenne) Marking
The CE Marking is an official mark required by the European Community for electric equipment regulated by the European health, safety, and environmental protection directives. A product usually requires this Marking if it is to be sold in Europe. Two companies included an assessment of their designs to meet the CE Marking requirements as part of their contractual agreements. Ascension/APC reviewed their new SunSine™ design to define the modifications that would be necessary to obtain the CE Marking, but will not include those changes in the first production run. PowerLight determined and made the necessary modifications to their system to meet the CE Marking requirements, and that Marking has been received.

3. CONCLUSIONS

Testing and verification are key elements to support the results of the PV Manufacturing R&D project. These tests are included as part of the contractual agreement, and they are defined in collaboration with the testing engineers and the manufacturer. Their purpose is to support industry in building advanced, reliable PV products and systems. The examples presented here illustrate the range of these tests. Manufacturers rely on these results to validate their products and to establish performance and reliability information. These results also provide data to support product warranties and guarantees. Testing to evaluate performance, qualify products for safety, and estimate product lifetimes will continue to be an essential part of this project.

4. ACKNOWLEDGMENTS

This work is supported under DOE Contract DE-AC36-99GO10337 with NREL, a national laboratory managed by Midwest Research Institute, and Contract DE-AC04-94AL8500 with Sandia National Laboratories, a multi-program laboratory operated by Sandia Corporation, a Lockheed Martin Company. Many people have contributed to developing and implementing the PV Photovoltaic Manufacturing R&D project and to the efforts carried out in this project. The authors acknowledge that this paper represents their work.

REFERENCES
[1] Witt, C. E., Mitchell, R. L., Symko-Davies, M., Thomas, H. P., King, R., and Ruby, D. S., 1999, "Current Status and Future Prospects for the PV Manufacturing Technology," *Proceedings, 11th International Photovoltaic Science and Engineering Conference (PVSEC-11)*, T. Saitoh, ed. (to be published)
[2] Witt, C. E., Mitchell, R. L., Thomas, H. P., Symko-Davies, M., King, R., and Ruby, D. S., 1998, "Manufacturing Improvements in the Photovoltaic Manufacturing Technology (PVMaT)," *Proceedings, 2nd World Conference and Exhibition on Photovoltaic Solar Energy Conversion*, J. Schmid et al., ed., European Commission, Ispra, Italy, Vol. 2, pp. 1969-1973.
[3] Thomas, H. P., Kroposki, B., Witt, C. E., Bower, W., Bonn, R., and Ginn, J., 2000, "Progress in Manufacturing R&D in Photovoltaic Components and Systems," *Proceedings, 16th European Photovoltaic Solar Energy Conference and Exhibition* (to be published)
[4] Thomas, H. P., Kroposki, B., McNutt, P., Witt, C. E., Bower, W., Bonn, R., and Hund, T. D., 1998, "Progress in Photovoltaic System and Component Improvements," *Proceedings, 2nd World Conference and Exhibition on Photovoltaic Solar Energy Conversion*, J. Schmid et al., ed., European Commission, Ispra, Italy, Vol. 2, pp. 1930-1935.
[5] Ginn, J.W., Bonn, R.H., and Sittler, G., "Inverter Testing at Sandia National Laboratories," Albuquerque, NM, Available at http://www.sandia.gov/pv (1999).
[6] Porter, D., Meyer, H., and Leang, W., *Three-Phase Power Conversion System for Utility-Interconnected PV Applications, A PVMaT Phase I Technical Progress Report*, NREL, Golden, CO, NREL/SR-520-24095, February 1998.
[7] Kern, G., *SunSine™300: Manufacture of an AC Photovoltaic Module, Final Report*, NREL, Golden, CO NREL/SR-520-26085, March 1999.
[8] Quarterly Highlight of Sandia's Photovoltaic Program, Volume 3, Dec. 1998, Sandia National Laboratories, Albuquerque, NM.

QUALITY CONTROL PROCEDURE FOR SOLAR HOME SYSTEMS

Miguel A. Egido, E. Lorenzo, E. Caamaño, P. Díaz, J. Muñoz, L. Narvarte
Instituto de Energía Solar – ETSI Telecomunicación
Ciudad Universitaria. 28040 – Madrid (Spain)
email: egido@ies-def.upm.es

ABSTRACT: The fieldwork in Photovoltaic (PV) rural electrification programmes shows that many technical problems are still present in the operation of PV systems and, also, that are mainly related with local aspects. In this paper, a PV stand-alone systems qualification testing procedure is proposed, which has been developed in combination with a previously technical standard, widely distributed. The resulting methodology can be reproduced in local environment and guarantees safety, reliability and good operation of Solar Home Systems. Now, our laboratory can certificate the fulfilment of the Universal Technical Standard for SHS in PV components and full systems and offers, until June 2001, free testing to all organisations or manufacturers interested.
Keywords: Solar Home Systems – 1: Qualification and Testing – 2: Rural Electrification - 3

1. TECHNICAL STATUS OF PV RURAL ELECTRIFICATION

During the last ten years, the number of rural electrification installations has been highly increased. At present, it could be roughly estimated in almost one million. This process has been accompanied by an important, although unequal, development of PV components: modules, charge regulators, batteries, DC/AC converters and charges. Nevertheless, this development, excepting the case of modules, has been reached in a poor co-ordinated way. While great attention has been paid to modules –maybe due to its technologically novelty product nature, which has lead to the development of a highly reliable product–, this has not been the case for the rest of the components, whose complexity of adjustment has been undervalued.

However, PV rural electrification observed through the field experience shows quite a different reality. Failures on the installations are linked not to the modules, but to the components and also, largely, to the installation. The reason of the bad operation is usually the malfunction of some of the components, although it is also frequent that the failure is associated to a bad installation or maintenance and, in some cases, a wrong system design. The rate of faulty SHS is taking up the confidence in this electrification method [1]. Even if all the elements worked properly and the system design was correct, the user would consider it to have failed when his energy demand was not covered. Frequently, the users' lack of awareness causes that expectations about their PV system energy production are not consistent with the power installed.

The improvement of the photovoltaic rural electrification situation is based on the analysis and comprehension of the characteristics of all the elements building up the installation. That is the reason why the standardisation task and consequently, the development of test procedures allowing the corroboration of the compliance of the standards, are a priority [2]. The results of the first phase of the *Programme Regional Solaire*, which have been evaluated quasi unanimously very positively by the scientific community, constitute an example of how this mechanism determines the success of a project [3].

The process of standardisation and certification is far from being innocuous. The development of a highly sophisticated normative with a low flexibility degree, either for standards or for certifications, can turn up the norm into a curb for the local markets development, which are, in fact, the main beneficiaries of the rural electrification programmes. Besides, sophistication does not necessarily imply reaching the aim of quality improvement. On the contrary, it increases the technological dependence of developing countries, which hinders the sustainability of electrification programmes. One of the key points on the technical success of PV rural electrification is a proper maintenance; although PV systems were considered not to need it, this old idea has been rejected as a result of the experience, proving that no large-scale project can be implemented without ensuring this point. Such maintenance is based on the availability of qualified technical staff, spare parts and affordable costs, as well as the necessary infrastructure to perform it. These three conditions could be seriously compromised if the local PV market is not encouraged as a consequence of a certification methodology where local actors are not involved.

From the photovoltaic material suppliers point of view, the certification of their products by accredited laboratories could imply a notable increase on their manufacturing costs. Several consequences are derived from this fact: increase of the electrification total costs, reluctance to include the certification of products as a contractual requirement, and boost on the free market of non standardised products. As a result of that, the dissemination of photovoltaic electrification could be sharply restricted.

This paper proposes a technical certification methodology, developed from the *Universal Technical Standard for Solar Home Systems* [4], which can be implemented without high technical skill observance on the countries using photovoltaic electrification. It has been intended that, within the adequate precision limits, the degree of observance of the PV system components specifications can be measured. Moreover, because of the increase of AC charges demand, DC/AC converters have also been included; being jointly developed their technical specifications, as well as the test procedures.

The aim is that all these procedures can be implemented on Electrotechnical laboratories in the same countries that are promoting the rural electrification development on disperse communities by means of photovoltaic solar energy, either from an institutional or private perspective. This strategy helps the technological development of local producers by the establishment of a feedback channel, thus making it possible the improvement of their products.

2. QUALITY CONTROL PROPOSED

2.1 Methodological aspects

The previous discussion has been focused in the process to attain PV systems more reliable. The main factors having influence on the way to reach this goal, are the recognition of dependency on local parameters, the importance of technical standards widely accepted, and the necessity to verify its fulfilment by technical quality control procedures. Several consequences can be drawn from this. Firstly, procedures to be developed (technical specifications, tests, etc.) should elude sophistication, in order to avoid that their application become restricted to a few laboratories from industrialised countries. Secondly, it should be considered that the proposed quality procedures should gain credibility by, for example, their validation against recognised, more sophisticated methods. The third consequence is that, in any case, quality assurance goes beyond technical aspects to reach quality of installation, servicing and service guarantees, which are local by essence.

Our proposal is based on the previous development of the standard for SHS [4], which could be extended for more general PV stand-alone systems. Requirements listed in the document intend to guarantee safety, performance and reliability of PV systems and were elaborated analysing all technical factors. For this, we considered PV components separately, but also the PV system as a whole. Within the frame of the present project, we have developed tests to check the accomplishment of the technical prescriptions.

In our opinion, it is possible to develop measurement procedures with high accuracy levels while reducing considerably the costs. This is more a question of philosophical principles and consideration of the socio-economical situation of the places where these systems are going to be installed, than of technical barriers. From the combination of this point of view with the experiences carried out up to now, the PV Systems Laboratory of the Instituto de Energía Solar (IES) proposes a set of tests for PV components and systems in stand-alone applications. The tests proposed intend to fulfill the following requirements:

1. Methodological simplicity, without loosing scientific rigor in their approach.
2. Based on simple low-cost instruments, making the methods applicable "in the field" (where PV systems will be installed).
3. Accuracy and reliability of results similar to other methods presently in use.

Detailed measurements procedures and instrumentation needed are collected in a document entitled: "PV Stand-Alone Systems Qualification Testing" [5]. In some cases, procedures have been taken directly from previous experiences (due to their low cost and simplicity), and in other cases our laboratory has developed them. All have been elaborated after a wide SHS campaign test with PV components coming from very diverse origin

(Europe, Argentina, Bolivia, Brazil, China, Ghana, Morocco,...)

2.2 PV Modules electrical test

The IES has developed a measurement procedure for quality control of PV modules that relies on rather simple outdoor I-V measurements. Outdoor tests have been questioned due to the possible uncertainties associated with the spectral mismatch and cell temperature determination. The way of obtaining such data is far from being straightforward: pyranometers, spectrum analysers, temperature probes and wind meters are the most typical instruments used, all of which require of auxiliary equipment, regular maintenance and, frequently, protective measures against noise. We have analysed the possibilities of simplifying the process, first of all, by means of using reference modules (previously calibrated in an independent laboratory having traceable sensors) as temperature and irradiance sensors (through the measurement of their short-circuit current and open-circuit voltage) and, secondly, by using as specific reference module one with similar technology to those under test.

All procedures proposed are based on the measurement of certain (i,v) points from the PV module characteristic curve, and their extrapolation to the internationally accepted reference conditions, the Standard Test Conditions (STC): 1000 Wm^{-2} of normal irradiance, AM1.5G solar spectral content and 25°C of cell temperature (IEC 60904-1:1987).

As results of uncertainty in different parameter estimations at STC given by the proposed method, the following are worth mentioning

— Short-circuit current estimation: within 3%.
— Open-circuit voltage estimation: within 2%.
— Fill factor estimation: within 1%.

Therefore, it is possible to characterise PV modules with an uncertainty below 5% in terms of maximum power, using methods fully accessible for basic-equipped laboratories. This makes the methods suitable for Quality control processes of PV modules, within the framework of Rural Electrification programmes undertaken in Developing countries [6].

2.3 Battery test

Three different tests are performed on batteries: discharge, charge and gassing. From the first one, initial and stabilised values of the battery capacity can be obtained. It is important to know its real capacity available, especially in sizing studies. The discharge curves are also used to establish the recommended deep-discharge regulation points that the associated charge controller should include. The test procedure consists on sampling voltages and currents when discharging the battery through a resistive charge (incandescent lamp) up to 10.8 volts.

The charging test gives information about how charged is the battery at different voltages. Related with the charge controller, it estimates the overcharge regulation points that protect the battery against excessive gassing and corrosion. This procedure consists of charging the battery at a constant current given by a power source, while voltage and current values are measured.

Finally, in the gassing test the loss current as a function of voltage and temperature can be extracted with a fully charged battery and a power source. In addition, it is useful to predict the water losses that the battery will have during its operation because this parameter is related to corrosion due to the formation of O_2 by dissociation of water.

The instrumentation used to execute all these tests is simple: a power source used as generator, a calibrated shunt to measure currents, some DC lamps and a PC with a data acquisition board to do the sampling. Nevertheless, it could be enough with manual sampling of voltage and current with a voltmeter every 15 or 30 minutes.

2.4 Charge controller test

The aim of charge controller tests is to analyse the device performance in normal conditions but also under extreme or anomalous situations that could appear in solar home systems. First, operation tests determine the energy losses inside the charge controller due to current self-consumption and internal voltage drops. Moreover, the regulation set points against overcharge and deep-discharge at a wide range of temperatures can be detected and compared with the ones recommended for the associated battery. Finally, its high current resistance, which has resulted to be of great significance. All these tests are done with a power source as generator, a battery and a load formed by some 12V lamps and optionally a potentiometer.

In a second group of tests, protections against anomalous conditions like short-circuit, reversed polarity, over-current and over-voltage situations are verified. The charge regulator should work correctly after that kind of events. Like in battery tests, very simple instrumentation is needed: a power source used as generator, a battery, 12V lamps, a calibrated shunt, a potentiometer and a voltmeter to measure voltages and currents.

2.5 Inverter test

The definition and specification for inverters (DC/AC converters) were not considered in [4] because normally they are not installed in SHS. Nevertheless, stand-alone PV systems, in general, require AC applications. The measure and characterisation of inverters and their application in photovoltaic systems has been analysed using commercial products and the scarce bibliography about the matter. This work has become, on the one side, a standard specification for them and, on the other side, a definition of the test procedures to check its compliance.

2.6 Lamp test

The following sequence of tests of luminaries (composed of ballast and fluorescent tube) is carried out electrical behaviour, protections, durability, extreme conditions and luminosity. This way, correct operation and user security is checked.

The electrical behaviour and luminosity tests give information about the efficiency and the general principles of operation. They are implemented at the different possible input voltages of a SHS, and in the two main regimes of operation of a lamp: the continuous operation and the switching on.

On the other hand, the protection tests allow knowing the response of the luminary when anomalous but possible circumstances occur. Protections against different risks when polarising the luminary or when the tube is damaged are checked in order to guarantee users safety, and protections against interferences are taken into account to allow the comfortable use of radios and TV sets. The durability and extreme condition tests are related to the

correct operation of the luminary when the number of cycles is big (that is to say, to check if the life of the lamp is going to be long) and when climatic conditions are hard. In this last case, the special conditions of the region where the lamp is going to be installed are taken into account.

The last step is to measure the illuminance behaviour. Normally, this measurement requires important test facilities and our objective was, once again, to develop a simpler procedure. The test equipment is based on a box with an inner cover made of a black material with low reflectivity (in fact, the cover is made using egg transport boxes black painted). This guarantees the non-interference of external lights and the independence of the reflections caused by the same lamp under test. Dimensions of the box have been optimised after extensive research. [7].

2.7 PV system test

The optimal situation to quality control is when the full PV system is available. Then, besides testing the PV components separately, it can be tested, also, other characteristics that involves various or, even, all components, i.e. the correct adjust of switching voltages in the charge regulator for the specific model of battery.

The test starts with a visual inspection to verify protections, labelling and features of the switches and cables. Then, it is calculated the mean daily energy production with the measured parameters of the components. After that, the autonomy of the system is tested with the battery full charged and without energy generation. Assuming that the PV system is delivered with the cables and the length of them, the wiring voltage losses are measured.

2.8 About the instrumentation

As it has been shown, instrumentation used in all tests is very simple, economically reasonable and easy to use with a basic technical training, which can be considered essential in any Electrotechnical laboratory. Table I shows the list of equipment needed for the measurement procedure for inverters (this is the one needing more instrumentation of all proposed).

Table I. Instrumentation for the tests of inverters

Instrument	Function
Digital oscilloscopes	Voltage and current measurements Waveform analysis Data storage
Digital multimeters	Voltage and current measurements
Voltage probes	Voltage measurements
Current probes	Current measurements
Calibrated shunts	DC and AC current measurements
DC power supply	Input power supply Input voltage setting
Batteries	Input power supply
Resistive and inductive loads	Efficiency measurements
Real loads	

Some of the test facilities have been specifically built looking for these requirements, for example, the black box mentioned in section 2.6. Or, as another example, the lamp cycling test unit consisting of a programmable timer and a relay: the former activates the latter causing the switching on and off of the luminaries.

The most expensive measurement instrument is the oscilloscope. The price that one has to pay for utilising one of the simplest and cheapest of the market is in terms of noise and defective insulation between the channels. We have overcome these difficulties by using opto-isolated or Hall effect probes: they do the insulation function that the oscilloscope lacks.

In any case, the price of the whole instrumentation equipment is equivalent to the price of 7-9 SHSs.

3. RESULTS

Table II summarises the PV equipment tested within the frame of the project.

PV Component	Units	Country of Origin
Module	4	Spain, France
Battery	9	Germany, Bolivia, Spain, France, UK
Charge regulators	16	Germany, Argentina, Spain, France, China, Morocco
Inverter	10	Argentina, Spain, France, Switzerland
Lamp	26	Argentina, Spain, UK, Brazil, Morocco
Full PV System	4	Spain, France

Table II. PV component tested to develop the Quality Control procedures proposed

Once compared the tests with the technical prescriptions [4] important deviations have been observed. As an example, following figures show some results particularly interesting. Figure 1 shows the low initial capacity of a battery after delivered by the supplier, clearly indicating an insufficient grow of the plates in the manufacturing process that can be easily solved with a deep charge, before of the installation.

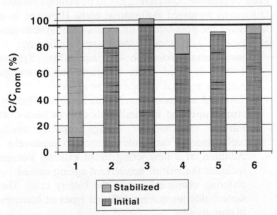

Figure 1: Tested capacity normalised with nominal capacity for six battery specimens. The solid line represents the minimum value required for the prescription.

Figure 2 represents the voltage output of three stand-alone inverters versus the input voltage variation. As it can be seen, in two of them the output voltage tolerance is surpassed.

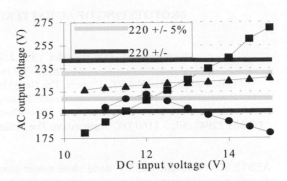

Figure 2: Voltage regulation on three inverters. The tests were made using a resistive load fixing the output power to the half of nominal power.

4. CONCLUSIONS

A critical review of PV rural electrification programmes implemented reveals that technical deficiencies are caused by the decentralised character of this electrification and not by the solar character of the application.

The Instituto de Energía Solar has developed several PV stand-alone qualification testing procedures that can be reproduced in a local context and allows certificating the fulfilment of the technical standard previously developed [4]. Also, a set of tests for inverters have been added together with the corresponding technical standard (still in draft phase).

Finally, as this action is being developed within the contex of a JOULE project, all organisations and manufacturers interested can request the certification of PV components and full SHS in our laboratory for free.

REFERENCES

[1] G.Foley, "Photovoltaic Applications in Rural Areas of the Developing World". The World Bank, 1995

[2] R.Posorski, B.Fahlenbock, "Technical Standards and Tender Specifications for SHSs and PV Supply of Rural Health Stations in Developing Countries", *Proc. 2nd World Conference and Exhibition on PV Solar Energy Conversion*, Vienna, 1998

[3] *Programme Régional Solaire*, Ed. Fondation Energies pour le Monde, 1996

[4] *Universal Technical Standard for Solar Home Systems*, Thermie B SUP 995-96, EC-DGXVII, 1998

[5] Solar Home System Testing Procedures. Download in http://www.ies-def.upm.es

[6] E.Caamaño, E.Lorenzo, "Quality control of wide collections of PV modules: lessons learned from the IES experience", Progress in Photovoltaics, March-April 1999, volume 7, No. 2,

[7] O. Perpiñán, "Desarrollo de un protocolo para la carcaterización fotométrica de lámparas fluorescentes", Master Thesis, ETSI Telecomunicación (UPM), 1999

ACKNOWLEDGEMENT: This activity is been possible thanks to the funding of Joule Programme (DG XII) in the project *Certification and Standardisation Issues for a Sustainable PV Market* (JOR-3-CT98-0275)

PROTOTYPING OF AC BATTERIES FOR STAND ALONE PV SYSTEMS

K. Burges[1], S. Bouwmeester[1], T.C.J. van der Weiden[1], J. de Rijk[2], M. Kardolus[2]
1) ECOFYS energy and environment
PO Box 8408, NL 3503 RK Utrecht, The Netherlands, tel +31 – 30 – 2808 300, fax +31 – 30 – 2808 301
e-mail: K.Burges@ECOFYS.NL
2) Mastervolt solar bv
PO box 22947, NL – 1100 DK Amsterdam, The Netherlands, tel +31 – 20 – 342 21 00, fax +31 – 20 – 342 21 69

ABSTRACT: Conventional PV stand alone power supply systems (SAPSS) mostly operate at low voltage and use directly coupled battery cells as energy storage. Disadvantages of these systems are, for example, serious restrictions in extension of the storage matching growing demand and short battery life resulting from imperfect management of individual battery cells. As a consequence, exploitation costs and reliability of electricity supply do not meet users expectations. An innovative system concept, using AC components only overcomes these problems. In the past AC PV modules already have been developed. However, AC batteries meeting the requirements are not yet available. In the framework of a European research project, development of AC batteries has been tackled. First tests with a number of experimental units have been carried out. The experience gained allows identification of the most promising topology (flexible AC / DC clustering). This knowledge and experience is a sound basis required for prototyping and testing of systems and, finally, industrial product development.
Keywords: Stand-alone PV Systems - 1: Battery Storage and Control - 2: Power Conditioning - 3

1. INTRODUCTION

Solar energy plays a major role in rural electrification at remote sites in Europe and in developing countries. During the last decade, more than a million solar home systems (SHS) has been installed all over the world, supplying electricity to households and community facilities [1].

The growing demand for stand alone electricity supply is generally recognised. To meet this demand, improvement of user acceptance, technology and cost effectiveness still is necessary.

2. STATE OF THE ART

2.1. Topology

Conventional stand alone power supply systems (SAPSS) using renewable energy consist of the following components (see **Figure 1**):
- Renewable generator
- Battery energy storage
- Charge regulator / controller
- DC loads or / and
- Inverter and AC loads.

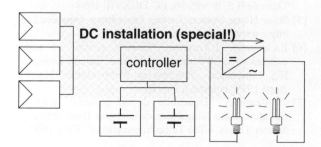

Figure 1: conventional topology stand alone PV system

For the battery energy storage mostly lead acid batteries are used.

The topology of larger systems (e.g. hybrid mini grid systems) to a certain extent differs from the simple design presented in **Figure 1**. Nevertheless, in both cases a cluster of directly coupled battery cells is used as energy storage.

2.2. Problems

The given topology faces the user to serious limitations:

1. Optimal match between demand and system capacity is exceptional.

In rural electrification programmes, regularly a growth of demand has been observed after the introduction of electricity supply. At the moment of design, system capacity is matched with the expected actual demand as good as possible. Oversizing has to be avoided because this means increased exploitation costs. However, in a later stage the possibilities for extension of system size and, especially, of storage capacity are very limited:

- Incremental extension is impossible. System voltage is fixed. Only complete battery strings with the nominal voltage can be added.
- To avoid accelerated ageing the number of battery strings operated in parallel should not exceed 4. Further extension of storage capacity needs replacement of the whole battery with larger cells.
- Addition of a new string to an existing storage includes the risk of accelerated ageing caused by differing characteristics of the battery cells. The same holds for mixing different types of batteries in one storage.

In practice, these limitations regularly result in an evident mismatch of energy demand and system capacity. Direct consequences are a significant reduction of battery life as well as poor performance and reliability of the electricity supply system.

2. Limited battery life

   The cells of the battery are coupled directly. Hence, the battery can be controlled only as a whole. Optimal management of individual cells or monoblocks is impossible. In practice, this results in a battery life which is significantly shorter than the design life specified by the battery manufacturer.

3. Low voltage DC appliances mostly are much more expensive than equivalent AC appliances. Additionally, they often are of poor quality.

4. DC system installation needs additional considerations and safety measures, especially in case of higher voltage and / or power output. Otherwise safety problems arise. The respective knowledge is not common and components are expensive.

Most of these limitations are related to the monolithic lay out of the battery energy storage. For the user these limitations result in increased exploitation costs, reduced performance and, hence, dissatisfaction.

3. THE IDEAL SOLAR HOME SYSTEM

3.1. Requirements

From the users point of view, a stand alone energy supply system should be characterised by the following features:

- Reliable, safe and easy in operation ➔ conventional AC installation and power quality;
- Comfortable ➔ use of common AC appliances;
- Flexible match of capacity and demand ➔ ease of extension of essential components (generator, energy storage) in appropriate steps achieved by modularity;
- Sustainable ➔ battery life according to specifications by optimal management and high quality components

3.2. Key features

Figure 2 shows the general concept of a system which meets these requirements. PV generator and battery storage consist of a number of independent modules. Power conversion devices are integrated in these modules. All components are connected to the conventional electrical installation using AC interfaces, making special DC connections and equipment unnecessary. Controlling the power conversion devices with a supervisory control mechanism, management of individual modules can be achieved in harmony with the required balancing of load and supply in real time.

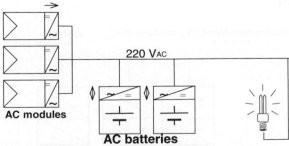

Figure 2: completely modular system using AC batteries

Summarised the key features of the storage modules are:

- completely modular;
- bi directional, parallel power conversion in island grids;
- management and control of individual units.

4. CHALLENGES

4.1. Earlier achievements

The described system concept means a drastic change of SAPSS technology. Nevertheless, a respective approach can profit from a number of efforts and developments performed in the past. Examples of earlier contributions to the elaboration of this concept are:

- Development and market introduction of AC-modules for grid connected PV systems [2];
- Design, implementation and exploitation of residential and professional hybrid SAPSS [3], [4];
- Development and market introduction of modular PV inverters for residential, grid connected PV systems (nominal capacity 2.5 kW, applied in the 1MW demonstration project in Amersfoort, NL, [5]);
- Investigation concerning the feasibility of AC battery concepts for SAPSS [6].

4.2. Results feasibility study

The results from [6] supported the assumption that the AC battery concept offers promising opportunities for stand alone power supply. Major conclusions are:

- Of course, comparing SAPSS using AC batteries with conventional DC systems, the new concept introduces extra losses related to the power conversion of the AC battery charger / inverter. However, other effects compensate these losses. Examples are:
  - The fact of finite storage capacity dominates the level of performance ratio to be achieved in stand alone systems (in both cases, when the battery is completely charged generation of electricity is reduced to the actual power demand and any surplus of energy has to be dumped);
  - Avoided mismatch losses in PV array;
  - Reduced ohmic losses due to higher voltage;

As a consequence the energy balance of AC battery systems shows little differences with conventional (DC) PV systems.

- Because of additional electronic components initial investments for an AC battery system are higher. Exploitation costs however can be competitive. Most SAPSS are faced to short battery life as a consequence of imperfect battery management. Battery replacement intervals of about 2 years are common. In those situations only a short extension of storage life achieved by optimum battery management allows compensation for the costs of the AC battery charger / inverter (see **Figure 3**).

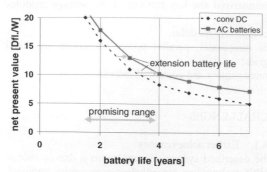

Figure 3: net present value of PV system using AC batteries compared to conventional topology (interest 6%)

In cases where the battery life in conventional systems is 5 years or longer the respective extension of battery life needed is not realistic. In these cases the AC battery concept does not promise direct reduction of exploitation costs of the power supply system itself. However, taking into account reduced costs for appliances and improved matching of system capacity and load, in many cases the concept still will be competitive.

- The market volume for AC battery systems is considerable. However, the immediate application in the low budget solar home system market is not considered as a promising strategy. Launching customers can be found in segments as recreation in industrial countries and electrification projects using hybrid mini grid systems. Nevertheless, after a period of successful demonstration and market development, other segments can be penetrated.

4.3. Progress needed

In other applications components are used which cover part of the functionality of the AC battery. Examples are uninterruptable power supply units e.g. for computers and control equipment in industry and a diversity of charger / inverter combinations for SAPSS and recreational applications. However, none of these products covers all three key features mentioned above:

- completely modular;
- bi directional, parallel power conversion in island grids;
- management and control of individual units.

There are more efforts to implement AC battery like concepts in SAPSS. Examples are the BacTERIE [7] and its successor the Sunny island unit of SMA [8], [9]. However, these approaches focus on systems and AC battery units in the kW range. With the given concepts producability and cost effectiveness for consumer markets are not likely.

In order to realise the idea of AC batteries in residential size SAPSS too, low cost solutions suitable for mass production have to be developed.

5. AC BATTERY DEVELOPMENT

5.1. Framework

Based on the requirements defined above, a consortium consisting of four European companies started the development of an AC battery unit. The parties involved and their respective major tasks are:

- ECOFYS (NL): co-ordination, concept development, energy management
- Mastervolt (NL): power conditioning
- TTA (ESP): monitoring and testing
- Sonnenschein (D): battery specification and testing

The work is carried out in the framework of the JOULE programme. The work programme covers prototype development, testing and monitoring of a number of units under real world conditions.

In the first two project phases, requirements and specifications of the AC battery have been defined, topologies have been developed and investigated and, finally, practical experiences with a number of hardware prototypes (see **Figure 4**) in bench tests have been evaluated.

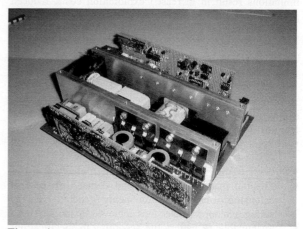

Figure 4: prototype AC battery power converter

5.2. Available topology options

To implement an AC battery concept, different topology options have to be considered. The alternatives are:

- Fixed AC / DC clustering versus
- Flexible AC / DC clustering.

Using fixed AC DC clustering each power converter is strictly detached to one battery compound (see **Figure 5**). Hence, power management (real time control of the power balance in the grid) and energy management (control of gross energy flows to individual batteries) are interacting closely. Nominal power in- and output of both parts of the power converters (DC↔DC and DC↔AC) have to be matched.

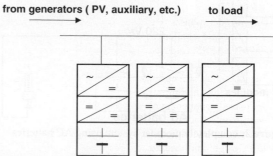

Figure 5: AC batteries, fixed AC / DC clustering

In case of flexible AC / DC clustering (see Figure 6) a local DC busbar connects the power conditioners (DC / DC converters) attached to the storage batteries. The DC busbar is connected to the bi directional inverters creating the grid interface.

from generators (PV, auxiliary, etc.) →   **to load** →

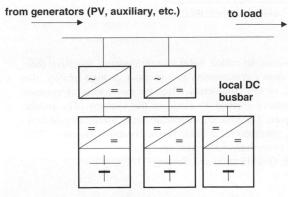

local DC busbar

Figure 6: AC batteries, flexible AC / DC clustering

In this option the number of DC / DC converters not necessarily has to be identical to the number of inverters. Coupling of control functions (grid support versus battery management) is less close than in the first topology. Nominal power in- and output of the clusters of converters has to be matched at system level.

5.3 Comparison

Flexible AC / DC clustering offers a number of important advantages compared to fixed AC / DC clustering:

- Separation of power modules opens the opportunity to match storage capacity (energy available) and power independently. Growing energy demand mostly requires extension of storage capacity but not necessarily means a need for increased nominal power. The user only has to acquire components which, consequently, are really used and, hence, this concept allows most effective investments.

- As a consequence of the separation of the control tasks (1. balancing power at the AC grid and 2. managing energy flow between the batteries), real time requirements for the control algorithm are more relaxed. Additionally, complexity of the control system is reduced which means a direct reduction of costs and effort for debugging and maintenance.

6. RESULTS

The project phases executed gave a clear idea of opportunities and challenges of the described AC battery concept for SAPSS. Important steps remaining are:

- Development of complete prototype systems;
- bench testing and optimisation;
- field testing and optimisation;
- industrial product development;
- market introduction.

Essential knowledge concerning the first steps has been gained. Design and testing of prototype units supplied practical experience concerning the most appropriate approach implementing the concept. Hence, a reliable basis

for successful continuation of this development path has been created.

7. CONCLUSIONS

A concept for modular, extendable stand alone power supply systems (SAPSS) has been developed. The key of the concept is the integration of AC batteries. This concept overcomes important shortcomings of conventional SAPSS. In that way the concept contributes to the further distribution of stand alone solutions for rural electrification. Residential PV systems as well as mini grid systems in community and professional environments will profit from the proposed concept.

First steps to implement the concept have been made. The knowledge and experience gained form a reliable basis for further elaboration of the approach.

ACKNOWLEDGEMENTS

The work reported has been supported by the European Commission (contract no. JOR3-CT98-0224) and NOVEM.

REFERENCES

[1] M.R. Vervaart, F.D.J. Nieuwenhout, "Solar Home Systems- manual design and modification of Solar Home System components", ECN, 2000.

[2] I-panelen – Feasibility study for AC modules (in Dutch), ECOFYS, Utrecht, 1992

[3] K. Burges et al.: Terschelling hybrid system research, final report Joule contract JOU2-CT92-0222, ECOFYS, Utrecht, 1995

[4] H. de Gooijer: "PV diesel hybrid energy systems on houseboats and barges in the Netherlands, Belgium and Germany, final report Thermie contract SE 41-93-NL-DE, ECOFYS, Utrecht, 1998

[5] Sunmaster 2500, Specifications and product information Mastervolt bv., Amsterdam, 1999

[6] L. Folkerts, K. Burges. T.C.J. van der Weiden: Feasibility study AC batteries (in Dutch), NOVEM contract 146.201-013.1, ECOFYS, Utrecht, 1997

[7] C. Schmitz, B. Willer: "Modular AC batteries in PV Power Supply Systems: Experimental Results of Field Tests and Laboratory Experiments", in 2nd World Conference and Exhibition on PV Solar Energy Conversion (Proc.), ISPRA, 1999

[8] A. Engler et al.: "Control of parallel working power units in expandable grids", ISET, Kassel, 1997

[9] B. Burger, G. Cramer: "Modularer Batteriewechselrichter für den Einsatz in Hybridsystemen", Kasseler Symposium Energie-Systemtechnik, ISET, Kassel, 1999

INDOOR TEST FACILITY FOR SOLAR HOME SYSTEMS

A METHOD TO VERIFY PERFORMANCE AND QUALITY OF SOLAR HOME SYSTEMS

Oliver O'Nagy, Claas Helmke

Renewable Energies Unit,EI, DG Joint Research Centre, I-21020 Ispra, Italy

Tel.: +49 170 6887 167, E-mail: oliver.onagy@svdesign.de

Tel.: +39 0332 789119, E-mail: claas.helmke@jrc.it

ABSTRACT: It is necessary to test small PV-stand-alone-systems, so called Solar Home Systems, to prove their reliability and operation performance. Even if each of the system's components is a product of high quality, the complete system often does not fulfil the user's requirements, or fail completely. Outdoor monitoring of systems operating under local environmental conditions is a time-intensive method to evaluate the systems. The results obtained in different regions of the world are difficult to compare. Unexpected meteorological conditions and non-reproducible results make it more difficult to obtain a general statement of a Solar Home System's function. This made it necessary to set up an Indoor Test Facility to verify test procedures for system approval.

Keywords: Solar Home System - 1: Stand-alone PV Systems - 2: Qualification and Testing - 3

1. INTRODUCTION

A small PV stand-alone system including PV-module, battery, charge regulator as well as several electric loads (lamps, radio, TV sets, refrigerator, etc.) is called a Solar Home System. A Solar Home System usually consists of components of diverse qualities: on one hand there are components of proven quality like the PV- modules and energy saving lamps, on the other hand there are components which are not approved or of a poor quality like cables, connecting elements, charge regulators and often the batteries. A test of complete Solar Home Systems is required to prove their reliability and operation performance. The reliability of a PV system is one of the most important information for a customer. An indoor test for Solar Home Systems provides reproducible measurements and results.

2. APPROACH

Six Solar Home Systems (SHS) are installed in a climatic chamber with a built-in solar simulator. They are connected to a data acquisition and to a programmable loads-control set-up. With the six-step steady state sun simulator any irradiance profile can be set. The climatic chamber operates in a wide temperature range, which meets the conditions of almost every region in the world. The programmable loads control makes it possible to apply any load profile on the Solar Home Systems by controlled switching of the appliances delivered with each system kit. This method allows the systems to operate under defined conditions. There is also a possibility to trace IV-curves of each of the systems' PV-modules/arrays besides the normal monitoring of operation.

3. THE SET UP

The set up of the Indoor Test Facility consists basically of three parts:

1. A controlled Environment
2. The Instrumentation
3. The Software

Figure 1: Six different SHS installed in the test chamber

3.1 Controlled Environment

A controlled environment is realised with a Solar Simulator integrated in a climatic chamber. The Solar Simulator is a six-step steady-state simulator which can apply stable irradiances of 250 [W/m²], 470 [W/m²], 640 [W/m²], 850 [W/m²], 1050 [W/m²] and 1200 [W/m²].

A variation of the distance between the lamps and the illuminated area can reduce the irradiance slightly if desired. The lamps are switched to a way that an area of about 12 m² is exposed to a homogeneous irradiance at every step. A window in front of the lamps separates the Solar Simulator from the test chamber thermically and contributes to an equal distribution of the irradiance (Figure 2). The irradiance can be changed remotely. The control equipment includes an interface for PC connection.

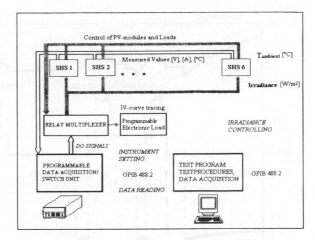

Figure 3 Summary of components and their functions

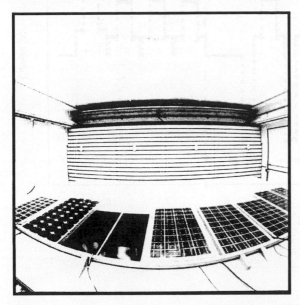

Figure 2: PV-modules and Solar Simulator in the climatic chamber

The size of the test area in the climatic chamber is about 24 m². The ambient temperature can be set in a range between − 40 °C and + 50 °C.

3.2 The Instrumentation

Two programmable data aquisition/switch units are used to scan simultaneously seven measurement points per Solar Home System. An electronic relay multiplexer is connected to a digital output line of a switch unit. The PV modules and the Loads (Lamps, Radio, TV set, etc.) are connected through this relay multiplexer to their charge regulators. Regarding the loads, the multiplexer has the functionality of a programmable timer switch. Six terminals of six separate controlled relays allow many ways to control or regulate the loads. The terminals which the PV-modules are connected to are used to separate the modules from their system and connect them to an electronic load. It can be set to trace IV-curves of the modules. The electronic load is a bipolar operating power supply which is used as a power sink in this case.

The Software

The testprogram is a LabVIEW® application with customised instrument drivers for instrument setting and data reading as well as new created subprograms to sort data, create new files and to control the relay multiplexer and the solar simulator. The whole program is built up modularly which makes it easy to implement modifications or new test procedures.

4. MEASUREMENTS

The following parameters of a Solar Home System are measured:

PV-module:	Voltage
	Current
	Temperature
Battery:	Voltage
	Current
Charge regulator:	Voltage
Loads:	Current

The first procedure ever was testing the systems' autonomy and the service of operation near the Low Voltage Disconnection mode (LVD) of the charge regulator. At the beginning of the test all batteries were charged with a C100 current. The ambient temperature of the climatic chamber was set at 25°C, all loads of the systems were connected.(figure 4, 1. Discharge).

After the last system LVD mode was reached, the first irradiation cycle started. During irradiation all loads were kept disconnected through the relay multiplexer and reconnected afterwards. It can be seen that only one of the systems – shown in figure 4 – exceeded the LVD mode again after every irradiation cycle. Charging and discharging cycles had to be repeated four times until the last system was out of the LVD mode again. (figure 4, 2. Cycling)

Recovery test: after discharging all systems, irradiance cycles were started again without connecting the loads. It is shown how many cycles are necessary to reach High Voltage Disconnection (HVD) mode again (figure 4, 3. Charge).

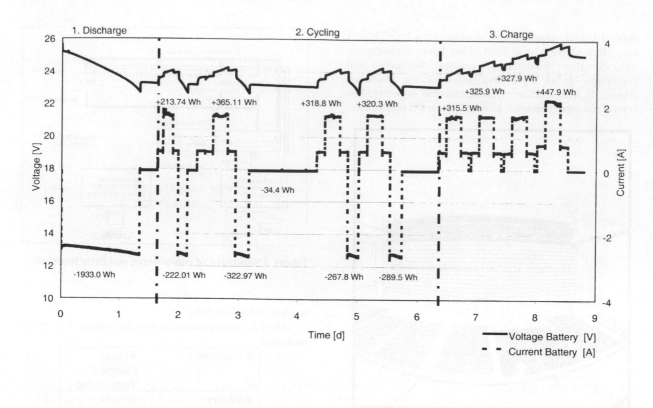

Figure 4: Battery Voltage and charge/discharge current of an SHS, as measured during the test procedure

In between two scans of the common measurements, it is possible to measure IV-curves of all the PV-modules, too. One module after the other is disconnected from its system, the Open circuit Voltage is measured, the module is connected to the electronic load and the program operates a calculated logaritmic voltage ramp from the Voc to the Isc point of the module.

In addition to an analysis of internal system losses (between module and battery and between battery and loads) it is shown, how close the PV modules are working to the the Maximum Power Point .Mpp. (figure 5).

5. CONCLUSION

The presented results of the first indoor test show, that any given test procedure can be followed under controlled ambient conditions.The results provide input to ongoing standardisation activities. New approaches can be analysed and verified.

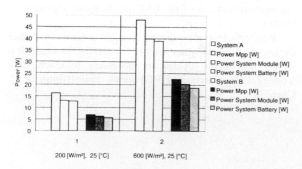

Figure 5: Power of module and battery charging power compared to the Mpp of the module. Two different sized but similar balanced systems are compared

RELIABILITY TESTING OF GRID CONNECTED PV INVERTERS

P.M. Rooij, J.A. Eikelboom and P.J.M. Heskes

Netherlands Energy Research Foundation ECN

corr. address : PO Box 1 1755 ZG PETTEN - The Netherlands

e-mail : rooij@ecn.nl - fax : +31 224 56 3214

ABSTRACT: Large numbers of grid-connected PV systems are being installed throughout the world. In a PV system where PV modules have expected lifetimes of up to 20 years it is desirable that the other system components have a comparable expected lifetime, or at least a predictable lifetime. The PV inverter is a key component with regard to reliability and lifetime. Procedures to test PV inverters for reliability are discussed in the paper and results from tests in the laboratory and the field are presented.

Keywords: Grid-connected – 1 : Reliability – 2 : Inverters – 3

1 INTRODUCTION

Concerning PV modules a vast body of data and experience has been built up in the past years and is available to the PV community. This has resulted in an internationally accepted standardisation of test procedures for PV modules. In these tests, e.g. IEC1215, the electrical, thermal and optical properties of modules are determined and experience has shown that these tests can give an indication of the expected lifetime.

With the growth of the annually installed volume of grid-connected PV systems there is a concern for the reliability of the Balance-of-Systems components, especially the PV inverter as a key component. As the financial investment is relatively large, the consumer understandably prefers to purchase a PV system where the components have comparable expected lifetimes. At this moment no international norms for design qualification or type approval of inverters exist. At national levels various requirements on various aspects have to be met by inverters. For the Dutch normalisation institute the Netherlands Energy Research Foundation ECN has developed a guideline for the design qualification and type approval for PV inverters, [1].

The types of PV inverters and the circumstances for which they are designed to operate vary widely. For small PV systems (< 10 kWp) until recently inverters were mainly housed indoors under mild temperature and humidity conditions. The development of AC modules, where the small inverter is coupled directly to a single PV module and mounted on the rear side, introduces new operating conditions. At the back of a roof-integrated PV system the temperatures range from well below zero Celsius up to 80 °C, with the entire range of humidity conditions. Under these conditions the electronics must be ready to operate for a period comparable to the expected lifetime of the PV module.

String inverters and the larger PV array inverters are available in special versions developed for housing in outdoor conditions.

The Netherlands Energy Research Foundation ECN is involved in various projects on the subject of reliability of inverters and on the monitoring of PV systems. In the first project two types of AC inverters have been subjected to outdoor tests and tests in climate chambers. In the second project extensive testing was performed on small inverters (i.e. power less than 300 W) in climate chambers and outdoors. A third relevant project is a noise barrier with 2160 AC modules in the Netherlands, where ECN is in charge of the monitoring of the PV system. The results are discussed in this paper.

2 Life-time testing of AC modules

2.1 Outdoor tests

As part of a research project two AC module inverters have been extensively tested. In order to determine the temperature variations under real conditions six AC modules have been placed in the outdoor test facility on a south oriented tilted roof. Three types of constructions were used: two AC modules mounted above the tiles, two AC modules with an isolated box at the rear side of the modules and two AC modules having a box with limited ventilation through the box. Temperatures of the ambient air and of the housing of the inverters were recorded for several months in the summer, see figure 1.

The housing temperature of the inverter can be described by the linear relation

$$\Delta T = k \times G_i + T_0$$

with ΔT the temperature rise of the inverter temperature in relation the temperature of the ambient, G_i is the incident irradiance and T_0 is a temperature offset. k is the temperature coefficient of the inverter, it is determined by the thermal properties of the inverter, those of the PV module and the prevailing meteorological conditions.

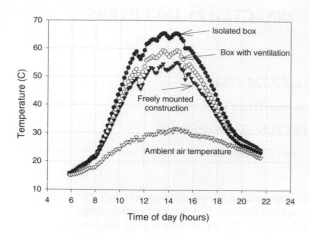

Figure 1 - Temperature distribution of ambient air and inverter housings in the three types of roof construction, as a function of time during a summer's day

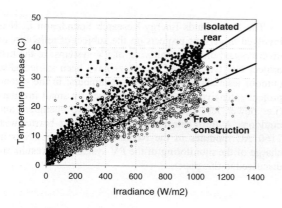

Figure 2 - Temperature increase as a function of the irradiance during a summer's month.

The k values for an inverter attached directly to the rear side of the PV module varied from 0.024 ($^{\circ}$C/(W/m^2)), for freely mounted constructions, to 0.035 ($^{\circ}$C/(W/m^2)) for constructions with an isolated rear side, see figure 2.

Limited ventilation shows k values close to those of the freely mounted construction 0.025 ($^{\circ}$C/(W/m^2)). These k values for inverter housing temperatures were little above corresponding k values for the PV modules. It should be noted however, that k values show an average thermal behaviour. Under conditions of absence of wind, for example, the inverter/module temperatures can be 15 $^{\circ}$C above the values predicted by the linear relation shown above, as shown in figure 2. The k values are used to compare ageing in accelerated lifetime tests to expected ageing under outdoor conditions.

Figure 3 shows the temperature distribution for a typical Dutch year for the ambient air and the inverter (for a specific k value of 0.035 ($^{\circ}$C/(W/m^2)); the number of hours for a bin of 3 $^{\circ}$C wide are shown. The inverter temperature is modelled to be higher than the ambient air temperature in proportion to the specific k value. As high temperatures correlate often with high irradiance, the inverter temperature distribution has a tail toward higher temperatures up to

70 $^{\circ}$C.

At elevated temperatures the ageing takes place at a faster rate. For electronic components the influence of the temperature can be described by the Arrhenius rule, for the activation energy the value of 0.6 eV is taken. If we chose the temperature of 21 $^{\circ}$C as a reference point, figure 3 also shows the effective ageing for the inverter. For this case the total number of hours of effective ageing is equal to 365x24=8760. Therefore we define 21 $^{\circ}$C as the annual mean effective ageing temperature for continuous operation of the inverter.

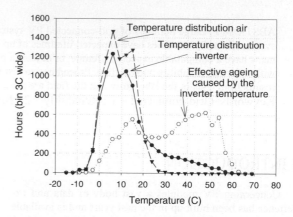

Figure 3 - Temperature distribution of the hour-average annual temperatures, for air and the inverters (for K=0.035) and ageing described by an activation energy of 0.6 eV

Temperatures of specific components in the interior of the inverters can be substantially different from the housing temperatures. Inverters, which are not directly attached to the rear surface of the module, will have lower k values than those discussed here. So, the results should be interpreted with care. The general conclusion is the inverter housing temperature can go up to 80 $^{\circ}$C on sunny days, with even higher temperatures in the interior. Ventilation is very effective in lowering the average temperatures.

2.2 Tests in the climate chamber

In the high-temperature-test eight inverters of both types are connected to an IV simulator and delivered energy to the grid at a DC power between 70 and 90 W inside a climate chamber providing an ambient temperature of 70 $^{\circ}$C. The power levels and the ambient temperature were chosen such that the temperatures in the interior of the inverter reached the allowed maximum temperature. Before and after these tests extensive electrical characterisation of the inverters took place. One of the most conspicuous components of the inverter is the relatively large electrolytic capacitor at the DC side of the inverter. For a single-phase inverter this capacitor is needed for energy storage and retrieval during the 50 Hz cycle. A large capacitance is beneficial for an optimal maximum power point tracking mechanism. Ageing of electrolytic capacitors can easily be quantified by measuring the Effective Series Resistance ESR of the capacitor. During the 2000 hours of testing increases of the ESR values up to 50% have been observed. The effect of this ageing on the electrical properties, such as inverter efficiency, is negligible. Continuation of these tests showed further increase of the ESR without any detectable deterioration of the inverter properties.

Ageing during 2000 hours of electrolytic capacitors under these conditions in a climate chamber can be compared to ageing that would occur under realistic outdoor conditions during 15 years. The acceleration is calculated on basis of the above defined effective ageing temperature of 21 °C. Although the capacitor is by no means the only inverter component to undergo significant ageing, it is generally thought to be a critical part with to the lifetime of the inverter. The fact that none of the sixteen inverters became defect shows that there is basis for confidence in the concept of an AC module inverter.

Besides through the choice of optimal electrical components the lifetime expectations of a PV inverter is determined by the reliability considerations concerning the design. Hot spots inside the inverter should be avoided by good thermal design, critical components such as the electrolytic capacitor should be positioned at a cool spot. Tests showed that the inverter should be thermally isolated as much as possible from the module surface. As the irradiance will heat the module surface to temperatures 40°C above the ambient temperature, it is preferable to provide a heat sink at a lower temperature, e.g. the module frame or the ambient air at the rear side.

For ageing of inverters, such as AC module inverters, tests developed for PV modules during which the ambient temperature and possibly the humidity is cycled from temperatures below freezing up to 85 °C have limited relevance. For PV modules these variations having ramp rates up to 200 °C per hour may test significant properties, for PV inverter it is hardly conceivable that rates like these will occur in the inverter ambient under outdoor conditions and inverters may not be designed for these circumstances. Although the inverter must be capable to operate in the entire module temperature range the rate of change in temperatures must be adapted to realistic levels, with maximum rates of e.g. 50 °C per hour. As the reliability is meant here to be measured by the period during which the inverter is operating according to its specifications, these cycling tests have more to do with the performance of the inverter than with the expected lifetime.

In a second research project on inverter reliability two further types of inverters for AC modules were tested. Several tests were carried out.

Environmental stress tests of both inverters were carried out in a climate chamber, where the inverters were submitted to a high-temperature-long-exposure test, a high-temperature-high-humidity-long-exposure test, a temperature-cycling test and a humidity-freezing test. Before during and after the tests the inverter efficiency and the electrolytic capacitors were monitored.

The first inverter passed all tests. After the two high-temperature tests two inverters failed out of eleven inverters. One inverter failed after 3720 hours and one after 6540 hours. Seven of the nine inverters, who were still operating, were tested for 6540 hours and the remaining two inverters were tested for 3910 hours. The electrolytic capacitors show an increase of ESR by a factor 4½, a decrease in capacitance by 5% while the efficiency remained stable over the 6540 hours of test time. The second inverter

appeared not to be suitable for accelerated lifetime tests at increased operating temperatures. As long as the inverters were operating within specs no problems were encountered. Tests show that several protection mechanisms inhibit the inverter to operate beyond temperature limits given in the inverter's specification. Therefore no test results are available. Both inverters passed the temperature cycling test and the humidity freezing test with good results. No failures occurred during these tests.

3 FIELD EXPERIENCES

A large photovoltaic system consisting of almost 2200 AC modules has been integrated into a one-mile long noise barrier along a highway in The Netherlands [2],[3]. One of the aims of introducing AC modules to this project was to obtain statistically relevant numbers of inverters for making estimates of the reliability of the concept of AC modules. AC module inverters by two different manufacturers were installed, with a respective share of 80/20% for the total number of AC inverters.

The system has been in operation about eighteen months (April 2000). The central monitoring PC daily contacts one out of every six AC module inverters, a total of 360 inverters. The energy yield as indicated by their internal kWh-counter is stored in the PC database. During the 18-months monitoring period the PC detected 11 inverters of one of the manufacturers which indicated no further increase of their kWh counters. This can indicate defect inverters, but it cannot be ruled out that defects occur in the monitoring prints of the inverter. The possibility of eleven defect inverters out of 360 in 18 months can therefore be considered a worst case situation. Assuming a similar and constant failure rate during the lifetime of an inverter would result in an expected mean-time-to-failure of around 20 years. For the inverter of the second manufacturer the kWh counters indicate that none of the 60 inverters failed during this period.

4 CONCLUSIONS

Inverters, especially inverters at the back of PV modules and those mounted outdoors, have to operate under harsh conditions for long periods of time. Field experiences and accelerated lifetime testing have shown that the electronics can be designed and manufactured in a way that reliable operation can be guaranteed for a period comparable to the expected lifetime of the PV module. Manufacturers are clearly learning from experiences and the reliability of modern inverters is steadily improving, [4].

Besides failing through ageing, inverters can fail for a great variety of reasons, e.g. problems linked to EMI, faulty installation, or 'infant mortality', which tends to disappear with ongoing improvements of the design and manufacturing processes. These causes of failure should be identified as soon as possible in order to be able to show the customers that PV inverters can be trusted to convert the solar energy to grid-power for as long as the PV module lasts.

5 ACKNOWLEDGEMENTS

Part of this work has been funded by the Netherlands Agency for Energy and the Environment Novem under contract 146.200-109.2.

6 REFERENCES

[1] A. Kanakis, N. van der Borg, Design Qualification and Type Approval of Inverters for Grid-Connected Operation of PhotoVoltaic Power Generators, ECN - C - 99-085

[2] N.J.C.M. van der Borg, E. Rössler, and E.B.M. Visser, A traffic noise barrier equipped with 2160 AC-modules, paper presented at this conference.

[3] C.W.A. Baltus, F.J. Kuiper, H.E. Oostrum, B.C. Middelman, E. Roessler, T. Nordmann, A PV data acquisition program for supervision monitoring of a large number of AC-modules, 2-nd World Conference on PV Solar Energy Conversion, Vienna, 1998, 2360-2363

[4] H. Haeberlin, Ch. Liebi, Ch. Beutler, Inverters for grid-connected PV systems: test results of some new inverters and latest reliability data of the most popular inverters in Switzerland, 14-th EPVSEC 1997, Barcelona, 2184-2187

EFFECT OF THE OPTICAL PERFORMANCE ON THE OUTPUT POWER OF THE EUCLIDES^{TM} ARRAY

I. Antón, G. Sala and J.C. Arboiro
Instituto de Energía Solar - UPM
E.T.S.I. Telecomunicación. Ciudad Universitaria.
28040 Madrid – Spain
Phone: +34 91 5441060 Fax: +34 91 5446341
e-mail: nacho@ies-def.upm.es

J.Monedero, P.Valera
ITER, Pol. Ind Granadilla
San Isidro
Tenerife - Spain
Phone: +34 922 391000 Fax: +34 922 391001
e-mail: iter@iter.rcanaria.es

ABSTRACT: Factors related to the mechanical and optical components can reduce the power output of a PV concentrator system. The EUCLIDES^{TM}-THERMIE plant of Tenerife has allowed the evaluation of all the aspects affecting the power output in a big concentrator system based on parabolic troughs. Causes of optical losses will be evaluated, providing new concepts and methods. Due to optical mismatch, the I-V curves show a significant false shunt resistance. Analysing the I-V curve, a mathematical approach based on statistical concepts will be proposed to characterise the illumination currents over the receivers. In order to reduce power losses, secondary concentrators have been installed with a power increase of 11%. The optical aperture and tracking requirements have been studied in the light of these results, concluding that the aperture angle is lower in power than in short circuit current. The accuracy of the EUCLIDES tracking system can be set up to ±0.05°, high enough to ensure a correct tracking.
Keywords: Optical Losses - 1: Concentrator - 2: Characterisation - 3

1.- INTRODUCTION

Photovoltaic concentrator systems have become one of the most important attempts to reduce the kW·h cost in the PV industry. Although the cost reduction seems to be possible, experiences such as the EUCLIDES^{TM}-THERMIE Plant of Tenerife [1] try to validate the technology and to demonstrate the real cost of energy in large generation PV plants. One important factor affecting the output of a PV concentrator system and which is not found in conventional flat systems are the losses associated with the opto-mechanical components. In these systems, there are many factors affecting the power output: reflectivity of the mirror, mirror dust, mirror orientation, mismatch due to several causes and receiver transmitivity. All them are less well known than equivalent topics in flat arrays [2-4] and usually the data comes from medium sized prototypes. Now we have the opportunity to study such factors on a large scale where components and the plant installation have been carried out by the final user.

A complete analysis of all factors of optical nature affecting the power has been carried out, providing new equipment and methods to evaluate each one. The dispersion of all of them causes a large optical mismatch. This results in a non-uniform illumination along the array, causing a significant tilt (false shunt resistance) on the I-V curve and thus, a decrease of current at the maximum power point. A mathematical approach based on statistical concepts will be proposed to characterise the non-uniform light.

2. FACTORS AFFECTING THE OUTPUT POWER

Before their installation, modules have been classified [5] to reduce electrical mismatch, so electrical losses caused by the different photo-electrical response are within 1%. The 1932 modules of the plant were sorted into five classes in accordance with their one sun short circuit current.

The optical performance of the EUCLIDES^{TM} array is affected by the following parameters.

2.1 Reflectivity of the flat surface [ρ_f].

The ratio of output to the input power light of the reflecting material. Three types of mirrors have been installed with different reflective and weather-resistant cover films:

1) Polished aluminium film covered with a transparent varnish coating (ρ_f = 78 %)

2) Silvered plastic film with a transparent weather-resistant acrylic film. (ρ_f = 89 %)

3) Silvered weather-resistant plastic film (ρ_f = 91 %)

The reflectivity of each reflective material with the corresponding weather-resistant material has been measured in the laboratory with a silicon solar cell as receiver and using AM1 as light source.

2.2 Efficiency of the shaped mirrors [ρ_m].

Defined as the ratio of light collected by the receiver to the input light of the shaped mirrors. It includes not only the reflectivity of the materials but the possible manufacturing defects of the mirror [6]. This parameter should be measured avoiding the edge effect due to both the collector and the receiver (a single calibrated cell receiving the light from the mirror surface). Values measured for the three types of mirrors are ρ_{m1} = 74 %, ρ_{m2} = 85 % and ρ_{m3} = 87 %.

2.3 Collector and receiver cleanliness [ρ_c] and degradation [ρ_d].

The amount of dust and dirt on the collectors and receivers depends on the site. Due to the salt and dust carried by wind, Tenerife has proved to be a hard environment in this sense. A calibrated receiving cell located at the southern mirror of one array of each type of collectors is used as both cleanliness sensor and film reflectivity degradation sensor. The reference value is the short circuit current measured at the moment that the completely cleaned collectors were installed. Comparing the output of the sensor before and after cleaning the mirrors, cleanness and degradation factors along time are obtained. The gradual changes of the cleanliness factor over days suggests a cleaning frequency of one wash every two weeks. The degradation factor measured over a long time period will give information about the life expectancy time for each class of mirror technology.

2.4 Edge effects and flaws in the mirrors

Depending on the position of the sun, the separation between mirrors (4 mm on average) may shadow only one or two cells of a 10 cells receiver and thus limit the illumination current of the module.

Other expected effect of the mirror is less angular aperture of the more curved end. Due to this region, certain amount of light, about 3 %, may be outside the active area of the receiver. Although this effect has been included in the optical efficiency of the mirrors, we would like to evaluated for a specific mirror if required.

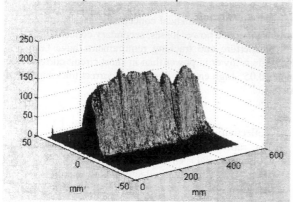

Fig 1: 3D plot of the incident light on the receiver, captured with a CCD camera.

A method based on a CCD (fig 1) camera to evaluate these effects has been developed and tested. The spot is projected onto a matt white surface and captured with a CCD. The overall irradiance on each cell region is integrated. The result is that the light outside the active area is less than 3%, but some located differences up to 8% has been found, caused by edge effects or flaws in the mirror.

2.5 Effective optical misalignment

The misalignment along the system is a function of the deformation and the array position [7]. The torsion due to wind and weight loads causes slight variations of them with the position of the array.

The alignment of the mirrors can be adjusted by tightening a pair of screws by operators and is carried out with the array tracking the sun. With the help of a simple optical device the spot is placed inside the active area of the receivers. The positioning of the 140 mirrors cannot be carried out instantaneously and the torsion of the structure during this operation may cause errors.

The power losses due to the misalignments have been evaluated by installing secondary concentrators and comparing the I-V curves (fig 2). The increase of power with secondary concentrators is of 11% and is caused, firstly, by an overcollection of light (3%) and secondly, by an increase of the optical aperture, which reduces the misalignment effect (8%).

2.6 Power loss analysis.

In figure 2 we have potted the I-V curve of an array with mirrors made of silvered plastic film with transparent weather-resistant film. An ideal I-V curve under the

detailed operating conditions is compared to I-V curves with and without secondary concentrators. In addition, an I-V curve with secondaries but discounting the I_{sc} increase has been plotted to evaluate the misalignment losses. The loss breakdown is shown in Table I.

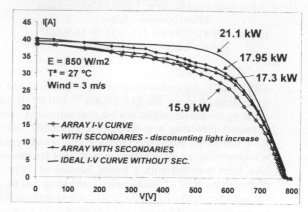

Fig 2: Experimental I-V with and without secondary concentrators. Power output analysis.

Table I: Summary of power losses measured in one EUCLIDES™ array

OVERALL POWER LOSSES	21%	Measured
Electrical mismatch	1 %	Measured
Misalignments	8 %	Measured
Cleanliness Mirror discontinuity Defects	13 %	Deduced

3. INFLUENCE OF THE OPTICAL MISMATCH ON THE PHOTOGENERATED CURRENT AND THE MAXIMUM POWER POINT.

First, a particular case, i.e., the EUCLIDES™ array will be studied in order to present the subject. One of the EUCLIDES™ arrays of Tenerife consists of 138 modules and 140 mirrors. Each module has 10 BP Solar LGBG cells in series connection and a by-pass diode. If all modules are connected in series, the output voltage at standard operating conditions (SOC: 800 W/m2, 20 °C ambient temperature and 2m/s wind speed) is 750 Volts at the maximum power point.

All the factors studied that can influence the efficiency of an EUCLIDES™ array are distributed along the 140 mirrors and 138 receivers, i.e., 252 square meters of mirrors or 168 linear meters of receivers. The number of variables affecting the power output makes it impossible to monitor all them in order to analyse their influence on the final I-V curve.

Thus, let us analyse the I-V curve to understand, at least in a statistical way, how many modules and receivers are properly illuminated.

3.1 Short circuit current analysis.

The first recorded IV curves of an EUCLIDES™ array showed a short circuit current value lower than the I_{sc} of the best illuminated module as expected in small PV generators. The mismatch in concentrators caused by non-uniform light is larger than in flat panels and the voltage drop across 137 by-pass diodes cannot be biased by only one module (about 6 volts).

Let us consider an array with N_t ($N_t=138$ in the EUCLIDES array) by-passed modules in series. Along the I-V curve, the fraction of modules with illumination current I_L over the array (I_{sc}) will be forward biased (fig 3). Reciprocally, modules with illumination current under the array current will have the by-pass diode forward biased.

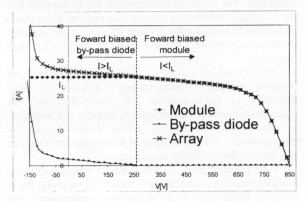

Fig 3: PSPICE simulation showing a module and by-pass diode bias in relation to an array I-V curve.

The number of modules forward biased is a function of the bias point, that will be called N(i). With the array in short circuit, the diodes of the worst illuminated modules are the load of the array. The equation of equilibrium at the short circuit point is approximately:

$$N(I_{sc}) \cdot V_{mp} = (N_t - N(I_{sc})) \cdot V_\gamma \qquad (1)$$

where V_{mp} is the maximum power voltage of a module, and V_γ is the voltage contribution of the diodes. If we consider that all modules are forward biased at the maximum power point of the array, this can be calculated as $V_{mp} = \dfrac{V_{mparray}}{N_t}$; V_γ can be obtained from the data sheets of the diode. In a EUCLIDES™ array, the number of reverse biased modules at the short circuit point, and thus, with higher I_L than I_{sc}, is 19 or 20.

3.2 Pseudo parallel resistance effect in the I-V curve.

All the factors responsible of power losses have a large dispersion along the array. This results in a large illumination mismatch and the I-V curves show a pseudo parallel resistance.

Considering the high number of modules, we shall assume that the illumination currents (I_L) of the modules follow a Gaussian distribution, centered on an average value of $<I_L>$ and with a standard deviation of σ_I. One method to calculate this parameters from the I-V curve will be proposed.

Let us consider the average value $<I_L>$ of the current distributions as the value in the middle of the flat part of the curve, taking into account the $(N_t - N(I_{sc}))$ reverse biased modules at the short circuit point:

$$V(<I_L>) = \frac{1}{2} \cdot \left[\frac{V_{mparray}}{N_t} \cdot (N_t - N(I_{sc})) + V_{mparray} \right] \qquad (2)$$

As the number of modules with an illumination current over the I_{sc} of the array is known (ec. 1), we can calculate the standard deviation of the illumination currents from the error function:

$$1 - \frac{N(I_{sc})}{N_t} = \frac{1}{\sqrt{2\pi} \cdot \sigma} \int_{-\infty}^{I_{sc}} \exp\left(-\frac{(i - <I_L>)^2}{2\sigma^2}\right) di$$
$$\Rightarrow \quad erf\left(\frac{I_{sc} - <I_L>}{\sqrt{2}\sigma}\right) = 1 - 2\frac{N(I_{sc})}{N_t} \qquad (3)$$

Equations (2) and (3) provide the values of the Gaussian pdf moments $<I_L>$ and σ. A Pspice simulated I-V curve has been obtained by using a population of 138 random illumination currents belonging to this Gaussian pdf. In figure 4, the measured I-V curve and simulated curve are shown to verify the approach. In addition, the number of reverse biased modules (number of active by-pass diodes) along the I-V curve has been drawn, calculated from the same simulation. Another PSPICE simulation with uniform illumination and the same I_{sc} has been also represented, with a 14.5 % power increase.

A minimum value of the best illumination current (I_{max})

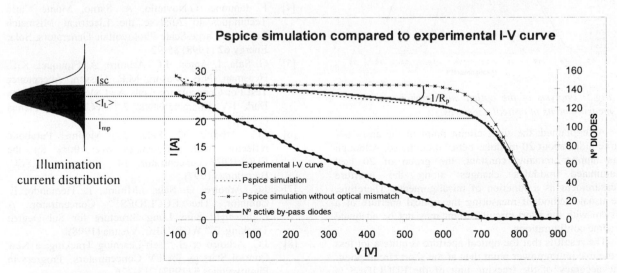

Fig.4: Experimental I-V curve compared to a PSPICE simulation with a population of illumination currents following a Gaussian pdf ($<I_L>=24.85$ A and $\sigma_I=2.09$, calculated with (2) and (3)); representation of the number of by-pass diode active along the I-V curve.

can be obtained taking into account the properties of the error function by:

$$\operatorname{erf}\left(\frac{I_{max} - <I_L>}{\sqrt{2}\sigma}\right) = 1 - \frac{2}{N_t} \qquad (4)$$

Therefore, assuming a Gaussian distribution of light there is at least one module with illumination current higher than I_{max}.

5. EXPERIMENTAL VERIFICATION OF THE MISMATCH CAUSES

Among the different factors affecting mismatch, only misalignments between modules and collectors along the 84 m of the structure justify this distribution and the values of power losses.

As the semiacceptance angle of the reflectors is 0.5°, misalignments greater than 0.5 ° causes important light losses over all the receivers, and thus optical mismatch. In addition, the accuracy of the tracking must be within the semiacceptance angle to ensure that all modules are illuminated normally.

Figure 5 shows the short circuit current of an array as a function of misalignments, compared with the same curve for the most shifted modules to the East and West. A maximum short circuit current tracking leads to point A, with large illumination current dispersion. At this point, although the I_{sc} of the array is maximum, the I_{sc} of the most eastely shifted module is only 50 %. On the other hand, point B has a slight lower short circuit current (95 % of the maximum) but also less current dispersion (worst illuminated module has 84 % of the maximum) and therefore, point B has the maximum power point.

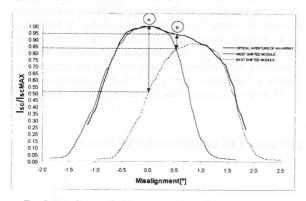

Fig.5: analysis of the optical mismatch based on measurements of optical apertures.

As explained, the short circuit point of the array I-V curve hides about 20 modules better illuminated. Although this number remains constant, the group of 20 best illuminated modules changes along the aperture measurement as a function of misalignments. Therefore, the usual method of measuring the optical aperture of a system with the short circuit current may not be adequate in some configurations.

The result is that the optical aperture is quite a lot less at the maximum power point than at the short circuit point. The accuracy of the tracking unit of the EUCLIDES is presently set to ±0.2 degree but can be increased easily to ± 0.05°, accurate enough to ensure a correct maximum power point tracking [8].

6. CONCLUSIONS

The factors affecting the power output of a concentrator system based on parabolic troughs has been identified. Among them, the illumination mismatch due to the opto-mechanical components has proved to be the most significant. The proposed mathematical model suppose that the statistical distribution of the illumination is gaussian and allows to calculate the average value $<I_L>$ and standard deviation σ_I from an I-V curve with the modules in series connection.

An increase of power of 11% can be achieved by installing secondary concentrators. Besides an over-collection of light causing 3% power increase, the secondaries provides a larger optical aperture and therefore, a reduction up to 8% of power losses caused by misalignments. Further designs of heat-sinks and modules will include the secondary concentrator.

The optical aperture of the system measured with the modules in series connection as receiver varies along the I-V curve, being quite lower at the maximum power point than in short circuit conditions. Therefore, the aperture, usually obtained from the short circuit current in the past, must be measured with the array polarised near to the maximum power point, at least with configurations with high number of modules in series.

REFERENCES

[1]   G. Sala, J.C. Arboiro, A. Luque, I.Antón, M. Gasson, N. Mason, K. Heasman, 480 kWpeak Concentrator Power Plant Using the EUCLIDES™ Parabolic Trough Technology, Proceedings. 2nd WCPVSEC, Vienna (1998).

[2]   D. Roche, H.Outhred, R.J. Kaye, Analysis and Control of Mismatch Power Loss in Photovoltaic Arrays. Progress in Photovoltaics: Research and Applications 3 (1995) 115-127

[3]   A. Luque, E.Lorenzo, J.M. Ruiz, Connection Losses in Photovoltic Arrays, Solar Energy 25 (1980) 171-178

[4]   F. Iannone, G.Noviello, A. Sarno, Monte Carlo Techniques to Analyse the Electrical Mismatch Losses in Large-Scale Photovoltaic Generators, Solar Energy 62 (1998) 85-92

[5]   G. Sala, I. Antón, J.C. Arboiro, A. Flouquet, K.C. Heasman, N.B. Mason, M.P. Gasson, Acceptance Test For Concentrator Cells And Modules Based On Dark I-V Characteristics, 2nd WCPVSEC Vienna (1998).

[6]   J.C. Arboiro, G. Sala, J.I. Molina, Parabolic reflectors with efficiencies over 90% for the EUCLIDES concentrator. 14th European PVSEC. Barcelona (1997)

[7]   J.C. Arboiro, G. Sala, I.Molina, L. Hernando, E. Camblor, The EUCLIDES™ Concentrator: A Lightweight 84m Long Structure for Sub-Degree Tracking. 2nd WCPVSEC, Vienna (1998).

[8]   J.C. Arboiro et al. "Self-Learning Tracking: a New Control Startegy for PV Concentrators" Progress in Photovoltaics 5 (1997) 213-226

The project has been subsided in the framework of the JOULE-THERMIE IV Programme.

EUROPEAN PHOTOVOLTAIC V-TROUGH CONCENTRATOR SYSTEM
WITH GRAVITATIONAL TRACKING (ARCHIMEDES)

F. H. Klotz, H.-D. Mohring, C. Gruel
Zentrum fuer Sonnenenergie- und Wasserstoff-
Forschung Baden-Wuerttemberg (ZSW)
Hessbruehlstrasse 21C, 70565 Stuttgart, Germany
Tel.: +49 711 7870 222, Fax.: +49 711 7870 230
e-mail: Fritz.Klotz@zsw-bw.de

M. Alonso Abella
CIEMAT - PVLabIER
Laboratorio del Instituto de Energias Renovables
Avda. Complutense, 22 - 28040 Madrid, Spain
Tel +34 91 3466 492, Fax +34 91 3466 037
e-mail: abella@ciemat.es

M. Gasson, J. Sherborne, T. Bruton
BP Solarex
Unit 12 Brooklands Close, Windmill Road
Sunbury-on-Thames, Middlesex TW 16 7DX
Tel.: +44 1932 765947, Fax.: +44 1932 765293
e-mail: BRUTONTM@bp.com

P. Tzanetakis
University of Crete
Department of Physics
71 003 Heraklion, Crete, Greece
Tel : +30 81 394116, FAX: +30 81 394101
e-mail tzaneta@physics.uch.gr

ABSTRACT: The general objective of the ARCHIMEDES project is the development of a photovoltaic system with passive tracker for highly efficient and long term reliable water pumping with remarkable cost advantages compared to conventional fixed flat plate systems. The system is based upon irradiation enhancement in the module plane by flat plate mirrors in V-trough configuration and elimination of losses from off axis incidence using a maintenance free solar tracking unit, the gravitational/thermohydraulic tracking system (GTS). The new ARCHIMEDES system can demonstrate up to more than 40% cost advantage over systems using fixed standard flat plate PV modules.
Keywords: Concentrator – 1: Tracking – 2: PV Pumping - 3

1. INTRODUCTION

PV V-trough concentrator systems with passive tracking for medium to large central power stations have been developed and demonstrated formerly [1-4]. The ARCHIMEDES project will exploit now passive tracking and V-trough concentration also for small decentralised systems, as PV pumping, through a modular and integrated design of the PV receiver.

In contrast to PV concepts with higher concentration, as the EUCLIDES system, the ARCHIMEDES system relies on standard LGBG cell technology, which is already now suitable for low concentration. Structural and tracking requirements are relaxed due to the high acceptance angle of the V-trough concentrator and allow thermohydraulic gravitational tracking. Therefore the ARCHIMEDES system is most likely to fullfill all requirements for low cost series production and reliable long term system operation.

The ARCHIMEDES project (EU-JOULE III Program) is dedicated to the development of system components followed by the field test evaluation of a complete PV pumping prototype system on Crete island.

In this presentation the ARCHIMEDES concept is described in detail and the first experimental results of the component tests are discussed.

2. OBJECTIVES AND APPROACH

The implementation of the following objectives will lead to achieve the full cost reduction potential:
- integral design of PV module (conventional c-Si cell technology) and V-trough concentrator (X = 2) for

high performance and low temperature effect, which implies a reduced excess operation temperature (Tmodule - Tambient) as a function of irradiance in the module plane compared to fixed flat plate modules,
- adaptation of the gravitational tracking system to the system area with a projected tracking accuracy of about ± 3 deg for clear sky condition,
- development of a light weight support structure, which is modular in size and designed for low cost production and flexible installation,
- PV pumping system optimisation using the most advanced pump analysis techniques,
- optimisation of the entire system by field tests and monitoring combined with component engineering of a complete PV pumping system at an end users site.

The output from these objectives will be the specific achievements:
- ARCHIMEDES PV power supply unit specifically adapted for decentralised small scale applications and optimised for pumping systems.
- ARCHIMEDES system will double the energy harvest of conventional LGBG cells in comparison to fixed tilted systems. Harvest gain factor is 1.8-2.2 dependend on fraction of direct solar radiation.
- Tracking, concentration and reduced temperature sensitivity of the modules will result in an annual array yield of up to 3000 kWh_{dc} per installed kWp PV module power in Europe.

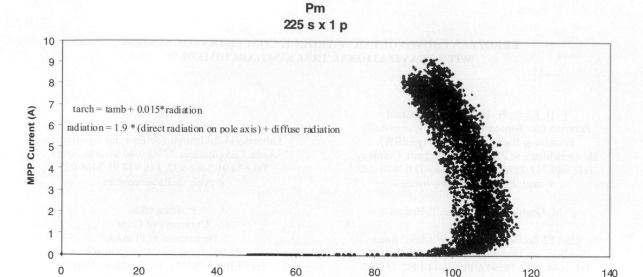

Figure 1: MPP current and voltage for two strings of 225 cells in series. Simulation with radiation data from Madrid. Inverter voltage window of 85 V to 125 V is fully matched.

3. ARCHIMEDES SYSTEM DESIGN CRITERIA

Several large PV pumping projects have been analysed. The results show that the typical village water supply system in the sun belt countries has an array rating of around 1 000 Wp with more than 75% of the systems operating in the range from 500 Wp to 1500 Wp

Therefore a system size around 1 kWp is regarded as optimum regarding water requirements and market situation. Moreover, from the technical point of view, the size of PV collector and tracking structure are most appropriate for a low cost production and in addition, several units can be combined to form a higher power system.

The market of photovoltaic pumps from different manufacturers world wide has been analysed. From the product range of each manufacturer a selection of potential systems suitable for operation with the ARCHIMEDES PV generator of about 1 kWp power has been made.

Four PV pump systems have been analysed in simulated operation with the ARCHIMEDES PV array output and have demonstrated their suitability for integration into the system.

Simulations have been performed using TMY of irradiance and ambient temperature for Madrid for the Grundfos SP5A-7 pumping system (with an inverter voltage operational range of 85 to 125 V) operating at ARCHIMEDES conditions. Figure 1 shows simulated current and voltage in the maximum power point of an ARCHIMEDES collector consisting of two parallel strings of 9 modules connected in series.

It is evident that a configuration with 225 cells in series has an excellent match to the Grundfos SA1500 Inverter. None of the relevant MPP points during any hour of the year has a voltage value lower than 85 V or greater than 125 V. Therefore a system of 18 modules is adapted to this pump/inverter type which will be presumably used for the field tests. Two further modules of the ARCHIMEDES system will be used for reference.

4. COMPONENT DEVELOPMENT RESULTS

4.1 ARCHIMEDES PV module

Geometry and modularity of the V-trough concentrator and ARCHIMEDES array are determined by size and electrical parameters of the used photovoltaic cell technology. For this project the Saturn cell technology of BP Solar is used.

The ARCHIMEDES PV generator will be in total composed of 500 half size one sun standard cells, in order to reduce series resistance losses (Figure 2). A single PV V-trough module (3.35 m length, approx. 252 mm width) is subdivided into 2 units each consisting of 25 half cells connected in series. Ten rows of PV modules are combined to form the V-trough array with dimensions 3.47 m length, 2.65 m width and up to 600 Wp installed PV power. Due to the modularity of the system the actual installed PV power can be adapted easily to specific pump needs. The subunits can be electrically connected according to the required voltage level of the pump.

Figure 2: Two PV submodules with V-trough concentrator

The back cover of the PV modules is an extruded Al tray to implement passive cooling (fins), back wiring and bypass diode and Standard IP67 electrical connectors.

The mirrors are mounted on triangular extruded aluminium profiles to support easy assembling. The supporting structure is developed for operation at different geographic latitudes (e.g. rotation axis inclination angle adjustable) and for easy mounting of the V-trough modules and the reflector profiles (Figure 3).

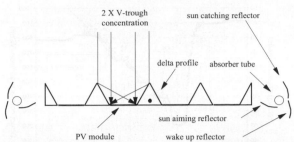

Figure 3: Principle arrangement of the ARCHIMEDES gravitational tracker and V-trough concentrator (PCT-patent pending).

4.2 V-trough concentration

V-trough geometry allows for 2X-concentration of direct beam sunlight. Effective concentration will be a little lower due to the imperfection of mirrors (reflection < 100%) and some additional Fresnel reflection losses on the front glass of the PV-laminate due to the somewhat higher incidence angles of the redirected sunlight (> 60°).

Remarks:
- 2X geometrical concentration means that about half of the aperture area is covered with PV modules (Figure 3).
- Polar-axis tracking in Europe increases the irradiance in the module plane of flat plate systems by a factor of 1.25 for Central Europe to 1.35 for South Europe depending on the fractions of direct and diffuse solar irradiance.
- Since the ARCHIMEDES system converts not only the direct sunlight but also the diffuse sunlight the effective irradiation enhancement of the ARCHIMEDES system is not 2X but about 1.5X to 1.6X depending on the site specific fraction of direct solar irradiance.
- In total the annual array yield (in kWh per installed kWp of PV modules) of ARCHIMEDES is about twice that of a conventional fixed tilted system. This means that with 2X geometrical concentration, the ARCHIMEDES system arrives at the same area specific energy harvest with half of the solar cells.

Thin glass reflectors glued onto the V-troughs demonstrated an effective reflectivity of > 90% (lab measurements showed 92 %). Irradiance levels in the module plane of more than 1800 W/m² have been obtained at the ZSW test field Widderstall. At the same time non-concentrated irradiance on a 40° tilted plane has been 1000 W/m² which corresponds to an effective concentration factor of 1.8X.

The first samples of PV modules have been manufactured and tested under one sun illumination and under V-trough concentration and the optical components of the system (flat mirrors for irradiation enhancement and shaped reflectors for the tracking system) have been characterised. Figure 4 shows current-voltage characteristics of a PV module without concentration and with V-trough concentration. The power gain obtained in the V-trough is in close agreement with the respective irradiance levels taking into account the temperature effect.

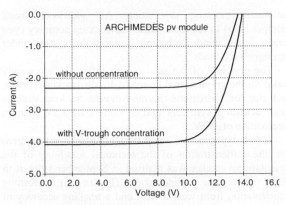

Figure 4: Current-Voltage characteristics of a single ARCHIMEDES PV module

4.3 Gravitational tracker

The functional principle is based on the thermohydraulic forced weight displacement in the absorber tubes on both side of the tracked array (Figure 5). For this purpose the gravitational tracker consists of two communicating absorber tubes equiped with a symmetric set of sun sensing reflectors.

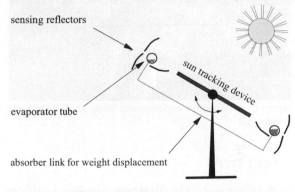

Figure 5: Functional principle of the thermohydraulic gravitational tracker.

The absorbers are filled with a volatile liquid. The absorber pressure is only dependent on it's individual absolute temperature. If the tracker is misaligned to the sun the absorber tubes will be irradiated unequally. This causes a temperature difference resulting in a pressure difference, which shifts the liquid portion until the pressure equilibrium is intalled again. With the displacement of liquid mass the structure is moved by a certain angle. In this way the tracker follows the sun in a self-regulating and energy self-sufficient way.

The sun sensing reflectors are consisting of wake up, sun catching and sun aiming reflectors. The wake up reflectors are active if the sun hits the collector from backwards (90° to 180° misalignment angle) and will reorient the collector back to the sun. The sun catching reflectors are active in a range of 20° to 90° from due sun and move the collector further towards the sun. The sun aiming reflectors are active in the range of 0° to 70° and finally „lock" the collector within a small tracking band of ± 3°. In this way a high tracking accuracy is obtained within a very large orientation range which allows for automatic morning orientation. An additional shock damper helps to realise a sufficient tracking accuracy even under lower wind loads and protects the system under stronger wind loads.

The principle of the sun sensing reflectors is not only valid for the gravitational thermohydraulic tracking system, but also for hydrostatic thermohydraulic tracking systems with actuating cylinders and is subject of a PCT-patent application of ZSW.

A first experimental setup (roughly 5 m² aperture area) for the demonstration of the technical feasibility of the gravitational tracker was realised in September 1999 at th ZSW test site Widderstall (Figure 6). Morning reorientation from west to east and a tracking accuracy of about ± 3° could be verified already in the very first full day test (see Figure 7), in spite of system oscillation due to thermal mass of the absorbers and stick-slip effect of the damping system.

Figure 6: Archimedes experimental setup at ZSW test site Widderstall.

5. NEXT STEPS

Three full scale ARCHIMEDES prototypes with 600 Wp installed cell power each (1.2 kWp effective power with tracking and concentration) are actually under manufacture and will be installed in May 2000 at the ZSW test site in Widderstall for a general proof of the function.

After the proof one of the prototypes will be installed on Crete island for a PV pumping field test evaluation and another will be installed at CIEMAT in Madrid for a systematic PV pumping system analysis using different pump technologies.

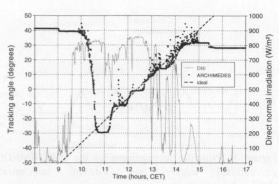

Figure 7: Thermohydraulic gravitational tracking system. Operational behaviour during a partial cloudy day (19.09.99).

Based on this experience BP Solarex and ZSW will develop an exploitation strategy with the aim to evaluate the economic advantage of the system for applications in Europe and the sun belt countries.

6. CONCLUSION

According to the results from component testing all the technical objectives regarding the PV module (efficiency, passive cooling), V-trough concentrator (reflectivity) and passive tracker (accuracy, reorientation) are reached or exceeded. The criteria for adaption of the ARCHMEDES system to a conventional pumping system are met.

REFERENCES

[1] EU-THERMIE programme, projekt no. SE165/92 DE, final report.
[2] EU-JOULE programme, projekt no. JOU 2 92 CT 0139, final report
[3] F.H. Klotz, G. Noviello & A. Sarno, "PV V-trough systems with passive tracking - technical potential for mediterranean climate", Proceedings of the13th European Photovoltaic Solar Energy Conference, Nice, France 23.-27. Oct. 1995, 372
[4] F.H. Klotz, "PV systems with V-trough concentration and passive tracking - concept and economic potential in Europe", Proceedings of the 13th European Photovoltaic Solar Energy Conference, Nice, France 23.-27. Oct. 1995, 1060

MANUFACTURING TECHNOLOGY IMPROVEMENTS FOR THE PVI SUNFOCUS™ CONCENTRATOR

W. R. Bottenberg, N. Kaminar, T. Alexander, P. Carrie, K. Chen, D. Gilbert, P. Hobden, A. Kalaita, J. Zimmerman
Photovoltaics International, LLC
171 Commercial Street
Sunnyvale, CA 94086, USA

ABSTRACT: PVI has developed and implemented an improved manufacturing technology for the SunFocus™ linear Fresnel lens PV concentrator with the goal of reducing the module manufacturing cost to less than one dollar per watt and increasing the manufacturing capacity to 50 MW/yr. PVI made substantial progress on these goals as part of a US DOE sponsored PVMaT project, including the development of a lens extrusion system, automated cell assembly station, EVA encapsulation system, an advanced collector design and a lowcost frame design. In addition, use of volatile organic compounds (VOC) was eliminated from the manufacturing process.
Keywords: Concentrator – 1: Manufacturing and Processing– 2: PV Module – 3

1. INTRODUCTION

The PVI SunFocus™ linear concentrator uses manufacturing processes identified as the low cost methods for each operation [1-4]. The linear-focus Fresnel lens is made by a plastic extrusion process. The collector assembly consists of the lens and an extruded aluminum heat sink; two roll formed aluminum sides and two stamped aluminum end caps. These components are joined together with inexpensive fasteners. The SunFocus™ uses solar cells manufactured using low-cost methods used for one-sun cells. Twelve modules are mounted on a stationary frame to move in unison for single-axis tracking, as shown in Figure 1.

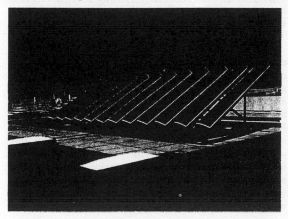

Figure 1: PVI SunFocus™ Panel

As part of its internal research and development program, PVI entered into a subcontract with the National Renewable Energy Laboratories to conduct research on manufacturing process improvement. This subcontract was funded by the US Department of Energy as part of the Photovoltaic Manufacturing Technology (PVMaT) program. This cost shared development program has made an important contribution to the ability of PVI to address the PV concentrator market.

PVI focused on four areas that were identified to be critical to the ability to achieve low-cost manufacturing.

The key components of the project were:
- Development of an advanced state-of-the-art lens extrusion system
- Development of a second generation automated receiver assembly station
- Development of low-cost roll-formed steel panel frame members
- Reduction of the use of volatile organic compounds (VOC) and hazardous materials in an automated module assembly process

2. DESCRIPTION OF CONCENTRATOR MODULE

The SunFocus concentrator module employs a line focus Fresnel lens to produce a concentrated image on a string of 36 cells located approximately 48 cm below the crown of the lens. The lens design produces a double humped image when the incident radiation is normal to the lens. As the angle of the sun varies from direct normal, the focal plane shifts up causing some beam spreading. A reflecting surface along the sides of the lens provides a secondary optical element to refocus the stray radiation towards the cell.

The cell is 48.3 mm by 96.5 mm in dimension. The aperture width of the cell is 44.2 mm since the side reflector covers the bus bars on the long side of the cell.

2.1 Module Characteristics

Table 1 shows the numerical values of the primary module characteristics.

Table 1: Module characteristics

Module Parameter	Value
Module Length, m	4.19
Module Width, m	0.51
Module Aperture, m²	2.13
Direct Mode Aperture, m²	1.77
Cell Area, cm²	42.66
Number of Cells	36
Total Cell Aperture m²	0.154
Geometric Concentration Ratio	11.5

Typical cells have 1-sun current densities of 0.0315

amp/cm². The modules produce 11.5 amps of current at standard conditions of 850 w/m² Direct Normal Irradiance. The average intensity at the cell surface can then be calculated to be about 8,600 w/m². The direct mode aperture is the active aperture area of the lens. Given the geometric concentration ratio of 11.5 and the direct mode aperture of 1.77 m² yields an approximate lens efficiency of 84%

Figure 2 shows a drawing of the module.

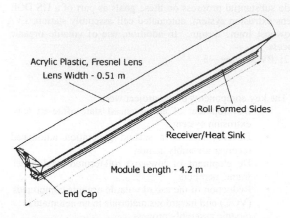

Acrylic Plastic, Fresnel Lens
Lens Width - 0.51 m

Roll Formed Sides

Receiver/Heat Sink

Module Length - 4.2 m

End Cap

Figure 2: Module Drawing

3.0 MANUFACTURING PROCESS IMPROVEMENTS

3.1 Extrusion Processing

At the beginning of the PVI PVMaT program, lens extrusion for PVI was done at an outside company that specialized in custom profile extrusion of acrylic parts. The normal instrumentation and controls found on commercial extruders are fine for most parts but are not sophisticated enough for the required high quality lens. Die design and production pressures at the outside extruder made lens development time too long.

During the project, we designed and installed a state of-the-art, computer controlled extrusion system. The computer acquires, stores, and displays data so that the operator can see the effect of changes. Lens transmission is measured as a process variable so that effects of changes are known in real time.

The extrusion system consists of an extruder proper with barrel; a material handling system; product cooling system; a tractor type puller that controls the line speed; a flying cut-off saw to cut the product to length; a product take off system that stores up to 5 lenses; a die and die manifold with heaters; and a control and data acquisition system. This system consists of a main and a subordinate computer, various sensors, power control systems, autonomous controls on peripheral equipment, and inline lens transmission tester. Peripheral equipment consists of a high-pressure air compressor, a low-pressure air blower, and a chilling tower.

The lens is formed as the plastic passes through the die which has the shape of the part machined into it designed to produced the lens. The lens continues to be formed as it

is pulled from the exit of the die. It is reduced in width by being pulled by the puller. The final outside radius of the lens is formed in the cold shoes and the final shape of the attachment features is made by after-extrusion forming tools. During this process the plastic is cooling and freezing to its final shape in the air rack. The die shape is adjusted to give the final shape of the part once cooled. Die design is based on theory, experience, and trial and error.

The most important feature of the PVI extruder is the sophisticated control system. Without this advanced control system, extruding quality lenses and sides would be impossible. The heart of the system is an industrial quality computer, which runs several programs concurrently, acquiring information from various sensors, and controlling various pieces of equipment. There are also microprocessor based blocks at various places that have immediate control over specific items, such as heaters and that acquire information from sensors. The inline lens transmission tester is the key part of process control.

With our extrusion system, we have been able to obtain lenses with 85% transmission at 11.5 to 1 geometric concentration ratio. This was obtained in a production setting of a hundred lenses per run with yields of 99%.

The first dies were designed for a line focus at the center of the cell. Normal inaccuracies in the extrusion process tended to spread out the flux over the cell. With the accuracy obtainable with our new system, we were able to focus to the point where fill factor was beginning to suffer. Our new die is designed for two foci at the quarter points of the cell active area width. We are also using a slightly longer focus length. We expect up to 87% transmission with these lenses and an improved fill factor in the module.

An important issue is the use of an acrylic lens as the cover for the solar cells. The question is whether the concentrated solar flux in the ultraviolet region is sufficiently high to damage the organic polymer encapsulation system employed on the receiver to protect the solar cell string. We provided the test group at the US National Renewable Energy Laboratory in Golden with a special lens/receiver test bed to measure the solar flux under outdoor test conditions. Figure 3 shows the irradiance in the UV and near UV region at the cell plane measured using a spectroradiometer. The sun conditions were in the range of 1000 w/m² global irradiance and 850 w/m² direct beam irradiance.

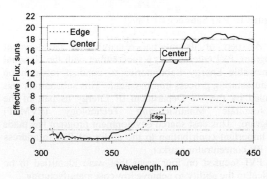

Figure 3: Ultraviolet flux at the cell plane

The data shows that in the UV damaging region between

350 and 360 nm the amount of flux, even at the center of the cell, is below 1.5. Below 350 nm, in the UVA range the UV flux is negligible. Based on these measurements we expect that the normal UV inhibitor additives found in commercially available ethylene vinyl acetate (EVA) should be adequate for protection of the encapsulation materials.

3.2 Receiver Fabrication: Heat Sink

At PVI we designed a new heat sink and module sides that are designed to assemble quickly, work together to dissipate heat, and eliminate solvent bonding. The new heat sink is a deep channel aluminum extrusion, which brings the module side attachment up higher, reducing the material usage in the sides. The upward extension of the heat sink also improves the free-air natural-convection heat transfer that is important on windless days. The sheet aluminum sides also help to dissipate heat. We found that only the bottom 6 inches of the module sides need to be highly reflective to achieve the full effect of the reflective secondary optics. With the new heat sink, a separate reflective piece can be used which eliminates having to make the entire module side reflective.

The new heat sink and aluminum module sides form the backbone of the module, eliminating the need for separate reinforcements as used in our previous module. This has greatly simplified the manufacturing and significantly reduced cost.

3.3 Receiver Fabrication: Automatic Lead Attachment

During the project we implemented an improved automated cell assembly station. This station solders the cell leads to the cells. The automated cell assembly station consists of the following features:

- lead punching from sheet material
- lead and cell transport by a robot
- automatic placement of insulating tape between the top and bottom leads
- employment of a two-position nest where the cell assembly is built and soldered
- automatic solder paste dispensing
- infrared lamp soldering

Solder joints obtained in the automated cell assembly station show a 50% improvement in pull strength over solder joints obtained by hand, and there was a 50% reduction in lead attachment series resistance losses compared to hand soldering. In addition, the new automated cell assembly station provides a soldered cell in about one-fifth the time taken by hand. The lead locations are consistent and accurate compared to the hand soldering case. The automated cell assembly station now fabricates a leaded cell in 24 seconds, yielding an annual machine capacity of greater than 3 MW.

3.4 Receiver Fabrication: EVA

Our previous encapsulation systems used either adhesive tape or a silicone encapsulant. The tape system was easy to apply but did not provide a void-free sandwich. The silicone encapsulant was expensive, contained a high fraction of VOC, and was a very hazardous material to use. Going to an EVA system eliminated both problems.

We developed custom tooling to use EVA material, based on a bag lamination process. The process uses the normal evacuation and heating sequences. We have been able to achieve void-free encapsulation of our full-length receivers using production equipment.

3.5 Collector Fabrication: Roll Formed Sides

The new aluminum sides are manufactured using a roll forming process. The forming adds the attachment features for joining the sides to the heat sink and the lens. Ribs formed in the process add stiffening to the module. The sides are structural elements for the module and are attached to the end caps using self-tapping screws. Each module side is one piece, which replaces 8 pieces each on the old design. Although the aluminum is more expensive than the plastic it replaces, the new sides are significantly lower cost when all the additional material parts and assembly labor are considered.

3.6 Collector Fabrication: Cast Aluminum End Caps

The new aluminum end cap is manufactured using a casting process. The end cap is dished which adds stiffness and strength with minimum material. A plastic bearing is pressed into the end cap for the pivot point and an adjustable pin is used for the tracker bar. These three simple parts replace over 25 parts on the old design, at a considerable cost saving. The bearing is designed to accommodate any misalignment. An adjustment pin allows each module to be adjusted for maximum output.

3.7 Panel Frame

The panel frame is roll formed from galvanized steel which is a significant cost saving over the previous extruded aluminum frame. The frame members have a cross section in the form of the letter C. The open part of the C is closed with a piece of galvanized steel sheet, forming a closed channel that can be used as an electrical conduit. The steel sheet is screwed to the C-section using self-drilling, self-tapping screws, which makes the channel into a closed tube for torsional rigidity. The frame design has also been greatly simplified to reduce manufacturing cost. The part count has been reduced by introducing multiple functions for parts and by using the same part in different places.

4 MANUFACTURING BENEFITS

The objectives of our internal program and the PVMaT project were been met by reducing material and labor costs and by improving production rate, product performance and product quality.

The cost has been halved over the previous design by production of our own lenses, elimination of 40% of the parts, using an EVA/fluoropolymer encapsulant system, and use of lower cost frame parts. Production rates for the extrusion process have increased by 200% over previous external fabrication, with the near term prospect of a further 250% increase. By eliminating extruded sides, we have freed up the extruder to make lenses. One extruder is now capable of 11 MW/year production rate.

Employment of roll formed sides and metal end caps

has greatly reduced module assembly time. Product performance has been improved by being able to produce better lenses and thus being able to tune our lens design to the cell design.

Quality and product utility has been increased during the project as a result of the lower weight and stiffer module design. The new design weighs 50% less than the previous design. The new weight of 30 kg permits installation of modules to be carried out easily by 2 persons with no special handling equipment.

The automated cell lead assembly station has significantly improved the accuracy and electrical quality of the product.

5. MODULE RESULTS

The new 51 cm aperture modules developed as part of this project produce twice as much power as the previous module based on a 38 cm lens. This is due to the improved collecting power of the lens, to the employment of more efficient cells and to improved heat dissipation in the heat sink design. Figure 4 shows a current-voltage curve for a module tested at near standard PVUSA conditions (PTC). These conditions are:

Direct Beam Irradiance: 850 w/m^2
Total Irradiance: 1000 w/m^2 AM1.5
Tambient: 20°C

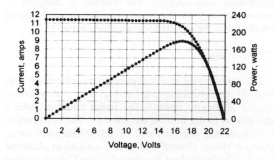

Figure 4: SunFocus™ Module Current-Voltage Curve

Correcting the module parameters to standard conditions we evaluated the module output to be 166 w.

6.0 CONCLUSIONS

We view the successful collaboration between NREL and PVI during our PVMaT project to have been extremely productive in providing the impetus to move forward in development of manufacturing technology for a superior product at low manufacturing cost. As a result of this project we are able to produce competitively priced product ready to be scaled to 50 MW/year rates.

ACKNOWLEDGEMENTS

This work was supported in part by NREL subcontract ZAF-6-14271-11. The authors wish to thank the technical review committee of R. Mitchell, A. Maish and B. Kroposki for their support and assistance. Special thanks to Daryl Myers of NREL who conducted the spectroradiometer tests on the receiver. The authors would also like to acknowledge the support and assistance of the other members of the PVI staff who supported this work.

REFERENCES

[1] Kaminar, N. R., and Curchod D., "Design and Construction of an Extruded, Curved Linear Focus Fresnel Lens," presented at the Twenty First IEEE Photovoltaic Specialists Conference, Kissimee, FL, May, 1990.

[2] Kaminar, N. R., McEntee, J. Stark, P. and Curchod D., "SEA 10X Concentrator Development Progress," presented at the Twenty Second IEEE Photovoltaic Specialists Conference, Las Vegas, NV, October 1991.

[3] Kaminar, N. R. et al., "Cost-Effective Concentrator Progress," presented at the Twenty Third IEEE Photovoltaic Specialists Conference, Louisville, KY, May, 1993.

[4] N. Kaminar, et al., "Manufacturing Technology Development of the Powergrid Linear Focus Photovoltaic Concentrator System", NCPV Photovoltaics Program Review, 16th Conference, AIP Conference Proceedings 462 (1998).

QUALIFICATION STANDARD FOR PHOTOVOLTAIC CONCENTRATOR MODULES

Robert McConnell and Sarah Kurtz
NREL, 1617 Cole Boulevard, Golden, CO 80401-3393, TEL: 303-384-6419, 303-384-6475, FAX: 303-384-6481, 303-384-6531, e-mail: robert_mcconnell@nrel.gov, sarah_kurtz@nrel.gov

William R. Bottenberg
Photovoltaics International, LLC, 171 Commercial Street, Sunnyvale, CA 94086, TEL: 408-731-1207, FAX: 408-746-3890, e-mail: wbotttenberg@pvintl.com

Robert Hammond
Photovoltaic Testing Laboratory, Arizona State University EAST, 7349 E. Unity Ave., Mesa, AZ 85212, TEL: 480-727-1221, FAX: 480-727-1223, e-mail: b.hammond@asu.edu

Steven W. Jochums
Underwriters Laboratories, Inc., Engineering Services, 413M, 333 Pfingsten Road, Northbrook, IL 60062, TEL: 847-272-8800 x42229, FAX: 847-272-9718, e-mail: steven.jochums@us.ul.com

A. J. McDanal
ENTECH Inc., 1077 Chisolm Trail, Keller TX 76248, TEL: 817-379-0100, FAX: 817-379-0300, e-mail: ajmcdanal@entechsolar.com

David Roubideaux
Amonix Inc., 3425 Fujita Street, Torrance, CA 90505-4018, TEL: 310-325-8091, FAX: 310-325-0771, e-mail: dave@amonix.com

Charles Whitaker
Endecon Engineering, 2500 Old Crow Canyon, Suite 22, San Ramon, CA 94583, TEL: 925-552-1330, FAX: 925-552-1333, e-mail: Chuckw@endecon.com

John Wohlgemuth
BP Solarex, 630 Solarex Court, Frederick, MD 2170, TEL: 301-698-4375, FAX: 301-698-4201, e-mail: JWOHLGEMUTH@Solarex.com

ABSTRACT: The paper describes a proposed qualification standard for photovoltaic concentrator modules. The standard's purpose is to provide stress tests and procedures to identify any component weakness in photovoltaic concentrator modules intended for power generation applications. If no weaknesses are identified during qualification, both the manufacturer and the customer can expect a more reliable product. The qualification test program for the standard includes thermal cycles, humidity-freeze cycles, water spray, off-axis beam damage, hail impact, hot-spot endurance, as well as electrical tests for performance, ground continuity, isolation, wet insulation resistance, and bypass diodes. Because we can't verify concentrator module performance using solar simulator and reference cell procedures suitable for flat-plate modules, the standard specifies an outdoor I-V test analysis allowing a performance comparison before and after a test procedure. Two options to this complex analysis are the use of a reference concentrator module for side-by-side outdoor comparison with modules undergoing various tests and a dark I-V performance check.
Keywords: Qualification and Testing – 1: Concentrator – 2: Reliability – 3

1. BACKGROUND

In 1997, an IEEE working group began developing a qualification standard for photovoltaic concentrator modules as a result of industry concern that the lack of a standard was affecting the marketing and sales of their products [1]. The first draft was based on evaluation tests developed in the late 1980s at Sandia National Laboratories and published in a Sandia report in 1992 [2]. It followed the general outline of tests in the IEC Standard 61215 and the IEEE standard 1262-1995 for flat-plate modules [3,4]. These flat-plate standards, the result of more than a decade of module and standards development, have contributed greatly to the present level of flat-plate module reliability such that manufacturers can back their products with guarantees as long as 25 years [5].

Since the publication of these earlier documents, photovoltaic materials technology and concentrator module development have advanced. Efforts in module reliability research have produced a better understanding of known and potential failure mechanisms associated with photovoltaic concentrator modules, especially regarding the effects of moisture ingress. The results of these efforts and experience gained have been used in formulating new tests and modifying earlier tests for a proposed concentrator qualification standard.

In following the outline of tests for flat-plate module standards, the working group identified some fundamental differences between the two technologies that required significantly different test approaches. Most flat-plate modules have a thin two-dimensional geometry, whereas concentrator modules usually have optics that are mounted away from the cells, forming a three-dimensional structure. As a result, one major difference between flat-plate standards and the proposed concentrator standard is that the concentrator standard has parallel test sequences for concentrator receivers and modules. A receiver is defined as an assembly of one or more PV cells that accepts concentrated sunlight and incorporates the means for thermal and electrical energy removal. A module is the smallest, complete, environmentally protected assembly of receivers and optics and related components, such as interconnects and mounting, that accepts unconcentrated sunlight. Figure 1 shows two linear concentrator modules and an associated receiver. Another significant deviation from the flat-plate standards arises from the difficulty in measuring performance after a test sequence because most concentrator modules cannot be accurately characterized with solar simulators. Later in the paper, we will discuss proposed power conversion efficiency tests, as well as two optional performance checks, to replace the use of indoor simulators.

All tests and procedures in this paper are under consideration by the IEEE working group. We plan to ballot on the proposed standard later this year, in hopes of reaching a consensus for the final standard. We have submitted an earlier draft for consideration by an International Electrotechnical Commission working group, as a proposed international qualification standard.

(a)

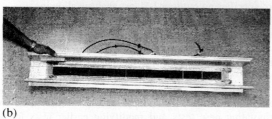

(b)

Figure 1. a) Two linear concentrator modules ready for qualification standard testing, at the Photovoltaic Testing Laboratory at Arizona State University East.   b) A receiver for the linear concentrator receiver tests.

2.   OVERVIEW OF QUALIFICATION TEST PROGRAM

Figure 2 shows the proposed test program in the most recent draft (April 2000) of the proposed standard. It requires 7 receivers, 5 modules, and, for a receiver design with inaccessible bypass diodes, 1 specially constructed receiver. Details for the temperature extremes, exposure duration, isolation determinate procedures, etc., are in the latest draft of the standard. In this paper, we will only discuss each sequence in general terms that will still, in many cases, contain tests similar to flat-plate qualification tests. Many of these tests are based on ASTM standards [6].

2.1   Baseline
Referring to Figure 2, we propose baseline tests for electrical performance, ground continuity, electrical isolation, and wet insulation resistance, along with visual inspections to determine the initial status of the modules and receiver sections.

2.2   Sequence A
In this sequence, we specify a test for bypass diodes and a thermal cycle test for two receivers.

2.3   Sequence B
Sequence B specifies thermal cycle tests different from that in Sequence A and humidity-freeze tests for two receivers and with parallel tests for two modules involving thermal cycling, humidity-freeze, electrical isolation, and terminations.

2.4   Sequence C
Here, we recommend damp heat exposure for two receivers, followed by a test for electrical isolation.

2.5   Sequence D
This sequence involves several module stress tests, including outdoor exposure, water spray, off-axis beam damage, hail impact, and hot-spot endurance.

2.6   Sequence E
For receivers in which the by-pass diodes are inaccessible, we specify a specially prepared receiver having access to diodes for a bypass diode test.

2.7   Final Test and Inspections
All modules and receivers are subjected to final tests of visual inspection, electrical performance, electrical isolation, wet-insulation resistance and ground continuity.

3.   ELECTRICAL PERFORMANCE TESTS

A critical question after many stress tests is the performance of the photovoltaic concentrator module or receiver. In the case of flat-plate photovoltaic modules, a solar simulator and reference cell provides a means to verify performance after a stress test. Photovoltaic concentrator modules, however, are critically dependent on the concentrated light incident on the receivers (See Figure 1). We have developed three possible tests to verify any degradation in the performance of modules and receivers. One of these is described in more detail in another presentation at this conference [7]. The testing organization conducts baseline outdoor performance tests under various temperature and solar irradiance conditions while measuring output currents and voltages to obtain an analytical expression for the module and receiver performance. After a stress test or series of tests, the module is again measured outdoors, and its performance, under conditions encompassed by the analysis parameters, is compared with that predicted by the analytical expression for the module's performance. It takes a lot of time to determine the analytical equation for the module's performance, so we explored two other possible determinations of any module performance degradation.

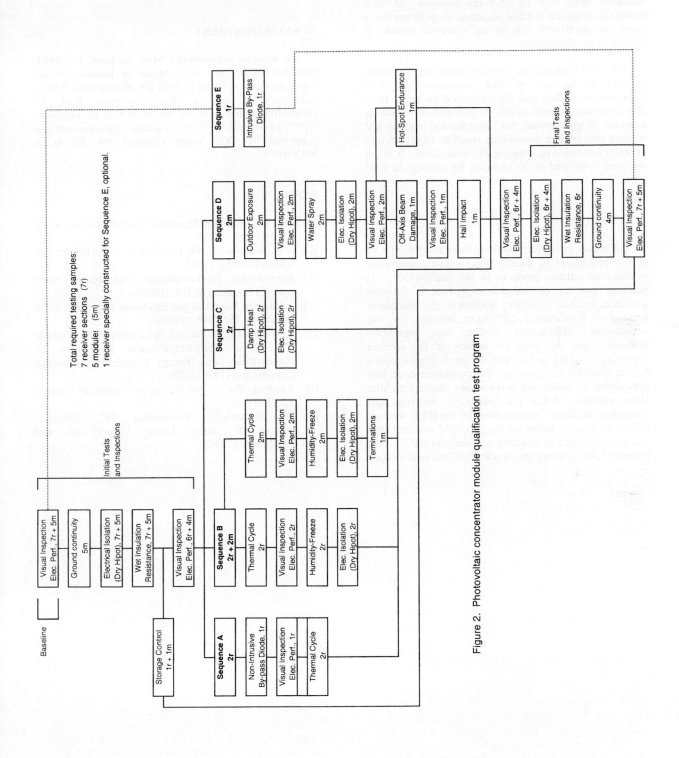

Figure 2. Photovoltaic concentrator module qualification test program

A second possible test for characterizing module performance uses a reference module. This module is not subjected to any stress tests, and its performance is measured along with all other test modules. After a module is subjected to a test sequence, its performance is again compared with that of the reference module to determine any degradation.

The third possible test for identifying degradation in modules is the use of dark current-voltage (I-V) measurements before and after intermediate stress tests. This procedure, as well as the use of a reference module, provides simpler means for determining module and receiver performance degradation resulting from increases in series resistance or breaking of connections. It is not intended to be used as the criterion for passing the final tests.

4.   CONCLUSION

For almost three years, we have been developing a set of test procedures to qualify photovoltaic concentrator modules as reliable products in the marketplace. In building upon the considerable experience in flat-plate standards, we had to explore alternative test procedures because of fundamental differences between the two photovoltaic technologies. The most significant differences are the need for testing of the receiver separately from the module and the much large size of a typical module, necessitating the development of new procedures for identifying performance degradation after stress testing. While the details of the tests and procedures are under consideration by the IEEE working group, we plan to ballot on the proposed standard later this year in hopes of reaching a consensus for the final standard. We have submitted a draft of this qualification standard for consideration by an International Electrotechnical Commission working group as a proposed international standard.

ACKNOWLEDGMENT

We wish to acknowledge Mark Jackson, previously with ENTECH, Inc., Alex Maish at Sandia National Laboratories, and Liang Ji from the Photovoltaic Testing Laboratory at Arizona State University East, for substantial contributions to the standard's development. NREL's coordination of the standard's development was supported by DOE under Contract No. DE-AC36-99GO10337.

REFERENCES

[1]   IEEE Concentrator Standards Working Group, IEEE 1513/Draft8, (2000)
[2]   J. R. Woodworth, M. L. Whipple, SANDIA Report, SAND92-0958, (1992)
[3]   International Electrotechnical Commission, International Standard 61215, (1993)
[4]   Institute of Electrical and Electronic Engineers, IEEE Standard 1262-1995, (1995)
[5]   J. W. Bishop, J. Sachau, W. Zaaiman, Proceedings 2nd World Conference and Exhibition on Photovoltaic Solar Energy Conversion, European Commission (1999) 1920.
[6[American Society for Testing and Materials, Section 12, (1999).
[7]   W. Bottenberg, Proceedings 16th European Photovoltaic Solar Energy Conference and Exhibition, (2000)

ULTRA COMPACT HIGH FLUX GaAs CELL PHOTOVOLTAIC CONCENTRATOR

C. Algora, J.C. Miñano, P. Benítez, I. Rey-Stolle, J. L. Álvarez, V. Díaz, M. Hernández, E. Ortiz, F. Muñoz, R. Peña, R. Mohedano and A. Luque

Instituto de Energía Solar – E.T.S.I. Telecomunicación (Universidad Politécnica de Madrid)
Ciudad Universitaria s/n - 28040 Madrid (SPAIN)
Phone: (34) 91.3367232; Fax: (34) 91.5446341; E-mail: algora@ies-def.upm.es

G. Smekens and T. de Villers
Energies Nouvelles et Environnement, Brussels (BELGIUM)

V. Andreev, V. Khvostikov, V. Rumiantsev and M. Schvartz
A. F. Ioffe Physico-Technical Institute - St. Petersburg (RUSSIA)

H. Nather and K. Viehmann
Vishay Semiconductors GmbH – Heilbronn (GERMANY)

S. Saveliev
Progressive Technologies – Sosnovy Bor (RUSSIA)

ABSTRACT: A new kind of concentrator suitable for operation at 1000 suns has been developed. The technological challenges related with this high concentration level have been undertaken and solved. More specifically, the main goals already achieved have been: a) design and manufacturing of concentrator GaAs solar cells with efficiencies over 26% (at 1,000 suns) for the best devices, b) design and manufacturing of optical concentrators with a current gain of 1,110 (geometrical concentration 1,256), an acceptance angle of ±1.5 degrees, an aspect ratio of 0.3 and an optical efficiency greater than 82% and, c) a temperature drop in operation from GaAs solar cell to ambient lower than 30°C.
Keywords: Cost reduction - 1: Concentrator - 2: Gallium Arsenide Based Cells - 3

1. INTRODUCTION

Photovoltaic (PV) technology has penetrated today's electricity market to only a limited level, mainly due to the relatively high costs of PV power systems. In areas of the world having a high proportion of direct irradiation one important way to reduce costs is to replace solar cells with lower-cost optical concentrators. Accordingly, the collector described in this paper could constitute a breakthrough for the generation of economical PV electricity thanks to its new both design and manufacturing philosophy.

2. GUIDELINES AND PHILOSOPHY

Consequently, the purpose of this work is the design and built of a PV concentrator based on a new optical design (RXI) and a specifically designed and developed GaAs solar cell. The whole device, called HERCULES, is capable of operating at 1,000 suns with an acceptance angle of ±1.5 degrees. Although a global efficiency as higher as possible is intended, the main goal of this work is to demonstrate the technological feasibility of operating at around 1000 suns with the following characteristics:

a) a GaAs solar cell efficiency around 25%,
b) a total concentrator thickness not greater than 0.5 its aperture diameter,
c) a cell-to-ambient temperature drop lower than 40°C.

Small GaAs solar cells and optical concentrators are key characteristics of this approach. This increases both the efficiency and the manufacturing yield of the GaAs solar cell, simplifies cooling, reduces the concentrator cost and increases modularity. The assembly cost concern has been solved by using the LED's (Light Emitting Diode) standard assembling techniques.

As a consequence, a PV plant cost of 2.74 Euro/W_p and an electricity cost of 0.104 Euro/kWh could be obtained based on a 10 MW_p production [1].

2. SET-UP

Figure 1 shows the cross section of the HERCULES concentrator unit. The optical device, called RXI [2], is made of a single dielectric piece in which the GaAs solar cell and part of the heat sink is embedded. Rays coming from the sun are first refracted at the upper surface, second reflected at the lower surface (that is metallised) and finally are reflected again at upper surface (by total internal reflection excepting at the centre of this surface, where a small metallic reflector must be placed).

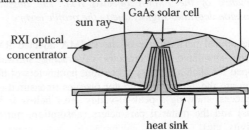

Figure 1. Set-up of the HERCULES concentrator. Darker regions of the RXI indicate the top and bottom mirrors.

3. OPTICAL CONCENTRATOR

A series of low-cost moderate precision RXI concentrators have been manufactured by PMMA injection moulding (see Figure 2). The mould is made using diamond turning technique (its accuracy is ±1 μm). The PMMA is pressure injected at high temperature into the mould. The material cooling after the injection produces a contraction of the moulded piece so, the manufactured piece has lower accuracy than the mould. These contractions are favoured because the thickness of the RXI is not uniform throughout its profile. The profiles of the concentrator have been measured with a contact profilemeter.

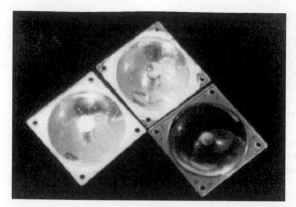

Figure 2. *Photography of the PMMA injected RXI concentrators with silver mirror.*

The metallisation of the lower surface and the front mirror of RXI has been made by evaporation of silver. A layer of 0.3 μm is deposited by e-gun evaporation. After this, the metallisation is protected with a coat of painting.

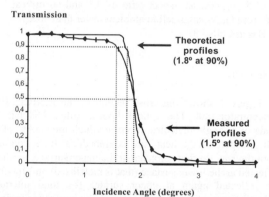

Figure 3. *Results of the ray-tracing simulation of the performance of the PMMA injected RXI. The angular transmission degrades slightly due to the profile errors.*

The optical performance of the PMMA injected RXI has been simulated by computer ray-tracing software developed for such a purpose. The input parameters of the program are the actual concentrator profiles (measured on some pieces, showing a peak-to-valley error below ± 30 microns) and the material parameters (absorption, mirror reflectivity, etc).

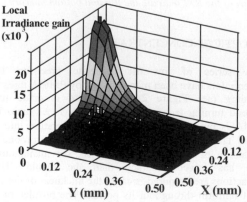

Figure 4. *Irradiance pattern on a quarter of the GaAs solar cell surface inside of the RXI.*

From the simulation, we deduce that the main effect of the manufacturing errors of the profiles is not in transmission drop, but on the acceptance angle. Thereby, Figure 3 shows that computer simulation of the acceptance angle predicts a value of ± 1.8° for an optical transmission of 90% of the maximum while the experimental profile shows an acceptance angle of only ± 1.5°.

Also, from computer simulation the irradiance target of light over the GaAs solar cell is obtained (see Figure 4). Light is concentrated onto the centre of the receiver and at this point, irradiance gain could be greater than 20,000. However, this calculation is pessimistic since the actual value should be lower because, in this case, the analysis considers neither optical losses nor dispersion at the optical surfaces. This non-uniform illumination could cause a drop of the *FF* of the GaAs solar cell.

4.   CHARACTERISTICS OF THE SPECIFICALLY DESIGNED CONCENTRATOR GaAs SOLAR CELL

Two different growth technologies have been used, the LT-LPE (Low Temperature Liquid Phase Epitaxy) and the MOCVD (Metal Organic Chemical Vapour Deposition).

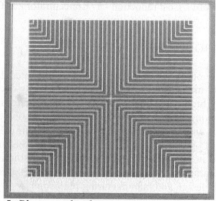

Figure 5. *Photograph of a concentrator GaAs solar cell. The total area inside the busbar is 1 mm².*

Considering the characteristics of the RXI concentrator, both a new GaAs solar cell structure and its related technological process have been specifically designed and developed [1]. Some its main novelties are:

1. The development of an accurate theoretical model of the GaAs solar cell. This model considers simultaneously the semiconductor structure, the antireflecting coatings and the ohmic contacts.
2. The determination of the optimum size of the solar cell by maximising efficiency (using the aforementioned model) and minimising both the cost of the whole concentrator system and the heating effects. The optimum size ranges from 0.5 to 1 mm².
3. The design of a front metal grid able to achieve a series resistance of around 3 mΩ·cm² with a simultaneous 1000-sun $J_{sc} \approx 26$ A/cm² (see Figure 5).
4. The improvement of ohmic contacts.
5. Semiconductor structure growth (LPE & MOCVD).
6. The optimisation of the antireflecting coatings (including the window layer) for homogeneous light inside a cone of 70 degrees.

Such a kind of solar cell based on a new design approach and on a manufacturing process close to Optoelectronics has been patent registered [3].

By using this process, many GaAs solar cells with efficiencies ranging from 23 to 26% under normal incidence and out of the RXI concentrator were obtained. The measurement of the best solar cells was performed at NREL. In order to calculate the efficiency a linear dependence between short-circuit current and irradiance was assumed. So, an efficiency of 26.2% at 1004 suns and a 25% at 1920 suns (see figure 6) is obtained.

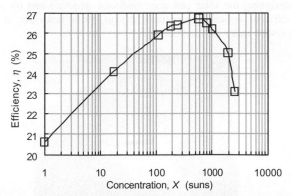

Figure 6. *Efficiency under concentrated light (normal incidence, AM1.5D spectrum) of the best GaAs solar cell measured at NREL. Efficiency has been calculated by assuming linearity of short-circuit current with irradiance.*

5. ASSEMBLING

Several of these solar cells were introduced in the RXI concentrator. Previously heat sinks were manufactured in a copper material by conventional machining. The calculations based on analytical approximations of the heat-transfer solution indicate a cell to ambient drop of 27.8°C. This value agrees well with the temperature drop fitted from the open-circuit voltage decrease which indicates something about 25-30°C.

The assembly process has been done using a laboratory-sized translation of conventional optoelectronic techniques for mass production. The cell is glued to the heat sink with a silver filled epoxy or solder. The front contact has been carried out using standard wirebonding. Finally, the encapsulation has been done by casting a transparent silicone rubber, which glues the GaAs cell to the concentrator. This method of encapsulation, based on elastomers, is currently being used for some high-efficiency LEDs in the optoelectronic industry. The selected silicone rubber has a high ultraviolet stability, needed for the 1000 suns light intensity near the cell.

6. CONCENTRATOR CHARACTERISATION

Several complete concentrator prototypes have been fabricated. Due to the low yield in this experimental phase, which is not a mass production one, no prototype has reached a high efficiency.

Optical efficiency, η_{opt}, of a concentrator is defined as the fraction of power incident on concentrator aperture that reaches the receiver's surface. For measuring this parameter, it has been decomposed as follows,

$$G_c = \frac{I_{SC-RXI}}{I_{SC-1sun}} \qquad (1)$$

with

$$\eta_{opt} = \frac{G_c}{C_g} * f_i^{-1} \qquad (2)$$

and

$$f_i = \frac{I_{SC}(\pm 10° - 70°, silicone, P)}{I_{SC}(normal\ inc., air, P)} \qquad (3)$$

G_c is the current gain of the cell inside of RXI, C_g is the geometrical concentration and f_i is a factor that takes into account that GaAs cell efficiency measurements are done under normal incidence in air while the cell within the RXI is almost isotropically illuminated and embedded in silicone rubber. Thus, this factor takes into account that Fresnel reflection losses on the cell's surface are different when we measure I_{sc} of the receiver at 1 sun of that one of the concentration measurement and that the change of the refraction index is more gradual (air-PMMA-silicone-GaAs for a cell inside of RXI, air-GaAs in the measurement at 1 sun illumination). Moreover, when the cell is inside of the concentrator the incidence angle of the light is between 10° and 70° over its surface (almost isotropic).

So, f_i is the ratio between the response of the receiver illuminated at normal incident angle and with air as incidence media and its response when it is illuminated under conditions of the concentration measurements. In Eq. (3), the short circuit currents that must be compared, are those caused for the same power over cell's surface. To specify this, "P" is put in brackets. The others parameters that specify the short circuit current in equation (3) make reference to the angular distribution of incident light on cell's surface and the incidence media (air or silicone).

Table I presents several I_{SC} measurements of the GaAs solar cells at 1 sun and inside the concentrator, I_{SC-RXI}. So, we can deduce the current gain produced by the RXI. Differences in current gain could be due to defects of surfaces (like scratch or dirt) that could increase losses in refraction, to little differences in mirrors quality (probably caused by similar reasons) and to differences in the antireflecting coatings of the solar cells that would produce a different behaviour with the light incidence angle. Measurements were made under natural sun irradiation. A polar axis tracking system to follow the sun during the process is used. A reference GaAs solar cell and a pyrheliometer (mod. Eppley NIP4) measure the direct irradiance of the sun at the moment of the measurement. The precision to point the tracking system to the sun is 0.2°. A black tube shadows the diffuse radiation over ± 2.7° that may illuminate the reference cell.

As an additional parameter of merit of a device, we can introduce the current density of the whole device under 100 mW/cm² , which is also shown in Table I. Observe that it coincides with the ratio G_c/C_g multiplied by the current density under 100 mW/cm² of the solar cell without RXI and divided by (1-0.058). Precisely, a 5.8% is the area of the front mirror that is an inactive area. Anyway, it has been excluded in the geometrical concentration calculation. Moreover, the last device in Table I is the same than the first one, but the aperture diameter has been reduced from 40 mm to 38 mm. This reduces the geometrical concentration and increases the efficiency.

As yet we haven't measured the value of f_i. The problem arises when trying to illuminate the cell under isotropic conditions within a dielectric media while controlling the amount of power being sent to the cell. In order to have a lower bound of the optical efficiency, we have measure this optical efficiency as the ratio of the short circuit current when the cell is in the RXI to C_g times the short circuit current outside the concentrator but immersed in a dielectric media. The illumination on both cases is 1 sun (indoors) with the angular spread of the sun. Thus the denominator of the short circuit currents is a normal incidence measurement. Since the cell+ARC has a maximum response under normal incidence, this definition of optical efficiency is pessimistic. These measurements gave an optical efficiency between 82-83% [4]. Now, if we consider the theoretical performance of the ARC we can estimate that optical efficiency is in the range of 84-87.8%.

If we consider global electrical efficiency, η, of the device as output electric power between power incident on aperture entrance, it can be decomposed as follows:

$$\eta = \frac{FF\ I_{SC}\ V_{OC}}{B\ A_E} = \frac{FF\ V_{OC}}{B} J_{SC(\text{withoutRXI})} \frac{G_C}{C_g} \quad (4)$$

where B is the beam irradiance, A_E is the concentrator entry aperture area, V_{oc}, I_{sc} and FF are the open circuit voltage, short circuit current and fill-factor of the whole device. The best values obtained for these parameters in Eq. (4) are given in Table II. As can be seen an expected efficiency of 19.3% is obtained. It is important to stress that the best values for these parameters could be obtained in a device of present technology (there is no trade-off between them).

The reason why no single device with an efficiency of 19.3% has been found is that the yield of the complete process, which is not a mass production one, is low. The fill factor degradation is the main cause to limit the efficiency. A value of 0.84 has been routinely measured on

GaAs solar cells at 1000 suns with uniform illumination under normal incidence, but 0.77 is the best one measured inside the RXI concentrator. This is due to the non-uniform illumination produced by the RXI on cell surface as predicted Figure 4. This prediction is experimentally confirmed as Figure 7 shows, where it is seen that light is concentrated producing a spot of high local irradiance gain.

Fortunately, this low fill factor can be improved by re-designing the solar cell semiconductor structure and by a different optical concentrator design. Both tasks are at present under development.

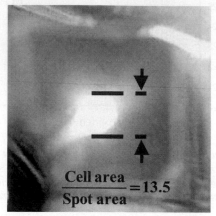

Figure 7: *Photograph of the sunspot produced on the GaAs solar cell inside the RXI.*

Table I: *Experimental measurements of current gain. The geometrical concentration is 1,256 in all cases, excepting the last one, for which is 1,134.*

GaAs solar cell	$I_{SC\text{-}RXI}$ (mA)	Irradiance (mW/cm^2)	$I_{SC\text{-}RXI}$ (mA) (@100 mW/cm^2)	$I_{SC\text{-}1\ sun}$ (mA)	G_c	G_c/C_g	$J_{SC\text{-}RXI}$ (mA) (@100 mW/cm^2)
T84x8	206.5	83.6	247.0	0.2225	1110	0.883	20.9
1-3000	195.8	90.2	217.2	0.2064	1052	0.837	18.3
6	207.7	86.2	241.0	0.2330	1034	0.823	20.4
T86xx	186.5	83.0	224.6	0.2190	1025	0.816	19.0
2	210.3	94.0	223.7	0.2188	1022	0.814	18.9
T84x8	192.0	84.3	227.7	0.2225	1023	0.902	21.3

Table II: *Best values obtained for the parameters shown in equation (4).*

Parameter	FF (at I_{sc}=191-195 mA)	V_{oc} (V)	$J_{sc\text{-}\ without\ RXI}$ (mA/cm^2) (@100 mW/cm^2)	G_c/C_g	η(%) expected
Value	0.77	1.15	24.2	0.902	19.3
Device	T8478, 1-3000	Dis7,1-3000	T8322	T84x8	-

SUMMARY AND CONCLUSIONS

This work has proven the feasibility of a 1000 sun operation PV concentrator. This is linked with the demonstration that the cost of PV energy can be reduced by using the new high concentration optical device together with high efficiency GaAs solar cells which are encapsulated according to optoelectronics industry production techniques. Thus, this work is pre-competitive. Demonstrated the success of the manufactured prototypes, an additional effort will be required in order to build commercial flat concentrator modules.

ACKNOWLEDGEMENTS

This work has been supported by the European Commission within the JOULE-THERMIE program under contract JOR3-CT97-0123 and by the Comunidad de Madrid under contract 07M/0260/1997. The authors would like to thank Keith Emery and Tom Moriarty of NREL for the concentration test of GaAs solar cells.

REFERENCES

[1] C. Algora, I. Rey-Stolle, E. Ortiz, R. Peña, V. Díaz, V. Khvostikov, V. Andreev, G. Smekens and T. de Villers (in this Conference).

[2] J. C. Miñano, J. C. González and P. Benítez, Appl. Optics 34 (1995) 7850.

[3] Universidad Politécnica - C. Algora, Registration number P200001088 (Spain)

[4] These measurements were carried out by Prof. V. Rumiantsev at Ioffe Institute.

BEHAVIOUR OF m-SI PLANT APPROACHING ITS 20-YEAR DESIGN LIFE

G. Travaglini, N. Cereghetti, D. Chianese, S. Rezzonico
LEEE-TISO, CH-Testing Centre for PV-modules
University of Applied Sciences of Southern Switzerland (SUPSI)
CP 110, CH - 6952 Canobbio
Phone: +41 91 / 940 47 78, Fax: +41 91 / 942 88 65, E-mail: leee@dct.supsi.ch

J. Bishop, A. Realini, W. Zaaiman & H. Ossenbrink
ESTI, JRC, I-21020 Ispra (VA)
Phone: +39 0332 789172, Fax: +39 0332 789268

ABSTRACT: The TISO 10 kWp m-Si PV plant was set up on 13th May 1982. It was the first plant in Europe to be connected to the electricity grid. It consists of 252 ASI 16-2300 Arco-Solar m-Si modules and is sited on the flat roof of the LEEE-TISO.
At first, the aim of the plant was to study possible technical and safety problems when connecting a plant to a public grid, whilst now, the aim is the study of mechanical and electrical degradation as well as that of the lifespan of PV modules. The detailed observations of the various PV components and, above all, of the modules supply important information on the critical spots of the PV plants. Observation of the life cycle of the modules makes it possible in particular to verify reliability and working life predictions that are also used for economic evaluation.
The initial electrical performance of the plant was measured in January 1983, and system operation has been monitored continuously. Furthermore, indoor measurements have been performed periodically at ESTI on a reference batch of modules. This combination of systematic monitoring and laboratory measurements provide a unique opportunity to study the system at the end of its 20-year design life. More recently, particular emphasis has been placed on the reliability of the modules: at present a special study is being performed to correlate field reliability with accelerated lifetime tests in order to assess PV module reliability.
Keywords: PV module - 1: Lifetime - 2: Reliability - 3

1. INTRODUCTION

The TISO 10 kWp m-si PV plant was set up on 13th May 1982. It is situated on the flat roof of *Laboratory of Energy, Ecology and Economy* (LEEE-TISO) of the University of Applied Sciences of Southern Switzerland (SUPSI) in Lugano. It was the first plant in Europe to be connected to the electricity grid.

Initially, the main aim of the project was the study of safety and technical problems posed by the connection of PV plants to the public grid. The plant is now used in the study of mechanical and electrical degradation and PV module lifespan. During all its 18 years in operation, the principal electrical parameters have been recorded.

Figure 1: TISO 10kWp m-Si grid-connected PV plant

The configuration of the plant has been modified twice for research purposes. The modifications have already been described in the preceding papers ([1] and [2]).

The substitution of the inverter (1992) caused a modification of the electrical wiring. Part of the plant remained disconnected because of the different number of modules per strings. The initial as well as the present configuration are outlined in table I.

MODULES		
Type of modules: Arco Solar, ASI 16-2300 m-Si		
Module power @STC: 37Wp		
PLANT, connection to the grid: 13 May 1982		
Configuration	Initial	Present, since 92
Nominal power	10.656 kWp	9.324 kWp
N° of modules	288	252
Strings, modules	24 str of 12 mod	12 str of 21 mod
Working voltage	200 V	± 350 V
Array tilt / No field	65° / 3	55° / 3
Inverter	Abacus, 10kW	Ecopower,15kW

Table I: Main features of the TISO 10kW PV plant

2. MEASUREMENTS

2.1 Performances at Standard Test Conditions (indoor)

A sample of 18 modules has been periodically observed and measured at STC during 18 years of operation. This allows, a correlation of field performance with indoor performance measurements, and helps define an energy rating scheme for PV modules. The STC measurements on the references modules have highlighted the stable electrical efficiency of this kind of module, whilst more recent modules either show a decrease after only one year of exposure, or don't even respect the limits of the guarantee already at time of purchase (see [3]).

The average power @STC of the 18 reference modules measured between 82 and 2000 remained virtually stable at 35.08 W, corresponding to 5.2% less than the manufacturer's given value. As figure 2 shows, 16 out of 18 modules did not suffer any obvious degradation outside the measurement accuracy limits (absolute measurement error 2.24%). The reduction in power in February 97 (average power –1.37%) is probably due to a systematic error of measurement since the decrease was recorded for all models.

Only 2 modules show a degradation, which after various fluctuations reached -8.1% and -8.5% respectively (difference in power with respect to that measured in '82); in both modules there is serious water infiltration and performance varies according to the level of dampness. The latest measurements, carried out in February 2000 after two months without rain confirm this hypothesis: power in one of the modules increased.

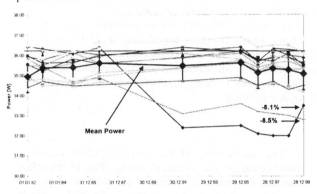

Figure 2: Power @ STC of 18 reference modules and mean power vs. time from 1982 to 2000. The two models which suffered degradation are also included in the mean value

2.2 Performances at real operating conditions (outdoor)

After 18 years the plant is still currently in operation and shows no electrical degradation.

During the course of the last 3 years of service, the plant produced on average 9,185 kWh/year (mean insulation of 1513.4 kWh/m^2.year) i.e. its Final Yield was 985 kWh/kWp.year against the Swiss average of 869. Peak production was reached in 1998 (9685 kWh): the rate of production was 1039 kWh/kWp; the Swiss average in 1998 was 858 kWh/kWp! In 1999 the plant produced 8696 kWh. The drop in production (-10.2%) with respect to the previous year is due to low insolation (7.3% less than '98) and a failure in one module. Nevertheless, the production rate of the plant was 933 kWh/kWp, and despite it being 20 years old and having an inclination which was not ideal (inclination 55°, orientation 7° east), it was still above the average Swiss value (869 kWh/kWp). It must also be said that the nominal power of the inverter (15kW), much greater than plant power, does not make high output possible.

The Performance Ratio of the plant shows seasonal variation, strongly influenced by the operating temperature of the modules. The oscillations occur around the same

value, thus confirming the performance stability of the modules and their reliability over time (see figure 3).

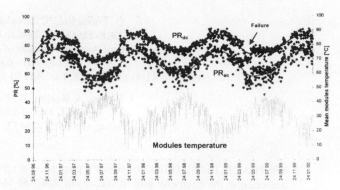

Figure 3: PR$_{dc}$, PR$_{ac}$, and daily mean module temperature of the TISO 10 kW plant

As in previous years, the difference between PR$_{dc}$ e PR$_{ac}$ is higher in summer than in winter. This could be due to a drop in performance of the inverter in summer due , for example, to the operating temperature of the inverter itself (negative influence of high temperatures on power elements), or to a different summer-winter distribution of the daily insulation (in summer the inverter operates for longer periods but at a lower load and so at lower efficiency). In 1999 the mean value of the PR$_{dc}$ was 76%, and the mean value of the PR$_{ac}$ was 65% respectively.

2.3 Comparison with new generation modules

In order to compare ASI module features with more recent ones, 2 16-2300 ASI modules taken from the plant -having thus been in operation for 17 years- underwent testing procedures together with cycle 6 modules (see [3]).

NOCT measurements. The NOCT values of 18 types of modules tested (cycles 5 and 6, [3]) vary from 42.0°C to 51.5°C (average 45.4°C). Leaving aside the Solarwatt modules (solar tiles), the ASI modules have the lowest NOCT (42.7°C); the advantage these modules have in energy production is due to their low A$_{activity}$/A$_{aperture}$ ratio.

Energy comparison. The energy production of ASI modules was excellent, 1,097.1 Wh/Wp with respect to declared power and 1,200.5 with respect to measured power. In the ranking for declared power, the PR for the ASI type (76.4%) is the highest. In the ranking for measured power, its PR (83.6%) is 4th overall. However, the relative difference with respect to the best module is of only 3.4%, whereas other more recent modules have a PR which is up to 16.7% lower than the best module.

Power stability. Despite this module having been exposed under real operating conditions for seventeen years, before the tests, only three c-si modules out of a total of 16 had better results than the ASI module in the 'declared power-power after exposure' field. It should be noted that after 17 years and after testing, this module still performed within the limits of the guarantee. This is not true of most recent modules (14 c-Si modules out of 15 tested), even at time of purchase (see [3]).

These results confirm the exceptional quality of these modules.

2.4 Thermic behaviour

All the ASI modules show local overheating of a cell linked to the terminal box averaging 4°C. In the plant, after 18 years of service, 56 cells show overheating of over 10°C more than the other cells (19.4% of the modules of the field, against 16.7% in '96) of which only one is not linked to the terminal box of the module. A large number of hot spots correspond to cells with oxidized grills. No module has more than one overheated cell. On the whole, temperature distribution inside the module is very regular unlike many models currently on the market.

The actual β coefficient of the ASI 16-2300 under real operating conditions was precisely determined. At 800 W/m², β= –2.132 mV/°C for the cell and –0.0746 V/°C for the module respectively. This value was confirmed (difference 0.9%) by measurements carried out indoors at Ispra on one of the aged modules (see ch 3.2). From this data, it emerges that, although these modules are not of the latest type, their thermic behaviour is similar to the most recent modules.

3. RELIABILITY OF ASI 162300 MODULES

3.1 Natural ageing (outdoor)

Damaged modules. These m-Si modules have shown their reliability: during the 18 years of operation only 8 modules broke down out of a total of 288 modules (2.8% of the modules). In 4 cases, the breakdown was caused by water infiltration of the cell linked to the terminal box.

Mechanical degradation of the modules - visual inspection. After 18 years of exposure under real conditions, in general the modules show numerous evident signs of ageing and mechanical deterioration (brownish colouring, air-bubble, oxidization of the conduction grills of the cell,...) and, in particular, a persistent increase of infiltrations. Nevertheless, the modules haven't suffered a drop in performance.

A visual inspection conducted in '96 brought to light the presence of infiltrations in 74% of the modules (461 infiltrations); in '98 a second inspection, showed a 40% increase in infiltrations (646 infiltrations). At the moment only 4.0% of the modules show no infiltrations.

The infiltrations usually start in the corners of the modules, and are, in most cases, very similar. The increase in infiltrations during the last few years may be due to the increase in the voltage of the system (from 200Vdc to ±350Vdc) resulting from the replacement of the inverter.

The brownish colouring on a number of modules found in the first years of operation of the plant, did not spread and the modules did not suffer any electrical degradation. The oxidization of the conduction grills of the cell linked to the terminal box, which occurred in some modules in the first years of service, has not affected their efficiency in the years that have followed.

3.2 Rapid ageing test (indoor)

3.2.1 Introduction and aim of the test

In collaboration with the ESTI-JRC Centre, rapid ageing tests have been carried out on the ASI modules of the TISO plant in service for fifteen years. The aim of the tests was:

- to study of the final years of the design life (20 years) of the TISO 10 kW plant, in particular to predict the failure mechanisms which will determine the actual lifetime of the PV modules installed on this system;
- to determine the accelerated degradation factors caused by the IEC 61215 qualification tests and investigate as to whether the degradation processes caused by the IEC tests are similar to those found in the field.

Moreover, by carrying out further IEC 61215 tests on a few TISO plant modules, a better overall view of the real working lifespan of PV modules and of the degradation processes involved, has been gained. The increased degradation should correspond to roughly 15-20 years of exposure.

3.2.2 Identification of modules and tests carried out

A batch of ten modules (M1..M10) was removed from the system and subjected to repeated sequences of the CEI/IEC 61215 type approval tests, as follows:

1M: Reference module (M1)
1M: Measurement of temperature coefficients only, as the modules had been already exposed for 15 years (M2)
2M: Reserve modules as laid down by the regulations (M3, M4).
6M: Various ageing cycles carried out (M5, M10). The most important tests were **UVE** (UV test: 15 kWh/m2, 1 week), **TC50** (Thermal Cycling test, 50 cycles of 4 hrs, from –42°C to +85°C), **HUF** (HUmidity Freeze test, 10 days at 85°C, 85% r.h.; **1 hr** per day at –42°C), **TC200** (Thermal Cycling test: 200 cycles of 4 hrs, from –42°C to +85°C), and the **DAH** (DAmp Heat test: 1'000 hrs at 85°C, 85% r.h.). Apart from some other mechanical tests laid down by the regulations, these modules also underwent the following:

- M5, M6:   UVE+TC50+HUF: 2 repetitions
- M7, M8:   TC200:   5 repetitions
- M9, M10:   DAH: 4 repetitions

Apart from these tests, all modules underwent the "Visual inspection / Performance @ STC / Insulation test" before and after each tests.

3.2.3 Pass criteria

A module design shall be judged to have passed the qualification tests, if all the following criteria are met:

a) the degradation of Pm @ STC after qualification tests does not exceed the limit of 5.6% (5% + 0.6% due to the repeatibility of the measurements).
b) no open-circuit or ground fault during the tests; there is no visual evidence of a major defect;
c) no dielectric breakdown or surface cracking; insulation resistance not less than 50MΩ at 500Vdc after the tests.

3.2.4   Summary of results

Electrical performances. No electrical performance degradation was observed after:

- 2 repetitions of the UVE/TC50/HUF cycle (M5&M6)
- 1020 thermal cycles (modules M7&M8)
- 4000 hours of damp heat (modules M9&M10)

Table II shows that average degradation for the 8 modules which had aged was 0.5%. Therefore, with respect to electrical performance, all the modules passed the test with distinction (test failed when degradation > 5.6%). Greatest degradation occurred on the two modules which underwent DAH (Damp Heat) cycles: their power degraded on average by 1.4%.

Module	Performance at STC Pm [W]		Δ Pm
	Before the tests	After the tests	[%]
M5	34.2	34.5	+ 0.9
M6	34.5	33.7	-2.3
Mean			**-0.7**
M7	35.1	35.4	+0.9
M8	34.3	34.4	+0.3
Mean			**0.6**
M9	34.8	34.1	-2.0
M10	35.2	34.9	-0.9
Mean			**-1.4**
MEAN	**34.7**	**34.5**	**-0.5**

Table II: Performance @ STC before and after the tests

Defects. The major defects were caused by the DAH tests –repeated 4 times- executed on modules M9&M10. Delamination of the external layer of the tedlar backsheet from the underlying aluminium foil was observed after 2,000 hours of exposure. Despite this major defect, both modules met the electrical performance and the electrical insulation test requirements. Apart from this defect, the only other major defect recorded was the partial detachment of the terminal box of module M7 after 650 thermal cycles (TC200). However, the delamination of the backsheet, with consequent exposure of a conductive surface (Al foil), might constitue a safety hazard, as the foil will, by capacitive coupling, be raised to a maximum potential above ground equal to the maximum system voltage.

Table III shows the variation of electrical variation which occurred during the tests. The M9 & M10 modules suffered an important reduction in electrical insulation which was much higher than the modules which underwent other tests. For these modules the leakage current increased by 7.7 volts at 0.5kV, and 11.4 volts at 2kV respectively (mean of 2 modules).

Mod.	Initial measure		Final IN measure		Current increase	
	I[μA] at		I[μA] at		Final/Initial	
	2kV	0.5kV	2kV	0.5kV	2kV	0.5kV
M5	4.62	0.29	11	0.84	2.4	2.9
M6	4.14	0.59	9.7	0.61	2.3	1.0
M7	3.5	0.4	6.0	1.4	1.7	3.5
M8	4.4	0.3	4.6	0.48	1.0	1.6
M9	3.6	0.6	42	4.6	**11.7**	**7.7**
M10	4.7	0.9	52	7.0	**11.1**	**7.8**

Table III: Insulation test (IN), before and after the tests

Broken cells during the tests. In module M6 one broken cell was found, but still working, after TC100. In module M10 one broken cell was found, also still working, in two points after DAH1000, and bubbles/delamination on another cell, after DAH4000 respectively.

Finally. Considering that the ASI 16-2300 modules were manufactured in 81/82, and had been exposed outdoors for 15 years, their resistance to the repeated CEI/IEC 61215 tests was remarkable. On the basis of these results, it is reasonable to assume that the modules could continue to provide useful electrical power for at least another 15 years. The DAH test seem to provide the most aggressive conditions. Modules undergoing this test show most electrical (greater reduction of Pm and electrical insulation), and mechanical (delamination of the multi-layer backsheet, broken cell) degradation.

4   CONCLUSIONS

On 13th May 2000 the 10kW LEEE-TISO plant, consisting of 252 ASI 16-2300 m-Si modules, will be 18 years old. In these 18 years only 2.8% of the modules broke down mainly because of water infiltration in the cell linked to the terminal box. At the moment, most modules show clear signs of mechanical deterioration, some brownish colouring, air-bubbles and oxidization of the conduction grills of the cell; 96% of modules show infiltrations. Nevertheless, the plant has the same performance as 18 years ago. Measurements @ STC on 18 reference modules do not show any drop in performance for 16 modules.

The CEI/IEC 61215 artificial ageing tests carried out on 10 modules exposed for 15 years has brought to light their remarkable resistance. After the tests the modules generally show clear signs of degradation but they are still working well, and don't show a reduction in power @ STC. On this basis, it is assumed that the plant can last another 15 years (18+15=33). Safety considerations, rather than degradation of electrical performance, seem to dictate the actual lifetime of the system.

REFERENCES

[1]   M. Camani, N. Cereghetti, D. Chianese and S. Rezzonico, Proceedings 14th EC Photovoltaic Solar Energy Conference (1997), p. 709

[2]   M. Camani, N. Cereghetti, D. Chianese and S. Rezzonico, Proceedings 15th EC Photovoltaic Solar Energy Conference (1998), p. 2058

[3]   G. Travaglini, N. Cereghetti, D. Chianese and S. Rezzonico, 18 types of PV modules under the lens, Proceedings 16th EC Photovoltaic Solar Energy Conference, Glasgow (UK), May 2000

ACKNOWLEDGEMENTS

This project is financially supported by the Swiss Federal Office of Energy and by the Azienda Elettrica Ticinese. The authors would like to extend particular thanks to Dr. Mario Camani.

100 kWp GRIDCONNECTED PV PLANT A13 IN SWITZERLAND
- 10 YEARS AND 1'000'000 kWh LATER

Luzi Clavadetscher, Thomas Nordmann
TNC Consulting AG, Seestrasse 129
CH-8810 Horgen, Switzerland
Phone: ++41 1 725 39 00
Fax: ++41 1 770 10 50
E-mail: mail@tnc.ch

ABSTRACT: 10 years ago this 100 kWp PV-Plant, built on top an existing sound-barrier structure along the A13 motorway in the Swiss Alps, went into operation. At the time this project was unique, as it was the first PV-plant along a motorway worldwide and the largest PV-plant in Switzerland. Also unique is the fact the plant performed extremely well over the whole period.

Monitoring is still carried out and detailed information about the technical operation is available. The plant produces on average 108'000 kWh net energy per annum at a specific annual yield of 1'035 kWh/kWp. This paper will focus on the operational experience over the past 10 years with information on:

Performance - Availability - Reliability - Component failure - Influence of dirt - Maintenance - Non-technical issues

KEYWORDS: Grid-Connected - 1: Performance - 2: Reliability - 3

1. INTRODUCTION

1.1 100 kWp PV Plant

Build in 1989 on top of a soundbarrier along the motorway A13 in the upper Rhine valley in the Swiss Alps, this 100 kWp PV plant has now entered its tenth year of operation. On the 31st of March 1999 the first million Kilowatt-hour was fed into the local grid. After eight months of planing the plant was constructed in eight weeks. In the initial six months the inverter showed some problems, mainly to do with the starting up in the mornings. This was then rectified by the manufacturer and the plant operates smoothly ever since.

1.2 The Project

The purpose of this project is to gain information on the longtherm behaviour of a large gridconnected PV-pant and its components under real operating conditions. Monitoring started at the same time as the plant went into operation. For the first three years the currents of all the 92 strings where also monitored. The monitoring and evaluation is carried out in accordance with the EU-Guidelines for PV Monitoring [1].

The life-span of a PV-system is thought to be about 20 to 25 years and this plant has now reached the halfway mark with its history recorded step by step. Monitoring data is published annually [1]. This paper represents a summary over ten years of monitoring.

2. OPERATIONAL DATA

2.1 Energy Produced

Since going on grid in December 1989 until March 2000 the PV-Plant A13 has operated for 38'157 hours and produced 1'105'712 kWh of net energy at a total availability for the PV-generator and the inverter of 94 %. Over the 10 year period the annual average insolation in the array plane is 1'403 kWh/m^2, the energy produced 109'442 kWh and the energy used 1'863 kWh.

Of the 1.7 % of energy used by the system, 0.4 % is used by the inverter and 1.3 % by the monitoring system. The favourable geographic location with clear sky conditions all the year around and the high reliability contributed mainly to

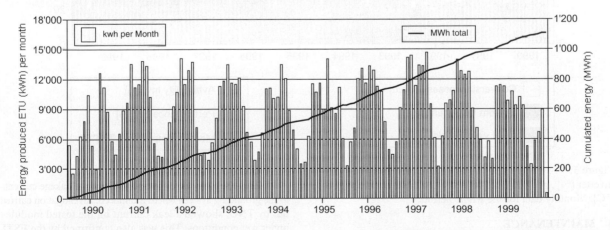

Figure 1: The energy fed into the local grid [kWh] in monthly steps and the cumulated energy [MWh] from December 1989 to March 2000.

the high average annual specific yield of 1'035 kWh/kWp. As most Swiss PV-plants are located in the centre of the country, the monitored Swiss average is 840 kWh/kWp.

2.2 Yield, Performance and Availability

Figures 2 and 3 and table 2 show the relevant operational data for the 10 year period.

The availability of the Inverter is 96 % and the PV-array 99 %. The average performance is 75 %. Figure 3 shows the monthly value of the yield and losses, the availability of the inverter, performance and the module temperature. Table 2 shows all the relevant operational data for each year of the 10 year period.

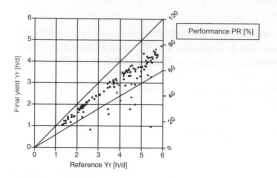

Figure 2:   Monthly performance PR [%] over the whole period, represented as final yield Yf [h/d] over reference yield Yr [h/d].

2.3 Detailed presentation

Figures 4 to 7 show plots of selected hourly values for the whole 10 year period. Each dot represents the mean hourly value at noontime when the plant was in operation. These plots clearly show the smooth operation of the plant.

10 years a total of 33 module had to be replaced. 12 modules where stolen and 21 damaged.

Table 1:   Major Events and Interventions 1989-1999

1989	Dec.	On grid
1990	Aug.	Power measurement
	Oct.	1 Module stolen, 2 Modules damaged
1991	Feb.	1 Module damaged
	Mar.	7 Modules damaged
	April	Module wettest
	Jul.	2 modules stolen
	Sept.	Exchange of all modules
	Oct.	Normal operation
1992	Feb.	1 Module stolen, 1 Module damaged
	Nov.	1 Module damaged
1993	Jan.	DC switch replaced
	Aug.	Rodent damaged datacable
	Sept.	1 Module damaged
1994	Aug.	1 Module damaged
1995	Jun.	"Solar Parcour" opened
	Jul.	Display damaged
	Oct.	All plants on the back removed
	Nov.	7 Modules damaged
1996	Oct.	8 Modules stolen
1997		No major intervention
1998	Aug.	Monitoring system battery replaced
1999	April	Electrical fire in arraybox 3
	Jul.	Maintenance monitoring system
	Aug.	Arraybox 3 repaired

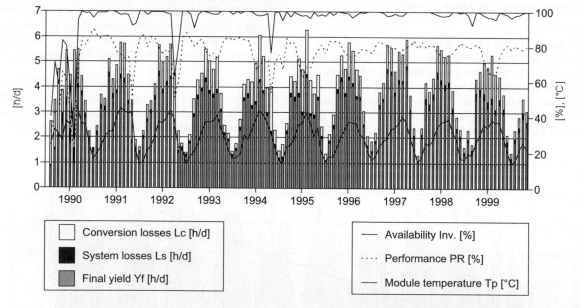

Figure 3:   Yield and losses [h/d] , availability of the inverter [%] , performance [%] and mean module temperature [°C]. Monthly data for the whole period.

3. MAINTENANCE

3.1 General Maintenance

The general maintenance is carried out by the local road department. It includes, the cutting back of plants and also the replacement of damaged and stolen modules. During the last

3.2 Exchange of modules

During the first year of operation, frequent leakage currents to the grounding system where monitored. Investigation carried out by TNC, showed a leak current on the tested modules under wet conditions. This was also confirmed by the ESTI, JRC in Ispra. Subsequently the problem was localised and recognised by the manufacturer which in turn agreed to replace all the modules. In October 1991 all 2'208 modules where replaced. The whole operation lasted for four weeks.

The plant was left running over the whole period and produced a reduced amount of energy.

Table 2:   Relevant operational data, annual values over the whole 10 year period.

3.4 Influence of dirt deposits

During the whole period the surface of the modules was never cleaned. In 1994 and 1995 special measurement where carried out on two strings (NEFF Project 656 [6]). Results showed no significant deposits of dirt caused by road traffic on an inclined surface two meters above the road level.

DOMAT 103.97 [kWp]	Data M —	O —	Operation [h]	Meteo H I [kWh/m²]	T am [°C]	Energies E IO [kWh]	to Grid [kWh]	cum. [kWh]	Yields and losses Y r	Y a	Y f [kWh/(kWp*d)]	L s	L c	PR —	Efficiencies Array —	Inv. —	total —	Availability Inv. [%]	Array [%]	Tp b [°C]	specific yield [kWh/kWp]
1989			200			5'410	5'410	5'410													
1990	1.00	0.26	2'929	1'436	9.7	84'151	82'184	87'594	3.96	2.47	2.23	0.24	1.49	**0.56**	0.067	0.904	0.061	74	98	34.8	790
1991	1.00	0.02	3'840	1'422	9.1	114'785	112'850	200'443	3.90	3.17	3.02	0.14	0.73	**0.78**	0.087	0.956	0.083	98	95	35.2	1'085
1992	1.00	0.05	3'747	1'357	9.6	107'697	106'049	306'493	3.71	2.96	2.83	0.13	0.75	**0.76**	0.086	0.957	0.082	95	100	32.3	1'020
1993	1.00	0.01	3'845	1'360	9.3	112'133	111'082	417'575	3.73	3.07	2.95	0.12	0.65	**0.79**	0.089	0.962	0.085	99	100	31.7	1'068
1994	1.00	0.03	3'852	1'372	10.7	104'602	103'496	521'071	3.76	2.87	2.76	0.11	0.89	**0.73**	0.082	0.961	0.079	97	99	33.4	995
1995	1.00	0.02	3'786	1'383	9.4	106'262	105'011	626'082	3.83	2.95	2.83	0.12	0.88	**0.74**	0.083	0.961	0.079	98	99	31.7	1'010
1996	1.00	0.01	3'858	1'429	9.1	116'189	114'949	741'031	3.91	3.17	3.05	0.12	0.73	**0.78**	0.087	0.963	0.084	99	100	29.8	1'106
1997	1.00	0.01	3'860	1'522	10.0	127'095	125'876	866'907	4.17	3.47	3.35	0.12	0.70	**0.80**	0.090	0.964	0.086	99	100	31.3	1'211
1998	0.98	0.01	3'732	1'396	9.5	113'530	114'808	981'715	3.89	3.23	3.04	0.19	0.66	**0.78**	0.089	0.941	0.084	99	100	30.5	1'079
1999	1.00	0.02	3'706	1'348	9.5	103'207	102'072	1'083'787	3.69	2.91	2.72	0.19	0.78	**0.74**	0.085	0.935	0.079	98	97	31.4	982
2000	0.25	0.01	802	269	2.4	22'202	21'925	1'105'712	2.89	2.48	2.30	0.18	0.41	**0.80**	0.092	0.927	0.085	99	100	21.7	211
total Nov. 89…Mar. 99			38'157			1'117'263	1'103'125	1'105'712													
mean 1990-1999	1.00	0.04	3'715	1'403	9.6	108'965	107'838		3.85	3.03	2.88	0.15	0.83	**0.75**	0.084	0.950	0.080	96	99	32.2	1'035

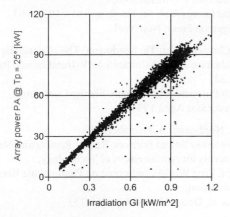

Figure 4:   The array power PA calculated to 25 °C module temperature over Irradiation GI, hourly values at noon over the whole 10 year period.

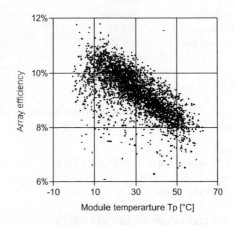

Figure 6:   Array efficiency to the module temperature Tp, hourly values at noon over the whole 10 year period.

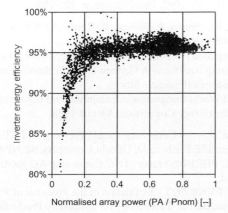

Figure 5:   Inverter efficiency, hourly values at noon over the whole 10 year period.

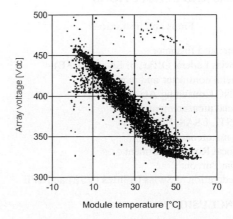

Figure 7:   Array voltage UA to the module temperature Tp, hourly values at noon over the whole 10 year period.

4. TECHNICAL DATA

Table 2:   Technical Data

Modules
Number	2208
Manufacturer	: Kyocera
Type	: LA 361J48, polycrystalline
V stc	: 16.7V dc
P stc	: 48 W

Array
Sub arrays	: $7\,^2/_3$
Strings / array	: 12
Strings	: 92
Modules / string	: 24
Length	: 800 m
Area	: 968 m^2
Nominal power	: 104 kWp
Operating voltage	: 401 V dc
Orientation	: 25 ° east
Inclination	: 45 °

Inverter
Manufacturer	: Siemens
Type	: Simatic
Controller	: Simoreg
Efficiency	: >95 %
Nominal power	: 100 kW

5. COSTS

Table 3:   Investment costs in Swiss francs at 1989 prices and in Euro converted at 1.6 Swiss francs.

Modules	39%			
	832'000	CHF,	520'000	EUR
Inverter	11%			
	229'000	CHF,	143'125	EUR
BOS	45%			
	963'000	CHF,	601'875	EUR
Turn key costs	100%			
	2'130'000	CHF,	1'331'250	EUR

6. FIRMS AND INSTITUTIONS

Table 4:   Firms and institutions

Owner and financier
    Swiss Federal Office of Energy (BFE)
General contractor and monitoring
    TNC Consulting AG
Special measurements
    ESTI, ESAS, JRC, Ispra
Site maintenance
    Local Road Department
"Solar Parcour"
    Federation of Swiss Utilities (VSE) and BFE

7. CONCLUSION

Since this plant was built many more PV noise barrier projects have been realised throughout Europe and Japan. Pilot projects and studies on the potential of PV and noise protection integration where realised [6], [7], [8], [9].

REFERENCES

[1] G. Blässer,
Guidelines for the Assessment of Photovoltaic Plants,
Document A, Photovoltaic System Monitoring,
Document B, Analysis and Presentation of Monitoring Data,
JRC, ESTI, Ispra, Italy 1991/92.

[2] L. Clavadetscher, Th. Nordmann,
100 kWp Grid-Connected PV-Installation along the A13 Motorway in Switzerland • Plant Monitoring and Evaluation • Operation and Maintenance,
Annual Reports for the Swiss Federal Office of Energy, 1990 to 1999,
Project Number: 32046.

[3] Th. Nordmann, L. Clavadetscher, R. Hächler,
100 KWp Grid-Connected PV-Installation Along Motorway and Railway,
Photovoltaic Demonstration Projects, pp 133 - 139, CEC Publication, Luxembourg, 1991.

[4] L. Clavadetscher, Th. Nordmann,
Prediction and Effective Yield of a 100 kW Grid-Connected PV-Installation,
Solar Energy Vol. 51, No. 2, pp 101 - 107, 1993, Pergamon Press, New York.

[5] L. Clavadetscher, Th. Nordmann, TNC Consulting AG,
Evaluation of Grid-Connected PV-Installations, Paper: 1B.32,
12th European Photovoltaic Energy Conference, Amsterdam April 1994.

[6] Th. Nordmann, A. Goetzberger,
 Motorway Sound Barriers: Recent Results and New Concepts for Advancement of Technology,
IEEE First World Conference on Photovoltaic Energy Conversion,
Hawaii, December 1994, p. 766-773

[7] Th. Nordmann, A. Goetzberger,
Motorway Sound Barriers: The Bifacial North/South Concept and the Potential in Germany,
13. European Photovoltaic and Solar Energy Conference and Exhibition,
23.-27. 10. 1995, Nice, France, p. 707-709

[8] Th. Nordmann, A. Frölich, K. Reiche, G. Kleiss, A. Götzberger,
Integrated PV Noise Barriers: Six Innovative 10 kWp Testing Facilities - A German Swiss Technological and Economical Success Story!,
2nd World Congress and Exhibition on Photovoltaic Solar Energy Conversion, Vienna 1998.

[9] TNC Energy Consulting GmbH Freiburg D, ENEA Rom, ISE Freiburg D, Utrecht University Nl, NPAC UK, PHEBUS France, TNC Consulting AG Switzerland,
EU PVNB POT - Evaluation of the Potential of PV Noise Barrier Technology for Electricity Production and Market Share,
EU Thermie B Project, 1999

EMC and Safety Design for Photovoltaic Systems (ESDEPS)

T. Degner[1], W. Enders [2], A. Schülbe[3], H. Daub[4] *et.al.*

(1) Project co-ordination: ISET e.V., Königstor 59, D-34119 Kassel, Germany
Phone: ++49–(0)561–7294–224, Fax –200 Email: abt-a@iset.uni-kassel.de
Project page: http://www.iset.uni-kassel.de/esdeps
(2) ÖFPZ Arsenal (arsenal research), Faradaygasse 3, Objekt 221, A-1030 Wien, Austria
(3) Elektrizitäts AG Mitteldeutschland, Monteverdistraße 2, D-34131 Kassel, Germany
(4) EnEtica S.L., Calle Nuestra Senora del Carmen 1 Ba A, E-35640 Corralejo, Spain

Abstract

Within the framework of the project ESDEPS (EMC and Safety Design for PV Systems) electromagnetic compatibility (EMC) and safety aspects of PV systems are investigated in detail. The findings from these investigations shall be the basis for the improvement and/or creation of standards concerning the EMC and safety of PV systems.

Topics covered by the project are investigations regarding the electromagnetic environment, like the effect of lightning on PV systems and the effect of transients on the mains on PV inverters, as well as investigations with respect to emissions from PV systems on the mains and DC lines and radiated emissions at radio frequencies.

Keywords: Safety - 1: Qualification and Testing - 1: Grid-Connected - 2

1 Introduction

The aim of this project is to make a contribution to improve the reliability, safety and quality of PV systems in order to achieve competitiveness in future electricity markets. Therefore the complementary parts of system design i.e. electromagnetic compatibility (EMC) and safety aspects are considered together to ensure compliance with all essential requirements. The results and conclusions of the project will be submitted to the relevant standardisation committees to further the European harmonisation process.

The project consortium consist of:

ISET e.V., Germany (Project co-ordination): A non-profit research institute, associated with Kassel University.
arsenal research, Austria. An independent demand oriented service enterprise for applied research and testing.
EAM, Germany. A central German energy provider.
EnEtica S.L., Spain. A SME which projects and installs solar and wind energy systems at the Canary Islands.

Currently the main activities in the project are the investigations regarding the effect of lightning on PV systems, the effect of transients on the mains on PV inverters, emissions from PV inverters into the mains and emissions at radio frequencies from PV systems.

2 Influence of the electromagnetic environment and immunity tests of PV-inverters

2.1 Influence of lightning

PV-systems may be effected by lightning in a very high degree. The parameters of the usual tests approach from measuring data which are about 40 years old and were made with devices with not enough bandwidth. But the electronic-components inserted in recent state of the art systems often are very sensible for high frequency inductions. For a systematic analysis of the effect of lightning strokes in PV-systems, arsenal research has installed a measuring station for the automatic recording of induced voltages and currents occuring in PV-systems and special PV-components in the

frequency range up to 50Mc. Fig. 1 and 2 give an impression about the experimental set-up. The station is located on top of the mountain "Gaisberg" (with an altitude of 1300m) near Salzburg. In a distance of about 150m there is an ORF-broadcasting station with a measuring system for the lightning current of a direct stroke into the antenna mast (with a height of 100m). These measurements are performed in cooperation between ALDIS (Austrian Lightning Detection & Information System, ÖVE) and TU Vienna.

Figure 1: Measuring station of arsenal research

The measuring station of arsenal research consists of some elements (e.g. PV-panels and antennas, an automatic high-speed camera oriented to the top of the antenna mast) mounted on a scaffold and some data recording systems located inside of a shielded cabin which is mounted on a trailer. Furtheron a lightning rod with a height of about 14m is mounted beside the scaffold. Recorded data are the voltage of some

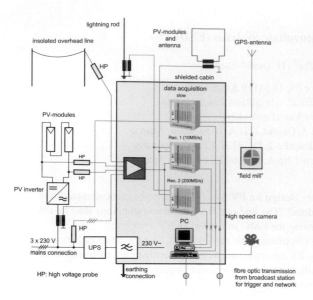

Figure 2: Measurement setup on the station Gaisberg.

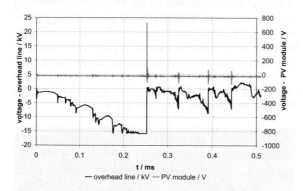

Figure 3: Measurement of induced voltages caused by a lightning stroke.

PV-generators connected with PV-inverters, currents of PV-panels in short circuit-operation (screened and not screened), voltage of an overhead-line and an underground-cable, E-field and B-field and the current of direct stroke into the lightning rod. The Arsenal-measuring-stations of ALDIS/TU Vienna and arsenal research are connected via fiberoptic-transmission lines. This gives, in the case of a direct lightning stroke into the antenna mast an unique opportunity of having recorded the inducing current in a known direction and distance and to the same time the induced voltages and currents in different elements which will give the base of better theoretical analysis of the relevant parameters of induced quantities. Fig. 3 shows an example for near lightning stroke induced voltages into a overhead line with a peak-voltage of 15kV and a voltage gradient of $1MV/\mu s$ and into a single PV-panel with a amplitude of 700V.

The aim of the tests will be the definition of immunity levels for PV-inverters, definition of test parameters and development of suited protection-concepts.

2.2 Influence of transients in the mains, caused by switching and failures

Utility interactive inverters are influenced in a high degree by transient phenomena in the grid. The most important are switching phenomena and disturbances caused by earth-faults and short circuits. A very high stress for the utility-connected inverters occur, if a short circuit in a connected branch is interrupted by a fuse or a circuit-breaker. In this case first the voltage will break down and the short-circuit-current in the utility will increase. In the moment of the fuse interruption the energy stored in the inductive components of the utility will be free and generate transient overvoltages. Recently in international standards there do not exist equiv-alent test-pulses for this kind of stress though it can happen relative probable and can generate defects in PV-inverters, as some tests in arsenal research showed. The energy of these transients depend on many parameters like the mo-ment of the short circuit, net impedance at the short-circuit-point, kind of the fuse or circuit-breaker. The experience showed, that fuses with "fast characteristic" cause the high-est steepness of the occuring transients. Arsenal research has developed a special test-setup for a systematic analysis of these phenomena with the aim of generation of new stan-dard test definitions. Fig. 4 shows the response of a small inverter (130W) to a short circuit in a device connected to the mains interrupted by a 6,3A fuse with fast characteristic. The voltage reaches a maximum of 750V and in this mo-ment the current flowing into the inverter has a maximum of 48A. But in this case it caused no inverter damage.

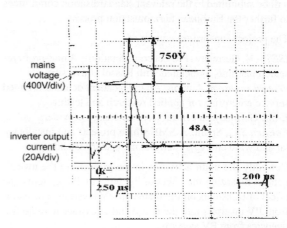

Figure 4: Response of a small inverter (130W) to a short circuit in a device connected to the mains interrupted by a 6,3A fuse with fast characteristic.

The switching of devices connected to the same utility as the inverter can lead to transient overvoltages and voltage dips. These transients can be simulated on the arsenal-PV-inverter-test-stand and the behaviour PV inverters of differ-ent construction-concepts were tested there under variation of different parameters like moment, duration and ampli-tude. The behaviour of the single inverters was very dif-ferent dependent of their individual construction concept. Some of the inverters were very robust against such distur-bances, but one type was so sensible, that even voltage dips of 5% lead always to disconnection, and another type of PV-inverters in the power class of 100W sinks a peak-current of 48A after a half-wave voltage-dip to zero (Fig. 5).

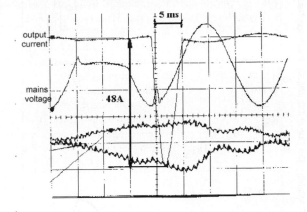

Figure 5: Response of a small inverter to a half-wave voltage-dip.

3 Emissions at radio-frequencies

One important topic of the project is the investigation of emissions from PV systems at radio-frequencies. These investigations aim at the developments of methods, measures and procedures to ensure the EMC of the PV systems. They imply (i) the investigation of the source of signals at radio frequencies (e.g. the PV inverter) and (ii) the investigation of the propagation mechanisms.

3.1 Sources of radio-frequency signals: EMC Tests on PV inverters

Comprehensive tests concerning the emissions of signals in the frequency range 150kHz up to 1 GHz have been performed on eight up-to-date PV inverters for grid parallel operation.

Table 1 gives an overview about the tests. The test have been performed according to the requirement for devices in household applications. For applications in an industrial environment the required tests are different:

- measurements of rf voltage and power on DC lines are not required.
- measurements of the rf field strength requires a greater distance (30m instead of 10m), so the results from the measurements cannot easily adapted to an industrial environment
- measurements of rf voltage on AC lines is required. Here the limit values are higher than for household applications.

The test results were rated according to the limit lines for household applications: Inverters below the limits lines were rated "good", less than 10dB above the limit lines "critical" and more than 10dB above the limit with "bad". The result is shown in Figure 6 and may be summarised as follows:

- many inverters exceed the limits.
- Only one out of eight inverters tested is below the limit lines for radio field strength.
- Half of the tested inverters exceed the limits for rf voltage on DC lines.

For an industrial environment the rating for the emissions on AC lines is: 5 inverters are "ok", none are "critical" and 3 are "bad". As mentioned above in an industrial environment a rating for the other emissions is not required (emissions on DC lines) or not easy to derive (radio field strength).

Other results from the tests are, that the amplitude of the emissions may show a strong dependency on the working point of the inverter.

The emissions on the DC lines of the PV inverter are very important, because the PV generator and the wiring may acts as an antenna and lead to radiated disturbances. This topic is subject of the next section.

freq. range	test	applied standard
0.15–30MHz	rf voltage on AC lines	EN50081-1
0.15–30MHz	rf voltage on DC lines	EN55014-1
30–300MHz	rf power on DC lines	EN55014-1
30–1000MHz	rf field strength	EN50081-1

Table 1: EMC tests performed on eight PV inverters

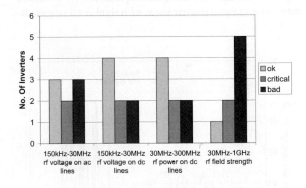

Figure 6: Summary of PV-inverter tests: The emissions are compared to the limit lines which apply for household applications. Rating:
"ok": below the limits
"critical": limits exceeded by less then 10dB
"bad": limits exceeded by more than 10dB

3.2 Propagation of radio signals in PV systems

The PV inverters tests presented in the previous section showed, that there are signals on the DC side of the inverters which may have a significant amplitude in a frequency range up to about 100 MHz. The question of how do these signals propagate in the PV generator and of how strong is the radiated signal is the topics of this section.

Therefore fundamental investigations of the rf impedance and radiation properties of PV cells and modules were performed first in order to characterise these components from an rf engineering point of view /2/. At present measurements on an experimental PV system (about $400W_p$ rated) are performed. The topics under investigation are here the radiation characteristics of the PV plant, e.g. the antenna factor of the system. The investigations are currently performed and the results shown here should be considered preliminary.

Fig. 8, top shows the radiated magnetic field measured in front of the PV generator at a distance of 3 m. The PV generator was connected to a PV inverter. The common mode disturbance current measured at the system is shown in the middle of the figure. For comparison purposes the radio frequency voltage measured on the DC lines of the inverter is shown. Here the inverter was connected to an artificial mains network with 150Ω rf terminating resistor.

It should be noted that the disturbance current measured at the PV system is in the same order of magnitude compared to the disturbance current calculated from the disturbance voltage at the artificial mains network. Disturbance current and radiated magnetic field show a frequency dependent relation.

The figures show, that if the signals on the DC lines exceed the limit lines, the limits for the radiated field must not be exceeded. However, the opposite is true also. Further investigations aim at the determination of the "antenna factor" for the PV system to enable the estimation of radiated fields from the amplitude of the conducted signals.

Fig. 7 shows, that even at higher frequencies a PV generator may cause radiated electric fields with significant amplitude.

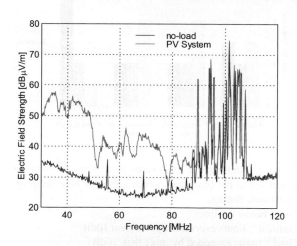

Figure 7: Electric field strength measured in front of our experimental PV system. The PV generator is connected to a commercially available PV-inverter. For comparison shown is a measurement with the PV system not in operation.

4   Conclusion

A significant high share of PV inverters currently available on the market produce signals at radio frequencies above the limit lines for household devices.

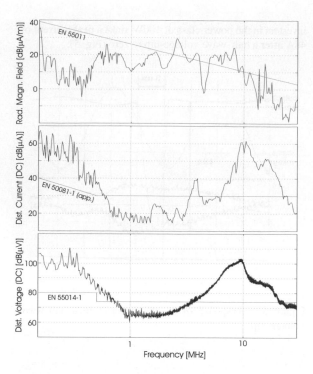

Figure 8: Top: Radiated magnetic field measured in front of a PV generator. Middle: Common mode disturbance current measured at the system. Bottom: Radio frequency voltage (inverter connected to artificial mains network)

Radio frequency signals on DC lines from PV inverters may lead to significant radiated electric fields.

The experiments on the lightning station in Austria will form the basis to define immunity levels for PV inverters, to develop suited protection concepts and to define test parameters.

The investigations at PV inverters with respect to the response to various transient current and voltage peaks will help to develop suited test set-ups and criteria for PV inverters.

References

/1/ H. Wilk, G. Schauer, H. Harich, W. Enders: "Testing Inverters for utility interactive operation". 2nd Photovoltaic World Conference, Vienna, 6.-10. July 1998

/2/ N. Henze, J. Kirchhof, B. Rothauge. "Hochfrequenzeigenschaften von photovoltaischen Generatoren - Die Ausbreitung von Funkstörspannungen in Photovoltaikanlagen". 15th Symposium Photovoltaische Solarenergie, Staffelstein 2000.

Acknowledgement

This work is funded in part by the European Commission in the framework of the Non Nuclear Energy Programme JOULE III under contract No. JOR3–CT98–0246.

LIGHTNING AND OVERVOLTAGE PROTECTION IN PHOTOVOLTAIC (PV) AND SOLAR THERMAL SYSTEMS

H. Becker, W. Vaaßen, F. Vaßen

TÜV Immissionsschutz und Energiesysteme GmbH, Am Grauen Stein, D-51105 Cologne, Germany
Phone: +49/221/806-2476, Fax: +49/221/806-1350, e-mail: beckerh@de.tuv.com

Ivan Katic, Miroslaw Bosanac
Danish Technological Institute – DTI, Gregersensvej, P.O. Box 141, DK-2630 Taastrup, Danmark
Phone: +45/4350-4567, Fax +45/4350-7222, e-mail: Ivan.Katic@teknologisk.dk

ABSTRACT: Photovoltaic as well as solar thermal systems employed for the exploitation of solar energy are subject to the climatic effects of the given surroundings in a manner depending on the system in question. Most importantly, because of their frequently exposed location on roofs or as free-standing installations in unsheltered areas, these systems risk damage from direct or indirect lightning.

Through measures for the protection against lightning and overvoltage, the planned operating times of the installed systems can be assured and economic expectations fulfilled. In this article, we explain the origin and possible effects of lightning. On the basis of existing standards and previous experience, we describe measures for lightning and overvoltage protection, and illustrate these measures by examples.

The complete results of the studies on this topic are presented in a brochure available from the European Commission Directorate General for Energy and Transport or from the authors.

Keywords: Lightning, Overvoltage - 1: Safety - 2: PV System - 3

1. INTRODUCTION

The utilisation of regenerative energies has greatly increased within the European Union since the 1980s. This development has been supported by the extensive promotion programmes not only of the EU but also of the Member States. Especially in the sector of one-family houses and duplex houses as well as in the commercial sector, many systems have been established for the generation of electrical power from sunlight (photovoltaic systems) and for water heating (solar thermal systems).The preferred locations for the installation of these systems are roofs or facades. Consequently, the systems are often very exposed and therefore endangered by direct strikes of lightning or the induction of overvoltage following near-strikes. Nevertheless, the establishment of a photovoltaic or solar-thermal system does not in itself necessitate setting up a lightning protection system (LPS). On the other hand, the investment value of photovoltaic and solar-thermal systems may be high enough to make their protection advisable. Lightning and overvoltage protection measures should therefore be carried out according to the risk of damage from lightning and according to the value of the systems to be protected.

2. THE FORMATION OF THUNDERSTORM CELLS AND THE PROCESS OF LIGHTNING DISCHARGE

The basis of any thunderstorm activity is the formation of thunderstorm cells. For the formation of these cells warm masses of air with sufficiently high humidity must be transported to the appropriate heights.

During this process, ice particles and water droplets are electrostatically charged through friction. For physical reasons, the negative charges migrate to the lower portion of the cloud and the positive charges to the upper region.

The charge separation causes the thunderstorm cell to become electrostatically charged.

Besides this charge separation within the thunderstorm cell, the lower negative cloud charge causes all negative charge carriers on the ground to be displaced. As a result, the surface of the earth is positively charged.

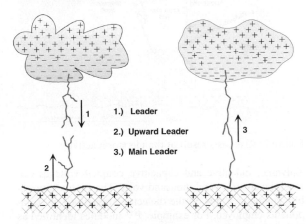

1.) Leader

2.) Upward Leader

3.) Main Leader

Figure 1: Negative downward flash

If the local electrical field strength in the space between the charges increases to several 100 kV/m, then lightning discharge can occur. More than 80 % of lightning discharges take place within the thunderstorm cells – cloud-cloud-flashes. Lightning discharges to the earth mainly originate in the middle, or negative, section of the thunderstorm cloud. At a proportion of about 90 %, these negative downward flashes are the most frequent type of discharge to the earth.

A so-called „leader" moves from the negative charge centre of the cloud towards the earth. The forward expansion of the leader moves in jerks of several 10 m with a velocity of approximately 300 km/h. When the leader has come within a distance of 10 m to 100 m from earth, the field strength between the head of the leader (negative potential) and protruding peaks on the earth (positive potential) increases so greatly that an „upward leader" extends from the earth towards the leader.

The point of strike of the lightning flash is then determined. Once the leader and upward leader have met, the main discharge occurs from the earth to the cloud at a speed of approximately 100,000 km/s for a period of up to several 100µs. The spark channel can be heated up to 30,000 °C, so that the pressure of the heated air is increased to as much as 100 bar. Thunder is caused by the explosion of the spark channel.

3.   THE   OVERVOLTAGE   EFFECT (GALVANIC, INDUCTIVE AND CAPACITIVE COUPLING)

Each thunderstorm affects its surroundings not only through the electric field of the thunderstorm cells but also through the electromagnetic field of the actual lightning discharge itself. These electric and magnetic fields induce voltages and currents in electrical systems and electrically conductive constructions. Depending on the tripping influencing variable of the overvoltage, we speak of inductive, capacitive or galvanic coupled overvoltage.

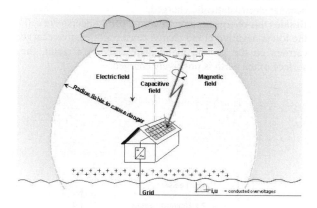

Figure 2: Risks as a result of thunderstorm activity

Galvanic, inductive and capacitive coupled voltages can reach levels of several thousand volts. The extent of possible damage depends on the dielectric strength of the components employed. For example, PV modules identified as Safety Class II Equipment are tested at a voltage of 6000 V. Consequently, as a rule they are sufficiently surge-proof against coupled voltages.

Galvanic coupling: Direct lightning strikes at PV systems or solar thermal systems drive currents through the mechanical components of these systems towards the earth. If, as a consequence of the lightning discharge, the insulation of the electrical equipment fails, then parts of the lightning strike current will also flow through this electrical equipment in the direction of the earth. We then speak of galvanic coupling.

Inductive coupling: Each lightning discharge is accompanied by the creation and subsequent collapse of a magnetic field, with the field surrounding the flash channel of the

discharge path and all conductors that carry the lightning strike current. This change in the field induces hazardous voltages in building constructions and all electrical installations. The level of the voltage is basically determined by the rate of increase of the lightning current, the distance to the discharge path and the magnitudes of the conductor loops of the technical equipment. Not only direct strikes but also near-strikes induce voltages in all conductor loops that can destroy the electrical equipment of the PV system.

Capacitive coupling (PV generator)
If a PV generator is located within the electrical field of a thunderstorm cloud, the freely moving charge carriers in the metals and semiconductors of the PV generator are displaced; a charge separation occurs. The charge transfer continues until the electrical field no longer exists at the interior of the PV generator. The electric field that collapses at the moment of lightning discharge causes a charge transfer. The charge then flows through all conductors connected to the earth in the form of a transient wave. Depending on the system design, these are the active conductors or the earth conductors of the PV generator frame.

4.   DECISION ON THE NEED FOR A LIGHTNING PROTECTION SYSTEM (LPS)

To assess the need for protecting a construction, the acceptable frequency of direct lightning strikes is compared with the number of expected lightning strikes. The calculation proceeds according to the standard ENV 61024.

Acceptable frequency of direct lightning strikes

$$N_c = A \cdot B \cdot C \cdot D$$

A,B,C,D are individual factors such as
- investment value of the solar system
- investment value of any equipment also damaged
- demands on availability
- other consequential damage

Probability of direct lightning strikes:

$$N_d = N_g \cdot A_e \cdot C_e \cdot 10^{-6}$$

N_g the average density of the ground flashes per km²
A_e the equivalent air terminal area
C_e a coefficient determined by the surroundings of the object to be protected

Example: The fewest thunderstorms are registered in the coastal regions of northern France, Holland, Germany and Denmark. For example, we statistically expect 1 stroke of lightning every 858 years ($N_d = 0.0012$) for a terraced house with the dimensions $L \cdot W \cdot H = 10 \cdot 8 \cdot 8 \; m^3$ in these regions.

The estimations of the acceptable annual number and the anticipated annual number of direct lightning strikes can then be used as the basis for deciding on the necessity of an external lightning protection system.

Comparing the two variables N_c and N_d, we can state the following:

$N_d \leq N_c$: external LPS not needed

$N_d > N_c$: external LPS is required

The components of an external LPS:
Air-termination systems occasionally are simple lightning rods, but usually are a combination of smaller lightning rods and taut wire meshes mounted directly on the main body of the object to be protected. This arrangement creates a protection zone where, with a certain statistical probability, no direct lightning will occur. The efficiency of an LPS is defined by the number of lightning flashes that can reliably be intercepted and diverted, and is standardised in the ENV 61024-1:1995 in terms of Safety Classes.

Safety Class	Efficiency
I (extreme)	0.98
II (high)	0.95
III (normal)	0.90
IV	0.80

Figure 3: Safety Classes and efficiencies of lightning protection systems

The task of a **down-conductor system** is to divert a coupled lightning current from the air-termination system to the earth-termination system.
The **earth-termination system** should safely divert and distribute the lightning current over the earth. In most cases earth-termination systems are designed as ring earth electrodes or foundation earth electrodes.

5. METHODS OF OVERVOLTAGE PROTECTION

The fundamental idea of modern overvoltage protection is to equalise all different potentials to a common potential at the instant of overvoltage, thereby preventing flash-arcs. Equipotential bonding is accordingly the most important overvoltage protection measure. To achieve equipotential bonding, all electrically conductive components of the object to be protected must be connected to one another, to the bonding bar and to the earth electrode. Generally these are system components such as metal supporting structures and mounting structures, piping or operational earthed conductors of the electrical systems. Naturally active conductors cannot be directly included in equipotential bonding, as this would be equivalent to a short-circuit. Lightning arresters have therefore been developed that have sufficient resistance during normal operation and that induce only a brief short-circuit once a certain voltage difference between the back-to-back insulated potentials has been exceeded.
Lightning arresters of quality Class B, also known as lightning current conductors, are designed for the down-leading of direct lightning currents. Models exist for lightning strike test currents from 25 to 100 kA. Depending on the diverted lightning strike current, lightning current conductors limit the overvoltage to values under 4 kV. Lightning arresters of quality Class C limit coupled overvoltage,

overvoltage from remote strikes and switching overvoltage. They need only divert currents with distinctly lower peak values, charges and specific energies. These arresters have a discharge capacity of at least 15 kA at a current waveform of 8/20 μs and limit the voltage to well under 2 kV.

6. APPLICATION EXAMPLES OF PV SYSTEMS

6.1 Roof-mounted, grid-connected PV systems
without protection against direct lightning strikes

Several dangerous overvoltages within the DC circuit must be expected in a grid-connected PV system during a target service life of several years. The generator electric circuit should therefore be protected by lightning arresters of Quality Class C, located as close as possible to the PV modules. If a wire length of approximately 15 to 20 m between these arresters and the inverter is not exceeded, then lightning arresters are not necessary directly in front of the inverter.

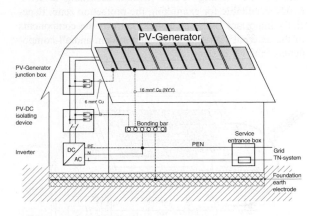

Figure 4: Roof-mounted, grid-connected PV system without protection against direct lightning strikes

Should these arresters be necessary in individual cases, then the DC isolation point is suitable for installation of the arrester. The DC isolation point must then be located in the vicinity of the inverters.

6.2 Roof-mounted, grid-connected PV systems
protected against direct lightning strikes

With PV systems protected against direct lightning strikes, the external lightning protection system must be designed so that all PV components are located within the protection zone. As a rule, this is possible only by means of appropriately located lightning peaks.

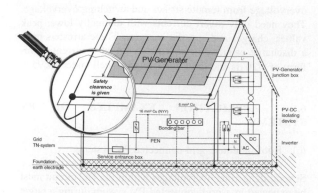

Figure 5: Roof-mounted, grid-connected PV system protected against direct lightning strikes (with safety clearance)

If the roof incline is too low or if the PV generator juts from the roof, then usually smaller peaks are also required on inclined roof areas in addition to the air termination network. The rolling sphere method according to ENV 61024 is suitable for examining the protection zone. Especially important is the clearance between the components of the external lightning protection system and all components of the PV system.

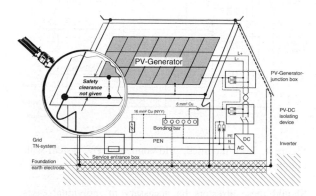

Figure 6: Roof-mounted, grid-connected PV system protected against direct lightning strikes (without safety clearance)

Arc flashovers must be safely prevented. If the clearance is more than 1 m it may be assumed that the clearance is sufficiently large. In all other cases the requisite minimum clearance can only be determined through calculations according to ENV 61024. As the comparison of figures for "safety clearance is given " and "safety clearance is not given" shows, the exposure problem affects the equipotential bonding and its cross-sections. Safety clearance is always desirable as a means to prevent harmful partial lightning currents in the PV system.

If a wire length of approximately 15 to 20 m between the arresters at the PV generator and the inverter is exceeded, additional lightning arresters must be installed directly in front of the inverter. The DC isolation device is suitable for the installation of the arrester, which must be located nearby the inverter, however.

6.3   Mountain huts as an example of PV stand-
        alone systems

Any PV component (PV generator, BCR, battery inverter) can be damaged by lightning strikes and overvoltage. These components can be protected using the same methods as already described for the other applications:
- Due to their exposed locations, an external LPS is usually required for mountain huts.
- All PV components should be located within the protection range of an external LPS.
- Protection of the active conductors by interconnection with lightning arresters: if the sum of the consumer-side wires does exceed an order of magnitude of approximately 10 m, then this consumer-side should also be protected by lightning arresters.
- Equipotential bonding.

7.   SUMMARY

Lightning and overvoltage protection of solar systems is feasible both technically and economically for the various system configurations. The extra cost (approximately 1% of the system cost) is of secondary importance, given the increase in operating safety and availability.

8.   ACKNOWLEDGEMENTS

This publication is based on results obtained in a project which was founded by the EUROPEAN COMMISSION Directorate General for Energy and Transport.

9.   REFERENCES

[1] F. Vaßen und W. Vaaßen:
Bewertung der Gefährdung von netzparallelen Photovoltaik-Anlagen bei direktem und nahem Blitzeinschlag und Darstellung der daraus abgeleiteten Maßnahmen des Blitz- und Überspannungsschutzes. VDE/ ABB-Blitzschutztagung, Neu-Ulm, November 1997

[2] F. Pigler, VDB-Info 1: Blitzschutz von Photovoltaik-Anlagen
Edition 9/94, Druckerei Hans Zimmermann, Cologne

[3] H.Häberlin and R.Minkner: Tests of Lightning Withstand Capability and Measurements of Induced Voltages at a Model of a PV System with ZnO Surge-Arresters".
Proc. 11th EC PV Conference, Montreux 1992, pp. 1415-1418

[4] H.Häberlin and R.Minkner: A Simole Method for Lightning Protection of PV Systems.
Proc. 12th European Photovoltaic Solar Energy Conference, Amsterdam 1994

REMOTE MONITORING OF PV PERFORMANCE USING GEOSTATIONARY SATELLITES

Richard Perez & Marek Kmiecik
ASRC   251 Fuller Rd.
Albany NY 12203
perez@asrc.cestm.albany.edu

Christy Herig & David Renné
NREL   1617 Cole Blvd.
Golden CO 80401
christy_herig@nrel.gov

ABSTRACT: Satellite remote sensing could potentially span entire continents, in real time, with high ground resolution. This paper investigates geostationary satellites to monitor the performance of ground-based PV arrays. A comparison between the actual output of photovoltaic (PV) power plants and satellite-simulated output estimates is presented. The paper shows evidence that the satellite resource can adequately monitor PV performance and detect potential problems and strengthen the credibility of ongoing remote PV monitoring efforts (e.g., [1,2])

1. APPROACH

2. 1 Simulation Models

Intermediate resolution (~10 km) images from the visible channel of the Goes East & West satellites are used as primary input [3]. Irradiances are modeled from these images following approaches previously developed by the authors and colleagues [4,5,6]. Ancillary model inputs include meteorological data (i.e., regional temperature and wind speed) readily available weather services.

PVFORM is used for irradiance-to-PV modeling [7]. We modified it to handle the custom multi-row 1-axis tracking procedure used in two of the considered PV systems -- this procedure, which prioritizes the elimination any row-to-row shading, involves non-ideal tracking and some backtracking in early AM and late PM hours.

2.2 Ground Truth

Six PV arrays from the UPVG program [8] were selected. Sizes range from 2.5 to 140 kW. Geometries include both fixed and one-axis tracking systems. Climates range from arid to humid continental (see Figure 1).

Ground truth data include plane-of-array irradiance (POAI) and PV-ac output. This information enables a

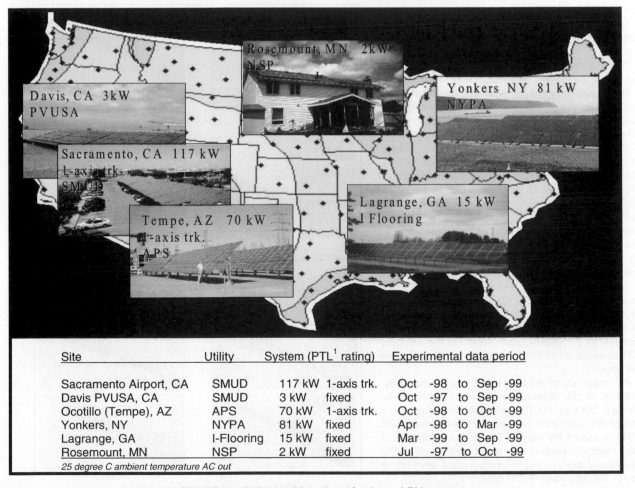

Site	Utility	System (PTL[1] rating)		Experimental data period			
Sacramento Airport, CA	SMUD	117 kW	1-axis trk.	Oct	-98	to Sep	-99
Davis PVUSA, CA	SMUD	3 kW	fixed	Oct	-97	to Sep	-99
Ocotillo (Tempe), AZ	APS	70 kW	1-axis trk.	Oct	-98	to Oct	-99
Yonkers, NY	NYPA	81 kW	fixed	Apr	-98	to Mar	-99
Lagrange, GA	I-Flooring	15 kW	fixed	Mar	-99	to Sep	-99
Rosemount, MN	NSP	2 kW	fixed	Jul	-97	to Oct	-99

25 degree C ambient temperature AC out

Fig. 1 Description and location of selected PV arrays

TABLE 1: Overall RMS and Mean Bias errors

	DAYTIME AVERAGE IPOA W/sq.M	MBE %	RMSE %	DAYTIME AVERAGE PV-OUT PER KW	MBE %	RMSE %	Satellite PV Minus Satellite POAI	UPVG simulation Minus Measured PV
Sacramento	453	8%	20%	407	22%	36%	14%	14%
Davies	452	-8%	22%	367	16%	50%	27%	39%
Ocotillo	639	4%	21%	610	0%	22%	-3%	1%
Yonkers	329	9%	26%	328	13%	32%	5%	9%
Lagrange	362	9%	26%	312	26%	44%	16%	14%
Rosemount	368	-4%	38%	328	11%	50%	16%	18%

precise accounting of errors between the satellite-to-irradiance and irradiance-to-PV models.

2. 3 Performance Evaluation

The performance validation benchmarks include short-term errors (hourly RMSEs) and long-term periodic errors (e.g., monthly MBEs). The former is relevant to applications involving a real-time knowledge of PV output (e.g., grid interaction, peak load management, outage mitigation analyses), while the latter is relevant to the focus of this paper -- performance feedback and remote metering applications.

3. RESULTS

3. 1 Overall Simulation performance

Table 1 reports the overall MBEs and RMSEs observed at each site and for both POAI and PV output.

POAI biases remain 10% for all sites. This is a little higher than recent global irradiance validations by the author and others [4,5,6], but still very reasonable, especially given the fact that we did not account for site-specific ground-reflectivity and local obstructions in the field of view of the arrays. RMSEs are of the order 20-25% except for Rosemount, MN. These RMSEs are consistent with past evaluations [4,5,6]. Note that we had previously shown that much of this RMSE is attributable to the comparison between the time-integrated pin-point measure vs. instantaneous spatially extended pixel [4]; the effective, or intrinsic pixel-wide accuracy of the satellite has been estimated to be of the order 15%. The large RMSE in Rosemount may be explained on the following grounds: (1) Minnesota represents one of the worst possible case for the satellite model with winter ground snow cover affecting model accuracy, (2) the same snow cover may also have affected ground measurement accuracy, and (3) a much smaller average irradiance at that site translates in a higher relative error for the satellite, while absolute errors remain roughly unchanged from other sites.

PV errors are larger than POAI's but, remarkably, as shown in the rightmost columns of Table 2, the difference between the PV and POAI biases precisely matches the difference between UPVG-simulated and UPVG-measured PV output. This indicates that much of the observed satellite-bias is traceable to either PV performance shortcoming or overestimated array rating (see site specific observations below).

The scatter plots in Figure 2 provide a graphic illustration of the satellite's short-term accuracy. The plots on the left compare satellite (y) against measured POAI (x). The plots on the right compare satellite-derived and measured normalized PV output. These plots are consistent with earlier satellite model validations [9]. Again, much of the scatter is a direct result of the pixel-pinpoint comparison [4]. Some of the POAI under- or overestimating trends may be explained on the grounds of remaining (but eventually correctable) satellite uncertainties, such as regional turbidities, as well as unaccounted-for site-specific ground reflectivity and close/distant field-of-view obstructions.

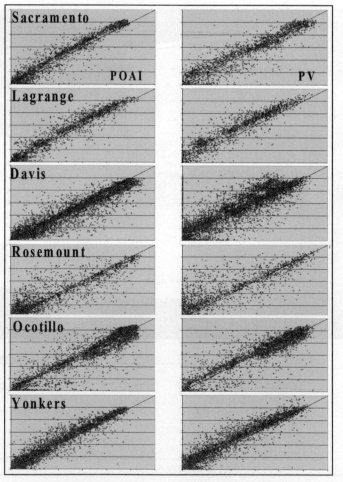

Fig. 2: Satellite (y-axis) vs. ground measured (x-axis) POAI (left) and PV output (right)

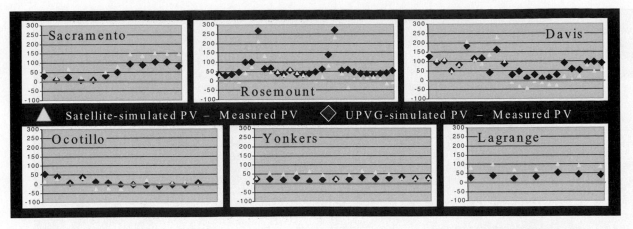

Fig. 3: Monthly MBE (day-time W per installed kW) for satellite predicted and measurement based UPVG PV output estimates

The PV plots are similar to the POAI plots with the stronger tendency for overestimation discussed above, more scatter, particularly for Davis, Sacramento and Rosemount – an indication of performance shortcomings discussed below.

3.2 Focus Satellite PV Performance Monitoring

Figure 3 shows the monthly evolution of bias errors -- satellite simulated PV minus measured PV; and UPVG-simulated PV minus measured PV. UPVG simulations are based upon measured POAI values. Each point represents one month of data.

The trends of the satellite and UPVG biases over time are remarkably similar. High UPVG overestimates, indicative of potential system problems are matched one-on-one by high satellite overestimates. A site-specific discussion follows:

Sacramento: Biases are small until June 1999 when the satellite's MBE reaches and remains around 150 Wm-2. A ground-based simulation MBE of about 100 Wm-2 matches this MBE jump. The ground-based discrepancy is indicative of some PV underperformance starting in June 1999. Yet undocumented causes for this shortcoming include a slight mis-tracking of one of the array's nine trackers [10], and dry season soiling which has been observed to account for as much as 20% performance shortfall in some cases [11]. The first two plots in Fig. 4 illustrate the measured and satellite-simulated POAI and PV output for a few days in July 1999. Satellite predictions are right on for POAI, but PV output falls short from expectations by about 20%.

Rosemount: Both satellite and UPVG biases remain slightly positive throughout. However during January 1998 and 1999 and to a lesser extent in the surrounding months, the satellite bias reaches very high values. Ground-based simulations closely match this pattern. These large biases are indicative of a major PV underperformance that is most likely the result of a persistent snow cover on top of the PV array, as no system malfunction was reported [12]. Snow cover is also the likely source of a greater winter discrepancy between the UPVG and satellite simulation biases – by affecting the POAI pyranometer as well as the satellite model. The bottom two plots in Fig. 4, showing POAI and PV profiles for a few days in January 1998, confirm that the PV system remained down for several days at a time. Look for instance at the eighth day in the time series: this is a very clear day with good POAI agreement between the pyranometer and the satellite; PV output gradually picks up in the late afternoon of that day after having been flat for three days, probably after the sun's

action on the roof got the snow to gradually slide off the modules

Davis: Both satellite and UPVG PV simulations substantially over predict the system's output until September 1998 – the 12th point on the graph. Discussions with the PVUSA operators [13] confirmed that the array did not operate properly until the fall of 1998 for a variety of reasons including inverter, array, and data acquisition problems. This array was in fact "field-rated" in the fall of 98 after all the problems had been resolved. Both satellite and ground biases remain acceptable until late spring 1999 where a tendency toward overestimation is again noted. This second set of over prediction, although not as strong as the first, corresponds to the over prediction at the nearby Sacramento array, indicating a possible effect of dry weather array soiling. Note that the satellite and UPVG trends are about 50 Wm-2 apart but follow very similar patterns over time.

Ocotillo: Very good agreement is observed for both the ground-based and satellite-based simulations. This power plant operated well during the considered period.

Yonkers array: The persistent small positive satellite bias is matched to a significant extent by the UPVG simulation bias, possibly indicating a minor – but well within specs – system underperformance. However, the bias trend remains roughly constant throughout the period and reflects the reliable performance of this array.

Lagrange: As in Yonkers but more so, the satellite bias is matched to about half by the UPVG bias, indicating some underperformance. This bias remains fairly stable through the considered period, indicating no performance problem besides a small over-rating. The possible reasons for the difference between the UPVG and satellite biases were discussed above and include unaccounted for field of view obstructions, and an underestimated turbidity in the satellite model (a problem that will eventually be correctable).

4. DISCUSSION

Results indicate that the satellite resource is capable of operationally detecting PV array performance shortfalls. The resource provides two performance evaluation tools: (1) large discrepancies (> 100 Wm-2) between satellite-simulated and actual PV output reliably point out to major

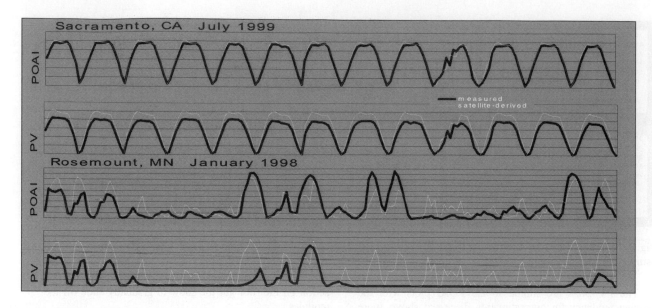

Fig.4: Satellite-derived (white) and measured (black) POAI & PV for 2 output shortfalls periods in Sacramento & Rosemount

array shortcomings; (2) the satellite bias trends over time, even if prediction is not always right on provide another troubleshooting tool. Any disruption in the bias timeline trend can be used to identify smaller problems.

By providing checks on the reliability of PV performance, the satellite resource could also be used in some aspects of system metering. Of course, remote energy production metering is out of question, because this depends on the reliability of the PV system -- some form of ground readout is necessary (e.g., monthly meter readout). But the satellite could be used to meter capacity credit assignable to a PV installation if its energy output matches the simulated expectation (the issue of capacity credit assignment for non-demand installations is discussed in [14]). Likewise, the satellite could be used for real time pricing metering purposes by providing a real time/site specific "dimension" to monthly energy production meter readings.

In other respects, the analysis of short term errors confirmed the contention that time/site specific PV simulations from satellites can be used with confidence to conduct regional PV-grid investigations including inquiries on peak shaving, grid penetration and outage mitigation [15].

ACKNOWLEDGEMENTS: This article includes results from research projects supported by NREL (Contracts No. XAD81767101 & XAH51522201). Many Thanks to Dan Greenberg (Applied Power), Bill Brooks and Tim Townsend (PVUSA), Gill Duran and Rick West (UPG), Guy Sliker (NYPA), Mark Rogers (NSP) and Tom Lepley (APS) for their help.

REFERENCES

[1] Reise, C. et al., (1999): Remote performance check for grid connected PV systems using satellite data. Proc. 2nd Satellite/Solar Resource Workshop, NREL, Golden Co.

[2] 2. Perez et al. (1999): Ongoing remote performance assessment of Astropower arrays deployed in the NYSERDA PV residential program. ASRC, the University at Albany, Albany, NY, 12203, USA.

[3] Internet Data Distribution System (IDD). Unidata-UCAR, Boulder, CO

[4] Zelenka, A., et al., (1999): Effective Accuracy of Satellite-Derived Irradiance. Theo. & Appl. Clim. 62, 199-207.

[5] Ineichen, P. and R. Perez, (1999): Derivation of Cloud Index from Geostationary Satellites and Application to the Production of Solar Irradiance and Daylight Illuminance Data Theoretical and Applied Climatology Vol 64, 119-130

[6] Hammer, A., et al., (1998): Derivation of daylight and solar irradiance data from satellite observations. 9th EMETSAT/AMS Conf. on Sat. Meteorol. and Ocean., Paris, 25th-29th May 1998, pp. 747-750.

[7] Menicucci D.F., and J.P. Fernandez, (1988): User's Manual for PVFORM. Report # SAND85-0376-UC-276, Sandia Natl. Labs, Albuquerque, NM

[8] Utility PhotoVoltaic Group (1999): Team-up Grid-connected PV Installations. UPVG, 1800 M Street, Washington, DC 20036-5802, USA.

[9] R. Perez et al., (1994): Using Satellite-Derived Insolation Data for the Simulation of Solar Energy Systems Solar Energy Vol. 53, 6; 7 pp.

[10] Eyewitness accounts (1999), Sacramento airport.

[11] Townsend, T.U. and P.A. Hutchinson, (2000) Soiling Analyses at PVUSA, Proc. ASES-2000, Madison, WI

[12] Rogers, M., (2000): Personal Communication, NSP, Minneapolis, MN

[13] PVUSA, (2000): Photovoltaics for Utility Scale Applications, Davis, CA

[14] Lampi M. and R. Perez, (2000): Assigning a capacity value to distributed renewable resources in restructured electric markets – the case of New York State, Proc. ASES-2000, Madison, WI

[15] R. Perez, R. Seals, H. Wenger, T. Hoff and C. Herig, (1997): PV as a Long-Term Solution to Power Outages. Case Study: The Great 1996 WSCC Power Outage. Proc. ASES Annual Conference, Washington, DC.

LATENT HEAT STORAGE ON PHOTOVOLTAICS

Tobias Häusler, Harald Rogaß

Brandenburgische Technische Universität Cottbus, Lehrstuhl für Angewandte Physik, Germany

Postfach 101344, D-03013 Cottbus, ☎ 0049-355-692315 / 📠 0049-355-692103 / E-Mail: tobias.haeusler@tu-cottbus.de

ABSTRACT: The use of photovoltaic elements in building facades allows besides the generation of electrical energy additionally the utilization of thermal energy caused by the absorption of solar radiation. But in most cases thermal energy must be stored until there is real need of it. Storing requires big volume and complicated systems to transfer the heat from the absorption layers to the storage. A solution of that problem is the usage of phase change material for storing. This material can store a high amount of thermal energy in a small volume. On that way a compact storage of phase change material is arranged directly behind the photovoltaic module. The system can run as passiv building facade with temperature balance effect or as heat storage for temporary use by active components. Experimental investigations on several developed PV hybrid elements with latent heat storage are described in this contribution.
Keywords: Thermal Performance - 1: Storage - 2: PV Module - 3

1. THERMAL ENERGY BY PV MODULES

The aim of the investigation was to combine a photovoltaic facade with a flat plate water collector. The black monocrystalline photovoltaic cells are quite good absorption layers and without cooling the temperature can achieve values up to 60 °C and more in realized photovoltaic facades.

On the one hand the result of high temperatures is lower electrical efficiency and a thermal problem in the building (overheating). On the other hand thermal energy for heating or personal using is required in every building. Without storage the use of the thermal energy is often not practicable. Therefore a latent heat storage system with high capacity on a small volume is integrated directly behind the photovoltaic cells in the facade (Figure 1).

Because of the melting point of 24 °C both the photovoltaic cells and the wall or room behind can not be overheated. The phase change material holds a constant temperature level for the outdoor wall of a building by storing energy during the day and transfer it into the building during the night.

After a sunny day the thermal energy of the storage can also be used by pumping water through the system and use the warm water into a heating system or to a heat exchange system. Only in that case the system needs active components.

An estimation shows that about 62 % of the overall solar radiation can be collected as thermal energy by the system. In addition about 7 % of the solar radiation energy are used as electrical energy. On the described way photovoltaic facades can be much more usefully without complicated systems or great space needs.

PV HYBRID ELEMENT FOR TEST 3

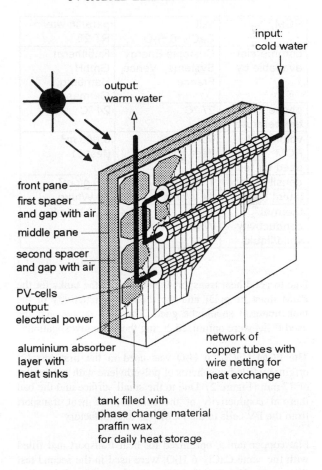

Figure 1: The PV cells are laminated directly on the aluminium absorber in the newest test system

2.   COMPARISON OF DIFFERENT SOLUTIONS

Phase change material (PCM) can store a high amount of thermal energy in a small volume. That is reached by using the effect of energy storage during the phase change from the solid to the liquid phase. In the change from the liquid to the solid phase there is an output of that stored energy [1]. This effect is well known by water, but the low melting point of 0°C is not suitable for storing solar energy. The optimized phase change temperature for the described application depends on following terms:

- can be achieved by solar radiation
- can be used by stabilization of room temperature
- can be used for a heating / heat exchange system
- no overheating of the PV cells.

Therefore the phase change temperature should be from 20 °C to 35 °C.

A high thermal conductivity for the PCM is very important for fast charge and discharge of the heat storage. Unfortunately the thermal conductivity of the most PCM is not very high.

PCM	salt CaCl₂·6 H₂O	paraffin wax RT 25
commercial available by	Cristopia Energy Systems, Vence, France	Rubitherm GmbH, Hamburg, Germany
phase change temperature	27 °C	24 °C
volume of PCM used for test	0.0155 m³ (test 1 and 2)	0.046 m³ (test 3)
density / solid	1.5 g/cm³	0.8 g/cm³
latent heat	44 Wh/kg	39 Wh/kg
thermal conductivity solid/liquid	1.09 / 0.54 W/mK	0.20 W/mK

Due to good heat transport the surface of the tanks for the PCM must be great and the thermal conductivity of the tank material should be good. The tank design and the used PCM were optimized during the three investigations.

The PCM $CaCl_2 \cdot 6\ H_2O$ was used in the first test [2] originally filled in spheres of polyethylene with a diameter of 77 mm (Figure 2). Due to the small surface and the bad thermal conductivity of the spheres, the heat transport from the PV cells to the PCM was dissatisfactory.

Flat copper tanks, optimized for heat transport and filled with the same $CaCl_2 \cdot 6\ H_2O$, were used in the second test (Figure 3). The PV module is the front side of the glass „aquarium". But the corrosive and salty PCM and the pressure caused by the volume change from solid phase to liquid phase destroyed the hermetic tanks in some weeks. A quite good heat transport from the PV cells to the PCM was the advantage of test 2.

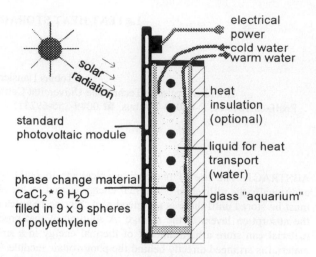

Figure 2: The first test system with PCM in spheres

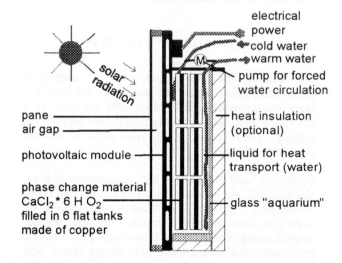

Figure 3: The second test system with PCM in flat copper tanks

Paraffin wax RT 25 is used in test 3, the newest investigation (Figure 1). This PCM is not corrosive and not hygroscopic. Therefore the used tank has at the top a little hole for pressure compensation. The whole tank is made of aluminium with special heat sinks on the absorber side. In contrast to conventional PV elements made of glass layers, the PV cells are laminated directly on the aluminium absorber layer separated only by a thin plastic film for electric isolation. There are two panes and two air gaps in front of the PV hybrid element for good heat insulation to the absorber.

The capacity of the tank with paraffin wax is optimized for the storage of the energy of one sunny day. The storage temperature can not reach values over 24 °C, if there is daily use of the stored energy. The size of the PV hybrid element is 1 meter in the length, 0.5 meter in the height and 0.15 meter from the front to the back. The system needs heat insulation on the back side, if it is not integrated in a facade.

3. RESULTS OF TEST 3

The temperature of the storage strongly depends on radiation. There is only a small influence by the outdoor air temperature in cause of the two panes in front of the system and the heat insulation around the tank.

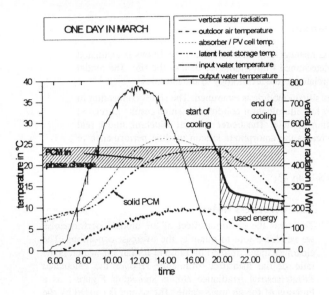

Figure 4: Temperature curves for the newest test system

The temperature of the latent heat storage increases quickly up to 20 °C after the first sunshine hours (Figure 4). Because of the melting point in the range of 24 °C, there is a temperature plateau. Also the temperature of the absorber and the PV cells can not increase over approximately 26 °C.

The use or storage of the thermal energy can lead to a cooling effect for the photovoltaic module. During the sunshine hours a normal photovoltaic module will be warmer than the module with latent heat storage. According to the theory the electrical output of the crystalline photovoltaic cells increases by 0.4 % per 1 K temperature decrease. That means an increase of electrical power by 15 % - 30 % in comparision to very hot conventional PV facades. But there is also a decrease of electrical output because of the additional pane in front of the PV element.

The use of stored thermal energy starts in the evening. The difference between the output and input water temperature shows the used energy. A thermal energy of 1.44 kWh can be stored only by the melting process of the PCM, until the temperature is increasing over 24 °C.

Due to the heat insulation the stored thermal energy is protected from high losses during times without radiation. Therefore a high temperature difference between the storage and the cold outdoor air is measured also some hours after the solar radiation. On this way a flexible use of the energy is guaranteed.

4. ENERGY BALANCE

The determination of the thermal efficiency of the investigated system according to the test procedure for flat plate water collectors (EN 12975-2) is not possible, because of the heat storage. Therefore the efficiency was determined after daily charging of the system by solar radiation by measurement of the input and output water temperature and the water flow, always at the same time in the evening. That procedure simulates the real operation of the PV hybrid element. The efficiency depends also on the distribution of daily solar radiation as there are heat losses during hours without high solar radiation and cold air temperature.

The determination of the efficiency is not finished. The solar radiation depends on season. The spring and autumn are signed by a relative big amount of solar radiation on the investigated system. There are fewer sunshine hours per day than in summer, but the angle of solar radiation on the vertical south facade is advantageous. The aim are results for all seasons. First results were got for several days in March and April (Figure 5).

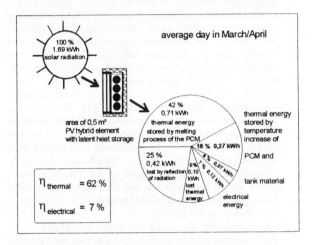

Figure 5: Energy balance for the newest test system

The averaged solar radiation was 1.69 kWh on an area of 0.5 m² with PV hybrid element. About 1.05 kWh of the incoming energy was stored in the system as thermal energy and additionally 0.1 kWh was changed into electrical energy. The low absorber temperature leads to good thermal efficiency. The electrical efficiency is quite low as the PV cells occupy only 65 % of the absorber area.

The investigations on the system will be continued to get results for all seasons and perhaps with another PCM.

REFERENCES

[1] Manz, H.: Sonnenstrahlungsbeladene Latentwärme-speicher in Gebäudefassaden. Diss. ETH Zürich Nr. 11377, (1996).
[2] Häusler, T; Rogaß, H.: Photovoltaic module with latent heat storage-collector. Proceedings of 2nd World Conference on Photovoltaic Solar Energy Conversion, volume I, pp.315-317, Vienna 1998.

Assessing the outdoor performance of single junction and double junction amorphous silicon solar modules

J.Merten, J.M.Asensi, C.Voz, J.Andreu

Universitat de Barcelona, Departament de Física Aplicada i Optica, Av.Diagonal 647, E-08028 Barcelona,

Phone:..34/934021134, Fax:..34/934021138, E-mail: jmerten@fao.ub.es

Abstract:

We propose here a performance model where the spectral mismatching under overcast conditions is evaluated from an empirical clearness index, which can be easily determined from irradiance data of the site. The model allows to determine the site dependend yields and losses related to the technology of the modules.

As an example, a single and a double junction amorphous silicon module is examined. The series connection in the double junction modules leads to a higher loss under low illumination conditions, which results in related yearly losses of 7% instead of 4% for the single junction module. The better spectral matching under real operating conditions than under standard conditions increases the amorphous silicon module performance in Barcelona by about 5%.

Keywords: Evaluation –1; PV-Modules - 2; Amorphous Silicon - 3;

1. Introduction

The accurate determination of the performance of photovoltaic modules is crucial for the design of PV systems and their economic evaluation. This performance is usually specified under standard testing conditions (STC) which are irradiance H_{STC}=1000W/m^2, air mass AM=1.5, and a temperature of T_{STC}=25°C. Under real operation conditions, however, these parameters are quite different. Therefore, models have been developed, which determine the site dependend module performance under real operation conditions (ROC) [1][2]. They require, however, spectral response measurements of the modules. We propose here a performance model where the spectral mismatching under overcast conditions is evaluated from an empirical clearness index, which can be easily determined from irradiance data of the site. The model allows one to determine the site dependend performance yields and losses related to the technology of the modules on a monthly or yearly basis.

As an example, we compare here two amorphous silicon solar modules, a single junction and a double junction. In a first experimental section, we describe the outdoor measurement of these modules, and develop a novel empirical clearness index q. In the following section, we develop the model, reviewing the effects affecting the performance of the modules. We then use this model to estimate the yearly energy production and to evaluate the different gains and losses with respect to standard conditions.

2. Experimental Procedure

2.1. Outdoor Performance Measurements

We present experimental data from outdoor measurements of commercial modules with single junction and double junction structure. These modules are placed outdoors (tilt angle: 44°) in a closed box in order to avoid humidity degradation problems. This lead to high operation temperatures, which were monitored by a platinum resistance sensor glued on the back of the module with thermally conductive silver paste. All efficiency values are evaluated with a Haenni pyranometer with a flattened spectral response. All data shown here have been normalised to 25°C using measured temperature coefficients for each module.

2.2. Clearness Index

To assess the seasonal effect of the module performance, we needed a tool to quantify the overcast conditions. The clearness index defined by Liu and Jordan [3], which is the ratio of the measured irradiance H over the calculated extraterrestrial irradiance $H0$, is shown in Figure 1 as a function of the air mass value. The scatter is caused by the different overcast conditions on the site. The upper evolvent H_{max}/H_0, drawn by the line, should empirically represent the clearest weather conditions at the testsite. We have found that H_{max}/H_0 depends on the detector used. We therefor divide the data of Figure 1(a) by their upper evolvent in order to obtain an empirical detector-independent clearness index q:

$$q = \frac{H / H_0}{H_{max} / H_0} \tag{1}$$

This index is one for clear irradiance conditions, and below one for overcast conditions. Figure 1(b) shows this index, which furthermore does not depend on the air mass AM.

3. Modelling the module performance under real operation conditions

The outdoor performance at the maximum power point of amorphous silicon module depends on four factors, which are briefly reviewed here, using our experimental data. We

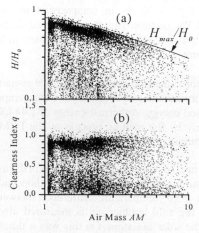

Figure 1: The empirical clearness index (see text).

thereby develop several correction factors, which then give the model of the module efficiency η_{ROC} under real operation conditions.

3.1. Thermal Anneal Effects

Higher temperatures of amorphous silicon modules lead to a slight improvement of the material properties of amorphous silicon, for example the mobility lifetime product (Staebler-Wronski effect). This may lead to positive temperature coefficients [4]. However, the evaluation of the temperature effect is beyond the scope of this study. So, the modules were exposed at relatively high temperatures (noon temperatures were 60°C in winter and 80°C in summer), and no significant changes have been found for the mobility lifetime product [5].

3.2. Temperature Effect

The temperature dependence of the module performance is usually modelled with following temperature correction factor:

$$F_\eta(T) = 1 + \beta_\eta \cdot (T - T_{STC}) \tag{2}$$

The temperature loss of efficiency β_η has been measured to be -0.3%/K for the single junction module [5], and -0.4%/K for the double junction module. In order to assess the remaining effects, we show in this publication only data which have been already normalised to 25°C.

3.3. The angular effect

Apart from the so called "cosine effect", a high incidence angle between module normal and sun may lead to increased reflections at the glass surface. The related additional loss reduces the carrier generation and should affect the module performance with a reduced circuit current. Examples for the measurement of the angular loss function can be found in [2][6], where the angular loss function becomes significant for angle from about 50°. The energy provided at such high incident angles is low and has been neglected in this study for simplicity.

3.4. Irradiance Effect on V_{oc} and FF

The open circuit voltage V_{oc} and fill factor FF of the modules depends on the irradiance level which is presented in Figures 2(a)-(d) as a function of the short circuit current. The low scatter of the data suggests that the spectral effect

is a second order one. Furthermore, it has been shown, that the spectral influence on the open circuit voltage for a given carrier generation (short circuit current) can be neglected, when there are buffer layers between p and n-layer of the junction [7]. We therefore assume that V_{oc} and FF depend on the global level of the carrier generation only. This dependence has been discussed in detail in [8] and can be using variable illumination measurements (VIM). Partial results of these measurements are the data shown in Figures 2(a)-(d), which will be used now to determine the model correction factors F_{Voc} and F_{FF}.

The irradiance effect on the open circuit voltage is empirically adjusted with:

$$F_{Voc,fit} = P_{0,Voc} + P_{e,Voc} \cdot \log(e) \tag{3}$$

where $e = I_{sc} / I_{sc,STC}$ is the carrier generation referred to standard conditions (STC). This form can be also directly evaluated from the equation of the carrier injection diode of the module. The model fit can be seen in Figure 2(a)-(b), and the parameters in the Table 1. Note that $P_{e,Vo}$ of the double junction module is **two** times the value of the single junction module. This fact is caused by the series connection of the two diodes in the case of the double junction module (which therefor suffers a stronger loss of voltage under low illumination levels).

The irradiance effect on the fill factor is modelled as:

$$F_{FF,fit} = P_{0,FF} + P_{1,FF} \cdot e + P_{2,FF} \cdot exp(-e / P_{3,FF}) \tag{4}$$

The last term have to be introduced in order to take into account the strong decrease of FF of the double junction module, which is caused by its relatively low parallel resistance.

Note the different signs of $P_{1,FF}$. Whereas the fill factor for the single junction module decreases with increasing carrier generation, the fill factor of the double junction increases.

The modelled irradiance correction $F_{voc} \cdot F_{FF}$ can be seen in Figure 3(a): there is only a weak dependence in the case of the single junction module, whereas the double junction module tends to losses.

3.5. Spectral Effect

In contrast to the V_{oc} and FF-data, the I_{sc}-data in Figure 2(e)-(f) exhibit a strong scatter. This scatter is

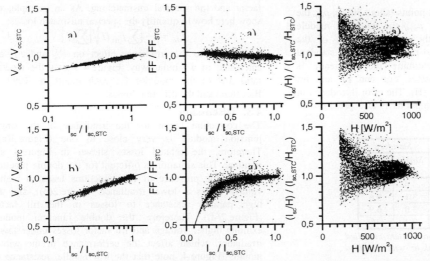

	single	double
$P_{0,Voc}$	1,00	1,00
$P_{e,Voc}$	0,136	0,256
$P_{0,FF}$	1,041	0,966
$P_{1,FF}$	-0,066	0,032
$P_{2,FF}$	—	-0,862
$P_{3,FF}$	—	0,152
$P_{0,Isc}$	1,364	1,423
$P_{q,Isc}$	-0,298	-0,363
$P_{AM,Isc}$	-0,378	-0,341

Figure 2: *Examination of the characteristic module parameters during one year of outdoor measurements. Each point (there are 8000) represents one measured I(V) curve. All parameters have been normalized to the values from standard conditions. Top figures: single junction, bottom figures: double junction module. The lines are fits according to the simplfied performance model. The wide scatter of the short circuit current data is mainly caused by fluctuations of the incident irradiance spectra.*

Table 1: *Parameters of the empirical performance model for the single and double junction module.*

caused by deviations of the irradiance spectrum from the AM1.5 standard irradiance spectrum. These deviations are caused by variations of the Air Mass value, which is the number of times the sun has to cross the atmosphere. High sun position means low Air Mass values and less losses of blue light in the direct irradiance. Another factor with influence on the spectrum is the overcast, expressed by a clearness index q less than one.

We model the spectral correction factor F_{Isc} for the short circuit current from a multiple linear regression of the data in Figure 2(e) and (f) as a function of clearness index q and the logarithm of the airmass AM:

$$F_{Isc} = P_{0,Isc} + P_q \cdot q + P_{AM} \cdot \log(AM) \qquad (5)$$

The accuracy of the multilinear model proposed (R^2 only 0.6) is limited. Cutting off data with incidence angles more than 50° did not result in a better fit of this model to the data. However, the wide scatter of I_{sc} (Figure 2(e)-(f)) could be reduced to less than 10%. The dependence of F_{Isc} can be seen in Figure 3: For the dependence on the AM, there is no significant difference between the two modules. The negative sign of P_{AM}, indicates that the models perform **better** under overcast ($q<1$); there, the double junction module seems to be more performant.

3.6 Modelling the module performance

The model of the module performance takes into account the factors developed above:

$$\eta_{ROC} = \eta_{STC} \cdot F_{FF}(e) \, F_{Voc}(e) \, F_{Isc}(AM,q) \, F_\eta(T) \qquad (6)$$

This model requires as input parameters:
- calibrated module data, measured under standard conditions: FF_{STC}, $V_{oc,STC}$ $I_{sc,STC}$
- uncalibrated module data, measured with variable illumination measurements (VIM): $FF(e)$, $V_{oc}(e)$,
- module data measured outdoors over a large period: $I_{sc}(AM,q)$

With these data, the simple model allows the prediction of the module performance for any H, AM, and q.

To assess the validity of the model proposed we present the statistical error of each parameters (p) determined with:

$$Error = \sqrt{\frac{\sum_N H \cdot \{(p_{Exp} - p_{Mod})^2 / p_{exp}^2\}}{\sum_N H}} \qquad (7)$$

where N is the number of data points (about 8000), p_{exp} the measured module parameter, p_{mod} the modelled module parameter. Table 2 shows the error for each of the parameters using the presented model. Assuming standard operation conditions to determine the efficiency is equivalent to assume a constant Voc and FF and Isc proportional to the irradiation (H). The error introduced in the parameters with this assumptions is also presented in Table 2..

p	Present Model		Standard	
	Single %	Double %	Single %	Double %
I_{sc}	6.5	5.9	11	9.1
V_{oc}	0.6	0.9	3.1	5.6
FF	2.2	3.6	3.0	6.9
η	6.1	5.9	9.4	15

Table 2: *Error of the parameters modelled with the present model or with standard operation..*

The last line of Table 2 show that the model prediction of η is better than assuming constant module efficiency. We conclude that the model significantly improves the

prevision of module performance under real operation conditions.

4. Yearly Module Performance

We now use the performance model to evaluate the site dependent performance of the two modules in Barcelona. This requires of regular spaced irradiance data. From these data, the clearness index q and the air mass value AM has is evaluated.

4.1. Solar Irradiance in Barcelona

As an example, we use solar irradiation data measured 1992 in Barcelona on a horizontal plane. The total irradiation in this year was 4.0 kWh/(m²·day), which is the equivalent of four hours of full sunshine per day. The characteristic histograms in Figure 3 show how this energy is distributed with the irradiance H, the clearness index q and the air mass AM. The most remarkable result is that most of the energy is achieved at clearness indexes close to one. For the performance model, this means that the accuracy of the model with respect to the overcast conditions is not critical.

4.2. Performance Analysis

For the estimation of the yearly energy output of a PV system, one usually takes the product of the specified module power at STC, the yearly irradiance, and the performance ratio PR. This latter correction factor takes into account losses on system, array, and module level. For the performance analysis of the modules examined, we evaluate the module ratio MR, which is a similar parameter, however, taking into account losses on module level under real operation conditions ROC:

$$MR = \eta_{ROC}/\eta_{STC} \qquad (8)$$

According to the performance model in equation 6, the module ratio is:

$$MR = F_{FF}(e) \cdot F_{Voc}(e) \cdot F_{Isc}(AM,q) \cdot F_\eta(T) \qquad (9)$$

Note that MR only depends on the irradiance H, the air mass AM and the clearness index q. For a yearly performance analysis, the average MR is weighted with the irradiance H ($F_\eta(T)$ is set to one here, see above):

$$MR = \sum (F_{Voc} F_{FF} F_{Isc} H) / \sum H \qquad (10)$$

where the sums are executed over all data of the averaging period (month or year). The performance model also permits to assess the yearly losses of the voltage, the fill factor and the spectral mismatching. As an example, we show here how to quantify the spectral mismatch losses:

$$L_{Isc} = 1 - \left[\sum F_{Isc} H\right] / \left[\sum H\right] \qquad (11)$$

Losses due to the irradiance effect on V_{oc} and FF are calculated in a similar way. Figure 4 shows the module ratio MR of the modules for each month of 1992 in Barcelona and the different losses.

4.3. Discussion

The correction factor for the irradiance of the single junction module is very close to one (Figure 3(a)). Therefore, the related losses shown in Figure 4 are minimal. This situation is different for the double junction module: the presence of **two** junctions leads to higher voltage losses at lower irradiance (Figure 2(b)), and the high parallel resistance to losses in the fill factor (Figure 2(d)). Therefore, the double junction module exhibits higher voltage and fill factor losses under lower irradiance, which affect the performance in the winter months (Figure 4, note that the low parallel resistance of the double junction module is a special feature of the module examined and not typical for the technology).

The correction factors for the spectral mismatching are in most of the cases higher than one ($AM<1.5$ and $q<1$,

Figure 3). Therefore, there is a **gain** in I_{sc}, especially in the summer months (Figure 4), which is more pronounced for the double junction module. This gain is 4..5% for both modules.

5. Outlook

Further studies have to show, whether the clearness index q and the correction factors P_q, P_{AM} developed here are independent of the site. The methodology developed here may be extended for any PV module technology.

6. Conclusions

A performance model has been developed, which allows to attribute site dependent losses to the technological properties of the modules. The spectral correction of this model is determined from outdoor measurements of the module behaviour and an empirical clearness index. It does not require spectral response measurements.

The series connection of junctions in amorphous

silicon modules leads to a higher loss under low illumination conditions. The related yearly losses are 7% The better spectral matching under real operating conditions than under standard conditions increases the amorphous silicon module performance in Barcelona by about 5%.

7. Acknowledgements

The authors would like to thank Prof. G. Lorente for providing irradiance data. This work was supported by the DGICYT of the Spanish Government and Generalitat de Catalunya.

8. References

[1] K.Bücher, 13th Europ.PV solar energy conference, 1995, pp2097-2103
[2] D.L.King, J.A.Kratochvil, W.E.Boyson, 26th IEEE PV specialist conf. 1997
[3] B.Y.H.Liu, R.C.Jordan, Solar Energy, 7, 53 (1963)
[4] G. Friesen, W. Zaaiman, J. Bishop. Proceedings of 2nd World Conference on Photovoltaic Solar Eneergy Conversion. 1998, pp.2392-2397
[5] J. Merten, J. Andreu, Solar Energy Materials & Solar Cells 52 (1998) 11-25
[6] R.Preu, G.Kleiss, K.Reiche, K.Bücher, 13th European PV solar energy conference, 1995,pp. 1645-1468
[7] J. Merten, C. Voz, A. Muñoz, J. M. Asensi, J. Andreu, Solar Energy Materials & Solar Cells, 57 (1999) 153-165
[8] J. Merten, J.M. Asensi, C. Voz, A.V. Shah, R. Platz, and J. Andreu, IEEE transactions on electron devices, Vol. 45, No.2, February (1999) pp. 423-429

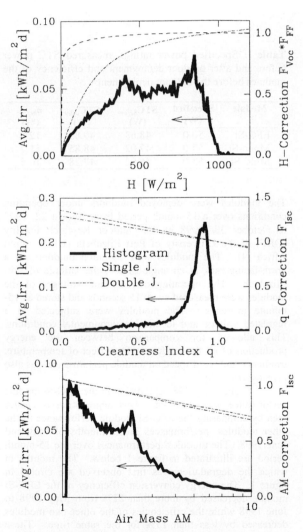

Figure 3: *Histograms of the irradiance H, the clearness index q and the Airmass in Barcelona 1992: The thick lines show is the yearly energy measured on a horizontal plane (classwidth H: 10 W/m², q: 0.01, 1/AM: 0.01). The thin lines show the model correction factors assuming the remaining illumination parameters to be as STC.*

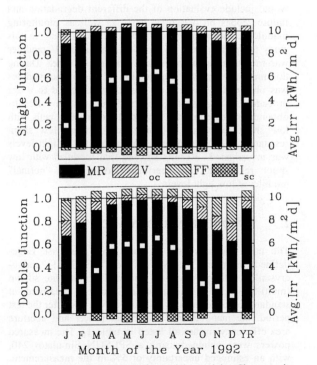

Figure 4: *Performance analysis of the modules. Shown is the module ratio MR, losses in I_{sc} due to spectral mismatching, and losses due to the irradiance effect in V_{oc} and FF. The squares indicate the average daily irradiance.*

DEGRADATION ANALYSIS OF SILICON PHOTOVOLTAIC MODULES

E. L. Meyer and E. E. van Dyk

Department of Physics, University of Port Elizabeth, PO Box 1600, Port Elizabeth, 6000, SOUTH AFRICA,

Tel: +27 41 504 2579, Fax: +27 41 504 2573, E-mail: phbelm@upe.ac.za

ABSTRACT

The continuing quest to reduce photovoltaic module degradation and improve performance requires a complete set of techniques to evaluate the module's entire cells' parameters. The foremost technique to observe module degradation is by monitoring the module performance under actual operating conditions. In this study the performance of silicon photovoltaic modules was monitored over an extended period of time. Modules comprising cells of three different technologies were used in the investigation. The cell technologies are Edge-defined Film-fed Growth silicon, mono-crystalline silicon and multi-crystalline silicon. During the monitoring period degradation in the performance of one of the modules was observed. An analysis of this degradation showed that low cell shunt resistance caused the observed degradation. It is also shown that these shunt paths reduce a module's efficiency when operating under low light levels. Hence, verifying the fact that shunt paths caused the observed degradation.

Keywords: Monitoring – 1: Degradation – 2: Shunts – 3

1. INTRODUCTION

Photovoltaic (PV) modules are renowned for their reliability. However, some modules do degrade and can fail when operating outdoors for extended periods of time. Since there is an ever-increasing deployment of PV as a form of energy in South Africa [1], it is desirable to have a failure reporting and analysis program to support the local industry as illustrated in [2,3]. A failure analysis program would include evaluation of the different degradation and failure modes on a regular basis as well as monitoring module performance under operating conditions. In this study the performance of three silicon-based PV modules was monitored over an extended period of time. During this time, performance degradation of one of the modules was observed. This module was then subjected to a test procedure to analyse the observed degradation. Results obtained indicate that the degraded module had cells with low shunt resistances. The effect of these low shunt resistance cells on module performance at low light levels was investigated. Results showed that modules with low shunt resistance cells perform worse than "normal" modules at low light levels.

2. MODULE PERFORMANCE

The three modules used in this study comprise Edge-defined Film-fed Growth silicon (EFG-Si), mono-crystalline silicon (mono-Si) and multi-crystalline silicon (multi-Si) cells. The modules' powers measured at standard test conditions (STC) both before and after the test period, the manufacturer's specified power and aperture area efficiency, η, are all listed in table 1. The measured powers were obtained using a SPI-SUN Simulator 240, with an estimated uncertainty of 5% in the measurement. The efficiency corresponds to the power measured before deployment.

Table I: Specified power ratings, measured STC power before and after outdoor deployment and efficiency of the modules before the outdoor deployment.

Module	Specified (W)	STC$_{before}$ (W)	STC$_{after}$ (W)	η_{before} (%)
EFG-Si	50.0	48.63	49.83	12.5
Mono-Si	53.0	47.02	48.85	11.7
Multi-Si	53.0	46.02	47.28	10.5

The modules were deployed outdoors under operating conditions over a 15-month period from August 22, 1997 to October 30, 1998 at the Outdoor Research Facility (ORF) of the University of Port Elizabeth (UPE), South Africa [4]. The modules were mounted outdoors on a north-facing rack at an angle of 34°, the latitude of Port Elizabeth. The operating currents and voltages of the modules were measured every 15 seconds and stored as 15-minute averages. The modules were subjected to a regulated voltage and identical environmental conditions. This allowed for comparison between the energy production of all the modules. The effect of temperature, irradiance and wind speed on module performance was also investigated [5,6].

From June 1998 degradation in the EFG-Si module performance was observed. This degradation is clearly seen by comparing the EFG-Si module performance to the other modules' performances in the months of May and June 1998. The modules' performances over the 15-month period are illustrated in figure 1 below. The months in which the degradation was first observed are circled in figure 1. The energy conversion efficiency of the EFG-Si module decreased by more than 22% from May 1998 to June 1998 while the efficiencies of the other two modules decreased by less than 13% for the same time. These percentages were calculated relative to the energy measured during May 1998. This was due to the lower, total daily irradiance-levels. The average daily irradiance for June 1998, was approximately 4500 Wh/m²/day. From June to July 1998 the efficiencies of all the modules increased by about 13% indicating that the EFG-Si module has degraded in efficiency by approximately 10% from May to July 1998. Energy production of the EFG-Si module in months after July 1998 indicates that the observed degradation was permanent.

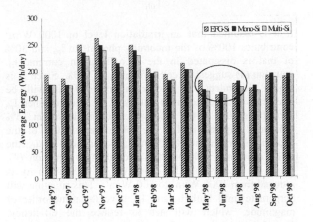

Figure 1: Average monthly energy per day produced by each module. Note the performance degradation of the EFG-Si module.

3.   DEGRADATION ANALYSIS

A photovoltaic module or cell can degrade for a number of reasons. To effectively analyse any observed degradation when a module operates outdoors, a complete set of techniques is required. Such a set of techniques was formulated for this study. These techniques include measuring light and dark current–voltage characteristics, electrical evaluation, hot-spot investigation and measuring individual cell parameters non-intrusively.

The EFG-Si module was subjected to these testing techniques in order to analyse the observed performance degradation. No visual defects on the cells of the module were observed. No electrical fault or cell mismatch was observed. Since the module conformed to all the external tests' requirements, it was suspected that the degradation was caused by inherent cell defects.

These inherent cell defects may be caused by four factors, namely, an increase in the cell's series resistance, a decrease in the cell's shunt resistance, antireflection coating deterioration, and discoloration of the encapsulant. The module's total series resistance as obtained from light I–V measurements was 0.96 Ω. No antireflection coating deterioration or yellowing of the encapsulant was observed during visual inspection.

A two-terminal diagnostic technique was then used to measure the individual cell shunt resistances, non-intrusively, without de-encapsulation [7]. Flawed manufacturing processes and material defects cause a shunt path in a PV cell (cell having low shunt resistance). The results of measuring the individual cell shunt resistance of the modules under test are presented and discussed in the next section.

4.   RESULTS

Results obtained for the individual cell shunt resistances are shown in figure 2 for the EFG-Si module and in figure 3 for the mono-Si module as measured at the time of the observed degradation.

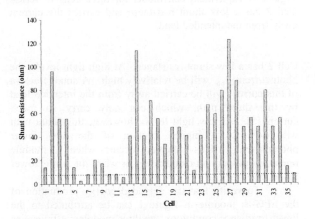

Figure 2: Shunt resistance of individual cells in the EFG-Si module after the observed degradation. Cells 6 and 12 will detract from the module output.

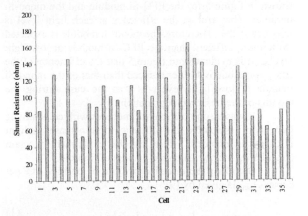

Figure 3: Shunt resistance of individual cells in the mono-Si module. This module had no cells with low shunt resistances.

The multi-Si module had a similar trend as that of the mono-Si module (fig. 3). The two crystalline modules had no cells with low shunt resistances. The EFG-Si module on the other hand, had a few cells with low shunt resistances as can be seen from figure 2. The dotted line in figure 2 corresponds to 7.5 Ω, the shunt resistance of all the cells when fully illuminated. This means that cells with shunt resistance close to or less than 7.5 Ω will detract from the module output under low light levels.

Although the EFG-Si module did not decrease in maximum power at STC (table 1), its total energy production decreased. To establish whether the low shunt-resistance cells were responsible for the observed energy degradation, consider the equivalent cell model for three cells in series (figure 4).

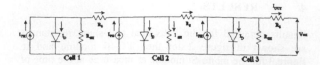

Figure 4: Equivalent cell model for three cells in series. Cell 2 has a low shunt resistance and carries the current away from the intended load.

Cell 2 has a low shunt resistance. At high light levels, the photocurrent, I_{ph}, will be relatively high. A small fraction of this current will be carried away from the intended load by the shunt path, which can only carry so much current [8]. At low light levels, however, this shunt path will dissipate a bigger fraction of the now lower photocurrent. This situation occurs when a module operates outdoors on cloudy days or when the sun is lower in the sky.

To verify that the degraded energy production of the EFG-Si module in figure 1 can be attributed to the lower irradiance conditions, the three modules' efficiencies were characterised as a function of light intensity. This was done by placing the modules in the solar simulator and covering it with masks of different opacities, attenuating the light levels accordingly. The module aperture area efficiency was then determined for each light level and is shown in figure 5 for the EFG-Si module and the mono-Si module. The error in the efficiency at each light level is less than 0.2%. Therefore, the mono-Si module is expected to be more efficient than the EFG-Si module at low light levels. It is evident from figure 5 that the efficiency of the EFG-Si module is further reduced than that of the mono-Si module at low light levels as was the case during the months of the observed degradation.

It was also noted that for modules with cells having high shunt resistances, the efficiency first increases as light intensity is reduced. This can be explained by again considering the equivalent cell model of figure 4.

By taking the direction of I_{ph} to be positive, the net current I_{out}, is given by [9]:

$$I_{out} = I_{ph} - I_D - I_{sh} \qquad (1)$$

The dark recombination current I_D, which is formed by recombination of electron-hole pairs within the semiconductor, detracts from the I_{ph}. The rate of recombination depends, among others, on the availability of charge carriers and subsequently on carrier lifetime. Using the fact that the I_{ph} is directly proportional to the incident irradiance, we can define the efficiency of the cell to be:

$$\eta = \frac{I_{out}}{I_{ph}} \qquad (2)$$

and from equation (1):

$$\eta = 1 - \frac{I_D}{I_{ph}} - \frac{I_{sh}}{I_{ph}} \qquad (3)$$

If the cell's shunt resistance is sufficiently large so that its shunt current, I_{sh}, is negligibly small, equation (3) reduces to:

$$\eta = 1 - \frac{I_D}{I_{ph}} \qquad (4)$$

If it is assumed that an irradiation level of 1000 W/m² contributes 100% of the incoming photons to I_{ph} and 30% of that is dissipated in the recombination current, I_D, equation (4) suggests a cell efficiency of 70%. Since I_D is highly dependent on carrier lifetime [9] it will be drastically reduced at lower values of I_{ph}. Hence, when I_{ph} constitutes 50% of the incoming photons, less than 15% will be dissipated in I_D. If, for instance, 10% of the incoming photons at 500 W/m² are dissipated in I_D, the cell's efficiency will be 80% according to equation (4), hence an increase in the efficiency as light intensity is reduced from 1000 W/m² to 500 W/m². This situation will perpetuate until both I_D and I_{ph} have the same order of magnitude, which will act to reduce the efficiency. Therefore, modules that first show an increase in efficiency as light intensity is reduced, are expected to have no cells with low shunt resistances. If, on the other hand, the cell's shunt resistance is small, the shunt current, I_{sh}, in equation (3) cannot be neglected. The efficiency in this case is only expected to decrease as light intensity is reduced.

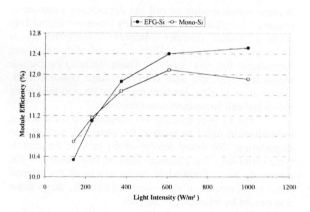

Figure 5: Peak-power efficiencies as a function of light intensity for the EFG-Si module and the mono-Si module.

5.   CONCLUSIONS

The outdoor performances of 3 silicon, PV module were monitored over a 15-month period from August 22, 1997 to October 30, 1998. Degradation in the performance of the EFG-Si module was observed since June 1998. This emphasises the importance of monitoring the performance of PV modules deployed outdoors in order to detect degradation modes.

To analyse these degradation modes, a complete set of techniques is required. Such a set of techniques must be able to evaluate all cell parameters.

The EFG-Si module had cells with low shunt resistances causing shunt paths. These shunt paths lead the current away from its intended load. These shunt paths cause a module to be less efficient at low light levels than "normal" modules. This was verified by measuring the module efficiency as a function of light intensity.

Results showed that all modules operating under reduced light intensity have a significant decrease in their efficiency. It also showed that if cells in a module have low shunt paths that divert current away from the intended load, the module's efficiency will be even further reduced under these lower light conditions. The EFG-Si module exhibited this behaviour. Knowledge of individual cell shunt resistance of an encapsulated module is therefore imperative to identify cells that are degrading module performance.

6.   ACKNOWLEDGEMENTS

The authors wish to express their gratitude to Tom Basso of the National Renewable Energy Laboratory, Golden, CO for useful communications in building the shunt resistance measurement system. The financial assistance of the South African National Research Foundation is also acknowledged.

7.   REFERENCES

[1]   Swanepoel R. (1998). "An appropriate energy source for Africa". *S. Afr. J. Sci.* **94**. pp. 46.

[2]   Dumas L.N., Shumka, A. (1982). "Photovoltaic module reliability improvement through application testing and failure analysis". *IEEE transactions on reliability, vol. R-31, No. 3.* August.

[3]   Forman S.E. (1982). "Performance of experimental terrestrial photovoltaic modules". *IEEE transactions on reliability, **vol. R-31**, No. 3.* August.

[4]   Meyer E.L. (1999). *Investigation of properties and energy rating of photovoltaic modules.* M.Sc. dissertation. University of Port Elizabeth, South Africa. January. pp. 74 – 82.

[5]   van Dyk E.E, Meyer E.L., Scott B.J., O'Connor D.A., Wessels J.B. (1997). "Analysis of photovoltaic module energy output under operating conditions in South Africa". *26th IEEE PVSC.* pp. 1197 – 1200.

[6]   Meyer E.L. and van Dyk E.E. (2000). "Development of an energy model based on total daily irradiation and maximum Ambient Temperature", *Renewable Energy Journal 21/1.* pp. 37 – 47. 20th April.

[7]   McMahon T.J. (1995). "Cell shunt resistance and PV module ratings". *13th NREL photovoltaics programs review, Lakewood, CO.* May.

[8]   McMahon T.J., Basso T.S. (1995). "Two terminal diagnostics for cells in series connected photovoltaic modules". *ASME International Solar Energy Conference, Lahaina, HI,* March 19–24.

[9]   Mazer, J. A. (1997). *Solar cells: An introduction to crystalline photovoltaic technology.* Kluwer Academic Publishers. pp. 98 – 99.

OUTDOOR EXPOSURE TESTS OF PHOTOVOLTAIC MODULES IN JAPAN AND OVERSEAS

Masakatsu Ikisawa, A. Nakano, and T. Ohshiro
Japan Quality Assurance Organization Solar Techno Center
HIC BLDG. 2F, 4598, Murakushi-cho
Hamamatsu-shi, Shizuoka-ken 431-1207 Japan
Phone : +81-53-484-4101, Fax : +81-53-484-4102

ABSTRACT: Outdoor exposure tests of photovoltaic (PV) modules have been conducted in Japan, Australia, and Oman to understand the long term stability of PV modules and the effects of climate conditions on efficiency. From the test in Japan, it was found that the efficiencies of crystal silicon (c-Si) and amorphous silicon (a-Si) module showed little degradation tendencies even after about 9 years of exposure. The tests in Australia showed that a-Si module efficiencies were strongly influenced by climate conditions especially by temperature. The efficiencies showed very slight degradation tendencies (2-3% per year) for about two years of exposure. The spectral distribution of solar irradiance were measured and the results are discussed in terms of short circuit current of a-Si. The test in Oman for a-Si related modules (single type of a-Si, two layers type of a-Si/a-SiGe, and three layers type of a-Si/a-Si/a-SiGe) showed initial degradation of efficiency which were a little different among the types of a-Si modules.

Keywords: Degradation - 1: PV Module - 2: Evaluation -3

1. INTRODUCTION

Japan Quality Assurance Organization (JQA) has started long term outdoor exposure test of various kinds of photovoltaic (PV) modules including crystal silicon (c-Si) and amorphous silicon (a-Si) modules at Kitami site in north of the Japan since 1991 and at two more sites in Japan since 1992 and 1993, respectively. Some interesting results have been obtained so far from these outdoor exposure tests, for example, the degree of seasonal fluctuation of a-Si module efficiency increased as the latitude of the site increased [1]. In this report, the efficiencies of c-Si and a-Si module for about 9 years of outdoor exposure are shown and their seasonal changes and long term tendencies are discussed.

As far as the performance of PV module efficiency is concerned, that of a-Si module is the most interesting. In addition to the initial degradation of efficiency known as Staebler-Wronski effect, a-Si modules show seasonal fluctuation of efficiency due to thermal annealing effect and spectral change of the sun light during the year. In order to understand the long term stability of the modules and the influence of the climate conditions on a-Si module efficiency more deeply, another outdoor exposure test have been started since December 1996 at three sites (AliceSprings, Darwin, and Perth) in Australia whose climate conditions are very different from those in Japan. In this report, the efficiencies of a-Si modules for about 2 years of outdoor exposure are shown. The relationship between the climate conditions and the characteristics of efficiencies at each

site is discussed by the results of temperature fluctuation of the year and the measurement of the spectral distribution of solar irradiance at three sites.

In addition to these two outdoor exposure tests in Japan and Australia, the other test has been started since 1999 at Muscat in Oman. The climate conditions at Muscat is very unique for its little rainfall, extremely high temperature in summer, and extremely high humidity. It is very meaningful to understand the characteristics of PV modules in the long term in this unique climate conditions at Muscat. In this report, the efficiencies of single, two layers, and three layers of a-Si type modules for about 100 days of outdoor exposure are shown and the results are discussed.

2. EXPERIMENTAL

Kitami site is located in north of the Japan with its latitude of 43.8 ° and it is in sub-frigid zone. In winter, there is a snowfall and the minimum air temperature reaches as low as about -20 ℃. Since 1991, c-Si and a-Si modules have been exposed together with other types of modules at Kitami site.

Three sites in Australia are AliceSprings, Darwin, and Perth. AliceSprings is located in the center of Australia and it is in desert region with very little rainfall and low humidity. The difference of air temperature between day and night and that between summer and winter is very large. Darwin is located in north of the Australia and it is in tropical rain forest region with much rainfall and high humidity in rainy summer season. Air temperature is high throughout the year. Perth is located in south west of the Australia close to Indian Ocean and it is in temperate region with mild climate conditions. Since December 1996, a-Si modules have been exposed together with other types of modules at three sites in Australia.

Muscat in Oman is located on the tropic of Cancer and its average total irradiation of the year is as much as that at the sites in Australia. Maximum average air temperature is higher (about 37 ℃) than that at AliceSprings (about 30 ℃), and its humidity is as high as that at Darwin (about 80%). Since 1999, single, two layers (a-Si/a-SiGe), and three layers (a-Si/a-Si/a-SiGe) of a-Si type modules have been exposed together with other types of modules at Muscat site in Oman.

At all sites of these outdoor exposure tests, I-V curves of the exposed modules were measured automatically once a day at around noon. Irradiance was measured by the pyranometer tilted with the same angle of each site latitude and the temperature of the modules was measured by the thermocouple attached on the back of the modules. These two values were used for the correction of the obtained raw efficiencies into the ones at standard conditions (1 kW/m^2, 25 ℃). The spectrum distribution of solar irradiance and its seasonal changes were not taken into account for the correction of raw efficiency. The data with irradiance of less than 0.8 kW/m^2 and apparently abnormal values were omitted for analysis. The measurements of module efficiencies by solar simulator were also conducted twice a year at AliceSprings and Darwin, fourth a year at Perth and sometimes at Kitami. The spectrum distribution of solar irradiance at three sites in Australia on one day were measured to evaluate the influence on short circuit current of a-Si.

3. RESULTS AND DISCUSSIONS

3. 1 Tests in Japan

The efficiencies of c-Si and a-Si modules measured automatically on site and by solar simulator for about 9 years are shown in Fig. 1. The average efficiency of c-Si (about 10%) was about twice as that of a-Si (about 5%). The seasonal fluctuation of efficiency of a-Si was much larger than that of c-Si because of the thermal annealing effect and seasonal change of spectral distribution of solar irradiance. The efficiencies measured on site and that by solar simulator were almost like the same both for c-Si and a-Si. The efficiency of c-Si in this long term of exposure may look decreased from about 7 years of exposure but this trend is not so clear. In the case of a-Si, the efficiencies from the on site measurement may look decreased from about 7 years of exposure but the efficiencies measured by solar simulator did not show

this kind of decreasing tendency at least from 3 to 9 years of exposure. From the results of the outdoor exposure test at Kitami site in Japan, it can be said that c-Si and a-Si showed little or almost no decreasing tendencies of their efficiencies for about 9 years of exposure.

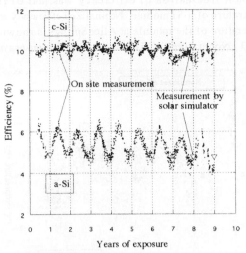

Fig. 1 Efficiencies of c-Si and a-Si modules exposed at Kitami site in Japan

3. 2 Tests in Australia

Normalized efficiencies of a-Si modules measured on site and those by solar simulator at three sites (AliceSprings, Darwin, and Perth) in Australia are shown in Fig. 2. As far as the normalized efficiencies for

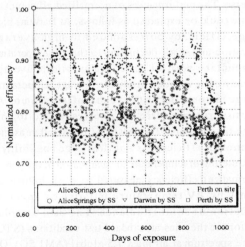

Fig. 2 Comparison of normalized efficiencies at three sites in Australia

about 1000 days of exposure are concerned, it is difficult to recognize clear trend from on site data. But from the data by solar simulator, normalized efficiencies showed very slight decreasing tendency. The degrees of decrease were about 2-3 % per year after initial degradation at all the three sites. Whether this rate of decrease in efficiency will continue or saturate is not known at this present.

Average efficiency was highest at Darwin, middle at AliceSprings, and lowest at Perth. Normalized efficiencies showed seasonal fluctuation, which were different from site to site. The degree of seasonal fluctuation of efficiency was large at AliceSprings, and small at Darwin and Perth. One of the reasons for seasonal fluctuation of efficiency was that of the temperature of a-Si module. Normalized efficiency and temperature of the module at AliceSprings is shown in Fig. 3. Normalized efficiency and PV temperature

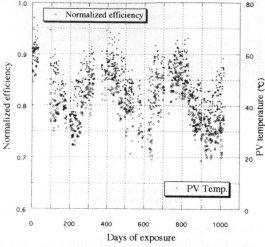

Fig. 3 Normalized efficiency and PV temperature at AliceSprings

corresponded to each other very well; normalized efficiency was high when PV temperature was high and vice versa. The performances of normalized efficiency at each site could be explained as follows. At Darwin, high average efficiency corresponded to high average temperature and small fluctuation to that of temperature throughout the year. At AliceSprings, middle average efficiency corresponded to middle average temperature and large seasonal fluctuation to that of temperature. At Perth, low average efficiency corresponded to low average temperature. The reason of the small seasonal fluctuation of efficiency is not clear. One possibility is large scattering of temperature in each season of the year.

Normalized efficiency measured on site is influenced not only by PV temperature but also by spectral distribution of solar irradiance. The data by solar simulator are the ones at standard test conditions (STC) with the spectrum at air mass 1.5 global (AM1.5G). On the other hand, the measurement of spectral distribution of solar irradiance at each site and the correction of the obtained data into the ones at STC were not conducted for on site data. As can be seen from Fig. 2, the difference in efficiency between on site data and that by solar simulator was large at Darwin and small at Perth. Darwin is located in low latitude and has less air mass throughout the year, which causes more shorter wavelength components in sun light. This pretends to increase the efficiency because the spectral response of

a-Si is larger in shorter wavelength range. It is not clear why the differences in efficiency between on site data and that by solar simulator were small at Perth. Seasonal fluctuation of spectral distribution of solar irradiance at Perth is expected to be larger from the calculation of seasonal change of air mass due to its larger latitude.

The spectral distribution of solar irradiance were measured in summer (at 12:30 on 8th Dec. '99 at AliceSprings, 11:48 on 10th Dec. '99 at Darwin, and 12:22 on 14th Dec. '99 at Perth) and the result is shown in Fig. 4. Irradiance was large at AliceSprings, Perth,

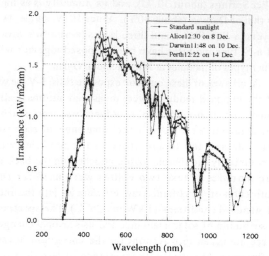

Fig. 4 Spectral distribution of solar irradiance at three sites in Australia

and Darwin in this sequence. The irradiance at AliceSprings was larger than that of standard sunlight for all the measured wavelength (300-1100nm) but that at Perth and Darwin were larger than that of standard sun light in shorter wavelength range (about less than 640nm) and smaller in longer wavelength range (about more than 640nm). This means the spectral distribution of solar irradiance at Perth and Darwin pretends to increase the efficiencies of a-Si because photoresponse of a-Si is larger in shorter wavelength range.

In order to clarify this result more clearly, integrated spectral irradiance and calculated short circuit current of a-Si for these spectral distribution of solar irradiance are shown in Fig. 5. Integrated spectral irradiance ratio (Er in the figure) is defined as the equation (1) and is an indicator to show the ratio of integrated spectral irradiance at each site compared to that of standard sun light.

$$Er=((\ \Sigma\ Ei-\ \Sigma\ Estd)/\ \Sigma\ Estd \qquad (1)$$

where, Er is integrated spectral irradiance ratio, Ei is spectral distribution of solar irradiance at each site, Estd is spectral distribution of standard sun light. The range of integration was 300-1100 nm. From this result, integrated spectral irradiance was about 10% larger, 2% smaller, and 2% larger than that of standard sun light at AliceSprings, Darwin, and Perth, respectively. Short circuit current ratio (Isc r in the figure) is defined as the

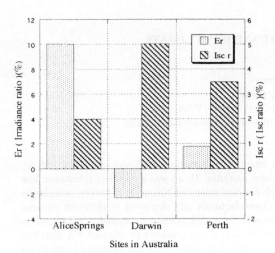

Fig. 5 Characteristics of solar irradiance
at three sites in Australia

equation (2) and is an indicator to show the influence of the spectral distribution of solar irradiance on short circuit current of a-Si by no spectral correction.

$$\text{Isc r} = \frac{(\Sigma \, Qa^*Ei/\Sigma \, Ei - \Sigma \, Qa^*Estd/\Sigma \, Estd)}{\Sigma \, Qa^*Estd/\Sigma \, Estd} \quad (2)$$

where, Isc r is short circuit current ratio, Qa is photoresponse of a-Si, Ei is spectral distribution of solar irradiance at each site, Estd is spectral distribution of standard sun light. The range of integration was 300-1100 nm. From this result, short circuit current was about 2%, 5%, and 3.5% larger than that of standard sun light at AliceSprings, Darwin, and Perth, respectively.

3. 3 Tests in Oman

The normalized efficiencies of three types of a-Si modules (a-Si, a-Si/a-SiGe, a-Si/a-Si/a-SiGe) measured automatically on site for about 100 days are shown in Fig. 6. All of the normalized efficiencies showed sharp

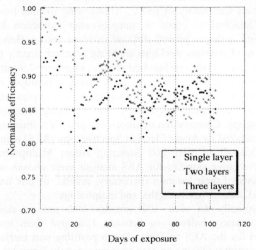

Fig. 6 Normalized efficiencies of single, two layers,
and three layers of a-Si modules

decrease after about 30 days of exposure, which could be recognized as initial degradation. The degrees of this initial degradations were large for single, two layers, and three layers in this sequence. The normalized efficiencies increased at about 50 days of exposure and were large for three layers, two layers, and single in this sequence. After about 50 days of exposure, all of the normalized efficiencies looked stabilized in a certain range (85-90% of initial efficiencies). As this is the result for a short time (about 100 days) of exposure, the evaluation of the performance of efficiency and their prediction is a task for future study.

4. CONCLUSIONS

Outdoor exposure tests of PV modules have been conducted in Japan, Australia, and Oman. From the test in Japan, it was found that the efficiencies of c-Si and a-Si modules showed little degradation tendencies for about 9 years of exposure. The tests in Australia showed that a-Si module efficiencies were strongly influenced by the climate conditions at each site, for example, seasonal fluctuation of their efficiencies well corresponded to that of temperature. The efficiencies showed very slight degradation tendencies (2-3% per year) for about two years of exposure. The spectral distribution of solar irradiance measured at three sites in Australia were favorable ones for a-Si short circuit current as they had larger irradiance in shorter wavelength range. The test in Oman for a-Si related modules (single type a-Si, two layers type a-Si/a-SiGe, and three layers type a-Si/a-Si/a-SiGe) showed initial degradation of efficiency. Performances of the efficiencies were a little bit different among the type of a-Si modules.

ACKNOWLEDGEMENTS

This work was supported by NEDO (New Energy and Industrial Technology Development Organization) as a part of the New Sunshine Program under Ministry of International Trade and Industry, Japan.

REFERENCE

[1] M. Ikisawa, et al, Renewable Energy vol. 14, Numbers 1-4, pp 95-100, 1998

PV MODULE DURABILITY IN HOT AND DRY CLIMATE

Neelkanth G. Dhere
Florida Solar Energy Center
1679 Clearlake Road, Cocoa, FL 32922-5703, USA
Phone: (407) 638-1442, Fax: (407) 638-1010, e-mail: dhere@fsec.ucf.edu

ABSTRACT: Average adhesional shear strength at the Si/EVA interface in a module deployed in the hot and dry climate in Arizona was ~2 MPa or ~35% of that (~5.7 MPa) in new modules. There was a correlation between low adhesional strengths and low surface concentrations of C and high concentrations of Na and P. A similar correlation has been observed in c-Si PV modules fabricated by different manufacturers and deployed in different climatic conditions. Chemically assisted diffusion of Na and P in samples from field-deployed modules may lead to formation of sodium phosphates and hydro-phosphates. Precipitation of reaction products at silicon/EVA interface resulted in loss of adhesion. Thus PV module durability can be improved by controlling the sources of impurities and their diffusion.
Keywords: PV Module Durability - 1: Hot Dry Climate - 2: C-Si EVA - 3

1. INTRODUCTION

The world Photovoltaic (PV) module and cell production has grown 31.5% from 153.2 MW in 1998 to 201.5 MW in 1999. With a growth of 63%, the Japanese production of 80 MW has surpassed the slowly increasing US production of 53.7 MW. The lions share (~80%) of PV module production goes to the crystalline silicon (c-Si) technology.

Pottant or encapsulant is a critical component for structural support, optical coupling, physical and electrical isolation, and thermal conduction. Over the years, polyvinyl butyral (PVB), silicone rubber, ethylene-vinyl-acetate (EVA), and more recently, other proprietary encapsulants have been used. Among these, EVA copolymer is used predominantly by the PV industry [1,2]. Therefore, research at FSEC PV Materials Lab has concentrated on the durability of c-Si/EVA PV modules. Because of their potential of both higher module efficiencies and lower cost, the production volumes of amorphous silicon, $Cu(In,Ga)Se_2$ and CdTe thin film technologies are expected to rise sharply during the next decade. Hence the study of the durability of thin film PV modules is also being carried out.

Until recently, relatively little effort had been made to evaluate the degradation mechanisms in field-aged modules, because of the difficulty in dissecting the laminated modules and also because of the lack of established diagnostic procedures. FSEC has developed techniques for dissecting the modules as well as for establishing some of the diagnostic techniques, in collaboration with the Sandia National Laboratories.

Delamination, resulting from the loss of adhesion must be addressed in order to achieve 30-year module lifetime. Delamination has been observed to occur in a small percentage of modules from all manufacturers, to varying degrees. The causes for this failure mechanism are not well understood. Most of the delamination observed in the field has occurred at the interface between the encapsulant and the front surface of the solar cells in the module. The delamination is more frequent and more severe in hot and humid climates, sometimes occurring after less than 5 years of exposure. Unfortunately, typical accelerated-ageing tests have not been effective in accelerating the mechanisms responsible for delamination, making laboratory investigation of the phenomenon more difficult.

The problems of loss of adhesion, accumulation of impurities at various interfaces, corrosion of metallic contacts in PV modules deployed in hot and humid climate as well as deterioration of mechanical properties of the encapsulant in hot and humid and hot and dry climates have been addressed in earlier studies [3-6]. This paper presents the study of loss of adhesion and impurity precipitation at Si/EVA interface in a PV module deployed in a hot and dry climate in Arizona during 1985-93 together with a summary of adhesional strength and impurity segregation data.

2. EXPERIMENTAL TECHNIQUE

The sample extraction process developed by the Sandia National Laboratories was further improved at the PV Materials Laboratory of FSEC. Unscrewing of nuts from bolts attached to cored Si samples after extraction by applying a torque proved difficult in some cases. Hence small 6-32 stainless steel nuts are being attached directly to the samples. Over one hundred samples (diameter 7.9 mm) were extracted from three cells from a module deployed in the hot and dry climate. Morphology of the samples was studied by optical microscopy and scanning electron microscopy. Composition of silicon cell and EVA interface was analyzed by Auger Electron Spectroscopy (AES) using a Perkin Elmer PHI 660 Scanning Auger Multiprobe. Representative samples for AES analysis were chosen on the basis of their location (periphery, middle, or bus line region), adhesional strengths, and morphology.

Typically an area 102 μm by 102 μm in size, approximately halfway in between two grid lines was chosen for the AES analysis. Depth profiling was carried out by sputter-etching an area of approximately 3 mm by 3 mm in size with an argon-ion gun at a rate equivalent to ~100 Å min^{-1} of silicon. Auger line scan was carried out over a distance of 127 μm at an intermediate depth. Chemical compound formation was inferred from correlation between AES line scans of different elements.

3. RESULTS

All the cells had partial to moderate delamination around bus lines and were beginning to delaminate between bus lines in a ring pattern. Several of the cells were brown in color and had light to dark blue coloration around ~1 cm of the perimeter. The worst cell was # 13 from the positive terminal. It showed evidence of a significant amount of delamination near one of the bus lines and moderate delamination around the other bus line. Sporadic areas of delamination were observed throughout the cell # 13. The best cell was # 28, with evidence of bluish color along one edge. The average torque values required for extraction of samples in the three regions were: middle 0.205 N m; periphery 0.207 N m; buss 0.372 N m. The adhesional strength of samples from module deployed in Arizona was ~2.0 MPa or ~35% of that of samples from a typical new module of ~5.7 MPa.

The results of analysis of the sample #13delam extracted from a delaminated area and hence having adhesional strength of 0 MPa are provided in the following. AES survey of as received sample to kinetic energy of 2050 eV is shown in Figure 1. It revealed the presence of silicon at 95 eV, phosphorous at 113 eV, carbon at 270 eV, titanium at 383 and 416 eV, oxygen at 511 eV, and sodium at 949 and 989 eV. The surface atomic concentrations calculated using the peak-to-peak height from the AES survey and relative Auger sensitivities of the elements are as follows: silicon 1.8 at. %, phosphorous 4.2 at.%, carbon 17.4 at.%, titanium 18.1 at.%, oxygen 43.8 at.%, sodium 14.7 at.%.

A depth profile was obtained by sputter etching ~3 x 3 mm^2 area at a rate of ~100 Å min^{-1} equivalent of silicon for 25 minutes (equivalent to 2500 Å). The depth profiles of peak heights and atomic concentrations at different depths (times) for the different elements are shown in Figures 2 and 3 respectively. The peak heights obtained by the three-point method and relative Auger sensitivities were used to calculate the concentration of the respective elements. A comparison with the more accurate atomic concentrations obtained using peak-to-peak heights and elemental sensitivities for an equivalent sample gave similar values.

The depth profiles in Figures 2 and 3 show an initial increase in the concentrations of the anti-reflection coating element titanium and then a gradual decline beyond depths of 1600 Å. Beyond the anti-reflection coating of TiO$_x$, the concentration of silicon increases continuously with increasing depth. The concentration of oxygen is high at the surface, remains high through the anti-reflection TiO$_x$ layer, and then decreases more gradually than that of titanium. The low relative Auger sensitivity of silicon results in lower values of peak heights for silicon and hence a larger scatter in the concentration values. Concentrations of carbon falls rapidly from the surface up to a depth of approximately 50   Å. Later the carbon concentration decreases gradually.

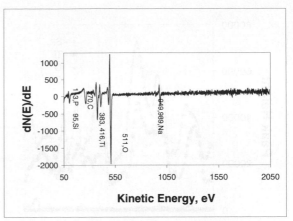

Figure 1: AES survey of a sample #13delam from a delaminated area of cell # 13.

The high surface concentrations of sodium and phosphorous of 14.7 atomic % and 4.2 at.% respectively decreased gradually with depth. The initial high concentrations of sodium, phosphorous, and oxygen point to a compound formation between these elements (Fig. 3). Atomic concentration of elements calculated from an AES survey at the depth of 2500 Å were as follows: silicon 85.0 at. %, phosphorous 1.1 at.%, carbon 0 at.%, titanium 4.6 at.%, oxygen 9.3 at.%, and sodium 0 at.%. As can be seen, the concentration of phosphorous remained significant to the depth of 2500 Å.

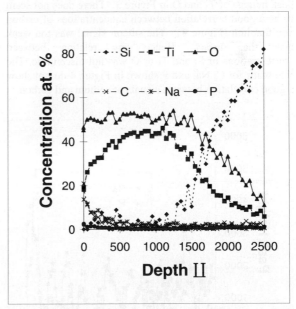

Figure 2: AES depth profile showing elemental peak heights of sample #13delam.

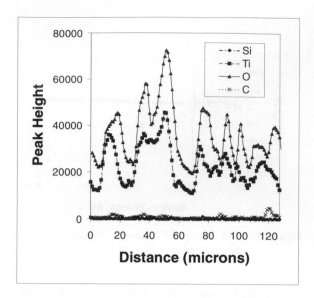

Figure 3: AES peak height line scan for O, Ti, C, and Si of sample #13delam

The depth profiling was paused at the depth of ~300 Å and an Auger line scan was obtained over a distance of 127 μm. The plots of peak heights of AES line scans divided in groups of Ti, O, Si, and C; and C, Na, and P, for ease of comparison, are provided in Figures 3 and 4. An almost direct correlation can be verified in the line scans using peak heights of Ti and O in Figure 3. There does not seem to be a good correlation between concentrations of carbon and titanium (Figure 3). The silicon signal was too weak near the surface. Hence any relation between concentrations of Si and Ti or O was not discernible. The line scans for C, Na, and P shown in Figure 4 do not show a good correlation between signals of sodium and carbon.

The phosphorous signal is too weak for inferring any correlation. The maxima in the sodium concentration can be seen to be located at the minima in the Ti concentration (Figures 3 and 4). The morphological irregularities i.e. orientation differences can be seen to be smoothed out in plots of elemental concentrations. However, relatively constant proportions between titanium and oxygen, as well as the anti-correlation between titanium and sodium can be verified. Higher concentrations of oxygen are observed in regions corresponding to higher concentrations of sodium, providing an additional evidence for compound formation between Na and O. The correlation between titanium and oxygen results from titanium oxide. Maxima in sodium concentration at minima in titanium concentration show regions covered with a sodium compound. AES analysis of samples from the middle, peripheral, and bus line regions gave similar results except for a better correlation between line scans of sodium and phosphorous. All the samples that showed low surface concentrations of carbon had low adhesional strengths. The carbon signals arose from EVA clinging to the surface after sample extraction.

It is interesting to note that a similar correlation has been observed between the adhesional shear strength and impurity concentration at Si/EVA interface in c-Si modules fabricated by different manufacturers and deployed in hot and humid and hot and dry conditions [3-6]. Compilation of the adhesional strength and concentration of carbon, sodium and phosphorous of samples extracted from new modules and modules deployed in hot and humid and hot and dry ambient is given in Figure 5. It clearly shows that the adhesional strength is higher when carbon concentration at the Si/EVA interface is high and the concentration of sodium and phosphorous is low and vice versa. In a few cases, there is an exception. In these rare cases, bulk properties of EVA deteriorate and a weak layer is formed within EVA. Consequently the adhesional strength is low while the carbon concentration is high.

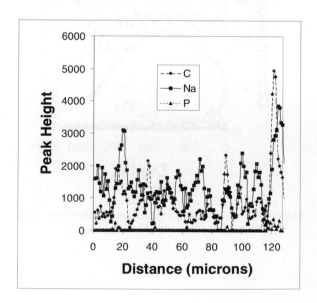

Figure 4: AES peak height line scan for C, Na and P of sample #13delam

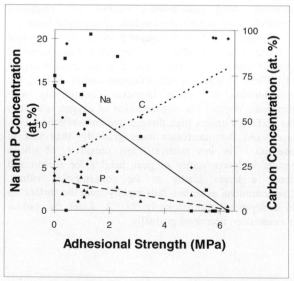

Figure 5: Correlation between adhesional strength at Si/EVA interface with concentration of C ◆, Na ■ and P ▲.

4. DISCUSION

Surface concentrations of sodium and phosphorous in the samples from the modules used in Arizona were very high. Often significantly high sodium and phosphorous concentrations were observed up to depths of a few thousand angstroms equivalent of silicon. The AES line scans from these samples usually showed correlation between concentrations of sodium and phosphorous. Often these were accompanied by proportionately larger concentrations of oxygen after discounting oxygen in the antireflection titanium oxide layer. This points to a strong probability of the formation of compound such as sodium phosphates and hydro-phosphates. The high sodium and phosphorous concentrations both at the surface and throughout the depth of a few thousand angstroms always correlated with low adhesional strengths. As seen above there is a correlation between low adhesional strength on the one hand and high sodium and phosphorous concentrations and low carbon concentrations on the other. This clearly shows the durability of PV modules could be improved if precautionary measures are taken to control the sources of impurities into the structure of PV modules as well as if the diffusion of inadvertent impurities is limited.

The origin of the phosphorous surface concentrations is in the n-type dopant, phosphorous. The sodium may originate from the atmospheric sodium containing aerosols or from soda-lime glass. Sodium in samples from the Arizona module may have originated from the front soda-lime glass. Chemically assisted diffusion seems to be responsible for the excessively high sodium and phosphorous concentrations.

The diffusion of active species such as sodium and phosphorous would begin satisfying some of the bonds at the EVA and silicon surfaces. This surface passivation would, in turn, reduce the strength of the adhesional bonds. Diffusivity of oxygen, water vapor, and other atmospheric species is high in EVA because it is an amorphous copolymer. Sodium and phosphorous at the Si/EVA interface will react with oxygen and water vapor and form sodium phosphates and sodium mono- or di-hydro phosphates. Such impurity precipitation would deplete the sites of adhesional bonds between EVA and silicon cell surface. Sodium and phosphorous concentrations in samples from delaminated regions were amongst the highest. The high concentrations of precipitated impurities may have resulted in complete delamination. Conversely, voids resulting from the delamination provide a preferential location for accumulation of moisture and precipitation of active impurities. These impurities greatly increase the possibility of corrosion failures in metallic contacts.

Several manufacturers are providing 20-25 year warranties for their PV module. An improvement of service lifetime to 30 years will enhance the value of the modules by 20-50%. It will also validate the economic viability calculations that assume a 30-year lifetime.

5. CONCLUSIONS

Average adhesional shear strength at the Si/EVA interface in modules deployed in the hot and dry climate in Arizona was ~2 MPa or ~35% of that (~5.7 MPa) of samples from new modules. Samples having low surface concentrations of C and high concentrations of Na and P had low adhesional strengths and vice versa. A similar correlation is shown to exist between the adhesional strength and the concentrations of carbon, sodium and phosphorous in samples extracted from modules fabricated by different manufacturers and deployed in different climates.

Chemically assisted diffusion of Na and P in samples from field-deployed modules may lead to formation of sodium phosphates and hydro-phosphates. Precipitation of reaction products at silicon/EVA interface resulted in loss of adhesion.

The durability of PV modules can be improved by controlling the sources of impurities in PV modules and by limiting the diffusion of inadvertent impurities.

ACKNOWLEDGEMENTS

This work was supported by the US Department of Energy through Sandia National Laboratories (SNL) Contract # AP- 7660. The author is thankful to Kaustubh S. Gadre for assistance in sample extraction and analysis, Michael A. Quintana and David L. King of SNL for useful discussions and Eric Lambers, of MAIC, University of Florida for assistance in AES analysis.

REFERENCES

[1] A.W. Czanderna, F. J. Pern, Solar Energy Mater. Solar Cells **43** (1996) 101.

[2] W.H. Holley, S.C. Agro, J.P. Galica, R.S. Yorgensen, Proc. 25th IEEE PV Specialists' Conference, Washington, DC, (1996) 1259. N. G.

[3] Dhere, K. S. Gadre, and A. M. Deshpande, Proc. 14th European Photovoltaic Solar Energy Conference, Barcelona, Spain, (1997) 256.

[4] N. G. Dhere, M. E. Wollam, and K. S. Gadre, Proc. 26th IEEE Photovoltaic Specialists' Conference, Anaheim, CA, (1997) 1217.

[5] N. G. Dhere and K. S. Gadre, Proc. 2nd World PV Solar Energy Conf. Vienna, Austria, (1998) 2214.

[6] N. G. Dhere, K. S. Gadre, N. R. Raravikar, S. R. Kulkarni, P. S. Jamkhandi, M. A Quintana, and D. L. King, NCPV Photovoltaic Program Review, Proceedings of the 15th Conference, Denver, CO, Sept. 1998. AIP Conference Proceedings 462, (1999) 593.

PHOTOVOLTAIC MODULES "KVANT" WITH INCREASED SPECIFIC POWER.

M. Kagan, V. Nadorov, V. Rjevsky and V. Unishkov
"KVANT", 129626, 3-d Mytischinskaya 16, Moscow, Russian Federation.
Phone: 7 095 287 97 42, fax: 7 095 18 71.

ABSTRACT: The unified series of PV modules "KVANT" having power 20, 30, 40, 60, 80 and 100 W based on the base module with power 10 W are presented in this report. The following approaches to increase the efficiency of the base module are investigated:

1. Application of combined (ion-diffusion) method to receive n^+-p-p^+ structures allowed to increase the average production efficiency of the solar cells up to 15-16% (AM1.5, T=25°C).
2. Application in the structure of modules of the solar cells with bifacial (double-sided) sensitivity or transparent in IR region of spectrum has brought to the additional power or to a decrease of the module operating temperature.
3. Application of fluorine co-polymer film in combination with the thermal plastic layer EVA type instead of the glass on the front surface of the base module brings to the increase of current and output power on the account of the optical loss decrease that was present in the glass.
    As the result the specific electric power of the unified series of PV modules under the standard conditions of illumination has increased up to 144 W\m².

Keywords: Combined method - 1; double-sided - 2; optical loss - 3.

1. INTRODUCTION

The efforts on perfecting technology of the silicon solar cells and research of a capability of application of new structural materials were undertaken with the purpose to increase the efficiency of PV modules of "KVANT".

Combined method to receive p-p^+ and n^+-p junctions was used to optimize n^+-p-p^+ structure of the solar cell and to increase voltage (V_{oc}). At the same time p-p^+ junction was created as before by method of ion doping by boron and dividing n^+-p junction was created by thermal diffusion from the solution composition containing phosphor.

2. RESULTS

Researches on the influence of phosphor concentration in a strongly doped layer and therefore on the value of open circuit voltage of the doping composition content were conducted. Simultaneously the experimental improvement of diffusion temperature and time for obtaining a required profile of additive distribution in a doped layer was conducted. The profiles of distribution of an additive in a doped layer was measured on the device A P-100C. It is shown that the depth of location of p-n junction makes $0.36 \div 0.4\mu$, surface concentration of an additive $(2 \div 5).10^{19}$ cm^{-2} under the absence of plateau and maximum. In figure 1 the typical volt-ampere characteristic of the solar cell having size 50×50 mm made by the indicated method under standard conditions of illumination AM 1.5, 1000 W\m² and T=25°C is shown.

It is visible, that the voltage V_{oc} has increased on $10 \div 15$ mV in comparison with the solar cell made by the standard method. FF has increased as well due to the better quality of the strongly doped layer and at surface area. In figure 2 the spectral sensitivity of the solar cell made by the improved technology is submitted.

From a curve it is visible, that, if the short-wave part practically is not changed on a comparison with the standard solar cell, then a maximum (0.7A\W under λ=0.95μ) and a long-wave part of spectral characteristic has increased on 10-15%, that indicates to the better

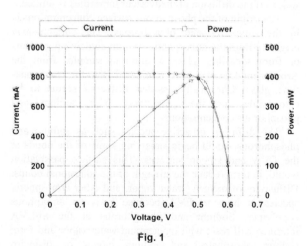

Volt-ampere and volt-watt characteristics of a Solar Cell

Fig. 1

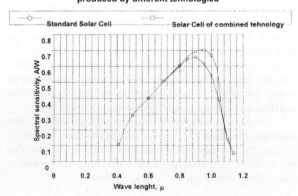

Comparison of spectral sensitivity of Solar Cell produced by different tehnologies

Fig. 2

gathering of carriers and consequently to the larger efficiency of the solar cell.

The investigation of the optical characteristics of the solar cell with application of films made of silicon nitrogen is conducted. Calculation and the experimental researches have shown that the reflection coefficient from a solar cell with the improved structure of brightening coatings in the field of spectrum 0.58-1 μ has made less than 7%.

The application of the solar cells made with the back contact drawing ensuring double-sided sensitivity in a structure of PV modules results in obtaining additional power or reduction of equilibrium operation temperature in a version of a design with carrying back glass. So for a base module consisting of 36 solar cells by the size 50×50 mm the addition to power makes 11% therefore under standard conditions of illumination the power of such module is 12.9 W. The volt-ampere characteristic of such module is shown on figure 3.

Volt-ampere and volt-watt characteristics of a PV module

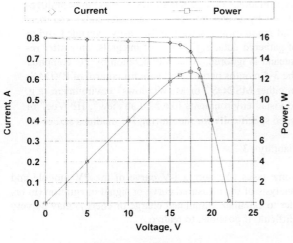

Fig. 3

The comparative volt-ampere characteristics of different modules are submitted in figure 4, where 1 is the module with protective glass on the frontal side, 2 are modules with polymeric layers on a frontal surface and 3 are the modules according to item 2 measured in conditions, when the additional illumination from the back gives an increase to power about 11%.

Volt-ampere characteristics of different PV module

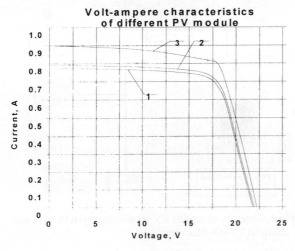

Fig. 4

As the field test shown, in conditions of a six-hour solar day, such module will ensure the additional energy 30÷45 W \ hour.

The important problem is to minimize the optical losses arising at the expense of reflection of radiation from a glass surface. In a report the version of a design of a base module is offered in which the hardened glass is only a carrying substrate and the frontal surface is made as a package from fluorine co-polymer film and film of EVA type. The replacement of a frontal glass on fluorine co-polymer film has allowed to minimize optical losses and to increase efficiency of PV module. The results of experimental researches of such module have shown, that at the expense of reduction of optical losses the current is increased on 4-6% as contrasted to the same module, when the hardened glass faces the radiation. Thus the effect of double-sided sensitivity of a module is saved. The researches of relation of short-circuit current and open circuit voltage of a module from temperature were conducted. The measurement gave the following results: under the illumination from the frontal side K_v = -0.31%\grades, K_j =0.06%\grades; under the illumination from the backside K_v= -0.30%\grades, K_j=0.09%\grades.

Based on the above-mentioned it is possible to make the following conclusion.

3. Conclusion

Improvement of technology of solar cells and application of new materials in a base module have allowed to increase the electrical characteristics of a unified series of modules by power from 10 up to 100 W, in particular, specific power up to 144 W\m² under standard conditions of illumination only from a frontal surface of a module.

PVDA-MODULES: COST EFFECTIVE AND EASILY INSTALLED ALTERNATIVE FOR USE IN AUTONOMOUS PV MONITORING SYSTEMS.

T.Żdanowicz, H.Roguszczak, T.Zawada

Wroclaw University of Technology, Institute of Microsystem Technology, SolarLab
ul.Janiszewskiego 11/17, Wroclaw, Poland
phone/fax +48 71 3554822 e-mail zdanowic@sun1000.pwr.wroc.pl

ABSTRACT: The concept of autonomous PV monitoring systems consisting of many low-cost self-contained specialized functional modules based either on 8-bit PIC (RISC type) or AT89C52 family microcontrollers is presented. Each of the PV data acquisition modules (PVDA-Modules) has been designed and programmed for specified application using the concept of EOP (Event Oriented Programming). All types of modules may communicate with PC in both directions via RS485 bus. Since each PVDA-Module may be accessed only by its own unique two character address so there is no practical limit for the type and number of used modules and monitoring system may be easily expanded and/or modified when necessary. PVDA-Module starts transmission of gathered data, e.g. values of integrals, only after receiving properly addressed request from the PC, otherwise command is ignored. After initialization by PC computer PVDA-Modules can work autonomously and collect data over the 3-4 months period. Software for control of PVDAS-Modules may be easily written in any programming language in either MSDOS® or MSWindows® environment (e.g. Borland's Pascal, C, C++, MS Visual C++) or using specialized visual environments like NI LabView®, HP Vee® or TestPoint®. More detailed description of the IVModule-A designed specifically for monitoring PV modules and/or arrays is presented.
Keywords: Monitoring – 1: Qualification and Testing – 2 : Reliability - 3

1. INTRODUCTION

It is well known truth that reliability and performance of a complex system is as good as is reliability and performance of its individual components and connections between them. This situation is especially well experienced in the case of the PV systems which are naturally designated to work over many years in changeable natural conditions and hence the predictability of the behaviour of the PV modules and other PV systems components in outdoor conditions may be very helpful for the estimation of the long time performance of the overall PV system.

For this reason some PV laboratories design and install advanced monitoring systems performing measurements and collecting data over prolonged periods. However, complexity of installation, specialized electronics and usually high cost of the commercially available general purpose data acquisition equipment make such systems very costly and hence practically nonavailable for many potentially interested PV systems installers and/or users. Another disadvantage of the complex monitoring systems may be that signal conditioning and measurements are often carried out in a single localization which may be quite distant from the signal source, e.g. in a special data acquisition unit or data acquisition card installed in PC, resulting in a complicated cabling of the system. What more, voltage signals may be easily interfered and in the case of high current signals cables must be thick enough to ensure allowable voltage drop. To overcome at least some of these problems commercially available data acquisition modules for field applications may be used. They normally have RS485 (or at least RS232) output for data transmission but they are quite costly and their functions are limited basically to voltage or low current signal measurements. Using such modules in PV monitoring systems will still require units for signal conditioning and

some tasks like taking I-V curve of the PV module and analysis of the measured curve or signal integration (in order to calculate cumulated charge or energy) may be very difficult, if possible, to realize.

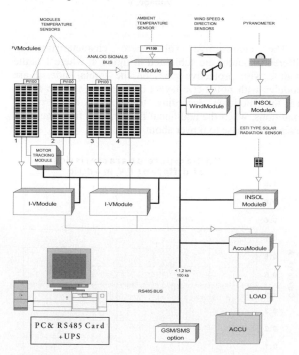

Fig.1 General concept of PV monitoring system with specialized PVDA-Modules.

PV Data Acquisition Modules (PVDA-Modules) developed in Solar Lab combine advantages of both solutions: those of advanced data acquisition units for the specified functions (e.g. they carry out signal conditioning and AD conversion) together with simplicity of installation of general purpose data acqustion modules for field applications [1]. They can be installed as close as possible to monitored objects and connected to common data bus for data transmission. Since they are microcontroller based they can perform complicated tasks, collect data and send them to PC only when required. They are low-cost and they have been specifically oriented for PV monitoring applications. Besides of low cost the another advantage of the PVDA-Modules is extreme simplicity of their installation. Monitoring system consisting of the appriopriate PVDA-Modules after initiallization may work without any communication with PC. All data (mainly integrals) are periodically stored in the EEPROM memory of the microcontroller and will not be lost even if the the power has been cut off.

Using suitable PVDA modules even very advanced monitoring system may be relatively simply assembled. Measured and collected data may be periodically transmitted to PC station via RS485 bus which, contrary to RS232 bus, allows for control of monitoring system from distance as long as 1.2 km at 100kB baud rate for data transmission. General concept of PV monitoring system with PVDA-Modules is shown in Fig.1.

2. EXAMPLES OF PVDA-MODULES DEVELOPED IN SOLAR LAB

Basing on the experience gained from the installation of several monitoring systems [2],[3] the following PVDAS-Modules have been developed so far or are currently under development in SolarLab [1]:

IVModule-A – ready-to-use multifunctional module for measurement of I-V curves of the two PV modules/arrays up to about 1kWp each with MOS power transistors as active load and additional 4 auxiliary channels for other measurements. More detailed description of this module may be found in the following section.

IVModule-B - module functionally similar to IVModule-A but with capacitive load enabling to measure PV arrays up to 20kWp (this module is currently under development).

InsolModule-A – module for insolation measurement with use of commercial pyranometer, e.g. K&Z CM11.

InsolModule-B – module for measurement insolation and temperature using ESTI type sensor or any other silicon sensor. Both types of InsolModules have built-in integrator function for cumulated incident solar energy calculation. Calibration constants must be written into module's memory using PC.

AccuModule – module playing both the role of charge controller for standard lead-acid battery (only 12 and 24 V at present) working in PV system as well as test unit measuring charge/discharge currents with integrator calculating real charge coming to or out of the battery and load; typical treshold voltages are stored in

module's EEPROM but they may be changed using PC and then corrected according to real battery temperature.

TModule – module for temperature measurement. Presently only module for use with Pt100 thermoresistors (8 channels) has been developed.

Software for the overall control of the monitoring system may be written in any programming language under either MSDOS® or MSWindows® control or with use of specialized software like like NI LabView®, HP Vee® or TestPoint®

2.1 Multifunctional IVModule-A

Multifunctional IVModule-A is the most advanced of all modules developed in Solar Lab so far. It can work alone as small PV monitoring system. Together with a laptop computer it can be very powerful instrument for PV engineers servicing PV systems installed in remote locations.

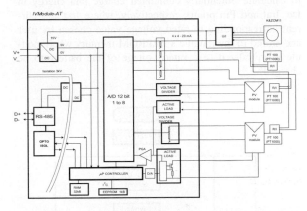

Fig.2 General scheme of multifunctional PVDA Module, type IVModule-A.

Fig.2 shows general scheme of the IVmodule-A. Program controlling operation of the microprocessor was written using concept of Event Oriented Programming (EOP). Its simplified flowchart is shown in Fig.3. User may communicate with the module using either PC with RS485 card or via RS232 port using commercial RS485/RS232 converter. As may be seen in Fig.3 only command addressed to a specified module and with a proper syntax and parameters will be executed otherwise it will be ignored or module return an error value. Module's address is two character long and it must be unique in the installed system.

IVModule-A has two independent four wire channels for measuring full I-V curve of the PV module and/or PV array up to about 1 kWp of the maximum power. Current is measured as voltage drop on the 8 mΩ manganin resistor using high-quality programmable gain instrumentational amplifier (PGA) with three automatically set ranges (0.4, 4.0 and 40 A, respectively). Voltage range is preselected manually by the module's installer as either of 25 V 50 V values, independently for each channel. Since the MOS transistor has channel resistance about 11 mΩ in saturation state so total voltage 'loss' at the short-circuit point should not exceed ~800 mV at the maximum measureable current. To minimize the amount of the energy dissipated in the transistor during I-V measurement in the case when bigger PV array is monitored there is a special procedure adjusting step of the voltage coming from 12-bit DA converter which biases gate of the MOS transistor. Practical example of how the procedure works may be seen on Fig.4 where I-V curve with maximum power point exceeding 400 Wp for the array conisting of 24 PV modules is plotted.

Additionally there are four auxiliary measuring channels which can be used for either voltage (0..4 V) or current signal (0..20 mA) measurement. Each of these inputs can be easily adopted to either temperature or insolation measurement. Another unique feature of the IVModule is possibility of calculating integrals using three independent 32-bit microcontroller's timer/counters. This has been used for calculating the integrals of some parameters independently for each I-V channel and one of the auxiliary channels (selectable). Integration time step may be selected for each channel independently from the range 10..2550 sec with the resolution 10 sec. This option allows to calculate potentially cumulated charge and energy in Isc, Irat and Pm points, respectively, where Irat is a current corresponding to rated voltage value chosen by user and stored in the module's EEPROM memory when module is initialized (e.g. rated voltage of the battery to be charged).

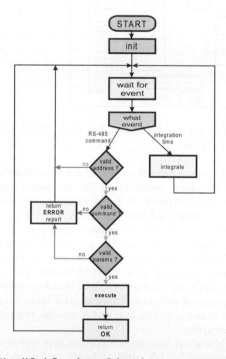

Fig.3 Simplified flowchart of the microprocessor program controlling operation of the IVModule-A.

There is more than 20 commands that IVModule-A will accept and execute. Some of the more important commands are: perform I-V curve measurement, send last measured I-V curve, measure and send Isc, Voc, Irat and Pm values. Commands Init, Start, Stop, Continue deal with integrating procedur. These commands are independently addressed to either of channels for I-V measurements and process of integrating may be launched also for one of auxiliary channels.

Many default parameters, like integration interval and rated voltage for each channel, are stored in EEPROM memory and they can be easily accessed and changed using PC. Some of these parameters can be used to preadjust the module according to special needs of the user.

Built-in watchdog function controls the autonomous work of the module. In for some reason microcontroller 'hangs' the module will reset after few seconds and start work with the initial parameters read from EEPROM.

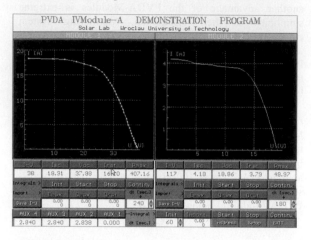

Fig. 4 View of the PC screen for IVModule-A demonstration program showing examples of two I-V curves measured for different PV systems:
Module 1: ≈1 kWp (array of 24×40 Wp modules in series/parallel connection) and Module 2: 80 Wp module with slightly damaged cell (two bypass diodes installed).

3. SUMMARY

Concept of modular PV monitoring systems has been presented. It is based on specialized, PC oriented data acquisition modules designed for field applications. Such alternative will simplify design and installation of even advanced monitoring systems at relatively low costs. Some of the features and possibilities of most advanced IVModule-A designed in Solar Lab mainly for PV modules and/or array tests have been described.

ACKNOWLEDGEMENTS

This work was supported by Polish State Committee for Scientific Research contract PBZ KBN 05/T11/98.

REFERENCES

[1] T.Zdanowicz, H.Roguszczak, European Conference on Photovoltaics, Krakow, Poland (1999) to be published in Optoelectronics Review.

[2] T.Zdanowicz, H.Roguszczak, Proceedings 13th EC Photovoltaic Solar Energy Conference, Nice (1995) p. 2322.

[3] T.Zdanowicz, T.Rodziewicz, M.Waclawek, Proceedings North Sun'99 8th International Biannual Conference on Solar Energy in High Latitudes, Edmonton, Canada 1999.

Flexible, Polymer Encapsulated Solar Modules – A New Concept for Cu-In-S Solar Devices Produced by the CISCuT Technology

R. Güldner, J. Penndorf, M. Winkler and O. Tober

Institut für SolarTechnologien, Im Technologiepark 7, 15236 Frankfurt (Oder), Germany; Tel.: +49 335 5633 0; Fax: +49 335 5633 150; e-mail: gueldner@ist-ffo.de; internet: www.solarzentrum-ffo.de

ABSTRACT: A new module assembly concept has been developed, based on quasi endless, flexible solar cells. The solar cells are fabricated in a base line by the CISCuT technology. It consists of a sequence of five distinct roll-to-roll batch processes. Within the module assembly, solar cell stripes are interconnected in series to strings by the so-called „roof-tile" principle. Any number of strings then can be interconnected in parallel by bus bars. Finally, the absorber is encapsulated in a stack of polymer foils. An assembly line has been developed, integrating these steps in an automated continuous process in order to realise a full base line for module manufacturing.

The technology allows fabrication of flexible modules, adaptable in voltage, current and power output as well as in shape and size. Thus, the module fits easily to different applications.

Keywords Glasgow Conference – 1: Module manufacturing, – 2: CuInS$_2$, CuIn(S$_x$Se$_{1-x}$)$_2$, – 3: Cost reduction

1. INTRODUCTION

In order to minimise the costs for photovoltaic energy conversion, a production line is necessary, which promises high throughput and yield as well as low investment costs. With respect to this, CIS is one of the promising candidates, used as a thin film material for photovoltaic applications. In the past years a relatively wide range of fabrication processes has been developed to deposit the thin film absorber layer and manufacture solar cells and modules [1, 2].

A new technique has been introduced that generates a Cu/In/S absorber layer continuously in different roll-to-roll processes on a Cu-tape (CISCuT) [3]. Using this approach, a base line concept has been developed, fabricating quasi endless solar cells in at least five distinct roll-to-roll processes. Encapsulated into polymer foils, flexible modules are built by „roof-tile" interconnection of solar cell stripes in series and subsequent interconnection of cell strings in parallel in a highly-automated assembly line.

This approach consists of relatively few technological steps. In contrast to other cell manufacturing technologies, the interconnection of solar cell stripes is shifted to the module assembling process. Thus, scaling up of the cell production processes is not necessary and fabrication of solar modules in a wide range of power output, shape and size is made possible. The concept should be highly productive and appears to be promising for the fabrication of low cost solar cells and PV modules.

In this paper we will present this approach in general with a deeper insight into the module assembly process as well as the present status of our investigation.

2. MODULE DESIGN CONCEPT

The module concept begins with flexible stripes of solar cells, typically 1 cm in width. A defined number of these stripes are electrically connected in series by attaching the back side contact of the first stripe to the front side contact of the following solar cell stripe. In this so-called „roof-tile" interconnection the longitudinal edges of the stripes overlap ca. 0.5 mm. Current collection grids at the transparent front side contact are unnecessary, due to the narrow width of the stripes and the "roof-tile" interconnection principle.

The voltage output of the module will be fixed by the number of the interconnected stripes. The variation of the quantity of stripes easily adapts the voltage output of the module to various applications under market conditions. To change the current output of the solar modules, several strings of series interconnected stripes can be connected in parallel by two bus bars on the back side of the cells. For this, the back side contacts of the first stripe of every string are electrically connected with the first bus bar. Respectively, the second bus bar connects the back side contacts of the last stripe of every string. Between the bus bars and all other back side contacts, a dielectric layer is deposited for electrical insulation. A well defined current output can be realised by modification of the length of the stripes as well as the number of the interconnected strings in parallel. The device power output can be placed at any corner of the module.

In Figure 1, the front and back side view of the absorber shows schematically the interconnection principle of cells and strings.

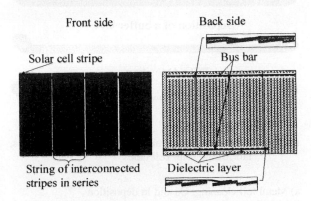

Fig. 1: Solar cell stripes attached in series and strings of them interconnected in parallel by bus bars

Flexible PV-modules are fabricated by laminating a transparent polymer encapsulation on the front side and a polymer encapsulation foil on the back side of the interconnected strings.

By the explained concept, flexible modules can be fabricated, suitable for several applications in voltage, current and power output as well as variable in shape and size.

3. PRESENT STATUS OF THE WHOLE FABRICATION PROCESS

3.1. The solar cell fabrication process

A solar cell tape, presently up to 250 m long, is fabricated by a sequence of at least five distinct roll-to-roll batch processes, schematically described in Figure 2. The concept starts from a metal foil 10 mm in width and 30 μm to 100 μm in thickness. This metal foil serves simultaneously as the mechanical carrier of the solar cell as well as the electrical back contact of the device. First the precursor is formed on the cleaned metal foil, followed by coating the foil edges. Then the precursor is sulfurised and converted into the Cu/In/S absorber layer. After removing a binary $Cu_{2-x}S$ phase at the top of the absorber, a buffer layer is deposited on the surface. Finally, a transparent front contact has to be added.

For all processes a roll-to-roll technique has been employed. Excepting the TCO deposition, where we use the sputter technique, only non-vacuum processes have been developed.

Solar cells, prepared with this technique, show efficiencies up to $\eta = 5$ % so far. Recombination processes and shunting problems are the main reasons for the efficiency losses.

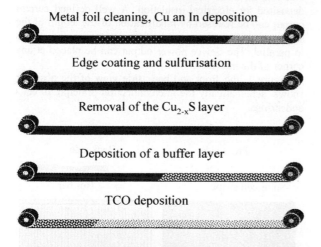

Metal foil cleaning, Cu an In deposition

Edge coating and sulfurisation

Removal of the $Cu_{2-x}S$ layer

Deposition of a buffer layer

TCO deposition

Fig. 2:   Sequence of roll-to-roll batch processes, fabricating solar cell tapes

a) Metal foil cleaning, Cu and In deposition

In the first roll-to-roll processing step, the metal foil is chemically and electrochemically cleaned. After some rinsing processes, the foil is electrolytically covered with a Cu alloy followed by a one sided electrolytical Indium deposition step. The Cu alloy has a thickness of typically 1μm to 5 μm, the thickness of In is in the range of 0.5 μm to 1 μm.

b) Edge coating and sulfurisation

In the first part, the borders of the foil are coated by wet-chemical deposition, using a silicate-like solvent. The covered areas do not react with sulphur gas in the following sulfurisation step. There, a sequence of several Cu/In/S layers is formed by sulfurisation and subsequent conversion of the precursor into the absorber. Inside the layers an internal p/n-junction has grown [4, 5].

c) Removal of the $Cu_{2-x}S$ layer

Currently this layer is removed by the well known KCN etch with some rinsing and drying processing steps in a roll-to-roll equipment.

d) Deposition of a buffer layer

Depending on the properties of the absorber layer, a p-type material is required as a buffer layer. An alcoholic CuI solution is continuously sprayed onto the surface of the etched precursor. After drying, a well defined CuI buffer layer is obtained in a thickness of about 100 nm. Presently, thin layers of CuS are also being investigated.

e) TCO deposition

As a transparent front contact, ZnO:Al is deposited by DC-sputtering. For this, a roll-to-roll equipment has been developed and built up in the base line.

3.2. The module fabrication process

A highly-automated assembly line, that is about to be completed at our institute, should fulfil the above explained module design concept and contribute to a low cost module production. For this, development was focused on simple and automatic handling of cell and module components, continuous processing and reduction of manual labour.

The construction principle is based on existing foil web converting systems. In our system in particular (Figure 3), the front side foil is first unwound and drawn into machine direction X (a), passing the following process steps (=machine units). In the following, the cell stripes are laminated onto the front side foil (b) and are electrically connected to modules (c). Subsequently, the back side foil is laminated onto the front side foil (d), representing the encapsulated module that is finally cut in machine direction X and transversal direction Y (e). Optional, the module adhesive has to be cured and the module can be equipped with the frame.

The absorber width is limited by the assembly line width and can vary between 200 and 1000mm. The absorber length theoretically has no limits. All process steps are adapted to the feed of the front side foil that can be adjusted between 10 and 100mm/min.

a) Front side foil transport

The front side foil is unwound and drawn from a stock coil with a controlled constant feed and furthermore positioned in transversal direction Y by a guiding control. The pull-roll pair behind the back side foil lamination maintains a constant web tension. The control of these parameters is a requirement for the quality of cell overlapping in the following cell laying process.

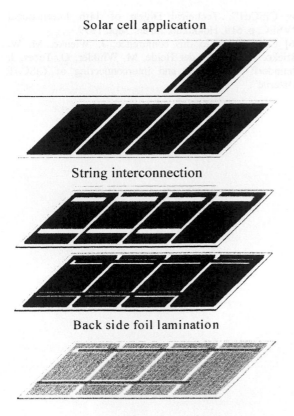

Solar cell application

String interconnection

Back side foil lamination

Fig. 3: Process steps of solar module fabrication

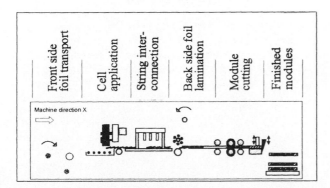

Fig. 4: Scheme of the assembly line for fabrication of flexible solar modules

b) Cell application

In preparation for the cell lamination, the front side foil is pulled over a heating plate that adjusts temperature and consistency of the adhesive layer situated on the top side of the foil web. Stripes of solar cells are now laminated into the adhesive, crosswise to the feeding direction X of the front side foil. As a result, the absorber side of the cell is embedded into the adhesive, the cell backside faces upwards.

Process (b) is realised by an X-Y-axis drive controlled laying head, where the cell is unwound from a stock reel and subsequently heated and pressed into the adhesive layer by a laying wheel. Finally, the cell stripe is cut to a specified length. Simultaneously the cell is "roof–tile" interconnected with the previously-laid stripe. A constant overlapping of the stripes is achieved by speed

synchronisation of the corresponding driving axis X to the front foil feed. In order to get a reliable mechanical and electrical connection in the overlapping zone, an adhesive has to be dispensed locally onto the overlapping edge, either before or within the cell laying process. Interconnection methods, using soldering processes, are also investigated [6].

The control interface of the machine allows programming of various process parameters like cell heating temperature, laminating pressure, laminating speed and overlapping width, in order to adapt the technology to the cell interconnecting process, to different cell carrier and foil materials, as well as to the production speed. Programmable module parameters, like the number of cells per string/strings per module, stripe length, string and module margin, etc., guarantee the flexibility of geometrical and electrical module features mentioned above. Prepared for the utilisation in a base line, the manufacture of a quantity of different module types can be programmed in a batch file. This will be executed when the cell laminating process is started.

c) String interconnection

The front side foil enters this section with a defined number of strings and cells per string which will be connected electrically on the cell back side by soldered bus bars. The machine concept permits an interconnection of any number of strings to a module. Furthermore, the bus bars can be applied onto the string back sides (Figure 1) or alternatively beside of the absorber.

Two application systems, each of them equipped with a set of devices, independently carry out the interconnection cycle of two bus bars. Required module parameters and cell positions are received from the cell laying unit. Interconnecting parameters can be programmed via the control interface.

First, lines of an insulating layer are sprayed to avoid short circuits between bus bar and the cell stripes below. Solder paste is then dispensed onto those cell stripes, which will be later soldered with the bus bars. The conducting stripes can now be placed in this manner that the two external device connections have a variable distance and can be situated at any corner of the module. Finally, application of solder paste and punching of two holes into the back side foil prepares the mounting of the junction box that completes the module fabrication.

d) Back side foil lamination

The module is encapsulated by laminating the heated back side foil web onto the front side web. Both are pulled through between laminating rolls by a pull-roll pair placed behind the laminating unit. Adjustment of laminating gap and adhesive temperature of the back side enables bubble free and smooth laminates. An integration of a module reinforcing layer into the back side foil is a possibility to regulate the module flexibility.

e) Module cutting

In the edge- and cross-cutting machine, the module margin Y can be defined manually, adjusting the edge cutter positions. The controller for cross-cutting the module panels from the laminated endless web gets the important dates from the cell laying unit and the incremental shaft encoder of the front side foil.

Using EVA as a cell embedding material results in the utilisation of a subsequent curing process. As one possibility, a standard vacuum laminator could be employed.

Depending on the embedding material, the adhesion between the module components and the field of module employment, the panel needs a certain level of edge protection against vapour diffusion and mechanical wear. A frame can fulfil these requirements and simultaneously ease the panel installation by integrated fixing points.

4. CONCLUSION

A new concept has been developed, fabricating flexible solar modules in an automated process. Presently, the module assembly line is still under construction. First investigations in single technological steps show promising results to manufacture solar modules according to this concept. For continuing investigations, a new project, supported by the German Government, has been started in order to developed the whole assembly process. In the next few years prototypes will be produced to demonstrate the main objectives of the whole base line concept. For that, further investigation will be done to optimise the solar cell production process. The activities are especially focused on the performance increase of the solar devices, based on the CISCuT technology.

ACKNOWLEDGEMENTS

The authors thank G. Barth, B. Färber, P. Rau (IST) and I. Perschke (FHTW) for technical assistance. We would also like to thank the "Ministerium für Wirtschaft" des Landes Brandenburg (LMWi) for its financial support of the technological equipment. Thanks are also addressed to the European Commission, supporting this work in the PROCIS-project JOR3-CT980159.

REFERENCES

[1] T. Negami, T. Satoh,Y. Hashimoto, S. Nishiwaki, S. Shimakawa, S. Hayashi: "Large Area CIGS Absorbers Prepared by Physical Vapour Deposition". Technical Digest of 11th International PVSEC, p. 633, Hokkaido, Japan, Sept. 1999.

[2] B. Dimmler, M. Powalla: "Scaling up Issues of CIGS Solar Cells". European Material Conference Proceedings, O-VIII.2, Strasbourg, France, June 1999.

[3] K. Jacobs, J. Penndorf, D. Röser, O. Tober, M. Winkler: "CISCuT - A non Vacuum Roll to Roll Technique for Preparation of Copper Indium Chalcogenide Absorber Layers on a Copper Tape". Proc. 2nd WCPEC, p. 409-412, Vienna, 6-10 July 1998.

[4] M. Winkler, O. Tober, J. Penndorf, K. Szulzewsky, D. Röser, G. Lippold, K. Otte: "Phase Constitution and Element Distribution in Cu-In-S based Absorber Layers grown by the CISCuT-Process", thin solid films, Vol. 361-362, (2000), p. 273-277

[5] I. Konovalov, O. Tober, M. Winkler, K. Otte: "Characterisation of Cu-In-S Absorber Structure obtained by CISCuT". Technical Digest of 11th International PVSEC, p. 619, Hokkaido, Japan, Sept. 1999.

[6] Contribution to this conference: J. Wienke, M. W. Brieko, A.S.H. van der Heide, M. Winkler, O. Tober, J. Penndorf: "Contacting and Interconnecting of CISCuT Material".

THE PVUSA FIELD WET RESISTANCE TEST PROCEDURE

C. M. Whitaker, A.B. Reyes, T.U. Townsend
Endecon Engineering, 2500 Old Crow Canyon Road, Suite 220, San Ramon, CA 94583.
+1 925 552 1330, fax= +1 925 552 1333, info@endecon.com
D. L. King
Sandia National Laboratories, Bldg 848 Dept 6219 1515 Eubank SE Albuquerque, NM 87123-0752
+1 505 844 8220, Fax = +1 505 284 3239, dlking@sandia.gov

ABSTRACT: The Field Wet Resistance Test (FWRT) is a procedure used to evaluate the integrity of a PV array's insulation. A field version of the Wet Insulation Integrity test in IEEE 1262 and IEC 60146, this test evaluates wiring, connectors, and junction boxes in addition to module encapsulation. A decade of experience at PVUSA shows that arrays nearly always pass easily or fail miserably. The test deftly locates nicked wire insulation, loose junction box covers, screw driver-penetrated module backsheets, inadequate connector seals, hairline cracks in module junction boxes, and other potentially unsafe and unreliable conditions. These mechanical problems are typically indicated as a short or near short to ground in the presence of an aqueous surfactant. Even without these problems, there are often large variations—factors of 2 to 10— in readings throughout the array. In this paper, we investigate a number of variables that affect the measured wet resistance of a PV array. Sorting through these variables will lead to more accurate measurement and diagnosis of impending failures. Results indicate glass temperature and settling time have a much greater influence than previously anticipated. Conversely, surfactant concentration and array voltage have very little effect.
Keywords: Field Testing – 1: Safety - 2: Reliability - 3

1. INTRODUCTION

Since 1989, Photovoltaics for Utility Systems Applications (PVUSA, www.pvusa.com) has performed a variety of field tests to evaluate the performance, reliability, and safety of installed PV systems. One of these, the Field Wet Resistance Test (FWRT), is used to assess the quality and health of a PV array's electrical insulation system. Performed more commonly in the US than in Europe, the FWRT is the field companion to the factory Wet Insulation Resistance Test in IEEE 1262 [1] and the Wet Leakage Current Test described in IEC 60146 [2]. Recently, the test was standardized by the American Society for Testing and Materials ASTM [3].

Readings of greater than 36 $M\Omega$-m^2 are considered adequate, however, field experience shows that arrays either pass easily or fail miserably. The test regularly locates cut wire insulation, loose junction box covers, module backsheets cut by over aggressive use of screw drivers, inadequate connector seals, hairline cracks in module junction boxes, and other potentially unsafe and unreliable conditions. PVUSA performs this test on all systems installed at its Davis

California site as part of performance verification and commissioning testing [4 lists a variety of test procedures used by PVUSA and others]. The question arises, can the accuracy and utility of the FWRT be improved through a better understanding of the factors influencing test results?

PVUSA and Sandia National Labs coordinated an evaluation of the test procedure to investigate alternate implementations (megohmmeter attached to plus, minus, and shorted segment leads), variations in surfactant concentration, effects of array open circuit voltage and temperature, and other parameters.

2. TEST DESCRIPTION

The basic test procedure involves connecting a megohmmeter between earth ground and one leg of an open circuited PV module or array segment, elevating the segment well above ground voltage. Next, the module(s) are sprayed with a mild aqueous surfactant solution. The insulation leakage resistance is then measured with a megohmmeter (Fig. 1). The megohmmeter works by providing a voltage (nominally

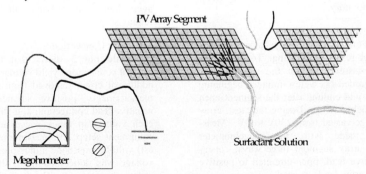

Figure 1 Field Wet Resistance Test Setup

500 or 1000V, sometimes higher) to the array segment, measuring the leakage current with an internal shunt resistance, and calculating and displaying the resistance. When connected as shown, the voltage generated by each module in the array segment is added to the meter voltage thus increasing the leakage current and overestimating the array leakage resistance. For an array segment with a voltage, V_{SEG}, the overall leakage resistance, R_{SEG}, can be expressed as a function of the megohmmeter voltage, V_M, its internal shunt resistance, R_M, and the meter reading, R_{Read}:

$$R_{SEG} = \frac{R_{Read}(2V_M + V_{SEG}) - R_M V_{SEG}}{2V_M} \quad \text{(Eqn. 1)}$$

As an alternative to the arrangement in Figure 1, the leads of the array segment can be shorted together, thus reducing V_{SEG} to zero and providing a direct reading of the segment leakage resistance. Shorting the segments requires a switching device of sufficient voltage and current (load-break) rating. In large arrays where many segments are tested sequentially, it's often more convenient to leave the segments open and correct the readings.

3. EVALUATION OF VARIABLES

A number of parameters will impact the results of the FWRT. Some of these parameters affect the leakage resistance (temporarily or permanently), while others affect the measurement itself. Using the variety of systems present at PVUSA, specific tests were performed to evaluate the effects of the following variables on the FWRT results:

- Array Voltage
- Array Temperature
- Surfactant Concentration
- Module Construction
- Array Age

Our approach was to perform the tests outdoors using full-size PV array segments (10 to 150 modules, areas of 9 to 62 m^2, with allowable leakage resistances of 0.6 to 4.2 MΩ). This approach made it difficult to precisely control each variable and to separate the simultaneous effects of multiple variables. However, it also provided a more realistic environment and insight that would not be apparent by testing single modules in the laboratory.

3.1 Array Voltage and Temperature

As shown in Eqn. 1, array voltage has a linear impact on the leakage resistance reading. To verify this, wet resistance measurements on the subject array segment were recorded every few minutes beginning just before sunset and ending after the segment open circuit voltage reached zero. During the test, array open-circuit voltage varied from 230V to 0V. Measurements were made with the megohmmeter connected to the array segment three ways: open-circuited to negative lead, open-circuited to positive lead, and with the array leads shorted.

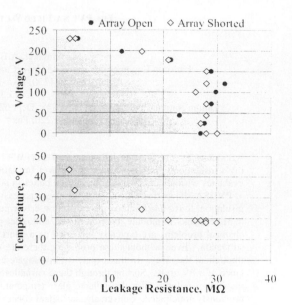

Figure 2 Leakage Resistance versus Segment Open-Circuit Voltage and Temperature

The upper plot in Fig. 2 shows the segment leakage resistance as the array voltage dropped to 0V. Measurements with the segment open-circuited (and corrected using Eqn. 1) and with the array shorted are shown. Though there are variations between the two methods, they follow the same general pattern, with resistance apparently rising with decreasing array voltage until the segment reaches roughly 150 volts, at which point, the resistance stays relatively constant. Of course, with the array shorted, there should be no impact due to array open-circuit voltage.

The lower plot in Fig. 2 shows the same shorted segment leakage resistance data points as in the upper plot, this time plotted against array temperature. Here, we see that, as expected, the leakage resistance is inversely proportional to temperature. Note that array temperature measurements are not terribly accurate due to the dynamic effects of solar absorption, module heat capacitance, and surfactant evaporation. Nonetheless, these measurements corroborate the notion that changes in leakage resistance indicated above 150V are actually related to temperature. Below 150V, readings were taken at a relatively constant temperature and show no obvious correlation to voltage, better identifying the expected voltage relationship.

3.2 Meter Connection

While Fig 2 shows that the open-circuit and shorted connections yield similar results, there was an additional effect removed from that plot. The three measurements—positive segment lead connection, negative lead connection, and shorted—were made at each voltage in quick succession. Because of the equipment setup, the measurement at the first (highest) voltage was done minus-plus-shorted; the second voltage was done plus-minus-shorted, the third was minus-plus-shorted. The Eqn. 1 corrected measurements for both the "plus" and "minus" readings are

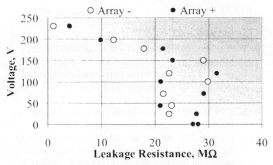

Figure 3 Example of Settling Time (1ˢᵗ reading low)

plotted in Fig. 3. This plot shows that the first measurement tended to result in a lower resistance measurement (the shorted measurements at each voltage tended to match the higher of the two open-circuited readings).

When performing the FWRT, there is a noticeable "settling time" where the measured resistance drops or rises asymptotically over a period of 30 to 90 seconds from the initial application of the meter voltage. Typically, the test is performed as quickly as possible to keep labor costs down. Also, leakage resistance will rise as the array dries and as it cools, which it often does simultaneously. However, the effect here appears to be more of the capacitive charge/discharge type.

3.3 Surfactant Concentration

The surfactant acts as a wetting agent to reduce surface tension and ensure the solution sheets. It also helps the solution penetrate cracks. The surfactant concentration does not have to be very high—a 500:1 (or 0.2%) water to surfactant concentration is specified in most standards. However, plain water is often not sufficient., FWRT testing using plain water, performed by one manufacturer at the time of installation of a 300kW array, showed no problems. Subsequent testing by PVUSA using the same equipment but adding a gallon of LiquiNox surfactant to the 500-gallon spray tank uncovered a systematic problem with a large percentage of module junction boxes.

A sample array segment was tested with a range of surfactant concentrations at night with the array cool

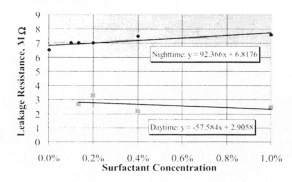

Figure 4 Effect of Surfactant Concentration

and thermally stable (20.5 – 23 C) and during the day with higher, more variant temperatures (35 – 40 C). Fig. 4 shows that under these two different sets of conditions, there is little variation in leakage resistance with large variations in surfactant concentration. In fact, the plot would indicate that plain water is not substantially different than a concentration 5 times stronger than recommended. However, the plot only shows bulk property effects and does not demonstrate the surfactants ability to penetrate, a necessary characteristic for locating many mechanical problems.

3.4 System Age

Dating back to the first system installation in 1989, PVUSA has performed the FWRT as part of the commissioning test on each system and periodically thereafter. Over the years, there have been changes in the procedure, the equipment used, the personnel performing the test, and other factors making direct then-and-now comparisons difficult. However, a qualitative evaluation of the historic and current records indicate that arrays using modules that have passed appropriate qualification tests are able to consistently pass the field wet resistance test. There is less certain, though persuasive, anecdotal evidence that systems exhibiting generic initial failures that were field repaired are likely to continue to show problems in subsequent years. Field "band-aids" are not an adequate substitute for manufacture process controls and proper installation procedures.

4. CONCLUSIONS

Test results show significant variations in FWRT readings—20 to 50% differences were not uncommon—under apparently similar conditions.

Specific tests indicate that these variations in readings are not related to differences in surfactant concentration or variations in array voltage.

Module temperature, however, has been shown to have a significant impact on leakage resistance. It is recommended that for the pass/fail criteria to be effective, standardized procedures include a module temperature range specification, and that the effect of temperature be described.

Settling time also has a considerable influence on results and appeared to be more of an issue with the array open-circuited than shorted.

PVUSA has traditionally performed the FWRT with the test segment open-circuited to avoid problems and safety concerns related to shorting the array. However, this testing suggests several reasons for using the shorted array approach. The shorted method simplifies data analysis by providing a direct measurement of leakage resistance (as opposed to having to correct the megger readings). It also appears that problems/errors associated with settling—possibly related to capacitive discharge—are reduced with the array shorted. Thirdly, though it did not occur during this testing, Endecon Engineering personnel experienced a megohmmeter failure caused by overvoltage within the device when the array and megger voltages added. This risk is significantly reduced if the array is shorted thereby not providing an additional external voltage to the megger.

The open circuit method may still be preferred when

♦   the array short circuit current is high
♦   the positive and negative legs of the test segment are far apart (necessitating long leads to the shorting device
♦   the module manufacturer places restrictions on when the array may be shorted

No specific results of this testing suggest that the traditional pass/fail criteria should be changed or, for that matter, that it is appropriate for the purpose of the test. At this point in time, every module construction tested can pass the $40M\Omega\text{-}m^2$ criterion if there are no specific problems with the insulation system. However, as manufacturers attempt to reduce module costs, by using a thinner glass superstrate for example, this test criterion may limit their abilities to do so.

PVUSA has rejected entire systems because of high failure rates observed with the FWRT denoting generic problems in some aspect of module construction or system installation. In several cases, these failures indicated improper or incomplete application of the factory wet resistance test. Conversely, we have seen several more recent multi-hundred kilowatt installations where failures were limited to a single loose module junction box cover or a module interconnection wire caught under a mounting bolt.

It appears some additional theoretical analyses or a more thorough review of previous theoretical analyses may be necessary to determine if the present selected criterion is meaningful. Also, should the criterion be specified at a different temperature for the outdoor test?

The FWRT can be rather labor-intensive and therefore expensive. It may be sufficient to test random samples of individual (small) systems or segments of larger systems. This is especially true of systems that have significant field performance history, and which have been installed by a competent, qualified installer who has installed this system previously without incident.

The test remains an extremely effective tool for locating existing and potential safety and corrosion problems.

REFERENCES

[1] IEEE Std 1262-1995, "IEEE Recommended Practice for Qualification of Photovoltaic (PV) Modules."
[2] IEC 60146, "Thin-Film Terrestrial Photovoltaic (PV) Modules - Design Qualification And Type Approval"
[3] ASTM E2047, "Test Method for Wet Insulation Integrity Testing of PV Arrays".
[4] IEEE P1373, "Recommended Practice for Field Test Methods and Procedures for Grid-Connected Photovoltaic (PV) Systems", Draft Standard.

AN INVESTIGATION OF THE ENVIRONMENTAL IMPLICATIONS OF THE CHEMICAL BATH DEPOSITION OF CdS THROUGH ENVIRONMENTAL RISK ASSESSMENT

Kathleen M. Hynes and John Newham
Northumbria Photovoltaics Applications Centre, University of Northumbria
Ellison Place, Newcastle upon Tyne, NE1 8ST, UK
Tel. +44 191 2274555, Fax. +44 191 2273650, Email. Kathleen.hynes@unn.ac.uk

ABSTRACT: Cadmium sulphide is widely used as the window material in cadmium telluride and copper indium diselenide thin film modules. The layer is usually deposited in a hot chemical bath, a cheap technology producing good quality material. Experimental analysis has shown the existence of an aerosol, which is located above the liquid surface of the bath, containing both volatile and non-volatile components from the bath. A hazard assessment has been undertaken to highlight the possible environmental risks which may occur due to the deposition process itself and the associated transport, storage and handling of the materials used in and generated by the production process. This paper will use the technique of environmental risk assessment to identify the major hazards within the CdS CBD process and assess their significance to human health and the environment.
Keywords: CdS – 1: Environmental Hazard – 2: Carcinogen -3

1. INTRODUCTION

Cadmium sulphide (CdS) is a commonly used window material in cadmium telluride (CdTe) and copper indium diselenide thin film PV cells. Other materials have been considered as alternative window layers but none have yet provided the same level of technical performance that is exhibited by CdS [1].

Chemical bath deposition (CBD) is a widely used technique for the deposition of the CdS window layers. It is a very simple process to perform, giving cheap, reproducible material. CdS deposition baths are heterogeneous in composition, and contain a variety of soluble inorganic and organic substances, some ionic, together with reactive components that form a fine precipitate. An auxiliary air flow is used to maintain high dissolved oxygen levels and to create vigorous stirring of the bath contents. The aerosol located above such a bath will contain the volatile components present in the bath, but little information exists regarding the release of non-volatile components, which are generally considered to remain in the solution. Aerosol concentrations will be dependent on the various transfer processes which bring each component into the gas phase. Information concerning the aerosol is an input to the environmental risk assessment for the production of the CdS layer.

This paper will use the technique of environmental risk assessment to predict where environmental risks may arise within the CdS CBD process and assess their significance. This should enable the operator to manage the risks and thus ascertain how effective are the existing or potential control measures in eliminating the impact and/or its likelihood.

2. ENVIRONMENTAL RISK ASSESSMENT

There is a growing interest in the use of environmental risk assessment as a tool to help organisations identify priorities and allocate resources for environmental improvement. It was referred to in the Conventions agreed at the Earth Summit in Rio de Janeiro in June 1992 and has formed the basis for EC legislation on the prioritisation of new and existing chemical substances. Expressing effects on the environment in terms of risk can have many advantages. It allows many environmental problems to be measured and compared in common terms. Risk

assessment and risk management have formed the basis for the regulation of human health risks within the Health and Safety Executive in the UK for many years. They form the underlying control of hazardous substances in the workplace as set out in the UK Control of Substances Hazardous to Health Regulations 1999 (COSHH). These were introduced to safeguard the health of, and risks to, workers in their place of work and define hazardous substances as including substances used directly in work activities, substances generated during work activities and naturally occurring substances.

Environmental risk assessment is a methodological approach to allow for an informed decision to be made on the precautionary measures required to protect workers, the general public and the environment from the impacts of hazardous substances. The methodological approach used in this work conforms to Commission Directive 93/67/EEC on principles for assessment of risks to man and the environment of substances notified in accordance with Council Directive 67/548/EEC. It comprises the following: hazard identification, dose-response assessment, exposure assessment, risk characterisation, regulatory guidelines for hazardous chemicals, recommendations for risk reduction. Hazards associated with the properties of these materials are identified in order to design appropriate storage and handling procedures. The assessment of risks and health effects considers issues of toxicity, potential for irritation, mutagenicity, carcinogenicity and teratogenicity. The environmental assessment also includes eco-toxicity and bioaccumulation potential data.

3. HAZARD ASSESSMENT OF THE CHEMICAL BATH DEPOSITION OF CdS WINDOW LAYERS

The hazard assessment has to consider all the various procedures and associated materials involved in the window layer production process. The materials are listed in Table 1, along with the major hazard information. The various procedures are as detailed below:

3.1 Transport and storage

The possible hazards associated with the transport and storage of the bath constituent materials (sodium hydroxide, cadmium chloride, thiourea and ammonia) and hydrochloric acid, used in the cleaning of the bath, have been identified. The Chemical Hazard Information and

Packaging for Supply (CHIP) Regulations (1994) which are a classification and labelling system, cover the transport and storage of chemicals. Possible hazards may be caused by improper packaging and spillage that could occur by the incorrect implementation of the regulations or as the result of an accident during transport. Specific precautions are required for ammonia (gaseous emissions), thiourea (flammability) and cadmium chloride (spillage).

3.2 Preparation of processing materials

This involves the preparation of the processing solutions from the basic materials. Hazards at this stage could be due to spillage, which could lead to impacts on human health if workers were exposed to dust or fumes, and to impacts on the environment if the spillage disposal is unsatisfactory.

3.3 Chemical bath deposition process

The deposition of the CdS layer takes place in a chemical bath, which is heated to, and maintained at, a temperature of 70°C. The exact reaction mechanism taking place within the bath and its consequent reaction products has not been fully identified.

A programme of experimental work was undertaken to investigate the emissions from the bath and attempt to clarify the chemistry of the bath solution [1]. A discussion of the outcome of this work is given in Section 4 and the results from the work are incorporated into the risk assessment reported in Section 5.

The baths are composed of potentially hazardous reagents. Aerosol emissions from open baths form the most common escape route, and are enhanced by the use of an air pulse which is passed through the bath in order to homogenise the bath contents during film deposition. The bath components are ammonia (1.5M), sodium hydroxide (23.5.mM), cadmium chloride (10.0 mM) and thiourea or thioacetamide (10.0mM). Initially only ammonia, sodium hydroxide and cadmium chloride are added, and are heated to 70^0C prior to the addition of thiourea or thioacetamide which then react with cadmium chloride to produce a fine orange-brown precipitate of cadmium sulphide (part of which forms a thin film deposit on a glass substrate), cyanamide, and urea. Cyanamide is a highly toxic gas but is considered to decompose in aqueous media above 40 ^{0}C to yield carbonate ion. Apart from ammonia all reactants and products possess low volatilities, although for thiourea and thioacetamide significant volatility is likely (for thiourea, cyanamide and cadmium cyanamide, few physical data are available, and in the case of cadmium cyanamide the structure itself remains unknown).

The properties of the major species of interest in the CBD are summarised in Table 1 together with the various transfer processes which bring them into the gas phase aerosol. Hydrogen sulphide is formed during acid dissolution of the sulphide precipitate during waste disposal operations and could pose an environmental hazard.

3.4 Cleaning of deposition vessel

This hazard assessment has considered the possible emissions from the cleaning of the deposition tank after the waste solution has been removed. Treatment and/or recycling of the solution from the chemical bath has not been considered in this work to date. Hydrochloric acid is used to clean the bath, and reacts with the deposits in the vessel to produce hydrogen sulphide.

4.   TRANSFER PROCESSES IN THE AQUEOUS AEROSOL

In this paper, evidence from natural environmental aerosols is applied to aerosol formation above chemical baths. Hence, it is postulated that the aerosol above a hot aqueous chemical bath forms by two main processes, designated below as *Types I and II.*

I   Thermal Aerosol Condensate (TAC)

This is formed by the thermal vaporisation of water from the bath solution and its subsequent condensation onto dust particles, which are normally present in the ambient atmosphere at densities ~10^4 - 10^5 cm^{-3} [2]. During condensation, this aerosol simultaneously absorbs gaseous substances that previously desorbed from the bulk solution. Equilibrium compositions can be treated using classical thermodynamics based on Henry's law and related expressions, and provide estimates of maximum emissions.

II   Dispersed Solution Aerosol (DSA).

A second aerosol is considered to form as a result of physical and thermal agitation of the solution interface. Both thermal energy input and the sparging action from the passage of air through the solution provide effective generation processes. Thermal buoyancy allied to air flows passed into the bath entrain this aerosol and carry it into the gas phase, a process similar to the well known formation of airborne marine aerosols from ocean wave action at ambient temperatures [3]. The aerosol composition will initially resemble that of the bulk solution from which it is formed, but will be mediated following its formation due to further transfers with the ambient gas phase.

Normally the TAC component will dominate, with relatively minor contributions due to the DSA. Initially these two aerosols will differ in composition, but rapid intermixing will occur, while the high aerosol surface areas will promote rapid kinetics for processes such as absorption and desorption which occur at the droplet-gas interface. The various discrete transfer processes associated with these two major aerosols are identified below and the appropriate processes for the materials involved in these transfers are detailed in Table 1.

Processes involved in formation of the TAC component.

1   Evaporation and condensation of water vapour at the surfaces formed by both the bath and the aerosol.

2   Thermal transfer of volatile substances from the bath solution into the gas phase (which may be assisted by any auxiliary air flow).

3   Uptake of volatile species from the gas phase into the aerosol condensate.

Processes involved in formation of the DSA component

4   Entrainment of solution droplets containing the volatile solutes by thermal buoyancy and by auxiliary air flows accompagnied by transfer of these solutes from the bulk solution into the aerosol.

5   Entrainment of soluble aquo ions and molecules, and insoluble precipitates by the the same mechanisms as for (4) above.

Additional processes contributing to changes in aerosol chemistry

6   Chemical reactions of substances within the aerosol condensate.

7   Release of volatile solutes from the aerosol liquid phase.

Other processes, such as the dissolution of precipitates within the aerosol condensate, may be considered unlikely given the low concentrations and slow rates of dissolution likely to be found.

Table 1: Hazard data on materials involved in the CBD of CdS

Material	Role	Hazard Information					
		Carcinogenicity	Toxicity	OEL*	Volatility	Main Transfer Process	Other Hazards
Ammonia	Major reactant	None	LD50 350 mg/kg oral rat Highly toxic to fish	18 mg/m^3	High	2, 3, 4	Low toxicity. Poison by ingestion and inhalation. Asphyxiant. Corrosive
Cadmium Chloride	Key reactant	Suspected human, animal carcinogen. Mutagen, teratogen.	LD50 88 mg/kg oral rat Highly toxic and eco-toxic.	0.025 mg/m^3	Non-volatile	6	Poisonous and an irritant. High bioaccumulation potential.
Cadmium Cyanamide	Reaction product	Possibly a volatile carcinogen	Highly toxic	N/a	Non-volatile	5, 6	
Cadmium Sulphide	Reaction product	Human carcinogen. Human mutation data reported.	Highly toxic.	0.03 mg/m^3	Non-volatile	5, 6	Corrosive, Irritant, may cause systemic effects. High bioaccumulation potential.
Carbonate (as ammonium salt)	Reaction product	None	LD50 96 mg/kg intr. mouse	0.03 mg/m^3	Non-volatile	5, 6	Low toxicity. Irritant to eyes and respiratory system.
Cyanamide	Reaction product	Unknown	High toxicity LD50 125 mg/kg oral rat	2 mg/m^3	High	4, 5	Moderate irritant to the eyes, skin and respiratory tract. Combustible. Possible systemic effects.
Hydrochloric acid	Cleaning reagent	None	LC50 5 mg/l inh rat	7 mg/m^3	Low	-	Strong acid. Corrosive to eyes, skin and respiratory tract
Hydrogen sulphide	Product	None	High toxicity LC50 444 ppm inh. rat	14 mg/m^3 STEL	High	2, 3	Human poison by inhalation. Highly toxic to fish. Severe irritant to eyes and mucous membrane. Human systemic effects by inhalation. Fire hazard on heating.
Sodium (with OH⁻ ion)	Key reactant	None		1.2 mg/m^3 STEL	Non-volatile	5, 6	Irritant. Corrosive to eyes and skin. Harmful to aquatic organisms.
Thioacetamide	Alt. Key reactant	Carcinogen. Experimental teratogenic data	LD50 301 mg/kg oral rat	No data	Low	2, 3, 4, 5	Poison by ingestion and intraperitonial routes. Harmful by inhalation. Irritant to skin and eyes
Thiourea	Key reactant	Carcinogen. May cause mutagenic and teratogenic effects.	Highly toxic. LD50 125 mg/kg oral rat Toxic to aquatic organisms.	0.3 mg/m^3 STEL (Russia)	Low	2, 3, 4, 5	Poisonous by ingestion. Possible systemic effects. Irritant to respiratory tract, eyes and skin. May damage blood and bone marrow.
Urea	Reaction product	Evidence of mutagenic effects.	Low toxicity. LD50 8471 mg/kg oral rat	N/a	Non-volatile	6	Moderate irritant to the eyes and skin.

Notes: Volatiles classified as: High – gaseous at 273.2 K, Volatile – low boiling liquid at 273.2 K, Non-volatile – solid at 273.2 K
* OEL values are for Long –term exposure limit (8 hour TWA ref. period unless other value is noted i.e. STEL – Short term exposure limit

Processes 2 and 3 will lead to increased airborne concentrations of volatile chemical substances. Maximum airborne concentrations are subject to the constraints imposed by chemical equilibrium. In contrast, the physical entrainment processes 4 and 5 are dependent on actual site specific circumstances, such as the thermal energy input into the bath and the volume air flows, and hence are not subject to equilibrium limitations with regard to distribution of these substances within an aerosol. The flux transferred may be determined experimentally. Processes 1, 6 and 7 modify the concentration profiles but do not in themselves bring concentration enhancements in the gas phase. Volatile solute release from the aerosol condensate should be especially significant for Type II aerosols. Information on the total aerosol composition has been

gained by experimental observations based on sampling and measurement [1]. The data obtained have then been used to determine the emission flux for each analyte.

5.   ENVIRONMENTAL RISKS ASSOCIATED WITH CdS DEPOSITED BY CBD

The data presented in Tables 1 and 2, indicate that environmental risks do exist within this production technique. The three major hazards associated with this are as follows:
• Ammonia storage in bulk
• Gaseous emissions from the hot bath
• Hydrogen sulphide generation whilst cleaning the bath

Table 2: Gaseous emission flux from a Cd deposition bath

Sample No.	Air Input Litres	Gaseous Emission Flux			
		Ammonia mg m^{-2} min^{-1}	Thiourea μg m^{-2} min^{-1}	Cadmium μg m^{-2} min^{-1}	Sodium μg m^{-2} min^{-1}
6	0.0	4.2	14	15	1.6
7	0.5	36	16	4.5	1.0
8	6.00	85	18	2.1	1.9
Means		42	16 +/- 2	7.1	1.5 +/- 0.5
Bath Solution Concentrn [1]		25 560[1]	761.2[1]	1124 [1]	540.5 [1]
(1) Bath solution concentrations are expressed in mg dm^{-3}.					

Storage of large quantities of ammonia could lead to the possibility of gas escape into the workplace or surrounding residential areas. In order to minimise this risk "just in time" supply procedures should be used to keep quantities stored to a minimum.

A major risk from this process arises from the gaseous emissions of Cd and thiourea/thioacetamide from the bath. Measurements of airborne concentrations of ammonia, thiourea, cadmium and sodium have been used to calculate an emission flux for each analyte from the hot bath at 70°C, using a first order expression for the desorption rate constant k. For the bath reactants this takes the form:

$$k = (1/At) \ln (c_o/c)$$

where A is the solution area in m^2, t is the sampling time in minutes, and c$_o$, and c are the reactant concentrations at zero time and time t respectively; an analogous equation applies for the reaction products. Under the standard conditions used in this study, with reaction times of 20 to 30 minutes, the removal of analytes due to desorption is relatively low and never above 0.5%, so that bath concentrations undergo significant change only as a result of chemical reaction. The gaseous flux generated for the site B measurements is summarised in Table 2.

Ammonia emissions show a steady increase with the volume of air fed into the bath solution, consistent with earlier evidence for the stripping action of the auxiliary air flow as it passes through the bath solution. A report [4] on the desorption of ammonia from an ammonium chloride bath at 25°C and pH11 suggests a flux of 1.1 mg m^{-2} min^{-1}, showing satisfactory consistency with the Table 2 values for the higher temperature. Thiourea and sodium flux values are approximately constant at 16 and 1.5 μgm^{-2} min^{-1} respectively, matching their relative volatilities. Assuming that all of the sodium is associated with the DSA component enables the DSA and TCA thiourea flux to be estimated at 2.1 and 13.9 μgm^{-2} min^{-1} respectively, confirming that major thiourea emissions arise from both thermal desorption from the bath solution and direct solution dispersion yielding the DSA.

Cadmium data show similar characteristics although subject to greater variability, possibly as a result of fluctuations in precipitation kinetics for cadmium sulphide. Given the similar very low volatilities of cadmium chloride, cadmium sulphide and sodium hydroxide, cadmium cyanamide remains the candidate to account for the variable emissions. In view of the carcinogenic properties noted for thiourea and for the cadmium species, the flux values in Table 2 constitute a significant environmental hazard. Droplets of 20μm diameter are small enough to be entrained with the DAS aerosol, and on evaporation of host water would form particulates comprising ~5pg cadmium salt or 3.2 pg thiourea. On release into the environment these nuclei possess long residence times in dry atmospheric conditions and high aveolar penetration if inhaled. Hence emissions from hot deposition baths are shown to form a hitherto unidentified source of environmental carcinogens, which can make undoubted contributions to urban hazards. In order to control this risk, all bath emissions should be extracted and filtered, according to current legislation and health and safety guidelines.

Similar measures should be introduced to control the emission of hydrogen sulphide, a highly toxic gas to both humans and the aquatic environment.

6.   CONCLUSION

Due to the nature of the materials involved in the deposition of CdS by CBD, environmental risks could occur at any stage of the process. However, it is particularly at the deposition stage that safety procedures should be strictly observed. The introduction of carcinogens into the environment within the production facility or the local area would be highly dangerous due to the long residence times in dry atmospheric conditions and the danger to human health if inhaled.

If followed correctly, existing regulations as to the storage, transportation and handling of the various materials, should ensure that risks to the environment and human health will be minimal.

ACKNOWLEDGEMENT

This work is supported by the Commission of the European Union under JOULE Contract NO. JOR3-CT97-0124. The authors would like to acknowledge the other contract partners for information provided and the assistance of Mr. M. Rodgers, Dr. M. Deery and Environmental Analysis Ltd. in the experimental analysis work

REFERENCES

[1] Final Report on Joule Contract No. JOR3-CT97-0124, to be published.
[2] S.K. Friedlander, Smoke, Dust and Haze, Wiley, 1977
[3] S.E. Gryning and N. Chaumerliac, Air Pollution Modeling and its Applications XII, Plenum Press, 1998
[4] A.M. Wachs, Y. Folkman and D. Shermesh, in Applications of New Concepts of Physico-Chemical Wastewater Treatment, American Chemical Society, 1972

RESOURCES AND PROCEDURES FOR THE SYSTEMATIC STUDY OF THE PRODUCTION OF A PHOTOVOLTAIC FIRM.

V. Martínez, M.J. Sáenz, J.C. Jimeno, A. Lizarduy, J. Ruiz* and J. Alonso*

Teknologia Mikroelecktronikoaren Institutua / Instituto de Tecnología Microelectrónica
Euskal Herriko Unibertsitatea / Universidad del País Vasco
Escuela Superior de Ingenieros de Bilbao
Alameda de Urquijo s/n, 48013 Bilbao. Spain.
Phone: 34.944.396.466, fax: 34.944.396.395, e-mail: jtpmasav@bicc00.bi.ehu.es

*ISOFOTON, S.A.
Polígono Industrial Sta. Teresa. C/ Caleta de Vélez, 52, 29006 Málaga. Spain

ABSTRACT: The need for procedures to improve the supervision of production and quality control systems that can be implemented in the photovoltaic industry, is well known. In this regard, this paper describes the resources and procedures developed by our group in order to systemise the study of production. These resources and procedures have been taken from the classification process. We consider that these may be used satisfactorily to improve quality control and, consequently, production.
This method takes advantage of the special characteristics of the classification system developed by our group, which enables I-V measurements to be made at different levels of illumination and/or temperature, in addition to the measurement carried out in the classification process.
The ultimate objective of this work is to enable production to be studied in two ways. On the one hand, the possibility of varying the classification margins and even the definition of new classification procedures in order to obtain panels that operate more uniformly within the same type. On the other hand, to define a measuring system that allows the correct supervision of production as a whole, using a few well-selected cells for this purpose.
Keywords: Manufacturing and Processing - 1: Evaluation - 2: Experimental Methods -3

1. INTRODUCTION

On numerous occasions the need has been identified for analytical methods and systems for the supervision and control of production within the photovoltaic industry [1], and particularly nowadays, with an increasing production level, extensive use of automation is estressed. In this sense, our group has performed several studies of the company Isofotón, S.A. and the European consortium MONOCHESS-II.

The basic idea is to take advantage on the fact thal all cell production go through a classifying procedure. At this point, at the end of cell production, only one measurement is usually done to carry out appropriate class for the cell under test. The use of an automatic versatile measuring system together with appropriate control and analysis programs could carry out extra information. It might be drawn not only statistical information of normal production (e.g.: all currents measured while classifying), but also enough data to analyze what is wrong, and why, when something fails.

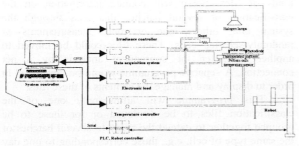

Figure 1: Measuring system sketch.

This article described resources and procedures for the systematic study of production, by using the classification process. We consider that these may be

satisfactory for improving quality control and, therefore, production.

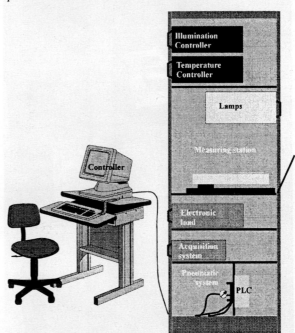

Figure 2: Measusing system layout.

2. THE CLASSIFICATION AND MEASURING SYSTEM

The classification and measuring system used for this study can is sketched in figure 1. It consists of a base controlled thermostatically by thermoelements and an illumination system which makes use of dichroic halogen lamps. Both the temperature and the illumination can be modified, although 25°C and 1 sun are the typical values used for the classification process. In order to polarise the solar cell at the required operating point, there is a 4 quadrant supply source that allows a wide range of variation in I-V values. For measuring, there is also a multichannel data acquisition system which records both the voltage and current as temperature and the illumination level at the time of measurement. A programmable PLC is responsible for controlling the different mobile elements in the system and communicates with a computer via an RS-232 connection. The computer communicates with the data acquisition system via a GPIB / IEEE-488 bus, the four quadrant supply source and the radiation and temperature controllers [2]. A layout of the system can be seen in figure 2,. and a photograph in figure 3.

Figura 3: Photograph of the actual measuring system.

The classification process performed in the plant is totally automatic, using a robotised system for placing and extracting the cells from the measuring system. In this way up to 12 thousand cells per day are classified.

During this procedure, only one I-V point measurements (required for the classification) are performed at this time, and certain statistical values on the process are recorded.

In order to carry out a more exhaustive control of one cell in particular, i.e., an I-V curve, it is necessary to stop the classification process.

3. SYSTEMATIC STUDY OF CLASSIFICATION

In order to obtain more information on the cells that pass through the classification system, and in order to avoid having to stop the classification in order to carry out exhaustive measurements, the control programme code could be modified by introducing a series of routines that in principle only have a slight effect on the speed of the classification process.

The first variation is in the recording process. In addition to the aforementioned statistical data, more data on the classification measurement can be had. The subsequent analysis of the contents of the resulting files will allow us to observe how currents of the same type, or in a complete classification batch, are distributed, assess the performance of the classification process itself, and check this against recorded statistics.

On the other hand, it could be arranged that one or several I-V characteristics were measured for a certain number of cells. The cell to be measured would be chosen randomly, and random selection could be different for different classes. Those measured cells would generate a file containing all the information on the measurement made. Likewise, a mark could be added to the previous classification record in order to indicate which cells have been measured.

The routine, a simplified flow diagram for which is shown in figure 4, could be parameterised outside the control programme in several aspects, both with regard to the data stored during classification and how many and how the measurements would to be made. In this way it would be possible to test several recording modes and make measurements without having to make continuous modifications in the main program.

4. DATA FILE MANAGEMENT PROGRAMME

The routine described above would give rise to two kinds of files. Some will contain information on the classification of several batches that pass through the system and, others, one or more measurements of randomly-selected cells. These files could be moved to another computer for subsequent analysis by a net connection. A separate programme would be developed in order to display and manage the contents of these files.

This programme will allow only one of the classification files to be displayed or for these to be grouped together to extend the study to several batches of the same type of cell, e.g., those corresponding to one day or month of production. The file, or set of selected files could be converted to MS-EXCEL tables (also recovered by SPSS), or ASCII files for treatment with other computer programmes.

In order to extract the parameters of the incorporated cell models, files compatible with MultIV.2 [3] could be generated for measurement files.

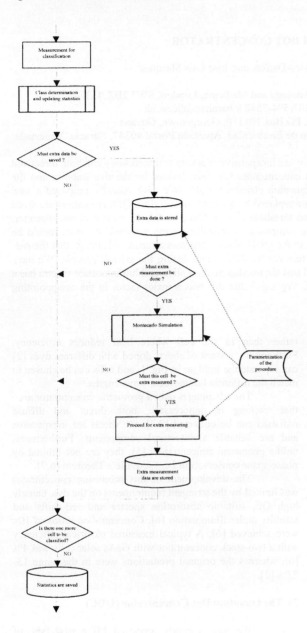

Figure 4: Flux diagram.

5. CONCLUSIONS

Within the automated classification procedure used in the plant, the recording of data corresponding to classification as such could be systemised, together with the incorporation of complete I-V measurements in randomly-selected cells. These procedures, incorporated within the classification system control programme, might not cause significant delays and, on the other hand, would allow a more complete set of information to be obtained on production, without the need to interrupt the classification process in the plant.

Furthermore, in order to be able to test different combinations, the routines introduced could be

parameterised outside the control programme and there were no need to make continuous adjustments to this.

The introduction of systems and procedures like that described in this paper would enable production to be studied in two ways: On the one hand, extensive estatistical information would be drawn so normal production will be well known, but the possibility of varying the classification margins and even the definition of new classification procedures in order to obtain panels that operate more uniformly within the same type could be an added possibility.

On the other hand, it could be defined a measuring procedure that allows the production, as a whole, to be supervised correctly, using a few well-selected cells for this purpose.

Futhermore, the capability for varying light conditions from dark to some extend over 1 sun, as occurs with the measuring system of this paper, could be useful in order to test cells that were going to work under concentration.

ACKNOWLEDGEMENT

This work has been made under the sponsorship of the Universidad del País Vasco / Euskal Herriko Unibertsitatea and its contract UP-147-375-TB-040/96.

REFERENCES

[1] Summary Discussion Sessions. Eigth Workshop on Crystalline Silicon Solar Cell Materials and Processes, 1998.

[2] V.E. Martínez, M.J. Sáenz, J.R. Gutiérrez, J.C. Jimeno and E. Perezagua, "A Flexible Measurement System for R&D in Industry". 10th European Photovoltaic Solar Energy Conference, pp. 689-691, 1991.

[3] V. Martínez, J.C. Jimeno, M.J. Sáenz and V. Rodriguez, "Multi-fit, Fitting Strategies and Batch Fitting in MultIV 2", 14th European Photovoltaic Solar Energy Conference, pp. 2427-2430, 1997.

MODELLING THE QUANTUM DOT CONCENTRATOR

Martin Williams[1], Keith Barnham, Ned Ekins-Daukes and Jose Luis Marques[2]

Physics Department , Imperial College of Science, Technology and Medicine, London, SW7 2BZ, U.K.,
Phone: (44)-207-594-7565; FAX: (44)-207-594-7580 k.barnham@ic.ac.uk
[1]Department of Physics, University of Guyana, PO Box 101110, Georgetown, Guyana.
[2]Instituto Universitario de Technologia Region Capital, Departamento de Electricidad, Apartado Postal 40347, Caracas, Venezuela.

ABSTRACT: Luminescent concentrators have the advantages that they are inexpensive, tracking is unnecessary and both direct and diffuse radiation can be collected. The development of dye-doped concentrators has been limited by the dye stability and the requirements of suitable absorption spectrum, red-shift and high quantum efficiency (QE). We have recently proposed a new approach, the quantum dot concentrator (QDC), in which the dyes are replaced by quantum dots (QDs). QDs are nanometre-sized crystallite semiconductors. The first advantage is that the absorption threshold can be tuned simply by choice of dot diameter. Secondly, high luminescence QE has been observed. Thirdly, being composed of crystalline semiconductor, the dots should be more stable. We describe the application of a thermodynamic model to the QDC which illustrates a further advantage that the red-shift is *quantitatively* related to the *spread* of QD sizes, which in turn can be determined during the growth process. We have extended this model to allow for the effects of re-absorption. We find that the overlap of absorption and luminescence spectra has a strong effect on the system efficiency, even in the case of unit QE. We argue that this was a major factor in the disappointing performance of early dye concentrators.
Keywords: Concentrator - 1: Modelling - 2

1) Introduction

Concentration can reduce considerably the cost of electricity from solar cells. However, conventional high concentration techniques require solar tracking, which is expensive, and utilise only the direct component of radiation. In the late 1970s a novel approach, the luminescent (or fluorescent) collector, was proposed [1,2,3], consisting of a transparent sheet doped with appropriate organic dyes. The sunlight is absorbed by the dye and then re-radiated isotropically, ideally with high quantum efficiency (QE), and trapped in the sheet by internal reflection. The trapped light is converted at the edge of the sheet by a solar cell with band-gap just less than the luminescent energy i.e. the cell is at optimum efficiency.

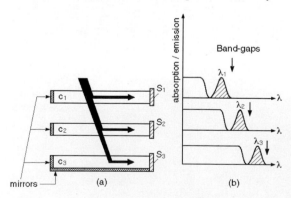

Figure 1: a) Schematic of a three-stack luminescent concentrator. Each layer in the stack absorbs and re-emits light of longer wavelength with ideal absorption coefficient and luminescence spectra as in b).

The excess photon energy is dissipated in the collector by the luminescent red-shift (or Stokes' shift)

rather than in the cell where heat reduces efficiency. Furthermore, a stack of sheets doped with different dyes [2] can separate the light, as in Fig. 1, and cells can be chosen to match the different luminescent wavelengths.

The advantages over a geometric concentrator are that tracking is unnecessary, both direct and diffuse radiation can be collected [3,4], the sheets are inexpensive and are suitable architectural components. Furthermore, unlike geometric concentrators [5], they are not limited by phase-space conservation i.e. Liouville's Theorem [6,7].

The development of this promising concentrator was limited by the stringent requirements on the dye, namely high QE, suitable absorption spectra and red-shifts and stability under illumination [6]. Concentration ratios of 10x were achieved [6]. A typical measured electrical efficiency with a two-stack concentrator with GaAs solar cells was 4% [6], whereas the original predictions were in the range 13-23% [2].

2) The Quantum Dot Concentrator (QDC)

We have recently proposed [8] a new type of luminescent collector, the quantum dot concentrator (QDC), in which these problems are addressed by replacing the organic dyes with quantum dots (QDs). QDs are nanometre-sized crystallite semiconductors, which can be produced by a variety of methods [9]. We believe that the colloidal [10], or single-molecule precursor [11] growth techniques, are particularly suited to mass-production and low costs.

The first advantage of QDs over dyes is the ability to tune the absorption threshold simply by choice of dot diameter. Colloidal InP quantum dots, separated by dot size, have thresholds which span the optical spectrum [10]. Colloidal InAs dots sample infra-red wavelengths [12]. Secondly, high luminescence QE has been observed at room temperature. Colloidal CdSe/CdS heterostructure dots have demonstrated luminescence quantum yields above 80% at room temperature [12].

We will discuss the importance of limiting the overlap between the luminescent peaks and absorption thresholds. Large red-shifts have been observed with QDs but their origin is a matter of some controversy, particularly as they depend on many factors. We present a thermodynamic model which is in reasonable agreement with the shape and red-shift of the luminescence in the remarkable room-temperature data of Ref. 10 under "global" illumination (well above absorption threshold). The model demonstrates quantitatively that the red-shift is determined primarily by the *spread* of dot sizes, which in turn can be optimised by choice of growth conditions [13]. This is a significant improvement compared to the dye-concentrator. We discuss below how the model can be extended to allow for the affects of re-absorption.

3) Thermodynamic Model for the Red-shift.

Yablonovitch [7] applied a detailed balance argument to the original dye-concentrator to relate the absorbed light and self-absorbed trapped light to the spontaneous emission in terms of the QE of the luminescent process, the chemical potential μ and the frequency dependence of the absorption cross section $\sigma(v)$. Assuming that μ does not depend on x, the luminescent intensity I_2 at an arbitrary position x is:

$$I_2(v,x) = \frac{8\pi n^2 v^2}{2c^2} e^{(\mu - hv)/kT} \frac{\Omega_2}{4\pi}\left(1 - e^{-N\sigma(v)x}\right) ....(1)$$

where n is the refractive index of the medium, Ω_2 is the solid angle of the internally reflected luminescence and N is the

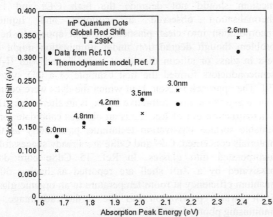

Figure 2. Data on the global red shift (absorption peak – luminescence peak energy) for illumination well above threshold, taken from Fig.1, Ref. 10. Crosses indicate the predictions of eqn.1 assuming $\sigma(v)$ is a Gaussian which fits the threshold in the relevant absorption spectrum.

density of the luminescence molecules. This is essentially the generalized Planck equation used to determine solar cell efficiency limits. Yablonovitch was applying it to fluorescent molecules but it should apply to any system interacting reversibly with a photon field providing the thermal equilibration is fast. The mechanisms responsible for energy relaxation in QDs are a matter of some controversy. However, experiments on mixtures of two dot sizes suggest very effective energy transfer mechanisms operate in colloids and solid films [14].

We find that this thermodynamic approach can be extended to the global illumination of an ensemble of QDs. The shape and position of the luminescence peak given by I_2 in eqn. 1, depend critically on two factors, the shape of the absorption cross section $\sigma(v)$ near threshold and the temperature T. They do not depend strongly on μ or the *absolute* value of $\sigma(v)$. We assume that the frequency dependence of $\sigma(v)$ at threshold is Gaussian as expected for dots with δ-function density of states and Gaussian distributed diameters. The luminescent line shapes in the global illuminated QD data of Ref. 10 can be quite well approximated using eqn. 1 if the width of the Gaussian describing $\sigma(v)$ is somewhat narrower than the absorption data [8]. If the width of the Gaussian is forced to fit the threshold behaviour of $\sigma(v)$ in the data of Ref. 10, then, though the shape of the luminescence is not particularly well described, the global red-shift as a function of absorption threshold (related to QD diameter) can be well reproduced as shown in Fig. 2. These fits suggest that this thermodynamic model can predict the dominant red-shift due to the distribution of dot sizes and therefore help to determine the absorption-luminescence overlap.

4) Modelling the effects of Re-absorption.

The model can be extended to allow for the effects of re-absorption [7] by addition of a second term:

$$\frac{dI_2(v,x)}{dx} = -N\sigma(v)I_2 + \frac{8\pi n^2 v^2}{c^2}\frac{N\sigma(v)}{e^{(hv-\mu)/kT}-1}\left(\frac{\Omega_2}{4\pi}\right) ....(2)$$

Note that in eqn. 2 μ is now a function of position x measured from one end of the concentrator and we have used the full expression for the Planck function. We solve

$$\int \sigma(v)I_1(v)dv + \int \sigma(v)I_2(v)dv =$$
$$\int \frac{1}{Q}\frac{8\pi n^2 v^2}{c^2}\frac{\sigma(v)}{e^{(hv-\mu)/kT}-1}dv\left(\frac{\Omega_1}{4\pi} + \frac{\Omega_2}{4\pi}\right) ....(3)$$

this equation numerically for I_2 and μ, in finite steps along the length of the concentrator, subject to the overall detailed-balance constraint:

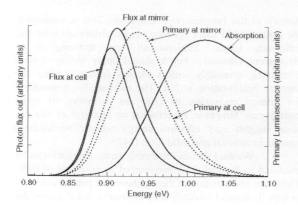

Figure 3: Calculated primary luminescence (dotted lines) near the mirror and near the solar cell for a single-slab concentrator (e.g. one layer in Fig. 1a)). Shape of the trapped flux at the same points following re-absorption (thin lines). The curves are numerical solutions of eqns. 2 and 3 allowing for re-absorption assuming the absorption as in thick line.

In eqn.3 I_1 is the incident light intensity (assumed to be AM1.5), Q is the quantum efficiency for the primary luminescence and Ω_1 is the solid angle corresponding to the escape cone.

We find that for typical concentrations, e.g. length 10x depth, a mirror on one end as in Fig.1 and for illumination on the top surface with an AM1.5 spectrum, there is a significant fall in the value of μ between the end mirror and the solar cell. This is illustrated in Fig. 3 where we show the primary photon intensity due to luminescence at the mirror and at the solar cell and also the photon flux I_2 at the same positions (in different units). It can be seen that the μ variation means that the primary luminescence falls, while the re-absorption results in a shift of the collected photon flux to longer wavelengths as well as a fall in intensity.

The fall in intensity and μ between the mirror and the cell is more marked for a quantum efficiency Q less than unity. This is expected due to non-radiative losses following re-absorption. However, we also observe a significant intensity fall for 10x concentration even with Q = 1. This is because, following re-absorption, the resulting secondary luminescence is again emitted over 4π and the fraction emitted *upwards* within the critical cone Ω_1 is lost. (The fraction emitted downwards into this escape cone can be collected by a second sheet as in Fig.1). The authors of the early studies of luminescent concentrators were aware of the importance of avoiding re-absorption [2]. The significance of this thermodynamic model is that it provides a quantitative method to calculate the effect of re-absorption. Furthermore one could possibly determine the primary Q from the variation in shape and height of the trapped flux depending on the position of the primary excitation.

The fits summarised in Fig. 2 further suggest that the shape of the QD size distribution can be chosen to

optimise the red-shift and minimise the overlap between absorption and luminescence. This is a significant advantage for the QDC compared with the dye-concentrator as the spread in QD sizes can be determined by the choice of growth conditions [13].

We have extended the model with re-absorption to a three slab geometry allowing for the downward re-radiation from the first two sheets to be collected by the lower ones and assuming a mirror on the lower-side of the third. We allow for the effect of reflection at interfaces where the refractive index changes and assume material attenuation factors similar to those in Ref. 2. The latter, however, are relatively unimportant compared with the re-absorption losses discussed above. We find that three solar cell band-gaps in the optimum efficiency configuration for an AM1.5 spectrum are 0.79 eV, 1.25 eV and 1.8 eV. If we minimise the effects of the re-absorption then the overall system efficiency (AM1.5 to electrical power) approaches that calculated for a three slab system in Ref. 2, namely 22%. However, introducing realistic re-absorption overlaps like those used in the fits in Fig. 2, results in a major fall in efficiency to 13% for slab length to depth ratios as low as 10x, even for Q = 1.

We conclude that the disappointing results obtained with 10x concentration in dye-concentrators [6] were probably mainly due to the effects of re-absorption. We note that the thermodynamic model will provide us with a way to explore this problem quantitatively. Furthermore the model suggests that the shape of QD size distributions can be tailored to reduce this overlap. A full description of this work is in preparation.

5) Practical QDC Design.

Ideally the incorporation of the dots into a suitable medium should not degrade the high QEs and fast thermalisation observed in colloids and liquids. Incorporation into clear plastics does not appear to be a problem though degradation under illumination might be less in glass or silicon dioxide. Glasses doped with II-VI semiconductors formed the first example of a QD system [9]. The approach of Ref. 3, in which the dyes were coated onto a glass of similar refractive index, is an alternative. The heterostructure dot of Ref. 12 is an obvious candidate for a suitable surface passivation technique, particularly as the materials concerned, CdS and CdSe are already successfully incorporated into glasses. In Ref. 15 CdSe core dots passivated by a ZnS shell are reported as having 60% quantum efficiency at room temperature in an organic glass. Such surface passivation has the further advantage of minimising photo-degradation effects.

Anti-reflection (AR) coatings at the interfaces between the low refractive index wave-guide and high n cell will be more effective than in conventional concentrators as the spectral range is much smaller. Light emerging at large angles can be effectively transmitted by multi-layer AR coatings [16].

Finally PV cells themselves have made great progress over the past two decades. Direct band-gap cells based on III-V systems have external QEs near unity just above threshold. The quantum well solar cell (QWSC) is particularly suitable as one can tune the band-gap by changing the well width [17]. The band-gaps near the optimum values discussed in Sec. 4 are possible with III-V QWSCs.

6) Conclusions

Many properties of QDs, in particular the size-dependence of their absorption threshold, red-shift and luminescent efficiency, make them ideal to replace organic dyes in a new type of luminescent concentrator. We have shown that the thermodynamic model of Ref. 7 quantitatively predicts the way that the global QD red-shift depends on the spread of QD diameters, which can be tailored by the growth conditions. The model provides a way to assess the effects of re-absorption, it explains how the spread in the QD sizes determines this overlap and suggests that by optimising the spread during growth a QDC can be tailored to minimise the re-absorption. This overlap was probably responsible for the disappointing performance of dye-concentrators.

Providing that the dots can be incorporated in suitable transparent media and retain their high QE, the possibility to minimise the effects of re-absorption makes it possible to foresee QDCs which perform in the upper range of efficiency predicted for the original dye-concentrators, i.e. around 20%.

REFERENCES

1. W.H.Weber, J.Lambe, *Appl. Opt.* **15** (1976) 2299.
2. A. Goetzberger, W.Greubel, *Appl. Phys.* **14** (1977) 123.
3. Charles F.Rapp, Norman L.Boling, *Proc. 13th IEEE Photovoltaic Specialists Conference* (1978) 690.
4. A.Goetzberger, *Appl. Phys.* **16** (1978) 399.
5. W.T.Welford and R.Winston, *High Collection, Non-Imaging Optics.* (Academic Press, San Diego, 1989).
6. A.Goetzberger, W.Stahl, V.Wittwer, *Proc 6th European Photovoltaic Solar Energy Conference*, (1985) 209. V.Witter, K.Heidler, A.Zastrow, A.Goetzberger, *J. of Luminescence*, **24/25**, (1981) 873.
7. E.Yablonovitch, *J.Opt.Soc.Am.*, **70**, (1980) 1362.
8. Keith Barnham, Jose Luis Marques, John Hassard, Paul O'Brien, Appl.Phys.Lett. **76** (2000) 1197.
9. U.Woggan, *Optical Properties of Semiconductor Quantum Dots* Springer-Verlag, Berlin, (1997).
10. O.I.Micic, H.M.Cheong, H.Fu, A.Zunger, J.R.Sprague, A.Mascarenhas A.J.Nozik, *J.Phys.Chem.B*, **101** (1997) 4904.
11. Tito Trindade, Paul O'Brien and Xiao-mei Zhang, *Chemistry of Materials* **9** (1997) 523.
12. A.P.Alivisatos *MRS Bulletin*, February 1998, 18.
13. O.I.Micic, C.J.Curtis, K.M.Jones, J.R.Sprague, A.J.Nozik, *J. Phys. Chem.* **98**, (1994) 4966. D.J.Norris M.G.Bawendi, *Phys.Rev.B*, 53 (1996) 16,338.
14. C.R.Kagan, C.B.Murray M.G.Bawendi, *Phys. Rev. B*, **54** (1996) 8633. O.I.Micic, K.M.Jones, A.Cahill A.J.Nozik, *J.Phys.Chem. B* **1998**, **102**, (1998) 9791.
15. S.A.Blanton, M.A.Hines, P.Guyot-Sionnest, *Appl. Phys. Lett.* **69** (1996) 3905.
16. V.Diaz Luque, C.Algora del Valle I.Rey-Stolle Prado, *Proc. 2nd World Conference and Exhibition on Photovoltaic Solar Energy Conversion*, (1988) 8.
17. K.W.J.Barnham, B. Braun, J.Nelson, M.Paxman, C.Button, J.S.Roberts, C.T.Foxon *Appl. Phys. Lett.* **59** (1991) 135.

MEETING THE CHALLENGE OF NEW GENERATION SOLAR CELL TESTING - LARGE AREA LONG PULSE I-V-CURVE TRACING

Tilman Thrum[*], David M. Camm[*], Dean Parfeniuk[*], Dirk Slootweg[+]
[*] Vortek Industries Ltd., 1820 Pandora St., Vancouver, B.C., Canada V5L 1M5
[+] Fokker Space B.V., P.O. Box 32070, 2303 DB Leiden, The Netherlands

ABSTRACT: A new approach in I-V-curve tracing based upon a continuous water-vortex high-pressure arc lamp is presented in this paper. Some key characteristics of this system are: customized spectrum (e.g. AM0 or AM1.5 compliant), continuously adjustable irradiance levels, suitability for cleanroom environments and vacuum applications. Pulse widths from 250 ms to continuous are achieved over a target area of a typical size of approximately 3 m x 4 m. It is shown that both spectral as well as spatial and temporal characteristics comply with the high standards that are common in the space industry.

The new measurement technique is made possible by using a flexible long pulse to continuous radiation signal. The ability to vary pulse lengths avoids limitations encountered by traditional I-V-curve measurement methods. Data is taken using a specifically designed load cell that holds constant voltage across the sample while modulating the intensity around the requested cell irradiance. Current and voltage are analyzed as a function of irradiance to extract accurate I-V-data at the requested irradiance level. Transient effects due to cell capacitance as well as interconnect capacitance and inductance can be measured by altering sweep rates. Results for a number of different types of solar cells/panels are presented.

KEYWORDS: solar simulation - 1: characterization – 2: capacitance - 3

1. INTRODUCTION

Historically testing of solar panels was performed by means of xenon flash lamp testers. A very intense flash maintained a constant irradiation level for 1 to 5 ms with a spectrum that approximates for example AM0 conditions. The I-V-characteristics of the test object were measured by scanning the cell voltage range during this constant irradiation time. The advent of new generation solar cell technologies such as thin film or multi-junction cells has raised concerns as to how reliable I-V characteristics for these cell types can be determined when measured by means of a flash tester [1]. Two major sources of errors have been identified: false measurements due to transient electrical effects in the cells ("capacitance effects") and errors due to spectral mismatch of flasher spectrum and quantum efficiency of multi-junction cells.

The Vortek VLASS system (Vortek Large Area Solar Simulator) addresses and overcomes both problems by using a different approach for both the lamp technology as well as the I-V-curve data acquisition. Core of the VLASS solar simulator is a high power Vortek arc lamp, the most powerful continuous radiation source worldwide in the UV-VIS-IR range. It contains a water-vortex stabilized Ar high-pressure plasma typically 20 cm in length. Fig. 1 demonstrates the lamp operation principle.

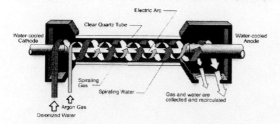

Fig. 1: Vortek arc lamp – Principle of operation

The lamp can be operated with 250ms long radiation pulses up to continuous operation. Due to the high radiative output, target areas up to the order of 15m² can be

illuminated at up to +/-2% spatial uniformity. The original spectral output of the lamp is that of a typical Ar high-pressure discharge at around 11000 to 12000 K. By applying thin film filter coatings onto the quartz envelope, which surrounds the actual discharge, the spectral output can be tailored to satisfy various standards and specific cell quantum efficiency curves. Since the lamp can run in a continuous mode it can also be used for thermal testing, e.g. radiating into a vacuum chamber through a quartz window. The selectable pulse-width allows for adjusting the solar panel illumination duration depending on the actual substrate material eliminating any transient cell effects.

2. SPECTRAL OUTPUT

A major problem in matching the sun's radiation by means of high-pressure plasma light sources operated in a continuous mode is that the discharge temperature (e.g. 8000 K to 13000 K) is typically lower than that of a flash lamp (e.g. 18000 K and hotter). The reason for this is mainly due to the input power limitation for operating the respective plasma in order to keep the quartz envelope from softening or melting. Since the plasma temperature correlates directly to the line-to-continuum emission ratio it

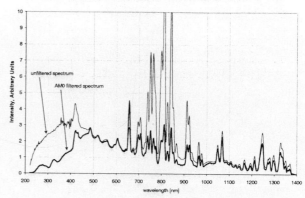

Fig. 2: Spectrum Vortek lamp - unfiltered/filtered for AM0

means that a continuously operated high-pressure discharge always tends to have a more pronounced line contribution in the emitted spectrum. In contrast a flashlamp spectrum shows a smoother spectrum due to a higher continuum emission contribution and so matches the sun's spectrum closer than a continuously operated unfiltered high-pressure arc lamp. To overcome this obstacle a special filter technology is used to adjust the spectral output of the Vortek arc lamp [2]. A multi-layer thin film filter is coated onto the outside of the quartz envelope. The actual wavelength-dependence of this filtercurve is customized depending on the specifications the spectral output of the lamp must comply with. Since the quartz envelope is simple to replace the system's spectral output can be modified easily for different standards and/or requirements. Fig. 2 shows the comparison between a raw Vortek lamp spectrum and a filtered spectrum, adjusted for AM0 requirements. The figure demonstrates the vast improvement the filter causes by drastically dropping the

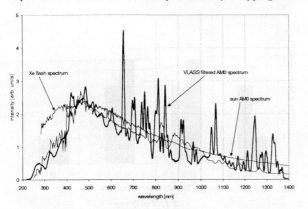

Fig. 3: comparison of different spectra

IR line contribution and smoothing the overall spectral output. Fig.3 compares the VLASS spectrum modified for AM0 compliance against a Xe-flasher spectrum and the AM0 sun spectrum defined by WMO. From this graph it is obvious that the filtered and smoothened VLASS spectrum provides a far better match to the AM0 requirements especially in the UV part of the spectrum. This closer match is especially of significance for multi-junction cells. The filter applied to the raw spectrum giving the results shown in Fig. 3 is designed to comply with the AM0 sun

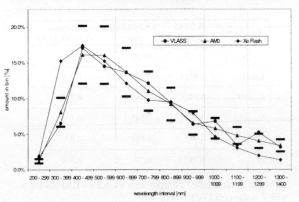

Fig 4: 100 nm binning for VLASS, Xe-flasher and AM0

spectrum binned in 100 nm bins as is typical for characterization of solar cell response. Fig. 4 shows the binned spectral results. The graph shows the superior

match of the VLASS system compared to the flash-tester spectrum especially in the UV region from 200 nm to 400 nm and the IR region above 900 nm. Utilizing the same filter technology for matching the terrestrial AM1.5 sun spectrum, shows equally good results.

3. I-V-CURVE TRACING

A second major source of error in "traditional" I-V-curve determination for state-of-the-art solar panels by means of flash testers arises from the very short pulse duration of around 1 to 5 ms and the fast voltage-sweeping rate the panel is exposed to. This sweeping rate during a flash is too fast for most modern solar panels to reach their DC operating conditions due to the transient rearrangement of the internal charge carrier distribution, also commonly described as cell capacitance. This effect causes a false prediction of the cell/panel current-voltage DC characteristics by either over- or underestimating the panel current depending on the voltage sweep direction [1].

The VLASS system eliminates this problem by using a new data acquisition approach. Instead of continuously sweeping the load voltage and measuring the respective load current while the flash lamp discharge maintains uniform irradiation on target the VLASS electronic load sets a constant load voltage and measures the panel current and voltage while the irradiation intensity is modulated around the requested irradiation level. This irradiation

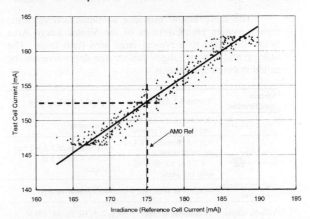

Fig. 5: Determination of actual panel current

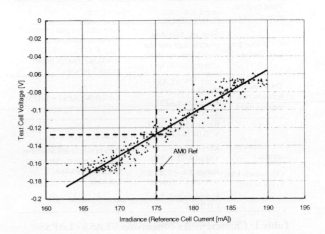

Fig 6: Postprocessing of actual panel voltage

modulation, which is in the 200 – 500 Hz range and hence slow enough to not influence the DC characteristics by any transient cell effects, causes small variations in the panel current and voltage around the nominal value for the requested irradiation level. Each I-V-data point is determined by measuring 200 to 300 solar cell/panel current-voltage data points together with the simultaneous reference cell outputs. This data is stored for subsequent postprocessing before stepping to the next load voltage level. In the data-postprocessing a linear least square fit for both the panel voltage and the panel current over the reference cell output is calculated and the panel's current-voltage values are determined by using the values at the requested irradiation level. Fig. 5 and Fig. 6 demonstrate this process for a plain silicon test cell and a silicon reference cell.

Using the described method the entire I-V-curve scanning process from V_{oc} to I_{sc} typically takes 2 to 3 seconds. Experimental results show no impact due to panel heating.

Due to the fact that multiple data points are taken for each investigated I-V-curve point during the above described irradiation modulation the system also allows for a confidence prediction and error estimation of the derived panel I-V-curve based on statistical data analysis. The described data acquisition technique is also found to be highly reproducible due to collecting this many data points.

4.   COMPARISON OF VLASS TO FLASH TESTERS

The table below provides a comparison of some typical system characteristics of the Vortek Large Area Solar Simulator to a typical large area flash tester. The voltage and current ranges can change depending on the actual system requirements.

System specification	Vortek VLASS system	Typical LAPSS system
Test plane area	Approx. 3 m x 4 m Starting 0.75 m off ground	5 m diameter
Test plane uniformity	< ± 1.5 % inside 1 x 1 m centre of beam < ± 2.0 % for 99 % of total test plane	< ± 1.5 % inside 2.5 x 2.5 m < ± 3.0 % inside 5 m diameter
Current/voltage resolution	0.003 % in all ranges	0.025 % in all ranges
Voltage ranges	by customer request	1, 2, 5, 10, 20, 50, 100 V
Current ranges	by customer request	0.1, 0.2, 0.5, 1, 2, 5, 10, 20 A
Pulse to pulse repeatability	< ± 0.5 %	< ± 1.0 %
Cycle time for single pulse	Start arc, take data, process data and plot with 2 minutes (from system idle 30 sec)	Take data and plot corrected IV curve, including I_{sc}, V_{oc}, P_{max}, V @ P_{max} and current at selected voltage points within 3 minutes
Total error between two measurements	< ± 1.5 % VLASS – VLASS < ± 2.0 % LAPSS – VLASS	< ± 2.0 % LAPSS – LAPSS
Measurement accuracy (All values refer to full scale)	Range 150 V - 10 A V: 0.04 % - I: 0.01 % Range 12 V – 2 A V:0.01% - I: 0.01% Range 5 V – 0.25 A V:0. 1% - I: 0.01% Reference cell 2 A: I: 0.012%	Data presently not available, for some manufacturers only full scale resolution available

Table 1: Characteristics comparison VLASS - LAPSS

One noticeable difference is that some flash tester manufacturers do not provide any measure of confidence for the acquired data sets (measurement accuracy category). Together with the above-described spectral match to required standards like AM0-compliance, this table shows that the overall system performance of VLASS is more accurate and gives a superior reproducibility.

The high level of accuracy is also confirmed by measurements taken during various stages of production, shipment and set-up of a VLASS system for Fokker Space B.V. A plain silicon DVT[1] sample (total surface size: approx. 0.06 m^2) from a commercial satellite solar panel project was used for all of these tests. Ten tests on seven different days covering a time-span from November 1999 through March 2000 were performed. During the tests I-V curves of the DVT sample were taken, including the characterization of V_{oc}, I_{sc}, P_{max} and other parameters. In between the tests, the system had been disassembled, shipped and reassembled. During this time the VLASS

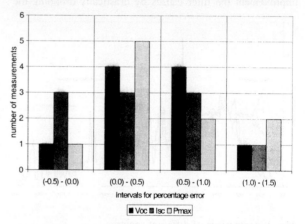

Fig. 7: Error distribution for repeated VLASS system test

system was also subject to maintenance work as well as multiple changes of the arc envelope and electrodes (to do experiments with different filter coatings on the envelope). The percentage error shown in Fig. 7 is a total system error including all uncertainties of the system, including potential data acquisition errors, mechanical tolerances etc. The percentage error is calculated by deriving the relative error of a measured data point from the data value averaged over all experiments. Fig. 7 suggests that the overall error distribution should eventually converge towards a Gaussian

Parameter	VLASS system	LAPSS flash tester	Difference VLASS/LAPSS
V_{oc}/[V]	10.905	10.90	+0.2 %
I_{sc}/[A]	0.8315	0.839	-0.7 %
P_{max}/[W]	7.142	7.182	-0.7 %
V_{max}/[V]	9.017	9.045	-0.3 %
I_{max}/[A]	0.791	0.794	-0.8 %
FF	0.787	0.790	-0.3 %

Table 2: Performance comparison VLASS - LAPSS for a BSFR solar cell DVT

[1] DVT: Design Verification Test Coupon

distribution. Further testing and measurements are underway and will be taken continuously to monitor the actual system performance. These results will be published at a later date.

A direct comparison between a LAPSS flash tester and the VLASS system was performed by Fokker Space B.V. using a commercial satellite program DVT comprised of a plain silicon solar cell type with BSFR. The results of this test are shown in Table 2. Since this type of cell is not expected to be subject to any transient effect the results of this comparison are a good indication of the relative spectral match between the systems both optimized for AM0 conditions. The results in Table 2 indicate that the differences between the measurements are of a very small nature and fall consistently into the range of total error between two measurements as given in Table 1.

Fokker Space B.V. performed a further comparison between the Vortek VLASS solar simulator and a European manufactured large area pulse flasher simulator. The results of this test are documented in Table 3.

Parameter	VLASS system	LAPSS flash tester	Difference VLASS/LAPSS
V_{oc}/[V]	18.5927	18.42	+0.6 %
I_{sc}/[A]	0.914	0.908	+0.9 %
P_{max}/[W]	13.01	13.03	-0.15 %
V_{max}/[V]	15.10	15.226	-0.83 %
I_{max}/[A]	15.10	0.856	+0.11 %

Table 3: Performance comparison VLASS-LAPSS for a LAPSS manufactured in Europe

This table shows again that the measurement differences between the systems are within the limits of the reproducibility error of Table 1. This implies at least an equally good representation of the required AM0 spectrum

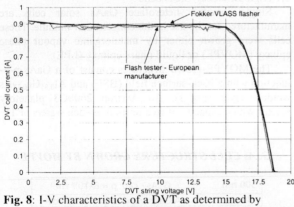

Fig. 8: I-V characteristics of a DVT as determined by VLASS and a flash tester

by the VLASS system over the entire illuminated target area. Fig. 8 finally compares an I-V-curve of another commercial satellite solar panel DVT as measured by VLASS and by ESA using a standard large area flash tester. The data is corrected for temperature. As for all above described tests a space-qualified plain silicon cell was used as a reference cell in this test as well. It is suggested that the more pronounced ripple on the flash-

tester curve is due to an increased noise impact on the measured signal during data acquisition.

After demonstrating that the VLASS system shows identical results and better measurement performance for cell types which are not expected to be affected by transient effects, further performance comparisons between the VLASS system and traditional flash testers are currently underway. These are focused on simulator performance for new cell types (e.g. multi-junction or thin film cells). Results of these tests will be published at a later date.

5. CONCLUSION

The Vortek VLASS system has proven its capability as a mature and robust large area solar simulator in comparison to traditional large area pulse solar simulators (LAPSS). In addition the system demonstrates improved versatility due to adjustable irradiation pulse duration to overcome transient measurement effects as well as adjustments to various spectral requirements by simply exchanging the coated arc envelope. Furthermore the system's radiative output is not affected by envelope blackening due to the proprietary water wall technology. The system is easy to service and can be adapted to various test conditions, e.g. vacuum chambers.

6. ACKNOWLEDGEMENTS

The authors would like to thank Mr. Steven McCoy of Vortek Industries Ltd. for the careful performance of numerous spectral measurements and calibration of the Vortek spectral measurement facility against NIST-traceable standards. The authors also acknowledge the support of Mr. Daniel Richard of ESA-ESTEC during the qualifying tests of the VLASS system.

99. REFERENCES

[1] H. Ossenbrink et al., "Errors in current-voltage measurements of photovoltaic devices introduced by flash simulators", 10th European PVSEC, Lisbon, pp. 1055-1057, 1991

[2] Dean A. Parfeniuk, David M. Camm, Tilman J. Thrum et al., "Liquid cooled high intensity radiation apparatus and method", Jan. 2000, US-Patent, application pending

CONCENTRATOR ARRAY BASED ON GaAs CELLS AND FRESNEL LENS CONCENTRATORS

V.D. Rumyantsev*, M. Hein[#], V.M. Andreev*, A.W. Bett[#], F. Dimroth[#], G. Lange[#], G. Letay[#],

M.Z. Shvarts*, O.V. Sulima [#]

[#] - Fraunhofer Institute for Solar Energy Systems, Oltmannsstr. 5, D-79100 Freiburg, Germany

Phone: (+49) 761-4588257, Fax: (+49) 761-4588250,

* - Ioffe Physico-Technical Institute, Polytechnicheskaya 26, 194021 St. Petersburg, Russia

E-mail of the corresponding author: hein@ise.fhg.de

ABSTRACT: A high-efficiency concentrator PV array with an aperture area of approximately 1 m^2 and a geometrical concentration ratio of 120x was fabricated. An all-glass design was chosen because of big advantages for many reasons. Glass was used as a superstrate for the concentrating lenses and as backside and sidewalls of the module-housing. Additionally, a very short focal length of 75 mm made the modules extremely compact. In this paper the manufacturing of the module is described and first results are presented. Long-term outdoor tests are foreseen.

Keywords: Concentrator - 1: Gallium Arsenide Based cells - 2: III-V - 3

1. INTRODUCTION

Photovoltaic (PV) conversion of concentrated sunlight is a recognized way to reduce PV electricity costs and to reach very high conversion efficiencies both for PV cells and modules. As the price of concentrator PV cells is reduced by the concentration factor, they can be fabricated with sophisticated structures and from relatively expensive materials.

Several manufacturers have built concentrator systems using point-focus Fresnel lenses and two-axis tracking [1-3]. The module housings are made of aluminum or some synthetic material, while the frontside of the module is represented by acrylic Fresnel lenses. These can be made by compression molding or polymeric lens film solvent laminated to an acrylic superstrate. In any case, it is aspired to keep the module sealed to prevent dust or humidity from entering it.

Reservations are still made against this technology for several reasons. Different thermal expansion coefficients of the lenses and the rest of the housing lead to thermal stress. This may result in leakiness after some thousand thermocycles, what can be expected to happen during lifetime. Entering humidity or dust will reduce optical efficiency or even short-circuit the entire module.

Moreover, acrylic lenses as well as sidewalls made of aluminum are relatively flexible. Hermetically sealed modules suffer under pressure variations. A strong supporting structure is necessary to prevent the bowing of lenses, which could change the focus and lead to major losses. The supporting structure is also necessary to avoid mechanical stress on lenses. Likewise other environmental effects on acrylic lenses like soiling, UV radiation or hail storms can lead to damages of the lenses [4].

The purpose of this work is to fabricate a high-efficiency concentrator PV array and to perform a long-term outdoor test of it. This array is based on GaAs cells and point focus Fresnel lenses and consists of several submodules with a total area planned to be approximately 1 m^2. We realized a short focal length, which is far below that of comparable systems. Therefore, our modules are extremely compact and easy to handle.

A key-step in fabrication of the all-glass modules is to laminate the Fresnel structure directly on a glass superstrate. This way, the surface of the concentrator module does not distinguish from flat plate modules. To exclude all possibilities of thermal stress, glass was also used for side walls and back cover. This material is cheap, and calculations show, that its thermal conductivity is sufficient.

In this paper the manufacturing of the all-glass submodules is described. First results of the submodules operation are presented. Thus the feasibility of the developed technologies is demonstrated.

2. MANUFACTURING OF POINT FOCUS FRESNEL LENS MODULES

2.1 High-efficiency GaAs cells

Single junction heteroface GaAs solar cells are manufactured for the use in the concentrator module. These cells were grown either by metalorganic vapour phase epitaxy (MOVPE) or liquid phase epitaxy (LPE).

The MOVPE-grown structures consist of a GaAs p-n-junction with a back-surface field (BSF) and $Al_{0.85}Ga_{0.15}As$ window layer (Fig. 1). An Aixtron 2600G3 planetory reactor with a capacity of five times four-inch wafers or 15

GaAs CELL STRUCTURE GROWN BY MOVPE

500 nm	GaAs - cap p = 8x10^{18} cm^{-3}
30 nm	AlGaAs-window p = 8x10^{18} cm^{-3}
30 nm	GaAs - FSF 1 p = 7x10^{18} cm^{-3}
800 nm	GaAs - emitter 2 p = 2x10^{18} cm^{-3}
2.5 µm	GaAs - base n = 2x10^{17} cm^{-3}
300 nm	AlGaAs - BSF p = 3x10^{18} cm^{-3}
1 µm	GaAs buffer n = 5x10^{18} cm^{-3}
350 µm	GaAs substrate n =1-4x10^{18} cm^{-3}

Figure 1: Scheme of the MOVPE-grown AlGaAs/GaAs concentrator solar cell with back and front surface fields.

times two-inch wafers was used. Our current wafer design consists of 37 solar cells with a diameter of 4 mm made on a two-inch wafer. Therefore, 555 solar cells can be fabricated per one epitaxial run.

Solar cell structures grown by isothermal LPE method are essentially simpler (Fig. 2). No BSF is used. The p-n junction is formed by a Zn diffusion from a Ga-based melt simultaneously with the window-layer growth [5]. In a specially designed crucible up to ten 2.5×2.5 mm^2 substrates can be processed in one epitaxial run. Sixteen solar cells can be fabricated from one substrate.

The front metallization is optimized for concentration ratios in the range of 120-180x. In order to minimize shadowing losses, silicone prismatic covers are used on the top of the solar cells.

GaAs CELL STRUCTURE GROWN BY LPE

0.1 µm	graded-x window-emitter layer $Al_xGa_{1-x}As$
2-3 µm	p-GaAs:Zn diffused emitter $p_{max} = 5 \times 10^{18}$ cm^{-3}
350 µm	n-GaAs:Si substrate $n = 2-4 \times 10^{17}$ cm^{-3}

Figure 2: Scheme of the LPE-grown AlGaAs/GaAs concentrator solar cell.

One should note, that the structure of MOVPE-grown GaAs-based concentrator solar cells is not fully optimized yet and has a potential for further improvement. The LPE-grown cells achieved efficiencies nearly as high as MOVPE-grown solar cells (Fig. 3). The crucial advantage of MOVPE-grown structures is, that the reproducibility is much higher. Therefore, for LPE-cells a higher effort in selecting cells is necessary to reach the same module efficiency.

CELL-EFFICIENCY UNDER CONCENTRATION

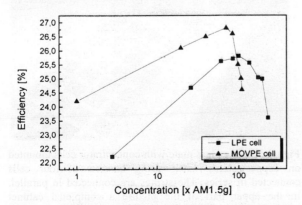

Figure 3: Efficiency versus concentration of two GaAs cells grown by MOVPE and by LPE; both cells were equipped with prismatic covers and had a diameter of 4 mm.

2.2 Cell mounting

A 4×4 cm^2 copper plate, which was 2 mm thick, served as a substrate for the cell and as a heat spreader. A ring made of copper-insulator-copper printed circuit boards (PCB) with an additional Ni-Au frontside metallization was used as a busbar for upside p-contacts. Both cells and PCB ring were soldered on the copper plate in one step. The interconnection of this busbar with the cell metallization was performed through the bonding of gold wires (Fig. 4).

The mounted cells were tested by measuring the IV-characteristic while running the cell as a diode. As criterion of choice we took a voltage at a current, which corresponded to the current of the solar cell under operation. The overall yield of cells was more than 90%. The data in Fig. 5 gives an insight in the variation of cell efficiency.

CONCENTRATOR CELL WITH HEAT SPREADER

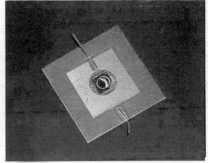

Figure 4: Solar cell mounted on a heat-spreading copper plate.

DISTRIBUTION OF CELL-EFFICIENCIES

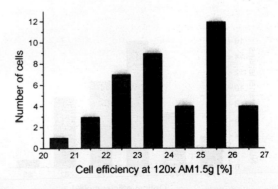

Figure 5: Cell efficiency for a number of 40 GaAs cells from different MOVPE-epitaxy runs. The concentration ratio is 120x (±4x) assuming a one-sun illumination of 1000 W/m^2 AM 1.5g; measurements were not calibrated.

Fig. 6 shows a cell-array which is intended to be used for 48 concentrator cells. The copper plates with cells are glued on the glass plate, which serves as backside of the module cabinet. For the accurate positioning of the cells beneath the lens panel the cell is illuminated with parallel light. The same procedure of adjustment was undertaken before sealing the entire cabinet.

ALLIGNMENT OF LENS ARRAY

Figure 6: Cell array and lens panel during fixing of concentrator solar cells; an optical device was constructed, which was used to position each cell accurately in the focus of the Fresnel lens. The procedure of solar cells positioning under the Fresnel lenses is shown.

2.3 Development of high-quality Fresnel lenses

The Fresnel lens panel in each submodule consists of a 4 mm thick silicate glass superstrate and a profiled silicone refractive layer polymerized directly on the rear side of the glass. Acrylic molds are used for the fabrication of the refractive layer. So far, each of these 48 acrylic molds are handmade leading to a wide distribution in optical quality. However, the cost of such a lens panel can be lower compared to acrylic lenses. Moreover, one can expect a very high long-term stability of such glass/silicone lenses.

The Fresnel structure is divided into two sections. The focal length of the inner section is larger than that of the outer one. This procedure reduces losses through chromatic aberration and results in a more homogenous light distribution on the cell.

DISTRIBUTION OF OPTICAL EFFICIENCIES

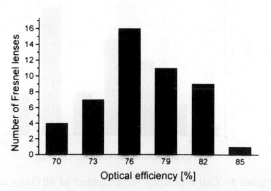

Figure 7: Optical efficiency of a 4x12-lens-panel; the average value was 76.6 %.

In Fig. 7 the result of testing a Fresnel lens panel is shown. The optical efficiencies were measured outdoor on a sunny spring day. For the determination of the optical efficiency a GaAs cell of the same type like those in the submodules was used. Therefore, the measured optical efficiency is related to the spectral response of utilized GaAs cells. Values as high as 84 % have been reached. The

difference in optical efficiency is solely caused by the varying quality of the handmade molds.

2.4 Assembly of submodules

The submodules are assembled as all-weather installations. Each of the 50x18 cm^2 submodules includes 48 pieces of 4x4 cm^2 Fresnel lenses and 48 GaAs-based concentrator solar cells with a designated illumination area of 0.13 cm^2. The geometric concentration ratio is 120x.

ASSEMBLY OF CABINET

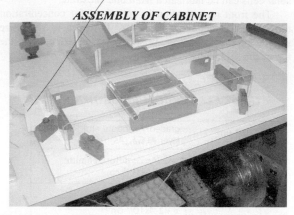

Figure 8: Manufacturing of an all-glass cabinet; the height of the glass walls was chosen after testing of ready Fresnel lenses and concentrator cells.

In Fig. 8 the building of the module cabinet is shown. In the center two aluminum-tubes for stabilizing the walls can be seen. The chosen distance between lens panel and cell array and therefore the height of the glass walls was 75.5 mm. Building the entire submodule from glass has one important advantage: Thermal stress through different thermal expansion coefficients is avoided and therefore the danger of leakiness. In addition the material is very stiff.

CELL-ARRAY ON GLASS PLATE

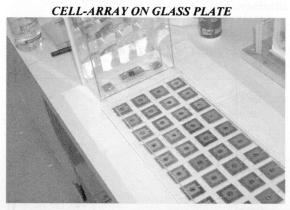

Figure 9: The glass plate with concentrator cells mounted on copper plates. Each string consists of four cells connected in series. The strings are connected in parallel. In the upper part of the picture a completed cabinet including Fresnel structures and electrical port can be seen.

Consequently no arching or variation in distance between cells and lenses through pressure and temperature changes are expected.

The submodules are filled with dry nitrogen and hermetically sealed by use of silicone rubber. The same material is used to fix the copper plates which carry the cells on the glass plate. Fig. 9 shows the cabinet before closing. After sealing the cabinet two brass screws are the only connection to the interior and thus acting as electrical ports.

3. OUTDOOR MEASUREMENT

A two-axis sun tracker from the company EGIS is used [6]. The sun tracking accuracy is 0.2° according to the company's specification. Each submodule can be adjusted separately on the tracker.

Several submodules were finished and mounted on the tracker. Fig. 10 shows the IV-curve of one submodule measured on the EGIS-tracker, where altogether 12 submodules are being installed. The resulting module efficiency was 17 %. The direct insolation during measurement was 880 W/m² measured by a pyrheliometer on a separate tracker.

BALANCE OF SYSTEM

Table I: Summary of losses for one sample submodule

Cell efficiency @ 120x	24 %
Optical losses	23 %
Losses through temperature raise	2 %
Losses through series connection	5 %
Module efficiency	17 %

IV-CURVE OF A SUBMODULE

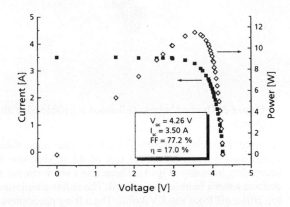

$V_{oc} = 4.26$ V
$I_{sc} = 3.50$ A
FF = 77.2 %
$\eta = 17.0$ %

Figure 10: IV-curve of one submodule installed on the EGIS-tracker; the direct insolation during measurement was 880 W/m², and the ambient temperature 13.1°C.

The relatively low fill factor of 77.2 % gives reason for a closer view to the sources of losses. On the basis of the IV-curve and the measurement of cell efficiencies and optical efficiencies of lenses, an estimation of losses in the module was performed.

We assume that the medians resulting from Fig. 5 and Fig. 7 are also valid for the submodule. Starting from a cell efficiency of 24 %, the relative percentages for the losses

through imperfect optics, heating of cells and current matching have to be subtracted (Table I).

The determination of the temperature raise of the solar cells in the submodule was done indirectly through measuring the decrease of V_{oc}. The submodule mounted on the tracker was shadowed until it reached ambient temperature. Under illumination the V_{oc} was lowered by 125 mV corresponding to 20 K in the case of four cells connected in series and a temperature coefficient of 1.6 mV/K for a GaAs cell at 120x.

The losses through series connection are set equal with the percentage, that is missing to obtain the measured value of module efficiency.

Much higher module efficiencies can be reached, if the efficiency of the best lenses is realized for the entire lens panel. It is obvious, that this technology bears the potential of more than 19 % module efficiency. Even higher efficiencies will be reached with tandem solar cells instead of single-junction GaAs.

4. CONCLUSIONS

A high-efficiency concentrator PV array with an all-glass design was presented. Glass was used as superstrate for the lenses and as backside and sidewalls of the module-housing. The geometrical concentration ratio is 120x. The short focal length of 75 mm made the submodules extremely compact.

First measurement of submodules with an aperture area of 770 cm² showed an efficiency of 17 %. This efficiency was measured under realistic outdoor conditions and includes all optical, electrical and thermal losses.

In order to determine the capability of the submodules to work at various environmental conditions, long-term measurements have to be performed. The installation of a datalogging-system for the long-term-testing of different kinds of submodules is in preparation.

ACKNOWLEDGEMENTS

This work was partly supported by the German Ministry of Economy and Technology (BMWi) under contract number 0328554D and by INTAS under grant 96-1887. We also gratefully acknowledge the support given to one of the authors (M.Hein) by the German Environmental Foundation.

REFERENCES
[1] C. Chiang and E. Richards, Proceedings of 21st IEEE PVSC. (1990) 861.
[2] D. Carroll, E. Schmidt and B. Bailor, Proceedings of 21st IEEE PVSC. (1990) 1136.
[3] V. Garboushian, D. Fair, J.A. Gunn, G. Turner, S. Yoon, Proceedings of 1st WCPEC. (1994) 1060.
[4] V. Garboushian and S. Yoon, Proceedings of 13th EPSEC. (1995) 2358.
[5] A. Baldus, A.W. Bett, U. Blieske, O.V. Sulima, W. Wettling., Journal of Crystal Growth **146** (1995) 299.
[6] EGIS Equipment Gesellschaft für intern. Elektronik Systeme GmbH, Flutstr. 34-36, D-63071 Offenbach, Germany

LONG TERM STABILITY OF PV-MODULES, DAMAGE CASES AND DAMAGE ANALYSES

Dipl.-Ing. Dr. Gerd Schauer
ÖsterreichischeElektrizitätswirtschafts AG (Verbundgesellschaft)
A-1010 Wien, Am Hof 6a
Tel.: ++43-1-53113-52439 (Fax: 52469), email: schauerg@verbund.at

Dipl.-Ing. Dr. Andreas Szeless
ÖsterreichischeElektrizitätswirtschafts AG (Verbundgesellschaft)
A-1010 Wien, Am Hof 6a
Tel.: ++43-1-53113-52425 (Fax: 52469), email: szelessa@verbund.at

ABSTRACT: More than 10 years ago Austrian electric utilities and in particular Verbund, the biggest electricity supplier in Austria, have begun to erect, operate and test photovoltaic plants /1/. Thus, today it is possible and valid to draw some conclusious regarding the long term stability and qualtiy of the PV panels supplied at that time. The described investigations are based on performance analyses of stand alone PV systems, grid coupled PV plants and module test installations directly or indirectly operated by Verbund and/or its partners.

Keywords: Stability of PV-Modules – 1: Module damage cases – 2: Damage analysis - 3

1. Introduction

Our oldest PV-plants are about twelve years old. During the years we obtained fundamental knowledg and collected experiences about design of PV plants. In the following, differnt types of damage cases are shown.

2. Laminat disintegration

2.1 Bubbles

Various disintegration phenomena of the back side laminations have been documented. Large size bubbles

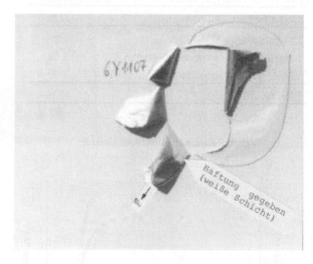

Figure 2: Opened bubble (adheason at bubble edge still given)

occurred (desolving and lifting of tedlar – aluminum sheet) which consequently led to the loss of the module due to penetrating humidity (Fig. 1). There was a smell similar to acetone when a bubble was opened. The tedlar-aluminium-foil lifted off from the EVA-foil. The lifting phenomenon started first with little bubbles which increased later. Routine checks of the electrical characteristics usually do not show the failure before the bubbles reach the module frame and cause faults to ground.

2.2 Small bubbles at soldering spots

Furthermore small size desolving of the laminate from the silicon cell along the soldering spots has been detected. Investigations concerning the exact failure causes are still going on.

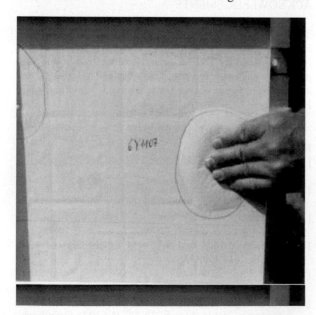

Figure 1: Large bubble

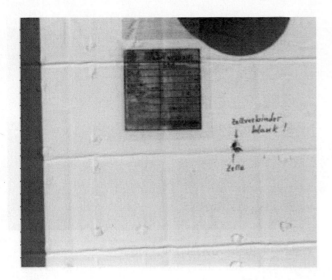

Figure 3: Small size desolving at soldering spots

2.3 Fissues

In PV plants located in the southern parts of Austria with high annual insolation various cases of fissures in the laminates were observed. Similar effects could also be investigated at visited plants in other countries (Fig. 4).

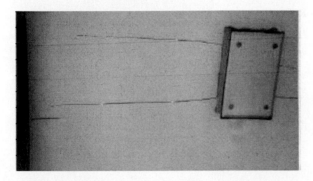

Figure 4: Fissures

Figure 5: Delamination of large glas/glas modules

2.4 Delamination (large glas/glas-module)

Another kind of damage, namely the delamination of glas/glas large size modules using epoxy type sealing technology (as used in the „house of future" of Energy Upper Austria Co.) has been studied (Fig. 5) /2/.

2.5 Hot-spot effects, brown decolouring

Local shadowing can lead to hot spots in individual cells (Fig. 6). For their detection current voltage characteristics of various types of cells were measured in order to determine their effect on the reliability of the modules. The occurrence of hot spots has been assessed and analyzed (bypass diodes).

During the course of the year the EVA and Tedlar sheets are exposed to high solar radiation intensities which can lead to brown decolouring of various degrees, thus resulting in power degradation of the modules. Different types of brown colouring were detected.

In one plant only the central parts of the cells were affected (Fig. 7) by the decolouring with no visible decolouring of the outer parts and edges. In another case brown decolouring occurred between the cells (back side laminate) of a large size module.

Figure 6: Local browning

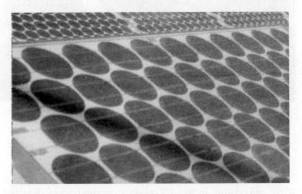

Figure 7: Browning in center part of cell

3. Degradation

A 3 kW alpine PV-pilot plant with a-Si modules at 2050 above sealevel is situated at the top of the upper dam of the hydro power plant Mooserboden/Kaprun. Degradation of the module voltage, short circuit current and the maximum power shows that the modules behave according to (after degradation) the expected nominal data /3/. Fig. 8 shows the distribution of module output power at the beginnig and after degradation (one, two and three

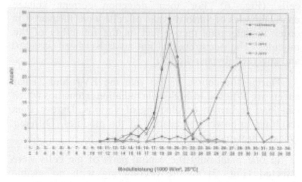

years).

Figure 8: Stabilized degradation of a-Si modules

4. Mechanical damage

Mechanical stability and rigidity of the modules and module frames are important quality criteria. To this respect a wide variety of damages has been recorded and investigated, damages originating by unsuitable choice of module size, by deformations due to snow and ice loads (Fig. 9) at alpine weather conditions and at extreme wind speeds.

Figure 9: Damege caused by ice-load

Fig. 10 shows a PV-generator after a foehn storm. About 25 % of the modules were broken

Finally, corrosion phenomena of a-Si modules of the first generation in the PV-plant Leonding were detected and analyzed.

Last not least there were damages caused by vandalism and theft (PV-sound barrier of motorway, PV-plant on the Loser).

Figure 10: Broken modules after foehn

Summarizing the results of more than ten years of PV-plant operational experience and continuous observance and recording of module damages has put Verbund and other Austrian utilities into the position to feed back information to PV-module producers helping to improve design and manufacturing of PV-modules aiming at increasing their operational life time.

References:

/1/ SCHAUER, G., SZELESS, A., FRIESS, M., KORCZAK, P.: 10 Years Operational Experience of Joint Utility-Siemens PVProjects in Austria. 2nd World Conference and Exhibition on Photovoltaic Solar Energy Conversion, HofburgCongress Center, Vienna, 6 – 10 July 1998.

/2/ WILK, H.: Haus der Zukunft, teilsolares Heizen mit Luftkollektoren, Systemtechnik und Energiebilanz 1997/98. Tagungsband 11. Int. Sonnenforum, Köln, 1998, Seite 137ff.

/3/ SCHAUER, G.; HUBER, J.; KORCZAK, P.; VAN DEN BERG, R.: Betriebsergebnisse der alpinen Photovoltaikanlage Mooserboden/ Kaprun. 10. Symposium Photovoltaische Solarenergie, Staffelstein, 15. - 17. März 1995.

MODULE DESIGNS WITH IMPROVED ENERGY COLLECTION PROPERTIES: A COMPREHENSIVE EXPERIMENTAL STUDY

H. Schmidhuber[*], P. Koltay, R. van der Zanden[†], C. Hebling
Fraunhofer-Institut für Solare Energiesysteme ISE
Oltmannsstr. 5, 79100 Freiburg, Germany
[†] Shell Solar Energy BV
P.O. Box 849, 5700 AV Helmond, Netherlands

ABSTRACT: The optical design of PV modules plays an important role in improving module performance and energy yield. Various designs have been studied in the past which lead to reduced reflection, enhanced light trapping and improved outdoor performance (see for example [1, 2, 3]). In this paper the impact of various front cover materials on the optical properties and especially on the outdoor performance of PV modules will be examined. Therefore a comprehensive experimental study has been performed in which the optical properties of modules laminated with commercially available front cover materials like different glass types and plastic foils, as well as materials with novel types of anti reflection coatings reported in [4] are determined.
The results for the commercial materials indicate, that especially for façade applications the choice of the cover material can influence the calculated energy yield by more than 8%. This result shows the importance of the optical design of PV modules (especially the impact of the front cover material) and the relevance of the characterisation of PV modules at non perpendicular incidence.
Keywords: PV Module – 1: Optical Losses – 2: Measurements – 3

1 INTRODUCTION

Many photovoltaic (PV) systems exhibit a poorer performance as is to be expected from the installed peak power. Performance ratios of 'reasonable' systems have experimentally been determined in the range of 60%-85% of their performance at standard test conditions (STC). This deviation is caused by the variable meteorological conditions which will be referred to as realistic reporting conditions (RRC). Therefore PV modules should not only be compared and rated by their STC efficiency but also classified according to their behaviour under different meteorological conditions and installation types. The same is true for the optimisation of PV modules. A module which is optimised for STC will in general not have the optimal performance under RRC (described in [5, 6]. Specific optimisation efforts for different climatic zones and application types are needed.

2 MODULE DESIGNS

The impact of structured surfaces on the optical properties of PV modules has been verified by studying the module designs listed in Table I.

All of these modules consisted of one poly-crystalline cell laminated with standard technique using EVA as encapsulant. Except of the metalisation no other inactive surfaces have been added to the module (this corresponds to a packing-factor (PF) of 1).

module	front cover	structure
K1	glass, 3 mm	stochastic
K2	glass, 3 mm	inverted pyramids
K3	glass, 3 mm	smooth
P60-1	pc type1, 3mm	v-grooves 60°
P60-2	pc type2, 3mm	v-grooves 60°
P90-1	pc type1, 3mm	v-grooves 90°
P90-2	pc type2, 3mm	v-grooves 90°

Table I: Description of the samples with structured covers used for the experimental optical characterisation.

3 STRUCTURED SURFACES

The first quantity to be considered after encapsulation of a cell in a module compound is the ratio of short circuit current before and after encapsulation called the encapsulation factor EF.

$$EF := \frac{\text{short circuit current after encapsulation}}{\text{short circuit current before encapsulation}}$$

Though a high encapsulation factor is preferable typically the short circuit current is reduced after encapsulation, because of the additional reflection at the glass/air boundary. Nevertheless EF can be greater than 1 too, if light trapping by inactive surfaces inside the module (e.g. by structured interconnectors; see ref. [6]) is considerable. The values of EF determined for the different module designs are displayed in Table II.

The measured values from Table II show an improved short circuit current for all investigated module designs. While the gain obtained from module K2 covered with structured glass (inverted pyramid structure) is especially high, the improvement for the polycarbonate covered modules (P60, P90) is smaller than predicted. This is due to technological problems related to the fabrication of these structures which where custom made at Fraunhofer ISE by hot embossing technique. The surface roughness of these polycarbonate covers was too high compared to the roughness of the glass surface used for sample K2 to reach the predicted performance enhancement.

For further analysis of the test samples profile measurements of the structured covers where carried out as well as angular dependent short circuit current measurements. The results of these experiments are compiled in Figure 1. The periodic structure of the v-grooves and the inverted pyramids in b) and c) can be clearly identified. In part c) obviously the top of the

module	EF
K1	1,028
K2	1,049
K3	0,996
P60-1	1,011
P60-2	1,016
P90-1	1,013
P90-2	1,018

Table II: Encapsulation factors EF of the different samples at STC

grooves has not been transferred properly, resulting in a more parabolic structure than intended. This non ideal structure is also responsible for the low performance of the P60 samples. The 90° grooves shown in part d) have the desired sharp top,

[*] Corresponding author; Tel.: +49-761-4588-193,
Fax: +49-761-4588-320, email: helge@ise.fhg.de

but still don't perform as expected because of the surface roughness problems mentioned before.

Obviously the best performing samples K1 and K2 have not only high encapsulation factors but also an improved angular dependent behaviour as far as the short circuit current is concerned. For the polycarbonate covers (Figure 1 c) and d)) the absorption function has also improved for high incidence angles, but this improvement is compensated for by the reduced performance at medium incidence angles. To clarify the extraordinary good performance of the sample K2 the angular dependent short circuit current of this module has been simulated by ray tracing. A good fit to the measured values (see Figure 1 a)) could be achieved for an apex angle of 150°. This corresponds very well to the value of the apex angle $2\gamma=160°\pm20°$ determined from the profile measurements. In the other cases the simulations did not fit well the measured results due to the reasons discussed before.

4   SOL-GEL ANTI REFLECTION COATINGS

In this section the performance of nanoporous layers for reflection reduction will be studied, which is based on a different principle. So far anti reflection effects by multiple reflections

and light trapping have been studied. Nanoporous layers however profit from a graded index of refraction which leads to a quasi continuous change of the refractive index. By avoiding a sudden change of the index of refraction the reflection is reduced. Information about manufacturing and optical properties of such layers can be found in [4]. For our purposes mainly the total reflectivity and the angular dependence are important.

Because the ARC is applied to both sides of the plate also module designs can be considered which consist of an air/glass/air boundary. This means that in such a module the glass is not laminated to the cells, but just fixed in front of the cells leaving a small gap of air between cell and glass cover. We call such a design "inert gas design" after reference [7].

The considered modules consisted of a sol-gel coated glass and a poly-crystalline cell which was mounted approximately 5 mm behind the glass plate. The backside consisted of aluminium covered by standard white Tedlar. To enhance the reliability of the results two different cells have been combined with two glass plates (samples G42 and G44), so that 4 different samples could be studied in total. Measurements of the short circuit current have been performed with and without glass to determine the encapsulation factor. Also angular de-

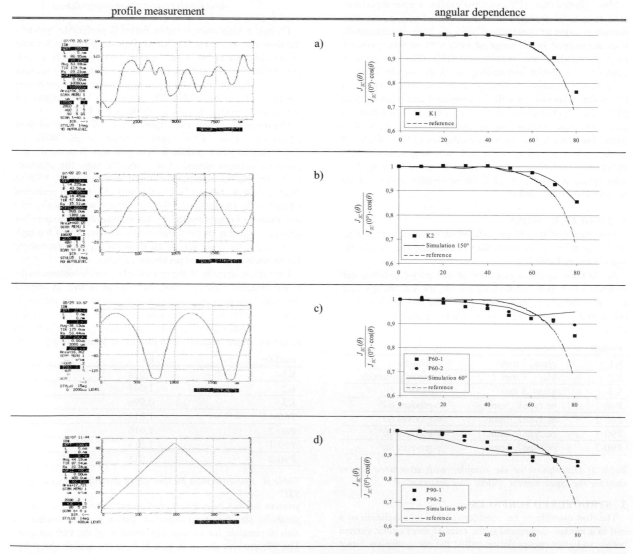

Figure 1: Cross section profiles of the front covers and the resulting angular dependence of the short circuit current of the different test samples: a) K1 b) K2 c) P60 and d) P90 (in part d) only the top of one groove is shown, the period in this case was about 1.5 mm). For comparison the angular dependent absorption function of the reference module (dashed line) and some ray tracing simulations are displayed.

pendent short circuit current measurements have been performed, which are displayed in Figure 2.

The angular dependence of the short circuit current corresponds approximately to that of the standard module, except of a stronger decrease at high incidence angles and the strong fluctuations of the data which can be observed around the mean value. These variations are due to the poly-crystalline surface of the solar cells, the reflection properties of which depend strongly on the incidence angle in combination with the fact, that the possible angles of incidence on the cell surface are higher for the inert gas design (up to 90°) than for the standard design (up to 41° if the index of refraction n=1.5) under the same conditions. The reduced light trapping properties of the design (caused by the double-sided ARC) further enforce the mentioned effect.

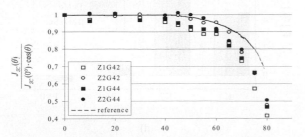

Figure 2: Normalised short circuit current versus incidence angle for 4 samples of the sol-gel coated inert gas design. (cell Z1 and Z2 combined with glass samples G42 and G44). For comparison also the angular dependence of the short cicuit current from the standard module is displayed.

The encapsulation factor of the inert gas design is (averaged over all 4 samples) at perpendicular incidence was about 96±0,1%, which corresponds exactly to the solar transmission of the glass plates in the considered wavelength interval. This means, that this design does not profit from light trapping at all!

Another possibility for using sol-gel coated glass for module construction is to laminate the glass with EVA like in an standard process. In this case the only difference with respect to the standard module is the additional ARC coating on the module surface. The coating on the inside has according to the following measurements no significant impact on the optical properties.

In Figure 3 the results of the angular dependent short circuit

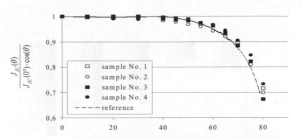

Figure 3: Normalised short circuit current versus incidence angle for 4 samples laminated with EVA and sol-gel coated glass. For comparison also the angular dependence of the short circuit current from the standard module is displayed.

current measurements performed on such laminated sol-gel coated modules are displayed. Again four different samples have been evaluated, each of them containing a single poly-crystalline cell. Like for the inert gas design also for the shown modules the angular dependence has not improved compared to modules laminated with smooth glass. Nevertheless at least the strong scattering of the data has decreased due to the index matching by the EVA, which reduces the influence of the surface reflectivity of the poly-crystalline cells.

Finally the encapsulation factors for the laminated cells are also poor, because light trapping is reduced by the ARC which greatly suppresses the total internal reflection at the glass/air boundary. Therefore in conclusion it can be stated, that sol-gel coated glass covers have no positive impact compared to smooth glass on module performance, neither if the standard design (lamination with EVA) nor if the inert gas design (gap of air between glass and cells) is considered. Therefor an evaluation of the RRC-efficiency has not been carried out.

5 OUTDOOR MEASUREMENTS

In order to verify the extraordinary performance of the glass with inverted pyramids (module K2 in Table 1) outdoor test have been performed at Helmond (The Netherlands) by Shell Solar. During the experiment four modules of type K2 and a reference module of type K3 where installed at latitude tilt and continuously monitored. The measured increase of the short circuit current compared to the module covered with smooth glass (type K3) is displayed in Figure 4. The crosses symbolise the individual data points, while the solid lines are the floating average of the data.

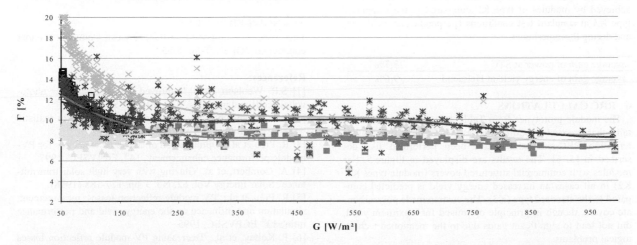

Figure 4: Results of long term outdoor measurements of the short circuit current of four identical optimised modules with structured glass cover at Helmond. The gain Γ displayed as function of the global irradiation refers to the short circuit current of a reference module with smooth glass.

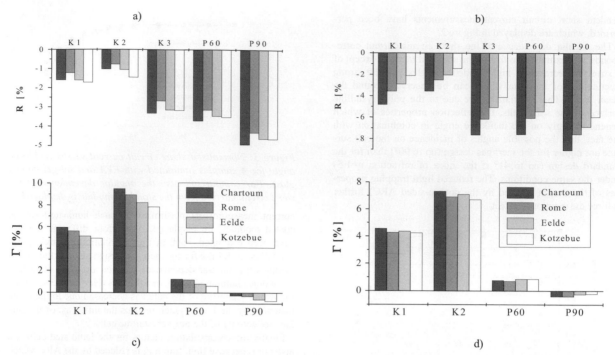

Figure 5: Reflection losses R of the modules from Table 1 at the locations Chartoum (Sudan), Rome (Italy), Elde (Netherlands) and Kotzebue (Alaska) for a) latitude tilted installation and b) facade installation. In c) and d) the gain factors Γ for latitude tilt and façade installation are depicted.

At low irradiation values the recorded gain is higher than at medium light levels, which is due to the dominance of the diffuse part of the irradiation at low light level conditions (overcast days, morning and evening hours etc.). At this operating conditions the individual modules perform different, because small changes in the environment (reflecting surfaces, different albedo etc.) can have a big effect on the amount of irradiation received by the module. Also differences in irradiation dependence of the efficiency of the samples can add to this effect explaining the different performance at low light levels. For medium and high light levels this effect gets negligible and the differences between the curves and the scattering of the data decreases.

All samples exhibit an at least 8% better performance (i.e. 8% higher short circuit current) than the reference module of type K3 at nearly all operating conditions. The following table finally shows the total averaged gain in power and energy yield achieved by modules of type K2 compared to the sample of type K3 at standard test conditions (perpendicular incidence) and during the experiment.

average gain of power at STC	3,7%
average gain of energy yield at Helmond	9,6%

6 RRC CALCULATIONS

The module prototypes from Table 1 have been experimentally characterised as outlined before. On basis of this characterisation RRC-simulations have been performed like described in [5, 6]. The results are displayed in Figure 5. For modules with commercial structured covers (module types K1, K2) in all cases an increased energy yield is predicted compared to the standard type (K3). The custom made polycarbonate covers, though in principle optimised for maximum yield, did not lead to significant gains due to the mentioned technological problems.

The gains have been achieved in all cases by two effects caused by the structured surface: First the power at STC (perpendicular incidence of the light) could be raised by reflection

reduction and light-trapping. Second the angular dependent absorption has improved leading to higher currents at oblique incidence.

7 CONCLUSIONS

The optical properties of various module designs have been studied experimentally with respect to their performance for terrestrial applications. The importance of the angular dependence of the short circuit current on the incidence angle could clearly be demonstrated. As the best result an optical optimised module design with 9% higher energy yield compared to the reference case has to be named. Outdoor measurements confirmed that the specific design increases the light absorption and leads to an improved energy yield.

Experiments revealed, that the considered reflection coatings based on a nanoporous layers on the module front side can not raise the optical performance significantly.

ACKNOWLEDGEMENT
This work was supported by the European Community under contract no. JOR3-CT-96-0095.

References:
[1] S.R. Wenham, et al. 'Improved optical design for photovoltaic cells and modules', 22nd IEEE PVSC1991
[2] J.M. Strohbach et. al. 'Specific module design for realistic working conditions', 2nd WCPVSEC, 1998
[3] R. Preu et al.; 'Optimisation of cell interconnectors for PV-module performance enhancement', 14th, ECPVSEC, 1997
[4] A. Gombert, et al. 'Glazing with very high solar transmittance', Solar Energy Vol. 62, No. 3, pp. 177-188 (1998).
[5] R. Preu et al.; 'PV module reflection losses: measurement, simulation and influence on the energy yield and performance ratio', 13th ECPVSEC, 1995
[6] P. Koltay, et al.; 'Decreasing PV module reflection losses under realistic irradiation conditions', 2nd WCPVSEC, 1998
[7] H. Nagel, et al.; 'PV-Module With Enhanced Optical Performance', 2nd World Conference PVSEC, Wien (1998)

VERIFICATION OF AN IMPROVED TRNSYS SIMULATION MODEL
FOR PV-ARRAY PERFORMANCE PREDICTION

Miroslav Bosanac, Ivan Katic and Erik Scheldon

Solar Energy Centre, Danish Technological Institute
P.O.Box 141, 2630 Taastrup, Denmark
Tel: +45 72 20 24 86, Fax: +45 72 20 25 00
Email: miroslav.bosanac@teknologisk.dk

ABSTRACT: Purpose of this work is to improve accuracy of already available simulation tools for prediction of a PV-array performance within grid-connected or stand-alone systems. The mathematical model applied is based on the double-diode PV-cell model. The model is incorporated in the TRNSYS program as one of the components. In that sense, already available components of the TRNSYS program may be used (solar air-collector and liquid collectors, building model, etc.) enabling simulation of hybrid systems including building-integrated PV-arrays. Special attention has been devoted to evaluate and identify the dependence of PV-energy output on (i) irradiation level, (ii) incident angle of irradiance and (iii) a part of diffuse irradiance in a hemispherical irradiance.

The accuracy of the simulation results has been analysed using measurement data. The verification has been carried out by measurement data of Solgården-system situated in Kolding, Denmark. Solgården system consists of 846 PV modules with approx. 90 kWp power roof mounted and 16.5 kWp integrated in the façade.

It is shown that the introduction of correction functions improves the accuracy of long-term energy predictions of PV-array performance. This may be especially relevant for PV-arrays situated in the climate zones (like e.g. northern part of Europe) with a considerable part of diffuse irradiation.

Keywords: Performance Simulation – 1: Silicon cell – 2:

1. INTRODUCTION

Along with a stronger development of the PV technology, a number of new software tools for prediction of photovoltaic systems performance are being available on the market. The majority of these tools deal with both grid connected and stand alone photovoltaic systems.

In general, these tools are either dealing with simplified methods for computation of energy yields or performing detailed simulation of PV system performance. The target groups for these tools are: (i) consulting engineering agencies and/or architects, (iii) dealers of the PV systems.

In principle, each program for PV system simulation consists of two basic parts: (i) estimation of solar irradiance at the PV array and (ii) computation of the output power of the PV system based on irradiance data, cell's technical data and the electrik-load profile. Nowadays, in the focus are the PV systems integrated in buildings. Therefore, the part of the program computing the solar irradiance on the tilted plane (roof) or vertical surface (wall) must be reliable as it directly influences accuracy of long-term energy prediction..

The TRNSYS [1] program contains the components for modeling of PV-thermal and PV-only systems made up of PV arrays, batteries, regulator-invertors and utility interfaces. The specific components are: Electric storage battery, Regulator/Inverter, Combined PV system and the silicon solar cell. Silicon cell model uses a set of empirical values to model the I-V characteristics.

The main advantage of the TRNSYS is availability of (i) the accurate models for computation of direct and beam irradiance on the tilted surface and (ii) simulation of hybrid solar systems.

2. MATHEMATICAL MODEL OF PV ARRAY

The mathematical model applied is based on the double-diode PV-cell model. The model is incorporated in the TRNSYS program as one of its components. In that sense, already available components of the TRNSYS program may be used (solar air-collector and liquid collectors, building model, etc.) enabling simulation of hybrid systems including building-integrated PV-arrays.

The model predicts the complete current-voltage characteristic over the entire operating voltage range of a silicon PV arrays.

Operating point is determined on the basis of irradiance, ambient temperature and operating voltage.

Subroutine simulates a maximum power tracking device as one of operation modes for a PV power system.

While a series resistance is taken into account, a shunt resistance is assumed to be infinite and thus neglected in the model.

An option is provided to evaluate the series resistance with the bisection method, if not given as input. A shunt resistance is assumed to be infinite.

To overcome convergence problems that appear when simulating direct coupled systems, a bisection method type of convergence promotion is included.

The user-friendly interface has been written in TRNSED [1] program to run the TRNSYS command (deck). A typical screen-layout of the TRNSED program is shown on Fig. 1:

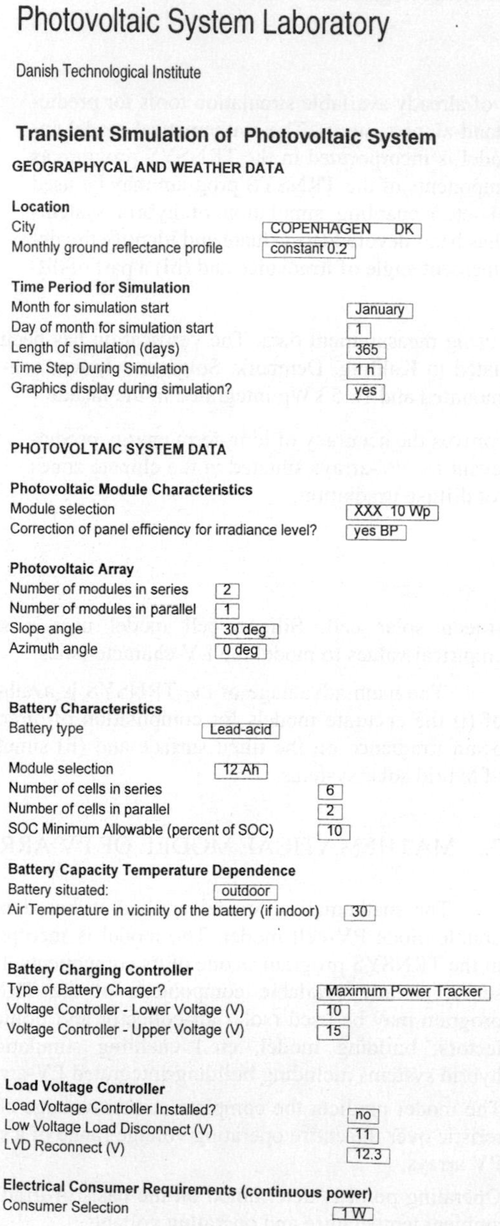

Photovoltaic System Laboratory

Danish Technological Institute

Transient Simulation of Photovoltaic System

GEOGRAPHYCAL AND WEATHER DATA

Location
City   COPENHAGEN   DK
Monthly ground reflectance profile   constant 0.2

Time Period for Simulation
Month for simulation start   January
Day of month for simulation start   1
Length of simulation (days)   365
Time Step During Simulation   1 h
Graphics display during simulation?   yes

PHOTOVOLTAIC SYSTEM DATA

Photovoltaic Module Characteristics
Module selection   XXX 10 Wp
Correction of panel efficiency for irradiance level?   yes BP

Photovoltaic Array
Number of modules in series   2
Number of modules in parallel   1
Slope angle   30 deg
Azimuth angle   0 deg

Battery Characteristics
Battery type   Lead-acid

Module selection   12 Ah
Number of cells in series   6
Number of cells in parallel   2
SOC Minimum Allowable (percent of SOC) 10

Battery Capacity Temperature Dependence
Battery situated: outdoor
Air Temperature in vicinity of the battery (if indoor)   30

Battery Charging Controller
Type of Battery Charger?   Maximum Power Tracker
Voltage Controller - Lower Voltage (V)   10
Voltage Controller - Upper Voltage (V)   15

Load Voltage Controller
Load Voltage Controller Installed?   no
Low Voltage Load Disconnect (V)   11
LVD Reconnect (V)   12.3

Electrical Consumer Requirements (continuous power)
Consumer Selection   1 W

Figure 1: The typical layout of the TRNSED (example for stand-alone system)

In this work we analysed influence of certain influences which are not of primary importance in computation of PV array energy yields.

The simulated PV-array energy output has been corrected taking into account the following variables:

(i)   Irradiation level.

        Correction, C_G, has been computed on the basis of the regression:

        $$C_G = a_o + b_o*G + c_o*G^2,$$

        where

        G is hemispherical irradiance in kW/m²

        a_o, b_o, and c_o are regression parameters:

        $a_o = 0.872$

        $b_o = 0.264$

        $c_o = -0.136$.

(ii)   Incident angle of beam part of hemispherical irradiance. The Ambrosetti equation gives dependence of correction factor, K_θ, on incident angle:

        $$K_\theta = 1 - \tan^a (\theta/2),$$

        where

        θ is incident angle of incoming (beam) irradiance

        a is parameter characterising the correction factor, i.e., for single-glass cover it's value is approximately 3.3.

(iii)   Fraction of diffuse irradiance in hemispherical irradiance.

        The diffuse irradiance is assumed isotropic. An ´effective´ incident angle of 60 deg has been applied for diffuse irradiance.

3.   SOLGÅRDEN SOLAR SYSTEM

Solgården is an apartment building which has been renovated in order to improve comfort and reduce the energy demand. A PV system was installed to supply the building with about half of its electricity demand.

The PV system consists of two parts:

• 757 m² of PV panels installed on the roof,

• 175 m² façade modules integrated in the balcony enclosures.

The total peak power is 89.5 kW on the roof and 16.5 kW on the façade.

Electrically, the plant is divided in 16 subsections, in order to handle the different irradiance levels.

Fig. 2 shows the balcony and roof-situated modules at the Solgården system.

Fig 2 Solar system at the Solgården

5. RESULTS

Typical results of comparison between simulated and measured energy yield is shown on Figs 3 to 5.

The measured DC energy output of the PV array has been compared for each sub-array of the Solgården system.

The error is considerable in February due to low level of irradiation available. The third column presents the results when correction due to irradiance level is applied. It is noticeable that error is reduced for winter months because the portion of impact of low irradiances is much higher in winter months than in summer. Thereafter, corrections for incident angle modifier and part of diffuse irradiation have been carried out. Results are shown in respective columns.

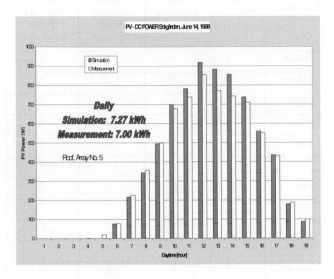

Figure 4. Comparison between simulated and the measured energy yield for balcony-subarray No. 5

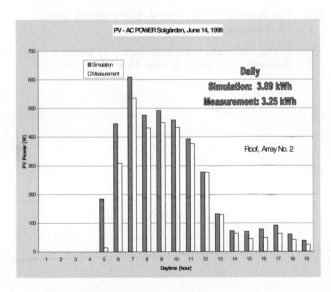

Figure 3. Comparison between simulated and the measured energy yield for subarray No. 2

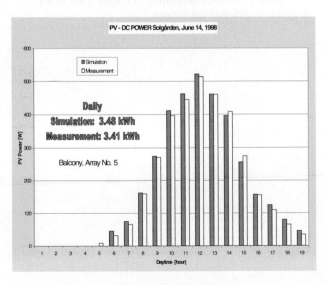

Figure 5. Comparison between simulated and the measured energy yield for roof-subarray No. 5

In the first run, no corrections were applied. The error for of the prediction of the DC energy output of the system is presented in the second culoumn of the Tab 1.

Month	ERROR IN SIMULATION OF THE PV ENERGY CONTRIBUTION			
	(Simulated PV Energy – Observed Energy) / Observed Energy in [%]			
	Without Corrections	Correction due to Irradiance Level	Correction due to incident Angle Modifier	Correction due to Fraction of Diffuse Irradiation
Januar	15	11	10	8
Februar	22	18	15	12
March	12	9	8	7
April	8	7	7	6
May	7	6	5	5
June	7	6	5	5
July	6	6	5	5
August	7	6	6	6
Septem.	7	6	6	6
October	9	8	7	7
Novem.	9	9	8	6
Decem.	11	10	8	7

Table 1. Error in simulation of energy yield of PV-array due to different corrections.

6.   CONCLUSION

It is shown that the introduction of correction factors improves the accuracy of long-term energy predictions of PV-array performance.

Results presented in Table 1 show that corrections are specially important for winter months because the corrections taken into account have higher influences than in summer months. Accordingly, it may be recommended to carry out the corrections for climate regions with higher portions of diffuse irradiation (like e.g. northern part of Europe).

7.   ACKNOWLEDGEMENTS

The support of the Danish Ministry for Science is gratefully acknowledged.

8.   REFERENCES

[1]   TRNSYS, Transient Simulation Program, Version 14.2, University of Wisconsin.

[2]   Test Reference Year, Dansk Meteorologisk Institut, Denmark

[3]   Duffie, Beckman, Solar Engineering of Thermal Processes, John Wiley & Sons, New York, 1980.

NEW GLASSLESS ENCAPSULATION MATERIALS FOR TRANSPARENT, FLEXIBLE, LIGHT WEIGHT PV-MODULES AND BUILDING INTEGRATION

Sabine Degiampietro[1], Albert K. Plessing[1], Peter Pertl[1], C. Bucci[2], R. van der Zanden[3]

Ad [1] :ISOVOLTA AG

Vianovastrasse 20, A-8402 Werndorf

Tel.: 03135-54314-48

e-mail: degiampietro.s@isovolta.com

Ad [2]: Eurosolare SpA

Via dÁndrea 6, I-00048 Nettuno, Tel.: +39-6-985-60501

e-mail: eurosolare@eurosolar.agip.it

Ad [3]: Shell Solar Energy

Lagedijk 26, 5700 JC Helmond, Netherlands, Tel.: +31 492 5086 08

e-mail: Ruud.R.vanderZanden@SI.shell.com

*Correspondence to Sabine Degiampietro

ABSTRACT: The encapsulation of solar cells in PV modules has several important functions. 80% of PV-power-module producer use laminated film material.

Current developments led to promising improvements of the main properties of film materials. A new system of encapsulation without glass – transparent film materials and rigid plates - has been investigated. The main characteristics and requirements on the encapsulant are adhesion to encapsulating EVA, layer-layer adhesion, poor heat shrinkage, optimal damp heat durability, UV resistance, high light transmission rate, good module laminating processability, long term stability in different climates, surface characteristics, low water vapour permeability, module stability and low costs.

To achieve all these goals, it is important to have high quality solar cells and also high quality encapsulating components. Results concerning technology and new encapsulation materials are presented here.

Keywords: Encapsulation - 1; Barrier Layer - 2; Cost Reduction - 3

1. INTRODUCTION

Standard PV modules consist of glass, EVA and laminated film material. All components have to seal and protect the solar cells from environmental impact. The front surface encapsulation has to allow the maximum possible amount of light transmission. In standard PV-module production the following composition is used:

♦ glass
♦ encapsulating EVA
♦ connected solar cells
♦ encapsulating EVA
♦ laminated film material

PV-module producer try to create highly efficient solar cells and long lasting modules. This paper reports on the activities and results for the development of new encapsulating materials; flexible film materials for front and back surface, and rigid materials for back surfaces, which can be used for the protection of high quality PV-modules.

1.1 Requirements and application of glassless PV-modules

The general requirements for glassless PV modules are the same as for standard modules. Flexibility, weatherability and electrical insulation are all important, not to mention the stability of the units regarding hail, UV rays, and climatic factors in general.

In order for PV modules to function for up to 25 years, rigorous durability tests had to be conducted, like the heat and humidity test as mentioned in chapter 5.2. These requirements have caused the investigations to achieve better performance and stability for film material for PV module production. Glassless PV-modules can be used in building integration, in low weight modules, cars, and for flexible photovoltaic applications.

Today, cost reduction is very important, but the product performance is not allowed to decrease.

2. MATERIAL FOR THE REAR SURFACE

In developing encapsulating materials depending on the requirements of a PV-module - long term stability, high performance and cost reduction – the following rear surface materials have been selected after many tests of various materials.

♦ Laminated film in thickness of 0,27 mm to 0,40 mm
♦ Laminated sheet of phenolic hard paper

Table 1: Types of rear surface material for glassless modules

Material	Thickness[mm]	Composition
Rigid, laminated sheet	8 stable	Phenolic hard paper
Laminated film	0,27 to 0,40 flexible	PVF/PET/PVF, treatment optional

MAX Exterior is a rigid plate based on phenolic hard paper and phenolic melamine press board. The main interest in phenolic hard paper is the processability in the PV module lamination. Therefore the adhesion to EVA had been tested as can be seen in chapter 5.4.

3. MATERIALS FOR THE FRONT SURFACE

The standard PV-module has a white, black or other colored back surface with a transparent front surface (normally glass), but it is also possible to use a transparent polymer film on the front side. The most important requirements for the front side of a PV module are the UV stability, mechanical stability, temperature stability, high light transmission, medium moisture permeability, and good adhesion to EVA, good adherability to adhesives and long term stability. It was necessary to get information on the technical properties of the film materials, which are exposed to severe climates.

It is well known that water vapour decreases the efficiency and life time of solar cell arrays. Therefore, different substrate films have been investigated on their suitability for barrier coatings of SiOx and Al_2O_3 to reach a WVP lower 0,1 g/m²d.

SiOx has a property which enables it to absorb UV light depending on the ratio of silicon to oxygen. The SiOx layer has a perfect optical performance and prevents rapid aging of solar cell arrays because of the barrier effect for moisture. Table 2 shows the films selected for SiOx coating:

Table 2: Substrates films for SiOx-coating

Material	Thickness[mm]
ETFE	0,10; 0,05
PET	0,05 to 0,10
PEN	0,25
PET/PEN	0,37

For front surface application the highly transparent ETFE film is the first candidate, followed by PET film. PEN has poor UV-stability and can only be used for back surface application.

PET film can be processed with PVF to a composite. The composite can be surface treated to achieve a higher adhesion to EVA.

SiOx coating trials were made by IVV (Fraunhofer Institut für Verfahrenstechnik und Verpackung) and LMP (Lawson Mardon Packaging) by a cost effective PVD process and different adjustments of the coating system. Significant improvements in higher adhesion , transparency and stability of SiOx layers could be achieved as can be seen in chapter 6.

4. INTEGRATED MATERIAL

At the beginning, external trials to laminate encapsulating EVA onto protective laminates had been made for identifying the correct adjustment parameters for the EVA extrusion line. A stable process could be developed to produce composites of EVA/film laminate.

Larger test quantities of integrated material with standard and fast curing EVA had been produced successfully without any problems in processability.

5. CHARCTERISTICS OF FILM MATERIAL

5.1 Water vapour permeability

A very important fact is the water vapour permeability (WVP) of the film materials. The value depends on the thickness and especially the kind of film. The objective in this work is to achieve a moisture permeability of < 0,1 g/m²d.

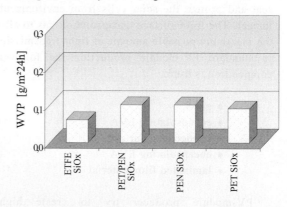

Figure 1: Water-vapour permeability SiOx layers coated on different substrate films

Excellent results of water vapour permeability < 0,1 g/m²d could be achieved by coating SiOx layers on a substrate film of ETFE, PET/PEN, PEN and PET.

The table 3 shows the WVP for encapsulating film composites, which are commonly used for PV module production:

Table 3 : Water vapour permeability film materials

Film Material	Thick-ness[mm]	Wat. Vap. trans-mission[g/m²24h]
PVF-PET-PVF transparent	0,27	0,63
PVF-PET-PVF opaque	0,17	1,3

5.2. Layer-Layer Adhesion

The layer-layer adhesion depends almost entirely on the type of adhesive system used. Tests are made on the amount of adhesive, thermal stability and adhesion strength.

An important test for long term performance is the "heat/humidity test". The materials are aged for at least 2000h in a hot and humid climate of 85°C and a relative humidity of 85% according to IEC 61215.

MAX Exterior passed 1000 hours, all other new developed materials passed 2000 hours in this test without failing.

5.3. Heat shrinkage

The heat shrinkage of polymer materials cannot be avoided, because of the internal stress of the macromolecules. This shrinkage is a problem in the manufacturing of PV-modules, but it can be minimized through the laminating process of the films. Typical shrinkage characteristics for film materials are:

Table 4: Typical shrinkage characteristics of encapsulating materials

Material	Heat shrinkage[150°C/30min.]		
	MD	TD	Method
ETFE SiOx	0,2	1,3	DIN 40634
PET SiOx	1	0	DIN 40634
PET/PEN SiOx	1	0,5	DIN 40634
PEN SiOx	1	0,5	DIN 40634
Exterior 701	0	0	DIN 40634

5.4. Adhesion to EVA

The adhesion of rear surface and front surface material in a PV-module to EVA is a very important property of the encapsulating system. PVF surface exhibits in poor adhesion compared to surfaces of treated PVF or SiOx-coated film material. The figure 3, 4 and 5 show the adhesion of front surface and rear surface encapsulation materials:

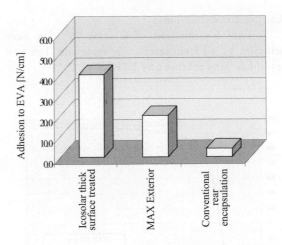

Figure 3: Adhesion to EVA rear surface materials

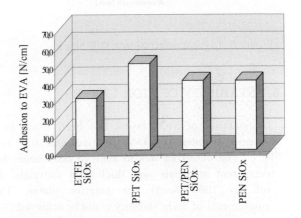

Figure 4: Adhesion to EVA front surface materials.

The adhesion to EVA does only decrease slightly after 2000 hours of aging in a hot and humid climate as shown in figure 5:

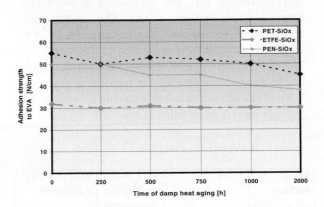

Figure 5 : Adhesion to EVA after aging in a hot and humid climate of 85°C/85% r.h.

5.5. Light transmission

A light transmission of at least 90% has been achieved by all transparent film materials. The best result has been achieved by ETFE film:

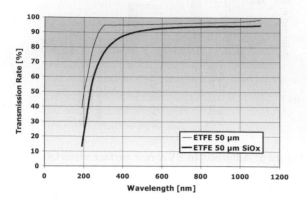

Figure 6: Light transmission ETFE 50μ

6.   STABILITY OF SIOX LAYERS

The stability of SiOx layers has been tested in a climatic chamber at 85% relative humidity and 85°C. The objective was to achieve 2000 hours for transparent materials and thick film materials, as well as 1000 hours for exterior plates. The requirements of SiOx stability could be achieved.

7.   COMPOSITION/APPLICATIONS OF GLASSLESS MODULES

The transparent materials and exterior plates are very interesting for the application in buildings, because of time and cost savings. Flexible materials are suitable for special applications, where PV modules need to be bent or altered.

Polymer substrates with transparent barrier layers can be processed with PVF for top quality products. Table 5 shows the evaluated materials for new front- and rear surface encapsulation materials.

Table 5: Materials for glassless PV modules

Front surface	ETFE	PET/PEN (+ PVF) transparent	PEN (+ PVF) transparent
Rear surface	PVF-PET-PVF thick	MAX-Exterior	
Integrated Material	EVA n.c./ laminated film	EVA f.c./ laminated film	

8.   TECHNOLOGY-LAMINATING PROCESS

All test modules had been laminated in a standard type vacuum laminator ICOLAM 060. The laminating cycle depends on the use of normal cure or fast cure EVA. Processing with ETFE had to be made by a slowed down temperature process to avoid wrinkling of the ETFE film.

9.   COST SAVINGS

High Quality and performance of new encapsulating materials will increase the reliability of PV modules. The cost of new materials can only be estimated.

A 4-layer composite could be replaced by an integrated material including barrier layers – this product could save approximately 23% of the costs.

A reduced process time of 50% could also reduce costs and will increase the productivity of PV module production. The shortened cycle time will lead to process automation and shortening the return on investments.

10. CONCLUSION –FUTURE PROSPECTS

New types of film materials and rigid sheets offer the possibility for new applications of glassless PV modules at lower production costs.

A new design for building integration has been investigated for special applications.

A very interesting development is the SiOx barrier layer, which reduces the water vapour permeability to a value of < 0,1 g/m²d. The high quality and excellent protection against water vapour make the new materials very promising candidates for encapsulation of thin film solar cells.

ACKNOWLEDGEMENT: This work was funded by the European Commission under the contract No. JOR3-CT97-0155. We also thank the participating companies BP Chemicals PlasTec, Shell Solar Energy, Fabrimex and Eurosolare for their contributions.

MEASUREMENT AND PERFORMANCE PREDICTION METHOD FOR PV CONCENTRATOR SYSTEMS

W. R. Bottenberg
Photovoltaics International, LLC
171 Commercial Street
Sunnyvale, CA 94086, USA

ABSTRACT: Accurate methods for measuring and predicting the performance of PV concentrator modules under real time circumstances are needed for qualification and rating of systems as well as for comparison to equivalently rated flat plate systems. Several groups have proposed energy rating models for flat plat modules. These rating methods require accurate models for module performance under a wide and complete variety of operating conditions. Extensions of these models to tracking systems and concentrators systems have been also considered. The purpose of his work is to propose a suitable method to measure and rate concentrator modules given that the modules cannot in general be measured indoors under controlled conditions because of their size and the unavailability of laboratory collimated light of sufficient intensity.
Keywords: Modelling – 1: Concentrator – 2: Performance – 3

1. INTRODUCTION

King et al. have proposed a model for measuring and rating PV modules that uses outdoor, real-time data [1]. This method has been extended in this paper to include, in particular, the case for linear, 1-axis tracking concentrators where annual declination angle effects on performance are important. The equations require a simple non-linear curve fitting method to obtain the required coefficients and functions.

In this paper we present the measurement methodology and compare measurements made of module performance on a 2-axis tracker with measurements made for modules mounted in a rooftop panel using the standard 1-axis tracking method. The non-linear curve fitting method allows the determination of the functions f_1 and f_2, which are the air mass and angle-of-incidence functions respectively. A coefficient C_6 measures the effect of collection of the diffuse insolation component by low concentration modules.

Module energy prediction models as reviewed by Whitaker and Newmiller and proposed by Marion et al. require indoor measurements of modules using heating pads and simulators [2,3]. The method for I-V curve translation, while effective, requires a linearization of the translation curve. The method discussed in this paper extends the methods of King et al. by the employment of non-linear curve fitting and use of module heat sink temperature as an indicator for temperature [1].

We have found that data reduction of several days of rooftop data performance can be made to yield accurate estimates of air mass and angle of incidence effects when compared to more controlled 2-axis measurements.

2. METHODOLOGY

Module performance is modeled using the approach described below. This approach requires a value for the cell temperature. If cell temperature is not measured, it can be computed using V_{oc} and I_{sc}, using Equation 8, or it can be computed using heat-sink temperature, using Equation 9. To extend this model for predicting module performance based on ambient environmental conditions, one can compute the coefficients in Equation 11, which relate heat-sink (or cell) temperature to ambient temperature, insolation, and wind speed. Changes made to the original flat-plate modeling approach are included to account for the different irradiance to which low concentration ratio concentrators respond versus flat plates. Concentrators respond to the direct-normal irradiance (E_{DNI}), rather than to global irradiance. Some low-concentration modules may also respond to a fraction of the diffuse irradiance (E_{DIF}), which is the plane-of-array irradiance minus direct-normal irradiance.

The approach is based on the premise that module I_{mp}, V_{mp}, and V_{oc} are functions of I_{sc} and cell temperature (T_c) only. This implies that the I-V curve is the same for any solar spectrum and angle of incidence for a given I_{sc} and T_c. Performance characterization of a module is then a two-part process. In the first part, the parameters and functions used in computing I_{sc} (1) are determined, and an "effective irradiance," E_e, is computed (2). The effective irradiance gives the response of the module short-circuit current relative to the reference conditions. At reference conditions, effective irradiance has the value 1. The parameters and functions used in computing I_{sc} are the absolute air-mass function, $f_1(AM_a)$; the angle-of-incidence function, $f_2(AOI)$; the short-circuit current at a reference operating condition, I_{sc_o}; the response fraction to diffuse irradiance, C_6; and the I_{sc} temperature coefficient, α_{isc}. In the second part of the analysis, I_{mp}, V_{mp}, and V_{oc} are related to E_e using Equations 3-5. The default reference operating condition is 50°C cell temperature, 850 W/m² direct irradiance, 150 W/m² diffuse irradiance, 1.5 absolute air mass, and 0° angle-of-incidence. Alternatively, a reference condition may be selected to correspond to required module or array rating conditions. When this procedure is be used for the purposes of a module qualification procedure, all measurements should be made at an angle of incidence of 0°, with $f_2(AOI) = 1$ to maximize output.

$$I_{sc}(T_c, E_{DNI}, E_{DIF}, AM_a, AOI) = \qquad (1)$$
$$[(E_{DNI}/850) * f_1(AM_a) * f_2(AOI) + C_6 E_{DIF}/150] *$$
$$\{[I_{sc_o} / (1 + C_6)] + \alpha_{isc}(T_c - 50°C)\},$$

where $I_{sc_o} = I_{sc}(50°C, 850$ W/m², 150 W/m², 1.5, $0°)$,

$$E_e(E_{DNI}, E_{DIF}, AM_a, AOI) \equiv$$
$$I_{sc}(T_c = 50°C, E_{DNI}, E_{DIF}, AM_a, AOI) / I_{sc_o}, \qquad (2)$$

$$I_{mp}(E_e, T_c) = C_1 + E_e * [C_2 + \alpha_{imp}(T_c - 50°C)], \quad (3)$$

$$V_{oc}(E_e, T_c) = V_{oc_0} + C_3 \ln(E_e) + \beta_{voc}(T_c - 50°C), \quad (4)$$

$$V_{mp}(E_e, T_c) = V_{mp_0} + C_4 \ln(E_e) + C_5[\ln(E_e)]^2 + \beta_{vmp}(T_c - 50°C). \quad (5)$$

The inclusion of the term $1/(1 + C_6)$ in equation (1) normalizes the effect of any diffuse irradiance in I_{sc_0} [4].

The sequence of analysis should be as follows:

a) Obtain or estimate a module-level temperature coefficient for $\beta_{voc} = \Delta V_{oc} / \Delta T$ for modules or cells. This can be done using the procedure presented in ASTM E 1036 Annex A2, which describes measuring V_{oc} at different temperatures under a solar simulator and computing the change in V_{oc} divided by the change in temperature at a reference temperature and illumination. Because β_{voc} varies with insolation and temperature, cell β_{voc} values need to be determined at the operational concentration levels, not at one-sun. As part of the analysis, a refined estimate of β_{voc} will be obtained, but an initial estimate is required to develop I_{sc_0} and to compute T_c from scan measurements. Because of the non-linear nature of Equation 1, values for α_{isc} cannot be assumed. The coefficients α_{isc}, α_{imp}, and β_{vmp} will be obtained as part of the analysis procedure.

b) Compute air mass (AM) and absolute air-mass (AM_a) using:

$$AM(Z_s) = \{\cos(Z_s) + 0.50572 [96.07995 - Z_s(degr)]^{-1.634}\}^{-1}, \quad (6)$$

$$AM_a(Z_s, P/Po) = (AM) * P/Po,$$
where $P/Po = e^{-0.0001184*h}$ (with h in meters). (7)

Air mass is the relative path length that the sun's rays traverse through the atmosphere before reaching the ground. Z_s is the zenith angle of the sun, which can be calculated given the site location, the day of the year, and the time of day. Absolute air mass is "pressure corrected" to compensate for altitudes, h, other than sea level. The air mass variable may be considered to be a proxy variable for spectrum effects.

c) Determine the cell temperature. If cell temperature is not measured, it can be computed using one of two methods. Equation 8 provides a method to compute cell temperature directly from the V_{oc} and I_{sc} measured in each I-V scan. It uses β_{voc} determined in Step a and the corresponding values V_{oc_0}, I_{sc_r}, and T_r taken at an arbitrary reference condition when the module is in thermal equilibrium. It is useful to pick T_r as the reference-cell temperature used for standard conditions. T_r is measured in degrees Kelvin.

$$T_c = \{[(1 / N)(V_{oc} - V_{oc_r})] + \beta_{voc}T_r\} / \{[(nk / q) \ln(I_{sc} / I_{sc_r})] + \beta_{voc}\}, \quad (8)$$

where:

N	=	Total number of cells connected in series in the array
I_{sc}	=	Measured array short-circuit current, A
V_{oc}	=	Measured array open-circuit voltage, V
I_{sc_r}	=	Array short-circuit current at the reference temperature, A
V_{oc_r}	=	Array open-circuit voltage at the reference temperature, V
T_c	=	Temperature of cells inside module, K
T_r	=	Arbitrary reference temperature for cells, K
β_{voc}	=	Temperature coefficient for V_{oc} for individual cell, V/°C
n	=	Cell diode factor (n=1 can be assumed for typical silicon cells)
k	=	Boltzmann's constant, 1.38066×10^{-23} J/K
q	=	Elementary charge, 1.60218×10^{-19} C

In the above definitions, the array or module values should be used as required by the situation. Because β_{voc} varies over wide temperature ranges, this method is probably not as accurate for wide ranges of T_c as a second method. This alternate method, given by Equation 9, computes cell temperature from a measured heat-sink temperature, T_{hs}. The differential-temperature constant, C_7, is computed using Equation 10 with measured values of T_{hs}, E_{DNI}, and E_{DIF}, and a value of T_c computed using Equation 8. This is done using data measured with conditions near the reference condition that was used to compute β_{voc}. Constant C_7 represents the temperature difference between the cell and heat sink per unit of insolation at 850 W/m^2 direct insolation and 150 W/m^2 diffuse insolation. For flat plate modules, C_7 is generally about 3 degrees C per 1000 W/m^2 total insolation. For concentrators, C_7 can range from 3 to 10 degrees C per 1000 W/m^2 insolation, depending on the module design.

For estimation of the initial value of C_6, set $C_6 = 1 / X$.

$$T_c = T_{hs} + C_7 * (E_{DNI} + C_6E_{DIF}) \quad (9)$$

$$C_7 = (T_c - T_{hs})/(E_{DNI} + C_6E_{DIF}) \quad (10)$$

Equation 11 provides a relationship between heat-sink temperature and ambient temperature, insolation, and wind speed. (Cell temperature can be substituted for heat-sink temperature in this relationship.)

$$T_{hs} = T_{amb} + (E_{DNI} + C_6E_{DIF}) * [C_8 + C_9 * \exp(-C_{10} * V)] \quad (11)$$

Constants C_8, C_9, and C_{10} are computed by performing a regression analysis of Equation 11, where V is wind speed in m/s.

d) The influences of the parameter C_6, the air-mass function $f_1(AM_a)$, and the angle of incidence function, $f_2(AOI)$, cause Equation 1 to be non-linear. A non-linear regression method is required to properly estimate these effects. For the functions employed in this model, the iterative, non-linear regression techniques found in Bevington and Robinson readily allow rapid estimation of the parameters [6]. A spreadsheet can be designed to perform the required iteration. An estimate of goodness of fit and of the parameter uncertainties can also be made. The air-mass function, $f_1(AM_a)$ and angle of incidence function, $f_2(AOI)$, can be expanded as a one- or two-term power series as follows:

e) Compute amz = $[AM(Z_s) - 1.5]$ for each data point. Assume the expansion of $f_1(AM_a)$ as follows:

$$f_1(AM_a) = 1 + C_{11} * amz + C_{12} * amz^2 \qquad (12)$$

The air-mass function is analogous to the spectral mismatch parameter. It is the ratio of the module short-circuit current to the output of the solar irradiance sensor (Eppley thermopile, broadband, normal-incidence pyrheliometer, normal-incidence pyranometer) as a function of air mass.

The functional form for $f_2(AOI)$ is assumed to be as follows:

$$f_2(AOI) = 1 + C_{22} * (AOI/90)^2 + C_{24} * (AOI/90)^4 \qquad (13)$$

The angle of incidence is either measured from the orientation of a 2-axis tracker measurement platform with respect to the sun or using local sun-module orientation calculations based on module deployment, location, and time data [2].

f) Perform a non-linear curve fit of the following function to determine the parameters C_6, C_{11}, C_{12}, I_{sc_0} and α_{isc}:

$$I_{sc}(T_c, E_{DNI}, E_{DIF}, AM_a, 0°) = [(E_{DNI}/850) *$$
$$(1 + C_{11} * amz + C_{12} * amz^2) +$$
$$C_6(E_{DIF} / 150)] *$$
$$[I_{sc_0} / (1 + C_6) + \alpha_{isc}(T_c - 50°C)] \qquad (14)$$

g) Compute $I_{sc}(T_c = 50°C, E_{DNI}, E_{DIF}, AM_a, AOI)$ using Equation 15:

$$I_{sc}(Tc = 50°C, E_{DNI}, E_{DIF}, AM_a, AOI) =$$
$$I_{sc}(measured) - \alpha_{isc}(T_c - 50°C), \qquad (15)$$

then compute E_e, the effective irradiance, for each data point using Equation 2.

h) Perform a regression on Equation 15 to identify C_1, C_2, and α_{imp} for Equation 3:

$$I_{mp}(meas) = C_1 + E_e * [C_2 + \alpha_{imp} (T_c - 50°C)]. \qquad (16)$$

Compute $I_{mp_0} = C_1 + C_2$.

i) Perform a regression on Equation 17 to identify V_{oc_0}, C_3, and β_{voc} for Equation 4:

$$V_{oc}(meas) = V_{oc_0} + C_3 \ln(Ee) + \beta_{voc}(T_c - 50°C). \qquad (17)$$

j) Perform a regression on Equation 18 to identify V_{mp_0}, C_4, C_5, and β_{vmp} for Equation 5:

$$V_{mp}(meas) = V_{mp_0} + C_4 \ln(E_e) + C_5[\ln(E_e)]^2 +$$
$$\beta_{vmp} (T_c - 50°C). \qquad (18)$$

k) Finally, compute P_{maxo} using Equation 19:

$$P_{max_0} = V_{mp_0} * I_{mp_0}. \qquad (19)$$

Once all the functions and coefficients are obtained, Equations 1, 3, 4, and 5 are used to compute I_{sc}, V_{mp}, I_{mp}, and V_{oc} for the measured data. These can then be used to compute the maximum power by multiplying I_{mp} by V_{mp}.

3. RESULTS

As an illustration of the method, a PVI 20" linear focus, Fresnel lens concentrator module was measured over several days on a 2-axis tracker at different orientations. Figure 1 shows $f_2(AOI)$ measured on the tracker.

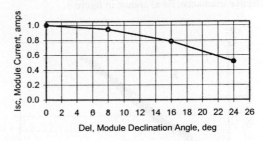

Figure 1: Module response function, $f_2(AOI)$

Table 1 shows the calculated coefficients for a similar module measured on the PVI rooftop array

Table I: Measured parameters for a rooftop module

Parameter	Coeff	Sigma
Isc_{oo}	10.74	0.07
C_6	0.037	0.002
a_{22}	-6.89	0.04
a_{24}	0.00	
a_{12}	-0.086	0.012
a_{12}	0.025	0.004
α_{isc}	0.00	

The tilt angle of the array was 30°, the local latitude 36.5° and the array points 11° to the west of south. The values from each assessment, test tracker vs. rooftop modules, compare well. The contribution from diffuse radiation is small. The coefficient of 0.037 corresponds to a diffuse contribution of slightly less than 0.6%. The temperature coefficient is zero. The angle of incidence values correspond very well. The data used in the non-linear calculation was filtered to remove points when the modules partially shadowed each other. For a complete formulation of the model a multiplicative shadowing term needs to be added to equation (1).

Figure 3 shows the calculated air mass response function, $f_1(amz)$, as a function of reduced air mass, amz.

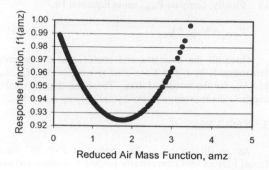

Figure 3: Module Isc as a function of effective irradiance

The short circuit current is shown to be linear in the effective irradiance, Ee as shown in figure 4.

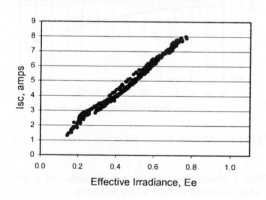

Figure 4: Module Isc as a function of effective irradiance

Finally in Figure 5 we show the measured and calculated values of Pmax. The calculated values were obtained using equation 19. The correspondence between measured and calculated values is good. The data for this analysis was collected in November and the module output peaks in the mid-morning around 10:00 local solar time when the angle of incidence is closer to 0 degrees. The fall-off in output during the afternoon reflects the high angle of incidence because of three factors: low sun declination, west of south orientation of the array and less-than-latitude tilt of the array. The model accounted for these three factors very well.

4.   DISCUSSION AND CONCLUSIONS

Improved prediction of concentrator array performance has been demonstrated. The methods used in this paper can be used to obtain required module characteristics to be incorporated into module energy rating algorithms and in module qualification methods.

Much further work remains to be done to qualify all aspects of the model, particularly the temperature components of the methodology.

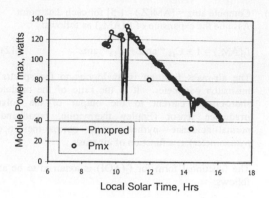

Figure 5: Measured and predicted values of Pmax for roof top module

ACKNOWLEDGEMENTS

The author thanks Alex Maish, Bob McConnell and Sarah Kurtz for valuable discussions in the development of the proposed model.

REFERENCES

[1]   D. King, et al., "Field Experience with a New Performance Characterization Procedure for Photovoltaic Arrays", 2nd World Conference and Exhibition on Photovoltaic Solar Energy Conversion Proceedings, Vienna, Austria (1998) pp 1947-1952.

[2]   C. M Whitaker, and J. D. Newmiller, "Photovoltaic Module Energy Rating Procedure", Final Subcontract Report, NREL/SR-520-23942 (1998)

[3]   B. Marion, et al. "Validation of a Photovoltaic Module Energy Ratings Procedure at NREL", NREL Technical Report, NREL/TP-520-26909 (1999)

[4]   ASTM Standard E 1036, American Society for Testing Materials (1990)

[5]   Personal communication from Alex Maish.

[6]   P. R. Bevington, and D. K. Robinson, "Data Reduction and Error Analysis for the Physical Sciences", McGraw-Hill, New York (1992)

AMORPHOUS SILICON MODULES ON FACADES WITH NON-OPTIMAL ORIENTATIONS

N.J.C.M. van der Borg and E.J. Wiggelinkhuizen
Netherlands Energy Research Foundation ECN
P.O. Box 1; 1755 ZG Petten The Netherlands
E-mail vanderborg@ecn.nl; fax +31 224 564976

ABSTRACT

An existing office building has been retrofitted with a PV-system on each of the four vertical façades. The façades are facing north, south, east and west. The applied system consists of single-junction amorphous silicon modules. The paper presents monitoring data obtained from August 1999 to April 2000. Although the monitoring period is not yet adequate for a complete assessment of the performance of the PV-systems, some conclusions can be already drawn.
Keywords: Amorphous silicon - 1: Vertical façades - 2: Monitoring - 3.

1. INTRODUCTION

An existing office building of ECN has been retrofitted with four identical PV-systems, each consisting of 42 amorphous silicon modules with a total nominal power of 500 Wp and a 700 W inverter. The PV-systems are positioned on vertical façades facing north, east, south and west respectively. The performance of the PV-systems is monitored since August 1, 1999. The aim of the monitoring programme is twofold: assessing the performance of the PV-systems and obtaining a set of data to be used for the validation of numerical models. This paper addresses the first item: the performance of the PV-systems.

The monitoring programme is supported financially by the Netherlands Agency for Energy and Environment (Novem).

2. PV-SYSTEMS

Each of the four PV-systems is made of 42 single-junction amorphous silicon modules connected as 6 strings of 7 modules. The modules are manufactured by Fortum (former Neste) and have a nominal power of 12 Wp. Each PV-system is connected to the grid through an SMA Sunny Boy 700 inverter with a nominal power of 700 W. Simple aluminium profiles have been mounted against the four vertical façades of the building in which the frameless modules are positioned. The total thickness of the construction is 9 cm. The façades of the building that are referred to as north, east, south and west have azimuth angles of -7, 83, 173 and 263 degrees with

respect to north. East of the building a neighbouring building is shielding part of the irradiance. Towards west the view is free from obstacles.

3. MONITORING SYSTEM

The monitoring system is based upon a measurement-PC connected to two groups of decentralised data acquisition units. One group of DAQ-units consists of five sub-units: one for data from the roof of the building and one for each of the four façades. These five DAQ-units are connected to the measurement-PC through a Local Operating Network (LON). The other group of DAQ-units consists of the four monitoring devices of the Sunny Boy 700 inverters, connected to the measurement-PC through an RS485/232 network.

The DAQ-unit of the roof measures meteo-data on the roof : horizontal global irradiance and its diffuse component (using two Kipp pyranometers of which one is equipped with a shadow ring), wind speed, wind direction and ambient temperature.

The DAQ-units of the façades measure the in plane irradiance (using a Kipp pyranometer and a reference cell), the module temperature and a number of wall temperatures. The reference cells are made of crystalline silicon with a filter glass (manufacturer Schott Nederland, type KG5). This combination of cell and filter results in reference cell with a spectral response similar to the spectral response of amorphous silicon modules. The reference cells have been calibrated in the ECN-laboratories.

The monitoring devices of the inverters measure the electrical quantities such as DC-voltage, DC-current and AC power.

All DAQ-units sample the measurement channels once per 10 seconds. The measurement-PC calculates the DC-power of the four inverters by multiplying the DC-voltage with the DC-current and then it calculates during consecutive measurement periods of 10 minutes the averaged values of all measurement channels. The 10-minute averaged data are stored for subsequent off-line evaluation.

4. MONITORING RESULTS

4.1. Irradiation

The monthly irradiation, obtained by integration of the irradiance data from the pyranometers on the façades and on the roof, is presented in figure 1.

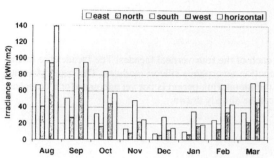

Figure 1 Monthly irradiation on the facades and on the roof (pyranometer data)

The irradiation values have been determined using the reference cells on the façades as well. The ratio between the results from the pyranometers and the reference cells are given in figure 2. This figure shows also the diffuse fraction of the horizontal irradiance.

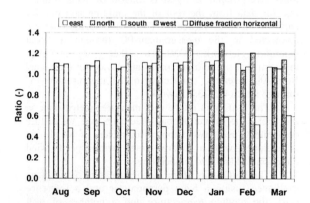

Figure 2 Ratio between the irradiance measured with the pyranometers and with the reference cells plus the diffuse fraction of the horizontal irradiance

Figure 2 shows significant differences in irradiation data from the pyranometers and the reference cells. These differences are caused by differences in the response of the reference cell and the pyranometer for non-normal incidence angles and for deviations from the Air Mass 1.5 spectrum. This is illustrated in figure 3, showing a time series of the irradiance on the west-façade on March 22. Figure 3 gives the irradiance measured with the pyranometer and the ratio between the pyranometer data and the reference cell data. A large difference is observed around noon, caused by the grazing angle of incidence on the west-façade. Also at the end of the day a large difference is observed. To investigate this the spectrum has been measured on the west-façade on March 22 on 13.30 h and on 17.45 h. The spectra, normalised to a total irradiance of 1 kW/m^2 below 900 nm, are given in figure 4.

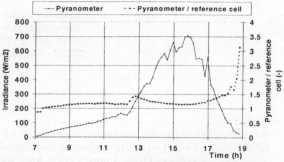

Figure 3 Irradiance on the west facade(22/3/00)

Figure 4 shows also the standard (AM1.5) spectrum and the response of a-Si in that spectrum. A shift of the spectrum towards red at the end of the day is observed. This part of the spectrum contributes to the irradiance measured by the pyranometer but remains unnoticed by the reference cell, having the main spectral characteristics of a-Si.

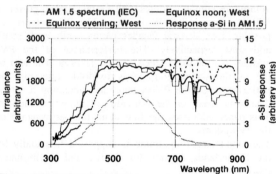

Figure 4: Normalized spectra

The observed difference between the data from the reference cell and from the pyranometer leads to the question which is the proper irradiance sensor to be used in monitoring programmes of a-Si systems. The answer to this question depends on the purpose of the monitoring. In case the monitoring focuses on spotting possible anomalies in the PV-system (such as partial shading effects, ageing of the modules, malfunctioning strings etcetera) a reference cell is more appropriate. In case the monitoring is intended to gather performance data to be used in the prediction of the energy production of similar systems on other locations a pyranometer will be more appropriate since irradiance data on various locations are (almost) always obtained with pyranometers.

4.2. Performance

The performance ratios of the four PV-systems have been calculated using monthly irradiation data from the reference cells and the monthly AC-energy from the inverters. The used value for the array power at STC is 435 Wp. This value has been obtained by measuring the IV-curve of the array on the south-façade (figure 5).

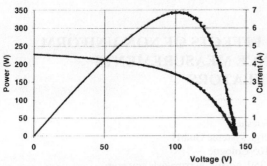

Figure 5 IV-curve at 42 º C and 820 W/m2

The PR-values are given in figure 6. This figure shows low PR-values for the east- and west-façades during the winter period and for the north-façade during all monitored months. This is caused by the low conversion efficiencies of the inverters due to the severe partial load operation.

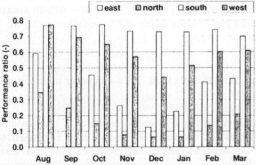

Figure 6 Monthly performance ratios

The efficiencies of the four arrays have been measured as a function of the irradiance. No significant differences have been observed between these four curves. Unfortunately the DC-current data obtained from the monitoring device of the inverters proved to have large measurement errors. A correction has been made for this by an in-situ calibration of the monitoring device of the inverter on the south-façade. After this correction the efficiency of the south-array has been determined as presented in figure 7.

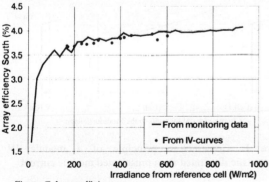

Figure 7 Array efficiency

Figure 7 shows also efficiency data obtained with some short measurement campaigns with an IV-tracer. With the efficiency curve of figure 7 and the monthly measured frequency distribution of the irradiance on the south- and north-façade the monthly DC-energies have been calculated. This resulted in the array-efficiencies in figure 8. If the inverter of the individual PV-systems would have been optimised to the actual power of the arrays, rather than chosen the same for all orientations, the range of the PR-values of 0.1 to 0.8 (figure 6) would be limited to a range of 0.4 to 0.8.

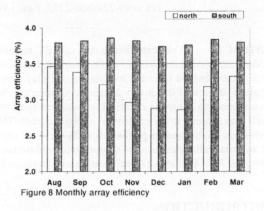

Figure 8 Monthly array efficiency

5. CONCLUSIONS

- A detailed set of measurement data is becoming available for the validation of models for irradiance and performance of a-Si modules on various façades

- Differences are observed between the response of the reference cells, designed for monitoring a-Si modules, and pyranometers. The differences are most significant during grazing incidence angles and at moments with spectra deviating from AM 1.5. The reference cells facilitate the assessment of the PV-systems with respect to anomalies and the pyranometers facilitate the prediction of energy production of similar systems at other places.

- The performance ratio for the PV-systems ranges between 0.8 (south) and 0.1 (north in the darkest month). A more appropriate choice of the inverters nominal power would improve the performance ratio very much.

6. ACKNOWLEDGEMENT

This work has been supported by the Netherlands Agency for Energy and Environment NOVEM.

MODELLING OF PV MODULES - THE EFFECTS OF NON-UNIFORM IRRADIANCE ON PERFORMANCE MEASUREMENTS WITH SOLAR SIMULATORS

W. Herrmann, W. Wiesner

TÜV Immissionsschutz und Energiesysteme GmbH

Am Grauen Stein, D-51105 Cologne

Tel.: +49-221-806-2272, Fax: +49-221-806-1350, e-mail: herrma@de.tuv.com

ABSTRACT: The performance data of a PV module are derived from the current-voltage characteristic measured under standard test conditions (STC). For this purpose, test laboratories as well as module production plants employ solar simulators which, depending on type and module size, display a more or less pronounced non-uniform distribution of irradiance in the test plane. In view of the quality assurance for solar simulator test facilities and of the measuring and manufacturing tolerances, it is important to know the effects of non-uniformity of irradiance on the module current-voltage characteristic, and to take the effects into account in the calibration of light sources. This study presents the fundamental effects of a non-uniform irradiance distribution on the module performance. By means of computer modelling the specific influences of the solar simulator system and of solar cell parameters are demonstrated and measurements methods for performance rating of PV modules developed.

Keywords: PV-Modules - 1: Qualification and Testing – 2: Simulation -3.

1. INTRODUCTION

The extent of different irradiance levels at different cell locations in the test plane of a solar simulator is described in terms of the degree of non-uniformity ω_G, which the Test Standard IEC 60904-9 on „Performance Requirements on Solar Simulators" defines as

$$\omega_G = \frac{G_{MAX} - G_{MIN}}{G_{MAX} + G_{MIN}} \cdot 100\%$$

G_{MAX}:   cell location with maximum irradiance
G_{MIN}:   cell location with minimum irradiance

The following questions are posed by the problems of performance measurements on PV modules by means of solar simulators:

a) How is the non-uniform distribution of irradiance related to the measurement accuracy for the performance parameters of the module?

b) Do there exist solar cell related parameters affecting the characteristic?

c) Which reference value of the irradiation is best suited for rating the performance of a PV module at known irradiance level distribution?

2. COMPUTER SIMULATION OF THE CURRENT-VOLTAGE CHARACTERISTIC

The current-voltage characteristic of a PV module can be formed from the current-voltage characteristics of individual solar cells, if the type of serial/parallel connection and protective switching with bypass diodes are taken into account. Crucial for the computer simulation is that the current-voltage characteristic of a solar cell is considered across the entire voltage range. Compared with the forward range, which for crystalline solar cells is limited by the open circuit voltage of approx. 0.6 V,

voltages of up to more than –20 V can occur in the reverse range /1/.

Fig. 1 illustrates the design principle of the module characteristic in the case of 3 solar cells connected in series. For a preselected module current I_M the individual cell voltages are summed according to the operating points on the solar cell characteristics, yielding the module voltage. Cell shading has a considerable effect on the total voltage if the module current I_M exceeds the short-circuit current I_{SC} of this cell and an operating point is generated on the reverse characteristic at this cell. This negative voltage can exceed the positive contributions from the other cells many times over.

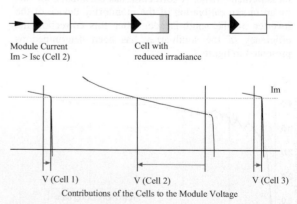

Figure 1: Construction principle of a module IV characteristic illustrated for a preselected module current

The level of the negative voltage contribution of a individual cell due to inadequate irradiation depends on two factors:

- The degree of reduced irradiance/cell shading: in the graph shown in Fig. 1, increasing shading corresponds to a parallel displacement in the direction of current. Thus a higher shading rate means a greater negative voltage.

• The slope of the reverse characteristic: a solar cell with a high blocking behaviour and thus a flat reverse characteristic, will cause a greater negative voltage contribution at the same degree of shading.

On the basis of this design principle, the computer simulation program COSIMO (COmputer SImulation of PV MOdules) was developed for calculating the current-voltage characteristic of a PV module for freely adjustable non-uniformity profiles of irradiation and module temperature. Fig. 2 shows the variables that must be taken into account when computing the current-voltage characteristic of a module. The present study is limited to the reverse properties of the solar cells and the irradiance level distribution (boxes 1 and 3).

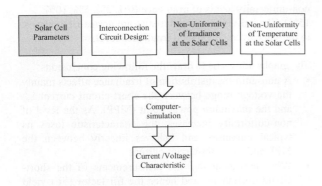

Figure 2: Computer simulation with COSIMO: different variables affecting the current-voltage characteristic of a PV module

The calculation algorithm of computer modelling is based on the following principles:

• The forward characteristic is described by the 2-diode-model of the solar cell. Beside the parallel resistance the modelling parameters are the same for all cells.

• The reverse operational behaviour of a group of solar cells cannot be uniformly described. As shown in Fig. 3, the reverse characteristics are subjected to scattering which is a characteristic of the type of cell.

Studies on the reverse operational behaviour of the most important commercially available solar cells have shown that the reverse characteristics in the range relevant to practical applications can be approximated by a straight line, the slope of which approximately corresponds to the parallel resistance of the cell /1/. For example, bypass diode protective switching across 24 cells will limit the reverse voltage to approx. -13 V. Computer simulation therefore does not require that an individual reverse characteristic be described in its entirety by means of a complicated function with 3 modelling parameters, as has been proposed in the literature /2,3/. It suffices to specify a set of parameters for the parallel resistance of the cells. For the types of solar cells considered, Fig. 4 shows the dispersion width of the measured reverse characteristic fields at a fixed reverse voltage of −10V. It can be clearly seen that the reverse behaviour among the different cell types varies greatly and consequently that the effects on the module characteristic must also vary in the case of non-uniform irradiance.

The special feature of computer simulation with COSIMO is that the individual reverse behaviour of the solar cells is taken into account and that each cell can be assigned its own irradiance level and temperature. The simulation program was experimentally confirmed by comparing the measurement and simulation results for preselected irradiance level distributions. The inaccuracy of the simulation is max. ±2%.

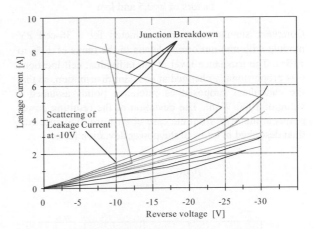

Figure 3: Reverse characteristic field of a batch of 10 solar cells of the same type. In the range down to −13 V relevant for the practical use of a module the reverse characteristics are approximately linear.

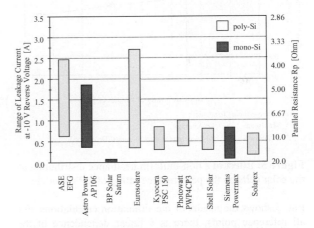

Figure 4: Comparison of the reverse operational behaviour of commercially available solar cells based on the dispersion width of the reverse characteristic field at a reverse voltage of −10 V.

3. RESULTS OF THE COMPUTER SIMULA-TION FOR NON-UNIFORM IRRADIANCE

3.1 REFERENCE POINT OF THE IRRADIANCE LEVEL

The quantitative description of the effects of non-uniform light on the output of the module raises the question of the reference point for the performance rating. Given an irradiance level distribution $(G)_i$ at the individual solar cells of the module, the following variables are candidates for the reference or effective value G_{EFF}:

a) $\overline{G}$:   average value of the irradiance level
                              for all cells
b) G_{MIN}:   minimum   irradiance   level   at   the
                              module surface
c) G_{MAX}:   maximum   irradiance   level   at   the
                              module surface
d) $\overline{G} - k \cdot \sigma_G$:   average value for all cells relative to
                              the standard deviation σ_G of the
                              irradiance level, with assumed linearity
                              factors of k=0.5 and k=1

Computer simulations were conducted for a 36-cell PV module with non-uniformity levels in the range of ±2% to ±15%. The irradiance level at the individual cell locations was proportionally changed at constant distribution, so that the candidates proposed as reference points assumed a value of 1000 W/m². The deviation of the resulting power points at the MPP from the case of uniform irradiance was then described in terms of the power deviation factor

$$PDF = \frac{P_{non-uniform}}{P_{uniform}}$$

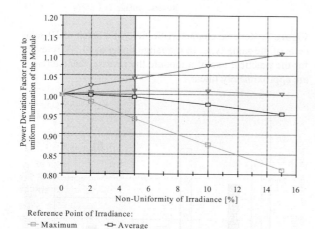

Reference Point of Irradiance:
- Maximum
- Minimum
- Average
- (Average minus Standard Deviation)

Figure 5: Effect of a non-uniform irradiance distribution at the solar cells on the performance rating of a PV module.

Fig. 5 shows the results of the simulation calculations. For all reference points, there is a linear dependence of the PDF from the level of non-uniformity. In the range ω_G <5% the average value of the irradiance $\overline{G}$ is the most suitable reference point for rating the performance of a module and hence for setting the light intensity of a solar simulator. The maximum and minimum at the module surface lead to a clear underrating and overrating, respectively, of the actual module performance.

3.2 EFFECT OF THE MODULE PERFORMANCE PARAMETERS

Fig. 6 displays influence of an increasing non-uniformity of irradiance on the module performance parameters based on computer simulation for a 36-cell PV module. As a result of the above considerations, all IV-characteristics have been normalised to $\overline{G}$ =1000 W/m² as the reference point of the irradiance level.

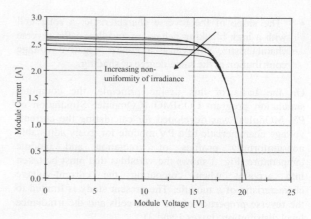

Figure 6: Modelled module characteristics for different non-uniformity levels of irradiance (0%, 2%, 5%, 10%, 15%). The individual characteristics are normalised relative to the same average value of irradiance level.

The graphs allows us to draw the following conclusions:
- A non-uniform distribution of irradiance affects mainly the voltage range between the short-circuit current I_{SC} and the maximum power point (MPP). As the level of non-uniformity increases, the characteristic loses its typical curvature and behaves linearly between the MPP and the short-circuit current.
- With increasing ω_G, the measurements of the short-circuit current I_{SC} and hence the fill factor (FF) yield false results. The short-circuit current is underrated and the fill factor consequently overrated.
- The open circuit voltage remains unaffected by the non-uniformity of the light.

3.3 EFFECTS OF THE REVERSE BEHAVIOUR OF SOLAR CELLS

Fig. 7 shows the simulated module characteristics of a 36-cell module, in which a single cell was subjected to a cell shading of 25%, while the shunt resistance was changed. The diagram clearly shows that the behaviour of the resulting module IV-characteristic is basically determined by the reverse behaviour of the cell. The better the reverse properties (greater shunt resistance), the greater is the effect on the module characteristic.

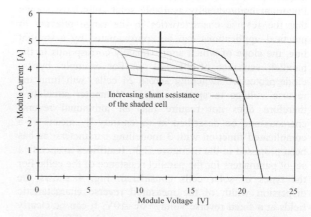

Figure 7: Effect of the reverse behaviour of a solar cell on the current-voltage characteristic of a 36-cell PV module with 25% shading

But what relationships obtain if not just one cell is shaded, but the states of illumination vary, as in the case of non-uniform irradiance? This question was considered for the cases of two cell types, *ASE-EFG* and *BP-Saturn*, which greatly differ in their reverse behaviour (see Fig. 3). Table 1 compares the basic reverse properties.

Cell type	I_{rev} (-10 V) lower limit	I_{rev} (-10 V) upper limit	Isc	$\Delta I_{rev}/I_{Sc}$
ASE-EFG	0.65 A (R_P=15.4 Ω)	2.45 A (R_P=4.1 Ω)	3.1 A	58.1%
BP Saturn	0.05 A (R_P=200 Ω)	0.1 A (R_P=100 Ω)	4.8 A	1.1%

Table 1: Reverse characteristic parameters as input variables for the computer simulation

Figs. 8 and 9 give the results of computer simulation calculations for different irradiance level profiles. In the simulation of the module characteristic with COSIMO, the correspondence (irradiance level at the cell location) $\leftrightarrow$ (parallel resistance of the cell) was determined by a random-number generator, with the R_P ranges from Table 1 being taken together to form the value range. We see that solar cells with high parallel resistance and consequently a flat reverse characteristic have a basically stronger effect on the module performance parameters compared with the case of uniform irradiance.

Whereas for cells with a large dispersion in the reverse characteristics (ASE-EFG) the deviations of the performance parameters for $\omega_G < 5\%$ lie in the range of 1%, the simulation for cells with extremely good reverse behaviour (BP Saturn) leads to deviations of up to 3% for short-circuit current and fill factor. In both cases the deviation of maximum power lies in the range 1%.

4. CONCLUSIONS

<u>Comparability of measurement results from different test facilities:</u> For rating the precision of a solar simulator test facility, the irradiance profile should be documented in addition to the level of irradiance when the performance of a PV module is measured. In the case of cells with good reverse behaviour, the short-circuit current and the fill factor are clearly affected by measurement errors even at low levels of non-uniformity.

<u>Application range of solar simulators:</u> The critical level of non-uniformity for performance measurements on PV modules by means of solar simulators can be specified as $\omega_G = 5\%$, which corresponds to the IEC specification for the classification of class B solar simulators. Below this value, the measurement error regarding the MPP is max. 1%, if the average value of the irradiance measured at the cells is used as a reference point for setting the light intensity. If a level of non-uniformity below 5% cannot be achieved (large modules), then qualifying measurements should conducted only with outdoor tests.

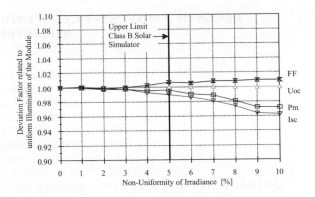

Figure 8: Change in the module performance parameters for cell type: ASE-EFG

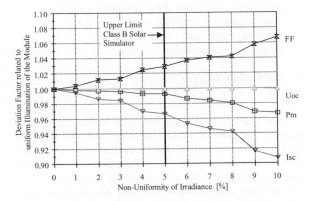

Figure 9: Change in the module performance parameters for cell type: BP Saturn

<u>Quality assurance in module production:</u> If a module has been calibrated by a test laboratory using solar simulator 1, and now serves as the reference module for quality assurance in the module production by means of simulator 2, then the calibration is valid only for modules of the same design with the same design of the cell interconnection circuit. Owing to the difference in lighting conditions, the irradiance level must be adjusted not to the short-circuit current, but to the calibrated module output at the MPP.

5. ACKNOWLEDGEMENT

This work was supported under contract No. 254 112 96 by "AG Solar", which is a financial support programme of the State of North Rhine-Westphalia in the sector of renewable energies and buildings.

REFERENCES

/1/ Herrmann W., Wiesner W., "Operational behaviour of commercial solar cells under reverse biased conditions", 2nd WCPEC, Vienna, 1998, pp. 2357-2359
/2/ Quaschning V., "Numerical simulation of current voltage characteristics of photovoltaic systems with shaded solar cells", Solar Energy **56**, 1996, pp. 513-520
/3/ Bishop J.W., "Computer simulation of the effect of electrical mismatches in photovoltaic interconnection circuits", Solar Cells **25**, 1988, pp. 73-89

DESIGN OF A NOVEL STATIC CONCENTRATOR LENS UTILISING TOTAL INTERNAL REFLECTION SURFACES

N.C. Shaw and S.R.Wenham
Centre for Photovoltaic Engineering,
University of New South Wales,
Sydney 2052, Australia.
Phone : +61-2-93855471   Fax:+61-2-93855412
Email: n.shaw@unsw.edu.au, s.wenham@unsw.edu.au

ABSTRACT: This work describes the operation a non-imaging static concentrator lens utilising refraction and total internal reflection with a geometrical concentration ratio of 2. Analysis with computer ray-tracing indicates a lens efficiency of 94% for collection of the solar energy in Sydney, Australia. The annual averaged optical concentration ratio is 1.88 for direct insolation within $\pm60^0$ in the East-West and $\pm25^0$ in the North-South directions, referenced to the normal of the lens. Peak optical concentration ratio for an ideal lens is 1.91. Analysis modelling indicates that non-imaging static concentrators utilising refracting and total internal reflecting surfaces can exhibit high optical efficiency and concentrate solar radiation over large acceptance angles.
Keywords: Concentrator – 1: Modelling – 2: Optical Properties – 3:

1. INTRODUCTION

The use of photovoltaic modules incorporating non-imaging static concentration has long been considered a possible solution for cost reduction in crystalline photovoltaic modules. Several designs have been proposed and developed [1,2,3] with reasonable success. These concentrators operate at low irradiance gain levels of 1 to 3 suns and posses high acceptance angles of solar radiation. Utilisation of low level optical concentration lenses in current crystalline photovoltaic modules can be beneficial; 1. The low irradiance gain obtained from this type of concentration allows the use of smaller high efficiency bulk crystalline devices, of which one type is already being commercially produced [4], 2. The range of angular acceptance of solar radiation for a low-irradiance-gain concentrator module need not be compromised as severely as high-irradiance-gain modules and 3. The low concentration levels also mean that active cooling of devices is not required to maintain efficiency.

Cusp type concentrators [1], are an example of a simple design with geometrical concentration ratio of 2. This design has been investigated for it possesses easy integration into current crystalline PV module technology [5,6]. Refractor type concentrators posses moderate irradiance gains of 2 to 3 suns, and are simple to fabricate [3]. Wedge-shape type concentrators [2,7] are more complex designs with higher achievable irradiance gains, 3 to 6 suns, but are more complex to manufacture. With the exception of [3], all of the above mentioned concentrators, utilise some form of metallised reflecting surface in their designs. Due to the non-ideal reflectivity of metallised reflectors, optical losses are unavoidably introduced, thereby lowering the optimum performance achievable. In addition, the long-term reliability of such reflecting interfaces becomes an issue.

The use of total internal reflection surfaces within non-imaging static concentrators provides an ideal and reliable reflection mechanism that can be used in low-irradiance-gain concentrators. This paper will outline the development of a non-imaging static concentrator lens using refracting and total internal reflecting interfaces, that could be incorporated into current crystalline photovoltaic module technology.

2. LENS DESIGN

2.1 Solar Radiation Distribution.

Investigation of the annual hemispherical distribution of solar radiation provides a mechanism to determine what the angular acceptance criteria of a static concentrator lens should be to operate efficiently. By collecting and concentrating the higher-energy, directly incident solar radiation and deliberately avoiding collection of the relatively lower-energy diffuse solar radiation, higher energy yields are obtainable. To set the acceptance angles for which the concentrator lens would operate in, the angular intensity of hemispherical solar radiation for Sydney was obtained [8].

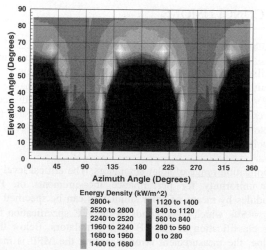

Figure 1. Plot of the hemispherical distribution of solar radiation (kW/m^2) that annually strikes a tilted plane in Sydney. The tilt angle of the plane is 34^0. The region of high-energy direct solar radiation extends over $\pm60^0$ in the East-West and $\pm25^0$ in the North-South axes from the plane's normal.

This distribution was chosen as Sydney's climate provides a large proportion of direct solar radiation over most days of the year, this type of climate is most suited to static concentration modules. The hemispherical distribution of solar energy is depicted in Figure 1

In this diagram, the collection plane's Azimuth Angle of 0^0, is orientated towards the equator, i.e. North in the Southern hemisphere, and the elevation angle component of the solar radiation distribution is referenced to the normal of the plane. Figure 1 indicates that the region of high-energy direct solar radiation is bounded by the angles of $\pm 60^0$ in East-West and $\pm 25^0$ in the North-South directions. These angles have therefore been set for the acceptance angles within which the lens must collect and concentrate direct solar radiation.

2.2 Design of a Trough Concentrator Lens.

Based on the angular acceptance criteria outlined in Sec 2.1, a solid trough-shape type concentrator with 2 dimensional light trapping was considered, Figure 2. The design exhibits an angular collection pattern to hemispherical solar radiation that is large in the longitudinal axis (East-West) and a smaller angular acceptance in the transverse axis (North-South).

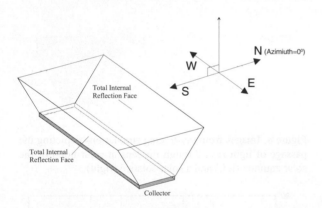

Figure 2. A solid trough-shape type concentrator lens indicating the orientation of the lens with respect to the incident solar radiation of Figure 1.

For confinement of all input light over the North-South acceptance angles, the orientation of the total internal reflecting interfaces is crucial. With reference to Figure 3, the angle θ must be chosen such that a ray of light striking the top surface of the lens at a declination angle of 25^0 (point M), will after refraction, strike the interface (point Y) at the critical angle for total internal reflection. For the case of the lens made from an optical material with refractive index of 1.5, such as glass or acrylic, this critical angle for reflection equates to 41.8^0, and θ was chosen to be 60^0.

2.2.1 Flat Entry Aperture.

Initially, the top entry aperture was designed to be flat and parallel to the collector surface, Figure 3. On subsequent analysis with raytracing simulations, it was discovered that rays striking the outer edges of this surface, at a declination angle of 25^0, were not confined within the lens. Figure 3, depicts 2 rays of light which are coupled

into the lens. Ray No.1, strikes the edge of the collection surface when reflected from the outer face of the lens at X. Ray No.2, and any other rays of light striking the top entry aperture between point L and the outer edge of this surface, will escape the lens near point Z. This loss mechanism occurred over 24.1% of the entry aperture area.

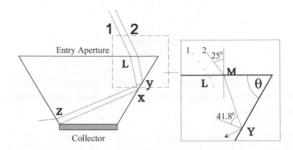

Figure 3. Cross section of the trough type concentrator lens indicating the total internal reflecting interfaces. The angle θ is chosen such that a ray of light striking the entry aperture at 25^0 to the normal will strike the reflecting interface at the critical angle for total internal refection.

2.2.2 Grooved Entry Aperture.

To increase the collection probability from the outer regions of the entry aperture, a structure is introduced which will re-orientate the incident light rays to strike the reflecting interfaces at an angle much greater than the critical angle. After subsequent reflection, these rays will successfully strike the collector surface. The initial modification to the entry aperture centred on incorporating asymmetric refracting grooves, extending along the longitudinal axis of the lens. A groove angle of 36^0, provides sufficient re-orientation to the input light. The profile of the grooves is depicted in Figure 4.

Due to the inclusion of this refracting structure, a different mechanism for loss of optical confinement within the lens was introduced. Light rays originating from a North-South acceptance angle boundary that strike the upper area of a groove, are refracted into the lens and strike the vertical wall of an adjacent groove. Here the rays are reflected by total internal reflection and redirected in such a way that they are coupled out of the lens, Figure 4.

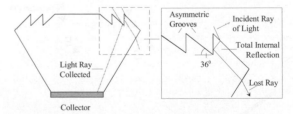

Figure 4. Profile of lens indicating the orientation of the asymmetric grooves and highlighting the loss mechanisms associated with the inclusion of asymmetric grooves.

A modification made to these grooves consisted of making the groove depth shallower, when the location of the groove is closer to the centre of the entry aperture. By choosing this design, grooves closer to the centre presented a smaller cross-sectional area for the reflection of light. This design reduced the severity of the side wall collection loss associated with the constant groove depth design. This is shown in Figure 5.

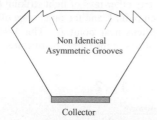

Figure 5. Modification to refracting grooves structure, indicating the varying depth with horizontal displacement of the grooves.

2.2.3 Curved Entry Aperture.

To remove the optical confinement losses associated with the grooved structure, a curved refracting structure was considered. The design of the structure possesses a quasi-curved surface formed from five flat surfaces, arranged along the longitudinal axis of the lens. The profile of the structure of depicted in Figure 6. The orientations of the facet angles are set to 36^0 and 19.94^0. Whilst not providing the low aspect ratio of the flat or grooved lenses, this surface provides complete confinement of input light rays over the North-South acceptance angles at an azimuth angle equal to 0 and 180 degrees.

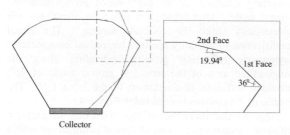

Figure 6. Quasi-curved entry aperture lens. This lens provides complete confinement of input light rays over the desired acceptance angles set in Sec 2.1.

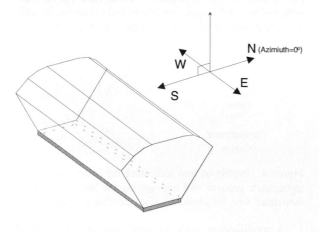

Figure 7. Quasi-curved refractor trough-type concentration lens indicating operating orientation with respect to the incident solar radiation distribution of Figure 1.

3. RESULTS OF COMPUTER MODELLING

3.1 Lossless Lens Modelling

Computer ray-tracing analysis was used to determine the hemispherical collection ability of the lenses with ideal optical transmission properties and complete absorption of light rays at the collector surface. The results of this modelling indicated that over the range of acceptance angles determined sec 2.1, the interaction of the quasi-curved refracting structure and the total internal reflection interfaces provided the highest collection and concentration ability of the four designs. Figure 8 shows images from ray-tracing simulations indicating the passage of light rays through the lens for normally incident light and for light striking the lens from a solstice condition at solar noon. The angular optical concentration ratio for this ideal lens in response to uniformly distributed hemispherical light was calculated, and the results are displayed in Figure 9.

When the distribution of solar radiation for Sydney (Figure 1) is used to determine the annual averaged irradiance gain, this lens possesses a concentration ratio greater than 1.80 for all direct solar radiation over the designed acceptance angles. Peak irradiance gain is 1.91 and occurs for normally incident direct light.

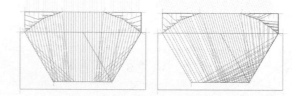

Figure 8. Images from ray-tracing simulations depicting the passage of light rays through the lens at solar noon of the solar equinox (left) and a solar solstice (right)

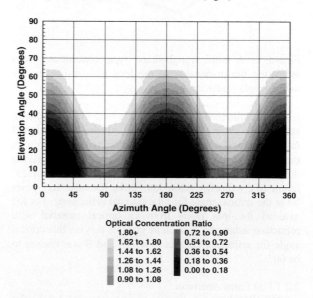

Figure 9 The angular distribution of optical concentration ratio for the ideal curved entry aperture lens of Sect. 2.2.3. This lens has an annual averaged irradiance gain of 1.88 over the acceptance angles set in Sect. 2.1.

3.2 Lossy Lens Modelling of Quasi-Curved Structure Lens.

The collection ability of this lens was considered for non-ideal optical transmission properties of the lens material. Suitable optical materials that the lens could be fabricated from, such as low iron glass [9] or acrylic polymers [10], posses average optical absorption coefficients between $(1.5 \times 10^{-3} - 5 \times 10^{-3})$ cm^{-1}, over 400nm to 1100nm wavelengths. Figure 10 indicates the averaged annual optical concentration ratio for this lens as the optical absorption coefficient is increased. The results displayed in Figure 10, were derived from the analysis of a lens with a depth of 50mm and a length of 200mm, these dimensions were chosen as they are comparable to current crystalline photovoltaic modules dimensions.

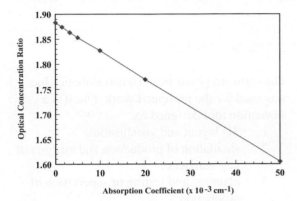

Figure 10. Optical concentration ratio for the curved entry aperture lens of Sect 2.2.3, as the optical absorption coefficient of the lens material is increased. Common optical materials, such as glass and acrylic, posses absorption coefficients of $(1.5 \times 10^{-3} - 5 \times 10^{-3})$ cm^{-1}.

The analysis concluded that the reduction in annual irradiance gain for this sized lens fabricated from glass or acrylic, was reduced from 1.88 to 1.85.

4. CONCLUSIONS

Non-imaging static concentrator lenses that operate without the need for metallised reflectors provide moderate concentration ratios with large acceptance angles. The work undertaken here indicates that over the acceptance angles of $\pm 60^0$ in the East-West and $\pm 25^0$ in the North-South axes, a refracting and total-internal-reflecting lens assembly with ideal optical properties possessed an annual optical concentration ratio of 1.88. Analysis further indicated that the irradiance gain is 1.85 when the lens size is compatible with current crystalline photovoltaic module dimensions and is fabricated from materials with non-ideal transmission properties such as glass or acrylic.

5. ACKNOWLEDGEMENTS

We would like to thank Jeff Cotter for providing the project with the raytracing software. The Photovoltaics Special Research Centre was supported by the Australian Research Council's Special Research Centres Scheme and by Pacific Power.

REFERENCES

[1]   A Geotzberger, Proceedings of the 20th IEEE Photovoltaic Specialist Confernece, Las Vegas, Ed (1988), 1333.

[2]   J.Parada, J.C.Miñano, J.L.Silva, Proceedings 10th EC Photovoltaic Solar Energy Conference, Lisborn, Ed (1991), 975.

[3]   K. Yoshioka, S. Goma, S. Hayakawa, T.Saitoh, Progress in Photovoltaics, 5, (1997), 139.

[4]   N.B. Mason, T.M. Bruton, K.C. Heasman, R.Russell, Proceedings 13th EC Photovoltaic Solar Energy Conference, Nice, Ed (1995), 2110.

[5]   U. Ortabasi. Proceedings 26th IEEE Photovolatic Specialist Confernece, Anaheim, Ed (1997), 1177

[6]   B.Mayregger, R. Auer, M. Niemann, A.G. Aberle, R. Hezel, Proceedings 13th EC Photovoltaic Solar Energy Conference, Nice, Ed (1995), 2377.

[7]   S.R. Wenham, S. Bowden, M. Dickinson, R. Largent, N. Shaw, C.B. Honsberg, M.A. Green, P.Smith, Solar Energy Materials and Solar Cells, 47, (1997), 325.

[8]   S. Bowden, Ph.D. Thesis, UNSW, 1996.

[9]   Bk-7 Data Sheet, Schott Glass Inc.

[10]   J.D. Lytle, G.W. Wilkerson, J.G. Jaramillo, Applied Optics, Vol. 18, No. 11, (1979), 1842.

3D SIMULATION OF SOLAR CELL PRODUCTION

Henrik Pettersson, Hovik Moosakhanian, Per Johander and Tadeusz Gruszecki
The Swedish Institute of Production Engineering Research (IVF)
Argongatan 30, S-431 53 Mölndal - SWEDEN
Tel. +46 31 706 60 00, E-mail: henrik.pettersson@ivf.se

ABSTRACT: We have illustrated and analysed a future large-scale production of dye photovoltaic cells for consumer electronics by simulating a full-scale production line in 3D. The dye solar cell technology has not yet entered production phase. However, since production methods have been developed, the 3D simulation of the designed production lines combined with thorough cost analysis creates a credible guideline for further development and investment for production of the dye photovoltaic cell. This simulation concept can be applied to other solar technologies and processes that are undergoing industrial development.
Keywords: 1: Simulation - 2. Fabrication - 3. Dye-sensitised

1. INTRODUCTION

Simulation enables the possibility to forecast problems and prevent them already in the planning phase of a production line. Today, simulation models and software which enable testing of products, production methods and manufacturing systems have been realised. New possibilities to optimise and improve throughput and quality of existing and designed products arise for companies that are able to apply simulations at an early stage in the production development process.

3D simulation of production lines is a tool for optimising the throughput and working routines for an existing production line. However, it is also an interesting method to foresee bottlenecks, find solutions to problems when addressing new techniques connected with processes that require unique solutions.

This possibility to illustrate and analyse a future production facilities has been applied to dye photovoltaic cells. This solar cell is not yet commercially produced, but manufacturing technologies for indoor applications have been developed and tested in manual work stations, and using in-line machines whenever they exist [1, 2]. The 3D simulation of the designed production line enables a visualisation of future production lines and visualise possible automated solutions to processes that are today manually performed and for which no known in-line production equipment exists.

2. THE SOFTWARE

The software *Quest* from Deneb Robotics Inc. was used for the presented work. *Quest* is a simulation tool designed for:

1. 3D layout and visualisation
2. simulation of production and analysis of key numbers
3. planning and testing of supervision of production

3. THE SIMULATION

Initially, 3D models of the required pieces of equipment were created, see Figure 1 and 2. The modules were then introduced into *Quest*. The production parameters of all equipment were set according to experimental data obtained and manufacturers specification. Finally, the different producing units were connected forming a production line (see Figure 3).

The simulated lines could then be used for studies of the influence of different parameters on the throughput. Due to the fact that many parameters were uncertain since there is no existing in-line machines for certain processes, the simulations served as illustrations rather than optimisations of the production flow.

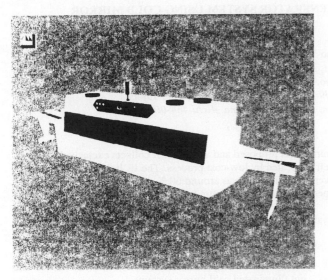

Figure 1   **3D model of washing machine**

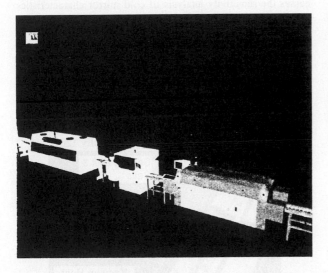

Figure 2   **Washing, screen-printing and drying processes**

- it served as an excellent illustration of automated solutions to manual processes
- it illustrated well processes for which no known in-line equipment exists
- it created a common visualisation for everybody involved in the project

It is important to be aware of the following possible pitfalls when dealing with 3D simulation of production lines for techniques that are under development:

- the simulation might simplify the state-of-the art and hide technological problems. Consequently, it might mislead people which lack the required background information
- modelling of machines may take a long time if there is no library for the equipment of interest

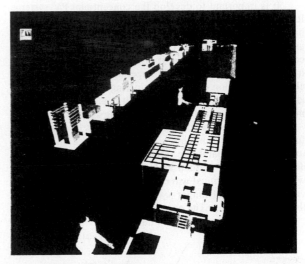

Figure 3   **Overview of one of the simulated production lines**

4. CONCLUSIONS

The simulation created, in combination with thorough cost analysis of the processes and materials, a credible guide-line for further development and investment for production of dye photovoltaic solar cells. This simulation concept can be applied on other solar technologies and processes that are undergoing industrial development. Especially the following points are worth mentioning:

- the simulation might dramatically cut down the costs and time for transfer from pilot line to full-scale production
- the 3D simulation provided very informative illustrations of non-existing processes

5. ACKNOWLEDGEMENTS

This work was partially supported by the European Commission, under contract number JOR-3CT97-0147.

6. REFERENCES

[1] European Commission funded project "Dye Photovoltaic Cells for Indoor Applications"

[2] A. Hinsch et. al., "Dye Sensitised Solar Cells: Large scale batch processing of mini modules for applications in consumer electronics", proceedings of the 11th International Photov. Science and Eng. Conference (Sapporo 1999)

ENERGY FLOW EXAMINATION ON PV GENERATOR SYSTEM USING COLD MIRROR

Kenji Araki, Masafumi Yamaguchi, Hisao Izumi*, Tetsuo Yamamoto**, Masao Hiramatsu***, Masanori Tsutsui***
Toyota Tech. Inst., 2-12-1 Hisakata, Tempaku, Nagoya 468-8511, Japan,
* IDEX., 297 Nakashinano, Seto, Aichi 480-12 Japan,
** Daido Steel Co., Ltd. 2-30 Daido-cho, Minami-ku, Nagoya, 457-8545 Japan,
*** Daido Metal Co., Ltd.Tendoh-Shinden, Maehara, Inuyama, Aichi 484-0061 Japan

ABSTRACT: Co-generator concentration system using cold mirror was proposed and analyzed. It collects exclusively visible sunlight and illuminates to newly developed Si concentrator cell fabricated by low-cost process. The IR light goes through the cold mirror and collected by vacuum collectors. The total conversion efficiency was around 60 % and excelgy generation was around 16 %. The energy output was stable in wide range of spectrum change due to sun height, as well as production tolerance.

Keywords: Concentrator system –1, Cold mirror –2, Hybrid system –3, Excelgy -4

1. INTRODUCTION

It is important to develop the concentrator PV systems for reducing the overall system cost and saving precious high-grade semiconductor materials and massive energy used to cell fabrication. One of the disadvantages of the concentrator system is that it collects the infrared-red (IR) irradiation and raises the cell temperature, which curtails the electric conversion energy. Considering the fact that IR flux is a part of the energy of sunlight, it is advantageous to remove this component before illuminating concentrator cells.

An idea was proposed to utilize this useless and harmful component of sunlight[1][2]. The system based on this idea uses a selective parabolic reflector (cold mirror). The incident light can be divided and concentrated into two parts; with less than a critical wavelength λ_c are used for PV conversion and that of longer wavelength above λ_c are used to thermal conversion. The overall system is shown in Figure 1.

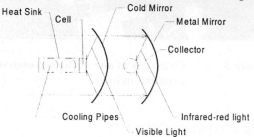

Figure 1 System overview diagram

It is apparent that this system utilizes the total sun power better than simple concentrator applications. However, it is not clear to what extent it excels the simple concentrator system. The purpose of this work is to examine the energy balance of the concentrator system with cold mirror and make it clear its advantages in view of energy conversion.

This paper deals with energy flow or balance of this hybrid system and discusses advantages and stability resulted from spectroscopic parallel process of energy conversion. This is the first research on quantitative examination of selective reflector concentrator PV systems. Any previous report did not do quantitative analysis on the power flow on selective wavelength concentrator systems. It also shows the accurate energy flow calculation method on the PV output illuminated cold mirror systems.

This work does not only quantitatively compare the cold mirror systems to the conventional systems. It also discusses the sensitivity analysis of cold mirror characteristics to mirror design and sun height.

2. SYSTEM OVERVIEW

Figure 2 is an overall image of the system. The gray long component is cooling unit and cells. The return water is also utilized as warm water. The typical temperature is 40 C. The vacuum heat collector, set behind the cold mirror and not seen in the figure 2, collects NIR light and raise the water or other heat media temperature typically 120C.

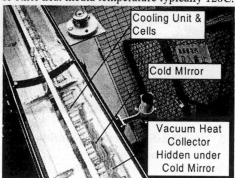

Figure 2: Photograph of the prototype system

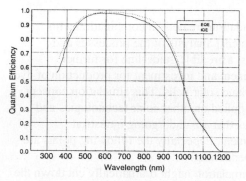

Figure 3: Quantum Efficiency of the newly developed cell

This system taken in the figure 2 was fabricated by Daido Metal Co., Ltd. and is debugging.

The cells used to this system are newly developed Si concentrator cells[3] as shown in Figure 3. They are made by low cost process with single photolithography step and from low-cost B-doped CZ silicon. The cell size is 48 mm X 12 mm. Typical efficiency is 17 % at X20 concentration.

Figure 3 is the most recent quantum efficiency plot of the cell. This data is used for the calculation throughout of this paper.

3. ANALYSIS

3.1 Optimization and sensitivity of mirror characteristics

One of the most important parameter is the bandwidth of the cold mirror. Obviously, the best bandwidth is the exactly the same range to the cell response. The problem is that the bandwidth of typical $(AB)^n$ structure optical multi-layer films is derived only by refractive index of A layer and B layer and not affected by the thickness and number of layers. Since these numbers are material properties, a limited range is industrially available. The typical number is 500 nm or 600 nm. Unfortunately these widths are not wide enough to capture full range of sensitive wavelength to solar cells and the impact of this limited bandwidth is significant. Therefore, the first step of calculation is to give a fixed bandwidth and to optimize cut-off wavelengths of reflection window.

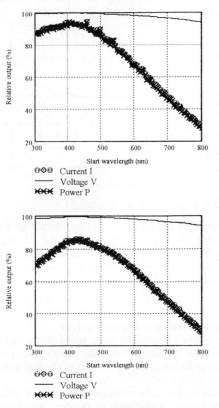

Current I
Voltage V
Power P

Current I
Voltage V
Power P

Figure 4: Photovoltaic generation vs. Mirror optical characteristics (Top: Bandwidth = 500 nm, Bottom: 600 nm)

The successive discussions are done by the combination of optimal cut-off wavelength. The next step is spectroscopic calculation of generated electric power, heat loss inside solar cells, loss in fixtures and optics, heat by useless illumination of infrared-red illumination, and PV generation. Figure 4 is the results of the sensitivity analysis and optimization. In this calculation, the losses of the cover glass and cold mirror are ignored for simplicity.

The photovoltaic output is less than the unity, because of the lack of deep-blue and deep-red region of the spectrum. The efficiency peaks when filtering characteristics of the cold mirror, spectrum response of the cell and spectrum of sunlight are matched. The peak appear at around 400 to 430 nm to both 600 nm and 500 nm bandwidths. Obviously, 600 nm bandwidth mirrors generate more electricity than that of 500 nm.

It is true that the total electricity is slightly reduced by the cold mirror. The conversion efficiency of the cell, on the contrary, increases due to the reduction of unnecessary spectrum component of the sunlight. Selective illumination by the cold mirror enables to boost the low-cost cell efficiency from 17 % to 24 % in the wide range of filtering characteristics.

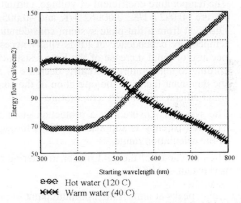

Hot water (120 C)
Warm water (40 C)

Figure 5: Heat generation sensitivity (bandwidth = 600 nm)

The filtering characteristics of the cold mirror also affects to energy distribution to hot water (vacuum collector) and warm water (cooling unit) section. Figure 5 represents the sensitivity to filtering wavelength region. The mirror collecting longer wavelength region raise the total energy illuminating to the cell and decreases the energy to the vacuum collector underneath the cold mirror. Thus, the heat generation of the warm water section increases and that of r section decreases.

The optimum parameters of a cold mirror are 600 nm of bandwidth and the reflectance range is from 420 nm to 1020 nm. The acceptable error is ± 50 nm. It is true that wider bandwidth always increases photovoltaic output, but is difficult to fabricate over 600 nm.

3.2 Excelgy optimization

Heat conversion and photovoltaic are completely different energy source and difficult to compare. The distribution of spectrum energy should be considered by appropriate weighting these two types of energies. The typical scale is thermodynamic effective energy or excelgy. Excelgy is defined as the following equation.

$$e_T = \int_{T_0}^{T_1} mC_p dT - T_0 \int_{T_0}^{T_1} \frac{mC_p}{T} dT \quad (1)$$

where, T_1 and T_0 are temperature of a heat media and environment, and m and C_p are mass and specific heat. Excelgy is a portion of heat energy that is theoretically converted to mechanical energy. Obviously, excelgy is less than the heat energy and reaches to energy value with the increase of temperature. In case of water as heat media, only 2.4 % or 13.2 % can be converted to mechanical energy, when the temperature is 40 C or 120 C. On the other hand, all photovoltaic energy can be converted to mechanical energy. Therefore, the quality of converted energy will be improved, if the sunlight is more distributed to photovoltaic, and secondarily, hot water rather than warm water.

Throughout this section, we compare the electrical energy and heat energy. The conversion efficiency of electricity and heat were modeled as follows.

1. Conversion efficiency of the vacuum collector and cooling unit are 90 % and 50 %. It is supposed to constant regardless of output temperature. These values were estimated by the first test run of the prototype system and very likely to be improved.

2. The input temperature of cooling water is 25 C it increases linearly by the output port.

3. The temperature coefficient of voltage, current, and FF are –0.0027 1/K, 0.00062 1/K and –0.005 1/K. Different from a flat-plate system, consideration of temperature dependence of FF is vitally necessary, because of rapid increase of the base resistance according to temperature. Actually, decrease of FF is the dominant factor in the operation of the concentrator silicon cell in high temperature.

4. The system is covered by a protection glass plate. The transparency is supposed to 92 % throughout of the calculation range.

5. The reflectance of the cold mirror in filtering range is constant and 95 %.

The excelgy peaks at almost the optimum point of photovoltaic generation. The optimum mirror is also best in view of energy quality.

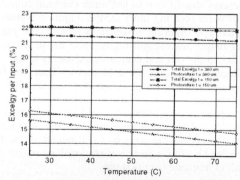

Figure 6: Excelgy vs. Cooling water temperature

Another required consideration is output temperature or flow rate of the water in cooling unit. Obviously, higher flow and lower temperature always improves electrical conversion and reduces heat generation output. The question is the output temperature to provide the best mixture between heat and electricity. Figure 6 indicates the calculation result of excelgy optimization. As it is expected, photovoltaic output simply decreases with the increase of output temperature of cooling water. However, excelgy value is almost constant,

because of compensation of the heat energy. It means conversion quality is unchanged regardless of convenience of the users for output temperature of the warm water.

3.3 Sensitivity to spectrum change –sun height

One of the advantages of the concentrator system is stability of output energy regardless of the sun height. It may produce comparable amount of energy in early morning or late evening, when any flat-plate system shows significant drop. This advantage may not be true to the system sensitive to spectrum of the input light, because color of the sunlight shifts to red in morning or evening.

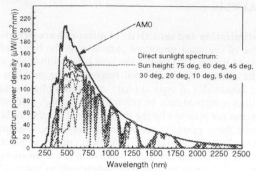

Figure 7: Spectrum changes by sun heights

Figure 7 is a synthesized direct sunlight spectrum in various sun heights. With the decrease of sun height, absorption of O_3 or H_2O is getting significant and shorter wavelength region is more scattered, leading to decrease of color temperature and intensity even to the normalized plane.

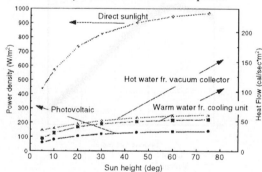

Figure 8: Energy output vs. Sun height

Figure 8 is the calculated energy output at 5 deg, 10 deg, 20 deg, 30 deg, 45 deg, 60 deg, and 75 deg of sun height. All kinds of output energy slightly drop when sun height is low. The dropping speed does not seem to be different among the three.

3.4 Implementation to the real system

The above calculations were based on rectangular approximation of the filtering characteristics of the cold mirror. This section discusses the difference to the real cold mirror system.

Figure 9 shows the measurement result of reflectance, transmittance and absorbance of the first prototype of the cold mirror for this system with 6 layers of optical films. The reflectance range, defined as 50 % reflectance points, was 384.4 nm to 947.8 nm. The bandwidth was 563.4 nm. Since it was made prior to optimizing work, the filtering range is located in much shorter wavelength region. However, it is not difficult to adjust to the target region and increase the bandwidth to 600 nm by increasing the layer number to 12.

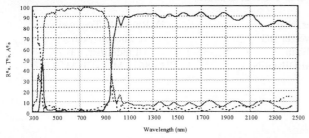

Figure 9: Reflectance, Transmittance, and Absorbance of the prototype cold mirror

Comparing to the rectangular filtering approximation, the most distinct difference is significant amount of absorption at the filtering band edge, UV region less than 330 nm, deep IR region more than 2200 nm. This gap leads to pessimistic influence to heat energy.

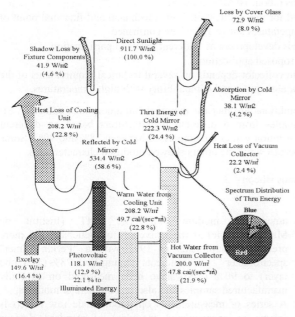

Figure 10: Diagram of Energy Distribution

Energy flow was calculated by the above data. The results are summarized in Figure 10.

A slightly lower photovoltaic efficiency was due to the error of the mirror design. The optimum design is supposed to raise photovoltaic conversion. Another option is to add UV reflecting film on the surface of the cold mirror and utilize shorter wavelength region to photovoltaic. The fixture efficiency defined as the ratio between the cell efficiency inside and outside of the fixture was modest comparing to reported systems. The advantage of our system is generation of warm water and hot water with keeping sufficient photovoltaic energy. Actually, when excelgy generation of heat collector was included as equivalent photovoltaic energy, the fixture efficiency reached about 93 %.

Total energy conversion efficiency was estimated as 61 %. The efficiency measured by independent organization to a reduced system had been reported as 59.5 %[4] and was close to the estimated value. These values may be improved after debugging is completed.

7. Conclusions

A cold mirror concentrator system is analyzed quantitatively. The main advantage is the utilization of the spectrum components out of the cell response range. IR and UV region are superlatively collected by a vacuum heat collector and generate heat energy. Energy loss of the cells are also collected and supply warm water.

Since this system is based on selective spectrum operation, adjustment on spectrum matching is important to keep high conversion efficiency. Numerical calculation revealed that optimal reflectance range of the cold mirror was 420 nm to 1020 nm. The acceptable error was plus or minus 50 nm. The design window was wide enough to production.

For optimization of output energy quality, excelgy was evaluated for various conditions of operation. The optimum point was almost the same point to the optimum photovoltaic operation point. The output excelgy was stable, in spite of the change of output water temperature. This means user can control water temperature or water flow in wide range while keeping the quality of output energy.

In spite of the fact that this system relies on spectrum separation, it was shown robust to spectrum change by sun height. This implies it can supply high amount of energy from early morning to late evening with the help of a tracker.

Detailed energy flow was examined with using a prototyped cold mirror. The photovoltaic efficiency was 12.9 % by newly developed low-cost Si concentrator cells with single photolithography process. The efficiency of the cell was increased to 22.1 % by selective illumination. The excelgy generation efficiency was 16.4 %, and total energy conversion efficiency was 61.1 %.

Acknowledgements

This work was supported by the New Energy and Industrial Technology Development Organization as part of the New Sunshine Program under the Ministry of International Trade and Industry, Japan. This work is partially supported by the Japan Ministry of Education as a part of studies of the Private University High-Tech Research Center Program.

Reference

[1] Ming-ju Yang, et al., "A 3 kW PV-Thermal System for Home Use", IEEE PVSC, 1997, pp 1313-1316

[2] Ming-ju Yang, et al., "A Hybrid PV-Thermal Concentrator System for Terrestrial Applications", 14th European Photovoltaic Solar Energy Conference, Barcelona, Spain, 1997, pp 329-331

[3] Kenji Araki et. al., "A Si concentrator cell by a single photolithography step", PVSEC-11, Sapporo, 1999

[4] Mikihiko Sato, et al., "Efficiency improvement of hybrid photovoltaic system", Research of Industrial Research Institute, Aichi Prefectural Government, No. 34, 1998

Toyota Technological Institute
2-12-1 Hisakata, Tempaku, Nagoya 468-8511, Japan

Tel: +81-52-809-1877
FAX: +81-52-809-1878
Email: araki@toyota-ti.ac.jp

A NEW GENERATION OF HYBRID SOLAR COLLECTORS

ABSORPTION AND HIGH TEMPERATURE BEHAVIOUR EVALUATION OF AMORPHOUS MODULES

P. Affolter
Ecole Polytechnique Fédérale de
Lausanne (EPFL)
Laboratoire d'Energie Solaire et de
Physique du Bâtiment
Bâtiment LESO
CH - 1015 LAUSANNE
Phone: + 41 (0)21 693 45 39
Fax: + 41 (0)21 693 27 22
Email: pascal.affolter@epfl.ch

A. Haller, H.-J.
Ernst Schweizer AG
Metallbau
CH - 8908 HEDINGEN

D. Ruoss, P. Toggweiler
Enecolo AG
Lindhofstr. 13
CH - 8617 MÖNCHALTORF

ABSTRACT: Although the idea of producing a solar collector for both electricity and hot water is not new, this new generation is based on a very innovative concept. In a feasibility study [1] we have collected a lot of information about important aspects (costs, technology, market, contacts with industries, ...) to prove that further development is worth being done. The results are encouraging since they show:

- A potential market does exist for several specific applications (about 10 MW in 2005)
- Photovoltaic (PV) thin film technology is likely to be suited for this application from a technical and financial point of view, provided that the long term stability of the cells at temperatures above 100°C are confirmed
- Several photovoltaic industries are ready to collaborate in this development at different levels of participation
- The several technical concepts proposed are suited for the proposed application.

The feasibility study showed that the competitiveness of a hybrid collector depends on several technical requirements of the integrated PV-module. The most prominent aspects are high solar absorption and compatibility with high temperatures.

The goal of the present study is to verify if these technical conditions are met for the available amorphous silicon (a-Si) technology. Measurements on commercial unencapsulated samples from 6 different manufacturers based on different substrates (glass, stainless steel, polyimide) were performed. Absorption values are comprised between 78% and 90%. Some samples heated up to 210°C for one hour kept their original properties while the others showed modified characteristics.

Keywords: Thermal-PV – 1:   Heat recovery – 2:   Amorphous silicon – 3

1.   INTRODUCTION

The results from the feasibility study phase show that the competitiveness of a water based photovoltaic-thermal (PV/T) collector depends on several requirements. They are:

- The absorption coefficient for high thermal efficiency (value of absorption over the whole solar spectrum) of the photovoltaic absorber should be higher than 80% (the photovoltaic absorber will be the photovoltaic device that replaces the absorber of a conventional thermal collector).

- The stability of a- Si cells at temperatures of 100-160 °C (stagnation situation of the thermal collector).

The information needed for a better understanding of these two points is available neither in technical documents and scientific literature nor by the PV companies with the best know-how. For this reason, the development of the PV/T collector must proceed with a series of measurements and more detailed work on these two subjects before product design is approached.

2.   METHOD

2.1 ABSORPTION ASSESSMENT

Few data are available from PV manufacturers concerning solar absorption; basic measurements in the feasibility study returned ~65 % absorption for a commercial module. 1996 Ernst Schweizer AG measured a solar absorption coefficient of ~72 % on a commercial Tefzel-EVA-

amorphous module [1]. The IMT (Institut de Microtechnique) at the University of Neuchâtel did more precise measurements in 1997 with a "Perkin Elmer" spectrometer [2]. Coefficients ranging from 78% (a-Si, one layer) to 90% (c-Si) have been measured on specially manufactured samples and also on commercial modules.

A series of measurements on photovoltaic raw materials (plates or encapsulated modules) of standard devices (glass-substrate, steel and polyimide) provided by six different manufacturers were performed (see Table 1) in order to show the effective reflections of the different optical layers of photovoltaic devices.

Sample code	Substrate	Junction type	Encap-sulation
A	glass	single	none
B	glass	tandem	resin
C	glass	single	none
D	glass	single	resin
E	glass	single	resin
F	polyimide	tandem	none
G	stainless steel	tandem	none
H	stainless steel	triple	none
I	polyimide	tandem	EVA / EVA
J	polyimide	tandem	none
K	stainless steel	triple	none

Table 1: list of samples available for the two tests

Two different measuring methods were used to offer a better understanding of the sample properties. The first was based on spectrometry. It yields a wavelength-resolution. This was subcontracted to the IMT. They used a Perkin Elmer UV/VIS/NIR spectrometer (type Lambda 900) to measure the reflection. The measurement is performed using an Ulbricht integrating sphere, therefore direct and diffuse reflection is summed up to yield the total reflection of the sample. The instrument baseline is taken prior to the measurement using a white standard of the material used for the interior of the Ulbricht sphere, assuming the reflectivity of this material to be 100%.

The second one was based on **calorimetry** and yields a global value. It was subcontracted to the IOA (Laboratory of Applied Optics of the Microtechnic Department of the Swiss Federal Institute of Technology/EPFL). The experimental apparatus consists of a thermal flux sensor (Episensor type A 02-050 or Peltier element of type TEC1-12714, both of a surface of 50x50 mm). As the Episensor type thermal flux sensor has an ill-defined surface, a copper plate of a thickness of 1mm was glued to the surface with a high thermal conductivity adhesive (Omegabond 200). The samples are then fixed on the sensitive surface with a heat conductive paste (HTC 35SL) and illuminated by the source. The illumination aperture has been provided with the upper surface constituted with mirrors, to avoid at maximum thermal radiation from the aperture frame to the cell under investigation. The illumination source is a dichroic halogen lamp (Philips, 150W nominal, type 13117), powered by a stabilized power supply under constant voltage (15.20V), at a distance of 400 mm, resulting in a fairly constant light intensity over the aperture of 20cm². The incident power has been measured with a pyroelectric wide band laser power meter (Ophir Nova) and yields 2.48W for 20cm² (1240 W/m²), respectively 1.99W for 16 cm² (1244 W/m²).

2.2 THERMAL STABILITY OF A-SI CELLS

Although some literature addresses the topic more or less directly [2][3][4][5], the weakness of data about stability of thin film PV-cells at temperatures above 100 °C makes proof of the suitability of such cells for an application in a PV/T collector necessary. Therefore long time temperature tests have to be performed.

Since there are no guidelines for such tests with amorphous silicon, it is first necessary to work out a testing program. Amorphous silicon is well known in the chip industry and thus we should also pursue a specific study of scientific literature (a-Si in general) in order to gain as much data and experience as possible for defining these tests. The literature study also has the task to prepare the final concept for the accelerated aging tests which will be done on laminated samples (PV cell and thermal exchanger) planned to be built in the next project phase.

Based on the results of the absorption measurements, two important parameters of the PV/T collector can be evaluated (modeled); these are the stagnation temperatures of a normal flat plate collector with this absorber type and its thermal efficiency.

The testing procedure has the following features:

- Raw materials are thermally stressed by thermal cycles with different durations (1-1000 h) and at different temperatures (100°C, 120°C, 140°C, 160°C, etc.)

- Testing place: automatic oven with a computer controlled temperature curve for heating (no active cooling)
- The samples are periodically measured with a constant lamp simulator (7 dichroic halogen lamps with good approximation of the solar spectrum: Philips, 150W nominal, type 13117)

3. RESULTS

3.1 SPECTROMETRY MEASUREMENTS

The transmission of all samples is equal to zero, due to metallization of the back surface. Therefore the absorption of the samples is given as $\alpha(\lambda)=(1-\rho(\lambda))$ where $\rho(\lambda)$ is the wavelengths dependant reflection.

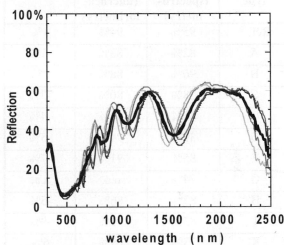

Fig. 1 shows a typical reflection spectrum as measured on one of the samples. Thin lines are measurements on different spots on the same sample; the thick line is the mean value at each wavelength.

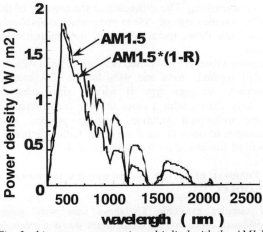

Fig. 2: this mean spectrum is multiplied with the AM1.5 spectrum. The total absorbed power corresponds to the surface under the curve, i.e. is obtained by integrating over all wavelengths.

Multiplication of the absorption spectrum $\alpha(\lambda)$ with the spectral power density in a standard AM1.5 solar spectrum and integration over all wavelengths yields totally absorbed power in the sample (see Fig. 1 and 2).

3.2 MEASUREMENTS BASED ON CALORIMETRY

Each sample has been measured several times and the mean value has been reported. No divergence larger than a few percents has been encountered. The heat absorption of the samples has been determined according to the following formula, by taking the nominal value of the black coated surface of the reference sample (absorption = 94.0%).

$$\alpha_{sample} = \alpha_{référence} (U_{epis} - U_{backgr})/(U_{ref} - U_{backgr})$$

Measurement results of the α-coefficients are presented in Table 2.

Sample type	IMT (spectro-metry)	IOA (microca-lorimetry)	Difference
Ref. *	95%	94%	1%
A	82%	83%	-1%
B	90%	88%	2%
C	78%	80%	-3%
C	78%	81%	-4%
D	89%	94%	-5%
F	85%	91%	-7%
G	74%	n.a.	n.a.
H	82%	86%	-5%
I	83%	91%	-9%
K	73%	69%	6%

Table 2: Absorption measurement results

* Reference: piece of stainless steel selective absorber, (courtesy of Energie Solaire SA, Sierre, Switzerland)

Comparison of the two methods shows that a very precise relation is resulting. The difference of the majority of the samples is in the range of –9% to +6%, which is absolutely tolerable and shows the reliability of both measurement methods.
Absorption values are comprised between 73% and 90% for the IMT method,. 69% and 94% for the IOA method respectively. Samples type B with a glass substrate technology shows, with a close to 90% α-coefficient, a perhaps well-suited solution for the project. Also technologies of types D and F are under further evaluation and will be considered for the prototype work.

3.3 THERMAL BEHAVIOR OF AMORPHOUS TECHNOLOGY SAMPLES

Most samples available for these tests were none-encapsulated. Since critical temperature was not precisely known and since the quantity of sample modules was limited, it was decided to separate the test modules in two groups:

• those to be tested at medium temperature during a long duration (first group)

• those to be tested at higher temperature during a short duration, up to electrical breakdown (second group).

Measurements were done as follows:

1. Measuring the sample (I-V measurements under halogen dichroic lamps at 300 W/m² in the centre of the probe table)

2. Thermally cycling the sample (warming up, plateau at a given temperature level for a given duration, passively cooling down)

3. Continuing with 1. until end of measurement or breakdown.

With the sample modules of the first group, the goal was to confirm that operation at 120°C for long periods was not a problem. Obtained results are presented in Fig. 3. Each curve corresponds to the average values of a series of measurements. Only the samples type "I" did not withstand these tests. Since the same none-encapsulated cells (type "J") passed the tests, we attribute this breakdown to mechanical problems due to thermal stress on the encapsulation. Another possible cause could be corrosion due to gazes emitted from the molten of the EVA.

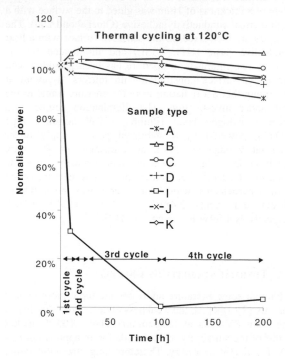

Fig.3: First thermal cycling series: "long durations" (4 cycles of 10, 10, 80 and 100 h at 120°C)

With the second group, the goal was to evaluate the critical temperatures with regards to breakdown (see Fig.4). We started therefore with a first cycle at 140°C during one hour, then 160°C during one hour, then 170°C during one hour. The fourth cycle was planned to have a plateau temperature of 170°C during 10 hours, in order to confirm the results of the third cycle with a longer duration. A bad operation of the temperature controller created a higher temperature: 210°C. The results of these series are nevertheless of prime interest. The results of this 210°C temperature stress series are the following:

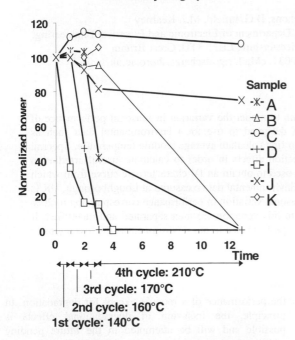

Thermal cycling up to

Normalized power

Sample

-✳- A
-△- B
-○- C
-+- D
-□- I
-✕- J
-◇- K

Time

4th cycle: 210°C
3rd cycle: 170°C
2nd cycle: 160°C
1st cycle: 140°C

Fig. 4:Second thermal cycling series:"high temperatures" (1h at 140°C, 1h at 160°C, 1h at 170°C, 10h at 210°C)

- the samples type "I" did not pass the first cycle,
- the samples type "D" showed a hard degradation after the second cycle
- the samples type "A" showed a hard degradation after the third cycle
- the samples type "B" showed a full degradation after the fourth cycle
- the samples type "J" kept their properties unchanged after the first cycle. One notices a reduction of 20% of the efficiency after the third test. The fourth cycle was not damaging
- the samples type "C" showed no degradation all along whole test.
- The samples of the type "K" showed no degradation after the third cycle. For technical reasons, they were not measured after the fourth cycle.

Since electromigration (transfer of molecules due to a current) is a possible cause of degradation of semi-conductors, we made further tests on the samples type "C" and "J" in order to check if the current increases the degradation. We forced currents close to the nominal current into the PV device. For a reverse current, no effect was stated for cycling of up to 170h. To force a direct current into the device means to load the parallel resistance with the nominal current. For the sample type "C", it was not possible since the voltage source was limited. For the sample type "J", the samples did pass the cycle at 140°C during 10h but were destroyed during the cycle at 170°C during 10h.

The best stability is obtained by type "C" samples which do not include any aluminium (contacts made of silver). The second sample with a good stability features a back contact made of aluminium. An explanation of this stability is the following: for a n-i-p structure, the Al-back contact is deposited first. Since natural oxidising is more important for aluminium than for amorphous silicon, the resulting

oxide layer between a-Si and back contact will be thicker. This oxide plays the role of an aluminium diffusion barrier [6]. We attribute the different behaviours of the other samples to the presence of a passivation layer on the a-Si thin films, that also plays the role of a diffusion barrier (good for zinc oxyde/ZnO, bad for tin oxyde/SnO_2 and ITO).

4. CONCLUSION

Average absorption values are comprised between 71%and 92%. The lower limit established in the feasibility study is overcome which means that several kinds of a-Si modules are suited for the hybrid application as far as absorption is concerned.

After hard thermal cycles, the comparison between the different technologies shows behaviours that are very different. The presence of aluminium increases the risks of pollution of the critical layers and contacts by diffusion processes. The best way of avoiding such problems is to use back contacts made of silver. One has in this case an excellent stability to thermal cycling (plateau of 210°C/10h). Another possible protection is to isolate back contacts from a-Si films with oxides. At least two technogloies show a very strong behaviour under high temperatures. The others could be used if several minor changes of the manufacturing process were made.

5. ACKNOWLEDGEMENTS

This work was financially supported by the Swiss Federal Office of Energy (OFEN/BFE) represented by Mr. J. Gfeller und Mr. S. Nowak (NET AG). We would like to thank also Mr. D. Fischer, M. Götz and R. Platz from the Institut de Microtechnique of the University of Neuchatel for their crucial help. Mr. T. Sidler (Institut d'Optique Appliquée/EPFL) gave an excellent confirmation of the absorption measurements. The expertise of Mrs. Anne Labouret (Solems) was particularly appreciated. A lot of help was provided by Mr. Robin Humphry-Baker (Laboratoire de Photonique et Interfaces/EPFL). Thanks to all of them.

6. REFERENCES

[1]   "New generation of solar hybrid collectors / Phase 1", P. Affolter (LESO/EPFL) and al, OFEN Project No 56360/1686, November 1997

[2]   " Long Term Behaviour of Passively Heated or Cooled a-Si :H modules ", C. Hof and al., Institute of Micro., Univ. of Neuchâtel, 25th IEEE PVSC, Washington DC, May 13-17, 1996

[3]   " Accelerated Life Testing of Solar Energy Materials", B. Carlsson and al., an IEA report of Task X, February 1994

[4]   " Belastbarkeitsgrenzen von PV-Modulen", Hermann Laukamp and al., Fraunhofer-Institut für Solare Energiesysteme ISE, Freiburg im Breisgau, April 1998

[5]   " Interface Degradation in a-Si:H Solar Modules", D. Peros and al., Universität-GH Siegen, Institut für Halbleiterelektronik, 14th EPVSEConference and Exhibition, Barcelona (Spain), 30.6 – 14.7 1997

[6]   "Solarzellen aus amorphem Silizium auf Aluminium: drei Wege, den Substrateinfluss zu beschreiben", M. Götz, Thèse No 1637, dép. Microtechnique. EPFL, 1997

MODELLING THE PERFORMANCE OF a-SI PV SYSTEMS

M.J. Holley[1], R. Gottschalg[2], A.D. Simmons, D.G. Infield, M.J. Kearney
CREST (Centre for Renewable Energy Systems Technology), Department of Electronic and Electrical Engineering,
Loughborough University, Loughborough, Leicestershire, LE11 3TU, Great Britain.
Tel.: 0044-1509-228140, Fax: 0044-1509-610031, eMail: r.gottschalg☐lboro.ac.uk

Abstract: This work describes the development of an algorithm to model the variation in seasonal performance of amorphous silicon (a-Si) photovoltaic devices. The model is designed to use local environmental data such as ambient temperature, solar irradiation and windspeed in order to first calculate average module temperature. Spectral irradiation data is then used together with module cover reflection effects in order to calculate 'useful' irradiation reaching the solar cell. An equivalent circuit specific to a-Si is used to obtain an IV characteristic curve from which maximum power point and hence efficiency can be calculated. Environmental data measured at Loughborough, UK is used to present a series of results that demonstrate an overall seasonal variation in performance corresponding to 20% of performance. The individual loss mechanisms contributing to this variation are then separated and quantified. It was found that the main contribution to the seasonal variation in performance in our environment is the low intensity of the light in winter time.

Keywords: amorphous silicon: 1; system: 2; modelling: 3

1 Introduction

Thin film solar cells promise lower PV costs for the future. The only thin film technology which has a significant market penetration to date is amorphous silicon (a-Si). This technology varies in its performance from the commonly used crystalline silicon technology in that a-Si exhibits a lower efficiency and that it degrades. It also shows an idiosyncratic seasonal performance variation, often resulting in maximum efficiency occurring in the summer. Several researchers attribute this to a change in the spectrum whilst others attribute it to a seasonal annealing effect.

Rüther and Livingstone [1] suggest that a major contributor is the seasonal spectral shift in the incident solar spectrum. A strong seasonal variation in the spectral region a-Si can accept light was reported for the location of Loughborough [2]. Hirata et al. [3] also show that the spectrum has an important role to play in contributing to higher summer efficiencies, despite the fact that the operating temperature is elevated in the summer months. Merten and Andreu [4] reported that the influence of the spectrum is in the range of 16% in the summer-winter comparison and that degradation plays a minor role only.

Nakamura et al. [5] suggest that seasonal performance can be modelled by accounting for degradation with a stretched exponential function. This is, however, in disagreement with the findings reviewed in the previous paragraph, where the main variation in performance is attributed to the change in spectrum.

A model taking the spectral changes into account and thus allowing a clear separation of all effects contributing to performance variations appears to be important for developing confidence in this technology. This paper describes a system model, which describes all effects on the performance of a module except for degradation. In principle, the inclusion of degradational effects is possible and will be attempted in the future pending additional research.

2 Model

Prediction of outdoor performance on the basis of empirical models was suggested by Nakamura et al. [5] and Kleiss et al. [6]. These models have the disadvantage that they are limited to the maximum power point and thus are not suitable for a full system model. The approach reported in this work is based on an a-Si specific parametric approach, which describes the whole I-V characteristic, thus allowing effects such as mismatch to be taken into account. An overview of the algorithm employed is given in Figure 1 and illustrates the three main contributions to the overall model: the thermal model, the optical model and the cell model. These components are described below.

2.1 Thermal Model

The thermal model used was originally developed by Fuentes [7] and predicts the average module temperature according to local environmental conditions. Windspeed, plane-of-array irradiation, ambient temperature and Installed Nominal Operating Cell Temperature (INOCT) are used as input parameters.

INOCT is defined as the cell temperature of an installed array at NOCT conditions (800 W/m^2 irradiation, 20° C ambient temperature and 1 m/s windspeed) [8] and therefore takes into account the module mounting configuration. For example, stand-alone rack-mounted modules will clearly possess different heat flow characteristics from those designed for building integrated installations. The model initially uses INOCT

---

[1] new address: PV Systems (EETS), 104 Portmanmoor Road Ind. Est , Cardiff, Wales, CF24 5HB, UK
[2] author for correspondence

to estimate heat gain and convective/radiative heat losses from both top and bottom module faces at NOCT conditions. These values are then modified according to the local environmental conditions. Thermal capacitance of the module is also incorporated to account for the natural temperature lag.

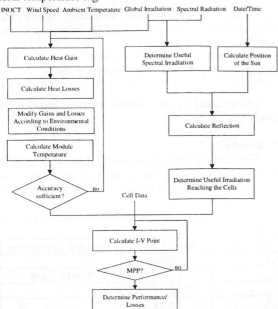

Figure 1: *Flowchart of the Algorithm.* The graph visualises the three main parts of the simulation, the thermal model, the optical model and the cell model, which are performed in the order shown.

The model is used with minor modifications from the original to account for the specific geographical location. A new tilt angle was incorporated for calculating the convective coefficient of the module top surface. Additionally, the actual clearness index is included for each data point while Fuentes assumes a single average clearness index value. The influence of this model is shown in Figure 2.

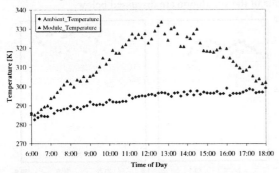

Figure 2: *Variation of Module Temperature in Summer.* The graph illustrates the relation of the ambient temperature and the predicted module temperature for a rack mounted system on a sunny day in summer 1999.

The appropriateness of this model was validated against the a-Si array installed at CREST and showed a reasonable agreement. Differences are due to the lack of module specific values for the thermal model, more accurate values would certainly give a better fit. The values used for the simulations are shown in Table 1.

Property	Value
INOCT	325 K
Thermal Mass	11000 J/m²K
Emissivity	0.84

Table 1: *Values Used for the Thermal Model.*

2.2 Optical Model

The optical model has to perform two tasks: it has to calculate the reflection occurring and adjust for the seasonal spectral shift. This shift is important because it changes the useful percentage of the incident energy drastically, as shown in the measurements shown in Figure 3, taken in Loughborough in the year 1999.

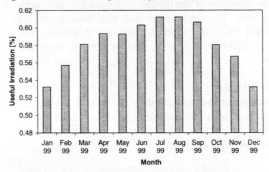

Figure 3: *Seasonal Spectral Shift.* The graph shows the variation of the percentage of the light which is in the range useful for a-Si at Loughborough.

Cells are typically calibrated with an AM1.5global spectrum. This spectrum is defined in the spectral range up to 4000 nm, while a-Si typically can only use irradiation below 790 nm. Radiation with longer wave lengths is not seen by the devices. As seen in Figure 3, the useful proportion changes over the year.

The approach chosen for including the spectral effect is to scale the incident global irradiation. This is done by comparing the energy in the useful range and calculating the global irradiation (with AM1.5 global spectrum) that contains the same amount of energy in the useful range. This can, as is illustrated in Figure 4, have a beneficial effect but can also have a detremental effect. This effect depends on different factors, e.g. humidity and number of particles in the atmosphere, so that it is much stronger in winter then expected from Figure 4.

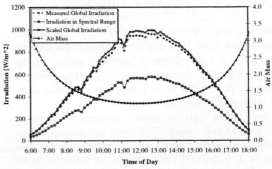

Figure 4: *Incorporation of the Spectral Variation.* The graph illustrates the effect for June 18th in 1999, and yields a beneficial result.

Furthermore, the reflection needs to be considered for a non-tracking system. The approach implemented uses the model presented by Erbs et al. [9] for the separation of diffuse and beam irradiation. The radiation on the elevated module is calculated following the standard approach given by Liu and Jordan [10]. The beam radiation reaching the surface is then reduced by the reflection occurring on the glass cover. This reflection is calculated following the Fresnel model as given by Duffie and Beckman [11]. The reflection is obviously dependent on the angle γ between module and the beam direction, which is calculated for a module facing due south as:

$$\cos\gamma = \sin\delta\sin(\phi-\alpha) + \cos\delta\cos(\phi-\alpha) + \cos\omega \qquad (1)$$

where δ is the declination of the sun, ϕ is the latitude, α is the elevation of the module and ω is the hour angle of the sun. The declination and the hour angle are calculated using the method given by Walraven [12], which is claimed to give an accuracy better then 0.01 degrees.

2.3 Cell Model

Amorphous silicon exhibits a voltage dependent photocurrent, thus the commonly used one diode model, as shown in equation (2), needs to be modified. The standard model is given as:

$$I = -I_{ph} + I_D(e^{\frac{e(V_j)}{nkT}} - 1) + \frac{V_j}{R_P} \qquad (2)$$

where I is the current through the device, V_j is the voltage at the junction, T is the temperature at the junction, e is the elementary charge and k is the Boltzmann constant. The behaviour of the solar cell is described through the photocurrent I_{ph}, diode saturation current I_D, diode ideality factor n, series resistance R_S and the parallel resistance R_P. The voltage at the junction is calculated as:

$$V_j = V - IR_S \qquad (3)$$

Models allowing for this were presented in recent times by Crandall [13] and Merten et al. [14]. A test of these showed that the Merten model is appropriate for the task given [15]. Merten et al. propose to modify equation (2) as follows:

$$I_{ph} = I_{ph}^{'}\left(\frac{1}{\frac{\mu\tau}{d_i^2}(V_{fb}-V_j)} - 1\right) \qquad (4)$$

where the fraction models an additional recombination term. It is described through the ratio of the mobility μ and lifetime τ over the square of the thickness of the intrinsic layer d_i. The voltage dependence is included by considering the relation of V_j to the flatband voltage V_{fb}.

The model gives good agreement between measurements and theory. The thermal variation of the equivalent parameters needs to be incorporated because otherwise a significant error will occur resulting in an incorrect representation of the behaviour of V_{OC} and I_{SC}.

It was found in our lab that series resistance as well as the $\mu\tau/d_i^2$ factor exhibit only small variation with temperature that can be modelled as a linear dependency. The diode ideality factor varies as:

$$n(T) = \frac{1}{Z} + \frac{1}{ZT^*}T \qquad (5)$$

where Z is a geometric factor which is needed to explain why the diode ideality factor is larger than two and T^* is a characteristic temperature. The diode saturation current varies as:

$$I_0 = I_{01}T^{\frac{6+\gamma}{2}}\exp\left(\frac{-E_G}{kT}\right) \qquad (6)$$

where γ is an empirical parameter, which was determined to be −1, and E_G is the band gap of the material.

The photocurrent is assumed to vary with temperature following a simple relation given as:

$$I_{ph}^{'} = (C_1 + C_2T)G \qquad (7)$$

where C_1 and C_2 are empirical constants and G is the irradiation.

3 Results

The model developed was used for the separation of the daily and seasonal variation of losses occurring in the operation of an a-Si PV system. All cells were taken as identical and of size A.

The parameters used for the simulation are given in Table 2.

Parameter	Base Value	Thermal Correction
A	100 cm^2	
I_{ph}	1.1×10^{-3} A	9.9×10^{-7} A/K
I_{01}	2.228×10^{-6} A	
E_G	1.6 eV	
Z	1.77	-
T^*	1.80×10^4 K	-
$\mu\tau/d_i^2$	1×10^7 V^{-1}	3.8×10^4 V^{-1}K^{-1}
V_{fb}	1.1xeV	-
R_S	5.25×10^{-2} Ω	-4×10^{-5} Ω/K
R_P	1.14×10^4 Ω	-1.95×10^2 Ω/K

Table 2: Electrical Parameters used for the Simulation.

It was found in our lab that these values give a good description of a single junction a-Si device in simulator measurements and are thus suitable for modelling a PV system as long as the degradation is neglected. The results of the simulation are discussed below.

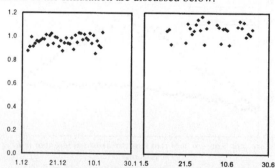

Figure 5: *Seasonal Variation of the Apparent Efficiency.* The figure shows the efficiency with respect to the global irradiation. 100 percent efficiency indicate the average of all simulated days.

3.1 Seasonal Performance

The obvious requirement has to be that the simulation is able to generate a performance pattern observed by researchers in the field. This is the case, as can be seen in Figure 5. The figure shows a normalised efficiency, the value one indicates the average of all points.

The seasonal performance variation is well observable without considering degradational effects. This agrees with the findings of Merten and Andreu [4], who claim that degradation plays an insignificant role after the initial degradation period of 6 months. However, differences to the findings of Merten and Andreu occur when one looks at the reasons for the decreased efficiency in the winter, as done in the following section.

3.2 Losses

The losses investigated in this section are presented as daily averages. Figure 6 gives the results in graphical form. The graph illustrates the impact on the electrical output of the device, not the losses of the input, i.e. it shows how much current is lost due to reflection rather than how much light is reflected.

As expected, the variation of the operating temperature leads to gains in efficiency in winter time and losses in the summer. The small impact of these variations could be expected because a-Si has a large band gap and an increase in temperature will result in a band gap that is narrower, thus closer to the value for optimum efficiency. This is offset, however, by a deterioration in the electrical parameters, thus yielding an overall loss in summer.

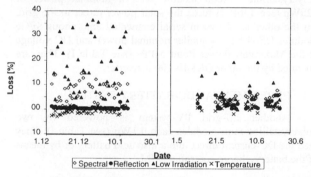

Figure 6: *Investigation of the Losses.* The graph compares the differences of the losses in the different seasons.

It appears surprising that the reflection losses in summer are higher than in winter. This is because the days are much longer and thus steeper angles are achieved in the course of a single day, reaching a point where the sun vanishes behind the module.

It is clear from Figure 6 that spectral losses have a significant impact on the seasonal performance. In areas closer to the equator this should be even stronger because lower air masses are achieved and thus the effect can result in net gains. Not so here, though, which can be attributed to the local climate. It has to be emphasised, however, that the values presented here are for a whole day. It is obvious that during certain hours of the day a net gain is achieved which is offset at other times of the day. The rather large scatter of low irradiation losses in the winter is explained by the fact that only measurements with more then $10 \ W/m^2$ were considered. For example, clear skies around midday followed by heavy cloud in the afternoon can result in only 2-3 h of data being considered (out of the 6 h daylight).

The low irradiation losses appear to be excessive in the winter months. However, this appears reasonable considering the level of irradiation. The *maximum* irradiation on 26.12.98, the worst day of all, is below $50 \ W/m^2$, which justifies such significant low irradiation losses. This unusually low value of irradiation was checked against CREST's thermopile, the integrated irradiance of the Spectroradiometer and the meteorological station operating on the campus of Loughborough University. These losses are higher than for other devices because a-Si has a voltage dependent photocurrent leading to an increased slope of the I-V characteristic for low voltages. Because of this, the I-V characteristic resembles more of a bend line then an exponential function for these very low irradiation levels, leading to increased low intensity losses.

4 Conclusions

The simulation shows that by using an appropriate equivalent circuit and modelling the appropriate operating conditions a seasonal change is observed in the range as expected in the literature. In the given climate of the United Kingdom, it can be concluded that the above model is sufficient to adequately model the output of an amorphous silicon module. The device performance is dominated in our climate by low irradiation losses.

5 References

[1] R. Rüther, J. Livingstone, Solar Energy Materials and Solar Cells **36** (1994) 29.

[2] R. Gottschalg, D.G. Infield, M.J. Kearney, UK-ISES Silver Jubilee Conference, UK-ISES (1999) 35.

[3] Y. Hirata, T. Inasaka, T. Tani, Solar Energy, **63** (1998) 185.

[4] J. Merten, J. Andreu, Solar Energy Materials and Solar Cells, **52** (1998) 11.

[5] K. Nakamura, S. Nakazawa, K. Takahisa, K. Nakahara, International PVSEC-5 (1990) 359.

[6] G. Kleiss, K. Bücher, A. Raicu, K. Heidler, Proceedings 11th EC Photovoltaic Solar Energy Conference, Kluwer (1992), pp. 578.

[7] M.K. Fuentes, A Simplified Thermal Model for Flat-Plate Photovoltaic Arrays, Sandia (1987).

[8] Jet Propulsion Laboratory, Block V - Solar Cell Module Design and Test Specification for Residential Applications, JPL (1981).

[9] D.G. Erbs, S.A. Klein, J.A. Duffie, Solar Energy, **28** (1982) 293.

[10] B.Y.H. Liu, R.C. Jordan, Solar Energy, **4** (1960) 1.

[11] J.A. Duffie, W.A. Beckmann, Solar Energy, Thermal Processes, Wiley (1974).

[12] R. Walraven, Solar Energy, **20** (1978) 393.

[13] R.S. Crandall, Journal of Applied Physics, **54** (1983) 7176.

[14] J. Merten, J.M. Asensi, C. Voz, A.V. Shah, R. Platz, J. Andreu, IEEE Transactions on Electron Devices, **ED45** (1998) 423.

[15] R. Gottschalg, M. Rommel, D.G. Infield, M.J. Kearney, Proceedings 15th EC Photovoltaic Solar Energy Conference, Stephenson (1998) 990

PHOTOVOLTAIC INTERNET LABORATORY

F. P. Baumgartner, R. Heule and M. Brechtbühl
University of Applied Sciences Buchs NTB
Werdenbergstrasse 4, CH-9471 Buchs, Switzerland
email: baumgartner@ntb.ch / Tel:+41-817-553377 Fax:+41-817-553377
http://www.ntb.ch/Other/PVwwwLab/

ABSTRACT: The purpose of the project, Photovoltaic Internet Laboratory PVwwwLab, is to develop an experimental system to perform basic PV experiments with real working PV sub-units via the Internet. Thus, each internet-user has the opportunity to measure the present performance of PV modules exposed to the sun, together with DC/AC converters and battery charge controller. According to the instructions of the internet-user the PVwwwLab at NTB provides immediately real measurements results on the Internet. The current interactive measurement-solution is based on pictures generated by the software tool *LabWindows™*.

Keywords: Monitoring - 1: Small Grid-connected PV Systems - 2: Internet Measurement - 3

1. INTRODUCTION

While PV power gain market share more and more people have to learn about the basic technique of photovoltaic not only on the cell level but also on the system level. Today the Internet becomes an integrated tool in education and information also on renewable energy. Some internet addresses are listed in [1], dealing with theory of solar cells, and impressive interactive-simulations on solar cell performance and sun path calculations. Several online reports of PV system-monitoring are also available on the Internet. As an early example, the performance of the NTB 18kWp grid-connected PV-system [2] is on the Internet since 1996. The aim of the presented NTB Photovoltaic Internet Laboratory is to assist this internet-teaching process on the experimental side and give the user the opportunity to work on-line with real outdoor PV-systems.

2. EXPERIMENTAL SETUP

The PVwwwLab provides access to two different types of experimental PV-systems: Grid-connected and DC-battery systems mounted on top of a NTB building (oriented south, inclination 35 degree).

2.1 EXPERIMENTAL GRID-CONNECTED SYSTEM

Two independent sub-systems are connected by the use of the OK4E 100 module DC/AC converter to the grid. Thus a polycrystalline Silicon module with a nominal power of 120Wp (see Table I) feeds the solar power direct to the grid. On the other hand two in serial connected amorphous Si/Ge modules US64 have a similar nominal power and DC voltage at the Maximum Power Point MPP (see Table I). They are operated by the other OKE4E 100 converter.

2.2 EXPERIMENTAL DC-BATTERY SYSTEM

Another off-grid PV-system is powered by two polycrystalline mini-modules each 9.1Wp (see Table I). They feed the DC-current direct into a lead acid battery or by means of the battery-charge controller SCM.

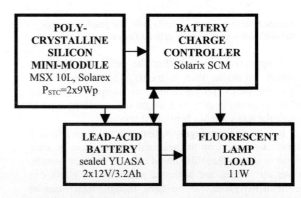

Figure 2: Block diagram of the DC-battery system.

Table I: Manufacture specification of the PV module at Standard Test Condition STC (1000W/m², 25°C, AM1.5)

module type	V_{MPP} [V]	I_{MPP} [A]	P_{STC} [Wp]	η_{module} [%]
GPV120P24	33.4	3.60	120	11.3
US64 a-Si/Ge	16.5	3.88	64	6.3
MSX10 poly	15.9	0.57	9.1	7.6

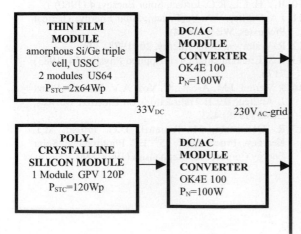

Figure 1: Block diagram of the grid-connected subsystem.

2.3 MEASUREMENT - HARDWARE - SENSORS

The measurement system of the PVwwwLab is outlined to obtain the relevant electrical values of both systems: DC-voltage/current, AC-voltage/current (Hall-current sensor, rectifier-voltage), DC/AC converter efficiency. The final module efficiency results from the immediate measured power from the sun by pyranometer and silicon reference solar cell (see Table II) under the present conditions of temperature and load-resistance. The modules current/voltage characteristics are performed by a variable electronic load. These measurement values are completed by an on-line temperature-profile measurement on the rear-surface of the 120Wp poly-crystalline and the amorphous module based on 15 PT1000 temperature sensors on each module.

Table II: Measured physical values, sensors and instruments

Physical value	Sensor / Instrument
Irradiation in PV-plane	Pyranometer CM11, Kipp&Zonen, silicon cell, ESTI reference sensor
Temperature-profile module back-surface	30 RTD Resistance Temperature detectors PT1000 are placed on the amorphous and crystalline module back-surface
Temperature ESTI	V_{oc} measurement of the ESTI cell
Voltage/current curve of the modules	electrical load, type PL306, Höcherl & Hackl, via RS232
DC-Voltage	smart link, RS485, Keithley, Instr., DCV12, 20 bit resol., 8 channels; DCV32, 16 bit resol., 8 channels
AC and DC current	Hall-current sensor LEM, HY10-P

2.4 FUNCTIONAL REQUIREMENTS

The internet presentation of the PVwwwLab is separated into the two basic systems: grid-connected and PV-battery system. While the performance data of the grid-connected system is monitored via internet the off-grid PV-battery system can be controlled inter-active by the internet-user.

- Monitoring: grid-connected PV-system including modules current/voltage curve with Maximum Power Piont (MPP) values and weather data

- Interaction: PV-battery system, different load types

The addressed user should not only be found among PV-specialists. Furthermore all internet-surfers are welcome, interested in the performance of real working PV-systems.

2.5 MEASUREMENT SOFTWARE

The measurement hardware (see Table II) is controlled by a personal computer via RS485 interface using the commercial software tool *LabWindows* (National Instruments Inc). Results of the measurements and system performance are shown in five different internet-pages (see Fig. 3-7). They are permanently updated and transferred to the internet server based on the TCP/IP standard protocol [4]. Another solution of conducting laboratory experiments over the Internet is given in [5] applied to semiconductor characterization by Java.

3. INTERNET PRESENTATION OF THE RESULTS

3.1 RESULTS GRID-CONNECTED SYSTEM

A PV generator in use supply about 10% less DC output-power than the specified STC value on the data sheet (see Table I). The loss of module efficiency is mainly due to the higher working temperature of about 50°C compared to the 25°C at STC. In Fig. 3 the performance of the real working two grid-connected subsystems are shown together with relevant weather data and temperatures. Thus the current performance of polycrystalline and amorphous modules and the DC/AC converters can be studied in the PVwwwLab. In combination with the measured module current/voltage characteristic the current MPP tracking of the DC/AC converter can be checked.

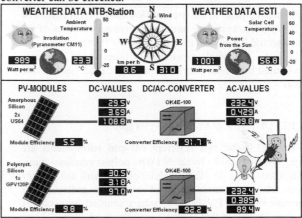

Fig. 3: Internet-page of the grid-connected PV-System monitoring (PVwwwLab measured at 14:30 / 2000-04-26)

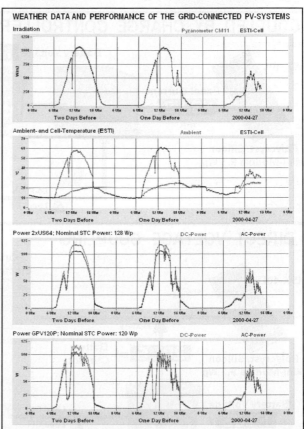

Fig. 4: Internet-page of the 3-day performance of the grid-connected PV system with an update rate of 10 minutes (last day 2000-04-27).

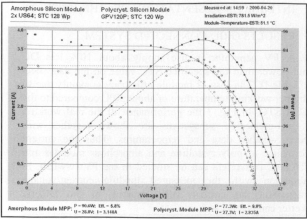

Fig. 5: Internet-page of the present IV-measurement results of the 120Wp poly-crystalline and the amorphous modules of the grid-connected PV system including the MPP values are given (see Table I). The whole measurement takes about 2 minutes and is limited by the used variable electronic load. (update rate 2 hours)

Fig. 6: Internet-page of the rear-surface temperature profile measurement of the 120Wp poly-crystalline and the amorphous modules of the grid-connected PV system. The temperature of the poly-crystalline module is in general higher but the variation across the surface is smaller compared to the amorphous module type.

3.2   RESULTS PV-BATTERY SYSTEM

In Fig. 7 the two different off-grid sub-systems are shown powered each by a 9.1Wp poly-crystalline mini-module (see Table I). The module current and the battery voltage are permanently updated and thus indicates the performance of the PV battery system coupled by a charge controller. By use of this charge controller the battery voltage is well below the maximum specified charging voltage of the battery. (overcharge is avoided)

While the second low cost PV battery sub-system is direct voltage coupled where permanently overcharging is not prevented. On this internet-page of the PVwwwLab the user is welcome to interact with the PV systems indicated by the "hands-on" symbols in Fig. 7. The direct voltage coupled sub-system can be loaded with a selected resistance by means of the electronic load. Additionally the immediate measurement of modules current/voltage curve can be requested.

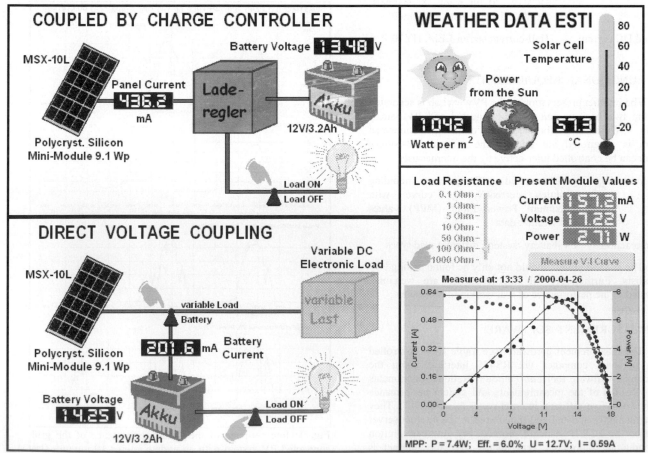

Fig. 7: Internet-page of the inter-active PV battery PV system. (PVwwwLab measured at 13:33 / 2000-04-26)

4. SUMMARY AND OUTLOOK

A public PV laboratory system has been developed conducting experiments and system-monitoring on outdoor PV-systems over the Internet. The permanently measured electrical values, the sun irradiation and the module temperature is presented on five internet-pages together with calculated performance values. Thus a small grid-connected and a small off-grid PV-battery system can be studied in detail under the present real working conditions. The extracted data like module and converter efficiency, MPP values can be compared to the manufactures-specifications by the internet-user.

One short-coming of the presented PVwwwLab is the reaction-time of the measurement system in the minute range and the transfer of the results as pictures over the Internet. In future alternative solutions will be tested, like applying Java to generate the figures at the user-side and only transfer the update measurement values via the Internet.

During the further optimisation the PVwwwLab will be extended by the following features:
☼ download list of measured data, monthly performance
☼ availability of simulated data of the different sub-systems
☼ accuracy calculation
☼ information on theory of PV-components (e.g. [6])
☼ related links (e.g. PV system-monitoring)
☼ web-camera

The PVwwwLab is not restricted to the educational sector. Furthermore, the future websites of PV companies can show their products in outdoor use, therefore giving the potential customer the opportunity to interact and watch the performance of real working PV-products.

REFERENCES

[1] Links to other interactive learning systems on photovoltaic, based on several simulation tools:
http://emsolar.ee.tu-berlin.de/~ilse/solar/index.html
http://www-lse.e-technik.fh-muenchen.de/
http://www.anu.edu.au/engn/solar/Sun/
http://www.ntb.ch/TT/Labors/EMS/WWW_links_erneuerbare_energien.html

[2] http://www.ntb.ch/Other/PV/Index.html
performance data of the 18kW PV plant at NTB;
S. Roth, E. Schönholzer, SEV/VSE Bulletin 10 / 1996

[3] NTB information on theory and application of different types of sensors: i.e. photodiode, pyranometer, ESTI, PT100 (language - German).
http://www.ntb.ch/Pubs/sensordemo/
http://www.ntb.ch/Pubs/sensordemo/pdf/NTB_23_ESTI.pdf
http://www.ntb.ch/Sensor/photodiode/Photodiode.html

[4] R. Heule, M. Brechtbühl; Diploma theses 1999 at Fachhochschule Buchs NTB, Switzerland
F. Baumgartner et.al. SEV/VSE Bulletin 1 / 2000

[5] http://nina.ecse.rpi.edu/shur/remote/
Automated Internet Measurement (AIM) Laboratory on semiconductor device characterization;
Hong Shen, et. al; IEEE Transactions on Education, Vol. 42. No. 3, August 1999

[6] NTB public information on solar cell technology
http://www.ntb.ch/TT/Labors/EMS/duenn/img001.gif

SENSOL® *global* – AN NEW MEASUREMENT SYSTEM FOR SOLAR RADIATION

Christian Bendel, Edgar Kunz, U.Rudolph, M.Schröder
Institut für Solare Energieversorgungstechnik, Verein an der Universität Gesamthochschule Kassel
(ISET e.V.), Königstor 59, D-34119 Kassel, Germany

Phone:(49) 561/7294224, Fax:(49) 561/7294200, Email: cbendel@iset.uni-kassel.de

ABSTRACT:

PV-plants and sites are assessed with solar radiation sensors. The partly very different physical qualities of the solar radiation sensors (spectral sensitivity, reflection, temperature dependency) lead to erroneous results. Thus different sites and plants are not comparable. Therefore, a new solar cell sensor SENSOL® was developed at ISET/Kassel. Based on SENSOL® a specific equipment SENSOL® *global* is created.

It measures the solar radiation in different directions as well as tilt angles. By a corresponding standardization of the measured values, a fast statement is possible for powergeneration, referring to the different directions and angles. The data transfer can take place via phone or radio transmission.

Keywords: Solar Radiation - 1: Solar Radiation Sensor - 2: SENSOL® *global* - 3

1. INTRODUCTION

The assessment of photovoltaic plants and sites occurs with insolation sensors. Normally these are pyranometer or silicon cell sensors. The partly very different physical qualities of the insolation sensors (spectral sensitivity, reflection characteristics on the surface and temperature dependency of the measurement signal) lead to erroneous results and should, if possible, take place with optimal solar radiation practice. High-quality pyranometers are not used in small photovoltaic plants and cheap cell sensors do not have the required long-term stability.

Thus different sites and plants are not comparable.

Pic. 1 PV-measuring station in Almeria

2. OBJECTIVE AND APPROACH

For the scientific site assessment, specific measuring modules are operated in the MPP (maximum power point). Pic.1 shows a measuring station in Almeria (in the south of Spain) which is in operation since 1993. This method of measuring is very practical but extensive and expensive.

Therefore, a new solar cell sensor SENSOL® was developed at ISET/Kassel whose physical qualities match well the solar modules on silicon basis.The following pictures (pic. 2 and 3) show the CAD-model and the first protype.

Pic.2 CAD- model SENSOL®

Since five month different SENSOL®prototyps with mono- and multicristalline cells are in operation at the ISET- test field. The test results up to now are promising.

The development of SENSOL® is especially significant for the solar market. A lot of solar firms and users wish a short-term market launch.

Pic. 3 First prototype of „ SENSOL®"

A new PV-measuring station has been developed especially for scientific purposes. Picture 4 shows a CAD-Model (design) of SENSOL® *global.*

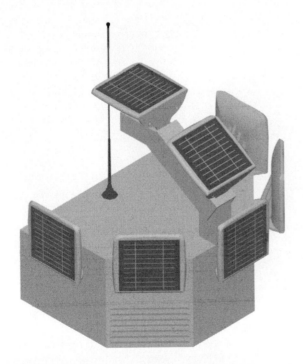

Pic. 4 CAD-Model of SENSOL® *global*

SENSOL® *global* consists of identical measuring sensors SENSOL® in different alignments. They measure the solar radiation in different directions (east, southeast, south, southwest, west) as well as tilt angles (0°, 32°, 90°). Simultaneously the cell temperature in every SENSOL® is measured with a sensor Pt 1000. The first preproduction model (pic.5) is located at present in the ISET test field The experimental results up to now are promising.The diagrams (pic. 6 and 7) show the typical solar radiation circumstances in the southern hemisphere.

Pic. 5 Preproduction Model „SENSOL®*global"*

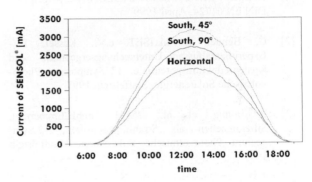

Pic. 6 Solar Radiation in different tilt angles

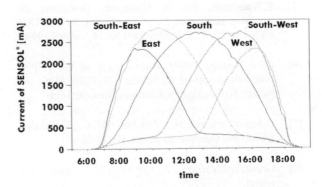

Pic.7 Solar radiation in different directions

3. CONCLUSIONS AND PERSPECTIVES

According to the observed deficits a solar cell sensor SENSOL® was developed, which meets the demands of the market. The sensor has the following specifications:
- high physical correlation with a real PV-generator,
- acceptable accuracy and instrumentation reliability,
- mechanically sturdy, weatherproof housing,
- any installation position possible,
- inexpensive structure,
- nice design.

On the basis of the SENSOL® a compact measuring equipment **SENSOL®** *global* was developed, by which unknown sites may be analysed with high quality, sufficient accuracy and measurement-technical reliability. In addition, **SENSOL®** *global* is inexpensive compared the use of test modules and has a nice design.

By a corresponding standardization of the measured values, a fast statement is possible for powergeneration, referring to the different directions and angles.

The complete measuring system is planned for supply by the grid and/or as an autonomous operation.

The data transfer can take place via phone or radio transmission. An integration into the internet is planned.

REFERENCES

[1]   Normen : DIN EN 60904-2 / IEC 904-2 , April '95; DIN EN 61724, April 1999

[2]   C. Bendel et al.,ISET e.V., Kassel, PV-Experimentalfassade-Untersuchungsergebnisse und Kostenreduktionspotentiale, 12. Symposium Photovoltaische Solarenergie, Staffelstein 1997

[3]   B.Grimmig et al., ISFH GmbH,Emmerthal, Siliziumzellen als Strahlungssensoren, 12 Symposium Photovoltaische Solarenergie, Staffelstein 1997

[4]   H. Ossenbrink et al.,ESTI, Ispra, DER QUASI-AMORPHE ESTI-SENSOR, 13. Symposium Photovoltaische Solarenergie, Staffelstein 1998

[5]   R.Wiegmann, Fa. if, Hannover, Gestaltung der Produkte nicht dem Zufall überlassen, INDUSTRIE DESIGN`99, VDI-Nachrichten Nr.14, 9.April 1999

[6]   Final Report, 1997, „The Results of the PEP`93 Intercomparison of Reference Cell-Calibrations and Newer Technology Performance Measurements"

[7]   Gebrauchsmuster Nr. 299 09 648.3, „Strahlungssensor und Gehäuse zu dessen Herstellung"

[8]   Geschmacksmuster Nr. 499 05 182.3, „Strahlungssensor"

[9]   Markenschutz Nr. 399 31 408.3/09, „SENSOL®"

One-Year Comparison of a Concentrator Module with Silicon Point-Contact Solar Cell to a Fixed Flat Plate Module in Northern California

P.J. Verlinden, A. Terao, S. Daroczi, R.A. Crane, W.P. Mulligan, M.J. Cudzinovic and R.M. Swanson

SunPower Corporation
430 Indio Way, Sunnyvale, CA 94085, USA
Tel: (408) 991-0900, Fax: (408) 739-7713
pverlinden@sunpowercorp.com
http://www.sunpowercorp.com

ABSTRACT: The objective of this paper is to report on a one-year field test comparison of a concentrator PV system to a fixed flat plate PV module. A concentrator PV module consisting of a Fresnel lens and a Point-Contact Silicon solar cell was mounted on a two-axis tracking system and tested over a one-year period in Northern California (Sunnyvale, CA). The results are compared to a multicrystalline Silicon flat plate module having a fixed southern orientation and tilt angle equal to the latitude (37 degrees). While one would expect that, during the winter months, the concentrator system would perform very poorly due to the small amount of direct sunlight, it appeared that the concentrator actually outperforms the flat plate module almost constantly. The total amount of energy produced by the concentrator is 37% greater than the energy produced by the flat plate on a kWh/m^2 basis.
Keywords: Concentrator - 1: Tracking - 2: High-Efficiency - 3

1. INTRODUCTION

In a recently published paper [1], R.M. Swanson studied the history of concentrator PV and discussed the reasons why concentrator technology has so far not taken a larger segment of the PV market. Concentrator PV appears to be the least expensive of all PV technologies for most of the locations in USA or in the world [1]. Also, concentrating photovoltaics has several other advantages over other PV technologies. It has the highest conversion efficiency demonstrated so far. It is not as dependent on the availability of low cost starting material (Silicon Feedstock) or the abundance of rare elements in the Earth's crust like Tellurium, Indium or Cadmium. The manufacturing of concentrator systems is, in a very large part, based on well established industries like steel, sheet metal and plastic. Therefore it is easily scalable up to several gigawatt per year of capacity.

Despite these advantages over the other PV technologies, the concentrator PV technology has had very little growth for the last 20 years, whereas the conventional crystalline Silicon PV technology has grown at an almost steady rate of 20% to 30% per year. Probably the main reason is that concentrating PV is only suitable for systems of 1 kW minimum or even greater in size. Therefore, concentrating PV has not received its chance yet to demonstrate its capability and its advantages.

There are many preconceived opinions about concentrator PV. One of them is that concentrators are reserved to very sunny locations, like the deserts in the Southwest US (New Mexico, Arizona) or North Africa, and that they do not produce enough energy in other locations where the amount of direct solar radiation is small compared to diffuse solar radiation [3]. The objective of this paper is to demonstrate that concentrators actually work very well, even in winter, in most locations. Over a one year period, from October 1998 to October 1999, we compared the energy produced by a Fresnel lens concentrator module to a commercial flat plate module (in

Sunnyvale, a location about 40 miles south of San Francisco, CA).

2. AVAILABLE SOLAR RADIATION

The annualized energy cost for a given PV system in a particular location is inversely proportional to the total amount of solar energy available per unit area at this location and to the conversion efficiency of such PV system. The concentrating receivers have a small acceptance angle and do not absorb diffuse light. Therefore, they have to track the sun. In theory, if the solar radiation were entirely direct, the gain in energy due to the 2-axis tracking would be about $\pi/2$, or 1.57. However, because of the low acceptance angle of the concentrator system and because the direct radiation is usually only 85% of the total solar radiation, the ratio between the direct solar radiation with 2-axis tracking and the global solar radiation with a fixed receiver is much lower than $\pi/2$, as Figure 1 shows. It varies between 0.61 and 0.97 in Seattle, WA, where the amount of direct radiation is quite small compared to diffuse radiation. In Albuquerque, NM, however, the ratio varies between 0.97 and 1.2 [2]. This means that, in order to produce, every month, more energy with a concentrator than with a flat plate module of the same area, the concentrator PV system has to be at least 3% relatively more efficient than a flat plate PV system in Albuquerque, but about 63% relatively more efficient in Seattle.

According to the National Solar Radiation Data Base [2], Seattle almost represents the worst case for concentrator systems in the contiguous 48-state US. Surprisingly, Seattle receives about the same amount of direct normal solar radiation over one year as Fairbanks, AK, i.e. 1074 kWh/m^2 and 1059 kWh/m^2 respectively. On the other hand, Albuquerque represents the best case for direct normal solar radiation with 2442 kWh/m^2 in average

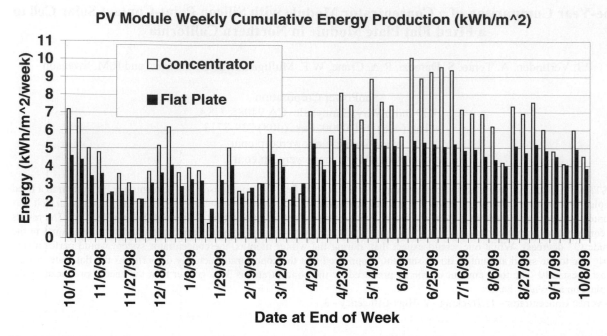

Figure 2: Weekly produced energy comparison of the Concentrator and fixed Flat Plate modules from October 1998 to October 1999 in Northern California (Sunnyvale), about 40 miles south of San Francisco. Over the year, the Concentrator module produce 37% more energy per square meter than the fixed Flat Plate module, and outperforms the Flat Plate module almost every week.

over 30 years. Finally, San Francisco and Minneapolis represent an average solar radiation profile that is comparable to most US locations within the contiguous 48 states [2].

Figure 1: Ratio of average monthly direct normal solar radiation (kWh/m^2) and global solar radiation for a fixed tilt angle equal to the latitude in four different locations [2].

Considering that the average flat plate module efficiency is around 12% (AM1.5, 25C), a concentrator PV modules should be at least 19.7% (AM1.5, 25C) in order to produce every single month more energy than a fixed flat plate module located in Seattle, WA. This efficiency value is already achieved by most high-efficiency, high-concentration PV modules. For example, Point-Contact Silicon solar cells manufactured in a production environment are up to 26.8% efficient at 100X (or 10 W/cm^2) and 25.5% efficient at 275X (or 27.5 W/cm^2, AM1.5, 25C) [4].

Optical and interconnection losses are in the range of 15% and 5% respectively, resulting in a typical module

efficiency of 21.6% and 20.6% at 100X and 275X respectively (AM1.5, 25C). We previously demonstrated such high efficiencies, up to 22.9% and 20%, for 88X and 250X concentration ratio respectively [4]. Therefore, with such efficiency, we should expect the concentrator module to outperform the flat plate module in any US location. The following experiment demonstrates it for one location, in Sunnyvale, CA, about 40 miles south of San Francisco.

3. EXPERIMENTAL SET-UP

A data acquisition system was used to record parameters from both concentrator and flat plate modules every 5 minutes over a one-year period, from October '98 to October '99. The measured parameters were: short-circuit current (Isc), open-circuit voltage (Voc), maximum power current (Imp) and voltage (Vmp), ambient temperature (Ta), direct normal incident power density (DNI), global incident power density at the same angle as the fixed flat plate module (GIP), backside temperature of the flat plate module (Tfp) and heat-spreader (Tsp) temperature of the concentrator module. From these measured data, efficiencies, power and energy produced were calculated. For the purpose of comparison, the output power and energy were normalized to the module area. The flat plate module is a 145 W rated module, commercially available, consisting of 42 multicrystalline solar cells of 225 cm^2 each. The concentrator module is composed of one 650 cm^2 Fresnel lens concentrating 27.5 W/cm^2 onto a Silicon Point-Contact solar cell of 1.56 cm^2 in active area. The concentrator cell is soldered onto a ceramic substrate and is passively cooled.

4. RESULTS

The measured efficiencies of the concentrator and flat-plate modules were 16.5% and 10.7%, respectively, at operating temperature, and estimated 19% and 12.5% respectively at 25C cell temperature. Over the 1-year period, the cumulative energy production of the concentrator was more than 37% greater than the flat-plate module (Fig. 4) on a kWh/m² basis. On a kWh/kW basis, i.e. the amount of produced energy per rated power, the concentrator outperforms the fixed flat plate by more than 24%. In other words, the capacity factor of a concentrator system in Sunnyvale would be more than 24% greater than the capacity factor of a fixed flat plate system. Of course, the comparison on a kWh/kW basis is difficult, and subject to discussion, because concentrators and flat plates are usually not rated the same way. Concentrator are usually rated at PVUSA testing conditions (PTC) or at 850 W/m² AM1.5D, 20C ambient temperature, 1 m/sec wind speed. On the other hand, flat plate modules are rated at Standard testing conditions (STC) or at 1000 W/m² AM1.5G, 25 C cell temperature. In order to have a fair comparison between the flat plate and the concentrator PV systems, it appears very important that both modules be rated at the same conditions. But, more importantly, the PV industry and scientific community need to agree on an "energy rating" of the modules instead of "peak power rating".

Surprisingly, even under relatively cloudy conditions, such as those of Northern California in the wintertime, the PV concentrator is superior to the flat-plate most of the time. Conventional wisdom says that concentrators will be effective only in geographic areas with exceptional direct solar resources, such as the desert southwest US. But testing in the relatively cloudy San Francisco Bay area showed that the PV concentrator outperformed the fixed orientation flat-plate module in all but 4 weeks of the year. On a monthly average basis, the concentrator outperforms the flat plate module every single month (see Fig. 3), and, on a daily basis, 269 days out of 365. The explanation of this remarkable result is that while PV concentrator output is indeed near zero in cloudy weather, the output of the flat-plate module under the same conditions, while non-zero, is also extremely low (see Fig. 5). If there are even a few sunny hours in any given week, the performance advantage of the PV concentrator will more than compensate for the small advantage of the flat plate system in low-light cloudy weather.

In the summer months, when demand for electricity is often highest, the concentrator outperformed the flat-plate by as much as 2 to 1 on a daily basis (see Fig. 6). The ratio of produced energy is greater than 1.65 in June.

Because of the tracking system, the power production of the concentrator is much more level throughout the day than a fixed orientation flat-plate system (Fig. 7). This not only leads to higher daily energy production and a higher capacity factor, but also has significant load matching advantages. For example, in sunny climates peak electric load due to air-conditioning occurs late in the afternoon rather than at solar noon.

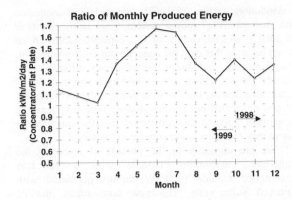

Figure 3: Ratio of produced energy (kWh/m²/day) by the concentrator and the flat plate module from October '98 to October '99 in Sunnyvale, CA (month #1 is January 1999). Compare Fig. 3 to Fig. 1, and notice that the efficiency of the concentrator module largely compensates for the small advantage of the flat-plate system in cloudy weather.

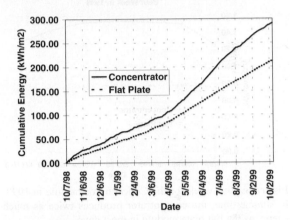

Figure 4: Cumulative Energy Production (kWh/m2) over one year period in Sunnyvale, CA of a concentrator and a flat-plate module. The concentrator module produces 37% more energy than the fixed flat-plate, on a kWh/m² basis.

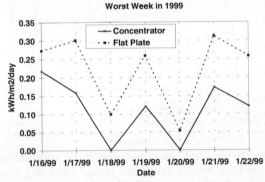

Figure 5: Worst week in 1999 for direct sunlight and for the concentrator module. But the flat plate module does not produce much electricity either.

Of course, the comparison of the performance of the concentrator and the flat plate modules may lead to different conclusions if the PV system is a grid-connected or stand-alone system. If the PV system is grid-connected with "net metering" billing system, what really counts is the total produced energy or the capacity factor on an annual basis. In this case the concentrator system wins with a large margin. And due to the efficiency difference, we estimate that the concentrator would even win in a cloudy climate like Seattle. On the other hand, if the PV system is not grid-connected but is a stand-alone remote system, the comparison leads to the conclusion that the concentrator system may require a larger capacity of battery.

We are currently in the process of replacing the single-lens concentrator module with a 12 lens concentrator module (120 W rating) populated with improved solar cells. The new concentrator module efficiency is now more than 18% at operating temperature. We will continue the comparison with the fixed flat-plate module. We expect the concentrator module will deliver about 50% more energy than the multicrystalline flat-plate module in Northern California, and even more in desert locations.

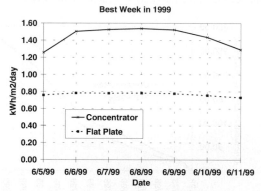

Figure 6: Best week for the Concentrator module in 1999. In summertime, the concentrator produces twice as much energy as the flat plate module in most days.

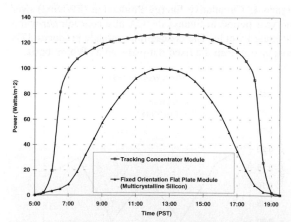

Figure 7: Electrical power produced by the concentrator and flat plate module during one typical sunny day. One can notice that the power production of the concentrator is much more level throughout the day than a fixed orientation flat-plate system.

5. CONCLUSIONS

There are many preconceived opinions about concentrator PV systems. One of them is that concentrator

PV systems do not perform as well as flat plate modules in cloudy conditions. The objective of this paper was to put this preconceived opinion under a very simple and practical test. A flat plate module on a fixed orientation and a concentrator module on a 2-axis tracker have been tested side-by-side in Sunnyvale, CA. This simple experiment just confirms what we could simply have calculated from the National Solar Radiation Data Base.

The concentrator module produces 37% more energy than the flat plate module, on a receiver area basis. The estimated capacity factor of the concentrator is more than 24% greater than the capacity factor of the flat plate module. The concentrator also outperformed the flat plate module every single month, 269 days per year and 48 out of 52 weeks. This outstanding result is explained by the fact that the high efficiency of the concentrator module largely compensates for the small advantage of the flat plate module in cloudy weather conditions. While the concentrator output is indeed near zero in cloudy weather, the output of the flat-plate module under the same conditions, while non-zero, is also extremely low. If there are any few sunny hours, the tracking system allows the concentrator to absorb the direct solar radiation and to produce often more energy than the flat plate.

It appears that the STC peak power rating of the PV modules is not adequate for comparing different PV technologies. The PV industry needs to agree on an energy rating of the PV modules, instead of the standard peak power rating.

ACKNOWLEDGMENT

We would like to thank other members of SunPower Corporation for their contributions to this work, particularly Matt Piper for his constant and much appreciated technical support.

REFERENCES

[1] R.M. Swanson, "The Promise of Concentrators", Progress in Photovoltaic Res. Appl., **8**, pp. 93-111, 2000

[2] "Solar Radiation Data Manual for Flat-Plate and Concentrating Collectors", NREL/TP-463-5607, April 1994

[3] A. Goetzberger and W. Stahl, "Global Estimation of available solar radiation and cost of energy for tracking and non-tracking PV systems, 18th IEEE Photovoltaic Specialists Conference, 1985, pp. 258-262

[4] P.J. Verlinden, R.M. Swanson, R.A. Crane, K. Wickham, J. Perkins, "A 26.8% Efficient Concentrator Point-Contact Solar Cell", 13th EC Photovoltaic Solar Energy Conference, Nice, October 23 -27, 1995, pp. 1582-1585.

FUTURA2 AND THE WSC - COMMERCIAL PV CAN BE EFFICIENT

C. Bucci [(+)], R. Nacci, G. Freda, R. Peruzzi, F. Ferrazza, E. Fioravanti
Eurosolare S.p.A, Via A. D'Andrea 6, 00048 Nettuno, Italy
tel +39 06 98560519, fax +39 06 98560234, email francesca.ferrazza @ eurosolare.agip.it
[(+)] present address University of Urbino, Italy

G. Coia, A. Crisante, R. D'Antonio, G. Izzicupo, G. Di Gregorio.
FUTURA TEAM c/o giuscoia @ tin.it
http://web.tiscalinet.it/futura2

ABSTRACT. The FUTURA team designed and realised a solar powered vehicle which participated to the DREAM Electro Solar Cup in Japan in August 1999, and to the World Solar Challenge in Australia, October 1999. FUTURA2 represents an evolution of the first model, based on the experience built up during the edition 1997. The PV power system was realised by EUROSOLARE S.p.A. in the pilot line of the R&D unit. An eight month programme was set up to improve the standard production process to fulfil the tight specifications of a solar car, that is high power, low weight, reliable encapsulation and easy assembly.
Keywords: Solar car – 1: Manufacturing and processing – 2: System – 3.

1. INTRODUCTION

The first improvement the FUTURA team required for generation 2 was a power supply of about 1200 W on the available surface of the car, which meant an encapsulated efficiency of at least 15.5%. Furthermore, the previous vehicle mounted commercial cells directly soldered by the team on the body of the car, a lengthy and difficult operation which also created isolation problems because the car is in carbon fibre. Therefore, a different solution was adopted by realising light, flexible, easily connected modules which the team could assemble in a short time.

The outcome of the programme and of the co-operation between FUTURA and EUROSOLARE until now has produced an improved solar cell pilot line process based on simple industrially scalable processes able to realise up to 16% efficient cells, and a convenient packaging process to produce light flexible modules. The cell process developed is an optimisation and an improvement of the existing screen printing based one, adds no extra passivation or selective emitter and is therefore particularly interesting because it provides a good compromise between moderately high efficient cells and low production costs.

The car has been used firstly in a challenge in Japan 1999, on the racing circuit of Suzuka. Although the structure of the vehicle was not really suited for high speed performance, having been designed for running long distances in the harsh conditions provided by the Australian desert, the final rank of this first exhibition was an encouraging 14th place out of 80.

The World Solar Challenge took place in Australia in October 1999, and the car ended in 24th position in the final ranking, slightly below starting expectations but largely in excess of the first outcome in 1996.

In any case the development work done in co-operation between FUTURA and Eurosolare has brought a number of advantages to both groups, and demonstrated that cells realised with upgraded industrial processes can be used in applications with stringent requirements, which can open new perspectives for commercial applications.

2. CELL PROCESSING

An example of the distribution of the cells realised for the solar car is reported below. The histogram refers to about 1200 single crystalline (commercial substrates) cells and is the early outcome of the programme. By the end of the study, all lots showed average efficiency values around 15.8%, as witnessed by the performance of the modules. A small number of cells with efficiency exceeding 16% resulted as well, which is one of the highest efficiencies reported for this kind of process. The cells were measured in standard conditions at Eurosolare using secondary references provided by certified institutions.

Class	Frequency
13,75%	1%
14,25%	2%
14,75%	5%
15,25%	31%
15,75%	56%
16,25%	3%
Aver. Eff.	15.7%
n. cells	1200

Table 1: Efficiency distribution of cells before tabbing and encapsulation

The performance of these cells is a remarkable improvement compared to that of a standard solar cell production line for terrestrial silicon PV, but is the result of only a slight complication in the process sequence. In particular no selective emitter was used, or any sort of intentional passivation. The cells had instead a pseudo-BSF Al layer which produced a partial gettering effect as well. The Al deposition and firing conditions were optimised for this programme, but are now available for the standard industrial process, which has benefited of the results of this work.

The flow chart of the process used is reported below. The only real difference with the standard production baseline is the greater accuracy in handling and all the chemical etching/washing sequences, and, most of all, the final cut of the outer edges with a mechanical saw. This is of course an unpractical step in a large scale production line, but came out as a necessity due to the constraints on power density for the car. The cells were therefore truly square single crystal CZ based, with a surface of about 100 cm^2.

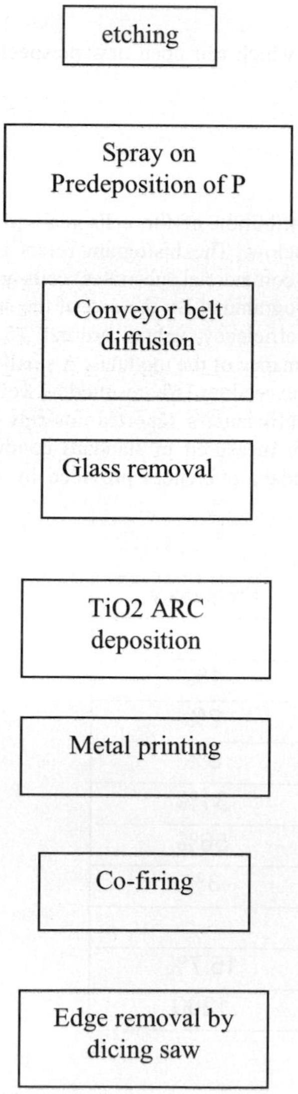

Figure 1: Schematic of process optimised for the cells

3.   MODULE FABRICATION

Another task of the R&D unit at Eurosolare was the realisation of modules suitable for mounting on the (existing) structure of the vehicle. The first Futura car in 1996, was in fact powered by the team by soldering the cells they had bought from a commercial distributor directly on the body of the car. This implied difficult and long operation, uncertain results because of the impossibility of checking the performance of cells after soldering, and little reliability of the structure due to the conductance of the car, in carbon fibre.

Therefore a number of plastic encaspulants were tested for application on relatively large modules, with an easy to handle "plug-in" connection system. The material was selected on the basis of electrical performance, ease to laminate, absence of aesthetic defects etc, and the modules were realised, tested one by one and sent to the FUTURA team for assembling.

The initial test of the system was the Dream Cup in Japan, were for the first time the compatibility of the car with its power supply were evaluated. The test was successful, and the modules were able to provide the required power to charge the batteries.

The second test was far more challenging, as the car participated to the World Solar Challenge in Australia, October 1999.

From the point of view of the module assembly and operation, little difficulties were observed both in terms of performance or in terms of durability of the PV components.

The plastic laminates were able to withstand the thermal excursions and the high peak temperatures with no delamination evidenced by the 3000 km race. In fact, the modules are still functioning and, thanks to the easy set-up developed for the car, are ready to be disassembled and used in a different application.

In our opinion, this field test witnesses the growing technical expertise in both materials and module assembly technology for applications which can readily be considered unconventional, but with an increasing demand from the public, in very different sectors (boating, leisure, spacecraft). This is in turn asking for a set of newly designed specifications for products which are evidently never going to sit 20 years on house roofs or in static power plants.

4.   CAR SPECIFICATIONS

4.1 Car data
Car n°   63
Team Name   FUTURA
Vehicle name FUTURA2
Country   Italy
Team Leader   Crisante Angelo
Cost ($US)   150.000 (100.000 vehicle)
Project time 18 months
N° team members   5

4.2.
Predicted Average speed (km/h) 60
Predicted Speed sol./ batt. (km/h) 70/90
Actual average speed 46 km/h

4.3 Dimensions
Vehicle weight (kg) (with battery) 240
Lenght (mm)   6000
Width (mm)   1960
Height (mm)   1350

4.4 Chassis / body
tubolar structure in ERGAL 70/20 / sandwich of carbon fiber and nomex

4.5 Aerodinamics
Frontal area (m2)   1.04
Drag coefficient   0.1
determined using constant velocity energy measurements

4.6 Solar Array
Cells: 792 Eurosolare mono silicon 98*98 mm 2 area, 6 g weight about
15.5 % efficiency

4.7 Encapsulation:
Cells laminated by EUROSOLARE between layers of ICOSOLAR, Tedlar and EVA,
and bonded to shell with silicon rubber

4.8 Array type: two array in parallel (Dx and SX) each
formed by 6 panels of 33 cells + 6 panels of 30 cells + 1 panel of 18 cells all in series

4.9. Array specifications:
total area (m2)   7.98
effective area (m2)   7.60
active area (%)   95.2
Voltage (V)   192
Weight (Kg)   15.72
Est. Eff (%)   15.68
Total Power (W)   1192

4.10 Motor and trasmission
Motor: MGM Motori Elettrici high efficiency asynchronous
Weight (Kg)   12
Voltage (V)   180
Maximum Speed (rpm): 3500
Cont. Power output (KW): 1.5
Peak power output (KW): 3.5

Wheels: MARCHESINI RUOTE magnesium 3/2.15"
Tyres: MICHELIN Solar Car Radial 65/80-16

5.   WHAT NEEDS OPTIMISING

Given the relatively limited funding used for the project, and the respectable ranking obtained in the prestigious World Challenge, the team is satisfied. However, improvements to the car structure, a better matching of the car design to the PV generator system (in this case, the car was built prior to the design of the PV array) can certainly be of great help in gaining position in a future race. What especially is worth considering is the heavier bank of batteries (for the same output power) FUTURA had compared to other competitive cars, which undoubtedly slowed the vehicle beyond demerits.

6. CONCLUSIONS

A development effort was carried out by the R&D group of Eurosolare with the aim of improving the performance of industrial - like silicon solar cells and of providing easy to handle - low weight modules to be used as power systems for the solar car FUTURA2, designed and realised by the Futura Team in Italy.

The development led to remarkable improvement in the efficiency of   low cost solar cells, and to encouraging performances of the car in official competitions.

It is worthy to note that module performance was not the major limiting factor to better performance.

On the whole the initiative, promoted by the FUTURA team with very limited funding available, and with the support form Eurosolare to improve the cell efficiency and realise the PV generator, can be considered quite successful, also in the manufacturing of light weight nodules.

7. ACKNOWLEDGEMENTS

Thanks to Dr. Plessing and S. Degiampietro of ISOVOLTA for extensive assistance in choosing and using plastic encapsulants for the modules.

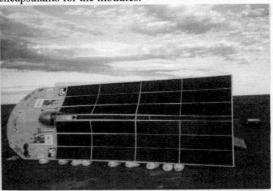

Figure 2: FUTURA 2 enjoying the morning sun

LOW COST SYSTEM FOR ILLUMINATION TEST OF HIGH CONCENTRATOR SOLAR CELLS

I. Rey-Stolle, L. Calderón and C. Algora
Instituto de Energía Solar; E.T.S.I. de Telecomunicación - U.P.M.
Avda. Complutense s/n, 28040 Madrid (SPAIN)
Tel.: +34 91 544 10 60; Fax: +34 91 544 63 41; Email: irey@ies-def.upm.es

ABSTRACT: High efficiency high concentrator (above 1000X) solar cells are emerging as a real alternative for dropping PV costs. The development of such a family of photovoltaic converters demands specific characterisation tools able to provide high illumination fluxes in an efficient accurate and simple fashion. The purpose of this work is to describe a very simple and low cost flashlamp based system for the characterisation of very high concentrator III-V solar cells at a wide range of concentration levels (25X – 4000X). Such a system produces results in excellent agreement with those of more elaborated techniques.
Keywords: Concentrator- 1: Calibration - 2: III-V - 3

1. INTRODUCTION

Concentration has been traditionally considered as an attractive way of dropping the global cost of a PV system. Particularly, high concentrator approaches where illumination intensities of 1000 suns or higher are involved together with very small converters (usually III-V cells) are a promising technology which has a clear potential for very low cost and has already shown technological feasibility [2]. Furthermore, the future prospects for the performance of high efficiency tandem solar cells under concentrated sunlight are of remarkable interest for terrestrial applications as demonstrated by the new absolute efficiency record for a PV device recently obtained by a triple junction ($GaInP_2$/GaAs/Ge) device under 50 suns.

It is obvious that the development of such a family of PV converters demands *ad hoc* efficient and accurate characterisation tools able to provide high illumination fluxes. To solve this task many systems have been proposed. Among the variety of possible choices, those based on flashlamps [1,3-4] are of widespread use due to their advantages regarding cost, power consumption, maintenance and controls involved. The purpose of this work is to describe a very simple and low cost flashlamp based system for characterising very high concentrator solar cells at a wide range of concentration levels (25X – 4000X). Such a system produces results in excellent agreement with those of more elaborated techniques.

2. SYSTEM DESCRIPTION

The main feature of the system herein presented is the simultaneous coincidence of high performance, low cost and simplicity. High performance is achieved through a low error level, a reasonable speed and a wide range of concentrations analysable: form 25X up to 4000X. Low cost and simplicity are accomplished by means of the utilisation of a simple electronic circuit (a voltage follower) as a fast, cheap and easily adaptable active load and the use of a low cost data-acquisition board (low sampling rate and no need of *sample and hold* capacity). Moreover, the system has the additional advantages of being greatly modular and therefore it can be facilely upgraded or adapted to different needs.

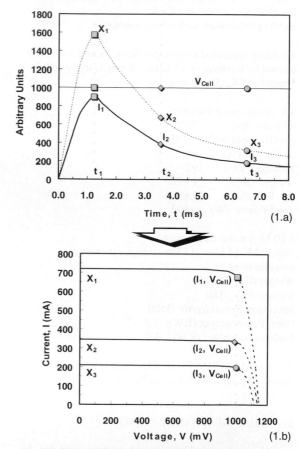

Figure 1: (a) Evolution of irradiance (dashed line), voltage (thin solid line) and photocurrent (thick solid line) during a flashlight pulse. (b) I-V curve building

2.1 Measuring Principle

The measuring principle was taken form [1] and is depicted in Figure 1. When the flashlamp is triggered the irradiance steeply increases up to a maximum value and then it decreases in a smoother manner down to zero again (dashed line in Figure 1). If the voltage drop across the measured solar cell is held constant during light pulse (thin solid line in Figure 1), the illumination current generated (thick solid line in Figure 1) will be that corresponding to the fixed voltage in the I-V curve for every concentration present in the light pulse. Therefore, given a known fixed value for the cell voltage (V_{cell} in Figure 1) if we take samples of the cell current (I_1, I_2 and I_3 in Figure 1) in different instants of the light pulse (t_1, t_2 and t_3 in Figure 1), we will get a point of the I-V curve (I_i, V_{cell}) for the corresponding concentration (X_1, X_2 and X_3). If we repeat the process for as many voltages as our precision needs demand we will be able to trace the whole I-V curve for a wide range of concentrations (Figure 1.b).

Of course, the latter procedure implicitly takes for granted that the solar cell is in a quasi-steady state during the measurement. In other words, we are assuming that the cell's response is much faster than the irradiance variation. This hypotheses restricts the validity of our system to fast direct-gap semiconductors (most III-V) where lifetime of generated carriers is in the range of nanoseconds while our light-pulse variation is in the range of miliseconds.

2.2 System Architecture:

The system architecture is depicted in Figure 2. As shown in that figure the system is formed by: 1) a flashlamp as high intensity light source; 2) a simple electronic circuit (a voltage follower) acting as a very fast active load to maintain constant the voltage drop across the cell during the light pulse; 3) a data acquisition board to control the reference voltage that has to be fixed by the active load and to measure the current generated by the cell;·4) a computer with specific software to process the signal obtained and to display the results and 5) a couple of DC power sources to supply the energy demanded by the flashlamp and the active load.

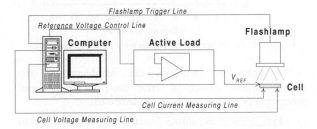

Figure 2: *System architecture (power sources for the flashlamp and active load are not included for simplicity).*

A main feature of the architecture used is that there is no reference cell present. The absence of such a common element in cell measuring systems is due to joint requirement of high concentration operation and low cost, and will be justified in detail later on. Let us first consider the particular characteristics of the main components of the system.

The flashlamp used is based on a conventional xenon bulb triggered by a current pulse generated by a R-C network. In order to achieve high irradiance the pulse is very short, being 90% of its energy condensed in the first ten miliseconds.

A simplified schematic for the active load is depicted in Figure 3. It is a classical voltage follower, formed by an operational amplifier (OA) and a power stage that provides/soaks the current demanded/generated by the solar cell. There is also a shunt resistor acting as a current sensor. The transistors in the power stage and the power supply voltage ($\pm V_{EE}$) can be chosen to match the current needs of the type of solar cell to be analysed. The card sets

the desired voltage on the negative input of the OA (V_{Ref}) and that voltage is forced on the cell. While in the dark, the load provides the polarisation current demanded by the cell (by means of the npn Darlington pair) and when the flash lamp is triggered soaks the generated current (by means of the pnp Darlington pair), which is measured through the voltage drop across the resistor R_{shunt}.

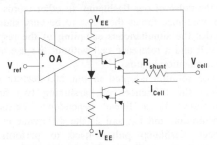

Figure 3: *Simplified schematic for the active load circuit.*

The data acquisition board is the component devoted to controlling and actually performing the measurement. The control functions are mainly to trigger, monitor the charge state, and reset the flashlamp; and to set the appropriate reference voltage (V_{Ref}) for every measurement. The measuring capabilities are of course related to reading the magnitudes of interest (i. e. voltage drop across R_{shunt}) with the maximum accuracy. The key parameter for that last function is the sampling rate (T_S), defined as the minimum period elapsed between two consecutive samples of the same signal. Such parameter determines the quality of the measurement, obviously the price of the board, and besides has a definite influence in our approach. So let us consider it more carefully. A general expression for the sampling rate is given in equation 1.

$$T_S = (T_{MUX} + T_{ADC} + T_{Save}) \cdot n \qquad (1)$$

where T_S is the sampling rate; T_{MUX} is the time elapsed by the channel selection (i.e. the time that multiplexer takes to switch between the different inputs); T_{ADC} is the time elapsed by the analog to digital conversion; T_{Save} is the time that takes to transfer the data from the Analog-to-Digital Converter (ADC) output to the memory; and n accounts for the number of signals being sampled at a time. For high-end data acquisition boards $T_S \approx n \cdot T_{ADC}$ as the multiplexer is very fast and usually the boards have some memory to hold the incoming data while the measurement is taking place (*sample and hold* feature). Conversely, in low-end boards all three times have to be considered as usually switching between channels is not so fast and the transfer of incoming data to the computer memory takes a non-negligible while. It is noteworthy that most manufacturers will only quote the T_{ADC} in their specifications but a real sampling cycle will always imply the selection of an input, the conversion of the data in that input and the saving of the results.

According to the foundations of discrete time signal processing [7] a signal has to be sampled at a frequency higher than twice the value of its maximum frequency component (f_{max}) in order not to lose information. This condition –known as the Nyquist Theorem– expressed in equation (2), determines the set of signals that with a given T_S is possible to sample without losing information.

$$T_S < 2/f_{max} \qquad (2)$$

The analysis of the current pulse produced by our flashlamp reveals that the maximum frequency component of that signal has 12 kHz, and thus should be sampled at a frequency ($f_s = 1/ T_s$) higher than 24 kHz (i. e. $T_s \approx 42 \ \mu s$). Usual low-end data acquisition boards have ADC conversion frequencies around 40 kHz ($T_{ADC} \approx 25 \ \mu s$), which just allows the correct sampling of a single signal ($n=1$) like the pulse of our flashlamp. In other words, achieving high irradiance forces the pulse to be very short and thus impedes the simultaneous sampling of the response of a test cell and a reference cell without the use of very fast data acquisition boards.

The solution adopted in our architecture was to do without the reference cell, assuming two fundamental hypotheses: 1) a linear dependence exists between concentration and I_{sc}; and 2) the difference in irradiance between flashlamp pulses used to perform a whole measurement is negligible. The above hypotheses allow the use of the cell under test itself to determine the irradiance of a measurement. In a first flashlight pulse short circuit current (I_{sc}) is measured by setting $V_{Ref} \approx 0$. Then equivalent concentrations are calculated, dividing the measured I_{sc} by the given one-sun short circuit current. For the subsequent flashlight pulses, performed at different values of V_{Ref}, the same concentration distribution is taken for granted (i. e. the pulses are almost identical and thus equivalent instants always correspond to the same irradiance).

2.3 Data Processing:

The raw data (i.e. data as read by the board) have to be processed in order to minimise the systematic and random errors that our system and procedure might cause. Among the main sources of error are:

- *Electrical noise*: Both the PC board and the active load are sensitive to electrical noise if placed in a noisy environment or if connected to a non-stabilised supply. Such noise would cause slight oscillations or *glitches* in the current signal measured.

- *Slight differences in consecutive light pulses*: Even though the flashlight has proven to be very stable, slight differences in consecutive pulses might appear mainly when the flashlight has been working for a long period.

- *Sampling limitations:* as referred in the previous subsection.

The issues related to noise and to flashlamp instability can be addressed by the repetition of the measurements and a subsequent statistical analysis of the results assuming a Gaussian distribution of the error for both phenomena. On the other hand, errors arising from sampling limitations have to be treated in a different manner. Let us take a closer look at the process. To trigger the flashlamp the computer closes a relay that allows the discharge of a couple of capacitors to supply the current pulse needed to excite the xenon flashbulb. The time elapsed in that process is variable mostly due to the relay closing time. To be sure that the whole light pulse is sampled, the *sampling window* starts just before the closing command is sent to the relay. Accordingly, the complete pulse is sampled but there is no absolute reference about the initial sample of the pulse since the relay lag is included within the *sampling window*.

As shown in Figure 4 such variable lag associated to the mechanical parts in the relay causes a delay (of three samples in the figure) between two consecutive measurements of the current generated by the flashlight pulse. Such large-scale delay has to be corrected and measured signals have to be aligned to an absolute reference. This is important to prevent a large error to occur since a hypothesis has been made regarding the equivalence between time and concentration, which is obviously violated by the pulses in Figure 4. In addition, yet another small-scale delay exists. If we consider the first non-zero samples of both pulses in Figure 4 (circled) evident differences in current between equivalent samples are observable. This again contradicts the assumed time-concentration equivalence. As a matter of fact, the first non-zero sample of a pulse can be randomly located in any instant of the first T_s seconds of the pulse (i.e. a *small-scale delay*). However, as both signals in figure 4 have been sampled observing the Nyquist theorem (eq. 2) full information about how to rebuild intermediate portions of the signal is present in both sequences. Thereby, as many intermediate samples as required can be interpolated to minimise the incidence of the *small-scale delay*.

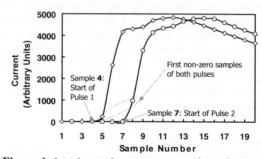

Figure 4: *Initial part of two consecutive pulses as measured*

In order to minimise the incidence of the aforementioned causes of error the following signal processing procedure has been implemented:

a) *Pre-filtering*: An initial filtering step is performed to detect correctable errors in the measurement (zeros caused by a too early reading of the ADC,...)

b) *Interpolation:* New samples are generated by low pass filtering the signal measured to minimise the *small-scale* delay.

c) *Origin determination*: The absolute alignment of the signals is performed by calculating the correlation of the measurement with an estimator. The sample where such correlation achieves its maximum value indicates the starting point of the pulse. This procedure has turned out to be much more robust than any other based on the sudden change in the signal value or its slope. The reason for this is that those methods are very sensitive to spurious noise. On the other hand, the correlation proposed deals with the signal general shape and thus it is insensitive to errors in single samples.

d) *Filtering*: Once aligned, the measurements are filtered using a five-coefficient FIR filter. This minimises the incidence of Gaussian noise. In addition, the signals are shortened to their *useful window*, i.e. the interval of samples that guarantees an acceptable signal to noise ratio.

e) **Final Acceptance or Rejection:** If after filtering the signal still shows excessive noise or specific irregularities (failure of the multiplexer, ...) the measurement is rejected, and has to be repeated.

The described data processing provides us with *clean* measurements of the cell current for different cell voltages. With such measurements the construction of I-V curves for as many concentrations as present in the *useful window* of the light pulse is straightforward. The measurement, the signal processing and the subsequent IV curve construction are processes fully automated by a software program.

3. RESULTS

The validation of the system designed has been carried out through the comparison of its output with the results obtained in other laboratories, namely Ioffe Institute (St. Petersburg, Russia) and National Renewable Energy Lab. (Golden, Colorado, USA). The formers are two institutions of great international renown in concentrator cell testing. Figure 5 and Table I show the results of the comparison for a typical LPE AlGaAs/GaAs concentrator solar cell. All comparison experiments showed that the results produced by the new systems are in excellent agreement with those provided by reference institutions.

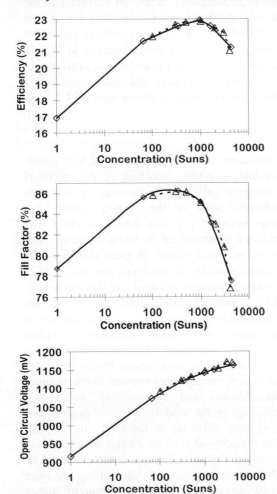

Figure 5: *System Results. Our measuring system (solid line, diamonds) compared to those obtained for the same device at Reference Institute (dashed line, triangles)*

Table I: *Results of our measuring system compared to those obtained for the same device by a Reference Institute*

Developed System Results				Reference Institute Results			
X (suns)	η (%)	FF (%)	V_{OC} (mV)	X (suns)	η (%)	FF (%)	V_{OC} (mV)
1	16.9	78.7	915	1	16.9	78.7	915
67.0	21.6	85.6	1073	103	21.9	85.8	1092
348	22.6	86.3	1121	305	22.6	86.2	1121
991.0	22.9	85.2	1142	1026	22.9	85.1	1148
1633	22.5	83.1	1151	2025	22.4	83.0	1155
4281	21.2	77.5	1163	3062	22.2	80.8	1172

Typical values to rate the system performance –mean relative errors– as deduced form the validation phase are:

- Error in Efficiency: ± 1.5 %
- Error in V_{OC}: ± 0.5%
- Error in Fill Factor: ± 1.5%

4. SUMMARY AND CONCLUSIONS

Summarising, a very simple and low cost system for measuring high concentrator solar cells from 25X up to 4000X has been described. The system is based on a flashlamp as a high intensity light source, a data acquisition PC board, a simple electronic load and a computer. The system architecture proposed allows the simplification of the hardware needed (and thus low cost) by overriding the signal analysis process. Accordingly, a sequence for signal processing has been proposed, including algorithms for pulse origin determination and filtering of the as-read data. As a result, efficiency, fill factor and open circuit voltage evolution versus concentration can be monitored and analysed with the aid of a specially dedicated computer program. The measurements obtained with the system show very good agreement with those of other labs obtained with different approaches involving much complex equipment.

ACKNOLEDGEMENTS

The Environmental Technologies Programme of the Autonomous Government of Madrid has financed this work under contract 07M/0260/1997. Mr. I. Rey-Stolle holds a PhD. scholarship from the Spanish Ministry for Culture and Education.

REFERENCES

[1] W. Keogh, A. Cuevas, *"Simple Flashlamp I-V testing of Solar Cells"*, 26th PVSC, Anaheim, Sept-Oct 1997

[2] M. Yamaguchi and A. Luque, *"High Efficiency and High Concentration in Photovoltaics"*, IEEE Trans. Elec. Dev., Vol. 46, No. 10, p.2139, 1999

[3] V.D. Rumyantsev, *"Technical Documentation on the Installation for measurement of load characteristics of concentrator solar cells"*, Ioffe Institute technical Report, St Petersburg (Russia), 1996

[4] F. Lipps, A. Zastrow and K. Bucher, *"I-V Characteristics of PV Modules with a msec Flash Light generator and a 2Mhz Data Acquisition System"*, 13th European PVSEC, Nice, France, 1995

[5] A. V. Oppenheim, R. W. Schafer and J. R. Buck, *"Discrete-Time Signal Processing"*, 2nd Ed., Prentice Hall International (London, UK), 1999

THE REDUCTION OF MODULE POWER LOSSES BY OPTIMISATION OF THE TABBING RIBBON

S. Roberts, K.C. Heasman, T.M. Bruton

BP Solarex, 12 Brooklands Close, Sunbury on Thames, Middlesex, TW16 7DX, UK

tel: +44 1932 765 947, fax: +44 1932 765 293, e-mail: roberts@bp.com

ABSTRACT: As the power of photovoltaic modules continues to increase due to advances in commercial crystalline silicon solar cell technology, power losses in the tabbing ribbon used for series interconnection of cells become increasingly significant. The shading loss is directly proportional to the surface area of the tabs, while the resistive (I^2R) loss is inversely proportional to the tab cross section. The use of a thick but narrow tabbing ribbon is therefore highly desirable to minimise both the I^2R and shading loss components. In the present work, the losses in a typical high power (85Wp) module were modelled as a function of tabbing ribbon dimensions and the results were confirmed by experiment. A comparison of continuous and spot soldering of the tabs and the effect on the I^2R loss component was also made. The suitability of commercially available copper tabbing ribbons was evaluated for use in a manufacturing process. The yield in the module assembly process was assessed and a number of modules were fabricated and tested. In an 85 Wp module of 36 cells, a reduction in the combined resistive and shading losses of 1.4 W was shown to be achievable by increasing the thickness of the continuously-soldered ribbon from 125 μm to 250 μm. In practice however, it was found that continuous soldering of the thicker tab caused an unacceptable degree of curvature in the wafer, due to mismatch of the thermal expansion coefficients of copper and silicon. The use of spot soldering reduced the curvature to an acceptable level, but contributed an additional resistive loss of 0.4 W. Therefore the net gain obtainable in practice by increasing the ribbon thickness from 125 to 250 μm was 1.0 W. In external lightsoak and accelerated environmental testing, the use of 250 μm interconnect ribbon was found to have no adverse effect on the module reliability. These results indicate that by the use of a tabbing ribbon of dimensions appropriate to the module power, a power gain of 1.2 % is obtainable.

1. INTRODUCTION

Silicon solar cells with 16.5 % efficiency and 150 cm^2 area have been commercially available for several years, eg laser grooved buried grid (LGBG) cells fabricated on 125 mm pseudosquare Cz wafers. A typical 36 cell module made using such cells has a Pmax in excess of 85 Wp and a maximum power point current (Imax) of 5A. The conventional method of interconnecting cells in the module is by two parallel copper interconnect ribbons (tabs), of width 1.5 to 2.0 mm and typical thickness (height) 125 μm. The tabs are soldered to the front metallisation busbars of each cell and the rear metallisation of each adjacent cell. This method of interconnection has changed little since it was first used in modules consisting of 100 mm round cells with efficiencies of 12 % and Imax of 2 A. Consequently, taking into account the increased current and cell size as solar cell manufacturing technology has progressed, the resistive loss (I^2R) in the interconnect ribbon has increased approximately 8 fold and is a significant factor in the design of high power modules. While reduction of the electrical resistance calls for an increase in the cross sectional area of the tabbing ribbon, the requirement for a low busbar shading loss places a constraint on the width of the tab. For an interconnect ribbon of given thickness, the optimum width is therefore the result of a trade-off between resistive and shading losses. The combined loss decreases as the tab thickness is increased, making a tabbing ribbon of high aspect ratio (thickness / width) desirable. However, the use of high aspect ratio ribbon has been limited by the difficulty of manufacturing copper ribbon to the required specification and by concerns about cell breakage in the module assembly process and long term reliability in the field. In the past, the majority of tabbing ribbon of low aspect ratio (eg 125 μm x 2mm) was manufactured by slitting copper sheet. While slitting has been used successfully to make ribbon of higher aspect ratio (180 μm x 1.5mm ribbon made by slitting is currently being used in commercial solar module production), further progress is likely to be limited by difficulty in controlling the dimensions of the ribbon as the aspect ratio is increased.. More recently, ribbon has been manufactured by the shaping of round wires by rolling; an approach which is more suited to the fabrication of ribbons with high aspect ratio.

The present work sets out to optimise the dimensions of high aspect ratio tabbing ribbon by modelling and experiment, fabricate modules using a commercial manufacturing process and evaluate their reliability by environmental testing. In addition to predicting the power gain theoretically obtainable by the use of high aspect ratio interconnect ribbon, the practical implications of using such materials are considered and the relative merits of continuous and spot soldering are discussed.

2. MODELLING OF POWER LOSS

The total resistive (I^2R) loss in the interconnect ribbon of a solar module is the sum of several series components. The simplest case, where the tab is continuously soldered along the whole length of the busbar on the front and back of each cell, is shown in Figure 1a. Each short section of ribbon between adjacent cells carries half of the module current (I) and contributes a resistive loss of $I^2\rho_1 l/4$, where ρ_1 is the resistance of the ribbon per unit length and l is the length of ribbon between cells. On the front of the cell, the tabbed busbar is a parallel combination of the busbar metallisation, tabbing ribbon and solder which has a resistance per unit length of ρ_2. The resistive loss contributed by each tabbed busbar is therefore $I^2\rho_2 L/12$, where L is the length of the busbar. The situation on the back of the cell is similar to that on the front, except that current flow in the back contact metallisation, in parallel to the current flow in the tabs, is also possible. However, because the sheet resistance of the back metallisation is normally much greater than that of the tabbing ribbon, the fraction of the module current flowing in the back metallisation is small and is neglected in the present analysis. For a fixed ribbon thickness, the total resistive loss is therefore inversely proportional to the width of the ribbon, w. In addition to the resistive loss, each tabbed busbar contributes a shading loss of $X\eta Lw$, where X is the irradiance and η is the efficiency of the cell. The shading loss is therefore directly proportional to the width of the ribbon, w (which is assumed to have the same width as the cell busbar).

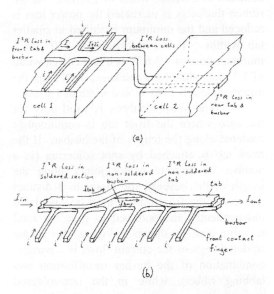

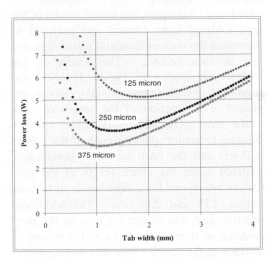

Figure 1: Diagrams showing the location of the power losses in the cell busbar and tabbing ribbon. (a) The shading and I^2R loss components in a cell string with continuously soldered tabs. (b) The current flow and additional I^2R loss components in the front busbar of a cell with spot soldered tabs.

Figure 2: The combined shading and I^2R tabbing losses as a function of tab width and thickness. (85 Wp module with 36 cells and 2 continuously soldered copper tabs per cell).

In Figure 2, the combined resistive and shading loss is plotted as a function of ribbon width for 3 different ribbon thicknesses, for the case of an 85 Wp module consisting of 36 cells (125mm square) connected in series. The minimum loss achievable with a ribbon thickness of 125 μm is 5.1 W, which is obtained at a ribbon width of 1.8 mm. As the ribbon thickness is increased the power loss is reduced and the minimum is shifted to smaller tab widths. For a thickness of 250 μm a minimum loss of 3.6 W is obtained at a width of 1.3 mm, while for a thickness of 375 μm the minimum loss is 2.96 W at a width of 1.1 mm.

The above analysis is valid only for the case where the front tab is continuously soldered along the length of the busbar. If the front tab is attached by spot soldering (ie a number of discrete solder joints along the length of the busbar), the losses in 3 distinct regions of the front busbar must be calculated independently and summed to give the total resistive loss, as shown in Figure 1b. In the soldered regions the current flows in a parallel combination of the busbar metallisation and tabbing ribbon, while in the non-soldered regions the current divides, with one component flowing in the tab and another component flowing in the busbar metallisation. Current entering the busbar from the front contact metallisation fingers has to flow through the busbar before reaching the tabbing ribbon. Because the resistance of the busbar is higher than that of the tabbing ribbon, an additional resistive loss component is introduced. An equivalent circuit model of the spot soldered front contact was analysed to determine the current flow and resistive power loss in each section. The results of this analysis are shown in Figure 3, where the combined resistive and shading losses are plotted as a function of the length of each solder spot. The number of solder spots is constant (5 solder spots for each of the 2 front busbars) and the back tabs are continuously soldered along the entire length of the cell. Curves are shown for 2 thicknesses of tabbing ribbon, 125 and 250 μm. The additional resistive loss due to current flow in the non-soldered sections of the front busbar is seen to increase as the length of the soldered regions is reduced and becomes a significant effect when the length of the soldered region approaches 5 mm, as is typically the case in modules fabricated using automated tabbing machines. Thus, the power gain of 1.4 W obtainable by increasing the tab thickness from 125 to 250 μm in the continuously soldered case is reduced to 1.0 W in the case where the 250 μm tab is spot soldered. This is an important factor in determining the optimum tab thickness in practice, as continuous soldering of thick tabbing ribbon causes curvature of the cell, which can be especially severe in the case of thin wafers.

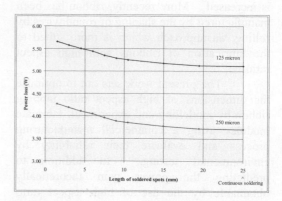

Figure 3: The dependence of the combined shading and resistive tabbing losses on the length of the soldered sections when spot soldering is used. (85 Wp module with 36 cells and two 1.5 mm wide copper tabs per cell, each soldered in 5 spots on the front and continuously on the rear).

3. EXPERIMENTAL DETERMINATION OF RESISTIVE POWER LOSS

To test the predictions of the model, 16 LGBG cells were tabbed using continuously soldered 125 μm x 1.5 mm copper ribbon and their light and dark IV characteristics were measured. 250 μm x 1.5 mm ribbon was then soldered on top of the original tabs, to increase the effective tab thickness to 375 μm and the cells were measured again. Connection to the back of the cell was made by vacuum contact with the base plate of the cell tester and was unaffected by the changes to the front of the cell. The average fill factor from light IV measurements was found to increase by 1.0 %. The dark IV measurements showed a reduction in series resistance consistent with the measured increase in fill factor. This increase in fill factor is due to reduction in the front tab resistance only, as the back contact remained constant. The fill factor increase predicted by the model for the same situation is 1.2 %.

4. MODULE FABRICATION AND ENVIRONMENTAL TESTING

The practical limitations to the thickness of high aspect ratio tabbing ribbon for use in solar module manufacturing are the stresses introduced into the cell by thermal expansion mismatch and by the pressure applied during vacuum lamination. The thermal expansion mismatch between the copper tabbing ribbon and the silicon wafer results in curvature of the cell during the soldering process and may also cause failure of the cell string after repeated thermal cycling. These effects may be alleviated by the use of spot soldering and soft annealed copper tabbing ribbon.

The suitability of 250 µm x 1.5 mm tabbing ribbon for use in standard glass-EVA-Tedlar module construction was assessed by fabricating 5 modules and subjecting them to accelerated environmental testing. The modules were made using 36 LGBG cells (125 mm pseudosquare) and had an average Pmax of 84.8 W. The tabbing ribbon was manually soldered at 5 spots of 5 mm length on each front busbar and continuously on the back busbars. A stress relief bend was formed in the ribbon between adjacent cells. A slight curvature was observed in the cells after front tabbing, but this did not cause any problems during the subsequent module assembly operations. No cells were cracked or broken during the entire module assembly process, from tabbing through to lamination. The 5 modules were exposed to natural sunlight in Madrid for 40 days, during which time measurements were taken on a Spire solar simulator. All 5 modules exhibited normal stable behaviour. There was no evidence of current mismatch effects which are indicative of cracked cells or of series resistance increase arising from failure of the solder joints. The average fill factor of the 5 modules changed by less than 1 % during the test, which is within the measurement repeatability (see Figure 4).

After external testing, one module was subjected to thermal cycling between -20 to +85 °C with a cycle time of 8 hours for 30 cycles, while a second module was exposed to UV radiation at a temperature of 45 °C and 100 % relative humidity for 280 hours. Table 1 shows the measured changes in the module parameters. The changes are within the experimental error of the measurement technique. While more extensive testing is required, these results are a good indication that the use of 250 µm x 1.5mm does not impair module reliability.

	Thermal cycling -20 to +80 °C 3 cycles / hr 30 cycles	Humidity and UV 45 °C 100 % RH 280 hours
Δ Isc (%)	+0.0	-0.6
Δ Voc (%)	+0.4	-0.2
Δ FF (%)	+1.1	+1.6
Δ Pmax (%)	+1.5	+0.8

Table 1: The changes in the measured parameters of the modules after accelerated environmental testing.

5. CONCLUSIONS

The effect of the dimensions of the tabbing ribbon on the power loss in an 85 Wp solar module has been evaluated by modelling and verified by experiment. While changing the tab dimensions from 125 µm x 1.8 mm to 250 µm x 1.5 mm reduces the power loss by 1.4 W in the case of continuous soldering, the greater mechanical stress introduced by the thicker ribbon necessitates the use of spot soldering. The additional power loss component introduced by spot soldering reduces the benefit from 1.4 W to 1.0 W. In practice the gain achievable by using the thicker tabbing ribbon is therefore 1.0 W (1.2 % of Pmax). No reduction in module assembly yield was seen in small scale production trials using a commercially available 250 µm x 1.5 mm ribbon and initial environmental test results show no adverse effect on reliability.

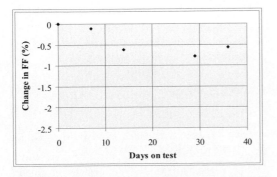

Figure 4: The change in the average fill factor the 5 modules fabricated using 250 µm x 1.5mm tabbing ribbon during ageing under natural sunlight.

6. ACKNOWLEDGEMENTS

The authors wish to thank G. Martinez for fabrication of the solar modules and J. Pascual for carrying out the environmental testing. The contribution of J. Woodcock to the work on power loss modelling is also gratefully acknowledged. The high aspect ratio tabbing ribbon was supplied by Neumeyer GmbH. The work was co-funded by the European Commission in the framework of the JOULE III Non Nuclear Energy Programme - (Contract number JOR3 - CT 98 - 0226, ASCEMUS Project).

CONTROLLED ATMOSPHERE PV CONCENTRATOR

F. Dobón Giménez
Freelance
Urb. Mayber, 107
38296 La Laguna
S/C de Tenerife-Spain
Tel. +34 922 26 01 75
E-mail: fdobon@mailcity.com

J. Monedero, P. Valera, M.P. Friend
ITER – Instituto Tecnológico y de Energías Renovables
Polígono Industrial de Granadilla
38611 Tenerife-Spain
Tel. +34 922 39 10 00
Fax +34 922 39 10 01
E-mail: iter@iter.rcanaria.es

ABSTRACT: This work pretends to solve some of the problems found in PV reflectors concentration techniques when some of its sensible parts are totally or partially exposed to aggressive agents.
The basic idea consists on enclosing in a box the most sensible components of the concentration system and eliminating when possible the components normally exposed to degradation. For this purpose the box should be completely sealed and the gas closed inside should not be aggressive for any of the internal parts.
The box should have a low absorption window to allow solar radiation passing through it. It has to be provided by a heat sink element that transfers the remainder heat to the out side, especially from the PV cells.
Keywords: Concentrator – 1: Cost reduction – 2: Optical losses – 3.

1. INTRODUCTION

The most common problems found in PV concentration systems are:

✓ Reflecting surface degradation due to moisture penetration, which destroys the silvered or aluminized layer, especially in the edges.
✓ Reflectance losses in the reflecting surface due to cleaning processes, which causes scratching on the reflecting surface and afterwards dust accumulation in the cracks.
✓ Optical mismatches due to reflectors surface deformation, especially if they are large size.
✓ Fast degradation of the photovoltaic module encapsulating organic elements (EVA, silicones, etc.) due to high solar irradiation concentration. Although the solar radiation is filtered, normally by a glass material window, an important part of the ultraviolet radiation is not blocked and reaches the cell. This, together with the high temperature reached on the cell's surface, accelerates the degradation process (browning of the encapsulating material).

2. DESCRIPTION

The controlled atmosphere PV concentrator is composed by the elements that follows (Fig. 1):

1- Window	4- Heat Sink
2- Reflector	5- Box
3- PV Module	6- Atmosphere

2.1 Window

The window is made of low absorption glass. To make it more impact resistance (hail, etc.), it can be made of glass sandwich (two layers of thin glass joint together with a non-degradable adhesive).

2.2 Reflector

The reflector has cylindrical-parabolic shape and is on the bottom side of the box. Different materials and techniques can be used in the parabolic mirrors manufacturing process such as glass strips or flexible adhesive layers.

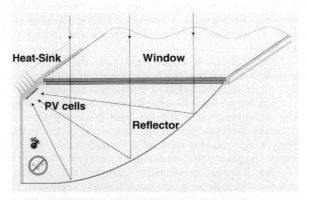

Figure 1: Concentrator schema

For serial manufacturing processes, direct vacuum surface metalization, such the one realized for the mirrors of astronomical telescopes could be used for cost reduction.

Any of the methods proposed will avoid using any further weather protection for edges or surfaces since they are content inside a non-aggressive environment.

2.3 PV Module

Different PV serial connected cells compose the PV module. On the one hand, the module backside is electrically isolated of the heat sink and on the other hand, it is thermally coupled with it using some of the materials developed for refrigerating power semiconductors.

The PV cells front side is directly exposed to solar radiation and does not use any encapsulating elements.

A low electrical resistance bus connects the PV cells between them and at the same time keeps their position over the heat sink.

2.4 Heat Sink

The heat sink can take part of the module and is made of aluminium. The cells are positioned in the inside surface and the backside is composed of aluminium sinks that expose a large area to the environment in order to transfer the residual heat to the outside.

2.5 Box

The box is sealed, structurally stable and has the geometry needed in the face where the reflecting surface is located.

Different procedures can be used for manufacturing the box.

Metal sheet manufacturing techniques may cause sealing and structural deformation problems, and is also sensible to bumps produced by mechanical impacts.

Best solutions include modelling procedure such as:

✓ One shot moulding using a thermo-plastic material.

✓ Fibre-glass and epoxy moulding, using an appropriate mould.

Any of these two moulding procedures guaranties that the box is made in just one piece, which avoids little labour, small dimensional tolerances and is corrosion-resistant.

2.6 Atmosphere

Vacuum is the ideal solution for the inside of the concentrator but it is not considered since it implies a lot of technical problems and a simpler concentrator concept is searched.

Two different solutions have been chosen for the controlled atmosphere. The first one consists in filling the box with an inert gas for the elements contented inside.

The second one is easier and consists in filling the box with atmospheric air which has been previously filtered extracting the most aggressive gases such as the water steam which corrodes most of the metals inside: silver, cooper, tin, plumb, aluminium and PV cells grid alloys.

The sulphur compounds damages the silvered surfaces. Although oxygen is originally very aggressive, it forms an oxide layer that protects from corrosion the inner side of the metal (unless water steam is present).

Fortunately, this problem is solved for other applications and commercial equipment developed for the telecommunication industry is available for this purpose.

The pressure inside the concentrator is slightly higher than the atmospheric pressure as a consequence of the gas dilatation due to the internal heat. A mechanism able to regulate the increase of pressure must be provided.

3.   FUTURE WORK

A two axis tracking system is a must in order to test the proposed concentrator. Following this research line, a tracking system has been designed for supporting one concentrators array (fig. 2). The tracker is in process of world patent.

In a second phase, a concentrator experimental prototype will be developed. Tests will be performed in order to evaluate different manufacturing techniques to detect faults and deviations from the original design characteristics. They will also help finding the most appropriated materials.

Finally, a short series of concentrators will be manufactured in order to test them working under severe climatic conditions. Internal and external parameters will be monitored during a one-year time period, which will help us find out the degradation suffered.

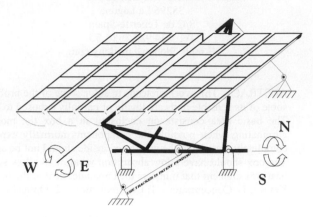

CONTROLLED ATMOSPHERE PV ARRAY

Figure 2: Controlled atmosphere PV array

4.   CONCLUSIONS

The proposed concentrator pretends the cost-reduction of the PV kWh since it uses less silicon per m² than the equivalent flat panel system, using standard materials and components with moderate prices and without complex manufacturing procedures.

The performance is similar to flat panel systems since the energy losses occurred in the glass window are similar in both cases. On the other hand, the additional energy losses produced in the concentrator mirror will be compensated by the concentrated energy in the cell (fig. 3).

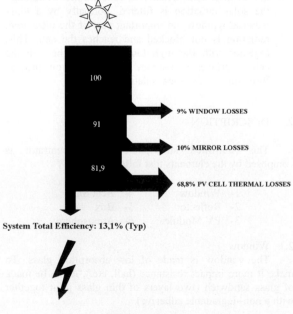

9% WINDOW LOSSES

10% MIRROR LOSSES

68,8% PV CELL THERMAL LOSSES

System Total Efficiency: 13,1% (Typ)

Figure 3: Losses

In contrast, a two axis tracking system is required in order to avoid the shadows projected by the box lateral walls.

The over cost of the tracking system will be highly compensated by the increase of energy collected by the system.

REFERENCES

[1] A. B. Meinel, M. P. Meinel,
     "Applied Solar Energy",Addison-Wesley
     Publishing Co., Inc. Reading-Massachusetts.

[2] E. Lorenzo,
     "Solar Electricity. Engineering of Photovoltaic
     Systems", Editorial PROGENSA. Sevilla.

[3] O. Knacke, I.N. Stranski,
     "Progress in Metal Physics - 6", Pergamon Press
     Limited - London.

[4] A. G. Guy,
     "Essentials of Material Science", Mc Graw-Hill Co. -
     USA.

[5] J.L. González Díez,
     "Materiales Compuestos", Universidad Politécnica de
     Madrid.

[6] Notes: "Mirrors Vacuum Metalization for
     Astronomical Telescopes", Instituto de Astrofísica de
     Canarias - Tenerife - Spain.

ON THE NOCT DETERMINATION OF PV SOLAR MODULES

M.C. Alonso, J.L. Balenzategui, F. Chenlo
CIEMAT, Departamento de Energías Renovables
Av. Complutense 22, E-28040 Madrid, Spain
Tel.: 34 91 346 66 72, Fax: 34 91 346 60 37
e-mail: carmen.alonso@ciemat.es, jl.balenzategui@ciemat.es

ABSTRACT: The determination of the Nominal Operation Cell Temperature (NOCT) of a module in a reliable way is important to predict the behaviour of the module in real operating conditions, being an useful parameter when comparing the performance of different module designs. Two IEC standards, IEC 1215 and IEC 1646 indicate the method to calculate NOCT in crystalline silicon terrestrial PV modules and thin film terrestrial PV modules respectively. Such standards have been applied to different types of PV modules, including especial designs for building integration applications and thin-film modules. The paper shows the results obtained when calculating the NOCT in accordance to the mentioned standards to different types of PV modules, including glass-glass, glass-tedlar, crystalline and thin film samples. Possible error sources that can bring about erroneous values of NOCT are analysed and points that might need clarification are suggested.
Keywords: Qualification and testing - 1: Characterisation -2: PV module -3

1. INTRODUCTION

NOCT is defined as the mean solar cell junction temperature within an open-rack mounted module in Standard Reference Environment (SRE): tilt angle at normal incidence to the direct solar beam at local solar noon; total irradiance of 800 W/m²; ambient temperature of 20°C; wind speed of 1 m/s and nil electrical load. It is an important parameter in module characterisation, since it is an indicative of how the module will work when operating in real conditions. Furthermore, in PV system design and simulation programs, many of the calculations are based in the determination of module temperature from ambient temperature and NOCT.

International Electrotechnical Commission has issued two standards that include the means to calculate NOCT, IEC 1215 [1] for crystalline silicon terrestrial PV modules, and IEC 1646 [2] for thin film terrestrial PV modules, being the method the same for both type of modules.

Within this work it has been tried to verify the suitability of the procedure in a country like Spain, with a Mediterranean type climate, and to analyse the results on the application of the standard to different type of modules, including special designs for building integration applications and thin film modules.

2. SUMMARY OF THE METHOD

The procedure to determine the NOCT of a PV module included in the standards IEC 1215 and IEC 1646 is based on the fact that the difference between the module temperature T_j and the ambient temperature T_{amb} is largely independent of the ambient temperature and essentially linearly proportional to the irradiance at levels above 400 W/m². The procedure calls for plotting $(T_j - T_{amb})$ against irradiance for a period when wind conditions are favourable. All data points taken during the following conditions must be rejected: irradiance below 400 W/m², wind speed outside the range 1 ± 0.75 m/s, ambient temperatures outside the range 20 ± 15 or varying by more than 5 °C, a 10 min. interval after a wind gust of more than 4 m/s and wind direction within ± 20° of east or west.

From the linear regression of the plotting of the difference between module minus ambient temperature against the irradiation a preliminary value of NOCT is obtained. This value will be corrected to 800 W/m² and 20° C depending on the average values of ambient temperature and wind speed during the test. The procedure will be repeated at least in another day.

2. EXPERIMENTAL DETAILS

3.1 Set-up

An outdoor facility with a data acquisition system has been set-up at Ciemat following the indications of the above mentioned standards. It consists of a structure to place the modules similar to the ones used in PV power plants and the necessary instrumentation to record the parameters used in NOCT determination:

- Irradiance sensors: Pyranometer Kipp-Zonen CM11 and m-Si ETC.
- Wind direction sensor: Micro Response vane of Wathertronic with a range of 0 to 360 DEG.
- Wind speed sensor: Micro Response Anemometer of Wathertronic. Range 0 to 100 mph, threshold 0.5 mph.
- Ambient temperature sensor: Aspirated Radiation Shield YOUNG. Range −50 to 50°C. Accuracy ±0.3°C.
- Modules temperature sensors: thermocouples attached in the back side.
- Data Acquisition System Fluke 2280 A PC controlled.

Irradiance, ambient temperature, wind speed, wind direction and module temperature have been recorded every 30 seconds during the testing periods. Besides, open circuit voltage of some the modules under tests has also been recorded. For the special cases of modules with an air camera between the front and rear side, several thermocouples have been adhered in both sides to evaluate temperature differences.

3.2 Sample description

Modules tested since the installation of the NOCT facility can be divided into two groups. First group was formed by 4 conventional m-Si PV modules of the same manufacturer whose main difference was the cell area and

Table II. Second group of PV modules under test. ASITHRU are amorphous Silicon semitransparent cells, while standard a-Si cells are opaque. EFG/multi are multicrystalline-Si substrate cells processed as EFG technology. Ref. and S-3 belong to the Spanish manufacturer Isofoton.
Used key: LI = Low iron glass; CG = Cerium-doped low iron glass; CT = Chemically Tempered glass; TG = tempered glass; TCO = glass with transparent conductor oxide; FG = float glass; PVB = poly-vinil-butilen adhesive.

Name	Ns.	Module Area (m²)	Cell Area (cm²)	Front Cover	Rear Cover	Cell	Encapsulant	Edge Seal
S-1	72	0.8996	100	LI (TG)	FG (TG)	EFG/multi	Resin	2mm tape
S-2	72	0.8996	100	CG (TG)	FG (CT)	EFG/multi	EVA	None
S-3	36	0.4400	104.5	LI (TG)	Transp. Tedlar	Mono-Si	EVA	Anodised Al frame
S-4	72	0.8996	100	CG (TG)	FG (CT)	EFG	EVA	None
S-5	72	0.8996	100	LI (TG)	FG (TG)	EFG	Resin	Hot melt
S-6	30	0.6000	174.6	TCO	TG	ASITHRU	EVA	None
S-7	30	0.6000	174.6	TCO	TG	a-Si	EVA	None
S-8	30	0.6656	174.6	HG/PVB/TCO	FG/air/TG	ASITHRU	PVB	Sealing material
S-9	57	0.6656	104.4	HG/PVB/TCO	FG/air/TG	ASITHRU	Resin	Sealing material
Ref.	33	0.3998	104.5	LI(TG)	White Tedlar	Mono-Si	EVA	Anodised Al frame

the number of cells per module (see table I). Front cover is glass, back cover is white tedlar, the encapsulant is EVA and the frame anodised aluminium. These modules will be denoted as F-1 to F-4.

Table I. First group of modules subjected to the test. Ns is the number of cells serially connected.

Name	Ns	Module Area (m²)	Cell Area (cm²)
F-1	36	0.96989	235.8
F-2	33	0.47601	117.9
F-3	36	0.63120	147.5
F-4	36	0.47601	117.9

The second group of modules, denoted from now S-1 to S-9, are modules specially designed for building integration under a JOULE III project, and they include m-Si, multi-Si and a-Si cells with different transparency degrees and encapsulated with different materials, as it can be seen in table II.

4. RESULTS AND DISCUSSION

4.1 Temperature measurements on special module designs

For modules listed in table II, a problem arose when determining module temperature. Samples S-8 and S-9 are specially designed for the integration into buildings, and count on an air camera in their back side, as it can be appreciated in figure 1.

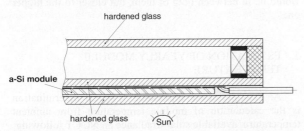

Figure 1. Special structure of the modules S-8 and S-9.

In these cases the temperature measured with a thermocouple in the back side of the module is not representative of the cell temperature. Several thermocouples were attached in the front and rear side of one of these modules, obtaining temperature differences of more than 15°C, as it could be though for a structure that is designed precisely to isolate the inner space of a building from environmental conditions (see figure 2).

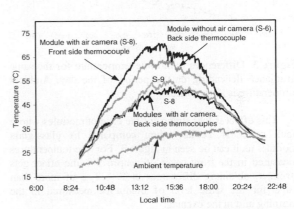

Figure 2. Differences in module temperatures along the day depending on the position of the thermocouples.

In this cases, or in general in non-conventional PV modules in which the temperature of the rear side of the module is not indicative of the cell temperature, a method should be stated to correctly determine module temperature in order to avoid incorrect predictions of module behaviour.

4.2 Effect "morning-afternoon".

When applying the procedure to the modules listed in tables I and II, first thing observed was that different periods along one day that could fulfil the conditions expressed in the Standard were possible to find. Figure 3 shows an example of a typical plot of T_j - T_{amb} versus irradiance for a clear sunny day. It can be observed that, even though at high irradiances the two lines seems to come near, for irradinaces below 700 or even 800 W/m² the lines separate with differences that can reach more than

5°C at the same irradiance. One of the specifications of the standard is that *ambient temperatures outside the range 20 ± 15 or varying by more than 5 °C should be rejected.* Paying attention to that, one might think that the temperature should not change in more than 5°C in the period of testing. If we apply that to the graph of figure 2, only a portion of it would be valid, furthermore, several different portions of it would be valid depending on the temperature range. Calculations on NOCT were made for different temperature intervals along the day in all the samples here tested. It was found that when NOCT is calculated with data measured only in the morning, their values were around 3°C lower than when it is calculated with data measured only in the afternoon, after midday. If all data points are taken into account, the value of NOCT obtained is an average, and is very close to the value obtained when only the portion of the graph with higher irradiances is observed (notice that the irradiance range should be of at least 300 W/m² according to the standard).

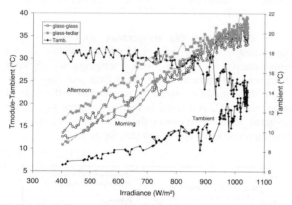

Figure 3. Differences in module temperature for the same irradiance depending on the period of the day. Ambient temperature is also plotted.

This effect is less strong for the case of modules whose back cover is Tedlar, when compared to glass-glass modules as it can be seen in figure 3. For the temperatures measured in the front cover of sample S-8, the effect gets stronger, obtaining differences in $T_j - T_{amb}$ of more than 10°C for the same level of irradiance measured in the morning and in the evening.

The values of NOCT that are presented in this paper have been calculated taking into account all possible data points, including morning and afternoon, as in this case the value is averaged and seems to be more representative of the behaviour of the module along the day. Nevertheless, special care must be taken when only a portion of the day is taken to calculate NOCT, i.e. for example, when only half of the day has stable irradiance conditions, as it could result in NOCT values over or under estimated.

4.3 NOCT values

Following are presented the values of NOCT obtained for the two groups of samples. For the second group of samples, NOCT values of S-8 and S-9 are not presented because, as it has been analysed in the preceding section, the measured module temperatures are not indicatives of the cell temperature, and the obtained values of NOCT would be erroneous. Sampled noted as "Ref." in table II is a conventional m-Si module that has been used as reference in all the measurements performed on table II samples.

All the calculations of NOCT have been performed in at least three days per sample, and the values presented are the average of the different values encountered.

In table III are presented NOCT values obtained for the first group of samples. No conclusion related to the module cell sizes can be inferred for the results showed in table II, except that samples F-2 and F-4, that have the same cell size, present very similar NOCT's.

Table III. NOCT values of sampled in table II

Sample	F-1	F-2	F-3	F-4
NOCT (°C)	48.1	49.0	50.3	49.1
	Test period: September F-1 March F2 to F-4			

In table IV are presented the results on the second group of samples, excepting S-8 and S-9. First thing that can be observed in table IV are the lower values of NOCT for the a-Si modules when compared to the other glass-glass samples. Besides, the semitransparent a-Si module presents a specially low value of NOCT. Glass-tedlar crystalline Si modules presented in table IV show lower values of NOCT than glass-glass modules.

Table IV. NOCT values of samples in table II.

Sample	S-1	S-2	S-3	S-4	S-5	S-6	S-7
NOCT	49.4	50.2	46.8	49.1	50.4	46.5	48.5
	Test Period April-May					Test Period: July	
	NOCT Ref.: 48.19						

Another point is the possible influence of the period of the year in the determination of NOCT. The "Reference Module" was tested together with S-1 to S-5 modules in April-May, and with S-6, S-7 modules in July. In general, values obtained in the testing days of April-May where lower than the values obtained in July. If we take into account that average ambient temperatures during the period of test in April-May ranged from 13 to 25°C, while in July ranged from 26 to 30°C, a possible combined influence of both ambient temperature and available irradiance could be consider, although more test in different periods of the year should be performed.

As it has been previously said, the temperature measured with a thermocouple is not indicative of the cell temperature for samples S-8 and S-9. Nevertheless, the same calculations were made from the measurements of the thermocouple in the front and rear side of sample S-8. The average values obtained were 36.20°C for the thermocouple adhered in the back side, and 50.20°C for the thermocouple in the front side. Probably the real NOCT would be in between both of them, but closer to the higher one.

5. ESTIMATION OF YEARLY MODULE TEMPERATURE

An example of the application of NOCT determination is the calculation of module temperature from ambient temperature, available solar irradiance an NOCT following:

$$T_m = T_{amb} + (NOCT - 20)\frac{E}{800} \qquad (1)$$

where T_m is the module temperature, T_{amb} is the ambient temperature and E the irradiance in W/m².

A simulation has been performed using eq. (1) and a program developed by Ciemat based in the SRADLIB library [3]. Typical Meteorological Year of Madrid has been used. The values of diffuse irradiance on horizontal and tilted surfaces has been calculated by means of the models proposed by Macagnan [4] and Perez [5]. Ground radiation is calculated from the isotropic model [6].

Simulation has been performed for vertical and sloped systems with different tilted angles. In figure 4 is shown an example of the maximum temperatures that can reach a module with NOCT = 47°C along the year when it is located on a south oriented vertical or tilted 35° surface.

Another example is shown in figure 5, where the differences in the temperature that the same module would acquire when it is located in a 20° tilted plane oriented to the Southwest or Southeast are presented. It must be noticed the higher temperatures in the Southwest orientation when main insolation of the module takes place during the afternoon. This is in accordance with the results obtained previously of higher temperatures of the module in the evening than in the morning for the same level of irradiance.

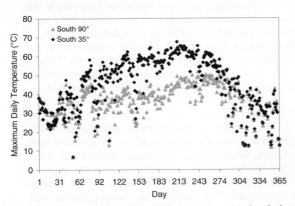

Figure 4. Maximum module daily temperature simulation in a south oriented vertical surface and south oriented 35°C tilted surface in Madrid. NOCT of the module = 47°C.

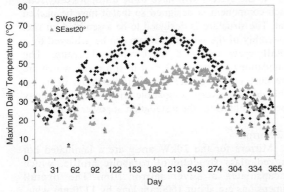

Figure 5. Maximum module daily temperature simulation in a Southwest and Southeast oriented surface tilted 20° in Madrid. NOCT of the module = 47°C.

The same simulations were calculated with modules with other NOCT, obtaining differences of the same order of the differences in NOCT. For example, when increasing NOCT from 47 to 50°C, the module temperature increased during the year a mean of ~3°C for south oriented surfaces

tilted 20 and 50°, and ~2°C for vertical Southwest and Southeast surfaces. Maximum differences for this increase of NOCT were of ~4°C for the two south oriented surfaces and ~5°C for the two vertical surfaces. An example is shown in figure 6.

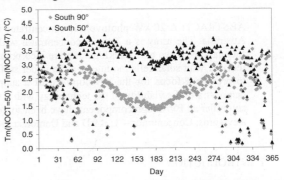

Figure 6. Differences in maximum daily temperature between a module with NOCT=50°C and another one with NOCT=47°C

6. CONCLUSIONS

Main conclusions inferred from the application of the International Standards IEC 1215 and IEC 1646 to determine the NOCT of a PV device are the following:

− When we consider the temperature of a PV module during the day, a heating of the module is observed during the afternoon-evening. This makes differences of T_{mod}-T_{amb} higher in the afternoon than in the morning for the same values of irradiance, causing differences of about ~2-3°C in NOCT. This would increase the error in the final value of NOCT.

− The method as it is proposed in IEC 1215 and IEC 1646 is not the most adequate for modules integrated into buildings: modules located in vertical or horizontal surfaces, modules closed to walls or isolated in their back side, or window modules that count on air cameras between the two external glasses, as their working conditions will be very different to the conditions expressed in the Standard.

− A method for the correct determination of module temperature (for example through V_{oc} measurements) should be defined for window modules thermally isolated in order to determine their NOCT.

ACKNOWLEDGEMENTS
This work has been partially supported by the European funded Project JOR3-CT97-0182.

REFERENCES

[1] International Standard IEC 1215. 1993-04
[2] International Standard IEC 1646. 1996-11.
[3] J.L. Balenzategui. "SRADLIB: a C library for Solar Radiation Modelling". ITC-904. Ed. Ciemat. Madrid 1999.
[4] M. Macagnan, E. Lorenzo, C. Jiménez. International Journal of Solar Energy 16 (1994) 1-14.
[5] R. Pérez, P. Ineichen, R. Seals, J. Michalsky, R. Steward.. Solar Energy 44 (5), 271-289 (1990).
[6] M. Iqbal. "An introduction to solar radiation". Academic Press, Toronto, 1983.

THE ANU 20kW PV/TROUGH CONCENTRATOR

J. Smeltink, A.W. Blakers and S. Hiron
Centre for Sustainable Energy Systems,
Australian National University, Canberra 0200, Australia
Tel. 61 2 6249 4884, John.Smeltink@anu.edu.au

ABSTRACT: A 20 kW photovoltaic concentrator has been constructed at Rockingham near Perth in Western Australia. The array is a Trough concentrator with a total collection area of 150 m². The system comprises foundations, a two axis tracking structure incorporating mirrors & supports and an aluminium passive heat sink receiver with solar cells mounted on the under surface. The primary goal for the construction of the array was to provide a focus for the development of a supplier base for the PV/Trough concentrator system. It also demonstrates the viability of the technology with a view to marketing arrays. With the exception of the tracking controller all PV/trough components have been sourced from Australian companies.
Keywords: Concentrator – 1; R&D and Demonstration Programmes – 2; Tracking - 3

1. INTRODUCTION

Conventional PV systems generate electricity at a cost of US$0.40 c/kWh or more. Costs must be reduced in order for the photovoltaic industry to expand further into the diesel fuel replacement, water pumping and grid connected markets. The cost of conventional PV technology is driven mainly by the cost of silicon wafers. This leads to a need to decouple PV systems from dependence on a large numbers of silicon wafers if costs are to be reduced. It also means that the silicon in use should be used efficiently.

Concentrator systems have the potential to greatly reduce the cost of photovoltaic power by using a large area optical system to focus sunlight onto a much smaller area of cells. In principle, most of the expensive cell area is replaced by a cheap focusing system. A recent paper provides an excellent summary of activity in the area [1].

A PV/Trough system, such as is being developed at ANU, consists of a parabolic reflective trough, concentrating light onto a line of cells. The few remaining solar cells in the system (at the focal line of the trough) are a relatively small part of total system cost. Thus expensive but efficient cells can be used without economic penalty.

The ANU has been actively engaged in the development of PV concentrator technology since 1995. Having gained experience and confidence with smaller systems the ANU has designed the prototype of a system which promises commercial viability. A system recently constructed in Rockingham, Western Australia (which is 50 kilometres south of Perth) has a power generation capacity of 20kW from a total collection area of 150 m². Installation of the array was carried out by Solahart Industries, which is a major solar water heater manufacturer.

2. A 20kW PV/TROUGH ARRAY

The 20kW array consists of 80 trough modules. Each trough module consists of a parabolic mirror with an area of 1.92 m², a receiver heat sink with solar cells and associated supports. A diagram of the system is shown in figure 1. This demonstration array will form the basis for commercialisation of ANU PV/Trough technology, act as a test bed for all system components and provide a means of evaluation for interested parties.

From the start of the project the ANU has maintained an awareness of the functional requirements for the system. The 20kW array is connected directly to the grid, thus avoiding the need for an energy storage system. Anticipating a harsh service environment in the Australian outback, the system is designed for a long service life with a minimum of maintenance. Passive heat sinking was chosen for this reason. With the exception of the tracking controller all PV/Trough components have been sourced from Australian companies, which assists with sourcing of spare parts.

2.1 Support structure

The geometry of the support structure of the 20kW system features the widespread use of standard steel sections, which are fully galvanised for long life. It has been optimised with a view to limiting the mass of material used. The distributed mass for the support structure including supports and tracking mechanisms is approximately 29 kg/m² of aperture. By comparision the one axis EUCLIDES™ system has a distributed mass in excess of 38kg/m² [2].

As the system will have to be transported by road each component is designed so that it may be packed flat. The structure is designed to be assembled on site. Assembly of the whole structure can be achieved by a small crew using a truck-mounted crane. The foundations consist of 11 cast in-situ piers, which are low cost, easy to install and offers flexibility in sizing. This design is very compatible with the methodology used by utilities for the installation of power poles.

2.2 Mirrors

Mirrors for the 20kW array are a laminated hot sagged glass structure of parabolic profile, which encapsulates a silver mirror film. The overall dimensions are about 1600mm long by 1170mm wide. The mirrors have a reflectivity of 91% and a focal region intercept of 82%. Physical characteristics of this mirror design are:

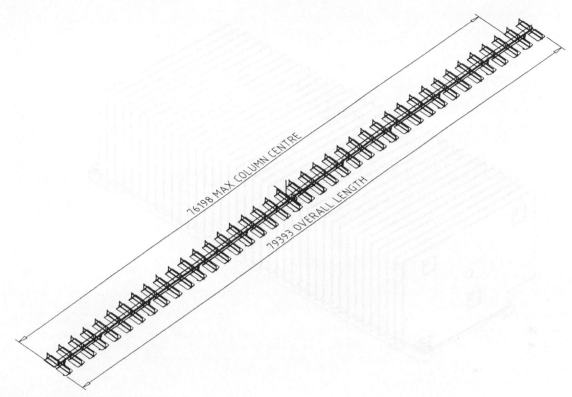

Figure 1: 20kW ANU PV/Trough Array

1. Impact resistance: the mirrors are sufficiently thick to withstand hail damage, handling and transport.
2. Abrasion resistance: the glass surface can withstand sand blast in dust storms and cleaning. This is an important requirement in the Australian outback.
3. Rigidity: sufficient to withstand large wind loads.
4. Corrosion resistance: the mirror design is similar to a car windscreen, in which the problem of ingress of water between the two panes of glass has been entirely solved.

2.3 Receiver

Solar cells fabricated at ANU are mounted directly on the receiver, which also provides a cooling mechanism via an integrated passive heat sink. The solar cells are assembled into a cell package which incorporates two bypass diodes per cell attached to the copper tabs that later form the cell interconnection. The cell packages are bonded to the receiver and joined in series by interconnecting cell tabs. Once assembled the cell string is encapsulated to exclude moisture and covered with glass for protection.

PV cells work at an optimal level when kept cool. Passive cooling has been developed for this array because it exhibits great reliability and low maintenance. The cells are kept at a temperature of 30 to 40 °C above ambient, a temperature similar to that experienced by cells in a flat plate system. This convective cooling system is constructed by joining aluminium fins directly to an extruded aluminium base (Fig. 2). The profile of the base allows cells to be

bonded to it directly. It incorporates a clipping feature to attach glass retainers, which also act as heat shields for the cell tabs. The ANU has applied for patent protection for the heat sink design.

Each heat sink is 1.6m long and has a total surface area of 4.54 m^2 and a distributed mass of 4.2 kg/m^2. By comparision the heat sinks developed for the EUCLIDES^{TM} system have a distributed mass of 8.75 kg/m^2 of aperture and the fully extruded ENTECH units are 16.5 kg/m^2 [3].

The heat sink mass can ultimately be reduced to 2.7kg/m^2 through the combination of using thinner fins and a wider mirror. The fin material can be reduced from 1.0mm to 0.7mm with minimal loss of performance. This material was unavailable at the time of manufacture. The heat sink is suitable for mirrors up to 1.5m wide.

2.4 Solar Cells

An elegant fabrication technique for high performance concentrator silicon solar cells has been devised. A total of 2,500 cells of area 20 cm^2 were fabricated at ANU on 100mm wafers for the 20 kW project. The minimum acceptable efficiency is 20%, but the average cell efficiency is close to 22% at 20-30 suns. The cells are of a conventional design with an electroplated silver metal grid on the front surface.

A novel flash tester has been constructed for cell measurement. A reliable measurement can be made of a cell's complete current voltage characteristics at five to ten different light intensities over a decade range in a

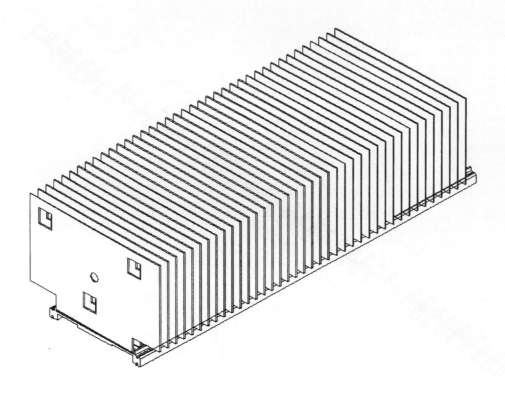

Figure 2: Section of an ANU Heat sink which incorporates fins attached to an extruded base.

few seconds. The complete measurement, including mounting and dismounting of the cell, takes about 30 seconds. The system automatically computes and displays all of the important cell parameters.

An important aspect of cell fabrication is environmental testing to ensure cell performance stability. Accelerated life testing of the cells is being carried out for humidity and UV resistance, air-ageing at elevated temperatures, temperature cycling and salt spray resistance. Similar tests are being performed on finished receivers.

Mirror Aperture	1170mmx1600mm
Number of Mirrors	80
Total Reflector Area	150 m^2
Concentration Factor (Geometric) (Actual)	30:1 23:1
Cell Efficiency	22% under concentration
Power Output per Module	250 Watts(peak) (SOC)
Power Output of System	20kW
Tracking Mechanism	2 Axis (accurate to 0.5°)
Total System Efficiency	14.8%

Table 1: System performance for 20kW array
(SOC: DB: 900W/m^2 Amb:20°C Wind:1m/s)

2.5. Array Tracking

The 20kW system features two axis continuous tracking. All trough modules are mechanically linked so that one motor actuates the tilt and another actuates the roll. Both motors are controlled by a time based open loop central processing controller via a motor driver interface and position feedback system. The range of motion for the array is 180° about the tilt axis and 130° about the roll axis. It has been calculated that the array will have an average intercept of 87% of the direct beam at a latitude of 35°.

4. CONCLUSION

The ANU PV/Trough technology has been developed with a view to creating a commercially viable system for outback Australia. A 20 kW demonstration array has been constructed, and the project has reached the commercialisation phase. Economic analysis shows that the PV/Trough project can produce electricity for US$0.20/kWh if appropriate manufacturing economies of scale are achieved. There is an awareness that demand already exists in Australia for this type of technology. The next step will be to take the technology to full commercialisation.

5. ACKNOWLEDGMENTS

The support of Solahart Industries Pty. Ltd., Western Power, the Australian Greenhouse Office, the Australian CRC for Renewable Energy, Alternative Energy Development Board of Western Australia, the NSW Department of Energy, the Australian Research Council, the Energy Research Development Corporation and the Power and Water Authority is gratefully acknowledged.

6. REFERENCES

[1] R.M. Swanson , "The promise of concentrators", Progress In Photovoltaics, Vol 8, pp 93-111, 2000

[2] J.C. Arboiro, G.Sala, J.I. Molina, L. Hernando, E. Camblor, "The EUCLIDES™ Concentrator: A Lightweight 84 m Long Structure for Sub-Degree Tracking"-Proceedings of the 2nd World Conference and Exhibition on Photovoltaic Solar Energy Conversion, July 1998, Vienna, Austria.

[3] G. Salo, A. Luque, J.C. Zamorano, P.Huergo, J.C Arboiro, "Lightweight Heat Sinks for the EUCLIDES Concentrator Array"- 14th European Photovoltaic Solar Energy Conference and Exhibition.

[4] A.W. Blakers and J.Smeltink "The ANU PV/Trough Concentrator System"-Proceedings of the 2nd World Conference and Exhibition on Photovoltaic Solar Energy Conversion, July 1998, Vienna, Austria.

16th European Photovoltaic Solar Energy Conference, 1-5 May 2000, Glasgow, UK

NEW STATIC CONCENTRATOR FOR BIFACIAL PHOTOVOLTAIC SOLAR CELLS

M. Hernández, R. Mohedano, F. Muñoz, A. Sanz, P. Benítez and J. C. Miñano.
Instituto de Energía Solar E.T.S.I. Telecomunicación, Universidad Politécnica de Madrid, Ciudad Universitaria s/n, 28040, Spain, Phone (3491) 544 10 60,Fax (3491) 544 63 41, e-mails: maikel@ies-def.upm.es, ruben@ies-def.upm.es, fernando@ies-def.upm.es, amsanz@ies-def.upm.es, pablo@ies-def.upm.es, minano@ies-def.upm.es.

ABSTRACT: Two new static nonimaging designs for bifacial solar cells are presented. These concentrators have been obtained with the Simultaneous Multiple Surface design method of Nonimaging Optics. The main characteristics of these concentrators are: (1) high compactness, (2) linear symmetry (in order to be made by low cost extrusion), (3) performance close to the thermodynamic limit, and (4) at least one non-shading sizable gap between the cell edges and the optically active surfaces. This last feature is interesting because this gap can be used to allocate the interconnections between cells, with no additional optical losses. As an example of the results, two designs for an acceptance angle of ±30 degrees gets a geometrical concentration of 4.8× and 4.3×, with an average thickness to entry aperture width ratio of 0.24. The three dimensional ray-tracing analysis of the concentrators is also presented.
Keywords: Concentrator - 1: Bifacial - 2: Building Integration – 3.

1. INTRODUCTION

The interest of static concentration in photovoltaics (PV) resides on several aspects. First, the concentration reduces the area of solar cell, which is by far the most expensive component of a flat PV module, and thus cost reductions seem possible. Second, as the cost of the module components is more distributed in concentration systems, they are less affected by the fluctuations of the market price of silicon wafers. Third, when compared with tracked concentrators, the static ones do not need higher maintenance than flat modules to operate. And four, with a correct design its possible to compensate the power loss due to the inclination angle when the modules are placed at the facades in building applications [1].

The purpose of this work is the development of two new static linear nonimaging designs for bifacial solar cells. It has been done under the framework of the "Venetian Store" project, funded by the European Union via the JOULE program. This project is leaded by the Spanish PV manufacturer ISOFOTON (which has already a bifacial cell technology), and aims to develop a commercial static concentration product for building integration at competitive cost with the esthetic of a venetian store, whose slats will be the concentrators.

The PV static concentrators are usually designed for bifacial solar cells, that is, cells whose two faces are active. The bifaciality doubles the possible ray trajectories that can reach the cell, which implies that the concentration limit is also doubled. This type of cells are not practical in high concentration systems (as tracking ones), because there is no inactive cell face to fix a heat-sink to and the dissipation becomes more difficult.

The concentration limit can also be increased if an optically dense media surrounds the cell. If its refractive index is n, then the concentration limit multiplies by n^2, so n = 1.5 implies a factor of 2.25 (slightly higher than that of the bifaciality). However, excepting a few concentrator designs [2,3], most approaches have linear symmetry (i.e., that of a trough), which leads to a lower theoretical limit for the attainable concentration: the factor 2.25 reduces to ~1.73 [4]. This concentration limit reduction is possibly compensated with the simpler manufacturing of linear symmetric devices by extrusion techniques.

These two new concentrators have been obtained by the Simultaneous Multiple Surface (SMS) design method of Nonimaging Optics [5]. This method consists in the simultaneous calculation in two dimensions of two optical surfaces to couple the edge rays of the source with those of the receiver. The surfaces to design can be of the sequential type (dioptric or mirror), on which all the rays of the transmitted bundle impinge, and also non-sequential mirrors (such as that of the Compound Parabolic Concentrator or CPC), according to the nomenclature introduced in reference 6. For referring to the different designs, each concentrator can be named with a succession of letters indicating the order and type of incidence on the optical surfaces that the transmitted bundle encounters on its way from source to receiver. The following symbols are used: R = refraction, X = sequential metallic reflection, I = sequential total internal reflection, X_F = non-sequential metallic reflection, I_F = non-sequential total internal reflection (the subscript F refers to the coincidence of the non-sequential mirror with of the ray bundle).

The parameters of this design are: (1) the acceptance angle α, (2) the refractive index n, (3) the cell width, (4) the distance from cell bottom to mirror and the distance form the cell top to the dioptric, and (5) the geometrical concentration C_G. The geometrical concentration (since the whole input ray bundle is collected) is bounded by the limit $C_{MAX}=2n/\sin \alpha$. This paper is focused in two designs that have lower concentration that the maximum possible, to improve practical aspects.

Section 2 deals with the RXI_F concentrator: its design and 3D-analysis of its angular response. Section 3 presents the RX-RXI concentrator. Section 4 shows the irradiance distribution that produces at the cell surface in two designs. Thermal characteristics are presented in section 5. Finally in the last section wee will see brief conclusions.

2. THE RXI_F CONCENTRATOR

The RXI_F concentrator is made of a dielectric piece of refractive index n>1 with one mirrored face (see figure 1). The rays transmitted by the RXI_F are refracted on the upper dioptric (R), reflected on the back mirror (X) and may reach the cell directly or after being reflected again (one or more times) by total internal reflection (TIR) on the upper dioptric (I_F). Then, the upper aspheric has a double function: it acts as a dioptric for the incident light and as a non-sequential mirror by total internal reflection for the rays after reaching the cell. Thus the top edge of the cell is in contact with the dioptric, as usual in non-sequential mirrors.

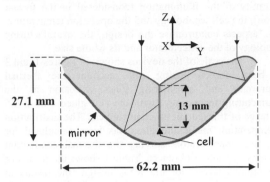

Figure 1. RXI_F concentrator designed for a cell
13 mm wide, $\alpha \pm 30^\circ$, n = 1.473, C_g=4.8x (0.812 C_{max}).

In figure 1 are shown some interesting geometrical characteristics of the RXI_F concentrator; the first one is a gap between the cell bottom and the mirror (this gap is interesting to place for the interconnections of cells of the module without raising optical losses). The second feature is that the cell top is in contact with the dioptric (entry aperture), which may complicate the environment isolation of the cell, and third one is that the concentrator has a kink at the entrance aperture center (which may accumulate dust).

2.1. Intercept factor of the RXI_F

A three-dimensional ray tracing has been done for the 4.8× RXI_F concentrator of the figure 1 in order to determine the intercept factor of the device, defined as the ratio of the power reaching the cell to the power which impinge on the entry aperture of the concentrator (neglecting the optical losses). Figure 2 shows the result of the lossless calculations, represented as a function of the direction cosines of the incident ray with respect to the axes x and y defined in figure 1. If the concentrator were placed with the x axis parallel to the east-west direction and tilted an angle equal to the latitude, then the sun will move along

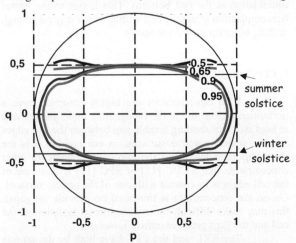

Figure 2. Intercept factor for the 4.8x RXI_F neglecting the optical losses. The variables p and q are the direction cosines of the ray before entering the concentrator.

the year within the area of the p-q plane enclosed by the lines q = ± sin 23.5° (which correspond to the solstices). The concentrator has a high geometrical collection within such area, where the energy density is higher due to the beam radiation. This transmission is properly adapted to the annual sun irradiation in most places, when considered as an inhomogeneous source [7].

The optical efficiency vs. wavelength was also calculated (at normal incidence) for glass as dielectric taking into account the different optical losses. The result obtained is a 72.5 %, integrated over range 400 - 1100 nm, being absorption in the glass the main contributor to this low value (if excluded, the efficiency raises to 91.3 %). The high absorption is due to the glass type selected for the simulation, which on the other hand presents excellent mechanical properties and is suitable for extrusion.

3. THE RX-RXI CONCENTRATOR

The RX-RXI concentrator is also made of a dielectric piece with one mirrored face (see figure 3). Conceptually, it is a combination of the well-known RX and RXI concentrators [8,9]. These two last characteristics, of the RXI_F, depend on the design parameters and the ray assignation for the design has been done to improve these aspects.

The rays transmitted by the central portion (which is the RX one) are refracted on the upper dioptric (R) and reflected on the back mirror (X) towards the cell. On the other hand, the rays transmitted by the outer portion (which is the RXI one) reach the cell after being reflected once more by total internal reflection (TIR) on the upper dioptric (I_F). Then, as in the RXI_F, the upper aspheric has again a double function, as a dioptric and as a sequential mirror by total internal reflection. As no non-sequential mirror is used this time, both edges of the cell are not in contact with the optical surfaces.

Figure 3 shows the RX-RXI concentrator and

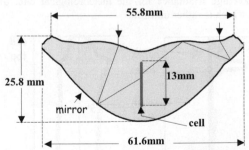

Figure 3. RX-RXI concentrator designed for a cell
13 mm wide, $\alpha \pm 30^\circ$, n = 1.473, C_g=4.3x (0.728 C_{max}).

we'll see that the undesirable features of RXI_F have been improved. The drawback is the concentration decrease (4.8× to 4.3×) for a design with the same dielectric volume.

3.1. Intercept factor of the RX-RXI

A three-dimensional ray tracing has been done for the 4.3× RX-RXI concentrator too. Figure 4 shows the results of intercept factor of the lossless calculations, like in section 2. This transmission is also properly adapted to the annual sun irradiation in most places.

Like in the previous design the efficiency vs. wavelength was calculated (at normal incidence) taking

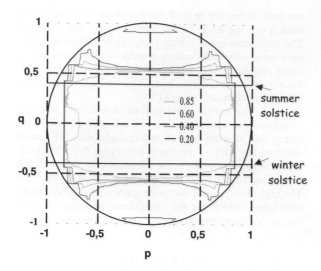

Figure 4. Intercept factor for the 4.3x RX-RXI neglecting the optical losses. The variables p and q are the direction cosines of the ray before entering the concentrator.

into account the different optical losses. The result obtained is a 70.2 %, integrated over range 400 - 1100 nm. This low value is due to the same thing that for RXI_F.

4. IRRADIANCE DISTRIBUTION

The three-dimensional ray tracing analysis (with optical losses) has been also used to estimate the irradiance distribution at the cell surface. Figure 5.a shows the instantaneous irradiance gain pattern that is produced the summer solstice at noon by the $4.8\times RXI_F$ concentrator on each face of the cell. The local concentration reaches values higher than 10 (both irradiance distributions have a similar profiles for the RX-RXI). Figure 5.b shows the annual average irradiance for the meteorological data of Madrid, which is rather uniform.

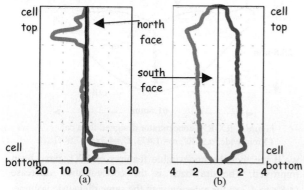

Figure 5. Ray tracing results on the irradiance distribution at the cell surface for (a) the solstice at noon for the RXI_F concentrator, and (b) annual average. Both distributions are for Madrid climate.

The instantaneous high local concentration factors often degrade the cell fill factor (exceptions depend on cell parameters; further information can be found in reference 10), and the design of the contacting grid lines of the bifacial cell has to be done to maximize the annual energy production.

5. THERMAL ANALYSIS

The complete technical design of the concentrator must take into account not only optical aspects, but also electrical and thermal ones. The concentration factor that provides maximum efficiency is affected by the non-uniformity of the illumination (considered in the former section), the cell technology and the operation temperature. These aspects compromise the design, the manufacturing technology of the concentrator and its whole size.

The scale of the devices shown in figures 1 and 3 has been chosen taking into account the thermal compromise and considering glass extrusion as the manufacturing technology (with the cell glued with a 0.5 mm layer of transparent silicone rubber). The distribution of temperature inside the glass has been calculated for these two concentrators solving the thermal conduction problems by finite differences. Table 1 shows the results of the thermal analysis for the RXI_F design (the results in table 1 are for the same volume of glass and plastic in RXI_F analysis).

Illumination profile	Δ Temperature (in °C)	
	Uniform	Punctual
RXI_F, Glass	32.6	33.8
RXI_F, Plastic	80.2	80.7
Flat module	23.0	-

In both designs, the vertical position of the cell inside the dielectric material makes the heat removal more difficult than in other design [2]. In order to keep the operation temperature at reasonable levels, acrylic as dielectric material (whose thermal conductivity is lower than that of glass) require smaller concentrator sizes. For a $50^{\circ}C$ temperature drop from the cell to ambient the concentrator aperture should be smaller than 28.5 mm.

It shows that the cell temperature under uniform illumination would be 9.6°C higher than that of a flat module under the same operating conditions (Irradiance = 800 W/m^2, wind speed =2.3 m/s (front) – 0 m/s (back), cell efficiency = 10%). It has been also found that the non-uniform illumination does not increase appreciably the average cell temperature (1.2°C for a 56× pill-box type illumination at the cell bottom). This is due to the lateral heat conduction along the cell width, which is good enough at these concentration scale levels.

6. CONCLUSIONS

Both concentrators are highly compact, have a performance close to the thermodynamic limit, and provide at least one non-shading sizable gap between the cell edges and the optically active surfaces, in our opinion it is the most important feature of the two concentrators. In other concentrators like a CPC [11] or SPC [12] at least one of the cell edges is in contact with one of the optical surfaces. Unless the concentrator is truncated (to provide the gaps), this may make difficult the environmental isolation of the cell and the gaps produce optical losses.

The RXI_F and the CPC have both by design one edge of the cell far from an optical surface. However, they are not equivalent, because, on the contrary to the CPC, in the RXI_F no ray of the bundle crosses the plane of the cell trough the gap. This implies that putting the cell interconnections outside the cell at this plane will produce negligible shade in the RXI$_F$, but not in the CPC case (though less than in the SPC). Note that while the cell width is reduced by the concentration factor, the tabs do not scale proportionally, and the shading is higher.

However, both RXI$_F$ and the CPC still have an edge in contact with an optical surface, and thus the gap forced at this edge will produce some optical losses as in the SPC. The RX-RXI design is advantageous with regard to the RXI$_F$ because the two lossless gaps are provided by design.

ACKNOWLEDGEMENTS

This research was supported by the European JOULE program under contracts JOR3-CT98-0252, ERK5-CT1999-00012 and ERK56CT1999-00002, contract TIC1999-0932-CO2-01 from the Plan Nacional de Investigación y Desarrollo with the Spanish Comisión Interministerial de Ciencia y Tecnología and contract 07T/0032/98 of the Consejería de Educación y Cultura of the Spanish Comunidad Autónoma de Madrid.

REFERENCES:

[1] R. Mohedano, P. Benítez, J.C. Miñano. "Cost reduction of building integrated PV's via static concentration systems". *Proc. 2nd World Conference and Exhibition on Photovoltaic Solar Energy Conversion. Viena, Austria*, 1998, pp. 2241-2244.

[2] S. Bowden, "High Efficiency Photovoltaic roof tile with Static Concentrator", *23rd IEEE Photovoltaic Specialist Conference, Procc.*, pp.1068-1072, (1993)

[3] K. Yoshioka *et al.* "Performance simulation of a Three-dimensional lens for a Photovoltaic Static Concentrator", *13rd European Photovoltaic Solar Energy Conference, Procc.*, pp. 2371-2376, (1995)

[4] J.C. Miñano, A. Luque, "Limit of concentration for cylindrical concentrators under extended light sources", *Appl. Opt.* **22**, pp. 2751-2760, (1983)

[5] J.C. Miflano, J.C. Gonz lez, "New method of design of nonimaging concentrators", Appl. Opt. **31**, pp. 3051-3060, (1992)

[6] P. Benítez *et al.* "Design of CPC-like reflectors within the Simultaneous Multiple Surface design method", *Nonimaging Optics: Maximum Efficiency Light Transfer IV*, Roland Winston, Editors, Proc. SPIE 3139, pp. 19-28, (1997)

[7] J.C. Miñano, A. Luque, "Limit of concentration under non-homogeneous light sources", *Appl. Opt.* **22**, pp. 2751-2760, (1983)

[8] J. C. Miñano, P. Benítez, J. C. González. "RX: a nonimaging concentrator", *Appl.Opt.* **34**, pp. 2226-2235, (1995)

[9] J. C. Miñano, J. C. González, P. Benítez. "RXI: a high-gain, compact, nonimaging concentra-tor", *Appl. Opt.* **34**, pp. 7850-7856, (1995)

[10] P. Benítez, R. Mohedano, "Optimum irradiance distribution of concentrated sunlight for photovoltaic energy conversion", *Appl. Physics Letters*, **74** (17), pp.2543-2545, (1999)

[11] R. Winston, H. Hinterberger, "Principles of cylindrical concentrators for solar energy", *Solar Energy* **17**, pp. 255-258, (1975)

[12] D.R. Mills, J.E. Giutronich, "Ideal prism solar concentrator", *Solar Energy* **21**, pp. 423-430, (1978)

OPTIMAL CLASSIFICATION CRITERIUM TO MINIMISE
THE MISMATCH LOSSES IN PV SERIES ARRAYS

Pablo Benítez, Fernando Muñoz, Juan C. Miñano
Instituto de Energía Solar, E.T.S.I. Telecomunicación,
Universidad Politécnica de Madrid, 28040 Madrid, Spain
Phone: (+34 91) 544 1060 - Fax: (+34 91) 544 6341

ABSTRACT: A new classification criterium that provides the minimum mismatch losses is presented. Its main innovation is the fact that this novel criterium, which can be deduced from variational principles, breaks with the generalised idea that links classification and less dispersion. To compare with other classification criteria a computer simulation has been carried out with 1000 PV cells modelled by a single-exponential equation with three parameters: short-circuit current, open-circuit voltage and series resistance.
Keywords: System– 1: PV array –2

1. INTRODUCTION

The dispersion of parameters of PV individual generators (cells or modules) causes that the maximum power that can be obtained from an array composed by them is smaller than the sum of the maximum individual powers. This subject has been an active field of research, not only from theoretical but also experimental approaches [1]-[6], with the objective of analysing and predicting the mismatch losses.

In many applications the solar cell can be considered an electrical power source of moderate to high current but low voltage. For this reason the series interconnection of cells, which provides the increase of voltage keeping the current level, is of special interest. Even when higher output current levels are needed, as in the case of large-scale PV plants, it is common to make first strings of series connected modules that provide the plant output voltage, and then connect the strings in parallel to provide the plant output current. In this configuration, the mismatch losses due to the final parallel connection are negligible with respect to those due to the series connection [3], and therefore, studying the mismatch losses in the series arrays would characterise the performance of the whole field.

Reducing mismatch losses has been usually considered as a goal. In order to get this reduction, in module manufacturing and in large PV installations it is not unusual to classify the generators into categories with the general idea that the dispersion of each category should be smaller than that of the whole set. This general idea have led to several classification criteria that have been useful in practice, but no guidance about how close to optimal these criteria are has been reported.

In this work we present the classification criterium that provides the minimum mismatch losses in series connected arrays. On the contrary to conventional ones, this criterium is not based in the aforementioned general idea that links classification and less dispersion.

Section 2 presents some of the criteria conventionally used by installers of modules in large PV plants and by manufacturers of cells to build modules. Section 3 is devoted to the definition of a generic criterium, which will be useful later in Section 4, in which the optimal criterium is presented. In Section 5, this criterium is compared with the conventional ones introduced in Section 2. Section 6 brings into question the suitability of minimising the mismatch losses as the goal to reach. Finally, in Section 7 practical aspects are discussed and in Section 8 some conclusions are given.

2. CONVENTIONAL CLASSIFICATION CRITERIA FOR SERIES ARRAYS

Considering the minimum mismatch losses in the series array as the goal, a criterium of classification of PV generators has been proposed [1][8], which is based on the measurement of the current I_m at their maximum power point (MPP). It consists in: (1) defining as many ranges of the parameter I_m as categories, (2) classifying the PV generators with the criterium that a generator belongs to a category if its I_m value is inside the range corresponding to that category. The ranges of I_m are usually selected with a given constant pitch, which is previously chosen. This criterium has been used, for instance, in the installation of Toledo-PV plant [7]. Another close criterium consists in using the short-circuit current I_{sc} instead of I_m. This criterium, which might be theoretically less effective, has got a practical advantage: the parameter I_{sc} is obtained from a single measurement, while I_m needs several measurements to be estimated.

A different criterium used by several module manufacturers classify the cells according to the value of the current given by the cells at the average voltage $\langle V_m \rangle$ of their MPP. This criterium also needs a single measurement to be implemented (plus the estimation of $\langle V_m \rangle$).

There is no general proof about which of the aforementioned criteria is better. However, the first one may seem superior because, as the series connection makes all the PV generators work at the same current, if all of them had the same current I_m (although there is dispersion in any other parameter), no mismatch losses would appear, because I_m would be also the current at the MPP of the series array.

3. DEFINITION OF A GENERIC CRITERIUM

Let us consider a set of N generators and assume that there is dispersion on the parameters which define the generator operation. Without loss of generality, let us consider, as an example, that the generator can be characterised by a model that depends on three parameters only; for instance, the short-circuit current I_{SC}, the open-circuit voltage V_{OC}, and the series resistance R_S.

In the three-dimensional space of parameters, a generator is represented by a point (I_{SC}, V_{OC}, R_S). The whole set of generators form a cloud of points in that space, as shown in Figure 1.

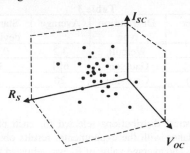

Figure 1. A PV generator is represented by a point of the space of parameters.

A criterium of classification can be defined in the space of parameters as a family of surfaces that do not cross one to another, and a category is given by the region of the space between two adjacent surfaces. These surfaces can be mathematically defined by the equation:

$$H(I_{SC}, V_{OC}, R_S) = C$$

where different values of the parameter C provide the different surfaces.

This representation of a generic criterium in the space of parameters is not completely general. For instance, a random criterium of sorting cannot be represented in this way, because the topological boundaries of the regions corresponding to each category do not define surfaces in this case. However, this representation is valid for all the criteria included in this paper.

As an example, in the case of two categories, a single surface is needed and the parameter C can be chosen as C=0. The classification of the PV generations is then done with this generic criterium as shown in the Table 1.

Table 1

Category 1	Category 2
$H(I_{SC}, V_{OC}, R_S) < 0$	$H(I_{SC}, V_{OC}, R_S) > 0$

The generators of each category are located at different sides of the surface H, as shown in Figure 2.

4. CRITERIUM FOR MINIMUM MISMATCH LOSSES IN SERIES ARRAYS.

Let us consider now that the N generators are classified into n categories, and that the generators of each category are connected in series, as shown in Figure 3. The maximum power generated by this configuration is given by:

$$P_{T,m} \equiv P_{S1,m} + P_{S2,m} + \ldots + P_{Sn,m}$$

where $P_{Sk,m}$ denotes the power at the MPP of the series array k (i.e. formed by generators of category k).

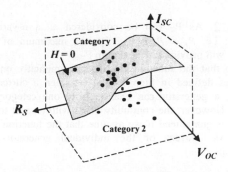

Figure 2. A generic criterium for two categories is defined by a surface H=0 in the space of parameters.

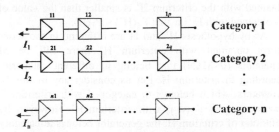

Figure 3. The merit function is defined from the series connection of the PV generators of each category.

Minimising the mismatch losses of these series arrays is equivalent to maximise the function $P_{T,m}$. Different classification criteria will distribute the individual generators differently, and then they will get different values of $P_{T,m}$. In the next section we find the classification criterium that makes $P_{T,m}$ maximum.

4.1. Definition of the optimal criterium

The criterium that maximises $P_{T,m}$ is defined in two steps.

Step 1. Assume the values of the currents I_1, I_2, ... and I_n of the series arrays are given. Let us define the function:

$$P_T \equiv P_{S1}(I_1) + P_{S2}(I_2) + \ldots + P_{Sn}(I_n)$$

where $P_{Sk}(I_k)$ denotes the power generated by the series array k at the bias current I_k. As proven below in Section 4.2, the classification criterium that makes P_T maximum is:

A generator belongs to the category k if, and only if, generates more power at the current I_k than at any other current I_j, $j \neq k$

For the particular case of two categories, this criterium says that a generator belongs to the category 1 (or 2) if and only if it produces more power at I_1 (I_2) than at I_2 (I_1), as expressed in Table 2.

Table 2

Category 1	Category 2
$P(I_1) > P(I_2)$	$P(I_1) < P(I_2)$

In the space of parameters, this criterium says that the surface H=0 is composed by the points fulfilling $P(I_1) = P(I_2)$.

Step 2. As P_T can be considered as a n-variable function, i.e. $P_T \equiv P_T(I_1, I_2,... I_n)$, the maximum of this function will provide the maximum $P_{T,m}$.

Note that the function P_T depends implicitly on the categories defined in Step 1, that is, when a current I_k changes, a generator can change from one category to another. However, the continuity of P_T with respect to the currents is guaranteed if we assume that the function $P(I)$ (power vs. current of the individual generators) is continuous.

4.2. Proof of Step 1

Consider a criterium H' different from the optimal H defined in Step 1. We will prove that the value of P_T obtained with the criterium H' is smaller than the value of P_T obtained with H, i.e. that $P_T(H') < P_T(H)$.

As by hypothesis H' and H are different, in the series arrays obtained with criterium H' there will be M generators (0<M≤N) that belong to different categories according to criterium H. Let us consider any of this M generators, which belongs to category j with criterium H and to category j' (≠ j) with criterium H'. However, by definition of criterium H, the generator belongs to category j of H because it generates more power at the current I_j than at any other current, in particular at $I_{j'}$. Therefore, the power provided by this generator in the configuration of criterium H' will be smaller than the power provided in the configuration of criterium H, and this applies to the M generators aforementioned.

The demonstration of Step 1 can be also stated in the framework of Variational Calculus. Although that proof is somewhat more complex than the one given above (it is also long enough to be excluded from this paper), as the variational formulation is more general, can be useful to find in the future the optimal classification criteria for other merit functions (see Section 6) or for mixed series-parallel configurations.

4.3. Dual criterium for parallel arrays

For the case of parallel arrays, it is possible to define the optimal criterium for minimum mismatch losses analogous to that of Section 4.1. In this case, the N generators are classified into n categories, and the generators of each category are connected in parallel, and the function P_T in Step 1 is defined as:

$$P_T \equiv P_{P1}(V_1) + P_{P2}(V_2) +...+ P_{Pn}(V_n)$$

where $P_{Pk}(V_k)$ denotes the power generated by the parallel array k at the bias voltage V_k. The classification criterium that makes P_T maximum for given voltages V_1, V_2,... and V_n is:

A generator belongs to the category k if, and only if, generates more power at the voltage V_k than at any other voltage V_j, j≠k

5. COMPARISON

In order to compare the optimal criterium presented in Section 4.1 with some of the conventional criteria defined in Section 2, a computer simulation has been carried out with 1000 PV cells modelled by a single-exponential equation with the three parameters: I_{SC}, V_{OC} and R_S [9]. A pseudo-

random number generator has been used to define the parameters of the cells with the statistical characteristics shown in Table 3.

Table 3

Parameter	Distribution type	Average	Standard deviation
I_{SC}(A)	Weibull	3.2	0.16
V_{OC}(mV)	Gaussian	601	30
R_S(mΩ)	Gaussian	20	2

The statistical distributions selected for each parameter are in accordance with the experimental results obtained in reference [2]. The average value of R_S was adjusted to get an average cell efficiency of 15%.

The cells have been sorted into two categories according to three criteria: (1) the optimal presented in Section 4.1, (2) according to the I_m value, taking the average value $<I_m>$ of the cells as boundary between categories, and (3) according to the current $I(<V_m>)$, taking analogously the value $<I(<V_m>)>$ as boundary.

The results, shown in Figure 4, have been expressed in terms of the global *connection efficiency*, defined as:

$$\eta_C = \frac{P_{T,m}}{\sum P_{ij,m}}$$

where the $P_{T,m}$ was defined in Section 4, and the denominator is the sum of the maximum powers delivered by the individual generators at their respective MPP.

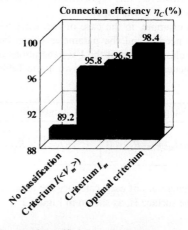

Figure 4. Connection efficiency obtained in the simulation with the different classification criteria.

6. OTHER MERIT FUNCTIONS MINIMISING THE MISMATCH LOSSES.

Up to this point we have considered as the goal to minimise the mismatch losses, expressed as maximizing the function $P_{T,m}$ defined in section 4. However, this is not the most suitable merit function in specific cases where other constrains appear. For instance, if the series arrays have to be connected composed by the same number of generators (to connect them in parallel later), this additional condition should be included in the problem (in Figure 3, p=q=...=r). The optimal classification in this case can be obtained in a similar way (with steps 1 and 2) to that provided in Section 4. For instance, for the case of two categories, teh classification is given by the surface H=0 that fulfills

$$P(I_1) - P(I_2) = \lambda$$

where λ must be calculated in Step 1 to provide the same number of generator in both categories. The proof of this step is analogue to that given in Section 4.2.

These merit functions are possibly adequate for the final customers, who would like to get the maximum power from their installations. However, this is not the case, for example, of the module installer, who must be concerned about what it is specified in the sale contract (which of course does not mention the mismatch losses). Another case is that of the module manufacturers, who usually sell at the same price, for instance, modules of nominal power 50 W ±10%, and thus any watt over the 45 W boundary is given to the customer for free.

As a simple example, let us define the appropriate merit function for a module manufacturer, assuming that it is going to sell a single type of modules, and that no cells/modules are rejected in the manufacturing process. Then, the merit function should be to *maximise the power generated by the worst module*, because this is going to fix the sale nominal power of all of them (with a ±10% tolerance, the nominal power would be $P_m^{worst}/0.9$, where P_m^{worst} is the nominal power of the worst module). Then, instead of the connection efficiency, the manufacturer could use the following *practical efficiency* η_P:

$$\eta_P = \frac{P_m^{worst}}{n < P_{cell,m} >}$$

where n is the number of cells per module and $< P_{cell,m} >$ is the average of the power at the MPP of the cells.

The optimal classification criterium in this case should provide how to select the cells to make modules in such a way that the worst module produces maximum output.

7. PRACTICAL IMPLEMENTATION

The practical implementation of the optimal criteria presented is not as direct than the conventional ones, but can be also done with some additional work. For instance, for a large PV installations, it would be necessary: (a) to have an model for the PV modules with sufficient accuracy, whose parameters for an individual generator can be obtained from few simple measurements, (b) to do such measurements for all the modules, and (c) calculate with a computer program the optimal classification according to the two steps given in Section 4.1, and (d) sort and connect the models according to the classification obtained. The main difference between this process and the conventional ones is that the module sorting cannot be done when they are measured in Step (b), but the computer processing of Step (c) is needed. However, note that this step would be

also needed in conventional criteria if the boundaries between categories, which are presently selected arbitrarily, were optimised.

In the case of the classification of cells when building modules in the factories, we have not found the optimal criterium (but only the merit function for a specific case). However, note that for practical implementation other constrains should be included, as the sequential and temporal nature of cell manufacturing, the possibility of reject cells, the interest of selling two or more types of modules with different tolerance and price per watt, etc.

8. CONCLUSIONS

Results show that optimal criterium really optimises the classification of cells. The simulated mismatch losses are 4.2% with criterium A (sorting by the current produced at the mean value of the voltage of the MPP of the cells), 3.5% with criterium B (sorting second by the current of the MPP of the cells) and 1.6% with the optimal one.

This criterium does not classify the generators into categories with the general idea that the dispersion of each category should be smaller than that of the whole set.

The reasons given in section 7 about practical implementation point to the optimal criterium may not be the best one for industry.

9. ACKNOWLEDGEMENTS

This work has been done under contracts TIC96-0725-C02-02 and TIC96-0825-C03-02 with the Plan Nacional de Investigación y Desarrollo with the Spanish Comisión Interministerial de Ciencia y Tecnología, under contract 06T/027/96 with the Consejería de Educación y Cultura of the Spanish Comunidad Autónoma de Madrid, and under contract JOR3-CT97-0123 with the JOULE program of the EC.

REFERENCES
[1] A. Luque, E. Lorenzo, J.M. Ruiz, *Solar Energy* 25, pp.171-178, (1980).
R. Zilles, E. Lorenzo, *Int. J. Solar Energy* 9, pp.233-239, (1991).
R. Zilles, E. Lorenzo, *Int. J. Solar Energy* 13, pp.121-133, (1993).
F. Iannone *et al.*, *Solar Energy* 62, pp.85-92, (1998).
L.L. Bucciarelly, *Int. J. Solar Energy* 9, pp.233-239, (1991).
D. Roche *et al.*, *Progress in Photovoltaics* 3, pp.115-127, (1995).
R. Zilles, Doctoral Thesis, UPM, (1993).
T.J. Lombarski *et al.*, *15th IEEE Photovoltaic Specialists Conference*, (1981).
E. Lorenzo, *Solar Electricity*, Ed. Progensa, (1994).

SKIAMETER SHADING ANALYSIS

M. Skiba[1], F.R. Faller[1], B. Eikmeier[2], A. Ziolek[2], H. Unger[2]

[1]Shell Solar Germany, Überseering 35, 22297 Hamburg, Martin.M.Skiba@ope.shell.com

[2]Department for Nuclear and New Energy Systems, Faculty of Mechanical Engineering,
Ruhr University of Bochum, Universitäts-Strasse 150, 44780 Bochum

ABSTRACT: When estimating energy yields of planned solar technical systems, size-specific effects have to be considered. During operation, shading effects can lead to considerable losses of solar irradiation, which in turn would lead to significant discrepancies between the energy yields predicted by the sales company of the solar system and the energy yields actually obtained.

In this paper a precise but yet cheap and fast method to evaluate shading effects on the energetic use of solar radiation is presented. An essential part of this procedure is the Skiameter (skia, greek = shadow), a device for opto-electronic surveying of the obstructed horizon. The practical application and results realised by using the Skiameter are discussed by analysing actual photovoltaic systems. Additionally, a validation of the whole procedure was done by comparing the data to measurements from the german 1000-roof-program. Losses of solar irradiation onto PV systems due to shading effects were calculated from Skiameter measurements with an average deviation of 3.3% compared to direct measurement.

Keywords: Shading -1: Solar irradiation -2: PV-Systems -3

1. INTRODUCTION

In the ideal case a photovoltaic system is installed such that (a) it is facing directly south, (b) is tilted ideally depending on the geographical latitude of its location and (c) no shading effects of nearby objects obstruct the solar irradiation. Practically, however, these requirements can be met only in part. When estimating energy yields of planned solar systems effects from the deviation from the ideal case have to be considered. In this paper shading effects are analysed since they can lead to significant losses of solar irradiation which in turn may cause significant discrepancies between the energy yields predicted by the sales company of the solar system and those actually obtained.

A procedure to evaluate the effect of shading on the overall performance of a photovoltaic system has been developed. An essential part of this procedure is the Skiameter (skia, greek = shadow), a device for opto-electronic surveying of the obstructed horizon. The data from the skiameter is further processed by computer programs specifically written for this purpose.

In this paper the practical application and results obtained by using the Skiameter with eight actual photovoltaic systems are discussed. In addition, a verification of the whole procedure was done by comparing the Skiameter data and subsequent calculation with values obtained from the same photovoltaic systems by direct measurement (I-MAP data-bank from the 1000-roof-program of the federal government of Germany).

2. HARDWARE

The method used for our Skiameter setup was chosen from a variety of possible methods; it it based on a concept developed by Tonne [1]. Figure 1 shows the principle components and functions of the setup used. The image of the surroundings is projected from a planoconvex, highly reflective lense onto the chip of a CCD camera through an objective which is placed directly above the lense. The CCD camera is

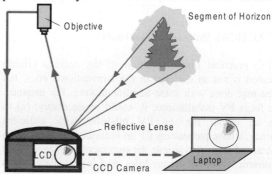

Figure 1: Principle of Skiameter measurement setup

located in the lower part of the system and stores the image directly as a digital picture. All optical components are able to project the entire hemisphere onto the CCD-chip of the camera, the actual image can be controlled via a LCD screen which displays online the viewed image. The camera can also easily be detached and used for other purposes.

The objective and its supporting holder as well as possibly people present at the time of the measurement may cause a loss of information by partly obscuring the hemisphere. However, the image can be taken twice with the position of the objective being different from the first picture. By superposition of the two images with suitable software the disturbing effects can be eliminated. Only the image of the front-side of the objective remains to be seen. Figure 2 shows a photograph of the Skiameter, Figure 3 depicts a 360° image made with this setup. In this example the picture was taken from a 32° tilted roof facing directly south (49.5° geographical latitude, 6.7° altitude). The measured solar irradiation onto the 2.1 kW$_p$ photovoltaic

Figure 2: Skiameter with objective at top as well as lense and digital camera

system that was actually installed on this roof was then compared to the value obtained through the procedure described in this paper.

3. CALCULATION

The simulation of the solar irradiation including shading effects is based on a multi-step procedure. For the calculation the following input data is needed:

- 12-monthly locally specific average daily sum of the global- and diffuse-irradiation ('macro-climatic characteristics')
- reflecting, transmitting or shading properties of objects in the surroundings ('micro-climatic characteristics')
- properties of solar system resulting from its geometry and alignment

In a first step the daily sums of the solar irradiation onto a horizontal non-obstructed surface are calculated by means of a stochastical model [2] using the monthly values of the global irradiation. However, in order to consider the local characteristics completely, another stochastical model to compute hourly sums of the global irradiation was developed; it is based on time-resolved data taken over many years.

An evaluation of the effect of a restricted or obstructed horizon onto the solar irradiation is only possible if the fractions of the diffuse and direct irradiation are known. Therefore, a deterministic algorithm was developed, that computes the diffuse fraction of the global irradiation dependent on the day-time, monthly sum of the diffuse irradiation as well as the cleanness-factor of the air. For the calculation of irradiation onto tilted surfaces a comparison of

various models was made and the one of Perez [3]

Figure 3: Example Skiameter image taken on a rooftop. In the top part the electrical grid connection to the house is visible, in the bottom part the horizon as viewed from the roof is to be seen. The Skiameter objective is still visible in the middle, the holder was eliminated.

was selected and adjusted for german climatic conditions.

The modelling of the shading effects is done separately for direct and diffuse irradiation. In the case of direct irradiation shaded areas can obviously be extracted time-resolved from the sun's daily path. For diffuse irradiation (which can be split in circum-solar, isotrope or near-to-horizon irradiation) the modelling can be done by determination of the view-factors after the unity-sphere-method of Nusselt [4]. The equations to compute the specific fractions of the diffuse irradiation can then be expanded in order to take shading effects into consideration. The corresponding view-factors can be determined from the Skiameter image; further details in [4].

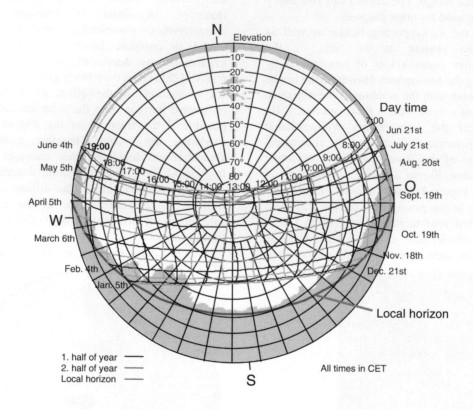

Figure 4: Compiled data from Skiameter with the sun's yearly trajectory

4. SOFTWARE

A specific simulation tool was developed (S3/S3⁺) which is divided in different modules for the interpretation and processing of the Skiameter image, determination of the irradiation, etc. The main tasks of the image processing are to extract the obstructions of the horizon from the raw photograph, possible corrections of single elements in the picture (e.g., elimination of the objective) and the determination of the view-factors and hourly values of the maximal possible duration of sunshine. It is also feasible to correct for

season-depending effects, like the presence of leaves on trees. Resulting from this process one obtains a digitised picture of the horizon and its local obstructions, as shown in Fig. 3 and 4.

For the further evaluation, the monthly average irradiation sums for the relevant region were taken

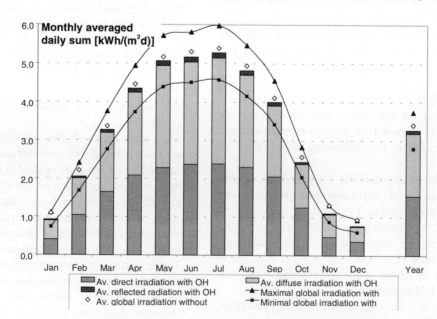

Figure 5: Results for 2.1 kWp photovoltaic system. 'OH' = Obstructed Horizon.

from the *European Solar Radiation Atlas* and were used as input for the calculation of the hourly sums of irradiation onto a horizontal or tilted surface over a year. In order to implement the effects of the obstructed horizon the digitised Skiameter image data together with the above mentioned hourly sums are processed and yield the solar irradiation under local conditions. The comparison of the solar irradiation with and without obstructed horizon allows for quantification of the losses. For the 2.1 kW_p system chosen as an example in this paper the results of the evaluation are shown in Figure 5. Separately shown are the direct and diffuse irradiation and the reflected radiation as well as the monthly average day-sums together with global irradiation excluding and including losses caused by obstructions of the horizon at this location. In this example the yearly loss of irradiation was about 50 kWh/m² at a total of about 1200 kWh/m².

5. VERIFICATION

In order to verify this method of determining the yearly losses of solar irradiation due to obstructions of the horizon the final results were compared to irradiation data measured directly over a period of three years at the module level of the photovoltaic system.

A more detailed description of the overall verification method is given in [5].

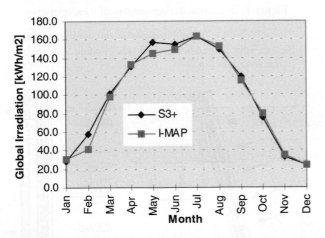

Fig. 6: Comparison of the global irradiation

Figure 6 shows the comparison of the global irradiation (S3⁺ vs. I-MAP) for the example roof chosen here. The average relative error is 6.6%. More significant are the yearly values of the global irradiation which were 1197 kWh/m² calculated by

S3⁺ compared to 1170 kWh/m² from the I-MAP data. The relative error is only 2.3%, standard deviation is 6.5 kWh/m².

6. CONCLUSION

A method to quantify the possible losses of solar irradiation onto a solar system through shading effects was described. Eight actual examples of photovoltaic systems which were thoroughly monitored through the german 1000-roof-program were chosen to compare the results of our method to the data obtained from direct measurement. The overall results of our method lead to deviations of the yearly global irradiation from I-MAP data of only 0.4-5.5%, average of all eight PV systems chosen was 3.3%. The described method can therefore be recommended to evaluate shading effects and to provide the annual global irradiation.

ACKNOWLEDGEMENT

We gratefully acknowledge the financial support by the 'Arbeitsgemeinschaft Solar' of state of North-Rhine Westfalia, Germany. We would also like to thank the 'Deutsche Wetterdienst' for providing data on the solar irradiation.

REFERENCES

[1] Freymuth, H., Lenz, H., Lutz, P., Schupp, G. *Licht, Luft, Schall.* Forum Verlag Stuttgart, 1977

[2] Heidt, F.-D. *Vergleich stochastisch simulierter täglicher Globalstrahlung.* 7. Int. Sonnenforum, DGS, Frankfurt 1990

[3] Perez, R., Seals, R., Ineichen, P., Stewart, R., Menicucci, D. *A new simplified version of the Perez diffuse irradiation model for tilted surfaces.* Solar Energy, Vol. 39, p. 221, 1987

[4] Skiba, M., *Ein Verfahren zur Standort-evaluation energetischer Nutzungsmöglichkeiten solarer Einstrahlung.* VDI-Fortschrittsberichte, Reihe 6, Nr. 375, 1997

[5] B. Eikmeier, A. Richei, A. Ziolek, U. Unger. *Validierung des Verfahrens SASKIA durch Messdaten aus dem I-MAP des Bund-Länder-1000-Dächer-Programms.* Final Report RUB E-I-253, Ruhr-Universität Bochum, January 2000

INFLUENCE OF TEMPERATURE ON CONVERSION EFFICIENCY OF A SOLAR MODULE WORKING IN PHOTOVOLTAIC PV/T INTEGRATED SYSTEM

E.Klugmann [1], E.Radziemska [2], & W.M.Lewandowski [2]

1- Solid- State Electronics, 2- Apparatus & Chemical Machinery Department / Technical University of Gdańsk
80-952 Gdańsk, Narutowicza 11/12, Poland

ABSTRACT: Experiments are conducted on a photovoltaic single-crystalline silicon solar module, integrated with a flat heat collector. It has been shown that the performance of the module degrades with rising temperature about 0.08%/K. Several advantages associated with water cooling of the module from 60°C to 25°C have been observed: the output power increase of 23%, the open circuit voltage increase of 18% and PV conversion efficiency increase of 3%. The energetic conversion efficiency of the PV/T integrated system is 13% higher than the efficiency of a separate heat collector.
Keywords: Silicon solar module - 1 : Solar heat collector - 2 : Conversion efficiency - 3

1. INTRODUCTION

Solar radiation arriving at the surface of the earth, at clean atmosphere, covers ultraviolet, with 9% of the radiation energy, visible radiation with 44% and infrared with 47% in a broad range of heat radiation wavelengths (0.75 - 3.0 µm.). The basic radiation with 98% of the whole radiation energy inside the atmosphere cover the wavelengths from 0.2 to 3.0 µm.

Due to the absorption of electromagnetic radiation, the heat is produced in a solar module. Apart from heating, radiation entering a solar cell can set free electrons from its atomic bond creating electron - hole pairs. In order to generate the electron-hole pair, the proper photon energy must be provident to the cell; at least equal to the material band-gap energy, and by means of a built-in "potential barrier" of the p-n junction the electrons are separated from holes, generating a photoelectric current and Joule`s heat in the series resistance of the module and electric power in the load circuit as well.

The thermodynamic analysis of an integrated PV/T system was given in previous papers [1-4].

The aim of this paper is to present the quantitative evidence that the temperature increase degrades significantly the output power and conversion efficiency of a silicon solar module. Efforts are being made to improve the conversion efficiency by reducing the temperature increase using a flat heat collector integrated with the PV module.

2. EXPERIMENTAL

Measurements of insolation were made at the latitude of 54N by means of a radiometer during June-July, at quiet insolation as seen in Fig.1. Consideration has been given to hours of sunshine during which the insolation changes only a little, e.g. between 11 AM and 1 PM. The current-voltage characteristics I(U) were measured in order to determine the fill-factor (FF), the electric output power and the conversion efficiency of the module. The electric current and voltage were measured by means of a digital multimeter METEX at the stabilized temperature of the module.

As a temperature sensor, a thermocouple copper - constantan was used.

A module of ASE-100-DGL-SM type was joint together with a heat collector, working as a 12 liter water capacity cooling and thermoregulating unit. The unit is an Al frame construction with the total surface of 0.88 m^2, covered by 72 single-crystalline silicon solar cells with the surface of 0.72 m^2.

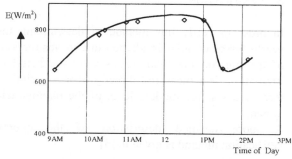

Figure 1: Insolation during one day of measurements.

The unit could be turned toward the sun in order to direct the sun light perpendicular onto the module. The mobile frame is seen in Fig.2.

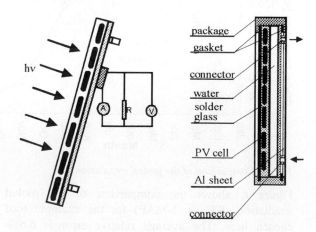

Figure 2: A schematic diagram of the experimental stand.

3. ENERGY LOSSES AND CONVERSION EFFICIENCY

The major phenomena that limit PV efficiency of the module are:

- light reflection from the module surface,
- solar light energy too little or too much of that which is needed to separate electrons from bonds,
- recombination of electrons and holes before they could contribute to electric current,
- series resistance to current flow,
- self-shading resulting from top-surface electric contacts,
- performance degradation at non-optimal operating temperatures.

In this paper we give some experimental data supporting the last sentence.

Generally about 55% of the solar radiation power is lost before the photons could be converted into electron-hole pairs. In case of silicon, 23% of photons is not energetic enough to separate electrons from their atomic bonds, and 32% has extra energy beyond that is needed to separate electrons from bonds.

Other losses are due to direct and indirect recombination of electrons and holes, due to reflection of the radiation from the cells surface (up to 5%) and due to self-shading from contacts.

The conversion efficiency is given as a ratio of the maximum electric and radiation power

$$\eta_{PV} = \frac{P_m}{ES} \qquad (1)$$

where S is the surface of the solar cells and E the radiation rate (W/m^2).

Introducing the fill-factor FF of the current-voltage characteristic I(U) as:

$$FF = \frac{I_m U_m}{I_{sc} U_f} \qquad (2)$$

we find:

$$\eta_{PV} = (FF) \frac{I_{sc} U_f}{ES} \qquad (3)$$

where I_{sc} is the short circuit current limit and U_f the open - circuit voltage (photo-electric voltage).

The heat collector efficiency:

$$\eta_{coll} = \frac{C \, m_{av} \, (T_{out} - T_{in})}{ES_M} \qquad (4)$$

where: m_{av} is an average flux of water, C- its specific heat, S_M-the surface of the module (heat collector).

The energetic conversion efficiency of the integrated system:

$$\eta_{PV/T} = \frac{C \, m_{av} \, (T_{out} - T_{in})}{E \, S_M} + (FF) \frac{I_{sc} U_f}{ES} \qquad (5)$$

The efficiency of a solar heat collector usually is within the range of $\eta_{coll} \sim 60\%$, whereas that of the silicon single-crystalline solar module $\eta_{PV} \sim 12 - 15\%$.

3.1. Performance degradation at higher temperature.

All semiconducting materials loose its efficiency at higher temperature, quantitatively described by a coefficient of the relative thermal decay of the conversion efficiency:

$$\beta = \frac{1}{\eta_{PV}} \frac{d\eta_{PV}}{dT} \qquad (6)$$

where the short circuit current limit of silicon changes with temperature only a little and the adequate coefficient [5]:

$$\gamma = \frac{1}{J_{sc}} \frac{dJ_{sc}}{dT} = 3 \cdot 10^{-4} K^{-1} \qquad (7)$$

At a small temperature increase ΔT near 300K :

$$\eta_{PV}(T) = \eta_{PV}(300K) \cdot (1 + \beta \Delta T), \qquad (8)$$

where $\Delta T = T - 300K$.

Solar cells and solar modules work best at certain temperatures, according to their material properties. At normal and lower temperatures, 25°C, silicon is a good material, but at high temperatures, 200°C for instance, the silicon efficiency droppes to 5%, e.g. the silicon efficiency temperature coefficient [6]:$\beta = -4,7 \cdot 10^{-3} K^{-1}$. It means, performance of silicon degrades with rising temperatures.

Solar cells without cooling fail in high temperature applications like concentrated sunlight or in space near the sun. In this case another solution is to adopt the cell material to the higher operation temperature and use gallium arsenide or cadmium sulfide. But these materials are much more expensive.

The temperature attained by the module depends on its construction, ambient air temperature, cooling conditions (wind speed) and insolation level. The intensity of sunlight directly affects the temperature of a typical average module [7].

4. RESULTS

In Fig.3 the current-voltage characteristics are shown at 60°C (without cooling) and at 25°C with cooling by using a 12 liter capacity water heat collector integrated with the module and a rotameter for indicating the rate of the flow.

As seen in Fig.3 the voltage and current from a PV source change with temperature.

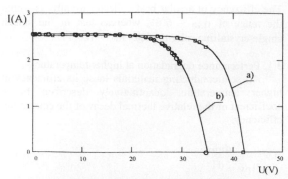

Figure 3: Current-voltage characteristics of the module
at the insolation 840 W/m^2 for temperature:
a) 25^{0}C and b) 60^{0}C.

Fig. 4 and 5 shows the output power against the load
resistance R :
a) $P_m = I_m U_m = 79.6$ W at $T = 25^\circ$C , and $R = R_{opt.,}$
b) $P_m = 61.3$ W at $T = 60^\circ$C , and $R = R_{opt.}$
The fill-factor :
a) FF = 0.742,
b) b) FF = 0.690,
and the conversion efficiency :
a) $\eta_{PV} = 13.1$ %,
b)   b) $\eta_{PV} = 10.1$ %
decrease with increasing temperature of the module.

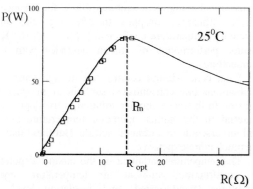

Figure 4: Output power against the resistance at 25^{0}C.

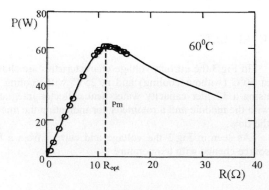

Figure 5: Output power against the resistance at 60°C.

The parameters of the PV module working in the PV/T
integrated system are shown in Table 1.

Table 1: Parameters of the PV/T unit

T($^\circ$C)	25	60
U_f(V)	42.18	34.75
I_{sc}(A)	2.545	2.555
P_m(W)	79.60	61.28
FF	0.742	0.690
η_{PV}(%)	13.1	10.1
$\eta_{PV/T}$(%)	75.8	72.8
η_{coll}(%)	-	62.7
$Q_{PV/T}$(W)	-	464
$Q_{PV/T}/S_M$(W/m^2)	-	527

As follow from the data of table 1, reduction of the
temperature from 60 $^\circ$C to 25 $^\circ$C results in the:
- increase of electric output power, 23%,
- improvement of PV conversion efficiency, 3%,
- improvement of PV/T conversion efficiency,13%,
- gain of heat flux from the cooling unit, range of 0.5
  kW/m^2. This heat can be applied to useful purposes.

5. CONCLUSIONS

The physical effects determining the efficiency
relationship with the temperature are quite complex. For
the most part, two predominate in causing efficiency to
drop as the temperature rises:
- increase of the thermal lattice vibrations, leading to
  electron-phonon scattering and decrease of charge
  carriers mobility,
- reduction of the p-n junction ability to separate
  electrons from holes in the pairs generated by
  photons,
The first effect severely degrades the silicon
performance, even at room temperatures. The second
effect does not occur until the temperatures of about
300^{0}C for silicon. In order to diminish these effects, it is
useful to decrease the module temperature by removing
the heat produced by:
- non-active absorption of photons, that do not
  generate pairs,
- recombination of electron-hole pairs outside the p-n
  junction region,
- photo-current (Joule heat generated in the series
  resistance of the p-n junction) and parasitic currents.
The temperature coefficients of our module, which
are: β= - 6.4 10^{-3} K^{-1} and γ = 1.1 10^{-4}K^{-1} agree well with
that, given for silicon in the literature [5], [6].
The results show that the energetic conversion efficiency
of the integrated system (76%) is 13% higher than the
efficiency of a separate solar heat collector (63%) with
the same dimensions.

REFERENCES

[1] H.P. Garg, R.K. Agarwal, Energy Conversion and Management **36** (1995) 2.

[2] M.A. Hammad, Renewable Energy **4** (1994) 8.

[3] W.M.Lewandowski, E.Radziemska, T.Wilczewski, H.Bieszk, Proceedings „Small Energetic 97" Conference, Zakopane, Poland, 18-20.09.1997.

[4] E. Radziemska, E. Klugmann , Proceedings XXIII Conference of the International Micro-electronics and Packaging Society, Poland Chapter , Technical University of Koszalin, Poland, 21-23.09.1999. Ed. (1999) 97.

[5] R.N. Hall, Solid - State Electronics **24**(1981)595.

[6] E. Klugmann, E. Radziemska, Alternative Energy Sources. PV Energetic Ed. by: Ekonomia i Środowisko, Białystok, Poland, (1999) 155.

[7] Basic Photovoltaic Principles and Methods, Ed. Van Nostrand Reinhold Comp. Inc. Solar Energy Research Institute, New York (1984) 199.

REMOTE CONTROL OF THE I-V CHARACTERISTICS OF SOLAR CELLS IN THE ENCAPSULATED MODULES

S.V.Litvinenko[1], L.M.Ilchenko[1], S.O.Kolenov[1], A.Kaminski[2], A.Laugier[2], E.M.Smirnov[1], V.A.Skryshevsky[1]

[1]Radiophysics Department, Kiev Shevchenko University, 64 Vladimirskaya, 01033, Kiev, Ukraine, Fax: +38-044-2656744, e-mail: litvin@uninet.kiev.ua

[2] LPM, UMR 5511, INSA de Lyon, 20 av. Albert Einstein, 69621 Villeurbanne Cedex, France, Fax: +33-472438531, e-mail: Anne.Kaminski@insa-lyon.fr

ABSTRACT: The possibility of monitoring solar cell encapsulated into batteries is considered in present work. Our approach is based on LBIC-like measurement when the module is loaded in galvanostatic operation mode. Periodical testing illumination of a selected solar cell results in the signal output of the module that is proportional to the derivative dV_m/di of the examined solar cell. This signal does not depend on the presence of other cells in the module. The $dV_m/di = f(i)$ functions can be obtained for all cells of the module by 2-D scanning and variation of the current. The integral of these functions gives the I-V characteristics, and the comparison of "dark" and "illuminated" $dV_m/di = f(i)$ functions determines the short circuit current of each tested cell of the module. The experimental verification was performed for a commercial encapsulated module.

Keywords: characterization – 1; batteries – 2; monitoring - 3

INTRODUCTION

The world-wide trends in PV research are to improve the conversion efficiency, reduce the material consumption and total cost. The large area multicrystalline Si and thin film solar cells (SC) are considered as promising devices to achieve these goals. Owing to the efforts of researchers and essential financial support during the last years the industrial scale production of 14 –16 % efficiency multicrystalline based SC are available. There is no doubt that the ratio efficiency/cost will be yet improved by using larger ingots and slices, by reducing the waste of semiconductor material, and by the implementation of new technologies. It is well known that the large active area of the solar cell provokes its inhomogeneity and the dispersion of its parameters.

The power output of PV batteries is contributed by several factors such as separate solar cell (SC) efficiency, module assembly quality and conditions of the operation. The efficiency of the module is somewhat lower than the average efficiency of the cells owing to the dispersion of their parameters. Thus the contribution of each cell to the total module efficiency in various conditions should be taken into account to minimise the energy conversion losses and the cells should be choose by criterion of minimum energy conversion losses before their assembly into the module or battery. Being encapsulated they are exposed to harmful influence of solar irradiation, dust, temperature and humidity impact. Besides that, the assembly process itself can lead to undesirable and uncontrolled variation of the SC characteristics. Usually after the encapsulation the direct access to every cell is not already available. The possibility of the SC characteristics monitoring is desirable because it allows to detect the origin of PV conversion energy losses in the module to make the forecast of the module reliability and to optimise the SC technology.

The aim of the present work is to introduce an original technique for testing the SC I-V characteristic parameters in modules which are already encapsulated. Though the rather informative technique such as Laser Beam Induced Current (LBIC) is widely used in SC study [1], it is not directly applicable to the mentioned task.

Really, the separate illumination of a certain cell of the module loaded by voltage constant circuit (potentiostatic mode) or simply by resistance results in a PV signal dependent on the SC photocurrent and on the parameters of other cells of the module. Thus the interpretation of such signal and extraction of the information about the tested separate SC is questionable [2].

EXPERIMENT AND DISCUSSION

In our approach the LBIC-like experiment is performed with SC module loaded to a galvanostatic mode Periodical illumination of a certain SC, for instance, by modulated laser beam, results in the periodical shift of its I-V characteristic along the current axis [3,4]. This well known "shift approximation" is valid for most of rather high efficiency solar cells, wide intervals of incident light intensity and applied voltage bias. Since the workpoint during the testing light impulse jumps to this shifted curve and the total current through the module circuit remains constant, the corresponding periodical voltage signal appears on the module output.

Let us consider the properties of this signal. First, it is proportional to the local slope of the tested number m cell I-V characteristic, i.e. the derivative of dV_m/di. Besides that, this signal does not depend on the rest of the cells of the module provided the input resistance of the measuring device is rather high. In the comparison with the case of the constant voltage or simple resistive loading these other cells are in the conditions of the invariable current and illumination and thus they "don't know" about the appearing of the alternative signal. Finally, this signal is obviously dependent on the photocurrent of the examined cell, i.e. on the mentioned shift of it's I-V characteristic during the testing light impulse.

Thus it is possible to get the $dV_m/di = f(i)$ functions for all the cells of the module by 2-D scanning and variation of the current. To recover the cell I-V characteristic the integral $V_m(i) = \int_0^i V'_m(i)\,di$ (where $V_m(i)$ is a voltage drop across the cell number *m*, $V'_m(i) = dV_m/di$ is a derivative in absolute units) should be calculated. As the current $I = 0$ at $V = 0$, the constant of integration $C = 0$ for dark conditions. In the common illumination conditions the constant of integration is equal to the corresponding open circuit voltage

$$V_{OC(m)} = \int_0^{I_{PH(m)}} V'_m(j)\,dj,$$ where $I_{ph(m)}$ is the short

circuit current of the cell number *m*. It is proposed to determine the last value by the comparison of "dark" and "illuminated" functions $V'_m(i)$. Fig.1 shows a typical experimental dependence of the discussed induced signal on the module current in dark and illuminated conditions. The curves on this figure differs by the shift along the

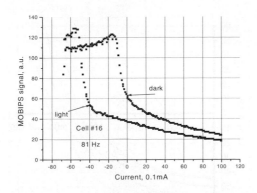

Fig.1. Dependence of the Modulated Optical Beam Induced PV Signal (MOBIPS) on the module current in the dark and illumination conditions

current axis. The values of the $V'_m(i)$ functions in absolute units are calculated by taking into account the dV/di derivative of I-V characteristic of the whole module.

The possibility of the evaluation of $V'_m(i)$ derivations from the measured induced signal is based on the following suggestions. (i) Firstly, the induced signal U'_m determines the derivative in the absolute unites $V'_m(i)$ taking into account the photocurrent of the measured cell and the amplification coefficient **k** equal for all the cells:

$$U'_m = k\,i_{ph(m)}\,V'_m; \qquad (1)$$

(ii) the voltage applied to the whole module is equal to the sum of each cell voltage drops, therefore the derivative of the whole module I-V characteristic is equal to the sum of each cell derivatives:

$$\sum_m V'_m = V'_{module}. \qquad (2)$$

The mentioned goal is easily achieved from these two expressions:

$$V'_m = \frac{U'_m V'_{module}}{i_{ph(m)} \sum_m (U'_m / i_{ph(m)})} \qquad (3)$$

In the experiment the originally designed laser scanning technique was used. It allows to examine in such way the modules and batteries of different size, from various distances and exploitation conditions. The user can choose the velocity of scanning, the size of the scanning area, frequency of the light beam modulation. The last value should be set low enough to avoid the influence of the capacitance conductivity and high enough to make the experiment faster. The obtained data storage is performed both in the mode of graphical (bitmap) files and in ASCII codes, that allows to observe and evaluate the data.

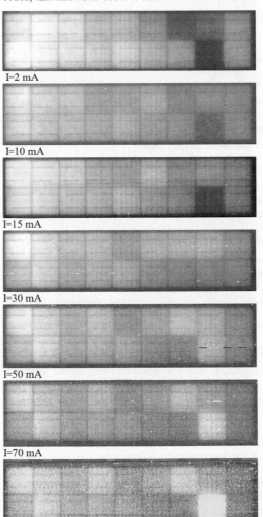

Fig.2. Distribution of MOBIPS (LBIC-like distribution) for (2*9=18)-cell commercially available module at different currents in dark conditions.

We investigated the commercially available batteries. A sample of the mentioned signal distribution for different current in the module's circuit is shown on the Fig. 2.

The contrast among different cells is often bigger than the contrast inside the cell area. According to the previous discussion the contrast among the cells corresponds to the derivatives $V'_m(i)$ that characterize each cell as a whole, while the contrast inside the cell reflects the photocurrent inhomogeneity. The contrast ratio among different SC changes depending on the applied voltage and illumination conditions. Let us consider whether the observed image contrast corresponds directly to any of the solar cell parameters. The MOBIPS image contrast between any two cells is proportional to i_s^{-1} (i_s : saturation current) when the voltage is low enough ($V < nkT/e$), since for the common I-V characteristic

$$V \approx \frac{nkT}{e}\frac{i}{i_s}$$ and the corresponding derivative is

$$\frac{dV}{di} \approx \frac{nkT}{e}\frac{1}{i_s}.$$ In the case of higher voltage

($V > nkT/e$) the ideality factor $n = \frac{e}{kT}\frac{dV}{di}i$.

Therefore the image contrast displays the cell-to-cell distribution of the ideality factor.

To obtain the numerical data which allows the recovering of the I-V characteristics, the distributions like shown on Fig.2 were measured and stored in ASCII format. The data corresponding to the coordinates of the certain cell were averaged by special software application in order to represent each cell as a whole by it's values of $V'_m(i)$.

The values of photocurrent were calculated from the current shift between the dark and illuminated $U = f(i)$ curves for each cell. I-V characteristics of the module were measured directly by the common technique and the V'_{module} derivatives were calculated numerically. Then, according to (3), the $V'_m(i)$ values in the absolute unites were calculated and I-V characteristics in the form of $V_m(i)$ dependencies were recovered by integration taking into account the constants of integration as discussed above.

The set of such I-V characteristics are shown on Fig.3 for dark conditions in the semilogarithmic scale and on Fig.4 for low intensity illumination conditions (about 2% of AM0) and in linear scale. Though the conversion efficiency of the module is rather high (13% at AM1) the valuable dispersion of the cells parameters is observed at low currents and voltages. This factor does not deteriorate the module efficiency but it is beneficial to display the possibilities given by the present approach.

The most of the cells have common quasi-exponential dark characteristics $i=i_s(\exp\frac{e\,V}{nkT} - 1)$, with the dispersion of ideality factor n and saturation current i_s. The set of the illuminated I-V characteristics displays also the dispersion of the short circuit currents. It is necessary to remember that the induced by the module PV signal in the

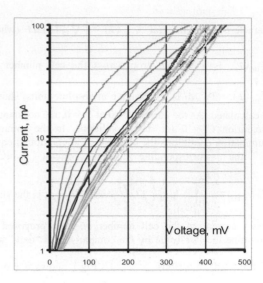

Fig.3. Set of I-V characteristics for the cells of the studied module in the dark conditions

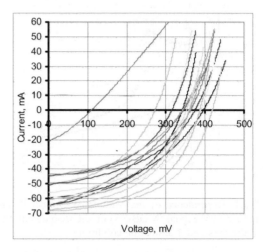

Fig.4. Set of I-V characteristics for the cells of the studied module in the conditions of the constant low illumination (2% of AM0).

described experimental conditions as well as the contrast of the LBIC-like image does not correspond directly to the examined cell photocurrent or to its contribution to the module efficiency as one can suggest from the experience of the single-cell LBIC experiment. One of the cells shares out by its poor loading characteristic, relatively high shunt resistance (8th in the second line of the cells on the MOBIPS images, Fig.2). According to the previous discussion, the small signal from this cell at low currents (dark area on the image) displays its big shunt currents, while the relatively high signal from this cell at higher currents (bright rectangular corresponding to the cell position on the image) can be explained by the small slope of its I-V characteristic.

Having the full I-V characteristics of the cells, it is possible to estimate their efficiencies and their contributions to the conversion efficiency of the module. Such calculations are represented by Fig.5.

The dark columns on the figure correspond to the conversion efficiency of each cell in its maximum power point. This efficiency can not be realized for all the cells of the module simultaneously because their the most

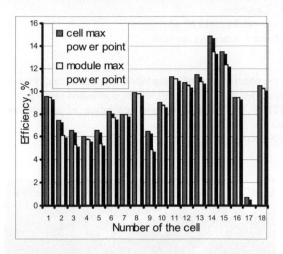

Fig.5. Conversion efficiency of the cells at each cell maximum power point (dark columns) and at the maximum power point of the module.

beneficial currents are different but the actual current is equal for all the cells. The current of the module's maximum power point has been taken for calculation of each cell contribution to the whole module power and efficiency (bright columns). The decrease of total module efficiency due to the cells parameters dispersion is calculated on the base of the obtained data. The contribution of each cell that is a product of the total I-V characteristic has been shown to depend on the level and homogeneity of the illumination, loading current, method of the assembly etc.

CONCLUSIONS

The photovoltaic signal induced by modulated incident optical beam on one cell of the series connected module has been shown to give an information about the dV_m/di derivation of this illuminated cell. The I-V characteristics are recovered by integration of the $dV_m/di = f(i)$ functions. The values of photocurrent are calculated from the current shift between the dark and illuminated f(i) curves for each cell. This information necessary for estimation of the PV module quality is extracted without disassembling the module. The ability of the proposed technique has been proved experimentally for commercially available modules.

We hope the proposed approach will help to monitor the parameters of cells and modules at their operation, minimise the energy conversion losses due to the dispersion of different cells, increase the reliability of the modules. It gives also an effective tool to investigate the integrated (single-substrate) modules [5].

REFERENCES

[1] Marek J. Light beam induced current characterization of grain boundaries. J. Appl. Phys., 1984, Vol. 55, # 2, p. 1621 - 1625.

[2] I.L. Eisgruber, R.J. Matson, J.R. Sites, K.A. Emery. Interpretation of laser scans from thin-film polycrystalline photovoltaic modules. Proceedings of First WCPEC; Dec. 5 – 9, 1994, Hawaii, p.283 – 286.

[3] S.M.Sze, Physics of Semiconductor Devices, 2nd ed., Wiley, N.Y., 1981

[4] S. V.Litvinenko, L. Ilchenko, A. Kaminski, S. Kolenov, A. Laugier, E. Smirnov, V.I. Strikha, V.A.Skryshevsky, , Mat.Sci.Eng. B, 71, pp.238-243, 2000.

[5] H.Forest. Efforts on commercialization of amorphous thin film technology. Proc. 14th European photovoltaic solar energy conference, Barcelona, Spain, 1997, p. 1365 – 1368.

BEHAVIOUR OF TRIPLE JUNCTION a-Si MODULES

N. Cereghetti, D. Chianese, S. Rezzonico and G. Travaglini
LEEE-TISO, CH-Testing Centre for PV-modules
University of Applied Sciences of Southern Switzerland (SUPSI)
CP 110, CH - 6952 Canobbio
Phone: +41 91 / 940 47 78, Fax: +41 91 / 942 88 65
Internet: http://leee.dct.supsi.ch, E-mail: leee@dct.supsi.ch

ABSTRACT: Initially the purpose of this work was to study the phenomenon of degradation induced by exposition to sun-light (Staebler-Wronski effect) on triple junction a-Si modules. Then the modules were connected to the grid. The aim of this plan is to compare the behaviour of old (TISO 4kW plant) and new generation of amorphous modules observing the cycles of degradation and regeneration.
Keywords: a-Si - 1: Evaluation - 2: Small Grid-connected PV Systems - 3

1. STUDY OF STAEBLER-WRONSKI EFFECT

1.1 Approach

8 triple junction a-Si modules of the firm Canon have been exposed under the following working conditions:

- 4 modules at MPP (of which 2 in parallel)
- 3 modules on Isc
- 1 module on Voc

In order to study light induced degradation (Staebler-Wronski effect) 4 of the 8 modules (2 at MPP in parallel and 2 on Isc were tested with measures of I-V @STC after various intervals of exposure (outdoor) to light. The other 4 modules were always exposed.

According to the experts at Canon, the power of these modules should settle at around 85% of the initial value after an exposure of 1000 kWh/m2 and the degradation should happen mainly in the first 50 kWh/m2.

1.2 Results

As illustrated in figure 2 at the LEEE-TISO centre a strong degradation between 100 and 200 kWh/m2 has been ascertained.

The difference from that declared by the manufacturer is probably caused by the different exposure conditions; at LEEE-TISO the modules have been exposed in real environmental conditions and therefore with irradiance and temperature in continuous change.

The measured medium power of the 4 modules, which have always been exposed, after the initial degradation was 58 W with respect to the initial 67 W; the degradation has been -13.4% and it approaches the value declared by the manufacturer (- 15%).

According to the manufacturer the module should have, after degradation, a nominal power of 62 W; the measured power, probably after an incomplete degradation (385 kWh/m^2), was therefore 6.5% less than the declared value.

Table I: Nominal power .

P_n	P_0	P_A (385 kWh/m^2)	$\dfrac{P_A-P_0}{P_0}$	$\dfrac{P_A-P_n}{P_n}$
[W]	[W]	[W]	(%)	(%)
62	67	58	-13.4	-6.5

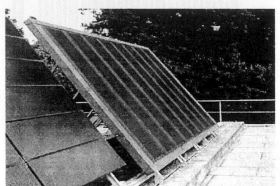

Figure 1: TISO 0.5kW triple junction plant.

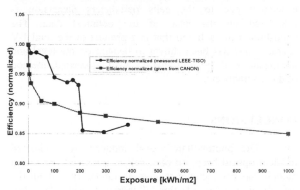

Figure 2: Efficiency versus exposure.

For the two modules at MPPT, the energy produced during exposure has been measured. In figure 3 the PR (Performance Ratio) over a period of about one year for these modules is illustrated; initial degradation during the first two months of exposure is significant. After this initial reduction, the PR value remains fairly constant at around 0.83 (with reference to the nominal power of the manufacturer).

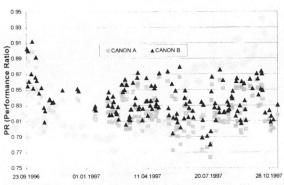

Figure 3: Performance Ratio versus Time.

2. TISO 0.5kW a-Si TRIPLE JUNCTION PLANT

2.1 Structure of the plant

In July 1998 a plant with these 8 modules was installed at the LEEE-TISO centre near the "old" TISO 4kW plant.

Table II: Main features of the 0.5 kW a-Si plant.

Connection to the grid:	July 1998
Nominal power @STC:	0.496 kWp
Type of modules:	Canon a-Si triple-junction
Module power @STC:	62Wp
N° of modules:	8
Configuration:	2 strings of 4 modules
Array tilt:	55°
Inverter:	Dorfmüller DMI500, 500W

2.2 Energy Production

During the first year of operation the plant produced 484 kWh (976 kWh/kW) with a daily peak of up to 2.7 kWh.

In the following graph the monthly energy production with the estimated (simulation with PVSYST) value is presented..

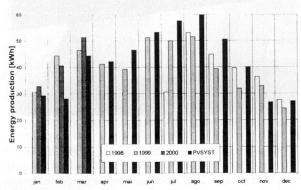

Figure 4: Energy production.

The difference between actual and simulated production ids due to the fact that the actual power (58W) of the Canon modules is about 6.5% less than nominal power.

2.3 Efficiency

The operating efficiency of the 2 strings oscillates between 6 and 6.8% and the variations depend strongly on the outdoor temperature. The two strings behave in a similar manner and their average is illustrated in figure 5.

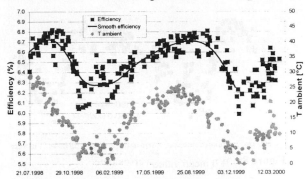

Figure 5: Efficiency of the TISO 0.5kW plant.

2.3 Thermal behaviour

Figure 6 allows a better understanding of how temperature affects plant production.

3 clear tendencies can be observed.

- From 0 a 10°C linear increase in efficiency (coefficient +0.6%/°C).
- From 10 to 22°C maximum level of efficiency reached.
- Above 22°C slight reduction in efficiency.

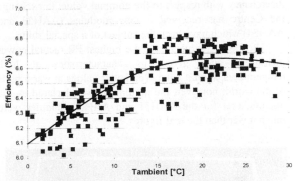

Figure 6: Efficiency versus ambient T.

2.4 Comparison old- new generation a-Si

Until now the measurements carried out don't show a decrease in the performance of the plant whereas the measurement performed on the old generation of a-Si modules show a slow but persistent degradation tendency ([1]).

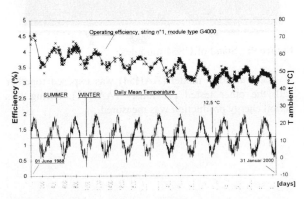

Figure 7: Efficiency of the TISO 4kW plant.

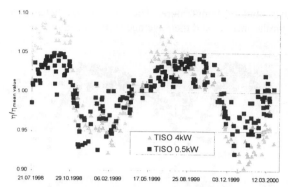

Figure 8: η/η mean value versus time

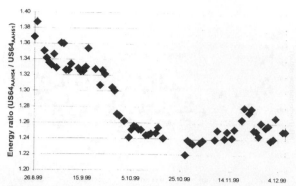

Figure 10: Energy ratio versus time

Figure 8 shows the normalized efficiency of the TISO 4kW single junction and TISO 0.5kW triple junction plants using their mean value.

2.5 US64 a-Si triple junction Module tests

2 US64 (a-Si triple junction) modules belonging to the Uni Solar [2] Company have also been tested at the Centre.

These modules show high initial degradation. After having seen the results of the two modules tested (AAH51 and AAH52), the manufacturer asked for one to be returned in order to analyse the reasons for the large discrepancy with respect to the nominal value; in exchange the Centre has received 2 new modules (AAH53 and AAH54) which have been the subject of a special study.

The US64 modules have the highest PR (actual power measured) and, given its characteristics at high temperatures, should have even better values in warm parts of the world; however, the nominal power declared by the manufacturer for the AAH51 and AAH52 modules were much lower than the real figures.

Figure 9: Stand of US64 modules

Later a 'new' model (AAH54) was exposed together with the other tested previously (AAH51); during this period energy production was measured.

Figure 10 shows that after initial degradation of the new module, the energy production ratio settled at around 1.25; this means that the AAH54 module produces 25% more energy than the module tested previously.

Initial degradation of the AAH54 modules more or less corresponds with that declared by the manufacturer (15%). The results of this test show that the AAH51 has a much lower nominal power than that declared (64W); the manufacturer's own measurements on the module returned (AAH52) confirm this conclusion (Pm @STC=52.1W).

3. INFLUENCE OF THERMAL INSULATION ON A-SI SINGLE JUNCTION MODULES

The efficiency of the a-Si modules during the seasonal degradation and regeneration cycles is affected by temperature above all. In order to ascertain in more detail the influence of temperature on the a-Si modules, the backs of series 7 of the TISO Plant were thermally insulated (May 1999). The temperatures measured on the back surfaces showed a subsequent average increase of 20°C.

Figure 11: Thermographic photograph of series 5, 6 e 7

Figure 11 shows the efficiency ratio between some of the non-thermally insulated series (S1, S5 and S6) and the insulated series 7. Insulation work on the series was carried between the broken line and the continuous red line. The increase in temperature has caused an increase in efficiency in the series 7: after 7 months of exposure this series with thermal insulation shows an efficiency which is between 6 to 9% higher that the non-insulated ones.

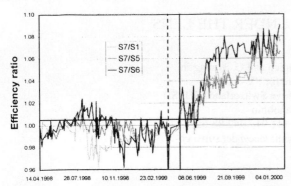

Figure 12: Efficiency ratio versus time.

During the first period of thermal insulation, the efficiency of series 7 did not increase much. With the arrival of summer heat (June-July), efficiency increased.

During the summer months efficiency results differ between series, whereas during the winter months the series maintain a fairly constant ratio.

The differences between the series are probably due to the different dynamics of the series themselves.

4. CONCLUSIONS

- Initial degradation of the a-Si triple junction modules induced by light (Staebler-Wronski effect) is around 15%; it occurs (outdoors) mainly during the first 200kWh/m² of exposure. After this degradation the PR settles at around 0.83.
- .The operating efficiency of the TISO 0.5kW plant triple junction oscillates between 6 and 6.8% and the variations depend strongly on the outdoor temperature.
- The efficiency of this plant seems to be constant unlike that observed with the s-Si single junction plant.
- At ambient temperatures between 0 and 10°C there is a linear increase in efficiency (coefficient +0.6%/°C).
- The US64 modules tested at the Centre also showed a high PR and stable power.
- During the project, the effect of thermal insulation on a series of the a-Si single junction TISO 4kW plant was also studied. An increase in efficiency from 6 to 9% with respect to the other 7 series was observed.

REFERENCES

[1] M. Camani, N. Cereghetti , D. Chianame and S. Rezzonico, Proceedings 15th EC Photovoltaic Solar Energy Conference (1998), p. 2058

[2] G. Travaglini, N. Cereghetti , D. Chianame and S. Rezzonico, 18 types of modules under the lens, Proceedings 16th EC Photovoltaic Solar Energy Conference, Glasgow (UK), May 2000

ACKNOWLEDGEMENTS

This project is financially supported by the Swiss Federal Office of Energy and the AET (Azienda Elettrica Ticinese).

The authors would like to extend particular thanks to the staff of the ESTI of the Joint Research Centre (JRC) in Ispra and to dr. Mario Camani.

18 TYPES OF PV MODULES UNDER THE LENS

D. Chianese, N. Cereghetti, S. Rezzonico and G. Travaglini
LEEE-TISO, CH-Testing Centre for PV-modules
University of Applied Sciences of Southern Switzerland (SUPSI)
CP 110, CH - 6952 Canobbio
Phone: +41 91 / 940 47 78, Fax: +41 91 / 942 88 65
Internet: http://leee.dct.supsi.ch, E-mail: leee@dct.supsi.ch

ABSTRACT: At the Testing Centre for photovoltaic components (TISO-LEEE) the modules, chosen among the ones most frequently used in the PV power grid connected plants in Switzerland, undergo a series of tests in order to verify their characteristics, reliability, medium and long-term performance and high voltage reliability.
The most important aspect of the tests is to compare energy output of modules exposed in identical real outdoor conditions. The aim of these tests is to answer the questions which need to be posed when planning a PV plant.
At the moment of purchase 1/3 of the modules are no longer under guarantee, while after a year of exposure under real environmental conditions and at MPP, 2/3 of the modules have a power which is lower than the limit stated in the guarantee.
Not only do the amorphous-Silicon modules undergo initial degradation, but also a number of crystalline-Silicon modules show significant degradation. Comparison between c-Si modules exposed to different high voltage conditions demonstrate that, in the short term, a correlation between the degradation observed and the voltage applied does not exist

Keywords: Energy Rating - 1: Degradation - 2: Qualification and testing - 3:

1. INTRODUCTION

The nominal electrical parameters supplied by the manufacturers normally refer to typical parameters recorded during the manufacture of the modules which use specific or even calculated measurements. Project planners and installers are increasingly asked to provide production and behaviour estimates for the systems they install, in particular when they involve Solar Grants and Contracting. They therefore ask the following questions:

1. Is the electrical data supplied by the manufacturers useful for our purpose?
2. Do the modules degrade over time? In what way?
3. Is there a difference in energy production (Wh/Wp) between the different types of modules?
4. What is the actual energy output of the different modules?
5. Does the high voltage between frame and cells of a module in a plant influence its efficiency?

In order to answer such questions the LEEE-TISO testing centre for PV components has, since 1991, carried out systematic tests, under real operating conditions, on the most important modules currently on the market ([1] to [5]). At Lugano-Trevano, the modules for each cycle of tests are exposed on stands tilted at 45° and 7° south of azimuth and kept at maximum power point (MPP). Each module is equipped with a Maximum Power Tracker adapted for its power range.

In more recent times, 18 modules divided into two cycles with 11 (cycle 5) and 7 (cycle 6) modules respectively have been test. They were selected from those most commonly found on the market or which had interesting innovations. In order to guarantee impartiality and neutrality regarding measurements, the modules were purchased anonymously unknown to the manufacturer. Two examples for each kind of module were acquired.

Figure 1: View of the LEEE-TISO test facility with the 18 modules under the lens.

2. TESTING PROCEDURE

During normal testing procedures [1], the electrical characteristics of the modules were measured, at regular intervals, at standard test condition (STC) at the ESTI laboratory of the Joint Research Centre in Ispra (I) as described in the following table:

Table I: Electrical measurements during the tests.

P_n	**Registration** of data of manufacturers
P_0	**Before exposure:** electrical behaviour @STC
	Exposure under real environmental conditions at MPP during 1 YEAR
P_6	**After 6 months:** electrical behaviour @STC
	Exposure under real environmental conditions at MPP for the next 6 month.
P_{12}	**After 12 months:** final tests @STC

3. ELECTRICAL CHARACTERISTICS OF THE MODULES

3.1 Initial Power P_0

The results of the measurements carried out at LEEE-TISO show real initial power@STC of the modules (P_0) differ from the nominal power of the manufacturer (P_n) (see table II,1st column). This is not surprising since the nominal value P_n is a mean indicative value, while the value of each single module should fall within the variance of the production parameters. What is surprising is that, even before exposure, a 1/3 of the modules were under the minimum limit of the guarantee (-10%). The technology used (mono or poly-crystalline or amorphous-silicon) has no bearing on the failure to respect limits. This means that either the manufacturers don't know the variance with respect to their products or their system of measurement differs from from the World PV Scale (WPVS) [13], adopted by JRC at ISPRA where measurements @STC were carried out.

Table II: Difference between nominal power (P_n) and measured power (P_0, P_6, P_{12}) of module type tested (average of two modules). Modules no longer under guarantee, before and after one year of exposure, on dark background.

Type of module		ΔP_{0-n}	ΔP_{6-0}	ΔP_{12-6}	ΔP_{12-0}	ΔP_{12-n}
	ASE-50-PWX	-4.7	-7.3	-1.6	**-8.8**	**-13.1**
	BP580	1.4	-2.2	-0.4	-2.6	-1.3
	GPV75M	**-13.1**	-3.1	-1.9	-4.9	**-17.3**
	KC80	**-10.8**	-4.3	-2.2	-6.4	**-16.5**
CYCLE 5	MST43MV	4.4	-10.7	-0.5	**-11.1**	-7.2
	PWX500	**-13.6**	-7.2	-1.0	-8.1	**-20.6**
	SDZ34-10	-4.1	n.a.	n.a.	0.3	-3.8
	SF75	**-13.7**	-1.1	0.3	-0.8	**-14.4**
	SP75	-3.2	-5.5	0.0	-5.5	-8.5
	Sunslates ①	**-18.4**	-6.9	-3.2	-9.6	**-26.4**
	US64 ②	n.a.	n.a.	n.a.	n.a.	**-18.6**
	ASI16-2300	-8.4	-0.9	-0.3	-1.2	**-10.8**
	H800A	-8.5	-3.0	-1.8	-4.8	**-12.9**
CYCLE 6	I55	**-13.1**	0.0	-1.7	-1.7	**-14.5**
	M30-16GEGK	**-15.0**	1.6	-3.9	-2.4	**-17.0**
	PL800	-6.5	-6.8	-0.2	-7.0	**-13.1**
	RSM100	-6.8	-6.7	-1.4	-8.0	**-14.3**
	SFM36Bx	-7.9	-3.4	-1.1	-4.5	**-12.0**

① average of 6 modules without significant reduction of power; ② value of P_{12} measured by manufacturer. **Bold type** : Significant difference.
Measurement error @STC: absolute ± 2.4% of P_0; repetitivity ± 0.85% of P_6 e P_{12} with respect to P_0.

In cycle 6, an old generation ASI16-2300 module, with 17 years of outdoor exposure, was installed for comparison with new generation modules.

3.2 Initial degradation and high voltage tests

PV solar modules with c-Si cells also show a degradation in performance when exposed to light at real operating conditions ([1] to [5]).

During the first periods of measurements, the modules were exposed to real environmental conditions at MPP and with high voltage at 1.2kV applied between the frame and the active part of the module. In order to separate voltage effect on initial photodegradation, a test on a sample of 6 identical modules (c-Si) exposed at different voltages (0, 300Vdc, 600Vdc, 900Vdc, 1200Vdc, 1800Vdc) was carried out. The results demonstrated that no short-term correlation exists between the voltage applied and degradation (see figure 2).

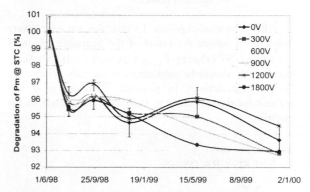

Figure 2: Degradation of Pm, measured @STC and normalised, for 6 modules **exposed under different voltage conditions.**

Degradation of these modules is mainly due to a reduction of Isc (from −5.5% to −7.5%) whilst Voc and Fill Factor (FF) didn't undergo any significant changes (-1.8% and +0.6% respectively).

In the new measurement procedures, applied to cycles 5 and 6, PV modules are not submitted to high voltage but only to normal operating conditions at MPP. Photodegradation in the performance of the modules when exposed to light (table II, 2nd column) was also observed in cycles 5 and 6.

As reported also by other authors [9,10,11], during the first hours of exposure initial degradation occurs; this is principally linked to a decrease of carrier lifetime in the bulk material. Part of this degradation also occurs if the modules are stored in the dark over a time of some weeks, but final stable module efficiency after photodegradation is equal even after predegradation in the dark [9].

Storage time and subsequent predegradation of the modules purchased is not known so the initial P0 value could correspond to the power of the m-Si modules which have undergone the degradation described above.

However, apart from the predegradation discussed above, significant degradation can not only be seen in 11 c-Si modules out of 15 (table II, 3rd column: ΔP_{6-0}), but also over the whole period in 13 c-Si modules out of 15 (table II, 4rd column: ΔP_{12-0})

In the standard modules with c-S cells, the average reduction after one year of exposure was −4.8% peaking at -9.6%

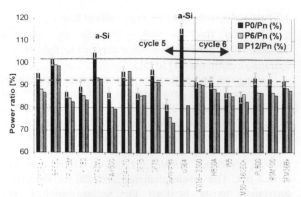

Figure 3:   Power ratio of measured value (P0, P6, P12) vs. nominal power.

After a year of exposure 13 modules out of 17 were found to be outside the limits of the guarantee (figure 3 and table II, 5rd column: P_{12-n}). In the case of PV plants the modules should be substituted or partially reimbursed by the manufacturer. In practice, however, it is still difficult for the consumer to verify the real power of the modules.

4. ENERGY RATINGS

Two modules of different types but of equal power can have different levels of energy production even though exposed in the same way, at the same time and under the same metereological conditions. The principal parameters which influence energy production are the temperature coefficients, behaviour at low irradiation, spectral response and reflections [10] [11].

4.1 Technical comparison

Figures 4 and 6 show energy ratings in the form of Performance Ratio (PR) calculated on the basis of real measured power values (Peff.=(P6+P12)/2) for cycles 5 and 6 respectively.

The relative percentage differences in energy production of the different types of modules are calculated with respect to the first module on the left.

In this type of comparison, the SDZ34-10, Sunslates and M30-16GEGK modules are at a disadvantage. They are designed to be laid on a sloping roof and therefore the test conditions (open rack) do not respect their real operating conditions. Moreover, during cycle 5 a Sunslates tile broke while another lost half its power; for this reason the relative difference reached 24.6%. Energy comparisons between cycle 5 modules and cycle 6 modules are not possible in that they took place in two different periods under different metereological conditions.

Energy ratings calculated using Peff. refer therefore to a technical comparison of the different types of modules tested.

The US64 module has the highest PR (calculated only with P12 measured by the manufacturer) and bearing in mind its characteristics at high temperatures, should have even better values in warm areas of the world;

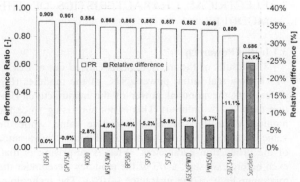

Figure 4:   Energy production (cycle 5) with respect to the power measured (Peff.).

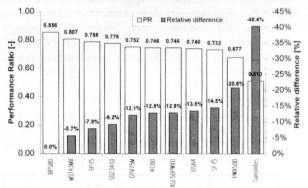

Figure 5:   Energy production (cycle 5) with respect to the manufacturer's given nominal value. (Pn).

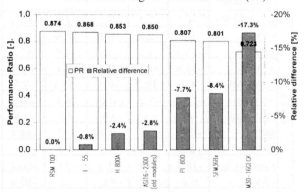

Figure 6:   Energy production (cycle 6) with respect to the power measured.

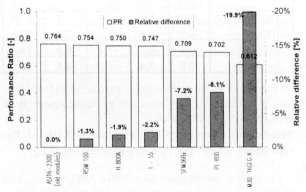

Figure 7:   Energy production (cycle 6) with respect to the manufacturer's given power value.

unfortunately, however, the manufacturer's nominal power is much lower than the actual figures (see chap.4.2).

From a technical point of view, the production differences for standard modules reach a maximum of 8.4%.

4.2 Consumer viewpoint comparison

Figures 5 and 6 represent energy ratings in the form of Performance Ratio calculated on the basis of nominal power (Pn) declared by the manufacturer for the cycle 5 modules and for the period 1998-1999, that is to say for a one year period after the first four months of expusure during which initial degradation of power took place (important for the a-si modules) as well as for cycle 6 (May 1998 - May 1999). This graph shows energy production for the powers actually purchased by the consumers, which, for the most part, as seen in chapter 3, are below the limits of the guarantee.

The relative differences as a percentage of energy production of the various types of modules are calculated with respect to the first on the left. Apart from the Sunslates modules, these differences reach values of up to 20%. This is due to the fact that the nominal powers, which are explicitly used in PR calculations, are, in fact, not 'contained' in the modules, as can be seen in the P0/Pn ratio in figure 3.

The PWX 500 and ASE-50-PWX-D modules, produced by the same manufacturer but with a different label, have a similar PR with respect to Peff.; but, since Photowatt (PWX500) declares a higher nominal power, there is a difference in production of 8% between the two modules. In cycle 5 the BP580 module shows better performance due to a correct declaration of power. The good energy rating of the old ASI16-2300 module with respect to other cycle 6 modules should be noted: after 18 years of exposure, and with nominal power being equal, this module produces more.

5. CONCLUSIONS

The photovoltaic solar modules, produced using crystalline-silicon cells, show a degradation in performance when exposed to light.

The initial degradation of c-Si modules takes place during the first period of exposure to light. This is often followed by further degradation during the following months irrespective of the season of exposure. The causes of this further degradation are still being studied.

Stabilized power after 12 months of exposure to light is lower than the guarantee limits for 12 out of 17 modules purchased. Definition of nominal power values given by the manufacturers for photovoltaic modules must therefore be revised. Measurement of electrical characteristics for every module which leaves the production line would only partly prevent non-respect of nominal power values since the modules still have to undergo initial photoinduced degradation.

Comparison between c-Si modules exposed at different high voltage conditions show that no short-tem correlation exists between initial degradation and voltage applied.

The choice of a module for a specific application must not only take into account performance (or power) of a module but also other factors linked to energy rating under different metereological and construction conditions. Measurements show a mean relative difference of PR (Performance Ratio) in an open –rack system tilted at 45° and based on real measured power values, of –4.6%, with a maximum difference of 8.4% (modules which are integrated in the roof as tiles are not taken into consideration).

REFERENCES

[1] M. Camani, N. Cereghetti, D. Chianese and S. Rezzonico, Comparison and behaviour of PV modules, 2nd World PVSEC, Vienna, July 1998.

[2] M. Camani et al.: Tests of reliability on crystalline and amorphous silicon modules, 14th EPVSEC, Barcelona (S), June 1997.

[3] M. Camani et al.: Vergleich und Beurteilung der PV Module, 13. PV Symposium, Staffelstein (D), March 1998.

[4] M. Camani et al. : Centrale di prova per componenti e sistemi per progetti nel campo della tecnica fotovoltaica, periodo IV: 1994-1996, Final Report.

[5] G. Travaglini, N. Cereghetti, D. Chianese, S. Rezzonico e M. Camani; Leistungsverhalten von PV Modulen; Simposio nazionale fotovoltaico, Zürich (CH), November 1999.

[6] D. Chianese et al.; Degradation of crystalline silicon modules, Simposio nazionale fotovoltaico 1999, Zürich (CH), November 1999.

[7] S. Sterk, K.A. Münzer, S.W. Glunz, Investigation of the Degradation of Crystalline Silicon Solar Cells, 14th EPVSEC, Barcelona (S), July 1997.

[8] J. Schmidt, A.G. Aberle, R. Hezel, Investigation of Carrier Lifetime Instabilities in CZ-grown Silicon, 26th IEEE PVSC, Anaheim (US), October 1997.

[9] A. Moehlecke et al., Stability Problems in (n)p$^+$pp$^+$ Silicon Solar Cells, 2nd World PVSEC, Wien, 1998.

[10] D.L. King, PV Module and Array Performance Characterization Methods for All System Operating Condition, NREL/SNL PV Program Review Meeting, November 1996.

[11] Bücher et al. , OPTIMOD project contract n°JOR3-CT96-0095.

[12] N. Cereghetti et al., Behaviour of Triple Junction a-Si modules, this conference.

ACKNOWLEDGEMENTS

This project is financially supported by the Swiss Federal Office of Energy and the AET (Azienda Elettrica Ticinese). The authors would like to extend particular thanks to the staff of the ESTI of the Joint Research Centre (JRC) in Ispra and to Dr Mario Camani.

OPTIMISED PHOTOVOLTAIC CONVERSION CHAINS IN INHOMOGENEOUS RADIATION CASES

Corinne Alonso, Mohamed-Firas Shraif, Augustin Martinez
Laboratoire d'Analyse et d'Architecture des Systèmes - CNRS,
7 avenue du Colonel Roche, 31077 Toulouse, cedex 04
Tel : 33. 5 61 33 62 00. Fax : 33. 5 61 33 62 08. Email : alonsoc@laas.fr.

Abstract: Today, photovoltaic (PV) conversion chains are generally designed much more as prototypes than as industrial products. As a consequence, design and implementation of such systems are very expensive. To reduce the cost and converge towards industrial processes, we ought to design tools that will help to obtain easily optimal PV chains adapted to a wide range of power and loads.

To achieve this goal, it seems to be more logical to use a distributed approach of the power conversion between PV generators and loads compared to one inverter with its efficiency optimised for the maximum power delivered.

This principle leads us to design an elementary conversion chain that we have optimised through simulations and that we are currently implementing as a real prototype.

In this article, we describe our chain and focus on its PV generator modelling. We illustrate the validity of this model in different normal and degraded radiation conditions. Finally, we show the behaviour of our PV conversion chain in inhomogeneous radiation cases.

Keywords: Defects - 1: PV system - 2: Modelling - 3.

1. INTRODUCTION

The main goal of our researches is to design optimal photovoltaic (PV) energy conversion chains with high efficiency and low manufacturing cost. In that context, we are developing an elementary optimal chain that is designed to be the basic module to build optimal chains adapted to a wide range of power and loads [1].

In this article, we describe our PV generator modelling. Its main advantage is its ability to represent classical behaviours but also some fault cases occurring in real systems.

2. MODELLING

Our elementary conversion chain is composed of a PV array, a DC-DC Boost converter driven by a MPPT control and a resistive load as shown on Fig. 1.

Each of theses elements is modelled separately by equivalent electrical circuits. These models are then integrated in ELDO which is a simulation software tool.

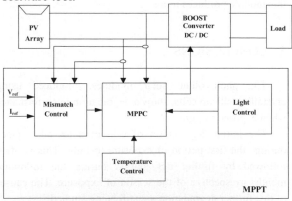

Figure 1: Block diagram of the elementary conversion PV chain.

A key point to achieve correct representation of the full chain behaviour in normal but also in degraded conditions is to use an accurate enough PV generator model.

As a consequence, we have decided to focus in this paper on the modelling of the PV generator and its validation. The rest of the modelling is detailed in [1].

2.1 PV modelling

To be able to represent inhomogeneous solar radiation effects on the PV subsystem, we decided to model individually each PV cell.

Theses cell models are then associated in series and/or in parallel like in reality to represent a global PV generator. We chose to represent only static electrical behaviour. Issued from literature [2], the static electrical behaviour of an individual photovoltaic cell can be described by the simplified equation :

$$I_{cell} = I_{CC} - I_S \left(exp\left(\frac{Vj}{\eta V_T} \right) - 1 \right) - \frac{Vj}{R_{sh}} \qquad (1)$$

$$\text{with} : V_{cell} = V_j - R_S I_{cell}$$

2.2 Validation of the PV model in classical working conditions.

To realise a first validation, we compared the simulated behaviour of a solar module constituted by 36 mono-crystalline silicon photovoltaic series-cells with data-sheets issued from the PVSYST software version 2.1 [3] from which, we extracted parameters I_{cc}, I_s, η, R_s, and R_{sh}. We see on figure 2a the static characteristics of the module ARCO M55 and on figure 2b, the simulation results which match exactly the data sheets information.

2. 3 Model ability to present classical defaults.

The first scenario we have studied corresponds to a 36 series-connected cells PV generator without protection where most cells are irradiated at 1000 W/m2 level while three are shadowed. This type of default corresponds for example to dust or leaves felt on the PV modules obscuring several cells.

We can see on the simulation results (Fig. 3) as described in literature [5] that this type of default has a dramatic effect on the power effectively delivered by the PV generator. A small localised default can induce easily 70 % power lost.

To go further, in the model validation, we also simulated different associations of PV modules (series and parallel) with and without protection diodes.

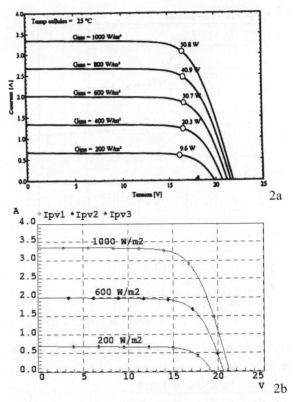

Figure 2 : Validation of the PV generator model. (a) Data sheets extracted from Pvsyst of an array referred ACRO M55. (b) Simulation results with ELDO.

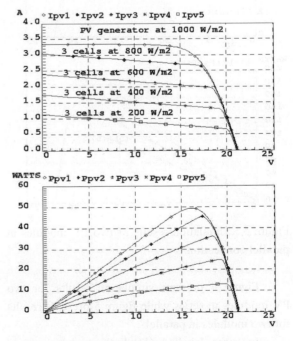

Figure 3 : Simulation results in cases of inhomogeneous radiation.

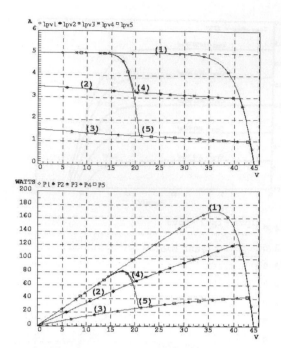

Figure 4 : Simulation results of two PV modules in series connection.

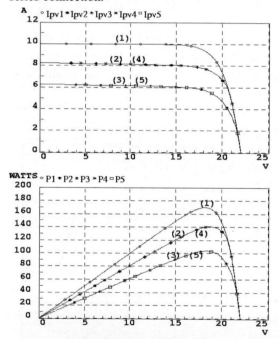

Figure 5 : Simulation results of two PV modules in parallel connection.

Figure 4 presents simulation results for two PV modules in series while figure 5 presents results for two modules in parallel.

In cases labelled (1), both modules are in normal conditions of 1000 W/m2 solar radiations. In series, we obtain the same current as one module while the open circuit voltage is doubled. Dually, in parallel, it is the current that is doubled.

Curves labelled (2) and (3) represent the behaviour when one of the modules has one cell less radiated than others (respectively with 600 W/m2 and 200 W/m2) without any protections.

Curves labelled (4) and (5) correspond respectively to (2) and (3) conditions but with by-pass diode.

In parallel configuration, the addition of by-diode has no effect. In series, this diode allows to obtain at least the power of the normally working module.

2. 4 Practical tests.

To verify these simulated results, we recently made physical tests with real PV modules (BP 585L). In practice, these modules are selves protected half by half. Experimental results are presented for two PV modules in series in normal conditions (Fig; 6a) and then with one partially obscured (Fig. 6b). They are compared to simulation results in same conditions (Fig. 6c).

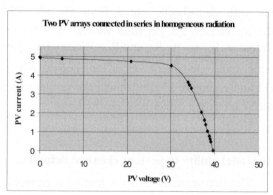

Figure 6a : Experimental results of two PV modules in series for a 1000 W/m2 radiation and a temperature of 25 C in normal conditions.

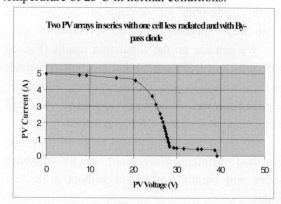

Figure 6b : Same conditions as Fig. 6a with one cell of one module obscured.

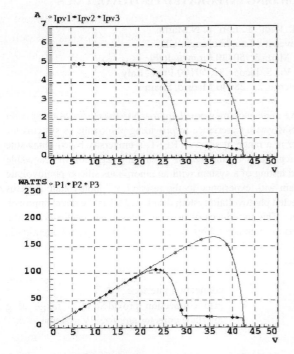

Figure 6c : Simulation results of two PV modules in series protected half by half.

3. ELEMENTARY CONVERSION CHAIN BEHAVIOUR IN INHOMOGENEOUS SOLAR RADITION CASES.

Our modelling approach permits to simulate some degraded working points of the PV generator and theirs consequences on the global chain yield.

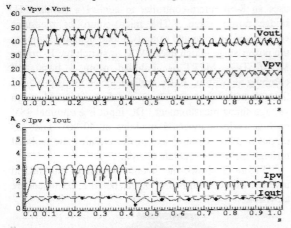

Figure 7 : Simulation Results of I_{input} and I_{output} Currents, V_{input} and V_{output} Voltages of the BOOST converter.

Test conditions: Resistive load = 50 Ω, Boost frequency = 100 kHz, Crystalline silicon PV panel ARCO M55 [3], for a solar radiation =1000 W/m2. At t = 0.3s, three cells are enlightened with 600 W/m2.

Figure 7 present our elementary chain behaviour before, during and after an inhomogeneous solar radiation variation happens (like in Fig 3). It represents the PV generator Voltage and current (Ipv and Vpv) and the boost output Voltage and current. We can see that our system recovers rapidly form the perturbation and that it reaches again a new maximum power point.

CONCLUSION

We have proved through experimentation that our PV generator model was accurate enough to simulate correctly classical deficiency cases. We have then shown in simulation that our elementary chain was robust even in such conditions. These results leads us to continue to validate and finalise the prototype of our elementary conversion chain.

BIBLIOGRAPHY

[1] M. F. Shraif, C. Alonso, A. Martinez. "A simple and robust maximum power point control (MPPC) for ground photovoltaic generators." Proc. of the International Power Electronics Conference IPEC – Tokyo 2000, pp 158-163.

[2] A. Duffie, W. A. Beckmann. "Solar engineering of thermal processes". 2nd Edition. 1991. John Wiley and Sons, Inc.

[3] A.Mermoud, PVSYST (manuel de l'utilisateur) version 2.1.Université de Genève.

[4] A. S. Kislovski "Power Tracking methods in photovoltaic applications". Power Conversion 1993 Proceedings, pp 513-528.J.

[5] J. M. Rolland, S. Astier, L. Protin. "Static device for improving a high voltage photovoltaic generator working under dusty conditions". Solar Cells n° 28. 1990, pp 277-286.

OPTIMISATION OF CONVERTERS FOR BUILDING INTEGRATED PHOTOVOLTAICS

A.D.Simmons[x], D.G.Infield[x], P. Redi[+], L. Lori[o] & N.Martin[^]
[x]CREST, Loughborough University, Loughborough, UK
[+]University of Florence, Via S. Marta 3, 50139 Firenze, Italy
[o]SEI-Sistemi Energetici Integrati srl, Via S. Jacopo,32, 59100 Prato, Italy
[^]CIEMAT-DER. Avda. Complutense, 22. 28040 Madrid, Spain

ABSTRACT: This EC funded JOULE project had the objective of developing a novel approach to the electrical integration of photovoltaics into commercial buildings. SEI, an Italian SME with experience in providing photovoltaics systems for the Italian railways, have developed a grid connected inverter in a range of sizes. CREST, a university based renewable energy laboratory in the UK with growing experience and reputation in the area of electrical integration of renewable energy, have developed the grid interface and conducted field testing of a system with an amorphous silicon photovoltaic array. CIEMAT, an independent research laboratory in Spain with experience in the research of photovoltaic systems provided bench testing of a prototype system. The grid connected photovoltaic system developed takes the novel approach of separating the inverter and the grid interfacing transformer. This introduces distinct advantages and challenges in the engineering of large scale grid connected photovoltaic systems for commercial buildings. In addition, special attention is paid to the use of amorphous silicon photovoltaics, in particular the MPP tracking requirements.
Keywords: Grid Connected - 1: Building Integrated - 2: Inverter - 3:

1. INTRODUCTION

InGrid is a new concept in grid connection for safe, efficient and cost-effective Building Integrated Photovoltaic Systems:

- Low AC & DC voltages with array centre grounded ensures safety
- Islanding protection provided separately by inverter and grid interface
- Benign impedance measurement methodology for islanding identification
- Modular system for the rapid deployment of any size of grid connected PV in the built

environment

Figure 1: InGrid Inverter

Typical Applications
- Domestic grid connected PV systems
- Commercial roof and façade PV systems
- School grid connected PV systems
- Motorway sound barriers

General Characteristics
- 38Vac 50Hz ac output voltage
- 48Vdc nominal dc input voltage
- 250W & 500W unit size
- MPP tracking algorithm for amorphous silicon
- high conversion efficiency for this inverter typology
- high efficiency and power factor
- connection of multiple units to single grid

connection interface
- grid connection provided by a range of grid interface units for any size PV system

2. SYSTEM DESIGN

The InGrid system conists of many inverters connected to a single grid interface.

2.1 Inverter Design

The purpose of the inverter is to convert the DC output of a PV array to 50Hz AC while optimising the operating point through Maximum Power Point Tracking (MPPT).

SEI have designed and developed an inverter which uses an 8-bit micro-controller to control a standard H-bridge power stage. The micro-controller measures current and voltage and outputs a load demand signal to the power stage which uses PWM to switch the H-bridge MOSFETs. MPPT is included using a basic hill climbing algorithm. Basic grid side detection is provided through under & over voltage and frequency and rough voltage shape measurement. DC input is a nominal 48V or four series connected modules at 40-80V and AC output is 38V. Control and power circuitry are isolated through the use of separate boards.

Figure 2 shows the internal layout of the InGrid inverter. The control board; power board & terminals are all visible.

Figure 2: InGrid Inverter

Figure 3 illustrates normal and islanded operation of the inverter, the upper line is the injected current and the lower line is the 'grid' voltage at the transformer terminals. Without the strong grid impedance the effect of the grid interface transformer is clearly seen as a significant distortion of the current and voltage waveforms.

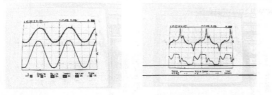

Figure 3: Normal & islanded operation

Hence, through measuring the shape of the 'grid' voltage waveform the islanding condition can be easily determined using this benign impedance methodology.

2.2   Grid Interface Design

The purpose of the grid interface is to connect the paralleled low voltage output of many inverters to the utility grid. The grid interface monitors the condition of the grid and connects one of two transformers: a 10VA monitor for night time a main transformer rated at full output for generation. A fast 16-bit controller is used to enable more accurate islanding detection in addition to voltage and frequency level measurement. In addition, potential exists for communications between inverters and grid interface and remote PC for sophisticated diagnosticis and data logging.

Figure 4 shows the internal layout of the InGrid grid interface. The small & large transformers; control board and terminals are visible.

Figure 4: InGrid Grid Interface

3.   SYSTEM TESTING

Testing of the complete system of inverters and grid interface was conducted in various phases through the development project.

SEI carried out basic testing of the inverter as development progressed and also to provide separate identification of system performance.

CIEMAT performed bench testing of a system to confirm operation consistent with European regulations for grid connection and also to determine basic operating performance. In addition, CIEMAT performed a limited field test with two PV arrays, two inverters and a transformer for grid connection.

CREST performed extended field testing of a system consisting of four inverters and a full grid connection interface as shown in figure 5. This shows the test equipment including various AC & DC instrumentation boxes.

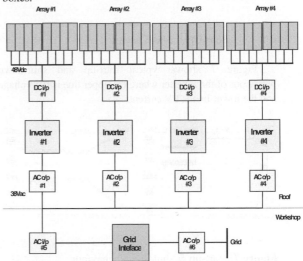

Figure 5: Field test configuration

Figure 6 shows the field test hardware installed on the roof of CREST's research laboratory, clearly seen are irradiance sensors, one of the amorphous silicon arrays and one of the inverters installed.

Figure 6: Field test hardware

Up to four InGrid inverters (3 x 250W & 1 x 240W), were tested with one grid interface and four PV arrays (3 x 352Wp amorphouse silicon & 1 x 240Wp poly-crystalline). The system specification is given in table III.

Table III: Specification of system used in CREST testing

	amorphous silicon	Polycrystalline silicon
Number of arrays	3	1
Peak Power, Pp (Wp)	352	240
Area, A (m^2)	9.52	2.21
Efficiency (%)	3.7	10.8
Voltage Range (V)	40-90	40-80
Inverter Power, Pi (W)	250	320
PV:Inverter Match, Pp / Pi	1.4	0.75

4.   PERFORMANCE RESULTS

The performance of any grid connected PV system is covered by three aspects: on-off behaviour; power quality and efficiency.

4.1   Start-Up & Shut-Down

Figure 7 shows typical start-up and shut-down behaviour of the inverter where the upper line is PV voltage and the lower line is PV current.

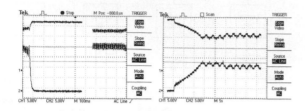

Figure 7: Start-up & shut-down behaviour

When starting up the inverter ramps the current up from zero and the PV voltage down from open circuit to a nominal operating value. This operating point is then adjusted up and down in order to keep track of the maximum power point. Clearly seen in the figure is the ramping up/down, the adjusting up and down and the settling to a new operating point after three adjustment cycles. Time to start-up from no load to full load is variable and dependant on irradiance, i.e. PV operating point, and is typically of the order of 15 seconds. Operating point adjustment is on a cycle of approximately 3 seconds.

When shutting down the inverter unloads the PV and the current collapses to zero as the PV voltage rises to its open circuit value within approximately 40ms or two grid cycles.

4.2   Power Quality

Power quality if defined as the total ahrmonic distortion of the current being Injected into the grid and its power factor. Figure 8 shows the inverter current waveform at 6% and 86% of peak power. Switching harmonics are visible but small in magnitude, especially at high power.

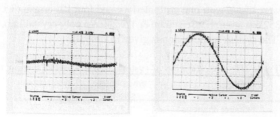

Figure 8: Current waveform at 6% & 86% of Pmax

4.3   Efficiency & Power Factor

The efficiency and power factor of two inverters, total maximum power of 500W, was measured by CIEMAT. Two mono-crystalline PV arrays, each rated at 300Wp, was operated over the full range of irradiance. The results are shown in figure 9.

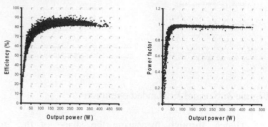

Figure 9: Efficiency & power factor vs. power

Inverter efficiency is good, although not outstanding, at over 80% for power delivered greater than 20% with an average of 85% and a peak of 90%. Power factor is greater than 0.9 above 10% of peak output.

4.4   Typical Operation

Figure 10 shows a time history of a sample day of operation of two of the inverters: one with an undersized polycrystalline PV array and the other with an oversized amorphous silicon PV array. Once radiation levels and PV voltage have reached a minimum threshold values the inverter starts to operate and produces PV current which modulates the PV operating voltage.

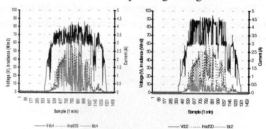

Figure 10: PV Voltage & Current

The wide excursions of voltage between 60V and 80V (90V for aSi) with a mean of 70V (75V for aSi) are coupled with on-off variations in current illustrating imperfect MPP tracking of variations in irradiance.   Operation with amorphous silicon is markedly worse than with poly-crystalline despite the greater array size.

5.   SYSTEM ANALYSIS

The InGrid system separates the DC-AC inverting from the grid connection, thus benefiting from the distribution over the array of the string inverter and the single point of connection to the grid of the larger array inverter.

An important aspect of grid connected PV system design is the matching of the PV array to the inverter. Mismatch will result in the system under-performing. The performance of the grid connected PV system depends on the efficiency of the various components over the expected power range: PV, MPPT, inverter, transformer. In standard inverters the transformer is not a separate component unlike in the InGrid system.

A typical InGrid system consisting of four inverters and a grid interface is illustrated in figure 11 and the efficiencies of the various components are shown.

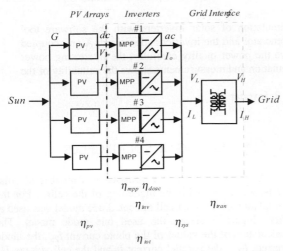

Figure 11: System efficiencies

PV efficiency is determined by the PV response to environmental conditions, peaking at:15% for mono-crystalline; 12% for poly-crystalline and 4% for amorphous silicon. Inverter DC-AC efficiency is determined by the H-bridge components, specifically the 'on' resistance which is related to cost and speed and part-load operation and the MPP tracking efficiency. Transformer efficiency is determined by its relative size and average system output. PV, inverter and transformer efficiencies are thus determined by system design and component sizing. MPP tracking efficiency is a measure of the effectiveness of the control system in tracking accurately and quickly changes in environmental conditions.

Figure 12 shows total inverter average efficiency, transformer efficiency and total system efficiency versus irradiance.

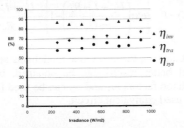

Figure 12: Total efficiencies

Average inverter efficiency is 85-90% across the range of irradiance from 200W/m^2 to 1000W/m^2. Transformer efficiency at 68-72%, is low giving a low overall system efficiency of 58-65%. The low transformer efficiency is due to the operation of the multiple inverter system and in particular the dynamic MPPT control. Figure 10 illustrated the difference between tracking MPP for amorphous silicon versus polycrystalline silicon.

Discontinuous behaviour such as this results in lower average output and hence lower average efficiency. To illustrate this further, figure 13 shows the efficiency of the four inverters tested versus irradiance. Inverter number 1 was operated with poly-crystalline and the other three were operated with amorphous silicon.

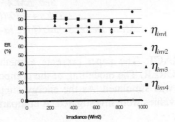

Figure 13: Inverter efficiencies

Inverter number 1 is clearly superior in operation, at an average efficiency of 90%, than the other three, which have average efficiencies of 75-85% .

Inverter efficiencies are partly constrained by the matching of the PV array peak power to inverter rated power. Neither amorphous nor polycrystalline silicon arrays were well matched to the inverters they were tested with. Typically, a PV array:inverter over-rating of 1.2 is the best compromise between maximum power operation and good low power efficiency.

6. CONCLUSIONS

Building integration of photovoltaics is advancing rapidly and the design of components and systems is developing options for greater optimisation of energy generation. Amorphous silicon presents a serious challenge for MPPT design due to its higher voltage, lower efficiency, dependence on temperature and spectrum and its flatter I-V characteristic.

The InGrid system of grid connection interface and multiple inverters has been developed and tested with crystalline and amorphous silicon PV. The InGrid system more than satisfies standard requirements for grid connection of PV. The InGrid system is more complex than standardand hence testing has identified options for optimisationof system operation. InGrid presents the BIPV system designer with a new approach to maximising energy generation for the client.

Improvements to the basic system design will address the dynamic behaviour and its impact on average system performance. In particular, a 32-bit DSP will replace the 8-bit micro-controller enabling far greater precision and accuracy of control.

Finally, it should be stressed that the performance, efficiency and cost of energy generated, of any PV system is a combination of the components in the system. System design is as important as component design.

SHORT-TIME SIMULATION OF SOLAR INVERTERS

Detlef Schulz and Rolf Hanitsch

Berlin University of Technology · Institute of Electrical Power Engineering
Sec. EM 4 · Einsteinufer 11 · D-10587 Berlin · Germany
Tel.: ++49/(0)30/314-24538 · Fax:++49/(0)30/314-21133
eMail: Detlef.Schulz@iee.tu-berlin.de Berlin · http://www.iee.tu-berlin.de/personen/schulz

ABSTRACT: The paper presents a method of short time simulation of solar inverters with the software tool SIMPLORER. The modelling of the components for the solar generator and the inverter is described. A grid coupled system was simulated. As the energy suppliers have to observe the power quality carefully, the interesting power quality values were investigated. A comparison between the simulation and measurements shows the suitability of the models.
Keywords: Simulation – 1: Small Grid connected PV Systems – 2: Inverter - 3

1. MOTIVATION

To receive information concerning yield and economy of solar plants in advance long time simulations are done. These calculations are based on daily or monthly insolation data. Thus unfortunately, no reliable results can be achieved for the short time domain. The analysis of the current shape in the interval of 20 to 200 milliseconds is of interest because the power quality values are calculated from this time interval. Grid coupled photovoltaic systems are common in Europe. Electrical energy becomes recovered in these plants with an inverter and is then fed into the grid. For the admittance of the PV plants the guidelines of the utility concerning power quality must be taken into account. The power quality regulations of the grid have to observed. For this purpose the software SIMPLORER is able to deliver exact results within the mentioned time domain.

2. SYSTEM OVERVIEW

The investigated system consists of the solar generator, the inverter and the connection to the public grid. Figure 1 gives an overview of the grid connected system.

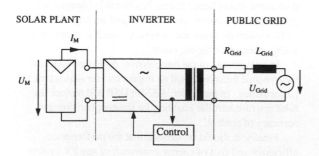

Figure 1: Overview of the grid coupled system

For the three parts solar generator, inverter and public grid the modelling will be described and the basic equations will be given.

3. MODELLING OF SOLAR CELLS

As the solar generator consists of a number of solar cells, we started with the modelling of the cells. . For the simulation of the solar cell the one diode model was used at first. Figure 2 shows the used one diode model. The calculation of the values of the photo current I_{Ph} , the diode current I_D , the parallel current I_P and the cell voltage U_C was made as described in [1] [2] and [3].

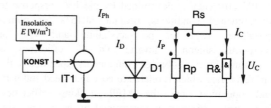

Figure 2: One diode model for solar cells

The resistance R& stands for the time-dependent value of the variable load, with this resistor the characteristic curves were obtained.
With the diode parameters m, U_T and I_S the cell current I_C was calculated:

$$I_C = I_{Ph}(E) - I_S \cdot \left\{ \exp\left(\frac{U_C + I_C \cdot R_S}{m \cdot U_T} \right) - 1 \right\} - \frac{U_C + I_C \cdot R_S}{R_P} \quad (1)$$

Later on, the two diode model was applied. With this model better results can be obtained, because two different diode factors m_1 and m_2 can be used [1]. Figure 3 shows the used equivalent circuit.

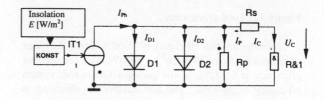

Figure 3: Two diode model for solar cells

For the two diode model the cell current I_C was given as:

$$I_C = I_{Ph} - I_{S1} \cdot \left\{ \exp\left(\frac{U_D}{m_1 \cdot U_T}\right) - 1 \right\} - I_{S2} \cdot \left\{ \exp\left(\frac{U_D}{m_2 \cdot U_T}\right) - 1 \right\} - \frac{U_D}{R_p} \quad (2)$$

To be able to work with this models in the simulation furthermore, module equivalent parameters for the series connection of cells were defined. With help of these parameters the modelling of complete solar modules or even solar generators was possible.

4.   MODELLING OF SOLAR MODULES

The obtained equations for the one diode model was adapted for the series connection of cells. This results in equivalent module parameters for the temperature voltage of the diode U_{TM}, the series resistance R_{SM} and the parallel resistance R_{PM} of the one diode model:

$$I_C = I_{Ph} - I_S \cdot \left\{ \exp\left(\frac{U_M + z \cdot R_S \cdot I_M}{\underbrace{m \cdot z \cdot U_T}_{U_{TM}}} \right) - 1 \right\} - \frac{U_M + z \cdot R_S \cdot I_M}{\underbrace{z \cdot R_p}_{R_{PM}}} \quad (3)$$

Considering the series connection of the solar cells the current I_C is equal to the current of the module I_M :

$$I_M = I_{Ph} - I_S \cdot \left\{ \exp\left(\frac{U_M + R_{SM} \cdot I_M}{m \cdot U_{TM}} \right) - 1 \right\} - \frac{U_M + R_{SM} \cdot I_M}{R_{PM}} \quad (4)$$

The module with its equivalent parameters was connected to the inverter model.

5.   INVERTER MODEL

For the inverter model the power electronic components were simulated. The used model is shown in figure 4. The full controlled bridge is connected to the grid via transformer.

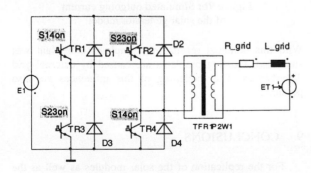

Figure 4: Inverter model for simulation

With the grid model, consisting of the grid impedance and the grid voltage, the simulations for different grid conditions are possible. The incoming voltage replicates the voltage of the solar module.

6.   SYSTEM CONTROL

Various control methods are checked with Petri nets known from the automation engineering. Therefore, for the test of the control approach no lavish programming is necessary. With this simulation approach unsuitable control methods can be avoided. The Petri nets consists of state variables and defined conditions for the change of the states. Figure 5 shows an example for a current control loop.

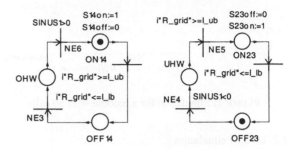

Figure 5: Control with Petri nets

The test of different parameter changes was made with this control method.

7.   SIMULATION RESULTS

7.1 Simulation of cells and modules

The simulation for the solar cells was done by changing the value of the connected resistor. To obtain the voltage and current for different levels of insolation E, the value of the current source according figure 2 was changed. The cell curves for different insolation conditions shows figure 6.

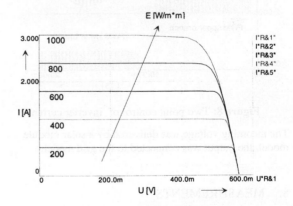

Figure 6: Simulation of solar cells
at different radiations

For the simulation of solar modules the use of the one diode model was sufficient, because the properties of the two diode model could be applied in the module equivalent parameters. Simulations with the one diode model with module parameters delivered good results similar to simulations the two diode model. As an example the series connection of 30 cells was simulated with this model.

The result is shown in figure 7. The simulation results agree with the measurements.

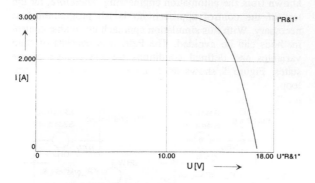

Figure 7: Simulation for a module with 30 cells

7.2. Inverter simulation

The inverter as described in figure 4 was controlled by Petri nets as shown in figure 5. With a set point current I_{set} of 1.5 A and a tolerance band of $0.2*I_{set}$ the simulation results are shown in figure 8.

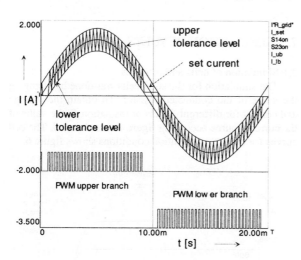

Figure 8: Two point control of inverter current

The incoming voltage was delivered by a solar module model, the output was connected to the grid model.

8.   MEASUREMENTS

To be able to judge the results of the simulation a comparison with existing real data from plants is necessary. For this purpose measurements at an existing solar generator were made. The comparison of the outgoing current was chosen to evaluate the functionality of the model.

Figure 9a shows the measured current shape, Figure 9b gives the result from simulation with a small tolerance band of $0.04*I_{set}$ and a switching frequency of 24 kHz. The measured current has a distorted shape, because the voltage shape in the grid was not fully sinusoidal. This phenomenon is a topic for further investigations. Except from this distortion, the simulation results agree well with

the real behaviour of the existing solar generator. The extension of the model with consideration of distorted grid voltages is another demanding task for further simulations.

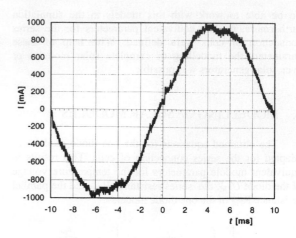

Figure 9a: Measured outgoing current of the existing solar generator

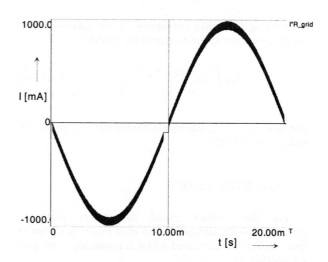

Figure 9b: Simulated outgoing current of the solar generator model

With this obtained results the complete solar plant was simulated. Existing problems are caused by external grid disturbances. The modelling of this influences are also possible.

9.   CONCLUSIONS

For the replication of the solar modules as well as the inverters models are necessary with which the real events can be studied well. The software SIMPLORER offers possibilities for the modelling and delivers good results. A model was developed for feeding of electricity into the grid using current control for the inverter. This inverter model can be coupled to the solar cell model without problems. With the overall model, consisting of solar cells, inverter and supply grid the behaviour of the solar generator connected to the public grid can be simulated fully.

From the simulation information is available of the electrical parameters in the milliseconds interval for variable radiation conditions e.g. the current quality for grid connected systems.

A solar plant was simulated, which is situated on the roof of the faculty building of the Electrical Engineering Department at TU Berlin and for which data are available permanently. The simulation leads to results which agree well with the measurement results of the existing plant. SIMPLORER is a well proven tool for system developers or users.

REFERENCES

[1]   Wagemann, H.-G.; Eschrich, H.: Grundlagen der photovoltaischen Energiewandlung. Stuttgart: Teubner 1994

[2]   Quaschning, V.: Regenerative Energiesysteme, München Wien: Carl Hanser Verlag 1998

[3]   Schulz, D.; Hanitsch, R.: Kurzzeitsimulation von Solarwechselrichtern. Tagungsband. 15. Nationales Symposium   Photovoltaische   Solarenergie. Staffelstein 2000. S. 149- 154

[4]   SIMEC GmbH Chemnitz: Reference Manual SIMPLORER Version 4.2.2000

NOVEL MICROPROCESSOR CONTROLLED REAL TIME MPPT
FOR PHOTOVOLTAIC CHARGING APPLICATIONS

Ljubisav Stamenic
Photovoltaic Energy Applied Research Lab, British Columbia Institute of Technology
3700 Willingdon Ave., Burnaby, BC, V5G 3H2, Canada

Matt Greig
Photovoltaic Energy Applied Research Lab, British Columbia Institute of Technology
3700 Willingdon Ave., Burnaby, BC, V5G 3H2, Canada

Eric W. Smiley
Department of Engineering Science, Simon Fraser University
8888 University Drive, Burnaby, BC, V5A 1S6 Canada

Joe Newton
Photovoltaic Energy Applied Research Lab, British Columbia Institute of Technology
3700 Willingdon Ave., Burnaby, BC, V5G 3H2, Canada

William Dunford
Department of Electrical Engineering, University of British Columbia
2356 Main Mall, Vancouver, BC, V6T 1Z4, Canada

ABSTRACT: The purpose of maximum power point tracking (MPPT) is to match the current and voltage characteristics of the load to the PV module's maximum power point, automatically for any given illumination. There are several ways to accomplish MPPT. One way is to have a large database containing the PV panel's output voltage and current values at different illumination levels at the maximum power point stored in the MPPT module.

A second way to accomplish MPPT is to use a dynamic system that tracks the maximum power point continuously. This system tracks the maximum power point by use of short-term memory. In this method the relative maximum power level of the PV module is dynamically tracked. This control method will constantly move the operational power point around the maximum power point. This method has several advantages and was chosen by the PEARL research team.

The PEARL research team designed and developed a digital (microprocessor controlled), as well as an analog version of a novel MPPT battery charging controller. Both systems use a unique dynamic-relative-maximum algorithm. A high power buck switcher was used to alter the apparent impedance of the load as seen by the PV module. The power from the module is calculated using analog circuitry and is then digitized for the microprocessor. The micro-controller then increases and decreases the PV module output voltage to maximize the power delivered by the module. Both systems obtained efficiencies of approximately 94%.
Keywords: Maximum Power Point Tracking - 1: Buck Switcher - 2: Analog - 3

1. INTRODUCTION

Photovoltaic Energy Applied Research Lab (PEARL) designed and developed two PV battery charging controllers with built in true maximum power point trackers. The two designs covered both analog and digital solutions. The digital solution used a microprocessor to implement a dynamic relative maximum tracking algorithm. The analog solution achieves the same results using an analog circuit.

Both units use a buck mode switching power supply. This power supply was used due to the performance requirements (high efficiency and low cost).

These units were developed as an interface between the PV array and the battery bank. These units also have charging regulator functions built in to allow proper co-operation between the MPPT circuitry and the battery charging process.

If an MPPT module is connected to a conventional charge controller, conflicts may arise. A MPPT module will always try to maximize the charging current to the battery. If a charging controller attempts to stop a fast charge and switch to trickle charge, the MPPT module may drastically increase its output voltage in an attempt to re-maximize the charging current. This conflict may damage the battery or the charging regulator. For these reasons the MPPT module and charging controller are integrated into one unit and have hardware and software to allow co-operation of these two components.

2. MAXIMUM POWER POINT

The maximum power point is the point on the I-V curve of a solar panel that defines the largest possible rectangle under the curve. The slope of the I-V curve at the maximum power point is approximately equal to negative one.

$$dI/dV \approx -1$$

The voltage of this point varies a little with changes in the illumination level. The current available at this point varies greatly with changes in the illumination level. Panel temperature and age will also change the maximum power point.

Another method to track the power point is to plot power with respect to voltage. The peak of this curve is the maximum power level. At this point of the curve the slope is zero.

$$dP/dV = 0 \qquad , \qquad P = I * V$$

So, what is the reason for Maximum Power Point Tracking? When a conventional 12 volt solar module is used to charge a 12-volt battery it may be that the panels rated power is not being achieved. This is due to the fact that the maximum power point voltage may be as high as 17 volts on some 12-volt panels. This means that the battery, which has a maximum voltage of about 14.5V, is forcing the panel to run at a lower voltage then it should. To correct this situation an MPPT module will convert the panel's higher voltage to a lower voltage, while at the same time increasing the charging current.

For loads like pumps a MPPT module can greatly increase the performance of the pump. The module will allow earlier starts in the morning and afternoons and increased performance during the day, as well as during cloudy days. The MPPT module greatly helps with loads that have very low resistance. With these loads increases in power of up to 60% have been observed.

An MPPT module will increase the power to any load, especially when the sun is rising and setting or if the light incident angle is poor.

3. METHODS OF POWER POINT TRACKING

There are several ways of power point tracking. One method is to have an extensive database of maximum power point voltages at all light levels. This control algorithm has several disadvantages:

- Specific to a particular PV panel;
- Data base must be maintained as PV panel degrades;
- Light level must be sampled;
- No method of temperature compensation.

This database method also requires a large amount of peripheral parts. A large memory storage location and light intensity measurement device would be needed. This method of power point tracking is specific only to a particular solar panel or array.

Another method is dynamic relative maximum tracking. This method uses an algorithm that continually searches for the maximum amount of power either from the source or to the load. Most of these units either try to maximize the current or voltage to the load, this method is the best way to accomplish MPPT. If you try to track the power from the panel you will require hardware or software to actually calculate the power level and then track that value. The multiplication of current and voltage will cause more problems then it is worth. By tracking current into a load or battery, you are effectively tracking the power into the load or battery. This control algorithm has several advantages being:

- Adaptive system, input can be anything with a maximum power point
- No extensive data base is needed
- System operation not effected by other outside variables i.e. temperature
- System will adapt to any load or radiation level

The method of measurement used in both of the MPPT modules developed at PEARL uses a dynamic relative maximum algorithm. These units track the current output of the buck switcher used as the power core of the MPPT module.

4. BUCK SWITCHER POWER CORE

Power conversion for the MPPT prototypes was accomplished with a buck switcher. Due to the fact that the application is battery charging, and that a battery's charging voltage is below the maximum power point voltage of the solar panel, this type of switcher was used. A switching power supply was used to make sure that as much of the panels power would get to the battery or load.

A buck switcher uses a high side MOSFET switch, which is either a P channel MOSFET or an N channel MOSFET driven by a high-side driver. It is preferable to use an N channel MOSFET as they tend to cost less for a particular current handling ability. Also N channel MOSFET have a lower on state resistance.

The high side MOSFET is switched at a high frequency, and employs a pulse width modulation (PWM) technique. This allows the output voltage to be changed from a higher voltage to a lower voltage, with a boost in available current. The chopped PWM power is then passed through a low pass inductor and capacitor filter. There is also a diode in the power supply to allow proper current flow during operation.

The PWM power supply can approach efficiencies of 95% or greater. The high efficiency allows for significant gains in power delivered to a load or battery in solar applications.

In cases where a high voltage is required for a load or large battery bank a boost switcher can be used to increase the voltage to the load.

Switching power supplies can also be used to design power supplies with isolated outputs, negative voltages or other rare power supply applications.

5. METHODS OF ELECTRICAL PARAMETER MEASUREMENTS

There are several ways to measure the 'power' from a solar PV module. There are two parameters that can be measured -- current or voltage. Power can also be measured by multiplying these two parameters. In early prototypes developed at PEARL it was found that using a microprocessor to multiply these two parameters was not a feasible method. When a switching power supply is used, noise is generated. If measured values are multiplied with noise in them the result will have a greater amount of noise. This large amount of noise corrupts the power data and makes peak power tracking difficult, if not impossible.

The other prototype uses an analog multiplier. If the power level was measured on the solar panel it was found that both parameters must be measured, and than multiplied together. It was best to monitor output power parameters. When monitoring output parameters, only one parameter had to be tracked. This is because any useable power is sent into a resistive load. After some research it was discovered that it is best to use a current sensor for the battery charging application.

Measuring current using a resistor was inaccurate. Resistors have inductive properties, which cause too much noise for accurate measurements. It was decided to use an open loop hall-effect current sensor. These sensors are more expensive, but are necessary for proper operation. An advantage of these sensors is that they often have a frequency range that allows direct current measurement, but ignores the high switching frequency.

6. SUMMARY

The PEARL research team designed and developed two prototypes of PV charging regulators with built in maximum power point trackers. One uses a microprocessor to implement a dynamic relative maximum algorithm. The other prototype uses an analog circuit to do the same thing. This new circuit dramatically reduces the cost of the unit.

Both units use an open loop hall-effect current sensor to measure the battery charging current.

These units use a buck switcher power core to transform the DC voltage. These switchers are 94% to 96% efficient therefore they can give gains in deliverable power to a load of up to 94% under some circumstances.

The two units have additional circuitry inside to allow proper battery charging regulation. These novel charging regulators support various types of deep-cycle lead acid batteries. They will also support multiple battery configurations.

The developed units are built as a single unit to allow co-operation between the battery charging regulator and the MPPT circuitry. This allows the MPPT circuitry to shut off when it is not needed.

REFERENCES

[1] In-Su Cha, Jang-Gyun Choi, Gyun-Jong Yu, Myung-Woong Jung, MPPT for Temperature Compensation of Photovoltaic System with Neural Networks, 26th PVSC, 1997, Anaheim, CA, USA (p. 1321).

[2] Y.J. Cho, K.S. Lee and B.H. Cho, Inductor Current sensing Peak-Power-Tracking Solar Array Regulator, The 34th Intersociety Energy Conversion Engineering Conference, 1999, Vancouver, BC, Canada (1999-01-2444).

[3] Y.J. Cho, B.H. Cho, A Novel Battery Charger-Discharger of the Regulated-Peak-Power-Tracking System, The 34th Intersociety Energy Conversion Engineering Conference, 1999, Vancouver, BC, Canada (1999-01-2445).

ENERGY MANAGEMENT FOR PV STREET LIGHTING

G. Goulas, T. Romanos, S. Tselepis
CRES, Centre for Renewable Energy Sources
19th km Marathonos Ave., 19009 Pikermi, Athens, Greece
Tel: +30 1 6039900, FAX: +30 1 6039905
E-mail: gegoula@mailandnews.com

ABSTRACT: The PV modules are ideal for use with batteries for stand-alone systems for street and parks lighting, etc. The energy management unit developed for these systems (PV-panels, batteries and lamp) is performed by a microcontroller system in order to optimise the energy use and the provided lighting service. The control unit (CU) is responsible for the lighting profile and lamp illumination and is programmable according to the needs through a user-friendly Windows communication software. A universal dimmer is responsible for the illumination of the lamp. The work was carried out during the course of the CRAFT-JOULE project "JOR3-CT98-7017".
Keywords: Stand-alone PV Systems - 1: Monitoring - 2: Expert-System - 3.

1. INTRODUCTION

The penetration of photovoltaic street lighting rapidly increases, especially in low traffic streets, squares, parks, etc. These systems are stand-alone and consisted of a PV panel, a battery, a lamp and a control unit. According to the energy management used, a longer battery life can be achieved resulting in less maintenance costs. Particularly in streets with low night traffic, the use of such systems reduces the energy consumption. For example, a significant low illumination is used and only when a move is detected the illumination increases resulting in an extremely high energy saving. A considerable energy saving can be achieved in squares by setting the illumination of the lamp at 50% late at night, without compromising the effectiveness of lighting. A similar case is the advertising signs where an illumination of 60% can be used after e.g. 2:00 a.m. without having any effect in the visibility of the sign, but reducing the energy consumption by 40%. An energy management, which satisfies the above cases, is developed.

2. HARDWARE DESCRIPTION

The PV Street Lighting System consists of a Control Unit (CU), a charge controller, a universal dimmer, a battery and a PV panel as shown in figure 1. The CU is the heart of the energy management system because it controls the energy of the battery and the way it is spent on the different parts of the system.

The prototype CU is made up of a box consisting of the CPU card, the Power Supply and sensors card. Figure 2, shows the Control Unit box. The installation procedure of the Control Unit is considered to be simple. The prototype Control Unit developed by CRES is based on the Motorola microcontroller MC68HC11F1. The maximum of its resources are: 8 analog inputs, 8 digital inputs, 8 digital outputs and 3 PWM inputs/outputs. The software-controlled power-saving modes make the MC68HC11 family especially attractive for battery-driven applications.

The universal dimmer is capable of switching and dimming all kind of lamps, e.g. incandescent, fluorescent, sodium, iodine, etc.

The charger is responsible for the charging and the protection of the battery while discharging. The operating voltage of the system may be 12 or 24V.

The total consumption of the control unit and sensors is less than 4W.

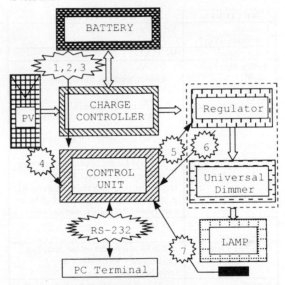

Figure 1: Block diagram of the PV Street Lighting System

Table I: **Monitoring values**

	Description	Type	Measuring range
1	Battery Voltage	Analog	0 - 50Vdc
2	Battery Temperature	Serial	-55 - +125°C
3	Battery Current	Analog	-5 - +5Adc
4	PV Voltage	Analog	0 - 50Vdc
5	Reference output	Analog	0 - 5Vdc
6	Output Voltage of Regulator	Analog	0 - 200Vdc
7	Luminous measurement	Analog	0 - 5000Lux

Figure 2: The prototype control unit

Table I shows the monitoring values of the Control Unit, while table II shows the specifications of the Control Unit.

Table II: Specifications of the Control Unit

FEATURES	SPECIFICATIONS
External Communication	Serial through RS232
Power Supply	9 - 36 VDC
Analog inputs	- 0 - 5V
	- Conversion type: Successive approaches
PWM output	- 0-100% (1kHz frequency)
Memory	- 1kB static RAM for temporary storage
	- 32 kB non-volatile RAM for raw data collection
	- EPROM for the operational software
Real Time Clock	Real time clock on the CPU card, registering time with the measurements in the stored data files (backed up by a battery with 10 years power autonomy)
Operation Temperature	0 to 70°C
Data File Output	ASCII format (to download data) that can be used with commercial spreadsheets
Sensors (provided with CU)	-LEM Hall effect transducers for Voltage -NT Hall effect transducer for Current -Digital Thermometer for Temperature

3.   ENERGY MANAGEMENT

The illumination output of the lamp has four levels of intensity (0%, 30%, 50% and 100%) which are controlled by the output reference of the CU. When the reference is set to 0% then the lamp is off. By having different levels of intensity the best energy saving can be achieved without compromising the effectiveness of lighting. The levels of intensity are programmable and can be ranged at any value between 0 and 100%. The lighting schedule can also be programmed based on lighting demand instead of the prevailing local conditions. The built-in real time clock is responsible for the scheduled operating condition. The default lighting schedule is from dusk to dawn, which is the most common way of operation.

The default lighting schedule is as follows:

At dusk when $V_{pv} < 3.5V_{DC}$ for 60 sec the lighting intensity is set at 30% for 10 minutes, then at 50% for 1 hour and then at 100% for 3 hours. Thereafter, the reference is set at 50% until dawn when $V_{pv} > 9V_{DC}$ for 60 sec or until low voltage disconnect is reached (if $V_{bat}<23$ V_{DC} then the lighting intensity is set to 30%,if $V_{bat}<22V_{DC}$ then the lighting intensity is set to 0%) as shown in figure 3.

This way the longest time of night operation can be achieved and the battery life can be prolonged since the low voltage disconnect alert prevents the battery from being damaged, resulting in less maintenance costs.

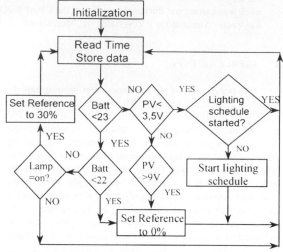

Figure 3: Logic diagram of the Energy management

The universal dimmer is capable of switching and dimming all kind of lamps, e.g. incandescent, fluorescent, sodium, iodine, etc. providing different solutions proportional to the needs.

4.   MONITORING AND PROGRAMMING OF THE CONTROL UNIT

The Control Unit monitors the battery voltage, temperature and current, the PV voltage, the luminous of the lamp and the input voltage of the lamp. The state of operation of the lamp is also stored, i.e. in which stage of the lighting schedule the lamp operates.

By combining the data together, the operating

Figure 4: The visualisation software used for monitoring and programming of the Control Unit

condition of the system can be defined. Data is stored every 15 minutes, which is the mean value of 1-minute measurements except the temperature, which is the moving average of 2 continuous minutes. Data is stored in non-volatile memory together with the time and date, so that the measurements can be corresponded to real time.

The CU is connected to a PC serially through RS-232 for monitoring and reprogramming of operations. By having a cable connected to the CU there is no need of physical access to the CU, which is significant for places that are difficult to approach.

CRES has developed a menu driven software for monitoring and programming the Control Unit that provides user friendly communication between a PC and the CU, which is shown in figure 4.

Each option represents a function and it is activated by two different ways. Either by its number (keyboard) or by pressing the button next to the option (mouse).
The available options are:
1. Delete: Deletes the data collected in CU memory.
2. Upload: A data file is created in the PC, where the collected data is saved.
3. Download: Defines a number of parameters concerning the CU's operations, e.g. customised intensity levels and lighting schedules.
4. Time & Date: Checks and reprograms the time and date kept by the CU's real time clock.
5. Calibration: It applies correction values to the data being uploaded from the CU to the PC.
6. On line view: Displays the measured parameters
7. Exit: Exits the communication program.
On the main MENU the user has also the option to select the communication port (Port Setup button) of the PC being used.

5. THE EXPERIMENTAL INSTALLATION AT CRES

At CRES an experimental PV Street Lighting System was installed in order to evaluate its performance. At the present the simplest case is examined where on/off control is applied to a fluorescent lamp, together with a STEKA charger. A 24V battery was used which was already tested in other experiments.

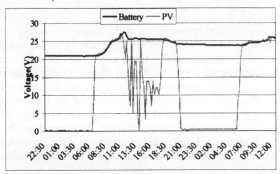

Figure 5: PV and Battery voltage

Figure 5 shows the charging of the battery according to the PV voltage. The fluctuation appearing is due to the charger. When the battery reached its maximum voltage, the PV voltage was cut off to prevent the battery from overcharging. When the battery voltage became low then

Figure 6: The experimental PV street lighting system installed at CRES

the PV voltage was turned on again and the charging of the battery started from the beginning. The graph has also this sharp fluctuation because the values displayed are mean values of 1-minute measurements. Figure 6 shows this installed system of the PV street lighting at CRES

6. CONCLUSIONS

Energy management is very important for the PV street lighting system. Bad energy management means a discharged battery, which means less night operation, and greater maintenance costs. A low cost control unit for energy management is developed. It can drive a universal dimmer for a variety of PV street lighting applications.

SolSim – A Software Tool for Simulation of Solar Hybrid Systems

C. Schaffrin, I. Knoblich
Konstanz University of Applied Sciences
Institute for Applied Research: Solar Systems
Brauneggerstr. 55, D-78462 Konstanz, Germany
Tel.: +49 7531 206-406, Fax: +49 7531 206-400, e-mail: schaffrin@fh-konstanz.de

ABSTRACT: The software package *SolSim* has been developed for simulating renewable hybrid energy systems. These systems can consist of units converting all sorts of renewable energy into electric energy. At this stage of development *SolSim* provides the following components: PV generator, wind energy converter, battery, DC/DC converter, DC/AC inverter (used for grid coupling), biogas/biomass combined heat power station and engine generator. Furthermore there are modules allowing an economic analysis and optimisation of the tilt angle of a solar generator. The software program *SolSim* is used not only for designing energy systems but also for developing strategies of energy management within those systems to satisfy the technical as well as the economical demands.
Keywords: Simulation – 1: Hybrid – 2: Sizing - 3

1 INTRODUCTION

Today the use of renewable energy for meeting the energy demand is already profitable within island systems (not coupled with a public grid), even considering only the operating efficiency. This island situation is given e.g. to the electrification of villages in India or other countries without a covering electrical grid. Since only in exceptional cases the energy demand can be adapted to the energy supply a high or even total supply coefficient is indispensable. Renewable energy is subject to statistical variations. Therefore a supply coefficient of 100 % at every time is not possible. This may be pretended by considering yearly sums, but a detailed analysis by time steps will unveil time periods of energy deficit.

The technical task of designing such energy systems is to minimize the periods and the amount of an energy deficit without increasing the amount of lost surplus energy (energy which cannot be used because of a fully loaded energy storage unit). This aim can be achieved by combining different sorts of energy conversion devices, e.g. PV generator and wind energy converter, and by adjusting the size of the energy converters and storage units. In this way a hybrid renewable energy supply system can compensate the seasonal variations.

Designing such hybrid systems merely according to technical criteria is not trivial, but regarding economic criteria additionally makes the problem even more complex. Satisfactorily working island systems may be supported by an intelligent energy management the rules for which should be individually optimized.

The described task can only be solved by means of computer simulation. Therefore already in 1988 the development of a simulation software has been started by means of which both any energy supply system can be optimized according to practical operating conditions, and the strategy for the energy management can be developed, considering technical as well as economic needs.

2 SIMULATION SOFTWARE

The simulation program called *SolSim* allows the modelling, the analysis and the optimisation of autonomous or grid coupled multivalent solar energy systems. Solar here is meant in its widest sense and includes radiation, wind ,water power and biogas/biomass.

2.1 User surface

During the development special attention was paid to practical every day usefulness and simple operation. All system components are stored in different libraries. This prevents long input sessions for the user. Nevertheless the program includes the option to define specific component parameters to be stored in the library. A graphic user surface with integrated context sensitive help function assures a simple and easily learnable operation.

2.2 System components

SolSim provides a wide range of energy supply components which individually can be connected to the main transmission line according to the system configuration (fig. 2). *SolSim* allows the user to change all component parameters. So not only the components available on the market can be used for simulation, but any component can also be generated individually in order to optimise the whole system, and stored in the corresponding library. In co-operation with a manufacturer important impulses may arise for future developments.

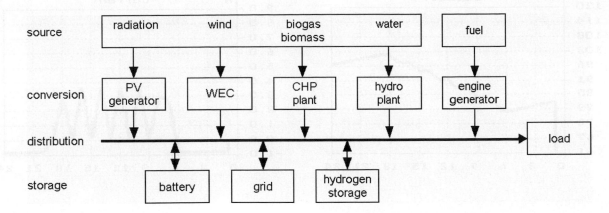

Fig. 1: Structure of the software package

The PV generator model determines the energy yield of a stationary and arbitrarily inclined or tracked (one axis or two axes) panel using the actual sun position and the diffuse and direct insulation. The albedo is also considered by the simulation algorithm. For operation in the maximum power point an MPP tracking alternatively can be switched on. An optimisation algorithm determines the most favourable degree of inclination taking into consideration the shadowing of the panel by a given horizon profile. Tracking can be continuously or discrete in time steps.

For the wind energy converter (WEC) the power output (as hourly mean value) is calculated from the power characteristics of the unit, the roughness of the ground and the actual weather data (wind speed, wind direction, Temperature and air pressure). The rotor power can be controlled by stall or pitch.

The battery model simulates the terminal voltage depending on the state of charge. It is based on characteristic data from the data sheet and data to be determined empirically.

The DC/AC inverter is used for energy conversion in both direction to and from the grid. The model is based on an efficiency curve obtained by mathematical interpolation.

An extensive software module for simulation of biogas and biomass plants has been developed. The bio-energy is converted into electric energy within combined heat power plants (CHP). The thermal energy is summed up, as well.

The fuel engine generator is required in case the supply coefficient must be 100 %. It can be operated in a static power profile and switched on and off in an intelligent controlling mode, in which a least running time, the battery state of charge and the power of the consumer are considered. The operational characteristics can be defined by the minimum running time of the engine generator and the state of charge with priority of the renewable energy sources. By varying the criteria the strategy of the energy management is optimised.

The energy consumer data are described by load profiles (current, voltage or power as functions of time). Periodically repeated load cycles are generated automatically.

The integrated economic analysis related to operational economy makes the comparison of the specific energy costs of different solar system designs possible. In addition a software module regarding the external social costs is also available.

2.3   Simulation results

The main simulation results such as surplus and deficit energy, performance ratio and plant efficiency etc. as well as hourly, daily, weekly or monthly values for each component will be displayed on the screen (fig. 2 and 3) after a simulation run.

All hourly simulation data are stored in files automatically for further calculation or publishing purposes.

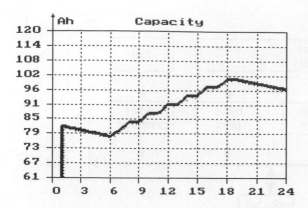
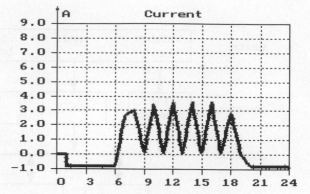

Fig. 2: Graphical output of component results (example)

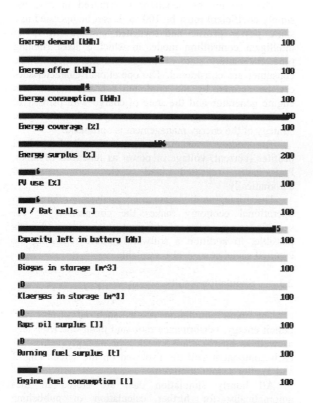

Fig. 3: Output of the main system results

3   SYSTEM DESIGN EXAMPLE

Independent of the used system components for converting all forms of renewable energy into electricity the aim of every solar plant is to maximise the available renewable part of the energy supply. For achieving this the user of *SolSim* can set different priorities. Depending on the individual case these may be the supply ratio, the economy, the renewable energy ratio or others.

For demonstrating the performance power of SolSim in the following example an energy supply system is designed for a farm not connected to the public grid. The basic part of energy supplied by a CHP unit converting the renewable energy from biogas and biomass produced by the cattle. A PV generator and a WEC have been added while a battery compensates the different time profiles of energy supply and demand (fig. 4). The demand peaks are caused by the milking machine.

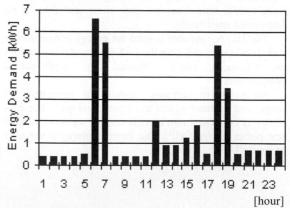

Fig. 4: Energy demand profile

Table 1: Parameters of the system components

component	parameter
PV generator	module: MQ 36 (53 Wp), total power varied installation: azimuth angle 10°, tilt angle 30°
WEC	power varied min. wind speed: 2 m/s nom. wind speed: 10 m/s max. wind speed: 25 m/s
biogas (CHP)	number of big cattle units (CA): varied fermentation unit: temperature 28° C, surface 63,7 m^2 gas storage: 70 m^3 mean energy loss: 0,7 W/m^2 K CHP: electr. efficiency 25% spec.energy of biogas 21,6 MJ/ m^3
battery	lead acid accumulator 120 V, capacity varied
energy demand	33,8 kWh/d
economic analysis	lifetime 30 years (battery 10 years) annuity 6 %

The main parameters of the system components are summarised in table 1. The situation examined here seems to be typical for the electrification in rural areas.

Different system variants have been simulated over one year. The results of some of them are given in table 2, while fig. 6 shows a bar graph for variant no. 13 with the monthly energy sums for PV, WEC, biogas, consumed and demand energy.

Table 2 also shows the relative specific energy costs where 100 % correspond to variant no.11 describing the reality at the moment (24 CA).

It should be pointed out that the simulation results show a battery state of charge not dropping below 50% in the variants with supply rate near 100% and a battery capacity of 900 Ah. This guarantees a long lifetime of the battery.

Table 2: Supply rates and specific energy costs of different system variants

Variant	PVGenerator: Power	WEC: Power	CHP: El. Power	Battery Capacity	Supply Rate	Surplus Energy	Surplus Energy	Deficit Energy	Relative Specific Costs
	kWp	kW	kW	ΔU	%	%	kWh	kWh	%
1	21,2	0	0	1800	81,7	72,0	8899	2264	131
2	6,4	10	0	1800	89,8	20,1	2487	1258	117
3	6,4	10	3 (8)	1800	97,9	44,8	5540	255	83
4	4,7	8	3 (8)	900	88,8	14,9	1840	1383	96
5	4,7	6	5 (12)	900	95,4	16,6	2055	563	94
6	4,7	8	5 (12)	900	96,6	30,3	3752	425	90
7	3,7	6	6 (15)	900	96,7	15,6	1930	405	92
8	2,7	6	7 (17)	900	97,5	14,3	1772	308	91
9	1,6	6	8 (20)	900	98,1	13,4	1659	234	89
10	0,0	6	10 (24)	900	95,8	16,6	2055	518	56
11	0,0	0	10 (24)	900	76,0	0,0	5	2965	100
12	4,7	0	10 (24)	900	96,8	15,3	1887	396	92
13	4,7	6	10 (24)	900	100	52,8	6532	0	83
14	1,6	6	10 (24)	900	99,6	24,2	2986	53	85

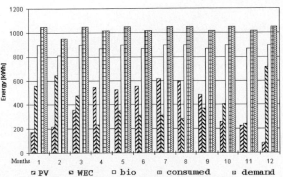

Fig. 6: Monthly energy sums [kWh] for variant no. 13

4   CONCLUSION

The software package *SolSim* is used for designing multivalent renewable energy supply systems (which means sizing of the different components) and for optimising the strategy of energy management. *SolSim* is a technically sophisticated and very flexible tool for engineers, as it allows entering a large amount of very specific data to adjust the simulation to a certain project. Nevertheless it is very comfortable as it uses component libraries and a simple surface structure. It provides all renewable energy sources to be combined with each other and with a back up generator. The system includes all major losses what enables *SolSim* to simulate close to reality. Also the very down into details analysis of the economic performance of the system is a good help to find out the optimal layout with respect to technical and economic criteria and to judge a solar hybrid system.

REFERENCES

[1]   Knoblich, I. (1998), Selbstoptimierende Softwaretools zur Simulation von Solar-Hybrid-Anlagen, Thesis, Konstanz University of Applied Sciences

[2]   Seeling-Hochmuth, G. (1997), A Combined Optimisation Concept for the Design and Operation Strategy of Hybrid PV Energy Systems, Solar Energy Vol.61, Elsevier Sience Ltd.

[3]   Winkler, S. (1991), Simulationsprogramm für Solaranlagen, Thesis, Konstanz University of Applied Sciences

[4]   Fengler, K. (1991), Entwicklung eines Programmbausteins für die Auslegung von Windkraftanlagen, Thesis, Konstanz University of Applied Sciences

[5]   Hertweck, M. (1989), Optimierung der Aufstellwinkel eines Solargenerators, Thesis, Konstanz University of Applied Sciences

NAVIGATIONAL AID SYSTEM WITH SATELLITE MONITORING

Christer Nyman
Soleco Ltd
Nygård 45 FIN-06100 Borgå, Finland
Email christer.nyman@avenet.fi
Phone: + 358 400 458790, +358 19 543209 fax: +358 19 543209

The first PV-powered navigational aid system in the world which communicate through the commercial Orbcomm -satellite has been presented and installed. A new technology, energy efficient buoy lantern has been developed.

The buoy lantern has been developed for use as a light signal on floating devices in harsh marine environment. The light source and the optical system of the unit is designed for up to 10 years of unattended operation - which is a great improvement compared to incandescent lamps. Special attention has been drawn to reliability, cost efficiency and energy consumption as well as performance in extreme temperature areas.

In remote areas where there is no cellular phone network coverage, the flasher communicator can be supplied with an Orbcomm modem. Utilizing the new Orbcomm message service, flasher reports are sent via a satellite communication link which converts the messages into standard email messages that can be received anywhere in the world.

The complete performance data of a small navigational system has been monitored since February 95 and the aim is to get long time performance data.

Keywords: 1: navigational aid, 2:satellite, 3: PV

1. BACKGROUND

The navigational aid system consists of the lantern, flasher and controller unit, communication unit, batteries charged by PV. The complete communication system is PV powered, like the satellite.

The buoy lantern has been developed for use as a light signal on floating devices in harsh marine environment. The light source and the optical system of the unit is designed for up to 10 years of unattended operation - which is a great improvement compared to incandescent lamps. Light Emitting Diodes (LEDs) are used as a light source of the unit. Highest possible light intensity and right signal colours are achieved by using the latest LED technologies. The high performance optical lens of the lantern has an unparalleled light output at extremely low input power. Compared to standard 85mm lanterns, the light output of the red and the green signal at same input power is more than 5 times greater.

Sabik Oy has developed a remote monitoring system with a modular design that enables all kind of lighted aids to navigation to be monitored and controlled. Special attention has been drawn to reliability, cost efficiency and energy consumption as well as performance in extreme temperature areas. [1] [2] [3]

On the lowest monitoring level, the monitoring information and control functions are integrated in the flasher, thus enabling monitoring of the smallest Aids to Navigation equipment simply by use of a telecom unit (Flasher Communicator) connected to the flasher serial port

Originally the system hierarchy was designed for only the lowest level, and data was to be collected by a mobile data collection unit board, a Buoy Tender or from a helicopter

When the first stage was completed and implemented, developing of the second level was initialized, aiming for a cost efficient way to monitor larger geographical areas using a central base approach. A system hierarchy was developed enabling a single outstation to report to a local hub or directly to central base depending on telecommunication options. By introducing local hub's, isolated geographical areas can be monitored through a single communication point from the base station, and local communication is handled using point-to-point UHF-radio communication.

2. SYSTEMS COMPONENTS

2.1 Solar powered lights and buoys

The simplest system can be used on small solar powered lights and solar powered buoys with a maximum solar power system of 100W. The system consists of only three components:

- SmartFlasher 6-28 intelligent flasher
- FC flasher communicator motherboard
- Orbcomm Subscriber Communicator, with built-in GPS receiver

The flasher incorporates all functions needed to operate a small PV powered aids to navigation, and is capable of monitoring the most vital functions of the light:

- Battery condition
- Lamp condition
- Panel production

2.2. Light emitted diod (LED) Buoy Lantern

This buoy lantern has been developed for use as a light signal on floating devices in harsh marine environment. The VP-4/LED lantern is characterised by a robust mechanical design, high reliability, long lifetime and high visibility at extremely low wattage.

The light source and the optical system of the unit is designed for up to 10 years of unattended operation - which is a great improvement compared to incandescent lamps. Light Emitting Diodes (LEDs) are used as a light source of the unit.

Highest possible light intensity and right signal colours are achieved by using the latest LED technologies.
External connector for serial communication to flasher is on the outside of the lantern.

Figure 1. The LED light buoy

The light source, which is formed by using several LEDs, has a high number of parallel connections to minimize the light output degradation of the signal due to a failure of a single LED. The current of the LED's are controlled with high accuracy by built-in adjustable controller. All LED's used are classified according to luminous intensity and colour chromaticity (within IALA preferred colour zones).

The high performance optical lens of the lantern has an unparalleled light output at extremely low input power. Compared to standard 85mm lanterns, the light output of the red and the green signal at same input power is more than 5 times greater. The vertical divergance is equivalent to or better than standard 85mm and 155mm lenses on the market. The lens is moulded as an single piece in UV stabilised Lexan™ polycarbonate which ensures durability of light output with minimal fading or discoloration during aging. The lens also exhibits good resistance to mechanical impacts and abrasion. The lens is also characterized by a good uniformity of the light output in the horizontal plane.

The base is manufactured of corrosion resistant marine aluminium alloy, which is finished externally with a high quality hot baked epoxy.

2.3 Smartflasher 6-28

The light is controlled by the SmartFlasher 6-28 that feeds the LEDs with constant current independent of variations in supply voltage and voltage transients. The LED driver can be adjusted to a wide range of output current thus enabling the output light intensity to be adjusted from 20% to 100% of nominal max. fixed intensity. [4]

2.4 Flasher communicator

The Flasher Communicator has three communication options:
1. UHF radio modem
2. GSM Cellular phone
3. Orbcomm satellite modem

3. USING THE FC WITH AN ORBCOMM MODEM

In remote areas where there is no cellphone network coverage, the FC can be supplied with an Orbcomm modem. The Orbcomm modem communicates through a satellite communication system based upon a number of low orbital communication satellites. The Subscriber Communicator (SC), as the Orbcomm modem is called, connects to the serial port of the
FC in the same way as the UHF modem and the GSM Cellphone.

When the Orbcomm SC unit is used, the outstation communicates directly with the Base Station computer.

When the FC-Orbcomm wants to send a message to the Base station (daily/weekly report or failure), the report is send to an Orbcomm satellite passing the site using VHF frequency. When the Orbcomm satellite has received the message from the SC, the report is immediately forwarded to the Orbcomm Gateway Earth Station where the message is converted into an email message and send as standard email to the address configured for each SC unit (=Base Station Computer email address).

The communication works the same way when messages are send from the Base Station to the outstation. The command is send as an email from the Base Station computer, then transmitted from the Gateway Earth Station to the closest Orbcomm satellite, where the SC will receive the message trough the downlink from the satellite.

The Orbcomm SC unit may also incorporate an optional GPS receiver, and preferred GPS window for buoys can be configured. If the buoy moves outside this preffered window, an OLI (Off Location Indicator) report will be generated immediately, and repeated at a preset intervals, thus enabling operator to track a drifting buoy.

4. ORBCOMM SHORT MESSAGE SERVICE

This new communication system makes global remote monitoring of aids to navigation easy and affordable. Utilizing the new Orbcomm message service, flasher reports are sent via a satellite communication link which converts the messages into standard email messages that can be received anywhere in the world.

In addition to status reports, the FC-Orbcomm GPS communicator can also monitor the location of the buoy using integrated GPS and generate an alarm report (OLI) when the buoy has moved outside a set location.

The Orbcomm modem communicates through a satellite communication system based upon a number of low orbital communication satellites. The Subscriber Communicator (SC), as the Orbcomm modem is called, connects to the serial port of the FC in the same way as an UHF modem ar a GSM Cellphone.

When the FC-Orbcomm wants to send a message to the Base station (daily / weekly report or immediately upon alarm or warning), the report is send to an Orbcomm satellite passing the site using VHF frequency. When the Orbcomm satellite

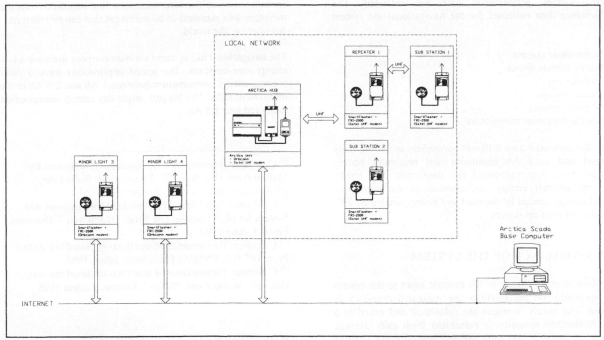

Figure 2. A complete system.

has received the message from the SC, the report is immediately forwarded to the Orbcomm Gateway Earth Station where the message is converted into an email message and send as standard email to the address configured for each SC unit (=Base Station Computer email address).

The communication works the same way when messages are send from the Base Station to the outstation (=polling). The command is send as an email from the Base Station computer, then transmitted from the Gateway Earth Station to the closest Orbcomm satellite, where the SC will receive the message trough the downlink from the satellite.

The Orbcomm SC unit (KX-G7101) also incorporates a GPS receiver, and preferred GPS window for buoys can be configured. If the buoy moves outside this preferred window, an OLI (Off Location Indicator) report will be generated immediately, and repeated at a preset intervals, thus enabling operator to track a drifting buoy.

5 .LIGHTHOUSE / COMMUNICATION HUB

If an outstation cannot be polled directly from the Base Station in a point-to-point communication network (outside UHF radio range, outside cellphone network coverage), a second level containing local Hub's for geographical areas can be used.

5.1 Arctica Field Computer

The Arctica unit is configured for usage as a hub, performing as the local controller in a geographical area. The Arctica is a field computer for control and data acquisition, specially designed for stand-alone usage in remote areas.
Normally, the Arctica unit is installed in a lighthouse acting as the lighthouse main controller,
• controlling the solar panel production,

• controlling the main light,
• controlling the emergency light
• controlling the main sound
• controlling the emergency sound
• monitoring the fog detector
• monitoring the racon
• monitoring the main battery
• monitoring the auxiliary battery
• monitoring the main light
• monitoring the photocell (several photo sensors can be used)
• monitoring main sound
• monitoring the solar panel production

In addition to controlling and monitoring the lighthouse, the Arctica unit can also log the data collected by the unit. The operator can program the Arctica to log actions or measuring data together with date and time stamp for any of it's I/O channels. There is room for up to 50.000 logged data's, and logging of data can be trigged by change in I/O or at preset time intervals.

6. BASE STATION

The base station has a PC computer with Windows NT operating system and ArcticaScada Windows™ software. The communication to the site is via Internet connection *or/and* dial up modem *or/and* Satel 1As UHF-modem.

7. MEASUREMENT OF THE TEST

The performance of the system is monitored to verify the currents of the whole control unit including regulator, timer and flasher and its applicability as controller of automatic light systems. The sizing of the PV-system used for this application can be verified by this test.

A test station for photovoltaic applications has been built where outdoor performance data are collected. The performance data collected for the navigational aid system are:

1. Photovoltaic current
2. Battery current in/out
3. Load current
4. Battery voltage
5. PV panel voltage
6. Panel and outdoor temperature

From the measured data different parameters including daily charged and used Ah maximum and minimum battery voltages have been calculated and monitored. The results give the internal energy consumption of the controller, typical energy needed by the load and energy needed by PV-modules for a set-up system.

8. PERFORMANCE OF THE SYSTEM

The daily average energy of the module input to the system and insolation on the modules are measured every 15 th second. The hourly averages are calculated and saved in a file. A monthly summary is calculated from daily average values. In figure 1, the daily average energy flows (Ah) of load, module and battery are presented.

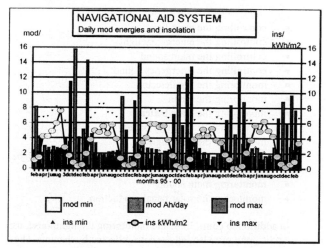

Figure 3. Daily average energy flows of module and insolation with daily maximum and minimum.

The average daily loads vary from 1 Ah in June to 2.5 Ah in wintertime. The maximum load of a particular day is about 3.5 Ah

9. CONCLUSIONS

Sabik Oy has developed a remote monitoring system for visual Aids to Navigation with a modular design that enables all kind of lighted aids to be monitored and controlled. Special attention has been drawn to reliability, cost efficiency and energy consumption as well as performance in extreme temperature areas.

This new communication system makes global remote monitoring of Aids to Navigation easy and affordable.

Utilizing the new Orbcomm message service, flasher reports are sent via a satellite communication link which converts the messages into standard email messages that can be received anywhere in the world.

The navigational aid systems for multipurpose use have a low energy consumption. The tested applications have a daily average energy consumption between 1 Ah and 2.5 Ah in the different months. The biggest single day energy consumption measured was 3.5 Ah.

[1] C. Nyman, Photovoltaic Navigational Aid System for Multipurpose Use Proc 13th Photovoltaic Solar Energy Conference, Nice, France, October 1995.
[2] C. Nyman, L. Mansner Photovoltaic Navigational Aid System for Multipurpose Use Proc   North Sun 7, Otaniemi, Finland, June 1997.
[3] C.Nyman, Performance of small navigational aid system, Proc. 14th E.C. PVSEC; Barcelona . Spain 1997
[4] C.Nyman, Performance of small navigational aid system, Proc. 2nd World Conf. PVSEC; Vienna, Austria 1998

System information:

Sabik Oy
Österby L 34 A
FIN-07370 Pellinki
Finland

Tel +358-19-540 719
Fax +358-19-540 810
sales@sabik.com
www.sabik.com

A SELF-OPTIMISING TRACKER FOR SMALL PV CONCENTRATORS

G.R.Whitfield, C.K.Weatherby, B.M.Towlson, L.Rahulan and R.W Bentley
University of Reading, Dept. of Cybernetics, Whiteknights, PO Box 225, Reading, RG6 6AY, UK
Phone 0118 9318223, fax 0118 9318220, e-mail cybgrw@cyber.reading.ac.uk

ABSTRACT: This project is part of a continuing programme to reduce the cost of solar photovoltaic electricity, by developing small concentrating pv systems for stand-alone systems in remote areas. The tracker uses a cheap microcontroller with a tracking head and drive motors; it is autonomous and self-aligning. For setting-up, the collector is roughly aligned and the tracker switched on. The tracker scans the sky until it finds the sun, and tracks it for a while, recording both the collector's position and the calculated position of the sun. From these the parameters of the collector's axes and position sensors can be calculated, and used subsequently for open-loop tracking. The microcontroller also controls a dc-to-dc converter to hold the solar cell array at its peak power point, and from time to time makes small adjustments to the direction of the collector to maximise the array output; this corrects for initial and slowly varying alignment errors in the collector and drift in the clock. No human intervention is required to set up or run the tracker, so the user can avoid the expense of a qualified engineer.
Keywords: Tracking – 1: Stand-alone PV Systems – 2: Cost Reduction - 3

1. INTRODUCTION

This project is part of a continuing programme to reduce the cost of solar photovoltaic electricity, by developing small concentrating pv systems that are sufficiently close to practical production to interest potential manufacturers. A recent EC-supported programme [1,2,3] examined a wide range of small concentrating pv systems, and built four 2 m² prototypes (Figs 1,2). It was concluded that several types of small pv concentrator, manufactured in quantity, would be cheaper than conventional planar arrays by a factor of 2 or 3. Such pv systems would be appropriate for stand-alone systems, used for domestic electricity supply or water pumping in remote areas. In this type of application the system must be easy for the user to install and commission; it will not be possible to pay for the services of an installation engineer.

Cheap convenient trackers are an important part of any such practical system. Conventional closed-loop trackers, using a sensing head and amplifier to drive the tracking motor are simple, and work well when the sun shines. But their accuracy depends on the alignment and cleanliness of the tracking head, and there is no automatic capability for morning re-set, wind-stow or handling intermittent cloud cover. Open-loop systems generally rely on a clock and computer to calculate the position of the sun, and point the collector in the calculated direction. The performance depends on the accuracy of setting of the collector's rotation axes and position transducers, the clock, and the latitude and longitude of the site. The setting-up procedure is quite complex.

The system now being developed combines the advantages of both open-loop and closed-loop trackers. It is based on a cheap microcontroller, which performs all the tracking functions, and also automatically sets up and optimises the tracking system. At no time in the setting-up or subsequent adjustment is human intervention required, so the user can commission his system without the need for a qualified engineer.

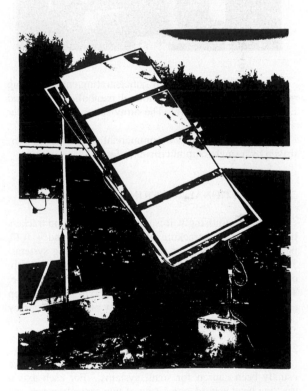

Figure 1: A prototype concentrating pv collector using thirty two point focus Fresnel lenses mounted in gimbals. The outer gimbal rotates about a polar axis. The inner one rotates in declination about an axis perpendicular to the polar axis.

This prototype uses a commercial closed-loop tracker; the tracking head is half way up the collector on the right.

Figure 2: A prototype pv concentrating collector using an array of eight offset cylindrical parabolic mirrors and eight lines of solar cells. The array rotates about a polar axis.

This prototype uses a commercial open-loop tracker, driven by a single-chip microcontroller.

2. CONVENTIONAL TRACKING SYSTEMS

Perhaps surprisingly, it is quite easy to develop trackers that give nominal tracking accuracies of around ± 0.1°, better than is needed for most pv concentrating systems. The difficulty lies in developing a system that is cheap, easy to set up, and that maintains its accuracy all the time. Trackers are usually divided into two types, closed-loop and open-loop systems. A good review of tracker accuracy and difficulties is given by Maish [4].

2.1. Closed-Loop Tracking Systems.

These are the simplest and cheapest systems, and have usually been chosen for small systems. For each axis a tracking head is used, consisting of two optical sensors and a shadowing plate which normally shadows half of each sensor; as the sun moves, one sensor is shadowed less, and the other more, and the difference between their outputs is the error signal which is amplified to drive a tracking motor. This is easy to make and set up, but suffers from the following disadvantages:-

It moves the collector away from the sun when clouds pass.

Its accuracy depends on the alignment of the tracking head with the collector, which can be easily upset, and on the cleanliness of the two sensors.

It has no automatic capability for subsidiary functions such as wind-stow and morning reset.

Being a feedback system, it has a tendency to hunt, unless properly damped.

All these problems can be overcome, but at additional cost and complexity.

2.2. Open-Loop Systems

These are in principle more complex, and have usually been reserved for large systems. They rely on a clock, usually driven by a computer or microprocessor, to establish the direction of the sun, and to point the collector in this direction. They are of course independent of cloud, and morning reset is automatic. But they have their own problems:-

The collector axes must be rigid and accurately aligned.

The collector needs position sensors.

The clock needs occasional updating to counteract drift.

Commissioning usually involves a complex manual set-up phase, in which the computer is fed with date, time, latitude, longitude, and the parameters of the collector axes and sensors.

3. THE PRESENT PROJECT

3.1. The Aim

The overall aim is to design, develop and test a tracker that combines the best features of both closed-loop and open-loop tracking, at a cost that is affordable for small systems, and that can be set up easily by unskilled users; in the applications envisaged, small free-standing systems in remote areas, there is no money to pay for the services of an engineer. The tracker will take advantage of the low cost of modern single-chip microcontrollers, and the fact that complex digital processing costs very little once the relevant programs have been written. The tracker is autonomous and self-commissioning; all the user has to do is to install the collector, roughly aligned, on a reasonably rigid mounting, and switch it on on a sunny day. The tracker will scan the sky, using its tracking head to find the sun, and will then track the sun in closed-loop mode. For a while it will record the sun's position in collector coordinates, for comparison with the sun's position in hour angle and declination calculated from the date and time, and the longitude of the collector. Hence the parameters of the collector's axes and position sensors can be found, after which the system will switch automatically to open-loop tracking.

The microcontroller will also be used to control a dc-to-dc converter between the solar cell array and the load, so that the array always operates at its peak power point. Since it now has a measure of the power output, it can make the tracker scan a small region around the nominal pointing direction, to maximise the array output. This will compensate for small alignment errors in the collector, and if regularly repeated will correct for other errors such as clock drift and sinking of foundations.

The system is suitable for two axis trackers, with any choice of axes that allows the whole path of the sun in the sky to be scanned, or for a single axis tracker with an appropriate choice of axis. For convenience, the development hardware has polar and declination axes (Fig. 1), but the algorithms being developed are universal.

3.2. The Tracker Hardware

The collector in figure 1 uses point focus Fresnel lenses, mounted in gimbals. The outer gimbal rotates in hour angle about a polar axis, driven by a commercial tracking unit, comprising a dc motor geared down by a large factor. There are cam-operated microswitches that serve as end stops, and a magnet on an intermediate gear operates a reed relay, to provide a pulse every revolution. Collector position is determined by counting these pulses, starting from the East end stop. The inner gimbal moves in declination, driven by a screw jack, with a similar dc motor, end stops and reed relay pulse generator. The laboratory tracker is similar, using the same commercial unit for rotation in hour angle, and a small geared motor for rotating a small Fresnel lens collector in declination. The motors are driven by electronic switches, controlled by digital outputs from the computer; the collector moves so slowly that there is no need for analogue speed control. A two-axis tracking head, with four small solar cells and a shading plate is mounted with the collector.

The collector (3 solar cells and 3 lenses in the laboratory test rig) drives an electronic load, controlled by the computer, which sets the array output voltage. Array voltage and current are fed to the computer via its A-to-D converter.

Inputs to the computer are:-

Digital input from the reed relay for counting to obtain position in hour angle
Digital input from the reed relay for counting to obtain position in declination
Analogue difference signal from the tracking head in hour angle
Analogue difference signal from the tracking head in declination
Analogue sum signal from the tracking head, to determine whether the sun is shining
Analogue array voltage
Analogue array current

Outputs from the computer are:-

Two digital outputs to each motor controller, "drive/stop" and "direction"
Analogue output to the electronic load, to set the array voltage

3.2. The Functions of the Controller

The program is broken down into a number of subroutines or (in C) functions. The principal functions are described below.

3.2.1. Functions for Open-Loop Tracking

Sun Position. This calculates the position of the sun, in hour angle and declination, from the date and time, the longitude of the collector, and astronomical data about the sun. The basic equations are taken from Maish [4].

Sun Position (Collector Coordinates). This calculates the values of the hour angle count and declination count corresponding to the sun's calculated hour angle and declination. It uses the information about the collector axes and position sensors calculated by *Find Conversion Parameters* (Sect. 3.2.3)

Set Count Zero. There are two functions, one that sets the hour angle count, by moving the collector to the East stop and setting the count to zero, and another that similarly sets the declination count to zero at the South stop. Whenever the collector is moved, the counts are appropriately incremented.

Move Collector. This moves the collector to the specified position, defined by the specified hour angle and declination counts.

3.2.2. Functions for Closed-Loop Tracking

Sun Seek. This moves the collector in a line scan pattern, from fully East to fully West and back, stepping 5° in declination between successive lines. When the sun is found, sensed by the tracking head's sum channel, the mode changes to *Sun Track.*

Sun Track. This moves the collector in hour angle and declination, until the error signals from the tracking head fall to zero.

3.2.3. Functions for Setting Up the Tracker

Store Track Data. While closed-loop tracking, under *Sun Track,* the function stores at intervals sets of calculated sun hour angle and declination, and the two collector position counts.

Find Conversion Parameters. This uses the stored track data to determine the directions of the collector axes, and the origins and scales of the collector position counts. These parameters are required by *Sun Position (Collector Coordinates).*

3.2.4. Functions for Optimising the Performance of the Tracker

Peak Power Tracker. This function adjusts the output voltage to the electronic load, so as to maximise the power output of the solar cell array.

Maximise Output. This moves the collector around its nominal pointing direction, to find the direction that gives the greatest output from the solar cell array. The necessary positional corrections are recorded, and added to the nominal directions in subsequent tracking. A similar system has been used with the EUCLIDES collectors at Madrid and Tenerife [5].

3.2.5. Supervisory Functions

In the development prototype, all supervisory functions are manual; in each mode, appropriate outputs and control menus are displayed on the computer's VDU, and the operator selects which mode to use. So it will be easy to test a range of control strategies. In the final microcontroller version of the system, the displays will be omitted, and a supervisory program will select the mode of operation and call the necessary functions.

4.   THE PRESENT STATE OF THE PROJECT

The required functions are being developed in C, on a PC with an I/O card having both analogue and digital interfaces.   So far, sun position calculations, the sun finding routine, and open-loop and closed-loop tracking are operating, using a small laboratory hardware test rig.   The peak power tracker circuit for the cell array is running, and the program for it is almost finished.   The higher-level routines for setting up the system, and for adjusting the collector position to maximise its output have still to be written.   At present, to facilitate testing, switching between functions is done manually.

The next stage is to adapt the program for a lap-top computer, for field testing, after which the supervisory programs can be written to replace the manual controls. This system will be used to control one of the collectors shown in Figures 1 and 2.   Then the whole system will be rewritten for a microcontroller, probably a PIC 16F876, which has enough digital and analogue inputs and outputs to make separate interface boards unnecessary.

5.   ACKNOWLEDGEMENT

The authors are grateful to the EPSRC for the grant which is supporting this work.

6.   REFERENCES

[1]. G.R. Whitfield, R.W. Bentley, C.K. Weatherby, A. Hunt, H-D. Mohring, F.H. Klotz, P. Keuber, J.C. Miñano and E. Alarte-Garvi. *The Development and Testing of Small Concentrating PV Systems*. Presented at the ISES Solar World Congress, Jersusalem, Israel, July 1999, Solar Energy (In the press).

[2]. G.R. Whitfield, R.W. Bentley, C.K. Weatherby, H-D. Mohring, F.H. Klotz, J.C. Miñano and E. Alarte-Garvi. *The Development of Optical Concentrators for Small PV Systems*. Final Report, EC JOULE contract JOR-CT96-0101, 30th November, 1998.

[3]. G.R. Whitfield, R.W. Bentley, H-D. Mohring, F.H. Klotz and J.C. Miñano. *Development and Testing of Optical Concentrators for Small PV Systems*. Proc. 2nd World Confr. on PV Solar Energy Conversion (15th EU PV Solar Energy Confr.), Vienna, July 1998, Vol. II, pp 2181-2184. EC, Luxembourg, 1998.

[4]. A.B. Maish, *The Solartrak Solar Array Tracking Controller*, Sandia Report, SAND90-1471.UC-275, Sandia Nat. Labs, Albuquerque, New Mexico, U.S.A., 1991.

[5]. J.C. Arboiro and G Sala, *A Constant Self-Learning Scheme for Tracking Systems*, 14th EU PV Solar Energy Confr., Barcelona, H.S. Stevens, Bedford, 1997, pp 332-335.

TRAXLE™ THE NEW LINE OF TRACKERS AND TRACKING CONCENTRATORS FOR TERRESTRIAL AND SPACE APPLICATIONS

V. Poulek
Poulek Solar Co. Ltd.
Kastanova 1481, CZ 250 01 Brandys n.L., Prague East District, Czech Republic
Tel/Fax: +420 202 804017, e-mail: poulek@telecom.cz, URL: www.traxle.cz

M. Libra
Czech University of Agriculture, Technical Faculty
Kamycka 129, CZ 165 21 Prague 6, Czech Republic
Tel: +420 2 24383284

ABSTRACT: A very simple solar tracker and tracking concentrator are described in the paper as well as the tracking strategy which enables high collectible energy surplus at medium tracking accuracy. A new low cost tracking soft concentrator combined with the solar tracker can double PV energy harvest in comparison with fixed panels and substantially reduce price of PV energy.

1. INTRODUCTION

The low cost solar tracker was described in details in the paper [1]. The apparatus enables backtracking within 5 minutes. It is important advantage in comparison with popular passive trackes where backtracking times as long as one hour are usual. The tracker works also at low temperatures down to -40°C with average accuracy ±5°. The low cost solar tracker is based on a new arrangement of solar cells. Solar cells both sense and provide energy for tracking. Costly and unreliable electronics has been completely eliminated.

2. THE TRACKER

Fig.1 shows the principle of the tracker. Two antiparallel sensing/driving solar cells are connected to reversible DC motor, the transmission is self-locking. As the sun moves from the east to the west, the angle of incidence β of solar radiation on sensing/driving cells increases until the power of the driving DC motor is high enough to move solar collectors. Then the collectors start to move and the angle of incidence β starts to decrease until the power of DC motor is lower than what is necessary to move solar collectors. Additional antiparallel solar cell enables backtracking of the tracker in the morning. In Fig.1b it is seen that the sun can move the collectors from any position.

Theoretical calculation of the energy surplus in the case of tracking collectors is as follows: We assume, the maximal radiation intensity $I = 1100 \text{W.m}^{-2}$ is falling on the area which is oriented perpendicularly to the radiation direction. We assume the day length $t = 12 \text{ h} = 43200 \text{ s}$ and we compare the tracking collector which is all the time optimaly oriented to the sun with the fixed collector which is oriented perpendicularly to the radiation

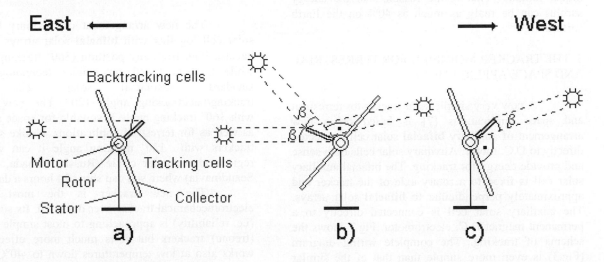

Figure1: The principle of the tracker.

direction only at noon. We mark the collector area S_o.

a) For fixed collector: The projection of this area on the area which is oriented perpendicularly to the radiation direction is equal $S = S_o \cdot \sin\varphi$ and the angle φ is changing in the interval $\varphi \ni \langle 0^o ; 180^o \rangle$ during the day. The angular velocity of the sun mooving cross the sky is equal $\omega = 2\pi / T = 7,27 \cdot 10^{-5} \text{ s}^{-1}$ and the differential of the falling energy is equal $dW = I\,S\,dt$. When we don't consider the atmosphere influence, we can calculate the energy, which is fallen on the collector area $S_o = 1 \text{ m}^2$ during one day:

$$W = \int_0^{43200} I\,S_o \sin\omega t\, dt = I\,S_o \left[\frac{-\cos\omega t}{\omega} \right]_0^{43200} =$$

$$= \frac{2\,I\,S_o}{\omega} = 3,03 \cdot 10^7 \text{ W.s} = 8,41 \text{ kW.h}.$$

b) For the tracking collector which is all the time optimaly oriented to the sun: When we don't consider the atmosphere influence, we can calculate the energy, which is fallen on the collector area $S_o = 1 \text{ m}^2$ during one day:

$$W = I\,S_o\,t = 4,75 \cdot 10^7 \text{ W.s} = 13,2 \text{ kW.h}.$$

We can see, the energy surplus is 57% when we don't consider the atmosphere influence. We would realy obtain this surplus for example on the Moon surface and in the space applications. On the Earth surface the sun is schining cross thick atmosphere layer after sunrise and before sunset. In the morning and in the evening the radiation intensity falling on the area which is oriented perpendicularly to the radiation direction is much lower than at noon is. On the other hand the day can be longer than 12 h at higher latitude. That is the reason, that the energy surplus can be realy as much as 40% on the Earth surface.

3. THE TRACKER MODIFIED FOR TERRESTRIAL AND SPACE APPLICATIONS

A new very simple solar tracker for terrestrial and space applications [7] is based on a new arrangement of auxiliary **bifacial** solar cell connected directly to D.C. motor. Auxiliary solar cells both sense and provide energy for tracking. The bifacial auxiliary solar cell is fixed to a rotary axle of the tracker and approximately perpendicular to **bifacial** solar arrays. The auxiliary solar cell is connected directly to a permanent magnet D.C. electromotor. Fig.2 shows the schema of tracking. The complete wiring diagram (Fig.3) is even more simple than that of the similar system with monofacial cells.

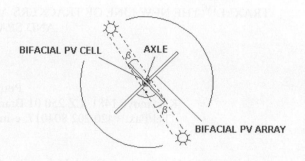

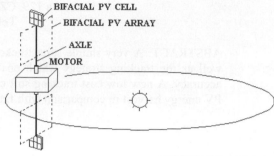

Figure 2: The schema of the solar tracker modified for 360° tracking.

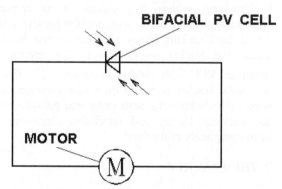

Figure 3: The complete wiring diagram of the solar tracker modified for 360° tracking.

The new arrangement of auxiliary **bifacial** solar cell together with **bifacial** solar arrays enables backtracking from any position (360° tracking angle) while trackers based on similar technology with standard monofacial solar cells have tracking/backtracking angle 120°. The new design with 360° tracking angle is suitable for space systems as well as for terrestrial applications. Unlike standard trackers with 120° tracking angle it can work in regions above polar circle (Russia, Canada, Alaska, Scandinavia) where sun can shine 24 hours a day.

The new tracker is the most simple electromechanical tracker ever designed. Its simplicity (i.e. reliability) is approaching to most simple passive (freone) trackers but it is much more effective. It works also at low temperatures down to -40°C where passive trackers doesnot work at all.

4. TRACKING CONCENTRATOR

A new tracking soft (C=1.6 -1.7) concentrator [8] can double PV energy harvest (in comparison with fixed panels). The new system combines simple low-cost tracker [1] with flat booster mirrors but unlike V-trough concentrator [2,3] by the new ridge concentrator the "outer" mirror has been eliminated (Fig.4). On single axis trackers, the mirror have to be extended beyond PV panels to ensure uniform illumination of panels at seasonaly variable elevation of the sun. On polar axis trackers with seasonaly adjustable slope of the axle the extended mirror is not needed.

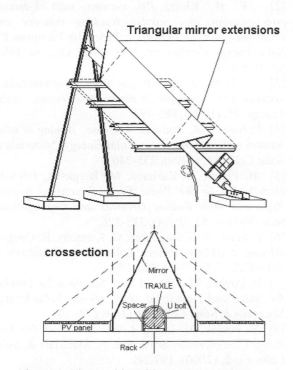

Figure 4: The tracking ridge concentrator.

It is advantageous that soft concentrators for photovoltaics does not need highly specular expensive mirrors. Weather resistant (at least 10 years) mirrors with high total reflectance are needed. The mirror can be made of rolled stainless steel sheet with special surface finish [4], of rolled aluminium alloy sheet (plated with pure aluminium) protected by a weather resistant polymer (PVF) film [4,5], of silver coated polymer (acrylic) film [6] or sheet, of aluminium coated polymer (acrylic) film [4] or sheet, of silver coated hardened glass.

The new tracking soft concentrator (Fig.6) is very compact and simple (reliable). Proven existing tracker hardware is used. Unlike V-trough concentrators no additional mirror supporting structures are needed and wind induced torque is strongly reduced. It can be used on polar or horizontal single axis trackers, two axis trackers as well as on $360°$ trackers for space and terrestrial applications [7].

Concentration ratio 1.6-1.7 reduces temperature of PV panels (higher efficiency) and avoids degradation of the encapsulant. The new design also improves (against V-trough) air flow around PV panels (improved cooling). Concentration ratio 2-2.4 of standard V-trough concentrators frequently caused browning of the EVA encapsulant while elevated temperature reduces efficiency of PV panels.

The new tracking ridge concentrator can double annual energy harvest (100% energy surplus in arid climates like e.g. northern Africa, Arabia, Arizona or western Australia, 70% in central Europe) in comparison with fixed panels and pumping capacity surplus can be as high as 150% [2].

One year comparison of energy production between fixed tilt PV panels and and PV panels mounted on polar axis tracking ridge concentrator was started in May 1999 (Prague region, $50°$ N). Very first results show that e.g. on clear (6.8kWh/m2) day in June 99 the energy surplus of 107% was observed.

5. WIND LOAD

The international norm P ENV 1991-2-4 „The fundamentals of designing and loading of the constructions, part 2-4 Loading of the constructions by wind" speaks about resistance of constructions to wind. The whole territory of Europe is never subject to speeds exceeding $v = 160\text{km.h}^{-1}$ near ground level. That is why our device was tested in a wind tunnel up to this particular wind speed.

Our stand is protected against wind by a self-locking transmission with a maximum torque of $M = 500 \div 3500$ N.m (it depends on the size), and is designed in a way that lets is withstand wind of more then up to $v = 160\text{km.h}^{-1}$. We performed a factual test of durability in a large wind tunnel with a diameter of 3 meters (Fig.5), which is located in the Aircraft Research and Testing Institute in Prague. This institute has the authorization to perform aerodynamic testing. The whole solar system was placed sucessively into four positions relative to the direction of wind. These directions were: perpendicular to the front, perpendicular to the back, perpendicular to the side, and sideways to the back with and angle of $45°$. In each positions we slowly increased the speed of the wind up to $v = 160\text{km.h}^{-1}$, and then let it affect the solar tracker for the time $t = 3\text{min}$.

The whole structure was very stable and had minimal vibrations during the tests that we described above. The air currents around the solar systems also behaved calmly. The system kept it's stability, even when we raised the wind speed to $v = 180\text{km.h}^{-1}$ for a short time. The tests did not damage the solar tracker in any way. We judge this from the fact of it's working properly after the test.

Table I: Economical calculation of the solar pumping system with concentrator.

System with fixed stand	Price (US$)	System with tracking stand	Price (US$)
10x 120W PV panel	8400	6x 120W PV panel	4200
1x fixed stand	780	1x tracking stand	1590
1x water pump	1770	1x water pump	1770
Total	10950	Total	7560

6. ECONOMICAL CALCULATION

Maybe, the most effective application of the solar tracking systems is the usage for water pumping. The pump is then much longer time over the liminal value in comparison with the solar system with fixed stand and that's the reason that the pumping capacity surplus can be as much as 80% when we use the tracking solar system without concentrator and as much as 150% when we use the tracking solar system with soft concentrator. The economical calculation is in Tab.1. There is one example of price comparison of fixed and tracking concentrator solar pumping systems with equal pumping capacity. We can see that the profit is approximately 3390 US$ in the case of tracking system.

7. CONCLUSION

The described newly designed and patented device for automatic orientation of the solar energy collectors with the soft concentrator of radiation can double the solar energy harvest and substantially reduce price of PV energy. It has range of advantages in comparison with existing devices. First of all its simplicity has positive influence to the reliability and price. In the near future we will use our device for the construction of the solar pumping systems which can be used above all in agriculture for irrigation, feeding and so on. We expect the pumping capacity surplus up to 150% which corresponds with our experiments.

There are more informations and pictures at http://www.traxle.cz.

8. REFERENCES

[1] V. Poulek, M. Libra, *New Solar Tracker*, Solar Energy Materials & Solar Cells, 51, (1998), 113-120.

[2] F. H. Klotz, *PV systems with V-trough concentration and passive tracking concept and economic potential in Europe*, Proc.13th European PV Solar Energy Conference, Nice 23-27 October 1995, pp.1060-1063.

[3] S. Nann, *Potentials for tracking photovoltaic systems and V-troughs in moderate climates*, Solar Energy, 45, (1991), 385-393.

[4] P. Nostell, A. Roos, B. Karlsson, *Ageing of solar booster reflector materials*, Solar Energy Materials & Solar Cells, 54, (1998), 235-246.

[5] B. Perers, B. Karlsson, M. Bergkvist, *Intensity Distribution in the Plane From Structured Booster Reflectors With Rolling Grooves and Corrugations*, Solar Energy, 53, (1994), 215-226.

[6] P. Schissel, G. Jorgensen, C. Kennedy, R. Goggin, *Silvered PMMA reflectors*, Solar Energy Materials & Solar Cells, 33, (1994), 183-197.

[7] V. Poulek, M. Libra, *A Very Simple Solar Tracker for Space and Terrestrial Applications*, Solar Energy Materials & Solar Cells, 60, (2000), 99-103

[8] V. Poulek, M. Libra, *A New Low Cost Tracking Ridge Concentrator*, Solar Energy Materials & Solar Cells, 61, 2, (2000), 199-202

Figure 5: The test in the wind tunnel.

Figure 6: The tracking concentrator.

TYPE TESTING OF INVERTERS FOR GRID CONNECTION OF PHOTOVOLTAICS

J.M. Thonycroft[1], R.J.Hacker[1], W.He[2], T.Markvart[2], R.J.Amold[3]

[1]Halcrow Gilbert,	[2]School of Engineering Sciences	[3]Siemens plc,
Burderop Park, Swindon,	Engineering Materials,	Sir William Siemens House
Wiltshire SN4 OQD United Kingdom	University of Southampton	Princess Street
Tel +44 (0)1793 814 756,	Southampton SO17 1BJ	Manchester
Fax +44 (0) 1793 815020	Tel: +44 (0)23 80593783	Tel. +44(0)161 4466351 email
Email hackerrj@halcrow.com	Fax +44 (0)23 80593016	arnoldr@plcman.siemens.co.uk
Thornycroftjm@halcrow.com	email: t.markvart@soton.ac.uk	
	ty.he@soton.ac.uk	

ABSTRACT: This paper presents the results of experimental research to develop a type testing procedure for inverters in the United Kingdom. The project is funded jointly by the Department of Trade and Industry, the PV industry and the electricity supply industry. The primary purpose of the project is to establish the performance requirements of PV inverters that will allow them to be accepted by the electric utility without further site commissions tests.

The results have important implications for the acceptability of PV as an energy source in the UK and for the likely costs of installing such a system, and will help to standardise the planning process, as well as the installation and approving grid-connected PV systems. This is essential if PV is to be widely seen on buildings within the UK.

Keywords: grid-connected-1, islanding-2, qualification and testing-3

1. INTRODUCTION

As the possibility of a proliferation of grid-connected PV systems in the UK over the next few years has emerged so too have concerns regarding safety, power quality and the impact on the network. There is an immediate need to develop standards that are acceptable to the electricity supply industry while not imposing excessive restrictions, and hence cost, on the manufacturers and installers of PV systems. A draft Engineering Recommendation (G77) has been developed providing guidelines for the grid-connection of small PV generators in the UK. While not legally binding this will be accepted as a mandatory standard within the UK when it is published. As part of the consultation between the electricity supply industry and PV industry a consensus was reached that companies would probably accept certification by manufacturers if a standard test procedure for inverters was developed. Engineering Recommendation G77 includes, as an appendix, an outline definition of the checks and tests required to establish whether an inverter can be classified as 'Approved'. This was used as a basis for the research along with the background of test developments in other countries and proposed international standards.

This paper describes the work carried out at the Solar Test and Reference Facility (STaR) at the University of Southampton, building on previous research at Southampton and elsewhere [1,3,4], with the aim to define an appropriate test procedure to be used in the type approval of single-phase photovoltaic inverters in the power range up to 5kVA. To this end, a circuit was developed to test the inverter's performance against the limits proposed for voltage band, frequency band, loss of mains, harmonic distortion, power factor and dc injection. A number of commercial inverters were tested for compliance to ensure that the tests that have been developed correspond to practically meaningful situations. The testing procedure and suitability of these limits were discussed with the PV industry and electricity supply industry.

2. BACKGROUND

The connection small PV generators to the distribution system at low voltage level poses two principal issues of concern to the distribution managers - the safety of personnel and protection of equipment. It is these issues that must be addressed in any type approval tests which are to accompany the G77 guidelines.

When the source of power from distribution system is disconnected from the network section to which the inverter is connected, the inverter must shut down automatically within a given time. If it does not, the resulting island operation can critically affect the safety of both electricity supply staff and the public as well as the operation and integrity of other equipment connected to the network. A satisfactory, safe and reliable method is therefore required for the detection of "island" operation.

Two key technical issues at the AC interface concerning power quality are power factor and harmonic distortion. In general, equipment operating at significantly less than unit power factor is unlikely to be acceptable for grid-connected applications. The VAR demand of both loads and inverters may become acceptable if the utility charges money to the owner of the co-generator for reactive power consumption. Since current regulations do not permit this for residential customers, the utility requires a power factor close to unity at the converter output.

Another AC interface issue of considerable importance is the possibility of DC injection by the inverter into the network. Consumer and utility fault clearing devices are incapable of safely interrupting DC fault currents. Furthermore, DC current is likely to saturate the distribution transformer and cause disruption of service to other consumers, both those on the same transformer and those on the same feeder since some primary fault clearing devices will then operate.

The development of the type test circuit and test procedures to fulfil the requirement of the G77 are described below.

3. TECHNICAL UNDERSTANDING OF ISSUES AND TEST DESIGN

The principal issue in the test design is to design a circuit which will simulate the operation of the inverter as part of the utility supply system. To deal adequately with the possibility of island operation it is important, in particular, to represent accurately the operation of inverter on the same line as another inverter, or when the load is a rotating machine. These two situations have been found especially significant in inducing island operation in some commercial inverters [1,2]. In particular, experimental results show that an inverter which will not island on its own may island when in parallel with another inverter which will island on its own.

These phenomena require analysis of parallel operation of two inverters, as shown in Fig. 1. Inverter 1 (INV1) and inverter2 (INV2) are in matched load condition, that is power generated by INV1 just consumed by load R1 and the same is true for INV2, and R2. Therefore, there will be no energy flow to/from the mains. Assuming both inverter operate together in the island mode after the switch is turned off, no energy is exchanged at point A but the two inverters continue to generate reference signals for each other.

In the type tests, it is more convenient to replace inverter INV2 by a resonant circuit with resonant frequency equal to the operation frequency of INVV1. The circuit theory then shows that there will be no energy exchange at point A. but inverter INV1 will 'feel' an identical voltage reference signal at point A as in the case of two inverters operating in parallel. Thus, the test can be carried out using a resonant circuit whose frequency is equal to inverter

operation frequency to simulate an islanded inverter (Fig. 2).

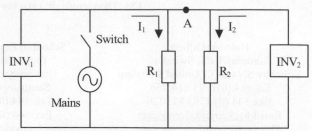

Fig. 1 Parallel operation of inverters in island mode

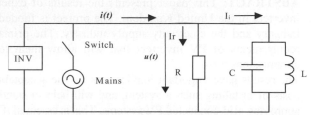

Fig. 2 Parallel operation with a resonant circuit

To analyse the experimental circuit in more detail needs some theory of the islanding phenomenon. To this end we should recall that, if an inverter is to manifest islanding, the active and reactive power produced by the inverter must be exactly balanced by the load. If the load is represented by resistance R in parallel with an inductor L and capacitor C, the active and reactive power consumed by the load is given by

$$P_{act} = \frac{U^2}{R} \quad \text{and} \quad Q_{react} = \frac{U^2}{\omega L}\left\{1-\left(\frac{\omega}{\omega_r}\right)^2\right\} \quad (1)$$

where ω_r is the resonant frequency of the circuit without damping. If θ denotes the phase angle between the voltage U and current I,

$$\tan\theta = \frac{\omega C - \omega^{-1}L^{-1}}{R^{-1}} = Q\frac{\omega_r}{\omega}\left(\frac{\omega^2}{\omega_r^2}-1\right) \quad (2)$$

the solution of (2) is:

$$\frac{\omega}{\omega_r} = \frac{\sqrt{1+(4Q^2-1)\cos^2\theta} + \sin\theta}{2Q\cos\theta} \quad (3)$$

where

$$\omega_r = \frac{1}{\sqrt{LC}} \quad \text{and} \quad Q = R\sqrt{\frac{C}{L}}$$

Equation (3) determines the relationship between the frequency and phase angle (and, therefore, also the power factor) for an inverter that injects current with a fixed power factor. If the mains is absent, ω and θ represent the islanding frequency and phase angle with respect to voltage of current injected by the inverter. This relationship can be conveniently represented graphically, as shown in Figure 3.

Returning now to the type test circuit, the phase angle (or the power factor) in Fig. 3 can be understood to indicate a perturbation of the inverter-load system under realistic operation. It is now seen from Fig. 3 that for small Q, a very small perturbation results in a large frequency shift which can be easily detected by the loss of mains protection system of the inverter.. To attain a reasonable level of protection, therefore, a sizeable value of Q must be used. It was found from experiments [1,2] that Q=0.5 represents a satisfactory limit for this purpose whilst ensuring a truly resonant behaviour.

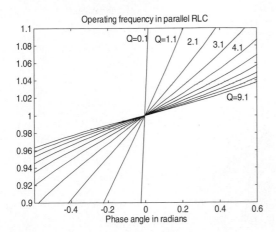

Fig. 3 Inverter frequency and phase angle for different quality factor of RLC load

4. EQUIPMENT AND TEST CIRCUIT

The set up for the tests follows directly from the analysis in Sec 3, and the circuit is shown in Figure 4. The PV array or PV simulator supplies the input power to the inverter, and P1 and P2 indicate monitoring points. L, C comprise the resonant circuit and R, L1 is matched load.

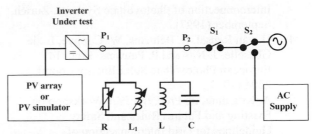

Figure 4 The test circuit used in the type tests

Since the frequency of the resonant circuit is 50Hz and quality factor Q is larger than 0.5, the values of R, C and L are

$$R = \frac{U^2}{P}$$

$$C = \frac{Q \times P}{2\pi f \times U^2} \geq 3 \times 10^{-8} \times P$$

$$L = \frac{U^2}{2\pi f \times Q \times P} \leq \frac{336.8}{P}$$

where P denotes the power output of the inverter.

5. CONCLUSION

A test circuit and procedure for the type approval of photovoltaic inverters to accompany the UK G77 guidelines have been developed. Examples of test results for eleven commercial inverters are shown in the Appendix. Despite difference in detail, the test circuit resembles the circuit proposed as part of the US [1] (the stricter conditions implied by the higher Q factor of the circuit should be noted) or the circuits used in the research work [4,5]. The results have important implications for the acceptability of grid-connected PV as an energy source in the UK and for the likely costs of installing such a system. When they are endorsed they will help to standardise the whole process of planning, installing and approving grid-connected PV systems. This is essential if PV is to be widely seen on buildings within the UK.

6. REFERENCES

1. W.He, T. Markvart and R.J. Arnold, Proc. 2nd World Conf. on Photovoltaic Solar Energy Conversion, 2772 (1998)

2. W.He, T. Markvart and R.J. Arnold, in Halcrow Gilbert (eds), Co-ordinated Experimental Research into PV Power Interaction with the Supply Netwrok – Phase I. ETSU S/P2/00233 (1999).

3. H. Haeberlin, J. Graf and Ch. Beitler, in: Proc. Int. IEA Workshop on Existing and Futures Rules and Safety Guidelines for Grid

Interconnection of Photovoltaic Systems, Zurich, September (1997).

4. H. van Reusel, R. Belmans, W. Coppye, L. de Gheselle. J. Nijs and P. Pauwers, Proc. 14th European Photovoltaic Solar Energy Conmf, 2204 (1997).

5. A. Kitamura, in: Proc. Int. IEA Workshop on Existing and Futures Rules and Safety Guidelines for Grid Interconnection of Photovoltaic Systems, Zurich, September (1997).

APPENDIX: TEST RESULT

TABLE1 INVERTER TYPE TEST ROUND 1 (at STaR 02-March-1999)

Inverter Type	Power range	Voltage Band (V)		Frequency Band (Hz)		THDi (%)	Power Factor	DC Injection (mA)	Loss of mains at load match (Seconds)	Reconnection Time (Sec)
		Low	High	Low	High					
INV1A	50%	200	256	49.7	50.3	23.9	0.99	±9	0.18	5
	100%					22.5	1.0	±11	Islanded	5
INV1B	50%	194	251	47.9	52.7	22.5	0.98	±6	0.02	2
	100%					N/A	0.99	±8	0.02	2
INV1C	50%	198	252	N/A	N/A	N/A	1.0	N/A	0.12	60
	100%					N/A	1.0	N/A	0.14	60
INV1D	50%	209	261	49	51.1	1.5	0.98	±1	10	8
	100%					1.1	0.99	±2	10	8
INV1E	50%	190	235	49.5	50.6	2.6	0.99	±1	0.03	5
	100%					2.4	1.0	±1	0.04	5
INV1F	50%	200	235	47.9	51.9	35	0.97	±1	0.08	3
	100%					42	0.99	±3	0.06	3

Notes:
1. THDi was measured under AC power supply with 0.5% THD of voltage.
2. Tests were carried out with DC power supply to simulate PV array.
3. Loss of mains tests were carried out with actual network.
4. Shading indicates the results outside G77.
5. Frequency band and THDi of INV1C is not available because it will not operate under AC power supply.
6. THDi of INV1B under 100% output power is not available as the test could endanger AC power supply.
7. THDi of INV1F is extremely high because the output AC current is distorted by input from DC simulator.
8. "±" sign used in DC injection is because sign and value fluctuate. Maximum value is shown.

TABLE2 INVERTER TYPE TEST ROUND 2 (at STaR 11-October-1999)

Inverter Type	Power range	Voltage Band (V)		Frequency Band (Hz)		THDi (%)	Power Factor	DC Injection (mA)	Loss of mains at load match (Seconds)	Reconnection Time (Sec)
		Low	High	Low	High					
INV2A	50%	207	253	49.5	50.5	2.3	0.98	−1	0.4	180
	100%					1	1	−3	0.5	180
INV2B	50%	213	259.5	49.5	50.5	7.5	0.99	0.4	0.38	180
	100%					2.2	1	-0.8	0.5	180
INV2C	50%	207	253	49.5	50.5	2.5	1	1	0.28	180
	100%					0.7	1	2.3	0.5	180
INV2D	50%	208	250	49.8	50.2	0.8	0.99	-0.8	0.07	210
	100%					1	0.99	-1	0.07	210
INV2E	50%	196	255	49.8	50.2	12.0	0.99	±6	0.3	30
	100%					2.4	0.99	±13	>5	30

Notes:
1. THDi was measured under AC power supply with 0.5% THD of voltage.
2. Tests were carried out with DC power supply to simulate PV array.
3. Loss of mains tests were carried out with actual network.
4. Shading indicates the results outside G77.
5. The voltage band of INV2B is outside G77 after setting Uac=196 to 253 V via communication channel.
6. "±" sign used in DC injection is because sign and value fluctuate. Maximum value is shown.

DESIGN, REALIZATION, TESTS, COMPARATIVE ANALYSIS OF LOW ELECTRIC CONSUMPTION PHOTOVOLTAIC COOLING SYSTEMS

COSTIC - E.MICHEL - F.BONNEFOI - Rue A Lavoisier - Z.I Saint-Christophe - 04000 DIGNE - FRANCE
Phone : 33.4.92.31.19.30 - Fax : 33.4.92.32.45.71 - Email : costic.digne@costic.com
CIEMAT - F.ROSILLO - 22 avenida complutense - 28040 MADRID - SPAIN
Phone : 34.1.346.6675 - Fax : 34.1.346.6037 - Email : rosillo@wanadoo.fr
ALPES FROID - P.RICHARD - avenue de la gare - 04700 ORAISON - FRANCE
Phone : 33.4.92.78.62.09 - Fax : 33.4.92.78.68.74
APEX - A.MINE - E. LAGET - 4 rue de l'industrie - 34880 LAVERUNE - FRANCE
Phone : 33.4.67.07.02.02 - Fax : 33.4.67.69.17.34 - Email : apex@mlrt.fr
UNIVERSITY OF ATHENS - A.ARGIRIOU - P.O BOX 20048 - GR-118 10 ATHENS - GREECE
Phone : 30.1..342.1270 - Fax : 30.1.346.4412 - Email : thanos@env.meteo.noa.gr

ABSTRACT :

The aim of this project funded by the European Commission is the following one : determine low electric consumption cooling systems well adapted for photovoltaic and for application in developing countries.
Two different systems were intended : evaporative air-cooler, buried pipe system.
The efficiencies in sensible heat, quite high for these systems (6 to 10) in application with 220VAC, have been optimised for photovoltaic. The tests showed that these coefficients could exceed values of 20.
The innovative aspect of the studied systems are :
- low consumption motor for fans
- adapted fan for different systems
- range of device defined in order to provide cooling power necessary for premises from an office to a little shop
- cooling "along with the sun"

Keywords : Ventilation - 1 : Stand-alone PV Systems - 2 : Developing countries - 3

1. PRESENTATION OF THE COOLING SYSTEMS

1.1.Evaporative air cooler
- **Principle :**
The hot air is humidified in contact with water so that to get cool by losing as much sensible heat as latent heat of vaporisation it gives to water.
The contact air-water may be made by a humid section, or by atomisers.
The air evolution can be represented on a psychrometric diagram : it is characterised by a constant wet bulb temperature. Thus, the air goes out of the air cooler with a dry bulb temperature lower and a relative humidity higher.
The efficiency of those air coolers is :

$$\varepsilon = \frac{\theta e - \theta s}{\theta e - \theta h}$$

θe : dry bulb temperature of the air at the inlet
θs : dry bulb temperature of the air at the outlet
θh : wet bulb temperature of the air

The sensible condensing unit capacity in [W] is :

P = 0.34 Q (θe - θs)
with Q : volumetric air flow rate in [m³.h⁻¹]

The coefficient of performance is : COP = Pf / Pel
with Pel : electric power absorbed in [W]

We have to distinguish two types of evaporative systems : direct and indirect.
- **Direct evaporative air-cooler**
In those devices, the air to blow in premises is directly in contact with water.

The exchange air-water can be made by several ways :

Plane Pad
The evaporative air cooler contains humidification plates (in wood or PVC) on which water, brought by a pump, flows.
A fan (in general, a centrifugal one) pumps in the air through plates, and blows it in the premises. The efficiency can reach 80%.

Rotary Pad
The pad has a drum shape : it is humidified and washed, turning in the water tank. The air goes through the drum thickness and is cooled. Thus, the water pump is removed.

Pulverisation
The air goes through water spray towards an evaporative and filtering plate and after towards another plate which traps the humidity obtained. The efficiency can reach 80%.
This type of direct evaporative air cooler is well-adapted for dry and hot climates. However, we must be careful with stagnant water, because of risks of legionellosa.

• Indirect evaporative air-cooler
Principle
The air is cooled by direct evaporation. Then, it cools new air through an exchanger air-air.
Thus, we obtain cool air, without supplies of humidity. Moreover, the risks of legionellosa are limited, since the air in contact with water is thrown back outside.

In this case, the efficiency is : $\varepsilon = \dfrac{t_{1e} - t_{1s}}{t_{1e} - \theta_{2e}}$

With :
t_{1e} : dry bulb temperature of the primary air at the inlet
t_{1s} : dry bulb temperature of the primary air at the outlet
θ_{2e} : wet bulb temperature of the secondary air at the inlet
We get efficiencies between 40 and 80%.

The use of that kind of air cooler can be broadened to countries less hot and less dry, for cooling only, or in complement with a classic air conditioning system, to reduce the load of the refrigerating compressor and thus, to reduce the costs of exploitation.

1.2. Cooling with buried pipes
• **Principle :**
The temperature of the ground can be lower that the temperature of the air.
This is the case for the Mediterranean climate. The same principle is used in the USA and Canada, to warm up the air in winter.
The exterior air is blown in a buried pipe and is cooled by forced convection and then it is blown in the premises.
The ground is an interesting source of freshness because its temperature decreases while the depth increases.

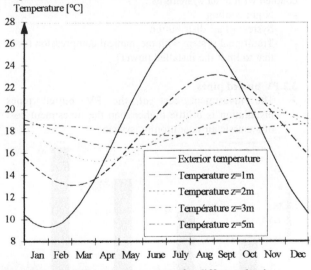

Figure 1 - Evolution of temperature for different depths into the ground (Athens - GREECE)

Furthermore, the temperature of the ground less and less depends on the daily variations of the exterior and the peaks of temperature don't take place at the same moment.
However, the performances of such a system can decrease during an extended use, on account of the heat accumulation in the ground.
So as to avoid this problem, we can improve the heat transfer by laying sand around the pipes.

We also can let the installation running all night long, if the exterior temperature is low enough to disperse the heat stored during the day.
Studies brought the following conclusions :
- The more the pipe is long, the more the temperature at the outlet decreases (beyond a certain value, the length doesn't bring improvements anymore).
- The more the radius of the pipe is long, the more the temperature at the outlet increases (because the coefficient of the convective transfer is lower).
- The more the air speed is big, the more the temperature at the outlet increases.
- The more the depth is big, the more the temperature at the outlet decreases (beyond 6.30m, the decrease in temperature isn't significant).
-
Naturally, those parameters depend on the nature of the ground, particularly on its humidity and on its diffusivity.
On the other hand, the choice of the pipe diameter must take into account the losses of head it generates and thus its influence upon the fan sizing.

2. TESTS IN LABORATORY AND IN REAL CONDITIONS

All partners have contributed to develop two different direct evaporative air cooler prototypes with rotary pads and vertical pads and one buried pipes prototype supplied with photovoltaic (PV) energy. These prototypes have been studied to run along the sun thanks to an impedance adapter. The photovoltaic system sized by the partner APEX in collaboration with COSTIC, based on the design information provided by the University of Athens, coupled with the fans specifically designed for the needs of the current project by the partner ALPES FROID, showed a reliable and robust behaviour throughout the experimental period. No failures were observed and the system operated without any interruptions.
In laboratory and in real conditions, tests results were conclusive. They showed that the prototypes had reached the wanted objectives.
Performances are the following ones :

2.1. Test results of the evaporative air cooler

➤ The evaporative air coolers designed by COSTIC and ALPES FROID can cool the air until 80% of the gap between the dry bulb temperature and the wet bulb temperature.

➤ The average performances (COP) for a sunny day can reach 40. The average COP of the system is about 30 for a working during summer in Madrid.

Photos 1 and 2 - PV evaporative air cooler prototype

2.2.Test results of the buried pipes

➢ The average COP of the system in real conditions is about 12 (majority between 5 to 25) for a length of tube equal to 20m. For the optimum length 30m, we can improve the COP to 18.

Photo 3 - PV buried pipes prototype in Athens

➢ The dampers designed in order to maintain the air flow across the exchanger close to 3 m.s⁻¹, operated as expected since the major part of the velocity values recorded were between 2.5 and 3.5 m.s⁻¹.

➢ The obtained temperature difference across the exchanger is satisfactory.

➢ The ground, due to its important thermal inertia, can provide cooling throughout the whole summer. This is because the ground temperature takes its highest value at the end of the cooling season, when the cooling requirements become lower.

➢ The performance of the system is not affected for distances between exchanger tubes greater than 0,5 - 1 m. Closer distances were not considered, since it is not likely to encounter them in practice.

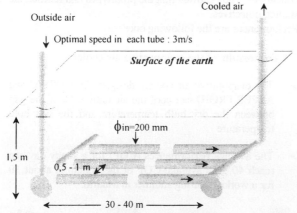

Figure 2 - Optimal data for a buried pipes installation

3.   ECONOMICAL AND TECHNICAL ASPECTS

Several simulations on TRNSYS allow to define the performance of the prototypes and its influence on the thermal behaviour of different typical buildings (office and dwelling). Analysis of simulation and experimental results allow to draw some conclusions about the performances, economical aspects.

3.1.PV evaporative air cooler

➢ For building with medium heat loads located in hot and dry climate, the PV evaporative air cooler can reduce the indoor temperature by 3 to 5 K in comparison with the same building without cooling.

➢ The air temperature of the indoor space should be around 2 K higher than the discharge air temperature and its relative humidity should be below 70%.

➢ The resulting temperature of the indoor space is 4 to 7 K below the outdoor dry bulb temperature.

➢ Concerning the costs, the PV evaporative air cooler is more competitive in term of cost in comparison with classic systems. The cost of functioning is very low in comparison with a conventional system. But we have to add the water consumption, 50 l/day (case for a dwelling of 100 – 150m²). This consumption can be penalising in certain regions where water is rare.

PV evaporative air cooler offers an interesting alternative for cooling. It can provide cooling, just consuming the necessary electricity for the operation of air fans and for a small water pump. This is much less compared to the electricity required for the operation of conventional air-conditioners.
In case of building with high heat loads, this system can be coupled with other systems as :
- Night cooling
- System of air distribution
- Traditional system with mechanical compression (so that to limit the installed power).

3.2.PV buried pipes

➢ The investment cost of the PV buried pipes installation is a little higher than the investment cost of a classic air conditioning system.

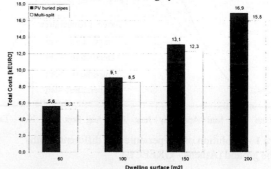

Figure 3 - Comparison between the total investment costs (Tax free) of PV buried pipes and classic air conditioners (multi-split) for different dwelling surfaces.

➢ Energetic costs : very low with alternative current, free with photovoltaic supply.

➢ Maintenance costs : very low. There are no major operating and maintenance costs associated with buried pipes except for an occasional cleaning of the system. The life span of the system is about equal to the buildings ones themselves.

➢ In case of very hot climate, the main advantage of this system is that it reduces peak demand for cooling and heating. This not only reduces energy costs, but also overall equipment and installation costs, since smaller dimensions may be used.

➢ Economical : low electrical consumption with photovoltaic supply or AC supply.

➢ Environmental friendly : it doesn't use freon fluids, nor compressor.

➢ Performance : coefficients of performance can reach 20.

➢ Maintenance is limited.

➢ Within the context of a photovoltaic solution, the system run along the sun.

PV buried pipes system is attractive for cooling and preheating ventilation air for many buildings in mild climatic regions. The resulting temperature of the indoor space should be about 3 to 5 K below resulting temperature of the indoor space without cooling.

The technology is already in successful use in several commercial, school and residential buildings. Maybe, we can hope to see the development for smallest buildings : dwellings, little shops, little offices...

4. CONCLUSIONS

These products can be used as an air conditioning systems in developing countries and Mediterranean countries. With these systems, it could be possible to cool premises such as little office buildings or dwellings. It is not possible to ensure the conditions of comfort all along the hot season as these systems only allow to reduce the temperature inside the premises of 3 to 4 °C compared to the situation without any conventional air conditioning systems. For instance, in the south of France it is possible to reach these conditions during about 80 to 90 % of summer.

At the moment these premises can be treated by an air conditioning split system with a heat pump as cold generator. A financial study, undertaken during the study, showed that costs of conventional systems compared to the costs of the new devices developed during the programme are rather equal. But these new products present the benefits to use a renewable energy which means that the working costs are low.

These systems present a disadvantage when the heat loads are important : it is not possible to ensure conditions of comfort. So there is an important work to do so that explain, to potential customers, the limit of the systems in order to avoid any bad references. Thus, it would be interesting to obtain , through a demonstrating programme,

some examples of successful installations to convince potential users.

RÉFÉRENCES

G.R.E.T : Bioclimatism in tropical zone, June 1984.

E.MICHEL - F.ROUESNE : Rapport d'étude d'un climatiseur évaporatif adapté au climat sahelien, 1989.

ASHRAE Handbook, HVAC systems and equipments, Evaporative air cooling, 1992.

P.BLONDEAU, M.SPERANDO, L.SANDU : Potentialités de la ventilation nocturne pour le rafraîchissement des bâtiments du sud de l'Europe. Conférence européenne sur la performance énergétique et la qualité des ambiances dans le bâtiment, November 94.

M.SANTANOURIS , D.ASIMAKOPOULOS : Natural Cooling Techniques, May 1995.

A.MARTIN, J.FELTCHER : Night cooling strategies, cooling options, Building services journal, August 96.

NEW BATTERY CHARGE CONTROL FOR PV SYSTEMS

COSTIC - E. MICHEL, V. BAYETTI - Rue Lavoisier - Z.I. de Saint Christophe - 04000 DIGNE- FRANCE
Phone: + 33.4.92.31.19.30 - Fax: + 33.4.92.32.45.71 - Email: costic.digne@costic.com
TENDANCIEL - B.MONVERT, P. LAGARDE - 15 rue Carnot - 91170 VIRY-CHATILLON - FRANCE
Phone: + 33.1.69.46.22.00 - Fax : + 33.1.69.46.00.44 - Email : tendanciel@worldnet.fr
HYPERION - S. Mc CARTHY - Main Street - Watergrasshill - CORK - IRELAND
Phone : + 35.321.889.461 - Fax : +35.321.889.465 - Email : hyperion@indigo.ie
CIEMAT - N. VELA, M. ALONSO - Avenida Complutense, 22 - 28040 MADRID - SPAIN
Phone : + 34.91.346.67.45 - Fax : +34.91.346.60.37 - Email : vela@ciemat.es
OLDHAM - P. LENAIN, M. MARCHAND - ZI Est - Rue Alexandre Flemming - BP 962 - 62033 ARRAS CEDEX
Phone : + 33.3.21.60.24.38 - Fax : + 33.3.21.60.25.74 - Email : jpatrick.smaha@oldham-france.fr

ABSTRACT: The common bad charge of PV batteries is, for us, due to the use of battery voltage as the main charging regulation parameter. We propose to use an intrinsic parameter : the gassing of lead-acid batteries when charged.
- Gassing appears when the battery is charged, whatever the "environment" parameters of the battery.
- Proposed gassing detection is simple and reliable.
The main results obtained in this programme are the following :
- Development of a mathematical model for the overcharge process
- Study, design of gassing detection sensors
- Study, design controllers
Algorithm of the control has been developed and control card s have been designed.

KEYWORDS : Battery storage and control – 1: PV system – 2: R&D and Demonstration Programmes – 3

INTRODUCTION

To develop a new PV regulation principle, regarding the charge of lead acid storage batteries : Why ?

The common bad charge of PV batteries is, for us, due to :

- The use of battery voltage as the main charging regulation parameter,
- This parameter is not an intrinsic one for the level of charge : it depends of a large quantity of other parameters, charging current, age of battery, temperature, electrolyte gravity, gravity stratification, state of charge before charging, "memory" of battery...
- This parameter is not, within the PV systems general working conditions, a good and reliable parameter to decide if a battery is well charged or not...

We propose to use an intrinsic parameter : the gassing of lead-acid batteries when charged :

- Gassing appears when the battery is charged, whatever the "environment" parameters of the battery.
- Proposed gassing detection (already tested on a prototype in a laboratory) is simple and reliable : storage of gas in a small receptacle installed in the electrolyte with electrodes inside : when gas has filled the receptacle, the electrodes are in gas and not in electrolyte and the electric resistance between these electrodes become infinite instead of zero, giving an information to the "outside" word of the battery (the receptacle has a small calibrated hole to evacuate slowly the gas).

- The use of this binary information can be taken into account in a "PV regulator" which decides when the battery is charged, for example if "we can count x electrodes-drying for each battery element of the set within y minutes".

The use of this binary information can also be used to check the good health of battery elements, for example if "the time to count x electrodes-drying of one element is y times greater than the average time for the set of the battery, the involve element has certainly some problems !".

1. : DEVELOPMENT OF MATHEMATICAL MODEL FOR THE OVERCHARGE PROCESS

The overall charge process has been divided into two parts, the properly called charge and the overcharge. The criteria to separate them are :

- *charge process :*
  from the voltage corresponding to charge starting (t= 0) (0 % SOC) to the voltage corresponding to initial gassing (Vg) (intermediate SOC)

- *overcharge process :*
  from the voltage corresponding to initial gassing (Vg) to the constant final charge voltage (Vfc) (100% SOC).

As a good approach we have calculated the initial gassing voltage (Vg) by means of the graphical method described in the following picture.

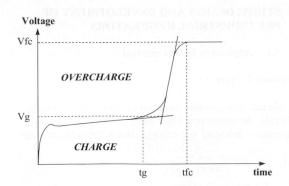

Figure 1 Graphical determination of Vfc, Vg, tfc and tg

In previous works we have developed a theoretical model for the discharge and charge processes based on external parameters easily obtained from laboratory measurements: current rate, voltage and temperature. This model includes the influence of key parameters like operation temperature and state of charge (SOC) evolution. To calculate the parameters included in the model equations we have developed in LabView (National Instruments) for Windows a software based on the Marquardt algorithm. This software allows single or multiple curve fitting, giving as result a set of optimised parameters suitable for the set of curves considered. The tests consisted of charges and discharges at different current rates (from C5 to C100) and temperatures (from 5 to 45°C) on a set of stationary lead-acid batteries commonly used in stand-alone PV installations (12V, 100-350 Ah). The test sample includes different types of batteries available in the European market: vented, sealed, cells and monoblock.

The mathematical equation proposed for the charge (+) and discharge (-) is:

$$V = \left[Vo + k \frac{Q}{\frac{Ct}{1+aI^b}(1+\alpha_c\Delta T + \beta_c\Delta T^2)} \right] \pm I \left(\frac{P_1}{1+I^{P_2}} + \frac{P_3}{\left(1-\frac{Q}{C_T}\right)^{P_4}} + P_5 \right) (1-\alpha_r\Delta T)$$

A set of optimised parameters for the whole set of experimental charge and discharge curves has been obtained for each battery. In general a good agreement between experimental and modelled curves is observed

This charge equation is only valid from the voltage corresponding to charge starting (t= 0) to the voltage corresponding to initial gassing (Vg).

We have considered the overcharge process from the voltage corresponding to initial gassing (Vg) to the constant final charge voltage (Vfc). The proposed model approximate the overcharge curve to two lines:

$$V = Vfc \quad for \quad V > Vfc$$
$$V = Vg + m(t - tg) \quad for \quad V < Vfc \text{ and } V > Vg$$

Where the gassing voltage, Vg, the slope, m, the gassing time, tg, can be written as:

$$tg = (A_{tg} + B_{tg} * I^{c_{tg}})(1 + \alpha_{tg}\Delta T)$$

$$Vg = (Ag + Bg\ln(1+I))(1 + \alpha_g\Delta T)$$

$$m = (am + Bm * I^{cm}))(1 + \alpha_m\Delta T)$$

$$Vfc = (Afc + Bfc\ln(1+I))(1 + \alpha_{fc}\Delta T)$$

A set of empirical parameters have been derived for different battery models and in general good agreement between the models and the experimental data have been found.

2. TESTS OF THE PROTOTYPES OF GASSING DETECTION SYSTEMS

2.1. Tests performed by OLDHAM

Measure of the voltage and the gassing evolution on vented lead-acid batteries during recharges at different constant currents to establish the relation between the voltage and the gassing evolution in charge according to the state of charge.

The voltage and the gas flow evolution are measured during recharges at different constant currents after discharges at 50% of the C10 capacity

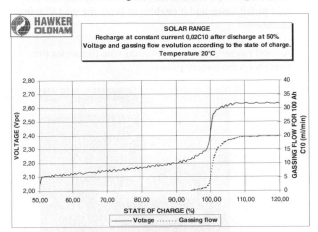

Figure 2 Voltage and gassing flow function of state of charge at 0.02C10

2.2. Tests performed by COSTIC

A procedure of testing has been analysed and testing system developed.

Photo 1: general view of test facilities

Tests objectives

The envisaged tests are the following :
- verification and balance of acquisition board
- charge at different temperature and charge current

- discharge at different temperature and discharge current

All along charge and discharge tests battery state of charge is controlled by the measurement of electrolyte density. Same tests will be performed several times to insure the good reproduction of results.

Tests results will allow :
- to draw characteristics as voltage / state of charge, voltage / charge time, voltage / discharge time for constant currents and temperatures
- to calculate « gas flow » detected by sensors and to correlate it with gas flow parameters necessary to controller development and to regulation algorithm.

First a problem appear : wetting and hole calibration. This problem was resolved by removing the bottom of the low electrode from the sensor body. It helps the last drop of acid to fall down and it enables to break the acid film in the chamber.

A good running has been obtained with the sensor put directly on the battery plates.

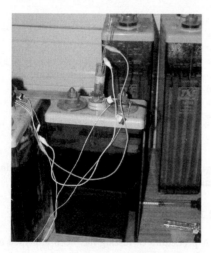

Photo 2: Sensor in the battery

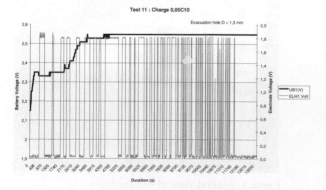

Figure 3 : test 11 for a charge of 0.05C10 with glass sensor with evacuation hole Ø= 1.3 mm

3.   STUDY, DESIGN AND DEVELOPMENT OF PRE-INDUSTRIAL REGULATORS

3.1.   Algorithm for the control

Functioning cycle :

A cycle has an adjustable duration of ten minutes. During one cycle, the micro-processor make 300 acquisitions of temperature, intensity and voltage, and it stores the average

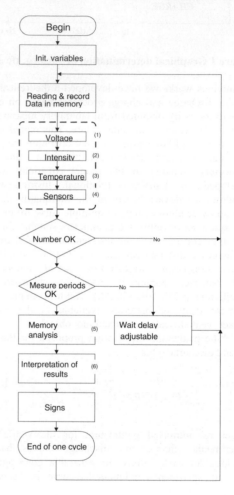

Voltage function :

This part of program must survey the voltage level of four accumulators element on the six of a battery. If more than the half of average measurement is in alarm state, the program control the alarms treatment card.

Intensity function :

This part of program must survey the intensity level incoming the battery. If the average measurement is in alarm state, the program control the alarms treatment card.

Temperature function :

This part of program must survey the temperature level of the battery. If the average measurement is in alarm state, the program control the alarms treatment card.

Sensors function :

The main idea of this research program is that the regulation can be done by detection of the gas flow at the end of the charge of the battery. This flow is detected by flooding/unflooding on the electrodes of the sensors. So we could have used the flooding/unflooding frequency as the basic information for regulation. But the sensors, even if they are more or less calibrated, have a very huge difference of frequency for a same flow. So the frequency, for the moment, is not the best information to use to detect the end of charge. We decided to use the fact that there is a big variation in the flow at the end of the charge (see Oldham tests), and to measure the variation of the frequency of flooding/unflooding. With this method, the detection is not affected by calibration problems. The other advantage is that this method can work for any battery without a new gauging of the sensors.

This part of program do the acquisition of flooding/unflooding frequency.

Analysis and interpretation of results :

This subroutine compute the variation of the frequency of flooding/unflooding and determine the state of battery charge.

3.2. Hardware for the controller

Three cards compose of controller:
- The acquisition card
- The central unit card
- The visualisation card

Photo 3: The acquisition card

The acquisition card, developed, by Ainelec, convert the analogic signals in numeric. It is the interface between the sensors and central unit.

Photo 4: the central unit card

The central process unit includes the software, it runs with a micro controller 68HC11A0.

The visualisation card cuts or connects the PV generator, the battery or the use, according to the battery state.

Photo 5: Prototype of visualisation card

The visualisation card must signalise by some diodes the different states of the controller :

Alarms
- Maximum temperature level
- Maximum intensity level
- Low electrolyte level
- Overcharge security
- Deep discharge

Process
- Charge process
- Discharge process

Battery
- Charge battery
- Discharge battery

4. CONCLUSIONS

The mains results are the following :

- Development of a mathematical model for the overcharge process

- Study, design of gassing detection sensors

- Study, design controllers

In conclusion

We have verified the principle of using the gassing of lead-acid batteries when charged and elaborated prototypes of detection sensors.

We have established algorithm for controller and designed the control cards.

It will be necessary to go on laboratory tests of controllers to verify the good running of algorithm.

ADVANCED OPERATION CONTROL CONCEPT FOR STAND-ALONE PV HYBRID SYSTEMS CONSIDERING BATTERY AGEING

Mohamed Ibrahim[*] Martin Rothert[**], Philipp Strauß[**], Peter Zacharias[**]
[*] Dept. of Electrical Energy Supply Systems (EVS), University of Kassel (IEE-GhK)
Wilhelmshöher Allee 73, D-34121 Kassel
e-mail: mibrahim@iset.uni-kassel.de
[**] Hybrid System Group, Institute for Solar Energy Supply Technology (ISET),
Königstor 59, D-34119 Kassel, Germany

ABSTRACT: This work focuses on the operation control of stand-alone hybrid systems and presents a recent development of system operation. The basic objectives of the developed concept are to achieve consistent technical and economical operating conditions for the system components, especially for the battery and the back-up unit. This concept is flexible enough to account for the actual status of the components (e.g. -battery state of charge (SOC), -stratification level in the battery, -renewable energy availability, -diesel loading ratio, etc.) and decide hour-to-hour the most efficient energy dispatch strategy. To consider the continuous changes of the components status, (e.g. battery ageing) the dispatch decision module is designed on the bases of event driven mechanisms, in which the system transfers from one state (mode) to another prescribed state based on the actual components status. Also, the constraints imposed on the components operation have been taken into account. Simulation results illustrate that longer lifetime of the battery is rather achieved via the developed operation control concept than via traditionally used ones. Continuous checking of the battery state of health and suitable sizing of the back-up unit with efficient use of its characteristics keep the battery from quick ageing and reduce the total system replacement, operation and maintenance costs.
Keywords: Stand-alone PV Systems – 1: Battery Storage and Control – 2: Lifetime - 3

1 INTRODUCTION

Due to the modular hybrid system concept and standardisation of the PV system components a drastic reduction of the system costs is achieved [1]. However, there is still a strong potential to improve the operation of the system in a way which guarantees a high level of supply reliability and reduces the kWh rates.

The life-cycle costs of an energy system are based on the investment and the operation and maintenance costs, including replacement cost. The investment cost mainly depends on the system configuration and the sizes of the components. In contrast, the operation and maintenance cost is a function of the components performance characteristics and the implemented operation strategy. Figure 1 introduces a simple representation of the life-cycle cost elements for stand-alone PV hybrid system.

Lead-acid batteries are the dominant energy storage in the PV stand-alone systems. Nevertheless, uneconomical lifetime of the battery is one typical feature of such systems. During the last decade, the attention of system developers has concentrated on analysing the reasons of this premature battery failure. Sensitivity of the battery and severe operating conditions of the PV systems stimulate the battery ageing mechanisms and result in rapid degradation of its capacity. Therefore, the battery always suffers from short life-cycle under stand-alone PV systems. That makes the costs of the battery in decentralized PV systems constitute more than 40 % of the total annual costs of the system [2]. However, there is some technological development on the component level in form of gel-, flies- and special solar-batteries, which have relatively longer life-cycle under PV operating conditions. These technologies are still expensive and rarely available in the current markets of developing countries.

On the other hand, operation control plays an important role in the performance and economics of PV hybrid systems. Although nowadays components used in PV systems have their own controller to ensure safe operation, the optimal operation of the entire system can only be achieved via the supervisory controller which receives its instructions from the operation control strategy [3].

That directs the attention towards utilizing the available system characteristics (components and operation strategy) to alleviate the stresses which the battery suffers from under decentralized PV systems operation conditions. Consequently, improvement of the battery lifetime and reduction in the €/kWh are warranted. In this work, an advanced and flexible concept for an operation control strategy which is based on the battery state of health and efficient use of the diesel generator will be introduced.

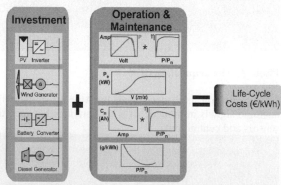

Figure 1: Life-cycle cost structure of stand-alone PV hybrid system.

2 OPERATION CONTROL

2.1 Tasks and limitations

The operation control for the stand-alone PV hybrid system refers to the strategy that manages the interaction between the system components, scheduling and execution of actions in a way that improve the system performance and reliability. Furthermore, the operation strategy takes actions to ensure that system operating goals are fulfilled and maintained despite different operation uncertainties. Also, it can prevent critical states of the system and alleviate or reduce consequences of failures in system components [4]. That signifies the influence of the implemented operation control strategy on the system energy availability and cost.

However, the functionality of the control strategy is constrained by boundaries which do not allow unlimited flexibility in its decisions. Such factors are; - *environmental*, - *cost*, - *energy availability* and - *quality of supply*.

Limited control over the renewable converters due to the continuous variation of their resources of energy makes it difficult to consider them as main elements in the energy dispatch process. In contrast, the energy of the fossil converter and the battery storage are more controllable, since they can deliver exactly the amount of energy needed and at the desired times. That makes them qualified to be the main components in many of the implemented energy dispatch strategies [5].

2.2 Towards an Optimal Operation Strategy

Practice to date in operating decentralized PV systems has shown that fixed operation concepts are inadequate for such complex hybrid energy systems because they concentrate on using operation decisions which do not consider the actual system components status. Ideal operation control may be fully achieved if it can:

- Consider the history of the system components and forecast the operating conditions (pre-dispatch activities).
- Take optimal measures to utilize the available energy in order to cover the demand reliably and optimize the costs and components lifetime, (dispatch activities).
- Evaluate and improve the operation decision procedures and continuously adjust the operation boundaries and criteria (post-dispatch activities).

Different operation concepts have tried partially to implement these criteria in order to reach a high performance level with the hybrid system. Although it is difficult to consider all these features in one operation control concept, this is an essential step towards the development of a reliable operation. Indeed, operation control moves towards a horizon where many information about the actual status of the system components, the weather and the demand have to be available before the phase of the energy dispatch decisions.

Battery ageing: The basic issue in this context is the battery state of health and the possibility to consider it during the dispatch decisions. Actual SOC, intensity of the acid stratification, time in deep discharge conditions, etc. are all necessary information for the operation strategy to carry out efficient operation actions.

Forecasting: Weather- and demand-forecasting supports the operation control either to start an extra conventional generator or to use the running engine to charge the battery. Many of these compromising decisions specify the ability of the operation control concept in order to enhance the system performance. On these bases the development of an advanced operation control is introduced.

3 DEVELOPMENT OF AN ADVANCED OPERATION CONTROL STRATEGY

Defining the rules and the boundaries of operation for both the battery and the diesel generator are important steps in developing an efficient operation control strategy. Another important consideration, which has to be integrated in modelling the dispatch strategy, is the actual system components status. In this way the enhancement of the energy dispatch decisions is guaranteed [6].

3.1 Features of the developed operation concept

The main goals of the developed strategy are concluded in the following two items:

- To achieve consistent operation conditions for the battery and the back-up unit and
- To optimize energy availability and total system costs.

Many optimization criteria are inherent in these two tasks. For example, consistent operation conditions include operating the diesel generator as efficient as possible, implementing regimes for alleviation and/or slowing down the degradation process of the battery capacity (acid stratification, grid corrosion, etc.). Moreover, optimizing the system operating cost includes ensuring high level of energy availability, minimizing the spilling of renewable energy, minimizing the environmental consequences, etc.

The developed concept is characterised by the following three measures:

1) Instantaneous identification of components operating conditions and consideration for forecasting.
2) Account for actual components status (e.g. the battery) and the constraints imposed on their operation.
3) Flexibility in deciding for the dispatch strategy.

Consequent results of these approaches are more reliable hybrid systems and cost effective of the produced energy.

In order to study the proposed operation concept a computer model has been constructed for the system configuration shown in Figure 1. Also, an ageing model to estimate different degradation mechanisms of the lead-acid battery has been developed and integrated in the simulation model.

3.2 Structure of the operation control concept

The main outlines of the developed operation control system architecture and its parts can be seen in Figure 2. In this figure, the developed operation control system is divided into three modules, where each module carries a specific operation responsibility. The communication interface (e.g. INTERBUS) is used to transfer the data between the components and the supervisory controller. Operation control receives data from sensors, component-controllers and internet to be processed. As the dispatch strategy is defined, the communication interface transfers the energy dispatch decisions to components controllers and actuators. The system data is processed by the operation control via the following modules:

- *Operation constraints module* includes optimization measures and components setting parameters. Such measures are the operation-domains and –limitations (e.g. min. SOC, min diesel load ratio, max. acid stratification level, etc.) which are recommended by the components manufacturers or gained from operation experience. The values of these parameters are a main key to high operation performance of the system.

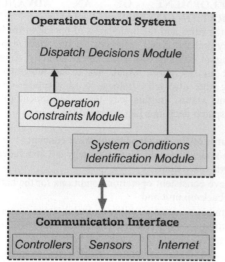

Figure 2: *Schematic diagram of the operation control system structure.*

- *System condition identification module* carries the responsibility of interpretation and evaluation of the information available from the input interface. This information is converted into parameters which identify the current status of the system. Also, in this module forecasting of the renewable resources is considered. The present operation strategy does not need exact forecasting of the renewable sources. However, vague information in the form of cloudy- or windy-hours is enough to estimate how the renewable potential will be in the next operation period. Forecasting in this work assumes periodical downloading of information about the weather situation for the site under consideration via the INTERNET. This information is further processed and supplied to the operation control in form of parameters which estimate the renewable potential. Upon these parameters the energy dispatch decisions are supported and operation of the back-up unit and the battery storage are optimized.

- *Dispatch decision module* determines the best control strategy that meets the design objectives without violation of the operating constraints. This includes elaboration of the operation scenario in form of decisions to choose the system mode of operation (i.e. which components are to be switched On/Off and energy allocation between the different sources and destinations).

3.3 kernel of the operation concept

To ensure the flexibility necessary for the functionality of the operation control, the dispatch decision module is designed on the bases of finite state mechanism [7]. This technique is a representation of an event driven system. In an event-driven system, the system transitions from one state (mode) to another prescribed state, provide that the condition defining the change is true. Accordingly, the activity or inactivity of a state changes dynamically, based on the actual operating conditions of the system. In this way the actual system operating conditions leads the strategy to a specific state. Correspondingly, energy dispatch decisions are generated.

3.4 System states identification

Seven states have been defined for the operation control as can be seen in Figure 3. In this figure examples of the transition conditions and the developed dispatch decisions of each state (i.e. operation strategy) can be noticed. The basic differences among the states are based on the value of the net-load[1] and the interaction between the battery and the diesel generator. The latter considers for the load ratio of the diesel and the battery actual state of heath. Whenever necessary the diesel generator, with its flexibility in delivering power, is used to prevent or alleviate the critical degradation mechanisms of the battery. That provides a better operation environment for the battery and ensures a longer life-cycle.

4 SIMULATION RESULTS

Typical Mediterranean weather data have been used. System components capacities are:-Storage capacity of 15 hours of the average load, PV and wind turbine each has a nominal capacity equal to the maximum load. Two conventional operation strategies (*battery-4-weeks charging* and *diesel-full-power*) are compared with the developed concept as can be seen in Figure 4. The results shown in Figure 4 are chosen at zero losses of load probability. The conventional strategies implement fixed-time operation decisions. In the four-weeks strategy, the battery is to be charged to 100% SOC every four weeks (i.e. consideration for the battery state of health). From Figure 4 it can be noticed that the battery capacity loss develops quickly, since the four-weeks full charging is either started too late (i.e. long time elapsed and the battery is in severe condition such as high acid stratification or deep discharge conditions) or very early (i.e. the battery is in a good condition due to light operation in the previous period). The full-power strategy operates the diesel engine at full load as soon as it is switched on. Accordingly, the engine fuel consumption and start/stop frequency are minimized. However, this strategy wears out the battery active material due to frequent charge and discharge cycles. In contrast the developed operation concept shows great difference in the battery degradation behaviour. This indicates the flexibility of the developed operation concept in choosing between different dispatch decisions so that the utilization of the battery energy and the diesel power is improved. Also, the developed strategy indicates its ability to take the necessary decisions for alleviation of the battery degradation mechanisms at the right times. A clear interdependence between the diesel relative rated power and the battery degradation behaviour appears when implementing the developed strategy, as can be seen Figure 5. A very slight increase in the fuel consumption (ca. 4.5 %) which is negligible if it is compared with the reduction in the battery degradation (ca. 33 %) is the result of the developed concept as can be seen in Figure 6. A preliminary cost comparison has shown that a 20 % reduction in kWh cost can be achieved by the developed operation concept.

Definitions:N_L= net-load, P_{B_ch}, P_{B_dis} = power charged and discharged from the battery respectively, P_{D_n}, P_D= rated and operation power of the diesel generator respectively and Δt = simulation time step.

---

[1] Net-load is the difference between demand and renewable power

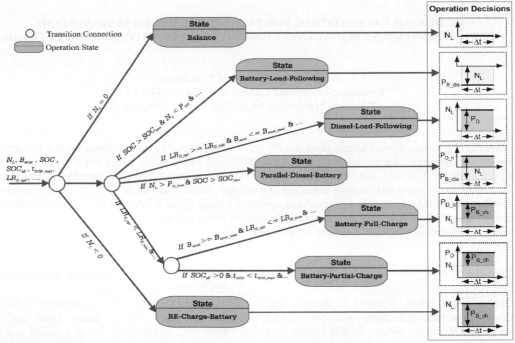

Figure 3: *Structure of system states identification.*

5- CONCLUSION

Continuous checking of the battery degradation mechanisms and slowing down their propagation helps in prolonging the operation lifetime of the battery. Also, efficient loading of the available diesel generator and rational use of the battery energy ensure a high level of energy availability. Another essential issue in developing the operation control strategy is the flexibility in dispatch decisions and implementation of the experience gathered from operating different hybrid systems.

REFERENCES

[1] W. Kleinkauf, O. Hass, F. Raptis, J. Schmid, P. Strauss, P. Zacharias, 2nd World Conference on PV Solar Energy Conversion, Vienna, Austria, (1998) 2864.

[2] G. Bopp, H. Gabler, D. Sauer, A. Jossen, W. Höhe, J. Mittermeier, M. Bächler, P. Sprau, B. Willer, M. Wollny, 13th EC Photovoltaic solar energy conference, Nice, France, (1995) 1763.

[3] "Entwicklung von PV-Versorgungsanlagen mit modularer Systemtechnik", Abschlußbericht, ISET (1998).

[4] A. Pereira, H. Bindner, P. Lundsager, O. Jannerup, Proceedings EC Wind Energy Conference, Nice, France, (1999) 960.

[5] C. D. Barley, C. B. Winn, Solar Energy **58**, (1996) 179.

[6] M. Ibrahim and Peter Zacharias, 11th Photovoltaic Solar Energy Conference, Hokkaido, Japan, (1999).

[7] David Harel; Science of Computer Programming **8**, (1997) 231.

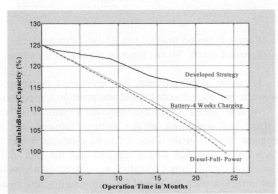

Figure 4: *Development of the battery capacity losses under different operation control concepts*

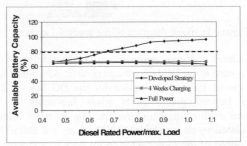

Figure 5: *Available battery capacity under different control strategies*

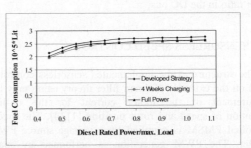

Figure 6: *Diesel fuel during 5 years of operation under different control strategies*

DC BUS VOLTAGE CONTROL FOR PV-POWERED WATER PUMP SYSTEMS
USING A PERMANENT MAGNET SYNCHRONOUS MOTOR WITHOUT SHAFT SENSOR

G. Terörde, A. Woyte and R. Belmans
Katholieke Universiteit Leuven; Dep. E.E./ESAT-ELEN
Kardinaal Mercierlaan 94, B-3001 Leuven, Belgium

ABSTRACT: The system studied in this paper is a PV powered water pump system, consisting of a PV array, a low cost inverter, a permanent magnet synchronous motor (PMSM) and a water pump with a water storage. The PV array has a peak power of 4,32 kW. The overall control of the system consists of a current controller, a speed controller without a shaft sensor and a main control consisting of a voltage controller and the MPPT. The voltage control varies the speed of the PMSM to stay in the calculated optimum voltage given by the MPPT. Measured results of the different control units are presented to demonstrate the stability of the system.
Keywords: PV Pumping - 1: Implementation - 2: Stability - 3

1. INTRODUCTION

The entire control system is shown in figure 1. The inverter operates as a variable frequency source (PWM) for the PMSM driving the pump. Since a PMSM in open loop is unstable, a field oriented control with feedback of speed and position, estimated by an extended Kalman filter (EKF), is proposed. Furthermore, the control system is equipped with a maximum power point tracker (MPPT) and a voltage control guaranteeing a balanced input/output power ratio in the DC bus.

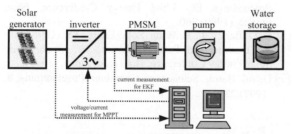

Figure 1: Block diagram of the analyzed system

The most important control loop for the stability of the entire system is the DC bus voltage control. All control and measurement units are supplied by the DC bus. A DC voltage beyond given limits leads inevitably to an undesired crash of the entire system. The optimum voltage is calculated by an overlaid MPP-Tracking and controlled by the speed of the motor. Due to the lack of a storage in the DC bus, the power of the PV array must be used immediately to accelerate the PMSM. Therefore, the voltage controller has to guarantee a balanced input/output power ratio in the DC bus.

2. SENSORLESS SPEED CONTROL

The structure of the implemented sensorless control is based on the extended Kalman filter theory using only the measurement of the motor current for the on-line estimation of speed and rotor position [1]-[4]. The speed-controlled PMSM is supplied by a voltage source PWM inverter. The PWM generation is performed by space vector modulation. The motor voltages required for the

Kalman algorithm are calculated with consideration of the non-linearity of the inverter.

Figure 2 shows the experimental result of a speed reversal using the estimated speed and position as feedback. Additionally, the real speed and position are measured and compared.

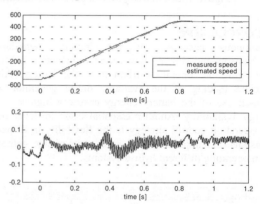

Figure 2: Speed reversal test Above: Measured and estimated speed; Below: Error of the angle estimation

It can be seen that there is a very good coincidence between real and estimated speed and position respectively.

At low motor speed ($\omega \Rightarrow 0$) the equations of the PMSM are simplified as the voltage induced by the magnets is very small. Thus, no more predication can be made on the position of the magnets and the EKF fails. Since at standstill only DC-values are given, the necessary flux variation must be forced by impressing a test signal into the system. A signal, which can be implemented easily, represents an additional sinusoidal reference current in the d-axis of the motor, using the d/q axis-symmetrie of the rotor to estimate the real position. In all presented experimental results the following d-axis reference current is used:

$$i_{d,ref} = i_d^* + 5\text{A} \sin(2\pi\, 200\frac{1}{\text{s}} t) \cdot \left(1 - \frac{|n|}{500\text{RPM}}\right) \quad (1)$$

whereby i_d^* results from the speed control. The generated reluctance torque is compensated by a complementary q-axis current.

The electromagnetic torque T_e can be expressed as:

$$T_e = p \left[\Psi\, i_q - (L_q - L_d) i_d\, i_q \right] \qquad (2)$$

The optimal control of the motor takes advantage of the reluctance torque by introducing a negative ($L_d < L_q$) direct axis current component. In figure 3 a comparison is given of motor control with feedback of the estimated speed and optimum d-axis current, motor control with feedback of the estimated speed and no d-axis current and motor control with feedback of the measured speed and position (FOC). The maximum torque using the EKF is smaller due to the tiny error of the angle estimation. This leads to a higher acceleration of the motor with feedback of the measured position. However, the bandwidth of the speed control with the EKF is comparable to the common control due to the dropped filter for speed measurement.

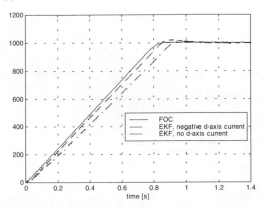

Figure 3: Speed step with feedback of the estimated speed and position (EKF), optimum d-axis current, no d-axis current and with feedback of the measured speed and position (FOC)

3. VOLTAGE CONTROL

The most important control loop for the stability of the entire system is the DC bus voltage control. All control and measurement units are supplied by the DC bus. A DC voltage beyond given limits leads inevitably to an undesired crash of the entire system. The optimum voltage is calculated by an overlaid MPP-Tracking and controlled by the speed of the motor. Due to the lack of a storage element (e.g.: battery) in the DC bus, the power of the PV array must be used immediately to accelerate the PMSM. As the irradiance increases, resulting in a higher output power of the PV array, the input power of the DC bus is higher than the output power. The voltage control must immediately accelerate the PMSM to stay in the MPP of the PV array. With decreasing irradiation, the power of the PV array is smaller than the output power in the DC bus. The difference comes from the capacitor. This is the most critical condition of the system. The DC bus may collapse, if this condition remains. Hence, the inverter must slow down the PMSM. Therefore, the voltage controller has to guarantee a balanced input/output power ratio in the DC bus. Figure 4 shows the energy flow within the system.

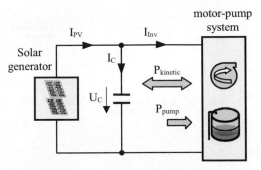

Figure 4: Energy flow of the total system

The dynamic behavior of the voltage control is determined by energy equations. The energy of the motor-pump system can be divided into kinetic and pump energy. Only the kinetic energy can be used to fed back energy to the DC bus and control the DC voltage.

Normally, the motor speed of a conventional drive supplied by a regular grid is completely independent of the DC voltage. Here, a PV array is the source. A relation between the motor speed and the DC bus voltage can be obtained by linearization. The kinetic power $P_{kinetic}$ changes faster than the pumping power P_{pump} and the current I_C faster than the voltage. Assuming a constant pumping power and a constant current I_{PV} of the PV array for a short time, the linearized relationship between DC voltage U_{DC} and motor speed ω is described by the following equation.

$$P_{pump} \approx U_C I_{PV} \approx const \qquad (3)$$

$$P_{kinetic} = J\,\omega\,\frac{d\omega}{dt} \approx -U_C I_C, \quad J = \text{inertia of the drive} \qquad (4)$$

$$\Rightarrow U_C \frac{dU_C}{dt} = -\frac{J}{C}\,\omega\,\frac{d\omega}{dt} \qquad (5)$$

The resulting transfer function with the reference speed N^* as input and the DC voltage U_C as output is defined by

$$\frac{U_C(s)}{N^*(s)} = -\frac{2\pi}{60} \sqrt{\frac{J}{C}} \cdot \frac{1}{sT_{speed} + 1} \cdot \frac{1}{sT_{Vf} + 1}, \qquad (6)$$

where T_{speed} is the equivalent time constant of the speed control and T_{Vf} the time constant of the voltage measurement. In Eq. (3)–(6), the efficiency of the drive is not taken into account, because of the opposite influence at acceleration and deceleration. During practical investigations the best results has been obtained using a common PI controller for the voltage control.

3.1 Experimental Results of the Voltage Control

In Figure 5 the voltage response for a step of the voltage reference from 225V to 125V and back to 225V is shown. The speed of the motor changes until the reference voltage is reached. It must be noted, that without voltage control the voltage area below the MPP (~185V) is unstable.

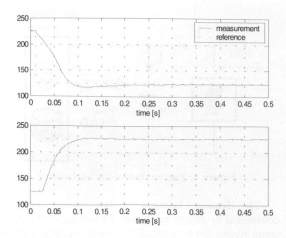

Figure 5: Step of the voltage reference
Above: 225V ⇒ 125V; Below: 125V ⇒ 225V

Figure 6 demonstrates the independence of the pump characteristic. The pumping head changes from a half meter to ten meters and back to a half meter. Above the response of the voltage control is plotted, below the speed. The speed of the motor changes until the reference voltage is reached.

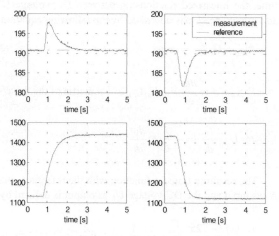

Figure 6: Variation of the pumping head
Left: 0,5m ⇒ 10m; Right: 10m ⇒ 0,5m

One of the most important features of the voltage control is the robustness during power interruptions, occurring at instantaneous decrease of irradiance (e.g.: passing clouds).
This property has been tested using a regular grid as power supply and applying a voltage dip on all three phases for a short time (~2 sec). The implemented regenerative braking scheme allows the inverter to keep its DC bus voltage at a minimum level (330 V), expanding the time in which supply voltage can be reapplied without the time-consuming DC-link capacitor recharging cycle.
Figure 7 shows the experimental result of a short-time three phase voltage dip. The deceleration of the motor during the power interruption is small, because no load is applied.
Figure 8 shows the experimental results of a comparable voltage dip but with load torque applied to the motor. The applied load amounts to 75% of the rated

torque. Due to the load, the deceleration is much faster. However, the power needed to keep the voltage at a minimum level is the same, as can be seen at the small negative q-axis current during the voltage dip. In fact, this power (~10 W) generated by the PMSM compensates the inverter losses.

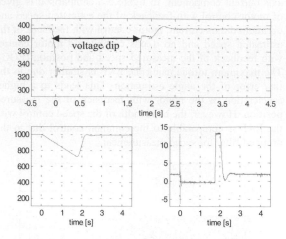

Figure 7: Voltage dip without load torque

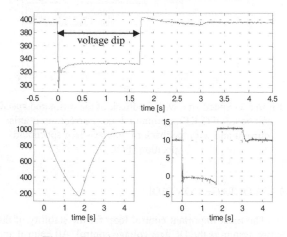

Figure 8: Voltage dip with load torque
(75% of the rated torque)

Keeping the capacitor well charged has the additional advantage of the control electronics being powered over a longer period of time, avoiding a time-consuming restart. With the implemented voltage control, a kinetic buffering during a total energy drop is possible for longer than one second.

4.   MAXIMUM POWER POINT TRACKING

The voltage reference is calculated by an overlaid maximum power point tracking. The Maximum Power Point (MPP) is characterized by the voltage, where the PV array generates maximum output power. The characteristics of the PV elements are affected by the irradiance and the cell temperature. To reach the point of maximum power at rising irradiance, the current in the DC bus must be increased while the DC voltage remains nearly constant. The voltage at the MPP changes with the array temperature and the current is almost unaffected. At lower cell temperature the MPP characteristic is situated in a higher voltage range. Thus, the optimum output voltage of the PV array is not constant and moves as conditions vary.

The MPP-Tracking delivers the reference quantity of the voltage control loop. First, a default voltage and search range must be given. After the default voltage is reached, the voltage is varied around this point. The quantity of the variation is given by the search range. During this variation, the power generated by the PV array is measured and the voltage linked with the maximum power is stored during the respective searching procedure. The new optimum voltage and the new search range are calculated from these actual measurements and in its history-stored values by an adaptive controller. With these quantities the controller starts a new searching procedure to find the MPP. Figure 9 shows the flow chart of the MPP-Tracking.

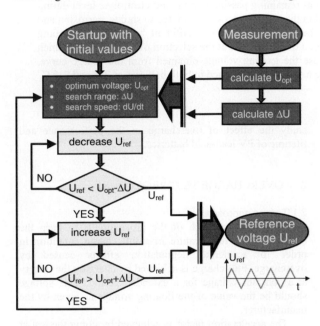

Figure 9: Flow chart of the MPP-Tracking

The previous values are very important for the calculation of the new optimum voltage and search range. If, e.g., the new calculated optimum voltage during a searching procedure with rising voltage is situated higher than the last optimum voltage, the MPP-voltage seems to change. But this can also indicate an increasing irradiance. If the second condition occurs, the controller should not change the new optimum voltage. Otherwise, the calculated voltage drifts away from the MPP. The same considerations are also valid for a decreasing irradiance. Thus, the adaptive control must be able to distinguish between a changing MPP and changing conditions, respectively. The search range depends on the variation of the calculated optimum voltage. If the calculated MPP is situated in the half of the past voltage range, the search range is reduced, otherwise it is increased.

Measurements of the implemented MPPT are plotted in figure 10 showing the power of the PV array as a function of the DC voltage. Three different starting conditions (a-c) are shown to demonstrate the ability of reproduction of the MPPT. The characteristic c exhibits the MPPT starting

with an unbalanced input/output power ratio on the DC bus.

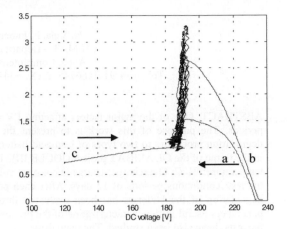

Figure 10: Measured results of the MPP-Tracking

5. SUMMARY AND CONCLUSION

The paper presents the design and the implementation of a speed based DC bus voltage control for PV-powered water pump systems using a PMSM without a shaft sensor. Practical investigations are done to demonstrate the stability of the DC bus voltage control, the independence of the pump characteristic and robustness during power interruptions Results of the dynamic and steady state behaviour of the sensorless speed control (extended Kalman filter) are given. A MPPT control without using an additional DC-DC converter is explained. The implemented MPPT is realized by feeding back the DC voltage and current to the controller, adjusting the speed of the PMSM and keeping the system operating at its MPP. The measured results of the MPPT exhibit the ability of reproduction and the stability of the entire control system.

ACKNOWLEDGEMENT

This research is supported by the Ministry of Economy of Flemish Region (IWT-Vliet). The authors are also grateful to the "Fonds voor Wetenschappelijk Onderzoek Vlaanderen" for its financial support and to the Interuniversity MicroElectronics Centre IMEC for its practical support of this work.

REFERENCES

[1]   Rajashekara K S, Kawamura A, 1994, "Sensorless Control of Permanent Magnet AC Motors", <u>IEEE IECON Proceedings</u>, pp. 1589-1594.

[2]   Bose B K, 1997, "Power Electronics and Variable Frequency Drives", IEEE Press, New York.

[3]   Brammer, Siffling, 1994, "Kalman-Bucy Filter, Deterministische Beobachtung und stochastische Filterung" R. Oldenbourg Verlag München, Wien.

[4]   Strejc V, 1980, "State space theory of discrete linear control", John Wiley & Sons.

ACCELERATED OVERCHARGE TEST PROCEDURE FOR A QUICK ASSESSMENT OF LIFETIME OF PV BATTERIES

N. Vela, F. Fabero, J. Cuenca and F. Chenlo
CIEMAT – Departamento de Energías Renovables
Avda. Complutense, 22. 28040 Madrid. Spain
Tel.: +34-91-3466745 FAX: +34-91-3466037 e-mail: nieves.vela@ciemat.es

ABSTRACT: One of the major factors affecting PV batteries life-time is their operation in overcharge for extended periods. The purpose of this work is to present the results obtained after applying a new accelerated overcharge degradation procedure to two types of lead-acid batteries used in PV installations. This procedure has been developed in the frame of the QUALIBAT project (JOULE III). The following conditions hold for the overcharge test: Electrolyte temperature: 71°C; Overcharge: charge at constant voltage (floating voltage specified by the manufacturer); Length of the test: continuous periods of 12 days. After each period of 12 days the battery capacity is measured at 25°C. The degradation of the batteries has been evaluated through the capacity evolution along the overcharge tests. Other parameters useful in the characterization of the process, such as remaining passing current and electrolyte level during the tests, have also been studied. The main damages observed in the tested batteries are related to sludge, corrosion and self-discharge, leading to gradual capacity losses. Results here presented show a different behaviour concerning degradation between both batteries. Following this we present a discussion about the selection of the voltage to which the degradation test should be performed. We propose to use the gassing voltage obtained from the charge curve. Additionally, at the end of the overcharge tests, self-discharge have been determined in both batteries at 25° and 71° C.
Keywords: Batteries - 1: Degradation - 2: Lifetime -3

1. INTRODUCTION

During the final part of the charge process (overcharge), when the battery state-of-charge (SOC) is high, the quantities of active materials decrease. For this reason, a part of the passing current is used to transform active materials but the other part is used to electrolyse water. This last phenomenon is called gassing. The main effects of gassing during overcharge are:
- Loss of water.
- Evolution of gas (H_2, O_2).
- Corrosion of the positive grid (reduction of battery life-time).
- Elevation of battery temperature (acceleration of corrosion rate).
- Reduction of battery efficiency.

Moderate gassing could help to prevent stratification by stirring the electrolyte but severe gassing can dislodge active materials from the plate with the consequent capacity loss.

The gassing voltage is dependent on temperature (the higher temperature the lower starting gassing voltage) and charge current (the higher charge current the higher starting gassing voltage).

Once the battery is gassing the rate of the process depends on:
- Temperature: the higher temperature the higher gassing rate.
- Operational voltage: the higher voltage the higher gassing rate.

Lead-acid batteries operating in stand-alone PV systems should be charged enough to assure a high SOC allowing a moderate gassing to prevent stratification but avoiding extended periods in overcharge state. However some problems in sizing (oversize the PV array, undersize the battery system) or in operation (less use than expected) in a given PV installation can lead to operate batteries in overcharge conditions for extended periods. The purpose of this work is to develop an accelerated test procedure to study the effect of overcharge in the performance and lifetime of PV lead-acid batteries.

2. OVERCHARGE TEST PROCEDURE.

2.1 Testing approach.

The basic approach of the procedure is to keep the battery at high temperature in a full charge condition in order to accelerate degradation effects caused by overcharge. Full charge is achieved by charging the battery at a constant voltage for a extended period. This voltage should be the value of the floating voltage as given by the manufacturer.

The acceleration factor is achieved by doing the test at a high battery temperature, according to the Arrhenius Law. To calculate test duration at different temperatures the following expressions should be used:

$$L_t = L_d / K \qquad K = \cfrac{1}{\left(\cfrac{1}{2}\right)^{\left(\frac{T-20}{10}\right)}} \qquad (1)$$

where:
L_t = Test duration
L_d = Battery design life
K = Ageing acceleration factor
T = Test temperature

All the overcharge tests presented in this work have been done at a battery temperature of 71°C (i.e., K = 34.3). This, according to expressions (1), gives a test duration of 106 days for a battery design life of 10 years .

2.2 Description of the testing facility.

Overcharge and capacity tests have been performed using two DIGATRON UBT 6-50 Universal Battery Tester units. These units allow to control and measure both voltage and current during the tests.

Electrolyte temperature have been measured using a T-type thermocouple protected by a Teflon cover inmersed in the battery electrolyte. Batteries have been heated inside two water circulating thermostatic baths. Two layers of polypropylene balls were placed on water surface in order to minimise water thermal losses and evaporation. A HYDRA Data Logger (FLUKE) have been used to record (every half an hour) the following parameters: date, time, voltage, current, electrolyte temperature and water bath temperature.

2.3 Sample description.

BATTERY 1. Lead-acid vented 12V monoblock battery. Tubular positive plates. Totally new.
Nominal capacity = 174 Ah (C10).
Floating voltage = 13.50 V (2.25 V/cell). Data given by the manufacturer.

BATTERY 2. Lead-acid vented 2V cells battery. Tubular positive plates. Totally new.
Nominal capacity = 325 Ah (C10).
Floating voltage = 13.38 V (2.23 V/cell). Data given by the manufacturer.

2.4 Test description.

a)   Pre-conditioning cycling test.

This has been necessary in order to avoid possible interference in measurements caused by memory effects and to prepare batteries to have an identical starting point before any test.

	BATTERY 1	BATTERY 2
1. Discharge	17.5 A, 5 h	32.5 A, 5 h
2. Pause	30 minutes	30 minutes
3. Charge	17.5, 5 h	32.5 A, 5 h
4. Pause	30 minutes	30 minutes
5. Repeat steps 1, 2, 3 and 4 ten times		

b)   Initial capacity test.

Determination of the initial capacity at an electrolyte temperature of 25°C according to the following sequence:

	BATTERY 1	BATTERY 2
1. Charge	17.5 A until 14.5 V	32.5 A until 15.6 V
2. Charge	14.5 V - 4 h	15 V - 5 h
3. Pause	30 minutes	30 minutes
4. Discharge	17.5 A until 10.8 V	32.5 A until 10.8 V
5. Obtain the capacity		
6. Repeat steps 1 to 5 until capacity remains constant (5%)		

The initial capacities obtained were: 149 Ah (Battery 1) and 299 Ah (Battery 2).

c)   Overcharge test.
- Charge the batteries at an electrolyte temperature of 25°C according to point b), steps 1 and 2 above.
- Heat the battery to an electrolyte temperature near 71°C.
- Maintain the batteries at a constant voltage chosen equal to their corresponding floating voltage V_{float} (manufacturer):

Battery 1. V_{float} = 13.50 V
Battery 2. V_{float} = 13.38 V
- Control the electrolyte temperature to 71°C.
- Keep batteries 12 full days in these conditions, then stop the charging process and leave the batteries cool until an electrolyte temperature of 25°C.
- Obtain the capacity following the corresponding point b), step 3 above.
- Repeat overcharge test eight times (Total duration of the test: 96 days).

3.   RESULTS

3.1 Capacity.

Figure 1 shows the evolution of the capacity (normalized) of each battery after their corresponding overcharge tests. In both batteries a capacity increase after overcharge test no. 1 can be observed. This can be due to the deep charge imposed by the conditions of the overcharge test. After this first overcharge test a continuous capacity decrease is observed for Battery 1. In the case of Battery 2 the capacity remains in an approximated constant trend.

After overcharge test no. 4 *anomalous* data in the capacity sequences were obtained. This can be explained by the fact that, in these cases and because of experimental problems, an extended storage (2 days for Battery 1 and 14 days for Battery 2) at 71°C happened after finishing the overcharge process, delaying capacity determination. This indicates large self-discharge processes occurring in the batteries, that have been probably accelerated because of both internal degradation and high temperature. For the remaining overcharge tests the capacity measurements were done immediately after finishing the overcharge period. Additional results concerning self-discharge will be presented in a later section of this paper.

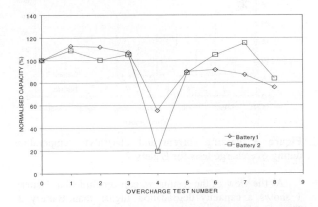

Figure 1: Evolution of the normalised capacity after each overcharge test.

3.2 Remaining current during overcharge.

Because the overcharge test has been performed at a constant voltage an useful parameter to characterize battery degradation could be the remaining current passing through the battery during the test. Figures 2 and 3 plot both remaining current and electrolyte temperature for Battery 1 and Battery 2 respectively. In the case of Battery 1 a continuous increasing of the remaining current is observed among the 8 overcharge tests. In the case of Battery 2 the remaining current only shows a narrow fluctuation around 6 A. Concerning electrolyte temperature it results that for Battery 1 many sharp variations could be observed because of the very frequent additions of water (at room temperature) necessary to keep the water level into allowable limits. This is a consequence of several factors: small water reserve in this battery, severe gassing and high temperature. For Battery 2 water refilling was practically not required during every single test mainly because of its very large water reserve and lower gassing rate compared to Battery 1. As a general rule it is strongly recommended to have large water reserve in PV batteries that can be subjected to extended overcharge periods.

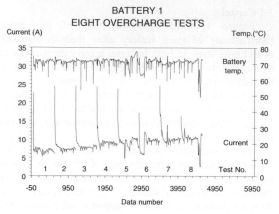

Figure 2: Passing current and electrolyte temperature during overcharge tests for Battery 1.

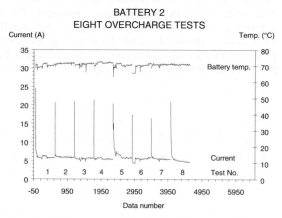

Figure 3: Passing current and electrolyte temperature during overcharge tests for Battery 2.

At the view of the results we can conclude that Battery 1 shows a capacity degradation higher than Battery 2. These could be due to the voltage selected to perform the overcharge test, i.e., the floating voltage indicated by the manufacturer.

3.3 Gassing voltage selection.

Gassing voltage Vg is affected by both charge current rate and temperature. In a previous work [1] we have determined gassing voltage values as a function of charge current rate and temperature for batteries equal to those ones used in this work. Every gassing voltage was calculated graphically [2] from a complete experimental charge curve at constant charge current rate and temperature. Gassing voltages obtained by applying this

method are presented in Table I for batteries 1 and 2.

Table I: Gassing voltages Vg at different current rates and temperatures for Battery 1 and Battery 2.

	I (A)	T (°C)	Vg (V)
Battery 1	1.7	27.0	2.21
	3.5	24.0	2.23
	9.0	24.2	2.25
	17.3	26.3	2.27
	9.0	38.9	2.25
	17.3	38.9	2.28
Battery 2	13	23.7	2.27
	30	24.2	2.31
	45	25.8	2.33
	10	42.7	2.26
	20	42.3	2.29
	33	43.7	2.29

Gassing values can be modelled as a function of charge current and temperature by the following expression:

$$Vg = (A + B \ln (1+I)) (1 + \alpha \Delta T) \qquad (2)$$

where:
I = charge current
$\Delta T = T-25°C$
A, B, α empirical parameters

The corresponding empirical parameters obtained by applying this expression are shown in Table II. The very low values of mse indicate a good accuracy to reproduce experimental values of Vg.

Table II: Empirical parameters obtained for Battery 1 and Battery 2 following expression (2) (mse=mean square error).

	A	B	α	mse
Battery 1	2.179	0.03188	-0.00008	1.24E-05
Battery 2	2.210	0.03142	-0.00045	2.80E-05

Table III presents a comparison between the values of Vg obtained from (2) and the floating voltages V_{float} given by the corresponding manufacturers. In the case of Battery 1 both values are very similar but in the case of Battery 2 its Vg is clearly higher than the corresponding V_{float} given by the manufacturer.

Table III: Vg modelled and V_{float} given by the manufacturer for Battery 1 and Battery 2.

	Vg (25°C, C10) (model)	V_{float} (manuf.)
Battery 1	2.27 V	2.25 V
Battery 2	2.32 V	2.23 V

Consequently both batteries are operating during the overcharge test in very different conditions concerning gassing, being Battery 1 operating at more drastic conditions than Battery 2. This situation suggests that the voltage selection criteria should be redefined in order to unify test conditions for different batteries, e.g. by using a gassing voltage value obtained from experimental charge curves (near 25°C) instead of the floating value given by the manufacturer.

3.4 Physical damages.

Photos 1 and 2 show some details of physical damages observed in the batteries during the overcharghe tests.

Photo 1: Physical damages observed in Battery 1 after some overcharge tests: large sludge.

Photo 2: Physical damages observed in Battery 2 after some overcharge tests: particles from the positive plate.

The main damages are related to sludge and corrosion in positive plates. Sludge is considerably larger in Battery 1 that in Battery 2, and from degradation test number 4 sludge partially covered the bottom part of the plates in Battery 1.

3.5 Self-discharge test

As it has been shown before, a storage at high temperature happened after the overcharge test no. 4 in both batteries, resulting in two *anomalous* data in Figure 1. In order to analyse this effect, a self-discharge test was applied to both batteries after 8 overcharge degradation tests. The self-discharge test was performed at electrolyte temperatures near 25°C and 71°C. Previous to the self-discharge test 10 charge-discharge identical cycles were applied to determine battery capacity. Test procedures are detailed in tables IV and V. Table VI presents the capacity values obtained for both batteries after self-discharge tests. These results show:

- Self-discharge is largely enhanced at high temperature
- Self-discharge is larger in Battery 1 than in Battery 2
- After 10 cycles Battery 1 shows a decrease in capacity at high temperature, indicating strong degradation.

Table IV: Test procedure for self-discharge test (Battery 1)

	25°C	71°C
1. Charge 17.5 A	until 14.5 V	until 13.35 V
2. Charge 4h	14.5 V	13.35 V
3. Pause	2 h	
4. Discharge 17.5 A	until 10.8 V	until 9.8 V
5. Pause	1 h	
Repeat 1 to 5 ten times		
Pause	180 h	
Discharge 17.5 A	until 10.8 V	until 9.8 V

Table V: Test procedure for self-discharge test (Battery 2)

	25°C	71°C
1. Charge 32.5 A	until 15.6 V	until 14.1 V
2. Charge 5h	15.0 V	14.1 V
3. Pause	2 h	
4. Discharge 32.5 A	until 10.8 V	until 9.8 V
5. Pause	1 h	
Repeat 1 to 5 ten times		
Pause	180 h	
Discharge 32.5 A	until 10.8 V	until 9.8 V

Table VI: Capacities (Ah) obtained for both batteries after self-discharge tests.

	Battery 1		Battery 2	
	25°C	71°C	25°C	71°C
Initial (10 cycles)	141	126	270	348
Self-discharge (180h)	93	11	167	136

CONCLUSIONS

A new accelerated test procedure for overcharge degradation has been applied to two different PV lead-acid batteries. Results show a different degradation between both batteries probably because of the selection of the voltage to which the test has been performed. As an inital approach we suggest to do the test at a voltage equal to the gassing voltage value obtained from experimental charge curves (near 25°C) in order to have similar testing conditions for different batteries. Otherwise, large self-discharge has been observed in both batteries after the overcharge tests.

ACKNOWLEDGEMENTS
This work has been partially supported by EU through the JOULE III research project QUALIBAT, contract number JOR3-CT97-0161.

REFERENCES

[1] N. Vela, F. Fabero, J. Cuenca, F. Chenlo and M. Alonso-Abella. Proceedings 14th EC Photovoltaic Solar Energy Conference, Barcelona, Spain (1997), 1688-1691.
[2] J. B. Copetti. Modelado de acumuladores de plomo-ácido para aplicaciones fotovoltaicas. PhD Thesis. Universidad Politécnica de Madrid (1993).

DEVELOPMENT, SIMULATION AND EVALUATION OF DIGITAL MAXIMUM POWER POINT TRACKER MODELS USING PVNETSIM

T. Schilla, R. Hill, N. Pearsall, I. Forbes, G. Bucher*
Marienstr. 18, 73262 Reichenbach, Germany
NPAC, School of Engineering, University of Northumbria, UK
*Polytechnic Heilbronn, Max-Planck-Str. 39, 74081 Heilbronn, Germany
Tel.: +49 7153 950995, Fax +49 7153 950994, eMail: thomas.schilla@schwaben.de

ABSTRACT: This paper discusses the development of an inverter model for grid connected PV systems. The model reflects the behaviour of a digitally controlled inverter and, in particular, the inclusion of a maximum power point tracker (MPPT). The model simulates the operation of an inverter in combination with a wide range of photovoltaic generators and includes all loss mechanisms that influence the performance of the system. The transient behaviour of a photovoltaic inverter is considered including oscillation of the operating voltage around the maximum power point. The model includes the digital maximum power point tracker and a DC to AC conversion efficiency model.
Keywords: Simulation - 1: Standard Circuit Simulator -2: Software -3

1. INTRODUCTION

The inverter model is a virtual electrical circuit that is able to reflect the dynamic behaviour of a digitally controlled inverter and, on the basis of detailed solar generator performance data, determine the transient behaviour of the inverter. It was developed within the program environment of the Design Center programs, PSpice, Schematics and Probe. Parts of the model have been implemented with the help of analogue behaviour modelling features of PSpice. All other parts of the developed circuit utilise standard PSpice components. Irradiance dependent loss mechanisms can be simulated and do not need to be estimated in the form of factors or assumptions. Losses due to the connection of system components such as cable losses, tracking losses, the influence of diodes or inhomogeneous temperature and irradiance profiles, as well as shading effects that directly influence the maximum power point, are included in the computation of the current operating point. Manufacturer tolerances of the solar cells or modules and the resulting different electrical influences on the I-V characteristic of the components are included in the solar cell or solar module model.

2. SIMULATION OF DIGITAL COMPONENTS

The digital components and models that have been developed in the course of this work were designed with the help of the Design-Center-Programs Schematics, PSpice and Probe. The model was then added to the model library in PVNETSIM [1] to allow the simulation of complete systems. PSpice is a mixed analogue/digital electrical circuit simulator that can calculate the behaviour of analogue-only, mixed analogue/digital, and digital-only circuits with speed and accuracy. PSpice reads the circuit files set, generated by Schematics, which includes a netlist of device declarations. PSpice calculates voltages and currents of the analogue devices and nodes, and states of the nodes connected only to digital devices, subject to a specified PSpice analysis [2].

3. DIGITAL MODEL OF A MAXIMUM POWER POINT TRACKER

In the case of the digitally controlled MPP tracker (DMPPT), probably the most widespread tracking method forces the operating point to oscillate around the MPP. Such systems mostly influence the PV generator performance by the means of a DC/DC converter either by varying the operating voltage or current. The control unit of the MPP tracker evaluates either the operating current or power to analyse the instantaneous operational status (see Figure 1). The DMPPT developed in this work is based on these design principles, evaluating the power and regulating the operating voltages via a controlled voltage source. The system consists of D/A and A/D converters, a memory unit, the logic control, a counter and a voltage controlled voltage source. The PV power is converted and compared with the value of the last measurement which is stored in the memory. The "Direction of change of VMPP" component in the block diagram depends on the output of the "Comparator", the "MPP voltage control" and the "Maximum power control". It increases or decreases the operating voltage with the next signal of the clock. The MPP voltage control checks if the limits of the MPP range are exceeded and the maximum power control evaluates whether the maximum power capability of the inverter has been reached. If the input power is too high, the MPP voltage control forces the operating point to move to smaller voltages along the I-V characteristic until the power is reduced to the maximum allowed power level.

The counter is a very important element of the circuit since it fixes the operating voltage at any time during simulation. At every clock signal, the count of the counter is converted to an analogue voltage level. This voltage controls a voltage source and this is directly connected to the PV generator. When a simulation run commences, the solar generator is operated at the smallest possible MPP voltage V_{mppMin} until the system is switched on by an internal signal and the voltage is set to V_{Start}. As soon as the power level P_{ON} at V_{Start} has been reached, the MPP tracking is activated and the operating voltage of the system moves towards the true MPP. The voltage level of V_{Start} must be between V_{mppMin} and V_{mppMax}. The system cannot operate at voltage levels below V_{mppMin}.

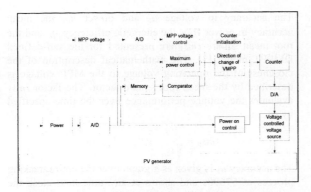

Figure 1: Block diagram of the digital inverter model developed in PSpice and implemented in PVNETSIM

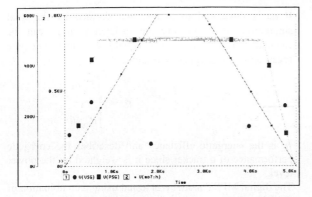

Figure 2: P_{max} tracking of the digital controlled MPP tracker model. The curves are: the operating voltage (V(VSG)), the DC power (V(PSG)) and the irradiance level (V(Mo7:h)). The module temperature was fixed to 25°C for the entire 5000s simulation time.

Furthermore the MPP limits specify the maximum resolution of internal voltage step width V_{sw} which is defined by $(V_{mppMax} - V_{mppMin})/2^8$. An 8-bit resolution of this voltage interval is sufficient to simulate the internal step width of most grid-connected PV inverters.

All inverters have an upper DC power limit. This is taken into account in the MPPT algorithm by the parameter P_{max}. As soon as the DC power level of the PV generator tries to exceed this power limit the operating voltage of the generator is moved towards smaller voltages. This operational mode is active until the power level falls below the limit again. An example is given in Figure 2, where P_{max} was set to 500W and the generator would normally provide up to 2.2kW from 2000 to 3000 seconds. The transient behaviour until 750s is described later in this paper.

During night time operation, when the solar generator power level has fallen short of P_{ON}, the counter is stopped. Under this condition, the simulation speeds up during phases of inactivity of the PV-system.

It is essential that appropriate adjustments are made to the MPP voltage limits (V_{mppMin} and V_{mppMax}) where under- or oversized grid-connected PV systems are simulated. If this is not the case, the MPP voltage will attempt to go beyond the relevant limits. In such a case, the model is unable to

accurately reproduce the performance of a real system and the operating voltage will be fixed to the MPP voltage limit. If the system voltage is at one of these limits, the DC input power can exceed the level of P_{max}. All parameters of the DMPPT are summarised in Table 1, which also includes values for a typical grid connected inverter.

Name	Description	Value
V_{Start}	The initial voltage of the system before the tracking starts	140V
P_{ON}	As soon as the power falls below this level the DMPPT stops. Usually this parameter is identical with the self-consumption.	4W
t_{sw}	Time between two successive regulation steps (clock interval)	5s
V_{sw}	Voltage value by which the operating voltage is increased or decreased. In case of an 8-bit resolution the minimum voltage step is $(V_{mppMax}-V_{mppMin})/2^8$	2V
P_{max}	Maximum allowed input power including the overload capability	1650 W
V_{mppMin}	The lower limit of the MPP voltage range	125V
V_{mppMax}	The upper limit of the MPP voltage range	250V

Table 1: Parameter of the DMPPT model and parameter values taken from a standard grid connected inverter

4. DC/AC CONVERSION EFFICIENCY MODEL

Such an exact simulation of the inverter behaviour requires careful consideration of the DC to AC conversion efficiency model. This was achieved by making the conversion efficiency dependent on the input power and the operating voltage. Since such a characteristic is specific to an inverter, it requires detailed information from the manufacturer. The efficiency model utilises the look-up table function of PSpice to compute the efficiency during simulation. The model computes a linearly increasing function of the form z = f(x , y), where z is the efficiency of the power output, x is the power input and y is the voltage at the maximum power point. This model is applicable to inverters in general. So far it has been adapted to the manufacturer data provided by SMA for their SunnyBoy 1500 inverter.

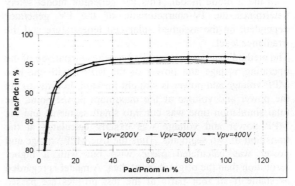

Figure 3: Voltage dependent inverter efficiency (SunnyBoy 1500)

5. VERIFICATION OF THE DIGITAL MPPT MODEL

In order to verify the electrical and functional behaviour of models of PV system components, test profiles containing data for ambient temperature, wind speed and global irradiation were developed. The test profiles have two columns of data. The first column contains the simulation time in seconds and the second column the input quantity separated by a space. The test profiles are designed in such a way that each quantity reaches its maximum and minimum levels and changes with the maximum rate of change during simulation. Furthermore, the data of the test profiles are synchronised in such a way that all possible combinations of these states can be investigated. Table 2 shows the key parameters that formed the basis for the development of the test profiles. The test profiles are subdivided into four different parts. Part I represents the data of a normal sunny day with changing irradiance and ambient temperature levels and low wind speed. In parts II, III and IV various combinations of maximum and minimum values and rates of change are investigated. The condition in which all input values are zero is also checked. Part I of the test profiles starts from 0s and ends after 57600s. After 86000s the end of part IV is reached.

Input quantity	Maximum	Minimum	Maximum rate of change
Irradiance $[H]$=W/m^2	1300	0	15 W/(m^{2}s)
Ambient temperature $[T_d]$=°C	55	-20	0.1 °C/s
Wind speed $[v_w]$=m/s	55	0	10 m/s^2

Table 2: Parameter of the test profiles including limits and the maximum rates of change.

The assessment of the MPPT algorithms is separated into three different aspects, the static behaviour, the dynamic behaviour and the stability of the transient simulation. For this a PV generator was selected that consists of two parallel connected strings each with 16 series connected solar modules. Each 50W module comprises 36 mono-crystalline solar cells whose characteristics are calculated with the 2 diode model. This PV generator model solely determines the IV-characteristic of the PV generator dependent on the assigned solar cell temperature and the irradiance level.

The method that has been applied for the comparison of the operating voltage and power of the MPPT with the true MPP voltage and power is straight forward. In order to get the power and voltage at the maximum power point, the total simulation time was cut into small segments and the MPP power and voltage was exactly calculated at the transition of two segments. Thus the true MPP voltage and power was calculated piece wise but with a higher resolution than the original input data. A model is regarded as stable if all four parts of the test profiles are passed successfully when the final simulation time was set to 90 days.

The accuracy in voltage a_V and power a_P, the mean accuracy in power $\eta_{A,P}$, the energetic efficiency η_E and the root mean square rms_V are presented for the self-defined DMPPT model [3]. The mathematical description of the closeness of the operating voltage to the MPP voltage is described by the mean root square factor. The factor rms_V evaluates the voltage performance over the time specified by t_{sim}.

$$rms_V = \sqrt{\frac{1}{t_{sim}} \int_0^{t_{sim}} (V_{SG} - V_{mpp}) dt}$$

The accuracy a_V is given as a graph over the entire tracking process of 57600s and shows at any time how close the voltage is to V_{mpp}. The instantaneous accuracy in power and voltage is given by:

$$a_V = \frac{V_{SG}}{V_{mpp}} \qquad a_P = \frac{P_{SG}}{P_{mpp}}$$

If a_P is integrated over time and divided by the total simulation time, the average efficiency $\eta_{A,P}$, can be calculated. The value of η_A is equally weighted over time regardless of the irradiance or power level.

$$\eta_{A,P} = \frac{1}{t_{sim}} \int_0^{t_{sim}} a_P dt \qquad \eta_E = \frac{\int_0^{t_{sim}} P_{SG} dt}{\int_0^{t_{sim}} P_{mpp} dt}$$

η_E is the energetic efficiency and describes the energetic performance of a tracker since it is weighted by the power level.

The digital MPPT model was tested using a combination of the PV generator as described earlier and the parameters listed in Table 1. The model successfully passed all criteria for reliable and stable simulation and showed a very good performance as can be seen in Table 3. The duration of the simulation of the first 57600s on a Pentium with 166MHz and 64 Mb memory was 1324s. When the clock frequency $1/t_{sw,}$ was decreased but the ratio of V_{sw}/t_{sw} kept the same, much lower simulation times were reached (see Table 3).

V_{sw}	t_{sw}	rms_V	$\eta_{A,P}$	η_E	t_d
2V	5s	9.5V	0.993	0.9994	1324s
4V	10s	11.2V	0.990	0.9985	485s
8V	20s	12.5V	0.987	0.9959	342s

Table 3: Simulation result of the 10-bit DMPPT model at different parameter settings over the time range of part I of the test profiles.

The operating voltage of the inverter is adjusted in steps of V_{sw} every time t_{sw} elapsed and the algorithm uses a 10-bit AD-converter and comparators for the evaluation of the operating power. This means that the difference between the two successive measurements must be at least $|P_{max}/2^{10}|$ for the system to detect an inversion point of the operating voltage. This resolution is independent of the operational conditions of the solar generator and therefore the operating point moves far away from the true MPP in the case of a flat PV-characteristic. This behaviour is reflected in Figure 4 and Figure 5 after start up and before shut down. In Figure 4 it can also be seen that the operating point drifts further away from the true MPP if the V_{SG} is smaller than V_{mpp}. This is because the absolute value of the slope of the PV-characteristic, at a specific power level is higher on the right side of the MPP than on the left. A voltage step away from the MPP voltage with a fixed step width V_{SW}, as in the case of the DMPPT model, therefore results in a bigger power step on the right side of the MPP where $V_{SG} V_{mpp} > 0$.

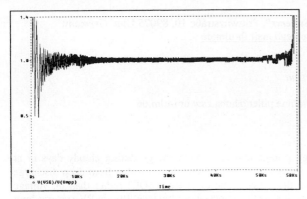

Figure 4: Accuracy a_V of the 10-bit digital MPPT model over the range of part I of the test profiles.

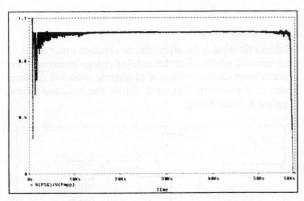

Figure 5: Accuracy a_P of the 10-bit digital MPPT model over the range of part I of the test profiles.

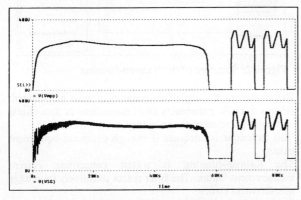

Figure 6: V_{SG} (V(VSG)) and V_{mpp} (V(Vmpp)) of the digital MPP tracker over the full range of the test profiles.

A demonstration of how well the model is able to follow the MPP is given in Figure 6 and Figure 7 which show the

V_{mpp}, V_{SG} and P_{mpp}, P_{SG}. The operating voltage is oscillating equally around V_{mpp} and can even follow when the irradiance level changes at its maximum rate of 15W/(m²s) or the temperature changes by 0.1°C/s. The tracker follows the MPP values without significant fluctuations. Even in part II, III and IV of the test profiles, when the quantities change at their maximum rate of change, the MPP values are also very good.

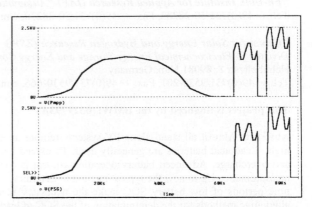

Figure 7: P_{SG} (V(PSG)) and P_{mpp} (V(Pmpp)) of the digital MPP tracker over the full range of the test profiles, the curves are mostly identical.

6. CONCLUSION

The inverter model demonstrates that the simulation of digitally controlled components in combination with a long-term (90 days) simulation of PV system components is both possible and reasonable. Of course the simulation time is longer than for standard PV simulation software. However the results, in terms of the electrical system behaviour, are much more detailed, accurate and transparent at any part of the circuit.

This digital model can be regarded as the beginning of a series of implementations of digitally controlled components into PVNETSIM. In addition to the maximum power point tracker model, the simulation of digitally controlled load management systems or charge controllers is also possible.

REFERENCES

[1] Schilla, T., R. Hill: PVNETSIM a versatile PV simulation software based on the Design Center Programs Schematics, Spice and Probe. 14th European Photovoltaic Solar Energy Conference, Barcelona 1997: H. S. Stephens & Associates

[2] MicroSim PSpice A/D Reference Manual, Version 7.1, October 1996 MicroSim Corporation, 20Faibanks, California

[3] Jantsch, M., Measurement of PV maximum power point tracking performance, 14th European Photovoltaic Energy Conference, 1997, Stephens & Associates

STATE-OF-CHARGE DETERMINATION FOR LEAD-ACID BATTERIES IN PV-APPLICATIONS

Ch. Ehret[a], S. Piller[b], W. Schroer[a], A. Jossen[b],

[a] *Fh-Ulm, Institute for Applied Research (IAF) "Automation Systems",* Prittwitzstraße 10, 89075 Ulm, *Germany*
Tel: ++49(0)731/50-28282, Fax: ++49(0)731/50-28270, e-mail: ehret@mail.fh-ulm.de

[b] *Center for Solar Energy and Hydrogen Research (ZSW)*
Division 3: Electrochemical Energy Storage and Energy Conversion
Helmholtzstr.8, 89081 Ulm, Germany
Tel: ++49(0)731/9530-201, Fax: ++49(0)731/9530-666, e-mail: sabine.piller@huba.zsw.uni-ulm.de

This project is financed by the BMWi (0329793A)

Abstract: Almost all stand-alone PV-systems require an energy storage unit to provide energy during cloudy days or at night. Lead acid batteries are generally used. To extend battery lifetime, the battery must be protected from deep discharge and overcharge. Advanced battery operation control is based on the battery state of charge (SOC) rather than on voltage measurements. Up to now the typical method to determine battery SOC is the ampere-hour balance. But, in PV-systems long time periods of low battery SOC make the necessary recalibrations for this method difficult. This paper presents three alternative methods for the determination of the SOC of lead acid batteries.
Keywords: - 1: Batteries - 2: Modelling - 3: Lifetime

1 SOC DETERMINATION

To extend battery lifetime, the battery must be protected from deep discharge and overcharge. For operation control the battery state-of-charge has to be known. All presented methods use relatively easy accessible measurements (battery voltage, current and temperature) to determine the SOC. The methods are based on the following battery model represented as an electrical equivalent circuit [1]:

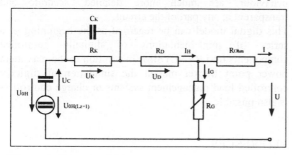

Figure 1: Battery model

with U_{OH}: rest potential, U_K: diffusion potential,
U_D: charge-transfer potential, I: current, U: voltage,
I_G: gassing current

In particular, the approaches are:

➢ **Kalman Filter:** an algorithm to estimate inner states of dynamic systems.
➢ **Artificial Neural Network:** an Artificial Neural Network can be applied if training data from batteries of similar type are available
➢ **Linear Model (parametrical or nonparametrical):** applies a linear equation to calculate the SOC.

1.1 Kalman Filter

Kalman filtering is an algorithm to estimate inner states of a dynamical system. For the state-of-charge assessment the battery open circuit voltage is of interest, which is an inner state of the battery. Figure 2 shows the structure of the applied Kalman-Filter.

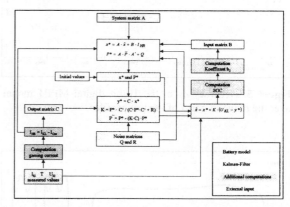

Figure 2: Structure of the Kalman-filtering

Before starting the algorithm parameters must be declared. Battery specific parameters from manufactures data sheets and noise parameters are necessary.
This technique is adequate if enough computer equipment exists.
The Kalman-filtering is robust concerning current measurement failure. The filter reacts sensitively to voltage measurement errors

1.2 Artificial Neural Network (ANN)

An Artificial Neural Network can be applied if training data (current, voltage, temperature and SOC reference) are available. For the training of this ANN, the SOC, which was calculated by the Kalman Filter, is used as reference. From this data the weighting factors of the ANN are obtained. The application of a backpropagation neural network showed good results for the SOC determination. The input data for this determination are voltage, current, temperature measurements and the fed back SOC, the output is the SOC difference between two time steps (see Figure 3). The overall SOC is given by the integration of the SOC-difference.

The trained ANN can be applied to other battery types and to batteries of different age phases.

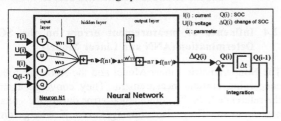

Figure 3: Structure of the fed back ANN

1.3 Linear Model

The linear model applies a linear equation based on the battery model to calculate the SOC. From the model two linear models were developed:

1. A **parametric linear model**: the parameters are determined by separating measurements of a battery if no reference is available. The measured parameters of the battery are: rest potential, diffusion resistance and charge-transfer resistance.

Input data are voltage, current measurements and the fed back SOC (see Figure 4).

2. A **nonparametric linear model**: the factors of the linear equation are determined from reference data through the method of the least-mean-squares-algorithm. If reference data are on hand, this model is an easy method for SOC-determination

To 1.: If no reference is present, a **parametric linear model** with the follow equation can be used:

$$\Delta Q(i) = \alpha_0 + \alpha_1 \cdot U(i) + \alpha_2 \cdot U(i-1) + \alpha_3 \cdot I(i) + \alpha_4 \cdot I(i-1) + \alpha_5 \cdot Q(i-1)$$

In this equation, the α-parameters are calculated by the following terms:

$$\alpha_0 = U_{OH}\Delta t / R_K \quad \alpha_1 = \left(-\tau_K / R_K - \Delta t / R_K\right)$$
$$\alpha_2 = \tau_K / R_K \quad \alpha_3 = \left(- R_D \tau_K / R_K - R_D \Delta t / R_K\right)$$
$$\alpha_4 = R_D \tau_K / R_K \quad \alpha_5 = \left(-\Delta t / CR_K\right)$$

R_K, R_D and C_K are the components of the underlying equivalent circuit (see Figure 1). The single elements can be determined via a separating measurement procedure.

To 2.: The **nonparametric linear model** is a simpler model, which is characterised by high robustness in relation to measurement errors and wrong initial conditions. It is important to notice that the β-factors of the model do not resemble with the pysically derivable α-parameters. The β-faktors are calculated by the least-mean-squares-algorithm. Therefore, reference data is necessary in form of voltage and current measurements and a reliable SOC-reference. The equation of the model is:

$$\Delta Q(i) = \beta_0 + \beta_1 U(i) + \beta_2 I(i) + \beta_3 Q(i-1)$$
$$Q(i) = Q(i-1) + \Delta Q(i)$$

with

Q(i): SOC, ΔQ(i): SOC-difference, U voltage and I current measurements

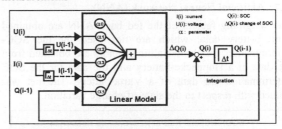

Figure 4: Structure of the linear Model

The linear model can be applied to other battery types and to batteries of different age phases.

2 DISCUSSION AND RESULTS

In Figures 5, 6, 7 the SOC curves of the three models are shown, in comparison with the SOC reference. The reference for the SOC is taken from [4].

The algorithms show low errors. The average absolute error lies in the range of 4 to 9%. Error extremes are observed during long periods of charging and discharging with high constant current (capacity-tests). Further simulations were done with batteries of different types which were exposed to other operation regimes (i.e. non-cyclic data).

2.1 Kalman-Filter

To apply this method initial values and parameters must be specified. The initial values are not crucial. Critical are however the stochastic defaults for the filter.

The parameters of the Kalman-Filter were adapted to the tested battery. If the parameters for a cycle and a battery are optimised, very good results could be achieved as shown in **Figure 5**. In simulations with data of similar structure the error in comparison to the reference remains under 7 %, even if parameters are used which were optimised for another battery. When the data is extended to include long charging/discharging phases with (for PV-systems) high currents ($\geq I_{10}$) maximum absolute errors of 20 % were observed during this high current periods. But, the overall mean error still stays under 7 %:

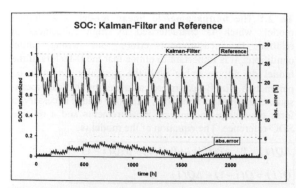

Figure 5: SOC curves of the Kalman-Filter, the reference and the error between the two curves

2.2 Artificial Neural Network (ANN)

The weighting factors of the fed back ANN are obtained from battery data, which are not used for the SOC-determination, i.e. the weighting factors are obtained from the data of a Hoppecke-battery and the data for the SOC determination are data of a Varta-battery. The net was trained with respect to the state-of-charge variation.

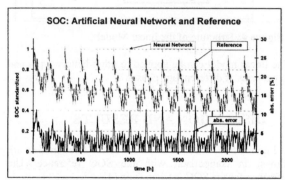

Figure 6: SOC curves of the Backpropagation Artificial Neural Network, the reference and the error between the two curves

2.3 Nonparametric Linear Model

With this linear model the factors are also computed from a cycle of a battery (here a Hoppecke battery with nominal capacity 150 Ah) .Then the state-of-charge can be estimated for other battery types and for batteries of different age phases (here a Varta battery with nominal capacity 180 Ah) with sufficient accuracy. The average error stays under 10% with the linear model although the β-factors were calculated by the battery data presented here.

In order to reduce the error for batteries, which have a greater or smaller nominal capacity, the β-factors have to be multiplied by a factor. Also for different sample rates a factor have to be considered. Due to the simple linear equation a technical implementation of the algorithm on a processor (8 bits) is possible. This algorithm was already integrated into a charge indicator for testing in co-operation with the firm Steca.

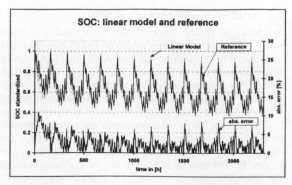

Figure 7: SOC curves of the Linear Model, the reference and the error between the two curves

2.4 Influence of measurement errors on the SOC Determination (ANN and Linear Model)

The Nonparametric Linear Model and the ANN are robust concerning current measurement. They converge against the reference SOC by coincidental error during the voltage measurement. They also converge even if wrong initial state-of-charge, defined measurement errors (voltage and current) against the SOC reference.

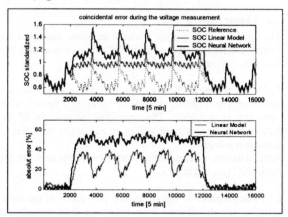

Figure 8: Coincidental error of the voltage measurement

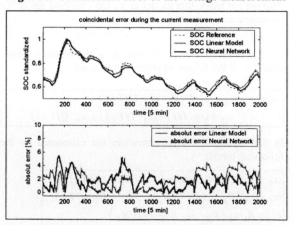

Figure 9: Coincidental error of the current measurement

2.5 Parametric Linear Model

The mean error of the SOC determination with the Parametric Linear Model is about 15%. Possible reasons for this are:

➢ The gassing current is not considered in the model.
➢ The diffusion resistance and the charge-transfer resistance could not clearly separated by the measurement.
➢ The acid stratification is not considered.

Deduction:

If in further work the mean error of the parametric model is reduced a standardized, industrially usable measuring procedure could be developed. The advantage of this method would be that for the model no reference data have to be produced.

3 CONCLUSIONS

➢ The algorithms show low mean errors (range 4 to 15%)
➢ The Kalman-filtering is adequate if enough computer equipment exists. It has the advantage to relay on electrochemical background. But the expenditure for optimal parameter estimation is high.
➢ The linear model is very suitable for technical implementation in simple charge indicators (e.g. with 8 bit processor).
➢ Alternatively to the purely empirical model adjustment by reference data, it is also possible to determine the parameters by measurements at the battery (the parametric linear model). The purely empirical model becomes a physical parametric model.
➢ The ANN is a simple non-linear model. The technical implementation is somewhat more difficult than that of the linear model.

4 Literature

[1] W. **Burkner**, *Verfahren zur Ladezustandsbestimmung von Blei-Batterien in Photovoltaikanlagen*, IfE Schriftenreihe, Heft 27, Lehrstuhl für Energiewirtschaft und Kraftwerkstechnik, Technische Universität München (1994)

[2] C. **Ehret**, M. Ivanov, W. Schroer, P. Adelmann. *Nichtparametrische Modelle zur Bestimmung des Ladezustands*. Tagungsband, 14. Symposium Photovoltaische Solarenergie, OTTI-Technologie-Kolleg, Staffelstein, 171-176. (1999).

[3] S. **Piller**, V. Späth, A. Jossen. *Kalman-Filter zur Ladezustandsbestimmung von Bleibatterien in PV-Anlagen*. Tagungsband, 14. Symposium Photovoltaische Solarenergie, OTTI-Technologie-Kolleg, Staffelstein, 454-458. (1999).

[4] **ZSW** (Hrsg.), ISET, ISE.: *Referenzdaten zum Test von Verfahren zur Ladezustandsbestimmung von Bleibatterien*, CD-ROM. (1999).

"PV-EMI" - DEVELOPING STANDARD TEST PROCEDURES FOR THE ELECTROMAGNETIC COMPATIBILITY (EMC) OF PV COMPONENTS AND SYSTEMS

S. Schattner[1], G. Bopp[1], T. Erge[1], R. Fischer[2], H. Häberlin[2], R. Minkner[2], R. Venhuizen[3], B. Verhoeven[3]

[1] Fraunhofer-Institut für Solare Energiesysteme ISE
Oltmannsstr.5, 79100 Freiburg, Germany
phone: ++49 761 45 88 313, fax: ++49 761 45 88 217, email: sschatt@ise.fhg.de

[2] Berner Fachhochschule, Hochschule für Technik und Architektur Burgdorf
Jlcoweg 1, 3400 Burgdorf, Switzerland
phone: ++41-34-4266853, fax: ++41-34-4266813

[3] KEMA T & D Power
Utrechtseweg 310, P.O. Box 9035, 6800 ET Arnhem, Netherlands
phone: ++31 26 3 56 35 81, fax: ++31 26 3 51 36 83

ABSTRACT: The DC side of a PV system acts as an antenna. In combination with an inverter this antenna is able to radiate emissions down to relatively low frequencies of 150 kHz disturbing appliances such as radio and TV sets. So far no standard considers this quite PV specific EMC problem. On the other hand PV systems are exposed to the impacts of direct and, more often, indirect lightning strokes. Until now no reliable knowledge exists on the induced overvoltages of indirect lightning strokes which stress module insulation and require overvoltage protection for the whole PV system. This, too, is a matter of EMC. Both mentioned EMC aspects - emission and immunity - are now jointly treated in the course of the PV-EMI project which is partially funded by the EU and will finish in June 2000. This paper presents its major outcome.
Keywords: Electromagnetic Compatibility (EMC) - 1: PV System - 2: Inverter - 3

1. INTRODUCTION

With photovoltaics becoming more and more widely spread the EMC of PV components and systems represents a quality characteristic of increasing importance. PV systems are indeed able to cause electromagnetic interferences. Radio and TV sets as well other radio appliances can be affected, even interferences of telephone systems occured.

Particularly the DC side of a PV system, consisting of the PV generator and the DC main cable, is problematic to manufacturers of PV power electronics, since from a radio frequency (RF) point of view it behaves like an antenna being supplied by the inverter or a switching charge regulator and, by this, causing electromagnetic field disturbances. Manufacturers often feel uncertain about applicable limits and measuring procedures which ought to help ensure the EMC of their product.

EMC always refers to two points: in a particular electromagnetic environment a device must not produce too much RF emissions. On the other hand, it must show sufficient immunity against impacts resulting from this environment. This also applies to the DC side of a PV system. Not only does it produce RF emissions in an antennalike manner, it is on its part exposed to impacts of lightning strokes. Lightning strokes produce transient magnetic fields which can induce overvoltages in nearby loops. PV generators form such loops. This requires overvoltage and lightning protection for the PV system.

Both phenomena are still not considered in standards concerning EMC and lightning protection [1]. The authors' work, presented in the following, as well as an EU project carried out in parallel [2] should provide a sufficient basis for filling this standardization gap.

One can see from Figure 1 that several electromagnetic coupling paths are relevant to a PV system (in this case a grid-connected system). But in contrast to the DC side and its specific EMC behaviour the AC side represents a common EMC problem which can be treated by existing standards.

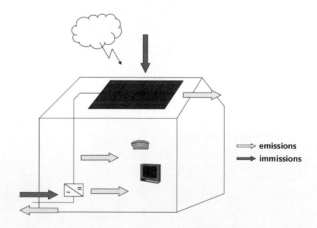

Figure 1: Electromagnetic coupling paths of a grid-connected PV system

2. EMISSION TESTS

2.1 Principal approach

In order to fill the standardization gap concerning RF emissions two values had to be determined in the frequency range of 150 kHz - 30 MHz by emission tests:

- the antenna factor H/V of the DC side, defined as ratio of the magnetic field strength H (measured at a distance of 3 m) and the disturbance voltage V (measured for example at the DC inverter input), and
- the input impedance Z of the DC side (the antenna impedance).

As these two items depend on several system parameters, a reasonable worst case assessment has to be done in the end. Once the antenna factor is determined and maximum magnetic field strengths are determined as well, one can derive the maximum disturbance voltages which then serve as EMC limits specified for the DC terminal of inverters or charge regulators. As maximum magnetic field strengths the limits given by the former VDE 0878-1 [3] (Figure 2) seem somewhat to high, but appropriate enough for giving a certain orientation. Current standards with more reasonable limits do not exist.

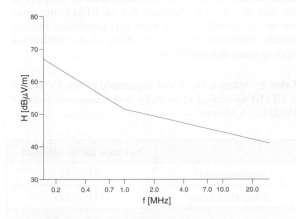

Figure 2: Magnetic field strength limits specified in the former VDE 0878-1 (measurement distance: 3 m)

Appropriate impedances can be considered in an impedance simulation network (ISN), shown in Figure 3. This ISN allows relatively comfortable RF voltage measurements in the laboratory under well defined conditions.

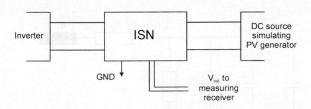

Figure 3: Scheme of an EMC test set-up for conductor-borne RF emissions on the DC side

In fact, four (namely two times two) different values have to be determined, since two modes of conductor-borne electromagnetic disturbances are to be distinguished: differential mode and common mode. Differential mode disturbances cause RF currents flowing in an antiparallel way, whereas common mode disturbances cause RF currents flowing in a parallel way. Usually, installed AC cables are bundled, so that differential mode disturbances are of very minor importance, since they do not cause any significant electromagnetic fields. Such fields then result from common mode disturbances. But this does not apply to the DC side of PV systems. The DC main cable normally is bundled, but the PV generator itself forms a loop. In other words, the inverter (or the charge regulator) does not only supply a (common mode) dipole antenna but also a (differential mode) loop antenna (Figure 4). In theory, each is able to cause electromagnetic disturbance fields.

Therefore, the following values had to be determined in the course of emission measurements: the common mode antenna factor, the differential mode antenna factor, the common mode impedance, the differential mode impedance.

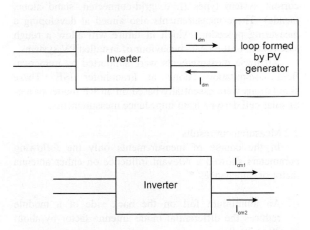

Figure 4: Differential mode (dm, above) and common mode (cm, below) antenna model

In order to determine these four values, emission measurements as well as numerical field computations were made. RF emission measurements were done in mainly two different ways.

Tests under defined conditions were done at the KEMA premises. First, tests were made on the basis of single modules, hereby investigating the antenna factor and the impedance of each module type. The modules differed in several parameters such as cell material, existing/non-existing frame, existing/non-existing aluminium foil on the back side and cell interconnection. Subsequently, arrays of modules were investigated, again determining the antenna factor and the impedance. Figure 5 shows the array used for this purpose. It was supplied by a defined either common mode or differential mode RF signal.

Figure 5: Module array and loop antenna at the KEMA premises

In addition, on site measurements were made by Fraunhofer ISE and HTA Burgdorf considering various current system types (i. e. grid-connected, stand alone, façade). These measurements also aimed at developing a measuring procedure which in future will allow a rough assessment of the EMC behaviour of installed PV systems.

All these measurements were supported by numerical field computations done at Fraunhofer ISE. These simulations were essentially based on an RF model of a c-Si solar cell derived from impedance measurements.

2.2 Measurement results

In the course of measurements only the following parameters showed a relevant influence on either antenna factor or impedance:

- An aluminium foil on the back side of a module reduces the differential mode antenna factor by about 20 to 30 dB.
- A frame reduces the differential mode antenna factor only by about 10 dB.
- A shielding of the DC main cable may reduce electromagnetic fields significantly (at one PV site up to 50 dB!), but only in case the shield is grounded on both sides with (from an RF point of view) a very good connection to earth (at the PV plant mentioned above: aluminium cladded flat roof). Shielded DC main cables grounded on both sides are the best choice for optimal lightning protection, but are not used very often due to the increased cost.

Regarding EMC behaviour the worst case PV system is therefore represented by a system without any shielding of the DC main cables and with modules having no aluminium foil. Metallic module frames which are very common do reduce the differential mode antenna factor. On the other hand, grounded module frames can increase the emission of common mode signals of a PV system (see 2.3 Simulation results).

Filter measures turned out to be quite simple. At PV plants and inverters which were dominated by common mode disturbances two common mode capacitors of a few µF and a common mode choke of about 1 mH were

sufficient in order to suppress electromagnetic fields almost down to the ambient RF noise. Figure 6 exemplifies this showing the affected magnetic field strengths at one PV site.

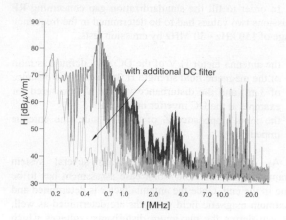

Figure 6: Magnetic field strengths measured at a PV site in Burgdorf (with and without DC filtering)

The measurement results gained at KEMA are listed in Table 1. When interpreting these results and particularly when comparing them to the results gained by simulations one has to take into account that at KEMA the array consists of modules with a back side aluminium foil. As already mentioned, this has an effect on the differential mode antenna factor of the array.

Table 1: Antenna factor and impedance of the PV system at KEMA consisting of modules with aluminium foil (f = 150 kHz .. 5 MHz)

	antenna factor (dBS/m)
common mode	-91 .. -82
differential mode	-111 .. -124
	impedance (W)
common mode	1150 .. 100
differential mode	20 .. 30

2.3 Simulation results

Figure 7 shows the AC equivalent circuit of a c-Si solar cell. The different capacitances are due to the behaviour of majority and minority charge carriers within or beyond the depletion region.

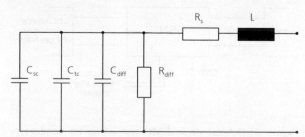

Figure 7: AC equivalent circuit of a c-Si solar cell

For a c-Si solar cell of 100 cm² the series resistor R_s typically amounts to about 10 m$\overline{\text{W}}$ and can be considered as constant. In contrast, the RC element ($R_{diff} \parallel C_{total}$) is dependent on the cell's bias voltage, the irradiation, the temperature and the frequency. Accordingly, these parameters were varied systematically in the course of impedance measurements. For bias voltages between 350 and 450 mV, irradiation between 500 and 1000 W/m², temperatures between 30 and 75 °C and frequencies around 100 kHz Table 2 lists the determined values of C_{total} (= C_{sc} + C_{tc} + C_{diff}) and R_{diff}.

Table 2: Values determined for C_{total} and R_{diff} (f = 100 kHz)

	f = 100 kHz
C_{total}	10 µF
R_{diff}	10 .. 400 m$\overline{\text{W}}$

Taking into account that $Z = 1/2pfC_{total}$ and R_{diff} decrease with increasing frequencies, the impedance of the RC element can be neglected in the entire frequency range from 150 kHz to 30 MHz. The cell impedance then results to R_s plus the inductance L which primarily is determined by the geometry of cell interconnection.

In a model of the PV module R_s was subsequently considered as a plain with a finite specific conductivity. The inductance L is generated automatically by CONCEPT [4], the applied software. Figure 8 shows as an example the model of a PV system consisting of 6 strings, each with 4 standard modules in series.

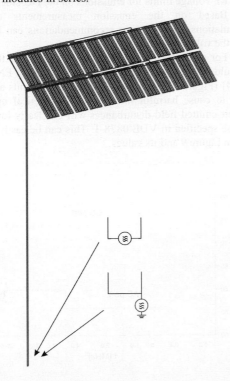

Figure 8: Model of a PV system as considered during numerical computations

The influence of the following parameters was investigated when doing simulations:

- height h of the PV generator
- number of PV modules in series (affecting the size of the differential mode loop)
- type of PV system (n parallel strings, each with m modules in series)
- the influence of a frame
- in case of an existing frame: type of grounding (either near the DC main cable or in quite a distance)

When interpreting the results, one must take into consideration that above a certain frequency (mostly around 5 MHz) the investigated antenna becomes resonant and therefore no general tendency can be recognized there. However, for frequencies below 5 MHz the following main results were obtained.

- **common mode**:
  - The common mode impedance is mainly dependent on the grounding of existing frames (which is quite common). A connection to earth, regardless at which location, reduces the common mode impedance by a factor between 2 and 3.
  - The common mode antenna factor is mainly influenced by the height of the PV generator. With increasing height (h = 1.5 .. 7.5 m) the common mode antenna factor also increases, in maximum up to 10 dB.
- **differential mode**:
  - The differential mode impedance is strongly dependent on the type of PV system. It decreases with increasing number of strings in parallel and it increases with increasing number of modules in series (that is, with increasing loop size). A PV system with 6 x 4 modules (6 Strings with in total 24 modules) has a differential mode impedance which is about 5 times lower than a system with 1 x 24 modules (1 string with in total 24 modules).
  - The differential mode antenna factor is only affected slightly by the parameters mentioned above.

The simulation results were obtained for the worst case (without an aluminium foil). Taking into account the effect of the aluminium foil for the differential mode antenna factor, these results are in satisfying accordance with the results gained through measurements by KEMA. Figure 9 and Figure 10 show reasonable worst case assessments regarding antenna factors and impedances.

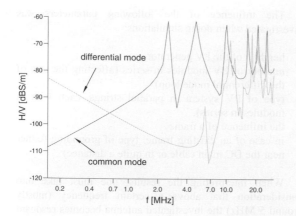

Figure 9: Antenna factor worst case assessment (based on simulations)

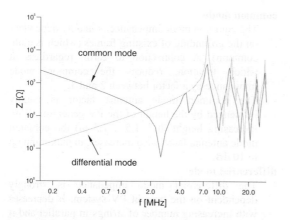

Figure 10: Impedance worst case assessment (based on simulations)

2.4 DC-ISN

As in general common mode disturbances are dominant for frequencies higher than a few 100 kHz, already in the beginning of the project a DC-ISN for 1000V and 100A with a common mode impedance Z_{cm} of 150 W was developed and sucessfully tested by HTA Burgdorf.

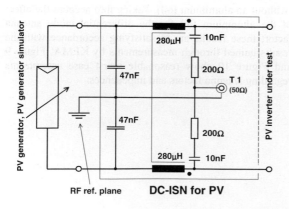

Figure 11: Simple DC impedance stabilisation network with Z_{cm} = 150 Ω (T1 terminated with 50 Ω); a voltage signal (-9.5 dB) is available at T1

Based on the emission measurements and the simulations now the following recommendations can be given for the impedances of a universal ISN:

- common mode impedance Z_{cm}: 100 W to 300 W
- differential mode impedance Z_{dm}: 100 W to 300 W

In cooperation with a nearby EMC company (Schaffner AG) HTA Burgdorf is developing a modified DC-ISN to meet the new values and to offer a simple attenuation factor for practical measurements. Like the LISN's used on the AC side, the new modified ISN takes into account not only common mode emissions, but also differential mode emissions.

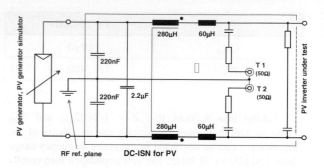

Figure 12: Modified DC impedance stabilisation network with Z_{cm} and Z_{dm} between 100 W and 300 W (T1 and T2 terminated with 50 W); exact values to be defined at final meeting in May 2000; voltage signals (attenuation depending on resistor values) available at T1 and T2

2.5 RF voltage limits for emissions on the DC-side

Based on the emission measurements and the simulations the following recommendations can be given for the voltage limits:

For conducted RF voltages on the DC side, the limits should be between the AC and the DC limits of EN 55014-1 [4] (Figure 13). Inverters meeting these limits are likely not to cause harmful interference in practical operation. Their emitted field disturbances will be clearly lower than those specified in VDE 0878-1. This can be easily derived from Figure 9 and its values.

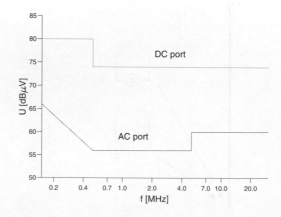

Figure 13: AC and DC port voltage limits specified in EN 55014-1

A final decision on the exact recommendations for the impedance values of the ISN and the applicable voltage limits will be made in the course of May 2000. These recommendations then will be submitted to the responsible IEC committee.

3. IMMUNITY TESTS

In the course of the PV-EMI project, HTA Burgdorf carried out also immunity tests at PV modules and models of PV arrays. Figure 14 shows the test generator developed at HTA Burgdorf. With this impulse current generator impulse currents up to 125 kA with maximum di/dt values of up to 40 kA/µs can be produced. Such rise velocities are representative for typical lightning currents. The usable array test area is 2.2 x 1.2 m².

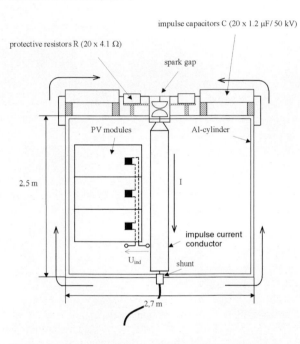

Figure 14: Impulse current generator at HTA Burgdorf

The following parameters were examined so far:

- module type (mainly differing in the number of cell rows and the wiring method)
- influence of the frame (on module and array level)
- influence of aluminium foil on back side of module
- bypass diodes
- type of grounding and location of grounding wires

Figure 15 shows the results for a single KC60 module (typical for a standard module: four cell rows) and a di/dt$_{max}$ of 25 kA/µs. Induced voltages were measured by varying the distance between the first cell row of the module and the lightning current up to 630 mm. Results above 630 mm were gained by calculations.

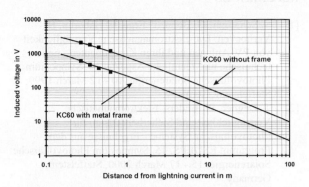

Figure 15: Induced voltages for different distances from simulated lightning current (KC60 module, di/dt$_{max}$ = 25 kA/µs)

An important result from these measurements is certainly that a frame reduces induced voltages by a factor of 3 to 5 which is in good accordance with previous experiments gained with clearly lower di/dt values [6]. This also applies to arrays of modules when all frames are connected. An aluminium foil on the back side of the module causes an even higher reduction (measured factors 7 to 9), which can be cumulated with the reduction due to the frame. These results agree well with the reductions for differential mode emissions (see section 2.2). Further results are:

- Some modules were tested with their bypass diodes. As some manufacturers use diodes with only 100 V break-down voltage, some of these diodes were destroyed during the tests. For immunity reasons bypass diodes with much higher breakdown voltages should be used.
- Grounding of frames has to be done by **bundling** the ground lead and the DC main cable close together.
- Varistors in the PV generator junction box, at the DC inverter input and at the AC output are necessary and sufficient for overvoltage protection.

Immunity measurements are still in progress. Final recommendations concerning measures for lightning protection and specifications for surge measurements will be given at the end of the project in June.

4. CONCLUSION

In the course of this project emission tests as well as immunity tests were made on the DC side of PV systems and on single components. Emission tests have already come to an end, so that soon detailed recommendations can be given for an EMC standard. Immunity tests will be carried out further. The project partners expect as a further outcome recommendations concerning measures for lightning protection and for a standardized surge test.

REFERENCES

[1]   S. Schattner, G. Bopp, T. Erge, H. Häberlin, B.
       Verhoeven: Die elektromagnetische Verträglichkeit
       von PV-Anlagen und ihren Komponenten - neue
       technische und rechtliche Aspekte, 14. Symposium
       Photovoltaische Solarenergie, 10 - 12 March 1999,
       Staffelstein, Germany
[2]   T. Degner, W. Enders: Elektromagnetische
       Verträglichkeit und Sicherheitsdesign für
       Photovoltaische Systeme - Ergebnisse aus dem
       Programm ESDEPS, 15. Symposium Photovoltaische
       Solarenergie, 15 - 17 March 2000, Staffelstein,
       Germany
[3]   DIN VDE 0878 Teil 1, Funk-Entstörung von
       Anlagen und Geräten der Fernmeldetechnik -
       Allgemeine Bestimmungen, December 1986
[4]   CONCEPT II, Version 8.0, November 1998,
       Technical University Hamburg-Harburg, Department
       of Theoretical Electrical Engineering
[5]   EN 55014 (CISPR 14:1993), Suppression of radio
       disturbances caused by electrical appliances and
       systems, Limits and methods of measurement of
       radio disturbance characteristics of electrical motor-
       operated and thermal appliances for household and
       similar purposes, electric tools and similar electric
       apparatus, December 1993
[6]   R. Möck: Einkopplung von Überspannungen in
       Solarmodule, practical course, University of applied
       sciences Augsburg in cooperation with Fraunhofer
       ISE, 1992

This work was funded in part by the European Commission
in the framework of the Non Nuclear Energy Programme
JOULE III.

PV SIMULATION AND CALCULATION IN THE INTERNET - THE ILSE TOOLBOX

Volker Quaschning *, Rolf Hanitsch **, Mike Zehner *** and Gerd Becker ***

* DLR PSA Plataforma Solar de Almería · Apartado 39 · E-04200 Tabernas (Almería)
Tel.: ++34-950-387906 · Fax: ++34-950-365313
Volker.Quaschning@psa.es
** Berlin University of Technology · Institute of Electrical Power Engineering · Sec. EM 4
Einsteinufer 11 · D-10587 Berlin · Tel.: ++49-30-31422403 · Fax: ++49-30-31421133
Rolf.Hanitsch@iee.tu-berlin.de
*** FH Munich University of Applied Science · Electrical Engineering Department · SE Laboratory
Dachauer Str. 98 · D-80335 Munich · Tel.: ++49-89-12651451 · Fax: ++49-89-12651299
zehner@e-technik.fh-muenchen.de

http://emsolar.ee.tu-berlin.de/~ilse

ABSTRACT: Since 1996 the Interactive Learning System for Renewable Energy (ILSE) is accessible in the Internet. ILSE is one of the first interactive applications in the Internet about renewable energy. This learning system offers lessons about solar energy, wind energy and energy policy. The users can test their knowledge with interactive multiple choice test. Besides the lessons and tests for learning purposes there is an extensive toolbox with multiple possibilities for online calculations and simulations. The toolbox contains tools for the calculation of characteristics of solar cells, PV modules and inverters as well as complex economic analysis and performance analysis of grid connected PV systems. The ILSE learning system and the toolbox are for free for everybody using the following Internet address: http://emsolar.ee.tu-berlin.de/~ilse.

Keywords: Simulation - 1: : Software - 2: Education and Training - 3

1. INTRODUCTION

In 1995 we started to develop a totally new learning system for renewable energy. The Interactive Learning System for Renewable Energy (ILSE) should present information about risks of the present energy supply and different types of renewable energy conversion. Lessons about solar energy, wind energy or energy policy give a general idea about renewable energy conversion, and there is a German and an English version of ILSE, so users can switch directly between both languages.

The aim was to develop a learning system that does not only offer static texts and graphics. One main emphasis during the development of ILSE has been interactive components for calculation and simulation. Since 1996 the ILSE learning system is accessible in the Internet.

Since the ILSE learning system has got very positive resonance and the access rates are increasing continuously we have enlarged the whole system. For this we have developed new powerful online simulation and calculation tools, which attend the lessons and we have enlarged the toolbox.

2. LANGUAGES OF INTERACTIVE COMPONENTS

One of the first computer languages that has been used in the World Wide Web (WWW) was HTML (Hypertext Markup Language). In the first version of HTML Internet applications were only able to present information in texts and graphics. An interaction with the user was only possible in a very simple way.

In 1995 the Sun Microsystems Company presented the new computer language Java. This object oriented language is similar to the C++ computer language. Wit Java it is possible to develop programs that can be executed directly from the Internet. Java Internet applications, the so-called applets, can be embedded into a HTML Internet page. The Java applets can be loaded by a Java compatible Internet browser such as Netscape Communicator or the Microsoft Internet Explorer.

The Java compiler produces Java applets in a machine independent byte code. For running Java programs it is not longer important what kind of operating system is used. A Java compatible browser that is available for most common operating systems is sufficient.

Another possibility to develop interactive Internet applications is the script language Javascript. This language is also available since end of 1995. The Javascript program code can be integrated directly into a HTML page. The Internet browser interprets and executes the script program. Script programs are not so fast as Java applets and they offer fewer opportunities. Hence, the Liveconnect concept from Netscape offers possibilities to combine Java and Javascript programs and to exchange data.

Microsoft has developed the new Internet languages Active X and Visual Basic Script that can be only used with Microsoft operating systems. In the first time there was a lack of security at the Microsoft languages.

By these reasons we have only used Javascript and Java for the development of the interactive components. The ILSE tools run with the Netscape Communicator or the Internet Explorer under different operating systems.

3. THE STRUCTURE OF ILSE

We have equipped ILSE with a comfortable navigation system and an attractive graphic design using frame technology. The three main subjects energy policy, solar energy and wind energy are presented in several lessons. Photovoltaics is one of the main topics. Fig. 1 shows an example of a lesson about principles of photovoltaic energy conversion.

The lessons and tools are available in two languages. The users can switch between the English or the German version every time by clicking on icons at the left frame.

All links of ILSE are relative. This is necessary to install the learning system in any directory. The users can download the whole ILSE system from the Internet and install it on a local hard disk. This possibility saves expensive online time when working with the ILSE system. When choosing the download option from the main menu the user will get a zipped version of ILSE (about 2.2 MB). After unzipping it to the local hard disk it can be started with a browser such as Netscape Communicator even in the off-line mode.

For most lessons there are multiple choice tests to check the knowledge and simulation tools for a better

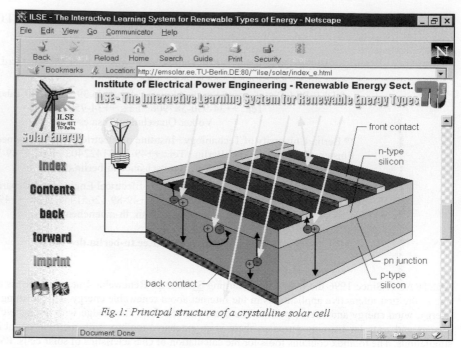

Figure 1: *ILSE* lesson about photovoltaics

understanding of complex correlations.

4. THE ILSE TOOLBOX

All calculation and simulation tools of the several lessons are unified in the ILSE toolbox. Due to the use of Java and Javascript there is no restriction to a certain operating system and all tools run directly in the Internet. The users can change parameters and achieve the results immediately online. So they could understand a complex correlation interactive through these applets and even better than it is possible by reading in a book. Besides you can not only use most of the tools for learning purposes. They are also usable for practical applications.

4.1 Interdisciplinary Tools

Concerning the subject energy policy, there are interdisciplinary tools that can not only be used for photovoltaics. One of the most powerful tools is the Energy tool for complex economic calculations and emission estimation (Fig. 2). This tool calculates economic data such as production costs of two alternatives that can consist of two different components. The impact of increasing fuel prices and carbon dioxide

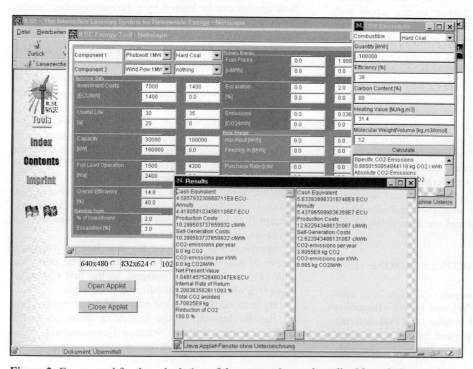

Figure 2: Energy tool for the calculation of the economics, carbon dioxide emissions and influence of taxes at photovoltaic systems and competitors

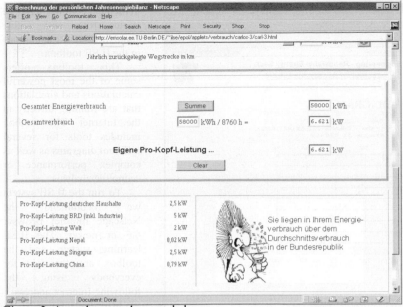

Figure 3: Annual personal energy balance

taxes can be included in the calculations. An emission tool compares the CO_2-emissions of different conventional and renewable generation types.

Another useful tool is an energy unit calculator that converts different energy units such as Joule, Wh (Watt-hours) or Btu (British thermal unit) into each other.

The annual personal energy balance tool calculates the total energy consumption due to electricity consumption, heating or transportation and it compares the total result with mean values of other countries in the world. A comic interprets the result (Fig. 3).

4.2 Calculation of Diagrams

A main emphasis in the ILSE toolbox is put on diagrams. Most of these tools show more complex correlations of equations explained in the learning lessons. The users can change parameters and get online new diagrams calculated with the new parameters.

In the toolbox there are the following diagrams about photovoltaics:

- I-V curves of different types of photovoltaic cells
- impact of temperature changes on I-V-curves
- characteristic diagrams of photovoltaic

modules
- characteristic curves of inverters
- calculations of the sun position and sun orbit diagrams

The inverter tool calculates the characteristic curve of the efficiency over the relative performance with only three characteristic points. Another result of this tool is the European efficiency (Fig. 4).

With the sun orbit tool the user can obtain a sun orbit diagram for every place on earth. The location can be chosen from a given list of international locations or it can be defined by latitude and longitude. The tool calculates the sun orbit diagram for every day of the year as well as the minimum and the maximum curve (Fig. 5).

Besides the tools about photovoltaics there are other diagrams in the renewable energy sector such as the solar thermal collector efficiency, the Rayleigh distribution for wind speeds or characteristics of asynchronous generators for wind turbines.

4.3 Performance Analysis

Another interesting tool gives a performance analysis for grid connected PV-systems. After defining the location, the usable area, the orientation and the system quality; results for the installable size of the PV-generator, the annual energy gain and the costs are predicted (Fig. 6).

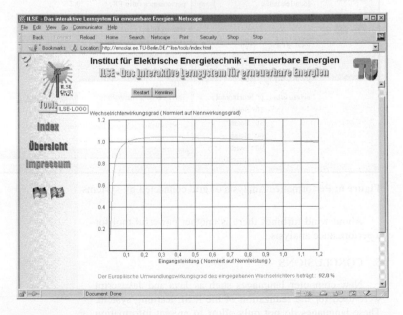

Figure 4: Characteristic curves of the inverter efficiency over the performance

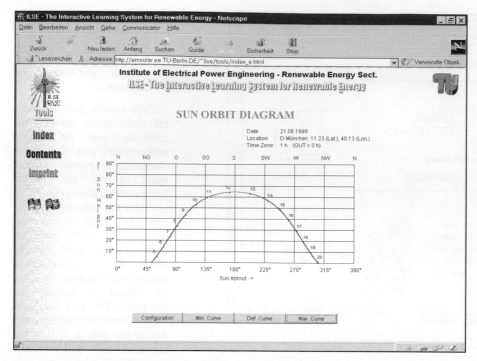

Figure 5: Calculation of sun orbit diagrams

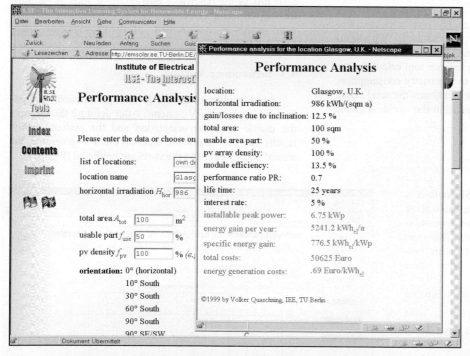

Figure 6: Performance Analysis of grid connected pv-systems

About wind turbines there is another powerful tool for a performance analysis.

5. CONCLUSIONS

New computer languages such as Java and Javascript were used for the realization of the ILSE learning system. These languages do not only allow to present information in form of texts and graphics but also make it possible to implement online multiple choice tests and online simulations and calculations in the ILSE toolbox.

This toolbox offers some of the most powerful calculations and simulations that are running directly in the Internet. The toolbox includes tools for several different diagrams as well as complex performance or economic analysis.

To run the ILSE system we recommend the Netscape Internet Browser 4.6 or newer. The ILSE learning system and the toolbox are for free for everybody using the following Internet address:
http://emsolar.ee.tu-berlin.de/~ilse

ACKNOWLEDGEMENTS

The authors would like to thank Carsten Anders, Christoph Blaschke, Martin Lemke, Carlitos Martini, Bruno Schneider-Schnettler and Berhard Schwarz who were involved in the development of the *ILSE* learning system.

DEVELOPMENT OF A VRLA BATTERY WITH IMPROVED SEPARATORS, AND A CHARGE CONTROLLER, FOR LOW COST PHOTOVOLTAIC AND WIND POWERED INSTALLATIONS

A.J.Ruddell[1], M.Fernandez[2], N.Vast[3], J.Esteban[4], F.Estela[4]

[1] Energy Research Unit, Rutherford Appleton Laboratory (RAL), Chilton, Didcot, Oxfordshire OX11 0QX UK
Tel : +44 (0)1235 445551 Fax : +44 (0)1235 446863 email : alan.ruddell@rl.ac.uk
[2] S.E. Accumulator Tudor, S.A., Ctra N-II, km 42 (P.O. Box 2), E-19200 Azuqueca (Guadalajara), Spain
Tel: +34 949 263316 Fax: +34 949 262560 email: lab.investig_tudor@gdl.servicom.es
[3] Bernard Dumas S.A., Creysse, 24100 Bergerac, France
Tel: +33 5 53232105 Fax: +33 5 53233713 email: bernard.dumas@wanadoo.fr
[4] ATERSA, C/Fernando Poó 6, 28045 Madrid, Spain
Tel: +34 91 4747211 Fax: +34 91 4747467 email: atm@bitmailer.net

ABSTRACT: There are many applications for small standalone photovoltaic (PV) and wind powered systems, in areas where it is not possible or economical to connect to the utility electrical network. The battery is ideally required to be low maintenance, have good cycling performance, and be resistant to acid stratification. The maintenance-free valve-regulated lead-acid (VRLA) battery is suitable for use in such applications. This paper describes the development of a VRLA absorptive glass mat (AGM) battery with improved cycling performance, through the use of a newly developed separator material with improved characteristics. This significantly reduces acid stratification, thus prolonging battery lifetime and reducing the overall system lifetime cost. A further innovation is the development of an adaptive charge controller which optimises the recharge time by taking the previous operating conditions of the battery into account, and ensures that the battery is fully recharged whenever possible. Prototype batteries and charge controllers have been designed, tested, and incorporated in a demonstration PV/wind system. The results of laboratory and application testing demonstrate the increased cycling capacity of the batteries, and the improved battery recharge achieved by the charge controller. Bench tests of the new battery indicate a 50% increase in the expected lifetime.
Keywords: Stand-alone PV Systems – 1: Batteries – 2: Battery Storage and Control – 3.

1. INTRODUCTION

The background to this research project is the provision of power in isolated areas for systems such as telecommunications repeaters, navigational beacons, environmental monitoring systems, as well as home light and power. These systems often have modest power requirements, and in many cases it may not be possible or economic to connect to the utility electrical network. The utilisation of renewable energy sources such as solar photovoltaic (PV) and wind turbine generators can provide an economic alternative source of power.

This paper describes the development of an improved low cost PV/wind power system, consisting of three main subsystems: solar photovoltaic (PV) or wind turbine generators, a battery bank for storing the generated energy, and a charge controller and power conditioner. The anticipated charge/discharge profile experienced by the battery depends on the system sizing calculations at the design stage, taking account of the meteorological conditions and available power from the generator(s), the load profile, and the required reliability of supply. Clearly it is desirable to minimise the overall system lifetime cost, including the battery installation, maintenance, and replacement cost, and this can be achieved by proper system sizing and specification of the key components.

The specific problem limiting the battery performance and lifetime in PV applications is the progressive development of acid stratification (different acid densities from the bottom to the top) in the battery. This is mainly due to the fact that in solar PV/wind applications, due to the random nature of the resource, a complete recharge of the batteries cannot always be guaranteed, and the battery may be required to operate for long periods at low SOC. The development of electrolyte stratification leads to the following problems: lower energy capacity available from the battery, high sulphation of the plates, and high corrosion and growth of the positive plates. The result is a severe reduction of the expected life.

Valve regulated lead-acid (VRLA) batteries, which hold the acid in either a gel or an absorptive glass mat (AGM) are inherently resistant to acid stratification, and have the additional important advantage of being maintenance free. On the other hand, flooded cells will suffer from acid stratification unless they are periodically over-charged, or alternatively provided with a means of electrolyte stirring. Therefore VRLA batteries are a suitable choice for low-cost standalone applications. In order to further improve their performance, a new absorptive glass mat (AGM) separator material has been developed for the VRLA battery. The new separator material includes special filling materials which modify the structure of the glass microfibre, in order to achieve the desired properties. This significantly reduces acid stratification, thus prolonging battery lifetime and reducing the overall system lifetime cost.

VRLA batteries require careful recharging, because overcharging can provoke water consumption, and eventually thermal destruction due to high recombination efficiency and associated heat generated. Most small charge regulators have two regulation modes (bulk charge, and taper charge), and include temperature compensation. A further innovation of this project is the development of an adaptive charge controller which takes the previous operating conditions of the battery into account. The SOC history for the previous two weeks is used to modify the voltage set-points, ensuring that the battery is fully recharged whenever possible, while avoiding overcharge.

Prototype batteries and charge controllers have been designed, constructed, tested, and incorporated in a demonstration PV/wind system. The results of laboratory and application testing are presented, to demonstrate the increased cycling capacity of the batteries, and the improved battery recharge achieved by the charge controller. Bench tests of the new battery indicate a 50% increase in the expected lifetime. The expected system operational benefits are presented.

2. DEVELOPMENT OF IMPROVED VRLA BATTERY

2.1 Development of a new improved separator material

Several different filling materials including anionic and cationic silica, and ceramic fillers, were selected and tested for inclusion in the new separator material. Laboratory samples of a proposed range of separator materials were prepared and extensively tested to determine and measure the most important parameters relevant to battery performance improvement, as shown in Table 1.

Characteristics	Standard	Low filler content	High filler content
Specific surface area (m^2/g)	1,17	7.28	14,83
Pore size distribution (µm)			
Minimum	2.54	2.40	2.25
Maximum	15.05	8.77	3.05
MFP	4.78	3.29	2.84
Capillary rise (mm)			
2 min	65	56	45
24 hours	620	780	830
Tensile strength (daN/25.4 mm)	2.44	1.70	1.20

Table 1: Characteristics of two new separator materials compared with the standard material.

The variation of characteristics due to the inclusion of low and high levels of filler are clearly shown in Table 1. There is a great increase of surface area, pore size distribution shifted to lower values, and an increase of capillary rise at long times (these last two properties are essential to eliminate acid stratification). The decrease in pore size has another important benefit, that is the ease of developing short circuits by the growth of lead dendrites is reduced. Magnified views of the modified and standard materials, shown in Figure 1, illustrate the increased specific surface area, and decreased pore size.

The inclusion of filler in the structure of the separator modifies the interlocking of the fibres, decreasing the

tensile strength of the material. This places additional requirements on the manufacturing process in industrial conditions.

These new separator materials were expected to give improved battery performance, and were suitable for assembly in battery prototypes, and battery cycling performance tests.

Figure 1: SEM photographs of the standard (without filler) and modified (with filler) separator material.

2.2 Prototype VRLA battery specification

The inclusion of the new separator material in a new VRLA battery offers longer life and increased reliability. Initial prototypes of the new battery were assembled and tested. The main characteristics of the new battery are:

- Technology : ....Valve regulated lead-acid.
- Dimensions (LxWxH) :.................. 244 x 190 x 280 mm
- Weight : .... 33 kg
- Voltage : .... 6 V
- Capacities 100h : .... 220 Ah
  5h : .... 180 Ah
- Stored energy   100h rate : ....1350 Wh
- Specific energy 100h rate :............................40.9 Wh/kg
- Expected life : ....~800 cycles at 60% DOD.
- Special features : .... New separator material to avoid acid stratification

2.3 Results of cycle life testing

Three groups of batteries were assembled using the standard separator material (Group I), the material with low filler content (Group II), and the material with high filler content (Group III). The three Groups were submitted to cycling tests, similar to the expected duty in PV applications, according to the following profiles.

a) Cycling with 20% depth of discharge (DOD).
Batteries are cycled continuously between 60% and 80% state of charge (SOC), with complete recharge every 25 cycles and a capacity test every 50 cycles.

b) Cycling with 60% DOD.
Batteries are cycled continuously between 20% and 80% SOC, with complete recharge every 25 cycles, and a capacity test every 50 cycles.

The results of cycling at 60% DOD show a significant improvement in cycle life of 50% with respect to the standard battery, see figure 2. Tests in light cycling conditions of 20% DOD have already reached 1400 cycles, and do not yet show differences in capacity between the three groups.

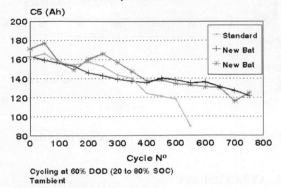

VRLA 6V/220Ah BATTERY FOR SOLAR APPLICATION
Joule Project JOR3-CT97-0131

Cycling at 60% DOD (20 to 80% SOC)
Tambient

Figure 2. Cycling at 60% DOD (20% to 80% SOC) for three groups of batteries.

3. DEVELOPMENT OF ADAPTIVE CHARGE CONTROLLER

3.1 Adaptive features of the new charge regulator

The new charge regulator utilises a microcontroller and MOSFET technology to optimise the charge control efficiency, and in addition provides an information display and keyboard for the user.

The charging cycle, illustrated in Figure 1, includes two phases, deep charging and float charging. Following a

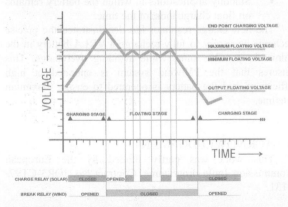

Figure 3. Illustration of charge control

discharge, the regulator allows the available charging current to flow to the battery without interruption, until the battery reaches the end point charging voltage, when the battery is almost fully charged. The end point is calculated as a function of temperature, charging current, and battery capacity. Float charging is made within the limits of a 'dynamic floating band', where the maximum and minimum voltages are calculated as a function of the SOC during the previous seven days (higher voltages after lower SOC). The charging current is pulse width modulated

(PWM) in a limit cycle mode, and when the battery reaches full charge the battery voltage response to charging current becomes fast, automatically reducing the duty cycle and reducing the mean charge current. The adaptive setting of the 'dynamic floating band' ensures that the battery charge is completed in the minimum time, and the battery is automatically equalised after a period of low SOC, while minimising overcharge after a period of higher SOC.

Other features of the regulator include :

Load disconnect :
Disconnection of the supply to the load when the battery SOC is low, to avoid deep discharge of the battery.

Status indicators :
Battery low and over voltage; Load under-voltage; Float and deep charge.

Alarm systems :
Battery low and over-voltage.

Protections :
Short circuit, over-voltage and reverse polarity.

Technical data :
- Consumption.. 40 mA
- Maximum PV charging current............................... 50A
- Maximum wind charging current............................ 10A
- Maximum load current.. 25A

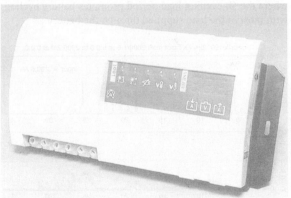

Figure 4. The completed charge regulator

4. APPLICATION TESTING

4.1 PV/Wind power system

A PV/Wind system was implemented to demonstrate and test the new battery and regulator. The system is installed in the UK and was sized to supply a load of up to 30Ah/day, see Figure 5.

Conventional energy balance techniques were used to size the system, and computer simulation was used to investigate operation for a period of one year. The combination of PV and wind generators is particularly advantageous in climates similar to the UK, where the solar resource in winter is rather low, and where the seasonal PV and wind resource tend to be complementary[1][2].

The main characteristics of the system are:

- PV array:............................120Wp, two ATERSA A-60
- Wind turbine:............... 50W at 10 m/s, Marlec FM 910
- Charge regulator: ....new ATERSA 12V/50A
- Battery: ....two new TUDOR 6V/220Ah
- Load:.................8 x 15W nominal, computer controlled

The PV/wind system is implemented with instrumentation and data-logging for generator, battery, and load voltages and currents, solar panel and battery temperatures, solar irradiance on the PV panel plane, and wind speed. The data acquisition system is PC-based, incorporating a National Instruments AT-M10-16E-10 DAQ board with a custom LabVIEW program. Data analysis following acquisition is by Matlab programs.

Figure 5. The PV/wind standalone power system used for demonstration and test.

4.2 Results

Battery capacity and coulombic efficiency tests were made before the operating period (capacity 213Ah, energy 2580Wh @ 40h rate, coulombic efficiency 95%, watt-hour efficiency 89%). A test procedure using a variable daily load cycle was developed to test the system over a range of operating conditions in a short operating period of 35 days. The PV, wind, and load currents are shown in Figure 6, where the 5h evening load period promotes battery cycling, with most of the load supplied through the battery.

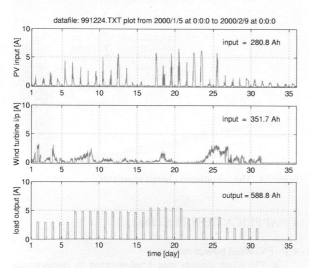

Figure 6. PV, wind, load currents for a period of 35 days.

The performance of the battery during this period is shown in Figure 7. The battery SOC has been calculated using Ah-counting, utilising the measured coulombic efficiency. It can be seen that the load cycle subjected the battery to a deep cycle, with superimposed daily cycles. The results confirmed the efficiency of the battery, and the adaptive charge control, ensuring that the battery is fully recharged, and is not overcharged.

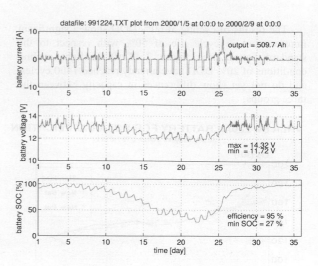

Figure 7. Battery performance for a period of 35 days.

5. CONCLUSIONS

The new VRLA battery, incorporating an improved separator material to reduce acid stratification, has already demonstrated a significantly increased cycle life. Tests conducted so far show an increased lifetime from the present value of around 500 cycles, to figures of the order of 750 cycles, for cycling at 60% DOD (20% to 80% SOC). Testing at lighter cycling levels continues. The battery is expected to have increased lifetime in all applications where acid stratification tends to develop in standard VRLA batteries, including standby, light cycling, and deep cycling applications. The main areas of use are expected to be :

- Small and medium-sized solar PV installations
- Applications requiring deep and repetitive cycles, such as light motive power applications.
- Standby applications in which the battery remains on float charge most of the time.

The new charge regulator has demonstrated the precise recharge necessary to fully recharge a VRLA battery in the minimum possible time, while avoiding overcharge. This ensures that the PV/wind system is operated at high efficiency, and the battery is operated to ensure maximum lifetime.

6. ACKNOWLEDGEMENT

This work was partly funded by the European Commission Joule programme, Contract No. JOR3-CT97-0131.

7. REFERENCES

[1] J.W. Andrews, Energy storage requirements reduced in coupled wind-solar generating systems, Solar Energy, pp73-74, 1976.

[2] T. Markvart, Sizing of hybrid photovoltaic-wind energy systems, Solar Energy Vol.57, No.4, pp277-281, 1996

DIRECT-COUPLED VERSUS BATTERY BACKED PHOTOVOLTAIC PUMPING SYSTEM

S.Bouwmeester[1], B. Van Hemert[1], G. Loois[1], H. De Wit[2]
1) ECOFYS energy and environment
PO Box 8408, NL-3503 RK Utrecht, The Netherlands, tel. +31 – 30 – 2808 300, fax +31 – 30 – 2808 301
2) VOPO Pompen- en machinefabriek bv
PO Box 6, NL-1483 ZG De Rijp, The Netherlands, tel +31 – 299 – 671312, fax +31 – 299 - 673358

ABSTRACT: In 1999, Staatsbosbeheer, the Dutch National Forest Service, commissioned a redesign of the nature reserve "De Steendert" which had to content with severe drought. A 5ha storage pond was made for irrigating the area. Accepted fluctuations in the pond allow for several day's water buffer. Ecofys supervised the construction by VOPO of a dual photovoltaic pumping system, which controls the water level in the pond. The system consists of two 45m^3/h pumps, one powered by an 1kWp photovoltaic array with 230 Ah battery bank and one direct powered by an 1kWp photovoltaic array. Measurement equipment is installed in order to measure and compare both systems characteristics. The aim of this research is to compare the technical and economical feasibility of both pumping systems. Expected achievements are: reduction of BOS and maintenance costs and ecological friendly system design by elimination of the battery bank. The energy generated by the photovoltaic modules covers the demand of the pumping system during a reference year. Direct-coupled pumping systems are expected to be both economically and technically competitive with their battery-backed counter parts.
Keywords: PV Pumping - 1: Cost Reduction - 2: Monitoring - 3

1. BACKGROUND

In 1999, Staatsbosbeheer, the Dutch National Forest Service, commissioned a redesign of the nature reserve "De Steendert". This nature reserve is located in the municipal Ophemert, about 7 km from the town Tiel. The reserve is about 80ha and mainly consists of wetlands, and some groups of trees. The area is provided with water from the nearby river Waal. Certain types of plants do not grow in this area because of relative drought.

The water household of the reserve is altered by constructing two 5ha ponds, one which collects the percolating water from the river Waal and the other which functions as a water storage reservoir. The second pond provides the area with sufficient water so an appropriate water level is obtained.

The water level in the percolation water pond fluctuates between +2.00m NAP[1] and +2.20 NAP, while the water level in the storage pond must fluctuate between +3.40m NAP and +3.70m NAP. A pumping system lifts the water from the lower into the higher pond.

2. EXPECTED ACHIEVEMENTS

Conventional autonomous photovoltaic systems are battery backed to buffer electrical energy. This buffer enables the system to continue to operate during periods of reduced solar irradiation. A storage pond is advantageous in the sense that it stores energy, thus, making a battery bank redundant. The efficiency of the storage pond is higher than that of a battery bank because no energy conversion takes place. For example, a mere one centimetre increase in water level equals a whole day pumping. A

---

[1] Nieuw Amsterdams Peil, the Dutch reference for altitude measurements.

battery bank is not necessary and the pump can be directly coupled to the photovoltaic array. Moreover, BOS and maintenance cost are reduced and system design is ecological friendly by elimination of the battery banks. Direct-coupled pumping systems are expected to be both economically and technically competitive with their battery backed counter parts [2,3,4,5].

3. DIMENSIONING

The monthly water demand in the 80ha reserve is estimated using a simple compartment model. In our model, we assume that the reserve is only irrigated from the water which stored in the storage pond. The appropriate water level in this pond is maintained by pumping water from the percolation water pond. The amount of percolated water is assumed to be sufficient to irrigate the 80ha reserve. Furthermore, it is assumed that the water is only drawn from the reserve by evaporation.

Table I. Estimation of the demanded pumping volume and corresponding demanded electrical energy during one year.

Month	Evapotranspiration (mm)	Rain fall (mm)	Pumping volume demand (m^3)	Electrical energy demand (kWh)
January	7.7	65.7	–	–
February	15.4	47.8	–	–
March	30.9	62.9	–	–
April	54.1	51.5	2240	37
May	81.9	61.2	17850	290
June	89.7	68.3	18460	300
July	89.3	74.9	12420	200
August	78.3	70.9	6380	100
September	49.8	66.7	–	–
October	27.9	72.0	–	–
November	10.9	81.0	–	–
December	6.2	79.5	–	–

Table I represents the amount of water needed for each month [1]. It can be seen that pumping is only necessary during the months April through August. For example, in the month June the pumping system has to transport about 18,460 m³ water in order to maintain an appropriate water level. Assuming a duty cycle of 7 solar hours per day, the pumping system is designed to lift 90 m³ water per hour. The electrical energy need for pumping water over an average 1.5 meters height, assuming a pump efficiency of 25%, is given in the last column of Table I.

A plot of the electrical energy demand by the pumping system and energy generation by the photovoltaic modules during one year reveals a strong relationship, see Figure 1. Assuming a performance ration of 0.7, the energy generated by 2.16kWp photovoltaic modules covers the energy demand during a reference year.

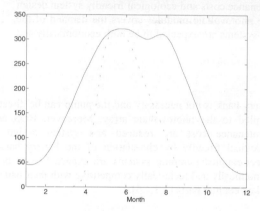

Figure 1. The energy generated by the photovoltaic modules (solid line) completely covers the energy demand of the pump (dashed line) during a reference year.

Connecting the pumping system to the main grid requires an 1km long electrical cable, which costs about €37,000. The photovoltaic modules costs €14,000. Moreover, "The Steendert" is a nature reserve and digging a cable into the ground should be avoided. Thus, construction of an autonomous direct-coupled photovoltaic pumping system is both economically and technically attractive.

4. OBJECTIVE

The aim of this research is to compare the technical and economic feasibility of the direct-coupled and battery-backed pumping system. The following questions concerning the feasibility will be answered:

- Can the system maintain the water level in the storage pond between +3.40m NAP and +3.70m NAP?
- Do the direct-coupled and battery backed pumping system transport an equal amount of water on a monthly basis?
- What is the overall efficiency of both direct-coupled and battery backed pumping system?
- What is the utilisability of the direct coupled system?
- Is the simple compartment model useful in dimensioning a pumping system?

- What is the estimated installation cost of the direct-coupled, battery backed and grid connected pumping system?
- What is the estimated maintenance cost of the direct-coupled, battery backed and grid connected pumping system?

5. IMPLEMENTATION

Ecofys supervised the construction of a dual photovoltaic pumping system by VOPO, which controls the water level in the storage pond. The system consists of two 45 m³/h pumps, one powered by an 1kWp photovoltaic array with a 230 Ah battery bank and one direct powered by an 1kWp photovoltaic array. Figure 2 shows a photo of the system. In order to increase the life time of the battery bank a smart pump control algorithm is devised. This algorithm determines if the battery backed pump should be switched on depending on the water level in the storage pond and state-of-charge of the battery bank. The state-of-charge is estimated by averaging the battery voltage during 15 seconds. The decision criteria are given in Table II.

Figure 2. The ditch from the pond with percolation water, two individual pump systems, the storage pond, pump control cabinet and photovoltaic modules can be seen.

Table II. The battery-backed pumping system is switched On or Off depending on the water level and battery voltage.

		Battery voltage (V)		
		≤22	22/24	≥24
Water level (m)	≤3.40+	Off	On	On
	3.40/3.70	Off	Off	On
	≥3.70+	Off	Off	Off

Measurement equipment is installed in order to measure and compare both systems characteristics. Figure 3 depicts the schematic layout of the system.

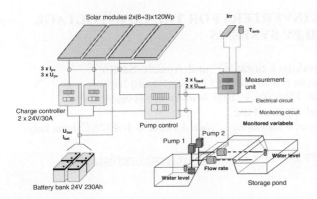

Figure 3. Schematic layout of the system; dashed lines indicate monitoring circuit and monitoring variables are given in bold typeface.

6. CONCLUSION

Direct-coupled pumping systems are expected to be both economically and technically competitive with their battery backed counter parts.

Pumping is only necessary during the months April through August. Assuming a duty cycle of 7 solar hours per day, the pumping system is designed to transport 90 m³ water per hour. The energy generated by 2.16kWp photovoltaic modules covers the energy demand of the pump during a reference year.

The aim of this research is to compare the technical and economic feasibility of the direct-coupled and battery-backed pumping system.

Final analysis of the monitoring results shall be available at the end of this year.

ACKNOWLEDGEMENTS

The work reported is supported by NOVEM, The Netherlands.

REFERENCES

[1] Koninklijk Nederlands Meteorologisch Instituut, Normalen van neerslag voor het tijdvak 1961-1990: Supplement bij het maandoverzicht neerslag & verdamping in Nederland, De Bilt, November 1992, ISSN 0925-3009.

[2] Van Hemert, B. *et al*, Autonomous PV-generators in agriculture and water management in The Netherlands, Germany, Spain and Finland, Technical Reports no. 1-6, Ecofys energy and environment, 1996-1998.

[3] Van Hemert, B. *et al*, Autonomous PV-generators in agriculture and water management in The Netherlands, Germany, Spain and Finland, Final Report, Ecofys energy and environment, 1999.

[4] Hilmer, F. *et al*., Investigation of a directly coupled photovoltaic pumping system connected to a large absorber field, Solar Energy, **61**, pp 65-76 (1997).

[5] Argaw, N., Optimum Load Matching in Photovoltaic Water Pumps Coupled with DC/AC Inverter, Int. J. Solar Energy, **18**, pp. 41-52 (1995).

ENEA'S FACILITIES TO TEST SMALL-SIZE INVERTERS FOR THE LOW-VOLTAGE GRID CONNECTED PV SYSTEMS

Michele Guerra*, Riccardo Schioppo*, Teodoro Contardi[1] and Angelo Sarno[1]#

(1) Enea Centro Ricerche. Località Granatello P.B. 32. I-80055 Portici (Napoli) Italy
*Area Sperimentale Enea di Monte Aquilone S.S.89 km 174+700 . I-71043 Manfredonia (Foggia) Italy

Salvatore Raiti and Marco Tina

Università degli Studi di Catania Dipartimento Elettrico, Elettronico e Sistemistico. Viale A.Doria 6. I-95125 Catania Italy

(1)# ph. +39 081 77 23 202 fax +39 081 77 23 344 e-mail: sarno@epoca1.portici.enea.it

ABSTRACT: It is common opinion that the diffusion on large–scale of the photovoltaic application, based on the installation of small-size grid-connected PV systems, even if promoted and supported by governative initiatives, will be successfull if technical and non-technical barriers are overcome.

In the past, in Ital, because of some regulation restrictions, the barrier, concerning the low-voltage grid connection and interface device, caused many uncertainties so that national industries weren't encouraged in developing inverters for this kind of application.

In order to promote the development and optimization of national products, a dedicated test-facility has been realized by Enea (Italaian National Agency for New Technologies, Energy and the Environment) at Enea Monte Aquilone Test-Site.

A detailed description of the test-facility, designed to be able to assess the inverters technical main features and to test them in simulated and operating conditions, is reported.

In the paper, preliminary experimental results will be presented and discussed in order to compare different commercial products and draw some conclusions about design criteria to optimize this kind of component.

Keywords: Diffuse generation – 1: Inverter – 2:Islanding – 3: Power Conditioning

1. INTRODUCTION

The large-scale diffusion of small size grid-connected PV systems, expected in Italy owing to the "10.000 PV roofs" national program /1/, will be successful if technical and non-technical barriers are overcome.

In the past, in Italy, because of some regulation restrictions, the barrier, concerning the low-voltage grid connection and interface device, caused many uncertainties so that national industries weren't encouraged in developing inverters for this kind of application.

Enea (Italian National Agency for New Technologies, Energy and the Environment), in according to the new contents of the latest edition of the CEI 11-20 standard (issued by the Italian Electrotechnical Committee in Dec.1997), has started experimental activities about small-size grid-connected inverters, by realizing, at Enea Monte Aquilone Test-Site, located in the South Italy, a dedicated test-facility to assess the inverters technical main features and to test them in simulated and operating conditions.

The experimental results, which will be presented in the paper, are related to some inverters available on the european market, having rated power ranging from 1500 to 2500 kVA, based on different design criteria. They report the measurement of significative parameters, characterizing the inverter under steady-state or transient operating conditions. In general, the experimental results show a value of the efficiency lightly lower than that one, reported by the data sheet. For all the inverters under test, an output current Total Harmonic Distortion,TDH, greater than 5% has been found at 30-40 % of the rated power. Regarding the behaviour following to a sharp grid cut-off, over-voltage values around 350-400 V and 50-100 ms long have

been measured, while the inverter shut-down time ranged from 0.2 up to 3.5 s. The analysis and discussion of the experimental results come to suggest some design criteria to optimize this kind of component.

2. TEST FACILITY DESCRIPTION

The experimental facility was designed to be able to measure significative parameters characterizing the inverters behaviour under steady-state or transient grid and load conditions, in order to assess the inverters performance and to test them in operating conditions.
It consists of
- a test-bench, powered by either a 3 kWp PV generator or a DC power supplier in according to the experimental requirements, in order to test up two components in connection with the grid, at the same time;
- a test field based on some grid-connected (1 – 3 kWp) PV systems operating in connection with the grid;

In both cases the connection with the utilit MV grid can be realized or directly through the local LV grid or by means of a dedicated LV-MV transformer.

The PV generator powering the test-bench consists of 10 strings (about 300 Wp each) divided into two subarrays. To meet desired voltage and current values, the PV generator electrical arrangement can be modified by connecting or unconnecting strings or parts of them..

Figure 1 shows the block scheme of the measurement chain of the test-bench, designed to measure the following characteristics:

- inverter efficiency (efficiency curve, maximum and weighted average (European efficiency), MPPT efficiency, voltage and current Total Distortion Harmonics ,(TDH), and DC current ripple;
- DC and AC EMI (Electromagnetic Interference)
- Single and mutual "islanding effect", due to a grid cutoff, under specific load and electrical lay-out conditions.

The measurement chain is able to measure current up to 20 A and voltage up to 1000V ca and 300 Vdc.

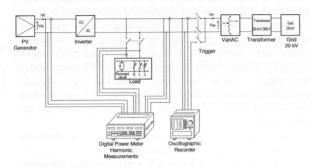

Figure 1: Measurement chain block scheme.

3. EXPERIMENTAL RESULTS

The inverters tested have been selected among those ones available on the european market, presenting different technical solutions and having a rated output power ranging from 1500 to 2500 W.

Table I: Inverter main technical features

Main Technical Features	A	B	C
Hardware Configuration	PWM multi-stage Transformless	PWM multi stage High frequency transformer	PWM – Full bridge with transformer
Nominal Input Power	1600 W	1800 W	2500 W
Nominal Output Power	1500 W	1600W	2200 W
Power control system In operation	< 15W	20W	15W
Power control system Stand-by or night	<<1W	0W	3W
Nominal input voltage	400V	100V	96V
Range input voltage	125-500 V	80-130 V	72-145V
Range output voltage	196-253V	196-253V	195-256V
Nominal output current	8A	8A	10A
Frequency	50Hz (49.8-50.2Hz)	50Hz (49.8-50.2Hz)	50Hz (49-51Hz)
Commutation frequency	20Hz	20Hz	30Hz
DPF	1	1	1
Insulation industrial frequency Transformer	-	-	1
Starting voltage	156 Vdc	82Vdc	82Vdc

Table I reports for each inverter under testing, identified by letters "A", "B", "C", detailed information about the hardware configuration and the main electrical characteristics, as listed in the commercial data sheet.

All the components are self-commutated inverters based on PWM technique with current control, without internal voltage or current reference, considered by the regulation as "systems not able to maintain voltage and current during a grid unavailability".

The purpose of the tests is to verify on one hand the values declared by the manufacturer, when given, and on the other if the operating behaviour, with particular attention to the quality of the output energy and to the "islanding"effect, is in according to the regulations.

3.1 Efficiency

The DC/AC conversion performance was measured in terms of the inverter efficiency.The efficiency curve has been calculated , by means of the interpolation method based on the minimum square technique, on the basis of the following equation, that takes into account the different loss mechanisms:

$$\eta = 1 - k1 - ko/y - k2\,y$$

where $y=Pdc/Pn$ is the nominal power fraction, whereas ko, k1and k2 are related to the losses, that can be ascribed to effects, respectively, independent on the power (as magnetic, control system energy), dependent on the power (conduction, switching) and on the square of the power (Joule).

ko, k1and k2 have been estimated for each component on the basis of a data base obtained, carrying out specific experimental campaigns during many clear sky days and collecting data every 5 sec. The values are reported in the following table II.

Table II: Loss coefficient

Inverter type	Ko= ko* Pn	K1 = k1	K2 = k2 / Pn
A (Oper. Volt. 207 V)	15.27	0.019	3.02E-5
A (Oper. Volt. 414V)	16.31	0.008	2.7E-5
B	36.95	0.022	4.62E-5
C	15.17	0.055	1.48E-5

Figure 2 shows, for each kind of inverter the efficiency, and the maximum efficiency and the weighted average.

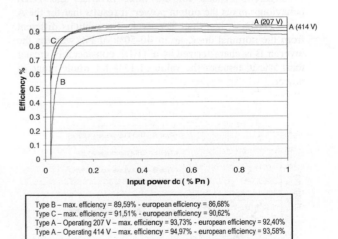

Type B – max. efficiency = 89,59% - european efficiency = 86,68%
Type C – max. efficiency = 91,51% - european efficiency = 90,62%
Type A – Operating 207 V – max. efficiency = 93,73% - european efficiency = 92,40%
Type A – Operating 414 V – max. efficiency = 94,97% - european efficiency = 93,58%

Figure 2: Inverter efficiency curve

By comparing the different efficiency curves, it results that the presence of both high frequency rectifier circuit and final polarity inverter influences negatively, in terms of efficiency, the hardware configuration based on multistage circuit, (inverter B) in comparison with the solution transformerless (type A) and that one based on the presence of an industrial frequency transformer (type C). It is worth noting that the increase of the joule losses together with a higher energy consumption by the control unit leads to lower efficiency values at low power zone.

Testing the inverters "A" and " B", it has been found that the absence of the insulation transformer allows to reach better efficiency (about 5% more) over the whole power range (10 – 100%).The comparison between the model "A" (without the transformer) and "C" (single stage with transformer) shows an improvement in terms of efficiency at high power values: it is possible to operate at high voltage (400-500 V) with corresponding reduction of joule losses.

By analysing the loss coefficients and their effect on the efficiency curve and the european values, it results that a decrease of k0 involves an increase of the efficiency at low-medium power and a gain in terms of european efficiency: A reduction of k1 causes a light translation of the efficiency curve without a significative variation in the european value; finally a decrease of k2 leads to an increase of efficiency at high power values with an light improvement of the european value.

3.2 Output energy quality

To characterize the inverters as regards the quality of the ac output, the low frequency (less than 2 kHz) harmonic components of the ac voltage and output current and the corresponding Total Harmonic Distortion (TDH) have been measured, in order to verify that at the nominal conditions the values are less than the values foreseen by the CEI EN 61727 (CEI 82-9): 1% for the voltage single harmonics, 2% for THDv , and 5% for THDi

The THD tests have been carried out in real operating conditions: the inverters are connected to the grid by means od a dedicated LV/MV transformer and powered by the PV generator. For all inverters an average THDv less than 2% has been measured, as well as it has been verified that in any case the current harmonic spectrum is in agreement with the limits imposed by the CEI EN 61000-3-2.

Figure 3 reports for the three inverters the THDi behaviour versus the output power: it results that for the A and C THDi is less than 5%, even at conditions different from the nominal ones, up to 40-50% of Pn, whereas the inverter B is characterized by a THDi considerably greater than 5%; it assumes the value of 15% for whole range of power.

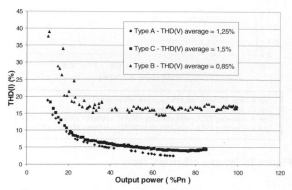

Figure 3: THDi versus output power

Only the kind B showed a vulnerability to the MV grid stability, probably due to final thyristors bridge, in terms of THDv.

It has been verified that for this type of inverter, on the contrary of the other two, the quality of the power fed in the grid (THDi/THDimax) gets worse as THDv increases, that occurs mainly at conditions far from the nominal ones.

3.3 Islanding effect

Finally the inverters have been tested to analize their behaviour when "islanding effect" conditions might occurr. In general, the regulations or utility specifications, in order to assure grid safety, obblige the inverters, to be connected to a first category grid, must be equipped with particular protections based on voltage and frequency relays (OVR/UVR over/under voltage and OFR/UFR over /under frequency). These devices, classified as passive protections, have the duty to shut down the inverter when the voltage and/or frequency assume values out of the allowable range.

Table III reports the values imposed by different regulations in force in different countries.

The purpose of the test was to verify the right operation of the protections, in terms also of NDZ (Not Detection Zone) and DPF (Deplection Power Factor), in such a way that it was possible eventually to evaluate the necessity to install additional protection devices.

Table III: Passive protection limit values

Interface protections threshold	Enel DV 1606	IEEE P929	DIN VDE 0126
UVR	0.8 Vn	0.92Vn	0.8Vn
OVR	1.2Vn	1.10Vn	1.15Vn
UFR	49.7Hz	59.5Hz	49.8Hz
OFR	50.3Hz	60.5Hz	50.2Hz

To simulate the loads variation on the grid, a RLC variable circuit parallel connected to the grid has been utilized. The tests involve the measurements of the shut down time following the grid cutoff, in different load conditions: no-load; fixed load; perfectly coupled load.

In absence of loads, following up a grid cut-off, for all the inverters an overvoltage less than 600 V have been measured. For the inverter type "C", equipped with the protections based only on the voltage and frequency relays it has been found that at an active power variation greater than 20% the relays worked correctly and the inverter shut down time was less than the value imposed by the regulations. For active power variation less than 20% the shut down time reached values around 3s, figure 4.

In the condition of perfectly coupled load, the tests have been carried out at different values of the capacitance. In this case, following up the grid cutoff, the shut down time was dependent on the capacitance and reached values greater than 3.5s, too long for the safety grid, figure 5. That means the passive protections could not be effective in some cases, then it could be necessary to install additional devices, as active protection, based on the capacity of altering the steady state conditions.It has been verified the effectiveness of the solution based on the monitoring of the grid impedance, utilized by the inverters types "A" and "B", even if it, resulting particularly sensible to the local load starting transient, could involve many undesired outages.

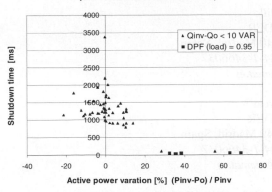

Figure 4: Shutdown time versus active power variation

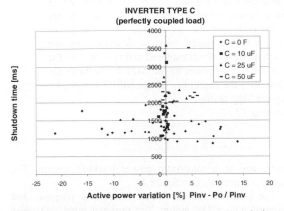

Figure 5: Shutdown time versus active power variation

Finally, at the beginning of islanding conditions, some changes in the voltage harmonics, in particular to the third harmonic, as reported by the figure 6. That means there is the possibility of avoiding the islanding phenomenon by means of continous monitoring of the third voltage harmonic.

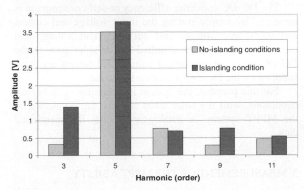

Figure 6: Changes in voltage harmonics at the beginning of islanding conditions

4. CONCLUSIONS

The characterization of the small-size inverters for the low voltage grid connection has been focused on two main aspects: the performance in terms of efficiency and energy quality, and the safety aspects related to itself and the grid.

The efficiency analysis has confirmed the advantage, presented by the hardware configuration without the insulation transformer, that permits to work at higher voltage values with an improvement of the operating efficiency average.

Regarding the energy quality the experimental data show that for the inverters under test, exception made for that one based on final stage, both the single harmonics and the THD at nominal conditions are in good agreement with the present day regulations.

Finally as concerns the islanding effect the protections based on a voltage and frequency relays present, in some cases, shut-down times too long to assure the grid safety. The installation of additional protections, as the ENS system, (grid impedance monitoring), showed itself very effective when on the grid there is one inverter. In the case of more units connected on the same grid branch the ENS effectiveness could be compromised.

By analyzing some voltage harmonic components it resulted that the islanding effect could be avoided by monitoring the voltage harmonics. The real effectiveness of this kind of active protection is under investigation.

REFERENCES

/1/ S.Castello, M.Garozzo, S. Li Causi and A.Sarno, Progress on Ongoing Activities of the Italian 10.000 Rooftop PV Program, to be presented at 16th European Conference.

/2/ H.Haeberlin, F. kaeser,Ch.Liebi and Ch.Beutler, Results of recent performance and reliability tests of the most popular inverters for grid connected PV systems in Switzerland, European Photovoltaic Solar Energy Conference, Nice 23-27 October 1995.

/3/ M.Begovic, M:E:Ropp, A.Rohatgi, A.Pregelj, Determining the sufficiency of standard protective relaying for islanding prevention in grid-connected PV systems, Proceedings of the 2nd Conference and Exhibition on Photovoltaic Solar Energy Conversion 6-10 July 1998, pagg. 2519-2524.

/4/ H:Haeberlin,J.Graf, Islanding of grid-connected PV inverters:test circuits and some test results, Proceedings of the 2nd Conference and Exhibition on Photovoltaic Solar Energy Conversion 6-10 July 1998, pagg. 2020-2023.

ON THE MODELLING OF A MAXIMUM POWER POINT TRACKING SYSTEM

Félix García Rosillo
CIEMAT-PVLabDER
Avda. Complutense 22, E-28040 Madrid, Spain
Tel: 34-1-3466672, Fax: 34-1-3466037
E-Mail: f.rosillo@ciemat.es

M. C Alonso García
CIEMAT-PVLabDER
Avda. Complutense 22, E-28040 Madrid, Spain
Tel: 34-1-3466672, Fax: 34-1-3466037
E-Mail: carmen.alonso@ciemat.es

ABSTRACT: Under a JOULE DG XII project, it has been developed the modelling of a prototype of maximum power point tracking system (MPPT). The aim of this work is to show that when describing a MPPT system or any device that include the capability of finding the maximum power point, it could be necessary to pay attention to the ability of the MPPT for finding the maximum power of the solar array. An example of simulation of this situation is given. This example suggests that a MPPT system must be described for two conceptually different kinds of efficiency. The first efficiency is that commonly considered and related with the self-consumption of the MPPT system (described as the rate of the output power of the MPPT system to the output power of the solar PV array). The second efficiency describes the capability of the MPPT system for finding the maximum power of the array. We call this efficiency the Mpp ability.
Keywords: DC-DC converter - 1: Modelling- 2: Power conditioning - 3.

1 INTRODUCTION

The purpose of this paper is to show a possible way for modelling a MPPT system, taking as example a specific system.

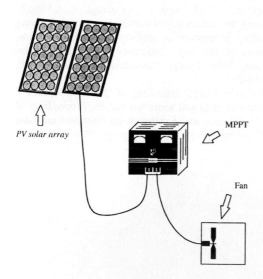

PV solar array

MPPT

Fan

Figure 1: Scheme of the PV system, with solar array, DC-DC converter and fan.

The system analysed consists of a DC-DC converter that interfaces a PV array and a load. The purpose of using the DC-DC converter is to use it as a power conditioning system and as a maximum power tracking system. The system is represented in figure 1.

The load is a fan used for climatization purposes and equipped with a DC motor. The nominal input voltage is 24 V and the nominal power is 210 W.

The PV array consists of two sets of panels in serie of two panels in parallel. The nominal al power of each solar panel was 53 W.

The DC-DC converter has a nominal output voltage of 24 V.

2 MEASUREMENT OF THE DC-DC CONVERTER EFFICIENCY

The DC-DC converter efficiency or self-consumption is measured by simply storing the input voltage and current and the output voltage and current. Input voltage and current are those values supplied by the solar array to the DC-DC converter and output voltage and current are those supplied by the DC-DC converter to the fan.

For this purpose, the system is operating as in normal conditions, with the solar array and the fan connected to the MPPT system. A datalogger records the mentioned data.

3. MEASUREMENT OF THE MPPT ABILITY

For the measurement of the MPP ability, the power supplied to the DC-DC converter from the solar array is varied by changing the tilt angle of the solar array.

At a certain value of the array tilt angle and consequently of the solar irradiation, the input and output voltage and current of the DC-DC converter in measured with the datalogger.

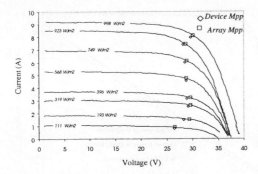

Figure 2: IV curves of the PV array at different values of the solar irradiation . The maximum power points of the solar array are designed as Array Mpp. The label "device Mpp" refers to the point of power selected for the DC-DC converter at each solar irradiance.

Immediately after this measurement is performed, the solar array in disconnected from the DC-DC converter and their IV curve is measured with a Stella Field array tester. This measurement permits to find the real MPP of the solar array, which must have been found for the DC-DC converter. The result of these measurements is illustrated in figure 2.

4 MODELLING OF THE EFFICIENCY OF THE DC-DC CONVERTER

A modelling for the DC-DC converter is presented in figure 3. The DC-DC converter efficiency is defined as the rate of the power consumed by the fan to the power consumed by the array. In our model, the DC-DC converter efficiency is found as a function of the power supplied by the array.

Equation 1, presented also in figure 3, is obtained using mathematical criteria, without intention of finding any physical meaning to the coefficients that match the

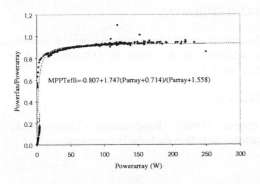

Figure 3: DC-DC converter efficiency versus array power. Dots represent the experimental data and continuous line represents the model of the curve.

experimental curve. Commercial software is used for

finding the value of the parameters that match the experimental data . As a starting point of the possible mathematical formulise that could be useful in the modelling, we pay attention to the Keating model for inverters (1):

$$\frac{Powerfan}{Powerarray} = -0.807 + 1.747\frac{Powerarray + 0.714}{Powerarray + 1.558} \quad (1)$$

The DC-DC converter efficiency, described as the rate of the input power to the output power, is due to losses of power associated to the conversion of the energy. This is conceptually important in order to distinguish this loss of energy of the ability of the DC-DC converter for finding the maximum power point of the array.

5 MODELLING OF THE MPP ABILITY OF THE DC-DC CONVERTER

The Mpp ability is related to the capability of the DC-DC converter for finding the maximum power point of the solar array. We define that parameter as the rate of the measured power that inputs the DC-DC converter at certain conditions of solar irradiance and temperature to the real maximum power point of the solar array. Figure 4 shows the modelling of the curve in our specific system.

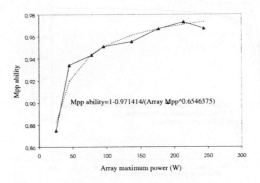

Figure 4: Mpp ability vs. the maximum power that the array could ideally generate. It is presented also the mathematical model of the curve.

Equation 2 shows the mathematical fit of the experimental data:

$$Mppability = 1 - \frac{0.971414}{ArrayMpp^{0.655}} \quad (2)$$

As in the case of the DC-DC converter efficiency, the Mpp ability is modelled with mathematical criteria that would permit to match the model in a program, without giving physicall sense to the parameters of the equation.

6 MATCHING OF MPP ABILITY AND MPP EFFICIENCY

For the calculation of the power that reaches the load, the first step is to calculate the power point chosen for the DC-DC converter. That value is as given in equation 2, evaluated at the theoretical MPP of the array at certain value of irradiance and temperature, named ArrayMpp.

The product ArrayMpp*Mppability(ArrayMpp) is the power that enters the DC-DC converter.

If we calculate the efficiency at this value of power as defined in equation 1, we will find the real efficiency. This efficiency multiplied for ArrayMpp*Mppability(ArrayMpp) is the output power.

Figure 5 shows the results for the calculation of the whole efficiency taking into account the Mpp ability of the DC-DC converter. The figure also shows another simpler model. In this model the efficiency is calculated with equation 1 and evaluated in ArrayMpp, which means that the Mpp ability has been neglected. The figure shows the percentual difference of both models.

In our case, for the specific DC-DC converter analysed, the higher differences between considering or not the phenomenon of the Mpp ability are observed at relatively low values of the output power. At for example a value of the Array Mpp of 50 Watts, the difference between the two models of efficiency is of only –0.25 %.

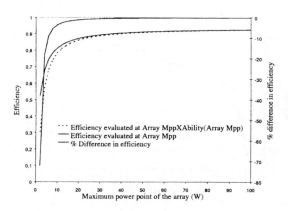

Figure 5: Efficiency of the DC-DC converter, calculated from the modelling formulas. The efficiencies are calculated supposing the existence of the Mpp ability and neglecting it. The percentual difference is also shown.

This difference between the two efficiency models, as it can be appreciated in figure 5, is quite low. Nevertheless we have to take into account that the input power would be in general different in each model, precisely for the effect of the Mpp ability, that is neglected in the simpler model. Consequently, the output power will be different too. When we compare the output power of both models at the same values of the irradiance and temperature, we find a difference of –8.4 % at 50 W.

Depending on the behaviour of the load with relation to the power, the model that includes the Mpp ability will improve considerably the description of the system.

Figure 6 shows the results for the calculation of the output power taking into account the Mpp ability of the DC-DC converter and without taking it into account. The percentual difference in power is also shown.

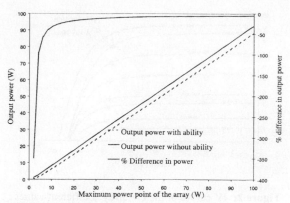

Figure 6: Output power of the DC-DC converter, calculated from the modelling formulas. The output power is calculated supposing the existence of the Mpp ability and neglecting it. The percentual difference is also shown.

7 CONCLUSIONS

A method for improving the calculations of the efficiency of a power conditioning system with maximum power point tracking has been described.

The conceptual difference of the Mpp ability and the conventional efficiency has been explained.

For a particular case, it has been described the method for combining the Mpp ability and the conventional efficiency.

It has been evaluated the effect of modelling in a particular system, considering two methods: one that takes into account the existence of the Mpp ability and other method that neglects the Mpp ability.

The importance of including the Mpp ability in a specific application will depend of their magnitude and the sensitivity of the load to the input power.

ACKNOWLEDGEMENT

The authors would like to acknowledge the contribution to the fulfilment of this work to J.Olivares, M.A.Ariza, J.Cuenca, and F.Chenlo.

REFERENCES

[1] Ciemat (1998); Fundamentos, Dimensionado y Aplicaciones de la Energía Solar Fotovoltaica.

OPTIMISATION OF AN INCIDENT ENERGY ON A STEP TRACKING PV ARRAY

W. Tayati and J. Watana
Department of Electrical Engineering, Chiang Mai University, Chiang Mai
THAILAND 50200, Tel/Fax : ++ 66 53 221 485, Email: wtayati@loxinfo.co.th

ABSTRACT: Optimisation of an incident energy on a step tracking PV system with equatorial installation was considered in this study. Calculation of incident energy on the tilted PV array was conducted using a latitude and longitude of the installation and a 45-day average of the clearness index for Thailand's sky obtained from previous studies by others. A uniform sky was assumed and surrounding reflection was neglected.

Two to six steps per day tracking modes were investigated. In each mode, optimisation of the incident energy was carried out to determine appropriate time and angle to adjust the array. The incremental energy obtained as compared to that of a fixed (and tilted) array was used as a criteria for optimum steps per day.

The calculations were verified by field measurement of PV module short-circuit currents. Two sets of PV module were installed as the fixed and the two step tracking installations. A microprocessor-based data acquisition system was employed and data analysis was carried out on a personal computer.

Result of the study shows that with two steps per day tracking, an increase in incident energy of 22 % is attainable. The energy increases as a number of step increases but the incremental is less.

KEYWORDS: Solar Tracking – 1: Optimisation of Incident Energy – 2

1. INTRODUCTION

This study is carried out with support from the National Energy Policy Office (NEPO) under Energy Conservation Promotion Fund to investigate an optimisation of incident energy on PV array installed in remote area for water pumping and PV electrification. Initially, a step tracking system with equatorial installation which requires simple structure and ease of operation is contemplated. Baltas [1] evaluated power output of step tracking PV array with azimuth installation using solar irradiation data obtained from measurement at 26 weather stations in the US.

2. ANALYSIS OF SOLAR RADIATION ON A STEP TRACKING PV ARRAY

2.1 Calculation of Solar Radiation on Horizontal Surface Under Clear Sky

Direct solar irradiation on a surface perpendicular to the sunlight under clear sky condition can be determined as follow.

$$I_{NC} = I_m \cos Z \tag{1}$$

where I_m is derived from a numerical function comparison with Exell's power series [2] and has a value

$$I_m = 1250 + 173 \cos \frac{360(n+6)}{365} - 50 \cos \frac{360(n-75)}{198} \tag{2}$$

where n is number of day in a year having value of 1 to 365
Hence, the direct solar radiation is

$$I_{HC} = I_{NC} \cos Z = I_m \cos^2 Z \tag{3}$$

and the diffuse solar radiation is

$$D_{HC} = CI_{NC} \tag{4}$$

where as $C = 0.120$ is derived from the above mention comparison. Therefore the total solar radiation on horizontal surface is

$$G_{HC} = I_{HC} + D_{HC} \tag{5}$$

2.2 Solar Radiation on Horizontal surface Under Cloudy Sky

The solar radiation under cloudy sky can be determined from that of clear sky when the clearness factor of the installation is known.

$$G_{HD} = k_d G_{HC} \tag{6}$$

where k_d (clearness factor) is the ratio of incident energy under clear sky to that under the cloudy sky. From Exell's study [3], the clearness factor for Chiang Mai was determined using daily sunshine hour and Angstrom's regression [4], [5].

$$k_d = \frac{H_{HD}}{H_{HC}} = a + b \frac{S}{S_m} \tag{7}$$

Exell [2] also calculated the two radiation components, direct and diffuse, necessary for calculation of solar radiation on a tilted surface by determining the ratio of direct to diffuse radiation (r_D) from the following equation

$$r_D = 1 - M(1 - r_C) \tag{8}$$

where $M = 0.733 k_d{}^2 + 0.267 k_d{}^4$

It follows that D_{HD} and I_{HD} can be calculated as

$$D_{HD} = r_D G_{HD}$$

(9)

$$I_{HD} = (1 - r_D) G_{HD}$$

(10)

2.3 Solar Radiation on Tilted Surface Under Cloudy Sky

A direct solar radiation on a tilted PV array in terms of a horizontal radiation is

$$I_{PD} = I_{HD} \frac{\cos\theta}{\cos Z}$$

(11)

And a diffuse solar radiation on a tilted surface can be determined using the model developed by Lui-Jordan [6] with an assumption of isotropic sky.

$$D_{PD} = D_{HD} \frac{1 + \cos\beta}{2}$$

(12)

Consequencely, the total solar radiation on the tilted surface is

$$G_{PD} = I_{PD} + D_{PD}$$

(13)

Using relationships from equations (1) to (12) and substitute in equation (13) gives

$$G_{PD} = Mk_d I_m \cos Z \cos\theta$$
$$+ k_d I_m \cos Z (C + (1-M)\cos Z) \frac{1+\cos\beta}{2}$$

(14)

2.4 Solar Radiation on an Array with Equatorial Installation

Array with equatorial installation is described as an array with polar axis lies along North-South axis and tilted angle equals to the location latitude as illustrated in Figure 1.

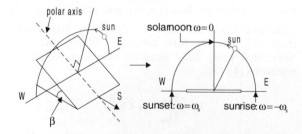

Figure 1 Equatorial Installation and frame of reference

Using vector representation, one can calculate zenith angle, incident angle and tilted angle of equatorial installed PV panel from the following equations.

$$\cos Z = A_1 \cos\omega + A_2$$

(15)

$$\cos\theta = B_1 \cos(\omega - c) + B_2$$

(16)

$$\cos\beta_c = C_1 \cos c + C_2$$

(17)

where
$$A_1 = \cos\phi\cos\delta \qquad A_2 = \sin\phi\sin\delta$$
$$B_1 = \cos(\phi-\beta)\cos\delta \qquad B_2 = \sin(\phi-\beta)\sin\delta$$
$$C_1 = \cos\phi\cos(\phi-\beta) \qquad C_2 = \sin\phi\sin(\phi-\beta)$$

Then $\cos Z$, $\cos\theta$ and $\cos\beta_c$ are substituted into equation (14). The result obtained is an equation for calculating solar radiation on an array with equatorial installation and is adjusted with an angle c along the polar (hour) axis.

$$G_{st} = Mk_d I_m (A_1 \cos\omega + A_2)(B_1 \cos(\omega - c) + B_2)$$
$$+ k_d I_m (A_1 \cos\omega + A_2)[C + (1-M)(A_1 \cos\omega + A_2)]$$
$$\times \frac{1 + C_1 \cos c + C_2}{2}$$

(18)

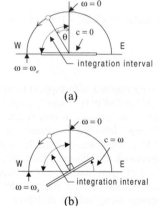

(a)

(b)

Figure. 2 Fixed and Continuous Tracking Array

Equation (18) can be applied to fixed and continuous tracking installations. In a fixed installation as shown in Fig.2(a), the angle c is equal to zero. Similarly, for a continuous tracking installation as shown in Fig 2(b), the angle $c = \omega$.

2.5 Incident Energy on Equatorial Installed PV Array

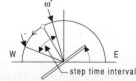

Figure 3. An Equatorial Installation with Angle "c"

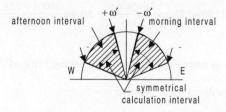

Figure 4. Symmetrical Calculation Time Interval

A time interval where the angle is c can be defined in terms of hour angles between ω' and ω'' as shown in Fig.

3. And the incident energy for this interval can be determined by integrating the solar radiation i.e.

$$H_{st} = \frac{24}{2\pi} \int_{\omega'}^{\omega''} G_{st}(\omega,c)d\omega$$

(19)

When the angle adjustment of a step tracking is designed in such a way that solar radiation on the array during the morning and afternoon are symmetrical as shown in Fig.4, the total energy for the symmetrical intervals becomes

$$H_{sb}(\omega',\omega'',c) = 2 \times \frac{24}{2\pi} k_d I_m \{M$$

$$\times [\frac{1}{2}A_1 B_1 (\cos(\omega''+\omega'-c)\sin(\omega''-\omega') + (\omega''-\omega')\cos c)$$

$$+ 2A_1 B_2 (\cos(\frac{\omega''+\omega'}{2})\sin(\frac{\omega''-\omega'}{2}))$$

$$+ 2A_2 B_1 (\cos(\frac{\omega''+\omega'-2c}{2})\sin(\frac{\omega''-\omega'}{2}))$$

$$+ A_2 B_2 (\omega''-\omega')]$$

$$+ C\frac{1+C_1\cos c + C_2}{2}[A_1(\sin\omega''-\sin\omega') + A_2(\omega''-\omega')]$$

$$+ (1-M)\frac{1+C_1\cos c + C_2}{2}[A_1^2(\frac{1}{2}(\omega''-\omega')$$

$$+ \frac{1}{4}(\sin(2\omega'')-\sin(2\omega')))$$

$$+ 2A_1 A_2(\sin\omega''-\sin\omega') + A_2^2(\omega''-\omega')]\}$$

(20)

Similarly, one can determine the incident energy on a fixed and continuous tracking array by integrating corresponding solar radiation from $\omega' = 0$ to $\omega'' = \omega_e$ (for a fixed array) and from $\omega' = 0$ to $\omega'' = \omega_s$ for a continuous tracking respectively.

2.6 Daily Incident Energy on an Array with m step per day tracking

For any m step per day tracking system (Fig. 5) there are n intervals to calculate the symmetrical energy (Fig. 6).

$n = (m+1)/2$ when m is an odd number

$n = m/2$ when m is an even number

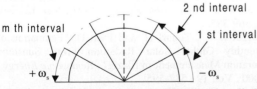

Figure 5 Time Interval of m step tracking installation

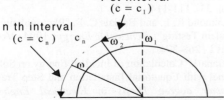

Figure 6 Energy calculating interval

Total incident energy on a m step per day tracking array is calculated from the sum of symmetrical incident energy as shown below.

$$H_{ajm} = H_{sb1}(\omega_1,\omega_2,c_1) + H_{sb2}(\omega_2,\omega_3,c_2) + ...$$
$$+ H_{sbn}(\omega_n,\omega_{n+1},c_n)$$

(21)

where

$\omega_1 = 0$, $\omega_{n+1} = \omega_s$ and $c_1 = 0$ for m is odd number $\omega_1 = 0$ and $\omega_{n+1} = \omega_s$ for m is even number

and there are $m-1$ independent variables.

3. OPTIMISATION OF A STEP TRACKING EQUATORIAL INSTALLED PV ARRAY

3.1 General Data

This study is conducted using location of Chiang Mai province of latitude 18.78 N and longitude 98.98 E. The date is November 7, 1999 (mid interval of $n_p=7$) and $\bar{k}_d = 0.765$ where $\bar{k}_d$ is an average of k_d obtained from Exell [2].

3.2 Step Tracking Optimisation

Equation (21) is subject to optimisation process to determine the panel angle and the hour angles for two to six step tracking. The objective of the optimisation is to obtain maximum incident energy on the panel.

Table 1 Hour and Panel Angle for $n_p = 7$
($H_{fx} = 4995$ Wh/m^2).

Installa Tion Steps.	ω_1	ω_2	ω_3	ω_4	c_1	c_2	c_3	En. Wh/m^2	Inc. En % H_{fx}
Fixed	0.0	84.3	-	-	0	-	-	4995	-
2	0.0	84.3	-	-	21.5	-	-	5270	5.51
3	0.0	24.2	84.3	-	0.0	33.4	-	5350	1.61
4	0.0	36.9	84.3	-	12.1	41.0	-	5383	0.66
5	0.0	14.9	44.8	84.3	0.0	20.3	46.1	5400	0.33
6	0.0	25.1	50.2	84.3	8.3	26.4	49.9	5409	0.19

It is shown in Table 1 that the 2 step tracking gives the maximum incremental energy of 5.51%.

Table 2 below shown details optimisation for other n_p's.

n_p	$\bar{k}_d$	c_1	H_{aj2} (Wh/m^2)	ΔH (% H_{fx})
1	0.770	± 21.6	5342	5.62
2	0.718	± 20.4	4734	4.65
3	0.681	± 19.2	5639	3.94
4	0.578	± 14.6	4693	2.17
5	0.536	± 12.8	4318	1.58
6	0.650	± 17.6	5019	3.21
7	0.765	± 21.5	5270	5.51
8	0.779	± 21.5	4895	5.61

It is seen in Table 2 that k_d has direct influence on incremental energy. That is the step tracking will be less effective on a cloudy day.

4. EXPERIMENTAL FIELD MEASUREMENT OF INCIDENT ENERGY ON PV ARRAYS

To validate the optimisation results, field measurement of PV arrays short circuit current were carried out every 10 minutes using data logger and personal computer. Short circuit currents of fixed and 2 step tracking array had been monitored for 90 days during October 16, 1999 to January 13, 2000. The data is analyzed to obtain incident energy and results obtained are given in Table 3.

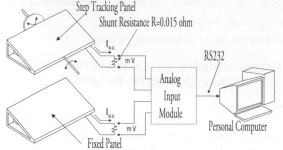

Figure 7 Experimental Measurement Setup

Table 3 Incident Energy from Experimental Measurement

Installation	Energy $n_P = 7$		Energy $n_P = 8$	
	(Wh/m²)	% H_{fx}	(Wh/m²)	% H_{fx}
Fixed Panel	4194	-	4680	-
2 Step racking	4734	12.88	5673	21.22

5. DISCUSSION

From the experiment it is found that for $n_P = 7$ and $n_P = 8$ the two step tracking array energy increases are 12.88% and 21.22% respectively compared to 5.51% and 5.61% obtained from the analysis. It is noted that during $n_P = 7$ the average sky condition was cloudy which must have affected the incident energy. While during $n_P = 8$, the sky was less cloudy. For this reason, the incremental energy during $n_P = 8$ is higher. It is also noted that the measured energy increases are much higher than the analysis due to isotropic and no surface reflection assumptions.

6. CONCLUSIONS

Optimisation of step tracking PV array using incremental energy under isotropic sky assumption has an advantage in function simplicity. The result obtained does differ from measurement. However, the discrepancy is more quantitative, as indicated by Mean Percent Error (MPE), than function representation, as indicated by correlation coefficient (R) which approaches unity. It is thus concluded that the optimum angles obtained from the analysis are correct. Evaluation of incident energy, taken into account the an-isotropic sky and panel surface reflection with the optimum angles derived, gives resultant incident energy well within 5% of the measurement [8].

7. ACKNOWLEDGEMENT

The authors would like to express their gratitude to the National Energy Policy Office (NEPO) for the financial support to carry out this work.

8. NOMENCLATURE

Z	zenith angle
ω	hour angle
ω_e	sun rise and sun set on tilt surface hour angle
ω_s	sun rise and sun set hour angle
θ	incident angle
β	tilt angle
c	panel's adjust angle
S	sunshine duration
S_m	maximum number of day light hours
a,b	coefficient of Angstrom's regression
D	diffuse solar irradiation
I	direct solar irradiation
G	global solar irradiation
H	daily solar radiation
I_m	maximum direct solar irradiation (sun at zenith)
H_{ajm}	daily solar radiation for m steps per day adjust panel
H_{sb}	solar radiation from morning-afternoon symmetrical calculation interval
H_{sb1}	solar radiation from 1 st calculation interval
H_{sb2}	solar radiation from 2 nd calculation interval
H_{sbn}	solar radiation from n th calculation interval

Subscripts

H	horizontal surface
P	panel surface
N	normal to beam surface
C	clear sky
D	cloudy sky

9. REFERENCES

[1] Baltas P. , Tortoreli M. and Russell P. E. "Evaluation of Power Output for Fixed and Step Tracking Photovoltaic Arrays" , *Solar Energy* , 1986 , Vol. 37 , No. 2 , 147-163.

[2] Exell R. H. B. "Simulation of Solar Radiation in a Tropical Climate with Data for Thailand" , *AIT Research Report* , 1980 , No. 115 , 9-16.

[3] Exell R. H. B. and Saricali K. "The Availability of Solar Energy in Thailand" , *AIT Research Report* , 1976 , No. 63, 8-10.

[4] Rehman S. and Halawani T. O. "Global Solar Radiation Estimation" , *Renewable Energy* , 1997 , Vol. 12 , No. 4 , 369-385.

[5] Togrul I. T. , Togurl H. and Evin D. "Estimation of Monthly Global Solar Radiation From Sunshine Duration Measurement in Elazig" , *Renewable Energy* , 2000 , Vol. 19 , 587-595.

[6] Klucher T. M. "Evaluation of Models to Predict Insolation on Tilted Surfaces" , *Solar Energy* , 1979 , Vol. 23 , 111-114.

[7] Hammond R. L. and Backus C. E. "Photovoltaic System Testing" , *Renewable Energy* , 1994 , Vol. 5 , Part I , 268-274.

[8] Watana J. "Calculation of Incident Energy on Solar Cell Panel with Equatorial Installation and Step Tracking", *Master degree Thesis in Electrical Engineering,* Chiangmai University, March, 2000.

DEVELOPEMENT OF A UNIVERSAL MIXED SIGNAL ASIC
FOR MODULAR INVERTERS

Bettenwort, G.; Gruffke, M.;
Ortjohann, E.; Voß, J.
Universität Paderborn
Elektrische Energieversorgung
Pohlweg 55 - Gebäude N
33098 Paderborn (Germany)
Telephone: 49 - 5251 - 60-2304
Fax: 49 - 5251 - 60-3235
email: bettenw@eevmic.uni-paderborn.de

Hüschemenger, M.;
Schönwandt, U.
Fachhochschule Köln
Abteilung. Gummersbach
Labor für Leistungselektronik
und Antriebstechnik
Am Sandberg 1
51643 Gummersbach

Magaritis, B.; Pohlmann, R.
Ascom Energy Systems
Senator-Schwartz-Ring 26
59491 Soest

ABSTRACT: A modular system concept for grid-connected photovoltaic systems is being developed within the scope of a research project of universities with an industrial partner. This project is supported by the Arbeitsgemeinschaft Solar Nordrhein-Westfalen. An important goal of the project is the low-cost production of modular inverters. Apart from devising a highly efficient RF power section, at present a unit is being designed for open and closed loop control on the basis of an Application-Specific Integrated Circuit (ASIC). For this purpose, first of all a special evaluation board (EB) emulating the entire functional scope of the ASIC was developed.

Highly flexible parameter configuration is a primary criterion for the design of the integrated circuit, so that the control unit can easily be adapted to changing boundary conditions. This paper introduces the development steps, the functional principles and the structure of the ASIC.

Keywords: AC-Modules - 1: Inverter - 2: Grid-Connected - 3

1. MOTIVATION AND GOALS

The state of technology for grid-connected photovoltaic systems is still oriented on the principle of central energy conversion, with all the consequences of this concept with respect to:

- Protection engineering

- Installation effort and

- System behaviour.

The electrical behaviour of the PV generators makes reliable short circuit detection impossible, because the short circuit current and the operating current are almost equal. Therefore the direct current collective line must be installed with protection against ground shorts and load circuit shorts. The effort which this entails is quite acceptable, on account of the high system power efficiency factor in the case of a simple configuration of the generator. Due to the complex system structure of this variant, which has to be defined unambiguously during the projecting phase of the plant, the flexibility of such PV plants with respect to subsequent modification is relatively poor.

The situation is different when the generator is to be integrated on surfaces which are poorly accessible, for example on the facade of a building, or if the solar radiation casts partial shadows.

Therefore in recent years a trend towards modularisation of the system technology has developed. The chained inverter constitutes a preliminary stage, with which - in agreement with the concept of a central inverter - a sufficient number of PV modules are connected in series to ensure that the generator voltage level lies above the grid voltage level, in as far as concepts without transformers are pursued. Otherwise a low frequency transformer usually provides voltage matching to the grid voltage level.

The smallest possible unit which makes sense on energy considerations consists of the combination of an inverter with a PV module to give a so-called AC module. System technology completely modularised in this manner largely avoids the disadvantages of conventional technology and can at the same time increase the power efficiency of the plant as a whole. The basic idea behind the modular system structure is to convert the PV direct voltage at the earliest possible stage into the grid-conforming alternating voltage.

The modularisation permits arbitrarily scalable PV-plants which can be constructed more simply and with greater dependability in the electrical engineering aspect, compared with a plant with central inverter.

Furthermore, efficiency losses through partial shadowing can be minimised because each AC module is operated with maximum power.

A disadvantage of such systems is that the power electronics as well as the open and closed loop control devices must be provided for each module in the same form as for a central inverter. The specific system price for modular inverters available on the market is usually higher than for plant variants with central inverter or string inverters. Therefore new development approaches are required particularly for modular inverters. A very promising concept is to use integrated circuits for open and closed loop control of the modular inverter, in order to reduce production costs.

2. REQUIREMENTS FOR THE DESIGN OF THE CONTROL UNIT

The rapidly increasing offer of power semiconductors can produce a situation in which power section topologies which have already been discarded in the course of this project, may within the foreseeable future become candidates once again for utilisation in modular inverters. For this reason the system concept has been devised such that the integrated circuit can be utilised in as wide a range of different inverter topologies as possible.

Criterion 1: Topology-independent control unit

A similar approach is required with regard to the effects of frequently changing feed-in directives, which also still feature large country-specific differences. In general, the monitoring devices, which too are part of the open and closed loop control system, must be designed such that adaptation to changed boundary conditions is possible with minimised effort.

Criterion 2: Adaptability to national standards

In recent years the offer of solar modules has continually increased with the high growth rate of the market for photovoltaic devices. This is true in particular for modules developed for utilisation in grid-coupled PV plants [1]. So far it is not known definitely, whether or not modular system technology will completely replace central converter systems, or whether both kinds of system will co-exist in future. A future-oriented control unit should therefore be flexibly adaptable to various voltage and power ranges.

Criterion 3: Suitability for most commercially available PV modules

Criterion 4: Suitable for inverters in various power rating classes

There is no longer any impediment for industrial mass-production, if and when it has been made possible to implement the requirements explained above nearly completely in an integrated circuit, i.e. when it is possible to realise a universally adaptable ASIC for open and closed loop control of a PV-inverter. Falling prices and competitive capability with respect to conventional PV-plants would then be the consequence.

3. STUCTURE

Figure 1 visualises the structure of the chief basic functions and their interaction.

The system-conditioned low input voltage implies that relatively little energy can be stored in the input capacitor. With otherwise the same control structure as for central inverters [2], the requirements imposed on the control dynamics of a modular inverter are significantly greater. The incident radiation interruptions, after previous high incident radiation level, give rise to problems. If the ambient temperature is very high at the same time, as can be expected for applications on the facade of buildings, switchover to an adaptive module voltage regulator can be contemplated.

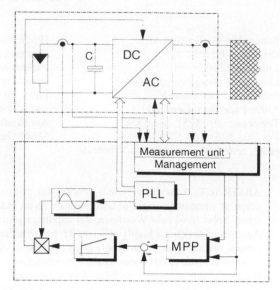

Figure 1: Structure of the inverter

Selection of a suitable maximum power point tracking is of central importance for full exploitation of the advantage of modular system technology, namely to operate each module at the optimum power output point. For this purpose a simulation model [2], [3] of the inverter was set-up, with which numerous known procedures can be verified. An important result of these investigations is that the efficiency difference between the various methods is very small, so that only the implementation effort needs to be considered as selection criterion. Multiplying methods should generally be avoided for the MPP-tracking [4].

Modular inverters have the decisive advantage compared with all other concepts, that the electrical parameters of the PV module can be measured directly. This makes it possible to monitor not only the state of the inverter, but also the state of the respective solar modules. For this purpose the control unit is equipped with a data communication interface, via which status and operating state protocols of the respective modular inverters are sent at regular intervals to a central monitoring unit which can monitor up to 32 AC-modules when fully equipped.

4. DRAFT DESIGN STRATEGY

The draft design of the ASIC is subdivided into four phases according to Figure 2. Phase 1 and 2 are used to check the basic feasibility according to technical as well as economic considerations. A discrete circuit construction was made on the basis of a structural analysis subjected to continual control with the help of simulation calculations. This made it possible to detect and remedy weak points already at an early stage of the project.

In cooperation with a system house for microelectronics, the procedure was then defined for the subsequent phases 3 and 4. An evaluation board (EB) was first of all constructed on account of the comparatively high complexity of the circuit.

Figure 3 shows the basic construction of the EB for the PV inverter. All specific functions can be emulated therewith and tested together with a power electronics section.

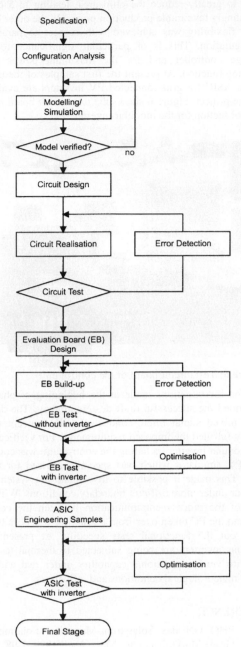

Figure 2: Draft design phases

The EB consists of a field programmable gate array (FPGA), a micro-controller (µC) and various A/D and D/A converters. The FPGA contains all time-critical functions such as:

- Grid monitoring (frequency and voltage)
- Internal fault monitoring
- Take-over of external faults from the power section
- The digital filter stages for signal processing
- The synchronisation mechanism for automatic grid connection of the inverter.

The heart of the system is a phase locked loop (PLL). The PLL performs grid synchronisation, drive of the power transistors in the unfolding stage of the power section and also provides the trigger signals for equidistant sampling of the measured parameters.

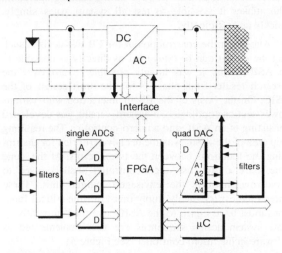

Figure 3: Basic construction of the evaluation board

To fulfil the demand for parameter configurability, the micro-controller contains the higher-level operational steering function as well as the maximum power point tracking function and the voltage regulator. The FPGA and the micro-controller exchange data via a serial universal asynchronous receiver-transmitter (UART) interface. The EB is controlled via a further serial communication computer interface which connects the EB to a PC. Figure 4 shows a section of the control menu of the EB.

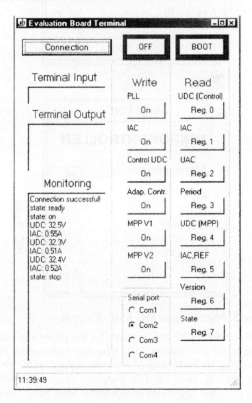

Figure 4: Control menu of the EB

For this purpose special control registers are set in the FPGA to place the system into the respective operating modes. This procedure permits step by step commissioning. Furthermore, status and measurement readings can be read from the registers via a PC during running operation. This makes it possible to test all system states simply, quickly and efficiently.

Altogether the construction of the EB was devised such that the circuit can be converted as directly as possible to an ASIC. Thus it was largely possible to take-over the research results from phase 3 for the draft design of the ASIC. On the basis of the EB in interaction with various power section topologies, a system variant was developed consisting of an ASIC and a micro-controller. The resulting chip area of the ASIC was the decisive criterion for distributing the functions between the two circuits. At the same time, only a low-cost micro-controller shall be used to avoid overshooting the envisaged budget limit. These criteria almost inevitably imply that fast time-critical functions must be assigned to the ASIC, and the comparatively slow system action sequences are best implemented as software in the micro-controller (see Figure 5).

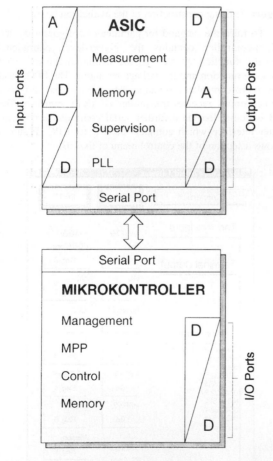

Figure 5: Mixed Signal MC/ASIC-Konzept

Integration of the driver stages for the power transistors was rejected for reasons of production engineering. The special ICs for switching-type power supply units, which are at present used for this purpose, can be integrated within the ASIC only with unacceptable large effort. So the driver stage for the power transistors remain separate components.

5   RESULTS

With this system concept it was possible on the one hand to greatly reduce the chip area, leading to a correspondingly favourable production price. On the other hand, great flexibility was achieved with respect to parameter configuration. This is of particular significance for the voltage controller and for the maximum power point tracking function. At present the first samples of the mixed signal ASIC for grid-connected PV inverters are available for the project. Figure 6 shows the test circuit board of the control section for the modular inverter.

Figure 6: Test circuit board of the control unit

The investigations made at the beginning of phase 4 confirmed the successful draft design strategy. The developed mixed signal micro-controller ASIC concept completely fulfilled the imposed requirements. For verification, the modular inverter including the control unit was coupled to a PV simulator which was specially devised for these tests. This made it possible to investigate the system behaviour under more difficult boundary conditions. With the help of the process communication between the control unit and the PC taken over from phase 3, it was possible to carry out the functional tests speedily. At present the modular inverters are being subjected to thermal tests in order to verify operational capability under real and extreme meteorological conditions and influences.

REFERENCE

[1]   PHOTON das Solarstrom-Magazin: 'Solarmodule Große Marktübersicht', März-April 1998, 2-98

[2]   Wasynczuk, O; Krause, P. C.: 'Computer Modeling of the American Power Conversion Corporation Photovoltaic Power Conditioning System', Sandia Contract No. 01-2860, March 1987

[3]   Ortjohann, E., Voges, B., Voß, J.: 'Dynamic Performance of Grid-Connected PV-Inverters', 12th European Photovoltaic Solar Energy conference, Amsterdam, Netherland, April 1994

[4]   G. Bettenwort, J. Bendfeld, C. Drilling, M.Gruffke, E. Ortjohann, S. Rump, J. Voß: 'Model for evaluating MPP methods for grid-connected PV plants', to be published in 16th European Photovoltaic Solar Energy conference, Glasgow , UK, Mai 2000

DESIGN OF DC INTERFACE UNIT FOR PV-GRID CONNECTED ROOF TOP SYSTEM

P R Mishra[a], E V Saktivel[a], J Panda[a], J C Joshi[b]
[a]Electronics Research Development Centre of India, Noida.
A-5, Sector-26, Noida-201301.
e mail: mishra_r@.hotmail. com.
[b]Centre of Energy Studies, Indian Institute of Technology Delhi-110016.

ABSTRACT: This paper presents design of DC interface unit for PV -grid connected roof top system. The proposed design takes care of safety standard given by Electro-Tecnical Commission in IEC-364 together with reliability and efficiency aspects. The DC interface unit is based on H-bridge topology and is operated at 100 KHz frequency. The reliability of the system is high due to use of embedded controller 80196 chip. Conventional pulse width modulation (PWM) technique can not be used because it results in poor efficiency at light load condition due to cross modulation and electro magnetic interferences. In addtion it is very difficult to operate the system at maximum power point (MPP) due non-availability of enough steps to provide effective modulation. These problems are overcome by using new technique i.e.Time Division Modulation. The design is very relevant because it enables the software control to implement MPP operation at very high frequency which resulted in 1KW power from 8 cubic inch transformer . In addition the energy conversion of DC interface module is 90% from quarter load to full load.
Keywords: Small Grid Connected PV system-1: DC-DC Converter-2: Roof Top-3

1. INTRODUCTION

The use of photovoltaic system as a power source is becoming popular due to easy deployment, urgent need of reducing emission of green house gases and peaking power demand. And so more PV-Utility connected roof top systems are getting installed around the world. Diferent types of Grid Connected Inverter are available and also given in literature [1-2]. These are based either on transfomerless or with isolated transformer designs. The problem with transformersless design is high earth leakage current and high input voltage of PV array. High earth leakage current and high input voltage is a safety hazard to human being. Therefore the International Electromechnical Commission specification IEC 60364 is interpreted for roof top photovoltaic system as [3]

i) Open circuit voltage of PV-array should be less than 120V.

ii) Utility should be isolated from PV-Array

These safety problems can be eliminated by designing a grid connected system with nominal voltage 64Volt and with isolation transformer. But the isolation transformer for 2Kwatt system at 10% no load current result in loss of 2.4 KWH energy per day for 10 hr operation. Taking average solar radiation as less than 5 sun, the loss of energy will be as high as 20%.

The other important criterias are greater packaging density and higher mean-time between failure. The packaging density can be increased by increasing the switching frequency. This permits a great reduction in the size and weight of the transformer, which reduces both wire and core losses.[4-5]

The packaging density can be increased by increasing the switching frequency from 60Hz to 100KHz. This results in the factor of 1666 reduction in size. This permits a great reduction in the size and weight of the transformer, which reduces both wire and core losses. At the same time faster switching speed of MOSFETs has permitted increase in switching frequencies from 50 to 200KHz or upto 400KHz. But the switching losses in MOSFETs with ferrite transformer increases with the increase in switching frequency. The trade-off between losses and density of converter is near 100KHz [6]. The other important aspects in the design of the DC interface unit is reliability and output voltage regulations. To achieve high reliability level, the Mean Time Between Failure i.e. MTBF should be very high which in-turn depends upon number of components used in the system. Therefore the components count should be bare minimum.

MPP tracking is generally provided by closed loop linear feed back control system and which normally leads to oscillation problems. To overcome this digital control with state space approach alongwith positive feedback control technique have been employed[7]. Both the above criteria can be effectively used with microcontrollers. Therefore in present work 16-bit Microcontroller 80196 is used to implement switching pattern for MOSFETs and also to provide feedback control. For MPP Operation generally PWM technique is employed with positive feedback control. PWM technique resulted in cross-over modulation and also reprogramming of registers whenever line/load varies. Above all the MPP Operation is not effective because enough resolution is not available at such high frequency. Further the upper limit of using PWM port of microcontroller is 33K Hz[8] To overcome this in present work Time Division Modulation has been developed and successfully used. The detailed operation is given in next section.

2. PRINCIPLE OF OPERATION

TheDC interface unit is basically H-Bridge DC to DC Converter and is shown in Fig.1. The incoming power from PV array is converted from 64V DC (nominal) to 325VDC (nominal) and output of DC Interface Unit is fed to the transformerless inverter input.

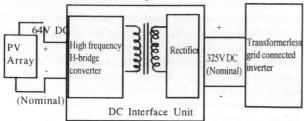

Fig.1 Block Schematic diagram of DC Interface Unit of a roof-top grid connected system.

The basic principle of H-bridge DC to DC converter is given below. The sequence of operation of H-bridge converter is given in Fig. 2.

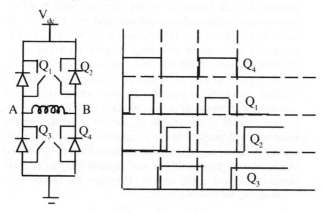

Fig. 2 H-BRIDGE AND WAVEFORM OF DRIVE PULSES

For power conversion, switches diagonal to each other should be simultaneously ON i.e. Q1 and Q4 or Q2 and Q3. For providing output regulation, the duty cycle of the switches is varied. The switches i.e. Q1 and Q3 or Q2 and Q4 in one arm should not be switched ON at any point of operation, otherwise shoot-through/ cross-modulationb occurs. The switches on the lower side (i.e) Q3 and Q4 should be switched ON before the operation is transfered from one arm to another arm i.e. during dead time. Otherwise residual energy in the transformer creates electromagnetic interference and disturbances in drive signals. The sequence of operation is clearly illustrated in Table 1.

To implement the above sequence at high frequency, High Speed Output (HSO) section of 80196 is used. High speed output can be operated up to maximum of 150 kHz. To switch ON and OFF the HSO the Content Access Memory (CAM) is to be programmed once and is locked. This is an independent section and can work without CPU intervention [9]. The maximum limit of clock frequency of 80196 is 20MHz. At this clock frequency 0.8 μs is minimum time resolution of the Timer. HSO operates with reference to Timer. With this time resolution, the cumulative time spent in avoiding cross-modulation and removing residual energy is 4μs.

Table 1

STEPS		Q1	Q2	Q3	Q4
1	RE	OFF	OFF	ON	ON
2	CM	OFF	OFF	OFF	ON
3	PT	ON	OFF	OFF	ON
4	CM	OFF	OFF	OFF	ON
5	RE	OFF	OFF	ON	ON
6	CM	OFF	OFF	ON	OFF
7	PT	OFF	ON	ON	OFF
8	CM	OFF	OFF	ON	OFF

PT-Power transfer stage
CM-overcome cross-modulation
RE-removing residual energy

So the maximum available time cycle is 6μs or 3μs per arm. Therefore the available resolution for output voltage regulation with PWM is only three steps. This resulted into poor line and load regulation.

Again there is ringing problem at 1μs level. This resulted in very low efficiency at light load conditions. Further the problem is aggaravated as the content in CAM register has to be deleted and reprogrammed for each value of PWM. For this duration the DC-DC converter operation has to be switched OFF.

The problems are overcome by new technique and coined as Time Division Modulation (TDM). In this technique, the H-bridge is operated with fixed or maximum duty cycle for predefined time period and the entire bridge operation is switched off. And the output voltage regulation is acheived by varying the On time of the bridge operaion while keeping total ON time and OFF time constant. Since in this technique the switching pattern is fixed, there is no need of reprogramming the CAM register, And ON and OFF time is controlled by the general purpose port of Microcontroller. The response time of MPPT is 250μS.

2.1 Feedback Control

The feedback control has been employed in two modules. For feeding maximum power to the grid positive feedback technique has been employed based on assumption that the utility voltage will not change during MPP tracking. In this technique at first phase angle is increased, if current fed to grid increases then the phase angle is further increased or vice-versa.

During above operation if inverter module voltage increase above the given range then the TDM ON period is decreased or vice-versa. This is based on digital state space technique.

3. DEVELOPMENT OF A 100 KHZ DC-DC CONVERTER

The design and development of DC-DC converter using TDM with following specification is discussed in this section:

Input Voltage Range	45-66 Volt DC
Output voltage	325 Volt DC
Output Current	5ADC (Max)
Switching Frequency	100KHz
Full Load Efficiency	>90%
Half load Efficiency	>85%

The first step in design is to generate the driver pulses of the, switches from the HSO. TO operate the HSO, the CAM register is to be programmed. Each HSO will need two registers for operation. For implementing this eight registers are required. Two additional registers are required i.e. one for resetting the timer and second for generating 4KHz for implemting the TDM. Against required ten numbers of CAMregisters, only eight registers are available.To overcome this, only three HSO is used for switching Q1, Q2, Q3,. The drive pulse for Q4 is generated with external hardware circuitry. The second step is to provide drive pulse with adequate current delivery/sinking power to charge and discharge gate to source capacitor of MOSFETS. This has been achieved by IR MOSFET driver IR 2110. The third step is to provide feedback regulation. Digital control is provided through using 10 bit in-built Analog to Digital converter(ADC) in Microcontroller. The response time off feed back control loop is 5 milli second. The port for controlling H-bridge operation can be updated by this time period.

The above technique is suited for power conversion in 1-5 Kwatt range. The system has been designed modular of 1 Kwatt to reduce mismatch losses between deffeent strings of PV array [10].

4. DISCUSSION AND RESULTS

The DC to DC converter using TDM technique and microcontroller has been designed and tested in the laboratory. The switching pattern of the DC to DC converter is given in Fig. 3(a), Fig 3(b) & Fig 3(c) for high input power & high load, medium inputpower & medium load and low input power and low load conditions respectivily. These switching pattern has been measured and recorded using 150MHz Digital storage Oscidoscope HP54602B.

The efficiency has been calculated by measuring input current & voltage and output current & voltage by Thurlby thander multimeter (TTi 1750 and Fluke 73 multimeter. The results are given in Table 2.
From table 2 , it can be seen that the overall efficiency of DC interface unit is near 90% even near quarter load. The size of the transformer is only 8 cubic inch for 1 KW module. The no load current is 310 milliAmpere at 60 Volt which is less than 2%. The overall gain of energy is around 10% due to varying

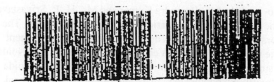

Fig. 3(a) TDM FOR HIGH INPUT POWER/HIGH LOAD

Fig. 3(b) TDM FOR MEDIUM INPUT POWER/ MEDIUM LOAD

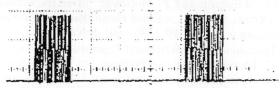

Fig. 3(c) TDM FOR LOW INPUT POWER/LIGHT LOAD

Table 2

Input			Output			Efficiency
Volt	Ampere	Watt	Volt	Ampere	Watt	(in%)
47.9	2.32	111	309	0.315	97.3	87.3
63.7	6.53	416	365	1.03	376	90.4
58.5	9.07	531	333	1.43	476.2	89.7
52.7	10.4	548	300	1.65	495	90.3
45.3	11.9	539	263	1.89	497	92.7

magenetisation current requirement from no load to full load. The switching pattern and test results corroborates our design criteria. The only draw back of the system is that the size of output filter capacitor is to be increased to reduce the ripple voltage.

The feedback control also works properly for wide value of utility voltage, ranging from 180V to 300Volt, the system operated near MPP.

5. CONCLUSIONS

The proposed new technique will find wide applications in small PV grid connected system.. The benefit of proposed technique are that CAM registers of miicrocontroller has to be programmed only once and effective duty-cycle can be achieved from 0% to 100%. The main achievement of the proposed technique is that microcontroller can be easily used for 100 KHz switching frequency resulting in low MTBF i. e. high reliability, better MPPT operation and high packaging density. In addition the above technique will be useful for power conversion form one DC level to another DC level i.e. in application like DC power supplies, UPS etc with minor modification in feedback control circuit.

6. ACKNOWLEDGEMENT

The authors would like to express their thanks to

our co-workers Mr. Rajendra Kumar and Mr. H. P. Srivastava for the work during fabrication and testing of the unit. The authors gratefully acknowledge the constant encouragement for the work by Mr. R. K Verma, Director, ER&DCI, Noida. The support and equipment provided by ER&DCI, Noida are greatly acknowledged. The authors also wish to thank Ministry of Non Conventional Energy Sources (Govt. of India) and especially Dr. B. Bhargava, Director MNES for sponsoring the research work.

REFERENCES

[1] S. Krauthamer, K. Bahrami, R. Das, T. Malic, W. Rippel, Photovoltaic Power Conditioning Subsystem: State of the Art and Development, jet Propulsion Laboratory, DOE publications, California, 1984.

[2] M. Meinhardt and P. Mutschler, Inverters without transformer for grid connected photovoltaic applications proceeding fo the EPE, PP 3.086-3.091, Selvilla, Spain, 1995.

[3] International Electro-technical Commission IEC 60364-7-712 : Electrical installation of buildings- Part 7 : Requirementt for special installation or locations section 712 : Photovoltaic Power supply systems (draft standard), 1997.

[4] N. Mohan, T. M. Undeland, Robbins. Power Electmnics .Applicaiions and Design. John Wiley & Son New york, 1994.

[5] N. W Bums, Power Electronics in the minicomputer Industry, Proc IEEE, Vol.76 No.4, 1998, PP 311-324

[6] Alex Goldman, Modern Ferrtie Technology, Van NosTrand Reinhold. New York, 1992, PP 259-256.

[7] R. J. Vaccaro, Digital Control, McGraw-Hill Inc. New york, 1995.

[8] H. Shinohara, K. kimto, T. Itami, Development of residential use, utility interactive PV inverter with high frequency isolation, Solar Energy material and Solar Cells Vol. 35, P. P429-436, 1994

[9] Intel, 16-bit Embedded Controller Hand Book, intel Mt. Prospect, 1L, 1997.

[10] M. Calasis, V. G. Agelidis & m. Meinhardt, Multilevel converters for single phase Grid Connected photovoltaic Systems : A overview. Solar Energy Vol. 66, No, 5, PP. 325-335, 1999.

PVSYST 3.0: Implementation of
an Expert System Module in PV Simulation Software

Dr. André Mermoud
Group of Applyed Physics (GAP),
University of Geneva,
Battelle, bât. A, 7, route de Drize
CH 1227 Carouge (Geneva)
Switzerland
Tel (+41) 22 705 96 61, Fax (+41) 22 705 96 39
e-mail : andre.mermoud@gap-e.unige.ch
http://www.unige.ch/gap-e/
http:// www.pvsyst.com

Christian Roecker,
Jacques Bonvin,
LESO-PB, EPFL,
CH 1015 Lausanne,
Switzerland.
Tel (+41) 21 693 43 41,
Fax (+41) 21 693 27 22
e-mail : christian.roecker@epfl.ch
http://lesowww.epfl.ch

Abstract: Version 3.0 of PVSYST has been developed by the University of Geneva in collaboration with the Swiss Federal Institute of Technology in the framework of IEA Task VII "Photovoltaics in the Built Environment" of the PVPS program (Photovoltaic Power Systems).

PVSYST is a sizing, simulation and analysis tool for grid-connected, stand-alone and DC-grid photovoltaic systems. It offers a three levels approach: "Preliminary Design" for quick pre-sizing of PV systems (monthly data based), "Project Design" for complete sizing and design (hourly detailed simulation), and "Tools" which proposes a deep validation process of measured data on existing systems. This level also includes the PV components and meteo database management, with a comprehensive description of each components behaviour, as well as general solar energy tools (solar geometry, PV-array behaviour under partial shadings or mismatch, etc.)

In the "Project Design" part an expert system guides the user in the layout definition of the PV-array and system, taking the component operating conditions into account. This level also includes a 3D CAO tool for partial shading studies. PVSYST also performs an economic evaluation of the defined system, in any currency.

Keywords: - PV design software – 1: PV system study – 2: Shadings – 3

1.- Objectives and Approach

PVSYST is an already well established simulation and calculation program for the study of photovoltaic systems. The previous version 2.2 has been evaluated by a Britannic team along with several other PV design tools [1], and was found as one of the most useable among the available PV software, with the ability to model the PV-system and its components in great detail. However it's main defaults were reported as the rather complex user interface, and the high level of PV expertise required from the user.

Therefore, a collaboration between the University of Geneva and the Swiss Federal Institute of Technology (LESO-PB at EPFL), recast it with the following objectives:

- Offer an ergonomic approach by introducing several project design steps, a "greenline" and several "variable depth" levels by presenting only needed dialogs or parameters at a given stage of the project.

- Implement user's guides ("expert system"-like approach) for the system design.

- Join an efficient and complete contextual "Help" system, allowing architects and not PV-specialists to use the program for pre-sizing at the earlier stages of the project.

- Use of a modern development platform (i.e. Delphi), which is expected to stay compatible with the next generations of running operating systems (from Win'95 and NT). Indeed, the program seems to give no problems with Win 2000 and NT5.

This upgrade translation in Delphi not only improved the user-friendliness of the whole interface, but also results in much more reliable operation.

As for the earlier versions, the development of this new release is financed by the Swiss Federal Office for Energy (OFEN/BEW).

2. – Program Structure

The homepage directly gives access to the three project design levels:

2.1. - Preliminary Design

It is the pre-sizing step of a project. For **grid-connected** systems, and especially for building integration, this level is *architect-oriented*, asking for available area, nominal power or energy yield desired. Further parameters are general properties about PV technology (mentioning colours, related to technology; transparency, etc), mounting disposition, ventilation. For **stand-alone** systems this tool allows to size the required PV power and battery capacity, given a user's load profile and the acceptable probability that the user will not be satisfied (equivalent to the desired "solar fraction").

In this mode the system yield evaluations are performed instantaneously in monthly values, using only a very few general system characteristics, without specifying specific system components. A rough estimation of the system cost is also available.

2.2. - Project Design

This engineer-oriented part is aiming to perform a thorough system design using detailed hourly simulations.

In the frame of a "project" – which mainly holds location and meteo - the user can perform and compare different system simulation runs. He has to define the plane orientation (with possibility of tracking planes, double-orientation or shed/sun-shields mounting), and to choose the specific system components. He is assisted in designing the PV array (number of PV modules in series and parallel) given a chosen inverter model or battery pack.

In a *second step*, he can specify more detailed parameters and analyse fine effects like thermal behaviour, wiring and mismatch losses, incidence angle losses, horizon (far shading), or partial shadings of near objects on the array, an so on.

The **3-D CAO tool** for near shading studies, which was described elsewhere [2, 3], has been noticeably improved, and provides now a means of evaluating electrical effect of partial shadings.

Results include several dozens of simulation variables, which may be displayed in monthly, daily or hourly values, printed or transferred to other software. A detailed **economic evaluation** can be performed using real component prices, additional costs and investment conditions, in any currency.

A summary of each simulation run is readily available on the printer, including all involved simulation parameters and most significant result plots and tables. According to additional features asked during the simulation process, other result sheets are also proposed for printing (shadings, economic evaluation, etc).

2.3. – Tools: Measured Data

When a PV system is running and carefully monitored, this part allows to **import the measured data** (in almost any ASCII format), to display tables and graphs of the real performances, and to perform **comparisons** with the simulated variables.

Simulations parameters can then be varied until obtaining close agreement between measured and simulated values. This provides a means of analysing and understanding the real parameters of the system, and identify even very weak running defects. In daily or monthly data accumulations, chosen faulty data can be eliminated (for example system failures, snow on collectors, etc.), in order to compare system performance in normal conditions.

2.4. – Tools: Database

This "Tools" part also includes the database's management, with the detailed specification of each element, a comprehensive behaviour graphical assistance (PV-module or battery characteristics, a variety of meteo graphical presentations, etc.) and printing of an exhaustive "data sheet" for any component.

PVSYST avails of about 200 geographical sites over the world, with **monthly meteo** data (irradiation and temperature). Further locations can be easily included by hand (monthly values), or read from the Meteonorm software [4]. A programmable interpreter allows for importing meteo data files in **hourly values** (e.g. own measured, TMY or DRY files), with almost any ASCII format. If hourly data are not available, the "project" will automatically call the **generation of synthetic hourly data** facility, using the monthly meteo values of the site.

The database of commercially available PV-components includes 300 PV modules, usual grid inverters, batteries, regulators. Of course any new or custom device can be easily added using manufacturer's specifications. Note that PV-module I/V characteristics is modelled through the usual one-diode model [5].

2.5. – General Solar Tools

This part also includes some specific tools useful when dealing with solar systems:
- generation of synthetic hourly data,
- quick meteo computations (transposition on tilted plane, horizon, sheds, sun-shields, IAM effects) in monthly values,
- plane orientation optimisation,
- tables and graphs of solar geometry parameters,
- irradiation under clear sky model,
- shed and sun-shields optimisation and shadings,
- PV-array electrical behaviour under partial shadings or module mismatch, etc.

2.6. - Help

A general contextual "Help" system is available from anywhere in the software. It includes program use assistance, as well as general information about system design, models used, validations.

3. – "Expert system" Features

These are essentially implemented in the second level **"project design"** part, after defining the project

3.1. – Grid System Design

The user has just to enter the desired **nominal power**, to **choose the inverter** and the **PV module types** in the database.

Then the program proposes the number of required inverters, and a possible array layout (number of modules in series and in parallel). This choice has to be performed taking the engineering system constraints into account: the number of modules in series should produce a MPP voltage compatible with the inverter voltage levels window, and should not exceed the maximum allowable voltage on the inverter input even in worst conditions (i.e. at low temperature, -10°C by default).

The user can of course modify the proposed layout: warnings are displayed if the configuration is not quite satisfactory: either in red (serious conflict preventing the simulation), or in orange (not optimal system, but simulation possible). The warnings are related to the

and the plane orientation. They are aiming to facilitate the PV system layout definition.

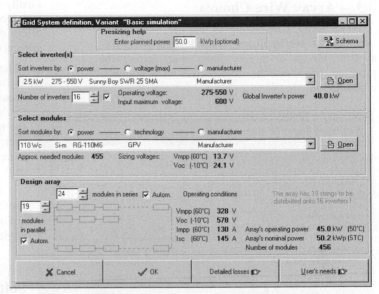

inverter sizing, the array voltage, the number of strings by respect to the inverters, etc.

3.2. – Stand Alone System Design

The procedure is similar, but requires a more refined sizing process, which in fact uses the same monthly optimisation as in the "pre-sizing" level. The user should first quantify the system load (choosing domestic-appliances, powers and operating times). Then he has to decide the **acceptable LOL** ("loss of load probability", i.e. the duration when load cannot be satisfied), and the required **battery autonomy** in days. According to the load level, the program also advises a system (battery) **voltage**.

Then the program computes the required storage capacity and PV array power. The user has just to choose the battery model and the PV module in the database, and the program determines the PV array and battery pack configuration.

In the same way as above, the user can adjust this configuration. The "Next" button gives access to the remaining of the system design (regulator choice, back-up generator, battery temperature, etc). Again, warnings are displayed according to the system design consistency: PV array sizing by respect to the battery pack, right choice of the regulator (voltage, input or output currents), etc.

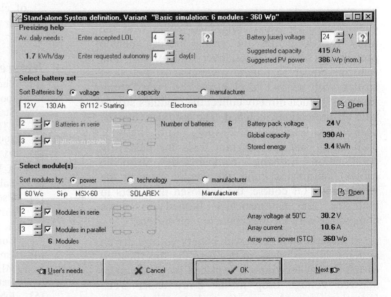

The pre-sizing of the system results from a very fast simulation process in monthly values. Storage capacity is simply evaluated for satisfying the load demand without solar yield. For the PV-array power determination, after computing the incident available irradiation in monthly values (using "average day" algorithm, which involves a dynamic computation of one average day each month, and provides accuracy of a very few percent on tilted plane transpositions), the program generates a 365 days stochastic series according to standard models [6], and computes the daily charge/discharge balance for estimating the "loss-of-load" probability. PV array power is adjusted in order to match the required LOL by a successive approximation method.

3.3. – Array Wire Choices

The array wire sizing is another example of user's assistance approach when designing the PV-system. Given a target ohmic loss factor, the program proposes the best suited wire diameter combination.

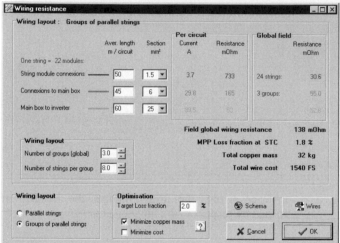

This can be applied to three common array cabling layouts: usual connection boxes near the array, which gather the string cables, include protections (diodes, fuses, over-voltage, etc), and transmit power to remote inverters; or groups of strings with second level boxes; or even "string inverters" which directly collect the modules strings with all protections included. The user specifies the desired loss level limit and average wire length for each level, and the program finds the best wire combination (among commercially available cables), optimising either the copper mass or the wiring cost. Of course, the proposed diameters in listboxes are automatically limited to wires compatible with the maximum branch current.

The same tool holds with battery voltage driven systems. Nevertheless, in this case the specification of the wiring loss ratio is replaced by a maximum allowed voltage drop in the array under STC. Indeed, wiring ohmic voltage drop will have no effect when the PV voltage is higher enough above the battery voltage (i.e. by low temperatures), and could be dramatic when this goes above the I/V elbow. Overall loss can only be evaluated by the detailed simulation.

4. – Conclusions

PVSYST 3.0 is much more than a simulation program. It yields now a real PV-system sizing and developing assistance. It offers the ability to model the PV system and its components in great details. Finally it also helps analysing measured data and real system behaviour or defects. Moreover, it offers comprehensive facilities for studying many aspects of solar energy use.

The involvement in the design team of people from field of architectural integration of PV has brought new points of view and new approaches to the user interface. This helped to overcome the main defaults which were reported by some users about the earlier versions, namely the complex approach and needed expertise of the user.

Therefore, it becomes a valuable professional and pedagogical instrument, for architects, students, as well as for confirmed engineers or researchers.

Software Availability

PVSYST V3.0 can be downloaded from the web site http://www.pvsyst.com, or ordered on CD-ROM from the author. It runs in evaluation mode during ten days with full capabilities (free of charge), and then turns in a "Demo" mode. Further use requires user's code.

References

[1]   Photovoltaics in Buildings: A survey of design tools. D. Lloyd Jones, P. Ruyssevelt, M. Standeven, J. Bates, C. Matson, D. Price. ETSU Report No S/P2/00289/REP, London, Nov. 1997.

[2]   PVSYST V3.0: Ergonomics and Accessibility. A. Mermoud, C. Roecker, J. Bonvin, 15th European Solar Energy Conference, Vienna 6-10 July 1998.

[3]   Use and Validation of PVSYST, a user-friendly software for PV-system design. A. Mermoud, 13th European Solar Energy Conference, Nice 23-27 october 1995.

[4]   Jan Remund, Esther Salvisberg, Stefan Kunz. METEONORM Version 4.0 (1999): global meteorological database. www.meteotest.ch.

[5]   John A. Duffie and W.-A. Beckman, Solar Engineering of Thermal process. Wiley and sons, N.-Y. 1991.

[6]   R.J. Aguiar, M. Collares-Pereira and J.P. Conde. Simple Procedure for Generating Sequences of Daily Radiation Values Using a Library of Markov Transition Matrices. Solar Energy Vol 40, No 3, pp 269-279, 1988.

A PV-MODULE ORIENTED INVERTER, FEEDING A LOW VOLTAGE AC BUS

Björn Lindgren, Chalmers University of Technology, Department of Electric Power Engineering,
SE-412 96 Göteborg, Sweden, E-mail: bjorn.lindgren@elkraft.chalmers.se

ABSTRACT: The development of photovoltaic inverters has recently moved towards smaller units in order to increase modularity and energy yield in shaded areas. This paper presents the design of a module-oriented inverter of 110 W. The current controller is implemented in an inexpensive 8-bit micro controller, which uses a hysteresis technique also known as bang-bang control. Emphasis has been put on high energy-efficiency (over 90 %) and low cost components.

Keywords: Inverters, Photovoltaic power systems, Solar power generation, Fault tolerance, Bang-bang control, Current control, MOSFET circuits.

INTRODUCTION

To make use of the expensive photovoltaic modules, it is important not only to position the PV-modules in an optimal way but also to consider the electrical behaviour of the cells [1]. Production losses due to shading effects are extensive in many PV-plants: For example, in the German 1000 roofs PV-programme, 10-15 % of the losses were due to shading [2]. Every measure should be taken to minimise shading, although sometimes it cannot be avoided. In such cases, a more shade-tolerant system should be employed. The fewer PV-cells connected to one converter the higher the shade-tolerance obtained. The disadvantage of such a system is its higher auxiliary power consumption and therefore lower power efficiency. However, the energy yield of a decentralised configuration can still be higher than that from a conventional system if shading is present. Additionally, the system becomes more modular, which makes it easy to exchange modules or expand an existing system with new modules.

An evaluation of the PV-plant at Sankt Jörgen, Göteborg, Sweden has shown that energy loss up to 30 % could be derived from shading effects even though only a very small fraction of the PV-cells was shaded [3]. This paper presents a module-oriented inverter that decreases this loss drastically and will be installed at Sankt Jörgen PV power plant in a near future.

CONVERTER TOPOLOGY

To select an efficient and cost effective converter a study of nine different converter topologies has been performed [4].

As a result of the study the configuration in Figure 1 with a low voltage AC-bus and decentralised inverters has been selected. It has relatively high energy-efficiency and low cost, few components in the main circuit, which implies long mean time between failure and robustness. The central transformer can to some extent filter harmonics generated by the inverters.

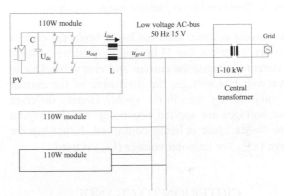

Figure 1. A full bridge inverter loads each PV-module optimally.

Each inverter operates its PV-module at the maximum power point and feeds a sinusoidal current into the low voltage AC-bus. A single-phase or a three-phase transformer boosts the voltage up to the grid level of 230/400 V. The transformer can be considered to be very reliable, so the system becomes fault-tolerant as the modules work independently of each other. If one fails the others will keep on operating.

Some drawbacks of this configuration are relatively high currents in the bus, and the control circuit, the sensors and the power supply are somewhat more complex to realise as compared with dc-topologies. The topology has a minimum input voltage level, 25 V at U_{grid} = 15 V, which together with the iv-cure for the PV-module set limits to energy production at low irradiation. Also the input capacitor must be quite large to limit the 100 Hz voltage ripple on dc-link.

CURRENT HYSTERESIS CONTROL

The current controller uses a hysteresis technique, also known as bang-bang control. It is a method to shape the output current which is relatively easy to implement in a microcontroller. A look-up table is used to produce a sinusoidal curve shape. This is then multiplied by the mean current value to form the reference current. Around this reference, the current controller sets an upper and a lower limit for the current as shown in Figure 2.

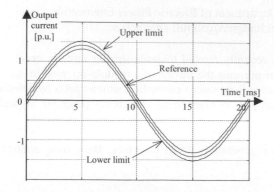

Figure 2. Upper and lower limit for bang-bang control.

If the current is lower than the lower limit, the current controller switches the H-bridge to positive voltage that starts to increase the current and vice versa. Due to the varying bus voltage, the derivative of the current will vary during the 50-Hz cycle. Hence, different output voltages are applied depending on what region of the 50-Hz cycle is being controlled. Either bipolar voltage ($\pm U_{dc}$) or unipolar voltage (U_{dc}) is used.

CRITERION TO CHANGE SWITCHING LEVELS

A full bridge can apply three voltages to its load, u_{out}: $+U_{dc}$, 0 and $-U_{dc}$. If control is done by applying $\pm U_{dc}$ the controllability becomes very asymmetrical around the peak grid voltage. Therefore unipolar voltage is used at these regions and bipolar voltage in-between. Another issue of this strategy is to minimise the power loss due to ripple. The regions with bipolar voltage ($\pm U_{dc}$), should be kept as narrow as possible as they tend to increase the current ripple. Although, the controllability is lost at high dc-voltage if the region is too narrow. A criterion for determining where to change the switching voltage levels (t_1 and t_2 in Figure 3) could be the points when the ratios of possible positive and negative voltage before and after t_1 are equal.

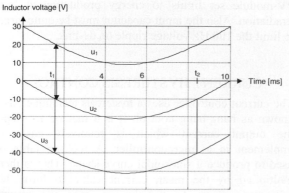

Figure 3.   Possible inductor voltages with U_{dc}=30 V and $\hat{u}_{grid}$=21 V.

In Figure 3 the three possible voltage levels across the inductor are shown during half a 50-Hz period (10 ms). The inductor voltages are

$$u_L = u_{out} - u_{grid}(t) = \begin{Bmatrix} u_1 \\ u_2 \\ u_3 \end{Bmatrix} = \begin{Bmatrix} U_{dc} \\ 0 \\ -U_{dc} \end{Bmatrix} - \hat{u}_{grid}\sin(\omega t)$$

(1)

Due to the control of the current, both positive and negative voltages must be available. For positive voltage, u_1 is used and must be greater than zero (i.e. $U_{dc} > \hat{u}_{grid}$). For negative voltage, both u_2 and u_3 can be used although u_2 gives a very low voltage close to t = 0 and t = 10 ms. Hence, u_3 is used there. The controllability of the current is optimal if the available positive and negative voltages are as equal as possible (the ratio positive voltage divided by negative voltage is close to one). This cannot be achieved but if the ratio of possible positive and negative inductor voltage is equal before and after t_1 an equalised controllability is acheved. This criterion is noted:

$$\frac{|u_3|}{|u_1|} = \frac{|u_1|}{|u_2|} \Rightarrow$$

(2)

$$\frac{-u_3}{u_1} = \frac{u_1}{-u_2} \Rightarrow \quad u_2 \cdot u_3 = u_1^2 \Rightarrow$$

$$(-u_{grid}) \cdot (-U_{dc} - u_{grid}) = (U_{dc} - u_{grid})^2 \Rightarrow$$

$$u_{grid} = \frac{U_{dc}}{3}$$

(3)

This means that the time for changing from bipolar voltage control to unipolar control, t_1, or vice versa, t_2, should occur when the grid voltage is one third of the dc voltage.

SIMULATIONS AND MEASUREMENTS

In theory, bang-bang control is an excellent method for shaping the output current since a certain fundamental frequency of the current is attained with a predetermined ripple. In practice however, there are disadvantages. For an analogue controller, there is always noise that will disturb the controller. In the case of a digital bang-bang controller, the execution speed and analogue to digital conversion time will limit the performance of control.

Simulations have been made with the simulation programme Saber to investigate what performance can be achieved with the digital control circuit and the main circuit selected.

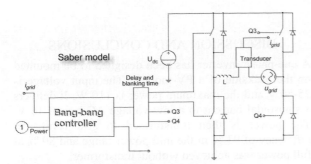

Figure 4. The scheme simulated in Saber.

In Figure 4 the simulated circuit is shown. The full bridge is modelled by a dc-source, four ideal switches, four ideal diodes, an inductance, a sinusoidal voltage source and a current transducer for measuring i_{grid}. The small rectangular block in the middle handles propagation delay and blanking time, which are set to 1 and 3 µs respectively. The block labelled 'bang-bang controller' performs the control described in section 'Current Hysteresis Control'. With the selected microcontroller (Atmel AT90S2313), one control loop takes approximately 34 µs so the implemented sampling frequency is 29.4 kHz. Further, the action from one sample is delayed one sample as it takes one sample time in reality to make the calculations.

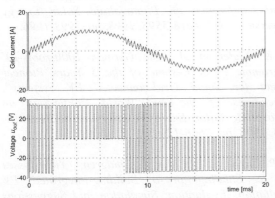

Figure 5. Output current and voltage due to simulation.

The grid current and the bridge voltage output are shown in Figure 5 from the simulation and Figure 6 from measurements. The achieved switching frequency is in average 3-4 kHz and the total harmonic distortion of the output current is 10 %.

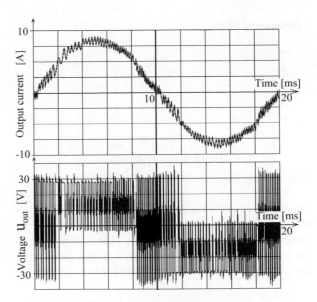

Figure 6. Measurements from the prototype.

As seen in the Figures 5 and 6 above, good agreement between simulation and measurement was obtained. However, the main reason for developing a model in Saber was to study internal parameters that are not easy to extract from the real inverter. Simulation environments are very helpful tools for developing new switching strategies and investigate what effects changes in the controller have.

REALISATION AND EFFICIENCY

The second generation of the digital controlled inverter is shown in Figure 7. It is surface mounted except for a few components including the grid inductor (not in the picture).

Figure 7. The 110-W inverter without the inductor.

The obtained efficiency of the PV-module oriented inverter is shown in Figure 8. In the mid power range, where most of the energy is captured, an efficiency of 93 % was obtained. At full load it decreases to 89 %.

Efficiency [%]

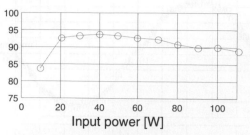

Figure 8. The measured efficiency.

The auxiliary power is less than 2 W and is included in the measured curve. An estimated 60-80 % of the losses is dissipated in the inductor. This is partly due to the relatively low switching frequency and therefore a high ripple current attained with the control algorithm.

FUTURE WORK

Since the bang-bang control algorithm described gives a fairly low switching frequency (3-4 kHz with a 10 MIPS-microcontroller) and therefore high ripple losses in the filter and wires, the intention is to try other control strategies.

The idea is to set up a table of the switching period during a 50-Hz cycle. The table can be designed to equalise the current ripple during the period. Thus, the switching frequency is not fixed, but varies as in the case of the bang-bang controller, *but* the switching frequencies are predefined and may be up to five times higher with the same microcontroller.

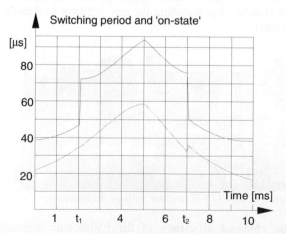

Figure 9. Required switching period to gain constant ripple.

Figure 9, top curve, shows a calculation of how the switching period may vary over half a 50-Hz period. The discontinuities at t_1 and t_2 are due to the changing of voltage levels. The lower curve is the required on state (voltage u_1) at the nominal power and input voltage.

This control strategy is to be developed and tested in Saber in the nearest future and later implemented in the prototype.

DISCUSSION AND CONCLUSIONS

A one-phase inverter has been designed to be mounted on the backside of a PV-module. The input voltage is 25-50 V and the maximum power is 110 W. It delivers a sinusoidal current to a low voltage AC-bus of 15 V. The power is then transformed to grid level. An efficiency of 93 % in the mid power range and 89 % at full power was achieved without transformer.

The control algorithm used is hysteresis control, which forms the output current. If different voltage levels are used during the 50-Hz period, a criterion for when to change is suggested.

ACKNOWLEDGEMENT

The Research Foundation at Göteborg Energi AB is gratefully acknowledged for its financial support.

REFERENCES

[1] V. Quaschning, R. Hanitsch, "Increased Energy Yield of 50 % at Flat Roof and Field Installations with Optimized Module Structures", Proceedings of the 2nd World Conference on Photovoltaic Solar Energy Conversion, ISSN 1018-5593, 1998, pp. 1993-1996.

[2] K. Kiefer et al "The German 1000-roofs-PV Programme — A Résumé of the 5 Years Pioneer Project for Small Grid-connected PV Systems", Proceedings of the 2nd World Conference on Photovoltaic Solar Energy Conversion, ISSN 1018-5593, 1998, pp. 2666-1270.

[3] P. Carlsson, L. Cider, B. Lindgren, "Yield losses due to Shading in a Building-integrated PV Installation; Evaluation, Simulation and Suggestions for Improvements", Proceedings of the 2nd World Conference on Photovoltaic Solar Energy Conversion, ISSN 1018-5593, 1998, pp. 2666-1270.

[4] B. Lindgren, "Topology for Decentralised Solar Energy Inverters with a Low Voltage AC-Bus", Proceedings of the 8th European Conference on Power Electronics and Applications, Lausanne 1999.

DESIGN AND PERFORMANCE OF A MODULAR STEP-UP DC-DC CONVERTER WITH A FAST AND ACCURATE MAXIMUM POWER TRACKING CONTROLLER.

Dr J A Gow, Dr J A M Bleijs.
Dept. of Engineering,
University of Leicester,
University Road, Leicester LE1 7RH.
U.K.

ABSTRACT: This paper describes a fast, accurate and efficient modular d.c.-d.c converter and maximum power tracking controller for use in medium to large scale photovoltaic generating plant. The modules are so designed that any number of these converters, each rated at 3kW, may be connected to a common d.c. bus in order to provide the required total plant power without significant changes to either the converter power chain hardware or the controller in order to support the parallel operation. The converter makes use of a fast and accurate maximum power tracking controller implemented in a digital signal processor (DSP) thereby providing localized maximum power tracking for each 3kW subsystem.
Keywords: -1 -2 -3

1. INTRODUCTION

Owing to the characteristic of photovoltaic (PV) arrays (Figure 1) being considerably different from the 'stiff', low impedance sources usually associated with power generation, considerable attention must be paid to the control of the power electronic conversion systems used with them. The PV array characteristic exhibits a maximum power point, the position of which is dependent upon both the amount of solar radiation incident on the array, and on the temperature [1] and for a given values of these environmental parameters it is only at this specific value of current and voltage that the array is operating optimally.

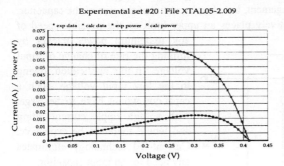

Experimental set #20 : File XTAL05-2.009

Figure 1 Sample PV array current/voltage and power/voltage curve

Given the considerable capacity for change of the position of the maximum power point with respect to the environmental parameters, it is necessary that the power electronic conversion systems are able to follow the changes in the position of this maximum power point to ensure that efficient operation of both the array and the converter is maintained. The converter must therefore be able to track maximum power.

1.1 Modular structure

Two aspects of photovoltaic operation are apparent. Firstly, while arrays are subjected to light the output will show a voltage - this can not be prevented. Therefore, in the interests of protection of plant maintenance engineers, it is desirable that the array terminal voltages be maintained within the safety extra-low voltage (SELV) band, giving a maximum terminal voltage of 55V.

This in turn limits the power rating of the converter. For a given power throughput, low input voltages will result in high input currents. These higher currents will then lead to higher losses. To this end the converters were designed with a maximum rating of 3kW.

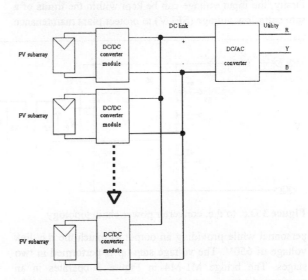

Figure 2 Modular structure of PV plant using these converters

To produce higher powers, the converters are designed as modules, each with its own controller (Figure

2). These modules permit the outputs of several converters to be connected in parallel and to feed a common d.c. link. A single high-power inverter can then be used to provide the utility interface. Such an arrangement has a number of advantages in that in a large plant, the total plant surface-area of array can be divided up into smaller modules, each of which has its own maximum power tracking controller. Therefore each module is independent from the others and can thus operate in a manner optimal to the sub-array. Individual subsystems can be taken off-line for maintenance without compromising the rest of the plant. The module controller is arranged in such a manner as to facilitate parallel operation without requiring any changes to the controller or converter hardware. Furthermore, the availability of reconfigurable high-power four quadrant vector controlled inverter systems as part of existing commercial motor drives allows the implementation of the inverter using readily available commercial drives. The current investigation is making use of a 20A Alspa GD4020 drive produced by Alstom Drives and Controls [2]. This inverter is configured to regulate the common d.c. link by changing the level of power throughput.

With a module power rating of 3kW, standard power components can be used at much lower cost than an equivalent d.c. to d.c. converter working at the total plant power.

2. CONVERTER STRUCTURE

The converter is based around current-fed bridge technology (Figure 3) and employs a high-frequency transformer - this topology offers a number of advantages within the design of the power chain in several areas. Firstly, the input voltage can be kept within the limits of a safe extra-low voltage (SELV) to protect plant maintenance

however, to limit the transformer leakage reactance and winding capacitance in order to avoid a high amplitude 'ring' as a result of the two combining to form a high-Q resonant circuit. A small snubber, across the primary of the transformer suffices to eliminate any residual ringing at the primary of the transformer.

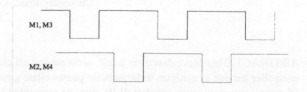

Figure 4 Bridge switching pattern

The design is fault-tolerant since input currents can not exceed the array short circuit current under fault conditions. Providing the switching devices are rated for this current, no condition can exist where the switching silicon is subjected to excessive levels of current (including shoot-through situations, which are no longer catastrophic). The converter becomes effectively current-sourced.

A further advantage, very important in photovoltaic conversion systems, is that the circuit can be designed to draw continuous input current over the input voltage design range. Some designs ([3],[4],[5],[6]) make use of converters such as buck/half bridge forward converters [3] with a capacitor across the array input. This configuration, in the opinion of the authors, is bad practice as it not only negates the inherent fault-tolerance of the converter by once again reverting to a voltage-fed arrangement, but also, by virtue of the size of the capacitor, effectively places an upper limit on the possible speed of any maximum power tracker by limiting the rate of change of input voltage. In this application, owing to the low input voltage, such a capacitor would have to be large, with a high ripple current rating and would therefore be expensive. Current-fed technology eliminates any need for an input capacitor.

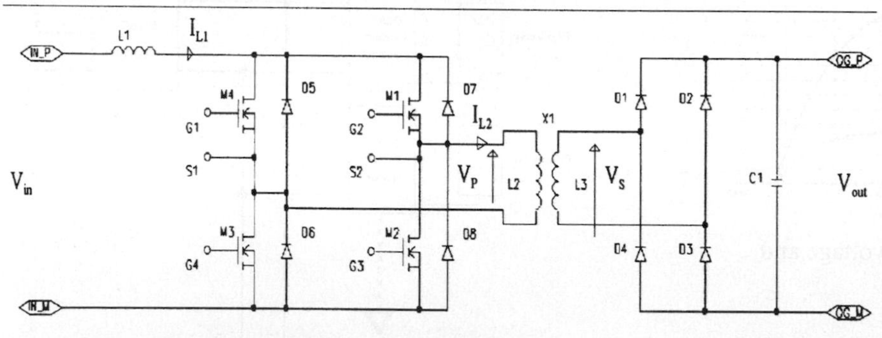

3. CONVERTER CONTROL

3.1 Hardware

Control is afforded by a digital control subsystem comprising a TMS320C50 digital signal processor (DSP) as the computational core. A large field-programmable gate array contains all required logic for timing, decoding and pulse-width modulation generation, thus removing the burden of PWM generation from the DSP. The PWM outputs from the gate array drive the bridge MOSFET gates using a commerical gate driver module. Analog to digital conversion is performed at 12 bit resolution by three

Figure 3 D.c. to d.c. converter power chain topology

personnel while providing an output to match the d.c. link voltage of 650V. The voltage step-up is performed in two stages. The bridge M1-M4 in Figure 3 operates in an overlapped pattern providing an initial voltage boost at the primary of the transformer. A further boost in voltage is provided by the transformer itself. Secondly, effects of transformer leakage reactance are minimized, avoiding the need for large snubbers and improving efficiency by reducing the associated dissipation. It is still necessary, channels based around readily-available devices and the outputs are buffered in order to render the initiation and control of the conversion cycle transparent to the DSP. The control algorithm is interrupt-driven and updates the PWM value register in the gate array at the beginning of each switching cycle, allowing for very fast responses to changes in the characteristics seen at the input to the converter. The algorithm itself is based upon a recently developed system making use of the characteristics of the power chain itself to derive the necessary slope information required to deduce the position of the converter operating point upon the array power curve.

3.2 Firmware

The control algorithm sets the converter up to act as a power source. Apart from protection mechanisms designed to shut down the converter in the event of a d.c. link overvoltage, there is no mechanism required for regulating the d.c. link voltage. Regulation is performed by the plant load, usually a d.c.-a.c. converter for grid connection. The d.c. link voltage is maintained within limits by regulating the power drawn from the link.

For testing purposes, it is necessary to provide a source with a maximum power point which has a known, and adjustable characteristic. The maximum power source can be approximated in two ways. The first of these makes use of a simple resistive potential divider constructed from a pair of rheostats to provide a source with a maximum power point within its operating range. The second method involves the use of a bipolar transistor as a current source. Such an arrangement produces a characteristic very close to that of a photovoltaic array, and the higher power can be provided with the aid of a fast analog servoamplifier.

3.3 Feedforward signal

In this plant arrangement it is possible for environmental parameters to change relatively rapidly , for example cloud cover. It is therefore possible for the voltage on the common d.c. link to change quickly. This change may occur quickly enough to cause a large excursion of voltage on the d.c. link before the inverter has time to adjust the power throughput. To prevent this larger d.c. link capacitors may be used, but this will also slow down the rate of change of power export to the utility. A better method is to employ a feedforward signal which not only facilitates rapid changes in exported power, but also allows the use of smaller d.c. link capacitance. Each d.c. to d.c. converter module generates a signal which is related to the distance the current converter operating point is from the module array maximum power point. These signals from each converter are averaged and passed to the inverter, which is capable of rapidly altering its power throughput based upon this signal. This prevents any significantly large excursions on the d.c. link from occurring.

3.4 Maximum power tracker

The maximum power tracking controller is based upon a fast algorithm that makes use of the characteristic of the converter topology itself to analyze the I/V curve of the array and is thus very fast. Samples of array current and voltage are taken immediately prior to the turn-on and turn-off points, and from these the slope information required to determine the proximity of the maximum power point can be computed. The algorithm operates on a per-cycle basis at the 20kHz switching frequency of the power chain,.

4. SIMULATION RESULTS

Simulation studies has shown the algorithm to work very well, and it is expected that a hardware prototype will behave in a similar manner. Figure 5 shows curves of the operation of the converter in response to a step decrease and increase in array power. These curves serve very well to demonstrate the speed with which this type of maximum power tracker can operate, with an acquisition time at increase of 2ms, and on decrease of less than 1ms. The operation of the feedforward signal can clearly be seen, and it is this signal that is used by the inverter to facilitate rapid changes in the exported power level.

5. HARDWARE PROTOTYPE RESULTS

Figure 6 shows an early result from a recently-completed hardware prototype. Curves of array voltage, array power, array current and converter output voltage are shown for a step input power increase of 740W. A source with a maximum power point was simulated using a high-current power supply and and a resistive divider. The inset shows the manner in which the maximum power tracking operates - during each effective switch on-time and off-time a portion of the I/V input curve is traced, with an aim to maintain both start and end points at the same value. The result is the 'double hump' ripple as shown in the inset. Owing to the algorithm parameters in this prototype not yet being optimal the algorithm is not as fast as that shown in the simulation results, but is still quite quick, with the re-acquisition occurring in about 40ms.

6. CONCLUSION

A modular d.c. to d.c. converter system has been developed which is aimed at medium to large scale photovoltaic power plant. These converters can be connected in parallel without any modification, readily allowing expansion of plant capacity limited only by the maximum power throughput of the inverter system used. Fast maximum power tracking algorithms have successfully been employed and combined with the feedforward signal provide a convenient and efficient means of handling photovoltaic array inputs in a fashion which makes for easy deployment.

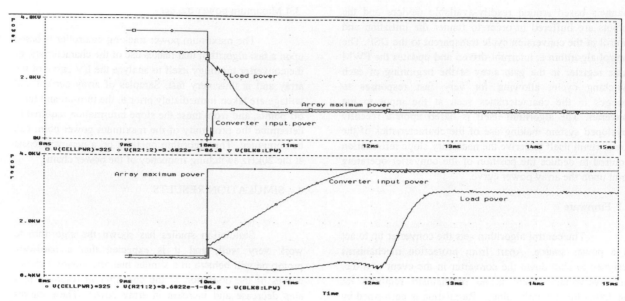

Figure 5 Response of maximum power tracking controller to a step increase in array power

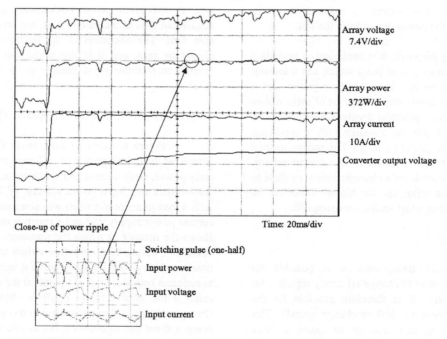

Figure 6 Early results from a hardware prototype for a step increase in load

REFERENCES

[1]   J A Gow, C D Manning, Development of a photovoltaic array model for use on power-electronics simulation studies. IEE Proc. Electr. Power Appl. Vol 146, No. 2, March 1999.

[2]   Alstom Drives and Controls Ltd. Kidsgrove, U.K. Alspa GD4000 Technical Manual

[3]   A Cocconi, S Cuk, R D Middlebrook, High Frequency isolated 4kW photovoltaic inverter for utility interface, Proceedings of the Seventh International PC'93 Conference, September 13-15, 1983, Geneva, Switzerland.

[4]   J D van Wyk, J J Schoeman, On total loss minimization in power controllers for photovoltaic systems. Proceedings of Mediterranean Electrotechnical Conference, Athens, Greece 24-26 May 1993 vol 2.

[5]   P Savary, M Nakaoka T Maruhashi, Novel type of high frequency link inverter for photovoltaic residential applications. IEE Proceedings vol 133 part B, No. 4, July 1996

[6]   J Howard Speer, Microprocessor control of power sharing and solar array peak power tracking for high power (2.5kW) switching power converters.

APPLICATION OF NEURAL NETWORKS IN THE SOLAR IRRADIATION FIELD: OBTAINMENT OF SOLAR IRRADIATION MAPS

L. Hontoria[*], J. Riesco[**], J. Aguilera[*], P. Zufiria[***]

* Grupo Jaén de Técnica Aplicada. Dpto. de Electrónica. E.U.P. de Linares.
Universidad de Jaén
Alfonso X el Sabio. 23700 Linares (Jaén). Spain.

** Dpto. de Electrónica e Ingeniería Electromecánica.
Escuela de Ingenierías Industriales. Universidad de Extremadura.
Avenida de Elvas s/n. 06071 Badajoz. Spain.

*** Grupo de Redes Neuronales.
Dpto. de Matemática Aplicada a las Tecnologías de la Información.
E. T. S. I. Telecomunicación. Universidad Politécnica de Madrid.
Ciudad Universitaria s/n. 28040 Madrid. Spain.
e-mail: hontoria@ujaen.es // jriesco@unex.es //aguilera@ujaen.es // pzz@mat.upm.es

ABSTRACT: In this work a methodology for obtaining solar irradiation maps is presented. This methodology is based in a neural system [9] called Multi-Layer Perceptron (MLP) [5, 8]. For obtaining a solar irradiation map is necessary to know the solar irradiation of many points wide spread the zone of the map where it is going to be drawn.

For most of the locations all over the world the records of these data (solar irradiation in whatever scale, daily or hourly values) are no existent. Only very few locations have the privilege of having good meteorological stations where records of solar irradiation have being registered. But even in those locations with historical records of solar data, the quality of these solar series is not as good as it should be for most of the purposes.

In addition to, for drawing a solar irradiation map, the amount of points of the maps (real sites) that it is necessary to work with makes this problem difficult to solve. Nevertheless, with the application of the methodology proposed in this article, this problem has been solved and solar radiation maps have been obtained for Jaén province, a southern province of Spain between parallels 38° 25' N and 37° 25' N, and meridians 4° 10' W and 2° 10' W.

Keywords: 1. Hourly solar radiation. 2. Neural networks. 3. Solar radiation maps

1.- Introduction

A very important and necessary element for the solar systems designer is the disposability of information about the solar irradiation in the zone where he is going to install the solar system. This information, in case that it exists, can be disposable, in several ways. The most common one is by means of several tables with a lot of very useful information, usually large solar sequences, but extremely difficult to handle. Nevertheless, another way can be by means of different solar irradiation maps of the zone where the installation is going to be made. If the purpose of the designer is to look for the best zone to install the solar system, this second way would be more efficient and easy to handle.

For obtaining a solar irradiation map is necessary to know the solar irradiation of many points wide spread the zone of the map where it is going to be drawn. These solar data can be disposable in several time scales. Usually there are two time scales in which the data are available: daily and hourly scales. For some applications the daily scale is appropriate, but in many other applications, where it is necessary more accuracy, a daily scale could be insufficient. In these cases it is necessary to work with

hourly solar irradiation data. In this work the data that have been necessary to work with have been hourly solar data.

Nevertheless, for most of the locations all over the world the records of hourly solar irradiation data are no existent. Only very few locations have the privilege of having good meteorological stations where records of solar irradiation have being registered. Also, most of the meteorological stations have daily solar irradiation records, but to find hourly solar irradiation registers is more rare. Even in the locations with historical records of hourly solar data, the quality of these solar series is not so good [2]. Usually there are many missing hours, mistaken hours, etc.

So the problem of obtaining a solar irradiation map in a certain zone is difficult to solve, due to the fact that it is necessary to know the solar irradiation of a great amount of sites.

In this work a methodology for obtaining solar irradiation maps is presented. The methodology proposed makes use of a neural system called Multi-Layer Perceptron (MLP). This neural network has been already presented in previous works [6, 11]. In those works it has been demonstrated the capabilities of the MLP to generate synthetic hourly solar irradiation series. What is more, in the comparison with other methods for generating synthetic sequences of solar data, the MLP improves [6, 7] the results obtained by those methods.

In this case, an improvement of the neural system has been made, with the objective of the application of this neural supervised model for drawing solar irradiation maps. This methodology, as it will be seen, is easily extensible and applicable to any other region of the world.

The article is divided in 7 sections. Section 1 is this introduction. In section 2 the problem of obtaining solar irradiation map is presented. Sections 3 and 4 propose solutions to the problem of section 2. In section 5 the methodology for obtaining solar irradiation maps is describe. Finally the conclusions and future actions are presented in sections 6 and 7.

2.- Problem to solve

As it has been mentioned before, the preparation of a solar irradiation map requires the previous knowledge of a lot of information, which can be summarised as follows:

1) To make a grid of the zone which is going to be represented in the map. Each point of the grid corresponds to a certain real site which will have some particular features as its latitude, its longitude and its altitude over sea level.

2) For each point of that grid knowledge of long sequences of solar irradiation (hourly preferable) is needed.

3) Once the series of solar irradiation are disposable for all the grid, it is necessary to obtain some particular statistical values for drawing the maps. For instance, it is interesting to summarised all that information in 12 maps, one for each month of the year. So, in this case, it is necessary to obtain the mean monthly values of the solar data.

A crucial problem, as it has been mentioned in the previous section, is found in step 2: usually there is no solar data available. Particularly in the province of Jaén, a southern province of Spain, where solar irradiation maps are being prepared that problem exists. Only in the city of Jaén records of solar irradiation are disposable. The Grupo Jaén de Técnica Aplicada, a research group of Jaén University is doing measures of hourly global solar irradiation since 1.995. To make that measures there is a pyranometer CM-11 Kipp & Zonnen installed in the terrace of the Escuela Superior Politécnica de Jaén. These measures are register by a data acquisition system called Meteodata and lastly send to a PC where the data are treated and saved.

With all of this data and training a MLP it is possible to generate long sequences of solar irradiation data which are indistinguishable of the real series, as it has been demonstrated in previous works. Nevertheless, for the generation of hourly solar irradiation series in other location of the province of Jaén two problems arise. These

two problems can be expressed by means of these two questions:

a) Is it possible to apply the MLP trained in certain location to generate solar series in other places different from? For instance, if a MLP has been trained in London, is this MLP valid for Glasgow?

b) If in a location where there is no information related with solar irradiation, as instance monthly mean solar irradiation values, number of sunshine hours, number of cloudy days, etc., is it possible to include in the MLP an other information available?

For answering those two questions two other studies were proposed. For answering the first question whether it is possible to use the information of a MLP trained in a certain location for the solar generation in another different place a first study was done. This study is described in section 3. As it will be demonstrated the answer to this question is yes.

In second place, for trying to answer the other question, a second study was done. With the first study, the problem of using a MLP trained in a certain place but used in another different place for the solar generation was solved. Nevertheless, for doing this, it is necessary to dispose of some solar irradiation information (direct or indirect) in the new place where the solar generation is going to be done. If this information is no available it is necessary to look for another information. This new information must be a distinctive feature of the new place. This new item is the latitude of the place and the altitude upon sea level. If a MLP is trained with these two other entrances it will be able to find out the relations between this new information and precedent information. This study is described in section 4.

3.- Universality capabilities of perceptrons

A first study trying to demonstrate universality capabilities of perceptrons was done. Three different locations, with different features where took. In table I these three sites, with some of their characteristics are shown.

Table I: Locations in study

Site	Weather	Latitude	Altitude	Years
Oviedo	Atlantic	43.35 °	348 m	1977-1984
Madrid	Continental	40.45 °	664 m	1978-1986
Málaga	Mediterranean	36.66 °	7m	1977-1984

As it can be seen three different cities has been chosen. The three ones are cities of Spain with different climates. Also these three locations have different altitudes upon sea level and their latitudes vary from 43.35° to 36.66° almost 7 degrees.

Once the hourly solar irradiation data of these cities is available the process followed is this one. First of all, three different MLPs are trained for each city. The way of doing this training is described in previous works [6,7,11]. The MLP structure is shown in figure 1.

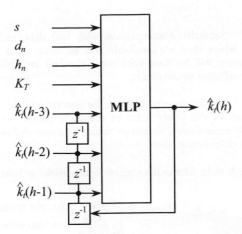

Figure 1: MLP Architecture for Clearness Indexes
Prediction

Secondly, with the MLP trained in Madrid, hourly solar irradiation data for the other two cities are generated with this MLP.

Thirdly, new synthetic hourly solar irradiation years are obtained by other classic generation methods. The classical methods employed are the Graham & Hollands method (GH method) [3, 4] and the Aguiar & Collares-Pereira method (AC method) [1].

Finally, with all of this information a comparative study is done. As a quality measure, the Mean Relative Variance (MRV) [6, 7] was used. This parameter, commonly employed in the digital signal processing community, quantifies the relative error, and it is defined as the quotient between the prediction error signal power and the AC power of the signal to be predicted:

$$ MRV = \frac{\sum_h (k_{th} - \hat{k}_{th})^2}{\sum_h (k_{th} - \bar{k}_t)^2} $$

The results are shown in table II.

Table II: MRV. Real data against data generated by MLP, AC and GH.
(* Trained with Madrid Data)

Method	Location	MRV
MLP	Madrid	0.1022
GH	Madrid	0.1611
AC	Madrid	0.1574
MLP	Oviedo	0.1723
MLP*	Oviedo	0.1761
GH	Oviedo	0.2481
AC	Oviedo	0.2567
MLP	Málaga	0.1255
MLP*	Málaga	0.1364
GH	Málaga	0.2418
AC	Málaga	0.2384

As it can be seen the MRV values are better for the data obtained by the MLP than those ones obtained by the classical methods (GH and AC). Even for the data obtained with the MLP trained with Madrid data but applied in the two other locations the MRV values are better than those ones of GH and AC method. With this study it can be concluded that the MLP presents universality capabilities, that is, a MLP trained with data of a place can be used for the generation of data in another different place.

4.- Latitude and altitude

Once a MLP is trained in a certain site it can be used for the hourly solar generation in a different place. Nevertheless, for the hourly generation, some solar information in the location where the generation is taking place is needed. For instance, this solar information can be solar global irradiation monthly mean values, number of sunshine hours, distribution of cloudy days, etc.

For drawing solar maps in Jaén province the only available solar information was the five years hourly solar irradiation data records in Jaén city. A MLP with this data was trained for Jaén city. Nevertheless, for the solar generation in some other sites of the province for obtaining the solar maps no other solar information was available. So with this problem it was necessary to look for other features of the sites, which were particular and different for each one. This new information is the latitude and altitude upon sea level. This two items were included to the MLP as new entrances.

A unique MLP was trained with the Madrid, Oviedo, Málaga and Jaén solar data, modifying the structure of figure 1 with two new entrances corresponding to the latitudes and altitudes of these four sites. (See table 1. For Jaén city Latitude = 37.73° & Altitude = 573 m.). This MLP is used for the hourly solar generation of each site of the grip in the solar maps. In next sections this procedure is described.

5.- Solar irradiation maps

For obtaining solar irradiation maps for Jaén province the procedure followed is this one. Firstly, a grid with more than 250 points has been prepared. These points vary from parallels 38° 25' N to 37° 25' N, and meridians 4° 10' W to 2° 10' W, and has been chosen in intervals of 5'. For all of this sites solar hourly values must be generated.

Secondly, as it has been mentioned above a unique MLP has been trained. To train this MLP, hourly solar irradiation data of Madrid, Oviedo, Málaga and Jaén have been used. This MLP is the tool used for the generation of solar hourly values in all the sites of the grid of the map. The solar information used for the generation is the Jaén solar data, and the particular information of each site is the latitude and altitude.

Finally the solar global monthly mean values [10] are obtained for each point of the grid, so twelve solar irradiation maps are possible to draw (one for each month). Next two figures present two maps obtained by this methodology. One map is for January (winter) and the other one is for June (summer).

Figure 2: Solar irradiation map of Jaén province in January.

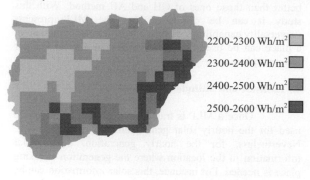

2200-2300 Wh/m²
2300-2400 Wh/m²
2400-2500 Wh/m²
2500-2600 Wh/m²

January

6.- Conclusions

A new methodology for the obtainment of solar irradiation maps has been development. For obtaining a solar irradiation map of a certain zone it is necessary to know the solar irradiation of a huge number of sites wide spread the zone. This information is usually no existent, so this problem has been solved with the methodology proposed,.

The methodology proposed is based in a neural system called Multi-Layer Perceptron (MLP). The MLP is a neural systems that needs a previous training and afterwards it can be applied in different places, varying only its entrances. The MLP proposed has been trained with hourly solar irradiation data of four sites of Spain. Each of these four sites has different features as climate, latitudes, and so on.

Once the MLP is trained a solar generation can be done in all of the sites of the grid which form the zone map. This generation is extremely simple and takes less time than the same generation than classical methods of solar generation.

Besides, the classical methods are unable to generate solar irradiation series in places where no solar information is available. Nevertheless, the methodology proposed is able to do this, is more versatile than the classical methods, and so is able to drawn maps of the zone.

This methodology is easily extendible to other places. The only requirement is the knowledge of the daily solar irradiation of only one site of the zone where the map is going to be drawn.

7.- Future Actions

As a first future action maps of different provinces, regions or even countries will be prepared. The extension of this methodology to other areas is, as it has been demonstrated quite easy. So drawing maps of more extended areas will be drawn.

Secondly a comparison with real data, in the places where they are available will be done. Also the comparison will be done with other classical methods of solar irradiation generation.

An statistically study of the generation data will be done. In this study a lot of statistical parameters as monthly mean values, variances, autocorrelations and so on will be taken into account.

Figure 3: Solar irradiation map of Jaén province in June.

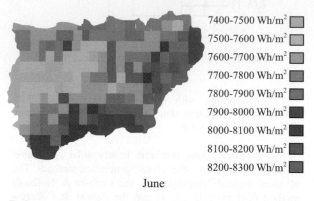

7400-7500 Wh/m²
7500-7600 Wh/m²
7600-7700 Wh/m²
7700-7800 Wh/m²
7800-7900 Wh/m²
7900-8000 Wh/m²
8000-8100 Wh/m²
8100-8200 Wh/m²
8200-8300 Wh/m²

June

References

[1] R. Aguiar and M. Collares-Pereira. TAG: A Time-Dependent, Autorregressive, Gaussian Model for Generating Synthetic Hourly Radiation. Solar Energy, Vol. 49, No. 3, pp. 167-174, 1992.

[2] R. Gansler, S.A. Klein, W.A. Beckman. Assessment of the accuracy of generated meteorological data for use in solar energy simulation studies.

[3] V. A. Graham, K. G. T. Hollands and T. E. Unny. A Time Series Model for Kt with Application to Global Synthetic Weather Generation. Solar Energy, Vol. 40, No. 3, pp. 269-279, 1988.

[4] V. A. Graham and K. G. T. Hollands. A Method to Generate Synthetic Hourly Solar Radiation Globally. Solar Energy, Vol. 44, No. 6, pp. 333-341, 1990.

[5] S. Haykin. Neural Networks. A Comprehensive Foundation. Macmillan Publishing Company, 1994.

[6] L. Hontoria, J. Riesco, P. Zufiria, and J. Aguilera. Improved generation of hourly solar irradiation artificial series using neural networks. EANN99. Varsovia. 1.999

[7] L. Hontoria, J. Riesco, J. Aguilera and P. Zufiria. Generación de series sintéticas de radiación solar horarias usando redes neuronales. IX Congreso Ibérico. Córdoba Marzo 2000.

[8] K. Hornik, M. Stinchcombe and H. White. Multilayer Feedforward Networks Are Universal Approximators. Neural Networks, Vol. 2, no. 5, pp. 359-366, 1989.

[9] R. P. Lippmann. An Introduction to Computing with Neural Nets. IEEE ASSP Magazine, pp. 4-22, April 1987.

[10] M. Iqbal. An introduction to solar radiation. Academic Press. 1.983.

[11] P. J. Zufiria, A. Vázquez, J. Riesco, J. Aguilera and L. Hontoria. A Neural Network Approach for Generating Solar Irradiation Artificial Series. Proc. of the (IWANN'99). 4-5 June, 1999. Alicante (Spain).

"GLASSY, GRID FRIENDLY" PV-IGBT-INVERTERS FOR HIGHEST CUSTOMER BENEFITS

Dipl.-Ing. P. Kremer, C. Mainka
Siemens AG, A&D SE S31, Wuerzburger Str.121, 90766 Fuerth, Germany
phone: +49 911 750 4050 / fax: +49 911 750 2246 / e-mail: peter.kremer@fthw.siemens.de

ABSTRACT: Siemens has developed high capacity 3-phase power conditioner units in a power range from 20kVA up to 1.2MVA (4x300kVA master/slave) for utility scale application. The inverter units SINVERTsolar are already constructed for the future demands of our customers concerning economy, protection equipment, operation, monitoring, visualisation and grid compatibility. The inverter units are based on the Simovert Masterdrives series, an AC drive converter from Siemens. High quality and reliability as well as a fair market price are some of the outstanding features of this IGBT-based inverter. Another advantage is the variety of types, which enables to offer a wide power range. Further features of the SINVERTsolar inverter units are the new 3-phase MSD (Mains Switching Device) to prevent islanding, insulation monitoring for PV generator with sequential fault location, phase shifter, VAR control, active THD filter and standardised operating and monitoring systems (PowerProtectsolar and PV-WinCC).
Keywords: Inverter - 1: Photovoltaic - 2: PV power plants - 3

1 INTRODUCTION

Siemens has developed high capacity 3-phase power conditioner units in a power range from 20kVA up to 1.2MVA (4x300kVA master/slave) for utility scale application. The inverter units SINVERTsolar are already constructed for the future demands of our customers concerning economy, protection equipment, operating, monitoring, visualisation and grid compatibility.

The inverter units are based on the Simovert Masterdrives series, an AC drive converter from Siemens. High quality and reliability as well as a fair market price are some of the outstanding features of this IGBT-based inverter. Another advantage is the variety of types, which enables to offer a wide power range.

Further features of the SINVERTsolar inverter units are the new 3-phase MSD (Mains Switching Device) to prevent islanding, insulation monitoring for PV generator with sequential fault location, phase shifter, VAR control, active THD filter and standardised operating and monitoring systems (PowerProtectsolar and PV-WinCC).

Important Features:
- Standardised series product
- Self-commutated IGBT-based inverter
- Silnusoidal modulation with highest dynamic
- Reactive power control (phase shifter)
- Active THD filter
- Mains monitoring with three-phase MSD
- Insulation monitoring
- Bus communication via RS232/RS485/Profibus DP
- Remote monitoring via telephone modem
- Fax alarm
- Visualisation via PC with PowerProtectsolar
- I&C with PV-WinCC

2 CONCEPT

The power conditioner units SINVERTsolar are modular based on the function units and are delivered ready for connection in the power range from 20kVA up to 1.2MVA (figure 1).

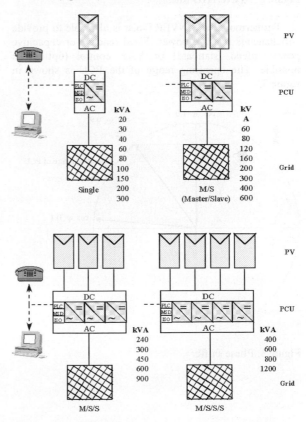

Figure 1: Master/Slave-concept

3   GRID FRIENDLY SINVERTsolar

The active THD filter is available as a standard in the SINVERTsolar. An active THD filter generates negative THD currents to smooth the sine-wave of the voltage (figure 2)

SINVERTsolar analyses separately in each phase the voltage harmonics distortion and feeds in the current of the same harmonics in damped phase opposition (the damping factor can be parameterised via the software PPsolar).

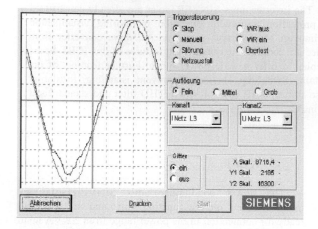

Figure 2: Active THD filter

Furthermore the SINVERTsolar is also able to provide simultaneous reactive power. Fixed reactive or capacitive power infeed (standard) or VAR control (option) is possible. The operating range of the PCU is shown in figure 3.

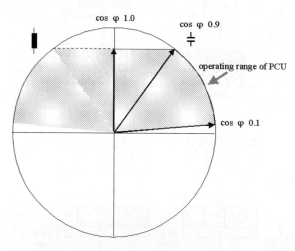

Figure 3: Phase shifter

4   GLASSY SINVERTsolar

A CD-ROM with the software PowerProtectsolar (PPsolar) for maintenance, service, measuring data storage and remote control via modem belongs to the volume supply of SINVERTsolar.

A fax modem is able to pass break downs on at once. The user is also able to read directly the most important system information on the control panel of the SINVERTsolar without PC.

Optional a professional information and control (I&C) system PV-WinCC is also available.

4.1 PowerProtectsolar

Software for visualisation, monitoring, service and remote control via PC

PPsolar on CD-ROM is the software solution for visualisation, monitoring, maintenance and service. On a direct or via modem connected PC (Windows 95/NT) there are a lot of functions available, which inform continually about the status of the PV plant.

The menu-driven user interface allows access to:

- Control panel
- Active system diagram (figure 4)
- Oscilloscope (figure 5)
- Event memory
- Process data
- Data storage (figure 6)
- Analysis

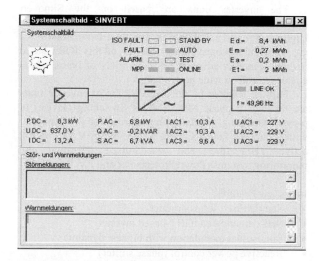

Figure 4: Menu - system diagram

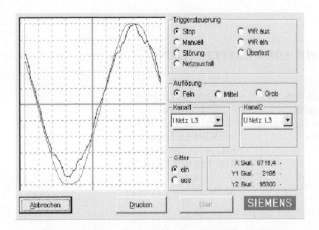

Figure 5: Oscilloscope for grid and load analysis

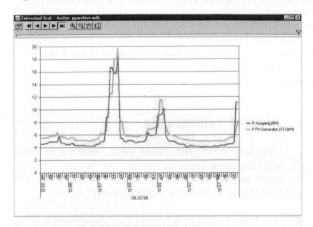

Figure 6: Archived data of the plant

4.2   Visualisation in a new dimension

PV-WinCC, the information and control (I&C) system of the professional power plant standard.

PV-WinCC (Photovoltaic Windows Control Center) is the professional software for visualisation and control of the PV plant (figure 7). It is capable of bidirectional communication with process and power plant control systems.

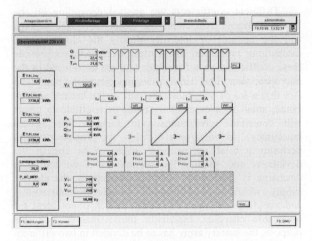

Figure 7: Clear arranged plant overview

It shows a really simple handling on basis of the operating system Windows 95/NT and the graphical user interface WIN 95 GUI.

The many configuring functions allow the implementation of almost any plant size, and subsequent expansion is straightforward.

PV-WinCC makes the energy flow of the plant transparent: it displays the latest statuses, messages, measured values and reports. All data occurring in the plant are automatically organised and stored. You have access at all times to the reliable Sybase SQL Anywhere data archive, and thus have continuous evidence of all events and measured values in the plant.

Communication between the PV plant and the WinCC terminal takes place quickly an reliably via Industrial Ethernet or Profibus.

The open system architecture and the integral OPC (OLE for process control) interface also allow incorporation in complex automation systems (figure 8).

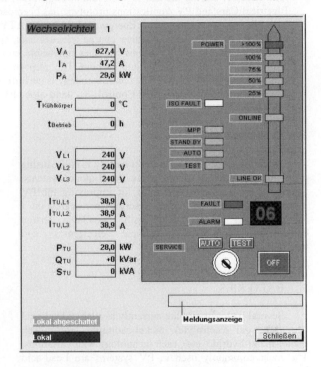

Figure 8: Copy of the control panel

STATIONARY BATTERY CONTROL FOR PV APPLICATIONS

J. Monedero, P. Valera, D. Baussou, A. Lugo, A.
Sánchez,, M.P. Friend, M. Cendagorta
ITER – Instituto Tecnológico y de Energías Renovables
Polígono Industrial de Granadilla
38611 Tenerife-Spain
Tel. +34 922 39 10 00
Fax +34 922 39 10 01
E-mail: iter@iter.rcanaria.es

F.Dobón Giménez
Freelance
Urb. Mayber, 107
38296 La Laguna
S/C de Tenerife-Spain
Tel. +34 922 26 01 75
E-mail:
fdobon@mailcity.com

ABSTRACT: During the end of 1998 and through 1999, the design and development of a Battery Control System for Electrical Cars and PV Applications prototype has been carried out by ITER's stuff and F. Dobón within the frame of MINER-Spanish Minister of Energy and Industry ATYCA program. The prototype was originally ideated for controlling electric vehicle battery charge (electric cars, electric pallet lift-truck, etc). This is an adapted version for stationary battery in PV systems or other renewable energy sources in stand-alone systems, where special energy requirements have been taken into account. The purpose is to monitor in real-time the charge remaining in the battery and its degradation in time. The control strategy is performed using a microprocessor.
Keywords: Battery Storage and Control– 1: Stand-alone PV Systems – 2: Storage– 3.

1. INTRODUCTION

Knowing the charge estate of a battery is still causing problems in most of photovoltaic stand-alone systems. Solving these problems is a must for improving the PV industry reliability. This work describes a research on a low-cost, user friendly and easily installable charge controller design, using simple electronics and microprocessors technology.

2. BATTERIES

Several battery types are currently available: lead-acid (liquid or gel electrolyte), nickel-cadmium, iron-nickel, nickel-metal-hydride, etc., each technology with variations. The most commonly used in PV systems are Lead-acid batteries.

The battery storage capacity depends on several parameters such as:

➢ Manufacturing and design factors.
➢ Charge/discharge current intensity and speed.
➢ Electrolytic temperature.
➢ Cut off voltage.
➢ Open circuit time.
➢ Charge/discharge cycles.

To know the charge estate from the outside of the battery, the minimum set of physical parameters that should be monitored are:

➢ Charge and discharge current.
➢ Battery temperature.
➢ Battery Voltage.

The electrolytic density is another important charge indicator but for practical reasons is not monitored.

One of the tasks that a charge controller must realise is to prevent the batteries from overcharging and from excessive discharge, which could cause damage and reduce the batteries' life.

3. BATTERY CONTROLLER

This is an intelligent system which measures in terms of charge (amperes times hours) the total balance between the output and the input battery current. The accumulator's real time charge is shown in a user's display (Fig.1).

Figure 1: Battery Control Prototype

Different coefficients are applied depending on the battery galvanic efficiency, the temperature, the voltage and the amount of intensity during the charging and discharging processes. The battery manufacturers provide these coefficients in a first stage and are introduced in the microprocessor. This data is often inaccurate and due to the fact that a lot of variables take part in the battery storage estate, small measurement mistakes or coefficients errors will contribute continuously to increase the system inaccuracy. In order to avoid this, the proposed system performs a total accumulator discharge to correct the deviations and modify the parameters if needed. The discharge is carried out without interrupting the electric supply (which is not admissible for telecommunications) and avoiding the waste of energy (which is not ideal for renewable energies, especially photovoltaics). For this purpose, the battery array has to be divided in two different sets. When one set of batteries is being discharged, its energy is delivered to the other group of batteries, which assumes the energy supply. This energy transfer is accomplished using a DC/DC converter. The discharge is performed during a long time-period (10 or more hours) to avoid the active substance detachment as a consequence of the volumetric variation in the plate. This avoids reducing the batteries life. If lead-acid batteries are being used, they must be charged immediately after the discharge for avoiding the lead sulphate crystallisation in the plate.

Although the electrochemical process inside a battery is mainly reversible, a small amounts of the active material does not reaction and produces continuos worsening in the accumulator storage capacity. This worsening effect is also taken into account by the system applying the corresponding coefficients.

On the other hand, the system is also provided by a charge/discharge cycle counter that allows to know the battery historical behaviour during his lifetime.

During the development stage, changes in the hardware configuration and software improvements have been carried out in order to obtain the best results. This has been especially difficult to do in the discharging current measurement since it is usually realised in an unpredicted way.

4. HARDWARE

The hardware works as an interface between the battery and the microprocessor's decision-making.

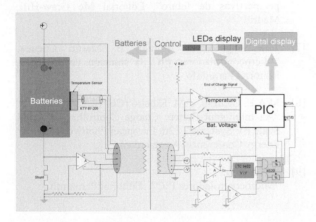

Figure 2: Main Scheme

The function scheme is shown in Fig. 2 and the main components are:

➢ Batteries for PV applications.
➢ Current and temperature sensors.
➢ Double wave asymmetric rectifier.
➢ Voltage/Frequency converter.
➢ Up/Down counter.
➢ LEDs' or digital display.
➢ PIC Microprocessor.
➢ PC for monitoring (optional).

The controller can be provided by two different charge visualisation systems: a digital display that shows the Ah accumulated or a set of LEDs that show it graphically.

5. SOFTWARE

The software is implemented in a PIC-Microchip microprocessor (16C84 or 16F84 versions) whose main function is to control the estate of the battery. For this role the PIC microprocessor (Fig. 3) saves in a register the pulses proportional to the value in amperes hours (Ah)

which charges battery and in another the ones that discharges it. To know the exact charge estate, different coefficients are applied as it was pointed in the preceding item. The charge shown in the display (or in the LED's set) is the difference between the accumulated charge and discharge values. If the LEDs' display is incorporated, the charge is expressed as a percentage of the nominal battery storage capacity.

Figure 3: Software Development Kit

Among the procedures included in the routines we have the following functions and characteristics:

➢ Voltage and temperature comparisons with data from manufacturer.
➢ Charge and discharge Ah.
➢ Total discharge Ah accumulated (to determine the number of cycles realised).
➢ Different coefficients for Voltage/Frequency converter are applied during the charge/discharge process.
➢ Analogic and digital user friendly interface.

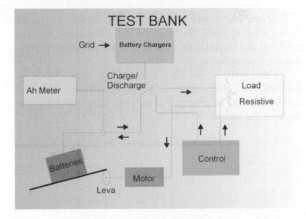

Figure 4: Test Bank Diagram

6. WORK TESTS

A series of tests were performed in order to verify and validate this electronic device. The tests carried out include:

➢ Battery charge/discharge cycles
➢ Floating charges
➢ Charge after full discharge
➢ Fatigue test in electrical car and PV installation

A test bank (fig. 4 and 5) was developed for this purpose where the batteries where connected to resistive loads and the battery control informs about their estates. The results were satisfactory.

Figure 5: Test Bank

7. CONCLUSIONS

A new and versatile battery charge control electronic device ideal to fulfil the needs of PV stand-alone systems has been developed. The system is low-cost, user-friendly and easily installable.

The system can be adapted to any kind of accumulator. The system decision-making and the coefficients can be changed in order to adapt the system to any specific battery.

REFERENCES

[1] E. Witte,
     "Acumuladores de plomo y acero", Editorial Gustavo Gili. Barcelona.

[2] E. Lorenzo,
     "Solar Electricity. Engineering of Photovoltaic Systems". Editorial PROGENSA. Sevilla.

[3] L. Lejardi,
     "Acumuladores de electricidad. Manual práctico". Editorial PROGENSA. Sevilla.

[4] J. Fullea,
     "ACUMULADORES   ELECTROQUIMICOS:
     Fundamentos, nuevos desarrollos y aplicaciones", Editorial Mc Graw-Hill. Madrid.

[5] G. Clereci,
     "Acumuladores eléctricos". Ediciones técnicas REDE. Barcelona.

[6] J. Fullea, F. Trinidad, J. C. Amasorrain, M. Sanzbero "El vehículo eléctrico. Tecnología, desarrollo y perspectivas de futuro". Editorial Mc Graw-Hill. Madrid.

[7] J.M. Angulo, I. Angulo, E. Martín Cuenca, "Microcontroladores PIC, la solución de un PIC". Editorial Paraninfo.

[8] M.A.E. Galdino, C.M. Ribeiro (CEPEL),
     "An Intelligent Battery Charge Controller for Small Scale PV Systems". 12th European Photovoltaic Solar Energy Conference. Amstaerdam.

[9] C. Tavernier,
     "Microcontroladores PIC". .Editorial Paraninfo.

INTERCONNECTION OF HETEROGENEOUS CONTROL AND MONITORING DEVICES AT THE EUCLIDES SOLAR PV CONCENTRATION PLANT IN TENERIFE

F. Pérez, P. Witsch[1], M. Friend, M. Cendagorta
ITER – Instituto Tecnológico y de Energías Renovables
Polígono Industrial de Granadilla
38594 Tenerife-Spain
Tel. +34 922 39 10 00
Fax +34 922 39 10 01
E-mail: iter@iter.rcanaria.es
[1] Uni-Hannover. Germany

I.Luque
INSPIRA
Chile 10, Ed. Madrid 92
28230 Las Rozas- Madrid-Spain
Tel. +34 91 630 45 34
Fax. +39 91 630 40 87
E-mail: inspira@inspira.es

ABSTRACT: The EUCLIDES PV Concentration Plant is a grid connected system that delivers 480 kW$_p$ output by means of an array of inverters. In order to control and assess the operation of the plant, as well as to enhance the local grid signal quality, a large network sensors, transducers, microcontrollers and PCs has been set up. All of them are linked together by a field bus with a specifically developed communications protocol aimed at smoothly interconnecting very heterogeneous devices. All elements of the EUCLIDES plant are connected to the EUCLIDES field bus, allowing operational control of the plant as well as the observance and recording of several different working magnitudes. The whole system covers the following: Grid Connecting Inverters, Tracking Systems, Weather Station & Central Command and Monitoring Station.

Keywords: Virtual machine organisation– 1: Connection of heterogeneus devices– 2: Multiple layer protocol– 3.

1. INTRODUCTION

The Euclides protocol has been designed in order to have the capability of connecting a set of heterogeneous devices The communication among devices connected to the EUCLIDES bus takes place at 3 different layers:
- Physical Layer
- Logical Link Layer
- Network Layer

The **Physical Layer** is responsible for the transmission of the electrical signals that codify the information, or explained in a different way, the one to guarantee that a logical 0 or 1 reaches their target device as such. The Physical Layer specifications are related to this task, referring to the characteristics of the signal carrying medium an its particular performances. The EUCLIDES Protocol Specifications do not affect this layer in any way, i. e. any transport medium can be used. Therefore, we will not discuss this layer later.

The **Logical Link Layer** is responsible for the safe transmission of byte groups, which we call *frames*, among devices on the bus. Frames have a special format, which will be defined furtheron, that makes transmission error detection easier. This is one of the most important duties of this level. The other important role is to assure correct packet addressing. The EUCLIDES Protocol does not demand any specific requirements of this layer.

The **Network Layer** is in charge of managing the access to the transmission medium, making flow control, and taking decisions in case of transmission errors. The groups of bytes that have been sent from one virtual device to another in a network level link will be called *packets*, in order to differentiate them from the byte groups of the logical link layer.

2. LOGICAL LINK LAYER

Communications on the EUCLIDES bus are based on a "master-slave with polling" network concept. Any given slave device, subsequently, may not transmit until it has received an explicit invitation to do so by its master. The master device, however, is allowed to send one or more frames to one or more slaves whenever it wishes to.

Frame Format
In this specific implementation of the protocol for the EUCLIDES plant in Tenerife, the format of the frames sent by the master matches the following pattern (each box represents exactly one byte):

Device Address	Frame Length	Data	...	Data	CRC High	CRC Low

DEVICE ADDRESS: Indicates the referenced slave device or devices.
FRAME LENGTH: Indicates the length of the whole frame.
DATA: These are the data bytes (usually an NL-FRAME).
CRC:16 bit cyclic redundant checksum.

The EUCLIDES Protocol Specifications do not demand this format. It has been adopted only for this specific implementation of the protocol, and is thus only an example.

The master always has device address 0. This drives slave devices to have addresses from1 to 255. However, this identifier must not necessarily be unique, at least as far as the Logical Link Layer is concerned. The restriction is that only a single device can make use of the bus at a time. This management is left in hands of the Network Layer, which identifies all single devices in the bus in a unique form.

During a transference from the master, the Device Address byte is sent with its 9th bit set to 1 (addressing mode) and only the device or devices which have this address identification will listen to the rest of the frame, which will be sent with this bit set to 0. This technique allows to save CPU overhead time for communication purposes in the slave devices.

3. NETWORK LAYER

The Network Layer is supported by the Logical Link Layer and NL-Packets are sent from point to point inside LLL-Frames. This operation is completely transparent to the Network Layer. The Network Layer Packet Pattern is shown below:

NL Address	NL Address	Packet Type	DATA	...	DATA

NL Address: Is 16 bit address, defining a device or a group of devices that are meant to receive the packet.
Packet Type: Indicates the type of packet.
DATA: Is the information which the packet is carrying.

No checking is performed, as it is assumed that the Logical Link Layer is efficient enough to detect a transmission error and inform the Network Layer of this occurrence.

NL-Addresses must be unique to each device, as they identify the devices which are connected to the bus. Group targeting is possible only by means of broadcasting methods. Together with its **device number**, every element that is connected to the bus must have a **group number**, identifying the cluster of devices it belongs to.

The structure of the 16 bit NL-Address is:

15	14	13	12	11	10	9	8	7	6	5	4	3	2	1	0
B	G	A_{13}	A_{12}	A_{11}	A_{10}	A_9	A_8	A_7	A_6	A_5	A_4	A_3	A_2	A_1	A_0

B: Is the broadcasting bit. When set to 1, the address bits A0 to A13 are ignored, addressing all the devices connected to the bus.

In this way, a broadcasting address targeting all devices on the bus will be 10xxxxxxxxxxxxxx.

G: Group broadcasting: When set to 1, A0 to A13 identify a target group. All devices belonging to this group are addressed.
A_{13}-A_0:These are the address / group number bits.

Bits B and G combinations:

BG	Meaning of A13-A0
00	Target *device* address
01	Target *group* address
10	General Broadcast (all devices)
11	Reserved

Table I. Broadcasting bits combinations

4. VIRTUAL MACHINE ORGANIZATION

The main philosophy behind the EUCLIDES protocol is that any device connected to the bus is regarded as a virtual machine with the following elements:
- Digital Input Ports   (Read Only)
- Digital Output Ports   (Write Only)
- Analog to Digital Converter   (Read Only)
- Digital to Analog Converter   (Write Only)
- Internal Memory Space   (Read / Write)

The ports and converters area called **channels**, which are identified by 8, 16, 32 or 64 bit numbers, while the internal memory is accessible through a 8, 16, 32 or 64 bit linear addressing scheme. This allow small to broad addressing ranges, enough for even larger systems.

As channel numbers and memory addresses are virtual, they do not necessarily have to refer to actually implemented port or address space identifiers. The manufacturer or a given device can present a virtual image of the machine, where an internal translation table can build up the correspondence between virtual and actual, physical identifiers. Memory space could even be split up, giving full 2^{64} elements linear virtual address space for reading access, as well as another 2^{64} for writing (or 2^{32}, 2^{16}, 2^8, or even all of then together, depending on the packet type sent). Each virtual memory word has 8, 4, 2 or 1 byte(s) of length, depending on the implementation. As process identifiers are present in the addressing scheme of a device (3 bits), ups to 8 virtual memory and channel spaces can be implemented, allowing multiple virtual device configurations on a single machine. This helps to avoid interferences among the different processes that are running in a given moment.

Virtual memory space is also a support for accessing variables and constants on the devices connected to the bus. A translation procedure at each device can relate special variables or constants to specific virtual memory addresses.

Of course, not all modes have to be implemented at a time. Usually it will only make sense for a given device to implement a given subset of the addressing and organisation schemes. If this device is accessed in a not implemented mode, it will answer back with a special packet which indicates that the specific mode is not available on this specific slave device.

These packet types can be defined in a systematic way using three different packet type byte formats. They have been designed to be very similar to each other, as this leads to a more straightforward implementation on both the software and the hardware levels.

5. OVERVIEW OF PACKET FORMATS

Masters can carry two types of operations:
- Control Operations (CONTROL format)

These packets consist in a byte in which the two higher bits must be 0 to declare the packet as a CONTROL Packet and the four lower bits stand for the type of operation (Reset Device, Data Block Inquiry, Cease Operation…)

- Port or Memory Access Operations (ACCESS format)

In these packets the two higher bits stand for the kind of accessing operation we need at a given time.

(00)	(Control Packets)
01	Digital Access
10	Analog Access
11	Memory Access

Table II. Access type bits

The other bits denote if we are in a reading or writing operation, the addressing format (length in bytes of Port /Memory Address) and the data format (length of the data word)

Slave packets can carry four types of answers:
- ACK packets
- NAK packets
- Data packets
- Status Information packets

The higher two bits stand for the type of packet, the third one is set to 1 if there is an alarm status on the device, the fourth is the result of a reading operation if it only involved one bit and the other four stand for the operation specified. When the packet is a Data Packet, these bits specify the type of word length returned (length in bytes of the information).

REFERENCES

M. Cendagorta, M.P. Friend, F. Pérez, P. Valera, S. González, V. Sánchez, F. Dobón, R. Galbas. "Differential MPP Controling"

G. Sala, J.C. Arboiro, R. Arcía, A. Luque, T. Bruton, D. Cunninghan. "Description and Performance of the Euclides Concentrator Prototype" Barcelona (1997)

G. Sala, J.C. Arboiro, I. Antón, A. Luque, M.P. Gasson, N.B. Mason, K.C. Heasman, T.M. Bruton, E. Mera, E. Camblor, P. Datta, M. Cendagorta, M.P. Friend, P. Valera, S. González, F. Dobón, F. Pérez. "480 kWpeak Concentrator Power Plant using the EUCLIDES™ Parabolic Trough Technology". Viena (1998).

G. Sala, I. Antón, J.C. Arboiro, A. Flouquet, A. Luque, K.C. Heasman, N.B. Mason, M.P. Gasson. "Acceptance Tests for Concentrator Cells and Modules Based on Dark IV Characteristics". Viena (1998).

J.C. Arboiro, G. Sala, J.I. Molina, L. Hernando, E. Camblor. "The EUCLIDES™ Concentrator: A lightweight 84 m Long Structure for Sub-Degree Tracking". Viena (1998).

K. Mukadam, F. Chenlo, M.C. Alonso, M. Alonso-Abella, A. Matas & P.Valera "Analytical Monitoring Results of the 1MWp PV Plant". 13th European Photovoltaic Solar Energy Conference. Nice (1995).

E. Lorenzo, C. Maquedano & P. Valera " Operational results of the 100 kWp Tracking PV Plant at Toledo-PV Project". 13th European Photovoltaic Solar Energy Conference. Nice (1995).

M.C. Alonso, G. Blaesser, F. Chenlo, H. Edwards, F. Fabero, K. Mukadam and P Valera. "Power Measurements at the 1MW Photovoltaic Plant Toledo".

RESEARCH AND DEVELOPMENT OF DEVICES FOR TAKE-OFF AND CONVERSION OF PHOTOVOLTAIC MODULE POWER

S.M. Karabanov, V.V. Simkin
JSC "Ryazan Metal Ceramics Instrumentation Plant"
55 Novaya St., Ryazan 390027, Russia
Tel./Fax: 7-0912-216334
E-Mail: rzmkp@org.etr.ru

ABSTRACT: The range of illuminance change in the limits of which a solar cell is able to give off power in load makes up thousands units and more. In order to provide the mode of giving off the maximum power in load it is necessary to change the value of the load current or resistance in the same limits. The paper presents the problem of construction of the extreme regulator providing the mode of giving-off PV module maximum power, and of the switching inverter within the framework of a united structure, when the significant part of its components is used simultaneously in both devices. Such an approach allows not only to reduce the hardware costs, but also the quantity of pulse transformations of energy, and, therefore to increase energy efficiency of modules, to simplify their operation and to make it more reliable. The given analysis allows to take into account the features of module current-voltage characteristics in the structure of the considered extreme regulator rather completely, to estimate the necessary deviation value of the regulation parameter at destabilizing factors.
Keywords: PV Module - 1: Extreme - 2: Inverter - 3

1. INTRODUCTION

The efficiency of solar cells application is determined to a considerable extent by two interconnected factors: the degree of matching the cell output resistance with the load and the change of illuminance range of a cell within the limits of which the matching conditions are provided. In turn, the load requires certain voltage value and its form. The extreme power regulator allows to decide the problem of matching load with the module resistance [1]. The inverter solves the problem of providing the voltage value or form. The joint use of these devices within the framework of the united structure imposes certain limitations on their structure and principle of operation. In the power regulator it is expedient to use the principle of searching for the power maximum on the increment of the output current value with the pulsed way of taking-off power in load [1]. Such a principle is easily integrated with a pulsed way of the output voltage formation in the switching inverter. The offered converter is based on these two types of devices.

Generally I-V characteristics of a PV module can be expressed by the ratio [2]:

$$I = I_s\,(e^{\beta U} - 1) - I_L,$$ (1)

where I_L represents excitation of unbalanced carriers of solar radiation and at U=0 determines the value of the short-circuit current I_{sc} of a solar cell; U is the output voltage of a cell; I is the cell output current; I_s is the diode saturation current in the equivalent circuit of a cell. $\beta=q/kt$ factor is determined by the value of an elementary charge q and radiation thermal energy kt.

According to (1), the point coordinates of the power extreme value at the output of a cell, I_m, U_m are given by the following equations:

$$J_m = -J_s\,\beta U_m e^{\beta U_m} \approx -J_L(1 - \frac{1}{\beta U_m}),$$ (2)

$$U_m = U_{oc} - \frac{1}{\beta}\ln(1 + \beta U_m),$$ (3)

where U_{oc} is the open-circuit voltage.

Ratios (1), (2), (3) show that I-V characteristics of a cell have two sloping sectors: one is about the axis of abscissas, the other –the axis of ordinates. The sectors are integrated by the inflexion region at which the extremum of the cell output power falls. The cell illuminance change results in change of I_{sc} and I_m values over a wide range. The U_{oc}, U_m values change in a rather small range, however, they can drift due to destabilizing factors, for instance, due to change of the cell surface temperature. In such conditions the principle of searching for the extremum on increment is expedient to use in the extreme regulator. This method is based on moving the operating point of an object in the extremum region and on determination of the quality function. The regulator structure can be essentially simplified if taking into consideration features of the cell I-V characteristics: almost constant position of U_m ordinate when changing double-parameter control mode to single-parameter one according to the load current value. In this case the extremum point position can be checked on the voltage U_{oc} recalculating it respectively.

2. EXPERIMENT

The considered regulation-conversion method was realized according to the functional circuit (Fig. 1).

The device is connected directly to a PV module (PVM). It includes a switching inverter (SI) with the control circuit (CC) and a pulse transformer (PT). In the structure the extreme regulator is shown by a scanning block (SB), a multiplier (M), a current-voltage converter (CV), a control circuit (CC) common with the inverter.

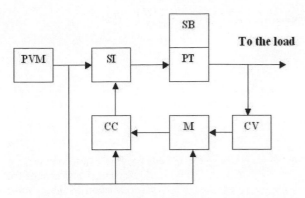

Figure 1: The regulation-conversion method.

In the regulator pulse multiplication of current and voltage on the inverter basic frequency (40 kHz) is applied. The result of multiplication is used immediately in the pulse form in order to generate the signal of pulse-width modulation in the control circuit. Scanning effect of the adjustable parameter is carried out on basic frequency of the inverter by the transformer TU owing to change of the load current, i.e. a current-voltage converter is realized immediately on the inductance of the transformer. Main logical functions of the control circuit were realized on PIC16C73A microcontroller.

3. RESULTS AND DISCUSSION

The considered combinative construction of the regulator-inverter functional circuit has allowed to provide high parameters of power (specific energy - 300 W/dm^3, efficiency - 0,93), weight and overall dimensions at comparatively low hardware costs. Deviation from MPP did not exceed 3% at the range of illuminance change in 1000 units. In the structure practically momentary entry (for 3-5 periods of the pulse voltage) in tracking mode at discontinuous change of the adjustable parameter is provided.

In order to provide normal operation of the extreme regulator the value of deviation ΔU_m from the extremum point U_m should completely overlap the region of possible changes of the output parameter both due to the illuminance change and effect of destabilizing factors. The value of deviation ΔU_m occurring due to destabilizing factors is managed to check only by indirect method on the value of deviation of the open-circuit voltage ΔU_{oc}. To determine the functional dependence $\Delta U_m = f(U_{oc})$ it is possible to use Lagrange formula:

$$\Delta U_m = \frac{dU_m}{dU_{oc}} \Delta U_{oc} , \qquad (4)$$

Distributing (4) on the whole interval of ΔU_m possible values you can estimate the maximum shift value of the extremum point coordinate:

$$\Delta U_{m\,max} = \left| \frac{dU_m}{dU_{oc}} \right|_{max} \Delta U_{oc\,max} , \qquad (5)$$

where $U_{oc\,max}$ is the maximum possible deviation of U_{oc} at destabilizing factors.

The function of U_{mmax} is defined implicitly and in general can be expressed by the following equation:

$$F(U_m, U_{oc}) = U_m - U_{oc} + \frac{1}{\beta} \ln(1 + \beta U_m) = 0 . \qquad (6)$$

The derivative of $U_m(U_{oc})$ function is expressed by partial derivatives:

$$\frac{dU_m}{dU_{oc}} = -\frac{\partial F / \partial U_{oc}}{\partial F / \partial U_m} = \frac{1 + \beta U_m}{2 + \beta U_m} . \qquad (7)$$

Then the increment value of ΔU_{mmax} will be determined by the product:

$$\Delta U_{m\,max} = \left| \frac{1 + \beta U_m}{2 + \beta U_m} \right|_{max} \Delta U_{oc\,max} . \qquad (8)$$

At destabilizing factors the maximum value of the first multiplier is provided at $U_m = U_{mmax}$. The value of $U_{m\,max}$ can be determined from (3), expanding the latter into a power series:

$$U_{m\,max} \cong U_{oc} - \frac{1}{\beta} \left[2\frac{\beta U_{m\,max} - 1}{\beta U_{m\,max} + 1} + \frac{1}{\beta U_{m\,max}} \right] .$$

Taking into account that $\beta U_{mmax} \gg 1$ we obtain for the approximate value of U_{mmax}:

$$U_{m\,max} \approx U_{oc} 2 / \beta . \qquad (9)$$

Substituting this ratio in (8) we determine the maximum deviation value:

$$\Delta U_{m\,max} \approx \frac{\beta U_{oc} - 1}{\beta U_{oc}} \Delta U_{oc\,max} . \qquad (10)$$

To provide the required deviation value of U_{mmax} in the extreme regulator structure it is necessary to generate the appropriate increment of the load current. The output current increment value of ΔI_m can be given by Lagrange formula:

$$\Delta I_m = \left| \frac{dI_m}{dU_m} \right|_{max} \Delta U_{m\,max} .$$

According to (2) we obtain the following for the approximate value of ΔI_m:

$$\Delta I_m \approx \left| -I_L \frac{1}{\beta U^2_m} \right|_{max} \Delta U_{m\,max} . \qquad (11)$$

Taking into account (9), (10) the maximum necessary increment value of the load current can be defined:

$$\Delta I_{m\,max} \approx I_L \frac{\beta U_{oc} - 1}{\beta^2 U_{oc}(U_{oc} - 2/\beta)^2} \Delta U_{m\,max} \qquad (12)$$

For practically important case of $\beta U_{m\,max} \gg 2$ we obtain:

$$\Delta I_{m\,max} \approx I_L \frac{\Delta U_{m\,max}}{\beta U_{oc}^2}. \qquad (13)$$

Or for the relative increment value of the output parameter it is possible to state:

$$\frac{\Delta I_{m\,max}}{I_L} \approx \frac{\Delta U_{oc\,max}}{U_{oc}} \frac{1}{\beta U_{oc}}. \qquad (14)$$

The necessary increment value of the output current referred to the open-circuit current value of the cell under consideration is determined by the relative deviation value of the open-circuit voltage due to destabilizing factors and depends on the cell activity (βU_{oc}). Each parameter is definitively specified by the mathematical model of current-voltage characteristics of the used cells.

4. CONCLUSION

The obtained results of the whole set of studies can be summarized as follows:
1. The new approaches to design of the devices of taking-off power from PV modules are developed.
2. Such an engineering solution increases essentially the power capabilities of PV modules, expands the area of their application, simplifies the modules operation and makes its more reliable.
3. The device produced according to the given technique had good parameters of weight and overall dimensions, high efficiency, high accuracy of power regulation, high response and huge dynamic range on the illuminance.
4. The given design procedure allows to determine main regulation parameters at destabilizing factors.

REFERENCES

[1] Karabanov S.M., Simkin V.V., Patent № 2117983.
[2] S.M. Sze "Physics of Semiconductor Devices", John Wiley & Sons NY, (1981).

ENERGY STORAGE MANAGEMENT STRATEGIES IN HYBRID SYSTEMS

C. Trousseau, D. Mayer, F. Neirac
Ecole des Mines de Paris, Centre d'Energetique
BP 207, 06904 Sophia Antipolis Cedex, France
Phone : + 33 4 93 95 75 99 - Fax : + 33 4 93 95 75 35
eMail : celine.trousseau@cenerg.cma.fr

ABSTRACT: The aim of the paper is to analyse and improve existing management strategies applied to renewable energy based stand alone hybrid systems including battery storage and conventional back-up supply and to evaluate the gain brought by modified strategies. A system including controllable energy supply and not only based on variable energy sources allows a greater number of freedom degrees in the energy dispatching point of view. This increased flexibility is of particular interest in order to better control the battery state of charge evolution with time. Focusing on a rational use of both renewable and conventional energy and since the grid back-up is the only power source to be controlled, different strategies of operation can be described taking into account the following parameters: starting time, power of operation and duration of operation from starting time. Examples of strategies of grid back-up used for actual hybrid systems are given and their advantages and drawbacks are highlighted. We define an improved management strategy of the energy stored in the battery under the particular criteria of maximising the use of available renewable energies and minimising energy taken from the grid.
Keywords: Hybrid - 1: Strategy – 2: Simulation - 3

1. HYBRID SYSTEMS AS RELIABLE SUPPLY

Electric systems here under consideration are in the power scale of a few kW and use a DC bus connection between components. The conventional AC load is supplied through a solely central inverter.

Stand-alone photovoltaic / lead acid batteries systems are widespread, but they have some drawbacks; indeed their reliability is not perfect since the load supply is interrupted when we have at the same time a too weak solar irradiation and a too low battery state of charge. On the other hand the issue cannot be solved with an increased storage capacity for reasons of battery maintenance and cost.

Hybrid systems including both renewable energy sources (photovoltaic and wind turbine generator) and conventional sources such diesel gensets or more seldom a grid back-up are known for allowing a highly reliable power supply, because this extra energy can be provided to the system at any time, thus avoiding loss of load. Furthermore the resources complementarity between solar and wind can play a role to tend towards autonomy.

2. PURPOSE OF WORK AND METHODOLOGY

2.1 Freedom degrees of controllable power source

With the possibility of adding a controllable power source some choices have to be done; indeed we have at our disposal three parameters which define the way this extra power source is used: starting time, power of operation and duration of operation from starting time. Contrary to the case of a hybrid system with a diesel genset, we have with a grid back-up supply much less constraints, especially regarding power of operation, which should not be less than 40% of rated power genset [1].

2.2 Existing energy management strategies

An example of hybrid system with grid supply is given in reference [2]. This system includes both solar and wind power sources, a conventional engine driving a generator, and a grid connection. But for reasons of subsidies to renewables, the optimisation of system operation is optimised under economic considerations: the daily energy exchange between system and grid is chosen so that the system is autonomous during the peak-tariff period; during off-peak energy can be imported from the grid to ensure an enough charged battery for the following peak-tariff period.

On the other hand, we can consider to use grid in the same way diesel gensets are usually used, that is to say as a back-up only for security of supply, so as to avoid to frequently resort to the grid. In this operation mode, the grid is switched on only when the battery reaches a low voltage threshold. Considering we use the grid for both load supplying and battery charging, we have therefore to choose the battery state of charge from which the grid is switched off.

2.3 Interest of simulation

Running simulations rather than carrying out field experiments with actual electric systems is cheaper and quicker. Moreover keeping in mind that we want only to compare operation strategies, with simulation we can reproduce strictly identical meteorological and load situations.

3. MODULAR SIMULATION TOOL

Existing simulation tools (SOMES, ARES, HYBRID2) provide a good proper modelling of DC bus hybrid systems, but did not appear to offer enough flexibility to meet requirements needed by the implementation of various customised operation control strategies.

3.1 Short models review

The models chosen are steady-state because the objective is the study of operation strategy influence on log-term performance.

Models used here are well-known. The photovoltaic array is modelled with the one-diode electric equivalent scheme which takes into account both solar irradiance and cell junction temperature to calculate current-voltage characteristic.

In the lack of thorough electrical and mechanical data on the wind turbine, the simple model of the manufacturer power curve, which gives measured output DC power (when connected to the battery bank) in function of wind speed in ten-minutes means, was used. For this last reason this behaviour model is convenient for ten-minutes time step calculations.

In the developed software a battery model [3] taking account for sharply increasing voltage during gassing phase was used because the implemented operation control strategies forced periodically the battery to have gassing for maintenance reasons.

3.2 Software organisation

The modular software was implemented under Matlab Simulink environment. Each component is represented by a block. Due to the lack of a Maximum Power Point Tracker the feedback between battery voltage and photovoltaic current-voltage characteristic determines the operation point.

Regarding reversible inverter, it has to be noticed that, in the case where the grid back-up is switched on at maximum power, the flowing AC current feeds the AC load, and the remainder is converted into DC current through the inverter used as a battery charger. Therefore during this operation mode the load is fed without conversion losses.

The energy management strategy is implemented as a independent separable block. Therefore it is quickly and easily replaced.

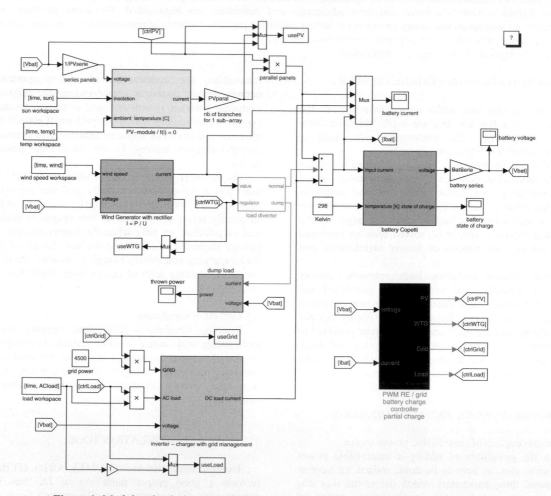

Figure 1: Modular simulation tool (the black box represents one possible operation strategy)

4. CASE STUDY

Calculations were carried out on the application case of a television re-broadcasting station in Uzbekistan, close to Tashkent. This system is composed of a photovoltaic array, a wind turbine generator with its dedicated dump load, a 48 V battery bank and an inverter which is reversible and thus be used as a AC-DC battery charger when back-up

supply from the grid is necessary. The load is composed of 3 retransmitters of 1.8 kW each (called afterwards T1, T2 and T3), operated during a part of the day. The following table gives a survey of the main system features.

SYSTEM FEATURES

component	rated power or daily energy
photovoltaic generator	6 kWp
wind turbine generator	3 kW (rated output)
battery bank	1120 Ah C10 equivalent to 54 kWh
reversible inverter	4.5 kW
retransmitter T1	34.2 kWh/day
retransmitter T2	23.4 kWh/day
retransmitter T3	18.0 kWh/day

Table I: Components size and daily load consumption.

Meteorological data were available only in mean monthly values for the concerned site, for daily horizontal irradiation as well as for wind speed. Nevertheless time-series data were necessary for running simulations. Existing algorithms suited to synthesis of statistically meaningful data time-series have been implemented.

Regarding solar data a first method [4] using Markov transition matrices allowed to synthetise daily values of clearness index for each month. Then a clear sky model [5] was applied to calculate the hourly profile for each day. Last, the anisotropic radiation model [6] known as "HDKR model" was used to convert horizontal values to tilted values. The tilt angle of the photovoltaic generator was assumed to be changed manually twice a year as following: 20° from April to August and 45° from September to March. Thus irradiation on array plane is improved at each season in comparison with a fixed generator without resort to a sophisticated and expensive sun tracking system.

Regarding wind data some assumptions had to be made due to the lack of further information for the concerned site. Obviously the mean wind speed value could not be taken as an input for the wind turbine power curve because of the non-linearity of wind turbine output with wind speed. We resorted to the long-term Weibull distribution on which wind energy specialists [7] are agreed to be a good fit for mean hourly wind speed values. A missing parameter describing the shape distribution was chosen among common values. Then a Monte-Carlo random simulation was used to generate one hourly time-serie for each month. But feedback from the system monitoring will allow in next months to check and retrieve further information such self-correlation.

To ensure a straightforward comparison between strategies only, the random synthesis was run one solely time, thus the meteorological inputs were strictly identical for all simulations.

Simulations were run along one year using a ten-minutes time step, using linear interpolation to convert the hourly meteorological inputs.

5. COMPARISON OF OPERATION CONTROL STRATEGIES

5.1 Considered strategies

In all cases the grid was switched on when the battery reached the low voltage threshold of 1.9 V per cell. In case of a boost charge mode, the grid is operated at inverter rated power until a voltage of 2.5 V per cell is reached. Then grid power is possibly reduced so that voltage does not increase more than 2.5 V until 3 hours are spent, so the grid is switched off. In case of a partial charge mode, the grid is operated until a defined percentage of battery C10 capacity was recovered, with a counting of Ah.

In all strategies, the boost charge occurs anyway one time in each 15-days period for reasons of battery maintenance.

In the basic strategy called afterwards "full", the boost charge mode is allowed each time grid back-up is switched on. In the modified strategies called afterwards "partial", the boost charge mode is not allowed more than one time in each 15-days period, otherwise the partial charge mode is used with 40 or 20% recovery charge threshold.

5.2 Comparison criteria

We considered the following energetic criteria: the minimisation of grid back-up supply and the rational use of renewable energies, defined as the ratio of the energy provided to the system to the energy theoretically available at these generators' output.

5.3 Results

Simulations were performed for 5 different load profiles combining one or several retransmitters; the following table shows the yearly required energy in each case, compared with yearly energies available at the renewable generators' output. Even from this raw balance we can see that the system could be autonomous with retransmitters T3 or T2.

ENERGY BALANCES

component	yearly energy in kWh
photovoltaic array	8 791
wind turbine generator	4 866
sum of renewables	13 657
load profile T3	6 570
load profile T2	8 541
load profile T1	12 483
load profile T2+T3	15 111
load profile T1+T3	19 053

Table II: Yearly available renewable energies with regard to load consumption.

Simulations showed that the system could run as stand-alone with, either retransmitter T3, or retransmitter T2. Thus no major issue of operation control arise with these both load profiles.

For the three other load profiles, the grid back-up was necessary to ensure a perfect reliability. The resource complementarity between sun and wind is effective; indeed, solar resource reaches its peak value in sommer, meanwhile wind resource reaches its peak value in winter, and simulations revealed two critical periods in spring and fall, when both solar irradiation and wind speed are weak.

YEARLY ENERGY TAKEN FROM THE GRID

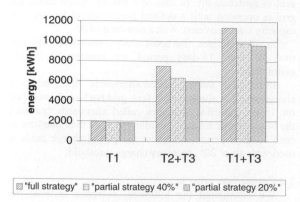

Figure 2: Comparison between back-up energy saving for three load profiles with three different strategies.

RATIONAL USE OF RENEWABLE ENERGIES

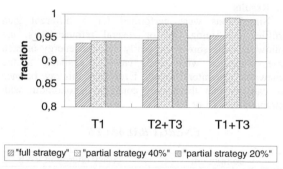

Figure 3: Fraction of available renewable energies provided to the system for three load profiles with three different strategies.

From these results we can notice that the "partial strategies" bring only a neglectable benefit in the case of the load profile T1. But much more energy is saved with the T2+T3 retransmitters (20.4%) and with the T1+T3 retransmitters (15.6%). As a preliminary explanation we might establish a parallel between this benefit and the grid contribution to the energy provided to the system.

**GRID CONTRIBUTION
WITH RESPECT TO THE LOAD PROFILE**

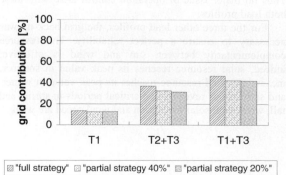

Figure 4: Grid contribution to the energy provided to the system (including load and losses).

This parameter is sensible but it would be worth to be investigated through other simulations.

6. CONCLUSION AND OUTLOOK

A modified energy management strategy consisting of a partial recharge coupled with a periodic boost charge for battery maintenance appeared to save until 20.4% of energy from grid in comparison with the "full strategy".

The rational use of renewable energies is improved, but is a less sensible criteria, because the fraction was already good, above 93%, with the basis "full strategy".

For these simulations we tested a same system layout with several load profiles under three operation strategies. The load profile appears to have an influence on the benefit induced by improved strategies. Nevertheless the relationship between components size / load profiles and operation control strategy requires further investigations, particularly from the generators sizing and the load profile energy and shape.

ACKNOWLEDGEMENTS

This work was performed with the help of a European Commission partially funded project Inco Copernicus.

REFERENCES

[1]   R. Hunter, G. Elliot, Wind - Diesel systems, A guide to the technology and its implementation, Cambridge University Press, 1994
[2]   D. Child, I. R. Smith, Modelling of the Energy Distribution within a Grid-connected Wind-Solar-CHP-battery system, International Journal of Ambient Energy, Vol. 16, No. 3, July 1995
[3]   J. B. Copetti, E. Lorenzo, F. Chenlo, A General Battery Model for PV System Simulation, Progress in Photovoltaics : research and applications, Vol. 1, 1993
[4]   R. J. Aguiar, M. Collares-Pereira, J. P. Conde, Simple procedure for generating sequences of daily radiation values using a library of Markov transition matrices, Solar Energy, Vol. 40, No. 3, pp 269-279, 1988
[5]   M. Collares-Pereira, A. Rabl, The average distribution of solar radiation - correlations between diffuse and hemispherical and between daily and hourly insolation values, Solar Energy, Vol. 22, No. 2, pp 155-164, 1979
[6]   J. A. Duffie, W. M. Beckman Solar, Engineering of Thermal Processes, 2nd edition, Wiley, New York, 1991
[7]   I. Troen, E. L. Petersen, European Wind Atlas, 1989

BATTERY ENERGY PAY-BACK TIME

Patrick JOURDE

CEA/SSAE bâtiment 238 CEA Cadarache 13108 Saint Paul lez Durance France

Phone : 33(0)442 25 21 52 Fax : 33(0)442 25 73 73 E-mail : patrick.jourde@cea.fr

ABSTRACT : A frequent issue is the module energy feed back ie how long it takes to a photovoltaic module to give back the energy necessary for the full process of its manufacture. The same question is applied to batteries in stand-alone PV systems. This paper presents an evaluation the energy necessary for the full process of making the batteries. In the best conditions, battery energy pay-back time is quite similar to modules pay-back time : 4,5 years for original batteries and 2,5 years and half for recycled batteries. But tubular plate life-time is more than three times shorter than module life-time, and SLI automotive or deep cycle batteries life-time 6 up to 15 times shorter. The apparently cheapest solution ie SLI automotive batteries, is by far the worst solution in terms of energy pay-back time. It is also the most expensive one, whatever the approach ie economical or ecological. However, for two main reasons, it could remain the most popular solution in developing countries. The initial cost is lower and the local availability much higher. High quality and long life-time batteries such as tubular plate stationary batteries could be recommended, especially in remote areas and even for small SHS application.

Keywords : 1 : Battery - 2 : Life cycle - Energy pay-back

1. ENERGY IN A BATTERY INDUSTRIAL PROCESS

The amount (and type) of energy which is necessary to make, install and maintain a battery include the energy in each component (plates, separator, box), is calculated as from the original mine or from the raw materials up to the final stage of the component's implantation into the battery. Two kinds of leads are compared : recycled and not recycled lead.

Since many stand-alone photovoltaic systems are installed in remote areas, various kind of transportation are taken into account, for instance one way and return for recycling, between factory and installation.

An average lead acid battery is considered. Then some variations related to types of batteries are taken into account. Three types are considered : SLI automotive car batteries, deep cycle batteries, tubular plates stationary batteries.

1.1 Raw material (1)

Lead :
Extraction from the mine and washing : 1 kWh/kg
Fusion and fining : 2,65 kWh/kg
including coal : 1,72 kWh/kg
    electricity : 0,93 kWh/kg

1.2 Recycled lead

Full process : 1,95 kWh/kg
including gas : 1,46 kWh/kg
    electricity : 0,49 kWh/kg

1.3 Recycling ratio

The battery recycling ratio is increasing , and is today : 70% in USA, 60 % in Europe, 50 % in Africa, 30 % in Asia. These ratios include all kind of batteries. It is much easier to collect a small SLI automotive car battery in a garage than to collect a heavy solar battery in a very remote area. Altogether, the recycling ratio of solar batteries are probably the worst. For recycled batteries various kind of transportation (car, truck, boat) are taken into account one way and return.

In many cases, the recycling in remote areas is too expensive, and return to the factory should be paid by the user. A ratio of 20 % is considered for solar application in developing countries (Fig.1).

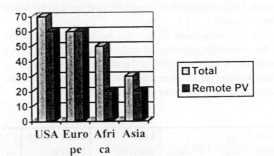

Figure 1 : Recycling ratio (%)

Therefore, the energy in lead is :
3,65 x 0,8 +1,95 x 0,2 = 3,31 kWh/kg
In Europe, a 60% ratio leads to 3,65x0,4 +1,95x 0,6 = 2,63 kWh/kg

1.4 Factory process

1.4.1 Lead (2)
Electricity : 0,95 kWh/kg
Gas : 0,584 kWh/kg
Water : 3 l /kg

1.4.2 Electrolyte.
Sulfuric acid requires a very small energy. It is an exothermic reaction from by-product SO_2 or sulfur combustion. The so-called distillated water is filtered by resins.

1.4.3 Case and cover
T case and cover are made of polypropylène or S.A.N[2] manufactured from natural gas liquids or petroleum. The energy required is approximately 22 kWh/kg, in the form of oil and gas (3)

If recycled, the energy requirement is approximately 4,5 kWh/kg

1.4.4 Separator (fiberglass) :

Sand, limestone, soda ash are melted at high temperature. Energy required is approximately 7,2 kWh/kg, in the form of oil and gas. The energy requirement in recycling is nearly the same.

1.5 Weight of various battery components

The tubular plate stationary battery such as Oldham TUS 5 (350 Ah 2 Volts) component weights is given in the first line of the table 1. This evaluation could be subject to large variations. For the same kind of battery (TUS 2 up to TUS 7) the ratio between lead and other components will vary. In the same case and with nearly the same amount of electrolyte, 5 plates (2 positives, 3 negatives) will be installed in the TUS2 and 15 (7 positives and 8 negatives) in the TUS7. This leads to :

Table 1 Type/Weight (kg)	Lead	Electrolyte	Case and cover	Separator	Other	Total (kg)
TUS 5	17	8,1	2	0,6	0,3	28
TUS 2	9,25	9,3	2	0,25	0,2	21
TUS 7	22,9	10	2	0,8	0,3	36,2

For an average battery, such as TUS 5, for each kg of lead, there is :
Case and cover : 2/17 = 0,117 x 22 = 2,6 kWh/kg (new batteries) or 0,53 kWh/kg (recycled batteries)
Separator : 0,6/17 = 0,035x 7,2= 0,25 kWh/kg

1.6 Battery transportation

A battery could be transported assembled or not, dry or not. Different types of transportation have to be taken into account :
- from the component factory (in industrialised country) to the assembling factory (in industrialised or developing country). A non assembled battery takes up twice as much volume as an assembled one. But the weight is the same, and this is the main financial factor.

- from the assembling factory to the wholesale dealer
- from the wholesale dealer to the shop or the consumer, and back when recycled.

An average specific transport will be by truck on few hundred kilometres. The transport to the shop or in small quantities will be also the most expensive one. The evaluation, subject to large variations, is :
Transport (total) :
0,3 kWh/kg without recycling, or
0,5 kWh/kg with recycling

1.7 Energy spent versus energy stored :

The total energy spent in kWh/kg is (Table 2) :

Table 2 Type/ Energy (kWh/kg)	Lead Raw material	Lead Fact. process	Case and cover	Separator	Transport	Total kWh/kg	Remark
New batteries	3,65	1,53	2,6	0,25	0,3	8,33	
100% recycled batteries	1,95	1,53	0,53	0,25	0,5	4,76	
Recycled batteries developing countries	3,31	1,53	2,19	0,25	0,34	7,62	20% recycled
Recycled batteries industrialised countries	2,63	1,53	1,36	0,25	0,42	6,20	60% recycled

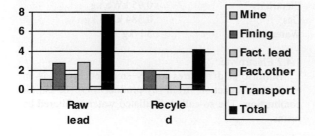

Figure 2 : Energy (kWh/kg) in battery

The following chapters will take in account the average value of **7,62 kWh/kg (or 27,4 MJ/kg)**. This

amount is similar to what was calculated in some studies for electric vehicles (4), and in environmental studies by Vattenfall (5) and (6).

The specific energy densities of lead acid batteries are about 40 Wh/kg for deep cycle, stationary or SLI automotive car batteries. Therefore :
=> **190 Wh spent per Wh stored** (average)
=> 208 Wh for new battery per Wh stored,
=> 119 Wh for recycled one per Wh stored

The energy density is different in other lead acid technologies :
35 Wh/kg for gelled batteries
=>217 Wh spent per stored Wh
45 Wh/kg for Active Glass Matter (AGM) batteries
=>169 Wh spent per stored Wh

Conclusion :the energy will be given back after roughly 190 full cycles. This value is not easily applied to PV systems

For a **12 volts 100 Ah battery** (1200 Wh) :
228 kWh are spent for the full process of making a new (or virgin) battery,
143 kWh are spent for the full process of making a recycled battery

1.7 Energy per type

For new batteries, coal, gas oil, and electricity are important. For recycled ones, gas is the most important (Fig.3)

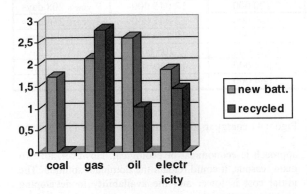

Figure 3 : Energy per type (%)

2 ENERGY SUPPLIED IN BATTERIES LIFE TIME

2.1
This chapter is devoted to the evaluation of the energy finally supplied by the battery for some typical applications (telecommunication, sea-lights, ...) or to private consumers. For the same amount of energy being delivered, the battery sizing can be quite different. It all depends on the type of application, the autonomy, the depth of discharge, the battery life time, the battery efficiency, the charge controller settings, etc.... The results of GENEC studies on the relation between deepness of discharge and battery life time are used (7).

Three kinds of batteries are considered :
Tubular plate batteries in stand alone house applications and professional applications : sold typically for 1500 full cycles.
SLI Automotive Car batteries used in small Solar Home Systems (SHS): sold for floating charge, and no reference to full cycles.
Deep cycle flat plate solar batteries used in small Solar Home Systems : sold typically for 500 full cycles.

2.2 Autonomy

The stand-alone PV system autonomy is an important factor for battery sizing, and, consequently, for pay-back time. This autonomy depends on meteorological conditions, on application and on expert or software for sizing. Between the tropics, or in Mediterranean countries, an average of one week could be applied for family applications, and two weeks for professional ones.

2.3 Tubular plate batteries in stand alone systems (comfortable house)

The first example takes in account the optimal conditions : latitude 20°, low sunshine annual variation : 4,5 kWh/day in winter, 6,5 kWh/day in summer. Batteries will be connected to a 1kWp - 24 V photovoltaic generator. The sizing to reach nearly a full autonomy (97%) leads to 12 two volts 1000 ampere.hour tubular plate batteries. Such battery's life-time is 10 years in best conditions.

This photovoltaic generator (modules +controller + batteries) will supply an average daily energy of 3,3 kWh from the battery. It does include all generator losses (modules, controller, wiring, batteries efficiency) out of one loss : the energy lost during overcharge controller disconnection. This could account for more than 30% in summer time for example

The overall energy supply by batteries in these optimal conditions is :
3.3 kWh/day X 3650 days = 12 000 kWh during their life-time
This batteries of 24 V x1 000 Ah = 24 000 Wh requires 190 X 24 000 = 4 560 kWh to be made.

The pay-back time is 10 years x 4 560 / 12000 = 3,8 years, and the energy is pay-back 2,63 times during battery life-time.

2.4 SLI automotive car batteries or deep cycle solar batteries in solar home systems

SLI battery life-time used in small SHS is still discussed. Ground experiences varies from 6 months up to 5 years. An average life-time with a good maintenance and load management is 3 years or less.

Deep cycle solar batteries, or deep cycle batteries life-time is 3 years or more. For the present evaluation, 3 years is an average for SHS using flat plate batteries.

A typical SHS includes a 50 Wp module and 12V 90 Ah batteries (autonomy: 6 days). With the best conditions as describe in 2.3, the daily available energy will be 165 Wh, and in 365 x 3 = 1095 days, 180 675 Wh will be produced.
This batteries of 12 V x 90 Ah = 1 080 Wh requires 190X 1080 = 205 200 Wh to be manufactured.

In these good conditions, batteries pay-back time is :
3 years x 205 200/180 675 = 3,4 years or 1 244 days. It is quite similar to batteries life-time (1095 days).

2.5 Tubular plate batteries in professional application (telecommunication, radio beacon,..)

Compared with 2.3, battery capacity is double to guarantee a 100% autonomy. Production will be the same : 12 000 kWh, but the 24Vx 2 000 Ah needs 9 120 kWh to be manufactured.

The pay-back time is 10 years x 9 120 / 12000 = 7,6 years, and the energy is pay-back 1,3 times during battery life-time.

2.6 European countries

In European countries, installations such as 2.3 or 2.4 will often be designed for winter and oversized for summer. 2 .3 will often be hybrid, and therefore designed for summer and under sized for winter. Diesel will provide the additional energy in winter

Two typical climates are studied : Southern Europe (Mediterranean), and Northern Europe. The batteries size doesn't change much from tropical, Mediterranean or Northern Europe climate. Twice times less energy production in Northern countries give with the same battery size twice as much autonomy required by long periods without sun.

The recycling ratio in Europe is 60% (172 Wh/ stored Wh).

Results are summarised in Table 3 :

Table 3	Daily produc. Wh Winter/summer	Battery size Wh / autonomy	Energy to make batteries Wh	Global production Wh	Pay back time
SHS tropics	150 / 180	1 080 / 6,5 days	205 200	180 675	3 years 148 days
SHS South Europe	90 / 190	1 200 / 6,3 days	206 400	153 300	4 years 14 days
SHS Northern Eur.	50 / 120	1 200/ 10 days	206 400	93 075	6 years 238 days
Profess. Appl. Tropics	3 000 / 3 600	48 000/14,5 days	9 120 000	12 045 000	7 years 208 days
Large House Tropics	3 000 / 3 600	24 000/ 7,2 days	4 560 000	12 045 000	3 years 286 days
Large House South EU	1 800 / 3 800	24 000/ 6,3 days	4 128 000	10 120 000	4 years 14 days
Large House North EU	1 000 / 2 400	24 000/ 10 days	4 128 000	6 205 000	6 years 238 days
Hybrid house South	1 800 / 3 800	12 000/ 3 days	2 064 000	10 120 500	2 years 7 days
Hybrid house North	1 000 / 2 400	12 000/ 5 days	2 064 000	1 861 500	3 years 119 days

Hybrid can appear as a good solution for pay-back time. It is not true. In hybrid systems, 1/3 up to ½ of the energy will be produced by another solution such as a diesel engine. The efficiency of such engine is quite low due to their small size, but this engine run at full charge. The average consumption is 0,3l of fuel per kWh.

2.5 Accuracy - Real daily energy production

These results are likely to be worse in reality. They are also average evaluations, due to the fact that batteries characteristics (i.e.1.4), systems sizing, load management, etc.. widely varies. For example, on summer afternoons, the battery is often overcharged and part of the production is lost. An average of 20/30% is lost in such a way, even if the controller and load management are more and more sophisticated.

CONCLUSION

The energy pay-back time in good conditions is approximately **4,5 years for new batteries and 2,5 years for recycled batteries**. This is quite similar to module pay-back time. But batteries life time is between 2 and 10 years and modules life time is 3 to 15 times more.

The conclusion shows a large scope of battery energy pay-back times. In some of the applications, the energy usefully delivered by the battery never reaches the amount of energy requested to make it.

The apparently cheapest solution ie car batteries, is by far the worst solution in terms of energy pay-back time. It is also the most expensive one, whatever the approach ie economical or ecological. However, for two main reasons, it could remain the dominant solution. The initial cost is lower and the availability in developing countries much higher.

High quality and long life-time batteries could be recommended, especially in remote areas. Another important factor in favour of high quality batteries comes from the low interest rate now being applied in many countries. The high initial cost can be transformed into reasonable monthly payments.

Recycling could be encouraged too, as it is since few months in India. Battery transport back to the factory for recycling is expensive in remote areas, and this could not be directly paid by the users.

Figure 4 : energy in manufacturing a 100 Ah 12V battery

REFERENCES

(1) information from Metaleurop : M. Paul Martin
(2) information from CEAC-Exide : M.Laurent Torcheux
(3) information Atoglass. M.Roberto Pardi
(4) Energy and environmental impacts of electric vehicle battery Linda Gaines Margaret Singh Total life cycle conference 1995
(5) PV cells environmental studies Vattenfall Utveckling Caroline Setterwall 96/99
(6) Battery guide for SAPV systems . Bengt Perers Bo.Andersson- to be published by IEA Task 3
(7) Ageing mechanisms in lead acid battery for photovoltaic systems. Florence.Mattera CEA GENEC 11/99

DESIGN OF A SIMPLE STRUCTURE FOR THE D-SMTS CONCENTRATOR

Rubén Mohedano, Pablo Benítez, Francisco J. Pérez and J.C. Miñano
Instituto de Energía Solar, E.T.S.I. Telecomunicación
Universidad Politécnica de Madrid, Ciudad Universitaria s/n, 28040 Madrid, Spain
Phone (3491) 544 10 60 - Fax: (3491) 544 63 41 - email: ruben @ ies-def.upm.es

ABSTRACT: This work addresses the designing of both the structure and the heat sinks of the D-SMTS photovoltaic concentrator, aiming simplicity and lightness. This mirror-based linear system offers a noticeable increased acceptance angle with respect to classical line-focus solutions for the same concentration level. Accuracy is then feasible to be relaxed at some assorted levels, owing to this high acceptance angle. The design of a novel light structure that, for instance, uses heat sinks as supporting beams and mirrors as heat sink fins has been carried out.

After a brief review of the D-SMTS main features, sections 2 and 3 are devoted to the design of the heat sinks and the structural analysis, respectively, of an approximately 3×3 meters array with four D-SMTS of geometrical concentration 30x. The calculation of the yearly-averaged optical efficiency predicted for this system is showed in section 4 and conclusions will be highlighted in section 5.

Keywords: Concentrator - 1: Cost reduction – 2: PV System - 3

1. INTRODUCTION

The D-SMTS (which stands for Dielectric-Single Mirror Two Stage) photovoltaic concentrator [1] is a trough-type system suitable for one-axis tracking operation and specially thought to work in the medium concentration range (C ≈ 30-100 suns) using silicon solar cells. This design, carried out in the non-imaging optics framework [2][3], has reflective primary optics and reflective/refractive secondary optics that increases the acceptance angle of this system (without losing concentration) with respect to its predecessor (the SMTS [4][5]), a concentrator that already showed outstanding capabilities with respect to conventional parabolic throughs, whose technology is very similar. This technology, widely used in solar thermal applications, has recently showed suitability also in the PV field, both from technological [6] and maybe cost-effectiveness (at large scale production) points of view [7] with the EUCLIDES™ approach.

Like the SMTS concentrator, the D-SMTS has a single mirror per cell row (Fig. 1), which acts as a primary for one of the cells rows and as a secondary for the other row (unlike the former, these two stages are overlapped in the mirrors of a D-SMTS). Deeper explanations about this and other features can be found in Reference [1].

Even though considering the annual averaged corrected sun-shape [8] instead of sun spread angle (0.265 degrees), the achieved high acceptance angle (±2.44 degrees in the 30x concentrator -compare at ±0.8 degrees of the parabolic through-) may appear as excessive unless we take into account allowance for errors at some levels. Among them, relaxing the optics, structure, tracking, orientation and mounting accuracies. For instance, ray tracing shows good performance even with gaussian slope errors (σ=0.5°) in the mirror profiles (the dioptric is nearly insensitive to errors due to the high local acceptance in its points[9]), a result unattainable with the parabolic shape.

It will be shown that an ultra-light and very simple structure supporting the concentrator cutting edge optics barely deteriorates its accuracy.

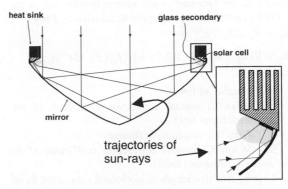

Figure 1 Two overlapped stages co-exist in each mirror of the D-SMTS, which form a single piece with the glass secondary, the cells and the heat sinks. After a first reflection on primary mirror, sun rays are cast onto the cells after either single refraction, or multiple-reflection plus refraction combinations.

2. DESIGN OF THE EXTRUDED ALUMINUM HEAT SINKS.

The main innovation is the use of heat sinks as supporting beams for the system. Thus, the extruded heat sinks have to meet both cooling and structural demands. The only restriction we impose are the current available geometries in the extrusion of aluminum: the ratio d/w (fins-length/gap between fins) must be bellow 8. Figure 2 shows the cross section of the designed 5-fins heat sink and its equivalent circuit (note its small dimensions permit its analysis as isothermal). The mirror, that uses a sheet of aluminum as substrate (thickness D=1.5mm), behaves as an additional fin. Because of its length (L=42.5cm), we cannot consider this extra-fin as isothermal, but an equivalent isothermal fin of shorter effective length L'(D,L) (see calculation below) has been used in the model.

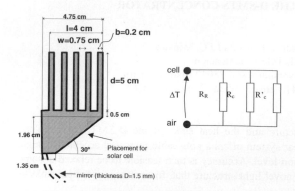

Figure 2 The geometry and dimensions chosen for the heat sinks meet extrusion available aspect-ratios. The equivalent circuit for thermal modelling is also showed. Notice this is not a scale drawing.

The thermal resistance R_T is the parallel of three resistances representing the radiation and convection phenomena, predominant in our system (conduction in the air will be considered as negligible to all effects). Thus:

$$R_T = R_R // R_c // R'_c \Rightarrow$$

$$R_T = \left(((n-1)h_c + h_{co})(2d+b) + h_R(d+l) + 2h_{co}L' \right)^{-1}$$

being:

- n=5 (number of fins)
- h_c (W/cm^2°C, natural convection coefficient of the interior sides of fins)
- h_R (W/cm^2°C, radiation coefficient)
- h_{co} (W/cm^2°C, natural convection coefficient of an isolated isothermal fin)
- L' (cm, effective length of isothermal equivalent fin of mirror)

The temperature drop from cells to air is:

$$\Delta T = P_{dis} R_T$$

The dissipated power per unit length, P_{dis}(W/cm) depends on the amount of sun power impinging on the system, and on the optical and conversion efficiencies. In this work, Pdis≅2.06 W/cm for a heat sink anodized in black (thus enabling heat removal by means of radiation) and Pdis≅1.58 W/cm if the heat sink is not anodized.

The interior and the exterior sides of fins do not share the same natural convection coefficient (this explains the utilization in this model of two different resistances for convection, R_c and R'_c). For the latter, the coefficient corresponding to an isolated isothermal vertical fin, can be used [10]:

$$h_{co}(W/cm^2 °C) = 4.4138.10^{-4} \left(\frac{\Delta T(°C)}{d(cm)} \right)^{1/4}$$

Notice h_{co} is almost insensitive to ΔT (which is raised to a power of 1/4). Thus, we consider h_{co}=7.1 10^{-4} (W/cm^2°C) (ΔT=33°C and d=5cm) constant and ΔT independent. The same number is used in the mirror, that, as we mentioned above, behaves as an isolated isothermal fin of effective length L'<L, which depends on mirrors length L and thickness D [10]:

$$L' = \frac{Dm\ th(mL)}{2h_{co}\rho} = 18.9\ cm$$

being ρ the thermal conductivity of the aluminum and m=(2ρh_{co}/D)$^{1/2}$. In this work, L'=18.9cm.

Proper thermal analysis should more carefully consider the less efficient convection in the gap between fins, caused by the difficult circulation of both incoming fresh air and outgoing warm air. The convection coefficient in this case, h_c, is deduced with the method of reference [11] after an extension to a 3-dimensional model for polar axis tracking. The result is the following expression for h_c:

$$h_c = \frac{4K}{w}\left(1 - \frac{d}{2d_a} \right) \qquad d < d_a$$

$$h_c = \frac{4Kd_a}{w2d} \qquad d > d_a$$

with 4K=10^{-3} and:

- $d_a = 0.2585w^4 \cos\gamma\Delta T$
- w = 0.75cm (gap between fins)
- $\cos\gamma = \cos\phi\cos\theta$ (being θ the time angle and φ the latitude)

Since h_c is obtained from two different expressions which depend on l_a and this latter is function a of ΔT, that, again depends on h_c, the value for the convection coefficient have to be found by means of numerical calculus, which leads to h_c=3.54 10^{-4} (W/cm^2°C).

Finally, regarding the contact resistance in the mirror-heat sink interface $R_{contact}$, it is worth stressing that, for a surface roughness of 1.65μm(rms) -quite reasonable in polished surfaces-, and a contact pressure of 10N/cm^2, a value of $R_{contact}$≅2cm^2/W°C may be expected[10] (for aluminum Al75S-T6). Considering a contact area (per unit length) of 1cm^2, and P_{dis}<2W/cm, the temperature drop in the interface will not exceed 1°C.

The result is an estimated temperature drop from cells to air of ΔT≅33.4° with not-anodized heat sinks (R_T≅20.5°C cm/W) and ΔT≅41.3° for the anodized heat sink (R_T≅20.06°C cm/W). Therefore, cooling by means of radiation is useless, due to the increased heat we have to remove in this case.

3. ANALYSIS OF THE STRUCTURE

As aforementioned, the supporting structure takes advantage of the fact that the whole of the optical components, the solar cells and the heat sinks form a single piece in a DSMTS. From the mechanical point of view, the system consists of four identical aluminum profiles (mirrors) placed in parallel with the aluminum heat sinks acting as main supporting beams, and only two more crossed steel beams that provide stiffness in the East to West axis (Fig. 3). Since polar axis tracking will be used, the system is tilted an angle φ identical to the latitude using two pillars, placed at both ends of the whole set N-S symmetry axis (which coincides with the system axis of rotation). Owing to simplicity, these two pillars have been considered as infinitely stiff in the structural analysis.

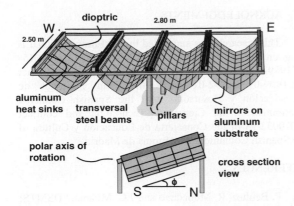

Figure 3 Four identical D-SMTS are attached in this approach system. The heat sinks are also the main beams of the supporting structure and two crossed East to West beams must be added to provide enough stiffness. Since polar axis tracking has been chosen, two pillars are used to tilt the system an angle identical to latitude.

The finite-element software ANSYS has been chosen for the structural analysis of such system. The creation of a model for the structure, basically consisting of a set of points and links between them plus boundary conditions like, for instance, degrees of freedom in those points, is needed to complete the program input. A detailed description of the materials involved is a must as well. This 2.8×2.5 meters array of about 24 Kg/m^2 (without pillars) has been submitted to gravitational (system's own weigh) and wind loads. The application of loads results in a deformation of the model that can be analyzed via nodal-displacements. Obviously, nodal displacements cause angular misalignment, which depends on direction of incoming sun rays besides.

Let x be the coordinate indicating the position along the polar axis of rotation, whereas z and y have been chosen perpendicular to the concentrators entry aperture at noon and parallel to the E-W axis, respectively. In this fixed global coordinate system, p, q, and r stand for the direction cosines referred to the x, y and z axes respectively. Since $r^2 = 1 - p^2 - q^2$, the direction of incoming sun rays **sun**(p,q) at a certain instant can be expressed as a function of only p and q.

The evaluation of misalignments will be carried out locally in a few sections $x = x_i$ of each concentrators entry aperture (Fig. 4). Let the section x_i be the one defined by two nodes x_{ki} and x'_{ki} of a concentrator k, (note the straight line linking these nodes is parallel to the E-W before application of any load, including the system own weight). The unitary vector $\mathbf{a}_{ki}$, the rest of the coordinates of nodes x_{ki} and x'_{ki} after displacement, is the key for the calculation of misalignment ζ_{ki} of section x_i. Indeed, the misalignment is the angle between vectors $\mathbf{a}_{ki}$ and $\mathbf{p}_i$ (a vector perpendicular to **sun** and contained in the plane of sun rays that passes through x'_{ki}). Therefore

$$\sin(\zeta_{ki}) \cong \zeta_{ki} = \cos(90 - \zeta_{ki}) = \mathbf{a}_{ki}\,\mathbf{sun}$$

Because of the symmetry, the misalignments in three sections (at the top, middle and bottom) of each trough have been found from simulation of only a half-day.

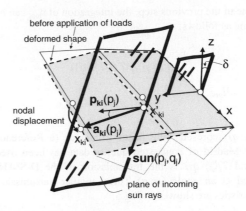

Figure 4. The application of loads result in a deformation of the shapes that can be analyzed through nodal displacements. The evaluation of misalignments will be carried out locally in a few sections $x = x_i$ of the deformed shape. Notice nodes do not remain necessarily at the plane of sun-rays (δ = declination) after displacement.

As an example, three systems with only difference the thickness (2, 3 and 4 mm) of steel on transversal hollow beams of square section (4 cm on a side) have been checked. We found that, at equinoxes, the maximum misalignment of the lighter system (#40.2) never exceeded $0.412°$ under gravitational loads (the worst case occurs when the concentrators face noon). The system has also been submitted to adverse wind loads in order to check if it remains within operation limits even then (wind velocities of 30Km/h), and in this case ζ_{ki} is always below .52 degrees (the most adverse conditions occur with wind impinging on the left-hand side of mirrors when the system faces morning). These angles are negligible compared with the acceptance angle of the D-SMTS, thus we can conclude that they are the thermal nor the structural demands who impose the geometry of these heat sinks. For operation conditions are widely assumed as more restrictive, dimensioning against break-down under heavy loads has not been carried out to date.

4. YEARLY-AVERAGED OPTICAL EFFICIENCY.

This contribution proposes the following two steps to predict the system performance over a whole year:

- For a given number of tracking positions p_j, calculate the vector $\mathbf{a}_k(p_j)$ that leads to maximum misalignment in each concentrator. Taylor $\mathbf{a}_k(p)$ through interpolation of resulting data. Notice the biggest misalignment will be often a consequence of the biggest nodal displacement, although the former depends also on **sun**(p,q). Also, notice that we are assuming that the less-illuminated cell limits the performance of each concentrator.

- Let the yearly-averaged optical efficiency η_{opt} be the ratio of sun power impinging on cells to the sun power impinging on the apertures of the four concentrators. As an example, this approach considers parallel connection between concentrators and also between the two cells rows of a concentrator. Once we have tailored a dependence of the misalignments with time in the previous step, the integration of η_{opt} can be done as follows:

$$\eta_{opt} = \frac{\sum_{k=1}^{4} \int_{p,q} R(p,q)T(\zeta(p,q))}{4 \int_{p,q} R(p,q)}$$

$R(p,q)$ represents the beam irradiance (a Reference Year of Madrid of contrasted quality data has been used [12]) and $T(\zeta(p,q))$ is the transmission of the D-SMTS evaluated at an angle $\zeta(p,q)$ Three angular transmission characteristics are showed in Fig.5.

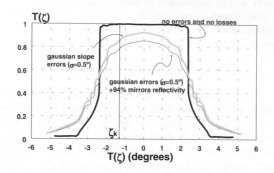

Figure 5. Ray tracing is used to determine the transmission characteristics of the D-SMTS. Gaussian slope errors of $\sigma=0.5°$ in all profiles and reflectivity losses (6%) on mirrors have been considered where indicated.

The calculation of η_{opt} with the loss-less and error-free shapes is meaningless. However, the assumption of errors and losses on profiles permit a more realistic calculation and, furthermore, shows how tolerant to errors is this system. Indeed, the predictions are an optical efficiency of about 93.1% for the case of surfaces with gaussian slope errors ($\sigma=0.5°$) on optical profiles and this value decreases down to 86.08% if we also consider 94% efficiency of mirrors reflectivity.

5.   CONCLUSIONS

In the D-SMTS, the duties of the structural, thermal and optical elements merge with the main consequence of improvements in both simplicity and performance. The high tolerance to errors allow for low-accurate mirrors. A work addressing the appropriateness of elastic conformation of these profiles has already commenced (supported under contract A9901 with the Universidad Politécnica de Madrid).

The lightness of the system also promises low-cost. However, the perception is that only installations in the range of 100kW-10MW can compete with the flat panel[13]. This suggest that a study addressing the cost-effectiveness of the D-SMTS approach should involve larger-sized systems.

6.   ACKNOLEDGEMENTS

This research was supported by the European JOULE program under contracts JOR3-CT98-0252, ERK5-CT1999-00012 and ERK56CT1999-00002, , contract TIC1999-0932-CO2-01 from the Plan Nacional de Investigación y Desarrollo with the Spanish Comisión Interministerial de Ciencia y Tecnología and contract 07T/0032/98 of the Consejería de Educación y Cultura of the Spanish Comunidad Autónoma de Madrid.

REFERENCES

[1]   P. Benítez, R. Mohedano and J.C. Miñano, "DSMTS: a novel linear PV concentrator", *Proc. of the 26th IEEE Photovoltaic Specialists Conference*, Anaheim 1997.

[2]   W.T. Welford, R. Winston. *High Collection Nonimaging Optics*, Academic Press, New York, 1989

[3]   J.C. Miñano, Juan C. González. "New Method of Design of Nonimaging Concentrators", *Applied Optics*, 31 (1992), pp. 3051-3060.

[4]   P. Benítez, J.C. Miñano, J.C. González, "Single mirror two stage concentrator with high acceptance angle for one axis tracking PV systems", *Proc. Of the 13th European Photovoltaic Solar Energy Conference*, 1995, pp. 2406-2409.

[5]   Elisa Alarte, Pablo Benítez, Juan C. Miñano, "Design, construction and measurement of a Single-Mirror Two-Stage (SMTS) photovoltaic concentrator", *2nd World Conference and Exhibition on Photovoltaic Solar Energy Conversion, Proc.*, pp. 2245-47, Viena, Austria, (1998).

[6]   J.C. Arboiro, G. Sala, J.I. Molina, L. Hernando, A. Luque "Parabolic Reflectors with Efficiencies over 90% for the EUCLIDES Concentrator", *14th European Photovoltaic Solar Energy Conference, Proc.*, pp. 1340-43, Barcelona, Spain, (1997).

[7]   G. Sala, J.C. Arboiro, A. Luque, I. Antón et al., "480 kWpeak EUCLIDES™ Concentrator Power Plant Using Parabolic Troughs", *2nd World Conference and Exhibition on Photovoltaic Solar Energy Conversion, Proc.*, pp. 1963-68, Viena, Austria, (1998).

[8]   A. Rabl, *Active Solar Collectors and Their Applications*, pp. 202-208. Oxford University Press, New York, 1985.

[9]   Pablo Benítez, Rubén Mohedano, Juan C. Miñano, "Manufacturing tolerances for nonimaging concentrators", *Nonimaging Optics: Maximum Efficiency Light Transfer IV*, Roland Winston, Editors, Proc. SPIE 3139, pp. 98-109, San Diego, US, (1997).

[10]   Alan J. Chapman, *Transmisión del calor*, (Ed. Bellisco, Madrid, 1990)

[11]   Antonio Luque, "Passive cooling of the Euclides", Internal report, IES, (1994).

[12]   M.H.Macagnan, E. Lorenzo and C. Jiménez, "Solar radiation in Madrid", *Solar Energy*, Vol. 16, pp. 1-14 (1994).

[13]   R. M. Swanson, "The promise of concentrators", *Prog. Photovolt. Res. Appl.*, **8**, pp. 93-111 (2000).

IMPACT OF LOAD CONTROL ON OPTIMAL SIZING OF STAND-ALONE PV SYSTEMS

A. Calì   S. Conti   G. Tina

Dip. Elettrico, Elettronico e Sistemistico,
Università di Catania
Viale A. Doria n. 6, 95125 Catania, Italy
E-mail: sconti@dees.unict.it

ABSTRACT: The utilisation of the solar energy source in domestic applications requires a reduction and a regulation of electricity consumption; in this perspective, especially in off-grid PV systems, the Load Control can contribute significantly to limit system costs reducing, for example, the required PV surface and the capacity of the battery banks.
The aim of the present study is to assess the influence of the application of Load Control techniques to optimal sizing of stand alone PV systems with energy storage. To evaluate the benefits coming from the proposed methodology, a Fuzzy Logic based multi-objective optimisation is developed and applied to a practical case.
The optimisation procedure is set out to find the best compromise between the total cost and the long term average performance of the PV system.
Keywords: Stand-Alone PV System - 1: Sizing - 2: Demand-Side - 3

1. INTRODUCTION

Many economical and environmental advantages, such as the reduction of green house effect, lead to a continuos increase in the number of civil applications, totally or partially based on renewable energies.

However, power systems supplied exclusively by stand-alone generating systems based on renewable sources cannot rely on a high level of availability like grid-supplied power systems do. Because of the intrinsic variability of some kind of renewable energy sources (e.g. wind and solar radiation), the energy generation is necessarily discontinuous: the use of diesel/wind/photovoltaic hybrid systems (only where the climate allows it), or, more commonly, the use of electrochemical energy storage systems is a possible solution to this problem.

Ideally, each system component as well as the energy supply system as a whole should be designed to meet continually the peak load. Obviously, this is not an economical solution as the actual power demand varies in a quite wide range of values, especially in distribution systems of relatively small size (with a low number of customers).

More over, besides the stochastic nature of the load pattern, the power supplied by the renewable sources undergoes stochastic changes too. In this context the possibility of "moving" non time-critical loads in order to match the load profile to the stochastic supply pattern of the renewable sources is very attractive, especially in the application area of optimal design procedures.

Although Load Management has been mentioned as an important research and development issue in literature [1], in recent years only limited research efforts have been made on this subject, whereas, in the 80's, the importance of Load Control, in particular, has been addressed by various studies [2] [3] [4], most of which are mainly focused on a particular kind of Load Control technique called Load Scheduling.

Nowadays, in view of a further enormous progress in electronic controls, Load Management techniques should not be overlooked to assure an efficient operation of renewable energy systems and cost-effective load control.

The aim of the present study is to estimate how the combination of different Load Control techniques (*scheduling, duty-cycle limiting, set-point moving, interlocking, demand-clipping, demand-regulating* and *load factor managing*) can influence the sizing optimisation of stand alone PV systems with energy storage.

A wide-ranging review of the issues related to Load Management is carried out in Section 2.

The general procedure proposed to perform the present study is described in Section 3, showing the interactions between the routines designed to implement the optimal sizing process and the load profile reshaping, respectively.

In Section 4, the routine developed for load reshaping is presented; the considered routine takes as inputs typical load profiles that are modified as a result of the application of Load Control.

In Section 5, the reshaped load profiles are used as inputs to a Fuzzy Logic based multi-objective optimisation procedure, that determines the optimal Solar Cell Array (SCA) power output and the storage system capacity required to supply electric energy to the load trough an autonomous Solar Energy System (SES).

Finally, in Section 6, a numerical example of SES optimal sizing, based on different load profiles, is presented to show the cost-effectiveness of the proposed approach. The results are, then, commented in Section 7.

2. LOAD MANAGEMENT STRATEGIES

The general term *Demand Side Management* (DSM) can be defined as design and implementation of all those activities aimed at modifying customers' energy demand in order to achieve the greatest advantages for both utilities and customers. In some cases the owner of the distribution system and the customer can coincide as in the case of some industrial distribution systems or of stand-alone renewable energy systems. *Load Management* (LM) can be considered as one of the various activities included in DSM. Normally, LM is carried out by power utilities to reshape load profile in order to obtain matching between customer's cyclic power demand and utilities' generation resources, as well as transmission and distribution resources. *Load Control* (LC) can then be considered as a subset of the activities included in LM. It is intended to act directly on customers' appliances so as to modify their load curve.

The daily load curve of a small power installation (such as a domestic one) results from the sum of the load curves of different appliances. Based on the feasibility of different load control actions, appliances can be classified as follows:

- *thermo-dependent appliances* (e.g. air-conditioners, electric water-heaters, refrigerators, freezers);

- *fixed cycle appliances*, (e.g. dish-washer and washing machine);

- *service appliances* (e.g. TVs, ovens, cleaning machines, lights, PCs).

Where LC is feasible, it is possible to obtain the following global load shape modifications: *peak clipping, valley filling* and *load shifting*. In Fig. 1 a schematic explanation of the aforesaid LC strategies is shown.

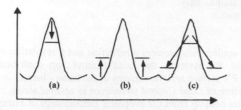

Figure 1: Types of load shape modification: (a) peak clipping, (b) valley filling, (c) load shifting.

Most systems required to implement LM techniques are installed at the customers' facilities to control the operation of one or more customer applications. Different types of LC are possible:

☐ *Direct Load Control*: the electric utility exercises a remote control on customer loads connected to the electric distribution system;

☐ *Local Control Concept*: local control actions are made by the controller at the customer's using only locally acquired information. There is no communication system connecting the utility and the customer.

☐ *Distributed Control*: in a distributed control system (also known as hybrid control) a local controller responds not only to local conditions but also to signals received from the utility. The utility retains the ability to activate the control action, while the customer retains the option to override or modify the action.

In order to provide a proper evaluation of LM techniques, their characteristics, costs and involved technologies must be taken into consideration along with system load typology and utility fixed and variable costs.

3. THE GENERAL OPTIMISATION PROCEDURE

In order to carry out the present study, some routines have been specially designed to implement the sizing optimisation process and the load reshaping. Their interactions are sketched in Fig. 2.

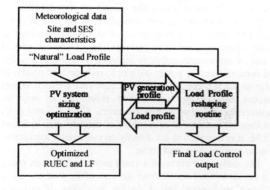

Figure 2: Optimisation procedure - Interactions between optimal system sizing and load reshaping routines.

The generation of a typical and significant load profile, required as initial input, is included in the proposed procedure. The obtained "natural" load profile, that is based on the natural customer power demand, is provided, at the beginning of the procedure, to a Fuzzy Logic based multi-objective *optimisation routine*, that determines optimal design parameters (control variables for the optimisation problem), i.e. the Solar Cell Array (SCA) power output and the storage system capacity required to supply electric energy trough an autonomous Solar Energy System (SES).

The resulting PV generation profile is an input for the *Load Profile Reshaping Routine*; the current load profile is modified to agree (as much as possible) to the generation pattern as a result of the application of Load Control. The reshaped load profiles are used as a new input to the optimisation program, that determines new values for the design parameters and, consequently, a new profile generation. The procedure carries on with data exchange between the sizing optimisation routine and the load reshaping routine, as long as significant improvement can be obtained in the objectives: *Load Fraction* (LF) and *Relative Unit Electricity Cost* (RUEC) (as will be explained in Section 5).

4. LOAD PROFILE RESHAPING

The basic idea is that the more the load profile overlaps the power generation profile of the PV field, the more the cost of the PV system decreases. Generally, an optimal sizing procedure deals with given load profiles that can not be modified and sometimes it is possible to specify a statistical variation of load conditions.

In the present study, load profile is introduced as a sort of new control variable thus increasing the problem complexity. To be more specific, load profiles are subject to an optimisation procedure performed by a dedicated routine for load profile reshaping; this routine is distinct from the main sizing optimisation procedure to keep more visible the result of the load profile optimisation and to have a direct control on constraints. The PV generation profile is an input to the *Load Profile Reshaping Routine*; it implements a non-linear optimisation in which some constraints are posed on the maximum relocated energy and on the numerical difference between the reshaped profile and the pattern; the current load profile is then modified to agree (as much as possible) to the generation pattern as a result of the application of Load Control. The reshaped load profiles are then used as a new input to the main optimisation program.

5. SES OPTIMISATION METHOD

In practical cases, the optimisation problem requires the minimisation of a set of Objective Functions (OFs). This is a complex problem because of the competition among the various objectives and also because they do not require the same priority and accuracy level. In general, it is not possible to define *a priori* a global performance index.

Fuzzy Logic has proven to be a very powerful tool in the definition of decision making schemes; it can be applied to obtain a fuzzy global quality index which is maximised using a scalar optimisation procedure, leading the evolution of the control variables towards the best solution.

As described in [5], the automatic Fuzzy Logic based multi-objective optimisation procedure, applied in this study, determines the optimum power output required from Solar Cell Array (SCA) and the storage system capacity to supply electric energy to consumers by an autonomous SES.

The objectives of the optimisation problem are the maximisation of the supplied *Load Fraction* (LF) and the minimisation of the *Relative* (compared to the cost of grid supply) *Unit Electricity Cost* (RUEC), whereas the control variables are:

the SCA surface (Ac), that strongly influences the overall cost of the SES;

□ the tilt angle (β), on which depends the best compromise between the amount of supplied energy and its uniformity during the year;

□ the storage system capacity (Bc), that has a strong influence on both the overall cost and the fraction of supplied load.

The optimisation process requires the following inputs:

□ site and SES characteristics;
□ load demand;
□ meteorological data;
□ economical data.

The calculation of the supplied LF is carried out using the *Utilizability Method* [7], which relates the solar radiation variability during the day and the year to the power that can be produced. This method is based on systematic experimental studies over the last forty years and allows to make a long-term forecast.

The RUEC evaluation involves the calculation of the *Life-Cycle Costs* (LCCs), given by the sum of all the system costs in its lifetime, expressed in today's money.

This is performed by the Life-Cycle Costing Method [6] in which, as said before, both initial costs and overall future costs (relative to the whole operational life of the PV system) are considered; these costs are then annualised, i.e. are expressed in terms of a cost per year, in order to obtain a comprehensive electricity cost per kWh.

As for the definition of a decision making scheme, to each design objective a *fuzzy* set is assigned, whose membership function (MF) associates to each actual value y of the OF, a normalised value μ(y), which expresses the degree of satisfaction of the considered objective. This procedure allows the various objectives to be described according to different scales of *fuzzy* values.

In the present work, piecewise linear functions are chosen to shape the MFs. They are characterised by two values (y_{acc}, y_{rej}): y_{acc} represents the full acceptability lower limit; y_{rej} represents the unacceptability upper limit (Fig. 3). If the values of each objective are below y_{rej} they are considered unsatisfactory and then they have to be rejected; on the contrary, if they are beyond y_{acc} they are totally acceptable. This two values are fixed according to the design requirements.

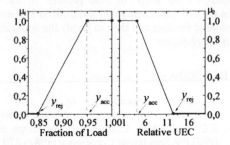

Figure 3: The membership functions employed in the reported example.

6. NUMERICAL EXAMPLE

To show the effectiveness of the procedure developed in this study, the proposed method is applied to a practical case for the optimum design of a PV system sited in Crotone, Italy. The characteristics of the system are given in Table I.

The load profiles provided as input to the optimisation algorithm are typical of a stand-alone system supplying residential loads; this load profiles have been built up by a

dedicated program based on various parameters such as social and economical conditions, habits of the family members, type of domestic appliances and so on, in order to take into account the dependence between load profiles and the kind of concerned users and appliances; the *family-customer* is composed by four members: an employed, a housekeeper, a boy or girl, a child. As for the installed appliances, two significant cases related to the kind of installed lamps are considered: low efficiency lamps (Case1) and high efficiency lamps (Case2), as shown in Fig. 4 and Fig. 5, respectively.

Table I: Site and SES characteristics.

Parameters	Values
Reference array efficiency	R = 0.12
Temperature Coefficient	$\gamma = 0.004$ [K^{-1}]
Cover transmittance	$\tau = 1.0$
Absorptance	$\alpha = 0.88$
Efficiency of PCS equipment	$\eta_{pc} = 0.88$
Battery storage efficiency	$\eta_b = 0,85$
Ground reflectance	$\rho = 0.2$
Array tilt range	$20°<\beta<60°$
Location	Crotone (Italy)
Latitude	39° 04' N
Distance from the grid	d = 1000 [m]

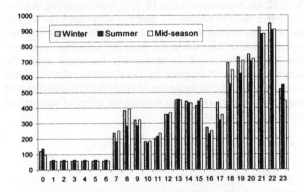

Figure 4: Load profiles for residential loads (Case 1).

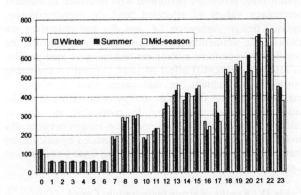

Figure 5: Load profiles for residential loads (Case 2).

Meteorological data concerning temperature and solar radiation at the considered site are shown in Fig. 6.

In Table II the economical data required by the performed evaluation are also provided.

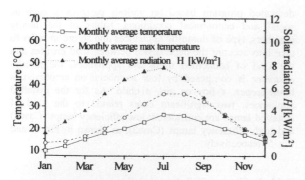

Figure 6: Meteorological input data.

Table II: Economical data

Parameters	Values	
Period of analysis	20	[years]
Excess inflation	0%	
Discount Rate	10%	
Battery price	100	[€ / kWh]
PV Array price	600	[€ / m²]
Electrical energy price	0.2	[€ / kWh]

The characteristic values of the membership functions (MFs) defined for the two optimisation objectives (LF and RUEC) are reported in Table III. It is worth pointing out that designers have to choose y_{rej} and y_{acc} very carefully: in fact, in order to obtain the best results it is important that the optimisation system works far from the saturation points of the MFs. This remark is particularly important when the acceptable level of supplied LF (load related reliability) has to be defined. The designer has to study the characteristics of the load and has to take into account the effect of automation system on the power demand.

Table III: Characteristic values of the MFs

	y_{rej}	y_{acc}
LF	0.85	0.95
RUEC	12	2

Tables IV and V summarise the results of the optimisation procedure performed with reference to different percentage values of relocated load. In correspondence, optimal design parameters (A_c and B_c) and objectives are shown. The simulations have been carried out assuming the possibility to perform a complete reshaping of load profiles to assess the potential effect of Load Control on the sizing optimisation. Actually, in real cases, there is always a maximum amount of feasible energy relocation due to the kind of installed appliances. In particular, in case of residential loads, the feasible relocation ranges between 6% and 20% [8]. In this study, we have estimated by calculation that the maximum load shifting is 15% in Case 1 and 20% in Case 2.

Table IV: Results of sizing optimisation: Case 1.

Relocated Load	A_c [m²]	B_c [Wh]	LF*10	RUEC
99%	43.23	2464	9.37 (+3.09%)	4.96 (-29.69%)
50%	43.91	11202	9.31 (+2.46%)	5.39 (-23.60%)
30%	48.67	12929	9.23 (+1.51%)	6.04 (-14.40%)
20%	57.34	5220	9.12 (+0.39%)	6.80 (-3.66%)
15%	57.52	5914	9.11 (+0.30%)	6.85 (-2.85%)
0%	58.03	10776	9.09 (Base value)	7.05 (Base value)

Table V: Results of sizing optimisation: Case 2.

Relocated Load	A_c [m²]	B_c [Wh]	LF*10	RUEC
99%	36.34	4304	9.39 (+3.14%)	4.78 (-30.70%)
50%	42.50	4653	9.28 (+1.93%)	5.59 (-18.96%)
30%	42.56	5017	9.28 (+1.92%)	5.63 (-18.38%)
20%	44.89	5384	9.23 (+1.32%)	6.00 (-12.91%)
15%	45.10	7131	9.21 (+1.17%)	6.11 (-11.42%)
0%	51.71	5091	9.11 (Base value)	6.89 (Base value)

In the first case, concerning the use of low efficiency lamps, the minimum amount of relocated load that leads to substantial improvement in the objectives is 30%, that is much higher than the assessed limit (15%). However, a relocated load of just 15% would provide a cost saving which is enough to acquire a small control system. In this perspective, the overall PV system cost would remain unchanged but the system reliability would be improved (+0.3%). On the other hand, in the second case with high efficiency lamps, the 15% of relocated load is enough to provide technical (LF = +1.17%) and economical (RUEC = -1.42%) benefits that justify the use of a load control system.

7. CONCLUSIONS

In this work the impact of load control on optimal sizing of stand-alone PV systems has been investigated. This complex task has been accomplished developing a Fuzzy Logic based multi-objective optimisation procedure for the design of PV systems with battery storage in order to determine the best compromise between cost and reliability.

More over a specific routine for optimal load reshaping has been developed. The overall procedure has shown to be flexible enough to cope with the intrinsic uncertainties related to PV systems and to load demand.

The application of the proposed method to a practical case has allowed us to verify that load control applied to PV systems is indeed cost-effective.

It is to be highlighted that the minimum amount of relocated energy able to provide economical benefits depends on the shape of load profiles. In the considered cases the estimated minimum threshold is low enough to be considered feasible and then to justify the application of load control techniques.

8. REFERENCES

[1] T. Schott, Proceedings of the 12th European Photovoltaic Solar Energy Conference (1994) 1996.

[2] R.C. Cull, A.H. Eltimshay, Proceedings of the 16th IEEE PV Specialists Conference (1982) 270.

[3] P.P. Groumpos, G.D. Papageorgiou, Proceedings of the 17th IEEE PV Specialists Conference (1984) 647.

[4] P.P. Groumpos, K.Y. Khouzam, L.S. Khouzam, Proceedings of the 20th IEEE PV Specialists Conference (1988) 1164.

[5] G. Tina, G. Adorno, C. Ragusa, Proceedings of the 33rd Intersociety Engineering Conference on Energy Conversion (1998) ID IECEC-98-I162.

[6] T. Markvart, Solar Electricity, John Wiley & Sons Ed. (1994).

[7] D. R. Clark, S. A. Klein, W. A. Beckman, Journal of Solar Energy Engineering, Vol. 105 (1983) 281.

[8] M. Rehm, G. Bopp, E. Rössler, 13th European Photovoltaic Solar Energy Conference (1995) 938.

EFFICIENCY MEASUREMENTS AS AN ADDITIONAL CRITERIA FOR BATTERY SELECTION

S. METAIS, D. DESMETTRE, J.L. MARTIN, F. MATTERA, Ph. MALBRANCHE

GENEC, CEA Cadarache, 13108 Saint-Paul-Lez-Durance, FR. Email : philippe.malbranche@cea.fr,
phone : 33 4 42 25 66 02, fax : 33 4 42 25 73 65.

ABSTRACT : In PV systems, efficiency is an important criteria. A lot of research projects are carried out in order to increase efficiency of PV modules or the whole system. What about batteries ? Lead-acid batteries, the more commonly used technology, have fluctuating faradaic (ranging from 90 % to 95 %) and energy (ranging from 80 % to 88 %) efficiencies. In relation to these variations, efficiency measurements could be another criteria to choose the battery which can restore stored PV energy as best as possible. At present time, few efficiency test procedures exist : one, for example, is included in the French standard NFC 58-510 procedure. However, in this paper, only one measurement is performed. The procedure which has been developed consists in deep cycles (50 % of the initial capacity) at low state of charge (between 0 % SOC and 50 % SOC) in order to avoid gassing, which decreases efficiency value. Moreover, low state of charge is the operating range for which efficiency is of paramount importance. The number of cycles was selected in order to avoid electrolyte stratification phenomena and to achieve plates formation. The results cover various types of lead-acid batteries.
Keywords : Glasgow Conference - 1: Storage - 2 : Batteries - 3 : Test procedure

1. DEFINITIONS

Generally, two types of efficiencies are considered :

$$\text{Faradaic effciciency (Ah efficiency)} = \frac{\text{Discharge capacity (Ah)}}{\text{Recharge capacity (Ah)}}$$

$$\text{Energetic effciciency (Wh efficiency)} = \frac{\text{Disch. capacity(Ah)} \times \Sigma \text{Disch. potential}}{\text{Rech. capacity(Ah)} \times \Sigma \text{Rech. potential}}$$

In this work both faradaic and energy efficiencies have been calculated.

2. EXPERIMENTAL DETERMINATION OF A REPEATABLE EFFICIENCY TEST PROCEDURE

2.1. Preliminary tests

This efficiency test procedure has been performed in low state of charge, because of the following points :

- Gassing of the electrolyte takes place at high state of charge, this phenomenon is not taken into account in this test.
- The recharge conditions of photovoltaic batteries are linked to the prevailing weather conditions, bad weather conditions will prevent the battery from being recharged leading to a low state of charge of those batteries (near 50 % of state of charge). It is then important to obtain "low state of charge" efficiency values at these conditions.
- The efficiency depends on the previous history of the battery ("memory effect"), several cycles must be performed in the

same conditions in order to have access to the real efficiency values at a given state of charge of the battery.

The efficiency test procedure chosen at the beginning if the experiment is as follow :

- Initial cycle :

- recharge until 100 % of SOC,
- discharge at 0.1 C/10 (= initial capacity), until 1.8 V per cell (= 100 % of SOC).

- Cycling :

- recharge until 50 % of the initial C/10 capacity value,
- discharge at 0.1 C/10 (initial capacity) until 1.8 V per cell.

This cycle is performed several times. The proposed efficiency test procedure is presented Figure 1.

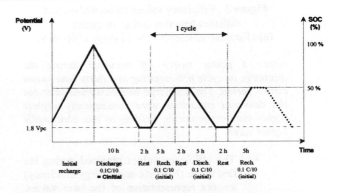

Figure 1 : Cycling conditions of the efficiency test procedure

This test procedure has been enhanced in our laboratory facilities.

2.2. Refinement of test conditions

Definition of cycle number

The efficiency test procedure has been applied to four batteries over ten cycles in order to be able to later define the minimum number of cycles required. Results are shown in Figure 2(a) and Figure 2(b).

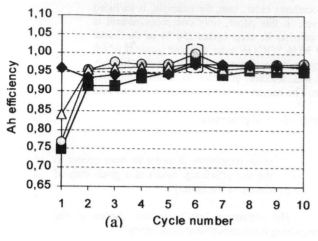

(a)   Cycle number

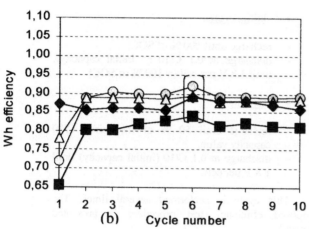

(b)   Cycle number

Figure 2 : Efficiency values calculated on four different batteries during 10 cycles.
(a) : Faradaic efficiency ; (b) : Energy efficiency.

Note : A power switch off occurred during the discharge on cycle n°6 resulting in a thirty hours open circuit period. The discharge was continued later but the delivered capacities were consequently slightly higher than other cycles, leading to this abnormally higher value.

- The efficiency values obtained during the first cycle (Faradaic and Energy efficiency) are not representative of the later values. Therefore, an efficiency characterisation is not possible in a one-step measurement and several cycles must be made.

- Efficiency values are nearly stable from the 2nd cycle. It is not necessary to perform all ten cycles, efficiency values will therefore be calculated between the 2nd and the 5th cycle only.

Efficiency test reproducibility

The efficiency test procedure has been applied to four identical flat plates batteries. The faradaic and energy efficiency curves are shown Figure 3(a) and Figure 3(b).

When deleting the first cycle efficiency value and extending the scale of the axis, the efficiency diagrams is as follow :

The batteries have the same behaviour regarding faradaic and energy efficiency values. With this optimised test procedure, a comparison has been be made between one battery of each type.

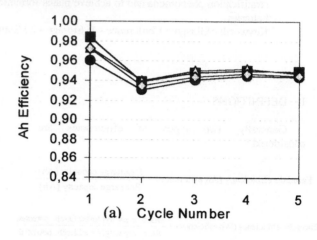

(a)   Cycle Number

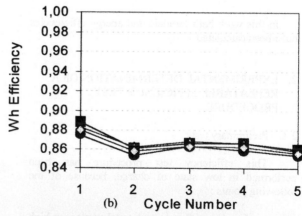

(b)   Cycle Number

Figure 3 : Efficiency values of four identical flat planes batteries (cycle n°2 to n°5).
(a) : Faradaic efficiency ; (b) : Energy efficiency.

3. EFFICIENCY TEST PROCEDURES

Different results have been obtained by applying the low state of charge efficiency test procedure on several types of batteries.

Description of batteries

This efficiency test procedure has been carried out on 17 different batteries :
- various types of flooded (14 batteries, 8 flat plates and 6 tubular plates batteries),
- AGM batteries (2 batteries),
- one gelled electrolyte battery.

These batteries are composed of tubular or flat positive plates and are made by several manufacturers.

Flat plates

The efficiency test procedure has been applied to ten batteries composed of flat plates, eight batteries are flooded and two batteries are of AGM type. Their rated capacity at I=C/10 are between 80 Ah and 100 Ah and between 82 Ah and 278 Ah at I=C/100 (*these values are given by the manufacturers*). The faradaic and energy efficiency values are shown Figure 4(a) and Figure 4(b).

values remain reasonably constant (difference less than a few %).
- Faradaic efficiency values lay between 0.93 and 1.0.
- Energy efficiency values are between 0.84 and 0.94.
- Both AGM batteries show high efficiency values (between 0.97 and 1.0 for faradaic efficiency values and between 0.91 and 0.94 for energy efficiency values).

Tubular plates

The efficiency test procedure has been applied to seven batteries with tubular plates. Rated capacities at I=C/10 are between 50 Ah and 360 Ah and between 140 Ah and 400 Ah at I=C/100 (these values are given by the manufacturers). The faradaic and energy efficiency values are shown respectively in Figure 5(a) and Figure 5(b).

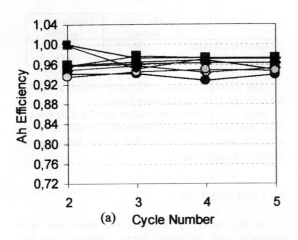
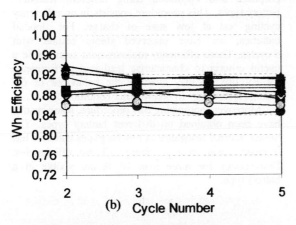

Figure 4 : Efficiency values of flat plates batteries (Cycle n°2 to n°5).
(a) : Faradaic efficiency ; (b) : Energy efficiency.

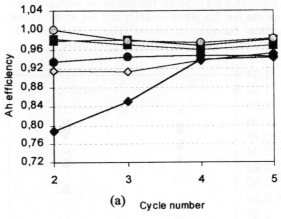
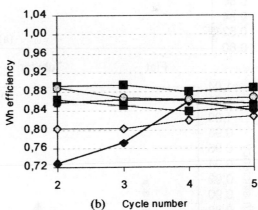

Figure 5 : Efficiency values of tubular plates batteries (Cycle n°2 to n°5).
(a) : Faradaic efficiency ; (b) : Energy efficiency.

- Faradaic and energy efficiency values are almost stable for the most of the batteries, except for one, whose values steadily increased. In such cases, additional cycles have to be performed until two consecutive

- Faradaic and energy efficiency values are almost stable for all the batteries, except for one battery.
- Faradaic efficiency values lay between 0.91 and 1.0, only one battery exhibited lower gains in both Ah and Wh efficiencies. The efficiency values of this battery begins at 0.79 (2nd cycle) and increased up to 0.93. This low value may be related to the active material

which is not entirely formed at the first cycles.

- Energy efficiency values lay between 0.80 and 0.89, except for one batteries. The efficiency values of this battery are steadily increasing from 0.73 to 0.84.

Comparison between flat and tubular plates

In order to compare the faradaic and energy efficiencies of flat and tubular battery plate designs, an average efficiency value has been calculated between the 4th cycle and the 5th cycle. This range has been selected because the corresponding efficiency values are mostly stable. The cycling conditions of this efficiency test (between 0 % and 50 % of SOC) prevents gassing and this eventually lead to stratification of electrolyte after a high number of cycles, therefore, we propose to perform only four efficiency test cycles. Furthermore, this low number of cycles will allow the test to be completed in a short time (14 h x 4 = 56 h). The efficiency average values of the ten flat plates batteries and the seven tubular plates batteries are shown Figure 6(a) (faradaic efficiency) and Figure 6(b) (energy efficiency).

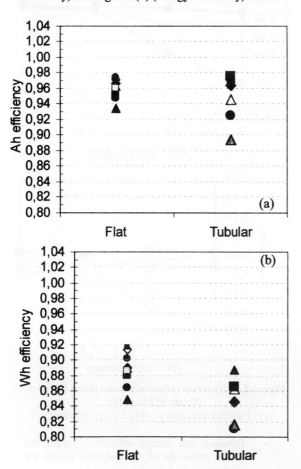

Figure 6 : Efficiency average values (between the 4th and 5th cycle) of flat and tubular plates batteries.
(a) : Faradaic efficiency ; (b) : Energy efficiency.

- Faradaic efficiency values lay between 0.93 and 0.97 for flat plate batteries. Lower faradaic efficiency values are generally obtained on tubular plates battery (between 0.89 and 0.98).
- Energy efficiency values are between 0.84 and 0.92 for flat plates battery. Lower energy efficiency values are obtained on tubular plates batteries (between 0.81 and 0.89).
- The lower efficiency values obtained on tubular plate designs is due to the higher ohmic resistance of this type of battery. The tubular plate battery will then reach the discharge threshold earlier than flat plates batteries, leading to a lower discharge capacity and hence to lower faradaic efficiency value.
- The average values for both types of batteries are shown Table 1.

Table 1 : Average values of efficiency recorded between the 4th and 5th cycle.

	Flat plates batteries	Tubular plates batteries
Faradaic efficiency	0.96	0.96
Energy efficiency	0.88	0.85

The average faradaic efficiency values is similar for both flat plates and tubular plates.

Both energy efficiency values show a difference of 3 %, the higher internal resistance of tubular plates batteries compared to the flat plates batteries can explain this difference.

4. CONCLUSIONS

In this work, a new test procedure has been proposed and optimised using different lead-acid battery types. This procedure consists in an efficiency cycling test at low state of charge. For flooded batteries, we have observed stable results with 5 cycles. These efficiency measurement seem to be a useful criteria to characterise lead-acid performance for PV use : it may evaluate the behaviour of the storage technology in bad weather conditions.

Moreover, optimising this procedure, some results have been obtained on different battery types. The higher internal resistance of tubular plates batteries has induced lower average energy (3 %) efficiencies. Discrepancy for more than 10 % are visible in a battery type.

NEW POWER CONDITIONING UNIT INCORPORATING CHARGE CONTROLLER, ENERGY FLOW MONITOR, DATA LOGGER, DC/AC CONVERTER FOR STAND ALONE AND COMBINED PV-DIESEL OPERATION

Stefan Krauter, Rodrigo Guido Araújo
Laboratório Fotovoltaico, UFRJ-COPPE/EE, Caixa Postal 68504
Rio de Janeiro, 21945-970 RJ, Brazil, http://www.solar.coppe.ufrj.br
Tel.: +55-21-2605010, Fax: +55-21-2906626

ABSTRACT: PV often has to be operated in conjunction with conventional generating units as diesel generators. AC output is often required due to lack of DC consumer equipment. In many locations a connection to the public electrical grid exists, but supply is unstable. Therefore the power conditioning unit should be able to synchronize to the grid when it is working and switch immediately to an autonomous operation mode (supplied by a battery), as soon as supply is getting unstable or is lacking. The purpose of this work is to combine different units (MPPT, advanced charge controller using charge balance, an energy flow monitor with a data logger and a single phase DC-AC converter which works under stand-alone and in combination with other generating sets) to one system. MPPT and charge control are already built and tested, the integrated inverter is in the stage of development .

To adapt the unit to different conditions and applications, a computer control for MPPT, charge controlling, energy balance monitoring and data logging was chosen. The computer also is also serving as a direct control of the power electronic devices. The design is relatively simple to ensure low costs and high reliability, but also features high efficiency. The 300 Hz PWM switching is carried out by a single IGBT device. The voltage of the DC bus at the output of the MPPT is controlling the energy flow into and from the battery. A fuzzy control strategy is compared to conventional approaches.

Keywords: Hybrid system –1: MPPT – 2: Charge control –3

1. INTRODUCTION

Due to higher production and lower prizes of PV modules, the market for PV applications is increasing, especially in remote locations. Still PV often has to be operated in conjunction with conventional generating units as diesel generators, due to high peak demands or periods of low irradiance. AC operation is very often required due to the lack of equivalent DC consumer equipment.

In many locations a connection to the public electrical grid exists, but supply is unstable and blackouts occur very often. For these cases the power conditioning unit should be able to synchronize to the public grid or to the local grid (e.g. consisting of a diesel generator) when it is working and switch immediately to an autonomous operation mode, as soon as supply is getting unstable or is lacking. Additionally the possibility of PV grid injection during times of grid connection helps to prevent grid overload and the probability of a blackout.

Batteries are the most sensible component of stand-alone PV systems. Most charge controllers used are just considering voltage as an indicator for the state of charge, which is not very appropriate for an optimal operation and a long lifetime of the battery. Therefore an adequate charge controller and/or control strategy should be used.

Often the different power conditioning units are installed separately and are from different manufacturers with different specifications. That increases costs and failure rate of PV systems. Therefore all components should be integrated and monitorable. Nevertheless for the extension of the system, power electronics should be modular and exchangeable while keeping the control units.

2. PROJECT

The purpose of the project is to combine the different units (MPPT, advanced charge controller using charge balance and/or fuzzy control, an energy flow monitor with a data logger and a single phase DC-AC converter which works under stand alone, in combination with other generating sets and as a grid injector) in one system. An intelligent master control unit has to find the best operation mode to achieve a safe, ecological and economical power supply.

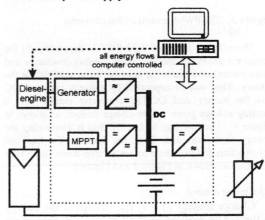

Figure 1: Schematic layout of the system with a DC bus.

To get a versatile construction making it adaptable to different conditions and applications, a computer control for MPPT, charge controlling, energy balance monitoring and data logging is used. This serves also as a direct control of the power electronic devices (IGBT, MOS-

FET). The design is relatively simple to ensure low costs and high reliability, but also features high efficiency.

The use of single units as a standard is limiting the number of external parameters contributing to the system performance, reduces possible faults of installation and makes measurements of different systems more comparable. This helps to quantify photovoltaic energy resources of different locations accurately and ensures electricity supply in remote areas.

2.1. MPPT and DC-DC converter

The 300 Hz PWM switching is carried out by a single IGBT device.

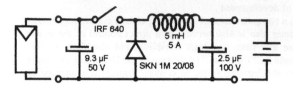

Figure 2: Circuit of converter.

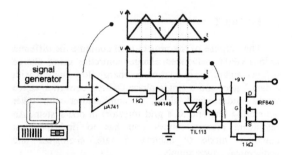

Figure 3: The PWM control of the converter.

The voltage output of the MPPT is depending on the actual conditions of the PV panel (mainly irradiance and temperature) and has to be adapted to be suitable for the battery. This was incorporated by using a common DC bus for battery and DC-DC converter output with a floating voltage given by the charge control, as shown in Figure 4. The voltage of the DC bus is controlling the energy flow into and from the battery. To facilitate construction and to reduce losses a "step-down" converter was used as shown in Figure 2 and Figure 3.

2.3 Charge control

Charge strategy is I-V, the voltage at the battery is limited in general to 2.42 V per cell to prevent excessive gassing, which may damage the electrodes. This threshold voltage is a function of electrolyte temperature, of charging current, and of the construction of the battery and has to be adapted. Optional is a charge control by using the dV/dt method [1].

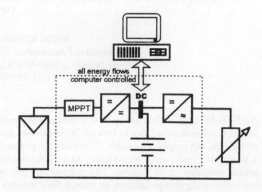

Figure 4: Internal DC bus with variable voltage given by charge control.

The most accurate results are obtained by a density measurement of the electrolyte, which is tested by using an optical refraction method. While the density measurement requires manipulation on the battery and is difficult to handle, a charge balance method is going to be implemented. Under development is an intelligent control strategy, which generates actual properties of the battery (as actual capacity and favorable depth of discharge) by using information from the history of the battery (operation temperature, discharge currents, depth of discharges, charging, cycles, self discharge). Acid density will be measured using an optical refraction method.

2.4 DC-AC converter

The DC-AC converter is in the stage of development. At the moment different topologies are investigated and compared. While a huge amount of preliminary work was already carried out for DC-AC converters either for grid connection only or for autonomous operation only, the combined operation mode is poorly investigated.

2.5 Combined operation

A preliminary study was carried out by Carlos Iriarte [3] to investigate the potential to reduce costs of power electronics used in a hybrid system. That work presents considerations about PV coupling with an AC source, as AC occurs at the diesel generator or at grid connection. While the AC output of the diesel generator (which is in general much more powerful than the PV generator) does not have to be rectified and adapted to the DC bus and then again re-converted to AC (as shown in Figure 5 in comparison to Figure 1), a simplification and cost reduction could be expected. The tested prototype has proven its ability to couple both sources in parallel or independently, without influencing voltage and frequency of the common bus. It turned out that the complexity of the control system increased due to the need to handle phase shifting and frequency variation at different load conditions. Also the layout is difficult to extend for more sources of energy supply (e.g. a wind generator). Therefore it was decided to apply the more versatile layout with a common DC reference bus.

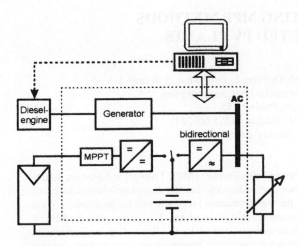

Figure 5: Design with common AC bus.

2.5 Operation control

To determine the threshold to start the diesel generator a fuzzy logic algorithm was used with success [2]. It uses the actual irradiance as a parameter for the decision and allows to save about 20% of operation costs and reduces the wear out of the mechanical parts. While the original layout used battery voltage as a decision parameter only the new system will also use charge balance to determine the state of charge.

3. EXPERIMENTAL RESULTS

The MPPT, the DC-DC converter and a simple charge control are already working. The experimental results of theses devices are shown in Figure 6 using fast changing irradiance conditions to test the dynamic response.

After the initiation period, the system reacts quickly to changing. Due to the change of irradiation also the operation temperature of the module is changing, which causes a shift of the voltage where the maximum power point is occurring. The system did not have any difficulty to follow this change.

Efficiency of the device is that high it could not be quantified with equipment available at the lab. Further investigations with a highly accurate power meter will be carried out soon.

4. CONCLUSION

The system designed offers significant advantages and will lead to a considerable cost reduction of applied combined PV system. Power electronics is kept simple using a minimum of components. Lifetime of battery will be extended due to adaptation of the possible level of discharge to the actual capacity, determined by an intelligent control, using the history of the battery with all its relevant parameters. The versatile construction allow easy extension to the system as well as multiple energy sources.

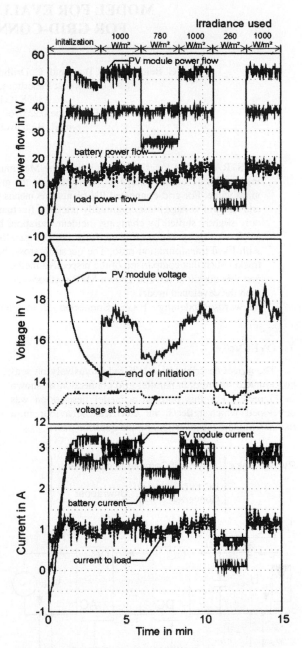

Figure 6: Dynamic response of power flow, voltage and current of PV module, battery and load for three different irradiance levels.

REFERENCES

[1]   A. Araújo, M. Sc. thesis, UFRJ-COPPE-EE (2000).
[2]   R. Rhomberg, Ph.D. thesis, UFRJ-COPPE (1999).
[3]   C. A. Iriarte, M.Sc. thesis, UFRJ-COPPE (2000).

MODEL FOR EVALUATING MPP METHODS
FOR GRID-CONNECTED PV PLANTS

G. Bettenwort, J. Bendfeld, C. Drilling, M. Gruffke, E. Ortjohann, S. Rump, J. Voß
Universität Paderborn, Elektrische Energieversorgung
33098 Paderborn, Pohlweg 55
Telephone 05251/60-2302, Telefax 05251/60-3235
e-mail. Bettenw@eevmic.uni-paderborn.de

ABSTRACT: The functional principle of the Maximum Power Point Tracker (MPP Tracker) is based on matching the PV generator voltage to the voltage on the grid take-over side such that the output power has its maximum value. For grid-connected PV plants this means that the transformation ratio between the generator voltage and the grid voltage must continually track the fluctuating meteorological conditions. It is known that the MPP is only slightly shifted by changing incident radiation, but it is strongly shifted by the fluctuating temperature. A MPP tracking system is efficient if the variable transformation ratio produces greater energy output than a system with fixed transformation ratio. The question arose, how to quantify the efficiency boost of the respective MPP tracker such that comparison of the respective methods is possible under load. This paper is devoted to answering just this question. Several known MPP methods are mutually compared with the help of simulating calculations with the developed model.
Keywords: Modelling - 1: Grid-Connected - 2: Inverter - 3

1. OVERVIEW

The model described here is based exclusively on self-commutated inverters in parallel grid connection as shown in Figure 1. For this purpose a model description was developed which reflects the electrical behaviour in a manner suitable for investigating the question posed.

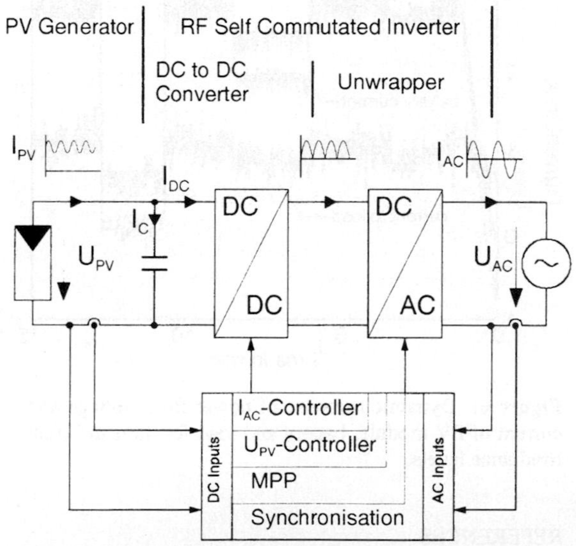

Figure 1: Basic components of a PV inverter

The chief components for evaluating the MPP methods are the models of the PV generator, of the inverter including the current control loops as well as the superimposed control loop for PV voltage maintenance. It is remarkable that we here have a model of the inverter which does not require specific details regarding the construction of the power electronics circuits.

Only the different variants for the MPP tracking must be modelled individually according to the type of method.

Figure 2 shows the structure of the PV inverter model described below.

To permit comparison of the methods under conditions of partial load too, the efficiency factor is calculated by simulations for altogether 6 different irradiation intensities and compared with the theoretically possible maximum power.

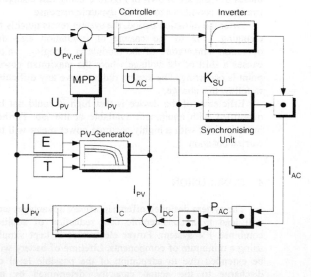

Figure 2: Simulation model of a PV inverter

According to the definition of the European power efficiency factor, the individual power efficiency factors are weighted according to the frequency of occurrence of the corresponding solar energy supply, and then summed. The losses appearing in the inverter are thereby deliberately ignored in the calculation of the balance. This procedure permits direct efficiency assessment of the various methods according to concrete figures.

2. MODELLING

2.1 PV generator

For modelling the PV generator, the two-diode model according to [1] is simplified so that the output variable I_{PV} can be determined from the input variables incident radiation E, temperature T and PV generator voltage U_{PV} without resorting to an iterative method.

2.2 Power section of the inverter

The model of the inverter is decisively simplified by ignoring the resulting effects of the switching actions [2]. This assumes that the current controller and the power section of the inverter implement the setpoints virtually proportionally at all times. The justification for this assumption is that the energy storage device of the power section is designed for the range of the switching frequency. This leads to negligibly small time constants in relation to the much smaller grid frequency.

2.3 Synchronising unit

On account of the missing reactive power capability, the output current of the inverter must always be in phase with the grid voltage and thus must correspond to the grid voltage in the stationary case too. For the application considered here, the grid frequency and phase are taken to be static, so that simulation of the dynamic response of the synchronising device can be omitted.

2.4 Transformer

A possibly existing transformer for matching to the grid voltage level is for simplicity assumed to be ideal and taken into consideration in the model according to the transformation ratio by adaptation of the corresponding parameters.

2.5 Self-commutating PV inverters

Figure 2 shows the remaining structure for the PV inverter model. It is essentially reduced to the power balance of the storage capacitor at the inverter input and that of the power section. Furthermore, the structure contains a number of non-linear elements such as the characteristic curve families of the PV generator and the multipliers for the power balancing.

2.6 Control

The functional principles of the overall system are based on the cascade structure consisting of the control loops

- MPP tracking

- PV module voltage regulator and

- Grid current regulator

whose respective time constants lie far apart (cf. Section 2.2). For the module voltage regulator this means that the underlying current control loop is in nearly settled state. However, the reference variable $U_{PV,ref}$ of the voltage regulator, which is at the same time the output variable of the MPP-tracking, should be changed only slowly. These setpoint changes should not be made so strong that an oscillation with the sampling frequency of the MPP tracking is superimposed on the grid current current.

On practical considerations this means that the PV voltage controller must be designed such that any operating point on the characteristic of the PV generator of the inverter can be set permanently. A selected method of MPP tracking can be investigated only when the stability within the defined input voltage range is ensured for each possible combination of temperature and incident radiation. This is true for the simulation as well as for implementation of a PV inverter.

It is possible in principle to omit a PV voltage regulator only if the power fed into the grid is always kept smaller than the power rating of the PV generator. When this prerequisite condition is fulfilled, the actual operating point never goes outside the stable range. However, fulfilment of this condition entails comparatively large implementation effort which can be avoided with a suitably designed PV voltage regulator.

The MPP methods introduced below are based on the previously made assumption that the PV voltage regulator operates stable. As explained at the outset, the MPP voltage essentially depends on the temperature, so that adjustment of the setpoint of the PV voltage must be very slow. This feature can be exploited in many cases to significantly reduce the circuit complexity for the MPP tracker.

However, this leads to significantly longer simulation times, which were able to be reduced with the model here involved, to the extent that nevertheless numerous simulation runs were made possible and thus power efficiency optimisation of a chosen procedure can be carried out.

2.7 MPP tracking with reference to two examples

The procedure selected below is intended to make clear by example that in addition to the power efficiency as most important criterion, some other criteria also decisively affect the choice of a suitable MPP tracking procedure. Typically the behaviour of the respective procedure can be modified specifically. By varying corresponding parameters, the dynamic behaviour and the precision can be adapted to fulfil the desired specifications.

The resulting implementation effort must be taken into consideration too. For an inverter in the kilowatt rating range, the costs for implementing MPP tracking tend to play only a subordinate role, but for a low power inverter these costs may well bring a decision in favour of a different procedure costing less.

For practical reasons a MPP tracking should operate on the basis of measuring the PV generator voltage. The operating point can then be moved specifically to avoid undesired operating states. An example of this is to shift the operating point towards the open circuit voltage, to avoid otherwise possible overload of the inverter.

Example 1: ΔP/ΔU method

The power characteristic in Figure 3, valid for a pair of values of temperature and incident radiation, serves to explain the procedure.

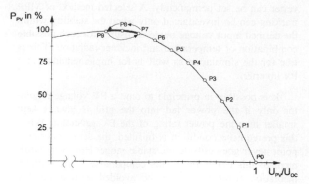

Figure 3: Search procedure

Starting with the operating point P0, the PV generator is in the no load state. When the inverter is connected, specifying a starting point for the voltage setpoint (operating point P1), the MPP tracker starts to maximise the power fed into the grid.

This method checks at fixed intervals whether the module power has changed due to a setpoint variation of the PV voltage (forced disturbance). In this way the operating point is continually shifted closer to the MPP. At the same time intervals, an observation is made whether the module voltage has increased or decreased. By calculating the differential quotient

$$\Delta P/\Delta U = (P_{\text{new}} - P_{\text{old}})/(U_{\text{new}} - U_{\text{old}}) \qquad (1)$$

it is recognised whether the operating point has shifted closer to the open circuit state or closer to the short circuit state. Passage of the operating point through the power maximum (P12) is recognised by the MPP tracker as change of sign of the differential quotient ΔP/ΔU. If the actual operating point lies on the falling branch with the inequality

$$\Delta P/\Delta U < 0$$

the setpoint is lowered accordingly. Otherwise the operating point lies on the rising branch for which the inequality

$$\Delta P/\Delta U > 0$$

holds. As reaction. the module voltage setpoint is lowered. In the vicinity of the MPP the sign of the differential quotient alternates in rapid sequence. This is a stable boundary cycle (P7 => P8 => P9 => P7 ...).

Example 2: Characteristic curve method

An advantage compared with the method in the first example is that the search for the MPP is traced back to direct measurement of the generator current and voltage. The complication of multiplication for power calculation is obviated. Commencing with a starting point, for example the idle state point, the PV voltage is reduced by a factor k which must lie between 0 and 1. Then the PV voltage is

increased until the module current has reduced to k times the previous value. The tracker repeats these two steps continually, whereby the operating point moves progressively closer to the MPP [3].

This procedure is implemented by the simulation model shown in Figure 4.

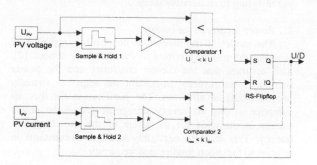

Figure 4: Model of Characteristic curve method

To explain the functional principles of this model, it is first of all assumed that the module voltage becomes smaller, i.e. that the output U/D carries logic level LOW. Therefore the RS flipflop is cleared so that the sample and hold circuit of the voltage channel is in "hold" state. When the module voltage becomes smaller than k times the stored voltage, the comparison is fulfilled and the RS flipflop is set. This makes the output signal U/D take logic level HIGH, so that the module voltage increases and stores the module current value just determined by the sample and hold circuit. The procedure now runs analogously for the current channel. Figure 5 shows the transient settle-down behaviour of the characteristic curve method for k = 0.8 and 0.95.

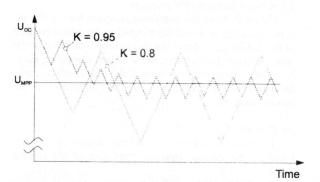

Figure 5: Transient settle-down behaviour

A feature of this method is that the deviation from the MPP is the smaller, the closer the constant comes to the value 1, but this increases the length of the settle-down time. The extent to which these deviations reduce the yield can be assessed completely with the model presented here, only when the partial load behaviour is taken into consideration appropriately. This is the subject of the next section.

3. ASSESSMENT CRITERION

Especially with the low irradiation conditions frequently found in central Europe, the PV plant should be optimised not only with respect to the nominal power rating, but should also exploit the available energy in the partial load range as completely as possible. As already mentioned above, the European power efficiency factor

$$\eta_{EU} = \quad 0.03\,\eta_{5\%} + 0.06\,\eta_{10\%} + 0.13\,\eta_{20\%}$$
$$+ 0.10\,\eta_{30\%} + 0.48\,\eta_{50\%} + 0.20\,\eta_{100\%} \quad (2)$$

is therefore used to assess the energy yield [4]. In order to achieve consolidated results in the determination of the respective power efficiency factors, only the steady state condition is taken into consideration. For this purpose, irradiation and temperature are held constant for each simulation run. The course of the PV power as a function of time is plotted in Figure 6 for visualisation.

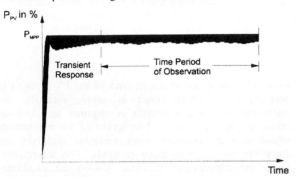

Figure 6: Time function of the module power

For calculating the average power, the energy yield obtained during the time period of observation is related to the interval duration T and compared with the theoretical maximum power at the MPP.

To demonstrate the performance of the model, an important parameter, namely the capacitance of the buffer capacitor, can be varied. The theoretical MPP value is taken as setpoint for the PV voltage. The coupling to the AC grid has the effect that no power can be delivered to the grid in the region of the zero transition, i.e. the output energy must then be buffer-stored in the capacitor. This has the consequence that the PV voltage pulsates cyclically at twice the mains frequency. This undulation leads to deviations from the MPP which become smaller with increased capacitance of the capacitor, i.e. smaller undulation of the PV current. Figure 7 shows these conditions for an inverter with 150 W nominal power rating.

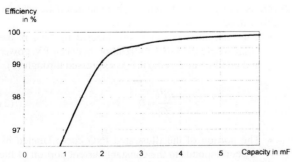

Figure 7: European power efficiency factor as a function of the capacitance of the buffer capacitor

The European power efficiency factor is decisively affected by the choice of capacitance up to 2500 μF. Further increase of the capacitance brings further increase of the power efficiency factor of only less than 0.15%.

4. RESULT

Table 1 summarises the simulation results with the MPP methods discussed in this paper.

Table 1: Measured parameters and the European power efficiency factor

	ΔP/ΔU method	CC method k = 0.8	CC method k = 0.95
PV voltage	yes	yes	yes
PV current	yes	yes	yes
PV power	yes	no	no
η_{EU} in %	98.9	95.4	98.7

The developed model is a suitable means for evaluating the different MPP variants. Furthermore, the MPP trackers can be optimised with respect to energy yield. This is evident from the characteristic curve method: For different values of the parameter k the resulting European-Efficiencies can be compared (see Table 1) and the optimal value can be found. The design of the PV voltage regulator can be mentioned here as further application of the model. It is even possible to detect malfunctioning by smart modelling of the MPP tracker and voltage controller.

REFERENCES

[1] M. Wolf, H. Rauschenbach, Series resistance effects on solar cell measurements, Solar Cells, IEEE Press, New York 1976, pp. 146-170

[2] Ortjohann, E., Voges, B., Voß, J.:
Dynamic Performance of Grid-Connected PV-Inverters, 12th European Photovoltaic Solar Energy conference and Exhibition, Amsterdam, Netherland, April 1994

[3] Boehrimger, A.:
Self-adapting dc converter for solar spacecraft power supply.
IEEE Transactions on Aerospace and Electronic Systems 4, 1968, S. 102-111

[4] H. Häberlin, H.R. Röthlisberger, PV converters from 1KW to 3 KW for grid connection: results of extended tests, 11th Photovoltaic Solar Energy Conference, 1992, Montreux, Switzerland

CONTROL INTEGRATED MAXIMUM POWER POINT TRACKING METHODS

Lícia Neto Arruda[*]
liciaarruda@bol.com.br

Braz J. Cardoso Filho[*]
cardosob@cpdee.ufmg.br

Selênio Rocha Silva[*]
selenios@cpdee.ufmg.br

[*]Universidade Federal de Minas Gerais
Av. Antônio Carlos, 6627 – Pampulha
31270-901 – Belo Horizonte – MG – Brazil
Phone: +55-31-499-4841

ABSTRACT: This paper introduces and describes two new Maximum Power Point Tracking (MPPT) methods: The Cross-Correlation Method and The Conductance Variation Method. The proposed methods take properly advantage of the direction of the conductance growth and the behavior presented by the photovoltaic (PV) system when operating at the constant current or voltage portion of the IxV characteristic curve. The proper evaluation of the effects of a small perturbation signal applied to the PV system is the basis of the proposed methods. The nature of the perturbation signal distinguishes the introduced methods from the conventional ones in the sense that only the resulting variation of the PV system voltage and current is required as the input variables of the control loops. The methods do not require either a fix step perturbation signal or look-up tables. These information are extracted from the PV system itself and are properly manipulated by the introduced control loop, which demands only an integral controller or a first order low-pass filter. No additional dedicated converter is required and the interaction with inverter control is straightforward.
Keywords: Tracking – 1: Inverter – 2: PV System – 3

1. INTRODUCTION

During the last two decades many researches have been performed aiming at reducing the cost of PV systems as well as the corresponding pay back time for the costumer. In this sense, many MPPT methods have been proposed in the literature [2][4][5][6][7][8]. It is believed that a very efficient MPPT method can result in an energy gain greater than 25% [3], which can markedly contribute to decrease the pay-off time of PV systems.

Most proposed MPPT methods are based on the application of a perturbation or an exploratory signal, which helps to determine the actual operating conditions of the PV system. This information is used to evaluate the location of the maximum power point. Some kind of logical decision is then taken in order to make the PV system converge toward its maximum power operating condition. These methods are named *Perturb and Observe* (P&O) [1].

Despite the proved functionality of the above-mentioned MPPT methods, some drawbacks can contribute to deteriorate their application possibilities:

- Some methods require a specific converter to perform the MPPT function, which can decrease the system total efficiency and reliability and increase the costs.
- Look-up tables are used in some of the proposed methods. The inherent fixed step size of the correction signal as well as the time required to go through the length of the table can compromise the convergence rate of the system toward a new maximum power operating condition.
- The decision about the convergence direction of the PV system toward the maximum power point aggregates, sometimes, many unnecessary logical testing steps. Besides that, these methods usually require the application of the already mentioned fixed step size correction signal.

This paper introduces and describes two new MPPT methods: *the cross-correlation method* and *the conductance variation method*. The knowledge of the general behavior and characteristics of the PV systems is properly used, which result in simple methods. No additional dedicated converter is required and look-up tables are not necessary. The nature of the exploratory signal and the required input variables highlight the flexibility of the introduced methods. The information about the convergence direction toward the maximum power point is extracted from the system itself and the generation of the correction signal is intrinsic to the proposed methods. Simulation results are presented. A complete description of the tuning of the controller gain as well as the experimental results are due on a later paper.

2. CROSS-CORRELATION METHOD

For each specific atmospheric condition, the maximum power point, which is a global maximum as indicated in fig. 1, is well defined as the point where the relationship between PV power and voltage is expressed through (1).

$$\frac{dP}{dV} = 0 \qquad (1)$$

The well-known characteristic PxV curve can be redrawn in order to obtain a more useful curve, as illustrated in fig. 2. Some remarkable characteristics can be readily extracted from the PxG curve:

- The maximum power point, labeled by "b", is the point where the cross-correlation between PV power and conductance vanishes, as expressed through (2).

$$\frac{\partial P}{\partial G} = 0 \qquad (2)$$

- The portion of the illustrated PxG curve labeled by "a" corresponds to the constant current portion of the IxV characteristic curve. In this portion, the slope of the PxG curve is negative.

- In a similar way, the labeled "c" portion of the illustrated PxG curve, whose slope is positive, corresponds to the constant voltage portion of the well-known IxV characteristic curve.

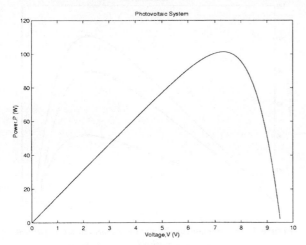

Figure 1: Characteristic PxV curve.

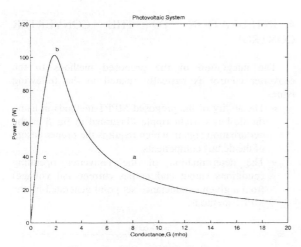

Figure 2: Characteristic PxG curve.

The features mentioned above can be properly used to guarantee the convergence of the operating point toward the maximum power point, as illustrated in the block diagram of the *cross-correlation method* presented in fig. 3. When the system is being operated on the "a" portion of the characteristic PxG curve, the product ΔP•ΔG is negative, which causes the conductance signal generated by the integral block to decrease. Similarly, the operation on the "c" portion of the PxG curve causes an increase of the conductance of the PV system. As the conductance grows from the constant voltage portion to the constant current portion of the characteristic IxV curve, the method causes the conductance of the PV system to converge toward the maximum power point. At this point the relationship between power and conductance vanishes, as expresses in (2), and the system stabilizes.

The *calculation block* in fig. 3 is just used to illustrate that the method requires two distinct values of power and conductance: (P₁;G₁) and (P₂;G₂). These values can be readily obtained by the application of a known perturbation signal Ǧ. It must be stressed, however, that the method

does not require the knowledge of the perturbation signal, since the variation of power and conductance are the input variables. In this sense, there is the possibility of use the dc bus current ripple as the perturbation signal, which demands the characterization of the dynamic behavior of the solar panel and proper design of the dc bus components.

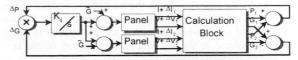

Figure 3: Block diagram of the *cross-correlation method*.

Simulation results illustrating the proposed method are presented in fig. 4. The curves illustrate the convergence of the operating point toward the maximum power point. The atmospheric conditions were suddenly varied in order to monitor the behavior of the PV system. Table I presents the atmospheric conditions used in the simulation.

Table I
Atmospheric conditions for the simulation results

Curve	Temperature (K)	Solar Irradiation (W/m²)
a	323	1000
b	323	800
c	308	400

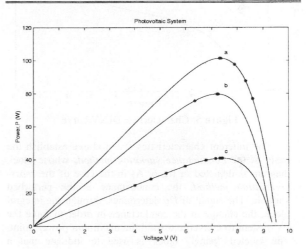

Figure 4: Simulation results for the *cross-correlation method*.

3. CONDUCTANCE VARIATION METHOD

The *conductance variation method* is based on equation (1), which can be rewritten as a function of voltage and current, as stated by (3).

$$\frac{\partial I}{\partial V} = -\frac{I}{V} \tag{3}$$

From (3), one can properly define a new variable expressed by (4) whose behavior can be advantageously used to guarantee the convergence of the operating point toward the maximum power point.

$$Dif = \left| \frac{\partial I}{\partial V} \right| - \frac{I}{V} \qquad (4)$$

The *conductance variation method* can be understood from the *DifxV* curve plotted in fig. 5. This curve highlights some important features, which are properly used in the introduced MPPT method:

- As the signal of *Dif* in the labeled "a" portion of the illustrated curve is negative, the conductance can be lead to a lower value when the PV system is being operated in the constant current portion of the IxV characteristic curve.
- Similarly, as a result of the positive values presented by *Dif*, the PV system conductance can be made large when the operating point is in the constant voltage portion of the IxV characteristic curve.
- The maximum power point is uniquely defined by the null value presented by *Dif*.

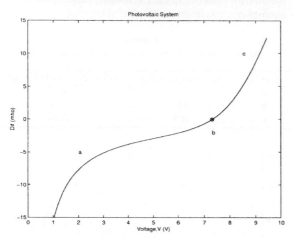

Figure 5: Characteristic DifxV curve.

The inherent characteristics stated above establish the basis of the *conductance variation method*, whose block diagram is depicted in fig. 6. As in the case of the *cross-correlation method*, the conductance is the perturbed variable. The signal of *Dif* determines, through the integral block, the change in the conductance in order to force the PV system to converge toward the maximum power point. The labeled "panel" block is used to indicate that a conductance signal is somehow sent to the inverter control loop (e.g. the conductance signal can be used to determine the amplitude of the current or voltage reference vector), which, in turn, actuates to extract from the panel the correspondent voltage and current values. Additionally, it must be stressed that the proposed method does not require the use of known perturbation signals ($\check{G}$), since voltage and current variations introduced by this signal are used as input variables.

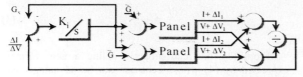

Figure 6: Block diagram of the *conductance variation method*.

Fig. 7 presents the simulation results of the introduced *conductance variation method*. The atmospheric conditions are the same described in Table I.

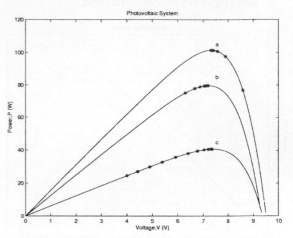

Figure 7: Simulation results for the *conductance variation method*.

4. INTEGRATION TO THE CONVERTER CONTROL

The integration of the proposed methods to the converter control is basically related to the following issues:

- The ability of the proposed MPPT methods in using the dc bus current ripple illustrated in fig. 8 as the perturbation signal, which requires the proper design of the dc bus components.
- The determination of the converter operating conditions (input and output current and voltage) from a given conductance set point generated by the MPPT methods.

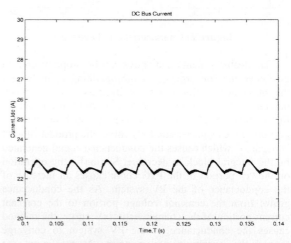

Figure 8: DC bus current ripple.

The above mentioned topics highlight the need to determine the relationship between a given conductance set point, *G*, generated by the MPPT methods and the modulation index, *m*, defining the converter switching function. The conductance command, as illustrated in fig. 9, defines the relationship between current and voltage at the input of the converter, as expressed in (5).

$$G = \frac{I_{dc}}{V_{dc}} \qquad (5)$$

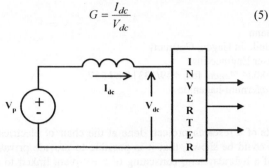

Figure 9: Simplified dc link.

For a CSI type inverter, the modulation index is defined in (6), where $I_{L,max}$ is the reference peak value for the ac line current.

$$m = \frac{\sqrt{3}}{2} \frac{I_{L\,max}}{I_{dc}} \qquad (6)$$

Assuming lossless converter and from (6) one can readily obtain the expression for V_{dc} as a function of the modulation index, where $V_{\phi\phi rms}$ is the line-to-line rms voltage.

$$V_{dc} = m\sqrt{2}V_{\phi\phi rms} \qquad (7)$$

Substituting (6) and (7) into (5), one obtain the relationship between modulation index and the conductance, as expressed in (8):

$$m = \pm \sqrt{\left(\frac{\sqrt{3}}{2}\right)\left(\frac{1}{G}\right)\left(\frac{I_{L rms}}{V_{\phi\phi rms}}\right)} \qquad (8)$$

Expression (8) illustrates the generation of reference signals to the control loop of the inverter using the conductance signal synthesized by the proposed MPPT algorithms. It must be said, however, that when calculating m, some restrictions must be observed:

- The calculated m value can not lead to V_{dc}, as defined in (7), greater than the open-circuit voltage of the solar panel.
- The short-circuit current must be observed as the limit of the dc link current expressed from (6).

The restrictions mentioned above limit, for a given conductance command, the possible values for m and I_L, as illustrated in fig. 10 by the shaded area. This figure relates control variables m and G with the power injected in the grid. However, this relationship as expressed in (8) is not suitable for practical implementation. A better solution is obtained solving (5) and (7) for the modulation index (9). Fig. 11 illustrates the integration of the proposed MPPT methods with the control of the inverter, where G is the conductance reference signal generated by the MPPT methods. Notice that the constant conductance panel loading is obtained by changing the modulation index and V'_{dc} according to the current supplied by the PV modules.

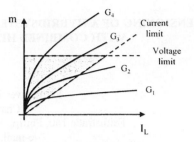

Figure 10: Operational restrictions.

$$m = \frac{I_{dc}}{G\sqrt{2}\,V_{\phi\phi rms}} \qquad (9)$$

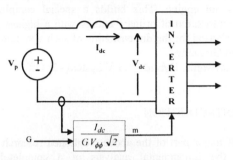

Figure 11: Integration between the introduced MPPT methods and the inverter.

5. CONCLUSIONS

Two new MPPT methods were introduced. The intensive and proper use of the solar cell characteristic curves resulted in extremely simple control structures. The integration of the proposed methodologies with the control of the inverter is straightforward, adding more attractive features to the MPPT methods presented is this paper. Additionally, the methods do not require look-up tables and the usual decision blocks are not present.

The use of the dc bus current ripple is advantageous since it eliminates the need for extra power and control hardware. Further study is still required to clarify all design issues

REFERENCES

[1]   C. Hua and C. Shen. Comparative study of peak power tracking techniques for solar storage system. In: *APEC IEEE*, Anaheim, CA, February 1998, pp. 679-685.

[2]   J.H.R. Enslin and D.B. Snyman. Combined low-cost, high-efficient inverter, peak power tracker and regulator for PV applicaiotns. *IEEE Transactions on Power Electronics*, vol. 6, no. 1, pp. 73 – 82, January 1991.

[3]   J.H.R. Enslin, M.S. Wolf, D.B. Snyman and W. Swiegers. Integrated photovoltaic maximum power point tracking converter. *IEEE Transactions on Industrial Electronics*, vol. 44, no. 6, pp. 769 – 773, December 1997.

[4]   K.H. Hussein, I. Muta, T. Hoshino and M. Osakada. Maximum photovoltaic power tracking: an algorithm for rapidly changing atmospheric conditions. *IEE Proc.-Gener. Transm. Distrib.*, vol. 142, no. 1, pp. 59 – 64, January 1995.

[5]   M. Calais and H. Hinz. A ripple-based maximum power point tracking algorithm for a single-phase, grid-connected photovoltaic system. *Solar Energy*, vol. 63, no. 5, pp. 277-282, 1998.

[6]   M. Matsui, T. Kitano, D. Xu and Z. Yang. A new maximum photovoltaic power tracking control scheme based on power equilibrium at dc link. In: 34th *IAS Annual Meeting*, 1999.

[7]   S.J. Chiang, K.T. Chang and C.Y. Yen. Residential photovoltaic energy storage system. *IEEE Transactions on Industrial Electronics*, vol. 45, no. 3, pp. 385 – 394, June 1998.

[8]   Z. Salameh and D. Taylor. Step-up maximum power point tracker for photovoltaic arrays. *Solar Energy*, vol. 44, no. 1, pp. 57 – 61, 1990.

DIMENSIONING OF A HYBRIDSYSTEM CONSISTING OF A PV-GENERATOR AND A STEAM ENGINE WITH COMBINED HEAT AND POWER FOR PRIVATE HOUSEHOLDS

K. Brinkmann
FernUniversität Gesamthochschule in Hagen, Germany
Chair of Electrical Power Engineering
Feithstraße 140, Philipp-Reis-Gebäude, D-58084 Hagen, fax: +49/2331/987 357,
e-mail: klaus.brinkmann@fernuni-hagen.de

ABSTRACT: This paper presents the main part of the results of a research project, done at the chair of electrical power engineering of the University of Hagen / Germany. It could be shown, that it is possible to provide private households self-sufficient with heat and power by the help of a hybridsystem, consisting of a PV-plant linked to a steam engine. This builds a special complex combined heat and power system, which enables a maximum exploitation of primary energy with a highest possible portion of renewable energy. This contribution presents the criterion for the dimensioning of such a system and as a concrete example the data for an average private household in Germany.
Keywords: Hybrid - 1: PV System - 2: Off-Grid - 3

1. INTRODUCTION

A major part of the above mentioned research project was the fundamental analysis of a completely self-sufficient energy-supply system for summer/winter or mixed conditions, including a photovoltaic system and a piston-type steam engine with combined heat and power [1], as shown in figure 1.

To reach this aim, this hybridsystem has been examined with the help of experimental investigations [2]. Additional theoretical analysis [3] and computational simulations [4] were helpful to build up a process scheme with all necessary components and to determine their sizes.

Energy-Supply System for Summer and Winter

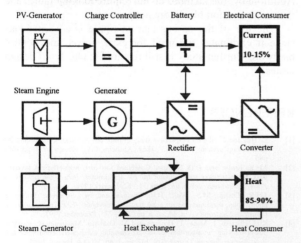

Figure 1: simplified process scheme of the hybridsystem

May be at the first view, the idea to realise the combined heat and power with the help of a piston-type steam engine seems to be curious. This technology has the disadvantage of the missing psychological valuation as a 'high-tech' solution.

But especially technical and scientific developments concerning energy-supply technologies, which are well conditioned to the fundamental rules of the nature and its ecology, ought not to be seen under a pure technical point of view, as it is usual for modern industrial product developments, where the attribute 'high-tech' serves as an important marketing element.

This does not mean, that the technical standard of the beginning of the industrial area has to be taken over. It is self-evident, that the hybridsystem as shown in figure 1 needs a 'modern' steam engine with respect to today's possibilities of the mechanical engineering.

An impressing example for such a development, which encourages to snatch up older technologies, is the success-story of the modern wind energy converter.

The combination of a PV-Plant to a combined power helps primarily to save fuels effectively.

In future, an increasing amount of biomass has to be used, because of its CO_2-neutrality; with respect to efficiency, as far as possible without preceding preparations. This implies shortest possible transportation and in the most cases solid biomass like chopped wood.

These arguments support decentralised energy-supply systems with external combustion. A hybridsystem with a PV-plant linked to a piston-type steam engine with combined head and power could be one possible option.

Additional considerations concerning other important properties like longevity, simple construction, rigidity, low susceptibility to trouble, the remaining necessary degree of efficiency and controllability of the ratio of power to heat, supported the decision to think about a PV-hybridsystem with a steam engine to provide private households.

2. PHYSICAL FOUNDATIONS

The today's annual ratio of electrical to heat energy consumption of average private households in Germany is nearly one to ten.

Essentially for the determination of the dimensioning criterion for the hybridsystem is the correlation of the time dependent consumption functions for electrical power as well as for heat [5].

It is necessary to correlate the effects of the three stochastic parameters current consumption, heat consumption and insolation. Therefore extensive long-term measurements in addition to available publications concerning these parameters for average values, as well as for the dynamic behaviour, were made and systematically examined regarding the seasonal variations.

An average household in Germany for 2,2 persons with 80 m² consumes annually 3146 kWh. The seasonal consumption in winter is greater than in summer. It is possible to approximate the daily consumption as a cosines-function, with the whole year as period.

The daily power consumption in a typical week in summer is 7,439 kWh and for a typical week in winter 10,123 kWh. The following annual ratios are valid:

$$\frac{power}{heating} \approx 0,11 \quad and \quad \frac{power}{process-heat} \approx 0,56. \quad (1)$$

It is also possible to approximate [2],[5] the daily heat-consumption as a cosine-function with sufficient accuracy. For a general private household, the following scaling could be used:

$$P_{general} = z \cdot P_{average} \quad with \quad (2)$$

$$z = \frac{annual \quad power-consumption}{3146 \quad kWh} \quad (3)$$

With these formulas for power and heat, an approximated daily ratio of power to heat could be given:

$$\sigma(d) = \frac{P_{electrical}(d)}{P_{heat}(d)} \quad (4)$$

The installation of a PV-Plant for the household results in an additional power supply and reduces the residual ratio of power to heat, which is left for the combined heat and power system:

$$\sigma(d)_{PV} = \frac{P_{electrical}(d) - P_{PV}(d)}{P_{heat}(d)} \quad (5)$$

With respect to the maximum possible degree of efficiency η_{max} of the combined heat and power, the realisable ratio of power to heat is limited to be smaller than υ:

$$\sigma_{PV} \leq \upsilon \quad with \quad \upsilon = \frac{\eta}{1-\eta}. \quad (6)$$

The usual electrical efficiencies of realised combined heat and power systems for the usage of biomass with piston-type steam engines are today about 16% [2].

These information lead to a formula to determine the PV-plant for the hybridsystem. The following condition allows to adjust the PV-Plant to the steam engine:

$$P_{PV} \geq P_{electrical} - \upsilon \cdot P_{heat}. \quad (7)$$

This condition has to be fulfilled for every day of the year.

3. TEST SETUP

For experimental investigations a test setup with a real steam engine for approximately 2.5kW from the beginning

of this century has been built up and connected to a PV-Generator-Simulator, which could be co-ordinated to the results of the photovoltaic plant of the FernUniversität building.

The installed PV-Generator-Simulator with a power output of maximum 2,5 kW enables an independence of the momentary weather conditions and a time-lapsed investigations.

For the storage of electrical energy, four batteries with a capacitance of 120Ah respectively were implemented in series, with a voltage of 48V totally.

Figure 2: piston-type steam engine test setup

Unfortunately, it seems that nowadays there exists no industrial producer of piston-type steam engines for smaller machines up to 5 kW. So the only way to make progress in this research project was to recourse to a historical model, as shown in figure 2.

For simplicity concerning installation and measurements, the steam generator works with electrical energy and produces maximum 18 kW saturated steam at 6 bar and 159°C.

Two control valves split up the steam flow in two directions, one of them to the steam engine and the other bypassed directly to the condenser. This gives the possibility to regulate the ratio of power to heat and enables to produce only heat without the steam engine, which is necessary if the batteries are sufficiently charged.

The self-made condenser consists of tube heat exchanger, which are implemented in two different heat circuits, one for a radiator and the main part for the production of hot water in a 50 l water vessel.

A pump in the condense water pipe was installed to reduce the back pressure, which results in a greater efficiency of the steam engine [2].

The shaft of the steam engine is coupled to an electrical generator via a cone belt. The nominal power of the generator is 1,3 kW and it is connected to a 3-phase bi-directional converter. This converter allows to switch between motor and generator mode. Because a steam engine with only one piston is in the most cases not able to start by itself, the motor mode can be used to give the first rotations.

The implemented measuring technique is based on 4-20 mA signals, which are used by a self-made control box with a μ-controller.

4. DIMENSIONING OF THE HYBRIDSYSTEM

The process scheme in figure 3 shows the functional connections of the above mentioned components.

In order to determine the dimensioning criterion of the whole system as well as of the single components and their accommodation mutually, a computer program for a system simulation has been developed.

In addition to the experimental experiences with the test setup, this programme is a valuable tool to prove a chosen dimensioning on the basis of mass flow and energy balance calculations.

This simulation was realised with the help of 'LabView', whereas each component can be mathematically described by characteristic curves. The time dependent changes and results of the calculations are readable in a process visualisation [4].

Most important for the determination of the dimensioning was the analysis of the consumption characteristics of private households in Germany as well as for electrical power and for heat [2], which had to be correlated to the expected energy production of the PV-plant. This energy production could be estimated with the help of extensive available meteorological data.

Hybridsystem "Photovoltaic + Combined Heat and Power with Steam Engine"

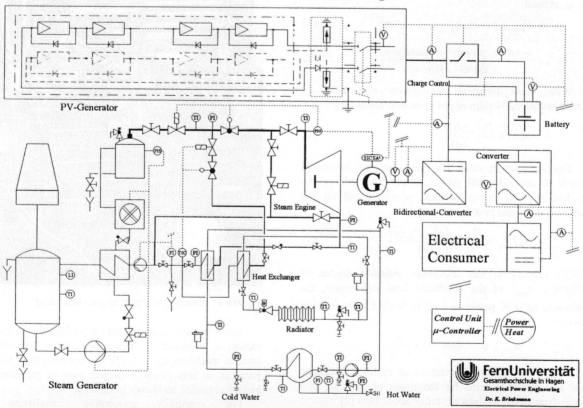

Figure 3: process scheme of the test setup

The upper line in the scheme above symbolises the photovoltaic generator with the charge controller and the connection to the batteries.

Under this, there can be seen three different functional areas, on the left side the steam generation, in the middle the heat exchange circuits and on the right side the electrical components for the power supply and the control unit.

Apart from some details for the heat exchange circuits, one can use this process scheme as basis for a pilot plant.

In order to allow a simple conversion to individually different local conditions, the average time dependent insolation characteristic in Germany has been standardised to an annual total global insolation of 1000 kWh/m².

This characteristic could be described with the help of a numerical approximated function, which can be used in formula (7) for the estimation of P_{PV} and therefore to determine the dimension of the PV-plant [2].

With respect to the piston-type steam engine, the parameter υ was carefully chosen to be 0,1. This value can surely be seen as a lower limit, with respect to the possibilities of modern mechanical engineering.

Figure 4: front view of the test setup

Including the influence of the performance ratio PR of a PV-plant, we obtain the following dimensioning for a general average private household in Germany:

$$PV - plant \cong \frac{z \cdot 2}{PR} kWp . \tag{8}$$

For the steam engine, it could be demonstrated, that the mechanical power has to be about

$$P_{mech} \cong z \cdot 1,7kW . \tag{9}$$

Consequently follows for the steam generator with respect to υ:

$$P_{steam} \cong z \cdot 17kW . \tag{10}$$

This energy flow is a measure for the dimensioning of the condenser and heat exchanger.

These are with sufficient accuracy the main values, which are realised for the dimensioning of the test setup, as shown in figure 4 above. For further details see [2].

With respect to the dimensioning of this hybridsystem as a self-sufficient energy supply the capacity of the storage batteries should be

$$C_{Battery} \cong z \cdot 52kWh . \tag{11}$$

5. CONCLUSION

This paper presents the main results of a research project concerning a hybridsystem consisting of a PV-generator linked to a piston-type steam engine, a process scheme and the essential information of dimensioning.

REFERENCES

[1] K. Brinkmann
"PV-Generator linked to a Piston-Type Steam Engine with Combined Heat and Power as a Hybridsystem for a completely self-sufficient Energy-Supply"
2nd World Conference and Exhibition on Photovoltaic Solar Energy Conversion, Vienna Austria, July 1998.

[2] "Systemtechnische Untersuchung eines Hybridsystems bestehend aus Photovoltaikanlage und Dampfmaschine mit Kraft-Wärme-Kopplung"
K.. Brinkmann, Dissertation 1999
FernUniversität Hagen / Fachbereich Elektrotechnik.

[3] K. Brinkmann
"Physical Analysis of a Hybridsystem consisting of a PV-Generator linked to a Piston-Type Steam Engine with Combined Heat and Power for a completely self-sufficient Energy-Supply"
11. Internationales Sonnenforum 1998, July 1998.

[4] K. Brinkmann, R. Taubner
"Simulation eines Hybridsystems bestehend aus Kraftwärmekopplung mit Dampfmaschine und Photovoltaikanlage"
14. Symposium Photovoltaische Solarenergie, Staffelstein 1999.

[5] K. Brinkmann
"Kriterien zur Auslegung von photovoltaischen Hybridsystemen mit Kraftwärmekopplung zur autarken Versorgung privater Haushalte"
15. Symposium Photovoltaische Solarenergie, Staffelstein 2000.

PREDICTIVE CONTROL STRATEGY FOR STAND ALONE PV HYBRID SYSTEMS USING EUROPEAN INSTALLATION BUS (EIB).

T. Romanos, S. Tselepis, G.Goulas
Centre for Renewable Energy Sources,
19th km. Marathonos Ave., GR-190 09, Pikermi, Greece,
Tel. +301 6039 900,
Fax +301 6039 905,
E-mail : troman@cres.gr

ABSTRACT: The paper focuses on the design and development of a Predictive Control System for stand alone PV Hybrid Systems using the European Installation Bus (E.I.B.). This work was partly performed during the course of realization of two JOULE projects (JOR3-CT97-0158 and JOR3-CT98-0244).
The Control System minimizes the production costs through an on-line optimal scheduling of the power units, taking into account the technical constraints according to the State Of Charge (S.O.C.) of the battery Bank, as well as forecasts of the load and the Photovoltaic energy contribution several hours ahead.
Keywords: Stand-alone PV Systems –1: Hybrid – 2: Expert-System - 3

1. PREDICTIVE CONTROL FOR STAND ALONE HYBRID SYSTEMS

A Control Strategy for Hybrid Systems is being developed, based on k-step ahead control horizon coupled with on-line system identification procedures.

The Hybrid System may consist of a Diesel generator unit, a Photovoltaic generator unit, a battery inverter and a load simulator for a typical small Greek island. The Diesel generator model is based on a typical deterministic linear model.

The estimation of the PV contribution, depending on the PV penetration phase, can be decided arbitrarily and it corresponds on an average day of a month in the Aegean Islands. A deterministic model has been developed for the state of charge (SOC) of the battery bank. A representative deterministic model can be used for the simulation of a two way battery inverter. Linear parametric models, like ARMAX model, can be derived for simulation, using data from load simulator correlated with the data of the PV inverter.

Predictive control strategy is developed for stand-alone Hybrid Systems. The advantage of predictive controllers is that they are relative easy to tune. They can be derived for and applied to multi-input, multi-output (MIMO) processes. That means that they can be adapted to the principles of the PV-hybrid's concept. Predictive controllers can also be derived for nonlinear processes. Predictive control is the only methodology that can handle process constraints in a systematic way during the design of the control strategy. Predictive control is an open methodology. That is, within the framework of predictive control there are many ways to design a predictive controller. Feed-forward action can be introduced for compensation of measurable disturbances and for tracking reference trajectories.

Unavoidably, predictive controller design has some drawbacks. Since predictive controllers belong to the class of model-based controller design methods, a model of the process must be available. In general, in designing a control system two design stages can be distinguished: modeling and controller design. Predictive control provides only a solution for the controller design. A second drawback is due to the fact that the predictive control concept is an open methodology. This means that different control strategies can be derived and so different behavior of the Hybrid System will be yielded. Therefore, a unified approach to predictive controller design is needed, which allows treatment of each problem within the same framework and results in significant reduction in design costs. Such a unified approach is developed and it unifies the strategies for stand-alone PV Hybrid Systems.

2. THE PREDICTIVE CONTROL CONCEPT [1]

Usually, predictive controllers are used in discrete time. The predictive controllers will operate the hybrid system with two inputs (the energy from the diesel and the battery inverter) and one output (the delivered energy to the loads minus the energy from the PVs). As it is illustrated in Figure 1, the time scales in parts a, b, c, d, e and f are time scales relative to the sample k, which denotes the present. The time scales shown at the bottom of Figure 1 are absolute time scales. Consider first, Figure 1 a, c and e and suppose that the current time is denoted by sample k which corresponds to the absolute time t. Furthermore, $u_d(k)$, $u_b(k)$ and y(k) denote the diesel unit output for stand alone system, the battery inverter output and the station unit output at sample k, respectively. These functions are expressed mathematically as follows:

$$U_d=[u_d(k),....u_d(k+H_p)]^T \quad ,U_b=[u_b(k),....u_b(k+H_p-1)]^T \quad ,$$
$$\hat{y}=[y(k),...,y(k+H_p)]^T$$

where: H_p is the prediction horizon and the symbol ^ denotes estimation. Then, a predictive controller calculates such a future controller output sequence u_d and u_b (shown in Figure 1 d and f), where the predicted output of the hybrid system corresponds to the load profile of a small island for a typical day.

Rather than using the controller output sequence determined in the above way, in order to control the process in the next H_p samples, only the first element of this controller output sequence (=$[u_d(k),u_b(k)]$) is used to control the process. At the next sample (hence, at $t+1$), the whole procedure is repeated using the latest measured information. This is called the *receding horizon* principle and is illustrated by Figures 1 b, d and f, which show what happens at time *t+1*. Assuming that there are no disturbances and modeling errors, the predicted process output $\hat{y}(k+1)$, predicted at time *t* is exactly equal to process output y (k) measured at t+1.

Now, again, a future controller output sequence is calculated such that the predicted process output is solved under minimization of energy consumption. In general, this controller output sequence is different from the one obtained at the previous sample, as it is illustrated in Figure 1 c and e. The reason for using the receding horizon approach is that this allows us to compensate for future disturbances or modeling errors. For example, due to a disturbance or modeling error the predicted process output $\hat{y}(k+1)$ predicted at time t is not equal to the process output y (k) measured at t+1. Then, it is intuitively clear that at time t+1 it is better to start the predictions from the measured process output rather than from the process output predicted at the previous sample. This way, the predicted process output is now corrected for disturbances and modeling errors. A feedback mechanism is always activated. As a result of the receding horizon approach, the horizon over which the process output is predicted shifts one sample into the future at every sample instant. The process output is predicted by using a model of the process to be controlled. Any model that describes the relationship between the input and the output of the process can be used. Hence, not only transfer-function models can be used, but also step-response models, state-space models and nonlinear models. Further, because the process is subject to disturbances, a disturbance model will be added to the process.

In order to formulate mathematically the way to minimize the fuel consumption, a criterion function is used.

The criterion function is a function of $\hat{y}$, u_b, and u_d.

The criterion function is:

$$J = \sum_{i=1}^{H_P} (u_d(k+i))^2 = \sum_{i=1}^{H_P} (\hat{y}(k+i) - u_b(k+i))^2$$

where the SOC constraint may be: $40\% < u_{SOC} < 90\%$ where u_{SOC} is the estimated SOC of the battery bank.
and the allowable power variation per time step may be: $0\% < d\,(u_d, u_b) < X\%$
the magnitude of X should be defined by a well-experienced operator.
Now the controller output sequence u_{opt} over the prediction horizon is obtained by minimization of J

with respect to **u**: $u_{opt} = \arg\min_{u_d} J$

Then, **u**$_{opt}$ is optimal with respect to the criterion function that is minimized taking into consideration the deterministic modes of diesel unit and battery bank.
The control strategy, together with the various components developed, is going to be implemented and tested in the pilot plant at CRES, in Greece.

3. THE EXPERIMENTAL PILOT PLANT AT CRES

A pilot plant serving all the options of the Control Strategy is necessary. For realistic simulation and testing purposes CRES has set-up a Hybrid system at its premises with

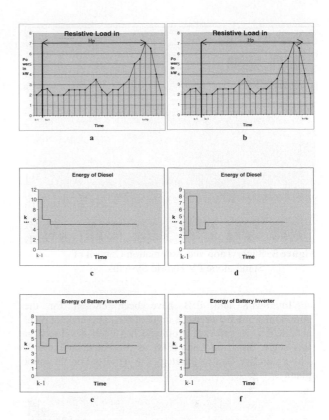

Figure 1: Receding horizon predictive control for the hybrid system. Parts a, c and e denote the situation at k sample, while parts b, d and f denote the situation at k+1 sample.

energy producing and load units connected to an AC grid. All components are connected to a single twisted pair wire bus. The system may be composed of a 4.4 kWp PV array, mounted on one axis sun tracker, a 9 kVA four quadrant battery inverter (96 VDC-220VAC), a 12 kVA Diesel electricity generator, ohmic, capacitive and inductive loads. An operational control unit which communicates with the local control units, controls voltage, frequency and power and operates as a data acquisition unit. The block diagram of the Experimental Pilot Plant layout is shown in Figure 2.

The aim of the experimental application was a standardized applicable system technology which will allow a variety of different hybrid islands supply situations to be covered. Combined with a standard communication bus for operational control, a basis is defined for future development of a set of fully compatible units, which can be integrated into various hybrid systems in islands.[2]

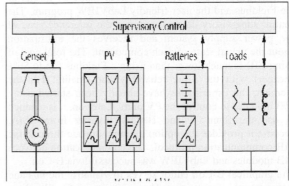

Figure 2: The block diagram of the Experimental Pilot Plant at CRES

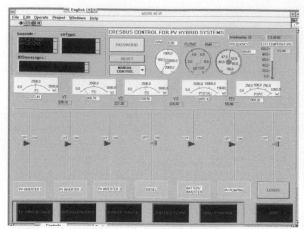

Figure 3: Supervision and Automation Control of Stand-alone PV Hybrid Systems

Interbus-S and EIB have been chosen for the communication between the units of the experimental plant at C.R.E.S. A software prototype has been developed at this stage, based on LabVIEW, as presented in figure 3. The Predictive Control Strategy can be tested on this pilot plant.

Figure 4: Photovoltaic array of 4.4kW$_p$

The Supervising Control has been made by means of the Fieldbus and the user-friendly LabVIEW program. The central supervisory control unit sets the reference state of the power supply system and receives the information about the actual state of each component. The total cost of using EIB is much lower than using a Fieldbus (up to 80% cheaper) with the same reliability and quality. For the moment, the experiments have already been started using EIB modules to control the PV units. Later on, it is planned to use the Sunny Island Battery Inverter from SMA, because it provides the option to communicate through the EIB communication protocol. The communication between EIB modules and LabVIEW was successful via B-Con. The graphical B-Con editor serves to visualise the bus and management functions. The purpose of all these

experiments is to simulate a stand-alone small Greek island as a Hybrid System. Figure 4 presents the photovoltaic array of the PV-Hybrid system and figure 5 depicts the distribution board, which measures and controls the power from photovoltaics. Notice that Interbus-S modules are used to measure the power from photovoltaics, while EIB modules switch on or off the SMA inverters.

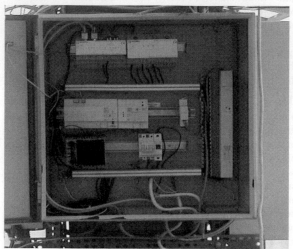

Figure 5: Distribution board of Interbus-S and EIB modules for the photovoltaic array

In the following figures, the experimental results are presented between May 9 and May 18 1999. The figure 6 presents the power flows of the hybrid system over a time period of 9 days. It is noticed, that during the first day the batteries are being charged and the loads have been switched off.

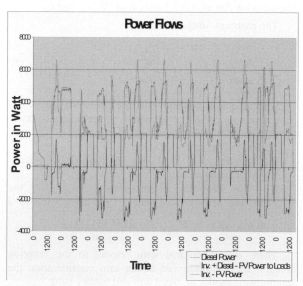

Figure 6: Power Flows

The negative values of power correspond at the times when the batteries are being charged. The PV power is contributing directly on the loads and reduces the load presented to the inverter and diesel generator by the value of its instantaneous power. The maximum instantaneous AC power offered by the two PV inverters together was 1 kW.

Figure 7 presents the battery bank state, in terms of voltage and flow of Ahrs to and from the batteries. The negative flow of Ahrs means that there are being charged and the positive that they are discharged. The voltage became temporarily zero, when the inverter was turned off.

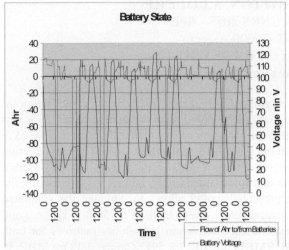

Figure 7: Battery State

By presenting, in the figure 8, the power flows of one day, the picture about the operation of the pilot plant becomes clearer. Notice that the effective load during the daytime morning hours is much higher than the resistive load, due to charging of the batteries.

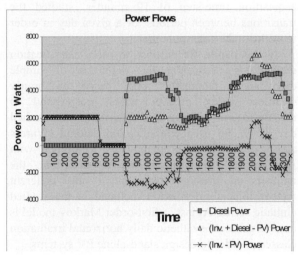

Figure 8: Power Flows

In the last figure 9, the energy produced daily by the diesel generator is presented next to the energy provided to the resistive loads without counting the energy contributed in a day by the photovoltaics, which in any case it is of the order of 5 kWh.

Therefore, during the period of 10 days we had 542.59 kWh of energy produced by the diesel generator, while the energy consumed by the loads, excluding the energy provided by the photovoltaics, was 475.14 kWh. Therefore, the energy efficiency of the system is obtained by a simple division of the two numbers, 475.14/542.59 = 87.5%.

Furthermore, during the same time period the light diesel fuel consumption was monitored. It was noted that, 305 liters of fuel were used for the production of 542.59 kWh of electric energy. Therefore, the energy consumption of the diesel generator was 562 ml/kWh or 469gr/kWh.

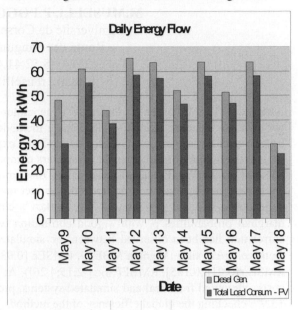

Figure 9: Daily Energy Flow

4. CONCLUSIONS

The Predictive control strategy developed here provides a worthy energy saving for a stand-alone PV Hybrid system. EIB technology seems to be the optimum choice for communication interface between the units of the hybrid system because it can combine the operational control in Hybrid Systems with home automation. The use of E.I.B. technology opens the way for compatibility of products from different manufacturers of PV components and Home Automation, opening the market in both directions.

REFERENCES

[1] Soeterboek R. (1992), "Predictive Control. A Unified Approach", Prentice Hall International, Englewood Cliffs, N.J.

2] W. Kleincauf, F. Raptis, P. Romanos et.al, "Modular Systems Technology For Decentral Electrification", 13th European PVSEC, Nice France, October 1995

A MARKOV CHAIN MODEL FOR DAILY IRRADIATION CLASSES RECORDS IN ORDER TO DESIGN PV STAND ALONE SYSTEMS

M.MUSELLI, P.POGGI, G.NOTTON, A.LOUCHE
Université de Corse - URA CNRS 2053 - SPE
Route des Sanguinaires - F-20000 Ajaccio
Tel : (33).4.95.52.41.41 - Fax : (33).4.95.52.41.42
E-mail : muselli@vignola.univ-corse.fr

ABSTRACT : In order to elaborate a generator of "typical-days" sequences necessary for the sizing of photovoltaic systems, this paper develops a methodology to class daily global solar irradiation based on a first order Markov chain model. The method has been experimented for two stations in Corsica (France) and we have limited the number of discriminant parameters computed from the hourly clearness index kt(h). The classification is based on the Ward's method checked by discriminant analyses. Previous works have shown that the different days could be clustered in 3 groups with distinct mean values and lower standard deviations. After checking the dependence and the stationary of the Markov's chain, the presented model allows to compute the simulated marginal probabilities p_i with a good correlation with the experimental ones (MBE$\in$[-0.154;0.151]). A good correlation has been observed between the simulated and experimental transition probability matrices for both meteorological sites (Vignola MBE=0, RMSE$\in$[0.035;0.056], RMBE(%)$\in$[0.17;3.59] ; Campo del'Oro MBE=0, RMSE$\in$[0.016;0.105], RMBE(%)$\in$[-2.15;4.26]). At least, sizing of photovoltaic systems has shown that the kWh costs, computed from real and simulated systems, present weak relative errors belonging to the range -13.6% and +3.9% checking the global efficiency of the method.

Keywords: Global Irradiation - 1: Markov Model - 2 : Stand-alone PV Systems - 3

1. INTRODUCTION AND AIMS

The photovoltaic solution often represents the best economical and technological choice to electrify remote areas. In this study, we develop a methodology of global irradiation "typical" day classification in order to elaborate a generator of synthetic daily global irradiation profiles in view of PV system sizing. This method has been experimented for two Corsican (France) and coastal meteorological stations (Vignola 41°55'N, 8°39'E, alt : 70 m and Campo del'Oro 41°55'N, 8°44'E, alt : 20 m) distant to about 10 kilometers.

Many papers have been dedicated to the problem of typical days classification. Aranovitch et al. [1] have elaborated classifications using 3 groups of days : clear sky, partially clouded and clouded (opaque) in order to synthesize daily tilted global irradiation profiles and to study PV system performances in Ispra (Italy). A study investigated in Trappes and Carpentras (France) [2] based on the integration of PV systems in houses used the parameters previously described combined with the partial vapor pressure, the wind direction and the sky nebulosity. This work conduced to 10 complex classes. The proposed methodology is difficult to use because it needs several recorded parameters that are rarely measured. Fabero et al. [3] distinguished 9 "typical" day periods by decomposing each studied day in 3 periods. This study (Madrid, Spain) using global horizontal irradiation recorded with an

acquisition time-step of 10 minutes, studied the transitions between periods for a given day in order to evaluate the solar potential.

In this paper, we purpose to use a classification methodology requiring the definition of simple discriminant parameters only based on hourly clearness index profiles which correspond to meteorological information describing the studied day. Then, an usual classification method (Ward's aggregation) on these parameters, combined with Discriminant Analyses allows to determine the number of clusters representing the different meteorological information included in the studied climatic series. At last, a first-order Markov model is used to generate synthetic daily horizontal irradiation classes in order to design stand-alone PV systems.

2. DESCRIPTION OF THE DATA AND THE PARAMETERS

The discriminant parameters are constructed from hourly horizontal irradiation data collected at both stations on the same period (1448 days i.e. 34752 hourly data). These data are computed on tilted planes for angles equal to 30°, 45° and 60° using classical models [4,5]. Horizontal (and tilted) discriminant parameters are the following :

- *Hourly clearness index kt(h)* computed from the hourly global horizontal irradiation I(h) and the hourly global extraterrestrial one I_0(h) :

$$kt(h) = I(h)/I_0(h) \qquad (1)$$

Moreover, we use the hourly clearness index on a tilted plane $kt_\beta(h)$.

- *Daily clearness index KT(d)* calculated from the daily global horizontal irradiation H(d) and the daily extraterrestrial one H_0(d) :

$$KT(d) = H(d)/H_0(d) \qquad (2)$$

Moreover, the daily clearness index on a tilted plane $KT_\beta(d)$ is constructed.

- *Integral of the squared second derivative of hourly clearness index profile S2* estimating the perturbation state of the hourly clearness index curve during a day ; it represents the integral of the squared second derivative of the hourly clearness index profile :

$$S2 = \sum_h \left[kt(h+2) - (2 \times kt(h+1)) + kt(h) \right]^2 \qquad (3)$$

This parameter is equally computed from the tilted clearness index $kt_\beta(h)$.

- *Monthly mean clearness index ktm* representing the mean value of the hourly clearness index for a given month m and a given angle β :

$$ktm = \overline{kt(h, m, \beta)} \qquad (4)$$

The discriminant parameter called *Sl* represents the number of hours for a given day where the hourly clearness index is inferior to the threshold *ktm*. This parameter is normalized by the total number of acquisition hours for the given day.

3. CLASSIFICATION METHODOLOGY AND RESULTS

We propose to cluster the global irradiation days on both sites using Ward's method, using the Euclidean distance to compute the difference between samples. This method allows a graphical representation of the progressive classification. The levelized histogram allows to locate the breaking points of the level indices characterizing the classification in the sense of the inertia. This method is validated using discriminant analyses (Fig. 1) which ranges the day samples into groups in order to verify the dendrogram obtained by the Ward's classification.

The methodology is applied to each site and each angle for a daily time-step. The optimal classification chosen between 3, 4 or 5 clusters is based on the derived discriminant functions (Table I). It lists the two highest scores among the classification functions for each of the 1448 observations used to fit the model corresponding to each cluster (polynomial

regression utilizing the discriminant parameters allowing to determine daily sequences of the typical days [6]). This procedure indicates the percentage of observations correctly classified by each clustering process.

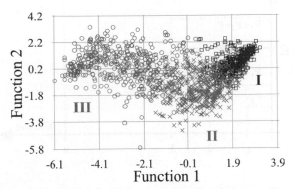

Figure 1: Example of discriminant analysis (Vignola 0°, 1448 days).

Table I: Percentage of the event number correctly classified derived discriminant functions.

Site	Campo del'Oro			Vignola		
Angle Nber of classes	3	4	5	3	4	5
0°	92.2	90.5	89.5	88.7	87.7	86.6
30°	93.3	92.4	92.2	94.1	89.3	87.1
45°	93.0	88.8	92.6	94.5	92.8	90.9
60°	92.6	88.8	92.5	95.6	88.8	91.4

The classification based on 3 clusters presents in all cases the best percentages and we chose to work with this solution. The 3 groups presenting distinct mean values and weak standard deviations, characterize different classes corresponding to various meteorological information : Cluster I for clear sky conditions, Cluster II for overcast sky conditions and Cluster III for cloudy sky. The comparison between both meteorological stations is impossible because the class scales of the different parameters (mean ± standard deviation) are not identical. However, the main conclusion is that two sites only distant of 10 kilometers can be subjected to the same climate but their statistical properties present clearly some differences.

4. FIRST-ORDER MARKOV MODEL AND RESULTS

A first-order Markov chain model was used to synthesize daily global irradiation sequences. This simulator performs a random variable in the range [0;1] and this random number is compared to the conditional probability matrix p_{ij} constructed from experimental data in order to determine the daily corresponding cluster (Fig. 2).

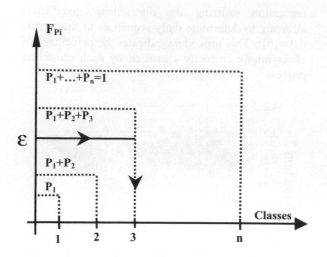

Figure 2: Markov Process (radom variable ε).

After having checked the dependence and the stationary of the Markov chain, the simulated marginal probability p_i (Figs 3a and 3b) are computed and have presented a good accordance with experimental data (Vignola MBE ∈ [0.059;0.090]; Campo del'Oro MBE ∈ [-0.154;0.151]).

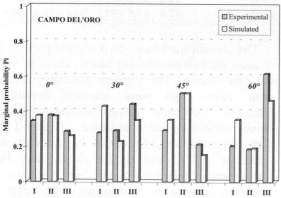

Figure 3a: Simulated and experimental marginal probabilities (Campo del'Oro).

Moreover, Table II presents the simulated transition probability matrix p_{ij}. A good accordance with experimental data is noted : Vignola MBE = 0, RMSE ∈ [0.035;0.056], RMBE(%) ∈ [0.17;3.59] and Campo del'Oro MBE = 0, RMSE ∈ [0.016;0.105], RMBE(%) ∈ [-2.15;4.26].

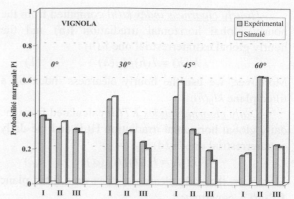

Figure 3b: Simulated and experimental marginal probabilities (Vignola).

Table II: Real and calculated transition probabilities.

Site	Experimental			Simulated		
C (0°)	0.723	0.217	0.060	0.750	0.224	0.026
	0.179	0.486	0.335	0.150	0.478	0.372
	0.100	0.416	0.484	0.147	0.424	0.429
C (30°)	0.985	0.015	0.000	0.989	0.011	0.000
	0.012	0.721	0.267	0.012	0.905	0.083
	0.002	0.173	0.825	0.006	0.048	0.946
C (45°)	0.850	0.136	0.014	0.863	0.127	0.010
	0.072	0.697	0.231	0.076	0.723	0.201
	0.036	0.533	0.431	0.065	0.631	0.304
C (60°)	0.918	0.079	0.003	0.927	0.073	0.000
	0.075	0.748	0.177	0.110	0.732	0.158
	0.005	0.049	0.946	0.009	0.055	0.936
V (0°)	0.671	0.182	0.147	0.668	0.263	0.069
	0.265	0393	0.342	0.187	0.438	0.375
	0.146	0.378	0.476	0.187	0.359	0.454
V (30°)	0.770	0.160	0.070	0.770	0.200	0.030
	0.272	0.427	0.301	0.220	0.480	0.300
	0.140	0.364	0.496	0.250	0.280	0.470
V (45°)	0.822	0.143	0.035	0.845	0.140	0.015
	0.220	0493	0.287	0.230	0.508	0.262
	0.108	0.448	0.444	0.204	0.419	0.377
V (60°)	0.745	0.246	0.009	0.800	0.200	0.000
	0.062	0.752	0.186	0.043	0.756	0.201
	0.016	0.508	0.476	0.042	0.541	0.417

5. PV SYSTEMS DESIGN AND COSTS

Lastly, the complete methodology was checked on PV systems designs. The comparison between experimental and simulated (constructed from synthetic "typical day" sequences) global energy on the global period calculated for each day by Eq. (5) has shown a good accordance (Fig. 4).

$$E(kWh)_{period} = \sum H_0(d) \times \overline{KT_i} \quad (5)$$

for i = {I, II, III}.

The global energy for each site and each angle on the whole period was constructed from :

Case 1 : the experimental data recorded at both stations ; the energy is called E{d},

Case 2 : the experimental "typical-days" sequences computed by the Ward's classification using the discriminant parameters defined in this paper ; the energy is called E{d,C$_{(real)}$},

Case 3 : the simulated "typical-days" sequences computed from the Markov's process elaborated in this work ; the energy is called E{d, C$_{(sim)}$}.

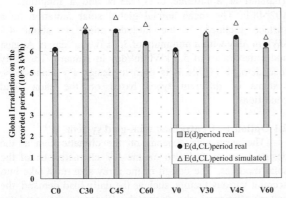

Figure 4: Real and Simulated Energy.

Then, using classical simulations of stand-alone PV systems and economical procedures [7], the comparison between experimental and simulated kWh prices produced by small PV systems (0.1 kW PV and 1.44 kWh Battery supplying small loads) has shown weak errors (Table III) with the RMBE(%) in a range -13.6% and 3.9%.

6. CONCLUSION

In this paper, a methodology is elaborated in order to size stand-alone PV systems in remote areas from global solar irradiation typical days on horizontal or tilted planes. This method utilizes the coupling of a classification process and a first order Markov chain model.

Table III: PV electricty costs computed from experimental and simulated solar data and relative mean bias error obtained from the comparison between the different cases.

	PV Electricity Cost $US/kWh			Cases 2/1 (%)	Cases 3/1 (%)
	Case1	Case2	Case3		
C(0°)	1.28	1.29	1.33	0.8	3.9
C(30°)	1.14	1.14	1.09	0.0	-4.4
C(45°)	1.14	1.13	1.03	-0.9	-9.6
C(60°)	1.25	1.24	1.08	-0.8	-13.6
V(0°)	1.30	1.30	1.35	0.0	3.8
V(30°)	1.16	1.16	1.15	0.0	-0.9
V(45°)	1.19	1.19	1.08	0.0	-9.2
V(60°)	1.28	1.25	1.18	-2.3	-7.8

After having defined the discriminant parameters and the Ward's classification used in a previous paper, this work has characterized the Markov chain (dependence, stationary) and a simulator has been elaborated and validated in order to construct typical days sequences. A good accordance has been noted between experimental and simulated marginal and transition probability matrices at both meteorological stations and for all angles. At last, using classical models computing the kWh cost, we have shown a good correlation between real and simulated kWh costs produced by the PV system. In conclusion, we have defined a process which is able to conserve and to reproduce the statistical and stochastic properties of the experimental global irradiation data on horizontal and tilted planes and which could be use to synthesize global irradiation in order to utilize them in studies of PV systems running.

REFERENCES

[1]. Aranovitch, E., Gandino, C. and Gillaert, D., *Colloque Météorologie et Energies Renouvelables*, Valbonne, 1984, 315-338.

[2]. Sacré, C., *Colloque Météorologie et Energies Renouvelables*, Valbonne, 1984, 615-630.

[3]. Fabero, F., Alonso-Abella, M. and Chenlo, F., *Proc. of the 14th European PV Solar Energy Conference*, Barcelona, 1997, 2299-2302.

[4]. Hay, L.E. and Davies, J.A., *First Canadian Solar Radiation Data Worksh.*, 1980, **59**, 59-72.

[5]. Orgill, J.F. and Hollands, K.T.G., *Solar Energy*, 1977, **19-4**, 357-359.

[6]. STATGRAPHICS Software. Technical Documentation, 1996.

[7]. Notton, G., Muselli, M., Poggi, P. and Louche, A. *Renewable Energy*, 1996, **7-4**, 353-369.

AN OPTIONAL DESIGN OF PV SYSTEMS FOR A THAI RURAL VILLAGE

S.Hiranvarodom, R. Hill[*], P. O'Keefe[**] and N.M. Pearsall[*]
Department of Electrical Engineering, Faculty of Engineering,
Rajamangala Institute of Technology, Pathumthani, Thailand.
[*]Northumbria Photovoltaics Applications Centre, School of Engineering,
University of Northumbria, Newcastle Upon Tyne, UK.
[**]Department of Geography and Environmental Management, University of Northumbria, Newcastle Upon Tyne, UK.

Abstract : A sample village with 100 households in a rural area of Thailand is selected for design of a solar home system, a photovoltaic (PV) battery charging station and a centralised mini-grid system. This paper focuses on the design and analytical evaluation of three main PV system types for a Thai rural village. This is based on climatic data at the design location, daily load demand, I-V characteristic of a solar module and so on. The daily electrical energy needs of a village can be broadly split into three categories, namely (a) for each household, (b) for community centre and (c) for public use. Analytical comparisons of each PV system type in terms of system cost and possible problems for using in typical rural villages of Thailand have been addressed.

Keywords : PV System - 1: Rural Electrification - 2: Life Cycle Cost - 3

1. INTRODUCTION

Rural electrification is one of the main applications of photovoltaic systems. Stand-alone PV generators are suitable for use in rural houses or villages that are located far away from the national utility grid. Since extension of the grid system into rural villages is expensive, there are many rural villages in Thailand that are without access to electric power. These villages are spread in rural areas through the country. Rural people use kerosene lamps and candles for lighting applications and there are no facilities for community entertainment. In some areas, the roads are in very poor condition and transportation of humans and materials is a problem. Hence, PV systems are a valid alternative source of energy to use in these rural areas.

PV systems have been widely used in typical remote villages in Thailand. Most systems are PV battery charging stations [1]. The results of this study are able to provide useful information for rural electrification planners and PV engineers to choose the optimum system for installation in a particular rural village in Thailand.

2. ASSESSMENT OF THE DAILY LOAD DEMAND

The first step of the design requires knowledge of the daily load demand in each household. The design approach is to calculate the array size based on solar irradiation at the design location and system parameters. The hourly load profile of the village is one of the most important data sets. Four load profiles are selected for analysis in this paper, namely: (a) night-time load, e.g., indoor lighting, street lighting and a TV set; (b) daytime load, e.g. water pumping; (c) constant load demand, e.g. refrigeration; (d) variable load demand, e.g. radio. The system designers should be given the correct data concerning the load profile in the village by the Head Villager or the Committee of the Village, who are a responsible for making a decision to install the system. Typical loads for design of a PV system in a rural village of Thailand with 100 households are shown in Table I.

3. EXAMPLE FOR OPTIONAL DESIGN OF PV SYSTEM IN A THAI RURAL VILLAGE

A typical Thai rural village, with approximately 100 households and far from the utility grid, is selected for the sample design. It is located in the Udon Thani province of Thailand at a latitude of 17°23'N and a longitude of 102°48'E. The mean annual global solar radiation on a surface inclined at the latitude angle is about 4.5 $kWh/m^2/day$ with a minimum of 3.8 $kWh/m^2/day$ in August and a maximum of 5.3 $kWh/m^2/day$ in January. The daily load demand (Table I) is used to design PV stand-alone systems of different types based on the same load conditions.

2.1 Option 1: Centralised mini-grid system

The array size depends on the climatic data at the design location and also on the I-V characteristic of the solar modules. The design of the array size must take into account various factors, such as the daily load demand, the battery efficiency, the mismatch efficiency, the regulator efficiency, the inverter efficiency and the line loss factor of the system. By using data from weather stations in similar climate zones, some estimate of the insolation can be made. This needs to be modified by a "variability factor" to allow for the variation from year to year, both in the mean and worst cases. By taking a large variability factor of 10%, it is reasonable to work with mean monthly insolation, as storage capacity will be high and so system performance is unlikely to be seriously affected by short-term fluctuations. The minimum array size is given by

$$N_A = (N_S \times AH_{(load)} \times \varnothing)/(K \times VF \times A_M \times D_M) \quad (1)$$
$$K = \eta_B \times \eta_M \times \eta_L \times \eta_R \times \eta_I \quad (2)$$

N_A is the total number of modules connected in a PV array and N_S is the number of modules in a series string. $AH_{(load)}$ is the total daily load in Ah, $\varnothing$ is the reference insolation intensity in kW/m^2, A_M is the current in amperes and D_M is the solar insolation in $kWh/m^2/day$. VF is the variability factor specified, η_B is the battery charging efficiency, η_M is the mismatch efficiency, η_L is the line loss factor, η_R is the regulator efficiency and η_I is the inverter efficiency.

$$N_S = (V_B + V_D + V_W) / V_{mp} \qquad (3)$$

where V_B is the nominal battery bus voltage, V_D is the forward voltage drop of the blocking diode, V_W is the total wiring voltage drop between the solar modules and the battery and V_{mp} is the voltage at the maximum power point and at operating temperature.

The most important factor influencing the selection of battery capacity is the occurrence of periods of continuous low insolation. These are likely to arise during the rainy season in Thailand. Choosing the available reserve days depends on solar radiation data and on battery types including the maximum depth of discharge (DOD). For a typical battery used for storage energy of PV power systems, it is recommended that it should not be discharged over 50 % DOD. The battery capacities required can be found from the following equations [2]:

$$E_b = (x E_n + y E_d)/[\eta_L \times V_D \times (DOD/100)] \qquad (4)$$
$$xE_n = x_1 E_{(radio)} + x_2 E_{(lighting)} + x_3 E_{(street lighting)}$$
$$\quad + x_4 E_{(TV)} + x_5 E_{(video)} + x_6 E_{(refrigerator)}$$
$$\quad + ... + x_n E_{(other loads)}$$
$$yE_d = y_1 E_{(radio)} + y_2 E_{(refrigerator)} + y_3 E_{(pumping)}$$
$$\quad + ... + y_n E_{(other loads)}$$

where E_b is the battery capacity expressed in ampere-hours, E_n is the total energy of load demand during night time in watt-hours, E_d is the total energy of load demand during daytime in watt-hours and $x_1, x_2, ..., x_n$ are the days of autonomy provided for different loads during the night. On the other hand, $y_1, y_2, ..., y_n$ are the days of autonomy provided for different loads during daytime. V_D is an average voltage of discharge of the batteries in volts.

For a centralised mini-grid system, the selected distribution voltage in the village is 220 VAC, 50 H$_z$. The input voltage of the battery bank is 240 VDC. It is recommended that PV systems over 5 kW$_P$ should use 240 VDC [3]. The village requires water of 20 m^3/day for pumping. As a result, the AC loads in Table I will be needed for design.

In practice, the efficiency of power conversion produced by all passive components in the AC circuit must be considered. It is supposed that there is a measuring system or equipment operated in a control room housing. Accordingly, the daily load demand is approximately 165 Ah at 240 VDC (power conversion efficiency is 85% and energy demand for measuring systems is 4800 Wh). For an overall efficiency factor of about 0.55 and based on a solar module type # BP 585 (at 1 kW/m^2, 25°C, AM 1.5, I_{mp} = 4.72 A, V_{mp} = 18 V), the equations 1 through 4 provide the PV array and battery size as shown in Tables 2 and 3.

A computer programme has been developed to calculate the daily state of charge (SOC) of the battery. The inputs to the computer programme include array and battery size, load profile, daily global solar radiation on an inclined surface at Udon Thani province of Thailand [4] and the I-V characteristic of the solar module. It has been found that, even in the case that the climatic data is reduced by 10%, the best relationship between array size and battery capacity corresponds to the system designed previously. The system needs a back up diesel generator of

15.5 kW and an inverter size of 16 kW with a battery charger. Furthermore, the conductor size and type of this mini-grid system are calculated. It is supposed that the length of cable from the control room housing to the farthest household is 0.5 km and lightning protection is included. This design is based on a maximum allowable voltage drop of 2% and a power factor of 0.85. Thunderstorm days and mean lightning flashes per km^2 per year are 80 and 4.7 respectively [5]. It can be concluded as follows: (i) the suitable cable size for the mini-grid system is 120 mm^2 (single phase system), (ii) for electrical equipment in a control room housing, the number of air terminators is 3, the number of ground conductors is 2 and the area of the copper conductor is 50 mm^2 and (iii) a grounding system should use 16 mm^2 (bare copper cable).

3.2 Option 2: Battery Charging Station System (BCS)

To design a BCS is less complicated than a centralised mini-grid system. The array size depends mainly on the number of charging points (units), ampere-hour loads that need to be charged and local solar radiation. Considering all DC loads (100 households) in the village, a BCS provides electrical power for charging the batteries used in each household. Other loads, such as public lighting, water pumping, refrigerator/freezer and community facilities are designed as PV individual systems. It is supposed that a BCS in the village has 25 charging points and each battery is normally discharged for 4 days with daily load demand. Based on the same module type and each rated battery of 120 Ah 12 V, the design of a BCS and other loads in a sample village at Udon Thani province of Thailand is shown in Table IV.

3.3 Option 3 : Solar Home System (SHS)

A typical load for SHS is lighting. By reducing the amount of power used for lighting, the householder can also use a small black and white television (B/W TV). Hence, household loads in Table I are slightly changed. They consist of a radio (6W) and a fluorescent lamp (8W) that are both used for 2 hours/day, and a fluorescent lamp (18W) and a small B/W TV (10W) that are both used for 3 hours/day. All loads are operated at 12 VDC using 130 Wh/day. Considering the same module type and battery rated in previous topics, the SHS for a sample village at Udon Thani province of Thailand is also shown in Table IV.

4. LIFE CYCLE COST (LCC) ANALYSIS

The LCC analysis is the appropriate method for comparing different electrification options. The present worth of different PV systems can be compared and the least cost option selected. A LCC analysis is based on the key assumptions (year 2000) shown in Table V. The costs of installation and operation and maintenance are estimated by multiplying the capital cost of PV arrays with 0.2 (20%) and 0.02 (2%) respectively. Using the data from the previous design above, the LCC analysis for the different PV systems is shown in Table VI.

5. CONSIDERATION OF POSSIBLE MERITS AND PROBLEMS FOR USING EACH PV SYSTEM TYPE

5.1 Centralised mini-grid system

When the public loads and community facilities are considered to operate under the control of a single system, a centralised mini-grid can be designed to cater for the overall village energy needs. It is usually designed to provide 220 VAC that feeds the village distribution grid. The power station consists of a large PV array, a control room housing the electronic load controllers, a large inverter and switching gears, battery storage and a back up diesel generator. Nevertheless, the investment cost of a project is extremely high. Experiences from other countries have been shown that, shortly after a centralised mini-grid was installed in a village on an energy need basis, the demand greatly exceeds installed capacity. This is because extra unauthorised loads, such as TV sets, and other electrical appliances are used. Another problem is that the equipment is very sophisticated and repair and maintenance are also costly.

5.2 Battery charging station system:

A fairly sophisticated load control system is required to regulate the charging of each of the batteries. An over-discharge protector should be installed in each household to protect the battery from over-discharging. In addition, all users or battery owners have to transport their batteries to the station and pick them up after the batteries are fully charged. This process may involve expenditure of time and money.

One of the big problems is that system failures cannot be technically repaired by a skilled person in the village. The waiting time for repair is long and this causes disillusionment amongst the villagers and a reluctance to use the PV station. Since skilled manpower is so scarce in remote areas, any systems having other than very simple operating or maintenance requirements are generally unsuccessful. The local small population, especially hill-tribe people, cannot provide skilled manpower and cannot usually keep those who may have been trained as part of a PV project. The system costs or sizing will vary greatly depending on local solar radiation and the number of charging points (units), which should be sufficient to avoid long queues.

5.3 Solar Home System

The advantages of a SHS are associated not only with improved lighting, but also with the elimination of dry batteries for radio/tape cassettes (whose cost usually represents a significant part of the annual earnings of an average rural family), and with the possibility of access to a television. Users can also work at night, adding income for their families, and children can have quality of studying, but the system sizing will be larger as well. Furthermore, a SHS usually requires DC appliances that have not been widely available in some rural shops, especially in Thailand, and are more costly than AC appliances. The investment cost of a SHS is too high for a rural family to pay as one single payment, but this problem can be overcome through appropriate financing schemes.

6. RESULTS AND CONCLUSIONS

According to data in Table VI, a SHS is the least-cost option compared with the battery charging station and kerosene mantle and wick lamps for lighting and rechargeable batteries used for operating small applications. The investment cost of a centralised mini-grid system is extremely high. Even though it closely mimics a conventional electrification system, it is much more restricted in the loads. Seemingly, interest in centralised mini-grid systems has been declining in the light of the experience gained to date.

A strategic model for PV dissemination in Thailand is necessary. Central PV systems are not an appropriate basis for a commercial market while individual PV systems such as SHS have already been proven to be commercially viable in some countries. It seems sensible therefore to move towards a commercial market for SHS with government provision on PV systems for public services in health and education.

In general, household PV systems are the least-cost option for a village with fewer than 200 connections and the PV mini-grid system is the least-cost option with 400 household connections and 100 households per km^2 [6]. Based on the analysis of existing projects in rural areas of Thailand, the SHS system should be encouraged more strongly. It seems clear that these systems provide a suitable level of service for households in rural areas at a monthly cost which is comparable to their present monthly outlay on lighting and small power. This contrasts with the situation for battery charging stations where the capital cost is beyond the capabilities of almost all individuals or local companies.

7. REFERENCES

[1] S. Hiranvarodom, R. Hill and P. O'Keefe, A Strategic Model for PV Dissemination in Thailand, *Progress in Photovoltaics: Research and Applications*, **7** (409-419) 1999.

[2] H. Saha, Design of a Photovoltaic Electric Power System for an Indian Village, *Solar Energy*, **27** (103-107) 1981.

[3] M.G. Imamura, P. Helm and W. Palz, *Photovoltaic System Technology* : A European Handbook, H.S. Stephens & Associate, England 1992.

[4] S. Hiranvarodom, *Estimation of Solar Radiation on Inclined Surfaces for Thailand*, Internal Report, NPAC, University of Northumbria, 1999.

[5] M Johansson, *Lightning Protection of Building and Their Contents: a review of current practice*, Technical note in 1/94, BSRIA, 1994.

[6] A. Cabraal, Mac Cosgrove-Davies, Loretta Schaeffer, *Best Practice for Photovoltaic Household Electrification Programs: Lessons from Experience in Selected Countries*, World Bank Technical Paper 324, Washington, D.C., 1996.

Appliances	Quantity	Power (w)	hrs/day	Remarks
FL small	100	8	3	loads for each
FL medium	100	18	6	household
Radio/tape	100	10	2	
FL 36W	10	36	5	loads for a
Television	1	120	5	community
VCR	1	40	5	centre, school
R/F	1	320	10	
LPG sodium	20	26	6	loads for
Pumping	1	600	8	public use

Table I: Typical loads for design a PV system in a sample village with 100 households

(all loads are AC appliances for a PV mini-grid system)

FL = Fluorescent Lamp, R/F = Refrigerator/Freezer,
VCR = Video Cassette Recorder, LPG = Low Pressure Gas,

Table II: Array size for a centralized mini-grid system in a sample village

Nominal system voltage (Vdc)	Modules in a series string	No. of strings in parallel	Maximum current (I_{mp})	Maximum voltage (v_{mp})	Maximum power (kW$_P$)
240	15	16	75.52	270	20.4

Table III: Storage battery parameters and sizing designed for a centralized mini-grid system

$x_1, x_2, x_3,$ x_4, x_5, x_6, x_7 (days)	y_1, y_2, y_3, y_4 (days)	V_D (Vdc)	DOD (%)	$x E_n$ (kWh)	$y E_d$ (kWh)	Capacity Designed (Ah)	Battery rated each (Ah)@12V	No. of batteries in series	No. of batteries in parallel
5	0,5,2,5	240	50	10.35	29.32	1600	200	20	8

Table IV: PV system sizing of different types based on the same load conditions in Table I in a sample village with 100 households

System Types	Array Size (kW$_P$)	No. of Modules	No. of Batteries
Battery Charging Station	10.62	125	100
Public Lighting	2.80	40	20
Refrigerator/Freezer	0.68	8	6
Community Facilities	1.61	19	18
Pumping System	1.02	12	-
Solar Home System	8.50	100	100

Table V: Key assumptions used in estimating cost in US$ (lifetime)

Life cycle period 20 years	Regulator $7 / amp (5)
Discount rate 10 %	Charger $6 / amp (10)
Inflation 5 %	Inverter cost $650 / kW (10)
Module cost $6 / kW$_P$ (20)	Fluorescent lamp $3-5 / lamp (2)
Battery cost $150 / kWh (5)	LPG sodium $25 / lamp (2)

Table VI: Comparison of LCC analysis for economic indicators of different system types in US$ (based on 100 households)

System Types	Capital Cost ($)	Levelised Energy Cost ($/kWh)
Centralized Mini-Grid	262,240	2.88
Battery Charging Station	108,860	8.54
Public Lighting	31,460	3.37
Refrigerator/Freezer	8,110	3.38
Community Facilities	17,570	2.54
Pumping System	12,500	1.26
Solar Home System	102,750	2.95

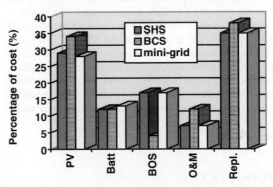

Figure 1: Comparison of life cycle cost of 3 PV different system types (all loads are included): solar home system, battery charging station and centralized mini-grid

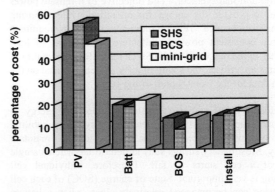

Figure 2: Comparison of capital cost for solar home, battery charging station and centralised mini-grid system

A Reliable Hybrid PV Power Architecture
for Telecommunications

Elias K. Stefanakos*, Paris Wiley and Sudharma Wijegunawardana
Clean Energy and Vehicle Research Center
University of South Florida,
4202 East Fowler Avenue, ENB 118, Tampa, FL 33620, USA.

ABSTRACT: The selection of a reliable and technically attractive power source is of great importance to the telecommunications industry especially where such equipment operates at remote unmanned sites. The utilization of a stand-alone hybrid PV power system is usually the solution in such cases. Most hybrid PV systems use batteries for energy storage and batteries become the weakest link that affects the reliability of the system because battery strings often start to deteriorate even due to failure of a single battery or cell. This paper is based on findings of the utilization of a hybrid PV power system as a reliable power source for Telecommunication applications through proper system management and battery (cell) equalization. The system is set up at the Clean Energy and Vehicle Research Center at the University of South Florida. It was found that the individual battery, generator, and PV system status could be determined in advance so that corrective actions can be taken without loss of reliability which is very important in this type of applications. The system was field tested under constant and variable load conditions to establish its performance.

I. INTRODUCTION

To assure reliability, stand-alone hybrid PV power supplies used as power sources for telecommunication systems require remote monitoring and management. The objective of a remote power system is to provide the energy for operation while requiring minimum maintenance. Telecommunication systems require a ripple free consistent power architecture with a loss of load probability in the range of 0.01%. Since reliability and quality of the systems are critical, the sizing of a PV system is based on the radiation data for the worst month of the year.

The performance of stationary batteries supplying energy to these remote applications often determines system reliability. Battery systems do need regular monitoring to maintain reliability. Most battery strings start to deteriorate when a single battery or a cell starts to fail. Therefore, individual cell monitoring is very important. State of charge (SOC) of each cell in the string can be managed relatively close, when a battery equalization method is implemented.

With the objective of providing a reliable power source for telecommunication applications, researchers in the Clean Energy and Vehicle Research Center at the University of South Florida have designed and field-tested a hybrid PV system that includes an intelligent management system.

The design of the system is based on the energy balance principle that guarantees the availability of the supply in the worst case. Fig. 1 is a schematic diagram of the stand-alone hybrid PV system, which is self-explanatory. The energy needed for the load is supplied by a PV array, and a generator is available under inclement weather conditions. The Battery string stores the excess energy and supplements the load during nighttime or during periods of reduced solar availability. A data acquisition system integrated in a personal computer controls the entire application. The monitoring system not only monitors the status, but it can also transmit information and alarms to a remote monitoring station.

II. EXPERIMENTAL SETUP

The overall system was set up using part of an existing 20kW solar installation at the University of South Florida. The load was sized according to the installed capacity of the solar array (variable load up to 288W (maximum); battery capacity of 140Ah; array capacity of 660W at a radiation of 1000 W/sq-m; system voltage of 48V). The charge control functions are performed using the energy monitoring and management system.

The energy monitoring system consists of sensors attached to all measuring points; a multiplexer and D/A board connected to a PC for data acquisition and control; and a switching module to switch charging modes. A PV charge controller controls the array voltage. A PC at a central monitoring station can communicate with the controller via a modem to receive alarms, to transmit controls and to download data for analysis. Fig. 2 shows the functions associated with the battery monitoring and management system.

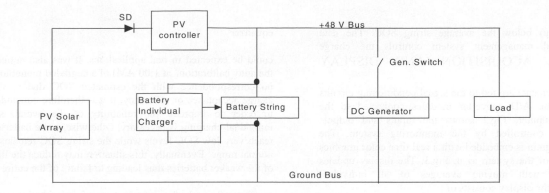

Fig 1: Block Diagram of the Hybrid PV System

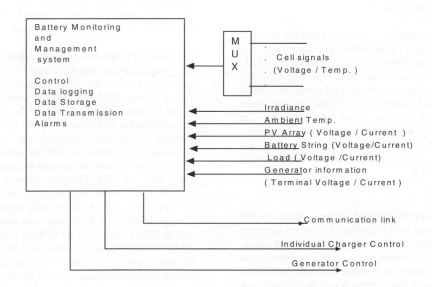

Fig 2: Battery Monitoring and Management System

III. EXPERIMENTAL PROCEDURE

The charge control is one of the essential functions expected from the energy monitor. The voltage and state of charge (SOC) of the battery string and the individual monoblocks are continuously being monitored at preset time intervals in order to decide what control method to follow. Special algorithms based on the discharge voltage profiles of the lead acid batteries have been used to determine the SOC of the batteries. Each day at 1:00 AM, the SOC of each battery was recalibrated to eliminate any accumulated errors in the SOC estimation.

Energy requirements are met either by the solar array, battery string or DC generator. The DC generator is simulated by a 48V/15A-power supply, which can charge the battery string from 40% to 80% SOC within a reasonable time. Once the battery string SOC is below a threshold (i.e. 40% SOC-arbitrarily set) and sufficient solar energy is not available, the generator is automatically connected to the DC bus. As a result the battery is not drained during insufficient insolation. When the battery string reaches 80% SOC, the generator is

automatically disconnected. At very low insolation levels the charge controller disconnects the solar panels from the system. The charge controller maintains the load voltage within ± 5% of 48V. Digital outputs from the D/A board have been used to change the operating modes as instructed by the charge control algorithm. The battery string is the most unreliable component of the hybrid photovoltaic system, since the reliability of the battery system often depends on the weakest battery of the string. Due to capacity differences, identical batteries may vary in performance from each other. The reliability can be improved significantly, by identifying and addressing the weakest cell or monoblock that is prone to failure.

In this study, a non-dissipative charger (equalizer) is used in order to boost the weakest battery module up to the average charge level of the remaining batteries. The single battery (cell) equalizer consists of a muliplexer, alternative switching bridge and a bus-isolated dc-dc converter. With a suitable switching arrangement, the weakest battery is charged using the energy available on the main bus. In order to activate the individual charger, the SOC of the weakest battery must be at least 10% (an arbitrary setting) below the average string SOC. The data acquisition and management system controls the charge

IV. DATA ACQUISITION AND DISPLAY STATUS

System measurements are fed to the signal conditioning circuits and then to the A/D converter modules connected to the monitoring computer. Fig.2 shows the inputs and outputs supervised or controlled by the monitoring system. The monitoring program is embedded with a real time color graphics screen display of the system as in Fig.3. The display updates every 5 min with moving averages of all individual measurements. the display consists of :

- Individual battery voltages, SOC and temperatures
- Voltages and currents of the battery sting, PV array, load and generator
- Solar Radiation
- PV array power output and string SOC
- Special alarms, date and time

The display indicates the status of the system and the outcome of the fault check. The display is organized in a such a way that attributes a change in color from GREEN > YELLOW > RED under alarm conditions for each individual item. Under normal conditions , all items are displayed green. Red alarms are critical and should be attended immediately. Some of the alarm conditions are displayed on the screen in text form, for convenience in trouble shooting. Other than displaying alarms, the management system is configured to collect all data and store them for analysis. The central station can download and analyze the data as required.

V. RESULTS

The hybrid PV system was field tested in three phases. In the first phase it was operated with a constant telecommunications load but without a battery equalization. In the second phase, individual charging was introduced under constant load conditions. In the third phase, a variable load was introduced consisting of three different load current levels throughout each day. Subsequently, the findings from those three phases were analyzed.

In the first phase, with the exception of the batteries, the balance of the system operated reasonably well. Battery charging was fully controlled and the possibility of overcharging was negligible. It was noticed that the charge controller controls the voltage from the array whereas the monitoring and management system controls battery overcharging and undercharging.

It was found that the SOC of individual batteries varied significantly from each other; this is especially true for the weaker batteries. A screen capture of the monitoring system in Fig.3 (real time display) illustrates this well. Consequently, alarms were triggered for weaker batteries. Alarm information could be made available to the Central Monitoring station alerting trained personnel of the potential problems with batteries. It is important to note that although few batteries have triggered alarms, the battery string SOC remains in the normal range.

Later in the experiment, it was found that battery #7 was defective and was then replaced by a good battery. Due to electro-chemical differences in batteries, such characteristics

equalizer

could be expected in real applications. It was also noticed that the auto_calibration(at 1:00 AM) of a degraded monoblock had no correspondence with the estimated SOC due to it's low charge acceptance or efficiency. It was therefore concluded that in order to improve the reliability of the overall system, individual charging is necessary. Otherwise, some batteries may reach very low SOC levels while the string SOC remains in the normal range. Eventually, this situation may reduce the life time of the weaker batteries thus leading to failure of the entire string.

In phase 2, with the introduction of individual monoblock charging, individual charging took place in the case where the weakest monoblock SOC differed from the string SOC by at least 10 %. The individual charger would stop charging when the charged monoblock SOC equals the string SOC. Unless dealing with a defective battery, the chances of battery failure due to long term discharge are greatly reduced and the alarming situation displayed in Fig. 3 is avoided.

Fig.4 shows the analysis of sample data obtained during a period of 172 hours in the second phase. the results are shown in two segments. During the first segment, the battery string SOC varied between 95% and 54% and individual charging of battery #7 took place in two occasions. The SOC error corrected by auto-calibration is greater for battery #7 as compared to other batteries. Generator charging may occur after long term discharge under inclement whether conditions. The second part of Fig.5 shows that the generator started charging the battery string SOC of 40%. Charging stopped when the string SOC reached 80%. In the last phase, i.e. with the introduction of a variable load , new batteries were installed. The variable load was specified at the following three current levels: 2.5A (8:00 AM to 6:00 PM), 1.5A (6:00PM to 12:00 AM) and 1A (12:00 AM to 8:00 AM). It was found that the hybrid PV system with variable load is a good and reliable power source for telecommunications under proper management.

VI. CONCLUSIONS

Results presented in this paper have clearly demonstrated that a stand-alone hybrid PV system can be used as a reliable power architecture for telecommunications especially in remote applications where power quality is of essence. a suitable monitoring and management system is needed to provide information on the status of all the system components including: PV system and controller, auxiliary generator, batteries and other control hardware and battery(cell) equalization. Experience has shown that batteries may be the weak link affecting the reliability of the hybrid system as a power source.

Monitoring the status of individual batteries in a battery string is necessary because a bad battery(monoblock) may effect the entire battery. The method used in this work for charging(equalizing) the weakest monoblock in the string has provided an acceptable solution that eliminates differences in the quality and electrochemical behavior of batteries.

The remote information about the power supply reduces the frequency of maintenance, increases the reliability and provides a convenient tool for troubleshooting in this type of application.

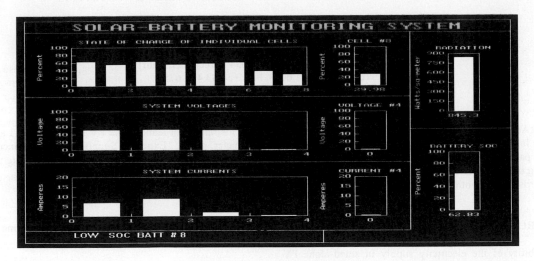

Fig 3: Hybrid System - Screen Display
Large deviations in monoblock SOC are clearly indicated by means of different colors and alarms

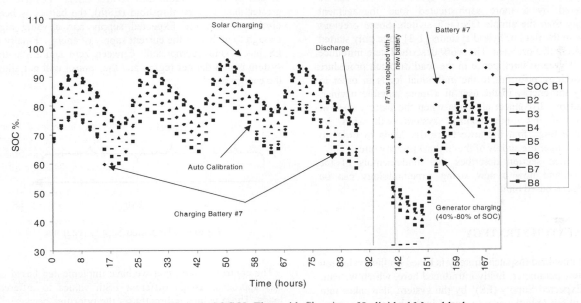

Fig. 4.   Phase 2: SOC Vs Time with Charging of Individual Monoblocks

ACKNOWLEDGEMENTS

We would like to thank Mr. Thomas Smith and Mr.G. P. Dissanayaka for their help in this project. The financial support provided by the Aboly Foundation is greatly appreciated.

REFERENCES

[1] A handbook of recommended design practices, Sandia National Laboratories, Albuquerque, New Mexico, USA, March 1990

[2]. Photovoltaic System Design. Laboratory Manual, Florida Solar Energy Center, Cocoa, FL, USA. 1988

[3]. D. Berndt, Maintenance Free Batteries, Handbook of Battery Technology, Second Edition, 1997

[4]. E. Stefanakos, G. P. Dissanayake, P Wiley, Monitoring of a Hybrid Photovoltaic System for Telecommunications, 2nd World Conference and Exhibition on Photovoltaic Solar Energy Conversion, Vienna, July 1998

A NEW LOAD DISPATCH STRATEGY FOR STAND ALONE PV SYSTEMS

A. Moreno, J. Julve, S. Silvestre and L. Castañer
Universitat Politècnica de Catalunya
Campus Nord, Mòdul C4, 08034 Barcelona (Spain)
Tel. +34 93 401 6773, Fax. +34 93 401 6756,
E-mail: amoreno@eel.upc.es

ABSTRACT: The aim of this work is to develop a new load dispatch strategy for stand-alone PV systems. A first attempt to include load priority in the control system is presented. A new parameter, named 'Expected Supply, ES' is introduced as key control parameter which is evaluated in real time out of monitoring results. Experimental and simulation results are presented showing that the effective loss of load probability can be improved for the highest priority loads
Keywords: Plant Control - 1: Expert System - 2: PV System - 3

1. INTRODUCTION

Reliability of the electricity supply in stand-alone PV systems is the result of several factors, among them the incertitude with the user consumption is known. Generally indiscriminate use of the energy while it is available leads to supply failure of all loads at a given time. This can be improved by a more sophisticated load management strategy than the simple ON/OFF switch. Some previous works in the field of hybrid systems [1-3] have only started to address this problem. The work described here introduce some degree of intelligence in the load dispatch procedure by assigning priorities to the individual loads in order to supply the energy to the user in a more suitable manner. Instead of disconnect all loads when the battery state of charge is low, our control system will start by disconnecting first the lowest priority loads in order to extend longer the supply of the highest priority ones.

The next section describes the fundamentals of the strategy and how a new sizing methodology can be introduced based on it.

2. DISPATCH STRATEGY

The new load dispatch strategy is based on the evaluation of a new parameter, firstly introduced here, which we call: the 'Expected Supply (ES)' by the system, that takes into account not only the energy stored in the battery at a given time (E), but also the instantaneous balance between the PV array generated power (P) and the power being delivered to the load (L) at the same time. The Expected supply is defined as follows:

$$ES(h) = \frac{E(Wh)}{L(W) - P(W)} \qquad (1)$$

where it can be seen that the ES is given in units of time, namely in hours if the energy available is given in Wh and the Power in W. This parameter is a measure of the energy state of the system in terms of the time it would supply the current power requirements if the radiation was the same. If ES is positive implies that the consumption is greater than the PV power, the system is loosing stored energy, however, if ES is negative the energy stored by the system

is growing and ES has the meaning of the time need to double the energy stored at the battery.

Three working areas of the system are identified with the help of Figure 1. In the first region where ES is negative, the battery is actually being charged, so the current supply of energy can be authorised. When ES is positive but greater than a given threshold (ESth), the battery is been discharged but the Expected supply has a value high enough to authorise the current supply of energy. Finally is ES has a value comprised between zero and ESth, the system is in a danger region and the control will not allow the energy supply.

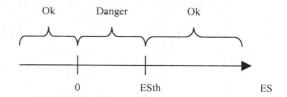

Figure 1. Expected Supply reegions

The control system, that we have implemented based on this approach, assign different ESth values to different loads. These values indirectly set the priorities because, a higher priority implies lower ESth value. The way the control works is schematically shown in Figure 2. A sequential procedure is started in real time by detecting which of the loads are demanded by the user at a given time. The systems reads the value of ESth assigned to lowest priority load among the loads demanded. The system then computes the value of ES in Equation (1) from the instant values of the Battery stored energy, power delivered by the PV array and the load demanded.

A comparison between the computed value of ES and the threshold of the lowest priority load and the system checks if the state falls into the danger region, if this occurs, the system takes action and disables the load.

Immediately the same procedure re-starts with the second priority load in the list until the value of the computed SE falls in a safe region as defined in Figure 1.

As can be seen at any time the last loads to be disconnected are the ones assigned with the higher priorities.

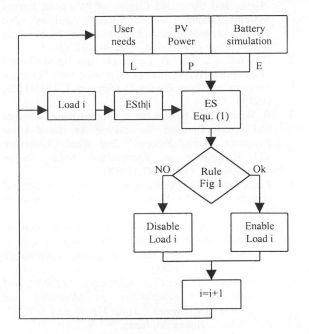

Figure 2. Load Dispatch Strategy

A key issue in this procedure is how to know the value of E, this is the energy stored in the Battery at a given time. The developed battery model is based on the equations described by Coppetti et al. [4], the model parameters contemplated are: Initial SOC, C10, number of series cells, and two empirical constants depend on the battery characteristics. Direct implementation of the analytical equations of the model has been programmed in Matlab taking into account the possible different operation modes of the battery. A function has been developed to iteratively solve the equations governing the PV system, taking into account: The PV module and battery model equations and the measured irradiance and temperature profiles. The implemented algorithm is based in the Newton-Rhapson method. In order to extract the final values of the battery model parameters a second function has been implemented modifying the internal LINFIT function of Matlab based in the Levenberg-Marqdart algorithm [5].

The inputs of the extraction function are: Voltages and currents of the system obtained by monitoring, temperature profile, and initial values of SOC and C10. The results obtained are the final values of SOC and C10.

This developed battery model has been contrasted successfully with previous models developed under Spice environment [6].

3. RESULTS

The new control system has been validated firstly by simulation of a PV system with a normalised capacity of 0.882 for PV modules and 2.0 for battery. The load is composed by four 150 W individual loads that consumes

all time. This implies a total daily load of 14400Wh. The simulation time is one full year with a simulation step of one hour in which the control system has been included. From this simulation, suitable values of ES thresholds have been identified in order to fulfil a given value of the systems loss of load probability (LLP). In fact, and this a feature of our method, the LLP concept can be applied to every individual load and not only to the system as a whole. Moreover changing the ratio of ES thresholds among the loads, produces changes in the individual values of LLP's as can be seen in Figure 3. An index ftu is used such that $SEth0 = 0*ftu$, $SEth1 = 1*ftu$, $SEth2 = 2*ftu$ and $SEth3 = 3*ftu$, for every load (C0, C1, C2 and C3). We can conclude that as the threshold values space out, the LLP of higher priority loads (C0 and C1) improve.

Finally, we have also implemented this strategy in a prototype system, deliberately undersized to be able to monitor failure of supply situations.

The PV section size is 550 Wp and the battery 740 Ah at 24 V. The load has four bulbs of 100 W with a total daily load of 4800 Wh. Figure 4 shows the results of six days of august 1999 in Barcelona (Spain). The upper graph shows the load profile demanded by the user. We have four loads from higher to lower priority (C0,C1,C2 and C3). The bottom graph shows the power effectively supplied, by the system. It is clear that not all loads required are served.

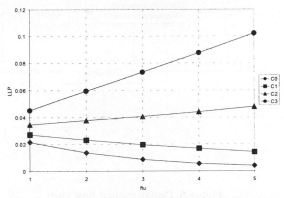

Figure 3. Dependence of the LLP on the threshold values

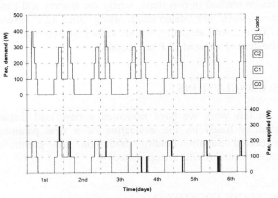

Figure 4. Experimental results

The implications of using advanced control in PV stand-alone systems in the sizing of the array and battery size are important and new methods of system can be introduced. The LLP concept should be revisited due to the individualisation of loads and separate control of each of them. Convergence with domotic systems is a clear way to implement these methods and control systems.

4. SIZING IMPLICATIONS

In order to do the design process easy with this new control, we have also developed a new design method that depart with the number of loads, the LLP desired for each one of them and location. The method gives the battery and panels dimension and the threshold values, SEth for every load. The method is inspired in the Numerical Iso-realibility method [7] that iterates the simulation of one year until it arrives at a good solution.

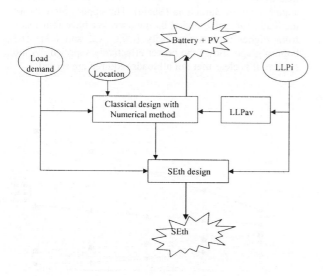

Figure 5. Design method flow chart.

In the Figure 5 we can find the flow chart showing the design method in two steps. Firstly the Battery and panels dimensions are determined using the LLP average (LLPav). And secondly, the set values of Seth that verify the desired value of LLP are founded.

5. CONCLUSIONS

In this work we have presented a new load dispatch strategy based on the evaluation of a new parameter: Expected Supply. Finally simulation and preliminary experimental results are presented

ACKNOWLEDGEMENTS

This work was supported by the CICYT TIC97-949 project

REFERENCES

1. C. Dumbs, N. Maizi, F.P. Neirac and D. Mayer, 'Sizing and Optimised Control of PV-Diesel Energy Systems Desing and Performance Analysis', 2nd Word Conference and Exhibition on Photovoltaic Solar Energy Conversion, 3, 3234-3237, (1998).

2. B. Wichert and W. B. Lawrance, 'Intelligent Control of Modular Hybrid Energy Systems', 14th European Photovoltaic Solar Energy Conference, 1, 1026-1029, (1997).

3. B. Wichert and B. Lawrance, 'Photovoltaic Resource and Load Demand Forecasting in Stand-Alone Reneware Energy Systems', 2nd Word Conference and Exhibition on Photovoltaic Solar Energy Conversion, 2, 3194-3197, (1998).

4. J.B. Copetti, E. Lorenzo and F. Chenlo, "A General Battery Model PV System Simulation", Progress in Photovoltaics, vol. 1 nº4, 281-292, (1993).

5. William T. Vetterling, William H. Press, Saul A. Teukolsky, Brian P. Flannery. "Numerical recipes in C. The Art of Scientific Computing.", Cambrige University Press, (1988).

6. S.Silvestre, D.Guasch, A.Moreno, J.Julve and L.Castañer. *"A comparison on Modelling and simulation of PV systems using Matlab and SPICE"*, *11th* PVSEC, Hokkaido, Japan, 901-902, (1999).

7. M. Egido and E. Lorenzo, " The sizing of stand-alone PV-systems: a review and a proposed new method", Solar Energy Materials and Solar Cells, 26, (1992).

POTENTIAL USE OF RENEWABLE ENERGIES FOR WATER AND WASTE WATER TREATMENT IN RURAL ZONES

E. MICHEL, A. DEVES
COSTIC - Comité Scientifique et Technique des Industries Climatiques
Rue Lavoisier – Z.I. de Saint Christophe - 04000 Digne les Bains – France
Phone: 33.4.92.31.19.30 - Fax: 33.4.92.32.45.71 - Email: costic.digne@costic.com

ABSTRACT: Drinking water and waste water treatment is an essential concern for rural communities. Power supply of installation is one of the main problems to solve in rural zone where water treatment plants are often remote from the grid. The use of renewable energies could be an interesting way to ensure the growth of these installations.
As concerns drinking water, some surveys show that needs are still important in rural zone, even if many realisations supplied by renewable energies already exist. Regarding purification, few installations have been achieved, although renewable energies could allow to fit with adapted systems some rural communities or could improve the working of some existing systems.
KEYWORDS: PV system – 1: Rural electrification – 2: Remote – 3

INTRODUCTION

Needs for treatment of drinking water and for purification are very important in rural zones. In many places no systems exist or the existing installations are not satisfactory.

To solve the problem of drinking water supply two main directions of action should be developed simultaneously. On the one hand it is necessary to treat more systematically the water intended for human consumption, and on the other hand it is necessary to find the means of effectively treating waste water to avoid contamination of the natural environment, and then of the water reserves.

In urban zone it is often only a financial problem, while in rural zone the problem is more recent and presents more obstacles. Indeed, in addition to financial means, it is necessary to take account of specificities of the small communities and to find adapted techniques. The systems must be thus technically adapted (low flow rate, not very charged effluents in the case of the waste water treatment) and must have reduced exploitation constraints and reduced operation costs (limited maintenance and possibility for unskilled personnel to carry out this maintenance).

There is sometimes an additional problem: the power supply. Indeed, many rural zones are far away from the electrical grid and cost of connection can be expensive (about 26 000 EURO/km of line). That is why the production of decentralized electricity using renewable energy can be proved to be profitable or even necessary.

This paper presents a study realised for French Agriculture Department (Ministère de l'agriculture et de la pêche - Direction de l'espace rural et de la forêt - Sous-direction de l'aménagement et de la gestion de l'espace rural - Bureau des infrastructures rurales et de l'hydraulique agricole).

1. ANALYSIS OF THE NEEDS

1.1. Drinking water
A study carried out in a rural and mountain zone [1] showed that:
– Only 28% of water collected are treated;
– 12% of the networks have a bad bacteriological quality and 46% a medium quality.

These data confirm that even in mountain zone, a good quality of water is not always naturally obtained.

With regard to the potential installations that can be achieved by using renewable energies, it can be concluded that:
– The distance from the grid and the low capacities of investment of the rural communes make that a power supply of the treatment systems by renewable energies seems to be the best solution in about 50% of the cases;
– In mountain zone, a hydroelectric pico-power station could often produce energy necessary to the treatment for there is an important difference of level between the collecting and the tank.

1.2. Waste water treatment
In rural zone, installations are often small sized (many stations with capacities lower than 100 Population Equivalent (PE)). The most used type of installations is trickling filter. The use of renewable energies for the supplying of small sized purification plants would make it easier to implement them in remote areas. Moreover to supply by renewable energies systems like automatic screening plants, with few maintenance, could improve the working of some existing systems.

1.3. The existing installations
Some installations supplied by renewable energies have been listed in France. At the moment the use of renewable energies for the treatment of drinking or waste water is mainly limited to installations of chlorination (90% for chlorination, 5% for UV treatment and 5% for purification). Solar energy is the most used. No example of use of the wind power to supply with electricity a system of treatment of water has been found (93% of the installations are supplied by photovoltaic panels and 7% by hydroelectricity pico power plant).

2.   POSSIBLE APPLICATIONS

2.1.   Treatment of drinking water [2]

For drinking water, various solutions quite well adapted to the needs have been already implemented. Three main systems exist:

- Chlorination by soda hypochlorite solution measuring pump (the most widespread application)
- Gaseous chlorination
- UV sterilisation.

These systems can be either supplied by some electric batteries charged by photovoltaic panels or by a pico-power station.

Intermediate chlorination supplied with a pico-power station is a spreading application. Indeed in the case of very wide water supply networks with only one point of chlorination, it is sometimes difficult to remain in acceptable standards concerning the remanent chlorine amount. It is then useful to add intermediate chlorinations.

Finally, an application which should grow in the next years is the treatment by turbine action of collecting water, in the case of a source, in order to supply the system of treatment (chlorination or UV treatment).

Table 1 - Drinking water treatment with renewable energies

	TREATED FLOW RATE m³ / day		ELECTRIC NEEDS Wh / day		PHOTOVOLTAIC PEAK POWER According to location Wp		PICO-HYDROPOWER PLANT INSTALLED W
	Min	Max	Min	Max	Min	Max	
Hypochlorite solution	0	150	25	75	40	90	
	1 000		500				100
Gazeous chlorination	150	500	75	250	90	300	
	2 000		1 000				200
UV	2		350		150	300	
	75		1 425		800	1 250	
	200		2 200		1 200	1 800	500
	200	350	7 200				1 000

2.2.   Example of application

This application is located in Savoie (France). It allows to treat 2 000 m³/h with a hypochlorite solution measuring pump supplied by a pico power station. The power of the turbine is 400 W. the total cost of this installation is about 20 000 EUROS.

Photo 1 - Hydroelectric pico-power station, alternator

Photo 2 - Hypochlorite solution measuring pump

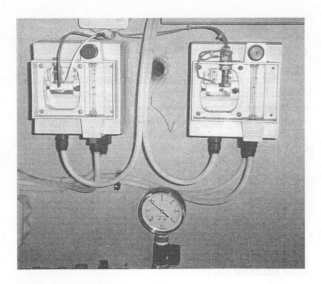

Photo 3 - Chlorometers

2.3. Waste water treatment

2.3.1. Supplying of systems which require electricity

Some installations could be supplied by renewable energies according to their low energy consumption:

- Dry toilets (toilets that don't need water or toilets with compost) are often supplied by photovoltaic panels. Indeed, the only electric consumption comes from mechanical ventilation. This system is widely used in mountain sites remote from the grid;
- Rotating biological contactors could be supplied by renewable energies. There is already an installation supplied by photovoltaic panels in Austria. Nevertheless the supplying of such systems by solar energy remains limited to small capacities;

- Trickling filters whose power consumption is about 10 Wh/inhab/d. It is one of the main systems installed in rural zones. A wide number of potential installations supplied with renewable energies thus exist. For instance, on the major part of the French territory, photovoltaic panels could supply stations up to approximately 500 PE. Nevertheless, according to zones, the use of wind power would be preferable for it would reduce the operation costs;
- Natural lagoons are sometimes slightly ventilated to improve their effectiveness. The energy needed is only some Wh/inhab/day;
- Finally, prefabricated stations of purification (trickling filters or activated sludge plants) for small capacities (about 50 PE) can also be supplied by renewable energies.

With regard to other techniques, like activated sludge plants and ventilated lagoons, it would initially be necessary to work on the process and on the savings in energy before considering the possibility to supply these systems by renewable energies (wind generator for example). It would be then necessary to evaluate the electrical needs with a margin for the future evolution of the station.

2.3.2. Supplying of some annex facilities to improve the effectiveness of the systems

- Screening;
- Raising stations to allow connection to the collective network of located downwards dwellings;
- Stirring machines for sludge silos where sludge is stored until agricultural spreading;
- Alarms and remote transmissions.

Table 2 - Waste water treatment with renewable energies

	POPULATION EQUIVALENT PE	ELECTRIC NEEDS Wh / day	PHOTOVOLTAIC PEAK POWER According to location Wp		PICO-POWER PLANT W
			Min	Max	
Ventilation of dry toilets	10 000 times used	60	50		
Lagoons ventilation	400	5 000	2 500	5 000	1 500
Rotating biological contactors	100	6 750	3 000	7 000	1 000
Trickling filters	400	3 500	2 000	4 000	500
	1 000	9 200			1 000
Screening plants	500	300	200	500	300
	1 000	800	500	1 000	400
Raising pump of waste water (mean values – vary according to the site)	400	4 000	1 700	3 500	1 000
	1 000	8 000			2 000
Stirring machines	1 000	4 000	2 000	4 000	4 000

2.4. Example of application

This application is located in Herault (France). It allows to ventilate the first lagoon of a purification plant which capacity is 800 PE (Population Equivalent). The peak power of the photovoltaic array is 4 080 Wc. The installation works "along the sun". The total cost of the installation is about 40 000 EUROS.

Photo 4 - The photovoltaic panels

Photo 5 - Aeration units

Photo 6 - Biological supports

CONCLUSION

Regarding drinking water treatment, it appears that renewable energies suit well with needs of local communities. Many realisations already showed the possibilities offered by such an energy supply system.

As concerns purification, two research orientations can be defined:
–   First, to seek the less energy consumer systems, likely to be supplied by renewable energies;
–   Second, to improve the operation of rustic processes that don't require energy by using some applications supplied by renewable energies.

Anyway local communities have interest to integrate renewable energies in their reflection when they are working on the realisation of their drinking and waste water treatment plants.

REFERENCES

[1]   DDASS des Alpes de Haute Provence, service santé environnement, Qualité des eaux destinées à la consommation humaine Bilan 1993 – 1994 – 1995 (avril 1997)

[2]   Michel E. Application de l'énergie photovoltaïque à l'alimentation en eau potable en zones rurales (1995) – documentation technique FNDAE n°12, Ministère de l'Agriculture et de la pêche

[3]   Michel E., Deves A. Application des énergies renouvelables à la potabilisation et à l'épuration des eaux (submitted) – documentation technique FNDAE n°23, Ministère de l'Agriculture et de la pêche

STAND-ALONE AC PV SYSTEMS AND MICRO GRIDS WITH NEW STANDARD POWER COMPONENTS – FIRST RESULTS OF TWO EUROPEAN JOULE PROJECTS "PV-MODE" AND "MORE"

Ph. Strauß[*], R.-P. Wurtz[*], O. Haas[**] M. Ibrahim[**],
[*]ISET e.V. / [**]University Kassel,
Königstor 59, D-34119 Kassel
Tel: (49) 561/7294-144   Fax-100
pstrauss@iset.uni-kassel.de

D. Mayer, C. Trousseau
ARMINES, Ecole de Mines
F-06904 Sophia Antipolis
mayer@cenerg.cma.fr
celine.trousseau@cenerg.cma.fr

P. Romanos, S. Tselepis
CRES / 19th km Marathonos Av.
GR-19009 Pikermi
tromanos@cres.gr
stselepi@cres.gr

J. Reekers
SMA Regelsysteme GmbH, Hannoversche Str. 1,
D-34266 Niestetal
reekers@sma.de

F. Perez-Spiess
ITER, Parque Eolico
E-38594 Granadilla Tenerife
spiess@iter.rcanaria.es

M. Bächler
WIP, Sylvensteinstraße 2
D-81369 München
renewables@mail.tnet.de

ABSTRACT: Stand-alone PV and hybrid systems can efficiently supply power for small remote consumers. Besides applications with a low power demand of up to some 100 W which can be covered by the well known solar home systems there is a demand for a new generation of standard island systems which are able to deliver single-phase or three-phase AC power of high quality. These systems should be able to supply power above about 2 kW e.g. for small workshops, small clinics and establishments for tourism. The produced power should be comparable to standards fulfilled by interconnected grids and the systems should also have the capability to form micro grids. The realisation of this task with the Modular Systems Technology (Kleinkauf 1999) requires further development and adaptation of appropriate electronic inverters which are able to form inverter dominated grids. The necessary features and adaptations for the power components battery inverter, PV-inverter and diesel genset are analysed in the two projects which are described here. Laboratory test concerning the grid quality of such AC PV battery systems and PV diesel hybrid systems are evaluated and compared to utility grid standards and to standards for diesel genset power quality. Further investigations include the comparison of different operational control strategies applying a newly developed modular simulation tool. Another aspect that is examined is the applicability of the European Installation Bus (EIB) for control purposes. In the experiments the EIB control is not only applied on the loads side but also on the generators side.
Keywords: Stand-alone PV Systems – 1: Hybrid – 2: Kythnos – 3

1   INTRODUCTION

The main objective of the work is to improve standard low-cost AC PV systems technology for small stand-alone applications and thus force the decentralised electrification with renewable energies. Simple expandability and compatibility to public grids will allow - in a later phase – adaptations to increasing consumer needs and also interconnection with extending public grids

The feeding of solar power into interconnected grids applying industrial PV inverters is successfully performed for several years now. But there is considerable R&D need for setting up small stand-alone inverter dominated grids where the PV generators feed in the power decentralised on the AC side. Market available PV inverters for interconnected grids can be adapted for this purpose if the properties of the island grids are comparable to the properties of interconnected grids. Principally such small grids which have a very high penetration level of PV power need PV inverters which can be limited in power output. Further items that have to be regarded include safety aspects, grid stability, grid quality as well as grid control and communication structures. Besides these technical issues which can be investigated in the laboratory the two projects help to identify problems which can be uncovered in the field tests like systems operation management, remote operation and adaptation of systems operational control according to specific consumer needs.

The projects which are described have the following technical and industrial objectives:

- Development of design rules for stand-alone AC PV systems and minimisation of expenditure for design, installation and maintenance

- Development and implementation of control and communication techniques for the PV inverter-dominated island systems with effective communication technology.
- Improvement of the functionality of components and operational control in laboratory plants.
- Verification of the technology by setting up and operating first end-user systems.

2   TECHNOLOGY

For supplying small single consumers up to a total power of a few kW, 230V/50Hz single-phase technology is suitable, whereas for larger systems including systems supplying bigger rotating loads, 400V/50Hz three-phase technology can be used. In the following the three basic system types which are assembled applying a set of only a few basic components are under investigation:

- single-phase AC PV-battery system
- three-phase AC PV-battery system
- three phase AC PV-battery-diesel system

For the first two system types the consequent parallel AC coupling of all components leads to inverter-based PV stand-alone systems with the battery inverter as the grid master. The third system type which is a hybrid configuration can operate in two operation modes: a) the battery inverter is the grid forming unit and b) the diesel genset is the grid forming unit.

The following steps for reaching the above mentioned aims are necessary:

- Improvements of components in the laboratories,
- Development of a modular simulation tool,
- Installation of end-user systems in Greece and

- Remote monitoring for long-term assessment.

The technology for the AC PV-battery systems without rotating generators has special requirements concerning the power reserve in the inverters e.g. for starting machines and tripping electrical circuit-breakers (Burger 2000). For the hybrid systems which also have a diesel generator set special technical aspects for the diesel integration are investigated. These include new concepts of operation and new maintenance strategies for such hybrid configurations as well as necessary communication interfaces.

3   TECHNICAL REALISATION

3.1 Single-Phase PV-Battery System

The smallest extension step consists of one 3.6 kW battery inverter and a 700 W PV inverter. The battery bank connected is dimensioned according to the users requirements.

Figure 1: Single-phase AC PV system in ISETs DeMoTec centre including one 3.6 kW battery inverter, three 700 W PV inverters and a contol and monitoring unit which can be controlled via GSM-technology

A similar system will be installed at a farm house in the mountains of Kythnos applying 2.2 kWp PV and a 60V / 490 Ah battery bank. The loads to be supplied include standard AC household loads and a small water pump for irrigation.

3.2 Three-Phase PV-Battery System

Three synchronised single-phase battery inverters are interconnected to form a three-phase system. The battery inverter on Phase 1 is the master and the others provide an output with the same voltage and frequency but with a phase difference of 120° and 240° respectively. The nominal power output of this system is 11 kW. Unbalanced loads can be supplied if the total load on each single phase does not exceed the limitations given for one battery inverter i.e. 3.6 kW. All three inverters are connected to the same battery bank.

The described three-phase application is a possible extension step for end-users who are already have the single-phase PV-Battery-System described in above. Two additional battery inverters and a software reconfiguration

permit a stepwise plant extension with the same standard components. The total energy of the PV-inverters connected to the AC-bus should not exceed the battery inverters nominal power.

In the frame of the PV-MODE project the three-phase PV-Battery System will be installed for the electrification of some houses in a remote area on Kythnos. The houses will be interconnected with a micro grid. Due to the system structure it is possible to feed in PV power at any point reached by the grid. This allows the integration of several PV generators at optically or functionally optimal places. All the known PV integration options like e.g. roof integration or shadowing facilities are possible. First discussions with the end-users have shown that a smooth integration of the PV is of great interest.

3.3 Three-Phase PV-Battery-Diesel Hybrid System

The integration of diesel gensets into small AC coupled PV-battery micro grids is investigated in the MORE project (see Figure 2). Here two principle operation modes are possible: in one operation mode the battery inverters form the electrical grid and in the other operation mode the

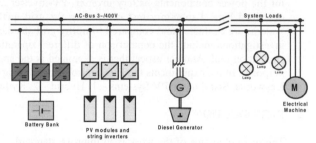

Figure 2: Structure of a three-phase HPS

diesel genset is the grid master. The first operation mode is similar to the three phase PV-battery system. In this mode the diesel genset can either be switched off or work as a controlled current source in grid parallel operation. In the second operation mode the diesel is forming the grid and all inverters operate in grid parallel operation.

Several aspects have to be regarded for the integration of small diesel gensets into such automatically controlled systems. According to guidelines of the diesel genset manufacturers the engine should not run under less than 40% partial load. Also frequent start-ups and shut-downs of the diesel engine diminish the generators life expectancy and generally increase the maintenance effort. Furthermore, a minimum surveillance of the gensets is required which can be performed automatically but especially the small gensets normally do not supply the appropriate sensors as a standard.

In the special application of the MORE project, an 11 kW HPS will be installed for the electrification of about six houses in a small remote valley on Kythnos. The system will supply electricity for about six houses with a three phase 400 V micro grid (see Figure 3).

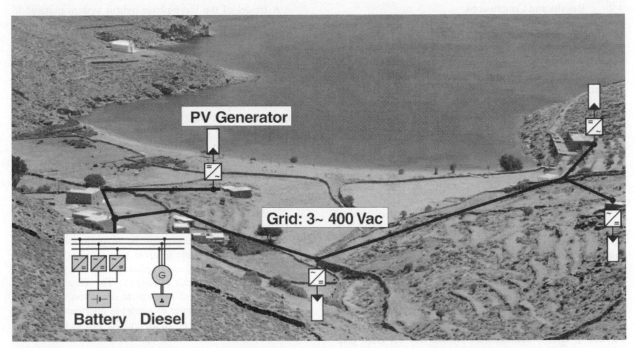

Figure 3: Integration possibility for a modular hybrid power supply system with three-phase micro grid

4   COMMUNICATION TESTS WITH EIB

4.1 EIB Test Environment

In order to examine the applicability of the European Installation Bus (EIB) for communication tasks in modular AC PV systems and hybrid systems a test environment was established. With personal computers supply components like e.g. PV generator or battery storage unit can be modelled. Furthermore, Supervisory control tasks and visualisation can be performed. The PCs are interconnected exclusively via EIB using a standard EIB/RS232 adapter.

Additionally, standard home-installation components like EIB-controlled switches and a 4-channel input module were integrated. The EIB-actors perform the tasks of main switch and load switches. The 4-channel analog input module transmits the measured grid values voltage, frequency and load current to the supervisory control unit. The "real world" consists of three ohmic loads, the main switch to the AC grid and the grid data acquisition unit.

The simulation environment was configured to simulate a small, single-phase, autonomous PV-battery power supply system with three resistive loads. Its schematic structure is depicted in Figure 4.

The main purpose of the test environment up to this time is to develop a communication interface based on the EIB standards for superimposed control and visualisation tasks. Therefore, the following functionality was implemented:

- supervisory control, using a unified state model for all supply components to describe its states of operation and the corresponding switching commands
- demand of reference values for active power
  monitoring of the active values for voltage, frequency and load current

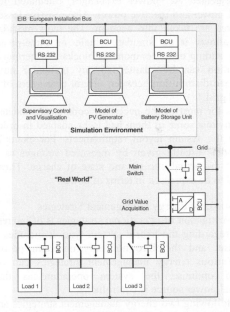

Figure 4: Structure of the EIB Test environment.

- visualisation of component specific values like state of charge, dc voltage and solar radiation
- error messages for monitoring and diagnosis purposes

For a safe and reliable communication it is necessary to apply the bus specification. The maximum transfer rate of the EIB twisted pair media is 9600 bd. Hence, the smallest time interval for cyclic data transfer is determined as 1s. All functions which do not have to be performed with the cyclic repetition are controlled by event.

4.2   Results and Conclusions

Regarding the bus specification, the applicability of EIB for supervisory control tasks was shown with the implemented test environment. All the implemented functionality described above was tested and the system worked reliably. In order to easily control multi-manufacturer systems in the future there is a demand in defining an unified component description. This will open the way to a standardised communication interface for PV systems and hybrid systems.

MODULAR SIMULATION TOOL

The development of a new simulation tool dedicated to AC bus photovoltaic hybrid systems was undertaken because already existing simulation software did not present the aimed features and flexibility. Components (generators, inverters, loads) as separate graphic boxes can be retrieved from libraries, which include also blocks with weather data processing algorithms.

For modelling and calculations the grid is considered stable in voltage and frequency as the tool is intended for the use in energy management studies and not from grid stability point of view. Components are interconnected with implemented AC power exchanges, calculated for both active power and reactive power. Also resistive losses on the wires are taken into account. The tools modular character allows it to simulate single or three-phase systems by adding extra components blocks. Furthermore, the simulation can be adapted to any particular non-balanced load distribution between several consumers of a small local grid.

Models for most of the necessary components existed already. Anyhow some of them had to be adapted in order to fit with the system layout requirements. For example battery models usually given by measured voltages as a function of the actual current and state of charge. These models had to be expressed in terms of power.

4.3 Comparison of Energy Management Strategies

The investigation of operational strategies is of interest especially regarding the hybrid system type. It includes a diesel genset and therefore not only uncertain power sources. Various energy management strategies can be applied to optimise the system operation as this conventional power source is controllable.

In the following two energy management strategies are described which were compared applying the simulation tool. In both strategies the diesel genset is used as a back-up power source, that means the diesel is started and operated at full power when the system would normally initiate a system power-down due a low battery state of charge. The threshold for the genset start was chosen to a battery state of charge at SOC = 40%. When the battery is full and the battery voltage upper threshold is reached, the photovoltaic inverter's output is reduced in order to protect battery from overcharge. The battery is completely recharged once a week for reasons of maintenance.

In the basic strategy called "full", the genset operates until the SOC ≈ 100% if is started. In the second strategy called "partial", the genset runs until SOC = 80% is reached.

A model of the three-phase hybrid system planned to be set up in Gaidouromandra on the Greek island Kythnos was built using blocks from libraries described above. The used load profile corresponds to an energy demand of 74.5 kWh per day and the battery storage capacity is 48 kWh. The genset has a rated power of 7 kW. A 5.76 kWp PV generator would be able to produce 9 MWh per year at PV inverter's output. It should be noticed that the load consumption of about 27 MWh per year cannot be covered with photovoltaic generator and the genset will necessarily provide the major part of energy consumed by both loads and losses. A second model differing from the original only in photovoltaic size was built using a 11.52 kWp PV generator, which can provide 18 MWh per year. Even this cannot completely cover loads and losses.

4.4 Simulation Results

After the simulation had been performed, energy balances have been made in order to assess both strategies considering the rational use of photovoltaic energy, the fuel consumption and overall efficiency. The diesel genset is not damaged because of inadequate operation; indeed the minimum operation power of genset is uncritical as it runs at 97% of rated power in mean, and the genset is always at least 7 hours in continuous operation. The results of the simulations are given in the following tables.

Table 1: Share of available PV energy which is used for different PV generator sizes

PV power	strategy "full"	strategy "partial"
5.76 kW$_p$	99 %	99 %
11.52 kW$_p$	75 %	81 %

Table 2: Fuel consumption for different PV generator sizes

PV power	strategy "full"	strategy "partial"
5.76 kW$_p$	8.8 m^3	7.7 m^3
11.52 kW$_p$	8.0 m^3	6.3 m^3

For both PV array sizes the second strategy allows to save fuel. The exploitation of photovoltaic energy is best with a smaller photovoltaic generator. The application of a bigger photovoltaic generator requires changes in the energy management strategy e.g. on the demand side.

The simulation tool allows an easy variation of operational strategies and of the component sizing. This is of great advantage for the system planning because both aspects cannot be planned independently for searching an optimum solution. The storage strategy of the energy management should depend on the ratio between renewable uncertain power generator and conventional controllable power source.

The bigger the PV generator in a system, the more attention has to be put to the selection of an adapted energy management strategy in order to exploit more renewable energy and to avoid high fuel consumption.

ACKNOWLEDGEMENT

This work was partially funded by the European Commission under the Joule 3 Programme PV-Mode CT98-JOR3-244 and MORE CT98-JOR3-215

REFERENCES

[1] W. Kleinkauf, F. Raptis, O. Haas: Electrification with Renewable Energies – Hybrid Plant Technology for Decentralised, Grid-Compatible Power Supply, Excerpt from Themes 96/97 Solar Energy Research Association, Germany, ISSN 0939-7582, Köln 2/97.

[2] B. Burger, G. Cramer, W. Kleinkauf, P. Zacharias, Hybrid Systems – Easy in Configuration and Application, 16th EPVSEC, 1-5 May 2000, Glasgow, UK.

[3] J. F. Manwell, A. Rogers, G. Hayman, C. T. Avelar, J. G. McGowan, Hybrid2 – A Hybrid System Simulation Model – Theory Summary, National Renewable Energy Laboratory, Massachusetts, 1997.

[4] Skarstein, Uhlen, Design Considerations with Respect to Long-Term Diesel Saving in Wind/Diesel Plants, Wind Engineering, Vol. 13, No. 2, 1989.

[5] J.A.M. Bleijis, C.J.E. Nightingale, D.G. Infield, Wear Implications of Intermittent Diesel Operation in Wind/Diesel Systems, Wind Engineering, Vol. 17, No. 4, 1993.

Remote Performance Check for Grid Connected PV Systems Using Satellite Data

Christian Reise[1], Edo Wiemken[1], Peter Toggweiler[2], Vincent van Dijk[3],
Detlev Heinemann[4], Hans Georg Beyer[5]

[1] Fraunhofer Institute for Solar Energy Systems ISE, Oltmannsstraße 5, D–79100 Freiburg
Phone +49-761-4588-282, Fax +49-761-4588-132, e-mail pvsat@ise.fhg.de
[2] Enecolo AG, Lindhof 235, CH–8617 Mönchaltorf
[3] Dept. of Science, Technology and Society, Utrecht University, Padualaan 14, NL–3584 CH Utrecht
[4] Dept. of Energy and Semiconductor Research, University of Oldenburg, D–26111 Oldenburg
[5] University of Applied Sciences Magdeburg, Breitscheidstraße 2, D–39114 Magdeburg

ABSTRACT: In this paper, we describe a remote performance check for grid connected PV systems. No additional hardware installation will be necessary on site. The site specific solar irradiation data will be derived from satellite images rather than from ground based measurements. On the basis of monthly irradiation time series, monthly values of PV system yield will be calculated and distributed automatically towards the system operator. He or she may then compare the estimated production to the real production meter reading. First tests reveal an overall accuracy of about 10 % for that period of a year offering 90 % of the annual solar irradiation.
Keywords: PVSAT - 1, Performance - 2, Reliability - 3

1 INTRODUCTION

A large number of small grid connected photovoltaic (PV) systems is in operation in Europe today, and a strong increase is expected for the near future. Today, the installed PV power of small systems increases with remarkable rates in some countries, e. g. with some 10 MW$_\mathrm{p}$ per year in Germany. Generally, these PV systems in a power range from 1 to some 10 kW$_\mathrm{p}$ do not include any long term surveillance mechanism. As most system operators are not PV specialists, system faults (component failures) or decreasing performance (e. g. due to increasing shading by growing vegetation) will not be recognized. At least two negative effects would be related to a bad performance of numerous PV systems: the overall energy production (and thus the saving of CO_2 emissions) would be reduced, and the individual plant operator will see financial losses. Regarding the increasing pay-back rates for PV energy (0.99 DM/kWh in Germany from spring 2000 onwards, similar initiatives are foreseen in other European countries), the last point becomes more and more important both for the plant operator as well as for the PV industry.

Therefore, there is a need for methods which allow for a cheap and reliable yield check of small grid connected PV systems. Hardware solutions (small irradiance sensors, intelligent monitoring devices) are available, nevertheless, extra devices will cause extra costs and require additional maintenance effort.

The EU JOULE III project PVSAT will set up a remote performance check for grid connected PV systems. No additional hardware installation will be necessary on site. The site specific solar irradiation data will be derived from satellite images rather than from ground based measurements. A target yield will be estimated for each individual PV system on a monthly basis. It will be reported to the system operators to allow for a comparison of targeted and real yield values.

2 HOW DOES PVSAT WORK?

The PVSAT procedure is based on three main components:
o A data base of PV system configuration data
o A satellite image processor
o A generic PV system model
The interaction between these components is depicted in figure 1 and will be explained in the following.

A PC-based data base contains geographical, component and operator related data for each individual system. The entries in the data base cover the following details:
o Addresses of PV system site and of operator
o Geographic coordinates of the site
o Orientation and tilt angle of PV system
o Horizon obstruction at the site
o Manufacturer and type of PV modules
o Size and wiring scheme of PV generator
o Mounting technique of PV generator
o Manufacturer and type of inverter
These data are to be collected once for each participating system. For new systems, the system supplier might aid the data acquisition process, as he will be acquainted with all technical details of the system at the moment of its installation. Nevertheless, also sys-

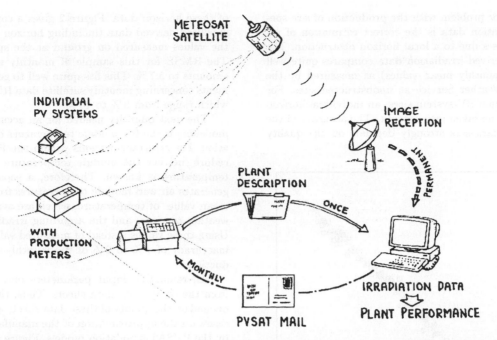

METEOSAT
SATELLITE

INDIVIDUAL
PV SYSTEMS

IMAGE
RECEPTION

PERMANENT

PLANT
DESCRIPTION

ONCE

WITH
PRODUCTION
METERS

MONTHLY

IRRADIATION DATA
⇩
PLANT PERFORMANCE

PVSAT MAIL

Fig. 1: Overview on the PVSAT procedure. For each individual PV system participating in PVSAT, geographical, component and operator related data are entered into a data base of configuration data. The continuous reception of METEOSAT images allows for the production of site specific time series of solar irradiation for each of the locations. At the end of a month, individual yield values are calculated for all PV systems. For this purpose, a generic system model is fed with plant data and corresponding irradiation time series. The results are mailed or faxed to the operator, he or she may then compare the estimated production to the real production meter reading.

tems already running for a long time may participate in PVSAT.

The continuous reception and processing of ME-TEOSAT images (done by one of the project partners) allows for the production of site specific time series of solar irradiation data for each of the locations. For Mid-European countries, METEOSAT images offer a resolution of 2.5×4.5 km^2. They are available in 30-minutes intervals. Global horizontal irradiation on ground is estimated from the pixel values in the visible channel using the concept of the HELIOSAT method [1]. This method has been recently modified [2], and the SATELLIGHT team [3] made further improvements to it. While single irradiance values (in 30-minutes steps) may differ by up to $\pm 25\%$ from ground measurements, the statistics of local irradiation climates are reproduced very well by the satellite derived data. In consecutive steps, the irradiation on a horizontal plane is converted to the tilted plane irradiation, and a local horizon obstruction is taken into account.

At the end of each month, individual yield values are calculated for all PV systems. For this purpose, a generic system model is fed with the configuration data and according irradiation time series. The model has been set up using the simulation system INSEL, which has been developed by the Oldenburg group several years ago [4] and which has been used for several system simulation and evaluation purposes [5]. The model incorporates numerous effects:

o PV IV-curve according to a 2-diode-model
o Temperature dependency of IV-curve
o Losses due to dust or soil on modules
o Ohmic losses in wiring
o Module mismatch losses
o MPP-tracking inaccuracy
o Inverter losses
o Inverter power limitation

The interface between the data base (running on MS-Windows) and the system model (running, as the satellite image processor, on Unix machines) is based on e-mail. Therefore, several distributed data bases may access the PVSAT server without interference. The results of the model calculation are transferred back to the data base, from where they are distributed (mailed, faxed or e-mailed) to the individual system operator. He or she may then compare the estimated production to the real production meter reading.

3   ACCURACY OF THE PROCEDURE

A main task of the project work programme has been concerned with the assessment of the accuracy of the PVSAT routine, both for the image processing and for the system modelling section. Basis of the investigations are PV system operation data acquired within the German 1000-Roofs-Programme [6]. Irradiation data on horizontal and tilted planes are available as well as system performance data, both with a time resolution of 5 minutes.

A major problem with the production of site specific irradiation data is the correct estimation of radiation losses due to a local horizon obstruction. The satellite derived irradiation data compares quite well to data (monthly mean values) as measured by the German Weather Service at unobstructed sites. For the individual PV system sites, an individual horizon line has to be taken into account. The accuracy of the radiation data sets strongly depends on the quality

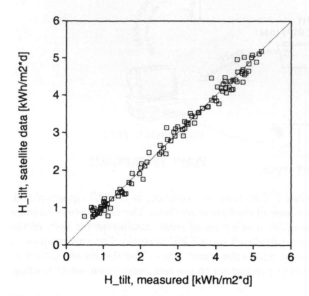

Fig. 2: Comparison of satellite derived to ground measured irradiation data. Each dot represents a monthly mean value of daily irradiation sums, the plot comprises one year of data for 9 PV system sites in Germany.

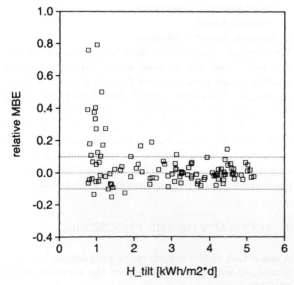

Fig. 3: Relative error of estimated production values versus observed meter readings, calculated on a monthly basis. The error values ([estimation − observation] / observation) are plotted versus the mean daily irradiation values for each month. The plot comprises one year of data for the same 9 PV systems as in figure 2.

of these horizon data. Figure 2 gives a comparison of satellited derived data (including horizon effects) and the values measured on ground at the specific sites. The RMSE for this sample of monthly mean values amounts to 5.7 %. This fits quite well to general statements concerning monthly satellite data RMSE values, which range from 5 % to 10 %.

The next quantity needed for an accurate system modelling is the PV module temperature during operation. For arbitrary systems within the PVSAT procedure, neither the module temperature nor the air temperature is known. Therefore, a separate model generates air and module temperatures from monthly mean values of temperature, which are available from weather services, and the according irradiation data. Using these values instead of measured values adds an inaccuracy of 1 % to 2 % of the monthly energy production.

All remaining input parameters may be derived from the component data sheets. Thus, they strongly depend on the quality of these data sheets and, in some cases, on the approximation of the manufacturer data by the PVSAT simulation models. Figure 3 shows the overall results for 9 PV systems and a test period of one year. For nearly all months with an average daily irradiation of more than 1.5 kWh/m^2, the differences between estimated and real PV production are smaller than 10 %. About 90 % of the annual energy production is related to these months.

For the remaining time (3 or 4 months during the winter period), the accuracy still needs some improvement. The main reasons for the large deviations during the winter period are:

o Increasing errors in the irradiation calculation from satellight images during winter (especially in case of snow covered ground)
o Inaccurate horizon lines which show a stronger effect at lower solar elevation angles
o Poor knowledge of component behaviour under part-load conditions

The first item may be improved after the transition to the next generation of European geostationary meteorological satellites in 2001. The last item is subject of further efforts of the PVSAT team, while a solution for the second problem is still under discussion.

4 TEST PERIOD

A semi-public test period of PVSAT has been started in May 2000 and will last for one year. This test phase is supported by associations of solar energy users and of system suppliers. It will provide information on several questions:

o What is the real-world accuracy of the procedure as a whole?
o Does the automated communication between the data bases and the PVSAT server work properly?
o What kind of information and aid do the PV system operators need?
o How do system suppliers interact with PVSAT?
o How would future service models look like?

After the test phase and its evaluation, PVSAT will be available to the public. The concepts of operation (by system suppliers, associations or neutral institutions) is currently under discussion.

5 CONCLUSION

Up to now, the results of the prototype PVSAT procedure are encouraging. The probability of calculating the monthly yield with a deviation less than 10 % is greater than 90 % for all months with an average irradiation of more than 1.5 kWh/m²d.

Considering this accuracy, PVSAT is not directly suitable for the concept of "guaranteed results", which could establish legal aspects between the system supplier and the owner or operator. Nevertheless, the goal of PVSAT is the early detection of partial system faults, and this goal will be achieved for all standard systems.

While PVSAT is designed for the estimation of monthly yield values (in order to support an early fault detection), the overall accuracy on an annual basis is considerably higher and reaches values of about 5 %.

ACKNOWLEDGEMENT

This work is funded in part by the European Commission within the JOULE-III programme (JOR3-CT98-0230).

REFERENCES

[1] D. Cano, J. M. Monget, M. Albuisson, H. Guillard, N. Regas, L. Wald: *A Method for the Determination of the Global Solar Radiation from Meteorological Satellite Data.* Solar Energy **37** 1986, pp. 31–39

[2] H. G. Beyer, C. Costanzo, D. Heinemann: *Modifications of the Heliosat procedure for irradiance estimates from satellite images.* Solar Energy **56** 1996, pp. 207–212

[3] A. Hammer, D. Heinemann, A. Westerhellweg, P. Ineichen, J. Olseth, A. Skartveit, D. Dumortier, M. Fontoynont, L. Wald, H. G. Beyer, Ch. Reise, L. Roche, J. Page: *Derivation of Daylight and Solar Irradiance Data from Satellite Observations.* Proc. 9th Conf. on Satellite Meteorology and Oceanography, Paris 1998

[4] J. Luther, J. Schumacher-Gröhn: *INSEL – A Simulation System for Renewable Electrical Energy Supply Systems.* Proc. 10th CEC PV Solar Energy Conference, Lisbon 1991 pp. 457–460

[5] H. Gabler, M. Raetz, E. Wiemken: *Analytical Evaluation of the Performance of Realized Photovoltaic Systems.* Proc. 12th CEC PV Solar Energy Conference, Amsterdam 1994 pp. 879–882

[6] Th. Erge et al.: *The German 1000-Roofs-Programme – A Resume of the 5 Years Pioneer Project for Small Grid-Connected PV Systems.* Proc. 2nd World Conf. on PV Solar Energy Conversion, Vienna 1998, pp. 2648–2651

PAYBACK TIMES FOR ENERGY AND CARBON DIOXIDE: COMPARISON OF CONCENTRATING AND NON-CONCENTRATING PV SYSTEMS

Anne Wheldon[1], Roger Bentley[2], George Whitfield[2], Tamsin Tweddell[3] and Clive Weatherby[4]

[1] (Author for correspondence), Energy Group, Department of Engineering, The University of Reading, Whiteknights, Reading RG6 6AY, UK. Tel: +44(0) 118-931-8756 Fax: +44(0) 118-931-3327 email: a.e.wheldon@reading.ac.uk

[2] Department of Cybernetics, The University of Reading, Whiteknights, Reading RG6 6AY, UK

[3] Max Fordham and Partners, 42/43 Gloucester Crescent, London NW1 7PE, UK

[4] Solar Century, Unit 5 Sandycombe Centre, 1-9 Sandycombe Road, Richmond, Surrey, TW9 2EP, UK

ABSTRACT: The time for the energy "embedded" in the materials and manufacture of a PV system to be "paid back" during use is an important measure of environmental viability. Energy payback times were estimated for PV systems using lens and mirror concentrators, and compared with those for multicrystalline (mc-Si)and amorphous (a-Si) silicon flat plates, using solar data for three contrasting sites. Both concentrators gave significantly shorter payback times than the mc-Si flat plate at all sites (1.0 to 2.7 years, compared with 2.1 to 3.6 years), and shorter or comparable times to the a-Si flat plate. Carbon dioxide payback times are very similar to those for energy. The systems will save between 1.5 and 3 tonnes of carbon dioxide per m^2 aperture over a lifetime of 20 years. Key assumptions are discussed in the paper.

Keywords: concentrator-1: energy payback-2: environmental effect-3

1. INTRODUCTION

The main current motivation for the use of renewable energy is to offset the environmental impacts associated with conventional energy use. Renewables also extend the lifetime of finite resources. It is therefore important to consider the net environmental and energy budgets related to both the production and use of renewable energy technologies. One simple measure for the evaluation of these technologies is the time for the energy "embedded" during manufacture to be "paid back" from the conventional energy which is offset in use.

For crystalline PV, the majority of the embedded energy is associated with the production of the silicon wafers. There is uncertainty in how this should be calculated, particularly since most silicon used at the moment is a byproduct from the microelectronics industry. The estimated payback times for encapsulated cells therefore range from about 3 to 8 years depending on the assumptions used [1].

The use of thin films minimises the amount of silicon (or other semiconductor) required, and the associated uncertainty in allocating embedded energy to the silicon becomes less important. The estimated payback times are shorter, typically 1 to 4 years [derived from data in 1].

Another way to reduce the amount of silicon required is to use a large area of optical concentrator (mirror or lens) to focus solar energy onto a smaller area of crystalline PV cells. Concentrators have been shown to be cost effective, because the financial cost of the concentrator is much lower than that of the avoided PV cells [2].

The present study determines whether the use of concentrators can also reduce the embedded energy in PV systems, and therefore reduce the energy payback time, compared to conventional flat-plate technologies.

2. METHODOLOGY

2.1 Systems studied

A previous study considered in detail the performance of a range of designs of PV concentrator systems, with likely financial costs based on estimates of materials used and manufacture [2]. Four of the most cost-effective designs were constructed as 2 m^2 prototypes and their performance was assessed in laboratory and long-term field trials. Two of these prototypes, shown in Photographs 1 and 2, are considered in the present study.

Photograph 1: Prototype 2 m^2 concentrator using point-focus Fresnel lenses

Photograph 2: Prototype 2 m² concentrator using line-focus mirrors

One concentrator uses point-focus Fresnel lenses made from acrylic, and requires 2-axis tracking to focus solar radiation on the cells. The other uses line-focus mirrors made from silvered acrylic film, and requires 1-axis tracking. Both systems use single crystal silicon (sc-Si) PV cells.

The flat plate systems considered in the present study are assumed to be made from multicrystalline (mc-Si) and amorphous (a-Si) silicon, with modules framed and attached to a fixed mounting. For comparative purposes it is assumed that all four systems are mounted in such a way that they could be bolted to a rigid horizontal substrate, either the ground or a flat roof.

2.2 Embedded energy

Table 1 gives the main subsystems and materials used in the four systems, with estimates of the embedded thermal energy. The encapsulated cells and trackers are regarded as "subsystems", and the energy estimates given include materials and manufacture of these subsystems. Other values are for bulk materials. The electrical energy used in material production has been converted to thermal energy assuming a European average efficiency of 35% [3].

TABLE 1: EMBEDDED ENERGY IN SUB-SYSTEMS AND MATERIALS			
	Thermal energy	**Units**	**Comments**
Subsystems (energy includes bulk materials and manufacture)			
PV module: sc-Si	1667* 889**	kWh/m²	All values are for frameless encapsulated modules
mc-Si	1222* 722**	kWh/m²	*near term, estimated from [1]
a-Si	306* 250**	kWh/m²	**medium term (2007) quoted by [1]
Tracker - 1-axis	55	-	kWh per m² of collector aperture. Assumed, based
- 2-axis	95	-	on similar devices
Bulk materials (energy does not include manufacture)			
Aluminium	45	kWh/kg	Data from [3], assumes European average of 32% recycled aluminium [www.eaa.net]
Galvanised steel	14	kWh/kg	Data from [3] + 40% for galvanising, no recycling
Acrylic lens, 3.6 mm	44	kWh/kg	Assumed, based on similar plastics [4]
Silver-acrylic film	14	kWh/m²	Estimate, based on material plus processing cost
Glass sheet, 3 mm	52	kWh/m²	Data from [4]

Table 2 combines these values with the amounts of materials, to estimate the embedded thermal energy, per m² of aperture, in the different PV systems, and results are shown in Figure 1. Manufacture is assumed to require an additional 40% of the energy of the bulk materials. Note that no allowance has been made in these estimates for the recovery of energy from, or re-use of, the materials at the end of their working life.

2.3 Energy output

Predictions of the annual electrical output were made for all four systems, using solar data from three contrasting sites. These are: Almeria, Spain with average irradiation 1734 kWh/m² per year on the horizontal, of which 33% is diffuse; Manfredonia, Italy, 1580 kWh/m², 38%; and Widderstall, Germany, 1058 kWh/m², 54% [5]. The electrical outputs, shown in Table 2, assume that each system is maintained at its maximum power point, and they do not take into account any losses in power conversion. This is valid for comparing the systems.

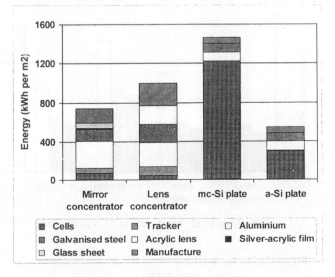

Figure 1: Embedded energy for the four systems

TABLE 2: EMBEDDED ENERGY, ENERGY OUTPUT, ENERGY PAYBACK AND CARBON DIOXIDE SAVINGS FOR THE FOUR SYSTEMS					
	Mirror conc.	Lens conc.	mc-Si flat plate	a-Si flat plate	Units
Embedded energy					
Encaps. PV cells	71 (10%)	46 (5%)	1222 (83%)	306 (56%)	kWh$_{th}$ per m^2 aperture (%)
Tracker	55 (7%)	95 (9%)	-	-	
Aluminium	280 (38%)	251 (25%)	91 (6%)	91 (17%)	
Galvanised steel	121 (16%)	187 (19%)	89 (6%)	89 (16%)	
Acrylic lens	-	192 (19%)	-	-	
Silver-acrylic film	15 (2%)	-	-	-	
Glass sheet	52 (7%)	-	-	-	
Manufacture	152 (20%)	230 (23%)	62 (4%)	62 (11%)	
Total embedded	746	1001	1464	548	
Annual energy output					
Almeria	242 [691]	296 [846]	242 [691]	141 [403]	kWh$_e$ [kWh$_{th}$] per m^2 aperture
Manfredonia	209 [597]	257 [734]	222 [634]	130 [371]	
Widderstall	108 [309]	132 [377]	143 [409]	84 [240]	
Thermal energy pay back (≈ carbon dioxide payback)					
Almeria	1.1 {1.0}	1.2 {1.2}	2.1 {1.4}	1.4 {1.2}	years near- {medium-} term
Manfredonia	1.3 {1.2}	1.4 {1.3}	2.3 {1.5}	1.5 {1.3}	
Widderstall	2.4 {2.3}	2.7 {2.6}	3.6 {2.4}	2.3 {2.1}	
Carbon dioxide savings over 20 year lifetime					
Almeria	3.0	2.5	2.4	1.4	tonnes CO$_2$ per m^2 aperture, near term
Manfredonia	2.6	2.1	2.1	1.3	
Widderstall	1.2	1.0	1.3	0.8	

It is assumed that the electrical output of the PV system offsets grid electricity, and a 35% conversion efficiency for thermal energy to electricity is used, as above. The annual figures for offset thermal energy are also shown for each system in Table 2.

2.4 Energy payback

The energy payback time for each system was calculated in years, as the total embedded energy divided by the annual output, both expressed in terms of thermal energy. These payback times are shown in Table 2 and Figure 2.

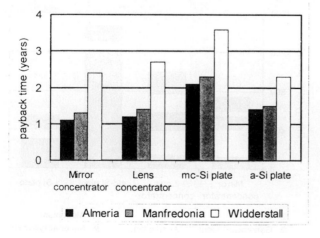

Figure 2: Energy payback times for the four systems (near-term)

2.5 Carbon dioxide payback and saving

Information on the "embedded" carbon dioxide and other pollutants is also available for various materials [3]. Because the mix of fuel sources does not differ greatly for the different materials in the systems used, a constant ratio of carbon dioxide:energy can be used for a first approximation. This implies that the payback times for carbon dioxide are about the same as those for energy, with the same ranking between systems. The net carbon dioxide savings were calculated over a 20-year lifetime, using an average emission of 0.2 kg of carbon dioxide per kWh$_{th}$ and these results are shown in Table 2.

3. RESULTS AND DISCUSSION

3.1 Embedded energy

The absolute amount, and the division, of the embedded energy differ significantly between the four systems (Figure 1). The energy in the silicon cells is the dominant factor for both the flat plate systems, particularly when crystalline cells are used. In contrast the cell energy represents 10% or less of the total in both the concentrator systems. In the mirror system, aluminium is the dominant factor because it is used for the combined housing and support for the mirror film. In the lens system the contributions of aluminium, steel, acrylic and manufacturing are all similar.

3.2 Energy output

The output from the a-Si flat plate, per m^2 of aperture area, is always less than the mc-Si plate, because of the lower efficiency. The concentrator systems perform worse than mc-Si, per m^2, at Widderstall but the same or better at Almeria. This is because they concentrate only the direct beam component of the radiation and therefore perform relatively better when the amount of diffuse radiation is lower.

3.3 Energy payback

All systems, at reasonably sunny sites, show encouragingly short payback times of between 1 and 2.5 years (Figure 2). The energy payback times for the flat-plate systems are at the low end of published values [1]. There are two reasons for this. Firstly, a near-term value has been used for the embedded energy in the cells, and this is lower than many values used for past and current PV production. Secondly, the output has been taken at peak power, and neither the losses, nor the embedded energy, in the power conversion system have been included.

The payback times for the concentrator systems are always shorter than for mc-Si flat plates, in particular at Almeria where the use of concentrators would halve the energy payback time. The concentrators are also slightly better than the a-Si flat plates, except at Widderstall which has the most diffuse radiation.

When medium-term estimates are used for the embedded energy in the cells, the payback times obviously decrease. The effect is most significand for the mc-Si flat plate since this has the highest relative contribution of the cells to the total embedded energy.

3.4 Carbon dioxide saving and payback

Carbon dioxide payback times are similar to those for energy. Table 2 shows that the systems studied provide carbon dioxide savings of between 1.5 and 3 tonnes per m^2 aperture, over an assumed 20 year lifetime. The concentrators provide similar savings to the mc-Si flat plate, and significantly greater savings than the a-Si flat plate.

3.5 Effect of assumptions

Power conversion If power conversion is included it will increase the absolute payback times, but not effect the relative ranking which this paper set out to establish.

Embedded energy There are many assumptions in estimating the embedded energy in the materials, which have been discussed elsewhere [1, 4, 6]. For comparative purposes it is important to recognise how the division of embedded energy differs between the four systems. For instance, if a substantially larger proportion of recycled aluminium can be used, this will make a significant reduction in the payback time of the concentrators, in particular the mirror system, but will have relatively little effect on the flat-plate systems. The relative contribution of different materials should be regarded as a significant factor when developing new renewable energy technologies of all types

Mounting All the systems have been assumed to mounted in such a way that they can be fixed to a rigid substrate. If systems are compared for a specific site then the method of fixing should also be considered in the energy budget. For instance, if a system is secured to the ground using about 0.2 m^3 of concrete then the additional (literally!) embedded energy is about 200 kWh per m^2 of aperture area. This makes a significantly different *relative*

increase in the embedded energy of the four systems – only 14% for the mc-Si flat plate but nearly 40% for the a-Si.

The flat plates can more easily be integrated into buildings, which could improve their payback compared to concentrators.

End-of-life The aluminium, steel and acrylic which form a significant component of the concentrator systems can be recycled, with energy recovery, at the end of the system design life. It may be more difficult to recycle the encapsulated PV cells, because of the difficulty in separating materials. Inclusion of end-of-life recycling would therefore improve the payback of the concentrating systems compared to the flat plates.

Replaced energy It has been assumed that the energy "paid back" is grid electricity. If the PV system is providing energy to, say, replace candles or dry-cell batteries, then the energy payback time will differ. If the fuel mix for electricity generation changes over the 20-year lifetime of the systems, then the carbon dioxide savings will differ.

4. CONCLUSIONS

Financial payback is the measure on which different energy systems are usually compared. However, the payback in terms of energy, and the lifecycle savings of carbon dioxide, are more significant in environmental terms and a better guide to sustainability. This study has shown that, particularly in the near term, the use of concentrators can reduce the environmental impact of PV.

REFERENCES

[1] E.A. Alsema, P. Frankl, and K. Kato. *Energy Pay-Back Time of Photovoltaic Energy Systems: Present Status and Future Prospects.* Proc. 2nd World Congress on Photovoltaic Solar Energy Conversion, Vienna (1998) 2125 - 2130.

[2] G.R. Whitfield, R.W. Bentley, C.K. Weatherby, A. Hunt, H-D. Mohring, F.H. Klotz, P. Keuber, J.C. Miñano and E. Alarte-Garvi. *The Development and Testing of Small Concentrating PV Systems.* Presented at the ISES Solar World Congress, Jersusalem, Israel (1999).

[3] R. Frischknecht, P. Hofstetter, I. Knoepfel, E. Walder, R. Dones and E. Zollinger. *Environmental Life-Cycle Inventories for Energy Systems - An Environmental Database.* Swiss Federal Energy Office (1994).

[4] U.K. Science Research Council. *Making the Most of Materials.* ISBN 0 900193 76 X (1979).

[5] G.R. Whitfield, R.W. Bentley, C.K. Weatherby, H-D. Mohring, F.H. Klotz, J.C. Miñano and E. Alarte-Garvi. *Development of Optical Concentrators for Small PV Systems.* Final Report, EC JOULE contract JOR-CT96-0101 (1998).

[6] T. Tweddell. *An Analysis of the Photovoltaic Roof on the Engineering Building at The University of Reading.* MSc. thesis, University of Reading (1999).

ANALYSIS OF RELATIONSHIPS BETWEEN PHOTOVOLTAIC POWER AND BATTERY STORAGE

FOR SUPPLYING THE END OF FEEDER – A CASE STUDY IN CORSICA ISLAND

V.Acquaviva, P.Poggi, M.Muselli, A.Louche

Université de Corse - URA CNRS 2053

Systèmes Dynamiques Energétiques et Mécaniques

Route des Sanguinaires-F-20000 Ajaccio

Tel :+334 955 241 41 – Fax :+334 955 241 42

E-mail : acquaviva@vignola.univ-corse.fr

ABSTRACT : The idea of this study comes up as a back-up system for a rural zone that is also end of line, and where the quality of the electricity is far from good: so frequent voltage drops appear all the day long. Grid-connected PV generation has several characteristics which are well suited to power systems applications. The main is that voltage drops occurring on heavily loaded lines would be reduced, minimising the need for voltage improvement. The option to decrease the voltage drops, in the evening or during low irradiation, is to integrate batteries into the PV system. With the availability of advanced batteries, it is now possible to store large amount of energy during off-peak periods of the day and use during the peak periods for load levelling.

Keywords: Grid-connected - 1: Voltage drops – 2: Battery - 3

1. INTRODUCTION

A system to support the utility in the hours with highest charge has been designed and the consumption curve peaks decrease. A case study is presented for an end of line in Corsica Island. The value of PV-electricity in grid-connected residential systems is studied through hourly simulations with meteorological data collected in Ajaccio (41°55'N, 8°39'E) and load data. The simulation studies are led in two different configurations of the power system : firstly, without storage, secondly, with capacitive energy storage. One of the principal achievements of this project is laminating the demand power curve with a small scale PV system and a storage unit. The effect of PV-inverter-battery will reduce in totality the voltage drops.

So by means of the PV system behaviour simulation, using a systemic approach, and taking into account the efficiency of each sub-system, we determine the set of configurations which allow us to reduce voltage drops. The use of the kilowatt-hour cost produced by the system and its minimisation constraint leads to the most optimised system.

2. VOLTAGE DROPS

2.1 Definition and voltage drops limitations

In Europe, typically, service for residential consumers is maintained within a range of 230 Vac – 10 % and + 6 % [1]. Voltage drop becomes unacceptable, when it exceeds 10 % and in this case, the utility has to increase voltage at the end of the line. An approach to solve this problem, is to satisfy the demand from a local point of view by investing in decentralised production units. Distribution generation facilities are strategically sited to deliver electricity where the needs are. These systems which are commonly called grid-connected rooftop PV systems can relieve voltage drops [2].

2.2. Simulation of voltage drops

The characteristics of end of the feeder [3], the seasonal load pattern (Fig. 1) and the formula of the voltage drop [4] (Eq. 1) allow to simulate the voltage drops for each section of the end of the feeder studied. The hourly relative cumulative voltage drop (%) for the segment n is equal to [5]:

$$\left(\frac{\Delta U}{U}\right)_n = 10^5 \times \left[\frac{(R + X \times tg\varphi)}{U^2}\right] \times \sum_{i=1}^{n}\left(P_i \times L_i\right)$$

(Eq. 1)

where $P_i = \sum\limits_{j=i}^{N} P_j$ is the cumulative load power (kW). N is the total number of sections, L_i is the load (kWh), R is the resistance (Ω/km) and X is the reactance (Ω/km).

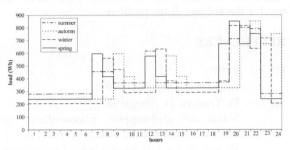

Figure 1: Load patterns.

We proposed to bring these drops under 10 % by using PV rooftop systems when the drops appear during the day. Thus, the hourly relative cumulative voltage drop (%) is:

$$\left(\frac{\Delta U}{U}\right)_n = 10^5 \times \left[\frac{(R + X \times tg\varphi)}{U^2}\right] \times \sum_{i=1}^{n}\left(P_i' \times L_i\right)$$
(Eq. 2)

where $P_i' = \sum\limits_{j=i}^{N}\left(P_j - P_{INV,j}\right)$ and $P_{INV,j}$ is the output power of the inverter (kW) computed from the PV power P_{PV} (kW) on segment j. If $P_j - P_{INV,j} < 0$ then $\frac{\Delta U}{U} = 0$.

For the month of January and for several PV surfaces, we plotted a graph (Fig. 2). The results shows that small scale PV systems decrease voltage drops during sunny hours, but the problem persists in the evening and in the early morning. So, grid-connected systems are not sufficient to level peaks of load all the day long. A PV system with battery would be necessary.

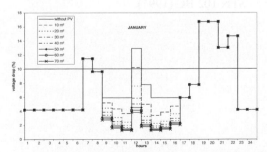

Figure 2 : Voltage drops with and without PV.

3. BATTERY PARK – METHODOLOGY

We used a method based on a system energy balance and on the storage continuity equations [6]. This system must laminated the load. Such a constraint leads to an infinity of possible system configurations. The addition of other physico-technical constraints allows us to reduce this system characteristic set to a more physical domain. The use of he PV-battery kilowatt-hour cost consumed and its minimisation constraint leads to the most optimised system and allow to find the couple (Surface, Capacity) (S,C) needed.

3.1 Sizing

Sizing a suitable battery adequate for shaving the peak demand hours in every month of the year requires the knowlegde of the peak of load period and the wanted reduction. An important decrease is useless because the comfort of the consumer is assured if the voltage drops are under 10 %. We chose to reach 9 % to resolve the problem of voltage drops and to limit the capacity of the battery. The figure 3 shows the energy required for this reduction.

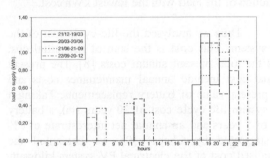

Figure 3 : Energy required to reduced voltage drops.

Operating modes: The following steps describe how the combined operation of the PV and the battery system is envisaged:

First operation mode:
- During the early morning hours, the battery State of Charge (SOC) is down to a low level. Therefore, constant power must be applied to charge the battery to a high level as possible before the discharge cycle begins. The charging power is composed of photovoltaic power generation available only after sunrise. The PV energy is converted with a constant charged efficiency ρ_{ch} equal to 85% [7]. If the PV array energy output is more than the load there are two possibilities. If the capacity of the battery does not reach its nominal capacity, the surplus of energy is converted in chemical energy, otherwise, the surplus is fed to the grid.

- All the photovoltaic generation must be applied to the peak of load. If the PV energy is less than the load, battery supply consumption (with a discharge efficiency ρ_{dch} taken equal to the previous charge efficiency [7]) but it does not reach the DOD, otherwise there is voltage drop.

- During charge periods, if the battery SOC reaches 100%, the PV output supply directly the load demand.

Second operation mode :
- Energy must be stored in batteries from the electrical network during the hours of the night (to 1:00 am to 6:00 am) in C (nominal capacity), C/2 (nominal capacity/2), C/3, C/4, C/5. During these hours, there is no voltage drops and the price of electricity is lower. At 6:00 am, the battery reaches is nominal capacity.

- if battery SOC reaches 100 %, the PV energy is directly injected into the grid.

System cost estimation: We determined the best configuration, introducing as a new minimisation constraint the consumed kilowatt-hour cost. Then, we will consider that the optimised PV system is the system which allows the desired reduction of the load with the lowest kWh cost.

Thus we analysed the life-cycle cost of the PV system. This cost is the sum of the initial cost plus the total present annual costs [8]. The present annual cost include annual maintenance costs and the present value of battery replacements. Taking a PV system life cycle cost of 20 years (m), a battery life of 5 years and an annual net interest rate of 6% as suggested by Chabot [9], we computed the actualised cost of the electrical PV system kilowatt-hour.

The kWh consumed cost is calculated by considering the life cycle cost and the yearly PV energy consumed by the load *Ec* (kWh). We present in table 1 the different couple (S,C) corresponding to the two operation modes. Charging during the night batteries to their nominal capacity seems to be the better option.

	Module surface area(m²)	Storage capacity (kWh)
PV only	39	66,0
nominal capacity/5	4	40,5
nominal capacity/4	3,5	33,1
nominal capacity/3	3	25,4
nominal capacity/2	2,5	17,8
nominal capacity	2	9,6

Table 1. Configuration of the optimal PV system to solve the daily voltage drops

4. CONCLUSION

In this paper, we applied an optimised method for an autonomous photovoltaic system in a case study in Corsica, taking into account the temporal distribution of the energy input (solar radiation) and output (electrical load). A grid connected systems with a battery park allow to resolve the voltage drops problem.

REFERENCES

[1]   EDF
       Internal publication (1997).
[2]   D. Travers, D. Shugar D,
       Value of grid-support photovoltaics to electric distribution lines, Progress in Photovoltaics : Research and Applications **2**:293-306 (1994).
[3]   Electricité de France,
       Technical documentation (1997).
[4]   M. Muselli,
       Electrification de sites isolés – Dimensionnement de systèmes hybrides à sources renouvelables d'énergie, Phd Thesis, University of Corsica (1997).
[5]   Centre de recherches : ARMINES, CONPHOEBUS, RAL, INESC, CIEMAT et CRES,
       Integration of Renewable Energies for Decentralized Electricity Production in Regions of EEC and Developing Countries. Project SOLARGIS. Rapport Final (1996).
[6]   J. Kaye,
       Optimising the value of photovoltaic energy in electricity supply systems with storage, Proc. 12th EPSEC Conference, Amsterdam, 431-434 (1994).
[7]   Oldham France S.A.,
       Technical documentation (1992).
[8]   P. Groumpos, G. Papageorgiou,
       An optimal sizing method for stand-alone photovoltaic power system, Solar Energy, **37**: 341-351 (1987).
[9]   B. Chabot,
       Abaques de coûts du kWh produits par énergies renouvelables. Rapport AFME, STN 102 BC (1991).

PHOTOVOLTAICS IN CONJUNCTION WITH ADVANCED SMALL HYDROPOWER

D. Reismayr
Institut für Elektrische Maschinen und Antriebe
Universität Stuttgart   Pfaffenwaldring 47   D – 70569 Stuttgart
Tel. ++49(0)711 685 7823   Fax ++49(0)711 685 7837
eMail reismayr@iema.uni-stuttgart.de

ABSTRACT: As an example of conjunction of photovoltaics and advanced small hydropower there are described two plant versions at a fixed load curve: For adapting the renewable energy supply curve (PV and hydropower) to the load curve, i.e. storing the renewable energy, either a reservoir or a battery is used. To assess these versions their applicability and energy costs are compared.
Keywords: Hybrid – 1: Small Grid-connected PV Systems – 2: Pricing – 3

1. INTRODUCTION

Not only e.g. Germany's about 10 000 stand alone properties, many of them situated on streams, but especially cases in developing countries led to the idea of investigating a photovoltaic system in conjunction with an advanced small hydropower generator at variable frequency in a power range of up to 50 kWp PV and 100 kW hydropower. This range covers a large number of plant versions for meeting the various load requirements. Notice: Due to a drastic reduction of investigations referring to these versions there can be given details only for one example.

2. OBJECTIVE

For a stand-alone plant in two versions and at a fixed load curve a comparison of applicability and energy costs will be given.

3. SYSTEM MODEL

3.1 Components

The investigated system consists of the main components photovoltaic generator, hydropower generator with rectifier (both feeding a DC-circuit and inverter to

the AC-consumption) and a propeller water turbine. The two versions are formed by using as an energy storage either a reservoir or a battery. See Fig. 1 without or with battery.

3.2 System data

Load curve private house oriented, peak load at about 18.00 hrs. (9 p.m.) see Fig. 2, annual load energy 298 560 kWh/a

Version 1, reservoir 430 000 m³, head 50 m, installed maximum hydropower 100 kW, PV 30 kWp spread over roofs of e.g. private and farm houses

Version 2, battery 1090 kWh, DOD 60 % installed maximum hydropower 60 kW, canal, head 30 m, PV 30 kWp

Maintenance costs 2.5 % of investment costs per year, except batteries at 5 %.

Interest rate 4 %, life times: Structures 70 years, machinery 40 years, photovoltaics 25 years, batteries 10 years.

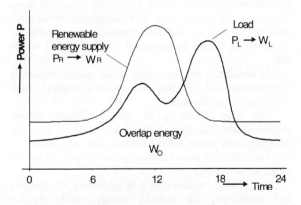

Figure 2: Daily energy curves: Renewable energy production P_R resp. W_R (PV and hydropower), load P_L resp. W_L, immediately used overlap energy W_O

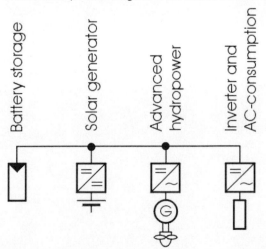

Figure 1: Scheme of stand-alone plant

4.   APPROACH

4.1   Energetic consideration

Looking upon Fig. 2, if there is $P_R > P_L$ a surplus of renewable energy is delivered and if $P_R < P_L$ a lack is occurring. According to [2] for grid-connected plants the autonomous renewable supply degree and the total renewable supply degree are defined as to

$$\rho_{aut} = W_O / W_L \qquad \text{(Equ. 1)}$$

and

$$\rho_{tot} = W_R / W_L \qquad \text{(Equ. 2)}$$

For minimising energy costs at a certain configuration of components and energy selling and buying prices it has to be obtained $\rho_{aut} \rightarrow 1$ and $\rho_{tot} \rightarrow 1$,
i.e. the overlap energy has to be maximised leading to an overlap degree

$$\kappa = \tfrac{1}{2} (W_O / W_R + W_O / W_L) \rightarrow 1 \qquad \text{(Equ. 3)}$$

In the present case of a stand-alone system ρ_{aut} is forced to be equal to 1. That means the energy production curve has to be adapted to the fixed load curve. If there shall be no loss of renewable energy and no lack in meeting the energy demand an energy storage has to be provided for balancing the system. But the adaptation by additional technical means will increase the absolute level of energy costs in spite of having $\kappa = 1$.

4.2   Solutions

The energy storage can be realised by a water reservoir or a battery.

For the first version of the present example a reservoir and a hydro generator normally running at 50 kW are provided. When there is a surplus of renewable energy , especially at insolation, the hydropower generation can be reduced by decreasing speed and water throughput for saving water. But that means on the other hand for meeting the peak load demand in the evening the maximum power of the generator has to be rated at 100 kW.

In addition to that depending on the special conditions the reservoir and the dam may cause a relatively high impact on the environment.

The plant seems to be able to operate almost autonomously. Therefore maintenance costs are assumed to be relatively low at 2.5 % of investment costs per year.

For rush peak power control purposes a small battery might be necessary.

The second version uses a large battery for charging and discharging the balance energy electrically for obtaining the congruence of energy supply and demand. This leads to a lower maximum power of the hydro generator of 60 kW. Instead of the reservoir a canal for deviating the stream has to be provided. 'The environmental impact by the structures would be much lower.

The battery requires a higher amount of maintenance, therefore its maintenance costs are assumed to be at 5 % of investment costs per year.

4.3   Energy costs

Specific (load) energy costs are calculated as follows.

$$c_L = (\Sigma A_{inv} + c_m) / W_{La} \qquad \text{(Equ. 4)}$$

if there is no supply resp. delivery from resp. to the grid due to the stand-alone system, taking into account the sum of the annuities A_{inv} of the components of the plant and the maintenance costs c_m.

The overall specific investment costs are:

Version 1: Hydropower plant (100 kW)   5700,- EUR/kW
Version 2: Hydropower plant (60 kW)   7500,- EUR/kW
      Battery system (1090 kWh)
      incl. DC/DC-converter and
      structures   280,- EUR/kWh
yielding to specific energy costs:
Version 1:   0.167 EUR/kWh
Version 2:   0.299 EUR/kWh

5.   RESULTS AND CONCLUSION

Concerning the applicability of the two versions of the present example it can be pointed out:
Version 1 with reservoir and dam can be maintained in a simpler way at lower costs due to its robust almost maintenance-free operation, on the contrary to version 2 with its battery.

Lower energy costs of 0.167 EUR/kWh of version 1 in comparison to 0.299 EUR/kWh of version 2 lead to the recommendation of version 1 if the environmental impact of the reservoir system is almost negligible.

System costs as for a stand-alone system are relatively low but due to small numbers of manufactured pieces still too high for developing countries.

REFERENCES

[1]   M. Kaltschmitt, A. Wiese (Editor), Erneuerbare Energien, Springer 1997
[2]   D. Reismayr, Photovoltaik und flexible Kleinwasserkraft für Gebirgstäler, 15. Symposium Photovoltaische Solarenergie, Staffelstein 2000

PHOTOVOLTAICS THE BACKUP FOR QUALITY POWER

B.Ghosh and R.Hill*

School of Energy Studies, Jadavpur University, Calcutta 700 032. INDIA
Tel: +91 33 4732853, Fax: +91 33 4725823, Email: sebsju@giascl.01.vsnl.net.in
*NPAC, School of Engineering, University of Northumbria, U.K.

ABSTRACT: The central electricity generating stations in most countries are located far away from the sites of domestic and commercial loads.. With the increase in population, along with the up-grading of life style, there is an increase in demand for electricity. Conventional distribution systems in many countries are now inadequate to meet this demand. The quality of the power supplied is degraded to such a level that it is often unable to meet the load requirement. It is not uncommon for the line voltage, frequency or the power quality to be outside the legal limits, producing problems for customers. The conventional solution to this problem is to up-grade the transmission lines, introduce a set up transformer at the feeder point and use an individual voltage stabiliser. This creates a problem at the off-peak period and generates a significant no load loss.

A parallel connection of a constant voltage source to the service line can stabilise the network voltage and maintain the quality of power. This can be done through embedded power generation using photovoltaics.

These issues, using preliminary data recorded at a commercial office complex in a typical city and a suburban residential complex, are discussed in the text of the paper.

Keywords: Grid-connected -1: PV System -2: Rooftop – 3

1. INTRODUCTION

India is a very large country with several states and central territories. Each state and territory have their own electricity board (SEB) and government controlled private limited bodies for supplying electricity to the cities. The tariff is not uniform throughout the country. Each state has its own tariff system depending upon several factors and almost all states and territories supply power to the consumer at a subsidised rate. The power generating stations are mostly thermal based and they are situated far away from the centre of demand. The power is transmitted and distributed through an old network. The growth in connected loads degrades the service voltage as well as the quality of power, particularly at the peak hours of demand. The service voltage sometimes falls below the legal limit and this can affect the life time of the equipment and can threaten human life when hospitals are affected. This problem demands appropriate solutions for the social needs of the population.

The supply of quality power is a challenging problem for the engineers, technologists and economists. The problem has two major dimensions, one from the engineering and technological point of view and the other related to the economic aspects [1,2].

In order to elucidate these aspects this paper reports a preliminary study on the demand, supply and the quality of power in two localities. One locality is a commercial office complex in the city and the other is a residential suburban area, both in an eastern part of India.

2. THE PRESENT SYSTEM

The usual service supply by the electricity authority is 220 Volts at 50Hz. A survey was carried out over six months on the variation of network voltage and the load demand with time.

Studies showed that the service voltage in the residential areas went down well below the usable limit between 6 p.m. to 9 p.m. with the increase in load demand. The line voltage steadily went up after this period and again rose above the usable limit during the off peak hours as depicted in fig.1.

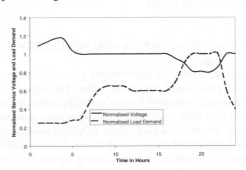

Figure 1: Variations of Normalised Service Voltage and Load Demand at Typical Residential Area

In the commercial office complex this picture is reversed. The demand increases after 9 a.m. and continues up to 5 p.m. and as a result the line voltage and quality of power degrade. This may be due to the unlimited load connected to the service line not accounted during the time of installation of power supply. The variation of this complex is depicted in fig.2.

A survey was carried out to find how the consumer dealt with this situation and it was found that the problem is partially solved through the use of individual voltage stabilisers in general and uninterruptable power supplies (UPS) for sophisticated devices. There is therefore a considerable market for these items. The survey also studied the real and apparent consumption of power and it

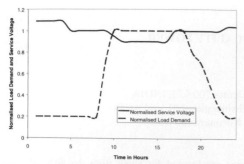

Figure 2: Variation of Normalised Service Voltage and Load in a Commercial Office Complex

was found that these items were consuming a reasonable amount of power at no load condition.

3. THE PROPOSED SOLUTION

In order to overcome these problems, it is proposed that the embedded power from photovoltaics may be fed to the input of the service line through electronic regulation. The design of such a system is carried out so as to account for the demanded load and availability of sunshine throughout the day. The power generated from the photovoltaic system may be fed to an inverter for generating 220 Volts at 50Hz. The output of the inverter is fed to the service network through a regulator, the function being to regulate the superposition of voltage coming from the two sources. The regulator also protects the feedback of power from the service network to the photovoltaic plant at the higher voltage period.

4. ECONOMIC ANALYSIS AND FEASIBILITY

Implementation of photovoltaic power is encouraged by the government of India. The Ministry of Non Conventional Sources of Energy (MNES) is the authorised body to deal with such matters. Various state government agencies are also working under MNES's umbrella. The Indian Renewable Energy Development Agency (IREDA), a financial institution, is also working for funding energy projects. IREDA is providing low interest loans and the government also institute other incentives such as tax free purchase on renewable energy consumer items in order to boost the applications and market for photovoltaics. In the commercial office complex the owners can choose the option for saving tax. In a residential area both tax saving as well as low interest loan schemes can help in implementing programmes. These programmes will help in resource generation, employment opportunity and energy conservation. A typical example can be projected in the light of the present studies. In an actual case, a commercial office complex containing a 100 units was surveyed and this revealed that, when the sun was at its peak, the usable load was also at its peak due to the enhancement of load demand. As a result there was a lowering of the voltage.

The adjustment of voltage can be done using the embedded power generation through photovoltaics. The roof of the commercial complex can be used for this purpose. The survey indicated that during peak hours each unit of the commercial complex needed about 50 Watts of

power to match the voltage level. Hence the commercial complex would therefore require about 5kW of power for this purpose. It is reported that, for a photovoltaic house in India, the average cost for a centralised photovoltaic power is about INR360 per Watt which is about US$8.4 per Watt. The cost of such a 5 kW system would be about US$42,000. The cost per unit would be US$420, which is about INR18,000. The unit owners agreed that this amount was easily affordable by them. If one looks at the long term benefit, at the present moment the cost of a UPS system is about US$400. The life of the UPS is certainly less than the life of the photovoltaic modules and the UPS is consuming power at the no load condition. The embedded power generated through photovoltaics is certainly more acceptable than the conventional UPS system. The same calculations are also applicable to the residential system. The calculations show that additional investment is needed for transformers and they would require replacement at least five times during the life span of the photovoltaic modules. It should also be noted that the reactive and resistive losses of the transformer increase the cost of the total system which is much more than the cost of the photovoltaic power system.

5. DISCUSSION AND CONCLUDING REMARKS

The present studies on embedded power generation, using photovoltaics, can be a substitute for meeting the overload demands in the utility distribution system. There are other benefits of a strategically located photovoltaic system, such as electrical loss savings, voltage support, generation of quality power and higher reliability. Each of the benefits described above can be verified using field demonstration.

The promotion of photovoltaic power is encouraged by the government and many financial institutions are interested. The advantage of tax benefits certainly add support for photovoltaic power promotion. Today in India there is a market for photovoltaics in the rural remote areas. Nothing so far has been tested for supplying power to the conventional grid in the urban areas. The present study may bring an opportunity for photovoltaics in the utility market. It is imperative that the photovoltaic manufacturer and the system engineer develop reliable and high performance products to meet the needs of the utility market.

This preliminary study shows that photovoltaic embedded generation could be a cost effective solution for Indian utilities and could form a large new market sector for the photovoltaic industries. The same picture may be observed in many other countries

ACKNOWLEDGEMENT

This work has been dedicated in the memory of Late Prof. Robert Hill who was one of the authors of this article. It is very sad that before the completion of this work he passed away. He had a belief that photovoltaics has a enormous opportunity as the back up for supplying quality power to the utility grid. Financial support from DST, Government. of India and British Council to help carry out this work is also gratefully acknowledged.

REFERENCES

1. R. Hill, S. Fifth, Deb Memorial Lecture, School of Energy Studies, Jadavpur University, Calcutta, INDIA December 11 (1995).
2. D.S. Shugar, Proc. 21st IEEE PV Specialist Conf. (1990) 836 – 843.
3. A.Kvora, Proc. International Workshop Eco Friendly Electricity in Off Grid Areas with Grid Quality Power **18-20** (1999) 77 – 83.

PHOTOVOLTAIC SYSTEM FOR EXPERIMENTATION

V. Martínez, M.J. Sáenz, M. García
Teknologia Mikroelecktronikoaren Institutua / Instituto de Tecnología Microelectrónica
Euskal Herriko Unibertsitatea / Universidad del País Vasco
Escuela Superior de Ingenieros de Bilbao
Alameda de Urquijo s/n, 48013 Bilbao. Spain.
Phone: 34.944.396.466, fax: 34.944.396.395, e-mail: jtpmasav@bicc00.bi.ehu.es

The aim of this paper is to describe the 1.6kWp photovoltaic system installed by the Instituto de Tecnología Microelectrónica. The system has been designed to operate as a stand-alone unit.

The special feature of this system is not its size but the selection of panels installed and their management and control. Panels of different nominal powers and voltages have been chosen in order to provoke the maximum level of mismatching possible in the system.

On the other hand, the connection and control system used allows individual access to each of the panels that make up the generation system, it being possible to connect/disconnect these to and from the system (giving rise to a variable size system of between 94Wp and 1.6kWp), as well as to modify the interconnection between the panels that make up the series branches of the system, or even connect one of the system panels to a measuring line, taking it out of the generation system temporarily, for individual measuring.

The management system gathers and stores on a periodic, exhaustive basis, several pieces of data relating to production, configuration and control status.

It is hoped that this system will be the support for other work being done in the Instituto de Tecnología Microelectrónica relating to the reliability of stand-alone photovoltaic systems and the minimisation of losses due to the mismatching of panels making up the same system.

Keywords: Stand-alone PV System - 1: Reliability - 2: Performance - 3

1. INTRODUCTION

The Instituto de Tecnología Microelectrónica, TiM, has designed and implemented a photovoltaic system specially conceived for experiments relating to the reliability of stand-alone photovoltaic systems and the minimisation of losses due to the mismatching of panels in any photovoltaic system. The field of panels has been installed on the roof of the new building of the Escuela Superior de Ingenieros de Bilbao.

The special characteristics of this system are set out in this article.

2. GENERAL DESCRIPTION OF THE SYSTEM.

The system has been designed to operate as a stand-alone unit, separate from mains. It is designed to be fitted with a total of 16 panels with powers of around 100Wp of nominal power, reaching a total of about 1.6kWp. These have been connected to a modular battery charge regulator

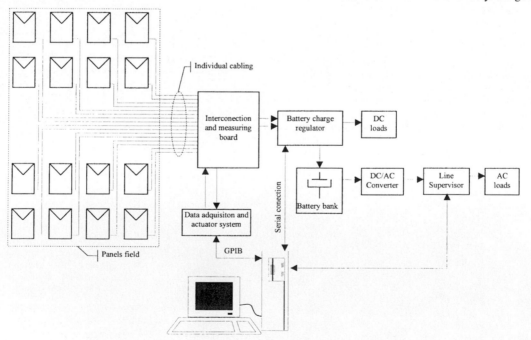

Figure 1: General description of the photovoltaic system.

with a limit of 30A per input module and for this reason the 16 modules have been laid out in two arrays, one for each of the available modules. The regulator has an RS-232 output for communicating the different statuses of the system, as well as the different measurements it makes. The battery bank is made up of 12 2V cells of acid lead connected in series, giving rise to a system with 24V of nominal voltage, with a capacity of 625Ah. The system is complete with an inverter with a 24Vcc input and output via a 220Vac transformer and a nominal power of 2.5kW, for the alternating current loads that are expected to be supplied. The entire system is controlled and supervised by a computer. Figure 1 shows the general layout of the system.

3. MAXIMUM MISMATCHING. INDIVIDUALISATION OF PANELS. VARIABLE SIZE SYSTEM.

The first special feature of the system is that the 16 panels that make it up have been chosen in order to attain the maximum level of mismatching possible. The panels have powers of between 94Wp and 110Wp, and are of both 24V and 12V for making serial/parallel connections within the system.

Figure 2: Connection cabinet.

The second major feature is that in this system, each panel has been wired individually and both terminals are taken to a connection cabinet where each input position corresponds to a position within the field of panels. The position within the field of panels, marked P1, P2, etc., will be linked to one of the parallel-connected arrays provided in the system, being connected in series or parallel within the array. The specific panel that is placed in these positions may be one of those referred to above, or changed for others at a later date. The idea behind this layout is the

individualised treatment of a specific panel which can be assigned an alias in order to refer to this. In this system, each panel is treated individually, different from the others, although it may be of the same type within the range provided for by the manufacture.

Once inside the connection cabinet, which can be seen in figure 2, each of the above inputs passes through a series of relays. These relays allow the connection/disconnection of a parallel branch whereby the generation of the system may vary from a minimum of 94 Wp up to a maximum of 1.6kWp. Furthermore, other relays allow different connections between the serial branch panels, increasing the mismatch, and still others allow a panel to be switched from the generation system to a measuring line.

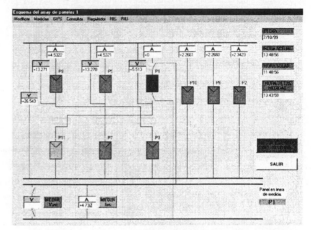

Figure 3: One of the screens of control. A panel conected to the measuring line can be seen.

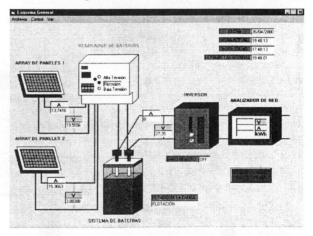

Figure 4: Main screen of the control and supervision program.

These features make the system especially appropriate, on the one hand, for studying the reliability of autonomous photovoltaic systems under different circumstances. On the other hand, it is the appropriate support for work relating to mismatching between panels.

4. CONTROL AND SUPERVISION SYSTEM

The photovoltaic system is complete with a full management and supervision programme running on a PC. This allows the status of the generating system to be displayed as well as each of the component branches (see figures 3 and 4), storing all the data acquired, together with the status of the panel connections in a number of databases. These databases can be used later to generate files compatible with MS-Excel and also in ASCII format for subsequent treatment by other computer programmes.

The management programme also records the configuration of the system and the modifications made to it. It also controls each and every one of the connecting relays, allowing the user to modify these. This can be seen especially in figure 3 in which one of the panels has also been diverted to the measuring line to determine its short-circuit current.

Figure 5: Database output.

5. RESULTS AND FUTURE WORK

The results of this system depend upon the information accumulated in the databases, figure 5, which is still insufficient to draw conclusions. Nevertheless, we hope that this system will allow us to implement work being done at this time on the minimisation of losses due to mismatching between the panels in a generating system, by means of the use of circuits that offset any differences in behaviour, and also due to the defining of new classification strategies, it being possible to check for changes in operation by comparing the databases generated.

It also represents a useful tool for studying the reliability of stand-alone photovoltaic systems, a field in which work is being done at present in the Institute by means of simulation techniques.

An photograph of the modules layout could be seen in figure 6.

ACKNOWLEDGEMENT

This work has been made under the sponsorship of the Universidad del País Vasco / Euskal Herriko Unibertsitatea and its contract UP-147-375-TB-040/96.

Figure 6: Actual layout of PV modules.

NEW SYSTEM FOR THE SUPPLY OF PHOTOVOLTAIC ELECTRICITY

Alfonso de Julián, Jesús García
Luis Alberto Calvo, Estefanía Reolid (CDT-Renewable Energy Area Technical Managers)

IBERDROLA
Hermosilla 3
Madrid, 28001
SPAIN
Tel: +34 91 577 65 00 Ext. 41440
Fax: +34 91 578 20 94
www.iberdrola.es
cdt@iberdrola.es

ABSTRACT: IBERDROLA developed a 3.3 kWp mobile unit for the supply of photovoltaic energy in order to demonstrate the versatility and applicability of solar photovoltaic energy in isolated areas. After three years of performance, in which operation conditions and manoeuvrability were studied, a new model has been made in order to improve some aspects of the first prototype. The new project consists of a 1.5 kWp portable photovoltaic system in which one of the most important aspects is its design, which facilitates its transfer by any conveyance. Due to its small dimensions and weight, this system is suited to be moved even by plane. This system has been developed in order to allow electricity to be supplied to remote customers under service conditions equivalent to those provided by the grid, this being accomplished in the shortest time thanks to the ease with which the system is transported and installed. In this way, IBERDROLA is able to contribute to the development of renewable energy sources and to care for the environment.
Keywords: Remote - 1: Stand-alone PV System - 2: Marketing - 3

1. DESIGN

The system consists of a container which is perfectly protected for its transfer. When this container is installed in the chosen place, it can be opened, unfolding the solar field which is made up of 18 BP-585 photovoltaic modules. This operation is designed to be performed without effort by only one person since the system is carefully balanced. When the two lateral structures are closed, a seal is formed. However the surfaces of panels are separated by an air channel that guarantees correct ventilation.

Figure 1: General view of the portable PV unit

The weight of the container is less than one thousand kilograms and the structure of it is made of aluminium, whose qualities are lightness and high corrosion and weather resistance.
The tallest side of the container folds down and becomes available for each use of the system. Opening this compartment, one has access to the inside of the container, in order to keep equipment of each application, for example a refrigerator for medicines, foods, etc…It is possible to have a set of bars and canopy which in a few minutes can be mounted to form a shelter of 6 m² surface and 12 m³ of volume.
Inside the container, batteries and also the necessary elements to use the system are located. On one side of the container, it possible to access the emergency energy source (diesel generator).

2. TECHNICAL CHARACTERISTICS

The main components of the unit are as follows:
- Photovoltaic generator
- Energy storage system
- Power conditioning system
- Emergency energy source
- Data acquisition system

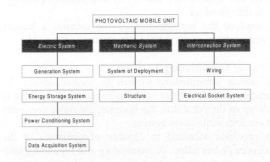

Figure 2 : Main system components

2.1 Photovoltaic generator:
The main function of the solar field is to generate photovoltaic solar energy for charging the batteries or direct supply via the control and regulation console.
The solar field is made up of 18 BP-585 mono-crystal silicon modules, each one with a nominal power of 85 W. The total power of the photovoltaic generator is 1,53 kWp

and its area is 11.32 m2. It is mounted in a structure which may be manually deployed. The structure is tilted 30º. The technical characteristics of BP-585 modules are:

Table 1: Technical characteristics of BP-585 modules

Typical peak power (Pmax)	85 W
Voltage maximum power (Vmp)	18 V
Current maximum power (Imp)	4.72 A
Short-circuit current (Isc)	5.00 A
Open-circuit Voltage (Voc)	22.03 V

Figure 4 : Back view of the PV portable unit

Figure 3: Frontal view of the PV portable unit

Some panels are exposed when the panels are not extended. The function of this field is to maintain battery-charging levels during periods in which the system is not operative, and therefore is not receiving energy from the main field.

2.2 Energy Storage System:
The function of the energy storage system is to store the energy produced by the photovoltaic generator to be used during the same period of time or during the hours of no insulation. The energy storage system has been designed to provide the system with an autonomy for 1.25 days of operation under zero sunlight conditions.

The energy storage system is made up of 8 FGB BLOC OPzV 6/200 stationary gel elements, each of 6 V and with a capacity of 200 Ah (100 h). These batteries need neither maintenance nor care of the electrolyte. That is why this type of batteries are adequate for applications in remote places. The batteries are located in a special compartment inside the container

To improve the performance of the inverter a battery working voltage of 48 V has been chosen.

2.3 Power Conditioning System:
The power conditioning system is located inside the container. A wall cabinet contains the assembly, which is made up of the regulator, the inverter, the dc/dc converter and the data acquisition system.

The regulator installed is capable of operating at the maximum power value, of monitoring and regulating load operation and battery discharge, and of controlling a possible auxiliary power system.

The dc/dc converter offers a 12 V output voltage as to cover some DC specific consumption's, such as mobile telephones.

The TRACE SW inverter is a single-phase self-switching unit with a power level of 1 kW. The direct input voltage is 48 V D.C., with a 50Hz, 220 V sinoidal wave on the alternating current side. Its maximum consumption is 0,5 W.

2.4 Emergency energy source :
As a complement to the photovoltaic generator, the system presents a diesel engine in order to accomplish battery charge in missing-sun periods or for direct 220 Vac energy supply .

The diesel generator is placed in a small compartment in one of the container laterals.

Figure 5: Emergency energy source (diesel engine)

2.5 Protection and Grounding System:
This system is made up of electronics protections against short-circuiting at the outlet to the user. The input to the equipment is protected by fuses and breakers.

2.6 Monitoring:
The data acquisition system has an information recording capacity for more than 12 months. The data may be downloaded to a PC via a series port, or via modem if the site is equipped with a telephone line. The user is provided with a display panel, which may be installed at the point of energy consumption, from where the most important operating parameters of the system may be viewed.

3. SUMMARY OF MAIN SYSTEM FEATURES

Table 2: Main system features

PORTABLE PHOTOVOLTAIC UNIT	
Location	Remote users
Connection mode	Isolated from the grid
Installed power	1.53 kWp
Type of modules	BP-585 (85 W)
Number of modules	18
Degree of inclination	30º
Total weight	1069 kg
High	2.00 m
Wide	1.99 m
Long	2.82 m
BATTERIES	
Direct current voltage	48 V
Type of battery	Gel elements (200 Ah)
Total weight	344 kg
Autonomous operation	1.25 days
INVERTER	
Power	1000 kVA
Input voltage	48 V
Alternating current-voltage	220 V mono-phasal

4. CONCLUSION

This photovoltaic system is one more application of how photovoltaic solar energy can cover the most basic and also the most important needs for people who do not have electric power. Furthermore it will be very useful for rural countries and areas and even NGOs, since they could use electrical energy from photovoltaic origin wherever the system is located. Therefore, this photovoltaic system will favour and help development of the countries and areas that need it.

PROGRESS IN PHOTOVOLTAIC COMPONENTS AND SYSTEMS

H. Thomas, B. Kroposki, C. Witt
National Renewable Energy Laboratory
1617 Cole Boulevard
Golden, CO USA 80401-3393

W. Bower
Sandia National Laboratories
1515 Eubank SE
Albuquerque, NM USA 87185-0753

ABSTRACT: The Photovoltaic Manufacturing Research and Development project is a government/industry partnership between the U.S. Department of Energy and members of the U.S. photovoltaic (PV) industry. The purpose of the project is to work with industry to improve manufacturing processes, reduce manufacturing costs, and improve the performance of PV products. This project is conducted through phased solicitations with industry participants selected through a competitive evaluation process. Starting in 1995, the two most recent solicitations include manufacturing improvements for balance-of-system (BOS) components, energy storage, and PV system design improvements. This paper surveys the work accomplished since that time, as well as BOS work currently in progress in the PV Manufacturing R&D project to identify areas of continued interest and product trends. Industry participants continue to work to improve inverters and to expand the features and capabilities of this key component. The industry also continues to advance fully integrated systems that meet standards for performance and safety. All participants included manufacturing improvements to reduce costs and improve reliability. Accomplishments of the project's participants are summarized to illustrate the product and manufacturing trends.
Keywords: Inverter- 1: Balance of Systems- 2: Components- 3

1. INTRODUCTION

The Photovoltaic (PV) Manufacturing Research and Development (R&D) project is a government/industry partnership between the United States (through the U.S. Department of Energy [U.S. DOE]) and members of the U.S. PV industry. The purpose of the project is to work with industry to improve manufacturing processes, reduce manufacturing costs, and improve the performance of PV products. This project is conducted through phased solicitations with industry participants selected through a competitive evaluation process. Since the project was implemented during 1991, there have been five solicitations for R&D on manufacturing process problems. Starting with the fourth solicitation in 1995, the topics were expanded to include manufacturing improvements for balance-of-system (BOS) components, storage, and system design improvements. These topics are included in the fifth solicitation initiated in 1998, and this work is now in progress. Accomplishments in the area of PV module manufacturing resulting from the PV Manufacturing R&D project are described in other publications [1,2]. This paper reviews specific progress and trends in PV systems and components.

2.0 TRENDS IN PV COMPONENT ADVANCES

Work in the area of BOS and PV systems has been in progress under the PV Manufacturing R&D project since 1995 [3]. These advancements and trends are illustrated by the work of the various participants in the Manufacturing R&D project and are described in the following sections. All project participants included manufacturing improvements to reduce costs and improve reliability as a part of their overall effort. A survey of the work accomplished over this period illustrates areas of

particular interest to industry and indicates general trends in product improvements. These accomplishments are described in the following sections.

2.1 Trends in Completed Projects

Thirteen awards were made in 1995. Of these 13, 8 awards were in BOS and system-related topics. Of these 8, three worked specifically to improve inverters, three addressed both power conditioning and integrated PV systems, and two worked on approaches to integrate systems and PV system manufacturing improvements. The following sections describe this progress, with participants grouped generally by topic. The participants and the focus of their work are listed in Table I.

2.1.1 Inverters

Three companies worked specifically to improve inverters. These three illustrate the trend for inverters to provide additional features in addition to basic power conversion. Companies working in this area used advanced electronics, new approaches to power bridge design, and complex software to improve performance, reduce costs, and provide monitoring and improved communications options. In alphabetical order, the first company described in this group is *Advanced Energy, Inc.,* which designed a hybrid inverter to interface with PV, diesel, and wind power sources. Advancements demonstrated by a 60-kW prototype included digital control, optimized magnetics, and smart-power components. Advanced Energy also included a communications interface. The second company in this group, *Omnion Power Engineering Corporation* (now a division of S&C Electric Company), improved their 100-kW inverter, implementing a highly integrated power-switching bridge, air-core transformers, and condensed printed-circuit boards (PCB). Omnion reports labor is

Table I: List of participants and work in PV Manufacturing R&D components and systems completed in 1998.

Company or Team	R&D	Inverters (Power Conversion Systems)	Advanced Components & Integrated Systems	Manufacturing and System Integration
• Advanced Energy, Inc.	Hybrid inverter	X		
• Omnion Power Engineering, Inc.	Advanced grid-tied inverter	X		
• Trace Engineering	Modular inverter	X		
• Ascension Technology, Inc., with ASE Americas, Inc.	Module-scale inverter & integrated PV product	X	X	
• Solar Design Associates, with BP Solarex & Advanced Energy, Inc.	Integrated PV array mounting, module-scale inverter	X	X	
• Utility Power Group, Inc.	Grid-tied PV system	X	X	
• Evergreen Solar, Inc.	Alternative encapsulants, new module mounting approaches			X
• Solar Electric Specialties, Inc.	Integrated PV products			X

reduced by half, and manufacturing costs and parts count are reduced to one-third that required for their earlier model [4]. The final company included in this group, *Trace Engineering,* developed a modular, bi-directional, 2.5-kW Series PS2 inverter designed for mass production. The PS2 is capable of grid-tied operation, or stand-alone with a battery, and can function in parallel with other inverters through a Sinewave Inverter Paralleling kit. Trace reports the PS2 has 35% fewer parts, is 40% smaller, and requires 42% less assembly labor than their previous SW series units [5].

2.1.2 Advanced Components and Integrated Systems

Three companies focused on power conversion improvements and PV product integration. *Ascension Technology, Inc.* (now a division of Applied Power Corp., or APC/AT), designed and completed pilot manufacturing of the SunSine™ 300 AC PV Module. This fully-integrated PV product incorporates their module-scale, 250-Wac SunSine™ 300 inverter attached to the back of the ASE Americas, Inc., large-area PV laminate (260 to 300 Wdc). The SunSine™300 inverter includes passive and active anti-islanding protection and maximum-power-point tracking [6]. *Solar Design Associates, Inc.,* led a team with Advanced Energy and BP Solarex to address several aspects of PV systems [7]. Advanced Energy enhanced their 250-Wac module-scale inverter to include software controls, and BP Solarex developed a low-cost, pre-engineered assembly that is now part of their pre-packaged Millenia® PV arrays. This assembly integrates module framing, mounting, array wiring, grounding, and lightning protection. The third company in this group, *Utility Power Group, Inc. (UPG),* established a manufacturing process for integrating PV laminates into modular panels that are pre-assembled into subarrays and installed in array strings of up to 15 kW each [8]. An Integrated Power Processing Unit (IPPU) manages single-axis tracking and power conversion for each string. The factory-assembled IPPU combines all the necessary power conversion/control electronics, array-tracking control

electronics, source-circuit protection hardware, and DC and AC switchgear for utility applications.

2.1.3 PV System Manufacturing Advancements

Two participants completed manufacturing and product advancements for fully-integrated systems. *Evergreen Solar, Inc.,* completed several manufacturing improvements, including the development of an air-cured transparent encapsulant for manufacturing PV modules using a continuous, non-vacuum lamination process. The company also identified a PV-laminate backskin material. The unique properties of the backskin resulted is a mounting system design for rapid panel installation in the field [9]. *Solar Electric Specialties, Inc.* (now part of APC), developed two fully-integrated PV products to demonstrate the benefits of product integration, including lower costs for materials and labor and higher overall quality and reliability. Two product lines resulted from this work: 1) the Underwriters Laboratories (UL)-listed, stand-alone Modular Autonomous PV Power Supply consisting of PV modules and an enclosure containing solid-state power control and sealed batteries; and 2) the mobile Photogenset, which integrates a generator, PV array, and battery storage to provide off-grid AC power [10].

2.2 Trends in Projects Now in Progress

The trends of increased sophistication, improved manufacturing, and streamlined PV systems continue to be demonstrated in BOS subcontracts initiated during 1998 with the fifth solicitation. Work now in progress includes APC/AT, which has redesigned their module-scale inverter to improve performance, enhance reliability, and reduce costs. With these enhancements, the product is named the SunSine®. PowerLight Corporation has undertaken a broad range of product improvements and manufacturing steps with the goal of achieving a $3.05/$W_p$ installed cost for their PV system. Omnion is developing a sophisticated design for 1- and 2-kW grid-

Table II: The APC/AT SunSine™ demonstrates significant advancements over the original SunSine™ 300.

Item	Original SunSine™ 300	Revised SunSine®	Relative Improvement
Power rating	251 Wac	275 Wac	9.6%
Efficiency	87%	91%	4.7%
Communications	Signal	Infrared data link	Multi-functional

tied inverters. UPG is developing a fully-integrated residential PV system that includes pre-packaged storage. As a group, these projects represent industry's continuing drive to reduce overall system costs and improve inverters. Details of this work are described in the following sections.

2.2.1 Manufacturing and Performance Improvements to the SunSine™ 300

The work now in progress by *APC/AT* follows the successful development of the SunSine™ 300. The advanced 275-Wac SunSine® incorporates significant operating enhancements, additional features, and streamlined manufacturing processes. Product improvements include a proprietary Zero Voltage Switching (ZVS) soft-switching approach and a modified transformer design to increase the unit's power and efficiency. Revisions to the PCB optimize performance and reduce overall size and costs. The PCB parts count has been reduced by 34%, non-standard parts are eliminated, and the board is redesigned for automated assembly. The SunSine® also has an infrared data link compatible with the IrDA™ interface used by PalmPilot™ and other electronic products. The SunSine™ advancements in performance and reduced costs as summarized in Table II.

2.2.2 Manufacturing Improvements for PowerGuard® PV System

The overall goal of *PowerLight Corporation*'s work in this subcontract is to reduce costs. Their approach is to introduce incremental improvements to their PowerGuard® PV system components and automated manufacturing methods.

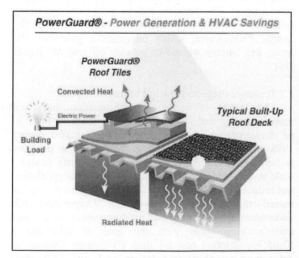

Figure 1: Schematic illustrating the structure of the PowerGuard® tile.

The PowerGuard® system is an approach that incorporates a proven roofing technique with PV. The system comprises electrically connected strings of PowerGuard® tiles, held in place, with no roof penetrations, by a ballasted curb (Figure 1). Over the past 18 months, PowerLight has designed an automated tile manufacturing facility and completed pilot manufacturing to demonstrate a capacity of 10 MW of PowerGuard® tiles/year. They have also reduced PowerGuard® tile production costs by 30% and reduced overall PowerGuard® system costs by 15%. The company is now refining automated processes to achieve production capacity of 16 MW per year. PowerLight also worked with lower-tier subcontractor Trace Technologies to modify their inverter to incorporate a data acquisition system and dial-up communication capability for audit-worthy verification of PV system performance. Trace also upgraded the design to digital circuitry and reduced total parts count by 25%.

2.2.3 Omnion Utility-Interactive Inverter

Omnion's objective is to produce 1- and 2-kW single-phase inverters for utility-interconnected applications. With modified design, the product's input voltage range will be 100 to 400 Vdc. The design will include provisions for optional or standard metering and data-acquisition connections, and interface and near-zero input voltage ripple due to active control to maximize array utilization. An inverter enclosure that meets UL requirements, and provides for a "plug together" inverter and base plate, will facilitate installation and service. Anti-islanding and ground-fault detection and protection are also planned.

2.2.4 UPG Integrated Residential PV System

UPG is completing development of a fully-integrated residential PV system that incorporates three distinct elements—a rooftop PV Array, a Power Unit, and an Energy Storage Unit. The modular PV Array comprises factory-assembled, modular subarrays, expandable to 4 kW, of either crystalline or thin-film PV modules. *Trace Technologies* is working as a lower-tier subcontractor to UPG to develop a Power Unit for this system. This bi-directional inverter supplies 120/240 Vac at 12 kVA continuous and 19 kVA of peak power, to manage battery charging, power-source switching (PV, utility, or battery), and utility interconnection. The unit's higher switching frequency increases power density, decreases the cost and weight of magnetic components, and increases conversion efficiency. The power-conversion topology also eliminates additional boost circuitry with its associated losses and expense. The modular Energy Storage Unit provides about 13 kWh of maintenance-free battery capacity. It interfaces with the Power Unit, communicating through a serial data bus. Each unit

Table III: Advantages of Utility Power Group, Inc., PV system compared to conventional residential rooftop PV systems.

Conventional Systems	Utility Power Group Residential System
Cost: $US 5.75 – 7.50 per watt	Cost: $US 4.75 – 5.50 per watt
Low inverter efficiency (85% – 93% at full power)	Efficiency 94% at full power
No integrated energy storage	Fully-integrated energy storage capacity (20 hour rate): 12.7 kWh

includes: 1) 32 valve-regulated lead-acid batteries, 2) overcharge protection circuitry for any given battery, and 3) a thermally insulated container with "smart" forced-convection cooling. The unit also incorporates software to monitor battery use. More units can be added if desired. The system improvements are compared to conventional systems and summarized in Table III.

3. CONCLUSIONS

The PV Manufacturing project has conducted R&D with industry in the area of BOS and systems since 1995. During that time, industry has demonstrated continued improvements in a broad range of inverter designs. These improvements have emphasized sophisticated power electronics, software, and other design improvements to enhance the capabilities of the inverters, as well as to reduce costs and simplify manufacture. Industry also continues to work toward integrated PV systems to reduce overall installed costs and to meet a variety of applications for PV products. These trends are expected to continue as new electronic sub-components become available and public demand for integrated systems continues to expand.

4. ACKNOWLEDGMENTS

This work is supported under DOE Contract DE-AC36-99GO10337 with NREL, a national laboratory managed by Midwest Research Institute, and Contract DE-AC04-94AL8500 with Sandia National Laboratories, a multi-program laboratory operated by Sandia Corporation, a Lockheed Martin Company. Many people have contributed to developing and implementing the PV Photovoltaic Manufacturing R&D project and to the efforts carried out in this project. The authors acknowledge that this paper represents their work.

5. REFERENCES

[1] Witt, C. E., Mitchell, R. L., Symko-Davies, M., Thomas, H. P., King, R., and Ruby, D. S., 1999, "Current Status and Future Prospects for the PV Manufacturing Technology," *Proceedings, 11th International Photovoltaic Science and Engineering Conference (PVSEC-11)*, T. Saitoh, ed. (to be published)

[2] Witt, C. E., Mitchell, R. L., Thomas, H. P., Symko-Davies, M., King, R., and Ruby, D. S., 1998, "Manufacturing Improvements in the Photovoltaic Manufacturing Technology (PVMaT)," *Proceedings, 2nd World Conference and Exhibition on Photovoltaic Solar Energy Conversion*, J. Schmid et al., ed., European Commission, Ispra, Italy, Vol. 2, pp. 1969-1973.

[3] Thomas, H. P., Kroposki, B., McNutt, P., Witt, C. E., Bower, W., Bonn, R., and Hund, T. D., 1998, "Progress in Photovoltaic System and Component Improvements," *Proceedings, 2nd World Conference and Exhibition on Photovoltaic Solar Energy Conversion*, J. Schmid et al., ed., European Commission, Ispra, Italy, Vol. 2, pp. 1930-1935.

[4] Porter, D., Meyer, H., and Leang, W., *Three-Phase Power Conversion System for Utility-Interconnected PV Applications, A PVMaT Phase I Technical Progress Report by Omnion Power Engineering*, NREL, Golden, CO, NREL/SR-520-24095, February 1998.

[5] Freitas, C. (1999). *Development of a Modular, Bi-Directional Power Inverter for Photovoltaic Applications: Final Report*, NREL, Golden, CO, NREL/SR-520-26154, March 1999.

[6] Kern, G., *SunSine™300: Manufacture of an AC Photovoltaic Module, Final Report*, NREL, Golden, CO, NREL/SR-520-26085, March 1999.

[7] Strong, S., Wohlgemuth, J., and Kaelin, M., *Development of Standardized, Low-Cost AC PV Systems, PVMaT Annual Report, September 1995-November 1996*, NREL, Golden, CO, NREL/SR-520-23432, June 1997.

[8] Stern, M., Duran, G., Fourer, G., Whalen, W., Loo, M., and West, R., *Development of a Low-Cost, Integrated 20-kW AC Solar Tracking Sub-Array for Grid-Connected PV Power System Applications, Final Technical Report*, NREL, Golden, CO, NREL/SR-520-24759, June 1998.

[9] Hanoka, J., Chleboski, R., Farber, M., Fava, J., Kane, P., and Martz, J., *Advanced Polymer PV System, Final Report*, NREL, Golden, CO, NREL/SR-520-24911, July 1998.

[10] Lambarski, T., Minyard, G., *Design, Fabrication and Certification of Advanced Modular PV Power Systems, Final Report*, NREL, Golden, CO, NREL/SR-520-24921, October 1998.

[11] West, R., Mackamul, K., and Duran G., *Development of a Fully-Integrated PV System for Residential Applications, Phase I Annual Report*, NREL, Golden, CO, NREL/SR-520-27993, March 2000.

Database of European Manufacturers and Distributors and Tests of Components for Solar Home Systems

H. Bonneviot[5], G. Bopp[1], M. Hankins[4], T. Paradzik[6], K. Preiser[1], D. Nuh[3], B. Rouvière[2]

[1] Fraunhofer-Institut für Solare Energiesysteme ISE, Oltmannstrasse 5, D-79100 Freiburg
Phone: ++49-7 61-45 88-2 40, Fax: ++49-7 61-45 88-217, e-mail: bopp@ise.fhg.de
[2] APEX, 4 rue de l'Industrie, F-34880 Laverune
[3] EETS, 104 Portmanmoor Road Industrial Estate, UK-Aandiff CF2 2HB
[4] ESD, Overmoor Farm, UK-Wiltshire SN13 9TZ
[5] IED, 46 rue de Provence, F-75009 Paris
[6] PSE, Christaweg 40, D-79114 Freiburg

ABSTRACT: Solar Home Systems (SHS) are small Photovoltaic systems consisting of a PV module, a charge controller, a battery, and directly connected DC appliances. They represent the majority of stand-alone PV systems in remote regions. To allow an optimum operation of the SHS, it is necessary that all components work optimally under the typical conditions of PV systems.
To get a view over the different components in the power range of a SHS and to have the possibility to compare the different products among themselves, a European market survey was realised. Over 500 manufacturers and distributors, which offer products like inverters (up to 400 W), charge controllers (12 V, up to 15 A), PV modules (up to 30 Wp) and DC lamps (12 V, up to 18 W) were written down. The information is prompted in a MS-Access database containing today 250 different components and is commercially available as a CD-ROM from the Fraunhofer ISE. Beside the standard characteristics like address of the manufacturer and/or distributor and type with nominal data, the information in the database comprise detailed information about electrical and physical properties, like self consumption and efficiency dimension and weight. Information about price, protection class and check digits are also available.
Keywords: market overview - 1: data base - 2: Solar Home System - 3

1. INTRODUCTION

To supply remote rural areas with energy, especially in developing countries, photovoltaically powered systems are often the only solution. Simply built Solar Home Systems (SHS) are very well suited to the special demands as has been shown over the last years,. Typically a SHS consists of one to two solar modules, a battery (12 V) and a directly connected load such as lamps, radio, and black and white television.

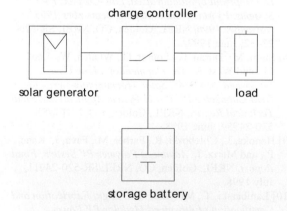

Figure 1: Construction principle of a SHS.

For the use of a SHS, it is essential to ensure reliable function. This will increase both the users acceptance and the dissemination of such installations. To enable optimal function of the SHS, all components must function under typical operation conditions for PV systems /1/. Furthermore, the single components of the installation, as well as the system parts, must be compatible with each another and optimised. Then the components as well as the whole system will operate efficiently /2/.

2. PROJECT/PARTNERS

Within the scope of the »ALERT-PV« project, which is supported by the European Commission, the demands on the components for SHS from completed projects in Kenya, Uganda, Mali and Europe were collected and analysed.
Single components were selected from an European market overview and tested according to these demands. Improvements in the products of the involved partners will be made in a last step.

3. MARKET OVERVIEW

Due to the participation of French, English and German partners it was possible to make a European market overview available. Over 500 inquiries were sent to manufacturers and distributors. The content of the 247 brochures and data sheets from approximately 35 manufacturers and distributors entered into a Microsoft Access data base.

Country	D	GB	AUT	CH	N	S	B	F
Manufacturer	7	3	1	3	2	1		1
Distributor	12	1		3			1	
Charge controller	22	3		14	1			2
Inverters	15	1		10	3			
Electronic ballast	49	22		21		3		
PV module	43	2	5	18	4		9	

Table 1: Information available in the data base for manufacturers/distributors and components.

Only products which are important for a SHS with a nominal voltage of 12 V were selected. Since there are other market overviews /3/ for PV modules with higher power, attention was focused on PV modules below 30 W. The maximum of the charging and discharging current of the charge controller is limited to 15 A. The nominal power of the inverters is up to 400 W and the power of the lamps is below 18 W.

Apart from the standard data such as address of manufacturer and/or distributor, type with nominal data, detailed information on electrical and physical specifications, (e. g. self-consumption, efficiency, dimensions and weight), the data base also gives information about prices, protection classes and possible testing certification.

4. COMPONENT TESTS

From the information collected in the data base, 15 charge controllers, 15 lamps with electronic ballast, 6 PV modules and 4 inverters were selected. These components were or will be tested regarding specific functions and details at Fraunhofer ISE. In the following, only the tests of the charge controllers and electronic ballasts are considered in detail, as the tests of the other components have not been finished yet.

5. REQUIREMENTS ON CHARGE CONTROLLERS AND ELECTRONIC BALLASTS

The following table shows an overview over the most important requirements on a charge controller and an electronic ballast in a SHS:

- high electrical efficiency
- little self-consumption
- charging and discharging thresholds adjusted to the battery
- behaviour under changing temperature and air humidity conditions
- safety measures, e.g. against reversed polarity of the module and battery, overload, removing of the fluorescent tube, etc.
- operation under abnormal conditions, e.. extremely low battery voltage, low temperature, etc.

6. RESULTS OF THE TESTED CHARGE CONTROLLERS AND ELECTRONIC BALLASTS

In the following the most important test results of 8 selected charge controllers and 7 electronic ballasts from Germany, France, USA and Kenya are shown.

Figure 2 shows the maximum self-consumption. It can be seen clearly that there are enormous differences between the charge controller. In three of the 8 cases the self-consumption current was significantly higher than the 5mA defined in the list of specifications /4/. In one case, it was 160 mA, which is ¼ of the current generated by a 50 W

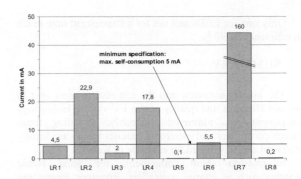

Figure 2: Self-consumption of the tested charge controller

SHS over a whole year. Figure 3 shows the threshold voltages of the deep discharge protection and the possible adjust ranges. The recommended threshold is 11,4 V /4/.

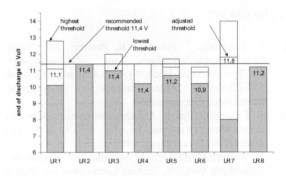

Figure 3: Highest and lowest allowable settings

It can be seen clearly that some charge controllers have an extremely wide adjustment range. On one hand, it helps the threshold to be adjusted to an optimum value, but on the other hand it can lead to incorrect settings due to unauthorised or inexperienced staff.

Nevertheless, the efficiency of the controller, which results from multiplying of the efficiencies of the charging and discharging cycles, plays an important role. The tests showed a result of 90 % and more at maximum currents. Compared to an investigation in the mid 90's /4/, the results show that there has been a significant improvement, particularly in the efficiency.

The electrical efficiency of the electronic ballasts is not so good, see fig. 4.

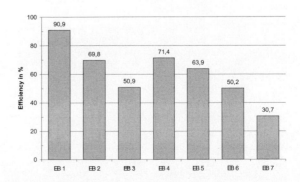

Figure 4: Electrical efficiency of the tested electronic ballasts.

Only one for LEDs and not for a fluorescent tube reach an efficiency of 90 %.

7.   SUMMARY

The market overview and the test results show that in the last 10 years, the quality of the PV system technology components has improved noticeably, including those charge controllers and electronic ballasts which are locally produced and developed for example in Kenya.

Information about components for Solar Home Systems such as charge controllers, inverters, lamps and PV modules is stored in a Microsoft Access data base. In addition to the technical data (nominal data from the data sheets), addresses of manufacturers and distributors are available. The data base makes it easier to search for components and gives an overview of the current market. The data base can be ordered from Fraunhofer ISE for a fee of 22 Euro plus 7% VAT.

REFERENCES

[1]   J. Kuhmann, T. Paradzik, K. Preiser, D.U. Sauer: "Herz-Kreislaufstörungen im PV-Inselsystem, Fallbeispiel Solar Home System in Indonesien", 13. Symposium Photovoltaische Solarenergie, Staffelstein, Germany 1998

[2]   K. Preiser, O. Parodi, J. Kuhmann: Quality Issues for Solar Home Systems, 13. EC PV solar energy conference, Nice, France, 1995

[3]   P. Welter: Marktübersicht Solarmodule, Photon das Solarstrom Magazin, Aachen, Germany, 1998 Issue 2

[4]   K. Preiser, J. Kuhmann, E. Biermann, T. Herberg: Quality of charge controllers in Solar Home Systems, results of detailed tests, ISES solar energy congress, Harare, Zimbabwe, 1995

Line Side Behavior of a 1-MW-Photovoltaic Plant

Gerd Becker *, Georg Maier **, Mike Zehner *

* Munich University of Applied Sciences · Electrical Engineering Department · SE Laboratory
Dachauer Str. 98 · D-80335 Munich · Tel.: ++49-89-12651451 · Fax: ++49-89-12651299
becker@e-technik.fh-muenchen.de

**Bayernwerk AG · Nymphenburger Str. 39· D-80335 München
georg.maier@bayernwerk.de

http://www-lse.e-technik.fh-muenchen.de

ABSTRACT: The world′s largest rooftop photovoltaic power plant, consisting of 7812 solar modules with a peak power of 1.016 kW, is situated on the New Munich Trade Fair Centre. The generated solar DC power is transferred into AC on the 0,4-kV-low voltage level by only one inverter unit consisting of three 330 kVA inverters. Measurements have been carried out to determine the line side behaviour of this 1-MW-PV-Generator on the AC-Side of the inverter. It can be seen that no limits of the standard EN 50160 are exceeded and that in times of high PV-Generation, the quality of the line side voltage is improved by the operation of the inverter unit.

Keywords: 1: Large Grid-Connected PV Systems - 2: Large Central Inverter Unit – 3: Harmonics

1. INTRODUCTION

The New Munich Trade Fair Centre, which was commissioned in February 1998, is currently one of the most modern trade fair centres in the world. On the roofs the world′s largest rooftop photovoltaic power plant is situated, consisting of 7812 frameless solar modules, each with 84 monocrystallin solar cells. The output of one module is 130 watts, which gives total module peak power of 1.016 kW. It was the aim of Bayernwerk, a large German electric power utility, who managed the project, to build the world's largest and most modern PV plant on six roofs of twelve of this new Centre and demonstrate it to an international public /1/.

Photovoltaic power systems connected to the grid via inverters may cause line side problems. Investigations have been carried out about the reactions of a great number (675) of 1 kW$_P$-photovoltatic systems in the feeding power system /2/. For a PV-Plant of 1 MW the reactions are unknown, therefore measurements have been carried out to determine the line side behavior of this large photovoltaic power plant.

The manufacturer of the inverter had stated that – under certain circumstances – the inverter would improve the quality of the line side voltage. Thus it was another goal of the measurement series - each for the duration of about one week - to verify this statement.

2. STANDARDS FOR LINE SIDE BEHAVIOUR

As a Standard for mains supply quality EN 50160 defines the voltages supplied by utility low voltage networks.

On the base of EN 50160 it was the aim to show the behavior of the following quantities depending on the load of the photovoltaic system.

- Harmonics - typical and non typical – in a frequency range up to 1 kHz – described by the THD = Total

  Harmonic Distortion $THD = \sqrt{\sum_{v=2}^{40} u_v^2}$

  with u$_v$ = Harmonic Voltage of the order v
- Flicker described by the Short Term and Long Term Flicker Intensity P$_{ST}$ and P$_{LT}$
- Unbalance of the voltage described by the negative sequence system
- Interharmonics

3. INVERTER UNIT

The Munich Megawatt solar generator with its arrays is feeding into a central DC bus bar, with DC voltages of 320 V – 435 V in the working point. To this bus only one central inverter unit is connected.

This selected central solution, with one 990 kVA inverter unit, is consisting of three 330 kVA standard units from a UPS (UPS = Uninterruptible Power Supply). The three inverters operate in master-slave mode with a rotating master. This way produces the optimum regarding technology, energy and cost. It also requires the least space and exhibits the best efficiency curve, in particular in the part-load range.

These inverters equipped with transistors, operate with pulse width modulation of 2,5 kHz. The inverters are capable to vary the form and thus the harmonics of the line side current. In this way, the harmonics of the line side voltage can be influenced, with the inverter operating as an active filter.

4.   GRID CONNECTION

As illustrated in **Figure 1**, the solar power from the PV-Generator is transferred through the inverter unit to the 0,4-kV-bus on the AC side of the inverter unit. Further it is transformed to the 20-kV-level of the internal medium voltage grid via one 1-MVA- transformer. This transformer is connected to the 20-kV-bus, which supplies the other load of the Trade Fair Centre, which can range from a few Megawatts up to 50 MW. The medium voltage bus is connected to the 110-kV-high voltage grid of the local electricity board via two 40-MVA-transformers, which means that a high short circuit power is available.

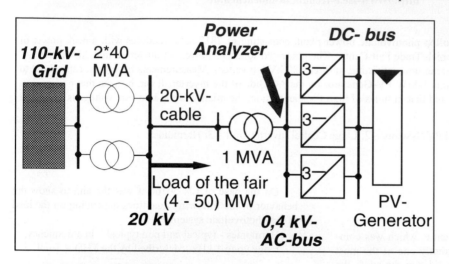

Figure 1: Connection to the Grid

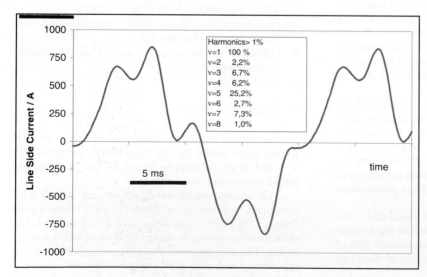

Figure 2: Line Side Current of one Inverter

5. MEASUREMENTS

A Power Analyzer – designed for the analysis of the quality of the power System due to EN 50160 - was applied, which permits the analysis of up to eight voltages and currents for a time range up to a few weeks. The measured values are stored on a harddisk. After measurement, they may be transferred to a PC for evaluation in a special analysis software, Excel or other evaluation software.

In **Figure 1** the arrow indicates the point where the line side currents and voltages have been measured.

5.1 LINE CURRENT OF ONE INVERTER UNIT

As an example of the time behavior of one of the three inverter units **figure 2** illustrates the time behavior of the line current of one line side current with its harmonics. Due to the master-slave principle the three inverters of the central unit operated at 150 kW, 163 kW and 159 kW.

The current contains lots of harmonic, especially the harmonic of the order 5 is very high. Therefore it was interesting to measure their reactions on the voltage.

5.2 HIGH LOAD OF THE FAIR

When there is a fair - up to 50 Megawatts are required.

In this case there are various reactions in the feeding power system from other electrical devices such as fans, elevators, illumination etc. The question was if the limits given by EN 50160 were exceeded and whether the inverter unit was capable to improve the THD.

Figure 3 on the next page shows the time behaviour of the THD of the line side 0,4-kV-voltage and the generated photovoltaic power during one week from the 18th March to the 25th March in 1999. The THD ranges from values greater than 2 % up to 5 %, it shows smaller values during the night, which means that some of electrical devices causing harmonics are only operated during daytime.

It can also be seen that the limit of 8 % due to EN 50160 is not exceeded. Furthermore it is obvious, that the value of THD decreases slowly on high PV-Power values (18/3/99, 21/3/99 and 24/3/99). During times with smaller values of generated power the THD value does not drop down. Investigating more details, it can be stated, that especially the value of the harmonics of the order 5 and 11 drop down. The value of the 5th order harmonic was on the 18/3/99 on PV peak power at 3 %, the 11th order harmonic at 1,5 %.

The other results can be summarised:

- A Small Long Term Flicker Intensity $P_{LT} = (0,1 - 0,3)$ occurs on frequently changing PV power
- The negative sequence system lies between 0,5 an 1 %. During times of high PV power these values drop.
- Very small values of interharmonics of the order 2, 4 and 6 can be detected in a range of about 0,02 %.

None of these values hits the limits of EN 50160.

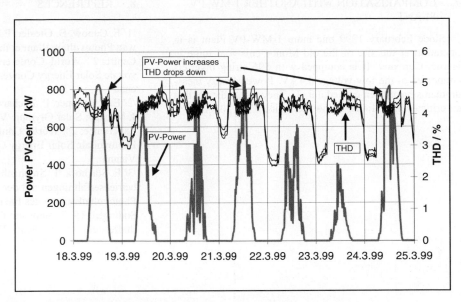

Figure 3: THD and PV-Power during one week of high load

5.3 LOW LOAD OF THE FAIR

As **Figure 4** indicates, the THD-Level is a little bit lower than during the high-load-time, with a minimum of 2 % . In the evening a peak is obvious. It occurs at about 20 h, and as at this time no solar power is generated, it cannot be caused by the PV-Plant.

The main results may can summarized as follows:

- There is no reduction of the THD, even though the PV-Power is high
- Minimal harmonics of the order 4 and 6 occur within a range of 0,02 % - 0,06 %
- The Long Term Flicker Intensity is nearly constant with $P_{LT} = 0,2$
- The negative sequence system shows values of 0,6% to 0,8%. These values drop slightly on high PV-Power.

Even of these of these values none hits the limits of EN 50160.

6. CONCLUSION

As one most important result it can be stated, that in this case - with the selected inverter unit and grid connection - no problems occur in the frequency range up to the order of 40. All the values of harmonics, flicker, negative sequence system and interharmonics remain within the given limits of the standards. On the other hand it can be seen that the operation of the inverter unit can improve the

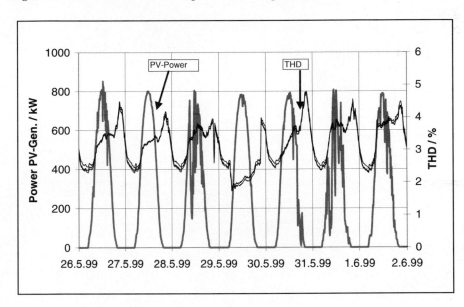

Figure 4: THD and PV-Power during one week of low load

values of the 5th order voltage harmonic produced by other devices, when the photovoltaic generation is high.

It should be emphasised that only a frequency range up to about a few kHz has been considered. As the inverter unit is operating at 2,5 kHz, disturbances in higher frequency ranges may be expected. Measuring the effects in the feeding power system in higher frequency ranges would be a very interesting task.

During the measurement series described, no transients have been considered. It is well known, that more harmonics occur when switching on or off the inverters. The transient behaviour of the inverters would be one more interesting object of investigation.

7. COMPARISATION WITH ANOTHER 1-MW-PV PLANT

Since February 1999 one more 1-MW-PV Plant is in operation at the site of the old Mont-Cenis coal mine in Herne, Germany. It is equipped with 596 string inverters connected to the low voltage 0,4 kV. Power quality measurements have been carried out /3/. They also indicate, that all critical values lie far below the limits of EN 50160.

8. REFERENCES

/1/ E. Cunow, B. Giesler, P. Hopf , G. Maier ; One Megawatt Photovoltaic Plant at the new Munich Trade Fair Centre; 2nd World Conference and Exhibition on Photovoltaic Solar Energy Conversion; 6 – 10. July 1998, Vienna

/2/ E.D. Spooner, P. Zacharias, D. Morphett, G. Grunwald, J. Mackay; Solar Olympic Village, Design and Testing Experience; 2nd World Conference and Exhibition on Photovoltaic Solar Energy Conversion; 6 – 10. July 1998, Vienna

/3/ E. Stachora, T. Stephanblome, T. Pierschke; Erste betriebserfahrungen mit der dachintegrierten 1-MW$_P$-Photovoltaikanlage des Energieparks Mont-Cenis in Herne Sodingen; 15. Symposium Photovoltaische Solarenergie, Staffelstein, 15 –17. März 2000

A TRAFFIC NOISE BARRIER EQUIPPED WITH 2160 AC-MODULES

N.J.C.M. van der Borg; Netherlands Energy Research Foundation;
P.O. Box 1, 1755ZG Petten, the Netherlands.
E-mail vanderborg@ecn.nl, tel +31 224 564401, fax +31 224 564976
E. Rössler; Fraunhofer Institut für Solare Enegiesysteme, Germany
E.B.M. Visser, NUON International/Renewable Energy, the Netherlands

ABSTRACT: A PV-system consisting of 2160 AC-modules has been integrated on a noise barrier along a highway. The AC-modules are clustered into 12 groups, connected in 12 cabinets, and feed in on a low voltage power line. Two types of inverter technologies are applied. The monitoring results show detailed information on the performance of the PV-system and its components.
Keywords: AC-modules - 1: Monitoring - 2: Noise barrier - 3

1. INTRODUCTION

A large photovoltaic energy system has been integrated into a 1650-meter long noise barrier at the A9-highway near Ouderkerk-aan-de-Amstel in The Netherlands. The realisation took place in the frame of a demonstration project, financially supported by the European Commission and the Netherlands Agency for Energy and Environment (NOVEM). The following partners carry out the project.
- NUON International / Renewable Energy (co-ordinator and owner)
- Netherlands Energy Research Foundation, ECN
- TNC Energy Consulting GmbH
- TNC Consulting AG
- Fraunhofer Institut für Solare Energysysteme

The main subcontractor was Shell Solar Energy who manufactured the PV-modules and installed the entire PV-system turnkey. The PV-system was put into operation on December 1st, 1998. On that date a two-year monitoring programme started in co-operation with Fraunhofer ISE. This paper presents the monitoring results obtained until April 1st, 2000.

2. PV-SYSTEM AND MONITORING SYSTEM

The PV-system consists of 2160 AC-modules, each with a nominal power of 95 W. Each of the AC-modules feeds in on a low voltage grid through 12 cabinets, as depicted below.

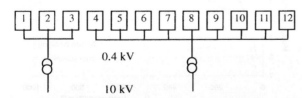

In each cabinet 180 AC-modules are connected to the grid. Two different types of AC-modules are used:
- Cabinets 1, 2, 11 and 12: per cabinet 180 modules with inverters of type A
- Cabinets 3 through 10: per cabinet 180 modules with inverters of type B

The monitoring system is based on decentralised data acquisition facilities (ref. [1]). The decentralised data acquisition facilities consists of the following categories.
- Global data acquisition units consisting of kWh-counters with digital data output in each of the 12 cabinets, measuring the total energy per cabinet.
- Supervision data acquisition units consisting of 360 Wh-counters, integrated in the inverters of 360 AC-modules, forming a statistical sample of 1 out of 6 modules and yielding information on the reliability of the individual AC-modules.
- Analytical data acquisition units positioned near the cabinets 2, 5, 8 and 11 (for irradiance, temperatures of the modules and ambient air and for electrical quantities of the modules and inverters)

The decentralised data acquisition units are connected via one single twisted pair cable to a measurement PC. The quantities from the analytical data acquisition units are measured continuously with a sample period of about 6 s and the results are condensed into values of the average, maximum, minimum and standard deviation of each measured quantity, based on periods of 10-minutes.
The remaining data are retrieved from the 12 kWh-counters and 360 Wh-counters once a day and stored, together with the analytical data in the memory of the PC. The measurement results are transported via a telephone line to ECN every night for subsequent evaluation.

3. GLOBAL MONITORING DATA

In this paragraph the results are presented for a one-year period from April 1, 1999 to April 1, 2000. In this period the number of days that the monitoring system operated correctly was 337 resulting in a monitoring fraction of 92 %. The final yield Y_f, as defined in IEC 61724, has been calculated per cabinet by dividing the energy production during the monitoring period obtained by the 12 kWh-counters by the nominal power (180 * 0.095 kW). The corresponding reference yield Y_r, as defined in IEC 61724, has been calculated per cabinet by numerical integration of the irradiance measured with the four in-plane reference cells near the cabinets 2, 5, 8 and 11 after division by the reference value of 1000 W/m^2. The result of the first reference cell has been used for the first 3 cabinets and so on. The performance ratio, again as defined in IEC 61724, has been calculated by dividing the final yield by the reference yield per cabinet. The results are given in figure 1. The figure shows significant differences in the performance of the AC-modules per cabinet. This will be evaluated further in the following paragraphs.

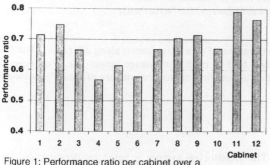

Figure 1: Performance ratio per cabinet over a one year period

4. SUPERVISION MONITORING DATA

Using the supervision data acquisition units the energy production of 1 out of 6 inverters has been determined per month and per cabinet. The monthly energy production of the 30 individual AC-modules in each cabinet has been compared with the averaged value of these 30 AC-modules. In case the production of an AC-module is less than 50 % of the average, it is concluded that the unit (AC-module or its monitoring device or both) is malfunctioning. The number of malfunctioning units out of 30 are given in figure 2 per month. Due to a monitoring problem with cabinet 8, the data of cabinet 8 is the number of malfunctioning units out of 15. Although figure 2 shows a significant variation of the number of malfunctioning units along the various cabinets, the data of figure 2 cannot account fully for the variation of the performance ratio in figure 1.

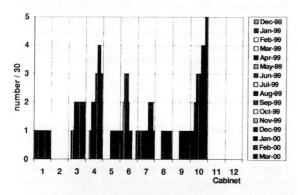

Figure 2 Number of malfunctioning devices

5. ANALYTICAL MONITORING DATA

5.1 Module efficiency
The module efficiency has been determined for 4 AC-modules using the DC-power data and the reference cell irradiance data obtained in the 10-minute measuring periods. The efficiency values are based on the net module area (72 cells of 10 x 10 cm^2) and are corrected to a module temperature of 25 °C using a temperature

coefficient for the power of -0.4 % / K. The module efficiency of the module near cabinet 8 and of the module near cabinet 11 is presented as a function of the irradiance in the figure 3. The module efficiency is not only a characteristic of the module but it is also influenced by the DC-voltage as set by the inverter. For this reason figure 3 shows also the corresponding DC-voltage, corrected to a module temperature of 25 °C using a temperature coefficient for the voltage of -0.4 % / K. Figure 3 shows no significant difference in the efficiency of the modules equipped with the two different types of inverters, although the applied voltages are different.

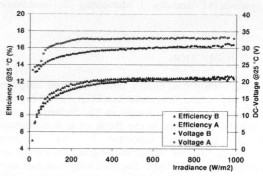

Figure 3 Module efficiency and voltage of the two types of AC-modules

5.2 Cleaning
During the monitoring the in-plane irradiance is measured using reference cells, made of the same cells as the PV-modules. In this way the response of the reference cell is as close as possible to the response of the modules regarding cell sensitivity, reflections of the glass, accumulated dirt, etcetera. In the end of May 1999 the 4 reference cells were cleaned. The modules remained uncleaned. This resulted in an apparent decrease of the efficiency of the four modules with a similar magnitude (figure 4). From these data it can be estimated that the dirt on the modules caused by the traffic on the highway accounts for a loss of energy of about 8%.

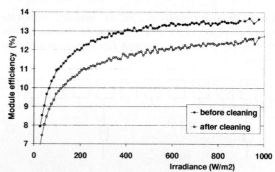

Figure 4: cleaning the reference cells

5.3 DC / AC efficiency
The efficiency of the conversion from DC-power to AC-power by the two types of inverters has been determined as a function of the DC-power. The results of the inverters near cabinet 8 and 11 are given in figure 5. This figure shows a significant difference in the efficiencies in

the low power range. The difference in the AC-energy caused by the observed difference in the conversion efficiency has been estimated using the frequency distribution of the DC-power, measured in the period from April 1, 1999 to April 1, 2000. This resulted in a difference in the annual AC-energy production of about 4.5 %. This accounts partly for the higher performance ratios of the cabinets 1, 2, 11 and 12 in comparison with the other cabinets.

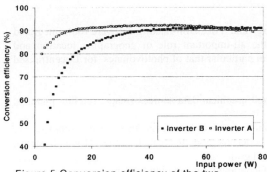

Figure 5 Conversion efficiency of the two inverters

5.4 Grid interference

It was observed that the AC-module near cabinet 5 frequently showed low values of the power at high irradiances. The 10-minutes averaged values of the DC-voltage and corresponding DC-current of this module is shown in figure 6. This effect was examined in a couple of dedicated measurement campaigns using laboratory equipment. It was demonstrated that the inverters located at a large distance from the 10 kV-transformer switched themselves off as a result of grid interference. After recognising this problem the manufacturer proposed a remedial action which will be implemented soon. It is expected that hereafter the differences in the performance ratio (paragraph 2) of the 12 cabinets will be much smaller.

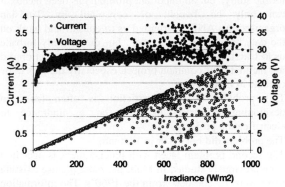

Figure 6: 10-minutes averaged data of DC-voltage and DC-current

6 CONCLUSIONS

The monitoring showed the following:
- The performance ratio defined for the 12 individual cabinets on a one-year period ranges between about 0.6 and 0.8
- The number of malfunctioning units (AC-modules or their internal monitoring devices) of the individual cabinets ranges between 0 and 5 out of 30 after a monitoring period of 16 months
- The DC-performances of the modules, equipped with the different types of inverters show no significant differences
- A rough estimate of the reduction of the annual energy production caused by the accumulated traffic dirt on the modules is 8 %.
- The conversion efficiencies of the two types of inverters are different in the low power range. This results in a difference in the annual energy production of about 4.5 %
- Due to grid interference, part of the PV-system has a sub-optimal performance. Remedial actions have been planned. It is expected that this will have a positive effect on the values of the performance ratios and on the number of malfunctioning AC-modules.

7. ACKNOWLEDGEMENT

This work has been supported by the European Commission and the Netherlands Agency for Energy and Environment NOVEM.

8. REFERENCES
[1] C.W.A. Baltus et al., Proceed. 2nd World Conference and Exhibition on Photovoltaic Solar Energy Conversion (1998), 2360 - 2363

DECISION-MAKING TOOL FOR THE CHOICE OF PHOTOVOLTAIC SITES :
THE TAKING INTO ACCOUNT OF THE GEOGRAPHIC LOCATION AND DISTRIBUTION OF HOUSEHOLDS

Corinne Lamache
ADEME / EDF
1 Avenue du Général de Gaulle, B.P. 408, 92141 Clamart Cedex, France
Tel : 33 1 47 65 42 77 - Fax : 33 1 47 65 32 18
e-mail : corinne.lamache@edf.fr

ABSTRACT : The subject of this paper is to present a methodology and a decision-making software tool, integrated into a Geographic Information System (GIS), that brings out the all-important role of geographic location and distribution of households in the choice of electrification systems, in particular that of photovoltaics, for Decentralised Rural Electrification.

The methodology and the tool can be discribed in three steps :
1. acquisition and extraction of households location from different graphical data sources (maps, satellite pictures, etc.),
2. the taking into account of the households' spatial distribution and creation of homogeneous « density zones »,
3. comparison cost of three electrification solutions for each « density zone » created : Solar Home Systems, generator set connected to a local grid, connection to the national grid.

Keywords : 1 : Rural Electrification - 2 : Households Density - 3 : Geographic Information System -

In rural electrification projects the choice between Solar Home Systems (SHS), multi-user electrification systems, and connection to the central grid, is made more and more frequently using adapted tools and methodologies. To make this choice, a number of technical, social and economic parameters allows economic comparison of different electrification techniques. Most often, this economic comparison does not take into account geographic parameters, in particular the households' distribution.

The method and the tool presented in this paper are using the households to be electrified as the central factor to make the comparison of different electrification solutions.

1. LOCATION OF HOUSEHOLDS TO BE ELECTRIFIED

The first step of the method consists on getting data of households' location. It is a difficult, time consuming and costly operation to do. This is a reason why this data is not often used in electrification projects.

To execute this first step, different data sources, from small survey areas in Tunisia and Vietnam, have been used. The following data sources have been compared : GPS surveying, maps, aerial pictures and satellite pictures.

1.1 Data sources
GPS surveying.
The GPS survey consists of taking longitude and latitude coordinates of houses with a measuring apparatus. This is the most precise way to get location information, nevertheless, it's very time consuming and costly.
Since a socio-economic study is often carried out for electrification projects, a GPS survey won't then be an extra cost. This survey could be made systematically when a socio-economic study is planned.
In Tunisia, within a research project with the National Agency of Renewable Energy (ANER), a GPS survey has

been done on 80 households in two different isolated zones.
In Vietnam, the Foundation « Energie pour le Monde » [1] is in charge of electrifying three zones of six villages. In order to validate the methodology and the tool, they carried out a GPS survey while they were doing the socio-economic investigation.

Precise maps and aerial pictures
The problem of precise maps in developing countries is that they are not often up to date and not easy to acquire. In Tunisia, the national cartography office is presently putting up to date all the country maps from aerial pictures and topographic surveys. This precise data was necessary for the study, but an important problem has been faced : it can only be bought or consulted by Tunisian organisations. ANER permitted us to consult and work on the data bought for the project ; but they strictly forbade us to take this data out of the country. This restriction did not facilitate our common project, especially the part to be execute in France.
This problem of data restriction is often faced in developing countries and can be a hindrance to electrification projects.
The problem did not occur in Vietnam.
The second problem faced with precise maps is age. In a zone previously studied in Mali, the more precise existing maps were 1 : 200 000 from the 1960's. The information on households' location was then no longer available.

Satellite pictures.
The main problem of satellite pictures is the cost.
For the Tunisian study one Spot picture was bought. Accuracy of Spot pictures is 10 metres, all the households under this size are not visible (except if the courtyard is identifiable). Nevertheless, the rapidity and the easiness for extracting this data make this source very interesting.
1.2 Extraction of households' information

---

[1] an NGO specialized in electrification in developing countries

The graphical data sources set out previously are of two different types : punctual information and picture information. Only the punctual one is of any interest for the GIS (Geographic Information System) tool. The GPS survey is a punctual type, it can be then straight introduced in the tool. The other data sources will have to be processed to extract punctual information on households. Appropriate image analyses software can extract automatically or manually this data.

GRAPHICAL DATA SOURCES

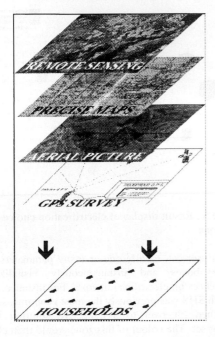

Figure 1 : Households data acquisition. Different graphical data sources allow to get information on households : remote sensing, precise maps, aerial pictures and GPS survey.

This household's information is integrated in a Geographical Information System as a new punctual lay.
The GIS will allow to :
- associate attributive information to this punctual information on households (incomes, energetic needs, etc.),
- carry out spatial and attributive requests (selection of non electrified households, selection of households distant from less than 3 kilometres to the national grid, etc.),
- carry out graphical analysis.

2. TAKING INTO ACCOUNT OF HOUSEHOLDS' DISTRIBUTION

2.1 Length of local grid
In rural electrification projects, three electrification systems can be found :
- Individual systems : each client has its own energy source. In developing countries these individual systems are usually SHS.
- Multi-user electrification systems : a group of clients is connected to a common energy source with a local electric grid. For this system, generator sets are generally used.

- Connection to the national grid : villages, or groups of households, are connected to the national grid.
The local electric grid is needed both for Multi-user electrification system and for the connection to the national grid. Its cost will determine whether an individual system is better than a collective system. The length of the local grid determines its cost, and therefore induces the choice between individual and collective system. Often, in electrification projects, the population average and the surface of administrative villages are used to determine the local grid length. This solution neglects the heterogeneity of households' dispersal. Therefore it induces errors in the choice of electrification system. To calculate the precise length of local electric grid, the real households distribution is needed.

2.2 « Density zones » creation
From the households' distribution, a tool developing in a GIS groups the households together according to their spatial distribution. The households close to one another will be grouped in a same « density zone ». To generate these zones, the user will have to specify two parameters : the maximal distance between two households and the minimum number of clients in a « density zone ».
In the GIS tool, this new data appears under two different forms :
- a graphical information, which appears as a lay in a « view »,
- attributive data, in a data base associated to the graphical information

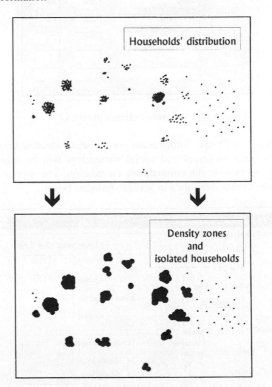

Figure 2 : « density zones » generation

The households outside of the zones will be considered as dispersed households and will be electrified with SHS. The tool calculates automatically the area, the number of households and the local grid length for the « density zones ». These zones will be used to compare the economic viability of different electrification solutions.

3. CHOICE OF ELECTRIFICATION SOLUTIONS

In each « density zone », the tool will compare the economic viability of three electrification solutions :
- Solar Home Systems,
- generator set connected to a local electric grid,
- connection to the national electric grid.
The precise identification of electrical needs will be used as a departure point for the calculation. First of all, as shown in the previous paragraph, the location of each household to be electrified will be taken into account.
Also, the energy profile of each potential client to come has to be given. Three groups of profiles are identified :
- domestic clients,
- communal clients,
- productive clients.
Several energetic profiles will be identified in each group. For instance, the communal group will include a school, a mosque and a health centre ; with an information on daily energy and peak power for each profile.

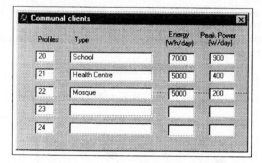

Figure 3 : Communal clients profiles.

Added to this information on client identification, economic, technical and social parameters will be taken into account in the comparative calculation. The user will have to enter these data in several dialogue boxes.

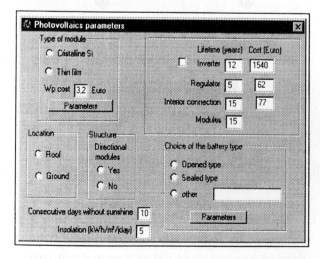

Figure 4 : Photovoltaic Parameters dialogue box.

The results of the calculation will be visualized in three different ways:
- in the attributive table joined to the « density zones »,
- on the graphic representation of the « density zones », with a colour code indicating the cheapest solution for each zone (SHS in yellow, generator set in red and grid connection in grey),
- in an Excel file, with an Excel spreadsheet corresponding to each zone.

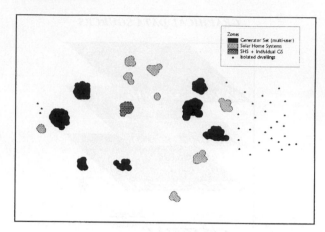

Figure 5 : Result display of electrification choice for each zone.

The tool user can modify one or many parameters in the dialogue boxes and instantaneously visualize the consequences on electrification costs. For instance, a zone for which SHS was previously the most economic solution would then tilt over to a collective solution with a generator set. The colour of this zone would then changed from yellow to red.

CONCLUSION

The adaptability of the tool allows the user to work even when all the data is not available :
- If the precise location of each household is unknown, the villages position could possibly be used.
- If energy profiles are not determined, it's possible to use approximate percentages for each group of profiles (for instance : 80% of domestic clients, 15% of communal clients and 5% for productive clients).
In these cases, however, the results may be less accurate.
The use of such a decision tool can be very useful for rural electrification projects. It allows to take into account the spatial dimension in the comparison cost of the different electrification solutions. In particular, it shows the determining role of the households' location in the choice between individual and collective solutions.

A SIMULATION PROCEDURE TO ANALYSE THE PERFORMANCE OF AN OPERATING GRID CONNECTED PHOTOVOLTAIC PLANT

A. Abete, F. Scapino, F. Spertino, R. Tommasini

Dip. Ingegneria Elettrica, Politecnico di Torino, C.so Duca Abruzzi 24, 10129 Torino (Italy)

Tel. +39-11-564.7105   Fax +39-11-564.7199   eMail spertino@athena.polito.it

ABSTRACT: A simulation procedure, based on suitable daily profiles of irradiance, allows to analyse grid connected PV systems with both crystalline and amorphous silicon modules. These profiles of irradiance are identified by a shape factor which takes into account deviations from sinusoidal profile. As a test bench, the simulation is applied to an operating grid connected PV plant (12 kWp). The matching between experimental data and simulation results, on daily basis, is adequate for DC and AC energy (5%). Therefore, good accuracy can be obtained for the determination of the performance ratio, because the losses due to over-temperatures on PV modules and due to DC/AC conversion are calculated taking into account the peak value of irradiance, besides the mean value.

Keywords: Simulation - 1: Grid-Connected - 2: PV system - 3

1. INTRODUCTION

The grid connected PhotoVoltaic (PV) plants are centralised to share peak power requests or decentralised for low power demand, e.g. to supply a dwelling house. Centralised PV systems can be a good choice for electric Companies aimed to shave the peak load during the midday hours, without a further upgrade of the distribution system. Really, these PV plants can be switch on/off very quickly and provide energy in the same place where it is delivered to the load.

A centralised PV system essentially consists of the PV generator, the DC/AC converter (Power Conditioning Unit, PCU) and the protection device (over/undervoltage, over/underfrequency and islanding).

To verify the efficiency, reliability and ageing of amorphous silicon (a-Si) modules, an Italian electric Company (AEM S.p.a.-Torino) has installed, within the European Thermie Programme, a small centralised PV system. It is constituted by 396 modules (12 kWp) and connected to 230V-50Hz three-phase grid by 3 single-phase PCUs.

The paper presents a simulation procedure, based on suitable daily profiles of irradiance, to analyse the performance of grid connected PV systems. These profiles of irradiance are identified by a shape factor which takes into account deviations from sinusoidal profile. In particular, as a test bench, the simulation is applied to the previous PV plant.

2. SIMULATION PROCEDURE

The starting point for the simulation procedure are the daily meteorological data (irradiance G(t) and temperature $T_{PV}(t)$ of the PV modules) obtained by proper models.

The following expression of irradiance [W/m²] is derived for clear days:

$$G(t) = G_{max} sin\left(\frac{\pi}{h_l}(t-t_0)\right)\left\{1+s\left[1-sin\left(\frac{\pi}{h_l}(t-t_0)\right)\right]\right\} \quad (1)$$

$$G_{max} = c_G \frac{\pi}{2} G_{mean} = c_G \frac{\pi}{2}\frac{H_d}{h_l} \quad (2)$$

$$s = \left(\frac{1}{c_G}-1\right)\bigg/\left(1-\frac{\pi}{4}\right) \quad (3)$$

where G_{max} is the peak value of the irradiance, h_l are the daylight hours and H_d is the daily irradiation in Wh/m². Moreover, $t = t_0$ represents the sunrise and $t = t_0 + h_l$ represents the sunset.

Differently from [1], here the coefficient c_G is defined as shape factor; it is the normalised ratio between the peak value of the irradiance G_{max} and the mean value $G_{mean} = H_d/h_l$. The value of c_G, for the same G_{mean}, determines the shape of the waveform: if $c_G = 1$, the waveform is sinusoidal; if $c_G < 1$, i.e. G_{max} is lower than sinusoidal peak value, the waveform is flat; if $c_G > 1$, the waveform is sharp (fig. 1). Different values of c_G, for the same G_{mean}, determine different values of cell temperature T_{PV}, of DC and AC energy. Hence it is important to determine the right value of c_G.

Then, the cell temperature T_{PV} with respect to the ambient temperature T_a is evaluated by a linear function of the irradiance G:

$$T_{PV}(t) - T_a(t) = \frac{(NOCT - 20)}{800}G(t) \quad (4)$$

where NOCT is the Nominal Operating Cell Temperature (G = 800 W/m², T_a = 20 °C and wind speed 1 m/s).

The current-voltage I(U) characteristics of the PV generator are determined by a typical equivalent circuit of the solar cell. The parameters of the one-diode model are: ideal current generator I_{ph}; reverse saturation current of the junction I_0; quality factor of the junction m; shunt resistance R_{sh} and series resistance R_s.

In this model [2], only the photovoltaic current I_{ph} and the reverse saturation current I_0 depend on G(t) and $T_{PV}(t)$:

$$I_{ph} = I_{ph}\big|_{STC}[1+\alpha_T(T_{PV}-298)]G/1000 \quad (5)$$

$$I_0 = I_0\big|_{STC}\left(\frac{T_{PV}}{298}\right)^3\frac{e^{-E_g/kT_{PV}}}{e^{-E_g/k298}} \quad (6)$$

where $I_{ph}\big|_{STC}$ and $I_0\big|_{STC}$ are calculated at the Standard Test Conditions STC (G = 1000 W/m² and T_{PV} = 298 K) and α_T is the temperature coefficient of I_{ph}, besides the band-gap E_g and the Boltzmann constant k.

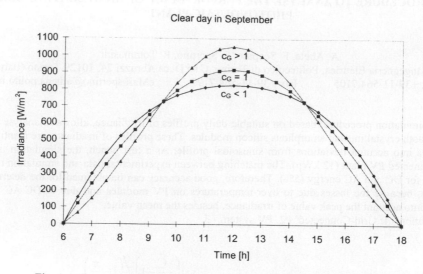

Fig. 1. - Three waveforms of irradiance with equal G_{mean} and different G_{max}.

The parameters of the solar cell equivalent circuit are obtained at STC by the manufacturer I(U) characteristic of one module.

The Maximum Power Point (MPP) is calculated on the PV generator I(U) characteristic corresponding to the actual irradiance and cell temperature. It is worth noting that MPP is the operating point at the DC side (P_{DC}) because the PCUs are usually equipped with MPP trackers.

Finally, by a model which takes into account the losses of the PCUs under open circuit (P_0) and load conditions ($a_P P_{AC}^2$ with a_P load loss coefficient), the real power P_{AC} and energy E_{AC} are obtained. Generally, the PCUs for grid connection operate at unity power factor ($\cos\phi = 1$).

$$P_{AC} = \left[-1 + \sqrt{1 + 4a_P(P_{DC} - P_0)}\right]/2a_P \qquad (7)$$

3. TESTING OF THE PROCEDURE ON THE OPERATING PV PLANT

The models of irradiance G(t), cell temperature $T_{PV}(t)$ and the equivalent circuit of the PV generator are tested by experimental data of the operating PV plant.

The PV generator of AEM, sited in mountains at 45° latitude, is formed by 3 sub-fields: each one comprises 66 parallel connected strings with 2 series connected modules. All the PV modules are constituted by 57 series connected a-Si cells (p-i-n/p-i-n). Each sub-field supplies a single-phase PCU operating at unity power factor ($\cos\phi=1$). Fig. 2 shows the scheme of the PV system.

A data logger monitors the following system parameters:

- solar irradiance G(t), ambient temperature $T_a(t)$ and cell temperature $T_{PV}(t)$;
- voltage $U_{DC}(t)$, current $I_{DC}(t)$ and power $P_{DC}(t)$ in MPP at the DC side and real power $P_{AC}(t)$, energy $E_{AC}(t)$ at the AC side.

The application of the simulation procedure requires the value of the coefficient c_G which minimises the sum of the square deviations of the function (1) with respect to the experimental data (fig. 3). Notice that the irregular profile of the experimental data is due to the obstruction of mountains in the morning. The cell temperature T_{PV} is evaluated by (4): the comparison with experimental data shows that the calculated values are very close (1-3%) for high irradiance levels (fig. 4).

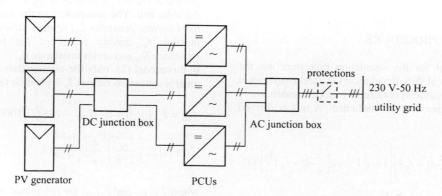

Fig. 2. – Scheme of the PV plant connected to the three phase grid.

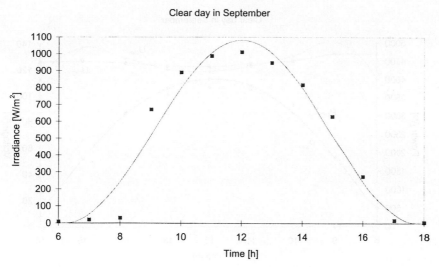

Fig. 3. - Comparison between experimental data (dots) and least square fit curve ($c_G > 1$).

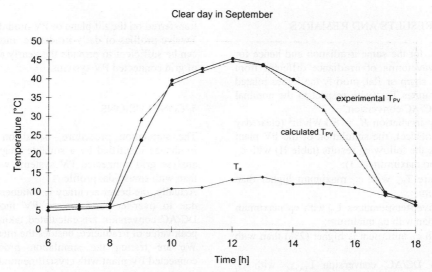

Fig. 4. - The actual temperature of PV modules with respect to the calculated one.

Table I shows the parameters of the equivalent circuit of the mean PV cell at STC; moreover the temperature coefficient $\alpha_T = 8\times10^{-4}$ K^{-1} and the band-gap $E_g = 1.6$ eV.

Table I: Parameters of the mean-cell equivalent circuit

I_{ph} [A]	I_0 [pA]	m	R_s [Ω]	R_{sh} [Ω]
0.58	1.6	2.3	0.35	14

Below 500 W/m², the experimental MPP and the simulated one by the equivalent circuit are different. In order to match the two points, the parameters of the equivalent circuit must be changed and, as a first approximation, R_{sh} must be increased ($R_{sh} = 24$ Ω). Really, at high irradiance levels, the Staebler-Wronski effect reduces the performance of the a-Si modules.

In order to calculate the AC power and energy, the parameters of the PCU model are $P_0 = 40$ W (open circuit losses) and $a_P = 2\times10^{-5}$ W^{-1} (load loss coefficient).

Fig. 5 shows the variations of DC voltage and power during a clear day, for one sub-field of the PV generator. The matching between the experimental and the calculated data is adequate for energy (5%), but not for voltage and power (2-20%). It should be noted that only energy is the parameter required for calculation of the system performance ratio.

This comparison is repeated for several days in the first month of operation with similar results. During the next months, the degradation of the performance, typical of the a-Si modules, begins to be important and the parameters of the equivalent circuit of the mean PV cell must be changed.

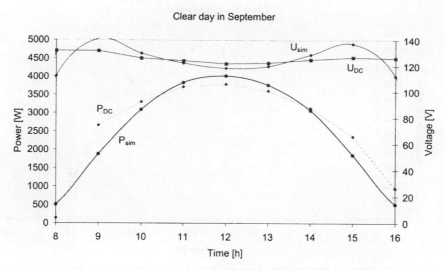

Fig. 5. – Test bench: experimental data and simulation results.

4. SIMULATION RESULTS AND REMARKS

As previously said, for the same irradiation and hence for the same G_{mean}, waveforms of irradiance different from sinusoidal one, i.e. sharp or flat, modify the losses related to the over-temperatures T_{PV} (with respect to the nominal 25°C) and to the DC/AC conversion.

For the same daily irradiation $H_d = 8$ kWh/m² (clear day close to summer solstice), the simulation of the PV plant behaviour provides the following results (table II) with c_G minimum (0.85) and maximum (1.3):

- peak temperature T_{PV} with c_G maximum higher than with c_G minimum;
- losses due to over-temperatures L_T with c_G maximum higher (28%) than with c_G minimum;
- DC energy with c_G minimum is higher (2%) than with c_G maximum;
- losses due to DC/AC conversion $L_{DC/AC}$ with c_G maximum higher (13%) than with c_G minimum;
- AC energy with c_G minimum is higher (3.4%) than with c_G maximum.

In summary, the daily Performance Ratio (PR) of the PV plant is 87% in the best case and 84% in the worst case.

Table II: Comparison between max. and min. c_G

c_G	T_{PV} [°C]	L_T [kWh]	E_{DC} [kWh]	$L_{DC/AC}$ [kWh]	E_{AC} [kWh]
0.85	57	2.1	31	2.2	28.8
1.3	68	2.7	30.3	2.4	27.9

Since the influence of temperature on the electric performance is lower for amorphous silicon with respect to crystalline silicon, sharp and flat waveforms of irradiance give higher differences, for crystalline silicon, on the DC and AC energy.

Three input data, i.e. daylight hours, daily irradiation and peak value of irradiance, define the analytical profile of daily irradiance.

In order to obtain the shape factor c_G required for the proposed procedure, the daily irradiation and the peak irradiance, given on the horizontal plane, must be transferred on the tilt plane of PV modules [3].

Twelve profiles of daily irradiance, monthly mean values, can be sufficient to provide the yearly energy performance of grid connected PV systems.

5. CONCLUSIONS

The simulation procedure, based on daily profiles of irradiance identified by a suitable shape factor, allows to analyse grid connected PV systems with better accuracy than with sinusoidal profile.

Actually, the better accuracy is obtained because the losses due to over-temperatures on PV modules and due to DC/AC conversion are calculated taking into account the peak value of irradiance, besides the mean value.

We are testing the simulation procedure on a grid connected PV plant with crystalline modules.

REFERENCES

[1]   IEC Standard 61725, Analytical expression for daily solar profiles (1997).

[2]   W. Knaupp, "Power rating of photovoltaic modules from outdoor measurements", 22nd IEEE Photovoltaic Specialists Conference, Las Vegas (1991), pp. 620-624.

[3]   UNI Standard 8477, Solar Energy - Assessment of irradiation on a tilt plane. (1983)

SUCCESSFUL TRIALS UPDATE FOR THE APPLICATION OF SOLAR REFRIGERATION IN FOOD TRANSPORT

A. S. Bahaj and P. A. B. James
Sustainable Energy Research Group,
Civil & Environmental Engineering Department
University of Southampton, Highfield,
Southampton, SO19 1BJ - United Kingdom
email: serg@soton.ac.uk - URL: http://www.soton.ac.uk/~serg

ABSTRACT: Solar photovoltaic driven refrigeration represents an application which warrants a major attention as a market opportunity, especially in northern latitude countries and as a means to achieve CO_2 reduction in the transportation area. The objective of this application is the replacement of the diesel generator with a photovoltaic system sized in such away as to provide enough power to allow operation under the same conditions for the two technologies. This paper describes an update of the results of operation of the solar powered, refrigerated delivery trailer (Solar trailer) operated by J Sainsbury plc. A month by month analysis of operation is described detailing the energy requirements of deliveries and the level of operation that is sustainable. The performance of the trailer is shown to achieve its design brief of achieving a daily delivery of chill produce throughout the year.
Keywords: Refrigeration - 1: Stand alone PV System - 2: Solar Cooling

1. INTRODUCTION

Solar photovoltaic (PV) driven refrigeration represent a novel approach in expanding PV into prominent applications which will have both environmental and market impacts. The utilisation of PV in transport refrigeration can also reap economic benefits and in many instances offers an enhanced viability in operation of the delivery regime.

The basic principle of the solar trailer is that

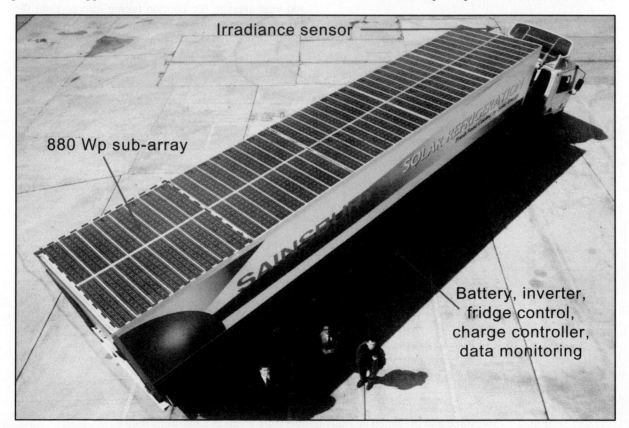

Fig. 1   The solar-Trailer. Roof mounted array and other systems positions shown.

the diesel engine that powers a transport refrigeration unit can be replaced by solar photovoltaic power. Photovoltaic (PV) modules or especially manufactures laminates can anchored into the roof of the trailer generate dc electricity. The system is controlled through a charge regulator coupled to a battery which stores the generated energy. The battery was sized to cope with overcast days and could also be used for night operation.

In general the energy requirement of the refrigerator is in phase with the seasons. During the summer months the high ambient temperatures mean that the refrigerator has to work hardest. The high energy load is however, more than offset by the available PV power from the long, bright days. Conversely, during the winter months although the days are shorter and the sun lower, in an often more overcast sky, the lower ambient temperature means the refrigerator's load is significantly reduced.

The solar trailer was commissioned in May 1997. It was designed to chill produce such as fresh fruit and vegetables en route to supermarkets. The refrigeration system was optimised by proper sizing of components and electronic control so that a reduction in energy demand can be achieved when compared with diesel driven units [1].

2. PV SYSTEM

The PV system, classed as stand alone, comprises the PV array, a especially designed and built charge regulator, battery and an inverter. The charge regulator designed and built to specification by Naps (Finland).

Appropriate 55 W modules were selected so that they can be nested structurally and aesthetically on the roof of the trailer.

Electrical wiring was achieved using standard marine quality cable and connectors. The array was configured into eight sub-arrays each having ten electrically matched monocrystalline silicon modules. The total nominal power output of the array under standard condition is approximately 4.4 kWp. An inverter was used to supply ac power to the refrigeration system.

2.1 Monitoring System

The monitoring system [2] was hooked up in a sealed container to the underside of the Solar Trailer. The onboard data logging system performing the collation and some mathematical manipulation of the data of temperature, refrigeratpr pressure, PV parameters and energy data was designed to be interrogated remotely through a cellular telephone link.

3. RESULTS AND DISCUSSION

Figure 2 shows the measured monthly temperature levels at Charlton, London determined from the monitoring system on the Solar-Trailer. The predicted values, shown in the figure, are based on a ten year average for London taking into account the effects of dust and smog in the atmosphere. The graph shows that the ambient temperature values measured for the winter months were up to 7 °C warmer than normal.

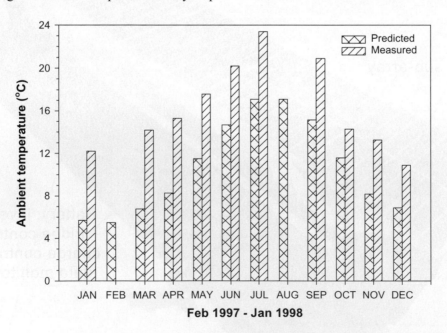

Fig. 2   Measured and predicted monthly temperature levels at Charlton London.

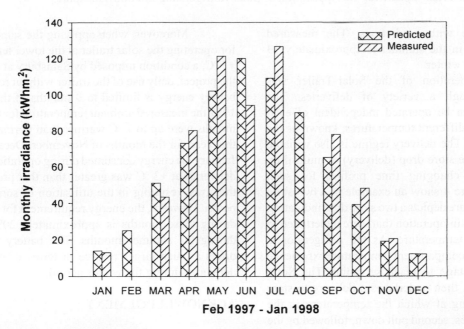

Fig. 3   Irradiance level received by calibrated sensor anchored in horizontal plane
at the front of the Solar-Trailer

century and this is clearly illustrated in the figure. This
means that the refrigerator was operated under harsher
conditions than those predicted by historical data. Data
for the months of August and February were incomplete
due to maintenance performed during both months.

The irradiance level received by calibrated
sensor anchored in horizontal plane at the front of the
Solar-Trailer is shown in Figure 3. The data collated is
broadly in line with predictions. The measured
irradiance level in the summer is approximately 13

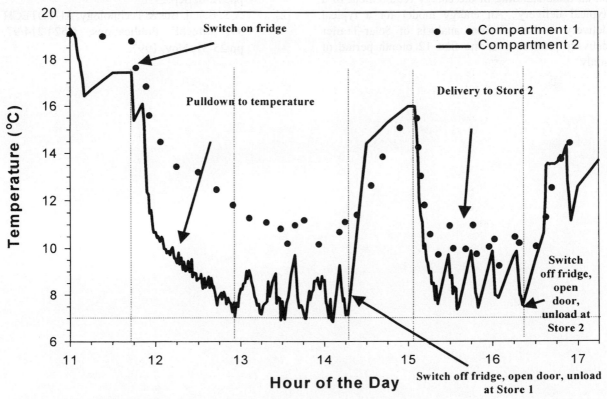

Fig. 4   Two store drop (delivery) indicating the variuos steps in operation that were undertaken to achieve the set
temperature.

broadly in line with predictions. The measured irradiance level in the summer is approximately 13 times that of the winter.

The operation of the Solar-Trailer was monitored through a variety of deliveries. The compartment can be operated independent of each other and set at different temperatures, known as the set-temperature. The delivery regime is also variable with either single store drop (delivery) or multi-drop resulting in a changing time profiles for such deliveries. Figure 4 show an example of a two store delivery. The figure depicts a two store drop indicating the various steps in operation that were undertaken to achieve the set-temperature. As the refrigerator is switched on, the compartment temperature experience the pull-down stage to reach set point. The Solar trailer is loaded, then undertakes a delivery to store one, door opening at which the temperature in the compartment rises, second pull down, followed by the delivery to store two.

Analysis of such results indicate that the Solar-Trailer can be operated, under UK weather conditions, at the designed temperature profile of +7 °C and usage throughout the year. Furthermore, such deliveries can be undertaken on a monthly basis with no net loss of energy from the battery. The number of deliveries that can be undertaken each month is based on an understanding of the energy requirement of a 'typical delivery'. An energy model for a typical delivery is derived from analysis of Solar-Trailer delivery profiles over the first 12 month period of study.

Moreover, when applying the same analysis for operating the solar trailer at the lower temperature +3 °C, a condition imposed by Sainsbury at the end of the project, daily use of the trailer without reverting to battery energy is limited to 9 months of the year. In 1997 the measured ambient temperature for the winter months were up to 7 °C warmer than normal. During this year, for the months of November, December and January the energy consumed during operation of daily deliveries at +3 °C was greater than that produced by the array, resulting in the utilisation of some battery power. However, the energy requirement for a delivery during these months is approximately 20% of that during the summer months. The battery therefore, offers a much larger reserve in terms of number of deliveries during the winter period.

ACKNOWLEDGEMENT

The authors wish to thank and acknowledge J. Sainsbury plc and Southampton University for financial and in-kind support for the Solar-Trailer Project.

REFERENCES

[1]   A. S. Bahaj, Renewable Energy, vol 15, pp572-6, Sep.1998

[2]   A.S. Bahaj, Bus & Technology, AUTOTECH 97, IMechE Publications, C524/214/97, pp.25-31, Nov. 1997.

STUDY OF A PHOTOVOLTAIC-THERMAL HYBRID SYSTEM

F. Serrano-Casares, A. M. Mateos-Medina

Instituto Andaluz de Energías Renovables. Universidad de Málaga.

Campus de El Ejido Málaga. España.

Tel.: +34 952 13 14 27. Fax: +34 952 13 24 09. e-mail: fserranoc@uma.es

ABSTRACT: Integration of photovoltaic systems in buildings offers a very important scope to expand the photovoltaic market by contributing, moreover, to a reduction in the use of fossil sources of energy. Photovoltaic-thermal hybrid systems generate electric and thermal energy, and they can increase photovoltaic systems efficiency. In order to evaluate this increase in the efficiency, it has been designed and built an hybrid system formed by a photovoltaic module and a thermal collector, both placed in parallel planes. In the gap between the planes there will be an air layer flowing up due to natural convection producing a cooling effect in the photovoltaic module and increasing its efficiency. To evaluate the increase in the efficiency, real-sun measurements have been made with the hybrid system. The results have been analyzed using different indicators of the system thermal behaviour.
Keywords: Hybrid – 1: Module Integration – 2: Energy Options – 3.

1. INTRODUCTION

At the present time, our theoretical knowledge about photovoltaic conversion and thermal solar energy have reached a stage that real progress in this matter will be obtained when practical solutions to give to these systems more uses will be found. In this way, a photovoltaic-thermal (PV/T) hybrid systems have been studied and built.

The systems of this type usually built [1,2] produce simultaneously electricity and heat ("solar cogeneration system") placing PV cells under the cover of a flat solar collector for air as shown in figure 1.

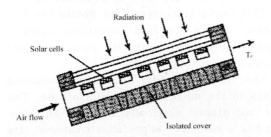

Figure 1: Conventional hybrid system

When we need to obtain high temperature heat in this hybrid panel, PV cell temperature becomes high and its efficiency falls down [3].

In an exergy analysis of the conventional hybrid collector shown in Figure 1 carried out in ref. [4] they conclude: for the exergetic efficiency of the thermal collector to be equal or greater than that one of the PV module, the difference between the environmental temperature and the average heat collecting temperature of the thermal collector for air (which matches the PV cell temperature due to the hybrid panel design) must be equal or greater than 42.8°C (for silicon conventional PV cells). At such high temperature the efficiency of PV cells decreases substantially, therefore, to get a good conversion efficiency in PV cells is necessary to obtain heat at low temperatures.

2. MODIFIED HYBRID COLLECTOR

It has been built the PV/T hybrid collector, shown in Figure 2, by placing in parallel planes a PV module and a thermal collector separated by a gap. The objective is to uncouple the thermal behaviour of the PV module respect to that of the thermal collector. This design of hybrid panel is based in one proposed in ref. [1]. It is aimed that part of the radiation passes trough the PV module and reaches the thermal collector. For that the PV module was built leaving free spaces between the cells and was packed with transparent Tedlar in its rear side. Thus, the thermal collector is heated and the air in the gap flows up in contact with the PV module, obtaining the requested cooling effect which decreases cell temperature and improves PV module efficiency.

Figure 2: Modified solar hybrid collector

In the model of ref. [4], they came to the conclusion that the mean air speed in the gap follows the next expression:

$$U_m = \frac{S^2 \cdot g \cdot \beta \cdot \tau_c \cdot \mathrm{sen}\,\theta}{12 \cdot v}$$

The meaning of each parameter is:
U_m: mean air speed between the parallel planes.
S: width of the gap.
g: acceleration of gravity.
ß: air expansion ratio.
τ_c:(air in the gap temperature) – (environmental temperature).
θ:angle between horizontal line and solar collector.
v: air kinematic viscosity.

Looking at the former expression, it can be seen that the air speed grows with τ_c and with θ because both factors increase the buoyancy effect. Moreover, when the gap width rise up, the air speed rises too, but there is an upper limit value for S which cancels the free convection effect.

If we consider a total energy efficiency and a total exergy efficiency as addition of individual efficiencies of PV module and

thermal collector, the hybrid collector analyzed improves the results respect to the hybrid traditional one used to heat air. This improvement is due to the cooling effect produced by the convective air motion in contact with the solar cells.

3. EXPERIMENTS.

In order to verify the conclusions established in the former analysis of the PV/T collector, real conditions measurements have been made with the hybrid collector of Figure 2, which was designed and built to this goal.

The hybrid system is composed by a PV module and a flat solar collector for water, both placed in parallel planes tilted 45° and leaving an air layer between them. Before their assembly, both devices (collector and module) have been subjected to tests in order to characterize their individual behaviour.

The PV module has been specially designed and built so that a part of the solar radiation pass trough it and reachs the thermal collector. It is composed by 32 solar cells packed with transparent Tedlar as back cover. Solar cell surface is 0.3395 m^2, and the cells are uniformly distributed on the total PV module surface of 1206x530 mm^2 , thus, solar radiation can pass trough the 44,2% of the total PV module area that is transparent (the remaining 55,8% of the area is occupied by the cells and the frame).

The PV module was characterized measuring its I-V characteristic curve under real sun conditions, using a tests procedure based in the standard method proposed in ref. [3]. This curve is corrected to obtain the I-V characteristic in standard conditions using a computer program properly designed and described in ref. [5].

The thermal solar collector mounted belongs to the model Garol-I of Isofoton. The tubes are placed vertically and the absorber plate has a selective superficial treatment. Its surface is 2x1 m^2 and the surface not covered by the PV module area will be considered as the transparent area of the PV module. Hence, the thermal collector area covered with PV cells is the 17% of its 2m^2 total area. The collector has been characterized to obtain its energy efficiency; in this way, real sun measurements have been made, following the standard ASHRAE tests procedure [6], with the protocol used in ref. [7].

After this individual analysis of the devices, three sets of tests have been made to the PV module in order to analyze its behaviour under different conditions and to find out if there are improvements when it is used in the hybrid system.

In the first set of tests, measurements have been made to obtain the efficiency of the PV module when it is completely exposed to the environment, even by its rear side. This configuration will be termed the isolated configuration. The isolated module has been mounted on a test-bed formed by an adjustable structure oriented to the South and on a plane fixed with an inclination of 45°. During a period of some days, measurements of the next magnitudes have been made each 30 seconds: voltage and current generated by the PV module, environmental temperature, PV module temperature and solar global radiation in the plane of the module. These measurements were recorded in a computer by a data acquisition system in order to allow a later treatment and analysis.

In the second set of tests, the same measurements have been made with the PV module, but covering its rear side in a disposition similar to that of an element of a photovoltaic facade or a photovoltaic roof. These measurements were recorded in the same way.

Finally, the third set of tests have been made placing the PV module as a part of the hybrid panel, as described previously. Again, during some days values of voltage and current generated by the module, environmental temperature and PV module temperature have been registered. Moreover, the thermal collector inlet and outlet temperatures, the circulating water flow and the thermal collector surface temperature in contact with the flowing air in the gap between the thermal collector and the PV module have also been measured.

4. RESULTS.

In figure 3 we show some of the results obtained on the efficiency yield by the PV module in the three different configurations previously described.

As it can be seen, the best results of energy efficiency are obtained in the isolated module and the worst results are obtained when the rear side is covered as in a photovoltaic roof. When the PV module is mounted in the hybrid panel, the results are improved in an appreciable way respect to those with the rear side covered.

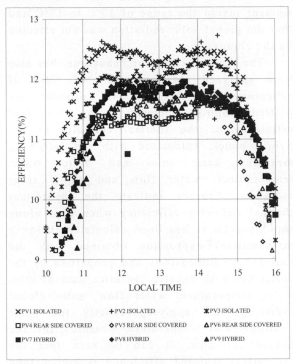

Figure 3: PV module efficiency

In figure 4, the PV module temperatures in the three situations are shown.

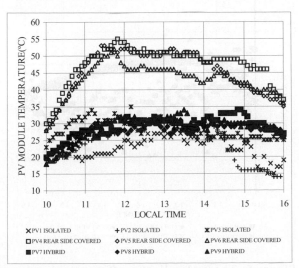

Figure 4: PV module temperature

It can be observed how the temperature affects to PV module efficiency. The lower temperatures are obtained with the isolated module and the higher ones belong to that with the rear side covered. When the PV module is mounted in the hybrid collector, its temperature is almost as low as in the isolated module. It is important to mention that environmental temperature was in every moment inside the range of 10°C to 17°C and that the global solar radiation was not affected by the clouds.

The thermal collector behaviour has also been analyzed. In figure 5 two types of efficiency for the thermal collector are plotted. Spots correspond to the real efficiency of the thermal collector in the hybrid panel, calculated with the following measured data: inlet and outlet water temperatures, water flow and global solar radiation. Crosses indicate the theoretical thermal collector efficiency when it is alone (not hybrid); it has been calculated (using a mathematical expression obtained with the real physic dimensions and properties of the collector) with exactly the same data of inlet water temperature, water flow, solar global radiation and applying exactly the same environmental temperature measured for the hybrid collector. It can be seen that the theoretical isolated efficiency is higher than the real hybrid efficiency, but in a relatively small percentage. In fact the mean of the real

values plotted is 70%, and the mean of theoretical values plotted is 73%. The disperssion in the real efficiency is produced by thermal transient phenomena.

As a conclusion, the results obtained confirm the improvement in the behaviour of the PV module when it is used in the modified hybrid collector respect to the conventional hybrid collector or to the roof-mounted module. Moreover, we have also found out that the thermal collector efficiency in the hybrid system scarcely decreases respect to that when it is alone.

REFERENCES

[1] T. Takashima, T. Tanaka, T. Doi, J. Kamoshida, T. Tani, T. Horigome, T. IEE Japan 115-B (1995) 430-435.

[2] R. Verlius, H.A.L. van Dijk, Proceedings 2° World Confference and Exhibition on Photovoltaic Solar Energy Conversion, (1998) 2542-2545.

[3] M.A. Green, Solar Cells, Pub. University of New South Wales, Australia (1992).

[4] T. Takashima, T. Tanaka, T. Doi, J. Kamoshida, T. Tani, T. Horigome, Solar Energy 52 (1994) 241-245.

[5] Aguilera, O., P.F.C., E.T.S.I. Industriales, Univesity of Málaga, (1998).

[6] ASHRAE Standard, 93-77 (1986).

[7] R. Porras, P.F.C., E.T.S.I. Industriales, University of Malaga, (1999).

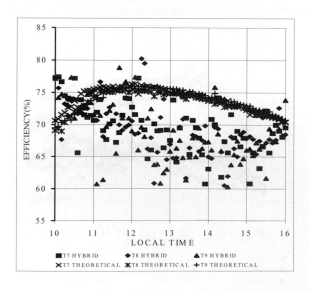

Figure 5: Thermal collector efficiency

MONTAGUE ISLAND PHOTOVOLTAIC/DIESEL HYBRID SYSTEM

R. Corkish a), R. Lowe a), R. Largent a), C.B. Honsberg a), N. Shaw a), R. Constable b), P. Dagger b)
a) Centre for Photovoltaic Engineering, University of New South Wales, Sydney 2052 Australia.
Email: rlowe@cheerful.com, r.corkish@unsw.edu.au, r.largent@unsw.edu.au,
c.honsberg@ee.unsw.edu.au, n.shaw@unsw.edu.au, r.corkish@unsw.edu.au
b) New South Wales National Parks and Wildlife Service,Princes Highway, Naroooma, 2546 Australia
Ph. +61-2-44762888; Fax. +61-2-44762757
Email: ross.constable@npws.nsw.gov.au, phillip.dagger@npws.nsw.gov.au

ABSTRACT: The diesel-electric power supply to the Montague Island nature reserve has been replaced by a photovoltaic/diesel/battery hybrid system, resulting in 87% decrease in fuel use with attendant reduction in environmental risk. Thorough electrical and environmental monitoring is carried out to enable improvements and for educational purposes. System performance and load management are reviewed.
Keywords: Hybrid – 1: Performance – 2: Remote – 3

1. INTRODUCTION

Montague Island is an ecologically sensitive, 82 ha nature reserve 9 km from Narooma off the eastern Australian coast (latitude 36°S). It is home to a wide variety of wildlife, including the most northerly colony of the Australian fur seal (*Arctocephalus pusillus*) in Eastern Australia, a significant fairy penguin (*Eudyptula minor*) rookery and other seabird species. There are a number of Aboriginal sites and a lightstation complex (construction commenced 1878) that is culturally significant. The island is managed by the National Parks and Wildlife Service (NPWS) of New South Wales, which restricts public access. The main structures are a photovoltaic-powered automatic lighthouse, three residences, originally built for head and assistant lighthouse keepers, a workshop, sheds and radio relay stations for various communications networks. A stand-alone photovoltaic (PV)/battery power system, separate from the power system that is the subject of this paper, supplies the lighthouse. A NPWS staff member and family occupy one of the residences, another is used intermittently by up to 15 visiting researchers simultaneously and the third is now an unpowered museum.

In 1996 NPWS, for both environmental and financial reasons, decided to install a PV-diesel hybrid system to replace the diesel generators which had previously powered the island (see Table 1). Around 100 200-litre fuel drums had been transported by launch from the mainland annually, a journey which involved crossing a dangerous sand bar at the mouth of the Narooma River. Electricity generation on the island was estimated in June 1995 to cost A$23,500/year. Aims included the reduction diesel running time to less than 10% (average 2.4 hour/day) in order to reduce fuel costs and risk of spillage of fuel and lubricants. Wind power was not favoured since the requirements of avoiding turbulence from existing structures and minimising the visual impact severely restricted the available sites.

University of New South Wales staff evaluated power requirements, set tender specifications and helped NPWS to evaluate tenders. The successful tenderer was Natural Technology Systems/Integral Energy. Electrical and meteorological data is collected for the purposes of system improvements, fault detection and correction, comparison with other systems, and to aid the specification and design of future remote-area power systems [1].

This paper describes and discusses the power and monitoring systems, load management, performance, and problems.

Figure 1: Lighthouse, photovoltaic arrays and nesting seabirds. The larger array to the right powers all the island's electrical loads other than the lighthouse.

2. SYSTEM LOAD

The island load consists of a family residence, housing for short-term visitors, facilities for tourists, communications systems, water pumps and a workshop. There are five radio communications repeaters belonging to NPWS, Surf Lifesaving Association, Marine Rescue Service, Royal Flying Doctor Service and Telstra. The main rainwater storage is underground cisterns beneath the courtyards on the residences. Overflow and some roof runoff is stored in above-ground tanks at 7m lower elevation. Pumps (760 W) transfer water from the overflow tanks to the underground cisterns when required (approx. 30 hours/year) and domestic pressure pumps supply residences from there.

In 1996 the entire island electrical load, apart from the lighthouse, was supplied by one 10 kVA and one 20 kVA diesel gensets, with the latter used only occasionally. Domestic hot water was heated electrically and electric space heaters were used. Large domestic appliances (refrigerators, freezers, washers dryers), water pumping, lighting and water heating accounted for 76% of the load and almost half the electrical energy was consumed in the occupied caretaker residence. Tourist facilities, which were used only intermittently, accounted for 15% of the total, principally due to the continuous supply of hot water for drinks (Table 1). The load was, and is, extremely variable due to intermittent use of large appliances.

Area	Average load (kWh/day)	Load fraction (%)
Caretaker's residence	13.5	47.5
Visitors' residence	7.2	23.5
Tourist facilities	4.1	14.6
Industrial	3.5	12.4
Total	28.3	100

Table 1: Analysis of 1996 load by area.

An earlier study of renewable energy options for the island [2] found a load of 110 kWh/day, including excess refrigeration capacity and many appliances and lights being deliberately left on in order to provide enough load for the oversized gensets to prevent glazing of the engine cylinders.

Figure 2: Part of the residential complex seen from the lighthouse. The large building in the centre houses the two occupied dwellings (caretaker and family and visitors). The building partly seen at right is the original keeper's residence, now an (unpowered) museum.

3. PV/DIESEL HYBRID SYSTEM

3.1. Load management

Significant changes were made to the load profile at the time of the installation of the hybrid power system. "Bottled" liquified petroleum gas is now used for water heating and cooking. This was preferred to solar domestic hot water heating in order to preserve the heritage values of the buildings. The need to artificially increase the load in order to protect the diesel engines was removed when the genset overcapacity was reduced. On the other hand, some communications and weather-station loads that were originally powered by independent PV/battery systems were moved onto the main island supply. There has been no attempt to urge the occupants to modify their lifestyle to reduce electrical demand.

3.2. Infrastructure

The larger of the two diesel gensets and two 20,000 l fuel storage tank and associated bunding were removed from the island and a 4,000 l fuel tank was installed within the existing generator building.

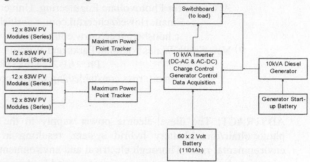

Figure 3: Block diagram of hybrid power system

The hybrid system (Fig. 3) consists of an array of four series strings of 12 BP Solar BP283 (nominally 83W) modules; two Australian Energy Research Laboratories maximum power point trackers (MPPT), each accepting the output of two strings of modules; a 10 kVA, 120VDC inverter/charger/controller manufactured by Power Solutions Australia; a pre-existing 10KVA diesel genset; and a 1101 Ah storage battery formed from 60 series connected BP Solar type BP 2P 1011 lead-acid cells. The array is approximately 50 m from the inverter and is steeply tilted (60°, latitude plus 24°) to avoid shading the adjacent array supplying the lighthouse and to boost winter performance. The inverter/controller is located in the same building as the genset and the battery is in a nearby building. In addition to managing the energy flows between the MPPTs, the battery, the inverter, the genset and the AC loads, the controller logs electrical system data at 15-minute intervals and produces daily summaries. The data are remotely accessible via modems.

4. ENVIRONMENTAL MONITORING SYSTEM

Since 1999 the controller's data logging has been complemented by a system to monitor three environmental parameters: insolation on a horizontal plane, ambient temperature and rear-of-module temperature [1]. The sensors are resistance temperature detectors (RTDs) and a pyranometer. Data are averaged and recorded every 15 minutes and may be downloaded via modems.

5. PERFORMANCE

5.1 Overall performance

The hybrid system has operated reliably except in the period immediately following a lightning strike in December 1999. The diesel genset ran continuously while repairs were made.

Records of the system performance stored by the controller for the hybrid system' lifetime of 865 days to October 1999 indicate that the genset had run for 2715.3 hours [1]. This equates to 13% of the period. In the absence of the hybrid system the genset would have run continuously. The PV array provides 23 – 24 kWh/day on cloudless days (results for March, May, July, August 1999), more than 90% of the average daily load.

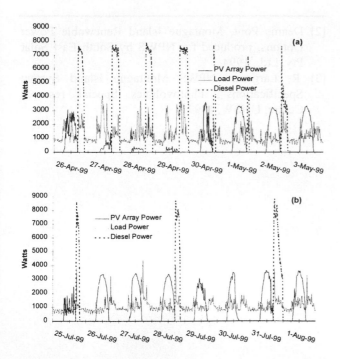

Figure 4: Array, genset and load variations (a) 26 April-3 May 1999, and (b)25 July–1 August 1999.

Fig. 4 is a plot of logged data for load, array and genset powers for (a) a partly-cloudy 8-day period in Spring 1999 when the load was high and (b) a mostly-sunny 8-day period in Winter 1999 when the load was smaller. The genset ran on all but one of the days in the Spring period, which included three days of very low insolation and ran for 4 − 6 hours following each of the four days of lowest sunshine. The genset started three times during the winter period when the load was smaller and the insolation more reliable.

A large fraction of the reduction in diesel running time since the installation of the hybrid system can be attributed to the addition of electrical storage to the genset. The genset still supplies more than half the load power while running time has been reduced by 87%.

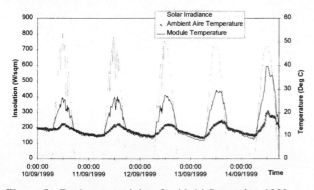

Figure 5: Environmental data for 10-14 September 1999.

The environmental monitoring system performed reliably until a lightning strike disabled it. A maximum rear-of-module temperature of 40 °C (for ambient 21 °C) was observed during September and October 1999 (Fig. 5). Such a cell temperature would be expected to result in 6.8% array power decrease relative to the standard temperature conditions

at which the array is nominally rated [1]. Temperature fluctuations that are not correlated with insolation variation may be due to wind cooling.

5.2 Load changes

There have been significant variations in the system load since the installation of the hybrid system. Firstly, caretakers (and their families) are changed periodically and power demands vary accordingly. Of particular note is the use of an electric clothes drier by one family with a baby. This was not anticipated in the system sizing [3]. Secondly, additional radio repeaters have been installed in the island. One of the original repeaters has since been provided with an independent PV/battery system. Thirdly, a desk-top computer which logs data as part of the automatic weather station on the island was (until September 1999) left with its monitor unnecessarily switched on continuously. This 2 kWh/day load had not been anticipated.

Nevertheless, a sample of load data for 80 days scattered throughout 1999 reveals a range of daily loads of 18.8 − 38.6 kWh/day and an average of 25.5 kWh/day. That average is less than the value of 30 kWh/day assumed in the design stage.

6 PROBLEMS AND ISSUES

There have been a number of operational problems which have affected system performance and data collection. It was noticed in August 1999 that for several months the controller's input value for battery size had been incorrectly set to 400 Ah. This had caused the controller to start the genset more frequently than was necessary. Another error, as yet unidentified, caused all the recorded values for array currents since September 1999 to be unrealistic values. Investigations are proceeding to identify the cause and to correct the data.

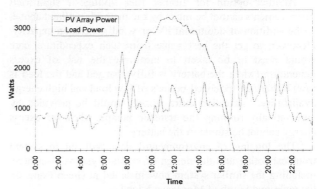

Figure 6: Load and array power profiles for 3 May 1999.

The PV supply is temporally unmatched to the typical daily load. Fig. 6 shows, as an example, the poor match between array supply and the morning and evening load peaks for a winter day. This mismatch results in an efficiency reduction due to the need for storage, with losses in the charger and the battery. Electrical energy that is produced by the genset and is stored in the battery undergoes both those losses and an additional one due to rectification.

A lightning strike on 8 December 1999 damaged on two PV modules, diodes on other modules and various components in the controller and data loggers (both electrical and environmental monitors). Local staff initially replaced the two 83W modules with 75W modules which were available, resulting in a 2.1% power loss. The orginal modules, with new diodes, have now been replaced.

The controller is currently scheduled to start the genset every 14 days to provide a gassing boost charge for the battery. This schedule is followed regardless of whether the generator has recently started for reason of low battery voltage. This would appear to be unnecessary with the current PV supply and load since genset operation is likely every few days anyway and some genset starts could be avoided if the controller was reprogrammed to give boost charges only when needed.

The baseline (overnight) load is typically 700-1000W, forming over 70% of the average energy load. This load comprises refrigerators and freezers for domestic and scientific research use, radio equipment, automatic weather station and computer, computers for visiting researchers and various domestic appliances in "standby" mode (two residences). It would appear to be important to try to minimise these constant loads in ways which do not significantly affect the comfort or lifestyle of the residents.

6. CONCLUSION

The replacement of a diesel-only remote power supply by a PV/diesel hybrid system has reduced the genset running time by 87%. The reduction is due to the combined effects of introducing energy storage into the system, supplementing the genset with PV, shifting some loads to LPG and some improvements in end-use efficiency. Further reductions in running time may possibly be achieved through careful analysis of the constant loads, use of more efficient equipment and appliances, switching off unnecessary devices, reprogramming of scheduled generator starts, and careful setting of system parameters in the controller.

Another option for further fuel saving, if significant improvements cannot be made by the methods outlined above, is the addition of additional PV or wind generation capacity. However, to get the best value from such expenditure, care would need to be taken to minimise the risk of excess generation (when the battery is fully charged and the load is low) if there are extended periods of low load and high energy availability. Effective extra storage could be provided by automatically running the transfer pumps whenever excess energy cannot be stored in the battery.

The logging of environmental data will aid in system troubleshooting and the design of similar systems elsewhere, including the similar system now installed at Green Cape, on the mainland South of Montague Island.

7. ACKNOWLEDGEMENT

The Photovoltaics Special Research Centre was supported by the Australian Research Council's Special Research Centres Scheme and by Pacific Power.

[1] Raymond Lowe, "Montague Island Photovoltaic System", B. Eng. Thesis, UNSW, Nov. 1999.

[2] Dennis Pont, Montague Island Renewable Power Options, produced for NPWS by South East Solar Pty. Ltd., 1994.

[3] R. Largent, "NPWS Montague Island System Specifications", Photovoltaics Special research Centre, UNSW, 1997.

INTERNATIONAL ENERGY AGENCY PVPS TASK 2:
ANALYSIS OF THE OPERATIONAL PERFORMANCE OF THE IEA DATABASE PV SYSTEMS

U. Jahn[1], D. Mayer[2], M. Heidenreich[3], R. Dahl[4], S. Castello[5], L. Clavadetscher[6], A. Frölich[6], B. Grimmig[7], W. Nasse[7], K. Sakuta[8], T. Sugiura[9], N. van der Borg[10], K. van Otterdijk[10]

[1] Institut für Solarenergieforschung GmbH (ISFH), Am Ohrberg 1, D-31860 Emmerthal, Germany
[2] Ecole des Mines de Paris, Centre d´ Energetique, Rue Claude Daunesse, F-06904 Sophia Antipolis, France
[3] Österreichisches Forschungs- und Prüfzentrum Arsenal, Faradaygasse 3, A-1030 Vienna, Austria
[4] Forschungszentrum Jülich GmbH, Projektträger BEO, D-52425 Jülich, Germany
[5] ENEA Casaccia, Via anguillarese 301, I-00060 Rome, Italy
[6] Thomas Nordmann Consulting AG (TNC), Seestrasse 129, CH-8810 Horgen, Switzerland
[7] Solar Engineering Decker & Mack GmbH, Vahrenwalder Straße 7, D-30165 Hannover, Germany
[8] Electrotechnical Laboratory (MITI), 1-1-4, Umezono, Tsukuba, JP-Ibaraki, 305, Japan
[9] Japan Quality Assurance Organization (JQA), 1084-2, Hatsuoi-cho, JP-Shizuoka-ken 433-8112, Japan
[10] Energieonderzoek Centrum Nederland (ECN), Westerduinweg 3, NL-1755 LE Petten, The Netherlands

ABSTRACT: As a part of the International Energy Agency (IEA) Photovoltaic Power Systems Programme (PVPS), the above mentioned participants of Task 2 are collecting and analysing operational data of photovoltaic plants in various system techniques located world-wide. The objective of this joint project is to provide suitable technical information on the operational performance, reliability and sizing of PV systems and their subsystems to the target groups. People working in the area of photovoltaics like manufactures, system designers, utilities, research laboratories, standardization organizations and vocational schools may benefit from the group of international experts working on the analysis of monitoring data and giving feedback to the PV system design. Task 2 participants have developed a database containing PV system data selected from different kinds of application e. g. professional and domestic stand-alone systems, distributed grid-connected systems and centralized PV power plants. A detailed analysis of more than 260 PV systems has been carried out using normalized quantities allowing any kind of cross-comparison between different systems. In case of stand-alone systems, new parameters have been introduced to better quantify the system behaviour from a technical point of view and to define a ranking procedure. This paper will summarize and present the most important results drawn from the performance analysis of the 260 different PV systems addressed in the IEA-PVPS Task 2 database.
KEYWORDS: Evaluation - 1: Performance - 2: Database - 3

1 INTRODUCTION

In order to state on the operational performance of PV systems and to develop guidelines for sizing and design optimization, the IEA-PVPS Task 2 has been launched in 1995 with the general objective to provide PV experts and other target groups with suitable information on the operation of PV systems and subsystems [1]. Task members have developed a database to accommodate the technical and operational data of different types of systems (grid-connected, stand-alone, hybrid).

At present the database contains more than 260 monitored PV plants with an installed capacity of about 12 MW_p operating at different climatic conditions and adapted to various applications (power supply, domestic uses, rural electrification, professional applications). Detailed system characteristics of selected PV plants as well as monitored data are stored in the database. The data are made available to the user through internal graphical displays and reports or by exporting the data into a standard spread sheet programme. This tool can also be used to check the operational behaviour of existing PV plants and to get a report on its performance expressed in standard quantities allowing any kind of cross-comparison between systems.

The implemented PV systems are located world-wide and operate therefore under different climatic conditions. Most of the monitoring data have been gathered under various national demonstration programmes in the IEA-PVPS member countries: e. g. Austrian Rooftop Programme, French Rural Electrification Programme, EU Thermie Programme, German 1 000-Roofs-Photovoltaic-Programme, Japanese Sunshine and Field Test Programme. There has been an increasing trend for demonstration programmes to focus on grid-connected PV systems. Particular complete statistics are available from 170 grid-connected PV systems ranging from small decentralized systems (PV roofs), dispersed systems (BIPV, sound barrier) to centralized systems (PV power plants).

France has focused its efforts on stand-alone systems (SAS) extracted from Ademe/EdF rural electrification programmes and Thermie ones. Nearly 40 PV systems from France and the overseas islands were collected representing off-grid domestic systems for rural electrification (isolated houses and Alpine huts) and off-grid professional systems (remote communication, control and protection devices).

A collection of such a variety of operational data can be considered as an unique tool for PV system performance analysis.

2 OVERVIEW OF PERFORMANCE INDICATORS

All system and subsystem performance data have been evaluated in terms of operational performance and reliability. The evaluation procedures are based on the European Guidelines, Document B [2], and the IEC Standard 61724 [3]. Additional parameters have been introduced for the analysis of SAS in order to better quantify the system behaviour from a technical point of view. Table I shows an overview of the derived parameters for performance evaluation.

Table I: **Overview of derived parameters for performance evaluation**

Parameter	Symbol	Equation	Unit
Array Yield	Y_A	$E_{A,d} / P_0$	h/d
Final Yield	Y_f	$E_{use,PV,d} / P_0$	h/d
Reference Yield	Y_r	$\int_{day} G_I \, dt / G_{STC}$	h/d
Array capture losses	L_c	$Y_r - Y_A$	h/d
System losses	L_s	$Y_A - Y_f$	h/d
Performance ratio	PR	Y_f / Y_r	—
Mean array efficiency	η_{Amean}	$E_A / \int_\tau G_I \cdot A_a \, dt \cdot 100\%$	%
Efficiency of the inverter	η_I	$E_{IO} / E_{II} \cdot 100\%$	%
Overall PV plant efficiency	η_{tot}	$E_{use,PV,\tau} / \int_\tau G_I \cdot A_a \, dt \cdot 100\%$	%
Annual irradiation, in plane of array	$H_{I,y}$	$\int_{year} G_I \, dt$	kWh/(m²·y)
Annual array yield	$Y_{A,y}$	$E_{A,y} / P_0$	h/y
Annual final yield	$Y_{f,y}$	$E_{use,PV,y} / P_0$	h/y
Annual reference yield	$Y_{r,y}$	$\int_{year} G_I \, dt / G_{STC}$	h/y
PV array fraction	F_A	E_A / E_{in}	—
Matching factor	MF	$PR \cdot F_A$	—
Usage factor	UF	E_A / E_{pot}	—

To compare PV systems, normalized performance indicators are used: e. g. energy yields (normalized to nominal power of the array), efficiencies (normalized to PV array energy) and performance ratio (normalized to in-plane irradiation). The most appropriate performance indicators of a PV system are:

The *final yield* Y_f is the energy delivered to the load per day and kW_p. The *reference yield* Y_r is based on the in-plane irradiation and represents the theoretically available energy per day and kW_p. The *performance ratio PR* is the ratio of PV energy actually used to the energy theoretically available (i. e. Y_f / Y_r). It is independent of location and system size and indicates the overall losses on the array's nominal power due to module temperature, incomplete utilization of irradiance and system component inefficiencies or failures.

To characterize the performance of stand-alone systems, two new parameters have been introduced: The *matching factor MF* is the product of the performance ratio and the array fraction (F_A) [4] and indicates how the PV generated energy matches the electrical load while using a back-up contribution (SAS) or energy from the grid (GCS). The matching factor is valuable for all hybrid systems (F_A less than one) and for grid-connected systems with a considerable contribution from the grid (F_A less than one).

The *usage factor UF* is the ratio of energy supplied by the PV array E_A to potential PV production E_{pot} and indicates how the system is using the potential energy. E_{pot} is a measured energy quantity, which differs from E_A for all SAS presenting PV array disconnection due to a fully charged battery.

3 SUMMARY RESULTS

3.1 Grid-connected PV systems

From the analysis of 170 grid-connected PV systems in the IEA-PVPS database, it was learnt that the average annual yield (Y_f) fluctuates only slightly from one year to another and has typical annual values (700 kWh/kW_p for Germany and the Netherlands, 830 kWh/kW_p for Switzerland and up to $Y_f = 1\,600$ kWh/kW_p for Israel). However, there is a considerable scattering around these average values for individual systems ranging from 400 kWh/kW_p to 950 kWh/kW_p (Germany) and from 400 kWh/kW_p to 1 400 kWh/kW_p (Switzerland).

Figure 1 shows the distribution of annual performance ratios calculated from 387 annual datasets of 170 grid-connected PV systems. The annual performance ratio (PR) differs significantly from plant to plant and ranges between 0.25 and 0.9 with an average value of 0.66 for 170 PV systems. It was found that well maintained PV systems operating well show an average PR value of typically 0.72 at an availability of 98 % (e. g. Switzerland). A tendency of increasing annual PR values during the past years has been observed.

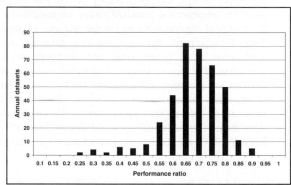

Figure 1: **Distribution of annual performance ratios for 170 grid-connected PV systems**

This section gives representative performance results for selected PV systems in different countries. Detailed results of all systems in the database are given in the IEA-PVPS Task 2 publication „Analysis of Photovoltaic Systems" [5].

Figure 2 shows the results in terms of performance ratio and availability of ten grid-connected PV systems in Switzerland operating between one and eight years. The monthly mean PR values are presented by a square dot, the annual PR by a circle. The small dots show the annual value of availability. Figure 3 shows the annual values of final yields as a function of reference yield for 120 annual datasets of 43 Swiss plants. The mean value of the reference yield is 3.3 h/d and for the final yield it is 2.3 h/d giving an average performance ratio of 0.70 at an availability of 95 % for all 43 Swiss PV plants.

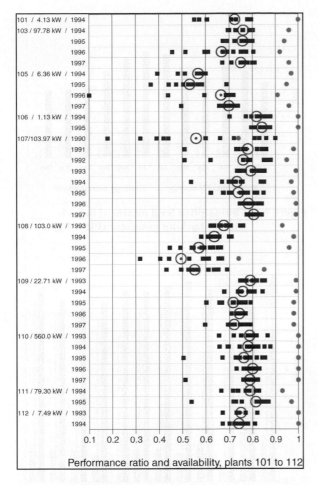

Performance ratio and availability, plants 101 to 112

Figure 2: Monthly and annual performance ratio and availability for 10 grid-connected Swiss PV plants

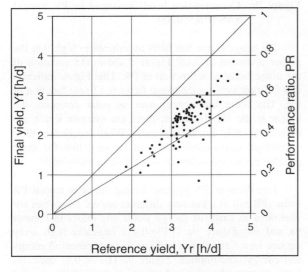

Figure 3: Reference yield vs. final yield using 120 annual datasets from 43 Swiss PV plants

A large grid-connected PV system (100 kW$_p$) in Italy is selected to present monthly performance results over a range of six operational years. Figure 4a shows monthly values of final yield, system and array capture losses presented in a stacked bar diagram. The annual Y$_f$ ranges from 1.62 h/d (1997) to 2.75 h/d (1993), while the annual

PR varies from 0.48 in 1997 to 0.65 in 1993 (Figure 4b). Besides the PR values, the capture losses L$_C$ are a good indicator for system problems in grid-connected systems. Because of the given definition (see [3]), a malfunction or inverter failure will result in a remarkable rise of L$_C$. This effect can be seen in Fig. 4a, where L$_C$ values are significantly high in June and July 1997 corresponding to very low values of monthly PR of less than 0.2 (Fig. 4b). The bad performance figures during that period in 1997 can be explained by severe failures of the inverter and other components. Using PR and L$_C$ as indicators, it is easy to detect system malfunction and failures in grid-connected systems.

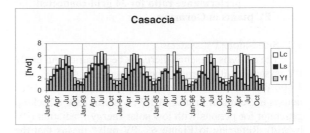

Figure 4a: Daily final yields (Y$_f$), array capture losses (L$_C$) and system losses (L$_S$) for an Italian plant from January 1992 to December 1997

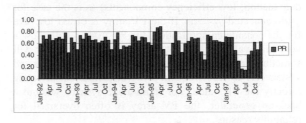

Figure 4b: Monthly performance ratios for the same Italian plant from January 1992 to December 1997

Figure 5 shows results of 34 grid-connected PV systems in Germany with respect to measured and nominal array efficiency. One can see in Fig. 5 that the measured array efficiency $\eta_{A,mean}$ significantly lies below the nominal array efficiency η_{A0}. The ratio of $\eta_{A,mean}$ to η_{A0} ranges between 60 % and 90 %. It has been noted that some manufacturers allowed systematic deviation of the measured array nominal power from the rated module power specified in data sheets.

As shown in Figure 5 the ratio of $\eta_{A,mean}$ to η_{A0} as a function of performance ratio indicates a linear relationship. Good operation of systems (PR > 0.7) can be expected from systems with a ratio $\eta_{A,mean}/\eta_{A0}$ > 80 %, although other parameters (e. g. inverter efficiency) and effects (e. g. shading) have to be taken into account. Thus the ratio of mean and nominal array efficiency is a good indicator for the evaluation of system operational performance for grid-connected PV systems.

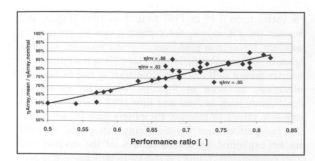

Figure 5: Ratio of mean array efficiency to nominal array efficiency as a function of performance ratio for 34 grid-connected PV plants in Germany

3.2 Stand-alone PV systems

The analysis presented here has been carried out on 27 PV systems with nominal power between 450 W_p and 1 500 W_p, which are used to provide electricity for isolated houses in Southern France. All PV systems have a back-up generator for periods with low solar energy irradiation (PV hybrid). Referring to Figure 6, „PV only" means that the back-up generator has not been used. The analysis in terms of performance ratio (Figure 6) shows that SAS present a wide range of PR which does not reflect the quality of a system from a technical point of view (component degradation, low efficiency components) as is the case for grid-connected systems. If the PV array is of too low size for the considered application, the PV system will show a very high value of PR, but at the same time the user will sometimes not be supplied with electricity. An oversized system has to face frequent array disconnection affecting directly the PR value. For stand-alone systems without a back-up generator, the PV array is often oversized for reliability reasons.

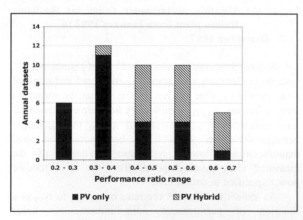

Figure 6: Distribution of annual performance ratios for domestic stand-alone PV systems in France

The value of PR depends on user consumption. If the consumption level is not correlated to the potential of the PV generator (Figure 7a), the PR will reach values which can be less than 0.2 on a monthly basis. Such a low PR is due to high array capture losses. Hybrid systems characterized by the use of a back-up generator, can show good performance if the consumption level matches quite well the potential of the PV generator (Fig. 7b).

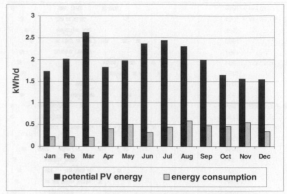

Figure 7a: Consumption level measured in SAS without back-up generator (PR = 0.2)

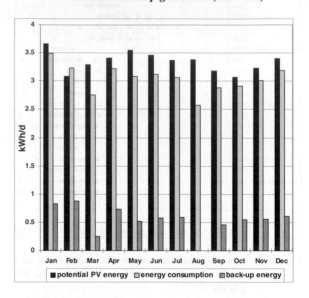

Figure 7b: Consumption level measured in PV hybrid SAS (PR = 0.65)

The usage factor has been introduced to highlight the proper operation of SAS. Figure 8 shows the variation of the usage factor as a function of PR. This Figure indicates that for most systems the usage factor is a linear function of PR. The better the system uses its solar potential, the higher is the PR. However, there are systems which are outside of this linear tendency. When analysing their operational characteristics, it can be seen that for these peculiar plants, the system losses are abnormally high.

Two different PV systems having the same annual PR value (PR = 0.31), but very different annual UF values are illustrated in terms of energy yields and losses in Figures 9a and 9b. Figure 9a (UF = 0.45) indicates high array capture losses and thus a bad matching of potential energy and energy consumption. Figure 9b (UF = 0.9) indicates high system losses and thus system failures and malfunction of components. These figures highlight the differences of operation of stand-alone systems and demonstrate that the usage factor allows to detect SAS with technical problems.

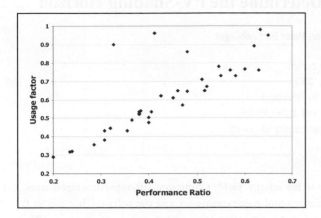

Figure 8: **Yearly usage factor and performance ratio for domestic stand-alone systems in France**

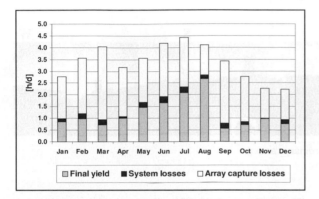

Figure 9a: **Indices of performance for one SAS with PR = 0.31 and UF = 0.45**

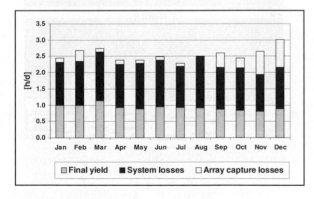

Figure 9b: **Indices of performance for another SAS with PR = 0.31 and UF = 0.9**

4   CONCLUSIONS

From the performance analysis of 260 PV plants in the IEA-PVPS Task 2 database the following annual performance ratios can be expected for the different types of systems:

- Grid-connected PV systems   PR = 0.6 - 0.8
- Stand-alone systems without back-up   PR = 0.1 - 0.6
- Stand-alone systems with back-up   PR = 0.3 - 0.6

The distribution of annual performance ratio calculated from 170 grid-connected PV systems shows that

the PR significantly differs from plant to plant and ranges between 0.25 and 0.9 with an average PR value of 0.66. It was found that well maintained PV systems operating well show an average PR value of typically 0.72 at an availability of 98 %. A tendency of increasing annual PR values during the past years has been observed. Despite good results, which have been obtained in many of the grid-connected systems, the investigation of the operational behaviour of the reported PV systems has identified further potential for optimization.

The performance analysis of data from stand-alone and stand-alone hybrid systems has revealed that operational performance is not only depending on the component efficiency, but also on system design and load pattern. Annual performance ratios range from 0.2 to 0.6 for off-grid domestic applications depending on whether they have a back-up system or not and from 0.05 to 0.25 for off-grid professional systems, which are often oversized for reliability reasons. The performance analysis of stand-alone systems in terms of performance ratio has shown that in contrast to grid-connected systems, the PR alone cannot be used to describe the proper operation of stand-alone systems from a technical point of view.

A more detailed analysis concerning the operation of stand-alone systems will necessitate:

- More detailed and more reliable monitoring campaigns, which are feasible even for small remote systems with the development of integrated data loggers.
- Several years of measurement to better appreciate the evolution of user behaviour over time.
- The use of simulation tools to evaluate the influence of new component sizes or new regulation strategies to increase the system performance.

REFERENCES

[1]   IEA Implementing Agreement on Photovoltaic Power Systems (PVPS):
       IEA-PVPS Annual Report 1999

[2]   Commission of the European Communities:
       Analysis and Presentation of Monitoring   Data
       Guidelines for the Assessment of Photovoltaic Plants, Document B, version 4.3, March 1997

[3]   International Electrotechnical Commission:
       Photovoltaic system performance monitoring - Guidelines for measurement, data exchange and analysis
       International Standard IEC 61724, Geneva, Switzerland, first edition, April 1998

[4]   A. Sobirey, H. Riess, P. Sprau:
       Matching factor - A New Tool for the Assessment of Stand-Alone Systems
       Proceedings 2nd World Conference and Exhibition (PVSEC 1998), Vienna, Austria, July 1998

[5]   IEA-PVPS Task 2 Report:
       Analysis of Photovoltaic Systems
       Editor: R. Dahl, Forschungszentrum Jülich, BEO Report IEA-PVPS 2-01:2000

A Fast, Efficient and Reliable Way to Determine the PV-Shading Horizon

Roland Frei, Christian Meier, Peter Eichenberger

energieburo zurich
Limmatstr. 230
CH-8005 Zurich, Switzerland
T +41 1 242 80 60; F +41 1 242 80 86
info@energieburo.ch, www.energieburo.ch

ABSTRACT: High horizons and shadows on PV-systems lower the energy yield. Chimneys, lift-superstructures, trees, buildings, mountains, etc. can hurt the performance. The 360°-camera-tool panoramamaster and the software horizon v1.0 allow the fast and precise photometric determination of the horizon. An expensive lens equipment is not necessary. The calculated horizon can immediately be processed in known simulation software (e.g. PolySun, MeteoNorm).

Keywords: Novel Devices – 1; Computing Devices – 2; Building Integration – 3

With HorizON, one can take full 360° panorama pictures with an ordinary camera

EASY TO USE

With the help of the panoramamaster,, any regular camera works as a true 360°-panorama-camera. Expensive lenses are not necessary anymore.

All one needs is a tripod, and any camera (analogue or digital) can together with the panoramamaster, produce precise 360-degree full circle panoramic pictures..

Load and merge your several pictures to a 360°-panorama and automatically calculate the horizon with the software horizon v1.0 (see demo-version).

The software horizON allows the fast and precise photometric determination of the horizon.

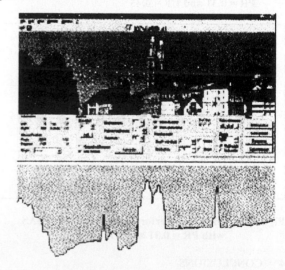

It produces a file, which can be processed with different standard simulation software (e.g. MeteoNorm, PolySun, etc.). For more information, please contact:
energiebüro Zürich, Limmatstrasse 230 CH-8005 Zürich
T +41 1 242 80 60   F +41 1 242 80 86
www.energieburo.ch   info@energieburo.ch

A DETAILED PROCEDURE FOR PERFORMANCE PREDICTION OF PV PUMPING SYSTEMS

M. Alonso-Abella and F. Chenlo, J. Blanco*
CIEMAT - Departamento de Energías Renovables
(*ABISYSA U. Las Lomas, 45, Olias del Rey, 45280 Toledo Spain)
Avda. Complutense, 22. 28040 Madrid Spain
Tel.: 34-91-3466492 Fax.: 34-91-3466037 e-mail: abella@ciemat.es

ABSTRACT: The main contribution of this work is obtaining the parameters that define the model of the subsystem converter/motor/pump. The procedure to predict the performance (water volume at a defined total head) of PV water pumping systems is based on experimental test data of a defined pumping system. The Flow-Array DC output power curves are measured in a laboratory facility and parametrized. A Typical Meteorological Year (TMY) of hourly global irradiance on horizontal surface and ambient temperature for a given location is needed as input parameter. A method to obtain the I-V array curves in any operating conditions is used. Solar irradiance on tilted surfaces is obtained using wide applied models. Performing hourly simulations the water volume at a given total manometric head can be obtained. The accuracy of this semi-empirical method is analysed and mainly depends on the accuracy of the solar irradiance models, the prediction of cell temperature from ambient temperature and the accuracy of the simulation method used to obtain the array I-V curves at defined irradiance and temperature conditions. The procedure here described can be applied to analyse the effects in Pumping systems of different parameters as: PV array configurations (module technology, number of cells in series and in parallel), loses due to threshold irradiances and loses due to non operation of the system due to out of range voltages.
Keywords: PV Pumping – 1: Simulation – 2: Modelling – 3

1. INTRODUCTION

The purpose of this work is to show an experimentally validated method to predict the performance of PV water pumping systems. This method is based on experimental data and simulation. Experimental data of the pumping system are used. The system is parametrized independently of the PV array technology and configuration obtaining in a PV pumping test facility the Flow-DC input power curves, *P-Q(h)*, at different total manometric heads, h. A model is proposed to parametrize these pumping curves in function of the head. Experimental data are also used to characterize the PV modules. Using 8 experimental parameters (obtained at STC or at any other reference irradiance (*Eo*) and temperature (*to*) conditions) and the analytical meted[2] the whole I-V curve can be obtained. A extrapolation method to obtain of the I-V curves at any irradiance and temperature is used. Hourly simulation is performed to obtain the irradiance on tilted surfaces from a TMY using wide applied models. The PV pumping system performance prediction method is mainly based on:

- The Flow-Array DC output power curves, *P-Q(h)*, are measured in a laboratory facility for a given pumping system at different total manometric heads. In this way the pumping system is characterized independently of the PV array configuration.
- A Typical Meteorological Year (TMY) of hourly global irradiance on horizontal surface and ambient temperature for a given location is needed as input parameter.
- A method to calculate the I-V curves of a given PV array configuration in different irradiance and temperature conditions from 8 PV module parameters: Isc°, Voc°, Im, Vm, slopes at Isc (PIsc) and Voc (PVoc), and temperature coefficients of Isc(α) and of Voc(β) at STC (or at other reference irradiance (*Eo*) and temperature (*to*) conditions).

Solar irradiance on tilted surfaces is obtained using wide applied models (as Erbs&Coll model to estimate the diffuse component on horizontal surface, and Perez model to predict the diffuse component on tilted surfaces).

The accuracy of the semi-empirical method proposed was analysed and mainly depends on the accuracy of the solar irradiance models, the prediction of cell temperature from ambient temperature and the accuracy of the simulation method used to obtain the array I-V curves at defined irradiance and temperature conditions.

2. PROCEDURE

One of the main objectives is the prediction of the performance of a PV pumping system (What is the flow or water volume for a given head in a given location?) in function of the PV array (serial x parallel configuration and technology). This information can be achieved in two ways: from direct measurements of the complete system (PV array + PV pumping subsystem) or from simulations, based on some experimental data.

The first way leads to very good and reliable results, we are measuring the complete and "real" installation, but is very time-consuming to have experimental data using different array configurations and technologies. Even it is difficult to have experimental results for different environmental conditions (irradiance and temperature). The simulation is less accurate (by definition), but allows us to predict results varying all possible situations (irradiance and temperature, array technology and configuration). We should take into account that a PV pumping systems operates in a defined range of voltage and current. The operational range should not be exceeded. In simulation, out of the operational range we will consider that the system does not work (Q=0). In real operation, the system could result damaged (if it has no protections) if operates out of this operational ranges. In the following we will describe the complete procedure to perform simulations in PV pumping systems. For this the only and necessary experimental data from a PV pumping system are the DC input Power (=array DC output power) versus water flow curves, *P-Q(h)*.

The necessary experimental data to perform a simulation of a PV pumping system are:

- The P-Q(h) curves for different and constant total manometric heads (also I-V curves for DC direct coupled)
- The *PV module reference parameters* that allow to simulate the array I-V curves in any environmental conditions (at reference conditions:E^o,t^o, Isc^o, Voc^o, Im, Vm, and the slopes of the I-V curve at Isc (Pisc) and at Voc Pvoc)), and the temperature coefficients α and β, of Isc and Voc

The necessary data inputs are: Irradiance data, ambient temperature data, total manometric head, PV array configuration (Number of PV modules in series and in parallel) and parameters of a PV module at STC (or at any other irradiance and temperature reference conditions) and Way of operation of the controller (MPPT or constant voltage).

Irradiance data can be an hourly record data file or can be obtained applying radiation models to obtain the irradiance on tilted surfaces. Ambient temperature data can be a record file for a given location.

With the characteristics curves P-Q(h=cte), the necessary information to obtain the volume pumped in any operational conditions (determined by the array serial x parallel configuration and a pumping Head) is done.

The PV array power to the pump is determined by the environmental conditions (irradiance, temperature), and has direct influence on the flow. The procedure to evaluate the pumped volume volume at a given total manometric head can be obtained either measuring directly with a Solar-generator-simulator or indirectly from the characteristics curves *P-Q(h)* performing hourly simulations by means the following steps:

1. Define the PV array, type of PV modules, by means of a set of 8 PV module reference parameters ($E^o,t^o,Isc^o,Voc^o,Im,Vm,Pisc,Pvoc,\alpha,\beta$) and serial x parallel PV array configuration (NsxNp).
2. The pumping Head, PV array tilt angle and orientation (tracking) are input parameters.
3. In plane irradiance, E, is obtained using solar radiation models from the horizontal TMY irradiance for a given location.
4. Cell operating temperature, tc, is obtained from the ambient temperature, ta, as tc=ta+te*E, where te is a module parameter variable in function of the TONC.
5. I-V curves are calculated at a given E and tc conditions in two steps, first using the analytical method[2] to generate the reference I-V curve (from the 8 module parameters in step 1) and second performing a geometrical translation method to extrapolate the obtained reference IV curve to the new Isc(Isc^o,E,tc,Eo,to,α,β) and Voc(Voc^o,E,tc,Eo,to,α,β) calculated at the new E and tc conditions .
6. The working point (I,V) of the Pumping system can be obtained, for AC pumping systems (or DC using matching devices). In MPPT systems the working point (I,V) is the MPP, in systems working at constant voltage, the current at this voltage can be obtained from the I-V curves of point 5. AC pumping systems (or DC using matching devices) operates most of times at a constant voltage or at MPP.
7. The flow can be determined in the P-Q(h) measured

curves, with the array power equal to the value determined in point 6 and the head value defined in point 2. The experimental P-Q(h) points have been modelled for each pumping system using a function with the head as parameter.

In step 6, for DC direct coupled pumping systems, it has been showed in previous work[4] that they can be characterized by a I-V motor curve in function of the pumping head; the cross point between the array and motor I-V curves gives the array power output. Also, for these DC systems, in step 6 the Flow-voltage curves can be used in place of the *P-Q(h)* curves, to obtain the water flow.

Repeating this procedure along the time, the hourly pumped can be obtained for every day, analysing the pumping capabilities of a PV pumping system in a defined place in function of the climatic conditions. Also, the best configuration serial x parallel of the PV array can be analysed for a determined motor/pump. The different parameters and models used in the procedure to estimate the performance of PV pumping systems in a given location and with a given PV array (some of them based in laboratory experimental data) are the following:

2.1 Solar Radiation

A Typical Meteorological Year (TMY) of *hourly* global horizontal irradiance data are used as input to the program. Calculations for different locations can be performed changing the TMY. A computer program can be used to obtain the hourly global irradiance on tilted surfaces. Models used in the program are: Erbs&Coll model for the calculation of horizontal diffuse irradiance and Perez[1] model for the calculation of diffuse irradiance on tilted surfaces.

2.2 Cell temperature

As in the case of solar irradiance, a Typical Meteorological Year (TMY) of hourly ambient temperatures are used as input to the program. Cell operation temperatures are calculated as:

$$tc = ta + te * E$$

Where *tc* is the cell temperature, *ta* is the ambient temperature, *te*≈0.025 (variable in function of the TONC) and *E* is the irradiance.

2.3 PV generator model

No interconnection, cabling or mismatch losses are considered, i.e., the voltage is calculated as the Ns times the voltage of one cell (or module) and the current as Np times the current of one cell (or module).

The method here proposed to obtain the I-V array curves in any operation conditions (E,t) is based on: 1) to use the analytical[2] model to obtain the I-V curve at the reference conditions (error<1%) from the the the 8 *PV module reference parameters* ($E^o,t^o,Isc^o,Voc^o,Im,Vm,Pisc,Pvoc,\alpha,\beta$), 2) to calculate the Isc and Voc at the new irradiance and temperature

$$Isc = Isc^o \frac{E}{E^o}[1 + \alpha'(t - t^o)]$$

$$Voc = Voc^o + mv_t \ln \frac{Isc}{Isc^0}[1 - \beta'(t - t^o)]$$

and 3) to apply a geometrical translation to the reference I-V curve calculated in 1) to obtain the new Isc and Voc values.

$$I = I^\circ + \Delta\, Isc$$
$$V = V^\circ + \Delta\, Voc \qquad \text{where} \qquad \Delta\, Isc = Isc - Isc^\circ$$
$$\Delta\, Voc = Voc - Voc^\circ$$

2.4 Converter/motor/pump model

We consider the converter/motor/pump as a "black box" with a given efficiency. The model is experimental and is based in the P-Q(h) curves measurements for each pumping system. Inverters for Pumping (or other kind of controllers) usually operates in two possible ways: at Maximum Power Point (MPP) or at a constant DC input voltage.

For centrifugal pumps, it is know that:

$$\left.\begin{array}{l} \dfrac{Q_2}{Q_1} = \dfrac{n_2}{n_1} \\[2mm] \dfrac{P_2}{P_1} = \dfrac{n_2^3}{n_1^3} \end{array}\right\} \Rightarrow Q \propto P^{1/3} \Rightarrow \left\{\begin{array}{l} Q = aP^{1/3} + b \\ a = a_0 + a_1 * h + a_2 * h^2 \\ b = b_0 + b_1 * h + b_2 * h^2 \end{array}\right.$$

Where Q is the water flow (l/min), P is the Array power (DC input power to the Converter) (W) and h is the total manometric head (m). The experimental data (Flow DC input power) can be fitted to the following function to obtain the coefficients **a (ao,a1,a2)** and **b (bo,b1,b2)**.

For positive displacement pumps similar polynomial fitting can be also performed.

The pumping starting threshold array power, Po, (Array output power), defined as the value for which the flow, Q, is >0, was evaluated as the root of this polynomial.

$$P_0 = \left(\dfrac{-b}{a}\right)^3$$

The simulation method here described can be applied to analyse the effects in Pumping systems of different parameters as: PV array configurations (module technology, number of cells in series and in parallel), loses due to threshold irradiances and loses due to non operation of the system due to out of range voltages; because of we obtain hourly values of all parameters.

3. PV PUMPING SYSTEMS USING FREQUENCY CONVERTERS

The use of standard frequency converters (FC) directly powered by a PV array (usually with a MPP voltage greater than 240VDC in most cases) and operating standard 200 V AC three-phase submersible centrifugal pump has been showed[6] as a wide PV pumping option: easy installation and operation, lower price compared with other equivalent PV pumping systems, wide range of powers available and use of standard equipments (independence of a defined supplier). These systems can also be modelled as showed in the previous point, but additional information is presented in order to predict performance and perform improvements in the voltage-frequency (VF) control of the FC to increase the system efficiency. Most FC have internal built in VF control algorithms that can be preselected to operate centrifugal pumps, but can also be reprogrammed. In the following we will present a procedure, alternative to the presented in point 2, in order to predict, from manufacturer supplied (or measured) H-Q and motor efficiency curves, the system performance.

It will be supposed hat: the head, h, for any number of pump stages is proportional to the one stage head, h1, and the motor sliding is constant. The input necessary data are

the curves (h_1, Q, η_p) for the pump and the (P_m, η_m) for the motor. Where Q is the water flow for one stage, η_b is the pump efficiency, P_m is the motor power and η_m the motor efficiency. If we define the following parameters: ho, the total system pumping head, ns the synchronous motor sped, no the motor speed at nominal frequency f_n (=50Hz), and e the number of pump stages, we can calculate in the following way: The head for a given number of pump states as $h = e * h_1$. If we define the constant fs as

$$f_s = \dfrac{n_0 f_n}{n_s}$$ the frequencies at given head

are $f_{ho} = f_s \sqrt{\dfrac{h}{h_0}}$. The pump power at nominal

frequency is $P_h = \dfrac{hQ}{0.367\, \eta_p}$ and the power at different

heads as $P_{ho} = P_{fn} \left(\dfrac{f_h}{f_s}\right)^3$ and the flow at ho as

$$Q_{ho} = Q\dfrac{f_h}{f_0} = Q\sqrt{\dfrac{h_o}{h}}.$$ The motor power at ho, P_{mho},

can be obtained interpolating (Fig. 1) from the motor efficiency curve (P_m, η_m) with $P_m = P_{ho}$ to give $P_{mho} = P_{ho}/\eta_m$.

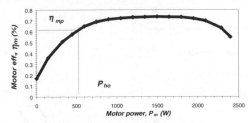

Figure 1: Motor pump efficiency interpolation.

Finally to obtain the necessary PV array power, P_{PV}, we can suppose that the FC has a defined efficiency, η_{FC}, (almost constant in all power range and >95%) and there

are a defined cabling losses, η_c. $P_{PV} = P_{mho}\dfrac{(1+\eta_c)}{\eta_{FC}}$

Using this procedure two graphs (Fig. 2) can be calculated (P_{PV}, Q) and (P_{PV}, f_{ho}). For a given array configuration and irradiance and temperature conditions, the PV array output power can be obtained and from these graphs, by interpolation, the flow, Q_{ho}, and frequency f_{ho} for the defined pumping system using an FC operating at a defined total head, ho.

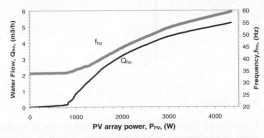

Figure 2: Flow and FC frequency in function of the PV array output power.

4. RESULTS

Three different PV pumping systems were characterized in the laboratory pumping test facility. Results were:

Table I: Parameters obtained fitting the flow-DC Power experimental values .

System	SP3A-10	SP5A-7	TSP-1500
a0	8.909	22.274	5.2201
a1	0.262	-0.371	0.1893
a2	-0.003	0.0047	-0.002
b0	-1.41	-77.26	12.976
b1	-3.583	-0.255	-2.527
b2	0.034	-0.009	0.0162
Po	$Po=14*H-73$	$Po=8.45*H-27.92$	$Po=10.27*H-16.752$

When the procedure here described ,in point 2, is performed for a given pumping system at various heads, the following monograms can be built in order to obtain the system performance in a given location and with a defined PV array configuration.

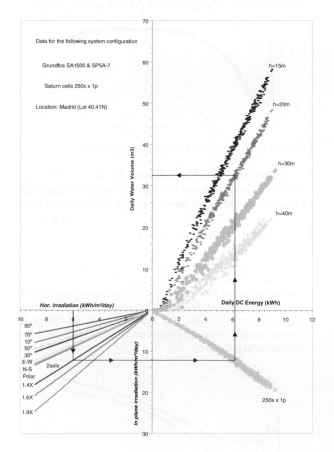

5. REFERENCES

[1]   R. Perez et Al., *A new simplified version of the Perez diffuse irradiance model for tilted surfaces*. Solar Energy 39, 221-231 (1987).

[2]   J.C.H. Phang et Al. *Accurate Analytical Method for the Extraction of Solar Cell model Parameters*. Electronic Letters V20 N°10 -1984.

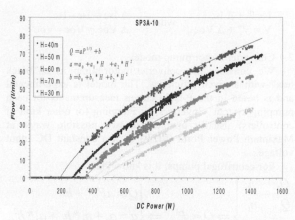

Figure 3: Example of fitting P-Q(h) for the grundfos SP3A-10 system.

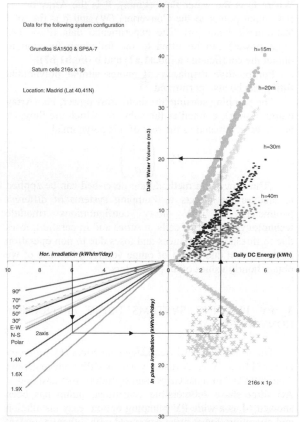

Figure 4: Example of simulation results for a PV pumping system. The same system with 2 different PV array configurations gives very different results (because of a bad coupling PV generator-Converter voltage operational range)

[3]   M. Alonso and F. Chenlo. *Testing Results of PV Pumping Systems for Stand Alone Installations*. 12th European Photovoltaic Solar Energy Conference, Amsterdam (1994).

[4]   M. Alonso and F. Chenlo. *Parameters influencing the efficiency of DC powered PV pumping systems*. 13th European Photovoltaic Solar Energy Conference, Nice (1995).

[5]   M. Alonso, F. Chenlo and U. Escudero. *Impedance matching and new use of speed controllers in PV pumping systems*. 14th European Photovoltaic Solar Energy Conference, Barcelona (1997).

[6]   M. Alonso, F. Chenlo, J. Blanco. *Use of standard frequency converters in PV pumping applications*. 2nd World Conference and Exhibition on Solar Energy Conversion, Wien (1998).

PREDICTIVE CONTROL OF PHOTOVOLTAIC-DIESEL HYBRID ENERGY SYSTEMS

B WICHERT and W LAWRANCE
Centre for Renewable Energy and Sustainable Technologies Australia
Curtin University of Technology
GPO Box U 1987, Perth 6845, Western Australia
Tel.: +61 8 9266 2960   Fax: +61 8 9266 3107   b.wichert@cc.curtin.edu.au

ABSTRACT: An advanced control strategy for PV-diesel hybrid energy systems has been developed, which reduces cycling of the battery bank, operates the diesel generator in its most efficient load range, and maximises the utilisation of the photovoltaic resource. The objective of this research project has been to develop an efficient, generally applicable control strategy, adapting to changing demand patterns, modular changes to the system configuration, as well as seasonal and short-term variations of the available solar resource. Photovoltaic resource and load demand forecasts constitute the knowledge base for the novel, predictive control strategy. Experimental results, as well as detailed simulations of the performance of several systems, show that only marginal fuel savings can be achieved due to the predictive control algorithm. However, battery cycling can be improved significantly, with estimated savings in lifecycle cost for battery storage exceeding 25%.

Keywords: Stand-alone PV Systems - 1: Energy Management - 2: Rural Electrification - 3

1. INTRODUCTION

Control methods optimising the interactive operation of multiple energy sources and storage devices promise significant improvements in system performance. Optimised control strategies aim to achieve quantitative performance improvements at the system level by implementing an 'intelligent' energy management system. Quantitative improvements represent efficiency gains, increased lifetime and reduced maintenance of components. Besides increased component efficiency and reliability, 'optimum'[1] operation of PV-diesel hybrid energy systems promises a significant reduction of the lifecycle cost.

The high battery replacement costs over the lifetime of hybrid energy systems has been the justification for extensive research, seeking to extend the cycle-life of lead-acid batteries. The initial cost of purchasing a motor generator is small compared with the price for renewable generators. However, fuel costs contribute significantly to the lifecycle cost of the system. Battery replacement and fuel costs promise significant opportunities for cost savings, justifying the development of advanced system control strategies [1,2].

2. FUNDAMENTAL CONTROL PRINCIPLES

Hybrid energy systems should be operated in such a way that the direct supply of energy to the load is maximised, thereby avoiding losses through additional power conversion. Furthermore, this results in reduced battery cycling, thus extending the lifetime of the battery bank. The utilisation of PV energy can be reduced as a result of a less effective control strategy. At times when the PV power exceeds the load demand and the battery is fully charged additional PV energy may have to be 'dumped'. This can be avoided by implementing an advanced battery management strategy, ensuring that sufficient battery capacity is available to store excess energy.

---

[1] The term 'optimum' describes the operation of a hybrid energy system, which allocates renewable and fuel-based resources in such a way that it results in minimum life-cycle costs, while also considering restrictive operational conditions. The optimisation problem includes sizing and controlled cycling of the battery bank, as well as sizing of power conditioning devices, the photovoltaic generator, and the motor generator.

Energy management requires the inclusion of planning as part of the control concept. This exceeds the ability of conventional strategies, which respond to changing operating conditions by controlling the power flow in the system based on fixed parameter settings. For the predictive control strategy presented in this publication strategic control decisions are based on the prediction of the daily photovoltaic resource and the load demand. The objective of the predictive energy management system can be described as follows:

Energy management is concerned with the optimum operation of multiple power sources and energy storage devices in hybrid energy systems, including planning in the strategic decision-making process. Additionally, appropriate measures for the user-interactive or automated management of the electrical load demand may be implemented.

Including forecasts of future operating conditions provides the additional information that is required to include 'time' as an additional degree of freedom in the decision-making process. Control actions are no longer restricted to instantaneous decisions but can include planning, therefore optimising control decisions. In the following section, theoretical performance criteria for optimum operation of PV-diesel hybrid energy systems are presented, constituting the formal basis for the predictive control strategy.

2.1 Optimum Control and Operation

For a given photovoltaic resource and load demand, as well as operational constraints of the motor generator and the battery, optimal *energetic performance* can be defined as operation of the system resulting in:

(a) Minimal fuel consumption of the motor generator;
(b) Minimal cycling of the battery bank;

Other important aspects of system operation, which contribute to the objective of optimal overall system performance, can result in sub-optimal energetic performance. Mainly the operation of the battery bank is of concern, since it requires a well-defined charge strategy to minimise battery ageing. Conceptually, optimum battery management is considered to be an operational constraint, as it conflicts with the solution for optimal energetic performance.

The time-series information for the photovoltaic power and the load demand has to be combined when defining a formal optimum for the operation of PV-diesel hybrid energy systems. The net load power (on the DC side) can be calculated as:

$$P_{NET} = \frac{P_{LOAD}}{\eta_{INV}} - P_{PV} \tag{1}$$

The net load can be described as the electrical power that has to be supplied from or is charged into the battery, assuming that the motor generator is not operational.

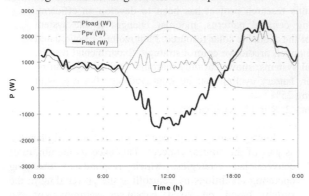

Figure 1: Typical profile of the daily net load demand

The inverter efficiency has a significant influence on the overall performance of the system, as inverter losses account for a substantial fraction of the total energy supplied by the system. However, to introduce this novel concept conversion efficiencies of the power conditioning devices and the battery bank are not considered explicitly in equations (2) to (4).

Essentially, the task is to schedule the operation of the motor generator so that optimal energetic performance is achieved. Formally, the objective of optimal supply side management can be expressed as:

$$\int_{t1}^{t2} |P_{NET} - P_{MG}| dt = \int_{t1}^{t2} |P_{BB}| dt \rightarrow \text{minimum} \tag{2}$$

The net energy flow to and from the battery bank can be minimised by using a controllable energy source, the motor generator. To ensure reliable system operation with a low loss-of-load probability the energy input and output of the system have to be balanced. Operation of the motor generator has to be scheduled to supply the difference between the load demand and the energy supplied by the photovoltaic array:

$$E_{MG} \approx E_{NET} = E_{LOAD} - E_{PV} = \int_{t1}^{t2} (P_{LOAD} - P_{PV}) dt \tag{3}$$

Operational constraints to be considered are the fuel efficiency characteristics of the motor generator and the limited storage capacity of the battery. Figure 2 shows how the motor generator is operated. Essentially, the condition for optimum energetic performance requires that the motor generator be scheduled to operate during the period of highest net load demand, as expressed with equation (2). If the motor generator is operated at nominal power, thereby maximising its fuel efficiency, the operating time of the motor generator can be calculated as:

$$t_{RUN} = \frac{E_{MG}}{P_{MGnom}} \tag{4}$$

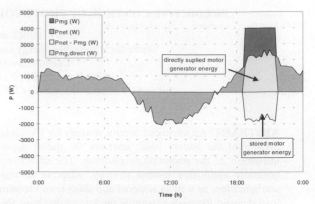

Figure 2: Example of motor generator operation at nominal power during period of highest net load demand

3. PREDICTIVE CONTROL STRATEGY

The following control decisions are addressed by the predictive control strategy on a daily basis:

(1) Whether or not the motor generator must operate to supply additional energy?
(2) For how long the motor generator must operate to supply sufficient additional energy?
(3) When the motor generator must operate to achieve optimum system performance?

At the end of each day (midnight 12am) the predictive control algorithm schedules the operating period of the motor generator for the next 24 hours. Control decisions (1) and (2) are based on the prediction of the daily photovoltaic resource and load demand, whereas decision (3) depends on forecasting the 3-hourly distribution of the net load demand. For the prediction of the photovoltaic resource and the electrical load demand, forecasting algorithms have been developed, which were described in detail in [2]. The daily net load demand is determined as follows, considering also conversion losses due to energy storage, as well as DC-AC conversion of the photovoltaic energy.

$$E_{NLf24h}(d+1) = E_{Lf24h}(d+1) - \bar{\mu}_{EPV} \times E_{PVf24h}(d+1) \tag{5}$$

The average conversion efficiency of energy supplied to the load from the photovoltaic array is calculated as:

$$\bar{\mu}_{EPV} = \frac{E_{PVdirect}}{E_{PV}} \times \bar{\mu}_{INV} + \frac{E_{PVstored}}{E_{PV}} \times \bar{\mu}_{BB} \times \bar{\mu}_{INV} \tag{6}$$

Considering energy losses due to conversion and storage of energy supplied from the motor generator, the required daily motor generator energy can be predicted as:

$$E_{MGf24h}(d+1) = \frac{E_{NLf24h}(d+1)}{\bar{\mu}_{EMG}} \tag{7}$$

The average conversion efficiency of energy supplied to the load from the motor generator is calculated as:

$$\bar{\mu}_{EMG} = \frac{E_{MGdirect}}{E_{MG}} \times 1 + \frac{E_{MGstored}}{E_{MG}} \times \bar{\mu}_{REC} \times \bar{\mu}_{BB} \times \bar{\mu}_{INV} \tag{8}$$

Assuming that the motor generator is operated at nominal power to maximise its fuel efficiency the required operating time over the next 24 hours can be predicted as:

$$t_{RUN}(d+1) = \frac{E_{MGf24h}(d+1)}{P_{MGnom}} \tag{9}$$

Based on forecasting the 3-hourly distribution of the photovoltaic resource and the load demand, the motor generator is scheduled to operate during the period of highest net load demand. To determine the optimum operating period for the next 24 hours, the control algorithm predicts the 3-hourly distribution of the net load demand for the next day (index i represents the eight 3-hourly periods $\{i=1..8\}$ for which predictions are made).

$$E_{NLf3h}(d+1,i:8) = E_{Lf3h}(d+1,i:8) - \overline{\mu}_{EPV} \times E_{PVf3h}(d+1,i:8) \quad (10)$$

Figure 3 shows a typical example of the 3-hourly net load distribution predicted on a daily basis. A search algorithm determines the optimum time period for continuous operation of the motor generator. The scheduled start time of the motor generator is selected as the beginning of the time period with the highest average net load demand.

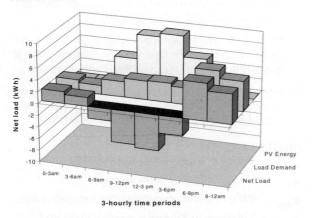

Figure 3:   Example of forecast for the 3-hourly distribution of the daily net load demand

4. EXPERIENCES AND RESULTS

In the following section we present experiences with the operation of a PV-diesel hybrid energy system using the predictive control strategy. The results are compared against the operation of a system using a conventional control strategy. To evaluate the feasibility of advanced control strategies a test facility was installed [3], which has been operated since February 1999 using the predictive control algorithm. Apart from analysing the performance of a small hybrid energy system under realistic operating conditions, the experimental data has been used to verify detailed simulation models using Matlab/Simulink.

Figure 4 shows the system performance over three days of normal operation using the predictive control strategy, compared against the performance using a conventional control strategy. For a system that is operated using the predictive control algorithm a larger fraction of motor generator energy is supplied directly to the load. Additionally, this strategy avoids operation of the motor generator during periods when the photovoltaic resource is sufficient to supply a significant fraction of the load. The motor generator is operated daily, typically for several hours in the evening. As a result of the fundamental control principles of the predictive control strategy, typical battery cycles are smaller and of shorter duration. This avoids the gradual discharge to a low state-of-charge (SOC) over several days, which can be observed for the system controlled by a conventional strategy. Additionally, the battery is not recharged regularly to a very high SOC, since the motor generator is turned off as soon as the required energy has been supplied. This results in extended cycling at intermediate SOC.

Laboratory testing reported by Newnham and Baldsing has shown that partial SOC operation can extend the cycle-life of gelled-electrolyte lead-acid batteries by a factor of three [4]. Due to their immobilised electrolyte, gel-type batteries do not experience significant electrolyte stratification. A regular recharge to a high SOC is therefore not required for this battery technology. This avoids charging at high voltage, effectively minimising battery ageing from increased material utilisation at the bottom of the plates, as well as corrosion.

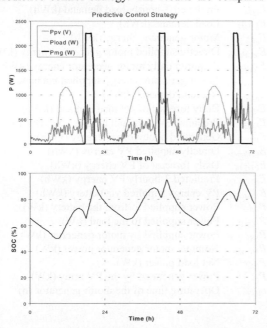

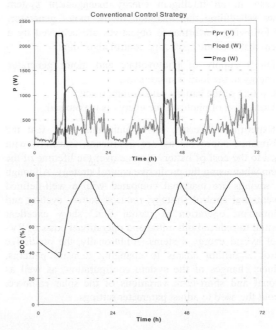

Figure 4: Power flow and SOC for the predictive control algorithm and a conventional strategy

Figure 5 compares the annual fuel consumption and energy supplied to the battery bank for different storage sizes. The potential for reduced fuel consumption is marginal for all simulations performed, typically ranging from 1% to 3%. However, reduced cycling of the battery bank is significant, with improvements ranging from 6% to 15% for a system operating with the predictive control algorithm. Combined with improved cycling at partial SOC, as discussed previously, the reduced reliance on energy storage can be expected to reduce the cost for battery storage by at least 25% over a 20-year lifetime.

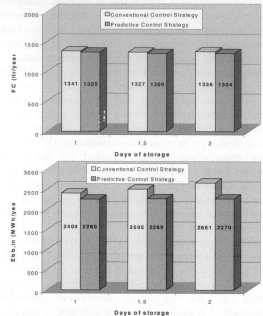

Figure 5: Fuel consumption and battery cycling for the predictive and the conventional strategy

5. CONCLUSION

The integration of photovoltaic resource and load forecasts in an intelligent energy management system allows scheduling the operation of the motor generator. The following performance objectives are achieved by a system that uses the predictive control algorithm:

(a) Direct supply of photovoltaic and motor generator energy to the load is maximised;
(b) Cycling of the battery bank is minimised;
(c) Utilisation of photovoltaic energy is maximised;

Experimental results and simulations show that the performance improvements are significant, especially with regard to the cost of battery storage over the lifetime of the system when using the predictive control strategy. Although fuel savings are marginal compared with a well-defined conventional control strategy, reduced battery cycling and predominant operation at partial SOC show excellent potential to reduce both initial and operational costs of PV-diesel hybrid energy systems. Additionally, the predictive control algorithm adapts to changing demand patterns, modular changes of the system configuration, as well as seasonal and short-term variations of the solar resource, without the need to adjust parameter settings.

ACKNOWLEDGEMENTS

The work described in this paper has been supported by the Australian Cooperative Research Centre for Renewable Energy (ACRE). ACRE's activities are funded by the Commonwealth's Cooperative Research Centres Program. Mr B Wichert has been supported by an ACRE Postgraduate Research Scholarship. The first author would also like to acknowledge CRESTA for providing the research facilities. The research work by Mr B Wichert has been supported by an Alternative Energy Development Board postgraduate scholarship, which is gratefully acknowledged.

REFERENCES

1. Wichert, B. and W.B. Lawrance. *Intelligent control of modular hybrid energy systems.* 14th EPVSEC. 1997. Barcelona, Spain.

2. Wichert, B. and W. Lawrance. *Photovoltaic resource and load demand forecasting in standalone renewable energy systems.* 15th EPVSEC. 1998. Vienna.

3. Wichert, B., et al. *Development of a test facility for photovoltaic-diesel hybrid energy systems.* World Renewable Energy Congress. 1999. Perth, Australia.

4. Newnham, R and Baldsing, W., *New operational strategies for gelled-electrolyte lead-acid batteries.* Journal of Power Sources. Vol.59, p.137-141. 1996.

NOMENCLATURE

$\bar{\mu}_{BB}$	Energy efficiency of the battery bank
$\bar{\mu}_{INV}$	Energy efficiency of the inverter
$\bar{\mu}_{EMG}$	Effic. of motor generator energy supplied to load
$\bar{\mu}_{EPV}$	Efficiency of PV energy supplied to the load
$\bar{\mu}_{REC}$	Energy efficiency of the rectifier
η_{BB}	Instantaneous battery efficiency
η_{INV}	Instantaneous inverter efficiency
η_{REC}	Instantaneous rectifier efficiency
d	Day of the year
$E_{Lf\,24h}$	Predicted daily load demand (kWh)
E_{Lf3h}	Predicted 3-hourly load demand (kWh)
E_{LOAD}	Load demand (kWh)
E_{MG}	Motor generator energy (kWh)
$E_{MGdirect}$	Directly supplied motor generator energy (kWh)
E_{MGf24h}	Required daily motor generator energy (kWh)
$E_{MGstored}$	Motor generator energy supplied via battery (kWh)
E_{NET}	Net load demand (kWh)
E_{NLf24h}	Daily forecast of net load demand (kWh)
E_{NLf3h}	Predicted 3-hourly net load demand (kWh)
E_{PV}	Generated photovoltaic energy (kWh)
$E_{PVdirect}$	Directly supplied PV energy (kWh)
E_{PVf24h}	Daily forecast of PV energy (kWh)
E_{PVf3h}	Predicted 3-hourly PV energy (kWh)
$E_{PVstored}$	PV energy supplied via battery (kWh)
P_{BB}	Power supplied to/from the battery (kW)
P_{LOAD}	Power supplied to the load (kW)
P_{MG}	Power supplied by motor generator (kW)
P_{MGnom}	Nominal motor generator power (kW)
P_{NET}	Net load power (kW)
P_{PV}	Power generated by the PV array (kW)
t_{RUN}	Operating time of the motor generator (h)

FULL AREA COVERAGE MEASURING NETWORK FOR RECORDING
THE DISTRIBUTION OF INCIDENT SOLAR RADIATION

J. Bendfeld, M. Gruffke, E. Ortjohann, T. Peters, J. Voß
Universität Paderborn, Elektrische Energieversorgung
Pohlweg 55 - Gebäude N, 33098 Paderborn (Germany)
Phone: 49 - 5251 - 60-2304, Fax: 49 - 5251 - 60-3235
email: gruffke@eevmic.uni-paderborn.de

ABSTRACT: Acquisition of incident solar radiation data is becoming increasingly important for electric energy supply projects and other technical applications. For this reason a new solar radiation sensor (Solar-Igel) has been developed at the University of Paderborn. With the help of 135 sectional sensors positioned radially on a hemisphere, the two-dimensional incident radiation distribution pattern is sampled twice each second. In the region around Paderborn a network of eight Solar-Igels has been set up over an area of 150 km^2 in order to extend the description of the incident radiation distribution at the site of a Solar-Igel for making temporal and spatial forecasts for the entire network. This paper first of all describes the utilised measuring technology. An important goal of the project is to make the data collected in the network available to interested users. Therefore the infra-structure is introduced here too, with which the data and results will be published on the Internet.
Keywords: Irradiation Measurement - 1: Pyranometer - 2: Software - 3

1. INTRODUCTION

Knowledge of the actual solar irradiation situation is very important in meteorology as well as for numerous technical application fields. Information which was formerly of subordinate importance is required today for active utilisation of available solar energy in optimised converter systems or in building control systems.

Sensors so far available on the market are based on measuring principles which are unable to make the required measurements with adequate accuracy or acceptable ratio of price to performance. Therefore the new measuring system called Solar-Igel has been developed within the scope of a research project supported by the Solar Research Group of the North Rhine Westphalia State (Arbeitsgemeinschaft Solar NRW). This novel sensor permits comparatively accurate description of the actual solar irradiation distribution in space at the operating site of the sensor.

In the region of Paderborn a measuring network has been set up consisting of eight Solar-Igels distributed over an area of 150 km^2, in order to extend the description of the distribution of incident radiation at the site of a Solar-Igel for temporal and spatial forecasting for the entire network. This paper first of all describes the utilised measuring technology. An important goal of the project is to make the data obtained in the network available to interested users. Therefore the infra-structure is here presented too, with which the data and results will be published on the Internet. Initial results are also presented here.

2. THE MEASURING SYSTEM SOLAR-IGEL

Instruments at present available on the market for making solar irradiation measurements are almost exclusively individual measuring devices for one of the three characteristic parameters of incident solar radiation (global, diffuse or direct radiation). Apart from an often delicate and therefore maintenance-intensive tracking

mechanism (for diffuse and direct radiation measurement), they have the disadvantage that they integrate the radiation distribution. Conversion of the irradiation readings by calculation to obtain the distribution of the radiation over the hemisphere of the sky is impossible or possible in only a very rudimentary form [1].

The Solar-Igel measuring system (Fig. 1) differs fundamentally from the conventional instruments. The sensor, based on the concept of [2], operates with separate response to parts of the celestial hemisphere. This solid angle resolution is achieved with 135 tubular optical systems each having an aperture angle of 7.1° (Fig. 1).

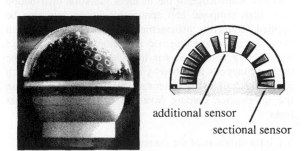

additional sensor

sectional sensor

Figure 1: Static solar radiation sensor Solar-Igel (photograph and cross-section).

The internal surfaces of the tubular optical systems are designed to suppress reflections. This ensures sharp delimitation of the individual angular regions of the sky. The 135 tubular optical systems (sectional sensors) are positioned radially oriented over the surface of a hemisphere in a configuration which minimises overlapping and missed regions of the sky.

The rotationally symmetrical construction of the tubular optical systems and their radial orientation give circular coverage areas on the virtual celestial hemisphere over the measuring location (Fig. 2).

An additional sensor is mounted on the zenith to avoid ambiguities which could otherwise arise when determining the global, diffuse and direct radiation parameters from the 135 individual sensor signals, due to overlapping and missed regions of the sky. The additional sensor's transducer is oriented parallel to the horizontal plane and has an angular response range of 180°.

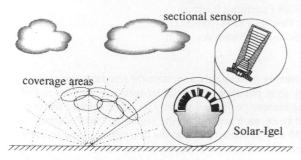

Figure 2: Coverage areas of the sectional sensors on a virtual celestial hemisphere.

3   EXTENSION OF THE MEASURING SYSTEM TO CONSTITUTE A NETWORK

Operation of a Solar-Igel provides high resolution temporal data describing the incident solar radiation. This measuring system is very suitable for supplementing conventional solar radiation measuring systems [3]. Furthermore, the system is suitable for extension to distributed measuring sites. The incident radiation data obtained can be processed centrally for mapping area coverage distribution of incident solar radiation. One conceivable application of this is the evaluation of a region for large area utilisation of solar energy converter systems, to predict the power and energy contributions of the converter systems. Knowledge of the incident radiation distribution with high temporal and spatial resolution can also be applied for facility management. For example, therewith systems can be implemented for automatic adaptation of the lighting conditions for interior rooms [4]. For the control functions it is thereby not necessary to equip each individual building with sensors. Instead it suffices to use the information coming from the network via the data lines.

3.1 Infra-structure of the measuring network

The sensor presented previously constitutes the basis for the measuring network. An individual station of the measuring network thereby consists of a Solar-Igel, an interface unit including radio controlled clock and a personal computer (PC) which controls the entire measuring operations and records the measured data delivered by the sensor (data logger function). The incident radiation data determined in the measuring cycles at half second intervals are transmitted to the data logger in serial data communication mode as a data telegram sent via the interface unit. The data logger processes the measured data and stores them for documentation. Evaluation can take place synchronously while measurements are being made (on-line) or later using archive measured data (off-line).

The technical signal link between the Solar-Igel and the data logger is established by the interface unit. It contains all connectors for data and control lines between the sensor and the PC as well as the power supply circuits

for the sensor. In addition thereto, a radio controlled clock module (DCF77) is integrated within the interface unit, as time reference source for the sensor and the data logger. This ensures that the incident radiation data picked up by two or more measuring systems located in different places can be mutually correlated correctly in time.

A network consisting of eight such decentralised individual measuring systems distributed over an area of 150 km^2 has been set up. The topology thereby established is shown in Fig. 3.

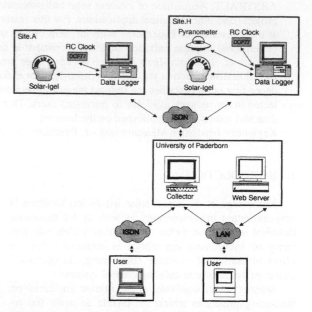

Figure 3: Topology of the measuring network taking two stations as example.

The overall system has the following features:

- Each decentralised station autonomously controls the daily measurements from sunrise to sunset and places the collected data (raw data = incident radiation distribution) on the data logger.
- Control and maintenance of the measuring routines take place centralised via the collector (Fig. 3).
- Parallel to the measuring routine, each station on request makes the current measurement results data available for central processing. For purposes of analysis and documentation, the data collected in the course of a day are sent out to a central processing system. This takes place at night outside of measuring hours.

3.2 Selection of the sites

The ISDN system which is now widely available is suitable as data transmission medium, because the individual sites are not specifically known in advance. Apart from ensuring that an ISDN connection was available, it was also necessary to make sure that a shadow-free location exists at each site. It should be a building with a flat roof, because this facilitates erection of the sensor.

To fulfil these requirements as well as possible, six schools participating in the project SonneOnline [5] were chosen as sites for the measuring network (Site.B to Site.G). These schools have identically constructed 1 kW$_p$ photovoltaic plants and an Internet connection via ISDN.

Two further stations were installed (Site.H = central station in the university and Site.A). Fig. 4 shows the locations of the selected sites.

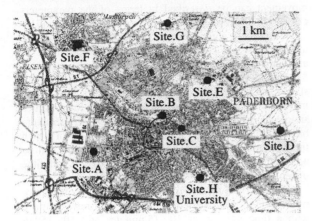

Figure 4: Location of the sensors in the measuring network.

3.3 Evaluation of the measured data and publication

The central processing facility automatically evaluates the measured data from the individual measuring systems. Thereby the data of the previous day are evaluated, which were transmitted to the central computer in the course of the night.

In the first stage of evaluation the data are checked to verify the availability of the individual sites (status display). In the second stage the standard radiation parameters global, diffuse and direct radiation are calculated. Thereby the calculated magnitudes of the measured parameters are compared with the readings of a reference pyranometer (CM21 from Kipp&Zonen). The final stage of evaluation is the calculation of meteorological parameters for characterising the day.

The results obtained in this way are automatically processed for publication as graphic display on a specially set-up world wide web (WWW) page [6]. Persons interested in using the data can download them free from there, already on the next day.

The following illustrations (Fig. 5 to Fig. 7) show some examples of the results of the evaluation described above.

Daylight duration in hours:	15:36:59
Sunshine duration in hours:	11:51:01
Relative sunshine duration in %:	75.33
Longest sunshine interval in hours:	01:00:01
Maximum global radiation in W/m^2:	927.92
Total energy in Ws/m^2:	203043.53

Figure 5: Characteristic meteorological parameters.

The status display (Fig. 6) provides various items of information which are respectively assigned to the individual bits of a status byte. At present the first 3 bits are interpreted as follows:

Bit 1: Measurement reading available for this site.
Bit 2: Sunrise / sunset.
Bit 3: Sunshine (direct radiation > 150 W/m^2).

This form of depiction gives a quick overview of the course of the general run of the measurements in the course of a day. For example, Fig. 6 shows a brief interruption of the measurements shortly after 10:00 a.m.

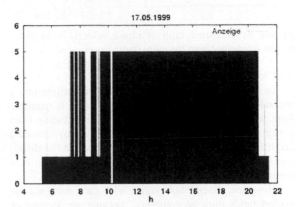

Figure 6: Status display.

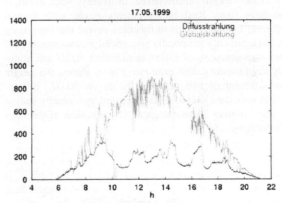

Figure 7: Global and diffuse radiation.

4 PRELIMINARY DISCUSSION OF FORECASTING INCIDENT RADIATION

Spatial and temporal fluctuations of solar radiation distribution are chiefly produced by cloud movement. Therefore it is possible to calculate the cloud movement parameters (cloud drift direction, cloud velocity, cloud height) from the measured radiation distribution. The measuring procedures have been described in [7].

4.1 Cloud drift direction and ratio of the cloud velocity and the cloud height

Measurement of the cloud drift direction and of the ratio of the cloud velocity v and the cloud height h are based on the time-offsets produced by a wandering radiation distribution between the temporal signals of two adjacent sectional sensors. Real measurements (Fig. 8) show that this procedure gives good results when there is only one cloud level in the sky. In this case the procedure also provides an assessment criterion n indicating whether the calculated parameters are reliable.

However, when cloud levels with different ratios of cloud velocity to cloud height are present, the measured parameters cannot be associated unambiguously with these cloud levels.

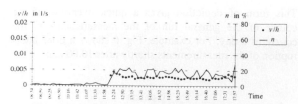

Figure 8: Measured ratio of cloud velocity v to cloud height h together with the assessment criterion n.

4.2 Cloud height

In order to be able to measure the cloud height too, a second Solar-Igel in a spatially different place is required. Both sensors thereby calculate the average radiation from a suitably positioned radiation field in many assumed heights h^*. The cloud height $h^* = h$ for which the difference of the two average radiation measurements is a minimum, is taken to be the actual cloud height.

However, several Solar-Igels with their assigned radiation fields must be combined, because on account of geometric boundary conditions it is not possible to cover all cloud heights encountered in reality with a single radiation field. Real measurements made with several different radiation fields in this case reveal the interesting fact that not only temporally sequenced measurements, but also measurements of different radiation fields are mutually confirmatory. For example, Fig. 9 shows the height measurement of two radiation fields on 30.07.1999 together with the cloud heights estimated by a nearby station of the German Meteorological Organisation (Deutscher Wetterdienst DWD).

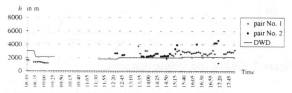

Figure 9: Measured cloud heights compared with the cloud height estimates of the weather station.

4.3 Algorithms for forecasting incident radiation

Apart from these parameters of cloud movement which are necessary input variables for forecasting incident solar radiation, further methods are being developed at present for separating the measured incident radiation distribution into a temporally and spatially slowly changing component (background model = diffuse background radiation), and a spatially and temporally quickly changing component (foreground model = clouds). Thereafter it is intended to shift the foreground model with the cloud movement parameters and then combine it again with the background model at the forecast location and time instant.

5. SUMMARY

The newly developed static solar radiation sensor Solar-Igel provides high resolution irradiation data in time and space. It responds to the incident solar radiation with 135 sector sensors mounted with radial orientation over the surface of a supporting hemisphere. From the output

signals of these 135 individual sensors, it is possible to calculate the basic solar parameters of global, diffuse and direct radiation [3]. Furthermore, using the new algorithms, the modulation of the irradiation caused by cloud movement can be described in terms of the parameters cloud drift direction, cloud velocity and cloud height [7].

Using these algorithms, it is intended in the next step to compile a full coverage irradiation description for an area of 150 km^2. For this purpose a measuring network of 8 individual stations (Solar-Igel and PC) has been installed and commissioned in the region of Paderborn. An important aspect of this project is to give third parties simple access to the data. For this reason an infra-structure has been developed and installed, with the help of which the data and results from the measuring network can be accessed during the measurements (on-line) and with time offset (off-line) [6].

On the basis of these data and the presented algorithms, the incident radiation forecast shall now be made with a time horizon of 10 - 15 minutes and 10 - 15 km against the cloud movement direction.

REFERENCES

[1]   Perez, R.; Seals, R.; Ineichen, P.; Stewart, R.; Menicucci, D.:
A New Simplified Version of the Perez Diffus Irradiance Model for Tilted Surfaces;
Solar Energy, Vol. 39, No. 3, pp. 221-231, 1987.

[2]   Appelbaum, J.:
A Solar Radiation Distribution Sensor;
Solar Energy, Vol. 39, No. 1, pp. 1-10, 1987.

[3]   Gruffke, M.; Heisterkamp, N.; Matzak, S.; Ortjohann, E.; Voß, J.:
Solar Irradiation Measurement With Silicon Sensors;
Proceedings 2nd World Conference and Exhibition on Photovoltaic Solar Energy Conversion, Vienna, Austria, 1998, pp. 78-81.

[4]   Bendfeld, J; Gruffke, M.; Ortjohann, E.; Voß, J.:
Der Solar-Igel: Neue Möglichkeiten der Steuerung solartechnischer Anlagen;
Bundesbaublatt,49. Jahrgang, 3/2000.

[5]   Skorka, I.; Kotzerke, C.; Bohlen, M.; Hoffmann, V.U.; Kiefer, K. :
SONNEOnLine? The Photovoltaic School Program
Proceedings 2nd World Conference and Exhibition on Photovoltaic Solar Energy Conversion, Vienna, Austria, 1998, pp. 2567-2570.

[6]   http://solarigel.uni-paderborn.de

[7]   Gruffke, M.; Heisterkamp, N.; Ortjohann, E.; Voß, J.:
Analysis of Cloud Movement by Reference to Solar Irradiation Data;
Proceedings 2nd World Conference and Exhibition on Photovoltaic Solar Energy Conversion, Vienna, Austria, 1998, pp. 2594-2597.

An Analysis of the Design of PV Systems in the Built Environment in The Netherlands in 1997-2000

J.W.H. Betcke, V.A.P. van Dijk,
Department of Science, Technology & Society, Utrecht University
Padualaan 14, NL-3584CH Utrecht, The Netherlands
Phone: +31 302537600, fax: +31 302537601
E-mail: J.W.H.Betcke@chem.uu.nl

ABSTRACT: The Dutch PV research programme NOZ-PV 1997-2000 states goals for the technical improvement of grid-connected PV systems. In the programme time span (1) performance ratios should increase from 65-70% to 77-81% (2) inverter efficiency should increase from 85 to 92% (3) the efficiency for tracking, ohmic and diode losses should increase from 85-90% to 93-95%. To determine these indicators by monitoring of systems in the field is costly and does not make a fair comparison possible. Here we present a protocol to analyse system designs by means of system simulation. The protocol gives the optimum performance of a system type (so without shading, soiling or system breakdowns). We apply the protocol to eight system designs, with various modules, inverter types, and mounting configurations, that were realised in the Netherlands between 1994 and 1999. The calculated performance ratios of these systems are in the range 74 to 84 %. We conclude that the NOZ-PV objectives can be achieved.
Keywords: PV System-1: Modelling -2: Evaluation -3

1. INTRODUCTION

Improvements in components and system design should lead to PV systems with a higher annual energy output. It is the intention of the Dutch national PV research programme NOZ-PV 1997-2000 [1] to stimulate such a development. In this programme, it is stated that for grid-connected PV systems on buildings in the Netherlands:
1) the performance ratio should increase from 65-70% in 1997 to 77-81% by 2000;
2) the inverter efficiency should increase from 85% in 1997 to 92% by 2000;
3) the combined efficiency for tracking losses, ohmic losses and diode losses should increase from 85%-90% in 1997 to 93-95% by 2000.

The performance of a PV system type can be evaluated by monitoring the performance of an installed system and then analysing the monitored data. However, this is a costly and time consuming method. Moreover, the measured performance may not the highest attainable with this system type. This is the case when all or part of the modules are not optimally oriented, or because of accidental conditions such as shading of the PV array by an obstacle, soiling of the array, an installation error or system breakdown. As an alternative, the performance of system types can be evaluated by simulating the performance of the system design with a simulation model and then analysing the simulated data. These results give the upper limit of the performance of the PV system type under realistic conditions. If it is required that results of evaluations of different system types can be compared, the procedure for the evaluation must be set up according to a fixed protocol.

In this paper we develop such a protocol and apply it to eight grid-connected PV systems in the Netherlands built in the period 1994-2000. The systems represent the development in components and design concepts, and include a system with a central inverter, decentral systems, AC module systems and a string inverter system. We analyse the results to study the progress towards the goals of the NOZ-PV.

We will present the protocol in section two. In section three we will apply the protocol to the eight PV systems and present the performance results. Section four will discuss the accuracy and limitations of the protocol and the observed trends in the system performance. Conclusions will be presented in section five.

2. THE PROTOCOL

2.1 Introduction

The protocol for the evaluation of the technical performance of grid-connected PV system in the built environment consists of a set of indicators to describe the technical performance of the PV systems, the method to calculate these indicators, the input data for the calculation and the selection criteria for the PV systems.

2.2 Indicators

The energy losses that occur in a PV system determine the performance of a PV system. Table I gives an overview of all energy losses in a PV system based on [2]. Usually, inverter manufacturers combine the static tracking losses and the inverter losses: we follow this convention. The performance ratio [3] is a measure for the overall performance of the system. We specify the energy losses as a percentage of nominal yield.

The NOZ-PV performance indicators are directly calculated (performance ratio) or derived from the energy losses.

2.3 The method of calculation

We use the simulation model SOMES 3.2 [4] to simulate the performance of the PV system. SOMES is an energy model with a time step of one hour. Component performance is calculated from efficiency curves, which give the output energy as a function of the input energy.

SOMES 3.2 contains the Sjerps-Koomen model [5] to calculate reflection losses and the Fuentes model [6] to calculate cell temperature. Low irradiation losses and the combined static tracking and inverter losses are calculated from respectively irradiation-efficiency curves and power-efficiency curves. DC cable losses, diode losses, mismatch losses and dynamic tracking losses are modelled as constant factors that reduce the module output energy (see Table I for the value of the factor).

2.4 Input data for the simulation

We use the Test Reference Year of de Bilt as meteorological input data (see Table II). We assume that the array consists of one type of modules that are oriented towards the South at tilt 45°. The following input data must be specified per system: the nominal power of the PV array, module characteristics (irradiation-efficiency curve, temperature coefficient), type of mounting (installed nominal operating conditions temperature); for the inverter the power-efficiency curve and the stand-by power consumption. When available we use data measured by independent institutes; else we use manufacturers' data.

2.5 Selection of systems

The PV systems are selected on the criterion of being state of the art at the time of realisation.

2.6 Acceptance of protocol

To make the results of the method widely accepted, the protocol has been defined in consultation with system suppliers in The Netherlands.

3. APPLICATION OF THE PROTOCOL

3.1 Introduction

Following the protocol, we analyse the performance of eight state-of-the-art grid-connected PV systems on buildings in The Netherlands, built in the period 1994-2000. We compare the performance of these systems to the targets mentioned in the NOZ-PV 1997-2000.

3.2 Description of systems

The main characteristics of the systems are described in Table III. The systems contain four different module types of three manufacturers. There are systems with a central inverter (Nieuw Sloten), a decentral inverter (Zandvoort, REMU-SCW, Thomasson Dura, Woudhuis), a string inverter (Eneco 2) and AC-module inverters (Arthur Andersen and Eneco 1). Modules are mounted in five different ways on the roofs. The modules that are roof-integrated with ventilation have different distances to the roof, which is taken into account in the simulation. We neglect the impact of channels on ventilation. The actual Nieuw Sloten system is made up of Shell Solar and BP modules: we simulate a system that has the same total peak power but is made up of Shell modules only.

3.3 Input data for the simulation

The irradiation-efficiency curves for the module were derived from manufacturers' data: from a graph on the data sheet (BP), from a fit with the two-diode model [7] to various IV curves (Siemens), and from one-diode model parameters (Shell). The temperature coefficient of the module efficiency and the nominal operating cell

Table I: Overview of energy losses in a grid-connected PV system and the method of calculation.

Indicator	Method of calculation
Irradiation losses	
Tilt conversion	Perez model
Reflection	Sjerps-Koomen model
Spectrum	Constant factor (99%)
Shading	*Not applicable*
Soiling	*Not applicable*
System losses	
Low irradiation	Irradiation-efficiency curve
Temperature	Fuentes model
DC cable losses	Constant factor (string inverters, AC modules: 100%; else: 98%
Diode losses	Constant factor (100% * (1-$\Delta V_{diode}/V_{string}$))
Mismatch losses	Constant factor (AC-modules: 100%; else: 98%)
Dynamic tracking	Constant factor (99%)
Static tracking and Inverter losses	Power-efficiency curve

Table II: Array characteristics in the protocol.

Tilt angle	Orientation	Location	Height
45°	South	De Bilt (52.1° N, 5.2° E)	6 m

temperature are manufacturers' data. For the NKF and Mastervolt inverters, we use power-efficiency curves measured by the Netherlands Energy Research Foundation ECN and manufacturers' data for the other inverters.

3.4 Results: energy losses

Table IV shows the calculated energy losses for the eight systems as a percentage of the nominal PV yield.

Reflection losses (3.2%) and spectrum losses (1.0%) are equal for all systems because all arrays are mounted at the same angle and orientation. The low irradiation losses are 6-7% for the Siemens and the Shell modules and 4.5% for the BP module. The temperature losses are in the range from 0-3.5%: the range is determined by mounting method and module type. DC cable losses are 0 for the AC modules and string inverter systems and 1.7% for the other system types. Diode losses are maximally 1%. Dynamic tracking losses are equal for all systems (0.9%). The static tracking and inverter losses are in the range 6.6-11.4%, with the best performing inverter types in descending order decentral inverter, string inverter, AC module inverter, and central inverter.

The total irradiation and system losses are in the range 19.7%-29.5%. There is a tendency that in the various system types, low values for certain losses are compensated by higher values for other losses. For example, compared to a decentral system, an AC module system has lower (zero) cable, diode and mismatch losses but higher static tracking and inverter losses. As a result, the total energy losses for the best performing central, decentral and AC module inverter system are fairly close: 26.2%, 24.0% and 23.7%. The string inverter system has the lowest total

Table III: Description of the analysed grid-connected system designs.

System name	Zero energy house	REMU /SCW	Nieuw Sloten	Arthur Andersen	Thomasson Dura	Eneco 1	Eneco 2	Woudhuis
City	Zandvoort	Amersfoort	Amsterdam	Amsterdam	Amersfoort	Rijswijk	Rijswijk	Apeldoorn
Year	1994	1995	1996	1996	1997	1998	1998	1999
Number of identical subsystems	2	8	1	272	19	5	1	70
Modules (strings x modules)	8x4 Shell IRS50 LA	15x18 Shell IRS50	248x18 Shell IRS50	1 Shell IRS95	10x3 Shell IRS95	1 Siemens SM 110F	10 BP585	3x8 Shell IRS95
Mounting con-figuration	roof integration	ventilated roof integration	Ventilated roof integration	stand-off	Ventilated roof integration	console	rack mount	roof integration
Inverter	Mastervolt Sunmaster 1800 (decentral)	2 x SMA 5000 master-slave (decentral)	SMA PV-WR-T 150 (central)	Mastervolt MV 130S (AC module)	ASP TCG III 2500 (decentral)	NKF OKE4 (AC module)	SMA Sunnyboy 850 (string)	Mastervolt Sunmaster 2500 (decentral)

Table IV: Calculated energy losses as percentage of nominal yield in the analysed system designs (%).

Energy Losses	Zandvoort	REMU-SCW	Nieuw Sloten	Arthur Andersen	Thomasson Dura	Eneco 1	Eneco 2	Woudhuis
Irradiation Losses								
Reflection	3.2	3.2	3.2	3.2	3.2	3.2	3.2	3.2
Spectrum	1.0	1.0	1.0	1.0	1.0	1.0	1.0	1.0
System losses								
Low irradiation	5.8	5.8	5.8	7.2	7.2	7.1	4.5	7.2
Temperature	3.5	1.6	1.9	0.3	1.7	2.4	0.0	3.5
Mismatch	1.7	1.7	1.7	0.0	1.7	0.0	1.8	1.7
DC cable	1.7	1.7	1.7	0.0	1.7	0.0	0.0	1.7
Diode	0.8	0.2	0.2	0.0	0.0	0.0	0.0	0.0
Dynamic tracking	0.9	0.9	0.9	0.9	0.9	0.9	0.9	0.9
Static tracking and Inverter	8.1	10.5	9.9	11.4	6.6	9.2	8.3	10.4
Sum of irradiation and system losses	26.7	26.5	26.2	24.0	24.0	23.7	19.7	29.5

Table V: Calculated values of the indicators of the NOZ-PV programme for the analysed system designs (%).

	Zandvoort	REMU-SCW	Nieuw Sloten	Arthur Andersen	Thomas-son Dura	Eneco 1	Eneco 2	Woudhuis
Performance ratio	77	77	77	79	79	80	84	74
Efficiency as a result of dynamic tracking, DC cable and diode losses	96	97	97	99	97	99	99	97
Static tracking and inverter efficiency	90	88	89	87	92	89	91	87

irradiation and temperature losses in the high-efficient rack-mounted modules are low.

3.5 Results: indicators of the NOZ-PV

Table V shows the calculated values of the performance indicators of the NOZ-PV programme. The objectives for the performance ratio by 2000 are met by all systems except Woudhuis. The Thomasson Dura system meets the objectives for inverter efficiency. The objectives for the efficiency as a result of tracking, DC cable and diode losses are met by all systems.

4   DISCUSSION

4.1 The protocol

The accuracy of the protocol depends on the accuracy of the simulation model and the accuracy of the input data. Based on previous experience with SOMES [8], we estimate the accuracy of SOMES 3.2 for the calculation of the performance ratio at 3% (assuming accurate input data). The accuracy will be higher for the various energy losses. In addition, the accuracy depends on the component data. It is hard to obtain reliable data on module performance, especially at low irradiation. We estimate the error in the annual module output at 3%. This would give a total accuracy of 4% for the performance ratio.

Comparison of results obtained for different systems by means of the protocol must be done cautiously:

1) The results of the simulation are representative for a system type on the condition that the component data are reliable and accurate. The monitoring report of the Nieuw Sloten system shows that the inverter performs better than specified [9]. As a result, the values that we calculated for the Nieuw Sloten system are too pessimistic.

2) The array mounting configuration has a large impact on the performance ratio: we observed a range of 0-3.5%. In practice the type of roof (horizontal/sloped) on which the array is mounted limits the number of possible mounting configurations. So, the roof type should be taken into account when comparing performance values of different system designs.

3) The system design is not representative for a system type when the size of the PV array is constrained by the available roof area: in that case, the system performance can be sub-optimal. This can more easily occur for a central or decentral system than for a string or AC module system.

4.2 Trends in system performance

For the eight analysed system designs, the performance ratio gradually increases from 77% to 84% from 1994 to 1998. However, the most recent system (Woudhuis) has a low performance ratio of only 74%, because of high static tracking and inverter losses. The Sunmaster 2500 has a higher peak efficiency than the predecessor Sunmaster 1800, which is reached at a higher percentage of the nominal input power. This design feature pays off only if the system is properly designed (the PV array must be oversized with respect to the inverter)

For each system type, for which more than one system design was analysed, the trend is towards a higher performance. For decentral inverter systems this is achieved by better mounting systems, fewer diodes, and more efficient inverters. For AC module systems this is achieved by better matching of module and inverter power.

5   CONCLUSIONS

We developed a protocol for evaluating the performance of a PV system by simulation of the system design. In the protocol, the array has a fixed tilt and orientation; it is assumed that no shading, soiling or system breakdowns occur. As a result, the protocol gives the optimum performance of a system type. The protocol can be used for comparison of the performance of different system types, provided that limitations of the method are taken into account (the input data must be reliable and accurate, the impact of the mounting configuration, constraints on the system design). Furthermore, the protocol does not take into account all design features of a system type (such as the better ability of AC module or string systems to deal with shading).

We applied the protocol to eight PV systems in the Netherlands that were installed between 1994 and 1999. In general, these results show that the system designs, which are built from existing components, can achieve the goals of the NOZ-PV programme. However, the sizing of the components is a critical factor for a well-performing system.

The protocol only investigates the technical performance of a system design. For an actual PV system, it must also be determined whether a system design is cost-effective.

ACKNOWLEDGEMENTS

This work was supported by Novem, the Netherlands Agency for Energy and the Environment, under contract no. 146.220-028.1. We would like to thank ASP, BP Solar, Mastervolt, NKF, Shell Solar, Siemens and SMA for supplying data. We would like to thank ECN, Ecofys, Ekomation, Nuon and REMU for supplying system details.

REFERENCES

[1] Report RC 4156, Novem, Utrecht (1997).
[2] A.M.H.E. Reinders, V.A.P. van Dijk, E. Wiemken, W.C. Turkenburg, Progress in photovoltaics: Research and Applications **7** (1999) 71.
[3] C.W.A. Baltus, E.A. Alsema, A.H.M.E. Reinders, J. Boumans, R.J.C. van Zolingen, B.C. Middelman, report RC 4154, Novem, Utrecht (1997).
[4] V.A.P. van Dijk and E.A. Alsema, report NWS92010, Department of Science Technology and Society, Utrecht University (1992).
[5] E.A. Sjerps-Koomen, E.A. Alsema, W.C. Turkenburg, Solar energy **57** (1996) p. 421.
[6] M.K. Fuentes, report SAND85-0330 • UC-63, Sandia National Laboratories (1987).
[7] J.A. Eikelboom, A.H.M.E. Reinders, Proceedings 14th European Photovoltaic Solar Energy Conference, H.S. Stephens & Associates (1997) p. 293.
[8] V.A.P. Van Dijk, Thesis, Utrecht University (1996).
[9] A.J. Kil, M. van Schalkwijk, H. Marsman, T.C.J. van der Weiden, report E2004, Ecofys (1999).

UNIVER Project.
A 200 kWp Photovoltaic Generator Integrated at Jaén University Campus.
First Experience and Operational Results.

P.J. Pérez; J. Aguilera; G. Almonacid; P.G. Vidal.
Grupo Jaén de Técnica Aplicada (Dpto. De Electrónica). Universidad de Jaén.
Avda. de Madrid, 35. 23071 Jaén.Tel: +34.953.212430. Fax: +34.953.212400.
e-mail: pjperez@ujaen.es

ABSTRACT: The Univer Project consists in the installation of a grid-connected photovoltaic system, with a total power of 200 kWp, at Jaén University Campus. This project is carrying out under the Thermie Programme (SE/00383/95/ES/UK).

The main objective is the integration of a medium scale PV plant using different architectural solutions. This project presents two innovative aspects: on the one hand, on development of the necessary technology that is able to implement medium–high scale PV plants in crowded places, mainly focused on safety and protection systems; on the other one, to develop and analyse different architectural solutions to integrate PV generators using constructive structures easily replicable.

The participating institutions are: Universidad de Jaén (coordinator), Isofotón S.A., Solar Jiennense S.L., Instituto de Energía Solar (UPM), Newcastle Photovoltaic Aplications Centre and Compañía Sevillana de Electricidad.

Keywords: Grid-Connected - 1: Safety - 2: Building Integration - 3

1. INTRODUCTION.

The Univer Project is developed under the Thermie Programme (SE/00383/95/ES/UK) of the E.U., with a budget of about 1.8 M. of euros. This project consists in the installation of a grid-connected photovoltaic system, with a total power of 200 kWp, at Jaén University Campus. The participating institutions are:
- Grupo Jaén de Técnica Aplicada (University of Jaén).
- Instituto de Energía Solar (UPM).
- Newcastle Photovoltaic Aplications Centre (UK)
- Solar Jiennense, S.L..
- Isofoton, S.A.
- Compañía Sevillana de Electricidad (CSE).

The total energy generated is estimated about 280 MWh/year. The annual electric consumption at University Campus is about 1,400 Mwh/year, which means that the PV Plant is able to cover approximately between the 15 and 20% of the total consumption. The pv generation coincides in peak hours to the electric demand at University, so most of this energy will be self-consumed. With these considerations, it is estimated an economic saving of around 30,000 euros/year[1]. Currently, the total power installed is about 70 kWp (PV subgenerator 1).

2. TECHNICAL DETAILS OF THE PV SYSTEM.

The Project consists in the installation of four photovoltaic subgenerators connected to the low voltage grid at Jaén University Campus. The installation is fully integrated in the existing buildings at University. This project presents two innovative aspects, on the one hand, to develop the necessary technology able to implement medium–high scale PV plants in crowded places, mainly focused on safety and protection systems. On the other one, to develop and analyse different architectural solutions to integrate PV generators using constructive structures easily replicable.

Apart from all the elements necessary to ensure the people and installation protection, the system includes a data acquisition equipment that will allow us to evaluate the system performance. In this sense, we pretend to use Internet to show the working data of the PV Plant in real time (telemonitoring) and to offer information about the project (http.//solar.ujaen.es).

Photo 1: PV generator (70 kWp) located in the Parking at University Campus.

2.1. Photovoltaic System 1.

It is integrated in one of the parking covers at University Campus. It consists in a photovoltaic generator with 70 kWp nominal power and a 60 kW inverter. The photovoltaic generator consists of 640 modules of model ISOFOTON I-106, 80 modules series connected each one, and 8 parallel arrays.

The inverter is made by Enertrón, a Spanish company.

The quality control of the PV modules has been carried out by the Instituto de Energía Solar (IES) at Madrid Polytechnic University. The result shows how the mean power measured by the IES varies about a 1,1% from the power measured by the manufacturer and within a 5% of nominal power. Thus, the power of the photovoltaic modules is within the range specified in the contract with the manufacturer.

For the integration of the photovoltaic generator, we use the existing parking covers at this University, which are almost totally free of shadows and with a 30° southeast orientation and tilted 7,5°.

2.2. Photovoltaic System 2.

It is the same as Photovoltaic System 1, and it is located in a cover parallel in the same parking.

Civil work (wires, grounds, etc.) for the installation of the two PV subgenerators located in the parking of the University, have been carried out. In this sense, the ground electrical characteristics of both PV subgenerators have been measured and the obtained value of earth resistance is lower than the expected, due to the good quality of the parking ground. This fact improves the safety aims of the system.

2.3. Photovoltaic System 3.

This PV generator is integrated in the Connection and Control Building of the project. In this building, the inverters, the data acquisition system and the safety and protection system are located.

The PV system consists in a photovoltaic generator with a 20 kWp of nominal power, and 20 kW string oriented inverters. As the two previous systems, the installation includes all the necessary elements to ensure people protection.

Photo 2: Connection and Control Building

One of the aims of this integration is getting a shady area, very useful either for this part of Spain, or for students, so that they can enjoy a nice place.

2.4. Photovoltaic System 4.

This PV generator is integrated in the south façade of the building located close to the Connection and Control Building. It consists in 40 kWp PV polycristalline modules and 40 kW string oriented inverter.

With this integration, we aim two objectives; on the one hand, to analyse the performance system to evaluate the potential of façades as an integration element in the south of Spain, and on the other one, to get a great visual impact in all the visitors coming at University.

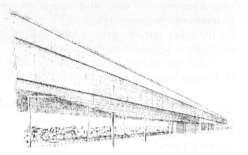

Photo 3: Picture of PV subgenerators 3 and 4

3.   ASPECTS RELATED TO SAFETY AND PROTECTION IN THE INSTALLATION.

With the aim of keeping people protection, the installation includes passive and active measures that avoid direct contacts with the active system parts (earth grids and an permanent insulation controller to detect the earth faults of the generators) [2,3].

This is one of the most outstanding aspects of the Project, and it has also been the most studied because of the high number of students at this Campus. The studies carried out about the installation safety and protection have been developed from two points of view: on the one hand, from the installation itself, and on the other one, from people safety. In this sense, it is important to point out the lack of a legal regulation related to such aspects in this type of installations in Spain.

In general, the risks that can affect an electric installation are due to overvoltages and overcurrents, although in our particular case, and because of the phovoltaic system working, we will only be affected by overvoltages as a consequence of the eventual presence of atmospheric discharges, induced inductions, etc. In this sense, the installation includes voltage limiters that reduce it to a value under the insulation level required to the equipment. These limiters are placed at the inverter input and output, at the DC junction general cupboard, and at the junction boxes of the different generators arrays.

As for people protection, the installation includes the necessary elements to avoid possible direct and indirect contacts. In this sense, there are three levels of protection: floating system, insulation control and earth connections.[1]

Photo 2: Inside view of the Connection and Control Building

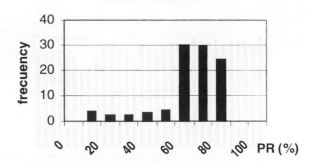

Figure 1: Daily performance ratio histogram

4. OPERATING EXPERIENCE.

The subgenerator 1 (70 kWp) was starting up on August of 1999. Energy generated by the system has been 30.315 kwh until March of 2000

In following table is shown the monthly values of the Performance Ratio. Also it is offered a histogram of the daily values of the Performance Ratio (Figure 1).

Table I: Energy Generated, Global Radiation and Performance Ratio.

	Energy Generated (kWh)	Gd (α,β) (kWh/m^2)	PR (%)
1999			
AUG	8.373	7,01	57
SEP	4.177	5,28	39
OCT	0	3,47	0
NOV	2.175	2,89	37
DEC	2.014	1,94	50
2000			
JAN	1.599	2,43	28
FEB	2.919	3,28	45
MAR	5.628	4,47	60

During the first months, the protection system of the utility company disconnected the PV plant many times (two or three per day). These disconnections were due to the poor fitting of the protection system values and made some inverter capacitors get broken. When these values were well fitted, the disconnections disappeared. Except for these disconnections, the PV plant was working without any problem worth a mention.

In September, we noted a serious problem with a computers room located in a building close to the Connection and Control Building of the PV plant. When the inverter was working, there were many problems with the computers. We analized the AC voltage and current and we tested the existence of a current harmonic of 5 kHz in the grid. The interferences in the computers were generated by this current harmonic. In October, Enertron company designed a capacitors filter, which eliminated this current harmonic and the grid interferences.

Another different problem appeared in the PV plant in winter time. The people safety system of the Univer PV plant is based on the use of an Permanent Insulation Control (PIC) device that is able to detect the loss of floating system. When this fail appears, the PV field is shortcircuited and connected to the ground. In some cold and wet mornings, the PIC device has detected a high decrease of the insulation resistance (about 5 kΩ) of the PV system, when the usual value of this resistance varies between 40 kΩ and 700 kΩ., depending on the environmental conditions. This problem is due to drops of condensation in connection boxes of the modules. At present, we are developing a functioning method of the control system to avoid these disconnections. Figure 2 shows the electrical parameters monitored of the PV installation in a typical day in Autumn. It is possible to see the disconnections of the inverter in the morning due to

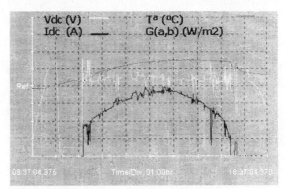

Figure 2: Electrical parameters of PV subgenerator 1 Disconnections of the inverter in the morning due to drops of condensation.

drops of condensation.

Figures 3 and 4 show the daily irradiation and energy produced by the PV plant in August. The total energy produced during this month is 8.373 kWh and the Performance Ratio is about 57%.

Daily Irradiation (kwh/m2)

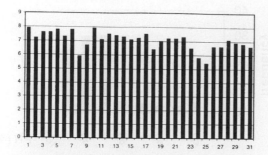

Figure 3: Daily Global Radiation on August

Daily Energy Generated (kwh)

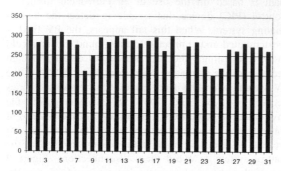

Figure 4: Energy Generated on August

5. FUTURE ACTIONS.

Next months, it must be achieved the followings future actions:

- Measurement of the PV generator real power and analysis of its working.
- Installation of a 70 kWp PV system on the cover of parking 2.
- Installation of a 20 kWp PV system at the pergola of the Control and Connection Building.
- Instalation of 40 kWp PV system integrated at the building located close to the Control and Connection Building

6. REFERENCES.

(1) G. Almonacid et al. "Univer project. A 200 kWp Photovoltaic Generator at Jaén University Campus". Proc. 14th EC PV Solar Energy Conference. Barcelona, 1997.

(2) S.A.M. Verhoeven et al. "Lightning induced voltages measurements in a Solar Array to improve the grounding structure". Proc. 13th EC PV Solar Energy Conference. Nice, 1995.

(3) H. Häberlin. "A simple method for lightning protection of PV systems". Proc. 12th EC PV Solar Energy Conference. Amsterdam, 1994.

NOVEL MONITORING CONCEPTS FOR STAND-ALONE PV SYSTEMS
INCLUDING OBJECTIVE AND SUBJECTIVE PARAMETERS

Jens Merten, Xavier Vallvé
Trama Tecnoambiental S.L.
Ripollés, 46 - 08026 Barcelona - Spain
Tel. 34 93 450 40 91 - Fax 34 93 456 69 48
tta@retemail.es

Philippe Malbranche, Pascal Boulanger
GENEC - Bat 351 CEA Cadarache
13108 St Paul lez Durance Cedex , France
Tel. 33 4 42 25 66 02 – Fax 33 4 42 25 73 65
Philippe.malbbranche@cea.fr

ABSTRACT: The objectives for monitoring of stand alone PV systems move away from system and component evaluation towards a *quality control of service* provided to the system user, and the detection of *correct interaction* of the user with the system. In this article, a novel monitoring concept is developed by analysing data from the existing monitoring scheme.

The user opinion concerning the global evaluation of the system is strongly correlated with the satisfaction of the energy needs. Users with energy needs not satisfied show a trend of not allowing the batteris to be fully charged. Users with adequate load management succeed to drawing more energy from their system without vitiating the batteries.

For the novel monitoring scheme, it is proposed to monitor the quality control of service by metering the energy provided to the user, and the time of low battery state of charge, which is found to be a critical parameter for the detection of system failures. The correct interaction can be assessed by the historical battery index, which indicates the level of load management performed by the user.

Keywords: Monitoring –1; Stand alone PV-systems –2;Quality of service –3

1. INTRODUCTION

1.1 Objectives for monitoring

The performance of stand alone PV-systems is well known from extensive monitoring efforts during the past years. Main objective was the evaluation of performance of the system, as well of the different system components employed.

Today the performance of components and systems is well known, and we are facing new objectives for the monitoring of stand alone PV systems. These are a control of the *quality of service* provided to the user, and the detection of *correct interaction* of the user with the system.

Another point of interest to be monitored is the satisfaction of the user. Analytical monitoring of system data in short time intervalls produces costs for data collection and treatment. Together with the new objectives for monitoring, novel monitoring concepts with reduced amounts of data should be developed.

For this purpose, we examine in Section 2 the subjective opinion of PV system users in Spain. In

Section 3, we examine the analytical monitoring data of TTA installations in Spain in order to extract parameters of interest for a novel monitoring scheme, which may be employed for smallest to medium scale PV systems.

2. THE USERS' OPINION

The opinion of about system users in the Catalonia region has been assessed by questionnaires [1]. In Figure 5, we show histograms of four aspects evaluated by 60 users. The *system reliability* and the *easiness of the system use* received good marks.

What can be observed, however, is a less satisfying trend in the marks concerning the *satisfaction of energy needs*. This reflects the difficulties of adequate consumption assessment in the design phase and of controlling the appliances that the user will connect. Furthermore, the final decision on system size is made on economical basis. The

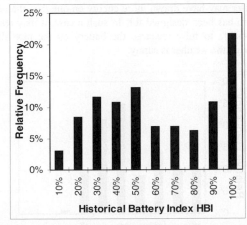

Figure 2: Histogram of historical battery index data occurred in 1999. The wide scatter of the data indicates the variety of user behaviours with respect to the battery. Reading example: for 14% of the monthly monitoring data, the HBI was between 40% and 50%.

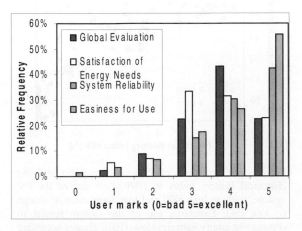

Figure 1: Marks given by the users in questionnaires.

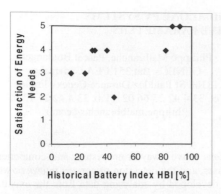

Figure 3: Correlating the historical battery index and the user satisfaction of energy needs. User with less satisfied energy needs tend not to allowing a full charge of the battery.

result is a tendency of the user to draw more energy from the system, than it has been designed for. In such a case, energy is systematically drawn until the user becomes disconnected due to low battery.

It is remarkable to note that the global satisfaction is strongly correlated with the satisfaction of energy needs, and not with the *system reliability* or *easiness of use*. This shows, that the satisfaction of energy is the main concern of the user, provided that the system is working properly and is easy to handle.

3. OBJECTIVE MONITORING

All data presented here come from installations with the *TApS-centralita* [2], which is an integrated power conditioning unit also containing a data acquisition system, which measures all relevant energy flows, battery voltage and other parameters in hourly intervals. We now analyse the data presenting monthly averages of characteristic parameters.

3.1 Battery Operation Data

A critical issue is to monitor whether the energy consumption is within the design limits of the system. As the autonomy of the batteries has been designed for four days, the user may have drawn more energy on one day, than the system has been designed for. In such a case, it may result impossible to fully recharge the battery on the next day, although the weather is sunny.

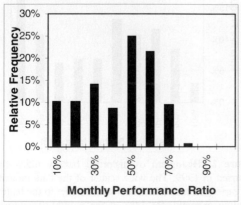

Figure 4: Histogram of Performance Ratio data occurred in 1999.

In order to better monitor this problem, a new indicator has been introduced, the *historical battery index HBI* [3]. *HBI* is the percentage of days within a month, on which the battery has become fully charged. High *HBI* values stand for batteries of systems, were an appropriate load management of the users allowed frequent charge. Low *HBI* indicate chronically empty batteries which will tend to sulphatation, it is furthermore an indirect indicator, that the user energy demands are not completely fulfilled by the PV-system. When *HBI* is too low, a red LED of the *TApS power conditioner* informs the user that the battery needs a complete charge.

Figure 6 shows the histogram of the *historical battery index HBI* for the installations examined, which is widely distributed. Only 19% of the monitoring data come from system months with a *HBI* exceeding 90%. Such data mostly come from periods where the user is not on site. The broad majority of the data tends to low *HBI* values. This shows that many users have a higher energy demand than the system has been designed for. This fact can be also underlined by the Figure 3, where it is shown that users less satisfied with the energy supply of their system allow less full battery charge.

3.2 Performance Ratio Data

The performance ratio is defined as the relation of the energy delivered by the system and its maximum theoretical achievable value:

$$PR = \frac{Energy/G}{P_{nom}/H_{STC}}, \tag{1}$$

where *Energy* is the energy produced by the system in the period examined (in kWh/month), G the measured irradiance during the period (in kWh/m^2/month), $H_{STC} = 1000$W/m^2 is the standard irradiance to measure the nominal power P_{nom} of the modules (in W).

Figure 5 shows the histogram of performance ratio data. There is a broad peak between 50% and 60%, which is a good typical value for stand-alone PV systems. High PR is related to strong energy demands, which entails in most cases a low state of charge of the battery, and thereby a good absorption of energy. On the other side, there is a broad spectrum of low performance ratios. In

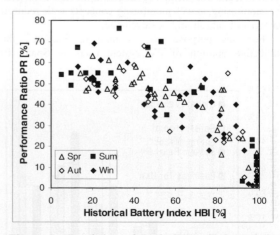

Figure 5: Correlation of Performance Ratio and the Historical Battery Index in 1999. Low use of the PV system, leads to high HBI and low acceptance of charge (= low PR). Excessive use of the system results in chronically empty batteries (low HBI), always accepting charge, which increases PR.

this case, not all the energy generated by the PV system could be used, as the battery has become fully charged. Such a reduced performance, however, is not caused by deficient operation of the system.

These observations are undelined by the strong correlation between the performance of a system and the historical battery index (Figure 8).

A relatively high historical battery index means that the battery is often fully charged. However, a battery with a relatively high state of charge does not absorb all the solar energy provided by the modules, which results in a low performance ratio. A relatively low historical battery index means that the battery is rarely fully charged. This is a bad factor for battery life, but increases the acceptance admit of the solar charge available, so the performance ratio will be the maximum value possible.

Note, that there is no seasonal influence in the data shown in Figure 8. The scatter in the correlation is caused by the different load management on site: user adapting their energy needs to the solar insulation, tend to use *more* energy without using the batteries. The result is a higher performance ratio and a high *Historical Battery Index HBI*. This observation underlines the need of user training and the provision of automated load management tools.

3.3 Energy consumption on site

The operator of the stand alone PV systems, SEBA, guarantees the user a fixed amount of energy, called *EDA* [4]. *EDA* is calculated from irradiance in the worst month on site and the PV-power installed. We present here the E*nergy Consumption Parameter ECP*, which is the ratio of the energy consumption divided by the guarantied energy *EDA*. The histogram in Figure 7 shows a broad spectrum of *ECP*. User may consume more than *EDA* in the summer months and achieve ECP up to 180%. Such high values are only achievable by appropriate load management in sunny periods. Lower values come from absence periods of the user.

High energy consumption reduces the possibilities for the batteries to obtain a full charge and therefore correlate with a low *historical battery index HBI* (Figure 8). The data points beyond this correlation, marked with a circle, come from situations, where on many days the energy consumption exceeds the energy generation. This is a sign

for a inappropriate load management (Figure 9), and indicates the need of further user training in these installations.

4. LESSONS LEARNED FOR THE OPERATION OF THE SYSTEMS

Better load management leads to increased performance ratio and allow the user to draw *more* energy form the system. In appropiate load management is reflected by a low historical battery index HBI and reduces the lifetime of the battery. Therefore, the increased involvement of the user in the oepration of his PV system enahces ystem performance and user satisfaction. However, the user should be equiped with adequate tools for this purpose. TTA provides remote visualisation panels indicating the complete system state in an easy understandable way, and remote controlled relais allowing load connection under certain system states only. In any case, the user has to be trained to understand the operation of the PV system and to understand the the tools provided.

5. A SET OF MINIMAL MONITORING PARAMETERS

For the monitoring of stand-alone PV hybrid system in the framework of large-scale rural electrification programmes, we propose a monitoring scheme, which contains electrical, electrochemical and subjective parameters.

5.1 Electrical data

We propose a reduced set of two variables for continuous monitoring in monthly intervals, which are cheap enough also for smallest scale systems. In order to check the *quality of service*, we propose to monitor the *Energy Consumption Parameter ECP*. This data does not require irradiance measurements on site. To verify the *correct interaction* of the system, we propose to monitor the *Historical Battery Index HBI*.

Analytical monitoring including other parameters and

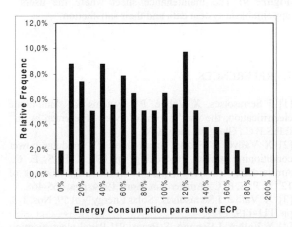

Figure 6: Histogram of the Energy Consumption Parameter. which is the ratio of energy consumption and the energy guaranteed by contract (EDA). High values come from sunny months, low values from low energy demand on site or lack of load management.

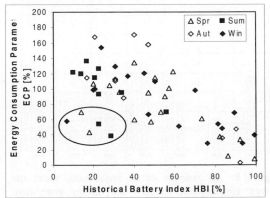

Figure 7: Correlation of the Energy Consumption Parameter with the Battery Historical Index. The Energy Consumption Parameter is the ratio of energy consumption and the energy guaranteed by contract (EDA). Points in the lower left corner (circle) come from installations with many days where the consumption exceeds the energy generation.

higher sampling frequencies may be further employed for the detection of hardware defects or for detailed studies of the system behaviour.

5.2 User data

We propose to combine the monitoring of the electrical data with a regular monitoring of data, which only can be provided by the user. For this purpose, a control card has been designed (Figure 9), which is filled by the user each month and sent to the system operator, four times a year. The aim of this card is to:
- Increase the user consciousness about the functioning of the system.
- Stimulate him to perform basic maintenance tasks
- Obtain electrochemical battery data

These are the data acquired by this card:
- The battery electrolyte: acid density and the water consumption. These data cannot be measured electronically yet.
- Battery operation data: operation of the battery, namely the *Historical Battery Index* indicator, the battery voltage and the *availability index* (related to the state of charge of the batteries). Although these data are measured by the monitoring system, they are requested here to increase the user sensibility about the weakest part of the PV system.
- Basic maintenance tasks: The user executes these tasks. It is checked, whether he has cleaned the PV-modules and the battery terminals. Also the state of the fire extinguisher is checked.
- Subjective user data: Important to know is whether the user plans to extend the energy consumption on site. The related mark serves as an alarm signal, which is then used to inform the user, that he may draw too much energy from his system. Finally, the subjective opinion of the user about the system operation is sought for.

6. CONCLUSIONS

Analysing the operation of PV systems in Spain, it has been found that the user opinion concerning the global evaluation of the system is strongly correlated with the satisfaction of the energy needs. Users with energy needs not satisfied show a trend of reduced *historical battery*

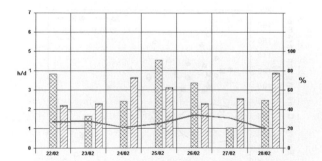

Figure 8: An example of an installation, where the energy consumption exceeds the energy generation. Shown is the generation (middle bars), and consumption (right bars). This graph is a standard output of the GIFA software developed at TTA for the management of stand alone PV systems. The data shown are normalised with the peak power of the PV modules installed (i.e. they are yields). They are expressed in units of equivalent sunshine hours (hours with 1000W/m2) per day.

index HBI, i.e not allowing the batteris to be fully charged. Users with adequate load management succeed in drawing more energy from their system without vitiating the batteries. The time of low battery state of charge is a critical parameter for the detection of system failures.

A novel monitoring scheme has been developed, which has the objectives to monitor the quality control of service provided to the system user, and the detection of correct interaction of the user with the system. The first point is covered by metering the energy provided to the user, and the time of low battery state of charge. The second point can be assessed by the *Historical Battery Index HBI*, which is a measure of the level of load management performed by the user. Both points are cross checked by regulary asking the subjective system evaluation of the user.

Figure 9: The maintenance sheet where the users appoint basic system data and their satisfaction.

7. REFERENCES

[1] J Serrasolses, X Vallvé, PV systems for stand-alone electrification: the user's point of view. Presented at 16th E.P.S.E.C., Glasgow, may 2000.

[2] X Vallvé, E Feixas, Stand-alone PV modular power conditioning station. Proceedings of 12th E. P. S. E. C., Amsterdam, April 1994, pgs. 1651-1654. Proceedings of 12th E. P. S. E. C., Amsterdam, April 1994, pgs. 465-468.

[3] X. Vallvé, J.Serrasolses, Solar Energy Vol 59, Nos. 1-3, pp. 111-119, 1997.

[4] X Vallvé, J Serrano, S Breto, PV Rural electrification program in Aragón (Spain): example of an implementation scheme involving users and local authorities. Presented at 16th E.P.S.E.C., Glasgow, may 2000.

PV-COMPARE: PRELIMINARY DATA FROM A DIRECT COMPARISON OF A REPRESENTATIVE RANGE OF PV TECHNOLOGIES AT TWO LOCATIONS IN NORTHERN AND SOUTHERN EUROPE

G.J. Conibeer[1] and D. Wren[2]

[1] Environmental Change Institute, Oxford University, 11 Bevington Road, OXFORD, OX2 6NB, UK
Tel: +44 1865 284690, Fax: +44 1865 284691, email: gavin.conibeer@ecu.ox.ac.uk
[2] The Solar Century, 1-9 Sandycombe Road, RICHMOND TW9 2EP, UK
Tel: +44 870 7358100, Fax: +44 870 7358101, email: dw@solarcentury.co.uk

ABSTRACT: PV-COMPARE aims to assess the comparative performance of a representative range of photovoltaic technologies. Eleven technologies are directly compared using eleven sub-arrays. There are two parallel project sites, with the same eleven sub-arrays, at Oxford and on a Spanish Mediterranean island, thus also comparing very different climatic conditions. Preliminary data are presented for the Spanish facility. Full annual data will not be available for another 6 months in Spain and another year in Oxford. However, when complete, they will provide valuable independent comparative data to the industry and to the user, that will enable informed choice of photovoltaic technology for a given application.
Key words: Evaluation - 1: PV Materials - 2: Grid-Connected - 3.

1. INTRODUCTION

The primary objective of PV-COMPARE is to directly compare a range of photovoltaic technologies at the same location. To this end each array consists of eleven sub-arrays of various technologies. The secondary objective is to compare these outputs at two very different locations. Hence, the parallel projects, in Spain and Oxford, consist of PV arrays that are very similar with the same peak power generation capacity.

Thus data are being collected that compare all of the technologies at the same location. This will give valuable information on diurnal and seasonal variation with respect to solar insolation and temperature; and on how performance is maintained over time. Their relative performance in low light conditions and at temperature extremes will be particularly interesting. In addition the data set from each location can be compared to the other at differing latitude and climatic conditions.

Preliminary data are presented here from the first 6 months operation of the Spanish array. The Oxford array has only recently been installed and so no data are available.

2. THE ARRAY OF ELEVEN TECHNOLOGIES

The total power of each array is 6.2 kWp, this being made up from the eleven sub-arrays each of about 550W peak. The support structure arrangements and electrical configuration have been previously reported [1], but the range of technologies employed is repeated here.

The eleven sub-arrays consist of:
Intersolar single junction a-Si, 36 Phoenix modules;
ASE double junction a-Si, 18 DG 30 modules;
Solarex double junction a-Si, 12 Millennia modules;
Unisolar triple junction a-Si, 8 US 64 modules;
Evergreen ribbon Si, 5 AC modules;
Siemens CIS, 14 ST 40 modules;
BP Solar CdTe, 10 50W Apollo laminates in Spain,
 14 40W Apollo laminates in Oxford;

ASE edge fed growth Si, 2 DG300 laminates;
Astropwer mc-Si on ceramic, 8 APX 80 laminates;
Solaex mc-Si, 10 MSX laminates;
BP Solar c-Si, 7 BP585 laminates.

With the slight exception of the CdTe laminates, the same range of technologies and connection arrangements are used in both Oxford and Spain.

3. AC INVERSION

Each sub-array is connected to inverters and feeds into the mains.

The voltage ranges from V_{mppt} @ 70°C to V_{OC} @ -10°C (the recommended temperature range for system sizing) were found to be large for all technologies. In the case of the a-Si technologies and the Siemens CIS this range was too large for the input ranges of the string inverters. For these sub-arrays module inverters were employed because this allowed more flexibility in the voltage inputs to the inverters because of greater choice in module connection arrangement. Even so the extreme voltages in the sub-arrays' range are not quite within the inverters' window, but as it is very unlikely for the sub-arrays to operate at maximum open circuit voltage when the temperature is as low as -10°C, or for temperatures to reach 70°C, this was a reasonable compromise.

Other considerations are also involved: the maximum string voltages and currents allowable in each case put further constraints on the configuration. Furthermore, power matching of each sub-array to its inverters such that the peak power is just above the maximum power of the inverters has had to be a very secondary consideration.

This difficulty in voltage and power matching highlights one of the findings, that the lack of standardisation between modules, whether in terms of power, voltage or size makes inverter matching a very difficult undertaking.

4. ARRAY SITES

The array in Oxford was installed in March '00. It is located on a building in the Materials Dept, Oxford University, just north of Oxford. It is a rooftop site with a pitched roof at 13° and orientation 5° W of S. The power generation capacity will be 6.2 kWp and in combination will generate about 6,000 kWh DC of electricity per year.

The other array is on a private estate on a Spanish Mediterranean island. Again it is a rooftop site, but at a pitch of 25° and orientation 15° W of S. This difference in pitch angle between Oxford and Spain is the biggest difference between the arrays. It is not expected to cause undue problems in comparative analysis. It is the same size as the Oxford array, 6.2kWp, but will provide about 10,500 kWh DC of energy per year, thus reflecting the higher levels of sunshine. This is about 6% of the electrical energy requirement of the estate on which it is established. Installation was completed in October '99.

(In both cases the AC energy generated will be less than these DC estimates because of non-unity inverter efficiencies.)

The interesting results will include how the comparison between the sites fares at various times of the year; for instance in the summer Oxford is likely to be much closer to parity with Spain because of the long summer days. Of particular interest will be the comparison of the technologies; for instance comparable performance in low light conditions and at different temperatures. Furthermore, in several cases novel module/inverter combinations are being tried. This will form an additional part of the experiment as well as contributing to the two buildings' electricity requirements.

AC performance is monitored for each sub-array. DC data are more problematic and have been calculated from AC in the data presented here, for the most part. Subsequent data will include a greater component of measured DC although still not for the module inverters.

Insolation and module and ambient temperature are also monitored. In Oxford this includes a four different spectral measurements, that will be used to assess spectral variation.

5. FIRST SIX MONTHS RESULTS FROM SPAIN

These are preliminary data, as there is not yet a full annual set for Spain, and no data yet for Oxford.

Figure 1 shows the performance of all eleven sub-arrays in terms of the DC energy generated daily, normalised to peak power for the sub-array, averaged over each month. (Equivalent to the peak sunshine hours per day, averaged for the month, *for each technology*.)

Two measurement anomalies in this data need explanation. Firstly the installation was commissioned towards the end of September '99. Hence the data for this month is averaged over only a few days and so will be affected more by extremes. The Evergreen and Solarex a-Si sub-arrays were in fact not commissioned until October and so result in zero values for output in September. Secondly, there was a data acquisition problem in February and March for the Astropower sub-

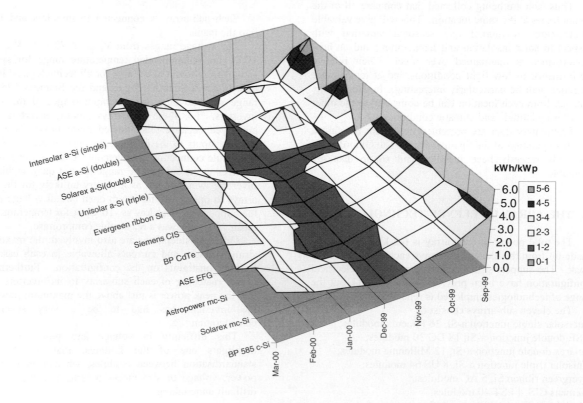

Figure 1: DC kWh / kWp per day for all the technologies averaged over each month.

array that resulted in very few days' data being collected, hence these two monthly average data points are also sensitive to extreme values.

Figure 2 shows the daily peak insolation as measured by a silicon reference cell in the plane of the array and the corresponding peak ambient temperature as measured by a shielded thermocouple.

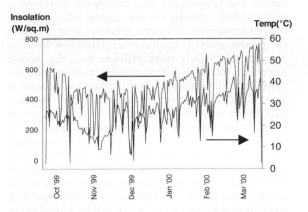

Figure 2: Daily peak insolation and temperature on site in Spain.

As would be expected the data show minima for both insolation and temperature in December. But they also show a relatively mild December overall particularly in comparison to the insolation, and also a sustained cool period in November. The spring months appear to have been sunnier than the corresponding autumn months, with sustained sunny periods at the end of January and the beginning and end of February. However, these are peak

value data and so do not reflect incident energy totals for each day, although the daily variation can be taken as partially indicative of the levels of direct sunlight. In subsequent publication, data on the daily variations of insolation and on module rather than ambient temperature will be presented.

The insolation data also do not reflect spectral variation. For the Oxford installation filtered sensors are being used that will give spectral information, with cut-offs relevant to the absorption edges of Si, CdTe and a-Si. These will give a better indication of the effectiveness of each sub-array for the part of the spectrum to which it is sensitive, in addition to its performance in the spectrum as a whole.

Figure 3 is a similar plot to Figure 1 but using inverted AC data on the monthly average daily peak sunshine hours.

AC performance is not a primary consideration of the project, but this is presented because the DC data is in the case of most data points, derived from the AC data, because of problems in measuring DC directly. A method has been established of calculating the DC values from the inverter efficiency with respect to power. The results from this have been checked against some directly measured DC values and this indicates a high degree of reliability for the method. Significant variation between the measured and calculated values was only found below about 5% inverter input power. At these values the amount of energy generated is a very small proportion of the total and the deviation in calculated values is in the direction that acts to negate the calculated values rather than that to give these low power values undue weight.

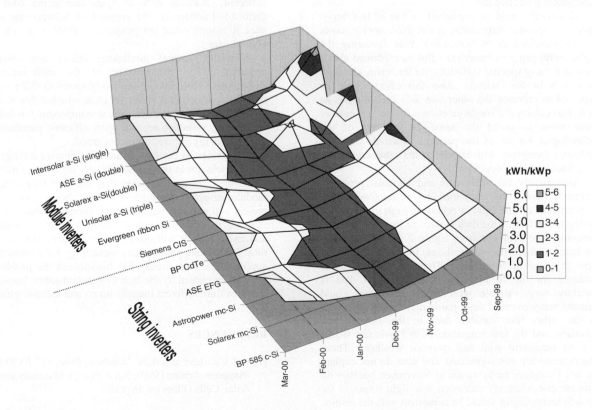

Figure 3: AC kWh/kWp per day averaged over each month.

6. DISCUSSION

It is no surprise to see the decrease in output in the winter values in figure 1. Similarly the fact that this decrease is less than the decrease in insolation indicated in figure 2, is amply explained by the lower temperatures in winter, again shown in figure 2, enhancing the efficiencies of all the technologies.

Of more interest is that the performance of all the technologies was lower in the spring than the autumn, despite higher insolation values, although again this is explained by the relatively low temperatures in the autumn for a given level of insolation. (Furthermore the caveat about the extra sensitivity of the September data mentioned in the previous section should be borne in mind, although this does not in itself negate the result.)

The comparison between sub-arrays indicates a general uniformity in performance between the technologies, particularly for the crystalline silicon, CdTe and CIS sub-arrays.

(The low values for Evergreen and Solarex a-Si in September and Astropower in February and March are due to the measurement anomalies discussed in the previous section.)

There are however some interesting variations. ASE a-Si (double junction) and Unisolar (triple junction) start with the highest values of kwh/kWp. The initial Staebler-Wronski performance degradation on light soaking explains this result, but does not explain the high values reached for ASE a-Si reached in March (or that of Solarex a-Si).

Furthermore the ASE a-Si and Unisolar both experience relative highs in November, whilst Evergreen experiences a relative low.

This result could be explained in one of two ways. Either the spectral distribution is weighted more towards shorter wavelengths in November, thus favouring the higher band gap a-Si materials. But as explained above there is not any spectral variation data for Spain, although there will be for Oxford. Also this effect would be expected to enhance the other two a-Si technologies as well, particularly the single junction a-Si. In fact both of these seem to about the same as the rest of the technologies for all of the period. It should also not disfavour Evergreen with respect to the other crystalline Si technologies for November.

Alternatively the ASE a-Si and Unisolar peaks in November could be explained by a relatively high proportion of diffuse light or relatively high temperatures having less effect on the dark current of the high bandgap material. The evidence from figure 2 does not support the latter of these two possibilities, in fact temperatures were at a low in November. There is however, some evidence for the former: the peak insolation data is certainly very variable with many lows that could indicate many overcast days, particularly when compared to the rather less variable data in the January and February, and the low temperatures in November would also be consistent with dull overcast weather. These observations are not significant and also do not explain the low Evergreen performance in November, but they do indicate that relatively enhanced low light response for the a-Si technologies would be consistent with the results. The subsequent analysis and publication of more detailed insolation and temperature data once a year's operation has been achieved will investigate these points more thoroughly.

The comparison of figures 1 and 3 yields another few points of interest even though AC performance is not of prime consideration. As explained in the previous section, figure 3 is the raw data from which figure 1 is derived and is presented even though the derivation is well justified. Overall the two graphs indicate a high level of consistency of inverter efficiency with respect to energy generation. It can just be discerned that the string inverters are slightly more efficient than the module inverters. This is in fact much more marked when the power data are analysed. This last point is because in both cases the inverter efficiencies start to drop significantly only at low power levels, and low power operation has little weight in determining the total energy values. In fact the efficiency profile for the module inverters remains high at lower percentage input powers than does that for the string inverters. The apparent poorer efficiency of the former is actually not due to the inverters themselves but rather due to their less well power matched application, which in turn is due to the wide voltage input voltage ranges for the a-Si and CIS technologies, discussed in section 3. (The Evergreen sub-array is an exception to this last point as the modules are designed for use with the module inverters.)

7. CONCLUSION

These are preliminary results and full analysis will not be possible until a full year's data have been collected. Autumn 2000 in Spain and spring 2001 in Oxford.) Furthermore the comparison between the two sites is naturally not yet possible as there is as yet no Oxford data.

However, several interesting effects are already apparent. The performance of the multi-junction amorphous Si technologies certainly seems to differ from the others, although it is not yet clear whether this is due to temperature, spectral or diffuse component insolation effects. AC inversion seems largely efficient, particularly as far as energy generation is concerned.

The project is the first to directly compare a range of PV technologies at two sites with very different climates. Additionally in several cases it is the first time certain module inverter combinations have been used. The project will generate independent information that will be valuable to manufacturers, systems designers, installers and end-users. It will enable informed decisions on the most appropriate photovoltaic technology for a particular application. And already has highlighted the problems that can be caused by the lack of standardisation between modules from different manufacturers and technologies.

REFERENCES

[1] G.J. Conibeer and A.W. Wilshaw, Proc. 11th PVSEC, (Sapporo, Japan; 1999), Solar Energy Materials and Solar Cells (Elsevier, in print).

OPERATIONAL RESULTS OF THE 250 KWP PV SYSTEM IN AMSTERDAM NIEUW SLOTEN

A.J. Kil[1], T.C.J. van der Weiden [1], J. Cace[2]
[1]Ecofys energy and environment, P.O. Box 8408, NL-3503 RK Utrecht, The Netherlands,
phone: + 31 30 28 08 300, fax: + 31 30 28 08 301, email: a.kil@ecofys.nl
[2]NUON Duurzame Energie, Arnhem, The Netherlands

ABSTRACT: In 1996 a 250 kWp grid-connected PV system was realised on 71 houses and apartments in Nieuw Sloten, a newly built city district of Amsterdam. Since then, the system has been operational and has been monitored. In this paper the results of one full year of monitoring are presented. The annual performance ratio of the total system was 77%, for subsections varying from 68% to 79%. Mismatch as a consequence of coupling of arrays with various orientations to a single inverter has led to a loss of annual energy yield of less than 1%. The annual yield of the PV system covers the electricity demand of approx. 85 households; the direct use of the generated PV power in the district has been 100% almost all the time. No losses due to inverter undersizing have been observed. A further reduction of the electricity production costs may be realised in similar projects by avoiding shading due to chimneys (3%) and by avoiding east and west oriented PV roofs (4%).
Keywords: Grid-Connected - 1: Monitoring - 2: Building Integration - 3

1. INTRODUCTION

The PV project in Nieuw Sloten was initiated by the utility company of Amsterdam, now merged with other Dutch utilities in NUON [1-4]. Ecofys participated in the project as a technical consultant and carried out the monitoring programme. The PV project has been financed by the EU Thermie Programme, Novem, the suppliers of PV modules, the utility company and the European subcontractors viz. the utility company of Copenhagen, Sermasa from Madrid and Euros from Genova. PV modules were supplied by Shell Solar Energy (The Netherlands) and BP Solarex (UK).

Nieuw Sloten was the first city district in the world where building integrated PV was demonstrated on such a large scale. The innovative integration aspects - mechanical as well as electrical - and the wide variety of system design characteristics within this project makes it a unique research object. The monitoring results can be used for future PV system design optimization which is needed for further cost reduction. Moreover the project offers data to assess the implications of large scale grid connected PV in the built environment for the local distribution grid.

This paper describes the operational results on the basis of monitoring data gathered in the period August 1997 - July 1998.

2. DESCRIPTION OF THE PV SYSTEM

2.1 Design considerations

The system was designed such that it is able to supply the electrical demand of approximately 100 households in Amsterdam. In reality, because of the very large roof areas, the 4400 PV modules were placed on the roofs and facade of 71 houses. Futhermore, operation should be easy, so a centrally placed control building was designed and built. A local underground DC grid connects the PV arrays to the inverter equipment in the control building.

Figure 1: On the left the PV roof of a single family houses is shown; on the background is the apartment building, with the PV facade and one of the penthouses.

2.2 Geometry

The PV modules have been integrated in the roofs of single family houses and in the roofs and facade of one apartment building. The PV integration technique was adapted from standard greenhouse building technology. The roofs of the single family houses are facing east and west under a tilt angle of 25° and south under 35°. On the apartment building the large roof and the roofs of two penthouses are tilted to 25°. The tilt angle of the PV-facade is 80°. All the PV areas on the apartment building are facing south.

2.3 Electric subsystems

The PV system is electrically divided into one large and nine small subsystems. The operating voltage of the large subsystem is 300 V DC. The roofs of the single family houses and the large roof of the apartment building, together 213 kWp, are connected to one SMA thyristor-bridge inverter of 150 kW (subsystem A).

The 21 kWp on the facade are connected to 3 SMA inverters of 5 kW in master-slave configuration (subsystem

B). The facade subsystem has a working voltage of 300 V DC as well.

The penthouses, in total amounting to 16 kWp PV, are connected to 8 Sunmaster 1800 inverters of 1.5 kW (subsystems C1..C4 and D1..D4). These inverters have an MPP-voltage of about 60 V DC.

Table I: Electrical characteristics of the subsystems

subsystem	kWp	inverter
A	213.7	SMA 150 kW
B - facade	20.8	3 x SMA 5 kW
C - penthouse	8.0	4 x Sunmaster 1800
D - penthouse	8.0	4 x Sunmaster 1800

Figure 2: Wide-angle photograph of the PV system, taken from the south. In the front the west-facing roofs of the single family houses are visible.

3. MONITORING

3.1 Monitoring goals

The system was monitored analytically according to JRC-guidelines and according to the Dutch guidelines for grid-connected PV monitoring [5], which are based on the JRC-guidelines. All electrical subsystems were monitored analytically, except the eight Sunmaster systems: they were monitored in two groups of four. The main goal of the analytical monitoring is to generate standard reports on the performance of the system.

Secondly, the monitoring was carried out for supervision pursoses. The design of the monitoring system was such that failure of one single string could be detected. Remote control of the monitoring system ensured a quick response time in case of possible failures.

Thirdly, a number of research issues were addressed:
- *Dependence of the PV yield on system characteristics* like orientation, tilt angle. Special attention was given to the (possibly negative) effect on the PV yield of coupling of arrays with various orientations to one inverter, as was done in subsystem A (SMA 150 kW).
- *Inverter field performance.* The nominal power of the various inverters lie in a broad range. Furthermore, inverters based on thyristor as well as on mosfet technology are present. Research was focussed on availability, efficiency, MPPT behaviour, inverter undersizing effects and temperature effects.

- *Correlation of PV yield with consumption patterns.* The extent to which the PV power was consumed locally was investigated in order to assess the maximum degree of penetration of PV in a similar residential area. Furthermore the electricity energy balance of the PV households was studied.
- *Thermal characteristics of the integrated PV roofs.* Is the ventilation underneath the PV roof sufficient to keep temperature losses of the PV yield down to a reasonable degree?
- *Shading*, by chimneys on the roofs of the single family houses and by nearby buildings. The chimneys were a part the architectural design and therfore unavoidable unless at high costs.

3.2 Monitoring system setup

Subsystems A and B were analytically monitored: DC array current, voltage and power as well as AC power (inverter output) was measured. Furthermore, ambient and cell temperature, horizontal irradiance (pyranometer) and in-plane irradiance (six reference cells, one for every different array orientation) were measured.

The eight inverters in subsystems C and D were monitored on the DC-side, but not on the AC-side. The total AC-output of C and D was measured instead.

The electricity power consumption of the total resedential area (322 households) was measured.

For supervision purposes the DC current of every group of four strings was measured continuously, as well as the status of the overvoltage protectors, fire alarms and isolation guards.

All measured physical quantities were averaged and logged on a ten minute basis. The data were logged by 11 dataloggers which were integrated in local string coupling cabinets. Communication from the dataloggers with the PC in the central control building was realised via a private telephone exchange. Free leads in the DC gound cables were used as telephone wires. This solution was chosen since it was the most cost-effective one.

Once a day all data were uploaded from the control building to Ecofys by telephone. Data were automatically processed. Monthly reports were generated automatically as well.

4. RESULTS

4.1 Analytical monitoring

Table II: Annual performance Aug. 1997 - July 1998: final yield Yf and performance ratio PR

subsystem	Yf [kWh/kWp/yr]	PR [%]
A	705	77.2
B	548	68.4
C	755	78.7
D	724	75.4
total	694	76.5

In Table II the performance over the total monitoring period is presented. The final yield of the total system was 694 kWh/kWp/yr. Normalized to the climatological average year at location De Bilt (NL) this amounts to 718 kWh/kWp/yr.

The penthouse systems C and D perform best with a final yield that is characteristic for state-of-the-art PV systems. The high performance of C was expected on the basis of the optimal orientation and tilt, the electrical design and the absence of shading. As expected, the facade system has the lowest yield. The main system A, a combination of various orientations, has an performance which may be considered as average for grid connected systems in The Netherlands.

In figure 3 the monthly final yield is presented. Also from this bar graph it is evident that in most months the penthouse system C performs best and B worse. However, in winter months the facade system B has the best performance, due to the high tilt angle.

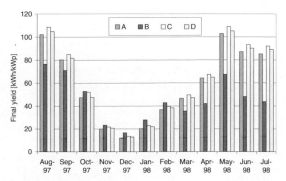

Figure 3: Monthly final yield of the four PV subsystems

4.2 Comparison of array performance

Application of PV on east and west oriented PV arrays, as was done in subsystem A, has led to a reduction of the annual yield of of the east and west arrays of approximately 20%. This is directly the consequence of the irradiation reduction of 20%. The annual yield of the system as a whole was due to this non-optimal orientation 4% lower than in case all arrays had been optimally oriented, i.e. south, with a tilt angle of 36°.

Mismatch due to coupling of arrays with different orientations to one inverter with a single MPP-tracker leads to a yield reduction of less than 1% on a yearly basis. This experimental evidence confirms earlier computer simulations [6].

4.3 Inverter field research

All inverters were 100% available during the monitoring period. The SMA 150 kW has the highest annual energy conversion efficiency of 95%. The monthly efficiency figures for the inverters are shown in figure 4. The SMA 150 kW and the Sunmasters have a clear seasonal dependence, presumable due to the temperature of the inverters. The SMA 5 kW inverters have a relative poor efficiency thoughout the year and do not show a seasonal dependence, because the input power (from the facade subsystem) is relatively low, even in summer.

Study of the MPPT-behaviour shows that the Sunmaster is better equipped than the SMA inverters to adapt the working voltage to variations of the cell temperature.

All inverters are undersized with respect to the installed PV power. No negative effects on the annual yield were observed in either of them. As far as the SMA 150 kW system (A) is concerned this is a consequence of the combinaton of various orientations, thanks to which the peak power at noon is lower than in a system with optimal geometry.

The Sunmaster inverters have a temperature control which reduces power if the inverter temperature becomes too high. The resulting annual yield loss was lower than 3%.

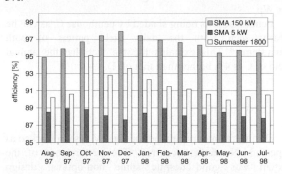

Figure 4: Monthly inverter efficiency for the three types of inverters applied.

4.4 Correlation with electricity consumption

The annual electricity production - in terms of kWhs produced - covers the electricity consumption of approximately 85 Amsterdam households. In figure 5 the monthly electricity production of the PV system is compared with the electricity consumption. It is evident that the strong seasonal effect of PV production is the cause of the observed PV surplus (or: a negative net electricity consumption) in summer and a deficit in the rest of the year (which is drawn from the grid).

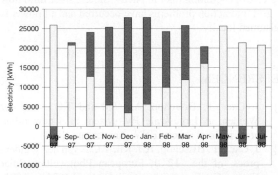

Figure 5: Monthly PV production (light colour) and net electricity consumption of the 71 PV households (dark).

The PV power was used locally for almost all of the time. Furthermore, the PV system did not cause any problems for the operation of the distribution grid.

From this is concluded that the maximum degree of penetration of PV can be much higher than now realised in Nieuw Sloten (71 out of 322 households), even without reinforcement of the grid.

4.5 Thermal aspects

The energy loss due to high cell temperatures lies for the various subsystems between 3 to 5% annually. The average heating coefficient of the cells was $0.05 \, K/(W/m^2)$. This value is characteristic for roof integrated PV systems with moderate ventilation.

4.6 Shading

Since strings were monitored individually, it was possible to investigate in detail the effects of shading by nearby objects. The annual yield reduction due to chimneys on the PV roofs was assessed to be 3-4%. Furthermore, the use of chimneys and dummy modules have resulted in a small number of shorter strings with a slightly lower yield per Wp.

Penthouse D is being shaded by penthouse C, as can be clearly seen from the daily yield patters of both subsystems.* This explains the the fact that the annual yield of C is 4% higher than that of D.

5. CONCLUSIONS

During the monitoring period the PV system worked as expected. No faults occurred. The performance of the system was good despite some non-optimal design characteristics. The losses due to east and west oriented roofs were more or less compensated by the high efficiency of the central inverter. In the given situation in which not all arrays can have an optimum orientation due to architectural constraints, the central electrical design has proven to yield good results in terms of energy, power and availability and therefore in terms of cost effectiveness as well.

A further reduction of the electricity production costs may be realised in similar projects by avoiding shading due to chimneys, avoiding east and west oriented PV roofs and further undersizing of the nominal inverter power.

Finally it is concluded that future PV systems in the built environment with a similar design can have a higher degree of penetration than in Nieuw Sloten, since until now no grid interaction problems have occurred and all power is consumed locally.

REFERENCES

[1]   J. Cace, F. Bisschop, A.J. Kil and T.C.J. van der Weiden: A large-scale PV system in a new urban area, internal report, ENW Amsterdam, 1994

[2]   J. Cace, F. Bisschop, A.J. Kil, T.C.J. van der Weiden, procs. of the 12 th European Photovoltaic Solar Energy Conference, Amsterdam, April 1994, p. 1091

[3]   J. Cace, J. Schlangen and A.J. Kil, procs. of the 5th Dutch Solar Energy Conference, Veldhoven, April 1995, p. 17

[4]   J. Cace, A.J. Kil, T.C.J. van der Weiden, procs. of the 14th European Photovoltaic Solar Energy Conference, Barcelona, July 1997, p. 698

[5]   C.W.A. Baltus et al., *Recommendations for the monitoring of grid-connected PV systems – version 02*, Novem, Utrecht, 1997 (report nr. RC 4154, in Dutch)

[6]   T.C.J. van der Weiden, A.J. Kil, procs. of the 1st World Conference on Photovoltaic Energy Conversion, Dec. 1994, Hawaii, p. 844

PV GENERATOR FEEDING A HYBRID STAND-ALONE / GRID CONNECTED SYSTEM IN GHANA

R. Eyras, O.Perpiñan
ISOFOTON, S.A.
Montalbán 9, 2º Izq.
28014 Madrid, Spain
E-mail:r.eyras@isofoton.es

ABSTRACT: This paper submits the design of a Centralised PV System to be installed in a building in Ghana, with the aim of promoting a large rural electrification project carried out in the country. The system, located at the Ministry of Mines and Energy Building, is devoted to allow continuos working of the Ministry staff, while frequent shut downs of the mains grid occur. In order to maximise the use of the solar energy, a Hybrid Stand Alone/Grid Connected system fed by a single PV generator has been designed.
Keywords: 1: Energy Options - 2: Hybrid - 3: Developing Countries

1   INTRODUCTION

Since 1998, a large rural electrification Project is being carried out by the Ministry of Mines and Energy (MM&E) from Ghana, financed under a FAD loan (Development Aid Founding) from the Spanish Government, and supplied and installed by ISOFOTON, SA. This project comprises stand-alone systems devoted to supply different rural applications, as solar home systems, schools, community centres, hospitals, battery charger stations, water pumping and streetlight, up to 230 kWp. The project has been concluded, with an important success in the rural areas, encouraging the MM&E to continue with this policy.

Table I collects the general characteristics of the installed systems

System	Nº	Generator (Wp)	Battery (Ah)	Uses
SHS A	750	50	100	Illumination Radio
SHS B	1150	100	150	Illumination Radio, TV
School	30	250	200	Illumination Audio-visual
Hospital	14	300	400	Illumination Refrigeration
Community Centre	12	250	250	Illumination Audio-visual
Streetlight	200	150	200	Public Illumination
Water Pumping	2	7200		Household Water Supply
Battery Charger	10	1000	100	Illumination Battery
University Kit	5	150	200	Educate

Table I: Characteristics of installed systems in the rural electrification project in Ghana

With the aim of promoting the photovoltaic systems, a 50 kWp centralised system was included in the project to be installed in the MM&E Building. This paper submits the final design of this system.

2   DESCRIPTION

2.1   Justification

Due to the almost total dependency on hydro power sources (hydro power from Akosombo central at Volta lake accounts for over 80% of the electrical energy supply), there are important problems with electrical power supply at urban zones of Ghana, having periods of insufficient generation with frequent interruptions of supply. Annual reliability is about 85-95%, though this is mainly dependent on climate conditions. For example, during 1998 there were poor rains in the Volta catchment area, and power supply had to be rashened, reducing reliability to less than 40%.

In this context, and with the aim of promoting the large rural electrification solar energy programme, the Ministry of Mines and Energy (MM&E) of Ghana planned the possibility of using photovoltaic solar energy in their own building, in order to continue normal job during the periods of supply break ("normally" these periods last two or three hours a day), which are cause of lots of lost hours of work.

2.2   Sizing

Below in Table II is the annual electricity consumption pattern at the MM&E building in 1997.

External Lighting	15.240 kWh
Internal Lighting	40.660 kWh
Air Conditioning	118.970 kWh
Office Equipment	42.070 kWh
Miscellaneous	21.150 kWh
TOTAL	238.090 kWh

Table II: Annual electricity consumption pattern at the MM&E building in 1997

The Photovoltaic Generator and the bank of batteries will supply energy to make internal lighting, office equipment and miscellaneous equipment work during periods of grid shut down. These equipment have an installed power of 36 Kw. Therefore, the size of the PV Generator and the capacity of the batteries are fixed by the electricity consumption of internal charge, except air conditioning, during periods of no less than three hours per day.

To satisfy this energy demand, a centralised system constituted by a 50 kWp PV Generator, a battery and a 50 Kw inverter (DGT model from Enertron) has been sized.

Battery capacity has been chosen for a 60% deep of discharge, with the aim of maximise the operational life, and for three hours of autonomy, which is the typical shut down period. The maximum voltage is 460 Vdc and the minimum voltage is 340 Vdc (these values are fixed due to the electrical characteristics of the DGT inverter).

With an installed power of 36 Kw during power supply break periods, it is needed a maximum current of:

$$I_{max} = \frac{36Kw}{340Vcc} \approx 105,88A,$$

and considering a battery yield of 0,9, the requested capacity is:

$$C_B = \frac{105,88 \cdot 3}{0.9 \cdot 0.6} \approx 588,2Ah.$$

It will be installed a system consisting in air blowers, which insurance quite good electrolyte homogenisation, avoiding the dangerous stratification phenomena and improving battery life. Besides, this system refrigerates the battery bank, whose temperature can be too high, because of the climate of Ghana and the discharge rate.

2.3   Design

Thus, to operate as a Uninterrupted Power Supply, the system is composed by the PV Generator (constituted by 504 modules of 100 Wp, connected in 18 groups in parallel of 28 modules in series each one), the 550 Ah battery bank, and a 50 KVA stand-alone Inverter type DGT (440Vcc/230Vac III), which supplies AC electrical energy for the MM&E building consumption.

A stand-alone Photovoltaic System like the UPS system, has the disadvantage that solar energy is lost when the battery is fully charged and the grid is working. As this is the more frequent situation during a typical year (the UPS works only during two or three hours a day), the use of a PV generator would be expensive and inefficient. Because of that, a hybrid system is proposed, combining a stand-alone system and a Grid Connected system with only one PV Generator.

The aim of the project proposed by ISOFOTON, is, on the one hand, to inject the energy generated by the solar panels directly into the mains grid (this will be the normal situation during most of the year), and on the other, to recharge a series of batteries which, during periods of supply failures, are responsible for satisfying the energy demand of the building. Thus, as it is explained below, the proposal combines a Grid Connected Photovoltaic System (GCPVS), during normal grid operation, and an Uninterrupted Power Supply (UPS), during power supply breaks.

Figure 1 reflects the global composition of the hybrid system, which consists of the previously commented PV generator alternatively connected to the two different subsystems:

UPS, as it has been explained, is composed by the PV Generator, a battery bank and a stand-alone DGT Inverter.

GCPVS comprises the PV Generator and three monophase Solete inverters whose electrical characteristics are collected in Table III.

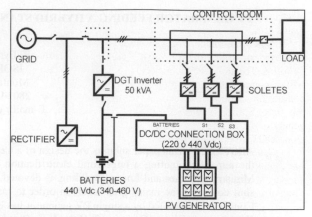

Figure 1: Scheme of the Hybrid Stand-Alone/Grid Connected System

Since standard inverters have been chosen, with different nominal input voltages (220 VDc for Soletes, 440 VDc for DGT inverter), a DC/DC Connection Box is needed. This Connection Box allows the connection of the PV generator subarrays in series or parallel, depending on which one of the subsystems is operating (series connection when UPS is operating and parallel connection for GCPS).

Another important function of the DC/DC Connection Box is to prevent loss of isolation through an Earth Failure circuit. When an isolation fault is detected, an alarm event is generated, so an authorised operator must commute the DC/DC Connection Box, which establishes a short-circuit in all the arrays, and connects them to earth.

Automatically switched with IGBT	
AC intensity control by hysteresis bands	
Nominal Power	14 Kw
Input	
Minimum Voltage in MPP	210 Vdc
Maximum Voltage in MPP	280 Vdc
Maximum open circuit Voltage	306 Vdc
Output	
Grid Voltage	220 V
Admissible Variations	±10%
Frequency	50 Hz±2%
Maximum distortion on AC injected into grid	<5%
Power factor	>0,98
Approximate yield	94%
Protection Mechanisms	
Automatic Circuit breaker on the input	
DC input contact	
AC output meter	
Manual output switch	
Electronic protection	

Table III: Electrical characteristics of the Solete inverter

Normal operation involves GCPVS, with PV generator connected to the three Solete inverters. In this case, the DC/DC Connection Box establishes a parallel connection of the subarrays of the PV generator (36 arrays of 14 I-100 modules in series), in order to feed Solete inverters with their nominal 220 Vdc. Each one of the monophase inverters is connected to one phase of the electric grid bars, delivering generated current into the mains grid.

When the mains grid shut down, the UPS subsystem is started in order to supply partial load of the MM&E building. To start UPS subsystem, an authorised operator will disconnect the mains grid, the three Soletes and the air conditioning equipment, commute the DC/DC connection box into Series Way (*batteries*), in order to allow PV generator to charge the batteries, and turn on the stand-alone DGT inverter.

When power supply from the grid is available again, the operator will turn off the stand-alone DGT inverter, connect the mains grid, and turn on the air conditioning. The PV generator will continue charging the battery until a voltage of 441 V is reached. At this moment, the control opens the charging line giving the end-of-charge information by means of an LED indicator. At this moment the operator will restart the GCPVS, feeding solar energy into the grid.

All of these tasks could be performed through an automatic mechanism with a control circuit. However, due to the extra cost this automatic mechanism would mean, the difficulties to manage the existing Air conditioning lines, and to the existence of a person who could undertake these tasks (permanent maintenance staff is available), a manual operation was selected.

Finally, the global system is completed by a grid-connected rectifier of 60 A nominal current, responsible for battery charging during normal grid operation. The rectifier is connected to the batteries to perform a floating charge, at a limited current of 15 A (in case of unavailability of energy from the PV generator, the rectifier has the possibility of raising the charging current to a value of 60 A, to allow full battery charge).

2.4   Architectural Integration

Initially, three different morphologies were going to be used with two principal objectives:

- To decrease installation costs (part of the PV generator was to be installed in an existing wooden garage)
- To achieve a large diffusion impact (part of the PV generator was to be on the terrace of the building, facing the city centre).

However, after some months of delay, the MM&E decided not to install any modules on the terrace of the building, thus the whole generator will be installed in the wooden garage, replacing the present cover by the PV modules, using the existing structure.

2.5   Expected Results

As it has been mentioned, the project is to be installed due to some problems with the MM&E, so no real results are available yet.

The PV generator is going to be on the roof of the garage, with North orientation a 15° tilted. Bearing in mind that the city of Accra, where the MM&E building is placed, is at 5,6° North latitude, the deviation from optimal orientation (South orientation) is supposed not to have a large effect in energy production. In fact, an simulation shows that the energy losses due to the orientation represent a 4,8% of the total generated energy during the year (see Table IV), but North orientation takes good use of an existing structure, reducing installation costs.

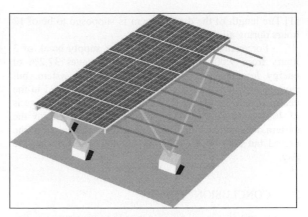

Figure 2: PV Modules on the garage roof of the MM&E building

South orientation		North orientation	
Month	E_{pv} *(kWh)*	Month	E_{pv} *(kWh)*
Jan	5.152,17	*Jan*	4.173,48
Feb	5.344,50	*Feb*	4.551,97
Mar	6.175,85	*Mar*	5.804,63
Apr	5.693,59	*Apr*	5.965,75
May	5.523,40	*May*	6.333,34
Jun	4.245,70	*Jun*	4.898,88
Jul	4.297,22	*Jul*	4.859,69
Aug	4.432,22	*Aug*	4.724,70
Sep	5.160,15	*Sep*	5.094,84
Oct	6.344,59	*Oct*	5.568,39
Nov	6.422,98	*Nov*	5.040,40
Dec	5.737,13	*Dec*	4.432,22
Total	**64.529,5**	**Total**	**61.448,3**

Table IV: Comparison between the generated electric energy (*kWh*) by a PV Generator south and north oriented, with 15ª inclination. A constant Performance Ratio of 72% is supposed.

If the system worked always as a GCPVS, part of this energy would be consumed at the MM&E building, and the rest would be feed into the electric grid. However, as it has been explained, the system depends on the way of working of the electric grid, so there will be periods with no energy exchange, since the system will not be connected to the mains grid. Nevertheless, during these periods the function of the system is to make working possible, that is, to save man-hours of job.

In order to evaluate the system working, the saved energy (because it is not consumed from the energy) and the saved man-hours of job can be calculated. Of course, the more saved man-hours, the less saved energy, because during power supply breaks no energy exchange with the electric grid exists.

With daily horizontal irradiation data it is possible to obtain hourly horizontal irradiation values with the conversion factor (Whillier):

$$r_g = \frac{G_{hm}(0)}{G_{dm}(0)}$$

which can be calculated through the Liu and Jordan equations. With these values it is possible to estimate the generated energy by the system during a períod in the day.

[1].The length of the day in Accra is supposed to be of 12 hours during all the year.

For example, supposing a power supply break of 3 hours from 9:00 to 12:00, the system saves 37,2% of energy less than an always grid-connected system, but saves 3 man-hours of job. If the period was located in the centre of the day (from 10:00 to 13:00), the percentage is of 40,9%, what means that the amount of the energy the system does not save depends not only on the length of the period but also on the location on the period during the day.

3   CONCLUSION

At this moment, the installation phase of the project is to be started, so no real results are available yet.

The project is expected to give solution to the energetic problems at the MM&E building, compensating the extra cost of hardware (since two different inverters instead of one and a DC/DC connection box are used) with the extra manpower hours saved. Besides, the hybrid configuration is expected to increase the global yield of the system taking good advantage of the available solar energy.

A large diffusion impact is expected, due to the localisation of the PV Centralised System in the garage in front of the Ministry of Mines and Energy Building.

REFERENCES

[1]   E. Lorenzo, "Solar Electricity", 1994 Progensa.

SPORE – AN EUROPEAN PROJECT FOR SOCIAL OBJECTIVES IN REMOTE PLACES IN ROMANIA

Dan I. Teodoreanu, Maria Teodoreanu,
Bogdan Atanasiu, Eduard Lauci
Research Institute for Electrical Engineering
New Energy Sources Laboratory
313 Splaiul Unirii, 74204 Bucharest, Romania
Phone/Fax: +401 3217247
e-mail: danteo@icpe.ro

Charis Demoulias
Alteren Inc.
Kassandrou 37a Str.
546 33 Thessaloniki, Greece
Phone: +30 31282528, Fax: +30 31 283725
e-mail: alteren@classic.diavlos.gr

Miklós Palfy
SOLART-SYSTEM LTD.
20 Gulyas Street, Budapest XI, Hungary 1112
Phone/Fax: +36 1 2461783
e-mail:solartsy@elender.hu

Alison Murray
Solapak Ltd, Cock Lane, High Wycombe,
Buckinghamshire HP13 7DE, UK
Phone: +44 1494 452941, Fax: +44 1494 437045
e-mail: solapak@intersolar.com

ABSTRACT: The project *"Small Power Photovoltaic (PV) Systems for social objectives in remote areas (SPORE)"* wishes to demonstrate the technical and economical competitiveness of PV systems for electricity supply in remote places, with the aim of improving the quality of life of people living in these regions and developing the mountain tourism. The development of standard PV systems for this kind of application is another goal of the present work. Ten PV systems have to be installed in different mountain locations not connected to the grid. The ten systems will supply with electricity: 6 schools – system A; 2 huts - systems B1 and B2; 2 cultural centers - system C.The sizing of PV generator is based on the monthly average solar radiation data (annual average value: 3.645 kWh/m^2/day, measured at the nearby meteorological station). The battery capacity is calculated for autonomy of 15 – 20 days. An innovation is the availability of the system to supply different kind of loads (12 V dc, 24 V dc, 230 V ac), with two important consequences for the improvement of system Performance Ratio: optimized efficiency of system components and reduced "non-availability to load" period.

Keywords: Stand-alone PV Systems – 1; Sizing – 2; Rural electrification – 3.

1. INTRODUCTION

From the statistics of the Romanian National Electricity Utility (CONEL), in Romania there are more than 63000 houses not connected to the grid and these are situated in 550 remote small villages. More than 30% of these small villages are located in the area of Apuseni Mountains (western Carpathian Mountains). The reports of the Ministry of Education show that more than 50 schools from this region are not connected to the grid. The mean distance to the grid is about 7.5 km and the estimated cost for electrical grid extension is more than 20000 ECU/km. In the Carpathian Mountain there are also more than 50 huts without electricity situated more than 10 km from the grid. If we suppose that the necessary funds for grid extension for these remote places are to be drawn from the energy taxes, we come to conclusion that the electrification of these places should last more than 50 years. The potential market for social objectives is estimated to be more than 1 MW and for rural electrification - to more than 5 MW. From this estimation, it is reasonable to develop first social energy applications taking into account also that the people living in these areas have to become confident about such things in order to invest individually later.

The paper presents the description of the PV systems, including the site locations, the insolation conditions and design parameters for loads, PV panels, batteries, inverters and other system components.

2. SOCIAL AND MANAGEMENT ASPECTS

The only solution for improving the quality of life for the inhabitants of these remote regions - in what it concerns their education, culture and access to information - is the use of renewable electrical energy sources. This is the main objective of the project. There are also other objectives of the proposed project, such as:

- to stabilize the rural population in these regions, to bring back people in villages, with benefits for agriculture, SME development, culture and tourism;
- to open and diversify a new market for clean and environmental friendly energy sources through electrical supply for social objectives;
- to create (through monitoring) a solar (and wind) energy database, for these regions, useful for the further replication of the project;
- to explain, demonstrate, educate people to optimize the use of the PV systems in accordance to their needs - the social dimension of the project;
- to choose the most reliable and cheap components for PV systems (including the production of some of them in the country) - the technical and economical dimension of the project;
- to complementary use the manual and automatically operation of the system - the intelligent management dimension (tilted mount, Ah meter, battery charger).
- an important aspect of the project management is the active end-user and local authorities participation in project implementation (installation, monitoring), thus

increasing the chances of further replicas of such kind of PV applications.

3. TECHNICAL DESCRIPTION

The project aim is the installation of 10 PV solar power sources for: schools - type A (6 systems), huts - type B (2 systems) and cultural centers - type C (2 systems).
The block diagram of the systems A and B1 are presented in Fig.1 and Fig.2 together with monitoring parameters.

3.1. Design considerations:
Loads. Estimating electric loads is very difficult for users who have access for the first time to the electricity. **Electronic equipment.** Robustness and reliability are more important than energy conversion efficiency for remote, isolated PV plants. We have to pay special attention to electronic equipment (controllers, chargers, and inverters) as the least robust PV system components. **Standards.** Special attention will be paid to the references [1-5] (CEI and other international and national Norms, Thermie and ISPRA-JRC recommendations, other JOULE projects).
Monitoring. Wind database is very important for rural and isolated sites (nonexistent or of poor quality at present).

3.2. General design criteria
    a.   The PV generators will be integrated in the houses both technically and architecturally;
    b.   The PV systems are sized for "stand-alone" operation under climatic condition corresponding to those in the mountain area (Deva meteorological station);
    c.   Description of the PV-houses (schools, huts);
    d.   The simplified system block diagrams and the power and signal distribution diagram permit the evaluation of the PV systems (A, B1, B2, C) and monitoring performances.

3.3. Brief description of the essential components:
PV Array is made of crystalline silicon PV modules, series and parallel connected. PV Modules have to correspond to requirements of the ISPRA/CEC-503 or IEC 1215 codes.
Batteries. The most suitable for our applications are deep cycle lead-acid batteries with tubular positive plates (flooded batteries).
Controllers/Inverters. The important properties of chargers are good mtbf (mean time before failure): (7-10) years and small standby consumption.
Inverter-General specifications:
- Type sine wave, interference free usage (for Na and fluorescent lamps, TV, VCR)
- Efficiency $\geq$ 90%, at Pnom (without charger)
- Efficiency $\geq$ 80%, at (5-10)% $\cdot P_{nom}$
- Distortion < 5% $\cdot P_{nom}$

Cables. The optimum voltage drop in electrical wires between PV array and battery is (2-3)%$\cdot U_{nom}$ and (4-5)% $\cdot U_{nom}$ between battery and loads. Battery's cables should be kept short and close together in order to avoid high values of resistances and inductances.
Fluorescent lights. Characteristics.
Indoor lighting:
- colour temperature: (2700 - 4000) K, warm white
- colour rendering index: $R_a > 60$

Outdoor lighting:
- Temperature range: - 15 °C ÷ + 45 °C
- Colour temperature: > 4000 K.
The expected lamp lifetime should be more than 5000 hours.
As an example, the detailed sizing of system B1 is presented in Table 2
The results of the sizing procedure are presented in the following table:

Table 1. The main parameters of the four PV systems

System	Array Power (W_P)	Battery capacity (Ah/V)
A	420	1000/12
B1	960	2000/24
B2	600	1400/24
C	480	500/24

All the systems will be monitored according to IEC and EC - Ispra Documents: analytic monitoring will be used for two systems (B1 and C) and global monitoring will be used for the other eight systems.

4. CONCLUSIONS

The design of the four different types of PV installations so as to standardise them from the user and operator point of view and to optimise the user needs taking into account the budget constraints was the main topic of this work. Some of these aspects are presented in this paper. Other tasks like: equipment procurement and manufacturing, measurement, installation and evaluation are in progress. Attention has been paid to the involvement of the end-users in the manufacturing of the mechanical structures transport and installation. As is projected at least 3 systems have to be installed this year: A, B1 and B2.

ACKNOWLEDGEMENTS

This work has been co-funded by EC within the INCO-COPERNICUS Programme (contract ICOP-DEMO-4008/98) and ANSTI - Romanian Agency for Research and Technology (contract 57-A49-5004/98).

REFERENCES

[1] IEC-1277/1995-02:Terrestrial photovoltaic (PV) Power generating systems - General Guide.
[2] IEC- 61836/1997-10: Technical Report - Type 2, Solar photovoltaic energy systems - Terms and symbols.
[3] IEC - 1194/1992-12: Characteristics parameters of stand-alone photovoltaic (PV) systems.
[4] IEC - 1829/1995-03: Crystalline PV array; On-site measurement of I -V characteristics.
[5] IEC-364-5-523/1983: Electrical installations of buildings.

Table 2: Detailed analysis of system B1 loads

Type:	System **B1**				
Location:	Fântânele Hut, Cibin Mountains				
Altitude:	1257 m				
Latitude:	46° N				
June - September Period					
DC Loads					
Load Type	Unit Power (W)	Number of Loads	Load Power (W)	Hours/ Day	Daily Energy Consumption (Wh)
Indoor Lighting in rooms	10	8	80	1	80
Outdoor lighting	8	2	16	10	160
Radio	25	1	25	7	175
Phone	20	1	20	2	40
Electronic Equipment	14	1	14		142
Monitor analytic	12	1	12		30
	Total		*167*		*627 DC*
AC Loads					
Load Type	Unit Power (W)	Number of Loads	Load Power (W)	Hours/ Day	Daily Energy Consumption (Wh)
Indoor Lighting in the Kitchen	14	1	14	6	84
Indoor Lighting in the Dinning Room	14	2	28	4	112
TV	55	1	55	4	220
Indoor lighting in Halls/other rooms	8	2	48	2	96
Refrigerator/Freezer	350	1	350		1700
	Total		*495*		*2212*
Correction for Inverter efficiency (×1.1): 2212 Wh ×1.1				**TOTAL**	**2433 AC**
Inverter Power 1 (W) *250* *Inverter Power 2 (W)* *2000*			*Battery capacity, (20 days autonomy): 2000 Ah/24V* *Total energy consumption, DC + AC Load:* **3060 Wh**		
October - May Period					
AC + DC Loads					
Load Type	Unit Power (W)	Number of Loads	Load Power (W)	Hours/ Day	Daily Energy Consumption (Wh)
Indoor Lighting in rooms (DC)	10	2	20	5	100
Indoor Lighting in	14	1	14	6	93
Indoor Lighting in Dinning Room (AC)	14	2	28	6	185
Outdoor lighting (DC)	8	2	16	14	224
Radio (DC)	25	1	25	7	175
TV (AC)	55	1	55	4	242
Phone (DC)	20	1	20	2	40
Electronic Equipment (DC)	12	1	12		58
	Total		**190**		**1117**

PV Array sizing, System B1	**SUMMER**		**WINTER**	
Energy consumption, E_L (Wh/day)	3060		1117	
Tilt array angle (°)	40	45	40	46
Solar irradiation in array plane, H_I (kWh/m²day)	5.68	5.53	2.25	2.34
PV array power, P_A , (W)	898	922	827	796
PV module power, P_M , (W_P)	60		60	
Number of PV modules: P_A / P_M, (pcs.)		15.5		
Nominal PV array rating: *16 pcs × 60 W_P (W_P)*		*960*		

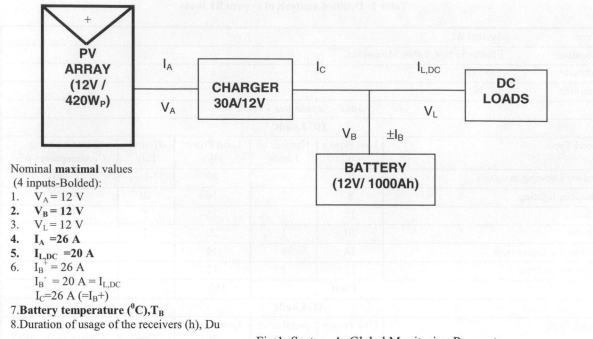

Nominal **maximal** values
 (4 inputs-Bolded):
1. $V_A = 12$ V
2. $V_B = 12$ V
3. $V_L = 12$ V
4. $I_A = 26$ A
5. $I_{L,DC} = 20$ A
6. $I_B^+ = 26$ A
      $I_B^- = 20$ A $= I_{L,DC}$
      $I_C = 26$ A $(= I_B^+)$
7.**Battery temperature (^{0}C),T_B**
8.Duration of usage of the receivers (h), Du

Fig 1. System A. Global Monitoring Parameters

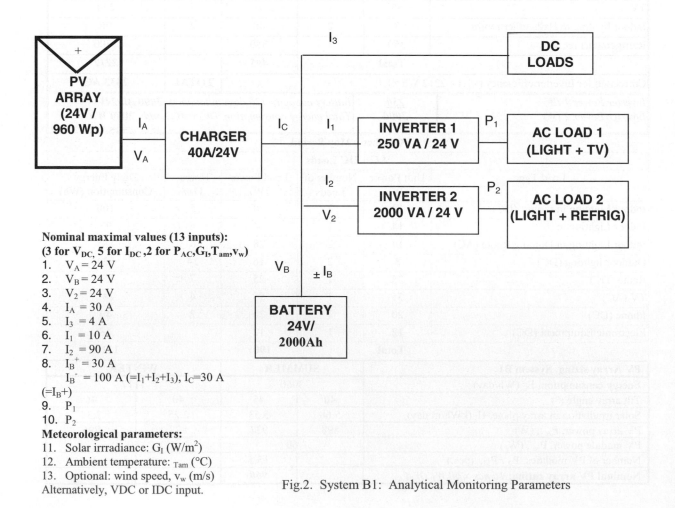

Nominal **maximal** values (13 inputs):
(3 for V_{DC}, 5 for I_{DC} ,2 for P_{AC},G_I,T_{am},v_w)
1. $V_A = 24$ V
2. $V_B = 24$ V
3. $V_2 = 24$ V
4. $I_A = 30$ A
5. $I_3 = 4$ A
6. $I_1 = 10$ A
7. $I_2 = 90$ A
8. $I_B^+ = 30$ A
      $I_B^- = 100$ A $(= I_1 + I_2 + I_3)$, $I_C = 30$ A
$(= I_B^+)$
9. P_1
10. P_2
Meteorological parameters:
11.   Solar irrradiance: G_I (W/m^2)
12.   Ambient temperature: $_{Tam}$ (°C)
13.   Optional: wind speed, v_w (m/s)
Alternatively, VDC or IDC input.

Fig.2. System B1: Analytical Monitoring Parameters

HIGH RELIABLE PV APPLICATIONS IN REGIONS OF THE EUROPEAN ALPS (EURALP)

P. Weber (1), E. Ehm (2), X. Vallve (3), K. Kiefer (4), E. Rössler (4)

DAV Deutscher Alpenverein e. V., Von-Kahr-Str. 2-4, 80972 München, Gemany

OEAV Österreichischer Alpenverein, Wilhelm-Greil-Str. 15, 6020 Innsbruck, Austria

SEBA / TRAMA TecnoAmbiental, Ripolles 46, 08026 Barcelona, Spain

Fraunhofer ISE, Oltmannsstr. 5, 79100 Freiburg, Germany

ABSTRACT: The purpose of the EURALP project is to show the great technical and economical progress made by photovoltaic hybrid systems which are now capable to support electric power with the same quality and service guarantee as a conventional power source even in very remote and critical areas. The project consists of 31 subprojects in the range of 1-5 kWp located in mountain regions all over Europe. First results of the comprehensive monitoring programme show good performance ratios and proof, that the systems are well designed and adapted to the users needs. This also results in a high degree of user's satisfaction.

Keywords: Stand-alone PV Systems - 1: Battery storage and control - 2: Hybrid - 3

1. INTRODUCTION

Within the EURALP project it was intended to demonstrate advanced and high reliable PV systems in regions of the European Alps, where many lightning strikes caused failure or even destroyed the main electronic parts of the system. Involved in the project are the German Alpine Cub (DAV), Serveis Energetics Basis Autonoms (SEBA) from Spain and the Austrian Alpine Club (OEAV). DAV owns about 350 alpine refuges and mountain restaurants in the German and Austrian Alps and carries out the project in close co-operation with the Fraunhofer Institute for Solar Energy Systems ISE. OEAV owns about 270 alpine refuges and mountain restaurants in the Austrian Alps and carries out the work in close co-operation with ATB/TBB. SEBA is a users association offering private users and local authorities in rural areas electric supply from stand-alone PV plants. It operates successfully PV plants through a concerted maintenance scheme with individual users and has meanwhile 130 associated systems. SEBA works in close co-operation with TRAMA Tecno Ambiental in realizing the systems.

Figure 1: "Brandenburger Haus". Highest located house of the "Alpenverein" (3277m). 2,6 kW$_p$ stand alone system.

The project consists of 31 subprojects in the range of 1-5 kWp located in mountain regions all over Europe.

This was to be reached by implementing an intelligent tele-monitoring system Since the 22 subprojects are different in their sizes, their power requirements and their natural resources it is impossible to design a standard hybrid system. For that reason it was planned to install at all sites the same advanced energy management system in combination with the monitoring equipment. The proposed total installed power was 65 kWp, the actual realized power finally will reach 68 kWp.

All the realised systems are stand-alone energy supplies for existing mountain lodges in very remote areas. Main objective of the project is the reduction of pollution and noise in natural reserve. Some of the existing diesel generator sets had been replaced within this project by rape-oil or liquid gas generator sets.

Figure 2: Subproject "Refugio de Estós" (Spain)

2. TECHNICAL DETAILS OF PV SYSTEMS

Even though the local conditions and technical and non-technical requirements at the different locations for the subproject systems had been quite different, at least basic standardisation for the system components and plant concepts had been one major target for the project. Therefore all the new energy concepts for the locations in the Alps were based on the guidelines of "Standardised PV-Systems", which had been worked out from a workgroup of experts in the beginning of the EURALP project. In these guidelines, 4 different categories of systems have been defined, which are characterised by different energy demands and typical operation schemes.

After the re-construction of the energy supply at each location, commissioning took place. For each systems an inspection booklet was prepared summarising all results of the comprehensive measurements and tests. During operation people responsible for the plant site regularly complete a so-called "Logbook", where besides meter reading defects and additional information on the plant operation are being documented.

It is not possible to give the technical details for all of the systems in this paper, only a few characteristic subprojects are to be introduced here.

2.2 Starkenburger Hütte

The installation of the PV system at the alpine refuge „Starkenburger Hütte" (Fig. 2) was finished in June 1997 when the opening season is starting.

Figure 3: Starkenburger Hütte with 4,95 kW$_p$ solar generator

The new electrical and energy concept is based on the PV generator supported by a fluid gas driven heat and power generation backup generator. The waste heat of the backup generator is used for hot water and central heating of the rooms. The 3-phase electrical power supply system is actually based on 6 PV string-inverters with a total power of 5 kWp and a 14 kW backup generator.

The solar generator was sized to total power of 4,95 kWp and consists of 45 solar modules type Siemens M110. Additionally the heat power generator can charge the 36 kWh storage battery at bad weather conditions. If no renewable energy is available the backup generator is charging the battery through 3 inverse working inverters and simultaneously supplying the 230 Volt AC to the refuge appliances.

The new solution within this concept is the totally AC coupling of all energy producing and using components. This means that even the solargenerators are feeding the AC grid via grid connected string inverters (SMA SunnyBoys). If there is excess energy at the AC grid the stand-alone inverters are working inverse and charge the batteries.

2.3 Schwarzenberghütte

Figure 4:   Schwarzenberghütte

The refuge „Schwarzenberghütte" (Fig. 3) is located at an altitude of 1380 m in Germany close to Hinterstein. The refuge is regularly open from middle of February to Easter and from June to October. The maximum capacity for overnight accommodation is limited to 44 visitors.

The new energy concept includes solar generator, liquid-gas driven power and heat generation plant, battery storage and inverters. The solar generator consist out of 24 solar modules type Siemens SM 110 with a peak power of 2,64 kWp. The solar generator is placed at the south roof of the refuge. If the liquid-gas driven generator is working the batteries will be charged by the charger simultaneously. At all other times the PV stand alone system with one 3,3 kW inverter is supplying the refuges electricity demand.

Figure 5:   Battery storage with 38,4 kWh capacity located in a separated room

The site received apart from the new PV system a data logger. This allows the control of the energy flows of the system and helps to check the plant performance over a longer timetable.

The data logger is located at the DC-distributor and stores the irradiance, the generated energy of the PV plant, the charge of the battery, the power of the inverters and the generated energy of the liquid gas driven generator.

The new energy concept now is based on a 48 V system voltage with one stand alone inverter of 3,3 kW for all standard 230 V AC-consumers. The battery storage capacity is about 38,4 kWh and in case of bad weather or deep discharge the backup generator is able to recharge the battery (Fig. 4).

The refuge received apart from the new energy supply system a visual display for the users. This allows the control of the battery and helps the users learning, how the systems is working.

2.4 Weilheimer Hütte

The refuge „Weilheimer Hütte" (Fig. 5) is located at an altitude of 1955 m in Germany close to Garmisch-Partenkirchen. It is regularly open May to October. Capacity for overnight accommodation is limited to 98 visitors.

The hybrid energy concept includes the solar generator, wind generator, gasoline driven backup generator, battery storage and an inverter. The solar generator consist out of 18 solar modules type Siemens SM 110 with a peak power of 1,98 kWp. The solar generator is placed at the south roof of the refuge. If the gasoline generator is working the batteries will be charged by the charger simultaneously. At all other times the wind-PV stand alone system with one 2,0 kW inverter is supplying the refuges electricity demand.

Figure 6: Solar generator "Weilheimer Hütte"

The new energy concept now is based on a 24V system voltage with one stand alone inverter of 2,0 kW for all standard 230 V AC-consumers. The battery storage capacity is about 28,8 kWh and in case of bad weather or deep discharge the backup generator is able to recharge the battery. If there is sufficient wind speed the 750 W Aerocraft windgenerator (Fig. 6) charges the battery additionally and thus reduces backup generator demand.

Figure 7: Windgenerator Aerocraft 750 W

3. MONITORING CONCEPT

According to the ISPRA monitoring guidelines a global monitoring for all systems below 5 kWp is recommended. Within the German subproject an analytical data logging system was installed additionally in some of the subprojects to investigate the new hybrid energy concepts. Data transfer had to be done by PCMCIA-Flash-cards or via GSM network (Global System for Mobile telecommunication) if available.

For the insolation a calibrated solar sensor and a special solar integrator counter (SIC100) were installed. A measurement device called "BAKO" to control the state of charge of the batteries was installed in each operating system. A display is informing about the percentage of the battery capacity.

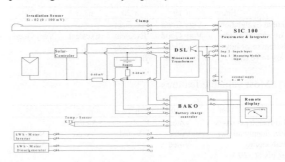

Figure 8: Monitoring concept

Within the Spanish subproject the PV-systems installed are subject to the service and maintenance program run by SEBA (a PV users association in Spain). This program requires the regular monitoring of the operation parameters of the PV-system. For this purpose, the complete monitoring system has been integrated into the TApS Power conditioning rack for stand alone electrification, which has been employed for all medium size systems of this project. Also a special software has been developed, which is needed to evaluate the proper operation and use of the PV-systems. Its name GIFA (gestión de instalaciones fotovoltaicas autónomas) stands for the management of autonomous PV-systems. GIFA has not only been designed for downloading and analysing monitoring data, but also for the overall management of the autonomous PV-systems, for example maintenance actions.

Once a system is operating, the following main data are automatically monitored:

- Maximum Battery Voltage [V]
- Minimum Battery Voltage [V]
- All relevant energy flows
- Availability of charge in the batteries [%]
- Charge cycle completed [yes/no]
- Battery charge cycles counter [1]
- full battery ratio [%]
- Disconnection time due to low battery [min]
- Battery temperature [°C].

Since most of the subproject installations have been operating now for only a few months, detailed long-term monitoring data are still not available and will be reported later.

4.   NON-TECHNICAL ASPECTS

A "Report on user's satisfaction" had been prepared by the Spanish project partners. At the end of the installation of the systems a questionnaire had been given to the user in order to be informed about his/her degree of satisfaction. Related to the Spanish subproject, there are available the questionnaires duly signed by the user related to 4 sites.

The analysis of the data gives the following conclusions:

a) Way of working of the installation

Site >	n°19	n°20	n°22	n°23
Global Evaluation	Very Good	Excelent		
Covering of needs	Very Good	Excelent	Excelent	Very Good
System reliability	Excellent	Excellent		
Easy to be used	Ecxellent	Ecxellent		

b) Most valuable features

Site >	n°19	n°20	n°22	n°23
Not fuel dependance	Voted		Voted	
Bec. Supply all day	Voted	Voted	Voted	Voted
Authonomy	Voted			Voted
Supply at 220 V				

c) Improvements that the users would do in their installations

Site	Improvement
n°19	Place a turbine
n°20	Install a SHD Water System
n°22	Install a SHD Water System
n°23	Install a SHD Water System and enlarge inverter power

d) Further conclusions

- Users normally show the installation to their visitors, this contributes to the diffusion of PV systems.
- Related to maintenance service, the general opinion is that the service is efficient when they repair malfunctions of the equipment. A feature that must be improved is the delay of time that sometimes happens to arrive to the site in question.
- Related to SEBA association, the most valuable features considered are the availability of guarantee of continuity of the installations, and seminars and papers related to teach the users and keeping them informed about SEBA activities.

5.   SUMMARY AND OUTLOOK

Within the EURALP project, 31 innovative PV hybrid systems have been realised on locations in the European Alps. These systems give a very reliable and environmentally friendly energy supply of high quality electricity for the users and at the same time establish objects of great tourist interest thus reaching a large public.

Elaborate measuring equipment at the sites and ongoing data acquisition allows a profound evaluation of the system operation and efficiencies as well as a quick detection of any malfunction. Comprehensive long-term measuring data are expected for the next years.

PV LIghting Systems Evaluation and rating methods (PLISE)

P. Boulanger [1] , P. Malbranche [1] ,T. Bruton [2] , L. Southgate [2] , S. Salvat G. Moine, J.C. Marcel[3]
F. Garcia Rosillo [4] , U. Hupach [5] , W. Vaassen [5] , C. Helmke [6] , H. Ossenbrink [6]

1 : GENEC, CEA Cadarache, 13108 St Paul Lez durance (FR)
2 : BP Solarex, Chertsey road, Sunbury on thames, Middlesex TW16 7XA (UK)
3 : Transénergie, 3 allée Claude Debussy, 69130 Ecully (FR)
4 : Ciemat, Avda Complutense, 22, 28040 Madrid (ES)
5 : TÜV Rheinland, Am Grauen Stein, D-51105 Cologne (DE)
6 : JRC ESTI, I-21020 Ispra (IT)

The PLISE project is founded in part by the European Commission (DG XII). The partners involved in this project are BPSOLAREX (UK), TRANSENERGIE (FR), CIEMAT (ES), TÜV RHEINLAND (DE), JRC ISPRA (IT) and GENEC (FR) who is the co-ordinator.

Abstract

This paper describes the work done after one year and a half within the framework of the PLISE project. This project aim at setting up a rating procedure for PV lighting systems, using mainly performance specifications in addition to component specifications. A rating method of the lighting system performance would provide a standardised information to potential customers and allow an objective comparative assessment of different designs. Initial studies showed that until now, no overall specifications nor procedures exist which address the performances of a complete system independently of the performances of the constitutive components. Three tests have to be performed to evaluate the performance of ten European solar home systems : components tests, definition and validation of indoor procedure using both a solar simulator or a programmable DC power supplied and definition and validation of outdoor procedure. Since the beginning, the procedure evolves and we present in this paper the procedure that will be proposed within the IEC TC 82 NWIP 82/218. First results are presented and allow to validate the different procedures established. Future work will give more results and allow a comparison between all the results obtained in indoor/outdoor conditions with data collected in the field.

1. Introduction

According to a market study conducted within the framework of this project many programmes have been launched throughout the world, supported by governmental or international agencies in order to impulse a free-driven market. The largest programmes deal with more than 100,000 or 1 million PV lighting systems (Indonesia, for example).

This situation obliges the manufacturers to propose a high quality product which competes with local, often lower quality products. If the choice is only driven by economic considerations, the global quality of the products will constantly drop. It is therefore very important to propose some rules and tools to enable the quality and performance of such systems to be compared.

The benefits for the PV industry are obvious : possessing a direct comparison scale for the performances of a system, in addition to the technical data concerning the components will enable the products to be more easily assessed, and will increase the level of market penetration for PV products to a wider public.

By adopting a simple and well-recognised classification scale, there will be two benefits :

1. For the final user, as for the main equipment specifiers, it will be easier to choose systems, since they will be better informed and therefore more satisfied with their purchase afterwards (presuming that the performances claimed are correct, which is not always the case).

2. For the PV system manufacturers, this rating will result in an increase in the average quality level, which will be beneficial to the user and will offer the possibility to easily prove that a product is a good product, thereby enabling a gain in time and trust for international calls for tender.

In this paper we present the objectives of the project, the technical approach adopted to address the problem and the main results obtained as for components tests, indoor tests and outdoor tests.

2. Objectives

The main goal of this project is to set up a rating procedure for PV lighting systems, using mainly performance specifications in addition to component specifications. A rating method of the lighting system performance would provide a standardised information to potential customers and allow an objective comparative assessment of different designs.

The additional goals of this project are:

- To develop a standardised representation of system features and performance, making the procurement process easier.

- During the work done for the development of the indoor and outdoor test procedures, several samples coming from different European manufacturers will be tested. These results will be a by-product of this project.

- To assess outdoor test procedures in some test centres located in developing countries, for comparison purposes under real conditions.

3. Technical approach

The main problem that must be solved is to develop an indoor rating procedure, which can fit several system designs and must be representative enough of most weather conditions and end-user needs. The selected approach consists of three steps:

1. A review of existing test procedures and standards, and at the same time, the supply of several systems, provided by the most representatives Europeans suppliers.

2. Three types of tests will be performed in parallel :
   - Component tests: performance of qualification tests for each individual component of the system.
   - Indoor system tests: testing of the completely assembled systems in a laboratory under managed conditions.
   - Outdoor system tests: testing of the completely assembled systems in outdoor conditions.

It is very interesting to see from this curve that operation is antagonistic. If the system offers high efficiency, a lot of solar energy will be lost, and the « performance ratio » will be low. To increase the performance ratio, which is often an important criteria for correctly dimensioning of a stand-alone system, it is advantageous either to increase the regulation threshold of the battery during recharging for a given battery capacity, or to increase the its capacity. But if the regulation threshold of the battery is increased, i.e. the battery is allowed to receive a greater carrying capacity, the risk is to encourage the segment A effect, which never completely recharges the battery. The situation is therefore increasingly that of mean and decreasing charge, which will favour the development of sulphating in the battery, and thereby reduce its life duration. Similarly, if the battery capacity is increased, the price of the system is also directly increased, which is not acceptable either.

4.1.c. Kits supplied

PV Kits	Wp	Ah	Volt	nb lights
Siemens AC	2*55	2*80	24	X
Free Energy	2x12	72	12	2
Total Energie	50	105	12	4
Total Energie	100	2*105	12	4
BP solarex	45	60	12	4
BP solarex	70	140	12	4
Shell Solar	50	100	12	3
Fortum	100	105	12	X

4.2. Components tests results

Each components of each systems has been tested separately with respect to their international procedure. In cases where no procedure exists at the moment (lamp, battery, charge controller) , a simple procedure is proposed. Components test results are inputs of overall system results to analyse their behaviour.

3.1. PV modules test results (IEC 61215)

PV modules have been tested according to IEC 61215 restricted to PV modules characteristics and IV curves.

PV Module	Rated Wp	Voc (V)	Isc (A)	μ (%)	Measured Wp
BPSolarex	70	21.02	4.26	10.1	63.45
BPSolarex	45	21.22	2.77	8.7	42
Photowatt	50	21.51	3.06	9.96	46.24
Shell Solar	50	NT	NT	NT	NT
Fortum	100	NT	NT	NT	NT
FEE	12	23.9	0.94	5	14.55
Siemens	55	21.6	3.2	12.1	51.5

3.2. Battery tests results (Capacity and efficiency)

Battery procedures for PV applications are manifold and several procedure address the evaluation of lifetime. The PLISE project is mainly focused on the instantaneous performances of a system. Once a ageing battery test duration is more than 3 months, it is not within the scope of the project to perform such a test.
We proposed to evaluate several capacity measurements :

- as soon as the battery has been delivered and filled up (C as delivered);
- after the first recharge (C initial)

- after 5 cycles (0-50%) to evaluate the battery efficiency.

Battery	C as delivered	C initial	Battery efficiency	C final	Rated C
Varta	99	108	95	103	105
Hoppecke	80	87	96	83	85
DETA	63	66	94	60	65
BP L60	42	62	94	50	60
Steco 3000	95	108	93	100	105
Solarbloc	132	141	92	138	140

The following graphs depict the battery efficiency obtained for each cycles. The efficiency procedure is detailed in [3].

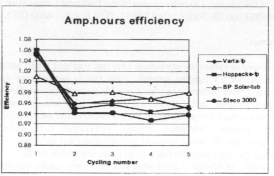

Figure 4 : Amp.hours efficiency

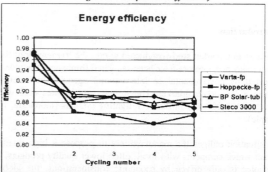

Figure 4 : Amp.hours efficiency

3.3. Lamp tests results (without luminaire)
Values measured for the reference sample at 100 h.

MODEL	LAMP NOMINAL LUMINOUS FLUX (lm)	LAMP MEASURED LUMINOUS FLUX (lm)	LAMP NOMINAL POWER CONSUM. (W)	LAMP MEASURED POWER CONSUM. (W)	LAMP NOMINAL EFFICACY (lm/W)	LAMP MEASURE EFFICACY (lm/W)	CURRENT CREST FACTOR
1	350	262.0	7.2	5.06	48.6	51.09	1.7
2	900	423.8	13	8.96	69	44.58	3.0
3	770	493.6	13	8.32	59.2	60.14	1.5
4	350	251.0	7.2	4.63	48.6	55.53	1.6
5	900	667.0	13	9.73	69	68.06	2.1
6	770	501.7	13	8.13	59.2	61.19	1.5
7	350	261.8	7.2	5.09	48.6	51.66	1.6
8	400	358.1	7	-	57.1	-	-
9	770	592.3	13	9.32	59.2	63.50	1.7
10	600	596.1	11	-	54.5	-	-
11	900	243.2	13	5.67	69	40.21	2.9
12	600	395.7	10	-	60	-	-
13	920	707.2	15	10.99	61.3	61.58	1.8
14	280	186.2	7.2	5.48	38.9	34.40	3.8
15	1040	558.2	20	9.83	52	55.20	2.3
16	400	313.6	7	4.69	57	67.42	4.69

3.4. Regulators tests results

Regulators or charge controllers will be tested according to the CENELEC PrEn 10/612 procedure. Here again, tests will be focused on performances and safety but not environment. Test are in progress.

3. Validation of the indoor rating and test procedure and use of this procedure within the appropriate standardisation organisations.

4. Results and discussions

After one year of work, the main results are summarised in the following tasks.

Task 1 (completed) : Critical review of existing procedures and standards is made, design simulation is made

Task 2 (completed) : Photovoltaic systems are supplied and monitoring system is defined

Task 3 (in progress) : Components test procedures are defined

Task 4 (in progress) : Indoor test procedure are defined, first tests are under progress

Task 5 (in progress) : Outdoor test procedure is defined, first tests are under progress, a simulation tool is developed

4.1. Critical review and kit supply

In order to better understand the behaviour of a small photovoltaic solar home system, the work has been divided into two parts :

- An analysis of existing test procedures
- An analysis of feedback from the field

4.1.a. Analysis of existing procedures

The following items result from this analysis of more than approximately 20 applied, or applicable references :

- None of the procedures analysed deal with the issue of evaluating the performances of a small domestic photovoltaic system - they all examine the problem from a « components » viewpoint, by systematically testing the components individually.
- Very little data and procedures are available to identify a solution to the problem raised. Most of the solutions proposed do not have any experimental foundations to validate them.
- The comparison between indoor and outdoor testing is never justified, and the procedures proposed are often very similar, not to say *too* similar. This has no significance in that the parameters from laboratory tests are controlled, whereas those conducted outdoor are closer to true conditions.

4.1.b. Analysis of field experiences

Our analysis has been mainly based on two major test campaigns, one in the laboratory [1], and the other on site [2]. The missing data to permit precise simulation of the behaviour of a system is the consumption profile. GENEC tests show that the behaviour of a kit can be affected by variability in consumption. The results presented in the data analysis from Indonesia show that the consumption profile is as shown in figure 1.

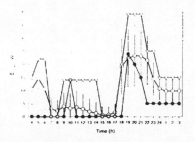

Figure 1 : Consumption profile example (from [2] - Utrecht, 1995)

The two independent references show that the performance of a system expressed such as quantity of electricity entering the battery depending on the average daily quantity of sunlight results in a characteristic graph shown in figures 2 and 3.

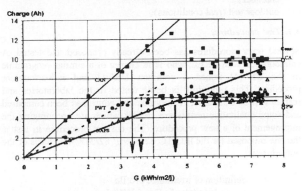

*Figure 2 : Characteristics of a PVLK
(GENEC data from laboratory tests)*

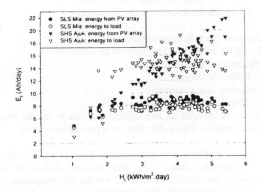

*Figure 3 : Characteristics of a PVLK
(data provided by the University of Utrecht - Indonesia).*

Such a graph is easy to analyse - the slope of the straight line segment is exactly in correspondence with the « global efficiency » of the system such as it is defined in the IEC 1724 standard.

The horizontal line corresponds to the limitation of carrying capacity of the system. It is useful to observe that a « perfect » photovoltaic system follows this curve exactly. A so-called « imperfect » system, due to the fact that additional losses are included in its efficiency, will be below the three straight sections described below and illustrated in figures 2 and 3. A short summary of a system behaviour is described in three segments :

- Segment A - the battery receives the given quantity of energy by the sun, affected by the efficiency coefficient. This corresponds to an incomplete recharge, on a day with little sunlight, for example.
- Segment B - the recharging of the battery is limited due to the fact that it is already at a high level of charge. The regulator comes into action and cuts out at a certain threshold (the threshold does not appear directly in Volts here). This corresponds to operation following several days of full sunlight.
- Segment C - the recharging of the battery is not limited by the regulator as the battery's charge level is low. The battery therefore recharges to a higher level. This will occur for example during the first sunny days after a period of bad weather.

4. System test results

Two types of tests have been performed :
- indoor tests (solar simulator or DC power supply + climatic chamber)
- outdoor test (real conditions)

4.1. Test procedures

Several tests procedures has been already proposed and tested. A collaboration with NREL, who is in charge of a similar programme in the US [4] issued a common US/EU and indoor/outdoor procedure. This procedure is being tested in the laboratories and results will be compared. A previous procedure has been compared and some qualitative results are similar which are encouraging the improvement of a new procedure. The current procedure in test is the one proposed for the IEC TC 82 NWIP 82/218. The basis of the procedure is :

- definition of irradiance profile
  "good day" : G > 5 KWh/m²
  "bad day" : G < 1.5 KWh/m²
- definition of two 7 days sequences
  Functional sequence : GBBGGGG
  Recovery sequence : GGGGGGG
- definition a complete test procedure

The purpose of the system performance tests is to
1. Ensure the system and load operate as expected.
2. Evaluate the system performance.
   Battery autonomy
   Battery time to recovery
   Overall system performances
3. Determine if there is any significant change in the usable battery capacity during the short-term performance tests.

Figure 1 is a graphical representation of the testing in terms of battery voltage and time. The time periods shown are only approximates. The procedure is divided into the following steps :
UBC_0 : Initial capacity test – After installing the system, charge and discharge the battery. Measure the usable battery capacity.
CB :   Recharge the battery before running the functional test.
FT :   Run the functional test to verify the system and load operate properly.
UBC_1 : Second capacity test – Charge and discharge the battery. Measure the usable battery capacity.
RT :   Recovery test – Determine the ability of the PV system to recharge the discharged battery.
UBC_2 : Final capacity test – Charge and discharge the battery. Measure the usable battery capacity.

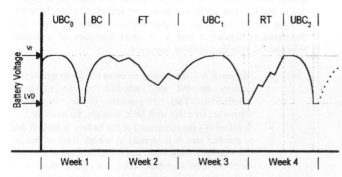

4.2. Tests results

JRC ESTI has tested several kits both in a solar simulator and in outdoor conditions. TÜV Rheinland tested the complete systems with an electronic power supply instead of PV modules, the others components of the systems being still installed in a climatic chamber. GENEC tested the systems in outdoor conditions.

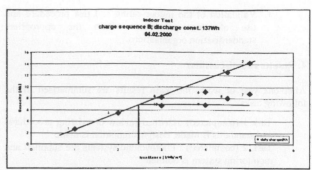

Figure 5 : Indoor performances test of an SHS (TÜV Rheinland)

First tests have been performed that validate the easy implementation of the procedure. Some difficulties have been encountered to performed instantaneous measurement in parallel for several kits which do not behave the same way at the same time. Some components have exhibited non co-operative behaviour.

5. Expected achievements and exploitation

It is expected that at the end of this project, the main results will be:

- the creation of a standardised representation of system features and performance, thereby simplifying the procurement process and improving the commercial marketability.
- a proposal made to the CENELEC and IEC TC 82 groups for a new standard.

6. Conclusions

This paper presented the objectives, the technical approach and the main results obtained within the framework of the PLISE project (JOR3-CT98-0258). The main goal is about to be reached and several tests results will be available at the end of the project. During the tests several aspects of SHS have been pointed out which can motivate a lot of innovative study to better understand how to properly size a SHS, the general quality control of the manufacturing process and how to evaluate some technical aspects of a SHS (influence of temperature, compatibility regulator/battery, failures, influence of the strategy management, etc ..).

Solar home system market and industry is booming especially in developing countries and a project such as the PLISE project intend to be a reference in the quality and performance evaluation specifications.

Références

[1] A.Reinders - Performance analysis of photovoltaic systems" - PhD thesis - Utrecht University - 1995
[2] JM Servant, JC Aiguilon - "Tests of PV lighting kits" - 11th PVSEC - Montreux (CH) 1992
[3] P. Malbranche, S. Metais, F. Mattera, D. Desmettre - "Quicker assessment of lifetime and of PV batteries"- 16th PVSEC - Glasgow (UK) - 2000
[4] "Draft Recommended Practice for Testing the Performance of Stand-Alone Photovoltaic Systems" - P1526/D1 - Draft 1 - NREL - April 2000

ELECTRICITY FROM A GOLDEN SOLAR GENERATOR FOR THE GOLDEN CITY PRAGUE

M. Grottke[1], P. Helm[1], A. Karl[1], J. Rehak[2], A. Kovar[2], F. Novak[3], A. Biasizzo[3], C. Panhuber[4]

1. WIP - Renewable Energies, Sylvensteinstraße 2, 81369 München, Germany,
tel.: +49 - 89 - 720 12 35, fax: +49 - 89 - 720 12 91, e-mail: wip@wip-munich.de
2. SOLARTEC s.r.o., 1. Máje 1000/M3, Roznov pod Radhostem, Czech Republic 756 64
3. Jozef Stefan Institute, Jamova 39, 1000 Ljubljana, Slovenia
4. Fronius Schweissmaschinen KG, Günter Fronius Str. 1, 4600 Wels-Thalheim, Austria

ABSTRACT: The objective of this joint demonstration project supported by the European Commission INCO-Copernicus programme, was to develop and install innovative, efficient and reliable photovoltaic PV modules together with the Czech photovoltaic industry. A unique feature of this PV façade extension is its gold coloured monocrystalline PV panels. This is the first coloured PV façade extension in eastern Europe. This paper presents an analysis of the first measurement data obtained. PV system efficiency is approx. 9% at STC and proofs that high PV module efficiencies can be obtained with the optically attractive coloured PV cell technology.
Keywords: Coloured PV Cells - 1: Grid-connected PV System - 2: Real-time Data Visualisation - 3

1. INTRODUCTION

In early April 2000 the golden colour PV façade extension was officially inaugurated in Prague in the presence of Czech ministers, decision makers, a representative of the European Commission and the active participation of CZ television and press. The grid-connected PV system is part of the façade of one luxurious hotel in Prague, the Corinthia Panorama Hotel (Fig. 1).

Figure 1: Part of the golden colour PV façade extension installed at the Corinthia Panorama Hotel in Prague.

It is the first publicly accessible and grid-connected PV system in the Czech Republic and the first larger-scale PV demonstration system installed by SOLARTEC s.r.o., the only manufacturer of PV cells in the Czech Republic.

PV system's performance is continuously monitored and evaluated by three independent data acquisition systems to allow a continuous operation control and a detailed PV system analysis and visualisation. Monitoring data and information on the PV system and the project partners is available via Internet at http://1fv-fasada.solartec.cz
The project was initiated and is co-ordinated by WIP.

2. TECHNICAL FEATURES OF THE PV SYSTEM

The 5,8 kWp PV system was installed in the South-facing building envelope of the Hotel. It consists of three grid-connected, independent sub-arrays with identical configuration: four strings in parallel with each 11 modules connected in series feeding one FRONIUS Sunrise Midi inverter.

PV MODULES

Manufacturer of PV cells	SOLARTEC, CZ
Solar Cell Type	Monocrystalline Si; golden active side
Size of PV cells	104 cm²
Number of PV cells per module	36
Size of PV modules	1,005 x 0,453 m²
Manufacturer of PV modules	TRIMEX Tesla, CZ
Peak power per module	44 Wp
Total number of modules installed	132

INVERTERS

Manufacturer	FRONIUS, AT
Type	Sunrise Midi
Pnom-dc	1,6 kW
Number of units installed	3

All three PV arrays are facing to South. However, for architectural reasons, the PV sub-arrays are not equally tilted. The line of PV modules installed closest to the hotel tower is tilted with 50°, all other PV modules are tilted with 35° (Compare with Fig. 5).

3. DATA MONITORING CONCEPT

The Prague PV system is remotely controlled by the plant operator SOLARTEC who is situated 350 km to the East of Prague, in Roznov pod Radhostem. The measurement concept was defined by WIP.

Continuous operation control is guaranteed via three independent Data Acquisition Systems (DAS):
- an inverter integrated DAS developed by FRONIUS for system operation control
- a high precision PC based Schlumberger DAS with Isolated Measurement Pods for detailed system analysis installed and operated by WIP
- a Real-time data visualisation system based on a PC-integrated data acquisition board for data visualisation on a monitor in the foyer of the hotel and Internet presentation. It was installed by JSI

3.1 Inverter-integrated System Operation Control

FRONIUS developed an inverter-integrated data acquisition system taking the most relevant inverter input and output data on a 15 min. avg. basis. A so called 'Modem Box' links inverters installed at different locations in the building and can transfer the monitoring data via modem to any remote PC equipped with a modem and a connection to a telephone line. Malfunctions of the PV system can immediately be detected.

This system is operating reliably. No failures occurred since the start of PV system operation. Measurement results (daily energy values) stay within appr. ±4% relative to the higher precision Schlumber DAS.

3.2 Schlumberger DAS with Isolated Measurement Pods

A PC based 16 bit Schlumberger DAS with Isolated Measurement Pods was installed and is operated by WIP to allow for an independent and remote in-depths analysis of the golden colour PV array performance and to check the quality of 'lower-cost' alternatives, such as the inverter-integrated DAS and the PC-integrated data acquisition board.

This system is operating to the full satisfaction of WIP.

3.3 Real-time Data Visualisation System Based on a PC-integrated Data Acquisition Board

Real-time data visualisation is based on a PC with integrated data acquisition board. The system is used to advertise for PV in the Czech Republic, (a) by a real-time data visualisation in the foyer of the hotel - Czech and English language - (Fig. 2) and (b) to advertise for the Czech PV industry in the world by supplying aforementioned information via Internet. On-site the visitor has the choice to check the current status of the system or / and to compare the current data with data measured in the past. He also has the possibility to learn about PV systems in general, to study the PV system design or to get in contact with the involved companies.

The concept of the Internet visualisation and the type of data to be monitored as well as the data evaluation concept was defined and designed by WIP. JSI programmed and installed the monitoring system in accordance with given system description and operation requirements. One of the Internet pages showing a simplified block diagram of the PV system is depicted in Fig. 3.

The access to PV system data and information via Internet turned out to be quite expensive. This is mainly due to the costs

of the leased telephone line providing permanent access to the web site.

Figure 2: Real-time visualisation monitor installed by SOLARTEC in the foyer of the Corinthia Panorama Hotel in Prague.

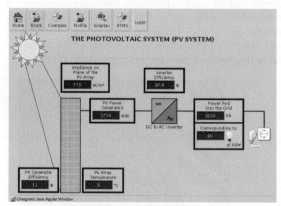

Figure 3: The simplified block diagram of the PV system in Prague – one of nine real-time data visualisation screens prepared for Internet presentation.

4. MONITORING RESULTS

Since the operation started in December 1999, the PV system is operating to the full satisfaction of the consortium. No major failures occurred and the energy production complies with the design targets as depicted in Fig. 4. A total yearly energy production of 4.600 kWh or 800 kWh/kWp of installed PV array power is expected, while the daily output will range between 5,0 kWh (Jan) and 26,0 kWh (May) in the monthly average or 0,9 kWh/kWp (Jan) and 4,5 kWh/kWp (May), respectively.

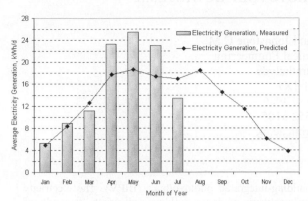

Figure 4: Measured and predicted PV array energy output since start of system operation. Results are depicted in kWh per day.

A more detailed evaluation of the PV array performance at a day with clear sky irradiance (Fig. 5) shows that PV sub-array output is different. Due to lower shadowing losses by the hotel tower (16%) and a lower PV sub-array tilt-angle (5%) PV sub-array #3 exhibits an energy output 21% higher than the energy output of the PV sub-array #1 installed close to the tower which is partially shadowed by the tower in the early morning and late afternoon hours. This reduction in power capability is the result of a compromise which had to be agreed-upon to allow for a visible installation of the PV array to the visitor of the hotel.

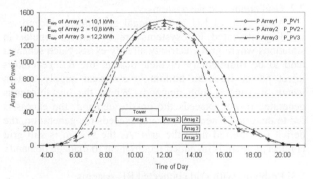

Figure 5: Measured PV dc power output of the three independent PV sub-arrays at a summer day with clear sky irradiance (June).

Efficiency of golden colour PV cells is lower than that of standard, dark-blue colour PV cells. Our measurements showed that the PV module efficiency reaches values between 8% and 9% (Fig. 6) which corresponds to about 75% of the power of dark blue-coloured, conventional PV modules, as supplied by SOLARTEC. Resulting PV cell efficiencies after encapsulation are appr. 11% at STC.

FRONIUS Inverter efficiency was calculated (Fig. 7) and reflects the characteristics of up-to-date transformer-technology inverters with 90% efficiency at 10% of nominal power; 93% up from 13% of nominal power and a maximum of 96% at 40% of the nominal power, respectively.

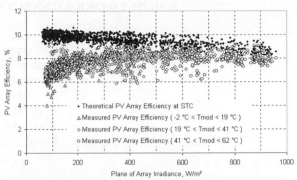

Figure 6: Measured PV array efficiency versus theoretical PV array efficiency at STC.

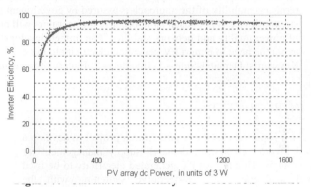

Midi inverter.

CONCLUSIONS

The target to install a PV demonstration system in the Czech Republic using innovative golden colour PV modules from national Czech production was achieved. PV system performance is up-to-date and the dissemination effect achieved at the official system inauguration (Fig. 8), via the real-time monitoring display in the foyer of the hotel and by Internet did fulfil the expectations of the consortium and will certainly contribute to boost PV industry in the Czech Republic and Europe.

Figure 8: Prague PV system inauguration.

ACKNOWLEDGEMENT

This project is co-financed by the European Commission through the INCO-Copernicus programme of the Directorate General for Energy and Transport and by the Ministry of Education, Youth and Sports of the Czech Republic.

DEVELOPMENT OF DEMAND AND SUPPLY MANAGEMENT CONTROL SYSTEMS FOR NETWORK CONNECTED PHOTOVOLTAIC SYSTEMS.

Clarke J A, Conner S J and Johnstone C M
ESRU, Dept. of Mechanical Engineering
University of Strathclyde
Glasgow G1 1XJ
Tel. +44 141 548 3986
Fax. +44 141 552 5105
Email: esru@strath.ac.uk

ABSTRACT
Network connection is often the preferred configuration for building-integrated photovoltaics (BIPV). The electricity generated by the system is exported to the local electricity supplier's network, while the building's electricity demands are separately imported. This approach places the responsibility for network stability on the local supply company. The short-term acceptability of the approach is assured by government subsidies. In the long term, the use of the supply network as a storage device is not sustainable because the operators will not be able to absorb the costs associated with the maintenance of power quality and network stability. The further development of BIPV will require approaches to power utilisation that avoid the need to export PV generated electricity to the grid.
Keywords: Demand-Side – 1; Inverter – 2; Battery Storage and Control – 3.

1. INTRODUCTION

This paper reports the results from a project to develop a smart inverter capable of demand/ supply management for PV systems. This allows PV generated electricity to be used when available, with co-operative switching with the electricity supply network when demand exceeds supply. The approach maximises the autonomous use of PV generated electricity while minimising network instabilities. It also reduces the costs associated with network connection, prevents contravention of the power quality/ compatibility legislation and allows for safe network maintenance by eliminating back feed of power to the network.

The paper reports laboratory results from power quality tests of inverter operation and describes the demand management control logic used by the inverter to enable co-operative operation of PV and supply networks.

2. BACKGROUND

2.1 Traditional RE systems

Building-integrated renewables have traditionally been arranged so that all the energy is used locally. A possible system might consist of several PV modules and small wind generators with DC outputs, fixed to the roof and connected to a battery. The power from the battery can be used directly by low-voltage DC appliances, or converted to standard AC mains voltage by a power inverter. These systems are usually found in remote locations where there is no grid electricity and the cost of installing power lines is high.

There are two drawbacks to this approach. Firstly, the incoming renewable energy is limited, as is the energy stored in the battery. If demand exceeds supply for long enough, the battery will discharge and the system will fail until sufficient renewable energy has been accumulated to recharge it. Careful system sizing and demand management is required to prevent this occurring. Secondly, the peak available power is restricted by current limitations in the battery and inverter. This can make it difficult to operate appliances with high starting currents such as microwave ovens and water pumps.

2.2. Grid connected RE systems

As a result of EU and national government incentives, building-integrated RE systems are becoming popular in situations where grid electricity already exists. Such schemes tend to be rudimentary in their design: the RE generators inject power into the building's electrical supply system where it cancels out demand. When the instantaneous generated power exceeds the on-site demand, the excess is fed to the grid. When the demand exceeds the generation, the deficit is imported from the grid. As the system is still grid connected, there is no added requirement to manage demand.

2.3. Problems with grid connected RE systems

This mode of operation contains hidden drawbacks. Most importantly, the electrical grid is a carefully balanced system [1]. Since electricity cannot be stored, generation must be planned to match demand. Grid-connected renewables represent additional generation that cannot be planned and has limited controllability. At present, this has little impact on network control and stability as renewables only form a small proportion of the total generating capacity.

3. INVERTER PERFORMANCE ASSESSMENT

Laboratory tests were undertaken to characterise a range of inverters used for both standalone and network connected applications. Experiments focused on the performance and part-load efficiency of hardware currently on the market.

The test procedure comprised a digital storage oscilloscope to capture the inverter AC current and voltage waveforms. The data was then subjected to an FFT harmonic analysis. The DC voltage and current were also measured to determine efficiency.

Tests were conducted on a low cost modified sine wave inverter, with a variety of loads imposed ranging from 10% to 100% of the rating. At each loading, the voltage and current waveforms were assessed for stability, frequency variation and harmonic distortion. Figure 1 shows the voltage profile for the inverter and Figure 2 shows the magnitude for the range of harmonic distortion measured.

A second series of tests were conducted for a network-connected inverter. This used microprocessor technology to deliver a true sine wave output, with maximum power point tracking and anti-islanding. Figures 3 and 4 demonstrate the voltage waveform and measured harmonic distortion respectively, while Figure 5 displays the efficiency against load characteristic.

Testing of the standalone inverter demonstrated that the waveform description, 'modified sine wave', is a marketing euphemism for a square wave. The usual modified sine wave has extra zero periods inserted in the rising and falling sections, but in this particular inverter these disappeared as the load increased.

The harmonic distortion associated with this waveform is high, causing supply sensitive appliances to malfunction, e.g. PCs, stereos etc. However, as a standalone unit, this inverter is not subject to power quality regulations. Testing of the network connected inverter demonstrated that the current waveform shows a reasonable emulation of a sine wave. Obvious defects were the missing sections near zero current, and flat tops, which correspond to a measured total harmonic distortion of 7%. The power factor was near unity at medium and high powers, and became more leading as the power decreased. The observed results suggested a constant reactive power of approximately 10VAr, presumably due to large capacitance effects on the output filter. Start-up, shutdown, and changes in power level, were effectively managed with no current surges or transients observed.

In network connected installations, the observed harmonic distortion of the line voltage depended on the harmonic distortion in the inverter line current and the impedance of the network. This impedance is usually relatively low so that the amount of current generated by typical PV/inverter installations does not result in large amounts of additional voltage distortion. When grid power was removed from the inverter/test load combination, to simulate an islanding incident, the inverter shut down within one cycle. The protection system proved robust under several power/ load scenarios.

4. ISSUES ASSOCIATED WITH GREATER RENEWABLE DEPLOYMENT

If future predictions for the increase in renewable based electricity generation are realised [2], severe problems in network stability will emerge.

One solution to this is to add centralised storage to the grid, either in the form of pumped-storage hydro or large capacity regenerative fuel cells. However, the time constants and conversion efficiencies associated with surplus electricity being transmitted to a central storage unit, stored, and then recovered and transmitted back to the demand site, will result in large losses.

Reducing these losses by introducing distributed storage would be effective and could be implemented using batteries in individual buildings, or fuel cells at the local/ community level. This would eliminate power being returned to the grid, and drawn from it. For this to function, effective demand side management is required.

5. ENABLING TECNOLOGY

There are many social, political and cultural hurdles to be overcome before a RE based society can be realised. The technologies required to undertake this have been developed for other types of operative configurations.

5.1. Inverters

State-of-the-art inverters are advanced. The task of converting DC to AC can now be undertaken with efficiencies greater than 90%, insignificant levels of harmonic distortion, and proven long-term reliability.

Low cost inverters: These are the original types primarily used in standalone applications. They deliver as power as required, until a safe limit is reached or the DC voltage falls below a certain level, indicating a discharged battery. In network connected applications, they extract as much power as possible from the source and supply it to the grid.

Smart inverters: These contain some rudiments of intelligence. Normally, grid-connected units can isolate themselves from the grid during a power failure and continue to supply the site, re-synchronising and connecting when power returns. Some units can operate in reverse as battery charge controllers, prioritising charging of the battery attached to the DC port when it is at a low level of charge.

Units of this type are a good example of co-operative switching. Although currently illegal in many countries, they are permitted in US network connections providing they comply with IEEE P929 [3].

Inverters of the future: These will possess greater levels of intelligence and will require open-ended possibilities for updating and expansion. One possibility is to remove the intelligence from the inverter itself and relocate this within a computer. The inverter then becomes a power converter acting on commands from the computer controller, with a minimum of on-board logic required to maintain safety and power quality. The computer would have the ability to communicate with several modules for maintaining load control with suitably enabled appliances, to form a smart power system. The main issue is the development of suitable control logic.

5.2. Controller

In the following, the inverter caters for low-level protection and synchronisation, and is permanently connected to the electricity supply network, being disconnected only in the event of a grid failure. This is in contrast to some co-operative switching systems where the grid is disconnected except when importing power. Response speed dictates that the switching should be controlled at a low level by the inverter's own logic.

The aim of such a high-level controller is as follows:
i)   to maximise utilisation of RE streams;
ii) to minimise grid power demand (real and/or reactive);
iii) to prevent back-feed to the grid; and
iv) to manage storage facilities subject to constraints (e.g. batteries must not be overcharged or discharged in deep cycles, and must not stay discharged for too long to reduce the risk of damaged).

These objectives may well conflict under certain conditions. For example, if the RE supply exceeds the on-site demand, aims i) and iii) require that the extra power should be stored on-site. However, if the batteries are already full, aim iv) will preclude any attempts to feed in more energy. Also, if the batteries have remained discharged for too long, aim iv) will result in a demand for energy to recharge them. If there is still insufficient RE for this purpose, power may have to be imported from the grid in conflict with aim ii).

More advanced systems incorporating demand side management would fulfil aim i) by switching on 'opportunistic' loads when the RE supply exceeded demand. It could also help towards aim ii) by switching off 'non-critical' loads when RE supply and stored energy is low.

However, this raises some complex issues when considering human interaction and user functionality. 'Opportunistic' and 'non-critical' loads would have to be defined to prevent crucial appliances ceasing to function when in use. Non-critical loads might include lights in an empty room, an idle computer, or a non-functioning stereo. Opportunistic loads might include a water pump feeding a large storage tank, or water heater. Critical loads might include lights in an occupied room, a washing machine in mid-cycle, an operating computer, or a TV in use

Clearly a given load will not always fall into the same category. For instance, a computer might be non-critical when lying idle, opportunistic when performing some low priority task, such as data compression, and critical when being used interactively. Lights can be non-critical or critical depending on whether the space is occupied and the level of daylight availability.

6. CONTROL LOGIC OPTIMISATION

From a mathematical viewpoint, conflict resolution is similar to optimisation. A weighting can be assigned to each unit of energy going towards a given objective, with positive weightings representing the cost of that energy, and negative weightings representing the benefit derived from it. For instance, there would be a high cost assigned to energy bought from the grid, but a still higher one for energy drawn from a discharged battery. Energy fed to lights in an empty room would have a near-zero benefit, while energy used for a computer in the middle of some important task would have a large benefit.

The problem is to find a way of sharing out the energy streams that minimises their weighted sum. This corresponds to maximum benefit delivered at minimum cost. Since there is one degree of freedom for each energy stream, this is a multi-dimensional optimisation problem. Some of these problems have a straightforward analytical solution, while other problems are complex in which case there are several possibilities for solution, such as simulated annealing, genetic algorithms, and heuristic or rule-based techniques.

Although the problem discussed is simple as optimisation problems go, a new solution will be needed every time the energy streams or weightings change. This may rule out methods such as genetic algorithms because of their slow speed.

6.1 Smart Loads

To operate demand management without user intervention would require some form of sophisticated technology. One way of implementing this is to equip each appliance with a communication device. This would not necessarily lead to extra data cables, since it is possible to transfer data over an existing power connection. The appliance would then report its status and demand for power to a central control program. Using this information, and data from other sources (e.g. infra-red occupancy sensors, battery management systems, voltage and current transducers), the program would compute an optimum course of action and return an on/ off command to a specific electrical appliance. This may require that operational state commands be sent to other equipment within the same network configuration. Although such control protocols may seem futuristic, the required technologies currently exist. Many appliances have a rudimentary 'brain' already and it would be possible to add the envisaged energy-related functionality. The 'central computer' would be nothing more than smart box discretely located in the building.

This raises issues associated with networking where the main obstacle to the success of a system would be compatibility. A networking standard would need to be agreed so that equipment from different manufacturers would co-exist. Networks are typically envisaged as layers.

For the physical transport layer, one of the many available types of power line modem could be used, e.g. the type used in the X-10 home automation system [4]. A short-range, low-cost wireless network such as DECT or Bluetooth might then be employed.

The transport protocol layer would need to support tens or hundreds of nodes and be able to handle with large numbers of small data packets. One of the Fieldbus type protocols used in building and plant automation might be suitable. There is already plans for a new generation of appliances each with its own IP address, although TCP/IP might be viewed as excessive due to its elaborate routing facilities.

A unified interface (know as an API) would also be required, with different degrees of implementation for different appliances. At the most basic level, the interface would indicate when the load was switched on and off, send a weighting and a rated power, and receive back an instruction to switch itself on/off. Such functionality might typically be located within a light switch.

More advanced appliances might require extra functions. A system which runs a well-defined cycle and should not be interrupted, such as a washing machine or microwave cooker, could benefit from being able to reserve a block of energy in advance. Opportunistic loads would also need to be told when they should proceed.

Such appliances might also have user interfaces to the power management system. For example, a washing machine might provide options such as "do the washing as soon as energy becomes available" or "do the washing immediately, even if it means importing energy" or even "notify me when there is enough surplus renewable energy for washing".

7. CONCLUSIONS

Using currently available technology, it is possible to build an integrated power management system, which can automatically minimise demand and maximise utilisation of renewable energy. The core problem is the definition of the time-varying costs and benefits associated with energy streams, and the determination of the alterations to the energy streams required to minimise the cost/benefit ratio. Optimisation problems of this nature have been addressed in other fields.

The most difficult aspect is to find a satisfactory way to define the cost/benefit weightings. The financial costs of energy, and the penalties associated with over-discharging batteries and back-feeding electricity to the supply network are well known. However, the human value related to energy demands is more subjective, e.g. the value that is assigned to a hot bath on a cold day. The configuration of benefits should be left to the system end-user.

References
[1] "Recommendations for the connection of embedded generation plant to the regional electricity companies distribution system" Engineering Recommendations G59/1, EA, UK, 1990.
[2] "NEW & RENEWABLE ENERGY, Prospects for the 21st Century" http://www.dti.gov.uk/renew/condoc/
[3] "Recommended Practice for Utility Interface of Photovoltaic (PV) Systems" P929 IEEE Standards Department, Piscataway, NJ, USA 1989.
[4] http://www.intellihome.be

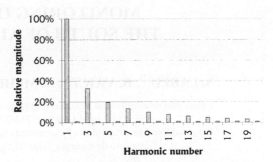
Figure 2: Magnitude of distortion for the range of harmonics measured

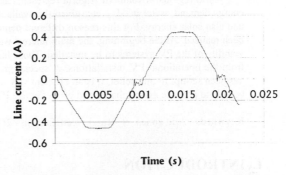
Figure 3: Voltage wave form for network connected inverter

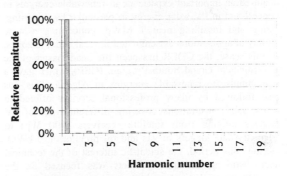
Figure 4: Magnitude of distortion for the range of harmonics measured.

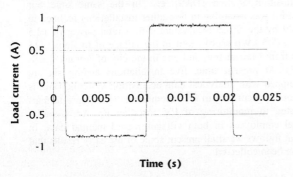
Figure 1: Voltage and current waveform for a standalone inverter.

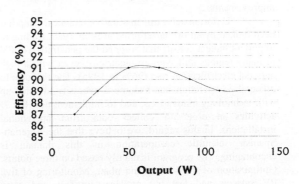
Figure 5: Efficiency varying with load for the network connected inverter

MONITORING OF FIVE PV SYSTEMS IN THE SOUTH OF ALGERIA: EARLY RESULTS

S. LABED[1], B. YAICI[1], A. MEHDAOUI[2], M. SADOK[2] and E. LORENZO[3]

[1] Centre de Développement des Energies Renouvelables (CDER) BP 62 - Bouzaréah, Alger (16340) - Algeria
Tél. 213/2/90.14.46 Fax. 213/2/90.15.60 E-mail: **LABED_S@Hotmail.Com**
[2] Station d'Expérimentation des Equipements Solaires en Milieu Saharien (SEES-MS) BP 478 Adrar, Algeria
[3] Instituto de Energia Solar (IES), Ciudad Universitaria s/n 28040, Madrid, Spain

ABSTRACT

Arid regions in southern Algeria represents an alternative to face many national problems: population growth, energy demand, water needs...etc, given the availability of many natural resources such as water reserves, oil, mines and also solar energy. For this reason different demonstration projects devoted to PV solar energy applications have been realized. At the beginning, the main interest was focused basically on the technical feasibility and the economic viability of the PV systems to be installed. These last years, a special attention has been accorded to data collection from the operational PV installations. A special program consisting in supplying 20 installations with data loggers has been set-up. In our belief, it is an efficient way to learn about the technical, economical and even social aspects to be able to propose more reliable and cost effective PV systems.

Keywords : Arid zones - 1: Rural electrification - 2: Monitoring - 3

1. INTRODUCTION

Since more than 10 years, the CDER has acquired an important experience in renewable energies in general and in PV solar energy in particular by designing, sizing and installing near 1 MWp, concerning a wide variety of applications disseminated in rural and arid zones. In this sense, the CDER has been the head master of the governmental Grand South Program (GSP), which gave a real impulse to renewable energies in Algeria. This was also followed by many professional services offered to many public institutions covering different types of PV systems ranging from satellite receivers (100 Wp) to telemetry data loggers of gas and oil pipelines (10 kWp). During this period, the scientific interest of the technical teams working on these projects was focused on the acquisition of know-how, demonstration of technical performances and reliability of the systems in such isolated regions, rather than collecting data necessary for future improvements.

Unfortunately, this qualitative technical jump has been braked by an oil barrel price drop in 1986, leading to important budget cut-off that stopped the penetration of PV in many regions still without electricity, even with a high national electrification rate (96% in 1995). The recourse to bilateral co-operation has become a necessity to catch-up with technology innovations, and to implement new R& D activities in order to collect data from operational installations. In this regard, we believe that the Algerian-Spanish scientific co-operation in this domain is encouraging. The program is mainly based on three points: Optimization of the Matriouane plant, Monitoring of five PV systems and last but not least module and array characterization (IV curve measurements).

This paper will try to focus on the two last points mentioned above, since the first has been achieved already. A special attention is accorded here to monitoring and data analysis. We believe that this last will help us in setting-up appropriate methods of: maintenance, technical and economic studies...etc. in favor of an open market (Sonelgaz national program 500 kWp), appearance of private operators,...etc. The objective of all this is to aware decision makers that fossil reserves will not last forever.

2. MONITORING & DATA ANALYSIS

The experience gained with the electrification of Matriouane as well as its monitoring and data analysis has been rich and very educational in different aspects : technical, social, management, test of equipment in harsh conditions...etc. This had helped the technical team in charge of the installation follow-up to identify the weak points and to propose many actions [1] such as: replacement of non-efficient equipment (inverter, light...etc.), installation of a PV DC supply for the Data acquisition System (DAS)...etc. In the same time our interest was extended to five other installations realized in 1989 by the SEES/MS(Adrar) with a total power of 12.5 kWp (2.5 kWp each) : four in the village of In Belbel (15 km from Matriouane) and one in the city of Adrar (SEES-MS). At present time, five installations among six are equipped with analytical data acquisition systems. Data is recorded according to EC recommendations (Ispra format) related to monitoring of PV installations. Additionally, local employees in both villages record manual data. In what follows we shall present some of the early results that have been collected.

2.1 Matriouane

Photovoltaics is synonymous of rational use of energy, efficient equipment ..etc. Results collected from the Matriouane installation, after replacing all the lamps of the village with low consuming lamps (11W) and changing the main inverter (with IGBT's)..etc. confirm this. In fact, if we compare consumption of January for both 1991 and 1996 as shown in figure 1 we can see easily the net difference. This saved energy reaches 30% (for the indicated month) of the total mean daily consumption, which means for the villagers more energy for agriculture (irrigation of green houses). This has always been their willing, since agriculture in this village has flourished with the coming of PV solar energy.

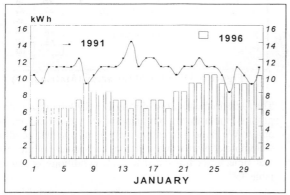

Figure 1 : Energy daily consumption - Matriouane

Much more than this is obtained in figure 2 where we can see the gradual increase (energy demand) from the starting of the installation in 1991 to the beginning of 1996.

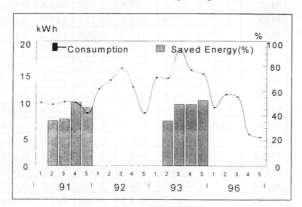

Figure 2: Daily consumption evolution & relative energy saving (reference year 1996)

It passed from 10.1 (1991), 12.2 (1992), 15.5 (1993) to drop to less than 7.5 kWh/day when efficient equipment have been connected. With this new configuration 50% of the total consumption can be saved as indicated in the same figure.

2.2 In Belbel

The village of In Belbel is 15 km from Matriouane and has been electrified by SEES/MS in 1989. Four PV stand-alone systems (2.5 kWp each) feed the whole village via four local grids with the same size of houses in theory. Three of these four installations are equipped with analytical data recorders since Dec.1995. This is the first time that data is recorded since 1989 and also an opportunity to learn about the operation of these systems.

Table 1: Mean monthly daily consumption (kWh)

System Id.	Dec. 95	Jan. 96	Feb. 96	Mean.
IB 02	4.25	4.06	4.26	4.20
IB 03	4.50	4.33	4.53	4.45
IB 04	5.66	5.33	5.56	5.51

A first remark can be made if we look at table 1, where we can note that even if the systems have the same size and feed the same number of houses, the daily consumption is different. This is basically due to the fact that villagers have connected other loads than those of 1989 as we noticed during our visits in Dec. 1995 and 1996. That can be easily seen in figure 3 especially in installation N°3 where the daily consumption can reach sometimes 3 times the mean consumption of the month. According to the operators of the installation this "attitude" was very frequent i.e. over consumption. As a consequence of this, and if we keep in mind the non-availability of distilled water (broken glass of water solar energy distillers), we can understand the bad situation of all the batteries. This explains the frequent shutdowns of the inverters because of the deep discharge of these batteries.

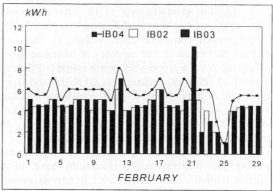

Figure 3. Daily consumption (February 1996)

To have a clear idea about the functioning of these systems, let's have a look at the energy balance of installation N°2 as shown in figure 4.

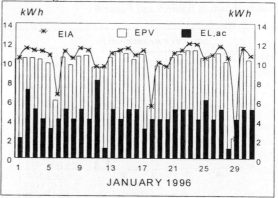

Figure 4 : Energy balance of installation N°2 (In Belbel)

We pushed our investigation with the available data in order to determine the indices of performance [2] of installation No. 3 as resumed in Table 2.

Table 2: Typical parameters and indices of performance (installation N°03)

	H_A	E_A	E_L	Y_A	Y_f	Ls	Lc
12/95	6.13	10.05	7.76	4.22	3.26	0.96	1.91
01/96	5.92	9.76	7.07	4.30	2.96	1.26	1.62
02/96	6.31	10.43	7.70	4.38	3.23	1.15	1.93
03/96	6.64	10.80	8.20	4.53	3.82	0.71	2.11
04/96	7.10	12.02	9.46	5.05	3.97	1.08	2.05
05/96	6.37	10.76	8.58	4.52	4.01	0.51	1.85
Mean	6.41	10.64	8.12	4.50	3.55	0.95	1.91

Irradiation as the first parameter (input) seems to be high even during the worst month (January) with a value of 5.92 kWh/m². For this reason the mean value of array yield 4.5 is fairly acceptable (if compared with the one of Matriouane near 5.0). On the other hand if system losses are kept under the value of 1.0, capture losses instead remain higher since they reach an average of 1.91.

3. IV CURVE MEASUREMENTS

On site measurements of IV curves (modules, strings, array...etc.) is a powerful complementary tool to the other physical and visual tests. To this end, many tests have been performed on the modules of Matriouane with an electronic load, a scopemeter and a portable PC in the framework of scientific exchange (CDER-IES and Ispra). The purpose of the measurements is to evaluate mismatch losses and to detect faulty connections or module's defection. Our main objective is to evaluate the degradation of modules in such harsh conditions.

The measurement procedure is well known nowadays [3]. Meteorological conditions have to be acceptable (radiation, wind speed ...etc.). After performing measurements and checking their quality, data is then translated to STC conditions after temperature correction for current and voltage. They can be filtered afterwards in case of fast measurements (<100 ms). In the case of Matriouane 20 ms was necessary to have an output quality signal as shown in figure 5 and figure 6.

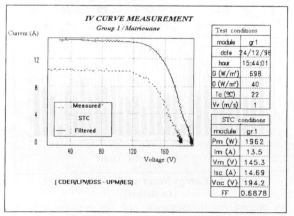

Figure 5. IV curve of Group1 (Matriouane)

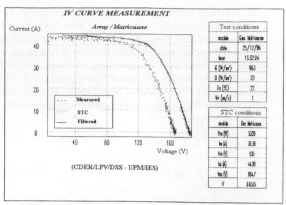

Figure 6. IV curve of the array (Matriouane)

The measurements done during Dec.1996 concern strings, groups (sub-arrays), combination of groups and the total array. We should mention here that weather conditions were note appropriate to perform individual module tests at that time. Data translated to STC conditions is shown in table 3.

Table 3 Summary of the electrical parameters (PV array of Matriouane)

Réf.	Isc	Voc	Pm	Im	Vm	FF
Stg1	3.048	192.3	373.7	2.688	139.1	0.6374
Stg2	3.018	197.0	397.6	2.731	145.6	0.6689
Stg3	3.070	197.9	406.9	2.755	147.7	0.6688
Stg4	3.088	196.2	419.7	2.765	151.8	0.6927
Stg5	2.926	195.0	414.4	2.898	143.0	0.7265
Stg6	3.137	188.8	395.2	2.762	143.1	0.6673
Stg7	3.014	191.3	379.1	2.672	141.9	0.6577
Stg8	3.001	191.8	388.6	2.732	141.9	0.6751
Stg9	2.979	194.0	386.8	2.672	144.8	0.6691
Stg10	3.084	190.7	393.9	2.790	141.2	0.6840
Stg11	2.938	192.4	388.0	2.694	144.0	0.6864
Stg12	3.039	175.3	334.0	2.581	129.6	0.6268
Stg13	3.002	186.0	364.0	2.654	137.2	0.6520
Stg14	2.984	188.7	348.0	2.685	129.6	0.6179
Stg15	3.080	189.1	369.5	2.620	141.1	0.6345
			5759			
GR1	14.7	194.2	1962	13.5	145.3	0.6878
GR2	15.3	187.8	1861	13.5	137.9	0.6476
GR3	15.1	188.5	1770	13.6	130.1	0.6196
			5593			
ARRAY	44.09	184.7	**5320**	39.4	135	0.6535

The visual inspection of the array and the analysis of measured data have indicated clearly the defection of two modules in string 12 and 14 respectively. On the other hand we also noticed that the underground cabling was in good state, whereas the other one (air) has been damaged by temperature variation (day and night) and also by sand erosion.

In this sense we believe that new electrical cables with appropriate sheathing should be used, otherwise it would not resist to these extreme conditions.

Table 4 Power losses - Matriouane PV array

Items	Losses(%)
Σ(Str1...Str5) & GR1	2.5
Σ(Str6...Str10) & GR2	4.4
Σ(Str11...Str15) & GR3	1.9
Σ(Str1...Str15) & Σ(GR1..GR3)	3.0
Σ(GR1..GR3) & Array	5.0
Σ(Str1...Str15) & Array	8.2

This campaign was also an opportunity for the team to evaluate the power losses of the array as shown in table 4. In this case also we noticed that inter-modules cabling especially in group2 were damaged. The quality of the cables used was not the same as those of group 1 & 3. The power losses between the strings and the array seem to be high and some actions should be taken : cable replacement, use of thermally isolated sheathing ...etc.

Finally we should mention the difference between the total PV array power measured in April 1993 (5.6 kWp) and the one measured in Dec.1996 (5.32 kWp). This represents 5% of the PV array . At the light of this we can affirm that degradation after a certain years of operation is a reality. As a mean value we can say that it is around 0.01%/module/year. This also means that after 20 years of operation we should expect a degradation of more than 30% of the hole PV array !!!.

4. CONCLUSIONS

Monitoring and data analysis is the only reliable way to know whether a system is operating correctly or no and to verify also if it was well designed to fulfill the initial requirements. A monitoring program is on due now for 20 different systems and applications all over Algeria. This paper has been an opportunity to present some early results on how these systems are running. If for the Matriouane installation frequent reports have been published, this is not the case for the In Belbel installations. In fact, this is the first time that we gathered data on it. We already noticed two main features for its installations: over consumption and deep discharge of all the batteries. We believe that batteries replacement is a necessity. We should also mention a positive point about inverters (AEG 2.5 kVA each), they are all (four) functioning up to now correctly.

On the other hand we also reported for the first time on site measurements of IV curves for different combinations of the Matriouane PV array. This technique is a powerful test for modules, which represent the main subsystem of any PV installation. This dissection let us in most of the time to detect any abnormality in any module , string, sub-array, array in addition to the detection of other problems: connections, cell or module defects, power losses...etc. Moreover if periodic IV scans are made for a PV array we can than detect any degradation, which may represent a barrier for PV penetration in a country like Algeria with an area of more than 80% of desert.

Finally all this effort to get a feedback from these installations is of prior importance to help researchers and also manufacturers in evaluating the performances and reliability but especially to propose to decision makers'

viable solutions. In our belief this is the only way to let PV solar energy compete fairly with other energy sources.

5. REFERENCES

[1] B. YAICI, S. LABED and E. LORENZO,
The PV plant of Matriouane (Algeria): Lessons learned 12th PVSECE Amsterdam April 1994

[2] Guidelines for the Assessment of PV plants
Document A PV system monitoring Issue 4.2 June 1993 JRC-Ispra

[3] G. BLAESSER, E. ROSSI,
Extrapolation of outdoor measurements of PV array IV characteristics to standard test conditions, Solar cells, 25 , pp 91-96, 1988

Acknowledgements :
The authors are grateful to: ICMA (Instituto de Cooperacion con el Mundo Arabe), CDER (Algeria), IES/UPM (Madrid-Spain) and also the Spanish Embassy delegation (Algiers) for their support and availability.

Syllabus for Solar Energy at School

E. Kuetz

Goethe-Gymnasium, 60325 Frankfurt/Main

Friedrich Ebert-Anlage 22
Fax-No. 0049/69-21 230 717

ABSTRACT: Goethe-Gymnasium Frankfurt (Main) founded a photovoltaic-working-group eight years ago to enable students of upper schools to attain a reasonable degree of competence in the current discussions in the field of renewable energies. In a common education and training project with an English school we have started to work out a syllabus for solar energy which is under permanent development. The project includes theory of solid state physics for photosensitive materials, experiments with solar cells and wind tubine-models, and excursions to commercial solar energy and wind turbine systems as well. Furthermore, students should become familiar with world-wide energy problems and get case studies for possible solutions of these problems.

Keywords: Education and training - 1: Experimental methods - 2.: Solar cell efficiencies - 3

1. Introduction

1.1 Background

The Goethe-Gymnasium was additionally awarded a solar power system of 2 KW peak by the Stadtwerke Ffm., for taking first and second prize in a competition featuring new ideas for the utilization of electricity. The idea to develop this education initiative was born by the impression that young people should necessarily become familiar with possibilities to install renewable energy sources (solar, water and wind power) and to assess the current maximum possibly to be substituted for fossil and nuclear power by these alternative methods. Obviously too many students leaving school with higher level certificates are talking about energy-problems and how to solve them without the necessary background in natural sciences. So our project has to include practical work and examples of application in renewable energies.

1.2 Objectives

The project was founded in 1992 in cooperation with a partner school in London and has been under permanent development ever since. 16 - 18 year old students take part in the programme, working two hours a week in addition to the basic physic lessons. Every year during the past decade the English and German partner groups met in London and Frankfurt/Main to explain each other what had been done in the field of renewable energies. One meeting took place in kind of a work-shop including experiments, theory and an excursion to visit a commercial photovoltaic system in Kobern-Gondorf near the river Mosel. The English group focussed their work on the investigation to produce a low price solar cell by a dip-method. These solar cells were given to the German group who determined the efficiency under laboratory-lamp-light using self-developed analysis methods as introduced below.

1.3 The module

a) introduces the current concepts for regenerative energies

b) ilustrates the behaviour of diodes in comparison to solar cells in the dark.

c) gives a survey over the theory of the band gap model for semi conductors, metals and non conductors

d) includes a collection of experiments with solar cells under laboratory light.

e) analyses the data supplied by the solar power system of the Goethe-Gymnasium

f) performs experiments with the model of a wind turbine

g) provides some excursions to a commercial wind turbinesystem and a solar system as well.

h) gives an outlook on the technical development to the future of projects supported by national governments and the EU Commission.[3]

i) is in extension to develop a dye solar cell under school laboratory conditions. These activities are supported by advice from several reseach institutes.[4],[5]

2. Contents and Results

In the first teaching area the students perform basic application experiments with LEDs, DIODES, LDRs and Solar Cells to awaken their interest.

After that they receive some insights in the process connected with conductivity of metals, non- and semi-conductors within the frame work of solid state physics.

The students establish voltage-current characteristics of diodes and solar cells in the dark to see their similarity and differences as well. Then the characteristic of a solar cell under light conditions is established. From this diagramme the maximum output is analysed. Then using a thermopile the input of the light source is measured. Determining efficiency from both the maximum output and the input of light energy presents a higher degree of difficulty for school physics. The output is first determined by taking the characteristic point by point with voltage current meters and after that it is demonstrated with an optimised experimental setup which connects an x, y-writer with the measuring device [2] (see picture 1).

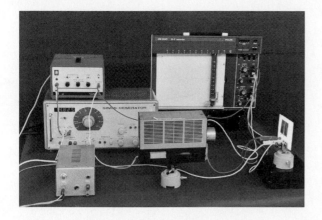

picture 1: Experimental setup

So we achieve an automatic determination of the characteristic of the solar cell. In a similar setup we have found a way to record the emission characteristic of the light source and to compare it directly with a graph of the solar cells sensitivity (see graph no. 1).

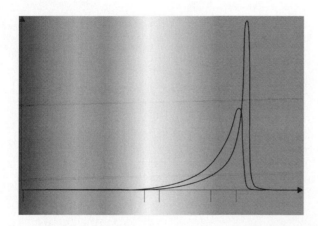

graph no 1: Spectral analysis

To measure the energy distribution of a light source in dependence of wavelength using a refraction prism causes certain difficulties under school conditions. Especially the area at the end of visible light to IR-light is not easy to measure with basic apparatus. Nevertheless it is the interesting area for the maximum sensitivity of silicon solar cells. We are proud to have solved this problem by support and advice of the Didactic Institut of Physics of the University of Gießen.[1]

3. EVALUATION

The experience of the past years with the photovoltaic group has shown increasing student interest in this field and is steadily growing in number of participants (16 - 20 students per group) in this project. On the one hand this is certainly due to the enhanced consciousness of the problems of energy consumption and renewable energies. On the other hand interest is also stimulated by the wish to participate in a pupils' exchange with our partner school or by the possibility to participate in an international energy congress. All these motivations, although secondary, should be worthy of support,

considering that the aim is to enable students of upper schools to attain, even before the I.B., a reasonable degree of competence in the current discussions in this field. These students, many of whom may perhaps not study physics at university, would otherwise not find the same level of self-assured competence in the political decisions pertaining to this field, which they will almost certainly be called upon to take part in, be it as voters or as otherwise politically active citizens.

4 REFERENCES

[1] Dr. Ganz
    Institut für Didaktik der Physik
    Justus-Liebig-Universität Gießen

[2] Dr. Bonnet, Dr. Richter
    ANTEC Kelkheim

[3] F. Jäger/A. Räuber
    Fraunhofer-Institut Freiburg

[4] Dr. A.J. Mc Evoy
    Laboratoire de Photonique et Interfaces Institut de Chimie Physique
    EcolePolytechnique Federale de Lausanne

[5] Dr. H. Wittkopf
    Firma Pilkington, Gelsenkirchen

Implementation, Strategies, National Programs and Financing Schemes

RENEWABLE ENERGY SOURCES IN THE 5TH FRAMEWORK PROGRAMME - FIRST QUALITATIVE AND QUANTITATIVE OUTCOMES

E Millich
European Commission, DG TREN
200 Rue de la Loi, B 1049 Brussels
Tel: +32 2 295 3625, Fax: 296 6261
e-mail: enzo.millich@cec.eu.int

W Gillett
European Commission, DG TREN
200 Rue de la Loi, B 1049 Brussels
Tel: +32 2 299 5676, Fax: 296 6261
e-mail: william.gillett@cec.eu.int

ABSTRACT: In the first call for proposals for the FP5 ENERGIE programme, almost twice as many proposals were assessed to be "worth funding" as the available budget could cover, so a large number of good proposals had to be rejected. The European Parliament's budgetary comments, that at least 60% of the budget should be allocated to renewables and of this 75% should be for demonstration, were exceeded with 70% allocated to renewables. A new call has now been issued, with a strong focus on tackling the near term EU commitments of reducing CO_2 (Kyoto) and increasing the contribution of renewables to the EU energy balance by 2010. There will be stronger links between these policy commitments and the projects, which are supported by DG TREN in the future. This is expected to result in fewer, larger projects, which have a greater impact on the market

Keywords: R&D and Demonstration programmes – 1: Strategy – 2: Dissemination – 3

1. INTRODUCTION

The European Commission's 5th Framework Programme, which began in 1999, is scheduled to run for four years until 2002. Within this major European RTD programme, work on renewable energy is funded within the ENERGIE sub-programme, which forms a part of the Energy, Environment and Sustainable Development programme.

The overall policy objectives of the ENERGIE programme are to

- develop sustainable energy systems,

- ensure security and diversity of supply

- provide high quality and low costs energy services

- improve industrial competitiveness

- reduce environmental impacts

The European Parliament issued budgetary comments in relation to the 1999 Call for Proposals, requiring that at least 60% of the budget should be allocated to renewables and, of this, 75% should be for demonstration.

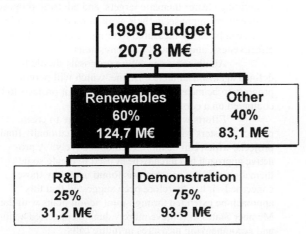

Figure 1 Budgetary criteria from European Parliament

They also required that the participation of SME's be greater than 10%. These requirements were met in this call, with an actual allocation of 70% to the renewable energy sector, and 15% participation by SME's.

2 QUANTITATIVE OUTCOME OF 1st CALL

Renewables are funded under two Key Actions in the ENERGIE sub-programme, namely Key Action 5 and Key Action 6.

The funding for each of these actions resulting from the 1st Call was similar, with 90 M€ for Key Action 5 and 104 M€ for Key Action 6, as shown in Figure 2 (below), where the split between the renewable energy sector including all relevant actions (RES) and the non-renewable energy sector (Non RES) can be seen.

Key Action 5	90	DEMO	60	R.E.S.	50
				Non R.E.S.	10
		R&D	29	R.E.S.	18
				Non R.E.S.	11
Key Action 6	109	DEMO	56	R.E.S.	42
				Non R.E.S.	14
		R&D	52	R.E.S.	24
				Non R.E.S.	28

Figure 2 Allocation to projects in 1st FP5 ENERGIE call

The proposals were submitted by over 3900 participating organisations from 20 countries. Out of these, it can be seen from Figure 3 that almost 1060 different organisations were retained as participants in the programme.

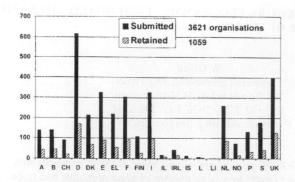

Figure 3　Participants in the FP5 programme by country

Out of the 802 proposals submitted to ENERGIE in the 1st Call, 347 were assessed to be "worth funding" and 228 were negotiated.

In terms of EC support, the programme provided 198 M€ of support to projects with a total eligible cost of 352 M€. A lot of good proposals had to be rejected because of budget limitations.

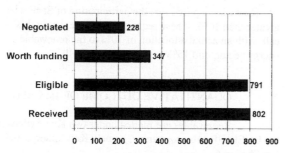

Figure 4　Proposals negotiated in FP5 ENERGIE 1st Call

In comparison with the funding provided for Renewables demonstration (demo') projects under the 4th Framework Programme, 59 M€ was provided in support of renewables demo' projects compared with an average of only 36 M€ in FP4. However, the total number of demo' projects supported was significantly less at 46 in FP5 compared with 53 in FP4. This trend to reduce the number of demo' projects supported was consistent across all of the RE technologies, apart from Biomass, where the number of supported projects increased (see Fig 5).

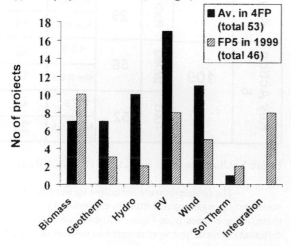

Figure 5　No of demo' projects supported by FP5 and Fp4

Nevertheless, whilst the number of demo projects was reduced, the level of EC support for demo' projects was increased in all sectors apart from geothermal and hydro (see Fig 6).

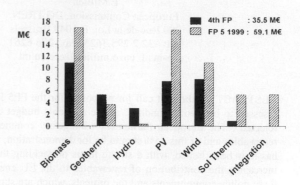

Figure 6　M€ support in FP5 1st call compared with FP4

3　QUALITATIVE OUTCOME OF 1st CALL

A qualitative assessment has been carried out to assess the coherence and consistency of the proposed actions with the expected results, in the context of the programme objectives.

In overall terms, the quality and technical coverage of the selected projects were good. However, more good quality proposals were needed in the areas of renewables, fuel cells and CO_2 abatement technologies.

Several of the selected projects lacked focus. It was concluded that this problem could be overcome by introducing more targeted and specific calls in future. Specific areas of weakness in the proposals were cost-effectiveness, socio-economic impact, and the definition of tangible deliverables.

Out of 26 "combined projects", only one was actually negotiated on a "combined" basis, and it is not yet clear how to address this issue.

The lack of sufficient good quality new and renewable energy proposals has now been addressed by issuing an additional call for proposals (3rd ENERGIE call), which specifically targeted these sectors (see below). Efforts are also underway to address the other weaknesses, by defining clearer thematic targets, and advising proposers to give more emphasis in their work to producing results which can be easily measured in terms of improved cost effectiveness, and reduced CO_2 emissions.

All objectives in future proposals should be defined against a suitable baseline, which will permit progress to be measured, and allow different projects to be compared on a common basis.

Efforts will also be made in future to create project "clusters", including both EC and nationally funded projects, which can be monitored collectively. A pro-active approach has already been used to create some thematic networks and this was found useful by those concerned. It has therefore been suggested that this approach be extended through joint action plans with the Member States to achieve more valuable concerted actions and accompanying measures in future calls.

There is evidence to suggest that the new FP5 "problem solving approach" has not been well understood by some proposers, especially those submitting research proposals. The key requirement in this context is "to find solutions to strategic problems", which for demonstration projects may involve what has been called "Business Demonstrations", which address all of the problems faced by a typical business when working to get its product into the market. Similarly, the FP5 objective of tackling socio-economic problems should be interpreted as "enhancing the capability to compete in global markets".

4   NEW CALLS FOR PROPOSALS

A new call for proposals (Energy third call [2]) has been issued with a deadline of 31 May 2000. This is focused on a limited number of topics, which have been selected with a view providing innovative energy technologies and solutions within a short to medium term time frame (less than 10 years). These topics are:

- Promotion of RES, including efficient integration into energy systems

- Fuel cells in mobile and stationary applications

- Energy efficiency in urban transport including the use of cleaner and alternative fuels

- Efficient energy services (heating, cooling, ventilation, lighting) and domestic appliances, and integration of renewables into buildings

- reduction of local and global environment degrading emissions

Fourth and fifth calls are currently foreseen in the first half of 2001 and 2002 respectively. Proposals submitted under the open calls for accompanying measures, thematic networks and concerted actions will from now onwards be assessed at the same time as the main calls, as well as on 1st September in 2000 and in 2001.

Proposers are recommended to include clear quantified targets in connection with their energy, efficiency and economic objectives, and similarly to specify clear socio-economic targets for the results of their work in terms of job creation, environmental protection, etc.

Proposers should also present a good technology implementation plan, showing how their work will help the EU to meet its commitments to CO_2 reduction - the Kyoto commitment - and to increasing the contribution of renewables to 12% of the EU energy balance by 2010 - the RE White Paper commitment [1]. Proposers are advised to note the trend towards larger projects, which are likely to achieve a greater market impact, and to "cluster" several projects together if necessary in order to achieve this goal.

Further information on FP5 calls for proposals can be found at

- Cordis web site: http://www.cordis.lu/fp5/src/t-4.htm,
- through National contact points,
- by e-mail to : helpline-energy@cec.eu.int or eesd@cec.eu.int,
- by fax to DG Research +322 296 6882 or to DG TREN +322 295 0577.

5   COHERENT SECTOR ACTIONS

The fusion of the previous Directorates General for Transport and Energy to form DG TREN, together with the arrival of our new Commissioner, have provided an opportunity for refocusing work in the energy sector. As a result, there is now a very strong emphasis in the new DG TREN on achieving the EU commitments and targets in relation to climate change and renewable energies.

EU Policy Commitments and Targets

**Directives and Legislation
Market support, Trading & Certification
Business & Technology Demo's
Promotion**

**Pilot Plants
Research**

Figure 7 DG TREN actions focus on policy commitments

To achieve these commitments, it is clear that a combination of legislative actions, market support with trading and certification, business and technology demonstrations, and promotion is needed. Further, this must be underpinned by a dynamic programme of innovative technology development, including pilot plants and other research activities.

DG TREN sees the three active energy programmes SAVE, ALTENER and ENERGIE as important tools, which must be used wisely to achieve its policy commitments and targets. Using these tools, it will continue to encourage strong team working, between [A] the public sector (EC and Governments) which are responsible for legislative and fiscal actions, and publicly funded actions, [B] Industry, which acts as technology developers and technology providers, and [C] the citizens of the EU, who are the users of the energy services and are also the main investors in the new technologies.

DG TREN also recognises the key supporting roles played by researchers, project developers, promoters, Bankers and consultants.

Figure 8 Teamwork to achieve EU policy commitments

6   PROPOSAL FOR A DIRECTIVE

The EC Vice President Loyola de Palacio has recently responded to the requests made by the Energy Council in December 1999 and by the European Parliament in March 2000, by preparing a proposal for a Directive on the promotion of electricity produced from renewable energy resources, ie: for green electricity.   The text of this proposal is not yet publicly available, but the *"major principles on which the Commission proposal for a Directive for the promotion of electricity from renewable energy sources will be based"* were presented by Mrs de Palacio in a speech at Santiago de Compostela on 10 April 2000.

The speech is being made available on the internet, and the following extracts may be of general interest :

- The draft proposal will refer in the recitals to an annex containing **quantitative indications** for the **targets** to be chosen by **individual Member States** with a view to assuring that the Community objective of 12% renewable energy sources penetration by 2010 as well as Community and national climate change commitments are achieved.

- The draft proposal will foresee that the **Commission monitors** the **compliance** of the national targets with the community objective of 12% and the Community climate change commitments. Furthermore, the Commission will have the right to present proposals to the European Parliament and to the Council with respect to national targets if they are inconsistent with the **Community objective**.

- The draft proposal will **abstain** from proposing **a harmonised Community wide support system** for electricity from renewable energy sources ("green electricity") as we do **not have sufficient experience** which would allow us to decide for one particular system. The **pragmatic approach** of the Commission thus will consist in further studying the **experiences** gained in Member States with the operation of the **different national support systems** in order to **make a proposal** for such **harmonised support system within 5 years**. This will be done on the basis of a **Commission report** assessing the various **support systems** in favour of electricity production from **renewable** as well as **conventional energy sources**.

- As concerns technical issues, the approach supported by the Commission in the past will be maintained: The draft proposal will oblige Member States to **assure** that **certification** of green electricity is both **accurate and reliable**; it will furthermore **oblige** Member States to **streamline and expedite authorisation procedures** applicable to installation of generation plants for green electricity. Also, Member States will **have to assure that electricity** from **renewable energy sources** has **priority access to the grid**. Furthermore, the **calculation of** costs of connecting new producers of green electricity should be **transparent** and **non-discriminatory**.

7   CONCLUSIONS

There is now a very strong emphasis in the new EC DG TREN on achieving the EU commitments to climate change (Kyoto commitment to reducing CO_2 emissions by 8% relative to the levels pertaining in 1990, and to doubling the share of renewables in the EU energy balance from 6% to 12 % by 2010.) To meet these goals, there must be much greater coherence between the different actions supported by the Commission and by the Member States. A smaller number of bigger projects and related actions is foreseen in future, with each project having a very much greater impact on the market.

REFERENCES

[1]   Energy for the future: Renewable Sources of Energy, White Paper for a Community Strategy and Action Plan (COM (97) 599 Final)

[2]   Third call for proposals for RTD actions, Part B Energy - Key actions 5 and 6, Official Journal of the EC, 2000 / C 73/10, 14 March 2000.

RECENT ADVANCES IN SOLAR PHOTOVOLTAIC ACTIVITIES
IN JAPAN AND NEW ENERGY STRATEGY TOWARD 21st CENTURY

Yoshihiro Hamakawa

Department of Photonics, Faculty of Science and Engineering, Ritsumeikan University

1-1-1 Nojihigashi, Kusatsu, Shiga ,525-8577 Japan

Tel: +81-77-561-2607   Fax: +81-77-561-2613   e-mail:hamakawa@se.ritsumei.ac.jp

ABSTRACT: The current state of the art in recent progress of Japanese photovoltaic (PV) activities is overviewed. Firstly, a brief discussion is given on **"the 3E Trilemma "** which is the most important issue for the 21st century's civilization life.   Secondly, a new strategy for the renewable energy promotion so called -Fundamental Principle to promote New Energy Developments and Utilization-, and its action planning for PV technology up to year of 2010 are introduced.   Some new strategies with tangible actions such as tax reduction, government financial support of 1/2 subsidy of PV system developments for public facilities namely as **"PV Field Test Experiments"**, and a 1/3 subsidy for the private solar house as **"PV Solar House Plan"** are demonstrated.

Thirdly, recent R&D efforts in the field of solar cell efficiency improvement technology are reviewed togather with progress of some new technologies such as multi-bandgap stacked solar cells, utilization of new materials; microcrystalline silicon (μc-Si), CIGS etc.   In the final part of paper, a prospect of future industrializations of the photovoltaics is discussed, then possible new roles to contribute to the global environmental issues by the PV system developments are introduced with some live technologies in progress, such as ashing of pollutant gas, cleaning of water and "greening" the desert, etc..

Keywords: Strategy-1: National Programme-2: Environmental Effect-3

1. INTRODUCTION

In the discussions of the Grand Design for the 21st century's civilization life, a big attention is focused on the 3E-Trilemma.   That is, in the regular way along capitalism, for the activation of economical development (E:**Economy**), we do need an increase of the energy expense (E:**Energy**).   However, it induces environmental issue (E:**Environment**) by more emissions of pollutant gases.   On the contrary, if the political option choses a suppression of pollutant gas emission, it inactivates the economical development.   This is **3E-Trilenma**.   Figure

1 illustrates the relation of the 3E-Trilenma [1]. According to the result of world energy trends analysis, the energy consumption per capita in a country is directly proportional to the country's annual income per capita [2] or its GNP (Gross National Product) [3].   On the other hand, the number of the world's inhabitants is steadily increasing as shown in Fig.2, and will reach about 6.1 billion by the year 2000, when it can be expected that worldwide energy demand will increased by multiples of population increase and another factor due to promotion of modernization. This positive increment in energy demand seems unavoidable in the near future even if energy saving

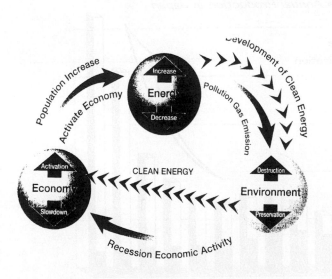

Fig.1. 3E-Trilemma, the most important task assigned to 21st century's civilization.   Only way to be solved is developing clean energy technology.

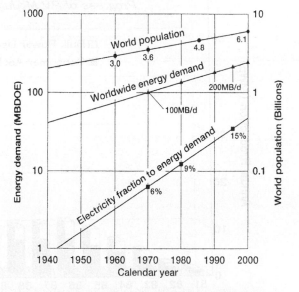

Fig.2. Transitions of world population, total energy and electricity demands.   Percentages are the ratios of electricity to the total energy

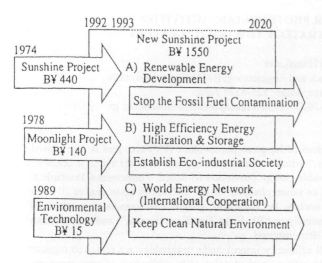

Fig.3.   Organizations and objectives of the
          New Sunshine Project

technologies in progress are applied to a moderate degree. For example, the rate of energy consumption per production unit in the heavy industries in well-developed countries might decrease, it will be completely compensated by the rapid increase of energy demand in the newly advancing countries such as China, India, Indonesia, Malaysia and Thailand.

Considering the two sided nature of the energy policy, that is, continuous growth of mass consumption of the limited fossil fuels one side, and becoming severe the global environmental issues on the other side, Agency of Industrial Science and Technology (AIST) in the Ministry of International Trade and Industry (MITI) in Japan has

decided to establish "New Sunshine Program" for the development of clean energy technology and environmental technology. Figure 3 illustrates a comprehensive structure of the new program and its relation to the Sunshine Project which was formulated in May 1973 prior to the first energy crisis, and started in 1974, the Moonlight Project which was initiated in 1978 for the energy saving technological development, and also environmental technology project progressed from 1989 [4]. The past injected budgets are also inserted in the figure. While the new program consists of three parts; a) Renewable Energy Development Technology, b) High Efficiency Utilization of Fossil Fuels and Energy Storage, and c) International Energy Cooperation, so-called WENET (World Energy Network) utilizing hydrogen fuel production technology by PV and distributions with a wide area energy utilization network system nicknamed "Eco-energy City".

2. KEY ISSUES FOR PV INDUSTRIALIZATION

The direct conversion of solar radiation to electricity by photovoltaics has a number of significant advantages as an electricity generator. That is, solar photovoltaic conversion systems tap an inexhaustible resource which is free of charge and available anywhere in the world. The amount of energy supplied by the sun to the earth is more than five orders of magnitude larger than the world electric power consumption to keep modern civilization going. Roofing tile photovoltaic generation, for example, saves excess thermal heat and conserves the local heat balance. This means that a considerable reduction of thermal pollution in densely populated city areas can be attained.

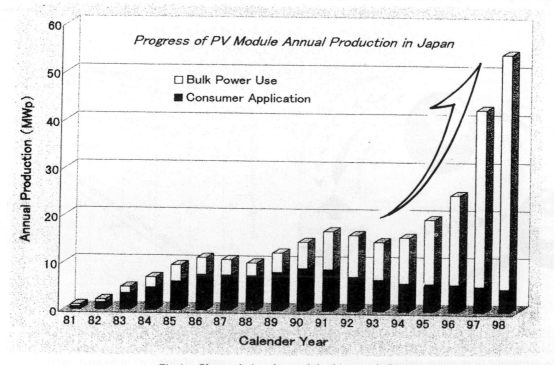

Fig.4.   Photovoltaic solar module shipment in Japan

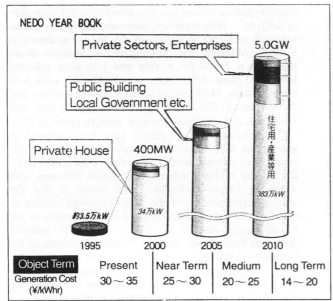

Fig.5.   NSS project mile stone for the PV system installation up to 2010

Figure 4 shows transitions of the solar cell module annual production in Japan since 1981 as surveyed by Optoelectronic Industry Technology and Development Association (OITDA) [5].   As can be seen from the figure, a remarkable increase of the annual production has been seen since starting of the New Sunshine Project in 1993.   In spite of various advantages in photovoltaic power generation as mentioned above, a big barrier impeding the expansion of large-scale power source application was the high price of solar cell module, which was more than \$30/Wp (peak watts) in 1974.   That is, the cost of the electrical energy generated by solar cells was very high compared with that generated by fossil fuels and nuclear power generation.   Therefore, the cost reduction of the solar cell is of prime importance.   To achieve this objective, tremendous R&D efforts have been made in a wide variety of technical fields, from solar cell material, device structure, and mass production processes to photovoltaic systems over the past 20 years.   As a result of recent fifteen years R&D efforts, about one order of magnitude price decrease has been achieved, and now, the module cost has come down less than \$4/Wp in a firm bid for the large scale purchase.   In December 1994, a new initiative of Japanese domestic renewable energy strategy. **"Fundamental Principles to Promote New Energy Developments and Utilization"** has   been   identified by the Cabinet Meeting.   Related action planning by law such as tax reduction, government subsidy etc. are approved by the Congress on April 10, 1997 [6].   The strategies are applied to not only whole ministries and government offices but also local government authorities and private enterprises.

With the government new policy, development and promotion of PV technologies have been complied as the most promised project.   An integrated installations of 400MWp PV modules by FY2000 and 5.0GWp by FY2010 for Japanese domestic use are scheduled as a mile stone in the program as shown in Fig.5.   A special regration of tax reduction for investment to the renewable energy plant, a government financial support of 1/2 subsidy on the PV

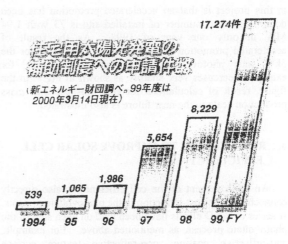

Fig. 6.   Number of accepted PV house monitor planning by the government subsidy

system for public facilities so called **PV Field Test Experiments**, 1/3 subsidy for the private solar houses as the **PV House Planning** of the field testing etc., are in progress.   Figure 6 shows number of accepted PV houses with the monitor plan by the government subsidy in recent six years.   As can been seen in the figure, number of government subsidy accepted PV house increases doubling year by year.   On the other hand, in the PV Field Test Experiment totals of 259 sites with 6.84 MW has been installed during 7 years since 1992.   A noticeable result

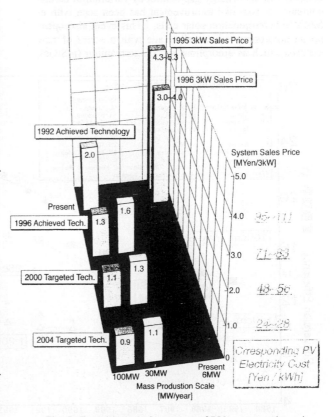

Fig. 7. Transitions and prospect of PV system sales price for 3kW private house monitor planning

in this project is that an accelerated promotion has been done.　In fact, number of installed site is 73 with 1.94 MW in only one year of 1998.　As the result of accelerated promotion strategy, in fact, sales price for the 3kW solar photovoltaic system for private house, for example, decreases very sharply as shown in Fig.7. In the figure, result of calculations for cost prospect with mass production scale in the near future is also illustrated.

3.　R&D EFFORTS TO IMPROVE SOLAR CELL EFFICIENCY

An improvement of the cell efficiency is also directly connected to the cost-reduction in the photovoltaic systems. A series of R&D efforts have been paid on each step of the photovoltaic process, as mentioned above.　For example, anti-reflective coating, non-reflective texture surface treatment, and heterojunctions with wide energy gap material have been investigated to get more efficient optical energy collections.　Through these technological progress, the cell efficiency of a single crystalline Si solar cell reaches 17–18% in the mass-production lines.　The polycrystalline and cast silicon solar cells show 14–16% on average and 15.7% in the best case in the mass-production line.　Some top records in the R & D phase are plotted in Fig.8.

In amorphous silicon solar cells, there has been noticeable progress in recent decade with the invention of amorphous silicon carbide (a-SiC), because a-SiC alloy permits not only valency control by impurity doping (p-n control), but also energy gap control by variation of carbon content.　A step-like improvement has been seen with a-SiC/a-Si heterojunction solar cells.　This event also opens up an amorphous silicon alloy age with a series of new materials such as amorphous silicon-germanium (a-SiGe),

microcrystalline silicon (μc-Si), amorphous silicon-nitride (a-SiN), etc.　With these new materials, amorphous silicon solar cell efficiency is improving day by day as shown in Fig.8.　Although the in-line mass-production phase efficiency is 8–9% for a wide area a-Si solar cell, the laboratory phase efficiencies are 13.4% for a-SiC/a-Si heterojunction, 14.3% for a-Si/a-SiGe stacked junction, and 15.04% and 21.0% for a-Si/poly-crystalline Si two- and four-terminal stacked junction solar cells, respectively [7].

Tremendous R & D efforts have been invested in solving the light induced degradation in a-Si solar cell which is a big weak point.　According to the results [8,9], the amount of light induced degradation increases with increasing i-layer thickness.　The reason is that the volume recombination of the photo-generated carriers in the i-layer becomes relatively large with a decrease of the lowest electric field in the i-layer [10].　The author's group has been shown from the theoretical calculation for electric field distribution in i-layer that the lowest electric field sharply increases with decreasing the i-layer thickness[10].　Therefore, for the purpose of suppressing this effect, the tandem type solar cell has been offered as a more reliable a-Si solar cell[10].　Along with this concept, wide varieties of experimental trials are in progress on the multi-bandgap stacked solar cells.　Recently Takakura [11] has made a systematic investigation of an optimum design of the two stacked tandem type solar cells.　The result shows that the best combination of a stacked junction material with a-Si would be poly-Si thin film from the viewpoint of theoretical efficiency, stability, plenty of natural resources and cost.　Then, a series of intensive investigations has been initiated elsewhere [12-15]. For example on the two-terminal a-Si solar cell stacked with poly-Si, a conversion efficiency of 15.04% with open circuit voltage Voc=1.478 V, Short circuit photocurrent density Jsc=16.17 mA/cm^2 and Curve fill factor FF=63%

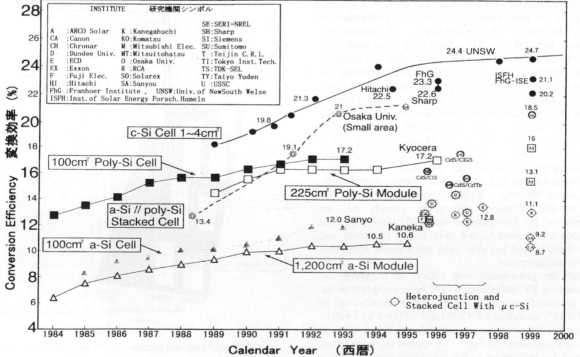

Fig.8. Transition of R&D phase solar cell and module efficiency

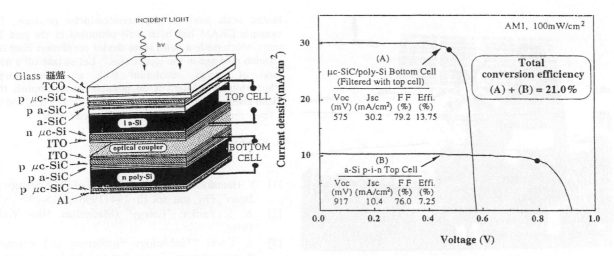

Fig. 9. Cell structure and V-I characteristics of a Si//poly-Si 4 terminal stacked solar cell

has been obtained under AM1 (Air Mass 1) illumination. With the same combination for a four-terminal cell, 21.0% efficiency has been obtained by Ma Wen et al [12] as shown in Fig.9.

In this cell structure, although the bottom cell polycrystalline silicon is employed with the conventional cast bulk silicon, a considerable attention has recently been gathered to the material combination of a-Si with poly-Si, so called "Honeymoon Cell" [13]. A number of intensive technological approachs to fabricate polycrystalline thin film growth on inexpensive substrates has been going on elsewhere. For example; "Molecular Beam Graphoepitaxial Growth (MBGE)" [14], "Solid Phase Crystallization (SPC) Method" [15], "Zone Melting Recrystallization (ZMR) Method" [16]. At the present, the best the data for the poly-Si bottom cell, 16.4% efficiency has been obtained by the ZMR method of the Mitsubishi group with 2x2 cm^2 cell area having the thickness of $50\sim60\mu m$ active layer [16].

In the course of R&D efforts for the honeymoon cell technology, an interesting byproduct has born. That is, a- or μc-Si (SiC) // poly-Si heterojunction solar cell [12]. Significant features of this kind of cell are, 1) a very simple fabrication processing, 2) low cost, 3) more than 17% efficiency is easily obtained. With a similar design concept, the Sanyo group [17] has recently reported 21%

efficiency with a structure of heterojunction with intrinsic photovoltaic active thin-layer (so called HIT). If the low cost poly-Si thin-film production in progress become a realistic technology, this kind of a-Si//poly-Si heterojunction solar cell will be one of the most promised technologies for the next generation of solar cell. Figure 10 shows an example of the best recent data on the a-Si// μc-Si stacked cell reported from Kaneka Group [18] having a total efficiency of 13.5% with the bottom cell active layer thickness of 0.6μm a-Si//2μm poly-Si produced by the plasma CVD method.

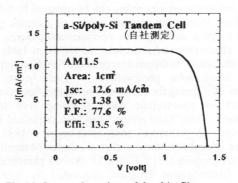

Fig. 10. Present best data of the thin film
Honeymoon cell   (presented by Kaneka)

Table 1. New contributions to the global environmental issues by photovoltaic technology.

Environment ← Local → Global	(1) Solar PV power generation	Clean sustainable Energy resource
	(2) Cleaning of air pollution	Ashing of pollutant gases by the glow discharge decomposition by PV
	(3) Cleaning of water	Electrochemical processing by PV
	(4) Generation of hydrogen energy	Electrolyze of water by PV
	(5) Stop of the desertification Greenery of the desert	PV Water pumping with implantation

Fig. 11   Gel Electrification Program
organized by NEDO

4. NEW ROLES FOR THE CONTRIBUTION TO ENVIRONMENTAL ISSUES

As has been descreibed in the former section, solar photovoltaic power generation is the most promising clean energy technology. Therefore, the penetration of PV technology into utility power generation might be of prime importance for market size expansion with use of scale merit of solar cells in the PV system development. With increase of feasibility onto semi-power applications and full use of maintenance-free with solar photovoltaics, many kinds of anti-pollution processing can be operated by solar photovoltaic power – for example, ashing of pollutant gas in the air by glow discharge decomposition and cleaning of water by electrochemical processing, as shown in Table 1. Mass production of hydrogen energy in the Sahara desert is planned using solar photovoltaics. There exists a possibility of stopping the desert expansion and "greening" the desert by photovoltaic water pumping with tree implantation. The "Gobi project" has been organized for this purpose. A preliminary study team began in 1990 to survey the natural conditions as part of a Japan-Mongolian cooperation program [19]. Figure 12 shown a photograph of "Gel Electrification Program" supported by NEDO Project in the cooperation program.

The result of these considerations indicate that the economically feasible age of photovoltaics will come unexpectedly early in the near future, if continuous efforts are made to promote R&D work and the market pull stimulation by means of government support and international cooperation. In any case, the most important emphasis should be placed on stopping fossil fuel contamination with a worldwide energy policy and to develop this newly born clean energy technology for the future benefit of mankind.

The new energy revolution from coal to oil had been accomplished only in 1/4 century years from 1950 to 1975. The reason of this short time transition for the energy revolution is due to a large scale merit of oil cost by massproduction with massprocessing in the petrochemical plant. With respect to the scale merit, solar cell in the massproduction line might be larger than that of oil at the stage of well developed PV utilization systems. The

largest scale merit in the semiconductor products, for example DRAM has been well identified in the past 20 years which made **a solid state device revolution** from the vacuum tube age in the electronics. Let us take off a new kind of energy revolution with **a clean energy Photovoltaics**. Might it be possible to accomplish **the clean energy revolution** in coming 25 years? It would be a fancy dream and let us enjoy our new challenge!

5. REFERENCES

[1]   Y. Hamakawa: "Recent Advances of PV Activities in Japan", Phy. stat. sol. (b) 194 (1996) pp.15-29.

[2]   K. S. Parikn: "Energy" (Macmillan, New York, 1976).

[3]   E. Cock: "Technology Forecasting and Research Planning Report" Exxon R & D, CRSIDG-73 (1973), p.5.

[4]   K. Miyazawa, K. Katho, and K. Kawamura: Report of Solar Energy Division Meeting, Technology & Industrial Council, MITI (1997) March 17.

[5]   "News Letter No.6", published by OITDA, March 1st. (2000).

[6]   Nihon Keizai Shinbun, News Paper Release (1997) April 11.

[7]   Hamakawa Y.: Proc. of WREC Regional meeting, Perth, Australia, (1999) p.37.

[8]   Y. Ichikawa, T. Yoshida, and H. Sakai: 1st WCPEC (Hawaii, 1994) 441.

[9]   Y. Hamakawa: Current Topics in Photovoltaics, ed. T. J. Coutts and J. D. Meakin (Academic Press, New York, 1985) pp. 111-167.

[10]   Y. Hamawaka and H. Okamoto: "Amorphous Silicon Solar Cells" Chapter 1 of *Advances in Solar Energy*, Vol.5 edited by K.W.Boer, Plenum Pub. Corp, (1989) pp.1～98

[11]   H. Takakura; Optimum Design of Thin-Film-Based Tandem-Type Solar Cells, Japan. J. Appl. Phys., vol 31 (1992)pp.2394-2399

[12]   Ma, W., T. Horiuchi, C. C. Lim, M. Yoshimi, H. Okamoto and Y. Hamakawa: Proc. 23rd IEEE PVSC (1993) 338.

[13]   Y. Hamakawa, Y. Matsumoto, Xu Chong-Yang, M. Okuyama, H. Takakura and H. Okamoto: Mat. Res. Soc. Symp. Proc. 70 (1986) 481.

[14]   Y. Hamakawa, and H. Takakura; Solar Cells, 12 (1984)pp.99-104

[15]   K. Kumagai: Report of 31st Solar Energy Tech. Promotion Committee, New Sunshine Project HQ, MITI, (June, 1994) 83.

[16]   A. Takami, S. Arimoto, H. Morikawa, S. Hamamoto, T. Ishihara, H. Kumabe and T. Murotani: Proc. 12th EC-PVSEC, Amsterdam (1994) 59.

[17]   K. Kumagai: Report of 31st Solar Energy Tech. Promotion Committee, New Sunshine Project HQ, MITI, (June, 1994) 87.

[18]   K. Yamamoto et al.: Thech. Dig. of PYSEC-11, Sapporo, Japan (1999) p225.

[19]   Nihon Keizai Shinbun, News Paper Release (1990) May 11.

PV IN EUROPEAN DEVELOPMENT AID

W. Palz

European Commission

Brussels, 200 rue de la Loi, JECL

ABSTRACT : The European Commission has in its external aid implemented only 3 relevant PV programmes i.e. the Sahel programme PRS and the Kiribati projects both from the European Development Fund and a new programme in South Africa from the EU budget. One is very far from a massive deployment strategy as proposed for instance in "Power for the World".

Most donors and the EU alike put currently their emphasis in development aid on eradication of poverty. PV promotion in future has to fit into this general strategy. Eventually PV has to find its place in the CSPs (Country Strategy Papers) which the partner countries in the Third World are drafting in co-operation with the European Commission and the EU member countries.

1. INTRODUCTION

It is generally accepted that decentralised electrification of the rural areas of the Third World countries is a huge potential market for PV. Unlike building integrated PV in the industrialised countries PV is cost effective in these markets today.

The question to be addressed is why this particular PV market is progressing only slowly. Furthermore, why are current aid programmes in the EU and those from other major donors virtually absent from this sector ?

*   Is it that current development aid concepts are old fashioned and not open enough to innovation ?

*   Does the European PV community not want to bother with social and humanitarian aid aspects ?

*   Is the PV industry immature ?

It is often claimed that finance is a major obstacle. However this cannot be the whole truth. From the European Commission's 8.6 billion € of external aid in 1999 [1] PV was virtually absent.

In any case it is striking to see that there is not much visible effort from the PV industry to help in levelling the playing field with other sectors in the aid programmes and to make global PV initiatives possible instead of focusing downstream on the project opportunities once they are open for contracts. One cannot reap the apples without having planted the tree !

2. PV PROJECTS AS OF TODAY : THE LESSONS LEARNT

The most relevant PV programme in the European Commission's international aid was PRS1, the regional Sahel programme for water pumping.

It is interesting to note that this programme was initiated at one of the previous European PV conferences [2] in 1986 when it was attended by the Commission's head of the Sahel aid programmes. At a total fund of 60 million € more than 600 waterpumps were successfully powered with PV and there was an equal number of home applications. After completion in 1998, the programme was extensively evaluated by the consultants Gauff/Cowi and major results published [3]. A follow up programme PRS2 is going to be decided this year. The budget will also be approximately 60 million € to be financed by the 8th EDF (European Development Fund). Thanks to recent cost decreases it is expected that several 1000 pumps can be equipped this time.

With a budget of 15 million € the second largest programme concerns PV for 1000 schools in South Africa. It has only recently been decided. In this case the EU budget bears 100 % of the project costs including the related appliances.

Still another example is Kiribati, a group of islands in the Pacific Ocean and part of the ACP countries. The funding from the 8th EDF is 4 million € for home systems. Although the budget figure looks low this project is quite meaningful as it represents 50 % of the overall Commission aid to this country.

In terms of technical lessons learnt a large wealth of experience has obviously come out of the Sahel project. It has been well documented by visiting experts that the plants are running as planned and the whole operation was a success.

This being said, it has to be reported that during incidental visits of decision makers from Brussels to whom one of the plants in the Sahel was shown a bad impression prevailed as the PV plant was found to be a playing ground for local kids with stones being thrown on the modules and thick covers of sand obstructing some of the modules. This leaves a bad reputation in Brussels and even a well done 1000 page thick expert report does not help to alleviate it.

We want to stress here rather the unambiguous role of the 3 EU projects here above from a financial and social point of view. Socially the projects are well integrated in national politics and the village environments. There is an incentive for people to look after their plant to have the benefit of the water and the electricity. They pay the financial contribution which is asked from them for the service without much problem. Most interesting is the fact that the fees cover only the cost of operation, maintenance (spare parts) and services but not the amortisation of the investment, the latter being covered by the European Commission's aid package.

3. STRATEGIC ENVIRONMENT

Commissioner Nielson in charge of development aid called recently for "Globalisation with a human face". We think this is exactly the right frame in which PV can fit perfectly.

The official frame for EU aid is Article 177 of the Amsterdam Treaty in which 2 of its 4 items encompass

- fostering sustainable economic and social development and

- campaign against poverty

In particular the poverty eradication target is linked with the World Bank/IMF sponsored Poverty Reduction Strategy. The same has become the key for the European Commission's policy agenda on development as laid down in the communication to the EU Council of Ministers and the European Parliament in the spring this year [1]. Eventually this policy has to be translated into practical measures and actions via the Country Strategy Papers (CSP) which the Commission is proposing for all countries that are benefiting from

its aid. This may be of particular interest for the ACP countries. After the signature of the next partnership agreement with the ACP countries at Suva, in the Fiji next June such SCPs will have subsequently to be put together by the ACP countries and Europe.

As far as PV is concerned the question arises to which extend it has a role to play in the new emphasis against poverty. Hitherto health and education programmes together with road construction and infrastructure for water and sanitation were key elements of EU aid programmes, in line with what the World Bank, the UN and most national donors are supporting. Also the new policy communication mentioned as ref. [1] maintains an emphasis on these subjects.

Indeed health and education of people is necessary for poverty eradication. But is it also sufficient ? Maybe not.
More than in the past, renewable energies in which PV is embedded have received new attention on the political level. Firstly there was the Commission's White Paper on Renewable Energies which calls for 500 000 PV systems for export by 2010 [4]. The new aid policy communication [1] mentions specifically : "Access to sustainable energy services has a key role to play in supporting social and economic development. The provision of energy services in particular through decentralised activities and the promotion of renewable energies is an increasingly important issue". A clear support for renewable energies in the ACP countries comes from the ACP-EU Parliamentary Assembly which adopted a resolution in its Assembly in Nigeria on   23 March 2000 [5]. The competent services of the Commission have, as far as they are concerned, recently produced an internal discussion paper which speaks out very clearly in favour of support for the renewables, too.

4. OUTLOOK

The overall target for PV implementation in the Third World is clear. The "Power for the World" concept has – if any – the merit to pinpoint the dimension of the problem which consists of providing electricity to a major part of the world's population [6]. How can there be progress without electricity ?

Probably there are now enough declarations. Time has come for drafting concrete programmes and projects with the involvement of all those concerned i.e. the donors, the beneficiaries (countries, civil society, villagers), the PV community at large and the industry, NGOs and service companies, agencies and banks …

In the following some rules and essential elements are proposed for such programmes.

*   In terms of financing, development aid funds should be used to alleviate the investment burden for the rural poor. Those in need of electricity are in most cases the poor. Current market sectors for PV where full payment is asked from the rural client should probably not be generalised and only address the better offs in rural society. As described here above, the Commission's PV projects are leading the way in this direction : PV investments are borne by foreign aid, public funds, utilities etc while only operation and maintenance costs are financed by the users.

    Eventually electricity may be considered as a basic commodity and every client in a country pays the same price for it. Aiming at the same prices in towns and villages alike, i.e. adjusting PV kWh costs in stand alone systems to prices from the grid would mean National Harmonised Electricity Prices for all electricity including PV.

    A precedent for the scheme is France where until now all electricity clients paid the same price wherever they lived in the country and whatever the origin of the generator. What is a proven concept for a rich country may fit also for less developed countries.

    The necessary competition for the PV sector in such a national frame of equal prices is eventually assured through tendering of the PV plant investments.

*   Information remains an essential issue. We constantly observe that there continues to be a big lack of knowledge about PV. Only the negative experiences are known by almost everybody.

*   PV projects should be done in co-operation between local companies and importers. As a rule half of the added value should be created locally. Local craft, industry, commerce, services and jobs must get new opportunities : that is what the alleviating of poverty is all about – activities created through the enabling of PV electricity comes as a second step.

*   European PV industry should take the opportunity to promote their case towards the decision makers of development aid in Europe and Europe's partners in the Third World.

    European PV industry has a lot of room for

doing more development work in view of the export markets. Standard products and families of products are not yet available as needed. There is also an urgency to conceive large project packages for electrifying wider areas so that decision makers can process them administratively and financially the same way it is done for conventional power generation or other larger projects.

5. CONCLUSION

At a time when donors of development aid are revisiting their concepts and focus more on results, PV may also have a new chance to get into higher implementation gears. PV fits well in the worldwide attempt to put poverty eradication on top of the development aid agenda. Co-operation with the ACP countries is a particular opportunity at present as the content of the future European Development Fund will be drafted in the year to come.

REFERENCES

[1]   "The Development Policy of the European Community" Brussels COM (2000) 212/11, 26 April 2000

[2]   Seventh EC Photovoltaic Solar Energy Conference, Sevilla, October 1986

[3]   The Regional Solar Programme, Brochure available from DG VIII, CILSS or Fondation Energies pour le Monde

[4]   "Energy for the Future : Renewable Energies", Brussels COM (1997) 599, 26 Nov 1997

[5]   "Intermediate Resolution on the Utilisation of the Renewable Energies in the ACP countries", ACP-EU 2885/00/fin

[6]   "Power for the World : A Global Action Plan, W. Palz, Journal of Solar Energy 14.3 p. 231 (1994)

THE INDIAN PHOTOVOLTAIC PROGRAMME

E.V.R. Sastry
Ministry of Non-Conventional Energy Sources
Block 14, CGO Complex, Lodi Road,
New Delhi 110 003, INDIA

ABSTRACT : The programme for the development of photovoltaics in India began in the mid-1970s and has grown to be one of the largest national programmes in the world. Early R&D efforts to develop the technology for the fabrication of silicon solar cells were followed by a programme for pilot production and development and demonstration of systems for various applications. Commercial production of photovoltaic products began around 1985. R&D projects presently being funded cover thin film technologies and system development. Specifications and test procedures have been developed for commonly used systems. Test facilities have been established at four locations in the country.

The Ministry of Non-conventional Energy Sources has been implementing a countrywide programme for the deployment of photovoltaic systems for a variety of applications. Among the systems being disseminated through a subsidy programme are solar lanterns, home lighting systems, street lights, water pumps and small power plants for villages. An experimental programme for installation of grid-connected power plants is also in progress. Among systems, which have reached a full stage of commercialisation, are those used in telecommunications, railways, oil & gas industry, broadcasting, etc. The Indian market is likely to develop into one of the largest markets for photovoltaic products in the world.

The paper gives an overview covering all aspects of the photovoltaic programme in India and discusses the potential for future growth.

Keywords : 1. Indian PV Programme 2. MNES

1. INTRODUCTION

India is endowed with a very vast solar energy resource. Most of the parts of the country have about 250 – 300 sunny days. There is an immense scope and potential for the use of solar photovoltaic technology in India. A large number of villages and hamlets, estimated to be over 80,000, are yet to be electrified. It is estimated that about 18,000 villages are not likely to be electrified by conventional means due to logistic, environmental or economic considerations.

The Indian solar photovoltaic programme covers the full range of activities from research and development, demonstration, industrial production and commercialisation to utilisation of technology. The Ministry of Non-Conventional Energy Sources (MNES) is responsible for the overall planning and programme formulation as well as overseeing the implementation of various activities.

2. RESEARCH AND DEVELOPMENT

Research and Development has been a major component of the Indian PV programme. The programme for development of photovoltaic technology and its applications in India was initiated as far back as 1976. Initial research work was focussed on development of solar cell technology . In 1980, a five year programme for bringing the technology to a stage of commercial production was taken up. The programme also included development and field demonstration of various PV applications. The programme led to commercial production of solar cells and modules in India based on crystalline silicon. A wide range of PV systems were developed and deployed for demonstration, field testing, evaluation and awareness promotion.

Besides leading to the establishment of indigenous manufacturing capability, continuing R&D efforts have led to reduction in costs, improvement in reliability and introduction of newer technologies. A pilot production line for high efficiency (> 16%) laser grooved buried contact solar cells was set up during the nineties.

During 1985-92, a major programme for the development of single junction amorphous silicon solar cell technology was implemented. At present a project on development of multi-junction amorphous silicon solar cell modules is progressing at the Indian Association for Cultivation of Sciences, Calcutta. The project is expected to lead to a production-worthy process. A set of projects on thin film solar cells based on cadmium telluride, copper indium diselenide and silicon thin films are also in progress.

A PV test facility has been set up at the Solar Energy Centre (SEC), Gwal Pahari, near Delhi to qualify PV modules to the recent IEC and Indian standards and testing of the BOS and complete PV systems to ensure product quality. Test facilities have also been set up at Bangalore, Calcutta and Trivandrum.

Performance specifications have been drawn for several PV systems including solar lanterns, solar home systems, street lights, solar pumps.

Training programmes on PV technology, PV system design etc. are being organised regularly. A set of training manuals developed recently under a World Bank supported programme were used in training courses on quality aspects in manufacture, testing, field installations and system design. More training programmes are proposed to be held during 2000 - 2001.

3. DEMONSTRATION AND UTILISATION PROGRAMME

A programme for development, demonstration and utilisation of PV systems was started in 1980s. There are numerous applications of photovoltaic technology which have been developed and demonstrated in the country. Among them are fixed and portable lighting units, water pumping systems, small power plants, power sources for telecommunication equipment, railway signalling, off-shore oil platforms, TV transmission and battery charging application. Several of these applications are commercially viable. During the past 15 years, the Ministry of Non-conventional Energy Sources has made sustained efforts for the deployment of photovoltaic systems in rural areas. A large amount of experience has been gathered on technical, economic, social and management issues. An analysis of the experience shows that solar photovoltaic technology can be a viable alternative to extension of grid lines to electrify villages, especially in remote and difficult areas. The overall reliability of PV systems is better than that of conventional power supply in rural areas.

A number of organisations are involved in the implementation of such a wide ranging programme. At the apex level, programmes are formulated by the MNES including the details of the schemes for central government assistance. The implementation of the programmes at the state level is through the state nodal agencies, which are specially set up to promote renewable energy programmes.

The Central Government supports the use of PV systems by individuals and non-conventional organizations by providing a subsidy of upto 50% of the cost of the systems. In some States, an additional subsidy is provided by the State Government. The procurement of photovoltaic systems by the state agencies is generally through normal tendering procedures of the concerned organisations. Specifications and guidelines issued by MNES are followed. Among the systems included in the scheme are solar lights, systems and small power plants.

Solar Lighting

Lighting is a basic need in households. Lighting is also required in community centres, clinics, adult education centres and in the streets. MNES has been promoting the use of PV technology to provide lighting in unelectrified villages and unelectrified homes for nearly 15 years now. Among the systems deployed are street lights, fixed home lighting systems, portable solar lanterns and larger community systems. During the last five years the emphasis is on large scale deployment of solar lanterns and home lighting systems.

Solar lanterns were developed and introduced about a decade ago. A solar lantern consists of a small 8-12 Wp PV module and a lantern unit which houses a 12 V, 7AH battery, electronics and a 5/7 watt compact fluorescent lamp. This design can provide light for 3-4 hours every day. Solar lanterns, which were conceived as a substitute for kerosene lanterns, have found wide acceptance as convenient portable lighting units indoors and outdoors. Field studies have shown that a lantern with a 7 watt compact fluorescent lamp gives a light output comparable to that of a petromax lantern. The solar lantern is a superior substitute for a kerosene hurricane lantern normally in use in rural households. Kerosene lanterns provide inadequate lighting, cause pollution and entail fire hazards. Solar lanterns are now being used not only in homes but in rural clinics, hostels, police stations etc.

A fixed type of home lighting systems comprises a 37/74 Wp PV module, a battery unit placed inside the house and two compact fluorescent lamps of 9 or 11 watts each. A typical home lighting system works for 4-5 hours each day. A small TV set or fan can also be powered by the system. Such systems are very popular in the states of Jammu & Kashmir, Rajasthan, Himachal Pradesh, Uttar Pradesh and West Bengal. 223 villages have been electrified in Rajasthan through provision of home lighting

systems. About 50 villages have been electrified through home lighting systems in Jammu & Kashmir including some remote villages in Leh and Kargil districts. Some of the projects were done through an NGO (SWRC, Tilonia) which also set up service facilities and trained local people in the repair and maintenance of PV systems.

Support is also provided for street lighting systems. A street lighting system consists of 74 watts of PV modules mounted on a pole, a battery at the base and a 11-watt compact fluorescent lamp with associated electronics. The lamp comes on automatically at dusk and switches off at dawn.

In several parts of the country it is found that PV systems of a small capacity are competitive in locations which are generally about 3- 5 km away from the grid when compared to the extension of the grid lines for that specific purpose. Stand alone PV power plants of 1 kW – 25 kW capacity have been installed in different parts of the country for village electrification such as Andaman & Nicobar Islands, Assam, Gujarat, Lakshadweep Islands, Rajasthan, Sikkim, Tripura, Uttar Pradesh and West Bengal. About 34 villages in Andaman, Nicobar and Lakshadweep Islands have been electrified through PV power plants and home lights. Some islands have been freed from dependence on diesel generators which required the fuel to be shipped from the mainland.

Several small PV power plants have been set up in West Bengal. Two power plants of 26 kW capacity each were set up in the Sagar 3 km to cater to two villages. These power plants are managed by local rural energy co-operative society. The monthly charges are collected through the local rural bank. Apart from home lighting, PV powered TVs and radios bring in information and entertainment to the villagers. Solar pumps are being used to provide drinking water supply. Recently three more power plants of 25 kW capacity each have been set up in the Sagar Island using a combination of grant and loan. Work on installation of three more plants is in progress.

So far abut 278,000 solar lanterns, 117,000 home lighting systems, 39,000 lighting systems and about 1 MW aggregate capacity of small power plants have been installed in different parts of the country.

Grid Interactive PV Systems

Besides the stand-alone power plants demonstration projects on grid -connected PV systems are also supported by the MNES. The programme envisages use of grid interactive PV systems for demand side management (DSM) or tail-end voltage support. The programme helps promote awareness among utilities. At present projects of 25-100 kW capacity are supported under this programme. Several projects for tail-end grid voltage support and DSM have been developed. A total of 1165 kW capacity of grid-connected systems have been installed so far, including some in the private sector. Projects of another 500 kW capacity are under implementation.

Water Pumping

Water pumping was one of the earliest applications of PV technology developed in India. Small pumping systems with an array capacity of 300-360 W were developed and tried during 1980s.

Based on the experience gained, a special programme was evolved in 1993 for the supply and installation of larger capacity PV water pumping systems for agriculture and related uses. Under this programme, users are offered a subsidy which is linked to the capacity of the PV array used in the pumping system and a soft loan towards 90% of the remaining cost of the system. Pumping systems with PV array capacity in the range 200-3000 W are covered under this programme.

A unique feature of this programme is that manufacturers are required to directly market the pumps to actual users. This has led to strengthening of the marketing networks of manufacturers and helped establish a direct link between the user and the manufacturer. About 3,320 pumps have been installed so far.

A separate study done in some agricultural universities and research institutions has brought out a report on types of crops that can be grown and area which can be irrigated with these systems, etc. A typical 900 watt array DC surface pump is found to be useful to irrigate 1.5 hactares of land with normal irrigation and upto 7.2 hactares using the drip irrigation technique.

4.   COMMERCIAL MARKET

The initial demonstration efforts proved that PV systems are commercially viable for several applications like power source for telemetry on off-shore oil platforms, railway signalling, low-power TV transmitters and microwave repeaters. A major expansion of the rural telecommunications network initiated in the early 1990s created a need for reliable small power systems for rural radio telephones and established a major market for the PV industry. Over 190,000 such systems have been supplied during the past five years. Such commercial applications are not supported by subsidy from MNES. During the last five years the market has significantly expanded to achieve a cumulative PV deployment of more than 44 MW in different parts of the country. In addition, export of PV products to other countries is also increasing. More than 13.5 MW of PV products have been exported during the last three years. Details of sector-wise use of PV cells and modules are shown in Fig. 1.

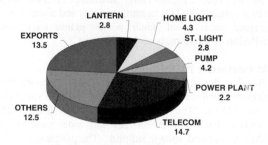

Figure 1 : Sector-wise use of PV Modules (57 MW aggregate capacity)

It is realised that there is a large potential for the commercialisation of several PV products if suitable financing arrangements are made available to individuals and commercial users and appropriate marketing efforts are made by the PV manufacturers. In order to promote the development of market for PV products, MNES has supported soft loan schemes through Indian Renewable Energy Development Agency (IREDA). Under these schemes soft loan assistance is available to individuals as well as commercial organisations for the purchase of solar photovoltaic systems. The loans can also be accessed by financial intermediaries, PV manufacturers, product distributors, cooperative societies, non-governmental organisations etc. These organisations are expected to market PV products and be responsible for timely return of loan to IREDA. IREDA provides loans for manufacturing

activities also. The World Bank has supported a US $ 55 million "SPV Market Development Project" for accelerating the commercialisation of PV products. IREDA has sanctioned projects of an aggregate capacity of more than 3 MW through soft loans under this facility. Among the projects being supported are small power plants, hybrid systems, lighting systems and water pumping systems. Recently, the IFC and GEF have launched a US $ 15 million project called "Photovoltaic Market Transformation Initiative" (PVMTI) to attract private sector investments in photovoltaic projects. Such projects are expected to accelerate the process for commercialisation of PV applications in India. The first loan under this project was approved recently.

Involvement of leasing companies in financing of PV systems has helped a number of private companies, including NGOs to install PV systems in rural areas. The leasing companies are able to offer PV systems on a one time lease payment basis by taking advantage of the 100% accelerated depreciation during the first year and soft loans offered by IREDA.

A grass-roots energy service company, in Karnataka, the Solar Electric Light Company (SELCO), has been successfully marketing and installing solar home systems and solar water pumps in several districts of the States of Karnataka and Andhra Pradesh through rural credit support schemes.

5.   INDUSTRIAL PRODUCTION

Considerable efforts made over the past two decades have made India one of the leading countries in the world in the development and use of PV technology. The expanding market in India for photovoltaic applications has led to the growth of domestic PV industry. The present domestic production is based on single crystal silicon solar cells. During the last five years many private companies have started commercial production and a number of joint ventures have been established.

There are 9 manufacturers of solar cells and 23 manufacturers of PV modules in regular production. About 60 companies manufacture PV systems, most of these are small businesses. There is also a manufacturing capacity of about 2 million silicon wafers. During 1999-2000 about 9.5 MW capacity solar cells and 11 MW capacity modules were produced. The total turn over of the Indian PV industry during 1999-2000 is estimated at about US $ 100 million. Fig. 2 shows the details of annual production of solar cells and modules in India.

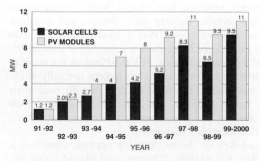

Figure 2: Annual Production of Solar Cells & Modules

Another important feature of the industrial development in this sector is domestic manufacture of key input raw materials and equipment required for processing of solar cells, modules, testing of cells, modules and systems.

6.   GOVERNMENT POLICY

There is a growing recognition that PV technology perhaps offers the best prospects for a breakthrough. Therefore, the policy is clearly directed towards a greater thrust on all aspects of PV technology and applications. The recent policy measures provide excellent opportunities for increased investment in this sector, technology upgradation, induction of new technologies, market development and export promotion.

The MNES provides subsidy upto 50% of the ex-works cost of certain PV systems covered under the demonstration and utilisation programme.   Besides the subsidies, there are a host of benefits available to both manufacturers and users of PV systems. These include: (i) 100% depreciation for tax purposes in the first year of the installation of the systems; (ii) No excise duty on manufacture of most of the finished products; (iii) Low import tariffs for several raw materials and components; and (iv) Soft loans (2.5% - 7.5%) to users, intermediaries and manufacturers.

7.   PROSPECTS FOR FUTURE

A major expansion in the deployment levels have been envisaged during the 9th Five Year Economic Plan (1997-2002). It is estimated that by 2002 a cumulative deployment of about 100 MW capacity PV systems would be achieved. 70 MW capacity PV systems are likely to be added during the 9th Plan period. During this period about 18 MW capacity PV systems comprising 300,000 solar lanterns, 200,000 home lighting systems, 4,000 solar pumps and about 3.6 MW of PV power plants are expected to be installed under the programmes of MNES.   It is expected that by the end of the Ninth Plan about 2000 villages may be electrified   through the PV route. The commercial applications and exports are likely to account for another 52 MW.

It is realised that at present there is a niche market for photovoltaic systems in unelectrified areas away from the grid. Applications like solar lanterns are useful in even those areas which have intermittent electricity supply. A large number of applications pursued so far have contributed in a significant way to the process of commercialisation of photovoltaics in India. The present demand of PV products in the country is about 6 - 8 MW / year. By 2002 it is expected to grow to a level of about 12 - 15 MW/year.

The potential market offers avenues for investment in technology and commercial production. High volume production and induction of new and improved technologies should lead to reduction in costs, contributing to further growth in the market. A major expansion in the deployment levels would be possible if the PV module prices can be brought down to US $ 2- 3 per watt or lower. This would also lead to opening of new markets in urban areas.

AUSTRALIAN GOVERNMENT SUPPORT FOR RENEWABLE ENERGY

Authors: David Rossiter, Gaelle Giroult[1]
Sustainable Energy Group, Australian Greenhouse Office
King Edwards Terrace, Parkes ACT 2600, Australia
GPO Box 621, Canberra ACT 2601, Australia
Tel: 61 2 6274 1392, Fax: 61 2 6274 1884

Abstract

Concern about climate change and concerted international action to reduce greenhouse gas emissions are powerful new drivers for renewable energy. The production and use of non-transport energy is the source of 62% of Australia's net greenhouse gas emissions. Emissions from non-transport energy increased by 13% between 1990 and 1997 [1].

Australia has progressively adopted a series of policies relating to the climate change issue with the initial *National Greenhouse Response Strategy* of 1992 being added to in 1997 by a suite of policies under the banner *Safeguarding the Future* and developing into the *National Greenhouse Strategy* in 1998. A further policy package was announced in 1999 under the heading *Measures for a Better Environment*.

Renewable energy has featured strongly in these policies and in total Australia has committed nearly A$1 billion towards the climate change issue – possibly one of the highest per capita commitments by any OECD country. Photovoltaic power systems are well known in Australia and substantial installations have been made in what may be one of the best near term market opportunities in a developed country. The high levels of insolation in the country together with the remote nature of many power supplies and the requirement of low maintenance facilities draw solar power closest to a market solution for power systems.

Australia has a lead agency devoted to coordination of climate change issues called the Australian Greenhouse Office (AGO). The AGO is a world's first in that the office is dedicated to greenhouse policy matters. The AGO reports to a Ministerial Council which includes the Ministers from the main Departments affected by greenhouse issues. This arrangement, which is unique in the OECD, facilitates policy development and enables coordinated greenhouse responses to be made more readily.

Keywords: National Programme Australia-1 Funding and Incentives – 2 Green Pricing - 3

1. BACKGROUND AND INTRODUCTION

On 20 November 1997 the Prime Minister of Australia in his statement *"Safeguarding the Future: Australia's Response to Climate Change"* announced a A$180 million package of measures designed to improve the greenhouse performance of Australia's highly competitive energy-dependent sectors. The package was designed to stimulate the uptake of renewable energy by encouraging increased market demand, by measures such as requiring large buyers of electricity to source additional electricity from renewable sources by 2010, and providing support to develop renewable energy industries.

As part of this package the Australian Greenhouse Office (AGO) was established in April 1998 as the lead Australian agency on greenhouse policy and programs. The AGO has responsibility for the coordination and delivery of the Commonwealth Government's climate change package and in particular the renewable energy programs.

The AGO provides a 'whole-of-Government' perspective and reports to a Ministerial Council comprising the Ministers for the Environment and Heritage, Industry Science and Resources, Agriculture Fisheries and Forestry, as well as the Minister for Finance. Other senior Government Ministers, such as the Minister for Foreign Affairs and Trade and the Minister for Transport and Regional Services (currently also the deputy Prime Minister), also regularly attend the Ministerial Council meetings.

In 1998 the *National Greenhouse Strategy* was signed by the Commonwealth and all States and Territories and it lays out an integrated package of 86 measures to reduce Australia's greenhouse gas emissions. It builds on the earlier 1992 *National Greenhouse Response Strategy*, incorporates the measures from *Safeguarding the Future* and coordinates activity at all levels of government and within industry. The AGO has responsibility for its coordination.

On 31 May 1999, as part of its revised tax package, the Australian Government announced an additional package called *"Measures for a Better Environment"* and committed a further sum of about A$800 million to support greenhouse abatement programs. These measures are designed to assist Australia in playing its part towards meeting its Kyoto commitments including through increasing the use of renewable energy.

Australian Government support for renewable energy now includes:

- Mandatory Additional Renewable Energy for Electricity Supplies;
- Greenpower scheme for electricity purchasers;
- Photovoltaics Rebate Program (PVRP);
- Renewable Remote Power Generation Program (RRPGP);
- Renewable Energy Commercialisation Program (RECP);
- Renewable Energy Showcase;
- Renewable Energy Equity Fund; and the
- Greenhouse Gas Abatement Program.

These packages together with previous commitments form a comprehensive and complementary range of measures which

[1] The authors wish to gratefully acknowledge assistance in writing this paper from members of several teams within the Australian Greenhouse Office Sustainable Energy Group including Philip Harrington, Martin Walsh, Stephen Bygrave, Gillian McDonald, Karla Wass, Caroline LeCouteur, James Stuart and David Liversidge.

now provide strong support for renewable energy and its development within Australia through market mechanisms, grants and other programs. This paper gives an overview of the renewable energy programs including their early stage impact on the Australian energy industry and in particular the photovoltaic industry sector.

2.   GOVERNMENT   SUPPORT   FOR RENEWABLES   THROUGH   MARKET MECHANISMS

The Australian Government has created two main market mechanisms to support the increase in uptake of renewable energy through a mandated additional renewable energy requirement for large electricity buyers and providing national support to a renewable energy accreditation scheme for electricity consumers called GreenPower.

2.1   Mandatory Additional Renewable Energy for Electricity Supplies

Large buyers of electricity such as wholesaler purchasers and retailers are collectively required to source an additional 9500 GWh of their electricity from renewable energy sources by 2010 relative to 1997. This takes Australia from around 16,000 GWh per annum of renewable energy in electricity in 1997 to about 25,500GWh per annum in 2010, or an increase of about 60% in that period.

This market mechanism reverses the decline in renewable content of electricity that has been occurring in recent years and will raise the renewable energy content of electricity to about 12.5 per cent by 2010.

This requirement is expected to liberate A\$2 to A\$3 billion of additional investment in renewables and will accelerate the uptake of renewable energy in grid-based electricity and provide a larger base for the development of commercially competitive renewable energy.

The market mechanism used to create this change is achieved through a scheme involving the use of tradeable renewable energy certificates.   Under this scheme, renewable energy certificates are produced from accredited renewable energy power stations and identified liable parties are required to hold specific numbers of these certificates on an annual basis. Certificates can be traded through a market system which ensures traceability of certificates through a central registry.

2.2   GreenPower

Under this measure electricity consumers are offered the choice of purchasing their normal electricity or paying a small premium to ensure their electricity has been partly or fully generated from renewable sources. A scheme called GreenPower has been operating in the State of New South Wales for several years to ensure that this power is genuinely generated from renewable sources. Under the *National Greenhouse Strategy* this scheme is being expanded to provide uniform standards of accreditation throughout participating States and Territories within the Commonwealth.   This scheme is being coordinated by the Sustainable Energy Development Authority (SEDA) of New South Wales.

Launching what it called the "World's Biggest Green Electricity Scheme" SEDA has included photovoltaic generation as a 'green energy" source.   There have already been a number of small systems of a few kW each connected to the grid in Newcastle, Sydney (5), Wollongong and Nimbin.   Energy Australia has called tenders for and constructed a 200kW$_{peak}$

photovoltaic plant in the Hunter Valley as part of its portfolio of green electricity generation.

Initially the majority of "green" generators connected to the grid were based on existing small hydro stations owned and operated by the NSW distributors.   However the SEDA rules require that by 31 December 1999, 60% of the total "green" electricity (in kWh) sold through green schemes must come from new green electricity generators, that is green electricity generators commissioned after 1 January 1997.   It is likely that a significant portion of this new "green" generation will come from grid connected PV systems, particularly if the cost of PV systems can be brought down.

Recently (March 2000) four States and Territories, New South Wales, Queensland, Victoria and the Australian Capital Territory (representing over 80% of Australia's population) have agreed to operate under the GreenPower accreditation guidelines.   The schemes had at last report (March 2000) about 60,000 consumers subscribing to them reflecting a strong interest in this voluntary market mechanism to increase renewable energy uptake.

3   GOVERNMENT   SUPPORT   FOR RENEWABLES THROUGH GRANT AND OTHER PROGRAMS

The Australian government has provided several opportunities for support of the renewable energy industry through grants and other programs. Under these schemes specific technologies are targeted, or programs are set up to support specific outcomes and capacity building is encouraged.   Some of these measures are described below.

3.1   Photovoltaic Rebate Program (PVRP)

Commencing on 1 July 2000, A\$31 million is provided over 4 years to householders and community use building operators who provide new photovoltaic power systems meeting specified criteria. The rebate scheme applies to rooftop and building integrated PV systems and provides up to half the system capital costs at the rate of A\$5.5 per peak watt of the installed system. Systems need to be at least 450W with household systems limited to a maximum of 1500W for the full subsidy and community use building applications limited to 5000W. The program is delivered by the Commonwealth with the assistance of the States and Territories.

3.2   Renewable   Remote   Power   Generation Program (RRPGP)

This measure provides A\$264 million over 4 years commencing on 1 July 2000 to replace off-grid diesel generation with renewables for existing and new systems. The program provides up to 50% of the capital cost of the renewable components of remote area power systems and will be delivered in consultation with the States and Territories.

3.3   Renewable   Energy   Commercialisation Program (RECP)

This A\$56 million program seeks to support and/or promote the demonstration and commercialisation of innovative renewable energy equipment, technologies, systems or processes that have strong potential for widespread commercialisation in Australia and/or overseas; and offer the prospect of significant abatement of greenhouse gas emissions over the longer term. Applicants need to provide at least 50% of the project costs from commercial sources and need to be an incorporated company, organisation, institution or consortium located in Australia.

Grants are normally within the range of A$100,000 and A$1 million and over a period of up to 3 years. Payments are made upon achievement of agreed performance milestones.

Successful projects to date include support for manufacture of photovoltaics, a solar sailor photovoltaic powered commercial sailing vessel using wind and solar power, building integrated solar photovoltaics, battery and inverter systems, demand side load management, integrated renewable energy control systems, solar thermal boosting dishes for thermal power plants, sugar bagasse dewatering systems, gasification systems for biomass, hot dry rocks, landfill gas and windpower systems.

A small part of this fund has recently been set aside to assist with Renewable Energy Industry Development (REID). A$6 million has been allocated for institutional strengthening and industry support in relation to renewable energy projects.

3.4   Renewable Energy Showcase
A$10 million has been allocated for some leading edge renewable energy "showcase" projects in areas of biomass, solar thermal power and wind energy. Four of the five projects will be providing "green power" to a major electricity grid. All the projects have potential for widespread use in Australia and overseas.

3.5   Renewable Energy Equity Fund
The Fund will manage a total of A$30million equity funds with A$20 million provided by the Australian Government and the remnant matched from industry money. The funds will be used to support renewable energy projects. The fund manager has recently been appointed and the fund is expected to be operating within a month or two.

3.6   Greenhouse Gas Abatement Program
This program provides A$400 million over 4 years commencing in 1 July 2000 and is designed to further assist Australia to meet its Kyoto commitments. The Australian Greenhouse Office will advise the Government on the development of options that have maximum carbon reduction or sink enhancement capacity. The Government will take into account the potential for employment growth, potential as catalyst for new private investment, the potential for export and the involvement/development of new technologies and innovative processes. Particular attention will be given to opportunities in rural and regional Australia.

4   POTENTIAL MARKETS FOR AUSTRALIA'S RENEWABLE ENERGY INDUSTRY

Australia has recognised that there are significant benefits to be obtained from the strategic positioning of Australian industry to develop a renewable energy market both internally and as an export market.

4.1   Internal Markets
Australia has one of the world's largest near term commercial (as distinct from government subsidised) markets for renewable energy systems such as photovoltaics, solar thermal electricity and for wind energy. A robust sector in this market is the displacement of diesel fuel used to generate electricity in remote towns, mining camps and homesteads by solar generated electricity. This can be a commercial market, unlike the residential programs in Europe, Japan and the USA, and hence is less subject to changes in government policy.

Such electricity grids range in size from 10 kW to 80 MW. Solar energy can substitute for much of the diesel fuel burnt in these places if the cost is less than the combined cost of diesel fuel, fuel transport to remote areas and associated diesel maintenance. This type of market does not exist in other developed countries because electricity grid coverage in those countries is nearly complete [2].

The combined capacity of diesel mini-grids in Australia is over 500MW. These are generally in areas with excellent insolation. Typical operating cost of diesel mini-grids (mostly fuel cost) is in the range 12-35Ac/kWh, which is 3-10 times higher than from coal fired power stations. In general, the smaller and more remote a grid is, the more costly is the electricity produced by it. A market penetration of 5% of the diesel electric market in Australia alone would represent a potential installed capacity of 22.5 MW and a value of A$90 million.

4.2   Export Markets
South East Asia is a large potential market close to Australia with a population in excess of 445 million. Most of the ASEAN member countries (population 326 million) have had GNP growth rates consistently between 6 and 11% per annum over the last five years. India is another large potential market, where government programs are in place to assist the marketing of renewable energy systems.

For example, tens of millions of rural dwellers on Indonesian and Philippine islands have little prospect of being connected to central electricity grids for many years to come. Sixty million Indonesians live on the outlying islands and do not have access to low cost electricity. They will rely on diesel systems with solar supplementation for the forseeable future.

Typically 70% or more of the population in these regions live in rural areas. Of these, more than 70% do not have access to the national grid. At least 50% of this group is unlikely to be connected to the grid within the next 30 years. In most of the countries in the region, grid extension is simply uneconomic because of the distances involved and the low energy requirements of the rural populations. Typical rural household electricity needs in the region would be between 300 and 500 kWh/year.

Assuming average households of around 5 people, a potential household market of about 40 million houses, worth in excess of A$4 billion, exists in the South East Asian region alone.

5   STATUS OF THE AUSTRALIAN PHOTOVOLTAIC INDUSTRY

5.1   Industry
The two main photovoltaic manufacturers in Australia were Solarex and BP Solar, and they have recently merged to form BP Solarex. In 1996, Solarex produced about 4MW of solar panels using polycrystalline silicon cells and BP Solar about 5MW of solar panels using single crystalline silicon cells. The merged companies have plans for increasing production capacity.

It is estimated 10 to 5 MW of PV electricity are installed in Australia for commercial applications, such as communications, navigation systems, private dwellings, etc.

5.2   Photovoltaic Demonstration Systems

Table 1 shows some of the solar renewable energy systems in Australia.

Table 1: Some Solar Energy Systems In Australia

Capacity	Technology	Location	Completion date
20 kW	Tracking PV flat plates	Kalbarri, WA	1993
10-15 kW	Fixed PV flat plates	Gylkminggan, NT	1995
Defunct, was 14kW	Solar Thermal	White Cliffs, NSW	1982, Retrofit PV 1999
50 kW	Solar Thermal- Big Dish	ANU, Canberra, ACT	1996
1.8 kW	PV/Trough	ANU, Spring Valley, ACT	1997
20 kW	PV/Trough	Perth area, WA	1999 REIP Grant A$300,000
50 kW	Fixed PV flat plates	Queanbeyan, NSW	1999
50 kW	PV flat plates	Dubbo, NSW	1997
40 kW	Concentrator technology	Dubbo, NSW	1999
100 kW	Fixed PV flat plates	Wilpena Pound, SA	1998
200 kW	Fixed PV flat plates	Newcastle, NSW	1999
> 50 kW *	Fixed PV flat plates	2000 Olympic Village, NSW (*)	2000
	Titania Cells manufacturing facility	Queanbeyan, NSW	2001? RECP grant A$1,000,000

(*) "The Solar Village will be some 33 times the size of any Australian solar project to date and will be the largest residential solar development in the world" according to Pacific Power. Pacific Power is purchasing 50 1kWp PV systems, including inverter, meters, switch gear, frames and flashing to be integrated into the roofs of the first 50 houses of the Solar Village.

Further details of installed and proposed renewable energy power stations, mainly those above 2kW capacity, are shown on the Australian Greenhouse Office website at www.greenhouse.gov.au

5.3   Photovoltaic Research And Development In Australia

Table 2 shows some of the research and development work in Australia.

Table 2: Some Photovoltaic Research In Australia

Company/Institution	Technology
Australian National University	Single c-Si thin film by Epilift; PV/trough system; high efficiency c-Si cells
University of NSW	Buried contact solar cells, c-Si thin film, high efficiency c-Si cells
Pacific Solar, Sydney, NSW	Si thin film by CVD on glass; joint project between Pacific Solar and UNSW
STA, Queanbeyan, NSW	Dye Doped Titania (TiO_2) electrochemical cell
Murdoch University, Perth WA	hydrogenated a-Si thin film

6   CONCLUSIONS

The Australian Government is strongly committed to greenhouse gas abatement with about US$600 million allocated to policies and programs. On a per capita basis this is one of the highest levels of funding in the OECD.

The Australian Government has an extensive support system for renewable energy projects and specifically has programs directed at encouraging photovoltaic installation.

The funding has been directed towards encouraging new market mechanisms, specific programs and grants.

Uniquely a dedicated office called the Australian Greenhouse Office has been created to coordinate all greenhouse policy matters. The AGO is structured to streamline government processes by directly including all relevant government departments within its governance council.

REFERENCES

[1]   Australia's National Greenhouse Gas Inventory 1997, September 1999.

[2]   G.Giroult-Matlakowski, U.Theden, A.Blakers, "Photovoltaic Market Analysis" Proceedings of the WCEPSEC'1998 Conference in Vienna, p3403-3406

Comparative impact of publicly funded Photovoltaic RD&D programmes (USA, Japan, Germany, France, Italy: 1976-1998).

Alain Ricaud

Cythelia

Savoie Technolac, BP 319, 73 375 Le Bourget du Lac Cedex, France

Tel. +33 4 79 25 31 75, Fax. + 33 4 79 25 33 09, email:ar.cythelia@wanadoo.fr

Abstract: This paper gives a comparative and independent review of the impact of publicly funded Photovoltaic RD&D programmes of five countries (USA, Japan, Germany, France and Italy). In 22 years, about 4 400 M€ were spent for PV RD&D by the public bodies of those countries (43% in the USA, 31% in Japan, 16% in Germany, 5% in Italy and 5% in France).We report our best estimates for the public funding in each country, as well as the ratio of public funding divided by the local industrial turnover, which characterises how public investments have served to share the risk with industry during the early period and the state of maturity of the photovoltaic industry in the recent period. All countries except Germany have managed to considerably reduce this ratio over the period.

We report our best estimates for the cumulated installed domestic capacity by segment (Remote residential, Remote industrial, Grid connected buildings and Central power plants) in the five countries. By the end of 98, out of the 902 MWp of PV modules world-wide cumulative shipments, only 269 MWp (30%) had been installed within the 5 countries in reference, this share increasing rapidly since a few years. We derive the preferred market segments by country. We show how the various national RD&D programmes have strongly contributed to the performances of commercial PV modules. We give details on efficiency, cost, reliability, energy pay-back time and volume produced. We review the performances and cost of cells, materials, modules, systems and storage. Finally, we review the average selling price of power modules throughout the period and derive the value 0.82 for the learning curve coefficient of the PV modules industry. We propose to use this coefficient to predict future cost reductions.

Keywords: R&D and demonstration programmes – 1: National programmes - 2: Funding - 3.

1. STATISTICS ON RD&D PROGRAMMES IN EUROPE, USA, AND JAPAN

1.1 Public funding versus research activity.

Since it has been considered as the most important and consistent event world-wide in 22 years, we have reviewed by field and by category the scientific papers presented over the last 15 EU Photovoltaic Solar Energy Conferences, including the 2nd world PVSEC –Vienna – 1998. We have sorted the topics as follows: Fundamentals, Material, Solar cells, Modules, Systems, Measurement, Applications, Demonstrations, Economics, and Programmes. Reporting the number of scientific communications per topic and per country over the period, we derive some interesting trends in terms of specificity of the research by country and in terms of usage of the public money.

In the 15 EPVSEC conferences, 5 644 papers were published by the PV community. In order to eliminate the parasitic effect of the German supremacy in all fields and the limited presence of Japan, we have analyzed for a given country, the relative number of published papers by topic. This ratio gives the natural inclination of a country for one of the 10 research topics. As a result, in an ideal world, applying improperly the economic theory of Ricardo to the PV RTD, the topic "Fundamentals" would go to Australia, "Materials" would go to France, "Solar cells" to Japan, "Modules" to Spain, "Systems" to Germany and "Measurements" to Italy, "Applications" and "Programs" to Switzerland, "Demonstrations" to the UK, and the topic "Economics" to the USA.

1.2 Public funding versus industrial activity

Altogether in 22 years, from 1977 till 1998, about 4 400 M€ were spent by public bodies for the PV RTD of the 5 top countries of the world PV club. In Table 2, we report our best estimates for the public funding in USA [1], Japan [2], Germany [3], France [4] and Italy [5] by 5 years periods.

Topic	USA	Germany	France	Italy	Japan	All countries
Fundamentals	81	142	75	24	23	565
Material	77	104	92	29	16	459
Solar cells	299	505	261	143	172	2100
Modules	34	70	20	27	11	275
Systems	47	178	38	54	14	575
Measurements	27	30	17	37	7	206
Applications	93	169	74	62	17	882
Demonstration	32	33	11	17	8	236
Economics	18	3	4	1	1	33
Programmes	39	39	20	20	15	313
Total	747	1273	612	414	284	5644

Table 1: Papers statistics covering the 15 EU PVSEC

Funding	USA	Germany	France	Italy	Japan
1976-80	629	21	34	25	80
1981-85	510	158	66	49	244
1986-90	208	208	53	61	245
1991-95	319	259	44	62	315
1996-98	185	105	24	36	457
Cumul	1 851	751	221	233	1 341

Table 2: Public funding by period and by country in M€

In Table 3, we report the calculated turnover of the cells and modules manufacturers having plants producing in the country in reference.[6] The cumulated turnover (4 700 M€) is barely above the cumulated public funding.

Sales	USA	Germany	France	Italy	Japan
1976-80	70	13	9	3	5
1981-85	368	45	34	32	250
1986-90	355	44	52	45	541
1991-95	601	91	65	125	580
1996-98	696	34	96	55	498
Cumul	2 090	227	256	260	1 874

Table 3: Industrial turnover by period and by country M€

We then derive the ratio Funding/Sales (more precisely, Public funding divided by Local industrial turnover) which in fact characterizes rather well the state of maturity of the photovoltaic activity in each country as it is judged by the public bodies in charge of the subsidies.

Funding / Sales	USA	Germany	France	Italy	Japan
1976-80	9,0	1,6	3,8	8,3	16,0
1981-85	1,4	3,5	1,9	1,5	1,0
1986-90	0,6	4,7	1,0	1,4	0,5
1991-95	0,5	2,8	0,7	0,5	0,5
1996-98	0,3	3,1	0,3	0,7	0,9
Cumul	0,9	3,3	0,9	0,9	0,7

Table 4: Public funding / Sales by period and by country

In Table 4, we find a ratio around unity (Japan: 0.7; USA: 0.9; France: 0.9; Italy :0.9) except for Germany (3.3) which has a very peculiar way of funding its PV activities. In 22 years the German public bodies (essentially the federal BMBF) will have spent about 750 M€ for an industry which delivered only 227 M€ in cells and modules sales during the same period.

1.3. Public funding versus dissemination

In Table 5, we report our best estimates for the cumulated installed domestic capacity by segment in the 5 countries for the period 1976-98. We first notice that, out of the 902 MWp cumulated cells and modules shipments throughout the world, only 269 MWp (30%) have been installed within the 5 countries in reference and most part of the 70% left in developing countries.

Cumul installed MWp	USA	Germany	France	Italy	Japan
Remote residential	47,5	0,7	4,9	6,5	2,0
Remote industrial	39,7	1,7	1,1	6,5	19,0
Grid con. buildings	16,5	24,8	0,3	0,5	74,8
Central power plants	7,7	3,3	0,0	9,0	3,0
Total (1998)	111,4	30,4	6,3	22,5	98,8

Table 5: Cumulated installed domestic PV capacity [7]

With the recent political push in the "On-grid distributed" segment in Germany, then in Japan, and probably next in the Netherlands and the USA, we are now on the way of disseminating PV in the most proximate and general usage of energy: private homes and commercial buildings.

Table 5 gives also a good idea of the preferred market segments by country : Germany is truly engaged into the route of grid connected buildings. The USA started early on with central power plants in California, and have now a mix scenario fitting well with the average world-wide market segmentation. France is strong in the segment of remote habitations through the SAHEL programs in Africa (1991-1996) and the more recent DOM-TOM (oversee territories) programme with EDF. Italy through ENEL has focused most PV investments in central power plants. Japan is now following the German route with heavy public and private investments in grid connected buildings and private homes.

2. COMPARISONS, SYNTHESIS AND PROSPECTIVE

We review the various reasons for the countries to invest in PV research and development, how their R&D was organised and funded, what the achievements and failures have been, and where PV R&D is currently heading. Since it is by far the leading country in Europe, we detail the German activities in comparison with those of Japan and the USA.

2.1 National Policies

Underneath the official language, the main reasons for public investment in PV R&D are quite different in the countries studied: in Germany, R&D has been driven by politics; in Japan by the Government; and in the USA, it seams that the solar lobby sets the agenda. In France, the nuclear lobby (CEA-EDF) is so strong that it has de facto, banned the usage of PV in the country for the last 20 years. Japan and more recently, The Netherlands, are the rare countries having explicit national energy plans, to the extent of being able to set targets for solar cells price and efficiency, as well as installed capacity of renewable energy technologies.

In Germany, it is politically accepted that PV R&D is a long-term task in line with the activity character of base-funded public research establishments, similar to aerospace or environmental research. The most important source of funding is provided by the Federal Ministry of Education, Science, Research and Technology (BMBF).

The basic philosophy of Japan's energy policy is harmony among the three E: Economic growth, Environmental protection, Energy security. .Japan believes that there is a great possibility for the country to use Photovoltaic solar energy, and also it is considered as another tool for future commercial penetration abroad. The Japanese PV program started as part of the Sunshine Project, which was set up in 1974. In 1992, the New Sunshine Project absorbed the Sunshine Project, the Moonlight Project (energy saving project, started in 1978) and the Environmental Technology programme (started in 1989).

In the US, the national interest in renewable energy grew out of the oil supply disruptions of the early 1970's. It was considered as a tool for national security. The PV program was first the responsibility of the National Science Foundation, then transferred to the Energy Research and Development Agency. Then the Jet Propulsion Laboratory was assigned technical and management responsibilities in January 1975 for the program, based on their experience in spacecraft PV power system. In 1977, the Solar Energy Research Institute national laboratory opened in Denver to develop new technologies that could blunt the effects of rising oil prices; now, after two decades of oil drop, priorities have changed and the new NCPV in NREL is now more and more channeling some of its efforts into commercial ventures and profitable products.

2.2 RTD programmes and co-ordination

There are several ways of co-ordinating the RTD programmes. Only countries with a strong political will can publish targets, state a time frame and set up an organisation to achieve their objectives. It is clear that Germany and Japan have this long range view.

In 1998 in Germany, out of 800 people working in Solar Energy R&D, there were more than 350 researchers and scientists involved in the PV R&D programmes in seven Institutes and various Universities under the umbrella of FVS (The German Solar Energy Research Association), established in 1990 as a cooperative structure. Solar energy R&D is considered as an essential national responsibility since the tasks are long term, interdisciplinary and in the public interest. Consensus is that no existing or newly established national R&D entity could satisfy these requirements as the relevant technical and scientific tasks are extremely heterogeneous and because solar energy research itself has an especially decentralized character.

In Japan, from 1974 to 1983, the R&D program mainly focused on low cost solar cell production including mass production of raw Si material, poly Si, and amorphous silicon. The second decade of the program (1984 - 1993), in addition to material R&D efforts, mass production technologies and system technology were intentionally pursued, together with new cell technology such as a-Si based stacked solar cell and flexible substrate solar cell. In the New Sunshine Project, since 1993, new programmes have been organised focusing on large public promotion and cost decrease of PV systems, e.g. field test of PV systems for public facilities and a monitor plan for private houses with government subsidy and new regulation of tax reduction. Officials of MITI are renowned for their zealous efforts in finding new avenues to refine the government-industries cooperations. They are able to persuade the five leading photovoltaics producers to suspend their normally intense sense of rivalry in order to achieve common research objectives.

The USA have been pioneering terrestrial PV since 1973: the first programme was named Low Cost Silicon Solar Array Project, later to be called Flat-Plate Solar Array Project (FSA). In 1977, the government programme was expanded to cover all aspects of PV R&D with efforts initiated at the newly formed Solar Energy Research Institute (SERI) and the Sandia National Laboratory. These additional efforts included concentrator modules, PV systems, and research on the newly developed thin films. Under the DOE PV programme, the government-industry-partnerships initiated cost-shared subcontracts which were championed by NREL. The US having a long tradition of good process engineers and manufacturing engineers, contrarily to Europe, manufacturing has always been an integral part of the US RTD strategy. Since 1988, the national PV program has sponsored the Photovoltaïc Manufacturing Technology (PVMaT) programme and the Thin Film Partnership (TFP).

2.3 Budgets

Germany, with an annual federal budget of more than 35 M€, and a cumulated funding of 750 M€ in 22 years has the second PV budget per capita behind Japan. Compared to the cumulated national industrial turn-over, Germany has the poorest result (20 € of public subsidy per Wp produced). If we add together the cumulated budgets spent in 20 years in Germany (750 M€), France (220 M€) and Italy (230 M€), we end up with 1 200 M€ of public funding for 743 M€ of industrial turn-over (122 MWp produced), that is an average European public subsidy of 6.1 € / Wp produced.

The cumulated public budget spent in Japan during the period 1978-98 is about 1 340 M€. Compared to the cumulated national industrial turn-over (1 875 M€), with no surprise, Japan has the best efficiency of   public spending (with 4.8 € of public subsidy per Wp produced). There have been two major impulse, the first one peaking in 1983 has led to the watches and pocket calculators market supremacy. The second impulse that started in 1994 is now in phase with the national implementation of the 70 000 roofs programme.

The cumulated public budget spent in the US during the period 1978-98 is about 1 850 M€. Compared to the cumulated national industrial turn-over (2 090 M€), the US are among the best in terms of efficiency of their public spending (with 5.1 € of public subsidy per Wp produced). The strong impulse of the Carter's administration was followed by a temporary decline during the Reagan administration, decimating the least profitable industries, and followed by a robust industrial growth starting in 1988 which has not stopped since then.

2.4 Performances

The various national R&D programmes have strongly contributed to the performances of commercial PV modules, in terms of efficiency, cost, reliability, energy pay-back time and volume produced.

In 1978, industrial crystalline silicon cell efficiencies were in the range of 10-11% ; they are now in the range 14.5 to 18%. Pilot line cell efficiencies of 19% are confirmed, and are expected to be introduced in the industry in the coming 2 to 5 years. What was believed to be the theoretical limit of silicon crystal solar cells in 1978 (23.5%) has been in effect ovecome on small laboratory devices at the UNSW (Australia) a few years ago.

In 20 years, the average selling price of PV modules in current US dollars has decreased from 18 $/Wp in 1978 down to 6.8 $/Wp in 1988 and 4.2 $/Wp in 1998. During the same time, power modules for large scale applications have seen their price decrease from 13 $/Wp down to 3.7 $/ Wp. Due to progresses in encapsulation, the life time of the modules has been increased from 5 years to 30 years. Most leading manufacturers are now guarantying their modules for 20 years.The energy pay-back time of the early terrestrial solar cells was calculated in 1978 to be about 20 years. Recently, the energy pay-back time has been estimated at 5 years (multi c-Si) in continental climates, and 1-2 years for thin film modules [8] From 2 MWp shipped in 1978 essentially from the US, the total PV world shipments have now reached the level of 202 MWp in 1999, more equally divided between EU, US and Japan. The average growth rate of the world market over the period 1978-98 period is 23%. To our knowledge, there is no other example of an industry enjoying such growth over a so long period of time. Obviously for the years to come much more is to be expected from the down stream, i.e. the utilization of PV in a large variety of new systems serving new applications.

2-5 Learning curves:

Fig. 1 shows photovoltaic modules price history as a function of the cumulative shipments. The coefficient derived from past history is 0.82 for crystalline modules. To predict the future, the formula that has been used is:

$Ln(P_1/P_0)=(Ln\ a/Ln\ 2).Ln\ (V_1/V_0\ )$, where a is the learning curve coefficient. we keep the same coeff of 0.82 for c-Si and take 0.78 for the Thin Films technologies. So far the thin-films technologies, have essentially penetrated the market of consumer modules and most recently, a significant part of the market of building integration. The segment of grid connected (solar roofs and façades on private and public buildings) was almost inexistant eight years ago. With the current political push, through larger and more diversified demonstration programmes extended to other countries, we can expect a yearly growth of 26%

have gone to considerable lengths to support R&D. Then in the 80's the fundings went to demonstration in the remote applications of the developing countries. In the 90's, much has been learnt from the implementation of the pilot projects and demonstration programmes in industrial countries. The growing market of BIPV installations has helped module manufacturers, roofs integrators, inverters producers, architects and numerous local electricians to gain tremendous experience from the field. Methods have been developed to enable an accurate determination of final payments in the framework of "guaranteed results" contracts. Now, the PV systems work, the technical rules are in place, but the process for the utilities to accept grid connection from randomly distributed individual producers will need more work on the concept of dispersed power generators connected to a national grid.

ACKNOWLEDGMENT

Informations and figures presented in this paper are derived from a study conducted by Cythelia and funded by EU DG XII Contract # JOR 3-CT97-2007.

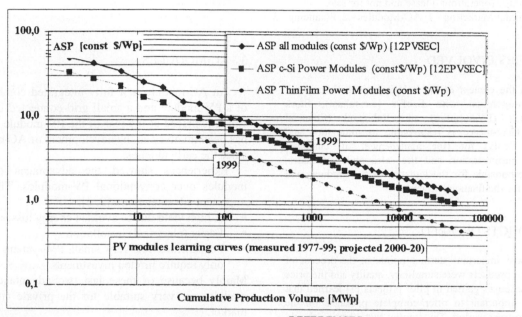

Fig. 1 Photovoltaic modules price history as a function of the cumulative shipments and projections 2000-20.

maintained constant for 10 years; then by 2010, the shipments would reach 2 100MW, (with a breakdown of 1 000 c-Si and 1 100 TF) the cumulated shipments would then be 10 400 MW (6 500 c-Si and 3 900 TF), the yearly turnover 4 500 M€, and the average selling price, would be as low as 1.1 €/Wp for TF power modules, 1.6 €/Wp for c-Si power modules and 2.2 €/Wp for the mix (all values in constant € 1999).

CONCLUSION

Although industrial engagement is indispensable for realising marketable solutions, past experience proved it insufficient to establish renewable energy utilisation in the absence of profitable markets. Both the public and private sectors had underestimated the difficulties of PV systems dissemination and the time frame needed for development and market introduction. In the late 70's, Public Bodies

REFERENCES

[1] Thomas Schott, BMBF Bonn, «Strategies for R&D of PV Japan-Europe in comparison» (Berlin, Germany, October 8-10 1997).

[2] Hamakawa, Proceedings 10th EPVSEC, Lisbon, 1991, p.1376

[3] K. Wollin, Procedings 10th EPVSEC, Lisbon,1991, p.1381

[4] Michel Rodot, Procedings 2nd EPVSEC Berlin, 1979, p.1135

[5] C. Messana & R.. Gilson, Procedings 9th EPVSEC, Freiburg, Sept 1989

[6] Derived from PV News (1980-1998)

[7] IEA statistics, Ref : Report IEA PVPS Ex. Co\TI 1997:1

[8] J.M.Woodcock, H.Schade, B.Dimmler, A.Ricaud, Proceedings 14th EPVSEC, Barcelona, Spain, June 1997, pp. 857-860.

SOLARIS: A MARKETING SUCCESS STORY WITH PRIVATE CONSUMERS

B. de Wit[1], S. van Egmond[2], P.S. Zaadnoordijk[3], D. Dijk[4]

1. Ecostream Renewable Energy Marketing Services, P.O. Box 8408, 3503 RK Utrecht, The Netherlands;
tel. +31 30 3002400; fax. +31 30 3002499; e-mail B.deWit@ecostream.nl
2. Greenpeace Nederland; 3. Stork Infratechniek; 4. Rabobank Nederland

ABSTRACT: To accelerate the market introduction of (small) PV-systems for private consumers, Greenpeace launched the Solaris-campaign in The Netherlands. The overall objective of Solaris is to realise 20,000 AC-modules on residential houses. The first phase of 5,000 AC-modules started in August 1999. By the end of 1999, about 2,500 AC-modules had been realised.

The AC-modules (100 Wp modules) are offered as Do-It-Yourself packages for € 442,50 (975 NLG) with a maximum of 4 AC-modules per address. This low price, which is about half the normal price, can be realised by purchasing the modules on a large-scale and by using subsidies of both the Dutch government and utilities and by applying favourable tax regulations. As a consequence of the use of these tax regulations, the AC-modules are offered for (operational) lease and not for sale.

Keywords: Marketing - 1: AC-Modules - 2: Financing - 3

1. PARTIES INVOLVED

Within the context of the Solaris-project, Greenpeace started co-operation with Ecofys (consultant), Stork Infratechniek (PV-supplier) and Rabobank (financial services). Greenpeace is responsible for promotion and publicity, Ecofys for (the management of) customer handling, administration and the sales and after sales. Stork is responsible for the logistics of the products and Rabobank for the financial schemes.

2. PRODUCTS OFFERED

The most important starting points in the process of defining the products were simplicity, quality and the price level. Experiences gained in pilot projects [1] learned that it is very important to offer complete packages on the private consumer market. This implies that attention should be paid to ease of use, for example when installing the PV. Furthermore, it is important to offer quality. An important spin-off effect of the Solaris-campaign is aimed for, so in order not to frustrate any future efforts on the private consumer market, the first experiences of private consumers with PV should be positive. Greenpeace targeted a price level of at most € 454 (1,000 NLG), which is about half the normal price.

The packages offered are complete Do-It-Yourself packages, consisting of the following elements:
- 1 Shell AC-module of 100 Wp, including cabling and a plug;
- 1 supporting construction for flat roof, sloping roof with tiles or façade;
- installation manual;
- a guarantee of 10 year on the module; 5 years on the inverter and 1 year on other components;

As a consequence of Dutch guidelines for grid connection, up to a maximum of 4 AC-modules per address can be ordered. A yield meter is offered for an extra € 102 (225 NLG).

The prices mentioned are net prices, in which subsidies and tax benefits have already been discounted.

3. AC-MODULES

An AC-module is a fully integrated combination of a PV-module and a small grid-connected inverter, which is placed at the rear of the PV-module. Ecofys has been involved in the development of AC-modules since the early nineties.

Experiences showed the advantages of AC-modules over conventional PV-modules. The most important advantages are [2, 3]:
- A high yield due to avoided energy losses
- Easier (electrical) installation
- Possibilities for very small PV-systems, which only require limited investments

Mainly because of these last two advantages, AC-modules are very suitable for the private consumer market.

Figure 1: AC-modules on flat roof of private dwelling

4. FINANCIAL SCHEME

In order to realise the price of € 442,50 (975 NLG) per package, it was not only necessary to purchase the modules on a large-scale, but also to use subsidies of both the Dutch government (Novem) and utilities and to make sophisticated use of favourable tax regulations.

Novem granted a subsidy for the first phase of 5,000 AC-modules. As a rule, private consumers can not directly obtain subsidy for PV from the Dutch government, but by grouping the buyers, subsidy could be made available.

Besides subsidies, also favourable tax regulations are used to bring down the price. These tax regulations can only be used by profit-making companies, which mean that private consumers can not benefit directly from the regulations. In order to be able to use the tax regulations any way, an (operational) lease construction is used. This lease construction is offered by De Lage Landen, a full subsidiary of Rabobank. All instalments are paid in advance at the start of the leasing period. After a leasing period of 9 years, the owner automatically becomes owner of the AC-modules, without extra efforts or additional payments.

5. LOGISTICS

The packages are assembled under supervision of Stork Infratechniek and then packed in boxes. The postal services take care of the distribution, delivering the packages at home.

6. MARKETING SERVICES

Customer handling, administration, sales and after sales are carried out by Ecostream. Ecostream is part of the Econcern holding, to which also Ecofys belongs. Ecostream offers renewable energy marketing services and is specialised in developing and implementing renewable energy campaigns in which private consumers are the target group. For the implementation of the campaigns, Ecostream possesses the newest techniques for customer handling, using amongst others a modern call-centre, an advanced and professional database-system and the latest web-techniques.

Ecostream offers the following services:
- Information dissemination via a special information-number and the website;
- Mailing the information materials;
- Reporting and sales via information number and website;
- Follow up of prospects;
- Watching over the contact with (possible) clients
- Co-ordination of the logistics during all the stadiums of sales;
- Administration for the sales, delivery, subsidy handling and after sales of the product.

7. RESULTS

In August 1999, Greenpeace made available about 3,000 addresses of people in principle interested in buying an AC-module. Greenpeace had already filtered out the people interested in buying 4 or more AC-modules and living in the service area of the utilities ENECO, ENW, EWR and REMU. For those people could participate in the SunPower-project (formerly known as PV-Growth project, see [1]), in which systems are installed consisting of 4 up to 10 AC-modules.

During the rest of year 1999, these 3,000 people received information materials about the products, including an order form. About 32% of all people addressed placed an order, ordering on average 2.6 AC-modules. This resulted in a total number of about 2500 AC-modules sold in 1999.

The yield meters, offered as an additional gadget, showed to be quite popular: almost 60% of the orders for one of more AC-modules also order a yield meter for € 102 (225 NLG).

Table 1 provides an overview of the number of AC-modules ordered per order as a percentage of the total number of orders placed. In table 2, the same is done with regard to supporting construction.

Table 1: Overview of orders qua number of AC-modules

	% of total
1 AC-module order	18 %
2 AC-modules order	35 %
3 AC-modules order	14 %
4 AC-modules order	33 % *
Average number of modules per order	2.6

* Leads for 4 AC-modules in the area where the SunPower-project is active were handled by SunPower. Thus, this percentage does not correctly reflect the interest in 4 AC-module systems.

Table 2: Overview of orders qua supporting construction

	% of total number of orders	% of total number of AC-modules
Flat roof system	43 %	48 %
Sloping roof system	53 %	48 %
Façade system	4 %	4 %

The project continued in 2000, be it with a delay of nearly 3 months. As a consequence of new tax regulations in The Netherlands per January 1st of 2000, new leasing contracts had to be developed, resulting in minor modifications in the concept (for example a reduction of the lease period from 108 months to 102 months).

8. PROBLEMS ENCOUNTERED

A very important challenge in the preparation phase of the project was the price level. In order to reach the target price, it was necessary to use tax regulations, which are in principle only directly available for companies subject to tax. It was the first time that these tax regulations were made available to private consumers, using an operational lease construction. However, being the first time that such a financial scheme was developed and thus the lack of experience of all parties with the implementation of the concept, it took quite some time to work out the concept and to have it approved by the relevant authorities.

Another important delaying factor were the subsidy procedures. It was the first time that a project of such size and impact and with private consumers as target group was submitted for subsidy.

As a consequence of these factors, the actual kick-off of the project could take place not earlier than August 1999, about 5 months before the official end of the project and over a year after the announcement of the project. This had several important consequences. First, many leads had turned 'cold': people who were interested a year ago, were not interested any more. Secondly, the timing was wrong. Because the systems were offered as DIY-packages, the best timing to start a project like this is early spring and not late summer when autumn is about to start. Thirdly, Greenpeace decided not to start a large-scale campaign to promote Solaris as the project was officially almost finished and the qualitative goal had been reached. Indeed, PV had been made available for private consumers and more and more parties got interested in acting on the consumer market, marking a significant break-through for PV in The Netherlands.

9. FUTURE PERSPECTIVE

8.1 Solaris

The other parties involved in the project (Stork, Rabobank and Ecofys/Ecostream) decided to continue the project in 2000. After a calm period of waiting for the new lease contract, the mailing to interested people has started again in March. Since then, almost 1,000 information packages have been sent to interested people. In order to secure continuation of the project after finalising the first phase of 5,000 AC-modules, subsidy was applied for in January 2000.

8.2 Other initiatives

Besides PV-panels, Ecostream is also involved in a project offering solar hot water systems to private consumers. This project, called SOL*id, is a co-operation with amongst others World Wildlife Found and a national network of certified solar energy installation companies.

The marketing for Solaris and SOL*id is combined as much as possible, promoting solar energy instead of only PV or SHWS. For both projects, the same special telephone number is communicated: 0900-BELDEZON (translation: CALL THE SUN). For the promotion of solar energy, the promotion possibilities of all parties involved are used, for example the communication opportunities of WWF and Rabobank.

Figure 1: AC-modules and a solar hot water system on sloping roof of private dwelling

10. CONCLUSIONS

The Solaris-project is the first large-scale marketing campaign on the consumer market in Europe, carried out by commercial partners. It is therefore a unique project to gain experiences with marketing concepts for renewable energy. The lessons learned within Solaris are indeed already applied in the SOL*id project for solar hot water systems.

The use of an (operational) lease construction is very new for The Netherlands and unique for Europe. Advantage of the lease construction is access to favourable tax regulations. Lease constructions therefore offer great possibilities for the financing of PV-systems, not only for private consumers, but also for e.g. public institutions.

REFERENCES

[1]   B. de Wit, T.C.J. van der Weiden, et al., *The development of marketing strategies for AC-modules*, Proceedings 2nd World Conference on Photovoltaic Solar Energy Conversion, Vienna, 1998.

[2]   C.P.M. Dunselman, T.C.J. van der Weiden, et al., *Feasibility and development of PV modules with integrated inverter: AC-modules*, Proceedings of the 12th European Photovoltaic Solar Energy Conference, Amsterdam, 1994.

[3]   L.E. de Graaf, T.C.J. van der Weiden, et al., *Third generation AC-modules: an important PV launching product*, Proceedings of EuroSun, Freiburg, 1996.

EXPERIENCE IN FINANCING SOLAR HOME PV SYSTEMS IN THE PHILIPPINES

EUFEMIA C. MENDOZA
Development Bank of the Philippines
Gil J. Puyat Ave., Makati City
Philippines
Tel No. (632) 893-4444 Fax. No. (632) 893-5380 E-mail: ecmendoza@finesse-dbp.com

ABSTRACT: In the Philippines, several government-initiated financing programs for NREs exist, namely: DOE –Energy Resources for the Alleviation of Poverty; NEA – Rural Photo-Voltaic Electrification Program; DBP – Window III Lending Facility/New and Renewable Energy Financing Program.
In addition, private initiatives are getting headway, some suppliers find selling PV Solar system on installment terms like any other appliances workable.
The Development Bank of the Philippines' (DBP) experience demonstrates that there is no single approach in financing solar PV for rural electrification system and it takes more time and effort in coming up with the appropriate loan package. The financing schemes vary from one project to project because of the economic profile of the target groups, fund sources, composition of the project implementors, etc. Learning from the experiences of previously financed projects and the successes and failures of other projects, other financing schemes evolve. DBP's innovative loan packages are done thru its W-III Lending Facility, the basic objective of which is to develop and implement financing packages which are responsive to the requirements of the target group beneficiaries who do not possess the traditional bank requirements of collateral and equity. In its desire to be at the forefront of NRE Lending, DBP is supported by UNDP's TA Grant to strengthen its technical capability to evaluate and manage NRE loan portfolio.
Keywords: Financing – 1: Rural Electrification – 2: Solar Home System – 3:

1. DBP – THE PREMIER GOVERNMENT DEVELOPMENT BANK

After 53 years of playing a major role in shaping the economic development of the country, the Development Bank of the Philippines emerged as a premier government development bank. Through its various lending programs, it provides vigorous support to the government's pump priming activities accelerate the economy of the Philippines from the adverse effects of the Asian economic crisis. This support takes the form of project financing to agriculture, industrial activities, housing, transportation and telecommunications, power and energy, health care and education. The provision of medium and long-term funds are the special niche of DBP in the Philippine banking system.

The DBP is predominantly a wholesale bank and is the largest conduit of official development assistance (ODA) funds in the Philippines from multilateral and bilateral agencies such as the World Bank, ADB, JBIC, OECF and KFW, among others. These ODA funds are onlent in pesos to accredited financial institutions which in turn relend to investment enterprises.

The Bank's retail lending operations, on the other hand, provide financing direct to investment enterprises to serve as catalyst to stimulate other banks to finance development projects that cannot normally access medium to long-term credit facilities.

While the Bank continuously build up resources to ensure its continued viability and financial strength to fulfill its development mandate, it integrates and implements environmental considerations into all aspects of its operations and services, loan portfolio management and credit decisions.

Financials	1998	1999
Total Resources	PhP115 (US$2.88)	PhP138 (US$3.45)
Networth	PhP14 (US$0.35)	PhP15 (US$0.38)
Loan Portfolio	PhP72 (US$1.80)	PhP82 (US$2.05)
Networks		139 PFIs / 76 Branches
Wholesale / Retail Lending Operations		59 : 41 Ratio

The various awards accorded to DBP by both local and international organizations are indicators that the Bank has been able to fulfill its development-banking role in an exemplary manner. The most recent was a Recognition of Excellence by the World Bank for the efficient implementation of WB-funded LGU-Urban Water and Sanitation Program last February 2000. DBP is the highest rated bank in the Philippines with a triple "Bpi" rating.

2. DBP – WINDOW III LENDING FACILITY

The reason why the Bank has been able to fulfill its role in an exemplary manner is its approach of looking beyond the problem in framing strategic lending programs and their administration. It does much more than what its conventional contemporaries do. In 1988, the Bank embarked on an important innovation in development financing by setting aside a portion of its profits of up to 30% annually to finance projects which are considered non-bankable under conventional terms but have great potential in contributing to the growth of the economy because of their characteristics... pioneering in the locality where they are located, using indigenous raw materials, delivering basic services to the remote areas. This is called Window III Lending Facility – the centerpiece of DBP's retail lending operations. Credit programs are formulated in consultations with the appropriate government and private institutions as well as the target beneficiaries, and as the Bank implements these credit programs, modifications are continuously made to enhance their responsiveness to the funding requirements of the target beneficiaries. Innovative financing packages are likewise developed on a per project basis to enable small enterprises to grow into a viable and sustainable business in the long run.

Through this lending facility, the Bank embarked on financing renewable energy projects in 1988 with the Photo Voltaic Solar Home System installed in 100 households in Burias Island as its first project.

3. DBP's EXPERIENCE IN FINANCING PV SOLAR HOME SYTEM PRIOR TO FINESSE

DBP recognized that rural electrification through non-conventional/renewable energy could efficiently be undertaken given the development support it needs. It pursues the development of the PV-SHS in support of the government's rural electrification programs in identified, sparsely populated, remote rural areas which remain unelectrified as the low electric consumption make the operation of conventional electric systems inefficient and too expensive, with the end in view of making improvement on the quality of life of Filipinos by appropriately addressing their specific energy needs utilizing renewable and environment-friendly, alternative sources.

The Bank's first assistance was given to Photo Voltaic Solar Home System (SHS) Project in Burias Island, San Pascual, Masbate in 1988 jointly with the Department of Energy. The Bank partially financed the acquisition and installation of 100 solar home systems in cooperation with the German Agency for Technical Cooperation – Special Energy Program (GTZ-SEP). The project essentially aimed to prove the technical maturity and market competitiveness of PV Technology against conventional electrification in specific areas of the country. The problems encountered in the implementation of the Burias Project provided learning lessons for the subsequent PV-project of the Bank. The following were the identified causes of problems:

- Poor financial management and accounting system.
- Unavailability of wearable parts and competent technicians to maintain the system. Replacement parts were sourced in Manila, which is some 400 kilometers away.
- The low level of awareness of the beneficiaries on the maintenance and

operation aspect of solar home system especially on load management side resulting in the overuse of the system and subsequent breakdown.
- Non-enforcement of the project's rules and regulation to protect the political interest of the head of the corporation who happened to be an elected local official.
- The inferior quality of the local component of the system.
- Inaccessibility of beneficiaries outside Burias Island, especially during rainy season; and
- Lack of understanding of the terms and conditions of the loan on the part of the key officers of the corporation

Learning from the experience in Burias Island, the Bank approached the BELSOLAR Project in a mountainous village in the north, differently, though, both are on a lease-purchase agreement with the end-users through the borrower corporation/cooperative. The following table summarizes the two different approaches the Bank instituted:

TABLE 1

	BURIAS 100 Households	BELSOLAR 100 Households
Project proponent	New corporation just organized for the project headed by the LGU Chief Executive	A 13 year old credit cooperative with high collection rate on its credit services headed by a Parish Priest
Start of DBP's involvement in the project	When financing is the only missing component of the project	At the start of the training and information dissemination of SHS operations to the identified target groups
Credit/Background investigations conducted	Limited to the officers of the corporation	All officers and the 100 end-users
Fund Sources	Equity 25% Subsidy 9% Loan 66%	Equity 27% Loan 73% (Specific uses of funds-Equity for BOS while loan is for panels and BCU
Interest Rate	DBP-Corp. 12% Corp.-End Users 16%	DBP-Coop. 15%with 3% PPR Coop.-End Users 15% with PPR
Repayment Term	3 years exclusive of 2 months grace period payable quarterly	5 years payable-annually
Collaterals	CHM on the Solar Home System	CHM on the Solar Home System Assignment of: Buy-back guarantee agreement of supplier Lease purchase agreement of End-Users Insurance coverage of SHS
Release of loan fund	Proof of equity	Upon the installation of the SHS

The mechanics of projects implementation in the loan package were not clearly specified in the BURIAS projects. In the case of BELSOLAR, these were spelled out and formed part of the loan documentation. The actual implementation of the mechanics could however, not be perfected. Even with several safeguards, there were hitches along the way surfaced. The following are the causes of problems identified in BELSOLAR:

- High maintenance cost because of battery replacement every two years and the PL Lamps every six months.
- The trained technicians looked for other opportunities with better compensation package leaving the project without technical support.
- The technical design is not appropriate because of the cloudy condition of the area. The Philippine standard is 5 watts per square meter but in Northern Philippines it can attain only 3 watts per square meter.
- Misinformation and undue interventions by other foreign nationals who visited the area and gave the information that a 50-watt panel is already obsolete and is being phased out in other countries.

Despite all these problems encountered, DBP remained determined in its commitment to promoting the NRE sector, mainly on the belief that it has been the lack of innovation existing loan packages that underscored the weaknesses of the previously funded projects rather than unviability of the project.

In its desire to be at the forefront of NRE lending, DBP is supported by the UNDPs Technical Assistance grant to strengthen its technical capability to evaluate and manage NRE projects to achieve high quality NRE loan portfolio.

4.   FINESSE PROJECT IN DBP

FINESSE or Financing Energy Services for Small Scale End Users was initiated in 1991 in Kuala Lumpur to develop innovative financial instruments for the commercialization of mature and market ready alternative energy systems to address a major constraint to the development and widespread use of New and Renewable Energy (NRE) Resources, which is the lack of funding / limited access to credit by project developers and end-users.

FINESSE recognized DBP's position that to address the situation, a necessary first step is to provide technical expertise to financing institutions to better evaluate and manage NREs, and in this light DBP was granted by UNDP with funding from the Dutch government in December 1997.

Under FINESSE, innovative mechanisms responsive to local conditions ad specific needs of project developers and end-users alike will be adopted. Specifically, the expected end-result is an appraisal

System, which focuses on project evaluation and loan processing guidelines and operating procedures specific to NRE investments.

To test the effectiveness of these guidelines, a pipeline of NRE electrification projects is currently being built up and the first is the solar PV home systems project in Palawan, in Southern Philippines which was already approved for financing using Window III funds.

5.   PALAWAN AND NRE: A PERFECT FIT FOR THE FIRST PROJECT UNDER FINESSE

Palawan is among the provinces considered in the Rapid Rural Appraisal (RRA) conducted by the Department of Energy in coordination with Palawan Associated Non-Conventional Energy Center and the rural electric cooperative in the area in line with the Energy Resources for the Alleviation of Poverty Program (ERAP). It is the country's second largest province composed of 247 barangays, out of which only 201 or 47% are considered electrified. Most of the electrified barangays use a combination of mini-diesel grids and kerosene for their power requirements. The low level of electrification can be attributed to the large number of small islands (1,768) and the highly dispersed household populations in many of these islands. Population in 1995 reached 635,033 and may double by year 2000 due to migration because of the bright economic prospects of the province. It is the host-province of the Malampaya Deep Water Gas-to-Power project being developed by Shell Philippines Exploration BV., a flagship project of the Estrada administration.

The local leadership is a strong advocate of the environmental conservation and protection. The province garnered the Presidential Awards for the Cleanest and Greenest LGU for three consecutive years 1996-1998 giving the province the Hall of Fame Award.

5.1 Site Selection

Barangay Cabayugan and Barangay Bahile of Puerto Princesa City were chosen among the 40 barangays identified in the RRAs of DOE's ERAP Program which constitute the Phase-I of the multi-phase renewable energy-based electrification program for areas not likely to be connected to the grid.

A comprehensive field survey conducted established the power load requirement and a high level of willingness and ability to pay for energy services from NRE systems such as PV Solar Home System based on the following:

Table 2

	BAHILE	CABAYUGAN
Load Requirements		
>125 watt hours per day	<38%	>51%
>250 watt hours per day	>29%	>20%
Monthly Income		
<P4,000.00	48%	51%
P4,001.00 - P10,000.00	40%	40%
>P10,000.00	12%	19%
Expenditures on Energy Uses		
<P100.00	33%	53%
P101.00-P200.00	31%	18%
P201.00-P300.00	17%	8%
>P300.00	19%	21%
Willingness to Pay		
P.101.00-P200.00	>46%	>26%
P201.00-P300.00	>28%	>32%
P301.00-P500.00	12%	>15%

Based on the need and payment capability of the two areas with 400 potential households, three different sizes of PV Solar System were determined. The following table summarized the features of each system size:

Table 3

SYSTEM SIZE	ESTIMATED EQUIVALENT (WH/day)	ALLOWABLE LOAD (WH/day)	EXAMPLES SIMULTANEOUSLY USEABLE LOADS
36 Wp class	123	86	Two 9-w fluorescent lamps (3.5 hrs, each) One 6-w radio (4 hrs.)
55 Wp class	188	132	Two 9-w fluorescent lamps (4 hrs. each) One 6-w radio (4hrs.) One 32-w B&W TV (1 hr.)
75 Wp class	257	180	One 36-w TV (1 hr.) One 45-w VHS (1 hr.) One 6-w radio (4 hrs.) Two 9-w lamps (4 hrs. each)

5.2 Basic Approach To The Financing Scheme

The Local Government Unit (LGU) which supports the installation of 400 PV Solar Home System in the two identified communities is the project proponent and owner of the systems. The operation and maintenance will be carried out by the rural electric cooperative (REC) in the area, under a management contract with the respective household heads. Wiring and appliances are owned by the households, which is basically the same concept as the grid-based electricity where the utility owns the facilities, power generation and distribution to the households.

The LGU as project proponent avails of DBP loan equivalent to 90% of the total project cost which includes the cost of Solar PV System, cost of bicycles for technicians' mobility, training expenses, brochures and information materials and community-based Solar PV (street light, school TV set). DOE infuse a portion of its Palawan-earmarked funds under ERAP Program as subsidy representing 7% of TPC while the households provide their equity of 3%.

The REC is the franchise holder of the area for power distribution and is rated as triple "A"

cooperative by the National Electrification Administration and has the following characteristics:

- Has good capability for billing and collection with 9% Efficiency Rate.
- Has the recognized authority to discount when fees are unpaid or if the systems are abused.
- Has a staff of competent technical people who can be trained in maintaining and repairing Solar PV System.
- Knows the local area best.

The household end-user fees based on system size installed fully cover the operation and maintenance cost of the REC. for a 36 Wp—PHP 200/month (US$ 5.0); 55 Wp—PHP 350/month (US$ 8.75); and 75 Wp—PHP 450/month (US$ 11.25).

5.3 Groundworks Undertaken Towards DBP Loan Approval And Project Implementation

- Site visit in April 1999 with the LGU and REC in coordination with DOE
- Comprehensive field survey in May 1999 in coordination DOE, NEA and REC
- Pre-feasibility study by DBP consultants in June 1999
- Coordination with other NRE players in Palawan (US-NREL, UNDP-GEF-SGP, Shell Solar)

6. STATUS OF ELECTRIFICATION IN THE PHILIPPINES

NEA reported that rural electrification at the barangay level has reached 72% of 36, 065 potential barangay. Electricity connections, however, have only been extended to about 65% or 5.013 million of the 7.686 million potential household connections as shown in the following table:

TABLE 4

Region	Barangay Level			Household Level		
	Total target	Total to date	Coverage (%)	Total target	Total to date	Coverage (%)
Luzon	15,534	12,856	83	3,377,000	2,606,673	77
Visayas	10,972	7,908	72	2,091,000	1,202,462	58
Mindanao	9,569	5,846	61	2,218,000	1,204,326	54
Total	36,075	26,610	74	7,686,000	5,013,461	65

The data indicates that a substantial number of Filipino are still "living in the dark" thus have limited opportunities to engage themselves on other productive endeavors that could only be made possible through the provision of adequate and continuous supply of electricity. Majority of them are living in off-grid, isolated islands and hinterlands. The ERAP Program aims to achieve a 90% barangay level of electrification by 2004 which translates to additional

5,875 barangays of which 1,400 cannot be connected to the grid and can be electrified by utilizing new and renewable energy such as the PV Solar Home System. And this is the challenge that DBP is taking to be at the forefront of NRE lending to enable every Filipino to enjoy the benefits of having electricity.

REFERENCES

[1]   DBP Annual Reports 1998 and 1999.

[2]   DBP-W III Monitoring Reports 1989-1999.

[3]   DOE-Energy Resources for the Alleviation of
       Poverty Program Framework, March 1999.

[4]   Project Document-UNDP-TA Grant to the DBP-
       FINESSE.

[5]   DBP Credit Policies on Window III Lending.

THE CHALLENGE OF MARKETING PHOTOVOLTAIC-POWER IN REMOTE RURAL AREAS

Gernot J. Oswald
Siemens Solar GmbH
Frankfurter Ring 152
80807 Munich

ABSTRACT: The potential of the PV-market in remote rural areas is rapidly increasing and almost unlimited. The actual development of this segment is disappointing because of well known barriers. Patience, time and co-operation of many parties are required to overcome the challenge of a market whose potential is several hundred times bigger than the current yearly production of the PV-Industry.
Keywords: Glasgow Conference - 1: Marketing; 2: Rural Electrification; 3: PV-Market

Photovoltaics, the art of converting sunshine into electricity, has been the subject of emotional discussions for a long time. While "solar freaks" have developed visions of turning the earth's deserts into gigantic solar power plants, their opponents have a hard time to accept the fact, that already today in many remote and off-grid applications solar power is the only practical and affordable means to provide electricity.

The most widely accepted idea for off-grid PV-application is the electrification of remote rural areas in developing countries. Almost everybody agrees that the quality of life of billions of people could be dramatically increased, if we would be able to provide them with solar power. The **potential** of this market segment is enormous. More than 2 billion people at 15 Wp per head represent a **market potential of 30.000 MWp** compared to a total World Market of 175 MWp in 1999.

It has resulted in expectations of business developments, which by far exceeded reality (Fig. 1).

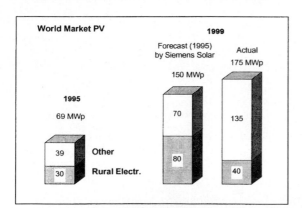

Fig. 1: PV-Market Hope and Reality

This widening gap between hope and actual business development has been subject of numerous meetings and conferences. The main barriers related to the different players have been identified. (Fig. 2)

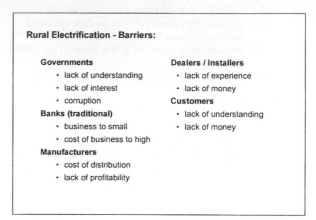

Fig. 2 Barriers

The financial crisis in Asia and the economic development in Latin America and Africa are additional reasons for the disappointing development of these markets. (Fig. 3)

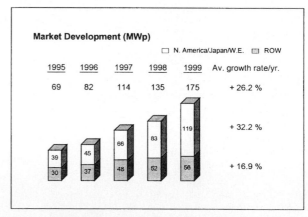

Fig. 3 Industrial- vs. RoW-markets

We have all learnt our lessons (Fig. 4) but it seems that we have not found convincing general solutions.

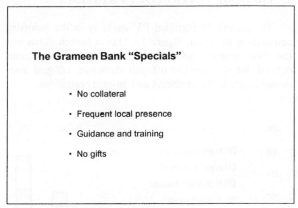

Rural Electrification - Lessons learned

Technology
- high quality products are available
- system design and maintenance create problems

Marketing
- dealer network financially and technically weak
- each country needs tailor made approach
- grants in form of products are detrimental
- cost of hardware is minor problem

Financing
- traditional banks are unable to do the job
  - need security
  - are too far from the client

Fig. 4 Lessons learned

But there is always an exception proving the rule. The approach of the Grameen Bank in Bangladesh (Fig. 5) has been quite successful, but it has also perfectly demonstrated that the development of PV markets in remote rural areas requires a lot of patience and takes much more time than we want to believe.

The Grameen Bank "Specials"

- No collateral
- Frequent local presence
- Guidance and training
- No gifts

Fig. 5 The Grameen Bank "Specials"

The co-operation of the governments of developing countries in the electrification of remote rural areas is essential but many times it is limited or expensive to secure.

The **role of the private sector** (Fig. 6) is easy to define but difficult to perform.

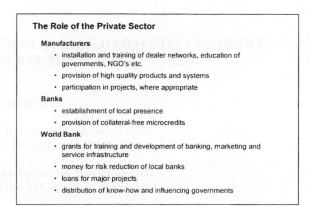

The Role of the Private Sector

Manufacturers
- installation and training of dealer networks, education of governments, NGO's etc.
- provision of high quality products and systems
- participation in projects, where appropriate

Banks
- establishment of local presence
- provision of collateral-free microcredits

World Bank
- grants for training and development of banking, marketing and service infrastructure
- money for risk reduction of local banks
- loans for major projects
- distribution of know-how and influencing governments

Fig. 6 The Role of the Private Sector

An obvious "problem" results from overheated market developments in industrial countries. Subsidy programs like in Japan or Germany make it difficult to stay focused on remote rural areas. It is much easier and cheaper to sell and install a 3 kWp roof top system in Germany compared to 60 – 100 solar home systems in the jungle. The surprising result of these "artificial" market conditions is the unexpected boom in grid connected applications (Fig. 7) which are concentrated in industrial countries.

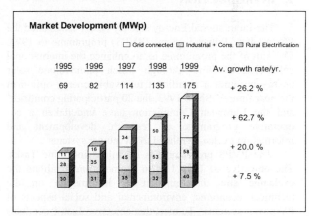

Fig. 7 Market Development / Application

The PV-industry nevertheless needs to and will overcome the challenge of marketing photovoltaic-power in remote rural areas for the benefit of both mankind and its own long term business.

TRENDS IN PHOTOVOLTAIC APPLICATIONS IN SELECTED IEA COUNTRIES: RESULTS FROM THE FOURTH INTERNATIONAL SURVEY

D. K. Munro[1] & J. M. Knight[2]

[1]ESD, Overmoor Farm, Neston, Corsham, Wiltshire, SN13 9TZ, UK
Tel +44 (0)1225 812102, Fax +44 (0)1225 812103, email: donna@esd.co.uk

[2]Halcrow Gilbert, Burderop Park, Swindon, Wiltshire, SN4 0QD, UK
Tel +44 (0)1793 814756, Fax +44 (0)1793 815020, email: knightjm@halcrow.com

ABSTRACT: The International Energy Agency initiated the Photovoltaic Power Systems programme in 1993. As part of this programme, the IEA conduct a periodic survey of PV power applications and markets. This paper summarises the results of the Fourth International Survey. It presents information on installed PV capacity, module production and production capacity, system prices, national budgets for R&D, demonstration and market stimulation and governmental and private initiatives. The paper provides a comprehensive and up-to-date review of the status of the PV industry and highlights trends in the industry over the period 1992-1998.
Keywords: IEA – 1: PV Market – 2: Module manufacturing - 3

1. INTRODUCTION

The International Energy Agency (IEA) established the Photovoltaic Power Systems (PVPS) programme in 1993. The aim of the programme is 'to enhance the international collaboration efforts through which photovoltaic solar energy becomes a significant renewable energy option in the near future'. Since 1993, the 20 participating countries[1] and the European Commission have undertaken a co-operative programme of research, development and information exchange related to PV power systems[2].

The PVPS programme is organised into nine Tasks. The objective of Task 1 is to promote and facilitate the exchange and dissemination of information on the technical, economic, environmental and social aspects of PV power systems. To meet this objective Task 1 conducts a series of surveys on the status of PV power systems applications and markets in the countries participating in the programme.

This paper presents the results from the fourth survey. It presents an overview of PV power system applications and markets as of the end of 1998 and analyses the trends in PV power systems implemented between 1992 and 1998. Data was drawn from national reports, which were supplied by each country.

The data was collated and analysed by Halcrow Gilbert as technical writer for the International Survey Report. Information is presented on installed PV capacity, module production and production capacity, national budgets for R&D, demonstration and market stimulation and new initiatives for promoting PV systems.

2. IMPLEMENTATION OF PV SYSTEMS

The growth in installed PV capacity in the reporting countries is shown in Figure 1. This is broken down into the four primary applications for PV power systems defined for the survey: off-grid domestic; off-grid non-domestic; on-grid distributed and on-grid centralised.

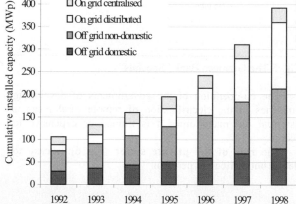

Figure 1: Cumulative installed PV power by application area in the reporting countries

Between 1992 and 1998 the total installed capacity increased from 106 MWp to 392 MWp[3], with an annual growth rate that varied between 20% and 28%. If growth continues at an average of 25% per year the installed power will be 613 MWp in the reporting countries by the end of the year 2000.

---

[1] Australia (AUS), Austria (AUT), Canada CAN), Denmark (DNK), Finland (FIN), France (FRA), Germany (DEU), Israel (ISR), Italy (ITA), Japan (JPN), Korea (KOR), Mexico (MEX), the Netherlands (NLD), Norway (NOR), Portugal (PRT), Spain (ESP), Sweden (SWE), Switzerland (CHE), the United Kingdom (GBR), the United States of America (USA)
[2] PV power systems are defined as all nationally installed (terrestrial) PV applications with a PV power of 40 Wp or more.

---

[3] The worldwide installed power will be significantly higher than the 392 MWp installed in the reporting countries.

Figure 1 shows that the proportion of the PV power that is grid-connected has increased rapidly. In 1992, only 30% of the installed PV power was on-grid; by 1998 it was 46%. This is due, largely, to a rapid growth in the on-grid distributed systems. These are a relatively recent application of PV where a PV system is installed to supply power to a building or other load that is connected to the utility grid. There are a number of perceived advantages for these systems: distribution losses are reduced because the system is installed at the point of use, no extra land is required for the PV system and costs for mounting systems can be reduced if the system is mounted on an existing structure. Compared to an off-grid system, costs are saved because energy storage is not required, which also improves system efficiency. Such systems are likely to become commonplace in the new millennium.

Table I shows the breakdown of installed power between the reporting countries. In Austria, Canada, Finland, France, Israel, Korea, Mexico and Sweden, over 90% of the total installed PV power is for off-grid applications. In contrast the majority of systems in Germany and Switzerland are grid-connected. Other countries with more than 50% of installed PV power for on-grid applications were Austria, Denmark and Japan. In most countries with a high percentage of on-grid systems, the majority of on-grid applications were for distributed PV power, with the exceptions of Italy and Spain where 89% and 71% respectively of the on-grid PV power was centralised.

59% of the total installed power is accounted for by just two countries - Japan and the USA, although Switzerland has the highest installed power per capita.

Table I: Cumulative quantity of PV power installed in reporting countries as of the end of 1998

Country	Off-grid domestic (kWp)	Off-grid non-domestic (kWp)	On-grid distributed (kWp)	On-grid centralised (kWp)	Total installed power (kWp)
AUS	5960	15080	850	630	22520
AUT	487	674	1630	70	2861
CAN	1378	2825	257	10	4470
DNK	35	140	330	0	505
FIN	2100	300	46	30	2476
FRA	5600	2260	140	0	8000
DEU	2900	6300	37300	7400	53900
ISR	88	200	6	14	308
ITA	5210	5100	780	6590	17680
JPN	450	52200	77750	2900	133300
KOR	306	2410	266	0	2982
MEX	9789	2195	2	0	11986
NLD	2476	954	3050	0	6480
NOR	1470	180	0	0	1650
PRT	384	102	17	0	503
ESP	5010	910	600	1480	8000
SWE	1823	433	114	0	2370
CHE	2210	190	7630	1470	11500
UK	108	254	328	0	690
USA	32000	40200	15900	12000	100100
Total	**79784**	**132907**	**146996**	**32594**	**392281**

3   INDUSTRY AND GROWTH

3.1 PV module production and production capacity

Figure 2 shows the trends in module production and production capacity between 1993 and 1998.

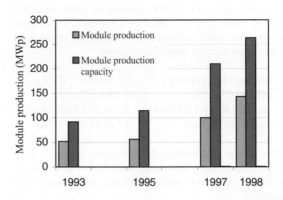

Figure 2: PV module production and module production capacity in the reporting countries

Total module production increased by 43% in the one year between 1997 and 1998. The module production capacity increased significantly in a number of countries, including Germany, Japan and Spain with increases of 6.2, 23.0 and 5.5 MWp respectively between 1997 and 1998[4]. The production of PV modules remained concentrated mainly in Japan (36%) and the USA (38%) although the European countries accounted for 19% of production and 28% of production capacity.

Commercial module manufacture is still based primarily on crystalline and amorphous silicon technologies as shown in Table II. However, other technologies such as copper indium diselenide, cadmium telluride and edge fed growth cells are starting to be commercially produced in Germany, Japan and the USA.

Table II: PV module production in the reporting countries in 1998

Module Production	MWp	%
Crystalline silicon	132.0	92
Amorphous silicon	8.1	6
Other	2.7	2
Total	142.7	100

3.2 Balance of System Components

A specialised industry exists manufacturing PV balance of systems (BOS) components such as battery charge controllers and inverters. As would be expected, those countries whose installed power is predominantly off-grid

---

[4] In order to avoid 'double counting', modules are considered to be manufactured in a country only if the encapsulation takes place in that country.

tend to have a BOS industry geared towards the manufacture of batteries and charge controllers, whereas in countries with predominantly on-grid power the industry tends to be geared towards the manufacture of inverters suitable for grid-connection.

3.3 System Prices

The average price of PV systems in 1998 remained similar to those in 1997 although a number of countries saw significant price reductions in some of the application sectors. System prices ranged as follows:

- off-grid (40Wp-1kWp)   7.8 to 24.0 USD/Wp
- off-grid (>1 kWp)   9.0 to 28.6 USD/Wp
- on-grid (40Wp –10kWp)   5.5 to 25.2 USD/Wp
- on-grid (>10kWp)   4.8 to 21.4 USD/Wp

The prices associated with on-grid systems were generally significantly lower than those for off-grid systems because no batteries and associated components are necessary.

4   BUDGETS FOR MARKET STIMULATION, DEMONSTRATION AND R&D

The total budget for market stimulation, demonstration and R&D spent on PV in the reporting countries is shown in Figure 3. The trend has been for the funds spent on market stimulation to increase whilst the proportion of funding spent on R&D has decreased. Funding for demonstration projects stayed fairly constant between 1994 and 1998.

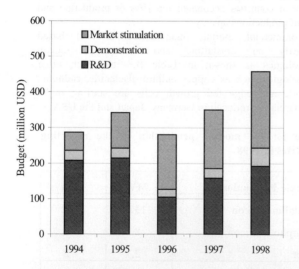

Figure 3: Budgets for market stimulation, demonstration and R&D in the reporting countries between 1994 and 1998

Of the total budget of 458 million USD available in 1998, 225 million USD was in Japan alone. Of the remaining countries the largest budgets for market stimulation, demonstration and R&D programmes were in the USA, Germany and the Netherlands (65, 56 and 38 million USD respectively).

5   NEW INITIATIVES

A large number of new initiatives were reported by the participating countries, including an increasing number of green electricity schemes, net metering, specific tariffs for PV and government grants for installations. In general, the utilities are showing an increasing interest in PV with utilities in several countries involved in the installation of demonstration projects and large scale implementation programmes.

Many of the new initiatives focus on building integrated PV systems, as illustrated below:

- In Japan, two residential PV system programmes resulted in the installation of 9244 residential PV systems between 1994 and 1997 with a further 8229 houses signed up for 1998. Japan has set a target of 5000 MWp of PV by 2010.

- In Germany, the 100,000 Roofs programme was announced in December 1998. It offers a low interest loan for 10 years with no repayments in the first 2 years. An installed power of 300 MWp is expected within 6 years.

- In the USA, the Million Roofs Initiative has commenced. This is a 10 year plan to stimulate the deployment of 1 million PV and solar hot water systems through loans and government grants.

- In Switzerland, 150 new installations were constructed in 1998 under the 'solar stock-exchange' which enables consumers to buy PV at a price which covers production costs of electricity from new PV installations.

Market deployment initiatives are beginning to replace national demonstration programmes in a number of countries, particularly where it is felt that the market has matured sufficiently and a particular application of PV is ready for stimulation.

6   SUMMARY OF TRENDS

By the end of 1998, the total installed PV capacity in the participating IEA countries was 392 MWp. The PV industry continues to grow at an average rate of 25% per year. If this trend continues there will be 613 MWp installed in the reporting countries by the end of the year 2000. An increasing proportion of this installed capacity is on-grid distributed PV systems (mainly building-integrated). The market is being strongly influenced by demonstration and market incentive programmes being carried out in a number of countries, especially Japan, Germany and the USA. Off-grid applications remain an important sector of the PV market particularly in countries such as Austria, Canada, Finland, France, Korea, Mexico and Sweden.

REFERENCES

[1] Report IEA-PVPS 1-07: 1999, 'Trends in Photovoltaic Applications in Selected IEA Countries between 1992 and 1998'

The Scolar Programme for Photovoltaics in Schools in the UK

P Wolfe[1], R Hill[2], G Conibeer[3]

[1]Intersolar Ltd, Cock Lane, HIGH WYCOMBE, UK
[2]Newcastle Photovoltaics Applications Centre, University of Northumbria, NEWCASTLE, UK
[3]Environmental Change Institute, Oxford University, 11 Bevington Road, OXFORD, OX2 6NB, UK
Tel: +44 1865 284690, Fax: +44 1865 284691, email: gavin.conibeer@eci.ox.ac.uk

ABSTRACT: The Scolar Programme for Photovoltaics is installing 100 grid-connected photovoltaic systems in schools and colleges in the UK. Participating schools gain an architecturally integrated PV array (average 650Wp), an invertor, a dedicated computer and an Internet link via the Scolar website. Here all the performance data for the 100 schools will be available to all the others. Access is also provided to a comprehensive educational package; which covers a wide range of information on renewable and solar energy as well as project based material aimed at various key stages of the UK national curriculum.
NB: Bob Hill has been an invaluable asset to the initialisation and the implementation of this programme. His death is a severe blow to it and to the PV community as a whole.
Key words: Education and Training - 1: Dissemination - 2: National Programme - 3

1. INTRODUCTION

Under the Scolar Programme for Photovoltaics, a total of one hundred PV systems are being installed in schools and colleges throughout the UK. All of these PV systems are connected to the local electricity distribution network. This has the advantage that any electricity that is not directly used on the premises as it is produced, is automatically fed into the utility's distribution network thus eliminating the need for any energy storage. Automatic performance monitoring using a Personal Computer based data acquisition system including a display unit is part of each installation. The PC also provides access to the internet.

Four different types of PV array are being installed: 35 walkway systems with modules on a canopy structure; 10 louvre systems located over windows; and 50 wall-mounted systems of two types, one of crystalline and the other of amorphous silicon modules (see figures 1 to 4 respectively). Finally 5 bespoke systems are also included bringing the total to 100.

Teaching material on PV has also been made available under the Scolar programme and each school can access performance data from installations at other schools through the internet connection to the Scolar website.

2. SYSTEM OVERVIEW

Figure 5 illustrates the layout of the Scolar PV systems. The two main components are the PV modules, which convert the sunlight into electricity, and the inverter, which feeds the electricity generated into the local distribution network. The PV modules are mounted on a support structure, which will vary according to the system chosen, and are connected to the inverter, which is located inside the building.

The system also comprises monitoring and display functions. In addition to the DC and AC voltage, current and power, the ambient temperature and the irradiance are monitored and the measured values are stored on the PC hard disk. The PC displays all monitored parameters on screen; and the values of energy generated to date and the current power output are also shown on an external display board. In addition to system monitoring, the PC facilitates access to the internet.

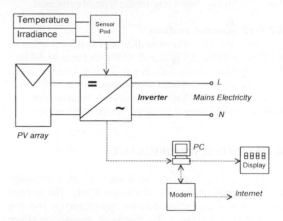

Figure 5: Schematic of a Scolar system.

3. TYPES OF SYSTEM

There are four types of Scolar design. A canopy system, designed to go in front of a doorway; a louvre system designed to go over a window and provide shading as well as electricity; and two wall mounted designs. The canopy, louvre and one wall mounted design use crystalline silicon PV modules, the other wall mounted design uses thin film single junction amorphous silicon modules (see again figures 1 to 4).

In designing the systems for the programme consideration was given to achieving a fair degree of architectural integration, whilst at the same time maintaining a reasonable amount of modularity in design, in order to minimise material and installation costs. It was also decided to include thin film amorphous silicon modules in one of the designs so as to represent the growing importance of thin film technologies. This has resulted in designs that are suitable for wall mount (both crystalline and thin film silicon); installation over windows (louvre); and free standing (canopy). Hence meeting the needs of schools which have a very wide range of building types and construction, whilst necessitating only four system types out for the total one hundred installations.

3.1 Modules for Canopy and Louvre Systems

Each PV module contains 48 PV cells, which have been encapsulated between two sheets of 4 mm toughened glass. The modules are thus semi-transparent and their construction is certified to IEC 61215. Module dimensions are 2155 mm x 500 mm. The PV cells are pseudo-square crystalline silicon cells of approximately 125 mm x 125 mm. The cells have an average output of 2.36 Watts when tested under Standard Test Conditions (1000 Wm^{-2}, AM 1.5, 25 °C).

Within each module, the PV cells are connected in series to provide an average output of at least 96W per module under standard test conditions. The maximum power point voltage of each module is 24V and the maximum open circuit voltage is 35V.

Both the Canopy and the Louvre Systems consist of six modules per system. The combined peak power of each is between 576Wp and 680Wp measured at Standard Test Conditions, depending on the type of cells used.

3.2 Wall mounted modules

The crystalline silicon wall mount system consists of 8 BP Solar 585s, at a power of 85Wp and area of 5.75m^2. The thin film silicon wall mount system consists of 12 single junction amorphous silicon modules at 50Wp each and a total area of 17m^2. (Each module is an assembly of four Intersolar Phoenix modules on one sheet of glass.)

4. INVERTER SPECIFICATION

The inverter cabinet is made of an electrically insulating material, which also meets IP54. The internal dimensions of the cabinet box are approximately 600 mm x 600 mm x 250 mm. To minimise misuse/vandalism there are no controls on the outside of the box. The DC and AC compartments within the inverter box are separated by a physical non-conducting barrier. The positive and negative DC inputs are physically separated by an isolating partition.

In addition to the mechanical covers, the electronics are protected from an over current or overvoltage with fuses and voltage suppressors. The voltage suppression is designed to cope with lightning surges on the AC side. The AC compartment components are earthed to the building earth. A residual current detector is provided on the output for additional protection.

A dual pole circuit breaker is installed between the input terminals and the lightning protection equipment. The maximum current is 15 Amps. The lightning / overvoltage protection equipment consists of two varistors in series with a mid-point earth connection. In addition, a fast semiconductor transient voltage suppression device (such as a Transorb) is connected in parallel.

Each inverter operates over a voltage range of 48 to 70V and at a peak power of 700W, designed to run with a power factor of unity. Inversion efficiency is 91% at 100% load and 78% at 10% load with a microprocessor based algorithm for maximum power point tracking.

4.1 Monitoring system

The PC receives data from the inverter via a serial communications link and displays it on the screen and saves it to disk. The data is transmitted on a fibre optic cable with the RS232 protocol of 9 600 baud rate, 8 bits, 1 stop bit and no parity. The program reads all parameters once every ten seconds, and averages the readings over one hour. Only the hourly averages are stored in the data file.

Sensors inside the inverter measure the following parameters: DC & AC current by current transducer; AC current by current transformer; DC voltage by potential divider and differential amplifier; and AC voltage by isolation amplifier. In addition a separate monitoring pod measures irradiance with a silicon reference cell and temperature by thermocouple. This pod is mounted beside the PV array and is connected to the inverter. These connections, to and from the invertor, are shown in Figure 6.

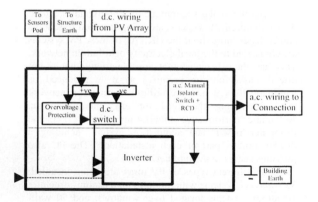

Figure 6: Inverter connections.

4.2 Display Unit

The PC also controls the display unit, which displays instantaneous power and energy generated to date. This is typically mounted in an entrance hall or similar place where it is visible to people passing by. It displays the instantaneous power output of the system in watts and the electricity generated to date in kWh (kilowatt hours).

4.3 Uploading to the Internet

The hourly averaged data is uploaded to the Scolar internet website on a weekly basis.

5. EDUCATIONAL DESIGN

The principle aim of the programme is educational. Pupils at the school gain hands on experience of photovoltaic technology. On the display panel they can see the power being generated at any moment as well as the total energy generated by the system since installation. The dedicated computer logs this data, which can be used for classroom based projects on the availability of solar energy; the efficiency of conversion to DC electricity; and the efficiency of the invertor in converting this to AC and of feeding it into the mains.

Once a week the data is uploaded to the Scolar internet site: **www.scolar.org.uk**
Here all the Scolar schools have access to the data from the other 99 schools for further project work comparing variations across the UK, these can also be correlated with seasonal variation and weather patterns.

Furthermore the website hosts an "educational resource package", which has a very extensive amount of information on solar energy. This is arranged in sections, one for each of key stages 2, 3 and 4 in the UK national curriculum. Each section includes modules on how PV cells work, what they are made from and what are their applications, each written appropriately for the key stage level. In addition there is an "Advanced resource pack" developed from the Open University course on Renewable Energy. This has a lot more information on photovoltaics and other renewable energy resources. It has a spreadsheet on which data from the school's system can be entered and analysed. There are also data sets from other European locations and worked examples for designing photovoltaic systems for particular applications. Finally there is a set of teacher notes to help in the management of the educational resource. All the relevant files can be downloaded from the website.

Scolar systems can be used in several categories within the national curriculum, as detailed in the resource pack on the website. These include the physics of photovoltaics, the technology of renewable energy application and the geography of solar resources across Europe.

6. SCOLAR CONSORTIUM AND FUNDING

Scolar is 30% funded by the UK Department of Trade and Industry, Technology Foresight Initiative and 30% by the companies and universities that make up the Scolar Consortium. This level of subsidy reduces the funding contribution from the school to £3,350; £3,850; or £4,500 from the full costs of £7,550; £8,750; or £11,515 - for

wall mounted crystalline silicon; wall mounted thin film silicon; and canopy and louvre systems respectively.

The members of the Scolar Consortium are all UK companies and universities involved in PV or related, research, manufacture or engineering. It consists of:
Beacon Energy; The Co-operative Bank; The Earth Centre; ECU (Oxford University); Open University; Schuco International; Building Research Establishment; CREST (Loughborough); Eastern Electricity; Halcrow Gilbert; Intersolar Group; Ove Arup; Sollatek; Colt Group; Dulas Limited; E.E.T.S. (Cardiff); I.T. Power; N.P.A.C. (Northumbria); RedNet; Southampton University; National Energy Foundation; Active Cladding; Microtech.

7. CONCLUSION

Scolar is numerically the largest PV installation programme in the UK. It is starting to provide hands on experience in PV and renewable energy to school pupils, through the systems themselves the extensive educational resource package and the internet link to other Scolar schools. In addition it is providing valuable experience in standardisation for the UK PV industry and breaking new ground in connection agreements to regional electricity companies.

At the time of writing, April 2000, 35 systems have been installed. Another 21 schools have signed contracts and are scheduling installation. There are a further 42 further live applications that have been submitted and that are being processed. Hence, currently there are an almost sufficient number of applications, the final few are expected over the next few months. Site visits and installation are well underway with completion of the 100 systems expected in the second half of 2000.

REFERENCES

[1] J.M. Wolfe and G.J. Conibeer. Proc. 2nd World Conf. on Photovoltaic Energy Conversion (Vienna, 1998).

PV ON VOCATIONAL COLLEGES IN SWITZERLAND
SEVEN YEARS EXPERIENCE IN TRAINING AND EDUCATION

Th. Nordmann, A. Frölich, Th. Bähler

TNC *Consulting AG, Seestrasse 129, CH-8810 Horgen, Switzerland*

Phone: ++41 1 725 39 00 , Fax: ++41 1 770 10 50, E-mail: nordmann@tnc.ch

ABSTRACT:

Within the framework of the Swiss Photovoltaic Programme for Vocational Colleges, 21 of the 60 vocational school centres in Switzerland were equipped with a photovoltaic installation. 4'300 apprentices or 42% of all electricians being educated in 1998 attended these vocational colleges. This means that the electricians programme for vocational colleges reaches almost 50% of all future electric installations professionals in Switzerland. Specific know-how and applications experience are introduced and integrated on a step-by-step basis into the normal theoretical curriculum. Continual checking of results and a measurement campaign document the high technical level of this application of PV. This user-group is further supported by an information network consisting of yearly experience-exchange conferences, a newsletter and an internet website (www.pv-berufsschule.ch).

Keywords: Education and Training - 1: Monitoring - 2: National Programm - 3

Fig. 1: The 50 kWp Installation on the roof of the Lucerne vocational college.

1. THE SWISS VOCATIONAL TRAINING SYSTEM

In Switzerland, practically all young persons who do not go on to receive an engineering or academic education go through an apprenticeship. The dual-apprenticeship system is built up on practical experience and part-time theoretical learning. Such apprenticeships are arranged within the framework of special apprenticeship contracts. Depending on the particular profession being learned, these run over three to four years. Between 60% and 80% of the apprentice's working hours are spent in the firm where he or she is employed. Here, the apprentice is trained for his or her later professional occupation in a specially structured training and work programme. All apprentices with the same profession take the accompanying part-time classes in a regional vocational school. These take up around one to one and a half days per week (depending on profession). About the half of all young people in Switzerland go through such an apprenticeship. After successfully passing the final apprenticeship examinations, the apprentices receive a state certificate of proficiency in the profession learned. This is the basis necessary for further professional education – e.g. going to a higher technical college - and, after appropriate practical experience, is also the prerequisite for taking the examination for the master craftsman's diploma.

No. of supported projects	24	
Power Range	1.8 .. 52.5	kWp
Total Power installed	193	kWp
Average power per project	8.04	kWp

Used cell types	
mono	16
poly	6
mixed poly and mono	2

Mounting	
Flat roof	16
Tilted roof	1
Building Integration	2
Tracking	2
Shed	3

Types of school	
Vocational schools	21
Other related schools	3

Tab. 1: Project summary by end of 1999.

In the photovoltaic market diffusion model [Ref. 1], it was shown that further development and dissemination of photovoltaics technology is dependent on well-trained professionals, especially in the area of electrical installations. Future professional electricians should as a group be specially looked after. Of the approximately 173'000 apprentices learning a profession in Switzerland in 1998, 11'150 were learning to be electricians and attended classes in one of the around 60 vocational colleges for electricians in Switzerland.

As was done in the recent past in the development of informatics, the PV-vocational colleges programme concentrates its development and dissemination activities in a similar way on these vocational colleges. This complements and extends Swiss efforts in photovoltaics research and development at universities and higher technical colleges.

2.THEINFORMATIONNETWORKFOR VOCATIONALCOLLEGESFORELECTRICIANS INSWITZERLAND

The aim of these pilot and demonstration activities is to introduce, over a wide geographical area, a new series of various types of PV installations on the roofs of Swiss vocational colleges, which produce outputs in the range between 3 kW to 50 kW

Efforts are being made to spread knowledge about photovoltaics as widely and as quickly as possible throughout the vocational training centres in cantons and local communities. This allows experience gained from the implementation of environmentally friendly renewable energy sources to be multiplied with a positive effect.

By actively assisting the students in the planning and construction of an on-grid photovoltaic installation, the technology for generating electricity from sunlight will be put within reach of these future professionals.

Exciting opportunities to investigate the efficiency of the system by altering various parameters are made possible by equipping the photovoltaic installation with a comprehensive, standardised data recording system and making energy flows more transparent. In this way, the photovoltaic installation will remain an interesting means of giving future generations of students the possibility of gaining experience in photovoltaics.

3.ACTIVITIESCOVEREDBYTHEPROGRAM

The program is being set up and conducted on behalf of the Swiss government by TNC Consulting AG, who also take over the role of project management. The project manager's job is to look after the whole program, from providing support for new installations in the planning and construction stages, via the procurement and setting up of measuring equipment, through to the evaluation, interpretation and cataloguing of data. The job also includes all co-ordination and organisational tasks, such as arranging the experience-exchange conferences, publishing newsletters, making programs available for evaluation, and setting up Internet links. In addition there are also activities related to the project as a whole, such as preparing data for the IEA database (see Fig. 1).

4.WHATTHEGOVERNMENTPROVIDESFOR THEPROJECT

The government supports the installations by contributing financially to construction costs, providing up to a maximum of 27% of the cost of the installation. One of the conditions for these financial incentives is that the participants must promise to make the data collected from the installation freely available.

As well as financial support, the government also provides the projects with a comprehensive, standardised monitoring

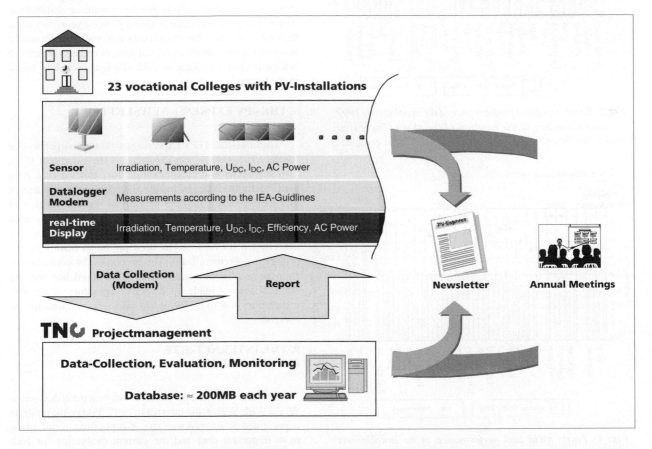

Fig. 2: The various components and activities of the programme are forming a network. The initial financing of the projects is accompanied by a monitoring campaign. Today the physical completion of the projects is almost finished. Now the information and experience-exchange activities are becoming more and more important.

system. The monitoring system consists of a data collection unit for measuring meteorological and installation values, a data logger for long-term recording and software for the evaluation and presentation of the data. This is used to prepare energy audits covering insolation, module yield (DC) and inverter output yield (AC), so that comparisons can be made between the reliability of photovoltaic installations and the output of individual components. All the sensors are fitted externally, independent of the components of the installation itself, thus making the measuring equipment very easily accessible, especially in the event of an installation breakdown.

5. DISPLAY PANEL (PRESENTATION FUNCTION)

As a further incentive, the government provides the vocational colleges with a standardised display panel. This makes it possible to display current weather and PV-installation

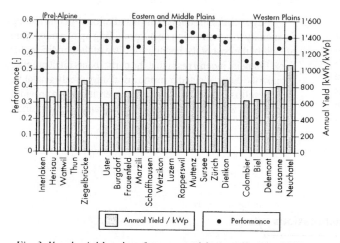

Fig. 3: Yearly yield and performance of the installations 1999, grouped according to the location of the plants. There is no significant difference in the average yield of these groups.

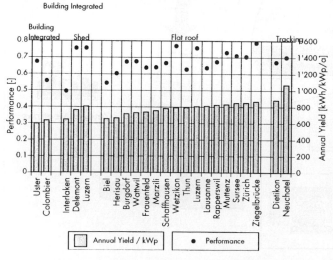

Fig. 4: Yearly yield and performance of the installations 1999, grouped according to the type of the plant. As expected, the building integrated installations show lower yields (facades and shed with partial shading). The yield of the Tracker installations is in the average 10 to 20% higher than those of fixed fields.

data, such as insolation, ambient temperature, power delivered, efficiency and other electrical operational data at a convenient location (e.g. in the college foyer, coffee-bar etc.). This presentation focuses the attention of students and teaching staff, and shows them clearly how solar energy can be used. The display panel can be driven locally or remotely via a modem. The local version shows present values whereas the remote panel shows hourly averages.

6. DATABASE AND EVALUATION

The stored data is sent each week via modem to the project management team and stored in a database. The records are all in the same format so that standardised evaluations are possible.

The data is evaluated and presented by the project management team in accordance with EU and IEA Directives. This means that the findings can be compared directly with other Swiss or European projects. The reports, together with a short interpretation of the data, are sent to the project participants every 2 months.

7. EXPERIENCE-EXCHANGE CONFERENCES

An important aim of the program is to encourage individual participants and the project management team to keep in contact with each other and share practical experience. This is the purpose of the conferences, which are always held in a college that has recently begun to operate an installation. Although these events take place only once a year, they serve their purpose well. The main topics covered are information about new installations and discussion of recorded results. It is hoped that more education-related questions will be raised in the future.

8. THE «PV EXPRESS» NEWSLETTER

The first issue of «PV Express» - an informative publication for the target group - was published in the autumn of 1996. This newsletter will be appearing 1 to 2 times a year. As a platform for the colleges and the project management team it tackles interesting subjects relating to the construction of installations and the interpretation of data. There is also space for presenting new installations and dealing with teaching matters. Up till now, it has been produced mainly by the project management team. The proportion of contributions from the installation operators is to be increased, however. An editorial team - made up of active operators of the PV-installations - is to be formed to take responsibility for the content.

9. THE INTERNET-SITE (WWW.PV-BERUFSSCHULE.CH)

Information on the programme was first published on the World Wide Web in the Internet in 1997. Today, the program is presented on its own web site. It is possible to call up the most important data and the current evaluation for each college PV-installation, at the moment in the form of a bi-monthly overview (address: www.pv-berufsschule.ch). Results and analyses, as well as the educational tools, are now available for viewing and can be downloaded.

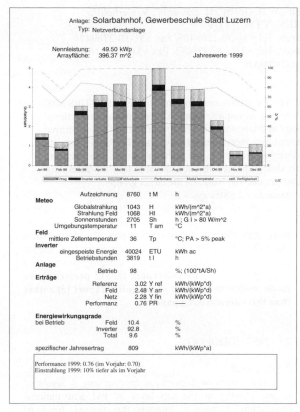

Fig.5: An example of the Monitoring Report which is sent to the schools regularly.

10.CONCLUSION

The programme for vocational colleges for electricians, combined with further activities in the educational area (Competence Centres in engineering colleges and the general PV Promotion Programme for Schoolhouses with a total of approx. 1 MW) appears to be a good idea and has produced a lasting impact.

11.REFERENCES

[1]   Th. Nordmann: Success Stories of Photovoltaic Financing in Europe, Fourteenth European Photovoltaic Solar Energy Conference, Barcelona, Spain, 30 June - 4 July 1997

[2]   Th. Nordmann, A. Frölich, L. Clavadetscher: The Swiss PV 200 kWp Pilot and Demonstration Programme for 26 Electrical Colleges, 14 th European Photovoltaic Solar Energy Conference, Barcelona, Spain, 30 June - 4 July 1997

[3]   Th. Nordmann: Grid-linked PV installations on vocational college buildings, implementation strategy of the Swiss photovoltaic promotional programme, Eighth National Symposium on Photovoltaic Solar Energy, March 1993, Staffelstein, Germany.

MARKET IMPACT OF THERMIE PV DEMONSTRATION PROJECTS

L.A. Verhoef[1], E.W. ter Horst[2], I.T.J. de Jong[1], B.Yordi[3], W.B. Gillett[3], H. Edwards[4], V. Gerhold[5],
[1]VSEC, [2]Novem, [3]EC-DG TREN, [4]ETSU, [5]TÜV
VSEC, Korte Jansstraat 7, 3512 GM Utrecht, the Netherlands, Tel +31-30-2369469, Fax +31-30-2369471
Novem, PO Box 8242, NL-3504 RE Utrecht, the Netherlands, Tel: +31-30-2393493, Fax +31-30-2316491
Web site: www.vsec.club.tip.nl

ABSTRACT: An effort has been made to assess the impact of 112 PV demonstration projects from the THERMIE programme 1992-1997. The IMPACT is assessed on 6 themes: Industrial impact, Markets, Programme interactions, Awareness, Clustering, co-operation and linkages, and Technology and innovations. This paper mirrors the executive summary of the report of that assessment, including lessons learnt and recommendations. The THERMIE projects have had a significant impact on the market penetration, the industry development, the technical innovations, and the clustering and co-operation between actors. The IMPACT methodology proved to be a valuable tool for extending monitoring of projects and programmes beyond the realm of proposers into the world of the target groups.
Keywords: Economic Analysis - 1, R&D and demonstration Programmes - 2 , Strategy - 3

1. INTRODUCTION

The THERMIE PV Demonstration Programme [1], which ran from 1990 to 1998, built on the experience of the earlier EC PV Demonstration Programme, which had started in 1978. These programmes together have contributed to more than 160 demonstration projects across the Member States of the EU, and to the installation of more than 10 MWp of PV, which represents approximately 10% of the PV power installed in the EU. The THERMIE programme has also supported a wide range of accompanying measures in the PV sector, including studies, workshops, publications and conferences.

The influence of these THERMIE A projects on the increased co-operation between European PV producers and users and on the dissemination of know-how is has been provisionally described [1]. In that analysis, the possible synergy, parallelism or divergence with National programmes of Member States and input of the main users and target groups of the programme was not yet included. The 5th framework programme is continuing support for demonstration projects, but with a different scope [2].

This "Impact project" has been carried out by experienced PV experts from NL, DE and UK, who have studied the archives of the THERMIE programme and carried out more than 145 interviews and on-site visits. These interviews involved a wide range of actors, including THERMIE project contractors, PV equipment suppliers, beneficiaries of the projects, researchers, national PV programme managers, and other related interest groups. The interviews produced a wealth of detailed information about the impacts of the THERMIE programme, from which the most important highlights are presented below.

An effort has been made to probe into the target groups of those demonstration projects, and not to limit ourselves to the experience and views of contractors only. A new methodology was developed called IMPACT by which a set of projects (or a programme) is assessed on 6 themes: **I**ndustrial impact, **M**arkets, **P**rogramme interactions, **A**wareness, **C**lustering, co-operation and linkages, and **T**echnology and innovations.

This paper is largely equivalent to the executive summary of the THERMIE action STR-1696-98-NL report [3]. It represents an assessment of the impact of

more than 100 PV demonstration projects contracted in the 1990-1997 THERMIE calls. Assessment of market impacts of demonstration projects and progrrammes is a complex process. **Any references should preferably be made within the context of the entire report [3] rather than this summarising paper**

2. METHODOLOGY

2.1 Approach

The survey was executed approaching 4 levels of actors, starting at the top level of EU and national programme managers, through the second level of multiple proposers and innovation experts, and the third level of proposers and co-proposers and project managers of the THERMIE PV projects. These resource persons were approached in such a way that the interviews have a mutual benefit. Finally as an important fourth level a selection of the relevant actors (not participating in the THERMIE PV projects) in the information dissemination and/or marketing/distribution channels of the proposers were approached to get real target group comments, experiences and feedback, see table 1 and figure 1.

Table 1. Levels of actors in PV demonstration programmes and data gathering activities

Resource persons and organisations	Method of data gathering
1. National/EC Programme Managers	Personal Interviews and workshop
2. Multiple (industrial) proposers and PV innovation experts	Questionaires and Personal interviews
3. Proposers and co-proposers	Personal interviews
4. Dissemination channels and target groups (not involved in projects)	Trade fair Interviews and Regional cluster analyses

2.2. Participation and PV volume

The 112 THERMIE projects were executed by 269 companies and institutes. 196 of these actors were involved as (financially contributing) proposer. The others supplied materials, components, or services. Approx. 70% of the proposers were SME's. The average size of participants was 1100 employees.

The PV power contracted by THERMIE between 1992 and 1997 amounts to 9,5 MWp, approximately 10% of the total power installed in the member states of the EC until 1998.

3. RESULTS

3.1 Industry, economics and jobs

During the course of the THERMIE Programme, the PV sector has matured from a largely R&D driven community to a growing world-wide business. The majority of the PV companies interviewed stated that the THERMIE demonstration projects had made an important contribution to their growth and improved commercial viability. However, in such a fast moving and complex process of market evolution, they found it difficult to identify exactly how many new jobs and additional income had been generated as a result of their involvement in the THERMIE programme.

Figure 1. Five levels of actors relevant to the impact of demonstration projects and the resource persons approached in this study.

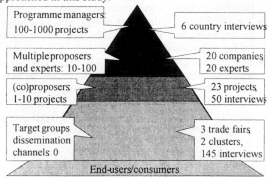

World-wide PV system and component prices have fallen significantly during the 8 years of the THERMIE programme, and the costs paid for PV components and systems within the EC THERMIE programme have fallen by similar amounts. This shows that the EC has succeeded in overcoming the well known problem of subsidy programmes, which traditionally tend to cause prices to remain higher than necessary.

3.2. Markets

The THERMIE programme has encouraged major EU utilities to become involved in its demonstration projects. This has played a major role in opening up the PV **utility markets**, and has allowed them to explore a wide range of both technical and non-technical innovations. Of particular importance have been demonstrations of different approaches to grid connected PV project ownership and financing, and of comparisons between individual grid connected inverters on each building and the use of centralised inverter systems for groups of buildings.

Major progress has also been achieved by the THERMIE programme in demonstrating innovative

approaches to PV **building integration**. As a result, a growing number of companies has now become established across the EU with specialist skills in PV integration in buildings, and an expanding range of new PV building products is now commercially available in the EU - several of which are already being exported.

In the field of **rural electrification**, the THERMIE projects have begun to include a number of "second generation" PV system components, such as intelligent systems controllers with facilities for data storage, which permit PV electricity service providers to offer a "guarantee of results". Historical data on system operation can quickly be downloaded from these controllers which allows the service technician to identify system faults that might previously have taken many days to understand. Whilst the markets for stand alone PV rural electrification within the EU are expected to become increasingly saturated in the next few years, such systems and components will form a major part of PV exports from the EU to developing countries for many years to come.

3.3. Programme interactions

During the past eight years, there have been increasingly close interactions between the THERMIE PV programme and the National PV programmes of the EU Member States. National experts advise the EC on the choice of the demonstration projects, and many are co-funded up to the ceiling of 49% using national programme funds. Moreover, a growing number of the larger THERMIE projects have spawned satellite projects, which have been supported by national funds. For example, satellite projects have build on the demonstration projects to explore socio-economic impacts in the region, or to measure component or system performance using additional techniques or procedures which are not included within the THERMIE monitoring guidelines.

3.4. Awareness and confidence raising

When working to develop and open up new PV markets, it is vital to raise the awareness and confidence of **financiers** and **decision makers**, and this has been greatly assisted by the involvement of well known companies and brand names in the THERMIE demonstration projects. For example, the following categories of well known organisations have been involved in THERMIE PV projects: major utility companies (RWE, EdF, NUON, Eastern Electricity, Union Fenosa, ENEL, PPC), major buildings companies (Pilkington, Schuco), major oil companies (BP, Shell, Agip, Repsol, Total), other famous names (Berlin Bank, Ford, City of Barcelona). Also, the awareness of PV projects was measured in target groups of the programme. We refer to [4] for details.

3.5. Clustering and partnerships

It has become clear from looking at the ways in which the PV markets in the EU have developed that a number of key linkages have been made which have resulted in clusters of important actors and businesses. Some of these clusters involve organisations in a local geographical region, such as the alpine clubs, isolated houses and local PV installers / artisans in the Alpine and Pyrenees regions. Other clusters have developed between complementary businesses, which have built up co-operative working relationships, each based on their individual strengths.

To a large extent, clustering occurred naturally within the THERMIE programme, but it has now been recognised a potentially valuable tool, which can be actively used to improve the effectiveness of demonstration programmes. Clustering is therefore being actively promoted in the Fifth Framework Programme.

Figure 5.1. Sources of information in five clusters

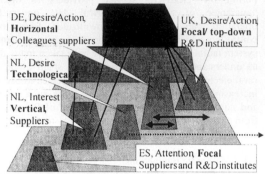

DE, Desire/Action,
Horizontal
Colleagues, suppliers

UK, Desire/Action,
Focal/ top-down
R&D institutes

NL, Desire
Technological

NL, Interest
Vertical,
Suppliers

ES, Attention, **Focal**
Suppliers and R&D institutes

3.6. Technical and non-technical innovation

Technical innovations: The main applications of PV in the EU have changed dramatically from small stand alone systems rated at a few hundred Wp to much larger grid connected kWp and MWp scale projects, with a strong emphasis on building integration. The THERMIE programme has played a major role in facilitating this maturing process, by supporting many of the first commercial scale examples of new applications, such as the first MWp grid connected generator in Toledo, and several of the first building integrated building facades in Germany. THERMIE has also supported demonstrations of other more technical innovations, such as module integrated inverters, multi-functional PV modules, and different coloured architectural PV modules. By encouraging demonstration projects, which involve partners from more than one EU Member State, the THERMIE programme has also stimulated detailed design modifications in grid connected string inverters. These allow devices, which were originally designed for use with the German electricity grid, to work also in France and the UK, where the grid control systems are still operated differently.

Non-Technical Innovations: It has been widely recognised in recent years, for example in the EU White Paper on renewables, that some of the most difficult barriers to the widespread implementation of renewable energy are non-technical. Through a combination of demonstration projects and accompanying measures in the PV sector, the THERMIE programme has supported the development and demonstration of a number of combined technical and non-technical innovations. Aogod example is the emerging renewable energy service companies (RESCO's) which have grown out of the user groups established by a cluster of PV rural electrification demonstration projects in the Pyrenean mountain areas of Northern Spain and Southern France. Another example are the clusters of grid connected PV generators on family houses in France. New SME companies have used these demonstration projects as an important starting point from

which to build their businesses, and they are now expanding their activities both within the EU and for exports.

4.   LESSONS AND RECOMMENDATIONS

Important lessons and messages can be drawn from this study. It should be noted, however, that the current impact assessment was based on finished and running projects, with a focus on positive feedback. **Predictions for future successful demonstration project design should therefor be done with care.**

• There is an important role for the EC to play in PV demonstration projects.
• Perfect projects are not the only way to achieve valuable PV demonstration, sometimes failures may have positive effects.
• Utilities are expected to be the major actors to be involved in countries with small PV efforts so-far. But for advanced PV countries other actors will become more important for hardware project success.
• Detailed knowledge on cluster structure is necessary for effective project execution and for successful targeted dissemination actions.
• Early identification of the most relevant actors and giving them leeway in their preferred project role (supplier or co-proposer) is recommended.
• The THERMIE PV demonstration projects have played an important and complementary role with national funded projects and vice versa.
• Both technological and non-technological innovations are recognised and adopted by the target groups in the dissemination channels of the project co-ordinators.
• It is recommended to place more emphasis on approaching and analysing end-user and target groups for any demonstration programme evaluation.

ACKNOWLEDGEMENTS

The authors wish to acknowledge the valuable input and warm hospitality of THERMIE project co-ordinators and co-proposers, PV industry, and national end European programme managers who were visited during this study. Also those who responded during our interviews on trade fairs are acknowledged. This project was executed with support from the EC Thermie B programme.

REFERENCES

[1] B. Yordi et al., "Lessons learned from the operation of PV demonstration projects in Europe", EC PVSC
[2] Vth Framework Programme, COUNCIL DECISION of 25 January 1999 adopting a specific programme for research, technological development and demonstration on energy, environment and sustainable development (1998 to 2002) (1999/170/EC);
[3] L.A. Verhoef, et.al., "Market Impact Assessment of THERMIE PV Demonstration Projects", Thermie-B STR-1696-98-NL, 2000, in preparation
[4] I. de Jong and L.A. Verhoef, "The impact of PV demonstration Projects in Not-involved target groups-Four Cases", Verhoef Solar Energy Consultancy, Utrecht, 1999

The Launch of European PV Marketing and Information Campaigns

A. Hänel, I. Weiss, P. Helm
WIP – Renewable Energies, Sylvensteinstr. 2, D-81369 München, Germany
Tel. +49-89-7201235, Fax +49-89-7201291, eMail: wip@wip-munich.de

M. Cameron
EPIA, Avenue Charles Quint 124, B-1083 Brussels, Belgium
Tel. +32-2-465 91 62, Fax +32-2-468 24 30, eMail: epia@epia.org

E. Ernst, G. Zähler
NHS Advertising Agency, Boschetsrieder Str. 71, D-81379 München, Germany
Tel. +49-89-74 888 20, Fax +49-89-74 888 222, eMail: ernst@nhs.com

E. Lutter, D. Chiaramonti, A. Grassi
ETA-Florence, Piazza Savonarola 10, I-50132 Florence, Italy
Tel. +39-055 500 2174, Fax +39-055 573 425, eMail: eta.fi@etaflorence.it

ABSTRACT: A high percentage of the population in Germany is in favour of solar energy, and PV in particular. However, the information level on PV is very low which constitutes one major hurdle for a purchase decision. In traditional market sectors it is common practice that manufacturers promote their products via nation-wide campaigns in one or more countries. In addition, there are numerous examples where enterprises pool their promotional activities (branch marketing). These advertising efforts are launched and co-ordinated by the respective industry association. The time is ripe to utilise successful product promotion strategies to fully tap the potential also for PV sales in Europe. Branch marketing is the best possible start for a nation-wide promotion of PV. It is a highly effective means to optimise the limited advertising budget of the European PV industry. The aims of an EC ALTENER-supported project which forms the basis for this paper are to elaborate nation-wide marketing strategies for two reference countries of the European Union with high potential for PV system sales (Germany and Italy), to launch a PV information campaign and to incorporate the national solar energy associations and a significant fraction of the European PV industry into this initiative. This paper presents the steps undertaken to create branch marketing activities in Germany and Italy and highlights the current status of PV campaign implementation in Germany.

KEYWORDS: Marketing-1, PV Information Campaign-2, PV Market-3

1. OBJECTIVES AND INTRODUCTION

As proven by representative surveys in Germany, a very high percentage of the population is in favour of solar energy, and PV in particular. These survey results represent the high level of environmental awareness in Germany. However, the information level in the general public on PV is very low which constitutes a major hurdle for a purchase decision. Even the existence of different solar energy applications (solar thermal and PV) are often unknown. A professional marketing of photovoltaic products, however, can help to overcome this information deficit.

Compared to the marketing of successful products such as computers or cars, the professional marketing of photovoltaics in Europe is still at an early stage.

In traditional market sectors it is common practice that European-wide acting manufacturers promote their products via nation-wide campaigns. In addition, there are numerous examples where enterprises pool their promotional activities (branch marketing) which are financed by common advertising pools (to which a certain number of enterprises financially contribute). Examples include the concrete, plastics and wool industries and even utilities, insurance companies and banks. These advertising efforts are launched and co-ordinated by the respective industry associations.

The European PV industry has not yet reached this level of professionalism. Advertising in PV is, in the best case, still an issue of local dealers or installers. No nation-wide campaigns exist for PV products to date.

The time is ripe to utilise successful product promotion strategies and to fully tap the potential for PV component and system sales in Europe. Branch marketing is the best possible start for a nation-wide promotion of PV: it helps to remove the information deficit and establishes a positive image of PV. It is a highly effective means to optimise the limited advertising budget of the European PV industry.

The aims of the EC ALTENER-supported project which forms the basis for this paper are, for the first time, to create a ready-to-implement strategy for the realisation of a PV information campaign financed by a common advertising pool of the European PV industry, the elaboration of nation-wide marketing strategies for two reference countries of the European Union with high potential for PV system sales (Germany and Italy), to launch a PV information campaign and to incorporate the national solar energy associations and a significant fraction of the European PV industry into this initiative. The participation of the European Photovoltaic Industry Association EPIA guarantees the transfer of know-how and lessons learnt to other European countries with high potential for PV campaign realisation.

This paper presents the steps undertaken to create branch marketing activities and briefly describes the conceptual design of the suggested information campaign. Both, Germany and Italy are currently in an excellent position to act as pilot countries for such an action as in both countries large-scale PV market introduction programmes are underway (the German 100,000-rooftops-programme and the new law on renewable energies - EEG) or planned (the Italian 10,000-roofs-programme) which urgently need to be backed-up by information campaigns to increase public awareness.

2. THEORETICAL BACKGROUND ON BRANCH MARKETING

Targets of Branch Marketing

Branch marketing is a widespread tool in traditional market sectors used for maximising the mutual benefits of participating companies and service providers. The targets of branch marketing are different to those of the marketing of well established products. Branch marketing serves basically to

- increase the degree of familiarity of product groups
- change an existing image or create a new and improved image
- professionalise young branches
- reduce marketing costs for the individual participating company

Type of Campaigns

Two classical campaigns can be distinguished which differ in the target group, the approach and the costs associated with the campaign.

An information campaign aims at reducing the information deficit on a product group. This type of campaign can be successful if the image on the products in the target group is positive. The target group just requires additional information in order to come to a purchase decision. The information carriers are classical advertising means such as an appropriate mix of brochures, flyers, internet presentation, etc. Less appropriate however are TV, radio, cinema spots, etc.

The costs of an information campaign comprise agency and production costs. As a further advantage of information campaigns a significant amount of information can be transported with a comparatively low budget.

An image campaign, however, aims at improving or readjusting the product's image and/or to increase the product's popularity. The target group differs depending on the product. However, most image campaigns address the part of the population who needs to be convinced of the particular advantages of a product or product group.

Past experience has shown that a well-balanced media mix of TV, radio, print media and internet provides the best basis for a successful image campaign.

The costs which will be required to achieve a change of image in the society and a long-lasting and well-balanced campaign will significantly surpass the costs of an information campaign.

Financing of Branch Marketing Activities

The financing of branch marketing activities is generally performed by the means of an advertising pool, a common source for funding to which a multitude of enterprises financially contributes.

A proper planning is required in order to identify the targets of the campaign and to exactly assess the financial needs. This must include, depending on the individual marketing plan and strategy, the costs of classical advertising, public relations, event marketing, sponsoring, etc.

The industry performing branch marketing may rely on its own financial resources. In the PV sector with limited financial resources however, public funding from national governments or the European Union will be sought. Examples have been identified in which the EU financially contributed to the promotion of e.g. olive oil in one of its member states.

In addition, ancillary industry and trade might be interested in the promotion and the support of particular branches. They should be included in the campaign from the beginning covering both financial issues and the conceptual design of the campaign.

Advantages and Disadvantages of Branch Marketing

Taking into account that even larger companies with significant advertising budget may be very often not in a position to reach the aims such as the change of the image of a product class, the cost aspects are of interest both for SMEs and larger enterprises. Effective marketing is possible with limited financial obligations.

On the other hand branch marketing offers the following disadvantages:

- branch marketing activities may also help competitors who do not contribute to the advertising pool
- the individual product or service of a company is not advertised
- the whole branch has to be motivated from a single associations

Implementation of Branch Marketing Activities

Typically, branch marketing activities are performed by the respective solar energy/industry associations. Independence from individual company policies is a prerequisite to be accepted by the whole branch. No individual company would be in a position to implement image and/or information campaigns.

In addition, sponsors and governmental organisations need one partner to support and to address. A prominent role in this context play the national associations, e.g. in Germany, the Deutsche Fachverband Solarenergie DFS and the Bundesverband Solarenergie BSE. These two pressure groups are the most important and influential in Germany.

EPIA can play the role of initiating and co-ordinating the launch of campaigns on a national level and transfer the know how and lessons learnt from one country to other solar energy associations in Europe.

Control of Campaign Success

The campaign approach chosen needs to be regularly assessed in order to be able to react to market needs and to consequently readjust the overall campaign. It is recommended to conduct a representative survey preferably once a year.

3. EXPERIENCE WITH BRANCH MARKETING IN OTHER BRANCHES

Starting point of the initiative of creating branch marketing for PV were the experience with branch marketing of products of traditional market sectors. In the framework of a non-exhaustive analysis pool advertisements in national press could be detected for the following branches in Germany:

- banks
- insurance companies
- wool industry
- plastics industry
- vine
- tourism
- concrete industry

It is particular amazing to see that even banks (on the occasion of the introduction of the EURO) and insurance companies (life insurance) were pooling their promotional activities in order to reach a common aim.

In Italy as well a non-exhaustive survey revealed advertising pools for

- food
- wine
- olive oil
- recycling
- oranges and
- heating by wood

As a mostly information based campaign entitled "Solar-na klar" was developed in Germany recently targeting potential clients of solar thermal systems. The strategy of this information campaign was, amongst others, to directly convey information from the specialist to the consumer. In order to convince and inform the specialists (i.e. the craftsmen) the campaign primarily targeted the craftsmen with information materials such as brochures, give-aways or an accompanying internet presentation and a hotline number from which information material can be obtained.

The results of this campaign of the solar thermal sector can provide significant input and lessons learnt for an adapted PV campaign.

4. PV MARKETING AND INFORMATION CAMPAIGN FOR GERMANY - PROJECT STATUS

In order to develop and launch branch marketing activities in the PV sector, the following groups need to be brought together for collaboration:

- PV industry
- national solar energy associations
- financing bodies

PV industry is not necessarily receptive to marketing efforts in times where PV is underrepresented in public opinion and when advertising budget is limited as most module producers claim to lose money with PV.

However, a change of opinion towards branch marketing could be detected in industry in Germany due to adequate financial support provided by the German 100,000-rooftops-programme and the new feed-in law (EEG) which offer the possibility of a faster return of investment and increasing turnover.

In order to attract PV industry two workshops with international participation were organised in Staffelstein, Germany and Glasgow, UK, respectively aiming at informing, sensibilising and attracting PV industry to join a marketing group on PV branch marketing. The workshops were well attended by numerous high-level PV industry representatives.

Following the experience with the solar thermal information campaign "Solar-na klar" the leading national solar energy associations DFS and BSE are in an excellent position to initiate and co-ordinate a PV campaign.

In the meantime, the steps to be undertaken to reach the target group via a PV information campaign in Germany were identified as follows:

- Step 1: Motivation of electricians and roofers to develop competence in the field of PV installations
- Step 2: Advice and motivation in the field of private house owners and builders
- Step 3: Motivation of decision makers, architects and house and town building companies
- Step 4: Initiate local holding companies for larger PV systems

The conceptual information campaign design in Germany was elaborated for step 1 and step 2 by the two participating advertising agencies Frido Flade and NHS AG, respectively.

Although the current situation (Mid 2000) in Germany seems to show no need for advertising taking into account that PV modules are sold out for months, long-term considerations show the need for marketing efforts as soon as module production capacity has been increased and first interest on PV has grown weary.

Step 1 is being carried out as a collaboration of the national solar energy associations and the national craftsmen organisations of electricians and roofers.

The starting date for step 1 of the campaign is scheduled for automn 2000, the start of step 2 is scheduled for mid 2001.

5. SUMMARY AND CONCLUSIONS

Branch marketing is an appropriate means for the PV industry to achieve maximum effect with minimum budget for the individual company. Branch marketing will help to further improve the image of PV and provide a maximum information on general aspects of PV.

As soon the information campaign has created sufficient demand marketing efforts of individual companies on their particular products can follow. If this stage has been reached PV is finally on the same level than typical products on the market.

With the help and initiative of EPIA the results and experience generated in Germany can then be transferred to other European countries.

ACKNOWLEDGEMENTS

PV Marketing and Information Campaign is supported by the European Commission through the ALTENER programme of the Directorate-General Energy and Transport under contract number XVII/4.1030/Z/99-492

CUSTOMER SITED PV –US MARKETS DEVELOPED FROM STATE POLICIES

Christy Herig & Holly Thomas
National Renewable Energy Laboratory
1617 Cole Boulevard
Golden, CO USA 80401- 3383
christy_herig@nrel.gov

Richard Perez
ASRC
The University at Albany 251 Fuller Rd
Albany, NY 12203
perez@asrc.cestm.albany.edu

Howard Wenger
AstroPowerWest
5036 Commercial Circle, Suite B
Concord, CA 94520
hwenger@astropower.com

Abstract: The customer sited PV market in the United States is dependant on State policies emerging from electric utility industry restructuring. These policies, most of which have appeared since 1996, reduce both the first cost and improve operating benefits. This analysis determines the breakeven turnkey cost of a PV system, from the customer ownership perspective, on a state by state basis. The results of this work are used by industry to target high-value markets, and by policy makers to identify options which will result in the greatest economic and market development. Still intangible external PV benefits, such as environmental value, are also analyzed and gauged against existing/potential policy actions.

1. INTRODUCTION

The US market for customer sited photovoltaics (CSPV) has historically been off-grid systems where the capitol cost of the distribution line extension offsets the CSPV cost. With residential energy prices ranging from ¢5-¢14/kWh, a consumer's values must extend beyond economics to make a grid-connected CSPV investment. PV system installed costs have declined from $6.21/W in 1996 to $3.90/W in 2000i with levelized energy costs of ¢17-¢12/kWhii, respectively [5].   Additionally, the PV industry has developed products targeted at the grid-tied residential market and developed financing packages to alleviate the up-front cost burden to the consumer. The gap between consumer value and cost for CSPV is close, but not close enough for most US consumers. However, as part of the electric industry restructuring, many states have included grid-tied CSPV market development policies for the purposes of resource diversity and economic development. The initial customer sited PV niche market analysis completed in 1996 [1] resulted in only 5 states with a breakeven turnkey cost (BTC) greater than $4 per watt. In 1999, 15 state had BTC's greater than $4 per watt, and four states were above $7 per watt. This increase in consumer market value is fully attributed to policies emerging from state restructuring activities. These include:

- 9 state income tax rebates 30 states with net metering[6]
- 12 buy-down or grant programs
- 11 property tax exemptions
- 2 state interconnection standards

Including these incentives in a life-cycle value analysis for CSPV, provides industry with geographic market targets.

2. APPROACH

The state-by-state database was developed to determine a breakeven turnkey cost for each state and presented in Table 1. The consumer breakeven turnkey cost (BTC), is the value per kW that a consumer can pay for a PV home energy system and neither gain nor loose money over the life of the system. The energy, tax, and policy benefits, as well as the capital (included in the home mortgage), operation and maintenance costs over the life of the system are forced to a net present worth of zero, using an 8% discount rate, by varying the initial cost of the PV system.

2.1 Assumptions

Many of the assumptions for the life cycle value analysis were more conservative for the 1999 analysis. The electricity price inflation rate has been lowered from 3.5% to 2%, consistent with market realities. This change decreased the BTC 5-10%. However, operation and maintenance costs are still inflated at 3.5%. An inflation rate was not applied to the environmental externality benefits over the life of the system, since this is still an intangible value. Also consistent with the 1996 analysis is the one kW installed PV system basis, taking advantage of full residential retail electric rate benefits. The mortgage financing is at 90% debt, 30year term, but the interest rate is set at 7% down from 8%, which results in an increased BTC.

2.2 Database Development

Next to capital cost reduction policy incentives, the analysis is most sensitive to changes in the residential rates. The current residential rates are based on annual residential revenue and consumption [3], resulting in lower more conservative rates. Of the 50 states, only 21 rates changed from the initial study by one- or two-tenths of a cent.

Net metering [6], property tax and sales tax [4] incentives are included in the table, but not in the analysis.   Full residential electric rate benefits are assumed, due to the BTC per kW installed basis.

The state buy-downs and grants are all new policies, which have developed since the original study. The authors have chosen to include two state (Florida and Illinois) buy-down policies and programs, which are not yet, but will soon be available. However, the actual buy-down may change upon availability. Additionally, the authors were unable to verify the availability of the Colorado Solar Energy Association 25% system cost rebate, but the rebate is included in the analysis.

Table 1: State-By State Attributes and Incentives

State	Res. Rank 1999	Res. Rate [3]	Res. Tax Credit [4],[5]	Net Meter [6]	Property Tax [3]	Sales Tax [3]	Buy Down, Grant[5], [7],[8],[9]	SOX- #/kW yr [10]	NOX #/kW yr [10]	CO2- #/kW-yr [10]	Cap Factor [11]	Res. BTC ($/kW) 1999	Ext. - NPV ($/kW) 1999	Res. BTC ($/kW) 1996
Alabama	37	6.7						16	8	2937	19	$2,497	$664	$2,440
Alaska	39	11.4						5	10	2644	12	$2,462	$413	$1,793
Arizona	14	8.8	25%/$1K	Y		Y		7	10	2957	24	$4,590	$456	$4,788
Arkansas	34	7.8						7	7	2791	19	$2,542	$434	$3,423
California	4	11.5		Y	Y		$3/W,50%	2	4	1389	22	$7,402	$286	$4,873
Colorado	10	7.4		Y			25%	10	16	4122	23	$5,196	$687	$2,789
Connecticut	19	12.1		Y				6	4	2062	18	$3,531	$319	$3,707
Delaware	21	9.2						28	10	3503	18	$3,497	$987	$2,910
Florida	11	8.1				Y	$2/W	14	8	2739	19	$5,016	$592	$2,544
Georgia	30	7.7						17	7	3090	19	$2,798	$687	$2,287
Hawaii	3	14.8	35%/$1750					15	9	4356	24	$7,911	$737	$7,500
Idaho	49	5.2						2	1	644	21	$1,690	$110	$2,189
Illinois	2	10.4			Y		60%/$5k	17	8	1962	18	$8,411	$612	$3,057
Indiana	29	6.9		Y	Y			24	18	4401	17	$2,815	$1,037	$2,231
Iowa	23	8.2		Y				16	15	3497	19	$2,995	$759	$1,915
Kansas	26	7.7						9	12	3350	21	$2,894	$576	$2,858
Kentucky	42	5.6						27	11	2976	17	$2,323	$930	$1,476
Louisiana	28	7.4						19	8	3206	20	$2,828	$735	$2,143
Maine	22	12.8		Y				8	3	2767	16	$3,462	$423	$3,158
Maryland	7	8.3		Y			$2.94/W	17	9	2876	18	$6,133	$693	$2,744
Massachusetts	13	11.6	15%/$1K	Y	Y	Y		9	5	2408	18	$4,647	$433	$4,321
Michigan	40	8.6						11	8	2198	16	$2,414	$481	$1,983
Minnesota	44	7.2		Y	Y	Y		7	10	2909	17	$2,217	$469	$1,885
Mississippi	36	7.0						11	9	3228	19	$2,502	$581	$2,737
Missouri	31	7.1						17	12	3165	19	$2,681	$730	$2,509
Montana	48	6.4			Y			3	6	2017	19	$1,919	$263	$1,771
Nebraska	43	6.4						8	11	2384	20	$2,222	$453	$2,066
Nevada	17	8.9		Y				8	13	3714	24	$3,610	$578	$3,021
New Hamp.	18	13.7		Y	Y			8	2	1230	16	$3,540	$294	$3,571
New Jersey	6	12.1		Y		Y	$2.94/W	4	6	1912	18	$6,719	$289	$3,608
New Mexico	16	8.9		Y				9	17	4447	25	$3,860	$684	$3,666
New York	1	14.1	25%/$3750	Y			50%	6	4	1540	18	$10,257	$289	$4,372
N. Carolina	5	8.0	40%/$1500	Y			$2.94/W	14	7	2409	19	$7,042	$563	$3,714
North Dakota	20	6.3	5%-3yrs	Y	Y			15	12	3630	19	$3,519	$710	$2,393
Ohio	25	8.6						27	11	2714	16	$2,956	$925	$2,354
Oklahoma	35	6.6		Y				9	12	3509	21	$2,538	$578	$3,429
Oregon	47	5.6	.40/kWh, $1K					1	1	415	18	$2,042	$55	$2,344
Pennsylvania	8	9.9		Y			$2.94/W	16	6	1854	16	$6,092	$548	$3,052
Rhode Island	15	12.1		Y			$1/W	1	9	2262	18	$4,564	$264	$3,541
S. Carolina	41	7.5						11	5	1738	19	$2,403	$416	$2,598
South Dakota	46	7.1			Y			5	4	947	19	$2,065	$215	$2,544
Tennessee	45	6.0						18	7	2223	18	$2,191	$652	$1,582
Texas	27	7.8		Y	Y			6	9	3166	22	$2,892	$453	$3,250
Utah	12	6.9	25%/$2K					3	15	4705	24	$4,907	$710	$2,588
Vermont	32	11.5		Y				0	0	293	16	$2,614	$21	$3,198
Virginia	9	7.8		Y			$2.94/W	12	6	2443	18	$5,753	$513	$2,744
Washington	50	5.0		Y				2	1	412	15	$1,020	$81	$1,084
West Virginia	33	6.3						29	12	3353	17	$2,605	$1,014	$2,105
Wisconsin	24	6.9		Y	Y		$0.5/kWh	14	10	2896	16	$2,994	$630	$1,770
Wyoming	38	6.2						8	16	4151	21	$2,477	$641	$1,887

The California SBC provides $54 million over 4 years for buy-downs of "emerging renewables", which include residential CSPV. The California Energy Commission administers the fund, which started in March of 1998. The buy-down provides $3/W up to 50% of the installed cost. It was designed to decline on an annual basis, but is currently still available at the $3/W value [12].

The Photovoltaic Buildings in Florida program will apply the major portion of $600,000 worth of funding from the Florida Energy Office / Department of Community Affairs towards system buy-downs. The residential CSPV buy-down is proposed at $2/W [8]. The program will be administered by the Florida Solar Energy Center.

The Illinois SBC will collect $5 million annually targeted towards renewable energy resources [5]. The Renewable Energy Resources Program under the Department of Commerce and Community Affairs is expected to administer grants to fund 60% of CSPV costs up to $5,000.

The Virginia Alliance for Solar Energy (VASE) is currently offering a $2.94/W buy-down for residential CSPV in five (Maryland, New Jersey, North Carolina, Pennsylvania and Virginia) states, through a request for proposal [7]. A minimum aggregate of 10kWac is required by the request for proposal.

The New York State Energy Research and Development Authority (NYSERDA) will administer the SBC fund

expected to collect over $234 million in the next three years [5]. Currently, NYSERDA has a program opportunity notice (PON) to deploy $1 million funds towards residential CSPV [9]. The PON limits the cost share at 50%.

The pounds per kilowatt-hour emission mitigation for SOX, NOX, and CO2 externalities were determined using the total industry generation and total industry emissions for each state [10]. Due to disclosure conflicts and externalities conflicts, six states, Arizona, Kentucky, Mississippi, Nebraska, North Dakota and Wyoming are calculated using utility generation and total industry emissions. The emissions mitigated for each kW of PV installed are then calculated using the state average PV capacity factor [11]. The value of the emissions mitigation by PV is based on the cost of control [15] versus the value of environmental damages.

TABLE 2: Emissions Cost-of-Control Values

	National[14]	CA [16]	WI [16]	MA 16]
SOX $/#	$2.03	$2.20-11.00	NA	$0.75
NOX $/#	$0.82	$4.50-$15.00	$1.35	$3.25
CO2 $/ton	$13	$9	$15	$22

FIGURE 1 Breakeven Turnkey Cost 1996 & 1999

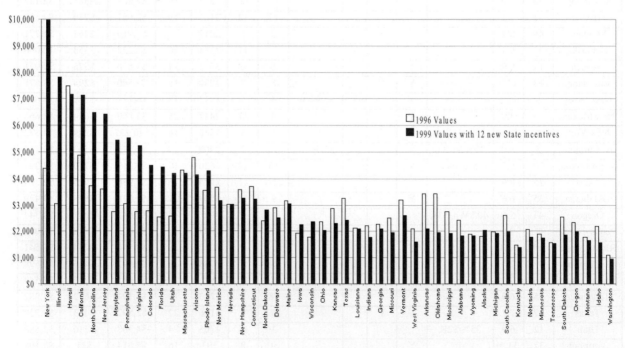

Figure 2: BTC, $/W, with Environmental Externalities and Real Time Pricing

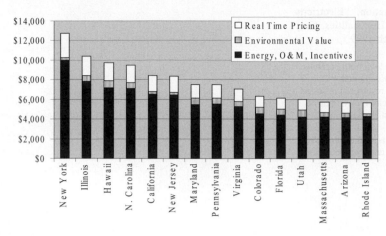

3 RESULTS

As shown in Figure 1, now fifteen states with BTC's above $4/W. In just four years, 12 new state policies, either buy-downs, grants or state income tax rebates, all of which reduce the initial cost of the CSPV system, increasing the BTC. Figure 2 examines the top fifteen states for other potential policies such as environmental externalities and real time pricing.

4 CONCLUSIONS

The economics indicate policies do effectively fill the gap between consumer value and price. The most active deployment areas in the country are the states with high BTC's.

Though many of the incentives used in the analysis have changed or sunseted. There is potantial for new incentives. Currently 23 States have initiated or implemented restructuring policies and 16 have renewables provisions. The system benefits charges (SBC) included in 13 state restructuring policies, are a source of funding for consumer incentives. Thus far, only 7 of the 13 state SBC's have been implemented leaving potential for more near term incentives. Additionally, 9 state renewable portfolio standards, the Million Solar Roofs initiative 40+ community partnerships working on consumer awareness, reducing infrastructure barriers and municipal policies, and the fifty utility green pricing programs either offered or under development are potentially new arenas for consumer incentives.

With the potential for market stimulation through policy incentives established by analyzing the increased consumer value, the next step is to determine the effectiveness as measured by participation in the various incentives programs.

REFERENCES

[1] Wenger, H., Herig, C., Taylor, R., Eiffert, P., and R. Perez, "Niche Markets For Grid-Connected Photovoltaics", IEEE Photovoltaic Specialists Conference, Washington, D.C., 10/96

[2] Osborn, D., "Sustained Orderly Development and Commercialization of Grid-Connected Photovoltaics: SMUD as a Case Example", A pre-print from Advances in Solar Energy XIV, 2000, Ver. 02/24/2000

[3] Energy Information Administration, "Electric Sales and Revenue 1997"

[4] Date base of State Incentives for renewable Energy-DSIRE, (1998). NCSC-IREC @ http://www-solar.mck.ncsu.edu/dsire.htm

[5] Sprately, W. A., "Consumer Charges Power Solar Financing", Public Utilities Fortnightly, Dec. 1998

[6] Starrs, T. (Feb. 1999), Personal Communications, "Summary of State Net Metering Programs (current)", version 2/15/99, Kelso Starrs and Assoc., Vashon, WA

[7] VASE Request for Proposal Phase 3, Virginia Alliance for Solar Energy (VASE) @ http://www.vase.org/index.html, (1999)

[8] Ventre, G. (1999), Personal Communication, Florida Solar Energy Center, Cocoa, FL

[9] Program Opportunity Notice (PON) No. 448-98, "Residential Photovoltaics Market Development", New York Solar Energy Research and Development Authority (NYSERDA)@ http://www.nyserda.org/448pon.html, 1999

[10] Energy Information Administration, "State Electricity Profiles", DOE/EIA-0629, March 1999

[11] QuickScreen software, Pacific Energy Group @ http://www.pacificenergy.com/software.htm
(12) Masri, M. (1999): Personal Communications, California Energy Commission, Sacramento, CA

[13] Osborn, D., "Commercialization and Business Development of Grid-Connected PV at SMUD", Proc. ASES Solar '98 Conference, Albuquerque, NM, June 1998

[14] Buchanan, C., P. Chernick, A. Krupnik, U. Fritsche, (1991) Environmental Costs of Electricity, Oceana

[16] Energy Information Administration, "Electricity Generation and Environmental Externalities: Case Studies", DOE/EIA-0598, March 1999

i These costs are the result of an aggregate long term purchase for the Sacramento Municipal Utility District pioneer PV program and represent the lowest reported residential installed costs. These are representitive of commercial PV systems of 30kW or more. The year 2000 cost was estimated from the committed contract price.

ii Levelized costs are for residential systems with 1st mortgage financing and retail rate compensation for energy production at ¢10/kWh.

ACTIONS TO BE TAKEN FOR A FURTHER DISSEMINATION OF PHOTOVOLTAIC IN GERMANY AND OBSTACLES FOUND IN AN EMPIRICAL SURVEY

M. Eichelbrönner, R. Hanitsch, D. Schulz
Technische Universität Berlin, Germany, Einsteinufer 11, D 10587 Berlin
T: +49.30.31422403, F: +49.30.31421133, Email: Rolf.Hanitsch@iee.tu-berlin.de

Abstract: The paper shows the results of an empirical survey on 162 photovoltaic operators out of a pool of about 900 operators of renewable energy technologies (RET). The survey describes their experiences with planning, permitting, financing and subsidising, installation and running photovoltaic-systems in Germany. From these answers conclusions were drawn, which actions should be taken for a further dissemination of photovoltaic systems in Germany.
Keywords - 1: Dissemination of photovoltaic systems – 2: Obstacles – 3: Actions for development

In the past several German and European studies have described in detail obstacles that hinder renewable energy technologies (RET). Two kinds of obstacles have been figured out: firstly the economical obstacles like high investment cost and low energy prices, secondly non economical obstacles like informational, legal, financial, fiscal, organisational and technical hindrances. It has been discussed widely how much the economical barriers do hinder the dissemination of photovoltaic.

In the years from 1994 to 1997 an "action plan to reduce obstacles which hinder renewable energy technologies (RET)", implemented by the German ministry of economics, was made to foster the further development of RET. Within this action plan empirical data of about 900 RET projects were gathered and discussed in an interdisciplinary panel of experts. A broad consensus was reached and a list of actions was developed how to bring RET into the market.

The following paper focuses on the results of the part photovoltaic. It presents the results of an empirical survey of 162 photovoltaic operators respective an actions' list. The survey describes the experiences in planning, permitting, financing and subsidising, installation and running photovoltaic-systems in Germany. From these answers and an additional panel of experts conclusions were drawn, which actions should be taken for a further dissemination of photovoltaic in Germany.

1 MOTIVATION AND INFORMATION

The photovoltaic technology provides the most expensive and advanced way to produce electricity. Therefore it is of high interest why operators invest in this technology. The survey gives qualified answers to this question, see fig. 1. Among all operators the spectrum of answers differs from - the most important - ecological reasons (mean value 1.2), followed by macro-economical reasons (mean value 1.9) and technology (mean value 2.0) to - less important - motives like prestige (mean value 4.0) or independency (mean value 3.3). The group of utilities seemed to be of specific interest. Their motivation starts with the highest interest in technical matters (mean value 1.4), followed by ecological reasons (mean value 2.0) or the aspect "to make something new" (mean value 2.4). The motives like advertising (mean value 2.6) and prestige (mean

value 3.0) within this group are of higher relevance as in the total number of operators.

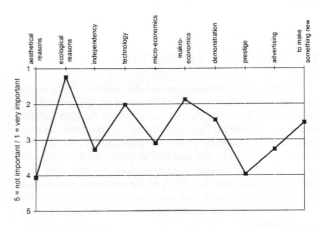

Figure 1: Operators' motives for investing in photovoltaic systems

The operators complain about a not contenting access to information about photovoltaics, esp. to economics, financing and legal procedures. Therefore they demand better information networks and specific brochures.

2 GENERAL AND LEGAL FRAMEWORK

In the past the general and legal situation has improved. Generally legal procedures for roof installations are not necessary but for larger stand-alone systems they are. Anyhow from decision to commissioning a mean time period of 24 months had to be noticed, see fig. 2. Asked for the most important obstacles the operators quoted the grant of subsidies, withstanding of utilities (by private operators) and technical problems. In some particular cases specific conditions were put on according to preservation of historical buildings. To prevent over-regulations and conditions the operators advice a qualification-programme for the representatives of public authorities.

According to the fast development of photovoltaics it is obvious that meanwhile these time periods were shortened significantly.

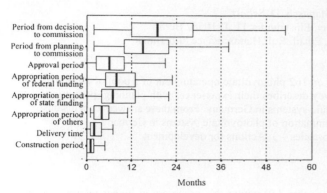

Figure 2: Range of time periods to realise photovoltaic-systems

3   OPERATORS' SATISFACTION

Of relevant importance seemed the question of the operators' satisfaction because this will be noticed by potential new installations. Therefore the operators' grades for their satisfaction are of relevance. Thus their rating differed from the mean value 1.7 for reliability to 2.0 for productivity and to 3.4 to economy. Obviously the amount of produced electricity did not reach the estimated output respectively the cost were higher than expected. Yet 97 % of the operators would invest into RET again. But in that case they would design a larger system or change the technology to solar thermal or wind-power systems.

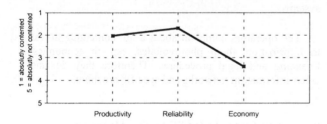

Figure 3: Operators' opinions of their photovoltaic-system

In general the operators quote a high technical reliability of 98,3 %. But 47 % of the operators complain about general technical problems and malfunctions of the inverter.

4   ACCEPTANCE AND IMAGE

Photovoltaics enjoys a very high acceptance in the public according to its high-tech-character. It stands as a synonym for RET at all due to an insufficient knowledge about its specifics. Therefore photovoltaic does not really have an image problem - it "only" has a cost problem.

On one hand it is to recognize that this technology itself has a very high acceptance for its sophistication and modernity. On the other hand photovoltaic systems have the highest cost for electricity production at all. This brings photovoltaic systems into a controversy. Recognizing this situation the survey researched the operators associates before and after the construction of the photovoltaic systems. The opinion of the craftsmen and planners before construction was already very positive but increased further with the realization. The highest improvement was found by friends (0.66 points), colleagues (0.45 points) and by neighbours (0.58 points). Utilities and financiers often refuse photovoltaic systems because of their high investment cost. Even the successful realization could not improve their opinion much so that utilities increased only 0.35 points and financiers 0.32 points, see fig. 4.

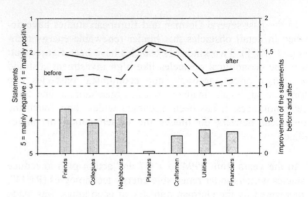

Figure 4: Opinions before and after the construction of the photovoltaic-systems

5   FINANCING AND SUBSIDISING

A further development of photovoltaics in Europe is only to be expected if there is a public and a political intention in terms of financial support. Because of its wide acceptance this should be not a major difficulty for the necessary amount of subsidies for a promising and successful market introduction programme challenges the willingness of the politicians.

In Europe, respectively Germany, grid connected photovoltaic systems have been built only if there is a significant grant or subsidy. Furthermore for an acceptance of these instruments a long-termed reliability is needed. Thus only under long-termed and reliable conditions operators as well as manufacturers can and will invest in new systems and in research or development of this technology. Therefore the operators propose a significant increase of the feed-in tariffs up to a cost covering value, distinctive development programmes or special low interest rates for investment credits.

It is of high significance that the German Government faced this situation and picked up the photovoltaic technology to launch a most ambitious market introduction programme. The first step was the "100 000-photovoltaic-roofs-programme" that finally was launched in the beginning of the year 1999. This programme gives a 0 % interest rate for investment credits for a time period of 10 years as a grant. The second step was to dismiss the "act to improve the renewable energies" in February 2000 which defines the feed-in tariff to 0,99 DM/kWh. With this approach a real break-through of photovoltaics might be foreseen. Referring to this fig. 5 shows a forecast of the possible effect to these instruments in the next years.

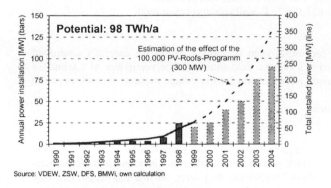

Figure 5: Development of grid-connected photovoltaic installations in Germany until 1998 and an estimation of the possible effect of the 100 000 PV-Roofs-Programme

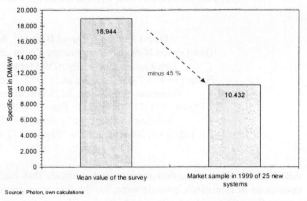

Figure 6: Development of the specific cost of photovoltaic systems form the early 90s (survey's pool) to the year 1999

7 RESULT

To bring photovoltaic forward into the future, the market introduction was emphasized by a new political framework. Therefore in the next years photovoltaic installations of more than 300 MW might be foreseen.

6 OPERATOR'S EFFORTS AND COSTS

After the installation in general a real maintenance or operation cots of the photovoltaic systems do not occur. However the survey turned up a significant operators' behaviour paying attention to their systems. Almost half of the operators quote the time spent to maintain the systems in average to 16,8 hours per kilowatt and year what is quite a high value. This is not to justify because of technical reasons. But taking into account the high technical interest of the operators and their support by the "1.000-photovoltaic-roofs-programme" (the first market introduction programme in the early 90s) related with a data monitoring obligation this attention seems to be understandable.

Nevertheless these operators were the pioneers of photovoltaics in Germany. They invested in the early 90s typical installed system cost of about 22.000 DM/kW. In the following years the system prices reduced continuously. Since that time an actual market sample in 1999 has recognized a price reduction of about 45 %, see fig. 6.

Regarding the actual situation, which means the "100 000-photovoltaic-roofs-programme" and today's system prices, production cost for electricity of about 1.3 DM/kWh is realistic. Taking into account the feed-in tariff of 0.99 DM/kWh a difference of only 0.31 DM/kWh remains to the operator.

8 LITERATURE

[1] M. Eichelbrönner; Erneuerbare Energien in der Stromversorgung - Errichtung, Anlagenbetrieb und Kosten auf Basis einer empirischen Situationsanalyse; Verlag Köster, Berlin 2000.

[2] J. Diekmann, M. Eichelbrönner, O. Langniß; Aktionsprogramm Abbau von Hemmnissen bei der Realisierung von Anlagen erneuerbarer Energien; Forum für Zukunftsenergien, Bonn 1997.

IMPLEMENTING AND FIELD TESTING A PV SYSTEM AT A PALESTINIAN VILLAGE

David Faiman[1,], David Berman[1], Michael North[2] and Ibrahim Yehia[3]
[1]Ben-Gurion National Solar Energy Center, Jacob Blaustein Institute for Desert Research,
Ben-Gurion University of the Negev, Sede Boqer Campus, 84990, Israel.
Phone: ++972-7-659-6933 Fax: ++972-7-659-6736 E-mail: faiman@bgumail.bgu.ac.il
[2]Greenstar Foundation, 6128 Blackburn Ave., Los Angeles, CA 90036, USA
Phone: ++1-323-936-9602 Fax:++1-323-936-7203 E-mail:mjnorth@greenstar.org
[3]Sol-Nur Ltd. Energy Systems, P.O.Box 300, Um-Alfahim 30010, Israel
Tel: ++972-6-631-0085/6 Fax: ++972-6-631-0087 E-mail:solnur@bayan.co.il

ABSTRACT: High-precision outdoor test methods that had previously been developed within the carefully controlled conditions of a research institute were, for the first time, applied to a genuine field situation. The venue was Al Kaabneh, a remote Palestinian desert village, which uses PV panels to provide electrification for the village mosque, the infirmary and the high-school.
Keywords: Evaluation : Rural Electrification: Developing Countries

1. INTRODUCTION

Al-Kaabneh is a small village on the edge of the Dead Sea desert, Latitude 31° 24' N, Longitude 35° 12' E, population 2000. Until the Greenstar Foundation began work there in December 1998, the village had no electricity, no pure water, little education or medical support, and no computer or communications. The village mosque now has a small solar panel on its roof, which powers an amplifier that announces the five-times-daily Muslim call to prayer, so that the people can hear the voice of the mosque for miles across the desert hills. In the center of the village, a new school and community center have been built. Eight large solar modules, supplied by ASE Americas of Boston for Greenstar, generate 2.4 kWp, at 48 volts.

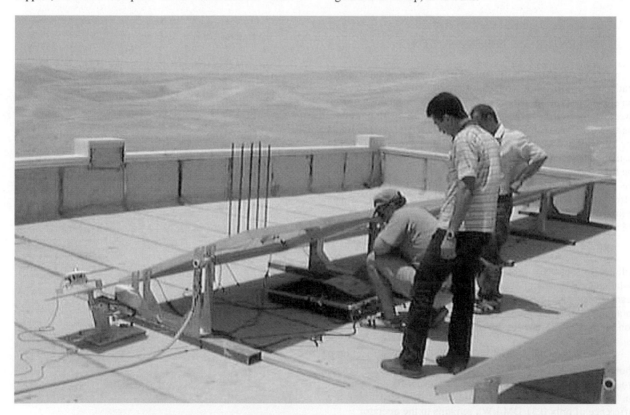

Figure 1: Measuring I-V curves for the eight 300 W PV modules on the roof of Al-Kaabneh school

During the day, a bank of batteries is charged by the sun; at night, the batteries and associated inverters deliver a stable supply of both DC and AC power to circuits wired throughout the building. One additional large module now also powers the nearby clinic, providing lights and cooling for vaccines and antibiotics needed by the village.

Greenstar contracted the services of Sol Nur Enterprises, of Haifa, to help install the solar array and supporting electrical components. Sol Nur's head instructed key people from the village on system maintenance and safety procedures. He visits Al-Kaabneh once a month to reinforce the training and conduct routine inspections. Greenstar's chief engineer instructed the villagers on the principles of solar power and electricity, how to optimize performance of their new solar equipment, and on basic conservation and preventive maintenance. Greenstar also contracted Ben-Gurion University's National Solar Energy Center to perform some on-site system tests at Al-Kaabneh. The results of those tests are reported here, together with some conclusions of a more general nature.

2. FIELD TEST METHODOLOGY

On July 18, 1999, field measurements were performed on the 2.4 kWp PV system that Greenstar had installed on the roof of the regional high-school. The system consists of two strings of four ASE 300-DGF/17 modules, each string having its modules connected in series. The modules face due south at a tilt angle of 25° to the horizontal. Cables connect each of the two strings to its own battery storage / inverter subsystem.

Measurement techniques were basically those developed, at the Ben-Gurion National Solar Energy Center, for high-precision outdoor module characterization [1-3]. Instrumentation, seen in Fig.1, included a Daystar I-V curve tracer, a carefully-calibrated Eppley PSP pyranometer mounted on a stand that allowed it to be orientated in any desired direction (specifically, at normal incidence, and parallel to the plane of array), thermocouples for measuring module temperatures, two small reference modules (one by ASE, made from similar materials to the modules that were to be tested and one by Solarex) and a Campbell micrologger.

Measurements were performed during the noon time period on a cloudless day. Several I-V curves were measured with the two reference modules held at normal incidence, in order to provide data for later comparison with our laboratory data set for these modules. Further measurements were performed with the reference modules orientated in the plane of the array that was to be tested. Because the latter measurements were not far from normal incidence they provided data that would be useful for correction purposes. Because of an acute water shortage at this desert location, it was considered inappropriate to clean all 8 of the 300 W modules that constitute the Al-Kaabneh array. Instead, measurements were performed with all modules in their "as is" condition and then, again, on 2 of them after washing.

3. CONTROL RESULTS

The precision that was possible, under these circumstances, is illustrated by the results shown in Table 1. These refer to all seven *in situ* I-V curves taken, at Al-Kaabneh, on one of our control modules - the Solarex SX-45. Four of these measurements were made at normal incidence and three were made with the reference module in the system plane of array. The table gives the observed range for each parameter, the corresponding STC-adjusted values and, for comparison purposes, the reference values that had previously been obtained from our laboratory measurements   [1].

Parameter	Observed range	STC-adjusted values	Reference value [1]
I_{sc} [A]	2.74 - 2.94	2.83 ± 0.02	2.86 ± 0.02
V_{oc} [V]	17.9 - 18.5	20.1 ± 0.3	20.7 ± 0.5
I_{pp} [A]	2.43 - 2.65	2.51 ± 0.03	2.55 ± 0.03
V_{pp} [V]	13.6 - 14.0	15.7 ± 0.3	16.5 ± 0.5
P_{max} [W]	33.7 - 37.0	39.4 ± 0.3	42.1 ± 1.1
FF [%]	67.3 - 68.1	69.6 ± 0.2	71.3 ± 0.2
H_{poa} [W m^{-2}]	951 - 1018	1000	1000
T_{mod} [° C]	46.3 - 56.8	25	25

Table 1: *In situ* measured values at Al-Kaabneh school, and reference values measured at the outdoor test laboratory at Sede Boqer, for the I-V parameters of a Solarex SX-45 control module

One sees that the agreement between our field results at Al-Kaabneh and the laboratory measurements at Sede Boqer, for the SX45 reference module, is sufficiently close to indicate that no serious systematic errors were made during the Al-Kaabneh field tests.

Measurements were also performed with a second control module: an ASE 50-AL/17, of similar type and construction to the modules under test, except smaller (a nominal 50 Wp compared with 300 Wp for the latter). These results were comparably similar to our previous laboratory tests and are, accordingly, not reported here. However, we used the temperature coefficients we had previously determined for this module, viz: $\alpha = 1.6$ mA K^{-1}, $\beta = -61$ mV K^{-1}, in order to arrive at appropriate values for the (6x larger) test modules, i.e. $\alpha = 9.6$ mA K^{-1}, $\beta = -61$ mV K^{-1}.

4. *IN SITU* STUDY OF THE ASE 300-DGF/17 TEST MODULES

Owing to the acute water shortage at Al-Kaabneh village it was deemed inappropriate to clean all of the PV modules prior to testing them. Instead, all modules were first characterized in their "as is" condition. Two were then washed and re-tested. Table 2 shows the measured I-V curve parameters, reduced to 1000 W m^{-2} irradiance and 25 °C cell temperature, for the 8 "as is" ASE 300-DGF/17 modules. Measurements were carried out on the modules sequentially and 3 experimental runs were performed. The standard deviations given in Table 2 correspond to the 3 measurements that were performed on each module.

Parameter	Module A	Module B	Module C	Module D	Module E	Module F	Module G	Module H
I_{sc} [A]	18.2	18.2	18.2	18.2	18.3	18.1	18.3	18.4
	± 0.12	± 0.09	± 0.11	± 0.12	± 0.10	± 0.10	± 0.09	± 0.09
V_{oc} [V]	20.3	20.4	20.5	20.3	20.2	20.5	20.5	20.5
	± 0.05	± 0.01	± 0.02	± 0.03	± 0.05	± 0.04	± 0.01	± 0.01
I_{pp} [A]	16.1	16.0	16.2	16.2	16.3	16.1	16.2	16.4
	± 0.05	± 0.10	± 0.05	± 0.12	± 0.15	± 0.03	± 0.10	± 0.09
V_{pp} [V]	15.9	16.0	16.0	15.8	15.8	16.1	16.0	16.1
	± 0.04	± 0.09	± 0.04	± 0.10	± 0.12	± 0.03	± 0.07	± 0.02
P_{max} [W]	256	256	260	256	257	259	259	263
	± 0.2	± 0.2	± 0.5	± 0.5	± 0.6	± 0.6	± 0.8	± 1.0
FF [%]	69.3	69.1	69.8	69.1	69.5	70.0	69.2	69.8
	± 0.23	± 0.26	± 0.28	± 0.23	± 0.11	± 0.12	± 0.14	± 0.08

Table 2: Measured I-V curve parameters, reduced to 1000 W m^{-2} irradiance and 25 $^{\circ}$C cell temperature, for the eight ASE 300-DGF/17 modules, *prior to cleaning* and at their *in situ tilt* angle of 25°

Two conclusions may be drawn from Table 2:
(1) All modules were found to be in satisfactory working order; (2) the standard deviations from the mean values of all parameters are unusually small: significantly less than ± 1% for Pmax, Isc, Voc and FF. The latter conclusion indicates unusually high quality control at the module production level.

5. EFFECT OF MODULE CLEANING

After the first three "as is" runs, modules "A" and "B" were cleaned. Each was then tested three times at the system design 25° tilt angle, and module "A" was subsequently tested four times at normal incidence.

Parameter	Module A "dirty"	Module A "clean"	Module A change [%]	Module B "dirty"	Module B "clean"	Module B change [%]
I_{sc} [A]	18.2	19.1	4.9	18.2	19.1	4.9
	± 0.12	± 0.03		± 0.09	± 0.01	
V_{oc} [V]	20.3	20.4	0.5	20.4	20.4	0
	± 0.05	± 0.01		± 0.01	± 0.02	
I_{pp} [A]	16.1	16.9	5.0	16.0	16.9	5.6
	± 0.05	± 0.17		± 0.10	± 0.07	
V_{pp} [V]	15.9	15.8	-0.6	16.0	15.8	-1.3
	± 0.04	± 0.14		± 0.09	± 0.07	
P_{max} [W]	256	268	4.7	256	266	3.9
	± 0.2	± 0.2		± 0.2	± 0.1	
FF [%]	69.3	68.9	-0.6	69.1	68.6	-0.7
	± 0.23	± 0.18		± 0.26	± 0.02	

Table 3: Observed change in I-V curve parameters (at 1000 W m^{-2} p.o.a. irradiance and 25 $^{\circ}$C cell temperature) after modules A and B were cleaned. Modules at tilt angle = 25°, non-normal incidence.

Table 3 shows a comparison of before cleaning and after cleaning results for modules A and B for measurements made at the 25° plane of array tilt angle. Not shown are the results of our 4 normal-incidence measurements on the cleaned module A as they reveal no significant difference between the STC-adjusted module parameters measured at a 25° tilt angle to the horizontal. This is not really surprising since both sets of measurements were made a few minutes on either side of solar noon, and at this time of year this coincides tolerably closely to normal incidence on a 25° tilted surface. [To be precise, solar noon was at 11:45, at which time the zenith angle was close to 10°. This means that the angle of incidence was approximately 15°, the cosine of which is 0.97].

Table 3 indicates that dirt on the modules was responsible for an approximately 5 % lowering of system performance. We have no way of knowing when, if at all, the modules had been cleaned prior to these measurements, however, the ease of access to the modules suggests that periodic cleaning should present no problem. We should also comment that the STC value of P_{max} = 270 W determined, from our normal incidence measurements

on the cleaned module A, represents a gross-area module efficiency of 11.2%.

6. OVERALL ARRAY PERFORMANCE

Some additional measurements were made within the school building in order to verify that the array had been re-connected in a satisfactory manner after our tests, and to provide some idea of the extent to which what was being generated on the roof was comparable to what was reaching the inverter/storage room. For the purpose of these tests we did not measure solar irradiance or module temperatures but these could be extrapolated with confidence from our previous measurements owing to the cloudless weather conditions. These final measurements revealed at worst, a 3% discrepancy between the sum of the outputs from all modules, as measured on the roof, and the total power delivered to the indoor inverters. In view of the stated uncertainties we may confidently conclude that the DC power supplied to the inverters coincided closely with what was being generated by the array.

7. CONCLUSIONS

This paper describes a self-consistent set of *in-situ* measurements which: (a) provide a detailed, high-precision characterization of all modules in the array; (b) indicate that all modules have rather similar characteristics to one another (with gross-area STC module efficiencies in excess of 11%); (c) indicate that the array modules have comparable performance parameters, *mutatis mutandis*, to an unused control module of similar construction; (d) provide some information of the effect of dust on the array; (e) provide
a quantitative set of initial parameters for future follow-up measurements.

On the day in question, we found the *array* to be functioning in a completely satisfactory manner in that: (a) no modules were defective; (b) all modules had similar characteristics to within ±1% of the mean; (c) the module performances figures were comparable to those of a control module of similar construction that we had taken along to accompany these tests; (d) the power arriving at the inverter was comparable to the amount being generated by the array.

At the scientific level, we may conclude from our results that *in situ* measurements of the kind described here can be almost as accurate as those performed under the most rigorously controlled laboratory conditions.

A small number of potential system problem points were discovered and brought to the attention of the system operators and these were quickly rectified. They included the need to move the array further from a parapet wall so as to prevent shading in winter, and the need to protect wiring from the chance encounter of humans.

An important point to emphasize here is that all of these changes were effected by the residents of Al-Kaabneh village themselves (who actually dismantled the array and placed it on a higher section of the roof). Such action was possible owing to the training program which had been provided by Greenstar and the chief engineer of Sol-Nur Ltd.

A suggested system design modification, which would make arrangements for the village to use electricity generated by the school array during the long summer school vacation, is being studied by Greenstar as part of their follow-up policy. Further information about the Al-Kaabneh PV project can be found on Greenstar's web page, http://www.greenstar.org

REFERENCES

[1]   D. Berman, S. Biryukov and D. Faiman, "EVA laminate browning after more than 5 years in a grid-connected, mirror-assisted, photovoltaic system in the Negev desert: Effect on module efficiency", Solar Energy Materials and Solar Cells vol. 36 (1995) 421.

[2]   D. Berman and D. Faiman, "EVA browning and the time-dependence of I-V curve parameters on PV modules with and without mirror-enhancement in a desert environment", Solar Energy Materials and Solar Cells vol. 45 (1997) 401.

[3]   D. Berman, D. Faiman and B. Farhi, "Sinusoidal spectral correction for high precision outdoor module characterization", Solar Energy Materials and Solar Cells vol. 58 (1999) 253.

TRENDS IN DANISH NATIONAL PV PROGRAMME

Bent Sørensen

Roskilde University, Institute of Mathematics and Physics, Energy & Environment Group.
email: bes@ruc.dk, www: http://mmf.ruc.dk/energy, fax +45 4674 3020
P.O.Box 260, DK-4000 Roskilde, Denmark, Phone +45 4674 2028

ABSTRACT: Developments in the focus of the Danish National PV Programme are presented and discussed, as an example of the considerations that a small country must make in order to obtain maximum dividend on its RD&D spending.
Keywords: R&D programme - 1: Photovoltaics - 2: Solar cells –3: Photo-electrochemical cells – 4.

1. INTRODUCTION

The Danish national PV programme started in 1991, although scattered support for PV projects had materialised earlier, within existing programmes. The new support was through the Renewable Energy Development Programme ("UVE") of the Danish Energy Agency (spending 2-4 MDKK/y on PV), although the basic Energy Research Programme ("EFP") additionally did have a slot for PV. The early phase focussed upon individual demonstration and knowledge dissemination projects (Katic et al., 1994). Two new dimensions were added in late 1999, where considerations regarding system development have expanded the scope of the programme in the direction of research and development, while at the same time demanding a programme extension into the area of market stimulation. One is a new PV programme, adding 10 MDKK/y to the existing funds and demanding specific R&D in the fields of novel devices, demonstration of building-integrated solutions, and basic information and quality assurance activities. The other is a recycling of energy taxes, first carried through in the electricity sector but being expanded to the gas sector, allowing R&D expenditures to be covered, if they are relevant for an energy transition strategy in tune with the official government planning (Danish Department of Energy, 1999). Industrial investments in building-integrated PV are supported by a 40% subsidy. In the future, the PV area may also carve itself a larger chunk of the basic research money ("EFP"), if promising new developments are materialising within the new programmes.

Historically, Denmark has contributed to crystalline silicon PV development, notably through the research performed at the Danish Technical University (see e.g. Leistiko, 1997). An offspring of this was a commercial cell production venture by the company "Solel", which, however, failed due to lack of market penetration. The problem was and is, that there is no significant niche market for photovoltaics in Denmark: Few vacation houses are so remote that they are not served by an electric power grid, interest in small-scale farm use is absent, and industrial "image greening" efforts have in Denmark been less willing to invest in building energy systems, compared to many other countries. Finally, public "showing the way" has been absent, as it is only this year (2000), that a suggestion of compulsory use of solar energy in public buildings

(schools, etc.) is being introduced to the parliamentary decision process. Today, the Danish business scene as far as PV is concerned consists of a number of small companies selling modules based upon imported cells, or other derived products. Newcomers include solar thermal collector companies who faces dwindling profits for their current products due to insufficient market penetration (presumably due to lack of market backing – sales of solar thermal systems quadrupled during one year, where the gas companies offered fixed price, warranty-backed solar systems to their gas customers). Now the solar thermal producers look to PV as a possible rescue technology, because they see public funding as creating at least a demonstration product market. The other types of newcomers are companies with no previous experience in renewable energy, typically building element producers, seeing PV integrated into windows and facade elements as an interesting addition to their product lines. These companies are interesting, because they usually have achieved high market visibility for their products, but they still likely to need some time to appreciate the technical problems of a technology considerably more complex than what they have been accustomed to.

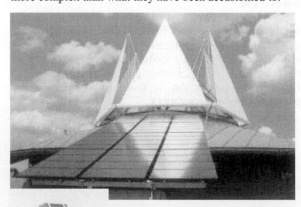

Figure 1 (above). Villa Vision at the Danish Technological Institute, Tåstrup (built 1993, with both 2.3 kW m-Si PV modules and also thermal panels).

Figure 2 (left). PV powered emergency phones at North coast of Sealand (Copenhagen Telephone Company, built 1997).

2. RESEARCH AND INDUSTRY PARTNERSHIP

The Danish Technological Institute showed an early interest in PV, resulting in the creation of a centre for information, testing of products and development assistance offered to industry. With some additional partners, this lead to the formation of a "Danish Solar Energy Centre". PV activity areas include

- Dye-sensitised cells
- PV/T system development
- Building-integrated components

Figure 3 (left). PV panels mounted on road protection barrier at Fløng (NESA, built 1999).

Figure 4 and 5. Brundtland Conference Centre in Toftlund (14 kW c-Si in roof and facade, built 1995). Interior (above) and outside view (below).

The testing and certification of solar products can develop into an advantageous synergism in product development, as it happened with the early collaboration between the Danish wind industry and the Wind Turbine Test Station near Roskilde. This has not quite happened in the solar case yet, perhaps because the solar industry (and particularly the solar thermal industry) has been less open to sharing bad as well as good experiences. The meticulous public dissemination of fault data from the Danish wind industry and its customers was essential in rapidly improving the product quality. The solar industry, on the other hand, seemed more interested in keeping problem information

away from marketplace exposure, and the removal of technical problems consequently progressed more slowly. The new generation of players in the field may ave a better understanding of the value of sharing experiences, at least during the first phases of development.

3. DEMONSTRATION ISSUES

Like in many other countries, the first demonstration projects in Denmark often had the flavour of plastering a solar cell panel on top of some building, with little attempt of integration or architectural considerations. Later, the importance of solar architecture and product integration has moved to the forefront (Sørensen, 1995), although the number of Danish architects educated in design of renewable energy systems is still exceedingly small. Energy basics were taught at the architecture schools for a few years during the mid-1970ies, and then removed. As a result, many architects still produce designs with miserable energy balances, if not North-facing solar panels, etc. Notable exceptions are the buildings shown in Figures 1, 4 and 5.

Danish building traditions have in the recent half century developed in the direction of accepting the cheapest possible raw buildings, but combined with large efforts to make the indoor decor as attractive as possible (a natural attitude in countries where people spent a large fraction of their time indoors, for climatic reasons). It is possible, that integration of solar energy devices into the building structure will heighten the concern over outside appearances, which would be a very positive contribution from solar energy.

Similarly, the recent interest in integrated solar products, which are strongly favoured in calls for public project support, has produced a lot of imaginative ideas, including (as it should be in this early phase) some doubtful concepts. Figures 2 and 3 show examples of taking advantage of off-grid PV opportunities, while Figure 6 shows a set of multicrystalline cell areas incorporated into a facade glass. In buildings where the architect uses huge glass facades in order to make the interior light and hot, the non-transparent PV may be a good idea, but for most buildings, the PV glass facade will just be replacing insulated walls, as the window area is already prescribed by building codes aiming at securing light penetration, to which the PV panels do not contribute, and thus the net energy gain by replacing an insulated wall by PV panels may well be zero (Sørensen, 2000b).

Denmark has used some of its modest public PV support money quite wisely, e.g. by establishing 30 PV houses at a location where they could all be connected to the same power substation, and hence give valuable answers on stability of large-penetration PV systems (Kristensen et al., 1999). A recent new project of this type entails is equipping 300 houses with PV panels, the locations being spread over a dozen villages (Encon, 2000, see Figure 7).

The new round of projects receiving support in early 2000 includes an emphasis on using PV in the apartment buildings of Building Societies, of which many in Denmark are part of the co-operative tradition. An example is a large building block at Skovlunde, where the roof shingles have to be replaced. The company Dansk Eternit has developed a new shingle incorporating PV modules (Figure 8), which will be used in the retrofit. There is an additional possibility of air cooling of the PV panel from the rear side, making this a PV-Thermal system (cf. Sørensen, 2000c).

Figure 6. Window integrated c-Si and m-Si solar cells at Skibstrup conference building (Midtglas, Gaia Solar, Peoples Renewable Energy Centre, built 2000)

Figure 7 (above). Single family dwelling furnished with 4 kW c-Si PV panels as part of the Solar-300 project (Encon, 2000).

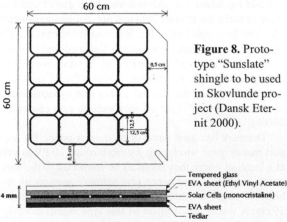

Figure 8. Prototype "Sunslate" shingle to be used in Skovlunde project (Dansk Eternit 2000).

Figure 9 shows an already completed apartment building renovation project incorporating PV panels in balconies, and Figure 10 a fringe project using PV on a hybrid luxury car, with plant oil diesel propulsion as its main source of energy.

4. ORGANIC SOLAR CELLS

The thinking behind this project is that organic solar cells (or photo-electrochemical cells, first suggested by O'Regan and Grätzel, 1991) might be more accessible for a country without any large microelectronics industry. Certainly the start costs are much lower than for production of PV cells, but the level of sophistication eventually characterising commercial organic solar cells is difficult to guess at this stage. Many other countries have started R&D efforts in this field, and there is certainly work to do, as only a few hundred organic dye materials have been tested (and all of fairly similar structure). Also, the precise engineering of the anode TiO_2 material is not yet fully optimised, and no suitable semiconductor cathode material has been found. Instead, an electrolyte Redox reaction with platinum catalyst is used, despite the much slower electron transport. Among the many alternatives considered are polymer electrode materials. This line of research (Bezzel et al., 2000) consumes about 15% of the recent addition to the Danish R&D budget (by political request).

5. CONCLUDING REMARKS

The Danish experience may be relevant for other countries with a strong interest in renewable energy solutions and greenhouse gas emission reductions, but without the industrial basis required for full-scale indigenous production of all required PV devices and components. The programme adopted aims at establishing high value added component production, but based on international collaboration and providing systems containing a fair range of imported components, combined in innovative products.

This work is supported by the Danish Energy Agency. The author was chairman of the Agency's Solar Energy Committee during 1999.

REFERENCES

[1] Bezzel, E., Lauritzen, H., Wedel, S., West, K., Lund, T., Sørensen, F. and Sørensen, B., 2000. Development of building-integrated photo-electrochemical cell. Danish Energy Agency: Programme for Building-integrated Solar Cells, Contract 51181/00-0028.

[2] Encon, 2000. Webpage http://www.encon.dk/sol300

[3] Katic, I., Sørensen, B. and Windeleff, J., 1994. Danish National Photovoltaic Programme. pp. 2302-2305 in "1st World Conference on Photovoltaic Energy Conversion, Hawaii", IEEE Electron Devices Soc., USA.

[4] Katic, I., Sørensen, H. and Sørensen, B., 2000. Solar cell systems with combined power and heat production. Energy Research Programme EFP2000 of the Danish Energy Agency, Contract 1753/00-0014.

Figure 9 (above). Balcony PV retrofits in Herning (2kW c-Si, installed 1997).

Figure 10 (below). "Connector 2001" hybrid solar cell-plant oil car (Toria, developed during mid-1990ies).

[5] Kristensen, P., Kristensen, F. and Ahm, P., 1999. Solbyprojektet. Final report (71pp). Encon, Brædstrup

[6] Leistiko, O., 1994. The waffle: a new photovoltaic diode geometry having high efficiency and backside contacts. pp. 1462-1465 in "1st World Conference on Photovoltaic Energy Conversion, Hawaii", IEEE Electron Devices Society, USA

[7] Sørensen, B., 1995. Fotovoltaisk Statusnotat 3. Final Report for Danish Energy Agency Project 51181/94-0002

[8] Sørensen, B., 2000. *Renewable Energy*, 2nd Edition, 912 pp., Academic Press, London.

[9] Sørensen, B., 2000b. The future Solar Landscape. Talk presented at DANVAC Solar Energy Seminar, Danish Technological Institute, 7. March.

[10] Sørensen, B., 2000c. PV power and heat production; an added value (contribution to this conference).

HOW MUCH DO I REALLY PAY FOR PV ON MY ROOF?
A COST COMPARISON AMONG SEVERAL EUROPEAN COUNTRIES.

Guido Haarpaintner

Cythelia, Savoie Technolac, BP 319, F-73375 Le Bourget du Lac Cedex, France

Email: guido@poboxes.com, Tel +33-6-6203 1610, Fax +49-89-609 1610, http://pro.wanadoo.fr/cythelia/

ABSTRACT: An overall cost model for solar-electricity- or photovoltaic- (PV) systems with normalisation to 1 kWp and prices in a single currency is proposed in order to facilitate cost comparisons across national borders while taking into account the different conditions for subsidy availability, electricity pricing, fiscal and technical regulations. PV systems from Germany, France and the UK are thus compared and the results show strongly varying cost structures. This is a measure of how far the EU is still from a real common market and suggests that future PV incentive programmes be coordinated on a European level in order to achieve a more equalising impact within the EU. Similar in depth studies of this kind will be a useful tool for renewable energy policy makers.

Keywords: Economic Analysis - 1: Pricing - 2: Small Grid-connected PV Systems - 3

INTRODUCTION

We are moving towards a common market in the European Union with progressive harmonisation of costs, prices, taxes, rules and regulations. At the same time there exists a multitude of subsidy programmes for PV systems from the EU, member countries, regions and even communes, while kilowatt-hours fed back into the electricity grid are remunerated at price levels varying from zero to one Euro. Technical regulations for grid connection of PV systems are still being defined on a national and applied and interpreted on a utility level. This includes requirements for special protections mechanisms (e.g. ENS in Germany) and metering arrangements (e.g. production versus grid feeding metering).

How do these geographically varying differences influence the real price tag for a consumer who wants to install a standard grid connected PV system on his roof?

In this study we compare overall PV system costs, which include equipment, and installation costs, taxes and tax relieves (e.g. recent VAT reduction in France and the UK), subsidies and pay back value through grid feeding. It is this cost, which the consumer perceives, which influences his decision to install PV or not and thus determines the success of the various PV market introduction programmes. Often, financial advantages can be cumulated, but in some cases they can exclude each other. Varying climate conditions and PV system design, operation, maintenance, repair and reliability makes a comparison of overall PV system costs on a geographical basis even more difficult.

In this paper we attempt to define a technical and financial normalisation of the PV systems from different locations. From this we create an overall PV cost model, which depends on the localised parameters representing the various factors, which influence the system cost, performance and pay back, thus making the overall cost comparison possible.

This will be an essential tool in evaluating the possible impact of the geographically varying financial support and market introduction programmes for PV systems.

1. PV SYSTEM COST MODEL AND NORMALISATION

The price tag for the end user of a PV system consists of a number of components which we group into 5 categories called (1) *installation costs*, (2) *running costs*, (3) *installation dependent subsidies*, (4) *production dependent revenues* and (5) *general parameters* describing economic, fiscal, meteorological and other local conditions. Note that we are treating only grid connected PV systems of a reasonable size for a private house here i.e. approx. 1-5 kWp without storage batteries.

1.1 Installation Costs

We define as *installation costs* the sum of all payments necessary so that the PV system can start working as planned. Thin includes material costs for the PV modules, DC and AC cabling, connectors and switches, inverter, support structures, housing and any other necessary equipment (e.g. an ENS imposed by German regulations which may or may not be included in the inverter price). Furthermore there are a number of services to pay for such as planning, transport, installation and connection of the PV system. Also the administrative tasks involved in acquiring subsidies, loans, planning and building permits, grid feeding contracts, etc. incur sizeable costs.

1.2 Running Costs

All expenses after the PV system is technically and legally ready to produce electricity fall in to the category of *running costs*. These include insurance, metering, maintenance and repair costs as well as loan repayment interests, rents and service charges. For example, some inverter producers now offer a lifetime warranty for a continuous service fee independent of the real efforts necessary to keep the inverter running. Lacking this, regular savings related to the expected inverter lifetime may be an option to cover the replacement costs in case of an inverter failure.

1.3 Installation Dependent Subsidies

We define as *installation dependent subsidies* all financial aides available purely linked to the installation of the system and independent of its subsequent operation and kWh production. This may also include indirect cost reductions such as income tax relief or the French VAT reduction for home repairs and improvements, which can be applied to installation work for a PV system.

1.4 Production Dependent Revenues

Production dependent revenue is all income related to the PV system due to its operation and kWh production.

This is mainly any compensation paid by the utility for electricity supplied to the grid but also primes or subsidies, which are made dependent on the PV system operability such as the 10th year payback relief in the German 100.000 roofs programme.

1.5 General parameters

All further information influencing for the overall cost calculation of a PV system we call *general parameters*. They are useful for comparing system costs across borders and otherwise different economic, fiscal and meteorological conditions. *General parameters* include interest rates, loan details, subsidy programme names, tax information (VAT, tax relief, etc.), PV System (location, orientation, size, equipment used, actual production), local grid operator, metering arrangements, kWh buyback rate and conditions, applicable laws and regulations (technical, fiscal, financial).

1.6 Financial and Technical Normalisation

Using the data from the overall cost model we need to render comparable PV systems, which have been installed and are operating under different conditions. Every owner will eventually be able to sum up all payments and revenues related to his PV system in his currency. The question is often whether the outcome corresponds to his initial expectations.

We suggest normalising PV systems by relating all data to 1 kW peak and the currency Euro. Furthermore, time dependent values should be given on a yearly equivalent basis as this also averages out any seasonal behaviour and simplifies financial interest calculations.

2. Examples from Germany, France and the UK

We have selected 3 sample systems to illustrate the differences one may encounter when comparing the financing of PV systems in the EU. The three countries lie relatively close together but have extremely different policies on PV subsidies.

2.1 Simplifying Assumptions

For an easy comparison we are assuming that the system is technically identical in the 3 cases. A house owner installs a 3.3 kWp PV system on his roof and feeds into the electricity grid. The system is designed to cover all of the owner's electricity needs averaged over the year, using the grid as intermediate storage. The system performs well and produces 900kWh per kWp installed in all three locations. The overall installation costs of the system are also equal before any taxes or subsidies are applied. In agreement with the German roundtable on PV system costs [1] we assume 7000 Euro/kWp for the key-in-hand system followed by annual maintenance costs of 1% and insurance of 0.5%. After subsidies, the financing source is the PV system owner's own capital on which he looses annual interest of 6.5%. After 20 years all investments should be entirely returned.

2.2 Parameters differing between countries

The differences between these countries lie in their respective VAT rates for system installation and operation, in the kWh buy back and sales rates, standing charges for grid connection and meter rental. The used values are summarised in Table 1.

Table 1: Differing parameters for the 3 examples

PV system location	Germany	France	UK
VAT (installation)	recoverable	5.5%	5.0%
VAT (operation)	16.0%	19.6%	5.0%
kWh (buy back)	0.496 €	0.079 €	0.042 €
kWh (sales rate)	0.130 €	0.079 €	0.101 €
Fixed grid charge	88 €	50 €	62 €
Grid feeding meter	30 €	0 €	45 €

The prices are taken from real cases in the respective countries. We have used 0.60 £/€ as the sterling exchange in April 2000. The low kWh buy back in the UK is taxed at the standard 17.5% VAT put for all other energy related transactions the reduced 5% VAT apply. Because the buyback rate in the UK is lower than the sales rate, any electricity produced is by preference directly consumed in order to avoid the costs of the sales rate. In the UK case, we assume that only 50% of the PV electricity is fed back into the grid.

2.3 Subsidies

In each country we have tried to find the optimum subsidy condition. Therefore in Germany, the system takes part in the 100.000 roofs programme [2]. 100% of the investment are covered by a 0% interest loan payable back in 10 annual rates, of which the first two and the last are waived on the condition that the system is still operating after 9 years. The high kWh buy back rate is defined in the new German renewable energy law [3] as 0.99 DM/kWh and is valid for 20 years.

In France, the system is purchased through the association Phébus [4] and installed in Rhône-Alpes region, which gives access to a total of 70% subsidies of the installation costs from the EU (via Phébus), ADEME and the regional government. Since summer 1999 it has legally become possible to feed PV-electricity back into the grid. A favourable VAT rate of 5.5% applies for the installation if done by an accredited installer on a building older than 2 years.

The UK also applies a reduced VAT rate [5] of 5% for installation and operation but no other subsidies are readily available.

2.4 The PV system Account

Taking into account all previously mentioned conditions, we calculate the balance of a PV system account on a yearly basis, through which all funds related to the PV system flow. The condition of the 20 year investment return is held by adjusting a regular annually cash flow into the account such that after 20 years the balance is zero. We call these 20 years the *loan period*. After this time the PV system has been entirely paid for and the annually constant cash flow into the account can be reduced such as to keep the account balanced at the end of each following year. The evolution of the account balance for the three cases mentioned can be seen in Figure 1.

The German curve clearly shows the pay back years of the 100.000 roofs loan while the French and British curves resemble a classic mortgage with different borrowing levels due to the 70% subsidy available in Rhône-Alpes. Note that the maximum debt is very similar in the German and French cases, they only occur at different times.

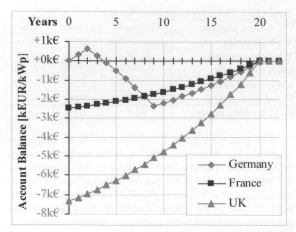

Figure 1: PV system account evolution during the 20-year loan period.

2.5   Average Electricity Cost

We compare the average electricity cost in Table 2 and Figure 2. For each of the three cases we plot the annual bill divided by the number of kWh used.

Table 2:   Average electricity costs in €/kWh.

Average kWh price	Germany	France	UK
without PV	0.19 €	0.11 €	0.13 €
with PV during loan period	0.08 €	0.39 €	0.92 €
with PV after loan period	0.18 €	0.14 €	0.18 €

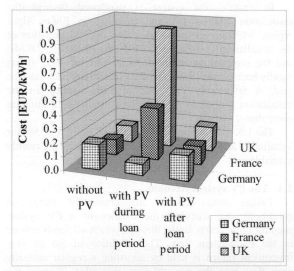

Figure 2: Average electricity costs without PV and with PV during and after the 20-year loan period.

2.5.1   Standard costs without PV

Without PV i.e. with a standard grid connection for the house the electricity costs are highest in Germany. This is mostly due to the high fixed charges for metering and basic connection. France has by far the cheapest electricity but this advantage is slightly reduced by very high taxes. Even at the strong Sterling exchange rate of April 2000 (0.60 £/€) the UK price is only slightly above France.

2.5.2   PV costs during the loan period

The tendency is totally inverted when looking at the PV system during the 20-year loan period. Now the UK rates are highest. Due to the lack of subsidies, UK residents need

to pay the real price of PV electricity. The cost approaches 1 €. This is in agreement with the *cost covering rates* [6] which have been defined by several German public electricity boards during the past years. The average cost in France is notably lower due to the strong investment subsidy but remains almost 4 times above the costs of conventional electricity. The currently very high subsidies in Germany, which can be cumulated, are able to reduce the average electricity cost for the PV system owner to less than 1/2 of the normal costs. This seems indeed like a very good incentive to install PV on your roof. However is must be made clear that the PV system owner is still not making a profit here. He is still paying for his electricity. It is merely a very low rate, which at 0.08 €/kWh is still above those rates paid for at the electricity market of the high voltage grid.

2.5.3   PV costs after the loan period

After the 20-year loan period, the average kWh price reverts back to reasonable values but which are still above the conventional kWh price in France and the UK. The reasons for this are the relatively high fixed costs for connection to the electricity grid and the maintenance and repair costs of the PV system. If one could further reduce the latter, a PV system can become financially interesting once it has been amortised.

2.6   Annual Cost Split

For our 3 examples, we have split up the annual costs for electricity with a PV system during its 20-year loan period in Figure 3. Note that by subtracting the kWh sales from the sum loan repayment, system operation, metering and electricity use, multiplying the result with the PV system size and dividing by the kWh consumed, one finds the middle column of Table 2 and Figure 2.

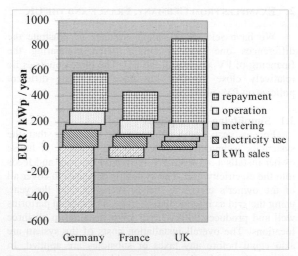

Figure 3: Average annual electricity costs during the 20-year loan period.

2.6.1   Loan repayment

In all cases the loan repayment is the most important part of the annual cost. In the German example the kWh sales almost offset the costs while they just compensate the costs of electricity use in the French and not even the metering charge in the British example.

2.6.2   Running costs

PV system operation is the second most important cost factor. These maintenance and repair costs need to include a regular saving for large repairs such as an inverter replacement. This is likely to happen at least once in the system lifetime until inverter technology catches up with the high reliability of PV modules. We have represented PV system operation, i.e. running costs as a regular expense here. One may indeed find a regular payment solution in form of a continuing warranty and maintenance fee. However in most cases one should look at it as a statistical value which may be close to zero for many years before suddenly rising, for example in case of an inverter failure.

2.6.3 Metering charges

One clearly sees that the fixed charges for the grid connection weigh much heavier in the German and the British case than in France.

3. FUTURE DATA COLLECTION AND ANALYSIS

The future aim of this study is to be able to compare in a similar way as presented in Section 2, a large number of PV systems from various locations and countries around the World, built under different financing conditions. Using such data, we hope to be able to make precise predictions about the cost of ownership for future PV systems dependent on their location, not only from a meteorological but also from a fiscal and subsidy related point of view.

The obvious task is to collect the relevant data from PV system owners who are willing to share their experience with others. If you have such information or know people who do, please contact the author of this paper.

The results from such a study are an important input for the design of further market introduction and incentive programmes for PV systems. This will be a useful tool for policy makers.

CONCLUSIONS

By analysing financing details for a technically identical PV system installed in Germany, France and the UK, we have shown that extremely differing cost structures result from the political, fiscal and economical situations in the respective countries.

This is a measure of how far we are still from the general strive to reduce cross border differences in the European Union. It is mostly so, because the PV market is still highly subsidy dependent and financial support destined for the end-user is very limited on the Union level

and varies greatly on a national and even on a regional level.

Another strong factor for the results is the differing pricing structure for electricity from the grid. It follows that the liberalisation of the European electricity market has not yet achieved the aim of creating a truly "common market" where former borders do not matter anymore. Further PV market introduction and incentive programmes should be co-ordinated on a European level in order to achieve a more equalising impact within the EU.

ACKNOWLEDGEMENTS

The Author wishes to thank in particular Wolf von Fabeck of the Solarenergie-Förderverein Aachen, Florence Barrault of the association Phébus, Jackie Carpenter of Eurosolar-UK, Irmgard Hoster, Ron Rumney, and Susan Roaf for their time, information, fruitful discussions and inspiration.

REFERENCES

[1] Wolf von Fabeck, „Wirtschaftlichkeitsberechung von Solarstromanlagen", SFV - Solarenergie-Förderverein e.V., D-52070 Aachen, http://www.sfv.de/, Solarbrief 1/2000, pp.11-16

[2] German 100.000 roofs programme, BMWi, http://www.bmwi.de/infomaterial/photovoltaik.html, Bundesanzeiger Nr. 13 vom 21. Januar 1999, Seite 770

[3] EEG – Gesetz zur Förderung der Stromerzeugung aus Erneuerbaren Energien, Deutscher Bundstag 25.02.2000, http://www.gruene-fraktion.de/

[4] Association Phébus, fiches d'information, F-69009 Lyon, France, http://perso.wanadoo.fr/phebus/

[5] "VAT Information Sheet 01/00", March 2000, Eurosolar UK / Energy 21, http://www.energy21.org.uk/

[6] W. von Fabeck, A.Kreutzmann. P.Weiter, „A new path to self-sustaining markets for PV - the Aachen model", Proc. 13th European PV Solar Energy Conference, Nice (1995)

ACTIVITIES OF IBERDROLA'S TECHNOLOGY DEMONSTRATION CENTRE ON PHOTOVOLTAIC SYSTEMS

Jesús García, Alfonso de Julián
Luis Alberto Calvo, Estefanía Reolid (CDT-Renewable Energy Area Technical Managers)
IBERDROLA
C/ Hermosilla, 3ª planta
Madrid, 28001
SPAIN
Tel: +34 91 577 65 00 Ext. 41440
Fax: +34 91 578 20 94
www.iberdrola.es
cdt@iberdrola.es

ABSTRACT: This paper describes the photovoltaic activities that IBERDROLA is developing in the Technology Demonstration Centre (TDC) on photovoltaic energy. As will be describes, it constitutes an essential support element towards innovative technological activities carried out inside IBERDROLA, and it represents an important competitive positioning in the scope of the new electricity generation liberalised market.
Keywords: Strategy - 1: Implementation - 2: Marketing - 3

1. TECHNOLOGY DEMONSTRATION CENTRE

Based on the strategic objectives that IBERDROLA has developed in its R&D Plan and meeting to the necessity of finding new technologies to satisfy the society's requirements, IBERDROLA, by means of its Generation Power Division, creates a purposeful instrument to help solve some of the problems of the electric supply: *the Technology Demonstration Centre*. This facility is believed to be a strategic tool addressed towards the client satisfaction and optimisation cycle.

The Technology Demonstration Centre can be considered as a common medium where, in a single facility, different technologies converge, among them PV solar energy. The research on PV technology, which needs a certain period of demonstration before its launch in the market, is one of the main purposes, in order to fulfil the objectives of technological management:

- Accessible technologies evaluation
- Technology initiation
- Technological transfer to and from the company
- Declining lagtime for market introduction

The Technology Demonstration Centre is located in San Agustín de Guadalix, 33,5 km from Madrid, in a small village near the sierra de Guadarrama (Km 33 of the Madrid-Burgos highway) with a very clean atmosphere.

This place receives a lot of visits from Universities, enterprises, and others public and private entities. That is why the TCD is an important help to disseminate photovoltaic solar energy.

It is divided in three Areas: Renewable Energy Sources, Energy Efficiency and Quality and New Power Generation Technologies. Within the Renewable Energy Area, here is a brief look at the systems and works that IBERDROLA has developed and is developing in the field of Photovoltaics.

Figure 1: Technological Demonstration Centre view

2. ACTIVITIES

The most relevant activities carried out in the TDC on photovoltaic solar energy respond to a strategy whose main objective is the enhancement of technological innovation. In the next figure, all activities related to photovoltaic solar energy can be seen:

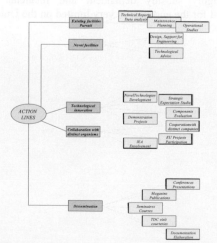

Figure 2: PV activities carried out in the TDC

Different photovoltaic activities carried out in the TDC are being achieved. Among them, the following activities can be highlighted:

- Development of Innovative Technology and Investigation projects, in collaboration with different organisms and national and international agencies.
- Design, construction and operation of novel PV technology-based facilities.
- Equipment and system analysis, design and development, related to PV energy.
- Promotion of all different PV technologies in the market, contributing to a high quality and environmentally benign energy supply in the near future.
- Technical and financial feasibility studies.
- Maintenance and operational system's design and adaptation with views to the improvement of efficiency and quality of energy processes.
- Equipment testing and rehearsal execution.
- Technological advice.

3. OFFER SERVICE

The TDC not only develops different and various innovative PV projects, but it also offers a series of services that contribute in enhancing nowadays market positioning, both to the Company itself and to different industries and institutions coming from the national and international sector.

These services are guaranteed by highly skilled personnel, who gives the transfer of all its knowledge and technological infrastructure to any one in need of it.

The service that the TDC provides to different industries is vary complete, from training to innovative system developments, passing through rehearsals and tests execution, and offering a variety of technological possibilities to the photovoltaic industry.

The next figure exhibits the most significant services offered by the TDC:

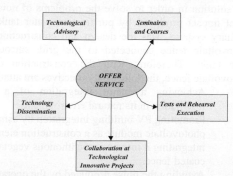

Figure 3: PV services offered by the TDC

The human team involved in this Area is believed to be highly qualified as to give technological advice to all sorts of companies and institutions in need of guidance. Taking this into account, the main services provided by this Area are the following:

3.1 Technological advice

Installation Design

Technological improvement reached in the photovoltaic energy field, as well as the existence of promotion and motivation programmes, both at national and European Commission level, contributes to increasing energy installations based on photovoltaic energy.

With this purpose, in the Renewable Area of the TDC, specialised technical counsel is offered for plants completion. This counsel includes events dealing with initial sizing of the facility and turn-key, looking for both energy optimums and economic feasibility during design and construction process.

The gained experience due to realisation of numerous projects becomes a guarantee in assuring service qualities offered in this field.

Technical-economic feasibility studies

To arrive at a society conscience point of meditation, in which photovotlaic energy is seen as an added element in the generation activity, different feasibility studies are made in the TDC as a step towards a further implantation of these systems.

From a technical point of view, a technological resource analysis is undertaken in order to verify its feasibility in plant construction, in terms of energy production. On the other hand, considering economic terms, studies of facility feasibility must be carried out to assume viability, taking into account existing economic aspects such as financing, funding income, inflation or interest rate.

At the study conclusion, applicant will be able to recognise the most suitable technology for its installation as well as the economic feasibility of the facility.

Preventive and corrective management plans design

Once the installation has finished, an important phase to assure its perfect operation, consists of the conception of a maintenance plan, preventing possible failures in normal plant operation and contributing with the necessary corrective measures in case this plan does not accurately perform.

Thanks to an exceptional experience in the numerous PV projects carried out in the TDC, a deep knowledge has been acquired related to photovoltaic plant operations. This permits maintenance plans, both for existing facilities and for those of new construction, guaranteeing a perfect performance during its scheduled operation.

Effective standards advise

Due to all current adjustments dealing with clean energy encouragement, the facility necessities seem to be growing in modern society. For this reason, organisations, institutions and interested persons will need standard legal advisory assigned to photovoltaic energy promotion.

From this point of view, the personnel responsible for this Area, consents to report about legal dispositions to be considered at time to process a technological energy facility. These dispositions also include aspects related to existing funding to be applied. For this purpose, all possible institutions and states promoting the massive use of PV energy are being contacted, in order to offer the most advanced and effective information to any interested party.

3.2 Seminars and courses

Among others, another service borrowed by the TDC consist of several courses in investigation centres and official agencies such as INEM or CIEMAT, at which the latest advances obtained in innovative developments are explained. This contributes to acquisition of different information and knowledge from people outside this Centre, which in other case would be very difficult to attain.

On the other hand, the TDC is a competent constituted facility that is able to carry out dissemination of information by means of different seminars and meetings for discussions of PV technology.

3.3 Test and rehearsal execution

The availability of different technical and human medium, as well as the wide available floor space, turns this Centre into an ideal place where to test and verify new PV components operation. To improve component and equipment contribution, a service provided by this Centre is the realisation of all sorts of tests assigned to test new prototype operation under real conditions. Last generation of Inverters, new photovoltaic modules, photovoltaic solar devices are some of the units usually tested and analysed in these installations.

By means of these tests, valuable information will be obtained about these new devices operation, proving itself useful for its incorporation into future facilities or for a possible commercialisation of such devices.

3.4 Technology dissemination

An interesting and important aspect of the TDC is that related to the possibility of offering its installations and terrain for the location of a number of various projects and equipment, not only with a view to testing and demonstration, but, as mentioned, as a teaching and assisting in those innovative and conventional PV technologies.

Personnel preparation, permanently sited in the TDC, will contribute to guarantee an adequate diffusion of all PV advances and innovations for companies, universities, schools and other institutions For this purpose, specific documentation of each of the possible applications is made in order to further disseminate it in specialised forums of this sector, the web or specialised publications, ands, all persons who visit these facility installations.

3.5 Collaboration at Technological Innovative Projects

One of the main objectives of the TDC is technological innovation of different PV systems. Another is reduction of Market Lagtime. Therefore, the realisation of different projects combining investigation, development and demonstration is one of the principal activities carried out in the TDC. To this effect, inside this Area, national projects such as PAEE, ATYCA, NATIONAL R&D PLAN, REGIONAL TECHNOLOGICAL PLANS or OCI-CIEMAT. International projects include: JOULE, THERMIE, FEDER, ESPRIT, BRITE-EURAM, or V EUROPEAN COMMISSION FRAMEWORK. In these various frameworks and together with different sorts of companies and entities, a technological development is being pursued in PV energy field, in order to eliminate barriers to market penetration.

The detailed explanation of the PV developed projects would be motive for another paper. Only a brief explanation of some of the most interesting projects developed in this Area is intended here.

1. **Photovoltaic School**. A 53 kW photovoltaic power generation system is integrated in the top of a

sportcentre of a special education school. It is a grid-connected system and will diminish energy consumption by means of a high quality supply.

2. **UMSEF. Photovoltaic mobile unit**.

Figure 4: Photovoltaic Mobile Unit

The system consists of a 3.3 kW photovoltaic power generator. The main objective is to supply electricity to remote customers under service conditions equivalent to those provided by the grid. The system has been designed for use especially at sites which are difficult to access and located at some distance from the grid. It is link to a battery energy storage system to guarantee five days of autonomy. IBERDROLA has developed another prototype of this system. This new project consists of a 1.5 kWp portable photovoltaic system, in which one of the most important aspects is the design of the system, to facilitate its transfer by any conveyance. Due to its small dimensions and weight, this system is suited to be moved even by plane.

3. **Photovoltaic Fence. Architectural and sonic barrier to diminish visual and sonic impact.**

This project consists of the construction of a 8.6 kWp photovoltaic fence in the TDC (Technology Demonstration Centre) which IBERDROLA has on the outskirts of Madrid, at San Agustín del Guadalix. The TDC is a place for testing and demonstration of new energetic technologies, from Renewable Energy Area. The best solution in order to solve the problems of noise and visual impact produced by purification water tanks and auxiliary systems, was the design and construction of a photovoltaic fence connected to the grid, surrounding these tanks. Therefore, with the construction of the photovoltaic fence, the following objectives are attained:

- Achieving a major integration of a TDC installation in its natural environment.
- Promoting PV building integration by using the photovoltaic module as a construction element, integrating it into an autochthonous vegetation coated fence.
- Avoiding the noise produced by the operation of the systems and the equipment described above.

4. CAPACITIES

The potential of TDC is based in three fundamental aspects:

1. - The team, distributed among central offices located in Hermosilla street and the TDC in San Agustín del Guadalix, is completely qualified to carry out all kinds of

advisory or consulting related to photovoltaic energy, both to the company itself and to external entities.

IBERDROLA has long been involved in developments dealing with PV technology. Since 1985 an experimental solar photovoltaic plant of 100 kW power is in operation, be coming the most important grid-connected photovoltaic power station at that moment in Europe. Participation in this project, and the continued work accomplished during following years in novel international and national projects related to PV technology, has endowed the TDC personnel with a complete knowledge of PV solar energy.

2. - The computerised and technical equipment available in the workplace in the TDC is complete and actualised to carry out, with great efficiency, works related to a technological demonstration centre, that wishes to publish as far as possible the results of testing.

This specialised equipment is capable of carrying out, among others, the following tasks:

- Solar photovoltaic facility control and monitoring, state verification and mending of panels, converters, and others.
- Data acquisition, analysis and treatment.
- Technical report elaboration.
- Solar photovoltaic plant state monitoring, as well as other projects via modem.
- Data exchange with other investigation centres, manufacturers and Universities
- National and international data analysis access.

3. - Availability of great floor extensions permits collaboration in novel national and international projects.

To conclude, The Renewable Energy Area in The Technology Demonstration Centre presents a great ability to carry out the PV activities and projects in which it is involved.

NEW STRATEGIES FOR THE INTRODUCTION OF PV ENERGY APPLICATIONS INTO UTILITY BUSINESS

Alfonso de Julián, Jesús García
Luis Alberto Calvo, Estefanía Reolid (CDT-Renewable Energy Area Technical Managers)

IBERDROLA
Hermosilla 3
Madrid, 28001
SPAIN
Tel: +34 91 577 65 00 Ext. 41440
Fax: +34 91 578 20 94
www.iberdrola.es
cdt@iberdrola.es

ABSTRACT: The main goal of this paper is to show the new strategies and technological priority lines that serve as a general framework to determine a series of strategic R&D objectives for the Power Production Division of IBERDROLA, and this way incorporate photovoltaic solar applications into the market as power generating technologies.
Keywords: Strategy - 1: Implementation - 2: Marketing - 3

1.INTRODUCTION

An open and world-wide market tendency is creating a new socio-economic world scenario, where technology and innovation will have an increasingly decisive role. Innovative societies, that is to say, those with a firm capacity of profitably generating and converting technological advances into final market products and services, will increasingly occupy the leading positions in the socio-economic future. Accordingly, those organisations with a permanent innovative attitude will increase their added value, creating more and more employment and wealth, as well as improving the society's quality of life.

2. POWER GENERATION DIVISION'S PLAN FOR RESEARCH AND TECHNOLOGICAL DEVELOPEMENT

The high level of technical advance attained in classic electric power generation and the overall restructuring- at both national and international levels- that the electric sector is currently undergoing, require the process of technological innovation to become much more selective with its investments. This selective approach will minimise the risks involves while maximising the returns yielded, by means of smooth-running mechanisms for the transfer and incorporation of the technological developments at IBERDROLA Power Generation installations.
That is why IBERDROLA has introduced a process of technological development capable of opening up new areas of expansion for the generation of electricity, through the optimisation of installations and facilities, thus they incorporate technological innovations and develop new technologies for generating electricity like photovoltaic solar energy.
The process of Innovation and Technological Development of the Power Generation Division (DIGEN) at IBERDROLA follows the general methodology defined at the corporate level, which is based on the referred points:

- Specifying a series of technological priority lines that serve as a general framework.
- Defining a co-ordinates set of programmes and actions within that framework.
- Developing these programmes and activities by integrating scientific resources and technologies, through active internal participation, and most suitable external co-operation.

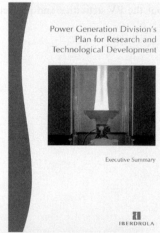

Figure 1: Plan for Research and Technological Development from IBERDROLA

2.1 General objectives
The strategy objectives are developed through the R&D Plan. One of the main objective of this Plan is to structure and ordered set of activities that allow the Power Generation Division's Process of Innovation and Technological Development to function, thus establishing a series of top-priority R&D lines.
The specific aims of these overall objectives are as follows:

- To set up formal channels making it possible for those technological demands or challenges that have arisen in the installation to be formalised and developed for later application in DIGEN,

- thus making their incorporation effective, while strengthening the emergence and development of new ideas.
- To establish R&D activities with a complete vision of the cycle: beginning and ending in the installations themselves, and focusing fundamentally on the needs and improvements of the existing installations; considering the incorporation of those that at some point in the future, and through the assimilation of new technologies, are considered by the Power Generation Division to be worth developing.
- To define a series of activity priority lines with the purpose of guiding the specific development needs and activities, together with the application of the results.

3. STRATEGY ON PHOTOVOLTAIC SOLAR ENERGY

With regard to photovoltaic solar energy, the main objective is to incorporate this type of technology as an electricity generator into the market. For this purpose, new photovoltaic opportunities and technologies must be exploited, and mainly those PV applications that, in a demonstration stage or as emerging technologies, have not been effectively establish in the normal energy use.

To apply this objective, IBERDROLA has established a strategy defined by areas of priority, which includes the needs and specific activities of development, and application of results. These areas are:

- Basic Research
- Characterisation of possible photovoltaic uses and existing applications related to technological developments, defining their state-of-the-art, and thus determining the final direction of this technology.
- Establishment of photovoltaic technology in the market, use of niches in the applications defined in the former stage.
- Planning of an introductory line of the defined applications in the market.
- Defined applications commercialisation.

These steps can be graphically visualised below:

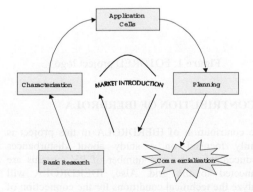

Figure 2 : Photovoltaic actuation lines

These steps are set up contributing to the specific objectives of each of these stages and directed towards their introduction into the energy market. These strategic areas will demonstrate photovoltaic solar energy applicability for different uses in our society, such as photovoltaic sound barriers, building integration, studies which allow eliminate the possible obstacles to entry of photovoltaic solar energy into the market.

Such strategy has been defined as to follow a main line, without losing the global objective perspectives, and have been assigned to demonstrate the applications of photovoltaic technologies.

Accordingly , IBERDROLA, with its Generation Power Division, creates an effective instrument to help solve some of the problems of electric supply: The Technology Demonstration Centre. This facility is believed to be a strategic tool for client satisfaction and the optimisation cycle.

The photovoltaic activities developed in the Centre as well as all related to Research and Technological Development, are developed in the Generation Business Area, both related to profitability objectives and opening new business opportunities (society strategic energy planning forecast), both made not only in conditions of maximum security, reliability, economy and quality, but also of innovation.

In this context, it seems clear that the innovative capacity of IBERDROLA in the photovoltaic field is a priority strategy to assure Spanish industry's competitiveness in the international market.

ACTIVITIES OF IBERDROLA IN FOTORED PROJECT

Jesús García, Alfonso de Julián
Luis Alberto Calvo, Estefanía Reolid (CDT-Renewable Energy Area Technical Managers)

IBERDROLA
C/ Hermosilla, 3ª planta
Madrid, 28001
SPAIN
Tel: +34 91 577 65 00 Ext. 41440
Fax: +34 91 578 20 94
www.iberdrola.es
cdt@iberdrola.es

ABSTRACT: The main goal of this paper is to describe the activities performed by IBERDROLA within the FOTORED project. FOTORED project is developed by a consortium of Spanish Companies and Institutions (IBERDROLA, ENDESA, CIEMAT, REE, UNED-FUE) and its main objective is to study the problems and propose solutions when building-integrated photovoltaic solar systems are connected to the grid on a large scale. The participation of IBERDROLA is mainly focused on a study of disturbances that appear on the grid due to connection of PV systems, and also technical conditions for the connection of pv systems to the grid.
Keywords: Grid-connected-1:Harmonics-2: Safety-3

1. INTRODUCTION

The current situation for photovoltaic solar energy is quite promising in Spain. The photovoltaic energy injected into the grid can be sold at the utility at advantageous prices thanks to a fix subside which is provided by the Spanish Government. Moreover, there are subsides for PV installations which decrease considerably their initial cost.

With these favourable conditions, it is supposed that the number of PV systems connected to the grid will increase in a near future.

The FOTORED project is been developing to analyse the consequences that could appear in the grid and in the final users when a large-scale of PV systems are connected to the grid.

2. FOTORED PROJECT

FOTORED project is achieved with the participation of the CIEMAT, ENDESA, REE UNED-FUE and IBERDROLA with the OCI-CIEMAT funding.

The main objectives of this project are:

a) To study the disadvantages and to propose solutions when building-integrated photovoltaic solar systems are connected to grid on a large scale.

b) To analyze the degree of economic profitability related to photovoltaic generation.

The main expected results of the project are:

a) Electrical Utilities will be able to know the advantages and disadvantages of promoting solar photovoltaic energy.

b) The results of the project will show which applications are the most competitives for grid-conected systems, and which costs must be reduced in order to use pv systems on a large scale.

c) The quality of the wave injected into the grid will be determined as well as the aspects that must be improved to avoid disturbances and interference.

d) According to the consumption profiles of different applications (air conditioning, commercial buildings, industrial buildings, etc.) we will be able to establish, through the results of the project, which applications are more competitive.

e) A general sizing programme for grid-connected pv systems will be obtained.

Figure 1: FOTORED project logo

3. CONTRIBUTION OF IBERDROLA

The contribution of IBERDROLA in this project is mainly focused on a study about disturbances produced when a large number of PV systems are connected to the grid. Also, IBERDROLA will analyze the technical conditions for the connection of the photovoltaic systems to the grid, and the security requirements both for the grid and the photovoltaic systems.

3.1 Disturbances analysis

PV arrays generate DC power. Before connecting a PV installation to the grid it is necessary to have a power conditioning system in order to convert DC power from the PV array to AC power compatible with the electrical utility. Besides, protection systems are required to prevent damage on the PV side due to alterations on the grid and vice versa. All these systems and equipment may produce disturbances and other problems on the grid. If small PV power generation systems become more common it will be necessary to investigate those effects that are not significant for single-inverter systems.

IBERDROLA will make a deep analysis in order to identify the main disturbances and disadvantages which could happen when photovoltaic solar systems are connected to the grid on a large scale. Also possible measures to prevent and/or to correct these problems will be analyzed.

From the point of view of the wave quality injected into the grid, this work will be focused on the following aspects:

- Harmonic distorsion due to grid connection inverters.
- Power factor
- Voltage variations
- Reverse power flow
- Electromagnetic interference.

3.2 Safety aspects

An increase of photovoltaic systems connected to the grid would mean that more people would have access to this type of systems since most of them would be installed in houses or buildings. That is why it is necessary to guarantee safety of the people, both system users annd maintenenance people, when connecting PV systems to the grid.

Issues like islanding operation of the PV system and the perfect isolation of all the active parties of the photovoltaic system that are accesible will be deeply analized. Also, countermeasures to solve the security problems that could happen will be proposed.

On the other hand it is necessary to guarantee the security of the equipment which form the PV array (PV modules, inverters..etc.) during possible variations on the grid and the security of the grid during a malfunction of the PV array. With this objective a study will be presented with the necesarry protections that must be considered when the connection of a PV system to the grid is made.

3.3 Technical conditions for the interconnection.

It is necessary to specify perfectly issues like the connection point, the grounding of equipment, the protections and the security both for persons and equipment when a PV system is connected to the grid.

Photovoltaic systems are a relatively new power generation technology. For this reason, in Spain, a specific regulation for the connection of photovoltaic systems to the grid does not exit. At this moment the Governement and the Energy Sector are preparing a preliminary draft with the guidelines for the interconnection of PV systems to the low voltage grid. After some discussions and modifications a final proposal will be presented and approved.

Within the work that IBERDROLA performs in FOTORED project, an important aspect is to compile existing regulation to connect PV systems to the grid.

4. CONCLUSIONS

With the FOTORED project, the main problems of PV systems grid connection will be identified, and this way, electrical utilities will position whith respect to the photovoltaic market, for the large scale connection of these systems.

Contribution of IBERDROLA is important in order to identify possible perturbances and variations on the grid, and to define the safety conditions when connecting PV systems .

The obteined results will contribute to speed up the use of photovoltaic energy in the Spanish market and to reach the objectives proposed in the White Book (12% RE in EU in 2010).

AN APPLICATION OF TECHNOLOGY DIFFUSION MODELS TO FORECAST LONG-TERM PV MARKET PENETRATION

Andrea Masini*, Paolo Frankl**

*INSEAD[1], Bvd. de Constance, 77305 Fontainebleau, France

tel: +33-1-6071.2529; fax: +33-1-6074.5564; e-mail Andrea.Masini@insead.fr

**Dip.to ITACA – Università degli Studi di Roma I "La Sapienza", Via Flaminia 70, 00196 Rome, Italy,

Tel: +39-06-4991.9014, fax: +39-06-4991.9013, e-mail: frankl@axrma.uniroma1.it

ABSTRACT[2]: this paper examines the diffusion process of PV in southern Europe in the next four decades. Towards this end, it first undertakes a detailed bottom-up analysis of two market segments that are appropriate for PV applications in five European countries. Then it simulates PV penetration under four macroeconomic scenarios. The analysis suggests that already today there are opportunities for PV diffusion in many islands of the Mediterranean region, which may trigger sufficient scale-economies to render the technology competitive in larger markets. The results also show that the diffusion process could be dramatically accelerated through the implementation of carbon-tax policies that support initial penetration. The environmental benefits (net avoided CO_2 emissions over the system life cycle) associated with the forecasted penetration are also evaluated.

1.INTRODUCTION

In spite of their numerous environmental benefits, photovoltaic systems are penetrating the European energy market at a lower rate than what predicted by some forecasts. Many top-down analyses identify the cost gap between the "average" kWh produced by means of non-renewable sources and that generated through PV systems as the primary culprit for this limited diffusion. Furthermore, these studies suggest that the gap is too large to be overcome either through subsidies or through dedicated policy measures aimed at supporting growth (see [1] for an exhaustive discussion of the issue).

The scenarios imaged by this rationale depict the establishment of a *vicious* cycle that hampers the growth of the PV industry. Because of the limited size of the market and of the pessimistic forecasts, module manufacturers suffer from the lack of scale economies and face high production costs. As a consequence, they typically set module prices far above the competitive threshold and force utilities to invest in different technologies. The market share of PV decreases even further, learning-by-doing does not occur and the cost difference between PV and non-renewable sources continue to increase.

Not surprisingly, policy makers at the national and international level are interested in devising strategies that may interrupt this vicious cycle and transform it into a *virtuous* one. The latter would take place when an initial penetration of PV spurs enough productivity increases at the manufacturing level to significantly reduce PV cost and further accelerate market diffusion. Indeed, there is large evidence that this virtuous cycle is already likely to occur if appropriate supporting actions are undertaken and a few inefficiencies that characterize present energy markets are removed. There exist important applications where PV is *already today* a competitive energy alternative and a few others where it is likely to be so in the short run. Thus, well-designed policies that exploit these opportunities may trigger a mechanism of self-sustained diffusion and accelerate PV growth. This sharp difference with the

results of other analyses is due to the fact that top-down studies typically compute the PV "cost gap" by evaluating the average energy cost in a market at a very aggregated level. In so doing they do not distinguish among specific segments with completely different cost structures that are somehow 'hidden" in the aggregated picture. Furthermore, they also often neglect the fact that a technology may become significantly cheaper as cumulative production increases.

Following these lines of thought, this paper aims to demonstrate the feasibility of policies to support large-scale PV diffusion and to quantify the expected rate of penetration of PV systems in southern Europe. Towards this end, it first undertakes a detailed bottom-up analysis of two market segments that are appropriate for PV applications in five European countries (Italy, Spain, Portugal, France and Greece). Then it examines how appropriate policies may be used as a leverage to favor the large-scale development of PV systems and it computes four scenarios of PV diffusion until 2040. Finally, it uses the expected PV penetration rates to compute the amount of net CO_2 emissions potentially avoided.

2.METHODOLOGY

2.1Theoretical background

Many analysts [2] expect PV systems to take over progressively, from those markets - or applications - where they are already a potentially viable technology (e.g. stand-alone applications), to those where only major techno-economic improvements may secure their success (e.g. bulk power production). Both the transition from one segment to the next and the progressive penetration within a single segment are mainly driven by the realization of scale economies and by the occurrence of learning-by-doing at the manufacturing level. As a result of the market growth, PV systems decrease cost and become economically attractive for a larger number of potential adopters

---

[1]The address of Mr. Andrea Masini at INSEAD should be used for all correspondence.

[2] The authors gratefully acknowledge Prof. Robert Ayres for his helpful support and advice.

Following this rationale, we examine the diffusion process of PV and the environmental benefits associated with it in a set of different markets under four different policy scenarios. We focus particularly on two market segments that are likely to be a fertile ground for PV diffusion in the short and medium term, namely: *i*) building integrated systems in small and remote islands with a local electricity grid; *ii*) building integrated systems for grid support.

2.2 Modeling PV diffusion

In each market segment the PV diffusion process is modeled as a sequence of individual adoption decisions made independently by the potential adopters and based on economic rationales that consider the following aspects:

- the current techno-economic performance of PV systems in the *local* market *vis à vis* those of the dominant energy systems in the same market;
- the effect of policy actions that may exogenously modify the relative cost of PV with respect to that of the old technologies.

A detailed discussion of the model is beyond the scope of this paper. A complete description can be found in [4]. In the following we just sketch the logical structure (Fig. 1).In each epoch the model performs the following operations:

- It estimates the cost of PV [1994$/Wp] from the PV learning curve (expected module and BOS cost as a function of cumulative worldwide production).
- It computes the cost of PV electricity [1994$/kWh] in different countries as a function of module and BOS cost, average insolation and possible incentives to PV;
- It estimates the expected PV shipments, based on the difference between the cost of PV electricity and that of the best available technologies for a similar application;
- It uses the cumulative shipments in *both* markets at the end of the year to "update" the learning curve (through a moving average model) and to compute a new PV cost for the following decision epoch.

The two segments retained in the analysis present important differences with respect to both the economic agents involved in the decision process and the technological constraints. Thus, step n.3 of the analysis is performed in a different fashion for the two cases.

Remote island

The potential adopters in this segment are local electric utilities that face a capacity expansion problem (as a result of a demand increase) and consider PV as a viable alternative to the purchase of new diesel generators (typical business-as-usual technology for these applications). PV diffusion is thus the results of a series of adoption decisions made independently by each utility on the basis of a rational evaluation of production costs. At each decision epoch and for each island the model:

- estimates the expected demand increase and the resulting additional capacity required;
- estimates the cost of electricity produced by means of diesel generators;
- computes the cost of PV electricity;
- compares the two costs and "adopts" the most convenient technology subject to grid-stability constraints (total PV less than 20% of total capacity).

The heterogeneous distribution of costs for both PV and diesel generators – due to geographical and economic differences across islands – generates a differentiated adoption pattern (i.e. utilities that are located in southern islands and that face high diesel cost are expected to be the earliest adopters). The analysis includes all the islands in the five target countries with a local grid and a population below 15000 inhabitants. The required data on energy cost, demand growth and local energy consumption were obtained from a previous study on European islands [5]

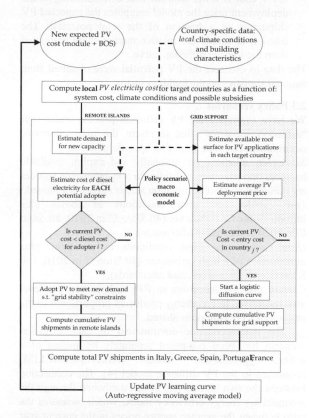

Figure 1: Logical structure of the diffusion model

Building-integrated systems for local grid support

The potential PV adopters for this segment consist of private end-users that use rooftop-integrated systems to satisfy their own electricity demand and eventually sell the surplus to the domestic grid. In this case, the diversity of potential adopters, the fact that they all are private agents who do not behave rationally and the limited diffusion of information render the use of rationale diffusion models meaningless. Thus, we have adopted a different approach.

The International Energy Agency [2] estimates that PV will start penetrating the grid support market when system cost approaches 4 $/W (average PV deployment price for this market), which corresponds to an energy cost of about 0.26 $/kWh[3] under the average Italian climate conditions.

Economic theory also suggests that any technology does not diffuse instantaneously. Rather it follows a typical S-shaped pattern that can be represented, for instance, through a logistic function [5].

---

[3] We refer to constant 1994 US $

Based on this rationale, we have used a pre-calibrated logistic diffusion model to describe expected PV diffusion in the grid support market. The process starts only after the cost of PV hits the IEA threshold for this market segment. Thus, at each decision epoch and for each country the model performs the following operations:

- It estimates available PV roof potential[4];
- It computes the average PV electricity cost;
- It compares current PV electricity cost to the electricity cost corresponding to the PV deployment price;
- If PV cost is lower than the cost corresponding to the deployment price the model computes the expected PV shipments as a fraction of the total potential. The fraction is simply the relative market share computed from the logistic diffusion curve.

The data to estimate the PV potential were obtained from national statistics on the building sector.

2.3 Policy scenarios

We have examined the PV diffusion process under four "diffusion scenarios" that represent the macroeconomic environment resulting from four specific policies.

No-regret: adopters act rationally and exploit available business opportunities. Subsidies to non-renewable sources are removed. Perfect information across agents is assumed.

Low carbon tax: same conditions as above, plus introduction of a carbon tax (40 $/ton C in 2010, 80 $/ton C in 2020 and constant afterwards);

High carbon tax: same conditions as no-regret, plus introduction of a high carbon tax (80 $/ton C in 2010, 160 $/ton C in 2020 and constant afterwards);

Carbon tax and incentives to PV: high carbon tax level plus incentives to PV energy producers proportional to the amount of CO2 emissions abated.

The macroeconomic environment of each diffusion scenario has been computed separately by means of a general equilibrium model (NEXUS-IMACLIM) connected to the PV diffusion model [4]. The connection between the two models has been established through the autonomous energy efficiency index, which measures the ratio between the average energy prices in the current year and the average energy prices in a reference year. In a "no-regret" scenario, the index is set equal to one and the market endogenously determines the expected diffusion pattern for the innovative technology. In a tax scenario, the index changes according to the macroeconomic fluctuations associated to the introduction of the tax. Thus, the micro diffusion models "see" the effect of a carbon tax as an increase in the cost of the business-as-usual technology.

2.4 Environmental analysis

The environmental benefits associated with the diffusion of PV are evaluated in terms of *net* avoided CO_2 emissions. These are computed as the difference between the expected emissions from the European electricity production mix (emission coefficient decreasing from 0.49 to 0.45 kg CO_2/kWh from 1998 to 2040) and the total CO_2 emissions of a PV system during its entire life cycle [6].

3. RESULTS

3.1 No-regret scenarios

As a first remark, this diffusion scenario suggests that already today and even without incentives, there exists a significant potential for PV diffusion in many islands of the Mediterranean region (up to 40 MW per annum in the next 5 years). Many utilities in this area would find economically convenient to adopt PV generators instead of diesel engines to meet new electricity demand. This potential is not currently exploited both because of imperfect information diffusion and because national governments tend to subsidy small utilities that operate in these islands[5]. This is unfortunate, as the diffusion in this initial "niche" is critical for the growth of the industry and the penetration in much larger markets. Indeed, if properly exploited, the potential embedded in Mediterranean islands may trigger enough scale-economies to render grid-support applications economically viable as early as in 2006. Under the same scenario PV is also expected to reach cost competitiveness in the peak power market. However this is not likely to occur before 2020 (Fig. 2 and 3).

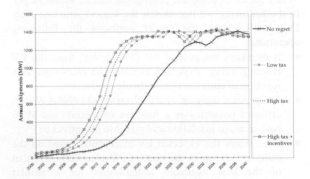

Figure 2: Expected annual PV shipments

3.2 Carbon tax scenarios

The analysis also suggests that the diffusion process could be dramatically accelerated through the implementation of dedicated policies that support initial penetration. The effects of this strategy are not obvious in the short term but they are impressive in the long run. Even a partial support helps the PV industry reach self-sustained competitiveness in the grid support market in the very early days of the new millennium. Scenarios of accelerated penetration clearly show that in this environment PV systems are able to "break even" much earlier than under a no-regret situation. (Fig. 2 and 3) The effect is partly due to the increased costs of the conventional technologies induced by the tax. But it is also driven by the accelerated development of the PV learning curve (Fig. 4).

As a result of the cumulative diffusion, in year 2040 the expected amount of energy produced with PV systems under the low tax scenario is almost three times bigger than under the no-regret scenario. It becomes almost seven times bigger under the high tax scenario.

---

[4] We have adopted a conservative approach and assumed that the installation of a PV system takes place either in a new building designed together with the generator or when an existing one is renovated for maintenance.

[5] The policy aims to attenuate the effect of geographical differences among end-users. By virtue of the subsidy the inhabitant of a remote island would pay the same electricity bill as a user in the mainland with the same consumption.

The environmental analysis provides similar and somehow amplified results, as PV systems implemented earlier exert their benefits for a longer period (Fig. 5).

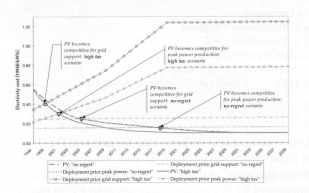

Figure 3: Expected evolution of electricity production cost: PV building-integrated systems vs. non-renewable

It is also worth noting that there exists only a small difference among the three tax scenarios retained, whereas they all differ significantly from the "no regret" case. This confirms the initial hypothesis that PV is already very close to competitiveness in the markets retained: a limited initial support is almost as beneficial as a policy that includes a high carbon tax and massive subsidies to the PV industry. Finally, further and unreported calculations suggest that, if electric utilities were to carry on spontaneously the initial additional investment necessary to favor PV diffusion, they could easily recover the capital through the savings of not paying the carbon tax.

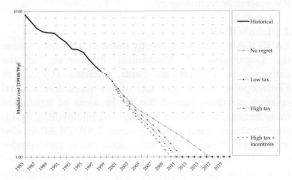

Figure 4: Historical vs. forecasted PV learning curve

4. CONCLUSIONS

This paper examines the expected penetration – and the associated environmental benefits - of PV systems in southern Europe. The analysis focuses on: *i*) building integrated systems in remote islands provided with a local electricity grid; *ii*) building integrated systems for domestic grid support and it is repeated under four different policy scenarios. The results clearly suggest that appropriately designed policies are fundamental to develop a virtuous PV cycle and maximize the long-term penetration of the systems. The results obtained have major economic and environmental implications for policy makers at the national and international level who aim at devising

optimal policies. They are also of interest for electric utilities that want to assess the future PV market potential.

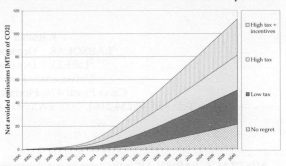

Figure 5: Expected net CO_2 emissions avoided over the system life cycle

5. REFERENCES

[1] "Photovoltaics", P. Frankl, in *Ecorestructuring*, R.U. Ayres, G.C. Gallopin, W. Manshard, R. Socolow, M. Usui, P.M. Weaver eds., The United Nations University Press, Tokyo, 1998.
[2] International Energy Agency: "World Energy Outlook to the Year 2010", IEA/OECD, Paris, 1993.
[3] Ayres R.U. et al., *Integrating technology diffusion micro Models for Assessing sustainable Development policy options*. Final report. EU project: env4-ct96-0292.
[4] Renewable Energies on Mediterranean Islands. APAS Contract RENA CT94-004. Final report, 1996.
[5] Fisher J.C., Pry R.H., *A Simple Substitution Model of Technological Change*. Technology Forecasting and Social Change, 3, 1971.
[6] Frankl P., Masini A., Gamberale M., Toccaceli D., *Life-Cycle Analysis of Building-Integrated PV Systems: Optimal Solutions for Reduction Of CO_2 Emissions*. Proceedings of the "XIV European Photovoltaic and Solar Energy Conference", Barcelona, Spain, July 1997.

SPREADING THE WORD ABOUT PHOTOVOLTAICS IN BRAZIL: "BRASIL SOLAR" AND OTHER PV DISSEMINATION ACTIVITIES PROMOTED BY LABSOLAR

R.Rüther[1,2] & M.M.Dacoregio[1]
[1]LABSOLAR – Departamento de Engenharia Mecânica
[2]LabEEE – Departamento de Engenharia Civil
Universidade Federal de Santa Catarina/UFSC
Caixa Postal 476, Florianopolis – SC, 88040-900   BRAZIL
Tel.: +55 48 234 2161   FAX: +55 48 234 1519   Email: ruther@mbox1.ufsc.br

ABSTRACT: LABSOLAR is the solar energy laboratory at Universidade Federal de Santa Catarina in Brazil. On top of the more fundamental research activities carried out on photovoltaics, LABSOLAR is also interested in the widespread dissemination of solar energy technologies. Since University students are future decision-makers, targeting these students appears to be a promising medium-term strategy for disseminating information on PV. In that context is that in November 1998, LABSOLAR promoted, together with GREENPEACE, a 4-hour PV-powered rock concert. The concert took place on campus and attracted an estimated 25 thousand people (entrance was free). The idea was to demonstrate the concept of both grid-connected and stand-alone PV systems to an audience largely unaware of the potentials of PV power. A 2kWp grid-tied a-Si PV system injecting energy in the public grid since 1997, and a 5kWp poly-Si stand-alone PV system installed next to the stage guaranteed a 100% solar-powered show. The concert started at 6PM while the sun was still high in the sky, and finished at 10PM at night, also to demonstrate that solar energy works even when the sun is not shining. Three bands played one hour each, and during 15 minutes intervals a large screen showed videos explaining the wonders of PV technology to the audience. Images of the concert are available at www.labsolar.ufsc.br. The event attracted considerable media coverage and was national news on television and newspapers, which suited well the objectives of LABSOLAR. The paper details this and other PV dissemination activities promoted by the laboratory, as well as the follow-up of these activities.
Keywords: Dissemination - 1: R&D and Demonstration Programmes - 2: Education and Training - 3

1. INTRODUCTION

While much progress has taken place in photovoltaic (PV) technology in the last decades, with considerable cost-reductions and relatively high annual production growth rates especially in recent years, market penetration remains limited. Total annual production volumes are still small (some 200MWp total world production in 1999 [1]), and economies of scale have yet to be reached. The main reasons for this limited penetration so far are the still high costs of PV and, perhaps more important, the unawareness of PV as a viable energy generation option by most of the potential users. Most people do not even know PV exists. Many others have a slight idea of what PV is about, that it powers satellites and remote communication systems, but are not aware that they could have a commercially available PV rooftop at their homes today, injecting surplus electricity into the public grid. A recent survey in Germany, where one of the most aggressive PV rooftop programs is in place, revealed that not even the financing agents for the German 100.000 Roofs Program know about photovoltaics [2]. Poor information dissemination is therefore a barrier to overcome in making PV a more widespread technology. School rooftop programs in Switzerland, the US and other countries, where grid-connected PV systems are installed on school rooftops with on-site monitoring, are targeting young students, informing future decision-makers that PV is an effective energy generating option. We describe an innovative approach showing how we are also targeting future decision makers, in this case University students in a direct way, and indirectly all the Brazilian society, disseminating information on PV technologies and acting as a nucleating agent for change towards the adoption of PV generation technologies.

2. LABSOLAR

LABSOLAR is the solar energy research laboratory within the Department of Mechanical Engineering at the Universidade Federal de Santa Catarina (UFSC) in Florianopolis, Brazil. On top of the more fundamental research activities carried out in the areas of irradiation measurement, solar water heating and photovoltaic (grid-connected and stand-alone) systems, LABSOLAR is also interested in the dissemination of solar energy technologies to a wide audience. Over 20 million people in Brazil do not have access to electricity, and rural electrification programs, some more and some less effective, are in place in the country, installing PV in solar home systems, battery charging stations and community centres. In urban areas, grid-connected PV has a considerable potential in a country with a huge solar resource and that will need to triple its generation capacity in the next 15 years (57GW to 170GW [3]). The Brazilian energy market is under considerable restructuring, and LABSOLAR intends to be a nucleating agent for change, encouraging the early adoption of clean solar energy technologies. Since University students are future decision-makers in our societies, targeting these students appears to be a promising medium-term strategy for disseminating information on renewables in general, and of PV in particular. In this context is that on Sunday, November 15th 1998, LABSOLAR promoted, together with GREENPEACE-BRAZIL [4] a 4-hour PV-powered rock concert, "BRASIL SOLAR".

3. BRASIL SOLAR

In 1998 LABSOLAR was working on a 5kWp stand-alone project for a hybrid PV-Diesel system to be installed on an island close to Florianopolis, where the University operates a historic site with a 18th century Portuguese fortress. The PV modules, batteries, charge regulators and inverters were already at the University warehouse ready to be installed, but clearance to install the system was delayed due to discussions with authorities concerned on whether the 50m^2 PV array would represent a considerable visual impact on the historic site. LABSOLAR was therefore with an idle stand-alone 5kWp system available, plus a 2kWp building-integrated, grid-connected system, which is operating on campus since 1997 [5-8]. The idea was born, to promote a PV-powered rock concert to showcase PV. GREENPEACE was considered the ideal partner to involve in such an event, and after being contacted, immediately agreed to take part, taking care of all the media coverage and seeking sponsorship to make the event possible. A pop band enjoying great popularity, especially among teenagers, called J-Quest [9], volunteered to fully sponsor the event, and so BRASIL SOLAR was made possible. The concert took place on the main campus at the Universidade Federal de Santa Catarina, which is located in a residential/commercial suburb in the sunny state capital city of Florianopolis (27° S), and attracted an estimated 25 thousand people (entrance was free). The idea was to demonstrate the concept of both grid-connected and stand-alone PV systems to an audience largely unaware of the potentials of PV power. The 2kWp grid-tied amorphous silicon PV system operating at LABSOLAR's building and injecting energy in the public grid since September 1997, and the 5kWp polycrystalline silicon stand-alone PV system installed next to the stage guaranteed a 100% solar-powered show. The concert started at 6PM while the sun was still high in the sky, and finished at 10PM at night, also to demonstrate that solar energy works even when the sun is not shining. Three bands played one hour each, with J-Quest closing the event at night with national TV coverage, and during 15 minutes intervals a large screen showed videos prepared by the LABSOLAR team, explaining the wonders of PV technology to the audience. Images of the concert are available at the laboratory's web site (www.labsolar.ufsc.br), and some are shown in figures 1 to 8 below. The event attracted considerable media coverage and was national news on television and newspapers, which suited well the dissemination objectives of LABSOLAR.

4. FOLLOW   UP,   OTHER   DISSEMINATION ACTIVITIES AND CONCLUSIONS

After the concert the laboratory has been receiving queries by people interested in the PV technology on a continuous basis, and television interviews have been given in many occasions. School science project students visit the laboratory and seek advice for their projects as a result of "Brasil Solar" and the media coverage. A tour LABSOLAR's facilities has also been included in the program of all first-year engineering courses at the University. The topic of PV now attracts more attention in the country and has been national night news on television again when a blackout left some 70 million people nationwide without electricity, evidencing the problems of

a centralised energy generation model. Reports and interviews about LABSOLAR's activities appeared in a number of magazines and newspapers, ranging from specialised building and architecture magazines to a Ministry of Education's publication distributed nationwide to all students in preparation for their University national entrance exam. However, most of the intentions by individuals inclined to install a rooftop PV system so far have failed to proceed due to financial barriers. Together with the further cost-reductions expected for PV, the aspect of financing will have to be addressed soon if this technology is to become popular in Brazil.

4.1 More BRASIL SOLAR

Brasil Solar has sown the first seed of what will hopefully become widespread practice in the near future. LABSOLAR has been approached by a number of Universities in Brazil to team up in promoting events like Brasil Solar at their campi. Students organising annual freshmen events at their Universities are seeking support to promote similar events, and an innovative approach was proposed to them. Each University will seek sponsors to provide funds in order to install 2kWp to 5kWp grid-connected PV systems somewhere on a highly visible spot on campus. One of the proposed system configurations is a PV-covered car park, designed by LABSOLAR in such a way as to enable the structure to be dismantled any time and be mounted side by side with a stage where PV-powered rock concerts will take place. After the show the system would be brought back to the car park configuration on campus, acting also as a long-term showcase of grid-connected PV. In this new concept, the grid-tied PV systems would inject power into the grid during most of the year, accumulating a credit in energy that could be used periodically to make 100% PV-powered events *ad infinitum*.

4.2 Displaying PV to Soccer Fans

Watching soccer games on television and at the stadium is by far the most popular leisure-time activity in Brazil and in many other countries. Until recently, players injured during a match were carried on canvas stretchers. Lately, small electric vehicles have replaced these stretchers in most situations, and here is another innovative way to put PV on display. In the wake of the Sydney 2000 Olympic Games, the so-called "green games", where PV is going to be put on display to all the world in a few weeks, LABSOLAR is aiming to further showcase PV to the wide audience of soccer fans. The idea is to fit the electric cars operating in the field with solar PV modules on their top, and also to fit the covered bench where coach and substitute players sit with BIPV covers that would recharge battery banks for these vehicles. Considering the impact this project can have in terms of dissemination of PV, and the relatively small investments needed to materialise this idea, especially in the multimillion $ world of soccer, LABSOLAR is confident to be able to further promote PV with this initiative.

Figures 1 – 8 (top to bottom, left column first): "BRASIL SOLAR" logo; mounting of the 5kWp stand-alone poly-Si PV system and the stage; part of the LABSOLAR team (undergraduate and postgraduate students), with the PV system and stage in the background; the PV system + stage, hours before the concert started; the second band on stage (day time); the audience at the beginning of the concert (day time); the third band on stage (night time); the estimated 25 thousand people that attended BRASIL SOLAR (night time).

4.3 PV-Powered Solar Electric Vehicles on Campus

Utility vehicles in Brazilian University campi are typically powered by small diesel tractors, to which small trailers are attached. These vehicles, called "tobattas", have a widespread use on campus to carry goods, furniture and trash. At the Universidade Federal de Santa Catarina they are also used to displace organic waste on a compost project carried out by the Department of Rural Engineering. Figure 9 shows one of these tobattas. The nature of the work performed by tobattas at UFSC is such that their diesel engines are kept running many hours every day, but the total distance they cover daily is typically under 20 km. Diesel engines are left running idle while workers load and unload their tobattas, and electric vehicles, powered by batteries fed by PV modules, seem perfect to do this job. Tobattas run around campus all day long, sharing pedestrian lanes with students and staff. All the academic community, being part of every day campus life thus sees these noisy and smelly vehicles. In this context the PV-powered electric tobatta will certainly serve well the aim of making PV more popular and widespread.

Figure 9: One of the nine Diesel-powered tobattas operating on campus at UFSC.

ACKOWLEDGEMENTS

R.Rüther wishes to acknowledge with thanks the Alexander von Humboldt Foundation (Germany), for sponsoring the 2kWp grid-connected PV system; the Brazilian Ministry of Mines and Energy for sponsoring the 5kWp stand-alone PV system; the Brazilian Research Council CNPq for sponsoring the author's attendance at the 2nd World Conference on Photovoltaic Solar Energy Conversion in July 1998, where the idea of promoting BRASIL SOLAR first occurred after watching a PV-powered event in downtown Vienna, and GREENPEACE-BRAZIL for so enthusiastically joining LABSOLAR in promoting the event.

REFERENCES

[1]P. Maycock, PV News, Vol.19(2), 2000, 1.
[2]A. Kreutzmann, Photon, Vol. 3-99 (1999) 3.
[3] www.mme.gov.br
[4] www.greenpeace.org.br
[5]R. Rüther, Proceedings of the 2nd World Conference on Photovoltaic Solar Energy Conversion, Vienna, Austria, (1998) 2655.
[6]R. Rüther, Proceedings of the International Solar Energy Society's Solar World Congress 1999, Jerusalem – Israel, in the press.
[7]R. Rüther & M. M. Dacoregio, Progress in Photovoltaics:Research and Applications, John Wiley & Sons, Vol. 8 (2000) 257-266.
[8]R. Rüther, This Conference (paper OD7.4).

Quality Assurance of Danish PV Installations

Ivan Katic, Bertel Jensen

Danish Technological Institute – DTI, Gregersensvej, P.O. Box 141, DK-2630 Taastrup, Danmark

Phone: +45/7220 2482, Fax +45/7220 2500, e-mail: Ivan.Katic@teknologisk.dk

ABSTRACT: Quality is considered to be an extremely important parameter for photovoltaic systems, first of all because the investment costs are high compared to other energy sources, and secondly because the depreciation of a system may take place over several decades. The Danish government has launched a support programme for PV systems on commercial buildings and recently also a program for apartment buildings and institutions. In order to ensure a good quality of the subsidised systems, it was agreed to set up a quality assurance scheme based on the experience with a similar model for solar thermal systems.

Keywords: Quality assurance, approval, education, certificates

1. INTRODUCTION

Denmark has a firm tradition for development of renewable energy sources, but when it comes to photovoltaics there has been very limited interest until recently. As a result of the global cost reduction and other factors, a certain political will to support PV is now a reality, and a number of demontration projects are supported. When initiating a national PV programme, it was natural to use the experience from the thermal solar energy, which has been supported for a number of years.

As in many other countries, the boom of the solar thermal industry in Denmark came to a sudden end because of a number of faulty systems and installations during the 80's. The situation was drastically improved with the introduction of a quality assurance scheme for thermal solar energy, first for components and since then for installations.

On the financial side, it was found that a blind support pr. system did not result in cost-effective solutions, on the contrary a fixed percentage reduction lead to artificially increased costs because the absolute payment from the state would then increase. As a consequence, the subsidy became variable and linked to expected system performance. This model has been running with success since 1993.

It was thus a natural step to introduce the same model for the PV market from the very beginning. In 1998 a PV quality assurance agreement was signed between the PV industry, installers and utilities with the main purpose to avoid the mistakes and bad reputation of the 80's caused through the malfunctioning of solar thermal systems

The *Quality Chain* may be used to illustrate how many mistakes could occur in the process:

Manufacturer
System house
Sales agent
Installer
Customer

Where is the weakest link?

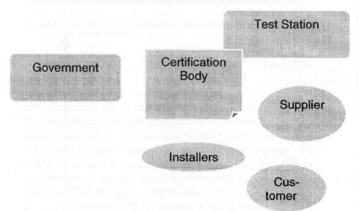

Fig.1. The key players in the Q.A. process.

2. APPROVAL SCHEME

The Danish approval and subsidy system for RE technologies is based on a close collaboration between the Energy Agency (government), Test Stations, manufacturers/ suppliers and installers. The system was set up to ensure that government money is only spent on quality products with a reasonable lifetime and performance. Another aim is to improve the products through a dialogue between manufacturers and researchers.

The scheme is divided into a component and an installation part:

Components: In order to gain approval, all inverters and modules must first of all show a 5-year guarantee and fulfil relevant standards. Only a few inverters with 5 years guarantee are
on the market today, however, it is just as important as ever to put pressure on manufacturers because the modules can now be bought with a 20-25 year guarantee. For a customer it is hard to see why the inverter should have a much shorter period of life. Components are checked with respect to efficiency and durability. The type-approved

components are listed in a publicly available catalogue as a service to the customers.

As a part of the component approval, the performance at various loads is found. In a Nordic climate the performance at low and medium irradiance gives a very important contribution to the annual yield as 50% of the energy stems from diffuse light. In order to compare the actual yield from different modules, the *module factor* is introduced. The module factor is calculated from the module efficiency at specific irradiance and temperature as well as the guaranteed minimum performance of the module. In this way manufacturers with a narrow performance interval are favorised.

A good *installation and users guide* in Danish is a demand in order to get the product approved. A technically perfect inverter or module may be damaged due to wrong installation or use.

Fig.2. Inspection of inverter installation.

be in the range 700-800 kWh/kWp. Most of the controlled systems have operated very well, but a few plants had fatal errors:

- Shading from overhang gave a very low performance in some facade-integrated systems.
- Inverter cut-outs gave a reduced number of operating hours because grid-impedance or voltage was outside the acceptance window

In order to improve the knowledge about PV, guidelines for architects and engineers have been issued. Seminars and workshops covering the special aspects of building integrated photovoltaics have also shown very useful.

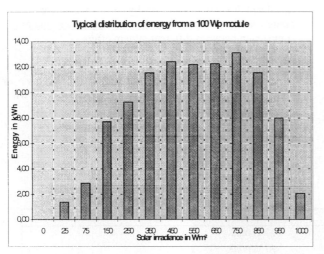

Fig.1. Typical energy distribution under DK operating conditions for a 100 Wp module

Installations: Good product does not automatically result in good installations. Factors such as mismatch, shading or wrong inverter settings may spoil a PV system completely. Therefore, Electricians who want to install PV systems with financial support must follow a 3-day course in a technical school. The course is a mix of theory and practise and is evaluated with a final examination. The compendium contains theory, examples and checklists for the commissioning of a final PV installation.

In order to have a system approved, it must be oriented within +/- 90° from south and have a minimum slope of 15° for self-cleaning by rain. (There have been examples of solar energy plants facing north!) Of course shading must also be avoided, so there should be a free look to the horizon between 12-58°.

From time to time the Solar Energy Centre will carry out spot-checks of installations (PV-police). A portable I-V curve measurement system is used to verify the performance in case of complaints from the customer. As a rule of thumb the annual yield in well-designed PV plants should

3.   SUBSIDY SCHEME

Unlike most other PV support programmes, the subsidies in Denmark are based on the expected performance, not a rated kW or m² number. The intention of the energy dependent subsidies is to push the development towards energy and cost effective PV systems. With a direct relationship between the expected yield and the subsidy, the market should ideally choose the systems with best price/performance ratio in the Danish climate. The expected yield (for a standard configuration) is calculated with a simple model:

Yield = Rated DC power * Module factor * Inverter factor [kWh]

The module factor can be considered as a local climate energy-rating factor. It is calculated as the weighted sum of energy contributions for all bins of solar irradiance and operating temperatures based on the reference year for a south facing roof. For the climate of northern Europe it is obvious that modules with a good efficiency at low insolation levels will have an advantage. It is a well-known fact that the efficiency stated by the manufacturer at STC (1000W/m² and 25°C) does not reflect the efficiency at typical operating conditions.

The module factor is not based on the rated power, but on guaranteed minimum power. This is important, because various studies have shown that the actual power is commonly some 5-15% lower than the rated value. It is also reasonable to use minimum power because the poorest module will determine the output of a string of modules.

$$Module\ Factor = \sum E_i * P_{Guaranteed} / P_{Rated}^{2}\quad [\ kWh/W_p\ pr\ year]$$

where E is the energy contribution for each bin of irradiance.

The inverter factor can be considered as the annual efficiency of the inverter if installed in a "standard system". If there is an internal consumption in dark hours, this is taken into account in the calculation. In order to get a high factor, an inverter for the Danish market should have a high efficiency at low and medium load.

$$Inverter\ Factor = \sum \eta_i * W_i - Relative\ self\ consumption$$

where W is an energy weight factor for each load range.

For AC modules, the test results may be used to calculate a combined module and inverter factor directly.

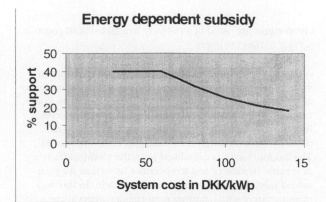

Figure 3. The support percentage is decreased with increasing costs

When the factors are found, the estimated yield can be calculated. It is not a scientifically correct value, but a rating value which can be used as a first estimate, and as a comparison between different system offers.

Currently the subsidy factor is 30 DKK/kWh, but with a maximum of 40% of the actual costs. The factor is currently adjusted, so that the most cost-effective systems wil just reach the maximum limit.

PV module data:							
Nominal power				100 Wpeak			
Guaranteed minimum power				90 Wpeak			
Power temperature coefficient				0,004 pr °C			
Irradiance W/m²	Reference year Hours	Air temp. °C	Module Temp. °C	Power Watt(25°)	Power, Temp.cor. Watt	Energy pr. year kWh DC	Relative Efficiency
0-0.1	4304	6,00	6,00	0	0,00	0,00	0
0.1-50	1297	6,70	7,70	1	1,07	1,39	40
50-100	542	9,10	12,10	5	5,26	2,85	67
100-200	622	10,10	16,10	12	12,43	7,73	80
200-300	416	11,40	21,40	22	22,32	9,28	88
300-400	361	11,20	25,20	32	31,97	11,54	91
400-500	302	12,50	30,50	42	41,08	12,40	93
500-600	244	13,30	35,30	52	49,86	12,17	95
600-700	209	13,30	39,30	62	58,45	12,22	95
700-800	194	14,50	44,50	73	67,31	13,06	97
800-900	152	14,80	48,80	84	76,00	11,55	99
900-1000	95	15,10	53,10	95	84,32	8,01	100
1000-1100	23	11,90	51,90	100	89,24	2,05	100
Total	8761					104,25 KWh	
				Specific yield:		1,04 kWh/Wpeak	
				Module factor		0,94	

Table 1. Module factor calculation for a typical 100 Wp module

4.   GRID CONNECTION REGULATIONS

Fortunately, the utilities have had a positive and pragmatic attitude concerning grid connection of PV plants since the first demonstration projects have shown that there are no practical problems with this.

There is no special law for grid connected PV inverters, but it is agreed to follow a simple set of guidelines until a more formal standard is established:

- The regulations are based on IEC 60364-7-712
- Grid impedance surveillance and external switch is not mandatory.
- Inverters must be connected to their own group in the fuse board.
- In general the sold electricity is paid with about 0,6 DKK per kWh

Practical experience has shown that disconnection due to high grid impedance is a reason for many cut-outs, unless the grid is very strong. It is therefore recommended to skip this function if it gives reason to problems.

The payment for sold electricity is the same as for wind turbines, which is far too little to make PV economically interesting. For most industrial PV plants, the internal consumption supersedes the production, so virtually no electricity is sold to the grid.

For institutions and private installations net metering is now allowed, avoiding costs and bureaucracy. However, a net-sale to the utility seen over a year will not be honoured. It can be expected that a number of pioneer-consumers will invest in small (AC-module) systems in the years to come, because the pay back time is not as important for private households as for commercial enterprises.

5.   SUMMARY

A subsidy and approval system for PV plants has been in operation for a few years now, with the aim to improve the quality and performance of the subsidised systems. It is still too early to see a clear effect of the programme, as the number of installed plants is too low. However, it is remarkable that the annual yield in most PV plants is in the range 700-800 kWh/kWp, which is in level with installations in central Europe. The few errors detected have mainly been in connection with the inverter impedance control circuit.

The financial support for PV plants in Denmark is still in its infancy, but there are several years of practical and organisational experience from the solar thermal sector. It is believed that the positive results can be copied, thus avoiding some of the mistakes that was done previously. Only by keeping a sufficient level of quality for components and installations, can a stable expansion of the PV market be assured.

6.   AKNOWLEDGEMENTS

This publication is based on results obtained under a contract with the Danish Energy Agency.

7.   REFERENCES

[1] B.Jensen & I.Katic : Typegodkendte solcelleanlæg i Danmark – Fabrikanthæfte. Solar Energy Center Denmark 1998

[2] Photovoltaics in Buildings, IEA 1996

[3] B.Jensen & I.Katic : Typegodkendte solcelleanlæg i Danmark – Installatørhæfte. Solar Energy Center Denmark 1998

[4] I. Katic, E. Scheldon, H.Sørensen & O.Rasmussen : Solceller i byggeriet. BPS Katagog 128. Jan.2000

SUCCESSFACTORS IN PV-HYBRIDSYSTEMS FOR OFF-GRID POWER SUPPLY: RESULTS OF A SOCIO-TECHNICAL INVESTIGATION IN GERMANY AND SPAIN

P. Schweizer-Ries, K. Preiser & M. Schulz
Fraunhofer Institute for Solar Energy Systems ISE
Oltmannsstr. 5, D-79100 Freiburg, Germany
Phone +49/761/4588-228, Fax +49/761/4588-217
e-mail: petra@ise.fhg.de

E. Ramirez, X. Vallvé & Ingo Vosseler
Trama TecnoAmbiental (TTA)
Calle Ribollés 46, E-08026 Barcelona, Spain
Phone +34-93-450 4091, Fax +34-93-4566948
e-mail: tta@retemail.es

ABSTRACT: For more than 15 years PV stand-alone systems have been installed in Europe, many of them financially supported by the European Commission. Our target was to investigate only successful ones and to find out what made them a success. The investigation team has studied 20 single and 3 village systems, their history, development, problems raised and solutions found. This paper summarises and explains the crucial success-factors for such PV stand-alone systems in four clusters: technical concept, infrastructure, interactions with its users and cost calculation with financing. Our approach combines technical, social and economic data.
Keywords: Stand-alone PV Systems – Implementation - Sociological

1 THE PROJECT: SUCCESSFUL USER SCHEMES

For more than 15 years, the authors have been installing and investigating PV stand-alone systems world-wide. A European project has now allowed to study some successful European installations. In this study factors have been analysed which made them successful from the point of view of the users, the technology and the economy. Therefore, ten individual installations in Germany and ten in Spain as well as three village power installations in both countries have been analysed with pre-formulated hypotheses and a socio-technical investigation design [1]. This included technical surveys of the systems as well as interviews with users, installers and suppliers.

1.1 What is a successful PV stand-alone system?

In this article we call all the systems PV stand-alone systems that base on standard alternating current (AC). These are often combined with other renewable energy sources or with auxiliary motor generators, that are then often called PV hybrid systems. PV stand-alone systems can supply single houses but can also provide village power supply. They can bring grid-quality power to remote areas 24 hours a day, 365 days a year.

We defined three criteria for the selection of successful PV stand-alone systems:
- The users are satisfied.
- The system works reliably.
- All costs are covered.

In most of the investigated PV stand-alone systems these criteria where fulfilled. In chapter 2 the factors of success, that were found in the successful systems, are explained and summarised.

1.2 Investigated systems

During the summer of 1999, different kinds of individual systems were investigated:
- private users in Germany (households and restaurants)
- remote restaurant of the German Alpine Club (DAV)
- users in Spain that belong to the users association SEBA (Asociación de Servicios Energéticos Básicos Autónomos)

Qualitative and quantitative data were taken from the users and technical measurements from the state of the system. Additionally three village systems have been investigated:
- Flanitzhütte constructed by the German Electricity Utility Bayernwerk
- Llabería belonging to the municipality of Tivissa
- La Rambla del Agua organised by SEBA and a local user association.

For the investigation of these community systems we applied the theories of using a community resource [2].

All the interviews with the users and other officials were recorded. Later on they were transcribed and analysed. Together with the questionnaire and the technical investigations they built the basis for the following results.

2 RESULTS

In the investigations different factors of success could be found, that are clustered in the following four sub-chapters.

2.1 Choose the right technical concept

All German systems have been designed to meet specific user requirements, e.g.:

- automatic start of the motor generator
- load management to avoid overload
- high performance lightning protection

Result: 99,99% reliability.

Most of the investigated Spanish systems have a compact, extensible central unit, where only generators, battery and load must be connected. The systems are easy to install and expand, the on-site engineering effort is low. The systems have an interface for communication with the users.

STANDARD SYSTEM LAYOUT

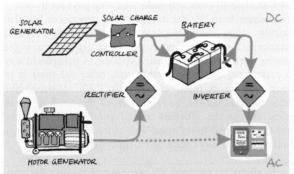

Figure 1: Standardised system layout with modular plug in technology as used in the Spanish systems.

The PV generator with a warranted lifetime of 20 years is the most reliable part of the system. The electronic components such as charge controller and inverter investigated in the European systems were of high industrial quality and efficiency. Lead-acid technology still provides the most economic energy storage. Batteries are easy to maintain. But they are susceptible to wrong treatment, e.g. low state-of-charge over a long time. The key issue is the battery's state-of-charge, which many low cost charge controllers can only estimate. More sophisticated devices feature high precision algorithms for state-of-charge estimation and reliable charge control with automatic operation control.

Looking at the back-up generator, the genset, with the accompanying battery charger, we have to distinguish: In small systems and systems, which are located in regions with high insolation, they need only an emergency generator, which does not necessarily be of industrial quality. In other larger hybrid systems, where the generator runs frequently, industrial quality is mandatory.

To promote a widespread use of PV stand-alone systems, the technical key issues are:

- standardised components
- improved user interface
- simple upgrading

2.2 Build up an infrastructure

All technical systems need maintenance and spare part supply. In the German PV stand-alone systems an average frequency of 0.5 to 2 incidents per system per year were found. The users were able to handle half of them, often by telephone support. Incidents can be reduced by high

technical quality and regular maintenance, but will hardly be reduced to zero. Therefore, a sustainable use of PV stand-alone systems needs an effective support structure.

Regular maintenance is necessary, there are two good reasons: (1) Prevention of faults, e.g. by regularly refilling of the batteries with distilled water or by readjusting charge controller threshold levels (2) Early detection of faults, such as regularly checking of the partial outage of the solar generator due to, e.g. a blown fuse or shading of modules. This enhances the quality of the energy supply and increases the lifetime of the system. Regular and professional maintenance has to be done at least once a year.

ORGANISATION OF SEBA

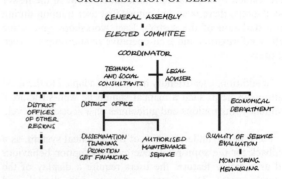

Figure 2: Organisation of the Spanish user association SEBA

Some systems allow remote control for maintenance and checks of recorded data. A technical hotline is available for the German and Spanish systems - a confidence feature users highly appreciate. If anything happens to the systems, the user can call and get the system repaired within a fixed and realistic time frame of 24 hours in Germany and 48 hours in Spain. This has to be included in the contract.

A good contract provides clarity and confidence between all partners. It makes clear who is the owner of the system, who is responsible for the energy service, who deals with maintenance and repair, what service has to be paid for, etc. In the investigated German systems, the energy supply and use is included in the rental contracts, e.g. of mountain lodges. The Spanish users have a standardised contract, which regulates issues such as financial support, warranty, insurance etc.

2.3 Work with the users

Energy demands and financial considerations determine the amount and the availability of power in PV stand-alone systems. The active decision on an appropriate and payable limit is very important. Not all users are willing to accept limits and not all suppliers are interested in making them accept limits. In contrast to grid based electrification, we see a different concept for the stand-alone systems. In successful PV stand-alone systems, a 'Stand-Alone System Culture' has emerged. In this culture, the energy use pattern is in line with the energy supply system. This means:

- The users have a basic knowledge about the technology and its use
- They have a favourable attitude towards the rational use of energy
- The technology is designed to be easily understood

and to communicate with the users, e.g. via displays

- The energy use behaviour complies with the "needs" of the technical system
- A sense of the technology, energy consumption and cost relationship is part of the user's outlook

If the consumers are willing and able to adapt their energy consumption as much as possible to the energy production in terms of time and total power, they can save themselves a lot of money and avoid frustration.

In all success-stories, the users knew their system well. They learned about it while living with the system and they were trained by a competent installation crew. In the newer installations, there is an obligatory first user training during the installation of the system. User workshops give some further information and a co-operative relationship between the users.

In addition, user-friendly handbooks adapted to the local culture can help: such a manual should be well-structured, with many illustrations and guidelines on a accessible level.

For a better understanding of the technical system, as a feedback for the appropriate energy consumption behaviour and as a safety feature, the users require a display of the system's state. The German systems use digital and analogue displays. Different consoles are installed in the control room and in the kitchen. The restaurants have additional displays for the guests. These show the momentary energy production and consumption. The Spanish systems have an extended display in the operation room and a small display installed in the living rooms. The users especially like the possibility of verifying at first hand that the system is running.

For the adapted energy consumption behaviour it is especially important to see:

- the current electricity consumption
- the battery's state of charge and the remaining energy
- the balance between production and consumption

BASIC DISPLAYS

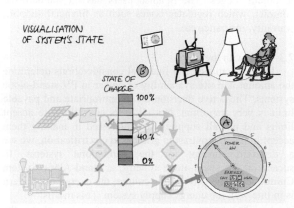

Figure 3: Basic display to support the rational use of energy. It shows the actual power consumption and the available energy

The system needs different levels of visualisation: one with only basic information for the 'normal' users. This must be easy to understand and shows that the system is running. In addition, another level of system monitoring with more measures for technically interested users and the maintenance staff is required.

Besides the use of energy-saving appliances, the users' behaviour is important. In all successful installations the users were aware of the energy consumption of the different appliances. They also knew the best times of energy production, for the Spanish systems these were the sunny hours. For the German systems these were the sunny hours and the running time of the diesel generator.

2.4 Calculate all costs and make sure who pays for it

Discuss from the beginning: Initial investment but also running costs like inspection, replacements, repairs, maintenance and financing costs. We really know now the running costs of the investigated systems. A transparent way to make total costs comparable with other solutions is to describe them in terms of available kWh. Although the latter can reaches the tenfold price of the tariff in urban grids, due to the efficient use of energy in stand-alone systems, 'off-grid users' need only a fraction of the energy amount of 'grid users' for a given energy service.

In all the investigated PV stand-alone systems the grid-connection could not be realised or would have been far more expensive than the system installed. In all of this installations the European Commission together with the local Governments helped to finance the installation costs; the owners paid the rest. As the running costs for maintenance at this remote sites are very high, the most appropriate way for covering this costs has still to be explored.

How to reduce costs

- Standardisation and wide dissemination of technology - regional projects with critical density of systems
- Operation of the system by trained users, resulting in long lifetime at low costs
- High-quality components are more expensive in the beginning, but lower in overall cost
- Good planning of the implementation process

The electrification of urban areas was supported by governments. Rural areas need the same! It is widely accepted to finance infrastructure like roads, telephone lines, electricity and sewage in regions where grid extension is viable. But often there are no concepts for complementing grid extension to supply isolated houses. In the investigated systems, sufficient funding was found for the installation and the running costs. This was possible due to governmental support and the efforts of the users who stood to benefit.

Subsidisation of energy systems can be realised in different ways, such as support on the investment, low interest loans, compensation of costs per kWh produced etc. Some programmes make financing available for the initial investment, e.g. the European Commissions, local governments and NGOs. Soft loans for financing exist as well. Here, private business and venture capital are becoming more and more involved.

In contrast, the running costs are rarely subsidised. For the establishment of rural electrification as a service, it is necessary to link subsidies with the development of sustainable markets. On this basis, rural electrification with solar energy becomes a cost effective alternative, system dissemination increases and costs decrease.

2.5 Village supply systems

All success factors mentioned for the PV stand-alone systems for single house supply are also valid for village power systems. In addition, there are some special tasks to solve in the world of PV stand-alone village power supply that is still evolving.

The most topical question is how to design the system in such a way that the community has the optimum benefit in sharing the electricity available. The synergy of different households can provide the possibility to design smaller systems than would be needed to supply the users with the same energy demand in combining single installations. In these systems, it is important to organise the users to share the energy in an optimal way. Each of the users should have a guaranteed amount of energy. In addition there should be the possibility of sharing energy and power with the neighbour.

We investigated three village power supply systems:
- Flanitzhütte that can demonstrate the virtually unlimited PV stand-alone power supply for a village in Germany. The disadvantage is the relatively high price of the kWh.
- Llabería that shows perfect running of the PV stand-alone system without auxiliary genset but high limitations to the users.
- La Rambla del Agua that has a flexible energy distribution system between the users. With 10 kWp photovoltaic power it can supply the electricity needs of 40 households.

All systems show that the power supply of a small village is possible with PV stand-alone systems. The distribution of information and energy among the users is the most challenging question for the future. Again, this is not only a technical question, but also depends on the user community and the available amount of money for installation and running costs.

3   SUMMARY AND CONCLUSIONS

No technical system is fault-free. The question is how quickly and cost efficiently faults can be detected, analysed and solved. Technical systems need high quality of components and optimised subsystems. Besides high quality of components they need a good infrastructure. Regular maintenance is required. This has to be organised with the users. Contracts can help to clarify rights and obligations of all involved.

The technical system and its users form a unit of energy supply and consumption. If the energy behaviour of the users can be adapted to the supply, the system can be optimised. This can be supported by technical assistance such as system monitors as well as user handbooks and

training. PV stand-alone systems can provide a viable and satisfying possibility for rural electrification.

To receive satisfied users, participation is crucial. This includes decision making as well as financing. But not only users have to be satisfied to make a new technology a full success. In addition to the direct PV users, there are planners, installers, investors and companies which provide the infrastructure supporting the systems.

When all costs are taken into account, PV stand-alone systems are often the cheapest solution for rural electrification. The costs must be clear from the beginning and include running costs. Money has to flow regularly from users, owners and other decision makers. In addition financing models are needed to support discriminated regions.

4   REFERENCES

[1] Preiser, K. & Schweizer, P. (1995) Interaction between PV systems and their users, PV Systems; FhG-ISE; COMETT book; Freiburg.
[2] Gardner, G. & Stern P. (1996) Environmental Problems and Human Behavior. Allyn and Bacon: Boston.

ESTABLISHMENT OF TEST LABORATORIES IN DEVELOPING COUNTRIES – A WAY TO ENSURE QUALITY OF PV-SYSTEMS AND TO SUPPORT MARKET DEVELOPMENT

Klaus Preiser, Jérôme Kuhmann, Tomislav Paradzik
Fraunhofer-Institut für Solare Energiesysteme ISE
Oltmannstrasse 5
D-79100 Freiburg / GERMANY
Phone: ++49-7 61-45 88-2 16
Fax: ++49-7 61-45 88-2 17
e-mail: preiser@ise.fhg.de

ABSTRACT Until only a few years ago, the discussion on photovoltaics for rural electrification was dominated by pilot and demonstration systems, which were financed by bilateral or multilateral co-operation. Today, the situation has changed fundamentally: Many countries have launched major programmes on rural electrification which are explicitly based on photovoltaic systems, e.g. Morocco, Indonesia, India, China or Argentina. The involvement of national and international investors such as the World Bank, oil concerns, electric utilities, banks and insurance companies has grown parallel to this development. Expected returns, risk minimisation, permanent energy services and infrastructure development are the topics which dominate the current discussion. Due to our interdisciplinary approach, which combines technical expertise with sociological and economic competence, we are able to support this market development in many different ways. The quality of photovoltaic systems and their components is a decisive factor for their sustainable dissemination. The market development can only be successful in the long term if investors gain confidence in the capabilities of this new technology. Definition of national and international standards, establishment of local expert centres, building up infrastructure for installation, maintenance and operation of systems, production quality control and authorisation measurement procedures on site - all these are aspects contributing to quality assurance.
Keywords: Qualification and Testing - 1: Rural Electrification - 2: Developing Countries - 3

1. INTRODUCTION

To accompany the currently running and future Photovoltaic activities in developing countries the establishment of centres of expertise is considered to be a key factor for a successful market deployment. To test and develop PV products that appear on the market with regard to their suitability for the foreseen application is hereby an important element. The partners in the market development, like the local PV industries, utilities, national institutions for standardisation, governmental organisations and institutions that are working in the field of solar energy research and last but not least national and international financing organisations and the final users of the technology will profit from the establishment of clear quality rules and quality control of locally produced and imported products and systems.

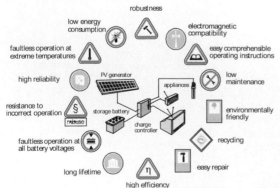

Figure 1: Quality of PV systems depends on various aspects [1].

International standardisation bodies have elaborated for many years norms and standards for PV-systems. These activities have been centred up to now to the PV module, to safety issues and to the definition of the PV specific nomenclature. During the last year, increased activities have been started towards the standardisation of PV-systems and their components, like IEC, CENELEC, PV GAP.

However, besides the institutional framework, the execution of quality control procedures has to be established. Of course there are already some test centres in the industrialised countries, but the same arguments, which speak for the local manufacturing of components are valid also for the test laboratories. Price and proximity to the market accelerate the acceptance of such an institution in the respective countries. The success of local test laboratories relies strongly on the reputation of the staff and the quality of the managers involved. Quality assurance of these institutions (accreditation) can be achieved through foreign testing bodies or international organisations, like IEC, TÜV or others.

To overcome the barrier of high price for low-income customers, commercial credits or financing systems are used all over the world. Therefore financing bodies will play more and more a key role in the dissemination of this new technology into rural areas. A basis for all decisions of financiers is the degree of risk they have to face when entering the playing ground. While national and international bodies have realised, that PV-systems for rural electrification may be good business in the future, today most financiers postpone the decision to get engaged with this new technology. To increase confidence, technical specifications and quality definitions, best represented by an internationally accepted quality seal, as well as positive experience in pilot projects are needed.

Fraunhofer ISE has been working in several countries in the accompanying and/or establishment of such test laboratories, e.g. in the Senegal, in Indonesia, in Morocco, in the Philippines or in Nepal. The detailed experienced gained in these co-operations is used by other countries or international donors.

2. TECHNICAL QUALITY CONTROL THROUGH STANDARDISATION

International standardisation bodies have elaborated for many years standards for PV-systems. These activities have been centred up to now to the PV module, to safety issues and to the definition of the PV specific nomenclature. During the last few years, increased activities have been started towards the and standardisation of PV-systems and their components. Today partly parallel work is done in IEC (International Electrotechnical Commission), CENELEC (Comité Européenne de Normalisation Electrotechnique), PV GAP (Global Approval Programme for PV).

The "Universal Standard for Solar Home Systems" (Universidad de Madrid for EU DG XVII) is an action undertaken to foster standardisation of Solar Home Systems. The World Bank's activities within their "Solar Initiative" firstly developed own specifications for Solar Home Systems and today is more and more co-operating and contributing to the elaboration of the PV GAP quality rules. In the following the different bodies and their activities in the field of photovoltaics are briefly summarised.

2.1 IEC - its role in the field of photovoltaics

IEC is the global organisation for electrotechnical standards. 49 national committees are elaborating continuously standards accepted in more than one hundred countries world wide. The standardisation work in the field of Photovoltaic components and systems has been done mainly in the Technical Committee TC 82. It is subdivided into three working groups:

- Working Group 1: "Glossary"
- Working Group 2: "Modules"
- Working Group 3: "Systems"

In WG 3 is currently circulating a first draft of a standrard with the title "PV Stand Alone Systems - Design Qualification and Type Approval". This draft refers to single house power supply with a maximum size of the PV generator of 1000 Wp and electrical appliances such as light, radio, TV, refrigerator and telecommunication systems. Test procedures shall be described, that will be used for the determination of the electrical and operational characteristics of PV-systems and their components.

Approved IEC standards are passed over to European norms (EN) and national norms, like the German DIN (Deutsche Industrie Norm) and VDE (Verein Deutscher Elektrotechniker).

2.2 CENELEC - its role in the field of photovoltaics

In the frame of the ALTENER programme of the European Commission DG XVII, CENELEC has started since 1996 a technical committee "Solar Photovoltaic Energy Systems", that today deals mainly with the standardisation of Solar Home Systems. The activities towards SHS are currently centred on charge controllers and electronic ballasts.

The structure of CENELEC standards implies, that the definition of standards and test procedures have to be organised under the following three titles:

- Safety
- Performance
- Electromagnetic Compatibility (EMC)

Principally CENELEC and IEC are working closely together, the agreed procedure is, that CENELEC hands over her activities to IEC when IEC steps into a field where CENELEC already is engaged. National standards of European member states like the German DIN and VDE are fed into CENELEC, while accepted CENELEC standards have to be accepted by the national comittees.

2.3 PV GAP - its role in the field of photovoltaics [2]

To accelerate the process of standardisation and to promote quality assurance for PV-systems globally, several manufacturers, financing institutions, like the World Bank and research institutions are joining in the PV GAP (Photovoltaic Global Approval Programme). The central office of PV GAP is located in Geneva, working closely with IECQ (IEC Quality Assessment System). The idea is, that on the basis of existing IEC standards for the qualification of components and systems, new standards are elaborated. For the time, that no generally accepted norms exist, interim "recommended standards" or "recommended specifications"(RS), that should rely wherever possible on national or regional standards, shall be published.

World wide test laboratories, in industrialised and in developing countries, will be identified, that can carry out the relevant tests of components and systems in a reliable and reproducible manner. A reference manual has been published in 1999 and will be further developed in the future.

The final aim of PV GAP is to develop a quality seal that will be given for PV components and systems, tested and approved under PV GAP conditions. This quality seal could be the pre-requisite for the acceptance of components and systems in international calls for bidding. Qualified and approved manufacturers, dealers and installers will be approved for carrying the PV GAP seal in their company logo.

2.4 "Universal Standard for Solar Home Systems" [3]

Under the leadership of the Universidad Politecnica de Madrid in Spain and funded in the European Energy Demonstration Programme THERMIE a proposal for an international standard for Solar Home Systems has been elaborated. Fifteen documents (calls for bidding, specifications and test requirements) have been used as the basis for the development of a catalogue of recommendations for Solar Home Systems. This work does not claim to be an international standard but has been created as a guideline for SHS manufacturers and installers. A discussion of SHS performances provides the line of argument to derive and to classify specifications. These specifications are formally arranged and presented in a way as to facilitate their use in bidding documents for PV rural electrification programmes. The following classification for the specifications has been made:

- Systems
- Components
- Installation

The „Universal Standard for Solar Home Systems" is available in English, Spanish and French through the Directorate General TREN in Brussels.

2.5 Adoption of quality rules in the respective countries

For the acceptance and the adoption of quality rules however, it is decisive to involve the local authorities in the countries themselves in the whole process of implementing quality rules for their rural electrification programmes. Here the basis has to be laid for the formulation of rules, the preparation of the infrastructure desired and the partner institutions to be involved. One example in this experience is our work with the authorities in Senegal, that we carried out on behalf of the GTZ Germany. First a seminar on quality rules and standards was held in Dakar, that led to an agreement on the quality rules to be established. Not until the responsible institutions had decided about the further steps in this process, the installation of the hardware and the training of the staff commenced.

Figure 2: Seminar on standards and quality rules in Dakar/Senegal

3. TECHNICAL QUALITY ASPECTS IN RURAL ELECTRIFICATION WITH PV-SYSTEMS

Certification of components and certification of systems is the needed basis for the qualification of PV stand alone systems for rural electrification. As described above various national and international activities are under way to find globally accepted standards for PV systems. For a successful implementation of PV systems in rural areas however further partners have to be involved. In the following the roles of local manufacturers, local test laboratories and financing bodies are described.

3.1 The role of local manufacturers

The main components of small PV-systems are the PV module, the storage battery, the charge controller, if necessary a DC/AC inverter, the appliances (lamps, radio, TV sets, refrigerators, fans etc.) and the installation material (safety boxes, cables, plugs, sockets etc.). While today and at least in the near future, the PV module will be imported from industrialised countries - in some developing countries the module assembly already started - all the other components are suited for local production. Of course these components have to fulfil the high quality standards as well which are needed for all components in a PV-system. This is the reason why in today's' larger PV programmes for rural electrification often components from industrialised countries are used.

However, there are several reasons for the assumption, that in the near future - and in reality this development has already started in many countries - good quality components will be produced locally. The proximity to the markets, i.e. close matching with the users' needs and desires, the lower price of the components due to the lower expenses for manpower and the better commercial situation (no import barriers), the ability to repair and replace faulty components very quickly and the promotion by local governments show clearly that there is a large potential for local production.

Today many local manufacturers work in joint ventures with industrialised countries which may bring benefits for both sides, quick establishment of a production line of high quality products for the local partner and increased market shares for the foreign partner.

If local producers make wise use of international and national quality rules and quality control bodies, they will increase their market competitiveness considerably.

3.2 The role of local test laboratories

The same arguments, which speak for the local manufacturing of components are valid also for local test laboratories. Price and proximity to the market accelerate the acceptance of such an institution in the respective countries. However, for the acceptance by local and foreign manufacturers, total independence has to be assured. The success of local test laboratories relies strongly on the reputation of the staff and the quality of the managers involved. Quality assurance of these institutions (accreditation) can be achieved through foreign testing bodies or international organisations, like IEC, TÜV or others. As mentioned before the complementation of international standards for local conditions is desirable and very often a prerequisite for the acceptance of these rules by manufacturers and users.

3.3 The advantage of quality control for financing bodies

PV-systems have a high investment price, while the running costs are normally lower than those of the conventional alternatives. To overcome the barrier of high price for low-income customers, commercial credits or financing systems are used all over the world. Therefore financing bodies will play more and more a key role in the dissemination of this new technology into rural areas. A basis for all decisions of financiers is the degree of risk they have to face when entering the playing ground. While national and international bodies have realised, that PV-systems for rural electrification may be good business in the future, today most financiers postpone the decision to get engaged with this new technology. To increase confidence, technical specifications and quality definitions, best represented by an internationally accepted quality seal, as well as positive experience in pilot projects are needed. Here also a wide field opens for organisations of international co-operation and international lending organisations, like the World Bank.

4. TESTING LABORATORIES

Testing laboratories are existing world wide to assure the quality of components and systems. While most of these test centres today are located in Europe or USA, it becomes more and more evident that local test centres for the accompanying of rural electrification programs are created at the place. Besides the lower costs for testing and certification of PV components and systems, the proximity to the local companies and users is one of the strongest arguments for this development. Testing laboratories may fulfil different roles:

PV Testing and Development centres assist manufacturers in the technical design of components and systems. Here quality requirements are elaborated and implemented on the basis of international standards and experience, under consideration, that the quality requirements have to be locally adapted and accepted. Based on agreed testing procedures, components and systems are analysed and recommendations for improvements are transferred to the manufacturer for re-designing.

Often launched by national activities, this class of testing centres has a strong background in pilot or demonstration projects. They realised the need of accompanying manufacturers and installers in their developments and they are today the „back bone" for many national or utility initiatives in rural electrification.

One example of this kind is the national centre CDER (Centre de Développement d'Energies Renouvelables) in Marrakech/Morocco. Closely linked with the national utility ONE (Office Nationale d'Electricité), they are in the centre of all discussions on running and future programmes for rural electrification in Morocco.

Figure 3: Training unit at the national test centre CDER in Morocco

Accredited testing laboratories are needed to certify components and systems according to international or nationally adapted standards. These independent bodies, equipped with high precision measurement technology, have to restrict themselves to testing and certification of components. In principal neither technical assistance nor own development of products and components should be performed to guarantee full independence of these entities.

In ISO 25 the structure, organisation and equipment of such laboratories is defined. During regular audits, the quality control of the laboratory is assured. As the requirements and restrictions of these centres are rather severe, probably only few bodies will be accredited.

Fraunhofer ISE together with the TÜV Rheinland and the Energy Laboratory LSDE in Indonesia are currently finalising their work in the set-up of an ISO 25 accredited test laboratory at LSDE in Serpong/Indonesia. Partly financed by the German government and the Worldbank, this centre will be the nucleus for all future activities in quality control in the huge Indonesian market for rural electrification with photovoltaics.

Figure 4: Training unit at the national PV test centre LSDE in Indonesia

5. SUMMARY AND CONCLUSIONS

Rural electrification with photovoltaic systems has reached a state of development where stable markets with high growth rates call for a professionalisation of the technology. A key role in the maturation of the market has the development of mechanisms for quality control of products, distribution channels and after sales services. On the basis of rules which are accepted in other fields of technology, on the basis of field experience with the new technology, international agreement on standards for PV-system components and on PV-systems is found today. These internationally accepted rules may have to be complemented by local rules which take into account local production and local application conditions.

These rules will be induced on the market through the conditions which energy planning authorities or electric utilities will write into their programmes for rural electrification. These rules will be induced on the market through the conditions which international or national development banks and private investors will impose on the companies which deliver, install or operate the new technology. The rules will have to be controlled through test laboratories which fulfil different functions. Test laboratories have to communicate the rules to planners, financiers, entrepreneurs and customer organisations. They have to control the adherence to the rules, and under certain conditions, they may have to make all the players fit for the game.

Quality control is not an objective in itself. In the end, all measures to maintain PV-system quality for rural electrification serve the aim to increase the satisfaction and the agreement of the final user. The user must get a good value, which means a reliable energy service, for his money. His satisfaction is the central condition for the sustainability of all rural electrification programmes.

REFERENCES

[1] Preiser, K.; Parodi, O.; Quality Issues for Solar Home Systems; 13. EU PV solar energy conference; Nice; France, 1995

[2] PV-GAP Reference Manual, PV GAP Secretariat c/o IEC Central Office, 1998, Geneva, Switzerland.

[3] "Universal technical standard for solar home systems", THERMIE B SUP 995-96, EC-DGXVII, 1998

PROGRESS IN THE UK DTI SOLAR PV PROGRAMME

Henry Parkinson
ETSU, (AEA Technology Environment)
Harwell, Didcot, Oxfordshire OX11 0RA, UK
Phone: +44 (0)1235 433462, Fax: +44 (0)1235 432331, e-mail: henry.parkinson@aeat.co.uk

This paper describes the current government activities relating to Photovoltaics as implemented in the Department of Trade and Industry (DTI) New and Renewable Energy Programme. To set the context the paper starts with a brief review of the UK Government's programme on New and Renewable Energy. Key activities in the programme are then described with a particular emphasis on the application of building-integrated photovoltaics (BIPV).
Keywords: National Programme – 1: Building Integration – 2: Grid-Connected – 3.

1. POLICY AND PROGRAMME BACKGROUND

In 1990, a review [1] of PV power technology commissioned by the then Department of Energy and undertaken by ETSU suggested that PV systems integrated into the fabric of new UK buildings could in the future provide an environmentally benign power source at a cost competitive with conventional sources.

Subsequent to this review of the technology, a small scale PV Programme was initiated by the DTI, implemented by ETSU, as part of the New and Renewable Energy Programme. In March 1994 Energy Paper Number 62 [2] was published setting out the Government's policy at that time.

In 1997 the new Government announced its proposals for renewable energy. As part of this review process a consultation paper [3] was published in April 1999. As a result of the responses received to this consultation paper a response document was published [4] in January 2000 setting out the proposed way forward for the New and Renewable Energy Programme. The aims are set out as:

- Assisting the UK to meet national and international targets for the reduction of emissions including greenhouse gases;
- Helping to provide secure, diverse, sustainable and competitive energy supplies;
- Stimulating the development of new technologies necessary to provide the basis for continuing growth of the contribution from renewables in the longer term;
- Assisting the UK renewables industry to become competitive in home and export markets and in doing so provide employment in a rapidly expanding sector;
- Contributing to rural development.

Some targets for deployment have also been set. Initially, the Government proposes that 5% of UK electricity requirements should be met from renewables by the end of the year 2003 and 10% by 2010. A 10% target for renewables electricity would be equivalent to an additional 2.6 to 3.0 million tonnes of carbon saving for the UK climate change commitments.

This policy fits in with the wider UK Climate Change Programme which was published [5] in draft in early 2000.

2. PROGRAMME CONTENT

2.1 Overview

In the light of these new policies, the full details of the new PV programme are still under development. However, recognising the large number of players with interests in PV, the programme aims are being pursued collaboratively with the utilities, the construction industry and their representative bodies, the UK PV industry, the programmes of the European Commission and relevant programmes of the International Energy Agency.

Work under the programme can be broadly grouped into four areas: building-integrated PV (BIPV), electrical network connection, supporting studies and industrial collaboration. The rest of this paper will focus mainly, but not exclusively, on the BIPV-related aspects of the programme.

As noted earlier, BIPV has been identified as one of the more promising routes towards the more widespread use of PV in the UK. A detailed consultation exercise with industry identified a number of areas for further work on BIPV:
- the need for tested and certified PV building products which are acceptable to the architectural profession;
- improvements in designer's and specifier's knowledge of PV system design, including the development of design tools;
- optimisation of the benefits of PV including the development and validation of systems which improve the overall energy balance of buildings;
- increased number of high profile built examples demonstrating a balance of innovation and best practice;
- production of information material, publicity and marketing strategies which will be developed at appropriate stages of the programme.

As a result of the above, the following broad objectives were formulated to address the specific areas of further work identified:

1. to encourage innovative research into, and the development of, PV building products and to improve the buildability of PV cladding systems including the establishment of testing and certification procedures.
2. to improve understanding of the integration of PV systems into building energy strategies to minimise energy consumption and dependence on fossil fuels.
3. to develop guidelines and design tools for the building industry.
4. to demonstrate PV technologies in a range of building types and disseminate the knowledge gained from the design, construction and monitoring process.
5. to encourage dissemination, information exchange, publicity and marketing of PV as a building component to home and export markets.

The results of this consultation have influenced the projects supported by the programme. (It should however be noted that this focus on BIPV is not to the exclusion of other applications or aspects of PV.)

2.2 Guides and Information

A number of studies have been undertaken or are underway exploring some of the key issues relating to the integration of PV into buildings. These have covered the following areas:

- Survey of Tools for the Design of Photovoltaics Power Systems in Buildings [6]. This concentrates on the available computer packages for the design of PV systems and makes a detailed assessment of those tools available on the market.
- Guidelines for the Testing Commissioning & Monitoring of PV Power Systems [7]. This guide covers the design of a PV system for the facilitation of testing, commissioning and monitoring, procedures, handover documentation for PV buildings with pro-formas included in the report, and guidelines for performance monitoring.
- Planning a PV Building - Guidance for Designers [8]. This aims to provide mainstream building professionals with the necessary information to assist in the planning and preliminary design of PV for new build projects.
- A Study into the Planning Issues Relating to the Integration of Photovoltaic Power Systems in Buildings [9] in consultation with a number of interested bodies including RTPI, the Urban Design Group, the Civic Trust, the House Builders Federation and RIBA.
- Construction (Design and Management) Issues Relating to Building Integrated PV [10]. The aim of this report is to assess the implications of the Construction (Design and Management) Regulations (1994) for BIPV systems and to provide Health & safety guidance for construction site managers.

It is intended that the production and dissemination of this series of guides will improve awareness of PV and promote best practice in its application, leading to an increased number of good quality, well designed installations.

2.3 BIPV Monitoring & Assessment Projects

Some of the most tangible expressions of recent achievements are the built examples of PV building integration. These are few in number at the moment and not all have directly benefited from government support although the research and experience accrued through the Programme and made available to the industry has contributed to their success.

This support has included assistance in the design phases of the project and monitoring of performance following construction. The objectives of Monitoring and Assessment projects vary in detail between projects but generally include the following:

- The sizing, selection and design of the PV system and its integration into the building envelope.
- Construction, integration with the utility grid, testing and commissioning.
- Monitoring of the PV output and other measured parameters.
- Evaluation of performance, cost effectiveness, reliability, customer satisfaction etc.
- Reporting and information dissemination

The following is a selection from projects which are or have recently been the subject of Monitoring and Assessment projects:

University of Northumbria. This building is one of the earliest large scale examples of BIPV in the UK. The 1960s building's southern facade has been reclad using PV modules angled slightly from vertical to maximise winter output. The 456 modules have a capacity of about 40 kWp. The project has been the subject of much publicity and is a landmark in the development of BIPV in the UK. Key results have been reported [11].

Oxford House. The south facing roof of this private house incorporates 48 modules in a 4kWp array. The aluminium section includes drainage and condensation channels and is deep enough to ensure ventilation to the back of the modules. The house also contains many other interesting low energy features and the owner uses an electric car charged from the PV system. Monitoring took place between September 1996 and September 1998 [12]. The system produces around 3000 kWh per year.

Ford Engine Factory. This is currently the largest PV power system in the UK. 26 rooflights covering 25,000 m^2 have been retrofitted to the existing roof of the factory. Each rooflight has clear glazing on the north side while the south side is covered by about 4kWp of large area laminates giving a total capacity of 100 kWp . The project was funded by the Ford Motor Company as part of its 'Factory of the Future' programme, with support from the DTI and the EC Thermie programme.

Doxford Offices. This prestige office development [13] has a 73 kWp facade and is one of the largest in Europe. It is estimated to produce 55,100 kWh per year. The striking design of this building won recognition by the UK Design Council a 'millennium product'.

Equinox building. It is planned to build a 300,000 m^2 development with a south facade incorporating more than 2000 m^2 of PV cells (around 194kWp). The PV facade will be the largest in the UK and will use a new glued structural double glazing incorporating the PV cells. The office development has been designed to be the first low energy fully air-conditioned building in London.

CREED. Growing public interest in a windfarm at Delabole in Cornwall has inspired the creation of a new visitor centre to be built at the site, the Centre for Renewable Energy, Education and Demonstration – CREED. Sympathetically designed to blend into the local countryside, the new building is partially earth-sheltered and incorporates state-of-the-art technology to ensure its minimal environmental impact. The roof structure is designed to integrate PV using both semi-transparent and opaque panels which together will provide 63 kWp. Warm air collected from the back of the PV panels will be used to supplement the buildings heated when required. Excess heat will be stored in a below-ground heat store for both space heating and domestic hot water.

Earth Centre. This project concerns the design and installation of a weather canopy which forms the central entrance point for the Earth Centre, a visitor centre focused on the twin themes of the environment and sustainable development. The canopy, supported on a timber space frame will consist of 1000 m^2 of mono-crystaline modules which are estimated to produce around 69 MWh per year [14]. This will meet around one third of the electricity needs of the adjoining building.

Monitoring is underway, planned or has been completed for 26 buildings covering a range of building types including houses, flats, offices, factories, school & university buildings and visitor centres. The integration has covered flat and pitched roofs, curtain walling, rain screen over-cladding and shading louvres. There is a reasonable geographic spread of monitored examples although there is a preponderance towards buildings in the south.

2.4 BIPV Design Studies

Design studies provide a means for learning about the design of building for the successful integration of PV but without the expense of actually proceeding to construction. A series of around 12 design studies has been undertaken (for example [15], [16], [17], [18], [19], [20], [21]). The studies are based upon the designs for real buildings that are either currently being designed or were recently designed or are subject to a major refurbishment. The general aims of such studies are to consider the following:
- Design - identify the ways that PV influences the design process
- Power Output - matching of the output to diurnal and seasonal demand
- Costs - estimate the costs and benefits of integrating PV into the building

- Building performance - design may affect building form and shading
- Aesthetics - PV may form a very visible feature on a building

The lessons and common themes from these design studies have been pulled together in a recent publication [22] which reviews the integration of PV into sixteen building designs.

2.5 Electrical Connection Issues

A substantial programme of projects has been undertaken in support of electrical connection issues. This work has benefited from collaboration with the electrical utilities and UK participation in Task V of the International Energy Agency Photovoltaic Power Systems Implementing Agreement.

One of the key outputs from this work is a connection protocol for small scale systems [23]. The protocol (Engineering Recommendation G77) is currently in final draft awaiting final approval for publication. These guidelines will make it much easier to understand and meet the technical requirements for small systems (<5 kW) wishing to be connected to the utility network.

3. NEW PROGRAMME INITIATIVES

Three new initiatives for the PV programme were announced in 1999 and these are now being enacted:

1. A call for proposals for projects to develop PV systems and components attracted considerable interest from UK industry and has resulted in the DTI allocating support of about £1 million towards these projects. The projects are led by industry who in most cases provide the majority of the funding.

2. A Domestic PV Field Trial is due to begin shortly and is being co-ordinated by a consortium led by the Building Research Establishment. The aim is to carry out a field trial of at least 100 domestic PV systems using the design, construction and monitoring of the installations as a learning opportunity for the utilities, building developers and other key players in the process. In this way, any significant barriers to the installation of domestic PV can be identified and supporting work to remove these barriers can be defined and undertaken as appropriate. Individual projects within this trial will be selected by a call for proposals with projects starting from the summer of 2000 onwards.

3. A Large Scale Building Integrated PV (BIPV) Demonstration Scheme has been defined with two main objectives:

- To establish best practice for large-scale BIPV in the PV and building industries.

- To use the buildings in the scheme to demonstrate industry capability in BIPV technology and design.

The scheme is anticipated to run for 3-5 years and will provide support for large BIPV projects. Broad consultation with the UK PV and building industry was carried out to define the scheme carefully to ensure that the aims were met and that funding support could be fairly distributed. The next step is to tender for a contractor to manage the scheme and to select the most appropriate candidate. The scheme is due to start later this year. Substantial funding of the scheme will enable significant levels of support to be provided to those BIPV projects that best demonstrate the technology. An initial budget of £3 million has been indicated for the scheme.

4. CONCLUSIONS

Renewable energy use in the UK has more than doubled since 1990, but the UK government hopes to achieve further substantial growth by 2010 and beyond. Initially, the Government proposes that 5% of UK electricity requirements should be met from renewables by the end of the year 2003 and 10% by 2010. Renewable energy is playing an increasing role in energy supply but it will continue to be introduced via a wide range of technologies and time scales depending on local circumstances. The integration of PV into buildings can make a contribution to their energy sustainability, functionality and appearance, particularly if projected reductions in cost are realised. Supporting activities for BIPV in the DTI New and Renewable Energy Programme are helping to prepare the way for the more widespread application of BIPV through the provision of information and development of data and experience. BIPV generation costs will need to reduce significantly before the technology can compete with conventional fossil fuels and some other renewable energy resources. However, unlike many other generation options, PV can be readily incorporated into the urban context and generate power at the point of consumption. The PV industry continues to work hard towards steady cost reductions and this together with the strong environmental signal which PV provides should see the increasing use of the technology in the UK.

ACKNOWLEDGEMENTS

ETSU manages the New and Renewable Energy programme on behalf of the DTI and the support from the DTI is gratefully acknowledged.

REFERENCES

[1] Review of Photovoltaic Power Technology, 1990, E H Taylor, ETSU-R-50

[2] Energy Paper 62, New and Renewable Energy: Future Prospects in the UK, March 1994, HMSO, ISBN 0 11 515384 5

[3] New and Renewable Energy: Prospects for the 21st Century, April 1999, DTI, (URN 99/744)

[4] New and Renewable Energy: Prospects for the 21st Century: Conclusions in Response to the Public Consultation, January 2000, DTI, (URN 00/590)

[5] Climate Change. Draft UK Programme. DETR 2000, Report Code 99EP0850

[6] Photovoltaics in Buildings: A Survey of Design Tools, ETSU S/P2/00289/REP, November 1997

[7] Photovoltaics in Buildings: Testing Commissioning & Monitoring Guide, ETSU S/P2/00290/REP, December 1998

[8] Photovoltaics in Buildings: A Design Guide, ETSU S/P2/00282/REP, 1999

[9] A Study into the Planning Issues Relating to the Integration of Photovoltaic Power Systems in Buildings, 1999, ETSU S/P2/00304/REP

[10] Photovoltaics in Buildings: Safety and the CDM Regulations, ETSU S/P2/00313/REP, 2000

[11] Architecturally integrated grid-connected PV facade at the University of Northumbria, 1997, ETSU S/P2/00171/REP and Investigation of operating characteristics of a PV facade, 1977, ETSU S/P2/00201/REP.

[12] Demonstration Project for a 4kW Domestic PV Roof in Oxford, 1999, ETSU S/P2/00236/REP

[13] The Solar Office: A Solar Powered Building with a comprehensive energy strategy by David Lloyd-Jones in 'Harnessing technology for sustainable development' CIBSE National Conference 1998, ISBN 0 900952 90 X

[14] Architectural Integration of a Large Scale Grid Connected PV Generator into the Weather Canopy of a National Exhibition Pavilion (the Earth Centre), 2000, ETSU S/P2/00293/REP

[15] Outline Design of an Advanced Solar Building for Loughborough University, 2000, ETSU S/P2/00329/REP

[16] BIPV Design Study For Renewable Energy Centre and Eco-Energy House, 2000, ETSU S/P2/00325/REP/1

[17] Building Integrated PV at Anglia Polytechnic University, 2000, ETSU S/P2/00324/REP

[18] BIPV Design Study: Kensall Green Gasworks, 2000, ETSU S/P2/00318/REP

[19] Design Study No 3: Girls' Boarding House, Haileybury College, 2000, ETSU S/P2/00312/REP

[20] Design Study for BIPV Installation, Yorkshire Artspace, Sheffield, 2000, ETSU S/P2/00307/REP

[21] Building Integrated Design Study: Morn Hill, Winchester, 2000, ETSU S/P2/00306/REP

[22] Photovoltaics in Buildings: UK Projects, ETSU S/P2/00328/REP, 2000

[23] Co-ordinated Experimental Research into PV Power Interaction with the Supply Network (Phase 2), 2000 ETSU S/P2/00233 REP/2

"SOLAR ELECTRICITY FROM THE UTILITY" IN SWITZERLAND – SUCCESSFUL MARKETING FOR CUSTOMER ORIENTED PV DEPLOYMENT

Erika Linder*, Stefan Nowak**,[‡] and Marcel Gutschner**

*Linder Kommunikation AG, Gemeindestr. 48, CH – 8030 Zürich, Switzerland
Tel: ++41 1 252 60 01, Fax: ++41 1 252 60 02
Email: zuerich@linder-kom.ch

**NET Nowak Energy and Technology Ltd., Waldweg 8, CH – 1717 St. Ursen, Switzerland
Tel: ++41 26 494 00 30, Fax: ++41 26 494 00 34
[‡]Email: stefan.nowak.net@bluewin.ch, corresponding author

ABSTRACT: "Solar electricity from the utility" is the name of an action within the Swiss National Action Programme Energy 2000, aimed at providing customers of utilities with the service of solar electricity. The fundamentals of the action can be described as a marketing approach towards both utilities and their customers in order to deploy the market for solar electricity for customers willing to buy this product at generation costs. After four years of operation, this action has achieved remarkable results: 90 utilities participate in the action as of spring 2000, half of the Swiss population now has access to this service, 3.4 MWp of photovoltaic power systems have been installed within this concept and 3.5 GWh of electricity are subscribed annually. A marketing survey shows that the market potential for this service is by far not yet saturated and highlights successful marketing strategies.

Keywords: Marketing – 1: Green Pricing – 2: National Programme – 3

1. INTRODUCTION

In the course of the present global liberalisation of energy and electricity markets, the utility industry is undergoing important changes in their way of operating and thinking. Customer orientation is rapidly changing the attitude of utilities towards the market. This represents a new situation for the deployment of photovoltaics, in particular for grid-connected applications. These changes may raise new barriers for photovoltaics but new opportunities may also emerge. Within this development, green tariff schemes, in particular for grid-connected photovoltaic power systems, are rapidly gaining importance [1,2] as they represent a model consistent with the present development in the utility sector.

In 1996, following the initiative and the experience of a few utilities, a new action under the title "solar electricity from the utility" was launched within the Swiss National Action Programme *Energy 2000*, together with the Swiss Electricity Supply Association (SESA). The purpose of this action is to provide utilities with a variety of possible financing models which serve their customers with electricity generated from photovoltaics. The goal of this action is to make solar electricity available to all customers in the service territories of the Swiss utilities. Thereby, an important contribution to the *Energy 2000* goals for photovoltaics is expected.

This contribution describes the principles of this action and the results regarding the deployment of photovoltaics achieved within this concept so far. Moreover, a recent marketing survey is described.

2. CONCEPT

The Swiss electricity supply industry is characterised by a large number of individual companies (a few large ones and many small, in total more than 1200), each of which traditionally served a limited geographic territory. Since 1989 a number of utilities have undertaken projects of grid-connected photovoltaic systems, ranging from small systems to the 560 kWp plant at Mont-Soleil. Many of the systems installed were built and financed by these utilities. A few utilities however started to address their customers by offering them possibilities to contribute to the construction of grid-connected photovoltaic systems. In 1996, 7 utilities provided some kind of solar electricity product or service to their customers.

At that time, the Utility of the City of Zurich introduced the concept of the "solar stock exchange", presented in more detail in [3]. This concept is characterised by

- buying solar electricity at actual generation costs from independent producers with long-term contracts of 15 or 20 years, thereby allowing for adequate upfront financing;

- providing solar electricity to the customers in the service territory of the utility on an annual basis, starting at small energy quantities, e.g. 50 kWh/y, and at actual generation costs;

- a strong marketing effort in order to motivate customers to buy solar electricity, and finally;

- the utility contributing by this marketing effort and taking over the long-term risk to find sufficient customers for the amount of solar electricity purchased from the producers.

Supported by a general federal subsidy programme, this concept proved to be very successful with a rapid deployment of photovoltaic systems in the City of Zurich.

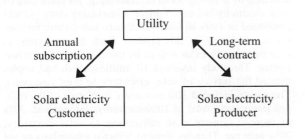

Figure 1: Concept of the "solar stock exchange"

The approach was subsequently extended towards a national initiative within the National Action Programme *Energy 2000*, supported by the Swiss Federal Office of Energy (SFOE) and the Swiss Electricity Supply Association (SESA).

3. MARKETING INITIATIVE

3.1 Approach

This national initiative is called "solar electricity from the utility" and basically represents a marketing approach, both towards the utilities and their customers. The purpose is to provide utilities with a variety of possible financing models which serve their customers with solar PV electricity. The overall goal is to make solar electricity available to all customers in the service territories of the Swiss utilities. The strategy consists in approaching the utilities and offering them instruments to

- evaluate their customers interest and willingness to pay higher rates for solar generated electricity

- identify suitable financing models according to their preference

- support their actions towards the implementation of these models

- support their publicity actions

An important element in the concept is to start with a representative survey of the customers interest in purchasing solar electricity. A comprehensive documentation package serves to support the utilities in this customer survey. Upon evaluation of this survey, the utility has an important decision element whether to start with a solar electricity concept. The next step is to define the appropriate financing concept (see below).

Marketing the product and establishing appropriate communication schemes from the utility towards their customers is a crucial element for success of this concept.

3.2 Financing models

When the utilities are approached, 5 different financing schemes are proposed in order to take into account their preferences:

- **Self-construction:** The utility owns the solar PV system and sells solar electricity at generation costs;

- **Stock exchange:** The utility buys solar electricity from independent producers and sells this electricity at generation costs (see above);

- **Solar pool:** Purchase and sale of solar electricity take place between different utilities and customers buy this electricity at generation costs;

- **Participation:** Sale of participation shares contributing to the investment costs, no sale of solar electricity;

- **Transfer:** High buy-back rates with no separate sale of solar electricity[1].

Mixed forms of these different models have been implemented in many cases.

3.3 Characteristics

The action "solar electricity from the utility" represents a shift from a technology oriented approach towards a market and demand driven operating mode. In this approach, motivation and marketing become the focus of the action. A number of innovative aspects are inherent to this approach:

- focus on customer preference

- the utilities learn to know their customers

- image gain of the utility as a customer-oriented service provider

- customers without the possibility to buy a solar PV system, e.g. tenants, obtain the opportunity to invest in PV

- the model allows to buy in at very moderate levels and is therefore accessible to anybody

4. RESULTS

The action has meanwhile been operational for 4 years. During this time, an increasing number of utilities has started to provide their customers with solar electricity according to one of the models described above. The evolution in time is presented in Figure 2.

---

[1] This financing model has recently been adopted by the new German Renewable Energy Law, setting a present buy-back rate for grid-connected photovoltaic systems of 0.99 DEM/kWh on a nation-wide basis.

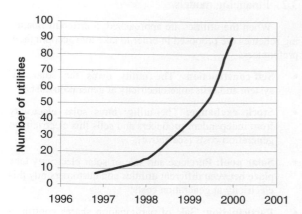

Figure 2: Evolution in time of the number of utilities providing solar electricity to their customers

Due to this growing number of utilities, an increasing amount of solar electricity has been subscribed by the customers in the service territories of the utilities involved. The evolution of the electrical energy subscribed is shown in Figure 3.

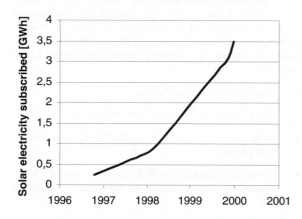

Figure 3: Evolution in time of the solar electricity subscribed within the action "solar electricity from the utility"

In four years of operation, the action "solar electricity from the utility" has achieved in a number of successful results:

- 90 utilities provide solar electricity to their customers (1996: 7 utilities);

- more than 3 Mio. customers (half of the Swiss population) are estimated to have access to solar electricity;

- 21'000 customers participate at a present rate of 0.88 – 1.2 CHF/kWh (0.6 – 0.9 €/kWh);

- 3.5 GWh of solar electricity are subscribed each year;

- 3.4 MWp of installed PV capacity has been built within this model.

5.   MARKETING EVALUATION

The action "solar electricity from the utility" is characterised by a strong focus on marketing. As marketing of solar electricity is a new subject, a marketing study [4] was performed in 1999 in order to discern key factors for successful marketing of this product. Both the marketing as seen by the utilities as well as by the customers was investigated. The study involved 10 utilities which had implemented the service of solar electricity to their customers. The survey included 100 customers from each of the utilities leading to a total of 1000 customers interviewed. 30% of these customers were subscribers to solar electricity, 70% were not. Thereby, reasons both for subscribing or not subscribing were addressed.

5.1. Market penetration

Market penetration is defined as the relationship between customers subscribing to solar electricity and the total number of customers. For the utilities surveyed, this factor varies between 0.1% and 4.4%, the majority being at about 2%. Different factors may affect this market penetration, such as:

- the offer of other renewable energies which may represent a competitive advantage

- the possible range of subscribed solar electricity

- rural or urban environments (urban environments tend to have a higher market penetration)

- the marketing and communication strategies followed

5.2. Marketing approach

There is no uniform marketing for solar electricity. While the basic idea is the same everywhere, product definition, distribution and communication differ considerably and reflect the variety of the Swiss electricity supply industry. The objectives of the utilities for this new energy service range between pronounced customer orientation, image promotion and increased demand or political pressure. A direct relationship between the marketing effort and the market penetration cannot be derived. This can be explained by different operational environments and differing general marketing efforts.

The product "solar electricity" is not uniform and depends on the kind of financing model applied (see 3.2). Large differences were observed between minimum order quantities, possible increments, length of subscription, etc.

Promotion and communication instruments applied differ strongly. The best impact was achieved either with direct mailings or with supplements to the electricity bill. Press communications are an important accompanying measure while communication through the internet cannot easily be linked to the success of the operation. TV and radio commercials are not judged to have contributed strongly so far.

The most important element is the intensity of the marketing. As the product is of "virtual" nature, frequent interaction with the customer becomes more important.

The introduction of the service of solar electricity is regarded as a very useful instrument for the transition towards more customer orientation and increased customer relationship.

5.3. Customers views

As stated above, direct and personal contacts are the most appreciated and successful ways of getting the customers attention for the subject.

The main reasons for the purchase of solar electricity are the consciousness for the environment and an ecological attitude. Further motivation is found in the sustainability, the potential danger of nuclear power, as well as the promotion of new technologies and workplaces. The main barrier is evidently the price of the product. Customers tend to know the product of solar electricity but, in spite of the fact that this product can be ordered in their service territory, 62% do not know where or how to purchase it.

Customers of solar electricity recommend the product further and almost half of them would be willing to increase their order, the main condition being a decrease in price. The importance of the marketing mix for the decision to purchase solar electricity is given in Figure 4.

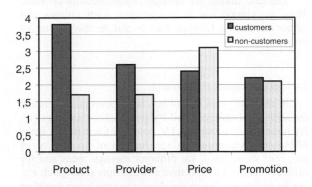

Figure 4: Importance of marketing-mix for customers and non-customers (1=not important, 4= very important)

5.4. Conclusions for marketing

The survey indicated that there is a further market potential for the product of solar electricity; many customers are not yet aware of the possibility to purchase the product. The market potential is estimated to be 25 GWh annually (3,5 Mio. customers, 3% market penetration, 250 kWh purchase volume); the present volume is 3.5 GWh. Successful marketing depends on several elements:

For the product and the customer these elements are

- the offer to **subscribe** to solar electricity

- the sale of a "**solar Swiss franc**" for something good

- a **free choice** of the amount subscribed

- an **annual subscription** with tacit prolongation

For the utility, important aspects are

- direct mailings to customers for information with regular invitation to increase purchase

- direct mailings or supplements to the electricity bill to non-customers

- present the offer on the internet

- regular information

- organisation of visits

- communicate future, sustainability, environment, etc.

6. CONCLUSIONS

The action "solar electricity from the utility" has proved to be a successful concept which is in agreement with the present changes in the utility sector. After 4 years of operation, the number of utilities has increased by more than a factor 10, the installed capacity is presently 3.4 MWp (about 30% of the present grid-connected capacity in Switzerland) and 3.5 GWh of solar electricity are subscribed annually. The marketing survey has indicated that there is further potential for this concept with the environment as main driver and the price as main barrier.

7. ACKNOWLEDGEMENTS

The support of the Swiss Federal Office of Energy (SFOE) and the Swiss Electricity Supply Association (SESA) is gratefully acknowledged.

8. REFERENCES

[1] S. Nowak, S. Rezzonico and H. Barnes, *Buy-back rates for grid-connected photovoltaic power systems: situation and analysis in IEA countries*; Proceedings 2nd World Conference on Photovoltaic Solar Energy Conversion; Vienna, 1998, p. 3365

[2] H. Gabler, K. Heidler and V.U. Hoffmann, *Grid connected photovoltaic installations in Germany – The success story of green pricing and rate based incentives*; Proceedings 2nd World Conference on Photovoltaic Solar Energy Conversion; Vienna, 1998, p. 3413

[3] B. Hürliman, A. Müller, P. Toggweiler and D. Ruoss, *Solar Stock Exchange in the City of Zurich*, Visual presentation VC2.56, 16th European Photovoltaic Solar Energy Conference, Glasgow, 2000

[4] S. Frauenfelder et al., *Erfolgsrezepte für das Solarstrom – Marketin*g, Swiss Federal Office of Energy, 1999

POTENTIAL AND IMPLEMENTATION OF BUILDING-INTEGRATED PHOTOVOLTAICS ON THE LOCAL LEVEL - CASE STUDIES AND COMPARISON OF URBAN AND RURAL AREAS IN SWITZERLAND

Marcel Gutschner* and Stefan Nowak

NET Nowak Energy & Technology Ltd.
Waldweg 8, CH - 1717 St. Ursen, Switzerland
Tel: ++41 26 494 00 30; Fax: ++41 26 494 00 34
* Email: marcel.gutschner.net@bluewin.ch

ABSTRACT: The economic and technical development of photovoltaics is predominantly shaped on the global level, whereas the implementation and the spread strongly depend on local policies and actions. Comparing the potential of available and solar-architecturally suitable roof area between the urban environment of the City of Zurich and the fairly rural Canton of Fribourg, it can be stated that the potential of electrical energy production by photovoltaics to the electricity supply is three times higher in the rural area. Nevertheless, the City of Zurich has deployed its potential area much better. The relative contribution of photovoltaics to the electricity supply is fifty times higher in the City of Zurich than in the Canton of Fribourg. This fact emphasises the local aspects of photovoltaics implementation.

Keywords: Implementation – 1: Building integration – 2: Marketing - 3

1. INTRODUCTION

1.1 Context

The economic and technical development of photovoltaics is predominantly shaped on the global level, whereas the implementation and the spread strongly depend on local policies and actions. Participants in the growing photovoltaics market are increasingly becoming aware of the decisive role of local factors and actors. Sound and comprehensible data are therefore needed for efficient actions and successful strategies.

1.2 Contents

This work presents two case studies - one in a typically urban context, the other in a typically rural context - as well as the comparison of the essential data about the potential of building-integrated photovoltaics (BIPV), the opportunities and barriers of the implementation of BIPV between rural and urban areas and their differences and similarities.

1.3 Objectives

The objective of this work is to provide extensive and concise data about the structure and texture of the existing building stock including specific and regional topics and taking into account technological, economic, financial, social and legal aspects. By means of a new tool to assess the solar-yield differentiated potential and by means of a detailed survey, analysing the local factors, and by contacting the local actors, a complete and precise data set is available.

The case studies are strongly implementation-oriented. They are carried out in co-operation with utilities and authorities in the City of Zurich and in the Canton of Fribourg.

1.4 Methodological Background

In the past years, a tool has been developed to assess the solar-yield-differentiated PV potential on buildings [1]. By means of this tool, the PV potential for different types of buildings (according to age, number of floors, type of use, roof shape, etc.), for different regional areas and for different categories of relative and absolute solar yield, etc. can be discerned. The regional building stock can therefore be precisely characterised taking into account the essential technical data for BIPV. The results are characterised by a high accuracy and a level of detail previously not obtained.

Social, legal, financial and other aspects are analysed and evaluated on the local level. Furthermore, local decision makers are contacted to understand their view and acting regarding the deployment of BIPV.

In this way, data can be collected for each region and then compared between the different regions. Conclusions can be drawn to set a successful framework.

This paper analyses the available and PV used roof area with very good solar yield (high yield: > 90 % of the maximum annual solar irradiation, good yield: 80 % < x < 90 % of the maximum annual solar irradiation) and explains thereafter the differences stated between the urban and rural area.

2. URBAN CASE STUDY IN THE CITY OF ZURICH

2.1 Objectives

This study examined the roof area potential for PV use in the whole building stock and its distribution for eleven different building categories in 1997. The technical and economic developments were considered and potential scenarios formulated taking into account the specific positive and negative factors prevailing in the local context of the City of Zurich [2].

2.2 Method

The study assessed the solar-yield differentiated BIPV potential. 13 typical zones were selected within the City of Zurich, the solar morphology of about 2'500 buildings was analysed in-depth. These extensive empirical data could be combined with very detailed statistical data from the local Office for Statistics. This data base allows almost any kind of analysis and extrapolations relevant for BIPV potential scenarios and future solar energy concepts taking into account factors like property structure, city planning zones, building types, roof shapes, installation size, energy yield, building periods, etc.

2.3 Characteristics of the Area Examined

The City of Zurich is the largest city in Switzerland with 360'000 inhabitants and 92 km^2 of land area (11 km^2 of ground floor area). The electricity consumption is 2.7 TWh/y.

2.4 Results

a) BIPV Roof Area Potential

The BIPV roof area potential is calculated for the existing building stock. Table 1 shows the potential for suitable areas with at least 90 % resp. 80 % of the maximum annual solar yield.

Table 1: BIPV (roof area) potential in the City of Zurich for fourteen building categories.

Building category	Number of buildings	BIPV (roof area) potential (m^2) with high solar yield	%	BIPV (roof area) potential (m^2) with good solar yield	%
Residential buildings (1 unit)	9'877	90'758	7	331'651	7
Residential buildings (> 1 unit)	16'631	952'464	38	1'747'631	38
Resid. + commercial buildings	8'016	374'694	16	758'050	16
Administrative buildings	1'013	197'067	5	236'845	5
Commercial buildings	3'119	416'707	13	595'995	13
Tourism buildings	486	37'904	1	58'996	1
Industrial buildings	1'952	272'730	8	376'673	8
Agricultural buildings	453	26'615	1	33'046	1
Buildings for cultural use	115	19'083	1	29'564	1
Buildings for sports	671	82'112	2	82'112	2
School buildings	722	90'780	3	140'395	3
Hospital buildings	230	10'979	2	99'090	2
Church buildings	234	24'591	1	49'441	1
Other buildings	3'488	71'609	2	78'585	2
All buildings	47'007	2'668'093	100	4'618'072	100

About 20 % of the gross roof area (13.7 km^2) is suitable and has a fairly high solar yield, another 14 % has a fairly good yield. 31 % of the roof area is unsuitable for construction elements and another 22 % for shading, the remaining roof area has to be eliminated purely for an insufficient solar yield.

b) Solar Electricity Potential

The solar electricity production potential with today's "common" technology is 0.27 TWh/y for the high yield roof area and another 0.17 TWh/y for the good yield roof area, that is 0.44 TWh. These values roughly correspond to 10 % resp. 16 % of the actual electricity consumption in the City of Zurich (2.7 TWh/y).

3. RURAL CASE STUDY IN THE CANTON OF FRIBOURG

3.1 Objectives

This study examined the roof area potential for PV use in the whole building stock and its distribution for eleven different building categories in 1998 [3].

3.2 Method

The study assessed the solar-yield differentiated BIPV potential. There are four steps to follow for the collection of empirical and statistical data:

- identification of individual building objects in the data base and on digitized land-registers,
- localisation of individual building objects on digitized land-registers and aerial pictures,
- sampling of roof elements of individual building objects by means of stereoscopes (in 3-D) and
- registration of the sampled empirical data in the statistical data base.

About 700 buildings were selected at random and integrated in a high performing extrapolation structure.

3.3 Characteristics of the Area Examined

The Canton of Fribourg is located in the middle of Switzerland between the mountainous regions of the Alps and of the Jura. Its area is about 1600 km^2 (18 km^2 of ground floor area) and is predominantly rural. The population is 230'000 inhabitants. The electricity consumption is 1.8 TWh/y.

3.4 Results

a) BIPV Roof Area Potential

The BIPV roof area potential is calculated for the existing building stock. Table 2 shows the potential for suitable areas with at least 90 % respectively 80 % of the maximum annual solar yield.

About 25 % of the gross roof area (22 km^2) is suitable and has a fairly high solar yield, another 15 % has a fairly good yield. 20 % of the roof area is unsuitable due to shading and another 15 % due to construction elements, the remaining roof area has to be eliminated purely for an insufficient solar yield.

Table 2: BIPV (roof area) potential in the Canton of Fribourg for eleven building categories.

Building category	Number of buildings	BIPV (roof area) potential (m^2) with high solar yield	%	BIPV (roof area) potential (m^2) with good solar yield	%
Public + administrative buildings	2'927	510'744	9	620'576	7
Residential buildings (< 4 units)	37'405	983'153	17	2'251'150	25
Residential buildings (> 4 units)	1'963	242'863	4	268'192	3
Resid. + commercial buildings	1'919	179'825	3	330'076	4
Agricultural resid. buildings	7'960	554'913	10	1'537'596	17
Other agricultural buildings	8'505	676'126	12	1'029'963	12
Buildings of transport infrastructure	959	109'962	2	118'447	1
Commercial buildings	1'750	517'628	9	517'628	6
Industrial buildings	3'548	989'234	18	1' 090'791	12
Tourism buildings	317	50'209	1	57'722	1
Other buildings	24'764	829'965	15	1'057'181	12
All buildings	92'017	5'644'624	100	8'879'321	100

b) Solar Electricity Potential

The solar electricity production potential with today's "common" technology is 0.56 TWh/y for the high yield roof area and another 0.30 TWh/y for the good yield roof area, that is 0.86 TWh/y. These values roughly correspond to a third resp. almost half of the actual electricity consumption in the Canton of Fribourg (1.8 TWh/y).

4. RESULTS AND COMPARISON

Some considerable differences can be stated by taking the figures concerning the potential (section 4.1) and actual (section 4.2) BIPV contribution to the electricity supply.

4.1 Potential BIPV contribution to the electricity supply

In the fairly rural Canton of Fribourg, the solar-architecturally suitable roof area could be used for the generation of 0.86 TWh solar electricity, which corresponds to 48 % of the actual electricity demand.

In contrast, in the City of Zurich, the roof area of the same quality could be used for the generation of 0.43 TWh solar electricity, which corresponds to 16 % of the actual electricity demand.

The potential BIPV contribution to the electricity supply (by simplifying other factors interfering in the distribution of solar electricity) is thus three times higher for the rural area. The most important reason for this is the ratio of ground floor area per inhabitant.

Table 3: Factors influencing the potential BIPV solar energy contribution to the electricity supply for the City of Zurich and the Canton of Fribourg (selection)

Factors influencing the potential BIPV solar energy contribution to the electricity supply	City of Zurich (urban area)	Canton of Fribourg (rural area)
Electrical use intensity (annual electricity use in relation with ground floor area)	243 kWh / m^2	99 kWh / m^2
Available ground floor area per capita	30 m^2/ capita	82 m^2/ capita
Utilisation factor I (architecturally suitable roof area with high solar yield per ground floor area)	0,25 m^2 / 1 m^2	0,30 m^2 / 1 m^2
Utilisation factor II (architecturally suitable roof area with good solar yield per ground floor area)	0,45 m^2 / 1 m^2	0,50 m^2 / 1 m^2
Maximum annual solar irradiation on best oriented surfaces	1167 kWh / m^2	1250 kWh / m^2

4.2 Actual BIPV contribution to the electricity supply

The actual BIPV solar electricity production is less than 20 MWh in the Canton of Fribourg, which corresponds to 0.001 % of the actual electricity demand.

The actual BIPV solar electricity production is around 1300 MWh in the City of Zurich, which corresponds to 0.05 % of the actual electricity demand.

Table 4: Favourable aspects for PV implementation in the City of Zurich (selection)

Favourable aspects for PV implementation in the City of Zurich
• Information / marketing
• Image / attitude with utilities, authorities, customers, etc.
• Financing schemes (e.g. solar stock exchange)
• Perception and conception of added value

The actual BIPV contribution to the electricity supply (by simplifying other factors interfering in the distribution of solar electricity) is fifty times higher for the urban area. Most important reason for this is the very favourable environment in the City of Zurich including the well-known solar stock exchange [4,5], pro-active marketing by the utility, hardly any legal construction restrictions, political and economical commitment and appropriate financing schemes.

5. CONCLUSIONS

Some conclusions can be drawn on the methodological and result level:

- in-depth analysis of the building stock with the newly developed tool to assess the solar-yield-differentiated BIPV potential yields highly accurate results
- evaluation of social, legal, financial and other aspects on the local level in order were analysed to discern particularly positive and negative factors for the implementation of BIPV
- comparison of the BIPV potential and implementation is made between two regions - one in a typically urban context, the other in a typically rural context
- useful data for the actions and strategies of the decision makers in and around the PV domain were provided

Two case studies were carried out in Switzerland, one in an urban context, the other in a rural region. These case studies allow to show similarities (possible thumb rules to determine the BIPV potential) and differences of the building stock and of the basic conditions between urban and rural areas and to bring out favourable and unfavourable basic conditions for the implementation of BIPV on the local level.

It can be concluded that e.g. the potential contribution of BIPV to the electricity supply of the Canton of Fribourg is three times as high as in the City of Zurich but the actual contribution of BIPV to the electricity supply of the City of Zurich is fifty times as high as in the Canton of Fribourg. These facts show clearly how much local factors and actors matter for the potential and, above all, for the implementation of BIPV. The results generated are finally of use for actions and strategies of decision makers in and around the PV field.

ACKNOWLEDGEMENTS

The support provided by the Utility of the City of Zurich and the Department for Transport and Energy of the Canton of Fribourg for the realisation of this work is gratefully acknowledged

REFERENCES

[1]   M. Gutschner, S. Nowak, *Approach to assess the solar-yield-differentiated photovoltaic area potential in the bulding stock;* Proceedings 2nd World Conference and Exhibition on photovoltaic solar energy conversion, Vienna, 1998

[2]   NET, S. Nowak, M. Gutschner, *Das Photovoltaik-Potential im Gebäudepark der Stadt Zürich, Teil 1: Analyse des photovoltaischen Flächenpotentials, Teil 2: Analyse der technischen, wirtschaftlichen und rechtlichen Indikatoren, Teil 3: Energetische Bewertung des Photovoltaik-Potentials;* October 1998

[3]   NET, S. Nowak, M. Gutschner, *Photovoltaisches Potential im Kanton Freiburg – Analyse des photovoltaischen, ertragskriterium-differenzierten Flächenpotentials im Gebäudepark des Kantons Freiburg;* October 1998

[4]   B. Hürlimann, P. Toggweiler, et al, *Solar Stock Exchange in the City of Zurich,* Visual presentation VC2.56 at 16th European Photovoltaic Solar Energy Conference and Exhibition, Glasgow, May 2000

[5]   E. Linder. S. Nowak, *"Solar Electricity from the Utility" in Switzerland – Successful Marketing for Customer Oriented PV Deploymen;* Visual presentation VC2.16 at 16th European Photovoltaic Solar Energy Conference and Exhibition, Glasgow, May 2000

AUSTRALIAN INITIATIVES IN PHOTOVOLTAIC ENGINEERING EDUCATION

S.R. Wenham, C.B. Honsberg, J. Cotter, T. Spooner, M. A.Green, M.D. Silver,
R. Largent, A. Bruce, A. Aberle and L. Cahill
Centre for Photovoltaic Engineering
University of New South Wales
Sydney NSW 2052, Australia
Ph: +612 9385 4018
Fax   +612 9662 4240
Email: s.wenham@unsw.edu.au

ABSTRACT: The Australian Government is responding to the growing concerns for the environment with a range of new initiatives. In the educational area, it has established a Key Centre for Teaching and Research in Photovoltaic Engineering with the primary new initiative being to develop and establish new undergraduate engineering programs in photovoltaics and renewable energy. The first new degree commenced in March 2000 at the University of New South Wales. The second undergraduate degree implementation is in Western Australia at Murdoch University. Despite these two institutions being located on the opposite coastlines of Australia, many of the educational units developed at these two institutions will be shared by making them available on-line via the Internet for cross-enrolments. Both programs introduce a range of teaching innovations including exposure to state of the art commercial and high performance solar cell technology, the use of interactive multimedia CD ROMs, Internet delivery, the implementation and use of a virtual production line for solar cell manufacturing and "hands-on" work with photovoltaic and other renewable energy systems. The programs will train engineers in all aspects of photovoltaic and renewable energy engineering. Excellent industry, end-user, government and host institution support (cash and in-kind), is being provided.
Keywords: Education and Training – 1: Tertiary Studies – 2: Engineering – 3

1. INTRODUCTION

The photovoltaic (PV) industry has been growing at a rate of 30% per annum [1], which is faster than the computer or telecommunications industries. Figure 1 shows the explosive nature of the PV industry growth. These soaring growth rates are predicted to continue as a new market, grid-connected PV on residential houses, develops. Government initiated plans throughout the world have already been formulated for the implementation of at least three million additional houses to be powered by solar cells during the first decade of the new millennium. 1.5 million houses are targeted for Japan, 1 million houses for Europe, and a further 1 million houses for America. Governments have demonstrated their willingness to offer whatever subsidies are necessary to ensure these targets are met, particularly in Europe and Japan.

In response to the booming PV market, many manufacturers globally are rapidly increasing their production capacity. Australian manufacturers currently enjoy almost 8% of the international market, a figure that is expected to increase in the future as state of the art Australian technology enters the market place.

The economies of scale that accompany the increased production of photovoltaics causes further reductions in the cost of PV, which in turn promotes additional growth. Figure 2 shows historically the relationship between the cost of photovoltaic modules and the corresponding installed capacity or market size. If this straight line relationship continues as expected, then the reduction of the photovoltaic module price to its apparent long term potential of under $1 /Wp, could lead to photovoltaic markets expanding by more than a factor of 1,000. International studies predict an expansion of more than a factor of 20 over the coming decade [2].

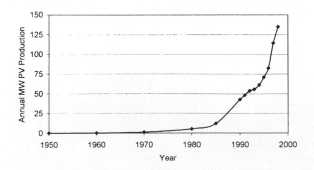

Figure 1: Annual production of PV modules since 1950.

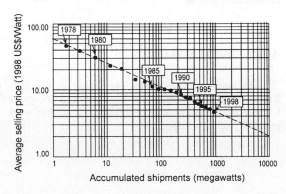

Figure 2:   Average selling price for PV modules as a function of cumulative installed capacity.

2. JOB CREATION AND EDUCATIONAL REQUIREMENTS

The rapidly expanding PV industry creates the need for PV engineers. International studies [2-4] indicate that approximately 50 new jobs are created for each 1 MW per annum increase in production capacity of photovoltaics. Based on present growth rates in the industry, this indicates that hundreds of thousands of jobs will be created in the photovoltaic sector alone during the next decade, with about 20% of these in manufacturing. In any booming industry, challenges are created for the educational sector due to the time lag associated with educating students through programs that often last at least four years. For the photovoltaics and renewable energy sectors, additional challenges arise due to the need to develop entirely new programs that do not currently exist. In Australia, the Federal Government and several universities are taking the initiative to ensure appropriate undergraduate engineering degree programs in photovoltaics and renewable energy engineering are developed.

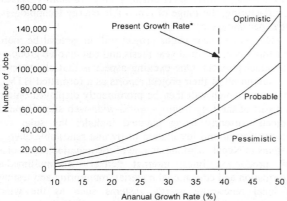

Figure 3: Predicted international job creation in the photovoltaics sector by the year 2004, when the first Photovoltaic Engineers graduate from UNSW.

To better understand the educational requirements, the most detailed international job study appears to be part of the 1996 European Green paper adopted by the European Parliament and subsequently expanded into a White paper. This paper cites a study showing that, for the photovoltaic sector alone, well in excess of 100,000 jobs will be created in Europe by 2010 [2], while for the broader renewable energy sector, hundreds of thousands of jobs will be created during the same time frame. In addition, in 1998 the Austrian Federal Minister for the Environment publicly announced the results of a study [3] that predicted 30,000-80,000 new jobs would be created in Austria alone in the photovoltaic sector by 2010. Similar types of studies have been carried out throughout the world, with similar types of conclusions drawn with regard to job creation. Using data from these studies, Figure 3 shows that the likely international job creation in the photovoltaics sector by the

year 2004, when the first Photovoltaic Engineers graduate from The University of New South Wales (UNSW), is about 50,000-60,000 new jobs [4]. Many of these will be engineering positions.

3. PHOTOVOLTAIC AND RENEWABLE ENERGY ENGINEERING JOBS IN AUSTRALIA

In New South Wales (NSW), Australia, the "Green Power" scheme provides electricity consumers with the option to pay a premium for their electricity but have it generated from "green" environmentally friendly sources such as photovoltaics or wind power. For example, energyAustralia allows consumers to pay a premium in the range of 0-40% for their electricity with a corresponding guarantee that 0-100% of that consumer's electricity will be effectively generated from environmentally friendly sources. In recent years, approximately 15,000 new subscribers per year have elected to join this scheme. This type of public support is also apparently experienced elsewhere throughout the world, providing significant opportunities for the PV industry to grow through applications which would not normally be considered economically viable.

A media release [5] in July, 1999 from the Minister for Energy, Mr. Kim Yeadon, announced that the growth in the *Green Energy sector* (which includes all aspects of photovoltaic engineering including the use of all renewable energy technologies and energy efficient building design), is outstripping that of the booming *Information Technology* industry in the state of NSW. In addition, the expected job creation in the *Green Energy sector* for the next 12 months is predicted to be approximately 1,200 new positions. Perhaps just as importantly, the same study revealed that 1,000 new jobs have already been created in the sector in the previous 2-3 years.

Unfortunately, as identified by many manufacturers and end users, limited educational opportunities exist for engineers to gain the necessary training and qualifications to suit the needs of the rapidly expanding photovoltaics and renewable energy sectors. For example, Western Power, who owns all the electricity grids in Western Australian, has found it impossible to find appropriately trained engineers for its rapidly expanding use of renewable energy technologies such as wind power and photovoltaics. Western Power has consequently taken the initiative to fund the establishment of the new undergraduate engineering program at Murdoch University specifically addressing this need. In NSW, photovoltaic manufacturing is particularly strong with almost all of Australia's manufacturing capacity being based in Sydney. A similar situation exists whereby appropriately trained engineers are unavailable. The local photovoltaics industry has been drawing heavily on graduates from Electrical Engineering at the University of New South Wales and then subsequently facilitating additional training for these graduates to equip them as photovoltaic engineers. Of the 137 graduates from Electrical Engineering in 1998 at UNSW, 3 of the top 5 students entered the local photovoltaics industry.

4. NEW ENGINEERING DEGREES IN PHOTOVOLTAICS AND RENEWABLE ENERGY ENGINEERING

In Australia, the Federal Government, through the Australian Research Council (ARC), has taken the initiative, following strong support from industry and end-users, to establish new undergraduate engineering degree programs in the photovoltaics and renewable energy areas. The mechanism for achieving these outcomes has been through the establishment of a Key Centre for Teaching and Research in Photovoltaic Engineering, one of only eight such Key Centres established Australia-wide across all disciplines. The ARC chose the UNSW team of academics as the ones to develop and implement the new educational programs. This team of academics has led the world for 15 years in terms of solar cell performance records but has also developed the most successfully commercialised new photovoltaic technology internationally over the same period. The new educational program at UNSW is the world's first undergraduate engineering degree in *Photovoltaics and Solar Energy* and commenced in March, 2000. The new engineering program at Murdoch University commences one year after the program at UNSW and takes advantage of many of the subjects developed at UNSW that are able to be offered for distance learning via the Internet. Strong industry and end-user support for these programs is also being provided, particularly from Western Power, BP Solarex, Pacific Solar, the Australian CRC for Renewable Energy, the Sustainable Energy Development Authority, the Alternative Energy Development Board, etc. These new programs give Australia an internationally leading position in the educational area.

5. COURSE CONTENT

The Photovoltaic and Renewable Energy Engineering programs cover a broad range of engineering tasks and disciplines, but can be summarised into five main areas.

These are:
♦ System design (computer based), modelling, integration, analysis, implementation, fault diagnosis and monitoring;
♦ Device and system research and development;
♦ Manufacturing, quality control and reliability;
♦ Policy, financing, marketing, management, consulting, training and education;
♦ Using the full range of renewable energy technologies including alternate electricity generating technologies (such as photovoltaics, wind, biomass, micro-hydro, solar thermal, geothermal, wave and tidal energy), complementary areas (such as balance of system components, fuel cells, batteries, interface electronics, etc) solar architecture, energy efficient building design and sustainable energy.

In the UNSW program, students also choose a second area of specialisation to study during years 2 and 3. In many cases, these second areas of specialisation can be built into double degree programs through an extra fifth year of study.

In general, the teaching in these programs contains a range of innovations. These include: exposure to state of the art commercial solar cell technology; unique insights into the design and fabrication of the world's most efficient silicon solar cells; the use of interactive multimedia CD Roms; internet delivery; the implementation and use of a virtual production line for solar cell manufacturing; and "hands-on" work with photovoltaic and other renewable energy systems.

6. UNDERGRADUATE PROJECTS

Major projects are taken during the second year and fourth year of the program. The final year thesis can be taken in virtually any area encompassed by the Photovoltaics and Renewable Energy sectors. In particular, the world class photovoltaic laboratories are well suited to thesis work in the device area. However, many students may prefer thesis topics encompassing system design, applications, device and system modelling, environmental issues, balance of system components, control electronics, policy, reliability issues, manufacturing, the range of renewable energy technologies, life expectancy, etc.

The second year major project will in general be more structured than the final year thesis and can involve group or individual projects. One exciting aspect is that all students will submit all of their project reports as a formatted HTML document, which will then be prominently displayed online as a part of the Key Centre's world-wide-web site. Typical projects currently being planned include: the solar car project; design and installation of pv and renewable energy systems in Nepal; working on one of the manufacturers solar cell production lines; internet delivery of multimedia information and educational material; monitoring and testing of local renewable energy systems such as the wind generator at Malabah or the pv powered light house at Montague Island; and involvement with Sunsprint, the state and national Model Solar Car Racing competition.

7. ENROLMENTS

The quota for the new program was set at 30 for the year 2000. High demand, however, has led to the enrolment of 41 students for the first year of the program, with most students having University Admission Index (UAI) scores well above 90.

8. CONCLUSIONS

The booming pv industry is creating jobs and the need for new educational courses to appropriately train engineers. The UNSW has initiated the world's first undergraduate degree in Photovoltaics and Solar Energy, with strong support from government, industry and end users. In March 2000, 41 students have enrolled into the first year of this four year program. Most of these students have University Admission Index scores well above 90 making this one of the more prestigious degrees offered by UNSW. This is indicative of the popularity of the new program. A similar program at Murdoch University, Perth, is being developed in Renewable Energy Engineering for implementation in 2001. This is being developed in conjunction with UNSW with many subjects to be offered to students in both institutions via the internet.

The new programs make use of many innovative teaching techniques and resources with approximately US$3 million being spent on their development. Course content covers all aspects of the photovoltaic sector as well as all other renewable energy technologies. Considerable emphasis is placed on students gaining "hands-on" experience such as through the design, installation and testing of pv systems in Nepal.

9. ACKNOWLEDGEMENTS

The Key Centre for Photovoltaic Engineering was established and is funded by the Australian Research Council in conjunction with the University of New South Wales. Other major support for the new educational initiatives is being provided by the Sustainable Energy Development Authority in NSW, the Australian CRC for Renewable Energy, Pacific Solar Pty. Ltd., BP Solar and numerous other organizations, companies and end-users of photovoltaic product. The new Murdoch program is also receiving strong support from Western Power and the Alternative Energy Development Board. The contributions of many staff of the University of New South Wales are also acknowledged, particularly the efforts of the support staff of the Key Centre for Photovoltaic Engineering and the Photovoltaics Special Research Centre.

10.   REFERENCES

[1] R. Curry, PV Insider's report, Vol. XVII, No. 2, Feb 98, p1

[2] European Commission, White Paper for a Community Strategy and Action Plan, COM(97)599 final (26/11/97).

[3] M. Bartensterin, Opening Adress at 2nd World Conference on Photovoltaic Solar Energy, Vienna, July 1998.

[4] S.R. Wenham, C.B. Honsberg and M.A.Green, ANZSES Solar '98 Conf, Christchurch, Nov 1998. Pp.532-538.

[5] K. Yeadon, Minister for Energy, NSW Government Media Release, July 1999.

FINANCING SOLAR ELECTRICITY IN THE U.S.:

AN INTRODUCTION

Michael T. Eckhart
Solar International Management, Inc.
1825 I Street NW, Suite 400
Washington, DC 20006
Mail: eckhart@solarbank.com

With contributions by Thomas J. Starrs, Keith Rutledge, Bill Brooks, David S. Carey, Chris Cook, Christy Herig, Thomas Hoff, Paul D. Maycock, Maurice E. Miller, John W. Stevens, and Steven J. Strong, plus oversight and inputs from Matthew Cheney, Mark Fioravanti, Roby Roberts, and Adam Serchuk

ABSTACT
This paper is the introduction to a paper being prepared under the co-sponsorship of the Utility Photovoltaic Group (UPVG) and the Renewable Energy Policy Project (REPP). The premise is that financing is the nexus of solar photovoltaic's (PV's) path to commercialization. Four things need to be done to bring the financial community into the solar electricity marketplace. First, financing criteria need to be factored more explicitly into the industry's marketing programs and the government's commercialization and incentive programs. Second, the underbrush of technical, regulatory and institutional issues needs to be cleared immediately for the marketplace to function. Third, there is a need to implement new financing programs for solar electricity. Finally, federal, state and local governments should be creating methods of monetizing the environmental benefits of PV, thereby bringing financial returns in line with societal goals and attracting financial resources to the PV marketplace.

1. INTRODUCTION

PV, the high tech material that converts sunlight directly into electricity, has gone through 25 years of research, development, demonstration, and commercialization, and now faces its real marketing debut at the dawning of the 21st Century. With existing equipment from the PV industry and increasing support from the public sector under the aegis of climate change and sustainable development, PV is now being used in many corners of the U.S. and world economy. The result is a high degree of acceptance by pioneering end users, but also an emerging realization that this capital-intensive clean energy option cannot be readily employed in mass markets without the companion development of an "installation, service, and financing infrastructure" that is suited for this product.

The next major hurdle in PV's path to success is establishing its "financeability." That is, if an end user wanted to purchase a PV system, and a lender wanted to extend financing for it, how likely is it to happen? It turns out that the answer is not very positive today, and is complicated by a rather daunting range of economic, technical and institutional issues involving system design, equipment availability, installation capacity, quality assurance, warranties, insurance, interconnection with the utility grid, regulatory treatments about metering and electricity rates, and economic incentives from federal, state, and local governments. This paper describes the issues and calls for a renewed effort by government and industry to resolve the issues and put them behind us, so that the widespread financing of the Solar Century [1] can begin.

1.1 Why PV?

It is becoming clear from market research that electricity consumers want this technology. In recent national surveys, [2], people ranked solar energy as their #1 choice of electricity, as shown in Figure 1. The second-ranked choice is windpower. Coal and nuclear are ranked last. Some of the utility industry's leaders are hearing the message from the public, and are forging new strategies that include use of the new technologies. Even so, many individuals and companies are not waiting for the utilities to adopt PV, They are doing it themselves.

Percentage Favorable Ratings

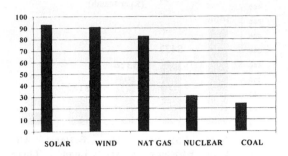

Fig. 1: Public Views About Electricity Options

People here and around the world seem to be drawn to PV out of an inherent understanding of its role in the sustainability of human development. PV might, in certain ways, be one of the most extraordinary of all technologies. Consider that humanity has invented a method of converting sand into a semiconductor material that can take in sunlight, absorb the energy and put it to work mobilizing a flow of electricity – our highest form of useful energy – and do it without motion, pollution, or making a sound. We will never run out of it. Its efficiency will only go up. Its cost will only go down. It is safe. Some people would ask, in terms of our energy future: why are we working on anything else?

Today's pioneering installations of PV cover a wide gamut. According to the PV industry:

Homeowners are putting PV on rooftops out of a concern for the environment, as part of a power reliability scheme, for high tech status, or to power off-grid homes.

Building owners are integrating PV into roofs and facades for reasons of economics, public expression of environmental consciousness, and looking modern.

Utilities are applying PV for off-grid applications such as signaling and lighting, and in new applications for power support at the end of long distribution lines. In all of these applications, there appears to be a common theme – that PV is used where the decision-maker is seeking a cost-effective

solution that also encompasses qualitative criteria such as the environment, appearance, and forward vision. This is an early market signal that the "value" of PV is greater than simply electricity cost.

In fact, surveys are discovering that people will pay more for clean and renewable energy. The National Renewable Energy Laboratory (NREL) recently completed a summary of utility market research concluding that, on average, a 70% majority of American people say that they are willing to pay at least $5 per month more for electricity from renewable sources. Some will pay even more: 38% said they would pay $10 per month more, and 21% said they would pay $15 per month more. If the average figure is true, then the American people appear to be ready to pay on the order of $10 billion more per year for cleaner electricity (5% of the nation's approximate $200 billion annual electricity bill).

1.2 Setting the Stage for Financing

Optimism about PV's future comes from the broadening realization that the energy-economic structure of the U.S. is "neither sustainable over the long-term future nor replicable by other countries around the world, if the global environment is to sustain life as we know it" as said in a speech to the American Solar Energy Society by Denis Hayes, in June, 1999.

But something is stopping the financial community from jumping into PV. Investors and bankers certainly hear the message but they have difficulty incorporating non-economic factors into today's financial decisions. To bring "financing" to the PV marketplace, industry and government sponsors have to make PV more appealing on the numbers, and more feasible.

2.   THE ROLE OF FINANCING

The decision to commit funds to a PV application is really the event for which all of the years of research, development, demonstration, and institutional work has been done. Financing of the end-use market is the nexus of commercialization, and should serve as a "focal point" for the industry's marketing programs and the government's commercialization programs.

2.1 Impact of Financing on Market Size

Current electricity rates for conventional power would nor be as "affordable" as they are today were it not for long-term and low-interest rate financing. The role of financing has been an essential ingredient in the development of the electric power industry for over 100 years. Indeed, the original structure of the Thomas Edison Company was an equipment supply division (now the General Electric Company), the Edison Operating Companies (Boston Edison, Consolidated Edison, etc.), and the Electric Bond & Share Company (EBASCO) for raising the necessary capital. The availability and cost of capital affects the economics of PV even more than other electricity generating technologies, because PV is the most capital intensive of all electricity generation options.

The provision of low-cost capital will be important to determining the size of the PV market. Information from around the world suggests that 2% to 5% of PV purchasers buy with cash, while the other 95% to 98% require some form of third party financing. Opinion in the PV industry suggests that credit programs will allow as much as 50% of all people in the market to purchase. If credit programs increase the addressable market from 2% or 5% of a population to 50%, then credit alone will have increased the market for PV by a factor of 10:1 to 25:1.

If monthly payments matter – and, in the U.S. they certainly do -- then Table 1 illustrates how the terms of financing can drive monthly payments on a PV system. Credit-card type debt on a $30,000 residential rooftop PV system would yield payments of $656/month. The monthly cost drops, however, to $182/month if this same PV system is included in a new home mortgage or financed with a low-interest loan. Such financing can be provided by banks, thrifts, credit unions, or even utilities. For example, if legislated by Congress, electric cooperatives might obtain 5% money from the Rural Utilities Service (RUS), mark it up to 6%, and make loans at those rates for PV installations and other environmentally-smart investments.

Table 1
Monthly Payments on a 3-Kilowatt Rooftop PV System

LOAN TYPE	TERM	RATE	PAYMENT
Credit Card	7 Years	18%	$656/mo
Credit Line	10 Years	12%	$442/mo
Home Equity Loan	15 Years	8%	$292/mo
Home Mortgage	30 Years	6%	$182/mo

2.2   Financing versus RD&D Programs

Looking at financing in the context of government policy shows that low-cost financing programs can actually have more impact on PV markets in the immediate future than technology research or manufacturing cost reductions. For example, Figure 2 illustrates the monthly payments for the 3-kilowatt grid-connected, residential rooftop PV system in

the U.S. The PV system is assumed to cost $10/watt ($30,000 total cost), with a total system efficiency of 10%. In the base case, financing is assumed to be a 7-year 18% loan, resulting in a monthly payment of $656. Then, three improvements are evaluated:

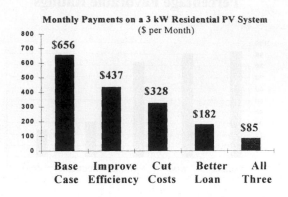

Monthly Payments on a 3 kW Residential PV System ($ per Month)

Fig. 2: Impact of Financing As A Policy Option

Research Programs to Improve Efficiency: By improving the efficiency of the system, its size can be reduced while still generating the same output, thus reducing the cost. For a 50% increase in system efficiency, the monthly payment can be reduced to $437 per month. It may take ten years or more, however, before technology research can be achieve these results in commercial products.

Cost Reduction Through Manufacturing Scale-up: By scaling up manufacturing, automating factories, and generally reducing the production cost per unit, total system costs can be reduced. Assuming a 50% reduction in system cost, the monthly payment of the sample system can be reduced to $328 per month. Achieving this result might take five years or more, as new factories are planned, built, and put into operation.

Better Financing: A third way of reducing monthly payments is to apply lower-cost, longer-term financing. Here, the 7-year, 18% loan is replaced with a 30-year, 6% loan. The monthly payments due to the change in financing alone come down to $182 per month, and this can be done immediately.

All three Initiatives: Of course the best policy is to accomplish all three cost reductions, but the immediate impact of lower-cost, longer-term financing is apparent. The provision of lower-cost, longer-term financing can have a more *immediate* impact on the affordability of solar PV systems than research and manufacturing programs. However, there is no question that the ultimate viability of PV depends absolutely on the success of future research and development programs, funded at a sustained and substantial level.

The bottom line is that financing is a key ingredient in determining the economic viability of PV. Whereas many people say, "the financing will be there when PV is more economic," the evidence also suggests that "PV will be more economic when low-cost financing is readily available".

2.3 How Much Capital is Needed?

How much capital is needed to finance the emergence of solar energy? Consensus industry forecasts indicate that annual shipments will increase from 150 MW in 1998 to about 2,000 MW in 2010 [3]. Preliminary analysis indicates that the capital required to support this growth will be $3.7 billion invested in PV manufacturing facilities, $3.8 billion invested in the distribution channels (financing inventory and receivables), and *$38 billion in end-user financing.* Clearly, the major challenge, by a factor of 10:1, is end-user financing.

Translating these figures to the U.S. market, assuming that the U.S. remains at approximately 10% of the world market, the capital requirement is nearly $400 million for PV factories, $400 million for working capital in the distribution channels, and $4 billion for end-user financing. An alternative calculation, based on the installation of solar PV on 200,000 rooftops (20% of President Clinton's Million Solar Roofs initiative), at an average cost of $20,000, also yields a $4 billion requirement for end-user financing.

However, $4 billion could prove to be a minimum figure, depending on how serious the U.S. gets about mobilizing clean energy technologies. Let's look at the scale of the issues. There are approximately 3,000 megawatts (MW) of conventional power plants installed around the world today. There are 75,000 MW to 100,000 MW of new conventional power plants being built each year, as compared to PV's relatively insignificant 200 MW of annual installations. In this context, one readily sees that PV needs to be installed at rates 10-times to 100-times the current rate to begin making a meaningful contribution to world environmental conditions. The magnitude of capital required would be 10-times to 100-times greater than the $4 billion cited above.

Likewise, in the domain of social development, there are 2 billion people lacking electric service in the world today, some 400 million family dwellings that could be served by PV if they could be reached by installation, service, and financing infrastructure. But today's rate of installation, at perhaps 400,000 solar home systems per year just scratches the surface. Again, one sees that the rate of PV installations needs to increase 10-fold to 100-fold before the technology is serving a meaningful role in world social development.

What kind of policies can have such an impact, increasing the market by 10-fold to 100-fold? Clearly, this is the domain of igniting "demand pull" and this, in turn, deals with the financeability of PV.

This paper recommends four actions: First, factoring financial criteria into the planning of the industry's marketing programs and the government's commercialization programs; second, clearing out the underbrush of technical, regulatory, and institutional issues so that end users, suppliers and financiers can transact business; third, implementing specific new financing programs for PV; and finally, creating a method to "monetize" the environmental benefits of a PV installation and return a share of the benefits to the owner, thereby bringing financial returns in line with our societal goals.

AKNOWLEDGEMENTS

This paper is the introduction to a larger work being prepared under the co-sponsorship of the Utility Photovoltaic Group (UPVG) and the Renewable Energy Policy Project (REPP). A group of expert co-authors, listed on the first page, is contributing to the paper, which should be published later in 2000.

REFERENCES

(1) Leggett, Dr. Jeremy, Oxford University, coining the term "Solar Century" while Director of Renewable Energy programs at Greenpeace Europe.

(2) Fahar, Barbara C., Ph.D., "Willingness to Pay for Electricity from Renewable Resources: A review of Utility Market Research." NREL Topical Issues Brief, July 1999.

(3) Maycock, Paul D., "Photovoltaic Technology, Performance, Cost, and Market (1975-2010), January 1999.

STAND-ALONE AND ISLAND APPLICATIONS OF PHOTOVOLTAIC POWER SYSTEMS: TASK III OF THE INTERNATIONAL ENERGY AGENCY PHOTOVOLTAIC POWER SYSTEMS PROGRAMME

A.R. Wilshaw[1], J.R. Bates[1], P. Jacquin (Task III Operating Agent)[2], K. Dongwhan[3], Y. Ishihara[4], A. Joyce[5], P. Malbranche[6], F. Minissale[7], F. Nieuwenhout[8], P. Krohn[9], K. Presnell[10], I. Stadler[11], D. Turcotte[12], O. Ulleberg[13], X. Vallve[14], M. Villoz[15],

1. IT Power, The Warren, Bramshill Rd., Eversley, RG27 0PR, UK;
tel: +44 118 973 0073; fax: +44 118 973 0820; e-mail: arw@itpower.co.uk.
2. PHK Consultants, 17 bis rue Jean-Marie Vianney, 69130 ECULLY, France;
tel: +33- 04 78 33 36 14; fax: +33- 04 78 33 38 08; e-mail: phkconsultants@compuserve.com.
3. Korea University, Seoul, Korea. 4. Doshisha University, Kyoto, Japan 5. INETI, Lisbon, Portugal.
6. GENEC, St. Paul-lez-Durance, France. 7. Conphoebus, Piano d'Arci, Italy. 8. ECN, Petten, The Netherlands.
9. Vattenfall Utveckling AB, Nyköping, Sweden. 10. Northern Territory Centre for Energy Research, Darwin, Australia.
11. Universität (Gh) Kassel, Kassel, Germany. 12. CANMET, Varennes, Canada.
13. Institute for Energy Technology, Energy Systems Department, Kjeller, Norway.
14. Trama Tecnoambiental SL, Barcelona, Spain. 15. Dynatex SA, Colombier, Switzerland

ABSTRACT: Task III of the IEA PVPS Programme began a new five year Work Plan in 1999. The focus of the work is on Quality issues of stand-alone PV, hybrid systems, storage methods, and load management. A selection of Case Studies will be chosen by each of the participating countries, which will serve to highlight the work of Task III, and also to provide field data for study. Task III is co-operating with Task IX and Task II on common areas of study. This paper presents the new Work Plan and summarises progress to date.
Keywords: Stand-alone PV - 1: Hybrid - 2: Storage - 3

1. INTRODUCTION

The overall mission of the IEA PV Programme is to encourage international collaboration efforts through which photovoltaic energy becomes a significant option in the near future. This paper describes the aims, objectives and progress to date of Task III of the International Energy Agency's Photovoltaic Power Systems Programme.

The objective of Task III is to improve the technical quality and cost-effectiveness of PV systems in stand alone and island applications. This is achieved through:
- the collection, analysis and dissemination of information on the technical performance and cost structure of PV systems;
- the sharing of knowledge and experience gained in monitoring selected national and international projects;
- the development of guidelines for improved design, construction and operation of photovoltaic power systems and subsystems;
- the further technical development of improved photovoltaic systems and subsystems.

The following 14 countries are actively participating in the work of the Task: Australia, Canada, France, Germany, Italy, Japan, Korea, the Netherlands, Norway, Portugal, Spain, Sweden, Switzerland, the United Kingdom.

Task III has been active since 1994, when its first Work Plan was agreed with the IEA. Work to date has dealt mainly with charge controllers and batteries for stand-alone PV systems, a survey of PV Programmes in developing countries, and a review of lessons learned in stand-alone PV system implementation across the Task III

member countries. Further details and results of these studies are referenced at the end of this paper [1] to [5].

In September 1999, Task III entered the second phase of its existence, with a new Work Plan and change in focus. The new Work Plan directs efforts on Quality Assurance of PV systems, PV hybrid systems, energy storage and load management.

This paper will present the new Work Plan and summarise the progress to date.

2. WORK PLAN

The Task III Work Plan is divided into two areas: Quality Assurance of PV Systems and Technical Issues. Experts participate in the Subtasks according to national experience and interests.

Subtask 1: 'Quality Assurance of PV Systems' aims to provide both end-users and programme managers with basic tools for ensuring the quality assurance of installed systems in the form of Recommended Practices or Guidelines. Subtask 2: 'Technical aspects of PV systems' intends to review the state-of-the-art of technologies used in PV-hybrid systems, energy storage, and load management. This will include detailed work on the design of PV-hybrid systems to provide technical information and guidelines to facilitate their deployment.

In addition to these subtasks, the Task will be undertaking some co-operative work with Task II (Operational Performance and Design of Photovoltaic Systems and Subsystems) and Task IX (Deployment of PV Power Technologies: Co-operation with Developing Countries).

The paper will describe the work plan in detail, and report on progress to date.

2.1 Quality Assurance of PV Systems

The primary aim of this subtask is to develop quality assurance schemes which will lead to installed systems with a guaranteed service life, at as low a cost as possible. This work is co-ordinated by the UK Expert. There are two aspects to this work: one is to ensure that the relevant frameworks are in place, and the other is to investigate how such schemes can be implemented in practice.

The issue of providing quality guarantees for PV systems is very complex. There are many stages in the design, supply, installation, operation and maintenance of a PV system, and each stage is a potential source of failure.

The starting point for this activity is a survey of existing guidelines, standards and Best Practices for stand-alone PV and hybrid systems. The purpose of this survey is to identify gaps in the existing material, and to propose appropriate guidelines to fill these gaps.

A number of complementary activities have been devised in order to identify the key issues that must be addressed with respect to QA schemes for PV systems. The first of these activities is to identify case studies in each participating country, which can serve as reference projects for the development of QA schemes. It is intended that the case studies will include a representative spread of stand-alone PV projects, and as many types of QA schemes as possible. The QA procedures will be reported for each project. Figure 1 shows one of the case studies identified to date.

Once the case studies have been selected, the 'field' experience of these projects can be evaluated. This is the part of the practical study work referred to above, i.e. to investigate which of the schemes actually work in practice. The case studies will be examined in detail so as to learn from the QA methods used to guarantee those projects.

Having evaluated the case studies for their efficacy, the schemes will be modified where necessary and used as input to the Task III QA methodology.

Figure 1: UK Case Study, PV-hybrid system, Scotland, UK (© The Solar Century)

In addition to the case studies, a number of work shops will be held in conjunction with Task III Experts' Meetings, where various target groups can be informed and consulted about the necessary elements to include in QA procedures for their particular sector of the market. These workshops will also be used to discuss and comment on the experience of QA procedures 'in the field'.

The following topics will be amongst the issues addressed as part of the work:

- Accreditation and certification
- User training
- Installer training
- Performance guarantees
- User satisfaction
- Monitoring guidelines
- Practical recommendations for performance assessment in the field

The QA methodology resulting from the above work will be presented at future international PV conferences, in order to encourage debate and comments on the proposed guidelines. The guidelines will also be presented to the relevant PV standards and QA organisations. It is intended that Task III experts will have ongoing input to the Standards committees in order to make timely recommendations. The channels of communication for this dialogue are currently being established.

2.2 Technical aspects of PV systems

There are three areas of work defined under Subtask 2. These relate to:

1. Hybrid systems
2. Storage
3. Load management and new appliances

Hybrid Systems

Most of the work on hybrid systems will address the relatively high cost of remote maintenance for hybrid systems. As a prime objective, the work on hybrids aims to facilitate large scale implementation of PV hybrid systems. This work is co-ordinated by the Australian Expert.

One of the objectives is to examine and develop component design such that hybrid systems are modular and easy to expand. This in turn will reduce the initial investment for project installation. A further aim is to develop remote monitoring of the systems, which will reduce maintenance costs.

In addition, optimum matching of the PV system to the auxiliary generator will be studied in detail, and ways of improving controllers and user interfaces will be examined with a view to reducing operational complexity.

The first activity is to construct a classification system, with which to categorise PV systems. This can be used to identify the areas where hybrid systems are appropriate, and to assist the selection of a range of case studies.

The alternative design procedures will be evaluated, and the performance of the systems will be compared to assess the most efficient design, in terms of both cost and energy generated. The evaluation will consider the entire process, from energy needs identification through to the final design engineered. It will focus attention on system modularity and matching of the PV and the auxiliary generator. Recommendations will be made with respect to the design of hybrid systems, based on the results of the evaluation.

In addition to the above, the systems will be assessed with respect to the monitoring strategies used. Based upon the field results, a recommended monitoring strategy for hybrid systems will be drawn up, with the aim to reduce maintenance costs of the installed systems.

Furthermore, the control and management of power from the systems will be analysed, with a view to optimising alternative strategies. Again, recommendations will be drawn up, based on the results of the evaluation.

The work undertaken will be presented at workshops and conferences, on a number of occasions during the course of the programme. These workshops have the aim of giving target groups and other interested parties an opportunity to discuss the results and input their experiences to the Subtask.

Figure 2: PV wind diesel hybrid system, Coconut Island, Queensland, Australia

Storage

The application of lead-acid batteries for electrochemical storage in stand-alone PV systems is a key influence on cost and reliability. Selection of the most appropriate battery for a given system and managing it correctly are important factors in optimising the design of stand-alone PV systems.

The Storage activity of Subtask 2 will address cost reduction of the storage function for PV and PV hybrid systems. This can be achieved from two perspectives: decreasing the investment costs or increasing the capacity and lifetime of the batteries. Both of these issues are being considered. In addition, there will be a watching brief on alternative forms of storage which are suited to stand-alone PV systems. This work is co-ordinated by the French Expert.

The work on lead acid batteries will use field data in order to obtain an overview of the batteries presently used in stand-alone PV systems. A number of case studies will also be selected in order to collect information about the lifetime of batteries installed as storage for stand-alone PV systems. This will enable the full cost of lead acid storage over the PV system lifetime to be ascertained.

The study of field data will therefore include work on battery lifetime with respect to the mode of operation of the PV system, as well as evaluating the impact of energy management strategies on the battery performance. The methods and devices currently used to assess state-of-charge and quality of the batteries will be considered, and recommendations on the minimum requirements will be made.

Specifications for monitoring and on-site testing will also be reviewed, and guidelines will be drawn up accordingly. Life-cycle cost analysis will utilise the data

from the case studies in order to assess the impact of different battery management strategies on the life cycle cost of a PV system.

Task III Experts will also conduct a brief review of alternatives to lead-acid batteries, for example new electrochemical couples, supercapacitors, fuel cells and fly wheels. Lastly, the Task III Experts working on Storage will make recommendations on new charge controller concepts and designs.

Once again, the work will be presented in conjunction with other relevant international events and workshops which will enable target groups to provide comments on this subject.

Load management and new appliances

The load profile is a very important factor in the design of a stand-alone PV or hybrid system. Consideration of load matching and the careful selection of efficient appliances therefore have a significant effect on the size of the PV generator, and thereby the investment cost of the system. A key consideration in sizing a system load is to allow for the natural tendency of the user to increase their energy needs with time.

The work aims to demonstrate the advantages of efficient load management strategies and well suited load appliances. It will look first at 'demand side management', i.e. how to manage typical user needs, how these needs are expected to change, and how to accommodate these changes. The work is co-ordinated by the Portuguese Expert.

In addition, the issue of whether to install d.c. or a.c. systems will be investigated. This will include an analysis of the problems related to the use of d.c. electricity with respect to safety of persons and equipment, due to the presence of high currents at low d.c. voltage. An assessment of the advantages and disadvantages of both electricity systems with respect to impact on overall system efficiency and cost will also be undertaken.

The issue of expanding an existing stand-alone PV system will also be addressed. Experts will consider methods of designing a stand-alone PV system which both satisfies present needs and allows for easy expansion when the user load increases.

The impacts of load matching and demand side management will be quantified in terms of their effect on the technical design and the investment costs of a stand-alone PV project.

3. WORK IN PROGRESS

3.1 Case studies

In order to maximise resources, case studies are being selected which may be used for the field studies of QA projects, hybrid systems and batteries, if possible. A questionnaire has been devised to describe the characteristics of each of the case studies.

To date, case studies have been identified in the UK, Spain, Canada and Australia. The UK system is an autonomous home and farm with 2.1 kWp PV and 11 kVA diesel generator. The warrantees for the system components were provided by the installer / supplier and the O&M is carried out by the owner. Two systems have been selected in Spain. The first of these is a residential 1.5 kWp PV system with a back-up butane gas generator, rated at 4.5 kVA. The second system is also residential, with a 1.3 kWp PV system and 4 kVA petrol generator back-up. Both systems are owned by a User Association, SEBA, which acts as an energy service operator. In both cases SEBA provides a 15 year energy service contract to the end user. The manufacturer provides SEBA with a guarantee of 90% of the rated power from the hardware. The system in Canada is a service application with 165 Wp PV and 740 Ah lead-acid battery back-up, providing power for natural gas well monitoring. Warranties for hardware were supplied by the manufacturer. There was no contractual warranty on the system sizing, apart from the designer's promise. The Australian system provides power to a community, with 3.2 kWp PV and a 15 kVA diesel generator. Both systems use lead acid battery storage. Hardware warrantees were provided by the manufacturers, and O&M is carried out by the owners.

The classification system which was devised as an initial activity of Subtask 2 is being used to select further case studies from participating countries.

3.2 Quality Assurance of PV Systems

Work to date on Subtask 1 has focused largely on the review of existing standards and QA schemes for stand-alone PV systems. An internal Task III document is being drawn up which details progress to date with respect to international and national standards and QA schemes for PV, batteries, and power management devices, as well as any available guidelines on stand-alone systems. It should be noted that, within this Subtask, the word standard also includes any guidelines and Best Practices which are in general use. Task III Experts will each contribute with information about the standards and QA schemes used by their respective countries, either at home or overseas.

This review aims to be as comprehensive as possible, and embraces the work of PVGAP, the IEC, and the World Bank ASTAE Unit. The purpose of the review is to detail the scope of each of the standards and QA schemes, so that any areas which require further consideration can easily be identified. Further information about this survey can be found in reference [6].

Task III Experts have also provided comments to PVGAP on the manual: Quality Management in PV Manufacturing, prepared for the World Bank ASTAE Unit.

3.3 Technical aspects of PV systems

The classification system for the case studies has been defined. Each of the case studies will be categorised according to this system.

A list of monitoring parameters has also been drawn up for the case studies selected. A summary of monitoring guidelines is being drafted to accompany this. This will enable a uniform monitoring protocol to be applied for each of the case studies. The purpose of these guidelines is to identify the best commercial practices for monitoring stand-alone PV systems, and to promote a common methodology which will facilitate effective comparisons. The document will also address protocol issues and include advice about monitoring equipment and methodology. Work is being contributed by various experts on monitoring experience in their countries.

In the storage area, a questionnaire has been devised to conduct a survey of field data from the participating countries. Work is also being carried out on the review of battery management strategies. Effort is currently being focused on an internal document which discusses the limitations of using voltage thresholds for battery management.

Other work includes preliminary data collection for Life Cycle Cost Analysis and a collection of example studies for battery specifications and test procedures.

The work on load management and new appliances is also proceeding to plan. A first draft of an internal Task III document on load management strategies will be completed towards the end of 2000, and a corresponding brochure on this subject will be available in 2001.

The work on appliances is focusing efforts on the typical problems encountered in practice with the application of new appliances in the field.

4. CO-OPERATION WITH OTHER TASKS

There are three main areas of co-operation between Task III and other PVPS Tasks.

Task III will co-operate with Task II by feeding in information collected from the case studies to the Task II database. In addition, Task II has been asked for suggestions on Quality criteria from existing systems for Subtask 1.

Task III will also be collaborating with Task IX, by feeding in technical suggestions which are relevant to the work on Task IX's Recommended Practice Guides for PV deployment in developing countries. In return, Task IX will request technical research from Task III Experts as and when such issues arise.

Finally, Task III and Task IX will be working together on guidelines for the certification and accreditation of training users and installers of stand-alone and hybrid PV systems.

REFERENCES

[1] E. Usher & M. Ross, Recommended Practices for Charge Controllers, CANMET, Varennes, Canada, 1999.
[2] J.R. Bates & A.R. Wilshaw, Survey of Stand-alone Photovoltaic Applications in Developing Countries, IEA PVPS Task III Report, IT Power, March 1999.
[3] Ph. Malbranche *et al*, IEA – PVPS Task III, 2nd World Conference and Exhibition on PV Solar Energy Conversion, Vienna, 6 – 10 July, 1998.
[4] IEA Task III, Stand-alone Photovoltaic Applications: Lessons Learned in 14 National Showcase Projects, James & James, London, April 1999.
[5] IT Power, Task III Slide Library, CD-ROM with 250 slides of stand-alone PV applications, 1999.
[6] A.R. Wilshaw et al, Review of Standards and Quality Assurance Schemes for PV Systems, 16th European PV Solar Energy Conference, James & James, 2000.

PHOTOVOLTAIC ENERGY STATISTICS OF SWITZERLAND 1999

Christian Meier, Christian Holzner, Peter Eichenberger

energieburo zurich
Limmatstr. 230
CH-8005 Zurich, Switzerland
T +41 1 242 80 60; F +41 1 242 80 86
info@energieburo.ch, www.energieburo.ch

ABSTRACT: In 1999 the total power of grid connected photovoltaic installations in Switzerland reached the „magical border" of 10 MWp. About 120 new systems with a total power of 1.8 MWp where installed during the last year. These data show that the positive growth trend of the PV power production in Switzerland was continued, but that the increase was not as high as in the years before.

Keywords: National Programmes – 1; System Reliability – 2; Building Integration – 3

1. A DECADE OF PV-MONITORING

The Swiss Federal Office of Energy together with the Swiss Utility Association runs since 1992 the project "Energy Statistics and Quality Control of Photovoltaic Installations" in Switzerland. In this period, the quality of all installations were monitored and the yield closely followed to insure a high quality of the installations and a satisfying performance yield.

year	number of PV instal-lations by the end of the year	nominal power [MWp DC]	annual electricity pro-duction [MWh]	annual yield [kWh/kWp]
1989	60	0.3	-	-
1990	170	0.8	400	-
1991	380	1.8	1100	-
1992	490	3.1	1800	800
1993	600	4.0	3000	810
1994	680	4.8	3500	800
1995	740	5.4	4000	815
1996	820	6.2	4700	825
1997	950	7.4	6000	880
1998	1100	9.1	7100	860
1999	1220	10.9	7700	770

Table 1: Contribution of photovoltaics to the electricity production in Switzerland since 1989.

A detailed analysis of the available data showed an average production of around 825 kWh/kWp. If this yield is compared with the available sunlight during the

Until the end of 1999 the overall installed power raised up to 10.9 MWp. For the calculation of the electricity production in the year 1999, the reduced yield of installations which were only built during the year of 1999 are considered. The column to the right shows the yield after comparison with different available sunlight in Switzerland.

To stay informed about the newly built photovoltaic installations and their corresponding power in Switzerland, every year a survey is done. All photovoltaic manufacturers and installers are asked to submit a list of their installations throughout the year. It was found that in 1999, 120 new

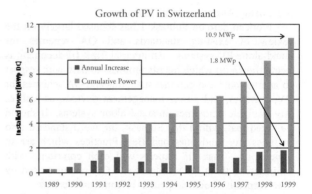

Growth of PV in Switzerland

photovoltaic installations where build adding up to over 1800 kWp power. In Switzerland, there are now about 1220 photovoltaic installations running, summing up to an overall power of 10.9 MWp. They produced in 1999 around 7.7 million kWh electricity (see Table 1).

2. SPECFIC YIELD

In 1999 about 1220 grid connected PV systems in Switzerland where producing 8 mio kWh of electrical energy. The cummulative power of these systems was 10.9 MWp.

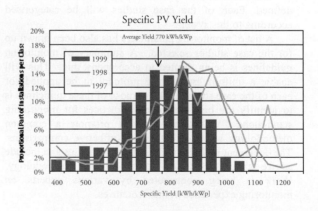

Specific PV Yield

The average yield of all reporting PV installations in 1999

was only 770 kWh per installed kWp. This low energy yield was caused by unfavourable weather conditions for photovoltaic power production. Meteorlogical measurements pointed ot that the solar iradiation in 1999 was about 5% below the long term mean

3. WORLD LEADERSHIP ENDANGERED

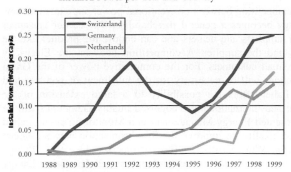

During the last decade, Switzerland had always the highest value of new installed PV power per year and inhabitant in Europe. This leading position is in danger because other european countries are more active in promoting PV systems.

4. GOOD QUALITY SYSTEMS

The quality of photovoltaic systems installed in Switzerland is on a high level. Investigations on inverter performance showed an availability of more than 97 %. In 1996 it was found also, that the inverters still run very satisfying. From surveys done by around 200 photovoltaic installations it was found that the biggest part of the inverters show a very good availability of around 98 %. It was shown, that in comparison to the previous years, some installations had longer down times due to inverter failure than in years before, but the overall performance yield of the inverters was still on a high level. At this point it has to be mentioned, that this data is available from surveys and not from measurements and therefore must be looked at with a certain caution. It is also imaginable, that some of the inverter failures, especially short ones or with only a slightly worse efficiency, have not been noticed and therefore not been reported.

5. OUTLOOK

The quality insurance and energy statistics will be continued. Because already for a number of years a significant number of installations has been monitored, it is expected, that for the future more detailed data can be found to determine the overall performance of photovoltaic installations over a time frame of ten or more years. To make the analysis of such data available, a structure is under development to incorporate the available data into a data bank.

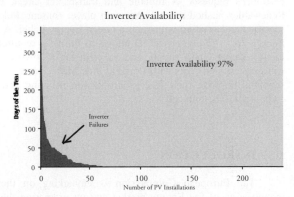

This work was made possible through the support of the Swiss federal Office of Energy and Swiss Electricity Producer and Distributor Association.

All questions regarding the preparation of the paper should be addressed to:

energiebüro Zürich
Limmatstrasse 230
CH-8005 Zürich
T +41 1 242 80 60   F +41 1 242 80 86
www.energiebuero.ch   info@energieburo.c

Energy Tariff Manager – Economic solutions
to Measure, Earmark and Track Electricity from Photovoltaic Systems

M. Heidenreich[1], E. Ploder[2]

[1] Österreichisches Forschungs- und prüfzentrum Arsenal, Faradaygasse 3, A-1030 Vienna, Austria
[2] PTS – Energie mit Strategie, Murfeld 14, A-8850 Murau, Austria

ABSTRACT: The liberalisation of the Energy market enables the competition between conventional power suppliers and new players such as Green Energy suppliers and dealers. Green Energy becomes a general merchandise. The earlier consumer supplied by the monopoly supplier is a today courted customer of suppliers and dealers and can choose the product like the Renewable or Fossil Energy technology, the production process and the supplier. A competition between the Electricity suppliers exists already, which decreases the prices for large and small customers. For the competitive power of the Green Energy suppliers apart from the economic criteria's quality Guidelines will be necessary for the fulfilment of non-standard customer's requests. A reliable and transparent Green Energy Label offers the possibility to a large extension of the Renewables pushed through the main player consent. Independent and internationally recognised quality Guidelines will be specified and verified with such a Green Energy Label. A now presented Energy Management and Monitoring device called *Energy Tariff Manager* (*ETM*) allows to fulfil the required product quality, to improve the Synchronisation between Green Energy demand and supply and to yield the following "Win, Win, Win" results: 1. The customers reduce their Energy consumption and costs with Demand Side Management and can choose the Energy source and quantity, 2. The Green Energy suppliers improve their Energy production planning and reduce expensive peak-load power and 3. The Greenhouse gas emission will be reduced.

1   INTRODUCTION

The European Commission is embarking on the liberalisation of the Energy market and on promoting the Renewables. The Directives of Electricity and gas market require initiatives of its members to progressively increase the free choose of suppliers of domestic and industrial customers and the White Paper pre-determines the ambitious goal to double the share of renewable Energy till 2010. The *Energy Tariff Manager* (*ETM*) allows to match the goals of these Directives by controlling the Energy consumption and by managing the demand side as well as by determining, by earmarking and by cost-effective tracking the quantity of the Green Energy through a device.

The *Energy Tariff Manager* (*ETM*) may give the cost-effective solution to some problems of a broad Green Energy market penetration are:

- How is it possible to guarantee the quality of the product "Energy"?
- How can the contribution of Renewable Energy be matched with Energy Efficiency measurements?
- How can the reserve capacity be managed?
- How can the supply and demand Synchronisation be optimised?

2   APPROACH AND MOTIVATION

The basic motivation for the *Energy Tariff Manager* (*ETM*) is to significantly increase the amount and proportion of renewable Energy generation and trading within the EU. Common rules and standards for "Green Electricity" supply will allow the optimum market penetration of the Renewables throughout the EU.

Promotion initiatives

Despite of the cost reduction efforts and continuos growth rate of world-wide PV industry Photovoltaic is definitely depending on promotion initiatives. Target of these initiatives is to trigger a learning process of the consumers and the suppliers. Voluntary customer contribution as well as innovative marketing strategies promote the use of PV systems or Renewable Energy in general. **Green pricing** has attracted broad attention up to now in The Netherlands, Germany, Switzerland and Austria. Within this program type utilities offer „green" Electricity – that is to say, Electricity generated mainly by PV and wind turbines – at a price that meets the generation costs over a long time period. The new **Renewable Energy Law in Germany** specifies the minimum price per kWh for the Renewables including Photovoltaics. The Return of investment of an independent PV power producer is now around ten. This year the European Commission has installed the **Green Electricity cluster**. The three clustered EU projects assess the costs, the benefits and the market potential of Green Energy by trading.

Green Energy Label

For the competitive power of the Green Energy suppliers apart from the economic criteria's quality Guidelines will be necessary for the fulfilment of non-standard customer's requests. The Green Electricity from PV systems is a valuable product. The active use of the Green Electricity contribute to the environmental protection measures concerning the Greenhouse gas reduction activities. This maximises the win of the society according to actual welfare criteria to reduce CO_2 and will establish a sustainable business of the PV systems and other Renewable technologies.

A reliable and transparent Green Energy Label offers the possibility to extend the Renewables pushed through the main player consent about what is Green Energy. One possibility to prescribe the frame and the joint principles of

Green Energy suppliers and customers and of an international Renewable Energy trading might be an European harmonised Green Energy Label. The targets of that Green Energy Label are: To give

- the customer the chance to choose the product either Green Energy, Conventional Energy or an Energy mix,
- the supplier and investor the needed market information (all relevant Data should be published periodically) and
- an independent control instrument and quality standards.

Independent and internationally recognised quality Guidelines will be specified and verified with such a Green Energy Label. EU harmonised quality criteria's for consuming Green Energy will improve the end-user perception. Beyond that the Green Energy label won't compete the existing national and European introduction on the market and advancement programs, but complete it.

Customer choice and protection

The earlier consumer supplied by the monopoly supplier is a today courted customer of suppliers and dealers and can choose the electricity from the Renewable or the Fossil the technologies, the production process and the supplier. The customer requirements for Green Energy within a liberalised Electricity market are:

- Trust to the market,
- the understanding how to choose the technology on the market and
- the understanding about the implication of its choices.

ETM is a powerful control instrument for gaining the consumer protection concerning Green Energy choice and quantity.

3 ETM - Energy Tariff Manager

For reducing costs the *Energy Tariff Manager* should be integrated in a still existing Energy meter junction box. An illuminated LC-Display will give the end-user an overview about its:

- Total Energy consumption (in kWh and national currency),
- share of Renewables (in kWh and national currency),
- +/- Consumption tendency (in case of power varied price limits) and
- the maximum Energy value per measurement period.

The amounts of the Energy and power Data is triggered via GSM communication from the involved utility. The consumption Data are collected, memorised and at any time local or remote readable with the measuring time and date.

Figure 1: The future device

Green Energy Tracking

The energy quantities from the Renewables such as Photovoltaic or wind energy can be determined with the innovative *Energy Tariff Manager* via a so-called "**Green Energy Tracking**". *ETM* allows to track the share of e.g. the PV Energy by counting the supplied Energy at the grid connection point and by communicating it via GSM to the *ETM* device placed on customer site. **The quantity of the Green Energy is determined and earmarked** through the device. This results an easy access to the information about the quantity and technology, e.g. Renewable Energy from a PV plant mixed with conventional fossil Energy. The present Energy source, the current load consumption, the actual costs per kWh and other *ETM* advises will be shown on the device display. Special appliances will be located on a good visible end-user place. The customer will be informed about *ETM* - *Energy Tariff Manager* activities anywhere in the building. A Green lighted LED will indicate the use of Green Energy, a red lighted LED the conventional Energy and the two signals the change from conventional to Green Energy.

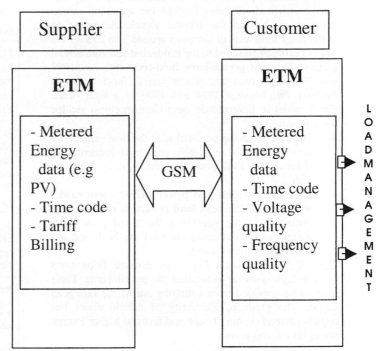

Figure 2: Green Energy Tracking via communication

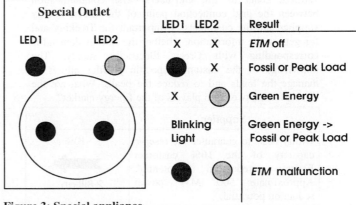

Figure 3: Special appliance

The requirement of the customer is covered with this stamped and contractually negotiated Green Electricity quantity. The end-user may control the quantities and resulted costs at any time.

Demand Side Management

The link between the active use of Green energy and the Rational use of Energy (RUE) is strengthened with *Energy Tariff Manager*. The consumption behaviour can be automatically matched with the fluctuated supply of the RES via the **load management**. A customer specified effective working time and power varied Tariff model will be implemented in a high integrated and cost-effective *ETM*. The device will record the consumption profile. Upon recognition of existing patterns, measures can be set to optimise energy consumption behaviour to reduce peak loading periods. This so-called "**peak-shaving**" produces energy savings and cost reductions for the end-users as well as suppliers. An optimised Tariff model based on the revised energy consumption profile will be designed to provide direct incentives for the user to reduce peak loading and find process improvements. The needless loads can be switched off either through the automatic energy manager or through the manual switch. According to preliminary field-tests in Germany around 5 to 25% of the yearly costs can be saved in the residential and commercial sectors. **The two preliminary field-tests** (between 1989 and 1990 of a time and power varied Tariff model in Freiburg and between 1994 and 1996 of a time varied Tariff model in Eckernförde, both German cities) yielded the following:

* The Energy saving potential with the time varied only Tariff model is less (6% peak-load reduction in Eckernförde, 11% peak-load reduction in Freiburg) than
* **the combined time and power varied Tariff model with 24% of the peak load reduction** (only used in Freiburg). In the interviews the asked participants announced their readiness to shift the loads and use the Energy more efficient.

The daily and seasonal Electricity demand fluctuations yield to high costs for generating the peak-lopping. Fuels have to be imported from countries out of the European Union. The produced Electricity of caloric plants has negative effects on our climate and leads to higher Energy prices for the end-users.

The benefit

The load management saves the Energy and reduces costs of the end-users. The communication between the grid connection point of the Green Energy supplier and the customers is important for Tracking and for getting the information, whether the Energy demand is corresponding with Green Electricity supply. The knowledge of the customer specific profiles allows to manage the loads and to reduce the peaks. What are the effects for the two main players of the Energy market?

Supplier	
Utility costs to guarantee the reserve capacity of the **10%** customer Energy reduction potential. Approximate 500 MW power reduction potential.	-10% ➡ **2 Billion**

Tab. 1: Techno-economic effects for the suppliers

The utility might use the *ETM - Energy Tariff Manager* to reduce peak-load cost shares. Investigation costs in gas-turbine plants for guaranteeing the peak-load capacity are quite high. **The specific construction cost of such a plant is around 4,000 €/kW (in Austria)**.

For the customers the following calculation is based on an average household's Energy consumption and includes the relevant specific technical requirements for measuring Energy consumption.

Customer	Estimated costs (€)
1. The **price for the devices** on the basis of an industrial mass production of around 100,000 sets per year.	**150,-**
2.1 Costs for implementation in existing installations of households, offices or industry will range from country to country.	218,- to 436,-
2.2 Costs for implementation in new installations	73,-
3. The economic benefit gained by the possibility of controlling the Energy use and consumption can be found in **the reduction of Energy costs**.	- 5% to - 10%
4. The yearly costs for Electricity consumption and heating in households in Austria shows an average of 872,- (5.000 kWh x EUR 0,1744) and 582,- (10.000 kWh x EUR 0,0581).	872,- 582,- 1454,-
5. A total of savings for Electricity and heating can be expected therefore in the range of 145,40 (-10%).	145,40 (- 10%)
Return Of Investment (ROI):	
6.1 Implementation in existing installations	**2.5 to 4 years**
6.2 New installation (devices and installation)	**1.5 years**

Tab. 2: Techno-economic effects for the customers

Based on the above mentioned calculations the benefit for the society lays in the CO_2 reduction possibilities by avoiding Greenhouse gas emission through gas turbines for guaranteeing the reserve capacity.

Society	
CO_2 **reduction** with 100,000 bought *ETM*	**95,000t/a**

Tab. 3: CO_2 reduction potential

If no communications (neither GSM nor ripple control) between the utilities and the customers exist ETM can be used with a fixed Tariff model. At fixed daily and seasonally time periods the loads can be managed as it is described above. An additional power limitation guarantees the observance of the contracted power quantity. This is very interesting for the SME like solarium operators, butchers or others. In Austria the Electricity costs are divided in a fix share and variable peak load depending power price. If the SME surpasses the contracted peak thresholds (integrated over 96h measurements through the involved utility) three times a year, you have to pay the much more expensive price over one calendar year. This daily and seasonally fixed Tariff model isn't as flexible and effective as the power and time varied model. But with the Energy profile information a customer specific Demand Side management with positive cost effects is always possible.

4   CONCLUSIONS

The two preliminary field-tests between 1989 and 1990 of a time and power varied Tariff model in Freiburg and between 1994 and 1996 of a time varied Tariff model in Eckernförde yielded that the acceptance of the participants to use a dynamic Tariff model with induced changes on the consumption behaviour is high. More than 80% of the interviewed persons in Eckernförde agreed, that cost-orientated Tariff models have environmental benign effects. The Energy save potential with the combined Tariff model with 24% of the residential peak load reduction. In the interviews the asked participants announced their readiness to shift the loads and use the Energy more efficient.

The above described *Energy Tariff Manager* allows a close link between the active use of Green Energy and the Rational use of Energy (RUE). The consumption behaviour can be controlled and automatically matched with the fluctuated supply of the of PV systems and other Renewables via the load management with positive effects on the market penetrations.

Strengthening the European industry know-how and learning process about the customer requirements will lead to significant improvements in terms of:

- The customer's acceptability
- The positive socio-economic effects
- The Energy supplier and trader perception and
- The technical design and dimensions of the PV systems.

5   REFERENCES

Michael HEIDENREICH, Hubert FECHNER, "Green Energy Labeling", Energieinnovation im liberalisierten Markt 2000, Austria

Michael HEIDENREICH," Successful Financial Incentives for a Dissemination of PV Systems – Evidence from Western Europe, CISBAT 99, Switzerland

T. MOROVIC, R. PILHAR and W. MÖHRING-HÜSER "Dynamische Stromtarife und Lastmanagement Erfahrungen und Perspektiven", Kasseler Symposium Energiesystemtechnik '97, Germany

AVALANCHE Server: Always Up-to-date Market Overview of PV Modules on the Internet
www.ret-market.org

Britta Buchholz[1], Lionel Ménard[2], Didier Mayer[2], Murray Cameron[3], Amy Francis[3], Bernard McNelis[3]

[1]Universität Gh Kassel
Wilhelmshöher Allee 73, 34121 Kassel, Germany
Phone: +49 561 804 6201, Fax: +49 561 804 6434, Email: britta@uni-kassel.de

[2]Centre d´Energétique, Ecole des Mines de Paris
B.P. n° 207, F-06904 Sophia Antipolis Cedex, France
Phone: + 33 493 95 75 99, Fax: + 33 493 95 75 35, Email: didier.mayer@cenerg.cma.fr, menard@cenerg.cma.fr

[3]European Photovoltaic Industries Association (EPIA)
Av. Charles Quint 124, B-1083 Brussels, Belgium
Phone: +32 2 465 9162, Fax: +32 2 468 2430, Email: epia@epia.org

ABSTRACT: The newly developed AVALANCHE Server uses an unprecedented approach to provide an always up-to-date market overview of renewable energy technologies, among them PV modules, on the Internet with minimum maintenance efforts. It is formed by an internet robot that searches individual HTML documents from PV module manufacturers with product information on PV modules. The concept is promising on the long term because it requires minimum efforts for maintaining the AVALANCHE Server and is based on the self-interest of PV module manufacturers to include standardised META tags into their own HTML documents. Simulation software can directly retrieve AVALANCHE data for its calculations.
Keywords: PV Market - 1: PV Module - 2: Simulation - 3

1. INTRODUCTION

Purpose of this work has been to improve the market transparency for PV modules, wind turbines, mini hydro turbines and biogas plants on the internet by developing a smart internet robot. This paper focuses on the section PV modules. The rapidly expanding internet offers a huge but disorganised amount of information on PV. On the one hand, most PV module manufacturers do have their own web-sites and present their products in their individual ways. On the other hand, conventional databases on PV modules are not always up-to-date and require high maintenance efforts. The developed AVALANCHE Server provides among others an overview of always up-to-date product information on PV modules with minimum maintenance effort.

This paper will introduce the functionality of the internet robot in general. It will further present the paths and META tags for the section "PV Modules" in detail. Next, it will show whose web-sites are AVALANCHE compatible today and which simulation software retrieves information directly from AVALANCHE.

2. FUNCTIONALITY OF THE SMART INTERNET ROBOT

2.1 General description

In the framework of the AVALANCHE project we have set-up a smart internet robot based on a tool called 'Compass Server' of the Netscape Communications Corporation company. This smart internet robot enables users to locate resources from the PV manufacturers that are spread on their web sites.

From the user´s perspective, this is a stand-alone index service, much like commercial web indexes, targeted on the renewable energies market including PV modules manufacturers.

The user will interact with the server by either typing in search queries using keywords or by browsing through a set of thorough categories. The server will respond with a list of items that correspond to the specified keywords or category.

From the administrator´s point of view the server is a database of information about resources available at manufacturers´ web sites. A program called a robot will find all relevant resources and generate information about them for the database. The robot works behind the scenes, invisible to end users.

Figure 1 presents a scheme of the whole process.

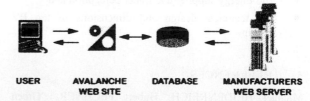

USER   AVALANCHE WEB SITE   DATABASE   MANUFACTURERS WEB SERVER

Figure 1: Scheme of the AVALANCHE functionality.

2.2 User´s side
On the AVALANCHE server users can
- Search by keywords;
- Browse by categories; and
- Combine searching and browsing.

So instead of having to know about and visit a number of sites, AVALANCHE provides a centralized, searchable database for renewable energies market where users can pinpoint and retrieve the resources they want.

2.3 Administrative side

The robot can index different kinds of documents. Among them we can find :

- Web pages (HTML documents)
- Plain text files
- Word-processing documents
- Other documents that can be converted into HTML
- dynamically generated documents such as database URL encoded documents
- FTP directories
- Meta information

For AVALANCHE, standardized meta-information is addressed which will be described in the next chapter. New categories can be added to hold any new manufacturer's document.

The robot can be scheduled to revisit all its sites to look for new documents and update descriptions of changed ones. This is an important feature that greatly decreases the maintenance effort while always providing users accurate and up-to-date information. It also promotes the idea based on the self-interest of PV-modules manufacturers to provide precise information on their web sites.

3. STANDARDIZED META TAGS FOR THE SECTION "PV MODULES"

3.1 Positioning of META tags

In order to be found by the AVALANCHE Server, web-sites of individual PV module manufacturers must be adapted to match certain newly defined standards. The AVALANCHE Server searches for standardised META-tags that are included in individual HTML documents that contain technical data on PV modules.

The basic structure of an HTML document is depicted in Figure 2. The upper side shows the source code of an HTML document and the lower side what is shown by the browser.

```
<HTML>

<HEAD>
<TITLE> This is the title of the
document </TITLE>
</HEAD>

<BODY>
This text is shown by the browser.
</BODY>

</HTML>
```

This text is shown by the browser.

Figure 2: Source code and browsed display of an HTML document.

Commands in HTML are called "tag". As the name tells, META tags contain information about an HTML document. Accordingly, they are located in the head of the document.

There exist a number of predefined names for META-tags, e.g. "description" or "keywords", which are included in most HTML documents. Search engines save time by only reading such META-tags instead of the full document text. Within the section "PV Modules", the AVALANCHE Server only reads the newly defined META-tags with the names "classification" and "data".

3.2 META tag "classification"

The META-tag "classification" reflects the categories in which the PV modules are classified in the AVALANCHE Server (Fig. 3), namely "Country of Origin", "Manufacturer" and "Power Rate".

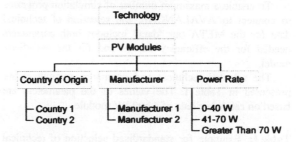

Figure 3: Detailed categories for the section "PV Modules".

For the visitor of the AVALANCHE Server, searching by each of the mentioned categories will result in a list of links that lead to the individual HTML documents of the manufacturer's web-site which contain technical data on the selected category of PV modules.

The content of the META-tag "classification" corresponds exactly to the three possible paths that the visitor of the AVALANCHE Server can choose in order to reach the HTML document under consideration. Its syntax is as follows:

```
<META name="classification"
content="
[Technology:PV Modules:Power Rate:
{0-40 W,41-70 W,Greater Than 70 W}]
[;Technology:PV Modules:Manufacturer:
{Manufacturer 1,Manufacturer 2}]
[;Technology:PV Modules:Country of
Origin:{Country 1,Country 2}]">
```

The AVALANCHE Server visits a selected list of HTML documents, i.e. PV module manufacturers must register their relevant HTML documents. It can be seen from the META tag "classification" that each data sheet for a PV module must be presented in one individual HTML document.

3.3 META tag "data"

The second step of META tag preparation aims at providing detailed information on the PV module that is necessary for simulating the energy performance, economic efficiency and environmental impact of a PV system containing the selected module. This is done by the META tag "data" for technical and economic product information. The AVALANCHE Server finds and extracts the content of this META tag into a database. The connection of simulation software to AVALANCHE will be described in Chapter 5.

The selection of technical data included in the META tag "data" is based on a survey on PV simulation software [2]. Input data requirements go hand in hand with the models behind the software. Mostly applied electric models include efficiency curves, one-diode-model and two-diodes-model.

Studies have shown that the one-diode model provides accurate results with a reasonable calculation effort [3]. Thus it should be preferred both to the two-diodes model with higher calculation effort and to the efficiency curves with a less accurate simulation. However, to date various simulation tools still employ the simple efficiency curves.

To enable a maximum number of simulation programs to connect to AVALANCHE, the selection of technical data for the META tag "data" includes both parameters needed for the efficiency curves and for the one-diode model.

The selected technical and economic parameters are presented in Table I. The values of the parameters are based on catalogue data of a real PV module.

Table I: Example for standardised selection of technical and economic data for the META tag "data" in the section "PV Modules".

Manufacturer 1 - Module 55 W - Technical data		
Parameter	Value	Unit
Cell type (amorphous/ mono-/ polycrystalline)	mono-crystalline	
Number of series cells	36	1
Area	0.43	m^2
Length	1293	mm
Width	329	mm
Maximum power rating	55	W_p
Rated current I_{mp}	3.15	A
Rated voltage V_{mp}	17.4	V
Short circuit current I_{SC}	3.45	A
Open circuit voltage V_{OC}	21.7	V
Fill factor	0.73	1
Rated efficiency	13	%
Number of bypass diodes		1
Diode blocking voltage		V
Dark Current		A/cm^2
Change of I_{SC} with temperature	+1.2	$mA/^\circ C$
Change of V_{OC} with temperature	-0.077	$V/^\circ C$
Technical Lifetime	20	years
Country 1 - Manufacturer 1 - Module 55 W - Economic data		
Price	175	
Currency	EUR	

The parameter „fill factor" is not needed by current simulation programs, but has been considered as relevant for the quick comparison of PV modules and has been added for completeness of the data. Life time, price and currency are needed for economic efficiency calculations. The currency is an extra parameter, because simulation programs must calculate with the value and thus cannot extract value and unit in one parameter.

In order to provide above data to various simulation tools, the META tag "data" must be included into the HTML document as described in Figure 4. In the META

tag, no units can be included. Thus, the values given in the META tag must correspond to the units given in Table I.

```
<HEAD>
<TITLE> Manufacturer 1 - 55 W Module -
Technical and economic data </TITLE>

<META NAME="classification"
CONTENT="
Technology:PV Modules:
Power Rate:41-70 W;
Technology:PV Modules:
Manufacturer:Manufacturer 1;
Technology:PV Modules:
Country of Origin:Germany">

<META NAME="data" CONTENT="
type=monocrystalline;
series_cells=36;
area=0.425397;
length=1293;
width=329;
Pmax=55;
Imp=3.15;
Vmp=17.4;
Isc=3.45;
Voc=21.7;
ff=0.73;
eff=13;
bypass_diodes=;
Vdb=;
dark_current=;
Isc_temp=1.2;
Voc_temp=-0.077;
life=20;
price=175;
currency=EUR">

</HEAD>
```

Figure 4: Standardised META tags "classification" and "data" for a PV Module.

4. WHOSE WEB-SITES ARE AVALANCHE COMPATIBLE?

Web-Sites of PV module manufacturers are compatible to AVALANCHE, if
- each PV module is presented in one individual HTML document within the web-site;
- there is preferrably an English product description.

A study was made of all the web-sites of the PV manufacturers who are members of EPIA as well as of some non-members to assess their compatibility with AVALANCHE. As shown in Table II, of the 15 PV module manufacturers, six have web sites currently in the correct format. Out of these, only three present their data in English language, namely SOLON AG, Photowatt International and Siemens Solar GmbH.

During February 2000 the Avalanche site was opened to the public. There are currently 10 manufacturers listed on the site under PV.

For those manufacturers who are currently restructuring their web-sites and integrating AVALANCHE META-tags, an interim demonstration version has been set up at the server of Kassel University. It contains HTML documents with the META tags "classification" and "data" that lead directly to the corresponding web-sites of the manufacturers. As soon as manufacturers have made their own web-sites compatible, this interim solution will be removed.

Table II: Internet presence of selected PV Manufacturers as regards their compatibility to AVALANCHE, i.e. presentation of each product on an individual HTML document and English language.

COMPANY	ONE HTML DOCUMENT/ PRODUCT	LANGUAGE
ASE GmbH Germany	PDF files	English
ATLANTIS SOLAR SYSTEMS LTD. Switzerland	Product information under construction	English
BP SOLAR England	Products not listed	English
EUROSOLARE SPA Italy	Yes	Italian
INTERSOLAR GROUP UK	No	English
ISOFOTON S.A. Spain	No	Spanish
PHOTOWATT INT. S.A. France	Yes	French English German
PILKINGTON SOLAR Germany	Products not listed	English
SHELL SOLAR Holland	No	English
SIEMENS SOLAR GmbH Germany	Yes	English Spanish German
FREE ENERGY EUROPE Holland	No	English
SOLARWERK GmbH Germany	Yes	German
SOLARWATT Solar-Systeme GmbH Germany	No	English German
SOLON AG Germany	Yes	English French German Spanish
WEBASTO Systemkomponenten GmbH Germany	Yes	English German

5. AVALANCHE COMPATIBLE SIMULATION SOFTWARE

Energy simulation software can retrieve technical data of PV modules directly from the AVALANCHE data base. This innovative internet feature is included in the simulation software PVS 2.000 by Fraunhofer Institut for Solar Energy systems and econzept Energieplanung GmbH. The functionality of this feature is fully integrated into the program. The only requirements for the user are the existence and the establishing of a connection to the internet.

PVS 2.000 is a powerful instrument to plan and dimension grid-connected and stand-alone photovoltaic systems[4]. With the program the user is able to compute the PV energy output, characterize and evaluate the operation of the system, compare different variants of the system or optimize system parameters such as the tilt angle or the battery capacity.

Options of connecting simulation software to AVALANCHE include URL encoding or an on-line tool programmed in JAVA. The latter has been done for the simulation of environmental impacts of PV modules within the AVALANCHE server.

6. SUMMARY AND OUTLOOK

The AVALANCHE Server is a very powerful tool. This paper showed the smart and innovative approach for providing a centralized information index of the renewable energy market with an internet robot, the Netscape Compass Server. Maintenance of the server requires very little effort.

Paths and META tags for the section "PV Modules" have been described in detail. They are currently integrated into web-sites of several PV module manufacuters. While the responses of all contacted manufacturers has been positive, only a few have adapted their web-sites already to AVALANCHE by integrating the META tags "classification" and "data". Others are represented on the server via an interim solution of a demo-server .

The simulation software PVS 2.000 retrieves information directly from AVALANCHE. In addition, environmental impacts of PV modules are simulated on the AVALANCHE server.

REFERENCES

[1] B. Bäurle, V. Nielsen, L. Ménard, Internet Renewable Energies Information System (IRIS), in: "Renewable Energy Databases: Fostering RE Market Expansion", Proceedings of the second European Workshop, 19-20 April 1999, Munich, Germany (1999).

[2] M. Voigt, Globale Simulation von PV-Anlagen im Internet, Diplomarbeit I, Universität Gh Kassel, Kassel (1998).

[3] V. Quaschning, Simulation und Abschattungsverluste bei solarelektrischen Systemen, Verlag Dr. Köster, Berlin (1996).

[4] V. Jung, B. Buchholz, PVS 2.000 – Ein Auslegungsprogramm mit Anbindung an die Internetdatenbank AVALANCHE für Solarmodule, in: Tagungsband 15. Symposium Photovoltaische Solarenergie, 15. – 17. März 2000, Staffelstein, Germany (2000)

The project AVALANCHE has been partly funded by EC-DGXII and Netscape.

PV R&D WITHIN THE DUTCH NOZ-PV PROGRAM

J.J. Swens, M. van Schalkwijk and F.M. Witte
Netherlands Agency for Energy and the Environment, NOVEM
P.O. Box 8242, NL-3503 RE Utrecht, The Netherlands, j.swens@novem.nl

ABSTRACT: On behalf of the Ministry of Economic Affairs of the Netherlands the Netherlands Agency for Energy and the Environment (NOVEM) carries out the Netherlands Development Program for Photo Voltaic Solar Energy (NOZ-pv). The governmental target is to have an installed pv power of 1450 MW in 2020. The efforts of the NOZ-pv program are divided over four clusters: 1. R&D on solar cells, components and systems, 2. Learning program pv in the build environment, 3. Tender program PV-GO and 4. Stand-alone pv systems: export and application in the Netherlands.

This paper only deals with the first subject, research and development on solar cells, components and systems. For solar cell R&D the focus is on multicrystalline silicon cells for the short term, amorphous silicon- and polycrystalline thin-film silicon cells for the medium term and organic solar cells for the long term. The R&D for components are split in two fields: electrical components and building integration components. For electrical components the R&D priorities are low cost ($< \varepsilon$ 0,5 /Wp) reliable inverters for the short term and international standards and electrical modularity for the medium term. A priority for the long term might be the development of products and components for use in larger DC networks. For building integration products the main targets for the short term are broadening of the product range, improvement of the quality and the establishment of national standards, for the medium term these are standardised pv building components and zero additional costs for BOS for building integrated systems and for the long term pv as standard building product.

Keywords: National Programme – 1; R&D and Demonstration programmes – 1; Funding and Incentives - 2

1.   INTRODUCTION

Solar power will be one of the most important sources of sustainable energy in the 21st century. To prepare the Dutch industry for a role in the future solar energy market, the Dutch Ministry of Economic Affairs has set up the Netherlands Development Program for Photo Voltaic Solar Energy (NOZ-pv). The NOZ-pv program is set out to create the conditions for the large-scale introduction of pv in the Dutch energy supply in the 21st century and ensures a broad approach [1]. The yearly budget is around ε 15 million. The Netherlands Agency for Energy and the Environment (NOVEM) is carrying out this program. The NOVEM efforts for the NOZ-pv program are divided over four clusters: 1. R&D on solar cells, components and systems, 2. Learning program pv in the build environment, 3. Tender program PV-GO and 4. Stand-alone pv systems: export and application in the Netherlands. The objectives of the program for the coming years are lower costs, a solid industrial position, a healthy market for pv systems, and a strong knowledge infrastructure.

The costs of solar electricity strongly depend on the combination of solar cells and balance of system costs like inverters and roof integration techniques. A continuous feed back from the experiments and applications in the built environment into cell, BOS component and system research is essential for making the right choices. Co-operation of market parties with universities, research centres and industry within technology clusters [2], is an important condition for a coherent and co-ordinated research that has to lead to cost price reduction and large-scale application of pv in the Netherlands.

The involvement of market parties in the application of pv is established by the PV Covenant Group. At this moment 28 parties have signed a combined effort agreement. Among the parties are the Ministry of Economic Affairs, power companies, pv industries, parties from the building construction sector, research institutes and NOVEM. The agreement lays down the role, contributions and obligations of the parties involved, as well as arrangements for mutual co-operation and transfer of knowledge. The targets set by the Covenant group will be reached by the end of 2000. It can in fact be seen as a successful application of the Dutch 'polder model' to the development of pv. Preparations for a new agreement from 2001 to 2007 have been started.

The ambition in the Netherlands is to built a solid pv infrastructure in order to become one of the five biggest pv producing countries in the world. Only by making the right choices and setting firm priorities a strong position can be created in a relatively small country as the Netherlands. The choices and priorities of the NOZ-pv program are continuously being monitored through market and stakeholders investigations and in brainstorm sessions in peer groups.

The subject of this paper is the R&D on solar cells, components and systems. The main goal on these fields is the support of fundamental research, industrial research and product development in order to create a strong pv knowledge infrastructure. This is necessary to play a major role in the fast growing international pv market and to be among the five strongest pv countries in the world.

2. R&D ON SOLAR CELLS

2.1 General

Fundamental solar cell research projects in the Netherlands mainly takes place in the Universities and Technology Centres. Some of these are supported by the leading Dutch pv industries. Industrial research is carried out in combined effort with the Universities and Technology Centres, but at least partly take place in the industrial environment. The estimated total public funding available for solar cell R&D ads up to ε 25 million for 2000, from which ε 4 million was will be supplied by the NOZ-pv program. The contribution from the Dutch pv industry for 200 will at least be ε 3 million.

2.2 Multicrystalline silicon

Short term R&D on the current generation multicrystalline silicon solar cells is targeted on metallisation, light confinement and passivation [3], [4],[5],[6]. Shell Solar and ECN mainly perform this work. Several technological developments and new production methods have the potential for considerable improvements within several years [7],[8]. It is expected that the efficiencies will increase from the existing 13% to about 17% for commercial mc-Si cells before 2005. New wafer materials (solar grade silicon), thinner wafers, new cell and module designs can lead to considerable cost reductions. It is expected that these developments, in combination with scaling-up and further automation, will lead to a three or four fold reduction of the production costs compared to the present (small) commercial production lines. In combination with the projected cost reductions for BOS this would result in

ε 2/Wp for the installed pv system. If indeed the current cell efficiencies are maintained and these cost reductions are realised within a few years, multicrystalline silicon will be a serious competitor for thin-film technologies on the mid-term. In conclusion, the R&D on multicrystalline silicon will ensure that mc-Si will be the leading pv technology in the next ten years.

2.3 Amorphous silicon

For the mid-term amorphous silicon is probably the most promising thin-film candidate for industrial upscaling in the Netherlands. The broad support of fundamental research at several Dutch universities has created a solid infrastructure for the development of this thin layer technology. Knowledge has been built up on (plasma) deposition, relations between material properties and deposition parameters, stability and device design. Increase of growth speed, the improvement of stability and the application of new (tandem) devices form the base for the desired high-speed large-area production of amorphous silicon solar cells. Promising is the Expanding Thermal Plasma deposition technology, which is developed at the Eindhoven University of Technology, that could lead to 50-100 times faster deposition of amorphous silicon compared to standard PECVD [9].

The industrial implementation of a-Si in the Netherlands has started within the Helianthos project of Akzo Nobel. Akzo Nobel has formed a technology cluster with the University Utrecht, the Delft and Eindhoven Universities of Technology, and the Netherlands Organisation for Applied Scientific Research (TNO). Pilot production of amorphous silicon flexible foil in a roll-to-roll process is planned in 2002. Commercial production of this thin-film technology is projected to start up five years from now. The price for the electricity from these solar cells should be comparable to the grid electricity consumer prices. This would mean a real breakthrough for the large-scale implementation for pv electricity production.

2.4 Thin-film polycrystalline silicon

Polycrystalline thin-film silicon solar cells are also a candidate for the mid-term. R&D on deposition, cell design and possible production processes must prove the perspective of this concept: high cell efficiencies on cheap carrier materials [10],[11],[12]. The Netherlands Energy Research Foundation (ECN) and the Delft University of Technology already proved the principles of this concept [13]. Closed polycrystalline silicon layers have been deposited on ceramics. Estimations of cost prices in combination with the developments of multicrystalline silicon must show if this technology could be the successor of the current crystalline solar cells. The breakthrough has to come from high-volume large-area production of these thin films.

2.4 Organic Solar Cells

Long-term research is mainly focussed on organic solar cells, as they bear the promise of becoming the most cost-effective type of solar cells in the future. Especially the large amount of freedom in the synthesis and choice of organic molecules and functional (conducting) polymers, and possible combinations with inorganic materials, make this approach attractive. R&D are focussed on the physics of energy collection, exciton transport, charge separation, synthesis of new materials, material stability and device design. Various types of cells like the Grätzel cell [14], a solid cell with a structured antenna layer and a bulk hetero-junction cell are being studied. Recently, a big R&D project, with university and industry partners, has started in the Netherlands to develop a stable bulk hetero-junction cell based on bucky-balls with 5% efficiency within 5 years from now.

3. R&D ON ELECTRICAL COMPONENTS

3.1 General

R&D on electrical components mainly focus on the improvement of inverters and energy measuring devices. The most important goal for the near future is the price reduction of these components. The R&D therefore mainly consist of product development and take place at the involved industries with sometimes the support from Technology Centres or Universities. The total effort, that will be put into the development of these products in the Netherlands in 2000 equals a budget of ε 2.7 million, from which the NOZ-pv program only supports ε 0,6 million (22%). The other ε 2.1 million is covered by the industry, showing a growing confidence in the future pv market.

3.2 Inverters

Several market introductions [15] and -investigations have come to the conclusion, that the first large-scale pv market in the Netherlands will be that of the small to medium sized roof integrated private solar systems. Based on that the Dutch inverter industry focuses its short-term R&D activities on low price inverters for systems of up to 4 panels. The price target set for the end of 2000 is < ε 0.5 per Wp. For larger inverters of 2000 Wp and more this target was already met in 1999. Here the efforts are aimed at DFP (Design For Production), at increased reliability and at adaptation to foreign markets. Furthermore to maintain the strong international position of the Dutch mini-inverters (single panel) for AC modules or simplified grid connection effort will be put into upgrading and cost reduction [16].

For the mid-term much effort is put into the target of the formulation of national- and international standards for electrical components for pv applications [17].

Another mid-term target is the development of fully electrical modular panels or so called 'plug-and-play' larger pv systems. For the long term the idea of large DC networks, requiring DC appliances, DC storage and AC-DC converters, has been launched, in order to open the discussion about future form of application of solar energy and the consequences for electrical components.

3.3 Energy measuring devices

In the field of energy measuring devices the price-quality ratio is the major problem. Presently both low price - low quality and high price - high quality devices are available [18]. Current developments show two approaches: 1. increasing the reliability of the low cost devices and 2. cost price reduction of the high quality devices. The target for 2000 is a sales price of < ε 50 for a reliable and sufficiently informative measuring device for systems up to 400 Wp.

For the mid- and long-term study groups are investigating the possible ways of monitoring addressing for example large system monitoring [19], and the possibility and necessity of short-term predictions for solar energy production.

4.   BUILDING INTEGRATION PRODUCTS

4.1 General

For building integration products the short-term target is to increase the number of products available, to decrease the costs of the products and to ensure their quality. The development of new products and the improvement of existing ones is taken up by a growing number of industries.

The available budget to support the industrial development of building components for pv within the NOZ-pv is approximately 0,5 million Euro per year. An additional 0,3 million Euro is invested by the NOZ-pv in the establishment of a system for quality control. As one of the first countries in the world the Netherlands have a standardisation committee for building integration of PV.

4.2 Product Development

During the last few years, and particularly in 1999, a number of products have been developed for the integration of pv in buildings. These products cover a vast range of integration applications, from simple flat roof systems to sophisticated tilted roof systems and facades [17],[21][23].

Active parties in this field are RBB-Lafarge (sloping roof integration systems), Laura Starroof (a-Si/Metal systems), NECO/ZND (a-Si/bitumen systems for flat roofs), Unidek and Ecofys (prefabricated subroofs suitable for many module types), Ubbink (support system for single and double panel systems for sloping roofs), Ecofys (modular flat and tilted roof systems) and Oskomera (pv façade systems). An interesting aspect is that most of the new systems are developed to have a watertight layer *underneath* the pv panels [20].

Cost prices are now in the order of 0,4 to 1 Euro per Wp, whereas still a considerable amount can be accounted for engineering and installation.

Challenges in the near and midterm future are twofold: integration systems should become available at cost prices comparable to conventional roofing material, and they should be easy to install.

A very important issue for building industry regarding cost price reduction of bipv components will be flexibility of choice of pv material for their product. A key aspect therein is standarisation of sizes between pv and building industry.

4.3 Standardisation

In order to facilitate the process of quality assurance for bipv products Novem has erected a standardisation committee with the NNI, the Netherlands Standardisation Institute. This committee, NC 353059 "Building physical aspects of solar energy systems in buildings", will start with the preparation of a national standard for building integration of pv (and solar thermal systems). The standard will describe which requirements a bipv system should meet in order to be in line with the national building act.

Furthermore Novem strongly supports initiatives for CE-marking of bipv products [3],[19], also carried out in the framework of the standardisation committee, in order to ensure that bipv products developed meet the European Construction Products Directive. This will be essential in a future European market, in which pv should be regarded as a standard building product.

Finally, the important issue of size adaptation between pv and building industry will be treated in this forum. It is anticipated, however, that the discussion should take place on an international level.

5.   CONCLUSION

The Dutch industrial and scientific activity in the field of pv solar development have shown a strong growth. The Dutch industry and research institutes have gained a strong international position. This could only be achieved through narrow collaboration of all market parties involved. In this collaboration the R&D cluster of the NOZ-pv programme played, plays, and will play a central co-ordinating roll.

For the future the NOZ-pv programme will maintain its roll as a collaboration stimulator and will keep organising open discussions with all market parties to develop a broadly supported Dutch vision on and approach towards new developments in solar energy.

REFERENCES

[1]   J.T.N. Kimman, E.W. ter Horst, L.A. Verhoef, E.H. Lysen, Proceedings of EPSEC-13, Nice (1995), p. 793-796.

[2]   J.T.N. Kimman, E.W. ter Horst, L.A. Verhoef, E.H. Lysen, Proceedings of PVSEC-9, Miyazaki (1996), p. 45-46.

[3]   J.H. Bultman, R. Kinderman, J. Hoornstra and M. Koppes, Proceedings of EPSEC-16, Glasgow (2000), paper VA1.56 (to be published).

[4]   A.R. Burgers and J.H. Bultman, Proceedings of EPSEC-16, Glasgow (2000), paper VA1.57 (to be published).

[5]   C.J.J. Tool et al., Proceedings of EPSEC-16, Glasgow (2000), paper VA1.58 (to be published).

[6]   L. Doeswijk et al., Proceedings of EPSEC-16, Glasgow (2000), paper VA2.20 (to be published).

[7]   W.J. Soppe et al., Proceedings of EPSEC-16, Glasgow (2000), paper VA1.55 (to be published).

[8] J.H. Bultman et al., Proceedings of EPSEC-16, Glasgow (2000), paper OD5.1 (to be published).

[9] B.A. Korevaar, R.A.C.M.M. van Swaaij and J.W. Metselaar, Proceedings of EPSEC-16, Glasgow (2000), paper VB1.20 (to be published).

[10] P.A.T.T. van Veenendaal, J.K. Rath and R.E.I. Schropp, Proceedings of EPSEC-16, Glasgow (2000), paper VB1.11 (to be published).

[11] J.K. Rath et al., Proceedings of EPSEC-16, Glasgow (2000), paper VB1.12 (to be published).

[12] F. Fung et al., Proceedings of EPSEC-16, Glasgow (2000), paper VD3.15 (to be published).

[13] A.J.M.M. van Zuphen et al., Proceedings of EPSEC-16, Glasgow (2000), paper VA1.53 (to be published).

[14] P.M.P. Sommeling et al., Proceedings of EPSEC-16, Glasgow (2000), paper OC8.5 (to be published)

[15] B. de Wit, S. van Egmond, P.S. Zaadnoordijk and D. Dijk, Proceedings of EPSEC-16, Glasgow (2000), paper OA4.5(to be published).

[16] M. Kardolus and H. Marsman, Proceedings of EPSEC-16, Glasgow (2000), paper VD1.23 (to be published).

[17] A. Kanakis, J.A. Eikelboom and P.M. de Rooij, Proceedings of EPSEC-16, Glasgow (2000), paper OB6.5(to be published).

[18] E.C. Molenbroek, A.J.N. Schoen, E. Vrins and P. Nuiten, Proceedings of EPSEC-16, Glasgow (2000), paper VC2.36 (to be published).

[19] E.C. Molenbroek, A.J. Kil, T. Schoen and F. Vlek, Proceedings of EPSEC-16, Glasgow (2000), paper VC1.50 (to be published).

[20] T.H. Reijenga and H.F. Kaan, Proceedings of EPSEC-16, Glasgow (2000), paper VC1.40 (to be published).

[21] J. Oldengarm, Proceedings of EPSEC-16, Glasgow (2000), paper VC1.69 (to be published).

[22] N.J.C.M. van der Borg and WE.J. Wiggelinkhuizen, Proceedings of EPSEC-16, Glasgow (2000), paper VC3.23 (to be published).

[23] P.C. Scheijgrond et al., J.C. Jol et al., Proceedings of EPSEC-16, Glasgow (2000), paper VC1.75 (to be published).

[24] M. van Schalkwijk et al., Proceedings of 2nd WCPVSEC, Vienna (1998), p. 2587-2590.

[25] J.C. Jol et al., Proceedings of EPSEC-16, Glasgow (2000), paper VC1.51 (to be published).

GUARANTEE OF RESULTS

FOR GRID CONNECTED PV-SYSTEMS

Gerd Heilscher[1], Robert Pfatischer[1],
J.Y. Quinette[2], E. Molenbroek[3], C. Meier[4]

[1]ist EnergieCom GmbH, Stadtjägerstr. 11, D-86152 Augsburg, Germany
Tel.: (+49) 8 21 / 3 46 66 – 0, Fax: (+49) 8 21 / 3 46 66 – 11,
eMail: heilscher@ist-energiecom.de

[2]tecsol (France), [3]Ecofys (The Netherlands), [4]energiebüro (Switzerland)

ABSTRACT: Grid-connected PV-systems were first introduced in demonstration projects where mainly technical aspects of the systems were examined. The price for the systems was based on the STC power. With rate-based incentives it will become important for an investor to be able to calculate the kWh-price from a given system. To link system price and system performance PV-system suppliers should give a performance guarantee. Guaranteed Results are a powerful marketing instrument. They establish confidence and give advantages both to customers and PV companies. Within an ALTENER-project a model contract for grid connected PV systems including performance guarantees is developed. Using this GRS^{PV} method will increase the quality of the systems. Confidence of investors will help to increase the PV market.

Keywords: Grid-Connected –1: PV Market – 2: Dissemination – 3:

1. INTRODUCTION

Besides the costs especially the unfamiliarity with PV stands in the way of large scale dissemination of PV technology. Although PV-systems technology has shown to be quite reliable, this is not enough to attract investors. For investors, a clear link between price and performance of a PV-system is necessary. Having PV companies give performance guarantees is a possibility to establish this link.

Giving guarantees is already a common marketing instrument in various fields. It establishes confidence and helps to convince the client.

Within an ALTENER-project a model contract for grid connected PV systems including performance guarantees is developed. Background to and details of the model contract are given in this paper.

2. PROJECT DESCRIPTION

The GRS^{PV}-Procedure is a contractual agreement between a PV company and a PV-client, which guarantees him a certain performance of the system.

The kind of guarantee is fixed in standardised or individual contracts. Both the client and the PV company

gain through the advantages of GRS$^{PV.}$ as you can see in table 1:

PV Company	Customer, Investor
gives the customer additional confidence	establishes additional confidence to the creditor
it is a quality label for professional operators	service and performance are guaranteed
it is a powerful marketing tool	simplified turn key solutions, it help's to compare offers
it provides a standardised framework for negotiations	guaranteed basis for calculations

table 1: Advantages of GRS^{PV}

Because of these advantages the GRS^{PV} concept is easy convincing to investors. However, some PV companies are afraid of the measures and costs which have to be taken into account to give such guarantees.

A standardised model contract, with detailed comments what is to consider at each point, should make it easier for them as well.

Apart from the suppliers, the customers have to be "educated" as well. They have to realise the added value may cost more. However, with GRS^{PV} issues like operation and maintenance are considered from the start, instead of being tagged on at the end as additional costs. Taking this into consideration, the added cost may not be as high after all.

In a PV project there is not only the relationship between the customer and the PV company. In different projects a lot of other connections are possible. The most important are shown in figure 1.

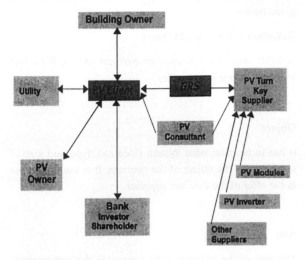

figure 1: project relationships

The PV Client is the middle point. He does not have to be the owner of the building or even of the PV system, but he is the moving force of the project. The project is financed by a bank or other investors, the energy is sold to a utility.

GRS^{PV} defines the relationship between the Client and the Turn Key Supplier. In some cases a PV Consultant helps to establish this connection. The services of the PV Turn Key Supplier are based on the suppliers or manufacturers of the components.

Regarding these points the presented project within the EU-ALTENER-Programme has the following aims:

1. Testing of the GRS^{PV} concept

2. Overcome the obstacles against using GRS^{PV} in PV industry

3. Definition of measures and costs of using GRS^{PV}

4. Gaining experiences in using GRS^{PV} in selected companies

In the first step the four project partners TECSOL (France), Ecofys (The Netherlands), energiebüro (Switzerland) and *ist* EnergieCom (Germany) analysed already finished projects. As a result a draft contract for using GRS^{PV} was developed.

In this model contract all influencing parameters regarding local and technical aspects and the performance guarantee have to be considered. This will be done in section 4. At first, an outline of the procedure is given in the following section.

3. GRS^{PV}-PROJECT PROCEDURE

A schematic diagram for the GRS^{PV}-Project Procedure is shown in figure 2:

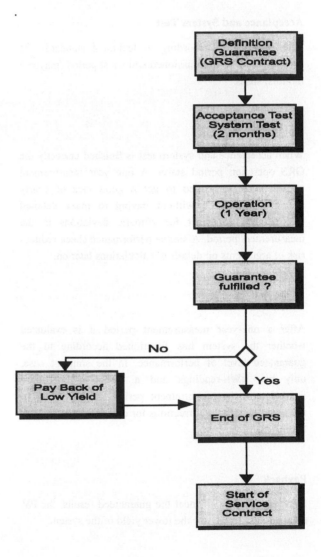

figure 2: GRS^{PV}-Procedure

Already in the tendering and offering phase of a PV-realisation project it should be discussed what PV-suppliers have to offer in terms of a performance guarantee. This might be difficult, because not all parameters that influence the yield at a given location might be known. For example, on a new housing location, it might be unsure what the shading situation will be. Or, performance estimates have to be made on the basis of preliminary sketches.

Definition Guarantee, GRS-Contract

Before the start of the project, negotiations between PV client and PV Turn Key Supplier define the contents of the GRS-contract. For completeness, all items from the model contract should be covered. A checklist of items for the contract is discussed in section 4.

Acceptance and System Test

After installation according to technical standards the system has to be commissioned and a test period (maybe 2 months) starts.

Operation

When acceptance and system test is finished correctly the GRS operation period starts. A one year measurement period is recommended to get a good idea of yearly system performance without having to make detailed calculations to correct for climatic deviations in the measurement period. A simple performance check reduces risk of arguments on details of calculations later on.

Guarantee Fulfilled?

After a one-year measurement period it is evaluated whether the system has functioned according to the guaranteed level of performance. In the simplest case, only two kWh-readings and a value of horizontal insolation in the measurement period are necessary. In more complex cases corrections for outages might have to be made.

Payback

If the system doesn't meet the guaranteed results, the PV company has to pay for the lower yield of the system.

End of GRS

In both cases the GRS period ends after one year. From that point it is the risk of the customer to operate the system.

Service Contract

PV systems should operate for about 20 years. So it is strongly recommended to start a service contract after the GRS period to provide long time operation without problems.

4.   ITEMS FOR THE MODEL CONTRACT

Naturally, details of a real contract are too project-specific to generalise. However, for the fulfilment of a performance guarantee, a number of items have to be covered in one way or another. A list of these items is given below.

Defintion of contract partners

In most cases the two contract partners are the PV-client and the PV turn key supplier.

Object

It has to be clear what system (location, type and size of system, ..) is the object of the contract. It is useful to refer to the offer of the turn key supplier.

Aim

The aim of the contract is to guarantee the performance and the quality of the system.

Duration

Duration and course of the different contract periods have to be defined. Important moments (start, end, test period, commissioning) have to be well defined.

Site Initial Conditions

The site initial conditions are very important for design and operation of the systems. They should be registrated in the contract. For example think about the available area (orientation and angle), the local climate and shadings (horizon, antennas, tubes,…).

Guarantee

The total amount of energy read from the kWh-meter is what counts. This is the value which is guaranteed, under conditions stated in other items in the contract.

Reference conditions

You need a meteorological reference for giving the guarantee and for checking it. So the meteo station to be used and possibly an interpolation method has to be defined.

Contract Limitations

The PV Supplier is not responsible of all errors. There are some limitations he can't be blamed for (grid failure, user faults, damages, theft, force majeur, overvoltage, ..). Also changing roof conditions can influence the performance of the system.

Correction Methods

For every parameter that influences the calculation method for correction of the guaranteed yield has to be cleared, e. g. deviation of the reference irradiation, grid outages, changing start conditions,... .

Failure Detection

Who is responsible of failure detection (e. g. interpretation of data)? What is the checking method (manually, automatic, periodically, …)?

Financial Aspects

In case the yield after correction is less than the guaranteed yield, the turn-key supplier shall pay an amount according to the lost production over the whole lifetime of the system to the PV client. Many variations on how to do this are possible.

Dispute Settling

In case a dispute between PV-turn-key supplier and PV-client arises on the execution of this contract, independent experts will be brought in to decide on such matters. It could also be stated who these experts will be.

Post Contract Measures

The GRS-contract should help to ensure that the PV-client gets the expected output from the PV-system. However, a GRS-contract only covers the first year after commissioning of the PV-system. The PV-system should work well for the whole depreciation period. The PV-client should have insight in the cost of maintaining the system throughout this period before the decision on making the investment is made. Therefore, an offer for a maintenance contract should be made by the PV-turn-key

supplier at the time of the offer for the PV-system. This should cover elements like yearly cost for operation and maintenance, availability of spare parts, necessity for overhaul of the system after a number of years. It could be decided that the PV-client or a third party will operate the system after the GRS-period. In this case, there should be training by the turn-key supplier (on operation of the PV-system as well as the monitoring system) to take over these tasks from the turn-key supplier.

Annex

Additional documents should be included in the annex of the contract, e. g. the design study, a site description, the commissioning report, the maintenance procedures, maybe a system log book.

5. CONCLUSIONS

By providing a framework for a GRS-PV contract, the project results will facilitate the use of performance guarantees by the PV turn key supplier. GRS^{PV} will lead to improved quality for the PV systems and ensures that clear agreements are made on operation and maintenance of the system. Also, the transfer of the model contract to an accepted quality standard is considered. With a proven contractual framework and the further utilisation of GRS^{PV} the confidence of potential investors and thus the PV market will be increased.

References:

/1/ J.Y Quinette et al.: "Guaranteed Solar Results for Solar Thermal Systems",1st ALTENER Conference –Renewable Energy entering the 21st Century, Proceedings, 25.-27. Nov. .Sitges, Spain 1996.

/2/ G.. Heilscher et al.: "Guaranteed Solar Results of Photovoltaic Systems", 1st ALTENER Conference – Renewable Energy entering the 21st Century, Proceedings, 25.-27. Nov. .Sitges, Spain 1996 p.279-287.

/3/ C. Meier et al.: "Excerpts from Draft Contract for PV-Energy Yield Guarantee", Zürich 1999.

/4/ A performance guarantee is part of the quality control programme in the 1 MW PV project in Nieuwland, Amersfoort , see also F. Vlek et all, 2nd World Conference and Exhibition on Photovoltaic Solar Energy Conversion, 6-10 July 1998, Vienna Austria, p 2492-2496.

PV SYSTEMS FOR STAND-ALONE ELECTRIFICATION: THE USER'S POINT OF VIEW

Jaume Serrasolses
Asociación SEBA
Mallorca, 210, 1r. 1a.
08008 Barcelona – Spain
Tel. 34 93 4463232
Fax 34 93 4566948
sebaasoc@suport.org

Xavier Vallvé
Trama Tecnoambiental S.L.
Ripollés, 46
08026 Barcelona - Spain
Tel. 34 93 450 40 91
Fax 34 93 456 69 48
tta@retemail.es

Pilar Chiva
Institut Català d'Energia
Diagonal, 453 bis
08036 Barcelona – Spain
Tel. 34 93 6220500
Fax 34 93 6220502
edificis@icaen.es

ABSTRACT: The PV users association SEBA carried on a survey to know the user's point of view about PV technology in stand-alone systems. In this paper, the authors present a summary of the most significant answers, the differences between the users groups and some conclusions about the user's believes referring to the role of the user's associations for satisfying system operation.

The clearest conclusion is a global positive assessment of the PV systems in three qualities: fulfilment of their electrical needs, system reliability and easiness of use. A significant majority of users would rather use the PV system than a possible electrical connection to the public grid, and they consider their system as a definitive and not a provisional solution.

A significant difference is that the members of a user's association in general are more satisfied with their PV system than the individual users.

Keywords: Rural Electrification, Evaluation, Sociological.

1. INTRODUCTION

SEBA is a PV system user's association who started its activity in 1989. Now it is managing about 275 stand-alone PV installations, mostly owned directly by SEBA, and has 400 members mainly distributed in the Northeast of Spain (Catalonia, Aragon and Balearic Islands) [1,2,3]. The growing number of PV systems managed by SEBA and the 10 years elapsed from the first installations, urge SEBA to initiate a global survey that would permit to evaluate the user's confidence in their systems.

This idea has been also supported by the Catalan Institute of Energy (ICAEN), interested in a larger survey, including also individual users of PV systems directly subsidised by the Catalan energy authorities.

Some PV users could think that they are taking part in a technical experiment and that they are necessary to improve prototype devices that in the near future will be used more extensively. Others could think that they are second-class citizens to which a low-performance electrification system is offered with the aim to reduce investment costs and they are waiting for the grid arrival to home. Or, maybe, the PV systems users believe they are using the most advanced technology, born in the space race, environmentally friendly, with performances similar or better than the conventional rural grid supply without the need to be wired to a electrical utility.

The questionnaire objective was to know or to answer some of these questions:
- what the users think about themselves?
- really they want to use this technology?
- is a real solution to their needs or a provisional system?
- are they confident in this technology and its permanence in the time?
- know they how to manage it?

- are the users kept under the umbrella of a user's association more confident with this systems than other individual users?

2. METHODOLOGY

To know with quantitative data the point of view of the stand-alone PV system users, SEBA carried out individual interviews during the lasts months of 1998 and the firsts months of 1999 to about 150 PV systems users. The three PV user's groups has been:
- members of the association SEBA
- members of two other local user associations
- individual PV user (not associated user)

In this paper, the dates from the second group have not been presented because of the little number of answers.

The interviews consisted in four types of questions:
- general data about the user, the installation equipment, and the electric consumption
- user's attitudes and assessment of their PV system
- functioning assessment
- operation level assessment

The following points summarise the numerical results of selected questions for the SEBA's users and the individual users. The first figure of percentage represents the results of SEBA's members and the second of individual users. In the final point, a general comparison between organised users and individual user is presented.

3. RESULTS

3.1 PV systems assessment

The first four questions ask a score from 0 (bad) to 5 (excellent).

3.1.1 Global evaluation

The 78.2% and the 59.4% respectively (SEBA/individual users) scores between the favourable answers (3 to 5), and only 11.6% and 18.7% scores between the negatives (0 to 2).

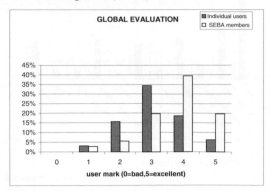

3.1.2 Satisfaction of energy needs

The 75% and the 57.6% respectively for both groups, score for a high level of fulfilment of electricity needs, and 10.2% and 30.3% a low level. SEBA members are more

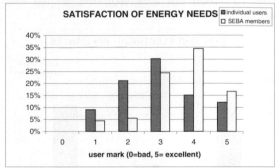

satisfied than the individual users. This can be explained by more PV power installed, but also by users training [6] and load management tools provided to this users.

3.1.3 System reliability

The 78% and the 69.7% score for a high degree of reliability and the 10.35% and the 18.2% for a low degree.

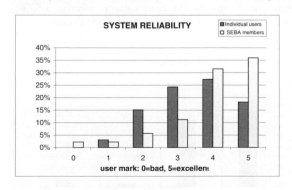

3.1.4 Easiness

The 79% and 74.2% scores for high easiness and only 4.8% and 6.4% for low easiness
So, the general assessment about the PV systems is favourable (77.5 % and 65.3% respectively) and only 9.2

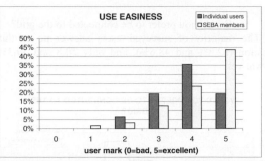

% and 18.4% scores for a negative level. It is remarkable to note that the global satisfaction is strongly correlated with the satisfaction of energy needs, and less with the system reliability or easiness of use. This shows that the satisfaction of energy is the main concern of the user, provided that the system is working properly and is easy to handle.

3.2. Advantages of the PV system

From several options, the most valued gains provided by the PV system are to dispose a permanent supply of electricity (40.2% and 57%), energy autonomy (29.9% and 14.3%), forgive the generator set (25% and 23%), and to enjoy a standard AC supply (23.7% and 5.7%) The most remarkable advantages see by the users are the permanent availability of electricity and the reduction of the operation time of the gensets.

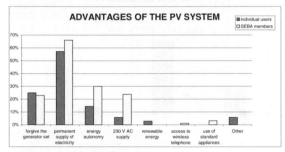

3.3. Why choose PV

When asked about the motive to opt for a PV system, most of them choose an economy reason (30% and 57% because was cheaper than the grid extension), or because was an option fit to their needs (29% and 20%) and at last because they prefer not to be wired with the electrical grid (12.3% and 11.4%), and there was no other alternative (4% and 2.8%).

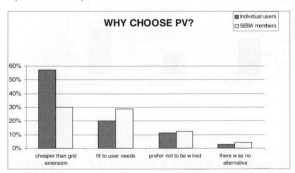

3.4.   Would you prefer to be connected to the grid?

The majority of surveyed users answer no to this question (62.8% and 48.6%), and only 20.8 and 34.3% answer yes.

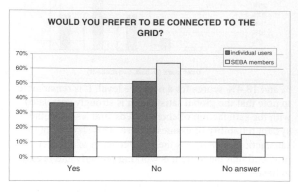

To specify more this question we asked them for two more opinions:

-would you negotiate the grid extension to your home?. The great majority answers no to this question (82.2% and 69.7%) and only 9.6% and 24.3% answers yes.

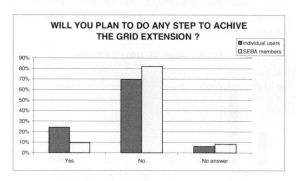

-do you consider definitive your PV system? The 15.3% and the 33.3% of the users consider the system definitive, 70.8% and 42% definitive if improvements are made, and only 9.7% and 18.2% provisional.

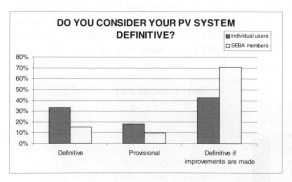

3.5.   Improvements to be made

From several options the demands proposed to improve the PV system are: enlarge the PV array power (40% and 48.6%), install a solar hot water system (25.7% and 22.8%), improve the information system (20.6% and 5,7%), enlarge de inverter power (16.5% and 28.5%), renew the battery set (11.3% and 22.8%), etc.

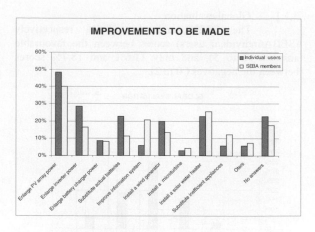

3.6.   Electrical appliances

The use of efficient electrical appliances is essential for a successful PV system. The answers to the question about the use of them was that 58.7% and 31.3% have got all the electrical appliances of the low consumption class; a 15.5% and 11.4% say that they will substitute the standard appliances for a more efficient; at last 3.1% and 45.7% (!) assert that these appliances are too much expensive to be used.

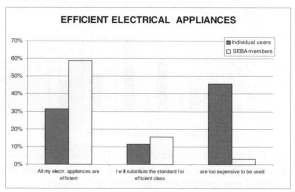

3.7.   Usual actions

An 81.4% and 74.3% of the surveyed users assets that they usually observe the energy gauge, whereas a 53.6% and 65.7% also look the battery electrolyte density.

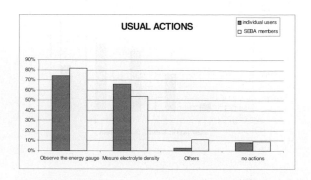

An other question focused on the user criteria to start the generator set and recharges the batteries. A 28 and 5.2% respectively work on this decision on the availability index (a battery energy gauge usual in the SEBA's PV systems), a 15.6 and a 13% work on the battery electrolyte density; a 18.8 and a 10.5% start the genset when the low battery indicator flash; and at last, a 10.4 and a 5.2% waits until the electricity is gone. Many people who have not answered this question haven't got any genset.

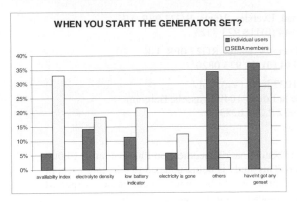

4. CONCLUSIONS

The clearest conclusion is a global positive assessment of the PV systems in three qualities: fulfilment of their electrical needs, system reliability, uses easiness. A significant majority of users prefer to use the PV system than a possible electrical connection to the public grid, and consider their system like a definitive one (despite the fact that many of them want some improvements) and not a provisional one. Despite the economy has been a powerful reason to choose the PV system, an important part of users choose it because was an option fit to their needs and they prefer not to be wired with the electrical grid. Very few were obliged to choose it because there was no other alternative.

As can be seen, a significant difference between the user's answers has been related to whether they are a member of a user's association or not: the members of a user's

association (SEBA) in general are more satisfied with their PV. The next table summarises the most significant differences between both groups (Table 1).

The results show the positive influence of a user's association like SEBA in terms of satisfaction of PV users and consequently a higher acceptance of PV technology.

In conclusion, this organised framework is decisive to get closer to achieve acceptance by the public that PV is also "real electricity" [4,5]

5. REFERENCES

[1] X Vallvé, J Serrano, S Breto, PV Rural electrification program in Aragón (Spain): example of an implementation scheme involving users and local authorities. Presented at 16th E.P.S.E.C., Glasgow, may 2000.

[2] Vallvé X, Serrasolses J, Design and operation of a 50 kWp PV rural electrification project for remote sites in Spain, in Solar Energy, vol. 59, Nos. 1-3, pp. 111-119, 1997, Elsevier Science Ltd.

[3] Vallvé X, Serrasolses J, PV stand-alone competing successfully with grid extension in rural electrification: a success story in Southern Europe. Proceedings of 14th E.P.S.E.C., Barcelona, june-july 1997, pgs. 23-26.

[4] Schweizer-Ries, P., et al. Success factors in PV-Hibrid systems for off grid power supply: results of a socio-technical investigation in Germany and Spain. Presented at 16th E.P.S.E.C., Glasgow, may 2000.

[5] Merten J., Malbranche P., Boulanger P., Vallvé, X. Novel monitoring concepts for stand-alone pv systems including objective and subjective parameters. Presented at 16th E.P.S.E.C., Glasgow, may 2000.

[6] Serrano, J., Ramírez, E., Vallvé, X.,, Vosseler, I. User training in pv rural electrification programs. Presented at 16th E.P.S.E.C., Glasgow, may 2000.

	Individual user (%)	SEBA user (%)
General assessment		
-Positive	59.4	78.8
-Negative	19	8.5
Fulfilment level		
-High	57.9	75.5
-Low	30,3	10
Would you rather to be connected to the grid ?		
Yes	36.5	20.8
No	51.5	64
Will you negotiate the grid extension to your home?		
Yes	24.2	9.6
No	69.7	82.2
Do you consider the PV system...		
Definitive +improvements	75.7	86.1
Provisional	18.2	9.7
I have got all the electrical appliances of the low consumption class	31.4	58.7
This appliances are too expensive to be used.	45.7	3.1

Table 1

PV 2010: AN UPDATE AND OBSERVATORY FOR THE MILLENNIUM

Bernard McNelis[1], Lara Bertarelli[2], Paul Cowley[2], Tayyab Shamsi[3], Umberto Tiberi[4]

[1] EPIA, c/o IT Power, The Warren, Bramshill Road, Eversley, Hants. RG27 0PR. UK
Tel: +44 (0)118 932 4417, Fax: +44 (0)118 973 7323

[2] IT Power, The Warren, Bramshill Road, Eversley, Hants. RG27 0PR. UK
Tel: +44 (0)118 973 0073, Fax:+44 (0)118 973 0820

[3] TEAM, Via G Marconi, 46 – 21027 Ispra (Va), Italy
[4] DGXVII, European Commission, 200, Rue de la Loi, B-1049 Brussels, Belgium
Tel: +32 2 295 22 81, Fax: +32 2 296 6283

ABSTRACT: PV 2010 was published in 1996, and has become a "bench-mark" document on Photovoltaics. 10,000 copies were printed and distributed by the European Commission. This study is being updated to take into account the many changes in the European and global PV markets and to bring it into line with the current and expected future PV projections - in terms of EU and national PV targets, PV industry, and end-users.

The key objectives of the updated PV 2010 report are in-depth analyses of the supply and demand structures of the PV sector, in order to create a greater understanding of, and attracting increased investment into, the EU PV industry.

As part of this action an Observatory will be created to support the European Commission and EPIA in the dynamic formulation and evaluation of strategic policies/scenarios. The aim of this Observatory is to build, sustain and increase the world-wide PV market share of the European PV industry through the continuous monitoring and evaluation of PV demand and supply.

In today's rapidly changing world economic climate, this is an appropriate time to consider refocusing *PV 2010* and obtaining more detailed information to enable the European PV industry to have a more accurate base for planning future strategies and direction.

Keywords: Strategy - 1: PV Market - 2: Modelling

1 BACKGROUND

In 1994, the European Photovoltaic Industries Association (EPIA) was commissioned to forecast the status of photovoltaics (PV) to the year 2010. The study, undertaken as part of the EC *ALTENER* Programme, took into consideration PV technology, market development and the status of the industry - both in Europe and the rest of the world. As a result of these investigations, a report was issued in 1996, which incorporated a strategic action plan devised to meet targets for European PV. It also identified the development of market capacity and the PV industry's capability to meet the stated targets.

This report, issued under the title of "**Photovoltaics in 2010**" (*PV 2010*) [1], was published in 1996 and is now considered to be a "bench-mark" survey of photovoltaics. Indeed, the European Commission drew on strategic options and figures developed in *PV 2010* for the photovoltaic component of its own "White Paper for a Community Strategy and Action Plan for Renewable Energy Sources", published at the end of 1997 [2].

Nevertheless, since 1995, when the research for the *PV 2010* report was undertaken, there have been many changes in the European and global PV markets. This has included major new programmes (e.g. USA's 1 million solar roofs programme, Germany's 100,000 rooftop programme, Italy's 10,000 roofs programme - all largely driven by climate-change and greenhouse gas emission abatement policies), restructuring of the electricity industry in many countries, new market players, changes in ownership of many of the PV companies, new products, new financing initiatives, etc.

As a result, a revision of the 1996 *PV 2010* report is required to bring it into line with the current and expected future PV projections - in terms of EU and national PV targets, PV industry, and end-users. Additionally, in the current rapidly changing world economic climate, this is an appropriate time to consider refocusing the report and obtaining more detailed information in order to enable the European PV industry to have a more accurate base for planning future strategies and direction.

1.1 Dynamic Observatory

As the past five years have clearly demonstrated, the PV industry and markets for the technology are dynamically linked. To ensure that future changes can be more readily accounted for, an important component of the PV2010 update is the establishment of an Observatory and dynamic market model to support the European Commission and the European photovoltaic industry in the formulation and evaluation of strategic policies/scenarios. The Observatory, currently under development, will serve as a tool for continuous monitoring and evaluation of PV demand and supply world-wide, which will give European Industry an inside track to help build, sustain and increase its share of the global PV market.

2 UPDATING PV 2010

The key objectives of the updated *PV 2010* report are in-depth analyses of the supply and demand structures of the PV sector, in order to create a greater understanding of, and attract increased investment into the EU PV industry.

In essence, the original PV 2010 report comprised the following components:

An analysis of the PV industry supply-side was undertaken to determine historical market development trends;

Similarly a historical demand profile was determined, based on analysis of industry shipments;

Future demand scenarios were then forecasted based on projection of the historical demand patterns together with analysis of demographic data, including certain socio-economic assumptions likely to affect the deployment of PV technology both in developing and industrialised countries (e.g. minimum basic energy service requirements and any incentive programmes). In addition, subsidiary benefits such as associated employment and avoided emissions of greenhouse gases were calculated;

The future demand scenarios, coupled with a system-cost reduction analysis, served as the basis for the definition of a strategy to ensure that European industry could retain a significant share of the predicted global market for PV technology;

A similar basic methodology is being employed to update the report.

Key players from the PV industry world-wide - including cell & module manufacturers and producers of balance of systems components - are being surveyed to establish current and future production capabilities, market development priorities, pricing structures and perceived inputs needed to achieve their growth predictions.

The demand structure is being analysed through the characterisation of the main types of buyers/users of photovoltaic technology and applications. This analysis will highlight the main players, as well as detailing international organisations that are having an important impact upon the market at a global, regional, national and/or local level, or who could do so within the next 10 years. The considerations for the supply-side assessment vary somewhat depending upon the "user" sector (e.g. national governments, international development organisations, utilities, industry or individual end-users), but key issues to address include:

policies and global strategies for the PV sector
demand trends for the next decade
incentives and financial support programmes
financial capacity and access to funds
operational sectors interested in PV technology
economic requirements from implemented projects

Given this base supply- and demand-side data, the project team aims to identify the relationships that exist between supply and demand in order to define the market for PV applications by broad application sectors, notably: Medium to large PV power plants; Grid-connected, small-scale applications (rooftops, facades and other urban components); Stand-alone applications (rural household electrification, community systems, solar home systems, water pumping, education, health, communication, etc.).

3 ESTABLISHING THE MODEL & OBSERVATORY

The key objectives of the development of the model and Observatory are:

In the short-term, to identify strategic indicators relating to world-wide PV supply and trends in demand. The European PV industry can use these indicators to evaluate its own strategies and positioning in the market, thus identifying new targets and business opportunities.

In the medium-term, the results of the model will be employed to develop a permanent Observatory at EPIA of the EU photovoltaic industry. The regular updating of the analytical data could result in an annual report on the status of and trends in the PV market. Such reports would provide the necessary indications for the characterisation of the market, covering both demand and supply aspects, with including data on policies and strategies in the public arena.

3.1 Defining the model

The demand-supply relationships identified during the updating of the main report will serve to define the fundamental parameters of the dynamic market model.

The model will put into context the development of PV in respect to historic supply/demand situations, and through appropriate algorithms will provide forecasts of the likely future market trends and technology directions. Consequently, it will serve to indicate to European operators in the sector the most suitable strategies to confront the competition.

The model will consist of three main sections:

i. *Evaluation of the competition*

Based on the supply data, this will allow the definition of the actual status and the trends of the PV industry around the world through the definition of the resources of the competition. These will be expressed through the following parameters:

- resources and technological capacity
- techno-constructive capacity and cost structure
- marketing capacity
- financial capacity
- nature and consistency of commercial activities

ii. *Evaluation of the strategic areas of the sector and the market*

Based on the demand data, this will allow evaluation of new demand tendencies for the various PV applications. The trends will be expressed through the following parameters:

- size of market application sectors by geographic area
- development rate of the sectors by geographic area
- main types of clients and estimate of actual purchases by geographic area
- growth forecast of the application sectors over various time spans by geographic area

iii. *Evaluation of various hypothetical strategies for entering new markets by European companies*

This evaluation will be carried out through the identification of significant competitive advantages and

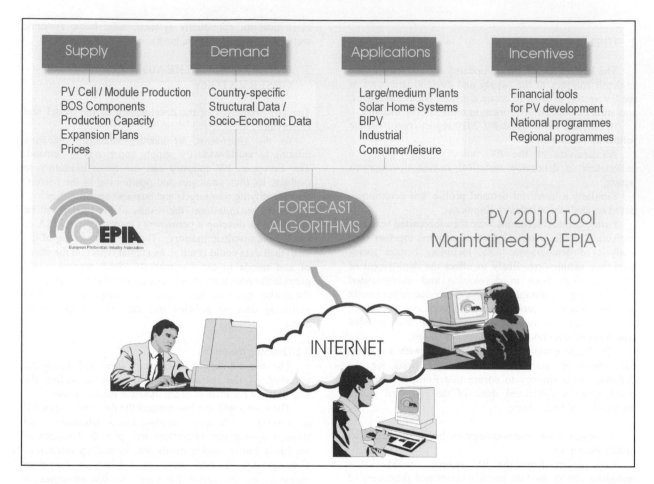

critical success factors.   The employed technique (and associated algorithm) will be based on a matrix of critical success factors and the strengths/weaknesses of companies, competitors and market profiles.

3.2 Maintaining the Observatory

The Observatory will be a long-term investment for the European PV industry, able to provide continuous updates on global market developments and analyses of competitors. It is envisaged that, once the model has been successfully implemented, it will be installed at appropriate European information providers (e.g. European Commission, EPIA).

Source material will be updated regularly in order to keep the results current. Key market indicators, regularly updated in line with the dynamic source data, will be made available via an on-line service to the European PV industry and other interested European parties. To facilitate this, the model itself has the following basic functionality:

i. *Provision for easy and continuous updating of market/product information:*

System utilities will be developed to facilitate the updating of the basic data. The purpose of these activities is to register new phenomena which disturb the behaviour of markets/products.

ii. *Continuous on-line data output.*

The system will provide a set of functions that can be activated periodically to extract basic market/product information strategies. These will be accessible via the internet using widely supported hardware/software environments.

The second function of the system will ensure that the model becomes a practical instrument to support users in taking new, strategic decisions. Two potential user groups can readily be identified:

i. Public entities: needing to make informed decisions on political interventions at European or single country level, based on the world-wide market trends

ii. Manufacturers and suppliers: for internal company strategic decisions

4   DISSEMINATING THE RESULTS

Aside from publishing the updated report and making key conclusion from the report and the outputs from the market model available on the internet, a number of workshops will be held to raise awareness of the outcomes of the study, and the availability of the decision-making tool.

The first series of workshops will be designed for government officials, and will be linked both to the findings of the report, and ways in which the EU target of 1 million PV systems installed by 2010 can be achieved. The programme for this type of workshop will include ways in which national and local Agenda 21 targets could be met using PV - with particular emphasis on integration into the urban environment.

The second type of workshop will be aimed at industry, and will be designed to inform existing players, as well as to broaden and deepen the sector by attracting new players and investments. Importantly, financiers will be considered within this target audience.

5 CURRENT STATUS

The preliminary collection of both supply- and demand-side data as outlined in section 2 above, is well progressed.

Analysis of the supply and demand structures, and of their inter-relationships, in order to define the dynamic market model is underway. Identification and characterisation of the parameters, describing the competition, and their correlation to the overall strength of the PV market have been defined. The calculation algorithms for the definition of projections up to 2010 relative to the supply and application sides are being completed and validated, while those relative to the demand side are under development.

The hardware and software architecture has been defined to give maximum system compatibility and a functional but user-friendly interface.

6 CONCLUSIONS

Major recent influences on the demand for and application of PV technology around the world have rendered the 1996 bench-mark report on photovoltaics - *PV-2010* - somewhat outdated. In recognition of this, the study is being re-visited. Furthermore, in anticipation of future significant developments in the sector, an observatory and market forecast model is being established to allow future dynamics in the supply and demand chains to be accommodated. The resulting PV 2010 tool will assist European decision makers from government, industry and finance to better formulate their strategic plans for future PV sector initiatives, with the objective of ensuring European industry retains a significant interest and influence in the global market for photovoltaic technology.

REFERENCES

[1] Photovoltaics in 2010, European Commission Directorate-General for Energy, ISBN 92-827-5347-6, ECSC-EC-EAEC, Brussels • Luxembourg, 1996

[2] White Paper for a Community Strategy and Action Plan for Renewable Energy Sources, Communication from the European Commission, Com (97) 599 Final 26/11/97

ECOWATT PROGRAM'S TECHNICAL EVALUATION AND USERS' SATISFACTION

Roberto Zilles and Federico Morante

Instituto de Electrotécnica e Energia, USP, Av.Prof. Luciano Gualberto 1289, 05508-900 SaoPaulo, Brazil
Fax: 55 11 8162878 e-mail: zilles@iee.usp.br

ABSTRACT – This study shows the results of the technical assessment visit made in November 1998 to Parque Estadual da Ilha do Cardoso (Cardoso Island State Reserve) in the County of Cananéia, State of São Paulo, in Brazil. The visit's aim was to evaluate the photovoltaic systems installed under the ECOWATT Program in the homes on that island. The visit was not intended to address problems, which is a task under the responsibility both the electrical services grantee and the company that won the bidding, but, instead, it was intended to detect and point out critical points, which were mentioned by users during several occasions in a dispersed way. During these visits we could record 75 systems (2 churches, 4 schools and 69 homes) and 56 interviews were made with users present at the time. Additionally, at some homes a test was carried out on performance and working conditions of the charge controller with the Solar Home System Tester and, during one year, the power consumption of two homes in the ECOWATT Program and of two homes out of the program was measured, with the use of Ampere/hour meters. The study includes comments related to the state of the facilities, users' satisfaction level, default level, measurement of batteries' capacity after two years utilization and about mistakes made during implementation. The results presented herein are intended to divulge the common mistakes made and which could have been prevented in order to ascertain the success of such programs.

Keywords: Solar Home Systems – 1: Rural electrification – 2: Utilities -3

1.   INTRODUCTION

In 1997, the program named ECOWATT, a name synthesizing the program's philosophy to conciliate commercial service and environmental preservation, was the first Brazilian commercial experience in serving clients with photovoltaic power [1]. *Cananéia, Iguape* and *Iporanga* villages were served by the program with 120 facilities. The invitation to bid for the supply of equipment items defined the final uses and it was up to each bidder to specify the system to be installed. The resulting set up contained two 70-Wp modules, two 54 Ah batteries, sealable box, brackets for fastening the modules, charge controller and Two 9 W lamps with electronic reactors and accessories for electrical installation in the homes.

The photovoltaic electrification on Ilha do Cardoso included the communities of *Marujá, Pontal do Leste, Enseada da Baleia, Canal do Ararapira, Itacuruçá* and *Perequê* and was decided by *Cia. Energética do Estado de São Paulo* (CESP) taking into account the peculiar situation of the residents on the island, subject to environmental law restrictions, which prohibits the use of equipment items that may cause any damage to the ecosystem. CESP's concern was focused on supplying electric power to all residences throughout the state, even those in the rural areas. That is how the ECOWATT Program was created to provide electrification to low-income consumers by using solar photovoltaic power.

Before taking this decision some studies were performed to verify the local population's low income, which, according to CESP's estimates, meant a low power demand that did not justify high investments to supply the area with traditional electrical lines. In sequence, there was a bidding process conducted by CESP, which *"privileged the establishment of final uses and it was up to each bidder*

to specify the equipment array to be supplied" [2]. Additionally, *"taking into account the local population's socioeconomical characteristics CESP thought it necessary to publish an equipment manual written in an easy-to-understand language. In addition to that manual, lectures were also offered under a didactic approach to the population"* [2].

CESP (currently ELEKTRO) considered an internal return rate of about 10% with a 20-year working life for the modules and the replacements of the batteries every four years. In accordance with those considerations a monthly tariff of approximately US$ 13,00 was calculated, regulated by a contract between the Company and each user. The battery maintenance and replacement every four years became the responsibility of ELEKTRO. As can be noticed, ELEKTRO took all the required steps to implement those facilities, however the problems presented afterwards were related to:

- the photovoltaic system type chosen;
- the way the installations were carried out, without taking into consideration the minimum caution measures recommended for these cases;
- the supervision quality during the assembly process; and,
- the management of the system after implementation.

In spite of the high investment made, time has shown that in addition to other questions related to the implementation of the photovoltaic systems none of the installations received much attention from the part of the Grantee Company, which focused on the implementation of a large scale program, therefore the technical precautions were not taken into account, thus compromising, in a general way, the systems' functioning.

2.   STATE OF INSTALLATIONS

The first observation, even before any technical visit for evaluation, is related to oversizing of the photovoltaic modules: 140 Wp for an accumulation capacity of 108-Ah (two associated 54 Ah batteries in parallel). This situation makes the user pay for an initial investment which they will not take advantage of in its entirety. The demand, to many users, could be supplied with a single 70 Wp module, so that the user could have the same service on a lower tariff paid to the electrical services grantee. In addition to this finding , it is worth mentioning that the overcharge control, according to the manufacturer's catalog, only actuates when the voltage on the battery terminals reaches 14.8 V. Checkings made with the Solar Home System Tester showed that the overcharge control only actuates when the voltage on the battery terminals reaches 16.0 V. An excessive value for the type of battery used.

Figure 1: Shaded photovoltaic generator.

There also where many installation mistakes, as those illustrated in pictures 1 and 2. On one hand, figure 1 shows a system where the installer chose the wrong way to locate the photovoltaic modules. The shading decreases the system's generation capacity, however, due to the fact that the generator is too big in relation to the accumulation system, in addition to the charge regulator being set to cut off at 16 V, in an indirect way this shading actuates as a protective system for the battery, since it reduces the power generated.

On the other hand, figure 2 illustrates a case where, because the charge regulator allows the system to reach voltage values above 16 V during the period in which there is solar generation, the parabolic antenna receiver causes interferences on the TV pictures. On noticing the problem only occurs on sunny days, the users found out that, if they put a piece of cloth on the generator it was possible to obtain a reduction of the interference or even to eliminate it, obviously because the piece of cloth causes the generation voltage level to decrease.

Figure 2: Installation of the ECOWATT Program where, by means of a piece of cloth, the electric generation is reduced.

There are many examples like this and, though some residents increased their charges without authorization from CESP, there was a large waste of generated power because the contracts permits the use of two 9 W compact fluorescent lamps only. In reason of this, the residents are forced to use candles, oil lamps and small gasoline-fueled generators, in order to meet their small power needs. Neither can they perform small maintenance tasks on the batteries, since the batteries are placed within a sealed box. Notwithstanding all these adverse conditions, ELEKTRO's invoice for the US$ 13.00 monthly fee comes in punctually. It should be mentioned that the systems maintenance became ELEKTRO's responsibility but, however, this aspect was abandoned. The final result of this cluster of factors was the complete discredit of the photovoltaic technology in that zone.

3.   USERS SATISFACTION LEVEL

In reason of all the problems found most users are not satisfied with the service delivered by photovoltaic systems and, as a consequence, this technology's credibility was compromised and is suffering a strong discredit in the region.

One of the ways to technically check the causes for this dissatisfaction was to understand the consumption habits in the households in that area; therefore, during one year some measurements using Ampere-hour meters were carried out [3].

In order to conduct our study, four homes in the Marujá community were considered. Two meters were installed in homes electrified through the ECOWATT Program (families 3 e 4). Another meter was installed in a residence that had not entered this program (family 1) and, additionally, the Photovoltaic Systems Laboratory of the Electrotechnique and Power Institute of the São Paulo University (LSF-IEE/USP) installed a pilot system for research activities in one of the homes in the village (family 2). The main purpose of that installation was to demonstrate functionality and efficiency of the photovoltaic technology and try to revert the people's negative opinion as caused by mistakes made in the ECOWATT Program. In table I the Ah meter monitored photovoltaic systems general characteristics are indicated.

Table I: Characteristics of the monitored facilities in the Marujá community.

	Family 1	Family 2*	Family 3	Family 4
Generator (Wp)	70	96	140	140
Battery (Ah)	136	190	108	108
Fluorescent lamps (# × W)	1 × 15 1 × 10 1 × 9	2 × 20 1 × 15 1 × 9	4 × 9	3 × 9 1 × 10
Incandescent lamps (# × W)	------------	2 × 2	-----------	1 × 2
Radio (W)	------------	15	10	----------
Fan (W)	------------	20	-----------	----------
Radio-transmitter	TX.20 W RX.8 W	----------	-----------	----------

*Has a 75 W DC/AC inverter.

Through the graph in figure 3 it is possible to confirm that the consumption levels of families 3 and 4, with facilities from the ECOWATT Program, are low in comparison with other families (1 and 2) with smaller photovoltaic generators (figure 4). If we take into account that most families having systems in the ECOWATT Program only use the two permitted lamps, we can have an idea of power waste level reflected in the very poor satisfaction level on the part of the users.

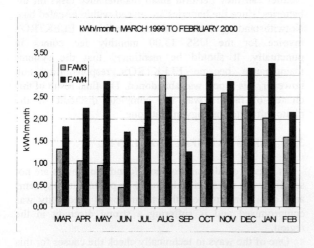

Figure 3: Consumption in kWh/month of families 3 and 4, ECOWATT Program, 140 Wp PV generator.

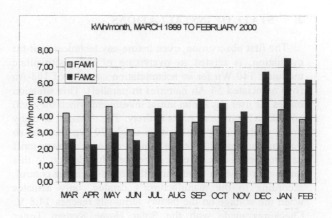

Figure 4: Consumption in kWh/month of families 1 and 2, 70 and 96 Wp PV generators, respectively.

During the evaluations carried out in November 1998, of the 56 users present at the time 38 informed they failed to pay the monthly fee to the electric power grantee due to the following reasons: dissatisfaction due to the frequent failures and lack of support in maintenance related issues.

4.   BATTERY CAPACITY MEASUREMENTS

The batteries at families 3 and 4 were replaced with new batteries so that the battery capacity test could be carried out. This replacement was made by the Photovoltaic Systems Laboratories that made available to the families new batteries in exchange for those initially installed by the ECOWATT Program. Two 54 Ah batteries were taken from each family and replaced by two identical new batteries. In the case of family 3, the batteries were changed after two years use and in the case of the family 4 after 2.5 years. In spite of the clear evidence that the batteries of many systems are technically dead there was not any action of the utility to substitute .Table 2 shows the capacity results as obtained after submitting the batteries to a charge cycle and a period of 24 hours to a 14.4 V constant voltage, discharge current of 2.75 A.

Table II. Measured capacities of the batteries of two families in the ECOWATT Program.

	Family 3	Family 4
Battery 1	6 Ah	4 Ah
Battery 2	7 Ah	5 Ah

The nominal capacity of the batteries in C_{20} is 54 Ah.

Figure 5 shows the discharging curve of Battery 1 of family 3 made right after the battery's withdrawal from the system. It can be noticed that only in three minutes the battery voltage reaches values lower than 10.8 V. Figure 6 presents the charge curve of this battery and the discharging curve obtained after keeping the battery for 24 hours under a constant 14.4 V voltage after the charge cycle.

The results indicate the deficiencies in the charge control equipment, the excessive overcharging allowed for the evaporation of the water from the batteries, which can not be filled up. Because the batteries do not permit this type of maintenance, they are car batteries without access for maintenance, but with a small orifice that permits evaporation. All the four batteries were weighed and showed 1 kg below their nominal weight. In other words, we can say each battery lost approximately 1 liter water, i.e., about 150 ml per cell.

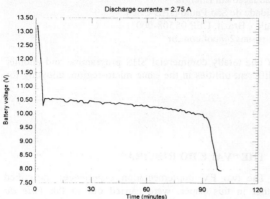

Figure 5. Discharging curve of battery 1, family 3, right after it was taken from the system.

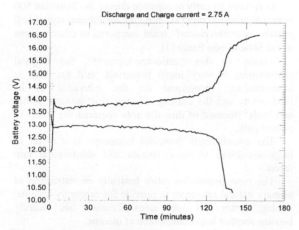

Figure 6. Charge and discharge curve of family 3 battery.

6.   FINAL COMMENTS

The observations presented herein reveal that at the time of preparation of the ECOWATT photovoltaic electrification program a series of factors was not taken into consideration, bidding specification, knowledge of the end user needs, training and schedule manteaning procedures which, doubtlessly, could have assured the final success of that program. To begin with, no quality control was performed on the equipment items intended for the project. In our experience, differently from conventional electrification, the engaging of the users in the new system is the only way for them to incorporate the new facilities into their way of living and, at the same time, to accept the cost arising from maintenance, replacement of lamps, reactors, etc. It was a very serious mistake that the ECOWATT original program has not considered either quality control procedures and a definition of a clear maintenance scheme.
The observed negligences don't find place in the universe of the conventional rural electrification where quality procedures are usual and they are part of the patterns of utilities service. For each new implementation careful bidding, technical supervision and equipment quality

controls are made, as well as maintenance plans, inspection routines and qualified personnel's training. The observance of those procedures in ECOWATT'S Program would drive to best results and certainly they would contribute in the diffusion of the photovoltaic technology. We don't know the reason that took the utility to disrespect those procedures.
In spite of the problems presented there is no reason for leave the installed systems abandoned, as is now. An answer to the users as for example, to leave clear where to obtain spare parts and the necessary technical aid concern the first compulsory step. On the other hand, the program deserves a revision of its objectives and a restructuring in the following aspects: change charge controls for new ones with better performance, replace batteries bank to match with PV generator take in account the user demand, revision of the electric installation and technical training in didactic conditions as foreseen in the original contract.

REFERENCES

[1] Zilles R., Andrade A. M. de & Almeida Prado F. "Solar Home System Programs in São Paulo State, Brasil: Utility and User Associations Experiences". *14th European Photovoltaic Solar Energy Conference*. Barcelona, pp. 931-933, 1997.

[2] Almeida Prado F. & Pereira O. S. "Programa ECOWATT - Uma Alternativa Comercial para Energia Solar Fotovoltaica", *III Congresso Brasileiro de Planejamento Energético*, São Paulo, pp. 216-218, 1998.

[3] Morante F. "Demanda energética em Solar Home System's". Master Degree Thesis, *Instituto de Eletrotécnica e Energia da Universidade de São Paulo*, Brasil, 2000.

Acknowledgements — This work was supported by FAPESP.

UTILILITY APPROACH TO PV RURAL ELECTRIFICATION: SOME COMMERCIAL EXPERIENCES IN BRAZIL.

Rosana Rodrigues dos Santos and Roberto Zilles
Programa Interunidades de Pós Graduação em Energia
Instituto de Eletrotécnica e Energia - Universidade de São Paulo (IEE-USP)
Av. Prof. Luciano Gualberto, 1289. São Paulo - Brazil. CEP 05508-900
Fax.: +55 11 8167828   e-mail: rosana2@uol.com.br

ABSTRACT: The paper aims at the description and analysis of one totally commercial SHS programme and another semi-commercial Battery Charging Station project lead by two different utilities in the same micro-region, allowing for geographically, economically and cultural consistent evaluation.

Keywords: Rural Electrification – 1: Utilities – 2: Implementation – 3

1. INTRODUCTION

A number of electricity utilities have been trying the use of PV rural electrification instead of grid extension to supply electricity to far-from-the-grid consumers within their concession areas.

In Brazil state electricity utilities were involved in the process of PV dissemination as implementing agencies and co-sponsors of national government and/or international agencies driven programmes [1]. However, the undergoing privatisation of the Brazilian energy sector changes this institutional framework, and the new institutional environment still lacks of proper regulation and share of responsibilities.

Besides the above described institutional path (external input), some utilities have decided to implement their own PV rural electrification programmes, as a least cost alternative to grid extension. From one utility to another, the programme implementation philosophy varies enormously: from pure donor/social to pure commercial.

The paper aims at the description and analysis of one totally commercial SHS programme and another semi-commercial Battery Charging Station project lead by two different utilities in the same micro-region, allowing for geographically, economically and cultural consistent evaluation.

After three years of functioning, the pure commercial experience lacks of sustainability. Users are generally not satisfied, tariff is high when compared to service, systems present signs of unskilled design and installation, the utility (now privatised) takes long time to perform maintenance, the contract between the utility and the user establishes the replacement of one battery whereas the tariff was calculated adopting the replacement of batteries along the entire life time of the system (four replacements), etc.

On the other hand, the semi-commercial Solar Battery Charging Station (SBCS) Programme had just a one-year life time. The utility decided - after this first year - for the installation of individual systems, and the charging station lies unused as part of the local landscape. The original SBCS project aimed at the creation of a micro-enterprise to manage the charging system and gather payments for each charged battery. This enterprise would live on this money and provide maintenance. The equipment was financed by the *Programa Nacional de Desenvolvimento Energético de Estados e Municípios - PRODEEM* together with the state utility COPEL.

The paper will present successes and failures of these two initiatives, paying special attention to the reasons why users are not satisfied and not committed to the solar technology. The whole analysis is carried out from their stand point, considering their difficulties and expectations.

2. THE "VALE DO RIBEIRA"

The two PV implementation experiences, described further in this paper, were carried out in the *Vale do Ribeira*, an environmental preservation area, only 250km far from São Paulo, the South-American largest economic centre.

In spite of its early occupation during the Brazilian 500 years colonisation process, the *Vale do Ribeira* region still remain "underdeveloped" when compared to other regions of the State of São Paulo [2].

Thanks to this "underdevelopment", the original environment is very much preserved, and the area is internationally recognised for the relevance of its biodiversity and the nursing of migrating animals, fishes and birds. Because of this, the area acquired the status of natural park.

The astonishingly beautiful landscape is a result of large mangrove swamps, islands and sub-tropical rain forest.

The rural population relies basically on extraction of primary natural resources to provide for monetary income: fishing, oyster and palm trees. Tourism has recently become another important source of income.

The extension of the national grid till the non-electrified communities may prove itself not viable, due to environmental considerations, scattering of users and rather low energy demand. The PV decentralised alternative could be a solution, if sustainability is taken into account.

3. RURAL "VALE DO RIBEIRA": ELECTRIFICATION AND THE INSTITUTIONAL FRAMEWORK

The *Vale do Ribeira* region lies part in the state of São Paulo and part in the state of Paraná. In São Paulo, the *Vale* belongs to the concession area of the electricity utility ELEKTRO (former CESP), and in Paraná, to COPEL.

These utilities are, in principle, responsible for extending the electricity service to every consumer within their concession area, according to the concession contract they maintain with the regulating body, the ANEEL[1].

However, from the utilities stand point, rural electrification has never been considered worth investing in, specially in areas such as the *Vale do Ribeira*, where environmental concerns, as well as access difficulties and low energy demand expectations are the constraints of the electrification problem.

In order to foster rural electrification under the newly privatised environment and also gather some political

---

[1] *Agência Nacional de Energia Elétrica.*

benefit, the Brazilian government has recently launched a R\$2.1x10^9 national electrification program, named *Luz no Campo*. By means of this program, utilities are enabled to apply for a loan (maximum of 75% of total investment; 5% interest; 1% administration tax and 7 years payback time) to extend the electricity grid to rural households within their concession areas [3].

Despite the *Luz no Campo* financing mechanism, electrification of the <u>seaside sub-region</u> of the *Vale do Ribeira,* named *Complexo Estuarino-Lagunar de Iguape-Cananéia-Paranaguá* after its dominant mangrove swamp natural environment, is still going to greatly rely on decentralised alternatives, either diesel or PV, due to the above mentioned environmental and consumer characteristics.

ELEKTRO, the electricity distribution utility resulting from the privatisation of CESP, is the legal responsible for the PV electrification project *ECOWATT* launched by CESP in 1997. There is not, however, any plan to expand the initial project.

COPEL, the public utility of the state of Paraná, has been carrying out a number of pilot experiences following the company's rural electrification policy. However, a field visit to the COPELS's project in the *Vale* proved the initiative to be rather unsustainable.

4. COST-RECOVERY IMPLEMENTATION EXPERIENCE

ECOWATT, the PV electrification program launched by CESP in 1997, was designed in a cost-recovery basis. Users where to pay monthly for the recovery of the entire cost of systems, including maintenance and battery replacement.

Institutional path

By means of a public bid, the utility CESP selected the supplier of the PV systems, which should also be responsible for the installation of the equipment and the training of local users, i.e., the utility signed a turn-key contract. After the installation period, the utility would take operation and maintenance of systems up. Electricity bill would be of approximately US\$13.

Successes and failures.

The ECOWATT program succeeded in installing 120 Solar Home Systems in the *Vale do Ribeira* region, of which 75 at Ilha do Cardoso[2][4], contributing therefore for the dissemination of the use of solar energy to local production of electricity.

However, a close look to the present situation of the project leads us to the conclusion that "something went wrong". After 3 years of functioning, users are definitively not satisfied. Diesel generators are still being largely used, and people claim for grid extension.

In order to analyse the reasons for such a situation, one should go back to the bidding process of the ECOWATT.

Instead of fostering potential suppliers to bid for equipment, CESP decided for the fulfilment of some requirements concerning electricity end-uses. In doing so, the utility lost control over the quality of system design, once the least cost rule was to be respected. The system design resulting from the winner proposal proved itself to

be unbalanced in pick power and storage capacity, causing batteries to die earlier from long operation under "gasing" condition[3].

Moreover, it is clear that with the same number of solar modules, one could provide a much better service, i.e., larger number of possible end-uses and greater use-period. For example, in figure 1 one can see four batteries being charged at the same time - in a completely irregular situation - due to the perception of the user concerning the extra amount of electric energy his oversized system could produce.

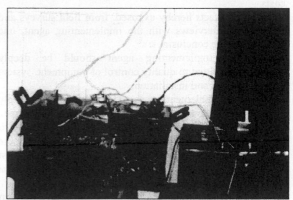

Figure 1: Four batteries being charged together.

The turn-key kind of contract forces the supplier to sub-contract installation and user training. The mechanism should work since efficient quality control is performed. Unfortunately, there are some SHSs within the ECOWATT project lacking skilled installation [5], as the one of figure 2.

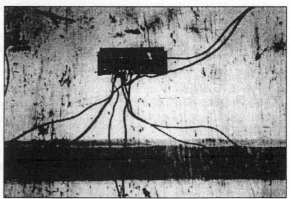

Figure 2: Unskilled installation.

Users claim that US\$13 is too much for such a low quality service. Moreover, the utility has rarely performed maintenance or battery replacement.

Because of access difficulties, monthly billing is not an easy task. Users say that, when they finally get their electricity bill, it is usually due, and they are therefore subject to late charges. They also mention that there is no DC appliances, balasts and lamps to be found in the proximity market Cananéia.

They complain as well about the contents of the contract they had to sign with the utility. According to this

---

[2] Ilha do Cardoso is an island located at the <u>seaside sub-region</u> of the Vale do Ribeira, surrounded one side by the sea and the other by the mangrove swamp. The whole are is a Natural Park, strictly protected by environmental legislation.

[3] Charge controller is not able to fully protect battery against over-charge, since its on-off operation allows current to flow into the battery when voltage level lies close to the over-charge disconnecting limit.

contract, the utility is only responsible for one battery replacement, although the monthly instalment was calculated considering four battery replacements. In addiction, the contract forbids customising of system use, under threat of loosing warranty rights.

On the other hand, the utility claims that decentralised PV systems are "much of a headache": people do not pay; access is difficult and time consuming; transport of spare parts is expensive and difficult; etc. Moreover, the public utility initially participating into the project is now the private utility ELEKTRO, whose guide line is to minimise costs.

From the facts hereby exposed, from field surveys and from some interviews with the implementing agent, one can reach some conclusions:

☐ The implementing agent should be deeply committed to quality control of equipment, system design and installation,

☐ The implementing agent should also be deeply committed to the project along its entire live-time,

☐ PV electrification is "peanuts" when compared to other priorities within the utility. Therefore, the decision to participate in any PV implementation as principal agent has to be a consequence of a company's long term policy,

☐ PV rural electrification means locally driven energy planning and project implementation,

☐ In order to determine end-uses prior to system design, a field research is useful.

5. SEMI-COMMERCIAL IMPLEMENTATION EXPERIENCE

COPEL, the public utility of the state of Paraná, started in the year 1996 a pilot PV electrification project in Barra do Ararapira, located 40min by boat south from the ECOWATT project. People there live basically from the merchandising of both fresh and/or salted fish. Apart from household electrification, their foremost energy need is refrigeration and freezing for fish.

The pilot project of 1996 aimed at the electrification of 35 households by means of a solar battery charging station (SBCS) [6]. Each of the 35 households received three fluorescent lamps and one 12V DC outlet and the battery kit, as the ones shown in figure 3.

Figure3: Battery kits: one charge controller and a maintenance free 95Ah C20 battery accommodated in a Fiberglas white box

Institutional path
People were to pay US$2 for each charging cycle.

There was a local responsible for the charging centre, whose salary was to come from the charging fees. However, these fees covered no battery or BOS replacement. Users were to bring in their " empty " battery and take home a charged one.

Financing of initial costs - solar array, battery building, battery kits and others - came from the PRODEEM[4] and COPEL. COPEL alone carried out the installation of the SBCS as well as household wiring.

Success and Failures
After one and a half year of functioning the service provided by the SBCS was replaced by 40 SHSs. The 1000Wp CdTe thin film solar array and the battery building were though not removed, as it can be seen in figure 4.

Figure 4: The CdTe solar array of the SBCS and a SHS to be seen on the backyard.

The foremost complaint of users concerns the frequent need to transport batteries to the charging centre. They claim they had to recharge the battery every 3 or 4 days in the beginning, and every two days by the end of the first year. The period between charging needs has shortened for SBCS batteries quickly lose Ah-capacity due to operation under deep discharging cycles.

It is worth mentioning that this fact opposes what many authors take for granted [7]: the task of having charged the battery every three days is NOT similar to the one of buying a new bottle of cooking gas every month or two.

One can say that the problem would be solved should batteries of larger capacity (or solar batteries) have been applied. Let us then consider some facts: larger batteries are heavier and bigger, therefore harder to be carried[5]; the loss of Ah-capacity due to deep discharging cycles would occur anyway; larger batteries are more expensive; deep cycle solar batteries are import goods and not to be found in any of the proximity markets.

A second round of complaints could be summarised in the following way: users do not own one specific battery, they take home any charged battery from the charging centre, otherwise he would have to wait at least one sunny day to have charged his battery.

There are a number of user behaviour resulting from this fact. Each household has different energy needs, imposing therefore different depth of discharge to the

---

[4] *Programa de Desenvolvimento Energético de Estados e Municípios*: a finance source to solar electrification of community uses, promoted by the Brazilian Ministry of Mines and Energy.

[5] Users are fishermen, they do not dispose of any other transportation mean apart from their boats. Moreover, some of the users are not able to carry battery themselves due to age.

battery, and consequently different charging periods.

This would not be a problem, should batteries be equaly charged in the SBCS. However, as battery loses capacity, people bring them in for charging more frequently and the charging unit does not match demand anymore.

People start to complain about the person responsible for the charging unit, blaming him for not fully charging batteries, or for selective handing of " bad batteries " for specific families. They do not pay the charging fee anymore, or they find other ways of doing so. The responsible for the station do not perceive the advantage in continuing the job.

There is, actually, no way out of this situation. In order to be economically attractive, SBCSs have to work with batteries submitted to deep discharging cycles. Moreover, frequent transportation of lead acid batteries is a fact, impossible to be overcome.

6. FINAL COMMENTS

Utilities are, to a certain extent, willing to participate in the process of PV electrification. They are an important implementing agent for their privileged position, concerning access to rural people, benchmarking, and information.

The major constraint to their effective participation is the centralised planning model they are used to: PV rural electrification means locally driven energy planning and project implementation.

Field experience proves that measuring the success of PV electrification projects by the number of Watt Picks installed means nothing. If an implementing agent is committed to rural electrification, he is committed to the electricity service, and not to Wp numbers.

Pilot projects are of great importance to the learning process of utilities and users. However, only sound technology and locally adapted implementing model should be applied, otherwise the project will negatively contribute to the dissemination of the use of solar energy to electricity supply.

REFERENCES

[1]   O.S. Pereira, **"Sources of Finance, Subsidies and Agents for Dissemination of Renewable Energy"**, Word Bank. August, 1997.

[2]   F. Morante, *"Demanda Energética em Solar Home Systems"*, MSc Degree Thesis, *Universidade de São Paulo*, April 2000. São Paulo, Brazil.

[3]   ELETROBRÁS, *"Programa Luz no Campo: manuais"*, Rio de Janeiro, Brazil. 1999.

[4]   R. Zilles, A.M. Andrade, F.P. Almeida, **"Solar Home Systems Programs in São Paulo State, Brazil: utility and users associations experiences"**, 14th European Solar Energy Conference and Exhibition", Barcelona, pp 931-933. 1997.

[5]   R. Zilles & F.Morante, **"Ecowatt Program Technical Evaluation and Users Satisfaction"**, 16th European Solar Energy Conference and Exhibition", 1-5 May 2000. Glasgow, UK.

[6]   COPEL, *"Energias Alternativas: utilização da Energia Solar na COPEL"*,

http://agencia.copel.br/copel/port/negocios-ger-energiasolar.html

[7]   World Bank, " **Solar battery Charging Stations: an Analysis of Viability and Best Practices**". February, 1999.

Acknowledgements — This work was supported by FAPESP.

DEMONSTRATION OF COMPETITIVE PV POWERED PUBLIC LIGHTING

G. Loois, H. de Gooijer,
Ecofys Energy and Environment, P.O. Box 8408, 3503 RK Utrecht, the Netherlands
fax +31 30-280 8301; e-mail G.Loois@ecofys.nl

High quality PV powered Public Lighting (PV-PL) is competitive to conventional lighting in dedicated markets. There, solar energy entails lower investments and running costs, mainly due to avoidance of grid extension and installation costs. The break-even point is from ten meters from the grid onward. In parallel, markets for grid connected PV lighting products emerge. This creates opportunities for a range of PV-lighting products for back-alleys, bus shelters, urban areas, rural roads, highway rest areas and railways. Currently over 350 locations in the Netherlands are lit through PV, to the full satisfaction of the end-users. The expected European potential for PV powered Public Lighting is over 10 MWp in the current decade.
Project developers, conventional and PV industry, governments, transportation companies, housing corporations and utilities co-operatively aim for over one thousand systems in the Netherlands within two years.
Market development of PV-PL currently profits from the availability of a range of reliable products, the feasibility of integration of PV, financing schemes and ways to embed PV-PL in regular lighting policy. For low cost quality control a new approach: *Econfirm* developed by Ecofys is applied.
Keywords: Strategy - 1, Stand-alone PV Systems - 2, R&D and Demonstration Programmes - 3. PV Market

Figure 1 *Blade*: Lighting with fully integrated PV system (design *Ecofys*, manufacturer *Stork Infratechniek*) for stand-alone and grid-connected applications.

1. INTRODUCTION

Public lighting is often required in locations where connection to the grid is only possible for a high price, due to large distances and difficult terrain for cabling. This is where PV power offers a cost-effective solution to public lighting needs. PV-PL is currently applied in those typical niche-markets.
In parallel a market for urban applications of grid connected PV-lights (fig 1) emerges, where municipalities or industries want to advertise their sustainable views and policies.
Ecofys aims to open up the conventional market of public lighting with both conventional and PV-related market parties.

This article addresses market potential, applications, experiences, design, cost ranges and quality control. The facts and figures presented in this paper describe the Dutch experience in the field of PV-PL.

2 PV-PL MARKETS IN THE NETHERLANDS

The current potential market for PV-PL systems is expected to be about 10 000 (over 1 MWp) in this decade. Extrapolation learns that the potential in Europe is over 10 MWp.
The main short term opportunities for PV-PL that are currently developed are:
- Service areas along motorways.
- Potentially hazardous locations on country roads: intersections, crossing places and dangerous curves.
- Railway tracks.
- Back alleys.
- Bus shelters.
- Public lighting in parks.
- Grid connected PV lighting on highly visible locations.
Furthermore, there is a market for lighting of locations by PV-PL due to their special character:
- Lighting of quays near bridges and locks.
- Remote working areas, e.g. along-side railway tracks.
- Security lighting of remote objects.
Demonstration of PV-PL in the niche-markets, has created confidence by the end-users and responsible authorities. Currently larger clients like utilities and municipalities are integrating PV-PL in their policies.

3. PV-PL: EXPERIENCES

The Dutch industry and end-users have gained experience with over 350 PV-PL systems in a range of applications. In comparison with conventional lighting systems, all stand-alone PV-PL systems described hereafter are, in their particular circumstances, competitive with conventional lighting, the accompanying grid extension, installation and maintenance costs. These products for the Dutch market are currently available in price range from € 750 (back-alley) to € 5,000 (design Blade).

This results in situations where PV-PL is competitive with 10 meters of grid extensional already (back alley). PV-PL streetlights can typically compete with 100 to 200 meters of cabling and feature benefits of independence and flexibility. All reliable PV-PL systems incorporate electronics for energy management -an absolute 'must' for cost reduction- in terms of limiting the size of the PV array and batteries.

The Energy management systems comprise:

- Customised lighting through adapted light levels, limited lighting periods through astronomic time switching and lighting on demand through motion sensing.
- Low-energy equipment: efficient lamps, luminaires (with catoptrics) and ballasts.

Basic specs for PV-PL are described by Ecofys in [2] and play an important role in defining reliable systems. To gain further acceptance in the market, well-designed systems and integration in the construction are currently made available by various industries.

Six typical applications and products of various industries are briefly described hereafter.

3.1 Lighting of back alleys

On an initiative of housing corporations back-alleys are lit for enhanced safety in Gouda, Maastricht and Sandpoort. First of all these systems feature independence of the grid, house owners, land owners and thus are easy to install. Despite short cabling distances of about ten meters, stand-alone systems are competitive due to simple mounting procedures and low hardware costs (PV arrays of 20 Wp only). Energy efficient lighting (11 W CFL) and 'lighting on demand' by presence sensing detectors achieve this. The systems are sized for one to two hours of light per night and feature seasonal storage of energy. A potential of many thousands of systems exists.

Figure 2   Lighting of back-alleys (*Stork Marine Solair*).

3.2. Hybrid PV-wind-battery streetlights

In the framework of a traffic security upgrade in the village of Rijssenhout one additional sides of the main street had to be lit. At 400 meters along the waterside soil conditions caused

high expenses for cable extension. At this windy location, 11 stand alone hybrid PV-wind-battery systems have been realised cost-effectively. The systems typically feature a PV array of 100 Wp, 80 Watt vertical axis wind turbine and 24 Watt energy efficient lights. Seasonal storage is featured.

Figure 3   Hybrid PV-wind-battery lights (*Logic Electronics*)

3.3 Lighting along railways

At Barendrecht PV-PL units have been installed to light foot paths between railway trackss for service personnel.

Figure 4   Lighting for service personnel (*Stork Infratechniek*)

Other typical railway applications are lighting of remote railway crossings and lighting of switch points. The systems incorporate a 100 Wp PV array, tuned lighting through a 7 and 11 W CFL and seasonal storage. Railway companies expect a potential for several hundreds of systems in the Netherlands

3.4 Lighting of rural roads
A Dutch commercial potential of about 10 000 systems exists for orientation lighting at curves and intersections of rural roads. Hundreds of systems have been realised. The systems shown form a part of a 10 solar streetlight project in the municipality of Barneveld. These systems feature 200 Wp, a 24 W dimmable CFL and seasonal storage. On short term leads to a thousand systems exist. Utilities prepare projects with a typical size of one hundred systems.

Figure 5   PVPL for rural roads (*Stroomwerk*)

3.5 Grid-connected and logo designs
A market for grid connected PV-PL emerges through municipalities that want to advertise their sustainable policies. This need is answered a/o by the well designed *Blade* (fig 1) that can be applied stand-alone and grid connected.

3.6 Lighting on demand in bus shelters
In the province of Flevoland, 37 bus shelters have been installed with a PV powered lighting unit. The systems feature a 100 Wp panel and a 11 W CFL and motion detection. The panel is invisible (roof integrated) for vandalism prevention.

Figure 6 a PVPL bus shelter, PV module integrated in the roof
(*Armada Outdoor International*)

Figure 6b PVPL bus shelter, PV module integrated in the roof
(*Armada Outdoor International*)

4. RESULTS OF MONITORING WITH *ECONFIRM*.

For quality control, Ecofys currently applies *Econfirm,* a low cost scheme for data monitoring. *Econfirm* is applicable in any type of battery operated system. This flexibility is needed in the broad PV-PL market.

Monitoring through *Econfirm* only addresses the essential question of end-users: 'What is quality of the energy service?' Scientific questions on the performance of the PV system are not addressed in order to save costs. The *Econfirm* scheme acquires data on the lighting period and the quality of the battery management, that are then evaluated automatically in an expert system. After a monitoring period of six weeks or a half a year the performance is evaluated and reported against end-user's requirements. The *Econfirm* report, concisely and accurately, answers the question whether the user requirements are met. An example is given hereafter. If not, the bottlenecks are revealed. In the Netherlands an *Econfirm* analysis typically costs 2000 € per assessment of six weeks or six months, preparation, installation and reporting included.

To illustrate the quality control with *Econfirm* the performance of the bus shelter lighting is presented hereafter.

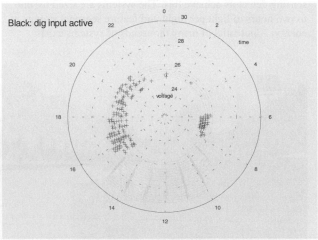

Figure 7   Data on cumulative 24-hour-clock: System voltage radial (22V-30V; lines) and 'light on' (crosses). Lighting in winter December 1999 to January 2000.

Figure 7 shows the system performance as it should be in a bus shelter lighting. The system voltage varies, even in winter, between 25 and 29 volts. The light is 'on' during dark and mainly when passengers are present, typically between 15:30 and 22:00 hours and in the morning from 6:00 to 8:00.

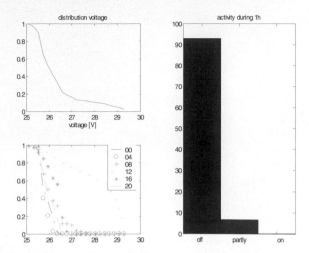

Figure 8   Voltage distributions (left) and (right) occurrence of light 'off'; 'partly off during measuring interval' and 'on' in winter December 1999 to January 2000

Statistical distributions of the system voltage (fig 8-left) confirm this performance. Fig 8-right shows that the lighting service is provided 7% of the time, shown by 7% of the measuring intervals with 'partly on' light. Absence of measuring intervals with 'light on' shows that the 'on time' of the light is always less than two hours.

In contrast to this perfectly operating bus shelter lighting, we present data of a bus shelter with a malfunctioning motion sensor.

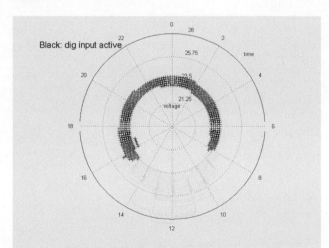

Figure 9   Data on cumulative 24-hour-clock: System voltage radial (20V-28V; lines) and 'light on' (crosses). Lighting in winter December 1999 to January 2000.

Two main aspects stand out in figures 9 and 10. Firstly the light is on all night between 15:30 and 8:00 (fig. 9 and 10-right). Secondly the system voltage ranges between 23 and 26 Volts (fig. 9 and 10-left).

Due to the defect sensor, the light was switched on all nights from sunset to sunrise. Consequenctly, the batteries were heavily discharged resulting in a structural low operating

voltage during the winter season. Due to the oversizing the voltage stays over the critical limit for freezing of the battery. Also from other analyses we have seen that the motion sensor is a very vulnerable component. However better components are available.

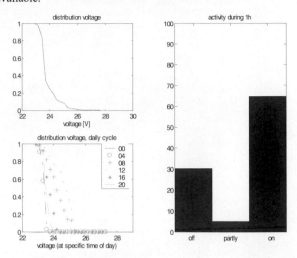

Figure 10   Voltage distributions (left); (right) occurrence of light 'off'; 'partly off during measuring interval' and 'on' in winter December 1999 - January 2000

6. CONCLUSION

In the Netherlands over 350 PV-PL systems have been realised, a Dutch commercial potential of over 10 000 systems is envisaged for this decade. This number will rise with the interest for grid connected PV-PL systems.

A range of PV-PL products is available, thus offering ranges in services, pricing and design. Such a range is an important tool in convincing and opening the market.

PV powered public lighting systems feature independence from the grid, landowners etc. and provide cost-effective solutions to lighting needs when the electricity grid has to be extended. Back alley lighting offers competitiveness for 10 meters of grid extension, streetlights typically compete from 100 meters onwards.

Energy conservation via lighting levels, lighting duration, energy efficient lighting, lighting on demand through motion detection has proven to be feasible in PV-PL systems.

The current interest and preparation of larger projects show prospect for a short-term market. Nicely designed systems to be presented in demonstration projects will stimulate this interest.

ACKNOWLEDGEMENTS

This article has been made possible by the support of CEC DG XVII - through the THERMIE programme and Novem (Netherlands Agency for Energy and Environment)

REFERENCES

[1] G. Loois, H. de Gooijer, et.al. Large scale market introduction of PV public lighting, Ecofys, Proceedings 15th EPVSEC, 3107-3110 (1998)

[2] H. de Gooijer, G. Loois, Requirements and recommendations for PVPL, E2521, Ecofys, NL 1997 (in Dutch)

A SIMPLE TOOL FOR CHECKING PV-PERFORMANCE

E.C. Molenbroek[1], A.J.N. Schoen[1], P. Nuiten[2], E. Vrins[2]

[1]Ecofys energy and environment, P.O. Box 8408, NL-3503 RK Utrecht, the Netherlands, tel. +31 30 2808 328, fax. + 31 30 28 08 301, E.Molenbroek@ecofys.nl

[2]W/E consultants sustainable building, Gouda - Tilburg, P.O. Box 733, NL-2800 AS Gouda, the Netherlands tel + 31 182 683434, fax +31 182 511296, w-e@w-e.nl, www.w-e.nl

A simple method was developed with which PV-system owners that are non-PV-specialists can check each month whether their system has been functioning well. It requires monthly kWh-readings, an estimate of the Performance Ratio, monthly insolation values of a nearby meteorological station and a tilt conversion factor. A field test with 36 PV-owners showed that the method worked well, in the sense that the owners could work with it and in the sense that failures were detected that very likely would have gone unnoticed otherwise. For large scale application of the method solar radiance and tilt conversion factors should be easily accessible, for example through newspapers or an internet site.
Keywords: Small Grid-connected PV Sytems – 1; Grid-Connected – 2; Plant Control – 3;

1 INTRODUCTION

As the amount of grid-connected PV-power installed increases from several kW per year to several MW per year, the need arises for simple tools to check PV-system performance. Failure to keep track of performance of PV-systems throughout its lifetime could lead to unnoticed system malfunctioning or failure. To this end, a simple method was developed with which PV-system owners that are non-PV-specialists can check each month whether their system has been functioning well[1]. Simplicity was required in order to make it a cheap method as well as to make it understandable for people who are not very familiar with PV-technology.

In this paper, details on the method of calculation and its accuracy are given. Furthermore, results of a field test with home-owners working with the tool are reported.

2 THE 'CONSTANT-PR' METHOD

2.1 Introduction

To check whether their PV-system has functioned well in a given month, the PV-owner or operator has to compare the measured monthly yield of the system with a calculated yield. The measured yield is equal to E_{fi}/P_{STC}, where E_{fi} is the ac-energy from the inverter (in kWh) and P_{STC} the nominal STC-power of the system (in kWp). For determination of the calculated yield Y_f it is assumed that the PR (Performance Ratio) is constant from one month to another, which reduces calculation of the monthly final yield Y_f to

$$Y_f = \frac{PR * H_i}{G_{STC}}$$

where

PR = the (estimated) PR of the system

H_i = monthly arrayplane insolation

G_{STC} = irradiance at STC (Standard Test Conditions) = 1 kW/m^2

The performance check consists of comparing the measured yield with the calculated yield. If differences larger than a certain percentage occur a fault situation is likely to exist. In total, the following data is needed for the performance check: (1) a reading of the kWh-meter that measures PV-production at the beginning and at the end of each month, (2) an estimate of the PR of the system, possibly given by the PV-supplier, (3) the monthly global horizontal insolation of the most nearby meteorological station and (4) a tilt factor for conversion of horizontal insolation to array-plane insolation. This requires knowing the tilt angle and orientation of the system.

2.2 Validation

The method and its accuracy have initially been tested on fourteen systems (of which five were identical, and all on the basis of crystalline silicon solar cells and with a central inverter) at ten different locations. A PR of 0.7 was used as starting value. It appeared that the ratio of the calculated to the measured yearly specific yield (normalized to nominal power) can vary 20% from one system to another. It appears that the initial estimate of the PR limits the accuracy of the method. However, once monthly yields have been calculated and measured for one year on a well functioning system, the PR can be adjusted to the experimentally found yearly averaged value, which makes consecutive monthly yield calculations considerably more accurate. If this is done, it turns out that the calculated monthly yield differs at most 10% from the measured yield in most, though not in all cases. This holds for the months March through October. In the winter months larger deviations are found. However, because only 10% of the insolation is received in the months November through February, an unnoticed fault will not cause large yield losses.

The 10% variation found is a sum of the following factors: (1) seasonal changes in PR, (2) spatial extrapolation error of monthly horizontal insolation meteostation to PV-location[2], (3) errors if kWh-readings are not taken exactly at the end of the month, (4) measurement error from the insolation measurement and (5) error in the calculated tilt conversion factor.

3   PERFORMANCE RATIO VARIATIONS

A follow-up study on a wider range of systems revealed that in some cases the 10% limit in monthly variation is exceeded without faults occuring. Therefore the limit was changed to 15%, which is the maximum deviation for all well functioning systems that were examined. In other words: for a system with a known PR (which can be calculated in the first year of faultless operation), if the monthly system yield is less than 85% from the calculated expected yield, there is probably something wrong with the system. The study included systems with non-ideal tilt angles (up to 80°) and orientation (up to east and west).

Naturally the assumption of a constant PR is an approximation. In reality, factors like roof integration method and temperature, non-linear behavior of modules and inverter with irradiation and input power, respectively, cause seasonal variations. In addition, specific shading situations can cause seasonal variations. This is illustrated in figure 1. However, we found that in all cases the yield for properly functioning systems did not drop below 0.85 times the yearly average.

From the above, it can be concluded that this simple 'constant-PR'-method could be a valuable tool for PV-owners to keep track of system performance. However, whereas to most PV-experts this method may seem extremely straightforward, it remained to be seen how easy it would be for PV-owners to work with. In the next section, preliminary results are described of a field test that was set up to examine this issue.

4   FIELD-TEST

4.1   Introduction
The field test is currently being held under 36 system owners in the Boegspriet project in Etten-Leur, the Netherlands. The Boegspriet project consists of 38 dwellings, with each 25,5 m^2 or 2.7 kWp PV panels. 36 owners of the 38 owners agreed to participate in the field test. The owners would first receive instruction on how to check the monthly system performance. Then they had to keep track of system performance for one and a half year. During this time it was checked whether the processing of the data and conclusions drawn from them by the home owners was correct. In addition, inquiries were held to assess the appreciation of the owners of the procedure.

4.2   Instruction of owners
First, the method was converted from a theoretical way of calculating system performance to an easy to use, easy to understand registration card. Second, an instruction booklet was written, describing the method and the required input data in everyday language. The participants received the registration card and instruction booklet during a personal visit. During this visit, the method was explained.

A few input parameters are needed to calculate the performance of the PV system. These are the readings of the PV electricity meter, the performance ratio, the monthly solar radiance, and the monthly conversion factor for the tilted plane. Given the electricity readings, users can calculate the realised monthly electricity yield. The other parameters are needed to calculate the theoretical yield. Comparing these two numbers indicates whether or not the system is functioning well. The tilt conversion factor should actually be determined each month, but they were given beforehand for reasons of simplicity, based on the Dutch Test Reference Year in De Bilt. Monthly figures for solar radiance were sent to the participants by mail. This number already incorporated the watt-peak power of the PV-system, so the actual figure participants received each month was (solar radiance*power) For the first year, the performance ratio (PR) was still unknown, but set at 0.7. The theoretical minimal yield to be calculated by the participants is then 0.8*PR*(solar radiance*power)*given tilt conversion factor. In this case, 80% of the calculated yield was used as a limit, because it was the first year of operation and the PR was not know yet. In consecutive years a value of 85% can be used, after the PR has been calculated based on results from the first year.

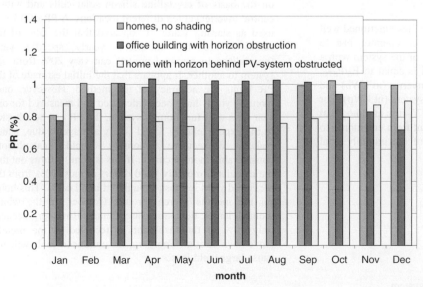

Figure 1 Examples of seasonal variation in the Performance Ratio.

After a full year of registration, the participants were asked to calculate the PR of their system.

4.3 System performance

The field test started in february 1999. From that moment on, the monthly electricity readings of the participants have been gathered. Figure 2 gives an overview of these readings. The rather large difference between the minimum and maximum yield is due to the fact that not all readings took place on the first day of the month.

In figure 3 these monthly yields are compared with the calculated yield, based on a PR of 0.7.
As explained above, a realised yield of at least 80% of the theoretical yield was considerd to be acceptable for the first year. A few systems turned out to score below this level. Figure 4 gives the number of systems that had an yield lower than (80% of) the theoretical yield, based on a performance ratio of 0.7.

After one year of monitoring, the performance ratio of the systems can be determined. For a system with an average yield of 2116 kWh, the PR is 0.72. As there is a large spread in annual yields, as can be seen in figure 5, there is a large spread in PR as well. The PR ranges from 0.63 to 0.77.

In several cases the registered electricity yield was lower than 80% of the theoretical value. In all these cases problems with (parts of) the system have been found, problems which would not have been found otherwise. Two times, there was a problem with the electricity meter. In two other cases the inverter broke down during the year. In one case, one of the strings wasn't installed properly, so it didn't contribute to the electricity yield. In the other cases the cause is unknown. System-owners became aware of these problems because they had to check their system's performance each month. The found problems are major failures of system components. An 80% limit on what is acceptable allows minor deviations in monthly yields without alarming the system owners.

4.4 Preliminary results of the field-test

During the instructional visit it became clear that not everyone is capable of calculating angles in general, which is necessary to determine tilt angle and orientation of the system, despite a detailed instruction on how to use a geometrical tool. This is not specifically related to this procedure, but more to a general lack of knowledge of mathematics.

A questionnaire was held under the participants of which at the moment of writing about 80% had been returned. The questionnaire shows that the registration card is easy to use. Filling in the card takes less than 10 minutes per month, which people consider to be not too much time. The card gives adequate information about the functioning of the system.

Determining the system performance is considered to be "not easy, not difficult", a bit easier for participants with a technical background. Only one participant made a wrong calculation.
Two-third of the participants state that they are more careful with their

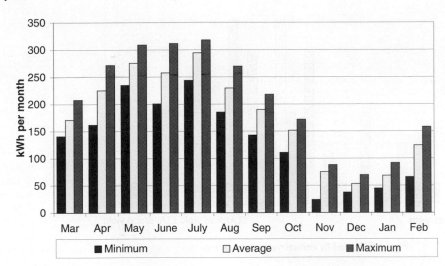

Fig. 2 Total monthly electricity yield for 36 systems, March '99 – February '00

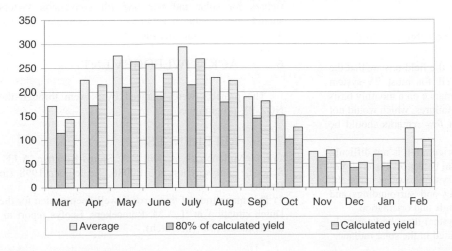

Figure 3 Total monthly actual and calculated yield for 36 systems, March '99 – Feb. '00, PR =0.7

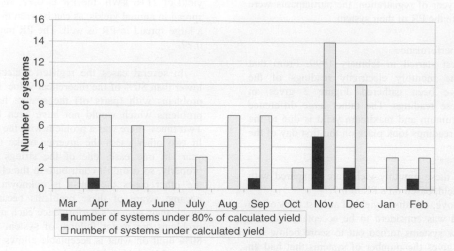

Fig. 4 Number of systems with lower than (80% of) the calculated yield.

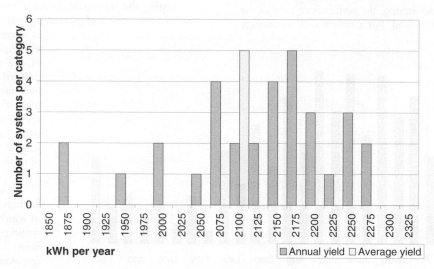

Figure 5 Spread in annual yield per system.

electricity use because filling in the card made them aware of it.

5   DISCUSSION AND CONCLUSIONS

Preliminary conclusions from the field-test are that the 'constant-PR' method works well for most PV-system owners. Checking system performance on a monthly basis directly shows major component failures, which would not be found otherwise. However, a few remarks should be made:
1) Calculation of tilt angle and orientation is not difficult, but for people with a non-technical background sometimes too complicated. Also, not all system owners have the necessary maps of their dwelling and surroundings and can therefore not determine tilt angle and orientation.
2) The watt-peak power of the system must be known. In this test, this figure was incorporated in the solar radiance, given to the participants monthly.
In both cases the installer could provide the necessary information to the owner.

To make this method work on a large scale, monthly figures for solar radiance and tilt conversion factors should be easily accessible, for example through newspapers or an internet site.

6   ACKNOWLEDGEMENT

The projects were financed by Novem through the NOZ-PV programme.

7   REFERENCES
1) E.C. Molenbroek, A. Kil, E.A. Sjerps-Koomen, A.J.N. Schoen, Ecofys report nr. E2086, September 1998 (in Dutch).
2) This magnitude of this error is well documented for the Dutch situation in 'L.A.M. Ramaekers, Ecofys report nr. E921, June 1996' (in Dutch).

PV IN BRAZIL

Stefan Krauter, Helmut Herold
Laboratório Solar Fotovoltaico, UFRJ-COPPE/EE, Caixa Postal 68504
Rio de Janeiro, 21945-970 RJ, Brazil, http://www.solar.coppe.ufrj.br
Tel.: +55-21-2605010, Fax: +55-21-2906626

ABSTRACT: Brazil has a very large potential for the use of photovoltaics: High irradiance levels at relatively small seasonal variations and a large rural population which does not have access to the public electricity grid (about 20 million people, 42% of the rural population). Despite this large potential the infrastructure on Photovoltaic supply, research and education is small. Several groups are working on the subject and try to improve the conditions.

A typical application for the non-electrified rural regions in Brazil is a little PV Solar Home System (SHS), which supplies energy for residential lighting and small home devices as radio or TV. About three thousand of these systems were implemented by a huge (70 mio Euro) governmental program called PRODEEM, by NGOs mainly working on rural development, and by many individuals. These SHS usually consist of one or two PV modules (20 to 60 W_p), a charge controller, a battery, and sometimes an inverter for powering conventional AC appliances. PV water pumping is increasingly getting popular, especially in the dry north eastern part of Brazil. Other typical applications are hybrid systems for "fazendas" (farms) consisting of a medium size PV generator (0.5-2 kW_p) and a diesel generator for more heavy loads (15-100 kW). PV grid connection is diminutive: only four photovoltaic grid injectors in Brazil are reported. A standardization of PV components, considering local conditions, is on the way.
Keywords: Brazil –1: PV applications – 2: Research –3

1. INTRODUCTION

Brazil has a size of 8.5 mio km², equivalent to continental US or 2.5 times the size of Europe. The population amounts 164 million in 1999 with an increase of 1.2% per year [1]. While the coastal areas are densely populated, inner regions suffer from extreme climatic conditions as dryness in the Northeast (Nordeste) or extreme humidity (Amazon) and lack of infrastructure and therefore loose population.

Figure 1: Map of Brazil with states and major regions.

Total energy consumption of Brazil is 228.8 mio TOE (9.58 EJ), 68.4% are from renewable energy sources, mainly hydro power and methanol from sugar crane, therefore Brazil has remarkable low specific carbon-dioxide emissions per capita of 1.7 t/a (Germany 10.5 t/a, USA 17.8 t/a) [2].

39% of total energy consumption is by electricity. While Brazil currently generates 92% of its electricity by hydropower, additional power plants may be driven by fossil fuels due to lack of water and limitations of suitable locations for additional hydro power plants. Growth of electricity consumption is 4% per year, mainly triggered by increasing commercial (+8.9%) and residential (+7.2%) consumption.

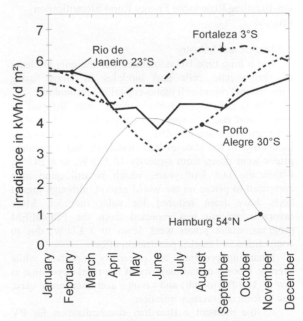

Figure 2: Daily irradiance on a inclined surface for different locations in Brazil in comparison to a location in Central Europe [3]. Angles of surface elevations are chosen equivalent to the latitude.

While the coverage of the electrical grid is almost total in the urban areas, a large part of the rural population (42%, about 20 million people) does not have access to the public electricity grid. The prerequisites for solar energy use are good: high value of yearly irradiance of 1600 to 2300 kWh/(a m²), while daily irradiance is not less than 3 kWh/(d m²) in average, even fat the least favorable locations and even in the worst month, as could be seen in Figure 2.

2. IMPLEMENTATION OF PV

2.1 Governmental activities

The Brazilian Government supplies various projects helping the development of poor rural regions located mainly in the North and Northeast of Brazil. Primary objective is to supply health, education, drinking water and energy for the rural population. One of the main steps of fighting poverty and providing development was the program "Brasil em Ação" (Brazil in Action). The total financing for the sector of energy was 4.2 billion Euro, while 70 million Euro have been spent on photovoltaics, mainly in the PRODEEM-Program.

The PRODEEM for the energy development of Brazilian states and communities was established in 1994 by the Ministry of Mines and Energy and is still going on. Prior objective is the energy supply for social, productive and development purposes. The program is coordinated by DNDE, which is acting for the Ministry of Mines and Energy. Total installed PV power exceeds now 1.7 MW_p. After phase four, which is actually in progress the number systems will exceed 5700 and further 2.8 MW_p of PV, thus making PRODEEM one of the most important PV electrification projects of its kind worldwide.

Other programs are carried out with foreign collaboration as Eldorado (German collaboration) or the US-Brazilian Renewable Energy Rural Electrification Program.

2.2 Market and Industry

For a long time Heliodínamica was the only producer of photovoltaic cells and modules in Brazil using conventional round-cell mono-crystalline technology.

The Brazilian photovoltaic market was distorted by high import taxes, to support local producers. Meanwhile a foreign manufacturer (SSI) set up a factory for module lamination in the state of Rio Grande do Sul. Consumer prices went down from typically 10 EU/W_p to 6 EU/W_p within the last four years, which is still quite high compared to prices on the world market, although import taxes have been reduced for solar modules. More favorable pricing was reported from the PRODEEM program, where prices went down to 3 EU/W_p due to competition and market penetration efforts.

Financing programs are extremely important, while average income of the majority of the rural population is low (< 100 EU/month) and savings accounts hardly exist, due to fears of currency inflation.

At the moment a Brazilian standardization for PV modules is in process at ABNT (Associação Brasileira de Normas Téchnicas) led by INPE (which was responsible for the PV panels of Brazilian space satellites) and is accompanied by a workgroup with members from UFSC, UFRJ-COPPE-EE CEPEL, INT and USP.

3. RESEARCH ON PV

3.1 Solar cells

Laboratório de Materiais e Interfaces, da COPPE. This group started in 1970 at the Metallurgy Department of the Federal State University of Rio de Janeiro, since 1982 research was carried out in amorphous thin film solar cells. Due to retirements activities stopped in 1998.

Laboratório de Microeletrônica da USP, ligado ao Departamento de Engenharia Elétrica da Escola Politécnica This group started in 1981 with production technologies for mono-crystalline silicon solar cells, including anti-reflection layers and contacting. Between 1983 and 1989 also activities in the multi-crystalline and amorphous area were reported.

Laboratório de sensores do INPE, Instituto de Pesquisas Espaciais de São José dos Campos. The "NASA" of Brazil, started its activities by testing and qualification solar cells for space applications. Is now also working on standardization of terrestrial PV modules.

Laboratório de Conversão Fotovoltaica, da UNICAMP. Group linked to the physics Institute is working since 1980 on photovoltaic materials and processes.

Laboratório de Microeletrônica e Células Solares, do IME. Group connected to the material science department at the Military University of Rio de Janeiro. Only group in Brazil which is working on Cadmium and Cooper Sulfite solar cells.

Also other groups worked partially on the development of solar cells and materials. Due to different reasons (discontinuation of projects, economic crises, lack of applied research, sell out of management) a lot of activities stopped.

3.2 Photovoltaic systems

A lot of groups are working partially on PV systems, often less on research, but on demonstration and monitoring.

Working groups

NAPER (Working Group for Renewable Energies)

IDER (studies on implementation of renewable energies in the state Pernambuco, where just 14% of rural properties are electrified, workshops for training, testing and development of solar equipment for 12V.

GEDAE (works since 1994 on installation of PV hybrid systems at *UFPA*)

UEES (Solar Energy Students Group - is working on information distribution about PV)

Aonde Vamos (amateur group working on popularization of renewable energies)

and many other smaller groups.

Universities

UFRJ-COPPE-EE Universidade Federal do Rio de Janeiro, PV-Labs. Started PV in 1994 with German GTZ and TU Berlin, now working on SHS, online monitoring, PV grid connection, system development, courses on PV-systems are given for graduate and undergraduate students [4], presently establishes an area for Renewable Energies.

UFSC Universidade Federal do Santa Catharina Works on PV building integration (see Fig. 3), PV grid connection, irradiance, also offers courses on solar energy for graduation and post-graduation).

Figure 3: Building integrated PV at UFSC (2 kW$_p$ a-Si).

USP Universidade de São Paulo. Works on SHS (installation and monitoring of several systems in São Paulo State), PV grid connection.

PUC-Minas, GREEN Reference and certification center (mainly for solar thermal components), simulation, remote education, applications of PV at electric fences, public lighting, water pumping, refrigeration. Courses on PV- and solar thermal systems.

UFPA – Universidade Federal do Pará. Works on PV-systems, organizes workgroup *GEDAE*.

UFPE - Universidade Federal de Pernambuco. Is working on SHS and applied PV systems.

UFCE Universidade Federal de Ceará. Works on solar thermal systems and SHS.

UFOR Universidade de Fortaleza. Is doing research on materials and irradiance measurement.

UFRGS Universidade Federal do Rio Grande do Sul. Research on irradiance, solar-thermal systems and photovoltaics, software development and planning

instruments (SolarCAD), courses are given for graduate and undergraduate students.

Electrical Utilities

CHESF / San Francisco, Recife. Test site, first combined PV-Diesel (1 kW) in Brazil financed by German GTZ, additional 10 kW system.

CELPE / Pernambuco. Installed 1000 PV systems for rural electrification

COELCE / Ceará. Set up 450 systems of residential illumination and water pumping.

CEMIG / Minas Gerais. 770 PV systems constructed, 4000 to be installed in collaboration with German GTZ.

CRESESB (Solar and Wind Energy Reference Center Sérgio de Salvo Brito) in Rio de Janeiro is publishing books and CD-ROMs on the area and supports a demonstration center, the Efficient Solar House, where seminars are held. Is linked to the research center of the utilities *CEPEL*.

Conferences on PV

Two to three national conferences a year on renewable energies, respectively solar energy, are also covering the area of photovoltaics and take place irregularly.

4. CONCLUSION

In relation to the potential and the quantity of activities carried out the resulting impact could be improved. A constructive, long-term program for scientific, educational and industrial development on the area would very favorable. Supervision, systematic monitoring and scientific analysis of the ambitious PRODEEM program could have led to significant progress in the area of applied SHS.

REFERENCES

[1] Instituto Brasileiro de Geografia e Estatística (1999).
[2] Almanaque Abril 2000 (2000).
[3] A. Krenzinger, Radiasol, UFRGS (1998).
[4] S. Krauter, R. Hanitsch, R. Stephan, Int. J. Electr. Enging. Educ., **35** (1998) 139-146.

THE POTENTIAL OF PV NOISE BARRIER TECHNOLOGY IN EUROPE

Thomas Nordmann, Andreas Froelich, TNC AG, Seestr. 129, CH-8810 Horgen, Tel+41/1/725 39 00, froelich@tnc.ch
Adolf Goetzberger, Gerhard Kleiss; TNC, Oltmannstr. 5, D 79100 Freiburg, Tel +49/761/4588-152, goetzb@ise.fhg.de
Georg Hille, Christian Reise, Edo Wiemken; Fraunhofer ISE, Oltmannstr. 5, D 79100 Freiburg, Tel +49/761/4588-0,
reise@ise.fhg.de
Vincent van Dijk, Jethro Betcke; Utrecht University, Padualaan 14, NL-3584 Utrecht, J.Betcke@chem.uu.nl
Nicola Pearsall, Kathleen Hynes; NPAC, Ellison Place, UK-NE1 8ST, Newcastle, +44/191/227-4595,
nicola.pearsall@unn.ac.uk
Bruno Gaiddon, PHEBUS, 1, rue de l' Oiselière, F-69009 Lyon, Tel. +33/4/78472947, bruno.gaiddon@ wanadoo.fr
Salvatore Castello; ENEA, Via Anguillanese 301, I –00125 Roma, Tel +39/630484339, email castello@casaccia.enea.it

ABSTRACT: Photovoltaics is expanding into new market segments. Photovoltaic noise barriers (PVNB) along motorways and railways permit today one of the most economic applications of grid-connected PV with the additional benefits of large scale plants (typical installed power: more than 100 kWp) and no extra land consumption. The aim of this study is to reveal the large potential that can be exploited for PV on noise barriers with the overall objective of raising the share of renewable energies for the EU´s electricity market. In contrast to many PV-potential studies published before, this proposal is focusing on PVNB only, as one of the cheapest ways to implement large scale grid-connected PV installations.
Keywords: PV Market- 1; PV noise barrier technology - 2; Grid-Connected- 3

1. INTRODUCTION

Photovoltaics (PV) is in moderate latitudes the only option to use the sun for electricity production to a larger scale. In the past various investigations have shown that large rated power in the GW range can be integrated in the existing house- and industrial-roof tops. Another area is very attractive due to its easy standardisation and legally clear properties: the transportation lines. These are highways and freeways as well as railway tracks.

Photovoltaic noise barriers (PVNB) along motorways and railways permit today one of the most economic applications of grid-connected PV with the additional benefits of large scale plants (typical installed power: more than 100 kWp) and no extra land consumption. The aim of this study is to reveal the large potential that can be exploited for PV on noise barriers with the overall objective of raising the share of renewable energies for the EU´s electricity market. In contrast to many PV-potential studies published before, this proposal is focusing on PVNB only, as one of the cheapest ways to implement large scale grid-connected PV installations.

Co-financed by the EU Commission DG XVII, TNC GmbH Germany co-ordinated the elaboration of the study together with the following partners:

- Fraunhofer ISE, Germany
- PHEBUS, France
- TNC AG, Switzerland
- NPAC, United Kingdom
- University of Utrecht, the Netherlands
- ENEA, Italy

2. METHODOLOGY

2.1 Data selection and preparation

The analysis of existing and planned noise barriers along rails and roads has been carried out by the national partners together with national authorities, which are experts and responsible for the required data. The methodical approach of this study includes the set-up of a grid with dimensions of 1 by 1 geographical degrees for Germany, Italy, France, United Kingdom and 0.5 by 0.5 degrees for the Netherlands and Switzerland. For each degree the length and orientation of rails and roads, the existing and planned noise barriers

are registered and grouped according to their orientations.
The solar radiation is based on data of a METEONORM data set [1]. This includes the solar radiation on horizontal orientation as well as various inclination angles for all possible orientations. Moreover, possible shading has been considered.
The following table 1 gives an overview over existing PVNB projects that have been installed in Europe:

Place Trafficway / Year	Peak kWp	Planner / Investor
Domat Ems, CH A13/1989	103	TNC AG / Swiss Office for Energy
Gordola, CH Railway/1992	103	TNC AG / Swiss Office for Energy
Seewalchen, A A1/1992	40	Oberösterreichische Kraftwerke
Rellingen, D A23/1992	30	TST (DASA)
Giebenach, CH A2/1995	104	TNC AG / Kanton Basel-Landschaft Swiss Office for Motorways
Saarbrücken, D A6/1995	60	Stadtwerke Saarbrücken
Utrecht, NL A27/1995	55	R&S und others
Ammersee, D A96/1997	30	3 Prototypes, TNC GmbH / Bayernwerk & BMFT
Zürich north, CH A1/1997-99	30	3 Prototypes, TNC AG / Swiss Office for Energy Swiss Office for Roads Zürich Utility EWZ
Ouderkerk, NL A9/1997/98	220	Shell & ENW / EU Commission

Table 1: Realised PVNB in Europe (status end of 1997)

The technical specifications of PVNB are based on the comprehensive knowledge of TNC GmbH and TNC AG with various plants realised [2],[3]. Technologies have been considered for both state-of-the-art and innovative concepts such as bifacial PVNB (see fig. 1). In bifacial PVNB the vertically mounted PV-module is light sensitive on both sides and is used at the same time as noise reflecting element.

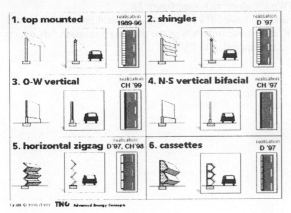

Figure 1: Schematic sketch of different photovoltaic noise barrier (PVNB) structures including integrated PVNB [4]

2.2 Definition of potentials
Installed PV power and produced electricity have been calculated for the following potentials:

Theoretical: Assessment of the maximal possible PVNB. All existing rails and roads (R&R) will be equipped with PV. PV is mounted on the support structure (which is not necessarily a NB) with the optimal tilt angle. A further development of the technology used is assumed. As shading is not considered all R&R will be equipped with PVNB.

Technical: All NB along R&R, planned today, will be equipped with PV. Moreover, the already installed NB will be upgraded with PV. A classification of the NB into all possible orientations is done for each 1 by 1 degree. The technology used is state-of-the-art. Shading is considered.

Short term: All NB planned today will be equipped with PV. A classification of the NB into all 1 by 1 degrees is done. The used technology is state-of-the-art. Shading is considered.

European extrapolated: In the progress of this study it became evident that France, Italy and United Kingdom have poor short-term potentials due to the lack of NB-planning. Consequently, a European extrapolated potential was defined: Length density within each degree was correlated with the average short-term potentials of the other countries investigated.

EU-member: For the EU-members not considered in this study an EU-member potential was defined: The average ratio of potential of electricity resp. power to the length of roads/rails installed was extrapolated to the other EU-members. Sweden and Finland have not been considered due to their low population density.

Anticipated: The anticipated potential is based on the analyses on the national basis according to the economic and political boundary conditions. This means the calculation of the economic competitiveness of PVNB in comparison with its alternatives. The relevant financial parameters for costs and revenues were considered. Moreover, the political willingness for a reinforced introduction of PV on NB were analysed.

2.3 Calculation and visualisation procedure
The calculation of potential is done in Excel 7.0. Special emphasis had been put on the visualisation of the results
Output data are used to produce maps, which are divided in coloured pixels of 1 x 1 degrees or 0.5 x 0.5 degrees. All colour levels represent the potential of PV power installed or PV electricity production "per pixel".
White pixels denote a zero potential or lie outside the area under consideration. The coloured pixels follow a near-logarithmic scale with 5 colour levels. These colour levels are identical for all maps, nevertheless, the same colour may denote different ranges for different maps.
In fact, the ranges change with the different types of potential (theoretical, technical, and short term). Moreover, colours are different for maps of installed power (green) and electricity production (red), but, the colour ranges remain constant for all countries. So, a map showing the theoretical potential for Italy will be comparable to the same type of map for Germany. On the other hand, two maps for France, showing the theoretical and the short term potential, will use different scales.

2.4 Technical process of map production
The maps are created from ASCII tables by the software package GMT. The output of GMT is one PostScript file per map.
The slides, consisting of a map on the right, the headline and the caption on the left, and several logos are formatted using LaTeX.
The following maps have been designed:
- 6 countries plus Europe as a whole,
- 3 types of potential (theoretical, technical, and short term/extrapolated potential),
- 2 quantities (installed power and electricity production),
- and all this for rails and roads

3. RESULTS

3.1 Theoretical Potential

Theoretical Potential		Countries						All six countries	Average of all countries
		CH	D	NL	UK	I	F		
Relevant roads	[km]	1868	11013	2701	10791	6830	12255	45458	7576
rails	[km]	1663	6652	3065	9967	4820	7850	34017	5669
roads	[MW$_p$]	2236	13183	3233	12917	8176	14669	54414	9069
rails	[MW$_p$]	1422	5687	2620	8522	4121	6712	29084	4847
rails&roads	[MW$_p$]	3658	18870	5854	21439	12297	21381	83498	13916

Table 2: Total of theoretical potential of expected installed power for each country

3.1 Theoretical Potential

The theoretical potential assesses the upper boundary of installable PV along traffic ways All existing R&R will be equipped with NB. The main potential is located in the European metropolitan areas London, Paris. Approx. 65% of the 75TWh/a annual production are related to roads, the remaining 35% to rails.

3.2 Technical Potential

The technical potential for the six countries investigated is encouraging: approx. 584 MWp PV along roads and 217 MWp PV along rails (see figure 2).

D and NL have 74% of the total technical potential, whereas UK and I have only 9% of the existing European NB which could be upgraded to PVNB.

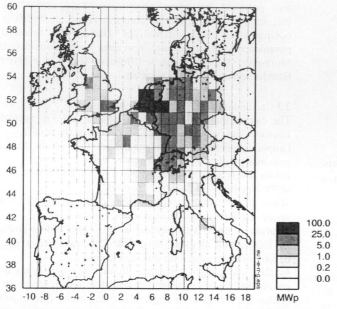

Figure 2: Technical potential of installed power of PV along rails & roads in all six countries

3.3 Short term/extrapolated potential

The short-term potential for Switzerland, Germany and NL is approx. 140 MWp PV along roads and 145 MWp PV along rails (see figure 3).

This results for all EU members to an extrapolated potential of 1145 MWp PV along roads and rails.

If the national policy changes in France, Italy and United Kingdom to a European extrapolation, the expected potential in France is 96 MWp, in Italy 170 MWp resp. in United Kingdom 385 MWp.

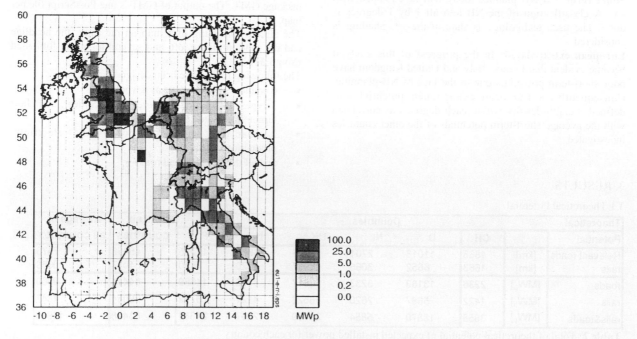

Figure 3: Extrapolated potential of installed power of PV along rails & roads in all six countries

3.4 The potentials for all EU-member countries

Not all countries of the European Union could be investigated in detail within the scope of this study. The following countries were not considered in this study: Belgium, Denmark, Finland, Greece, Ireland, Luxembourg, Portugal, Spain and Sweden

Consequently, an EU-member potential was defined: The existing road resp. rail lengths in these countries have been multiplied with the average ratio of the potential of electricity/power and road/rails in the six countries investigated.

The short-term – including the European extrapolated - potential of installed PV–capacities for roads will increase by 15%, for rails by 28% by including these EU countries. Spain is according to this assessment the most important country for a more detailed analysis

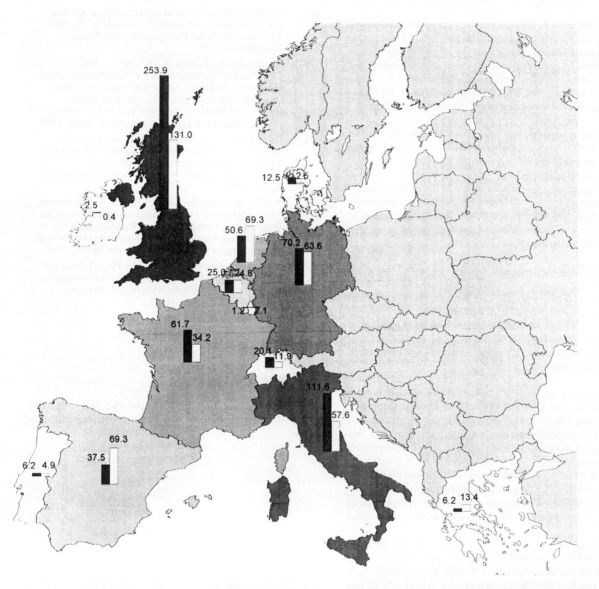

Figure 4: European extrapolation of the short term potential of installed power for PV in all EU-member countries. Data for CH, D, NL, UK, I and F have been calculated in detail; the potential is based on all noise barriers planned today (left bar: roads, right bar: rails)

3.5 Anticipated Potential

Switzerland is still the country with the highest installed PV power capacity in the world. The technology profits from a big interest and goodwill by many people. The idea of using PV on NB was lanced in Switzerland. NB are a fixed part of Swiss highways. A national plan for the construction of more than 105 km of NB along roads and 76 km along rails of NB in the next five years has been set up.

The interest for the construction of PVNB has grown very much during the past few years. Especially pressure groups of concerned people show interest, also communities and private persons.

Summarising, the boundary conditions in Switzerland are very favourable to realise a large share of the existing potential.

Germany has very favourable conditions for the realisation of further PVNB [5]:

- A high density of rails and roads
- An ongoing demand for a further construction of NB
- Reasonable to excellent tariffs for PV-electricity fed into the interconnected grid
- Favourable acceptance of PVNB by the public opinion

Consequently Germany is one of the key countries for a reinforced introduction of PVNB.

In **the Netherlands** attitudes of the parties involved in the erection of PVNB are neutral to positive. Obstacles to implementation are the large number of parties involved in the erection of noise barriers, the economic feasibility and the accessibility of noise barriers in urban areas.

The large rail projects offer the best perspective for short-term large-scale introduction of PVNB in the Netherlands.

In **United Kingdom**, the actual potential for the use of PVNB is rather low as a result of the low usage of noise barriers on roads and railways, the lack of formal incentives for PV electricity fed into the grid and the reduced programme of road building expected in the next few years. The attitude to the use of PV as a future source of electricity in the UK is generally quite positive, but demonstration of actual PV noise barriers would be required to make this application acceptable.

In **Italy**, the actual potential for the use of PV on noise barriers is rather low as a result of the low usage of noise barriers on roads and railways and the lack of formal incentives for PV electricity fed into the grid.

On the other hand the attitude to the use of PV as a future source of electricity is generally quite positive, but demonstration of actual PV noise barriers would be required to make this application acceptable.

In **France** the potential of PV noise barriers is rather low for the moment. The principal obstacles to their development are the very low price that the grid operator is disposed to pay for PV electricity and the lack of information about this new kind of noise barriers among specialised professionals and decision-makers.

The work to be done in order to overcome these obstacles is at first to bring higher pressure for increasing the feed-in tariff paid for PV electricity as a sustainable energy, free of greenhouse effect gas emission as well as of contaminated waste.

The second task is to inform and to convince decision makers of PVNB possibilities. For that purpose, we expect much from the achievement of SUNWATT's PVNB, as it will be the best demonstrative example in real size of this technology.

4. OUTLOOK

The result of this study confirms the current activities to implement PV on noise barriers as an important share in the PV market.

It is anticipated by the partners that the chances to realise large shares of the potentials are good in Germany, the Netherlands and Switzerland. In France, United Kingdom and Italy the realisation of larger PVNB is rather unlikely.

There are recent changes in the revenue rate for electricity of PV feed-into the inter-connected grid: In Germany 0.506 EURO/kWh for 20 years has been decided, in Spain 0.40 EURO/kWh are currently under discussion. In both cases (in Germany together with the soft-loan of the 100,000 Roof-Programme) the profitability of PVNB is strongly improved and reinforced actions to install further PVNB are envisaged.

REFERENCES

[1]   METEONORM - Global meteorological database for solar energy and applied climatology. Version 3.0. Bundesamt für Energiewirtschaft, Bern, Switzerland, 1997.

[2]   Nordmann, T.; Frölich A.; Reiche, K.; Kleiss, G.; Goetzberger, A.: Integrated PV Noise Barriers: Six Innovative 10 kWp Testing Facilities - A German /Swiss Technological and Economical Sucess Story!, 2nd World Conference and Exhibition on Photovoltaic Solar Energy Conversion, Wien, 1998.

[3]   Nordmann, Th.; Frölich, A,. Clavadetscher, L.: Eight Years of Operation Experience with two 100 kWp PV Soundbarriers, 14 th European Photovoltaic Solar Energy Conference, Barcelona, Spain, 1997.

[4]   Reiche, K., Goetzberger, A, Frölich, A,. Nordmann, Th.: Integrated PV-Soundbarriers: Results from the International Competition and Realisation of Six 10 kWp Testing Facilities, 14 th European Photovoltaic Solar Energy Conference, Barcelona, Spain, 1997.

[5]   G. Hille, K. Reiche: Potential der Photovoltaik auf Lärmschutzwänden an deutschen Verkehrswegen. 11. Symposium Photovoltaische Solarenergie. Staffelstein, Germany, 1996.

The full report can be found in the internet:
www.tnc.ch

ACKNOWLEDGEMENT

The authors thank EU DGXVII and the Swiss Federal Office for Education and Science for co-financing this study.

Financing Grid Connected PV Systems
in The Netherlands after the Year 2000

E.J. Koot M.Sc.[1], D.J. van Dulst[2], D. Gieselaar[2]
[1]Managing Director [2]Consultant

Ekomation Solar Energy Consultancy
P.O. Box: 29112, 3001 GC Rotterdam, The Netherlands
Tel.: ++31 10 280 72 64, fax : ++31 10 280 72 65
E-mail: e.j.koot@ekomation.nl, Internet: www.ekomation.nl

ABSTRACT: Ekomation has evaluated financial aspects for grid connected BIPV-systems beyond 2000, as preparation for the new Dutch PV Convenant. Several market segments are evaluated as well as channels of distribution are charted and market related (environmental) factors. Instruments for an accelerated market growth are also investigated. The study results in the definition of the most attractive market segments. For example companies profit the most from Dutch financial regulations and market circumstances.

Keywords: Financing -1: Funding and Incentives -2: Economic analysis –3

1. INTRODUCTION

1.1 Framework

In the Netherlands, the market introduction of photovoltaics is stimulated by a joint effort of various Dutch pv-organisations. This effort has been formalised in the so called *PV Convenant*, which is an agreement on realizing pv power to be installed and technological developments. Because the term of this very successful PV Convenant is ending in 2000, a new and more ambitious agreement for the period till 2007 is planned, in order to boost the Dutch pv market seriously. In the current periode this new convenant is being prepared, Ekomation has executed, for the PV Platform, where all convenant partners are represented, a survey on possible financial incentives for specific market segments for the Dutch BIPV-market.

2. OBJECTIVES AND APPROACH

2.1 Objective

The objective was to create an insight in the possible incentives for the most potential market segments in the Netherlands in the period 2000-2005. The study emphasised the needed and available financial incentives, caused by the most important obstacle for a grown up pv-market: the high price of the technology. It is assumed that in the concerning period, market players will not be capable of decreasing pv system prices in such manner solar kilowatt-hours will be price-competitive with conventional electricity yet. Therefore additional financial incentives are critical for a successful market introduction of photovoltaics.

2.2 Approach

Based on three criteria the Dutch pv market has been divided into several market segments. The segments which have the highest market potential in the period 2000-2005 are selected and the specific channels of distribution within these segments are defined. These channels are significant because they will be helpful with the determination of useful new incentives that need to be developed. Besides, the description of the specific channels of distribution results in a valuable insight in the organisations that are involved. Putting the channels together with the market segments, so called market configurations are being developed.

Furthermore all relevant external aspects for pv development in the Netherlands are charted. Subsequently an overview of possible instruments is presented.

3. MARKET CONFIGURATIONS

3.1 Segmentation

Market configurations are defined by putting segments together with the channels of distribution. Segments are defined based on three criteria:

1. *Size of the pv-installation*: Very small systems range up to 300 Wp and are mostly small single or double AC-modules. Small systems are (semi) integrated pv systems limited in their size up to 800 Wp due to grid interconnection regulation. Medium sized systems are in the range of 800 Wp till 3500 Wp. The latest class are the large systems with an installed pv-power of 3,5 kWp or more.

2. *Applications on new buildings or retrofit-applications*. This criterion is relevant because it influences system design, system size and the energy performance of a dwelling or building.

3. *The investor profile* is related to the fiscal scheme which is relevant for the investor: Private persons and non–profit organisations cannot utilise most of the fiscal incentive schemes whereas firms can.

Nr.	System size	Investor profile	Type of installation
1.	<300	Private households	Retrofit
2.		& Non-profit org.	New buildings/dwellings
3.		Profit organisations	Retrofit
4.			New buildings/dwellings
5.	200-800	Private households	Retrofit
6.		& Non-profit org.	New buildings/dwellings
7.		Profit organizations	Retrofit
8.			New buildings/dwellings
9.	800-3500	Private households	Retrofit
10.		& Non-profit org.	New buildings/dwellings
11.		Profit organizations	Retrofit
12.			New buildings/dwellings
13.	3500>	Private households	Retrofit
14.		& Non-profit org.	New buildings/dwellings
15.		Profit organizations	Retrofit
16.			New buildings/dwellings

Table I: Defined market segments

3.2 Channels of distribution

The channels of distribution for BIPV-systems in The Netherlands can be devided into direct channels and indirect channels. Direct channels of distribution concern direct delivery of the pv-system to its user by a so called pv manufacturer or manufacturers agent. An indirect channel concerns delivery via a market intermediair.

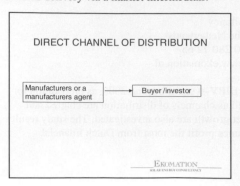

Figure I: The direct channel of distribution of BIPV-systems in the Netherlands

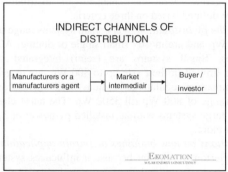

Figure II: The indirect channel of distribution of BIPV-systems in the Netherlands

Market intermediaries in The Netherlands are very divers and involved in various markets. The most important market intermediaries are:

- Property developers and housing associations
- Energy utilities
- System integrators (dedicated engineering companies)
- Electrotechnical installers
- Non governmental organisations (NGO's)
- Building component manufacturers
- Do It Yourself (DIY) stores

3.3. Configurations
        Based on the market segments and channels of distribution the most important market configurations can be defined for The Netherlands. The most of these configurations are now being developed by the early market players and it is to be expected that the next few years these configurations will expand rapidly.

Configuration 1: very small systems
        Very small systems can be sold by energy utilities (NUON's pv-prive) or NGO's (Greenpeace Solaris). It isn't unthinkable that future sales of very small systems will be realised by DIY's. These configurations have without exception indirect channels of distribution.

Configuration 2: Small systems, retrofit

Building integrated pv-systems between 200-800 Wp for existing buildings will likely be sold by energy utilities, electrotechnical installers or system integrators and maybe the building industry via DIY-stores. This configuration makes use of indirect channels of distribution.

Configuration 3: small systems on new buildings
        New buildings are mostly developed by property developers and sometimes housing associations (social dwellings). Property developers use small pv-systems in order to disthinguish their products (dwellings or commercial buildings) from others with a sustainable image. This configuration is based on the use of indirect channels of distribution, but there are also examples of delivering small scale systems directly by the manufacturer.

Configuration 4: Medium sized systems
        The configuration of medium sized systems for new buildings doesn't differ much from the configuration of small systems for new buildings. The same channels of distribution will be used with the exception of system integrators becoming more dominant. The segment of existing buildings (retrofit) can increasingly be interesting for the direct channel. But intervention of a energy utility or distributor of building components is thinkable.

Configuration 5: Large scale systems
        Sales of large scale systems in The Netherlands is mainly expected in new buildings like apartments and commercial buildings. Both types of distribution channels are used, whereas the energy utilities in the indirect channel are the most significant. They can use the generated solar power for their green energy products.

4. EXTERNAL ANALYSIS

4.1 General
        Next to the definition of the configurations it is important to know which external aspects influences these configurations. The most important aspects is the Energy Performance Ratio on new dwellings (called the EPC), the electricity market, the price of electric energy (per kWh) and the availability of subsidies.

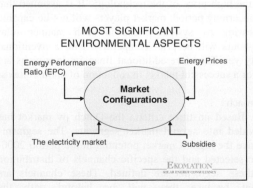

Figure III: The most important environmental aspects concerning pv market configurations in The Netherlands.

4.2 Energy Performance Ratio (EPC)
        The build environment is a potential surrounding were high results on energy saving can be realised. Therefore the Dutch authorities have created regulations for the energy performance of new buildings in a EPC. BIPV-systems have a positive influence on the energy (saving)

performance and can therefor be an incentive for property developers to invest in BIPV. The expected decrease of the EPC-standard will increase the value of pv as a measure to accomplish the standard.

4.3. The electricity market

As in many European countries, in The Netherlands the energy market will be liberalised and energy utilities privatised. As a result of this change the market is becoming rapidly international and more competitive. Former government owned utilities are now operating in a highly dynamic and hostile environment. The new Electricity Law dictates a phased liberalisation. The larger clients are since the first of January 2000 free of choosing its electricy supplier. Medium sized clients will be free to do so in 2002. Consumers and small enterprises in 2004.

It is to be expected that energy utilities will distinguish themselves more and more in order to create increasing market share. One of the possibilities is offering a wide range of energy products, including environmental friendly products like green electricity or pv-systems. Selling Solar power as a derivative of green electricity can also have some market potential, which schemes in other European countries already have been shown. Due to the internationalising energy markets, these products can also be offered by for example German utilities.

4.4 Energy prices

Compared with mobile communication tariffs the liberalised energy market will probably result in new tariff structures. Due to the uncertainty about these structure the influence on pv cannot be estimated. Pay back rates by energy utilities will be decreased, certainly for large scale systems. For (very) small systems it is not likely that the produced energy is going to be delivered back to the utility's mains. Therefor buy back rates will not influence (very) small systems sales. Prices for conventional electricity do. They have an influence on the attractiveness of small pv-systems because the higher the price, the higher the energy saving from the small pv-system is.

The price for energy in The Netherlands is artificial high, due to the so called Regulatory Energy Tax. The price of the kWh is increased by a regulatory tax which will be graduatedly raised in the coming years.

4.5 Subsidies

The Dutch energy utilities have agreed on an action plan in which they subsidise renewable energy projects, particularly pv-projects up to about 20% of the investment costs. This is an important incentive for projects realised in the Netherlands, but from 2001 this subsidy scheme will not be available anymore, so one of the most important pv-incentives will disappear. At this moment possible new incentives by the Dutch utilities are discussed, but aren't likey due to the liberalisation and commercial priorities of the utilities concerned.

Within the fifth framework of the European Union it is possible to receive fundings for cleaner energy systems and energy efficiency projects. The EU subsidy scheme makes it for innovative projects and systems possible to get a significant funding. But small projects or projects with a less innovative character will not be egible.

Fiscal schemes in The Netherlands will become increasingly important as an incentive for investments in BIPV's. The current fiscal regulations depend on the availability of funds fed by the ecotax. It is expected that these funds will be high enough for the coming years,

including a market boost. These fiscal instruments are only directly applicable for companies and are likely to remain for several years.

5.   POTENTIAL ANALYSIS

The study is based on a target of 50 MWp installed pv-power in 2005. The necessary financial support depends on various aspects like demand development, acceptation of pv in the build environment etc. When the demand for pv will rise, less additional financial incentives will be needed. When pv will not be accepted by the late majority, maintaining current financials incentives will not be sufficient, even with decreasing system prices.

6.   STIMULATION

6.1 General

Definition of necessary incentives depend on the the specific target for each market segment, which incentive is the most appropriate for the segment, the contribution of the incentive for each project and what the side effects are of the concerning incentive. These aspects were beyond the scope the study.

6.2 Accelerated market development

In order to accelerate the market to a level were the 50 MW target in 2005 will be exceeded, some kind of a shock therapy is needed. To create such a shock, incentives must create cost effectiveness for the investor. Main obstacle for such cost effectiveness is not the way these financial incentives will get to the investor, but the available fundings for such incentive. Selling solar electricity could be the approach for getting enough fundings for a cost effective incentive. The energy utility, or a third party, can offer this new product. Buyers will pay a higher price for a solar kWh, enabling the energy utility to invest in pv-power generation capacity. The price for the solar kWh to be sold will be higher than the price for conventional electricity, VAT and ecotax included. The current success of green electricity sales in The Netherlands show that there is a market potential for this product. The most important market segment for producing the solar electricity are small and medium enterprises which can effectuate most of the fiscal schemes on the pv-investment. They can sell the generated solar power to the Solar Power distributor.

Other measures for a accelerated market development in The Netherlands are an increased Energy Performance Ratio legislation, fiscal deductibility of pv-systems for private households and persons and increasing governmental subsidies.

7.   EVALUATION

7.1 Evaluation of potential incentives

For very small systems it is important that the incentive is addressed to the end user of the pv-system. Fiscal legislation could be an option. A threat could be the laboriouness of such an arrangement.
The provision of cheap loans has proved to be very difficult due to involved bureaucratic schemes. A direct subsidy can offer opportunities for this market configuration.

Small systems (<800 Wp) on existing buildings can be stimulated by a fiscal deductibility for private house owners. It can for example be combined with a cheap loan. An alternative is the combination of direct subsidy with a cheap loan.

Small systems on new dwellings and buildings can preferably be stimulated by a financial incentive like a direct subsidy. The concerning property developer is the right target group because he is always involved and the decision maker. Fiscal stimulation is sub-optimal due to the lack of effectiveness for project developers. Cheap loans will also be less effective because the are not beneficial for property developers.

Small systems (<800 Wp) for existing commercial buildings can be stimulated by continuation of existing fiscal schemes, development of cheap loans, and an additional direct subsidy.

Medium sized systems (<3500 Wp) for new dwellings can preferably be stimulated by high direct subsidies.

Medium sized systems for commercial buildings are most effectively stimulated by existing fiscal schemes and a direct subsidy.

Large scale systems (>3500 Wp) for new dwellings and existing commercial buildings can effectively be stimulated within the existing fiscal schemes, direct subsidies and for example sale of produced solar electricity as an individual product or as a part of green electricity. The necessary contribution is the price that consumers are willing to pay for this environment friendly product.

7.2 General

Companies profit the most from Dutch (financial) incentives and the current market circumstances. Large scale systems are more interesting than small pv-systems.

Small systems (200-800 Wp) are particularly interesting on new houses for private households due to the positive influence on the energy performance of dwellings and the high price of avoided kWh purchased from the local energy utility.

For new and existing buildings there is no difference in stimulating very small systems (<200 Wp) or small systems (200-800 Wp), due to the small effect of environmental factors.

Environmental factors make medium sized systems the most attractive for profit organisations, due to fiscal legislation.

The introduction of Solar Electricity as a product can accelerate market developments of BIPV-systems in The Netherlands.

REFERENCES

[1]   Energie-onderzoek in Nederland, advies van de algemene Energieraad aan de Minister van Economische Zaken, juli 1996

[2]   Milieu Actie Plan van de Energiedistributiesector, EnergieNed, Arnhem, maart 1994

[3]   Energiebesparingsnota, brief aan de Minister van Economische Zaken, Den Haag, april 1998

[4]   Energie voor de toekomst: Duurzame Energiebronnen, Witboek voor communautaire strategie en een actieplan, Commissie van de Europese Gemeenschappen

[5]   Nationale Energieverkenning 1995-2020, P. Kroon, O. van Hilten, ECN, maart 1998

EUROPEAN CREATIVITY WORKSHOP
PV MARKET STIMULATION

Drs. E.M. Koot
Ekomation Solar Energy Consultancy
P.O. Box: 29112, 3001 GC Rotterdam, The Netherlands
Tel.:++ 31 10 280 72 64, fax.:++ 31 10 280 72 65
E-mail: e.m.koot@ekomation.nl, Internet: www.ekomation.nl

ABSTRACT:
This paper presents the outcomes from a two-day European "PV market stimulation by training and creativity" workshop with participants from The Netherlands, Great Britain, Germany and Italy. In order to stimulate the application of PV, the market should not only be approached from a technological point of view but increasingly more with a marketing vision. Only a customised approach to market segmentation and a right positioning will trigger growth in the PV market. After the 'technology push' in the PV world the time is right for a 'marketing push' to realise stimulation of PV.
Keywords: PV market -1: Marketing-2: Building Integration –3

1. INTRODUCTION

In the framework of the Altener II-project "PV market stimulation by training & creativity workshops" a workshop took place in Amsterdam in April 1999. The main theme of the meeting was stimulation of the PV market (Photovoltaic solar energy) in technology, product development, finance, marketing and regulation. The two-day session was organised by Ekomation, project manager of the Altener-project, in association with partners from The Netherlands (ECN, Shell Solar, VSEC), England (Newcastle Photovoltaics Applications Centre), Germany (ISTEnergieCom) and Italy (Conphoebus). This workshop was also supported by Novem from The Netherlands. The intensive workshop involved presentations, brainstorm sessions and interactive cases. The participants are active in different fields of the PV market (the pv-industry, power supply companies, consultants, research institutes, etc.). The presentations mapped out briefly the recent developments in the PV markets in the four participating European countries. With brainstorm sessions in small groups concrete ideas were generated as a basis for the application of PV worked out in the interactive cases. This paper will present the main outcomes of the workshop.

2. PV AND MARKETING

One of the most remarkable aspects of the two day session was the growing attention for the marketing of PV. Two marketing experts expressed their vision on marketing of PV products and services in a presentation. This lead to new insights: in order to stimulate the application of PV, the market should not only be approached from a technological point of view, but increasingly more with a marketing vision.

Some other important aspects that emerged from the workshops will be described in this paper. Attention will be paid to the marketing approach of the so called 'expensive' source of renewable energy (pv-systems and solar energy), its positioning, the success factors and factors of failure of similar products.

After the presentation of professor Sicco C. Santema (Santema Consulting, marketing expert) "PV on a mission", a discussion took place among participants covering the following subjects:
- Is PV a product or service?
- Electricity from PV as an 'intangible' product.
- Image and perception of Quality of 'green products'.
- The value of eco-products in relation to availability (business-to-business marketing). What is the position of PV in the matrix?
- The 'eco-emotion' compared to the availability of these products (consumer marketing). What is the position of PV in the matrix?
- The difference between what is offered to the consumer and what consumers perceive (ServQual Model).

Santema concluded his presentation with 'the PV-mission':
'Products' like pv have product/ service effects, green effects, services appeal to customers, and therefor a (green) marketing future'.

Mister Rom Bult (Hügli Pollock Read, marketing expert in the field of construction) covered the second day 'The do's and don'ts of marketing and positioning in the built environment'. Examples were being discussed of success factors en factors leading to failure introducing other products in the built environment. Bult underlined the importance of the right positioning of the product, especially in its introduction phase. A new product with a high price should be positioned as high as possible. 'High' means 'in the top of the pyramid' (figure 1).

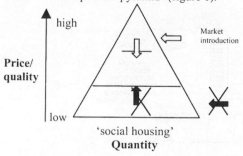

Figure 1: Positioning. (Rom Bult, Hügli Pollock Read)

In the case of PV positioning is 'top-down'; the product should be positioned in the introduction phase in the highest market segment. A 'bottom-up' introduction, through (larger) lower market segments to higher market segments, appears in most cases to be impossible.

Furthermore, the 'business chain' was being discussed. To be able to deliver a product in the right way to the end user, each link in the distribution chain has to be taken into consideration. Every party has specific needs, also in the PV market. The architect for example has demands relating design, the property developer wants to increase selling points of his houses and the end user has requirements in other areas (status, saving energy, etc.). Each market is a collection of different segments with specific characteristics, needs en desires. This is important in offering a product to a specific market; a total market, or total pv-market, is non-existent. Some other 'do's and don'ts' in respect to market stimulation that were mentioned during the workshop by Rom Bult were:

- Solving specific problems and meeting specific needs is only possible when the product is (the solution) is specific (customised).
- Demonstration is an essential tool in market stimulation.
- This demonstration needs to be in line with positioning, segmentation and long term marketing targets.
- Never start a market introduction of a particular product, when the product is not available in the short run.

The afternoons of the workshops were very intensive: groups of 5-6 participants brainstormed to generate new ideas for PV-projects. The method that was being used was the 'methodological analysis'. The participants were asked to make up examples and new ideas based on pre-set dimensions, for example 'technological application', 'building type', 'type of end use', 'promising target markets', all dealing with PV. On the basis of different combinations of these dimensions concrete ideas were generated. Ideas with the highest potential were transformed into action plans, that can eventually lead to new PV-projects. The second day participants were handed out 'additional bagage': 'do's and don'ts' with several good and bad examples of existing marketing concepts. These were used to develop concrete 'marketing plans' for existing PV product/market-combinations.

With the applied brainstorm method along with the creativity of the participants, over 50 ideas were being generated. Some of the concepts are:

'Polluter pays photovoltaics'
In relation to the amount of pollution, the polluter covers a certain part of its energy consumption PV solar energy.

'Sunshine Boulevard'
PV facades especially designated for the hotel and catering industry, in which each technology can be applied.

'Solar label on consumer products'
Each consumer product is provided with a hallmark or tag, when solar energy has played a role in the production. The tag or hallmark is controlled by an independent organisation/institute.

CONCLUSIONS

During the workshops the participants became aware of the fact that a custom tailored market segmentation of the PV market is necessary. Only after a clear market segmentation and right positioning chances exist to trigger growth in the PV market. This can also be concluded from the examples that were provided during the workshop by marketing experts. After the 'technology push' in the PV world, the time is right for a 'marketing push' to realise stimulation of PV in alignment with customer needs.

Following the Amsterdam workshops, a very successful workshop took place in London. A next workshop will be organised covering the same theme in Germany.

Project partners

Professor W.C. Sinke, ECN, The Netherlands
Ir. E.J. Koot, Ekomation, The Netherlands
G. Heilscher M.Sc., IST EnergieCom, Germany
Dr. N.M. Pearsall, NPAC (University of Northumbria), UK
Ir. J. Schlangen, Shell Solar Energy BV, The Netherlands

THE DUTCH GRID CONNECTED PV MARKET BEYOND 2000
MARKET INTRODUCTION STRATEGY FOR THE DUTCH GRID CONNECTED PV MARKET BEYOND 2000 BASED ON MARKET SEGMENTATION AND STIMULATION INSTRUMENTS.

E.J. Koot M.Sc[1]., D.J. Middelkoop[2]
[1]Managing Director [2]Consultant
Ekomation Solar Energy Consultancy
P.O. Box: 29112, 3001 GC Rotterdam, THE NETHERLANDS
Tel.: ++31 10 280 72 64, fax : ++31 10 280 72 65
E-mail: e.j.koot@ekomation.nl, Internet: www.ekomation.nl

ABSTRACT: More knowledge about the demand side of the Dutch PV market is needed as input for a new market introduction strategy. The new Dutch "PV Convenant" for the period of 2001 until 2007 will be based on this strategy for the Dutch grid connected PV market. Ekomation mapped out the market potential for grid connected PV systems and researched the most important market segments and effective instruments. A large part of the Dutch PV-sector prefer a market introduction strategy in which the sales of PV systems are stimulated adopting a "top-down" approach. A segmented approach to the PV market will give the advantage of an effective way of using instruments. The results of the research project "PV market after the year 2000" are shown in this paper. A market potential can be attained (with an unchanged policy) of 73-155 MWp in 2007, the potential of segments solar electricity and PV in Green Energy for the consumer and business market excluded. Market segments with the highest potential are PV in Green Electricity, individual house owners/ with PV on owner occupied property, PV property developers/ putting PV on new housing development and Solar Electricity as an independent product. The most important instruments are a generic subsidy and an agreement between the utilities tot set a minimum share of PV electricity in the (successful) Green Electricity products.
Keywords: Economic analysis -1: National Programme -2 : Funding and Incentives–3

1. INTRODUCTION

Preparing for a new PV Covenant for the period of 2001 until 2007 Ekomation was assigned by the PV Platform, in which the representatives of the 27 organisations that signed the current PV Covenant, to map out the market potential for grid connected PV systems in the most important market segments in the Netherlands.

The initiative to conduct research in the Dutch PV market was motivated by the changing phases of market development in the PV market, in which increasingly more attention is being paid to the demand side [1,2] instead of the supply side [3]. One of the questions that arose was if a 500 MWp cumulative installed PV capacity in 2007 is feasible, given the recent developments and existing instruments.

2. GENERAL OBJECTIVES & METHODOLOGY

2.1 Objectives

The main objective of the study was to create an insight in the most attractive market segments in The Netherlands in the period 2001-2007 and effective instruments for these segments. Also a proposal for a market introduction strategy is needed for achieving the objectives of the new PV Covenant.

2.2 Approach and limitations

It is assumed that in the concerning period, market players aren't capable of decreasing pv-system prices in such a manner solar kilowatt-hours are price-competitive with conventional electricity yet. Therefore additional financial incentives are critical for a successful market introduction of photovoltaics. Prices are expected to fall 6% per year [4].

In the study it is assumed that there will be no limitations for the financial governmental funds for generic investments subsidies and fiscal contributions on the actual levels [5]. Also the percentage of these instruments will not change. Current financial incentives in the Netherlands are not enough for realising the planned new goals. Unchanged policy implies that the amount of subsidy and sources of money available for fiscal arrangements are unlimited.

New technologies which will be introduced on the Dutch market and are expected to have little effects on the dominant position of multi-christallijn silicium (Si_{mc}). Some sub-segments can be interesting for other technologies but will not influence the results of this research.

The research is based on actual and future levels for regulated energy taxes.

Although some discussions in the Dutch parliament, an obligation for green energy for households is not taken as a variable for this research.

2.3 Methodology

In the scope of the project two workshops have been organised in which the most important parties (both demand and supply side of the market) were involved to discuss the most fitting market strategy until 2007, the market segments with the highest potential and desired instruments for market development.

Quantitative and qualitative market research of the market segments was done by desk research on literature and databases, (telephonic) interviews with representatives of sectors and other experts in the selected market segments.

Assumptions on the penetration of PV systems in market segments is done by use of earlier research [6,7] and quantitative research

3. RESULTS

3.1 Outcomes of workshops

All market parties involved support the proposed segmented approach of the PV market. A large number of participants prefer a market introduction strategy in which the sales of PV systems are stimulated adopting a "top-down" approach: first focus on qualitatively high market segments followed by the other segments. Qualitatively high means: high quality, aesthetic value and high appeal to other segments.
Eleven market segments are defined and chosen for further research based on the workshops.

3.2 Selected market segments

Quantitative and qualitative market research of eleven market segments show that with an unchanged policy a market potential can be attained of 73-155 MWp in 2007, the segments solar electricity and PV in Green Energy for the consumer and business market excluded. These results can be seen in table I.

Table I: Total PV Potential in Market Segments

Description market segment	
Market party / *PV Application*	Cumulative market potential
	MWp in 2007
1. Property developers/ *New housing development*	12-24
2. House owners/ *Owner occupied property*	31-61
3. Housing associations/ *Rented houses/apartments*	7
4. Private corporations & (General) Partnerships/ *New & existing property*	2-3
5. Government/ *Buildings & sites*	3-11
6. Office building proprietors/ *Office buildings*	8
7. Companies/enterprises/ *Buildings & sites*	6
8. Farmers Company/ *Ground based systems*	2-30
9. Infrastructure supervisors/ *PV Noise barrier*	2-5
Total (Solar electricity & PV in Green Energy excl.)	*73-155*
10. Households & corporate users/ *Solar electricity*	43-72
11. Households & corporate users/ *PV in Green Energy*	5-204

Solar electricity and PV in green energy are not included in the determination of the total market potential to avoid possible duplication. PV systems that generate solar electricity will be mainly realised in the prior mentioned segments.

3.3 Discussion Results for individual market segments

1. The segment private owner occupied property shows an interesting potential of around 30-60 MWp. Market research proves that especially in higher income classes approximately 100.000 to 200.000 households are willing to acquire PV systems of around 300Wp at the existing price level and with the present subsidy system [8].

2. Project based new housing and apartment houses/blocks offer a net potential of respectively 95 and 60 MWp. The DuBo Centrum (national centre of sustainable building development) and municipalities can play an

important role in developing this segment. The now lacking direct financial (fiscal) stimulation and insecure subsidies for property developers are crucial to market penetration, since another important driver in this segment is absent. The expected market potential for 2007 is 12-24 MWp.

3. Collective PV flat roof systems in apartment houses and individual systems for single-family dwellings lead to an interesting net technical potential in the segment rented houses/apartments of housing

associations. The first sub segment offers possibilities, because of the limited risks, attractive product pricing and financing options. Furthermore, opportunities exist for new large-scale collective project concepts involving a number of housing associations for systems on single-family dwellings.

4. In the segment of private corporations and (general) partnerships, the service companies can profit optimally from fiscal instruments. At times, financially very attractive PV systems offer perspective when used as marketing instruments and can moreover enlarge support because of the many business relations.

5. The technical potential for PV on government offices and buildings is high and mainly health care and educational facilities offer an interesting potential for the application of PV flat roof systems and PV facades. The expected penetration level, however, is low due to the absence of a real 'driver' and applicable fiscal instruments.

6&7. On office buildings a tremendous technical potential exists for large flat roof systems, an opportunity for solar electricity generation. New financial instruments are required to unlock the large technical potential and increase the expected market penetration. This can be achieved by increasing the familiarity with PV and indirectly by stimulation of solar electricity and Green Energy.

8. The gross technical potential for ground based systems used by farmers on agricultural land is very large. Market penetration is unknown, but is expected to be very low to zero because a 'driver' and financial resources lack and other more profitable alternatives exist (wind energy). The segment offers possibly opportunities for fast realisation of PV capacity when financed by third parties.

9. PV is largely supported at Rijkswaterstaat, the governmental organisation that is involved in the construction of noise barriers along roads. Although it seems a highly potential market [9], the net technical potential is fairly limited and market penetration is mainly dependent on the realisation of large railway projects (HSL, Betuwelijn). International contracts, a lacking driver of PV application in this segment and legal complexity of these projects are no stimulants for *fast* realisation of large PV capacity. In the long term (after 2010) the amount of noise barrier projects will decrease as well as the market potential.

10. PV as part of Green Energy has a very high net technical potential. When problems do not occur in the supply of Green Energy, with a wide support of energy suppliers (certification and/or agreements on minimum % PV) and with well aimed marketing campaigns, PV included in Green Energy can lead to a high market potential of PV. The segments solar electricity and Green Energy can compete with other segments, but can also stimulate other segments through wider acquaintance and support. The most attractive market segment in which a high volume of grid connected PV can be realised appears to be PV in Green Energy products. The announcement to open the Dutch market for renewable energy as of 01-01-2001 [10] together with the present turbulent growth of Green Energy [11], makes it rather

difficult to predict market growth. Tens of MWp PV capacity are feasible when the share of PV increases from 0,25% at the present to a few percent.

11. Looking at comparable experiences in the PV market abroad, especially the business market offers interesting possibilities to market the independent product "solar electricity". A high net potential can be taken advantage of using well aimed marketing campaigns, for example by new providers [12]. Additionally, solar electricity generating installations on company buildings possess likewise market value.

3.2 Instruments for the market segments

Some important instruments that can lead to a larger market penetration of grid connected PV systems in the Netherlands are:

1. Financial incentives: generic investment subsidy, fiscal contribution or rate based incentive. Important is the development of financial instruments that are easy accessible and are guaranteed to be available in the long run due to building planning.
2. Marketing campaigns to stimulate Green Energy combined with agreements to increase the amount of PV with energy utilities, for example in a PV Agreement (PV Covenant).[13]
3. Marketing campaigns to stimulate the use of solar electricity.
4. Broaden and enlarge applicability of existing fiscal instruments for several groups, particularly real estate developers, private persons and governments. A financial 'driver' for these and other groups increases the possibilities to unlock a large net technical potential as well as market potential.
5. Increase familiarity of PV and market value for companies. A viable option is mentioning or stimulating use of PV in agreements.
6. Stimulate and financially support application of PV by the government to set an example, while immediately accessing a large net technical potential.
7. Marketing activities aimed at specific groups of private owners of existing property.
8. A renewable energy obligation for households and business market can lead tot more PV capacity.

Confronting these instruments with the eleven market segments give the results as shown in table II.

Table II: Evaluation of effective stimulation instruments

Market Segment	Generic investment subsidy	Rate based incentive	Broadened fiscal instruments	Advantageous loans	Tax-deductions for private	Marketing campaigns Green	Increase regulated energy taxes	PV-obligation	Agreements/PV Covenant
1. Property estate developers; New housing development	++	0	0	0	0	0	0	++	+
2. House owners; owner occupied property	++	++	0	+	++	0	0/+	0	0
3. Housing associations; Rented/houses/apartments	++	+	0/+	+	+?	0/+	0/+	+	+
4. Private corporations & partnerships; New/existing	++	++	+++	+	++	0	+	0	0
5. Government; Buildings/sites	++	++	0/+	+	0	+	+	+	++
6. Office building proprietors; Office buildings	++	++	++	+	0	++	+	+	0/+
7. Companies/enterprises Company buildings/sites	++	++	++	+	0	++	+	+	+
8. Farmers; Ground based systems	++	++	+	+	0/+?	+	0		0/+
9. Infrastructure; PV noise barriers	++	++	+	+?	0	0/+?	+	0	+
10. Solar electricity	++	+	++	+	0	++	+	-	0
11. PV in Green Energy	++	+	++	+	0	++	++	-	++

4. CONCLUSIONS

4.1 Potential for Dutch market

With the current expected price decreases, PV systems will not be profitable in 2007. When the policy presently effective remains unchanged and with an objective set around 150 MWp in 2007, an average yearly subsidy is needed of 50 million guilders. It appears also possible to realise the objective or even 500 MWp cumulative capacity using the presently existing fiscal instruments, but without additional government subsidies. This can be done by stimulation of 'Green Energy' combined with agreements on the percentage of PV in Green Energy or a renewable energy obligation; stimulating the product 'Solar Electricity' or possibly an obligation to apply PV in new housing development. In this approach end users pay and more important, with 'Green Energy' and solar electricity, costs are being divided among a large number of users.

4.2 Market segments

Most potential market segments are PV in Green Electricity (segment 10), house owners/ owner occupied property (2), PV in property developers/new housing development (1) and Solar Electricity as an independent product (11). These four segments have a large market potential also a good image for other segments.

4.3 Important instruments

A generic investment subsidy will stimulate all market segments.

The most *fast* way to reach a high volume of PV is an agreement between utilities to set a (relatively high) minimum of PV electricity in the product Green Electricity.

A rate based incentive or an enlargement of fiscal instruments can be effective for reaching the great potential in segments 6 and 7.

Promotion campaigns which are focused on Solar Electricity and Green electricity will stimulate these market segments and will possibly have a positive effect on other segments.

4.4 Strategy

A segmented approach to the PV market will give the advantage of an effective way of using instruments. A large part of the Dutch PV-sector prefer a market introduction strategy in which the sales of PV systems are stimulated adopting a "top-down" approach.

If a rate based incentive system will be installed (like recently has been introduced in Germany) this instrument must be secured for a long term and funds must be large enough.

REFERENCES

[1]   S. van den Berg et al, Het marktpotentieel van netgekoppelde PV-systemen voor particulieren, Motivaction Amsterdam (October 1997)

[2]   VROM, Potentiële belangstelling voor CO2-gecompenseerde producten (August 1998)

[3]   KPMG Bureau voor Economische Argumentatie, Zonne-energie: van eeuwige belofte tot concurrerend alternatief, (July 1999)

[4]   W. Sinke et al. , PV-introductieplan, (1997)

[5]   Ekomation, PV Financiering na 2000, (May 1999)

[6]   S.A.H. Moorman et al. ,Het potentieel van PV op daken en gevels in Nederland, Centrum voor energiebesparing en schone technologie Delft (1997)

[7]   B.J.M. van Kampen en W.C.H.C. Jansen, Knelpunten bij inpassing van fotovoltaïsche zonne-energie in de gebouwde omgeving (herziene versie January 1999)

[8]   Geo-Marktprofiel, Socio-demografische en lifestyle analyse van Solaris aanvragers en Solaris kopers, July-august 1999.

[9]   G. Loois, H. Marsman, en Q. Sluijs), Haalbaarheidsstudie: Zonnestroom   uit geluidsschermen in Nederland, Ecofys Utrecht, (1998)

[10]   Toespraak minister van Economische Zaken tijdens NDEC, Noordwijkerhout, (November 1999)

[11]   Ministerie van EZ, Energierapport 1999

[12]   Arthur D. Little inc., Marktwerking in de leidinggebonden energiesector, (1999)

[13]   Nyfer, Keurmerk voor Groene Energie, (1999)

This study has been prepared for the PV Platform as a preparation new PV Agreement (PV Convenant) for the period of 2001 until 2007 in order of the PV Platform. Ekomation Solar Energy Consultancy has co-operated in this project with two Dutch consultants:

Dr. L.A.Verhoef, Verhoef Solar Energy Consultancy
R. Bult, Hügli Pollock Read Industriële Marketing

Further information about this paper can be obtained by:

Ekomation Solar Energy Consultancy
P.O. Box: 29112, 3001 GC Rotterdam,
The Netherlands
Tel.: ++31 10 280 72 64, fax : ++31 10 280 72 65
Internet: www.ekomation.nl
E-mail: e.j.koot@ekomation.nl,

The full text of the report "PV market after 2000" can be downloaded   of   the   website   of   Ekomation.

THE INTRODUCTION OF THE GST AND ITS IMPLICATIONS FOR THE FINANCIAL FEASIBILITY OF GRID CONNECTED PHOTOVOLTAIC SYSTEMS IN WESTERN AUSTRALIA: NEW OPPORTUNITIES

Barry Graham[1], Martina Calais[2], Vassilios G. Agelidis[3], William B. Lawrance[4]
[1]School of Accounting, Curtin University, GPO Box U1987, Perth WA 6845, Australia
[2]School of Engineering, Murdoch University, Australia
[3]Centre for Economic Renewable Power Delivery (CERPD), University of Glasgow, UK
[4]School of Electrical and Computer Engineering, Curtin University, Australia

ABSTRACT: This paper presents a financial analysis of the estimated cost and benefits of operating a 1.5 kW grid connected photovoltaic (PV) system in Perth, Western Australia. The analysis presents results for a typical residential electricity consumer and a small business consumer. The paper investigates the feasibility before, and after the introduction of the PV Rebate Programme for residential systems in January 2000 and the Goods and Services Tax (GST) in July 2000. For the financial analysis, Internal Rate of Return (IRR) and Payback Periods (PP) are used as project evaluation techniques. The results show, that the current PV Rebate Program is insufficient to make PV systems attractive for individual taxpayers on purely economic grounds.

Keywords : Small Grid-connected PV Systems - 1: Funding and Incentives- 2: Economic Analysis - 3

1. INTRODUCTION

In the past photovoltaic modules and system components such as inverters have been sales tax exempt in Australia. With the introduction of the Goods and Services Tax (GST) by July 2000, however, residential consumers are likely to face price increases of 10% for PV system components. However, even if that is the case, to stimulate the PV market, the Australian Government has introduced a "Photovoltaic Rebate Program" [1]. Specifically, since January 2000, this program provides rebates for the installation of residential stand-alone and grid connected PV systems. The rebate is based on the rated output of the PV component of the system, with its level currently being $ 5.50 per rated Watt of PV installed. The minimum system size is 450 W, there is no maximum size, but the rebate is capped at $ 8.250.00.

In this paper, a 1.5 kW grid connected PV system is chosen for the following economic analysis. Section 2 outlines the estimated system costs and energy generation when operated in Perth, Western Australia. Energy consumption and applicable tariffs are then described for the two potential operators, the individual taxpayer and a small business. Based on these figures, cash flow forecasts for the life of the system for both residential and small business electricity consumers are presented in Section 3. Section 4 describes the methods of investment evaluation and underlying assumptions and then presents the outcomes for different scenarios. Conclusions are presented in Section 5.

2. SYSTEM COST, ENERGY GENERATION AND CONSUMPTION ASSUMPTIONS

2.1 System Cost Assumptions

System cost assumptions are based on a one off purchase (not bulk purchase) with incidental operating and maintenance costs omitted. Table I presents the estimated costs for a 1.5 kW PV system based on quotes obtained from suppliers in Western Australia in April 2000. All prices are in Australian dollars. The installation cost is estimated on the basis of an average labour cost of $ 34/hour. Tilted roofs, requiring frame constructions for the installation of PV modules are assumed. A PV module price of $8.5/W is assumed when purchasing a significant number of modules, or a complete system, before the introduction of the GST.

Three different scenarios for both the residential and the small business case are considered in this investigation:

- Scenario A: Before Rebate and before GST
- Scenario B: After Rebate and before GST, and
- Scenario C: After Rebate and after GST.

The cost of the 1.5kW system for the three scenarios and both cases (residential and small business) is outlined in Table II. It should be noted, that after the introduction of the GST the system price rises by 10%. The rebate for the 1.5 kW system is $ 8,250.

Table I: Estimated costs of a 1.5 kW PV rooftop system before the introduction of the GST (all prices in Australian Dollar, exchange rate April 2000: AUS$ 1 = US$ 0.6).

Cost Item	Price ($)
1.5 kW PV array	12,750.00
Inverter	3,000.00
Installation	0.4/W+600.00=1,200.00
NPER engineer approval*	450.00
Total	**17,500.00**

*An application for a grid connected PV system in Western Australia must be endorsed by a Chartered Professional Engineer with NPER (National Professional Engineers Register) standing with the Institution of Engineers, Australia (IEAust).

Table II: System cost scenarios for a 1.5 kW PV rooftop system for residential consumers and small businesses.

Scenario	1.5 kW System Cost ($)	
	Residential	Small Business
A. Before Rebate & GST	17,500.00	17,500.00
B. After Rebate & Before GST	9,250.00 (=A-$8,250.00)	9,250.00 (=A-$8,250.00*)
C. After Rebate & After GST	11,000.00 (=A times 1.10-Rebate)	9,250.00 (=B**)

*Although the current rebate schemes only applies to residential systems the same rebate is included as a cost scenario for small businesses.

**For the small business scenario, the 10% GST is ignored as most small businesses will be able to claim a credit for GST on inputs against GST charged on sales to customers.

2.2 Energy Generation

Based on average hourly irradiance values [2] and average hourly temperature values [3] for each month, the annual energy output of a PV array tilted at the latitude angle in Perth, facing north can be calculated. The PV modules assumed are 12.6 % efficient crystalline silicon modules and have a power de-rating factor due to temperature of 0.005/°C. The annual energy output of the complete system is then calculated, based on typical PV inverter efficiency curves. Based on these constraints a 1.5 kW PV system in Perth generates approximately 2550 kWh/year.

2.3 Electricity Consumption – Residential Consumers

The analysis is based on electricity consumption for a family of four of 3900 kWh/year that is typical for a household with reticulated gas (main electrical loads include a fridge, freezer, TV, lighting, washing machine, air conditioner and microwave) [4]. Recorded data from a distribution feeder for a purely residential area in a southern suburb of Perth is used to determine average weekday and weekend daily load profiles for each month. An example of an average weekday load profile as well as the possible output of a 1.5 kW PV system is shown in Fig. 1a).

2.4 Electricity Consumption - Small Business Consumers

An engineering company in Bentley Technology Park, Perth, was chosen as an example of a small business consumer. The company has research and development facilities (electronic research laboratories and software development), a manufacturing section with workshops, and an administration and marketing section. The company employs 39 staff and occupies 600 m² of floor space. The annual electricity consumption is 125,078 kWh.

2.5 Applicable Tariffs - Residential Tariff

The local utility, Western Power Corporation, offers residential customers a Renewable Energy Buyback Scheme. In this scheme, customers can choose between two tariffs: (a) the standard residential Tariff A1, and (b) a SmartPower tariff. The A1 tariff is a flat tariff of 12.75 cents/kWh (23.39 cents/day supply charge) with no seasonal or daily variation. For the Smart-Power tariff, the energy charge is dependent on the time of day and varies between 6 cents/kWh and 18.5 cents/kWh as shown in Fig.

1b). In both cases, the renewable energy buyback rates are equal to the electricity purchase rates. Therefore, customers joining the scheme, are either billed for their net import or credited for their net export of electricity at the same rate over a billing period. The scheme is available to the first 100 "pioneer" residential customers who join before the end of the year 2000. Renewable energy systems (not only PV, but also wind turbines and micro hydro plants) between 500 Watts and 5 kilowatts are covered by the Scheme. Additionally required meters are supplied by Western Power, however, a one-off $100.00 Special Account Establishment Fee applies. When comparing the two tariffs [5] the Smart-Power tariff generally results in higher annual net cash flows, therefore this investigation will be based only on the Smart-Power tariff.

2.6 Applicable Tariffs - Small Business Tariff

For consumers where electricity is used for business purposes and at low voltage, the tariff L1 applies if the electricity consumption is less than 1650 kWh per day. This is a tariff typically used by small businesses and, for present purposes, we define a 'small business' in this way. Our example engineering company's average daily electricity consumption is well within this constraint.

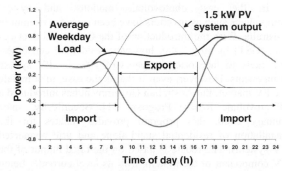

Figure 1 a): Average weekday load curves for a Perth household in February.

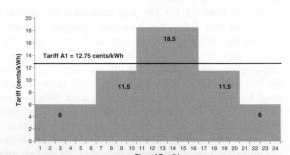

Figure 1 b): Weekday Smart-Power residential tariff for October to March.

Table III: Western Power Corporation's Tariffs

Tariff	Residential		Business
	A1	SmartPower	L1
Supply charge (cents/day)	23.39	22.54	24.31
Energy charge (cents/kWh)	12.75	18.5 (peak) 6.0 (off peak) 11.5 (high shoulder) 9.0 low shoulder	15.98

3. RESULTS

Table IV shows the estimated cost savings plus cash inflows arising from the installation of a 1.5 kW PV system for the residential (Smart-Power) and small business (L1 tariff) scenarios. In the case of the Residential/Smart-Power tariff scenario, it is assumed that consumers who install a PV grid connected system on their property have not previously paid their electricity bill according to the Smart-Power tariff. The item "Bill without PV" in Table IV therefore refers to the electricity bill of the household under Tariff A1. The Smart-Power Tariff, however, is also available for residents who are not part of the Renewable Energy Buyback Scheme. A one-off charge of $199 (new homes) or $329 (existing homes) then applies for the installation of the Smart-Power meter. Since the Smart-Power tariff favours residential load profiles the Smart-Power electricity bill for the investigated load profile is considerably lower ($503.55/year) than the flat rate tariff A1 bill ($583.34/year). The use of the $503.55/year in the following analysis would reduce the potential benefits to residential consumers of converting to a PV system.

With respect to the small business example, since the electricity consumption is greater than the energy generated, there is no export of electricity to the grid.

Table IV indicates a positive annual net cash flow due to avoided costs and/or the export of electricity to the grid for all scenarios. The question is whether these future cash flow benefits are sufficient to justify the cost of the investment in a PV system. This question is addressed in the following sections.

Table IV: Operating cash flow analysis for a 1.5 kW PV rooftop system.

	Residential	Small Business
Tariff	Smart Power	L1
A. Bill without PV ($/year)	583.34	20,075.00
B. Bill with PV ($/year)	308.08	19,666.80
C. Avoided cost ($/year)[A-B]	275.26	408.20
D. Income through export ($/year)	141.84	0.00
E. Net Cash Flow ($/year) [C + D]	417.10	408.20

4. METHODS OF INVESTMENT EVALUATION

The investment in a PV System is evaluated using two common investment evaluation models. Both models share the common purpose of seeking to determine whether an investment is economically feasible according to the decision criterion specific to each model.

The Internal Rate of Return (IRR) Method: This method seeks to discover the rate of discount that equates the future net cash inflows with the investment outlay. The IRR method also takes into account the time value of money. It is often simply described as the rate of interest at which the Net Present Value of a proposal equals zero. The decision rules for this method are: accept if the IRR exceeds the required rate of return; reject if the IRR is less than the required rate of return.

Payback Period Method: This method does not take into account the 'time value' of money. It simply calculates how many years it will take for the investor to recoup the investment. The payback period is compared with some maximum acceptable payback period in order to determine whether the investment is acceptable or not.

4.1 Assumptions Underlying the Investment Evaluation Analyses

1. Electricity prices are assumed to remain constant in real terms (year 2000 dollars) for the twenty-year period of analysis. From July 2000 the GST applies.
2. For the residential scenarios it is assumed that that the sale of electricity back into the grid does not constitute a 'business' for the purpose of making a profit. Therefore, no taxation consequences are considered for these scenarios.
3. For the small business scenario, the total cost of the PV system is considered depreciable. The depreciation rate used in this analysis is a straight-line rate of 10% for 10 years.
4. A company tax rate of 34% (which will come into effect from July 2000) has been used for the small business scenario. This rate is assumed to remain constant for the life of the project.
5. The period of analysis is equal to the assumed physical and economic life of the PV system of 20 years.
6. Maintenance costs and incidental operating costs have been ignored.
7. For the Internal Rate of Return analysis, a minimum "real" required rate of return of 5% is assumed. This is based on a real rate of return for a "risk-free" investment of 3% plus a "risk premium" of 2%.

4.2 Investment Evaluation Analysis - Residential Consumers

Table V presents a summary of the investment evaluation analyses for a 1.5 kW PV system for residential electricity consumers. The cost of the PV system is taken from Tables I and II and the future cash flow benefits are taken from Table IV.

Table V: Results of financial evaluation for a 1.5 kW PV system for residential use.

Evaluation Method	Scenario A	Scenario B	Scenario C
IRR	-6.21%	-0.96%	-1.95%
Payback	42.0 years	22.2 years	24.7 years

Table V demonstrates that the introduction of rebates on household PV systems dramatically improves the economic feasibility of a typical 1.5 kW PV system. However, the introduction of the GST from July 1, 2000 has a mildly adverse effect on this outcome. Even with the introduction of the rebate the investment decision is still negative for both evaluation methods:

- the IRR is negative in all cases, ranging from -0.96% to –6.21%;
- the payback period results show that the shortest payback period of 22.2 years exceeds, but only marginally, the projected 20 year life of the PV system.

Prior to the introduction of the rebate, effective January 2000, the major drawback to an investment in a PV system was that the expected stream of future cash inflows was nowhere near sufficient to financially justify the initial cost outlay (of course, individuals may have had non-financial incentives to make such an investment). The reasons for this include the present level of manufacturing costs of the components; the low efficiency ratings of PV cells at present; and the availability of relatively cheap fossil fuel energy sources in Australia. The rebate offers at least temporary relief during a period when the expected cost of PV cells declines as higher efficiency ratings are achieved and production costs are driven down globally.

4.3 Double Export Price of Electricity Policy

An alternative policy that rewards environment friendly energy generation using renewable energy sources is to pay a premium price for exported PV electricity. Rebate schemes which pay part of the initial costs of grid connected PV systems may lead to an increase in number of installations, however, the operation of these systems is not encouraged, only the installation. Schemes which give PV system operators a premium price on generated kWh encourage the operation of the system, not only the installation. An example is the German Renewable Energy Law, which guarantees individuals or businesses who invest into a PV system, 0.99 Deutsche Mark (DM)* per kWh over a period of 20 years. In combination with the German 100,000 roofs program it is now economically feasible to operate a PV system in Germany [6].

The analysis in Table VI and VII assumes that the export price for electricity is doubled. Two export scenarios are considered:
(a) the residential electricity consumer exports only the *surplus* electricity generated (Table VI), and
(b) the residential electricity consumer exports *all* of the electricity generated by the PV system (Table VII).

The 'export all' scenario is tantamount to making an investment in a PV system for the purpose of selling it back into the grid.

Table VI: Results of financial evaluation assuming a doubling of price for export electricity for residential consumers (export *surplus* scenario).

Evaluation Method	Scenario A	Scenario B	Scenario C
IRR	-3.94%	1.88%	0.62%
Payback	31.3 years	16.6 years	18.8 years

Table VII: Results of financial evaluation assuming a doubling of price for export electricity for residential consumers (export *all* scenario).

Evaluation Method	Scenario A	Scenario B	Scenario C
IRR	-2.36%	3.90%	2.03%
Payback	25.9 years	13.7 years	16.3 years

The introduction of a double export price policy considerably improves the financial outcomes thereby making a PV system installation more economically acceptable. In fact, when combined with the rebate policy, the IRR of the investment verges on acceptability and the payback period is less than the anticipated life of the PV system.

It is worth noting that a rebate policy favours smaller PV systems (such as the 1.5 kW system assumed in this analysis) due to the proposed upper limit on the rebate of $8,250. On the other hand, a policy of doubling the export price favours the larger systems due to their greater electricity generating capacity.

4.4 Investment Evaluation Analysis - Small Businesses

Table VIII presents a summary of the financial evaluations for the installation of a PV system by a typical small business for both of the investment evaluation methods. Once again, the cost of the PV systems is taken from Table I and II and the future cash flow benefits from Table IV. The rebate on component costs available to residential households (Scenario B) is not applicable to businesses under the proposed grant scheme by the federal government but is investigated for comparative purposes. The implications of the introduction of the GST are ignored on the basis that GST paid on inputs will be recouped on sales to customers.

Table VIII: Results of financial evaluation for PV systems for small business use. (Tax Rate 34%; Tariff L1)

Evaluation Method	Scenario A	Scenario B
IRR	-4.93%	-0.91%
Payback	30.8 years	21.9 years

Table VIII shows that the IRRs are negative, and payback periods exceed life of system, before the rebate introduction. The introduction of the rebate results in more favourable outcomes, including an almost break-even payback period, but not to the point of a PV system being acceptable in economic terms.

The outcomes for a PV system for a small business are more favourable than the residential scenarios presented in Table V due to the cost of a PV system being fully tax deductible for a business.

5. CONCLUSIONS

The results presented in the paper indicate that from the economic point of view rooftop PV systems have become more competitive due to Australian Government's Renewable rebate initiative. However, they still remain capital intensive on pure economic grounds with long payback periods and internal rate of return in relatively low negative range.

If certain targets are to be met, (the Australian Government has made the commitment to source 2% of its electrical energy from renewable sources by 2010) schemes such as the "double export price tariff", which is described in the paper, may also need to be introduced to send appropriate signals to consumers and hence encourage them to invest in PV systems.

ACKNOWLEDGEMENTS

The authors would like to thank Mr C. Hosking and Mr D. Harrison from Western Power and Mrs. S. Phillips from Advanced Energy Systems Pty Ltd for providing us with valuable information for this study.

REFERENCES

[1] Australian Greenhouse Office, "Photovoltaic Rebate Program," http://www.greenhouse.gov.au/renewable/initiatives.html#photovoltaic: 2000.

[2] T. Lee, D. Oppenheim, and T. J. Williamson, "Australian Solar Radiation Handbook, Attachment Five, Western Australia," Energy Research and Development Corporation, Canberra ERDC 249, 1995.

[3] Bureau of Meteorology, "Temperature Data for Perth," 1998.

[4] SECWA, "Domestic Energy Use in Western Australia," State Energy Commission of Western Australia Demand Management Paper No. 1, September 1991.

[5] V. G. Agelidis, B. Graham, M. Calais, and W. B. Lawrance, "A financial analysis of grid connected photovoltaic systems for Perth in Western Australia," In Proceedings of the Australian New Zealand Solar Energy Society (ANZES) Annual Conference, Solar '99, Geelong, Australia, 1999.

[6] A. Kreutzmann, "Ein irrer Erfolg, In German," *PHOTON das Solarstrom-Magazin, German Solar Electricity Magazine*, pp. 22-24, 2000.

* 1 DM=0.51 Euro, 1 Euro=0.91 US$ (May 2000).

PROGRESS ON ONGOING ACTIVITIES OF THE ITALIAN
10,000 ROOF-TOP PV PROGRAM

S. Castello[1], M. Garozzo[1], S. Li Causi[1], A. Sarno[2]
[1] ENEA, Casaccia Centre, Via Anguillarese 301 - 00060 S.Maria di Galeria - Rome (Italy)
[2] ENEA, Portici Centre, Località Granatello - 80055 Portici - Naples (Italy)

ABSTRACT: The Italian roof-top Program, jointly realised and defined by the Ministry of Industry, the Ministry of the Environment and ENEA, is intended to widely diffuse the roof-top technology in the Italian web, by means of the installation of a very large number (10,000) of PV systems for distributed generation in five years, with a total capacity of 50 MW. The Program, whose starting has been delayed to this summer, will be managed and monitored by ENEA and will be financed by the Italian Government.
In the mean-while, some fundamental preliminary and collateral activities have been started by ENEA, in order to make the Program effective as much as possible. Among them, technical and demonstrative actions, as well as training and advertising programs, are the most important ones. For instance, the Program guidelines have been defined and the preparation of a technical handbook is almost completed. In addition, specific demonstrative and experimental activities have been started, to disseminate pilot plants all over Italy. At present, 21 small systems, for a total power of 50 kW, are under installation in some important Italian cities, five out of them are indeed in operation.
Program preliminary and collateral activities, progress in overcoming recognised barriers and first results are discussed in the paper.
Keywords: National Programme – 1: Rooftop – 2

1. INTRODUCTION

Nowadays, very important national programs [1-4] have been world-wide launched, mainly devoted to promote the photovoltaic distributed generation. The Italian roof-top Program [5] is rather ambitious: 10,000 plants are expected to be installed in five years, with a total capacity of about 50 MW, on the basis of public incentives. In addition, some strategic benefits are expected during the Program. The more significant ones are:

- the diffusion of the photovoltaic technology and of the environmental awareness,
- the growth of the competitiveness of the Italian photovoltaic industry,
- the expansion of the market,
- a "driven" decrease of photovoltaic costs,
- the creation of job opportunities,
- the local development in unfavored regions.

Finally, some collateral activities have been scheduled, with the intent of maximising Program effectiveness. Among them, the realisation of several pilot plants, training activity for designers and technicians, advertising campaigns and a 12 years plants maintenance are the most significant ones.

The Program, jointly promoted by the Italian Ministry of the Environment and the Ministry of Industry and financed by the Italian Government, has been defined principally by ENEA, the Italian Agency for New technology, the Energy and the Environment. Program management and monitoring too will be performed by ENEA, in the framework of a dedicated agreement to be signed with the Ministry of the Environment.

The Program is aimed at disseminating all over Italy photovoltaic systems, with a rated power in the range 1-50 kW, to be operated in connection to the low voltage grid and preferably integrated in buildings (roofs, facades, shelters, sun-shadings, etc.). About 8,900 small plants having a peak power up to 5 kW and about 1,100 larger ones (5-50 kW rated) are foreseen.

From the operational point of view, a start-up announcement will be issued, to provide information about the main Program objectives and the relevant items. Moreover, yearly announcements are foreseen, in order to tune the Program on the evolution of the photovoltaic market and technology.

Concerning Program incentives, a contribution on capital cost has been chosen, principally due to the high present day costs of the photovoltaic technology. This should assure a massive participation of private citizens, to which the Program is mainly addressed. Moreover, because of a percentage of the investment cost has been assumed as public contribution, the upper limit of this cost has to be fixed. Therefore, both the upper limit and the percentage will be identified year by year, also taking into account the evolution and the trend of the international PV market. In conclusion, because of one of the main target is the reduction of the pay-back time of the share supported by the end-user below the figure of 10-12 years, an economic incentive equal to 70 % and an upper limit of 7 $/W have been fixed for the first year of the Program. In fact, they seem to be compatible with the present day major economical parameters in Italy.

Everyone (both private citizens and public or private companies) can ask the public incentive, by submitting a very simple application form, including a synthetic description of the plant to be installed on its own property.

2. ONGOING ACTIVITIES AND FIRTS RESULTS

Although the Program has been defined two years ago, it is not yet in operation, because of some bureaucratic problems related to its funding. At present, the Program launch is expected within the next summer.
Nevertheless, all the scheduled collateral activities have

been regularly started and some of them are almost completed. The principal items considered in the meanwhile regard: Program guidelines and handbook preparation, an effective involvement of ENEL (the major Italian Utility) and local Utilities. In addition, fiscal aspects, net-metering and new PV component development have been considered too. Among the above mentioned activities, the realisation of the experimental pilot plant is, of course, the most important one. Finally, it is worth mentioning that these collateral activities allowed to recognise some barriers, most of them already overcome.

The compilation of guidelines, jointly performed by ENEA and ENEL, turned out to be very useful to better define Program procedures and to stand out the most hidden barriers. Besides, main suggestions for plant designers and installers have been collected in a handbook, almost completed. This handbook, added to the didactic material under preparation, will result very helpful to carry on training activities together with local Authorities.

ENEL and Local Utilities have been constantly informed about the Program, mainly to obtain their shared consensus to the initiative and to overcame some technical barriers, such as grid interface devices, grid connection requirements and plant maintenance.

Regarding fiscal aspects, the simplification proposed by ENEA together with the Ministry of Industry in order to support the Program has been positively evaluated by the Italian Government. The first very important result consist in the abolition, for photovoltaic plants having rated power up to 20 kW, of both the license of operation (the related annual tax included) and State and local taxes on self-consumed energy.

As concern the net-metering, which will be allowed only for systems up to 20 kW, an effective co-operation between ENEA and the National Authority for Electric Power and Gas has been recently established. This co-operation is aimed at defining the application form of the contract which will regulate the technical and economic details of the energy exchange (in terms of kWh) between the end user and the local Utility. The text of this contract, to be applied by all Utilities in Italy, is almost completed.

In the framework of a co-operation with the Italian PV Industry, modules and low power inverters have been developed, to be used in the Program context. At present, three different kind of modules for facades and roof are available (large tiles, up to 2 m² double glass modules and double glazed window), as well as small single phase inverters, equipped with protection devices conform to the Italian safety regulations. Such components have been installed (ENEA Portici Centre) in some typical build integration and are currently under test.[6]

Finally, major progress occurred in demonstration and experimental activity, based on design, installation and performance analysis of several pilot plants. On the whole, 21 small pilot plants, for a total capacity of 50 kW, are under installation in some principal Italian cities. In this case, peculiar installation sites have been selected, with the main aim of experimenting the integration of the photovoltaic generator into the architectural design of roofs, facades and shelters, without neglecting to publicise this kind of applications. Besides, simplified protection devices are adopted in some of these plants, to test this new grid interface mode. In some other plants, recently developed (by the Italian Industry) modules and inverters are tested in real operating conditions. Of course, in order to analyse

plant performance and to verify the effectiveness of these technical solutions, all the pilot plants are monitored by means a dedicated remote network [7], already designed and used by ENEA to collect data from the most significant PV pants in operation in Italy.

Five of these plants entered in operation last year and their operational data are regularly collected and analysed. In the following, a brief description of the above five plants is given, together with their main performance.

Forte Carpenedo plant: the building concerned is the reception of Forte Carpenedo (Venice) exhibition area, a military fortification in the past century and, at present, under recovery and valorisation for cultural aims. The 3.1 kW array consists of 48 polycrystalline photovoltaic big tiles and represent a typical example of module integration into roof structures. The plant was connected to the grid at the end of May 1999. Measured data are not yet analysed in details, but by a quick view of these data, the resultant low values of both final yield (1.37 h/day) and performance ratio (about 40 %) seem ascribable only to the unavailability of the local grid in August-October 1999

Fig.1 Forte Carpenedo plant (Venice). Polycrystalline PV tiles (36 m²) are integrated in a traditional resort.

Fig.2 ENEL Research Building plant (Milan): two sub-arrays (1.2 kW on the roof and 1.7 kW on the facade) provide energy for low power utilisation in the building.

ENEL Ricerca plant: the photovoltaic array (2.9 kW) has been installed both on the southern facade of the ENEL building at Segrate (Milan) and on the horizontal surface of the roof. Because of the different module exposition and voltage operation of the two sub array, two separate inverters (1.5 kW each) have been used. From the architectural point of view, the array has been designed taking into account natural integration, visual pleasing and good composition between materials and colours. The size of photovoltaic modules has been choice to better fit the wall of the building, while green cells of polycrystalline silicon were adopted to harmonise with the logo colour. This pilot plant is the first one commissioned (4 May 1999). Up to now, no failures occurred and preliminary performance are very satisfactory. A yield value of 2.33 h/day is above the average expected in North Italy and represents a remarkable result for such module exposure.

Fig.3. Somaglia plant (Milan). Amorphous silicon modules compatible with traditional plates are installed on the roof side of a restaurant along an Italian motorway.

Somaglia plant: this installation has been realised in co-operation with the Autogrill Company, that generally manages the restorative area of service stations along Italian motorways. The 3.0 kW array has been installed on the pitch of Somaglia (Milan) service station, employing amorphous silicon modules. These modules correspond in size, appearance and mounting system to the traditional slates. The preliminary analysis of operational data (collected in the period May 1999 – April 2000) has showed a final yield about 15% less than the estimated one. Reasons for this are the low insolation level registered during last winter and a not optimal module exposure (south-west). On the other hand, a satisfactory value of performance ratio (67%) has confirmed a good plant reliability.

Fig.4 Bovisa plant (Milan). On the way to the brand new Malpensa International Airport, 64 polycrystalline PV modules serve the railway station with 3.1 kW.

Bovisa plant: the 3.1 kW array has been installed on the roof of Bovisa (Milan) railway station, very frequented because of the connection to International Malpensa airport. During the first ten months of operation, a good behaviour of the plant, both in terms of yield (2.44 h/day) and performance ratio (65%), has been demonstrated.

Fig.5. S. Gilla plant (Cagliari). In the shelters for shop trolleys, glass PV modules substitute usual plexiglass elements, providing 3.1 kW of PV power.

S. Gilla plant: the 3.1 kW plant has been realised at the shopping centre of S. Gilla (Cagliari, Sardinia). In this case, 32 polycrystalline modules have been utilised to partially substitute roof plexiglass elements, on two existing shelters for shop trolleys. Special emphasis has been given to constructional aspects, because of safety constrains. Plant performance are continuously evaluated by monitoring the most important parameters of the system. A detailed analysis of such parameters has provided value of yield (3.15 h/day) and performance ratio (78%) in very good accordance with the foreseen performance.

In order to quickly compare the performance of these plants, their final yield and performance ratio are shown in Table I.

Plant	In operation since	Final yield (h/day)	Performance ratio (%)
Forte Carpenedo	27/5/99	1.37	39.6
ENEL Research	04/5/99	2.35	70.9
Somaglia	13/5/99	1.92	67.3
Bovisa	28/5/99	2.44	65
S. Gilla	13/7/99	2.15	78.2

Table I: index of performances of the first five pilot plants

3. CONCLUSIONS

Although the photovoltaic option is not likely to represent one of the more significant contributions to both the energy need and the CO_2 emission reduction, it should be considered that its popular acceptance is sharply increased in Italy during these last years, as well as the environmental awareness. On the other hand, a wide interest to the roof-top Program of the PV operators and many Italian Utilities has been proved, while most of the fiscal, legal and technical barriers have been successfully overcome

In this framework, the Italian 10,000 photovoltaic roof-top

Program is ready to be launched, being fully defined all the management procedures and demonstrated the technical feasibility of the experimental solution adopted in pilot plants.

Therefore, a positive context has been established, providing the best condition for the success of the initiative.

REFERENCES

[1] J. Rannels, Implementation and Financing for President Clinton's Million Solar Roofs Initiative. Proceedings of the 2nd World Conference on Photovoltaic Solar Energy Conversion, Vienna, Austria, July 1998.

[2] M. Perez Latorre, Post kyoto Energy Policy Strategy The Res Strategy and Action Plan. Proceedings of the 2nd World Conference on Photovoltaic Solar Energy Conversion, Vienna, Austria, July 1998.

[3] R.Haas. Financial Promotion Strategies for Residential PV Systems – An International Survey. Proceedings of the 2nd World Conference on Photovoltaic Solar Energy Conversion, Vienna, Austria, July 1998.

[4] C. Cook. The Maryland Solar Roofs Program; State and Industry Partnership for PV Residential Commercial Viability Using The State Procurement Process. Proceedings of the 2nd World Conference on Photovoltaic Solar Energy Conversion, Vienna, Austria, July 1998.

[5] M. Garozzo, S. Li Causi, A. Sarno; The Italian 10,000 rooftops Program. Proceedings of the 2nd World Conference on Photovoltaic Solar Energy Conversion, Vienna, Austria, July 1998.

[6] F. Apicella, A. Sarno, S. Li Causi, S. Pietruccioli, G. Cesaroni; Architectural integrated PV systems: design and preliminary analysis. This Conference

[7] T. Contardi, F. Iannone, G. Noviello, A. Sarno, M. Valicenti; ENEA PV plants remote monitoring network: design and preliminary performance analysis. Proceedings of the 13th European Photovoltaic Solar Energy Conference, Nice, France, October 1995

All requests regarding the preparation of the paper should be addressed to:

S. Castello
ENEA, Casaccia Centre, Via Anguillarese 301 - 00060
S.Maria di Galeria - Rome (Italy)
Tel. +39 06 3048 4339
Fax. +39 06 3048 6486
e-mail: castello@casaccia.enea.it

A GLOBAL FRAMEWORK OF NATIONAL QUALITY PV PRACTITIONER CERTIFICATION PROGRAMS

Mark C. Fitzgerald
Institute for Sustainable Power, Inc.. P.O. Box 260145, Highlands Ranch, Colorado 80163-0145 USA
Phone: 303-683-4748. Fax:303-470-8239. e-mail: markfitz@ispq.org

ABSTRACT - Nearly every successful industry has some form of professional credential as a means of qualifying its practitioners. If the PV industry is to succeed in the long term, it must have a common set of quality knowledge and skills competency standards, along with a means to evaluate and implement them. Without such an objective means of qualifying practitioners, it will become increasingly difficult to engage investment and conventional financing. And, it is likely that, without international cooperation and coordination, development efforts around the world will tax their already-limited funds to simply reinvent the certification work that already exists elsewhere. The purpose of this work is the coordination and development of a framework of national programmes implementing a set of global standards for a number of quality PV practitioner certifications. The intent is that implementing an objective measure of competency through a third-party credential will improve the quality and reliability of fielded systems, encourage financing and investment, and create local, sustainable jobs. Cases reviewed here include projects sponsored by The World Bank (implementing quality training certification programs in India, Sri Lanka, and China) and the U.S. Department of Energy (supporting a PV training accreditation and certification system at the American Indian Tribal Colleges).
Keywords: Education and Training – 1: Qualification and Testing – 2: Financing – 3

1. INTRODUCTION

Sustainable energy technologies in general, and photovoltaics (PV) specifically, are particularly effective when implemented in a distributed fashion, whether integrated into conventional urban residential or commercial buildings, or at the remote home or village level. This results in the installation of many small systems, and requires a dispersed resource of field technicians and businesses to service this decentralized market.

Funding organizations and government agencies are very interested in accelerating the market development of sustainable technologies. However, they often find that there is not an adequate infrastructure in place to successfully market, install, and service the technical solutions that they wish to support. Training is seen as one component required for successful market development. However, until recently, there was not a broad-based effort to define and implement the wide variety of tasks, knowledge, and skills needed for the successful implementation of dispersed sustainable energy technologies, and this is one of the major factors preventing financing agencies from more aggressively.

The work of the Institute for Sustainable Power (ISP) is the development of a global accreditation framework of learning objectives and evaluation standards that can be met by training organizations. These training organizations would then be able to offer broadly recognized certifications at a number of professional and practitioner levels. This credentialing will help funding organizations assess the risk and potential of the projects.

From 1995 to 1999, the effort to develop this international framework was mostly organizational. However, beginning in 1999, with support from The World Bank and the U.S. Department of Energy, the ISP began to pilot the development of national programs and Master Trainer Accreditation, leading, in 2000, to the first internationally recognized certifications of photovoltaics (PV) installation and maintenance practitioners.

2. HISTORIC DEVELOPMENT

This project originated in October 1995, at a meeting sponsored by the Rockefeller Brothers Fund to bring together the PV community with the finance and investment communities. At this meeting, a senior investment manager said that he would consider investing substantial monies into a debt or equity fund, if only there was some means to assess the knowledge and skills of those accessing those funds. Without that basic risk assessment tool, though, he felt that he could not, in good conscience, make such an investment.

Based on this conversation, an effort was undertaken to survey the PV industry and finance/development community concerning the value of such a credential. Based on the positive outcome of this research, organizers incorporated a 501(c)(3) non-profit corporation, the Institute for Sustainable Power, Inc., to pursue the development of a global training accreditation and certification program for PV systems design, installation, and maintenance.

In 1998, the U.S. Department of Energy's (DOE's) National Renewable Energy Laboratory supported an industry-wide evaluation of competency standards for qualifying the installers of grid-tied PV installations. This was followed in 1999 by a project sponsored by the ASTAE unit of The World Bank, to develop and implement guidelines for establishing national framework programs for the certification of stand-alone PV systems installation and maintenance technicians in India, Sri Lanka, and South Africa. Also in 1999, the DOE supported the development of accredited vocational training programs to certify PV installation and maintenance practitioners at American Indian Tribal Colleges in the United States.

Today, the ISP has accredited a number of Master Trainers who have certified a number of installation and maintenance practitioners, and the global framework of training accreditation and practitioner competency standards is being implemented by organizations in a number of countries, and several countries are actively working toward the establishment of national programs to participate in the global framework.

3. CONTEXT

Just as PV is only one option in the mix of potential solutions to an individual's or community's energy needs, training quality is only one infrastructure component among a number of components that are critical to the success of the renewable energy industry. And, successfully implementing the infrastructure components that are critical for the long-term viability of the PV industry is as much an issue of timing as organization.

Standards for training and hardware are two infrastructure components (among others that include finance and distribution infrastructure) that must be closely aligned—the best hardware will not work if it is poorly installed and maintained, while the best-trained installers cannot make a system work with faulty hardware. In another issue of timing, organizing a training program before there are jobs for the graduates is a losing proposition. In the end, both hardware and training standards will be easier to implement with the financial driver in place. If programs such as PVMTI, the Solar Development Group (SDG), The Solar Bank, The World Bank, International Finance Corporation, United Nations Development Program, U.S. Agency for International Development, other bilateral and multilateral donors, government organizations, NGOs, and foundation programs specify certified hardware and installers, then the market will embrace them.

4. ACCREDITATION AND CERTIFICATION

4.1 Accreditation Components

Accreditation is the process by which an individual trainer or training organization wishing to receive recognition for meeting accepted standards applies for, is audited to, and meets those standards. In the ISP process, it accepts applications submitted by trainers and training organizations interested in accreditation. These organizations provide information on the levels of training that they offer, the candidate population they target, the experience of the training staff, and the physical and equipment facilities they have available for instruction programs. This application is reviewed by an auditor and then an audit team that would do an on-site assessment of the training organization. The report of this team (or individual, as appropriate) would then be given to a final auditor—one that did not participate in the on-site audit—who would make the recommendation either to accredit or not accredit. If accreditation was not granted, a full report would be supplied to the training organization, and that organization would be able to reapply.

This process would be carried out for accreditation for any or all of the available levels of certification. For the Master Trainer Accreditation or for accreditation of continuing education classes or workshops, a modified version of this system would be used.

4.2 Certification Components

Because the needs vary so much from country to country, and the range of practices in the PV industry is so great, there is a range of certification levels. In a country with a developed utility grid, the emphasis might be on the the Installer Technician III: Grid Tied Systems, while in a country where much of the population is off the grid, the emphasis might be on Maintenance Technician I/Installer Technician I: Solar Home Systems; Maintenance

Technician II/Installer Technician II: Large Stand-Alone Systems; or, Installer Technician IV: Hybrid Systems.

Once these certification levels are achieved, through testing for both knowledge and skills (experienced professionals may choose to take the tests without taking the training programs), the certification will be valid for a period of 3 years (though a country may choose to make this requirement more strict), at which time the candidate would have to show proof of the successful completion of a specified number of hours of continuing education (approved manufacturers' courses, conference workshops, academic classes, etc.) and a specified number of successful field installations to maintain the certificate. Without the required continuing education or field experience, the candidate would have to retake the knowledge and skills exam for the certificate.

The levels of certification either currently implemented or under development are

- Maintenance Technician I: Solar Home Systems
- Installation Technician I: Solar Home Systems
- Maintenance Technician II: Large Stand-Alone Systems
- Installation Technician II: Large Stand-Alone Systems
- Maintenance Technician III: Hybrid Systems
- Installation Technician III: Hybrid Systems
- Installation Technician IV: Grid-Tied Systems
- Installation Technician V: Master Installer
- Systems Designer I: Small Stand-Alone Systems
- Systems Designer II: Architects and Engineers
- Systems Inspector
- Small Business Practices: Dealers and Distributors
- Project Finance, Evaluation, & Assessment: Bankers, Investment Professionals, and Program Officials
- Certified Trainer

5. PILOT PROJECTS

Pilot programs for the accreditation of training programs and the certification of practitioners have been carried out with the support of the U.S. National Renewable Energy Laboratory, the U.S. Department of Energy, and The World Bank (with funding support from the government of The Netherlands). These pilot programs have included efforts in India, Sri Lanka, South Africa, and the United States (including programs at the Tribal Colleges of the American Indian nations). And, based on the success of these pilot programs, efforts are underway to expand the framework with programs and collaborations in Australia, Botswana, Canada, Costa Rica, Ghana, Kenya, and Zimbabwe.

5.1 The World Bank QuaP-PV Project

In 1999, the Asia Alternative Energy Unit (ASTAE) of The World Bank, with funding from the government of The Netherlands, implemented a program to improve the quality of PV components and systems design, manufacture, testing, and installation and maintenance, abbreviated QuaP-PV. These four programs were carried out by ECN, The Netherlands (for quality component design); the Global Approval Program for PV (PV GAP), Switzerland (for quality manufacturing; the Florida Solar Energy Center, the United States (for testing laboratory quality systems); and, the Institute for Sustainable Power, Inc. (ISP), the United

States (for quality training accreditation and installation and maintenance practitioner certification programs).

Under this contract, each of the four participating organizations held an initial pilot implementation workshop in Jaipur, India, in October of 1999. Building on the feedback and experience from this workshop, the four organizations revised the materials and held additional implementation workshops in other target countries. Each participating organization then held subsequent workshops and training programs in other target countries. The ISP held two additional workshops/trainings: a Train-the-Trainers workshop, in Sri Lanka, in February 2000, and a quality training framework implementation workshop, in South Africa, in February 2000.

For the pilot workshop, in Jaipur, India, the ISP developed a draft manual and workshop support materials. This workshop was aimed at the stakeholders interested in, or involved with, the development and implementation of a quality installation and maintenance training infrastructure. This target group included PV industry representatives, government officials, finance professionals, and standards experts. The intent was to use the manual and workshop materials to provide the participants with a number of information and implementation resources:

- Background on PV project development
- Background on practitioner training
- Information on quality systems
- Resource information on identifying stakeholders
- Guidance on the steps to develop a national PV training accreditation and practitioner certification framework, in line with the ISP's global PV training standards framework

The workshop, run for five days, was organized and coordinated for The World Bank by REIL, of India, which also identified and invited the workshop participants. The participants in the Quality Training workshop represented most of the target stakeholder groups. Through the course of the workshop, the participants and reviewers provided constructive feedback, as well as comments and suggestions on the content and format of the manual and workshop. They also provided valuable feedback on the relevance of the project to their needs, and unanimously indicated the importance of establishing such a quality program in India, both for the industry and for the professional development and standing of the participating industry individuals.

From the feedback of the participants and the input and guidance of the reviewers, the manuals and workshop materials were refined and adapted to better represent the material and to clarity the implementation guidelines.

Using the revised manual and workshop materials, The World Bank contracted with a South African expert in PV training and vocational standards to take the materials and offer the workshop to representatives of a number of African countries. The goals of this activity were to test the revised manual and workshop materials and to assess the potential for countries to use the materials without requiring a workshop instructor specially trained in the use of the materials. The hope was that the materials were designed in such a way that, through review and familiarization with the materials, an appropriate instructor in any country could conduct the training without outside assistance. As the development of the materials was ongoing throughout the course of the contract, an ISP representative was asked to join the workshop to review the use of the manual by an outside instructor and the gather first-hand review and assessment from the participants for use in the final version of the manual. However, the ISP representative was not to actively participate in the instruction of the workshop.

The workshop was attended by 13 experts representing five African countries (Botswana, Ghana, Kenya, South Africa, and Zimbabwe), all involved with PV projects and each with an interest in implementing a national-level program for the accreditation of training programs and the quality certification of the knowledge and skills of PV installation and maintenance practitioners.

In the course of the workshop, the participants moved quickly through the material, indicating that they felt that the manual provided sufficient guidance and detail to implement such a program on a national or regional level. And, rather than proceeding with the detailed steps of the workshop, they decided to move beyond the framework of the manual and work together to establish a pan-Africa working group to coordinate the development of framework standards for training quality and competence. By moving to this step, the participants felt that they could better share resources and lessons, mitigating the expense of developing such a program in isolation. The participants chose a coordinator who would represent them in the international training standards efforts, and together they agreed on a follow-up meeting, to be held in June 2000, to ensure that the momentum developed at the February workshop was not lost.

To test the training standards set out in the quality installation and maintenance guidelines, The World Bank contracted with Global Sustainable Energy Solutions (GSES), an ISP-Accredited Master Trainer from Australia, to hold a Train-the-Trainers workshop in Sri Lanka. Using the version of the Australian National PV Installer Certification training, adapted to the competency standards of the ISP, GSES held a 1-week training in Colombo, Sri Lanka, for 20 professionals from PV companies in Sri Lanka interested in becoming certified trainers, and one unaffiliated vocational training specialist, who would be able to provide qualified training to those concerned about conflicts of interest in taking training from companies that might be their competitors.

Under the contract, the training materials were translated into Sinhala, and the course was given by a GSES trainer in English, with the assistance of a translator who would provide the same material in Sinhala.

The participants took part in both classroom education and hands-on skills training, as the first phase of becoming certified as PV trainers for solar home systems installation. This first phase was to qualify the participants' at the level of certification at which they wished to be instructors (solar home systems installation). The second phase would be held at a later date, to qualify their instructor skills.

At the end of one week, the participants were evaluated through a knowledge examination and a hands-on skills evaluation. Of the 21 participants, 17 received passing marks and will participate in the second phase of the trainer certification program. In evaluating the four who were unsuccessful, it was determined that having the material presented primarily in English, even with a Sinhala translator available, was insufficient, especially for teaching technology-based subjects. This only emphasises the need for national-level programs to quality trainers, to avoid having to contract with outside trainers.

5.2 U.S. Department of Energy Tribal College Program

Also in 1999, the U.S. Department of Energy's Office of Economic Impact and Diversity funded a program to work with academic and vocational training programs at a number of American Indian Tribal Colleges and Universities to establish accredited training programs for certifying installation and maintenance practitioners.

Under this program, the ISP worked with the Native American Renewable Energy Education Program (NAREEP) to identify schools and instructors that were interested in and prepared to participate in a program to qualify their programs and instructors to offer training to certify PV systems installation and maintenance practitioners. NAREEP identified eight participants from six schools, and these instructors were asked to participate in a 2-week PV training course at D-Q University (a participating Tribal College), in Davis, California. The instructors for the course were from Solar Energy International, Colorado USA, and Native Sun/Hopi Solar, Arizona USA. Through the course, two schools and their instructors were identified as candidates for piloting the accreditation program: Crownpoint Institute of Technology, New Mexico USA, and the Lac Courte Oreilles Ojibwa Community College, Wisconsin USA.

In the ensuing months, the ISP arranged for the donation of PV hardware (from ASE Americas) to the two schools and began to work with them to review their programs and ensure that their instructors had the requisite resources and experience to provide certification training. This included reviewing the existing academic and vocational program, the resource library, job placement capabilities, linkages with industry, and instructor qualifications and experience.

On successful completion of this accreditation phase, the schools are prepared to begin training programs to certify their students. In addition, they will act as mentors and guides to the other participating tribal colleges and their instructors, to move these other programs toward full accreditation for PV training.

In a follow-on to this work, the DOE's Office of Economic Impact and Diversity has supported additional efforts to continue this work and to expand the scope of the support for the development of capabilities at the schools through the formalization of a number of Centers of Excellence at these and other participating schools.

6. CONCLUSIONS

There is a growing realization and consensus that some form of accreditation and certification to validate the knowledge and skills of industry professionals is critical to the long-term success of the PV industry.

It is the view of the development and finance communities that a broadly recognized professional credential will enable financial professionals to better assess their risk of lending money to PV businesses or for PV projects.

It is the view of the government an development communities that a formalized accreditation and certification program will promote the development of new, sustainable jobs; that systems they fund will be more likely to succeed; and that successful installations will encourage wider use of sustainable technologies, and reducing the emission of global climate change gasses.

And, finally, the PV industry believes that this type of training standardization will improve the quality of their employees; that their employees, dealers, distributors, and customers will stay more current with the state of the technology; and that this type of program would lead to more successful field installations, reducing their overall costs and encouraging market growth.

Through projects around the world, following the examples of the pilot projects described here, the interested stakeholders (e.g., PV industry, training community, standards organizations, government offices, utilities, community groups, NGOs, finance and development organizations, etc.) will be able to work together to improve the quality, reliability, and sustainability of PV projects through the implementation of common quality competency standards for training accreditation and practitioner certification.

ACKNOWLEDGMENTS

The author would like to acknowledge the support and guidance of the staff of the ASTAE unit of The World Bank and the staff of the U.S. Department of Energy's Office of Economic Impact and Diversity in providing the opportunity to develop and implement this work. In addition, the author would like to thank the staffs of the Lawrence Berkeley Laboratory, the Native American Renewable Energy Education Program, contributors, participants, and reviewers in the programs for their generous assistance and critical input.

PHOTOVOLTAIC SYSTEM INVESTMENT THROUGH INCLUSION IN MORTGAGES;
A COMPARISON AND ANALYSIS OF DIFFERENT MORTGAGES IN THE NETHERLANDS

M. Colijn

Shell Solar Energy B.V., P.O. Box 849, 5700 AV, Helmond, the Netherlands

Tel. +31-(0)492 – 508 608, Fax. +31-(0)492 – 508 600, e: Solarinfo @Si.Shell.com

Abstract

PV Systems are easily available for Grid Connection applications, but are hampered by a high up-front investment requirement for the end-user. Providing a clear, affordable finance option - without subsidies - for PV is therefore essential for the end-user market of Grid systems to grow. Including PV in the mortgage of the house is one readily available option. Given the multitude of mortgage types, and the benefit of interest payment returns to the end-user under the Dutch tax system, several mortgages are analysed for their suitability towards inclusion of PV. Three general outcomes show that inclusion of PV in the mortgage is a viable option: the savings mortgage is least costly per month, inclusion of PV in a higher mortgage is relatively cheaper, and high-income groups pay less for a PV system in their mortgage.

Keywords: Mortgage – 1: Solar Finance Scheme – 2: PV – 3

Introduction

The PV industry faces several challenges, of which the most important ones are efficiency, reliability and price. The perceived price of a PV system for application in the built environment, depends largely on the way the system is financed. In a European setting, where grid connected PV systems are often integrated into the roof of households, the end-user or house owner is confronted with either "lending out" his roof to a utility, or making a capital investment for which no subsidies are available directly. Including the cost of the PV system in a mortgage of the house, there where the system has been integrated as part of the house, is one method for spreading the cost of PV system ownership, and eliminating the need for up-front capital payment by the user.

Mortgages offer one way of including the capital cost of a grid-connected PV system in the monthly payments, without the need for subsidies and with a low monthly payment requirement. The Dutch mortgage system allows for repayment of interest to the end-user, thereby making both mortgages and the investment in PV more attractive.

To understand the effectiveness of PV inclusion the mortgage, several different mortgage types have to be compared for different sizes of grid-connected PV systems, for different income levels, and for different mortgage sizes. Additionally a varying interest rate is applied to create a clear overview of the implications of PV inclusion in this financing construction, all over an identical period of 30 years.

Types of Mortgages

There are many types of mortgage, offered by almost all finacing institutions, including banks, insurance companies, pension funds and also company finance schemes. In this paper, the institutions which offer the mortgage service are not considered, and out of the existing types of mortgage, the three most common types are used. These are the savings mortgage, the annuity mortgage and the linear mortgage.

The Savings Mortgage

In this type of mortgage, a loan is provided for which nothing is paid back during the mortgage period. Therefore, interest has to be paid over the entire loan amount, for the entire duration of the mortgage, which gives all the fiscal advantages of interest repayment. Apart from this, a fixed amount is saved monthly, over which many banks and insurance companies pay an equal interest rate as the mortgage taker has to pay for their mortgage loan. With this construction, the saved amount equals exactly the loaned amount that can be repaid at the end of the mortgage period.

By saving during the mortgage period, in combination with the possibilities offered fiscally for retirement savings that can be paid out tax free, the maximum tax deductibility is obtained for the interest payments. Additionally, tax-free savings are accumulated for the loan. Due to this construction, the savings mortgage has low net monthly payments that remain at the same level for the entire mortgage duration.

Linear Mortgage

The simplest form of mortgage is the linear mortgage. With this type of mortgage a loan sum is borrowed to finance the house purchase, over which interest is paid monthly, depending on the interest rate. Additionally, the mortgage is repaid in a pre-determined number of instalments, so that a fixed repayment sum is also paid each month. Therefore, the total monthly payment decreases linearly, as each month more of the loan has been repaid, and the interest on the remainder of the loan decreases as the mortgage debt decreases.

For low-income groups this mortgage is not attractive: the loan requires high monthly payments at the start, there where the salary is low (e.g. at the start of a career), and the fiscal advantages are also reduced because less interest deductibility can be claimed when in the lower tax groups.

However, for higher income groups wishing to increase a mortgage, or where the existing mortgage has been (mostly) paid off, the linear mortgage offers the possibility of a high tax deduction on interest payments, because the mortgage taker is in a higher tax group.

Annuity Mortgage

For an annuity mortgage, a loan amount is also borrowed for a predetermined time-scale, during which the loan amount is gradually repaid. However, in this form of mortgage, the combined gross amount of interest and loan repayment per month is fixed. In the beginning, when the loan amount is highest, most of the payment consists of interest, and only a small part is loan repayment. As time progresses, larger amounts of loan are repaid, and smaller parts of interest. This type of mortgage is of interest to those for whom the linear mortgage starting payments are too high. In the beginning the repayments are low, and the interest can be deducted, making the net repayments lower than in the case of the linear mortgage.

Fiscally this mortgage is not always advantageous, as more tax can be deducted from the annual income at the beginning of the mortgage period than at the end. Yet people tend to earn more at a later stage of the mortgage period, and would like to deduct more from their tax payments. Thus the net monthly payments increase during the repayment period. On the other hand, the home-owner knows exactly what the gross payments will be for the entire duration of the mortgage.

Base Assumptions and Comparisons

For the analysis performed here, several base assumptions have been made. The most important is the mortgage period, which was chosen at its maximum length of 30 years. The importance of this is that the interest rate is coupled to the mortgage period. Hence, a thirty year calculation can be made with one interest rate, and compared easily with a similar calculation in which the interest rate has been varied. This was done for the savings mortgage only. As variations showed to be low, only the 7% base was used for the annuity and linear mortgages. The interest rates currently available for this period vary between 6% and 7%, and calculations have been made on this basis.

The other cross-variation has been to differ the income levels, which changes the tax deductibility of the interest payments. Three base case annual salaries have been chosen to represent the largest set of income groups in the Netherlands, at Hfl. 50,000, Hfl. 75,000, Hfl. 100,000. For each income level, mortgage loans have been chosen, with the middle income group being applied to each of the mortgage levels. Each mortgage level for each income group had the two interest rates described above imposed on them. The mortgage levels used were Hfl. 200,000 for the lower two income groups, and Hfl. 300,000 for the upper two income groups.

From this base scenario, each mortgage was increased in two steps: the first step of Hfl 15,000, which represents 1 kWp of installed grid connect photovoltaic (PV) system, the second increase was for Hfl. 30,000, representing a 2.5 kWp installed grid connect PV system

Each set of combined income level, mortgage level, interest rate and PV system was applied to all three mortgage types, as described in the section above. The monetary figures used in the calculations were translated to Euro's in the results to make comparison easier.

	Mortgage Type			Mortgage Base Level		PV System	
Annual Income	Saving	Annuity	Linear	200,000	350,000	1 kWp	2.5 kWp
Hfl 50,000	X	X	X	X		X	X
Hfl 75,000	X	X	X	X	X	X	X
Hfl 100,000	X	X	X		X	X	X

Table 1, Income-Mortgage choices

Results

Three basic results have been obtained from the calculations made above:

1. The Savings Mortgage is the cheapest mortgage in terms of net monthly payments for a PV system over the entire period of the mortgage, for identical mortgage levels, income groups and interest rates.

2. The higher the income level, the lower the PV payments for an equivalent system, under comparable mortgage conditions;

3. The higher the mortgage amount loaned, the lower the PV payments for an equivalent system, under comparable income conditions.

The extra payment made for a 1 kWp PV system is lowest for a savings mortgage owner, with a medium to high income, and a high mortgage. This stands at Euro 27.2 per month, at 6% interest. The variation of payments over all three mortgage types, income groups and mortgage levels, ranges from the mentioned 27.2 Euro / month, to 51.7 Euro / month for the annuity mortgage at low income and mortgage levels.

In table 2, below, a comparison is given for the Hfl 50,000 income group, at 7% interest, and a mortgage of Hfl 200,000. The savings mortgage is cheapest per month.

The (L) indicates a lowest payment, with other payments increasing, up to double the amount mentioned.

Income also influences the payments made for the PV system. Table 3 above shows how the payments vary for a savings mortgage, at 6% interest, for a 2.5 kWp PV system. The higher the income level is, and related mortgage, the lower the payment requirements for the PV system.

Conclusion

When including a PV system capital cost into a mortgage, the monthly payment requirement offers an affordable finance option for the relative income levels. This is because the payment is spread out over a long period of time, and combined with a very secure finance method.

The success of applying PV in mortgages will depend on the effectiveness of marketing this option with end-users, and the willingness of finance institutions to adopt the PV mortgage in their packages.

The other advantage to using mortgages is that it simplifies financing arrangements for the end-user by eliminating the need for both the application and use of subsidies for Photovoltaics. The added benefit of reduced or eliminated electricity consumption from the grid supply has not been taken into account in this analysis, but should be added when considering the purchase of a PV system.

References

This paper was written using a variety of internet sites from institutions specialised in mortgage calculations:

Huis & Hypotheek - http://www.huis-hypotheek.nl/

ABN AMRO - http://www.abnamro.nl/wonen/financieren/stap_hypotheek2

Hypotheek Calculator - http://home.iae.nl/users/kunstein/hypotheek/

Ministerie van Financien - http://www.minfin.nl

Hypotheek Online - http://www.hypotheekonline.nl/welkom.htm

Telegraaf Hypotheek - http://krant.telegraaf.nl/krant/ditjaar/huis/teksten/huis.index.html

Mortgage Type	1 kWp (Net **Euro**/Month)	2.5 kWp (Net **Euro**/Month)
Savings	32.7	65.8
Annuity	39.9 (L)	68.1 (L)
Linear	35.4 (L)	54.9 (L)

Table 2 – (L) = lowest monthly extra, over thirty years.

Annual Income (Dutch Guilders)	Mortgage Level (Dutch Guilders)	Interest rate	Net Payment (**Euro**/Month)
50,000	200,000	7%	65.8
75,000	350,000	7%	65.8
100,000	350,000	7%	58
300,000	1,000,000	7%	49

Table 3 – 2.5 kWp PV system payments in a Savings Mortgage

USER TRAINING IN PV RURAL ELECTRIFICATION PROGRAMS

Jordi Serrano
Asociación SEBA
C/ Mallorca, 210, 1r. 1a. – 08008 Barcelona – Spain
Tel. +34 93 4463232 – Fax +34 93 4566948
sebaasoc@suport.org

Enma Ramírez, Xavier Vallvé, Ingo Vosseler
Trama TecnoAmbiental S.L. (TTA)
C/ Ripollés, 46 - 08026 Barcelona - Spain
Tel. +34 93 450 40 91 - Fax +34 93 456 69 48
tta@retemail.es

ABSTRACT: The evolution in the field of PV systems always has been regarded from a technological point of view improving components and technical system performance but until now very few steps have been undertaken towards a better assistance to the end user. This assistance has been limited to some advice and user manuals but in most of the cases the lack of user's knowledge about "his power plant" made it difficult to achieve an optimal operation of the PV systems.

For this reason, user training in the fields of operation of the PV stand-alone system as well as rational use and management of the energy available is a determinant factor for optimal use and enlarged lifetime of a PV stand-alone system. On the other hand, training and assistance leads to a higher degree of user's satisfaction with his PV system and consequently to increased acceptance of PV technology in general.

In this paper we present our approach in the field of user training as an important measure to increase performance and acceptance PV stand-alone systems.

Keywords: Education and Training – 1: Stand-alone PV Systems – 2: Rural Electrification – 3

1. INTRODUCTION

SEBA is a PV users' association, which offers energy service mainly in rural areas with PV stand-alone systems. SEBA offers 15 years service contracts including the services of maintenance, advice, user training and insurance. As owner of the systems and energy operator, SEBA had to find solutions for several problems, which arose during system operation in the past, as:

- Increased cost due to appliances with adverse effects on inverters (high power peaks etc.)
- Increasing energy consumption, which leads to chronical low batteries and reduces lifetime
- User satisfaction not optimal due to lack of transparency of system operation
- In general high service costs due to incidents and user uncertainties, which wouldn't appear if users had a better knowledge about their PV power plant
- Conflicts with users due to bad system use

As one step to tackle these problems obligatory user training was included in SEBA's service contracts as a determinant factor for better quality of service in the PV rural electrification programs driven by SEBA.

2. GENERAL OBJECTIVES

To avoid the situations described above user training has to address several aspects to influence users' behaviour in terms of energy consumption and responsibility for his system. So, user training was conceived with the following objectives:

- Transmit basic knowledge about the operation of PV stand-alone systems
- Transmit knowledge about basic maintenance and incentivate the users to care for their PV systems
- Increase satisfaction of users
- Create higher consciousness of rational use of PV energy
- Incentivate the user to interact with the technical system and to learn how to use it optimally ("learning by using")
- Incentivate the users to interact among each other and to learn from each other
- Induce interaction of the users with SEBA to profit from the support service of the association in case of problems

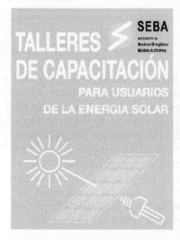

Figure 1: Official logo of user training courses in rural electrification programs of SEBA

3. APPROACH

To well address the training to the users different types of training courses were designed taking into account the type of system, the target audience and the special objective. The following courses are being held:

- Group training courses for new PV users of individual and multi-user systems
- Basic individual user training at the commissioning of the PV system
- Accompanying activities related to user training like special workshops for children and general workshops to refresh knowledge about PV technology

The training courses are performed by TTA, which has special know-how on this issue counting with an interdisciplinary team of sociologues and technical engineers.

Figure 2: Announcement of a training course for users of multi-user PV systems in the village of San Felices (Aragón, Spain)

4.   METHODOLOGY

The methodology of the user training courses depends on the type of activity. In the following we focus on the group training courses as the main activity to transmit the necessary knowledge to the users to be able to better operate their PV systems. These courses are mandatory when the service contract with SEBA is signed and thus are regularly offered. Basic individual user training is always done at the commissioning of each site. However the accompanying activities, like special workshops for children etc., depend on demand and are offered on irregular basis as extra service of SEBA for its members.

4.1 Group training courses

These courses are generally of a duration of one day in groups. This is not much time to transmit the knowledge about a complex system like it is in this case of the system "user - PV plant". Thus, the methodology cannot be a mere lecture about all the detailed aspects and problems that could occur. The methodology chosen tends to transmit the following basic issues:

- Consciousness about new situation
  Most of the people do not really know what it means to live with a PV stand-alone system, they are not only consumers but also energy producers and suppliers. Consciousness for this new situation has to be created.

- Possibility to tackle problems
  Basic relations of the behaviour of the technical system have to be shown in order to create transparency. The user shouldn't be afraid of the system, problems may occur but he has to know basic approaches to solve them. Confidence in the support service has to be created.

- Motivation to interact
  There is no use explaining the user what to do if he doesn't see the point why. So, clear reasons are to be shown in order to motivate him to interact with the system but as well with SEBA.

The training consists of three general phases, a theoretical phase in form of an interactive lecture, a practical phase and a final evaluation in form of a test and a validation of the course. In the theoretical part the following aspects are treated:

- General aspects
  Description of specific characteristics of stand-alone PV systems and presentation of the service offered by SEBA.

- Comprehensive description of components, functions and characteristics
  Description of the whole system with its subsystems in a metaphoric way using analogies with known subjects of daily life. In this way principles of the value "energy" and the difference to the value "power" is explained. In this context the specific signals and functions of the equipment used are explained.

Figure 3: Use of energy efficient appliances is crucial for good load management in PV-systems

- Importance of efficient energy management
  It is explained what are the contradictions between the optimum energy management and an "unlimited" energy provision like in the public grid. The user is motivated to do energy management by showing him the consequences of his behaviour. Means of improving energy management are shown, like energy efficient appliances, deferring big loads, avoiding stand-by consumptions, activating back-up genset, etc.

- Planning of load management in daily life
  Prevision of energy production and consumption during the week with examples.

- Basic maintenance
  - Clean the modules and avoid shading

- Check batteries: corrosion of poles, refill water, measure density
- Consult important indicators of the power conditioning unit (Historical Battery Index, Availability Index of battery)

• Safety aspects
This is related to the safety of the interior electric installations and wiring, handling acid and operating the genset.

Figure 4: One of the basic maintenance tasks is to clean the modules (Hospedería de Usón, Aragón, Spain)

In the practical phase exercises are done with the installed system components and typical situations are simulated and discussed together:

• Battery-density
• Equalisation charge
• Low battery cut-off
• Overload cut-off
• Limitation of energy entrance by the regulator – using the excess of energy
• Starting the back-up genset to charge batteries
• How to interpret the remote visualisation panel or how to use the energy dispenser / meter in multi-user systems
• Situations in case of using remote controlled relays for automatic load management

Figure 5: Explication how to handle sulfuric acid safely

In a final evaluation phase the knowledge acquired by users during the course is checked in a small questionnaire. Like this the critical issues can be detected. Finally a questionnaire about the quality of the course gives hints for development of the methodology and the user's opinion.

4.2 Special tools for efficient energy management
A very important point for a flawless system operation is the interaction between the user and the technical system. To help the user to do good load-management and to give him feedback about the system state, SEBA is using special user interfaces and control devices developed by TTA. These products are the outcome of the special experience of the authors working with users of PV stand-alone systems. The understanding of these tools is treated with special attention during the user training. The used devices are the following:

• TApS remote visualisation panel (see Figure 6): User interface, which gives the user feedback about the state of his PV-system helping him to do good load management and to give transparency about system operation.
• TApS remote controlled relays: Switches for automatic load management of special consumers allowing load connection under certain system states only.
• TApS energy dispenser / meter: User interface and energy limiter used in multi-user systems to fairly distribute the available energy among the users. It limits the individual energy consumption but assures a contracted amount of energy to each user and gives incentives for efficient load management [2].

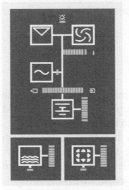

Figure 6: The TApS remote visualisation panel installed in every system gives the user feedback about the system state and helps to good load management (indications: energy input, in-out power balance, battery deliverable charge, hot water temperature in case of installed DHW and temperature of heating)

4.3 Basic individual user training
This type of user training is done at the commissioning of individual systems, the moment in which the system is handed over to the user officially. The following subjects are treated:

• General aspects
• Function of the inverter
• Function of the DC/DC converter
• Principle menus of the power conditioning unit
• Realisation of an equalisation charge
• Charging the batteries with the back-up genset
• Switch: PV-array – back-up genset
• Remote visualisation panel
• Basic maintenance and handing over a basic maintenance sheet (Hoja de Conservación Básica), which has to be filled in by the user and sent to SEBA each three months
• Safety of the interior electric installation
• Rational use of energy

Figure 7: Individual user training at the commissioning of the system: How to operate the PV system

4.4. Accompanying activities

General workshops are conceived to refresh knowledge about PV energy generation, to give the possibility to users to interchange experiences in operation of PV stand-alone systems and to create a relation based on confidence between the users' association and its members. The subjects to treat depend on the demand of the participants related to questions of operation of PV-systems and energy management.

Special workshops for children have been designed due to the fact that we found various families interested in this type of activity. The methodology adapted is "learning by playing", that means to offer games related with subjects of daily life in a PV system.

5.   LESSONS LEARNED

User training leads to increased interaction of the users with SEBA. So, possible problems can be discussed together and solutions are easier to be found.

In general user satisfaction increases as the system operation is more transparent to the user and he knows better how to manage the available energy. Thus, also motivation of users increases to care for their systems.

Monitoring of the systems of users, which already visited a training course, show their favourable behaviour. However, the user should be equipped with adequate tools for this purpose, which help him to do energy management.

The early education of children is important, as it is often them who want to watch TV or listen to music whenever they want and not when conditions are favourable. Experience shows that with proper education it is possible to make children being interested in the relation between energy consumption and the generation of the energy with the sun.

It is important to give the users more information on request. For this purpose SEBA offers a special user manual, which deals with detailed and comprehensive description of the operation and maintenance of PV-installations [1].

However, in many cases it is quite difficult to motivate users to participate in the training courses and to interact with the PV-system. This is due to the fact that most of the users did not choose the PV option because of their interest in this technology but because of PV being the most economic solution for the satisfaction of their urgent need of electrification.

6.   CONCLUSIONS

Special training methodologies have been developed, taking into account the principle issues to transmit to the users in order to increase system performance and user satisfaction. To reach these targets technical tools have been developed in order to help the user to better manage his system.

Analysing the evaluation of the training courses held up to now, it has been found that users appreciate this type of actions, on the one hand to get knowledge about their PV-plant and on the other hand to know other PV-users and to interact with SEBA.

Analysing the first experiences corresponding to user behaviour it can be concluded that user training accompanied by general information and transmission of knowledge about PV-systems is a basic condition for social efficiency as well as technological performance and evolution of these systems.

However, the analysis and especially the control of the system "user – PV-plant" remains difficult. So, to achieve what technology and science may not, we ask for support from heaven as is shown in Figure 8.

Figure 8: Blessing of the multi-user PV system in San Felices (Aragón, Spain) by the community priest

7.   REFERENCES

[1] J. Serrasolses, J. Serrano: Manual del usuario de instalaciones fotovoltaicas (SEBA), 1998

[2] X. Vallvé, J. Merten, K. Preiser, M. Schulz, Energy limitation for better energy service to the user: First results of applying "Energy Dispensers" in multi-user PV stand-alone systems, presented at 16th E.P.S.E.C., Glasgow, May 2000.

[3] X. Vallvé, J. Serrano, J. Campo, S. Breto Asensio, PV rural electrification program in Aragón (Spain): Example of an implementation scheme involving users and local authorities, presented at 16th E.P.S.E.C., Glasgow, May 2000.

PV RURAL ELECTRIFICATION PROGRAM IN ARAGÓN (SPAIN): EXAMPLE OF AN IMPLEMENTATION SCHEME INVOLVING USERS AND LOCAL AUTHORITIES

Xavier Vallvé
Trama Tecnoambiental S.L.
C/ Ripollés, 46
08026 Barcelona - Spain
Tel. 34 93 4504091
Fax 34 93 4566948
tta@retemail.es

Jordi Serrano, Javier Campo
Asociación SEBA
C/ Ramón J. Sender, 16, bajos
22005 Huesca - Spain
Tel. 34 974 244107
Fax 34 974 224241
sebaasoc@suport.org

Sergio Breto Asensio
Gobierno de Aragón – D.G.
Energía
Paseo María Agustín, 36
50004 Zaragoza - Spain
Tel. 34 976 714745
Fax 34 976 714723
sbreto@aragob.es

ABSTRACT: This program, called PAERA (Programa Aragonés de Energetización Renovable Autónoma), builds up on lessons learned and represents an evolution of PV rural electrification with a service scheme. After experiences in other regions, the user association SEBA started a new program in Aragón with the support of the Aragonese Government. It started in 1996 with a 4 year pilot phase and up to now it has electrified 8 small villages and 25 individual houses, farms and mountain huts with an installed PV-power of 60,7 kWp servicing 62 users. An implementation phase will follow to significantly reduce an estimated deficit of another 150 sites without electricity. The main innovative aspects are organisational, infrastructural and technological.

In the paper we present results from our experience with the PV rural electrification scheme in the following areas: Involvement of local authorities and users in the design phase and follow-up; Technical performance of the PV-systems; Effects of user training.

Keywords: Stand-alone PV systems-1 : Villages-2 : Implementation-3.

1 INTRODUCTION

Aragón is one of 17 Regions in Spain. Its territory has 47,556 km^2 divided into three provinces: Zaragoza, Huesca and Teruel. Total population is 1,2 million with a very uneven distribution with more than one half living in the capital city of Zaragoza.

Aragón has a total of 729 municipalities most of which have less than 1000 people and only a few have more than 5000. The average population density is 25 p/km^2, which is a low value compared to the average for Spain that is 78. This data is more significant considering the concentration in the capital. In the case of the province of Teruel, for example, the density is a mere 9 p/km^2.

At the same time it has high renewable energy potential: high isolation, water, wind and a large extension. The Government of Aragón, aware of this, has engaged an ambitious Renewable Energy Action Plan

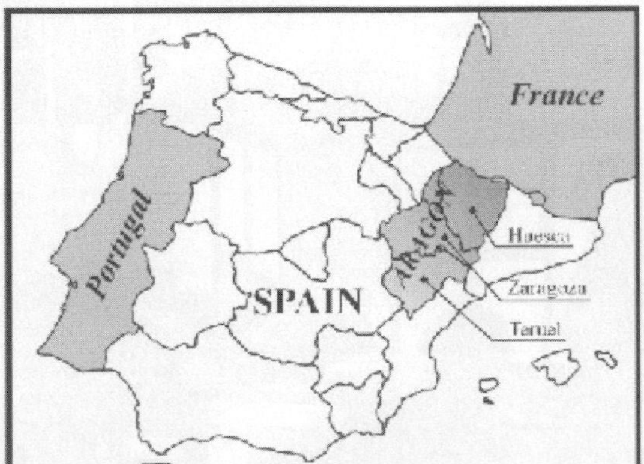

Fig. 1: The PAERA program is spreading from the North to the South of Aragón.

with a target to shift from an 11% share of R.E.S. of

energy consumption to 20% in 2005. Among other actions it has established a program through the Decree 68/1998 of the Government of Aragón to fund actions in energy saving and diversification, rational use of energy, use of local and renewable energy sources and rural energy infrastructure, with has been, in the last few years, one of the preferred technical areas for PV application.

Within this scope, SEBA's program PAERA is an evolution of PV rural electrification with a service scheme [1]. The program is driven by the users through the user association SEBA with support from the Directorate of Energy of the Government of Aragón. Project management is done by TTA and installation is done by local installers (SME's). Quality control of equipment and monitoring of most of the sites has also been done.

2 PROGRAM OBJECTIVES

2.1 Service aims

The overall mid term objective is to install PV-systems in all rural scattered sites where this alternative is more cost-effective than grid extension. The technical quality and security of supply should be comparable to the grid except for the energy load limitation.

2.2 Strategic aims

Lessons learnt from former projects showed the importance of involving regional energy planners in PV rural electrification [2]. For this project a close collaboration was establis-hed with the Aragonese energy planning authorities, to include also PV as a mature alternative, in rural electrification plans. The project has a first pilot phase, now completed, focused on target districts in Huesca. At the same time a survey on rural electrification deficit has been realised by the Directorate General for Energy.

2.3 Technological aims

It is estimated that a total power of about 300 kWp will be required until end of the year 2005. The sites can be either individual stand-alone systems or PV-hybrids with a multiuser minigrid, both offering standard electrical quality (230V/50Hz). Each user should have an energy deliverability assurance.

3   THE PILOT PHASE (1996-1999)

3.1 Technological aspects

Technical quality: All components met the highest state of the art for quality and efficiency. In the case of inverters high efficiency for very small loads was mandatory. Smallest systems could have a series or shunt regulators but in larger systems MPPT was specified.

Energy Deliverability Assurance: As a response to user requirements to be assured energy production has been developed. This concept, called EDA (Energía a Disposición Asegurada), is an assurance of energy deliverability on the worst month. In individual systems consuming more than the design load affects only that user if the service is disconnected, but in multi-user system this issue is of major importance. Load management was improved by the introduction of a user interface, which gives the user information on the system status. This helps the user to adapt his energy consumption to the deliverable energy. On the multiuser systems "Energy Dispensers" to control the individual were installed to fulfil this requirement individually [3]. Questionnaires have shown that users appreciate and use the devices as support for system operation.

Standard layout. There are three standard layouts that are adapted to each application:

Compact. For small loads of up to 25kW.h/mo.,3 compact systems were installed. The array power is in the range of 400 to 600Wp. Batteries are at 24V and 400Ah. Inverter power is of about 1 kW (Fig. 2A).

Fig 2: a) Compact system in "La Sardera". b) All modular PV-system in "Vizcarra y Salamaña".and c) Multiuser PV-hybrid plant and user training in the village of "San Felices".

All modular. For the medium and larger loads, from 33 to 100 kW.h/mo., 22 standardised SEBA state of the art [4] all modular layout was implemented with the aim to reduce maintenance costs and to allow an easy upgrading of the system. The average range of the PV-array of individual systems is about 500 to 2000Wp. Batteries are 48V with a capacity of 400 to 750Ah. Power conditioning is done by a modular rack, which combines MPPT-regulation and conversion with data logging and load management control. The typical inverter power is 2 to 3 kW (Fig. 2B). Upgrading of these systems is simple adding modules or inverter power. This has been necessary at two sites after a few months of operation.

Multiuser. During the project 8 villages were realised. This new need resulted in a new layout of centralised multiuser systems, economically more feasible than installing individual systems for each user. This, nevertheless, caused other social and organizational challenges that have had to be addressed. As an example for the multiuser micro grids, figure 2C) shows the PV-plant for the village power supply of "San Felices". It features 10,425Wp PV array, 180kWh batteries at 48V and 7,2kW inverter. The centralised generation requires an adequate distribution of energy on individual level and local involvement.

Components. PV modules were monocrystalline from two different European manufacturers meeting current standards. Layout is in strings of two or three modules at 55° inclination. These are comparatively the most standardized components of stand-alone PV plants.

Batteries are Pb-Acid tubular elements of several capacities with transparent enclosure.

The power conditioning and control rack used. For "all modular" and "multiuser" systems. It consists of a universal prewired main rack that can house standardized elements: Battery surveillance unit and data logger; PV MPPT regulator; Cascading inverter with sine wave in 1 kW steps.

3.2 Implementation aspects

Approach. The program is divided in two major phases. The first has been a pilot program with several subprojects to develop and validate the scheme. It started in 1996 and it has electrified 6 small villages and 26 individual houses, farms and mountain huts with an installed power of 60,7 kWp and a total of 57 users. It has concentrated in the northern province of Huesca to reach enough density of sites to make maintenance feasible and also to have a cultural impact at the locally. After adjustments on the scheme, an implementation phase (2000-2005) will extend the program trough the rest of Aragón.

Infrastructure. The association SEBA established a branch office in the city of Huesca to look after the service to the users. Local installers, trained participating in the installations, repair and maintain the equipment.

Local involvement. Users have been involved from the beginning, discussing their needs, the size of the systems, payment schemes, etc. Also, municipal authorities have been consulted and have played an active role, especially in multiuser installations.

Ownership. Rural electrification (grid or stand-alone) is a basic need that requires a significant public investment. In these case funding came mainly from the Regional Government and the EC but also from the users. The systems are owned by SEBA. The average investment

cost of each user serviced was 14,525€, of which 21% was paid directly by the user in the form of a deposit that can be partially recovered if the service contract is cancelled.

Contractual. Service contracts of 15-year commitment were developed which include the concept of EDA as well as other service obligations.

User capacity building. We have made mandatory user training, individually during commissioning, and in-groups in workshops. Users get acquainted in the operation of PV-systems, basic maintenance and rational use of energy. This helps considerably to reduce failures and increase lifetime of battery by good load management.

Nevertheless, it still results difficult to control the appliances that users eventually connect to their systems. This sometimes causes system failures due to inverter overcharge or low battery due to energy overloads.

Maintenance tasks. Are carried out at three levels:
- The user: inspection of operating parameters and level of electrolyte in batteries.
- Local technician: preventive maintenance, repairs and data collection.
- Engineering consultants: data analysis, advice on use of energy available and maintenance management.

Tariffs. A service tariff scale is applied to individual PV-users as well as users in a multi-user system. The fee is related to the "EDA" to pay for maintenance and repair costs as well as an insurance against theft, etc. Typical average payments are 20 €/mo. for contracted energy deliverability (EDA) of 65kW.h/mo.

4 RESULTS

4.1 Realised systems

Table 1. Summarises the systems realised to date to supply 62 users.

4.2 Quality of service

The quality of service is evaluated by subjective data from questionnaires answered and mailed by the users and from technical data. On larger systems hourly data is monitored; smaller ones have an energy meter recorded by the user.

Operating parameters are summarised in monthly average values for each installation as shown in Fig. 3.

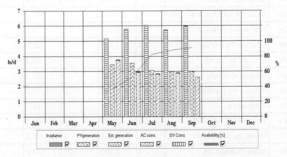

Figure 3: Monthly monitoring data such as irradiance (left), generation (middle), and consumption (right).

Fig. 4 summarises the average data available for most of 1999. For each site we show Performance Ratio; the Energy Load Ratio, that is the consumed load respect to the design load; Historical Battery Index, that is the fraction of days that battery equalisation has been reached; user satisfaction is evaluated from the questionnaires. In general the degree of satisfaction is high, and a wide range of PRis observed. This is often caused by low consumption periods due to seasonal activities. Also note that some users, -ie- HU98083, HU98084 and HU98088 exceed the design load resulting in high PR and low HBI. On the other hand see user HU99054 that is only using 30% of the EDA resulting in a low PR and high HBI.

5   OUTLOOK

The pilot phase realised mainly in the province of Huesca has been successful and lessons learned have resulted in general improvements 62 users have been electrified at a fraction of the investment cost of grid extension and are satisfied. A recent survey by the Government has identified more than 150 isolated sites without electrification. In the following phase as the service infrastructure is set up the PAERA program will be extended to the other provinces, Zaragoza and Teruel. One of the risks related to the low density of the region is

not being able to reach a critical density of use, to sustain the service without some additional support

6   REFERENCES

[1] Vallvé X, Serrasolses J, Design and operation of a 50 kWp PV rural electrification project for remote sites in Spain, in Solar Energy, vol. 59, Nos. 1-3, pp. 111-119, 1997, Elsevier Science Ltd.

[2] Vallvé X, De Juan J M, Comparative assessment of PV and grid extension to electrify a rural district in Southern Europe (Spain). Proceedings of 13th E. P. S. E. C., Nice, October 1995, pgs. 1175-1179.

[3] X Vallvé, J Merten, K Preiser, M Schulz, Energy limitation for better energy service to the user: first results of applying "Energy Dispensers" in multi-user PV stand-alone systems. 16th E.P.S.E.C., Glasgow, May 2000.

[4] Vallvé X, Serrasolses J, PV stand-alone competing successfully with grid extension in rural electrification: a success story in Southern Europe. Proceedings of 14th E.P.S.E.C., Barcelona, June-July 1997, pgs. 23-26.

Year	Site	N° users	PV Power (Wp)	Inverter Power (kW)	Battery Capacity kW.h	Design EDA (kWh/mo)
1996	Aldea de Pano	1	2.250	4	35.3	39
1996	Aldea de Caneto	1	675	2	16.6	50
1996	Refugio de Viadós	1	510	2	19.0	33
1996	Refugio de Estós	1	765	4	19.0	50
1996	Refugio de la Renclusa	1	1.020	3	26.4	67
1996	Molino de Yeste	1	765	2	19.0	50
1997	Casa Castillo en Ascaso	1	750	2	19.0	50
1997	Granja La Creu	1	600	2	15.8	42
1997	Refugio de San Úrbez	1	1.020	3	35.3	67
1997	Casa Lencina de Panillo	1	600	1	43.2	42
1997	Samper de Trillo	1	750	2	43.2	50
1997	P. la Coma y Villeros	1	375	0,8	43.2	25
1997	Granja Renales	1	1.125	2	16.6	75
1997	El Nadal	1	600	1	16.6	42
1997	Aldea de Escuaín	8	10.200	7,5	180.0	633*
1998	Aldea de Ascaso	3	1.125	3	35.3	83*
1998	Borda del Nogal	1	600	1,2	9.5	25
1998	Aldea de San Felices	8	10.425	7,5	180.0	803*
1998	Alm. Agríc. Sardera	1	450	1,2	9.5	25
1998	Hospederia montaña Usón	1	1.125	3	35.3	66
1998	Refugio de Lizara	1	1.125	3	35.3	58
1998	Refugio de Linza	1	900	3	35.3	66
1998	Refugio de Gabardito	1	1.350	3	30.2	58
1998	Montcalbós	1	450	1	19.0	25
1998	Vizcarra y Salamaña	2	2.025	4	35.3	117
1997	Morillo de Sampietro	5	2.250	3	61.4	140*
1998	Aldea de Mipanas	4	2.700	3,75	72.0	155*
1998	Refugio de Góriz	1	4.050	4	72.0	330
1999	Huerta la Sala	1	1.575	3	19.0	117
1999	Cámping de Los Baños	1	2.250	3	72.0	200
1999	Casa de Mingo	1	2.025	3	35.3	92
1999	Aldea de Mirabal	3	1.800	4	35.3	143*
1999	Aldea de Estaronillo	4	2.475	3,75	35.3	211*
		62	**60.705**			

Table 1. Systems realised during the pilot phase of the program.
* Total energy contracted to all users.

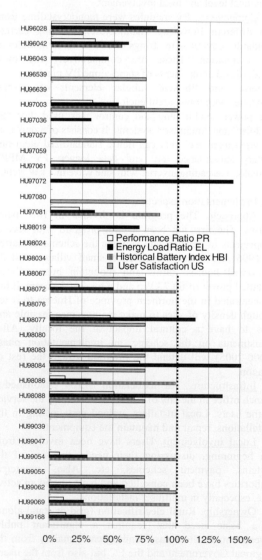

Fig. 4: Avg 1999 values of performance parameters

A PRELIMINARY ANALYSIS OF VERY LARGE SCALE PHOTOVOLTAIC POWER GENERATION SYSTEMS

K.Kurokawa[1], P.Menna[2], F.Paletta[3], K.Kato[4], K.Komoto[5], T.Kichimi[6], S.Yamamoto[7], J.Song[8], W.Rijssenbeek[9],
P.Van der Vleuten[10], J. Garcia Martin[11], A. de Julian Palero[11], G.Andersson[12], R.Minder[13], M.Sami Zannoun[14], M.Aly Helal[15]

[1]Tokyo University of Agriculture and Technology (TUAT), 2-24-16 Naka-cho, Koganei, Tokyo 184-8588, Japan
[2]ENEA, Centro Ricerche Portici, Localita Granatello, I-80055 Portici (Napoli), Italy
[3]CESI SFR-ERI, Via Rubattino 54 Milano 20134, Italy
[4]Electrotechnical Laboratory (ETL), AIST, MITI, 1-1-4 Umezono, Tsukuba 305-8568, Japan
[5]Fuji Research Institute Corporation (FRIC), Kanda-Nishiki-cho, Chiyoda-ku, Tokyo 101-8443, Japan
[6]Resources Total System (RTS), 2-3-11 Shinkawa, Chuo-ku, Tokyo 104-0033, Japan
[7]Photovoltaic Power Generation Technology Research Association (PVTEC), Otowa, Bunkyo-ku, Tokyo 112-0013, Japan
[8]Korea Institute of Energy Research (KIER), 71-2 Jang-dong, Yusong-gu, Taejon 3-5-343, Korea
[9]ETC-ENERGY, P.O.Box 64, 3830AB, Leusden, the Netherlands
[10]Free Energy International bv, P.O. Box 9564, 5602 LN Eindhoven, the Netherlands
[11]IBERDROLA, Hermosilla, 3, 28001, Madrid, Spain
[12]Electric Power Systems, EKC, KTH, S-100 44 Stockholm, Sweden
[13]Minder Energy Consulting, Ruchweid 22, CH-8917 Oberlunkhofen, Switzerland
[14]Ministry of Electricity and Energy, P.O.Box: 4544, Masakin Dobbat Elsaff-El-Hay Sades, Nasr City, Cairo, Egypt
[15]Mechanical & Electrical Research Institute, Ministry of Public Works & Water Resources, Delta Barage, Cairo, Egypt

ABSTRACT: It is expected that photovoltaic (PV) technologies will be one of major solutions of energy and environmental problems in the 21st century. Though solar energy is low density by nature, desert areas, where high-level insolation and large spaces exist, has high potential for introducing PV system on a large scale. In this study, potential of very large-scale photovoltaic power generation (VLS-PV) systems that have 10MW to a few GW capacity was preliminary evaluated under umbrella of IEA/PVPS Task VI. Review of existing studies for world deserts suggested that power generation by VLS-PV systems on deserts appeared to be economically promising and attractive. Preliminary analysis for socio-economic impacts of VLS-PV systems by using a methodology of Input/Output (I/O) table analysis showed that the VLS-PV systems would have a potential to promote the economic activities of the country where they would be installed. It was also found by rough life-cycle analysis for the VLS-PV system that it might have short energy payback time (EPT) and low CO_2 emission intensity. This preliminary study has expanded into a full-scale study on VLS-PV system since 1999 as a new Task VIII in IEA/PVPS.
Keywords: Energy Options – 1: Developing Countries – 2: Environmental Effects - 3

1. BACKGROUND AND OBJECTIVE

It is expected that renewable energy will be an important generation option for the 21st century as a response to global environmental problems. It is anticipated that with global population growth and economic growth the demand and supply of energy will be very tight, especially for developing countries (Fig.1)[1]. New energy sources and related technologies will have to be advanced with sufficient lead-time.

Before such long-term energy issues, global environmental problems are another major and urgent issue because increase in energy consumption causes CO_2 emissions almost directly. As pointed out at Kyoto COP-3, simple economical optimisation processes for world energy supply cannot be accepted any more to suppress global warming.

It is anticipated that solar energy will be one of the major clean energy sources in the future. A great deal of potential exists in desert areas around the world for capturing and converting the solar energy into electricity. If appropriate technologies can be found, they will assist in solving energy shortages in countries surrounding desert areas. Photovoltaic (PV) technologies are on track of becoming technically and economically viable for application at the very large-scale level.

This study was carried out under umbrella of IEA/PVPS Task VI for the purpose of examining and evaluating the potential of very large-scale photovoltaic power generation (VLS-PV) systems preliminarily.

2. CONCEPT AND DEFINITION OF VLS-PV SYSTEM

Presently, three approaches are being considered to encourage the spread and use of PV systems:

- Establishing relatively small-scale PV systems independent of each other. Currently two system sizes are used: stand-alone systems for private dwellings of several hundred watts; larger roof-mounted systems for dwellings and commercial buildings of two to ten kilo-watts and ten

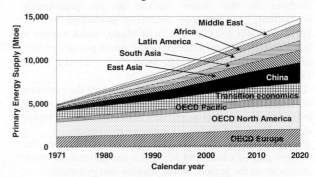

Fig.1 World Primary Energy Supply: business as usual[1]

to one hundred kilo-watts respectively. The smaller systems are mainly used in developing countries while the latter systems are used mostly in developed countries.

- Establishing 100 to 1000 kW medium-scale systems on vacant land on the outskirts of urban areas. This kind of system is in practical use at about a dozen sites around the world at present and these are expected to grow rapidly early in the next century.

- Expanding multi-megawatt systems on large areas of barren, unused land subject to high levels of sunshine. In such areas, systems greater than 10 MW in aggregate can readily be accommodated in relatively short periods. Many more systems can be envisaged when the cost is lowered beyond a threshold level as a result of mass production thus creating a beneficial cycle between cost and production. This may become one of the solutions for future energy and environmental problems and ample discussion of this is believed to be worthwhile.

The third category is called "very large-scale photovoltaic power generation (VLS-PV) system". The definition of VLS-PV system can be summarised as follows:

- Size may range from 10 MW to 1 GW or even a few giga-watts, consisting of one plant or an aggregation of multiple units located in the same region and operated on a collective basis.
- Amount of electricity generated by such plant is considered significant for the district, nation or region.
- System can be land-based (mostly arid or semi-arid regions) or water-based (lakes, coastal, open waters), although water-based system may not be considered in depth at present.
- System can be based in developing countries or developed countries, each having their special economic needs.

3. AVAILABILITY OF WORLD DESERTS

Solar energy is low-density energy by nature. To utilise the energy on a large scale, a huge land surface the earth is necessary. One third of the land surface is covered with very dry areas called "deserts", where high-level insolation and large spaces exist. It was estimated deserts had great potential for PV system, that is, the annual energy production would equal world energy consumption if a very small part of these areas, say 4%, was used for the installation of PV systems.

Rough estimation was made to examine desert potential by the assumptions of a 50% space factor for installing PV modules on the desert surface as the preliminary evaluation[2][3]. The total electricity production became $1,942 \times 10^3$ TWh ($=7 \times 10^{21}$J$=1.7 \times 10^5$Mtoe), which means a level almost 18 times as much as the world primary energy supply 9.2×10^3 Mtoe ($=107.5 \times 10^3$TWh $=3.9 \times 10^{20}$J) in 1995.

These are quite hypothetical values, ignoring the presence of loads nearby these deserts. However, at least these indicate high potential as primary resources for developing districts located in such a solar energy rich region.

Fig.2 also shows that the Gobi desert area between the western part of China and Mongolia can generate as much electricity as the present world primary energy supply.

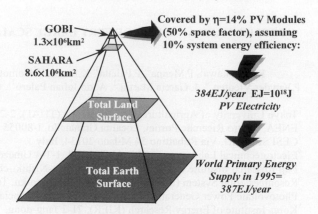

Fig.2 Solar Pyramid

4. REVIEW OF EXISTING STUDIES FOR WORLD DESERTS

4.1 Case Study on World Six Deserts

This case study was originally made by the Photovoltaic Power Generation Technology Research Association (PVTEC) Technical Committee on VLS-PV Systems which was organised in the PVTEC under the R&D contract with NEDO, Japan. It was first shown at PVSEC-9, Miyazaki, in 1996[4] and also presented at the International Workshop on VLS-PV Systems, Tokyo, in 1997[5]. The examples are a part of tentative results obtained by this work and are not the latest.

The purpose of the work was to show the huge potential of PV systems in the world and to study the feasibility of large-scale PV plants. The station would comprise 10 sub-units by 20 units of 500kW optimised size sub-units as shown in Fig.3. According to the tentative results of the work, possibilities are shown to realise electricity cost of 8-13 ¢/kWh for a 100MW plant located in any one of six desert sites in the world, Gobi/China, Thar/India, Sonora/Mexico, Great Sandy/Australia, Sahara/Mauritania and Middle East/Abu Dhabi, if system cost is assumed at

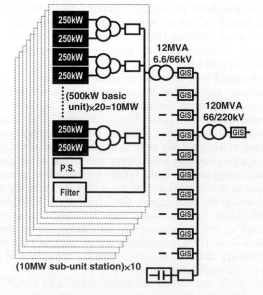

Fig.3 Schematic diagram of 100MW VLS-PV system

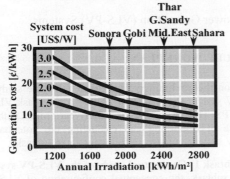

Fig.4 Indicative generation cost for different system costs and levels of annual irradiation

$1.5/W in consideration of site irradiation, local labour cost, and so on (Fig.4). In spite of the fixed flat plate, the cost can be maintained at a fairly low level.

4.2 Case Study on Sahara Desert[6]

ENEL made three case studies: i.e., I) a 1.5GW size, centralised PV power plant + 300 to 900km line, ii) 1.5GW produced by 30 PV plants of 50MW, iii) 1.5GW produced by 300 PV plants of 5MW.

The third case produced the most attractive results. Each of these plants should cover an area of approximately 10 hectare (=0.1km^2) and its power would be typically delivered through single a.c. MV lines (for example, 20kV). In this case the PV plants would be distributed within the coastal strip of North-African countries, placed less than 10km from the HV/MV substations and the distribution networks that feed the loads.

The model considered for calculating the transmission cost of PV energy is shown in Fig.5. This case considers just a 20kV line of adequate size and a new 20kV bay in the receiving substation, assuming that the output level in the PV plant is already at 20kV as it is in the Serre plant. The cost is estimated for 1, 5 and 10km lengths of the MV link, determined on the basis of ENEL's costs.

4.3 Case Study on Electricity Utilisation on Sahara Desert

The Southern part of Egypt has the highest solar energy. Annual variation of global solar radiation is shown for different zones. It reaches about 7kWh/m^2/day as an

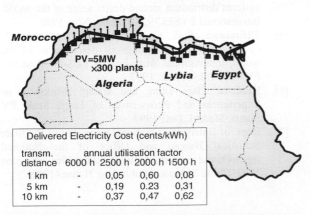

Fig.5 A case of small size PV systems along with network

average global radiation and about 8.9kWh/m^2/day average direct normal radiation at the zone proposed for implementing a pilot project. It is clear from all of these figures that the southern part of Egypt is the most suitable place to start our project.

Five sites were selected to make case studies concerning

electricity utilisation in the desert area. It was assumed in this study that irrigation wells and pumps were powered by PV systems at these sites.

5. VARIOUS ASPECTS OF VLS-PV SYSTEM

5.1 Institutional and Oganisational Aspects

Today there is a tendency with respect to emerging views on renewables and their importance in the world, and it now seems that a wider interest in renewables will be economically and environmentally benefiting them. Since this has created a large potential for large-scale PV applications, VLS-PV systems should be taken more seriously in contributing to the national and international environmental standards and agreements.

In developing countries, the major barrier for successful implementation of large-scale projects like VLS-PV systems seems to be the limited institutional capacity. Possibilities for legal embedding of VLS-PV systems should be taken into account as well. Most promising DCs are middle-income countries with a large population. Of the latter category, and in terms of technical Operation and Maintenance (O&M) -capability, India and China in the first instance offer the best possibilities.

The minimal O&M expertise and time involved in running a PV power plant, makes the BOT (Build, Own, Transfer) construction attractive for DCs. A drawback is the often weak institutional and legal framework, which then would favour applying the BOO (Build, Own, Operate) construction where ownership is maintained in the hands of the foreign investor. These options need to be further elaborated.

Financing options for VLS-PV systems includes project financing. The option of making use of CDM (Clean Development Mechanism) and/or CC (Climate Change) funds should be studied in more detail. These funds might be tapped successfully, although ensuring the eligibility of the project through cumbersome procedures might be a time-consuming factor when applying for subsidies.

Options for increasing the global market share of PV electricity include applying subvention of PV electricity and/or labeling it as 'green electricity'. An increasing number of potential consumers in OECD countries are willing to pay a higher price for this type of electricity. Moreover, allowing a certification system to be applied on buyers and/or end-users, offers scope for national governments and electricity distribution companies in meeting their targets with respect to the share of renewables in total energy production.

5.2 Socio-economic Aspects and Technology Transfer

In this study, preliminary analysis for socio-economic impacts induced by the installation of VLS-PV system was carried out by using a basic methodology of Input/Output (I/O) table analysis. Production of the system components, or PV modules, inverters, array structure and foundation as well as construction of the system were assigned to appropriate industrial sectors of the recent I/O table of China[7], and domestic production, employees and compensation for the employees that would be induced by the installation of 1GW VLS-PV system on the Gobi Desert were calculated. As a result, the induced production was estimated at two times as much as initial installation cost, and compensation for employees equal to 21% of the initial installation cost was also induced.

Although the case study used the I/O table of China in which industries were divided into only 33 sectors, the study suggested that VLS-PV system would have a potential to promote the economic activities of the country where it is installed.

5.3 Environmental Aspects

VLS-PV system is expected to act as a deterrent to further global environmental devastation such as climate warming, ozone depletion, acid rain, desertification and land degradation, and threats to biological diversity. Since it is indeed a very large-scale and long-range project new to us, the environmental issues throughout the life-cycle of the VLS-PV system must be discussed.

A methodology of life-cycle assessment (LCA) is becoming a useful approach to evaluate environmental impacts of various technologies. Since PV technology is one such energy technology, it is very important to discuss energy-effectiveness of the VLS-PV system as well as environmental aspects. Therefore, two kinds of indicators, i.e., "Energy Payback Time (EPT)" and "Life-cycle CO_2 Emission Intensity" should be considered.

In this study, a preliminary evaluation of the life-cycle of VLS-PV system on Gobi Desert was carried out to estimate EPT and life-cycle CO_2 emission intensity. Concerning the production stage of system components, primary energy requirements for PV modules, Inverters (with transformer), array structure, foundation and transmission line were taken into consideration. It was assumed that PV modules, inverters and transmission cables would be produced in Japan and transported to some installation site on the Gobi Desert. It was assumed that the remaining components were produced in China. Primary energy requirements for transportation were also considered.

As a result, EPT was calculated at around a year, which was much shorter than the expected lifetime of the PV system, and the life-cycle CO_2 emission intensity was estimated at slightly more than 10g-C/kWh, which was about 1/20 of that from the network in China. While the

Fig.5 Image of VLS-PV system on desert area

estimation was preliminary, it implied that the VLS-PV system has the potential to be energy-effective and reduce CO_2 emissions.

6. FUTURE WORKPLAN

Based on this preliminary analysis, a full-scale study on VLS-PV system has started in 1999 as a new Task VIII in IEA/PVPS. This task, which is a four-year project with the objective of examining and evaluating the potential of VLS-PV system, consists of following three subtasks:

- **Subtask 1: Conceptual study of the VLS-PV system**
In this subtask, the conceptual configuration of VLS-PV system will be developed by extracting the dominant parameters of the conditions in which the systems are technically and economically feasible while considering it from life-cycle viewpoint. The criteria for selecting regions suitable for case studies of the installation of VLS-PV will be identified and then the regions for case studies will be nominated.

- **Subtask 2: Case studies for selected regions for installation of VLS-PV**
Employing the concepts of VLS-PV and the criteria and other results produced under the Subtask 1, case studies on VLS-PV systems for the selected regions will be carried out and the effects, benefits and environmental impact of the VLS-PV systems will be evaluated.

- **Subtask 3: Comprehensive evaluation of the feasibility of VLS-PV**
The results of the case studies performed under Subtask 2 will be evaluated to summarize similarities and differences in the impact of VLS-PV system installation in different areas, and mid- and long-term scenario options that will enable the feasibility of VLS-PV will be proposed.

REFERENCES

[1] IEA, World Energy Outlook, 1998 Ed..
[2] K.Kurokawa: Conceptual considerations on advanced regional energy-supplying activities PV-AREA, JSES Conf., Tsukuba, No.23, Dec. 1993 (in Japanese).
[3] K.Kurokawa: Areal evolution of PV systems Solar Energy Materials & Solar Cells, **47** (1997) 27-36.
[4] K.Kurokawa et al., Case studies of large-scale PV systems distributed around desert areas of the world, International PVSEC-9, Miyazaki, Nov. 1996
[5] T.Hirasawa et al., Case studies of large-scale PV systems distributed throughout desert areas of the world, International Workshop on VLS-PV Systems, PVTEC, Tokyo, Mar. 1997
[6] F.Paletta, Plant siting and sizing, Workshop on Experiences and Perspectives of Large Scale PV Plants, Madrid, Dec. 1997
[7] Dept. of Balances of National Economy of the State Statistical Bureau and Office of the National Input-Output Survey, Input-Output tables of China 1992, China Statistical Publishing House (1996)

INTENSIVE INTRODUCTION OF RESIDENTIAL PV SYSTEMS AND THEIR MONITORING BY CITIZEN-ORIENTED EFFORTS IN JAPAN

Kosuke Kurokawa, Daisuke Uchida, Akihiko Yamaguchi,
Tokyo University of Agriculture and Technology
Naka-cho, Koganei, Tokyo, 184-8588 Japan/ phone: +81-423-88-7132/fax: +81-423-85-6729
and Ken Tuduku
REPP - Renewable Energy Promoting People's Forum
Higashi-Ueno, Taito-ku, Tokyo, 110-0015 Japan/ phone: +81-3-3834-2406

ABSTRACT

The authors have made sure that the adequacies of the evaluation of PV systems, with developed sophisticated verification (SV) method. The method provided analytical performance factors such as performance ratio, power conditioner efficiency, temperature factor, shading factor, incidence-angle-dependence factor, load matching factor and other array parameter still undefined factor. Using this method, the authors evaluated residential PV systems. However in their systems measuring point is only system-output without a part of sites, therefore we must substitute not measured data for the estimated data from AMeDAS. In this case, we suggest this estimated method and report the result that was evaluated for using that data. As a result, SV method is applied to this data taken at 39 systems for 5 months in Japanese REPP project. The mean system performance ratio K was identified 69.5% for 39 systems in FY 1999. Furthermore in the 29 systems that the estimated data was used system performance ratio identified 70.4%.

1. Introduction

Since April 1997, a Japanese NGO, REPP (Renewable Energy Promoting People's Forum) has aimed to improve social conscious of ordinary people for better environmental method of generation power, introducing PV systems into residences. On the other side, since there is no measuring instrument, it is difficult to evaluate residential PV systems without a part of systems. In this paper, the authors suggest the method of estimating irradiance data and temperature of PV module and clarifying operating characteristic of PV system. Therefore performance analysis of residential system was clarified by system performance ratio and system losses.

2. Measuring Point and items

There are 2 types of Simple type and Precise type in residential PV systems of REPP project. Each measuring items are shown in Table 2-1, and the distribution of their PV systems are shown in Fig. 2-1. Though Precise type have measured 4 points, Simple type have do only 1 point which is PV system output. Therefore last 3 items must be estimated from the AMeDAS data to use SV method, because it has been measured at 800 points.

Table 2-1 Measuring items of each type

	Precise type	Simple type
Measuring items	-In-plane irradiance -Temperature of PV module -Array output -PV system output	PV system Output
Number of site	10	85
Sampling time	10 minutes	30 minutes

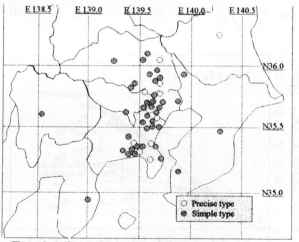

Fig. 2-1 distribution of PV systems in *Kanto* area, Japan

3. Evaluate method of residential PV systems
3.1 SV method

The SV (sophisticated verification) method has been developed by the authors [1][2]. The actual operational PV systems data divided into the loss factors by SV method: performance ratio K, power conditioner efficiency K_C, temperature factor K_{PT}, shading factor K_{HS}, load matching factor K_{PM}, incident angle dependence factor K_{PI} and other array parameter K_{PO}. SV method can evaluate from only 4 measuring point (irradiance, array output, system output and temperature) with externally available information.

Each loss factor is estimated as follows:
Temperature factor is able to estimate the formula (3-1).

$$K_{PT} = 1 + \alpha_{P\max}(T_{CR} - 25) \qquad (3\text{-}1)$$

The relation among incidence angle dependence factor K_{PI}, other array parameter K_{PO} and load mismatching factor K_{PM} are shown in Fig.3-1 and Fig.3-2. A scattered graph as shown in Fig.3-1 is the relation between array output and in-plane irradiance. An upper straight line corresponds to ideal energy production by array with its capacity P_{AS} under irradiance H_A. Scatted dots are all the hourly data divided by temperature correction factor K_{PT}. A lower straight line is drawn as the upper envelope of scattered points and no mismatch line is assumed along this line ($K_{PM}=1$).

The difference between upper line and no mismatch line is the incident angle dependent and other array losses and it means K_O. The difference between scattered dots and no mismatch line is shading and load mismatch losses.

$$K_O = \frac{E_{NM}}{E_S} \qquad (3\text{-}2)$$

The relation incident angle dependence and K_O is shown in Fig.3-2. In this figure method estimated incident angle dependence loss factor is shown. The line along maximum points of K_O is incident angle dependence line and the difference between scattered dots K_O is other array losses K_{PO}.

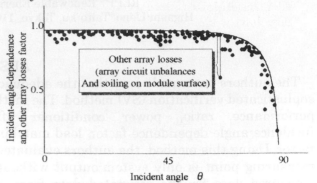

Fig.3-2 Identification of incident angle dependence factor K_{PI}

Therefore the influence of shadow can be estimated from Fig.3-3.

The processes identify the principle of shading effect detection as follow:

At first, irradiance pattern on a specific solar day representing a given month is calculated for each hour by a theoretical model considering array orientation and inclination angle, hourly monitored data for a certain site are plotted keeping hourly relation. Looking at a maximum value for each hour as a fine-day pattern for the month, the scale of given theoretical day pattern is adjusted to fit them as an envelope.

Second, supposing that the influence of a shadow doesn't change during the same month, it observed on the extracted maximum values can be as dip compared with the fit fine-day curve.

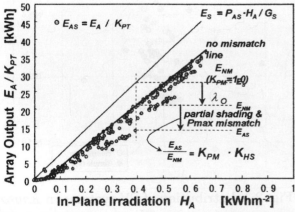

Fig.3-1 Identification of other array factor K_{PO} and load mismatching factor K_{PM}

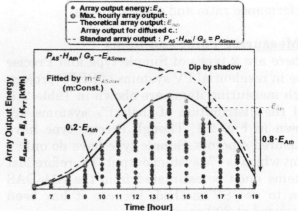

Fig.3-3 Fitting of clear-day power pattern and separation of shading

3.2 Method of estimate for insufficient data

The method of evaluation is different between Precise type and Simple type PV systems because the latter has not enough the measuring point for using SV method. Therefore in evaluation of Simple type the insufficiency data must be estimated from adopting AMeDAS data. Each PV system type's flowchart is shown in Fig.3-4 and Fig.3-5.

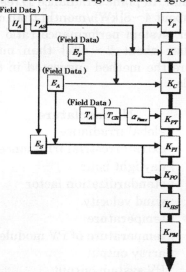

Fig.3-4 Flowchart of Precise type PV system

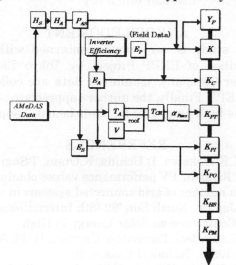

Fig.3-5 Flowchart of Simple type PV system

Irradiance data is estimated from the daylight hour of AMeDAS data.

$$H_S = CFh\{H_{S(n=0)} \cdot (1-n) + H_{S(n=1)} \cdot n\} \quad (3\text{-}4)$$

$$H_{S(n=0)} = a \cdot H_O + b \text{ (a,b:constant)} \quad (3\text{-}5)$$

$$H_{S(n=1)} = c \cdot H_O + d \text{ (c,d:constant)} \quad (3\text{-}6)$$

The temperature of PV module is also estimated from wind velocity of AMeDAS data. This method is estimated as follow:

$$T_{CR} = T_A + \Delta T \quad (3\text{-}7)$$

$$\Delta T = (-6.036 + 0.274 \cdot V + 0.071 \cdot V^2)$$
$$+ H_A \cdot (45.63 - 5.91 \cdot V + 0.333 \cdot V^2) \quad (3\text{-}8)$$

Besides array output can be estimated with inverter efficiency of each its type.

$$E_A = \frac{E_P}{K_C} \quad (3\text{-}9)$$

Their method of estimation can be evaluated Simple type's PV system.

4. Evaluation result of PV system of REPP project in Japan

Under the introduction of PV system in supporting Tokyo Electric Power Company, REPP has installed 95 PV systems in Japan since 1997. To demonstrate the applicability of SV method to actually monitored data, 10 systems of Precise type and 39 systems of Simple type are chosen as a part of the REPP Project in FY 1999. And the method of evaluate for applying to Simple type's PV systems was also applied in evaluate of Precise type's. This result was shown in Fig3-6. The appropriate method of the estimation for irradiance could be proved from this result. In System performance ratio and in-plane irradiance, estimate data is bigger than on-site data because there is possibility that pyranometer was soiled and it could not measured accurately. The result of evaluate their PV systems was shown in Fig.3-7. The average of system performance ratio K was 69.5%. The average of inverter losses λ_C was 5.5%, load mismatch λ_{PM} was 8.3%, efficiency decrease by temperature λ_{PT} was 3.8%, shading losses λ_{HS} was 7.5%, incident angle dependence λ_{PI} was 2.8% and other losses λ_{PO} was 2.6%. The losses related with inverter ($\lambda_C + \lambda_{PM}$) occupied half the whole of PV system losses. Therefore the main parameters are demonstrated in Fig.3-8, Fig.3-9 and Fig.3-10.

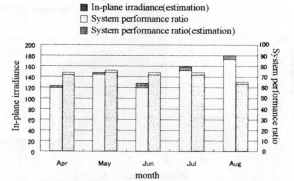

Fig.3-6 Appropriate method of estimate irradiance

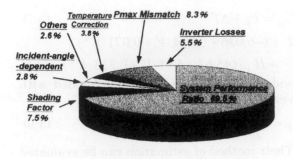

Fig.3-7 Average loss parameters in REPP
Project FY1999 Data

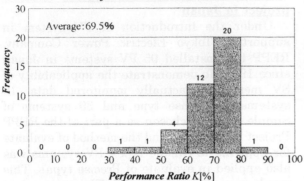

Fig.3-8 performance ratio of PV systems in
REPP Project

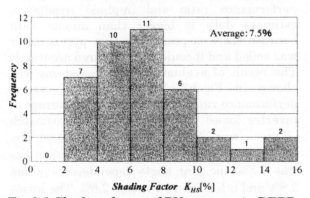

Fig.3-9 Shading factor of PV systems in REPP

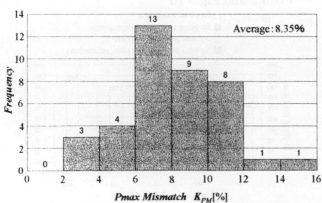

Fig.3-10 Pmax mismatch factor of PV systems
in REPP Project

5. Conclusion

The author developed SV (Sophisticated Verification) method and evaluated actual operating PV systems data in the REPP Project by SV method. In this study, the algorithm of estimating the method of the residential PV systems and of the PV system that monitoring data is insufficient was improved and certified. As a result, in-plane irradiance H_A estimated is bigger about 4~8[kWh/month] than on-site one. However, system performance ratio estimated is smaller about 2~4[%] than on-site one. Therefore the method suggested in this paper available.

Nomenclature

H_S : global irradiance
H_O : extraterrestrial irradiance
n : daylight hour
CF_h : standardization factor
V : wind velocity
T_A : temperature
T_{CR} : temperature of PV module
E_A : array output
E_P : PV system output
K_C : inverter efficiency

ANKOWLEDGMENT

This study is very much concerned with the activities of REPP Project by Tokyo Electric Power Company, monitored data are collected by REPP. Finally, the authors appreciate people in the both programs for their heartful support.

REFERENCES

[1] K.Kurokawa, D.Uchida, K.Otani, T.Sugiura: "Realistic PV performance values obtained by a number of grid-connected systems in Japan", North Sun '99 (8th International Conference on Solar Energy in High Latitudes), Edmonton, Canada, 11-14, Aug, (1999), Technical-Session 9.

[2] D.Uchida, K.Otani, K.Kurokawa : Evaluation of Effective Shading Factor by Fitting a Clear-Day Pattern Obtained from Hourly Maximum Irradiance Data, PVSEC-11, 11th International Photovoltaic Science and Engineering Conference, Sapporo, Japan, Sep.20-24, (1999), 22-C-1-5.

PV IN SCHOOLS AND PUBLIC BUILDINGS
A PILOT ACTION

I. Weiss, P. Helm
WIP
Sylvensteinstrasse 2
D-81369 München, Germany
Tel. +49-89-7201235
Fax +49-89-7201291

M. Sala, A. Trombadore
Univ. of Florence
Dept. of Architectural
Technology
via Pippo Spano, 15
Tel.+39-055-504 8394
Fax.+39-055-504939448

S. Gärtner, A. Grassi
EnergiaTA-Florence
Piazza Savonarola 10
50132 Florence, Italy
Tel. +39-055-5002174
Fax +39-055-573425

L. Bertarelli
IT Power Ltd.
The Warren
Bramshill Road, Eversley,
Hampshire, RG27 0PR
United Kingdom
Tel. +44-118-9730073
Fax +44-118-9730820

ABSTRACT: For the long-term promotion of PV, and its successful implementation, presentation and performance at schools and public buildings is considered to be essential. There have been some very promising initiatives started, on-going or completed in several European countries or regions, e.g. the "Sonne in der Schule" PV programme in Germany and the Scolar programme in the UK. These initiatives follow different approaches, have different initiators, different duration and were often limited to specific regions. The overall aim of this project is to elaborate and disseminate guidelines for a "PV in School and Public Buildings", which can be implemented in EU member states as a part of a wider consumer information campaign, in line with the proposal in the EC's White Paper on Renewable Energies. The results of the PV programmes investigated in Germany, UK, Switzerland and Italy and the activities performed which are 1) Survey of existing and completed programmes which aimed to promote and implement PV in schools and public buildings; 2) Identification of advantages and disadvantages of the different programmes, main issues and successful measures, and to define appropriate solution approaches for problems encountered; 3) Identification of financing schemes appropriate for the participating schools; 4) Development of guidelines which allow the implementation of a successful promotion and implementation campaign "PV at School" in EU member states by interested organisations and 5) A wide dissemination of the guidelines for multiple "PV in School" campaigns in EU member states This pilot action investigates "PV in schools and public buildings" programmes through-out the European Union as a means of compiling an overview of best practices. The project has been co-financed by the European Commission, DG TREN, in the framework of the ALTENER Programme.

Keywords: Grid-connected PV Systems - 1: Financing - 2: Dissemination - 3

1. OBJECTIVES AND INTRODUCTION

The overall aim of this project was to elaborate and disseminate guidelines for a "PV in School and Public Buildings", which can be implemented in EU member states as a part of a wider consumer information campaign, in line with the proposal in the EC's White Paper on Renewable Energies.

Within this project following activities were performed to achieve the result:
- to undertake a survey of existing and completed programmes which aimed to promote and implement PV in schools and public buildings;
- to identify the advantages and disadvantages of the different programmes, main issues and successful measures, and to define appropriate solution approaches for problems encountered
- to identify financing schemes appropriate for the participating schools
- to establish guidelines which allow the implementation of a successful promotion and implementation campaign "PV in School" in EU member states by interested organisations
- to achieve wide dissemination of the guidelines for multiple "PV in School" campaigns in EU member states.

This action investigated "PV in schools and public buildings" programmes through-out the European Union as a means of compiling an overview of best practice. Although individual project partners focused on dedicated target countries the data collection and analysis followed commonly defined and agreed criteria to facilitate trade-off analyses.

A wide dissemination of the elaborated guidelines focused on key decision makers in organisations best suited for the programme implementation (i.e. local and school authorities and utilities).

2. PROJECT RESULTS

2.1 Survey on existing/completed programmes

An investigation of existing and completed school programmes was performed in order to identify the specific advantages and disadvantages, the lessons learnt and the contentment of the users (i.e. schools, public buildings) of the various programmes existing and completed in Switzerland, Germany and United Kingdom. A brief description of the programmes is given below.

School programme in Switzerland

School roofs are a key focus for PV projects in Switzerland. The Ministry for Energy Conservation is providing financial support for this type of project so that schoolchildren will become familiar with PV electricity generation from an early age.

Between 1993 and 1996 a total of 155 primary, secondary, and high schools installed a combined PV capacity of 803 kW, thanks to a 30-percent subsidy provided by the Swiss government.

The Swiss have also installed grid-connected PV in vocational colleges to educate students studying in the electrical field about renewable energy. Here students install the systems themselves. In the past six years, systems have been put into service at 24 vocational colleges with a total installed power of 191 kWp.

PV in School programmes in Germany

The 'Sonne in der Schule' programme initiated in Germany, generally aims to demonstrate the innovative PV technology to young people, to motivate the pupils for the rational use of energy and to reduce the CO_2 emission. The programme has started in 1994 with different initiators such as the Federal or State government and utilities, supporting the schools nation-wide or on the regional level. Most PV systems are grid-connected applications but some utilities also offer stand-alone applications. The programmes in Germany last between one and ten years. The length of this period is determined by the period required for monitoring.

In total about 1400 PV systems have been installed in schools by the end of 1999 which amounts to a capacity of approx. 1.5 MWp. The PV systems support to meet the target of reducing CO_2 emissions, the rational use of energy and to demonstrate PV technology. Several of the programmes are still on-going and last of about 10 years including the monitoring phase.

Scolar Programme in United Kingdom

The Scolar programme launched in 1994 and supported by the Department of Trade and Industry, will install 100 interactive PV systems in schools throughout the UK. Another programme recently initiated in the UK is the Solar Network.

The Scolar Programme is part of the Government's Foresight Initiative. The Foresight panel allocated £1 million to the programme. The consortium members are matching these funds. Schools who wish to be part of the scheme will have to make a financial contribution of about £3,500 for the 1-kilowatt system.

Close to 20 British primary and secondary schools and technical colleges have already gone solar, and by the end of 2000 a total of 100 will be drawing electrical power from photovoltaic panels. Beside the hardware, the schools gain access to the Internet and real-time experience of PV power through a specially tailored teaching program.

In total more than 500 academic institutions have expressed interest in the programme, called Foresight Scolar. The Scolar schools will act as demonstration sites for PV technology to the next generation.

The programme is a collaboration between the government, industry, and universities, and is managed by the Intersolar Group, described as the oldest photovoltaic company in Europe. The government and industry are paying the majority of the cost, while participating schools contribute about one-third of the cost of each PV system — between 3500 – 4500 £ (depending on which solar electric designs is chosen).

2.2 Trade-off Analysis of the Various Programmes and Lessons Learnt

As a result of the investigation undertaken described before was resulting in the identification of the advantages and disadvantages of the different programmes, to identify major problems before, during and after the implementation of the programme(s), to identify successful measures and to define the appropriate solution approaches for problems encountered before, during and after the installation of the PV systems at the individual schools and public buildings.

Advantages and disadvantages of the various programmes are summarized in the following:

1) Pupils will benefit from:
- First hand experience of PV technology and a chance to learn about major technologies of the future.
- Increased environmental awareness through curricular and non-curricular education and activities.
- Links with other schools in their own country and across Europe. These links not only give a chance to share information between schools but can also be used as a means for cultural exchange and improving language skills, and of course fostering new relationships.
- Studying the benefits and performance of PV, often as a formal addition to science classes or after-school programmes.

2) Benefits to the School itself:
- Increases pride within the school.
- Enhances educational aims of the school.
- School becomes a centre for environmental education.
- Increases educational opportunities. Particularly in disciplines such as Mathematics, Physics, Environmental Science, Engineering, Geography and Languages.
- Increases public relations and profile within the community.
- Increases prestige compared to other schools.

3) Benefits to the Wider Community:
- Enhances integration of the community. Local authorities, organisations and businesses may be willing to co-operate in the initiative.
- Increases environmental consciousness among the wider community.

4) The initiators and Sponsors of the programme will benefit from:
- Increased public relations locally, regionally, nationally and internationally (for example through branding of educational and dissemination material).
- Increased awareness in the community about the goods/services of the sponsor.
- Improved environmental credentials.

- Contribution to the reduction of carbon dioxide emissions and the promotion and development of a clean energy technology.

5) National programmes
- Achievement of national or regional objectives with respect to the contribution of renewable energy.
- Dissemination of information on renewable energy technologies and in particular on PV to the larger community.

6) The Environment
- Reduction in carbon dioxide emissions.

It is essential that the potential beneficiaries of the programme are made clearly aware of the benefits they could accrue from joining a programme.
- To help with the development of new "PV in Schools" programmes across Europe, the main problem areas from completed and existing programmes have been compiled below.
-

Problem Indicators

Participation
- The project group received little support by the school director.
- The teacher's work was made more difficult by the fact that colleagues were not willing to discuss themes like "PV" or "Energy" in their lessons.
- Schools participating in the same programme and located near to each other were not aware that either was engaged in the same programme.

Planning
- The PV plant was installed and only later was the possibility for integrating PV within the curriculum assessed.
- Teachers complained that they were not acknowledged enough for their efforts.
- Public relations and marketing of the programme were neglected. The regional administrations of national programmes disseminated information on the programme with varying commitment, usually depending on the region's own commitment to the environment.

Management
- In most cases the success of the programme depended on the goodwill and co-operation between the school and the programme initiator. The support provided by the initiator was seen as crucial.
- Large consortiums involved in the day-to-day running of national programmes tended to work against the overall efficiency of the programme. Decision making and logistical arrangements between the consortium took long periods of time.
- The success of the programme was also dependent on the presence of a person within the school that championed the programme. In the case that this person was a teacher, PV as a topic had a good chance to be widely integrated in the curriculum. However, if the teacher left during the running of the programme the success of the project became at stake. If this enthusiastic person was a school-external the integration of PV in the curriculum became less probable.
- Participating schools that contributed to the total

system and installation costs were dissatisfied with long lag periods between application to the programme and the final installation of the system. This led to the schools' loss of enthusiasm, interest and support to the programme.

Education
- An educational concept was not included in the programme and this was felt to be a great disadvantage.
- The person responsible for the programme within the school had the task of finding out about PV, developing the educational material and integrating it within the curriculum. No support was provided by the initiator of the programme or the government, this was felt as a disadvantage of the programme and perceived as a huge burden to enthusiastic teachers.
- General information on renewable energy was provided, however, teachers would have considerably preferred material specific to PV and preferably also tailored to the different school age groups.

Installation
- The limited number of qualified installers could not meet the high demand for the PV systems fast enough. This led to an increased lagging period for the schools.
- The roof was not as robust as initially perceived, the carrying structure therefore had to be reinforced.

Monitoring
- The monitoring software was perceived as being too time-consuming to learn to use, teachers soon lost both interest and their motivation to work with the system and its components.
- Schools were not informed on how the monitoring data that they had to manually collect and submit to the initiator of the programme was then used. No analysis, graphs, nor the chance to compare their own system with any other was provided.

Other
- In some cases schools did not have any reason to install the PV system other than that the finance was made available to them. In other cases the only reason to install the PV system was that another school in the same city already had one, the "me-too" syndrome. In consequence, they had no concept about what to do with the system and no motivation to work with the system and integrate it into the curriculum.
- In some cases the failure of the inverter has caused system break-down. Pupils soon became disappointed about the PV system and lost interest.

Identification of Financing Schemes

The price of PV is still a barrier to its further use. In financing PV systems traditional funding channels and mechanisms may not be sufficient. Additionally, in recent years the budgets of schools have been severely cut, thus making it even more difficult for schools to participate in such programmes. External funds and subsidies are therefore required for schools to feel financially able to participate.

Several funding mixes should be envisioned, these

can include the following:

- Local government authorities;
- National State Funds available for renewable energy, environmental and educational projects;
- European Commission funding programmes available for renewable energy, environment and education (these will in most cases require at least two EU countries to be involved in the project);
- 'Green Electricity' funds operated by utilities;
- Utilities;
- Electricity generators;
- Energy industries (for example manufacturers and suppliers of goods and services);
- Private investment by businesses or private funds, for example large retail supermarkets;
- Sponsors interested in supporting the programme within a particular school, region or at the national level;
- 'Green Loans';
- Local Educational Authorities;
- Fundraising events organised by the school;
- Schools' own funds;
- Donations, both direct funding and consumables like PCs which could be supplied free of charge.

In Germany most of the initiators of the "PV in schools" programmes have been utilities. In these cases utilities have had an important role in providing funds and subsidising the systems. Utilities have often contributed 50% to 70% of the total system and installation costs and the rest of the funds have been provided by a wide range of funds including the state, the municipality, private sponsors and the school itself.

2.3   Guidelines for Action – PV in Schools

The present degree of uptake of photovoltaic (PV) systems in schools across Europe varies widely, and initiatives differ from one region to the next. It is therefore considered that guidelines are needed to facilitate the implementation of PV systems in schools across all of Europe.

The brochure provides a step-by-step guide to help and encourage initiators, be they government officials, utilities, administrators, schools, teachers or even pupils, to join an already existing programme or implement a new and successful "PV in School" campaign. These guidelines are the result of a one-year in-depth study, conducted by a European team of experts, on existing and completed "PV in School" programmes in Germany, Switzerland and the UK. The lessons learned from the studied campaigns have been instrumental for the development of this brochure.

A guideline for action has been prepared which includes **22 key elements** which any programme can adopt as a broad based methodology. The guidelines for action aim to

- provide an overview of the issues that need to be addressed in the development of a new 'PV in Schools' programme;
- provide an understanding of how to arrange a new programme logistically;
- ensure that all indicators of success and problems are fully addressed in a new programme;
- provide understanding of all key players that are needed in a new programme;
- highlight the importance of cross linkages with other on-going programmes at national and European level.

The lessons learnt and the main problem areas from completed and existing 'PV in School' programmes have been compiled to provide support with initiatives for new PV in School programmes.

The brochure is available as hardcopy and can be requested by any of the project partners. In addition, the guideline can be downloaded on the internet homepage: www.pvinschools.de

3.   CONCLUSIONS

The project result and the dissemination activities undertaken within this project will support to implement similar programmes for PV in schools not only European-wide also world-wide.

Pupils and students will be decision makers of tomorrow and it is important to prepare and provide them with the rapidly emerging PV technology. The installation of PV systems are sending a clear message to pupils that this technology is better, more environmental friendly and reliable.

INTRODUCTION OF PHOTOVOLTAICS IN POLAND

S. M. Pietruszko

Warsaw University of Technology, Institute of Microelectronics and Optoelectronics,
IMiO PW, ul. Koszykowa 75, 00-662 Warsaw, Poland
Tel.: +48-22-6607782, Fax: +48-22-628 8740, e-mail: pietruszko@imio.pw.edu.pl

ABSTRACT: This paper describes the barriers that exist or can be encountered in implementing photovoltaic technology in Poland. An obstacle to the development of photovoltaics is a lack of legal regulations, direct governmental subsidies, and low-interest loans. There are no incentives offered for manufacturing and the purchase of photovoltaic equipment by private investors. Another obstacle is the public's lack of knowledge about the potential of renewable technologies. This paper discusses future prospects and possibilities for developing photovoltaics in Poland. There are positive signs of PV development, e.g., the need and desire for new energy solutions are stated officially. A regulation on purchasing electric energy produced by renewable energy technologies is now in force. The demand for photovoltaics can also be stimulated by recent changes in the electricity tariff system in Poland. This paper also suggests ways to promote, disseminate, and deploy PV technology in Poland.

Keywords: strategy – 1: national programme – 2: implementation – 3

1. INTRODUCTION

Poland is currently one of the most dynamic economies in the world. Economic growth, as measured by the gross domestic product (GDP), has been at 5 %- 7 % in recent years. The economic growth in the last years was achieved by reducing the energy intensity of nation. Improved energy efficiency is the result of common-sense efforts motivated by higher energy prices and of ownership transformations in the economy. Poland, at present, is relatively self sufficient, with an indigenous energy source (mainly hard and brown coal) to cover its demand for power. The primary energy production has not changed in Poland for many years. In 1997,domestically produced coal, oil and gas made up 71.3%, 14.7% and 9.2%, respectively, of energy production. The reminder 4.8% include fuel wood, peat, waste fuels, hydro energy, and other renewable energy sources. The percentage of renewables such as wind, biogas, solar energy, and small hydro power is slowly increasing, but is still only around 2%.

Poland will have to increase the share of primary energy coming from renewable energy sources for many reasons. First, environmental protection is one of the constitutional obligations of the State. Bound by international agreements, Poland is obligated to reduce pollution. Poland signed the United Nations Framework Convention on Climate Change during the United Nations conference in Rio de Janeiro in 1992. Poland also became party to the UN Framework Convention on Climate Change in 1994. The forecast concerning the emission of greenhouse gases, particularly of carbon dioxide, indicates that in 2000 the emission of these gases should not exceed the 1998 level. Poland also signed the Kyoto Protocol, with the obligation to reduce CO_2 emission by 6% compared to the reference year 1998. Poland will join the European Union in several years and will be obliged to meet European standards on pollution and to significantly increase the implementation of renewable energy sources.

Another reason to wider use of renewable energy is to develop a more balanced mix of primary energy sources in Poland. Although coal reserves will last for the next 150

- 200 years, their share in primary energy will decrease in the future and gas and oil use will increase. However, Poland does not have significant resources of these two energy sources and must secure its future supply of energy and fuels.

2. LEGAL FRAMEWORK

The basic objectives of the Polish energy policy are laid out in "Poland's Energy Policy Guidelines by 2000" [1]. Compared to European Union member states, Poland consumes 2-3 times more energy to produce a single unit of gross national product. The higher cost of energy makes the Polish economy less competitive and contributes to the deterioration of the environment.

The new Energy Law, which entered into force in January 1998, offers the opportunity to change this situation. The Law enumerates the main targets of the new energy policy:

- secure a supply of energy and fuels,
- achieve efficiency in production, distribution, and use of energy and fuels,
- develop competitive conditions in the energy industries.

The new Energy Law initiates substantial restructuring and regulation of electricity and heat markets. It introduces a free market for energy and rate diversification by region. The Law fosters energy-saving innovative methods, supports clean technologies and recognizes renewable energy sources as one option to achieve environmental targets in Poland.

Two executive regulations related to the Energy Law are connected directly to renewable energy sources. The regulation of the Ministry of Economy deals with the obligation to buy electricity and heat from nonconventional energy sources. This executive regulation came into force in March 1999. It obligates energy enterprises and utilities to buy electricity and heat from nonconventional energy sources. However, energy enterprises are not obligated to buy energy from renewable energy sources in the following two situations:

1. enterprises dealing with trade in electrical energy, if the price of purchase energy is higher than the highest prices of energy paid by the end user,
2. enterprises dealing with trade in heat, if the price of purchase energy is higher than the highest prices of heat purchased from traditional (fossil fuels) source of energy.

In the present state, the regulation in some way favors the production and sale of electricity from renewable energy sources, but is a practical barrier to the sale of heat from them. The regulation needs to be modified, and the first part should be changed soon. In addition to the maximum prices of electricity, the minimum price level should also be defined. The second part of the regulation, which deals with heat, needs major modification, because it makes impossible to sell heat from renewable energy sources.

The second regulation of the Ministry of Economy deals with the need to obtain licenses energy for production. It concerns detailed rules for energy enterprises that are not obligated to have licenses for energy production. The rule concerns:

- electricity produced using sources with total installed capacity not higher than 50 MW;
- heat produced using sources with total installed capacity not higher than 5.8 MW.

However, the exemption from licensing does not concern renewable energy sources for which energy purchase is obligatory.

The Construction Law, which was brought into force in 1995, together with the later changes gives some indirect regulations that can be useful for developing solar systems, i.e., emphasizes that buildings should be designed, constructed, and utilised so as to ensure environmental protection and rational use of energy.

Unfortunately, there are no statements connected to the renewable energy technologies in the construction regulations. Of course, renewable technologies can be integrated in buildings if they comply with the general regulations. Although there are no official design principles, standards, codes, or guidelines on solar systems and their components, the following conclusions can be drawn concerning photovoltaic applications. Building-integrated photovoltaics and stand-alone PV systems can be used without any problems by the individual consumers (in commercial, housing, and service sector) if appliances are adapted for using electricity from PV. If a grid-connected PV system is used, an inverter would have to meet the requirements of the electric utilities concerning an acceptable level of harmonic distortion, quality of voltage, current output of wave forms, and level of electrical noise.

Another law important to implementing solar technologies is the Thermo-Modernization (Retrofitting) Law, which entered into force in January 1999. The law is accomplished by the executive regulations issued in May 1999. This law specifies rules for support of thermal modernization activities. One of the aims is a total or partial exchange of traditional fossil-fuels energy sources to nonconventional sources of energy. The law is oriented toward environmentally clean and energy-saving housing construction and technologies. The law defines the basis for supporting investments in thermal modernization and lays down the operating principles of the Fund for Thermo-Modernization of Buildings. In essence, it provides State guarantees for loans taken out for investments in energy efficiency and renewables, and it assists borrowers in paying off these loans.

There is no central authority to coordinate activities related to solar-to-energy issues, but Poland has several ministries partly responsible for renewable energy development and implementation that may be interested in applying PV technologies. Unfortunately, at present, the benefits arising from the use of renewable energy sources are not recognized as a priority in any of the ministries.

3. ECONOMIC AND SOCIAL ASPECTS OF IMPLEMENTATING PHOTOVOLTAICS

Use of renewable energy sources, including PV technology, in many countries has produced many benefits essential for sustainable development, from the points of view of the environment, economy and sociaty. For example, rural areas are a vital element in the Polish economy, with almost 40% of Poland's population living in rural areas. Although agriculture yields only about 6.3% of GDP, over 26.5% of the population works in this sector. At the same time, rural areas have a very high unemployment rate that may reach even 50%, underdeveloped technical infrastructure, and inefficient and often expensive and unreliable energy systems. Additionally, rural areas show a negative economic balance – a large part of agricultural incomes often flow to urban areas, e.g., with expenses related to energy costs.

Therefore, the creation of new jobs and the diversification of incomes seem to be among the highest priorities for sustainable rural development. It is particularly important in the rural areas, where unemployment is very high and where PV technologies and systems do not necessarily require highly qualified staff. Using local solar-energy sources to fulfil local energy needs creates employment and supports local markets, where additional incomes are supporting the local community. An indirect benefit arising from the use of PV systems may be to moderate the excessive migration of people from rural to urban areas.

Contrary to the above social *benefits* in the rural sector, PV technology can be perceived as social and political *obstacles* in other sectors. For example, the lack of knowledge of the potential of solar energy technologies, the risk of high unemployment in the coal industry, and the strength of the coal lobby will prevent the rapid change of the energy supplies structure.

The development of renewable energy sources is hindered by a lack of direct governmental subsidies and of low-interest loans. No incentives are offered for the manufacture and purchase of photovoltaic equipment by private investors. Because prices of fossil fuels are controlled by the government (especially prices of electricity), no subsidies for photovoltaic systems exist. The present low prices of energy obtained from fossil fuels squeeze potential profits of renewable energy installations. However, energy prices are increasing sharply. The government plans to allow energy prices to float, so they may reach the level of prices in Western Europe after 1999.

Note that the tax systems for installation equipment and components applying to renewables are not appropriate. Value-added tax (VAT) is charged in three bands: 22%, 7%, and 0%. For solar collectors, VAT is equal to 22% because they are classified as non-electrical ovens. PV modules are not classified at all. This situation a major obstacle to the wider use of solar systems in Poland, must be change quickly.

4.   PROSPECTS

There is a positive outlook for renewable energy in Poland. First, e.g., the need and desire for new energy solutions are stated officially (Junes 1999's resolution of the Polish Parliament on the Development of Renewable Energy Sources), high-level decision makers are becoming aware of the need to explore innovative energy systems, because the use of renewable energy sources, including PV technology, in many other countries has produced many environmental, economic, and social benefits essential for sustainable development. Furthermore, a regulation on purchasing electric energy and heat produced by renewable energy technologies is now in force (from March 1999). In the medium and longer term photovoltaics seems to be one of the most promising renewable energy sources in Poland. So, the substitution of solar energy for conventional energy sources is beginning to be considered worthwhile [2].

The demand for photovoltaics is also stimulated by recent changes in the electricity tariffs system in Poland. Many power plants and other components of the grid system, especially low-voltage transmission systems, need to be refurbished and upgraded. Most of the primary energy is produced in southern Poland, which results in substantial transmission losses (about 11% - 13%). Because Polish energy policy is determined to bring all energy prices to the market level, fossil fuels are becoming increasingly expensive. Electricity will become more expensive in rural areas with the introduction of the regionalization of energy prices. Local energy sources, including solar sources, are now becoming more attractive economically. These are the main reasons that renewable energy and particularly photovoltaics are perceived as a promising solution for rural electrification.

This situation meets a real interest and growing demand for energy self-sufficiency on the local level. In small municipalities, farms, and rural areas the power grid is often deficient and urgently needs greater capital. Proposed improvements are usually related to modern technologies that use local energy sources. Photovoltaics is considered a specific renewable energy option, that is suitable for changing the insufficient traditional energy supply system in agriculture.

Installing stand-alone PV systems in regions remote from an electricity supply grid will be the most profitable in the nearest term. Poland plans to develop a network of highways, and photovoltaic systems will be necessary for emergency phones and for lighting road lamps. Also, in telecommunications, PV systems can power communication systems such as microwave repeaters, television and radio transmitters or receivers, telephone systems, and small radios in remote areas. The recent rapid development of cellular phone systems (at present there are three providers) shows promise for developing photovoltaic systems.

Also in rural areas, the power required to drive small-scale agricultural machinery, such as small solar dryers or the ventilation of greenhouses, is lower than 500 W. A major advantage of such products to the PV industry is that the total cost of a very small PV power supply is easily affordable by consumers. The designer is able to match a PV system to the exact energy needs of the application, instead of being constrained to purchase a combustion engines generator. On the other hand, agricultural consumers not only pay for energy from PV systems; they also obtain better agricultural production and reduced labor requirements. In this area, PV systems offer reasonable potential in rural areas and particularly in agricultural production.

Poland should also follow the example of the industrially developed world, where investment in PV systems continues to grow, with most interest focused on the great potential for PV in buildings. The success of this so-called "Thousand Roofs" program in Germany has sprawned similar initiatives in other European countries (Switzerland, Austria), in the U.S. the Million Solar Roof Initiative and in Japan 70,000 Roofs Program.

In Poland, the production of solar cells and systems is possible because of a strong scientific and technical background. Polish industry is seeking new products and ideas, and offers medium-level process technology and a well-educated working force.

For effective promotion, dissemination and deployment of photovoltaics, it is necessary to establish a common strategy for different types of institutions (research, governmental, and NGOs) that considers specific conditions, social aspects, and consumer expectations in Poland. Considering all the potential social and economic gains that PV technology offers to end users, new political initiatives on the central and regional levels are needed to vitalize local economic activities and to create a long-term implementation strategy for PV systems, with necessary changes in the legal and institutional framework. Such activities will not be detrimental to the society, so they should be very attractive for all policy and decision makers. European Union experience can help to raise the credibility of modern legal, institutional, and political solutions on the central and local level in Poland.

Among different institutions involved in promoting and disseminating of photovoltaic technology in Poland, the most important are research institutions and non-governmental organizations [3]. Several Polish research institutions are active in research and development. They are involved in basic research projects concerning the design of appropriate technologies for applications, and they also develop indispensable research skills. These institutions include: Warsaw University of Technology, the Solar Lab in the Wroclaw Technical University, the Photovoltaics Laboratory in Kozy, IBMER in Warsaw, and the Foundation for Photovoltaics in Poland in Warsaw. The most important activity of these institutes, from the information and social point of view, was the establishment of a number of small demonstration projects (solar drying, ventilating of agricultural structures, supplying electricity in a shelter home and in a movable apiary house).

Difficulties with starting new development and demonstration projects in this area result from the type of donor organizations (e.g., KBN - the State Committee for Scientific Research) usually involved in funding and prioritizing the research institute activities. KBN started (March 2000) financing research which aims to develop technology for producing solar cells and modules. There are suitable technological and human resource conditions for such projects in all these institutions. They can select the most suitable projects, from the vantage of marketable applications, and can establish and monitor such systems.

NGOs are the other active group of institutions working in photovoltaics. The Polish Solar Energy Society (PTES-ISES), established in 1994, is active in information exchange and training seminars. The Society bridges the gap between research laboratories, suppliers, and potential end-users. It supports the development of local institutions for training. In this task, PTES is supported by other NGOs working more closely with local communities.

As shown above, a number of different institutions can provide specific services for solving specific problems and further implementing PV technologies. Which organizations should be involved in particular projects depends on the goal and existing local conditions. To successfully implement renewable energy technology, one needs concerted action and strong collaboration of different but complementary organizations (R&D institutions, technology suppliers, financial institutions, regional authority, local communities, educational institutions and groups of end-users).

For further commercialization of PV technologies, it is indispensable to establish a number of new demonstration projects. Potential users should be able to see and test the installations. Demonstration projects should be chosen in relation to their usefulness. Users are most interested in economic results (e.g., increased yield, quality, and market price).

The interaction between Poland and European Union countries is needed to rapidly develop photovoltaics in Poland. Some projects could be sponsored by the EU through programs such as the 5 Framework Programme, Altener, Eureka or by the U.S. Cooperation can be established in many ways: knowledge transfer; sponsoring organization or participation in topical workshops, international conferences and study visits; R&D cooperation to initiate new projects of common interest, working in mixed-nationality teams; demonstration projects; and exhibition sites.

5. CONCLUSIONS

Designing a comprehensive strategy for the use of photovoltaics technology can be one of the crucial steps to enhance the development of renewable energy technologies in Poland. The government has a major role to play in mitigating the barriers that hinder the wider and more successful adoption of renewable energy technologies, and particularly, PV technologies. If renewable energy technologies are to be accepted as viable and commercially attractive investments, the government must grant them the same benefits as those given to the established conventional energy industry. The creation of the situation in which photovoltaics and other renewables can compete fairly must be nurtured by institutional considerations and incentives.

Considering all the potential social and economic gains that PV technology offers to end-users and the real needs that exist in the Polish economy, new political initiatives on the central and regional levels are needed to vitalize local economic activities and to create a long-term strategy of PV systems implementation, with necessary changes in the legal and institutional framework. Such activities will not be detrimental to the society, so they should be very attractive for all policy and decision makers. European Union experience can help to boost the credibility of modern legal, institutional, and political solutions at the central and local level in Poland.

It must be mentioned that considerable work has already been done, compared to with the situation at the beginning of the 1990s. It is essential that the information campaign that has started is also joined by some changes in legislation. For the first time, some laws and regulations relate in a direct way to renewables. However, there is an urgent need to formulate the basis for a national Programme to develope, promote, and implement renewable energy technologies. It is therefore necessary to continue efforts to establish an adequate institutional and organizational framework to promote better use of solar energy in Poland.

ACKNOWLEDGMENTS

Part of this work was done thanks to support from the European Commission project "Building Integration of Solar Technologies", INCO-COPERNICUS-4080-98.

REFERENCES

[1] Energy Policy of Poland and the Draft Programme to the Year 2010, Report of the Ministry Industry and Trade, Warsaw, November 1992 (in Polish).
[2] S. M. Pietruszko, G. Wisniewski, D. Chwieduk, and R. Wnuk (1996), Potential of Renewable Energy in Poland, Renewable Energy, World Renewable Energy Congress, 15-21 June 1996, Denver, 1124 - 27.
[3] S. M. Pietruszko, The Status and Prospects of Photovoltaics in Poland, Renewable Energy 16 (1999) 1210-4.

Financing Solar Home Systems
Lessons from Swaziland

P.E. Lasschuit, C. A.Westra

Netherlands Energy Research Foundation ECN

P.O. Box 1755 ZG Petten, The Netherlands,

Tel. +31 224 56 4115, Fax +31 224 56 3214, e-mail: westra@ecn.nl

ABSTRACT: Many Solar Home System pilot projects have been implemented to date, but only few have resulted in a sustainable market development. 'Finance' or better the lack of it is often quoted as a major obstacle. Formal banks are reluctant to deal with rural households - the primary target group for SHS - which are generally considered as 'non-bankable'. This paper shows that in Swaziland as well as other countries in Southern Africa a latent demand exists, but actual demand for SHS is still too small for commercially sustainable market development. Sustainable finance scheme can help to reach the middle and higher income groups, but to reach a critical mass a lot of effort and time is needed. If SHS are to become within reach of the poor, a helping hand of the Government is needed.

Keywords: Solar Home System (SHS) -1: Finance -2: Southern Africa.

1. INTRODUCTION

Early 1997 the company Solar International Swaziland (SIS) was established, being a Joint Venture (JV) between an existing local company Swazitronix and Dutch parties. Since 1991 Swazitronix had tried to build a sustainable solar business, but failed to do so. Forced to diversify the business and on the brink of 'giving up' on solar, the JV - guaranteeing Dutch support - came at the right time and kept the solar spirit alive. With the necessary technical requirements already in place, the support concentrated on the establishment of a SHS credit scheme and creation of awareness. In the first year of its existence SIS managed to sell and install 170 SHS on credit. Swaztrionix used to sell an average of 20 SHS per year (between 1992-1996) on a cash basis. At first sight, SIS seems to have realised significant increase, but the numbers are still far below the projected 500 SHS. Sales remained rather constant in the following year and dropped to about 100 last year due to the fact that Triodos Bank stopped further loan disbursements to SIS and consequently SIS had to stop its onward lending to its customers.

In a country with a relatively high GDP per capita, and less than 5% of the rural households connected to the grid, one would expect a higher uptake of SHS. The analysis below tries to reveal some of the factors that may contribute to the discrepancy between realisation and projections.

2. SIS'S CREDIT FACILITY AND PERFORMANCE

SIS's credit facility

To make SHS affordable to more than just the high-income households, a credit facility was set up with the assistance of the Dutch Triodosbank. Initially local banks were approached to 'adopt' the SHS credit facility, but only with a 100% guarantee and unacceptably high service fees they were willing to do so. Due to this 'risk averse behaviour' of local banks, the credit was offered directly to SIS, who was prepared to carry the full responsibility for repayment of the loan.

A lump sum loan of US$ 150,000 was given at an interest rate of 14% and repayment period of 3.5 years. SIS in turn provided credit to its customers at 22% annual interest, which was slightly under prevailing 'formal' lending rates. The difference in interest rates (14% versus 22%) compensated for the low retail margin on SHS and was necessary to pay for SIS's overhead cost. The maximum repayment period was 3 years and a deposit of 25% of the purchase price was required. To keep administration and logistics simple, the credit facility started with one standard product costing US$ 675 and consisting of:

- 45 Wp poly crystalline panel
- 6 Amp regulator
- 96 Ah deep cycle battery plus storage box
- 4 energy efficient 9 W lights
- cabling, mounting
- transport and installation.

The design was based on prevailing rural energy demand for lighting, radio and television, generally met by candles, dry cell and car batteries [1].

Performance

To date about 400 SHS (45Wp or more) have been sold and installed, the majority on credit. Although most customers initially prefer to repay their loan in 3 year, in practise people tend to pay faster. The payments of the monthly instalments have been rather satisfactory. Sometimes, household budgets are tight due e.g. funeral expenses or school fees. In such cases SIS applied a flexible approach and allowed people to catch up with payments in the following months. So far about 40% of the customers have completely paid off their system. Among those that have entered their 3rd year of repayment, the discipline to pay their monthly instalments tend to have weakened, suggesting that a 2-year repayment scheme is likely to generate the best results. To date 5 SHS have been repossessed and 5 more are likely to be repossessed soon. With a default rate of 3% the credit scheme can be considered successful.

3.   REGIONAL EXPERIENCE WITH SHS CREDIT SCHEMES

Looking at similar SHS credit schemes in the region, more or less the same results can be noted.

Table 1: SHS Credit Schemes in Southern Africa

	Swaziland	Namibia	Botswana
grid electrified rural househ.	5%	10%	8%
project start	1997	1997	1997
price SHS*	US$ 700	US$1095	US$ 800
deposit	25%	20%	15%
interest	22%	5%	14%
repay. period (months)	60	36	36
no. installed SHS end '99	350	280	350
seed finance	US$ 150,000	US$ 650,000	US$ 900,0000
source of finance	Triodos bank	Nam.Gov/ USAID/ NORAD	Botswana Government
repayment seed money	repayment at 14%	repayment flows back in revolving fund	repayment flows back in revolving fund

*) In all 3 cases a SHS refers to a 4 light, 45-50 Wp SHS.
Source: [2], [3], [4].

Based on invested money and number of installations and considering the fact that the project was run without subsidies the project in Swaziland has done well. But like many other countries, no critical mass has been reached yet.

At present there are about 100,000 rural households in Swaziland that are not connected to the grid and have little or no prospect of being connected soon. So, why don't all these households line up for a SHS?

4. THE PRICE/QUALITY ISSUE

The low sales in the above examples seem to indicate a discrepancy between supply and demand

The SHS offered under the various schemes, are high quality systems with corresponding high price tags. Looking at the few cases in Africa where SHS have become a commercial success, notably Kenya, this was mainly attributable to small PV panels (around 10-15 Wp). The Kenyan solar market has been active for more than a decade during which 150,000 units have been sold commercially. Many of the systems are technically imperfect. They are often procured incrementally starting with a television powered by a car battery, charged in a nearby town, followed by a small solar panel and sometimes followed by another panel. Due to a mixed quality of components, incorrect installation and maintenance, about a third of the existing systems in Kenya are not fully operational [5]. A similar tendency can be noticed in Swaziland. The aforementioned end of the credit facility mid 1999 resulted in a significant reduction in the sale of 45 Wp systems. To counter balance this, small panels and PV kits have been offered on a cash basis. Demand is slowly picking up and more than 100 smaller systems/panels have been sold within a period of 6 months. It is not unlikely that also these systems will show a poor performance in due course.

The above highlights the present SHS market stalemate. One the one hand suitably sized and installed systems are preferred from a performance point of view, but are too costly for the majority. Only with a sustainable finance facility in place, good quality SHS are within reach of the middle and higher income groups. In the absence of such finance households tend to buy low quality/low price systems, which may fulfil a need, but often result in sub-optimal operations. For the low-income group electricity, be it grid or solar, will remain mere wishful thinking.

5.   SHS WITHIN THE POLITICAL CONTEXT

Rural electrification in developing countries in most cases can not be placed outside the political context. Grid electrification is highly visible and symbolises development. Electrification has a positive correlation with quality of life indicators such as life expectance, infant mortality, food availability and literacy [6]. As such the promotion of rural electrification creates political goodwill. But more often than not, rural grid electrification is too costly due to the dispersed rural settlements. The high expectations created by politician and the consequent 'wait and see' attitude of the rural population have in many cases been damaging for private initiatives. Large efforts, low margins and lack of volume have killed the spirit of many small solar entrepreneurs. Only a few involved in solar systems for professional applications, notably telecommunication, have financially performed well. The market for Solar Home System is generally regarded as too difficult and none rewarding.

On a higher level, however, there seems to be a general consensus regarding the role SHS can play in rural electrification. Evidence shows that SHS can improve rural life, but market development requires substantial efforts and success won't come overnight. Moreover, the commercial procurement of a SHS requires a certain level of income. It's an illusion to view SHS as the answer to electrifying to poor. The latter can only be achieved through direct intervention of Government or donor community.

6. CONCLUSIONS

The project in Swaziland shows that SHS seem to meet a demand, but only few have the financial capacity to acquire a SHS. The establishment of the credit facility has clearly shown its positive impact on the SHS market. By ending this credit facility a trend can be noticed towards low quality/low price systems, which may result in poor performance. Reviving and extending the credit facility will clearly help to keep the reputation of SHS high and will be beneficial for further market development. To date there are 100,000 rural households in Swaziland that still lack access to electricity. The many SHS publications that has appeared recently by World Bank and others often indicate a (commercial) market potential for SHS in developing countries for SHS of 10-20% of the rural population.

For Swaziland this would mean 10,000 to 20,000 households for whom SHS could become a reality. The project in Swaziland indicates that even without heavy subsidies results can be generated. They don't come easy and don't come overnight. Much more time and effort is needed, but it nevertheless poses a real challenge.

REFERENCES

[1] Lasschuit, P., *Rural household energy strategies in Swaziland*, Interfaculty Department of Environment Science, Amsterdam, 1994.

[2] Lasschuit, P., *Assessment of the PV market in Swaziland & Review of the Government Demonstration project*, Energy Research Foundation ECN, 1998.

[3] Müller, H., *Revolving fund for Solar Home Systems*, Experiences in Namibia, Windhoek, 1998.

[4] Government of Botswana, *Botswana National Development Plan 8 (1997-2002)*, Gaborone, 1997.

[5] ESMAP, *Implementation Manual: Financing Mechanisms for Solar Electric Equipment*, Washington, 1999.

[6] Menna, P., P. Paoli, L. Sardi, *Schemes for Photovoltaic Electrification in the rural Southern Mediterranean Regions,* 14th E.C. Photovoltaic Solar Conference, Barcelona, 1997.

The 'Solarstrom Stock Exchange' from the Electricity Utility of the City of Zurich (ewz)

Daniel Ruoss and Peter Toggweiler
Enecolo AG, Lindhofstr. 52, 8617 Mönchaltorf, Switzerland
Tel: 01 994 90 01 / Fax: 01 994 90 05
Email: info@enecolo.ch, http://www.solarstrom.ch

Bruno Hürlimann and Annina Müller
Elektrizitätswerk der Stadt Zürich (ewz), Tramstrasse 35, 8050 Zürich, Switzerland
Tel: 01 319 41 51 / Fax: 01 319 41 97
Email: bruno.huerlimann@ewz.stzh.ch, http://www.ewz.ch

ABSTRACT: The original ewz- model of the 'Solarstrom Stock Exchange' for the promotion of solar electricity has evolved to become a very successful project and the numerous positive feedback's and inquiries ewz has received from home and abroad are most encouraging.

Beginning of 2000, 1620 kW nominal PV power had been installed. 42 PV plants are under contract producing yearly a total solar electricity value of 1'200'000 kWh. 5700 ewz customers ordered solar power at a current price of 0.65 U$ / kWh. Negotiations with further suppliers indicate prices for solar electricity below the 0.50 U$ / kWh mark (this includes the federal contribution of approx. 1/ 3 of the installation costs). Cost reduction of around 20% for installed PV plants could be achieved due to the ewz 'Solarstrom Stock Exchange'- model, but in parallel the quality and the aesthetics of the installation was notable increased.

A first rough analyses shows that 60% of the installation are in the range 0 to +10% difference for the produced energy compared to the prognosticated figure. Some installations perform in the range of –12.5%, the main reason for this bad performance were malfunction of inverters and a three-phase safety disconnector, which is only manually recloseable.

Keywords: Green Pricing – 1: Marketing – 2: Strategy - 3

1. INTRODUCTION

A representative opinion survey carried out in July 1995 showed that there is a definite demand on the part of the electricity utility of the City of Zurich (ewz) customers for solar electricity. In the survey, 3,500 ewz customers were asked whether they would be prepared to accept solar electricity and to pay a higher price for it. The high response rate was surprising: 270 of those contacted – some 7% – responded positively to the idea of subscribing to solar electricity and indicated their readiness to purchase on average 70 kWh at 1 U$ / kWh per year. At this point, it was not clear to what extent intentions would be borne out by actual practice, i.e. how many customers would finally subscribe to the solar electricity scheme.

The second motivating factor behind the 'Solarstrom Stock Exchange' is contained in the ewz company profile. ewz is actively promoting the introduction of renewable energy, and this also forms one of the energy policy objectives of the City of Zurich. In 1991, ewz began providing financial support from the so-called 'electricity saving fund' for various plants, including those exploiting renewable resources. However, in an urban environment dominated by rented flats, this proved insufficient to encourage more solar plants to be built. The 'subsidisation' achieved only 51 kW to be installed. A new model was needed that would enable the majority of residents, i.e. the tenants, to support the building of solar plants, even though they were not property owners themselves.

Moreover, a Swiss government promotion program called 'Energy 2000' is presently in force that calls for 50 megawatts of photovoltaic generating capacity to be installed in the whole of Switzerland by the year 2000. For the City of Zurich, this translates to a PV capacity of 2,400 kW as the aim for 2000.

2. HOW DOES THE EWZ 'SOLARSTROM STOCK EXCHANGE' WORK?

In a word – the ewz purchases solar electricity on the open market, and offers it to customers at a flat rate corresponding to the average price of obtaining solar electricity, making no additional charge. In this way, ewz plays the role of mediator between producers of solar electricity on the one hand and customers on the other.

2.1 How does ewz negotiate with suppliers of solar electricity?

ewz offers producers an undertaking to purchase solar electricity, guaranteeing the purchase price (indexed to inflation) for 20 years. First, ewz issues a public tender to planning consultants, property owners, insurance companies, etc., to supply solar electricity at an economic, but nevertheless competitive, price.

Suppliers finance their projects via the capital market or from their own funds, and build and operate the equipment. Tenders are accepted from those suppliers with the lowest, yet still economic, prices. To encourage high-quality design, ewz issues technical guidelines.

An efficient and reliable plant is in the suppliers' own best interests, since frequent breakdowns would involve a financial loss for them.

'Solarstrom Stock Exchange' can only function properly with customers who are prepared to accept solar electricity at the present rate of five to six times the normal price. Thus, parallel to the tendering scheme, ewz is endeavouring to find customers who are willing to cover part of their electricity needs with solar electricity and pay the current rate of 0.65 U\$ / kWh for it. Customers order solar electricity in the form of a subscription to ewz. The aim is to convince customers that they do have a direct influence on whether or not solar plants are built in the City of Zurich. Subscribers undertake to purchase solar electricity for a period of one year, but this is automatically renewed for a further year if not cancelled in advance.

2.2 What happens if there is a difference between supply and demand?

An inbalance between supply and demand can occur in the long run if customer subscriptions for solar electricity – renewed on a yearly basis – should decline, ewz having committed itself to accepting solar electricity at an agreed price for 20 years. Assuming that the market volume over the next few years should expand to the 1 million kWh forecast, but that then demand should fall by (let us say) 10%, ewz would incur a loss – assuming a solar electricity price of 0.65 U\$ per kWh – of U\$ 65'000. This situation is, however, improbable, as the price of solar electricity is more likely to fall with increased market volume owing to reduced costs of production.

3. WHAT DOES EWZ ACTUALLY DO?

To publicise its solar program and increase public awareness in this area, ewz is performing active and intensive marketing [1]. Potential customers for solar electricity are directly or indirectly encouraged to take part via numerous different channels. ewz supplies customers with specific information designed to increase transparency and credibility. Further, new photovoltaic plants realised on customers' initiative are often inaugurated at an official ceremony. This is a good opportunity to invite customers to take part and to visit their plant. The rising number of customers (and plants) is making increasing demands on the book- keeping and energy-consulting resources, as well as on the engineers.

In coping with this, existing infrastructure and personnel are utilised, and, despite the increased

work- load involved, this does help to expand know-how for the future. The long-term success of models of this type is strongly dependent on the active and enthusiastic participation of employees, and on effective promotion at the highest level of management.

4. RESULTS ACHIEVED TO- DATE

ewz has met with a very positive echo on the part of the public. Also, co-operation with environmental organisations has been very constructive. To give an example, ewz has a joint program with WWF within the Living Planet Campaign with climate as the central theme.

Since May 1997, it has been possible for Zurich citizens to order solar electricity from ewz. By beginning of 2000, some 5700 customers had subscribed, including 200 companies or institutes, like the University hospital of Zurich, major Swiss banks, the public traffic sector of Zurich, etc. This means that over 2.9% of ewz customers now cover part of their electricity needs with solar electricity, paying five times the normal price in return for an environmental sound product – and more orders are coming in every day.

42 plants are already under agreement and further negotiation with possible suppliers will take place this year. The present 42 plants together produce a total of 1'200'000 kWh per year from three rounds of tenders.

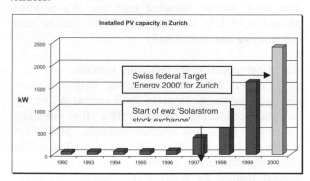

Figure 1: The development of the installed PV power in Zurich due to the 'Solarstrom stock exchange'

The present offered price of solar electricity is below the 0.65 U\$ / kWh mark. Additional price reductions will result from federal contributions, obtainable under the condition that they are passed on 100% to customers. A final price of less than 0.50 U\$ / kWh is negotiated with most suppliers. Cost reduction of around 20% for installed PV plants could be achieved due to the ewz 'Solarstrom Stock Exchange'- model, but in parallel the quality and the aesthetics of the installation were notable increased.

4.1 Installation ABZ in Zurich [2]

On two house rows a retrofit was done using a new integration system called SOLRIF (solar roof

integration frame).

Figure 2: 53.04 kW PV installation ABZ for the 'Solarstrom stock exchange' with SOLRIF (in this figure only one house row, 26.52 kW is shown)

Due to the very easy access to the PV installation and good insight, the client asked a perfect integration system. The system should be easy to mount (less construction work), in combination with the existing clay tiles usable and a good appearance should result. The goal was also to built a cost effective BIPV (Building Integrated PV) installation and a perfect appearance, due to the location and the usage of the buildings itself.

Since the SOLRIF system can be easily introduced in existing roofs in combination with any standard clay tile, a very esthetical BIPV installation was resulting. The final overall installed cost per Wp resulted around 6.50 US$, showing that BIPV can be cost competitive and aesthetically.

4.2   Cost reduction for mounting systems

Several other new mounting structures were designed for the 'Solarstrom stock exchange' leading to cost reductions around 50% for mounting structures. Following another example, SOFREL® was applied around 70% for flat roof installation. The mounting structure is a simple concrete element with special feature for holding the module and works with the weight foundation on the roof (no screws or any other mechanical connection to the roof skin is applied).

Figure 3: Most often applied mounting structure (SOFREL®) for flat roof installation within the 'Solarstrom stock exchange'.

SOFREL® is on the market for less than 0.35 US$ / Wp, including two elements and the needed stainless steel brackets. For flat roof installation system, cost less than 5.40 US$ / Wp did result.

4.3   Analyses of some installations

An analyses of almost all installations has started in April 2000. The first, although rough results are shown in this paper. So far 12 PV installations were evaluated and the energy data analysed. The distribution of the PV plants in view of the building area was as followed (Status 19.04.2000).

Flat roof installation	Produced (kWh/kWp)	Prognosticated (kWh/kWp)	Difference (%)
22.3 kWp PV plant no. 1	726	817	-12.5
12.8 kWp PV plant no. 2	750	846	-12.8
=> the difference is significant high, due to malfunction of the inverters			
33.7 kWp PV plant no. 3	863	845	2.1
50.4 kWp PV plant no. 4	825	847	-2.7
20.4 kWp PV plant no. 5	901	846	6.1
50.7 kWp PV plant no. 6	886	874	1.4
12.6 kWp PV plant no. 7	798	798	0.0
12.96 kWp PV plant no. 8	780	829	-6.3
73.5 kWp PV plant no. 9	884	856	3.2
88 kWp PV plant no. 10	861	846	1.7
Average (without no. 1 & 2)	**850**	**843**	**0.7**

Table 1: Evaluated flat roof installations

Sloped roof installation	Produced (kWh/kWp)	Prognosticated (kWh/kWp)	Difference (%)
6.6 kWp PV plant no. 11	800	815	-1.9
Average value	**800**	**815**	**-1.9**

Table 2: Evaluated sloped roof installation

Horizontal (4°) roof installation	Produced (kWh/kWp)	Prognosticated (kWh/kWp)	Difference (%)
81.6 kWp PV plant no. 12	763	721	5.5
Average value	**763**	**721**	**5.5**

Table 3: Evaluated horizontal (4°) roof installation

In all tables shown above is the PV nominal power, the produced energy yield, the prognosticated value and the resulting difference presented. The analyses covers the operating year 1999 (01.01.99 to 31.12.99).

Table 1 shows an average value for flat roof installation of 850 kWh / kWp. This value presents 8 installation including one plant (no. 7), which is in winter partly shaded, resulting in an overall energy reduction of approx. 6%. No. 8 was also not properly operating, due to an inverter breakdown on one phase for 8 weeks. These problems solved, an average value of 860 up to 870 kWh / kWp can be expected for the future. For comparison the average value for all Swiss installation [3] (incl. the alps the southern part of Switzerland, no facade installations and no shaded plants) is given with 830 kWh / kWp for the year 1999. As conclusion for the flat roof application, it might be pointed out that most installation are producing almost as prognosticated. The energy yield for the PV installation within the 'Solarstrom Stock Exchange' is 2% higher than the Swiss value. But nevertheless, the average value of 850 kWh / kWp has to be improved by better control of the inverter operation and fixing of the breakdowns within 2- 3 days.

For the sloped and horizontal application further data has to be collected and analysed. So far only two value are calculated.

A comparison is given in figure 4, showing all installation with the produced energy compared to the prognosticated value.

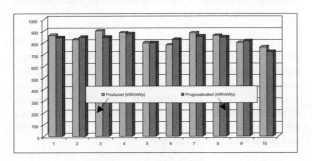

Figure 4: Comparison with the produced energy compared to the prognosticated value.

The difference has been calculated for all 12 installation and is divided into ranges.

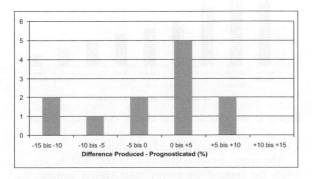

Figure 5: Difference of the produced energy compared to the prognosticated value.

The majority of the installations are in the range 0 to +10% difference for the produced energy compared to the prognosticated figure. Some installations

perform in the range of –12.5%, the main reasons for this bad performance were malfunction of inverters and a manual three-phase safety disconnector.

5. CONCLUSIONS AND OUTLOOK

The ewz- model of the 'Solarstrom Stock Exchange' for the promotion of solar electricity has evolved to become a very successful project and the numerous positive feedback's and inquiries ewz has received from home and abroad are most encouraging. Very positive echo was received from the public, press and environmental organisations, as well from related business partners, even in foreign countries.

The cost reduction achieved (new mounting systems and overall systems cost) are being reflected in the produced electricity cost too, making it possible to attract more customers and to hold the existing customers.

The produced energy is for most PV installation almost as prognosticated. But nevertheless, for the several PV plants the produced energy figure can be improved by better control of the inverter operation and fixing of the breakdowns within 2- 3 days. Some installations were not in operation over a period of 2-8 weeks. Would the energy loss be calculated with the contracted energy delivery tariffs (approx. 0.65 US / kWh), some owners and operators of installation would be heavily surprised about the sum, which is lost due to a not appropriate controlling and maintenance.
The analyses of the energy data will be extended for all installation including the mechanical condition of the substructure and the soiling of PV laminates and modules, especially. All failures so far will be noted and examined. A report will be available and will help to improve the PV performance for some installations.

Needless to say that ewz will increase their effort to promote and spread the 'Solarstrom stock exchange'. The Swiss federal target of the program 'energy 2000' 2.4 MW will be achieved. A further marketing communication campaign in order to gain more customers is planned.

REFERENCES

[1] Giger N. and Hürlimann B. (1998) Presentation of the ewz Solarstrom stock exchange
[2] Ruoss D. (1999) Presentation ABZ PV installation, IEA Task VII
[3] C. Meier and C. Holzner, Swiss PV Statistic Report 1999, VSE 10/2000

FULLY ECONOMIC MARKETS FOR PV IN THE UK:
THE GROWTH POTENTIAL FOR STAND-ALONE SYSTEMS

Oliver Paish

IT Power Ltd

The Warren, Bramshill Rd, Eversley, Hants, RG27 0PR, ofp@itpower.co.uk

ABSTRACT: A study [1] was completed by the author in January 1999 which aimed to identify and quantify the market opportunities for stand-alone renewable energy power supplies in the UK, covering solar PV, wind, and micro-hydro electricity-generating systems. The focus of the study was on small systems, probably charging a battery, and generally less than a few kW installed capacity. Emphasis on grid-connected renewables in the UK has drawn attention away from the fact that it is <u>off-grid</u> systems that can offer immediate and genuinely commercial markets for the renewables industry. There is likely to be a major increase in the worldwide market for renewable energy systems in the next 10 years, and renewed efforts are required to ensure that the UK is in a position to play a commercial role in that market. With targeted efforts to stimulate the market, UK industry believes that a potential market of over 6000kWp can be achieved and exceeded by 2010, with a business potential in excess of £60million for manufacture and installation.

1. RENEWABLE ENERGY SYSTEMS IN THE UK

While it is possible to obtain almost any renewable energy product in the UK, the number of actively marketed systems is limited, and there are large areas of the country without representation.

A literature review and survey of UK suppliers/installers was undertaken to establish the status of off-grid renewable energy applications in the UK. These were divided into the 3 categories in Table I, of which the 'well-established' applications account for the large majority (probably 80-90%) of the UK off-grid market to date.

Table I: UK applications using renewable energy systems

Well-established
Monitoring and data collection
Telemetry
Lighthouse power systems
Offshore beacon lighting
Mobile homes and boats
Remote homes & buildings
Demonstrated, not widespread
Ticket machines
BT telecoms repeater stations
Railway track greasers
Electric fencing
Warning signs and lights
Public information displays
Demonstrated as 1-offs
Area/street lighting
Emergency phones
Cell-phone network repeaters
Small water treatment units
Cooling in transport
Sluice-gate operation
'Intelligent' cats-eyes

2. WHAT MAKES A GOOD OFF-GRID SUPPLY ?

It is easily shown that, unless the mains grid is within a few 10's of metres, it is not worth installing a new grid connection for small levels of power demand. Furthermore, as Figure 1 illustrates, the smaller the energy demand, the more people are prepared to pay per unit of energy provided, especially if the solution brings other benefits such as reduced size and greater portability.

The key to the effectiveness of stand-alone supplies in remote areas is therefore not the unit cost of the energy supplied. More important is the ability to demonstrate that such systems can do the job not only cost-effectively, but also reliably, over a long timescale, and with the minimum of technical or other problems. Some of the main issues which dictate whether an off-grid power supply will be successful are summarised below.

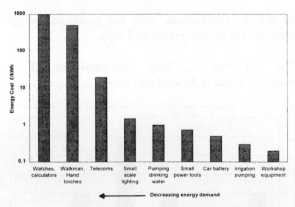

Figure 1: Acceptable cost of energy vs level of demand

2.1 Matching energy demand to energy supply

The most cost-effective RET system will be one in which the demand for energy most closely matches the variation in the natural energy resource; this keeps the need for energy storage to a minimum. Anyone wishing to market a particular renewable energy technology should be searching for applications which need the energy at times when it is readily available.

2.2 Use of energy efficient appliances

Until recently, mains appliances have rarely been energy efficient because it has been more important to get their capital cost down than for the consumer to save a small amount of cheap grid electricity. Attention to energy efficiency has reduced the energy requirements of certain off-grid applications by between 3 and 20 times. *Such improvements will act to greatly broaden the market for small-scale renewables, and needs to be a focus for future product development.*

2.3 Ease of maintenance

Renewable energy systems should have almost no maintenance needs, and have no need of repair or replacement for at least 10 years (excluding the batteries). This is one of their key selling points over petrol/diesel gen-sets, or battery-only systems. Such high reliability has been proven by numerous professional systems, but is by no means the norm.

2.4 Theft and vandalism

No one supplying an outdoor technology should underestimate the problems of theft and vandalism, which are endemic in many regions of the country. A system which can exhibit features which will deter removal of components by spanner or crow-bar, and resist damage from hockey sticks, flying bricks, etc. will be more successful.

2.5 Visual impact

Visual impact is a frustrating issue which cannot be ignored by system designers. They will enhance the marketability of their systems if they can develop aesthetically attractive products and arrange for the power supply to be flexible in where it is located, including merging the power supply system into the application hardware (most easily achieved with a PV-battery unit).

2.6 Consequences of loss of supply

It is important that the size and specification of a renewable energy system is appropriate to the importance and value of the load being supplied. The key technical issue is battery failure. By setting the charge controller to cut off the load whenever battery discharge reaches 40%, battery life can be extended for up to 10 years. *However this means that saving the battery is given higher priority than allowing the system to operate.* It is possible to programme more sophisticated controllers to reduce the load in steps, down to essential functions only, when the battery charge starts to get low.

3. MARKET BARRIERS

Off-grid renewables are already the most cost-effective energy solution in many situations, and their under-utilisation in the UK is not believed to be a particularly a price-sensitive issue. Instead, the market for RETs is predominantly held back by lack of awareness and confidence in the technology, lack of 'packaged' products, and lack of a conventional business infrastructure.

There remains a remarkable lack of awareness of RETs as a whole in the UK, and a resultant lack of confidence in their abilities. There is still a general belief that renewable energy is "something for the future", not a present reality and opportunity.

There are few dedicated commercial packages involving renewable energy systems. Most customers want a task carried out away from the mains grid, and don't want to get too involved in the technology. They need to be able to see attractive, reliable, off-the-shelf packages, with appropriate guarantees and with options for financing arrangements.

A sales and distribution network for RETs does not yet properly exist in the UK. This is in marked comparison, for example, with the USA, where specialised sales representatives, dealers, and catalogues are now widespread.

4. UK MARKET ESTIMATES

In order to develop order-of-magnitude estimates for the possible market opportunities over the next decade, system suppliers were asked to give an opinion on each of 25 applications, using 3 criteria:

1. overall potential as a commercial market for renewable power supplies
2. approximate size of power supply applicable
3. approximate total market size for the period 2000-10

The estimates were used to divide the 25 applications between *Promising today* (real markets available today), *Probable* (real markets emerging within 5 years), and *Possible* (markets emerging in over 5 years), and to infer:

- the total business potential (£), using an assumption on the cost per Wp of installed system.
- the clean energy generated per year by 2010, using an assumption of the kWh/year supplied by each Wp of installed capacity.

The market exercise 'boxed in' the realisable potential market for the period 2000-2010 between conservative high and low values, describing system sizes (whether PV, wind, or hydro) in terms of the equivalent Wp of PV. The results are summarised in Figure 2.

The 'moderate low' market prediction of around 1000kWp(equiv) for the next decade is believed to be similar to the figure for small wind and PV systems installed in the UK in the previous decade. Hence this is the 'no improvement' scenario.

The baseline estimate is three times this at over 3000kWp(equiv), and seems more realistic, though still conservative when considering that Holland installed over 3000kWp of PV alone in the decade to 1997. The 'moderate high' estimate is double the baseline figure, at nearly 6500kWp (equiv).

Figure 2: Total Market Size (2000-10) under 5 scenarios

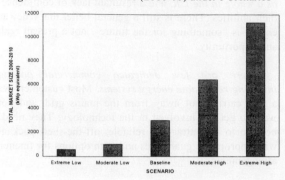

The baseline figure for energy supplied by 2010 is just over 1000 MWh/year. This amounts to the energy generated by an extra grid capacity of only 150kW operating all year round, and is roughly equivalent to saving 550,000 litres of oil-based fuel per year.

The low figure for energy generated stresses the point that the main benefits of stimulating the off-grid sector of the renewable energy market will not be in the consequent displacement of greenhouse gas emissions, but in:

- achieving tasks in off-grid locations which were previously economically unreasonable;
- strengthening the renewables industry so as to enhance the opportunities and capabilities to service (i) the much larger off-grid markets in the developing world and (ii) the grid-connected markets in industrialised countries (particularly for solar photovoltaics);
- highlighting to UK businesses and public sector organisations the many various opportunities that there are to achieve remote tasks with a renewable power supply;
- raising the profile, awareness, and support for renewable energy technologies with business and general public alike.

5. MARKET NEEDS

Ideas for possible Government initiatives to expand the market for RETs in the UK were compiled from industry responses, and can be summarised as follows:

1. Marketing assistance
- Support the industry in marketing renewables across different end-user sectors, for example through targeted information brochures and seminars.
- Support small businesses in marketing RETs at non-RET trade exhibitions.
- Encourage a positive attitude from electricity companies in becoming involved in off-grid system supply or system leasing.

2. Technical development
- Support the development of energy-efficient load applications to bring down RET system sizes.

- Assist research on improving the energy storage component of RET systems.
- Support product development and marketing of commercial packages.

3. Policy measures
- Reconsider planning regulations and guidelines towards the use of RETs.
- Remove VAT on RET hardware.
- Offer capital subsidies for domestic RET systems, along the lines of energy efficiency subsidies.

4. Improve general business and public awareness
- Circulate a glossy summary of the potential of off-grid RETs in the UK among relevant departments of government and industry.
- Assist the trade associations to develop the internet and network TV as information resources.
- Sponsor one or more design competitions for the innovative, commercial use of renewables.

Of the above measures, the most strongly emphasised were the need for marketing assistance and raised public awareness. These would provide a boost for UK companies who feel that they have sound products to deliver, but a very diverse and unknowledgeable market to reach.

6. CONCLUSIONS

In summary, the UK is behind many other European countries in the implementation of off-grid applications using a renewable energy resource. With targeted efforts to stimulate the market, UK industry believes that the 'moderate high' potential market of over 6000kWp can be achieved and exceeded by 2010, with a business potential in excess of £60million for manufacture and installation. With no action, then the industry will continue to a position between 1000kWp 'moderate low' and 3000kWp 'baseline' installed capacity by 2010.

REFERENCES

[1]   O.Paish, IT Power Ltd, New UK markets for off-grid renewable energy systems, ETSU K/BD/00186/REP, January 1999.

PHOTOVOLTAIC SOLAR ENERGY IN SPAIN'S RENEWABLE ENERGY DEVELOPMENT PLAN

C. Hernández Gonzálvez, J. Artigas Cano de Santayana, F. Monedero Gómez
Institute for the Diversification and Saving of Energy (IDAE). SPAIN
Paseo de la Castellana, 95, Planta 21. 28046 – Madrid
Tel. 34 1 4564900 Fax: 34 1 5551389 comunicación@idae.es - www.idae.es

ABSTRACT: The aim here is to summarize the role of Photovoltaic Solar Energy in Spain within Spain's Renewable Energy Development Plan. In 1998, photovoltaic solar energy in Spain totalled an installed capacity of 8.6 MWp. In recent years there has been a rapid rate of growth due to the expansion of the market, constant decrease in costs, broadening of the range of applications, and increase in the efficiency and reliability of the products and systems. The industrial sector comprises 73 companies, mainly manufacturers, installers and distributors. Spanish manufacturers offer high quality products and large generating capacity. Current capacity is 12.5 MWp/year and immediate capacity is 22.5 MWp/year. There is a privileged price treatment of energy generated by photovoltaic installations, with two levels corresponding to the limit of 5 kW. The elimination of barriers and providing incentives means the elaboration of priority lines of action, to achieve the increase of 135 MW in installed power (1999-2010 period).
Keywords: Energy planning – 1: Current situation – 2: Targets – 3

1. INTRODUCTION

The Plan has been developed as a response to the European Commission's White Paper, which sets as a target for renewable energy that it should meet 12% of primary energy demand in the European Union by 2010. In particular, it is foreseen that a level of 3,000 MW of installed photovoltaic solar capacity be achieved by 2010, which means multiplying installed capacity in 1995 by 100.

On 27 November 1997 Law 54/1997 on the Spanish Electricity Sector was approved. Its main aim was to deregulate the electricity sector and create a special regime for renewable energies, with guaranteed access to the grid and the possibility of a premium of over 90% of the average electricity price for photovoltaic solar energy. This Law requires a development plan for renewable energies, which is the one being presented here.

As part of the implementation of Law 54/1997, on 23 December 1998 Spain saw the appearance of Royal Decree 2818/1998 which established the administrative procedure for access to the special regime, the delivery conditions of electrical power and the premiums of 60 Pta/kWh (0.36 ECUs/KWh) for power produced by photovoltaic systems with an installed power of up to 5 kW(up to a total of 50 MW) and 30 Pta/kWh (0.18 ECUs/kWh) for the remainder of installations of this type. This implies privileged treatment of electricity generated by photovoltaic installations, with prices six times higher than for other renewable energies from other small installations and three times that from other sources.

The strategic character of the Plan is based on the intrinsic benefits of renewable energy, such as its independence from imports, the decreased environmental impact of its use, the strategic balance in the supply of energy and the creation of an infrastructure for sustainable use.

2. ELABORATION OF THE PLAN

Against this backdrop, and with the backing of the Spanish Ministry for Industry and Energy, the Institute for the Diversification and Saving of Energy (IDAE) has elaborated the technical details of the Renewable Energy Development Plan, which presents the current situation of renewable energy and the forecasts for the coming years, and the mechanisms and sources of finance supporting the growth targets (12% of primary energy consumption).

The methodology applied by the IDAE in the preparation of the Plan guarantees the participation and integration of the scientific, technical and industrial positions of the specialist representatives of central government, the autonomous regions, public bodies, social organizations, business associations and academic and social organizations. The Plan has become an exercise in reflection defining the basis for the future development of an energy sector capable of consolidating a new dimension in the industrial sphere.

The general benefits it is hoped will be obtained with this Plan are based on its structural nature –linked to specific prior RTD actions and the future opportunities that arise as the technology matures– through utilization of the new strategic definition of the EU structural funds and the providing of incentives for technologies and markets.

3. CURRENT SITUATION OF PV IN SPAIN

3.1 Installed capacity

In 1998, photovoltaic solar energy in Spain totalizes an installed capacity of 8.6 MWp, with vanguard technology, supported by an up-to-date industry, and with systems which offer over 20 years of useful life. In recent years the rate of growth has been rapid (in 1998 1.3 MW were installed, over twice as much in previous years) due to the expansion of the market, the continuous decrease in costs, the broadening of the range of applications and the increased efficiency and reliability of the products and systems.

In the figure 1 a distribution by Autonomous Communities of installed capacity using photovoltaic solar energy can be seen:

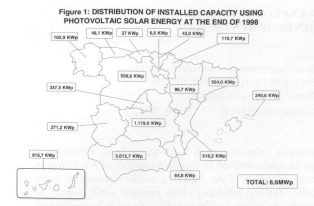

Figure 1: DISTRIBUTION OF INSTALLED CAPACITY USING PHOTOVOLTAIC SOLAR ENERGY AT THE END OF 1998

102,9 KWp   48,1 KWp   27 KWp   6,6 KWp   43,0 KWp   119,7 KWp

558,6 KWp

337,5 KWp   86,7 KWp   554,0 KWp

240,6 KWp

271,2 KWp   1.119,0 KWp

816,7 KWp   3.012,7 KWp   518,2 KWp

64,8 KWp

TOTAL: 8,6MWp

3.2 Photovoltaic manufacturing sector

The photovoltaic manufacturing sector in Spain comprises 73 companies, mainly manufacturers, installers and distributors. Spanish manufacturers offer high quality products and large generating capacity. Current capacity is 12.5 MWp/year and immediately available capacity is 22.5 MWp/year, without the need for major additional investments, rather by utilizing to the full the available equipment and extending shifts. Over 80% of Spain's output is exported (50% to the rest of the European Union and 30% to the rest of the world).

3.3 Environmental aspects

In terms of environmental aspects, photovoltaic solar energy regards the environment, offering local applications where generation and consumption are close to one another and where the environmental load in the usage phase is negligible. There is no significant impact on the physical environment or on the biota (air, soil, noise, hydrology, flora and fauna) and it avoids atmospheric CO_2 emissions. The main impact is produced by the extraction of the raw materials (by-products), manufacture of the components of the installation and in its disposal after it has completed its useful life (this is an area which is still under study). The main impact to avoid is the visual effect on the landscape, that is i.e. affecting the physical environment.

3.4 Costs

The cost of implementing photovoltaic systems remains high and basically depends on the type of installation (on- or off-grid), the type of technology, the size and development of the manufacturing processes, market conditions and company strategy.

There is a clear commitment on the part of the Spanish photovoltaic sector to achieve a reduction in costs by improving the technology through the use of fewer raw materials and less energy during manufacture, increasing the efficiency of photovoltaic cells (up to 18%), optimizing production processes, developing new technologies (CIS and CdTe) commercially, and improving concentration systems. By 2010 cost reductions of 20% are hoped for in off-grid systems and 25% in grid-connected systems.

4.   DEVELOPMENT'S   BARRIERS   AND   ITS SOLUTIONS

4.1 Barriers

The main barriers to the development of the sector which have been identified during the preparation of the plan concern a series of conditioning factors, such as:
-    The high initial investment,
-    The dependence on the price of silicon,
-    A lack of general information and the lack of legislation and, specific standards for components, installations, connection to the grid, installers and concerning the integration of installations in buildings.

Also, excessive rigidity has been detected in the systems to apply for grants or subsidies, a lack of coordination between the different government bodies and the limited involvement of councils, which are closest to the citizen.

4.2 Measures

The Renewable Energy Development Plan establishes the conditions for the elimination of these barriers by setting up a series of measures and incentives. In short, these measures take a variety of forms including:
-    the implementation of subsidies and financing of research and development activities;
-    tax deductions on investments;
-    tax regulation of electricity interchanges;
-    development of regulations for photovoltaic systems, administrative and technical conditions for the connection of photovoltaic installations to the electricity grid and regulation to obtain a photovoltaic system installer certificate; with official approval of "installer firms", and setting up of a photovoltaic system maintenance certificate;
-    running of public awareness-raising campaigns;
-    example activities using photovoltaic solar energy by government departments in public buildings;
-    financing of diffusion systems with specific lines of preferential finance; and,
-    promotion of pilot application projects.

The priority lines of technological innovation in the Plan are concentrated on a series of measures such as:
-    the development of panels and systems with high levels of integration in buildings,
-    the development and standardization of standard kits for small applications,
-    the research and development of thin-film technologies,
-    improvements in the development of inverters, and
-    the development of concentration technologies.

5. CONCLUSIONS

5.1 Forecast

The forecasts made by IDAE for the year 2010 are based on the following factors:
-    The market potential of solar photovoltaic energy is estimated at 2,300 MWp, broken down into two large groups depending on the type of application:
    - *Applications isolated from the grid*: 300 MWp (taking into account the requirements for electrification in dwellings and other types of isolated installations).

- *Applications connected to the grid*: 2,000 MWp (10%, which is the technically admissible percentage of generated peak power in Spain).
- Through the time scale of the Plan (1999 – 2000) industrial capability is set at the following levels:
  - *Present capability*: 150 MWp.
  - *Immediate capability* (which does not require additional investments, but using available equipment to the maximum and increasing shifts): 270 MWp.
- The estimate from Autonomous Communities is 41 MWp, primarily considering applications isolated from the grid.
- ASIF's (Asociación de la Industria Fotovoltaica. Photovoltaic Industry Association) forecasts amount to 120 MWp, of which 20 MWp are made up of isolated applications and 100 MWp of grid-connected applications.
- IDAE's forecast, in line with growth prospects in other EU countries adapted for conditions in Spain, is around 120-135 MWp for the year 2010. Of this figure about 20 MWp would relate to applications isolated from the grid and 115 MWp to connected applications.

The following figure (figure 2) represent the increase of installed photovoltaic solar energy in 1999 – 2000 period.

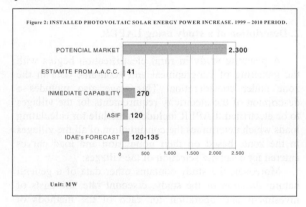

Figure 2: INSTALLED PHOTOVOLTAIC SOLAR ENERGY POWER INCREASE. 1999 – 2010 PERIOD.

POTENCIAL MARKET — 2.300
ESTIAMTE FROM A.A.C.C. — 41
INMEDIATE CAPABILITY — 270
ASIF — 120
IDAE's FORECAST — 120-135

Unit: MW

5.2 Targets

On the basis of the foregoing premises, the Plan sets the following targets for photovoltaic solar energy for 2010:

- Stand alone systems:
  20 MW
- Grid-connected systems > 5kW: 65 MW
- Grid-connected systems < 5kW: 50 MW
- Increase in energy generated: 203 GWh/year
- Emissions avoided (2010):   175,277 tCO_2

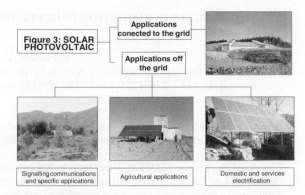

Figure 3: SOLAR PHOTOVOLTAIC

Applications conected to the grid

Applications off the grid

Signalling communications and specific applications

Agricultural applications

Domestic and services electrification

REFERENCES

(1)   "Promotion Plan for renewable energies in Spain". IDAE. Ministry for Industry and Energy.

(2)   "Special report for the Promotion Plan for Renewable Energies. Solar Photovoltaic Sector". Asociación de la Industria Fotovoltaica (ASIF).

(3)   "Renewable Energies in Spain. Balance and Prospects for 2000. '99 Edition". IDAE. Ministry for Industry and Energy.

(4)   "Solar Photovoltaic Energy. Renewable Energies Manuals". IDAE. Ministry for Industry and Energy.

(5)   "Solar Photovoltaic Installations". Alcor, E. PROGENSA.

(6)   "Fundamentals, Scope and Applications for Solar Photovoltaic Energy". CIEMAT. Ministry for Industry and Energy.

(7)   "The Environmental Assessment of Photovoltaics" Energy Technology Support Unit (ETSU) for Department of Trade and Industry.

(8)   "Atmospheric Contamination. Agents". Spanish Energy Institute.

(9)   "Agents and effects of Water Pollution". Spanish Energy Institute.

(10)   "Renewable Energies in Spain. Calendar of Projects 1997". IDAE. Ministry for Industry and Energy.

(11)   Consultative Commission on Energy Saving and Efficiency. Renewable Energies Working Group. Co-ordination Working Group. IDAE. Ministry for Industry and Energy.

(12)   Solar Radiation on Sloping Surfaces". Centre for Energy Studies. Ministry for Industry and Energy.

(13)   "Understanding and managing health and environmental risks of CIS, CGS and CdTe Photovoltaic module production and use: a workshop". P.D.Moskowitz., K. Zweibel and m.P. DePhillips. Report NBL-61480, 1994.

RURAL ELECTRIFICATION IN DEVELOPING COUNTRIES USING LAPER SOFTWARE

R. Fronius, V. Lévy, S. Garnier

EDF / R&D Division, 1 Ave. du G al de Gaulle, F-92141 CLAMART CEDEX - France

Tel : +33 1 47654205, Fax : +33 1 47654206, email : rainer.fronius@edf.fr, valerie.levy@edf.fr

ABSTRACT: Today, 40% of the world population need rural electrication (RE). The World Bank and other donors for RE-projects need a clear assessment of the viability and sustainability of the proposed electricity generation facility.

Electricité de France (EDF) and the French Agency for Environment and Energy Management (ADEME) have developed LAPER, a planning software using a GIS-based method for the electrication of vast regions. This program determines a masterplan using geographical data enhanced by socio-economic inquiries. After choosing the optimal means of electrication (MV-network, PV-panels, gensets, micro-windgenerators or micro-hydrogenerators) for each village, LAPER calculates an investment plan and an electrication schedule for the region, taking into account political, financial and geographical criteria.

The results are stored graphically and by tables in the GIS.

Keywords: Rural Electrication – 1: Developing Countries – 2: Energy Options - 3

INTRODUCTION

The electrification of rural areas in developing countries requires complex planning studies. The socio-economic and geopolitical situation in these regions implies that their electrification will not necessarily follow the same rules as those established for urban zones.

In cooperation with ADEME (the French Agency for Environment and Energy Management), EDF (Electricité De France) has developed a planning method adapted to the rural electrification of developing countries and the LAPER software to assist in implementing this method.

This software is described in this document. The first paragraph provides some background information on the tool and details the chronological development of a study using LAPER. This is followed by details on how calculations are made and a description of the technical aspects of the software.

1. WHY RURAL ELECTRIFICATION PLANNING SOFTWARE?

The benefits of electrification are recognised world-wide. However, even today, almost half of the world's population still does not have access to electricity. This is not only an issue for people who aspire to greater comfort and a higher standard of living, but also for local communities hoping to develop economically, for governments which see electrification as a way of limiting the rural exodus and, finally, for the international organisations whose objective is to even out the imbalances in the distribution of wealth.

The existence of all these actors and their converging interests means that it is vital to take into account the socio-economic and political dimensions when planning the electrification of a developing zone. The geographical characteristics must also be considered, all the more so since, pending further investigation, the choice between different technical solutions remains open.

A planning study in rural electrification therefore requires the processing of a large quantity of data from which different scenarios must be created and then compared. These facts naturally lead to the idea of computerising the process, with a tool that can fulfil these expectations, i.e. structure the data and study as many strategies as possible.

EDF and ADEME developed LAPER (Rural Electrification Planning Software) heading for this objective.

2. Description of a study using LAPER

A planning study in rural electrification begins with the gathering of geographical and economic data in the zone under consideration. This stage also includes a description of the electricity requirements for the villages to be electrified. LAPER includes a module for calculating loads which determines the consumption of all the villages in the zone, based on their population and load curves entered for a sample selection of the villages.

Moreover, the study contains other data of a general nature: duration of the study, discount rate, the costs of investment and operation for each of the methods of electrification compared, the budget allocated for each year of the study and the importance to be accorded to the socio-economic criteria taken into account in establishing electrification schedule.

The technical electrification solutions considered by LAPER are individual photovoltaic electrification, micro–hydraulic electrification, electrification by wind power plant, electrification by diesel generator and by network. To calculate the cost of the network it is necessary to know the lengths of the lines to be installed. This data will depend to a great extent on the geography of the zone and, where applicable, the location of any existing network. LAPER allows for a graphic input of the medium voltage network, which serves as a reference for the cost of electrification by network. It also indicates the optimum number of sources and outgoing circuits by source in the case of an "all network" solution.

All the data collected at this stage of the study constitutes what LAPER defines as a strategy. The planner can easily create other strategies by modifying one or more items of data. For each of these strategies, LAPER determines the most suitable method of electrification for each village and then an electrification schedule for all the villages over the study period.

The results of these calculations (assessment for the study as a whole, results by village and by year) are provided in the form of tables and charts. Tools are available to the planner to facilitate the creation of diagrams and layouts, making the study conclusions easy to read.

3. LAPER CALCULATION PRINCIPLES

3.1. Electrification target

Initially, LAPER determines the electrification target, i.e. the most appropriate technical solution for each village.

The method for determining the electrification target is as follows: for each village, the costs are compared for electrification by network, by generator, by wind power plant, photovoltaic and hydraulic electrification. If electrification using decentralised methods is cheaper for one village, LAPER verifies whether the increase in network costs for the other villages caused by its non-participation in the construction and maintenance of the medium voltage line concerned would exceed the earnings obtained through electrification of this village using a decentralised method. If it is established, for at least one village, that electrification by decentralised method is less expensive, this will necessitate a new iteration because the cost of the network would increase for the other villages. The comparison with the decentralised methods must then be repeated.

To take account of the dispersion of the settlement, LAPER enables you to distinguish between the centre and the periphery of a village. When determining the target, it is only the cost of electrification to the centre of each village which features. The periphery of the villages is electrified by individual solar panels.

The electricity requirements of each village can be entered in LAPER in different ways, enabling you to either give an complete description or, on the contrary, carry out the study even if your data is incomplete. Thus, residential customers can be distinguished from collective equipment and large customers can also be identified. Large customers are taken into account in the centre of the village. The presence of large customers rules out the photovoltaic electrification solution for the centre of a village.

The size of the installations under consideration is adjusted for the number of customers and the load level at the last year of the study period.

The low voltage losses are taken into account when calculating the cost of electrification by network or by a decentralised method linked to a low voltage network within the village (electrification by wind power plant, by generator or hydraulic electrification). The medium voltage losses are taken into account when calculating the cost of electrification by network.

Hydraulic electrification is automatically excluded for the villages located a long distance from a river with the minimal flow required for installing a system of this kind. Similarly, electrification by wind power plant is ruled out for villages situated in a zone where the wind speed is too low.

3.2. Electrification schedule

Once the electrification target has been established, LAPER determines the corresponding electrification schedule.

To do this, it determines an order of priority for the electrification of villages according to five criteria described below and weightings given to each of these villages by the planner.

The politics, solvency and development criteria require data on the villages. This data must be deduced from socio-economic inquiries made at the beginning of the study.

The financial criterion is calculated by LAPER. This relates to the villages supplied with electricity by the network. It enables the villages closest to the source to be electrified first. It does not penalise the villages electrified using a decentralised method.

The inter-regional equilibrium criterion is also calculated by LAPER. If a village has already been electrified, it gives priority to the villages situated in another area of the studied zone so that the areas identified by the planner are uniformly electrified.

Once the first priority village has been defined, it is electrified during the initial year if the assigned budget is sufficient. If the budget for a year is not sufficient it is brought forward to the following year and so on until the accumulated sum permits electrification. Priorities are then re-calculated for the remaining villages and the next village is electrified in the same manner as soon as the budget allows this.

4. TECHNICAL CHARACTERISTICS OF LAPER

LAPER operates on a PC in a Windows NT 4.0 environment.

LAPER has three components: the LAPER interface and the Village Type Manager which each have their own interface and the ER-SIG module which is an extension of the Arcview geographic information system.

The geographic information system enables data and results from the LAPER study to be displayed visually. Arcview has been personalised to facilitate the loading of the necessary data, the entry of a reference electricity network to establish the cost of centralised electrification, the mapped representation of the target and the basic development scheme determined by LAPER.

The Village Type Manager creates village types according to their consumption. The consumption of all the villages in the zone can then be determined by linking them to a particular village type.

The LAPER interface gathers the data for the study. The data on the villages, the zone and the reference network are loaded from Arcview. The LAPER interface determines the most suitable method of electrification for each village and an electrification schedule for all the villages over the study period. It offers the possibility of developing different strategies by varying one or more data sets. The calculations are made for each strategy.

The three components of LAPER exchange data and results (method and year of electrification, investment and operation costs, earnings, number of customers supplied, power and energy supplied, losses if appropriate). The following diagram (Fig. 1) will give an overall view of a study using LAPER and its information flows of information.

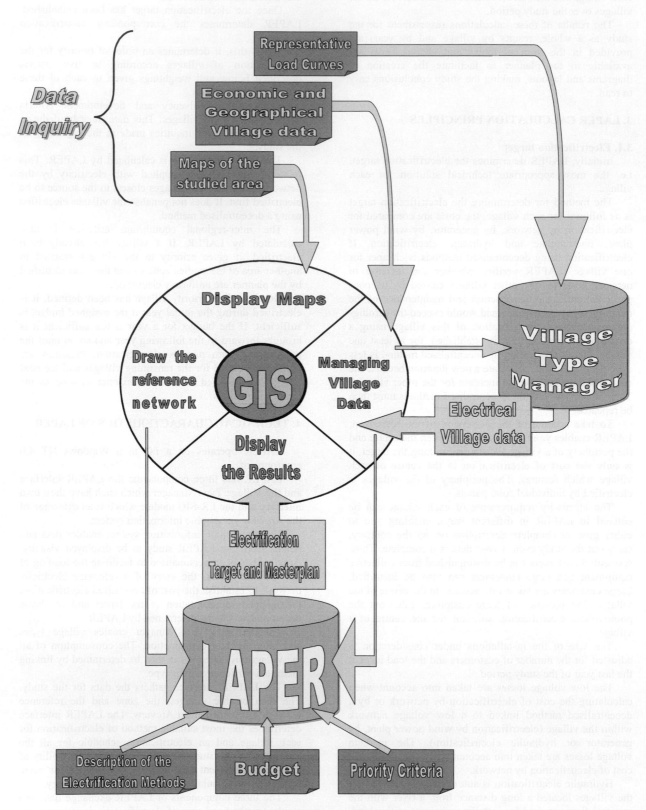

Figure 1

Market Deployment in Developing Countries

BUILDING ON EXPERIENCE: ASSURING QUALITY IN THE WORLD BANK/GEF-ASSISTED CHINA RENEWABLE ENERGY DEVELOPMENT PROJECT

Anil Cabraal, Senior Energy Specialist,
Asia Alternative Energy Program, The World Bank
1818 H Street, NW, Washington, DC 20433 USA
Tel: 202-458-1538   Fax: 202-522-1648   Email: acabraal@worldbank.org

ABSTRACT: The World Bank Group has almost a decade of experience with twelve projects around the world that use solar photovoltaic technology to provide basic energy services to rural households in developing countries. The Bank has learned that product and service quality is essential, if all participants in the supply and service chain are to make a profit or see a significant benefit -- a prerequisite for the widespread and accelerated dissemination of solar PV. The presentation highlights the comprehensive quality improvements that are being implemented throughout the supply and service chain by the Photovoltaic Market Development Component of the World Bank/GEF-Assisted China Renewable Energy Development Project. Initial results are promising and suggest that China's experience may well demonstrate that developing countries can not only compete internationally on the basis of quality, but also can be active contributors to ever-evolving global standards for solar PV.

Keywords: Solar Home System – 1, Qualification and Testing – 2, Developing Countries – 3

1. INTRODUCTION

The World Bank Group's experience with solar home systems began in 1992. Since then, it has approved twelve projects that use solar photovoltaic (PV) technology to provide basic energy services, such as lighting, radio, television, and operation of small appliances. These projects are focused on rural households that lack access to electricity grids. Solar home systems have been proven to deliver least-cost rural electrification and to be an effective supplement to grid-based electrification policies.

Users of solar home systems are able to avoid the costs of battery charging and LPG or kerosene purchases; they also enjoy PV's convenience and safety and the improved lighting and indoor air quality that it affords. Of additional significance are the environmental benefits associated with reduced CO_2 emissions.

More than 500,000 households in developing countries currently use solar home systems to meet their basic electricity needs. Installation targets for the World Bank's projects will increase that number by 750,000 by the year 2005. The China Renewable Energy Development Project is the largest renewable energy project the Bank has approved to date. It includes financing for 190 Megawatts of wind farms, up to 400,000 solar home systems, and technology improvement grant funds to support PV and wind.

The overarching objectives of the World Bank's solar PV projects are to provide electricity services to populations which cannot be reached in a cost-effective and timely way by lower cost options and, in doing so, to help develop markets for solar home systems and to overcome the main barriers to widespread and accelerated dissemination of solar home systems. The Bank's projects are still experimental. There is not enough accumulated experience to provide definitive answers about best practices, although the Bank's understanding or what does and does not work is evolving rapidly.

2. THE QUALITY CHAIN

My brief presentation will focus on what the World Bank has learned about the quality chain from its project involvement, in general, and from the China Renewable Energy Development Project, in particular. Those interested in broader information can obtain it from the paper we have just issued [1].

The Bank has found that there is no one correct method for PV dissemination. The approach must fit the country and market context. At the same time, one principle clearly holds true, regardless of the mode of dissemination. All participants in the supply and service chain must make a profit or see some kind of significant benefit. The supply and service chain includes component suppliers; system integrators; distributors and dealers; sales, installation and service organizations; as well as customers (see Fig. 1).

Product and service quality is central to the ability of anyone along the chain to make a profit or secure other important benefits. The World Bank promotes quality for its essential role in ensuring a sustainable and long-term service to consumers. We view quality as paramount to building confidence in solar PV and accelerating its widespread use.

The Bank's project experience in China offers a good example of both typical quality problems and some promising new ways of addressing them. A small but dynamic rural PV system distribution business is emerging in China, and sales have expanded quickly. At the same time, however, problems are evident in both quality of systems and components and responsiveness of after-sales service.

Several quality-related problems happen with enough frequency to be of particular concern. First, many smaller solar home systems are packaged and sold without charge controllers as a means of keeping costs down. This invariably shortens battery life and can considerably raise the life-cycle cost of the system to the user. Second, due to lack of after-sales service, when components fail or need routine replacement, the system is out of service until the user can obtain new or replacement parts from a supplier. Users must often go a considerable distance to obtain such parts, and their satisfaction can be considerably dampened by this inconvenience. Third, the quality of locally made and uncertified PV modules is uncertain, and their ratings may not be accurate.

Of course, such problems are by no means unique to China. They are similar to ones experienced in other countries and are ones that many people in this audience know well and are taking pains to address. There is growing awareness that, if such problems go uncorrected, they could mar PV's reputation and seriously inhibit market growth.

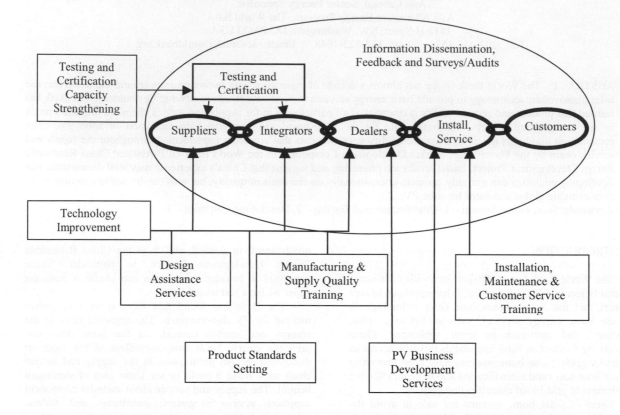

Figure 1 China Renewable Energy Development Project: Key Features of Quality Improvement along the Supply Chain

3. QUALITY IMPROVEMENT FEATURES IN THE CHINA RENEWABLE ENERGY DEVELOPMENT PROJECT

In order to address quality problems, the Photovoltaic Market Development Component of the China Renewable Energy Development Project has focused its attention and energies on five types of quality improvements.

First, it is improving the commercial capability of the participating companies to expand and sustain their PV businesses. Second, it is improving the quality of products and services available to consumers. Third, it is strengthening project management capabilities. Fourth, it is strengthening the capacity of testing centers and companies with regard to product quality assurance. Fifth, it is enhancing the access of suppliers, dealers and consumers to information.

A variety of capacity-building training and technical assistance activities are being conducted in China with the aim of bringing about these quality improvements. I will devote most of the remainder of my presentation to describing the nature and status of these activities.

Business development assistance is provided as part of the project. It takes the form of training and tailored assistance with business planning. It includes assistance with planning for the opportunities and demands of business expansion – extending sales and service networks, developing warranty documents, and setting up procedures for quality assurance. To date, 17 Chinese PV companies have received business development assistance, and more and more are requesting it. It is these companies that will lead the sales efforts under the project

The project has conducted a *market survey for PV systems* in four provinces in China [2]. The survey gathered information about the characteristics of PV purchasers and their preferred payment patterns. Its results are intended to help PV companies' design appropriate and effective marketing strategies. The results of this survey are on the project's Web site (http://setc-gef.newenergy.org.cn) in May.

The project has developed *technical product standards* for PV systems and for PV/wind hybrids [3]. Technical standards for quality products were prepared through a participatory process that involved key Chinese research and testing organizations and PV module and balance of systems manufacturers, as well as some assistance from international experts. The draft specifications underwent review at both the national and international levels [4]. The resulting revised draft technical standards were then disseminated to local and international companies.

It is noteworthy that two of the draft technical standards for inverters and controllers have since been accepted by the Global Approval Program for Photovoltaics (PV GAP) as PV Recommended Specifications and have been submitted to the IEC for consideration for adoption new global standards. The World Bank considers this to be an excellent example of how national and international standards for PV can be harmonized on win-win terms.

PV product quality improvement has been another project priority. A PV testing agency from the Netherlands trained technicians and experts at the Heifei University of Technology in ways to provide design assistance services to engineers working in Chinese design centers and PV manufacturing companies. The aim of the assistance is to improve PV product quality, in order to increase user satisfaction without increasing -- and even sometimes decreasing -- the cost to the manufacturer. The university is currently providing design assistance services to local suppliers.

The project has pioneered attention to *product certification* in China. Four Chinese testing institutes were helped to develop their capacity to test and certify key PV system components. They have since initiated product testing services, eliminating the need for Chinese PV module and component manufacturers to go to the time and expense of sending their products to distant countries for testing.

As an extension of the above, the project has also attended *to strengthening the capabilities of testing and certification institutions*. The current capabilities of the four aforementioned Chinese testing institutes were assessed by an international PV systems testing expert. A program was proposed to help them improve their capabilities to test and certify PV products, with the aim of helping the two that did not have ISO-25 accreditation to obtain it. This accreditation would give them globally recognized status as approved PV testing laboratories and would also assure that their test results accepted internationally. They are currently working towards achieving this accreditation.

Considerable project attention has been given to providing *training in manufacturing quality improvement.* Two Chinese experts attended a training program in India, funded by the World Bank, prepared by PV GAP and conducted by a team led by Rajasthan Electronic Industries, Ltd. The training addressed the key elements of quality manufacturing: (a) implementing an ISO 9000-compliant quality management system, (b) adhering to appropriate standards, and (c) submitting products for quality testing and securing recognized certification. This training was part of the World Bank's Quality Process for PV (QuaP-PV) Project, which promotes quality in four areas, in order to build confidence in solar PV. The four areas are quality design, quality manufacturing, quality testing labs, and quality installation and maintenance. Under the QuaP-PV project, four sets of training manuals and programs were prepared and field-tested in India. Subsequently training was given in Sri Lanka, China, the Philippines, and South Africa [5]. The Chinese experts, in turn, provided training in Chinese to 42 PV companies on establishing ISO 9000-based quality systems in their companies using the training materials prepared by PV GAP.

In another example of the useful cross-fertilization with the QuaP-PV Project, the China Renewable Development Project has also devoted attention to *training in installation and maintenance services*. The China National Institute of Standardization and Jike Energy New Technology Development Company trained more than 40 Chinese PV technicians from 32 companies. In the future, more technicians will be trained in quality installation and maintenance and customer service. Eventually, national PV installation and maintenance trainers and courses will be accredited through which practitioners can be certified. It is widely known that the life of solar PV systems can be reduced by years, due to poor installation and maintenance. The accreditation and certification infrastructure that is being put in place in China will help ensure that PV's reputation is upheld, that users' needs and expectations are met, and that access to financing is not constrained.

Lastly, the project has looked at the need for *information dissemination*. A project Web site has been established to disseminate information internationally of product and project requirements, qualified products and participating dealers. The project office has plans to conduct a public information campaign about PV systems and services, receive consumer complaints, and conduct surveys and audits to ensure that standards are met.

4. NEXT STEPS

All of the quality-enhancing activities I have just described were conducted during the project preparation phase over the course of approximately one-year. At present the China Renewable Energy Development Project is in the project implementation phase. During this phase, the project will provide continued assistance to manufacturers for business and quality improvement and continued information dissemination to consumers through local media and meetings. In addition, it will launch three financing initiatives. First, a technology improvement component will provide cost-shared assistance to PV manufacturers and dealers for product improvement and market development. Second, grants of US\$1.50/Wp will be made available to help reduce the initial cost of systems larger than 10 Wp. And, third, pilot financing schemes will be piloted to increase the affordability for poorer consumers who cannot afford the high front-end costs.

In conclusion, the China Renewable Energy Development Project has paid the most comprehensive attention to date in any developing country to PV product and service quality. It affords the World Bank and all those interested in removing barriers to the widespread and accelerated dissemination of solar home systems a unique opportunity to observe the effects of such attention on customer satisfaction, supplier profitability, and market and business development. Although it is still too early to draw any conclusions, the conceptual model of ensuring quality products and services all along the supply and service chain seems to be useful, and promising progress is being made as specific quality-enhancing activities are implemented. The World Bank hopes and expects that China's experience will demonstrate that developing countries can not only compete internationally on the basis of quality, but also can be active contributors to ever-evolving global standards for solar PV.

For more information on the China Renewable Energy Development Project see: http://setc-gef.newenergy.org.cn/. For more information on the Asia Alternative Energy Program of the World Bank see: http://www.worldbank.org/astae/.

REFERENCES

[1] Martinot, E., A. Cabraal and S. Mathur, World Bank/GEF Solar Home Systems Projects: Experiences and Lessons Learned 1993-2000, World Bank Informal Note, April 2000 (draft).

[2] Project Management Office, China RED Project, Assessing Markets for Renewable Energy in Rural Areas of Northwestern China, State Economic and Trade Commission, Beijing, China, May 2000 [forthcoming].

[3] Project Management Office, China RED Project, Solar Photovoltaic Systems and Photovoltaic/wind Hybrid Systems Specifications and Qualifying Requirements, State Economic and Trade Commission, Beijing, China, December 1998.

[4] Powermark Corporation, Chinese Photovoltaic Technical Documents and Standards Review, prepared for the World Bank, Phoenix, Arizona, USA, February 2000.

[5] (1) Energy Center Netherlands, Manual for the Design and Modification of Solar Home System Components, prepared for the World Bank, the Netherlands, February 2000. (2) Global Approval Program for Photovoltaics, Quality Management in Photovoltaics: Manufacturers Quality Control Training Manual, prepared for the World Bank, Geneva, Switzerland, February 2000. (3) Florida Solar Energy Center, Training Manual for Quality Improvement of Photovoltaic Testing Laboratories in Developing Countries, prepared for the World Bank, Cocoa, Florida, USA, February 2000. (4) Institute for Sustainable Power, Inc., PV Installation and Maintenance Practitioner Certification Infrastructure: Development Procedures, prepared for the World Bank, Highlands Ranch, Colorado, USA, February 2000.

DEPLOYMENT OF PHOTOVOLTAIC TECHNOLOGIES: CO-OPERATION WITH DEVELOPING COUNTRIES. TASK IX OF THE INTERNATIONAL ENERGY AGENCY'S PHOTOVOLTAIC POWER SYSTEMS PROGRAMME

J. R. Bates, B. McNelis, A. Arter[1], W. Rijssenbeek[2].
IT Power Ltd, The Warren, Bramshill Road, Eversley, Hampshire, United Kingdom, RG27 0PR. Tel +44 (0) 118 973 0073, Fax +44 (0) 118 973 0820, email: jrb@itpower.co.uk
1: ENTEC AG, Bahnhofstr. 5, CH-9000 St Gallen, Switzerland. Tel: +41 (0) 71 228 1020; Fax: +41 (0) 71 228 1030; email: alex.arter@entec.ch
2: ETC, Kastanjelaan 5, P.O. Box 64, 3830 AB Leusden, The Netherlands. Tel: +31 (0) 33 494 3086; Fax: +31 (0) 33 494 0791; email: office@etcnl.nl

ABSTRACT:

This paper describes the aims, objectives and progress to date of the latest Task (Task IX) of the International Energy Agency's Photovoltaic Power Systems Programme. The overall mission of the IEA PVPS Programme is to encourage international collaboration efforts through which photovoltaic energy becomes a significant option in the near future.

The objective of Task IX is to further increase the overall rate of successful deployment of PV systems in developing countries where PV is often the only viable option for remote electrification. Task IX is therefore exploring the possibilities for, and scope of, co-operation with developing countries and international financial institutions.

The following countries are actively participating in the work of the Task: Australia, Canada, Denmark, France, Germany, Italy, Japan, the Netherlands, Switzerland, the United Kingdom and the United States of America. The United Nations Development Programme and the World Bank also participate in Task IX.

Keywords: Developing countries - 1: Rural electrification- 2: Strategy – 3

1. INTRODUCTION

The primary objective of Task IX is to further increase the overall rate of successful deployment of PV systems in developing countries. This will be achieved by:

- Development of Recommended Practice Guides;
- Promoting improved techno-economic performance of PV in developing countries;
- Identification of areas where further technical research is necessary;
- Exchange of information with, and between, target groups;
- Workshops for, and information exchange with, donor agencies.

The Task IX Workplan was agreed by the PVPS Executive Committee in May 1999. The Task has been divided into 3 subtasks:

(i) Deployment Infrastructure;
(ii) Support and Co-operation;
(iii) Technical and Economic Aspects of PV Systems in Developing Countries.

The process of identifying a number of target countries has been initiated and these have been provisionally identified as: Argentina, Brazil, China, Dominican Republic, Ghana, Honduras, Kiribas, Indonesia, India, Morocco, Philippines, South Africa, Vietnam and Zimbabwe

2. DEPLOYMENT INFRASTRUCTURE

This subtask has been developed to identify and overcome the critical barriers to the widespread PV deployment and implementation through developing a series of Recommended Practice Guides (RPGs) to help promote the necessary infrastructure requirements in developing countries. These Guides will be both technical and non-technical in nature, covering issues such as quality and certification, operation and maintenance, training and accreditation, and financing mechanisms.

This work is investigating the various strategies for infrastructure development and deployment to ensure widespread and successful implementation of future PV programmes. Key issues relating to PV deployment strategies and initiatives are being investigated and collated to provide input to the preparation of the Guides.

A series of surveys are currrrenty underway on various aspects of PV deployment programmes. As far as possible, information is being gathered from existing sources and networks (e.g. local in-country expertise, PRESSEA, existing PV suppliers networks, results gathered from the IEA PVPS Task III Developing countries Survey Report [1]). If it proves necessary, in-country missions may also be undertaken. The surveys are collecting data on existing schemes and initiatives for each of the identified subject areas. Each survey will be then be reviewed by the Task experts and then used as input to the corresponding Recommended Practice Guide.

A framework for the development of the guidelines has been developed and will involve the following actions:

(i)   Preparation of an experts panel from Task IX and possibly with representation from target groups.
(ii)   Analysis of documents to identify relevant lessons learned.
(iii)   First Draft of the guidelines document.
(iv)   Workshop on the draft guidelines with the experts panel to improve it with their comment.
(v)   Finalising the guidelines document with the comments and improvements of the expert panel.

Guides relating to the following areas will be collated:

- government policy and RE planning
- financing mechanisms
- institutional development
- training programmes
- operation and maintenance of systems
- certification and accreditation
- systems planning
- infrastructure frameworks
- awareness raising.

The Guides will be published over the internet as this is both cost effective and easier than hard-copy publication. It will also facilitate easy and regular updating.

The Recommended Practice Guides will provide a comprehensive series of documents that will be disseminated via appropriate networks. The dissemination of the Recommended Practice Guides is crucially important to ensure that they are implemented and utilised by the organisations at which they have been targeted.

A key goal of this dissemination exercise is to ensure that the guides are implemented on a practical level in real implementation programmes.   In order to achieve this goal, relevant agencies will be encouraged to adopt the Recommended Practice Guides as an integral part of their rural electrification programmes.   A part of this process will involve a series of workshops and seminars, targeted at

relevant institutions in specified countries, as well as manufacturers, utilities, training organisations etc.

3.   SUPPORT AND CO-OPERATION

This second subtask aims to stimulate awareness and interest amongst multilateral and bilateral agencies, NGOs, development banks, on the technical and economic potential, opportunities and recommended practice of PV systems.   This will enable decision-makers to obtain the expertise and knowledge that is required to prepare PV programmes and appropriate PV system deployment.

The objective of the Subtask is to stimulate awareness and interest amongst the target sectors on the technical and economic potential, social implications, opportunities and best practice of PV systems and to establish a dialogue with multilateral and bilateral agencies and development banks.   The objectives will be met through two main areas of activity:

(i)   Support to Multilateral and Bilateral Donors and Development Banks
(ii)   Co-operation with IEA's Renewable Energy Working Party (REWP) and IEA / OECD

The programme of work for each Activity will take the form of:
- Educational seminars and workshops for donor agency, bank and client country staff;
- Information and dissemination services including publications;
- Review of publications.
- Co-operation with the IEA / REWP, IEA / non-member country committee and OECD Secretariats.

The dynamics of the interactions are are shown in Figure 1.

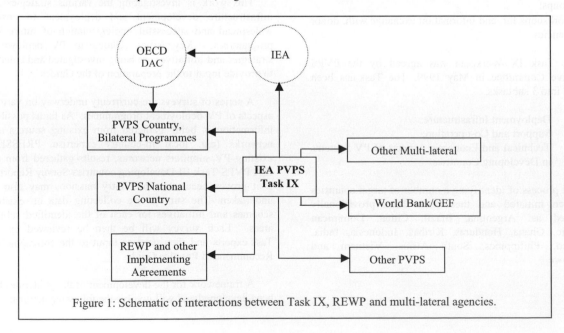

Figure 1: Schematic of interactions between Task IX, REWP and multi-lateral agencies.

Four workshops are planned over the next four years with staff of the multi-lateral agencies. The following agencies have been identified as potential host institutions:

- World Bank Group, Washington.
- United Nations Development Programme (UNDP), New York.
- Asian Development Bank (ADB), Manila.
- Interamerican Development Bank (IDB), in-country-office in a Latin American State.
- African Development Bank (AFDB), Abidjan.
- European Commission (EC), Brussels.

The proposed approach is to structure the workshops in such a way that invited speakers with relevant practical experience can bring their views to the workshop and assist the Task IX team in enlarging the know-how and experience in PV technology deployment.

As a part of these co-operative activiites, the ASTAE Unit of the World Bank has asked the PVPS Programme and Task IX in particular to review a series of QA manuals and associated training programmes. The ASTAE Unit has funded the development of QA documentation on:

- Quality Management in Photovoltaics: Manufacturers Quality Control Training Manual;
- Manual for design and modification of solar home system components;
- Training Manual for Quality Improvement of Photovoltaic Testing Laboratories in Developing Countries;
- PV Installation and Maintenance Practitioner Certification Infrastructure: Development Procedures.

Task IX undertook a peer review of the documents and the associated training programmes. The consensus within Task IX was that there was certainly a need for a quality standard of some kind for World Bank and other programmes and that these manuals went a long way towards addressing this.

Following further discussions with the ASTAE Unit, Task IX is to undertake a further review of the manuals prepared by ISP (PV Installation and Maintenance Practitioner Certification Infrastructure: Development Procedures) and FSEC (Training Manual for Quality Improvement of Photovoltaic Testing Laboratories in Developing Countries). These reviews will be undertaken with a view to further developing the manuals by Task IX.

Another organisation with which Task IX is collaborating is the Renewable Energy Working Party (REWP) of the IEA. Thhe REWP oversees the various Implementing Agreements on bioenergy, geothermal energy, hydrogen, hydropower, wind turbines, solar heating and cooling, solar thermodynamic power (SolarPACES) as well as PV. The REWP also advises the IEA Committee on Energy Research and Technology (CERT) and other IEA bodies on strategy. A Support Unit has recently been established at IEA Headquarters in Paris.

The REWP is paying particular attention to market deployment in developing countries, and therefore has a special interest in PVPS Task IX, as the first IEA task with this objective.

4. TECHNICAL AND ECONOMIC ASPECTS OF PV SYSTEMS IN DEVELOPING COUNTRIES

This subtask aims to investigate the techno-economic aspects and potential of PV systems, and the roles of utilities in developing countries. The work will identify areas of specific concern to developing country applications requiring further research and feed this into other parts of the IEA PV programme.

A Working Group led by Australia with input from Germany, Japan and the USA has been established to finalise the Workplan for the third subtask.

5. CO-OPERATION WITH DEVELOPING COUNTRIES

It is particularly important that Task IX co-operats with experts and institutions within developing countries in order to ensure that the needs of developing countries are properly addressed. In order to develop this co-operation the following activities have been proposed:

- Hold Task meetings in target countries in association with other related events
- Hold workshops in target countries in association with Task meetings
- Target country experts participate in special sessions of Experts meetings
- DC experts to undertake work - assist with surveys etc

6. CO-OPERATION WITH OTHER IEA PVPS TASKS

Part of the role of Task IX can be viewed as 'cross-cutting' the other PVPS Tasks. This 'cross-cutting' element is essentially in two areas:

(i) identifying technical areas of research that need to be addressed from a developing country perspective;
(ii) disseminating the results from the other Tasks to multi-lateral and bilateral agencies, NGOs, banks and to experts in developing countries.

The interaction of Task IX with the other Tasks is shown schematically in Figure 2. The relationship between Task III and Task IX is of particular importance. Task III is addressing relatively advanced technical issues, whereas Task IX will address information exchange and co-operation with international organisations in developed and developing countries, dealing with non-technical barriers in order to avoid duplication. Some of the Activities in Task IX will be undertaken in close co-operation with Task III, in particular the preparation of the Recommended Practice Guides relating to training, certification and accreditation. A number of publications may be published jointly with Task III.

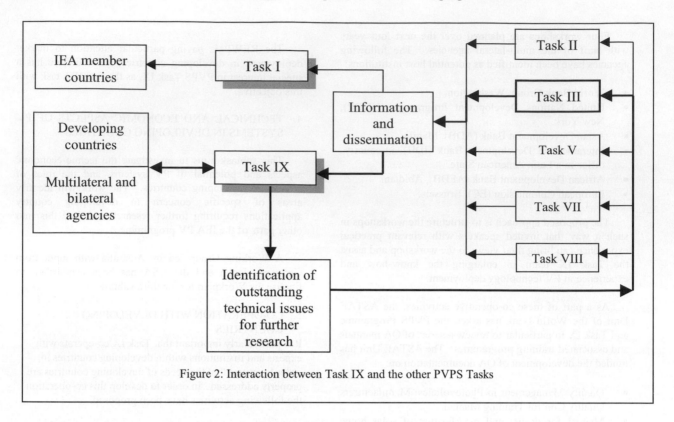

Figure 2: Interaction between Task IX and the other PVPS Tasks

7. CONCLUSIONS

This paper has highlighted the work to date and outlined the overall objectives of the newest IEA PVPS Task. Through the increased co-operation between the pool of expertise within the PVPS Programme and the various donor agencies, it is hoped that the impact of PV implementation programmes in developing countries can be maximised.

The IEA, by addressing the issues of co-operation with non-member countries (essentially the poorer nations of the world), is supporting establishment of the necessary dialogue.

It is hoped that the successful outcome of Task IX will deliver benefits to both the PV industry and the people of the developing world at one and the same time.

REFERENCES

[1] *Survey of Stand-alone Photovoltaic Programmes and Applications in Developing Countries in 1996*, J. R. Bates & A. R. Wilshaw, IEA PVPS Task III Report. September 1999.

THE OPTIMISED USE OF RENEWABLE ENERGIES IN DECENTRALISED RURAL ELECTRIFICATION : EDF'S APPROACH

A. SCHMITT[(1)], R. SOLER[(1)], G. MARBOEUF[(2)], C. NAPPEZ[(1)]

(1) Electricité de France, R & D Division,
1 Avenue du Général de Gaulle, B.P. 408, 92141 Clamart Cédex, France
Tel : 33 1 30 47 65 34 29/33 27 - Fax : 33 1 47 65 32 18
Email : alain.schmitt@edf.fr / robert.soler@edf.fr / christophe.nappez@edf.fr

(2) Electricité de France, International Distribution,
17, Place des Reflets, 92080 Paris La Défense Cédex 8, France
Tel : 33 1 49 02 88 71 - Fax : 33 1 46 92 88 30
Email : guy.marboeuf@edf.fr

ABSTRACT : Electricité de France is developping a global approach for decentralised rural electrification. This approach encompasses the institutional, organisational, technical, tariffs and financial aspects, and leads to the use of renewable energies in optimised techico-economic conditions. Ultimately the aim is to provide not only electricity but multiple services (electricity, water, telephone, ...) to wide rural areas. This method is especially adapted to developing countries in which the electrical system is still restricted to urban areas.

Among the research performed, a specific R&D effort has been done to define a planning software and also to enhance the performance of autonomous systems such as photovoltaic systems, wind energy systems, diesel or hydro micro-plants likely to be used in that context. The R & D effort is also aiming to develop a range of standardised systems.

Moreover, a particular emphasis has been put on financial aspects and especially the determination of tariff structures compatible with the low revenues of rural populations.

Key words: 1: PV System - 2: Rural electrification - 3 : Developing countries

INTRODUCTION

EDF is increasing its international activities and, at the same time, half of humanity is yet awaiting electricity, economical development and improvement of it's quality of life.

So a project called « Energy for all » aiming at building Rural Decentralised Electrification programmes is underway at EDF, in parallel with an R &D programme.

There are few villages remote from the grid in industrialised countries because of the high level of development of the electrical systems. The situation is very different in the rest of the world. To achieve this rural electrification it is necessary to choose between national grid extension and decentralised electrification using autonomous systems. The latter may be based on renewable energy sources. Total grid coverage is often only to be expected in the very long run due to the huge investments needed. Autonomous decentralised systems offer an alternate solution that could help achieving the goal within a few decades only and offer rural populations an access to electricity.

EDF has developed a global approach aiming at defining the electrification policy of a large area of a country or even a whole country. This approach encompasses the institutional, organisational, technical, tariffs and financial aspects. Ultimately the aim is to provide multiple services (electricity, water, telephone, etc.) to rural areas.

This method is especially adapted to developing countries in which the electrical system is still restricted to urban areas. Among the research carried out, a specific R&D effort has been done to define a planning software and also to enhance the performance of autonomous systems such as photovoltaic systems, wind energy systems, diesel or hydro micro-plants likely to be used in that context. Moreover, a particular emphasis has been put on financial aspects and especially the determination of tariff structures compatible with the low revenues of rural populations.

PLANNING RURAL ELECTRIFICATION

In order to develop an electrification policy for any given country, it is firstly necessary to take a long-term view (20 to 30 years ahead) of the electrification target which it is desired to achieve and then to put in place the means for achieving this target in the long run : the Master Plan.

In order not to arrive at an arbitrary separation of the two methods of electrification (decentralised and centralised), it is essential to adopt a global approach to such planning with the following objectives :

- to define an electrification target for the studied region. It consists in choosing between centralised electrification solutions (grid) and decentralised (collectives or individuals). It means to propose an optimal grid extension and to determine which decentralised solution is the least cost option wherever grid extension is too expensive ;
- to propose a time schedule for the electrification of villages in the region relying on criteria taking into account socio-economic, environmental budgetary and political aspects.

To meet these objectives a methodology has been defined. This methodology has been integrated into a computer tool called LAPER dedicated to the planning of rural electrification in developing countries. The latter works in relation to a Geographical Information System (GIS). The users (Ministry in charge of electrification, Financial Institutions, utilities, ...) are thus able to optimise the selection of the electrification modes, to evaluate the investments needed and their profitability and to prepare a work programme. The financial, economic, political and technical aspects are taken into account.

The electrification target can be presented today using geographical information systems (GIS). Each village is thus identified on a map of the country and colour coding can be used to indicate the method of electricity supply adopted.

The master plan, for its part, must enable the optimum date to be determined for carrying out the necessary works. The villages are classified by the decision-maker according to multi-criteria analysis methods (economic and political criteria etc.) for the purpose of planning the electrification works year by year or on the basis of five-year plans.

This entire process provides, period by period, a list of the villages to be electrified, the electrification method used (network, decentralised multi-user systems, decentralised single-user systems), the amount of investments required for each period and the number of structures to be built. These annual or five-year plans can be displayed by the GIS.

The figure 1 shows how the master plan is implemented year after year.

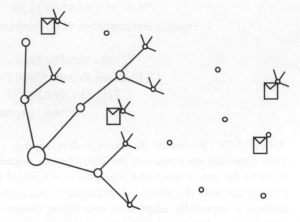

State of electrification at the end of year n

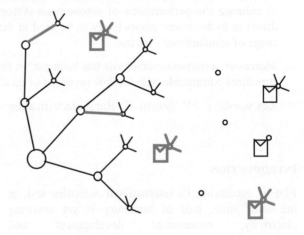

State of electrification at the end of year n+1

 Grid

○   Non electrified rural village

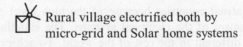 Rural village electrified both by micro-grid and Solar home systems

Rural village electrified by Solar home systems

Figure 1 : Electrification by grid and by renewable energies ; implementation of the master plan

R&D TO PROVIDE INNOVATIVE LOW COST SOLUTIONS

Network services

The triple-phase networks used in our European countries are powerful but much too luxurious and thus too expensive for the electrification of developing countries. EDF has developed a single-phase between phases (dual-phase) network concept, at least 30 % cheaper than the triple- phase system.

A comparison of this concept with North American-type networks. reveals that, for an equivalent or even lower cost, this dual phases network protection system is simpler and thus easier to operate and maintain, thus increasing its reliability.

Furthermore, in this system, transient single-phase faults, which are by far the most common faults, usually affect a large number of customers but are removed in a few fractions of a second by a very brief power cut, or even rendered imperceptible to the customer by means of a shunt circuit breaker. The disruption of networks by low-voltage periods is thus minimised.

Drops in voltage and losses are two to three times lower than in the North American-type networks.

The safety of equipment and persons is also significantly better, due to the greater sensitivity of the protective devices and the lower fault currents. Disruption of the neighbouring networks, e.g. telecommunications, is also lessened by the reduction in stray currents.

Decentralised services, defining a range of systems

The definition of a range of standardised systems and components both for micro-grids and autonomous systems was considered necessary in order to undertake large scale projects. Low cost distribution grid technology have been designed for highly populated areas and a range of standardised autonomous electricity production systems for villages remote from grids or scattered dwellings. The range is based on the use of diesel or hydro micro-plants connected to a local micro-grid or of photovoltaic and wind energy systems.

The range is geared to multi-user systems (Multiple-Home Systems or MHS) and single-user systems (Solar Home Systems or SHS).

The first of these is intended for densely populated rural areas. These systems comprise a micro-production station, a micro-distribution grid and user installations at the premises of consumers. This concept is particularly attractive as regards the pre-electrification stage. In fact, once the national grid

arrives, the production installations can be moved on to more isolated villages and the micro-grid connected to the national grid via a transformer.

Where the national grid is not scheduled to arrive in the village within the period covered by the study, the question then arises of choosing between a centralised multi-user system at village level and single-user systems. This question can be resolved by means of a comparative study of the net present value cost figures for both electrification solutions, but it must also take into account the sociological and cultural aspects. Depending on local mentalities, it may be necessary to opt for a micro-power station and micro-grid (communal) system or single-user systems of the Solar Home System type (individual).

Other considerations may influence the decision, such as the time period for which power is to be supplied. The simplest systems use small diesel engines and a micro-grid is then essential to distribute the energy amongst users. The diesel normally operates for limited periods during the day, e.g. from 7 p.m. to 12 p.m. The power requirement for productive uses, for instance in a village, may also be a criterion for choosing a micro-grid solution.

The use of hybrid micro-power stations allows an enhanced continuity of the power supply to be envisaged. Energy is produced by a photovoltaic field or a small wind-powered machine, and stored in batteries during the day. The energy may be available within the grid for much of the day, or even throughout the day. The diesel generator set is only used as a top-up supply, when the amount of electricity provided by renewable energies is insufficient. The figure 2 shows how a village could be electrified both by community and individual systems.

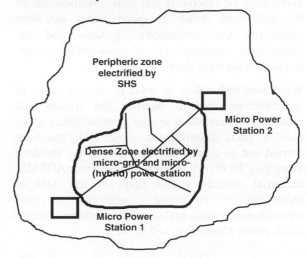

Figure 2 : Electrification of a village both by community and individual electrification systems

However, these systems are more costly in investment terms than those consisting solely of small diesel generators. Yet their net present value costs calculated over fairly long periods may give them an advantage, as they have a long life span, of approximately 15 or 20 years, whereas that of small diesel generators is fairly short (several thousand hours at most, i.e. two to three years of use). When the net present value cost is calculated over a long period and with low discount rates, e.g. over a period of 25 to 30 years with a discount rate of below 10%, it tends to favour renewable energies. If the calculation is made over a shorter period (several years), with high discount rates (15 or 20%), the diesel solution becomes inescapable.

Where the population is very scattered, the single-user system solution may become the automatic choice. Given the low amounts of energy normally involved, these systems can be small in size and their cost is then relatively low, especially if these systems can be produced in large numbers. Given the small amount of energy produced each day, it is absolutely essential to use appliances which consume very little electrical energy, such as low-energy lamps, for instance.

The purpose of the research under way is to determine a range of systems which are as standardised as possible, so as to reduce their cost, improve their reliability and meet the identified requirements of the greatest number. This assumes an in-depth study of the requirements of potential users located in rural areas in the developing countries.

SPECIFICATIONS FOR EQUIPEMENT AND STUDIES

The development of such a range of autonomous systems has needed an upstream study of users' requirements in isolated sites in parallel with the elaboration of functional and tests specifications of the equipment. A set of specifications has been elaborated for components, systems and the institutional environment of renewable energy production for rural electrification.

Work was undertaken in order to write a range of specifications able to improve the quality and reliability of renewable energy systems likely to be used in rural electrification projects. It has been carried out in almost 20 months and has involved more than 30 French experts, coming from ADEME, industrial companies and EDF, in the field of photovoltaic electricity generation, rural electrification, small-scaled wind machines, micro-grids, power electronics, and so on.

They have been accepted as IEC « Publicly Available Specifications ». A second edition is to be published in the second half of 2000.

ADAPTED TARIFF STRUCTURES

The success of rural electrification projects relies on the supply of services commensurate with needs and whose tariffs are in line with the means of the future customers. The tariff structure has to be adapted to make these services affordable to low-income end-users, and recover overall operating and replacement costs to meet debt service and reasonable equity return to project supervisors. While collecting those tariffs, a special care has to be taken regarding the possible losses due to clients not paying their due.

In several undergoing projects dealing with rural electrification in West Africa and Latin America, EDF in association with other partners is promoting such an approach aiming at sustainability in the long run.

CONCLUSION

The approach developed by EDF for rural electrification proposes to build a master plan and to choose the right electrification system for the right place. To compete with the grid solution, the solution of autonomous systems using renewable energies must be as professional and industrial as the conventional solutions. The electrification solutions for the rural population of developing countries must match their requirements and also their buying power but mustn't be handicraft solutions. So, it is necessary to undersize strongly the systems compared to the one used in developed countries, but to ensure a very good quality.

The first step for writing general specifications for the use of renewable energies in rural electrification programmes is now achieved. The DRE Specifications, proposed by the French renewable energies actors, could be a building brick for international standardisation, and may be a stepping stone to international certification. They have already been accepted as IEC « Publicly Available Specifications ».

We think that to ensure the quality of the electrical service provided to the user, it is compulsory to ensure the three main following key points all together : the quality of manufactured products, the quality of carrying out of projects, and the quality of operation, maintenance and renewal.

CLEAN WATER WITH CLEAN ENERGY
A Huge Market for PV - But Not Yet Explored

O. Parodi, K. Preiser, P. Schweizer-Ries
Fraunhofer-Institut für Solare Energiesysteme ISE
Oltmannsstraße 5, D-79100 Freiburg, Germany
Tel: +49-761-4588-281; Fax: +49-761-4588-217
e-mail: orlando@ise.fhg.de

ABSTRACT: World-wide there are 2 billion people living without electricity, and about 1.3 billion without adequate drinking water provision. But why does this happen, even though there are plenty of well known water treatment technologies, and photovoltaic systems can provide electric energy even to the most remote locations of the world?
In an EC-sponsored project, Fraunhofer ISE and its Spanish and Latin American partners investigated a new approach for PV-powered drinking water provision: The situation of rural water supply in 9 Latin American countries was characterised. In parallel, different disinfection technologies were analysed and units available on the market were tested. For 5 pilot villages in Mexico, Argentina and Morocco adapted PV-powered water treatment and disinfection systems were developed and implemented later-on into the communities. Special attention was paid to the careful analysis of the socio-cultural context in the pilot villages. The installations were accompanied by an education campaign and additional measures.
In the course of the project, some major barriers hindering the development of the rural drinking water market could be identified, but also recommendations were elaborated, how to overcome these barriers.
Keywords: Water Provision - 1: Remote - 2: PV Market -3

1. WATER AND ELECTRICITY IN RURAL AREAS

Statistics of the World Health Organisation WHO show, that 80% of all diseases in developing countries are water-born. 2 billion cases of diarrhoea are registered annually, and each year die about 5 million people, mainly children, due to the consumption of not adequate drinking water [1]. The United Nations report, that in 1998 1.3 billion people did not have access to hygienically adequate drinking water. On the other hand, around 2 billion people live in houses, which are not connected to a public electricity grid [2]. Few of them have the possibility to generate electric energy by themselves. This means, that about 17% of the world's population - mostly inhabitants of remote rural areas - actually do not have access neither to safe drinking water nor to electric energy.

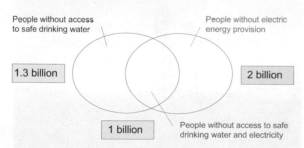

Figure 1: People living without electricity and healthy drinking water

2. "CLEAN WATER WITH CLEAN ENERGY"

The Fraunhofer Institute for Solar Energy Systems ISE in Freiburg, Germany, together with partners in Argentina[1], Mexico[2,3], Latin America[4] and Spain[5], has set up a project

with the aim to develop concepts for the provision of healthy drinking water to people in remote rural areas. The necessary electric energy is - wherever needed - provided by solar photovoltaics: For the provision of a limited amount of electric energy in rural areas this is usually the most suitable solution in terms of economics.

3. DRINKING WATER IN RURAL LATIN AMERICA

Figure 2: Drinking water in San Antonio was taken from an open channel

To develop new concepts for the provision and purification of drinking water, first the main problems existing in rural communities had to be identified. A survey with case-studies, carried out by RIER[4] in 9 Latin-American countries, showed, that surface water, which is one of the main sources for drinking water, regularly is contaminated with anthropogene germs. But chemical

[1] INTA - Instituto Nacional de Tecnología Agropecuaria, San Juan
[2] IIE - Instituto de Investigaciones Eléctricas, Cuernavaca
[3] CIRA - Centro Interamericano de Recursos del Agua, Toluca
[4] RIER - Red Iberoamericano para la Electrificación Rural
[5] IES - Instituto de Energía Solar, Madrid

pollution of rivers and lakes also rises, due to an increasing industry and agriculture. Groundwater and rainwater are normally of good quality, but not always available.

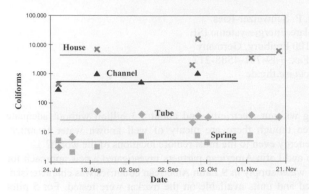

Figure 3: Contamination of the water rises on its way from the spring to the customer (San Antonio de Aguas Benditas, Mexico)

Problems arise while water is stored and distributed to the consumers, as the example of a village in Mexico, San Antonio, shows: While samples taken at the spring, which provides the village with drinkable water, only contain a small amount of coliforms, the contamintion at the end of the tube, which connects the spring with the village, is ten times higher. The water taken from an open channel, which passes through the village, contains 100 times more bacteria than the spring, and the analysis of the water stored in the containers in the houses shows a 1000 times higher level of bacteriological contamination than the original water.

Figure 4: The drinking water source of Nativitas (Mexico) is contaminated from the neighbour-village's waste water

There are many examples where people drink contaminated water, even though they know that this water can make them ill: In on Mexican village, people drink highly polluted surface-water, because it is sweeter than the ground-water, which is extracted from a depth of 150 m and in addition disinfected with chlorine. In Nativitas,

another Mexican village, only 10% of the interviewed people purified their drinking water before consuming it, even though 65% considered the water quality as "bad". The water source was contaminated by the waste water of another village located nearby.

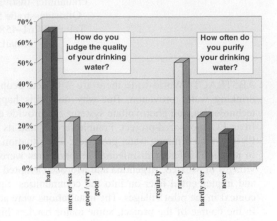

Figure 5: Although people from Nativitas do know, that their water source is highly polluted, few of them purify their drinking water regularly

So how can sustainable improvement of drinking water quality in rural communities be reached? The project-partners of the "Clean-Water"-project have developed a strategy, composed of centralised, mostly automatic water purification on the one hand and user integration on the other hand, combined with intensive training units and supporting measures. In the following an example for the chosen approach is shown:

4. ONE OUT OF FIVE: THE PILOT-VILLAGE BALDE DE SUR DE CHUCUMA

Figure 6: Wind-pumps provide water to the village of Balde de Sur de Chucuma, Argentina

The Argentinean village Balde de Sur de Chucuma is located in the desert of the province of San Juan. The settlement is composed of a central zone with 5 houses and a public place, were little children get a warm meal, of a few dispersal located houses and of another agglomeration of 3 houses, in which the school forms the central point.

Water is drawn out of 7 different wells. Some of them are equipped with a mechanical wind-pump. Although the water level is at more than 20 m depth, the extracted water is biologically contaminated. It must be disinfected.

4.1 A chlorinating-unit for the centre of the village

For the centre of the village chlorinating was chosen as most appropriate disinfecting technology. The heart of the system is a membrane-dosing-pump, which injects sodium-hypochloride to the water, which is pumped from the well to an elevated storage-tank. From that storage the water is distributed to the houses around. The dose of chlorine is controlled by a signal which is proportional to the water flow that feeds the tank.

Figure 7: The heart of the chlorinating unit is the dosing pump

Several dosing units were tested at the Fraunhofer ISE as well as at UPM-IES' laboratories in Madrid, to find out the most suitable one for the application in Balde de Sur de Chucuma. The specific energy consumption of the finally chosen disinfecting unit is 24Wh/m³ at an average flow rate of 170 l/h. This means, that around 100 Wh of electric energy are necessary to disinfect 4 m³ of water per day.

4.2 An UV-disinfecting unit for the school

Figure 8: Disinfection unit with buffer-tank, UV-reactor, control unit, magnetic valve and measurement devices

Another technology was chosen to disinfect the drinking-water for the school of the village: Compared to the application of chlorine, the treatment of the water with UV-light of 254 nm has the advantage of not altering the waters smell and taste. The disadvantage is, that there is no protection against re-contamination, once the water has been disinfected. Therefore, in the context of remote rural regions, this technology is suitable, if water does not has to be stored for a long time or transported over long distances after the disinfecting process.

Another disadvantage of the UV-technology is the fact, that the usually used mercury low pressure vapour lamps need some time (until some minutes) to become effective after having been switched on. Therefore for the school of Balde de Sur de Chucuma a concept with a small buffer-tank for disinfected water (capacity: 50 l) was developed, which is – controlled by a magnetic valve – from time to time filled up by water from a bigger storage tank (capacity: 1 m³), located on the schools roof.

At Fraunhofer ISE's laboratories extensive tests of UV-disinfecting units available on the market have been executed. For the selected device, electric power provision was changed to 12 V DC and an additional control unit was developed.

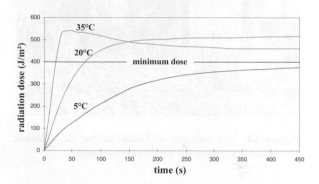

Figure 9: Switch-on curves of the employed UV-device show, that the time to become ready for operation depends strongly on the water temperature.

The disinfecting system finally installed in the school of Balde de Sur de Chucuma consists of a PV-generator of 50 Wp, a Battery-storage of 60 Ah/12 V, a charge-regulator and the UV-unit inclusive control devices. It is able to provide 1 m³ of drinking water per day. The real specific energy consumption is in the range of 160 Wh/m³.

4.3 Accompanying measures in the village

To assure a sustainable operation of the technology installed, a number of accompanying measures were carried out or induced into the community:

- A small PV-pump fills now the storage tank on the school's roof. The tank and the distribution network was restored and optimised.
- In the centre of the village, the availability of water was increased by adding a small PV-pump to the existing system.
- An elevated tank (capacity: 2m³) was included into the windmill's tower and the distribution network was renovated.
- The villagers themselves constructed an operation-house for the chlorinating-unit. In addition they will

build a small community hall. The PV- is designed to provide electric energy for both buildings. Technical data of the PV-system are: System voltage 24 V DC, 2 PV- modules à 50 Wp, 2 industrial batteries 100 Ah/12 V, 3 lamps with 11 W each and a high efficient DC/DC-converter 24/12 V.

- Intensive training units were executed with the villagers, to ensure a proper operation of the water provision and purification systems. Minor technical problems can now be solved within the village. Regular maintenance work can be done in situ as well. In case of major problems and for the control of water parameters a technician of the capital San Juan has to assist to the villagers.

- In a pilot-plantation of algarrobo-trees and in a demonstration garden pupils learn to improve their nourishment and at the same time manage the scarce water recourses.

Figure 10: The villagers of Balde de Sur de Chucuma constructed the house for the disinfection unit

5   LESSONS LEARNED IN THE PROJECT

In the course of the work in the pilot villages as well as in the evaluation of the case studies in the Latin American countries, some interesting experiences could be gained:

- Chlorination is a safe disinfection method, which provides protection against re-contamination - but chlorination is not allways accepted!

- Dosification of liquid chlorine (e.g. NaOCl) has the lowest specific energy demand of all tested technologies. Chlorine concentration of 0.4mg/l at the users tap is sufficant for safe disinfection.

- Where chlorination is not suitable or not accepted, UV-technology presents a more expensive, but feasible and well accepted alternative.

- Drinking water purification systems should be at least semi-automated. Periodic maintenance and controls from responsibles outside the village (e.g. health authorities, concessionaires) must be assured.

- Education and conciousness-forming in an early stage is crucial for long term user acceptance with regard to the introduced purification technology.

6   A MARKET FOR PV-DRIVEN DRINKING WATER PURIFICATION DEVICES

6.1 The major barriers

The market for drinking water purification devices driven by renewable energies still is in an early stage. In the project "Clean Water with Clean Energy" three major obstacles for a comprehensive diffusion of PV-driven water purification devices have been identified:

- Knowledge about the relevance of pure drinking water for health was found to be very poor in rural communities. Hence the health aspect of water is of low priority. Availability, smell, taste and temperature are considered as much more important.

- Norms for drinking water quality normally do exist, but as there is almost no control in rural areas the norms are hardly respected.

- Technical know-how is limited in rural regions as well as an infrastructure for maintenance and spare part provision. Complicated technologies can not be operated in this context. Drinking water purification devices available on the market are neither designed for rural regions nor for renewable energy systems.

6.2 Measures to explore the market

But in the project "Clean Water with Clean Energy" also strategies to overcome the obstacles mentioned above could be developed. Actions towards the exploration of the big potential market for decentralised drinking water purification systems are:

- In developing countries the key players in rural water provision (administration, health authorities, utilities, research institutions, private companies and user co-operatives) must be brought together to start discussion and to set up master plans for integrated infrastructure development.

- Local health authorities must be encouraged to extend their activities to rural areas. At the same time they should be made aware of the possibilities and advantages of decentralised water purification units operated by renewable energy sources.

- Purification units designed to attend the huge market of rural drinking provision must be developed and made available for a reasonable price. Those units must be energy-efficient, easy to operate and to repair, reliable and preferably semi-automated.

- The implementation of drinking water purification technology into a village should be accompanied by a process of awareness forming with regard to the relevancy of water for peoples' health. As this process may need a long time, actions to increase health-related water quality should be escorted by additional measures, which can be perceived as immediate benefits for the water customers.

Fraunhofer ISE is currently working on all these topics, in different projects and together with private companies, to explore the huge future market of rural drinking water provision.

REFERENCES

[1] World Health Organisation WHO (1993): Guidelines for Drinking Water Quality. Volume I: Recommendations, Geneva, Switzerland

[2] United Nations Development Report UNDP (1998): Human Development Report 1998, Goetzky, Bonn, Germany

ACKNOWLEDGEMENT

The Project "Clean Water with Clean Energy" was supported by the European Commission, DG XII, in the frame of the INCO-DC programme, contract number ERBIC18CT960104.

RESOURCE-CONSERVING IRRIGATION WITH PHOTOVOLTAIC PUMPING SYSTEMS

Andreas Hahn
Deutsche Gesellschaft für Technische Zusammenarbeit (GTZ) GmbH
P.O. Box 5180, D-65726 Eschborn, Germany

Reinhold Schmidt, Ariel Torres, Amador Torres
Universidad de Tarapacá, Centro de Energías Renovables
Casilla 6-D, Arica, Chile

ABSTRACT: In areas of developing countries with no access to grid power, diesel-driven pumps are commonly used for irrigation in farming and forestry. Diesel pumps, however, require regular maintenance and refuelling, and they place a burden on the environment. Wherever the failure of a diesel pump and the expense of its operation constitute an economic risk for the operators of small-scale irrigation systems, photovoltaic pumps (PVP) represent a reliable, alternative means of water delivery.

Keywords: PV Pumping - 1: Irrigation - 2: Developing Countries - 3

1. INTRODUCTION

The use of photovoltaic pumps for irrigation represents a promising option for using solar energy productively and for generating income.

However, the use of photovoltaic pumps for small-scale irrigation is still held back by a lack of information and practical experience [1]. This kind of information is very important for convincing people in the irrigation sector that it is worth considering the option in the first place. To bridge this gap, the pilot project entitled "Resource-conserving Irrigation with Photovoltaic Pumping Systems" was started early in 1998 with a duration of four years.

2. PVP IRRIGATION PILOT PROJECT

2.1 The Role of GTZ and its Partners

The pilot project, financed by the German Federal Ministry for Economic Cooperation and Development (BMZ), is being implemented by the Deutsche Gesellschaft für Technische Zusammenarbeit (GTZ) GmbH, in cooperation with national project executing organizations in the following countries:

- Ethiopia; Bureau of Agriculture and Natural Resources
- Chile; Centro de Energías Renovables, University of Tarapacá
- Jordan; Ministry of Agriculture (in preparation)

The interdisciplinary character of the pilot project calls for close cooperation with experts from diverse professional areas. With a view to introducing sustainable dissemination processes, the suppliers of PVP irrigation systems with their local structures, together with national and private-sector institutions, are being involved in the project activities.

2.2 Objectives

The project is designed to clarify whether photovoltaic pumping systems can be used to irrigate high-quality crops in a cost-effective and resource-conserving manner, and what management and technical requirements must be met in order to operate a PV-based irrigation system. In the course of the project, 10 pilot systems are being field-tested and intensively monitored at selected locations in Chile, Ethiopia and Jordan.

The field-testing of PV-based irrigation systems is intended to enable the users and operators of the pilot facilities to assess and evaluate the technology. Therefore the project places great emphasis on upgrading of project partners and training of system users.

2.2 Target Group

Due to the comparatively high initial investment cost (see 4.), PV irrigation is not a solution for the individual subsistence farmer.

The project focuses primarily on peri-urban small and medium-size farms that use energy- and water-conserving forms of irrigation to grow cash crops on up to 3 hectares of land, hence generating income that could be used to finance a PV-based irrigation system.

Up to now, four pilot plants (0.3-1.2 kW$_p$) for cash-crop production have been installed in the Atacama Desert in the northern part of Chile. In order to investigate the application of PV-based irrigation systems in forestry, three pilot plants (0.3-0.4 kW$_p$) have been installed in Ethiopia to supply tree nurseries with irrigation water. The aforementioned systems have been in operation for about two years. Another three systems for cash-crop production will be installed in Jordan in the year 2000.

3. TECHNICAL ASPECTS

3.1 Operating Principle

The operating principle behind any photovoltaic irrigation system is quite simple. A solar generator provides electricity for driving a submersible motor pump, which in turn pumps water into an elevated water tank (see fig. 1).

The water tank bridges periods of low insolation and supplies the pressure needed for the irrigation system. One major advantage of solar pumps is that they do not require batteries, which are expensive and need a lot of maintenance.

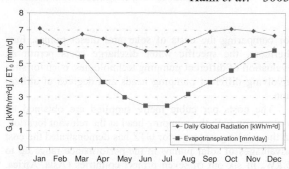

The maintenance of a PV-based irrigation system is restricted to regular cleaning of the solar modules. Depending on the water quality, the only moving part of the system, the submersible motor pump, has to be checked every 3 to 5 years.

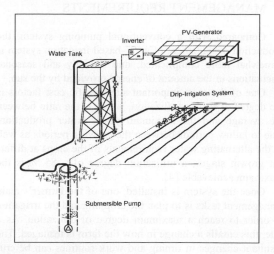

Figure 1: PVP standard system combined with a conventional drip irrigation system

Force of gravity causes the water to flow from the tank to the fields. The PV Irrigation Pilot Project is also testing units that feed directly into the irrigation system and therefore do not require a water tank.

Drip irrigation permits economical use of water, and its relatively low operating pressure makes it particularly well-suited for combination with photovoltaic pumping systems. However, it must be pointed out that such hi-tech irrigation systems are not necessary. Alternative irrigation techniques are possible, as long as they are water- and energy-conserving.

3.2 Technical Monitoring

In order to permit continuous monitoring of crucial operating parameters, all pilot systems are equipped with automatic data acquisition systems. In addition to the technical evaluation of the performance of the PVP irrigation systems employed, the recorded meteorological parameters (see fig. 2) facilitate management of the irrigation systems, and the water consumption data provide information on the degree of system utilization.

While the technical aspects of solar irrigation are generally regarded as adequately developed, a closer look reveals that there is still need for research at the laboratory and field-test levels. In that context, the design and testing of special low-pressure irrigation systems, including suitable filters and fertilizer injection devices, must be given special attention.

The Chilean project sites were equipped with locally available, low-pressure drip systems, suitable for operation with PVP equipment [2]. The field data confirmed that the systems are very reliable and operate at low system pressures of the order of 0.2–0.3 bar (tank height), thereby guaranteeing a uniform supply of water to the field (measured uniformity coefficient > 95 %).

Figure 2: Recorded daily global radiation (G_d) and evapotranspiration (ET_0) at pilot site Chaca, Atacama Desert, Chile

This presumes, of course, that all components of the drip irrigation system have been designed for such a low system pressure. The uniformity of water distribution must also be guaranteed for variable pressure conditions.

In PVP standard systems (see fig. 1) pressure fluctuations are compensated by the water tank. However, building such a tank involves much cost and effort, accounting for a correspondingly large share (35 %) of the capital outlay for a typical PVP system. On economic grounds especially, direct pumping of water to the field is advisable.

4. ECONOMIC CONSIDERATIONS

4.1 Investment Costs

Although the advantages of solar technology are evident, purchase decisions are often taken in favour of the competing diesel-powered systems. The comparatively high investment costs of the solar system are critical here. Figure 3 shows the price of a PVP system related to power generation.

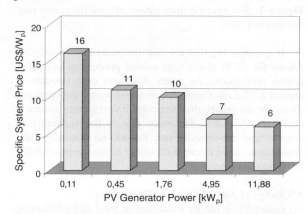

Figure 3: Specific PVP system prices (ex-works) without water tank, fence and surface piping (Source: Siemens Solar)

Today the operator of a ready-to-use solar pump pays about 3 times as much as would be needed for a diesel pump with the same performance [3]. However, it is frequently overlooked that after installation the solar system incurs only a fraction of the operating costs of a diesel pump. Consequently it does not make economic sense to compare different technologies solely on the basis of the investment costs.

4.2 Specific Water Discharge Costs

The specific water discharge costs [Euro/m⁴], covering both investment and operating costs, are taken as a basis for comparing the costs of solar and diesel pumps. Furthermore, the specific water discharge costs permit an evaluation of different pumping technologies, even for sites involving different pumping heads and degrees of utilization.

The costs per cubic metre supplied are obtained by multiplying with the pumping head at the relevant location. In the drinking water sector GTZ has demonstrated the cost advantages of solar pumps in the performance range up to 2 kW$_p$ in six out of seven project countries (Asia, Africa, Latin America) [3].

First results of photovoltaic water pumps applied in small-scale irrigation systems are promising, although after two years of operation the present economic data base is still insufficient for a final evaluation. However, it does suffice to identify a basic trend.

To illustrate this point, fig. 4 shows the latest results of an economic study at Vitor, one of the Chilean pilot sites.

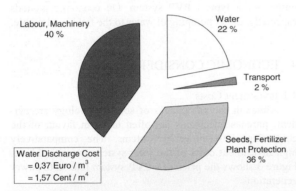

Figure 4: Site-specific distribution of cost for potato production in Chile

It is surprising that the cost for water supply only accounts for 22 % of the total annual production costs. For potato production, the water discharge cost amounts to around 1.57 Euro-Cent/m^4. Compared with a conventional diesel-driven pump, this particular PV-irrigation system is economically viable. However, due to the variability of country and site-specific cost factors, no generally valid conclusion can be drawn with regard to the overall viability of photovoltaic pumps.

4.3 Range of Application for PV Irrigation

Generally, it can be stated that as long as the following site-specific conditions apply, economic advantages of photovoltaic pumps over competing diesel pumps can be expected:

- arid/semi-arid climate
- no access to the public power grid
- problems with the maintenance of diesel pumps and the supply of fuel for their operation
- pumping head up to roughly 30 metres
- max. field size of 3 hectares
- cultivation of high-quality crops for secure markets
- use of water-conserving and energy-saving methods of irrigation (e.g. drip irrigation)
- high degree of system utilization through adoption of permacultures or systematic crop rotation

However, a site-specific analysis of the economic viability should always be carried out before a decision on investment is taken.

5. MANAGEMENT REQUIREMENTS

Compared with a conventional pumping system, the production management of a PV-based irrigation system is somewhat more complicated, due to daily and seasonal fluctuations in the amount of energy provided by the sun.

One of the most important site-specific cost factors is the degree of system utilization, which is the ratio between the average and the maximum yearly water production. Due to fallow periods between the growing periods as well as the alternating water requirements of the crops at different growth stages, a utilization degree of ≈ 85 % is the maximum achievable [4].

Once the system is installed, one of the farmer's main management tasks is to plan the cropping and the irrigation in order to reach a maximum degree of utilization. As a rule, this entails a change in how the farm is managed. The resultant changes in timing and work routines can be crucial for the acceptance of PVP technology.

One advantage of diesel-driven pumps is the direct availability of water after pump start-up. Some farmworkers seemed to have far more difficulties in getting used to handling a PVP-system than their employers. Accustomed to the "instant-power effect" and high pressure of diesel pumps, some farm-workers complained that the PVP system delivers too little water. In spite of initial doubts, the farmers and workers are now showing a high degree of acceptance.

The high level of reliability and low maintenance requirements of PV irrigation systems are the features most appreciated by Chilean farmers, who are used to facing daily problems with their diesel-driven pumps. That fact alone has contributed much to the acceptance of this new technology.

6. ENVIRONMENTAL IMPACTS

6.1 Life-Cycle Analysis of PVP and Diesel Pumps

Within the scope of the pilot project, the environmental impacts of photovoltaic and diesel-driven irrigation systems were investigated. Together with the *Institute for Applied Ecology*, GTZ has carried out a life-cycle analysis, comparing the two different technologies [5].

The study incorporates a calculation of greenhouse-gas emissions (as CO_2 equivalents), acid-air emissions (as SO_2 equivalents) and cumulated energy requirements (CER), as well as qualitative environmental impacts, such as the pollution of water and soil by diesel oil. In addition, energy productivity factors and energy payback times of three PV technologies (monocrystalline, multicrystalline and amorphous silicon modules) have been determined.

The life-cycle comparison is performed for conditions in sun-rich developing countries (assumed solar irradiation 2000 kWh/m²*a) and analyses the so-called "cradle-to grave pathway", including the manufacturing process, transport, operation and partial recycling of system components.

The results of the life-cycle comparison show that the greenhouse-gas emission balance of PV-pumps is approximately 10 times smaller than that of the diesel system (see fig. 5). For acid-air emissions, diesel and PV systems differ by a factor of at least 50.

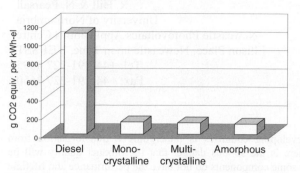

Figure 5: Life-cycle comparison of greenhouse-gas emissions

6.2 Energy Productivity Factors and Payback Times

The energy productivity factors and energy payback times are considered to be further indicators for the indirect environmental burden resulting from solar modules. The energy productivity factor is defined as the quotient of the total energy supplied by an energy system during its lifetime and the total manufacturing input for this system. It thus describes the energy productivity of the system.

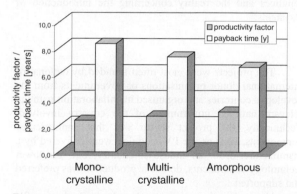

Figure 6: Energy productivity factors and energy payback times of PV modules

The energy payback time is calculated by dividing the energy-specific manufacturing input for an energy system by the amount of energy it supplies annually. This thus corresponds to the period in years after which an energy system has "paid back" the cost of its production through the energy it supplies.

As fig. 6 shows, both the energy productivity factors and energy payback times of PV modules are comparatively favourable in sun-rich regions.

In summary, PV modules are significantly less of an environmental burden than the diesel reference system, even for conservative assumptions regarding module lifetime, rack and frame construction. In the life-cycle comparison with fossil-fuel systems, the foreseeable developments of PV manufacturing technologies will further increase their advantages. Other aspects, such as contamination of soil and ground-water resources can be completely avoided when deciding in favour of the PVP option.

7. CONCLUSION

In conclusion, one can state that in principle, PV irrigation systems are suitable for small-scale irrigation purposes in farming and forestry. Photovoltaic pumps require little maintenance and no fuel and therefore often constitute the only reliable solution to the problem of irrigation water supply in remote areas. The adaptation of a conventional irrigation system to the photovoltaic pump still leaves scope for component improvement.

First results concerning economic efficiency confirm that PV-pumps are able to yield cost advantages over diesel-driven pumps, as long as certain site-specific conditions apply. However, the high initial investment costs are still the main obstacle to distribution of PV pumps. Therefore it is necessary to compensate for the high investment costs by providing loans on favourable terms via development banks or through other suitable financing models.

Besides the purely financial evaluation, additional criteria are needed for an overall evaluation of PVP-technology. Fuel and lubricants for diesel pumps often pollute wells, soil and groundwater. By contrast, photovoltaic pumps are an environmentally sound and resource-conserving technology. This fact, together with the high level of technical reliability, has contributed much to the farmers' acceptance, in spite of initial doubts.

REFERENCES

[1] Müller J., Internationaler Sach- und Wissensstand zum Thema Photovoltaische Bewässerung, GTZ, Eschborn (1999)

[2] Torres E., Schmidt R., Ovalle R., Flores C., Torres A., Hugo Escobar A., Implementación de un programa piloto para nuevas aplicaciones de riego tecnificado utilizando bombas solares fotovoltaicas, Congreso Internacional de Energias Sustentables SENESE X, Punta Arenas, Chile (1998)

[3] Posorski R., Haars K., The Economics of Photovoltaic Pumping Systems – Excerpt of Cross Section Study, GTZ, Eschborn (1994)

[4] Heitkämper K., Management and Profitability of Photovoltaic Operated Pumping Systems for Irrigation in Northern Chile, University of Hohenheim, Stuttgart (2000)

[5] Fritsche U., Life-Cycle Analysis of Photovoltaic and Diesel Pumps for Irrigation in Developing Countries, GTZ, Eschborn (2000)

Developing Countries - Developed Quality.
A discussion on the feasibility of locally produced quality products for solar home systems for rural electrification

C. Helmke & H.A. Ossenbrink
DG-Joint Research Centre
Environment Institute - RE - ESTI
I-21020 Ispra (VA), Italy
Tel: +39-0332-789 119
Fax: +39- 0332-789 268
Email: claas.helmke@jrc.it

R. Hill & N. Pearsall
University of Northumbria
Newcastle Photovoltaics Applications Centre
Ellison Place, Newcastle upon Tyne, NE1 8ST
Tel: +44-191-227 4595
Fax: +44-191-227 3650

The theme under which PV systems are installed in developing countries is undergoing a transformation from international aid programmes towards a commercial market. There is the serious danger that the product quality will be sacrificed for the low initial investment cost. There is concern that some components do not have the performance and lifetime which are required, damaging the reputation of PV systems in a number of countries and towards the donor organisations. The widespread adoption of PV will not occur if the customer and the client have bad experience of non-functioning systems. If the photovoltaic market is to mature, drastic improvements in terms of component and system quality, local involvement in selection and maintenance must be established as soon as possible. If industrial enterprises from developed countries use the LDC's as a market for low quality products, the entire PV market is in serious danger of completely losing its reputation both from the end users and the financing institutions, hence the commercial market will never become a wide-spread reality. In the authors' view, international standards for PV products, testing of PV systems, components, and the introduction of one common quality system are absolutely essential to ensure:
⇨ Continuity of PV electrification projects and growth of the commercial market
⇨ Reliability of systems
⇨ Local production of quality products at competitive cost
This paper describes the discrepancies between general prejudices and the reality concerning the introduction of international standards and quality systems in the field of solar home systems.
Keywords: Quality - 1, Rural Electrification - 2, Standardisation - 3

1. STATE-OF-THE-ART

All over the world rural electrification (RE) projects have been launched to provide electricity to inhabitants who have no access to public electricity supply using photovoltaic (PV) stand-alone power systems. Such systems usually consist of:

- a photovoltaic module
- a battery
- a charge regulator
- the loads (lamps, radio, TV)
- switches, cables and sockets
- the mounting structure

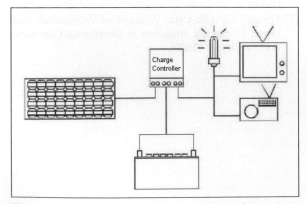

Figure 1: drawing of a stand-alone photovoltaic power system

The projects were most often founded by national or international donor organisations or governments from the developed countries and organised in collaboration with the relevant national government of the country involved. Technically, the project set-up was that sophisticated components, such as the PV modules, were delivered by a manufacturer from an industrial country, but, for lower technology components, domestic production was preferred and supported.

Until recently, no international standards for PV systems and a few standards for PV systems components were available. Hence, new technical specifications had to be designed for each single implementation project, considering the locally available technologies for the production of components on the one side and, on the other hand, accounting for already existing national codes and standards for the production of components.

The most reliable components of PV systems are the PV modules. International standards for the assessment of their quality have existed for many years and are well established. But the lack of global available and accepted quality standards for the testing of other components and the complete systems led to the result that the majority of PV systems do not match the performance criteria as requested in the contract.

The continuous failure of complete PV projects discredits the entire PV industry and limits the progress of providing electricity to developing countries by clean energy sources.

2. WHAT IS NEEDED

To give a clear statement in terms of what is needed to improve the situation described above, one certainly needs to define the base parameter:

What do we want ?

Surveys performed in the past [1] state clearly that performance and reliability problems can be overcome by the integration of a quality system, when implemented at an early stage of the project.

2.1 Definition of QUALITY

Many institutions are engaged in the field of quality, quality systems (QS), quality assurance (QA), quality control (QC), and total quality management (TQM). A large variety of ways in which quality influences the character of a company and the flow of information and processes are available, together with many descriptions on how and where quality can be implemented. Only a few definitions of what quality actually means are present or reproducible in a short and comprehensive way. One of these is provided by Duane A. Floyd [2]: 'Quality is the totality of customer desired attributes', with desired attributes such as:

⇨ Functionality
⇨ Reliability
⇨ Price
⇨ Delivery
⇨ Customer Service
⇨ Aesthetics, and many other things.

2.2 What do we want ?

Many RE projects are set up in a way that the initial investment costs of the PV systems are covered by national/international donor organisations, granting a loan to the end-user and receiving a monthly payback calculated on the amount the user would normally pay for alternative energy supply (kerosene, central battery charging etc.). These organisations, which are involved in creating and designing rural electrification projects, want a positive outcome of the project. The erected systems shall work after installation, and preferably shall work after a pre-defined period of time.

The longer the PV systems works, the higher is the opportunity to pay back the loan granted at the beginning of the project. That means project partners in rural electrification projects want PV systems which work reliably for a certain period of time.

The PV industry needs to maintain its reputation of producing and marketing reliable products. Even though the components which fail in RE projects are usually not the PV modules, but other electronic components and/or the battery, their failure comes back to the PV industry because most PV systems are marketed by the PV industry. That means the PV industry wants and needs reliable and adapted components to assemble reliable PV systems.

The end-user, who actually purchases the PV system, expects that this system will perform according to the specifications, covering his electricity needs, as he has to pay for this service on a monthly basis. He is certainly less willing to pay continuously for a system which does not provide the service promised.

The end-user wants a system which provides the specified service at the best price. It is clear that we all want a GOOD photovoltaic system. Combining the above points, we can define such a good system as:

"A good PV system delivers a specified service over a guaranteed period of time at the best price."[3]

Comparing the above statement with the previous definition of quality, it becomes obvious that we want a QUALITY photovoltaic system, as the definition for quality and the statement of what we want to obtain match very well:

Functionality	=	a good PV system delivers
customer service	=	a specified service
reliability	=	over a guaranteed period of time
price	=	at the best price

2.3 How to obtain a quality PV system?

For the PV industry, especially in developing countries, comprehensive manuals have been produced by the World Bank. These manuals provide all the necessary information and explanations regarding how a quality assurance program can be implemented and are available for the following areas.

⇨ manual on the quality assurance and approval process for PV manufacturers
⇨ manual on the certification requirements for testing laboratories in the field of PV
⇨ manual on quality control for installation and maintenance
⇨ manual for design modifications to improve quality

2.4 Obstacles for implementing a quality system

Manufacturers from developing countries are reluctant to implement a quality system due to a set of pre-existing prejudices. The majority of these prejudices are the result of a lack in understanding the necessity, together with a deficit of a clear definition of what a quality system is and what it means. If certification of the quality of products is not requested, this in the worst case opens the door for pure fraud.

Picture 1: Example of a zero-quality charge controller [4]

Picture 1 shows the most obvious example of such a product. Displayed is a locally produced product, which is sold as a charge controller to protect the battery from over charge and deep discharge. Dismantling the device it became obvious that no service or protection can be expected; the connections were simply wired together. This is fraud, which can not always be avoided. More damaging for the photovoltaic market are poor or non-performing products, which appear on the market with a good and serious intention.

2.4.1   The lack in understanding of the necessity of a quality system

The lack in understanding of its necessity results in statements like:

⇨   our products work fine, so we don't need a quality system, or even:

⇨   we know what we are doing, why waste time writing it down, and so on.

Reality shows that, if no proof of quality is requested by the client or customer, products appear on the market promising all the features needed, but which are just not able to fulfil them.

Another point is the cost effectiveness of quality products. Figure 2 provides a comparison of the costs of ownership of solar home systems over a period of 25 years, including initial investment costs. The solid line represents the costs of a quality system, whereas the dashed line characterises the cost development for a low quality PV system. Considering a total initial investment of 450 Euro for the bad system and 700 Euro for the quality system, the initial investment for the quality system is 35% higher, but:

⇨   already after two years, the quality system works out cheaper, considering the shorter life cycle of low quality products and hence their replacements costs,

⇨   after 12 years of operation, the bad system is double the price of the quality system, including the initial lower purchase price of the bad system.

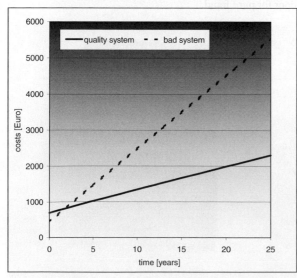

Figure 2: Analysis of the costs of ownership of SHSs

Table 1 summarises the assumptions for the purchase prices of the components and their life times. The components to create a PV system and the loads, like lamps and a TV set, are considered. Performing the calculation

without the costs for a TV set would lower the absolute investment and ownership cost, but the relation between the two systems would remain almost the same.

	Quality		no Quality	
	price [Euro]	life time [y]	price [Euro]	life time [y]
PV module (50 Wp)	300	25	250	5
battery (100 Ah)	120	6	80	1
charge controller	40	25	1	1
cables etc.	20	25	5	2
lights	60	4	10	0.5
TV-set	150	10	100	2
Total purchase cost	**690**		**446**	

Table 1: comparison of purchase cost and lifetime of quality and non-quality PV components

3.   THE DEFINITION OF A QUALITY SYSTEM

A simple approach to explain what a quality system is might be to say:

A Quality System is a tool to:

⇨   say what you do

⇨   do what you say

⇨   prove it

⇨   improve it

A quality system (QS) consists of five levels:
1.   Quality Policies are high level documents describing the company's commitment to quality. They correspond to sections from the Standard.
2.   Quality Procedures are more detailed documents describing the systems that have been set up to assure quality.
3.   Work Instructions are documents written in a step by step format so that the team can learn how to produce quality in each of their job processes.
4.   Quality Records are databases that create and maintain the records over time.
5.   Corrective Action Requests is a process of recording a problem, analysing the root cause of that problem, proposing a solution, approving that solution, implementing it, and then evaluating its effectiveness.

4.   PREJUDICES AND RECTIFICATIONS CONCERNING THE IMPLEMENTATION OF QUALITY

The previously explained obstacles, the lack of understanding, and the deficit of definitions, result in an excessive supply of prejudices against the implementation of quality and quality systems. Some examples and rectifications are listed subsequently.
1.   Standards are difficult to obtain

Fact is: standards for electrical components including PV can be obtained from the International Electrotechnical Commission. For cases where international standards are not yet available, recommended specifications for PV components and systems can be obtained from PV GAP, the photovoltaic global approval program. (www.iec.ch, www.pvgap.org)

2. Standards are too complex and difficult to understand

A set of documents has been developed by the World Bank, covering the most important aspects of implementing quality (www.worldbank.org):

⇨ training manual on the quality assurance and approval process for PV manufacturers

⇨ training manual on the certification requirements for testing laboratories in the field of PV.

⇨ training manual on quality control for installation and maintenance

⇨ training manual for design modifications to improve quality

3. Standards are not available

The process to produce an IEC standard is time consuming. Therefore the actual number of standards for PV applications is quite remarkable; currently, 25 IEC standards related to PV are available, of which 8 are related to PV systems. 7 IEC standards for stand-alone systems are in the pipeline

4. A quality system increases component, system and production cost

Fact is: the training of employees and the implementation of a QS costs money and the testing and certification of components and systems costs money as well, but once implemented,

⇨ a QS reduces the percentage of faulty products, which certainly reduces and not increases costs

⇨ "do it right the first time", a typical quality slogan, is a proven way for increasing profitability

⇨ a QS increases, instead of decreases, profitability through lower failure rates, higher reliability, and less warranty claims for repair

⇨ ensuring reliability and consistency of the products results in lower unit costs.

5. Requiring a QS excludes local manufacturers

The contrary is the truth,

⇨ strengthening the local market by quality products against global competition and avoiding the trap of local products only for local markets,

⇨ if the donor requests components and systems to be of approved quality, this will ensure that local companies have a market for quality products, and that they can compete and sell even outside the home country,

⇨ requesting the implementation of a quality system will help local companies to be internationally competitive, which will provide a major boost to the industrial development of the host country.

5. CONCLUSION

The local production of quality products is feasible, if:

⇨ a quality strategy for developing countries is developed, which must play an integral part of the project layout in future RE projects.

⇨ this ensures the internationalisation and hence a wider market for local produced quality products

6. REFERENCES

1 CESIS-PV, JOULE project of the European Commission, MTA report, not for publication

2 Duane A. Floyd, Member of the American Society for Quality Certified Quality Engineer and Certified Quality Auditor

3 Contribution during the IEC TC82 WG3 meeting, Albuquerque, US, December 1999

4 Picture © by University of Madrid

QUALITY CERTIFICATION FOR PV INSTALLATION AND MAINTENANCE PRACTITIONERS

Mark C. Fitzgerald
Institute for Sustainable Power, Inc.. P.O. Box 260145, Highlands Ranch, Colorado 80163-0145 USA
Phone: 303-683-4748. Fax:303-470-8239. e-mail: markfitz@ispq.org

ABSTRACT – In 1999, The World Bank's Asia Alternative Energy Unit, with funding from the government of The Netherlands, supported four organizations to develop and implement quality PV infrastructure programs. These programs, along with their implementing organizations, are (1) quality component design (ECN, The Netherlands); (2) quality PV component manufacturing (PV GAP, Switzerland); (3) quality system testing programs (Florida Solar Energy Center, USA), and, (4) implementing quality training accreditation and practitioner certification programs (Institute for Sustainable Power, USA). These programs were piloted in India, Sri Lanka, China, and South Africa. The focus of this paper is on the implementation of training accreditation and practitioner certification programs in India, Sri Lanka, and South Africa.
Keywords: Developing Countries – 1: Rural Electrification – 2: Reliability – 3

1. INTRODUCTION

Around the world, development organizations and government agencies are keenly interested in accelerating the market deployment of sustainable technologies for both environmental reasons (global climate change mitigation) and poverty alleviation. Unfortunately, experience has shown that deployment is not enough – without an adequate infrastructure (to successfully design, manufacture, test, market, install, and service the technical solutions), many systems have proved unreliable and unsustainable, and the sponsors have missed the opportunity to create local jobs.

With a significant financial commitment to implementing renewable energy projects and systems worldwide, The World Bank's Asia Alternative Energy Unit (ASTAE) identified quality infrastructure development and standards as key components to ensuring the long-term success of their renewable energy programs. To support the development of such an infrastructure, in 1998 the ASTAE staff applied to the government of The Netherlands to support quality infrastructure development as a component of the Dutch commitment to mitigating global climate change gasses (here, through the increased use of renewable energy technologies).

With the success of this proposal, in 1999 the ASTAE unit contracted with four organizations to each take one infrastructure component and develop appropriate quality guidelines that could be used and implemented by countries interested in improving the quality and reliability of their photovoltaic (PV) systems implementation projects, while at the same time creating local capacity. The ASTAE staff abbreviated the title of this quality program for PV as QuaP-PV.

2. ASSUMPTIONS

As the world becomes a more connected place, and as rural and remote populations become increasingly aware of the services and benefits available to those with access to services like electricity, there are growing social, political, and economic demands on governments to provide such services. This causes a dilemma. If electric service is not provided to these communities, social and economic pressures increasingly will drive people to leave the rural communities and migrate to the already-overstressed urban areas. Realistically, though, the option of providing electric

grid service to these communities is far beyond the financial wherewithal of any government, though many continue to install grid extensions to some and promise it to others. The real problem, though, is that if these governments and utilities could supply grid power to their constituents in rural and remote communities, their most likely choice would be conventional fossil-fuel generators, which bring with them significant environmental consequences.

With an increasing awareness that PV can provide a useful, environmentally benign electricity option for these rural and remote communities, development and donor organizations have sponsored and supported a number of large-scale projects to test the technical and practical aspects of the PV option in real-world applications. Unfortunately, over time, experience is showing that an inordinately large number of these projects are proving unsustainable [1]. On review, though, comparing the successful projects with the unsuccessful ones has provided guidance indicating that a few common, addressable issues are responsible for most of the failures – most of these related to issues of quality at any of a number of points in the product cycle, project development, and project implementation.

Based on this experience, and the need to improve the success of PV project implementation efforts, the staff of The World Bank's ASTAE unit believed that the implementation of quality systems at a number of key points in the project cycle could dramatically improve the viability and sustainability of PV projects and programs, in addition to creating local capacity. To implement this, the staff felt that it would be appropriate and important to develop resources that could be used at the organizational and national levels to implement and maintain these quality systems.

3. PROJECT FRAMEWORK

In reviewing both sustainable and unsustainable projects, along with the recommendations developed for its own document "Best Practices for Photovoltaic Households: Lessons from experiences in selected countries" [2], the Bank's ASTAE staff identified four aspects of the PV project cycle where better quality guidelines and practices might improve the projects and make them more sustainable and beneficial to the local communities. These were

- Product Design
- Product Manufacturing
- Component and System Testing
- System Installation and Maintenance

To address these, the staff outlined a number of resources, activities, and deliverables that would provide quality guidance both for the PV industry and the client countries. These included the development of guidance manuals for implementing and maintaining quality systems in each of the above four areas; a series of workshops to pilot test the manuals and implement the quality systems with the appropriate stakeholders; and, a review of the process and deliverables by clients and subject matter experts.

Initially, the staff targeted the testing and implementation of this program in three ASTAE client countries: China, India, and Sri Lanka. Later, it was decided that, as the scale of the PV industry and market was so different among these three countries, a fourth country (South Africa) would be added, and different infrastructure components would be tested in different markets.

To develop the quality guidelines, four contractors were chosen:

- ECN, The Netherlands, for Component Design
- PV GAP, for Quality Manufacturing
- Florida Solar Energy Center, for Quality Testing
- Inst. for Sustainable Power, for Quality Training

The ultimate deliverables of these organizations were to be a number of pilot and implementation workshops, along with formal manuals, and their associated workshop materials, for use by national and corporate entities wishing to implement quality programs for these important infrastructure components.

4. WORKSHOPS AND TRAINING COURSES

Under the QuaP-PV contract, each of the four participating contractors held an initial pilot implementation workshop in Jaipur, India, in October of 1999. These workshops provided the first opportunity to test the draft manuals and to obtain feedback from segments of the target communities in a real-world setting. The outcomes of these first pilot sessions were both the immediate feedback of the participants and the more formal comments and suggestions of the professional reviewers contracted with to provide specific and professional guidance. As a result of the Jaipur workshops, each of the four contractors re-evaluated the manuals and resource materials in preparation for subsequent workshops and implementation programs in the other participant countries.

Ultimately, the quality component design and the quality manufacturing workshops were translated into Chinese and held in China; the quality manufacturing workshop and the quality installation and maintenance training framework workshop were held in South Africa, in February 2000; and, an implementation of the quality installation and maintenance training framework was held in Sri Lanka, in the form of a "Train-the-Trainers" workshop, in February 2000.

5. TRAINING INFRASTRUCTURE

While all four components of the QuaP-PV project are equally important, the balance of this paper will focus on the specifics of the importance and development efforts of the quality installation and maintenance certification infrastructure.

Training is seen as one component required for successful market development. However, until recently, there was not a broad-based effort to define and implement the wide variety of tasks, knowledge, and skills needed for the successful implementation of dispersed sustainable energy technologies, and this is one of the factors preventing financing agencies from more aggressively funding distributed electrification programs.

Training involves the attainment of minimum levels of knowledge and skills, so that specific tasks can be performed to measurable standards. Historically, training efforts in sustainable technologies have been implemented without clear competence targets, as measured against third-party standards, so the effectiveness of training could not be properly judged. Programs did not build upon one another, and isolated training activities were undertaken around the world, often "reinventing the wheel" in terms of content, lab and field activity, methods, logistics, and other important dimensions. Modern learning theory and instructional design technology were not often among the strengths of the parties involved, and were often neglected. A worldwide acceleration in market development for decentralized, sustainable technologies will be impeded until training program accreditation and practitioner knowledge and skills certification against third-party standards emerge to help focus efforts and scarce resources. These will allow development and funding organizations to do conventional risk assessments for their investments and loans when reviewing project submissions; and, they will allow training organizations to concentrate on presenting their programs and products in the context of accepted standards for content quality and participant performance.

Under the QuaP-PV program, The World Bank contracted with the Institute for Sustainable Power (ISP) to develop guidelines and an implementation workshop manual for use by countries and companies wishing to develop and implement quality programs for the certification of PV installation and maintenance practitioner knowledge and skills competencies, as part of the larger global framework of training standards. Through this, there would be a common recognition of minimum competency, available as a guide for finance and development organizations evaluating potential funding recipients for PV projects.

6. RESULTS

For the Quality Training component of the QuaP-PV project, the ISP held, or participated in, three workshops – one each in India, Sri Lanka, and South Africa.

6.1 India
For the first pilot workshop, in Jaipur, India, held in October 1999, the ISP developed a draft manual and workshop support materials. This workshop was aimed at the stakeholders interested in, or involved with, the development and implementation of a quality installation and maintenance training infrastructure. This target group

included PV industry representatives, government officials, finance professionals, and standards experts. The intent was to use the manual and workshop materials to provide the participants with a number of information and implementation resources:

- Background on PV project development
- Background on practitioner training
- Information on quality systems
- Resource information on identifying stakeholders
- Guidance on the steps to develop a national PV training accreditation and practitioner certification framework, in line with the ISP's global PV training standards framework

The workshop, run for five days, was organized and coordinated for The World Bank by REIL, of India, which also identified and invited the workshop participants. The participants in the Quality Training workshop represented most of the target stakeholder groups. Through the course of the workshop, the participants and reviewers provided constructive feedback, as well as comments and suggestions on the content and format of the manual and workshop. They also provided valuable feedback on the relevance of the project to their needs, and unanimously indicated the importance of establishing such a quality program in India, both for the industry and for the professional development and standing of the participating industry individuals.

From the feedback of the participants and the input and guidance of the reviewers, the manuals and workshop materials were refined and adapted to better represent the material and to clarity the implementation guidelines.

6.2 South Africa

Using the revised manual and workshop materials, The World Bank contracted with a South African expert in PV training and vocational standards to take the materials and offer the workshop to representatives of a number of African countries. The goals of this activity were to test the revised manual and workshop materials and to assess the potential for countries to use the materials without requiring a workshop instructor specially trained in the use of the materials. The hope was that the materials were designed in such a way that, through review and familiarization with the materials, an appropriate instructor in any country could conduct the training without outside assistance. As the development of the materials was ongoing throughout the course of the contract, an ISP representative was asked to join the workshop to review the use of the manual by an outside instructor and the gather first-hand review and assessment from the participants for use in the final version of the manual. However, the ISP representative was not to actively participate in the instruction of the workshop.

The workshop was attended by 13 experts representing five African countries (Botswana, Ghana, Kenya, South Africa, and Zimbabwe), all involved with PV projects and each with an interest in implementing a national-level program for the accreditation of training programs and the quality certification of the knowledge and skills of PV installation and maintenance practitioners.

In the course of the workshop, the participants moved quickly through the material, indicating that they felt that the manual provided sufficient guidance and detail to implement such a program on a national or regional level. And, rather than proceeding with the detailed steps of the

workshop, they decided to move beyond the framework of the manual and work together to establish a pan-Africa working group to coordinate the development of framework standards for training quality and competence. By moving to this step, the participants felt that they could better share resources and lessons, mitigating the expense of developing such a program in isolation. The participants chose a coordinator who would represent them in the international training standards efforts, and together they agreed on a follow-up meeting, to be held in June 2000, to ensure that the momentum developed at the February workshop was not lost.

6.3 Sri Lanka

To test the training standards set out in the quality installation and maintenance guidelines, The World Bank contracted with Global Sustainable Energy Solutions (GSES), an ISP-Accredited Master Trainer from Australia, to hold a Train-the-Trainers workshop in Sri Lanka. Using the version of the Australian National PV Installer Certification training, adapted to the competency standards of the ISP, GSES held a 1-week training in Colombo, Sri Lanka, in February 2000, for 20 professionals from PV companies in Sri Lanka interested in becoming certified trainers, and one unaffiliated vocational training specialist, who would be able to provide qualified training to those concerned about conflicts of interest in taking training from companies that might be their competitors.

Under the contract, the training materials were translated into Sinalese, and the course was given by a GSES trainer in English, with the assistance of a translator who would provide the same material in Sinalese.

The participants took part in both classroom education and hands-on skills training, as the first phase of becoming certified as PV trainers for solar home systems installation. This first phase was to qualify the participants' at the level of certification at which they wished to be instructors (solar home systems installation). The second phase would be held at a later date, to qualify their instructor skills.

At the end of one week, the participants were evaluated through a knowledge examination and a hands-on skills evaluation. Of the 21 participants, 17 received passing marks and will participate in the second phase of the trainer certification program. In evaluating the four who were unsuccessful, it was determined that having the material presented primarily in English, even with a Sinalese translator available, was insufficient, especially for teaching technology-based subjects. This only emphasises the need for national-level programs to quality trainers, to avoid having to contract with outside trainers.

7. LESSONS LEARNED

In the course of developing the quality guidelines and materials for implementing a quality framework for training installation and maintenance practitioners, under The World Bank's QuaP-PV project, a number of lessons became apparent. Among these, it is imperative to involve a broad range of stakeholders and actors – in the development of the guidelines, in the review of the materials, and in the implementation workshops. Each segment of the stakeholder community (e.g., industry, government, finance, training, standards, etc.) brings a unique and critically important perspective to the project, identifying specific needs for their community and suggestions on how best to represent that material to the appropriate participants.

Another lesson was that many of the workshop participants were interested in a number of the workshops, and holding them in parallel precluded them from attending sessions they felt might be equally important. Because the program was constrained by the source of funding to a 1-year timeframe, it was difficult to spread the workshops out so that there was no overlap. In addition, with the workshops running a minimum of a week, it would have proved impractical to expect the participating experts to leave their work for additional weeks of participation in the framework workshops.

Another, perhaps obvious, lesson is the importance of providing the workshop participants with guidance, resource materials, workshop materials, and workshop goals and expectations sufficiently in advance of the event that they are able to properly prepare. Not only would this provide for a more efficient and productive workshop, but it will accurately represent the content and expectations of the workshop to the attendees, who need to prioritise their choices between work and participation in these events.

Finally, with the quality training accreditation and practitioner certification framework, the participants indicated a strong interest in working quickly to implement the standards at a national level. However, there was also a strong interest in identifying other national and regional programs with which they could coordinate, to gain from the existing experience and to share in the developmental process.

8. CONCLUSIONS

The development of a quality PV infrastructure project (QuaP-PV), by the ASTAE unit of The World Bank, was the result of a both a need for, and opportunity to develop, a means to improve the quality and reliability of PV systems implementation programs through guidelines for common quality standards. To do this, the ASTAE unit coordinated the development and testing of workshop manuals and guidelines for the implementation of quality standards and processes for four infrastructure components: Quality Component Design; Quality Manufacturing Processes; Quality Product and Systems Testing Procedures; and Quality Standards for the Training of PV Installation and Maintenance Practitioners. This was carried out by four contract organizations through the development and review of the materials, the piloting of the materials in workshops in India, followed by revision of the materials and full implementation of the workshops in a number of other participant countries.

The outcomes of the QuaP-PV project were workshop manuals, resource materials, and guidelines for implementing each of the four quality programs at the national level by countries interested in implementing quality PV programs. These materials are to be published by The World Bank, initially on CD-ROM, and made available for general distribution.

Ultimately the goals of the project are (1) the improved quality, reliability, and sustainability of PV systems, primarily in the rural and remote markets of developing and transition economies, and the concurrent development of local, sustainable jobs; (2) through this, to contribute to the mitigation of global climate change by encouraging the development of clean, renewable electricity generation as a viable alternative to fossil-fuel generation sources; and, (3) to provide the development and finance community with an improved means of risk assessment in evaluating funding opportunities and requests.

9. ACKNOWLEDGMENTS

The author would like to acknowledge the support and guidance of the staff of the ASTAE unit of The World Bank in providing the opportunity to develop and implement this work. In addition, the author would like to thank the contributors, participants, and reviewers in the program for their generous assistance and critical input.

REFERENCES

[1] P. Khuanmaung, S. Thepa, R. Sinklee, and K. Kirtikara, "Assessment of Photovoltaic Water Pumping Systems in Thailand," Proceedings of Solar '96: The 34th Annual Conference of the Australian and New Zealand Solar Energy Society, Darwin, Australia, October 1996.

[2] A. Cabraal, M. Cosgrove-Davies, and L. Schaeffer, Best Practices for Photovoltaic Households: Lessons from experiences in selected countries, World Bank Technical Paper Number 324, Asia Technical Department Series. The World Bank, Washington, DC. August 1996.

TECHNICAL EVALUATION OF A PV-DIESEL HYBRID SYSTEM RURAL ELECTRIFICATION PROGRAMME IN INDONESIA

J-C Marcel, G. Moine, P. Veyan [1]
H. Orengo [2]
P. Boulanger, P. Malbranche [3]
K. Preiser, P Schweizer-Ries [4]

1 : **Transénergie**, 3D allée C. Debussy, 69130 Ecully, France
2 : **Total Energie Indonesia**, Kuningan Plaza, JlHR RasunaSaid KAV, C11-14 Jakarta Selatan 12940, Indonesia
3 : **GENEC**, CEA Cadarache, 13108 Saint Paul lez Durance, France
4 : **Fraunhofer-Institut Solare Energiesysteme**, 5 Oltmannsstrasse, D-79100 Freiburg, Germany

ABSTRACT: In the framework of a project follow up led by Transénergie, dealing with the installation and the assessment of the operation of four hybrid photovoltaic-diesel systems in four villages in Indonesia islands, this paper describes the different tests and studies which has been conducted as a technical evaluation after one year of operation. This follow up is supported by E.C. Thermie B, French ADEME and the companies involved.
Three aspects have been covered by this assessment: a technical approach aiming at validating the state of the system one year after the commissioning stage and validating innovative aspects of the system (battery management, data satellite transmission, pre payment system), a sociological survey that can be used to evaluate the social impact of such a programme in order to better understand the expectations of the users and to try to anticipate in the future their new requests and an economical approach to set up a specific exploitation scheme in order to guaranty the durability and viability of the installation in the villages.
Keywords: Hybrid – 1 : Sustainable – 2 : Rural electrification – 3

1. INTRODUCTION

Rural electrification is a major concern for most developing countries. Two approaches are commonly used for decentralised rural electrification of villages : the use of individual small stand-alone photovoltaic systems, also called solar home systems (SHS) as they usually are installed in the houses, and at the opposite the use of a bigger centralised hybrid generator made of a PV generator and genset feeding in a mini grid.

The main challenges observed with those centralised hybrid system supplying tens or hundreds of houses are manifold :
- how to design and manage the complementary sources of energy once it has been decided to privilege the use of renewable ;
- how to implement an exploitation scheme and recovering costs that guaranty the sustainability of the system ;
- how to transcript in technical terms the social and economical aspects of such a system with sustainable use.

In order to address those aspects, a technical evaluation has been conducted in September 1999, in three of the four Indonesian villages where hybrid systems was installed. The programme and the installation is described in the first part of the paper. The second part of the paper presents some aspects of purely technical evaluation mainly focused on the battery state of charge evaluation. The investigations led for implementing an adapted exploitation scheme and data analysis process are addressed in the third part while the last part is devoted to the interim results of the sociological analysis after investigations hold end of 1999.

2. THE FRENCH INDONESIAN RURAL ELECTRIFICATION PROGRAMME AND TECHNICAL DESCRIPTION OF THE SYSTEM

2.1. Some inputs from the programme

Within the framework of a French-Indonesian co operation led by the Indonesian transmigration ministry, several villages have been created from nowhere to fight against rural exodus of the population. Among them, four have been selected to be equipped with a centralised hybrid photovoltaic-diesel plant as electricity supply system.

Three villages are located in Sulawesi island and one is located in Kalimantan island with a particularly hard access.

Each village is composed of three hundred up to four hundred individual houses. A family of an average of 5 people live in every houses (figure 1).

Figure 1 : Typical home

Each house is equipped by three light points and sometimes radio or TV set, more rarely fridge. The set up of the mini grid has been made by Indonesian people. Currently the plants belong to the Ministry of Transmigration with a contractual follow up by the French consortium, then a transfer to the local government is planned.

2.2. Technical description of the system

The system is set up with the following components (fig. 2):

- 23 kWp powered photovoltaic generator
- 40 kVA diesel generator ;
- 24 kW battery charger ;
- energy management cabinet ;
- battery storage (120 cells, 1400 Ah) ;
- 30 kVA inverter (three phases) ;
- a local mini-grid
- high efficiency lights.

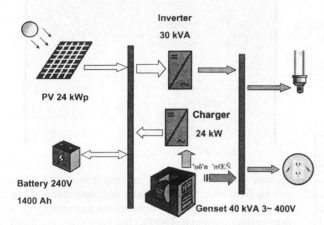

PV 24 kWp

Inverter
30 kVA

Charger
24 kW

Battery 240V
1400 Ah

Genset 40 kVA 3~ 400V

Figure 2 : Functional diagram of the system

2.3. Operating principle

The main characteristic of the system design is that 80% of the total energy required is provided by PV, diesel generator provides the 20 % left and could be used directly to the grid in case of dysfunction. During sunlight hours, photovoltaic generator charges the battery storage according to the available irradiation. The inverter, powered by the battery, is operating continuously and delivers electricity with respect to the load which is quite low during the day. At nightfall, the energy demand increases and if necessary the genset is turned on automatically so that the battery storage is preserved. The genset is switched off automatically when the battery state of charge has reached a sufficient level and/or after a programmed duration. Every day, the same operating cycle is running. If necessary the energy management system performs every three days a complete charge of the battery to warranty reliability and to extend lifetime. The running of the diesel generator is reduced to the minimum in order to limit exploitation costs, nuisances and to increase its lifetime.

To avoid energy waste, energy saver appliances are used.

In order to simplify the supervision, all the information needed to control the operation is available on the central control and dispatching panel of the electric cabinet or can be visualised on control displays of each component. At the same time, a monitoring and datalogging system performs several environmental and electric measurements (irradiation, DC and AC currents and voltages) and is able to transmit data to the abroad supervision centre in France by a satellite modem. Users finance a part of the maintenance and running costs, thanks to a prepayment meter in each house.

It assumes three functions :

- energy payment by code entered with keypad

- power limitation regarding the consumer subscribed contract;
- daily average energy limitation.

3. TECHNICAL EVALUATION

One year after commissioning, in order to assess the overall plant the evaluation covered several aspects of the system : PV array, cabinets, vending station, genset and battery inspection with later in France electrolyte and sludge samplings analysis in laboratory. The goals of the mission was also to evaluate and complete if necessary the competencies of the village operators and Indonesian devoted technician as well as to verify the consumption and stock of spare parts and consumables.

3.1. Overall system comment

For the villages, building, PV generator, genset, electronic cabinet, battery park, data collection and telecommunication set has been inspected.

The main observations are :

- overall good cleanliness of both building and components has been recorded thanks to the operators (who do not know our coming).
- the state of PV array and junction boxes was very good (see example figure 3)
- the genset was good maintained and cleaned
- in one village, rain water entered in the electronic cabinet but fortunately without consequences; this problem due to a non properly building draining system has been solved later by operators.
- few damages have been caused by reptiles and rats, therefore cabinets and wires trolleys have been hermetically closed with polyurethane bombs.
- some pre payment counters were by passed, some lights ballast's have been replaced (environmental aggression).
- taking advantage of the mission the data collection and telecommunication system has been upgraded and checked.
- the grid was in a good state.

Figure 3 : Very good general aspect

3.2. Battery measurements

For every battery storage, a complete measurement of voltage, density and temperature has been performed cell by cell. The measurement were done early in the morning so that the battery rest during the night. PV generator has been disconnected during measurement in order to be as closed as possible to standardised laboratory conditions.

The results demonstrate that the overall state of every storage bank is good, no weak elements have been detected which could jeopardy the system. (see example figure 4)

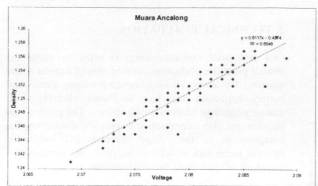

Figure 4 : One site battery – Nernst law

The results are summed up in the following table.

Data	Muara Ancalong	Kualumpang	Salopangkang
V tot)	249.57	249.39	252.18
V aver	2.079	2.078	2.107
Δ V	0.0041	0.0060	0.0054
Density	1.25	1.239	1.241
Δ Density	0.0036	0.0057	0.0054
Temp	29.21	27.43	30.93
Δ Temp	0.44	0.49	0.87

3.3. Sampling and analysis

For some cells, in each villages, sampling were stored in test tubes to be analysed back at GENEC. The main goal of this campaign was to evaluate if any sulfation has occurred after one year of operation. Samples were made at different depths in the battery electrolyte :
- upper plate powder (corrosion)
- electrolyte sampling at three depths (stratification)
- sludge sampling (sulfation)

(see figure 5 give the sampling practice with the "magic tube")

Figure 5 : The magic tube

The density measurement at different depths is very steady and demonstrates that no harsh stratification has already appeared in the battery storage. This conclusion is quite

satisfactory one year after the commissioning stage and it validates the energy management used in the systems.

Back to France, some samples which had a high amount of active material inside the test tube have been analysed via an electronic microscope after filtered with a silver membrane. Figure 6 exhibits one crystalline architecture observed from the samples.

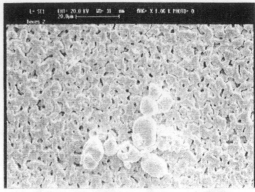

Figure 6 : small crystal observation

Crystals observed from one site was larger than those coming from others villages. From this observation, it can be notice that the battery incriminated has undergone a deep cycling constraint. This fact had been sawn, explained and solved on the field without consequences.

As a conclusion, those measurement are a powerful tools to analyse the state of a battery storage. and it is clear that they have an overall good "health and behaviour".

4. SOCIOLOGICAL SURVEY

In addition with the technical evaluation of the systems, a sociological investigation has been conducted by the Fraunhofer Institut Solare Energiesysteme with the University of Jakarta in order to evaluate how such a system is perceived by end user, in five different fields :
- the energy consumption
- the knowledge about the system
- the identification of the organisation inside the community
- personal judgements
- new needs for the future

From this survey, the following points raise:
the energy consumption is quite low, people are willing to consume more energy and also pay for it
- the knowledge about the PV-system is high and there exists an awareness by the fact that their system is powered by the sun
- by-passes are seen as unfair and the metering system is judged as fair and very good energy distribution system.
- until now electricity is mainly used for lighting, leisure activities; people enjoy to be connected to the city via radio and TV.

The two investigated systems where different. Whereas one of the villages is quite poor and far away from the urban area, the other one is richer. In the latter energy consumption will raise soon, people are more active and the solar organisation is very highly excepted. To integrate the systems into the life of the people further adaptations on both the technical and the social side are necessary.

5. EXPLOITATION SCHEME AND COSTS ANALYSIS

To perpetuate the use and the good health of an electric distribution system, whatever the kind of generator used, local technico-economic accompanying measures must be provided. The problem is most important when a renewable energy is used because people have no technical practical background with this new energy and because no networks exists to supply easily any spare parts and repairs.

The implementation of a durable exploitation scheme is based on the following steps :
- a monitoring/control system which allows to follow exactly the state of operation of the system, and enable to react as soon as a problem is detected;
- a local team involved in low and high qualification maintenance tasks
- a supervisor who is fully in charge of the complete system, technically and financially speaking, knowing exactly the complete life cycle costs
- a deep analysis of the end-users needs and forecasted evolution

All the systems are equipped with a datalogger ENERPAC, monitoring device that allows data to be collected on site and analysed everywhere in the world via a satellite telephone. Figure 7 depicts an example of a daily consumption and production profile.

The present exploitation of the system is given to a village committee which is in charge of :
- vending the electricity controlled by the pre-payment meters
- saving money for operators salaries and component replacements (genset filters for instance);
- realising the preventive maintenance routine of the system (cleaning PV array, genset sewage's, fill up distilled water in battery ...);
- making some repairs (e.g. light replacements);

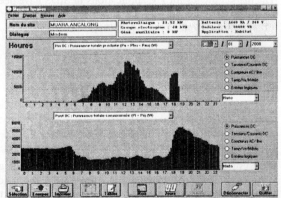

Figure 7 : A daily consumption and production profile

For that purpose, two technicians are employed as the first level maintenance team. Each six months or if necessary, a well trained Indonesian technician visits the site and do some high qualification tasks as the second level. The French company Total Energy Indonesia is currently in charge of the whole logistic for supplying new pieces, spare parts and solving technical problems if any. The data

are collected in Lyon (F) and also analysed in different laboratories. Transénergie is the basic technical support.

The technical evaluation, inspection of organisation and effective maintenance operations and the sociologic survey lead us to propose a new exploitation scheme which will be set up after the guaranty period as follows :
- local first level will remind unchanged, but exploitation procedures has to be more trained;
- second level will be the same in its principle, with a independent Indonesian company in charge of it (Fig.8)
- the third level will be transferred to the Indonesian ministry with the technical support of a public Indonesian PV test centre.

Figure 8 : The trained Indonesian technician operating

6. CONCLUSION

As a conclusion, the technical evaluation conducted by a motivated team (fig.9) in three a rural decentralised programme in Indonesia does not reveal any difficulty with those systems.

Some improvements in the system technical monitoring, the exploitation scheme and the social organisation have been proposed that could perpetuate the system operation, optimise the running costs and so satisfy and fix the population.

Figure 9: The Indonesian – French – German team

PV AS A TOOL FOR SUSTAINABLE DEVELOPMENT IN SMALL ISLAND STATES – AN OVERVIEW

Lara Bertarelli, Anthony Derrick

IT Power, The Warren, Bramshill Road, Eversley, Hants. RG27 0PR. UK
Tel: +44 (0)118 973 0073, Fax:+44 (0)118 973 0820, E-mail: lb@itpower.co.uk

ABSTRACT: Most small island developing states (SIDS) are blessed with abundant renewable energy resources, but they depend overwhelmingly on petroleum for their electricity production and biomass for the bulk of their energy consumption. Neither of these sources is about to be depleted completely, but access to both is limited, and the use of either has severe environmental and economic consequences.

SIDS have a high and relatively constant supply of solar energy but its use as high grade and clean energy source is largely under exploited. PV systems have been used on many islands, mainly in dispersed settings, for telecommunication transmission and reception, lighting, small medical refrigerators, and water pumping. PV can play an important role in improving the quality of life of people residing in small and remote island states.

This paper has the objective of providing the background to the underlying issues of SIDS, an overview of the PV applications and finally the challenges that PV in island states faces in the future.

The paper draws from a wide range of activities in rural electrification programmes undertaken by IT Power including a chapter co-ordinated by IT Power with contributions from Ecoenergy, Barbados and Tritech, Fiji, titled "Energy as tool for sustainable development in African Caribbean and Pacific Small Island Developing States". This study was carried out for UNDP and EC in 1999 and the report is now available from UNDP and EC offices.

Keywords: Strategy - 1: Sustainable - 2: Remote– 3: Developing Countries

1 BACKGROUND

There are many varying factors that distinguish individual island states from one another, that make any generalised paper on small island states a difficult one, however islands face a series of similar difficulties and have similar needs. These include:
- limited land availability
- high population density and population growth rates but low total populations
- rapid level of urbanisation
- urban unemployment
- deteriorating environmental conditions (waste disposal, pollution)
- impacts caused by climate change (enhanced level of natural disasters, sea level rise, land degradation, etc.)
- increased frequency of natural hazards
- isolation from the mainland
- remote and dispersed over large ocean areas
- rising energy prices for commercial fuels
- increased energy demand
- inefficient use, transmission and generation of energy
- high transportation costs (internal and international)
- expanding tourism industry which is unsustainable without a clean environment
- continuing dependence on foreign assistance
- vulnerability to changes in global energy prices, and external political and economic situations.

Common to most island states is also their abundant renewable energy resources. Island states have a high and relatively constant supply of solar energy.

The many and variable conditions that face dispersed island communities make PV systems an attractive energy source. PV could play a substantial role in the electrification of SIDS.

2 PV SYSTEMS IN SIDS

Renewable energy technologies like PV can have an important role to play in improving the quality of life of people in SIDS. A large proportion of the population in island states is without access to electricity and reside on small, remote and distant islands, stand-alone PV power systems can therefore be one of the most suitable options to provide them with the necessary electricity. The market potential for PV is substantial.

Although the share of PV to the total energy supply remains at a level significantly below its potential, the use of PV systems in SIDS is increasing due to the overall increased awareness on proven renewable energy technologies and the decline of capital, installation and operating costs.

The market niches for PV applications in SIDS have been identified as being:
- Solar Home Systems (SHSs)
- Solar lanterns
- Water pumping
- Bio-climatic interventions
- Community building power
- Desalination plants
- Refrigeration
- Marine and aviation navigation aids
- Telecommunications

PV systems in SIDS have been distributed to village households, village schools and community halls, for village water pumping and power for satellite earth stations. SHSs can provide reliable power more cheaply than diesel systems in small, remote villages for consumers with a limited number of appliances. Solar generators are used extensively in the Pacific region for remote applications where virtually every other Pacific Island country or traditional power producer/distributor has installed some solar generating capacity [1].

In Tuvalu, PV-based electricity is used in 40 percent of households in the outer islands for lighting. In New Caledonia and French Polynesia, solar generators are installed by local power utilities whenever grid connection is too costly [1]. Vanuatu has adopted an ambitious solar electrification policy.

In Kiribati PV systems have been installed in all the health centres on the 19 islands to provide power for vaccine refrigeration, lights and an emergency two-way radio. Water supply for 10 rural communities in Kiribati is provided by PV water pump systems and 10 more systems were to be installed in 1995. At present, a total of 310 solar home systems have been installed in Kiribati. Other PV power systems have been installed on many of the 26 inhabited islands of Kiribati for communication purposes.

PV systems have been installed in many islands of the Cape Verde archipelago and in Jamaica approximately 45 solar street lamps have been installed [2].

Table 1 provides some examples of PV applications in island states.

Solomon Islands	▪ SHS for 50 households in Sukiki Village ▪ SHS for 56 households & community high school & primary school
St Lucia	▪ Lighting for village school (400 Wp), campsite (1.2 kWp), illumination of fort (1kWp), PV training facility (250 Wp lighting kits).
Mauritius	▪ Hybrid PV diesel/petrol generator sets
Seychelles	▪ SHS for 5 households
Micronesia	▪ 242 SHS
Jamaica	▪ 45 solar street lamps
Marshall Island	▪ 3 kWp PV system
American Samoa	▪ 4 kWp PV system
Fiji	▪ Rehabilitation of 68 PV lighting systems at Namara village, Kadavu ▪ Naroi village on Moala island – 170 households received SHS (100Wp)
Tonga	▪ 350 homes have been wired up under the Vava'u's ▪ 50 PV lighting systems for Ma'onge'one, Ha'apai Group
Tuvalu	▪ 50 PV lighting systems for outlying islands, PV Vaccine (8) and domestic (7) refrigeration systems in outlying islands. ▪ Upgrading of 265 existing PV lighting systems. ▪ Earth Station PV Power Supply (1989)

Table 1: Examples of PV applications in islands

The awareness created by some of these projects has led to greater interest, co-operation and more requests from people and, in the case of the Solomon Islands, it has attracted support from the government to expand its programme. However, the majority of past PV programmes implemented in SIDS have had disappointing results, the reasons are varied and are discussed in the next section.

Donors are continuing to provide grant aid funding for renewable energy projects in the region but with varying approaches to their implementation. The Secretariat of the Pacific Community (SPC) has started implementing the largest renewable energy programme in the region, the AUD 3.45 million Regional Australia-France Rural Renewable Energy Project [RAFRREP]. The goal of RAFRREP is to advance the social and economic development of the Pacific Community through the use of sustainable renewable energy technologies. Its purpose is to increase the utilization of sustainable renewable energy

technologies, in particular PV and wind energy technologies, in island and rural communities.

In Kiribati the European Commission will be funding, later this year, a programme to install 1500 solar home systems and PV for 133 community buildings. The Kingdom of Tonga has also recently expressed interest in using solar for outer island electrification.

3 THE CHALLENGES

There are many and varied reasons for the disappointing results of renewable energy technologies promoted in SIDS, from a lack of sufficient, accurate renewable energy resource data; a lack of detailed understanding of the economic and technical viability of these technologies in the setting of SIDS; inappropriate institutions; scarcity of capital resources; to insufficient efforts in organising active participation of the local community at the early planning stage.

It is important that adequate training in system operation and maintenance be provided on a long-term basis, that assistance on energy efficient mechanisms and renewable energy technologies be provided and that support for local and regional organisations to plan, operate, maintain, finance and expand the use of the technology be continued until a truly sustainable basis is achieved.

Success depends on developing the right mix of regulatory, legal and institutional frameworks; energy and related policies; private and public sector involvement; and financing mechanisms. Many models are possible and there is no one package that can be universally applied.

Institutional capacities

The lack of energy policy-makers, energy planners, project organisers, energy economists, energy system managers, engineers, and technicians is a continuing problem in SIDS.

Putting sustainable energy into non-energy sector projects

Links should be developed between ministries and departments to inform and disseminate the use of PV in all other public sectors, such as education, health, agriculture, etc. The power industries also need to consider and develop decentralised and sustainable options. Renewable energy systems must be fully integrated into sectoral programme activities, and not treated as a low priority add-on.

Information, promotion and dissemination

Lack of knowledge, misunderstanding and bias have acted as deterrents to the wider utilisation of PV. Without adequate dissemination of information on sustainable energy technologies a number of misconceptions may arise. It is crucial that for building market confidence governments include clear signals that demonstrate a commitment to sustainable energy. Government promotional campaigns backed by good information are also needed to raise the profile and explain the benefits of PV to the business and finance sectors and also to households and the small-scale service sectors.

Financing

Experience has shown that to provide high quality and reliable energy supplies for SIDS, capital investments in renewable energy systems are essential. Sustainable energy development however requires affordable credit financing as many SIDS do not have the financial resources to afford the relatively high up-front costs of renewable energy technologies. In addition, traditional banks are averse to lending small amounts for unfamiliar purposes. Whatever financial mechanism is used, programmes must be financially sustainable if they are to expand and reach the rural population.

Mechanisms to mobilise financial resources are required and should include continued encouragement of private investment, small-scale grants and micro-enterprise loans. Financing methods depend to a large extent on the type of borrower and what security is provided by the lender. Different models and instruments (grants, soft loans, tax concessions, commercial loans, etc.) must be adapted to different needs and scales and should be explored to assess their suitability for the SIDS situation. In addition, it is important that international aid funds are made available.

In April 1999 170 households in the village of Naroi on Moala island, Fiji were equipped with a 100 Wp solar generator and prepaid meters. The metering scheme is attracting considerable interest within the region, and appears to be successful [1]. The local post office sells the codes for a flat monthly rate. The credit code is entered by means of an integrated keypad on the meter and the customers receive one month of service. The Fees collected will then be used to cover maintenance costs, battery replacement and installation of new systems as the village grows [1].

Project implementation

The dynamics of project implementation must be fully understood for effective and successful energy programmes.

- Participatory appraisal methods are necessary at all stages of the programme. There is a special role for all groups, including women, youth, senior, indigenous and local communities as well as private sector and non-governmental organisations [4]. Understanding the priorities of the community and whether the community is prepared to pay for the proposed system is crucial.
- Partnerships with local and regional organisations.
- Assessment of local resources including human resource capabilities. To ensure effective understanding and implementation of all projects suitable human resources should be available in country. Self-investment, self-construction and self-operation are all important elements [5].
- Assessment of most appropriate technology for local conditions. Decentralised systems have proven to be more effective in dispersed communities. Efforts need to be placed on ensuring, subject to national legislation and policies, that the technology, knowledge and customary traditional practices of local and indigenous people are adequately and effectively protected and that they thereby benefit directly on an equitable basis.

- Training/education at all levels. Training of local installers, maintenance personnel and users is important to the continued reliability of the systems. Simple measures, such as the provision of a users manual in the local language as well as a poster permanently fixed to a nearby wall or directly on to the system can be used to reinforce important messages.
- Establish sustainable infrastructures. Accessible after-sales services are needed and spare parts must be available at reasonable costs and relatively quickly. Suppliers only in urban areas leads to poor dissemination in rural areas.
- Installation, operation and maintenance support services. All systems require proper installation and good quality operation and maintenance support services, during and after the programme, if markets for the technology are to prosper and grow. Installation, operation and maintenance should be undertaken by qualified local technicians.
- Project assessment - follow-up. There is a need to ensure that the system runs successfully even after the programme has been contractually completed.

Technology transfer and adaptation

Co-operation with neighbouring islands and other SIDS to share experiences and expertise is essential for SIDS. A number of networks enhancing south-south co-operation on development and environmental issues have been developed, however there is still little focus on sustainable energy related factors. "Piggy-back" dissemination techniques using existing networks are particularly effective in rural areas.

New proven technologies usually cannot take off until there is a critical mass of: (i) customers who know about the technology and (ii) technicians who can support it. The private sector can be the major agent in the process of innovation/technology transfer.

It is crucial to note that no technical transfer can occur without consideration of non-technical issues that build on the knowledge and the experience of local people and their natural resources.

Reducing Costs

As in most other developing countries, electricity in SIDS is currently available mainly in cities and surrounding rural areas, and in most of these countries electricity is generated from fossil fuel-based systems, since that is the most economical option for power generation in remote areas. Prices of petroleum fuels landed in Pacific SIDS are typically 200-300% of international values. The cost of distributing the fuels within each country is added on, resulting in a very costly operation, especially as most generation systems are on a small-scale to very small-scale basis.

There are prospects for substantial reductions in the capital costs of PV as different markets develop and the policy environment is developed to facilitate their market penetration. Duties and taxes discriminate against sustainable energy technologies in many of SIDS. Removing these barriers will improve the viability of PV.

Policy

Each SIDS should have a sustainable energy policy and a Renewable Energy/Energy Efficiency Master Plan. Support for the preparation of Master Plans through existing regional organisations is a key area for technical and non-technical co-operation.

4 CONCLUSIONS

PV can play an important role in improving the quality of life of people in these small island countries however a number of challenges need to be addressed to ensure the future sustainability of PV projects. Developing an integrated project is time consuming, but the benefits in terms of reduced risk and likely success of the project are considerable.

Getting the right payment structure in place is one of the greatest ongoing challenges for PV rural electrification. Experience has shown that fee collection, even a symbolic fee, is fundamental to the success of solar projects – users must be made aware that solar equipment is valuable and service is not free. Payment methods and financing schemes need to be country specific because of cultural differences.

There are many lessons to be learned from past and present programmes. Co-ordination and co-operation between islands is a crucial component for the future of PV in island states. It is also critically important that donors' initiatives are well co-ordinated and are adopting a common consistent approach to the sustainable development of renewable energy in SIDS.

REFERENCES

[1] South Seas PVs – Metered Solar Electrification in Fiji. Renewable Energy World. January 2000.
[2] Headley, O. (1998). Renewable Energy in the Caribbean. The World Directory of Renewable Energy suppliers and Services.
[3] Solar energy: Lessons from the Pacific Island experience, Andres Liebenthal, Subodh Mathur and Herbert Wade, World Bank Technical Paper No. 244, Asia Technical Department Series, World Bank, Washington, ISBN 0-8213-2802-6, 1994.
[4] SIDS Programme of Action (SIDSPOA). WWW site. http://community.wow.net/eclac/SIDSPOA
[5] Cheatham, C. (1993). Overview of Energy Sector Issues and Problems of Power Development in the Pacific Island Countries. Hydroequip'93. Conference on Medium-Small Hydro Equipment, Hangzhous, China.
[6] UNDP/EC (1999). Energy as a tool for sustainable development for African, Caribbean and Pacific countries. ISBN 92-1-126-122-8.

Progress with the IFC/GEF Photovoltaic Market Transformation Initiative (PVMTI)

Anthony Derrick
IT Power Limited
The Warren
Bramshill Road
Eversley, Hampshire
RG27 0PR, UK
Tel: +44 118 9730073
Fax: +44 118 9730820
Pvmti@itpower.co.uk

Ian Simm
Impax Capital Corporation
Broughton House
6-8 Sackville Street
London W1X 1DD, UK
Tel: +44 207 434 1122
Fax: +44 207 434 1123
Pvmti@impax.co.uk

Vikram Widge
International Finance
Corporation (IFC)
2121 Pennsylvania Avenue, NW
Washington, DC 20433 USA
Tel: +1 202 473 1368
Fax: +1 202 974 4349
Pvmti@ifc.org

SUMMARY

The Photovoltaic Market Transformation Initiative (PVMTI) is an initiative of the International Finance Corporation (IFC) and the Global Environment Facility (GEF) to accelerate the sustainable commercialisation and financial viability of solar photovoltaic-based energy services in India, Kenya and Morocco. The investment phase of the initiative commenced in July 1998. By mid 2000 four investments had been approved or recommended for approval totalling more than $ 7m and with more deals under review, the investment management team is targeting IFC commitment on all US$ 25m of funds by mid 2001.

Background

The Photovoltaic Market Transformation Initiative (PVMTI) is an initiative of the International Finance Corporation (IFC) and Global Environment Facility (GEF) to accelerate the sustainable commercialisation and financial viability of solar photovoltaic-based energy services in India, Kenya and Morocco.

The IFC have appointed Impax Capital Corporation and IT Power as their External Management Team (EMT) for PVMTI. The PVMTI Initiative has a total of 25 million US$ of GEF funds available for investment by IFC in India (US$ 15m), Kenya (US$ 5m) and Morocco (US$ 5m).

A local partner organisation has been appointed in each country to work with the EMT namely IT Power India, Pondicherry, Pipal Ltd, Nairobi and Resing, Marrakech.

During 1997 an IFC/EMT group assessed the business flow and likely deal flows in the three countries. The investment phase commenced in July 1998 with the call for investment proposals issued in September 1998 for submission by the end of 1998.

By mid 2000 and following concerted due diligence, structuring and negotiation by the PVMTI External Management Team (EMT) on the most appropriate proposals received, IFC has approved or recommended for approval four PVMTI deals totalling US$12.3 million

Investments

The first PVMTI project approved in India involves a US$2.2m investment in Shri Shakti Alternative Energy Technologies. This project envisages setting up 300 "Energy Stores" across India, thereby increasing the visibility and availability of quality Solar PV products and providing reliable after-sales service.

The second project to receive IFC approval is a US$3.5m investment in SREI International Finance to co-finance the expansion of their solar loans operations into West Bengal. SREI is working together with the Ramakrishna Mission (a well-established local NGO) and Tata BP Solar, one of India's leading suppliers of PV systems. The project therefore has all the core elements required to encourage growth in the market - high quality PV products and supporting infrastructure and also availability of consumer finance.

In Kenya, the first two projects have also been approved - a US$1 million loan and grant to Kenya Commercial Bank (KCB) to co-finance the provision of loans for Solar Home Systems (SHS) through its widespread branch network, and a US$0.6 million investment in the Muramati Tea Growers SACCO for provision of solar loans to their members

With more deals under review, the EMT is targeting IFC commitment on all funds by mid 2001 including projects in Morocco, where the national utility (ONE) expects to launch its rural electrification tender for SHS

To date, PVMTI funds have been used to provide debt, equity and guarantees, as well as grant funding. IFC / EMT have identified several key themes.

- Business environment - each country is unique and "what works" in one may not work in another.
- Investment - flexible PVMTI investment terms have facilitated optimal structuring of projects
- Sponsors - a number of strong new entrants have joined incumbents thus increasing competition.
- Management - a strong and experienced management team is a vital component of the business plan.
- Profitability - several projects are likely to be commercially viable after only one round of concessional financing.
- Financing - third parties (e.g. banks) are committing funds to PV deals in the three countries.

Alternative Investment Instruments

Equity, debt and guarantees have all been utilised in PVMTI projects. Sponsors need to be aware of the relative advantages and disadvantages of these instruments.

Equity strengthens the balance sheet and facilitates subsequent raising of finance (e.g. bank debt). It is also usually provided for a longer period than debt finance, often with no fixed repayment date. Against this, PVMTI equity dilutes returns to sponsors and may involve costly legal documentation.

Debt finance (i.e. loans) increases sponsors' equity returns and may be tax efficient. However, debt inevitably involves a fixed repayment schedule, which could cause problems if the business struggles. In addition, US$ loans mean that sponsors face some foreign exchange risk.

A Guarantee from PVMTI can encourage a third party to accept a role in the project. Guarantees involve no foreign exchange risk for the sponsor. However, legal and monitoring costs may be high.

Quality Assurance

It is important that PV systems disseminated as a result of PVMTI investments are of a high quality and dependability. In all three countries activities have been initiated on quality issues and technical guidelines have been provided to investee companies.

In India, a survey has been undertaken to assess the capability of local laboratories to test PV systems and balance of system components. Research has also been carried out on the national and international standards relevant to PV systems.

In Kenya a local consultant was contracted to participate in a workshop on certification of PV training activities and in Morocco the CDER the PV certification programme and testing facilities were reviewed as a potential resource for PVMTI

STRATEGY FOR IMPROVING LIQUIDITY IN THE INDIAN PV TECHNOLOGY MARKET BASED ON THE MOBILISATION OF EXISTING FINANCIAL STRUCTURES

J. Stierstorfer, M. Cameron, P. Helm

WIP – Renewable Energies, Sylvensteinstr. 2, D-81369 München, Germany

Tel: +49-89-720 1235, Fax: +49-89-720 1291, wip@wip-munich.de

B D Sharma, S. Sinha, V. Bhatnagar

Tata Energy Research Institute (TERI), Habitat Place, Lodhi Road, New Delhi 110 003, India

Tel: +91-11-462 2246, Fax: +91-11-462 1770

ABSTRACT: The Indian PV market is not only one of the leading PV markets in Asia but also - through its size and potential - of significant importance for the European and the world wide PV industry. However, the exploitation of this potential faces numerous financial barriers that exist on both the customer side and on the supply side. Within the international initiative "PV Credit Mobilisation" existing financial barriers were identified, analysed. On the basis of these findings a strategy in form of a 10 Point Action Plan was developed to overcome these barriers as a means of enhancing the financial viability of the demand and supply side of the Indian PV market. The resulting strategy is presented in this paper. Since the barriers were not only identified within the financial sector itself the 10 Point Action Plan aims besides the financial sector also on the Indian PV industry itself and the national and state governments.

KEYWORDS: Financing-1, Stategy-2, India-3

1. OBJECTIVES AND INTRODUCTION

The Indian PV market represents a unique opportunity for photovoltaic technologies. Due to the favourable climatic conditions and the high share of villages that are not connected to the grid (approximately 80,000) the potential for photovoltaic applications is tremendously high. Besides rural areas, urban areas offer excellent conditions for PV since the public grid is usually weak and characterised by frequent black and brown outs. Under such conditions PV systems are attractive as a back up power supply. At the same time a national PV industry exists that already provides a wide range of products and services.

However, the exploitation of this potential is hindered by financial barriers to both the PV industry side as well as to end users. The main objective of the international initiative "PV Credit Mobilisation", co-financed by the THERMIE programme of the European Commission, is to develop a comprehensive strategy to overcome the identified barriers. The objectives of this work have been described in detail elsewhere [1].

2. FINANCIAL BARRIERS

A wide range of financial barriers were identified by a survey undertaken amongst banks and other financial organisations in India combined with an analysis of the current situation of the PV industry in India. In the course of the project work it became evident that the barriers are not only to be found in the financial sector but that several important barriers have their origin in the PV industry itself.

This had a significant impact on the strategy developed based on the findings of the survey work.

Specific details of these barriers have been described elsewhere [2].

Here we present an overview of the strategy which forms one of the key result from this work.

3. THE STRATEGY – A 10 POINT ACTION PLAN

The project consortium decided to present the developed strategy in a way that would emphasise the realisation of the recommended steps. A 10 Point Action Plan was considered as the best means to synthesise possible solutions and to demonstrate their potential to transform the recommendations into real actions. The following paragraphs describe the 10 Point Action Plan.

3.1 Fostering of regular round-table meetings between banks and PV industry

Intensifying the contacts between the financial sector and the PV industry improves the relationships and supports the development of mutual trust and the atmosphere of confidence. Through a better understanding of the needs and expectations for the future that currently characterise the PV industry, the financial sector can better react to these needs and gains a better understanding of the PV industry's situation.

Recommendations

Round–table meetings between the financial world and the PV industry will offer the opportunity to intensify the contacts between the two sectors. These meetings should take place on a regular basis at different locations involving not just senior personnel but also "frontline" representatives of the financial world and the

PV industry. These meetings should be moderated by a joint taskforce formed by a PV industry association and a financing industry association which could, in a neutral way, respect the interests and needs of both sides. A particular element of the literature provided at these meetings should be a handbook for financial organisations containing special issues relating to lending to the PV sector. the reliable business expectations of the PV industry

The launch of a 'PV Finance Forum' platform in the context of renewable energy conferences and exhibitions represent another means to intensify the contacts between the financial world and the PV industry.

3.2 Special training initiatives for larger national and international banks

One of the findings of the survey amongst the financial organisations was that 70% of the surveyed financial institutions have no technical understanding of PV systems. This clearly represents a barrier to the greater up-stake of PV in the financial world. Banks do not have to be technical experts on PV but should have some basic knowledge which would motivate them to become deeper involved in financing photovoltaic technologies. Having no knowledge at all means completely relying on somebody else's information and opinion, creating a feeling of dependence.

Recommendations

Training programmes targeting employees of the financial sector will help to improve their knowledge about PV and in this way the often negative image that PV still has amongst financial organisations will be improved.

The aim should be to establish a renewable energy unit in the headquarters of the larger banks and to have persons responsible for financing PV projects in each branch bank. It is important to offer employees responsible for PV projects training programmes on a regular basis in order to raise their awareness of PV and its market. Newsletters can be very helpful in order to provide the employees with important information beyond the trainee programmes. The training should be undertaken by a team of recognised experts.

3.3 Introduction of financial intermediaries while simultaneously cutting the distribution chain to reduce costs

In India the consumer products are generally marketed in a long chain and as a consequence the product lands up with the user at several times the ex-factory cost. Dealers are forced to accept lower margins and avoid stocking of PV systems to avoid holding costs. This results in poor takeoff and late deliveries.

The current PV market size in the rural areas is low and cannot sustain the usual margins of other consumer goods. The low volume and low margins do not attract reliable dealers and result in poor after sales service.

Recommendations

Instead of using the regular business channels, special marketing innovations are needed for marketing of PV products. The two points to be tackled are the reduction in direct costs to the end user and enough scope for the local entrepreneur to visualise a long term business by providing excellent after sales services. This can be achieved by adopting direct marketing strategies.

The introduction of a financial intermediary and a one point contact in the village managed by either a NGO or a local entrepreneur is essential. This mode of marketing would on the one hand cut down the costs which are associated with the several nodes of the supply chain while at the same time offering adequate scope to the local entrepreneur in terms of installation, commissioning, after sales service, collection of credits etc. This would result in better availability of systems, improved after sales service through detailed training of the local contact person, and higher end user confidence in the product due to the local presence.

3.4 Creation of project clusters in rural areas to reduce costs

Once PV systems fail, they generally lie unused especially in rural areas. Often due to incorrect use of the system the battery gets discharged below revival level and the repair/replacement costs become large. The manufacturers are unable to provide after sales service for just one or two systems at the frequently remote locations. Most of the banks feel that the recovery will be difficult as the systems are not likely to work due to lack of after sales services. Hence, they do not look at the PV market as a line of sustainable business.

Recommendations

All the stakeholders and promoters of PV technology including the manufacturers should deliberately plan their sales promotion in a way which leads to cluster marketing of products. The manufacturers could, for example, give free service in a particular location as an incentive. Similarly the government should select the areas which require PV systems and allocate incentives to promote clustering. The financial institutions and the funding agencies like World bank, GEF or projects like PVMTI should encourage projects with cluster concepts and associated after-sales service. In such locations special training should be provided to the local entrepreneurs.

3.5 Closer international co-operation on PV technology standards

Quality standards are one of the most important means in order to convince people that the product will perform as promised. The risk associated with a purchase is then minimised and made transparent through quality standards. Also questions of guarantee periods are closely linked to the quality standards.

There are still many quality standards issuing pending in India, not only in the field of photovoltaics. This lack of quality standards was identified within this project work as one of the major problems associated with the PV industry and is an important reason for the credit paralysis in PV sector in India.

Recommendations

The PV Global Approval programme (PV-GAP) is an initiative that aims at introducing world wide quality standards for PV products in order to strengthen the position of PV on the world market. India as one of the important producers of PV products should play an active role within this programme. Playing a major role in PV-GAP would increase the chances for Indian products on the world market. This would have a positive effect on the

- export opportunities of the Indian PV industry

- the quality of PV products in India

- the reputation of the photovoltaic technology in general in India and specifically in the Indian financial sector

Closer co-operation of the Indian PV industry on the international level could accelerate the process of introducing proper quality standards. The EC Joint Research Centre in Ispra (Italy) is an internationally well recognised institute that has developed quality standards in the field of photovoltaics that are world wide recognised. Co-operating with the EC Joint Research Centre as well as other such bodies, on promotion of standards would benefit the indigenous PV industry.

3.6   Closer European-Indian co-operation on PV demonstration programme policy and implementation know-how and best practise

Demonstration programmes for PV has proved to be a very important tool in Europe as a means of increasing the role of PV in Europe's energy portfolio. Still today many new PV applications and technologies prove their quality and reliability in demonstration programmes before they are ready to be self sustainable on the market. Europe's experience in the field of demonstration and dissemination of PV reaches back to the early 80ies. Many best practice stories arose from these programmes.

Recommendations

Within the European Commission's THERMIE Programme extensive experience was gained on all aspects of demonstration and dissemination of PV technology. The experiences gained in executing this successful programme might serve as a guideline when starting up a similar programme in India. The framework of such a programme would be different in India but the main results made in Europe could still be very valid for launching a PV demonstration and dissemination campaign in India.

A demonstration programme realising successful projects would prove to banks that PV projects can be very attractive for banks to finance. Most banks have not financed a PV project so far. If their first experience of getting involved in a PV project is positive, they will be willing to continue in financing PV projects in the future.

3.7   Sensitising electricity boards and local government organisations on the benefit of PV-Grid interaction

For several decades the electricity sector in India was exclusively dominated by the 19 Indian State Electricity Boards (SEBs) before the new private power policy was launched in 1992. The private sector plays an increasing role in electricity generation especially in fairly new technologies like power generation from wind energy. It is very clear that the power sector in India is moving towards a greater share of private initiatives in the field of power generation projects. But still today the SEBs have a very powerful position within the electricity market. This dominating role will not be altered within a short period of time. Therefore it is very crucial to convince the SEBs of the quality and the advantages of PV technologies.

Recommendations

Energy departments of the state governments as well

the SEBs should be targeted by a PV information campaign clearly emphasising the advantages of PV especially in remote areas and its potential to meet the steadily rising in energy demand.

The PV industry has to play an active role in proposing joint projects to the SEBs. Also the Indian State Governments should use their influence on the SEBs to support the role of PV in the SEBs' portfolios.

In order to start a successful co-operation the SEBs have to be convinced of the advantages PV offers by information especially tailored to their needs. The main argument in favour of PV should be the dramatic increase of potential customers that will be achieved by offering a PV based power supply in rural areas that are not connected to the grid.. A proper dissemination of information followed by some demonstration could lead to their active involvement in the promotion of PV. The following benefits could be presented to the SEBs:

- Peak load shaving capability of PV

- Tail end voltage support to weak grids / rural grids

- Long range benefits as the cost of PV electricity will be stable, while conventional costs will rise

- Modularity of PV systems, so it is easy to adapt the systems to changes in the energy demand

- Need to get familiar with PV installations as this will become a cost effective solution in the coming decades.

3.8   Take Advantage of Carbon Trading in the context of the global warming challenge

Photovoltaics is one of the important technologies for developing countries to reduce GHG (Greenhouse gas) emissions and help the global environment. The Clean Development Mechanism (CDM) one of the proposed vehicles to minimise the GHG emissions in developing countries and help meet the industrialised nations' commitment under the Kyoto protocol.

When the direct cost of carbon savings are calculated under the CDM, the cost per ton of carbon saved is quite high for photovoltaics and therefore does not attract the industrial nations to promote PV as a GHG abatement option since cheaper alternates are available. However, if one looks at the PV technology in the global picture, it offers global benefits in addition to local environmental and socio-economic benefits for developing countries.

Recommendations

If a proper initiative is taken by both the developed and developing countries to include PV as a prime GHG mitigation option under CDM, this would have a major impact on the PV market. This will allow the developing nations to effectively reduce PV programme costs by accounting the cost of carbon savings against the actual costs. This will represent an important impulse to the growth of PV in developing countries which can reap benefits of PV for socio-economical development. The new business opportunities created by carbon trading will encourage the financial world - big financial institutions as well as the small local branches - to get involved in PV as part of this new tool to minimise global warming.

3.9 Proposals to Governments to make PV more attractive without the use of subsidies

The Government of India has introduced several fiscal incentives to promote equitable socio-economic development especially in rural sector. These incentives help the particular sector of economy to grow as well as make the services affordable to rural people who actually need them. Although there are several such benefits available to sectors such as agriculture, power, housing sector and selected fossil fuels (such as LPG and kerosene). PV has not been allocated all of these benefits.

Recommendations

The Government should amend the policies so that the PV is put at equal footings compared to other similarly important technologies, both in terms of fiscal benefits and financing options, to the end-users. Some specific recommendations are:

• Government of India offers several benefits to the housing sector, namely long-term loans are available at a lower interest rate and the repayments and interest have certain tax benefits for the individuals. If Solar Home Systems are "declared" to be part of the house, the house owner will be able to get loan and tax benefits for PV and solar thermal system. This will also reduce the transaction costs as the loan will become part of home building loan transaction leading to very simple procedure .

• Government of India provides several benefits through local and import taxes for Mega Power Projects as a part of promotion of infrastructure facilities. PV, (although it is also a power source) however, is not considered as a power industry and these benefits are not available. The inclusion of PV in the power industry will entitle this sector to get the same benefits.

• Government of India, for the promotion of renewables, makes available subsidy or depreciation benefits. Depreciation benefits are available to companies and not individuals. Therefore the same PV system procured by a company is at effectively lower price than the one procured by individuals. Extension of depreciation benefits to individuals under the income tax act will encourage individual users to access this route and not subsidy route.

3.10 Set-up of a mechanism to exchange experience and expertise on photovoltaic solar energy between India and Europe

In a world with an increasing share of global players and increasing importance of the global market co-operation between different nations and between different continents is becoming more important.

Co-operation between Europe and India does exist in general in different economic areas but there should be a more specific co-operation in the field of photovoltaics.

Recommendations

European and Asian Photovoltaic Conferences take place on a regular basis with a period of 18 months. These conferences gather experts from scientific community but also relevant politicians and decision makers and therefore could be used as an excellent platform for discussing all relevant issues on solar energy. Subjects to be discussed could include political and fiscal boundary conditions as well as programmes that are able to foster the further expansion of photovoltaic technology.

These conferences are well established and known as high quality events. Integrating a European – Indian Forum would help to encourage the exchange of experiences and expertise while at the same time add value to the events.

Another channel of mutual exchange could be the fostering of greater co-operation between the European Photovoltaic Industry Association and its counterparts in India.

4. CONCLUSION

As presented, this 10 Point Action Plan is a concerted action addressing several target groups: the financial sector, the PV industry, the national and the state governments, NGOs and the state electricity board.

Realising the basis elements of this 10 Point Action Plan it will be important to co-ordinate the several action points in a way that will result in a maximum impact.

ACKNOWLEDGMENT

PV Credit Mobilisation is co-financed by the European Commission through the THERMIE programme of the Directorate-General Energy and Transport under project number STR-1078-96.

REFERENCES

[1] M. Cameron, P. Helm, J.C. Bonda, B D Sharma, *PV Credit Mobilisation: tapping latent financing opportunities to boost the penetration of photovoltaic technologies in developing countries*, Proceeding from the 14th European PV Conference , Barcelona, July 1997

[2] M.Cameron, F. Gostinelli, P. Helm, J.C. Bonda, B D Sharma, *Removing Barriers between Financial Institutions and the PV Market: Lessons from India*, Proceedings from the 2nd World Conference and Exhibition on Photovoltaic Solar Energy Conversion, Vienna, July 1998

LUZ SOLAR – AN ELECTRIC-POWER UTILITY PHOTOVOLTAIC PROGRAMME
FOR RURAL PRE-ELECTRIFICATION

A.S.A.C. Diniz, F. W. Carvalho, E. D. França, J. L.Tomé, M. H. Villefort,
D. Borges, M. Rezende, L. A Araújo, J.G.F.Rosa, C. A. Alvarenga[1] and J.A.Burgoa
Department of Commercialization and Demand Management,
Companhia Energética de Minas Gerais - CEMIG
CP 992 - Belo Horizonte, Brazil
Telephone: 55 31 299 4505
Fax: 55 31 299
e-mail: asacd@cemig.com.br

ABSTRACT: The first phase of the LUZ SOLAR Programme, designed as a larger-scale implementation of photovoltaics in rural pre-electrification in the Brazilian southeastern state of Minas Gerais, is described for this initial, pilot phase. The purpose of this paper is to describe the program, the experiences (technical and social), and the future plans for rural electrification with this utility. Several successful demonstrations have shown that PV can be reliable and cost effective in remote rural areas, and especially effective in improving quality-of-life for the users. The primary objective is to facilitate the access of lower-income people to education, lighting, communication, and (in some cases), potable water. A special feature is the close participation/collaboration of users and prefectures in the project in order to improve consumer acceptance and to ensure the effectiveness of the program. Experiences with system design improvement, performance, reliability, user training, maintenance, user-acceptance, financing, and reliability are presented. The model used for installation and maintenance is discussed, including the social, legal, and financial aspects to deploy the technology.
Keywords:1:Rural electrification – 2:Solar Home System – 3: Utilities

1. INTRODUCTION

Companhia Energética de Minas Gerais(CEMIG) is among the largest Brazilian electric-power utilities and supplies electricity for the population of the southeastern state of Minas Gerais. Since 1952, CEMIG has been supplying electricity to its consumers, mainly from its hydro facilities. In 1984, it extended its activities to new kinds of energy, such as gas, solar, wind, and biomass. Its installed generation capacity to about 5,500 MW, and its distribution grid supplies electricity to 5 million consumers in 7,000 localities, 300,000 farms, and 60,000 low-income rural homes. Urban areas are almost 100% supplied by the CEMIG grid, but there are more than 180,000 rural, and mostly low-income, households not grid connected. Generally, low-income rural consumers are in a minimal electricity consumption range (30 kWh/month) and, due to large tariff subsidies, they pay monthly rates just over US$1. In spite of state and municipal subsidies for grid installation, financial and operational costs make these consumers unprofitable to CEMIG. The Minas Gerais state government places a high priority on rural electrification and is commited to supporting geographically balanced development of rural areas by increasing the well being of the rural people and stimulating the growth of economic activities to meet economic, social and regional development goals. In order to balance shareholder interests with company social commitments to provide electricity to people, the shareholders have decided to invest 5% of annual company profits in social programs including *rural electrification.*

Throughout CEMIG (51% of the shares belongs to the state governments and 49% to the private sareholders), the aim is to benefit with electricity all the population of Minas Gerais state by 2003. To achieve these aims, the Lumiar Rural Development Programme was launched in 1999.

Depending on the distance from the grid and the population concentration of the communities, the PV alternative can be cost-effective compared to grid connection. As a result, a PV rural-electrication market was identified based on: (1) the existence of a substantial dispersed, non-grid-connected community population, and (2) the potential to provide electricity with costs below comparable grid extension options.

In its background, CEMIG has investigated the use of solar energy for many years, evaluating the state potential, carrying out pilot and demonstration projects, and accumulating experience and knowledge about the performance of solar equipment based on the various state climatic and social-economic parameters. These data clearly indicate that PV systems can perform well throughout the state, with the optimum areas situated in the northern and western regions based upon their global solar radiation averages over 5 kWh/m^2/day. Based upon these favorable conditions and several successful demonstrations, which have shown PV can be reliable and cost effective in remote rural areas, CEMIG (in collaboration with state and federal governments) launched the LUZ SOLAR—a larger scale PV rural pre-electrification programme that are based on quality-of-life improvement for users. The primary objective is to facilitate the access of low-income people to lighting and communication, and (in some cases), potable water.

A special feature of the Luz Solar is the close participation/collaboration of users in the project in order to improve consumer acceptance and to ensure the effectiveness of the program. Added benefits from this approach have

---

[1] now at SOLENER S.A

been the improvement of system design to meet users needs, and the requisite training of participants to guarantee sustainability of PV systems. The sustaintability model that is under implementation, as well as the model used for installation and maintenance, are discussed in this paper, including the social, legal, and financial obstacles to deploy the technology.

2. THE LUZ SOLAR PROGRAMME: PHOTOVOLTAIC RURAL-PRE-ELECTRIFICATION

Based on the sucessful installation of 140 PV systems in the demonstration phase completed in 1998, CEMIG's team concluded that the technology was potentially cost effective and sufficiently reliable for specific markets. Additionally, it was concluded that any PV rural pre-electrification programme in Minas Gerais should be centered on improvement of quality of life and meet required social and economic needs of the stakeholders to ensure their buy-in. It was determined that the new projects should include lighting and communications for houses, schools, health clinics, and community centres. The community characteristics and habits were mandated in planning stages and at the initiation of the project.

The **Luz Solar** is part of *CEMIG's Lumiar Development Program·* This 4-year program clearly reflects and demonstrates CEMIG's social mission. Depending on the quality of life of the municipality classification, CEMIG can provide 50%, 70% or 87% (the poorest ones) cost coverage, and the municipality (in which the community is located) covers the rest of the total investment cost. The programme includes some 185,000 low-income rural homes, with a mix of conventional grid and photovoltaics. Some 5800 of these new installations are or will be PV powered, with the particular decision to use this technology base on a comparison in cost to install a conventional, competing power source.

The LUZ SOLAR Programme has three subprogrammes:

- **Luz Solar - Rural School and Community Centre PV Pre-electrification Subprogramme:** In a cooperation among CEMIG, and federal and state governments, the aim is to electrify and to provide potable water to 800 isolated schools and 300 communal centers in municipalities that have the worst quality of life in the state.

The aim of this subprogramme is to implement TV education and adult evening classes in rural schools. The criteria for community selection include: at least 5 km away from the grid; no electrification plans for the next 5 years; population over 100 people and dispersed; local government participation and the school should not have less than 30 students. It should be emphasized that the selection of the schools to participate of the program is made by the State Education Secretary.

All public buildings in each community (i.e., schools, heath clinics, and churches) define the communal centers. In the first phase, implemented in 1999, 115 PV systems were installed in 90 rural schools in the northeast region (90 systems for lighting and communication and 25 for water pumping). The systems were partially funded by the Mines and Energy Ministry throughout PRODEEM (Program of Energetic Development of States and Municipalities), the Minas Gerais State's Secretary of Education, and the Minas Gerais State's Secretary of Energy and Mines. The basic installation included 360 -1000-Wp PV generator, which supplies electricity for AC loads (lighting, TV, VCR, satellite dish, and water pumping systems). The school and communal centres energy needs varies from 20 - 70 kWh/month.

The procurement was focused on the turn-key systems and also included 3 years of technical assistance—as well the creation of the infrastructure for the maintenance of the PV systems in the concentrated areas. This scheme has been chosen primarily to facilitate the maintenance requirements for this program, sharing technical staff, equipment, and knowledge. The school and communal centres energy needs vary from 20-70 kWh/month.

- **LUZ SOLAR - Rural Home Pre-electrification with Photovoltaics Subprogramme** will install 5500 SHS in rural communities. The basic criteria for selecting the communities are similar to the iLuz Solar - Rural School and Community Centre PV Pre-electrification Subprogrammeî and includes low-potential energy consumption, typically below 30 kWh/month/house, and community voluntary acceptance. The first 450 PV systems were installed in 1999, partially funded by the German Bank KfW. The systems were installed in 8 municipalities. Two basic 12 Vdc installations are defined: *Simple PV system (*50 Wp module, 1 maintenance-free battery and 1 load controller); and *Double PV system (*100 Wp modules, 2 maintenance-free batteries and 1 load controller) mainly for lighting, TV, and radio. Figure 3 shows a typical rural home pre-electrified by LUZ SOLAR programme in Minas Gerais State.

- **Photovoltaics Solar Energy Trainning Programme** - To ensure the sustentability and consequently the sucess of the LUZ SOLAR photovoltaic program in Minas Gerais, CEMIGís current issue is to set up a successful photovoltaic training program to avoid the lack of expertise among target-group personnel. This training has been focused on users, maintenance technicians from CEMIG, and the trainers. Through its Department of Commercialization and Demand Management, CEMIG has been working to establish and integrate this training within the electric-power utility training program. It involves two major components: technical education on PV from basics through system concepts, and hands-on training for maintenance. The initial target groups were the trainers (teachers or instructors) within the CEMIG PV Training Center.

As CEMIG's guideline, all the technical training for its staff is primarily conducted through its *School of Formation and Professional Improvement ñ EFAP*, in Sete Lagoas, 70 km from Belo Horizonte. To meet the training needs of Luz Solar Programme, a *CEMIG's PV Training Center* has been established inside EFAP. This facility has two functions: the demonstration centre, in which installed systems demonstrate the main PV applications in rural areas (such as lighting, communication, street lighting, water pumping, refrigeration and irrigation). The Centre includes a classroom equipped with relevant PV system components that are installed by the programme to give infrastructure for a complete photovoltaic training. Currently, the training course content for the personal involved in the Luz Solar Programme, which focuses on the prefectures electricians, trouble-shooters technicians, and regional CEMIG's electricians for technical support, is being refined and finalized. The training of these groups will be followed by the training of the project developers and regional technicians for marketing.

Customers training is very important for the sustainability model of the Luz Solar Programme, since costumers satisfaction with PV systems also depends on the user education provided. The training makes clear the capabilities and limitations of the PV systems, cover routine maintenance procedures, including interpreting control panel information, managing loads, modules clening procedures and replacing bulbs and other components.

3. THE SUSTAINABILITY MODEL OF THE LUZ SOLAR PROGRAMME

As the Luz Solar Programme is part the Lumiar Rural Development Program, it will follow the same guidelines for its implementation. The priority will be given for the poorest municipalities, since the subsidies to be given for the community is a function of municipality quality of life classification. This program clearly reflects and demonstrates CEMIG´s social mission. Depending on the quality of life of the municipality classification, CEMIG can commit R$800/customer, R$1100/customer, or R$1300/customer (U$1=R$1,80) and the municipality, in which the community is located, covers the remainder of the total investment cost. Actually from the 822 municipalities that compose the State of the Minas Gerais, 429 are classified as very poor and CEMIG will commit R$1300/consummer, more than 90% of the total investments costs for the simple PV systems type (these systems cost R$1398,00, when U$1=R$1 in the bid done in 1999). These are the municipalities, most of them situated at the northern and eastern regions of the state. The majority of the PV systems that will be installed by the Luz Solar Programme are in these regions.   It is critical is that the municipality must be located at optimum areas of global solar radiation, with averages over 4,5 kWh/m^2/day.

For the implementation of the Luz Solar Programme - Phase 1 in 1999, during which were installed 565 PV systems, in 100 comunities in 30 municipalities, a field socio-economic study was completed. From data analysis, the results demonstrates that the majority of households and prefectures are willing to pay the tariff (monthly bill), and the capacity to pay is influenced primarily by their income (average R$ 70, around U$50 monthly). They cannot pay downpayments, but they are willing to have CEMIG's bill them periodically. As a social value/investment, they are willing to pay the monthly to keep their system working well. Also from the field socio-economic study results, it is crucial to subsidize the downpayments due to the low monthly income of the public benefitting from and needing the programme.

The implementation model of the Luz Solar Programme obeys the following paths:

- the people from the rural communities require electrification to the prefectures;
- the prefectures negotiates with CEMIG's local office;
- this office chooses the best option based on the cost comparison analisys of grid versus PV;
- if the PV systems is the chosen option and if there is money still available (each local office have an amount of money for rural electrification), an order to install the systems is done by CEMIG´s local office;
- the Department of Commercialization and Demand Management, as the executing office, is responsible for all performance standards and technical specifications of the PV systems. It designs the systems and when an significant amount of system has been ordered by local offices it buys the PV turn-key systems (by national bid requiring that all the components follows international standards and passes in the tests required by the technical specifications) and coordinate all the installation, comissioning and training processes;
- once a community is selected as a project site the residents are chosen to receive a simple (50Wp) ou double PV systems (100Wp), based on the size of the house and their ability to pay the monthly installments. Potential consumers must be willing to purchase the PV

systems. Schools, communities centres (heath centres, clubs, children care places, etc. and houses) are automatically included in the prefecture agreement with CEMIG;

- the prefecture start to pay the as soon as Department of Commercialization and Demand Management defines the amount of PV systems for that specific community. CEMIG has 180 days after the total payment of the downpayment to install the PV systems in the community, including SHS and schools and communities centre systems. The systems have been bought in a volume, to get economy of scale in the procurement. The products are standardized to facilitate quality assurance. Only after the field tests are completed can the systems be installed.
- after the systems has been installed, the Department of Commercialization and Demand Management monitores the performance of the PV systems;
- after the installation, each municipality that received the school and community centre's systems will have a prefecture's electrician (prefecture's employee) trained at the CEMIG's PV Training. These technicians are trained mainly fot the routine maintenance of the schools and community centre PV's systems, helping with monitoring and overseeing the system's performance. The householders are trained to give routine maintenance for their systems;
- when (if) the PV systems fails, the users (prefecture or the householders) call the CEMIG´s maintenance regional office, which, in turn, orders the technical assistance to service the system (this is done by private electricians trained by the trrainer of the CEMIG's PV Training Centre). The idea is to have the availability of a many trained electricians, to select the best service. CEMIG´s maintenance regional office select the private electricians (small entrepenuers) and contract the electrician. This office also establishes the priority of maintenance, with the grid always has higher priority.

CEMIG's regional office (covering one or several muicipalities) is also responsible for inclusion of the rural communities in the programme, maintenance services, controlling the prefectures and householders payments, and enforcing any disconnections for payment defaults.   In the Luz Solar Programme, CEMIG sells energy services, but retains ownership of the system.   That is, the PV hardware (PV modules, charge controllers, inverters and batteries) is neither sold nor leased. CEMIG owns the PV system and provides complete installation and maintenance.

The consumers (householders, prefectures etc.) pay the monthly installment of R$237,5/MWh established by the ANEEL 289 resolution - July/2000), which represents part of the cost of battery replacement and maintenance of PV systems. For example the householders will pay a monthly PV fee (R$1,7 for the simple PV system and R$3,0 for the double one, around US$1 to US$1,80).

CEMIG subsidies the PV maintenance, similar to what it has been doing for rural consumers electrified with its distribution grid, as part of its social mission. The bill is due quarterly. If the users fail to pay two bills (that means 6 months of service), the modules will be decommissioned and replaced only after the payments of the bills and taxes of the administration services.

The philosophy of the Luz Solar Programme is a rural pre-electrification, consequently when the electricity demand of the PV pre-electrified community grows, the grid will become economically viable. Then the PV systems will be transfered to another isolated rural community.

Spare parts, such as fuses, light bulbs and other components that need to be replaced frequently, must be available locally. This is accomplished through a dealer in the municipality accessible to the users. CEMIG has worked very hard to ensure this accessibility and the expertise to maintain the installed system.

4. CONCLUSIONS

PV systems are an effective complement to grid-based power, which is often too costly for sparsely settled and remote area. They provide a cost-effective alternative to extending in the grid and also provide a social benefit to the users.

The idea of installing systems in concentrated areas has the main focus to establish suitable technical assistance, proper training, adequate local dealers and a viable maintence scheme. These factors together, between so many others, should ensure a success of the Luz Solar sustaintability model. Many early PV programs failed due to a several critical factors, including unreliable technical performance, poor system design, lack of ongoing, qualified technical performance, poor system design and unrealized user expectations. The Luz Solar Programme has benefitted from lessons learned from these experiences to avoid (or at least minimize) failure in the sustentability model.

There are no major technical problems with the nearly 700 PV systems aleady installed, their major components, or with users load management practices (despite the difficulties in user training).

The Luz Solar programme is especially important since it provides electricity to these schools for the *first time*. The academic environment is being enhanced with adequate lighting and with the ability to provide educational links through television and satellite communication. PV systems are safer and more convenient than kerosene lighting, diesel lanterns, and dry cells which are commonly used by the people in such rural communities. Consequently, this is also raising the living standard in these rural communities—having a positive and related effect on health-related issues.

Acknowledgements

The authors wish to thank specially all CEMIG's staff that also contributed to the success of the these PV programs. Also the authors acknowledge the unselfish staff participation of the Energy Ministry throughout PRODEEM (Program of Energetic Development of States and Municipalities), Minas Gerais State's Secretary of Education, Minas Gerais State's Secretary of Energy and Minas Gerais State's Planning Secretary, COPASA-MG and the German Bank KfW.

REFERENCES

[1] A.R. Lobo, C.A. Alvarenga, A.E. Prado, and M.S.C.C. Mendonça, Photovoltaic Rural Electrification and Electric Power Utility, IERE Workshop, Cocoyoc, Mexico, (1995) 146-160.

[2] C. A. Alvarenga, D. Costa & A. R. Lobo Proc. 4th World Renewable Energy Congress, Denver, U.S.A., Elsevier, London, (1996) 1152-1157.

[3] A.S.A.C. Diniz, A.E. Prado, M.S.C.C. Mendonça, F. Q. Al;meida, C.A. Alvarenga, 26th IEEE PVSC, Anaheim, U.S.A., (1997), 1317-1320.

[4] World Bank technical report - "Brasil: Proposta de Energias Renováveis para Eletrificação Rural no Nordeste", Workshop de Conceituação do Projeto, Rio de Janeiro, Julho/97.

[5] A. S. A. C. Diniz,., M. S. Mendonça,., F. Queiroz, D. Costa, C.A. Alvarenga
Progress in Photovoltaic: Res. Appl., 6 (1998) 99

[6] A. S. A. C. Diniz, C. A Alvarenga, J. A Burgoa
Second World Photovoltaic Specialist Conference (1998) 1560

[7] A. S. A. C. Diniz. R. Flora, J G. Rosa, R. Shroer., J. A Burgoa
Second World Photovoltaic Specialist Conference (1998) 580.

[8] A. S. A. C. Diniz., J G. Rosa, J. A Burgoa
CONLADIS – Congresso Latino Americano de Distribuição de Energia Elétrica (1999) 50

[9] C. Ribeiro, A. S. A. C. Diniz, O . Soliano
Congresso Brasileiro de Energia – CBE (1999) 380

MULTI-APPROACH ON THE DEPLOYMENT OF PV TECHNOLOGY IN THE COUNTRYSIDE OF BRAZIL.

O. Soliano Pereira
A. Souza Neto

Winrock International
Av. Luiz Tarquínio Pontes, 2580
Vilas Trade Center S/106 – A
Lauro de Freitas – BA – BRAZIL
42.700-000 – Phone: 55 71 379-1759
osoliano@winrock.org.br
asouza@winrock.org.br

ABSTRACT: Winrock's Brazil REPSO approaches the deployment of photovoltaic energy systems for Brazilian remote areas contributing at regulatory issues, installed systems and increased capacity building of key actors. The straight relationship between the National Regulatory Agency for Electric Energy (ANEEL), the Ministry of Mines and Energy (MME) and Winrock has lead to contributions to the regulatory framework strengthening the PV technology and extended to all renewable energy (RE) systems. In the same environment, Winrock has also helped in the development of permission holder to exploit areas inside utilities' areas. The partnership with MME has contributed to the installation of PV systems in community service centers such as schools and health centers. In addition, Winrock has participated and contributed to the national program for rural electrification created by the Government of Brazil (GOB), and efforts was made to assure funds for PV systems for rural electrification. Besides, there is a success story of the creation of a NGO network that increased the penetration of PV technology in remote areas after capacity building and technology transfer is provided. The overall idea of Winrock's work is to keep Renewable Energy "on the table" and increase its penetration in the countryside energy development.
Keywords: Rural Electrification – 1: Sustainable – 2: Developing Countries – 3

1. INTRODUCTION

This work presents the experience of Winrock International's Renewable Energy Project Support Office – Brazil REPSO in the deployment of photovoltaic energy systems for Brazilian remote areas. Winrock, with the support of the United States Agency for International Development (USAID), established a field office in Brazil four years ago. Since then a lot of work has been done to increase the penetration of PV systems in the remote areas.

Winrock has been working in many different approaches with partners such as the Brazilian Ministry of Mines and Energy (MME), the National Regulatory Agency for Electric Energy (ANEEL), local non-governmental organizations (NGOs), private and still state-owned power concessionaires, and potential permission holders which will be able to exploit areas within utilities' concession areas. These activities have placed the REPSO among the most important key actors in the Brazilian rural energy development scene.

Each Winrock's partner plays an important and different role in the Brazilian rural energy market. MME is responsible for the national energy policy, and coordinates the PRODEEM. ANEEL and the state regulatory agencies define the regulatory framework. The partner NGOs have credibility to assure the sustainability of their projects due to the serious work they are doing towards the sustainable development of the Brazilian society. Both public and private concessionaires are important players in the rural electrification process, either increasing its pace or slowing it down, according to state government policies. Finally, the new figure of the "permissionaire", recently established by ANEEL [1] can be the recipient of permissions bid inside the utility's concession areas, to explore the non-

electrified market, if utilities do not demonstrate appetite to do that.

Winrock International has a presence in over 40 countries and has other areas of action such as Forestry and Natural Resources Management, Gender and Agriculture, all of them with intersection with Clean Energy Group. That means that the renewable energy program, undertaken by Brazil REPSO, is developed through integrated solutions supported by the Institution's worldwide capability and experience coupled with the strong knowledge of local capability. The ultimate approach to a country like Brazil is that there are many solutions to the supply of energy to support the development of the rural areas.

2. SUPPORT TO GOVERNMENTAL ACTIONS

According to MME [2] there are 100.000 communities with no access to electricity supply. National statistics reports the existence of over 20 million people without access to electricity [3] and the National Agricultural Census mentions three million farms in a similar situation. This market would represent an energy demand of approximately 20 GW [4].

In order to overcome these challenges, two national programs were launched by the Ministry: the National Program for Energy Development of States and Municipality (PRODEEM), and Light in the Countryside National Program of Rural Electrification (Luz no Campo).

PRODEEM has been created through a federal Decree, in 1994, to assist the energy development of states and municipalities in the most remote localities of Brazil. Naturally, in these places, the cost of grid extension is prohibitive, thus PV becomes the more effective way to supply energy, in a substantial number of situations. PRODEEM's accomplishments can be seen in the Table I.

Table I: PRODEEM's results: Population benefited, installed PV capacity and resources invested.

Year	Communities Benefited	Population	Installed Capacity (kWp)	Resources (R$ x 10³)
1995	9	-	-	2.243
1996	116	34.403	187	4.348
1997	200	68.633	428	4.400
1998	1.776	351.200	554	7.498
Total	2.091	403.036	1169	18.389
1999 *	1120	800.000	1260	54.234

* Estimate for 1999.

Source: MME [4]

Winrock has a cooperative agreement signed with the Ministry to support the restructuring and help the implementation of PRODEEM. On the institutional front, Winrock has produced the first draft of the PRODEEM's Business Plan, which recommends the establishment of Regional Market Managers, and push the program from its current aid approach to a more market-based one.

Additionally to the institutional support, Winrock has implemented PV projects in rural communities jointly with members of a network of Brazilian NGOs, established by Winrock which provided technical capacity building and helped the implementation of sustainable mechanisms to assure the continuity of installed projects.

Luz no Campo was launched in December 1999. The U$ 1.7 billion program intends to bring electricity energy to 1 million rural families - half of them in the Northeast of Brazil, mainly grid connected. As a result of an effort by Winrock, which included meetings with the Federal and State Ministries officials, scenario analysis and estimation of remote market, a component of PV was included in the program. In the case of the State of Bahia, an amount around U$ 10 million will be destined to PV projects.

In the regulatory front, Winrock has been working with the National Regulatory Agency for Electric Energy (ANEEL), whose one of the key objectives is the provision of universal service in a continental country like Brazil, inside a new environment composed basically of private distribution utilities. Winrock has developed business plans demonstrating the feasibility of renewable energy business considering the regulatory incentives provided by the Institution. The final result was the evidence of the PV systems were the least cost alternative in the case of remote areas, and concessionaires will use them in the case the Agency decide to enforce the universal service compulsory need to the utility to invest in rural electrification.

ANEEL has promoted several public consultations and hearings to discuss new resolutions, and Winrock has participated in those, which any impact can be created to the diffusion of PV and other renewable sources of energy. The following procedures have been adopted by the REPSO:

- Daily survey the resolutions put in place and consultations opened to public participation, which could affect the implementation of renewable energy;
- Translate draft regulations expected to have a notable impact, and distribute the English text version to concerned partner institutions in the U.S. and other countries;

- Gather of all relevant contributions received, with their justification / rationale;
- Present alternative proposals to the agency and advocate for them during public hearings;
- Distribute final version of the regulations to key actors in both printed and e-mail form.

One of the implemented resolutions was *Resolution 112/99* that defines requirements for the registration/authorization for construction, upgrade and refurbishment of thermal, wind and solar power plants. ANEEL decided to simplify the process of registration for plants under 5 MW. Only plants over this limit requires previous authorization, and it was eliminated the requirement of authorization for feasibility studies.

Resolution 233/99 established a price cap (*Valor Normativo*) for energy freely negotiated between independent producers and concessionaires and permissionaires, establishing the limits, which are allowed to be transferred to end-user tariffs. The highest limits were given to PV: US$ 137/MWh, as opposed to US$ 32,7/MWh for competitive sources The establishment of these Standard Values attempts to provide cost coverage to make possible the of diversification of energy sources while keeping consumer protection.

Winrock has also contributed with comments to the draft of *Resolution 245/99* that create conditions to extension of a subsidy given to fossil fuel generation in remote localities, especially in the North of Brazil, to PV generation. The CCC fund (Fossil Fuels Consumption Account) amounts up to U$ 176 million that can be redirected to renewable energy projects. This promotes renewable energy as a least-cost option for rural electrification in many situations.

Another important resolution which Winrock has given a contribution during its discussion was the *Resolution 333/99* which created general conditions for the establishment of electricity installations of private use and the rules for operation of permissionaires, which can even operate inside the concession areas, after a public tender.

3. CONCESSIONAIRES AND PERMISSIONAIRES

COELBA, the private power utility in the State of Bahia, asked for Winrock's assistance to develop a model for investment evaluation in decentralized renewable energy. The model was part of a draft corporate business plan developed under business confidentiality conditions. An IRR (Internal Rate of Return) model was developed to assess the economic feasibility of PV projects, taking into account the above mentioned incentives given by ANEEL, and even the potential effect of the Clean Development Mechanism - CDM.

The developed model is been extended to other utilities (private and public) interested in the deployment of the rural market using RE technology, but because of the huge continental dimensions of the Brazilian territory the main barrier so far is uncertainty associated to O&M costs.

The same model has been applied to rural electrification cooperatives (REC) and agricultural cooperatives, which are the potential permissionaires. Existing REC are submitting their proposals to become permissionaires. These permissionaires can become more efficient and cost effective the supply of remote areas and are not biased to grid extension as the conventional concessionaires.

Winrock has been supporting a couple rural cooperatives and NGOs, potentially permissionaires, in the development of their business plans and implementation of these innovative projects. The business plans are especially important to NGOs that need to create, at least, a separated account or even a for-profit subsidiary to meet the requirements to become a permissionaire.

The concept of permissions for new areas is still under development by ANEEL, and will represent, in certain cases, the appearance of a new agent on the development of rural electrification, mainly using PV. It will represent a threat to the utilities, which, due to strategic reasons, will feel obliged to consider this least cost alternative.

So, Winrock is working on the development of mechanisms that could justify to the regulatory agency to bid those new areas to potential permissionaires while is promoting capacity building to institutions which can play this role in the rural market.

4.   NGO's NETWORK

Winrock put in place the idea of introducing renewable energy to the projects and programs of local NGOs working in different issues related to sustainable development. The twenty partner organizations work on areas such as rural development, health and nutrition, education, conservation and biodiversity protection, gender issues, among others. They generally involve local community in a participatory process and assure the sustainability of the project through the provision of a local support and the establishment of revolving funds, even generating for profit institutions.

The geographical area of influence of these NGOs varies from huge national parks in the middle of the Amazon region to small areas in the semiarid of Bahia. Thus, the approaches adopted do not privilege a single area but the diversity of the country and all his plurality.

Winrock normally analyses the other projects developed by the local NGOs, provides technical assistance, training opportunities, developed projects to be submitted to funding institutions or financial agent, provides cost-sharing to pre-feasibility studies and the development of business plans and occasional audits to their financial mechanisms and revolving funds.

The work that has been done in partnership with two different NGOs, the Projeto Saúde e Alegria (PSA) in the North of Brazil and the Associação de Pequenos Produtores do Estado da Bahia (APAEB) in the Northeast has produced out-standing results.

PSA, located in the city of Santarém, in the second biggest state of the Amazon region - Pará (PA), focuses its activities in health and nutrition aspects of small villages on the banks of the Tapajós and Arapiúns River. They have been working with many communities in order to improve the quality of life of these populations mainly by decreasing diseases and implementing actions towards a healthier way of life. These actions are oriented to nutritional aspects of the consumed food, water treatment or hygienic education but not loosing the focus on the conservation and protection of the environment.

PSA and Winrock have installed 19 PV systems (adding up to 11 kWp) in community centers of seven villages, another 6 is already signed up for the continuity of the project and over 80 villages has shown interest to participate in future phases. In this part of the Amazonian region, there are over 800 villages that use diesel generators for their electricity supply or have no electricity at all.

Those centers are the core of the communities' activities and the place in which villagers put in place community *spirit*. One of the observed results of the implementation of the project was the increase of the number of meeting in the community centers, numbers of students going regularly to school, access to information and health conditions improved.

The PV systems were installed in the following centers:

- 7 Community Centers (lightning and small appliances have created conditions to a cultural development and global awareness among population)
- 3 Health centers (lightning, communication radio and vaccine refrigerator are providing a much more reliable action against many diseases)
- 7 Schools (lightning for 14 classrooms, TV, VCR and Satellite Dish benefiting approximately 550 students that know have information of what is going on in the world outside their remote site.)

The whole process from selecting the villages and implementing the projects comprised many different stages and difficulties. Participation of the population, the local NGO and Winrock made the difference from other national programs that implement the same type of systems. The basic difference was the approach given since the beginning of the process. The success of these projects is simply based in asking what is that the population wants. This question has been forgotten in many PV projects worldwide.

In simple terms, the first steps were to determine what population wanted and their expectations and try to match this with their capacity and willingness to pay. A second step was the capacity building process: 25 members of different communities were trained on operation and maintenance of PV systems.

Another success history with PV technology in the countryside of Brazil is the implementation of over 500 SHS and a jointly implementation of 2-kWp PV system in a Family Agriculture School (FAS). All these systems where installed in the influence area of a local NGO called APAEB, in the semi-arid of Bahia. Unlikely the previous case, the environment here is dry and the average family income is lower than the Amazon population.

The state of Bahia, the biggest one in the Northeast of Brazil, has close to 780.000 rural properties without electricity and many of than situated in the driest part of the state. Grid extension in these cases is not cost effective leading to a huge amount of people unassisted by energy services and in very poor conditions.

The FAS is an agriculture school maintained by APAEB that teaches people on how to survive in the semi-arid region. The school has over 120 young students working and studying on a daily basis to improve quality of life in the driest area of Brazil. They used to use small PV systems basically for lighting.

The work done at the school provided capacity building to local technicians and improved the previous PV systems with a much more reliable and continuos supply of energy. Winrock prepared the PV project and submitted it for funding. The MME supplied the equipment and the other costs where shared between Winrock and APAEB. Today, the school operates totally on solar energy that powers electric fences to keep the goats confined out side the crops, water pumping both for human and animal use, lightning, computers, communication radios, refrigerators and many other appliances.

APAEB's records with solar energy precede Winrock Office in Brazil, although after the first contacts and jointly work, Winrock has helped APAEB in many different activities. Either by providing technical expertise to APAEB's projects, transferring technology or preparing PV projects to be submitted for funding by a third parties. In all these actions the main idea is to increase the penetration of PV technology as a solution to the semi-arid and to Brazilian countryside.

APAEB's long experience with PV energy has itself a unique approach and outstanding results. Over the last five years 560 SHS systems were installed and a revolving fund established as a model to facilitate the acquisition of the PV systems and guarantee the replacement of parts.

The creative approach was the facts that fees are indexed to their main product: goat meat. The goat-meat-equivalent-weight-price system converts system cost into kilograms of equivalent goat meat to be paid over eight years. The implication is a high level of consumers' satisfaction and a low non-payment rate.

One important issue that will really make a difference in the future deployment of PV is the effort that Winrock is putting in helping APAEB to become the first off-grid permissionaire in Brazil, a model with a high content of replicability. A business plan was developed to the establishment of a small subsidiary of APAEB dedicated to supply rural properties and communities in this area of Bahia.

Winrock has also helped APAEB's with its revolving fund and recently an audit has been done on it, preparing the institution to face the new responsibilities as a permissionaire. The expectation is that the market will find its own pace and institutions like APAEB will become important agents to implement wise energy solutions in the fields. Winrock role has been to assist and support these initiatives to archive a common objective: development of rural areas using RE technologies.

5.   AUDIT TO PRODEEM

Winrock has done, under a PRODEEM contract, a fieldwork to determine whether the PV systems installed by the government initiative over the last 5 years had accomplished their objectives. One of the conclusions was that the involvement of the population, which is fundamental to achieve the goal, was a missing part of the first round of pilot projects. No initial involvement to try to understand their demands has taken place.

Second observation: no information was provided on the technology, their technical limits, necessary maintenance practices. And above all no clear responsibility for replacement of parts, disposal of used batteries or recovery of the costs.

6.   CONCLUSIONS

Winrock's multi-approach is centered on the dissemination of new models for solar technologies, taking an initial assumption that in a country like Brazil, there is no unique solution, but a portfolio of them, including concessionaires, permissionaires, NGO's and private

sector, always under the umbrella of national policies and regulatory mechanisms.

The pioneer work with RE energy in Brazil has given Winrock the knowledge to carry on all kinds of projects for rural deployment and RE promotion. The PV market has a long way to go yet. The multi-approach adopted by Winrock in fact denotes the plurality of solutions and indicates many paths to take.

Hopefully these solutions will lead the country towards the improvement of the quality of life of the population by increasing their access to education and information, health and dignity.

REFERENCES

[1]   ANEEL Website – www.aneel.gov.br

[2]   E. M. M. Scheleder. "O mercado Invisível". DNDE/MME (1999)

[2]   IBGE "Pesquisa Nacional por Amostra de Domicílio – PNAD" (1996)

[3]   E. M. M. Scheleder. "O mercado Invisível". DNDE/MME (1999)

[4]   Ministry of Energy Website - www.mme.gov.br

[5]   DNDE/MME Website – www.mme.gov.br

[6]   Trade Guide – Winrock International (1999)

[7]   Luz no Campo Program (1999)

[8]   Energia Para Todos – Business Plan, Winrock International (1999)

EXPERIENCES WITH GIS PLANNING OF RURAL ELECTRIFICATION IN TUNISIA

Michel Vandenbergh
Centre d'Energétique - Ecole des Mines de Paris
BP 207, F-06905 Sophia Antipolis Cedex
Tel +33 (0)4 93957480, fax+33 (0)4 93957535
EMail Michel.vandenbergh@cenerg.cma.fr

Hédi Turki
Société Tunisienne de l'Electricité et du Gaz (STEG)
rue Ibn El Jazzar, 1002 Tunis Belvédère
Tel +216 1 841204, fax +216 1 842621,
EMail dpsc.steg@gnet.tn

ABSTRACT : This paper presents the general context of the rural electrification planning in Tunisia and the GIS based methodology applied to evaluate the market for solar photovoltaic home kits. The methodology, based on cost/benefit risk analysis has been applied within the frame of the European project IRESMED. First results for a selection of Tunisian villages are presented.
Keywords: Rural electrification - 1: PV Market - 2: Developing Countries - 3

I. INTRODUCTION

For many years, the Tunisian government has been financing social and economical development programmes in rural areas. Rural electrification is one of the key development factor. In ten years, the rural electrification level has been multiplied by 2.6 to reach 79% in 1997 (to be compared with 30 % in 1987). The ambitious objective of the 9th Plan (1997-2001) is to reach an electrification level of 85% in each region of the country. However, bringing the electric network to the remaining population is more and more expensive in remote areas due to the high cost of infrastructures and the low level of electricity consumption. Because of this high electrification cost, decision makers have come to electrify rural households with photovoltaic kits.

During the planning of the rural electrification with two technical alternatives (Other technologies like diesel mini-grids or hybrids are not considered to be adapted to the Tunisian situation), it has become rapidly evident that there was a need for new decision tools. Indeed, during the planning process, the Tunisian Electricity and Gas utility (STEG) has to answer many questions :

- How to optimise the use of renewable energy credits to electrify rapidly the maximum of households at the minimum cost ?

- How to choose between the two different technologies which are available for rural electrification (grid extension, photovoltaic kits)?

To solve the problem, STEG has built a national data base for rural electrification projects. The use of a Geographical Information System (GIS) was necessary to manage this huge amount of data. Indeed, taking into account the geographical factor is essential for an optimized midterm and long term planning.

The Renewable Energy department of "Ecole des Mines de Paris" and the Planning department of the Tunisian Electricity and Gas utility (STEG) are collaborating within the EC project IRESMED to define and apply the most adapted methodology for the integration of solar photovoltaic energy in Tunisian rural electrification programs. This paper presents the IRESMED project, the technology selection methodology and the calculated potential market for photovoltaic rural electrification in a selection of Tunisian regions. The methodology, which has been implemented on a GIS platform, is based on Cost/Benefit and Risk analysis.

II. THE IRESMED PROJECT

The IRESMED project aims at analyzing the integration of solar and wind energy for electricity supply in the rural regions of the southern Mediterranean countries. Research is carried out on the selection of the most appropriate sites to install the renewable energy systems, on the technologies most suited, on the cost/benefit analysis and on the possible financing schemes. A major issue is the competitivity of renewable energies vis-à-vis fossil fuel options and thus detailed selection of the adapted sites (where the solar or wind resources are highest and which are close to load centers) must be done based on cost/benefit analysis in order to select the least cost options. The external benefits related to social or environmental aspects must also be taken into account.

Another major issue in developing countries is the lack of available financing for renewable energy projects which can contribute to the social and economic development and the protection of the environment but are more capital intensive than fossil fuel based systems and also are to be used, in the case of rural electrification, and thus purchased by very poor rural populations. For these reasons adequate financing mechanisms must be developed which combine private sector financing alongside with financial support from the governments and international aid, otherwise the lack of financing will block the integration of renewable energies in these rural regions.

The project is the natural evolution of the process of north-south co-operation initiated with the INTERSUDMED which had identified one renewable energy project per country. IRESMED focuses on the identification of viable plans for the large-scale integration and market development of solar and wind technologies ; this requires a deeper understanding of the potential for these technologies along with the driving forces for the integration of renewable energies and the measures necessary to create local market demand. IRESMED also provides links to European industrial and financial participants that are needed for project implementation.

The research thus covers the various phases of an integration study on such a large scale: research on available wind and solar resources, study of potential sites for PV or wind power, evaluation of adapted systems, cost/benefit analysis including social and environmental aspects, methodology for financing schemes and market development.

The countries concerned are Morocco, Algeria, Tunisia, Egypt, Israel, Palestinian National Autonomy, Jordan and Turkey.

The IRESMED consortium is composed of: OME (France), IPTS (Spain), EDF (France), ENEL (Italy), ENDESA (Spain), CEEETA (Portugal), CIEMAT (Spain), CRES (Greece), DLR (Germany), ARMINES (France), ENEA (Italy), RISO (Denmark), TGI (Spain), MEKOROT (Israel), CDER (Morocco), SONELGAZ (Algeria), STEG (Tunisia), NREA (Egypt), PEC (Palestine), NEPCO (Jordan), EIE (Turkey).

III. METHODOLOGY FOR PV MARKET ANALYSIS

The methodology is based on comparing locally not only costs but also benefits of centralised and decentralised rural electrification solutions. Indeed, in Tunisia [1], the expected benefits for photovoltaic and central grid electrification are not equal. A

selection criteria based on the levelized electricity cost [2] [3] might therefore give unrealistic results.

The objective of cost-benefit analysis is to quantify in monetary terms the costs and benefits of an investment project. The main advantage of the method consists of the fact that it exhaustively systematises all the costs and benefits of a project, making at the same time transparent the set of assumed hypothesis in the analysis.

The methodology is based on six steps:

1. *Build the electrification projects database*. STEG has built a national data base for rural electrification projects. Every group of at least 10 households, which is not yet electrified, is described in the data base. Using this information, it is possible to localize and quantify the potential electrical loads.

2. *Analysis of the main supply network*. This concerns analysing the existing grid upstream of the zone to be supplied with electricity and defining the new equipment (Medium and Low Voltage lines lengths and substation sizes) needed to distribute electricity from the existing grid to the remote populations. The evaluation has been done by local STEG staff for each electrification project of the data base.

3. *Solar resource assessment* A map for the sizing solar resource of PV systems has been processed for the IRESMED regions using raster data from the new European Solar Radiation Atlas (ESRA). The ESRA global horizontal radiation data for the lowest resource month is first converted to global radiation on a tilted plane, using monthly correlation methods from Erbs and Klein/Theilacker [6]. The design parameters for the calculation are: Tilt angle of the PV panel = latitude + 10° (0° = horizontal); Azimuth of the PV panel = south; Ground albedo = 0.2. The choice of this particular tilt angle is related to the expected low seasonal variation of the load profile for small PV kits. It is thus important to maximize the production in winter time. Finally the raster data is averaged in each of the 282 Tunisian administrative entities called "délégation".

4. *Analysis of loads*: Socio-economic surveys from a sample of non electrified villages have been analysed to assess the spatial variation of the potential demand for electricity on the initial year of electrification and its evolution in time. Another output from these surveys is the social

level and the corresponding financial resources of the populations. The statistical distributions of the main socio economic parameters will be important inputs to the risk analysis.

5. *Analysis of uncertainties* Financial and economic analyses of electrification projects show considerable sensitivity to a number of key variables, including: load and load growth; cost of generation; willingness to pay; technical and non technical losses; capital costs. The uncertainties of each parameter are modeled by a simple triangular frequency distribution, determined by their pessimistic, most likely and optimistic values.

6. *Photovoltaic market evaluation* In order to determine the photovoltaic market, the costs and benefits are calculated in each electrification project of STEG for both photovoltaic and central grid technologies. To take into account the uncertainties related to many parameters, a risk analysis is done using the Monte Carlo simulation method. The probabilistic approach helps to evaluate the risk of overestimating the photovoltaic market.

IV. TUNISIAN PHOTOVOLTAIC MARKET

The methodology has been applied to study the market of photovoltaic kits in Tunisia. The selected regions (fig. 1) have a low level of rural electrification and/or high grid connection costs. Table 1 gives the 1994 population statistics for the 11 selected districts.

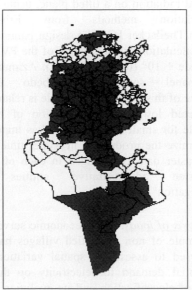

Figure1: Study regions

In the STEG rural electrification data base, the selected regions correspond to more than 2700 projects and 83000 households, with an average of 30 households per project.

STEG organized socio-economical surveys in 7 administrative districts (Gabès , Bizerte, Gafsa, Siliana, Kasserine, Zaghouan, and Sidi Bouzid). 38 villages were visited, among which 12 are connected to the national grid, 13 are partially electrified with diesel groups, 13 are not electrified at all. The head of the village and following the size of the village, 3 to 15 families were questioned.

Using statistics from the surveys, the electrification project database and past Tunisian experiences with PV projects, 10 parameters have been chosen for the risk analysis (table 2). The triangular probability density for each of these parameters is determined by the minimum, maximum and most probable values.

District	Population density inhab./km2	Electrification level
Kairouan	79	< 65 %
Zaghouan	52	< 65 %
Siliana	53	< 65 %
Kasserine	48	< 65 %
Sidi Bouzid	54	< 65 %
Bizerte	131	65% - 73%
Gabes	43	65% - 73%
Gafsa	34	65% - 73%
Le Kef	55	65% - 73%
Sfax	97	> 73%
Tataouine	4	> 73%

Table 1: Population statistics1994

Parameter	min	mean	max
Installed PV kit cost ($/Wp)	12	15	20
Lifetime of battery	2	4	5
PV annual maintenance cost ($)	10	15	30
Initial payment for PV kit ($)	0	110	110
Annual payment for PV kit ($/year)	0	17	34
Subsidies for zero emissions ($/kWh)	0	0	0.01
Production + distribution cost ($/kWh)	0.05	0.1	0.2
Demand level multiply factor	1	1	2
Lifetime.of.Grid Lines (years)	25	30	35
Lifetime.of. Grid substations (years)	25	30	35

Table 2: Risk parameters

Discount rate	0.08
Project duration (years)	20.0
Lifetime of kit (years)	20.0
Battery cost.($/kWh)	150.0
Investment for LV line ($/km)	8000.0
Investment for MV line ($/km)	9000.0
10kVA.substation cost ($)	3000.0
25kVA.substation cost ($)	3500.0
50kVA.substation cost ($)	4500.0
O&M cost for LV line ($/km)	147.0
O&M cost for MV line ($/km)	170.0
O&M cost for substation ($)	126.0
Initial payment for grid connection ($)	220.00
Fixed Annual payment ($/an)	5.00
Price of grid electricity ($/kWh)	0.07

Table 3 : Other parameters

The Monte Carlo algorithm (5000 iterations) is applied to evaluate the photovoltaic market by comparing the costs and benefits in each electrification project of STEG for both photovoltaic and central grid technologies. The resulting cumulated frequency for the rate of villages to be electrified by a PV kit is presented in Figure 2. The 10% and 90% points of the distribution establishes with an 80% confidence interval, that the photovoltaic market is between 4% and 19% of the total number of households.

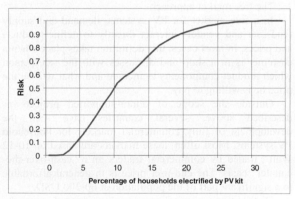

Figure 2: Risk of overestimating the photovoltaic market

V CONCLUSIONS

Within the frame of the IRESMED project, a photovoltaic market risk evaluation methodology has been adapted to the specific needs of the Tunisian gas and electric utility (STEG). The methodology is based on cost/benefit comparison of the decentralized and centralized electrification alternatives. The methodology has been applied to a set of 2700 projects (83000 households) from the STEG rural electrification data base.

The results of the risk analysis indicate a minimum value of 4% (risk = 0.1) and a maximum value of

19% (risk = 0.9) for the total number of households to be electrified with photovoltaic kits.

VI. REFERENCES

[1] N. Hammami et al., 1998 : "L'électrification rurale de base solaire en Tunisie - Approche et réalisation", AME et GTZ

[2] Neirac F.P., Vandenbergh, C. Andolina, F. Martin, G. Glinou, V. Miranda, J. Halliday, M. Rudden, 1997.: "SOLARGIS project : Integration of renewable energies for decentralised electricity production in regions of EU and developing countries " Proceedings of 14th European Photovoltaic Solar Energy Conference

[3] C. Monteiro, V. Miranda 1997: "Probabilities and fuzzy sets in the market evaluation of renewablme energies - the Solargis experience" - Proceedings of the PMAPS '97 Conference

[4] D. I. Banks 1998: "Criteria to support project identification in the context of integrated grid and off-grid electrification planning" - report of EDRC University of Cape Town

[5] W. Short, D. J. Packey, T. Holt 1995: "A manual for the economic evaluation of energy efficiency and renewable energy technologies" NREL / TP-462-5173

[6] Duffie J. A. and Beckman, W.A., 1991: "Solar Engineering of Thermal Process" John Wiley and Sons

MARKET DEVELOPMENT MODELS FOR HOUSEHOLD PV SYSTEMS IN DEVELOPING COUNTRIES

ir F. van der Vleuten-Balkema[1], drs T. Kuyvenhoven[2], prof.dr F.Janszen[2]
[1] Free Energy Europe, PO Box 9564, 5602 LN Eindhoven, the Netherlands,
tel: +31 40 290 1245, fax +31 40 290 1249, info@free-energy.net
[2] Erasmus University, Rotterdam, the Netherlands

ABSTRACT: Three main market development models for household PV systems in developing countries have been analysed: the greenfield project model, the system engineering houses model and the traders model. Although the traders model may have the best track record in serving end-users in developing countries, little is done to support and expand this model. The paper calls for a broader view on PV market development in developing countries, and for instruments to be developed to strengthen especially the free market for household PV products in developing countries.
Keywords: Rural Electrification – 1: Market development – 2: Policy – 3

1. BACKGROUND

Utility investment in rural electrification has been proven over the past three decades to be unable to reach an estimated target group of 2 billion people. A self-sustaining market for household PV systems seems to be the main option for basic rural electrification.

We have observed two different, not related PV market models that can currently be found in developing countries.

Table 1. Characterization of two existing market models.

	Project market	Free market
Defined by	Project developers	End-users
Main products	30-50W c-Si	10-15W a-Si /c-Si
Price level	400-1000 USD	50-200 USD
Distribution	(Semi)utility	Traders
Estimated annual installations	30.000-40.000	150.000-200.000
Interventions	Government and donor support	No support Distorted by projects

1.1 The project approach

The project approach is characterized by the fact that project developers and project managers are the principal actors deciding the products and prices in the market.

Defined by actors external to the market, projects have shown a tendency to a focus on singular and standardized market systems[1]. For instance, the World Bank has defined "best practices for Photovoltaic Household Electrification Programs" [2]. In Europe, the European Union has recently developed similarly narrowly defined "Universal technical standard for Solar Home Systems" [3].

Martinot [4] has evaluated twelve on-going World Bank group projects to promote Solar Home Systems and has come to the preliminary conclusion that "more project flexibility is needed for local partners to allow them to develop good business models". Notably the review suggests that customers desire a range of component

options and service levels and can benefit "from even small systems".

We estimate that the annual number of installations in project approaches is currently between 30.000 and 40.000 systems per year.

World wide experience with this project type has however shown that it has not created self-sustaining markets yet. Significant numbers of systems have been diffused in projects, but personal communications in former project areas time and again show that when a project has stopped, the market has generally largely stopped. Surprisingly, there is little evaluation of this failure of this project model to create sustainable PV markets.

1.2 The free market approach

In a free market for PV systems, demand and supply (end-user and vendor) interact directly to define products and prices. In most cases, this definition takes place when a customer is in a shop and discusses with the salesperson which product components to include and what will be the price he is willing to pay.

Outside the scope of donor-supported projects as described above, several countries have seen the development of fully commercial markets for household PV systems. Most often, these markets are based on 10-12 Watt , traded by commercial traders on a "cash-over-the-counter" basis. The end-user price is in general affordable to a significant part of the population at 60-200 USD.

Distribution channels generally depend on the traditional trading infrastructure in the market, and mostly build on existing distribution channels, micro-enterprises and the informal sector.

We estimate the market volume of commercially traded household solar energy systems at around 150.000-200.000 systems per year.

This market is already self-sustaining and depends on investment by importers and re-traders in developing countries, and equipment suppliers, such as Free Energy Europe.

1.3 Donor support for market interventions

Despite its proven potential for serving the end-user, the free market model has not received support from the international donors and development aid organizations. Sometimes it is even distorted through projects in their drive for strict technical standards that exclude the small

---

[1] in most cases the 30-50 Watt "Solar Home System", costing 400-700 USD. Due to the high price, substantial market is only seen in long term credit or fee-for-service dissemination [1][2].

systems from "market development projects". In Kenya, Morocco and China projects have even been financed with the explicit objective to transform the existing free market from its current state to a "better practice" high quality market. In these interventions, the better practice markets are defined by the project developers and managers, including implicit or explicit technical standards. Therefore we would characterize these interventions as replacing the free market by a project market.

Two main reasons appear to play a role behind this bias of the major international donors towards project markets:
1. Lack of awareness: international project developers and donors may be largely unfamiliar with the functioning of self-sustaining free markets in developing countries
2. Lack of capacity: International donors and project actors may not have adequate instruments for supporting dispersed free markets in which small enterprises operate. Their existing instruments may only fit a focus on large-scale operations such as nation-wide credit schemes and "fee-for-service" programs, inherently oriented towards project markets.

2. PURPOSE

In view of the above, the research presented in this paper has aimed to provide insight in the ways in which self-sustaining markets for household PV systems in developing countries can be created.

The objective of the research has been to identify and map the various market development models and their dynamic relations between actors and factors. The objective has also been to identify critical processes in these market development models.

3. METHODOLOGY

The research has started from contemporary innovation theory, dynamic business modeling and knowledge systems analysis.

The used research methodology is basically a knowledge management methodology. This means that the results from the study give insight in the present level of knowledge and insight in market development models by the combined actors involved in the study. The research has involved all supply side actors in the Netherlands and all demand side actors in South Africa. It has been carried out between September 1998 to July 1999.

4. MARKET DEVELOPMENT MODELS

Based on the interviews and secondary information, three typical PV-market development models (business models) have been identified. These models can be seen as very 'pure' business models for introducing PV to rural areas. In reality, several mixtures of these models will be found. Below, the basic characteristics (Organization(s), Applications, Market and Technology) of each model are described.

4.1 The project model

Organization

In this type of development, a donor, consultant or NGO defines a green-field project. The main purpose of most projects is the development of a new distribution infrastructure. The exposure of such projects is very large whereas the development process can be relatively slow (3-7 years).

Technology & Application

The main application in household electrification projects is a standardized solar home system with a crystalline silicon panel with a nominal power of at least 40 Wp. In most cases, this application is offered in combination with credit structures in order to increase the affordability.

Market

Projects are aimed at the rural population living in pre-selected areas in developing countries. To reach these actors a new distribution & financing, installation and after-sales structure is developed. Market development depends primarily on the success in building this new infrastructure.

4.2 The system house model

Organization

The central actors in this development model are 'systems engineering houses': organizations or companies that have the knowledge and know-how to put together components to tailor a solar home system to the needs of an end-user.

Most of the time solar energy is the main business activity of these companies.

These actors perform an engineering as well as a small-scale wholesale function and are, in both cases, quite flexible in their activities.

Technology & Application

The main applications is a solar energy system tailored to the needs of specific clients. SEH's use panels of different sizes, based on different PV-technologies (i.e. crystalline and amorphous silicon). They tend to have a strong focus on technical standards developed in projects.

Market

Based on the applications described above, SEH's often aim for niche markets, such as NGO installations and the telecommunications market. This causes a bias towards the project model. SEH's are also actually attracted by donors into their project model.

Market development depends primarily on evolutionary expansion of the business of the system houses.

4.3 The traders model

Organization

PV systems are imported and distributed by companies that have a well-developed commercial distribution infrastructure, that reaches the target group of PV systems. This can be most successful when PV complements trading in other products (for example black and white televisions and batteries). In general PV is only a small percentage of the turnover of the trader.

Application & technology

The main application is a standardized trading product that has been optimized in the market to fit the needs of the end-users in terms of product and price. For instance a 12W panel as battery charger, or a complete basic lighting system.

There is a strong focus on the end-user needs and choice of technology becomes a less important factor. As long as the needs are properly identified and fulfilled, IEC

certified as well as not certified, crystalline as well as amorphous silicon PV technology can be used.

Market

The market depends on the market for the complementary products or the existing distribution network from which one tries to profit.

Market development depends on the expanding market reach of the distribution network and the recognition of the product by vendors and end-users.

5. NON LINEAR MECHANISMS

The insights gathered from the interviewed and investigated actors (governments, donors, NGOs, utilities, equipment manufacturers, importers, distributors, consultants, systems engineering houses, financial institutions, project companies, retailers and end-users), have been explicated and linked together in dynamic business models in the indicated causal diagrams.

The mapping analysis has revealed several non-linear mechanisms that play a role in the relation between supply and demand and that can be critical processes in market development. These critical processes are summarized below:

5.1 Project model

A strong focus on affordability of expensive extensive design systems and the resulting development of credit structures causes a low level of product and price differentiation, which again increases the focus on financing for affordability.

PV projects create consumer awareness and initiate (free) market development, although the latter is a slow process. On a short time scale projects can also damage existing markets.

Where PV projects introduce extensive-design products, the free market will not develop and follow-up can only be in more projects.

5.2 System house model

The involvement of system houses and their focus on product engineering tends to reinforce the development of PV projects and vice versa.

System houses may have a temporary role in the market because of a shift from engineering to distribution.

5.3 Complementary traders model

PV sales by complementary traders can have positive effects on the PV market development process in three ways: combined use products, use of existing infrastructure and high local involvement.

Use of an existing distribution infrastructure makes quality assurance more difficult.

A lack of commitment by complementary traders can slow down the market development process.

6. CONCLUSIONS

The analysis has revealed that there are three market development models that may be the ingredients to a particular market development for household PV systems in a certain country.

It is possible to make some conclusions on the use of the basic approaches (development models) presented in this paper:

6.1 Projects

First of all, project development can lead to sustainable PV-market development, but only if a 'free market' follows-up on these projects.

We conclude that offering the end-user a larger enough differentiation of PV-products can play a critical role to attain sustainability. The end-user should be able to make a choice from a variety of products.

This implies that the focus on technology quality should not lead to over harmonization of options, such as suggested in [2] and [3].

Increased competition within projects can lead to more product differentiation, a better match with end-user needs and local habits resulting in an increased level of commercial activities i.e. sustainable market development.

6.2 System engineering houses

Secondly, the activities of system engineering houses can also lead to sustainable PV-market development. Creative system engineering houses can give the PV-business the product-differentiation, which it so badly needs.

There is however little incentive for system engineering houses to focus on product differentiation towards less extensive-design products. First of all, project customers mainly ask for extensive-design products and secondly the small distribution capacity of system engineering houses withholds them from focusing on simple-design products because they can be easily 'copied' by existing free market distributors, that can operate with higher financial and distribution capacity.

By combining their engineering function with a stronger focus on distribution, system engineering houses can very well initiate and maintain sustainable PV-markets in less developed countries.

6.3 Traders

Thirdly, traders of complementary products such as batteries, radio's and televisions can open up a way for sustainable PV-market development. PV-market development can benefit from the combined use of PV and complementary products, the use of an existing distribution infrastructure and a high level of local involvement.

This way has proven hard to combine with the development of PV-projects. It has to be prevented that PV projects undermine the market development through traders.

To assure best value for money for the end-user, easy to use products are to be offered in combination with very clear information and development of a certain level of technical capacity on the grass-root/market outlet level.

Although the conclusions described above are mainly based on information gathered in The Netherlands and South Africa, one can assume that the dynamics of these market development models have a more universal validity. Further research, in different countries and involving different stakeholders, can add to the insights.

7. RECOMMENDATIONS

The main challenge for donors and governments is to develop instruments to support the development of self-sustaining market models, tailored to the situation of a particular country.

Currently donors and governments focus only on the project model and more recently on supporting system engineering houses (with a bias towards projects). There is still no active support for the traders model, even though this may hold a large potential for market development.

We propose that the free market development in which the traders operate can be supported in the following ways:

➤ Assist the development of an effective and efficient PV trading infrastructure through supporting development of retailers and outlets;
➤ Remove market barriers such as standards that may prevent trade-able products from entering markets;
➤ Facilitate development of the inherent knowledge base in markets regarding realistic PV options, on the level of the vendor and the end-user.
➤ Facilitate the learning process of markets to distinguish between good and bad quality products, by arranging for consumer information feedback and independent product information services;
➤ Prevent all distortion of free market development by projects.

Finally, we propose to accept that best practices for market development can only be defined by markets, in which end-users and vendors are the main actors, rather than consultants, project developers or international donors.

REFERENCES

[1] Dlamini, S.N., 1999. Solar Home Systems: Insights from developing countries around the world - a literature survey of current issues and topics. Paper ECN-I--99-005. ECN, Petten.

[2] Cabraal, A., M. Cosgrove-Davies, L. Schaeffer, 1996. Best Practices for Photovoltaic Household Electrification Programs – Lessons from experiences in selected countries. World Bank Technical Paper Number 324. World Bank, Washington.

[3] EG-DG17, 1998. Universal Technical Standard for Solar Home Systems. Instituto de Energia Solar, Madrid.

[4] Martinot, E., A. Cabraal, S. Mathus, 2000. World Bank Solar Home Systems Projects: Experiences and Lessons Learned 1993-2000. Working draft paper – not for citation or quotation, Jan. 24, 2000, World Bank, Washington.

[5] Jacobsen, A., R. Duke, D. Kammen, M. Hankins, 2000. Field performance measurements of amorphous silicon photovoltaic modules in Kenya. Paper prepared for the ASES, Wisconsin.

[6] Kuyvenhoven, T., 1999. The innovation of photovoltaic applications in less developed countries: a dynamic perspective. MBA report. Erasmus University, Rotterdam (available in PDF format from Free Energy Europe)

[7] Van der Plas, R.J., M. Hankins, 1998. Solar electricity in Africa: a reality. Energy policy vol.26 No. 4 pp. 295-305.

SOLAR ENERGY IN YEMEN – AN ECONOMICALLY VIABLE ALTERNATIVE

Peter Ahm, Director, PA Energy Ltd., Snovdrupvej 16, 8340 Malling, Denmark
John S. Ijichi, Principal Cofinancing Advisor, The World Bank, 1818 H Street NW, Washington DC, USA
Dr. Towfick Sufian, Head of Coordination Unit, Ministry of Education, Sana'a, Yemen
Christer Nyman, Director, Soleco Ltd., Londbölentie 27, 06100 Porvoo, Finland

Abstract:
Of the 16 million people living in Yemen only a small minority - in particular in the rural areas - has access to electricity. Electrification through grid extension or local diesel powered mini-grids will be prohibitively expensive, take a very long time and often not provide a constant service.
Health and education are sectors fundamental to the social and economic development of rural Yemen. Access to a constant and reliable source of electricity widely expands the scope and effectiveness of these sectors allowing the use of modern technology
Yemen is endowed with an abundant and both over the regions as well as the seasons a pretty constant solar irradiation - in average 5.5 kWh/day/m^2.
Photovoltaics is a mature but yet relatively expensive technology directly converting light into electricity.
To be effective the health and education facilities require a relative small, but constant and reliable supply of electricity. The both technically and economically most viable source of electricity for off-grid health and educational facilities in rural Yemen have been found to be photovoltaic power systems. The technical replication potential for such systems is estimated to many thousands of units both in sectors of health and education. Access to electricity from a photovoltaic power system for the local clinic or school staff is seen as an important motivating factor for keeping staff in the field.
The local photovoltaic resource base in Yemen is in reality non-existing, and it is therefore a challenge to set up sustainable deployment of photovoltaic technology - even having initial adequate donor support.
The paper describes a novel, low cost, approach by the World Bank to the deployment of photovoltaic technology in Yemen, initially for the sectors of health and education but having a wider perspective in terms of off-grid electrification.
Keywords: Developing Countries - 1,: PV Market - 2: Strategy - 3

1. Introduction

Yemen is one the poorest countries in the world with per capita income less than $ 300 per year, which ranks 142 out of 174 countries in the 1996 Human Development Index. Its population is over 16 million with a population growth rate of 3.7 % per year, which means the population will be doubled in less than 19 years. The county's services indices are universally low: life expectancy at birth is 54 years, infant mortality at 96 per 1000 live births is about twice as high as the middle East average, adult illiteracy rate is 58 %, and gross primary enrolment is 70 % of school age population compared with the Middle East average of 96 %. Among the overall energy consumption 89 % relies on fuel wood whereas electricity occupies only 1.4 %.

Although the current democratically elected government has been implementing an austere structural adjustment program since 1995 with full support from the international community including the EU, Japan, the IMF and the World Bank, growth in GNP per capita between 1988 and 1998 shows an actual decline of 1 %. The political unification after the civil war in 1994 produced a democratic system including universal suffrage, most advanced in the Middle East region. However, economic damages caused at the time of the 1991-92 Gulf War by siding with Iraq and at the time of the Civil War between the South and North in 1994 are so wide spread, that it is estimated that it will take at least 20 years for Yemen to attain full recovery.

Against the background of more than 19 % of the absolute poverty level in the Yemeni economy,improvement in education and health services with special emphasis on the rural population is the key to economic development for the country. More than 70 % of the population lives in the remote rural areas, in villages of an average size of about 315 inhabitants. Only a fourth of the country is in principle covered by the national electricity grid, which itself suffers from inadequate generating capacity, lack of proper mainte-nance and human resources. Of the rural households only about 4 % have access to an electric grid, being the national grid or local mini-grids powered by diesel generators. Since most of the villages are situated on top of hills and small mountains in a generally rugged terrain extending the national grid to these villages is simply not an option due to very high cost. The diesel operated generators and kerosene based lighting are rapidly becoming economically unsustainable as the price of fossil fuels goes up as a consequence of the structural adjustment program. Moreover the physical maintenance cost is also rising to a prohibitive level to remote villagers.

Although the World Bank's energy policies stress the importance of economic viability and financial returns of energy investments, the Bank at the same time emphasizes the essential need of balancing these considerations with those of environmental compatibility and sustainability. In view of the high wood fuel ratio in Yemen, about 90 % of the total energy consumption, and the fact that this consumption of fuel wood is far beyond the level of sustainable wood fuel output, it is found to be imperative to promote alternative energy generation, but with a due regard to the economic viability of such solutions. It is under these circumstances tha the government of Yemen requested the World Bank to look into the possible introduction of solar energy based distant education facilities and health services for remote villages.

2. The Approach

Usually, when the Bank or other major donors move into a new sector, such as solar energy application in Yemen, it is done on a relative large scale including a broad and concerted set of actions intended to build the required capacity, to remove barriers where possible and eventually set in preparations for a kick-start of a commercially sustainable market.

However, the situation of Yemen calls for another approach. First of all very little PV technology can be found in Yemen, and the little to be found is in the very special niches of telecommunications and military applications. The Yemeni resource base in the field of PVs can be said not to exist, which means it is neither easy to identify targets for substantial capacity building nor for comprehensive barrier removal activities. A future commercially sustainable market situation for PVs can only be discussed in very general terms. Secondly the awareness of the potential benefits, the drawbacks and the general characteristics of applied PV technology is simply not present in Yemen. Thirdly information on household energy usage and spending – in particular in the remote rural villages – is at best fragmented.

The World Bank consequently decided to accommodate the request from the Yemeni government with regard to solar energy by launching a small scale combined demonstration and feasibility study. The objective being in the sectors of distant education and primary health care to:

·   investigate the technical, economic and social viability of Pvs;
·   set up a number of pilot-demonstration systems;
·   monitor the technical, economic and social performance and impact;
·   to develop options for further deployment of the PV technology;
·   to recommend next immediate steps – if any;

The rationale for this approach is, using little funds and combining demonstration and study, to try to establish a base for assessment of further interventions in the sectors of PVs for education & health so far not tried out in Yemen. The approach has been discussed by officials of the Yemeni government and the World Bank and found to make sense in the circumstances given.

Using financial support from Denmark and Finland the Bank launched the combined feasibility study / demonstration projects mid 1998. The results obtained so far are briefly outlined below.

3. The Project Methodology

When assessing the viability of introducing a new energy technology, in this case PVs, it is necessary to go through a screening process [1] including the following steps:

·   the resource must be adequate and with acceptable seasonal variations;
·   the application must be economically viable;
·   the application must be financially viable, if applicable;
·   the application must be institutionally sustainable;
·   the application must be locally replicable;
·   the application must be socially acceptable locally;
·   the resulting environmental impact must be acceptable;

3.1

The solar energy resource in Yemen is abundant and evenly distributed over the country and over the seasons. The average insolation is around 5.5 kWh/m^2. Average insolation varies about 10 % with location and seasonal variation is about 25 % with little variation as to location. Adequate design procedures including the orientation of the PV modules can reduce the seasonal variation considerably. Available data provides adequate base for design of small to medium scale solar energy systems.

3.2

Investigations have shown, that the basic demand for electricity at the schools and the clinics is in the range of 1 kWh/day. Including electricity for staff, a powerful motivating factor, may increase demand to a few kWh/day. In the Yemeni context taking into account the logistics of a reliable supply of fuel and adequate maintenance PVs simply constitute the least cost solution.

3.3

Financial viability only to a certain extent applies, when considering social facilities such as education and health. It is, however, mandatory that sufficient funds for maintenance and repair can be ascertained – by experience these funds should be generated/administered by the end-users or as close to as possible. In Yemen school fees and clinic fees have been shown to accommodate these costs.

3.4

Institutional sustainability is indeed a major issue when considering the introduction of PVs in Yemen. The resource base is very weak, and inadequate maintenance and service of prior installations have proved to be fatal. The present project has focused on the facilities of the Solar Energy Centre – a small unit at the Technical University of Sana'a. It is the hope, that this unit may develop into a commercially viable business entity in the field of PV installation and maintenance, either as a company or an NGO.

3.5

The issue of replicability is in Yemen simple: the number of off-grid schools and clinics is well into the thousands and the need of new facilities is enormous. The demand for electricity at communal facilities such as schools and clinics is small, but very critical. PVs constitute in off-grid areas the least cost solution. The local communities appear to be able and willing to pay for maintenance and service as demonstrated by a socio-economic study carried out as an integral part of the project. Donor/government assistance is needed for the initial investment.

3.6

The socio-economic study referred to above has amply demonstrated the social acceptance of PVs as source of electricity. However it is worth noticing, that a very common statement refers to the limited capacity of PVs ("we are not getting the same service as a grid hook-up would provide").

3.7

The environmental impact of PVs is in general regarded to be benign. Energy pay-back time is around 3-5 years with an expected lifetime of >>10 years. Life-cycle analysis of PVs demonstrates a very acceptable environmental impact, given responsible behaviour of producers, distributors and end-users. Stand-alone PVs are inevitably related to batteries, normally lead/acid batteries. Collection/re-cycling of lead/acid batteries is a must, if PVs - at a reasonable penetration – are not to have a negative environmental impact. In Yemen this situation is not imminent, but adequate precautions should be instigated at an early stage. The NGO or the commercial entity referred to in Para. 3.4 above would be made responsible for this to the extent they have handled the initial distribution, within a legal framework to be developed by the Yemeni government with assistance from the donor community.

4. Brief technical overview of pilot installations

The scope set by the available funds the following demonstration systems have been implemented in Yemen in the course of this project:

- a primary health care clinic system (200 Wp) to run a vaccine refrigerator and a light;
- a solar home system for the doctors house (50 Wp) to provide staff motivation;
- a secondary school system (200 Wp) to power TVs/VCR and lights;
- a solar home system for teaching staff (50 Wp) to provide staff motivation;
- a water pumping system at a secondary school (1200 Wp);
- a telecommunication system – for emergency use (200 Wp);

The refrigerator system includes a PV array of 200 Wp, a battery bank of 200 Ah, a charge controller and a vaccine refrigerator in compliance with WHO specifications. Standard mono-X Si modules are used with long life (>15 years) sealed lead/acid batteries.

The solar home system includes a single 50 Wp mono-

X Si module and an open tubular plate lead/acid battery of 75 Ah, and provides power for 4 lights and an outlet for a radio in the doctors house.

The secondary school PV system is the same as for the clinic. It provides power for two TVs and one VCR, plus 4 lights for evening classes.

The solar home system, identical to the above, was intended for a teachers home, but was converted to power a PA system in daily use at the school and a few extra lights.

The pump system consists of a PV array of 1200 W (mono-X Si standard modules), a DC/AC-inverter and a submersible pump (a Grundfos sub-unit). Water level is 23 m from ground level and tank height from ground level is about 11 m. A complete pipe system for a secondary school and 13 households, together with the water storage tank, were built by the inhabitants in the village. The pumping system provides in average about 8 cubic meters of water a day – a bit more than expected [2].

The emergency telecommunication system includes a set of two radio communication HF-SSB transceivers with antennas. One of the radio communication system is placed in the same village where the water pump is installed. This radio communication system is powered by a PV power pack, consisting of 4 modules of 50 W (mono-X Si standard modules) and batteries. The peak power consumption of the radio is 125 W, when operated. The other radio communication system is placed in a near by village where a health station with the access to grid electricity is located.

The first mentioned four systems are installed in the village of Al Udain, near Ibb some 200 km South of Sana'a, and were commissioned in November 1998.
The last two systems are located in the village of Al Jarrahi near Al Hodaidha on the Red Sea coast. The water pumping system was commissioned early in 1999; the telecom system has not yet been commissioned.

4. Conclusions

The approach to introduce a new and modern renewable energy technology, PVs, as now being implemented in Yemen by the World Bank and the Yemeni government, has shown great promise, in view of the enormous PV potential and the economic viability of PVs due to the scattered nature of the rural villages (more than 37,000 with an average size of 320 persons) in the rugged terrain. Alternatives – in particular extension of the national grid or diesel based mini-grid systems are under these circumstances prohibitively expensive.

The approach includes the following stepwise process:

- an initial, small scale but professional, action to demonstrate and verify technical, economical and social viability;
- identification of selection criteria and options

on how to proceed – if so;

· given donor assistance for the initial investment, identification of sustainable maintenance
and service mechanisms;

· initiation of an organic move towards relevant
capacity building and barrier removal;

REFERENCES

[1] P. Ahm, G. Foley, Proceedings ISES Conference 1998 on Renewable Energy for Rural
Development, Nepal 145.

[2] C. Nyman, First Experience of Water Pumping
System in Yemen, Paper for the 16th European
PV Energy Conference 2000, Glasgow.

RURAL ELECTRIFICATION WITH PV IN THREE REMOTE AREAS OF THE INDIAN HIMALAYAS, A PARTICIPATORY AND DECENTRALISED APPROACH.

M. Camps
Programme ASVIN - CNRS
Snow View - Papersely
Almora - 263601 U.P. - India
Ph. +91 5962 32336 - Fax +91 5962 31507 - Email: matthieu@nde.vsnl.net.in

ABSTRACT: Long term sustainability is the key feature of the project described, which revolves around the future users, and seeks their active participation. It aims at setting up decentralised and reliable networks in terms of institutional, technical and financial aspects in order to support the dissemination of solar energy (namely PV systems). Village committees collect and manage regular financial contributions from the users, which are deposited in bank accounts and generate interest in view of future replacement requirements. In each region, a solar workshop provides human and technical support. Unemployed young villagers are trained to become "Barefoot Solar Engineers", able to carry out all maintenance operations, either in the village, or in the workshop where spare parts are available. Employed by the project for its duration, they will, afterwards, become private entrepreneurs.
Keywords: Developing Countries - 1: Rural Electrification - 2: Sustainable - 3

INTRODUCTION

It is widely acknowledged that rural electrification is a basic social need and an important aspect of rural development. In India, while 86% of the villages are officially electrified, only 31% of the rural households have access to electricity. Out of 86500 villages not yet electrified, 18000 are considered "remote" i.e. of most difficult access, not likely to be ever electrified through grid extension.

This project operates in three areas of the Himalayas, where most of the villages fall under this classification. These areas are located in Ladakh (Jammu & Kashmir), Kumaon (Uttar Pradesh) and Sikkim. They all have a low level of infrastructure and services, with 80% of the work force engaged in agriculture, which remains at a subsistence level. Moreover, with few, or no industries, the emigration rate is high.

The Rural Electrification Corporation is reviewing its strategy and, instead of grid extension, would now favour non-conventional sources of energy for remote villages. India's potential for such sources of energy, and particularly solar photovoltaic, is great, and the policies of the Central Government have been trying, since 1993, to take into account the lessons of past experiences, evolving from a "technology push" to a "consumer pull" strategy.

Nevertheless, the rate of success of solar electrification in remote areas remains unsatisfactory (only 50% of solar PV devices were functional according to a recent survey [1]). There is, on the one hand, a growing thesis that the Government cannot do everything, develop markets or provide subsidies forever. On the other hand, private companies, manufacturers or financing intermediaries remain too far from the potential consumers and are not likely to establish close linkages with them considering the distances involved, and the extreme lack of communication networks. As a matter of fact, most potential users of solar PV remain ignorant about it, and existing users, most of the time, have no access to any "after-installation" service. The obvious consequence is that failure of even the cheapest component, is fatal.

The question is, in our opinion, "how to bridge this gap, ensure access to this technology, and its sustained use in the most under privileged and needy sections of the society?"

THE PROJECT

The partners

This project is implemented by the Social Work and Research Centre (SWRC), an Indian Non Governmental Organisation (NGO) and the ASVIN Programme (Applications of Solar Energy in Villages of India and Nepal) of the CNRS (National Council for Scientific Research—France), a research-cum-development programme. Its objective is to bring some answers or directions to the above question.

For more than a decade now, **SWRC** has developed a concept of decentralised solar workshops with "Barefoot Solar Engineers" (BSE). These are unemployed village youth trained to become local technicians, to fabricate parts of the systems, and assemble and maintain solar home systems (SHS) and lanterns. For their services, they are paid by the community. Successfully "demystifying" the technology in rural areas, SWRC has made it accessible to, understandable and maintainable by the villagers, decreasing their dependency on the outside. Their target has always been the "poorest of the poor", in very remote, almost inaccessible places, where people do not benefit from effective development programmes.

The ASVIN Programme, since 1981, has been studying the conditions of acceptability of community solar equipment (pumping, milling, lighting) in remote villages of India and Nepal in collaboration with local NGOs. Its original focus was on the social implications of the introduction of sophisticated innovations. After several projects, the main lesson learnt was that institutional, technical, as well as financial problems could multiply the social ones: only a comprehensive approach, tackling each of the issues can prove beneficial for the users [2 to 5].

Background

Each partner, in earlier projects, has been taking up the same challenge: to demonstrate the feasibility of collecting regular cash contributions from rural poor. For that, the same condition is applied, whether in India (SWRC) or in Nepal (ASVIN): the users had to be convinced of the usefulness and reliability of the service. It has been a long process. People are totally unfamiliar with the concept of reliable service, and their willingness to pay has been spoilt by subsidies and free distribution programmes.

In Ladakh, for example, where SWRC has more than 10 years of experience, the villagers were initially very reluctant to collect cash contributions to meet the BSE fees. Slowly, as the villagers understood that the process was under their control, with a minimal dependency on the outside, a regular practice was established. Rs 20 (€ 0,5) were collected per month, per family. It was already a revolution! Yet, the cash collected was only sufficient to cover the BSE salary, and the problem of maintenance and replacement became acute over the years. Now, former agreements are being re-discussed, on the lines of the present project, so that cash contributions can pay for new batteries, which, after an average of 7 years of functioning, require replacement. In a far away village, the record sum of Rs 100 (€ 2,5) per month has been agreed upon. In a place like Ladakh, where the inhabitants of Leh (district headquarter) pay a sum of Rs 40 (€ 1) per month per house (supposedly for 2 light points), such a contribution is unheard of.

Over the years, it became increasingly obvious that the contributions must be designed to facilitate the replacement of the batteries and panels in the long run: establishing a durable service remains the priority, but a sustainable structure must support it.

Strategy

Sharing their experience and vision, the two partners launched this project in 1998, based on a strong principle: the users, although rural poor with little education can become the masters of their own development process. It is possible to set up decentralised networks, with their active participation, in order to ensure long term sustainability. Three domains (Institutional ⇔ Technical ⇔ Financial) are interconnected, at three levels (Village ⇔ Region ⇔ National federation): this "alternative grid" is meant to operate in difficult conditions, and to evolve as a decentralised structure that can stand on its own (see fig. 1).

Generally, the hardware components are the focus of solar projects. Thinking in terms of "networks approach" brings a meaningful and indispensable complement: the

---

THE HARDWARE & SOFTWARE COMPONENTS
Hardware: the SHS, or the lantern, or any appliance (purchase, supply, and installation)
Software: the flow of information to the future users, their training, their technical and financial involvement, throughout the project life, the maintenance and replacement of the systems through a reliable network

Figure 2: Description of hardware / software components

THE NETWORKING

Institutional

The future users are involved at each step of the project, right from the planning phase. Easy to claim, but difficult to achieve! It is a time consumming process that lasted almost two years. It is the role of the promoting agency, SWRC, with its field centres (also referred to as **Regional NGOs**) and community workers to induce and facilitate the emergence of a local entity: **the Village Energy and Environment Committee** (VEEC).

For that, information and planning are two necessary tools. Transmitting fair information is not easy, especially in areas where the devices and the concept are totally unknown. Many meetings have to be organised, visits to places where people already use solar PV for lighting can be arranged. Moreover, it is important for the promoter not to lose sight of all the needs of the communities involved. The solar technology is the answer to one type of need only, and an integrated approach is necessary. The VEECs are trained to conduct micro-planning and surveys. Linkages with other agencies can help respond to other needs. The surveys, rather than being a source of scientific or accurate information, should be considered as a means of increasing the dialog with the community, and a way to know it more closely, as their findings can help generate useful discussions.

At a national level, SWRC head office represents a kind of federation of the regional groups, through which sharing of experiences, lobbying work can be carried out. In February 1999, for example, a national workshop was

---

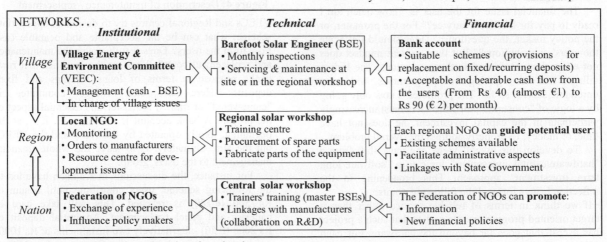

Figure 1: Networking: three domains, three levels

"software aspects" (see fig. 2).

Here, 850 Solar Home Systems (SHS), 300 solar lanterns and other appliances for income generation at village level such as crop, milk and wool processing will be installed, managed and maintained by the villagers.

organised to present the project and the work achieved during the planning phase. Most of the presentations were carried out by VEEC's representatives addressing high level Government officers, international agencies, or other NGOs.

Technical

The distance between the installed equipment and the "know how" must be as short as possible. One **Barefoot Solar Engineer**, selected by the villagers, and trained by the regional centre of SWRC is in charge of the technical follow-up, at village level. Monthly inspections, simple operations can be done at site, and he can conduct more complex ones in the **regional workshop**. There, a master trainer can, for sometime, supervise and guide the BSEs. The flow of spare parts is organised by the workshop, which can easily contact the manufacturers.

At the federation level, a workshop conducts the trainers' training (master BSEs), and develops a close relationship with manufacturers. A gap needs to be bridged, here also: an NGO cannot conduct effective research and development programmes -it does not have the scientific and financial means of a private company- but it gathers priceless information thanks to constant feed-back from the field. This information must go back to the manufacturers so that products can match the needs of rural users.

Once the project reaches a certain technical maturity and achieves a significant scale of operation with a sufficient number of systems installed, the concept of "private BSE" becomes feasible. In Ladakh, where more than 1000 SHS have already been installed by SWRC, the BSEs are going to form a co-operative, with their own workshop and management systems. Another way to increase the market of the BSEs, is to propose maintenance contracts to users of solar PV devices, installed by other agencies, but without appropriate service. Enlarging the competence of the BSE to included the repairing of other devices (radio, tape recorders, TV…) is one more way to secure his long term business at the village level.

In Kumaon, it has been estimated that once 1000 SHS are installed, if each user agrees to pay Rs 300 (€ 7) per year for a service contract, it can sustain a team of 10 BSEs (their salary Rs 1500 (€ 35) per month; their administrative and travel expenses).

Financial

The financial issue for the user is: "how much am I ready to pay for a reliable service?" For the promoter, or the policy maker, the question is: "how should I allocate the users' contributions?" But their present mindset does not allow them to formulate these questions. The Indian villager has been getting free programmes, with very unreliable service so far. The promoter is now only going by a logic of "consumer pull": the users must increasingly participate in the capital investment: he does not have time, or money, to properly address the overall problem.

To develop this overall view, one must shift from a "hardware/capital investment" logic to a "software/long term investment" approach. This broadening in the reflection brings forward the following points:
• If we think in terms of reliable service rather than a target oriented programme, we realise that such a project has a cost that goes far beyond the hardware. For each SHS installed, a comparable amount of money is invested in hardware and software (see Fig. 3).
• The setting up of this "alternative grid" must remind us of the equity factor: equity between the rural poor and the grid connected consumers. Should we ask the solar user in a remote village of the Himalayas to pay for the actual cost of generating electricity, and maintaining and replacing the SHS? As a comparison, in Jammu and Kashmir only 16,5% of the cost of supplying grid electricity is recovered from the domestic consumer (44%

COSTS OF HARDWARE / SOFTWARE			
For one SHS installed under this project:			
HARDWARE:			
SHS:	INR 10.850	€ 260	**➔ 51%**
SOFTWARE			
Replacement:	INR 5.700	€ 135	27%
Workshop:	INR 1.400	€ 35	7%
Training:	INR 1.150	€ 27	5%
Salaries BSEs:	INR 2.100	€ 50	10%
	INR 10.350	€ 247	**➔ 49%**
TOTAL	**INR 21.200**	**€ 507**	**100%**
➔ *Hardware & software costs are comparable*			

Figure 3: Costs of hardware and software components

in Uttar Pradesh, which includes both the hilly and plain areas) [6].
• Considering this equity question and keeping in mind that the systems installed have a finite lifespan, this programme has made a deliberate decision: to set up the systems for the users, as a grant, and to allocate their financial contributions only towards meeting the maintenance and replacement costs (see Fig. 4), without imposing a huge financial burden on people severely hit by poverty.

Conscious of the necessity of fair contributions, the

MAINTENANCE *AND* REPLACEMENT
"Maintenance" covers:
➔ up-keeping and minor repairs
➔ services of the Barefoot Solar Engineer
the costs are small, and the users can afford to meet them whenever they occur. They **are paid on the spot.**
"Replacement" covers
➔ PV panel replacement (20 years)
➔ Battery replacement (5 years)
the costs are relatively large and most users will be unable to pay for them in a lumpsum. To ensure their replacement, **regular provisions, by the users, from the beginning, are a must.**

Figure 4: Description of maintenance / replacement

VEECs and Regional centres try to design an appropriate scheme: what can be the acceptable and bearable cash flow from the users? Does it fully cover the maintenance and replacement costs? What are the facilities that rural banks propose in terms of loans, savings and fixed deposits? Here also, the village committee is "empowered" or trained to master this essential aspect of the project. A bank account is then opened. It is, at the beginning, jointly operated by the NGO and the VEEC, but the aim, ultimately, is to leave the whole financial management to the users' group.

For instance, the discussions in Kumaon have led to the following scheme: users contribute until a sum of Rs 7500 (€ 180) is reached. This will take care of batteries and panel replacement. An initial deposit of Rs 2000 (€ 50) -alternatively two instalments of Rs 1000- is invested in a fixed deposit for five years. In addition, the users have to pay the VEEC monthly instalments of Rs 90 (€ 2) which are invested in a recurring deposit scheme. The interest rate is 12%, and after approximately 5 years the user does not need to save more cash: the amount available is Rs 11000 (€ 260). This is sufficient to buy the new battery, and the balance is reinvested in a fixed deposit. The interest generated will continue to take care of the next battery replacement, and possibly the panel replacement too. The impact of the project there is already bringing many potential users forward. This can be explained as the State Government nodal agency for the promotion of solar energy is also supplying SHS in many villages, where candidate users have to contribute towards the investment and pay Rs 7200 (€ 170). But with no after sale service, and people having to travel for more than 100 km, to be told that spare parts can not be arranged, the life expectancy of the units is limited, though the amount paid by the users is comparable.

In Sikkim and Ladakh, the approach is slightly different, although the contributions too are meant as provisions towards the replacement of batteries and panels. With an initial investment of Rs 5700 (€ 135), these replacements are secured for at least 40 years. As this sum is out of reach for most users, a soft loan scheme has been organised by the project. The repayment schedule is tailored to the users' cash capacities. Rs 50 (little more than € 1) per month, invested during 3 years in long term fixed deposit is sufficient to recover the loan. After that, the users contribution should continue and go towards the fees of the BSE who, by then, will no longer be employed by the project.

Once the institutional, technical and financial aspects become clear for the VEECs, and the role and responsibilities of each partner is defined, a contract is signed between the future group of users and the regional NGO. A real commitment is agreed upon: the services or networks set up by the promoting agency are partly "handed over" to the community in return of their active participation.

CONCLUSIONS

• The approach described might appear "heavy" or "time consuming". Who can afford to spend two years before installing SHS, meticulously addressing all the issues described above? Neither the Government, which is ill equipped for that, nor private companies, which see no reason to spending money in this way. Therefore, the role of the NGO is necessary in this process, and if we look from a development perspective, the work achieved is a formidable investment for future collaborative activities of many different kinds. The village committees, that can manage their own energy production, are prepared to create and handle other new projects for development.

• Lighting is a social need, but except for a few cases, it does not generate income. It shows the possibility of generating electricity on site, under a process that is mastered by the users. On this basis, other applications can be discussed, taking into account the local economy, and the needs and competence available. The framework is then ready, confidence is raised, and further training aimed at enhancing technical skills and managerial capabilities can take place.

• From a global point of view, the question of solar PV in developing countries now revolves around the market deployment issues. In India, the Government authorities have a legitimate and ambitious programme in that sense. There are about 50 private companies now competing in that field, producing equipment locally. So, one can say, the policy is consistent with "global thinking". But is that sufficient to ensure a steady and sustainable dissemination of solar PV systems? Certainly not, if we look at the Indian reality in the remote corners of the Himalayas.

• There, one still wonders how potential users can avail themselves of such facilities? Through State Government programmes? Only the hardware aspects are taken care of. Directly from the manufacturers? Almost impossible! There is no dealer there. For people far away from the "power sources" whether electrical, political or economic, only local NGOs have established a credible relationship with them.

• Whether or not it is the NGO's role and responsibility to set up all the networking described in this paper remains an open question. In the absence of any alternative, SWRC and ASVIN have developed this strategy from the perspective of "development" and "sustainability". And, it must be underlined, it is a constant struggle. The NGO is here replacing both the public service and the private sector.

• But, the responsibility of any Government should be to ensure fair access to basic facilities to its whole population whether urban or rural, rich or poor. Rather than blind subsidies that destroy sustainable livelihoods, rural poor need well framed policies that deliver quality services. In future, one could imagine comprehensive schemes linking:
→ **NGOs** for the field work / development approach
→ **Manufacturers** for the setting up of decentralised workshop, training and constant improvement of the devices based on the feedback from the field
→ **Financing agencies** for providing soft loans for part of the initial investment, and the replacement of the major components in the long run
This could constitute an "alternative grid" and should be recognised and supported by the Government.

• Rather than suggesting itself as a model to be exactly replicated, this project tries to highlight the set of parameters that must be taken into account for successful rural electrification with PV in remote regions. As the implementation phase continues, there are still lessons to be learnt…

AKNOWLEDGEMENT

The project *"Development & Dissemination of Solar Energy Systems for Villages in the Himalayan Region - India"* is funded by the European Commission (DG DEV)

REFERENCES

[1] R.C. Pal, S.N. Srinivas and S. Dutta "Status of renewable energy technologies in Urjagrams in India" *International Journal of Ambient Energy, Vol. 20, No. 4 (10/99)*
[2] P. Amado "To whom it may not concern" *12th European PV Conference, 13/4/94, Amsterdam*
[3] D. Blamont "Les Structures Sociales : Un Facteur Non Négligeable" *Séminaire.sur l'Electrification Rurale Décentralisée par générateurs solaires en Asie du Sud-Est, 1-3/12/97, Hô Chi Minh-Ville*
[4] D. Blamont "Solar energy for production activities in India and Nepal: some problems" *International Conference on Role of Renewable Energy Technology for Rural Development, 12-14/10/98, Kathmandu*
[5] D. Blamont, P. Amado "Using photovoltaics for agricultural processing activities in Upper Mustang, Nepal" *ISES 1999 Solar World Congress, 4-9/7/99, Jerusalem*
[6] TEDDY (TERI Energy Data Directory & Yearbook) 1999/2000, table 31A, page 128, New Delhi

SOCIO-ECONOMIC STUDY OF DOMESTIC COLD PRODUCTION IN RURAL AREA IN MAURITANIA, SAHEL, WEST AFRICA

M. A. Kane
Cythelia
Savoie-Technolac, BP 319
F-73375 Le Bourget du Lac Cedex
Tel. : +33 4 79 25 31 75
Fax : +33 4 79 25 33 09
E-Mail : mak.cythelia@wanadoo.fr

ABSTRACT : Under the Sahelian climate, hot and dry, no productive agricultural activity can be developed without increasing the preservation capacity. Resolving the rural domestic cold problem is an absolute necessity in every rural decentralised electrification (RDE) project. Many technologies are available to produce cold but which one to pick to include in a RDE project ? In this paper we draw up an inventory of the various solutions and we attempt to define the best one for the specific project of DIAGUILI, Mauritania. The refrigeration technology can use various energy sources, solar, electricity, gas, kerosene. We compare and integrate each solution in our global RDE scheme. The comparisons concern the efficiency, the final costs and the conditions of socio-economic adaptation in the local environment. The gas solution appears the cheapest, due to the lower investment and reasonable recurring costs. However, in DIAGUILI hybrid generator mini-grid scheme, PV remains an interesting solution. But the most promising solution seems to be the production of domestic cold by thermal absorption. In the future, the demand will concern the cold-rooms and the freezers for productive activities, which can be best satisfied in good technical and economical balance by the absorption systems.
Keywords : Developing Countries – 1: Rural Electrification – 2: Refrigeration – 3

1. INTRODUCTION

The production of domestic cold in the African rural areas is a RDE major challenge. First, because of the relatively high level of demand for productive activities comparatively to the usual RDE demands. Second, because of the complexity of the required systems for a cold production in a RDE scheme. In this paper, we draw up an inventory and we compare the advantages of the different available technologies. We analyse the best solution for DIAGUILI, integrated into a RDE project within a mini-grid supplied by a generating set. Thus, we underline the priority of an integrated and sustainable development. We face also the different solutions at the conception of the project and we take into account the realism of the possible choices with a technical and socio-economic point of view.

2. DIFFERENT REFRIGERING SYSTEMS

2.1 Solar absorption and compression advantages

The cheapest the ratio of the energy costs (heat/electricity), the more interesting the process. The energy consumption of an absorption machine is ten times lower than a compressor consumption. Thus, a low cost of a heat source combined with a high cost of electricity will advantage absorption systems. One can assume that for a cooling temperature higher than 30°C, the absorption process is more advantageous than the compression. The advantages of a compressor are a high speed and a good-quality of the produced cold, the reliability, the mobility and a high-efficiency. An absorption system is more simple, requires lower maintenance and repairing costs, can easily work outside and works better for a partial charge.

2.2 Comparison between solar systems

There are three ways to produce solar domestic cold. If a heat source is available, one can use a tri-thermal machine (absorption). If an electrical source is available, compressor can be used.

There is also an absorption refrigerator working with electricity (12 VCC or 220 VAC) combined with gas. But a high level of consumption is the major drawback of this system.

In **Table I,** we report the standard characteristics of the solar refrigerators. The absorption process do not yet profit of low consumption refrigerators and requires a relatively big size of system when it is mixed with a gas system. Its other major inconvenient is its low-efficiency. The compressor with an auto-synchronous motor has a high consumption per unity of volume and is often low-efficient. The compressor with an open CC motor has often frigorific fluid losses through the joints. And the main disadvantage of the compressor with an hermetic CC motor is a short life due to the impossibility of its maintenance.

Type	Supply	Size	consumption	*LC system	drawback	Pack
Electrical absorption	12 VCC 220 VAC Gaz	50 l Big size for gas	high	no	**LE	yes
Auto-synchron	12/24 VCC	30 to 300 l	High	yes	**LE	recommended
Open CC motor	12/24 VCC or higher	30 to 300 l	Low	yes	Fluid losses	recommended
Hermetic CC motor	12/24 VCC	30 to 300 l	Low	yes	Short Life	mandatory-

Table I: Characteristics of electrical refrigeration systems.
*LC system : Low consumption system **LE : Low efficiency

In **Table II**, we report the investment costs per unity of frigorific power for a PV compressor and for a solar absorption refrigerator. The PV installation is 2.5 times more expensive even if the gap is decreasing with the market expansion (20% to 30% per year) and some innovative solutions like leasing systems. PV is not adapted to the high demands like the cold rooms. The optimised systems have now a high isolation, a high-efficiency, minimised losses during the openings...

Refrigerator System	Absorption	PV
Solar conversion	Heat at 70°C	Electricity
Energy efficiency	0.575	0.108
Exergetic efficiency	0.0754	0.108
Refrigerator		
Energy efficiency	0.248	4.45
Exergetic efficiency	0.100	0.236
Installation		
Energy efficiency	0.143	0.451
Exergetic efficiency	0.00754	0.0255
Flux ($Q_f = 3$ kW)	20.96 kW	6.24 kW
Investment		
Per unity of heat source	3.26 FF/W	-
Per Watt	-	139 FF/W
Per energy machine unity	24.9 FF/W	139 FF/W
Per frigorific Watt	13.1 FF/W	31.2 FF/W

Table II: Comparison of the solar absorption refrigerator and the PV compressor.

2.3 Other systems

The **gas refrigerator** is reliable and high-efficient with many advantages : easy use, easy maintenance, low spare parts and investment costs. However, it remains expensive and not so comfortable. The cold production is very slow and the working cost is high. Over all, gas supply can be a real problem for remote areas.

The **oil refrigerator** is not reliable. It has however some advantages : easy use, moderate spare parts cost and low investments. The disadvantages of this type of refrigerator are a bad quality of the oil in the rural African areas, a slow production and a bad quality of the cold, the difficulties of maintenance, a high working cost and the absence of a thermostat.

The **mixed refrigerator** (gas combined with a CC or AC electrical compressor) is likely the most promising. It is expected from the actual R&D results a 200 litres low consumption refrigerator.

2.4 Conclusion

For the domestic use, the gas or oil refrigerators are twice cheaper than a PV refrigerator.

The investment costs still remain the main obstacle when there is no real technical problem. For example, a good isolated freezer (from –22°C to –9°C) when it is started at –22°C, can be 72 hours autonomous and requires then only 6 or 8 hours of electricity per day. Thus, a high-efficient refrigerator with a separated freezer compartment consumes today less than 700 Wh/d.

The systems have now lower and lower energy consumption and the charge accumulators have been appreciably improved. Over all, one can expect for the coming years an important decrease of the panels cost which represents about 30% of the total cost of a PV refrigerator. One can assume that the working cost of the PV refrigerator, which is today similar to the working cost of a gas or oil refrigerator, may then appreciably decrease.

3   DIAGUILI RDE SCHEME

3.1 Mauritania context

Mauritania annual production of electricity is about 185 GWh for a consumption of about 145 GWh. The residential and tertiary sectors consume 82 GWh, Mines and Industries sector consumes 163 GWh and agricultural sector consumes only 0.6 GWh per year. Current energy prices are (with 1FF = 30 ouguiya) :
- Gas : 43 FF for butane charge of 12 kg
- Gas-oil : 3 FF/litre
- Fuel : 4 FF/litre

3.2 DIAGUILI context

DIAGUILI has a population of about 5,000 residents with 740 households regrouped into 340 concessions of about 15 persons.

There is almost one shaft per concession and three drill-holes with manual SE3 pumps. One pump is private and the two others are collective with a village commitee management and a turncock. The revenue of the water sales pay the repairing, the spare parts and the turncock salary.

The immigration is the first economic activity, concerning more than 26% of the total population and being the first income for the village, before agriculture (which concerns 21% of the population and represents the second income). The pensioners are 15% of the population.

There are 24 craftsmen : 1 wood joiner, 1 metallic joiner (the main private activity with 2 employees and 2 generating sets of 16 kVA and 9 kVA) who produces solar driers, 6 blacksmiths, 1 welder, 1 mechanic, 3 tailors. There are 84 stores, 1 restaurant, 2 standards, 6 mills and 3 huskers. There is also a daily market with many sailors coming from Mali.

The ancient dispensary is equipped by 6 PV modules of 50 Wp supplying the vaccine refrigerator. The annual working budget of this rural health centre is about 4,200FF including the salary of the volontaries. 40% of the benefits of the medicine sales serves to buy new medicines, 30% covers the centre working cost and 30% serves to pay the personnel.

3.3 Energy demand of DIAGUILI

The DIAGUILI population uses mainly paraffin lamps. However, they have more and more recourse to PV panels and generating sets. There are 13 generating sets of total cumulated capacity of 150 kVA. Two generating sets (55 and 20 kVA) have no grid to supply.

There are 3 light points per household that cost 42 FF per month. The half of this expense (21 FF) serves to buy dry cells. Audiovisuals cost 27 FF per month and remains the main households' demand. The total expenses for energy access (without cooking, which is mainly supplied by gathered wood) is thus 67 FF per month per household. The social organisation into concessions (3 households per concession) carries to a large economies of scale effect. The expenses for energy remain relatively low for the house sizes and the household incomes. The agriculture demand for pumping is about 94 MWh per year, mainly for the irrigation. The management, not the energy supply, is the main problem because there is an under invoicing and an under exploitation of the arable lands. Transforming the agricultural products requires an installed power of about 50 kW. A decentralised energy production is hence necessary. The handicraft (welding, millstone, saw) requires 10 kW of installed power. This demand is supplied by three generating sets of 45, 16 and 9 kVA. A grid connection requires an ongoing working of the generators.

There are two electricity sailors using the one battery chargers with a 50 Wp panel (2 batteries/day) and the other a micro-grid.

At the beginning of the project, there was 31 refrigerators and 2 freezers in DIAGUILI :
- 17 gas refrigerators for the trade (120 FF/month).
- 6 gas refrigerators for the domestic use (60 FF/month).
- 5 PV refrigerators (out of working cost).
- 3 electrical refrigerators connected to a generating set (unknown cost).
- 2 freezers connected to a private micro-grid with a working cost of 1,810 FF/month (for 4 households and 200 lamps).

3.4 Technical and economic analysis

With a fixed threshold of 34 FF/month, 83% of the concessions are solvent. The subscription for one mini-grid connection is 20 FF/month. The recovery is fixed to 4 FF/part (1 part for 1 lamp, 2 parts for a 80 W plug).

We studied five scenarios with two hypothesis : a high hypothesis which assumes an increase of 100% of the cold demand (62 refrigerators) and a low hypothesis which assumes an increase of 33% (42 refrigerators).

The calculations are done for an external temperature of 43°C with 20 minutes of daily opening. It represents 40 openings of 30 seconds that implies an opening every 15 minutes from 9 to 19 o'clock. It is a strong hypothesis for a domestic use but probably too low for a trading use.

3.4.1 Scenario 1 : All electrical - GS during 18 h/d

This scenario requires a continuous working for the generating set. For the high hypothesis, the cold produced is twice more expensive than the cold produced by a gas refrigerator. For the low hypothesis this rate increases to three. This solution is hence to avoid.

3.4.2 Scenario 2 : Mixed - gas (night) / PV (day)

The PV cost is calculated for K = 0.75 that takes into account the weather, the module depreciation and an efficiency of 90% for the batteries. This solution twice more expensive than a cold production by a gas refrigerator is to avoid too.

3.4.3 Scenario 3 : Mixed - gas (14 h/d) / Mini-grid (8 h/d)

This scenario requires a storage for a continuous working, a limitation of the number of connected refrigerators and a one-by-one connection. The low-efficiency of the refrigerators (cosφ = 0.85 due to the reactive energy) causes a cost over-run that requires a better repartition of fix costs by increasing the number of subscribers. One may also choose automotive batteries with a rapid discharge rate. Passing from a system to the other may cause management problems. The produced cold is lightly cheaper than the cold produced by a gas refrigerator.

3.4.4 Scenario 4 : Mixed - PV (day) / Mini-grid (8 h night)

This scenario allows to diminish the PV system cost by using low consumption CC refrigerators. It is the best long-dated and the most flexible solution. One can in fact increase the panels number with an increasing demand. It allows to increase the generating set life, to decrease the diesel consumption and over all to increase the availability of electricity. The cost over-run due to the converters, about 10,000 FF for 8 years is rapidly recouped by a repartition on the subscribers. It is the cheapest solution.

3.5 Conclusion

The generating set is expected to work 8 hours/day for the best scenario. It is assumed that an increase less than 8% of the power of the generating set causes no cost over-run. Producing the cold with a gas refrigerator remains attractive despite an expensive refrigerator cost. It remains the more advantageous solution for an easy supplying area when it is combined with a short storage of 2 or 3 days. Only the mixed system PV/mini-grid could be more interesting. A RE micro-financing scheme could increase the interest of this scenario. The low consumption refrigerators are recommended to diminish the energy demand. The results are summarised in **Table III**.

Scenario	Scen 1	Scen 2	Scen 3	Scen 4
GS power kVA	82	77	101	82
Gas bottle/month	0	0.54	2/3	0
Gas-oil extra consumption Litre/month	6,720	0	0	0
Electricity Wh/day	545	1,625	1,625	409
Cost/month ouguiya	5600	5850	2510	2336
FF	187	195	84	78

Table III-a : High hypothesis – 62 refrigerators

Scenario	Scen 1	Scen 2	Scen 3	Scen 4
GS power kVA	80	77	93	80
Gas bottle/month	0	0.54	¾	0
Gas-oil extra consumption Litre/month	0	0	0	0
Electricity Wh/day	545	1,625	1,625	409
Cost/month ouguiya	8300	5850	2510	2336
FF	277	195	84	8

Table III-b : Low hypothesis – 42 refrigerators

Table III : Cost comparison of cold production in DIAGUILI (domestic & trading).

4.   CONCLUSION AND PERSPECTIVES

To produce cold in the rural remote areas without supplying or mobility problems, the gas refrigerator is the best solution owing to its low investment cost. However, a PV solution combined with a mini-grid (a generating set with battery charger) could be an interesting solution. It is, in fact, a good technical and economic compromise (a maximum of energy at a low cost). The 12/24 VCC supplies then the low consumption refrigerators when the 230 VAC (with an inverter power < 2 kVA) supplies the concession.

The commercial efficiency is about 12% for PV panels and 90% for the auxiliary devices (accumulators, interfaces). The investment is 100 FF/Wp for a PV refrigerator and 1,500 FF/m^2 for a thermal refrigerator. The actual experimental systems combine a heat source (panels) with a mechanical power (1 kWp). Due to its low-efficiency, a plant of 30 m^2 is necessary for 1 kW frigorific power, corresponding to a cold-chamber of 5 m^2. With PV panels, 1 kW frigorific power requires 5 m^2, a too high investment. The future solution will probably be an absorption system, especially for the high demand of the cold-chambers and the freezers.

ACKNOWLEDGEMENTS

The Author wishes to thank Bernard Gay and Luc Arnaud of the Groupe de recherche et d'échanges technologiques (GRET) for their time and fruitful discussions and the personnel of the Alizés RDE programme in Mauritania for their help.

REFERENCES

[1]   L'électricité photovoltaïque
L. Chancelier, E. Laurent, GRET/Ministère de la coopération, Collection « Le point sur », décembre 1996

[2]   Electricité solaire au service du développement rural
M. Rodot, A. Benallou, IEPF / Réseau international d'énergie solaire (RIES), Collection « Etudes et filières » 1993

[3]   Manuel du frigoriste : Réfrigérateurs solaires photovoltaïques
Bulletin de L'Unesco février 1988

[4]   Bulletin africain n° 6 mai 1996

[5]   Le photovoltaïque en coopération et à l'export – PEV en Haïti
Rapport de l'Ademe 24 novembre 1992

[6]   L'électricité solaire photovoltaïque – Principes et applications
G. Moine, Publication Europe Information Développement, Octobre 1981

[7]   Maîtrise de l'énergie pour un monde vivable
B. Laponche, B. Jamet, M. Colombier, S. Attali, ICE Editions, Paris 1997

[8]   Manuel technique du froid,
Le nouveau Pohlmann, Verlag C. F. Müller

[9]   Le froid questions et réponses
M. E. Anderson

[10]   Production de froid en sites isolés,
H. Hamadou (CLER), P. Brule (ECEnR), Rapport de synthèse décembre 1997

[11]   Machines frigorifiques solaires,
J. Bougard & D. Vokaert, Publication Fac. Polytechnique de Mons et ULB.

PHOTOVOLTAIC ENERGY SUPPLY FOR RURAL HEALTH CENTRES IN MOZAMBIQUE

Jonas Sandgren, KanEnergi AS, Baerumsveien 473, N-1351 Rud, Norway
Phone: (+47) 67 15 38 59, Fax: (+47) 67 15 02 50, e-mail: jonas.sandgren@kanenergi.no
Mats Andersson, Energibanken i Jättendal AB, Sweden,
Phone: (+46) 652 134 24, e-mail: mats@energibanken.se
Boaventura Chongo Cuamba, Eduardo Mondlane University, Mozambique,
Phone: (+258) 1 49 33 76 e-mail: cuamba@nambu.uem.mz

ABSTRACT: Mozambique's Ministry of Health is implementing the project "Solar Energy for Rural Health Facilities". The first phase, consisting of systems for 40 health systems is just finished. In this paper the systems are described, and some preliminary observations on user views are presented. One of the systems is supplied with a data logger, and we give an example of how results from such detailed measurements may be used. Finally we sum up the lessons learned from the project so far, and we propose a model for how sector-wise PV-projects may benefit efforts for general PV-electrification.
Keywords: Stand-alone PV Systems – 1: Rural Electrification – 2: Dissemination – 3.

1 INTRODUCTION

Since 1995 Mozambique is implementing a Health Sector Recovery Programme with the main objective to increase the coverage and improve the quality of health services. In 1997 the Government of Mozambique received a grant from the Norwegian Agency for Development Co-operation (NORAD) towards the cost of a Health Sector Recovery Programme project,. The project consists of installation of photovoltaic systems at approximately 150 rural health facilities nation-wide. The project is carried out in two phases, Phase 1 and Phase 2, of which the first is under implementation. KanEnergi AS was contracted as Procurement Advisor to the Ministry of Health, and Energibanken i Jättendal AB provide PV-specialist services.

The project was initiated in the autumn of 1997 with fieldwork for defining needs and priorities, and tender documents were subsequently prepared. The technical specifications in the tender documents determined system design in terms of functional requirements defining the expected performance of the systems. Detailed system design was left to the suppliers to do.

Technical requirements and information on the individual pieces of equipment were used to safeguard minimum levels of quality and to ensure compliance with relevant standards. Systems for phase 1 were procured by international competitive bidding during the spring of 1998. Fortum AES Norway was awarded the contract, comprising systems for 40 health centres. Systems are supplied as turn-key installations, and training of maintenance staff and users is included in the delivery.

2 PV SYSTEMS DESCRIPTION

Each health centre is equipped with a clinic lighting system, a vaccine refrigerator system, and usually two staff house systems, although the number of staff houses may vary. Each of these systems is independent of each other. The clinic lighting system provides general lighting by fluorescent lamps, dedicated lighting for medical examination, and an external sodium-lamp. The staff house system is rather versatile, featuring fluorescent lamps for general lighting, a halogen reading lamp and a domestic refrigerator. In addition a portable solar lantern in included in the staff house systems. Table I lists the different systems.

	Lighting System	Vaccine System	Staff House (2)
PV modules	5x50 W	4x50 W	4x50 W
Batteries	2x100 Ah/10h	2x100 Ah/10h	2x100 Ah/10h
Lamps	5x18 W (general)	-	4x18 W (general)
	1x10 W (storage)		1x20 W (reading)
	1x18 W (outdoor)		Magic Lantern
	2x20 W (medical)		
Fridge	-	Dedicated vaccine	Domestic 60 l
Others	Meter unit on regulator		Meter unit on regulator
			2 double sockets

Table I. The standard installation at each health centre consists of four separate systems.

3   PROJECT IMPLEMENTATION

Installation work started in April 1999, and will finished in May 2000. The installation work suffered some delays due to tardy customs clearance. The installation work was carried out by one installation team consisting of two men. They were assisted by a maintenance technician from the local health administration and unskilled labour hired on-site. The installation team had not previously had experience with installing PV-systems.

Mozambique's Ministry of Health has a nation wide maintenance organisation, whose staff receives training in installation and maintenance of the systems. The training is given in two separate courses. The supplier developed special training manuals for this purpose, and the results of the training have been very satisfactory.

Figure 1. An example of a health centre building with the panels for the lighting and vaccine refrigeration systems on the roof. The installation on this building is aesthetically pleasing.

In July 1999 an installation monitoring survey was performed in order to check the quality of the installation work and commissioning procedures. At the time of the survey, installations had been made at 10 health centres, and eight of these were visited.

Figure 2. The fixtures used for mounting the panels are versatile and allow for adaptation to many different situations.

All systems inspected were working satisfactorily and, except for a few mistakes based on misunderstandings on behalf of the installation team, they were generally well installed. However, there is evidence to suggest that the lighting systems are being used beyond their design point.

Figure 3. Example of gable mounting of panels at staff house.

4   PRELIMINARY USER VIEWS AND OBSERVATIONS

The users' views on the systems were also sought during the installation monitoring survey. This was done by interviewing the staff on site. All health centre staff interviewed expressed great satisfaction with the **lighting system**, and the value of the medical examination lamps was particularly emphasised. The observed state of charge of the batteries was often rather lower than expected. It is not possible to ascertain the reason for this with certainty, but based on general observations and interviews we consider the most probable cause to be more intensive use of the lighting system than assumed in the consumption pattern used for design.

All 8 **vaccine refrigerators** inspected were stocked with vaccine. In marked contrast to the situation in the lighting systems, the batteries were without exception fully charged at the time of inspection, which was rarely before 11 o'clock in the morning. There are thus no concerns regarding the functioning of the vaccine refrigeration systems. Prior to the PV installation, the health centres visited had had kerosene refrigerators in varying working order, or no refrigerator. The vaccine refrigerator system is probably the piece of equipment provided under the project that gives the highest benefit by ensuring viability of vaccines.

Typically, health centres supplied under the project have two houses for the staff and their family to live in. These houses are also electrified in the project, as it is hoped that this will strengthen the motivation of the staff to take good care of the systems and make it easier for the Ministry to recruit qualified staff to isolated locations. The **staff house systems** provide lighting and are also equipped with a 60 litre domestic refrigerator.

Figure 4. An example of a spacious staff house. The PV modules could have been installed farther to the left in order to minimise shading in the late afternoon.

In all instances the inhabitants said they valued the staff house systems highly. In particular, they liked to have light that was free of charge and smoke- and smell-less. The appreciation and awareness of the benefits of the domestic refrigerator appeared to vary more. In most cases there was something in the fridge (e.g. fruit and vegetables, fish, a meal cooked the day before, water), and in two cases the fridges were fairly well stocked. It is possible that use and appreciation of the refrigerators will become more uniform once people have become more used to having them. It would therefore be appropriate to evaluate the impact of the presence of the domestic refrigerators after approximately three years. It appears that the portable solar lanterns are relatively little used.

The state of charge of the batteries in the staff house systems showed larger variations than in the systems installed in the clinic building. The most likely cause for the *very low* states of charge observed is certainly too high energy consumption. In two of the three cases the inhabitants had moved in after installation, and had thus not received any training and information.

5   SYSTEM MONITORING

In order to learn more about the user pattern of the systems, an in particular the lighting system on the clinic building, a data logging system was installed in December 1999 at the Maluana health centre, 70 km north of Maputo. The reason for choosing the lighting system was the intensive utilisation of this system, evidenced by a low state-of-charge of the batteries, observed during the installation monitoring survey. The Maluana health centre was chosen due to its relative closeness to Maputo, which made it feasible for staff working at the solar energy programme at University Eduardo Mondlane to collect the data. Already after the first few days of datalogging, we found that the Maluana health centre is used sparsely. This winter it has unfortunately been difficult to collect data due to the extreme weather conditions suffered by Mozambique in February and March, and consequently no thorough long-term evaluation has been possible.

However, one important observation has been made. It was observed during the system monitoring survey that the charging current the lighting system PV array did not reach the expected values as specified from the data sheet for the

module type used. This situation could be due to a number of reasons, e.g. sub-quality modules, high temperatures or operation far off the maximum power point.

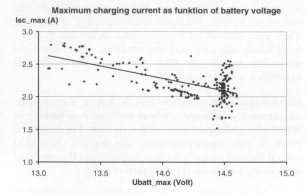

Figure 5. Normalised charging current plotted against battery voltage. In the figure is also inserted a least square fit straight line. The charging current appears to be influenced by the battery voltage at high voltages.

The logged data indicate that the likely reason is the long cable, approximately 20 meter, between modules and batteries. The cable does not have sufficient area to avoid too large a voltage drop over such a long distance. The data reveals that this mainly occurs at a high voltage of the battery and high cell temperature (figure 5). In such situations the working point on the IV-curve is shifted to a voltage where the PV array can not produce the expected current.

6   CONCLUSIONS AND LESSONS LEARNED

The first phase of the project is just concluded, and the Government of Mozambique and NORAD are now contemplating how to go about the second phase. The general conclusions drawn from the experiences made so far may be of value both for the second phase of this project and other PV projects in developing countries. These conclusions are presented in the following.

Procurement – Use of Functional Requirement in Technical Specifications
When the systems were procured, the tender documents relied on verifiable functional parameters rather than component specifications for describing the systems. We believe that this approach has proved to have advantages. First, it allows the suppliers to use their ingenuity to provide cost-effective and reliable solution, and second, it brings forth the skills and qualifications of the supplier quite clearly, thus facilitating the evaluation of the tenders.

System Design – Maintain an Even Standard of Quality
The staff house system consists of the same high quality components as the clinic lighting system, partly to suppress the need for storing additional spares. However, a portable lantern was included in the staff house system. Such lanterns have found market acceptance in some regions, but here the quality of the hardware and the hassle with charging them (connecting to panel, place panel in the sun, vigil against possible theft etc.) makes them appear inferior to the fixed installation, and they are thus little used. A low

rate of use – and thus low rate of charging – will make the sealed batteries to deteriorate quickly. It seems advisable to avoid to mix components with very varying quality in the same system, since they become sources of irritation and tend to discredit the system as a whole in the mind of the user.

Maintenance and Re-training of Staff - Institutional Framework

An important lesson learned has been that staff is exchanged relatively often. This is often due to the wishes of the occupants to get transfer to locations closer to their kin and similar reasons. This calls for re-training of the users at regular intervals or whenever needed in order to ensure that they know how to look after the system. In principle, the situation is of course not any different from that for other types of equipment, but in this case it is exacerbated by the fact that the PV-installations represent a totally new technology for *practically all* the users.

In order to ensure that the systems receive the care they need and are always kept in working order, it is necessary for the Ministry of Health to have in place routines for reporting on the state of the system, for carrying out preventive maintenance, and for responding to calls for corrective maintenance. It is also important the Ministry of Health has in place effective routines for management of stocks and other logistics. These issues should be addressed already during the planning of a project.

In this particular case, the organisation for carrying out re-training and maintenance is in place, but the existing routines need to be adapted to the needs of the PV systems. Presently these issues are under review in a special activity of the project.

Replication in Other Sectors – Taking Advantage of the Infrastructure Already Put in Place

Most places with a health centre also have a school. In Mozambique, and other countries as well, school facilities are in short supply, and provided that teachers and teaching materials of sufficient quantity and quality can be obtained, using the school building in the evening may significantly improve the access to education. This is particularly true for women, which often have to do work like wood and water collection, and farming in the day.

However, if the school building is to be an effective place for learning in the evening, good quality lighting must be available, and PV lighting systems offer excellent value for money in off-grid locations. PV would also make it possible to run a radio set on a regular basis, providing a tool for enhancing both the training of the teachers and the pupils. Finally, staff house systems for the teachers could be expected to have similar benefits as at the health centres.

Implementing a PV lighting projects in Mozambique would be greatly facilitated by the already existing health sector installations. First, the health sector project has put in place capabilities for installing and maintaining systems, and this capability is available in all provinces of the country. The Ministry of Education might therefore sign a maintenance agreement with the Ministry of Health. Second, important savings could be achieved by sharing stocks of spares.

The Role of Sector Projects in the Context of General PV Electrification

Through the health sector project, and possible future educational sector ones, practically everybody in a determined area will be exposed to PV technology. People may be expected to want to benefit from a system in their own house, and thus the sector projects may be said to have primed a market. Such a "own house" system might be either bought or leased.

Judicious efforts to build the commercial infrastructure necessary for realising this home-level PV electrification may with time achieve a full-fledged PV electrification of a village. Such efforts would consist in providing training for local "technicians", recruiting strategic customers, putting credit schemes in place, and motivating tradesmen to bring PV-adapted loads (lamps, radios, TVs etc.) to the market. In this context the sector projects have ensured that technical capabilities are already available on a nation wide scale. There are training concepts to build on, and technicians that might be qualified to give the training at the local level.

Figure 6 illustrates the process. The technology is introduced in the health and educational sectors. This introduces the technology, and provides for the establishment of basic technical capabilities in the area, albeit not necessarily *in* the village. Strategic "first customers" are then electrified. Programmes for developing the private sector – training for developing technical and commercial skills, credit schemes, awareness raising and information, enabling institutional arrangements, improving access to PV-adapted loads etc. – may then serve to put in place the commercial infrastructure for servicing a sustained PV market.

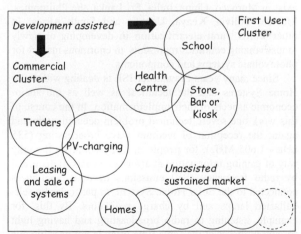

Figure 6. Schematic presentation of the proposed relationship between sector PV-electrification projects, the local private sector

7 REFERENCES

Sandgren, J. and Andersson, M. *Solar Energy for Rural Health Facilities. Installation Monitoring Survey Report.* KanEnergi, October 1999.

THE ELECTROMAGNETIC COMPATIBILITY (EMC) OF SOLAR HOME SYSTEMS

S. Schattner, G. Bopp, U. Hofmayer
Fraunhofer-Institut für Solare Energiesysteme ISE
Oltmannsstr.5, 79100 Freiburg, Germany
phone: ++49 761 45 88 313, fax: ++49 761 45 88 217, email: sschatt@ise.fhg.de

ABSTRACT: When dealing with Solar Home Systems (SHS) one is often confronted with EMC problems. Charge regulators as well as self-ballasted lamps are concerned. They often disturb medium wave (AM) radio reception which is wide spread in developing countries. Unfortunately, the useful field strength in these countries is rather weak. Besides this, TV sets and other radio appliances can be disturbed. The work described in this paper aimed at investigating disturbance sources, coupling paths and possible filter measures on the basis of a variety of marketable charge regulators and self-ballasted lamps. A simple measuring procedure ensuring the EMC of these components will be presented.
Keywords: Solar Home System - 1: Electromagnetic Compatibility (EMC) - 2: Rural Electrification - 3

1. INTRODUCTION

The electrification of remote areas in developing or newly industrializing countries is one of the most interesting markets for the application of photovoltaics now and even more in the future. According to estimations, up to the present approximately more than 1.000.000 Solar Home Systems were saled and installed worldwide. In the next years sales probably will increase further, due to several international and national funding programmes [1]. For instance, Argentina is seeking to supply 300.000 remote houses with electricity. It is estimated that 70 % of these systems will be photovoltaic systems. Indonesia is aiming at the electrification of in total 1.000.000 remote houses by photovoltaics. Further ambitious programmes exist in Morocco, China, India, Sri Lanka, the Philippines, Mexico, Bolivia, Kenya, Uganda and South Africa. In other words, rural electrification in developing or newly industrializing countries represents an enormous market for photovoltaic systems and components.

Since many years Fraunhofer ISE is dealing with Solar Home Systems and the technical as well as the socio-economic aspects of their implementation. In the course of this work one severe technical problem occured again and again: the reception of medium wave broadcasting (535 kHz - 1.605 MHz), for people in many countries the only way of gaining information at affordable costs, is disturbed by radio frequency (RF) emissions of other appliances connected to the Solar Home System, in particular by self-ballasted lamps and by charge regulators. By this, for example, listening to radio broadcasting and having light for any purposes at the same time often is not possible due to these EMC problems. But not only medium wave broadcasting is affected by RF emissions, also cases of disturbed TV sets and other radio appliances occured. It should be clear that the acceptance of a new technology like a Solar Home System suffers enormously through malfunctions like EMC problems of this kind.

For Fraunhofer ISE it was therefore a great concern to examine these EMC problems. The work described in the following particularly aimed at investigating disturbance sources, coupling paths and possible filter measures on the basis of a variety of marketable charge regulators and self-ballasted lamps. Finally a relatively simple measuring

procedure which allows to ensure the EMC of SHS components will be presented.

2. THE TEST SET-UP

A typical Solar Home System (Figure 1) was realised in the laboratory as a test set-up, hereby considering realistic conditions from an RF point of view [2]:

- Typical cabling lengths of a few meters were applied. All cables were bundled.
- The system was not connected to earth. This is typical for many countries in which earthing is very difficult due to arid climatic conditions causing a rather dry soil.
- A sun simulator consisting of 8 x 9 halogene lamps supplied an irradiation of approximately 800 W/m². Since the RF impedance of a PV module is dependent on irradiation, by this quite realistic conditions were ensured.

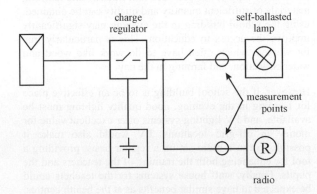

Figure 1: Schematic test set-up for conductor-borne RF emission measurements

The conducted RF emissions were measured by a current probe in the frequency range of 10 kHz to 30 MHz mainly at the input of the radio, but also at the input of the self-ballasted lamps. The radio was a very low-cost type, typical for usage in Solar Home Systems. It did not cause any significant RF disturbance itself. It was operated either with a rather weak medium wave broadcast station (1.33 MHz, signal level: 67 dBµV/m) or a rather strong broadcast station (870 kHz, 100 dBµV/m). The weak station resembles typical circumstances in developing countries.

Table 1 and Table 2 list the charge regulators, electronic ballasts and lamps chosen for EMC tests. The selected charge regulators cover the most wide spread regulators with puls width modulation (PWM) and typical switching frequencies of about 100 Hz. Two-position controllers which usually have clearly lower effective switching frequencies of a few Hz were not considered, since they presumably cause less RF emissions. When doing a selection of self-ballasted lamps, also a variety of current types was considered.

Table 1: Tested charge regulators

Charge regulator	Manufacturer
HLR 12-2	Helios Solartechnik
ProStar-12	Morningstar
Solsum 6.6X	Steca
SLR107	Uhlmann Solarelectronic

Table 2: Tested electronic ballasts and respective lamps

Electronic ballast	Manu-facturer	Respective lamp	Manu-facturer
n. n.	BP-Solar	F8T5/CW	Phillips
EVL14	Eckerle	CF-SI 7W/182 H48	Sylvania
Accutronic AT 7-9/12L	Osram	Dulux D/E 10W	Osram
n. n.	Photherm	FL .. 10 W daylight	SiBalec
Solsum ESL 7W	Steca	Solsum ESL 7W	Steca

3 RESULTS

3.1 The RF coupling paths

In principle, there are two ways for electromagnetic disturbances to cause interferences in radio sets. Disturbances can either interfere as conductor-borne disturbances or as field-borne disturbances. So at first, it had to be determined which coupling path is relevant for electromagnetic interferences in Solar Home Systems.

With switching frequencies of about 100 Hz and 10 kHz, respectively, charge regulators and self-ballasted lamps produce significant RF emissions in maximum up to a few MHz. Since the geometric dimensions of both

appliances are relatively small, from a physic point of view this RF power at first can only be emitted in the form of conductor-borne emissions. Significant cabinet radiation or radiation resulting from the lamp´s body does not take place. This even applies to lamps with lengths of a few 10 cm: when operating them near the battery-supplied radio, the radio functioned well in an acceptable distance above 1.5 m.

In principle, a conductor-borne disturbance can cause disturbance fields or even radiation which then may interfere as field-borne disturbance and affect medium wave radio reception. This strongly depends on the type of the existing disturbance modes.

In general, two modes of conductor-borne disturbances are to be distinguished: differential mode and common mode (Figure 2). Differential mode disturbances cause RF currents flowing in an antiparallel way, whereas common mode disturbances cause RF currents flowing in a parallel way. Because of this, common mode RF currents are likely to cause significant field disturbances, whereas differential mode RF currents do not produce any significant fields, as long as the two wires are bundled, by this neutralizing each other´s fields. It should be noted that this does not apply to wires forming a loop. In case of a loop, differential mode RF currents, too, are able to cause significant fields, similar to a loop antenna.

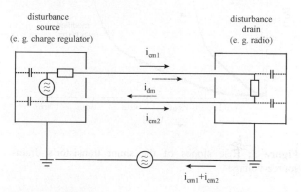

Figure 2: Conductor-borne differential mode (dm) and common mode disturbances (cm)

Since Solar Home Systems usually are not connected to earth, common mode currents do not exist. This was confirmed by RF current measurements. It represents an important fact: Even when firstly being emitted as conductor-borne disturbances, these disturbances can not cause any significant disturbance fields, since only differential mode disturbances exist.

This was in another way confirmed by disconnecting the radio and supplying it with batteries: The radio functioned well even at a distance of about half a meter from the PV system´s cabling.

To finally put it in other words: in case of no earthing and bundled wires the relevant coupling path is entirely conductor-borne. As long as all wires are bundled, no additional field-borne disturbances occur. This leads to one important requirement for practical SHS installations: wires should always be installed in a bundled way and never form some kind of loop.

3.2 Charge regulators

All tested charge regulators disturbed the weak broadcast station, none of them disturbed the rather strong station.

The implementation of EMC measures turned out to be relatively simple. Regarding the HLR 12-2 and the weak broadcast station a differential mode 1 μF ceramic capacitor between plus and minus at the consumer port was sufficient in order to achieve an acceptable reception. In case of the other charge regulators an additional measure was necessary. By adding a resistor in the transistor's gate driving the rise time of the slope was reduced to such an extent that radio reception was acceptable. Figure 3 shows the results for the SLR107 in comparison to the original rise slope of the HLR 12-2. The additional power dissipation caused by this measure lies in a range of 1 to 5 W, depending on the respective charge regulator. This might require additional measures concerning heat removal. Table 3 lists the successful EMC measures in detail.

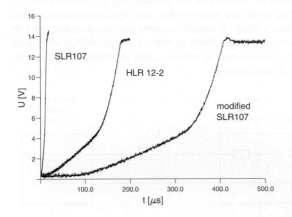

Figure 3: Rise slopes of the shunt transistor's drain-source-voltage

Table 3: EMC measures

Charge regulator	EMC measure
HLR 12-2	1 μF ceramic capacitor
ProStar-12	1 μF + additional 33 kΩ in the gate driving
Solsum 6.6X	1 μF + additional 2.2 kΩ in the gate driving
SLR107	1 μF + additional 110 kΩ in the gate driving

3.3 Self-ballasted lamps

In comparison to the charge regulators, the self-ballasted lamps caused quite severe disturbances. Both broadcast stations were disturbed by all electronic ballasts. As an example, Figure 4 shows the AC component of the input voltage of the Phototherm electronic ballast and

Figure 5 the corresponding RF current frequency spectrum, measured at the DC power supply input of the radio set.

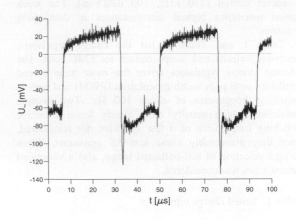

Figure 4: AC component of the input voltage of the Phototherm electronic ballast

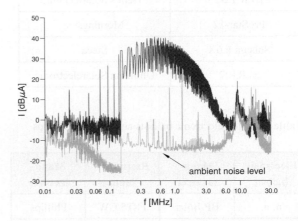

Figure 5: RF current frequency spectrum, measured at the DC input of the radio set with operating Phototherm electronic ballast

None of the EMC measures carried out were successful to such an extent that the quality of radio reception was tolerable. Figure 6 shows the results of an added low-pass filter. Due to its corner frequency of 22 kHz disturbances at relatively low frequencies still remain significantly high, whereas disturbances at relatively high frequencies are suppressed quite clearly almost to the level of ambient noise. It is therefore likely that disturbances at lower frequencies than about 300 kHz are responsible for the audible disturbances due to intermodulation and crossmodulation effects.

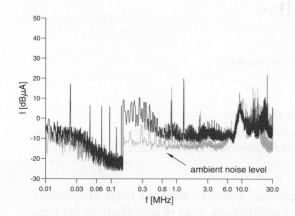

Figure 6: RF current frequency spectrum, measured at the DC input of the radio set with operating modified Phototherm electronic ballast (additional low-pass filter)

Suppressing even lower frequency disturbances requires high filtering efforts for the given electronic ballasts and their respective switching topology. It is doubtful whether these great efforts are justifiable for such a low-cost product. Under these circumstances, for a manufacturer of an electronic ballast likely to be applied in a Solar Home System it seems to be more reasonable to adjust the switching topology to the susceptible radio. Resonant switching topologies with specific filtering measures are imaginable, as well as higher switching frequencies.

4. PROPOSAL FOR AN EMC TEST SET-UP

Standards specifying EMC requirements for charge regulators (EN 55014-1) and self-ballasted lamps (EN 55015) do exist [3]. But it is rather doubtful whether these standards should be applied to the 12V/24V appliances such as treated here.

Firstly, regarding medium wave broadcasting, the electromagnetic environment of a developing country differs quite clearly from the environment of an industrialized country. As already stated, the useful signal for medium wave radio reception is significantly lower, this making the radio more susceptible against unintentional RF emissions. Furthermore, also the system impedances of a Solar Home System should differ significantly from the impedances specified in existing standards, always assuming a 230 V AC power supply. Further research seems necessary for creating reasonable EMC specifications for Solar Home Systems. As long as such specifications lack, it seems to be advisable to apply a very practical measuring method in order to insure the EMC of SHS charge regulators and self-ballasted lamps.

In a way similar to the test set-up described in 2. and depicted in Figure 1, the manufacturer should build up a typical Solar Home System in the laboratory. He should further connect a low-cost radio as a consumer, tune in a medium wave broadcast station (or even better, several stations) which has a rather weak useful signal level of about 60 to 70 dBμV/m and make sure that the operation of his charge regulator or his self-ballasted lamp does not interfere radio reception. According to the experiences gained at Fraunhofer ISE, this test set-up seems to be a very appropriate test method, quite easily to be realized by a manufacturer.

5 SUMMARY AND CONCLUSIONS

The EMC characteristics of four different charge regulators and five different self-ballasted lamps, typically applied in Solar Home Systems, have been examined. While the charge regulators seem less problematic regarding their EMC behaviour, the self-ballasted lamps produce severe electromagnetic interferences, made audible by a medium wave radio set. The coupling path is totally conductor-borne. Simple EMC measures such as low-pass filters were not succesful. Innovative switching topologies seem necessary to treat this EMC problem.

On the other hand, radio sets, typically used in developing countries, are of a very low quality. It is essential to EMC philosophy that not only maximum emissions, but also a minimum immunity must meet the requirements of a specific electromagnetic environment. These radios definitely lack this minimum immunity. Until now, there are no radios particularly designed for the use in Solar Home Systems, considering, among other things, also the specific EMC requirements. On the background of the dynamic market development, described at the beginning, in future this ought to change and actually might become a quite lucrative business.

REFERENCES

[1] K. Preiser, K. Reiche: Marktentwicklung für Photovoltaik-Systeme zur ländlichen Elektrifizierung, 15. Symposium Photovoltaische Solarenergie, 15 - 17 March 2000, Staffelstein, Germany

[2] U. Hofmayer: Elektromagnetische Verträglichkeit in Solar Home Systems, diploma thesis, Fraunhofer ISE, August 1999

[3] S. Schattner, G. Bopp, T. Erge, R. Fischer, H. Häberlin, R. Minkner, R. Venhuizen, B. Verhoeven: "PV-EMI" - developing standard test procedures for the electromagnetic compatibility (EMC) of PV components and systems, 16th European Photovoltaic Solar Energy Conference, 1- 5 May 2000, Glasgow, UK

This work was funded in part by the European Commission in the framework of the Non Nuclear Energy Programme JOULE III.

SPECIFICATIONS FOR THE USE OF RENEWABLE ENERGIES IN DECENTRALISED RURAL ELECTRIFICATION

A. SCHMITT, G. HUARD, C. NAPPEZ

Electricité de France, R & D Division,
1 Avenue du Général de Gaulle, B.P. 408, 92141 Clamart Cedex, France
email : alain.schmitt@edf.fr, guy.huard@edf.fr, christophe.nappez@edf.fr

ABSTRACT : In Electricité de France's (EdF) R & D programme, aiming to develop a range of optimised systems using renewable energies for Rural Decentralised Electrification, a set of specifications has been drawn up, in partnership with the French Agency for the Environment and for Energy Management (ADEME) and the French photovoltaic industry. This work has been carried out by working groups involving all the main French players in the field of renewable energies.

These specifications are designed to ensure the three key points for a sustainable and successful electrification : the quality of manufactured products, the quality of carrying out of projects, the quality of operation, maintenance and renewal. If only one of these three key points is missing, the electrification project is likely to be a failure. These specifications are dedicated to stand alone systems and are currently the legal reference in France and overseas departments, but also for the electrification projects which are currently carried out abroad by French actors. They have been accepted by the International Electrotechnical Commission (IEC) as Publicly Available Specifications and a new working group (Technical Committee 82) is being formed to have them converted to norms.

Key words : 1 : PV System - 2 : Rural electrification - 3 : Developing countries

1. INTRODUCTION

In EDF's research programme on the use of renewable energies in Decentralised Rural Electrificaztion (DRE), work was undertaken in order to write a range of specifications able to improve the quality and reliability of renewable energy systems likely to be used in such rural electrification projects. It was carried out in almost 20 months and has involved more than 30 French experts, coming from French Agency for the Environment and for Energy Management (ADEME), manufacturers and EDF, in the field of photovoltaic electricity generation, rural electrification, small-scale wind machines, micro-grids, power electronics, and so on.

The first edition was published in 1997. The second edition, revised, will be available in the second half of 2000.

2. WHY WRITE SPECIFICATIONS FOR DECENTRALISED RURAL ELECTRIFICATION (DRE) ?

It is generally the case that each DRE project promoter writes his own specifications and industrials must adapt their offer. To ensure the growth of the market and to improve the quality of the systems, it is then necessary to promote a common basis of technical specifications. This will make it possible to use off the-shelf systems for rural electrification and thus to reduce the cost of such programmes.

In 1993, EDF launched an R & D programme that aims at designing a range of electrification systems for rural areas, using renewable energies and especially solar energy. The DRE Specifications were are cornerstone of this programme.

3. WORKING METHOD

The work was carried out by working groups using the functional requirement study method, with each group having to reflect on the functions that each component and system must achieve. The groups had also to assess the level of performance to be reached for each function. The result is a functional specification with, for most cases, a technical specification associated.

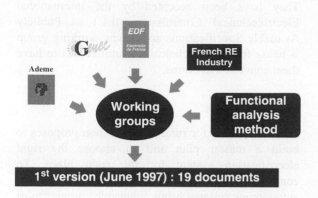

Fig 1 : The working method

4. STRUCTURE OF THE SPECIFICATIONS FOR DECENTRALISED RURAL ELECTRIFICATION (DRE)

The specifications for DRE's purpose is to help the introduction of renewable energies into rural electrification programmes. The structure of the DRE specification is designed to help a project supervisor to go from needs to product or to the right electrification system. This is what is shown by figure 2.

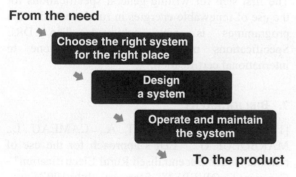

Fig 2 : The purpose of the DRE specifications

The three more important steps to be followed by a project supervisor are :

- first to choose the right system for the right place and the right user,
- second, to design the most cost effective system according to the site,
- third, to prepare the operation, maintenance and renewal of the installations.

The experience of numerous projects in the past shows clearly that every project omitting one of these three key points is a failure, and that after a few months or, in the best cases, after a few years, all the installations or systems are out of order. Furthermore, most experiences of management by

users' associations have proved to be failures too. For a sustainable electrification, the electrical systems must be operated by efficient professional companies as it is the case in industrialised countries.

The documents are classified, in the second edition, into three parts :

From the energy needs to the electrification system.

Rules for system design and operation.

Technical specifications.

Part A is dedicated to the quality of the project. It is clear that the quality of products is a necessary condition for a successful electrification but it is not the only one. The best industrial products have to be installed and operated in the best way to be fully efficient. Especially, a socio-economic study must be carried out by the project contractor to be sure to match the needs of the user and to install affordable systems.

Figure 3 shows the main actors involved in the realisation of a project, and their relationship framework.

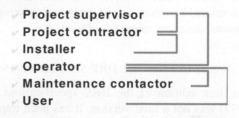

Fig 3 : the main actors involved in the realisation of a project

The project contractor is responsible toward the project supervisor for the quality of the installations, and for the quality of installed products. The project supervisor entrusts the installation to an operator who is responsible toward the user for the quality of the delivered service.

Part B deals with the system design and functional specifications.

When the groups began to work, they quickly realised that it was compulsory to build a classification of the systems. This classification is proposed by the B1 part and the following documents propose guidelines for system design, energy management, data acquisition, safety. All these guidelines are adapted according to the kind of systems. Figure 4 gives the summary of the architecture which is composed of eight types of systems. Diesel only systems are taken into account because they are widespread in rural electrification and they are in competition with the systems using renewable energies and sometimes complementary to them as it is the case in T5 type which is an hybrid one.

The classification is focused on three main kind of systems : battery-less systems, Individual Electrification systems (IES), and community Electrification systems (CES).

Battery-less REN	T1	❑ REN
IES *(Single-user)* Individual Electrification Systems	T2 T3	❑ REN + storage ❑ Hybrid + storage
CES *(Multi-user)* Community Electrification Systems	T4 T5 T6 T7 T8	❑ REN + storage ❑ Hybrid + storage ❑ Coupling REN / Diesel ❑ Diesel + storage ❑ Diesel

Fig 4 : the eight types of systems

Part C proposes technical specifications for the main components of the system, including generator and photovoltaic array, wind turbines, diesel generators, batteries, micro-grids and micro-power stations, inverters, energy managers, etc.

5. WHAT USE FOR THE DRE SPECIFICATIONS ?

The first edition of the DRE specifications (June 1997) was not a final version. It has been improved by feed-back from the field, and from various comments from around the world.

As far as EDF's rural electrification projects are concerned, several pilot projects are underway, located in France, West Africa and South America, involving the electrification of several tens of thousands of people. The electrification systems used for these pilot sites are both micro-power stations with micro-grids (CES) and solar home systems (IES).

These electrification programmes are a very good field experiments and are supplying a lot of information able to improve the current specifications.

But these specifications have also to be discussed internationally and to be improved by a feedback, as complete as possible, coming from the experience of international actors involved in the field of renewable energies and rural electrification. They could be useful for all initiatives aiming at the improvement of the quality of rural electrification using renewable energies.

They have been accepted by the International Electrotechnical Commission (IEC) as Publicly Available Specifications and a new working group is being formed (Technical Committee 82) to have them converted to norms.

6. CONCLUSION

EDF's approach for rural electrification proposes to build a master plan and to choose the right electrification system for the right place. To compete with the grid solution, the solution of autonomous systems using renewable energies must be as professional and industrial as the conventional solutions. The electrification solutions for the rural population of developing countries must match their requirements and also their buying power but mustn't be handicraft solutions. So, it is necessary to undersize strongly the systems compared to the one used in developed countries, but to ensure a very good quality.

We think that to ensure the quality of the electrical service provided to the user, it is compulsory to ensure the three main following key points all together : the quality of manufactured products, the quality of carrying out of projects, and the quality of operation, maintenance and renewal.

The first step for writing general specifications for the use of renewable energies in rural electrification programmes is now achieved. The DRE Specifications may be a stepping stone to international certification.

7. BIBLIOGRAPHY

[1] SOLER R., SCHMITT A., CAMEAU L., MARBOEUF G. : "EDF's approach for the use of renewables for Decentralised Rural Electrification" - Congress FLOWERS'97, Florence - July 1997

[2] EDF, ADEME : "Specifications for the use of renewable energies in rural decentralised electrification" - June 1997

INTEGRAL TRAINING PACKAGE FOR SOLAR HOME SYSTEM PROJECTS

Bernard van Hemert[1], Quirin Sluijs[1], Geerling Loois[1], Johan Willems[2],

1. Ecofys Energy and Environment, P.O. Box 8404 NL-3503 RK Utrecht,
phone: +31 30 2808387 / -397; fax +31 30 2808301;
e-mail: b.vanhemert@ecofys.nl. or q.sluijs@ecofys.nl

2. Shell Solar Energy, Lagedijk 30, 5705 BZ Helmond, The Netherlands
tel. + 31-492-508608, fax +31-492-508600, e-mail: Johan.W.G.M.Willems@SI.shell.com

ABSTRACT: The development of an integral training package for end-users, promoters, installers, technicians and trainers is described. After the definition of the relevant players, their baseline educational level and training needs were assessed. Materials were designed specifically for each target groups but in coherence with each other. Draft results were field tested in large scale Shell SHS projects in Bolivia and South Africa. The final training kit includes one promotion leaflet, 6 manuals, 2 poster sets and the Power Game, an exciting entertainment which teaches battery management to end users, whether literate or not.

Keywords Glasgow conference: Solar Home Systems – 1; Developing countries – 2; Training – 3.

1 BACKGROUND

Shell Solar Energy has a long trajectory of marketing Solar Home Systems (SHS). A bottleneck in the implementation of large-scale SHS projects has proven to be the lack of adequate knowledge with the various players at grass-root level. To overcome this problem, Shell Solar Energy in co-operation with ECOFYS has developed an integral training package for Solar Home System Projects. The package contains manuals and training materials for 5 central players in the implementation of SHS projects.

2 THE PROCESS OF DEVELOPING THE TRAINING PACKAGE

The process to develop the coherent set of training manuals and materials followed these steps:
1. All players in a SHS project that might benefit from some kind of standard capacity building were identified. This was done in consultation of several project managers of large-scale SHS projects. The result of this exercise is presented in section 3.
2. From all players the training needs were identified, as well as their assumed baseline educational level.
3. A set of draft materials were prepared. A specialist in 3-D technical drawings and a cartoonist were contracted.
4. Field testing took place in South Africa and Bolivia. This field-testing proved to be crucial, as is elaborated in section 5.
5. Based on the results of the field tests, the final package, briefly described in section 4, was produced.

3 ACTORS IN A SHS PROJECT

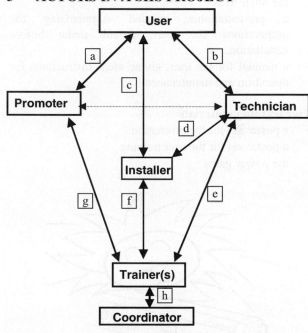

Figure 1: Basic interaction between parties involved in the project

The promoter introduces the solar home system The contacts between promoter and (future) user (a) are referred to as the promotion process.

The contacts between installer and user (c) consist of installation and commissioning of the SHS, including instructions during and immediately after the installation, and inspection and maintenance afterwards. The contacts between technician and user (b) take mostly place after the installation, when problems arise which cannot be solved by the installer, or in case of quality checks on installation work.

The contacts between technician and installer (d) are normally of a coaching or supervising nature.

Promoters, technicians and installers are capacitated and supervised by trainers (e, f, and g):

The co-ordinator will normally implement or guide the adjustments to the training materials of the kit. This will be in close co-operation with the trainers (h) in the project.

4   THE TRAINING PACKAGE

4.1   User

At the user level, the package comprises three manuals and three training tools:

User manuals
- a promotion leaflet, providing basic information on the Shell SHS,
- a pre-installation manual summarising the preparations the user should make before installation.
- a manual for the user, giving clear instructions for operation and maintenance.

User training materials
- a poster set for the promotion
- a poster set for the user training
- the power game

Figure 2: Straightforward black-and-white cartoons proved to be suitable for the user manuals

All these materials are written for a semi-literate public, with simple language and many self-explanatory illustrations.

The poster sets are used by the promoter and the installer as a guide for their user training sessions. They contain hardly any text, are cheap to reproduce and can be rolled and packed in a simple PVC pipe.

A qualitative sense for power management is taught by the innovative Power Game, where three players aim to get the batteries of their SHS full through proper load management. Specially designed dice define the weather conditions and the use of the various loads (lamps, TV, radio) over the day. Playing the game does not require any reading or calculating skills.

4.2   Promoter

The **manual for the promoter** deals with informing potential clients about the possibilities of the SHSs. This manual stresses not to rise false expectations, to assess whether a SHS is the best solution for the client's needs, to check his financial capacities, and how to convert the initial enthusiasm into a concrete deal. As a tool, the promoter can make use of the **poster set for the promotion**.

Figure 3: The Power Game proved to be a game which is not only exciting but also educative. Even people with no education level get a qualitative sense of battery management.

4.3   Installer

The **manual for the installer** focuses not only on the technical aspects of the installation and maintenance of the SHS, but also on the user training, as it is the installer who instructs the users on the do's and don'ts. Load management through the **power game** is a central topic in this practical on-spot training, while the **poster set for user training** helps to guide the discussion.

4.4   Technician

The **manual for the technician** is similar to that of the installer, only that the level is significantly higher. This allows in-depth analysis of topics such as system sizing and trouble-shooting.

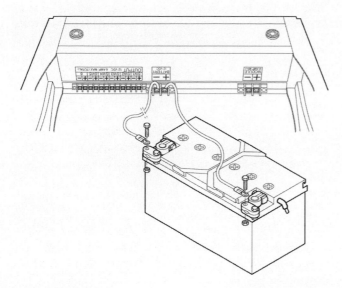

4.5 Trainer

The **manual for the trainer** provides a guide for the

Figure 4: Technical 3-D drawings can be easily understood by technicians and installers alike

training of promoters, technicians and installers, based on the respective manuals. The training methodology is participatory, with the field experiences of the participants as the starting point. Apart from proposed training programmes, the manual contains tips on communication skills, participatory training techniques and how to address gender issues

4.6 Co-ordinator

The **guide for de co-ordinator** discusses how to deal with the training package. Ideally, the different manuals should be adapted to the local situation:

- issues of ownership and (re)payments conditions are crucial for the successful implementation of a project, and have to be addressed properly. However, as these aspects are completely different for each project, they cannot possibly be included in a universal package.
- technical aspects such as tilt angle, irradiation, temperatures etc. may vary.
- local slang and customs may require textual adaptations to the user manuals
- the set-up of the specific project may cause different roles for the different players.

5 THE FIELD TESTS

In Bolivia and in South Africa, the materials were thoroughly field-tested. Trainings for installers and promoters were conducted, the real-life use of the manuals and training tools was evaluated, and the whole package was extensively discussed with project staff and field workers.

Field testing of the draft manuals proved of invaluable benefit, as it yielded some important and many small improvements:

- The educational level of the technicians appeared to be much higher in South Africa than in Bolivia. Therefore, it was decided to make a simplified version of the 'manual for the technician', the 'manual for the installer'. This manual has the level and presentation of the user manuals.
- Some drawings appeared to give an inaccurate message, and were adapted.
- The reproduction costs of the materials proved to be a more important factor than initially thought. As a consequence, the concept of hard covered manuals with coloured images was abandoned in favour of simple plain paper booklets with black-and-white drawings which can simply be photocopied.

6 CONCLUSIONS

Successful implementation of any large-scale SHS project requires an integrated capacity building strategy. The produced training package can serve as a basis for that. Precondition for its optimal use is that the provided elements be adapted to the specific project set-up and local conditions.

7 LITERATURE

During the production of the package, use was made of the following references:

1. TOOL Consult & GRUPO, Amsterdam / Lima, April 1996 :Small scale electricity generation with solar and wind energy – Technical training on the market development of renewable energy. Trainer's manual and Participant's manual
2. TOOL & KSCST, Amsterdam / Bangalore, 1995: Marketing and after-sales for renewable energy technologies – technical training for manufacturers, dealers and installers.
3. Mark Hankins, Commonwealth Science Council, 1995: Solar electric systems for Africa – revised edition
4. Rodríguez-Murcia, PNUD/OLADE, 1995 Training manual in Photovoltaic systems for rural electrification
5. Leitman, WB 1988 Some considerations in collecting data on household energy consumption
6. Shell Solar Energy, July 1997: Installation/operation/maintenance/inspection manual
7. Shell Solar Energy, November 1998: Training manual
8. SOLSISTEMA, manual de mantenimiento, n.d.
9. S. Rifkin & M. Johnston, London 1987: Health care together, training exercises for Health Workers in Community Based Programmes.

10. L. Vargas Vargas, Alforja, San José 1984: Técnicas
Participativas para la educación popular, tomos 1 y
2.

DEVELOPMENT PLAN FOR SMALL AUTONOMOUS RENEWABLE ENERGY SYSTEMS FOR THE DEPARTMENT OF LESVOS

Stratis Tapanlis, Lars Beuerman, Tom Scheer
University of Kassel - Department for Efficient Energy Conversion
Wilhelmshöher Allee 73, D-34121 Kassel, Germany
Tel.: +49-561-8046218 Fax: +49-561-8046434 email: stratis@uni-kassel.de

Dimitris Koniarellis
Union of local authorities of Lesvos
Iktinou 2, GR-81100 Mytilene, Greece
Tel.: +30-251-29570 Fax: +30-251-41463 email: tedklesv@otenet.gr

ABSTRACT: Scope of this work is the development of an integral solution for irrigation and cooling applications in agricultural areas, located far away from the public electric grid. By the use of renewable energies in windy or sunny areas a simple, cheap and reliable irrigation(pumping) or cooling(refrigerating) systems can be applied, avoiding the disadvantages of the already existed diesel-powered systems. Other applications e.g. solar powered road lighting or solar water purification systems will be also considered. Target area are the islands of the department of Lesvos, Greece. A database including meteorological data, system components and contacts to manufacturers and local sellers was set up. Financial and legal aspects were investigated and recorded. A pre-sizing software tool for small autonomous PV an wind plants was programmed.
Keywords: Rural Electrification – 1: Water pumping - 2: Stand alone PV systems – 3

1. INTRODUCTION

The Department of Lesvos consists of the three islands Lesvos, Lemnos and Agios Efstratios located at the north-eastern part of the Aegean Sea in Greece. The two major problems of the islands are considered to be the declining growth rate of the local population and the low per capita income. This decrease of the local population is attributed to the inadequate supply of employment, the low-level infrastructure and the inadequate interconnections with the continental country. The population of Lesvos is primarily located in rural areas (55% of the total population of Lesvos) and about 50% of the active population is working in the agricultural sector.

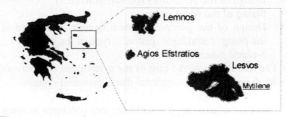

Figure 1: Map of the department of Lesvos

According the collected data of the National Meteorological Agency of Greece, the data of European Solar and Wind Atlas and self-collected meteorological data, there is a large potential on solar and wind energy available in the area of the department of Lesvos. The Union of the local authorities of Lesvos implemented in the years 1994 – 1996 measurements of the solar radiation and wind velocity on four sites of Lesvos, created a local wind and solar map of the island and developed a plan for the exploitation of wind energy in the department.

Scope of the present work is the development of an integral solution for irrigation and cooling applications in agricultural areas, located far away from the public electric grid. By the use of renewable energies in windy or

sunny areas a simple, cheap and reliable irrigation (pumping) or cooling (refrigerating) system can be applied, avoiding the disadvantages of the already existed diesel-powered systems. Object of this work is the removing of any kind of barriers and the penetration of wind and photovoltaic systems in the agricultural applications through simple sizing and designing of the systems. Even other stand alone applications such as solar powered road lighting or solar water purification systems will be considered.

Up to now, the common methods in case of irrigation in remote areas are as follows :
-There is no irrigation at all. That means only plants can be grown, which do not need much water (e.g. olive trees). The harvest is very low.
-The irrigation system consists only of a diesel motor (an old car or lorry motor) and a piston pump. The engine often has a 100%-2000% higher power as needed. The amount of pumped water is not controlled, so, too much water is used and the ground-water level drops. The motor is very inefficient (uses too much fuel), causes high cost and a high pollution, especially CO_2.
-A diesel motor-generator is combined with an underwater electric pump. This is the most efficient of the currently used methods. Due to the high power of the generators, this solution has high capital cost as well as the disadvantages of the other Diesel-systems (fuel cost, CO_2).

The same conditions are existing in the case of the provisional cooling of agricultural products (milk, vegetables etc.) until they be sold to the co-operatives or to traders. Farmers either cannot fridge their products so they cannot have enough production or they use diesel generators.

2.   DATABASE

Meteorological data (solar, wind, rainfall), needs of inhabitants of remote areas in energy (also water demand, kind of usual vegetation, water quantity per plant/hectare, frequency of irrigation, freezing of milk and vegetables etc.), RE-systems and components (panels, wind generators, inverters, water pumps, low energy DC-refrigerators etc.) and contacts to sellers and installers of such equipment are collected and set up to a database. The database is implemented in MS-Access and will be distributed to agricultural cooperatives municipalities etc.

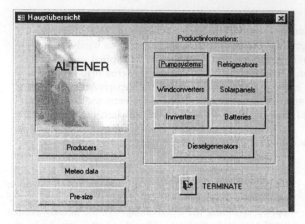

Figure 2: Front panel of the Access Database

Meteorological data includes informations from the National Meteorological Agency, European Solar Radiation Atlas, European Wind Energy Atlas and self-collected data of one pyranometer and four wind meters. The technical data of more than hundred PV panels, twenty five wind turbines and sixty inverters are included in the Database. The database includes also data about water pumps, DC-refrigerators, Batteries and diesel generators. Addresses, phones etc. of more than fifty manufactures or local sellers are included in the database too.

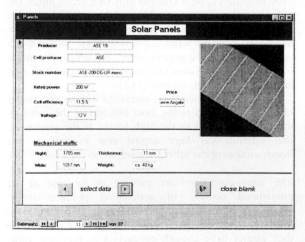

Figure 3: Screenshot of the database

The button "pre-size" starts directly the pre-sizing programme described below.

3.   PRE-SIZING SOFTWARE TOOL

A new software tool is programmed to help electricians and farmers to pre-size photovoltaic or wind energy autonomous refrigerating and pumping plants on Lesvos. The energy needs or the load profiles must be inputted. The software uses the technical and meteorological data of the database. New data can also be inputted and saved.

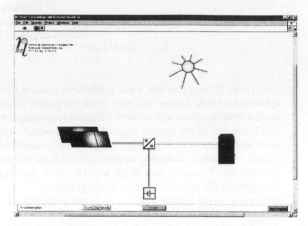

Figure 4: Screenshot of the pre-sizing software tool

If the energy consumption or the load profiles are not known, the tool calculates them from empirical data. For the pre-sizing of a pumping-irrigating system, can be used as input the needed water quantity or even the depth of the aquifer and the kind and number of the plants to be irrigated.

The programme will calculate the optimal design of the system based on the water needs, the solar and wind energy potential. The programme output will be:
-The design of the whole system :
- Selection of the type of the generator (wind, solar, both)
- Design of the generators (power, voltage...)
- Sizing of the water reservoir
- Design of the pumping system (power, size, type of the pump: centrifugal, piston..., type of the motor: DC, 3-phase synchronous AC ...)
- The calculation of the cost of the system and comparison with the cost of a conventional diesel powered system.

In the case of cooling, the software tool will have as input the needed refrigerating power and as output the design of the system and calculation of the cost similar to case of irrigation.

4.   FINANCIAL ASPECTS – DISSEMINATION

The legal aspects of autonomous energy production from RE sources are investigated. Ministry of agriculture and ETAL (programme Leader II) are identified as the more optimal way for financing of agricultural RE installations.

A small information brochure was made and two workshops took in April 2000 in Mytilene place involving and informing potential users, electricians, potential financiers and local authorities about the possibilities of installing small autonomous RE systems on Lesvos. The Database and the pre-sizing software will be distributed to the municipalities and agricultural cooperatives.

The first two pilot plants are already designed. An 0,35 kW PV-pumping plant and a 6 KW PV-Diesel-hybrid system will be installed at the end of this year on the island of Lesvos. Three other stand alone applications (solar road lighting, solar water purification) are planned too.

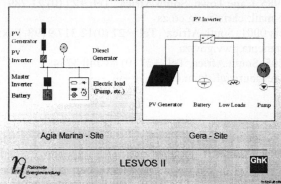

Small off-grid Photovoltaic-Diesel hybrid energy systems on the island of Lesvos

Figure 5: Systems to be installed on Lesvos in this year

5. AKNOWLEDGMENTS
This work has been supported through the European Commission, DG Energy and Transport, programme ALTENER II.

6. REFERENCES

[1] Dimensionierung einer autonomen Stromversorgung, sowie der Entwurf eines Energiekonzepts für ein Haus und ein Bewässerungssystem auf einer griechischen Insel, Lars Beuermann, Ghk-FB16-RE, 1999

[2] Entwicklung eines Software-Tools zur Vordimensionierung kleiner RE-Anlagen, Tom Scheer, GhK-FB16-RE, 2000-04-30

[3] Entwicklung einer Datenbank und Entwurf technischer Konzepte im Rahmen des Entwicklungsplans für kleine autonome Energieversorgungssysteme für eine griechische Insel, Stratis Tapanlis, GhK-E&U, 2000

[4] Energy in the north Aegean region, Regional energy agency of the north Aegean

PROGRAMME FOR THE ELECTRIFICATION OF 1 000 SCHOOLS IN EASTERN PROVINCE AND NORTHERN CAPE IN THE REPUBLIC OF SOUTH AFRICA.

J-Paul Louineau, J. R. Bates, C. Purcell[1], W. Mandhlazi[2], H. Van Rensburg[3].

IT Power Ltd, The Warren, Bramshill Road, Eversley, Hampshire, United Kingdom, RG27 0PR. Tel +44 (0) 118 973 0073, Fax +44 (0) 118 973 0820, e-mail: jpl@itpower.co.uk

1: Energy & Development Group, P O Box 261, Noordhoek 7985, Cape Town, South Africa. Tel: +27 (0) 21 789 2920, Fax: +27 (0)21 789 2954, e-mail: chris@edg.co.za

2: Department of Minerals and Energy, Private Bag X59, Pretoria 0001, South Africa, Tel: +27 (0)12 317 9198, Fax: +27 (0)12 322 5224, winnie@mepta.pwv.gov.za

3: Eskom NGE - TSI, Private bag X1, East London 5205, South Africa, Tel: +27 (0) 43 748 3734; Fax: +27 (0)43 748 1285, henry.jvrensburg@Eskom.co.za

Abstract

This paper describes the implementation of a project funded by DGVIII of the European Commission to provide electricity to 1 000 schools in remote areas of Northern Province and Eastern Cape Province in the Republic of South Africa using photovoltaic systems. The project is the EU contribution to an existing programme set up in 1995 under the Reconstruction and Development Programme to supply non-grid electricity to 16 400 schools in remote parts of the country. The project provides non-grid electrification as well as audio visual equipment for each school including PV power supply, control circuitry, AC lights and AC power points for the television, video, satellite dish and decoder. The systems have been designed to generate enough electricity to provide lighting for 3 to 5 rooms, as well as electricity for the audio-visual equipment.

The project is also training emerging contractors to install the equipment, as well as school staff to maintain the systems. Recurrent costs will be covered by the provincial departments of education which will be responsible for the upkeep of the systems.

This paper provides project background information and specific information on the role of the Technical Assistance Unit (TAU) focusing on work to date to ensure sustainable maintenance (both technical and financial) of the systems.

1. INTRODUCTION

The objective of the project is to provide non-grid electricity to 1000 schools in remote areas of Northern Province and Eastern Cape Province using photovoltaic systems. The project is implemented by ESKOM, the parastatal electricity utility under the authority of the Department of Minerals and Energy and with Technical Assistance from IT Power Ltd, UK and the Energy & Development Group, South Africa. The project is the EU contribution to an existing government programme (i.e. Reconstruction and Development Programme, RDP) set up in 1995 to supply non-grid electricity to 16 400 schools in remote parts of the country. To date, over 1 400 schools have been electrified with finance from the government's national RDP and from the Dutch government (300 schools) at a cost of approximately 12 000 000 Euro.

The EU funded programme has some unique features:

1. Installation of 1 000 schools with almost 1 MWp and as such is one the world's largest donor supported programmes (as a reference, the PRS programme in the Sahel region was 1.2 MWp).

2. Direct involvement of Eskom, one of the largest utility companies in the world, and government agencies (the Department of Minerals and Energy and the Department of Education).

3. Use of extension workers to increase community awareness of the potential benefits of the electricity for the schools, this includes the training of end-users.

4. Maintenance commitment: system installation cannot start unless the Provincial Departments of Education and the school governing bodies have committed themselves to pay for recurrent costs and take responsibility, ownership and management of the systems and ensure the security of the systems.

5. Training of emerging contractors in PV system installation and servicing.

2. IMPLEMENTATION ORGANISATION

The EU-funded project began in December 1998 at a cost of 15 million Euro. The Non-Grid Electrification (NGE) Unit of Eskom is managing and monitoring the programme, under the supervision of the Department of Minerals and Energy. This unit works directly with the Provincial Departments of Education (PDoE) and the Department of Public Works, the latter is in charge of the maintenance of government buildings.

The first stage in the implementation process was the identification of the beneficiary schools. Between December 1998 and May 1999, Eskom, in collaboration with the PDoE established selection criteria to choose the schools and subsequently identified 1 000 schools. Floor plans were completed and systems designed for each school. Tender dossiers along with technical specifications for PV hardware and for installations were drawn up as well. Extension workers were hired and received their first training.

The installation phase began in late 1999 and as of the 31st of April 2000, about 300 schools had been completed. The project is scheduled to continue until July 2001. In January 2000, a Technical Assistance Unit was appointed by the EU to provide extra input in project management and monitoring. It is unusual that the appointment of the TAU should be made at such a late a stage in the programme, however it has proved to be an effective measure for all stakeholders. The TAU works within the Department of Minerals and Energy.

3. PV SYSTEM TECHNICAL INFORMATION

The PV systems installed in this project differ from those previously installed under the RDP (from 1992 to 1995), not so much in size, but in concept. Under the RDP programme, systems made use of 12 volt improved car-battery banks and comprised a DC circuit for the lights and an AC circuit for powering audio-visual equipment. This design proved unsatisfactory in that it lead to high maintenance costs and frequent misuse and theft of the 12 volt batteries. DC lights proved less reliable than desired, and the inverters selected were prone to failure in the harsh rural conditions. Problems related to the design and commissioning inspection of the long runs of DC lighting cables (and switches and joints) exacerbated the problems of degradation over time, and lead to high voltage drops and low light efficiencies. The systems currently installed are completely AC. All systems are identical: 8 modules of 110Wp each, a 510Ah / 24 V stationery battery bank (2V cells), a 900W inverter, 14 luminaries of 36W each and a TV with VCR and satellite dish with decoder. The AC systems do not have the voltage drop problems experienced in the past, and therefore do not now require elaborate commissioning of the wiring, which is now standardised and certified.

The PV modules are mounted near the school buildings on a sturdy galvanised structure with tamper-proof bolts. The battery bank is housed in a dedicated galvanised battery enclosure with a strong locking system and located in the shade under the array. The control gear enclosure housing the charge regulator and the inverter is also mounted under the PV array. Finally, the AC feeder is buried under ground to reach the school building where a distribution board is installed. Electricity is then distributed to luminaries and 2 sockets. A special security enclosure has been manufactured to house the audio-visual equipment for safekeeping.

The PV modules, charge regulators and inverters are European made, along with 40 % of the batteries supplied. The rest of the batteries, the array support structures, battery enclosures and control gear enclosures are manufactured in South Africa. Local contractors, including contractor's previously trained by Eskom, are carrying out installations.

Each system is to be provided with a user's manual and a logbook. Installations are guaranteed for one year by the installation contractor against minor faults and poor workmanship. After this period, the Department of Education is responsible for the maintenance of the systems. PV modules are under warranty for 10 years, the inverters and regulators for 5 years and the batteries for 2 years. The PV hardware supplier also provides local after-sales services, this being a contractual obligation of the PV supplier.

4. THE SOUTH AFRICAN CONTEXT

It is important to describe the socio-political context in which the project is taking place in order to understand the challenges that the project is facing. The apartheid system was still in place less than 6 years ago and under that regime, schooling facilities for black people were appalling. Furthermore, schools in black South Africa were one of the oppressive tools used to reinforce apartheid until 1994. Consequently some schools are to this day, still regarded in a negative light by rural communities despite government efforts to improve education standards. Many rural communities have yet to reap the full benefits of such government policies however and still live in circumstances of severe poverty.

Given these circumstances the schools electrification programme faces many challenges due to the following factors:

1. School facilities are not always regarded as being part of the community. This is often compounded by the fact that school buildings can be physically isolated from the rest of the village.

2. Illiteracy levels remain very high, hence the population may not realise the full extent and benefit of better education opportunities. This reinforces the cycle of poverty.
3. People have more pressing needs such as safe water and food security.
4. Theft and vandalism are rampant in rural areas.
5. Access to isolated schools is difficult on poor roads.
6. As yet, there is limited community acceptance of the PV systems. People regard electricity in their homes as a priority.
7. The majority of school headmasters and teachers do not live near school facilities. Hence they are not in a position to reinforce security at the schools.

Ultimately, provincial governments will rely on the communities to ensure the success of the project in those communities will be partly responsible for recurrent costs of maintenance and upkeep as well as the overall management and security of the systems. This will be a heavy burden for those communities that are already struggling to meet their basic needs.

5. INCREASING THE USE OF THE SYSTEMS

Past experience has shown that there is serious under-utilisation of the non-grid power supply systems (i.e. PV) in schools. This is both a symptom and a consequence of the fact that the users have a low sense of ownership of their systems. Such a situation leads to system faults occurring much earlier than normal or quite simply systems being stolen or vandalised with little reaction from the community.

In principle, the system is designed to allow the use of electricity for lighting for the adult basic educational scheme (ABET), for TV to watch or record the South African Broadcasting Corporation educational programmes (broadcast every week day) and educational video tapes supplied by the Department of Education to schools.

The TAU with input from the implementing partners, is looking for ways to increase general use of the systems. PV power systems could be used for: public phones, a telephone for the school to increase communication with provincial authorities and maintenance contractors, use of the school as community centre for watching sports etc. The PV supply can also be used to power computers and other teaching aids and appliances. Computers with a phone link would also open the way for internet access.

Increasing the use of the systems will to a large extent, depend on the training of the end-users, i.e. school staff and community leaders. So far, training of the end-users is being carried out by installation contractors and by the extension workers. Thus far,

training consists mainly of teaching strict rules about system use (e.g. 4 hours of light per day in 3 to 5 classrooms, 2 hours of TV) and rules about first line maintenance (e.g. cleaning the PV array).

This training has three weak elements. Firstly, it does not follow a well-defined methodology which inevitably leads to gaps and inconsistencies in peoples' knowledge and understanding of the systems. Secondly, there is no standard procedure for evaluating the user knowledge level before or after training, making follow up training more difficult. Finally, the training focuses on providing rules for system use – leaving the users no better informed of the basic functioning and capability of the system. If for example, lighting is used less than 4 hours, users do not know that they can use the TV for much more than 2 hours per day. Indeed, the rules exclude all other uses of electricity and tend to frustrate attempts to increase energy usage and expand the range of applications, while the system can provide power for many other appliances which will be beneficial in the context both of the school and the community.

The TAU is in the process of developing a unique training package to train end-users, including a built-in evaluation methodology, based on similar methods used in other African contexts. This package will be used by installation and maintenance contractors, extension workers, Eskom commissioning officers and supervision officials of the Department of Education. This tool will ensure that the knowledge of users will be sufficient for them to use the systems to their fullest capacity, as well as allow for easy verification of users knowledge. The training will teach all stakeholders to decide for themselves whether they can use a system or not, and for what length of time depending on the appliance in use. Other possible apparatus could be: AM/FM radios, microscopes, overhead projectors, computers and printers, walkie-talkie chargers, alarms for the school etc.

6. REACHING THE COMMUNITY

In previous schools electrification programmes, whether grid or non-grid, the needs of the systems users or communities were poorly integrated into the project approach. Often the approach was top-down and driven by the need to meet self-imposed quotas. In order to improve community integration into this project, extension workers have been appointed to help facilitate a sense of ownership of the systems within the community. The extension workers are active before, during and after the installation of the systems. The scope of their work comprises:

1. To carry out needs assessment for power
2. To increase the utilisation of the audio-visual equipment through training designated school

staff on reception & recording of distance learning programmes

3. To support school management and governing bodies in the initial phase of the utilisation of the electricity during and after schools hours.
4. To increase awareness on the potential benefits of the system for the school and the community
5. To support school management and school governing bodies in improving the general security situation at the schools with particular emphasis on theft prevention.

Eight people have been hired full time for this role. Their profile is such that they have been, or still are, associated with the DoE; most are trained teachers. They have the advantage of having lived in the rural areas where they have been assigned and as such have been involved in community development. They are each provided with a vehicle and are responsible for a cluster of approximately 125 schools. They report their activities on a weekly basis to the NGE Eskom regional manager in each region

It is yet too early to measure the impact of the extension workers. Initial feedback however indicates some significant results such as the spontaneous construction of fencing around the PV system or the school by the community and one community which had welded an extra frame on a PV array to prevent further theft! Furthermore, extension workers are able to visit schools that were equipped under the previous RDP projects, and this at a marginal cost. Hence, they are able to gather feedback and report to the NGE Eskom unit on all aspects of the programme to date. Although extension workers received initial training, they will be trained again on various aspects of PV systems use and training of trainers. Finally, their impact is monitored by the project management team.

7. SUSTAINABLE MAINTENANCE

Sustainable maintenance will only be achieved through high quality installation work evaluated with sound commissioning procedures, high standards of maintenance work by systems users, high standards of servicing by technicians and good safe-keeping of the schools by the community. The TAU is working closely with Eskom to ensure high quality installation and servicing work. The TAU is proposing modifications to the original system design to reduce future maintenance costs (e.g. ensuring a higher ratio of switches per luminaries, from 0.1 to 0.5, better weather protection of electronic control gear). Although the present local contractors are able to install the systems in a rather professional way, initial findings indicate that there is a need for the TAU along with Eskom to ensure contractors receive advanced training in system servicing and user

training. Furthermore, the TAU is proposing modifications of the commissioning procedures and to train Eskom field staff accordingly. If reduction of recurrent maintenance costs is an initial objective, it must be complemented by a durable financing mechanism to ensure a budget is allocated for the maintenance. If this is not achieved, maintenance costs can rapidly become rehabilitation costs, which are usually substantially higher. This can happen in the space of a few years. The DoE has limited resources but has committed itself to dedicate what is necessary for the PV supply.

The beneficiaries of the on-going programmes include 400 000 schools pupils per year, 12 000 teachers and 50 000 to 100 000 adult learners. In order to have the same targeted beneficiaries every year, school systems need to be functional. The TAU and all stakeholders are focusing their efforts to ensure sustainable maintenance. Such a strategy will ensure that these systems remain viable and contribute to the Department of Minerals and Energy's efforts to integrate the electrification of schools (and clinics) into the newly defined National Electrification Plan (grid and non-grid). This move will give further insurance that community systems will be part of the global maintenance schemes of hundred of thousands of SHS planned to be installed and serviced by regional 'utilities' in the coming years in South Africa.

8. CONCLUSION

It is anticipated that the effect of the EU intervention along with the Technical Assistance Unit and Eskom will reach well beyond the installation of 1 000 schools. It should provide guidelines and implementation strategies and standards for the continuation of the non-grid electrification of schools programme. 15 000 schools or thousands of community facilities will need PV electricity in the coming years as the grid will not reach them for at least the next 20 years.

(1)

SURVEY OF STAND-ALONE PHOTOVOLTAIC PROGRAMMES AND APPLICATIONS IN DEVELOPING COUNTRIES

J. R. Bates and A. R. Wilshaw

IT Power Ltd, The Warren, Bramshill Road, Eversley, Hampshire, United Kingdom, RG27 0PR. Tel +44 (0) 118 973 0073, Fax +44 (0) 118 973 0820, e-mail: jrb@itpower.co.uk

ABSTRACT:

This paper details the key findings of a survey of PV programmes and applications in 21 developing countries undertaken as part of IEA PVPS Task III (Stand-alone and Island Applications of Photovoltaic Power Systems)[1]. The paper presents demographic and geographic data for each of the countries in the survey, including literacy rates, poverty levels - as given by the population on an income of less than 1 USD per day - and GDP per capita. Data on the status of national electricity generating capacity and electricity production was collected and summarised, as well as data on the population in each country without access to electricity services. Data on PV programmes and markets were also collected and summarised. All of the countries included in the survey had some experience with stand-alone photovoltaic power systems. For many of the countries, this was largely for social, health or educational applications, often funded by international agencies and installed as demonstration or pilot programmes. Most of the countries had plans with regard to the electrification of rural areas. A number of countries had policies and targets for rural electrification that explicitly included reference to the use of PV and/or other renewable energy sources. A summary of existing programmes on rural electrification and PV and the status of import duties and tariffs in the surveyed countries is given in the main report. One of the common features among the survey countries was the lack of finance available for the purchase of PV systems, either through cash sales or through affordable credit. This was especially problematic in rural areas, where the population was often reliant upon subsistence agriculture and informal employment. As this demographic group represented the largest market for stand-alone photovoltaic power systems, the problem of finance needs to be addressed in order to develop the potential market.

Keywords: Developing countries - 1: Stand-alone PV systems- 2: Rural electrification – 3

1. INTRODUCTION

This paper presents the results from a survey undertaken as part of the collaborative work of Task III (Stand-alone and Island Applications of Photovoltaic Power Systems) of the International Energy Agency's PVPS Programme. As a part of this collaborative work, a survey of PV programmes and applications in 21 developing countries was undertaken by the authors with assistance from the experts of the Task III participating countries.

This paper aims to provide a basic understanding of the state of the photovoltaic market in developing regions of the world at the end of 1996, and to highlight the perceived market barriers to the accelerated implementation of photovoltaic power systems. The key demographic and geographic data for each of the countries in the survey are presented in Table 1. These data varied very widely between the countries surveyed indicating the wide disparity of conditions in the various countries: land areas ranged from that of Brazil, encompassing 8 460 000 km^2 to Tuvalu, occupying 26 km^2. Population densities ranged from less than 2 people per km^2 in Mongolia to 380 people per km^2 in Tuvalu. Literacy rates ranged from an estimated 100 % in French Polynesia to 24 % in Ethiopia. Urban populations ranged from an estimated 79 % in Brazil to 12 % in the Cook Islands.

Poverty levels, as given by the population on an income of less than 1 USD per day, ranged from a high of 69.3 % in Uganda to less than 2 % in Morocco and Thailand. Consumer price indices ranged from over 640 % in Brazil (this figure has reduced significantly since 1996) to 1.5 %

in French Polynesia. GDP per capita in the surveyed countries ranged from 103 USD in Ethiopia to 7 554 USD in French Polynesia.

2. ELECTRIFICATION STATUS

Many of the countries had large, dispersed rural populations without access to conventional, grid or diesel based, electricity services. A reasonable correlation between per capita electricity production and GDP per capita can be seen from Figure 1. [The data point for the high electricity production per capita is for South Africa, which exports electricity to neighbouring countries. The data point for the high GDP per capita is for French Polynesia which receives significant amounts of French bilateral aid.]

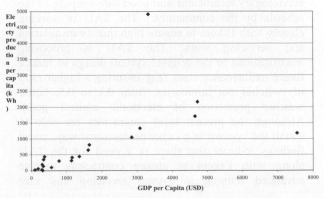

Figure 1: Electricity production per capita versus GDP

Table 1: Population details of surveyed countries

Country	Population (x10³)	Land area (km²)	Population density (inhabs per km²)	Population growth rate (1980-96)	Urban population (1996)	Literacy (%)
BRA	161 000	8 460 000	19.0	1.8 %	79 %	83
COK	20	240	83.3	1.08 % (1995)	NA	80
DOM	8 000	48 000	166.7	2.1 %	63 %	83
ETH	58 000	1 100 000	52.7	2.7 %	16 %	24
PYF	233	3 660	64.7	1.89 % (1995)	NA	100
GHA	18 000	228 000	78.9	3.1 %	36 %	60
IND	945 000	3 287 000	287.5	2.0 %	27 %	52
IDN	197 000	1 812 000	108.7	1.8 %	36 %	77
KEN	27 000	569 000	47.5	3.1 %	30 %	69
MYS	21 000	329 000	63.8	2.5 %	54 %	78
MNG	3 000	1 570 000	1.9	2.6 %	61 %	83
MAR	27 000	446 000	60.5	2.1 %	53 %	35
NAM	2 000	823 000	2.4	2.7 %	37 %	38
PHL	72 000	298 000	241.6	2.5 %	55 %	95
SEN	9 000	193 000	46.6	2.7 %	44 %	38
ZAF	38 000	1 220 000	31.1	2.0 %	50 %	76
TZA	30 000	884 000	33.9	3.1 %	25 %	46
THA	60 000	511 000	117.4	1.6 %	20 %	94
TUV	10	26	380.8	1.45 % (1995)	NA	NA
UGA	20 000	200 000	100.0	3.5 %	13 %	48
VNM	75 000	325 000	230.8	2.1 %	19 %	94

Table 2 provides the estimated electricity generation capacity, annual electricity generation, and per capita generation figures for the survey nations. Again, the per capita estimations must be taken in the light of the fact that there were a wide disparity of living conditions within a given country.

In an effort to develop a better understanding of the circumstances in the rural areas, estimates of the percentage of the population without access to electricity were collated. The figures provided were compared with the estimated labour force and estimated percentage of the labour force engaged in agriculture. Figures are shown in Table 3.

Table 2: Estimated electrical capacity, and generation

Country	Electricity Generation Capacity 1995 (GW)	Electricity Generation 1995 (GWh)	Electricity Generation per capita (kWh/capita)
BRA	63.77	275 000	1 711
COK	0.014	21	420
DOM	2.28	6 500	813
ETH	0.33	1 300	22
PYF	0.075	275	1 250
GHA	1.8	6 200	344
IND	81.2	415 000	439
IDN	11.6	61 200	311
KEN	0.73	3 700	137
MYS	8	45 500	2 167
MNG	1.25	NA	NA
MAR	2.4	12 000	444
NAM	0.406	994	925
PHL		29 700	413
SEN	0.215	900	100
ZAF	46	187 000	4 916
TZA	0.405	1 800	60
THA	10	80 100	1 335
TUV	0.0026	3	306
UGA	0.2	610	31
VNM		14 400	192

Table 3: Extent of rural electrification and employment

Country	Labour Force (1996)'000s	Agriculture (%)	Unelectrified (%)
BRA	72 000	23	12
COK	6	NA	NA
DOM	3 000	25	NA
ETH	26 000	86	90
PYF	76	15	12
GHA	8 000	59	60
IND	418 000	64	68
IDN	91 000	55	70
KEN	13 000	80	90
MYS	8 000	27	19
MNG	1 000	32	60
MAR	11 000	45	75
NAM	1 000	49	90
PHL	30 000	45	45
SEN	4 000	77	75
ZAF	15 000	14	33
TZA	16 000	84	96
THA	30 000	40	27
TUV	NA	NA	NA
UGA	10 000	84	95
VNM	38 000	71	80

From Table 3, it can be seen that a large proportion of the workforce in many of the countries was involved in agriculture. The correlation between the population engaged in agriculture and lacking access to electricity appeared to be clear, as can be seen from Figure 2. It can be assumed that the rural population made up the majority of both statistics.

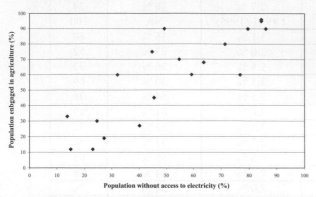

Figure 2: Population without electricity versus population in agriculture

There was also a correlation between the unelectrified population and GDP per capita as can be seen from Figure 3 suggesting that the poorer countries were more likely to have larger, unelectrified populations working in agriculture.

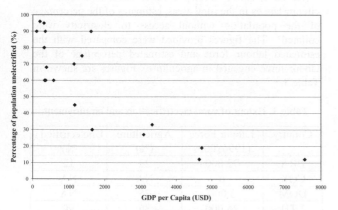

Figure 3: Population without electricity versus GDP per capita.

Furthermore, from Figure 4, a loose correlation between the unelectrified population and the urban population was apparent, providing further evidence that unelectrified populations tended to be those in rural areas.

3. PV PROGRAMME EXPERIENCES

All of the countries included in the survey had some experience with stand-alone photovoltaic power systems. For many of the countries, this was largely for social / health / educational applications, often funded by international agencies and installed as demonstration or pilot programmes. However, a number of countries had developed a commercial PV industry independently of large scale aid projects. Particularly important from this perspective were the industries in Kenya, the Dominican Republic and Namibia.

The installed PV power as of the end of 1996 is shown for each of the surveyed countries in Table 4. It was estimated that the total PV power installed in the surveyed countries was in the region of 53 MWp with over half of this installed in India. Figure 5 shows the installed power per capita. For reasons of scale, the data for Tuvalu have not been included in the Figure as it equates to 5.05 Wp/capita.

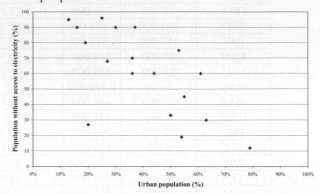

Figure 4: Unelectrified population versus urban population.

The countries with the highest total peak power installed at the end of 1996 were Brazil, Kenya, India, South Africa and Thailand, which accounted for 47 MWp between them.

Table 4: Estimated installed PV power.

Country	PV Power installed at the end of 1996 (kW$_p$)	Country	PV Power installed at the end of 1996 (kW$_p$)
BRA	2 000	MAR	1 000
COK	NA	NAM	800
DOM	225	PHL	133
ETH	NA	SEN	800
PYF	NA	ZAF	5 500
GHA	350	TZA	NA
IND	35 000	THA	2 500
IDN	1 800	TUV	50
KEN	2 000	UGA	150
MYS	640	VNM	100
MNG	80	**TOTAL**	53 078

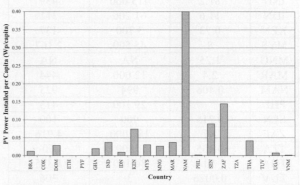

Figure 5: Installed PV power per capita in the surveyed countries

In India, PV had been used in many applications including water pumping, street lighting systems, solar home systems and solar lanterns as well as large scale (>100 kWp) PV power plants and telecommunications. PV in South Africa had been largely used in solar home systems, water pumping, electrification of schools and clinics as well as in professional applications such as telecommunications. In Thailand, PV had been installed for village battery charging stations, schools and clinics and water pumping applications. In Brazil, the first PV applications were for telecommunication relay stations but since 1992, PV had been used in solar home systems, schools and water pumping. In Kenya, PV had been largely used for solar home systems and solar lanterns, almost entirely in the private sector.

From Figure 5 it can be seen that the countries with the highest per capita figures were Namibia, South Africa, Tuvalu, Kenya, Senegal and Morocco. It is interesting to note that the governments of these countries were actively supportive of PV and, with the exception of Senegal, had an active private sector PV industry. In Morocco, Senegal and Tuvalu, large PV programmes funded by bilateral agencies had been implemented and these had been generally successful in ensuring continued government support. In Namibia and Thailand the PV industry had developed mainly through the active support of the national governments and in the absence of large aid programmes.

Table 5 shows the uses of stand-alone photovoltaic power systems in the survey countries. It must be noted that as information was often difficult to obtain there may be examples of particular applications that are not listed for a particular country. Every effort was made to ensure the accuracy and completeness of the data although there were inevitably areas where data were lacking.

Table 5: PV applications in the surveyed countries

Country	Domestic	Schools	Clinics	Water	Telecom
BRA	✓	✓	✓	✓	✓
COK	✓		✓		✓
DOM	✓				
ETH			✓	✓	✓
PYF	✓				
GHA	✓				✓
IND	✓	✓	✓	✓	✓
IDN	✓		✓	✓	
KEN	✓		✓	✓	
MYS	✓	✓	✓	✓	✓
MNG	✓				✓
MAR	✓	✓		✓	
NAM	✓	✓	✓	✓	✓
PHL	✓	✓	✓	✓	
SEN	✓		✓	✓	
ZAF	✓	✓	✓	✓	✓
TZA	✓		✓	✓	✓
THA	✓	✓	✓	✓	
TUV	✓				✓
UGA	✓		✓	✓	✓
VNM	✓		✓		✓

The use of PV water pumping and other agricultural applications had been piloted widely, as this was seen as a primary market for systems. The private sector purchase of systems for agricultural use was limited by the expense involved. Even as a community co-operative venture, the capital accumulated from subsistence agriculture was generally not enough to provide the down-payments necessary for developing a photovoltaic powered irrigation system.

Other agricultural applications, such as solar-powered electric fences were also being piloted, especially in areas of Brazil where animal husbandry was widely practised.

In the Pacific islands of French Polynesia and the Cook Islands, economic activity based upon photovoltaic electrification included pearl farming and 'eco-hotels', catering to environmentally conscious tourists.

Domestic lighting appeared to be the primary use of photovoltaic systems purchased in the private sector, mainly because these systems were the least expensive.

In urban areas, which were relatively affluent but had an unstable power distribution system, such as the Dominican Republic, photovoltaic power systems were used for back up power supplies.

4   CONCLUSIONS

Many of the surveyed countries had experience of stand-alone PV systems for remote service applications such as telecommunications, railway signal and switching devices, television relay stations, and coastal navigation devices. This market was largely a commercial market and operated without the need for direct subsidies. Solar Home Systems (SHS), vaccine refrigerators, school electrification, public street lighting, water pumping and desalination plants can all be considered to bring significant social benefits and represent a huge potential market for the PV industry. Many of the countries in this survey had experience with one or more of these applications, although they were often funded through multi-lateral or bilateral aid programmes.

The financing of photovoltaic systems was an issue to be resolved in each of the 21 countries surveyed. This was perhaps inevitable due to the low income levels of much of the population in these countries. However, the problems were not insurmountable and some of the countries had approached the problem of finance with some degree of success. Indonesia, India and Tuvalu each had some success at providing finance to end users. The Dominican Republic, Kenya and Namibia had also had considerable success in providing PV systems on a market basis in the absence of any government subsidies.

REFERENCES

[1] *Survey of Stand-alone Photovoltaic Programmes and Applications in Developing Countries in 1996,*
J. R. Bates, A. R. Wilshaw, IEA PVPS Task III Report.
September 1999.

---

[1] i Countries surveyed and corresponding ISO Country code:
Brazil (BRA), Cook Islands (COK), Dominican epublic (DOM),
Ethiopia (ETH), French Polynesia (PYF), Ghana (GHA), India
(IND), Indonesia (IDN), Kenya (KEN), Malaysia (MYS),
Mongolia (MNG), Morocco (MAR), Namibia (NAM), Philippines
(PHL), Senegal (SEN), South Africa (ZAF), Tanzania (TZA),
Thailand (THA), Tuvalu (TUV), Uganda (UGA), Vietnam
(VNM).

STRATEGY FOR REDUCTION OF WASTE FROM SOLAR HOME SYSTEMS

J. A. Verschelling[1], Dr. M. Takada[2] and T.C.J. Van der Weiden[1]

1. Ecofys Energy and Environment, P.O. Box 8404 NL-3503 RK Utrecht, ph/fax: +31 30 2808392/01 j.verschelling@ecofys.nl.
2. United Nations Development Programme, Energy and Atmosphere Programme, 304E 45th St, New York, NY 10017, USA, Ph/fax +1 212 906 5155/5148, minoru.takada@undp.org.

ABSTRACT:
Solar Home Systems (SHSs) are expected to provide a least-cost option for rural electrification and could supplement grid based electrification policies. Large scale application of SHSs is foreseen in the near future. However, SHSs contain substances that can be harmful to the environment and human health such as lead, mercury and other heavy metals, sulphuric acids and PVC. The issue of pollution caused by the waste of SHSs has not drawn much attention. In this paper, a strategy based on prevention, reduction and recollection and recycling is discussed to address the issue.
Keywords: Stand-alone PV-systems -1 Environmental Effects -2 Waste -3

1. BACKGROUND

Approx. 2 billion people live without access to modern reliable energy services and a large share of them without the perspective of a short term connection to the electricity grid. Solar Home Systems (SHS) are expected to provide a least-cost option for rural electrification and could supplement grid based electrification policies (World Bank, 1997).

Early 1999, approx. 1 Million SHSs were in use (ECN, 2000), and further rapid market growth is expected in developing countries. Multilateral development organisations are developing and testing new approaches for projects designed to develop markets for SHSs and to overcome the barriers to their widespread and accelerated dissemination. PV industry, donors, policymakers, NGO's and the private sector concentrate on the project financing, formulation and implementation issues. So far, there has been little attention for the post implementation phase and what should happen when (parts of) SHSs are disposed off.

2. SOLAR HOME SYSTEM WASTE

At present, the whole issue of the recycling of PV modules and other components, including design options, waste collection schemes, processing methods and environmental and cost impacts are still to be fully known (IEA, 1997). Among various PV applications, SHSs require special attention on waste issues for two reasons. First, SHSs market is growing where institutional bases are relatively weak. Second, BOS components for SHSs have relatively short life-time, vulnerable to mishandling, and as such tend to be a source of waste.

A typical SHS in use in developing countries consists of the following basic components: a PV module, a battery, a charge controller and one or more lamps, wires, connection and switch gear. All of these components have different life times, dependent on their quality, intensity of use and treatment during storage, transportation, installation, operation and maintenance. An example of the lifetimes of SHSs components is given in table 1.

There are two main issues associated with SHS waste. First, some of the components of the SHSs contain substances that can be harmful to the environment and human health such as lead, mercury and other heavy metals, acids and PVC. Both the extent of the potential spread of waste from SHSs into the environment and its

Component	Main substances	Approx. weight [kg]	Approx. lifetime [yr]	Approx. Waste after life [kg]
Module (Si, 40Wp)	Glass, Aluminium, Plastics, Silicon, Lead	5	20	5
Battery (45 Ah)	Lead, Sulphuric Acids, Plastics	15	3	100
Controller	Electronic Components, Plastics and Metals	0.5	10	1
Lamp armatures (2)	Electronic components, Plastics and Metals	0.3	10	0.6
Fluorescent lamps (2)	Glass, metals, Plastics, Mercury, Phosphore	0.25	1	5
Cables	Copper, Plastics	1.2	20	1.2

Tabel 1: Indication of waste of a 40 Wp Si SHS with a 20 year lifetime.

effects are hard to estimate and (will) remain largely unknown as the facilities for early detection are generally not available. As the 'incubation time' of the waste ranges from approx. 1 to 20 years after installation of the SHS, its scale may be unnoticed for some time before the effects of large scale projects show up. Absence of a legal framework in many developing countries or the capacity to enforce it, combined with the presence of more acute priorities, make this problem more difficult to address.

Second, attention is required for BOS components, which represent large fraction of waste in SHSs due to their shorter lifetime.

Given the fact that some kind of recycling (rather: down cycling) process already has been developed (BP solar 1994) for crystalline silicon PV modules (which accounted for 80% of the world production of PV cells in 1999 (PV news, 2000)) and that the battery and most of the other components of the SHSs can technically be processed after useful life, it is important to give attention to waste recollection and aspects of costs.

3.   A STRATEGY TO REDUCE WASTE

One important way to address the waste problem is to put appropriate product specification in place at the design stage. A case study (Ecofys, 1998) shows that good design can significantly reduce the amount of waste released at the end of the life cycle of a PV generator. Further research into design and waste of SHSs and an inventory of the toxic materials, the amounts and impacts should be performed.

Secondly, mechanisms for the recollection of SHS waste from developing countries markets should be evaluated. Following this, a framework should be put in place to promote the principles of Prevention, Reduction and Recollection & Recycling.

3.1.   Preventing harmful waste:

There is a need to prevent the use of chemicals and processes that will lead to hazardous waste at the end of the product lifecycle. Recently several efforts have been made, for example, water based pastes instead of pastes based on organic solvents for screen printing. International co-operation is needed to promote exchange of information and technology transfer on these aspects.

3.2.   Increasing the product's life time:

As an example, table 1 shows the amount of waste that is produced during the 20 year life of a 40Wp SHS. As can be seen from the table, the batteries contribute to the largest (physical) amount of waste. It is logical to start addressing the battery waste issue while the other parts of the SHSs are further investigated.

In larger SHSs projects the application of dedicated components (such as deep cycle batteries instead of car batteries) may be feasible. If the battery life can be extended from 3 to 5 years the amount of battery waste after life will be reduced accordingly by 40%.

In order to help extend the life time of SHS, establishment of appropriate standards and codes is needed both for SHS components as well as a SHS as a whole, which is being pursued by IEC and PV GAP. Furthermore, efforts are necessary for ensuring high quality products. These steps should include:

- Improved project and SHS system design
- Training for manufactures to improve quality in the design and manufacturing process
- Training for test laboratories
- Training for better quality installation and maintenance

The improvement on life-time of batteries will also have positive impacts on the Energy Pay Back Time (EPBT) because BOS components typically contribute about 50% of the EPBT for SHSs (Alsema, 1998).

3.3.   Recollection & recycling:

Technical potential and financial aspects of the recycling and processing of used PV modules must be further studied. Strategies for recollection and recycling of SHS materials should:

- provide a market mechanism based structure that links in with other waste flows such as that of car batteries, other electronic waste (such as old TV sets and radios), etc.
- call for capacity building for the central and local governments and the private sector
- provide feedback to the designers of SHS components and SHS projects.
- Set up a structure to incorporate issues of decommissioning as an integral part of training programmes.

3.4 Linking to international initiatives:

Most PV market development initiatives in developing countries are, more or less, linked to initiatives supported by international organisations. For example, World Bank/GEF pogrammes are promoting SHSs in many countries, such as 100,000 SHSs in Brazil, 200,000 SHSs each in China and Indonesia (Photon,

Figure 1: Batteries in a SHS.

1999). It is thus strongly suggested that these programmes should look into waste issues in light of the strategies discussed above.

4. CONCLUSIONS

Although SHSs are widely in use in many countries, there has been little attention given to waste problem. SHSs contain harmful chemicals, which could be a threat to people's health, and the local environment, if left entirely unaddressed. More attention is needed for this problem. An approach based on prevention, reduction and recycling is proposed. Since SHS market is cross-boundary, international co-operation is required to effectively address the waste issue.

5. LITERATURE

[1] J.A. Verschelling, K. Burges, "Dismantling of PV generators, a case study" (In Dutch), Ecofys 1998.

[2] M.R. Vervaart, F.D.J. Nieuwenhout, "Solar Home Systems- manual design and modification of Solar Home System components", ECN, 2000.

[3] J.G.M. Kortman, M.G.H. van Kampen "Chain analysis of PV powered and grid connected streetlight for public areas" (In Dutch), IVAM Environmental Research 1998.

[4] M. Schmela, 'Recycling comes later' (in German), Photon, Juli-August 1997.

[5] T. Bruton a.o., 'Recycling of high value, high energy content components of silicon PV modules' BP Solar, 1994.

[6] Evert Nieuwlaar and Erik Alsema, Environmental Aspects of PV Power Systems, IEA PVPS Task 1 Workshop, IEA, 1997.

[7] E.A. Alsema, P. Frank, and K.Kato, "Energy Pay-Back Time of Photovoltaic Energy Systems: Present Status and Prospects", Proceedings of the 2nd World Conference and Exhibition on Photovoltaic solar Energy Conversion, July, 1998.

[8] Photon (1), 1999.

A SIMPLIFIED ONE-AXIS SOLAR TRACKER FOR PV SYSTEMS FOR DEVELOPING COUNTRIES

Stefano Bechis, Pietro Piccarolo
DIPARTIMENTO DI ECONOMIA E INGEGNERIA AGRARIA, FORESTALE E AMBIENTALE
UNIVERSITÀ DEGLI STUDI DI TORINO
Sezione di meccanica via Leonardo da Vinci 44, 10095 Grugliasco (TO) – ITALY
tel. +39 011 6708589 fax +39 011 6708591 e-mail: bechis@agraria.unito.it

ABSTRACT:

A simple solar tracker has being developed, which is working on the principle of the solar clock. This tracker is designed to be directly built by the rural blacksmiths in developing areas. The design and the materials used for the tracker take into account the technical and equipment level of the average Sahelian blacksmith. This solar tracker should be adapted for the requirements of the micro enterprises that charge batteries for rural home power systems. This research is being carried out in the framework of a project of cooperation between the Faculty of Agricultural Science of Università di Torino and CNES, SONIES and PROFORMAR in Niger, with the cofinancement of Regione Piemonte.

Keywords: Tracking – 1: Developing Countries – 2: Solar Home System – 3

1. FOREWORD

The purpose of this research has been the assessment of a simple support for a PV panel, able to track the sun.

The solar tracker is designed for use in developing areas which can be built directly by local, rural blacksmiths. It works on the principle of a water clock.

The final users of this system should be the small enterprises that charge batteries for people living in villages without electricity.

For this reason, it was decided to provide it with a motion system based on a principle as simple as possible.

At present this system is still under development.

The name of this system is 'Girasole', the Italian for 'Sunflower'.

2. HOW THE SYSTEM WORKS

The mechanism that moves the support, in the present layout, is based on a water clock, in the future, it is proposed to study a solution that will use sand instead of water (sand-glass system).

In this tracker, the panel is fixed to two pivots, an upper and lower one (Fig.1).

Around the upper pivot, there is a pulley that is moved eastward or westward by a small rope.

The rope is linked to the top of two divergent bars that are fixed on a balance that is placed in the lower position (Fig.2).

When the balance is moved up and down (like a see-saw), the panel turns in the two directions.

The 'automatic' movement to the balance is given by a system that consists of a chain linked to one end of the balance, and a water tank fixed to the other end.

Figure 1: The Girasole support equipped with a 50W PV panel. In the middle of the figure the balance is visible with the tank on the left and the chain on the right.

Figure 2: Detail of the motion system. In lower position the hinge that fixes the bar to the rest of the frame. The small rope that links the top of the two bars making a turn around the pulley is visible.

The weight of the water tank and of the chain determine the position of the bar (and consequently of the panel).

As the tank continuously lets droplets fall down (at a rate between 34 and 42 drops per minute) its weight becomes lighter as the time goes by. The chain has a fixed weight per meter of length, and pulls the bar down a specific unit of length for a specific weight loss of the tank.

When the tank is full, the arm where it is fixed (left in Fig.1) is in the lower position, while the arm that holds the chain is in upper position and the panel is oriented towards East.

As the tank loses water droplets and the weight of the chain pulls the right arm down, the tank is lifted up, the two bars fixed to the balance move towards the chain (towards right in Fig.1) and the right bar pulls the small rope, thus making the pulley and the panel rotate.

This slow movement continues for all the day long, and by sunset time the panel looks West.

3. TESTS CARRIED OUT

The tests on Girasole have been aimed at two main objectives:
1. to assess the correct speed of rotation of the panel
2. to assess the quantity of energy received from a PV panel on Girasole, in comparison with a fixed PV panel

The first objective has been reached testing chains of different weights (and also springs, in a different design) with different droplet rates. Different maximum turning angles of the balance have been tested, as well as different lengths of the arms.

The second objective is being reached through comparative tests, using in parallel a PV panel mounted on Girasole and a PV panel mounted on a fixed support. In these tests the following parameters were measured:
- the incident radiation on the turning panel
- the incident radiation on a fixed panel (control)
- the currents and voltages produced by the two panels.

In this paper the results are given in term of incident radiation.

The tests have been carried out in Torino, Italy, at about 45 degrees of latitude North and 8 degrees longitude East.

During the tests the two panels have been oriented exactly in the direction of the Equator (South), for the turning panel it has been oriented in such a way to face South at the solar noon, and with an angle of 45 degrees to the horizontal.

4. TEST RESULTS

The first tests (April 2000) indicate that on an average day, with solar radiation up to 800 W/m^2 on the horizontal, with some clouds in the morning, the system is able to provide 27.8% more solar radiation on the rotating panel (Fig.3 and Table I).

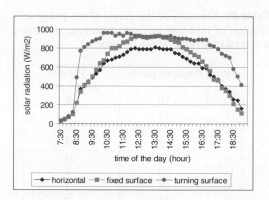

Figure 3: Behaviour of solar radiation on the horizontal, on the surface of the PV panel mounted on Girasole, and on the surface of the fixed PV panel.

The water clock still needs to be adjusted at least once during the day, because as the head in the tank changes, the rate of droplets becomes smaller, with a relative diminution in the number of drops per minute within 8% and 10% every hour.

Further developments of the system will try to eliminate this problem.

Table I: radiation incident on the fixed PV panel and on the rotating panel during a day (W/m^2)

Fixed panel	Rotating panel	Difference
7200	9200	+2000 (+27.8%)

With regards to the rotating speed, the system moves with small gaps, because every time it moves it has to overcome the static friction of the elements. In the Table II the measured angles of the pulley are reported. The starting angle of this test was 10 degrees and the arrive angle was limited to 153 degrees. At 90 degrees, the PV panel surface is exactly oriented South. In this tests this occurs between 13:30 and 13:45 legal hour, that is exactly the time when the solar noon for this location.

5. THE COST

The cost to build this mobile tracker is estimated being between 50% and 100% more expensive than a mobile support. All that is needed in addition to a fixed solution are a water tank, a chain, a pulley, a hinge and some iron bars. A skilled blacksmith should put together the parts in about one day of work.

The material to make the wooden mobile prototype costed in Italy about 25 Euros more than the material for the fixed support (control).

Table II: Variation of the Angle of the pulley as a function of time and difference between successive angles (displacement of the PV panel)

hour	Angle of pulley	Angle of displacement
8:00	10	0
8:15	13	3
8:30	13	0
8:45	15	2
9:00	19	4
9:15	20	1
9:30	23	3
9:45	25	2
10:00	27	2
10:15	30	3
10:30	35	5
10:45	36	1
11:00	40	4
11:15	45	5
11:30	47	2
11:45	53	6
12:00	58	5
12:15	61	3
12:30	65	4
12:45	74	9
13:00	77	3
13:15	82	5
13:30	84	2
13:45	92	8
14:00	95	3
14:15	100	5
14:30	101	1
14:45	106	5
15:00	109	3
15:15	111	2
15:30	121	10
15:45	121	0
16:00	123	2
16:15	130	7
16:30	136	6
16:45	141	5
17:00	145	4
17:15	148	3
17:30	152	4
17:45	153	1

6. FINAL REMARKS

A new type of solar tracker has been built. It works on a very simple type of clockwork motion system.

In the first tests it has allowed, in comparison with a fixed support, the collection of 27.8% more energy.

Some of the working parameters still have to be fixed at the moment. At this stage the system is under the form of wooden prototype.

The cost of this tracker is estimated to be low.

After the system will be optimised technical drawings will be made and distributed, to allow the construction in those countries where electric energy is still not available in large parts of rural areas.

The tracker is addressed mainly to the small enterprises that charge batteries for Single Home Power Systems (SHPS).

IMPROVED DESIGNS FOR SOLAR RECHARGEABLE LANTERNS AND THEIR DEVELOPMENT AND MARKETING IN DEVELOPING COUNTRIES

Kieron Crawley, Ray Holland, Stephen Gitonga

Intermediate Technology Consultants, Bourton Hall, Bourton-on-Dunsmore, Rugby, Warwickshire, CV23 9QZ
Tel.: +44 1788 661187, Fax.: +44 1788 661105, E-mail: kieronc@itdg.org.uk, Internet: www.itcltd.com/solar

ABSTRACT: Solar Home Systems of 20-50 Watts are not affordable for the majority of rural people in Africa, nor in many countries in Asia and Latin America. In Kenya, for example, 80 % of rural people possess kerosene hurricane lamps and spend between $3 and $8 per month on kerosene plus batteries for torches and radios, and candles. Battery charging is an increasingly common service in towns. For most of these homes a small power source for lighting using modern compact fluorescent lamps, power for a radio and possibly other electronic appliances would do much to improve living conditions if it was affordable. This paper describes the process of market research and the trade-offs during product development and testing to produce a lantern that is affordable and reliable. The requirements of the battery, panel, charge regulator and lighting circuit are described and the critical process of mass marketing in rural areas is described.
Keywords: Developing Countries – 1: Stand-alone PV systems – 2 : PV Market - 3

1. BACKGROUND

1.1 In many parts of Africa people who live in rural areas have no access to electricity.

Despite ambitious government plans, the constraints imposed by a scarcity of financial resources and the sheer practical difficulties of transmitting and distributing grid electricity over huge areas means that most people face the prospect of going without a connection for many years to come. As a result, most families are forced to rely on candles or Kerosene lamps to provide basic lighting in their homes. In Kenya 96% of householders use kerosene for lighting, while 70% also spend significant amounts of hard earned cash on dry-cell batteries for torches.

Successive studies have highlighted the potential for decentralised supplies of power for lighting at both community and household level, and advances in Photo-voltaic technology has resulted in the steady growth of sales in Solar Home systems over the last few years. Unfortunately the cost of installing even a moderate Solar Home System puts it out of reach of the majority of rural families in developing countries.

A recent World Bank survey carried out in Kenya pointed to the potential for solar rechargeable lanterns as a low cost and flexible lighting option for large sections of the rural community. The project identified seven existing lantern designs considered to be appropriate to the African environment and used between thirty and sixty of each (all imported to Kenya) to test customer demand for the products and to collect feedback on the technical performance of the samples. While the study demonstrated that there was a real demand for Solar Lanterns, customers highlighted a number of technical shortcomings with all of the products tested. Most of these shortcomings related to the poor construction of the lanterns, the quality of light and the relatively sharp drop off in performance after a period of months of use.

1.2 Why has an effective Lantern not been developed by the Private Sector in developing countries?

For manufacturing companies in developing countries, new product development for local markets is expensive and risky. This is particularly true when the product is targeted at rural mass markets where effective marketing and distribution techniques are still unproven.

Not only may the company have poor access to technical know-how, but the long term capital investment required for mass production tooling makes it an unattractive option compared to products for smaller but more accessible markets amongst higher income groups.

On a global scale however, the potential for solar lantern products is huge (2 billion people currently without access to electricity) and the potential for a commercially viable product would seem to be great.

As a result of this analysis, Intermediate Technology Consultants were able to secure funding from the Department for International Development (DFID) for a project to develop an improved lantern for use in rural households in developing countries. The main activities of the project were outlined as follows;

- using customer information as a starting point and working together with local manufacturers, provide technical "know how" to develop an improved lantern which meets all of the criteria demanded by customers,
- employ appropriate manufacturing and assembly techniques that would allow the product to be manufactured and assembled easily in developing countries,
- incorporate within the project a facility to overcome the constraint associated with capital outlay for mass production tooling,
- provide assistance with local marketing of the product using rural mass marketing techniques currently being developed in countries such as India through organisations such as International development Enterprises (IDE)

As a result the IT Solar Lantern has been designed as a low-cost alternative to a Solar Home System and is intended to allow rural African families to climb the first step on the "energy ladder".

1.3 Structure of the project

The project has been broken down into a number of stages, of which one to three are now complete;
1. Customer Research and product specification
2. Design and Development
3. Prototype production and household testing
4. Selection of manufacturing licensees

5. Tooling and setting up for production
6. Marketing

2. PROGRESS SO FAR

2.1 Customer Research

Market research which is carried out in countries of the south must take careful account of peoples lifestyles and cultural backgrounds. This is particularly important when dealing with communities in rural areas where conventional research techniques can easily fall short and yield inaccurate results.

Bearing this in mind, a number of studies have been carried out in Kenya by project partners Energy Alternatives Africa. These have attempted to identify which aspects of existing Solar Lantern designs are favourable to potential customers.

During the Lantern survey, focus groups were used to measure reactions to particular design details. They were also used to stimulate more general discussions where participants could air their viewpoints and develop their responses. Detailed results from question and answer sessions were recorded on paper while more general feedback from discussion groups (which often contained interesting and unexpected information) was recorded on tape for analysis later.

As well as gathering important design data, a number of interesting lessons have been learnt which may serve as important pointers in future surveys of this type.

The team discovered that survey groups which were well balanced, helped to engender an atmosphere where everyone felt comfortable about having their say. These groups tended to yield more accurate results than those where the discussion was dominated by a few knowledgeable individuals.

The location and time of day for group surveys was also important. Holding a survey after dark might at first seem to be the best way of demonstrating lanterns and their important characteristics to customers. In practice the team found that in rural areas with no street lighting, finding participants (women in particular) who were willing to travel from home at that time was more difficult.

The team also discovered that although participants drawn from slightly higher income groups are easier to access and interview, their spending patterns can differ from those of customers from lower income households at which the new products might be targeted.

2.2 Product Specification

As a result of the initial focus group study the project team were able to come up with a concise description of the "ideal lantern".

The most important features were identified were as follows;

Service characteristics;
- The price of the lantern should be between $75 and $100 if possible
- The lantern should provide light for up to 4 hours each evening.
- Customers should have access to affordable and readily available spares
- Customers expect an overall lifetime of the lantern of 6 years

- Customers expect a 12 months warranty for the product

Design characteristics;
- The lantern should give a 360 degree spread of light,
- The bulb enclosure should allow maximum transmission of light with minimum dispersion effects.
- The carry handle should be sturdy and comfortable.
- The preferred choice of bulb was (5w CFL type)
- The lamp should be portable and weigh no more than 2.5 kg.
- The lantern should be stable with a good base.

Some of the extra features that potential customers expressed a need for were;
- An indicator to show that lamp is charging,
- A warning light to show that the lamp is about to switch off when the battery is low,
- A power socket to allow a small radio to be connected to the unit.

The findings of this initial survey were used to form a design brief and as a result, the team produced a new design for a lantern which incorporates all of these features.

2.3 Manufacturing Options

There are a range of options in terms of manufacturing techniques that are available to the designer today, and the choice of the most appropriate comes down to considerations of scale of production and cost. As a general rule, individual component part costs come down as production levels increase. This is usually accompanied however by a higher level of capital investment in tooling.

Injection moulding is a well established and cost effective technology for the mass production of household items in the North, it is becoming increasingly available as an option for manufacturing in developing countries where it is used in the mass production of basic goods such as buckets, basins, tableware and packaging.

High Density Polyethylenes and filled Poly Propylenes are relatively inexpensive and robust materials which can be recycled using simple equipment.

Considering the technical requirements for the Solar Lantern and the projected production quantities, injection moulding was selected as an ideal technology for producing low cost, high quality components with the level of detail required for this product.

2.4 Battery technology

A crucial component for any rechargeable device is the battery. The project activities have included research into available battery technologies to identify a battery which;
- has the capacity to store charge sufficient for the required period of lighting,
- is suitably robust to withstand the heavy duty cycle required for daily charge and discharge,
- requires no customer maintenance (also spill and leak proof),
- has minimum impact to the environment if disposed of at the end of its life cycle,
- could be manufactured locally in the medium term in developing countries,
- provides a cost effective solution.

As a result a Valve Regulated Lead Acid (VRLA) battery with a gel electrolyte has been selected as the battery technology with which to prototype the lantern. Although this is available only through import at present, the project has established that with suitable investment the battery could be produced locally by manufacturers of traditional wet lead acid batteries.

2.5 Design and development - Rapid prototyping

It is vital during the development of any new product to show customers a sample and to listen to their ideas about it. This is especially important if fully-fledged production involves substantial investment in terms of tooling and machinery.

Until a few years ago designers could only produce "block models" of new injection moulded products. These were constructed by hand from wood and plastic and although they had the appearance of the final product they could not normally demonstrate any of the working characteristics. Today, computer aided design (CAD) software combined with rapid prototyping techniques and "soft-tooling" allow the designer to realise his new design in a matter of hours. A computer generated "electronic model" which contains all of the physical information about the size and shape of the new product is fed into a Rapid Prototype machine. The machine uses a filament of plastic and a moving nozzle to lay down successive layers of material (rather like icing on a cake) which are built up to produce a plastic replica of the design. A soft silicon rubber mould, which is created by pouring liquid rubber around the original and allowing it to set, can then be used in turn to produce a small batch of products.

This technique allows the design team to assess the design very quickly and more importantly allows manufactures to obtain feedback from potential customers, all before any significant tooling costs have been incurred. This technique which is not normally available to manufacturers in developing countries has been brought to bear in the project through facilities at Coventry University and has been used to produce a small batch of sample lanterns for use in field trials in rural households in Kenya.

2.6 Prototype production and household testing

At the time of writing, thirty fully working prototype lanterns have been sold to sample households where they have been used for a period of two months. Facilitators will shortly complete a series of households visits and use questionnaires to measure customers reactions to the new design. In addition, selected members from each household will be gathered together to form focus groups where information will be collected through more informal discussion about the lanterns. Information already gathered suggests that the lanterns have been well received and that the provision of thirty prototypes has stimulated demand amongst a large number of other potential customers who have come to hear about the project.

The main selling point of the lantern was identified by the focus groups as being the fact that it was "portable and compact". With further discussion other important features of the lantern compared to those of traditional kerosene lamps were that it was;
- economical,
- gave a better quality of light,
- cleaner,
- faster and easier to operate,
- it does not flicker nor is extinguished with the wind.

It was mentioned by the dealers and customers at Meru and Machakos that the inclusion of a good quality radio in the kit, made the whole package much more attractive to customers.

A few of the recipients in the test areas where lanterns were sold through dealers, already had solar home systems installed. They stated that they preferred to light their homes with the lantern, and use their existing panel to charge and use the other features of the system (most notably TV). They agreed that the portability and versatility of the lantern was an important aspect of why they chose it when they already had a solar home system. This demonstrates that there is also a significant market for lanterns amongst those who already have Solar Home Systems.

3. REMAINING WORK

3.1 Selection of manufacturing licensees - Investors prospectus

During the course of the project it has become apparent that the scope and potential for the Solar Lantern stretches far beyond Kenya. As a result the project geographical focus has been modified to encompass a number of other countries. The objective of this phase of the project will be to identify a number of regional centres where manufacturers are strategically placed to serve a number of high potential countries. Regions that have been identified so far are, Southern Africa, Western Africa, Asia and South America.

The Project team are currently working on developing an Investors Prospectus which will allow any potential manufacturer/distributor to assess the product in terms of the investment needed to commence local assembly and the likely returns in terms of sales to local markets.

Markets for new products in developing countries build up slowly and as a result manufacturers are naturally more cautious when it comes to investing in large production runs. It is likely that the project will facilitate the supply of component parts for lanterns to a number of manufacturers in regional centres. Injection moulding tooling held centrally in Europe or the Far East (funded and owned by the project), will be used initially to supply mouldings in batches to licensed manufacturers who are interested in local assembly. As local markets are developed, tooling could be leased and transferred for local production of component parts.

For local manufacturer this removes a major part of the risk associated with setting up for large scale production and for the project it allows the capital cost associated with tooling to be amortised using royalty payments from licensees in a number of countries.

3.2 Quality control

Quality control is an important issue that needs to be addressed if a new product is to gain a foothold in the market and build reliable reputation. This is no less true in developing countries where the Solar PV sector has suffered as a result of the introduction to the market of poor quality amorphous panels. A product that is produced, assembled and distributed locally in more than one country presents even more of a challenge. By licensing manufacturers to produce and assemble lanterns locally using components and mouldings supplied initially from a central facility, the product holders will have an important tool with which to ensure that all lanterns meet minimum quality standards.

4.   LESSONS LEARNT SO FAR

The project has broken new ground for ITDG in a number of areas and as a result has already produced a number of valuable lessons in development agency/commercial sector partnerships.

4.1   Commercial ownership and intellectual property

To ensure the success of the project, it has been essential at an early stage to involve potential manufacturers in the design and development process. This has thrown up interesting questions concerning commercial ownership and intellectual property rights. Ownership of a commercial product which has had design and development input from a number of different parties is potentially difficult to ascertain and agree upon. In addition, whereas development agencies are keen to disseminate and maximise the impact of their research, private companies have commercial interests which are focussed on maximising profit and protecting their share of the market. This can lead to problems when it comes the time for the technology to be disseminated.

With the Solar Lantern project, this potential hurdle was avoided by ensuring that all detailed technical input which had the potential to lead to an intellectual property asset, was carried out at arms length from the initial project manufacturing partners. These manufacturers although technically having no ownership of the resulting design, will be given first option to manufacture under license when the design process is complete. This ensures recognition of their input during the early stages of the project but provides the project team with the necessary flexibility and autonomy in selecting suitable licensees in other countries.

4.2   Development versus commercial interests

In the development sector the primary concern of agencies is to successfully deliver benefits to target groups (large numbers of poor people). In the commercial sector the bottom line is profit and continued growth of the business. In many instances these forces can work successfully hand in hand, and have real potential for mobilising change on a large scale. In projects which involve private sector/development agency collaboration, it is important however that these two goals are recognised and that there is a clear understanding of where they may cause project activities to diverge.

For products where there is a demonstrated demand across a variety of income level groups (eg. Solar powered lighting) the manufacturer has a choice in terms of pricing. A medium to high priced product will have a small market but will generate a relatively high margin and quick return on investment. These markets are generally easily accessible and are based around large centres of population. The same product, priced at a lower level will result in a lower margin but will give access to a much larger market with greater potential for overall profit. These markets are generally widely dispersed in rural areas and carry a larger risk in terms of return on investment for new products.

Whereas the latter generally contain the bulk of the beneficiaries targeted by development agencies, it is the former, lower risk group that manufacturing businesses are more comfortable in dealing with.

The marketing mechanisms needed to reach these customers are still being developed within countries of the South and the project has identified the need for a component which builds the capacity of local manufacturers to reach large rural mass markets through networks of rural marketing agents.

5.   CONCLUSIONS

Work on the development of the Solar Lantern has generated a considerable amount of interest amongst potential customers and manufacturers alike. It is clear from the experience so far, that the unique position of development agencies such as ITDG can serve to bring together the necessary ingredients and partners for the development of new products for mass production within the commercial sector.

There are still many lessons to be learnt however surrounding issues of ownership and control of such products and the best models for working relationships between the development and commercial sector.

How does it work?

The Solar lantern kit consists of a Photo-Voltaic Panel, and a lantern containing a high efficiency lamp, a rechargeable battery and a charge control circuit. The concept is a simple one – during daytime, sunlight falling onto the Photo-Voltaic Panel generates a small electrical voltage. This is used to charge the Lanterns battery so that the lamp can provide light during darkness.

The charge control circuit housed within the lantern is the "brain" of the unit. Not only does it ensure that the battery is charged and discharged correctly so that it gives a lifetime of maintenance free service, but it can also "decide" to give the battery an extra top-up charge if the panel has gone without its full quota of sunlight for a few days. It's on-board microprocessor will even store information (which can be downloaded later after "interrogation") on how the lantern has been used over a period of time. This information is extremely useful and will help the designers build a picture of how customers use their lanterns. This information will be used to design better lanterns in future.

Figure 1: The ITDG Solar Lantern

FINANCING SHS PROJECTS:
PRE-PROJECT LESSONS FROM THE PVMTI INITIATIVE IN MOROCCO

Jimmy Royer
PVMTI-Morocco

Myriem Touhami
PVMTI-Morocco

Mohamed Aboufirass
RESING

PVMTI-Morocco / Resing office: 17 rue Yougoslavie, No. 13, Guéliz-Marrakech, Morocco
Tel.: (212-4) 43.77.42 / Fax: (212-4) 43.76.25 / e-mail: pvmti@cybernet.net.ma

ABSTRACT: The Photovoltaic Market Transformation Initiative (PVMTI) is an initiative of the International Finance Corporation (IFC) with funding from the Global Environment Facility (GEF) to accelerate the sustainable commercialisation and financial viability of solar photovoltaic-based energy services in India, Kenya and Morocco. In Morocco the PVMTI program has attracted a number of Solar Home Systems (SHS) projects because of the existing favourable conditions for their implementation. This paper presents these conditions and shows how the PVMTI program intends to finance large-scale use of Solar Home Systems while making it a profitable business for dedicated private firms in Morocco.
Keywords: Financing - 1; Rural Electrification - 2; Stand-alone PV Systems - 3

1. BACKGROUND

1.1 The Moroccan Rural Electrification Program

In Morocco, there are about 28,000 douars or villages that are presently not connected to the national grid. A national rural electrification program called PERG (Programme d'électrification rurale global) was set up in 1996 to accelerate the introduction of electricity to these douars. Under this program, the national utility ONE (Office National de l'Electricité) is bringing the grid to rural villages at a pace of about 1 500 villages per year. This program intends to have electrified about 80% of all rural villages in Morocco by the year 2010 [1].

Under the same program, ONE recognises that many douars will not have access to grid electricity because they are too scattered or too remote. Bringing the grid to these villages will just be too costly. In these regions where ONE does not plan to extend the grid in the medium to long term (10 years or more), it has agreed to partially fund an individual PV Solar Home System (SHS) on each household. ONE thus expects to electrify 200 000 households with SHS bringing solar electricity to about 2 millions people in rural regions.

While it is easy to understand that ONE's contribution will make it more affordable for rural dwellers to pay for a SHS, its aim is not really to provide them with less expensive solar systems but, to provide them with an electricity service that is adequate and reasonably priced.

To have good electricity service coverage, at least equivalent in quality to what the grid can provide in rural areas, a SHS must be maintained adequately and replacement of specific components must be provided during its expected lifetime. These tasks not only require a large pool of technicians but also that they be accessible to remote villages in order to execute normal maintenance and respond rapidly to repair demands.

ONE rapidly realised that private firms, distributors and installers of these equipment where in a better position than itself to install and maintain these systems in the field. The presence of a number of experienced firms in Morocco has certainly motivated ONE towards this direction.

1.2 The PVMTI Program

The PVMTI program in Morocco came at the right time. Its aims is to co-finance private sector firms that are positioned to implement important PV projects. The level sought for a PVMTI investment in Morocco is between 500 000 to 1 000 000 USD with a co-financing leverage of 1 to 2 meaning that the smallest project needs to be of 1,5 million USD [2].

In investing on projects of relatively large importance, IFC wants to scale up the level of installations and go from individual projects to large-scale deployment program. In demonstrating that these projects can be good investments, the PVMTI program also wants to serves as a catalyst to get financial institutions involved in the PV field.

In Morocco, because of the existing PERG program and the desire from ONE to use private firms to install and maintain PV SHS, one of the most promising PV applications has been the deployment of SHS for decentralised rural electrification. With the advent of the PVMTI program, ONE is confident that it can successfully direct private firms to take a leading role in electrifying these remote douars.

2. THE REAL PRICE OF A SHS

The price of a solar home system is not limited to just a list price that a retailer will give a prospective buyer. The following points reveals that a SHS can be a lot more expensive than first estimated.

2.1 Purchasing a SHS

A rural dweller usually does not have the cash to buy a SHS upfront. He will need to take a loan or buy his system on credit so that he can spread the price of his equipment purchase over one to three years in recurring instalments that he can more easily afford. The interest charges of such a loan can be important and may add between 10 to 30 % to the original price.

2.2 Installing a SHS

The SHS beneficiary will usually need someone to install his system. This installation is normally not limited to installing the PV module on the roof and connecting it to a battery and a controller. Since the rural user is not familiar with electricity and PV systems, he will also request that his supplier wires the electricity network in his house. In Morocco, most companies include this installation in their list price. This usually adds another 15 to 20% to the PV system price.

2.3 Servicing a SHS

Finally, and more important, the user will need to maintain his system and replace some key components such as the battery and the controller. Since he is usually in a remote rural area, servicing his system can be costly. This price is usually the most neglected one and the most sensitive to the service life and expected lifetime of the system. Left without professional servicing, the life expectancy of a SHS is about 3 to 5 years as is attested by the number of dead systems lying in many households of Morocco and other countries.

Preliminary estimation in Morocco as shown that O&M costs is around 24 USD/year with an additional 15 USD for collection fee if the user is buying on credit or on a fee-for-service scheme. Battery replacement is estimated to be every 3 years and controller replacement every 5 years.

Figure 1 shows the total price of a system for different service periods. Note that all costs shown have been discounted to show net present values. The discount rate used was 5%, which is a low but acceptable rate. Note also that costs shown include a 30% profit margin on equipment and service costs.

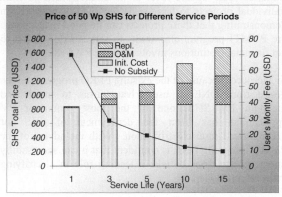

Figure 1: Breakdown of SHS prices for different service periods
(Prices shown are not from one specific firm but represent an aggregate of prices found in Morocco).

A first and obvious conclusion from this graph is that the longer the service period, the more expensive the total price of the system will be. However, on a yearly payment basis, the longer the service life, the cheaper the system. Thus, if a user can pay for his system over the service period provided by an operator, his monthly or yearly payment will be a lot less if the system his serviced for a long period.

What the graph also shows is that after 10 years, the difference in yearly payments is relatively small from year to year. This is mainly due to the fact that the total O&M and replacement costs increase significantly over the years and have taken a large proportion of the total price. At the

same time, the longer the payment period, the more risks that the operator faces on default payment from his client. A good balance between low yearly payments and low default risks would thus be a service period of 10 years.

Table 1 shows the breakdown of a SHS price for a service period of 10 years. Note that all calculations where done with a fixed discount rate of 5% over 10 years. Costs shown are retail prices charged to the user.

Price with	Lifetime	10 years	
Discount Rate	5% (yr-end)	7.722	
Initial SHS Price		867	60%
Total Equip. Price	650 $	650.0	45%
with 3-yrs loan	12% interest	87.0	6%
Installation Cost	130 $	130.0	9%
O&M Cost		301	21%
Operation Cost	15 $/year	115.8	8%
Maintenance Cost	9 $/year	69.5	5%
Collection Cost	15 $/year	115.8	8%
Replacement Cost		278	19%
Battery	91 $/3 yrs	205.2	14%
Controller	52 $/5 yrs	72.7	5%
Total Payment	$	1446	100%

Table 1: Breakdown of SHS price with service period of 10 years.

Table 1 shows that O&M and replacement costs will amount to around 40% of total price after 10 years. This is not cheap and certainly not in the 10-20% range commonly assumed by many PV project managers. In fact, a 50 Wp system that would normally retail for 780 USD (with installation) will have cost the user 1450 USD after 10 years of operation or more than double its original retail price.

3.   THE NEED FOR FINANCING

3.1 Making it More Affordable to the User

Figure 2 shows the yearly payments for the same PV system over the 10 years service period.

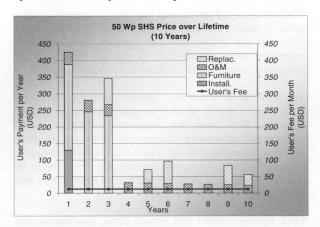

Figure 2 : Breakdown of SHS price with service period of 10 years.

The yearly payments (bars) in fig. 2 shows the importance of financing the user. In most cases, a rural household will not be able to afford the upfront cost of purchasing a PV system, nor will he be able to afford the recurrent cost of replacing a battery or a controller when it is needed. These needs can be covered by two forms of financing :

3.2 Micro-Credit Financing
Micro-credit financing basically permits a rural dweller to obtain a small line of credit over a short period of time so that he can purchase expensive equipment immediately. The main difficulties with micro credit financing lie with the solvency of the user and his access to this type of credit. Most national banks cannot afford to take on such low credit loans. A rural dweller is an unknown entity and to evaluate his solvency would be too expensive for them.

The Grameen Bank in Bangladesh is one successful example of micro financing where a company dedicated to micro-financing, the Grameen Shakti, was created to give loans of small sums of money, between 300 to 500 USD, to its members for the purchase of a SHS [3]. In Morocco, similar financial institutions also exist but they normally finance rural commercial projects and not consumer goods.

In the last few years, other micro-credit financing models for the purchase of SHS have been devised and/or reported extensively [4,5,6]. They range from hire-purchase contract, to peer-group lending, to leasing directly from the retailer or through an independent financial institution.

Most of these financing models address the immediate need of spreading the initial cost of equipment over a few years. Some of these models will take into account the O&M costs needed to service the system over the period of the loan. Very seldom will the models anticipate replacement costs and provision for them before they occur. They usually never take into account the cost of the infrastructure that will be needed to provide low-cost after sale service or replacement of equipment.

3.3 Financing the Service Firm, not the User
Because of its relative sophistication and high initial cost, a PV system cannot usually be purchased through traditional mercantile channels, i.e., from a local village merchant. Furthermore, the large-scale deployment of SHS installations in rural areas usually means that a private firm will have to set-up an infrastructure of regional and local offices from where its technicians will install the SHS and eventually service them. To be efficient, this infrastructure must be organised at the beginning of the project and can bring an unexpected demand for cash at that time.

Additionally, the SHS firm has to provide some sort of financing for replacement parts so that users do not need to wait to have enough cash to pay for the replacement part before they can enjoy their installation again when they fail. Financing of replacement parts should then also be planned from the beginning.

Thus, while many SHS financing schemes have focused in providing micro-financing to rural dwellers so that they can pay for the initial cost of the system, the real financing demand lies not only on the initial investment but also on the infrastructure that will provide both low installation

costs and low service costs of the systems.

Energy Service Companies or ESCOs are particularly well suited to respond to large-scale deployment of SHS systems. An ESCO does not sell a system, it sells a service. It normally retains ownership of the system for the period of service and it is responsible for its complete financing. The user buys the service of using the SHS system for a fixed monthly fee spread over the agreed upon service period (see user's fee line in fig. 2).

The financial requirements of an ESCO are similar to other service companies where large amount of cash is needed during the first few years of operation. For a SHS company this need is typically for the first four or five years when the company is installing SHS systems and building its service infrastructure. Once a sufficient number of systems have been installed and users have started paying their monthly fees, the company starts to make money (fig. 3).

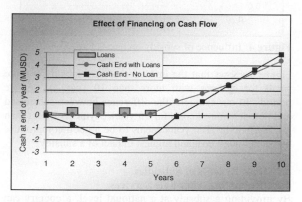

Figure 3: Influence of financing on cash-flow

While ESCOs can deliver low-cost, long-service SHS to rural dwellers, their presence is relatively new and financing them through normal financial institution is still not common. Its main disadvantages are with the high risks of long-term payments and the low solvency of its clientele.

The PVMTI program permits in part to solve the financial requirements of a service company by providing a subordinated debt or a guarantee fund to a financial institution working with this firm. The aim of PVMTI is to show that these firms can be profitable and sustainable in the long-term and are viable investments.

4. THE NEED FOR SUBSIDIES

4.1 Making it More Affordable to the User (bis)
As we have seen above, spreading the price of a SHS over part or all of the service period of the system can alleviate the burden of its high cost. However, it does not make the system cheaper, in fact, financing it increases its overall price. Under a national electrification program, where the aim is to give an electricity service to a majority of the population, subsidies are needed to lower the user's price to an acceptable level.

The Moroccan utility, ONE is providing such a subsidy for SHS through its PERG program. This contribution corresponds to the approximate cost of a module and a battery of a 50 Wp system or about 300 USD. It is substantially lower than the average amount that ONE would normally contribute to connect a rural household to the grid. This contribution must therefore not be viewed as a subsidy but more of a tool to equalise disparities of services given to the Moroccan population. ONE intends to give this contribution on a long-term basis so that those rural users can enjoy a quality electricity service provided by solar energy at a reasonable cost.

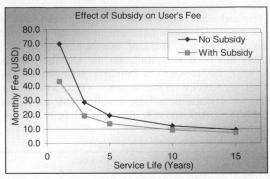

Figure 4: Influence of subsidy on user's fee

Figure 4 shows that the influence of this subsidy is quite small for systems with more than 10 years of service period. In our example, the difference between subsidised and non-subsidised costs for 15 years payments is less than 1 USD/month. Ten years seems to be a good compromise between lowest cost and highest impact of a subsidy.

4.2 Other Benefits of Subsidies
4.2.1 Promoting After Sale Services
By providing a subsidy at a national level, a country can oblige users to pay for the extra cost of service and thus make sure that the investment will be worth its money. Through its subsidy, ONE has in the past forced private firms to not only sell SHS equipment but to agree to a five-year service contract with the users. The latest initiative intends to have the private firms agree to a ten-year full guarantee service for the SHS.

4.2.2 Promoting a Program for Quality
By providing a subsidy, ONE can also enforce the use of quality equipment and make sure that installations are correctly done by trained technicians. In Morocco, the PVMTI program is also anticipating that a part of its funds will be used to train technicians to a high national standard. PVMTI is working with the Centre de developpement des energies renouvelables (CDER) and the french agency ADEME to develop a training course for PV technicians.

Another part of the grant may also be used by manufacturers to certify local products intended to be used on projects financed by PVMTI.

4.2.3 Imposing a Battery Recycling Program
The installation of SHS on a large scale will have a serious impact on the consumption and disposals of batteries. A simple calculation shows that to implement its solar program of the size that Morocco intends to have, there will be a need to replace at least 200,000 batteries every three years or at least 67,000 batteries per year once the program is completed.

Both ONE, IFC and GEF have recognised that this would have a serious impact on the environment if it were not dealt with from the beginning. ONE has recently included in its new Request for Proposals that each PV suppliers must have a recycling program for batteries if they are to install PV SHS under their program. The PVMTI program is also enforcing recycling of batteries within its projects. The grant component can be used to alleviate the costs of recycling batteries.

Fortunately, Morocco has a few battery manufacturers that have the capacity to recycle batteries. Preliminary discussions with them showed that they are willing to pay for the emptied batteries. In fact, it has been shown that this fee could pay for the transportation of the battery from the collecting depot out in the field to the recycling firm. Another firm that produces the battery electrolyte has also stated that it would neutralise any sulphuric acid brought to their plant.

Making the recycling of battery a mandatory clause to both a subsidy and a financing package insures that the beneficiaries will be obliged to recycle them.

5. PRELIMINARY RESULTS OF THE PVMTI PROGRAM IN MOROCCO

Significant results to date of the PVMTI program in Morocco are:
- It has helped PV firms position themselves in Morocco.
- It has permitted the establishment of international partnerships with Moroccan firms in the PV market.
- It has spurred the interest of Moroccan financial institution for the PV sector.
- It has helped ONE define a new full service approach to electrify the rural world with PV.
- It has evaluated national testing procedures for PV components.
- It is defining a training course for PV technicians.

The PVMTI program in Morocco is also involved in other types of PV applications. It is currently evaluating PV micro-grid, PV water pumping, tele-transmission and rural telephone projects.

REFERENCES:
1. Solicitation for Proposals for the PVMTI Program in Morocco, Marrakech, Morocco, Sept. 1998.
2. Programme d'Electrification Rural Global (PERG), Office National de l'Electricite, Morocco, April 1999.
3. Grameen Shakti - Development of Renewable Energy Resources through Micro-Power Companies, Bangladesh, 1998,
4. Gregory, J. et al., Financing Renewable Energy Projects: A guide for development workers; IT Publications, London, UK, 1997.
5. Cabraal, A. et al., Best Practices for Photovoltaic Household Electrification Programs: Lessons Learned from Experiences in Selected Countries; WB Technical Paper No. 324, Wash. D.C. USA, 1996.
6. Adib, R. et al., Financing Solar Home Systems: The Case of Indonesia, 2nd World Conference and Exhibition of PV Solar Energy Conversion, Vienna, Austria, 1998.

PRE-PROJECT LESSONS FROM PVMTI IN KENYA
"LINKING TECHNICAL SERVICE PROVIDERS TO FINANCIAL INTERMEDIARIES"

By PVMTI Team in Kenya
A. HAQ, *Country Manager*
A. NGIGI, *Deputy Country Manager*
F. MOHAMED, *Local Partner*

IFC/GEF Photovoltaic Market Transformation Initiative (PVMTI)
Purshotam House, Chiromo Lane, Westlands
P.O. BOX 42777, Nairobi, KENYA
Telephone: 254-2-742552/744973
Fax: 254-2-740687
E-mail: pvmti@iconnect.co.ke

There exists a huge PV market growth potential in Kenya, and Kenya represents one of the largest truly commercial PV markets. With a population of nearly 29 million, a grid connection of only 2% and a PV potential market penetration of less than 2%, the scope for market development appears great. Each newly installed system generates further demand, but limited financial resources and poor access to credit has hampered growth by confining traders to a 'cash-across-the-counter' approach, which mainly offloads components, not systems, into the market - with related technical problems. Interestingly though, Kenya has a very unique and successful Savings and Credit Cooperative Society (Sacco) Movement. Well spread into the rural regions where PV potential is highest, it caters for the livelihood of over 50% of the population, has a total annual turnover of about US$ 220 million and an individual membership of approximately 5 million Kenyans. Running in parallel to this 'movement' is a modern commercial banking sector, also with high potential for PV funding. Mainstreaming lending for PV in these two sectors portends a huge and sustainable PV market growth path, making *"Linking Technical Service Providers to Financial Intermediaries"* a key strategy for PVMTI in Kenya.
Keywords: Financing – 1: Rural electrification – 2: Kenya - 3

1. THE MARKET

There exists a huge PV market potential in Kenya;
1.1 Population of 28.7 Million (1999 census) with a 34% population increase over the last 10 years.

1.2 Over 75% of the population is rural-based.

1.3 Only 2% of Kenyan households are connected to the electricity grid.

1.4 Approximately 3.5 million Kenyan households with potential for electrification are currently off-grid, have no realistic chance of obtaining a grid connection in the near future, and currently use kerosene, dry cells, or car batteries for lighting.

1.5 Government sponsored rural electrification program exists, but has been slow, ineffectual and short of funds.

1.6 PV penetration to date is under 2% of potential PV market in Kenya.

2. THE DEMAND

The Demand is however mainly latent due to:
2.1 Bad reputation due to poor quality products and installations.

2.2 Failure of systems due to lack of after-sales service.

2.3 Limited consumer credit for Solar Home Systems (SHS) purchase, partly due to risks associated with poor technical quality.

2.4 Mainstream financial institutions view the sector with skepticism, due to associated systems' failure, default risk and general unfamiliarity with PV technology.

2.5 Available credit is very expensive - for example hire purchase finance at interest rates of 40 to 60 per cent and commercial banks currently lending at 25 to 35 per cent (generally 5 to 8% above base rate).

3. THE SUPPLY

There exists an established SHS supply sector; however this is small and entirely trader-oriented;
3.1 Fewer than 2% of off-grid dwellings have a PV connection, about 120,000 SHS.

3.2 Sale of individual components rather than systems with warranties / guarantees.

3.3 Most installations completed by untrained independent (mains) electricians with limited understanding of PV system installation requirements and service/maintenance needs.

3.4 Lack of technical Standards for PV systems/components and installation.

4. THE ACTION PLAN

The remedial and developmental plan is therefore:
4.1 To trigger and satisfy latent demand for consumer products based on PV technology by:

4.1.1Introducing financial institutions to leverage the existing PV distributor networks in both urban and rural areas.

4.1.2Employing standard approaches to consumer finance.

4.1.3Developing customer-oriented marketing and sales techniques.

4.1.4Introducing high-quality products through reputable PV product and technical service providers.

4.2 The critical action-plan is "strengthening and linking technical service providers - backed by international PV producers - to financial intermediaries, within project structures (including mandatory technical guidelines and requirements for staff training), to address identified markets, barriers and risks." The bottomline is therefore to broker commercially and technically viable, long-lasting and sustainable relationships in the industry.

4.3 The effort will steer the sustainable growth of key players at three levels in the Kenyan PV industry:

4.3.1Grass-Roots Financial Intermediaries.

The extensive network of savings and credit co-operative societies (SACCOs) in Kenya, which is an integral part of the co-operative financial system (dubbed the "Cooperative Movement" by the Kenya Government), provides a significant link to large groups of potential SHS consumers, often in contiguous geographic areas. This sector commands approximately US$ 220 million in lending activity, and is highly successful. It is a 'closed-loop' financial sector where members mobilise savings and utilise the same for member loans at interest rates determined by the members themselves. There are 9,000 societies which have been launched since 1908, when the first cooperative was set up for dairy and agricultural products, although the cooperative 'movement' started in 1940. The sector has about 5 million individual members who cater for the livelihood of over 50% of Kenya's population of 29 million.

4.3.2Mainstream Financial Intermediaries.

The stronger Kenyan local and international banks can play a critical role in aggregating lending to customers of PV products. They have historically lacked confidence in the performance of SHS in particular, and have avoided the PV market. Nevertheless, partnered with reliable technical service providers, they are willing to commit both management time and their own capital to the sector.

4.3.3Technical Service Providers.

The supply side of the PV market in Kenya has suffered from a chronic lack of investment of capital and management time. Most SHS importers and distributors have acted as traders rather than service providers, while international suppliers view Kenya as a high-risk market, and have avoided commitments. PVMTI is supporting the fixed and working capital needs and growth of PV businesses in Kenya in two distinct ways: (a) By providing PVMTI funds and (b) By soliciting the support of co-financiers interested in short to medium term investments in PV distributors and integrators.

4.4 These Partnerships (consortiums) are being designed to effectively address three key barriers to growth of the PV market in Kenya:

4.4.1Lack of Successful Business Models.

PVMTI projects will demonstrate how strong financial intermediaries can partner with established distributors to grow the SHS market. The market strategy involves matching Financial and Technical Partner sub-set of branches and service centres, chosen or launched for their proximity to likely markets and their sustainability in the long run.

4.4.2Managerial and Technical Skills.

Levels of technical competence among staff have been low, and management has been weak. By focusing support primarily on established and higher potential distributors, and securing on-going commitment from strong, committed suppliers, the Project aims to steer the growth of "high potential" relationships between industry players.

4.4.3Lack of, or expensive End-User Finance.

Potential providers of consumer finance for SHS in rural areas are typically deterred from lending by a lack of credit history of potential borrowers and a lack of adequate security. Those potential borrowers who could satisfy lenders are often deterred from borrowing by high, fluctuating interest rates. PVMTI will lend funds to financial institutions (both Saccos and commercial banks) to facilitate the marketing of solar loans to account holders and to bring down the cost of funds to the financial institutions. This will enable the institutions to initially offer interest rates to borrowers that are (1) below market (and therefore affordable), and (2) fixed at the time of borrowing.

5. THE PVMTI GRANT

5.1 PVMTI Grant funds will strengthen this effort by financing:

5.1.1 Market research, and

5.1.2 Training – technical/commercial/e.t.c.

5.2 As described in the PVMTI Project document, grant funds will be used in "non-commercial activities" believed critical to the success of the project.

6. THE FINANCING MECHANISMS

6.1 Sustainability of all project structures used is a key success factor. At end of the PVMTI project, all IFC financed projects must have returned funds lent to them and become totally commercial on their own merit.

6.2 The financing structures which will address the above are:

6.2.1Concessional Loans:

Availability of PVMTI funds will cater for lack of liquidity for PV lending, while concessional terms will counter the high cost of funds and high interest rates. Longer term financing (e.g. moratorium and phased repayments) will counter fluctuation of interest rates, by allowing fixation of consumer rates at the time of borrowing.

6.2.2 Lines of Credit:

Lines of Credit through strong banks who wish to limit their involvement to a custodian role, will avail funds at minimal add-on cost to the Grass-Roots Financial Intermediaries (mainly SACCOs) whose interest rate regime is much lower than conventional commercial banking.

6.2.3 Guarantee funds:

Guarantee funds will address lack of consumer credit due to credit risk, by sharing default with the financial institutions. Guarantee funds will also address the high foreign exchange risks associated with offshore borrowing in Kenya.

7. THE BENEFICIARIES

Who are the beneficiaries of these Partnerships?

7.1 Individuals, who are existing customers of the financial institutions (banks and cooperative societies) and have a regular source of income, for example teachers or civil servants;

7.2 Groups of customers, possibly aggregated by NGOs;

7.3 Small and Medium Enterprises and Institutions;

7.4 Participating sponsors who will benefit from PVMTI concessions, assistance and grants.

8. THE SECURITY

Expected Security for Loans from the financial intermediaries:

8.1 For individuals, a guarantee such as a personal pledge from a neighbour or colleague, or

8.2 A "check-off" mechanism which entails a letter from his or her employer guaranteeing to withhold interest and principal repayments from the individual's monthly salary.

8.3 For groups, a pledge of collective assets such as cash deposits and/or other security such as personal guarantees or NGO pledges.

8.4 For Small and Meduim Enterprises (SMEs) and institutions, standard security, for example, a pledge of assets.

9. THE RISKS

PVMTI deals in Kenya are particularly risky.

The Projects are most sensitive to the rate of consumer default and to the volume of sales. Consumer default is in turn highly influenced by the interest rate regime and economic (income continuity and stability) factors. The major risks are:

9.1 The management team fails to perform.

Management training partially funded by the PVMTI Grant will minimise this risk.

9.2 Customers do not respond well to the offering.

The PVMTI Loan will be disbursed in a series of tranches based on sales utilisation of previous tranches.

9.3 Customers do not repay.

This risk is exacerbated by the currently poor state of the economy (annual rate of inflation in excess of 10 per cent, swings of up to 40 per cent in both the KSh-US$ and the Treasury Bill rates, e.t.c.). Often, debtors are unable to meet debt service payments, hence the participating financial institutions will only lend to those individuals and companies who pass minimum tests of credit-worthiness and who offer adequate guarantees of repayment.

Other mitigating measures include short loan tenors and MOU (memorandum of understanding) with technical providers to ensure debtors do not default due to non-performance of system.

10. THE ENVIRONMENTAL IMPACT

At present, there is no policy on battery recycling in Kenya. To the knowledge of PVMTI, there exists only one lead recycling plant with no facilities at an industrial scale to recycle battery casings (mostly hard rubber). Sponsors are being encouraged to:

10.1 Source batteries from manufacturing facilities approved by Kenyan pollution-control authorities,

10.2 Design and implement a deposit/refund scheme for batteries to encourage user participation,

10.3 Use existing facilities to recycle lead, and

10.4 Monitor developments in the battery recycling industry and use environmentally sound recycling facilities, as they become available.

FIRST EXPERIENCE OF WATER PUMPING SYSTEM IN YEMEN

Christer Nyman
Soleco Ltd
Londbölevägen 27 FIN-06100 Borgå, Finland
Email: christer.nyman@avenet.fi
Phone: + 358 400 458790, +358 19 543209 fax: +358 19 543209

ABSTRACT: A photovoltaic water pump (PVP) was installed and monitored in a village at sea level in the east coast of Yemen. Economic study showed PVP to be competitive within the power range of small diesel pumps, where they often even constitute the least-cost option. Social study showed the high acceptance and better integration of PVP into project village. The water in Yemen is found in wells at depths of 15 m down to 100 m, and more. The village population usually lives on the top of the steep mountains and the well is down in the valley. The first pilot PVP was installed in a village at sea level with an existing well. The work is supported by World Bank financing and a co-operation with a Danish and Finnish CTF of World Bank.
Keywords: 1: PV Pumping 2: Developing countries 3: Villages.

1 Background

Yemen is classified as an LDC (low developing country) and it has among one of the highest MMRs (maternal mortality rates), IMRs (infant mortality rates) and illiteracy rates in the world.

Among the approximate 16 million people living in Yemen very few people have access to running water and/or electricity. There are multiple pockets in rural areas throughout Yemen without electricity. Access to electricity amplifies the quality of health and educational services and allows for rapid social-economic development.

Photovoltaic (PV) systems have recently been recognized as a possible viable source for electricity, for areas, which are off grid to the conventional electricity supply in Yemen. The PV systems are capable of converting solar light/energy into electricity, which offers a moderate reliable flow of electrical power adequate to supply electricity for distant educational and public health facilities, as well as for pumping water.

Presently in Al-Udayn (Ibb governorate) there are four PV system pilot demonstrations set up to supply electricity to the local educational facility and to the PHC (Public Health Clinic). In Al-Jarrahi (Al-Hodaidah governorate) there is a PV system set up to a water pumping and telecommunication scheme. The water pumping system is presented and discussed in this paper. A social study was done and is also presented here.

2. Objective of the PVP system

A reliably supply of clean drinking water is a basic need, but approximately one billion people in remote areas of developing[1] countries have no adequate access to clean drinking water. 78 % of all diseases in developing countries are regarded as "water induced". In order to demonstrate the technical maturity of PVP and to identify the conditions required for their dissemination the following objectives was set up by World Bank for this Yemen project:

- Investigate the technical and economical cost of PV operated water pumping compared to alternatives.
- Establish a Solar Energy Implementation Unit and provide training and expertise
- Select site for a pilot demonstration systems, design appropriate PV power system including appliances; procure and install pilot demonstration systems and monitor pilot demonstration systems
- Set up criteria for future selection of further sites of PV power systems in the fields of water pumping

The PVP system with water tank in the village

3. Pilot project characteristics

3.1 System design

The PVP system is designed on the basis of the findings from a local data survey. The solar energy resource in Yemen is abundant and evenly distributed over the country and over the seasons. The average insolation is around 5.5 kWh/m^2. Average insolation varies about 10 % with location and seasonal variation is about 25 % with little variation as to location. From the intended location field data as demand for water in the supply area, pumping head and geographical peculiarities as was needed.

- **Solar Array.** They are mounted on an old diesel pump house using a simple frame that holds the modules at a fixed tilt angle towards the sun. PV size 1200 W module = 24 times 50 W modules. With tilt angle of 15^0

- **Storage tank.** Direct-coupled PV pumps deliver water only when the sun is shining. This requires some type of water storage in order to satisfy the need when the sun is not out. Water level is 23 m from ground level and tank height from ground about 11 m, tank size is about 10 m^3, which gives a head of 34 m.

- **Pump.** The pump was originally sized for a head of 100 m and lift of 100m over ground level, but changed in last minute. Water pump yield: at solar radiation of 600 W/m^2 the pump gives 19,5 l/min, with means 1,2 m^3 per hour. The estimated pump yield will be in the range from 0,2 to 1,6 m^3 per hour giving a daily yield between 5 and 7 m^3. Pump type is a Grundfos SP1A-28, submersible.

- **Pump Controller.** Matching device used so systems will operate at optimum power, matching the electrical characteristics of the load and the array.

- **Inverter.** A 3-phase SA1500 inverter was chosen to output in a variety of voltages.

- **Data monitoring.** A weather station powered by a small 10 W PV module was included for solar radiation, temperature and humidity measurements. The water yield is measured by a simple water flow meter.

For the monitoring and follow up of the pump performance, an independent weather measurement station and a water flow meter was installed. Local engineers from the Solar Energy Centre in Sanaa supervise the performance and monitoring.

It is also important to include sociological factors in the planning process. The future users should be involved in the data-gathering process at the intended PVP site in order to make early allowance for their customs and traditions in relation to water.

3.2 Installation
About three complete days was needed for the installation and transportation of the pump system assisted by local engineers. 1 day was needed for testing, monitoring and giving instructions for the use of the pump system. A lot of dust will accumulate on the surface of the modules, during periods with strong winds. The wind is mainly occurred in January – February, lasting up to 1 month.

Inspection of installed PV modules

4. Water pump economics

4.1 PVP economics
Using a module price of 6 US$/Wp, PV pumping are usually competitive for village sizes of 1000 - 2000 people, hand pumps for up to 1000 people, and diesel driven pumps for more than 2000 people. The estimation applies to a 20 meter pumping head. Generally PV pumps are competitive with diesel pumps up to a total pumping head of 50 m.[2] [3]

This pilot system has a unit water cost of 0.54 US$/$m^3$ and the per capita capital cost is 3.9 US$ /person, based on 20 LCD (=per capita water consumption). This value 1.96 US$ /person based on water consumption of 10 LCD. The daily water output of the system is 9 m^3. The unit water cost is not high compared to other pumping projects in the world.[4]

4.2 Diesel economics
Different diesel water pump alternatives were calculated. With small water yield demand (10 m^3/day) the diesel pumps used in small villages in Yemen give a unit cost of 0.63 US$ /m^3 The unit cost for the Yemen calculation is high due the low usage of the pump. Water pumps are used only 10 –20 min per day and the daily water production is low. The lowest unit cost with general diesel pumps are about 0,32 US$/$m^3$ to 0,40 US$/$m^3$ if the diesel is used for large water production per day. With small water productions the PV pumps are viable. The water demand in the villages in Yemen is mostly small.

4.3 PV and diesel comparison
The economics of PV powered water pumps and diesel water pumps in Yemen give a water unit cost of US $0,80 /$m^3$ for diesel pumps and US$0,54 /$m^3$ for PV powered pumps.

With small and limited demand of electricity for water pumping and taking into account the present situation of rural electrification in Yemen there is technically no economical alternative to PVs when considering the service of electricity. The water pumping with small O&M and no fuel costs can only be made by PVs.

In villages with irregular access of grid powered by local diesel, even the national grid, one can considered an alternative system, where PV modules and solar energy are used as energy source and water pumping at day time is performed very safety every day. The storage is the water tank.

The chief value of grid-independent PV systems lies in their ability to provide a cost-effective alternative to expensive distribution line extensions. Grid-independent PV systems are therefore most likely to be cost-effective when they serve small or intermittent loads that would be costly to serve with a line extension. Although the length of a line extension is important, it is not the only factor involved in assessing the cost of a line extension. Also important is the cost of the transformer, the type of terrain to be crossed, and the need for trenching

Most utilities have calculated their average cost per km or "per span"—the distance between two distribution line poles—for line extensions. However, not all utilities calculate this cost in exactly the same way, and not all utilities charge their customers the actual cost of a required line extension. These differences were as much the result of different accounting methods employed by the utilities as of measurable differences in materials and labour costs. For example, one utility included only materials costs—poles, transformers, wires, etc.—in its line extension estimates, while another also included labour costs and an allocated portion of employee benefits and general utility management costs.

5. Technical performance of the system

The system have been will be monitored as to daily usage from February 99 to July 99 by the end users and an automatic mobile weather station powered by PV. The system have been visited and checked regularly by engineers from the Solar Energy Center (SEC).

Forms to take data for monitoring the PV-modules and the pump performance are given to the Solar Energy Centre. Regular visits of the SEC stuff will be carried out for maintenance and monitoring the systems.

The first measurement of the water pump site shows that the water yield is about 8 - 9 m^3 per day. The daily solar radiation (kWh/m^2) is also about 5.7 kWh/m^2 .The pump system has been monitored from February 99 to September 99. During this period the water output has been on average 8.1 m^3 / day, when the estimate was 7 m^3 / day for this system.

No special problems have been noticed. On the march 5th when the site was visited by local engineers, the modules had about 5% of sand coverage on. This may be seen on March 4th and 5th of the water yield.

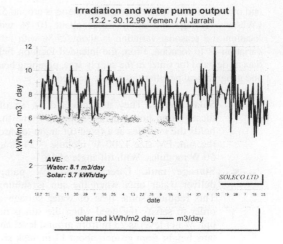

Daily water pump yield

6. Social impact study

The questionnaire was designed specifically to obtain substantial data, which would reflect the social impact the PV systems had in the target sites. The questions for the study were designed according to the outlined requests, as well as, other pertinent relevant questions. The study aimed at collecting data on the following issues:[5]

Social statistics: Gender, age and non-formally collecting information on the general overall social-economic circumstances of the community (including the general health and educational situation of the people).

Technical issues: End user understanding of PV system operation, daily use and system maintenance. End user satisfaction with supplier, service and their personnel, installation, quality of products, size adequacy, performance. Need for a range of standard sizes, easy extension.

End user attitude towards PV system electric services: Satisfaction of the system, what energy services the system replaced, advantages, disadvantages, recommend to others, others envious, communal system influence on tensions and willingness to pay a fee.

Due to a social assessment study of the pump system 84% of the village understand that the system needs maintenance with its cost sharing implications. However, 91% does not understand how the system produces power to operate the water pump. After setting up the system, regular follow up visits by technicians are essential to maintain local population's confidence in the system and commitment to maintenance cost sharing.

In conclusion the results of the PV system water-pumping scheme in village were successful. A community once without easy access to fresh drinking water, now has it pumped directly into thirteen homes two hours per day. The participants were all very satisfied with the access to water and encourage others to acquire about obtaining a similar system.   Notable points of the study were:

- Generally most of the participants did not have a good technical understanding of how the PV system operated nor of its daily use or maintenance, but this caused no negative impact on the project.
- All participants agreed that there were both benefits, as well as advantages to using the system.
- The majority of the people did not believe there were any disadvantages to the system, but that the system did have some "limitations" (low water pressure).
- There were both local tensions, as well as surrounding community tensions because of the existence of the system in the area, but fortunately there were no conflicts or problems as a result of these tensions.
- It is believed that a "communal PV system" would prevent tensions.
- Most of the participants stated that they were willing to pay for a system because it was so useful, but unfortunately "willingness" and affordability are not the same.
- No one, individually, could afford to pay for a system.
- Some of the participants stated that they could perhaps afford to pay an estimated 150 YR, equivalent to 1 US$ for a monthly service fee, in order not to loose the service.
- The access to water in their homes improved their general hygiene (regarding: personal, equipment, utensils, clothes and home) and it made for better over-all performance (of their household-work). The system also saved time and effort, especially for those who had previously had to fetch water for 2-3 hours daily.

This PV system would be beneficial for many communities similar to the village with the PVP system, where access to water becomes a very laborious, difficult and time consuming task. All the participants were very pleased with the project as a whole and consider it a blessing.

7. Conclusions

The approach looks at the implementation of a new technique as a socio -technical system where people and technique interact in the interplay of adaptation and adoption. The importance of including local engineers, social scientists and end users in the project is obvious, for the acceptance of new technology and making people satisfied.[6][7]

- A screening process has indicated the technical an economic viability of using PVs to provide electricity for water pumping and telecommunication stations in off-grid areas in Yemen
- The technical and economic viability of using solar PV power systems for water pumping

and telecommunication in rural areas has been highlighted partly through the desktop screening process and partly via demonstrations. An associated socio-economic study show that all stakeholders are very satisfied with the PV units implemented, and end users have expressed understanding and willingness to pay for maintenance and to some extent cost recovery. PV technology motivates staff and people to stay in remote areas, by having access to electricity and telecommunication.

- A high replication potential for PVs to villages, schools, clinics and irrigation sites have been identified
- A Solar Energy Implementation Unit has been identified in the form of the SEC and training has been provided; the SEC has been found to be a competent and appropriate choice as national focal point for PV installation, operation, maintenance and monitoring activities
- Deployment models and criteria for selection of further sites for PV power systems in the fields of water pumping and telecommunication have been developed.
- The PV resource base in Yemen is still very much in an embryonic state; however there is an increasing awareness of the technology in the private sector

References

[1] The World Bank, World development report 1994, Oxford University press (1994).

[2] N. Argaw, Evaluation of photovoltaic water pumps and estimation of solar energy availability in Ethiopia, Tamper University Publications 194, (1996).

[3] C. Nyman, N. Argaw, Simulation and measurement of water pumping systems in Ethiopia, 13th European PV conf, Nice, (1995)

[4] A. Hahn, Lessons learned from the international photovoltaic pumping program, 2nd World PV Conf, Vienna, 1998.

[5] J. Ijichi, T.Sufian, C.Nyman, P.Ahm, Solar Energy in Yemen - an Economically Viable Alternative, 16th European PV conf, Glasgow, 2000.

[6] P.Ahm, Small scale PV systems for primary health care and distant education in Yemen, World Bank final report 2000.

[7] C.Nyman, Small scale PV systems for water supply and telecommunications in Yemen, World Bank final report 2000.

PHOTOVOLTAIC ENERGY SUPPLY IN COMMUNITIES OF THE XINGÓ PROGRAM IN THE BRAZILIAN NORTHEAST

Elielza M. de S. Barbosa*; Chigueru Tiba*;
Ramiro P. da Silva Junior** Fabiana M. Ferreira**; Maria A. P. Carvalho ***
Grupo de Pesquisas em Fontes Alternativas de Energia - FAE
Departamento de Energia Nuclear - DEN Universidade Federal de Pernambuco - UFPE
Av. Prof° Luiz Freire, 1000, Cidade Universitária, 50.740540, Recife - PE, Brazil.
Phone/fax: 55 081 (271-8252) (227-1183) ; e-mail: elimsb@npd.ufpe.
*CNPq Grantees/Xingó Program
** Executive Coordination Xingó Program

ABSTRACT: The Xingó Program is a multidisciplinary initiative, developed jointly by the CNPq- Brazilian National Council for Scientific and Technological Development and CHESF- Hydroelectric Power Company of the São Francisco River. Its main objective is to promote the development of a semi-arid region through actions undertaken in different areas; more specifically, to seek energetic solutions on suitable techniques in the region and at the same time identify local demands and business opportunities that may lead to the introduction of enterprises in the region, principally focusing social and citizenship development. Eight rural communities located in the perimeter of Xingó program were selected for implementing the first pilot projects. This paper describes a technical and social diagnosis, and a conceptual project that were made for each community, considering the resources and the local available potentialities, prioritizing energy supply to schools, health centers and the supply of drinking water. These experiences, in spite of the successes and failures, could be deemed as a lesson that should be learned. The feeling of citizenship of the needy community inhabitants, in whom dignity lays hidden by illiteracy, malnutrition and lack of perspective, is recovered just because of the arrival of light and water.
Keywords: Stand-alone PV Systems-1: PV Water Pumping- 2: Sustainable-3.

1. INTRODUCTION

São Francisco River is the largest fully Brazilian river, and the most important one in the Brazilian Northeast region. With a length of approximately 3,160 km, it cuts through the northeast semi-arid region, and it is of utmost importance for the economy of the region, because of its significant hydro-electric potential. The Hydroelectric Power Company of the São Francisco River (Companhia Hidroelétrica do Rio São Francisco, CHESF) has an installed power of 11 MW, and over 15,000 km of transmission lines. Nine of its hydropower plants are found along the São Francisco River, and Xingó Hydropower Plant (3,000 MW) is one of them. However, these resources (power and water supply) are not available for a large number of people. These are the inhabitants of small villages located in the surroundings or in the islands of the São Francisco river, where the conventional power and water supply networks do not reach them.

Aiming at developing the Northeast semi-arid region around Xingó power plant, the Xingó Program was then given birth. It is a multidisciplinary initiative, developed jointly by the CNPq- Brazilian National Council for Scientific and Technological Development and CHESF. Its main objective is to promote the development of a semi-arid region by way of actions in different thematic areas: Agro-ranching, Aquiculture, Archaeology and Historical Heritage, Ecology-Biodiversity and the Environment, Education, Energy, Water Resources and Water Quality, Tourism and Hospitality. Specifically, the objective of the field of Energy is to seek energetic solutions on suitable techniques in the region and at the same time identify local demands and business opportunities that may lead to the introduction of enterprises in the region, mainly focusing social and citizenship development. From this perspective, 8 rural communities located within the coverage perimeter of Xingó program were selected for implementing the first pilot projects, during 1999. The photovoltaic (PV) systems installed were made available to Xingó Program by means of the Brazilian Government/Ministry of Mines and Energy, through the PRODEEM (Program for the Development of Sates and Municipalities).

2. CONCEPTUAL PROJECT

The Thematic Area Program (TA-P) foresees that: the PV electrification pilot projects should act, as a priority, in the fields of Education, Health, and Water Supply; local governments, as a counterpart, should take over the commitment of acquiring the basic equipment to be installed at school/health care units, as well as providing teachers and health care agents, and collaborating with the necessary physical refurbishment.

Within this focus, a Conceptual Project was designed, considering that in each one of the communities, the following modules should be developed, taking the most out of the locally available potential resources /3/:

M.1: **Education Module:** school building or refurbishment, and teacher allocation

M.2: **Health Module:** health center building or refurbishment, and visit of regular health care agents

M.3: **Power Supply Module:** photovoltaic electrification in the community (school, teacher's house and public illumination)

M.4: **Water Supply Module:** water pumping using photovoltaic energy

M.5: **Socio-Economic Module:** Literacy and/or vocational training for both young and adults, center for the breeding of small animals and/or plants adaptable to the climate of the Northeastern Semi-Arid region.

Unfortunately, it is not always possible to fully implement a Conceptual Project that is designed based on the ideal conditions that technology could provide. The reality of each region/situation overrides, therefore making necessary to make the conceptual project comply with whatever is possible to be accomplished.

Aiming at learning such reality, technical and socio-economic appraisals have been carried out in all pre-selected communities that were candidates to photovoltaic powering. The results achieved are the inputs for the specific project for the individual community, and for the socio-economic actions.

3. TECHNICAL AND SOCIO-ECONOMIC APPRAISAL

3.1 Socio-Economic Appraisal

This appraisal was carried out in a universe of 216 (two hundred and sixteen) families, including approximately 1,331 (one thousand three hundred and thirty one) persons. It comprises several aspects: historical, cultural, economic, educational, health, housing, infrastructure, main issues, and community claims – just to name a few.

In spite of some peculiarities, the locations present common features, such as: water and power supply shortage; high rate of illiteracy and school drop-outs in the diurnal shifts and high interest for the nocturnal shifts, almost all of their local adult population is illiterate; excessive use of child and adolescent labor; livelihood agriculture (beans, maize, watermelon, etc.); as far as health is concerned, we have detected high rates of malnutrition, anemia, conjunctivitis and pneumonia, among other diseases; as far as housing is concerned, almost all families live in loam huts, and their illumination is based on diesel-fueled rustic chandeliers, as diesel is cheaper than kerosene, in spite of their complaints regarding the dark smoke it produces, causing discomfort to their eyes and nostrils. The average expenditure per family is of 2.00 to 7.00 US$/month. Almost all of them did not know about Photovoltaic powering. The data were collected in 1999.

As an example, Tab. 1 shows the specific characteristics of three communities, regarding the theme of energy: **Gualter**, originated from a Landless settlement; **Monte Alegre Velho**, located in a municipality that presents one of the lowest human development indexes (HDI), and **Batida**, a Kantaruré Indian Settlement.

3.2 Technical Appraisal: Physical Conditions and Demand

The following was detected: the need of recovering the physical structure of the schools and water springs, an appraisal of the electric load (demand), the quality of the water of the available source, and the possibility of drilling wells in the region.

- Physical conditions

The adequacy of the conceptual project to the reality of the communities impaired the implementation of the development modules (M.1 to M.5) for most of them. Except for the Gualter community, to this date, it has only been possible to implement the powering of the schools (M.1 and M.3) in the remaining communities.

Gualter community's location offers excellent physical conditions for the development of the conceptual

Table-1: Main Characteristics of the Communities

Community	M. A Velho	Gualter	Batida
No. of families	17	37	17
No. of persons	102	296	102
No. of children	04 (average)	06 (average)	04 (average)
Health:center /local agent/ (hospital)	Not available, /01 agent/ (18 km away)	Not available, /Not available/ (37 km away)	Not available, /01 agent/ (33 km away)
No. of schools (student) /School level/	01 (46) / primary /	Not available,	01 (68) / primary /
Water: Resource (Quality)	Well (saline) Water-truck (importable)	Well (good) Water-truck (importable)	Well (good) Diesel-motor pumped.
Power for illumination: Type / usage and (monthly cost) (US$). [Family cost] (Average/ monthly)	Diesel-fueled *Chandelier* / 6 liters, (2.00); *Dry batt.* / 6 units, (1.50); Car batt/ 2 loads, (2.50) [US$2 - 7.00]	Diesel-fueled *Chandelier* / 4 liters, (1.80); Car batt/ 2 loads, (2.00) [US$ 2 - 4.00]	Diesel-fueled *Chandelier* / 8 liters, (3.00); *Dry batt.* 4 units (2.00). [US$ 3 –5.00]

project in its entirety: easy access, which makes possible a continuous follow-up; broad space, for the building of schools and socio-economic units; excellent quality well, with a depth of 60 m, an outflow of 6 m³/h, 671 ppm (parts per million), and 250 m away from the area reserved for the school; however, the lack of an appropriate social organization impairs the implementation of the actions. Different from the Monte Alegre Velho community, that has a reasonable social organization, a well with an outflow of 1.3 m³/h, and it is near the school with very poor water quality, 17,180 ppm. These rates are much higher than the levels acceptable for human consumption, (maximum accepted is 1,000ppm). Other locations present the same problem, but with lower salinity rates. Tab.2 shows the list of communities, specifying the development modules (M.1...M.5) implemented by Xingó Program during 1999, the first year of the execution of the project.

- Energy demands

In all units, the power system should meet a minimal demand: illumination + color TV + video + satellite dish for educational purposes. According to an agreement signed with the local administrations, which are responsible for acquiring the equipment, in some cases the demand corresponding to the use of some household appliances used in the preparation of the school food was added (whisker, refrigerator or a small freezer). The demand is within the range 400-1200 Wh/day. In some locations, the signed agreements have been observed, and some other city administrations are programming their resources in order to enable the acquisition of this equipment. Therefore, in some cases, the capacity of the installed system may seem to be inadequate to the current demand. Power and water consumption meters are in the process of installation, for a better compliance between the capacity of the PV systems and the real consumption. Tab.2 lists the specifications of both the predicted and the current power demand.

Table.02- Communities Powered by the Xingo Program until 1999.

Communities - inh	Initial status		Modules ****	Demand (Wh/d)		PV System	Ref
	School	Water*		Foreseen	Installed (current)	Generator; Battery	Fig.1
M. A. Velho-102	Yes	P.S (17,180ppm)	M.1, 3	1,200	600 (lamp11 x 20W)	(600W) ; 900Ah	(a)
Caqueiro-350	Yes	P.S (14,020ppm)	M.1, 3	600	460 (lamp 7 x 20W)	(200W) ; 650Ah	
Surrão-	Yes	P.S (15,688ppm)	M.1, 3	600	460 (lamp 7 x 20W)	(200W) ; 650Ah	
Uricuzeiro-180	Yes	P.S (3,538ppm)	M.1, 3	450	340 (lamp 7 x 20W)	(150W) ; 390Ah	
Turco-210	Yes	P.S (4,720ppm)	M.1, 3	800	660 (lamp13 x 20W)	(260W) ; 520Ah	
Batida-102	Yes	P.P (880ppm)**	M.1, 3	810	480 (lamp 8 x 20W)	(300W) ; 520Ah	(b)
Gualter-296	No	P.P (671ppm)***	M.1,3,5	1,600	1,500 (lamp11x20W + 4x40W + 02 TV)	(780W) ; 900Ah	(c)
			M.4	5 m³/d	12 m³/d (tank 15 m³)	TSP1000 SM041	
Caiçara-240	No	P.P (900ppm)***	M.4	-?	10 m³/d (tank 5 m³)	SCS 10/230	

*P.S= saline water well; P.P= fresh water well; ** diesel-fueled pumping, *** installed PV pumping; ****Implemented

(a)

(b)

(c)

Figuure.1:   Pictures of energy systems, original and recovered schools, PV pumping, improvements -a) M.A.Velho; b) Batida, Kantaruré Indian Settlement: c) Gualter

4. CHARACTERISTICS OF THE INSTALLED PV SYSTEMS

The installed electrification photovoltaic systems comprise: generator (PV ASE –50 AL/17 and 300 DG/17 modules); accumulator (130Ah CONCORDE PVX12105 and 300 Ah PVX12255 batteries); Charging and discharging (TRACE C40, C30-A and C30); inverters (PROWATT 800 and 1000). The pumping systems are of the following types: SOLARJACK (SCS 4/325 and 10/230) and TOTAL ENERGIE (TSP 1000 SM 041 and 021)

For designing the system capacity, the LLP (Loss of Load Probability) method was adopted with a deficit of 3 - 5%. In each school, in general, the generator was installed on the roof, the PV modules were connected in parallel, and the inverter transforms the voltage from 12 Vdc to 110 Vac. Each school built a small room where the batteries, controllers and inverters are installed. The illumination load is made up of 20-W fluorescent lamps. Higher efficiency lamps, although more appropriate, cannot be found in the local markets. Tab.2 lists the main characteristics of the installed systems, illustrated on the pictures of Fig 1.

5. ACTIONS FOR SOCIO-ECONOMIC DEVELOPMENT

During Xingó Program's first year of activities, we have further dedicated to the activities devoted to the installation of the systems, and to the planning of the activities to be implemented, which may also contribute for socio-economic development. However, it worths the while to mention the activities developed in Gualter location, where the following activities were successfully implemented, with a high degree of satisfaction on the part of the inhabitants: Leisure center (public square with a 29" color TV), School (250-m^2 area, sewage, power and water), Pumping and water supply system (PV system, tank = 15m^3, cap=12 m^3/day, fountains, laundries, ponds for animal drinking), Courses (sewing literacy for both young and adults taught at night at school). One of the classrooms was adapted with sewing machines.

The following activities are programmed for the year 2000: Training courses; basic maintenance in photovoltaic systems, and literacy programs for both young and adults – for all powered communities; basic courses on electricity, caprine breeding, pisciculture, apiculture and school vegetable-gardens – for the communities where correlated vocations were detected; Installation of irrigated planting systems - in Gualter and Caiçara; Energy supply of a flower house at Batida.

6. COMMENTS

One of the major challenges we have found in the photovoltaic powering projects in rural areas has to do with the planning of the activities, and the continuation of the institutional programs. What is usually not decided in the executive and/or technical levels.

Another major challenge, that is also related to the sustainability of these initiatives, is a greater lack of social organization in the "communities".

By the way, this is not exclusive to photovoltaic powering programs; other initiatives such as health care and education programs face the same dilemma. Usually, the target locations of these projects do not present the properties inherent to what means a *community*. Without a minimum of social organization and recovery of one's identity as a human being, no positive results will be achieved.

Examples: At Batida location - Kantaruré Indian settlement, where a relative social organization is found, there is a new pumping system, well installed in a good profile well, a tank of reasonable size, an installed water supply network, and all houses have toilet facilities and fountains. However the inhabitants endure with water shortage. The motor is diesel-fueled, and the institution in charge of the supply forgot their commitments. The monthly expenditure with fuel is approximately US$40.00. In spite of having some characteristics typical of a community (culture of their own, hierarchy and political leadership = tribal chief, and religious= shaman), the Kantururê settlement cannot get US$ 3.00/month per family, which would enable the water supply.

At Caiçara location, the PV pumping system was individually installed (September 1998) in a well located 400m away from the reservoir located at the side of the main road. No dwellers live near the well. The persons walk by the road, collect the water, and go away. There is no control on the distribution of pumped water among the users. Our technical team systematically checks the PV system, collects the data on the production, and on the water consumption, provided by simple meters installed in both the input and output of the tank. This system has not presented any failure or defect. For the users, everything works as if it were a "natural fountain".

At Gualter "community", the technical team is frequently called. Technical problems frequently emerge because of the interference of some user in the local administration of water or power (TV hours).

Gualter is a very new community, coming from the Landless encroachment movements onto an old farm approximately three years ago. After the land was conquered, basically no social work has been undertaken. Except for very few inhabitants, old dwellers, there is practically no socio-economic commitment/activity among the inhabitants that could promote the recovery of the sense of being a *person*.

This process of recovery is an objective of Xingó program. Joint integration activities among its thematic areas may help the creation of centers of communal and participatory development, and help improve life quality.

In this sense, during this first year of activities, some actions have been implemented, and we hope our projects may continue.

ACKNOWLEDGMENTS

The authors would like to thank CHESF, CNPq, SUDENE, PRODEEM, for their institutional support, the technical team of the Energy Thematic area for their availability, and the technicians of the Grupo FAE-UFPE for their collaboration and useful suggestions.

KEYS TO THE SUCCESS OF A PV UTILITY

W B Gillett
European Commission, DG TREN
200 Rue de la Loi, B 1049 Brussels
Tel: +32 2 299 5676, Fax: 296 6261
e-mail: william.gillett@cec.eu.int

G T Wilkins
AEA Technology – ETSU, Harwell,
Didcot, Oxfordshire OX11 ORA, UK
Tel: +44 1235 43 3128, Fax: 43 2331
e-mail: gill.wilkins@aeat.co.uk

Terubentau Akura
Solar Energy Company
PO Box 493, Betio, Tarawa,
Republic of Kiribati, Pacific Ocean
Tel:+686 26058 Fax: 26210

ABSTRACT: An independent evaluation of the EU's Lomé PV Follow-up Programme, which funded the installation of PV systems for rural households in the outer islands of Kiribati and Tuvalu in the Pacific Ocean, was carried out in 1998 / 99. It found that, after more than 5 years of operation, 95% of the solar photovoltaic (PV) systems installed by a solar utility in Kiribati were still working well. The users were delighted with their solar-powered household lighting, radios and cassette players, and the solar utility had created 27 sustainable new jobs, half of which were in the poor rural areas of the outer islands. The new solar utility had therefore contributed not only to social development, but also to poverty alleviation.

Studies of PV rural electrification schemes in other Pacific Island States and in rural Africa have not found the same levels of success. Although good quality PV system components are a basic requirement, our studies show that it is not simply the PV technology which lies at the root of a successful PV utility. Far more important is the institutional infra-structure which is put in place to ensure the long term sustainability of the PV Utility scheme.

Keywords: solar home system – 1: rural electrification – 2: developing countries – 3

1.   INTRODUCTION

1.1   Republic of Kiribati

Kiribati is an independent Pacific island state, located on the equator and accessible by plane from Fiji or Nauru.

It consists of 33 islands of which 17 are inhabited, with a total population of around 80,000. Its average GDP per capita is about 550 Euro. Typical households in the outer islands comprise 6-7 persons; typical household incomes in the outer islands are in the range 90–180 Euro per month.

The government of Kiribati is keen to provide electricity services to the outer islands in order to improve the quality of life for the local communities and hence to encourage them to remain living there. The main island of Tarawa is already experiencing serious environmental pollution from over crowding, so it is vital to discourage further migration from the outer islands to Tarawa.

Until now, there is no access to television in the outer islands, so typical households need electricity only for lighting, communications, radio, and music. Short term demands for electrical power to drive construction tools, photocopiers or film projectors are normally supplied by small motor generators.

1.2   Support from the European Community

The European Community (EC) has been providing funds for energy projects in the Pacific islands since 1984 under the Pacific Regional Energy Programme (PREP). The PV Follow-up Programme, which was an integral part of the PREP, was implemented by the Fiji-based Forum Secretariat and, more recently, by SOPAC. It involved a wide range of local training and capacity building activities, as well as the installation of PV systems in Fiji, Kiribati, Papua New Guinea, Tonga and Tuvalu.

In Kiribati, the EC supported the installation of 250 solar home systems on three outer islands, together with training and technical assistance. The training was led by the project consultant Herbert Wade, whose extensive experience undoubtedly contributed substantially to the project's overall success.

Training was given on how to specify and tender for good-quality PV modules, cabling, and high-quality (deep discharge) batteries. Training was also provided for technicians on installation and maintenance.

Technical assistance was provided for establishing the local manufacture of battery charge controllers and dc/dc converters, which could resist the highly corrosive environment in the Pacific islands. These locally manufactured units were not only used in Kiribati, but have also been supplied to PV projects in neighbouring Pacific island states.

The information presented here was gathered in 1998/99 as part of an ex-post evaluation of the PV Follow-up project, approximately 5 years after the systems had been installed. The evaluation was funded by the European Commission, and involved a six week mission to the region, during which detailed discussions were held with the responsible government organisations and their solar utility companies. On site investigations were also carried out on the outer islands of Kiribati and Tuvalu, where the PV systems had been installed. A new EC funded PV project for Kiribati is scheduled to begin later in 2000.

1.3   Solar Energy Company (SEC)

Solar home systems in Kiribati are installed and maintained by the Solar Energy Company (SEC), which is a government owned corporation, based in the capital Tarawa. SEC was originally set up in 1984 as a PV retail sales company, but was restructured in 1990 with a Board of Directors appointed by the Government. Since then, it has operated extremely successfully as a rural "Solar Utility", under a small but highly committed management team. SEC retains ownership of the solar PV systems, and charges the householders a monthly fee for their electricity. SEC also manufactures its own PV system controllers.

1.4 The PV systems installed in Kiribati

The 250 PV systems supported by the EC were installed on three islands (North Tarawa, Nonouti, and Marakei), which are each less than one hour's flight by small plane from the capital South Tarawa.

These systems typically provide power to three compact fluorescent lights, each rated at 7 or 11W, and a radio and/or cassette player. Two PV modules (total power ~100 Wp) are mounted on one large pole or four small poles beside the house, and charge a high quality deep discharge battery (~100Ah) via a locally manufactured charge controller.

PV systems also power lights and vaccine refrigerators in the local medical clinics, and Citizen Band radios for communications with the main island and fishing boats out at sea. Some small PV-powered water pumps have been installed to supply drinking water for children in the schools.

2 A SOLAR UTILITY

2.1 The concept

A "Solar Utility" (just like a conventional Utility) installs good-quality equipment, retains ownership of that equipment, and ensures that it is well maintained throughout its working life. Customers are charged a monthly fee for electricity services.

Selling PV systems directly to users was shown to fail in earlier EC projects in the Pacific. Even when given access to soft financing, poor households tend to buy the cheapest available, poor-quality PV modules and batteries, which are too small to meet their energy needs. As a result their PV systems typically fail within the first 1-2 years of operation.

PV systems, which are professionally installed using good quality hardware and managed by a Solar Utility, with well-trained technicians, normally provide a far more reliable and cost-effective service than those, which are bought by individual householders.

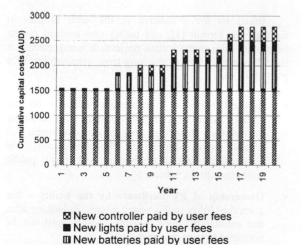

☒ New controller paid by user fees
■ New lights paid by user fees
▥ New batteries paid by user fees
■ User contribution to hardware costs
▨ EC contribution to hardware costs

Figure 1 PV system hardware costs spread over 20 years

2.2 PV Utility payment scheme

In Kiribati, all except 50 AUD (Australian dollars) of the initial PV system hardware costs were paid by the European Commission. (The 50 AUD was the initial downpayment paid by each user) It can be seen from Fig 1 that the EC contribution represents approximately 50% of the total hardware costs which will be incurred during the 20 year lifetime of the system. The remaining 50% together with all of the running costs are paid by the users through their fees to the solar utility.

Rural households, living on the outer islands of Kiribati, pay ~6-8 Euro per month for their PV electricity services, which is broadly comparable with their monthly tithes to the church. Discussions held with householders on the outer islands of Kiribati during the evaluation confirmed that this level of fee is affordable.

The evaluation found from SEC's accounts that such fees have been sufficient until now to cover the solar utility's operating costs, because SEC has cross subsidised its solar utility using income earned from other project activities. However, more systems are needed to make the utility economically viable. Independent calculations show that full economic viability of the utility will be achieved when it is able to spread its operating costs over at least 1000 PV systems, and this is now expected within the next few years, when further systems have been installed.

An approximate breakdown of the operating costs of the solar utility in Kiribati is shown in Fig 2, where it can be seen that the largest element is for the replacement of PV system components as they reach the end of their useful lives. Also key to its success is the fact that SEC invests an equivalent amount of its solar utility income in an interest bearing bank account to cover the future costs of replacing the PV system components.

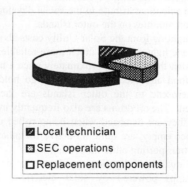

▨ Local technician
▥ SEC operations
☐ Replacement components

Figure 2 Breakdown of utility operating costs

2.3 Sustainability of PV Utility operation

The long-term sustainability of the solar utility depends on having satisfied customers who pay regularly, and on PV systems which work well. This sustainability is perceived in terms of job security by the local technicians, who collect the fees every month, and who are also responsible for cleaning the systems and topping up the batteries with water.

Each technician's salary is paid from his fees, so he is well motivated to visit each system regularly. Technicians also check that the users are not putting the lifetime of their expensive batteries at risk by taking too much energy (eg by making unauthorised connections to the battery which by-pass the charge controller).

3   BENEFITS OF SOLAR UTILITY APPROACH

3.1   Contribution to poverty alleviation

The establishment of the Solar Utility in Kiribati has already resulted in the creation of 13 full time and 14 part time jobs, and more will be created as the Utility expands. Jobs for field technicians have been created in the outer islands, to clean and maintain the PV systems. More skilled jobs have been created on the main island for senior technicians, for the local manufacture of controllers and for the management of the Utility business.

The availability of lighting and communications on the outer islands has made it more attractive for educated islanders to remain on the outer islands, bringing with them the incomes, which they earn though business activities on the main island. Similarly, islanders who go away for several years to work as seamen, are also more willing to retain their homes on the outer islands because they can remain in contact with their families though PV powered communications. The result of installing the PV systems is therefore a significant increase in the average level of income on the outer islands and an overall reduction in poverty.

3.2   Social development benefits

The provision of PV electricity to rural households has significantly raised living standards and the quality of life in the outer islands.

The availability of good quality and more reliable lighting is helping to discourage some of the better educated and more dynamic members of the outer island communities from migrating to the main island. As a result, they are able to contribute to the social development of their own communities on the outer islands.

PV lighting from the Solar Utility costs about the same as kerosene lighting, but is much more reliable. The ships, which visit the islands approximately once a month, cannot store food and fuel in the same cargo hold, and supplies of kerosene to the outer islands are therefore severely limited. The oil drums are also frequently in short supply, since they are damaged when being rolled up the beach and, once empty, are quite attractive for other uses (eg storing or transporting other materials).

The quality of lighting, produced by compact fluorescent lamps powered by solar electricity is far superior to kerosene lighting, and is comparable with that produced by conventional grid power. Islanders can therefore use it during the long evenings for reading, community meetings and work related activities such as preparing to go night fishing. It has also improved the working conditions in the local clinics, schools, and shops during the evenings.

PV lighting is available at the flick of a switch without the inconvenience and safety risks of having to light an oil lamp in the middle of the night. This is much appreciated by families who are looking after elderly or sick relatives, and by those with young children.

Benefits for both genders - Many of the benefits of PV systems apply to all members of the household, regardless of gender or age. However, the women in Kiribati benefit particularly from lighting whilst preparing food, dealing with the night fishing catch and tending to sick children at night. The men benefit particularly from lighting when preparing to go fishing at night and when getting ready to work on their land early in the morning (before dawn).

3.3   Environmental benefits

The introduction of PV systems has reduced the consumption of kerosene on the outer islands, which has in turn reduced the emissions of CO_2 and the risks of oil spillages, which damage the local environment.

Until the arrival of the solar lighting, families in the outer islands were totally dependent on wick or pressure lamps fuelled by kerosene, or on torches using dry cell batteries for lighting at night.

Kerosene is still used by those families without solar lighting, and has to be transported to the outer islands by ship in 200 litre drums, which are off loaded into small boats and rolled up the beach. The fuel is then sold to individual householders after being decanted into small plastic containers or bottles.

Whilst the benefit of the emission savings is relatively small in global terms, the local environmental benefits resulting from reduced transport of fuel to the islands are much more important. The risks of environmental damage caused by oil spillage to ground water supplies are very important as well as those involving damage to the coral and other marine life, that surround these tiny atolls.

PV systems blend well into the local environment. It is often difficult to see them – even from a short distance.

Damage is beginning to be caused to the local environment by discarded dry cell batteries, which are simply thrown onto the ground around the houses. Recycling schemes will soon be needed for these and for the lead acid batteries used in the PV systems, when they reach the end of their useful lives (after 5-7 years).

4   KEY SUCCESS FACTORS OF A PV UTILITY

Based on the evaluation of the solar utilities in Kiribati and Tuvalu [1], and taking into account results from other solar electrification projects, it is suggested that the following are key to the long-term sustainability of a PV Solar Utility:

- **Political support** - government officials on the Utility's Board of Directors can help with obtaining the necessary financing; a supportive government will reduce import duties on PV components; and government backed promotion will raise public confidence in the scheme.

- **Ownership of PV hardware by the Utility** - this gives the Utility staff a vested interest in looking after and maintaining their systems. Users must not be permitted to tamper with it.

- **Market demand** – users must be willing and able to pay an initial deposit for their systems (~30 Euro) and monthly Utility fees for their PV electricity services.

- **Adequate system sizing with good-quality**

hardware - Unlike a poor householder who is short of cash, a properly established Utility business should be able to invest in high quality certified PV modules, good quality (deep discharge) batteries, durable and reliable controllers, and well installed cabling (external quality if outdoors and underground quality where appropriate)

- **Monthly visits to each system by well-trained technicians** - if the technician visits every month to collect his fees, then he can ensure that the equipment is being properly cared for. He can top up the batteries, clean the systems and check for unauthorised connections to the battery.

- **Formal agreements with users, supported by a strict disconnection policy** - if users tamper with the PV system or do not pay fees on time, then it is essential that they be disconnected. Otherwise, the business will soon begin to experience financial problems.

- **Good business management** – a successful PV Utility must be run as a professional business, with monthly reviews of income and expenditure, and properly documented stock control. In addition, a senior technician should make regular checks (two or three times per year) on the work done by field technicians.

- **Well-managed investments** - the monthly fee income should be well invested in interest bearing accounts, to maximise the money available for buying replacement components in the future. A Banker on the Utility's Board of Directors is valuable.

- **Contacts and experience in international procurement** – the Utility needs to have staff who have international experience and can negotiate effectively in international markets, when purchasing replacement PV system components.

- **On-going training programme for administrative staff and technicians** - a rural Utility will inevitably have a turnover of staff, and needs to keep all of its employees up to date through regular training.

5 CONCLUSIONS

The Solar Utility concept has been shown to work well in Kiribati, and to be a far more sustainable approach to rural electrification than providing subsidised financing (soft loans or capital subsidies) for rural households to purchase their own PV systems.

By supporting the Utility rather than the user, international aid is used to purchase good quality hardware, which provides reliable electricity services, and to create long term sustainable jobs. It therefore contributes directly to poverty alleviation and to social development.

The Kiribati model has shown that, once established, a well managed Solar Utility can collect enough fees to cover all future component replacement costs, and its own operating costs. *No further subsidies are needed*, provided that fees are set at an affordable level (6-8 Euro / month), fee income is well invested, and

enough systems are installed (typically at least 1000 systems).

If a donor pays the initial hardware costs, which represent about 50% of capital costs for the first 20 year life-cycle, then users will pay the other 50% via the Solar Utility for replacement batteries (~5 years), and controllers and lights (~8 years).

In summary, the EC funded Kiribati project has demonstrated that using donor aid to establish a Solar Utility and pay for the initial hardware costs of about 1000 solar home systems *is an excellent example of sustainable development co-operation.*

6 REFERENCES

[1] Evaluation of the PREP Component : PV Systems for Rural Electrification in Kiribati and Tuvalu (7 ACP RPR 175) Final Report March 1999 AEA Technology – ETSU, (Available from the European Commission DG Development, Evaluation Unit)

[2] Bill Gillett and Gill Wilkins, Solar So Good – An EC funded solar utility succeeds in Kiribati, The ACP-EC Courier, No 177 October – November 1999.

ACKNOWLEDGEMENTS

The work reported here was financed by the European Commission, Directorate General for Development, through the European Development Fund.

The technical advice and guidance provided by Herbert Wade is gratefully acknowledged. The success of the PV Utility in Kiribati is largely due to Herb's professional skills and vision, coupled with the hard work of the staff in the Solar Energy Company.

ESTABLISHING A RENEWABLE ENERGY INDUSTRIES ASSOCIATION IN CHINA

Wang Zhongying, Zhu Junsheng, William Wallace, Li Junfeng, Frederic Asseline
GEF/UNDP Project Management Office
1811 Jinyu Mansion, 129 Xuanwumen Xidajie
Beijing 100031 P.R.C.
Tel and Fax: +11 86 10 6641 1401

Bernard McNelis and Jenniy Gregory, IT Power Ltd.
The Warren, Bramshill Road
Eversley, Hampshire, RG27 0PR, UK
Tel: +44 118 973 0073; Fax: +44 118 973 0820

Mathew Mendis, Alternative Energy Development, Inc.
8455 Colesville Road, Suite 1225
Silver Springs, MD USA 20910-3320
Tel: +1 301 608 3666; Fax: +1 301 608 3667

ABSTRACT: Renewable energy development in China is expanding beyond the capabilities of existing institutions to provide support for sustainable commercial development. China is currently in a transition phase in which the central and local governments are in the process of developing new policy incentives and environmental regulations impacting renewable energy technology dissemination. There is still a large gap between renewable energy resource potential and existing levels of technology dissemination in China; nevertheless, project development is increasing in scale and a broader investment base is becoming available for project financing. In this transition, renewable energy industry groups in China need services in the form of policy advocacy, business development assistance and training, facilitation of financing, and assistance in interfacing with the international renewable energy community. In 1999 the United Nations Development Programme (UNDP) and the Chinese State Economic and Trade Commission (SETC) initiated support for the establishment of the Chinese Renewable Energy Industries Association (CREIA). CREIA is a non profit and non-government organization that uses market-oriented approaches to interface with the domestic industry to promote renewable energy commercialization. CREIA will manage an Investment Opportunity Facility (IOF) to help connect projects with investors, and will play an advocacy role on behalf of industry with the government and other stakeholder groups. It will also provide services to members in form of market information, databases, trade missions, foreign exchanges, workshops, and training. This paper will summarize the establishment, activities to date, and future plans of the CREIA organization.
Keywords: Developing Countries - 1: Financing - 2: Sustainable - 3

1. INTRODUCTION

China is in a transition period in which the pace of renewable energy development is increasing. Although there is still a large gap between the potential of renewable energy based on available resources and current levels of market development, developers are in the process of increasing the scale and scope of current and near term projects. For example, in the case of photovoltaics (PV) the total installed capacity of PV systems in China at the end of 1999 was approximately 13 MW [1], distributed among telecommunications, industrial, agricultural, consumer, and rural electrification applications. However, in the fastest growing PV market sector of rural electrification, several large-scale projects are under development that will rapidly expand PV dissemination in China.

One example is a 10 MW solar home system program being developed by the World Bank for western China [2]. The Shell Company is implementing a large-scale solar home system project in the Province of Xinjiang [3]. In addition, several large-scale solar home system and village power projects are in the approval stage in the Brightness Program under the Chinese State Development Planning Commission [4]. Similar examples of increased scale of development can be found in the market sectors of large-scale wind farm development, solar water heating applications, biogas plants, and bagasse cogeneration.

With the increasing scale of renewable energy development comes new needs for support mechanisms to promote sustainable commercialization and market expansion. Policy initiatives for renewables formulated by central government agencies in China are a major activity in the Government's planning processes for the 10th Five-Year Plan (2001-2005). For example, the Chinese State Development Planning Commission, with foreign expert assistance, is assessing the potential for Renewable Energy Portfolio Standards (RPS), and other policy mechanisms, for promoting renewables. Provincial governments are also developing policy initiatives. For example, solar and wind renewable energy systems are integrated into the rural electrification expansion plans of most western provinces as one important option for village electrification.

The next stage of renewable energy expansion in China also has implications for business development and financing. Most of the domestic renewable energy business base in China has limited experience with international standards in project development and with commercial financing of large-scale projects. Technology transfer is also needed from the international and domestic marketplaces to promote the dissemination of state-of-the-art commercial technologies, achieve cost reductions, and improve the manufacturing base in China. Information exchange is needed to promote international best practices for project development and financing, facilitate business interactions between domestic and international organizations, and provide an interface between domestic industry groups and key government decision-makers and other stakeholders.

2.   RENEWABLE ENERGY STATUS IN CHINA

2.1   Potential of Renewable Energy

An indication of the potential for selected renewable technologies is given in Table I. Figures are expressed in terms of the cumulative installed capacity or production levels in 1998 vs. the levels projected for 2010 based on estimates from the Chinese State Economic and Trade Commission [1]. The potential for accelerated commercial development is especially significant for grid-connected wind and solar water heating technologies. Projections for solar PV installed capacity can be considered a lower limit.

Table I: Indication of potential of several key renewable energy technologies in China.

Technology	1998*	2010*
Solar PV (MW)	13	56
Solar Water Heaters (10^6 m^2)	15	171
Grid-Connected Wind (MW)	224	3000
Small Wind (MW)	17	430
Geothermal Power (MW)	25	100
Biogas Livestock Farms (10^8 m^3)	0.6	2.8
Biogas Industrial Wastewater (10^8 m^3)	3.2	24

*Cumulative

2.2   Industry Status

Table II provides estimates of numbers of companies operating in selected market sectors. Most market sectors are subject to heavy competition among a large number of small companies. Product quality is often poor and profit margins are low. In many market sectors a consolidation is occurring with larger successful companies beginning to increase market share and smaller companies merging or going out of business.

Table II: Characterization of industry groups in terms of number of manufacturing and distribution companies.

Market Sector	Large Companies*	Small Companies
PV	5	>45
Wind		19
Solar Water Heater	10	1000
Biomass	20	600
Geothermal		30

*Companies of sufficient size to take advantage of some economy of scale in manufacturing or distribution.
Source: Chinese State Economic and Trade Commission

2.3   Barriers to Commercialization

Factors contributing to commercialization barriers for renewable energy development in China include government policy and planning. At the national government level there is a lack of a systematic and comprehensive policy structure tailored for renewable energy development. Coordination among agencies responsible for renewable energy planning is weak, which

is reflected in the lack of a long-term strategic plan for sustainable renewable energy commercialization.

Other factors include the high cost of the domestic renewable energy technology base in some market sectors. There is a general lack of public and stakeholder awareness of the benefits and availability of renewable options. Assessment of renewable energy resources is incomplete and accessibility to resource databases is limited. There is also a need for the development and application of standards and certification testing to insure product quality and safety. Linkages from research and development to commercialization are poor due to lack of development capital and cutbacks in government support of R&D in China.

As the project development scale increases, financing barriers exist in the form of lack of access to information and lack of professional financial institutions familiar with the characteristics of renewable energy project financing. There is also a lack of intermediate organizations that can provide services to renewable enterprises in China for market development assistance.

3.   FORMATION OF CREIA

3.1   Supporting Organizations

In March 1999 the United Nations Development Programme (UNDP) initiated the project CPR/97/G31 entitled "Capacity Building for the Rapid Commercialization of Renewable Energy in China." The major objective of this project is to support the acceleration of sustainable commercialization of renewables by addressing policy, business, finance, and technical issues through capacity building activities. Corollary objectives include achieving a detectable increase in technology deployment through project activities, developing a strong interface with industry and investment communities, and promoting the transfer of international experience and technology to China.

The GEF/UNDP Project is funded at a level of $25.8 million (USD) over a period of five years. Core funding comes from the Global Environmental Facility (GEF) and the Chinese Government. Co-financing contributions come from the AUS AID Program of Australia and the Government of the Netherlands. The project is implemented by the Chinese State Economic and Trade Commission in Beijing and is executed by the United Nations Department of Economic and Social Affairs in New York.

One of the major activities in the GEF/UNDP project is supporting the establishment of the Chinese Renewable Energy Industries Association. Both the UNDP and the SETC will provide initial support of CREIA staff and facilities for a three-year period. After five years the organization should become fully self-supporting through membership fees and services. CREIA is viewed as a key long-term legacy of the GEF/UNDP project.

3.2   Establishment of CREIA

During the latter half of 1999, a series of consultative activities were conducted with a number of companies, industry groups, government organizations and investors in China. Consultations provided feedback from industry regarding their expectations and needs. Industry also provided recommendations for structuring the CREIA organization. A formal institutional workshop for the organization was conducted in January 2000, attended by over 30 key companies and trade associations in China, as founding members of CREIA. These companies are major representatives from the wind, PV, solar water heater, biomass, and geothermal business sectors.

During January 2000 CREIA was formally registered as CREIA Renewable Energy Ltd. under Chinese corporate law. Corporation status is temporary pending approval of CREIA's application with the Chinese government for trade association status, expected within several months. A Board of Directors and Secretariat has been appointed, as well as the first Director of CREIA, Mr. Zhu Junsheng.

CREIA is a non-profit corporation and although it receives initial support from the State Economic and Trade Commission in China, it is not formally associated with a government organization. The non-government status distinguishes CREIA from most existing trade associations in China, which are frequently departments of a government agency. As such, trade associations in China in the past have been more responsible to the government than to their industry membership base.

CREIA will conduct a membership drive in 2000 to attract a wide base of support from renewable energy companies. High priority market sectors for membership in the organization include solar photovoltaics, solar hot water heaters, biomass, grid-connected wind farms, small wind technologies, geothermal power, and hybrid systems (general rural electrification markets).

Other trade associations can also attain member status. Provision has been made for foreign companies to join CREIA as international affiliate members, allowing them to register in CREIA's company database and giving them access to information services. In the initial phase of CRIEA development, Chinese companies will be given full membership status.

3.3 Objectives

CREIA has been established with the specific objective to work with Chinese renewable energy industry groups to promote sustainable renewable energy development based on market-oriented commercialization principles. In implementing this objective, CREIA will provide several functions.

CREIA will serve as a bridge between regulatory authorities, research institutes and industry professionals, providing a forum to discuss renewable energy development at the national level. Subsequently, CREIA will play an advocacy role on behalf of industry to advise the Government of China on strategic policy formulation.

CREIA will be a window bringing together national and international project developers and investors. It will promote technology transfer and raise awareness of renewable energy investment opportunities. An Investment Opportunity Facility (IOF) will be developed and regional networking meetings and training activities will be conduits for information exchange.

CREIA will provide a communication network for its members from the Chinese renewable energy business community. This network will allow members to communicate across their respective subsectors and provide a platform to voice their concerns collectively.

CREIA will play an advocacy role on behalf of industry to raise the general profile of renewable energy technologies, applications, and industry as a whole. This includes information dissemination to the public and public education.

3.4 Industry Response

During 1999 and 2000 the level of interest shown in CREIA by Chinese companies has been high. International companies and trade associations have also exhibited a high interest in CREIA. There is a general recognition of the need for a trade association that will promote the needs of renewable energy companies to advance commercialization objectives.

Recommendations by companies have been given freely to CREIA in the formative stage. At the same time, expectations are high, and companies anticipate a high level of performance and effectiveness from CREIA.

4. CREIA STATUS

In 1999 a contract to support CREIA was awarded by the UNDP to a team consisting of IT Power in the U.K. and Alternative Energy Development, Inc. (AED) in the United States. The support team is working with members of the CREIA staff to prepare a five-year business plan and marketing strategy for the organization that will allow CREIA to become fully self-supporting in five years.

A series of business and management training activities for CREIA was developed that will be executed during May and June 2000. The primary objectives of the training program are to create renewable energy project identification skills, project assessment and preparation capabilities, and awareness of domestic commercial and international sources of financing for renewable projects within CREIA. Training will target CREIA Board members, operational staff, renewable energy project developers, financiers, and government decision-makers.

In April and May 2000 consultations with businesses, investors, and trade associations were conducted by CREIA staff and corporate members in the United States and Europe. These consultations included information exchanges between CREIA and experienced trade groups that was used to further develop a proposed inventory of information and services that CREIA will provide to its membership. Business contacts were also made between foreign and Chinese companies and with foreign investors and financial institutions.

5. INVESTMENT OPPORTUNITY FACILITY

5.1 Objectives

One of the key services that CREIA will provide to its membership base is the development and maintenance of an Investment Opportunity Facility (IOF). The principal objective of the IOF is to provide assistance to the domestic renewable energy industry to identify, develop and obtain financing for renewable energy projects.

Specifically, the IOF will work in close partnership with CREIA members to provide guidance, supporting information, and technical assistance to help members obtain investment partners and financing for the development of renewable energy projects. In addition, the IOF will help members recognize and capitalize on the growing opportunities for renewable energy investments to mitigate global climate change.

5.2 Key Elements of the IOF

The IOF will consist of three key elements. First, the *Investment Development Process* will pro-actively identify and solicit renewable energy projects for investment.

Second, the *IOF Database and Tracking System* will hold key information on potential projects and prospects. Information will be accessible to industry, potential financiers, and other key stakeholders.

Third, the *Investment Review Committee* will review investment opportunities and identify potential sources of available funding. Information will be accessible on the CREIA INTERNET website, which is in development.

5.3 Outcomes

It is expected that the IOF will play an important role in facilitating the linkage between projects and financing in China. In combination with the IOF information databases, CREIA will conduct investment opportunity forums to create opportunities for project developers, entrepreneurs, and companies to meet directly with investors and financial institutions interested in renewable energy projects. Active outreach to the financial community has already started.

A corollary outcome of the IOF is expected to be improvements in the quality of feasibility studies and project proposals. Part of CREIA's training services will be devoted to assistance to project developers in preparation of documentation meeting international finance standards.

6. INFORMATION SERVICES AND DATABASES

6.1 Databases

Membership and public services from CREIA will include availability of a series of databases that will be published on the CREIA website. Databases linked with the IOF will include a membership information base with company profiles and contact details, profiles of financial institutions and financing intermediaries, and information for presenting and tracking projects and proposals.

In addition, CREIA will post database information generated in collaboration with the GEF/UNDP project in China. Such databases will include renewable energy resource information, financing channels and methods, renewable energy technology information, and an expert and project developer database. There will also be an international affiliates database.

6.2 Workshops, Training, and Conference Support

CREIA will collaborate with the GEF/UNDP project in conducting policy, business, finance, and technical workshops and training events for promoting renewable energy commercialization objectives. For example, CREIA will be involved in the project's standards and certification activities, finance training for project developers, and regional workshops for technology and application information dissemination.

During November 28-December 1, 2000, CREIA will be a major sponsor and organizer of the 2000 China International Environment, Renewable Energy, and Energy Efficiency Exhibition and Conference, conducted in Beijing. During this period, CREIA will conduct two workshops directed at project financing for large-scale wind farms and renewable energy opportunities in China's western development initiative. During 2000, CREIA will also conduct a solar thermal standards and certification workshop and PV workshop (in October 2000 in Kunming, China).

7. SUMMARY

The establishment of the Chinese Renewable Energy Industries Association represents an opportunity to engage the domestic renewable energy enterprise base in China in a direct way to promote sustainable development and commercialization. CREIA will play an advocacy role on behalf of industry with the Government and other stakeholders, yet be independent of the Government and focused on the business sector and market-oriented principles for commercial development. Implementation of the Investment Opportunity Facility and maintenance of a series of website-based databases will help facilitate linkages of renewable energy projects with sources of financing. CREIA will also play a key role in interfacing the domestic industry in China with the global renewable energy community.

You may contact CREIA via the email address: creia@163bj.com.

ACKNOWLEDGEMENTS

The authors would like to acknowledge financial support from the GEF/UNDP Project CPR/97/G31 and from the Governments of Australia (AUS AID Program) and the Netherlands.

REFERENCES

[1] Zhu, J., Proc. Sino-USA Renewable Energy Business Workshop, Xian, P.R.C. China (Nov. 3 1999), No. 5.

[2] World Bank Report No: 18479-CHA CN-PE-46829, Project Appraisal Document (Dec. 2, 1998), p 7.

[3] Huang, X., Proc. GEF/UNDP Renewable Energy Business Development and Financing Workshop, Beijing, P.R.C. China (April 5-7 2000), to be published.

[4] Xu, H., Proc. Sino-USA Renewable Energy Business Workshop, Xian, P.R.C. China (Nov. 3 1999), No. 6.

Author Index

Author Index

Author Index

Author Index

Keyword Index

Keyword Index